GROUNDWATER: PAST ACHIEVEMENTS AND FUTURE CHALLENGES

PROCEEDINGS OF THE XXX IAH CONGRESS ON GROUNDWATER:
PAST ACHIEVEMENTS AND FUTURE CHALLENGES
CAPE TOWN / SOUTH AFRICA / 26 NOVEMBER – 1 DECEMBER 2000

Groundwater: Past Achievements and Future Challenges

Edited by

Oliver Sililo et al.

The Water Programme, Environmentek, CSIR, South Africa

A.A. BALKEMA / ROTTERDAM / BROOKFIELD / 2000

Photographs courtesy of:
British Geological Survey (top right)
C. Pointer (top left)
CSIR, South Africa (bottom)

The texts of the various papers in this volume were set individually by typists under the supervision of each of the authors concerned.

Published by
A.A. Balkema, P.O. Box 1675, 3000 BR Rotterdam, Netherlands
Fax: +31.10.413.5947; E-mail: balkema@balkema.nl; Internet site: www.balkema.nl
A.A. Balkema Publishers, 2252 Ridge Road, Brookfield, VT 05036-9704, USA
Fax: 802.276.3837; E-mail: info@ashgate.com

ISBN 90 5809 159 7

Printed in the Netherlands

Groundwater: Past Achievements and Future Challenges, Sililo et al. (eds) © 2000 Balkema, Rotterdam, ISBN 90 5809 159 7

Table of contents

2 *Groundwater modelling*

3 *Chemical and isotopic characterization of groundwater*

4 *Groundwater pollution vulnerability and remediation*

Groundwater: Past Achievements and Future Challenges, Sililo et al. (eds) © 2000 Balkema, Rotterdam, ISBN 90 5809 159 7

Foreword

Papers in this proceedings volume were refereed by at least two members of the Editorial Sub-Committee of the IAH XXX congress. The task of editing the more than 250 papers received was formidable, and I would like to thank the following for their time and effort: Steve Appleyard, Mike Barrett, Ebehard Braune, Willy Burgess, Lisa Cavé, Wen-hsing Chiang, John Chilton, Christine Colvin, Julian Conrad, Ingrid Dennis, Johan de Beer, Peter Dillon, Jane Dottridge, Matthias Eiswirth, Pannie Engelbrecht, Louise Fraser, Bjørn Frengstad, Ian Gale, Chris Harris, Simon Hughes, David Lerner, Reinie Meyer, Kym Morton, Ricky Murray, Roger Parsons, Mike Rivett, Irené Saayman, Nick Stevens, Siep Talma, Richard Taylor, John Tellam, Gideon Tredoux, John Weaver, Gerrit van Tonder, Balt Verhagen and Alan Woodford. Reviewers who read more than 10 papers are listed at the beginning of each topic in this volume.

Apart from the reviewers, a number of colleagues in the Water Programme, Environmentek, CSIR assisted in the final stages of editing by checking final papers submitted and where necessary made layout changes to meet the publishers instructions. In this regard, I would like to thank Lisa Cavé, Christine Colvin, Anthony Fielies, Irené Saayman, Louise Fraser, Reney Robyntjies and Kathy Pretorius.

Oliver Sililo
The Water Programme, Environmentek, CSIR
Stellenbosch, August 2000

Groundwater: Past Achievements and Future Challenges, Sililo et al. (eds) © 2000 Balkema, Rotterdam, ISBN 90 5809 159 7

Preface

This volume comprises over 200 papers of the proceedings of the 30th annual congress of the International Association of Hydrogeologists. The congress was held in Cape Town jointly with the Groundwater Division of the Geological Society of South Africa and hosted by the CSIR.

The overall theme of the congress, *Groundwater: Past Achievements and Future Challenges*, encouraged groundwater professionals to take stock of progress in our field during the last few decades and consider the priorities for future work. This volume represents the diversity of experience of the international community and the great range of application of hydrogeological expertise: from assessing fecal contamination of shallow groundwater in Kampala to characterising deep subsurface microbes in the Witwatersrand gold mines; from in situ treatment of Cr contaminated groundwater in Germany to studies in Japan of soil environment controls on VOCs. The field of hydrogeology remains an exciting and challenging area of theoretical study and practical application, pushing the boundaries of science and improving the quality of life for millions.

Professor Foster's keynote paper outlines some of the achievements in the field of hydrogeology in providing good quality, reliable water for domestic and agricultural use to hundreds of millions of people around the world. He also discusses the dramatic improvements in water management called for by the World Water Forum, held in March 2000. These included a major groundwater initiative to enable grassroots driven, sustainable aquifer management. The challenges currently faced by South Africa, discussed by Braune, present a microcosm of those apparent in our divided world. New water legislation in South Africa attempts to redress the legacies of apartheid by focusing on principles of sustainability, equity, transparency and efficiency. Groundwater will need to play a crucial role in the provision of water resources if the objectives of sustainable development are to be met.

The congress is arranged around 6 themes which form the chapters of this volume:
- Determination of aquifer parameters, resource assessment and evaluation;
- Groundwater modelling and water well issues;
- Chemical and isotopic and isotopic characterisation of groundwater;
- Groundwater pollution, vulnerability and remediation;
- Integrated water resources management;
- Geotechnical issues and mining.

I would like to thank the many hydrogeologists who willingly gave their time and expertise to produce this peer reviewed volume. I also wish to congratulate the senior editor, Oliver Sililo of the CSIR in South Africa, on coordinating the review process and thank him for his many hours of hard work. It would not have been possible to produce this volume without the backing of the Water Programme of CSIR, and they are gratefully acknowledged for their support.

Christine Colvin
Hydrogeologist, CSIR, South Africa
Stellenbosch, August 2000

Sponsors

CSIR

Your Technology Partner

The compilation of this volume was supported by the CSIR's Water Programme, South Africa.

The following sponsors of the IAH 2000 are also thanked for their support:

UNESCO

WRC

DWAF

CDG Carl Duisberg Gesellschaft e.V.

GCS

UMVOTO

CAPE METROPOLITAN COUNCIL

NRF

SRK Consulting
Engineers and Scientists

K L M Consulting Services
Specialists in Mine dewatering

E. MARTINELLI AND ASSOCIATES
CONSULTING HYDROLOGISTS & GROUNDWATER GEOPHYSICISTS

GSSA

IGS

Keynote papers

Groundwater: Past Achievements and Future Challenges, Sililo et al. (eds) © 2000 Balkema, Rotterdam, ISBN 90 5809 159 7

The mineralogy of microbial ecology

P.C. Bennett & J.R. Rogers
Department of Geological Sciences, University of Texas, Austin, Tex., USA

ABSTRACT: Mineralogy, microbial ecology, and mineral weathering in the subsurface are an intimately linked biogeochemical system. We have documented microbial silicate weathering driven by the nutrient requirements of the microbial consortium when the required elements are present as trace mineral inclusions. This relationship has been identified in an anoxic oil-contaminated groundwater, dominated by iron reducing and methanogenic bacteria. Here carbon is abundant but Fe and P are scarce, and even resistant feldspars are rapidly colonized and destroyed if they contain these elements. Addition of feldspar chips containing trace apatite inclusions to laboratory microcosms accelerates BTEX biodegradation and increases biomass. Chips of manufactured glass containing P are rapidly colonized, while similar glasses without P are barren. The subsurface distribution of microorganisms may in part be controlled by mineralogy, and an organism's ability to take advantage of mineral-bound nutrients. Conversely, mineral weathering may in part be a function of that mineral's nutritional potential, with microorganisms destroying only "beneficial" minerals.

1 INTRODUCTION

A variety of subsurface bacteria degrade organic contaminants, although the growth of the subsurface population is often limited by the availability of terminal electron acceptors and inorganic nutrients. The transformation of hydrocarbons by microorganisms, in turn, produces a wide variety of reactive byproducts that perturb geochemical equilibria, dissolving or precipitating minerals. These geochemical reactions are an integral part of subsurface microbial ecology, where the alteration of a mineral surface may offer an advantage to the population by enhancing attachment, buffering extremes of pH, altering habitat, or releasing limiting nutrients.

The geochemical ecology of a petroleum contaminated aquifer located near Bemidji, MN, USA was examined in collaboration with researchers from the U.S.G.S. (http://wwwmn.cr.usgs.gov/bemidji/). In this aquifer natural attenuation of the contaminating petroleum is proceeding in both aerobic and anaerobic saturated zones as well as the unsaturated zone. Here the subsurface microorganisms gain energy from oil oxidation; while O_2 offers the most energy for the aerobic community, energy can be conserved using other electron acceptors, such as $N^{(IV)}$, $Mn^{(II)}$, $Fe^{(III)}$, and $S^{(VI)}$ when oxygen is limiting (Ghiorse and Wilson 1988, Chapelle 1993). Most petroleum degraders also require macronutrients such as N and P, and strategies that increase their bioavailability enhance the survival of that microbial consortium. Different microbial consortia dominate in different aquifer regions, and each consortium produces distinct mineral-weathering phenomena.

2 APPROACH

To examine the interaction between microorganism and mineral surfaces, we use *in-situ* microcosms to characterize microbial colonization and mineral weathering in field settings, with adjunct laboratory-microcosm experiments used to examine substrate utilization and byproduct production. The field methods have been described elsewhere (Hiebert and Bennett 1992, Rogers et al. 1998), but in brief: clean sterile mineral chips are placed in permeable polyethylene chambers, and these chambers are suspended in wells, or buried, in the target aquifer for periods of months to years. Upon recovery, biological tissues are fixed using chemical critical-point drying (Vandevivere and Bevaye 1993). Fixed samples are inspected in the laboratory using conventional scanning electron microscopy (C-SEM) and environmental SEM (E-SEM) for microbial abundance, colonization pattern, and etch pit or mineral precipitate development.

2.1 *Field Studies*

The Bemidji aquifer was contaminated in 1979 when a petroleum pipeline ruptured and sprayed a light aliphatic crude over approximately 0.65 hectares of the land surface, resulting in both surface and subsurface accumulations of petroleum. This resulted in a region of groundwater contaminated by dissolved aromatic and aliphatic hydrocarbons, and these compounds are rapidly degrading by microorganisms coupled to iron (III) reduction (Lovley and Lonergan 1990, Baedecker et al. 1993) and by secondary methanogenesis.

The contaminated aquifer consists of ~20 m of calcareous silty sand deposited as glacial morainal and outwash material overlying a clay-till of unknown thickness (Bennett et al. 1993). The native aquifer material is composed of ~ 57% quartz, 29% feldspars, 4-6% carbonates 2% hornblende, <1% clays, and < 0.2 % organic carbon. Degradation of the contaminating petroleum has produced a plume of inert and reactive organic and inorganic solutes, including a variety of organic acids (Cozzarelli et al. 1990, Baedecker et al. 1993), and etching of native quartz and feldspar grains has been documented (Bennett et al. 1993).

Geochemical parameters, including dissolved carbon, nutrients, anions, cations, and field parameters have been determined annually since 1983 (Bennett et al. 1993). Total biomass on microcosm minerals and in the water was determined as the total lipid biomass, while microbial diversity on mineral surfaces was estimated from SEM imaged cell morphology, phospholipid fatty acid analysis, and 16SrDNA analysis. Diversity on mineral chips was also characterized using a modified Most Probable Number (MPN) method using PRAS (pre-reduced, anaerobically sterilized) prepared media (e.g. Bekins et al. 1999).

2.2 *Laboratory Microcosms*

In addition to the field approach, laboratory microcosms of native consortia were constructed using clean reference-mineral fragments inoculated with aquifer materials and groundwater from the contaminated aquifer. Cores were collected from the anoxic region of the aquifer using a frozen-shoe wire-line core barrel (Murphy and Herkelrath 1996). The core material and the selected mineral type were transferred to triplicate sterile serum bottles in an anaerobic chamber, finished with a 1:1 mixture of formation water and sterile anaerobic water, and amended with toluene. Dead and uninfected control samples were also constructed. Reaction progress was determined by the periodic withdrawal and chemical characterization of 1 ml sample aliquot.

3 RESULTS AND DISCUSSION

When oil is added to a carbon-limited starved habitat, the native microorganisms are given rich source of electron donor available for oxidation, as well as carbon for assimilation and growth. In the lightly contaminated sprayed-oil area, degradation occurs in the unsaturated sediments with molecular oxygen as the primary electron-acceptor. In the oil-saturated sediments near the pipeline rapid degradation occurs in both the unsaturated and saturated zones. Here degradation occurs faster than the inward flux of molecular oxygen, and the system is anaerobic. Aerobic conditions return down-gradient of the oil pool before the available carbon decreases back to the oligotrophic background conditions. In each of these biogeochemical zones, the byproducts of microbial oxidation of the components in oil perturb mineral-water equilibria over large areas.

3.1 *Carbonate Diagenesis*

Under the lightly contaminated soils up-gradient of the oil pool, petroleum degradation proceeds by aerobic beta-oxidation and oxygenlytic cleavage of simple aromatic compounds, depleting the reservoir of dissolved oxygen, while producing carbonic acid (H_2CO_3). The produced carbonic acid dissolves carbonate minerals, producing a plume of dissolved Ca^{2+}, Mg^{2+}, and HCO_3. Partial pressure of carbon dioxide increases from a background value of $10^{-2.5}$ atm to $10^{-0.8}$ atm at the edge of the iron-reducing zone as pH decreases to 6.6. Here the aquifer geochemistry is dominated by the carbonate dissolution:

$$CH_2 + 1.5O_2 + Ca_XMg_{1-X}(CO_3) \rightarrow XCa^{2+} + (1-X)Mg^{2+} + 2HCO_3^- \quad (1)$$

where CH_2 represents an average mid-chain alkyl carbon (i.e., hexane- C_6H_{14}) with a carbon oxidation state as ~C^{-II}, with a mixed Ca/Mg carbonate as the reacting mineral. The reaction occurs over wide areas of the aquifer, and does not require direct colonization of the mineral surface. There is no indication that the microbial consortium gains an advantage from the weathering of carbonates beyond pH buffering. No evidence of feldspar weathering is observed in the aerobic regions.

In the anoxic water under the oil pool, calcite dissolution was only occasionally observed, and *only* secondary to microbial colonization. Here the dominant reaction is calcite precipitation, and syntaxial overgrowths are observed on most surfaces. Changes in solute concentration in this region supports a hypothesis of calcite precipitation, driven by iron reduction, e.g.,

$$C_7H_8 + 36Fe(OH)_3 + 58HCO_3^- + 65\,Ca^{2+} \rightarrow 65CaCO_3 + 36Fe^{2+} + 87H_2O \quad (2)$$

Another possible reaction to drive the system toward calcite precipitation is bicarbonate consumption by methanogens, e.g.,

$$2HCO_3^- + 4H_2 + Ca^{2+} \rightarrow CaCO_3 + CH_4 + 3H_2O \quad (3)$$

3.2 *Microbial feldspar dissolution*

Under the oil pool the groundwater is extremely reducing, and dissolved silica and iron concentrations increase along the flow path, with parallel increases in dissolved Sr^{++} and K^+ concentrations. Dissolved silica increases only in the anoxic region under the oil and approaches saturation with respect to amorphous silica, while aluminum appears to be conserved in secondary clays. Results from the field microcosm experiments demonstrate that silicates are rapidly weathering, but only on surfaces colonized by microorganisms, with significant variability in colonization pattern, colonization density, and weathering intensity on different silicates.

Most silicate minerals are not colonized, and no evidence of significant chemical weathering was observed over periods of 3-14 months. Other silicates, however, are intensely colonized, with evidence of rapid weathering. With the exception of quartz (Hiebert and Bennett 1992), colonized silicates all contain minor phosphorus as apatite inclusions, whereas virtually identical feldspars without P are uncolonized and unweathered (Rogers et al. 1998).

The requirement of colonization to produce a sufficiently reactive microenvironment results in observed weathering behavior that is in apparent conflict with the Goldich weathering sequence (Goldich 1938), which predicts that silicates such as olivine and plagioclase are unstable at the earth's surface relative to microcline and quartz. In fact, "resistant" microcline dissolves much faster than olivine, if that feldspar contains nutrients of value to the colonizing microbial consortia.

3.3 *Hydrocarbon Degradation Rate*

The rate of toluene degradation in nutrient-limited groundwater by native consortia was examined using standard laboratory microcosms (described above). We found that adding some minerals to the microcosms dramatically increase degradation rate, while other minerals do nothing at all, and that the minerals that enhance degradation are different in different regions of the aquifer.

Most silicate and all carbonate minerals do not influence toluene and degradation in the anaerobic microcosms. Minerals that contain phosphorus, iron, or nitrogen, however, greatly enhance degradation over the control microcosms, and the results support the observations from the field experiments. Anorthoclase, containing both iron and phosphorus in minor inclusions, as well as simple magnetite, have the most dramatic effect, with some enhancement also seen with microcline and apatite.

The field and laboratory microcosm studies support studies by other researchers that the iron-reducing population dominates hydrocarbon degradation in this aquifer (Bekins et al. 1999), but also shows that the population is now limited by both iron and phosphorus. While these elements are present in sufficient quantities in the uncontaminated aquifer sediment, they have been scavenged from the contaminated aquifer after 20 years of intense microbial activity, and are now limiting. Addition of either P *or* Fe results in some enhancement of degradation, but adding both immediately accelerates toluene degradation in anaerobic groundwater.

4 IMPLICATIONS

The microbial degradation of petroleum in groundwater is a fundamental aspect of "natural attenuation", and most ground water professionals are familiar with the general processes. What is less commonly evaluated, however, is the relationship between the microbial community and the geologic framework within which natural attenuation progresses. Traditionally, hydrogeologists evaluate the effectiveness of natural attenuation simply by the rate of removal of the target compounds (for example, BTEX compounds). This fails to account for the influence of mineral-bound nutrients, the influence of microorganisms on mineral dissolution and precipitation, and the consequent changes in the hydrogeology and microbial ecology.

We have found that in oxic environments, aerobic heterotrophs rapidly oxidize hydrocarbons to carbon dioxide, producing mineral acidity that drives calcite and dolomite dissolution. This interaction is observable over a wide area of the aquifer up-gradient of the oil, and no evidence exists to indicate that it is restricted to the immediate vicinity of the microorganisms (except in rare cases). Weathering rate is controlled by the rate of microbial production of CO_2 or other acids or ligands, and by the kinetics of calcite and dolomite dissolution, and this reaction tightly buffers the pH to near-neutral conditions conducive to microbial growth. Over time the composition and permeability of the aquifer will change as calcite is removed, and eventually the buffer capacity of the aquifer will be eliminated, allowing the pH to move to more acidic regions. This is observed in other hydrocarbon contaminated aquifers where calcite is not present. The byproducts of calcite dissolution are advected downgradient under the oil, where the conditions change from calcite undersaturation to calcite supersaturation, due to the consumption of acidity by iron reduction and carbonate methanogenesis. Here, syntaxial calcite overgrowths form, resulting again in a pH buffered water, and a loss of porosity.

Under the oil, iron minerals and silicates are weathered, releasing limiting nutrients that locally stimulate microbial degradation rate. Silicate weathering occurs only near attached microorganisms, and only colonized silicates dissolve. The by-product of silicate weathering is the precipitation of clays that are observed on mineral surfaces as well as on the microbial membranes. Over time this results in the removal of oxidized iron minerals, and the selective weathering of primary silicate minerals that contain nutrients of value, while leaving behind "useless" feldspars , and complex clays. The production of clays also changes the character of the aquifer, forming surfaces where sorption and exchange reactions can occur.

Down-gradient of the oil pool, where oxygen is again available, iron (and silica) precipitate in the aquifer, reducing permeability, and potentially changing ground water flow paths. In this region acidity is again generated secondary to iron precipitation, and calcite dissolves. The microbial consortia in this region has not been characterized, however, and it is not known if there are specific nutrient limitations.

The biologically mediated destruction of silicate minerals to liberate limiting inorganic nutrients is a fundamental observation of subsurface microbial ecology, and may also impact our interpretation of the mineral weathering sequence in the rock record. A basic tenet of sediment diagenesis, the "Goldich Weathering Sequence"(Goldich 1938), states that the most unstable silicate mineral will weather (dissolve) first, with more resistant silicates taking progressively longer to dissolve, e.g. (from least to most stable), olivine < plagioclase < albite < anorthoclase ~ microcline < quartz. The observed weathering sequence of minerals in an anaerobic, microbially controlled system however, is almost opposite, and olivine is stable with respect to microcline, and the relationship between microbial colonization and weathering rate almost perfectly correlated. This suggests that in *some* environments the indigenous microorganisms may significantly alter weathering patterns as they aggressively scavenge limiting nutrients.

REFERENCES

Baedecker, M. J., I. M. Cozzarelli, D. I. Siegel, P. C. Bennett & R. P. Eganhouse 1993. Crude oil in a shallow sand and gravel aquifer-III. Biochemical reactions and mass balance modeling in anoxic groundwater. *App. Geochemistry.* 8:569-586.

Bekins, B. A., E. M. Godsy & E. Warren 1999. Distribution of microbial physiologic types in an aquifer contaminated by crude oil. *Microbial Ecology.* 37:263-275.

Bennett, P. C., F. K. Hiebert & J. R. Rogers 2000. Microbial control of mineral-groundwater equilibria - macroscale to microscale. *Hydrogeology. J.* 8:47-62.

Bennett, P. C., D. I. Siegel, M. J. Baedecker, I. Cozzarelli & M. Hult 1993. The fate of crude oil in a sand and gravel aquifer I. Inorganic geochemistry. *App. Geochemistry.* 8:529-549.

Chapelle, F. H. 1993. *Ground-Water Microbiology and Geochemistry*. John Wiley & Sons, N.Y.

Cozzarelli, I. M., R. P. Eganhouse & M. J. Baedecker 1990. Transformation of monoaromatic hydrocarbons to organic acids in anoxic groundwater environment. *Environmental Geology and Water Science.* 16:135-141.

Ghiorse, W. C. & J. L. Wilson 1988. Microbial ecology of the terrestrial subsurface. *Advances in Applied Microbiology.* 33:107-172.

Goldich, S. S. 1938. A study in rock weathering. *Journal of Geology.* 46:11-58.

Hiebert, F. K. & P. C. Bennett 1992. Microbial control of silicate weathering in organic-rich ground water. *Science.* 258:278-281.

Lovley, D. R., M. J. Baedecker, D. J. Lonergan, I. M. Cozzarelli, E. J. P. Phillips & D. I. Siegel 1989. Oxidation of aromatic contaminants coupled to microbial iron reduction. *Nature.* 339:297-300.

Lovley, D. R. & D. J. Lonergan 1990. Anaerobic oxidation of toluene, phenol, and *p*-cresol by the dissimilatory iron reducing organism GS-15. *App. Environmental Microbiology.* 56:1858-1864.

Murphy, F. & W. N. Herkelrath 1996. A sample-freezing drive shoe for a wire-line piston core sampler. *Ground Water Monitoring and Remediation.* 16:86-90.

Rogers, J. R., P. C. Bennett & W. J. Choi 1998. Feldspars as a source of nutrients for microorganisms. *American Mineralogist.* 83:1532-1540.

Vandevivere, P. & P. Bevaye 1993. Improved preservation of bacterial exopolymers for scanning electron microscopy. *J. Microscopy.* 167:323-330.

Groundwater: Past Achievements and Future Challenges, Sililo et al. (eds) © 2000 Balkema, Rotterdam, ISBN 90 5809 159 7

Towards comprehensive groundwater resource management in South Africa

E. Braune
Department of Water Affairs and Forestry, Pretoria, South Africa

ABSTRACT: Through the National Water Act of 1998, groundwater management in South Africa will be moving from its present low management interest to integrated water resource management, with national government as custodian. The Act allows for a phased implementation of groundwater use authorization through measures such as scheduled permissible use and general authorizations. Groundwater protection will operate in an integrated water resource management framework through prescribed tools like a resource classification system and a reserve, protecting basic human needs and aquatic ecosystems. The transition from a highly centralized situation towards groundwater playing its appropriate role in country-wide catchment management agencies and water user associations will have to be bridged with innovative strategies of sharing capacity, resources and information between all the relevant role-players in government and the private sector.

1 INTRODUCTION

As part of democratization, South Africa is also undergoing major changes to its water services and water resource management policies and practices. These are particularly significant for groundwater, because of the growing importance of groundwater and it's past neglect in South Africa. Significantly, the United Nations organization decided to devote World Water Day in 1999 to groundwater under the theme "Groundwater – the invisible resource". Groundwater is a major source, not only of safe water, but a key factor in local development and poverty alleviation. In fact, nearly half of the world's population depends on groundwater sources for drinking water supply and for other uses. The reasons are obvious – groundwater occurrence is widespread, it is generally safe to drink if protected against pollution and is affordable and close to the community who can manage it.

Management strategies are required which address the unique characteristics and roles of groundwater and at the same time take the socio-economic development situation in a particular country into account. The purpose of this paper is to highlight elements of an emerging groundwater management strategy within the new water policy and legislative framework in South Africa.

2 GROUNDWATER OCCURRENCE AND ROLE

South Africa has a largely semi-arid to arid climate and has a large and growing population, presently standing at 40 million. Using the United Nations definition of water scarcity as the available water resources per capita of a country, South Africa finds itself among the twenty most water-stressed countries in the world (United Nations 1995).

Historically groundwater had only contributed about 15% of total bulk water supply. Because of the very old geology in South Africa, groundwater occurrence in 90% of the country is in hard rock with only secondary openings. Groundwater is contained mainly in fractures, and to some extent also in pores in weathered rock. Depth of weathering and depth of open fractures is very variable, but is in general shallow. Groundwater recharge is largely a function of climate and varies between 200 mm per annum in the wetter parts of the country to less than 5 mm per annum in the driest parts. Only in the eastern and southern higher rainfall parts, does groundwater contribute to the flow in rivers and streams.

For many years groundwater had been regarded as a cheap source of water, requiring little or no planning or management. A lack of understanding of the occurrence, movement and recharge of groundwater, and its "private water" legal status, have however, in many instances led to unsustainable utilization of this resource. The consequent failure of boreholes in many instances has unfortunately promoted the view that groundwater is an unreliable source of supply which should be replaced as soon as possible by more reliable surface water supplies.

While emphasis of past governments had been on bulk water supply to the commercial agriculture and urban/industrial sectors, the democratization process following the 1994 election has put the spotlight on the poorest of the poor of society and the meeting of basic needs, including water supply and sanitation needs. These were addressed in the government policy paper of 1994 in which a minimum standard of 25l/person/day within a walking distance of 200m or less was set (Department of Water Affairs & Forestry 1994). Surveys indicated that over 12 million people, distributed over some 15 000 villages, did not have this most basic water supply.

Because of the highly distributed nature of water demand in these rural areas, regional schemes are in most instances not economically viable. Furthermore, at the local level river and spring flows are decreasing during the dry season and during drought periods and widespread problems of surface water pollution occur in both urban and rural areas. This provides groundwater, despite its low yield nature, with a completely new role as the most feasible option to meet the massive backlog in domestic water needs everywhere in the country. For the Eastern Cape Province, for example, planners now agree that groundwater has the potential to serve over 80% of the 5 700 communities there – a clear change in outlook from only a few years ago when groundwater's planned use was relegated to the dry western parts.

3 NEW POLICY FRAMEWORK AND STRATEGY CONSIDERATIONS

In line with the political priorities, the massive water services backlog was the first to be addressed with the above-mentioned policy paper and in 1997 with the Water Services Act. The purpose of the Act is to provide a regulatory framework for water services institutions and for the setting of national standards and of norms and standards for tariffs. The challenge of the coming years will be to devolve the water services function to local government, which is only just emerging in large parts of the country.

This was followed in 1998, after 3 years of policy making and very wide consultations, by the National Water Act.The purpose of this Act is to ensure that water resources are protected, developed, conserved, managed and controlled in ways which take into account other factors -

- meeting the basic human needs of present and future generations;
- promoting equitable access of water;
- redressing the results of past racial and gender discriminations;
- promoting the efficient, sustainable and beneficial use of groundwater in the public interest;
- facilitating social and economic development;
- providing for growing demand for water use;
- protecting aquatic and associated ecosystems and their biological diversity;
- reducing and preventing pollution and degradation of water resources;
- meeting international obligations

and for achieving this purpose, to establish suitable institutions and to ensure that they have appropriate community, racial and gender representation. As the public trustee of the nation's water resources national government must ensure that water is managed in a sustainable and equitable manner, for the benefit of all persons (National Water Act 1998).

Initially a groundwater chapter had been envisaged for the new water act. However, as the writing of the Act progressed, it became clear that the separation of resources and legislation into boxes would have to fall away and that the ideal to be striven for was integrated water resources management. Three of the principles on which the Act was based are particularly relevant in moving groundwater from its present neglected management and "private water" legal status towards integrated water resource management in South Africa (Department of Water Affairs & Forestry 1997).

Principle 2: All water, wherever it occurs in the water cycle, is a resource common to all, the use of which shall be subject to national control. All water should have a consistent status in law, irrespective of where it occurs.

Principle 3: There shall be no ownership of water, but only a right to its use (for environmental and basic human needs), or an authorisation for its use. Any authorisation to use water in terms of the water law shall not be in perpetuity.

Principle 5: In a relatively arid country such as South Africa, it is necessary to recognise the unity of the water cycle and the interdependence of its elements, where evaporation, clouds and rainfall are linked to groundwater, rivers, lakes, wetlands and the sea, and where the basic hydrological unit is the catchment.

Equity and sustainability are the highest level objectives of the Act and integrated water resource management the ideal to be striven for. At the same time the Act is very flexible and allows for a phased implementation. This is important for groundwater as there is usually an evolution of sustainability concerns in groundwater development (Gupta & Onta 1997), *i.e.*:

- providing a sustainable water use;
- protecting resource use from degradation, *i.e.* sustaining quality; and
- controlling environmental impacts of groundwater development, *i.e.* sustaining environmental diversity.

With progressive development it becomes more and more important to understand the functioning of the groundwater system and its interaction with both the aquatic and terrestrial environment. However, not only the scientific factors, but also institutional and social acceptance are crucial in addressing sustainability. These together are captured in three important operational management strategies as highlighted by Gupta and Onta 1997):

- understanding of resources availability should be enhanced;
- resource managers and decision makers should recognize groundwater as a crucial component of water resources and the environment; and
- knowledge and understanding of the groundwater system and the environment should be transferred to the groundwater users so that consensus-driven, sustainable management plans will be achieved.

While these will also be crucial strategies for sustainable groundwater development in South Africa, there remains the highest priority need here to make groundwater, the previously neglected resource, a cost-effective, generally accessible and reliable source of water for millions who have never even had a basic water supply.

4 IMPLEMENTATION OF ACT

4.1 *Groundwater use authorization*

The Act lays the basis for regulating water use which is defined very broadly and includes *inter alia*, taking and storing water, activities which reduce stream flow, waste discharge and disposal and activities which impact detrimentally on water resources and which can be declared controlled activities.

In general, a water use must be licensed unless it is listed in the specific schedule in the Act, is an existing lawful use or is permissible under general authorization. The authorization hierarchy obtained through these different instruments is illustrated in figure 1.

General Authorizations are intended to alleviate the burden of issuing licenses for common, but low impact water uses and form part of the migration to full implementation of the National Water Act, particularly for those water use categories that have never been controlled or regulated before.

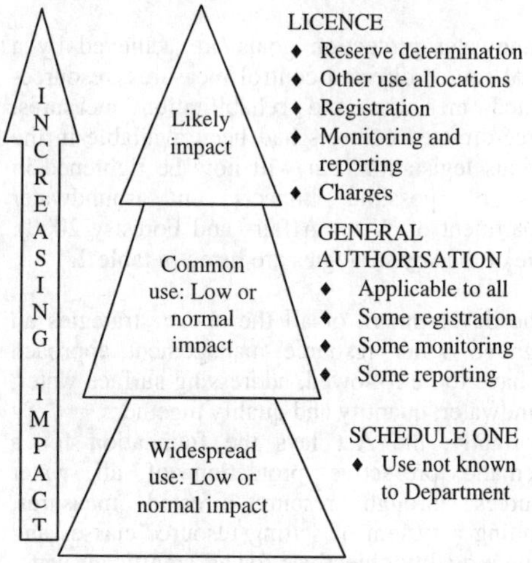

Figure 1. Hierarchy of water use authorization in the National Water Act

The first General Authorization for groundwater accepted 50% of "Groundwater Harvest Potential" (Seymore & Seward 1996)

for different regions as basis and was consolidated into 4 zones (table 1).

Table 1. General Authorization for groundwater in terms of the National Water Act

Zone	General Authorization (m^3/ha/annum)
1 (arid regions)	only Schedule 1 use
2	≤ 60
3	≤ 300
4	≤ 750

Areas that were already groundwater-stressed, were excluded from the general authorization. At the moment these are all the groundwater control areas that had been declared in terms of the previous water act.

A water use registration drive, to be completed within 2 years, is to provide the first overall assessment of water use. For groundwater the registration requirement has been set for uses of $10m^3$/day and more. This will form the basis for water resources planning, prioritization of licensing, adjusting of general authorizations, where necessary, and for the implementation of a water use charge system.

4.2 *Groundwater resource protection*

Groundwater protection goals are achieved by a mix of source-directed control measures, resource-directed measures and rehabilitation measures. Source-directed controls had been available in the previous legislation, but will now be tightened in terms of possible impacts on groundwater (Department of Water Affairs and Forestry 2000). Some of the key strategies are listed in table 2.

In the development of all the above strategies an integrated water resource management approach will have to be followed, addressing surface water, groundwater, quantity and quality together.
Importantly, the Act lays the foundation for a systematic pro-active protection of all water resources through resource-directed measures, including a system of setting resource classes and resource quality objectives for all significant water resources. Furthermore a Reserve must be determined for these water resources as soon as reasonably practicable after the class has been set. No water use can be authorized from a specific water resource before the Reserve has been set.

Table 2. Source-directed strategies for groundwater quality management

Measures	Implementation Actions
Groundwater abstraction, dewatering and artificial recharge	National Water Act: • Improve license conditions; • Recharge: develop guidelines for license evaluation.
Disturbance of aquifers by mining and related activities	Minerals Act: • Environmental Management Plans (improve groundwater component). National Water Act: • Develop a general authorization and new regulations.
Waste storage and disposal	Environmental Conservation Act: • Improve minimum requirements for groundwater; National Water Act: • Develop guideline documents and licence conditions; • Develop irrigation of waste as controlled activity.
Diffuse sources (urban and rural	• Assessment of different sources and development of procedures and regulatory instruments for diffuse source controls (Best Management Practices for different sectors).
Influencing land-use allocation and planning	• Water Services Act: o Development of appropriate regulations o Specific requirements for Water Services Development Plans. • Integrated Development Objectives (Local Government; National Consultative Forum for Dense Settlements).

The Reserve consists of two parts – a basic human needs reserve which provides for the essential needs of individuals served by the water resource in question, and an ecological reserve which relates to the water required to protect aquatic ecosystems of the water resource, and will vary depending on the management class that has been set for the resource.

Present thinking on the use of the different instruments provided in the Act is summarized in table 3.

Table 3. Resource-directed measures for groundwater resource protection

Measures	Objectives and strategies
Reserve	
• Basic Human Needs Reserve	Ensuring sustainability of community water supplies where groundwater is an important or sole source. (prevent over-pumping, pollution, intrusion of poor quality water, etc.).
• Ecological Reserve	Protecting groundwater-dependent aquatic ecosystems (springs, baseflow in rivers, wetlands, etc.)
Classification	• Protecting groundwater-dependent terrestrial ecosystems (vegetation, geological features, etc.); • Protecting aquifer integrity (aquifer matrix, e.g. sinkholes, subsidence, aquifer clogging, etc.); • Providing of classification framework for systematic setting of the Reserve.
Resource Quality Objectives	To specify the desired local groundwater water management regime *for all the above protection instruments* (in terms of groundwater levels, water level fluctuations, minimum groundwater discharge constraints, water quality objectives, etc.)

Resource Quality Objectives can also be used for the regulation or prohibition of land-based activities which may affect groundwater quality. Protection zoning, used extensively throughout the world (Keary 1993; National Rivers Authority 1992; U.S. EPA 1990) to protect important drinking water supplies from groundwater, can be introduced through this measure.

While the surface water protection focus will be driven by the Reserve, it is essential to develop the classification instrument to cater for pro-active protection of groundwater resources outside the present definition of "aquatic ecosystems". For this purpose the development of an "Eco-hydrological Region" classification system has been suggested which incorporates all water resource components and their water-dependent ecosystems (Hatton 2000) (see figure 2).

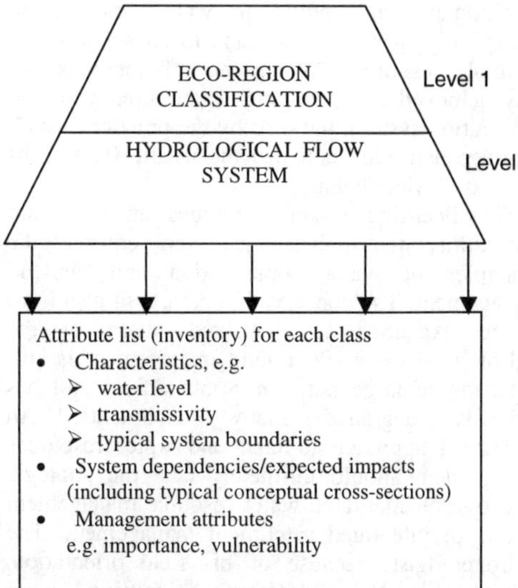

Figure 2. A proposed Ecohydrological Region classification system in support of groundwater resource protection (Hatton 2000).

Knowledge of the groundwater component and its interdependencies in such a regional system will be very limited in South Africa. However, the model of Tóth (1999) of a hierarchy of local, intermediate and regional flow systems as determined by composite topography, and each producing its own hydrogeologic conditions and associated natural phenomena, should, in general terms, also hold for the hard rock systems here. Of growing scientific and management importance are the ecotones or interfaces between aquatic and terrestrial ecosystems and between surface water and groundwater systems. Here the ecological processes are more intense and resources more diversified, but also more vulnerable to human impacts (Gibert et al. 1997). Typical examples in the South African context are the important groundwater-fed fresh water refuges in both our coastal lakes and in coastal zone sea waters.

4.3 *Groundwater conservation*

The National Water Act has shifted the water resources management attention in South Africa from the traditional primary focus of development of additional water sources to managing of water demand and to conservation of available resources. Water use efficiency is not only achieved by managing water already in the distribution system, but also by the practices which optimize and even enhance the usable fraction of the natural water balance.

The licensing system provides an important opportunity to spread and even enforce the principles of water conservation and demand management. The conservation ethic will also have to be expanded to the land phase of the hydrological cycle. Poor land management is still occurring in large parts of South Africa and has seriously degraded many catchments. An ecological approach to land and water resources management should ultimately take the strategic objective of integrated water resource management to one of integrated catchment management. The hydrogeologist, because of his field orientation, will be at the forefront of putting water conservation and catchment management into practice at the local level.

4.4 *Institution-building*

A major purpose of the Act is to achieve the establishment of suitable institutions for appropriate and participative management of water resources. This is particularly important for highly localized groundwater resources which cannot be physically managed centrally. The Act provides for three levels of management, namely national government, nineteen catchment management agencies as the new regional management level and water user associations acting cooperatively at the local level.

All three levels hold tremendous opportunities for improved groundwater management. Water user associations can practice conjunctive use of water resources and artificial recharge of aquifers, put catchment conservation into practice on their land and reverse over-exploitation and degradation of water resources. They can also become an opportunity for community development through the sharing of skills and resources.

The catchment management level with devolved management responsibilities is crucial for addressing groundwater in a planned and integrated way in the public interest. It will also have to provide the essential support for local level management, which had been completely lacking in South Africa for groundwater management.

The national level has for the first time got clear duties defined in the Act, which can be challenged in public, e.g. the national water resource strategy and the need to monitor and provide national information systems. While the authorization of water use will in future be devolved to catchment management agencies, the protection of water resources through the resource-directed measures will remain the responsibility of national government.

The biggest challenge for water management in South Africa will be the bridging period of the next ten to twenty years in which transformation from a highly centralized situation to new functional and highly participative institutions at different levels will have to take place.

5 THE PEOPLE-DIMENSION OF GROUNDWATER DEVELOPMENT

Sustainable water services depend on both the resource sustainability addressed in the National Water Act and the supply system sustainability addressed in the Water Services Act.

Audits of community water supply implementation in South Africa indicate most of the same problems that have influenced water development in other developing countries, ranging from non-payment, abandonment of schemes or elements thereof, pollution of water sources and lack of integrated development. These failures are costing the respective countries millions and are frustrating communities, who are to be served. What is still lacking, above all, is an active community participation in the full process which will lead to the design, installation, ownership and subsequent maintenance of water supply schemes. Throughout the world this people-dimension of water supply is now emphasized as the highest priority. The approaches and strategies adopted by developing countries at the 1990 Abidjan Conference also hold for South Africa and for water supply from groundwater (Palmer & Eberhard 1994):

- Investments should be based on effective demand and long-term sustainability;
- Priority should be given to the rehabilitation and maintenance of existing assets;
- Community participation should be an integral part of all project development and implementation;

- Technologies should be applied appropriately, taking into account, among others, social-cultural preferences, affordability and long-term sustainability;
- The strategies should aim at maximum coverage with a basic level of service before giving attention to upgrading existing adequate services;
- Water and sanitation investments should be integrated with primary health care strategies and solid waste and sewage disposal and stormwater drainage.

Particular challenges for the groundwater sector are a more effective planning and cost-effective development of hundreds of small, localized schemes from groundwater, the introduction of a wider range of low-cost, appropriate technologies for community supplies, the integration of different development needs and the massive awareness- and capacity-building required at both the community level and its support structures. Developing people clearly needs a much more long-term and planned commitment than developing water.

6 INFORMATION AND INFORMATION SYSTEMS

Information will be the cornerstone on which implementation of the new water policy and legislation will rest and on which new institutions can function effectively. New instruments in the Act, like catchment management plans, the reserve and controlled activities will be very information-intensive. The need for integrated water resource management will create the need for much greater integration of the relevant information systems, e.g. hydrology, geohydrology, water quality and land type, use and cover. The many new participants in a devolved management system will need to share in these information systems which are presently still very centralized. Widespread public participation will require information which is transparent and user-friendly and systems which can be used inter-actively to facilitate multi-user decision-making. The basis for these increased information needs will have to be a more intensive and highly systematic monitoring of water resources status and trends as well as of compliance with policy and regulations at all levels from national to local. A summary of proposed strategies and actions in this regard is shown below:

- A National Groundwater Information System (NGIS) being adapted from the Dutch REGIS system to be replicated in the 9 provinces and later in the catchment management agencies;
- Local data capture and decision-support systems, MuniWater and AgriWater, being developed for the two main groundwater user groups:
- Population of the NGIS through the regulation for the provision of borehole information and through achieving the participation of the groundwater industry in data and information collection and sharing;
- Establishment of a national groundwater monitoring, assessment and reporting system, together with regulations and guidelines for regional (catchment and aquifer) monitoring as well as for monitoring the local impacts of groundwater use.

Besides the essential need for use of high technologies like remote sensing, GIS and data transmission via satellite, the provision and sharing of information also requires the long-term commitment and investment in people development.

7 ENABLING STRATEGIES

The water industry in South Africa, and in particular the groundwater industry, is going through a period of major transformation. This will require enabling strategies that innovate, coordinate, facilitate and support implementation of a shared vision.

To be able to take a comprehensive strategy forward, national government will need to reassess its role, structure and budgetary resources as custodian for groundwater. All three functions which have suddenly gained importance, *i.e.* groundwater resource assessment, development support and resource management, need appropriate attention.

A major ongoing awareness campaign at all levels regarding the need for improved groundwater management will be necessary to achieve the desired understanding, participation and support for long-term transformation.

A much greater role for the whole groundwater industry, its skills, capacity, information and drive, will be necessary for the common objective of taking the country through the transitional phase.

Groundwater resource management should not operate in isolation, but should be part of the drive towards integrated water resource management. In this way synergies can be achieved and scarce resources shared. The national and catchment management strategies should become major vehicles for integration.

The new water resource management hierarchy and new and more integrated approaches to water resource management will require a strategic approach to research and development, technology transfer and capacity building. The networking of education and training providers and users, as well as researchers, as recommended by the UNESCO/WMO mission to South Africa in 1998 in their FET-WATER proposal (DWAF/UNESCO/WMO 1998), will need urgent and widely coordinated implementation.

Throughout this challenging period of implementation, linkages with the international groundwater and development community should be sought and strengthened, to exchange knowledge and skills, test ideas and approaches and achieve support in areas of critical need.

Wherever possible, regional rather than just national focus should be maintained in the implementation of the groundwater strategy and ties with the SADC water sector should be continuously strengthened.

To achieve the essential stakeholder buy-in, the groundwater resource strategy should have clear and agreed objectives and realistic targets on which progress should be reported regularly.

8 CONCLUSIONS

Bringing groundwater under an integrated water resource management umbrella will make the groundwater management task much more management- and information-intensive. Much greater understanding of resource availability and of the hydrological systems context will be required. A new emphasis will be necessary on knowledge transfer regarding the groundwater systems to the whole spectrum of groundwater users so that consensus-driven sustainable management plans will be achieved.

Besides the greater water resource management focus, a developing country like South Africa will need continued high priority attention to meeting the large remaining basic water supply need. This focus on developing people rather than water, in both the water supply and in the water resource management functions, will require a much more long term and planned commitment.

ACKNOWLEDGEMENTS

A groundwater resource management strategy is something that grows over time and in many different places. The direct and indirect contributions of many colleagues towards this growth and towards the contents of this paper are herewith gratefully acknowledged. The Department of Water Affairs and Forestry is thanked for allowing publication of this ongoing policy formulation.

REFERENCES

Department of Water Affairs and Forestry. 1994. Water Supply and Sanitation Policy. White Paper, Cape Town.

Department of Water Affairs and Forestry. 1997. White Paper on a National Water Policy for South Africa. DWAF, Pretoria.

Department of Water Affairs and Forestry. 2000. Groundwater Quality Management Policies and Strategies. DWAF, Pretoria.

DWAF/UNESCO/WMO. 1998. Mission on the Assessment of the Education and Training Needs of the Water Resources Management Services of the Republic of South Africa. Department of Water Affairs and Forestry, Pretoria.

Gibert, J., Fournier F. & Mathieu, J. (eds.) 1997. The groundwater/surface water ecotone perspective : state of the art. In *Groundwater/Surface Water Ecotones: Biological and Hydrological Interactions and Management Options. UNESCO International Hydrology Series.* Cambridge University Press.

Gupta, A.S. & Onta, P.R. 1997. Sustainable groundwater resources development. *Hydrological Sciences Journal 42, 565-582.*

Hatton, T. 2000. The Comprehensive Determination of Water Resource Reserve Groundwater Component. Review report to the Water Research Commission, Pretoria.

Keary, J.B. 1993. National groundwater protection guidelines for Australia. *AGSO J. Astraul. Geol. & Geophysics, 14(2/3), 303-306.*

National Rivers Authority. 1992. *Policy and Practice for the Protection of Groundwater.* NRA document. United Kingdom.

National Water Act. 1998. Act No. 36 of 1998. *Government Gazette Vol. 398*, Cape Town, Republic of South Africa.

Palmer, I. and Eberhard, R. 1994. Evaluation of Water Supply to Developing Urban Communities in South Africa. *WRC Report No. KV49/94.* Pretoria.

Seymore, A.S., & Seward, P. 1996. Groundwater Harvest Potential. Internal report. Department of Water Affairs and Forestry, Pretoria.

Tóth, J. 1999. Groundwater as a geologic agent: an overview of the causes, processes, and manifestations. *Hydrogeology Journal 7: 1-14.*

United Nations. 1995. Population and water availability. *Water. October, 1995.*

U.S.E.P.A. 1990. *Progress in Ground-water Protection and Restoration.* Office of Ground-Water Protection, U.S, Environmental Protection Agency.

Water Services Act. 1997. Act No. 108 of 1997. *Government Gazette Vol. 390*, Cape Town, Republic of South Africa.

Groundwater: Past Achievements and Future Challenges, Sililo et al. (eds) © 2000 Balkema, Rotterdam, ISBN 90 5809 159 7

Assessing pesticide pollution of groundwater: Current knowledge and remaining gaps

P.J.Chilton, M.E.Stuart, A.R.Lawrence & D.C.Gooddy
British Geological Survey, Wallingford, UK

R.J.Williams & A.C.Johnson
Centre for Ecology and Hydrology, Wallingford, UK

ABSTRACT: The paper summarises recent research on pesticides in groundwater in both temperate and tropical regions. Results of field, laboratory and modelling studies highlight the factors which determine the fate and behaviour of pesticides in groundwater systems. These include transport pathways from the soil to the water table and thence to supply sources, and the processes such as adsorption and degradation which can help to attenuate pesticide movement and reduce concentrations. Studies of degradation show that most compounds are likely to be much more persistent in aquifers than in soils, but below the water table the long travel times and potential for dilution may greatly reduce concentrations. The greatest risks are likely to occur in fractured aquifers, with their potential for much more rapid flow.

Important uncertainties and gaps in knowledge remain. Laboratory studies of degradation present difficulties of extrapolation to field conditions, and provide evidence of wide variations in half-lives spatially and with time; making the choice of values for transport modelling problematic. Further work is required to improve understanding of such variations. Studies also indicate that different degradation pathways can occur, and different pesticide metabolites can be produced, depending on environmental conditions. The occurrence and behaviour of metabolites in groundwater systems is also poorly known.

1 INTRODUCTION

Over the past 20 to 25 years, nitrate has remained the most important aspect of groundwater pollution from agricultural practices, and considerable research attention has been directed to this issue. Advances in understanding of the processes responsible for increased nitrate concentrations in groundwater led naturally to consideration of the risk of pollution by pesticides. While the potential for pesticide leaching to groundwater has been recognised for some years (Lawrence & Foster 1987; Foster et al. 1991), the equivalent research to characterise pesticide fate and behaviour in the subsurface has been only recently embarked upon. Such research was initially hampered by the technical sophistication and high cost of analysis to reach the detection limits required in relation to drinking water standards for the very wide range of compounds in common agricultural usage (Lawrence & Foster 1987).

Although analysis was indeed difficult and expensive, an increasing number of pesticide compounds are being detected in groundwater in the UK (DETR 1998), in Europe (Spliid & Køppen 1998) and in North America (Barbash & Resek 1996), as monitoring programmes have been developed, often in response to tightening drinking water standards. In Europe, the EC Drinking Water Directive sets a very stringent maximum admissible concentration of 0.1 μg/l for any pesticide, and compliance has produced significant costs for the water industries in member states. In the US, Federal and State EPA pesticide standards are toxicologically based as are those in the WHO Guidelines for Drinking Water Quality.

Pesticides are developed in the temperate countries and data relating to their use on unfamiliar crops and soils and to different climatic conditions are needed (Plimmer et al. 1998). In tropical regions, the growing demand for food produces increasing intensification of agriculture, often with irrigation and heavy applications of agrochemicals. In many areas, this highly intensive production of up to three crops per year occurs on soils directly overlying shallow aquifers which are used for potable water supply. Crop losses due to weeds, insects and pathogens in developing countries are still much higher than in the developed world (Figure 1). Great benefits to crop production can therefore be achieved from pesticide use, but these need to be balanced by knowledge of the potential impacts on health and the environment.

This paper provides a brief overview of published

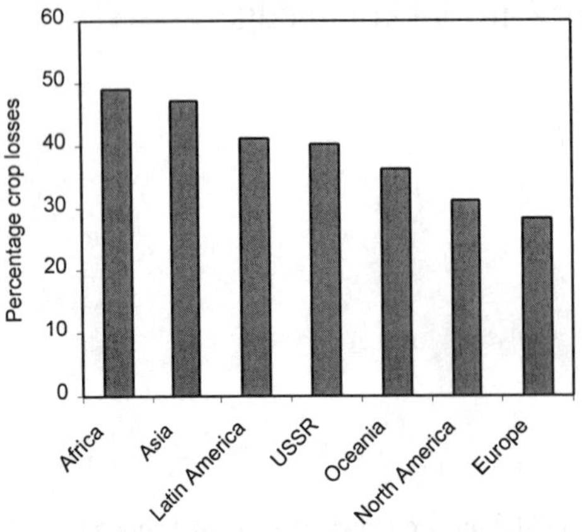

Figure 1. Losses to principal food and cash crops from weeds, insects and pathogens, 1988-1990 (from Plimmer et al. 1998)

findings concerning pesticide occurrence in groundwater, and the main factors which determine pesticide migration to and within aquifers. This is supported by the results from recent BGS projects which have investigated the fate and transport of pesticides both in UK aquifers and under tropical conditions.

2 PESTICIDE OCCURRENCE

In the UK, the pesticides most frequently encountered in groundwater reflect the widespread applications of pre- and post-emergent herbicides to autumn-sown cereal crops: isoproturon, mecoprop, MCPA and 2,4-D (DETR 1998). Atrazine and simazine are also found, although the triazines are not extensively used in UK agriculture. The occurrence of the former resulted mainly from its use as a total weedkiller on railway tracks, highways, airfields and other paved areas, a usage approval for which it was withdrawn in 1992, and the latter from similar usage in orchards and forestry.

The triazines together with various soil insecticides, especially the carbamates and chloropropanes, have been found in concentrations exceeding 1 µg/l in shallow aquifers in Europe and the USA (Jones et al. 1988; Rao & Alley 1993; Spliid & Køppen 1998; Carabias Martínez et al. 2000). Kolpin et al. (1997) found the herbicides prometon, atrazine and simazine to be the most frequently detected compounds in shallow urban groundwater. Most pesticide concentrations observed in these studies were in the range 0.1–100 µg/l, and it seems likely that concentrations significantly above this range can be attributed to local point source contamination close to the well or borehole.

Much less monitoring has been undertaken in tropical regions, largely as a consequence of the difficulty and high cost of analysis referred to above. However, there are now some published data appearing, for example Li & Zhang (1999) describe diffuse groundwater pollution from pesticides used in rice and cotton cultivation in China. In developing countries, some compounds, especially the organochlorines, are still used despite having been banned due to the adverse environmental effects resulting from their high persistence and extreme toxicity (Tonmanee & Kanchahakool 1999; Agrawal 1999; Matin et al. 1998).

In the United States, attention has begun to focus not only on the presence of pesticide parent compounds in groundwater but also their metabolites (Kolpin et al. 1996; 1997; Barrett 1996). The most commonly detected metabolites are derived from aldicarb, atrazine, carbofuran, DCPA and endosulfan and may be present in concentrations in excess of the parent compound.

3 PESTICIDE FATE AND BEHAVIOUR

The natural processes which govern the fate and transport of pesticides can be grouped into the following broad categories: leaching, volatilisation, degradation, sorption and, less significantly, plant uptake. The mode of application and action of the pesticide are important factors in relation to soil leaching since those targeted at plant roots and soil insects are more likely to be leached than those applied to the leaves.

The concentrations and timing of pesticide residues arriving at the water table are dependent on the mass applied, antecedent applications, the mobility and persistence of the compound and the hydrogeological conditions. Both field and laboratory studies are needed to determine the main factors that control the fate and behaviour in the subsurface. These include the transport pathways from the soil surface to the water table and thence to supply sources, and the processes such as sorption and degradation which can help to limit pesticide mobility and attenuate concentrations. Once in the saturated zone, dilution and degradation will further limit concentrations arriving at a point of abstraction.

Preliminary assessment of the transport of pesticides from soils to groundwater can be made, based on the physicochemical properties of the pesticides and on knowledge of groundwater flow and aquifer characteristics. Following this approach, Foster et al (1991) estimated the likely retardation in pesticide transport with respect to a conservative, non-reactive solute, using the partition coefficients of each compound. Although involving a number of simplifying assumptions (Foster et al. 1991), the approach showed that transport through the matrix of many aquifer materials is likely to be slow. However, the possibility of more rapid flow in the fractured sandstone and limestone aquifers of the UK could allow pesticide residues to reach the water table much more quickly, without adequate time for the attenuation processes mentioned above to operate. Further, once below the soil, pesticides enter an environment with much less clay minerals and organic matter and greatly reduced indigenous microbial populations. Sorption and degradation are, therefore, likely to be much less effective in attenuating pesticide concentrations. Thus, from a general consideration of the processes involved, pesticides could be much more mobile and persistent in aquifers than in the soil (Lawrence & Foster 1987). The relationship between and the factors governing the rates of transport and degradation have, therefore, been the key issues of concern in the pesticide studies summarised in this paper.

4 TEMPERATE AGRICULTURE – THE UK

One of the first studies in the UK which included the monitoring of groundwater for pesticides was in the Granta catchment on the Chalk of eastern England (Clark et al. 1995). A relatively few pesticides at low concentrations were found. Shallow groundwater near the river contained concentrations similar to those in the river. Monitoring of the unsaturated zone using suction samplers showed no consistent infiltration of pesticides through the soil profile under normal agricultural practice.

In another study a general strategy of porewater extraction from undisturbed core samples, similar to that used in earlier nitrate studies, was modified for pesticide sampling, particularly to take account of the larger volumes required for analysis (Chilton et al. 1993). Positive concentrations of atrazine were detected in two boreholes in the unsaturated zone of the Chalk aquifer beneath a maize crop, and persisted to at least 5 m depth (Figure 2). Similar results for atrazine were obtained by Clark et al. (1995), but little evidence of isoproturon or mecoprop beneath winter cereal crops was seen in either study.

Regular sampling of groundwater from shallow observation boreholes in the uppermost part of the Chalk aquifer at four sites indicated consistent but very low concentrations of atrazine, and few detections of isoproturon (Table 1). Since this might reflect the influence of dilution of localised infiltration in the broader lateral groundwater flow system, novel samplers were developed to try to collect the most recently infiltrated groundwater. Thus, a hanging sampler designed to take discrete samples from very shallow water during periods of rising water table was able to detect low concentrations of isoproturon (Figure 3). This suggests that some pesticide was able to reach the water table by a rapid flow mechanism.

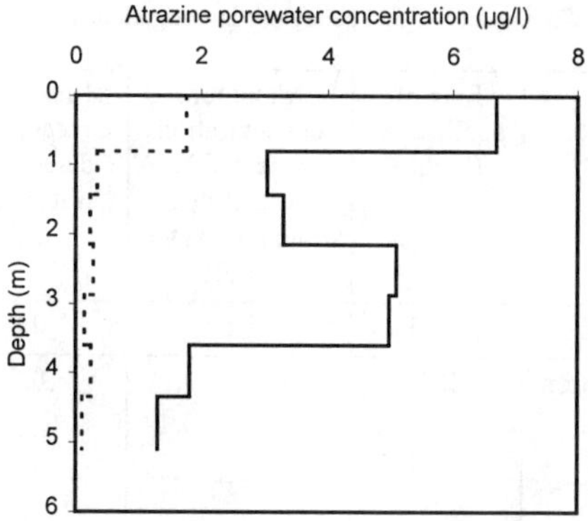

Figure 2. Profiles of porewater atrazine concentrations in the Chalk beneath a maize crop

This was confirmed by sampling at the water table during periods of winter recharge using an automatic borehole purging and sampling station installed in a shallow borehole. This was developed to allow programming to take samples frequently during periods of likely recharge. Significant concentrations of isoproturon were detected in the saturated zone following application, declining each year with the amount of residue available for mobilisation (Figure 4).

A study of the degradation of the herbicide mecoprop in the Chalk aquifer was carried out using laboratory microcosms (Chilton et al. 1993). This showed that there is potential for mecoprop degradation in the subsurface and that the majority of the pesticide was removed within 60 days.

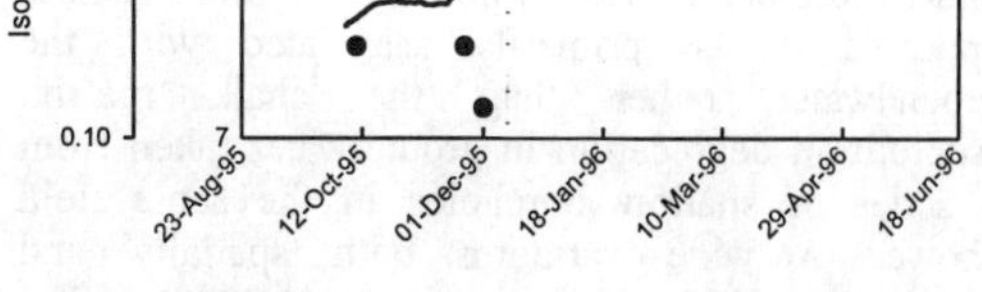

Figure 3. Detection of isoproturon (circles) at the water table after application during period of rapid recharge using a hanging sampler

Table 1 Summary of fate and behaviour investigations in the Chalk aquifer of the UK

Pesticide	Rate of application (kg/ha)	Maximum concentration detected in unsaturated zone porewater (μg/l)	Maximum concentration detected at water table (μg/l)	Half-life in laboratory microcosm (days)	Material used (depth, m)	Breakdown products detected
Atrazine	2.3 – 3.4	5.6	0.8	-	-	-
Isoproturon	2.4	0.1	0.5	< 7 - >15	Topsoil	
				7	Subsoil (0.5)	Mono-demethylated isoproturon?
				150-∞	Unsaturated zone (3-8)	
				7-∞	Saturated zone (4-20)	
Mecoprop	-		-	14-30	Unsaturated zone (3)	-
				30	Unsaturated zone (5)	
				∞	Saturated zone (19)	

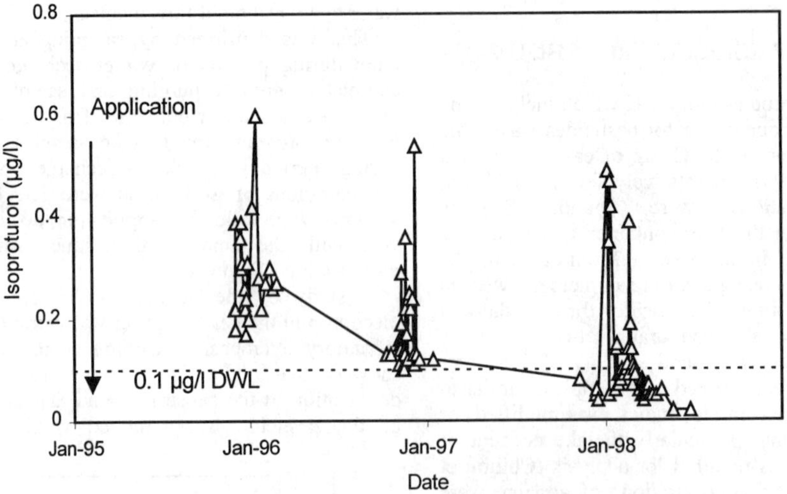

Figure 4. Isoproturon in the saturated zone detected by programmed borehole purging and sampling

Isoproturon appears to be more persistent, with some potential observed in very shallow chalk but very little in the unsaturated zone below 2 m (Johnson et al. 1998). The pesticide was not fully degraded under these conditions. In both studies the potential appeared to be primarily associated with the groundwater rather than the chalk matrix. Isoproturon degradation in groundwater taken from a series of shallow boreholes in the same field showed a wide variation both spatially and temporally (Table 2, Johnson et al. 2000). The results of our UK studies are summarised in Table 1.

Testing of an unsaturated zone leaching model against the observed porewater data and using the

Table 2. Isoproturon half-lives for groundwater samples taken from different boreholes in the same field but at different times (ND = not determined)

Borehole	May 1995	Nov 1995	Nov 1996	Jan 1997	Mar 1998
4	7	97	104	None after 149	None after 202
5	ND	362	ND	387	575
6	ND	ND	ND	None after 149	280
7	ND	ND	ND	174	228

measured isoproturon half-life data from the microcosms predicted that no isoproturon would be detected at the water table. Assuming that atrazine is twice as persistent as in the soil gave an atrazine profile falling between the profiles shown in Figure 2, with potential positive concentrations at the water table (Chilton et al. 1995a).

The results suggest that, in contrast to nitrate, it is unlikely that significant amounts of pesticide residues are accumulating and moving slowly downwards through the Chalk matrix to the water table. However, with the component of rapid fracture flow expected in the Chalk, transport of pesticides to the water table is possible. While the threat to groundwater from normal agricultural usage appears slight, further work is required to establish the potential for rapid transport.

5 TROPICAL REGIONS

Concern has been expressed about the impact on groundwater quality from intensification of agriculture in the developing world, often by means of irrigation, and the associated increases in pesticide use. Increases in pesticide usage are now more rapid than in the developed world but most of the published information on the persistence of pesticides refers to fertile, organic clay soils from temperate regions. There remains a great need for further information on pesticide persistence in groundwaters and for both soils and the deeper subsurface environment in tropical environments (Chilton et al. 1998).

Where intensive tropical agriculture is being carried out in regions overlying shallow aquifers, significant pollution of groundwater can be expected where soils are thin, well drained and permeable. The insecticide carbofuran is widely used on a range of crops including cereals, vegetables and bananas and there are many examples of leaching to groundwater (Walker & Porter 1990; Ritter 1990; Barceló et al. 1996; Diaz-Diaz et al. 1998). In Kalpitiya, Sri Lanka, monitoring of the leachate beneath a sandy experimental plot with a shallow water table indicated that 10% of the first application of carbofuran was leached to groundwater (BGS et al. 1992). However, leaching of the second application one month later was considerably reduced (1% of the applied rate) and this was attributed to enhanced microbial degradation in the soil. The following season the application was increased ten-fold and the percentage of the first application leached to the water table increased to 50%. Clearly the rate of leaching even at the same site can vary considerably and depends on the quantity applied and whether previous applications had been made recently.

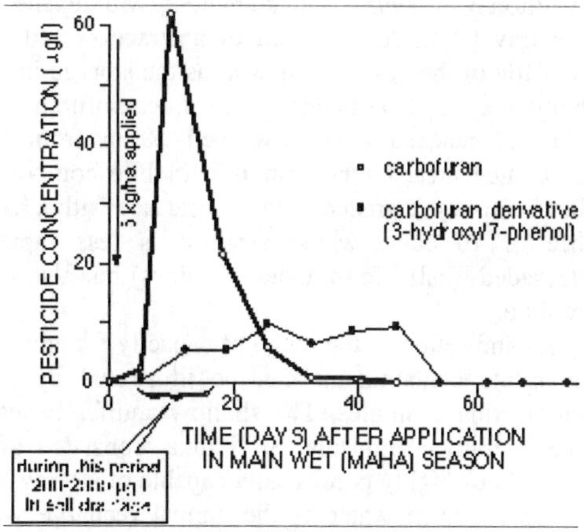

Figure 5. Leaching of carbofuran and a metabolite from irrigated cultivation in Sri Lanka

The method of application was also found to influence groundwater contamination, with furrow irrigation of bananas in the Canary Islands leading to greater leaching of carbofuran than drip methods (Diaz-Diaz et al. 1998.).

Pesticides have very different mobilities. In Sri Lanka, carbofuran arrived at the water table some 10–12 days after application, similar to a conservative tracer, whilst the herbicide alachlor was retarded by the aquifer matrix and appeared 75 days after application.

It is also important to establish the degradation pathway when assessing the risk to groundwater supplies, firstly because the metabolites produced could be of very different toxicities to the parent compound and to each other and secondly because the persistence of the different metabolites can vary.

In Sri Lanka (BGS et al. 1992) the degradation pathway for carbofuran in aerobic soil microcosms was initial oxidation to 3-hydroxycarbofuran followed by hydrolysis to 3-hydroxycarbofuran-7-phenol. This metabolite was also detected in shallow piezometers close to the cultivated plot (Figure 5). In contrast, in alluvial soils under paddy cultivation near Madras, India (CWR et al. 1993), where conditions are believed to be strongly anaerobic, the main metabolite produced was carbofuran phenol. Microcosm studies suggested that this metabolite may be more persistent in groundwater than the parent compound.

Once in the saturated zone, dilution and degradation are probably the most important processes attenuating pesticide concentrations

(Chilton et al. 1995). The attenuation will depend on the travel time to the point of abstraction and the half-life of the pesticide as well as the storage in the aquifer. In Sri Lanka, neither carbofuran nor alachlor reached a shallow well 20 m from the cultivated area. Carbofuran is a mobile compound but is rapidly degraded in the subsurface with a half-life of 10 days, whilst alachlor is less rapidly degraded (half life of about 200 days) but also less mobile.

An indication of the range of capacity for aquifers to dilute contaminants can be illustrated by two contrasting examples. The shallow aquifer beneath the Kalpitiya Peninsula, Sri Lanka, consisted of a 15 m thick highly porous sand capable of storing 5-8 times as much water as the annual recharge with considerable potential for contaminant dilution. The storage available in the fissured limestone aquifer of Barbados only represents once or twice the annual recharge with very limited potential for dilution.

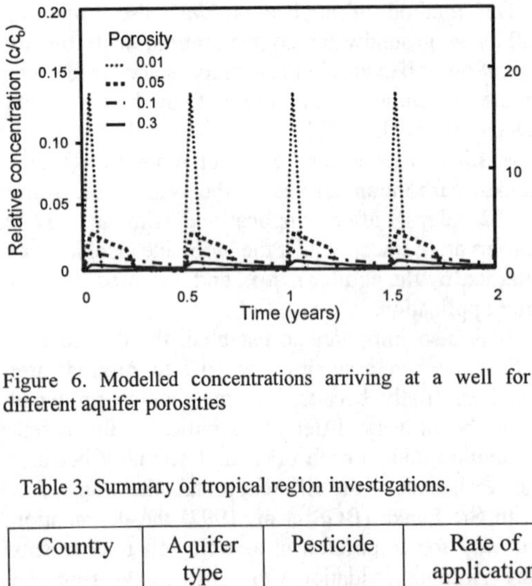

Figure 6. Modelled concentrations arriving at a well for different aquifer porosities

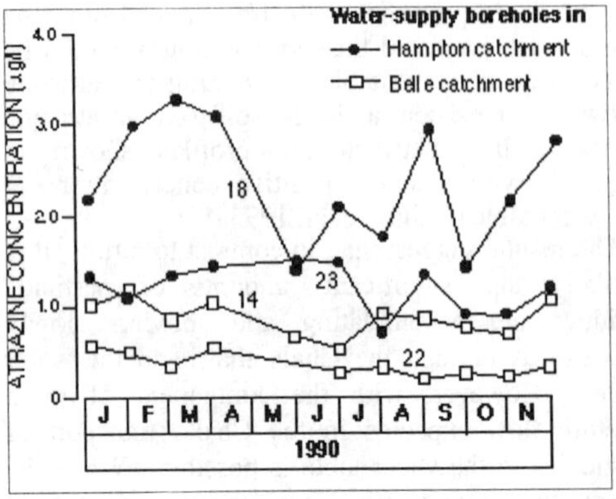

Figure 7. Groundwater atrazine concentrations in Barbados catchments under sugarcane cultivation

A simple model (Figure 6) can predict the impact of porosity on concentrations arriving at a well. The data in this figure assume a pesticide with a half-life of 100 days is applied twice each year to an annular area around and to within 5 m of the well (Barker & Lawrence 1992). The results shown in Figure 6 have important implications for monitoring, as infrequent, fixed-interval sampling of low-porosity aquifers such as fractured limestones might easily miss the sharp peaks in pesticide concentration and provide a falsely optimistic view of groundwater quality.

The persistence of relatively similar compounds can vary considerably and this may explain why atrazine is detected widely in the groundwaters of Barbados, but not ametryn. Both compounds are widely used and applied at about the same rate to sugarcane (4 kg/ha/a) and are considered from

Table 3. Summary of tropical region investigations.

Country	Aquifer type	Pesticide	Rate of application (kg/ha)	Breakdown products detected	Maximum concentration detected at water table (µg/l)	Range of concentration in abstraction (µg/l)
Sri Lanka	Sand	Carbofuran	12/crop	-	60	Not detected
				3-OH-7-phenol	10	Not detected
		Alachlor	4/crop	-	4	Not detected
India	Alluvium	Carbofuran	1.25/crop	-	-	Not detected
				7-phenol	Occasionally	Not detected
Barbados	Limestone	Atrazine	3.6/a	-	Not determined	0.5 – 3
				N-desethyl atrazine	Not determined	0.2 – 2
		Ametryn	4.3/a	-	Not determined	Occasionally

published data to be similarly mobile. However, whilst atrazine was detected at concentrations in the range 3-5 µg/l, throughout the year in all public supply boreholes tested (Figure 7), ametryn was rarely detected. It is presumed that ametryn is degraded to below detectable concentrations during migration through the unsaturated and saturated zones. A summary of our tropical studies is shown in Table 3.

6 NON-AGRICULTURAL USE

Weeds are a recurring problem in many situations, requiring control or eradication. The heaviest non-agricultural usage of herbicides is for total weed control of paved areas, highways, railway lines, car parks and airfields. This type of usage poses a particular threat to groundwater since, in many instances such areas are drained by soakaways. Pesticides may, therefore, enter the subsurface with much less scope for attenuation in the soil and unsaturated zone. The non-agricultural use of atrazine was banned in the UK in 1992 but it continued to be responsible for most pesticide exceedences in public supply from groundwater for some years.

The source of non-agricultural pesticides can be difficult to identify. Contamination of Chalk groundwater at a site used for market gardening was first brought to light by poor growth of glasshouse cucumbers. Concentrations of up to 100 µg/l of atrazine, 2,4-D and amitrole were detected in the farm borehole and others in the surrounding area. These were eventually attributed to weed control at a local airfield rather than spraying of an adjacent railway line since 2,4-D had not been used there (Chilton et al. 1993).

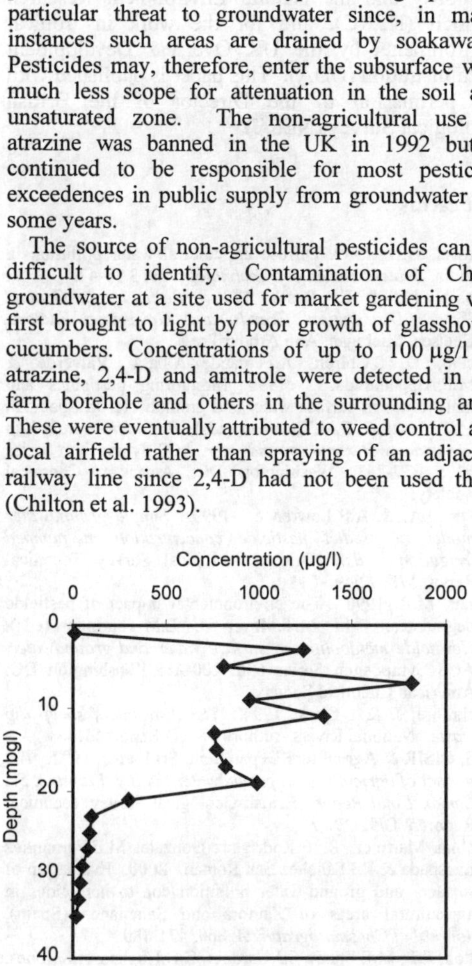

Figure 8. Concentration of methabenzthiuron in porewaters beneath a soakaway in the Chalk

The problem can also be persistent. Residues of atrazine and simazine were detected in two Chalk public supply boreholes abstracting from a dry valley containing a railway line more than two years after the application of triazines had ceased. Water utilities have been very interested to predict the length of time that troublesome concentrations might persist. Also, traces of diuron, one of the compounds which has replaced the triazines, were starting to appear.

Incorrect or careless handling or disposal can also lead to point source contamination. Elevated concentrations of a range of herbicides derived from a soakaway adjacent to a pesticide storage area into the Chalk of southern England were detected in porewaters beneath the site (Figure 8) giving rise to a groundwater plume extending some 3 km downgradient. The concentrations in the porewater profile were by far the highest seen during any of the BGS projects. The coincidence of high pesticide concentrations in a local public supply borehole and peak groundwater levels indicated very rapid saturated zone flow with limited attenuation by matrix diffusion.

Pesticides other than herbicides can also be responsible for point-source pollution. Diazinon, propetamphos and chlorfenvinphos, insecticides used in sheep dip at that time, were detected in groundwater at up to 5 µg/l downgradient of sheep dip installations (Blackmore & Clark 1994).

There is also evidence that pesticides are present in leachate from waste disposal sites. Liquid wastes from pesticide manufacturing were leached from a landfill site in the Jurassic limestone of eastern England, giving rise to contamination of the aquifer by a range of pesticides, including mecoprop (Sweeney et al. 1998). Concentrations of up to 30,000 µg/l were detected in the saturated zone, but little evidence was detected in the porewaters and it appeared that transport was mainly by the fissures of this heavily fractured limestone. The variability of redox conditions within the aquifer has been shown to play an important role in degradation of mecoprop (Harrison et al. 1998).

In all of these examples quantitative predictions of pesticide behaviour are precluded by a lack of knowledge of many aspects of the problem as well as the rates of degradation.

7 REMAINING UNKNOWNS

Important uncertainties in the understanding of pesticide behaviour in the subsurface still remain. There is a continuing need for further information on pesticide persistence in groundwaters in temperate environments and for both soils and the deeper subsurface environment in tropical environments. There are concerns that pesticide behaviour may be tested by the manufacturers under prescriptive 'standard' conditions which do not adequately represent either the soil types found on the outcrop of major UK aquifers or the conditions in tropical applications.

Laboratory studies of degradation can provide estimates of pesticide half-lives, but these can vary widely both spatially and with time, making the selection of parameter values for transport modelling problematic. Further work is required to improve our understanding of such variations.

Laboratory studies can also indicate that different degradation pathways may occur, and that different pesticide metabolites are produced depending on environmental conditions, but it is unclear as to how well such laboratory studies represent the actual conditions in the field. In particular, there is a need for determinations of degradation rates at realistic field temperatures and at the very low pesticide concentrations actually observed in field studies. Laboratory microcosms can never reproduce the distribution of active microorganisms and nutrients which exist in a fractured aquifer.

The public health implications of pesticide metabolites remain an outstanding issue. The occurrence and behaviour of metabolites in groundwater systems is also poorly known since they are not as yet normally included in legislation controlling pesticide concentrations in potable water. While the extensive screening study of Kolpin et al. (1996; 2000) showed that nine of the ten most frequently detected compounds in Iowa's groundwater were metabolites, there is as yet little routine monitoring for metabolites.

8 CONCLUSIONS

The research has shown that the risk to groundwater supplies from pesticides in typical UK diffuse agricultural applications appears to be limited, but that intensive tropical agriculture, non-agricultural applications and incorrect disposal of pesticides have the potential to produce serious deterioration in groundwater quality.

The greatest risk to groundwater supplies from pesticides will be where:

- aquifers are overlain by permeable soils and travel times to the water table are short
- the aquifer is of low porosity and consequently dilution is small
- the pesticide is relatively mobile and stable.

Pesticide behaviour may be tested by the manufacturers for product registration under conditions which do not adequately represent either the soil types found on the outcrop of major UK aquifers or the conditions in tropical applications.

The validity and applicability of results obtained from laboratory studies to the field scale remain unclear, particularly for consolidated fractured aquifers.

The occurrence, behaviour and significance of metabolites in groundwater systems are also poorly known.

Pollution of groundwaters by pesticides is a potentially serious and complex environmental problem and its management requires a balance between the need for increased crop production and the requirement to maintain drinking water quality.

ACKNOWLEDGEMENTS

The authors are grateful to their many colleagues and collaborators who have contributed to the work described here. Funding for the research carried out in the UK was provided by the National Rivers Authority and the Natural Environment Research Council (NERC), and for the work in tropical environments by the UK Overseas Development Administration (ODA). This paper is published with the permission of the Director of the British Geological Survey (NERC).

REFERENCES

Agrawal, G.D. 1999. Diffuse agricultural water pollution in India. *Water Science and Technology*: 39, 3 33-47.

Barbash, J.E. & E.A.Resek. 1996. *Pesticides in Ground Water: Distribution, Trends and Governing Factors.* Chelsea, Michigan: Ann Arbour Press.

Barceló, D, S.Chiron, A.Fernandez-Alba,A. Valverde & M.F.Alpendurada. 1996. Monitoring pesticides and metabolites in surface water and groundwater in Spain. In M.T.Meyer & E.M.Thurman (eds), *Herbicide metabolites in surface water and groundwater,* ACS Symposium Series 630: 237-253. Washington DC: American Chemical Society.

Barker, J.A. & A.R.Lawrence. 1993. *Simple groundwater model to predict pesticide concentration in pumped irrigation wells.* British Geological Survey Technical Report WD/92/49.

Barrett, M.R. 1996. The environmental impact of pesticide degradates. In M.T.Meyer & E.M.Thurman (eds), *Herbicide metabolites in surface water and groundwater,* ACS Symposium Series 630: 200-225. Washington DC: American Chemical Society.

Blackmore, J. & L.Clark. 1994. *The disposal of sheep dip waste.* National Rivers Authority R&D Report 11.

BGS, CISIR & Agriculture Department, Sri Lanka. 1992. *The impact of agriculture on groundwater quality, Kalpitya, Sri Lanka. Final Report.* British Geological Survey Technical Report WD/92/29.

Carabias Martínez, R, E.Rodríguez Gonzalo, M.E.Fernández Laespada & F.J.Sánchez San Román. 2000. Evaluation of surface- and ground-water pollution due to herbicides in agricultural areas of Zamora and Salamanca (Spain), *Journal of Chromatography A*: 869, 471-480.

Chilton, P.J., M.E.Stuart, S.J.Gardner, C.D.Hughes, H.K.Jones, J.M.West, R.A.Nicholson, J.A.Barker, L.R.Bridge & D.C.Goody. 1993. *Diffuse pollution from land-use practices,* National Rivers Authority R&D Project Record 113/10/ST.

Chilton, P.J., M.E.Stuart, D.C.Gooddy, R.J.Williams & A.R.Williamson. 1995a. The occurrence of herbicides in and the modelling of their transport to the Chalk aquifer beneath arable land in southern England. In A.Walker, R.Allen., S.W.Bailey, A.M.Nichol, A.M.Blair, C.D.Brown, P. Gunther, C.R.leake & P.H.Nicholls (eds), *Pesticide Movement to Water,* British Crop Protection Council Monograph 62: 111-116. Farnham.

Chilton,P.J., A.R.Lawrence & M.E.Stuart. 1995b. The impact of tropical agriculture on groundwater quality. In H.Nash & G.J.H.McCall (eds), *Groundwater Quality*: 113-122. London: Chapman & Hall.

Chilton, P.J., A.R.Lawrence & M.E.Stuart. 1998. Pesticides in groundwater; some preliminary results from recent research in temperate and tropical environments. In J. Mather, D.Banks, S.Dumpleton & S.Fermor (eds) *Groundwater contaminants and their migration*, Geological Society Special Publication: 333-345. London.

Clark L, J. Turrell, M.Fielding D.B.Oakes, I.Wilson & L.Taylor. 1995. Pesticides in major aquifers, Narional Rivers Authority R&D Report 17.

DETR. 1998. Report of the Chief Drinking Water Inspector. Department of Environment, Transport and the Regions.

Diaz-Diaz, R., J.E. Garcia-Hernandez & K Loague. 1998. Leaching potentials of four pesticides used for bananas in the Canary Islands. *Journal of Environmental Quality:* 27, 3, 662-572.

Foster, S.S.D, P.J.Chilton & M.E.Stuart. 1991. Mechanisms of groundwater pollution by pesticides. *Journal of the Institution of Water and Environmental Management*: 5, 2, 186-193.

Harrison, I., R.U.Leader, J.J.W.Higgo & G.M.Williams. 1998. A study of the degradation of phenoxy acid herbicides at different sites in a limestone aquifer. *Chemosphere:* 36, 6, 1211-1232.

Johnson, A.C., C.D.Hughes, R.J.Williams & P.J.Chilton. 1998. *Journal of Contaminant Hydrology*, 30: 281-297.

Johnson, A.C.,C.White and C.L.Bhardwaj. 2000. Potential for isoproturon, atrazine and mecoprop to be degraded with a chalk aquifer system. *Journal of Contaminant Hydrology*: in press.

Jones, R.L., A.G.Hornsby, P.S.C.Rao & M.P.Anderson. 1988. Movement and degradation of aldicarb residues in the saturated zone under citrus groves on the Florida ridge. *Journal of Contaminant Hydrology*, 1: 265-285.

Kolpin, D. W., E. M. Thurman & D. A. Goolsby. 1996. Occurrence of selected pesticides and their metabolites in near-surface aquifers of the mid-western United States. *Environmental Science & Technology,* 30: 335-340.

Kolpin, D.W, P.J.Squillace, J.S.Zorgorski & J.E.Barbash. 1997. Pesticides and volatile compounds in shallow urban groundwater of the United States. In P.J.Chilton et al (eds), *Groundwater in the Urban Environment*: 469-474, Rotterdam: Balkema.

Kolpin, D.W., E.M.Thurman & S.M.Linhart. 2000. Finding minimal herbicide concentrations in ground water? Try looking for their degradates, *Science of the Total Environment*: 248, 2-3, 115-122.

Lawrence, A.R & S.S.D.Foster 1987. *The pollution threat from agricultural pesticides and industrial solvents.* British Geological Survey Research Report 87/2.

Li, Y. & J.Zhang. 1999. Agricultural diffuse pollution from fertilisers and pesticides in China. *Water Science and Technology*: 39, 3, 25-32.

Matin, M.A., M.A.Malek, M.R.Amin, S.Rahman, J.Khatoon, M.Rahman, M.Aminuddin & A.J.Mian. Organochlorine insecticide residues in surface and underground water from different regions of Bangladesh, *Agriculture Ecosystems & Environment:* 69, 11-15.

Plimmer, J.R., K.H.Büchel, R.J.Hance & F.Führ. 1998. The future of pesticide use: the responsibility of developed nations toward lesser developed nations. In F.Führ, R.J.Hance, J.R.Plimmer & J.O.Nelson (eds) *The lysimeter concept, environmental behaviour of pesticides*, ACS Symposium Series, 699: 259-269, Washington DC, American Chemical Society.

Rao, P. S. C. & W. M.Alley. 1993. Pesticides. In W. M. Alley (ed) *Regional Groundwater Quality*. Van Nostrand Reinhold: 345-382, New York.

Ritter, W.F. 1990. Pesticide contamination of ground-water in the united-states - a review *Journal of Environmental Science and Health, Part B-Pesticides, Food Contaminants and Agricultural Wastes*: .25, 1, 1- 29.

Spliid N.H. & B.Køppen. 1998. Occurrence of pesticides in Danish shallow ground water, *Chemosphere*: 37, 7, 1307-1316.

Sweeney, J., P.A.Hart & P.J.McConvey. 1998. Investigation and management of pesticide pollution in the Lincolnshire Limestone aquifer in eastern England. In J. Mather, D.Banks, S.Dumpleton & S.Fermor (eds) *Groundwater contaminants and their migration*, Geological Society Special Publication: 347-360. London.

Tonmanee, N. & N.Kanchanakool. 1999. Agricultural diffuse pollution in Thailand. *Water Science and Technology*: 39, 3, 61-66.

Walker, M.J. & K.S.Porter. 1990. Assessment of pesticides in upstate New York ground water; results of a 1985-1987 sampling survey. *Groundwater Monitoring Review*: 30, 116-126.

Groundwater: Past Achievements and Future Challenges, Sililo et al. (eds) © 2000 Balkema, Rotterdam, ISBN 90 5809 159 7

Groundwater resources at the turn of the millennium: Taking stock and looking forward

S.S.D. Foster
British Geological Survey, Wallingford, UK & University of London, UK

ABSTRACT: In this paper the author makes a personal review of the rapid development of groundwater resources and hydrogeological science in the second half of the 20th century, and speculates upon the major challenges for the first decades of the 21st century, in the light of the recent World Water Forum, Vision Report and Framework for Action. A global perspective is taken, but the southern Africa context is also considered specifically.

1 TAKING STOCK OF 20TH CENTURY PROGRESS

1.1 *Groundwater Resource Development - Technological Achievement and Socioeconomic Contribution*

Groundwater has been the fundamental resource allowing the rapid and economical development of more reliable, improved quality, water supplies for a large proportion of the rural population across extensive areas of Africa, Asia and Latin America. This crucial, but formidable task began to gain momentum in the 1980s and has led to significant improvements in public health and life quality for innumerable rural communities. In the rural context also, waterwells have been of primary importance in the development of extensive livestock rearing in some semi-arid regions of Africa and Latin America.

Groundwater has also played a fundamental role in shaping the economic and social health of many urban areas in the developing world. No comprehensive statistics exist on the proportion of urban water-supply worldwide derived from groundwater. Nevertheless, more than 1,000 million urban dwellers in Asia and 150 million in Latin America (and somewhat less in Africa) depend directly or indirectly upon well, spring and borehole sources. On account of relatively low cost and generally high quality, groundwater has often been the preferred source for reticulated public water supplies and is widely exploited for private domestic and industrial uses. Much of this has occurred in the last 25-35 years, following earlier trends in Europe and North America.

During the same period, many nations have also witnessed an enormous increase in the exploitation of groundwater for agricultural irrigation. Comprehensive statistics are again not available but Table 1 provides some relevant data.

Table 1 : Selected national statistics on agricultural irrigation and groundwater use (from UN-FAO AquaStat database)

Country	Years	Irrigation		Origin of Water	
		area (kha)	water-use (Mm^3/a)	surface (%)	ground (%)
Bangladesh	1990/95	3,750	12,600	31	69
China	1993/95	48,000	407,770	78	18
India	1990/93	50,100	460,000	41	53
Pakistan	1990/91	14,330	150,600	66	34
Mexico	1995/97	5,370	61,200	73	27
Iran	1992/93	7,260	64,160	50	50
Saudi Arabia	1992/93	1,610	15,300	3	96

• the figures do not distinguish supplementary from near-continuous irrigation
•• conjunctive use, which is known to be practised in some areas, is also not specified

Groundwater resources have been underpinning the 'green revolution' in Asian agriculture, have permitted cultivation of high-value crops in a significant number of arid nations with inadequate food production and are offering security in numerous areas where irrigation with surface water resources has been found deficient during dry seasons and dry years. It is most important to record that the use of groundwater can be a major factor in stimulating increased irrigation water-use efficiency and productivity, because the scale of development allows individual control, greater flexibility and more incentives for efficiency of water use.

When reviewed in the broad historical perspective, the scale and rate of global development of groundwater supplies represents a major achievement with enormous socioeconomic benefits through a period of burgeoning world population. While groundwater exploitation is not new and wells have been excavated since the dawn of human history, the potential for groundwater exploitation has changed radically since the 1960-1970s and abstraction on the large-scale is a phenomenon of the late 20th century. This has been the result of technological advances in drilling and pumping, both of which owed as much or more to mechanical engineering than to hydrogeological science.

1.2 *Emergence of Hydrogeological Science Some Remarkable Advances*

While limitations in hydrogeological knowledge have not seriously constrained groundwater development, they will often, however, have increased its cost. Nevertheless, there have been some remarkable advances in hydrogeology in the second half of the 20th century, during which period, the subject evolved from a qualitative to quantitative science. However, it remains questionable in most nations whether such advances have received adequate application in terms of informing groundwater and land-use management practices.

It is beyond the scope of the present paper to enter into an extensive and detailed review of these scientific advances, but from the author's viewpoint the following achievements are particularly impressive:

- elaboration of hydraulic analyses for groundwater flow in porous and fractured media that permit approximation to a wide range of aquifer and ground conditions
- development of powerful software for numerical modelling of groundwater flow and contaminant transport, capable of simulating transient situations with complex boundary conditions
- identification of a suite of radio and stable isotopic analysis which provide powerful tools for the evaluation of groundwater flow rates and residence times, and the origin and fate of various contaminants.

2 GROUNDWATER IN THE WORLD WATER VISION PROCESS

The importance, current status and future needs of groundwater resources received prominent attention at the World Water Forum & Interministerial Conference in The Hague (Netherlands) in March 2000. The two key documents of the Forum were:

- 'World Water Vision - Making Water Everybody's Business' (WWC, 2000)
- 'Towards Water Security - a Framework for Action' (GWP, 2000).

These had as their overarching messages the need to mobilise broadly-based stakeholder involvement in water management and a large increase of investment in water infrastructure (including research and innovation), with improved water security as the goal. As the 'Vision Report' (WWC, 2000) puts it: "there is a water crisis - but the crisis is generally not about having too little water to satisfy our needs, it is a crisis of managing water so badly that millions of people - and the environment - suffer badly".

The documents contain numerous references to groundwater. The 'Vision Report' (WWC, 2000) calls for 'new techniques and institutional mechanisms for groundwater management'. The 'Framework for Action' (GWP, 2000) lists numerous targets and priorities to address the water crisis, including a number with specific or implied reference to groundwater such as:

- making water governance effective through establishing policies and institutions for managing river basins and aquifers
- promoting a major groundwater initiative to build the legal, institutional and technical basis from bottom-up for sustainable aquifer management
- integrating urban water sector planning and relating it with other urban planning processes.

A special half-day session of the World Water Forum on 'Groundwater Issues', jointly organised by the IAH and the IWMI (International Water Management Institute), provided input to the final forum session ('From Vision to Action') (Table 2).

An issue of serious concern was the reliability of the assertion in the 'Vision Report' (WWC, 2000) that the current global rate of aquifer overdraft (largely as a result of overabstraction for agricultural irrigation) is in the order of 200 km^3/a. The need for investments in groundwater management, especially in key countries (such as India, China, Mexico and of the Middle East), where groundwater levels in major aquifers show widespread evidence of continuous decline, was not questioned. However, the scientific basis for the above quantitative estimate is very weak and its uncertainty very high.

It should thus not be used as justification for major capital works schemes (involving new empounding reservoirs and long-distance aqueducts), and much more work on detailed investigation and local management of the aquifers involved is the first priority.

Table 2. Key global action points on groundwater resources from the World Water Forum

KEY ACTIONS	MAIN THRUST
Raising Political Awareness	• raise the profile of groundwater commensurate with its social and economic value, by improving the availability, reliability and accessibility of information for policy makers
Increased Management Capability	• shift emphasis from groundwater development to management, greatly increasing investment in management capacity, monitoring infrastructure and relevant research
Targeted Groundwater Development	• target controlled groundwater development on areas where exploitation levels are low and/or poverty alleviation is pressing
Strategic Groundwater Initiatives	• develop strategic initiatives that enhance awareness at operational level of the importance of groundwater resources, the significance of emerging problems and the practical management responses available to address such problems
Pilot Management Projects	• initiate pilot groundwater management projects and harvest lessons from on-going formal and informal management initiatives and interventions

The vision identified an urgent need to promote more effective and integrated water resources management. As regards groundwater there are two challenges to overcome:

- one technical - widespread inability to conceptualise the potential role of groundwater storage (often a major but undervalued asset) and to characterise aquifer systems adequately as the basis for management
- one institutional - common inability to be pro-active in controlling groundwater exploitation and constraining land-use for groundwater protection.

Amongst the key institutional changes widely needed, are improved definition of groundwater abstraction rights and tradability, mobilisation of the user-community at the appropriate scale to participate in aquifer management and progressive reduction or retargeting of subsidies which distort the natural economic constraints on groundwater use.

2.1 *Water for Food Production and Rural Development*

It is predicted (using population trends and agricultural 'business-as-usual' scenarios) that the world will experience a food security crisis within 15-25 years, as a result of continuing population growth, further losses of irrigated land through salinisation and continuing low crop yields in dry-land farming.

Groundwater resources play an important role in irrigated agriculture in numerous key countries and certain critical questions arise:

- evaluation of the significance to food production of past mining of groundwater storage in various major aquifers
- prediction of the growth of areas where food production is under threat from rising ground-water-table and salinisation of soil and groundwater (mainly as a result of excessive irrigation from surface water sources)
- potential for the development of relatively small groundwater supplies in areas of rainfed agriculture to act as a source of limited supplementary irrigation at critical times in the crop growth cycle, which greatly reduce agricultural productivity at present.

Amongst the key technical actions that need to be promoted in areas where groundwater plays an important role in agricultural irrigation are (Foster et al, 2000):

- demand management to reduce non-beneficial evaporation through more efficient irrigation techniques and increased water productivity through changes in crop type
- conjunctive use of surface and groundwater, so aquifer storage can be used to 'stretch the yield' and 'reduce the wastage' of surface water schemes
- more widespread artificial replenishment and water harvesting, to recharge overexploited aquifers where feasible
- ways of reconciling the use of groundwater for shallow domestic water-supply wells in areas subject to significant drawdown (within seasonal and longer-term climatic cycles) from irrigation wells.

2.2 *Water for Cities and Urban Development*

Groundwater is a major and economical source of potable public and private water-supply for in numerable cities, while the subsurface is at the same time the principal receptor for an increasing volume of human and industrial waste. The potential linkages between urban services and groundwater (Figure 1) needs to be more widely understood if potential conflicts between functions are to be avoided or minimised.

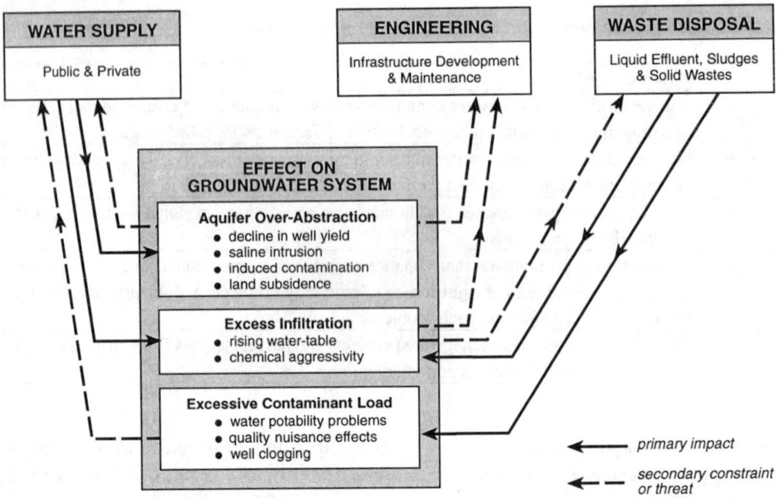

Figure 1: Linkages between urban services via the underlying groundwater systems (Foster et al, 1998)

Moreover, problems of counterproductive competition between municipal, industrial and private domestic groundwater abstractions needs to be addressed. Without appropriate abstraction controls and improved planning, self-supply from groundwater often mushrooms and leads to a chaotic resource utilisation situation and eventually to aquifer degradation, in which the smaller and poorer users suffer disproportionately. Given the frequent marked vertical variations in groundwater quality in urban areas, there is an urgent need to ensure that the better quality groundwater is preserved for potable use and to provide incentives for lower quality groundwater to be abstracted for non-sensitive bulk uses.

Amongst the key actions that are required to improve and rationalise the management of groundwater in urban areas are (Foster et al, 1998):

- systematic surveys of urban groundwater resources (their exploitation status, susceptibility to exploitation-related degradation, quality trends, pollution vulnerability, contaminant loading) in parallel with identification of all significant stakeholders
- incorporation of groundwater considerations into municipal planning to constrain groundwater abstraction and guide sanitation schemes, if the least sustainable of past practices are to be avoided
- using wastewater (the world's only increasingly available resource) from waterborne sewage systems in downstream areas more effectively, either directly for irrigation (while avoiding groundwater pollution) or indirectly through controlled aquifer recharge (effecting in-situ treatment) and repumping.

2.3 *Water for Nature: Environment and Ecosystems*

Groundwater plays an important role in sustaining many environmentally and economically important freshwater wetland habitats and coastal brackish water lagoons. Such environmental features, and the ecological systems that they house, may be threatened by groundwater development for urban and agricultural development.

Additionally, groundwater discharge forms the perennial baseflow of the upper reaches of many river systems and this has an important environmental and ecological role, which can be seriously perturbed by groundwater development.

To manage the situation effectively requires more widespread recognition of the linkages between groundwater systems and aquatic ecology, much improved understanding of minimum natural groundwater discharge requirements to maintain these features (even at reduced scale), together with well-informed stakeholder participation in water resources decision-making. Some starts have been made in this direction in a few countries worldwide, but much remains to be done to obtain an appropriate balance between the development and conservation of groundwater systems.

2.4 *Southern African Dimension*

On the occasion of the IAH XXX Congress in Cape Town (South Africa) in November 2000 it is appropriate to make special reference to the groundwater situation in the southern African (SADC) region. This region is characterised by rainfall patterns which show high variability both spatially and temporally, and includes:

- important, relatively arid, areas where groundwater is a vital resource underpinning human survival and all economic development
- extensive areas which exhibit high drought propensity, such that groundwater is very important for human water-supply and can provide a more secure basis for agricultural activity and industrial development.

The region possesses a number of sedimentary aquifers which contain a large groundwater storage resource and extensive areas of crystalline (igneous or volcanic) bedrock, overlain in some cases by alluvial or aeolian cover, from which modest water-supplies for rural development can be obtained.

The common groundwater resource agenda of the SADC countries includes:

- assessment of the (potentially fossil) aquifer storage resource in the major sedimentary aquifers, and definition of strategies for its planned mining in the arid regions
- improved evaluation of the rates of current infiltration and deep groundwater recharge, and their controls, in the semi-arid regions with rainfall between 400-600 mm/a
- further development of drought water-security plans through improved characterisation and mapping of weathered crystalline basement, other bedrock and thin surficial aquifers, in terms of their potential to meet both minimal rural domestic water-supply requirements and small-scale (garden) irrigation.

3 CHALLENGES FOR THE 21ST CENTURY

3.1 *Groundwater Resources : from Exploitation to Management*

Prior to advances in well drilling and pump technology, groundwater abstraction was very limited, and by comparison resources appeared infinite. In many cases, government organisations were formed to promote resource exploitation (rather than management) regardless of the status of the resource base. Well drilling and pumping energy have both been highly subsidised.

There is now rather widespread incidence of competition for available groundwater resources with the result of long-term aquifer water-level decline (in excess of 1.0 m/a in the more arid climates). This overdraft of aquifer storage can have quasi-irreversible side-effects, especially those involving the encroachment of saline water and/or the compaction of the aquifer and overlying aquitards.

There is also evidence of increasing social inequity where deeper, larger-capacity, irrigation boreholes lower the regional water-table and increase the cost of (or eliminate the access to) water-supply for users of shallow domestic wells. In the long-run, these processes call into question the sustainability of some existing agricultural and urban exploitation of groundwater.

The challenges inherent in this history are compounded by the increasingly critical role groundwater resources play in the livelihoods of individual users in many developing nations. These individual benefits make any subsequent constraints on groundwater use politically sensitive. In combination these factors represent both a cause of groundwater management problems and an obstacle to the implementation of effective responses.

Regulatory agencies in developing nations are often under resourced and weakly empowered in relation to the control of groundwater abstraction. However, increasing their financial budget and strengthening their legal provision alone will not necessarily improve the situation. Certain inherent factors also need to be addressed:

- the lack of public and political awareness of the consequences of excessive groundwater exploitation, and thus absence of an adequate consensus for action
- the need (in many nations) to define groundwater abstraction rights and mobilise stakeholder participation
- the uncertainty of resource evaluation due inadequate monitoring of aquifer response to abstraction and recharge.

The economic characteristics of groundwater have also played a major role in the emergence of management problems. Groundwater is generally undervalued, especially where exploitation in uncontrolled, and the exploiter (in effect) receives all the benefits of groundwater development, but (at most) pays only part of the costs. Undervaluation leads to economically-inefficient patterns of groundwater allocation and use. In addition, because in-situ values associated with groundwater are rarely reflected, undervaluation stimulates overexploitation, by reducing incentives for investment in water conservation and (more generally) in resource management.

While groundwater is normally easy and economical to develop, achieving sustainable and optimised aquifer management is amongst the most important, least recognised and highly complex of natural resource challenges facing society (IAH, 1999). It is a long-term multi-disciplinary and

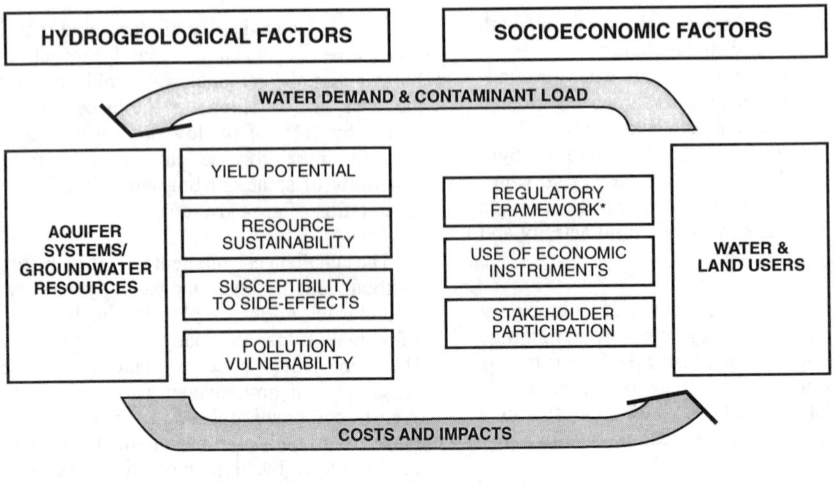

Figure 2. Conceptual framework for the effective management and protection of groundwater resources (Foster et al, 2000).

multi-facetted activity, which must recognise and respect hydrogeologic and socioeconomic diversity and that requires considerable investment in terms of institutional development (in the broadest sense) and systematic data collection (at field level) (Figure 2).

Key components that will widely be required to achieve these ends are (Foster et al, 2000):

- transformation of government organisations and enable them to facilitate the management process through strategic planning, legal provisions, groundwater databasing and aquifer monitoring
- raising political and public awareness of the need for sustainable management
- development and application of groundwater resource decision-support systems
- assignment and consolidation of groundwater use rights and introduction of realistic abstraction charges
- stimulation of stakeholder participation through the formation of aquifer user associations
- breaking the common resource-management paradigm that groundwater recharge is a fixed parameter, with recognition that it varies markedly in a temporal sense with long-term climatic cycles and changes, with surface water engineering works and with changes in irrigated agriculture and urban water-service systems
- recognition of the interaction and interdependence between groundwater and surface water resources, which often greatly favours their conjunctive use.

3.2 *Groundwater Sources: from Development to Protection*

The rapid growth in the development of groundwater sources for public water-supply, especially in the latter part of the 20th century, has not, in general, been matched by complementary actions in relation to groundwater source and aquifer protection. Even in places where there is relatively little pressure on groundwater resources in an absolute volumetric sense, there may still be a 'water resource crisis' in quality terms.

The threats to groundwater quality are manifold and increasing, but examples of successful experience in controlling them are still rather few. The major and (now) persistent concerns are:

- the growing volume and complexity of contaminant loads on the subsoil, as a result of inadequate handling, disposal and reuse of wastewater
- the very widespread diffuse contaminant load arising from the leaching of agrochemicals and their derivatives from permeable soils.

The need is to make real advances in protection through:

- identifying land areas where aquifers are vulnerable to pollution as a result of insufficient intrinsic pollutant attenuation capacity, especially in the capture areas of major groundwater supply sources
- controlling the subsurface contaminant load in such areas by exerting appropriate control over existing land-use practices and future land-use planning

- a more integrated approach to the management of groundwater and wastewater in urban areas.

Such actions will require justification in technical and economic terms.

3.3 *Hydrogeological Science - Future Needs and Probable Trends*

There are thus a series of well-defined pressures on groundwater resources which are likely to generate demands and opportunities in hydrogeological science, including the need for:

- tools for improved evaluation and prediction of groundwater interactions with the surface water environment
- realistic valuation (as opposed to evaluation) of groundwater resources, as a basis for more rational abstraction charging
- improved characterisation of groundwater storage in relation to assessing the role of aquifers in conjunctive use or integrated management schemes and for drought preparedness
- improved understanding of subsurface biochemical processes determining the natural contaminant attenuation capacity of aquifers in their vadose and saturated zones
- further development of natural and artificial tracer techniques to assess recharge rates and transit times of groundwater in aquifer systems, and thus their vulnerability to surface water contaminants.

There is little doubt that continued (and probably accelerated) technological development in the realm of hydrogeological data collection (field and remote sensing) and data processing (GIS-based approaches) will provide important opportunities, both for hydrogeological research and practical application.

However, amongst the key issues that will need to be confronted at the science-management interface is coping with (and communicating) geological and meteorological uncertainty in relation to groundwater resource decision-making.

Hydrogeology is still a relatively young, and inherently imprecise, science. Hydrogeologists still struggle to apply the growing knowledge-base effectively to the more efficient management of environmental resources and to improve public and political perception of groundwater. The two areas are inextricably linked. How well do we handle hydrogeological uncertainty and communicate risk, especially at the legal and financial interfaces? Approaches at present often range from the exceedingly precautionary to the widely optimistic. There is a tendency for some hydrogeologists (even in non- specialist circles) to dwell excessively on residual areas of uncertainty, while failing to make a clear statement about what is known with confidence. This is something we need professionally to remedy if we are to ensure that investment is directed towards water-supply protection (and natural attenuation) rather than treatment, to the optimal use of groundwater storage rather than building more surface reservoirs, and to efficient subsurface waste disposal rather than more costly (and perhaps more risky) alternative solutions.

4 ACKNOWLEDGEMENTS

The author wishes to acknowledge the contribution of many friends, both in the BGS, more widely in the groundwater community and in the World Bank's Water Resources Network, for stimulating his thinking on various aspects of this paper.

5 REFERENCES

Foster S, Lawrence A & Morris B 1998. Groundwater in urban development - assessing management needs and formulating policy strategies. World Bank Technical Paper 390.

Foster S, Chilton J, Moench M, Cardy F & Schiffler M. 2000 Groundwater in rural development - facing the challenges of supply and resource sustainability. World Bank Technical Paper 463.

GWP 2000. Towards water security - a framework for action. Global Water Partnership.

IAH, 1999. World water resources - a vision for groundwater. International Association of Hydrogeologists.

WWC 2000. World Water Vision - making water everybody's business. World Water Council.

Groundwater: Past Achievements and Future Challenges, Sililo et al. (eds) © 2000 Balkema, Rotterdam, ISBN 90 5809 159 7

Hydrogeology and waste management strategies

M.J. Knight
National Centre for Groundwater Management, University of Technology, Sydney, N.S.W., Australia

ABSTRACT: Hydrogeology is an important component of waste management strategies involving solids and liquids as they interface with land and groundwater. A successful strategy is one that produces sustainable technical, environmental, economic and social outcomes. This paper illustrates these principles in the cases of solid waste domestic landfills, cemeteries and the application of greywater to household land. The sustainable landfill recovers valuable land and water resources in an economic manner. Cemeteries produce few contaminants and burial sites can safely be used more than once. Reuse of greywater is possible in Australian cities, although there are some hydrogeological, economic and social constraints.

1 INTRODUCTION

Waste production and management are fundamental processes in the natural environment as well as being outcomes of people's creative or destructive activities. Some resources that are useful such as coal are the result of wastes (vegetative material) that are coupled with a disposal and a concentrating geological process. In general the waste management processes in natural systems often involve chemical consumption, disposal, recycling, storage, concentration, dispersion and detoxification (Knight, 1983). The key principle that emerges is one of holistic sustainability.

All civilisations down through the ages have produced wastes and have usually used some form of land based facility for disposal. Crude landfills within and outside city walls (now archaeological sites) were used for solid wastes. There is evidence in ancient Greek and Roman cities e.g. Ephesus and others where wastewater from households and runoff was used to flush toilets (latrines) with ultimate disposal to land for agricultural production (Crouch 1993). The filtering properties of gravel for wastewater treatment were also understood by the Romans and probably by the Greeks.

The present day problems of waste production and management are nothing new. Hydrogeology has much to contribute to the strategies used in order to maximise resource use, optimise the site where waste can usefully interface with land prevent groundwater contamination and ultimately produce a sustainable system. This positive approach needs to be developed further and overcome the lingering negative feelings that groundwater specialists have had in the past of waste management systems involving land, being all polluting.

2 WASTE MANAGEMENT INVOLVING HYDROGEOLOGY

Hydrogeology has made major contributions in developing complex technologies for managing hazardous waste materials such as high level radioactive waste (Fried 1979; Mukhopadhyay & Tsang 1999) and deep injection disposal of industrial wastes (Apps & Tsang 1996). Possibly of greater relevance to most countries including those in Africa for both rural and urban areas are the handling of solid wastes, and some wastewaters. To illustrate some key principles of sustainability and the role of hydrogeology, this paper examines domestic waste landfills, cemeteries and managing greywater (reusing) at the household level.

3 THE SUSTAINABLE LANDFILL

Sustainable landfills usually involve mixed domestic solid waste located on the appropriate hydrogeology and topography, as well as being designed, constructed and operated with the relevant hydrogeological and engineering principles. The concept was first presented some three years ago (Knight 1997) and discussed further by Knight & Dent (1998), Figure 1. Three phases are recognised. Correct hydrogeological principles are used in siting to

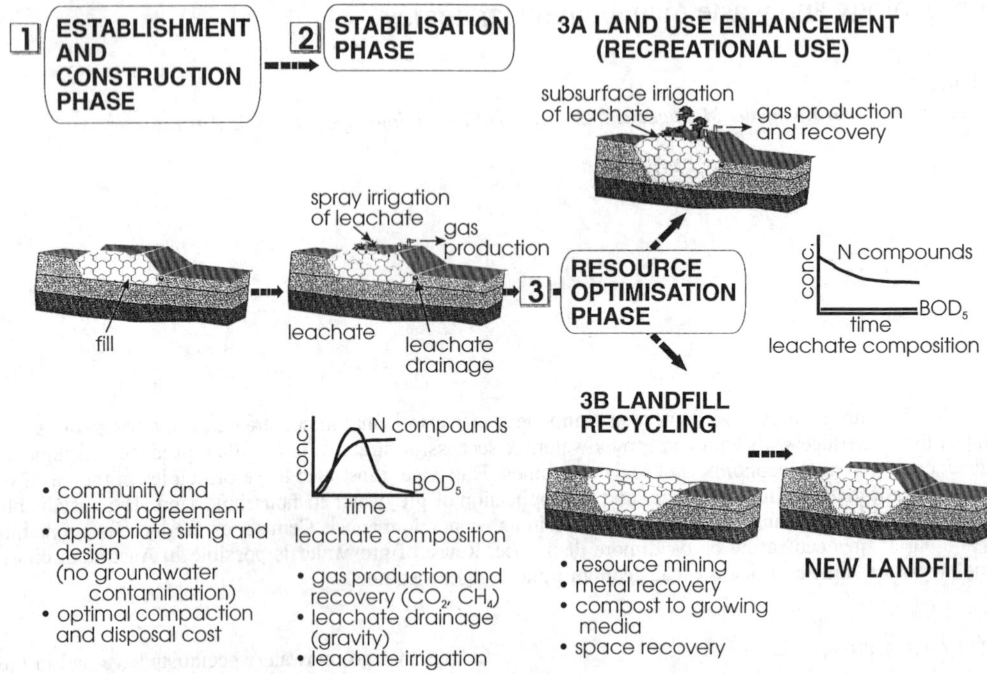

Figure 1. The sustainable landfill. (Data after Knight 1997 and Knight & Dent, 1998)

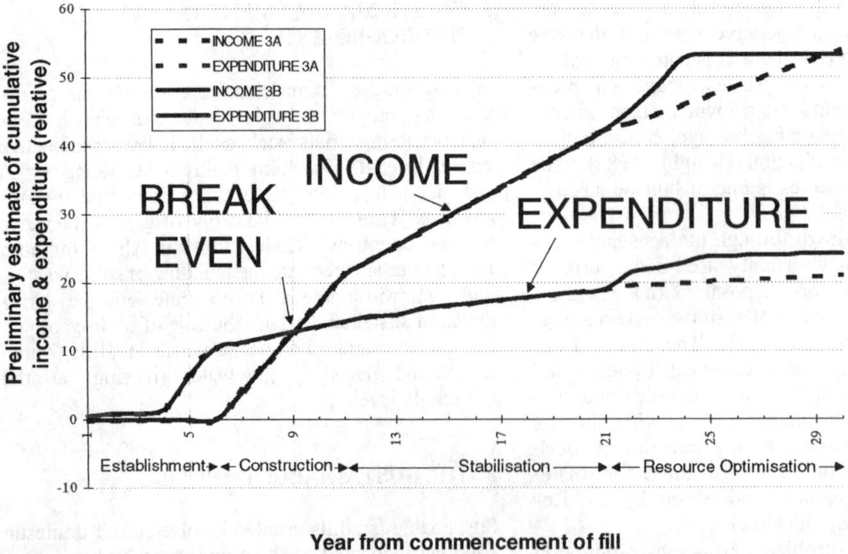

Figure 2. Estimates of cumulative expenditure and cash flows over the 30 years life of atheoretical 1 million tonne domestic waste landfill. The current 3A, 3B relate to landfill Phases. Data after Knight & Dent (1998).

avoid groundwater pollution (Phase 1), provide free leachate drainage (e.g. tabular sloping fill with gravity flow) and leachate recycling to maximise waste stabilisation (Phase 2).

The concept of a landfill bioreactor applies, (Knight & Beck 1987; Reinhart & Townsend 1998). Leachate is a resource (Knight et al. 1997) with useful nitrogen compounds that can be employed for irrigation during post stabilisation as land use changes to recreational uses e.g. sporting fields. Gas is a resource and is extracted for electricity and minimising greenhouse impact (Phase 3A). In some cultures and locations, the landfill could be re-mined for resources (Phase 3B). This has also been suggested recently by Reinhart & Townsend (1998).

Knight & Dent (1998) indicated that landfills need to be financially sustainable which result in positive cash flows in the medium to long term, Figure 2. Financial modelling should extend to 30 years or more. This has not been common practice in the past. In the establishment and construction stages negative cash flows are to be expected.

4 CEMETERIES

Dent & Knight (1998) pointed out that cemeteries are like domestic waste landfills. In Australia, 46% of the population prefer land interment for their remains (ACCA 1997). It is only in recent times that hydrogeologists have begun studying cemeteries in relation to processes and potential for groundwater pollution.

In a national study in Australia, Dent (2000 pers. comm.), and Dent & Knight (1998) found that the impact of cemeteries on groundwater is minimal even where an aquifer is below the burial zone. This is partly due to the low mass of contaminants distributed over a wide area and added to through time. A lean 70kg adult male human body contains: 16,000g carbon, 1,800g chlorine, 19g magnesium, 42g iron and water 70-74% by weight. Females are 2/3 to 3/4 of these amounts (Forbes 1987).

Some monitoring results for Cheltenham Cemetery, Adelaide, South Australia are indicated in Table 1. Cheltenham Cemetery is above the River Torrens fan aquifer (Aitchinson 1954) with the soils of the burial area comprising silty to sandy clays and clayey sands. There is a saline watertable below. Hydraulic gradients are low (0.0007 to 0.004) with hydraulic conductivity (K) about 10^{-4}m/day (Gerges 1996 pers. comm.). The data in Table 1 support the general observation of very low contaminant contributions with perhaps the exception of *Pseudomonas aeruginosa*. An important practice that occurs at Cheltenham (but not generally elsewhere in Australia) is called "lift and deepen". After 50 years (grave tenure) any remnant skeletal or artifactual material of the most recent person buried is placed in a small "ossuary" box. The grave is dug deeper, the box reinterred and then a new body placed in the space created. This saves considerable land space and cost. It is also consistent with the concept of sustainability for the site. There are no scientific reasons as to why this practice should not be encouraged.

5 GREYWATER REUSE ON HOUSEHOLD LAND

Greywater is wastewater produced in a household by the shower, bath, handbasin sinks and washing machines. It differs from "blackwater" that comes from toilets and the kitchen and is more contaminated. In most western urban areas, greywater and blackwater are disposed into the sewerage pipe system and rainwater runoff often flows to a separate wastewater system. In both cases there is a substantial water resource loss.

Currently there is extensive re-thinking of the urban water cycle in Australia. This process began five years ago (ASTEC 1995). An important outcome of the ASTEC (1995) study was the realisation that more emphasis was needed on water management at the household level rather than the traditional control by state or local authorities.

Table 1. Range of key cemetery analytes, Cheltenham Cemetery, Adelaide, South Australia.

Analyte	Near-boundary bores (g)	Internal bores (f)
Ph	7.5 – 7.8	7.3 – 7.9
NO_2-N	0 – 0.034	0 – 0.010
NO_3-N	0 – 0.6	0 – 11.4
Cl	52 – 1120	107 – 576
SO_4	22 – 255	52.5 – 179
NH_3-N	0.01 – 0.59	0 – 0.53
PO_4 (a)	0 – 7.0	0 – 6.2
TKN (b)	0.16 – 0.81	<0.05 – 0.61
TOC (c)	1.6 – 28.0	1.3 – 21.2
BOD	<2 – 15 (d)	<2 – 16
CO_2g	210 – 325	135 – 220
Total coliforms	0 – 1	0 – 17
Faecal coliforms	0 – 1	0
Faecal strepto-cocci	0 – 1	0
Pseudomonas aeruginosa (e)	0 – 1	0 - 40

Data after Knight & Dent (1998).
All measurement in mg/L except for pH units and bacterial counts which are per 100mL: (a) measured as orthophosphate; (b) total kjeldahl nitrogen; (c) total organic carbon; (d) 2 mg/L is the detection limit; (e) *Pseudomonas aeruginosa* was only sought from a second round of sampling; (f) internal bores are well within the cemetery; (g) Near-boundary bores are near the boundary but within cemetery grounds.

5.1 *Greywater use potential in Sydney and Brisbane*

In Sydney and Brisbane there is an ongoing study of hydrogeology of household land in relation to water reuse, its economic dimensions and peoples' attitudes (Knight et al. 2000).

Permeabilties of soils (lawns, garden beds) have been tested using a Guelf permeameter in both Sydney and Brisbane, Figure 3.

Figure 3. Location map

For a range of soils and weathered rock (regolith) the following hydraulic conductivities were obtained: Sydney (Sutherland) 1.7×10^{-4} to 4.4m/day; Brisbane 8.2×10^{-5} to 20m/day (Knight et al. 2000).
To function properly, slopes need to be less than 10° (less than 5° better) and rock should not be too close to the surface. In the Sutherland Shire of Sydney 20 to 50% of household properties are not suitable due to sandstone being too close to the ground surface.

Other constraints are low permeable soil horizons causing perched watertables. This can potentially cause spring discharge (ponding) runoff from downslope interflow where multiple households down a slope use greywater, Figure 4.

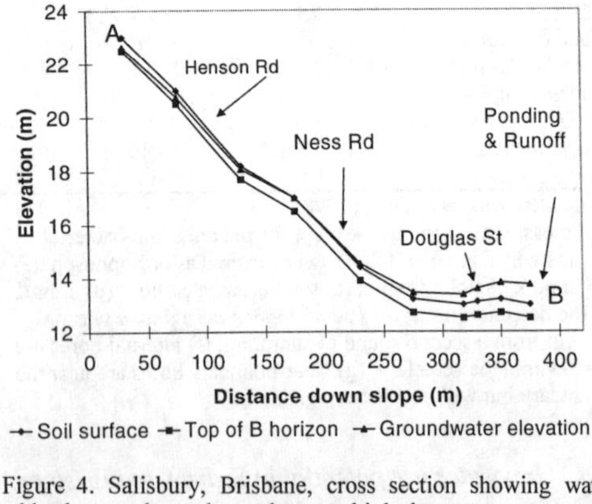

Figure 4. Salisbury, Brisbane, cross section showing water table changes down slope where multiple houses use greywater on podzolic soils with a low permeable B horizon. Data after Knight et al. 2000.

In such environments, application rates would need to be controlled and some form of drainage installed on the lower slopes.

Costing of greywater installations in Brisbane indicate that payback times resulting from capital costs offset against the value of water saved are about 20-40 years. Preliminary social surveys in both Brisbane and Sydney reveal that more than 5 years is unacceptable (Knight et al. 2000). To motivate householders in water reuse, government agencies would need to substantially raise the price of incoming clean mains water or offer some form of subsidy or rebate as an incentive. This strategy is clearly needed in the urban environment where hydrogeology allows. The result will be a step towards sustainable water resource management.

6 CONCLUSIONS

Hydrogeological principles are being increasingly applied to a range of waste management strategies. Wastes involved include radioactive, industrial, wastewaters and domestic solid materials that include household wastes and ultimately people buried in cemeteries. This paper has indicated that an appropriate waste strategy is one that draws on sustainability principles seen in the natural environment as well as applying appropriate engineering, finance, and social techniques to produce successful outcomes.

Useful landfills, unpolluting cemeteries and reused greywater at the household scale are possible in urban or rural areas.

REFERENCE

ACCA. 1997. Cremation in Australia. *News Winter*: 16-17. Australian Cemeteries and Crematoria Association.

Aitchinson, G.D., Sprigg, R.C. & Cochrane, G.W. 1954. The soils and geology of Adelaide and suburbs. *Bulletin Geological Survey SA* 32, 89p.

Apps, J.A. & Tsang, C.F (Eds) 1996. *Deep injection disposal of hazardous and industrial waste. Scientific and Engineering aspects.* San Diego, Academic Press.

ASTEC. 1995. *Curbing our thirst: Possible Futures for Australia's Urban Water System in the 21st Century.* Australian Science and Technology Council Future Needs 2010, Canberra.

Crouch, D.P. 1993. *Water Management in Ancient Greek cities.* New York, Oxford University Press.

Dent, B.B. 2000. Lecture, University of Technology, Sydney.

Dent, B.B. & Knight, M.J. 1998. Cemeteries: a special kind of landfill. The context of their sustainable management. In T.R. Weaver and C.R. Lawrence (Eds), *Proceedings International Groundwater Conference: Groundwater: Sustainable Solutions, Melbourne, 8-13 February*: 359-374. International Association of Hydrogeologists.

Forbes, G.B. 1997. *Human body composition: growth, aging, nutrition and activity.* New York, Springer Verlag.

Fried, S. 1979. Radioactive waste in geologic storage. *Proceedings of Symposium sponsored by the Division of Nu-*

clear Chemistry and Technology, 76th meeting American Chemical Society, Miami Beach, September 11-15, 1978. Washington American Chemcial Society.

Gerges, N.Z. 1997. *Senior Geologist*, Mines and Energy of South Australia.

Knight, M.J. 1983. Waste disposal in geological materials. *Current Affairs Bulletin* 60(7): 18-30.

Knight, M.J. 1997. Sustainable landfills – making environmental and economic sense of landbased waste management systems. *Proceedings of the International Symposium on Engineering Geology and the Environment, Greek National Group of International Association of Engineering Geology, Athens Greece, 23-27 June, 1997*. Eds. P.A. Marinos, G.C. Koukis, G.C. Tsiambaos and G.C. Stournaras, V4, Balkema, Rotterdam (in press).

Knight, M.J. & Beck, G. 1987. Modelling the long-term dynamic behaviour of selected groundwater quality parameters in a domestic waste landfill at Lucas Heights, Sydney. *Proceedings of the International Conference on Groundwater under Stress. Brisbane, Queensland, 11-16 May 1986*. Australian Water Resources Council Conference Series 13, AGPS, Canberra 475-490.

Knight, M.J. & Dent, B.B. 1998. Sustainability of waste and groundwater management systems. In T.R. Weaver and C.R. Lawrence (Eds), *Proceedings International Groundwater Conference: Groundwater sustainable solutions. Melbourne, 8-13 February*: 359-374. International Association of Hydrogeologists.

Knight, M.J., Sutton, B.G. & Yates, D.B. 1997. Landfill leachate – managing a valuable irrigation resource in the urban environment. *Proceedings of the XXVII IAH Congress on Groundwater in the Urban Environment, Nottingham, UK, 21-27 September, 1997*. Vol 1 *Problems, Processes and Management*. Ed John Chilton et al.., Balkema, Rotterdam 603-610.

Knight, M.J., Tomlin, J.S., Paffard, A.F. & Nicol, R.A. 2000. Potential of greywater reuse for irrigating household land in Sydney and Brisbane. *Paper 274, CDRom Xth World Water Congress, 12-16 March, Melbourne, International Water Resources Association.*

Mukhopadhyay, S. & Tsang, Y.W. 1999. Thermal – hydrological Modeling of the drift scale test at Yucca Mountain, Nevada. *Earth Sciences Division Annual Report 1998 – 1999*, 79-80. Lawrence Berkeley Laboratory, Berkeley, California.

Reinhart, D.R. & Townsend, T.G. 1998. *Landfill bioreactor index design and operation.* Boca Raton Lewis.

Groundwater: Past Achievements and Future Challenges, Sililo et al. (eds) © 2000 Balkema, Rotterdam, ISBN 90 5809 159 7

The use of monitored natural attenuation as a cost-effective technique for groundwater restoration

David N. Lerner, Steven F. Thornton & Ruth M. Davison
Groundwater Protection and Restoration Group, University of Sheffield, UK

ABSTRACT: There are many groundwater pollution problems caused by the spillage of fuels and industrial chemicals throughout the world. There has often seemed to be between strategies of ignoring the problem or going for a full scale and expensive cleanup. These extremes are now seen as simplistic, and there has been a growing acceptance of a risk-based approach to evaluating individual sites and designing appropriate strategies. This allows account to be taken of the natural processes which reduce concentrations along the pathway from source to receptor, particularly biodegradation. Taking formal account of these processes can often reduce or eliminate an engineered restoration, replacing it with a requirement that the plume is monitored in future. Protocols are available to assist with the assessment of such natural attenuation. They are especially useful for straightforward hydrocarbon plumes. Studies of two more complex cases, involving phenolic contaminants, are presented to show that natural attenuation can still as assessed on sites which do not fit within the protocols. In one case, natural attenuation has been successful. In the other, biodegradation is so slow that an alternative risk management strategy will be required.

1 GROUNDWATER POLLUTION PROBLEMS

Contaminated land and groundwater remain a large liability on the balance sheets of many companies, local authorities and nations. Groundwater pollution is particularly difficult to cleanup, and researchers have often argued that no severely polluted site has been successfully restored by engineered intervention (National Research Council, 1994). The underlying problem is that of inaccessibility of the pollutants. Heterogeneity causing greatly variable flow rates, diffusion of pollutants into low permeability materials and long term sorption, all mean that the time scales for cleanup are controlled by diffusion, a very slow process. This has been recognised in the USA, where it is now possible to obtain a "Technical Infeasibility Waiver" for some sites. Pump-and-treat, which is the standard approach when active restoration is undertaken, is best seen as a containment approach.

Until now, most "owners" of polluted groundwater in the UK have only dealt with the problems when they wish to sell land, or they have had such a serious incident that the Environment Agency becomes involved. This is going to change. Recent and imminent legislation will create a new contaminated land regime. This will require every contaminated site to be identified and evaluated for the risk it may cause to controlled waters. The Environment Agency is now empowered to require cleanup where groundwater is polluted but not linked to any contaminated land. The new IPPC legislation will require landholders to put baseline surveys of the environmental condition of their sites on the public register. Then there will be public information about which sites are polluted and powers to require cleanup – ignorance and inaction will no longer be options.

There is a potential conflict between the increasing pressure on landholders to act on groundwater pollution, and the acknowledged difficulties of engineered cleanup. Natural attenuation is a potential strategy to resolve this conflict for many sites – if natural processes can be demonstrated to be reducing risks to acceptable levels, then active cleanup will be unnecessary.

2 NATURAL ATTENUATION CONCEPTS

Natural attenuation of polluted groundwater refers to the suite of "….physical, chemical or biological processes that act without human intervention to reduce the mass, toxicity, mobility, volume, or concentration of contaminants in soil and groundwater……" (U.S. EPA, 1998). These in-situ processes include biodegradation, dispersion, dilution, sorption, volatilisation, and chemical or biological stabilisation, transformation, or destruction (U.S. EPA, 1999). It is also called *intrinsic (bio)remediation*, and the term *monitored natural attenuation* (MNA) is becoming standard as it emphasises that monitoring must be undertaken if MNA is to be accepted as a strategy for any site.

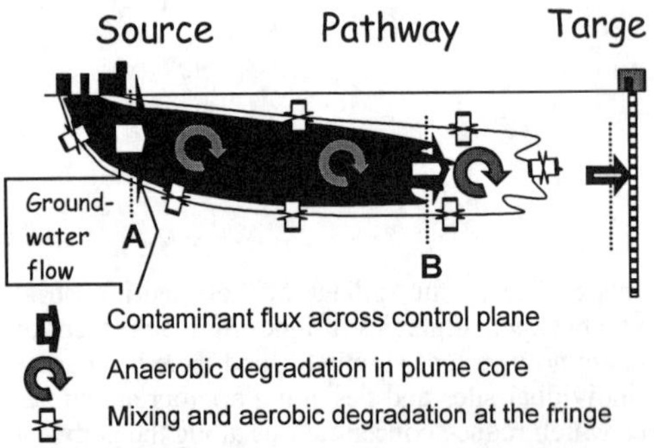

Figure 1. Concept sketch for natural attenuation of biodegradable contaminants in groundwater, in which fluxes are reduced during flow from source to receptor.

The processes of natural attenuation are shown schematically in Figure 1. With time, polluted groundwater moves from the source at A, through intermediate points in the plume (B), to the receptor at C. MNA will be successful if the flux and concentration of pollutants is reduced enough to make the risks to the receptor acceptably low. For organic pollutants, biodegradation is usually the most important process. Anaerobic degradation will occur inside the plume, but is slow and limited for many situations. Usually aerobic and nitrate reducing degradation are the most important reactions, but they can only take place at the fringe where background waters mix slowly into the plume. The plume will grow until the influx of pollutants at A is balanced by the rate of destruction of the combined fringe mixing-controlled and the slower internal reactions.

3 WHEN IS MNA APPROPRIATE?

The use of MNA has increased substantially over the last decade with, for example this approach implemented in 23% of US cases in 1997 compared with only 5% in 1991 (Kremer, 1998a). MNA was the preferred choice for the remediation of 46% of petroleum hydrocarbon-contaminated sites, and was also widely used for inorganic contaminants and chlorinated solvents (35%). The exploitation of MNA as a remediation technology is supported by the U.S. EPA "within the context of a carefully controlled and monitored site cleanup approach that will reduce contaminant concentrations to levels that are protective to human health and the environment within a reasonable time frame" (U.S. EPA, 1999). It requires sound technical justification within a site-specific decision process, that allows for contingency measures in the event that its predicted performance fails to meet regulatory cleanup criteria (Wilson, 1998). This implies a sophisticated site assessment.

MNA has many potential advantages, but also some possible limitations, when compared with conventional engineered approaches. It is not for those in a hurry, or for landholders who want to install equipment for PR reasons. Nor is it for all compounds. For example, chlorinated solvents only degrade anaerobically and incomplete degradation produces vinyl chloride which is more toxic than the parent compounds.

MNA is a viable option for compounds which do attenuate, and for sites with enough travel distance and time to allow the processes to operate. If these hurdles are passed, non-technical factors are considered, such as the institutional controls over long periods, the availability of adequate funding for reliable monitoring and performance evaluation, and public acceptance of the extended time scale required for remediation.

4 HOW TO DEMONSTRATE MNA

Natural attenuation is evaluated using multiple lines of evidence (Wiedemeier, et al., 1999). These distinct but converging lines of evidence are based on hydrochemical and microbiological data that document the occurrence and extent of in-situ contaminant removal processes. The nature of the information required to demonstrate MNA of organic contaminants includes:

- *Primary lines of evidence* such as historical trends in contaminant data showing plume stabilisation and/or contaminant mass loss over time;
- *Secondary lines of evidence* showing that *in-situ* hydrogeochemical conditions are suitable for biodegradation and that active biodegradation has occurred. This is based on geochemical indicators of naturally occurring contaminant biodegradation, such as the depletion of electron acceptors and donors, increased metabolic by-product concentrations, decreasing parent compound concentrations and increased daughter compound concentrations;
- *Tertiary or optional lines of evidence* that prove the processes and estimate rates.

Primary and secondary lines of evidence are obtained by sampling groundwater monitoring boreholes along the plume flowpath, including the source area, and uncontaminated zones of the aquifer. They are used to show that a plume is shrinking, stable or only growing at a rate slower than that predicted by conservative groundwater flow calculations. They also demonstrate that this behaviour is consistent with the hydrochemical environment. Tertiary lines of evidence include data from microcosm studies and are used to support an MNA assessment, when the other evidence is inconclusive, or when information is required on a specific degradation mechanism or on environmental factors that may limit biodegradation processes.

5 DEVELOPMENT OF PROTOCOLS FOR MNA

The emergence of MNA as a remediation technology has resulted in the development (which is ongoing) of technical protocols for its application. These are guidance documents which present good practice in site investigation and analysis of data required to implement and monitor MNA. The development of these protocols has been driven by American experiences, producing guidance for petroleum hydrocarbons (Wiedemeier, et al., 1995) and chlorinated organics (Wiedemeier, et al., 1997). Protocols have also been developed by the American Society for Testing and Materials (ASTM, 1998), and private sector companies in response to specific contaminant problems (e.g. Buscheck and O'Reilly, 1995). Significantly, there is little technical guidance for the application of MNA in many European countries, although this is presently under consideration in Denmark and in the Netherlands. A guidance document, commissioned by the Environment Agency, for the assessment of MNA in the UK is currently under development (Carey *et al.*, 2000).

The American protocols have proved very useful. They have introduced a common language and philosophy to MNA assessments. They have made studies more cost effective and enabled MNA to be widely accepted, reducing the costs of contaminated groundwater to society. However other countries are not the same as the USA. They have different legal and administrative arrangements, and different pressures on national resources. Often their hydrogeology is different. National protocols can be quickly developed from existing documents to take account of these factors. We strongly recommend that any protocol includes a requirement for peer review. More than anything, this will raise standards of investigation and interpretation, and share experience amongst the community. Without such review and transparency there is a serious risk that the public will see MNA as a do-nothing, money saving option which is justified in a secret report. MNA is more than this, it is a valuable addition to a repertoire of solutions to the massive problems of contaminated land and groundwater.

6 SITE INVESTIGATION AND MONITORING STRATEGIES

The site investigation programme needed to evaluate the performance of NA technology incorporates initial site characterisation and long-term monitoring (Kremer, 1998b). Site characterisation for NA is fundamentally different and more comprehensive and than that required for an active remediation scheme, because it requires greater understanding of processes affecting the contaminant plume and there is greater emphasis on data collected from within the plume. This initial monitoring phase should identify the location and extent of contaminant source area(s), the spatial distribution and concentration of contaminants, heterogeneity in the aquifer geological and hydrogeological characteristics and variations in the groundwater hydrochemistry. The objective of the initial site characterisation for NA assessment is to define the baseline conditions. This is necessary to ascertain whether NA is likely to be a viable remediation option and, if so, to provide a reference state from which its performance can be monitored over time. A typical network of observation boreholes used to evaluate NA of contaminant plumes is shown in Figure 2. At a *minimum* the borehole network needs to include wells that assess unimpacted background groundwater quality upstream of the plume (well A), the contaminant composition in the source area (well B) and groundwater quality along the plume flow path (wells C to G). To correctly delineate the plume, boreholes should also

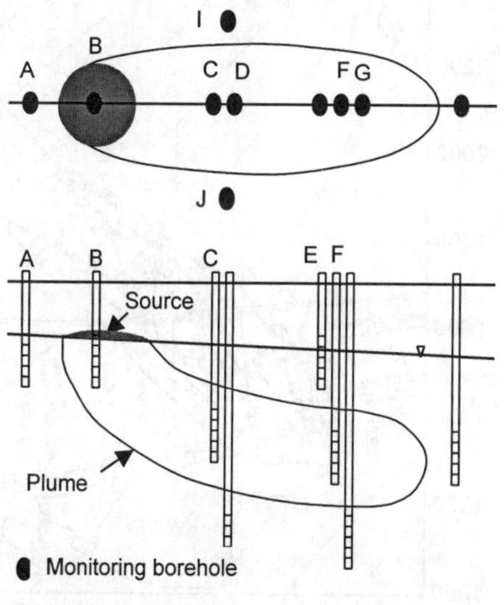

Figure 2. Network of observation boreholes used for assessment of natural attenuation (modified from Wiedemeier and Haas, 1999).

be positioned in the flow path ahead of the plume to define the downgradient extent of contamination (well H) and transverse to the plume to define the lateral extent of contaminant migration (wells I and J). The monitoring borehole network (e.g. wells A to J) is usually installed in phases since the extent of plume migration is unknown until these wells have been drilled.

Long-term monitoring (LTM) is required to assess the behaviour of the contaminant plume over time and to confirm that NA is occurring at rates which are protective of downgradient receptors. Additional wells are likely to be needed, their number, location, screened intervals and frequency of sampling being dependent on contaminant distributions, site stratigraphy, plume velocity, groundwater geochemistry and travel time to sensitive receptors (Barcelona, 1994; Wiedemeier and Haas, 1999).

7 CASE STUDIES

Of course real life is often not as simple as the conceptual models which are used to explain ideas, or as the protocols would sometimes have us believe. Two case studies below illustrate how natural attenuation can be assessed outside the protocols. In the first case it is probable that the risks have been reduced sufficiently, while the other shows such slow degradation that MNA is a not sufficient management strategy.

7.1 *Rexco plant, Mansfield, UK*

The site produced smokeless fuel, by the Rexco process, between 1935 and 1970 using coal from the on-site colliery. The site is on the Permo-Triassic Sherwood Sandstone, the second most important aquifer in the UK. There is no superficial protective cover to the aquifer, the water table is some 16-20 m below ground at the site, and groundwater velocity is about 60 m y^{-1}. Both ammonium liquor (quench water) and coal tar have contaminated the unsaturated zone but only the ammonium liquor has affected the groundwater. Ammonium liquor from a modern plant using a similar process was analysed and found to contain high concentrations of ammonium (~12800 mg l^{-1}), phenol (~7700 mg l^{-1}), chloride and sulfate. Extensive fieldwork (multilevel boreholes creating 109 monitoring points at 39 locations; Jones *et al.*, 1999) discovered a plume of ammonium (up to 400 mg l^{-1}) down gradient from Rexco but no organics were observed. A second plume was inferred, originating from a lagoon some 800 m east of the plant; this is the subject of continuing investigations (Davison and Lerner, this volume).

Phenol is expected to degrade in groundwater, although there are known to be inhibitory effects at high concentrations. An electron balance suggested that sufficient electron acceptors have mixed into the plume to degrade all the organics (Thornton *et al.*,

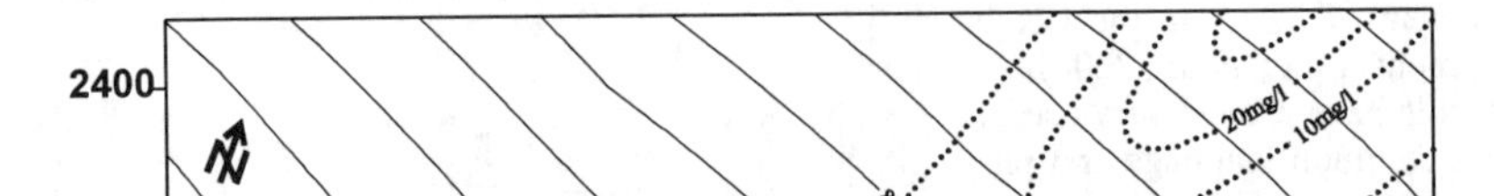

Figure 3. Simulated positions of phenol, ammonium and ethylbenzene plumes at Rexco in 1996, assuming no biodegradation.

1998). However there were no field data to confirm that complete degradation had occurred. The reactions would have been completed some time ago, allowing the geochemical and microbiological signatures in the aquifer to return to a more normal condition before the investigation took place. In addition, the still present ammonium undergoes similar oxidation reactions and so will confuse the evidence provided by field data. Another approach was required to evaluate whether NA had been successful for the phenol.

A groundwater flow and solute transport model was constructed of the site, and calibrated against the ammonium data (Davison and Lerner, in press). Figure 3 shows the model predictions for ammonium and phenol in the absence of biodegradation, and reveals that the phenol was likely to have moved off-site. It is probably beyond the locations of the observation boreholes, which had been drilled before the model was constructed. Thus the absence of phenol in the observation boreholes does **not** prove NA by itself, as it can be explained by the chromatographic separation of the pollutants during transport since the input stopped at the ground surface around 1970.

To differentiate between these explanations an organic compound that has a similar affinity to degrade as phenol and a similar retardation to ammonium was simulated. Ethylbenzene has such properties, and was known to be present in the source. When modelled with just advection, dispersion and sorption processes active, ethylbenzene was predicted to remain around the field site (Figure 3). In reality it is absent and so must have degraded, indicating that degradation is active at the site; NA has probably been successful at this site (but see Davison and Lerner, this volume).

7.2 *4A tar distillery, Wolverhampton*

The 4A plant has manufactured specialty organic chemicals since the site was opened in 1950. Originally it was a coal tar distillery, and it now uses feedstocks brought in from other chemical plants. As at Rexco, the plant overlies the important Permo-Triassic Sherwood Sandstone. Groundwater is about 4 m below ground, and the aquifer is 250 m thick in the vicinity of the site. Flow is towards the west, at about 10 m/y. In 1987, a pollution plume of tar acids and neutral aromatic compounds was found under the site. It extends about 500 m westward, and to a depth of 60 m when defined by the 1 mg l^{-1} phenol contour. The total concentration of organic compounds in the source area is currently 24800 mg l^{-1}. This includes about 12500 mg l^{-1} of phenol, substantial concentrations of cresols and xylenols, and lower concentrations of mono-aromatics and other compounds.

Consultants have provided two contrasting interpretations of the site data, one arguing that NA has been insignificant and that a substantial, engineered cleanup is required. This is estimated to cost £20 000 000. The other consultant has argued that substantial biodegradation has taken place and that no engineered clean up is needed. The difference between these views arises out of interpretation of the source term. As spillages have occurred over 50 years or so, and records are negligible, the amount spilt is almost impossible to estimate from site records.

Our approach was to investigate the plume in more detail with two multilevel samplers which provided sampling ports at 1 m intervals over 35 and 42 m deep boreholes (Lerner *et al.*, in press). Anions (F, Br, Cl, NO_2, NO_3, SO_4, PO_4), cations (Ca, Mg, K, Na, Fe, Mn, Al, Cd, Ni, Zn, NH_4), Si, S^{2-} total P, total S, organics (phenol, o-cresol, m/p-cresol, 2,3-xylenol, 2,4/2,5-xylenol, 2,6-xylenol, 3,5-xylenol, 3,6-xylenol, 4-hydroxybenzoic acid (4-HBA) and 4-

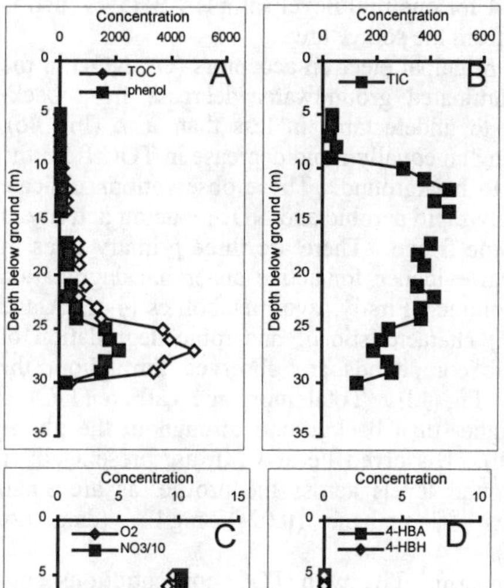

Figure 4. Selected vertical profiles from BH59 at 4A, showing (a) distribution of major phenolic pollutants, (b) concentrations of total inorganic carbon (TIC), (c) O2 and NO3 profiles, and (d) two metabolites of anaerobic biodegradation of phenol.

Table 1. Comparison of NA case studies at Rexco and 4A.

Nature and name of site	Rexco - coal carbonisation	4A coal tar distillery
Source of pollution	Aqueous ammonium liquor containing ~ 13 g phenol l^{-1}.	Non-aqueous phenolics mixture, leading to current in situ concentrations of 13 g phenol l^{-1}.
	Total spillage ~ 400 t of phenol, also NH_4, Cl, SO_4 and other organics in source.	Total spillage ~ 1000 t of phenol, also a substantial load of other polar organics.
	Active at surface 1955-1970	Active at surface 1950-1990
Unsaturated zone	20 m	3 m
Groundwater velocity	60 m y^{-1}	10 m y^{-1}
Changes in flow direction	Major changes up to 90° due to changing local abstraction on several occasions	Minimal
Current status	All organics believed to be degraded.	<10% of organic load degraded
	Substantial NH_4 present in aquifer	
Investigation and assessment	Large field investigation inconclusive. Solute transport modelling required to assess behaviour	Initial large field investigation inconclusive. De tailed interpretation of multilevel profiles, using isotope and geochemistry required
Natural attenuation outcome	NA has eliminated risk	NA will not eliminate risk

hydroxybenzaldehyde (4-HBH)), TOC, TIC, sulphur isotope ratios and wellhead parameters were measured. A small subset of the results are shown in Figure 4 for one multilevel sampler, BH59 which is 150 m from the source area.

The principal electron acceptors (O_2, NO_3) in the uncontaminated groundwater decrease from background to undetectable in less than 2 m (Fig 4c), mirroring an equally rapid decrease in TOC from the plume to background. These observations indicate relatively rapid aerobic and NO_3-reducing activity in the plume fringe. There are three primary lines of chemical evidence for active anaerobic degradation in the plume. Firstly, two metabolites (4-HBA and 4-HBH) characteristic of anaerobic degradation of phenolic compounds are observed throughout the profile (Fig 4d). Total inorganic carbon (TIC) is also higher than background throughout the plume (Fig 4b). Dissolved Fe and Mn are present above background levels across the profile, as are small amounts of methane (0.02-12 mg l^{-1}) (data not shown).

Comparing TIC with TOC concentrations suggests that a maximum of 1% of the TOC has degraded in the 15 y travel time from source to BH59. At a plume scale, the residual TOC is about 2000 t. Both an electron balance and a carbon balance suggest that the residual carbon is about 95% of the original spillage. At both local (BH59) and plume scales, NA has been slow and will not reduce concentrations sufficiently to protect the receptors at risk (Thornton *et al.*, 2000).

7.3 *Comparison of the case studies*

Both Rexco and 4A concern the same organic pollutants (phenolics) in the same aquifer (Sherwood Sandstone) and yet the outcomes are remarkably different. The similarities and differences are highlighted in Table 1. A number of factors came together at Rexco to help NA, such as the thickness of the unsaturated zone and greater mixing caused by higher and changing velocities. None of these factors were beneficial at 4A, with the consequence that NA has not been successful.

8 CONCLUSIONS

The severity of groundwater pollution problems is well known, at least to hydrogeologists. Similarly, the costs and difficulties of engineering clean-ups are well known, with the consequence that society cannot afford to take engineered action to restore many sites, and they have been left alone with little or no formal assessment. Rather than ignoring such sites, a more sophisticated approach is developing in which full account is taken of the natural processes which reduce pollutant fluxes and concentrations. This natural attenuation can sometimes reduce risks from the pollution sufficiently to allow a formal and well-argued decision that active restoration is unnecessary.

Assessing natural attenuation can be difficult, because of the complexity of the biogeochemical reactions involved. Protocols have been developed to help in such assessments, and they work well for straightforward cases, particularly of the common hydrocarbon fuel and chlorinated solvents spillages. Two more complex case studies have been used to show that NA can be assessed even when the protocols cannot be applied.

9 FURTHER INFORMATION AND ACKNOWLEDGEMENTS

NNAGS is the Network on Natural Attenuation in Groundwater and Soils. It is hosted by the Groundwater Protection and Restoration Group and funded by EPSRC to promote research in the field. NNAGS runs courses, workshops and an annual conference and is open for anyone to join by visiting the website, http://www.shef.ac.uk/~nnags/. Steven Thornton is the Environment Agency sponsored Research Fellow on Natural Attenuation, and the case studies were funded by the EU, EPSRC and the Environment Agency.

REFERENCES

American Society for Testing and Materials (ASTM) 1998. *ASTM Guide for Remediation by Natural Attenuation at Petroleum Release Sites.* ASTM, Philadelphia.

Barcelona, M.J. 1994. *Site Characterisation: What Should We Measure, Where, When and Why?* Proc. Symposium on Natural Attenuation of Ground Water: 20-25. U.S. EPA/600/R-94/162.

Buscheck, T. & O'Reilly, K. 1995. *Protocol for Monitoring Intrinsic Bioremediation in Groundwater.* Chevron Research and Technology Company, Health, Environment and Safety Group: 20.

Carey, M.A., J.R. Finnamore, M.J. Morrey, and P.A Marsland 2000. *Guidance on the assessment and monitoring of natural attenuation of contaminants in groundwater.* WRc R&D Technical Report P344.

Davison, R.M. and D.N. Lerner 2000. Proving natural attenuation at a site contaminated by ammonium and phenolics. *This volume.*

Davison, R.M. and D.N. Lerner (in press). Evaluating natural attenuation of groundwater pollution from a coal carbonisation plant: Developing a local scale model using Modflow, MODTMR and MT3D. *Water and Environmental Management Journal.*

Jones, I., D.N. Lerner & O.P. Baines 1999. Multiport socks samplers: a low cost technology for effective multilevel ground water sampling. *Groundwater Monitoring and Remediation,* **19**(1): 134-142.

Kremer, F. 1998a. *Trends in the Use of MNA.* Seminars on Monitored Natural Attenuation for Ground Water: 1-9 to 1-11. U.S. EPA/625/K-98/001.

Kremer, F. 1998b. *Framework for Use of MNA.* Seminars on Monitored Natural Attenuation for Ground Water: 1-15 to 1-18. U.S. EPA/625/K-98/001.

Lerner, D. N., S. F. Thornton, M. J. Spence, S. A. Banwart, S. H. Bottrell, J. J. Higgo, H. E. H. Mallinson, R. W. Pickup, G. M. Williams (in press). Ineffective natural attenuation of degradable organic compounds in groundwater: an initial overview. *Ground Water.*

National Research Council 1994. *Alternatives for Ground Water Cleanup,* National Academy Press, Washington DC, USA: 315.

Thornton, S.F., R.M. Davison, D.N. Lerner and S.A. Banwart 1998. Electron balances in field studies of intrinsic remediation. M. Herbert & K. Kovar (eds), *GQ 98 - Groundwater Quality: Remediation and Protection,* Proceedings of a conference held at Tübingen, September 1998. IAHS publication 250: 273-282.

Thornton, S.F., D.N. Lerner and S.A. Banwart 2000. Assessing the natural attenuation of organic contaminants in aquifers using plume-scale electron and carbon balances. Submitted to *Journal of Contaminant Hydrology.*

U.S. EPA (United States Environmental Protection Agency) 1998. *EPA Policy On Use of Monitored Natural Attenuation for Site Remediation.* Seminars on Monitored Natural Attenuation for Ground Water, 1-3 to 1-6. U.S. EPA/625/K-98/001.

U.S. EPA 1999. *Use of Monitored Natural Attenuation at Superfund, RCRA Corrective Action, and Underground Storage Tank Sites.* U.S. EPA Office of Solid Waste and Emergency Response Directive 9200.4-17P: 32.

Wiedemeier, T.H., & Haas, P.E. 1999. *Designing Monitoring Programs to Evaluate the Performance of Natural Attenuation.* Fifth International Conference on In Situ Bioremediation of Organic Compounds, Batelle Press, Columbus, Ohio, USA: 313-323.

Wiedemeier, T.H., Wilson, J.T., Kampbell, D.H., Miller, R.N. Hansen, J.E. 1995. *Technical Protocol for Implementing Intrinsic Remediation with Long Term Monitoring for Natural Attenuation of Fuel Contamination Dissolved in Groundwater.* U.S. Air Force Center for Environmental Excellence:. XX.

Wiedemeier, T.H., Swanson, M.A., Moutoux, D.E., Wilson, J.T., Kampbell, D.H., Hansen, J.E. & Haas, P. 1997. *Overview of Technical Protocol for the Natural Attenuation of Chlorinated Aliphatic Hydrocarbons in Ground Water Under Development for the U.S. Air Force Center for Environmental Excellence.* Proc. Symposium on Natural Attenuation of Chlorinated Organics in Ground Water: 37-61. U.S. EPA/540/R-97/504.

Wiedemeier, T.H., Rifai, H.S., Newell, C.J. & Wilson, J.T. 1999. *Natural Attenuation of Fuels and Chlorinated Solvents in the Subsurface.* Wiley & Sons:. 617.

Wilson, J.T. 1998. *Risk Management of Monitored Natural Attenuation.* Seminars on Monitored Natural Attenuation for Ground Water: 6-3 to 6-17. U.S. EPA/625/K-98/001.

Groundwater: Past Achievements and Future Challenges, Sililo et al. (eds) © 2000 Balkema, Rotterdam, ISBN 90 5809 159 7

Modeling groundwater flow and mass transport in heterogeneous media: Issues and challenges

Chin-Fu Tsang
Earth Sciences Division, Ernest Orlando Lawrence Berkeley National Laboratory, Calif., USA

ABSTRACT: The need for predictions of groundwater flow and contaminant transport in the subsurface—over large distances and long time periods—has imposed extraordinary demands on the field of hydrogeology. Such a need arises in assessing the safety of a geologic nuclear waste repository and in evaluating groundwater contamination and remediation designs. One of the main difficulties in modeling groundwater flow and mass transport is the heterogeneity of the flow system, both in terms of its characterization through *in situ* measurements and its conceptualization and simulation. This paper reviews some important issues and challenges in modeling flow and transport in heterogeneous media, and discusses approaches to address certain aspects of the problem. Topics discussed include dynamic flow channeling, tracer breakthrough curves, multiple scales for flow in fractured rocks, different scales in measurement, modeling, prediction and heterogeneity, and system characterization and analysis for predictive modeling.

1 INTRODUCTION

The need for predictions of groundwater flow and contaminant transport in the subsurface over large distances and long time periods arises in the safety assessment of a geologic nuclear waste repository or in the evaluation of groundwater contamination and remediation designs. These problems are of great concern in many countries. They require calculations of flow and transport in geologic systems over distances of tens to hundreds of kilometers and time periods of 10,000–100,000 years. This need imposes extraordinary demands on the field of hydrogeology. One of the main difficulties in modeling groundwater flow and mass transport is the heterogeneity of the flow system, both in terms of the system's characterization through *in situ* measurements and its conceptualization and simulation. This is also an important basic science problem in hydrogeology. This paper reviews a number of scientific issues and challenges in flow and transport in heterogeneous media and indicates possible approaches to certain aspects of the problem.

Heterogeneity of groundwater flow systems may be considered in terms of large, intermediate, and small features. The large features include major fault zones, layering structures, or regional compartmentation of areas with different hydraulic conductivity or permeability. The small features include permeability variations on a scale much smaller than the scale of measurement. Intermediate features are those of the same order as the measurement scale. It is quite apparent that large features would have a significant impact on flow and transport. They can be detected and characterized through a careful program of surface geologic and geophysical surveys, downhole logging at a number of locations, and short-term and long-term hydraulic tests. Once these large features are identified and characterized, flow and transport calculations may be made using numerical methods such as the finite-difference or finite-element methods. Continued progress in methodologies and techniques is being made to perform this kind of work, such as geophysical tomography and joint inversion of hydrologic and geophysical data (e.g., Alumbaugh and Neuman 1997; Smith et al. 1999 Wilt et al. 1995; Nekut 1994; Tichelaar et al. 1997; Walsh 1993; Vasco et al. 1997, 1998; Lee 1995; Hyndman et al. 1994; Rector 1995; Rubin et al. 1992; Karasaki et al. 2000).

For small-scale features whose dimensions are much smaller than the measurement scale, some kind of averaging can usually be done. One well-known example is the dispersion coefficient that can be used to describe solute-concentration spreading that results from flow velocity variations (caused in turn by small-scale heterogeneity). A number of authors have developed expressions for dispersivity in terms of permeability variance and correlation length (see Gelhar and Axness 1983; Gelhar 1986, Dagan 1984, 1986, 1990; Neuman et al. 1987; Neuman and Zhang 1990; Rubin 1990). However, for intermediate-scale features (typically 0.1 to 0.5 times the flow distance), the situation is not so simple.

We note that in principle we know how to handle the large and small features. With the former, we can use *in situ* surveys and borehole measurements, followed by deterministic modeling; with the latter, we can employ some kind of averaging and representation. Intermediate-scale heterogeneity is the main scientific challenge. A number of issues related to this challenge are discussed below.

2 DYNAMIC FLOW CHANNELING

Conventional approaches to flow and transport in heterogeneous media, even for intermediate-scale heterogeneity, assume the applicability of the well-known advective-dispersion equation (e.g., Javandel, et al. 1984), which may be written as

$$\frac{\partial}{\partial x_i}\left(D_{ij}\frac{\partial C}{\partial x_j}\right) - v_i\frac{\partial C}{\partial X_i} = \frac{\partial C}{\partial t}$$

where

$$v_i = -\frac{\bar{k}}{n}\frac{\partial h}{\partial x_i}$$

and

$$\frac{\partial^2 h}{\partial x_i^2} = 0$$

In these equations, C is the solute concentration, v_i is the average pore velocity in the direction x_i, D_{ij} is the dispersion coefficient tensor, n is the effective porosity, h is the hydraulic pressure head, $\bar{k}$ is the mean hydraulic conductivity, and x_i are the Cartesian coordinates.

For two-dimensional cases, if the flow is in the x direction with velocity v, and D_{ij} can be defined as D_L along the main direction of flow x and D_T transverse to it, we get the well-known solution for the spread of a tracer plume, which at t = 0 is a line concentration C_o with length 2a normal to the flow direction (Javandel et al. 1984, p. 19):

$$C(x,y,\tau) = \frac{C_o}{4(\pi D_L)^{1/2}}\exp\left[\frac{vx}{2D_L}\right] \cdot \left\{\int_0^t \exp\left\{-\frac{v^2\tau^2 - x^2}{4D_L}\tau^{-5/2}\right. \cdot \left[\operatorname{erf}\left(\frac{a-y}{2(D_T\tau)^{1/2}}\right) + \operatorname{erf}\left(\frac{a+y}{2(D_T\tau)^{1/2}}\right)\right]\right\} d\tau$$

This indicates a simple-body plume of tracers moving along the x direction with concentration smearing in the x and y direction, depending on the parameters D_L and D_T, respectively.

However, in a system with intermediate-scale heterogeneity under a pressure gradient, flow will seek out paths in proportion to their total resistance. The resulting tracer plume will not be a simple-body volume, but will follow a number of separate paths through the heterogeneous medium. These are paths of least total resistance, and such behavior is the *flow channel phenomenon* (Tsang and Tsang 1987; 1989; Moreno and Tsang 1994; Birkholzer and Tsang 1997). Since the main flow paths are dependent not only on the heterogeneity field, but also on the orientation of the pressure gradient, we may call it *dynamic channeling* (Tsang et al. 2000).

To illustrate this phenomenon and its dependence on heterogeneity, let us consider a heterogeneous field that may be generated by using a geostatistical method (the Geostatistical Library GSLIB, Deutsch and Journel 1992). The hydraulic conductivity k can be given by a lognormal distribution:

$$n(k)dk = \frac{1}{\sqrt{2\pi}\,\sigma \ln 10}\exp\left[-\frac{(\log k - \log\bar{k})^2}{2\sigma^2}\right]\frac{dk}{k}$$

where k is the value at each spatial location and σ is its standard deviation in log k, around a mean $\bar{k}$. The spatial variational structure may be described by an exponential covariance function given by

$$E\left[\log k(r_1) - \log\bar{k}\right]\left[\log k(r_2) - \log\bar{k}\right] = \sigma^2 \exp\left[-\frac{2|r_2 - r_1|}{\lambda}\right]$$

where E is the expectation value, r_1 and r_2 are two spatial points, and λ is the spatial correlation length. For the exponential variogram, the effective spatial correlation range is three times the correlation length (see deMarsily 1986). Let us define the ratio of the effective spatial correlation range to the total line flow distance as $\lambda' = 3\lambda/L$. Thus we can characterize the heterogeneous system by three parameters, $\bar{k}$, σ, and λ´.

Having generated the heterogeneous field, let us impose a pressure difference between the upper and lower boundaries, with side boundaries closed to flow. This can be solved straightforwardly by a finite difference or finite element method.

In Figures 1a–c, we show typical results for the 2-D case. Flow domain is discretized into 200 × 200 grid cells, and λ´ is 0.15. The lines show the tracks of the fastest 90% of the particles. For a small

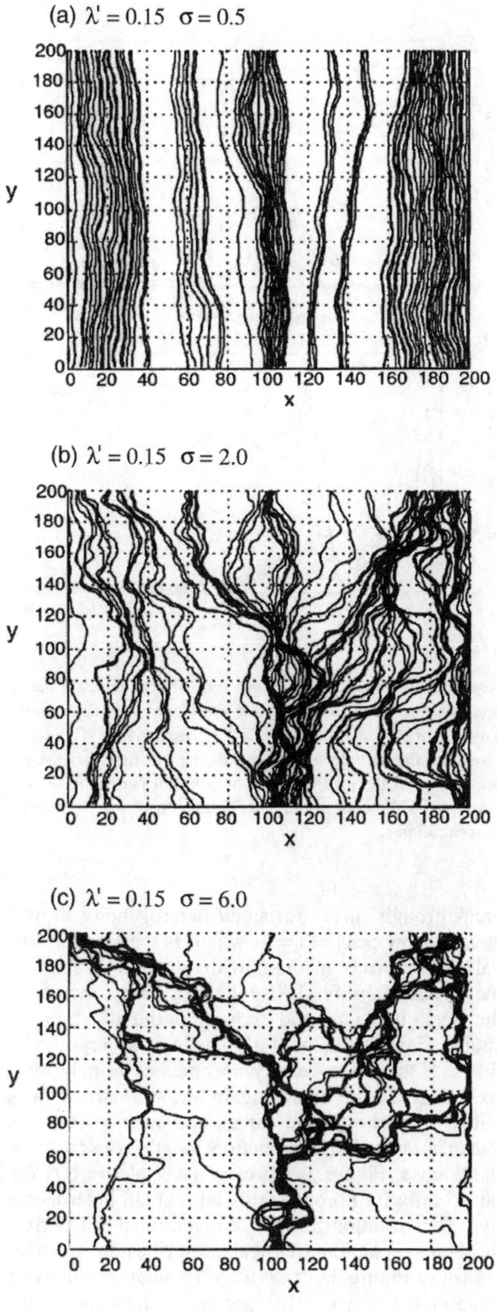

Figure 1a–c. Emergence of flow channeling, under a pressure step applied from the top to the bottom boundary, as a function of σ for a 2-D heterogeneous medium. Tracer flow paths are shown for $\lambda' = 0.15$ and $\sigma = 0.5$, 2.0, and 6.0 in a, b, and c, respectively.

standard deviation of natural log permeability, $\sigma = 0.5$, flow is essentially vertical. The travel-time contrast is small among all the flow paths. However, as σ increases, this contrast greatly increases, and flow becomes channelized (Figures 1b and 1c), with exit flow at the lower boundary concentrated at fewer and fewer locations.

Flow channeling also depends on the spatial correlation range as a fraction λ' of the flow distance. If λ' is small, channeling does occur, but the channels are closely spaced; there are many of them over the flow domain, and their effect is "averaged out." However, for λ' larger than ~0.1, it is found that flow channeling is important and is not very sensitive to the exact λ' value. Some typical results are shown in Figures 2a–c for the same σ value of 2.0, but λ' values of 0.015, 0.15, and 0.3, respectively.

According to these results, intermediate-scale heterogeneity gives rise to flow channeling, so that solute transport along these fast paths is much larger than the mean. Furthermore, a strong spatial variation of solute concentration exists along the low-pressure outflow area of the flow domain. How to deal with such variations in solute transport is an open question.

Similar results are obtained in 3-D systems (Morano and Tsang 1994, and Tsang et al. 2000). Further, flow channeling also occurs in unsaturated systems. In that case, it is also a function of the degree of saturation in the flow domain (Birkholzer and Tsang 1997).

3 TRACER BREAKTHROUGH CURVES

One characteristic feature of flow channeling is the highly asymmetric tracer breakthrough curves at the exit boundary of the flow domain. Let a pulse of solute be deposited at the upper inflow boundary (see Figure 1 or 2) at time t = 0, and let it be followed by particle or tracer tracking. The particles representing the solute are then collected as a function of time at the lower exit boundary, integrated over the lower boundary area. Summing the arrivals of the particles builds up the tracer breakthrough curves, which describe the arrival of the solute resulting from a unit release at the upper boundary. The results represent the case of negligible dispersion, if the particle steps are chosen to be very small relative to the heterogeneity scale.

Figures 3a and 3b show breakthrough curves for different standard deviations in the permeability distribution over a 3-D flow domain and for two different ratios of correlation range to travel distance. For the large correlation range ($\lambda' = 0.3$; Figure 3b), the

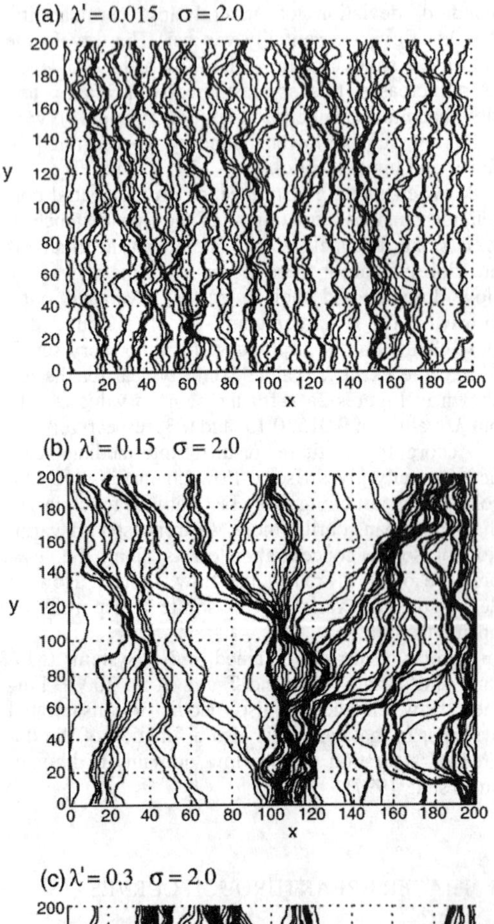

Figure 2a–c. Emergence of flow channeling, under a pressure step applied from the top to the bottom boundary, as a function of λ' for a 2-D heterogeneous medium. Tracer flow paths are shown for σ = 2.0 and λ′ = 0.015, 0.15, and 0.3 in a, b, and c, respectively.

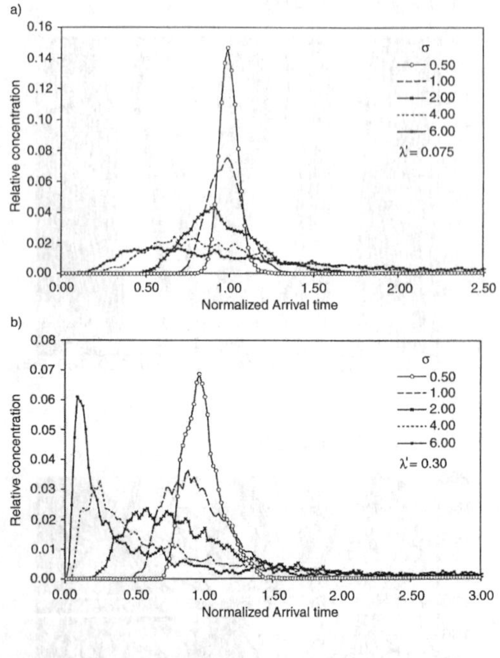

Figure 3a–b. Breakthrough curves for different standard deviation σ values and for a ratio of correlation length to travel length λ of .075 and 0.30 in a and b, respectively. The x-axis gives the arrival time normalized to (i.e., in units of) the expected time if the medium were of constant permeability. The y-axis gives the concentration as a fraction of the input pulse concentration.

breakthrough curves for small heterogeneity show a clear narrow peak at t = 1, which is the time of arrival for a porous medium with uniform geometric-mean permeability. When heterogeneity increases, the peak is wider and moves in the direction of shorter travel time, with a longer tail at large travel times. If the heterogeneity becomes extremely large (σ = 4 or 6), the curves again show a narrow peak but at very short travel times and with a long tail. When a very short correlation length is used (λ′ = 0.075 case, Figure 3a), a clear peak at t = 1 is obtained only for porous media with small heterogeneity. The solution of the conventional advective-dispersive equation accurately describes the situation shown in Figure 3a or 3b for very small σ values up to approximately σ = 1.0 (see curves labeled σ = 0.5 and σ = 1.0). The velocities peak around the mean flow velocity (with arrival time t = 1) and are symmetric on either side of the mean flow velocity. The spread of velocities about the mean increases with σ. However, as σ becomes much larger, with σ > 1.0, flow begins to be focused in a few channels, and the breakthrough curves show a much earlier peak, quite distinct from the t = 1 peak, and a long tail. The early peak arrives in as little as one-tenth the time of travel for a constant-permeability medium (see the $\sigma = 6.0$ case in Figure 3b). Such a phenomenon is believed to have been seen in a number of field experiments (see, for example, the review by Tsang and Neretnieks 1998).

These results have important practical significance. First, because of flow channeling, the peak arrival of a contaminant plume could be as much as an order of magnitude sooner than expected. This is an important concern for evaluating the potential migration of a contaminant plume, and its possibility needs to be accounted for in safety assessments. Second, because of the emergence of flow channeling at strong heterogeneity, the usefulness of the conventional advective-dispersive equation for analyzing this class of tracer breakthrough curves is very much in question.

4 MULTIPLE SCALES FOR FLOW AND TRANSPORT IN FRACTURED ROCKS

For a fractured rock in which flow occurs through a fracture network, multiple levels exist for the scale of heterogeneity. At the small scale, equal to or less than the dimension of a single fracture in the fracture network, there is the spatial range of fracture aperture variability. At that scale, flow and solute transport are affected by a spatial scale, λ_a, characterizing the aperture variations over the single fracture plane. The next scale is the mean spacing, s, between successive fractures in the fracture network. Beyond this is the variation of fracture density over space with a correlation range of d. This is illustrated by Figure 4, which shows schematically the dispersivity (assumed to reflect the heterogeneity scale) as a function of flow distance. When the flow distances are smaller than the single fracture dimension, the dispersivity of solute transport should be of the order of λ_a. When the flow distances are greater than the mean fracture spacing, the dispersivity will be of the order of s, meaning that solute may go through alternative nearby fractures at a distance of about s apart. When the flow distances are even larger, the dispersivity will reflect the spatial correlation range d of fracture density variation. Thus, the dispersivity as a function of transport distance should display a multiple-step structure and will level off at about d if no higher-scale heterogeneity exists. In practice, these multiple steps may merge with each other if the different scales, rather than being represented by single values of λ_a, s or d, are ranges of values overlapping with each other.

This concept is demonstrated by Nordqvist et al. (1996). In a detailed calculation, they developed a fracture network model that includes the effect of fracture aperture variation on each fracture in the network. One example of their results is shown in Figure 5, where the tracer breakthrough curves are shown for a number of transport distances. Fracture spacing s ranges between 4 and 8 m. Note the different x-axis scales for the different curves in this figure. At very small transport distances, less than 6 m, the breakthrough curves show a sharp peak with a very narrow spread, corresponding to λ_a. As the transport distance becomes larger than s (8–12 m), we see a two-peak structure. The separation between the two peaks corresponds to the effect of two alternative paths in the network separated by s. In these calculations, there is no spatial variation of fracture density. Thus, as the transport distance becomes larger than s, the breakthrough curves display a smooth single-hump structure, with its spread corresponding to the fracture spacing. These alternative tracer breakthrough curves have been noticed in field experiments (see the review by Tsang and Neretnieks 1998).

5 DIFFERENT SCALES IN MEASUREMENT, MODELING, PREDICTION, AND HETEROGENEITY

Predictive modeling of flow and transport in a heterogeneous medium involves a number of scales besides that of the heterogeneity itself—from the scale of measurements for obtaining mediumparameters, to grid cell size of the calculational model, to the sampling size or the scale for predictive quantities. These are illustrated in Figure 6.

In this figure, the heterogeneity scale S_H is the separation between the gray circular areas representing permeability heterogeneity. Measurements, with the measurement scale S_M, may then cover

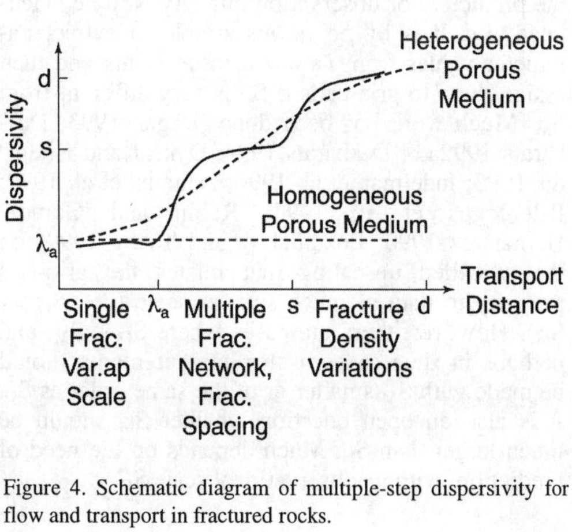

Figure 4. Schematic diagram of multiple-step dispersivity for flow and transport in fractured rocks.

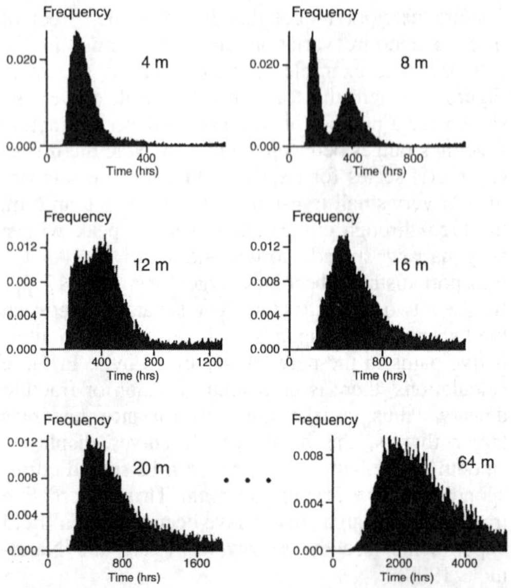

Figure 5. Tracer breakthrough curves for a fracture network, with variable apertures for each fracture. The fracture spacing ranges from 4 to 8 m. The transport distance is labeled on each plot.

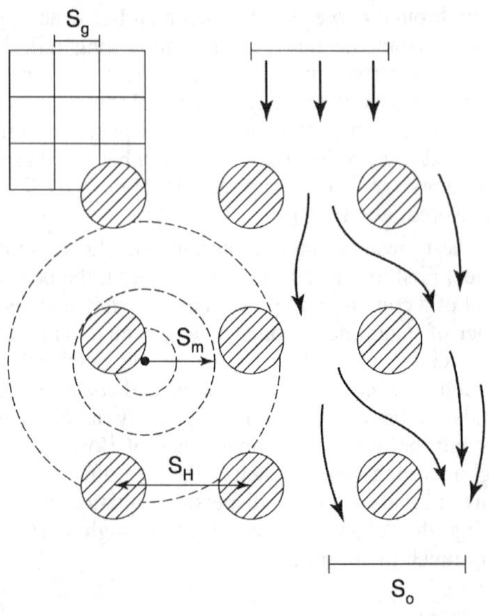

Figure 6. Schematic diagram of the different scales in predictive modeling: scale of heterogeneity S_H, scale of measurement S_M, scale of model grid S_G and scale of observation or predictive quantity S_O.

different regions of the system. If the measurement data are permeability data from cores, S_M will be on the order of core size, which is very small. If the measurement data are from well testing, S_M will be larger. If they are from long-term interference pressure tests, then S_M will be much larger. As can be seen, measurements with different S_M will sample very different permeability properties, depending on the relative value of S_M and S_H.

In performing predictive modeling, we need to account for two additional scales. The first is the scale of the grid cell for numerical calculations, S_G; the second is the dimension of the area over which the predictive or observation quantity is to be calculated, S_O. It is by no means simple to extract parameter values from *in situ* measurements and then assign them to grid cells if S_G is very different from S_M. Much work has been done (Dagan 1993; Desbarats 1992a,b; Desbarats 1994; Dykaar and Kitanidis 1993; Indelman et al. 1996; Neuman et al. 1992; Paleologos et al. 1996; Rubin and Gomez-Hernandez 1990; Pozdniakov and Tsang 1999) on the so-called upscaling that relates the grid-cell property to data of much smaller scale (i.e., $S_G >> S_M$). However, there are cases where $S_G < S_M$, and perhaps in such cases further measurements should be made with S_M smaller or of the same order as S_G. It is also an open question whether S_M should be much larger than S_H. Much depends on the need of prediction, with its observational scale S_O.

The lesson to learn is that in modeling, we need to be aware of these different scales—S_H, S_M, S_G, and S_O. To conduct modeling calculations with input parameters without understanding these four scales could be an important source of error, one that is very often overlooked. On the whole, how these four scales interplay and the proper way to deal with them in predictive models are yet to be determined.

One example to illustrate the importance of this issue is the use of dispersivity to calculate solute transport. Very roughly, dispersivity may be used to represent the effect of heterogeneity and corresponds to S_H. If we solve this problem with an advective-dispersive equation and use an observation scale S_O much smaller than S_H, the answer can be very wrong, because the advective-dispersive equation smooths out solute concentration over the S_H scale. In other words, solutions of the conventional advetive-dispersive equation are only valid if $S_O \gtrsim S_H$. It is possible to find examples in the literature where the advective-dispersive equation has been incorrectly used (i.e., when this condition does not hold).

6 SYSTEM CHARACTERIZATION AND ANALYSIS FOR PREDICTIVE MODELING

Prediction of flow and solute transport in a groundwater system requires characterizing the system to identify relevant features and processes and to obtain parameter values for model inputs. The required types of *in situ* measurements and observations depend on the predictive quantities to be calculated. The relationship was discussed in Tsang et al. (1994). It is not a simple problem because we must ensure that all the important features and processes are included in the model. In recent years, a number of countries have tried to develop procedures and methodologies for ensuring adequate characterization of a given site. These efforts have been made to enable modeling that includes a proper uncertainty estimate. The goal is predictions on nuclear waste repository safety over tens or hundreds of thousand years. Nuclear waste management organizations that have gone through such efforts include NIREX in UK (NIREX95 1995), SKI and SKB in Sweden (Site94, 1996 and SR97 1999), and Posiva in Finland (TILA-97 1999).

The procedures developed by these organizations can be broadly divided into a number of key steps:

(1) Identification of features, events, and processes (FEPs) relevant to the site under consideration.
(2) Evaluation of the interplay and influence of FEPs with each other.
(3) Geologic surveying and evaluation of the site, especially identification of major faults and fractures.
(4) Use of multidisciplinary data and information, such as geologic, geophysical, hydrological and geochemical input. Investigation of their consistency in developing site characteristics.
(5) Analysis of system evolution under different scenarios that may occur over the next 10,000 to 100,000 years (including changes in rainfall levels and possible glaciation over the site).
(6) Use of three kinds of models: (a) models used in data interpretation; (b) models used in observation of system behavior and estimation of effective parameters; and (c) regional flow models to be used for predictive calculations.
(7) Use of alternative models with different degrees of sophistication to analyze the system and compare results.
(8) Evaluation of uncertainties not only in parameter values, but also in FEPs that describe the system and in scenarios and external processes that impact the system.
(9) Use of "expert opinions" and a structured expert elicitation process for parameters and information that are not available for the particular site under study.
(10) Use of stochastic and probabilistic modeling methods.

As we can see from this list, it is a complex procedure, made necessary by the demand for defensible predictive modeling of flow and transport in heterogeneous field sites. Overall, these are preliminary efforts using techniques that are still improving and evolving. Much progress is needed before we can be confident in our prediction of system behavior and its uncertainty range.

7 SUMMARY AND CONCLUDING REMARKS

In this paper, I have discussed key scientific issues in the modeling of flow and mass transport in heterogeneous media, of which fractured rock is a particular class. Permeability heterogeneity on a scale comparable with the scale of the flow domain presents challenges, ranging from dynamic flow channeling, multipeak tracer breakthrough curves, and multiscale dispersivity for fractured rocks to the interrelationship between measurement scale, heterogeneity scale, numerical grid scale, and the scale for which predictive quantities are to be calculated. All these issues are unresolved and deserve further research.

Finally, note that for the safety assessment of a nuclear waste repository or remediation of a contaminated site, a system characterization and analysis is needed to make a defensible prediction of system behavior, with a proper estimate of uncertainties. Considerable work towards this end has been applied in different countries, and continued work is expected in the coming years. This work affords us further opportunity to study the issues presented in this paper and to attempt to meet the need of society for long-term and large-scale predictions of flow and mass transport in realistic geologic systems.

8 ACKNOWLEDGMENT

We thank C. Doughty and T.N. Narasimhan for their careful review of the manuscript. This work was jointly supported by the Japan Nuclear Cycle Development Institute (JNC) under a binational JNC/DOE Collaborative Agreement Annex with the DOE Office of Environmental Management, Office of Science and Technology, and by the Director, Office of Science, Office of Basic Energy Sciences, of the U.S. Department of Energy (DOE) under Contract No. DE-AC03-76SF00098.

9 REFERENCES

Alumbaugh, D.L. and G.A. Newman 1997. Three-dimensional massively parallel electromagnetic inversion: II. Analysis of a crosswell experiment. *Geophys. J. Int.* 128:355–363.

Birkhölzer, J. and C.F. Tsang 1997. Solute channeling in unsaturated heterogeneous porous media. *Water Resour. Res.* 33(10):2221–2238

Dagan, C. 1993. Higher-order correction of effective conductivity of heterogeneous formations of lognormal conductivity distribution. *Transp. Porous Media* 12:279–290.

Dagan, G. 1984. Solute transport in heterogeneous porous formations. *J. Fluid Mech.* 145:141–177.

Dagan, G. 1986. Statistical theory of groundwater flow and transport: Pore to laboratory, laboratory to formation, and formation to regional scale. *Water Resourc. Res.* 22(9):120S–134S.

Dagan, G. 1990. Transport in heterogeneous porous formations: spatial moments, ergodicity, and effective dispersion. *Water Resour. Res.* 26(6):1281–1290.

Desbarats, A.J. 1992a. Spatial averaging of the transmissivity in heterogeneous fields with flow toward well. *Water Resour. Res.* 28(3):757–767.

Desbarats, A.J. 1992b. Spatial averaging of the hydraulic conductivity in three-dimension heterogeneous porous media. *Mathl. Geol.* 24(3): 249–267.

Desbarats, A.J. 1994. Geostatistical analysis of aquifer heterogeneity from the core scale to the basin scale: A case study. *Water Resour. Res.* 30(3):673–684.

Deutsch, C.V. and A.G. Journal 1992. GSLIB—Geostatistical Software Library and Users Guide, Oxford University Press, New York and Oxford.

Dykaar, B.B. and P.K. Kitanidis 1993. Transmissivity of heterogeneous formation. *Water Resour. Res.* 29(4):985–1001.

Gelhar, L.W. 1986. Stochastic subsurface hydrology from theory to applications. *Water Resour. Res.* 22(9):135S–145S.

Gelhar, L.W. and C.L. Axness 1983. Three-dimensional stochastic analysis of macrodispersion in aquifers. *Water Resour. Res.* 19(1):161–180.

Hyndman, D.W., J.M. Harris, and S.M. Gorelick 1994. Coupled seismic and tracer test inversion for aquifer property characterization. *Water Resour. Res.* 30(7), 1965–1977.

Indelman, P., A. Fiori, and G. Dagan 1996. Steady flow toward wells in heterogeneous formation: mean head and equivalent conductivity. *Water Resour. Res.* 32(7):1975–1983.

Karasaki, K., B. Freifeld, A. Cohen, K. Grossenbacher, P. Cook, and D. Vasco 2000. A multidisciplinary fractured rock chaacterization study at Raymond Field Site, Raymond, California. *Journal of Hydrology*, accepted for publication.

Javandel, I., C. Doughty, and C.F. Tsang 1984. Groundwater Transport, *Water Resources Monograph* 10, American Geophysical Union, Washington, D.C.

Karasaki, K., B. Freifeld, A. Cohen, K. Grossenbacher, P. Cook, and D. Vasco 2000. A multidisciplinary fractured rock characterization study at Raymond Field Site, Raymond, California. *Journal of Hydrology*, accepted for publication.

Lee, D.S., V.M. Stevenson, P.F. Johnston, and C.E. Mullen 1995. Time-lapse crosswell seismic tomography to characterize flow structure in the reservoir during the thermal stimulation. In Rector, James W., III (ed.) Crosswell methods. *Geophysics* 60(3):660–666.

deMarsily, G. 1986. Quantitative hydrogeology, Orlando, Florida: Academic Press.

Moreno, L. and C.-F. Tsang 1994. Flow channeling in strongly heterogeneous porous media: A numerical study. *Water Resour. Res.* 30(5):1421–1430.

Nekut, A.G., 1994. Electromagnetic ray-trace tomography. *Geophysics* 55:371–377.

Neuman, S.P. and Y-K. Zhang 1990. A quasi-linear theory of non-Fickian and Fickian subsurface dispersion. 1. Theoretical analysis with application to isotropic media. *Water Resour. Res.* 26(5):887–902.

Neuman, S.P., C.L. Winter, and C.M. Newman 1987. Stochastic theory of field-scale dispersion in anisotropic porous media. *Water Resour. Res.* 23(3):453–466.

Neuman, S.P., S. Orr, O. Levin, and E. Paleologos 1992. Theory and high-resolution finite element analysis of 2D and 3D effective permeability in strongly heterogeneous porous media. In T.F. Russell, et al. (eds.*), Mathematical Modeling in Water Resources*, Vol. 2. New York: Elsevier, pp. 118–136.

NIREX95 1995. A preliminary analysis of the groundwater pathways for a deep repository at Sellafield. Volume 1: Development of the Hydrogeological Conceptual Model. Volume II: Derivation of Effective Hydrogeological Parameter for Regional Model III: Calculations of Risk. United Kingdom NIREX Limited, Science Report 5/95/012.

Nordqvist, A.W., Y.W. Tsang, C.F. Tsang, B. Dverstorp, and J. Andersson 1996. Effects of high variance of fracture transmissivity on transport and sorption at different scales in a discrete model for fractured rocks. *J. of Contaminant Hydrology* 22(1–2):39–66.

Paleologos, E.K., S.P. Neuman, and D. Tartakovsky 1996. Effective hydraulic conductivity of bounded, strongly heterogeneous porous media. *Water Resour. Res.* 32(5):1333–1342.

Pozdniakov, S.P. and C.-F. Tsang 1999. A semianalytical approach to spatial averaging of hydraulic conductivity in heterogeneous aquifers. *Journal of Hydrology*, 216, 78–98.

Rector, J.W., III 1995. Crosswell methods. *Geophysics* 60:627–920.

Rubin, Y. 1990. Stochastic modeling of macrodispersion in heterogeneous porous media. *Water Resour. Res.* 26(1):133–141.

Rubin, Y. and J.J. Gomez-Hernandez 1990. A stochastic approach to the problem of upscaling of conductivity in disordered media: Theory and unconductional numerical simulations, *Water Resour. Res.* 26(4):691–701.

Rubin, Y., G. Mavko, and J. Harris 1992. Mapping permeability in heterogeneous aquifers using hydrological and seismic data. *Water Resour. Res.* 28(7), 1809–1816.

Site94 1996. Deep repository performance assessment project. Volumes 1 and 2, Swedish Nuclear Power Inspectorate (SKI), Report 96:36.

SR97 1999. Post-closure safety. Deep Repository for Spent Nuclear Fuel. Volumes I and II. Swedish Nuclear Fuel and Waste Management Company (SKB).

Smith, T., M. Hoversten, E. Gasperikova, and F. Morrison 1999. Sharp Boundary inversion of 2D magnetotelluric data. *Geophysical Prospecting* 47:469–486.

TILA-97 1999. Safety assessment of spent fuel disposal in Hästholmen, Kivetty, Olkiluoto, and Romuvaara. Posiva OY, Finland.

Tichelaar, B.W. and P.J. Hatchell 1997. Inversion of 4-C borehole flexural waves to determine anisotropy in a fractured carbonate reservoir. Geophysics 62(5):1432–1441.

Tsang, C.F. and I. Neretnieks 1998 Flow channeling in heterogeneous fractured rocks. Reviews of Geophysics 36(2):275–298.

Tsang, C.F., L. Gelhar, G. de Marsily, and J. Andersson 1994. Solute transport in heterogeneous media: A discussion of technical issues, coupling site characterization, and predictive assessment. Adv. in Water Res. 17(4):259–264.

Tsang, C.F., L. Moreno, Y. Tsang, and J. Birkholzer 2000. Dynamic channeling of flow and transport in saturated and unsaturated heterogeneous media. Accepted for Geophysical Monograph 42, T. Nicholson (ed.).

Tsang, Y.W. and C.F. Tsang 1987. Channel model of flow through fractured media. Water Resour. Res. 23(3):467–479.

Tsang, Y.W. and C.F. Tsang 1989. Flow channeling in a single fracture as a two-dimensional strongly heterogeneous permeable medium. *Water Resourc. Res.* 25(9):2076–2080.

Vasco, D.W., J.E. Peterson, Jr., and K.H. Lee 1997. Ground-penetrating radar velocity tomography in heterogeneous and anisotropic media. *Geophysics* 62(6):1758–1773.

Vasco, D.W., K.K. Karasaki, and L. Myer 1998. Monitoring of fluid injection and soil consolidation using surface tilt measurements. *Journal of Geotechnical and Geoenvironmental Engineering*. 124(1):29–37.

Walsh, J.J. (Society of Exploration Geophysicists, 63rd Annual Meeting and International Exhibition, Washington, DC, United States) 1993. Fracture estimation from parametric inversion of SV waves in multicomponent offset VSP data. In: Society of Exploration Geophysicists; Expanded Abstracts with Biographies; Technical Program; 63rd Annual Meeting and International Exhibition. 63:140–142

Wilt, M.J.; D.L Alumbaugh, H.F. Morrison, A. Becker, K.H. Lee, M. Deszcz-Pan, 1995. Crosswell electromagnetic tomography; system design considerations and field result. *Geophysics* 60(3):871-885.

1 Determination of aquifer parameters, resource assessment/evaluation

Topic co-editors:

Johan de Beer
CSIR, South Africa

Richard Taylor
University of Surrey, UK

John Weaver
CSIR, South Africa

Gerrit van Tonder
University of the Orange Free State, South Africa

Willy Burgess
University College London, UK

Allan Woodford
Toens and Partners, South Africa

Groundwater: Past Achievements and Future Challenges, Sililo et al. (eds) © 2000 Balkema, Rotterdam, ISBN 90 5809 159 7

The applicability of DC resistivity in some geological surroundings in Sweden

J. Aaltonen
Division of Land and Water Resources, Royal Institute of Technology, Stockholm, Sweden

ABSTRACT: Direct current (DC) resistivity measurements have been used successfully for several years to delineate groundwater contamination from landfill areas, however the usefulness of the method differs due to the prevailing environments. This paper shows that the geology is the most common feature creating uncertain results. This is mainly due to irregular subsurface conditions, grounding problems for the electrodes, or difficulties to separate an anthropogenic effect from natural low resistivities. In Sweden, especially till, clay and peat areas cause problems in studies of leakage from landfill areas.

1 INTRODUCTION

As groundwater resources become more and more exploited the need increases for better resource evaluation techniques, as well as better monitoring tools. DC resistivity methods have been used for a long time, for groundwater investigation purposes, e.g. for delineation of contaminant plumes and mapping of subsurface geological conditions (e.g. Kalinski et al. 1993, Nobes 1995, Benson et al. 1997). However, DC resistivity methods have a number of limitations, for example the sensitivity to irregular subsurface conditions and the difficulty in adjusting the achieved physical model to a single geological model (e.g. White et al. 1984).

Different geological surroundings can be expected to also react differently as a result of seasonal variations in precipitation and temperature. It is important to be aware of these limitations and have sufficient knowledge of how the resistivity method reacts in different geological surroundings for a correct interpretation of the resistivity results achieved (Aaltonen 1998a,b).

This paper endeavours to show the applicability and discuss difficulties and problems of DC resistivity measurements in different geological surroundings in Sweden. This is done with a number of case studies, which have been aimed at delineating groundwater contamination due to landfill leakage.

2 FIELD EXPERIENCE

A number of geophysical investigations were carried out in different geological environments surrounding sanitary landfills in Sweden (Fig.1) from 1995 to 1999 (Aaltonen 1998a,b). The DC resistivity investigations were carried out with an ABEM Terrameter SAS300. Standard electrode configurations have been used in all of the investigation comprising Wenner for profiling and Schlumberger for the soundings.

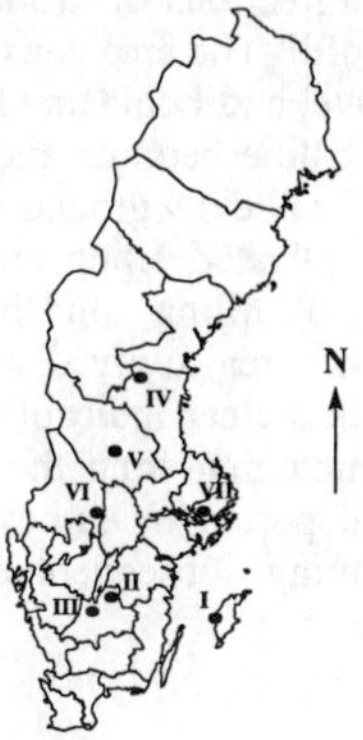

Figure 1. The field investigation areas in Sweden.

2.1 *Case I*

The aim was to verify leakage diffusion from a municipal landfill. The area comprise flat terrain, a thin overburden (< 0,5 m), karstic limestone with several fracture zones and groundwater levels between 1 and 10 m below ground surface.

The area was a complicated geological environment with unevenly distributed fractures in the karstic limestone and with a thin overburden which caused difficulties with electrode grounding. However, profiling delimited areas with lower resistivity and soundings revealed indications of leachate in the fractures. The analyses of soundings was complex with a multitude of layers, both thick and thin (Fig. 2). The contrast between the limestone and fractures with groundwater, was however detectable.

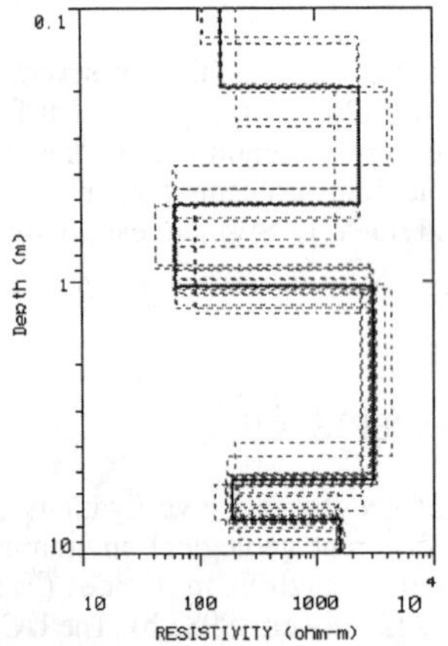

Figure 2. Soundings revealing the karstic structure, with fractures (low resistivity zones) at 1 and 8 m depth.

2.2 *Case II*

The aim was to clarify directions of groundwater flow from a sanitary landfill. The area consists of a hilly terrain of sand, gravel and boulders (15-35 m thick) underlain by crystalline bedrock and with a groundwater level 6 to 13 m below ground.

The hilly terrain complicated siting and interpretation of soundings. Profiling, on the other hand, gave an interpolated resistivity distribution over the whole area, with a clear indication of the affected groundwater, stretching from the landfill area. This result was supported by removing the topographical effects during subsequent analyses (Fig. 3).

2.3 *Case III*

The purpose was to investigate the use of DC resistivity as a complement to groundwater monitoring programs based on water chemical sampling. Thick layers of peat (approx. 5 m), underlain by fine sand and clay cover the area around the municipal landfill. Crystalline bedrock occurs at a depth of 5 m and the groundwater level is at a depth of 2 m.

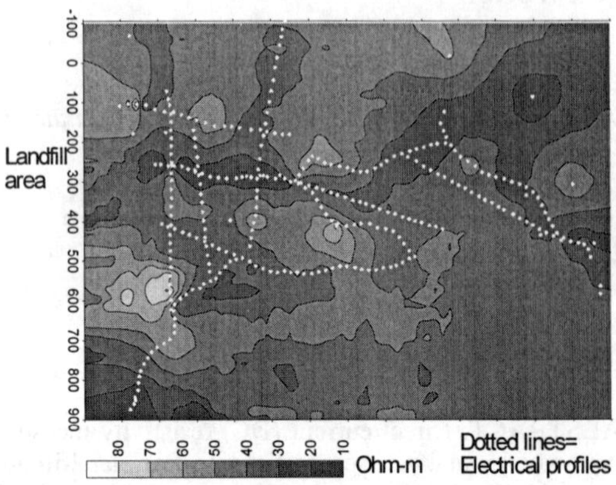

Figure 3. Resistivity distribution in the investigation area.

This geological environment was very suitable for both grounding electrodes and measuring DC resistivity due to the soft, flat terrain with a corresponding horizontal groundwater surface.

No clear indication of leachate was found outside the system of protective ditches, neither from chemical sampling nor from DC resistivity measurements. However, the low resistivity associated with peat and clay as well as the leachate can make it difficult to separate low resistivity measurements due to the presence of leachate from low resistivity measurments due to natural geolocial conditions. Interpretation of sounding curves measured between the ditch system indicated the presence of affected layers. This was also compared with measurements several years before, showing clear agreements. These results are important for future monitoring systems.

2.4 *Case IV*

The aim was to map possible leachate diffusion from the landfill (mining and waste rock) which could be a source of groundwater contamination. The study area is characterised by undulating topography, with boulder rich till (0.5 to 1 m in ø), spotwise covered with a thin layer of peat, characterise the area together with a groundwater level of 2 m depth.

It was hard to obtain good grounding for the electrodes in between the boulders and thereby difficult to ascertain the quality of the results achieved. The till can further make the resistivity vary both spatially and temporally over several decades due to the number of boulders and the heterogeneous texture in between.

Resistivity sounding proved more successful than profiling because it provided a more detailed interpretation of the different layers rather than profiling that depends on differences n apparent resistivity at a specific current electrode spacing. Even though the soundings gave results, they were comparably complex to interpret. It was possible, in several cases, to obtain an equally reliable model with a leachate layer as without (Fig. 4).

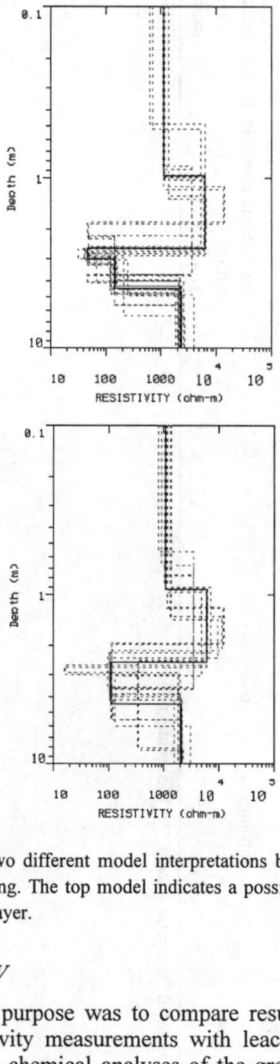

Figure 4. Two different model interpretations based on the same sounding. The top model indicates a possible leachate in the third layer.

2.5 *Case V*

The main purpose was to compare results of the DC resistivity measurements with leachate mapping using chemical analyses of the groundwater. The area around the landfill comprises thick layers of flat sandy material (sand, loamy sand and sandy silt), bordered by an extensive wetland area. The groundwater level is found 2-3 m below the ground and the sandstone bedrock 15 m below water level.

The uniform geology enabled good and clear results. The profiling showed good agreement with the chemical sampling. The geological properties could be considered fairly homogeneous so the variations in the apparent resistivity could be assumed as directly related to the ionic strength of the groundwater. The resistivity measurements in the adjacent wetland were almost impossible to interpret due to the water abundance and the naturally low resistivity values.

2.6 *Case VI*

The aim was to verify leakage from the landfill site (chemical wastes and sludge) and if possible to map the extent of this leakage. The area is dominated by till and silty clay with a bedrock level 10 m below surface. The whole investigation area is sloping with more clayey soils in the lower parts.

Even though the study area was rather small, the changing surface geology complicated the interpolation of profiling data. Similar low resistivity areas were obtained at different locations with only a few cases related to leachate presence. The other low resistivity areas were related to geological factors, as clay and a shallower groundwater table. A leachate plume could, however, be mapped out and confirmed by groundwater samples.

2.7 *Case VII*

The aim was to use DC resistivity for continuos monitoring of possible leachate from the landfill (industrial and construction waste). The area is characterised by two valleys of which one is filled with silty clay and the other with peat underlain by silty clay and sandy till. Sandy till and hard crystalline rocks on the higher elevations dominate the sides of the valleys.

Permanent electrodes for profiling have proved very successful during three year of monitoring (Fig. 5). The seasonal variation of resistivity in this hydrogeological environment seems to be within 15% and the effects of ground frost is avoided with the permanent electrodes. Problems have occured in the lower elevations, with separation of the naturally low resistivity values in the clay from an affecting leachate.

3 CONCLUSION

Table 1 is a compilation of the previous studies. As seen the geology is a prevalent factor creating difficulties for DC resistivity measurements aimed at delineating groundwater contaminants. Irregular subsurface conditions are one of the most important cause of interference on DC resistivity data (e.g. Greenhouse & Harris 1983, Stierman 1984). For example, the detection of contaminants can be difficult in lithologies, which is naturally of low resistivity, such as clays. A high contamination

Table 1. A compilation of the applicability of DC resistivity measurements for mapping groundwater contaminants in different geological surroundings of Swedish landfills. Level of evaluation: + better methods are available ++ satisfactory +++very good / o bad oo rather good ooo good (Modified after Aaltonen 1998b).

Field area	Geology	Usefulness of DC resistivity	Causes of uncertainty	Results achieved	Suitable improvements in combination with DC resistivity	Comments on seasonal variation	Considerations to make
I	Karstic limestone	++	Multitude of thin layers	oo	GPR**	Groundwater level varies considerably even over a short time	• Grounding of electrodes • Multitude of fractures, both horizontal and vertical
II	Sand/Gravel/ Boulders on crystalline bedrock	+++ / ++ (VES*)	Topography (VES)	ooo / o (VES)	GPR, EM***	Rather large (20%) reflecting the fluctuation in groundwater level	• Placing of soundings in hilly terrain
III	Peat/Clay/Fine sands on crystalline bedrock	++(+)	Geology	o(o)			• Separation of naturally low resistivity from an influence caused by a leachate
IV	Boulder rich till on crystalline bedrock	+	Geology	o	EM		• Grounding of electrodes • Boulders • Type of fine material
V	Sand/Loamy sand/Sandy silt on sandstone bedrock	+++ / + (wetland)	Water content (wetland)	ooo / o (wetland)	EM		• Groundwater level
VI	Silty clay/Till on crystalline bedrock	+++	Geology	oo	GPR, soil sampling		• Local variation in geology
VII	Silty clay/Sandy till on crystalline bedrock	+++	Geology	ooo / o(o) (in clay)	EM	Variation in resistivity is within 15 %	• Separation of naturally low resistivity from an influence caused by a leachate

*Vertical Electrical Sounding ** Ground Penetrating Radar *** Electromagnetical methods, like Slingram

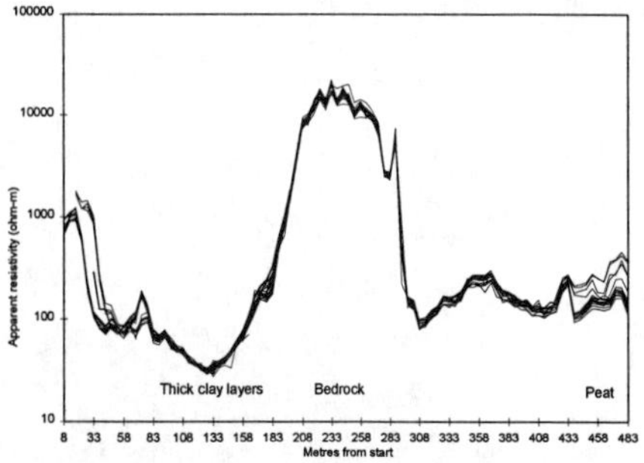

Figure 5. Results from monitoring. The different lines represent monthly resistivity values from 1996 to 1999.

level is needed for a significant and measurable difference (e.g. Baker 1990, Buselli et al. 1990).

The topography is also a cause for concern, especially for sounding data, where uniform horizontal layering is assumed for processing and interpolation. It can be quite useful to rotate the resistivity array to determine the azimuthal dependence of the apparent resistivity (e.g. Stierman 1984, Nobes 1996).

Another geological factor to take into account is the difference between groundwater level (i.e. the free water surface in an observation tube) and the capillary zone. The capillary zone may, in more fine grained materials, be interpreted as the groundwater zone in resistivity measurements, since the pore water is most often connected and thus results in a lower resistivity (Van Dam 1976, Aaltonen 1998a).

Some considerations also have to be made on how the present season is affecting the results achieved in variation of groundwater levels, ground moisture, for example (Buselli et al. 1992; Aaltonen 1998b).

Finally, it has to be emphasised that most often a single geophysical investigation technique is not satisfying enough so either the investigation should be complemented with other geophysical techniques (see Table 1) and/or with geotechnical investigations.

However, DC resistivity has in most of the presented cases proved useful and given solution to problems connected to groundwater contaminant mapping.

REFERENCES

Aaltonen, J. 1998a. *Landfill Investigations with Direct Current Resistivity Methods - A literature review*, KTH, Stockholm.

Aaltonen, J. 1998b. *The Use of Direct Current Resistivity Methods in Landfill Investigations - A study on glaciated terrains*, Lic. Thesis 2025, KTH, Stockholm.

Barker, R.D. 1990. Investigation of groundwater salinity by geophysical methods. In: Ward, S.H. (ed.). *Geothecnical and environmental geophysics*, II: 201-212.

Benson, A.K., K.L. Payne &M.A. Stubben 1997. Mapping groundwater contamination using DC resistivity and VLF geophysical methods - A case study. *Geophysics* 62(1): 80-86.

Buselli, G., C. Barber, G.B. Davis & R.B. Salama 1990. Detection of groundwater contamination near waste disposal sites with transient electromagnetic and electrical methods. In: Ward, S.H. (ed.). *Geothecnical and environmental geophysics*, II: 27-40.

Buselli, G., G.B. Davis, C. Barber, M.I. Height & S.H.D. Howard 1992. The application of electromagnetic and electrical methods to groundwater problems in urban areas. *Exploration Geophysics* 23: 543-555.

Greenhouse, J.P. & R.D. Harris 1983. Migration of contaminants in groundwater at a landfill: A case study. *Journal of Hydrology* 63: 177-197.

Kalinski, R.J., W.E. Kelly & I. Bogardi 1993. Combined use of geoelectric sounding and profiling to quantify aquifer protection properties. *Ground Water* 31(4): 538-544.

Nobes, D.C 1996. Troubled water: Environmental applications of electrical and electromagnetic methods. *Surveys in Geophysics* 17: 363-454.

Stierman D.J. 1984. Electrical methods of detecting contaminated groundwater at Stringfellow waste disposal site, California. *Environmental Geological Water Science* 6(1):11-20.

Van Dam, J.C. 1976. Possibilities and limitations of the resistivity method of geoelectrical prospecting in the solution of geohydrological problems. *Geoexploration* 14: 179-193

White, R.M., D.G. Miller, S.S. Bradwein & A.F. Benson 1984. Pitfalls of electrical surveys for groundwater contamination investigations. *Proceedings of the NWWA/EPA Conference on Surface and Borehole Geophysical Methods in Groundwater Investigations*, February 7-9, San Antonio, Texas.

Groundwater: Past Achievements and Future Challenges, Sililo et al. (eds) © 2000 Balkema, Rotterdam, ISBN 90 5809 159 7

The vertical head distribution in a coastal sand-dune aquifer

R.I.Acworth, R.Soriano & I.L.Turner
UNSW Groundwater Centre, University of New South Wales, Sydney, N.S.W., Australia

Bundled piezometers have been installed to 18 m depth in a 5 km long, 350 m wide coastal sand-dune aquifer at Hat Head in northern New South Wales, Australia. The ocean boundary of the aquifer is dominated by beach face wave runup, nearshore wave setup and tides which each have an impact throughout the aquifer. The inner boundary is a tidal creek without the impact of beach face wave runup. Vertical head variation has been measured at 3 locations across the aquifer using manometer boards to measure the head at 50 mm increments of depth. Vertical head variation in excess of 500 mm has been detected close to the creek boundary with a positive head in sands at depth. The vertical head variation has also been recorded at 60 minute intervals in shallow and deep piezometers at each of the 3 bundled piezometer locations. The observed distribution of hydraulic head and the fluid electrical conductivity distribution measured in the bundled piezometers are used to constrain conceptual models of groundwater flow within the aquifer.

1 INTRODUCTION

Groundwater discharge to the ocean through coastal aquifers has been identified as an area of major uncertainty (Buddemeier 1996) and is being investigated by the Land-Ocean Interactions in the Coastal Zone (LOICZ) core program of the International Geosphere-Biosphere Programme (IGBP). The importance of fresh groundwater discharge as a source of nutrients into coastal ecosystems has only recently been appreciated (Nordstrom & Roman 1996).

In New South Wales, disposal of sewage effluent (treated or untreated) is a major problem in many coastal settlements, particularly since offshore disposal has been banned by the State Government. The disposal of effluent into coastal sand dune complexes has been practiced for many years in some communities and many more are actively investigating this form of disposal. However, problems with the transport of nutrients and virus have been identified, and detailed investigations at some sites have shown that the initial simple conceptual model of a groundwater recharge mound in the aquifer, bounded by fixed head conditions at mean sea level, is not correct (Turner et al. 1996; Acworth et al. 1998).

Detailed investigations at Hat Head in northern New South Wales, Australia (Fig 1), have been undertaken to establish a more appropriate conceptual model of the vertical fluxes in these aquifer systems.

2 METHODS AND MATERIALS

Bundled piezometers have been installed to 18 m depth in a 5 km long, 350 m wide coastal sand-dune aquifer at Hat Head in northern New South Wales, Australia (Fig. 2). A transect consisting of three sites in line between the ocean beach and a tidal creek (Fig. 2) has been established to determine the flow mechanisms within the aquifer. Each site comprises 3 piezometers. The coastal site (HH15, 16 and 17); the middle aquifer site (HH21, 22 and 23); and the tidal-creek site (HH18, 19 and 20) are shown in Fig. 2, along with other piezometers previously installed.

Each site comprises:

1. A plain 50 mm piezometer to 18 m, used to provide access for geophysical logging.

2. A 25 mm piezometer with a 1 m intake between 17 m and 18 m. The hydraulic head is recorded using a DRUCK transducer and a HYDROKIT data logger. Mini-piezometers have been strapped to the outside of the 25 mm core pipe at 500 mm intervals (19 in total) with open areas of 20 mm each.

3. A 25 mm piezometer with a 1 m intake be-

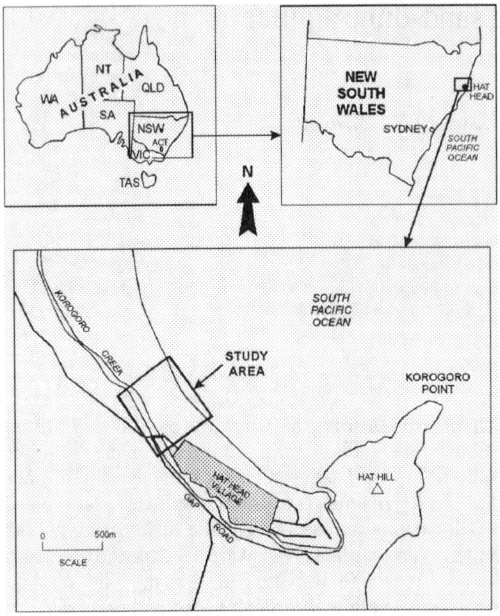

Figure 1: Location map for Hat Head

tween 8 m and 9 m is logged using a second pressure transducer and data logger and has mini-piezometers installed from the surface to 9 m depth at 50 mm increments.

The piezometers were installed using hollow-stem augers operated from a light-weight drilling rig equipped with tracks for moving over the sand dunes. The 50 mm piezometer at each site (HH17, HH21 and HH18) was logged using GEONICS EM39 logging equipment which included gamma-ray activity and bulk electrical conductivity sondes.

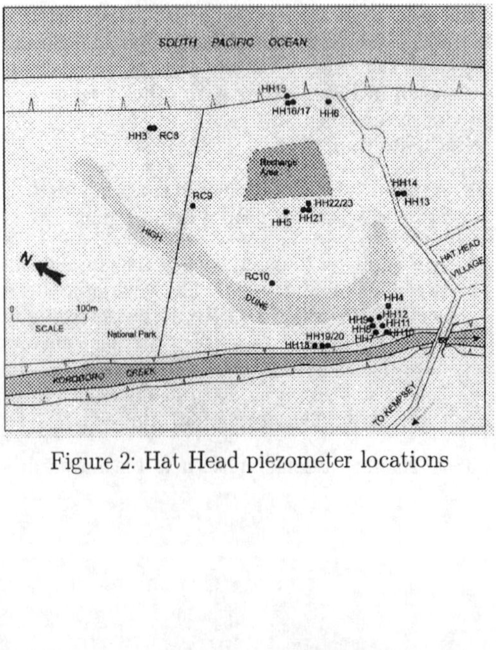

Figure 2: Hat Head piezometer locations

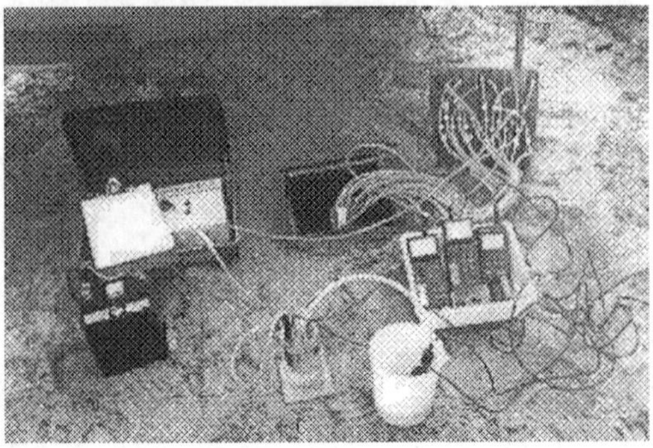

Figure 3: Well head sampling equipment at a bundled piezometer site

Figure 4: Multi-channel manometer board set up adjacent to HH19 and HH20

Water samples have been recovered from the bundled piezometers using a peristaltic pump. Measurements of fluid EC, temperature, pH and Eh were made using ORION meters at the well head (Fig. 3).

The effect of tidal variation on the vertical distribution of hydraulic heads in the mini-piezometers was established using specially constructed multi-channel manometer boards erected at each site (Fig. 4) during a period of king tides.

3 RESULTS

3.1 Geology

Fine to medium grained clean sand was recovered from all the bores during drilling. The sands have accumulated as dunes and form part of the near-shore dune complex.

Minimally disturbed cores were taken from one bore (HH17) and subject to grain-size analysis and

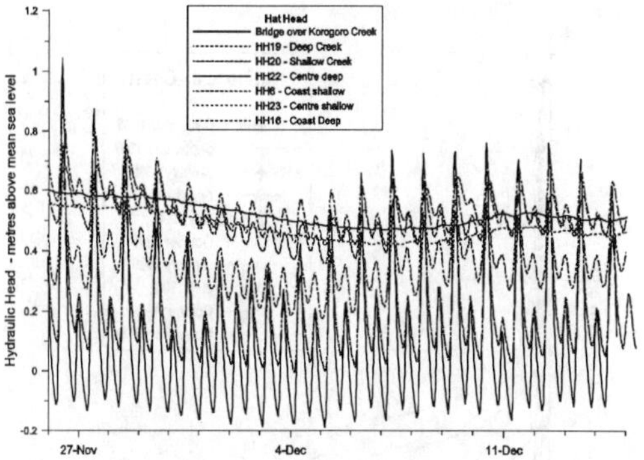

Figure 5: Hydraulic head variation at the 3 bundled sites and at Korogoro Creek

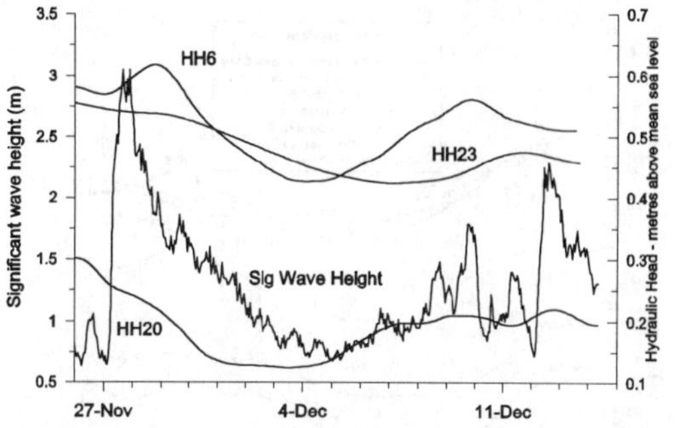

Figure 6: Comparison between the ocean significant wave height at Coffs Harbour and groundwater levels for HH6, HH20 and HH23

hydraulic conductivity determination using falling head tests. Values of hydraulic conductivity between 15 m/day and 35 m/day were measured.

A zone of cemented material, locally referred to as 'coffee rock' was recovered at various depths in the different bores. These layers had a reduced hydraulic conductivity (<1 m/day).

3.2 Vertical hydraulic gradients

Detailed hydraulic head data from the 3 bundled piezometer sites and the level recorded in Korogoro Creek at the road bridge (Fig. 2) are shown in Fig. 5. The levels are shown as metres above mean sea level. The impacts of beach face wave runup and nearshore wave setup at the beach (HH6 & HH16) and from near-shore wave setup in the tidal creek (HH19 & HH20) are clearly apparent beneath the dominant impact of the tides. The period shown in Fig 5 was predominantly dry with only 20 mm of rain recorded on 11 December. The response to tidal variation obscures longer frequency information in the record and a low pass filter (Thompson 1983) has been used to remove the tidal signal. The low pass filtered water level data is presented in Fig. 7. The significant results for this data can be summarised as:

1. There is a slight downward hydraulic gradient at the coast, with average differences in hydraulic head between HH6 (completed at an elevation of -1 m) and HH16 (completed at an elevation of -13 m) of 0.16m. There is very little variation in the gradient between the two dates. Higher heads during 28 - 30 November and again during 9 - 10 December are caused by a combination of beach face wave runup and near shore wave setup. Fig. 6 shows the correlation between the significant wave height measured at an ocean buoy approximately 60 km to the north of the site with the shallow piezometer at the ocean beach (HH6), a shallow piezometer at the creek (HH20) and the central piezometer (HH23). The correlation would be improved by data from a closer location.

2. There is a slight upward hydraulic gradient in the centre, with average differences in hydraulic head between HH22 (completed at -16.6 m depth) and HH23 (completed at -7.54 m depth) of 40 mm shown in Fig 7. Ground elevations for these piezometers were obtained using differential GPS and may be subject to slight error. The head difference is maintained throughout with heads at both depths equally impacted by changes at the boundary. There is a lag of several days between the impact of the changed wave climate on the coast and the head response in the centre of the aquifer, suggesting that this is not a pressure response but the result of groundwater flow. There was very little rainfall during this period.

3. There is a strong upward hydraulic gradient at the Creek between HH19 (completed at -17.2 m depth in clean low salinity water) and HH20 (completed at -7.3 m) of 0.36 m. The record for the deeper piezometer is very different, indicating that it represents a distinctly different part of the aquifer. The hydrograph for the shallow piezometer (HH20) closely follows that for the tidal creek with a constant head difference of approximately 100 mm. This indicates that there is almost a continuous discharge of groundwater to the creek. The colour of the groundwater between 2 m and 4 m in HH20 is a dark brown. The creek has the same colour during low tides and is only replaced by clear salt water at high tide.

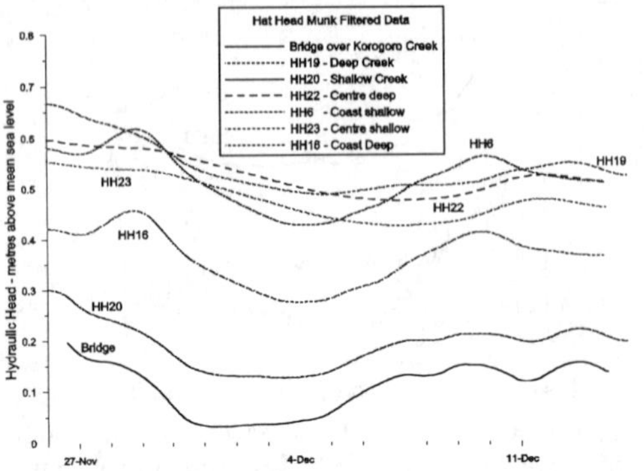

Figure 7: Low pass filtered water levels showing the vertical gradients

3.3 Coast - Geophysical logs and fluid EC variation

The geophysical logs for the coastal site show a fairly constant gamma-ray signature and a bulk EC signal reflecting the top of the saline interface at approximately 13 m. Fluid EC profiles for various times in the past 3 years have been taken at a bore installed at the edge of the ocean beach (HH3 in Fig 2). These have shown that the depth to the top of the saline interface has varied by as much as 3 m (Fig 9). Hydrogeochemical data obtained from samples of each of the mini-piezometers indicate that a zone of mixing occurs with an initial increase in fluid EC to approximately 12000μS/cm

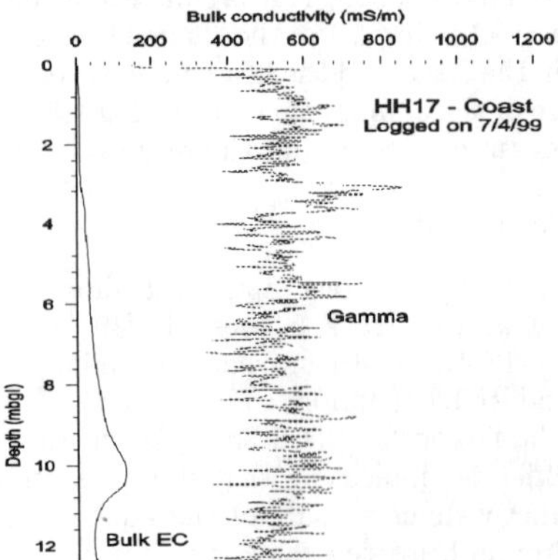

Figure 8: Geophysical log for HH17 (coast) showing the gamma ray activity and bulk EC variation

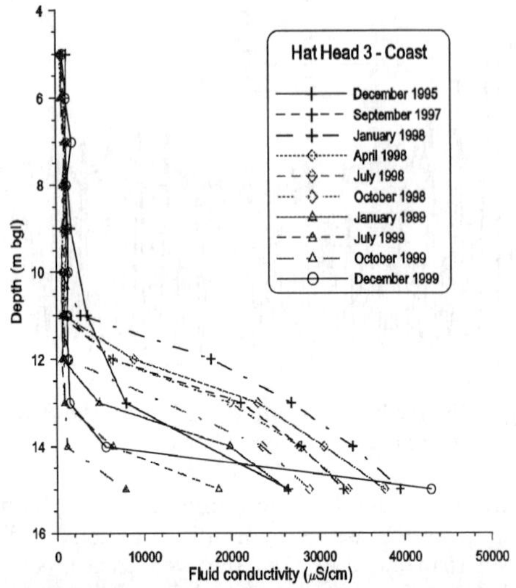

Figure 9: Fluid EC variation at the top of the beach (HH3)

at 10 m depth before freshening again to below 1000 μS/cm between 11 and 13 m and then increasing steadily to 46000μS/cm at 18 m. These features are also shown on the bulk EC log in Fig 8.

3.4 Centre - Geophysical logs, fluid EC profiles and manometer board data

The geophysical logs show the position of the top of the saline interface at the central bore at a depth of approximately 13 m. The depth to the saline interface has remained constant at this location. There is a gamma ray anomaly at approximately 7 m but this is not correlated with any other observation. The conditions in the centre of the aquifer at HH21/HH22/HH23 are the most stable. The interface between fresh water and saline water remains stable at approximately 14 m depth (Fig. 10). The water above the saline interface is strongly reducing with hydrogen sulphide gas noted in all the mini-piezometers sampled indicating the reduction of sulphate (Stuyfzand 1993). The manometer board data shows constant heads between the surface and 12.5 m depth with a decline in head to the base of the bundle at 18 m depth. Density corrections obtained using the fluid EC values (Stuyfzand 1993) indicate that this decline in head is the result of the change in fluid density and that there is very little hydraulic gradient with depth.

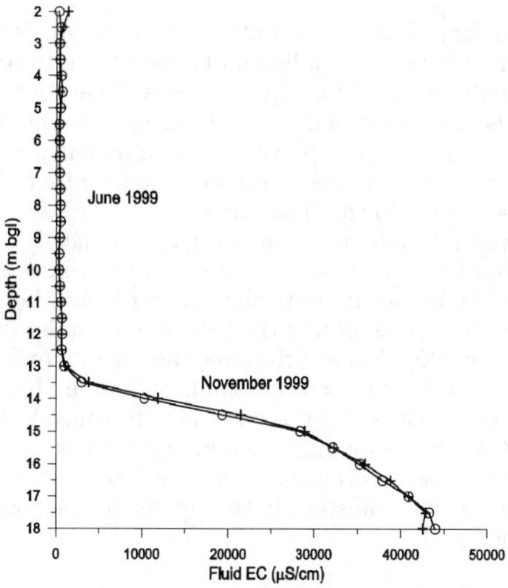

Figure 10: Fluid EC variation (HH22 & H23)

3.5 Creek - Geophysical logs, fluid EC profiles and manometer board data

The interface between fresh and saline water at the creek side is complex. Fresh water overlies saline water with a sharp boundary separating the saline water from deeper fresh water (Fig. 11). A gamma ray anomaly occurs at approximately 15 m depth. The fluid EC variation with depth and time (Fig. 12) shows that the magnitude of the saline

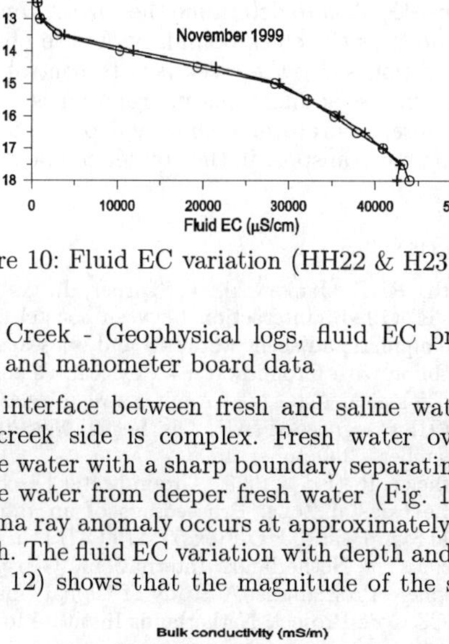

Figure 11: Geophysical log for HH18 (creek) showing the gamma ray activity and bulk EC variation

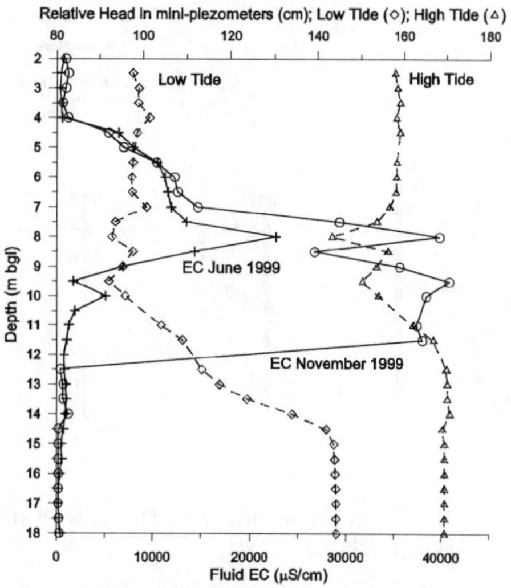

Figure 12: Fluid EC variation HH19 & HH20) and the hydraulic head distribution at high and low tides on 26/11/99 derived from the manometer board measurements

wedge adjacent to the creek varies considerably in time. The manometer board data confirms the head change with tide in the aquifer and demonstrates that the difference in head between the shallow and deep piezometers occurs at the same depths at which the fluid EC varies. The head loss is constant (0.30 m) throughout a tidal cycle in the clean water part of the aquifer below 15 m and constant (0.31 m) in the saline part of the aquifer at 9 m depth. The data in Fig. 12 have not been corrected for density variation as the base and top of the bundle have similar fluid EC and therefore similar density.

4 DISCUSSION

The filtered water levels reveal a continuous hydraulic gradient from the ocean to the tidal creek for the period of this study. This is the result of the tide falling more rapidly than the beach can drain (Raubenheimer et al. 1999) and is accentuated during additional beach face wave runup and nearshore wave setup (Nielsen 1989). The presence of approximately 12 m of low salinity water in the centre of the aquifer implies that there is not a continuous flow of water in response to this gradient. The significance of the upward gradient of low salinity water adjacent to the creek may be significant in controlling this flow. A conceptual model that shows the possible boundaries of the different water types is presented in Fig. 13. The angle of dip for the saline interface between HH16

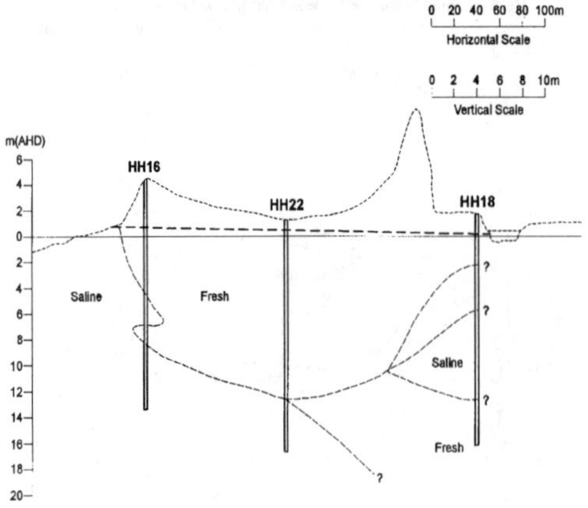

Figure 13: Conceptual model of the flow system based upon available data

and HH22 is surprisingly shallow. The conditions beneath the tidal creek require further investigation, particularly on the far side of the creek and beneath the creek. The water type at the base of HH18 is different to that in the upper part of HH22 and is also under considerable additional pressure. A lithological control may therefore exist separating these 2 water types and further work is underway to investigate this. The fresh water at depth adjacent to the creek may extend up to the creek bed leaving lobes of saline water on either side of the creek, or the interfaces may extend horizontally beneath the creek and may be lithologically controlled. Water in the creek is only saline for a few hours each day and may not be sufficient to contaminate the aquifer at depth.

5 CONCLUSIONS

The fate of contaminants in coastal and estuarine sand aquifers can only be determined if the characteristics of the boundary conditions can be accurately determined. The high hydraulic conductivity and small size of the aquifer means that groundwater flow and hydraulic head distribution will be determined by the boundary conditions (Frank & Reilly 1987). It is clear from the data presented that these conditions are very dynamic and significantly impacted by ocean waves and tides. The transmission of these effects inland by 200 m is surprising and may be result of the combined impact of the ocean boundary with the tidal creek boundary. If this is the case, the effects would be expected to decline as the width of the aquifer increases. The impact of waves and tides in aquifers less than 500 m wide can not be ignored if a tidal creek forms the second boundary in addition to the ocean boundary. The groundwater system responds to rainfall recharge rapidly and heads decline to pre-rainfall levels within 3 or 4 weeks. The position of the freshwater lens formed during these events may be sufficient to prevent saline intrusion occurring as a result of the continuous gradient from the ocean beach to the tidal creek.

Accurate measurements of hydraulic head, corrected for density, are necessary to determine the flow conditions. In particular, the conditions in the hyporheic zone around the tidal creek require further investigation to determine the correct conceptual model for the creek boundary. As the boundary condition along the creek acts to control flow systems in the aquifer, this information is necessary before accurate modelling and prediction of contaminant transport in the aquifer can be determined.

REFERENCES

Acworth, R. I., Jankowski, J., Turner, I., & Soriano, R. (1998). Interaction between coastel sand-dune aquifers, adjacent wetlands and sea water. In Gambling with Groundwater - Physical, Chemical, and Biological Aspects of Aquifer-stream relations: XXVIII Congress of IAH, Las Vegas, Nevada: 27 September - 2 October, 1998.

Buddemeier, R. (Ed.) (1996). Groundwater Discharge in the Coastal Zone: Proceedings of an International Symposium, P.O. Box 59, 1790 AB Den Burg - Texel, The Netherlands. International Geosphere Biosphere Programme: A Study of Global Change: LOICZ Core Project, Netherlands Institute for Sea Research.

Frank, C. L. & Reilly, T. E. (1987). The effects of boundary conditions on the steady-state response of three hypothetical groundwater systems - results and implications for numerical experiments. Water-Supply Paper 2315, US Geological Survey.

Nielsen, P. (1989). Wave setup and runup: An integrated approach. Coastal Engineering 13, 1–9.

Nordstrom, K. & Roman, C. (Eds.) (1996). Estuarine shores: evolution, environments and human alterations, pp. 287–310. New York: John Wiley.

Raubenheimer, B., Guza, R. T., & Elgar, S. (1999). Tidal water table fluctuations in a sandy beach. Water Resources Research 35(8), 2313–2320.

Stuyfzand, P. J. (1993). Hydrochemistry and hydrology of the coastal dune area of the Western Netherlands. Research and Consultancy Division, Groningenhaven 7, P.O. Box 1072, Nieuwegein, The Netherlands: KIWA. ISBN = 90-74741-01-0.

Thompson, R. (1983). Low-pass filters to suppress inertial and tidal frequencies. Journal of Physical Oceanography 13, 1077–1083.

Turner, I., Coates, P., & Acworth, R. (1996). The effects of tides and waves on water-table elevations in coastal zones. Hydrogeology Journal 4(2), 51–69.

Groundwater: Past Achievements and Future Challenges, Sililo et al. (eds) © 2000 Balkema, Rotterdam, ISBN 90 5809 159 7

Application of remote sensing and geographical information system to hydrogeological studies in the Upper West Region, Ghana

Bruce Banoeng-Yakubo
Department of Geology, University of Ghana, Legon, Ghana

Lilian Skjernaa
Geological Institute, University of Copenhagen, Denmark

ABSTRACT: A synergetic approach was applied to hydrogeological investigations in the Upper West Region, Ghana. The objective was the identification and delineation of lineaments from Landsat TM image and aerial photographs as a basis to determining the relationship between the fracture systems and borehole data in a GIS environment. Principal Component (PC) transformation of five of the bands produced decorrelated PCs that discriminated the structures. False Colour Component (FCCs) of the first three PCs enhanced the linear features distinctly exposing shear zones and fracture systems in well-defined patterns. A retransformation of the first three PCs distributed the image variance uniformly and so enhanced the discrimination of both structure and lithology. Two main shear zones were identified: C1 with a NNW direction is slightly oblique to the regional foliation; C2 ranging in direction from N to NNE. Quartz veins and joints also occur in these directions but dominantly trend ENE, NNW, and WNW. Buffer zones were created around lineaments of various orientations and attribute data were clipped from borehole records within these buffer zones. Analysis showed a strong positive relationship between yield and static water levels. High-yielding boreholes are located along NNE to E trending fractured veins and joints in granites. Yields decrease abruptly in boreholes more than 500m from the lineaments. Boreholes along fractures oriented between ENE and E have low average yields close to the lineaments but yields increase suddenly at a distance of 400m of these zones.

1 INTRODUCTION

The Upper West Region of Ghana is dominated by rural settlements that depend entirely on groundwater for potable water supply. The region is underlain by crystalline rocks comprising metasediments, metavolcanics and granitoids which have developed very highly heterogeneous aquifers resulting from localised distribution of fracture zones. This source of water supply is the most appropriate, in terms of cost and quality, as well as the ease with which it can be developed to serve the scattered communities. The water-bearing and yielding capacities of these rocks depend on the extent of development of secondary permeability resulting from fracturing and faulting.

The aim of the study therefore was to delineate the geology and structure of the area from satellite images and geological mapping as an important input in the development of an integrated approach to groundwater development. This was done in a GIS environment, using both ARC/INFO and ARVIEW softwares.

This study used remote sensing and ground geological mapping techniques to identify and delineate the geology and structure of the area. Hydrogeological data were then integrated in a GIS environment to capture, store, manipulate and analyse all data layers.

1.1 *Location of Study Area*

The study area lies within Latitudes 9°40'N-11°00'N and Longitudes 1°30'W-2°55'W. It covers an area of about 18,476 sq.kms. It is bounded on the north and west by Burkina Faso, south by Northern Region and east by Upper East region (Figure 1). It consists of five districts namely, Wa, Tumu, Lawra, Nadowli and Jirapa-Lambussie districts and has a population of about 620,000 people.

2 GEOLOGY

The Birimian System has been described to be geosynclinal deposits (Sanjara and Tkachev, 1964) deposited under quiet conditions with sporadic effusive activity. Banoeng-Yakubo (1987) believes that the Birimian was created from ductile thinning of the crust above an upwelling mantle. The age of the Birimian is between 2.25-2.05 Ga. Davis et al (1994) and Leubes et al, (1990) placed the Eburnean orogeny around 2.17 ± 65 Ga from Rb/Sr dating.

The Birimian in the study area occupies a N-S trending plateau in the west of the area and extends about 130 km northwards from west of Wa and has an

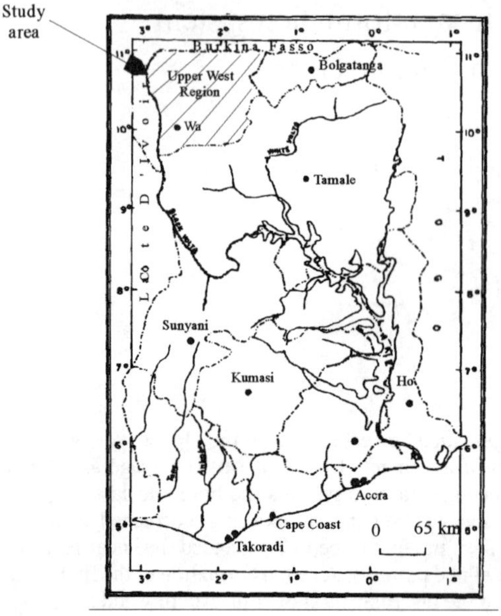

Figure 1. Location map of the Upper West Region, Ghana

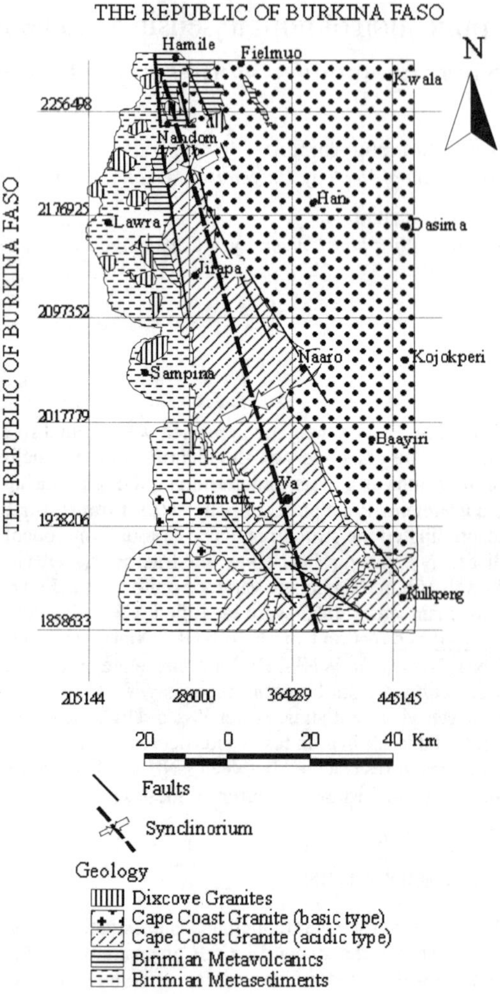

Figure 2. Geological map of the Upper West Region, Ghana

average width of about 20km to the border with Burkina Faso (Figure 2; Table 1).

The Lower Birimian is made up of tinted greenish, medium-grained biotite and sericite schist, greenish phyllite, sandstones and greywackes. Laminated fine to coarse-grained sandstone and black fine-grained schist forming the base of the Lower Birimian are common but not widespread. Some of these sandstone are tuffaceous with distinct massive to foliated structure. Typical in the area are microquartzites, normally massive thin-banded with fluidal structure of acidic lava (Roudakov, 1965).

The Upper Birimian series consists mainly of metamorphosed basic and intermediate lavas and pyroclastic rocks with some hypabyssal igneous rocks interlayered with phyllites and greywackes.

Table 1. Summary of regional geology of the Upper West Region, Ghana

System	Rock Type	Age
Upper Birimian	Basalts, tuffs, andesites, greywackes, phyllites	2.25 to 2.05 Ga
Lower Birimian	Graphitic phyllites, tuffaceous phyllites,schists greywackes, sandstones	2.25 to 2.05 Ga
	INTRUSIVES	
Dixcove Granitoids	Biotite granites, hornblende granites, granodiorites, quartz diorites	2106±59 Ma
Cape Coast Granitoids	Gabbros, gneissic granitoids, diorites	2216±72 Ma

The Cape Coast and Dixcove granitoids occupy over 80% of the Upper West region. The Cape Coast granitoids are thought to have been formed as a result of the reworking of the archaean basement. They consist of gabbro, gneissic granitoids, granodiorites and micaceous granitoids and are in tectonic contact with the Lower Birimian. Schists dominate the contacts between the Lower Birimian and both the Cape Coast and Dixcove granitoids due to extensive shearing at these contacts.

The Dixcove Granitoids usually intrude into the Birimian rocks and are younger than the Cape Coast granitoids. They consist of biotite granite, hornblende granite, granodiorite and quartz diorite.

3 SATELLITE DATA PROCESSING AND INTERPRETATION

3.1 *Methodology and Image Enhancement Techniques*

Five bands comprising TM bands 2, 3, 4, 5 and 7 covering the area were used for display. Band 1 has a lot of noise and was therefore not used. Band 6 is the thermal band with a spatial resolution of 120 m and not useful for this study. The image was geometrically rectified using ground control points located from a digital drainage map covering the area. The points are eastings and northings of the the Ghana local Transverse Mercator Projection system. In all 25 ground control points covering the whole area and uniformly distributed were used to compute a second order deformation model for the grid. The root mean square error was 3 pixels, each pixel having a spatial resolution of 30 m.

Various image-processing techniques were employed to enhance these five Landsat TM bands. These techniques include image rectification, contrast stretching, convolution filtering and principal component analysis (PCA).

The five bands of the image were inspected individually to identify the various rock types and structural entities. However, except major structures, it was difficult to identify and delineate the lithologies and all other structures in the area.

This necessitated the use of all available combinations of the various bands to obtain optimum data (Legg, 1992). Colour Composites of three bands comprising a combination of bands within the visible green, visible red and infrared were used to display the image. These bands were therefore combined as composites in Red, Green and Blue (RGB) filters. The infrared band was filtered through a red filter, with other bands filtered through green and blue filters to obtain False Colour Composites (FCCs) of the area (Richards, 1993). However, after a number of combinations of the bands, the contrast was not significant enough to discriminate lithology or structure. This combination, among others, does not distinctly discriminate lineaments and geology because of the extensive dimensionality of the image data resulting from the strong correlation among the bands.

Enhancement techniques such as contrast stretching and convolution filtering were performed on the FCCs to evaluate their lithological significance but they did not provide any useful lithological and structural contrasts in the image for study. In order, therefore, to decorrelate the various bands, it became necessary to use an appropriate multivariate statistical technique, such as Principal Component Analysis, to discriminate the geology and structure of the area.

3.2 *Principal Component Analysis*

Statistics of the images were computed using CHIPS FOR WINDOWS software. The statistics were used as input to perform a Principal Component Analysis of the image. The reduction of dimensionality separates non-random information, such as lithology and morphology, from random information, mostly noise (Dwivedi and Sankar, 1992; Kanea, 1997) making data analysis and interpretation easier.

The first three principal components, which captured most of the image scene variance, were re-transformed and all three new components designated as PCT1, 2 and 3 were displayed as band-triplets in RGB. Principal Components 1, 2 and 3 were enhanced using 3 x 3 convolution filters and then contrast stretched. This composite discriminates both linear features and lithology.

4 IMAGE INTERPRETATION AND DISCUSSIONS

4.1 *Geological Interpretation from Satellite Image*

A geological map of the study area was produced from the three re-transformed principal components.

The following rock types were discriminated from the interpretation:

- Sheared granitoids of the Cape Coast type
- Dixcove granitoids intruding into the metavolcanics and metasediments
- Mafic granitoids, mainly diorites, gabbros, amphibolites, hornblende granites of the Cape Coast type
- Metavolcanics of the Upper Birimian Series
- Metasediments of the Lower Birimian Series

The various rock types shown above were ground-truthed, the result of which is shown in Figure 2, the geological map compiled during this study. Various structural features such as faults were initially delineated as lineaments. Those that were proven to have been generated tectonically during the ground-truthing were delineated as faults. Other minor fractures, which could not be ground-truthed due to difficult terrain but were seen on the image to have tectonic significance, were also inferred as faults. A synclinorium was also inferred and confirmed from work carried out by Roudakov in 1965.

5 LINEAMENT AND STRUCTURAL INTERPRETATIONS

One of the main objectives of the study was to determine the relationship between lineaments and the geological structures mapped in the area. Lineaments can be discerned on an image variously as persistent and controlled trending of topography, well-patterned drainage system, linear alignment of vegetation as well as linear contacts due to tonal and textural changes in the image as a result of changes in the lithology (Sanders et al., 1997).

The lineaments range in orientation from 1 to 360 degrees with a standard deviation of 61 degrees (Figure 3). About 38% of the lineaments are within 315^0 and 360^0 which is comparable with the rose diagram of the attitudes of geological structures mapped in the field and presented in Figures 4a and b. A bird's eye view of the two rose diagrams show that the lineament data

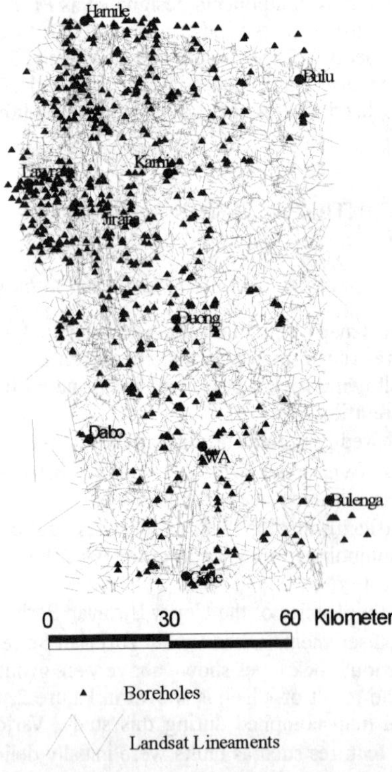

Figure 3. Lineaments delineated from Landsat TM images of the Upper West Region, Ghana. (Note the boreholes along the lineaments).

closely match that of the field data. The smooth peak of the azimuth is 165 degrees representing 21 % of the mapped data and 150 degrees in the case of the lineaments in the granitoids also representing 21.1%.

Most lineaments trending NW to NE are related to trends of foliation and C-surfaces. Similarly, most ENE and NNW lineaments are related to joints and quartz veins (Figures 4a, c, and d). The major very persistent NNW-SSE lineaments are major shear zones in the granitoids and those that act as contacts between the Birimian rocks and the granitoids of the Cape Coast type. These major lineaments tend to converge and present a synformal structure that contains the metasediments in the western limb and sheared granitoids in the eastern limb. There are two major shearing events in the study area as depicted by Figure 4d; the early shearing occurred in a NNW-direction swinging northward due to the heterogeneous stress regimes occurring during their formation, followed later by shearing in a NNE-direction accompanying the emplacement of the Dixcove granitoids. The early shear is denoted here as C1 and the later shear as C2.

The C1 shear planes are oriented, on the average, at an azimuth of 330^0 while the foliation trends 341^0. The shear planes are therefore aligned slightly oblique to the foliation. In shearing deformation, material lines rotate towards the shear plane as deformation progresses. The N to NNW oriented foliation therefore rotated in a sinistral sense into parallelism with the shear plane where especially, deformation is intense and the foliation coincides with the C-surface. The C-surfaces are manifested by anastomosing shear planes in the mylonites.

Quartz veins and joints are numerous and dominate the structure of the area. These veins are oriented in principally three directions. These are as follow: 1) ENE-WSW; 2) NNW-SSE and 3) WNW-ESE with a few disposed in a NNE-SSW direction (Figure 4e). Many of the veins in the Birimian rocks are oriented parallel to the two main shear zones and quartz appears to have crystallized in them during the deformation event accompanying granitoid emplacement.

6 GIS DATABASE ANALYSES AND INTERPRETATION

6.1 *Methodology*

The ARC/INFO GIS software was used to digitise and analyse the data that were generated from the satellite image as well as other spatial and attribute data. Some of the lineaments digitised were delineated using this software in the vector format and edited. In order to define explicitly spatial relationships, topology was created in ARC/INFO for all lineaments that were digitised using CHIPS FOR WINDOWS image processing software. This procedure was followed in order to facilitate the querying of the data and overlay of geographic features that would facilitate spatial analysis.

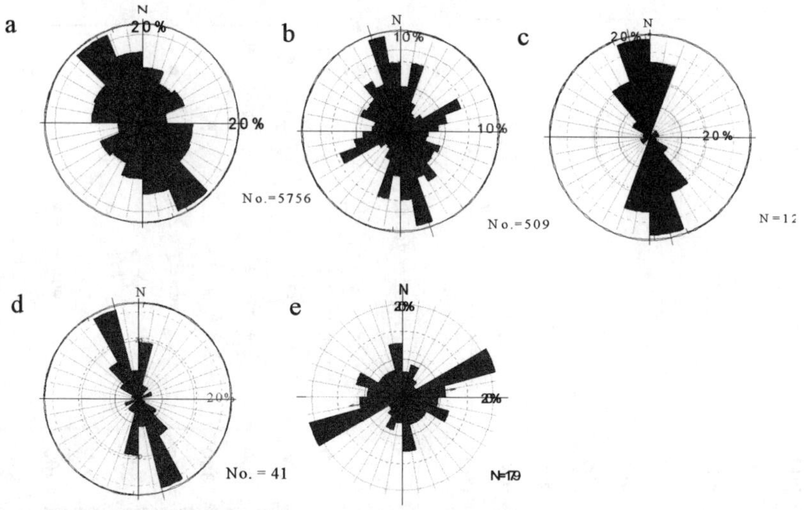

Figure 4. Rose diagrams of the orientations of (a) lineaments delineated from Landsat TM, (b) attitudes of structures mapped, (c) foliations, (d) shear zones, and (e) joints and quartz veins, of the Upper West Region, Ghana

The spatial data were created in ARC/INFO environment and the attribute data analysed in ARCVIEW. The analyses were carried out in the GIS in order to reference the data to positions on the ground. Similarly, spatial analyses were carried out between data layers or attributes.

During the processing and analyses of the data obtained for this study, the following data layers were overlaid in the GIS:

- Borehole parameters e.g. location attributes such as yield, depth etc.
- Digital topographical data e.g. contours, drainage, villages
- Drainage maps
- Geology Map of the area e.g. structure, rock types.
- Aerial photographs e.g. lineaments.

6.2 *Spatial statistical relationship of lineaments and Hydrogeological data*

6.2.1 *Methodology*

All lineament data generated from the satellite image and which had topology were imported into ARCVIEW. Similarly, all boreholes with spatial locations and accompanying attribute information in a database format were displayed as themes. Table 2 shows the descriptive statistics of borehole data used for the analysis.

Buffer zones were created in ARCVIEW GIS along all lineaments of various orientations. The buffer intervals are as follows; 100, 200, 300, 400, 500, 700 and 1000 m. Intervals of lineament orientations were obtained from queries. This procedure was used to separate lineaments of different azimuths. The azimuth intervals of lineaments used for the query are; 0-30^0, 31-60^0, 61-90^0, 91-120^0, 121-150^0, 151-180). Statistics of the various parameters were then obtained.

6.2.2 *Relationship between Lineaments and Hydrogeological data*

In granitoids, the shallowest SWLs are found in boreholes that have been sited on lineaments oriented ENE to E. However, as the buffer distance increases, the static water levels decrease. Boreholes sited on N to NNE oriented lineaments are very high-yielding but their yields fall sharply after a buffer distance of 500 m. These boreholes also have shallow SWLs which fall sharply after the same buffer distance (Figure 5).

The deepest static water levels are found in boreholes

Table 2 Descriptive statistics of borehole data

	N	Minimum	Maximum	Mean	SD
Depth	498	11.76	70.32	30.49	10.027
SWL	490	0.31	34.11	8.81	5.067
Depth	489	5.49	60.45	21.65	8.555
Yield	497	1.64	600.00	33.71	52.093
Valid (listwise)	477				

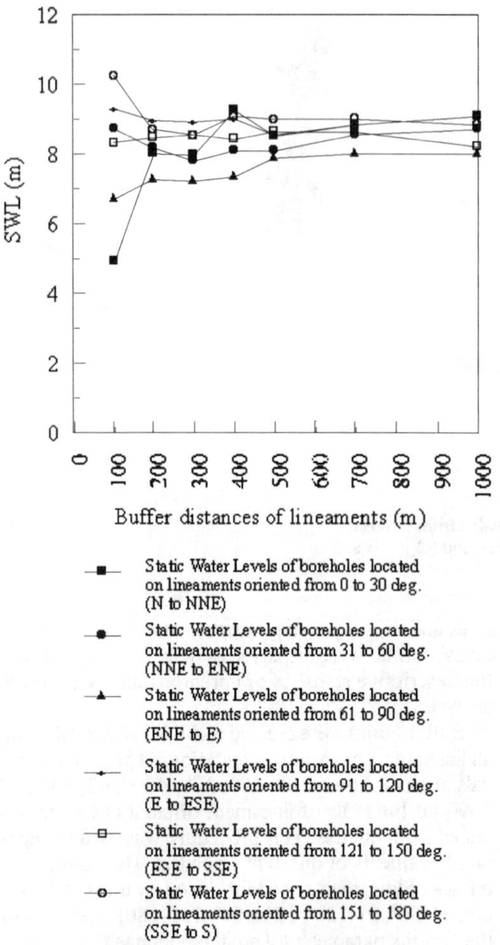

Figure 5. Graph of static water levels versus buffer distances of lineament orientations in the Upper West Region, Ghana

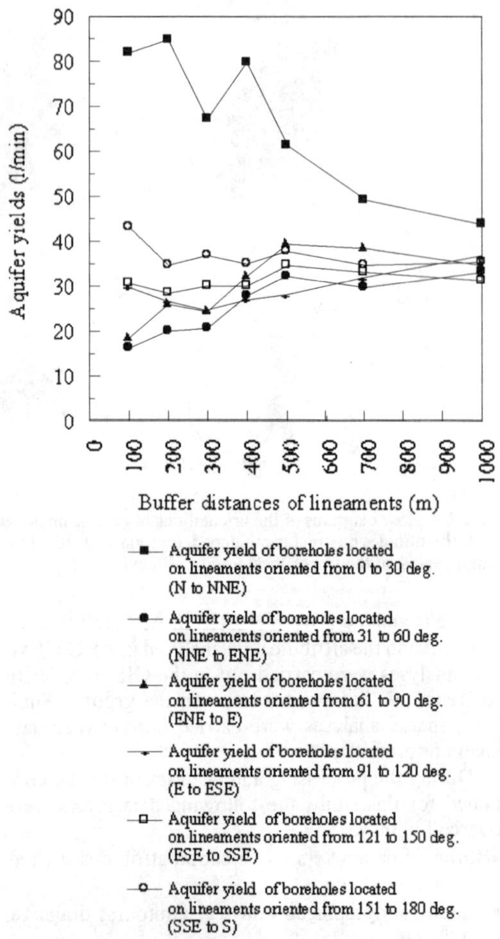

Figure 6. Graph of aquifer yields versus buffer distances of lineament orientations in the Upper West Region, Ghana

sited on N to NNW lineaments. These lineaments are related to reactivated shear zones, which contain very feldspathic and micaceous minerals, which tend to alter more easily and clog the permeable zones.

However, yields of boreholes located on NE and ENE lineaments increase with distance from the lineament. The most significant increase in yield was recorded along lineaments oriented ENE to E. It is worthnoting that fractures, quartz veins and joints enhance the development of permeability and therefore play a very critical role in groundwater occurrence in these rocks.

North to NNE lineaments have the highest average yields in the granites. (Figure 6). Since these lineaments are fractured, foliated rocks, shear zones and pegmatites and also relate to contacts, their ability to store and transmit water is enhanced especially where there are different generations of superimposed fractures which explain the higher yields. These yields decrease 500m from the lineaments, but still have higher average yields compared to the rest of the lineaments of different azimuths. Similar results were obtained from boreholes sited in the metasediments.

Lineaments in the NNW direction have higher average yields than the rest but lower than NNE trends. These yields decrease after 500m from the lineament. These lineaments have been reactivated shear zones with limited areal extent explaining the reduction in yield with distance from the borehole.

Yields of boreholes with azimuths between ENE and E initially have low average yields but the yields increase suddenly after a distance of 300m from the lineament. This observation is consistent with the fact that joints and quartz veins in this area increase the permeability of the rocks. However, it appears that areas close to the lineaments are densely sheared due to ductile deformation events rendering very low permeability near the lineament. Brittle-related deformation operated in the areas away from the shear zone enhancing the permeability of the rocks.

7 CONCLUSIONS

The study has shown that any exploration for groundwater in the Upper West Region of Ghana must necessarily involve a thorough understanding of the rock masses and structural setting. This is important in view of the fact that groundwater occurrence in this region is controlled by the rock types and structure.

Satellite image interpretation and ground geological studies identified and delineated metasediments and metavolcanics of the Birimian system and Cape Coast and Dixcove granitoids . These rocks which are crystalline in nature have low permeability for groundwater recharge except where secondary permeability has developed due to fracturing, quartz veining and weathering.

The remote sensing interpretations were very involving. To use this approach for a groundwater development project for any rural community elsewhere, the implementation of this phase should start far ahead of drilling phase. This would allow for a thorough ground-truthing of the lineaments and rock types identified during the remote sensing phase.

The lineaments and the various rock types from the satellite image interpretation formed the first input into the GIS. Topographical, geological and hydrogeological data were overlaid on the satellite data and analyzed spatially to obtain the ensuing information that formed the basis for this study.

Lineament analysis from remote sensing data in ARC/INFO and ARCVIEW GIS vividly demonstrated the utility of these tools as very important to groundwater exploration in crystalline terrain such as the Upper West Region. It is clear that the objective of establishing the relationship between borehole data and hydrogeological potential of lineaments was achieved. Structural analysis of the lineaments in relation to ground geological mapping established a similarity between them and identified the following:

The earliest foliation trends NNW with variations to N. That there are two major shearing events; C1, the earliest, in a NW to NNW directions and C2 which is a post intrusion shearing, trends in a N to NNE directions. This post-dates regional foliation.

That the earliest pegmatite intrusion trends N-S parallel to the regional foliation and predates NW to NNW-trending C-surfaces. ENE-trending pegmatites post-date regional foliation and NW to NNW-trending C-surfaces

That the dominant quartz veins trend ENE which cross-cut NNW quartz and pegmatite veins. ESE veins predate NNW trending joints that are as a result of shear zones reactivation and post-date ENE-trending joints.

In any future borehole siting, N to NNE-oriented lineaments should be targeted for high-yielding boreholes since they were forned from a major shear that post-dates the regional foliation and hence permeable enough for groundwater storage and transmission. Similarly, ENE to E-oriented lineaments which are commonly joints and quartz veins are good targets for prolific boreholes.

REFERENCES

Banoeng-Yakubo, B.K., 1987. Structure, petrology and gold-mineralization of the Birimian-Tarkwaian formations of the Asuokoo area, Ghana. Unpublished M.Sc. Thesis, University of Ghana, Department of Geology.

Davis, D.W., Hirddes, W., Schaltegger, U., and Nunoo, E.A., 1994. U-Pb age constraints on deposition and provenance of Birimian and gold-bearing Tarkwaian sediments in Ghana, West Africa. *Precambrian Research*, 67: 89-107.

Dwivedi, R.S., and Sankar T. R., 1992. Principal Component Analysis of LANDSAT MSS data for delineation of terrain features. *INT. J. Remote Sensing*, 13(12): 2309-2318.

Kenea, N.H., 1997. Improved geological mapping using Lansat TM data, Southern Red Sea Hills, Sudan: PC and HIS decorrelation stretching. *Remote Sensing* 18:1233-1244.

Legg, C., 1992. Remote Sensing and Geographical Information Systems: Geological mapping, mineral exploration and mining. Ellis Horwood , Chicester. 1st edition.

Leube A., Hirdes, W., Mauer, R., and Kesse, G.O., 1990. The Early Proterozoic Birimian Supergroup of Ghana and some aspects of its associated gold mineralization. *Precambrian Research*, 46: 139-165.

Richards, A., 1993. *Remote Sensing Digital Image Analysis; An Introduction*. Springer-Verlag, Berlin. Enlarged Edition.

Roudakov, V.M., 1965. Ghana Geological Survey Department Archive Report No. 50, Minerals Commission/GTZ Project

Sander, P., Chesley, M.M., and Minor, T.B., 1996. Groundwater assessment using remote sensing and GIS in a rural groundwater project in Ghana: lessons learned. *Hydrogeology Journal*. 4(3): 40-41.

Sanjara, L.A. & Tkachev, V.N., 1964-66. Geological structure and mineral deposits of the Northwestern part of Sunyani Sheet No.1. Unpublished report, Ghana Geological survey.

Groundwater: Past Achievements and Future Challenges, Sililo et al. (eds) © 2000 Balkema, Rotterdam, ISBN 90 5809 159 7

Groundwater reservoir characterization based on pumping test curve diagnosis in fractured formation

I. Bardenhagen
Department of Water Affairs, Windhoek, Namibia

ABSTRACT: In this paper methods and techniques for the diagnosis of pumping tests in fractured aquifers are reviewed and extended. These techniques, which include the comparison of drawdown and recovery data, straight-line analysis, special plots, derivatives, and pseudo-skin analysis are demonstrated on a continuous fractured model (double porosity case) and a discontinuous model (single fracture case) using theoretical and practical examples.

1 INTRODUCTION

Pumping tests are the most important tool for the aquifer investigation in the groundwater industry, as they are the only method that provides information on the hydraulic behavior of the well, and reservoir boundaries. These are essential information for an efficient aquifer and well field management and frequently, together with drilling results, are the only information available for the establishment of groundwater extraction schemes. To achieve reliable reservoir information in fractured aquifers, a sound understanding of the drawdown behavior is of high importance. A detailed diagnosis of drawdown data in combination with geological data can help to comprehend the reactions in a fractured aquifer. This paper will briefly review the two important and well-known models for fractured aquifers, the continuous model for double porosity or natural fractured aquifers and the discontinuous model for a single fracture or fault embedded in a homogenous matrix. These models will be used to demonstrate the diagnostic tools available to identify the different flow phases that can occur during a pumping test in a fractured environment. Field data will be presented to illustrate and discuss some of these cases.

2 THEORY

2.1 *Double porosity models*

The concept of double porosity introduced by Barenblatt et al. (1960) considers homogenous distributed conductive fractures embedded in a homogenous distributed matrix. Two matrix types are generally discussed: the spherical block matrix (Warren & Root 1964) used to represent aquifers like quartzite and the layered matrix (Kazemi 1969) adopted for example for sandstones with bedding planes. Commonly a pseudo-steady state block-to-fracture flow is assumed, which is a special case of transient flow restricted by skin, as illustrated by Moench (1984).

Moench (1984) considers that two representative elementary volumes (REV), one for the fractures and one for the matrix, are necessary for the mathematical description of the flow in such aquifers. The classic concept of double porosity cannot be applied if the cone of depression is not considerably larger than both REV because the influence of the single fractures is not negligible as demonstrated by Wei et al. (1998).

The typical drawdown curve in a double porosity aquifer is shown in Figure 1. Initially, mainly the fracture network releases water and the drawdown is characterized by a straight line in a semi logarithmic (semilog) plot. The flattening of the curve is originated by the ever-increasing additional contribution from the matrix. The later drawdown is the response of the matrix alone, which is represented by a straight line parallel to the initial one in the semilog plot. In a double logarithmic (log-log) plot, both the initial and the late drawdowns are characterized by Theis curves shifted horizontally from each other. The distance between the two parallels or two Theis curves depends on the fracture/matrix storage coefficient ratio.

Both the pseudo-steady state and transient models show an almost horizontal flattening when block-to-fissure skin is present. The flattening in the transient model without block to fissure skin is not horizontal.

It has a slope of half the slope of the two parallels (Fig. 1).

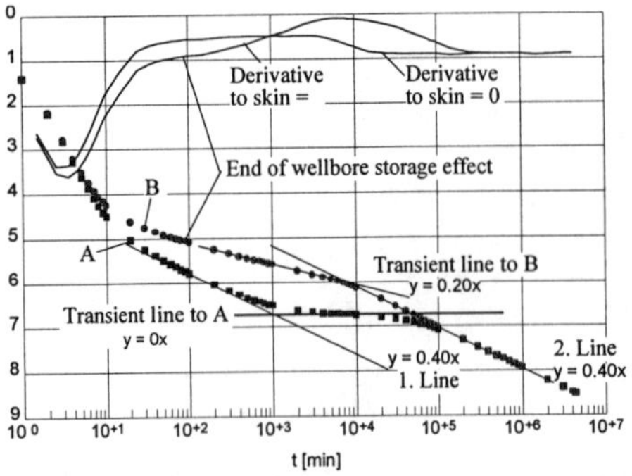

Figure 1. Drawdown in slab-shaped double porosity aquifer and transient block to fissure flow. A = fracture/formation skin-factor of 5. B = fracture/formation skin-factor of 0.

2.2 *Single fracture models*

In low permeability rock, features like fractures, faults, dykes, or bedding plains can represent zones of high permeability that act as conduits. Basically five models can be applied for these features:

- Finite vertical fracture with infinite fracture conductivity (Gringarten et al. 1974)
- Finite vertical fracture with finite fracture conductivity (Cinco-Ley & Samaniego 1978)
- Infinite vertical fracture with finite fracture conductivity (Boehmer & Boonstra 1986 dyke/formation model)
- Finite horizontal fracture with infinite conductivity (Gringarten & Ramey 1974)
- Finite horizontal fracture with finite conductivity (Valkó & Economides 1997)

Whether a vertical feature is finite or infinite conductive depends on the relative conductivity Cr defined as:

$$Cr = K_f \cdot w / (\pi \cdot K_m \cdot x_f) \text{ [-]} \quad (1)$$

where:
K_f = conductivity of the fracture [l/t]
w = fracture's width [l]
K_m = conductivity of the matrix [l/t]
x_f = half-length of the fracture [l]

Cinco-Ley & Samaniego (1978) demonstrated that for all practical purposes a single vertical fracture with Cr ≥ 100 could be regarded as an infinite conductive fracture that coincides with the Gringarten et al. (1974) infinite flux model for infinite conductive fractures.

If Cr < 100, then the fracture has finite conductivity and those models that considers the fracture conductivity must be applied (Cinco-Ley & Samaniego 1978, Boehmer & Boonstra 1986).

Additionally, Cr is used to classify the different flow phases that can occur while pumping from a vertical fracture as follows:

- Cr < 0.1, bilinear flow (after well bore storage or linear fracture flow), thereafter pseudo-radial flow
- 0.1 ≤ Cr < 100, bilinear flow, linear formation flow appears masked in the transition zone
- Cr ≥ 100, linear formation flow

When these single vertical fractures are represented in a log-log plot, the following characteristics can be depicted (Cinco-Ley & Samaniego (1981a):

- Linear fracture flow appears as a straight-line with slope 0.5 in the very early time data
- Bilinear flow is plotted as a straight-line with slope 0.25 in the early time data (Fig. 2)
- Linear formation flow shows a straight line with slope 0.5 in the intermediate time data (Fig. 2).

The radial-acting flow phase plots on a straight line in a semilog plot.

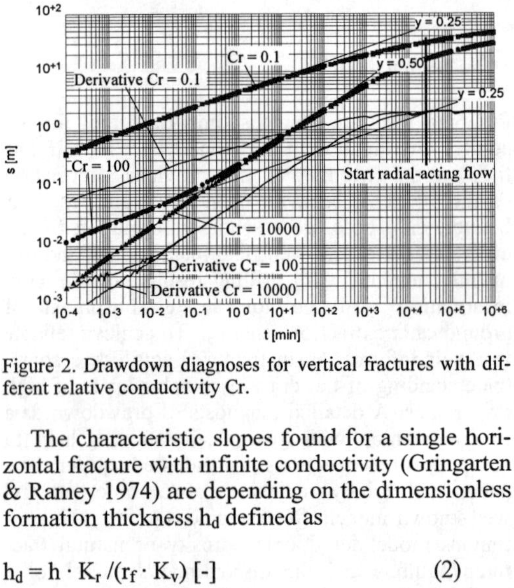

Figure 2. Drawdown diagnoses for vertical fractures with different relative conductivity Cr.

The characteristic slopes found for a single horizontal fracture with infinite conductivity (Gringarten & Ramey 1974) are depending on the dimensionless formation thickness h_d defined as

$$h_d = h \cdot K_r / (r_f \cdot K_v) \text{ [-]} \quad (2)$$

where:
h = reservoir thickness [l]
r_f = fracture radius [l]
K_r = radial fracture conductivity [l/t]
K_v = vertical formation conductivity [l/t]

When $h_d < 1$, the following characteristic flow phases are observed in a log-log plot:

- Very early time data shows a slope of 1 that indicates fracture storage flow
- Early time data shows a slope of 0.5 for linear flow
- Intermediate time data shows a slope of 1 again for transient flow influenced by the limited fracture reservoir

The flow phases for $h_d \geq 1$, when represented in a log-log plot, are reduced to a slope of 0.5 from very early time data to intermediate time data, indicating linear flow.

However, due to the very short early time period, this phase might be masked by wellbore storage or simply be missed. Valkó & Economides (1997) investigated the behavior of a horizontal penny-shaped fracture with finite conductivity, but did not find characteristic slopes over the dimensionless time t_d period in the range of 0.1 to 100.

In all these cases the radial-acting flow phase commences approximately at the dimensionless time $t_d \cong 15$, which is defined as:

$$t_d = T \cdot t/(S \cdot x_f^2) \quad [-] \qquad (3)$$

where:
t = Real time [t]
T = Formation transmissivity [l²/t]
S = Formation storage coefficient [-]

Only after the radial-acting flow phase is reached can the common analysis methods for primary aquifers be applied to determine the transmissivity T, e.g. Jacob or Theis. The storage coefficient S can be estimated with equation (3), if the fracture half-length x_f is known. Figure 3 illustrates the extremely large error made in the calculation of S when the Jacob straight-line method for time drawdown data is applied to the data measured in an observation well located near an infinite conductive vertical fracture with uniform flux. The straight-line method is only applicable when the distance of the observation well is at least 5 times that of the fracture half-length (which is in fact the REV of such a system).

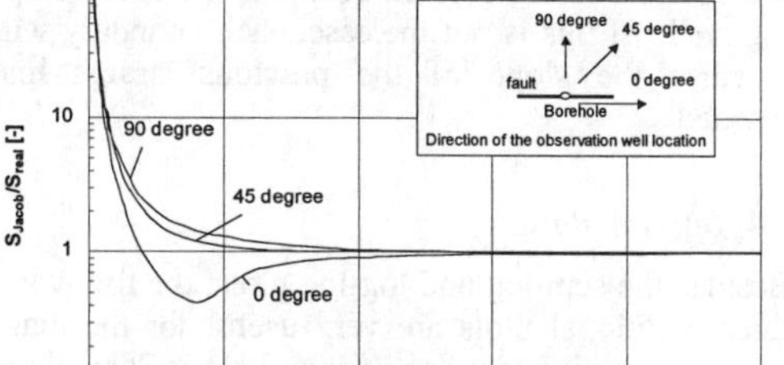

Figure 3. Deviation from the real storage coefficient calculated using the Jacob's straight-line method for data of the radial-acting flow phase in observation wells in the vicinity of a single vertical infinite conductive fracture with uniform flux.

One must bear in mind that the determination of the storage coefficient S is a function of its location and not the extraction time although the radial-acting flow phase in the observation well has been reached! However, Figure 3 can also be used to either determine the correct storage coefficient if the relative position of the observation well to the fracture is known or to determine the fracture half-length if the storage coefficient is known.

The influence of the fracture storage can be described with the relative storage capacity CDf after Ramey & Gringarten (1976) defined as:

$$CDf = S_f \cdot w/(S \cdot x_f) \quad [-] \qquad (4)$$

where S_f = Fracture storage coefficient [-].

Figure 4 shows the drawdown in a vertical fracture with infinite conductivity and infinite flux with CDf between 10^{-0} and 10^{-10}. All curves commence with linear flow (slope 0.5). All curves with CDf < 10^{-4} describe the drawdown in high storage capacity fractures and their behavior is significantly different to that of well bore storage. In these curves the slope of 0.5 indicates linear fracture flow and not linear formation flow and indicates that all discharged water is provided solely by the fracture. The curves show a transition zone characterized by an increasing slope from 0.5 to 1 with increasing CDf as a result of the leakage from the formation that gradually replaces the storage in the fracture. Once the radial-acting flow phase is reached, all water is provided by the formation.

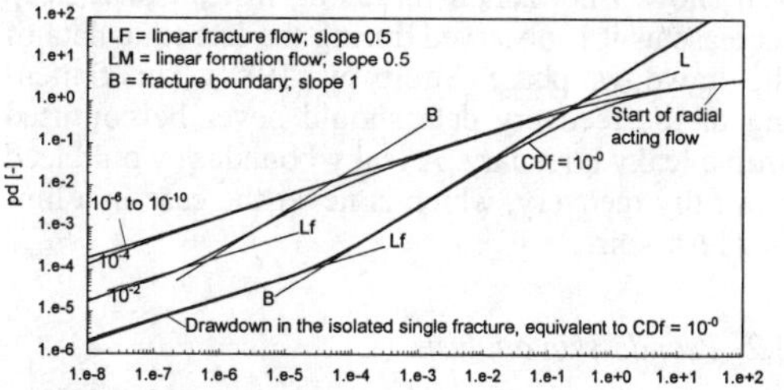

Figure 4. Drawdown in an infinite conductive vertical fracture (Cr = 10000) with various relative fracture capacity CDf (dimensionless drawdown pd = $\pi \cdot T \cdot s/(2 \cdot Q)$ [-] where: s = drawdown, Q = discharge rate, T = formation transmissivity)

The drawdown curve for CDf = 10^{-0} in Figure 4 coincides exactly with the drawdown curve observed in a isolated single fracture for the early time data and partly with the transient phase. It confirms the fact that the transition zone is caused by the limited extent of the finite fracture because the water pumped from such a isolated single fracture is provided only by the fracture storage. This observation leads to the interpretation that these effects i.e., wellbore storage, fracture storage, and the closed boundary reservoir can be described by the same principle of pumping from a limited reservoir.

3 DIAGNOSIS METHODS AND TOOLS

The following methods and tools can be applied for the diagnose of pumping test data in fractured aquifers:

- Comparison of drawdown and recovery data
- Pseudo-skin analysis
- Straight-line analysis in log-log and semilog plots
- Special plots
- First and second derivative of the drawdown data

3.1 *Comparison of drawdown and recovery phase*

The first check during a pumping test data evaluation should be the comparison of the drawdown and recovery behavior of the data if both are available. Due to the superimposing theory, both should show the same principal curve shapes. If a flattening is observed in the drawdown curve, the same typical flattening must appear in the recovery data. If this is not the case, the flattening in the drawdown was caused by:

- Variation of the discharge rate during the drawdown phase
- Discharge from a closed reservoir (limited reservoir)

In the first case, only the recovery data after a time correction (Birsoy & Summers 1980) can be used. In the second case, the late time recovery data will show a horizontal flattening instead of a steep increase as it is observed during the late time data of the drawdown phase (Streltsowa 1988). This flattening of the recovery data should never be confused with a leaky boundary. A leaky boundary would lead to a fully recovery, which is never the case in a limited reservoir.

3.2 *Pseudo-skin analysis*

A well located within or in the vicinity of a conduit (if the distance is below the REV) shows less drawdown than expected for wells in a homogeneous formation. This effect is known as pseudo-skin (Gringarten & Ramey 1974). The application of the skin determination methods as in wells situated in homogenous aquifers would lead to a negative skin-factor sf [-]. The skin factor sf and the drilled radius r_w are related to the effective radius r_{eff} as follows (Sabet 1991):

$$r_{eff} = r_w \cdot e^{-sf} \quad [l] \qquad (5)$$

The equation shows that a negative skin-factor (well in a fracture) would lead to an effective radius of the well larger than the real radius. In the case of a positive skin-factor (well in homogenous aquifer) the well is effectively acting with a radius smaller than the actual one.

However, the pseudo-skin effect can be used to determine whether a well is located in a fracture zone or not, due to the fact that, in principle, no negative skin factor or enlarged effective radius can be observed in a primary or continuous fractured aquifer. One useful application is for example to distinguish between the drawdown of a horizontal fracture with infinite conductivity and the drawdown of a well in a homogenous aquifer with wellbore storage. Unfortunately this holds true only for $h_d \leq 1$ where the skin-factor is always negative, because the skin-factor increases towards positive values with $h_d > 1$. Similarly, the method can be used for vertical features.

3.3 *Straight-line analysis*

As already discussed most of the flow phases that appear during a pumping test in a fractured aquifer show characteristic straight-lines either in a log-log or in a semilog plot. However, the log-log plot provides additional information as follows:

- Well bore storage shows a slope of 1
- Fracture storage shows a slope between 0.5 and 1 for linear flow and 0.25 to 1 for bilinear flow
- Two parallel closed boundaries show a slope 0.5. (Ehlig-Economides & Economides 1985)
- Three equidistant closed boundaries U-shaped show a slope 0.5
- Limited reservoir (four closed boundaries) shows a slope of 1.

The semilog plot provides additional information regarding the presence of negative boundaries:

- One closed boundary doubles the slope of the radial-acting flow straight-line
- Two perpendiculars closed boundary quadruples the slope of the radial-acting flow straight-line.

Only one straight-line will be observed when all the boundaries are located equidistant to the pumping well. If this is not the case, each boundary will increase the slope of the previous straight-line reached.

3.4 *Special plots*

Besides the semilog and log-log plots, the following three additional plots are very useful for the diagnose of pumping test data (Cinco-Ley & Samaniego 1981b):

- Linear drawdown versus square root of time
- Linear drawdown versus fourth square root of time
- Linear drawdown versus one divided by square root of time

The first plot is useful for the determination of the linear flow behavior. The drawdown data of linear and bilinear phases will plot on a straight line that commences at the origin of the diagram. In the second plot the drawdown data of the bilinear flow phase will plot on a straight line starting at the origin. The drawdown data of a spherical flow will plot on a straight line that commences in the origin in the third plot.

Cinco-Ley & Samaniego (1981b) demonstrated that the first two diagrams are very useful to determine skin effects in drawdown data of wells in single fractures with either linear or bilinear drawdown behavior. Such skin effects cause an additional drawdown that increases clogging phenomena and in extreme cases can even destroy the stimulation effect of drilling in a fracture zone (Economides & Nolte 1989).

Cinco-Ley & Samaniego (1981b) found that, due to skin effects, the early time linear flow data in a log-log plot are represented as an almost horizontal line that develops into the radial-acting phase (bilinear flow would plot similarly). Plotting the same data in a linear drawdown versus square root of time diagram will also show a straight-line, but will be shifted downwards from the origin. Both authors stated that, by data from the pumped well only, it is not possible to obtain a unique solution for the determination of the skin location, which could be located at the well, or between fracture and formation, or at both. However, Bardenhagen (1999) showed an unique evaluation method for the skin location for a single vertical fault with a relative conductivity $Cr \geq 100$ by using the plot linear drawdown versus square root of time for drawdown data of a pumped well and an observation well located in the same fault.

All these plots can also be applied for horizontal features as they are only related to the flow regime.

3.5 *Derivatives*

The additional use of curve derivatives is of great advantage because they are sensitive to small changes in the drawdown curve and are independent of skin effects, as illustrated in Figure 6. Usually, the derivative is plotted as ($\Delta s/\Delta t \cdot t$), which provides the following advantages:

- All characteristic straight-line slopes remain the same
- Radial-acting flow phase is plotted as a horizontal line, which eases the identification for the human eye

Unfortunately, derivatives applied to real data often show too much noise. Smoothing of the derivatives would overcome this problem, but it cannot be ensured that the applied mathematical algorithm would not produce misleading artifacts. However, recovery data are usually less noisy as they are not influenced by variations in the discharge rate and can be used. Nevertheless, with some experience even noisy derivative curves can be interpreted.

4 FIELD EXAMPLES

4.1 *Field example 1*

Two boreholes located 133.5 m apart were sited on a 15 km long, sub-vertical (77°S) fault zone crossing the Fish River in the southern part of Namibia, which might be a potential recharge source. The fault partly separates two low yielding formations composed at horizontal intercalated layers of claystone, siltstone and sandstone. Both boreholes intersect the fault at 27m below the surface. The water level in both boreholes rose immediately after the fault was struck at a level of 906.1 m amsl (8.3 m below surface in BH1 and 5.3 m below surface in BH2). The airlift yield was estimated at more than 100 m^3/h in each borehole. Screens with 0.5 mm slots were installed to avoid borehole collapse. Figure 5 shows the drawdown measured during one of the constant discharge tests. Only the drawdown in the observation well shows a slope of 0.5 indicating linear formation flow. However, the drawdown in the pumped well starts almost horizontal and developes at late time to radial-acting flow. This behavior is typical for a skin that is located at the well.

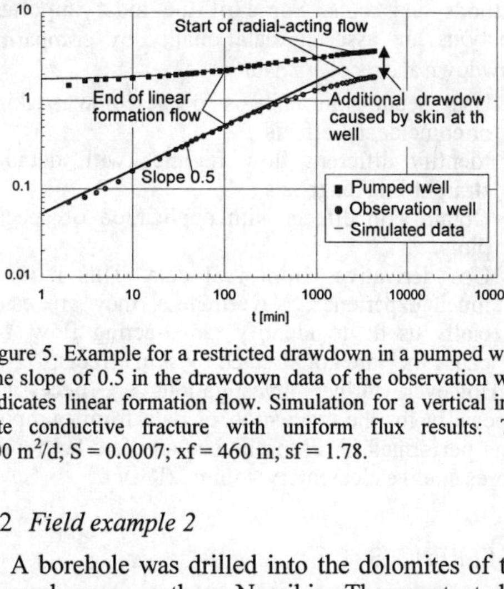

Figure 5. Example for a restricted drawdown in a pumped well. The slope of 0.5 in the drawdown data of the observation well indicates linear formation flow. Simulation for a vertical infinite conductive fracture with uniform flux results: T = 200 m^2/d; S = 0.0007; xf = 460 m; sf = 1.78.

4.2 *Field example 2*

A borehole was drilled into the dolomites of the Tsumeb area, northern Namibia. The constant discharge test indicated a short period of linear flux to a fracture in the early stages. Water strikes were recorded at depths between 14 m and 27 m below the water table. As soon as the water level in the pumped well fell to the level of the first water strike a significant increase in the drawdown was measured, with a further increase as the water level fell below to the second strike (Fig. 6). This behavior is a clear indication of over-abstracting at a rate of 20 m^3/h. However, due to unknown reasons an additional drawdown of 7 m can be determined from the special plot.

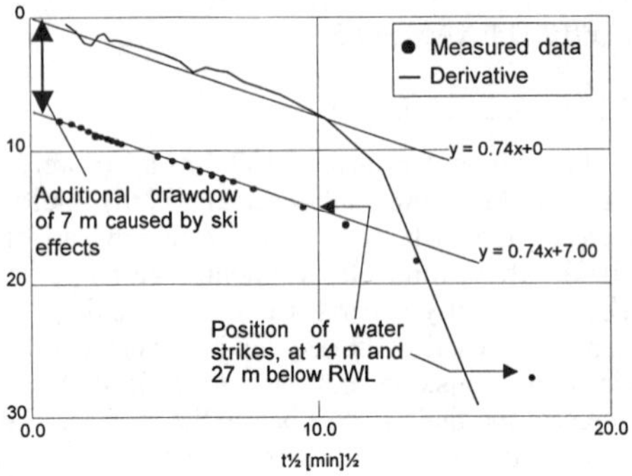

Figure 6. Graphical skin evaluation using the linear flow period of drawdown curve.

5 CONCLUSION

It is demonstrated that pumping tests can provide various types of useful information regarding the reservoir behavior of fractured rock aquifers, if reliable data and principal structural geological information are available. Furthermore, it is shown that only a detailed data analysis can produce reliable results, using a combination of diagnostic methods and tools. Some of the most important functions are assuring data quality by comparing drawdown and recovery data.

- Use pseudo-skin analysis to overcome masking or coincidental effects
- Identify different flow regimes with detailed straight-line diagnosis
- Obtain skin effects with application of special plots
- Use derivatives from real data. This requires much experience. Nevertheless, they are especially useful to identify radial-acting flow, because they are not affected by skin effects.

However, further investigations are necessary, especially for the evaluation of data from pumping tests performed in discontinuous aquifers below the representative elementary volume (REV).

REFERENCES

Bardenhagen, I. 1999. Skin Localization at Wells Drilled in a Vertical Fault Zone. GROUND WATER, Vol. 37, No 5: 764-769.

Barenblatt, G.I., Zheltov, Yu.P., & Kochina, I.N. 1960. Basic concepts in the theory of seepage of homogeneous liquids in fissured rocks. In Tatiana D. Streltsova (eds), *Well Testing in Heterogeneous Formations: 370-373. An Exxon Monography. 1988. New York: John Wiley & Sons, Inc.*

Birsoy, Y.K. & Summers, W.K. 1980. Determination of aquifer parameters from step tests and intermittent pumping data. GROUND WATER. Vol. 18: 137-146

Boehmer, W.K. & Boonstra, J. 1986. Flow to wells in intrusive dikes. Ph.D. Thesis, Free University, Amsterdam: 262pp.

Cinco-Ley, H. & Samaniego, V. 1978. Transient Pressure Behavior for a Well with a Finite-Conductivity Vertical Fracture. SPE 6014, Richardson, Texas: Society of Petroleum Engineers, American Institute of Mining, Metallurgical and Petroleum Engineers.

Cinco-Ley, H. & Samaniego, V. 1981a. Transient Pressure Analysis for Fractured Wells. SPE 7490, Richardson, Texas: Society of Petroleum Engineers, American Institute of Mining, Metallurgical and Petroleum Engineers.

Cinco-Ley, H. & Samaniego, V. 1981b. Transient Pressure Analysis: Finite Fracture versus Damaged Fracture case. SPE 10179, Richardson, Texas: Society of Petroleum Engineers, American Institute of Mining, Metallurgical and Petroleum Engineers.

Economides, M.J. & Nolte, K. G. 1989. Reservoir Stimulation. 2nd ed. Englewood Cliffs. New Jersey: Prentice Hall

Ehlig-Economides, C. & Economides, M.J. 1985. Pressure Transient Analysis in an Elongated Linear Flow System. SPE 12520, Richardson, Texas: Society of Petroleum Engineers, American Institute of Mining, Metallurgical and Petroleum Engineers.

Gringarten, A.C., Ramey Jr., H.J., & Raghavan, R. 1974. Unsteady-state pressure distribution created by a well with a single infinite-conductivity vertical fracture. SPE 4051, In Proceedings of SPE-AIME 47th Annual Fall Meeting, San Antonio, Texas, Oct. 8-11, 1972. Richardson, Texas: Society of Petroleum Engineers, American Institute of Mining, Metallurgical and Petroleum Engineers.

Gringarten, A.C., & Ramey Jr., H.J. 1974. Unsteady-state Pressure Distribution Created by a Well With a Single Horizontal Fracture, Partial Penetration, or Restricted Entry. Society of Petroleum Engineers Journal, Transaction Vol. 257: 413-426.

Kazemi, H. 1969. Pressure transient analysis of naturally fractured reservoirs with uniform fracture distribution. Trans. Soc. Pet. Eng. American Institute of Mining, Metallurgical and Petroleum Engineers AIME, 246: 451-462.

Moench, A.F. 1984. Double-Porosity Models for Fissured Groundwater Reservoir with Fracture Skin. Water Resources Research, Vol. 20, No 7: 831-846.

Ramey Jr., H.J. & Gringarten, A.C. 1976. Effect of high-volume vertical fractures on geothermal steam well behavior. In Proceedings Second U.N. Devel. And Use of Geothermal Resources, San Fransisco, U.S. Government Printing Office, Washington D.C., Vol. 3: 1759-1762.

Sabet, M.A. 1991. Well Test Analysis. Contributions in Petroleum Geology and Engineering, Gulf Publishing Company, Houston, Texas.

Streltsowa, T.D. 1988. Well Testing in Heterogeneous Formations. An Exxon Monography. New York: John Wiley & Sons, Inc.

Valkó, P. & Economides, M.J. 1997. Transient Behavior of Finite Conductivity Horizontal Fractures. SPE 38436. Society of Petroleum Engineers, SPE Journal, Vol. 2: 213-222.

Warren, J.E. & Root, P.J. 1963. The behavior of naturally fractured reservoirs. Trans. Soc. Pet. Eng. American Institute of Mining, Metallurgical and Petroleum Engineers AIME, 228, 3 (3): 245-255.

Wei, L., Hadwin, J., Chaput, E., & Swaby, P. 1998. Discriminating Fracture Patterns in Fractured Reservoirs by Pressure Transient Tests. Prepared for presentation at the annual SPE Annual Technical Conference and exhibition held in New Orleans, Louisana, 27-30 September 1998. SPE 49233. Society of Petroleum Engineers, Inc.: 609-618.

Groundwater: Past Achievements and Future Challenges, Sililo et al. (eds) © *2000 Balkema, Rotterdam, ISBN 90 5809 159 7*

Determination of aquifer characteristics of semi-confined aquifers from single well tests

W.K. Boehmer
Euroconsult, Arnhem, Netherlands

ABSTRACT: This paper describes a method to determine all the aquifer characteristics of semi-confined aquifers from data of single well tests only. They are always more abundant than aquifer tests with observation wells required at present. The method requires a semi-log time-drawdown or time-recovery data plot of a constant rate pumping test corrected for the well loss derived from a step-drawdown test. First the transmissivity (KD) of the aquifer is determined by the Jacob method. Then the product of aquifer storativity and the vertical resistance of the aquitard (SC) is determined by matching the data plot with a calculated time-drawdown plot of the transmissivity of the aquifer and the product of any assumed value of SC. The steady state drawdown of the data plot yields the values of the De Glee function Ko(r/L) and r/L from which the product of the transmissivity and the vertical resistance KDC can be determined. With above parameter combinations the transmissivity and storativity of the aquifer and the vertical resistance of the aquitard can be determined separately. Application of the method to single well test data without well loss correction and the application of Walton's type curve method are examined.

1 INTRODUCTION

Leaky aquifers or semi-confined aquifers are very common in many of the large groundwater basins in the world. Regional groundwater models of these basins become more reliable when more data on aquifer characteristics are available from pumping tests. The transmissivity *KD* and storativity *S* of the aquifer and the vertical resistance of the aquitard *C* may be determined by several pumping test methods summarised in (Kruseman and De Ridder 1990). All these methods require drawdown or recovery data from observation wells. This paper examines the possibility to determine all aquifer characteristics from drawdown or recovery data of single well tests only, which are far more abundant than aquifer tests with observation wells.

2 PREVIOUS WORK

(De Glee 1930) developed the first steady state method for determining the transmissivity of the aquifer and the vertical resistance of the aquitard of a leaky aquifer. (Hantush & Jacob 1955) developed an unsteady state method describing the drawdown around the well by the equation :

$$s = \frac{Q}{4\pi kD} W(u,r/L) \qquad (1)$$

where:
s = drawdown
t = time
r = distance from the pumped well or well radius

$$u = \frac{r^2 S}{4KDt} \qquad (2)$$

W(u,r/L) = the leaky aquifer function reduces to $2K_0(r/L)$ for steady state conditions.

$K_0(r/L)$ = modified Bessel function of the second kind and of zero order.(DeGlee 1930)

$$r/L = r/\sqrt{KDC} = \text{leakage factor} \qquad (3)$$

For steady state conditions equation 1 changes into the De Glee equation:

$$s_{max} = \frac{Q}{2\pi KD} K_0(r/L) \qquad (4)$$

Both methods require drawdown or recovery data from more than one observation well. (Hantush 1956) developed several methods for the analysis of pumping test data in leaky aquifers out of which the inflection point method applied to time-drawdown or time-recovery data of one observation well plotted on semi-logarithmic paper is generally used. (Walton 1962) by using the equations of (Hantush &

Jacob 1955) developed a curve matching method for matching time-drawdown or time-recovery data plotted on logarithmic paper with the best fitting of a family of type curves of $W(u,r/L)$ plotted versus $1/u$ for different values of r/L.

Until today single well test data could be used only for determining the transmissivity of the aquifer by applying:

- the Jacob method (Cooper and Jacob 1946) to time-drawdown or time-recovery plots. A more accurate determination of the transmissivity may be obtained by application of:
- The Hurr-Worthington method described in (Kruseman and De Ridder 1990).
- The Theis residual drawdown method (Theis 1935).

3 DETERMINATION OF ALL AQUIFER CHARACTERISTICS FROM TIME - DRAWDOWN OR TIME - RECOVERY DATA OF A SINGLE WELL TEST.

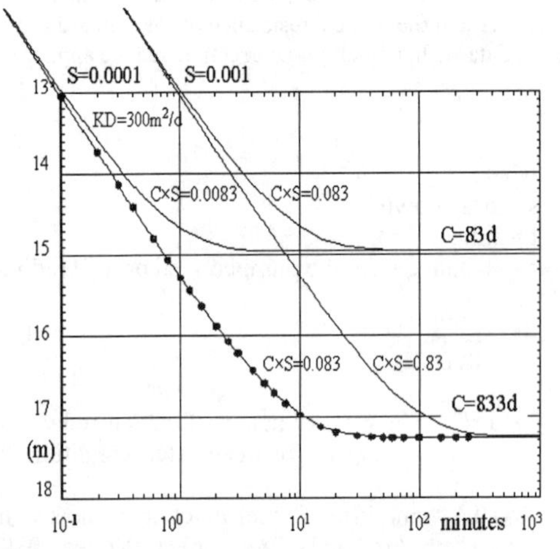

Figure 1: Semi-log time-drawdown graphs of a pumped well in a semi-confined aquifer for KD = 300 m^2/d and different values of CS.

Many of the drawdown and recovery plots of single well tests show early stabilization of the water level due to recharge by vertical leakage through over-lying and/or underlying aquitards indicating semi-confined conditions of the tested aquifers. The characteristic shape of the semi-log time-drawdown plot or time-recovery plot starts with an unsteady-state early time straight line section yielding the transmissivity of the aquifer when analyzed by the Jacob method. This straight line-section changes via a curved middle line-section into a late time horizontal line-section when changing from the unsteady state condition into the steady state condition. See Figure 1.

The following problems have prevented analysis of time-drawdown or time-recovery data of single well tests so far:

- The water level in the well is usually influenced by well loss and cannot be analysed by the above methods without a correction.
- The inflection point at half the maximum drawdown usually occurs at too early a time not shown on the time-drawdown plot. Backward extrapolation of the time-drawdown graph in order to find the position of the inflection point is generally very unreliable. This is because the slope of the time-drawdown plot is usually too small and yields too early values of t_i when extrapolated backwards resulting into too small values of the storativity S. This makes the results calculated with Hantush inflection point method very inaccurate.
- Early time-drawdown and time-recovery values may be effected by well bore storage and cannot be used for aquifer analysis.
- Another problem that may hamper the analysis is the unknown value of the effective well radius r. This may especially hamper a proper analysis of single well tests of semi-confined fractured rock aquifers. The form of the cone of depression in the immediate vicinity of such wells is often not in acccordance with the theory due to the influence of the local fracture pattern especially when the well taps a fracture in the aquifer. Even correction by well loss will not lead to the theoretical correct drawdown for the radius of the well in relation to the general aquifer transmissivity but usually to a smaller drawdown. Only at some distance from the pumped well will the cone of depression start to follow the theory. This means that observation wells will be required for the analysis of the aquifer characteristics. Only the transmissivity may be determined from single well test data by the Jacob method .

Figure 1 shows an equal shape for all four semi-log time-drawdown plots with the same KD but different C and CS values. This uniformity in shape of the semi-log time-drawdown plots and consequently the uniform shape of semi-logarithmic graphs of the Walton type curves of log $1/u$ versus $W(u,r/L)$ makes them unsuitable for separate determination of S and C by conventional matching techniques. A new matching technique of the single well test semi-log time-drawdown data plot with a calculated time-drawdown plot of the transmissivity determined by one of the above listed methods and assumed values of C and S will be developed.

The position of the time-drawdown graph of a semi-confined aquifer on semi-log graph paper may be described by the coordinates time t_i and drawdown s_i of the inflection point. The drawdown s_i is equal to half the maximum drawdown or $0.5s_{max}$ as described by equation 4. The equation for t_i may be derived from (Hantush 1956) equation:

$$u_i = \frac{r^2S}{4KDt_i} = \frac{r}{2L} \qquad (5)$$

After rearrangement Eq.5 may be rewritten as:

$$CS = \frac{4KDt_i^2}{r^2S} = \frac{t_i}{u_i} \qquad (6)$$

Equation 6 shows that *CS* is the ratio of t_i/u_i This relationship is valid for any value of u on the type curve sheet and its corresponding value of time on the data sheet after matching.

In Figure 1 following equations 6 and 1 the position of every time-drawdown plot can be described by the coordinates time *t* and drawdown *s* corresponding with a fixed value of u - W(u,r/L) = 1-1. A horizontal shift of the position of one of the time-drawdown plots with a known value of *CS* over a distance dt from position t_1 to t_2 changes the original *CS* value to $CS \times t_2/t_1$. In Figure 1 for example a shift of the upper left time-drawdown graph with CS = 0.0083 horizontally along the time-scale by one log cycle to the right yields a time-drawdown plot with a *CS* value of 0.0083 × 10 = 0.083. Only a vertical shift of this latter graph will be sufficient to match the time-drawdown data without changing the value of *CS*. Due to the absence of *r* in equation 6 determination of *CS* by this method is independent of the well radius and is very accurate.

Two pairs of graphs in Figure 1 show equal values of maximum drawdown s_{max} and consequently two pairs of graphs show equal values of C= 83.3 d and C=833 d following equation 4. Shifting the upper right graph in a vertical direction downward until it fits the drawdown data will not change *CS* but increase its maximum drawdown s_{max} by *ds* which will change the value of $K_0(r/L)$ for this graph in accordance with equation 4. The corresponding value of *r/L* can be calculated or read from a table of the function $K_0(r/L)$ yielding after substitution into equation 3 the product *KDC*. With *KD, CS* and *KDC* known *KD, C* and *S* can be determined separately.

4 PROCEDURE FOR DETERMINING *KD, C* AND *S* FROM THE DRAWDOWN OR RECOVERY DATA OF A SINGLE WELL TEST.

The following computerized matching procedure has been developed for determining all aquifer characteristics with a high degree of accuracy from data of single well tests. The computer program should be capable to analyze drawdown and recovery data of semi-confined aquifers on semi-log plots by regression techniques as well as by curve matching. The program should include the possibility to switch from "aquifer test analysis mode" into "drawdown calculation mode" without losing the calculated parameters. In this paper the program ECPUMP developed by the author has been applied for the analysis.

The procedure of the analysis described below will be explained and illustrated by application to the time-recovery data of production well H1 at Ho Chi Minh city Vietnam shown in Figure 2. Table 1 presents the residual drawdown and recovery data of this test corrected for fluctuations caused by a nearby well and well loss. The inner diameter of the well screen is 0.2m The depth of the well is 90m. The screened Upper Pliocene aquifer is composed of medium and coarse sand with lenses of fine sand and silt from 60 to 90m depth. It underlies a 12m thick semi-confining clay layer. The pumping rate during the test was 19.2 l/s.

The procedure for the determination of *KD, C* and *S* from single well test data is as follows:

(1) First (1) a step drawdown test and after full recovery of the water level, (2) a constant rate pumping test followed by (3) a recovery test should be carried out. Analysis of the step-drawdown test yields the well loss at the pumping rate applied during the single well test. The well loss should be subtracted from the drawdown and the recovery values measured during the constant rate pumping test. If required the skin effect and the effective well radius should be determined in accordance with methods described in (Kruseman & de Ridder 1990). In the example the corresponding well loss correction is 1.19m for the applied pumping rate.

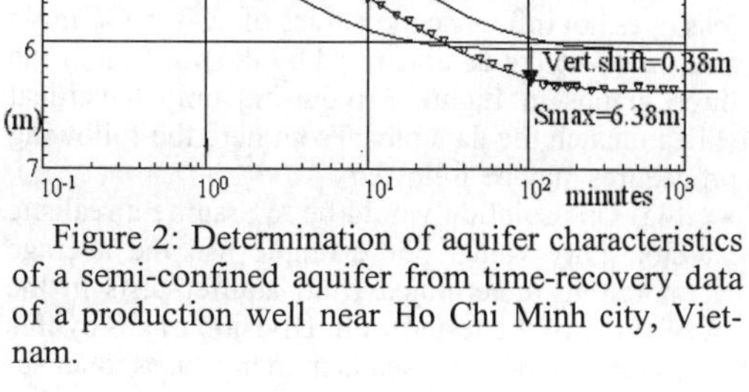

Figure 2: Determination of aquifer characteristics of a semi-confined aquifer from time-recovery data of a production well near Ho Chi Minh city, Vietnam.

(2) After the correction for well loss make a semi-log time-drawdown or time-recovery data plot of the pumped well and determine the transmissivity of the aquifer by using the Jacob method or the method of Hurr&Worthington. The Theis residual drawdown method may also be applied on residual drawdown plots. In the example presented KD=336 m^2/d.

(3) Calculate a time–recovery graph using the transmissivity of the aquifer (KD=336 m^2/d) and any convenient values of *S, C* and *CS*. For example S=0.0001, C=100d and CS=0.01 has been used in the example.

(4) Shift the calculated time-recovery graph until it matches the data plot and note the horizontal shift along the time-axis. Some reduction of the transmissivity may be required in order to obtain a better match of the calculated graph with the data plot. In this example the horizontal shift is 6.92 as shown on figure 2.

(5) Calculate the new values of *CS* by multiplying the original *CS* by the horizontal shift. In the example CS = 0.01 × 6.92 = 0.0692. Calculate a new time-drawdown graph with CS=0.0692. If correct this plot will require only a vertical shift to match the data plot.

(6) Calculate K_0(r/L) from the maximum drawdown s_{max} using equation 4 and read the corresponding value of r/L from a table of the function. In this example s_{max} = 6.38 m; K_0(r/L) = 8.217 corresponding with r/L = 3.35 × 10^{-4} After substitution into Equation 3 this yields the aquifer characteristics which are in this example: C= 262d S= 0.0692/262 = 2.64 × 10^{-4} and KD=336m^2/d.

5 PROCEDURE FOR THE ANALYSIS OF SINGLE WELL TEST DATA WITHOUT WELL LOSS DETERMINATION BY A STEP-DRAWDOWN TEST

In case no step-drawdown test has been carried out and consequently the well loss is unknown the above procedure may be applied with some adjustment. This will be illustrated by analysis of the time-recovery data of the previous example. Without correction for well loss the maximum recovery value $(s\text{-}s')_{max}$ = 7.57m. Steps 2 to 4 remain largely the same. The use of recovery data uncorrected for well loss does not influence the values of *KD* = 336 m2/d and *CS* =0.0692d as illustrated by the position of the three graphs in Figure 3 requiring only a vertical shift to match the data plot. From here the following procedures may be followed:

- (4a) One solution would be to assume a realistic storativity value. For example use the average storativity determined from aquifer tests in the vicinity of the tested well. Division of *CS* by this assumed storativity leads in many cases to an acceptable C-value although one must change the assumed storativity value in case it yields an unrealistic low *C*-value of the aquitard.
- (4b) Another solution would be to calculate time-drawdown curves using the new determined value of *CS* with estimated extreme values of the resistance of the aquitard *C* and the aquifer storativity S in order to find the "well loss" and an acceptable solution.
- (5a) In our example with CS= 0.0692d we start to calculate time-recovery graph (a) with an extreme low value of C=69.2d for the 12 m thick semi-confining clay layer and the storativity S=0.001. The maximum drawdown or recovery (s-s') = 5.85m. A vertical shift of 1.72m, representing the "well loss", is required in order to fit curve (a) with the data plot in Figure 3.

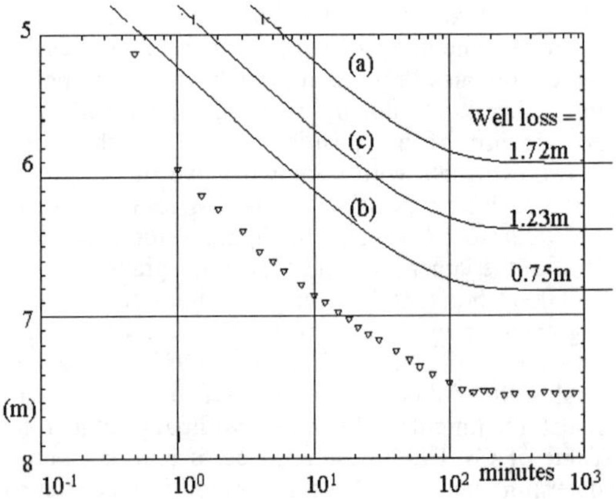

Figure 3: Time-recovery data not corrected for well loss of a production well at Ho Chi Minh city Vietnam and three time-recovery graphs calculated with extreme (curves a and b) and average C and S-values (curve c) for CS=0.0692d with the "well loss" to be subtracted in order to match the data plot.

Comparison with the real well loss of 1.19m learns that the extra 0.53m is caused by a selected too low value of *C*.

- (6a) Calculate another time-recovery graph (b) this time with an extreme low value of S=10^{-4} and the value of C= 692d to correspond with CS=0.0692d. The maximum recovery $(s\text{-}s')_{max}$ = 6.82m yielding a vertical shift or "well loss" = 0.75m to fit the data plot which deviates –0.44m from the real well loss.
- (7a) A similar calculation with an intermediate value of S=0.0003 and C=231d yields a time-recovery graph (c) with $(s\text{-}s')_{max}$ = 6.34 m. and an average "well loss" = 1.23m. This result is by chance very close to the real values of *KD, C* an *S* values as found in Section 4.

6 ANALYSIS OF SINGLE WELL TESTS BY MATCHING WITH WALTON's TYPE CURVES.

Both normal logarithmic time-drawdown plots of single well tests on semi-confined aquifers as well as the corresponding Walton type curves of *W(u,r/L)* versus *1/u* are sub horizontal due to small values of the well radius *r* and *r/L*. Therefore matching becomes usually very inaccurate, invalidating the Walton method for the analysis of single well tests. However figure 4 shows that exaggeration of the vertical drawdown-scale over the horizontal time scale in the logarithmic data plot by a factor 10 results in a steeper plot of the data as well as of the type curve. This greatly overcomes the matching problem and yields accurate determination of KD, S and C values from time-drawdown or time-recovery data of single well tests if corrected for well loss. The results presented in table 2 show that the best matching type curve type curve 2 with r/L = 0.00035 in Figure 4 yields with values of KD = 337 m2/d and CS = 0.0667 the same results as the semi-log curve matching method described in section 3.

Table 1. Recovery data from single well test on H1.

Time since stop Pumping (minutes)	Recovery* (m)	t /t'	Residual drawdown (m)
0	0	∞	6.38
0.5	3.49	2881	2.89
1	4.6	1441	1.78
1.5	4.94	961	1.44
2	5.04	721	1.37
3	5.20	481	1.18
4	5.35	361	1.03
5	5.42	289	0.96
6	5.49	241	0.89
8	5.58	181	0.81
10	5.66	145	0.71
12	5.71	121	0.67
15	5.78	97	0.60
18	5.83	81	0.55
21	5.89	69	0.49
25	5.935	58	0.40
30	5.98	49	0.39
40	6.06	37	0.32
50	6.12	29	0.26
60	6.17	25	0.205
75	6.21	20	0.17
100	6.26	15.4	0.12
125	6.325	12.5	0.055
150	6.34	10.6	0.04
175	6.33	9.2	0.05
200	6.35	8.2	0.03
250	6.38	6.7	0.0
300	6.38	5.8	0.0

* Corrected for well loss = 1.22m

Experience shows that the semi-logarithmic curve matching method of section 3 is easier to apply and yields more accurate results than the variant of the Walton method applied here.

Table 2: Aquifer characteristics determined by the Walton method from recovery data of well H1.

Curve No.*	'r/L $\times10^{-4}$	KD (m^2/d)	CS (d)	S $\times10^{-4}$	C (d)
1	2.16	357	0.0831	1.32	600
2	3.5	337	0.0667	2.8	242
3	5.42	322	0.0597	5.7	106

*Curve No. refers to type curve No. in Figure 4.

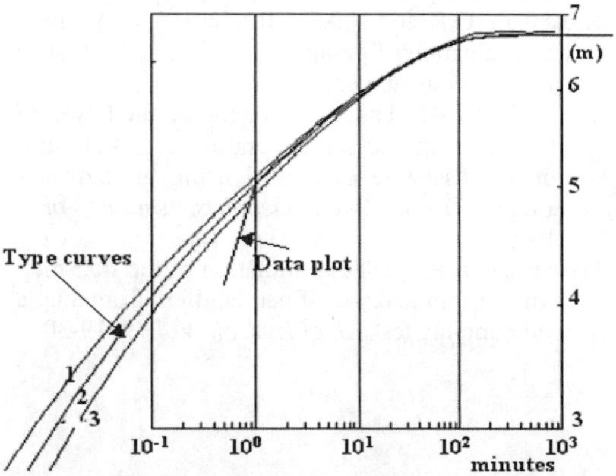

Figure 4: Time-recovery data of well H1 near Ho Chi Minh city Vietnam matched with Walton's type curves for the determination of aquifer characteristics of a semi-confined aquifer. (Y-scale of plot exaggerated for accurate matching)

7 CONCLUSIONS

1. The presented semi-logarithmic curve matching method of calculated time-drawdown plots with time-drawdown or time-recovery data plots of single well test yields all aquifer characteristics of a semi-confined aquifer accurately, provided that the drawdown or recovery data can be corrected for well loss determined by a step drawdown test.
2. Accurate values of aquifer transmissivity *KD* and the product of aquifer storativity and the hydraulic resistance of the aquitard *CS* can be determined from time-drawdown or time-recovery data of any single well test on a semi-confined aquifer without correction for well loss.
3. The semi-log time-drawdown matching technique is easier to apply and yield more accurate results than the conventional Walton curve matching method despite the development of an accurate matching technique for type curves with small values of well radius r and *r/L* as required for single well tests.

REFERENCES

Bierschenk W.H. 1963.Determining well efficiency by multiple step-drawdown tests. *Intern. Assoc. Sci. Hydrol. Publ.64*: 493-507.

De Glee G.J.1930. Over grondwaterstromingen bij water onttrekking door middel van putten. *Thesis J.Waltman* Delft (The Netherlands*)* 175pp.

Hantush M.S. and C.E. Jacob 1955. Non-steady radial flow in an infinite leaky aquifer. *Trans. Amer. Geophys. Union Vol.36* : 95-100.

Hantush M.S. 1956. Analysis of of data from pumping tests in leaky aquifers. *Trans. Amer. Geophys. Union Vol.37* : 702-714.

Kruseman G.P. and N.A.De Ridder. 1990. Analysis and evaluation of pumping test data. 2nd ed. *ILRI Publ.47* Wageningen.

Theis. C.V. 1935. The relation between the lowering of the piezometric surface and the rate and duration of discharge of a well using groundwater storage. *Trans. Amer. Geophys. Union Vol.2*: 519-524.

Worthington P.F. 1981. Estimation of the transmissivity of thin leaky-confined aquifers from single well pumping tests. *J. of Hydrol. Vol 49*: 19-30.

Groundwater: Past Achievements and Future Challenges, Sililo et al. (eds) © *2000 Balkema, Rotterdam, ISBN 90 5809 159 7*

The value of groundwater resources associated with carbonatites

F.S. Botha
Department of Water Affairs and Forestry, Pietersburg, South Africa
J.L. van Rooy
University of Pretoria, South Africa
J.J.P. Viviers
Southern Africa Geoconsultants, Pretoria, South Africa

ABSTRACT: The aim of this paper is ultimately to highlight the value of the carbonatite as a cheap sustainable alternative bulk water source and the value of regional groundwater exploration programmes. The Nebo Plateau forms part of the Lebowa Granite Suite, Bushveld Igneous Complex. Several geological events forming part of the structural setting may have resulted in excellent aquifer conditions and forming part of these geological events is the Spitskop Complex. The Spitskop Complex is the second target area in a regional groundwater exploration programme on the Nebo Plateau. Due to the genetic history and the mineral composition the complex serves as an excellent target for ground water exploration. Boreholes in the carbonatite have good blow yields with good expected pumping yields.

1 INTRODUCTION

The Reconstruction and Development Programme (RDP) of the South African Government has as one of its goals the provision of clean drinking water and sanitation to all its people. This is not a major problem in and around the cities and smaller towns. However, in rural South Africa the increasing demand for clean drinking water puts more pressure on one of our most valuable resources. Supplying all communities with surface water would be an impossible task in large parts of South Africa. The development of the necessary infrastructure for surface water supply schemes is a time consuming process which may take at least 10 –15 years to complete. It is therefore necessary that supplementary water supplies should be found and utilized as an immediate source. Groundwater is a quick and economic solution and is widely utilized in community water supply programs. Therefore groundwater may serve as an interim water supply and may come in useful during future dry periods. Some geological environments, however is classified as having a poor groundwater exploitation potential It is in these harsh environments that groundwater exploration programmes should be launched to investigated the occurrence of ground water and the sustainable economic exploitation potential. The Nebo Plateau, South Africa can be described as having a low groundwater exploitation potential (Figure 1). In an attempt to address the issue a research project was launched, aiming to understand the water regime and to find ways to utilise the available groundwater.

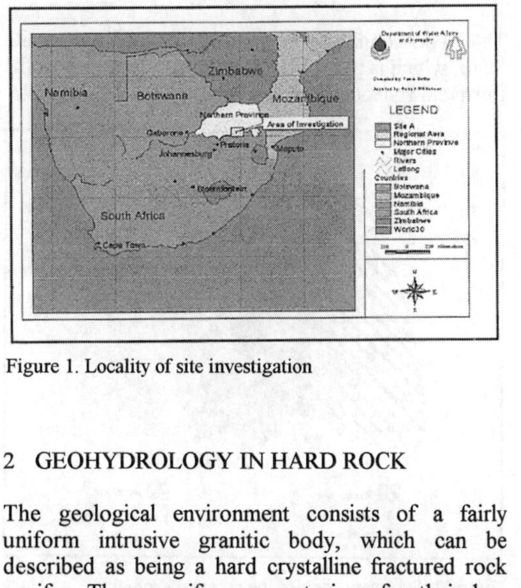

Figure 1. Locality of site investigation

2 GEOHYDROLOGY IN HARD ROCK

The geological environment consists of a fairly uniform intrusive granitic body, which can be described as being a hard crystalline fractured rock aquifer. These aquifers are notorious for their low groundwater potential and difficult groundwater exploration conditions. Groundwater exploration and development in hard crystalline rock areas is a difficult task and several perspectives on how to find suitable groundwater resources have been developed. Dijon (1984) found that in Precambrian rocks major fractures associated with faults weather along strike, forming deep basins. These weathered zones correspond with topographic lows and drainage systems. The topographical lows serve as water recharge areas and most successful wells are located in these areas. Taylor (1984) did work in the northwest of India on middle Precambrian granitic rocks, situated in a hot, semi-arid, Savannah-type climate. He found that borehole yield depends on the *number*, *width* and *spacing* of fractures and/or the relative thickness and permeability of the weathered layers. Larsson (1981) did some hydro-geological work on the Sardina granites in Italy, and concluded that hydrogeological investigations must include the setting up of rainfall gauges and constructing runoff stations and regional geological mapping. In *Project 8.6 of the International Hydrological Program-Ground water in Hard Rocks* (Larrsson, 1984), the use of satellite imagery and air photos is suggested, focussing on structural and geomorphological features. A methodology developed by Baweja and Raju (1984) suggested that groundwater development programmes should make use of geology, structure and groundwater conditions expected in the area. It is clear that the regional geological structures on the Nebo Plateau should be studied closely during a groundwater investigation.

3 REGIONAL GEOLOGICAL SETTING

The Nebo Granite forms part of the Lebowa Granite Suite, which is part of the later intrusive stages of the Bushveld Igneous Complex (Figure 2).

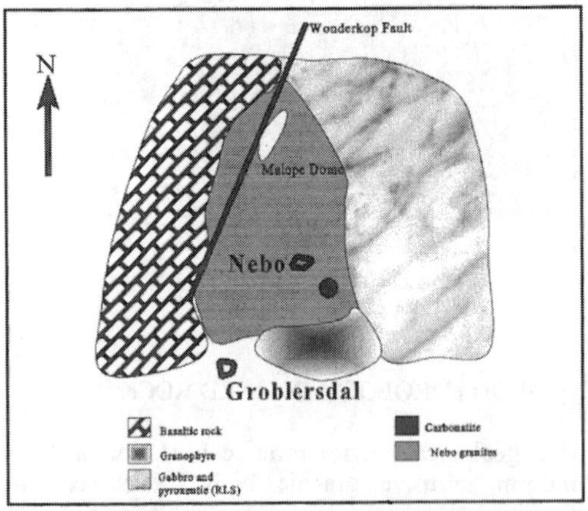

Figure 2. Schematic view of the regional geology

The Nebo Granite is a plate like intrusion resting on the Rustenburg Layered Suite (RLS), with the contact dipping 5°to the north- northwest. The thickness of the intrusion varies from 420m in the south to 2400m in the north. The Nebo Granite outcrop is surrounded by younger Karoo basaltic rocks in the west, which form the flat lying Springbok Flats with the Transvaal sediments in the north and rocks from the Rustenburg Layered Suite in the east. The ***Wonderkop Fault*** is the most prominent regional geological structure in the area. Intersecting the granite in the west, the Olifants River follows the fault, which extends into the Rustenburg Layered Suite, to the east. The Wonderkop Fault is a strike-slip fault and indications are that the fault extends beyond the dolomite of the Chuniespoort Group to the north-east (Schewllnue *et al, 1962).* The ***Steelpoort Fault*** and ***Sekhukune Fault*** are manifestations of two other major geological events which influenced the Lebowa Granite Suite. Odgers and du Plessis(1989) did a regional seismic survey in the north-eastern Bushveld Complex and postulated that the occurrence of the Wonderkop Fault and the ***Malope Dome*** in such close proximity are no coincidence. They found that the overburden thickness varies across the fault and estimate the material under the Malope Dome could have been less dense than the surrounding material making the material buoyant. The catalyst effect of the buoyant sub-layer under the dense overburden may have promoted diapirism. The rising diapir in turn induced further variation in the overburden and consequently induced further faulting in the overburden. The composition of the rising diapir (Malope Dome) is still not certain, but gravity surveys suggest that it is capped with a relatively thin layer of Transvaal rocks and RLS with the main body however consisting of material similar to the Nebo Granite. The **Sptiskop Complex** forms part of a suite of alkaline intrusions intruded into the Kaapvaal Craton during 1450 Ma and 1200 Ma (Harmer, 1999). It is an alkaline intrusion and can be classified as a dolomitic carbonatite. The complex is almost a round intrusion with a diameter of approximately 8 kilometres, covering a surface area of 50 km^2 . The complex consists of successive intrusions of pyroxenite, ijolite, nepheline syenite (foyaite) and carbonatite. The dolomitic carbonatite is much younger than the surrounding granites but due to its low silica content it tends to weather much faster forming a regional topographical low. The Complex are highly fractured and although this may be the result of the regional structural setting, it may also be associated with the introduction and mineralization of the complex.

4 SITE INVESTIGATION

The Spitskop complex was due for investigation forming part of the regional investigation. However, due to a community water supply programme close to the complex it was decided to drill much earlier than planned. Initially it was thought that dyke material was intersected, but the calcite and mica identified in the drilling samples indicated otherwise. The samples were identified as ijolite forming part of the Spitskop complex. The first borehole had a blow yield of more than 20l/s and seven more exploration boreholes gave good results. It was obvious that the

Spitskop Complex held must promise and therefore a full geohydrological investigation followed.

4.1 Siting and drilling of new boreholes

Initially conventional geophysical methods were used to site new boreholes but these methods failed. The conductivity of the cover material may be to high or the magnetic layers in the carbonatite may have altered the readings, eventually only normal magnetic profiles were used. Three targets were identified, (1) the contact between the carbonatite and the country rock, (2) the contact between the dolomite and the ijolite (3)structural features in the carbonatite. Magnetic profiles provide enough data to identify the contact between the complex and the country rocks. Typically the contact will have a magnetic low and boreholes were drilled across the contact (Figure 3).

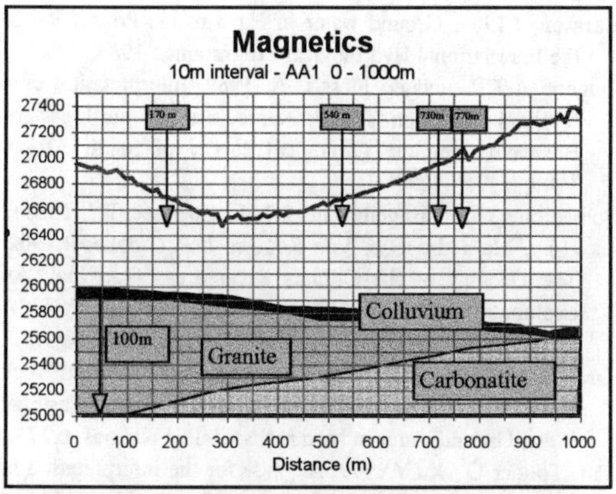

Figure 3. Typical magnetic profile showing magnetic low, with preliminary profile derived from drilling logs.

The best successes were near the contact, in the carbonatite where the ijolite are highly fractured. The contact area is highly weathered and in some places clay material were drilled to a depth of 78 m. Blow yields on the weathered contact area are between 1 and 3 l/s and blow yields in the fractured zone near the contact are as high as 23 l/s. The contact between the dolomite and the ijolite is well mapped and can be identified quite easily. Therefore siting was done mainly with the aid of available geological maps. Magnetic profiles were also placed across the proposed target area to verify exploration sites. All four sites identified were successful with blow yields in the order of 5 to 23 l/s. The boreholes sited on the radial geological features were also sited using both geological interpretation and magnetic profiles. The boreholes have fair blow yields ranging from 0.5 and 15 l/s.

4.2. Sustainable domestic water supply.

One borehole drilled during the exploration phase was pump tested. The borehole was tested at a constant yield of 22 l/s for 72 hrs and with the aid of the FC method (Van Tonder and Xu, 1999) recommended at 8 l/s for 24hrs. A T-early value of 338 and T-late of 93 m^2/d was calculated and a S-value of $5.3*10^{-3}$ given. An additional 15 boreholes will be tested and recommended. According to DWAF standards the ground water have a exceptional good quality. If the proposed boreholes will be used as an alternative bulk source it must be able to coupe with large abstraction volumes. Therefore a steady state water balance based on recharge were compiled using Bear's single cell example. The water balance (WB) therefore can be calculate as follows (Figure 4):

WB = (RI+RB+IA) - (WA+EV+RB+OA)

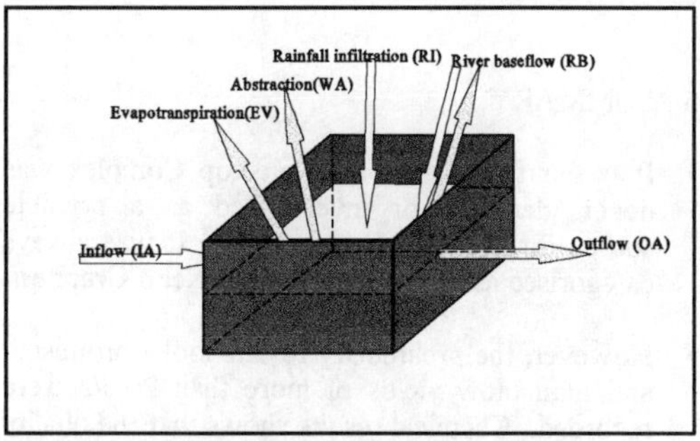

Figure 4. Single cell water balance based on Bear *et al.*

Based on the drilling results and available information the water balance were calculated. Two scenarios were taken in account making certain assumptions.

Table 1. List of assumptions

Scenario	A	B
Mean annual rainfall (mm)	719.2	601.5
Surface area (Km^2)	100	100
Recharge (%)	6	3
In-/outflow from aquifers (m^3/y)	0	0
In-/outflow from rivers (m^3/y)	0	0
Evapotranspirartion.(m^3/y)	0	0

Assuming that the in-/outflow from the rivers and the adjacent aquifers is equal and that A is our best case scenario and B our worse case. Then according to scenario A $3.58X10^6$ m^3/y can be abstracted and $1.80X10^6$ m^3/y according to scenario B. If the average aquifer thickness is 35 metres and the storativity is taken as $2X10^{-3}$ then there is approximately $4.67X10^{6}$ m^{3}/y in storage which may last 3 years if correctly managed. Therefore, taking the worst case scenario in account, 19 boreholes yielding 3 l/s@24 hrs can run for almost four years without any rain, suppling almost 200 000 people with clean drinking water at 25 l/p/d.

4.3 The value of the Spitskop Complex

The value of the ground water found due to the Spitskop Complex can be calculated. If the estimated cost of an alternative surface water supply is R10.00/m^{3} (Bardenhorst, 2000, *pers. comm.*) and the cost of the ground water supply is R1.50/m^{3} (Classens, 2000, *pers comm.*) then the value of the ground water can be calculated as the difference in cost multiplied by the volume abstracted. Therefore the value of the ground water can be calculated at almost R15 000 000 per year.

5 SUMMARY

- It is fair to say that the Spitskop Complex was never identified or investigated as a possible source for domestic water supply. It was always categorised as forming part of the Nebo Granites.
- However, the preliminary results looks promising and high blow yields of more than 20 l/s were recorded. Chemical results shows that the quality may also be promising and according to the F.C_method one borehole can already be recommended at 8 l/s for 24 hours.
- A single cell model was used to determine the recharge and to calculate the rate at which the aquifer can be pumped.
- It was concluded that 200 000 people could benefit from this source and that the source could be managed through a 4 year drought period.
- By using this source almost R15 000 000 can be saved per year.
- Ground water must not be ruled out as a water source on the Nebo Plateau. Boreholes should be carefully selected and exploitation recommendations should be done accordingly.
- The only way to promote ground water is to compile a water balance and calculate the ground water value and expressed it in a currency.

6 REFERENCES

Bardenhorst, B. 2000. Personnel communication.

Baweja, B.K., Raju, 1984. Ground-water potential of hard rock areas in the Noyil, Amaravati and Ponnani Basins, South India. Groundwater in hard rocks, Project 8.6 of the International hydrological programme, Unesco

Bear, J.1979. Hydraulics of Groundwater. Israel: Mcgraw--Hill. pp.442.

Classens, J. Personnel communication.

Dijon, R.1984. Ground-water exploration in precambrian rocks of Northern Cameroon. Ground water in hard rocks: United Nations, p 171.

Harmer, R.E. 1999. Petrogenetic association of Carbonatite and Alkaline Megmatism: Constrains of the Spitskop Complex, South Africa. Jrnl of Petrology, vol 49, no.4.

Larsson, I.1981. Ground water investigations in a granite area of Sandina, Italy., Report by UNESCO, Reg. Office for Science and Technology for South and Central Asia, New Delhi., 1981., pp 17 - 25

Larsson, I.1984. Ground water in Hard rocks, Project 8.6 of the International Hydrological Programme., 1984.

Odgers, A.T.R. and du Plessis, A. 1989. Interpretation of a regional reflection seismic survey in the north-eastern Bushveld complex, Geological survey of South Africa, 1989, p 125 –129

Schwellnus, J.S.I, Engelbrecht, L.N.J., Coertzee, F.J, Russel, H.D., Malherbe, S.J., Van Rooyen, D.P.,Crooke, R.1962. The Geology of the Olifants River area, Transvaal. An explanation of sheet 2429B (Chuniespoort) and 2430A (Wolkberg), Department of Mines, Geological Survey.

Taylor, G.C.1984. Ground-water utilization in hard rock aquifers of the Pali region, Jodhpur, Division, Rajasthan, India. Ground water in hard rocks: United Nations, p.174.

Van Tonder,G., Xu,Y.1999. A guide for the interpretation of pumping tests and estimation of sustainable yields of boreholes in fractured-rock aquifers. Technical Report no. GH3927, DWAF.

Groundwater: Past Achievements and Future Challenges, Sililo et al. (eds) © 2000 Balkema, Rotterdam, ISBN 90 5809 159 7

The study of karst spring recession in Slovenia

M. Brencic
Geological Survey of Slovenia, Ljubljana, Slovenia

ABSTRACT: In the article the theory of monotonic recession analysis of a karst spring hydrogram is presented. The recession curves of the hydrogram are analysed in the same diagram and described with descriptive statistic profiles that converge to one point. The profiles were described with the power and exponential law on the basis of which it is possible to calculate the parameters of convergence. The theory is illustrated by the hydrograms of two large spring systems, Vipava and Ljubija from the Dinaric karst in Slovenia.

1 INTRODUCTION

The analysis of a hydrogram and its recession part has long been recognised as an important tool in hydrology and has been the subject of extensive research. With this analysis it is possible to determine various model parameters. At the beginning of the 20th century Maillet (1905) published the results of hydrological studies of the river Vanne. He analysed various recession curves that were fitted with the exponential equation.

Before the Second World War new concepts in hydrogram analysis were established. The unit hydrogram was introduced and on the basis of this concept Barnes (1939) separated run-off into three components. Each component was described with the recession coefficient on the basis of the exponential outflow equation.

Recession analysis is used for solving various water management problems, such as low flow prediction, irrigation analysis, water supply, hydropower plant construction, precipitation run-off modelling etc. (Tallaksen, 1995).

In karst hydrogeology the analysis of a karst spring recession hydrogram represents an important method for the estimation of dynamic reserves, permeability, porosity and transmisivity in karst aquifers (Bonacci, 1993; Ford & Williams, 1989). The starting point for the analysis of a karst aquifer is the work of Schoeller (1962) who presents various analytical models under various boundary and initial conditions. Later Mangin (1970, 1975) developed new mathematical solutions of karst spring recession hydrograms.

In the literature various definitions of recession are given. They differ according to the needs and theoretical background (e.g. Chow, 1964; Linsley et al., 1988) and they are sometimes arbitrary and subjective. To avoid this we experimentally studied only the shape of the total hydrogram and its recession part without any other presumptions.

2 GENERAL SETTINGS

In Slovenia more than 40% of the area consists of karstic rocks which mainly belong to the well-known Dinaric karst. At the moment more than 7400 caves are registered, several of them are deeper than 1000 m and longer than a few kilometres. Among the significant karst phenomena there are also large karst springs. Well known are the springs of the Ljubljanica system (Gospodaric & Habic, 1974) and the springs with the recharge area in Trnovsko Banjska planota (Kranjc, 1997).

Karst in Slovenia is positioned in the area where the climate is under the influence of the Mediterranean and the Alps, with the average rainfall of more than 1500 mm/year, while in the higher parts the average is more than 3000 mm/year. The evapotranspiration is in the interval between 400 and 600 mm.

The recession of daily total hydrograms of two large systems of springs were studied. The first studied hydrogram is the one of the river Vipava. It consists of the hydrograms of several small springs flowing out from fractures. The river drains from the

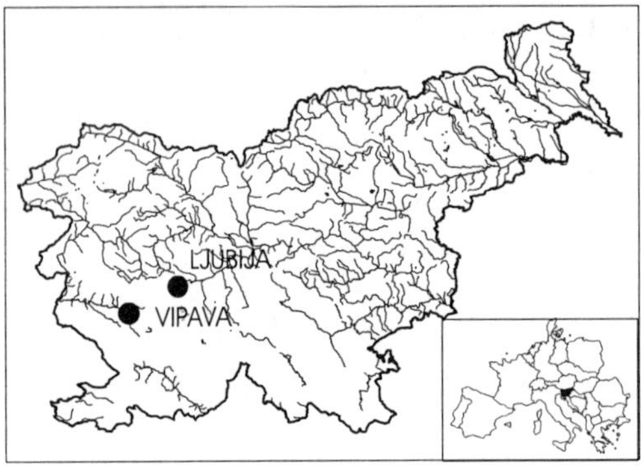

Figure 1. Position of springs

Trnovsko Banjska planota through the Upper Cretaceous bedded limestones. The springs are positioned at the bottom of the Vipava valley. They are the consequence of Eocene flysch at the bottom. The analysed total hydrogram covers the period from the beginning of the year 1960 to the end of 1995 over a period of 36 years with a total of 13149 observations. Average discharge is Q_{avg}=6,83m^3/s, standard deviation s=8,36m^3/s, $Q_{0,50}$=3,53 m^3/s, $Q_{0,25}$=2,10 m^3/s, $Q_{0,75}$=7,80 m^3/s, Q_{max}= 67,1 m^3/s, Q_{min}=0,73 m^3/s, Q_{max}/Q_{min}=92.

The river Ljubija, to which belongs the second studied hydrogram, is also represented by several smaller springs. Its recharge area is in the Jurassic and Cretaceous rocks of the Cerknica - Planina polje system. The analysed total hydrogram covers the period from the beginning of the year 1952 to the end of 1994 over a period of 43 years with a total of 15706 observations. Average discharge is Q_{avg}=6,87m^3/s, standard deviation s=5,07m^3/s, $Q_{0,50}$=5,48 m^3/s, $Q_{0,25}$=2,87 m^3/s, $Q_{0,75}$=9,55 m^3/s, Q_{max}= 25,80 m^3/s, Q_{min}=0,2 m^3/s, Q_{max}/Q_{min}=129.

The hydrograms were obtained from the database of the Hydrometeorological Survey of Slovenia.

3 THEORY

3.1 *Monotonic recession*

Sometimes in the modelling of hydrogeological phenomenon the system is described as a black box, which is defined with the input I(t) variable (e.g. rainfall, infiltration), output variable Q(t) in the form of discharge and with the system rules. In classical recession analysis we assume that the recession begins when I(t)=0. The point in time when I(t)=0 is usually chosen on the basis of the conceptual model of a real system or some theoretical suppositions.

In the karst hydrogeological system it is difficult to decide when, if ever, the input variable is zero. In order to avoid subjectivity and arbitrary decision in the determination of time point when I(t)=0, we experimentally studied the shape of the recession limb.

We define monotonic concentration as a part of the event hydrogram when

$$\frac{dQ(t)}{dt} > 0 \tag{1}$$

and monotonic recession as a part of the event hydrogram when

$$\frac{dQ(t)}{dt} \leq 0 \tag{2}$$

with this definition we separate the total hydrogram into N_{conc} monotonic concentration curves and N_{rec} monotonic recession curves. In the next step we are dealing with monotonic recession curves as a whole. In the preceding analysis all of the monotonic recession curves begin at the recession time t_i=1 day regardless of the position of the curve in the total hydrogram. The series of all monotonic recession curves of the same origin are defined as the cloud of monotonic recession.

3.2 *Description of recession cloud*

The cloud of monotonic recession is mathematically described with profile curves based on the calculation of descriptive statistics for every day of particular recession day t_i. For each vertical that belongs to particular recession day t_i with $N(t_i)$ observation discharges Q descriptive statistic where calculated, among them average, variance and percentiles. Calculated statistics for particular recession day t_i connected are treated as profile curves. We use average, variance and percentile profile curves.

3.3 *Model of profile curves*

Profile curves were described with two mathematical models:

a) Power model

$$Q_{P_i} = Q_{0P_i} t^{-\lambda_{P_i}} \tag{3}$$

b) Exponential model

$$Q_{P_i} = Q_{0P_i} \exp(-\lambda_{P_i} t) \tag{4}$$

Where λ_{Pi}, Q_0 are descriptive parameters of the model and index P_i presents the particular profile of descriptive statistics.

As a measure of goodness of fit of the modelled profile curve we calculate the correlation coefficient r.

3.4 *Convergence of recession cloud*

To describe the relationship between the descriptive parameters Q_{0Pi} and λ_{Pi} from equation (3) and (4) the following equation is choosen:

$$Q_{0P_i} = A\exp(B\lambda_{P_i}) \quad (5)$$

The equation (5) represents a convergence modelling curve. For the power model (3) from (5) follows

$$Q_{P_i} = A\exp(B\lambda_{P_i})t^{-\lambda_{P_i}} \quad (6)$$

To determine the point of convergence for the percentile curves of descriptive statistics we choose from (6) two different percentile curves Q_{P10} and Q_{P20} with different slope parameters λ_{P10} and λ_{P20}. At the point of convergence

$$Q_{P10}(\lambda_{P10}) = Q_{P20}(\lambda_{P20}) \quad (7)$$

and after a brief manipulation

$$\exp(B\lambda_{P10})t^{-\lambda_{P10}} = \exp(B\lambda_{P20})t^{-\lambda_{P20}} \quad (8)$$

$$B(\lambda_{P10} - \lambda_{P20}) = (\lambda_{P10} - \lambda_{P20})\ln t \quad (9)$$

$$t = \exp B \quad (10)$$

According to equation (10) the time of convergence for the power model is t_c=expB. In the simillar way we can show that other percentile curves of the monotonic recession cloud could converge at time t_c and that convergence discharge for power model is A= Q_c.

For the exponential model (4) from equation (5) follows

$$Q_{P_i} = A\exp(\lambda_{P_i}[B - t]) \quad (11)$$

After brief manipulation we can show that all modelling curves can converge only at convergence time B=t_c and convergence discharge A=Q_c.

4 RESULTS

The results of the monotonic recession cloud analysis of the karst spring systems Vipava and Ljubija are presented.

4.1 *Monotonic recession cloud of Vipava*

The monotonic recession cloud of Vipava is presented in Figure 2. It was modelled with the power model curves. 1633 monotonic recession curves with the average duration of 6,31 days were analysed. In the total hydrogram 78,37 % of the length is under the monotonic recession.

The average profile curve of the monotonic recession cloud and its correlation coefficient are

$$Q_{avg} = 12{,}87t^{-0{,}72} \qquad r = -0{,}9946 \quad (12)$$

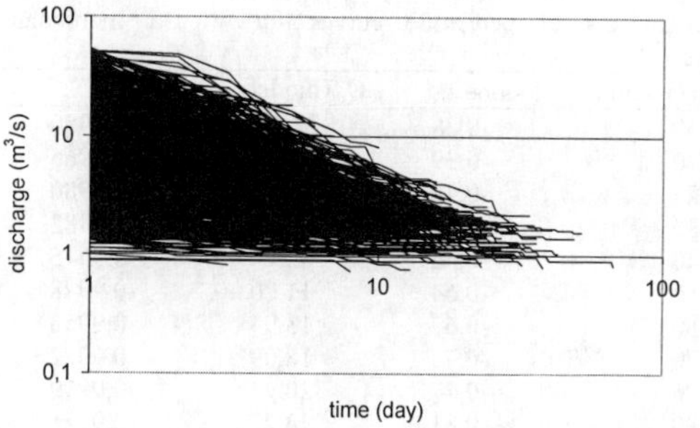

Figure 2. Monotonic recession cloud of Vipava in log log diagram

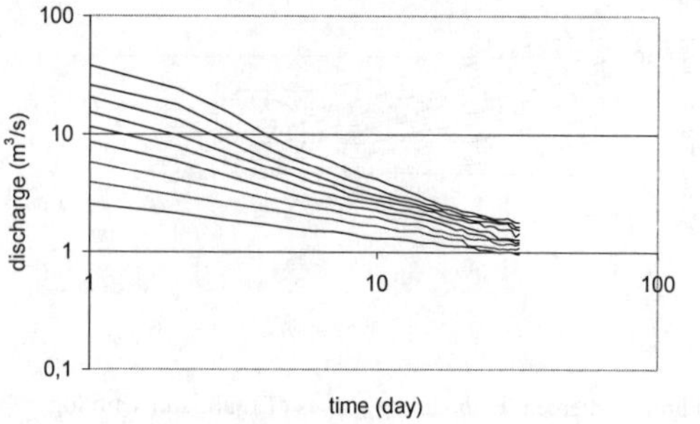

Figure 3. Percentile modeling curves of Vipava in log log diagram

The variance profile curve of the monotonic recession cloud and its correlation coefficient are

$$\mathrm{var} = 143{,}74t^{-2{,}33} \qquad r = -0{,}9932 \quad (13)$$

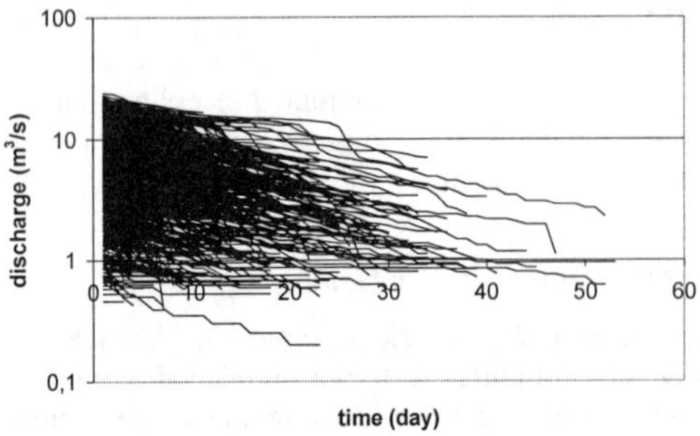

Figure 4. Monotonic recession cloud of Ljubija in semi log diagram

The results of modelling with percentile curves for the monotonic recession cloud are presented in Table 1 the empirical curves are presented in Figure 3.

Table 1. The descriptive parameters and correlation coefficients of percentile curves for Vipava monotonic recession

Percentile	slope λ	Q_0 (model)	r
10	-0,28	2,52	-0,9980
20	-0,39	4,00	-0,9965
25	-0,43	4,83	-0,9980
30	-0,47	5,66	-0,9982
40	-0,56	8,15	-0,9975
50	-0,64	11,20	-0,9978
60	-0,67	13,94	-0,9956
70	-0,73	18,09	-0,9922
75	-0,77	20,92	-0,9919
80	-0,81	24,27	-0,9921
90	-0,91	36,57	-0,9933

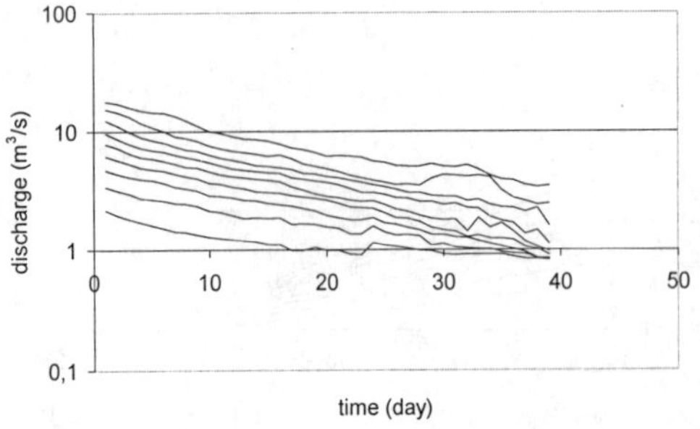

Figure 5. Percentile modeling curves of Ljubija in semi log diagram

The convergence modelling curve and its correlation coefficient are

$$Q_0 = 0{,}76\exp(4{,}27\lambda) \qquad r = 0{,}9994 \qquad (14)$$

From modelling curves follows that the convergence discharge of Vipava is Q_c=0,76 m^3/s and convergence time is t_c=72 days.

4.2 *Monotonic recession cloud of Ljubija*

The monotonic recession cloud of Ljubija is presented in Figure 4. It was modelled with exponential model curves. 1514 monotonic recession curves with the average duration of 8,22 days were analysed. In the total hydrogram 79,27 % of the length is under the monotonic recession.

The average profile curve of the monotonic recession cloud and its correlation coefficient are

$$Q_{avg} = 7{,}96\exp(-0{,}038t) \qquad r = -0{,}9928 \qquad (15)$$

The variance profile curve of the monotonic recession cloud and its correlation coefficient are

$$\text{var} = 30{,}47\exp(-0{,}077t) \qquad r = -0{,}9931 \qquad (16)$$

The results of modelling with percentile curves for the monotonic recession cloud are presented in Table 2, the empirical curves are presented in Figure 5.

Table 2. The descriptive parameters and correlation coefficients of percentile curves for Ljubija monotonic recession

Percentile	slope λ	Q_0 (model)	r
10	-0,025	1,77	-0,8906
20	-0,034	3,16	-0,9822
25	-0,038	3,89	-0,9930
30	-0,041	4,61	-0,9959
40	-0,044	5,97	-0,9952
50	-0,047	7,50	-0,9964
60	-0,044	8,76	-0,9950
70	-0,047	10,70	-0,9893
75	-0,050	12,08	-0,9902
80	-0,051	13,74	-0,9834
90	-0,047	17,33	-0,9855

The convergence modelling curve and its correlation coefficient are

$$Q_0 = 0{,}19\exp(82{,}53\lambda) \qquad r = 0{,}9730 \qquad (17)$$

From modelling curves follows that the convergence discharge of Ljubija is Q_c=0,19 m^3/s and convergence time is t_c=83 days.

5 CONCLUSIONS

1. Separation into monotonic recession and concentration shows very regular pattern, which

could be described with profile curves of descriptive statistics.

2. Each of the profile curves of descriptive statistics could be modelled with exponential and power curves with very high correlation. Only small discrepancies between empirical data and modelling curves are presented at the beginning of the recession.
3. Model curves converge at the convergence discharge Q_c and convergence time t_c.
4. Different convergence times and convergence discharges for various karst spring hydrograms are mainly the consequence of different karstification level in the aquifers that drains out through the springs.

REFERENCES

Barnes, B.S. 1939. The structure of discharge - recession curves. *Transactions of American Geophysical Union* 20: 721 - 725.

Bonacci, O. 1993. Karst springs hydrographs as indicators of karst aquifers. *Hydrological Sciences Journal* 36: 51 - 62.

Chow, V.T. (ed.) 1964. *Handbook of applied hydrology*. New York: McGraw Hill Book Company.

Ford, D. & Williams, P. 1989. *Karst geomorphology and hydrology*. London: Chapman & Hall.

Gospodaric, R. & Habic, P. 1976. *Underground water tracing. Investigation in Slovenia 1972 – 1975*. Ljubljana: Institute Karst Research.

Kranjc, A. 1997. Karst hydrogeological investigations in southwestern Slovenia. *Acta Carsologica* 26 (1), 1- 388.

Linsley, R.K., Kholer, M.A., Paulhus, J.L.H. 1988. *Hydrology for engineers*. London: McGraw Hill Book Company.

Maillet, E. 1905. *Mécanique et physique du globe – Essais d'hydraulique souterraine et fluviale*. Paris.

Mangin, A. 1970. Contribution à l'étude des aquifères karstiques à partir de l'analyse des courbes de décrue et tarissement. *Annales de spéléologie* 25: 581 - 610.

Mangin, A. 1975. Contribution à l'étude hydrodynamique des aquifères karstiques. Dijon, DES thesis Univ. Dijon.

Schoeller, H. 1962. *Les Eaux Souterraines*. Paris: Masson & Cie.

Tallaksen, L.M. 1995. A review of baseflow recession analysis. *Journal of Hydrology* 165: 349- 370.

Groundwater: Past Achievements and Future Challenges, Sililo et al. (eds) © 2000 Balkema, Rotterdam, ISBN 90 5809 159 7

Preliminary assessment of the groundwater potential of the Tsumeb Aquifers in northern Namibia

R. Bufler & P. Ambs – *GKW Consult, Mannheim, Germany*

T. Himmelsbach – *Bundesanstalt für Geowissenschaften und Rohstoffe (BGR), Hannover, Germany*

E. Tordiffe – *NamWater, Windhoek, Namibia*

R. Bäumle – *Department of Applied Geology, University of Karlsruhe, Germany*

ABSTRACT: This paper presents the preliminary results of investigations in the northern Otavi Mountainland and its flat foreland up to Oshivelo in northern Namibia. Comprehensive investigations, including hydrocensus, geophysical surveys and remote sensing interpretation have been executed. The geological and hydrogeological situation of the study area was assessed by hydraulic modeling with the goal to quantify the possible long-term sustainable groundwater abstraction and the short-term emergency bulk abstraction from the Tsumeb Aquifers. Different scenarios were delineated which allow a prediction of the impact of groundwater abstraction on the groundwater resources of the different aquifer systems encountered.

1 INTRODUCTION

Within the next decade the water supply for Central Namibia will face an urgent need for additional secure water resources. A feasible option might be to abstract more groundwater from the Tsumeb Aquifers in the Grootfontein-Tsumeb-Otavi area, to meet the growing water demand for the Central Area of Namibia.

In order to quantify the long-term sustainable abstraction and the possible short-term emergency bulk groundwater abstraction from the Tsumeb Aquifers, a comprehensive investigation program was initiated in September 1999 that will be finalized in April 2001. The investigation program consists inter alia of a hydrocensus in order to determine the groundwater level and its temporal variations and the impact of hydrological conditions on the groundwater resources. In addition it will allow the compilation of existing hydrogeological and hydraulic information, which serve as a basis for numerical modeling.

The main emphasis of the study lies in the determination of potential groundwater quantities, which may be abstracted from aquifers in the Grootfontein-Tsumeb-Otavi area on a sustainable and economical basis. This abstraction must however be considered in direct relation to water demands at the Kalahari area, where porous aquifers seem to be hydraulically connected to underlying fractured aquifers of the Grootfontein-Tsumeb-Otavi dolomite group. In a later phase the study will address the potential impacts of groundwater abstraction on the environment as well as socio-economical impacts on the study area.

2 DESCRIPTION OF THE STUDY AREA

2.1 *Geological frame*

The study comprises with an area of approximately 10,000 km^2 the northern Otavi Mountainland (OML) between Otavi, Grootfontein and Tsumeb as well as the flat foreland up to Oshivelo (see Fig. 1). The OML lies on the northern platform/shelf area of the Otjiwarongo branch of the Damara Orogen. Upon the granites and gneisses of the Grootfontein Basement Complex approximately 6000 m of sediments of the northern facies of the Damara Sequence have been accumulated (Hedberg 1979, Lombaard et al. 1986). The proterozoic Damara Sequence consists of a basal arenaceous unit (Nosib Group, up to 1500 m), a middle carbonate unit (Otavi Group, up to 3000 m) and an upper clastic unit (Mulden Group, up to 1700 m). The predominantly dolomitic rocks of the Otavi Group have been stratigraphically subdivided into the lower Abenab Subgroup and the upper Tsumeb Subgroup. The strata were moderately folded during the Pan-African Orogeny (680-450 Ma) into several synclines and anticlines generally trending west-east.

Four karstification periods of the dolomitic rocks of the Otavi Group have been delineated. The first and oldest period of karstification lasted presumably from 750 Ma to 650 Ma in a period of uplift and erosion (Lombaard et al. 1986, Plöthner 1997). A second karstification period interrupted by glaciation and sedimentation lasted from 450 Ma to 280 Ma. The third karstification period occurred in early Tertiary (Van der Westhuizen et al. 1988). The last ma-

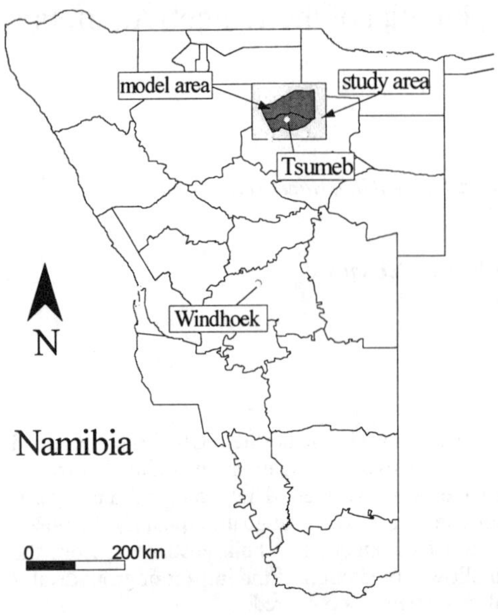

Figure 1. Location of the study area

jor phase of karst processes developed during the Pleistocene from approximately 34,000 to 14,000 yr BP according to Geyh (1995).

Dykes and dyke swarms with extensions of several kilometers in strike direction are present in the study area. These doleritic intrusive bodies of Karoo age roughly follow a SW-NE direction, indicating an inherent zone of weakness which is continuously rejuvenated (Hedberg 1979).

2.2 *Hydrogeology and hydrology*

The OML is a dolomitic massif dominated by a pronounced rugged relief with hills rising some 500 m above the surrounding plain. It is intersected by a series of major broad west-east trending valleys and only few minor lateral valleys. Below the main valleys either low permeable phyllites of the Mulden Group or low permeable metamorphic rocks of the Grootfontein Basement Complex occur. Within the investigation area three different aquifer types occur.

1. The Tsumeb Dolomite Aquifer as well as the aquifers in igneous and metamorphic rocks can be described as overall low to medium permeable fractured aquifers. At some places, however, the permeability of the Tsumeb Dolomite Aquifer may increase considerably due to intense karstification at contact zones and tectonic shear zones. The chimneys of the Tsumeb and Abenab mines, the Dragons Breath, Harasib and Ghaub Caves and the Otjikotoand Guinas Lakes are for instance major palaeokarst features.

2. The second aquifer system is formed by porous, unconsolidated to semi-consolidated Tertiary to Quaternary sediments of the Kalahari Group north of Tsumeb. The Kalahari is the flat and gently northwards dipping foreland of the dolomite mountains. The thickness of the Kalahari deposits increases towards the north. The western and northern margins of the Kalahari are the Etosha basin and the Ovambo basin.

3. The third aquifer type is formed by fractured dykes of different geological provenance. For instance, the dormant Tsumeb mines and the Abenab mines are situated near such dyke systems in the OML. Although the dyke systems seem to be not widely distributed, their influence on the entire hydraulic regime can not be neglected.

Table 1. Pumping test evaluation statistics.

	Porous Aquifer T [m^2/day]	Fractured Aquifer T [m^2/day]
Minimum	123	14
Maximum	9880	3200
Mean	3894	1840
Pumping tests evaluated	5	13

Reliable information on the aquifer parameters of the Tsumeb and Kalahari Aquifers is sparse. The transmissivities of the porous and fractured aquifers obtained from the analysis of former pumping tests range between 14 and 9880 m^2/d within the model area. These findings are explained by the heterogeneous nature of carbonate rocks of the Otavi Group with regard to structural and karst features (fractured aquifer) and sedimentological variations within the Kalahari Group (porous aquifer), leading to laterally and vertically varying transmissivity values. A similar variety can be observed for the evaluated permeability and storativity values.

The transmissivity of the dykes is unknown at this stage of the investigations.

The mean annual temperature at Tsumeb is 22 °C, and the mean annual potential evaporation at Grootfontein amounts to 2314 mm (Wrabel 1999). Maps of mean annual evaporation for the study area, however, show rates between 2600 mm and 2800 mm (Crerar & Church 1988), while the study area receives a mean annual precipitation of only 566 mm (1913-1994). The distribution of the annual rainfall is highly erratic (Plöthner 1997). Rainfall occurs mainly during summer thunderstorms between November and March

3 INVESTIGATIONS PERFORMED

During the desk study in the Phase 1 of the project all existing studies, the remote sensing investigation for lineament analysis and the determination of irrigation patterns were either compiled or reinterpreted. Additionally, the field survey for the assessment of the current groundwater abstraction and groundwater levels of the different aquifer systems was completed. Reconnaissance geophysical surveys with TDEM, FDEM, ground magnetic and gravity investigations were carried out for the delineation of the stratigraphical succession and the verification of the lineament analysis from satellite images. The generated and compiled data have been incorporated in a GIS database.

The synthesis of geological, geophysical and hydrogeological investigations leads to the delineation of two major dykes. One is located directly northwest of Tsumeb and the other one approximately 25 km east of Tsumeb. At the present stage of the investigations the dykes are considered to connect the fractured aquifer of the dolomite massif in the south and the porous aquifer of the northern foreland (see Fig. 2). Preliminary modeling of the geophysical data suggests that the SW-NE striking dykes are weathered and that enhanced hydraulic conductivities might be developed in their fractured margins. The interpretation of satellite images shows more regionally important, preferential flowpaths, which exist within the dolomites of the OML as N-S striking fracture zones. It seems very likely that along such structurally weak zones secondary aquifers exist due to locally enhanced karstification. The N-S striking fracture zones were generated during the Damaran Orogeny (580-450 Ma) and have been rejuvenated at least during the continental break-up of the African and American plates between approximately 180 Ma and 130 Ma.

According to Hedberg (1979) occurs the Karoo Group only at depth in the northern part of the study area. Isopach maps (Hedberg 1979) indicate that the Karoo Group commences approximately 40 km north of Tsumeb and increases towards the northern boundary of the study area to approximately 200 m in thickness. With the present studies, by way of the interpretation of a geoelectrical section extending northward from Tsumeb, it is suggested that Karoo strata may extend further south, closer to Tsumeb.

East of a north-south line, passing through Tsumeb, basalts of the Etendeka Formation occur within the Karoo Sequence according to Hedberg (1979). West of this line, only clastic sediments (Dwyka and Etjo Formations) are present (Hedberg 1979). Interpretation of the aeromagnetic data in the present study suggests that the basalts in fact extend further west, right across the northern portion of the study area.

It is interpreted from a line of TDEM soundings, conducted in the frame of the present geophysical investigations, that a series of sub-parallel conductors exists beneath the Kalahari Group sediments, presumably overlying the rocks of the Otavi Group.

This set of conductors in the northern study area occurs in depths between approximately 150 m and 300 m. It is interpreted that this set of conductors represents Karoo age rocks (high porosity rocks with saline groundwater or shales), particularly in view of the presence of overlying basalts as interpreted from the aeromagnetic data.

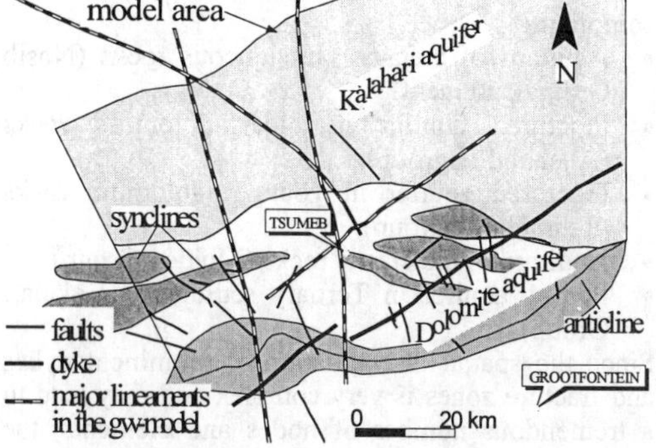

Figure 2. Simplified geological map with the major geological lineaments in the study area

The analysis of the irrigation patterns from satellite images (LANDSAT TM) indicates a substantial reduction of irrigation from 1984 to 1996 in the area of the Kalahari Aquifer. Thus, a reduction of groundwater abstraction for agricultural production can be assumed. According to the groundwater contour plan, the regional flow patterns are directed from the Tsumeb Dolomite Aquifers towards the northern Kalahari Aquifer. The hydraulic head lies between 1350 and 1460 m a.m.s.l. along the southern model boundary and amounts to approx. 1140 m a.m.s.l. along the northern model boundary. The major drop in hydraulic head occurs within the Tsumeb Dolomite Aquifer, i.e. in the south of the model area, whilst the hydraulic gradient in the Kalahari area is comparatively small.

4 DEFINITION OF THE MODEL

The finite element code of FEFLOW is used to simulate and predict the impact of groundwater abstraction on different aquifer systems. The main emphasis is laid upon the influence of preferential flow paths and how they might affect the entire hydrological regime, especially under the influence of irrigation affected by the respective geohydraulic and meteorological conditions.

First, 2-dimensional horizontal (2-DH) and vertical models (2-DV) of the Tsumeb Aquifers were established. The hydraulic boundaries of the model were defined according to the results from geophysical and geological investigations and from hydrocensus data. The model boundary types are:

- No-flow boundary (or flux boundary) in the south of the model area,
- Hydraulic head boundary in the north of the model area within the Kalahari Aquifer,
- No-flow boundaries to the western and eastern margins of the model area, which are aligned along prominent flow directions.

Five hydrogeological units have been delineated, comprising:

- Aquitard of metamorphic/igneous rocks (Nosib Group/Basement),
- Fractured aquifer in older dolomitic rocks (Abenab Subgroup),
- Fractured aquifer in younger dolomitic rocks (Tsumeb Subgroup),
- Aquitard of the clastic rocks (Mulden Group),
- Porous aquifer in Tertiary sediments (Kalahari Group).

Since the spatial discretisation of prominent dykes and fracture zones is very complex and may lead to a tremendous number of nodes and elements, the numerical model could incorporate only four discrete prominent lineaments (see Fig. 2). The existing numerous minor dykes, which occur as dyke swarms, are simulated therefore by anisotropic permeability tensors. With this approach the nature of recharge into the northern Kalahari Aquifer by deep lying dyke and fracture zones can be delineated.

The main problem for the definition of the models lies in the unknown vertical extent of different groundwater horizons. It can be outlined that the entire aquifer system is subdivided into different vertical, depth correlated horizons which are characterized by different mean residence times of the groundwater. Tectonic planes and ancient weathering surfaces bind the different horizons. An uppermost fast advection zone is underlain by a zone, which is characterized by a groundwater turnover-time amounting up to a few decades. The latter is underlain by deep aquifer levels, which show mean residence times of the groundwater of up to a few hundred years. Since the aim of the model is to predict the change of groundwater levels within a few decades, only the uppermost aquifer horizons, which are assumed to interact with surface-related hydrology, were incorporated in the hydraulic model.

An appropriate approach to define the bottom level of the vertical succession of the aquifer horizons can be derived from the evaluation of hydrochemical data. In the Tsumeb mine, variations of the sulfate concentration in the groundwater according to depth were monitored. These data show elevated sulfate concentrations down to a depth of approximately 750 m. From this level on the sulfate concentrations decrease with increasing depth. This fact indicates that at depths beyond approximately 1600 m sulfate reduction dominates. Accordingly, the zone of active convection seems to be restricted to a depth of not more than 1600 m. This depth is assumed to represent the top of the stagnant groundwater zone and is therefore chosen as the lower vertical model boundary.

5 MODELING RESULTS AND PRELIMINARY CONCLUSIONS

The aim of the numerical modeling during Phase 1 of the Project is to verify whether and how the northern Kalahari Aquifer is connected to the Tsumeb Aquifers. It is of particular interest whether the dykes and major fault zones act as conduits or barriers to the groundwater flow. This question is investigated with different horizontal and vertical model scenarios. The calibrated 2-DH and 2-DV models assist parameter and sensitivity studies in order to quantify and to test different hydrogeological hypotheses.

In general it can be concluded that the chosen approach with more than 8 different scenarios is advantageous/useful for the determination of a water balance for the Tsumeb Aquifers and of the potential abstraction impacts. The groundwater balance delineated for the different scenarios from the modeling is based inter alia on literature values from previous studies for the recharge and the hydraulic conductivity.

However, the modeling showed that discrepancies of the different estimations are apparently depending on the chosen boundary conditions. The modeling indicates that the calibrations are possible with either changes of the recharge or the hydraulic conductivity values. Overall, little changes in input parameters (recharge, hydraulic conductivity) show a major impact on the groundwater balance, as well as on the hydraulic behavior of different areas. Hence, recharge and hydraulic conductivity parameters have to be determined more accurately in Phase 2 of the Project when the 3D-model will be defined.

The Karoo Group possibly occurs between the Kalahari and the Dolomite Aquifer. According to the modeling results its distribution and lithology has a crucial impact on the groundwater balance. If the Kalahari Aquifer is separated from the underlying Dolomite Aquifer by the low permeable Karoo Group, the model scenario shows that a higher recharge factor can be allocated to the Kalahari area. Such a scenario indicates a strong increase of the groundwater potential of the Tsumeb Aquifers (see Fig. 3).

Without such a confining layer, the calibrations of the 2-D models indicate that recharge by precipitation in the Kalahari basin reaches only approx. 10 per cent of the recharge in the areas of outcropping

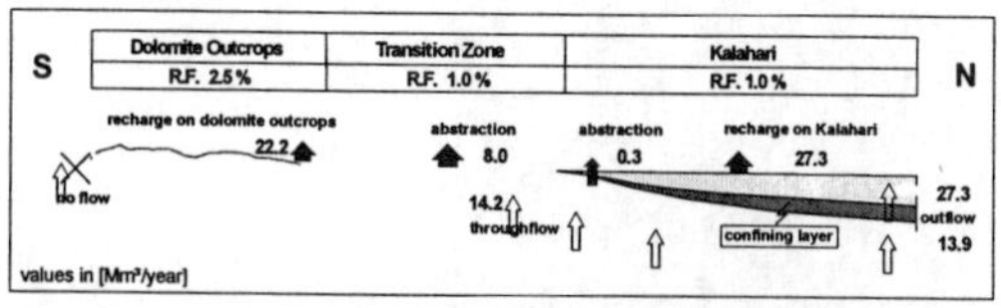

Figure 3. Groundwater balance components obtained from two different model scenarios. The lower scenario assumes the presence of a confining layer between the Kalahari and the underlying Tsumeb Dolomite Aquifer and a higher recharge factor (*R.F.*) in the Kalahari area.

dolomite within the OML. This implicates that the recharge area of the Kalahari Aquifer includes large areas of the OML and that all enhanced, long-term groundwater abstraction from the OML might, after a few decades, affect also the water abstraction in the northern Kalahari area. At this stage of the investigation, however, it is not yet fully understood whether the Kalahari Aquifer is mainly fed by discrete, highly permeable fault and fracture zones or by diffuse exfiltration of groundwater from the underlying fractured aquifer of the dolomite group.

It is emphasized that the presented preliminary groundwater balances for the model scenarios have to be refined with more reliable assumptions before they can be utilized for a determination of the sustainable groundwater potential in the study area. However, at the present stage of the investigations it already can be assumed that the Tsumeb Aquifers represent a substantial groundwater resource.

During Phase 2 the groundwater modeling will therefore include the findings from the upcoming detailed geological and geophysical surveys as well as the results from drilling, aquifer testing and the recharge study. In order to predict the reaction of the aquifer systems and the interaction between the Tsumeb Dolomite and the Kalahari Aquifers realistically, a 3-D model will be developed on the basis of the 2-DH and 2-DV models, which have been established during Phase1. The modeling of the aquifer responses for various abstraction scenarios will allow recommendations for a sustainable management of the groundwater resources in the study area will be provided.

Acknowledgments: The Consulting Group GKW CONSULT/BICON was appointed by the DWA (Department of Water Affairs, Namibia) to execute the KfW (German Development Bank) financed studies. Staff from DWA, NamWater and the Consultants jointly execute the study. We wish to thank the Director of Water Resources, Mr. P. Heyns (DWA) and the Senior Hydrogeologists I. Bardenhagen (DWA) and H. Grobbelaar (NamWater) for their support during the study.

REFERENCES

Crerar, S.E. & Church, J.T. 1988. *Evaporation map for Namibia.* Unpubl. Rep. (11/1/8/1/H1). Windhoek: DWA.

Geyh, M.A. 1995. Geochronologische Aspekte paläohydrologischer und paläoklimatischer Befunde in Namibia. *Geomethodica* 20: 75-99.

Hedberg, R.M. 1979. Stratigraphy of the Ovamboland Basin, South West Africa. *Bull. Precambr. Res. Unit Univ. Cape Town* 24

Loombard, A.F., Günzel, A., Innes, J. & Krüger, T.L. 1986. The Tsumeb Lead-Copper-Zinc-Silver deposit, South West Africa/Namibia. In C.R. Anhhaeusser &S. Maske (eds), *Mineral Deposits of Southern Africa* Vol.I & II: 1761-1787. Johannesburg: Geol. Soc. S. Afr.

Marchant, J.W. 1980. *Hydrogeochemical exploration at Tsumeb.* Unpubl. Ph.D. Thesis. Cape Town: Univ. SA.

Plöthner, D. 1997. *Hydrogeology and isotope hydrology of the Otavi Mountainland and its surroundings KARST_01 and KARST_02.* Unpubl. Rep. Hannover: Bundesanstalt für Geowissenschaften und Rohstoffe (BGR).

Van der Westhuizen, W.A.,Tordiffe, E.A.W., Loock, J.C. & De Bruiyn, H. 1988. Quaternary karstification and vanadium mineralization in the Otavi Mountain Land, South West Africa/Namibia. *Palaeocol. Afr.* 19: 391-403.

Wrabel, J. 1999. Ermittlung der Grundwasserneubildung im semiariden Bereich Namibias mittels der Chlorid-Bilanz-Methode. *Hydrogeologie und Umwelt.* 16.

Groundwater: Past Achievements and Future Challenges, Sililo et al. (eds) © 2000 Balkema, Rotterdam, ISBN 90 5809 159 7

Groundwater occurrence in hard rocks in the federal district of Brasilia – A sustainable supply?

J.E.G.Campos
Instituto de Geociências, Universidade de Brasilia, Brazil

U.Tröger
Institut für Angewandte Geowissenschaften II, Technische Universität Berlin, Germany

ABSTRACT: The demand of groundwater for urban supply in the greater area of Brasilia, the capital of Brazil, required an exploitation strategy to identify sustainable water resources. The design of a hydrogeological map and a vulnerability map was only possible after designing a new geologic/lithologic map without soil covers. A detailed tectonic map in combination with the others maps and information about aquifer systems was the base for the hydrogeological map. Both, protection and exploitation of groundwater as a sustainable resource is possible with management using the referred maps. This paper shows that the technical information forms a basis to turn the groundwater into a sustainable water supply for to the region.

1 INTRODUCTION

The Federal District (FD) is located in central Brazil and includes Brasilia the capital of the country. The planning of Brasilia in the late 1950s did not include any groundwater supply for the inhabitants of the FD. The population did increase in the mean time over one hundred percent of the planned number. New settlements in the surrounding of Brasilia are of two types: closed condominiums of more wealthy people and simple agglomerations of houses of poor people. Both have the same problem. In the first type the water supply is for a few who can pay more for the water, but the supply from dam sites is not feasible. The second type can pay only a little amount for the water. The only solution for the water supply for both groups is groundwater.

Recently a new geological map in a scale of 1:100.000 without soil cover was prepared. In combination with soil maps, geomorphological maps, satellite image interpretation, and the information from water wells a hydrogeological map and a vulnerability map were presented. The hydrogeological map shows that groundwater is more abundant in east-west striking structures and the specific yields are higher in the quartzite than in schist. Also it can be demonstrated that the strongly folded series with a dip of a higher angle have wider open fractures with better groundwater circulation conditions, when compared with series that dip with low angle.

1.1 *Soil*

In the area there are four main type of soils. The first group is represented by thick soils (up to 30 metres thickness), sandy and, reddish to yellowish-red in colour developed over the highlands. The second is similar to the first, but presents a clayey to silt texture. Another group is composed by clay material, varying from some centimeters to a maximum of two meters, are grey to white in color and is developed over the dissected portion of the landscape. The last soil class is almost absent, shows just some centimeters of thickness and contains some rock fragments.

1.2 *Geology*

The regional geology comprises low metamorphic grade Proterozoic folded rocks and is represented by four units: the Araxá, Bambuí, Paranoá and Canastra groups. The Canastra Group is constituted by phyllite, calciphyllite and fine grey marble; the Paranoá Group is defined by slate, coarse to fine quartzite, metarhythmic rocks (intercalation of quartzite and slate) and carbonate lenses; the Bambuí group is a siltstone and arkosic bearing sequence and the Araxá Group is composed of schist and fine micaceous quartzite (Freitas-Silva & Campos 1998). The Canastra and Paranoá are considered of Meso to Neoproterozoic age, while the Bambuí and Araxá are dated as Neoproterozoic.

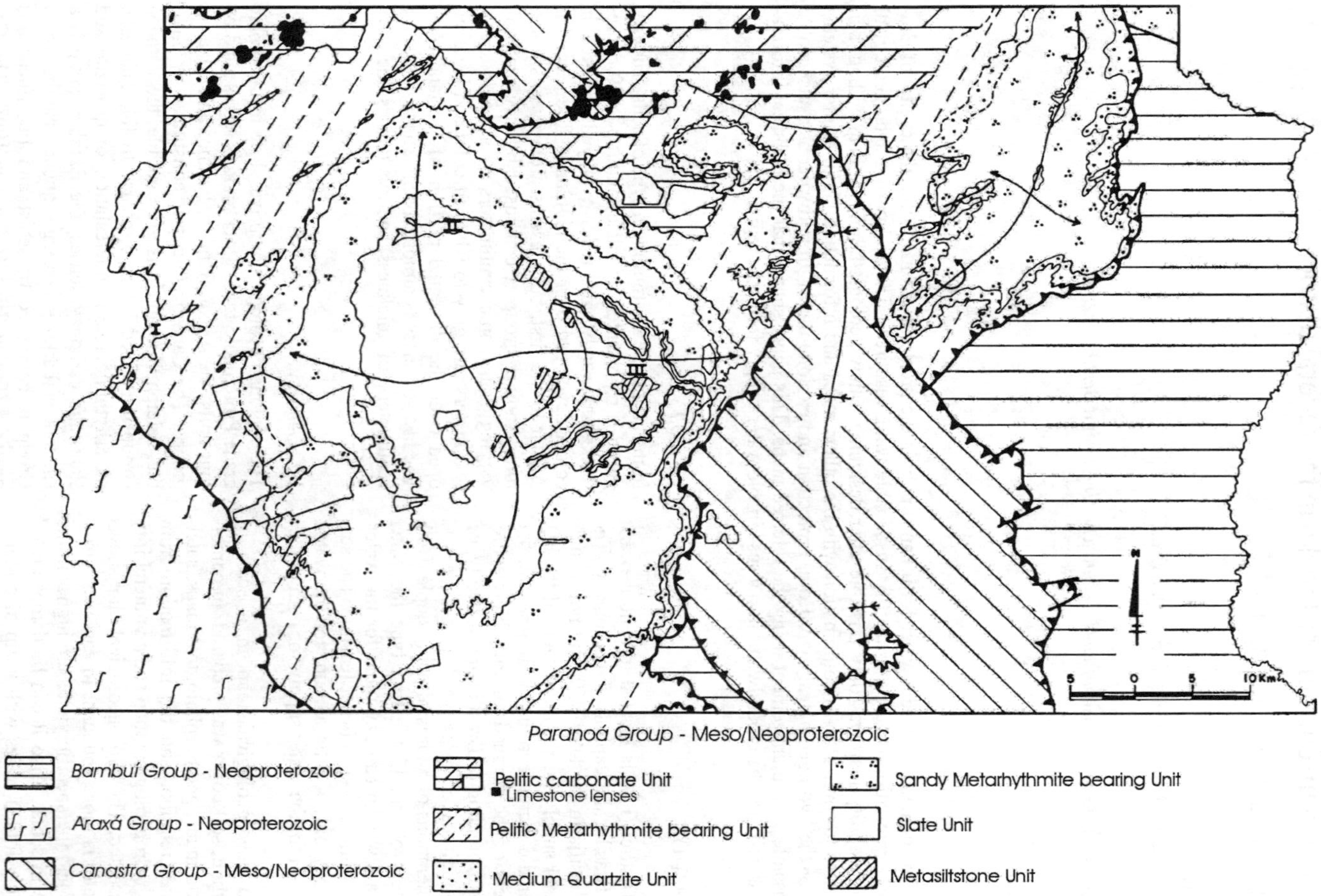

Figure 1- Simplified Geologic map of the Federal District (Brazil). I - Descoberto Dam; II - Santa Maria Dam and III - Paranoá Lake. (After Freitas-Silva & Campos 1998).

In terms of distribution the Canastra and Bambuí represent each 15% of the total area of the FD, the Paranoá Group is about 65% of the area and the Araxá represents a minor area of only 5%. The Paranoá Group is the most important (Fig. 1).

1.3 *Tectonic setting*

All geologic units are related by important thrust faults that evolved in the Brazilian Tectonic Cycle at the end of the Proterozoic. The complex tectonic model observed in the area is due to the position of the region, that is situated in the boundary of the external portion of the Fold Belt to the cratonic domain. In this situation there is a complex pattern of fold superposition which results in the dome and basin pattern.

The structural geology can be summarized in Table 1. The D_5 deformation phase is the most important in the point of view of groundwater, as it represents the ruptile control of the aquifers. The F_a fracture group are of extensional origin, the F_b is of shear or dilatational origin, while the F_c is the conjugated pair of the shear fractures. All groups of fracturing may be horizontal to vertical in dip, but the most important in the groundwater circulation control is the east-west trend and in part the conjugated pair. The east-west fractures may dip towards south or north, but the northward dipping are more common.

Another important control over the groundwater circulation is the reactivation of the fracture planes. This reactivation is attributed to the Cretaceous South Atlantic drift and also Tertiary neotectonic events, as can be shown by the young lateritic crusts faulting and small graben structures locally observed. Figure 2 shows the main apparent lineaments of the region.

Table 1. Summary of the main structural features of the Federal District region.

Deformation Phases	Main Structures	Mean Attitude
D_1	S_1 foliation, generated in a simple shear regime .	S_1 N15-27E/50NW
D_2	F_2 chevron asymmetric folds, with E2 axis and S2 axial plane foliation.	S_2 N56E/87NW E_2 N60E/15 and S60W/15
D_3	Open F_3 folds with E_3 axis and a sparse axial plane cleavage	S_3 E10-15W/70-90SW E_3 N10-15E/00-
D_4	Open F_4 folds with E_4 axis and a sparse axial plane cleavage	S_4 NS/80-90E E_4 E10W/10
D_5	Normal faults and generalized fracturing (three main groups of fractures F_a, F_b and F_c)	F_a EW F_b N10-15E F_c N45W and N45E

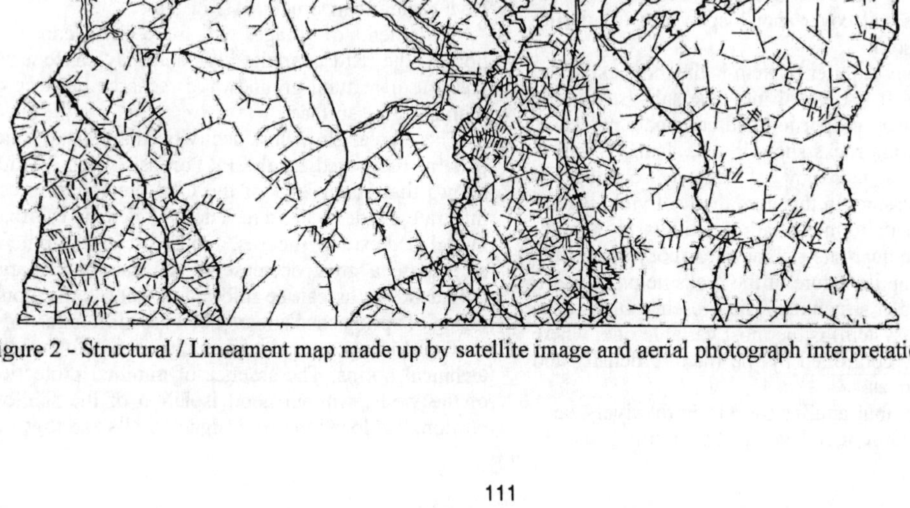

Figure 2 - Structural / Lineament map made up by satellite image and aerial photograph interpretation.

Geophysical studies (Lousada 1999) using eletromagnetic methods have shown good results in the subsurface mapping of the fractured zones.

2 HYDROGEOLOGY

All aquifer types, the porous, the fractured and the karst domain, are represented in the area. The porous is limited to the weathering zone with up to 50 meters thickness, including the sandy or silty-clay soils and weathered rocks. The fractured aquifers are represented by the fracture/fault/joint zones of the Proterozoic rocks. The karst aquifers are related to the limestone and marble lenses of the Paranoá and Canastra Group.

The porous domain is divided into four aquifer systems (P_1, P_2, P_3 and P_4 systems) based on the soil type and the total thickness of the weathered mantle. This domain, although presents low total water reserves is very important as a natural filter and a regulatory layer for the streams discharge. These aquifers are also important for the recharge of the underlying fractured and karst systems.

The fractured domain is the most important for groundwater supply and is divided in the Canastra, Paranoá, Araxá and Bambuí aquifer unit. The Araxá Aquifer is developed in low specific yield schist. It is the least productive aquifer in the FD area, with the lowest K, T and S values due to the unfavorable recharge conditions, resulting from low fracturing density, low mean foliation plunge dipping and shallow soil cover.

The Canastra Aquifer System is important because of the presence of the marble lenses which locally provide good wells with up to 100 m^3/h water discharge and is now important for public water supply. The importance of this aquifer system decreases outside the fractured zones of the marble lense occurrence. The vertical foliation dip and the thick soil cover control the aquifer recharge, due to the preferential planes to water percolation and to the water reserved in the soil cover (not lost by run off during the rainy season).

The Paranoá Aquifer System is the most exploited in the region. It is divided into five subsystems due to the clay/sandy proportion and carbonate presence. The clay bearing rocks show low K, T and S values as there is a tendency of fracture closing by lithostatic pressure. In this case the well yield is controlled by quartzite intercalation and quartz vein occurrence. The quartzite-rich aquifer subsystem is the best, as the ruptile nature of the rocks helps to maintain of the open structures. The presence of carbonate (PPC Subsystem) is the most variable one, where the yields are controlled by the fracture density and dissolution zones.

The Bambuí aquifer System is relatively unimportant because it is developed over siltstone and arkose. Where the arkose layers are more frequent the aquifers tend to show better circulation conditions and the best situation is observed where the Porous P_2 System covers the fractured aquifers.

Table 2. Summary of the classification of the aquifers domains, systems and subsystems and the mean discharge of the well (Campos & Freitas-Silva 1998).

Aquifer (system/subsystem)	Mean discharge m^3/h
Porous aquifers domain	
P_1, P_2, P_3 and P_4 systems	<0.8
Fractured aquifer domain	
Paranoá system	
S/A subsystem	12.5
A subsystem	4.0
Q_3/R_3 subsystem	12.0
R_4 subsystem	6.0
PPC subsystem	9.0
Canastra system	
F subsystem	7.5
F/Q/M subsystem	33.0
Bambuí system	5.5
Araxá system	3.0

3 GROUNDWATER VULNERABILITY

In general the groundwater is well protected and the effective risks of contamination are related to the fast occupation of the regional recharge areas which causes problems of sewage of waste water and in some cases the usage of fertilizer and agrotoxic products. The inadequate technical conduct of the well drilling increases the natural risk of contamination.

The map of vulnerability to contamination of the deep ground waters has been developed considering the DRASTIC and GOD methods (Aller *et al.* 1987 and Foster *et al.* 1987), applied to the local conditions, including the environmental protection areas and the recently growing urban areas, which do not have basic urban structures (Fig. 3).

Settlements of villages which are under construction on the hard quartzite highland may cause a decreasing of natural groundwater recharge because of sealing of the surface.

These areas show low declivity in a plateau relief pattern, recovered by the P_1 Porous System, which allows that about 20% of the total precipitation can infiltrate the deep aquifers. The high density of urban model occupation (houses, roads, urban equipment, etc) cause a great decrease of the natural recharge (about 80%), as can be inferred by long term monitoring of springs and stream decrease discharge.

The drilling companies usually do not follow the technical norms. The absence of minimal protection of the wells, without good isolation of the shallow portion, bad location and design of wells and low

Figure 3 - Vulnerability map to the contamination of the deep groundwaters of the Federal District (after Campos & Freistas-Silva 1998).

surface protection, often result in contamination of groundwater mainly by the seepage of waste water and local landfill leachates.

4 CONCLUSIONS

The hydrogeology of the FD can be summarised in two groundwater domains, defined by the porous aquifers (water contained in the soil covers) and the fractured aquifers (water contained in the Proterozoic rocks).

The importance of groundwater is increasing strongly in areas where surface water was dominating the supply of potable water. The demand of water in the more remote areas where condominiums are growing today can only be managed with groundwater supply.

An evaluation of the groundwater recharge helps to manage the important resource in the future. Calculation and modelling of exploitable and total reserves, will be important to define the safe yields of each aquifer system/subsystem.

As groundwater will be needed even more in the future, planning must take place. The practice of artificial recharge is viable in the highlands areas and is strongly recommended. Permeable sinks can be fed by rainfall water from the roof of the houses in the period from November to February, when almost of 50% of the annual rains precipitate.

Because of the low potential of the aquifers of the FD, as shown in the preview sessions, the groundwater must be considered as an strategic resource to complement public potable water supply or to supply little condominiums or rural communities.

However, the resource can be only a sustainable water supply if the protection of the recharge areas is respected.

REFERENCES

Aller, L., T. Bennet, L.H. Lehr, R.J.E. Petty & G. Hackett 1987. DRASTIC, a standardized system for evaluating groundwater pollution potential using hydrogeologic setting. U.S Environmental Protection Agency, Ada, OK. EPA Report 600/2-87-035:1-455.

Campos, J.E.G. & F.H. Freitas-Silva 1998. *Hidrogeologia do Distrito Federal.* In: Inventário Hidrogeológico e dos Recursos Hídricos Superfi-ciais do Distrito Federal. Brasília. IEMA-SEMATEC-UnB. Parte I. (Unpubl.).

Foster, S.S.D., M.E.Ventura & R. Hirata 1987. Contaminacion de las aguas subterráneas: un enfoque efecutivo de la situación en America Latina y el caribe en relación com el Suministro de agua Potable. CEPIS. Technical Report, OMS., OPS-HPE, CE-PIS, Lima, Peru: 42p.

Freitas-Silva, F.H. & J.E.G. Campos 1998. *Geologia do Distrito Federal.* In: Inventário Hidrogeológico e dos Recursos Hídricos Superficiais do Distrito Federal. Brasília. IEMA-SEMATEC-UnB. Parte I: 86p. (Unpubl.).

Lousada, E.O. 1999. *Estudos geológicos e geofísicos aplicados a locação de poços tubulares profundos em aquíferos fraturados na região do Distrito Federal.* Mcs Thesis. University of Brasilia: 107p.

Groundwater: Past Achievements and Future Challenges, Sililo et al. (eds) © 2000 Balkema, Rotterdam, ISBN 90 5809 159 7

Quantifying the heterogeneity of fracture systems: An index for fracture flow network representations at different scales

D. Cesano
Department of Civil Engineering and Engineering Mechanics, Columbia University, New York, N.Y., USA
Department of Civil and Environmental Engineering, Royal Institute of Technology, Stockholm, Sweden

A. Bagtzoglou
Department of Civil Engineering and Engineering Mechanics, Columbia University, New York, N.Y., USA

B. Olofsson
Department of Civil and Environmental Engineering, Royal Institute of Technology, Stockholm, Sweden

ABSTRACT: The heterogeneity of a fractured rock mass makes it difficult to represent its fracture network. Therefore, studying and predicting fracture flow and transport processes is challenging. However, fracture network representations might not be enough since the success of numerical modeling depends on the degree of heterogeneity of the fracture network, on the scale and on the amount of data available. This paper aims to go one step back to analyze and fully understand what heterogeneity really is by presenting an attempt to quantify the degree of heterogeneity of a fracture network. The computation of a heterogeneity index at different scales can show how the geological properties of the fracture network vary with scale, helping in selecting those scales where the rock mass seems to be more homogeneous.

1 INTRODUCTION

Hydrogeologists often claim the heterogeneity of a fracture network as the main source of uncertainty affecting flow and transport processes through fractured rocks as well as during the data collection campaign and the data analysis strategy. But what is really heterogeneity?

Typically, the heterogeneity that affects fracture flow and transport processes have equated with variations in hydraulic conductivity. Freeze and Cherry (1979) define heterogeneity as hydraulic conductivity "variations through space within a geologic formation". Greenkorn and Kessler (1969) defined a heterogeneous formation as one in which the probability density function of hydraulic conductivity is multimodal, that is it shows variations in K without maintaining a constant mean K through space.

From these and similar considerations it was assumed that the statistical description of the spatial variation in the hydraulic conductivity values of a site could allow the heterogeneity to be described and represented. It has been shown that the hydraulic conductivity follows specific distributions (Willardson and Hurst, 1965; Davis, 1969), through which the heterogeneity is nowadays represented in some modeling exercises (e.g., Monte Carlo simulations).

Debates on which type of representation should be adopted and on the importance of fracture network representations have been going on for quite some time. The Committee on Fracture Characterization and Fluid Flow in 1996 (Hsieh, 1998) stated "the hydraulic properties of rock masses are likely to be highly heterogeneous even within a single lithological unit if the rock is fractured. The main difficulty in modeling flow in fractured rock is to describe this heterogeneity". Further complications are also given by the so-called "scale problem", since heterogeneity also varies with scale. The problem is compounded by the difficulty in identifying the relation between the hydrogeological or hydraulic properties of a site, a process, and the scale upon which the properties are measured and the process occurs.

However, the heterogeneity of a fracture network still remains a vague concept. How heterogeneous can a fracture network be? Which is the most convenient approach to modeling a fracture network according to its degree of heterogeneity? Does the heterogeneity of the fracture network change with scale and how do such changes affect the flow? If until now no answer has been given to these and similar questions, it probably means that heterogeneity has not been fully and clearly understood yet. Therefore, complementary tools must be integrated in network representations to approach the problem.

The aim of the present paper is to introduce a preliminary and theoretical attempt to consider the heterogeneity of a fracture network as an intrinsic characteristic of the fracture system that can be quantified with an index. Such an index would indicate the degree of variability of fracture properties at different scales. This information can be then used to derive scaling characteristics of the fracture network and to indicate the scale on which the fracture characteristics and the corresponding hydraulically connected network are scale invariant.

2 DEFINING THE HETEROGENEITY IN FRACTURE NETWORKS

In rock masses, fractures are the only dominant pattern through which water can flow, with the overall effective porosity of the intact rock usually being very low. Each fracture has its own properties, which can be characterized and classified according to different criteria, such as those proposed by the International Society of Rock Mechanics (ISRM, 1978). Of all fracture characteristics that regulate fracture flow, the most important are the following:

Strike and dip: strike is the direction of a discontinuity in space measured from the true north, while the dip is the smallest angle measured from the horizontal plane. For fracture flow, strike and dip indicate the directions of flow within the rock mass. Strike and dip are two of the four factors together with spacing and trace length that define the connectivity of a rock mass. Connectivity is defined as the number of intersecting fractures within a given unit volume.

Fracture spacing: fracture spacing is the perpendicular distance between adjacent discontinuities. The larger the spacing is, the lower the degree of connectivity of the fracture network.

Trace length: trace length is the persistence of the fracture length on the rock wall and in real situations may give a crude measure of the areal extent and penetration depth of the fracture. The larger the trace length is, the higher the probability of having a high degree of connectivity.

Fracture aperture: fracture aperture is the distance between the two walls of the fracture and determines the possibility of water to flow through the fractures. Flow cannot occur where a fracture is completely closed.

Roughness: roughness is the shape and position of the fracture walls relative to the mean plane of the fracture. The roughness of a fracture itself can influence the rate and manner of flow under constant rates and it is an important factor for fracture flow since it is closely related to the geological and therefore hydraulic history of the fracture.

Thickness and type of mineral filling: mineral filling is the result of mineral precipitation or of mineral desegregation of the rock walls and can influence the permeability of the fracture.

Rock stress: significant rock stress conditions can influence the direction of flow by forcing water to flow along certain preferred directions. The magnitude of such stresses can be measured with in-situ techniques, while the direction of stress can be defined inexpensively and roughly by visual inspection of geological structures such as dikes and tension fractures.

If a rock mass has to be perfectly homogeneous from a hydraulic point of view, its fractures should be grouped into a number of classes according to their strike and dip.

In the ideal case of a homogeneous fracture network where a constant recharge is uniformly applied all over the upper boundary of the domain, water flow would be perfectly predictable in both direction and rate anywhere in the network. Figure 1 shows such ideal homogeneous conditions with two fracture sets of equal fracture characteristics but of mirror-image-like dip. Similar homogeneous conditions could be hypothesized for larger or smaller number of fracture sets.

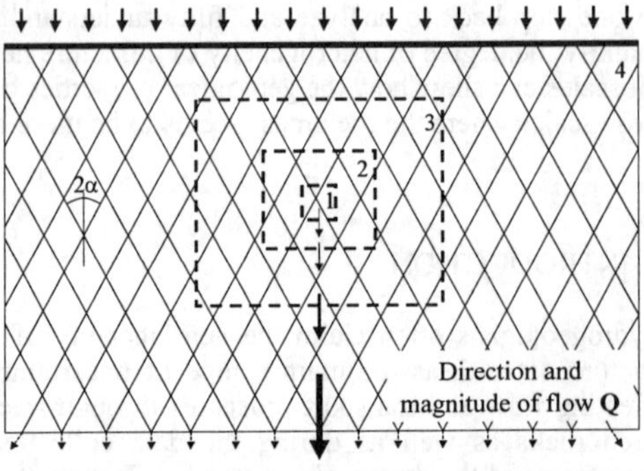

Figure 1. Example of homogeneous fractured rock composed by two sets of fractures of equal fracture and hydraulic characteristics (e.g., dip, roughness, aperture and spacing) and with a constant recharge. The dashed squares represent the three different scales denoted as 1, 2, 3 and the full scale 4. The vertical arrows in the fracture network represent the direction and the magnitude of flow (not in scale).

In the network represented in Figure 1, the overall amount of water Q [liters/second] flowing in a unit time through one horizontal section of given length is easily calculated by:

$$Q = (n * q_1) + (m * q_2)$$

where q_1 is the flow rate [liters/seconds] through one single fracture belonging to set 1 at any cross section, while q_2 [liters/seconds] is the flow rate through one single fracture at any cross section belonging to set 2, and n and m are the number of fractures crossed at each cross section by the two sets. However, such perfect homogeneous conditions are an ideal case and do not exist in nature. Fractures develop during different geological ages where regional stresses of varying magnitude and direction twist and tilt the rock mass forming rock blocks of variable size. Tectonic stresses cause shearing and

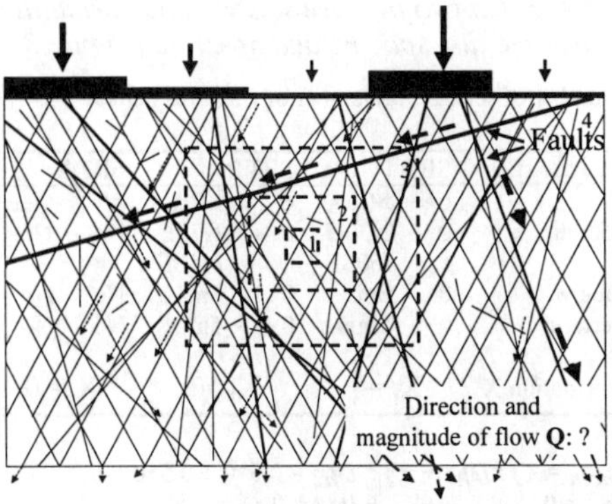

Figure 2. Real case of fractures at a hypothetical site. Different fractures are present, which can be grouped in a number of sets, each of which has hydraulic and fracture properties with distributions that vary within defined boundaries. It is theorized that for very large scales a well-defined mean and distribution description is reached. In real cases, recharge can also be heterogeneous, for example due to variation in soil composition or precipitation. However, the heterogeneity of recharge is not explicitly considered here.

tension that change the fracture characteristics not only between two different sets, but even within one single set and along one single fracture. The situation in a real case is therefore more likely to be the one represented in Figure 2, where the geological history of a site and the heterogeneous nature of the material produce different types of discontinuities(e.g., faults, fracture zones and different types of fractures) with different hydraulic properties. By assuming that every fracture belonging to each set has defined characteristics (e.g., strike, dip, aperture) that can be described by a well-defined frequency distribution, it becomes possible to think of the heterogeneity of a fracture network as a homogeneous network with a heterogeneous component. The comparison of the ideal homogeneous conditions with the real heterogeneity of the network will then give an index of how heterogeneous the fracture network is compared to ideal homogeneous conditions, where heterogeneity is here defined as the spatial variability of those fracture characteristics that regulate fracture flow.

3 QUANTIFYING THE HETEROGENEITY OF A FRACTURE NETWORK: THEORY

The aim of the index proposed here is to think of the heterogeneity of a fracture network as an intrinsic characteristic of the fracture system that can be quantified according to the principles described above. Such an index is based on attributing values to each fracture characteristic. An assumption made when calculating the index is that all fracture characteristics in question follow a normal distribution. This allows us to calculate the variability (e.g., variance) in fracture characteristics for both the homogeneous and heterogeneous components, which are then combined. The result is a measure of the heterogeneity of the network as a whole. However, since all fracture parameters are given in different units, it is necessary to transform all the parameters into a non-dimensional quantity of the same range (e.g., between 0 and 1) in order to compare and combine all the different variances.

Table 1 shows a generic table of results after computing the mean and the variance for each set. Then the non-dimensional variances σ^2 of the fracture characteristic of each set are combined to produce values X_i that represent the variability in the fracture characteristics within the specific set i. Finally, the heterogeneity index of a fracture network I_h combines all the different X_i, according to Table 1 and the following set of formulae:

$$\begin{cases} I_h = \sum_{n}^{i=1}(X_i \cdot D_{hi}) + (X_h \cdot D_{hh}) = I_{\text{hom}} + I_{het} \\ I_{\text{hom}} = \sum_{n}^{i=1}(X_i \cdot D_{hi}) \\ I_{het} = [X_h \cdot D_{hh}] \\ X_i = (\sigma^i_{s/d})^2 + (\sigma^i_s)^2 + \ldots + (\sigma^i_a)^2 \\ D_{hi} = \dfrac{N_i}{N} \end{cases}$$

Table 1. Variance σ^2 and associated values x of the fracture characteristics for various sets i. *Set h* is the set that groups all those fractures that do not belong to any of the previous groups, that is the heterogeneity component of the fracture network. Each X_i is also a numerical quantity.

Characteristics	SET 1		SET 2		…		SET i		SET h	
	σ^2	Mean	σ^2	Mean	…	…	σ^2	Mean	σ^2	Mean
Strike/dip	$(\sigma^1_{s/d})^2$	$X^1_{s/d}$	$(\sigma^2_{s/d})^2$	$X^2_{s/d}$	…	…	$(\sigma^i_{s/d})^2$	$X^i_{s/d}$	$(\sigma^h_{s/d})^2$	$x^h_{s/d}$
Spacing	$(\sigma^1_s)^2$	X^1_s	$(\sigma^2_s)^2$	X^2_s	…	…	$(\sigma^i_s)^2$	X^i_s	$(\sigma^h_s)^2$	x^h_s
…	…	…	…	…	…	…	…	…	…	…
Aperture	$(\sigma^1_a)^2$	X^1_a	$(\sigma^2_a)^2$	X^2_a	…	…	$(\sigma^i_a)^2$	X^i_a	$(\sigma^h_a)^2$	x^h_a
Final contribution	$X_1 = \Sigma\,(\sigma^1)^2$		$X_2 = \Sigma\,(\sigma^2)^2$		…		$X_i = \Sigma\,(\sigma^i)^2$		$X_h = \Sigma\,(\sigma^h)^2$	

where the σ^2 are the non-dimensional variances for each fracture characteristic, the X_i are the contribution of the different fracture sets to the overall heterogeneity (the homogeneous component), X_h is the contribution of all those fractures that do not belong to any set (the heterogeneous component) and D_{hi} is the measurement of the number of fractures that belong to the set *i*. The value of D_{hi} is the ratio of the total number of fractures *N* to the number of fractures N_i belonging to set *i*. The value of D_{hh} is computed in a similar way, but in this case it refers to the fractures that do not belong to any set (the heterogeneous component). The quantity D_{hi} can vary from zero (no fractures belong to any sets, which means maximum degree of heterogeneity) to one (all fractures belong to a fracture set). The relevance of such a factor in the overall estimation of the heterogeneity index is to take into consideration the amount of fractures that belong to the heterogeneous and homogeneous components. The larger the variability of the fracture characteristics is, the higher the value of the index, which indicates high heterogeneity in the network.

4 QUANTIFYING THE HETEROGENEITY OF A FRACTURE NETWORK: EXAMPLES

Three different cases are presented here to show the use of the heterogeneity index in practice. The main assumption, valid for all those cases, is a constant recharge over the whole area. Moreover, only four parameters have been considered, that is strike/dip, fracture spacing and aperture, for which variances are calculated according to a reclassified interval of 0 and 1, which means that all parameters are assumed to have the same importance in regulating the heterogeneity of flow through a fracture network. Such reclassification is also necessary to combine the variance of fracture characteristics that are, typically, given in different units (e.g., degrees in strike and dip and millimeters and meters in aperture and spacing). All the values of the parameters were random numbers sampled from a normal distribution of given mean and standard deviation. It was also assumed that fracture sets are displayed according to stereographic projection and cluster analysis of stereographic projections was conducted according to standard techniques (Robin and Jowett, 1986a;b and Robin and Jowett, 1988). In the cluster analysis presented here, different fractures are grouped together in pole clusters, and fractures falling within one pole cluster are assumed to belong to the same family.

4.1 CASE 1a: two fracture sets with no variability in strike/dip, spacing and aperture (Figure 3a)

Number of sets: 2. Number of fractures per set: 20.

Characteristics	SET 1		SET 2		SET h	
	σ^2	Mean	σ^2	Mean	σ^2	Mean
Strike/dip	0	N30E /60SE	0	N70E /45SE	No	No
Spacing	0	6m	0	6m	No	No
Aperture	0	5mm	0	5mm	No	No
Contribution X	$X_1=0$		$X_2=0$		$X_h=0$	

$I_{hom} = X_1 * D_{h1} + X_2 * D_{h2} = 0.5*0 + 0.5*0 = 0$
$I_{het} = 0$ $\quad I_{hom} + I_{het} = 0$

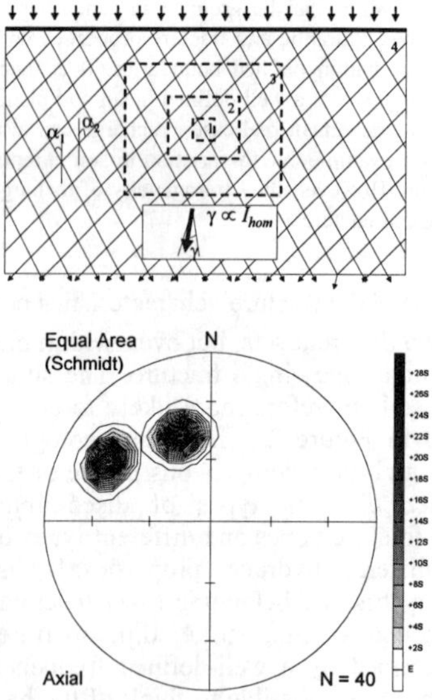

a

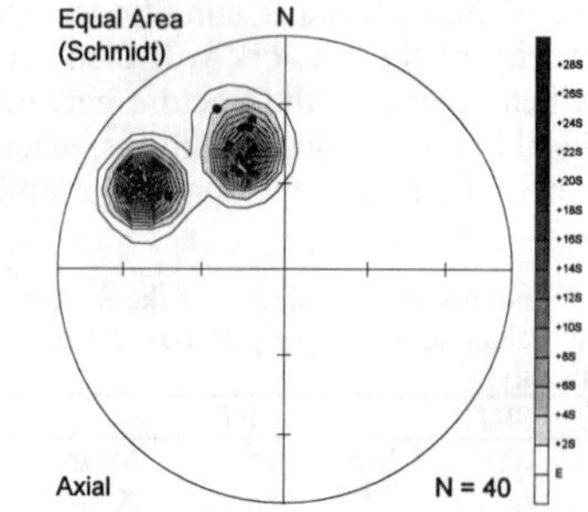

b

Figure 3. *Case 1*, subcases *a* and *b* and the corresponding stereographic projection. Only in case *a*, the angle γ that indicates the variability in the direction of flow is equal to zero. Strike and dip, aperture and spacing are equal everywhere. In case *b*, there is variability in strike/dip, aperture and spacing.

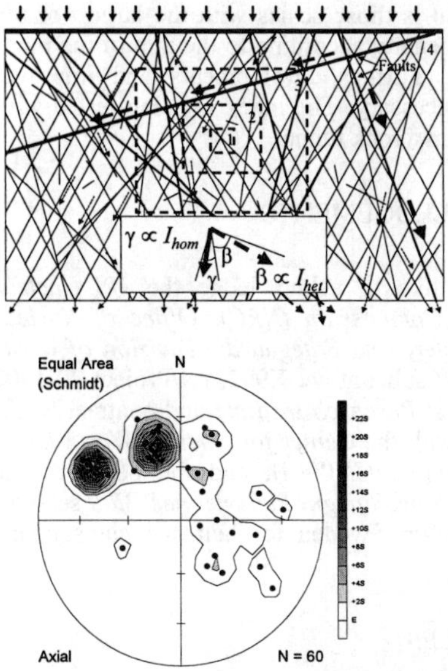

Figure 4. *Case 2*. This is a schematic representation of a fracture network formed by two fracture sets and one heterogeneous component. The two vectors represent the homogeneous and heterogeneous components of flow and the magnitude can vary in direction within an angle γ for the homogeneous component I_{hom} and β for the heterogeneous component I_{het}. Also the modulus of the two vectors varies according to the quantity I_{hom} and I_{het}.

4.2 CASE 1b: *two fracture sets with variability in strike/dip, spacing and aperture (Figure 3b).*

Number of sets: 2. Number of fractures per set: 20.

	SET 1		SET 2		SET h	
	σ^2	Mean	σ^2	Mean	σ^2	Mean
Strike/dip	0.007	N30E/ 60SE	0.014	N70E/ 45SE	No	No
Spacing	0.084	6m	0.071	6m	No	No
Aperture	0.055	5mm	0.095	5mm	No	No
X	X_1 =0.146		X_2 =0.18		X_h= 0	

$I_{hom} = X_1 * D_{h1} + X_2 * D_{h2} = 0.5*0.146 + 0.5*0.18 = 0.163$

$I_{het} = 0 \qquad I_h = I_{hom} + I_{het} = 0.163$

4.3 CASE 2: *2 fracture sets with variability in strike/dip, spacing, aperture and 1 set of random fractures (Figure 4)*

Number of sets: 2 homogeneous and 1 heterogeneous.
Number of fractures per set: 20.

	SET 1		SET 2		SET h	
	σ^2	Mean	σ^2	Mean	σ^2	Mean
Strike/dip	0.007	N30E/ 60SE	0.014	N70E/ 45SE	0.117	N28 W/39 SW
Spacing	0.084	6m	0.071	6m	0.084	6m
Aperture	0.055	5mm	0.095	5mm	0.055	5mm
X	X_1 =0.146		X_2 =0.18		X_h= 0.256	

$I_{hom} = 0.33*0.146 + 0.33*0.18 = 0.107$

$I_{het} = 0.33*0.256 = 0.084 \qquad I_h = I_{hom} + I_{het} = 0.191$

4.4 *Scaling property*

Plotting of the different indices versus the scale upon which they have been calculated gives the scaling property of the medium (Figure 5). The heterogeneity of the fracture network is expected to increase with scale until it reaches a peak, after which the index decreases and stabilizes.

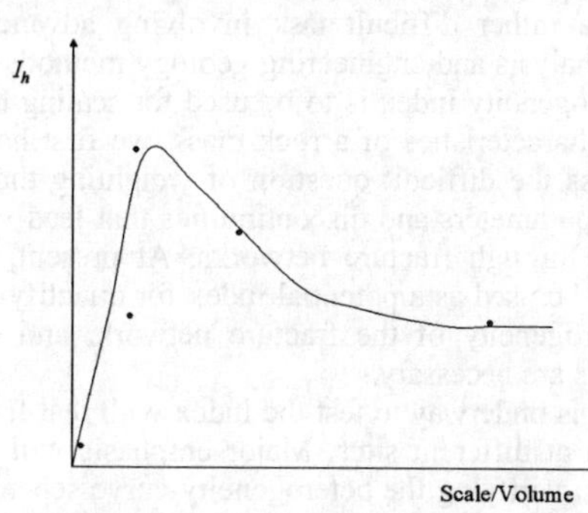

Figure 5. Hypothesized scaling property of a fractured medium.

5 DISCUSSION

Current prediction efforts are missing a quantity (an index) that is able to assess the degree of success of the prediction starting from the geological and hydrogeologic data that is available. The variability in rate and direction of flow is in fact directly affected by the variability in the fracture characteristics of the fracture network and of recharge. The heterogeneity index that has been presented here is an early attempt to quantify the heterogeneity of a fracture network from geological information and to derive scaling properties that can then be used while modeling the rock mass.

From the cases presented here, it is clear that the heterogeneity increases when a larger variability in the fracture characteristics is present. In the ideal case of just two poles formed by 20 fractures of fixed strike/dip (Case 1a), only two well-defined clusters were visible, where $I_h = 0$ since all fractures belonging to one set had the same fracture characteristics. If a small variability in strike/dip, spacing and aperture is present (Case 1b), the fracture characteristics change in space so that their variances were no longer equal to 0. When the variability in fracture characteristics increases (Case 2), the heterogeneity of the network increases accordingly. From a modeling point of view, high degrees of heterogeneity imply large difficulties in representing a fracture network due to such high spatial variability in the fracture characteristics.

However, there are a number of issues that must be addressed. The heterogeneity component and the variability in the fracture characteristics are influenced by the way in which fracture data are collected. Moreover, there is a subjectivity in collecting fracture data according to the classification method that is used in the geological characterization. In data processing, there is also subjectivity in grouping fractures into sets, where discerning fractures that belong to a cluster or that belong to no cluster can be a rather difficult task involving advanced cluster analysis and engineering geology methods. If the heterogeneity index is to be used for scaling hydraulic characteristics of a rock mass, we first have to address the difficult question of weighting those fracture parameters and discontinuities that lead water flow through fracture networks. At present, I_h can only be used as a potential index for quantifying the heterogeneity of the fracture network, and refinements are necessary.

Work is underway to test the index with real fracture data at different sites. Major emphasis will be placed on verifying the heterogeneity curve schematized in Figure 5. Another major concern will be to describe the heterogeneity of flow through a system of vectors at different scales, for which magnitude and variability in direction is estimated from the index I_h and its two components I_{hom} and I_{het}.

6 CONCLUSIONS

The present paper has presented an attempt to quantify the heterogeneity of a fracture network. The index has been developed to give a clear indication of the scale at which it becomes difficult to predict flow and transport processes through a fracture network. The index aims to be simple and meaningful, so that it can be derived from measurable parameters, which can be determined quickly and inexpensively. The index may be calculated and plotted at different scales to derive scaling properties of the fracture network. To increase the reliability of such scaling, we need to quantify the role of different fracture characteristics and discontinuities in regulating fracture flow.

This information can help to select the most suitable and convenient scales to represent a fracture network, that is those scales with the largest or the least heterogeneity according to the aim of the modeling effort. Further developments and improvements are necessary. The index is being tested with real data from a rock tunnel.

ACKNOWLEDGEMENTS

We would like to acknowledge the *US Nuclear Regulatory Commission (NRC), Office of Nuclear Material Safety and Safeguards, Division of Waste Management* subcontract X99219N/Project 20-1402 for *Southwest Research Institute* and contract NRC-02-97-009 with the *Center for Nuclear Waste Regulatory Analyses (CNWRA)*, and the board of the *Blanceflor Boncompagni Ludovisi neé Bildt* scholarship, Stockholm, Sweden, for their financial support.

REFERENCES

Davis, S.N. 1969: Porosity and permeability of natural materials. *Flow Through Porous Media*, 54-89, New York: Academic press, ed. R.J.M. De Wiest.

Freeze, R.A. & J.A. Cherry 1979: *Groundwater*. Englewood Cliffs, N.J.: Prentice-Hall.

Greenkorn, R.A. & D.P. Kessler 1969: Dispersion in heterogeneous non-uniform anisotropic porous media. *Ind. Eng. Chem.*, 61: 14-32.

Hsieh, P.A. 1998: Scale effects in fluid flow through fractured geologic media. *Scale Dependence and Scale Invariance in Hydrology*, 335-353, Cambridge University Press.

ISRM (International Society of Rock Mechanics), 1978: Suggested methods for the quantitative description of discontinuities in rock masses. *International Journal of Rock Mechanics and Mining Science & Geomechanics Abstracts.* 15 (6): 319-368.

Robin P-Y.F. and C. Jowett, 1986 a: Computerized density contouring and statistical evaluation of orientation data using counting circles and continuous weighting functions. *Tectonophysics*, 121: 207-223.

Robin P-Y.F. and C. Jowett, 1986 b: Characterization of superposed Fisher distribution from point density surfaces. *Geol. Assoc. Canada*, Annual Meeting.

Robin P-Y.F. and C. Jowett, 1988: Statistical significance of clustered orientation data on the sphere: an empirical derivation. *J. Geology*, 96: 591-599.

Willardson, L.S. & R.L. Hurst 1965: Sample size estimates in permeability studies. *J. Irr. Drain. Div., Amer. Soc. Civil Engrs.*, 91: 1-9.

Groundwater: Past Achievements and Future Challenges, Sililo et al. (eds) © 2000 Balkema, Rotterdam, ISBN 90 5809 159 7

Adequate depth of boreholes in hard rocks: A case study in Ghana

P.K. Darko & J. Krasny
Faculty of Science, Charles University, Prague, Czech Republic

ABSTRACT: In Ghana, groundwater supply from boreholes remains the only viable alternative for the numerous scattered rural communities. However, the bedrock consists mainly of hard-consolidated rocks. Boreholes are drilled to varying depths to tap the limited groundwater that occurs in the regolith and wet fractures within the bedrock. Since drilling cost is directly related to depth, quantitative appraisal of the relationship of potential yield to depth has been carried out to estimate adequate depth to which boreholes could be drilled for sustainable yield at minimum cost. For available drawdown of between 3 to 5m, adequate depths range from 34 to 50m for the Dahomeyan gneiss, 42 to 50m for the quartzite and sandstone of the Togo and Buem series, 50 to 59m in the Voltaian shale and sandstone, and 39-52m for granites. These depths are general guides for drilling in hard rocks for hand-pump rural water supply.

1 INTRODUCTION

During the long dry season in Ghana (November to April) many streams dry up and shortages of potable water, especially for the rural and urban fringe communities, are chronic. Though it is difficult to develop large concentrated supplies of groundwater due to the widespread distribution of hard, consolidated rocks of low to intermediate transmissivity (Darko & Krasny 1998), groundwater remains the most reliable alternative as a safe source of water supply. The increasing demand for potable water has led to the drilling of a large number of boreholes in many parts of the country. The boreholes are drilled to varying depths to tap the often-limited groundwater that has accumulated in the weathered zone and in wet fractures within the bedrock. The cost of drilling in such environments has been very high thereby putting a great strain on the ability to provide water to the numerous scattered rural communities. With drilling cost very much dependent on the depth of boreholes, it is necessary to study the relationship between hydraulic parameters (that characterise potential yield and well productivity) and the depth of boreholes, to ascertain adequate depths to which boreholes could be drilled in the various hard rock formations.

2 HYDROGEOLOGICAL BACKGROUND

About 45% of Ghana is underlain by a variety of Precambrian intrusive crystalline (mostly granites) and highly folded metamorphosed sedimentary rocks (Birimian and Tarkwaian formations, Togo and Buem series), as well as the Dahomeyan gneiss. Overlying these rocks are Paleozoic consolidated rocks of the Voltaian system, which forms an extensive sedimentary basin that includes about 35% of Ghana. The various rock components are given in Table 1.

In general, the rocks have little or no intergranular porosity. However, in several locations they are fractured and weathered to varying extent, imparting intermediate porosity that is responsible for storing and transmitting groundwater. The character of weathered and fissured zones of hard rocks in tropical regions has been described by a number of authors (e.g. Biscaldi 1968, Gill 1969, Chilton & Foster 1993). Many hydrogeological investigations are aimed at locating such zones that could provide appreciable yields. In Ghana, boreholes are usually of small diameter (165mm) and of shallow to medium depths (12.5-94.0m). The average yields in the Birimian schist and phyllite, the granites, and the shale and sandstone of the Voltaian system are 51.6, 58.7 and 108.7 l / min respectively. Table 1 gives a

Table 1. Selected borehole data in various hard rock formations in Ghana

Geological unit (rock types)	Number of Boreholes	Depth (m) Mean	Depth (m) Range	Yield (l / min) Mean	Yield (l / min) Range	Screen section (m)* Mean	Screen section (m)* Range
Granites	556	36.1	12.5-88.8	58.7	5-600	12.9	4-33
Birimian and Tarkwaian (phyllite, schist, quartzite, tuff)	66	34.8	22.9-52.0	51.6	8-413	8.9	1-18
Togo and Buem (quartz schist, shale, phyllite, sandstone, lava)	193	38.5	22.0-76.2	74.1	7-658	10.6	1-28
Dahomeyan (crystalline gneiss, migmatite, schist)	84	35.4	22.0-70.0	42.9	8-200	11.4	1-31
Voltaian (shale, sandstone, arkose, mudstone, limestone)	698	35.5	21.6-94.0	108.7	5-887	9.8	2-30

(* Screen section provided in place of aquifer thickness).

summary of the depths and yields of boreholes drilled in the various formations.

3 GENERAL CONSIDERATIONS OF TRANS MISSIVITY

Specific capacity and transmissivity are hydraulic parameters that characterise the water bearing and yielding potential of an aquifer. Specific capacity, the ratio of well discharge to drawdown in the pumping well, is a measure of the productivity of both the well and the aquifer. Besides giving us first approximations of groundwater abstraction, transmissivity distribution and variation also provide useful information on hydrogeological properties on regional scales (Krasny 1993). Measurements of specific capacity and available drawdown are neither difficult nor time consuming. Hence, to avoid the cost of long duration pumping tests, specific capacity values are often used to estimate aquifer transmissivity. Various authors have elaborately described the assumptions and limitations involved in this approach (e.g. Walton 1962, Razack & Huntley 1991). Though pumping tests remain the most reliable field method for aquifer parameter determination, pumping test analyses of fractured rock aquifers present unique problems due to anisotropy and heterogeneity in fractured areas.

3.1 *Depth related changes*

The decrease in permeability with increasing depth has been demonstrated from several places. This trend though not systematic, is attributed to a number of factors including the greater influence of surficial phenomenon at shallow depth (weathering and unloading) and the reduction in fracture disposition with depth. Several empirical relations have also been derived to determine the depth dependence of permeability (Black 1987, Lee & Farmer 1993, Carlsson & Olssen 1997).

3.2 *Optimum and adequate depth of boreholes*

The concept of optimum or adequate depth of boreholes is very relevant in hard rock aquifers. The depth at which an unsuccessful well should be abandoned to give the lowest average cost per well has been defined as the optimum depth (Singhal & Gupta 1999). The success or failure of a well may depend on the yield and water demand, the quality of water and its use, and cost of construction. Read (1982) provides a better idea by relating the optimum depth to an expected discharge. For rural, domestic hand-pump water supply systems in Ghana, a borehole with a minimum yield of 10 l / min is considered as successful. An adequate borehole depth is considered the depth at which the expected minimum yield could be tapped and thereafter, be sustainably withdrawn.

4 DATA ANALYSIS

4.1 *Available data*

Short duration (6 hours) pumping test data of more than 1500 boreholes spread all over Ghana have been used in this study. The principal information on this set of data is given in Table 1. Based on the age, petrography and occurrence, the data were grouped into five main hydrogeological units, which were taken as sample populations for statistical analysis. Selected results of previously statistical analysis carried out on transmissivity are listed in Table 2.

4.2 *Method of Analysis*

The specific capacity of the various wells was calculated and the index of transmissivity (Y) was used as

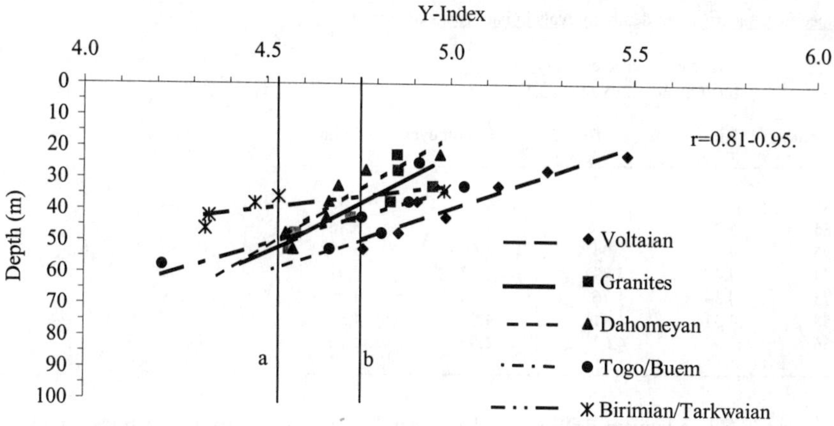

Figure 1. Transmissivity index in relation to borehole depth from hard rocks in Ghana. Vertical lines a & b are limits for 5m & 3m available drawdown respectively.

a comparative or regional parameter for the estimation of the transmissive capacity and yielding potential of the aquifers (Jetel & Krasny 1968). The index Y is expressed by the relation:

$$Y = \log(10^6 C) \qquad (1)$$

where C = specific capacity (l / s / m).

Table 2. Hard rocks in Ghana and their transmissivity (after Darko & Krasny, 1988)

Geological unit	Transmissivity (m²/day)	
	μ	$\mu \pm \sigma$
Granites	7.4	1.9 - 28.6
Birimian and Tarkwaian	3.9	1.0 - 15.4
Togo and Buem	10.6	3.4 - 32.8
Dahomeyan	6.0	1.8 - 19.3
Voltaian	11.9	2.1 - 67.1

Explanation: μ = sample mean, σ = standard deviation

From the index Y, the transmissivity coefficient, T (m^2 / day), was estimated by the conversion equation:

$$T = 86400(10^{Y-8.96}) \qquad (2)$$

From the available data, the Y index values for each of the five main hydrogeological units were averaged separately over a depth interval of 5m. These mean values were then correlated with depth. The coefficient of determination was high in all the hydrogeological units. It ranges from 0.81 to 0.95.

The expected minimum yield (i.e. 10 l / min) per unit available drawdown (i.e. difference between non-pumping water level and lowest practical pumping level) was used as the limiting specific capacity, and converted into Y. For hand-pump supply, an available drawdown of 3 to 5m is considered adequate for all practical purposes. "Adequate" specific capacity values (expressed as Y index) for this drawdown are 4.75 and 4.52 respectively. From the Y index plot (Fig. 1), the corresponding depths were estimated for the various hydrogeological units. These depths are regarded as adequate borehole depths at which the expected yield could be tapped and sustainably abstracted.

4.3 *Results*

There is a discernible decrease of transmissivity with depth for all the various hydrogeological units,

Table 3. Estimate of adequate drilling depths in hard rocks of Ghana.

Geological unit	adequate depth (m) at drawdown of	
	3m	5m
Granites	39	52
Birimian and Tarkwaian	36	39
Togo and Buem	42	50
Dahomeyan	34	50
Voltaian	50	59

Table 4. Mean specific capacity at 5m depth intervals in hard rocks of Ghana

Depth (m)	Mean specific capacity (as transmissivity index Y)				
	Granites	Birimian & Tarkwaian	Togo & Buem	Dahomeyan	Voltaian
22.5	4.85			4.97	5.48
27.5	4.86	5.32		4.77	5.26
32.5	4.95	4.98	5.04	4.69	5.13
37.5	4.84	4.47	4.89	4.67	4.91
42.5	4.73	4.34	4.76	4.66	4.99
47.5	4.58	4.31	4.81	4.55	4.86
52.5	4.56		4.67	4.57	4.76
57.5			4.21		

as shown in Figure 1. The adequate drilling depths are listed in Table 3 and the specific capacity of the rocks at different depths (in terms of Y index) is given in Table 4. The depths range from 34 to 50m in the Dahomeyan gneiss, 36 to 39m in the Birimian and Tarkwaian rocks, 39 to 52m in the granites, 42 to 50 m in the phyllite and quartzite of the Togo and Buem series. In the sandstone and shale of the Voltaian system, the adequate depth is between 50 and 59m.

The results obtained from this statistical averaging method are comparable to those obtained from hard rocks in different regions as reported by a number of authors (e.g. Read 1982, Wright 1992).

4.4 *Discussion*

Most of the data used for the statistical analysis lacked information on construction profile, especially on aquifer thickness and open sections. It is thus difficult to reliably estimate the saturated thickness in order to determine the hydraulic conductivity or permeability, or use any of the comparative regional parameters expressing permeability or specific capacity index. In all the various hydrogeological units, there was a decrease in transmissivity with increasing depth. This seemingly inconsistent trend can be explained by the fact that many of the wells were terminated just after enough yield was obtained. Therefore in deeper wells, it is probable that the overlying sections had very little flow, if any, and were of imperceptible permeability. Consequently, the average transmissivity index of deeper wells, as assumed in this analysis might be used as an indication of vertical permeability changes. A maximum available drawdown of 5m and the least expected yield of 10 l / min might then be used as limiting values for determining the potential yield. For a well fitted with a hand-pump for rural domestic water supply, the depth estimated using these values might be considered the upper limit.

5 CONCLUSION

Based on statistical analysis of available data, depths to which wells may be drilled for sustainable yields in hard rocks of Ghana have been estimated. The depths range from 34 to 50m in Dahomeyan gneiss, 36 to 39m in the Birimian and Tarkwaian rocks, 39 to 52m in granites, and 42 to 50m for the quartzite and sandstone of the Togo and Buem series. In areas underlain by Voltaian consolidated sedimentary rocks the depth is between 50 and 59m. Barring local differences in fractured rock aquifers, the estimated adequate depths could serve as general guides for the drilling of boreholes for rural domestic water supply in Ghana. From this statistical approach, only conclusions on regional scale can be made. In detailed hydrogeological studies on local scale, results might differ considerably from the represented averaged results.

REFERENCES

Biscaldi, R. 1968. Problemes hydrogeologiques des regions d'affleurement de roches eruptiv et metamorhiques sous climat tropical. *Bull. du BRGM* (deuxieme serie), section III, 2, 7-22.

Black, J. H. 1987. Flow and flow mechanisms in crystalline rocks, in J. C. Goff & B. P. J. Williams (eds.), *Fluid flow in Sedimentary Basins and Aquifers*. Geol. Soc. Special Publ. No. 34, 185-200.

Carlsson, A. & T. Olssen 1997. Variations in hydraulic conductivity in some Swedish rock types. *Proceedings 1st. International symposium Rock Store*, Stockholm, 301-307.

Chilton, P. J. & S.S.D. Foster 1993. Hydrogeological characterisation and water supply potential of basement aquifers in tropical Africa, in *Mem. 24th. Congress of IAH*, Oslo, part 2, 1083-1110.

Darko, P. K. & J. Krasny 1998. Comparison of transmissivity of hard rocks in Ghana and in the Czech part of the Bohemian massif, in R. Annau, S. Bender & S. Wohnlich (eds.),

Hard Rock Hydrogeology of the Bohemian Massif. 3rd. International Workshop, Windischenschenbach, Germany.
Gill, H. E 1969. A groundwater reconnaissance of the Republic of Ghana, with a description of geohydrologic provinces. *Geological Survey Water-Supply Paper 1757-K,* Washington.
Jetel, J. & J. Krasny 1968. *Approximative aquifer characteristics in regional hydrogeological study.* Vest. Ustr. Geol. 43 (5), Prague, 459-461.
Krasny, J. 1993. Classification of transmissivity magnitude and variation. *Ground Water,* 31(2) Dublin, Ohio, 230-236.
Lee, C. H. & I. Farmer 1993. *Fluid flow in discontinuous rocks.* Chapman & Hall, London, p. 16961.
Razack, M. K. & D. Huntley 1991. Assessing transmissivity from specific capacity in large and heterogeneous alluvial aquifer. *Ground Water,* 29(6), 856-861.
Read, R. E. 1982. Estimation of optimum drilling depth in fractured rock, in *Papers of the Groundwater in Fractured Rocks,* AWRC Conference, series No. 5, Canberra, 191-197.
Singhal, B. B. S. & R. P. Grupta 1999. *Applied hydrogeology of fractured rocks.* Kluwer, Dordrecht, the Netherlands.
Walton, W. C. 1962. Selected analytical methods for well and aquifer evaluation. *Bull. 49, Illinois State Water Survey,* Urbana, p. 81.
Wright, E. P. & W. G. Burgess (eds.) 1992. *The Hydrogeology of Crystalline Basement Aquifers in Africa.* Geol. Soc. Special Publ. No. 66, the Geological Society, London, p. 264

Groundwater: Past Achievements and Future Challenges, Sililo et al. (eds) © 2000 Balkema, Rotterdam, ISBN 90 5809 159 7

The use of borehole tomography to image the saline interface – A case study from New South Wales, Australia

G.R. Dasey & R.I. Acworth
UNSW Groundwater Centre, University of New South Wales, Sydney, N.S.W., Australia

ABSTRACT: Borehole electrical tomography is an efficient method to map the position of the saline interface if boreholes can be installed with bundles of electrodes in the surf zone. Preliminary bundle designs were tested in boreholes installed at the edge of a tidal creek at Hat Head on the Northern New South Wales coast. This paper details the methodology developed and describes the results of the initial studies. The RES2DINV program was used to invert the apparent resistivity data sets. An unexpected result of the study is that fresh water exists beneath the saline interface at the tidal creek. A layer of slightly cemented sand separates the deeper fresh water from the saline water. The tomographic results indicate upconing of the fresh water through the borehole annulus into the saline water, suggestive of a significantly higher hydraulic head in the deeper part of the aquifer.

1 INTRODUCTION

Borehole electrical tomography has been used as an investigation technique for a wide variety of applications. The uses of the technique have included the monitoring of steam injection into the subsurface (Ramirez et al 1992), monitoring in-situ vitrification tests (Spies & Ellis 1995), delineation of mine and other subsurface voids (Maillol et al 1999, Shi et al 1997), and to locate permafrost (Hauck & Muhll 1999). The surface use of direct current (DC) resistivity methods to locate saline soil and groundwater conditions is well established (Barker 1990, Acworth 1999).

Typical studies of the saline interface have been directed towards determining boundary conditions as part of resource development studies (Lebbe 1999). With increasing pressure around the world on groundwater resources the identification of boundary conditions for a wide variety of aquifer types is becoming increasingly important.

In this study borehole tomography using the DC resistivity method is used to map the saline interface near a tidal creek at Hat Head in northern New South Wales, Australia.

2 METHOD

2.1 *Piezometer Installation*

A total of six piezometers were installed near the tidal creek at Hat Head (Figure 2). Three piezometers (HH10, HH11 and HH12) comprised 50 mm uPVC casing to 15 m depth with a short section of screen at the base. These piezometers were designed for use with the Geonics EM39 logging equipment.

As a result of access restrictions, a track mounted drilling rig was used for installation of the piezometers. All holes were drilled using 100 mm diameter hollow stem augers. Water was used to maintain a positive head during drilling to prevent flowing sand entering the augers.

A further three piezometers (HH7, HH8 and HH9) were installed offset 4 m from HH10, HH11 and HH12 (Figure 1). The piezometers HH7, HH8 and HH9 were constructed using the following materials:

- 15 m casing comprising 25 mm uPVC.
- 1 m screen comprising 25 mm machine slotted uPVC.
- 27 mini-piezometers strapped to the outside of the 25 mm uPVC and terminated at 0.5 m intervals.
- Each mini-piezometer was terminated with an open area of 20 mm and filter sock.
- 27 electrodes consisting of a gold plated banana type plug crimped to copper wire and sealed with glue filled heat shrink.
- Electrodes were terminated immediately adjacent to each mini-piezometer.

HH7, HH8 and HH9 are located 3, 9 and 15 m from the creek edge respectively. During drilling a 0.4 m layer of 'coffee rock' was encountered at

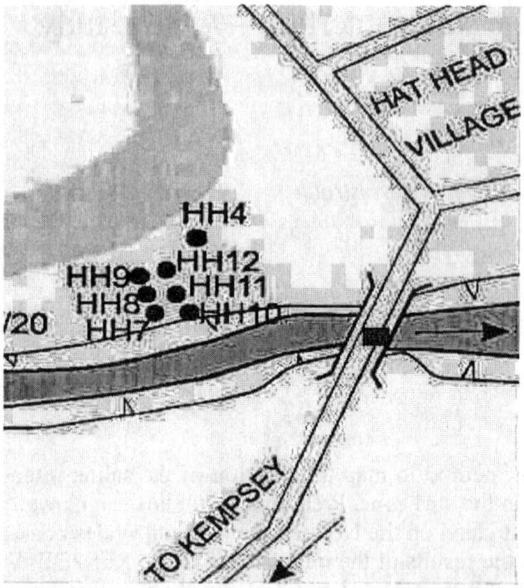

Figure 1: Piezometer Locations

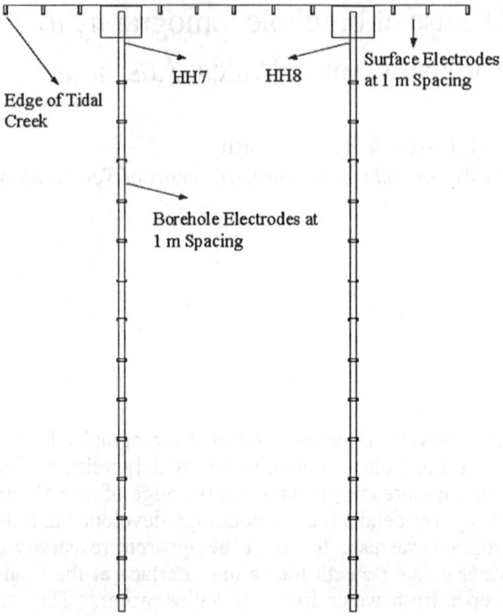

Figure 2: Electrode Locations Borehole Tomography

around 10.5 m. 'Coffee rock' is often described as a cemented sand layer and when brought to the surface is brown/black but can be washed clean leaving quartz sand. Above and below the 'coffee rock' fine to medium grained brown quartz sand was present.

2.2 *Resistivity*

The vertical four-electrode resistivity method utilized all electrodes for each piezometer. A patch board was used to connect an ABEM SAS4000 Terrameter to the electrodes. Combinations of electrodes were selected manually using banana type plugs. An inter-electrode separation of 1.0 m was used for all measurements.

Borehole tomography resistivity data was also collected using an ABEM SAS4000 Terrameter and four channel adapter. The minimum electrode separation was chosen as 1 m to limit the required measurements. Eleven surface electrodes (brass stakes) were installed along the survey line to increase resolution of surface features (Figure 2).

The pole-dipole array was used because of the difficulty with locating a second remote electrode at the site. The remote current electrode was positioned approximately 100 m in a northwesterly direction along the creek. The maximum N value used was 8, which is less than one tenth of the distance to the remote electrode.

To allow connection of appropriate electrodes to the Terrameter all wires were connected to a patch board. A series of simple banana type connectors were then used to sequentially connect the four measuring channels and current terminals to the relevant electrodes.

3 RESULTS

3.1 *EM39*

Piezometers HH10, HH11 and HH12 were surveyed using Geonics EM39 logging equipment. Figure 3 presents the bulk conductivity data for HH10. Note that HH10 is 4 m from HH7. The results for HH11 and HH12 are similar to that presented for HH10.

The dominant features of the bulk conductivity data are:

- Low electrical conductivity (EC) values to a depth of 4 m representative of dry soil conditions or soil saturated with low conductivity fluid.
- A sharp increase in bulk EC at 4.5 m with a plateau at 175 mS/m to 6 m.
- A slight decrease in values occurs at around 7.5 m followed by a rapid rise to a maximum value of 800 mS/m at 12 m.
- The rapid rise is immediately followed by a decrease to less than 200 mS/m at 14 m.

3.2 *Fluid Conductivity*

Fluid conductivity was measured for all mini-piezometers at HH7, HH8 and HH9. A peristaltic pump was used to purge and sample all mini-piezometers. The three bundled piezometers demonstrate similar variations with depth. Figure 3 presents the fluid conductivity values for HH7.

The variation of fluid conductivity differs slightly from that obtained with the EM39 logging tool. The EM39 demonstrates that the high salinity water is

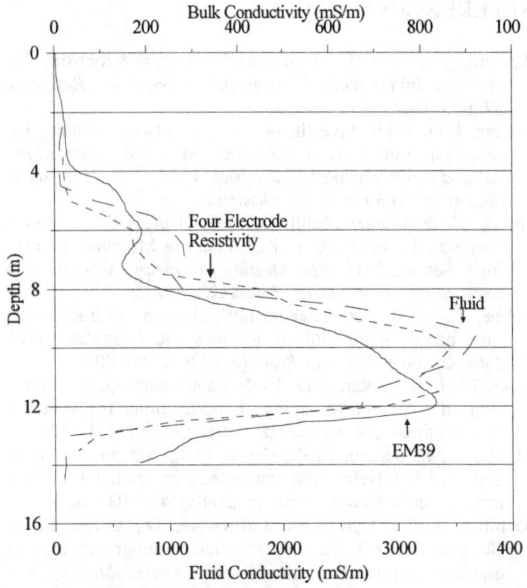

Figure 3: HH7 Measured Fluid Conductivity, Four Electrode Resistivity and EM39 (HH10) results

slightly deeper than indicated by the fluid conductivity. The EM39 was logged in HH10 while the fluid conductivity was taken at HH7 offset 4 m parallel to the tidal creek. The EM39 was also logged several days before the fluid conductivity was measured. Both these factors are likely to contribute to the slightly different saline interface location. The peak value of fluid conductivity was over 3500 mS/m at a depth of 9.5 m in HH7.

3.3 *Resistivity Measurements*

The four electrode vertical resistivity data is presented in Figure 3 for HH7. Note all values have been converted to units of milli-Siemens per meter to allow comparison.

Two tomographic lines were performed, the first from HH9 to HH8 and the second from HH8 to HH7. A total of 600 measurements were taken for each survey. Both forward and reverse measurements were taken to limit axisymmetric effects.

The inversion package RES2DINV was used to produce the images presented in Figures 4 and 5. The RES2DINV package is a commercial program developed by Dr M.H. Loke and is described in Loke and Barker (1996).

The root mean square errors for the two inversions are 10.4% and 5.6% for HH8 to HH7 and HH9 to HH8 respectively. Initial inversion attempts presented errors significantly higher than these values. Examination of the data revealed a number of points that contributed significantly to the error. The final inversions were undertaken with the outlier points removed.

The borehole tomography results of Figure 4 and Figure 5 demonstrate similar subsurface conditions. The common features of the images include:

- High surface resistivity values corresponding to dry-moist sand.
- Small increases in resistivity with depth to approximately 8 m where a large decrease in values occurs.
- An increase in resistivity at 14 m.

4 DISCUSSION

The simple single piezometer data presented in Figure 3 demonstrates the expected result that in an environment dominated by sand there is a direct relationship between the fluid conductivity and DC resistivity. An attempt at describing this relationship using an Archie type equation was unsuccessful.

In the area where the images overlap (HH8) there is good agreement in values. This result gives confidence in the inversion and data collection process.

The two features of most interest to the study are:

- The thinning of the low resistivity layer with increased distance from the creek.
- A high resistivity feature present below the low resistivity water.

Figures 4 and 5 demonstrate that the saline water decreases in thickness from approximately 9 m at HH7 to 6 m at HH9. A possible interpretation of this feature is that the tidal creek acts to recharge the sand aquifer. Infiltrating saline water at the creek will flow downward through the aquifer. As flow occurs away from the creek the density of the more saline water will concentrate near the base of the aquifer. The implication is that there must be a net flow from the creek into the aquifer. Currently it is not known whether the flow of saline water is a continuous or intermittent process.

The second notable feature in the study is the presence of relatively fresh water below the saline layer. Simple consideration of density and mixing factors indicate that a flow barrier must be present. The layer of 'coffee rock' is most likely acting as a barrier creating a two-aquifer system.

Where the piezometers penetrate the 'coffee rock' the images indicate an upward movement of water into the higher aquifer. A possible explanation is that there is a higher hydraulic head in the lower aquifer. The inference is that the 'coffee rock' is acting as a confining layer.

5 CONCLUSIONS

The results of the study indicate that borehole tomography techniques can efficiently map the saline interface and resolve small-scale detail. Surface

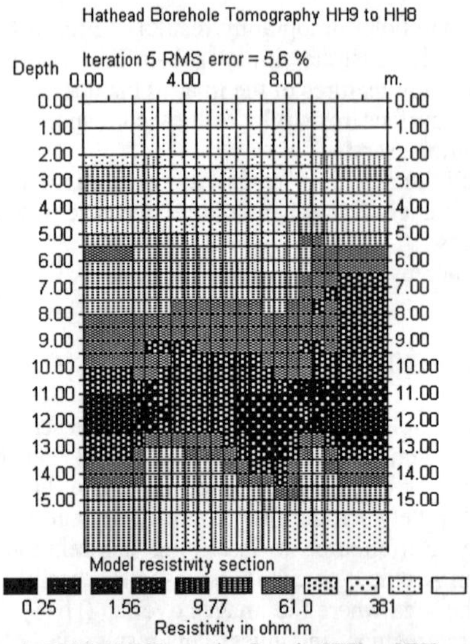

Figure 4: RES2DINV Inversion Results HH9 to HH8

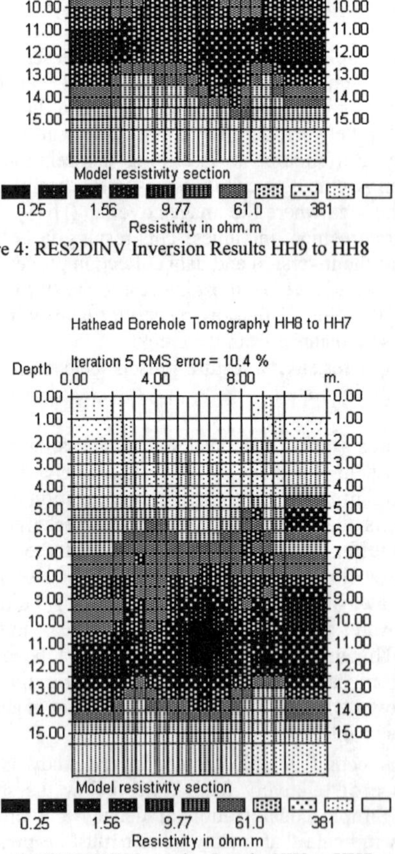

Figure 5: RES2DINV Inversion Results HH8 to HH7

techniques, such as electrical imaging, are unable to resolve the features mapped in this study. There was a close correlation between the measured fluid conductivity values and the inversions presented in Figures 4 and 5.

The resistivity analysis also provided additional unexpected information on the aquifer system. The presence of a flow barrier (inferred by fluid conductivity data) was identified using the technique. The flow barrier may also act as a confining layer as indicated by upconing of the higher resistivity fluid from the lower aquifer.

REFERENCES

Acworth, R.I. 1999. Investigation of dryland salinity using the electrical image method. *Australian Journal Soil Research* 37: 623-636.

Barker, R.D. 1990. Investigation of groundwater salinity by geophysical methods. In Nabighian, M.N. (ed.), *Geotechnical and Environmental Geophysics, Vol II, 201-221, Tulsa, Society of Exploration Geophysicists.*

Hauck, C. & Vonder Muhll, D. 1999. Using DC Resistivity Tomography to Detect and Characterise Mountain Permafrost. Session:2015. *61st Meeting European Association of Exploration Geophysicists Extended Abstracts.*

Lebbe, L. 1999. Parameter identification in fresh-saltwater flow based on borehole resistivities and freshwater head data. *Advances in Water Resources* 22(8): 791-806.

Loke, M.H. & Barker, R.D. 1996. Rapid least-squares inversion of apparent resistivity pseudosections by a quasi-Newton method. *Geophysical Prospecting* 44: 131-152.

Maillol, J.M., Seguin, M.K., Gupta, O.P., Akhauri, H.M. & Sen, N. 1999. Delineating uncharted mine galleries in West Bengal, India. *Geophysical Prospecting* 47: 103-116.

Ramirez, A. L., Daily, W., LaBrecque, D., Owen, E. & Chesnut, D. 1992, Electrical resistance tomography used to monitor subsurface steam injection. *Annual Meeting Abstracts, Society Of Exploration Geophysicists, 492-494.*

Shi, W., Morgan, F. D. and Wharton, A. E., 1997, Application of electrical resistivity tomography to image Harrison Caves in Barbados, West Indies. *Annual Meeting Abstracts, Society Of Exploration Geophysicists, 350-353.*

Spies, B.R. & Ellis, R.G. 1995. Cross-borehole resistivity tomography of a pilot-scale, in-situ vitrification test. *Geophysics* 60(3): 886-898.

Groundwater: Past Achievements and Future Challenges, Sililo et al. (eds) © 2000 Balkema, Rotterdam, ISBN 90 5809 159 7

Thermal analyses for groundwater flow assessment in a sedimentary basin of Japan

J.R. Dim & A. Miyakoshi
Graduate School of Science and Technology, Chiba University, Japan

Y. Sakura
Department of Earth Sciences, Chiba University, Japan

K. Fukami
Geological Survey of Hokkaido, Japan

ABSTRACT: In Ishikari lowland, temperatures measured in 22 observation wells (depth<200 m) were used to estimate direction and scale of groundwater flow. Groundwater flow systems of the area were determined from borehole temperature profile classification and, thermal gradient-temperature relationship. As a result, the regional groundwater flow system is recharged at the southwest highlands and discharged in the northeast lowland along Ishikari River; while two local flow systems appear in the shallow layers. The calculated heat flow of these terrains show high values of 140mW/m^2 in the northeast and 100mW/m^2 in the southwest and, low values of about 40mW/m^2 in the central part. Heat flow and thermal gradients generally increase from north to south creating an active geothermal area at the southwest mountains. However the anomalous northeast high temperature zone, results from heat surplus brought by the regional groundwater flow system.

1 INTRODUCTION

Knowledge of hydrogeologic structures is essential in the understanding of heat transport phenomena involved in the earth's upper crust as moving groundwater can cause advective perturbation to the conduction-dominated subsurface thermal regime. Sedimentary basins and active tectonic zones of high topographic relief appear to contain a significant area in which the thermal regime is controlled by advection (Smith & Chapman 1983). Heat flow measurements should be made deep enough to generate a high level of confidence that the temperature log has a representative constant-gradient zone below any water movement, however water movement at a still greater depth is always a possibility, and such water movement may be undetectably changing the heat flux within the logged zone (Mansure & Reiter 1979).

In this study, analyses of groundwater flow suggested by borehole temperature measurements of a sedimentary area in Ishikari basin are presented. The estimation of heat flow in this medium includes effects of groundwater flow. Generally, thermal gradients are thought to increase continuously from north Ishikari river to the southern mountains. A 2-D thermo-hydraulic field simulated numerically along a cross-section of the main water flow direction suggests effects of groundwater flow movements in the distortion of the geothermal regime.

2 SITE SETTINGS

Ishikari lowland is situated north of Sapporo city, Hokkaido in the northern part of Japan (Fig.1). The area, limited at the northern edge by the sea of Japan is composed by sedimentary formations of gravel, sand, silt, clay and peat. Its southwest highlands are made of late Pliocene-Pleistocene volcanic formations of andesites. 22 boreholes of 200 m depth or less were used to measure groundwater hydraulic head and temperature profiles. These temperature measurements were made with a thermister thermometer of 0.01 °C precision which was inserted in the hole and lowered every 2 meter-interval from the water surface to the bottom.

3 THERMAL ANALYSES

A wide geothermal gradient map of Hokkaido compiled by the geological survey of Hokkaido in 1995 shows a gradual increase of thermal gradient from the northeast Ishikari river to the southwest mountainous and volcanic formations. In our study, temperature profiles obtained were classified in three groups according to their shape:

1. concave upward profiles located in the southwestern mountainous formations with generally high thermal gradients.

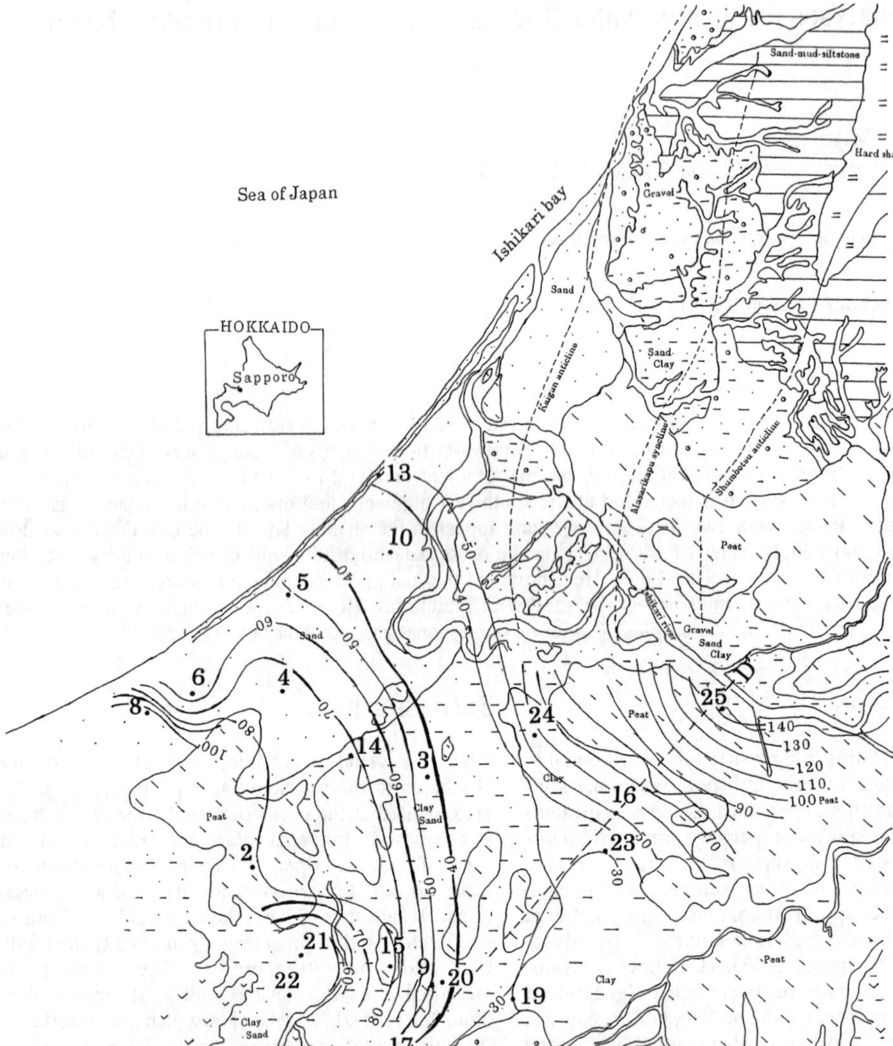

Figure 1. Location of the study site and Heat flow distribution (isolines are in mW/m^2)

2. intermediate profiles of the median or southern part of the lowland with medium to low thermal gradients

3. convex upward profiles of the northern part of the lowland, showing some high thermal gradients of 120 °C/km. Analytical methods in convection proc-

esses of heat transport pioneered by Bredehoeft & Papadopoulos (1965) have been applied worldwide. Mansure & Reiter in 1979 suggested that, a way to analyze the character of data exhibiting potential groundwater flow is by plotting dT/dZ versus T. The equation deduced stands as:

$$\frac{dT}{dZ} = (\beta/L)(T - T') - E/k \qquad (1)$$

Where $\beta = C_\rho VL/k$ is a dimensionless parameter positive or negative with the downward or upward water velocity V, E is the heat flow rate and is equivalent to the intercept of the curve with dT/dZ, T' is the temperature at which the internal heat energy is taken to be zero. To examine the degree of disturbance of the geothermal field by groundwater movements, this thermal gradient-temperature analysis was applied to our study field.

Results show that positive sloping curves dominate in the southern area and generally alternating negative-positive sloping types are encountered in the northern zone. An areal view of this distribution shows that groundwater moves within a regional and two local systems. The recharge and discharge areas of the regional groundwater flow system are respectively in the southwest mountains and the northeast lowland near Ishikari river. One of the local flow systems recharges in the southwest piedmont and discharges in the southern end of the lowland. The other local flow, recharges at the median lowland and discharges at the northeast end.

4 HEAT FLOW AND THERMAL TRANSPORT MODELING

A general equation of thermal transport can be expressed as:

$$\frac{\partial}{\partial x_i}\left[\left(\kappa + \frac{D_{ij}}{R}\right)\frac{\partial T}{\partial x_j}\right] - \frac{\partial}{\partial x_i}\left(\frac{v_i}{R}T\right) + \sum_{k1}^{N}\frac{Q_k(t)T_k(t)}{R\theta}\delta(x_k', y_k', z_k') = \frac{\partial T}{\partial t} \qquad (2)$$

$D_{i,j}$ is the hydrodynamic dispersion, v_i is the average linear groundwater velocity, x_i are the 3D spatial coordinates, κ is the aquifer thermal diffusivity, Q_k is the fluid volume flux for a source or sink located at (x'_k, y'_k, z'_k), θ is the porosity, T_k is the source temperature for an injection well. T is the unknown aquifer temperature, R is the thermal retardation defined as:

$$R = \frac{C_0 + L\omega\rho_w(\partial Wu/\partial T)}{S \cdot \theta c_w \rho_w} \qquad (3)$$

C_0 is the heat capacity of the porous medium, L is the latent heat of water, ω is the moisture content, W_u is the fraction of unfrozen moisture, ρ_w and c_w are respectively the density and specific heat of water. S is the degree of saturation.

The amount of heat flux in the area, evaluated with Bullard plot and interval methods include effects of groundwater flow and shows an average value of 58.8 mW/m^2 (1.4 μcal/cm^2.sec). Its areal distribution as illustrated on figure 1 shows high heat flow areas at both southwest (around 100mW/m^2) and northeast (reaching 140 mW/m^2) ends and, low concentrations at the median zone. These heat flow variations seem to differ from the general geothermal gradient tendency (continuous increase towards the southwest) of Ishikari basin. For these reasons, some explanations have been attempted and, simulated results of a vertical 2-dimensional numerical thermo-hydraulic model of a cross-section D-D' along the main water flow direction were used to assess the processes involved in heat transfer phenomena of these terrains. Groundwater flow simulation results obtained with MODFLOW were consistent with observed hydraulic heads and, served as initial values for heat transfer model.

For this coupled thermo-hydraulic model the finite element program developed by Molson & Frind (1983) was used. We assumed the following boundary conditions: (1) no water and heat flows across the lateral sides; (2) fixed hydraulic heads representing the water table and, a constant surface temperature of 9 °C at the upper boundary; (3) on the lower boundary, an increasing temperature with thermal gradient, in concordance with the general distribution of the geothermal map of the region. The modeled domain (16500 m length, 450 m depth) is formed of 6570 elements and 13468 nodes. The observed temperature distribution agrees with simulated ones. Obtained results combined to previous analytical interpretations show that the subsurface temperature greatly reflects the influence of groundwater flow systems. Convection processes highly distort isotherms of this geothermal field in the direction of flow as it is visible on Fig. 2 which is the thermal field of a southwest-northeast cross-section along the main water flow direction

In fact, the southwest mountainous terrains constitute a geothermal area where groundwater recharged in the high temperature zone, redistributes heat during its migration towards the regional discharge area therefore, creating the cumulative heat flux observed here. On the other hand, the low temperature zone in shallow layers of the central or south lowland is generated by the cooling effect of recharging waters of the local flow system.

5 CONCLUSION

In this study, analyses of field geothermal results have

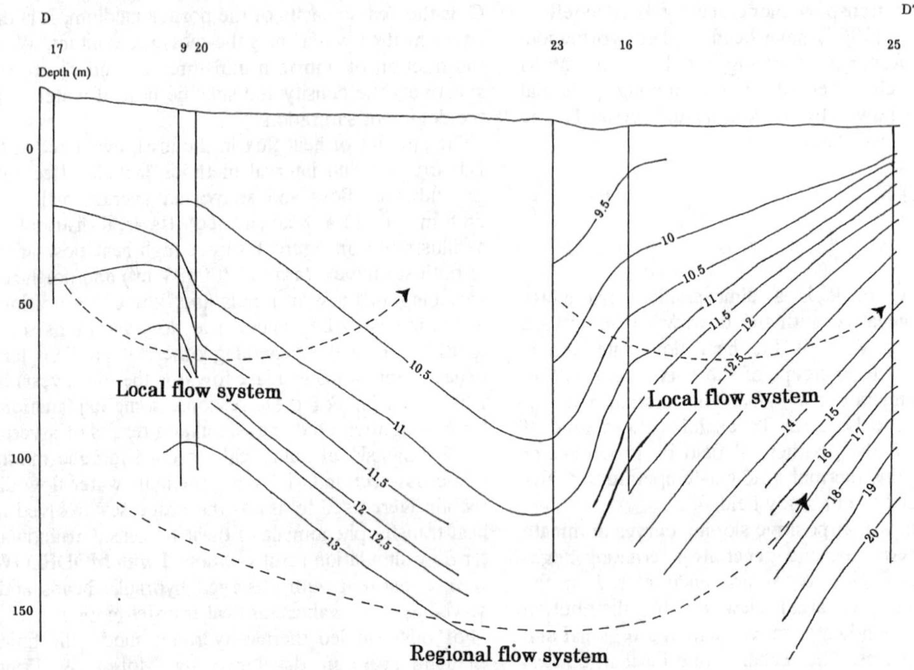

Figure 2. Thermal field on a southwest-northwest cross-section

shown the influence of groundwater flow in the redistribution of heat. Convective processes created by groundwater flow systems have highly distorted isotherms of this geothermal field in the direction of flow. This flow mainly operates from the southwest highlands to the northeast Ishikari river.

REFERENCES

Bredehoeft, J.D. & I.S. Papadopulos 1965. Rates of vertical groundwater movement estimated from earth's thermal profile. *Water Res. Research*. 1(2): 326-328.

Mansure, A. J. & M. Reiter 1979. A vertical groundwater movement correction for heat flow. *Geophysical research J.* 84(B7): 3490-3496.

Molson, J. W. & E. O. Frind 1993. *Heatflow version 2.0 Density-dependent flow and thermal energy transport model in three dimensions*. Waterloo Center for Groundwater Research, University of Waterloo. p. 65

Smith, L. & D. S. Chapman 1983. On thermal effects of groundwater flow. 1. Regional scale systems. *Geophysical research J.* 88(B1): 593-608

Groundwater: Past Achievements and Future Challenges, Sililo et al. (eds) © 2000 Balkema, Rotterdam, ISBN 90 5809 159 7

Application of the electrokinetic sounding technique for geohydrological investigations in a fractured rock aquifer system

F.D. Fourie, J.F. Botha, R. Grobbelaar & G.J. van Tonder
Institute for Groundwater Studies, University of the Orange Free State, Bloemfontein, South Africa

ABSTRACT: Since many South African communities depend on groundwater for their supply of potable water, efficient groundwater exploration techniques can make a significant contribution to the country's water budget. The vast majority of South African aquifers occur in fractured hard-rocks, the geometry and physical properties of which severely restrict the application of conventional geophysical techniques used in groundwater exploration, such as electromagnetic, magnetic and resistivity surveys. These techniques often fail due to the fact that groundwater exploration is usually done in an indirect manner whereby potentially suitable subsurface structures are identified. The Electrokinetic Sounding (EKS) technique is a relatively new technique that offers the advantage of direct groundwater exploration. The presence of groundwater in the subsurface is required for EKS signal to be generated. The recorded EKS signal can be processed to obtain first estimates of potential borehole yield. This paper describes the physical principles on which the EKS technique is based and evaluates its application in fractured hard-rock aquifers by means of a case study. Indications are that the EKS method may be able to resolve the localized occurrences of well-developed fracture zones.

1 INTRODUCTION

The new constitution of South Africa considers the availability of drinking water a basic human right. Considerable emphasis is therefore currently placed on the identification of sustainable water supplies. Groundwater is the sole source of potable water for many urban and rural communities in South Africa and efficient methods of exploration, development and management of groundwater resources are required.

Geophysical techniques, particularly magnetic, electromagnetic and resistivity methods, have been used for many years to locate groundwater in South Africa, but with varying degrees of success. The magnetic method is used almost exclusively to locate and delineate intrusive magmatic bodies and the contact zones between the intrusive bodies and the host rock are then considered to be targets in groundwater exploration. The electromagnetic and resistivity methods are used to map the subsurface conductivity distribution of the area under investigation. From the conductivity information inferences regarding the subsurface geology are made and geological structures are mapped.

Although highly weathered zones containing groundwater will in some instances produce large enough conductivity contrasts to allow direct detection by means of the electromagnetic or resistivity method, groundwater exploration by geophysical means is in general an indirect process whereby subsurface geological structures that may act as, or be associated with aquifers are identified.

The vast majority of South African aquifers are of the secondary type, that is, their water-bearing properties were developed through secondary processes such as faulting or fracturing. It is estimated that approximately 90% of South African groundwater occurs in secondary aquifers. More than 50% of South Africa is underlain by rocks of the Karoo Supergroup. These rocks are typically dense and not extensively fractured so that their permeabilities are low when compared with primary porous formations.

In the past groundwater exploration in Karoo rocks mostly targeted intrusive dolerite bodies. When such bodies do not occur in the area being explored, the magnetic method is of little use for groundwater exploration. The electromagnetic and resistivity methods on the other hand are also frequently unsuccessful in locating Karoo aquifers due to a lack of conductivity contrast between the fractured zones and the surrounding host rock.

The Electrokinetic Sounding (EKS) method is an electroseismic technique that offers the advantage of direct groundwater exploration since the presence of subsurface water is required for the generation of EKS signal.

2 PHYSICAL PRINCIPLES OF ELECTROSEISMIC TECHNIQUES

Electroseismic techniques are based on principles from various disciplines in the physical sciences, particularly colloid and surface science, and the theories of elasticity and electromagnetism.

The surfaces of the mineral grains of which rocks are composed usually display net electric charges due to the presence of unsatisfied chemical bonds (Fitterman, 1978), whereas all groundwater is electrolytic in nature. In fluid-saturated porous rocks an electric potential is therefore usually produced by the contact between the water and solid. As a result the electric charges within the fluid separate into an electric double layer (EDL).

A mechanical disturbance that propagates through such a water-saturated medium will cause a displacement of the solid phase relative to the fluid phase, thereby inducing an electric current and electromagnetic field in the medium. The possibility therefore exists that one may detect the presence of groundwater by observing the induced electromagnetic field on the surface of the earth.

2.1 *The electric double layer and zeta potential*

The EDL consists, as the name implies, essentially of two layers. The inner or Stern layer consists of ions adsorbed onto the solid surface through electrostatic and Van der Waal's forces, while the outer diffuse or Gouy layer in the water is formed by ions under the influence of ordering electrical and disordering thermal forces (Shaw, 1969). Since the adsorbed ions partially screen the effects of the charged solid surface off, the electric potential falls rapidly across the Stern layer, which is usually only one ion thick.

Any relative motion between the solid and the water will obviously cause a relative motion between the outer diffuse ions and inner strongly bound ions. The plane at which the ions of the inner and outer layers are sheared from each other, is called the slipping or shearing plane, and the electric potential in the plane the zeta (ζ)-potential. The zeta potential plays an important role in electrokinetic phenomena, since it is contained in the electrokinetic coupling tensor, which determines the existence and magnitude of the induced electromagnetic field.

2.2 *Propagation of elastic waves in a fluid-saturated porous solid*

The propagation of a disturbance in a porous medium saturated with a viscous fluid can be described mathematically by a theory developed by Biot (1956). According to this theory the disturbance will propagate in the form of two dilatational waves and one rotational wave. The two dilatational waves are commonly referred to waves of the first and second kind. Dilatational waves of the first kind correspond to the solid and fluid parts moving in phase, whereas waves of the second kind correspond to the solid and fluid parts moving out of phase.

Biot further shows that the dilatational waves of the second kind propagate at lower velocities than waves of the first kind. These waves are therefore frequently referred to in literature as the Biot slow and fast waves, respectively. The theory also indicates that the slow waves are dissipative in nature and die out rapidly with distance from the source. Slow waves are, however, again generated when fast waves intersect interfaces between dissimilar fluid-saturated porous media (Geertsma & Smit, 1961).

2.3 *Electroseismic coupling*

Various mechanisms exist whereby seismic and electromagnetic energies can couple in wetted porous rocks. However, Haartsen (1995) suggests that mechanically induced relative flow of charges is likely to cause the largest effects. This phenomenon is generally known as the electrokinetic effect, while the electric current caused by the effect is called the streaming current.

Pride (1994) derived from first principles equations that govern electrokinetic phenomena. These equations have the form of Maxwell's equations coupled to Biot's equations. Pride's work was extended to generalize the equations to include the effects of anisotropy (Pride & Haartsen, 1996). When compressional and rotational waves propagate in a homogeneous porous medium the streaming current is cancelled by an equivalent conduction current. The electric field is confined to the wave and has no extent outside of it. No independently propagating electromagnetic waves are therefore generated. However, when the wave is incident on an interface between dissimilar media, there is a variation in the streaming current density leading to a charge separation that varies with the time signature of the incident wave. The time-varying charge separation acts as an oscillating dipole source spread over the interface and generates an electromagnetic disturbance that propagates independently and can be recorded at the surface.

Even though Biot's slow dilatational wave causes the largest relative motion between solid and fluid, it is necessary to consider both the fast and slow waves, as well as the rotational wave to obtain a complete picture of the electroseismic conversion. Furthermore one needs to consider all seismic reflection and refraction events. A seismic disturbance can give rise to a number of different elastic waves propagating in the subsurface. In an infinite homogeneous isotropic medium, only dilatational and shear exist. However, if the medium is bounded, waves are generated that travel along the interfaces bounding the medium. Examples of such waves are Rayleigh and Love waves that travel along the surface of the earth. Stoneley waves are the subsurface equivalents of Rayleigh waves traveling along interfaces between rocks. Waves that are refracted at the critical angle are called headwaves. In fluid-saturated porous media, these waves cause relative motion between fluid and solid across an interface of different electrokinetic coupling properties, thus giving rise to electromagnetic disturbances.

Mikhailov et al. (1997) and Beamish (1999) described the different ways whereby seismic energy is converted into electromagnetic energy through the mechanism of electrokinetic coupling. Figure 1 is a graphical representation of a spherical spreading seismic wave incident on a boundary between two media at a depth d from the surface. As the wave traverses the boundary, successive Fresnel zones, i.e. radial zones along the interface for which the distance traveled by the seismic wave differs by less than half a wavelength (λ), intersect the boundary at later times and larger offsets. Each Fresnel zone is out of phase with the Fresnel zones preceding and following it and the contributions to the generated electrokinetic signal are therefore also out of phase. All the out of phase components of the generated signal cancel so that only the first Fresnel zone (with smallest offset) contributes significantly to the total signal.

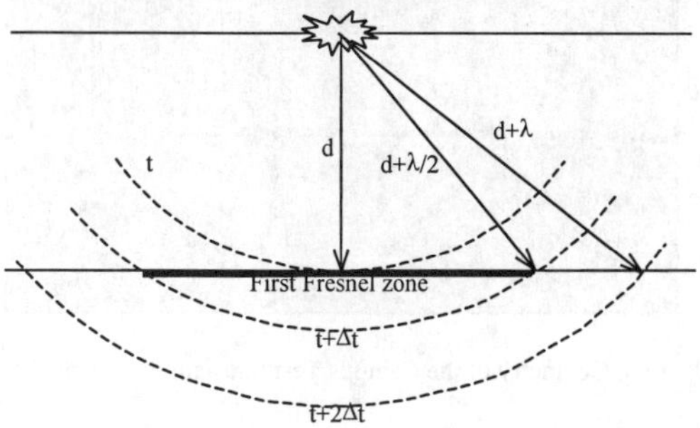

Figure 1. The Fresnel zone coupling mechanism. Only the first Fresnel zone contributes significantly to the total generated signal.

Fresnel zone coupling leads to near simultaneous electromagnetic arrivals at all antennae and the arrival time is approximately equal to the one-way travel time of seismic waves from the seismic source to the interface. Refracted head waves traveling along the interface give rise to electromagnetic fields that display time-dependent arrivals at an array of surface antennae. Figure 2 is a schematic summary of the electroseismic coupling occurring in the subsurface. In principle the Fresnel zone coupling mechanism may occur at any boundary between media for which the electrokinetic coupling tensor differs, whereas the refracted wave mechanism requires a difference in acoustic properties.

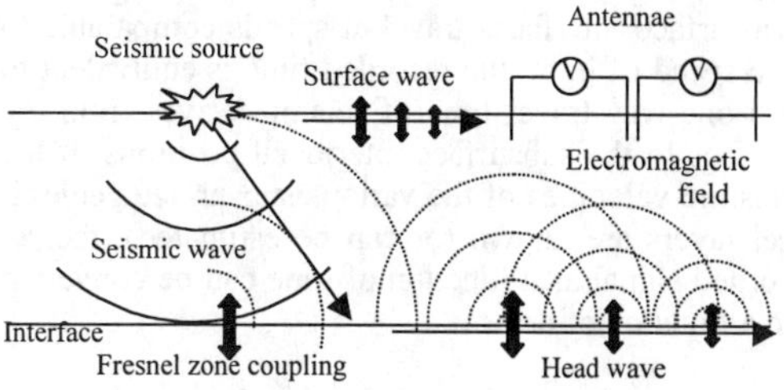

Figure 2. Schematic diagram of electroseismic coupling mechanisms. Vertical arrows indicate oscillating dipole sources.

3 THE ELECTROKINETIC SOUNDING TECHNIQUE

The electrokinetic sounding (EKS) technique employs an antenna geometry that facilitates the measurement of electrokinetic signal in a sounding mode. Antennae are positioned close to the surface source of seismic energy for the following reasons:

- In homogeneous isotropic media, Rayleigh and Stoneley waves exhibit radial symmetry. Once these waves pass beneath the positions of the surface antennae, they rapidly contribute less signal due to symmetry considerations.
- For the Fresnel zone coupling mechanism, the total electromagnetic signal has its origin mainly at interfacial positions directly beneath the seismic source.
- Calculations performed by Beamish (1999) to evaluate the electric field associated with a buried vertical electric dipole indicate a field that is sharply focused above the position of the dipole. A rapid spatial decay of signal amplitude can therefore be expected for antennae at increasing offsets. Using antennae located close to the seismic source allows larger signal-to-noise ratios.
- Working with small offset antennae reduces the 3D problem to a pseudo-1D problem whereby electroseismic surveying can be done in a sounding mode, measuring only signals generated vertically below the shot-point.

A typical electrokinetic sounding geometry is shown in Figure 3. The antennae generally employed consist of grounded electrode pairs instead of induction loops, for it has been shown (Fitterman, 1978) that due to symmetry reasons, the electrokinetic effect in a layered earth will generate no surface magnetic field. The use of two electrode pairs situated on either side of the shot-point allows data to be recorded on two channels, which allows distinction to be made between electrokinetic signal and incoherent noise.

The surface electric potential at the antennae is recorded as a function of time. Due to the fact that the electromagnetic effects of the oscillating dipoles at subsurface interfaces travel at speeds comparable to the speed of light, the recorded time is equivalent to the one-way travel time of seismic waves from the surface to the subsurface interfacial positions. If the seismic velocities of the various subsurface geological layers are known (or can be estimated), the recorded signal as a function of time can be converted to a function of depth.

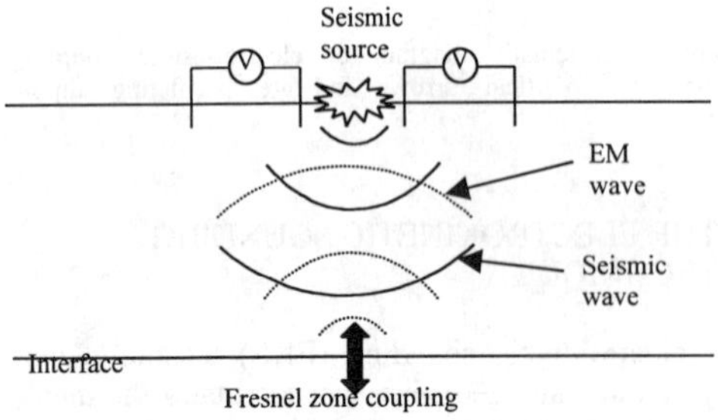

Figure 3. Electrokinetic sounding geometry.

Using the work of Chandler (1981), Millar & Clarke (pers. comm.) developed a method whereby the recorded EKS signal can be analyzed to obtain estimates of the vertical distribution of porosity and permeability beneath the seismic source. From these estimates a first estimate of potential borehole yield can be derived.

4 APPLICATION OF THE EKS TECHNIQUE IN KAROO AQUIFER SYSTEMS

Since rocks of the Karoo Supergroup underlie more than 50% of South Africa and much of the arid regions thereof, the successful application of groundwater exploration techniques in these rocks is of high priority.

Experience has shown that a borehole in a Karoo aquifer will only have a significant yield if it intersects a fracture. In addition, the water-yielding fractures in Karoo rocks are usually parallel to the bedding planes and have aperture widths of only a few millimeters. Although the areal extent of the fractures may be considerable, fracturing is not uniformly present. The physical dimensions of the fractures do not allow large quantities of water to be stored in the fractures themselves, and low storativity values are a rule. The main water supply to a borehole intersecting a fracture is the sedimentary rock matrix that surrounds the fracture. Fractures in Karoo rocks act as conduits that transport water from the sedimentary rock matrix to the borehole. Results of recent investigations performed by the Institute for Groundwater Studies indicate that transmissivity values of $1\ m^2\ day^{-1}$ and around $700\ m^2\ day^{-1}$ can be expected for the matrix and fracture, respectively.

4.1 *EKS survey on the Campus Test Site*

The Campus Test Site at the University of the Free State in Bloemfontein, South Africa was established in 1990. The test site covers an area of approximately $180 \times 190\ m^2$ and a total number of 30 percussion boreholes have been drilled at various positions on the site. Numerous hydraulic and geophysical tests have been performed on the on-site boreholes, and the geohydrological conditions are well known. The geohydrology at the Campus Test Site is presented schematically in Figure 4.

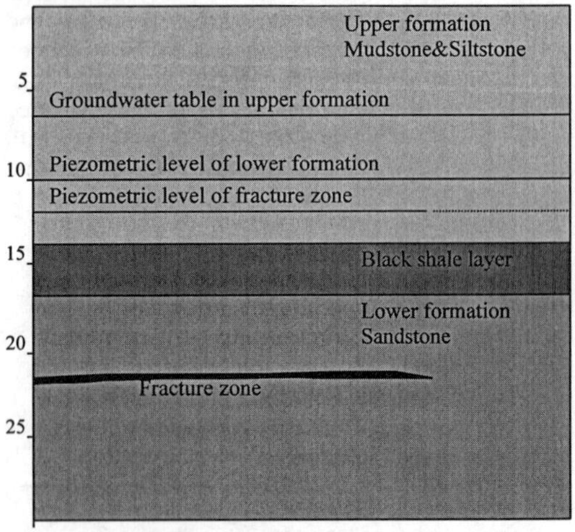

Figure 4. Geometry of the Campus Test Site aquifer system.

The geology consists of sandstone, mudstone and shale deposited under fluvial conditions. Core samples indicate parallel bedding plane fractures of which the frequency decreases downward from the more weathered zones at the top to thicker and more competent units below. The most significant bedding plane fracture is encountered at a depth of around 21 m. It is a horizontal fracture that follows the contours of the sandstone unit. The black shale layer encountered at 14-17 m acts as an aquitard. Boreholes intersecting the bedding plane fracture zone at 21 m are high yielding ($>3\ l\ s^{-1}$), whereas low yields ($<0.5\ l\ s^{-1}$) are observed in those boreholes that do not.

During 1999 an EKS survey was performed on the Campus Test Site. Commercially available EKS equipment from the company Groundflow Ltd. was used to conduct the survey. Surveying took place on a 10 × 10 m grid covering an area of approximately 5600 m^2. The grid had been established in the central part of the test site where the borehole density is high. The survey area contained both high and low yielding boreholes. A hammer-and-plate system was used as the source of seismic energy and stainless steel electrode pairs with dipole lengths of 2 m were employed as antennae. Triggering was achieved by means of an inertia trigger strapped to the hammer.

Figures 5 and 6 are examples of the EKS signals that were recorded along lines 10 and 30 of the grid (only one channel is shown). The only processing procedures performed on the data were 8-fold stacking and filtering to remove power line and high frequency noise.

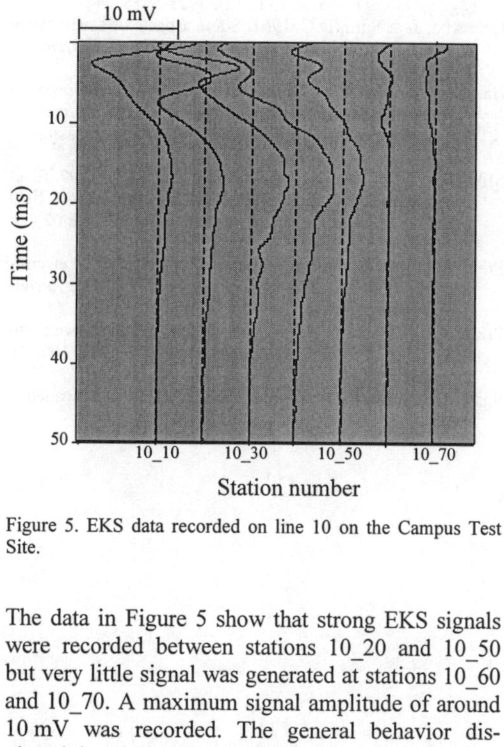

Figure 5. EKS data recorded on line 10 on the Campus Test Site.

The data in Figure 5 show that strong EKS signals were recorded between stations 10_20 and 10_50 but very little signal was generated at stations 10_60 and 10_70. A maximum signal amplitude of around 10 mV was recorded. The general behavior displayed by the recorded EKS signal consists of a strong negative component followed by a positive peak that gradually decays back to zero voltage. Coherent peaks are also observed between 3 and 4 ms. These peaks can probably be related to the EKS signal generated at the groundwater table, suggesting a seismic velocity of between 1500 and 2000 m s^{-1} for the shallow formations. This estimated velocity is in agreement with known values of seismic velocities for weathered Karoo rocks. Unfortunately the recorded EKS data were very noisy in early times, making it difficult to identify true EKS signal. The coherence of the early time peaks on traces 10_20 to 10_40 however suggests that the signal may indeed be due to the groundwater table.

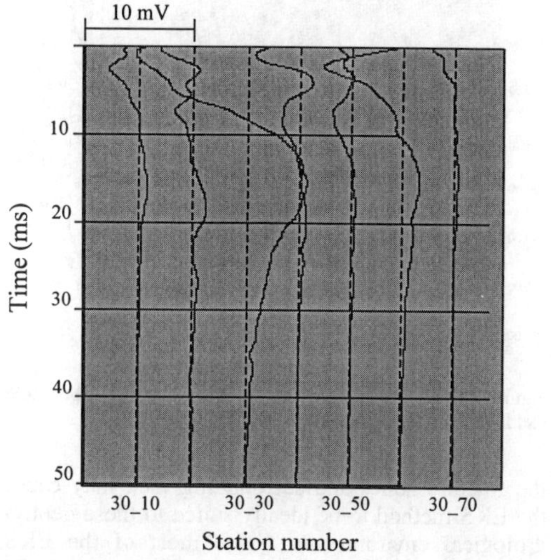

Figure 6. EKS data recorded on line 30 on the Campus Test Site.

The EKS signals recorded along line 30 display similar behavior to the signals on line 10, but strong signal is much more localized. The signal recorded at station 30_30 has a maximum amplitude of more than 10 mV, whereas little signal was recorded at its neighboring stations.

Figure 7 is a contour map of the normalized estimates of potential borehole yield over the survey area, as derived from the EKS data. The positions of all percussion boreholes are also shown. Boreholes intersecting a well-developed fracture zone at a depth of around 21 m are shown as a filled black circle, those intersecting a poorly developed fracture zone are shown as empty circles and boreholes that do not intersect the fracture zone at all are indicated by crosses. There is a general agreement between the fracture location as indicated by the drilling results and the EKS survey. The localized nature of fracture occurrence is also demonstrated by both the drilling and EKS results.

5 CONCLUSIONS

The results of the EKS survey performed on the Campus Test Site indicate that the EKS method may prove invaluable in groundwater exploration in Karoo rocks. The high permeability contrast between

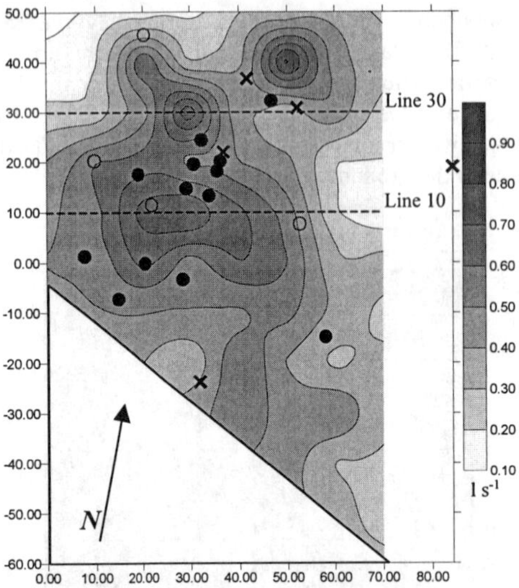

Figure 7. Contour map of the normalized estimates of borehole yield.

the fracture zones and surrounding rock may cause the EKS method to be ideally suited to these geohydrological environments. Since most of the EKS signal is generated at interfaces vertically below the survey position, indications are that the EKS method may be able to resolve the localized occurrences of well-developed fracture zones.

The high noise levels recorded during early times are problematic since the presence of EKS signals is masked. This problem can be addressed by reducing the noise associated with the source and the triggering mechanism. Due to the inconsistent nature of hammer blows, a shotgun source should be used instead of a hammer-and-plate source. Inertia triggers with very low noise content are commercially available and should be used to reduce noise levels. The use of shallow auger holes to introduce seismic energy at a depth below the highly weathered surface rock will reduce noise levels due to scattering and will allow more energy to penetrate to the deeper layers.

The long wavelength transient behavior of the EKS signal observed complicates the problem of defining an on-set time at which signal is generated. It is therefore difficult to relate the recorded signal to a specific depth. Considerable overlap of signal generated at different horizons is possible.

Due to the high permeability of the fracture zones, a seismic disturbance is likely to cause considerable mobilization of fluid in these zones. The resulting horizontal motion of fluid will give rise to an electromagnetic disturbance with behavior quite unlike that of a vertical oscillating dipole.

To understand the long wavelength behavior of the recorded signal and to predict the effects of horizontal flow on EKS signal, it will be necessary to apply the equations governing acoustic and electromagnetic coupling to the specific case of horizontal fractures. This is the topic of forthcoming research.

REFERENCES

Beamish, D. 1999. Characteristics of near-surface electrokinetic coupling. *Geophysics Journal. International* 137: 231-242.

Biot, M.A. 1956. Theory of propagation of elastic waves in a fluid-saturated porous solid. I. Low frequency range. *Journal of the Acoustical Society of America* 28 (2): 168-178.

Chandler, R. N. 1981. Transient streaming potential measurements on fluid-saturated porous structures: An experimental verification of Biot's slow wave in the quasi-static limit. *Journal of the Acoustical Society of America* 70 (1): 116-121.

Fitterman, D.V. 1978. Electrokinetic and magnetic anomalies associated with dilatant regions in a layered earth. *Journal of geophysical research* 83 (B12): 5923-5928.

Geertsma, J. & Smit, D.C. 1961. Some aspects of elastic wave propagation in fluid-saturated porous solids. *Geophysics* 27 (2): 169-181.

Haartsen, M.W. 1995. *Coupled electromagnetic and acoustic wavefield modeling in poro-elastic media and its applications in geophysical exploration.* Ph.D. thesis, Massachusetts Institute of Technology.

Mikhailov, O.V., Haartsen, M.W., & Toksöz, M.N. 1997. Electroseismic investigation of the shallow subsurface: field measurements and numerical modeling. *Geophysics* 62 (1): 97-105.

Pride, S.R. 1994. Governing equations for the coupled electromagnetics and acoustics of porous media. *Physical Review B* 50: 15678-15696.

Pride, S.R. & Haartsen, M.W. 1996. Electroseismic wave properties. *Journal of the Acoustical Society of America* 100 (3): 1301-1315.

Shaw, D.J. 1969. *Electrophoresis*. London: London Academic Press.

Groundwater: Past Achievements and Future Challenges, Sililo et al. (eds) © 2000 Balkema, Rotterdam, ISBN 90 5809 159 7

Specific surface area of glacial drift material determined by laboratory experiments

Å. Fransson
Department of Geology, Chalmers University of Technology, Göteborg, Sweden

ABSTRACT: The importance of, and difficulties associated with, groundwater flow and mass transport are reasons for studying the interaction between aquifer material, groundwater and solutes. A factor that influences both permeability and adsorption is the specific surface area and the aim of this study was to examine this linkage between fluid flow and solute transport. The strategy employed was a comparison of specific surface area determined by three methods: (1) analysis of particle-size data; (2) analysis of saturated permeability data from permeameter experiments; and (3) analyses of batch experiments and breakthrough curves for base-cations and chloride. Three materials, all sandy gravels with slightly different grain-size distributions were examined. When comparing the specific surface areas estimated from permeability data, breakthrough curves and particle-size data, the resulting areas were of the same order of magnitude.

1 INTRODUCTION

Groundwater is an important resource and its vulnerability depends upon factors such as the geological material, the groundwater flow and the resulting mass transport. The geological material is important since the mobility of the adsorbable solute is partly controlled by the specific surface area of the material, and the solute is delayed in its further transport, so called retardation. Theoretically there exists an expression describing groundwater flow that considers the influence of the specific surface area (Carman, 1937, 1956). Of interest is therefore to investigate how this specific surface area is related to the adsorbing area of a geological material, to increase understanding and the possibilities to link the different processes. Another reason for undertaking an experimental study also mentioned in Bjerg (1992) is the abundance of geochemical models describing groundwater flow and transport, and how these processes and the influencing parameters are not so often investigated experimentally. In Sweden, geological formations containing sand and gravel are of particular interest since they are commonly good aquifers and therefore especially vulnerable. Here, three sandy gravels with slightly different grain-size distribution curves were investigated. The adsorbing solutes utilised were chloride and the cations Ca^{2+}, Mg^{2+}, Na^{+} and K^{+} which are common in lechates from e.g. landfills and therefore their behaviour is also of interest.

2 THEORY

The central parameter in this paper is the specific surface area, defined as the area of a body divided by its volume. For a spherical grain of diameter, D, this area is:

$$S = \frac{A}{V} = \frac{\pi D^2}{\pi D^3/6} = \frac{6}{D} \tag{2.1}$$

The virtual size of the specific surface area is difficult to determine as it depends upon at what scale the area is studied, and none of the following investigated areas is therefore likely to be the total specific surface area.

To separate the specific surface areas determined by different methods, an index has in this paper been chosen for each one:

1. Sieving and particle-size data analysis, S_S.
2. Permeameter experiments and the Kozeny-Carman equation, $S_{K\text{-}C}$.
3. Batch- and permeameter experiments: retardation, S_R.

2.1 *Sieving and particle-size data analysis*

According to the definition of specific surface area, knowing the diameter should be enough to estimate a value. By sieving, the distribution of diameters for an entire sample is obtained and as a result a value of the total specific surface area can be determined (Gustafson, 1983):

$$S_S = \int_0^\infty \frac{\alpha f(D)}{D} dD \tag{2.2}$$

where, f(D), is a frequency function or a derivative of the grain size distribution curve, and α, is a shape factor. This factor, depends upon the shape of the grain and has both for spherical and cubic grains a value of 6, see Equation 2.1, where S = 6/D = α/D. For flakier grains, this shape factor was adjusted, see Fransson & Nordén (1996).

2.2 *Permeameter experiments: permeability*

The Kozeny-Carman equation, (Carman, 1937),

$$k = \frac{n^3}{k_0 \left(\frac{L_e}{L}\right)^2 (1-n)^2 S_{K-C}^2} \tag{2.3}$$

describes a semi-empirical relationship between permeability, k, porosity, n, and specific surface area, here $S_{K\text{-}C}$. The expression assumes laminar channel flow and the coefficients k_0 and $(L_e/L)^2$ take into account the shape of the channel and the tortuosity respectively. Tortuosity relates the actual length water is transported, L_e, to the theoretically shortest way, L. The coefficients k_0 multiplied by $(L_e/L)^2$ in the denominator of the Kozeny-Carman equation were through experiments found to have a value of about five (Carman, 1937). The porosity and the specific surface area are derived from the concept of channel flow where, as in hydraulics, the hydraulic radius is of major importance. Depending on the specific surface area in contact with water flow permeability changes. This is due to the friction in the water in contact with the material.

2.3 *Batch- and permeameter experiments: retardation*

In mass transport a distinction is made between the solutes following the water and those adsorbed onto the material, (e.g. Appelo & Postma, 1993). Adsorption, which is here assumed to be fast, reversible and described by a linear isotherm, results in a retardation of the solute. Retardation is expressed by dividing the velocity of water in the pores, v_p, by the velocity of the delayed front of solutes, v_s,

$$\frac{v_p}{v_s} = R = 1 + \frac{\rho_b}{n} K_d \tag{2.4}$$

where ρ_b is the bulk density and K_d the distribution coefficient: This coefficient relates the amount of solute adsorbed per unit weight of the material, C_s, to the amount still in solution, C_w, when equilibrium is reached,

$$K_d = \frac{C_s}{C_w} \tag{2.5}$$

Another way of describing a distribution coefficient, is K_a which is used in this work to relate the amount of solute adsorbed per unit area of the material, C_{sa}, to the amount still in solution, C_w,

$$K_a = \frac{C_{sa}}{C_w} \tag{2.6}$$

resulting in a retardation factor,

$$R = 1 + \frac{(1-n)S_R}{n} K_a \tag{2.7}$$

which increases when the specific surface area, S_R, does. The area related distribution coefficient, K_a, was estimated using batch-experiments and the specific surface areas from the particle-size data analysis, S_S.

3 EXPERIMENTS

3.1 *Sieve analysis*

For the experiments three materials of glaciofluvial sand and gravel, having different grain size distributions were taken in an end moraine, south of Göteborg, Sweden. Three samples from each material were taken out, washed and sieved. Sieving was carried out, first through square sieves and then through rectangular sieves combined with an ocular investigation in order to examine the flakiness of the grains. Grain size distribution curves derived from sieving were then used to calculate the specific surface areas of the three different materials, Equation 2.2. The total specific surface area is a summation of the areas of each fraction in a sample, and the distribution within these fractions is approximated by a logarithmic trapeze formula (Gustafson, 1983). The reason for having three different samples per material was to get representative values of the specific surface areas.

3.2 *Permeameter and batch experiments*

The permeameter experiment served two purposes; firstly to determine the permeability of the materials and secondly to study the breakthrough curves and the delay or retardation of five different solutes. Figure 1 shows the experimental set-up.

The permeameter was carefully packed to get as homogeneous conditions as possible and then slowly saturated from the bottom side. Finally, before starting the experiment the porosity, n, was determined. Since the experiment was aimed to determine

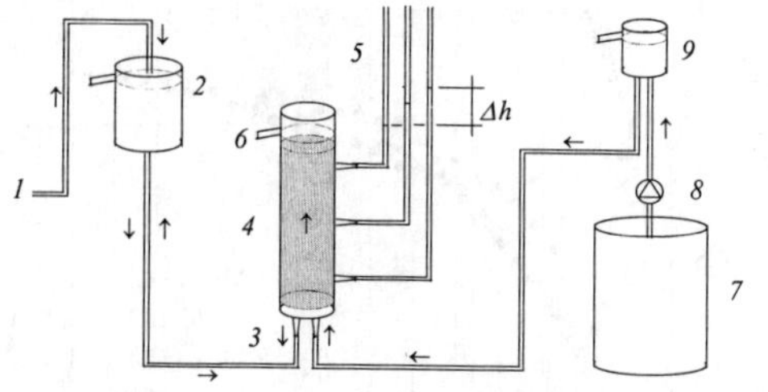

1 Inflow, tap-water
2 Container with overflow
3 Inflow permeameter, valves
4 Permeameter
5 Observation of pressure levels
6 Outflow
7 Container for solutes
8 Pump
9 Container with overflow

Figure 1. Experimental set-up to determine permeability and retardation

both permeability and retardation, a solution that contained chloride and one of the following cations: sodium, potassium, magnesium or calcium was prepared before starting. Chloride was used to indicate when the waterfront or a pore volume, V, had passed through the permeameter.

Initially, water was flowing from a container with overflow (2 in Fig. 1). During the experiment the flow, Q, the pressure-drop, Δh, over the length, L, of the sample and the temperature were measured to obtain the permeability, k (e.g. de Marsily, 1986):

$$k = \frac{QL}{A\Delta h}\frac{\mu}{\rho g} \tag{3.1}$$

where, A, is the cross-sectional area of the permeameter, g, is the force of gravity per unit mass, μ and ρ are the viscosity and the density of the fluid respectively. Temperature was measured continually and the experiments were performed at a temperature of 15-20°C. The solute was introduced by opening the right valve (3 in Fig 1.) reversing the flow between the water container (2) and permeameter (4) due to the higher position of the container for solutes (9).

To verify whether the adsorption was related to the specific surface area of the materials and to study the isotherm, batch experiments were carried out. The three different materials, that according to sieving experiments had three different sizes of specific surface area, were put in solutions of known volume and concentration. After about 20 hours, a chemical analysis of the remaining concentrations in solution was done and new solutions of higher concentration were then added to the three samples.

4 RESULTS AND DISCUSSION

In Figures 2 and 3, grain size-distribution curves and a diagram, which relates the concentration of potassium in water to the amount adsorbed, are presented for materials 1, 2 and 3. Among the studied solutes, potassium is presented here due to its more delayed breakthrough curves.

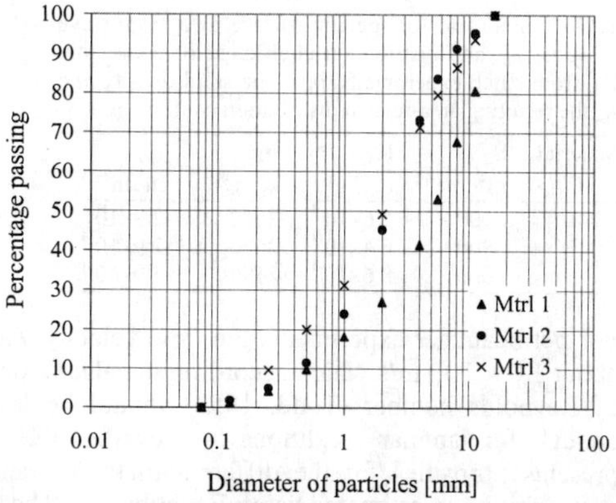

Figure 2. Grain-size distribution curves for material 1, 2 and 3.

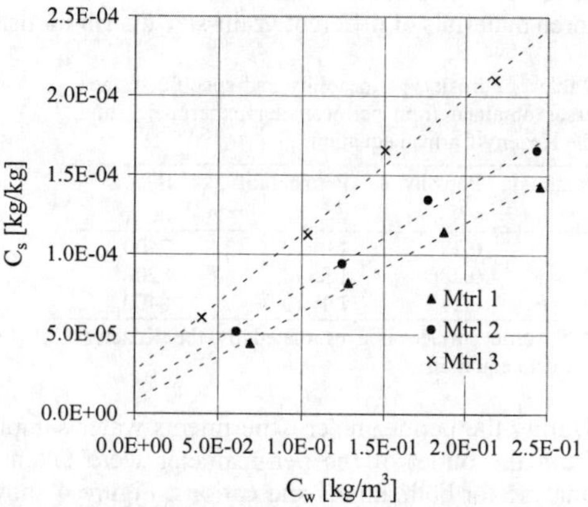

Figure 3 The amount of potassium in solution, C_w, and the amount adsorbed per unit weight of the material, C_s, obtained from batch experiments for materials 1, 2 and 3.

As shown in Figure 3, the isotherms appear to be linear and the amount of solute adsorbed and the value of the distribution coefficient for potassium, Equation 2.5, increases from material 1 to material 3. This is also true for the specific surface areas obtained from the grain size distribution curves, S_s, see Table 1. The areas are of the same order of magnitude as was presented in Aastrup *et al.* (1995), where gravel and silt had estimated specific surface areas of approximately 600 and 60,000 m^2/m^3. If utilising these areas and the solid density, ρ_s,

$$\frac{\rho_s K_d}{S} = K_a \tag{2.4 \& 2.7}$$

a constant value of the surface related distribution coefficient, K_a, is found. This indicates an area-related adsorption even though the specific surface areas are likely to be underestimated due to the assumption of spherical or cubic grains.

Table 1. Estimates of specific surface area from sieve analysis, S_S, and distribution coefficient of potassium, K_d, from batch experiments. ρ_s is the solid density and K_a the resulting area-related distribution coefficient.

Material	S_S m^2/m^3	K_d m^3/kg	ρ_s kg/m^3	K_a m^3/m^2
1	5100	5.6×10^{-4}	2670	2.9×10^{-4}
2	5900	6.4×10^{-4}	2670	2.9×10^{-4}
3	8000	8.6×10^{-4}	2680	2.9×10^{-4}

For permeameter experiments the flow velocity was about 1.0×10^{-4} m/s which according to calculations of Reynolds number (Todd, 1959) should be low enough for laminar conditions to prevail. Table 2 presents porosities for the different materials, and permeability, k, estimated using Equation 3.1. These figures, and the Kozeny-Carman equation (2.3), resulted in specific surface areas, here $S_{K\text{-}C}$, which describe the difference in internal friction between the three materials of different grain-size distributions.

Table 2. Porosity, permeability and specific surface areas obtained from permeameter experiments and the Kozeny-Carman equation.

Material	Porosity, n -	Permeability, k m^2	$S_{K\text{-}C}$* m^2/m^3
1	0.27	1.4×10^{-10}	7400
2	0.28	1.3×10^{-10}	8200
3	0.27	1.1×10^{-10}	8400

* Specific surface area estimated by the Kozeny-Carman equation

During the permeameter experiments water samples from the outlet of the permeameter were taken to analyse for both anions and cations. Figure 4 shows the water/chloride front and the retardation of potassium. Here, the y-axis presents the outlet concentration, C, divided by the concentration of the inlet water, C_0. The x-axis shows the number of pore volumes, V, passing through the permeameter, i.e. time divided by the time for breakthrough of the chloride front.

According to Figure 4, potassium does not reach its inlet concentration during the test. This has been observed for potassium earlier (Bjerg, 1992) and could be due to diffusion into the matrix of grains. However, this is a slow process and the retardation was evaluated from an early part of the curve where the relative concentration, C/C_0 is equal to 0.5. The retardation factors, Equation 2.7, for material 1, 2 and 3 are according to Figure 4 approximately 2.6, 3.6 and 3.6.

To estimate the adsorbing specific surface areas for the permeameter experiment, S_R, these retardation

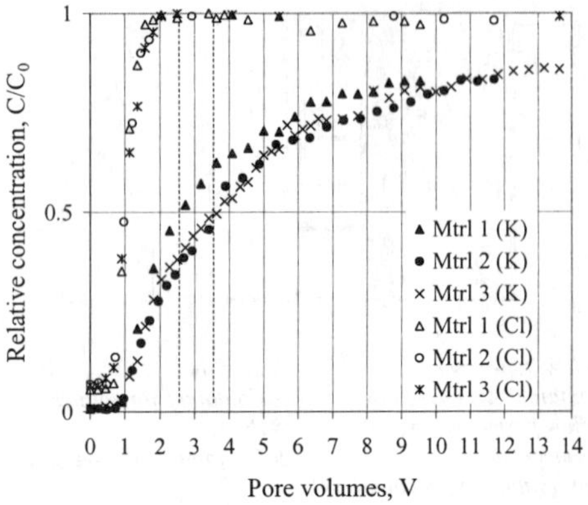

Figure 4. Breakthrough curves for chloride, Cl, and potassium, K, for materials 1, 2 and 3.

factors, the porosity and the surface-related distribution coefficient, K_a, Table 1, were used. Table 3 compiles these areas and those determined by sieving and analysis of grain-size distribution curves, S_S, and the ones from permeameter experiments and Kozeny-Carman's equation, $S_{K\text{-}C}$ The areas estimated from permeability- and retardation data are related to the areas estimated from sieve analysis (in parenthesis).

Table 3. Estimates of specific surface area from sieving and permeameter experiments. The areas estimated from permeability- and retardation data are related to the areas estimated from sieve analysis (in parenthesis).

Material	Specific surface areas		
	Sieving, S_s m^2/m^3	Permeability, $S_{K\text{-}C}$ m^2/m^3	Retardation, S_R m^2/m^3
1	5100 (1)	7400 (1.45)	2000 (0.40)*
2	5900 (1)	8200 (1.40)	3600 (0.60)*
3	8000 (1)	8400 (1.05)	3400 (0.40)*

* Values estimated from sieving, permeameter- and batch experiments.

As presented in Table 3, all methods give successively increasing values of specific surface area for materials 1, 2 and 3, with a small deviation for the ones related to the retardation. When comparing the specific surface areas from the permeability data and breakthrough curves to the areas obtained from particle-size data, the resulting areas from permeability were less than 150 % and the breakthrough curves gave estimated areas exceeding 40 % of these areas. Since the specific surface area from sieving is likely to be smaller than the actual area, the estimated area-related distribution coefficient gets a too large value. This results in an underestimated surface area from retardation. Consequently, an increased area from sieving would make all areas approach the values estimated from permeability data.

5 CONCLUSIONS

Specific surface areas estimated from grain-size distributions for three different sandy gravels were between 5100-8000 m^2/m^3, which is in accordance with results by e.g. Aastrup *et al.* (1995).

The batch experiments indicate an area-related adsorption for potassium and the shape of the potassium breakthrough curve can be due to a combination of advection, dispersion superimposed by adsorption and diffusion, where the latter is a slower process. The outflow concentration of potassium not reaching the one at the inflow, have also been observed by Bjerg (1992), and might be due to slow diffusion into the crystal lattice.

Permeability data estimated specific surface areas of 7400-8400 m^2/m^3 and breakthrough curves resulted in areas of 2000-3400 m^2/m^3 for the three materials. When relating the specific surface areas from permeability data and breakthrough curves to the areas obtained from sieve analysis, the areas calculated from permeability were higher (150 %) and the areas from breakthrough curves were lower (40 %) than the ones from sieve analysis. Thus, the order of magnitude of the specific surface area could be estimated from sieve analysis. However, the certainty and the implications of the results would increase if further investigations were performed including a larger number of samples having different grain-size distributions. Furthermore, an analysis of area-related distribution coefficients of other solutes would be interesting. Since both transport of water and solutes seem to be affected by the specific surface area this linkage could possibly be utilised when considering migration processes in glacial drift material having different grain-size distributions.

REFERENCES

Aastrup, M., B. Thunholm, J. Johnson, U. Bertills & A. Bernfell 1995. *Groundwater chemistry in Sweden*. Report 4416. Swedish Environmental Protection Agency. Stockholm.

Appelo, C.A.J. & D. Postma 1993. *Geochemistry, groundwater and pollution*. Rotterdam: A.A.Balkema.

Bjerg, P.L. 1992. *Attenuation of cations in groundwater*. Ph.D. thesis. Department of Environmental Engineering, Technical University of Denmark. Lyngby.

Carman, P.C. 1937. Fluid flow through granular beds. *Trans. Instn. chem. Engrs*: 15:150-166.

Carman, P.C. 1956. *Flow of gases through porous media*. London: Butterworths.

Fetter, C.W. 1993. *Contaminant Hydrogeology*. New York: Macmillan Publishing Company.

Fransson, Å. & C. Nordén 1996. *Hydraulisk konduktivitet och specifik yta vid vatten- och ämnestransport i sand och grus (In Swedish)*. Report B 427. Department of Geology, Chalmers University of Technology. Göteborg.

Gustafson, G. 1983. *Brunnssystem för värmelagring och värmeutvinning i akviferer (In Swedish)*. Report R39:1983. The Swedish Council for Building Research. Stockholm.

de Marsily, G. 1986. *Quantitative Hydrogeology. Groundwater Hydrology for Engineers*. San Diego: Academic Press Inc.

Todd, D.K. 1959. *Groundwater hydrology*. New York: John Wiley & sons.

Groundwater: Past Achievements and Future Challenges, Sililo et al. (eds) © 2000 Balkema, Rotterdam, ISBN 90 5809 159 7

Recharge assessment: Comparing tracers, micro-meteorology and soil water models

J.C.Gehrels

Faculty of Civil Engineering and Geosciences, Delft University of Technology & Netherlands Institute of Applied Geoscience TNO, Delft, Netherlands

ABSTRACT: Three methods were used to estimate the groundwater recharge rate below a natural grassland cover under temperate climatic conditions: (1) the micro-meteorological Bowen ratio energy balance (BREB) method, (2) tracer techniques using chloride and oxygen-18, and (3) soil water flow simulation with numerical and reservoir models. The BREB method resulted in recharge estimates over the short-term (one season). The use of oxygen-18 was based on the seasonal cyclicity that was traceable in the soil profile. Both the chloride and oxygen-18 vadose-zone profiles only detected the diffuse flow component. The recharge rates calculated with the two soil water models were very similar. The reliability of the recharge estimates was evaluated by comparing the independent outcomes. The short-term recharge estimates from the soil water models agree reasonably with those from the micro-meteorological measurements. For the long term, the agreement between the soil water models and the saturated-zone chloride mass balance was good.

1 INTRODUCTION

This paper describes a process-oriented study of soil water movement through the unsaturated zone. The study focuses on quantification of the groundwater recharge rate at the plot scale. Since the mid-1980s there has been an abundance of recharge studies, most of which have focused on semi-arid and arid conditions. There have been fewer studies, however, under more temperate conditions (e.g. Johansson 1987, Meinardi 1994). Environmental tracers, especially chloride and stable isotopes, have found wide application in arid regions, but much less so in humid areas (e.g. Saxena 1987, Gehrels et al. 1998).

The most common methods for estimating precipitation recharge through the unsaturated zone include lysimeter studies, tritium profiling, soil moisture studies, the zero flux plane approach, the chloride profile method, and application of the geothermal gradient and stable isotopes (Bredenkamp et al. 1995). The results of the individual methods, however, often contain substantial uncertainties that are difficult to quantify. It is therefore preferable to apply different techniques based on independent input data.

In the present study three different methods were applied, (1) micro-meteorological estimation of actual evaporation, (2) tracer profiling with chloride and oxygen-18, and (3) soil water modelling with a numerical model and a reservoir model. The usefulness of these methods is investigated under the prevailing temperate climatic conditions in the Netherlands. The results are compared for a site with a natural grassland cover to evaluate the reliability of the associated recharge estimates.

2 METHODS

2.1 *Micrometeorological BREB method*

Direct recharge may be determined indirectly by assessing total actual evaporation (interception, transpiration and soil evaporation) at the soil surface. In the present study actual evaporation is estimated using the Bowen ratio energy balance (BREB) method using standard deviations of temperature and humidity (Tillman 1972, Bink 1996). This method uses the fast fluctuations (at the scale of seconds) of temperature and humidity as a measure for changes in vertical water (vapour) fluxes. The theory is not new, but the method has been only sparsely applied to periods as long as an entire season. The method requires a limited amount of variables to be measured and is therefore an interesting alternative for the determination of actual evaporation. Given the actual evaporation, percolation recharge is estimated with a water balance approach including soil water storage changes.

The Bowen ratio β [–] is defined as the ratio of the sensible heat flux H from a surface to the accompanying vertical latent heat flux λE:

$$\beta = \frac{H}{\lambda E} \quad (1)$$

The BREB method was originally developed as a method based on the vertical gradients of temperature and humidity. More recently, the method has been formulated in terms of the standard deviations of temperature and humidity (Tillman 1972, Bink 1996). In contrast to the method based on gradients, this method requires observations on a single level alone. The method is based on the measurement of the turbulent temperature fluctuations. Turbulent temperature fluctuations close to the surface are related to vertical transfer of sensible and latent heat. Heat is transported vertically through the exchange of eddies with different temperatures. This implies that a vertical heat flux will be accompanied by a local variation of temperature. An appropriate measure of the intensity of the temperature fluctuations is the standard deviation of the dry-bulb temperature σ_{T_d} :

$$H = \rho c_p \rho_{wT} \sigma_w \sigma_{T_d} \quad (2)$$

where σ_w is the standard deviation of the vertical wind component and ρ_{wT} is the transfer efficiency of sensible heat, i.e., a measure of the net transfer of sensible heat by turbulent motions. ρ is the density of air ($kg.m^{-3}$) and c_p the specific heat capacity of air at constant pressure ($J.kg^{-1}.K^{-1}$). In a similar way, the standard deviation of the specific humidity σ_q is related to the latent heat flux:

$$E = \rho \rho_{wq} \sigma_w \sigma_q \quad (3)$$

where ρ_{wq} is the transfer efficiency of latent heat. Hence, if the transfer efficiencies for sensible heat and for latent heat have equal magnitudes, β can be calculated from σ_{T_d} and σ_q (Bink 1996):

$$\beta = \pm \frac{c_p}{\lambda} \frac{\sigma_{T_d}}{\sigma_q} \quad (4)$$

σ_q can be obtained from the standard deviations of dry-bulb and wet-bulb temperature according to Vugts et al. (1993).

Actual evaporation is then calculated from the standard deviations and the Bowen ratio with Equations (1) and (4). Finally, the recharge rate is estimated by combining the obtained actual evaporation with a water balance approach according to

$$\frac{dS}{dt} = P - E_a - R \quad (5)$$

where R is recharge, P is precipitation, E_a the BREB actual evapotranspiration at 2 m and dS/dt the net change in soil water storage.

2.2 *Tracing soil water with chloride and oxygen-18*

Of the tracer techniques available, chloride balance techniques appear to be the simplest, least expensive, and most universal for recharge estimation (Allison et al. 1994). Under steady-state conditions of piston flow, the flux of Cl at the surface is equal to the flux of Cl below the active root zone, and conservation of mass leads to:

$$P \cdot \mathrm{Cl}_P = R \cdot \mathrm{Cl}_R \quad (6)$$

where P and R are mean annual precipitation and recharge ($mm.y^{-1}$), and Cl_P and Cl_R are mean Cl concentrations ($mmol.l^{-1}$) in precipitation and soil water, respectively. Provided recharge and tracer inputs are steady state and there is no secondary recharge mechanism, tracer concentration should increase with depth until a constant concentration is reached.

In principle, the stable isotopes oxygen-18 and deuterium are preferable as tracers, as they are incorporated in the water molecules. In this study, the distinct seasonal character of oxygen-18 as present in precipitation is correlated with the cyclic pattern that is apparent from the observed soil water oxygen-18 profiles. In arid climates a clear cyclicity is usually absent, whereas a cyclic pattern is found to be more persistent under temperate conditions with higher recharge rates (e.g. Thoma et al. 1979, Saxena 1987).

A method is used here developed by Gehrels et al. (1998). The method translates the oxygen-18 ($\delta^{18}O$) precipitation record to a depth record using precipitation depth and soil water content to calculate the depth position of the isotope peaks. Depth d is written in terms of time t by introducing a mean soil water velocity $\overline{v}_{sw}$:

$$d(t) = (t - \varphi)\overline{v}_{sw} \quad (7)$$

where φ represents a phase shift. In addition to Equation (7), the translation from time to depth is rescaled by including the precipitation depth record and the soil water profile in the calculation of the depth, see Gehrels et al. (1998). The $\delta^{18}O$-precipitation curve is rescaled to a $\delta^{18}O$-depth curve by calculating the infiltration depth per time step from the precipitation amount and the soil water content. In addition, an exponential dampening is introduced to account for the dispersive attenuation of the isotopic fluctuations.

In this way, the $\delta^{18}O$ depth curve is calculated from the $\delta^{18}O$ precipitation record using precipitation depth, soil water content and introducing a dampening. The mean soil water velocity $\overline{v}_{sw}$ is estimated by minimizing the differences between the calculated and observed oxygen-18 profile using a least-squares optimization algorithm. Recharge is then calculated from the obtained $\overline{v}_{sw}$ and observed mean soil water content $\overline{\theta}$ as

$$\overline{R} = \overline{v}_{sw} \cdot \overline{\theta} \quad (8)$$

2.3 *Measuring and modelling soil water flow*

Groundwater recharge and water table fluctuations are modelled via two different approaches. The first approach uses a mechanistic physically-based numerical model (SWAP) based on the Richards equation, and the second approach applies a conceptual reservoir-type model (EARTH) based on simplified water balance equations. The simulations of soil water movement with both the SWAP and EARTH models were calibrated with soil water measurements. Soil water content was monitored with a capacitance probe (Bell et al. 1987), so that the water content could be determined at different depth intervals down to several metres below surface.

SWAP (Soil-Water-Atmosphere-Plant simulation model) is a one-dimensional soil physical finite-difference numerical model describing water flow in unsaturated or partly saturated soils based on the Richards equation, see Van Dam et al. (1997).

The reservoir model (EARTH) is a modified version of the EARTH model described by Van der Lee & Gehrels (1997). EARTH is a lumped parametric model for the simulation of soil water content, actual evapotranspiration, percolation and groundwater level fluctuations. The unsaturated zone is schematically represented by a number of subsequent modules, or reservoirs, see Gehrels (1999).

For both models, the boundary conditions at the soil surface are formulated in terms of rainfall, and interception, transpiration and soil evaporation. In the present study the interception loss is modelled with the analytical model according to Gash (1979). Potential evaporation is calculated using a reference evaporation (according to the Makkink formula, as is usual in the Netherlands), crop factors and soil cover fractions.

3 STUDY AREA

The study area, called the Veluwe area, is a sparsely populated natural area located in the central part of the Netherlands, see Gehrels (1999). A low-altitude continuous natural area of such size (~1250 km^2) is unique for the Netherlands, and even for the densely populated NW part of Europe.

The climate in the Netherlands can be classified as maritime. Annual precipitation in the central Veluwe is ~880 mm. The average annual reference evaporation according to Makkink is ~580 mm.y^{-1} at the main meteorological (KNMI) station of De Bilt.

The Veluwe area (up to ~100 m above mean sea level) forms a large infiltration area recharging a phreatic aquifer system. The Veluwe is for the larger part covered with forest, with a smaller share of heathland and an increasing share of grassland. In the central Veluwe area, the soils are generally podzolic soil profiles in poor sands. The subsurface can be characterized as subhorizontally bedded coarse fluvial sands and gravel, overlain by finer, well-sorted aeolian cover sand.

This paper concentrates on one site called 'Radio Kootwijk'. At the location of this site, the terrain is practically flat. The water table is at a depth of ~20 m below ground surface. The dominating vegetation type at Radio Kootwijk is purple moor grass (*Molinia caerulea*) and a decreasing fraction of heather (*Calluna vulgaris*).

4 FIELD MEASUREMENTS AND RECHARGE ESTIMATION

4.1 *BREB method*

The wet-bulb and dry-bulb temperatures were measured at a height of 2 m above ground surface at intervals of 2 s over a full year. From these measurements, the wet-bulb and dry-bulb standard deviations were calculated as well as the correlation coefficients between the wet-bulb and dry-bulb temperatures.

From these measurements, actual evaporation was calculated using Equations 2–4, see Figure 1. Precipitation rates were obtained by linear interpolation of the records of the surrounding meteorological (KNMI) stations. dS/dt was calculated from the soil moisture capacitance probe measurements carried out at the Radio Kootwijk site. The overall soil water storage change over 1994 was small: −18 mm over the 3 m profile of the capacitance probe access tube. The estimated percolation recharge was 656 mm.y^{-1} for 1994.

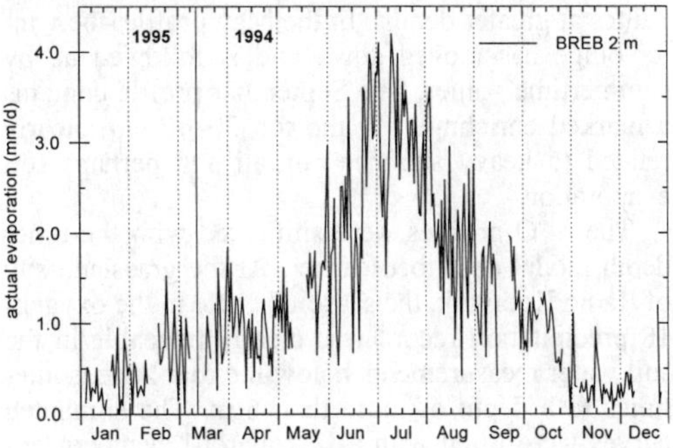

Figure 1. Daily BREB actual evaporation from 18 March 1994 (day 67) until 8 March 1995 (day 57). (*To clearly depict the seasonal cycle, the first 57 days of 1995 are shown on the left.*)

4.2 *Chloride and oxygen-18*

Chloride profiles were drilled bimonthly from January 1994 until January 1995. The concentration profiles generally showed a consistent pattern throughout the year. The top ~1 m is characterized by highly variable concentrations. The Cl content sharply drops to clearly lower values below ~1 m and remains fairly constant with depth. Below the effective root zone, the chloride profiles do not show distinct peak values or a distinct downward movement in time of a particular profile section. The relatively low Cl content below the roots is explained by mixing of diffuse flow with a preferential flow component of low concentrations that bypasses the bulk of the soil matrix (Gehrels 1999).

Recharge was calculated from the chloride profiles according to Equation (6). The results show a large variation in estimated recharge rates over the year, ranging from ~300 to ~700 mm.y^{-1}. The figures show a fairly consistent seasonal pattern. To arrive at a recharge estimate not biased by seasonal fluctuations, the average recharge rate was calculated over six profiles to obtain an average over a two-year period (as deduced from the calculated travel times). Estimated average (1993–1995) recharge rate at Radio Kootwijk was 510 mm.y^{-1}. (The associated actual evaporation for *Molinia* can then be estimated as $E_a = P_G - R_p = 550$ mm.y^{-1}.)

In addition, a long-term recharge estimate was deduced from the average chloride concentration in precipitation (0.1 mmol.l^{-1}), groundwater (0.4 mmol.l^{-1}) and the areal average, annual precipitation depth (ca. 900 mm.y^{-1}). Above forests (40% of the area), the ratio of total deposition to wet deposition was found to be ca. 2.5 (Gehrels, 1999). This leads to a groundwater recharge estimate of 360 mm.y^{-1}.

Oxygen-18 profiles were determined bimonthly from the same soil water samples as the chloride profiles. Figure 2 shows three examples of the observed $\delta^{18}O$ profiles. The profile drilled in January shows a clearly depleted (winter) profile in the first 2–3 m, quite sharply switching to heavier (summer) values at greater depths. In the May profile, the winter bulge has moved down and is followed up by summertime values. The September profile contains a marked enrichment in the top 30–40 cm, clearly related to heavy summer rainfall and perhaps soil evaporation.

The $\delta^{18}O$ profiles were simulated with the time-depth model described earlier. At the grassland site of Radio Kootwijk, the seasonal cycle in the oxygen-18 precipitation record was clearly traceable in the soil water measurements below the root zone, sometimes to a depth of more than 6 m. The simulated curves were fitted with a Marquardt-Levenberg optimization routine (Marquardt, 1963), resulting in an estimated mean soil water velocity (Fig. 2). From the combination of this soil water velocity and the observed average soil water content, the recharge rate is calculated using Equation (8).

Seasonally independent recharge rates were obtained from the average over the six oxygen-18 profiles giving a two-year (1993–1995) average value of 459 mm.y^{-1}. (The associated actual evaporation for *Molinia* can then be estimated as $E_a = P_G - R_p = 601$ mm.y^{-1}.)

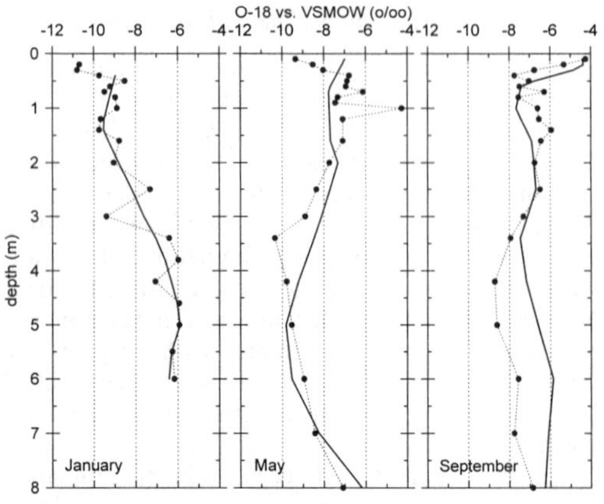

Figure 2. Observed oxygen-18 profiles (*symbols*) drilled at Radio Kootwijk in 1994 with simulated time-depth model (*solid*).

4.3 *Measuring and modelling soil water flow*

Input data of daily precipitation were obtained as described earlier. Daily reference evaporation according to Makkink was calculated from the mean daily air temperature and the daily sum of global radiation measured at the meteorological (KNMI) station of De Bilt. Potential evaporation was obtained as the reference evaporation multiplied by a crop factor. Gash interception parameter values were obtained from the literature. Soil physical (Van Genuchten) parameters were determined from samples taken at five different depths in the soil profile.

Soil water content was monitored throughout

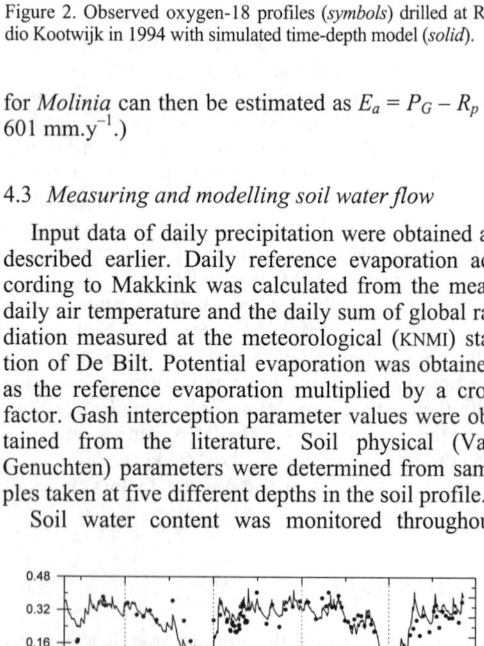

Figure 3. Observed (*symbols*) and simulated (*solid line*) soil water content (cm^3.cm^{-3}) at fixed depths for the Radio Kootwijk E4 site from September 1992 to December 1994. Simulation with SWAP. (*Note the different y-scales.*)

1993 and 1994, mostly at weekly intervals. Figure 3 shows time series of volumetric soil water content for a number of fixed depth intervals measured with the capacitance probe and simulated with SWAP.

The calculations with SWAP and EARTH were carried out on a daily basis. Equal upper and lower model boundaries were taken for both models, in this way being able to compare the results more clearly. With both models groundwater recharge was estimated over the period of 1955–1995. Calculated recharge over 1994 was 602 mm for SWAP and 591 mm for EARTH. The calculated water use of *Molinia* or actual evaporation E_a is around 480–500 mm.y^{-1}.

5 COMPARISON OF RESULTS

Table 1 compares the results obtained for *Molinia* grassland with the three independent methods.

The BREB method yielded estimates of actual evaporation that were comparable to figures for *Molinia* grassland reported in the literature (e.g. Schouwenaars 1993). The method was relatively easy to apply and the results were useful. However, as the measurement period was short (as is usually the case — typically one season), the recharge estimate should not be based on micro-meteorological measurements alone, but also on the monitoring of soil water content to account for storage changes, as was done in this study.

The SWAP and EARTH water balance estimates are practically identical. The annual evaporation totals were found to be markedly constant over the years, while the annual recharge totals were as variable as the precipitation totals. The total evaporation calculated with the models (i.e. actual plus interception evaporation) is ca. 50 mm.y^{-1} higher than the actual evaporation obtained from the BREB method (which includes interception as well). A difference of ~10% is not unrealistic, considering the combined range of (unknown) uncertainty of both methods. It is likely that a small overestimation is introduced via the parameters of the upper boundary in the water balance models, for example, via the crop factor or in the estimated interception rate of 50–60 mm.y^{-1}.

The tracer methods provide clearly higher estimates for E_a than the soil water models and BREB actual evaporation (Table 1). As a result of storage changes, the differences in recharge are slightly smaller than those for E_a. The difference with SWAP and EARTH is 40–60 mm.y^{-1} for chloride and 90–110 mm.y^{-1} for oxygen-18. The most plausible explanation for the differences in recharge is that with the tracer methods only diffuse recharge is detected and a possible preferential flow component is ignored.

A primary difference between chloride and oxygen-18 is that the chloride balance is based on measurements taken only below the root zone, whereas the oxygen-18 time-depth profile is fitted to measurements that include the root zone as well. As concluded earlier, in the root zone preferential flow is dominant and the diffuse flow component is probably small. The oxygen-18 estimate can therefore be regarded as a weighted average of root-zone and percolation-zone diffuse recharge components, whereas the chloride estimate represents the diffuse component in the percolation zone.

Based on the difference between chloride profiling and the water balance models, a difference between total percolation and diffuse percolation exists of ca. 10% of total percolation. This is in line with the conclusion that diffuse flow is dominant in the percolation zone. A tentative result is then that preferential flow below the root zone comprises of ca. 40–60 mm.y^{-1} or 4–6% of gross precipitation.

Table 1 furthermore shows that the long-term results for the soil water models and chloride are much better in line. The 'long-term' chloride estimate was deduced from saturated groundwater values and is compared with values over 1955–1995. The chloride values deduced from groundwater concentrations include both preferential and diffuse flow components through the unsaturated zone.

6 DISCUSSION AND CONCLUSIONS

Three principally different methods were used to estimate groundwater recharge: (1) micro-meteorological estimation of actual evaporation, (2) tracer techniques and (3) soil water flow models. All three methods suffer from uncertainties that are practically impossible to quantify. The reliability of the recharge estimates was therefore judged by comparing the independent outcomes.

The Bowen ratio energy balance (BREB) method

Table 1. Comparison of water balance results (mm.y^{-1}) obtained for *Molinia* at Radio Kootwijk from the methods treated in this paper. The percolation rate R_p deduced from the BREB method was calculated as $P_G - E_a - dS/dt$, where P_G is gross precipitation and E_a actual evaporation. For the tracer methods, E_a was calculated as $P_G - R_p$. Total evaporation for SWAP and EARTH is indicated as actual plus interception evaporation $E_a + E_i$. For chloride, the short-term (93–95) figures are obtained from unsaturated-zone profiling while the long-term (55–95) values result from observed saturated groundwater concentrations.

Meas. period		Bowen Ratio energy balance		Chloride mass balance		Oxygen-18 time-depth		SWAP2 soil water flow		EARTH lumped reservoir	
	P_G	E_a	$P_G - E_a - dS/dt$	$P_G - R_p$	R_p	$P_G - R_p$	R_p	$E_a + E_i$	R_p	$E_a + E_i$	R_p
1994	1072	434	656					482	602	481	591
1993–95	1060			550	510	601	459	487	551	486	574
1955–95	872			512	360			497	380	486	387

using standard deviations was relatively easy to apply and resulted in estimates of actual evaporation comparable to literature figures for *Molinia* grassland. However, the results are only available over short time spans.

Chloride was applied in both the unsaturated and saturated zones. The results for the saturated zone agreed well with the water balance estimates from the 1D soil water models. The chloride profiles in the unsaturated zone only detect the diffuse flow component. Hence, in cases where preferential flow occurs in the percolation zone (below the roots), recharge is underestimated. For shallow aquifers, this problem can easily be overcome by comparing estimates from the unsaturated and saturated zones. A problem arises in cases with deep water tables, such as in this study, as the figures obtained from the unsaturated and saturated zones are then incomparable because of the different time scales associated with both zones. The method should therefore be applied in combination with other methods in which case it provides additional information in the same way as it does for shallow aquifers.

Estimation of the downward percolation flux with the tracer of oxygen-18 was based on the seasonal cyclicity that is preserved in the unsaturated zone under the prevailing temperate conditions. However, its application as a tool for recharge estimation proved to be limited. This is mainly because application of oxygen-18 assumes diffuse infiltration in both the root and the percolation zones, as a result of which the oxygen-18 estimates were found lower than those from chloride.

The short-term and long-term recharge rates that were calculated with the SWAP numerical model and the EARTH reservoir model appeared to be very similar. The calculated recharge rates from the SWAP and EARTH models are obviously not independent, as both share the same top and bottom boundary conditions. The close agreement therefore indicates that, for conditions such as in the Veluwe, a fairly simple reservoir model such as EARTH is equally suitable to estimate recharge as the more complex SWAP model. The conclusion is that the recharge estimation is largely determined by the top boundary conditions and less by the more sophisticated root water uptake scheme included in SWAP.

The present study shows that only a combination of methods can result in reliable recharge estimation. The short-term recharge estimates (for 1994) from the SWAP and EARTH models agree reasonably with the results from the micro-meteorological measurements. For the long term, the agreement with the results from the chloride mass balance for the saturated zone is good.

REFERENCES

Allison, G.B., Gee, G.W. & Tyler, S.W. 1994. Vadose-zone techniques for estimating groundwater recharge in arid and semiarid regions. *Soil Sci. Soc. Am. J.* 58: 6–14.

Bell, J.P., Dean, T.J. & Hodnett, M.G. 1987. Soil moisture measurement by an improved capacitance technique, part II. Field Techniques, evaluation and calibration. *J. Hydrol.* 93: 79–90.

Bink, N.J. 1996. *The structure of the atmospheric surface layer subject to local advection.* Ph.D. thesis, Wageningen: Landbouwuniversiteit Wageningen.

Bredenkamp, D.B., Botha, L.J., van Tonder, G.J. & van Rensburg, H.J. 1995. *Manual on quantitative estimation of groundwater recharge and aquifer storativity.* Pretoria: Water Research Commission.

Gash, J.H.C. 1979. An analytical model of rainfall interception by forests. *Quart. J. R. Met. Soc.* 105: 43–55.

Gehrels, J.C. 1999. *Groundwater level fluctuations; separation of natural from anthropogenic influences and determination of groundwater recharge in the Veluwe area, the Netherlands.* Ph.D. thesis, Amsterdam: Vrije Universiteit Amsterdam.

Gehrels, J.C., Peeters, J.E.M., De Vries, J.J. & Dekkers, M. 1998. The mechanism of soil water movement as inferred from 18O stable isotope studies. *Hydrological Sciences Journal* 43(4): 579-594.

Johansson, P-O. 1987. *Methods for estimation of direct natural groundwater recharge in humid climates — with examples from sandy till aquifers in southeastern Sweden.* Ph.D. thesis, Stockholm: Royal Inst. Techn..

Marquardt, D. 1963. An algorithm for least-squares estimation of nonlinear parameters. *SIAM J. Appl. Math.*, Vol. 11: 431–441.

Meinardi, C.R. 1994. *Groundwater recharge and travel times in the sandy regions of the Netherlands.* Ph.D. thesis, Amsterdam: Vrije Universiteit Amsterdam.

Saxena, R.K. 1987. *Oxygen-18 fractionation in nature and estimation of groundwater recharge.* Ph.D. thesis, Uppsala: Uppsala Universitet.

Schouwenaars, J.M. 1993. De verdamping van pijpestrootje (*Molinia caerulea*) en veenmos (*Sphagnum papillosum*) in hoogveengebieden en haar betekenis voor het waterbeheer. *H_2O* 26(14): 376–382.

Thoma, G., Esser, N., Sonntag, C., Weiss, W., Rudolph, J. & Leveque, P. 1979. New technique of in-situ soil-moisture sampling for environmental isotope analysis applied at Pilat sand dune near Bordeaux. In: *Isotope Hydrology 1978*, Proc. Symp. Neuherberg, IAEA / UNESCO: 753–766.

Tillman, J.E. 1972. The indirect determination of stability, heat and momentum fluxes in the atmospheric boundary layer from simple scalar variables during dry unstable conditions. *J. Appl. Meteorol.* 11: 783–792.

Van Dam, J.C., Huygen, J., Wesseling, J.G., Feddes, R.A., Kabat, P., van Walsum, P.E.V., Groenendijk, P. & Van Diepen, C.A. 1997. *SWAP version 2.0, theory; simulation of water flow, solute transport and plant growth in the Soil-Water-Atmosphere-Plant environment.* Wageningen: DLO Winand Staring Centre.

Van der Lee, J. & Gehrels, J.C. 1997. Modelling of groundwater recharge of a fractured dolomite aquifer under semi-arid conditions. In: I. Simmers (ed.), Recharge of phreatic aquifers in (semi-) arid areas, *UNESCO / IAH Int. Contr. to Hydrogeology*, Vol.19: 129–144. Rotterdam: Balkema.

Vugts, H.F., Waterloo, M.J., Beekman, F.J., Frumau, K.F.A. & Bruijnzeel, L.A. 1993. The temperature variance method: a powerful tool in the estimation of actual evaporation rates. *IAHS Publ.* 216: 251–260.

Groundwater: Past Achievements and Future Challenges, Sililo et al. (eds) © 2000 Balkema, Rotterdam, ISBN 90 5809 159 7

VLF method applied to hydrogeological prospecting in the gneissic-migmatitic hard rock aquifer of Évora, South Portugal

A.Gomes & A.Chambel
Departamento de Geociências, Universidade de Évora, Portugal

J.Duque
Centro da Geologia da Faculdade de Ciências, Universidade de Lisboa, Portugal

ABSTRACT: The gneissic-migmatitic aquifer of Évora is an important hydrogeological system, defined in gneisses and some associated migmatitic and granodioritic rocks of Alentejo (South Portugal). The aquifer has an upper layer of 10 to 40 m of weathered rock that has transformed the primitive fractured aquifer into a porous medium. The rocks involved have 1 to 6 metres of the same type of weathering, in more slopy areas. The geophysical method of Very Low Frequency (VLF) profiling was performed in a small area (Álamo do Gavião Farm), in the eastern part of the hydrogeological system, in order to identify areas of high hydrogeological potential. The geophysical prospecting and hydrogeological research allowed the drilling of 13 boreholes, being 6 extraction wells and 7 piezometric wells. The productivity of the abstraction wells are between 4000 L/h and 20,000 L/h, high value ranking when compared with the regional yield.

1 INTRODUCTION

Alentejo is mainly a region of metamorphic and igneous Precambrian and Palaeozoic rocks, affected by the Hercynian orogeny, with special characteristics regarding to groundwater storage, flow, quality and yield.

The prospecting methods related to flowpath patterns and drilling methods are always improving in this type of hard rocks. This geological environment has a high fracturing network and flow random, which is always posing new challenges to hydrogeology. The VLF-EM (Very Low Frequency - Electromagnetic Method) is helpful to identify fractured deep structures in hard rock aquifers (Griffiths & King 1972, Astier 1975, Parasnis 1986, Milsom 1989). The main goal of this paper is to report on the application of VLF-EM in the Álamo do Gavião Farm area of the gneissic-migmatitic aquifer of Évora, in South Portugal.

2 GEOLOGICAL, GEOMORPHOLOGIC AND CLIMATIC SETTING

The gneissic-migmatitic aquifer of Évora is an important regional water storage, with a surface area of 226 km^2.

This region is placed in the middle of a domain called the Ossa Morena Zone (OMZ). This aquifer occurs in gneisses, migmatites, granodiorites and quartzodiorites with a weathered and highly fractured layer of 10 to 40 m deep, and a less fractured system under the 40 m depth.. The aquifer is surrounded by more impermeable rocks, like schists and other quartzodiorites, which have less than 6 m of the same type of weathering, in more slopy areas.

It occurs in a very flat area, which is very typical in the Alentejo region, and is near a divergence point between three main river basins: Tejo, (N) Guadiana (SE) and Sado (SW).

Geomorphologically, the Álamo do Gavião Farm is linked to the regional tectonics reflected in the field by a great fracture line which is coincident with the Degebe river, affecting the topography and drainage network.

The region is influenced by a Mediterranean climate, which is typical in the inner regions, with high temperature ranges (high temperatures in Summer and a very dry period between June and September).

The regional mean precipitation for a forty-year period for Alentejo is about 600 mm/year. However, only 4 to 5% of the total annual precipitation occurs in summer (Chambel et al. 1998).

Droughts are common in summer. The continental area of Alentejo is a typical semi-arid region with many desertification problems.

3 HYDROGEOLOGY

The gneissic-migmatitic aquifer system of Évora is a free type aquifer, highly permeable to contamination. The main groundwater flow direction is N-S, following the Albardão stream direction.

The interpretation of aerial photography shows that the possible fractures may be associated to four main directions: NE – SW, NW – SE, NNW– SSE, E – W.

The most important lineation correspond to NNW-SSE and NE-SW directions. The first ones are aligned with the Albardão stream, which cross the propriety. The possible fractures of the second direction seems to have less hydrogeological importance than the NNW-SSE direction. The intersection of the alignments can be of great importance to the investigation.

The aim of this research is to identify areas of high hydrogeological potential. The aim was to detect the most deeply weathered areas, which can reach 10 to 40 m depth in topographically depressed areas. This information was correlated to the interpretation of the possible deep fractures detected in the aerial photography.

4 VLF-EM PROSPECTION

The VLF-EM can be used to find steeply dipping structures that differ from their surroundings with regard to electrical resistance. This method is thus very well suited to water prospecting in fracture zones.

Detection is based on the secondary fields created by electrically conductive bodies in the ground excited by an electromagnetic primary field. This low frequency field is sent out from military radio transmitters, working generally in the frequencies between 15 and 30 kHz (Telford et al. 1976).

The radio waves can pass through the rocks. The depth of penetration (δ) depends on the transmitter frequency (f) and the resistivity (ρ) of the rock (Paterson & Ronka 1971):

$$\delta = 500 * \sqrt{(\rho / f)} \quad [\text{m}] \qquad (1)$$

Only a body having low electrical resistance can create a secondary field. The magnitude of the secondary field is normally obtained by comparing the signals in the horizontal and vertical directions.

Since the primary field is horizontal the normal reading will be zero. The reading will be zero even in the presence of a horizontal layer of material having low resistivity, such as clay or salt water.

Since the primary field is horizontal, it's possible to observe the amplitude and the phase of the secondary field relative to the primary. By this process it is possible to determine the position of the more conductive bodies in the surrounding rocks, also their dip and depth.

If a fracture has water, the system will behave as a conductive body. The fracture itself will have a response different from the surrounding rocks.

The software acquired from ABEM allows the interpretation of the profiles. For each group of profiles interpretative graphics were created. The first, here designated by A, puts side by side all the parallel profiles done in the same area. The anomalous zones appear as filled peaks. A sequence of anomalies in the various profiles indicates the presence of a geological structure, which might have hydrogeological value. It should always be confirmed in the field. Another type of graphics (B) involves, each time, only one of the profiles and it represents the actual data.

The graphics with the representative profiles of the filtered data are created in the same way. These graphics, designated by C in this text, are made up of two curves, corresponding to the real and imaginary components of the electromagnetic field.

The fourth type of graphics, here designated as D, represents each profile as a vertical cross-section. The darkened zones will correspond theoretically to anomalies of geological type, mainly when the patterns are asymmetrical.

4.1 Case study

Five groups of VLF profiles were observed in several places. They are marked as grid 1, 2, 3, 4 and 5 (Fig. 1). Locations were chosen to confirm in the field the existence of possible fractures identified on the aerial photography, but not evident in the topographic features.

Grid 3 is the profiles 1300 and 1320. The graphic of type A (Fig. 2) demonstrates the occurrence of several anomalies in both profiles, however the one which is more favorable corresponds to profile 1300.

The display of graphic type C leads to the identification of this anomaly at 95 m (Fig. 3).

The confirmation of this anomaly comes from graphic type D (Fig. 4). A borehole was drilled and the well (W3 in Figure1) reached 70 m depth with a yield (direct flow system) of 20,000 L/h. Another borehole was positioned on the same profile. The anomaly is located at 32 m from the start of profile 1300 (Figs 3 and 4), which corresponds to piezometer P6 (Fig. 1). This one has a instantaneous yield of 10,000 L/h.

A profile 1000, which is part of grid 1,an interesting anomaly was also detected at 150 m. This can be observed in the graphic of type C (Fig. 5).

This anomaly resembles a possible geological structure at depth (Fig. 6). The borehole was drilled to 52 m depth. This well (W2) has a yield (direct air flow system) of 20,000 L/h.

Another borehole was drilled at 65 m on this profile (piezometer P4 in Figure 1), corresponding to an

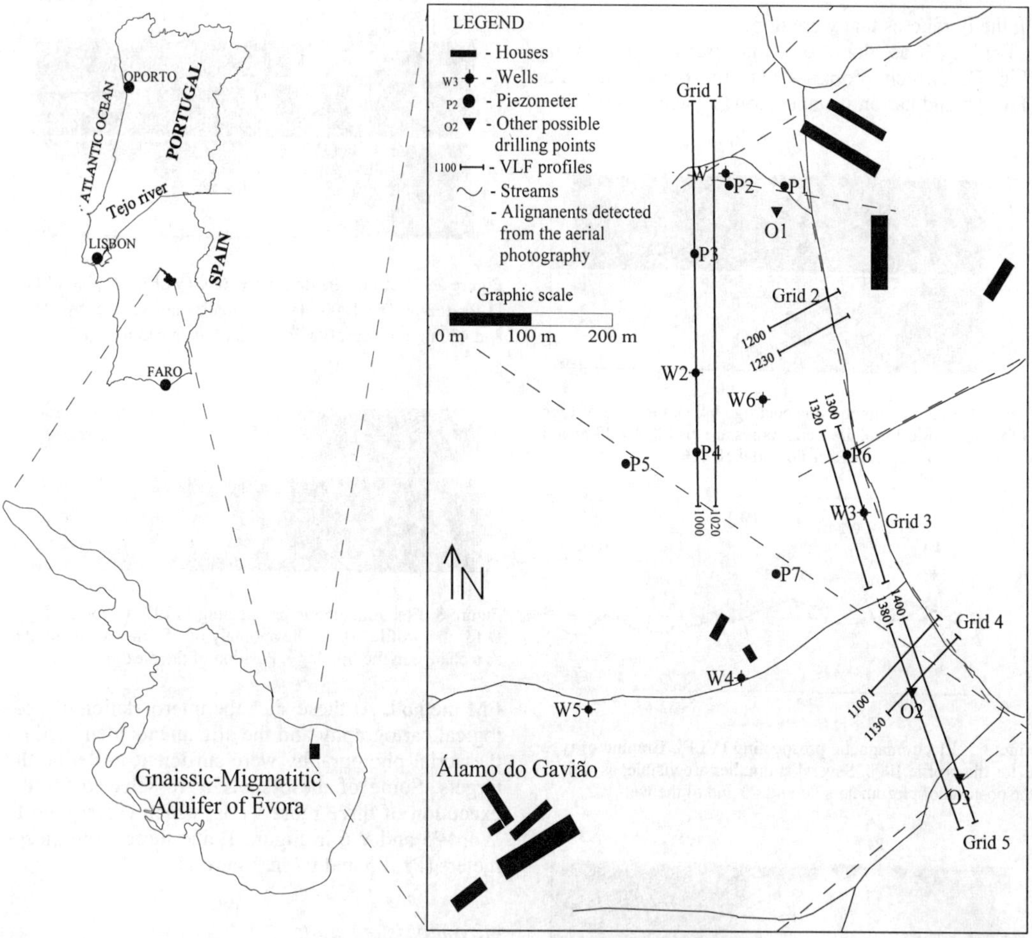

Figure 1. Localization of the study area. Also represented are the alignments defined by aerial photography, the VLF profiles, the piezometric and extraction wells and other possible drilling points in the same area. The profiles always begin in the point where are the profile numbers.

area where the geophysics shows a solid unfractured rock. The purpose was to confirm the geophysics interpretation.

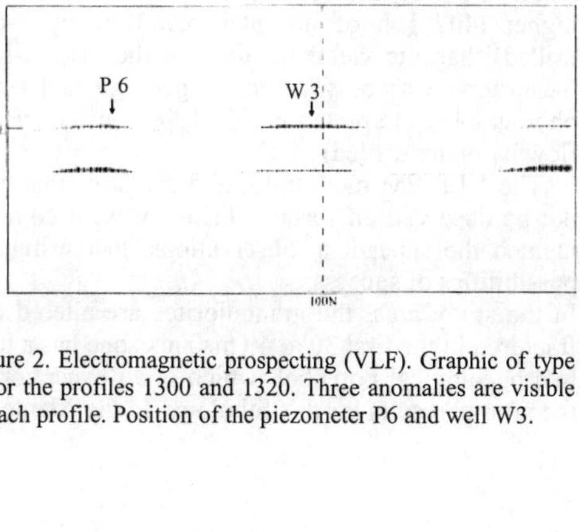

Figure 2. Electromagnetic prospecting (VLF). Graphic of type A for the profiles 1300 and 1320. Three anomalies are visible in each profile. Position of the piezometer P6 and well W3.

Figure 6 shows, a whitened zone starting at 10 m deep. This interpretation was confirmed by the results obtained in the drilling: after 10 m depth only hard rock was encountered

This piezometer has only got 3000 L/h of instantaneous flow and is clearly the most unproductive of

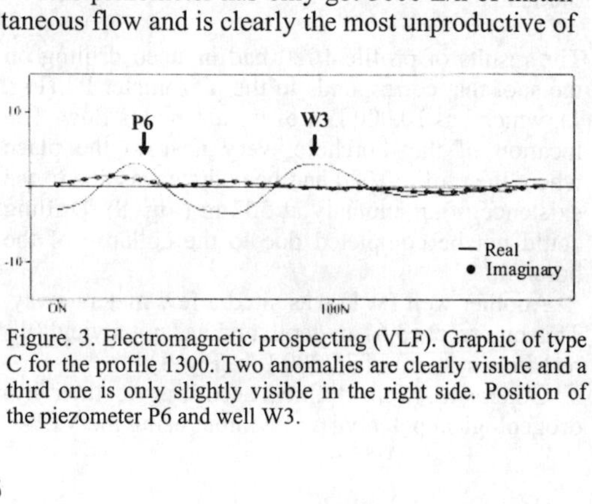

Figure. 3. Electromagnetic prospecting (VLF). Graphic of type C for the profile 1300. Two anomalies are clearly visible and a third one is only slightly visible in the right side. Position of the piezometer P6 and well W3.

all the boreholes that were done.

Drilling was done on this profile, at 278 m (Fig. 7), which corresponds to piezometer P3 (Fig. 1), and has an instantaneous flow of 5000 L/h.

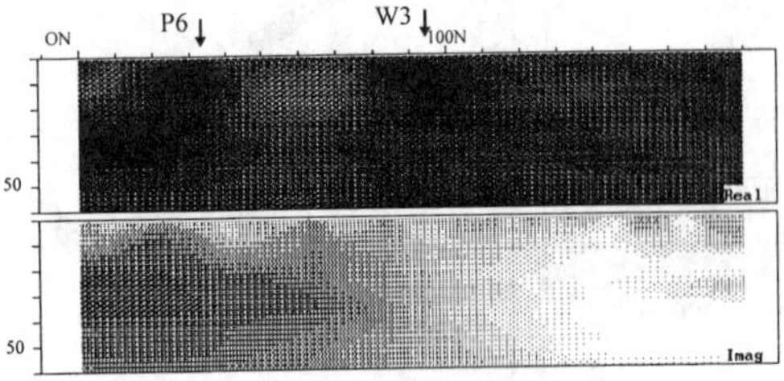

Figure 4. Electromagnetic prospecting (VLF). Graphic of type D for the profile 1300. Two anomalies are visible, by 30 m and 95 m. Position of piezometer P6 and well W6.

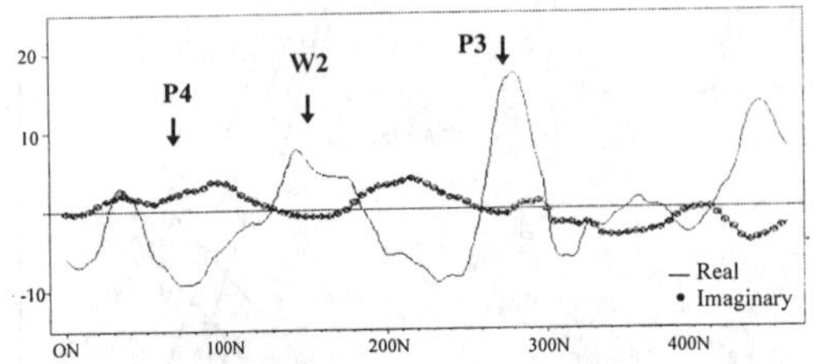

Figure 5. Electromagnetic prospecting (VLF). Graphic of type C for the profile 1000. Several anomalies are visible, as well as the position of piezometers P4 and P3 and of the well W2.

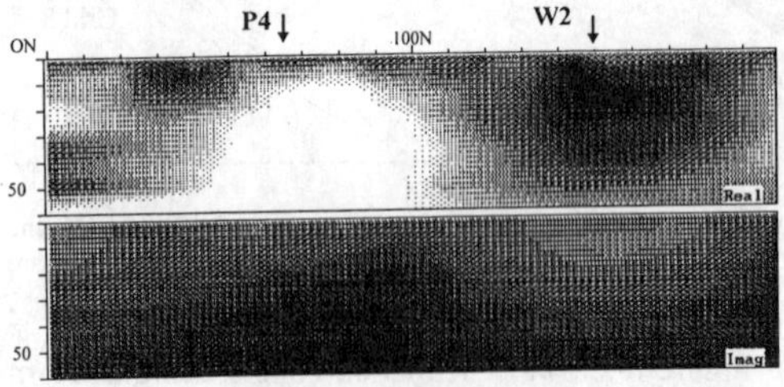

Figure 6. Electromagnetic prospecting (VLF). Graphic of type D for the profile 1000. Position of the well W2 (anomaly) and of piezometer P4, deliberately executed in a possible compact structure.

The results of profile 1020 had induced drilling on the spot that corresponds to the piezometer P2 (Fig. 1), which has 10,000 L/h of instantaneous flow. The location of this borehole, very near to the place where the profile 1020 had been done was due to the existence of an anomaly at 352 m (Fig. 8). Drilling could not be completed due to the collapse of the borehole.

Another well (W1) was sited a few meters away. This one reached 61 m depth and has got a yield (direct flow system) of 36,000 L/h (Fig. 1).

Other sites were defined according to their hydrogeological potential and without using the VLF-

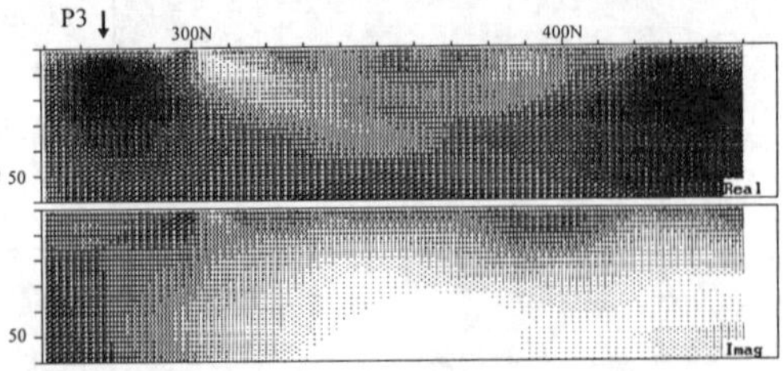

Figure 7. Electromagnetic prospecting (VLF). Graphic of type D for the profile 1000. Two anomalies are visible, by 280 m and 430 m in this profile. Position of the piezometer P3.

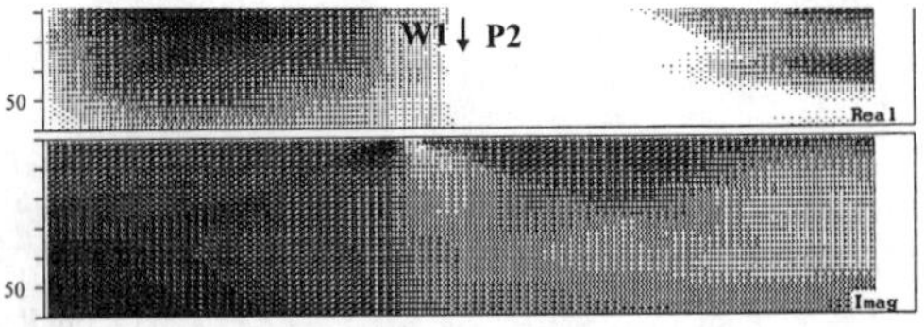

Figure 8. Electromagnetic prospecting (VLF). Graphic of type D for the profile 1020. The anomaly of 350 m was interpreted as a change in the lithology. Position of the piezometer P2.

EM method. At these sites the interpretation of geological cartography and the alignments identified on the aerial photography were sufficient to define the targets. Some of these spots were selected for the execution of three more wells, which correspond to W4, W5 and W6 in Figure 1, and three more piezometers (P1, P5 and P7 in Figure 1).

4.2 Analysis of results

The yields obtained in these wells are medium or above medium in terms of productivity in this lithology (Table 1). The instantaneous flows have to be seen as indicative of the productivity of these wells and not as abstraction values. The latter were reduced, in terms of abstraction, to 30 to 50%, in order to preserve the resource and the wells (Table 1).

The yields obtained per meter drilled in wells where the prospecting with VLF-EM was used are higher (407 L/h of instantaneous flow by meter drilled) than the yields obtained in the wells where the criteria was only based on geology and aerial photography interpretation (310 L/h of instantaneous flow by meter drilled).

The VLF-EM method located fractures that cannot be observed on surface. In this way, it complemented the superficial observations, increasing the possibilities of success.

In the study area, the granodiorites are altered and fractured in the first 30 to 40 m. This zone has a high hydrogeological potential, which is an exception to the normal results for this lithology. In the Alentejo

Table 1. Data related to the wells and piezometers executed in Álamo do Gavião Farm.

Well (W) / Piezometer (P)	Deepness of the wells (m)	Instantaneous yield (L/h)	Abstraction recommended yields (L/h)		Use of VLF-EM method
			30% of Inst. Yield	50% of Inst. Yield	
W1	61	36,000	10,800	18,000	Yes
W2	52	20,000	6000	10,000	Yes
W3	70	20,000	6000	10,000	Yes
W4	55	40,000	12,000	20,000	No
W5	79	21,600	6480	10,800	No
W6	55	12,000	3600	6000	No
P1	70	12,000	-	-	No
P2	15	10,000	-	-	Yes
P3	19	5000	-	-	Yes
P5	13	7000	-	-	No
P6	31	10,000	-	-	Yes
P7	52	8000	-	-	No
P4	19	3000	-	-	Yes, but not in an anomaly

region, the representative yields in igneous and metamorphic rocks are near 3000 L/h. The VLF-EM method, when used in hydrogeology, improves the knowledge of a region, especially when soils cover the structures. This equipment has a permanent source of energy and is accessible in any parts of the world. It emphasises the existence of heterogeneities like fractures and lithological contacts that cannot easily be identified by superficial cartography. It is also quick and easy to use. Its limitations to hydrogeology are the depth of penetration, which is small when the underground is very conductive, and its tendency to be highly sensitive to metallic objects e.g. fences and electric or telephonic cables, as well as to other electromagnetic sources, like thunderstorms.

5 CONCLUSIONS

The VLF-EM is a method that is simple to use, but must be used carefully, due to the multitude of answers that can be achieved; the specialist must know the geology and structural history of the region very well to interpret the results correctly.

It has given satisfactory answers to problems such as the detection of fractured structures under cover.

The practical results of drilling prospecting were good in terms of yields, even when compared with the boreholes drilled and the basis of visible structural features. Even better when considering that the places prospected with geophysics were the ones where the structures were not easily recognizable.

REFERENCES

Astier, J.L. 1975. *Geofisica Aplicada a la Hidrogeologia*. Madrid: Paraninfo.

Chambel, A. Duque, J. & Fialho, A. 1998. Groundwater in a semi arid area of south Portugal. In J. Brahana, Y. Eckstein, L. Ongley, R. Schneider, & J. Moore (eds.) Gambling with Groundwater-Physical, Chemical and Biological Aspects of Stream Relations - Proc. XXVIII Congress of IAH, Las Vegas, 5-9 September 1998: 75-80. Las Vegas: IAH.

Griffiths, D.H. & King, R.F. 1972. *Geofisica Aplicada para Ingenieros y Geólogos*. Madrid: Paraninfo.

Milsom, J. 1989. *Field Geophyics*. Chichester: John Wiley & Sons.

Parasnis, D.S. 1986. *Principles of Applied Geophysics*. London: Chapman & Hall.

Paterson, N. & Ronka, V. 1971. Five years of surveying with the VLF - EM method. *Geoexploration* 9: 7-26.

Telford, W.M., Geldart, L.P., Sheriff, R.E. & Keys, D.A. 1976. *Applied Geophyics*. Cambridge: University Press.

Groundwater: Past Achievements and Future Challenges, Sililo et al. (eds) © 2000 Balkema, Rotterdam, ISBN 90 5809 159 7

Hydraulic seepage within an astatic karst lake, North-Central Florida

J.D. Hirsch
Water and Air Research Incorporated, Gainesville, Fla., USA

A.F. Randazzo
Department of Geological Sciences, University of Florida, Gainesville, Fla., USA

ABSTRACT: This paper presents the results of hydraulic seepage testing performed within an astatic karst lake located within North-Central Florida to determine how hydrostratigraphy influences and/or controls the occurrence and magnitude of lake bottom seepage. Seepage meter results identified seepage outflow at all seepage meter locations except one, and determined that seepage magnitude and occurrence varied significantly throughout the lake. The highest seepage outflows were recorded within shallow littoral sections of the lake. In comparison, low and no seepage outflows were recorded in deep areas of the lake where lake bottom confining units had been breached by karst processes. Excavation seepage meter results, laboratory permeameter testing, and grain size analysis determined that impermeable organic muck layers, less than 60 cm in thickness, can restrict and/or prohibit seepage out of a lake. These occur in areas where the underlying intermediate confining layer has been breached and seepage outflow is expected.

1 INTRODUCTION

Scientists and regulatory agencies rely upon cost effective and accurate methods of quantifying and characterizing seepage flow in karst lake basins. Such information aids in determining the feasibility of augmenting lake stage levels during periods of drought by increasing inflow and/or decreasing outflow. Seepage measurement methods also are needed to aid scientists and planners in predicting and preventing the transfer of contaminants originating at the land surface to the underlying groundwater resources or surface water bodies (Fellows and Brezonik 1980) via seepage flow.

Characterizing groundwater seepage within an astatic karst lake is challenging because seepage patterns and magnitude are often controlled by subsidence features that breach confining units separating the lake and upper aquifers from deep sub-lake aquifers. Lakes that are underlain by leaking and/or breached confining units are susceptible to dramatic variations in seepage magnitude and patterns over short horizontal distances and time periods. Bathymetric maps, lithologic logs, seismic reflection profiles, potentiometric maps, aerial photographs, and lineament maps have all been employed to locate subsidence features and presumably areas of seepage outflow.

Results using these methods can be ambiguous. Seismic reflection and lithologic results can provide detailed lake bottom profiles that clearly show subsidence features and/or faults breaching a confining unit (SJRWMD 1992). While this information can identify potential seepage areas, it does not confirm the occurrence of seepage and/or provide seepage estimates.

Mass balance approaches can be used to calculate net seepage for this type of environment, but provide little information on the occurrence of seepage or its variation. Direct measurement using the seepage meter method eliminates the accumulation of errors inherent in mass balance methods and also minimizes assumptions on the homogeneity of the water shed substrate. This study was undertaken to test the hypothesis that the seepage meter method is an effective way to characterize seepage magnitude and distribution in astatic karst lake environments. Also, direct measurement of seepage, using the excavation seepage meter method, is an effective technique to characterize potential vertical variations in seepage within lake bottom sediments. To test this hypothesis, seepage meter and excavated seepage meter measurements were recorded within Brooklyn Lake. Located in North-Central Florida, Brooklyn Lake's sandy bottom, size and urban location make it ideal for residential and recreational activities. This lake experiences short-term fluctuations in lake stage elevations that cause it to divide into several individual pools. At maximum lake stage elevations, the lake has a surface area of 2,756,007 square meters, and at record low lake stage elevations, the area is reduced to 615,144 square meters.

During periods of sufficient rainfall and high lake stage elevations, the lake receives inflow from Alligator Creek and discharges to this creek along the north-central and west shorelines, respectively. Alligator Creek does not flow in or out of the lake during periods of low rainfall and low lake stage elevations. Previous non-empirical investigations have determined that the main factors causing low lake stages were lack of rainfall, regional declines of water levels in the Floridian aquifer, and lake bottom leakage (i.e., seepage outflow from the lake). Empirical methods for measuring seepage are needed to corroborate non-empirical seepage estimates and to better understand what factors control seepage.

Seepage measurements were made in littoral and non-littoral parts of Brooklyn lake and at locations where the underlying confining layer was determined to be breached or to be intact. Lake bottom core samples also were collected and analyzed to estimate the vertical permeability and grain size distribution of material near the water/sediment interface (i.e., upper 0.5 to 1 meter of sediment). Field and laboratory testing was conducted to determine how the lake's hydrostratigraphy influences and/or controls the occurrence and magnitude of seepage flow within the lake. Empirical measurement of the magnitude and distribution of hydraulic seepage within Brooklyn Lake is important because it will enable scientist to select target areas for detailed investigations more efficiently.

2 METHODS

Figure 1 shows a graphical representation of a seepage meter drum and baggie collector assembly (Lee 1997).

The method utilizes a bottomless cylinder (i.e., cut off end of a 55-gallon steel drum) that is inserted open end down into the lake bottom sediments for approximately 15 centimeters. The top closed end of the seepage meter is fitted with a threaded pipe nipple to which a plastic ball valve and 4-liter (L) plastic bag assembly are connected. At the lake surface, the ball valve and plastic bag assembly is filled with a measured volume of water, gas bubbles are excluded from the plastic bag and the ball valve is moved to the closed position. Subsequently, the ball valve and pre-filled bag assembly are connected to the submerged seepage meter drum.

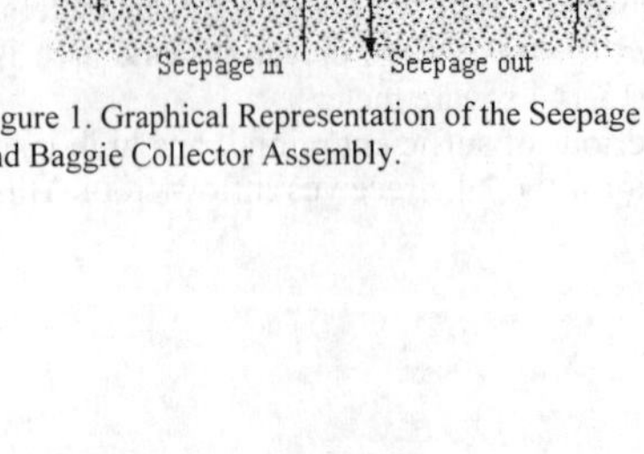

Figure 1. Graphical Representation of the Seepage Meter Drum and Baggie Collector Assembly.

Once a seal is formed between the drum and the bag assembly, the ball valve is opened and the time is recorded. Water enters or exits the plastic bag as a result of seepage flow. At the end of the test the ball valve is closed and the finish time and volume within the bag is recorded. At the surface, the contents of the bag assembly are poured into a 1,000-milliliter (ml) plastic graduated cylinder and measurements are recorded to the nearest ml.

Seepage flow is then calculated for the area of the lake bottom enclosed by the cylinder. Results of seepage meter measurements may be reported as either a flux (milliliters [mL] x m^{-2} x min^{-1}) or a velocity (centimeters [cm] x day^{-1}) across the sediment-water interface.

Excavation seepage meters utilize longer drum wall lengths and removable watertight lids. These modifications enable excavation seepage meters to be driven deeper into lake bottom sediment and allow the removal of sediments within the base of the seepage meter. Excavation of bottom sediments was accomplished using SCUBA equipment and a hand-held metal ice scoop.

Once the lake sediments were removed to a depth of approximately 60 cm below the lake bottom surface, excavation seepage drums were fitted with a watertight metal lid and seepage meter measurements were obtained as described above.

2.1 Field Procedures and Data Analysis

Because of inherent seepage measurement errors, seepage flux values varying by less than (+ or –) 0.5 L x m^{-2} x day^{-1} were considered indications of no seepage flow. Negative seepage values represent seepage outflow through the lakebed and positive values represent groundwater flow into the lake. Spatial and temporal seepage trends were analyzed by constructing an array of time- series plots.

During the initial seepage meter survey, 18 seepage meters were installed within an array of lake environments. These environments included: shallow sandy shoreline areas with no underlying collapse feature; shallow sandy shoreline with underlying or nearby collapse feature; sand/muck lake bottom, with no underlying collapse feature; sand/muck lake bottom with underlying or nearby collapse feature; and shoreline and lake bottom areas lacking prior hydrologic investigation. At a minimum, three seepage measurements were collected from each drum during the initial survey. Seepage meter locations are shown in Figure 2. Also shown in Figure 2, are the approximate locations of subsidence features mapped by the SJRWMD (1992).

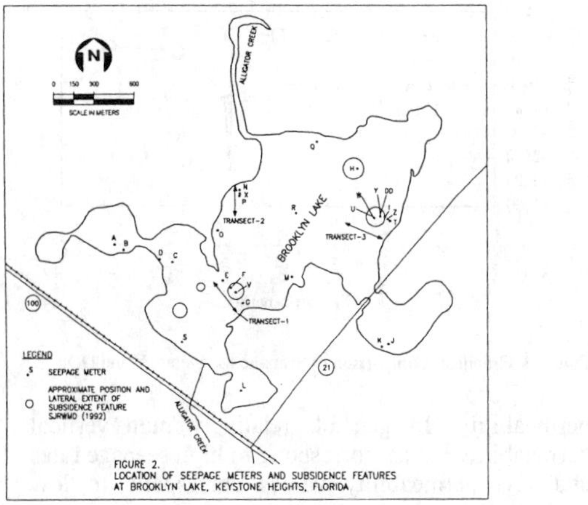

Figure 2. Location of Seepage Meters and Subsidence Features at Brooklyn Lake, Keystone Heights, Florida.

Special excavation seepage meters identified as W and DD were installed adjacent to meters U and Y, respectively to test how fine grained muck material at the sediment water interface affects seepage. Before sediments were excavated from within seepage meter DD, standard seepage measurements were recorded for two weeks. Subsequently, the inside of this drum was excavated, and additional seepage measurements were recorded.

After the initial seepage meter survey was completed, additional seepage meters were deployed to construct three separate seepage meter transects identified as 1, 2 and 3. Excavation seepage meters W and DD were installed in Transect 3. Figure 2 shows the location of seepage meters in Transects 1, 2, and 3. Seepage meter transects were established to monitor seepage variability over small distances and to compare seepage patterns and magnitude in areas with and without underlying collapse features.

2.2 Sediment Core Sample Collection and Laboratory Testing

Sediment core samples were collected adjacent to seepage meter locations A, BB, I, J, K, L, Q, T, U, Y and Z to obtain laboratory-derived vertical permeability values and sediment grain size distribution data. Core samples were identified by the letter of the nearest adjacent seepage meter. For example, core samples collected adjacent to seepage meter A were identified as A-1 and A-2. Core samples were collected by driving a clear, 6 cm inside diameter (i.d), PVC core barrel into the lake bottom for 60 to 75 cm . Once the core barrel was driven to its desired depth, a rubber stopper was pushed into the top open end of the core barrel and the sample was retrieved. Core samples were individually tested in the laboratory to determine vertical permeability using the falling head permeameter test method (Watson and Burnett 1995). Tests were generally repeated until three reproducible values were obtained. Samples requiring over five days to complete testing were only tested once. After initial permeameter testing was completed for core samples Y-1 and U-2, the top 37 cm and 40 cm, respectively, of organic muck sediment was removed from these samples. Subsequently, the remaining sandy bottom sections were tested to compare permeability between muck layers and sandy sediments.

After completion of permeability testing, grain size analysis by washing (*CRL Method Number 483*) was then performed on each core sample (EPA 1976).

3 RESULTS

Average seepage flux values for Brooklyn Lake are shown in Table 1. In general, seepage out of the lake and/or no seepage was recorded at all seepage meter locations except seepage meter M where seepage into the lake was consistently measured.

The magnitude and occurrence of seepage varied throughout the lake. Seepage magnitude and occurrence was determined to be highly variable in seepage meter Transect 3. As shown in Table 1, the highest average naturally occurring seepage outflow values were recorded at seepage meters located within shallow littoral sections of the lake. In comparison, low and no seepage outflows were recorded in deep areas of the lake underlain by subsidence features.

3.1 Excavated seepage meter tests

Because highly variable seepage patterns were recorded for adjacent meters within Transect 3, excavation seepage meters W and DD were deployed adjacent to meters U and Y respectively, to test how lake bottom sediments of variable permeability control seepage. These meters were installed in an area underlain by a large subsidence feature shown (SJRWMD 1992) to breach the confining unit underlying the lake (Figure 2). Only very low and no flow values were measured at meters U and Y, respectively. Prior to excavation, no seepage was recorded at meter DD. In contrast, the highest outflow values measured during the entire study were recorded at seepage meters DD and W after muck sediments were excavated to a depth of approximately 60 cm below the lake bottom within both of these meters.

Average seepage flux values recorded for seepage meters W and DD were -137.58 L x m^{-2} x day^{-1} and -1,014.49 L x m^{-2} x day^{-1} respectively

Table 1. Average Seepage Flux Values for Brooklyn Lake.

Sample Station I.D.	Seepage Meter Depth (m)	Average Seepage Flux (L x m^-2 x day-1)
A	8.54	0.07
B	1.37	-5.61
C	4.57	-0.56
D	1.52	-3.46
E	0.91	-0.04
F	5.79	-1.08
G	0.91	-0.02
H	6.10	-0.84
I	4.27	-14.64
J	6.10	-1.60
L	0.61	-12.40
M	0.91	0.44
N	1.22	-3.42
O	1.22	-0.96
P	6.10	0.10
Q	0.61	-89.86
R	1.83	-0.99
S	1.07	-3.88
T	1.68	-0.76
U	6.10	-1.88
V	6.40	-0.04
*W	6.40	-137.58
X	5.18	-0.09
Y	6.10	-0.07
Z	3.05	-0.07
*DD	6.10	0.18
**DD	6.10	-1,014.49

*Indicates Excavation Seepage Meter
**Indicates Seepage Values After Excavation

After sediment removal and prior to initially attaching the bag assembly to meter DD, particles in the water column were observed being sucked into the pipe nipple opening at the top of the meter. This occurrence became increasingly difficult to identify during later testing episodes. Similarly, seepage outflow values steadily decreased over time. It was hypothesized that sloughing of relatively low permeability material within the upper sediments of the meter into the excavated area underlying the meter was partially responsible for the reduction of seepage outflow over time.

Accumulation of fine particulate matter being sucked into the meter between sampling episodes could also have helped to reduce permeability and seepage outflow within the meter over time. Figure 3 shows this process graphically for adjacent seepage meters Y and DD.

3.2 Permeameter tests versus measured seepage

Figure 4 shows a graphical comparison of average daily seepage versus laboratory measured vertical

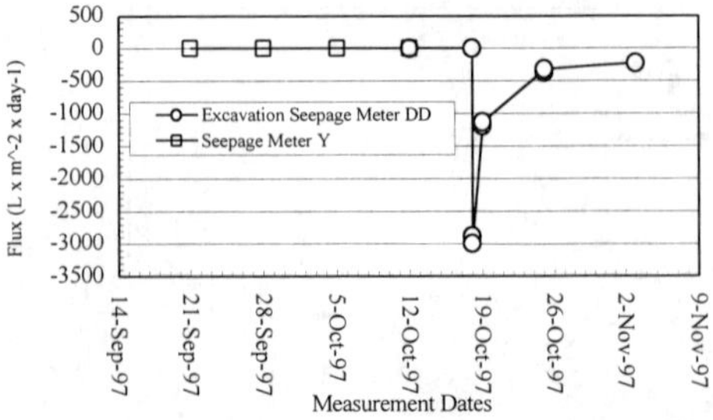

Figure 3. Graphical Comparison of Seepage for Meters Y and DD.

permeability. In general, relatively high vertical permeability values correspond to high seepage rates and low permeability values correspond to low seepage rates or no seepage. Permeameter tests conducted after the top 37 cm and 40 cm of organic muck was removed from core samples Y-1 and U-2 respectively, showed vertical permeability values had increased by two orders of magnitude. These results indicate that impermeable sediments at the water/sediment interface control the occurrence and magnitude of seepage in areas of the lake where the potential for seepage outflow is downward to the underlying aquifers. Exceptions to these trends include seepage meter locations T and Z where low and no seepage was measured, and high vertical permeability values were recorded for each corresponding core sample. These anomalous readings suggest deeper subsurface sediments of low permeability are restricting and/or prohibiting seepage in the area of these meters.

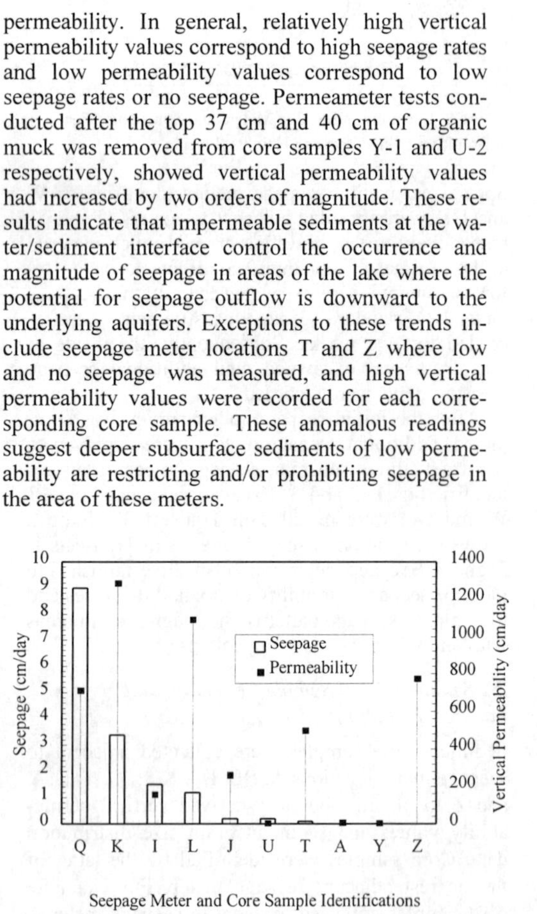

Figure 4. Seepage vs. Vertical Permeability.

3.3 Grain size versus permeability

Figure 5 shows a graphical comparison of percent material passing a No. 200 sieve versus average permeability results for core samples. High permeability results for core samples correspond to a low relative percent of material passing the No. 200 size sieve. Conversely, core samples with low measured permeability correspond to a relatively high percentage of material smaller than the No. 200 size sieve. Material passing the No. 200 sieve is categorized as silt/clay sized material.

As shown in Figure 5, an approximate 4 percent difference in clay and/or silt size material between core samples J and Y reflect a dramatic contrast in the vertical permeability of these two core samples. These results further support the hypothesis that low permeability, fine grained sediments (i.e., organic muck) at the water sediment interface play an important role in controlling the occurrence and magnitude of seepage within Brooklyn Lake.

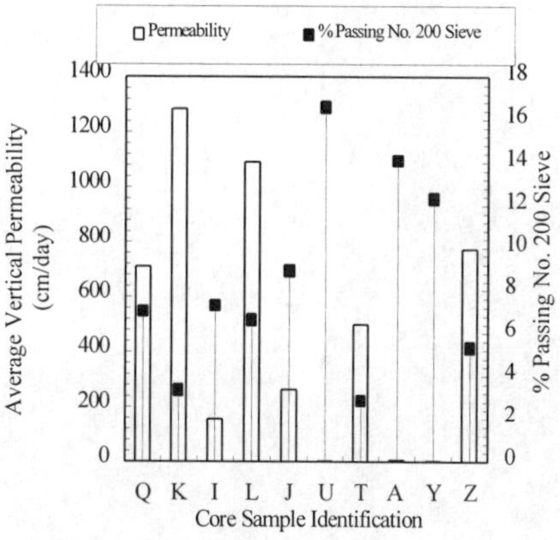

Figure 5. Average Permeability vs. Average Percent of Material Passing No. 200 Sieve.

3.4 *Annual Seepage Outflow*

Annual seepage outflow estimates using average and weighted average calculations are shown in Table 2. Annual seepage outflow estimates using average and weighted average calculations were 248.2 cm/yr and 211.8 cm/yr, respectively. Approximately 81 percent of all seepage outflow measured during the study occurred within the shallow littoral sections of the lake. The remaining 19 percent of seepage outflow occurred within deeper non-littoral areas.
Annual seepage outflow data derived from the seepage meter method support the range of water-budget derived seepage outflow values previously reported for the lake (Table 2).

4 SUMMARY AND CONCLUSIONS

Brooklyn Lake has experienced extreme and dynamic changes in lake stage elevations that reduced its area by half in less than five years. Seepage outflow is the largest outflow component of the hydrologic budget for Brooklyn Lake. No empirical seepage measurements have been reported for the lake prior to this investigation.

Table 2. Summary of Annual Seepage Outflow Values Calculated for Brooklyn Lake.

Source	Date	Seepage Outflow Estimate Methods	Seepage Outflow (cm/yr)
Hirsch (1)	1997	Seepage Meter Field Measurements	248.2
Hirsch (2)	1997	Seepage Meter Field Measurements	211.8
Annable + Motz	1996	Residual by Water Budget Calculations	210.6
Motz + Heaney	1994	"	185.2
Deevey	1988	"	90.17
Clark et al.	1963	"	426.72

(1) Simplified Average
(2) Simplified Weighted Average

Seepage meter results reveal that a much larger percent of outflow occurs within the permeable littoral areas of Brooklyn Lake compared to deeper non-littoral sections.

Excavation seepage meter results, and laboratory permeameter and grain size analyses determined that seepage variation is largely controlled by differences in the permeability of material at the water/sediment interface within Brooklyn Lake.

The seepage meter method is an effective means by which to measure the magnitude and distribution of hydraulic seepage within Brooklyn Lake. The results of this study can be used to model surface water and groundwater interactions within the lake. This information will also aid scientists and regulatory agencies in determining the feasibility of augmenting the stage level of Brooklyn Lake by increasing inflow and/or decreasing seepage outflow.

Brooklyn Lake and other seepage lakes will continue to fluctuate in lake stage elevation in response to precipitation trends. Future multiple year long deficiencies in rainfall can be expected to cause the stage of these lakes to drop dramatically, and year long periods of adequate precipitation will enable them to fill their basins once again. The degree to which these lakes respond to rainfall distributions will be a function of the thickness and lateral continuity of impermeable sediments at the water/sediment interface, potentiometric surfaces of the intermediate and Floridan aquifer systems, and karstic features underlying the lake beds.

5 REFERENCES

Annable, M.D. and Motz, L.H. 1996. *Investigation of the Lake and Surficial Aquifer Interaction on the Upper Etonia Creek Basin.* Final Report, Special Publication SJ96-SP15.

Report to SJRWMD, University of Florida, Gainesville, Fl. 75 pp.

Clarke, W.E., Musgrove , R.H., Menke, C.G., and Cagle, J.W., Jr., 1963. *Hydrology of Brooklyn Lake Near Keystone Heights, Florida.* Report of Investigations No. 33, Florida Geological Survey, Tallahassee, Florida, 43 pp.

Deevey, E.S., Jr., 1988. *Estimation of Downward Leakage from Florida Lakes.* Journal of Limnology and Oceanography, American Society of Limnology and Oceanography, Vol. 33, No. 6 (part 1), pp. 1308-1320.

Fellows, C. R., and Patrick L. Brezonik 1980. Seepage Flow into Florida Lakes. Reprinted from *Water Resources Bulletin, Vol. 16, No. 4.*, pp. 635-641.

Hirsch, J. D. 1998. *Characterization of Hydraulic Seepage within Brooklyn Lake, Keystone, Heights, Florida.* Masters Thesis, University of Florida, Gainesville.

Lee, D. R. 1977. A Device for measuring Seepage Flux in Lakes and Estuaries. *Journal of Limnology and Oceanography, American Society of Limnology and Oceanography, Vol. 22,* pp. 155-163.

Motz, L.H., and Heaney, J.P., 1994. Upper Etonia Creek Hydrologic Study. Special Publication SJ92-SP18. Report to SJRWMD, University of Florida, Gainesville, Fl. 177 p.

Subsurface Detection Investigations, Inc., 1992. *High Resolution Seismic Reflection Profiling in Selected Lakes in the SJRWMD.* Special Publication SJ92-SP13. Report to the SJRWMD. Miami, FL: Author.

Groundwater: Past Achievements and Future Challenges, Sililo et al. (eds) © 2000 Balkema, Rotterdam, ISBN 90 5809 159 7

CSAMT mapping of a deep hydrogeological structure in Mauritius Island

J.L.Join, B.Robineau & M.Courteaud
Laboratoire des Sciences de la Terre, Université de la Réunion, Saint Denis, France

M.Ritz
IRD, Dakar, Sénégal

ABSTRACT: Controlled Source Audiomagnetotelluric (CSAMT) sounding were done at several locations along 5 profiles in south-western Mauritius to map the resistivity and thickness of the upper layers of the volcanic sequence and to infer the aquifer system from the electrical interpretation. Several sounding were measured adjacent to wells to aid in correlating the geophysical interpretations to known lithologic logs. The sounding data were isotropic across the entire observational band width. One dimensional far-field modeling has been used to determine the resistivity stratification with depth. This study suggests that the main body of the aquifer is confined in a paleovalley containing young volcanic flows at depths ranging from 40 m to 180 m..

1 INTRODUCTION

Volcanic islands of the SW Indian Ocean are the aerial parts of large hot spot shield volcanoes. Their present shape and morphology depends both upon the volcanic morpho-structures (caldeira and rift zones) and the age of the eruptive events. In the highlands, aquifers can develop down to several hundred meters and prospecting needs deep investigation techniques. Application of electromagnetic (EM) methods for mapping subsurface structures and for groundwater studies (e. g. Ogawa and Takakura, 1990 ; Goldman et al., 1991) has become common over the last decade. EM soundings are used to estimate the electrical properties of the subsurface materials. As electrical properties of geologic materials are highly dependent upon porosity, saturation degree of rocks, and pore fluid resistivity , information about subsurface geologic structures, aquifer geometry, water quality and water content can be obtained. EM techniques have already provided useful results in volcanic environments (Ritz et al., 1996).

A controlled source audiomagnetotelluric (CSAMT) survey was performed which was directed at improving our deep hydrogeological knowledge of the southwestern part of Mauritius Island, as part of a statewide groundwater resource assessment. The Beaux Songes study area (Figure 1) was considered to be a major hydrogeological structure (Giorgi, 1997), draining part of the highlands groundwater system.

In this paper we report the results of CSAMT soundings carried out at Beaux Songes area. An interpretation of the data is presented here and compared with available drillhole data to test the feasibility of CSAMT in solving various hydrogeological problems.

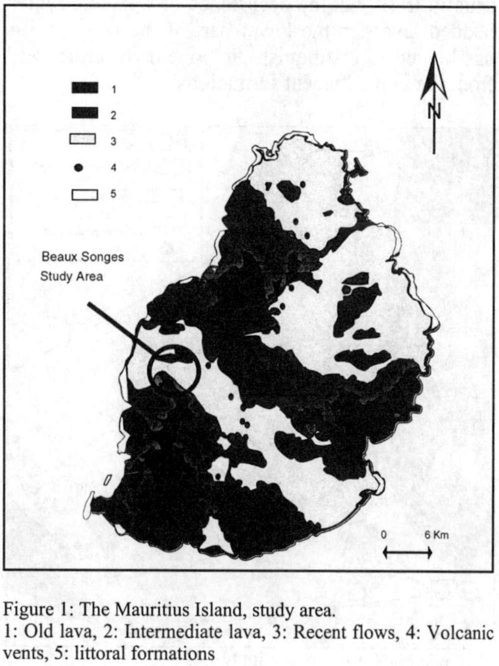

Figure 1: The Mauritius Island, study area.
1: Old lava, 2: Intermediate lava, 3: Recent flows, 4: Volcanic vents, 5: littoral formations

2 GEOLOGICAL AND HYDROGEOLOGICAL SETTING

Located in the SW Indian Ocean, Mauritius (57°35' E, 20°15' S) belongs to the Mascarenas archipelago. This low volcanic island (maximum altitude 823m has a tropical humid climate with a mean annual rainfall value of 1 500 mm.) Such a context is favourable to groundwater resources (Proag 1995, Jawaheer and Proag 1988).

The aerial part of Mauritius shield volcano was built due to the activity of a hot spot between 10 Ma and 0.025 Ma. Edification occurred during two magmatic stages separated by a long improductive period (5 Ma to 3.5 Ma). (. MacDougall and Chamalaun 1969).

The first stage (10-5 Ma) is characterised by the building up of a large shield volcano, more than 1 000 m high (Perroud 1982). By the end of this stage, the central part of the volcano collapsed to form a polylobe caldera depression. Old formations, the highest peaks of the island being their relicts, consist mainly in superimposed alkaline basaltic flows, pyroclastics, breccia, and a few late trachyte (intrusion) plugs.

The last stage appears as a repetition of rejuvenation phases after a 1,5 Ma long period of volcanic quietness and intensive erosion. Young formations are mainly nepheline basalts that issued from central pyroclastic vents aligned on a prominent N-S line. Flows have filled the central calderic depression and overflowed seaward in many places. The lavas are usually fresh, but the occurrence of weathered interbedded layers in the lower part of the young series has helped to distinguish "intermediate formations" from the upper "recent formations".

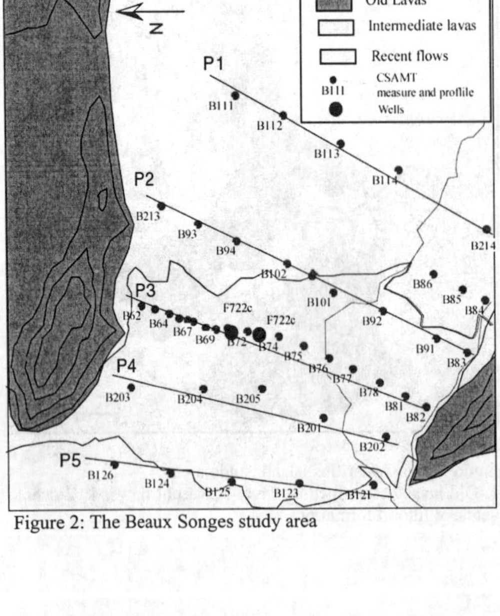

Figure 2: The Beaux Songes study area

The Beaux Songes study area, covers part of a large (2.5 km) and high (average altitude 250 m), E-W valley (Figure 2) gently deeping westward. The expected geological structure is a paleovalley in the old formations, outcropping on both sides (Mt St Pierre to the N and Mt du Rampart to the S), that was latter filled with the intermediate and recent formations. (Giorgi 1997). This structure may function as a main drain of the "Beaux Songes" aquifer. But its geometry and deepness was still unknown

3 CSAMT METHOD

Controlled-source audiomagnetotellurics (CSAMT) is a frequency-domain electromagnetic (EM) sounding technique which utilizes artificial EM fields to measure resistivity variations in the ground. Goldstein and Strangway (1975) introduced the use of a controlled source to overcome the weak signal levels associated with the conventional magnetotelluric (MT) technique (Cagniard 1953). The controlled source provides a dependable stable signal, resulting in higher precision and more efficient data collection. Typical CSAMT equipment usually consists of a transmitting station where the EM field is generated and a receiving station where both the horizontal electric (E) and magnetic (H) fields are measured at a range of frequencies for the computation of the tensor CSAMT (Zonge and Hughes 1991). In traditional CSAMT methods, one component of E parallel to a transmitter bipole, and the orthogonal H component are measured for the computation of the apparent resistivity and phase curve. However, the interpretation of CSAMT data is more complicated as it is known from MT. The reason is that the measurements of the EM fields often have to be carried out in the near-field zone of the transmitter, where the homogeneity of the EM field is violated. CSAMT allows plane wave interpretation if the transmitter-receiver distance is greater than 3.5 skin depths (Kaufman and Keller 1983). In the far-field zone of the source, MT equations derived for a plane wave source are applied in CSAMT data analysis. More recently advances in modeling have allowed near-field data to be included in the interpretation (Boerner et al. 1993 ; Lu et al. 1997).

In our study area, CSAMT measurements were recorded using the Stratagem tensor system of Geometrics. The transmitter operates in the frequency range from 500 Hz to 92 kHz. For the lower frequencies (10 - 1000 Hz) the system uses the natural-source EM fields as is done in MT. At the receiving station two components (Ex and Ey) of the electric field are measured with perpendicular dipoles, while two components (Hx and Hy) of the magnetic field are recorded with sensor coils. This leads to a tensor estimate of the impedance (Vozoff 1972).

4 CSAMT DATA AND INFERRED RESISTIVITY STRUCTURE

The CSAMT data were collected at 44 sites along 5 profile lines shown in Figure 2. The tensor CSAMT sites were usually spaced 500 m apart, with the telluric sensors oriented N-S and E-W at each location. The apparent resistivity and phases from the two directions of measurement are computed directly for each frequency by means of a built-in microprocessor. Impedance tensors (Vozoff 1972) were rotated to azimuths determined by the direction of maximum anisotropy at each frequency and for each site. Skews were evaluated and are low at all sites.

Most of the stations are characterized by a maximum (about 100 ohm.m) in apparent resistivity in the frequency range of about 2000 Hz to 500 Hz and display very little static shift (Jones 1988). Note that all the sites were free from near-field effects. Hence, the rotationally invariant determinant (Berdichevsky and Dmitriev 1976) apparent resistivity and phase data were inverted in the conventional manner using a one-dimensional (1-D) inversion routine (Interpex 1993). Along 5 profile lines, the resulting 1-D models reveal the electrical structure of Beaux Songes.

An interpreted resistivity-versus-depth cross section along profile P3 is shown in Figure 3. Two wells on this profile (see Figure 2) give information that is of use in evaluating the geophysical interpretation. The cross section shows a remarkable consistency from sounding to sounding and a pattern of smoothly varying structure.

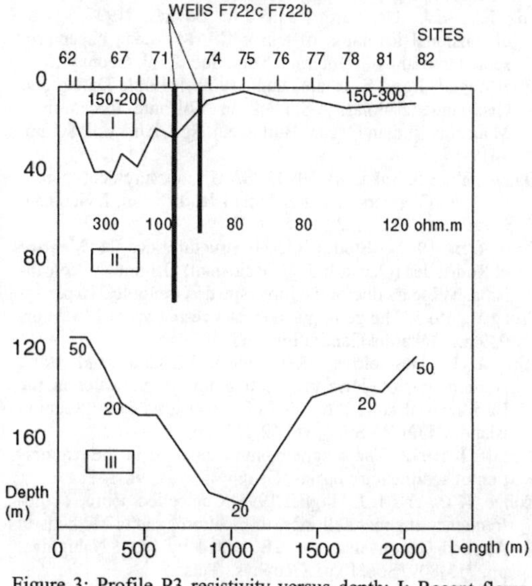

Figure 3: Profile P3 resistivity versus depth; I: Recent flows, II: Intermediate lava, III: Old lava

The CSAMT soundings indicate the following geoelectrical layers:

(1) an intermediate-resistivity (150-400 ohm.m) surface layer, 5-40 m thick. This unit is found with a sudden thickness jump from 5-10 m in the central part of the section to 20 - 40 m in the left part, suggesting the presence of an old drainage channel between sites S72 and S74. On the basis of well information, this layer is inferred to be recent basalt flows,

(2) a low-resistivity (less than 10 ohm.m) second layer 1-5 m thick, corresponding with bedded clay,

(3) an intermediate-resistivity third layer 100-180 m thick. This layer is up to 180 m thick in the central part of the section and its resistivity varies along the profile from 60 ohm.m to more than 300 ohm.m. This layer is also predominantly basaltic and variations in resistivity are probably indicative of the relative contribution of clay interbeds in the section,

(4) a low-resistivity fourth layer of older volcanic material corresponding to the conductive basement. A possible explanation for the extensive conductor is that it is a result of mixed layer clays and zeolites which lowers the resistivity of basalts (Benderitter and Gérard 1984).

5 DISCUSSION AND CONCLUSION

Constructed geophysical models have been correlated to available well lithostratigraphic logs. Complex interbedded units are seen as single layers with an average lithogical and weathering electrical response Hence, a good correlation has been found as follows.

The upper intermediate resistivy layer fits with a recent lava flow sequence, including its top soil, and water saturation in its lowest part . The resistivity at each site depends on the local influence of each component, and decreases towards he axial part of the valley.

CSAMT sounding were able to reach the thick electrical basement of the intermediate and recent formations across Beaux Songes valley. Considering the low resistivity values (20-30 ohm.m), the old formations outcropping on both sides of the valley and the lithology of the basement drilled in other places (Giorgi 1997),old volcanic formations are suspected, deeply weathered and water saturated. Geological sections of Beaux Songes paleovalley, eg. along profile P3, are now reliable.

The topographic surface of the basement, deduced from the top altitude of the last conductive layer at each site, gives the morphology of a paleovalley that was latter filled with intermediate and recent lava flows (Figure 4). The roughly E-W axis of the paleo-valley is dipping west ward from an altitude of 110 m upstream down to 25 m at Beaux

Songes site were a clear threshold is noticed on the map.

The geometry of the volcanic aquifer is deduced from the mapping of geoelectrical layers in the study area. Without mineralised interstitial water or temperature gradient, the various electrical units can be interpreted in term of resistivity – permeability relationship. Only the occurrence of clay or other secondary minerals like chlorite, serpentine or zeolite can explain the low resistivy of volcanic formations. Such minerals clog up the primary porosity, giving the formations a low permeability.

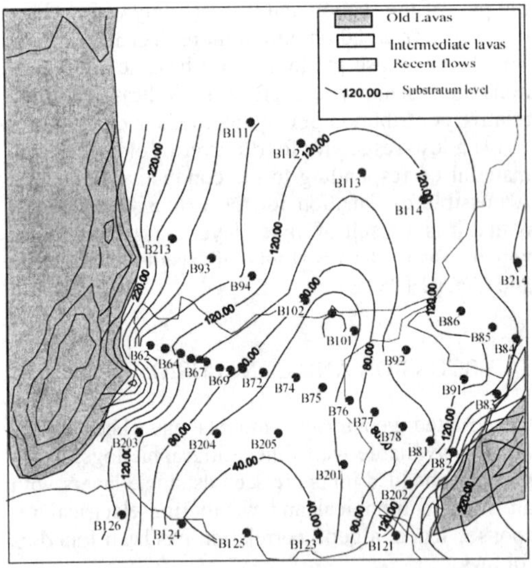

Figure 4: Topographic surface of conductive basement

The altitude of the conductive basement being always positive, the possibility of marine water intrusion is ruled out. Then, depending on thickness, low resistivity layers can be considered as semipermeable to impermeable formations

Two main characteristics define the Beaux Songes aquifer: (i) the thick poorly permeable basement represents the bottom of a deep aquifer. Its morphology is typical of a paleovalley filled with more permeable formations ;(ii) the thin conductive/impermeable middle clay layer, found at all sites, can stratify the aquifer. The upper part is free within the recent lava flows, when the thicker lower part is kept in less permeable intermediate volcanic formations. The aquifer total thickness varies from 40 m close to the valley flanks to 180 m at the paleovalley axis.

CSAMT is found to be feasible for mapping deep hydrogeological structures in a tropical volcanic environment. However, to realize the full benefits of surface geophysics, interpretation must be closed interprated with auxiliary data such as drillhole information. The survey results allowed the geometry of the deeper part of the aquifer to be defined.

REFERENCES

Benderitter, Y., and A. Gérard, 1984 Geothermal study of Réunion Island : audiomagnetotelluric survey, J. Volcanol. Geotherm. Res., 20, 311-332.

Berdichevsky, M.N., and V.I. Dmitriev, 1976. Basic principles of interpretation of magnetotelluric sounding curves, in Geoelectric and Geothermal Studies, edited by A. Adam, pp. 163-211, Akad. Kiado, Budapest.

Boerner, D.E., J.A. Wright, J.G. Thurlow, and L.E. Reed, 1993. Tensor CSAMT studies at the Buchans Mine in central Newfoundland, Geophysics, 58, 12-19.

Cagniard, L., 1953. Basic theory of the magnetotelluric method, Geophysics, 18, 605-635.

Giorgi L., 1997. Etat des connaissances géologiques et hydrogéologiques de l'île Maurice. Seminar on Management of Groundwater Resources in Mauritius. Proceedings, University of Mauritius.

Goldstein, M.A., and W.D. Strangway, 1975. Audio-frequency magnetotellurics with a grounded dipole source, Geophysics, 40, 669-683.

Goldman, M., D. Gilad, A. Ronen, A. Melloul, 1991, Mapping of seawater intrusion into the coastal aquifer of Israel by the time domain electromagnetic method, Geoexploration, 28, 153-174.

Interpex, 1993. EMIX MT user's manual.

Jawaheer K., V. Proag., 1988. Groundwater aquifers of Mauritius, Hydrogéologie, n°2, 1988, pp. 125-134, 7 fig.

Jones, A.G., 1988. Static shift of magnetotelluric data and its removal in a sedimentary basin environment, Geophysics, 53, 967-978, 1988.

Kaufman, A.A., and G.V. Keller, 1983. Frequency and transient soundings, 685 pp., Elsevier, New York, 1983.

Lu, X., M.J. Unsworth, and J.R Booker, 1997. Two-dimensional inversion of tensor CSAMT data, paper presented to 67th Ann. Internat. Mtg., Soc. Expl. Geophys..

MacDougall, I and F. Chamalaun., 1969. Isotopic Dating and Geomagnetic Polarity Studies on Volcanic Rocks from Mauritius, Indian Ocean, Bull. Geol. Soc. Am. Vol. 80, pp. 1419-1442.

Ogawa, Y. & S. Takakura 1990. CSAMT measurements across the 1986 C craters of Izu-Oshima Island, Japan, J. Geomag. Geoelectr., 42, 211-224.

Perroud B., 1982 – Etude Volcano Structural des Iles Maurice et Rodrigues (Ocean Indien Occidental) Origine du Volcanisme. Thèse de doctorat, Université de Grenoble 210 p.

Proag V., 1995. The geology and water resources of Mauritius, 939 p., Mahatma Gandhi Institute

Ritz M., M. Descloitres., B. Robineau, M. Courteaud. 1997, Audiomagnetotelluric prospecting for groundwater in the Baril coastal area, Piton de la Fournaise volcano, Reunion island, GEOPHYSICS, vol.62, N° 3, pp. 758-762.

Vozoff, K.,1972. The magnetotelluric method in the exploration of sedimentary basins, Geophysics, 37, 98-141.

Zonge, K.L., and L.J. Hughes,1991. Controlled source audiofrequency magnetotellurics, in Electromagnetic Methods in Applied Geophysics, vol. 2B, edited by M.N. Nabighian, pp. 713-809, Soc. Expl. Geophys., Tulsa.

Groundwater: Past Achievements and Future Challenges, Sililo et al. (eds) © 2000 Balkema, Rotterdam, ISBN 90 5809 159 7

Groundwater resource developments in the Northern Territory, Australia

P.Jolly, D.George, S.Tickell, U.Zaar & D.Chin
Natural Resources Division, Department of Lands, Planning and Environment, Palmerston, Plumstead, N.T., Australia

ABSTRACT: In the mid 1970's, the Northern Territory commenced a program of work aimed at providing towns and communities with a secure water supply. Due to the long "dry season" in the north and the aridity of the south, water supplies were sourced from groundwater. This was essentially completed by the end of the 1980's. In the 1990's, the emphasis of the groundwater program shifted to become the identification of groundwater resources with a significant regional development potential (eg. for irrigated agriculture). This change resulted in a work program being developed that was directed at providing relevant, up-to-date information on the groundwater resources of the Northern Territory.

A major objective of this program is to ensure that this information is presented such that it can be easily understood by landowners and land managers. The primary technique used has been coupling water resource maps with community education. CD-ROM's using interactive multimedia format have also been produced. Work is also well advanced in making most of the polygonal and point data on groundwater resources available over the Internet.

1 INTRODUCTION

The Northern Territory (NT) covers 1,346,200 km^2 of land with a population of approximately 196,000 of which 135,000 live in the two main population centres of Alice Springs and Darwin. The remaining population lives in small towns or remote aboriginal communities and pastoral stations (refer Figure 1 for locations). The climate ranges from the arid deserts of the south to the wet/dry monsoonal tropics of the north. Historically, all water supplies were sourced from springs, streams or shallow wells close to the towns and communities. Severe water quality, and sometimes quantity problems were experienced as the towns and communities developed. Subsequent water supply investigations have therefore primarily been focused on locating suitable groundwater resources, and groundwater supplies are now used throughout the NT, with the only major surface water retention dam being in the capital Darwin.

Prior to 1970, work was usually targeted at providing one or more production bores to meet the immediate needs of a community. No detailed studies of the water resources were undertaken. This often resulted in production bores being inappropriately sited.

Since 1970, water source development programs included groundwater resource evaluations. These have now been completed, and all towns and major communities now have a water supply capable of supplying 800 litres/person/day on an annual basis, and also able to sustain 1200 litres/person/day for periods of up to one month (Power and Water Authority 1996). Potable water quality guidelines, as set down by the National Health and Medical Research Council (1996), have also been met. Data and reports from these investigation programs provide the fundamental datasets on the groundwater resources of the NT. The reports form the basis for any future work and contain recommendations for both resource and borefield management (Jolly & Yin Foo 1994).

In 1989 the Government commenced a program of work, using the capacity of the recently introduced computer aided drafting system, to supply information on groundwater resources in a format that could be used by land owners, land managers, developers and others. A perceived shortage of water in two large mining regions provided the impetus for the commencement of this program (Domahidy 1990, McGowan 1989).

In 1992, representatives of the pastoral industry (the major primary producer in the NT) approached

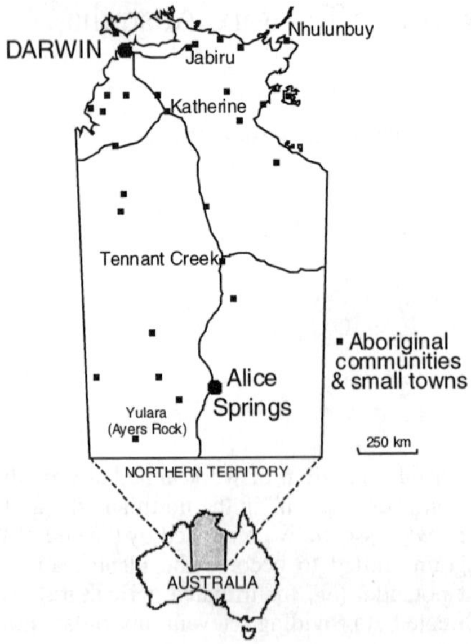

Figure1. Location map of the Northern Territory showing major population centres and aboriginal communities.

the Government to undertake similar work for them. Their requirement was maps and reports that could be used in the development of whole property plans. To achieve this, water resources information had to be integrated with information on soil, pasture and terrain, a technique that lends itself well to the present GIS technology. This work has been successful, and there are currently three projects in progress mapping the water resources of pastoral land (Mathews 1997, Sanders et al. 1994, Tickell 1994, Tickell & Rajaratnam 1998).

Approximately half of the NT is aboriginal owned and managed. The Government's regional mapping of water resources on aboriginal land commenced in 1995 with a joint program with the Australian Geological Survey and the Central Land Council, on a pilot project called the Western Water Study (Jamieson 1998, Wischusen 1998). This study mapped the groundwater resources beneath 6000 km^2 of land to the west of Alice Springs, and produced information in a format that aboriginal people could easily understand and use in their land management decision making (Gallagher 1998).

In 1996, the federal government jointly funded another pilot project on aboriginal land, the East Arnhem Land Water Study (Prowse et al. 1996, Zaar et al. 1999). The aim of this was to determine if the methodology being developed could be applied to the wetter northern regions of the NT. This project has recently been completed, and has been well received.

Both of these studies produced maps, posters, reports, a GIS and photographic database on the water resources. The East Arnhem Land Water Study also included aboriginal language names for rivers and other water related features, and are the first published maps in Australia to do so. Incorporating traditional aboriginal knowledge with regards to water resources into these products has been an important aspect of the studies, and has ensured acceptance of the value of the work. The success of these two pilot projects has resulted in funding of a further six regional mapping projects on aboriginal land.

Community consultation has occurred at all stages of these projects to ensure that the appropriate product and information needs have been met.

The evolution of the mapping program is illustrated in Figure 2. When the current mapping projects are complete, approximately 70% of the NT will be covered.

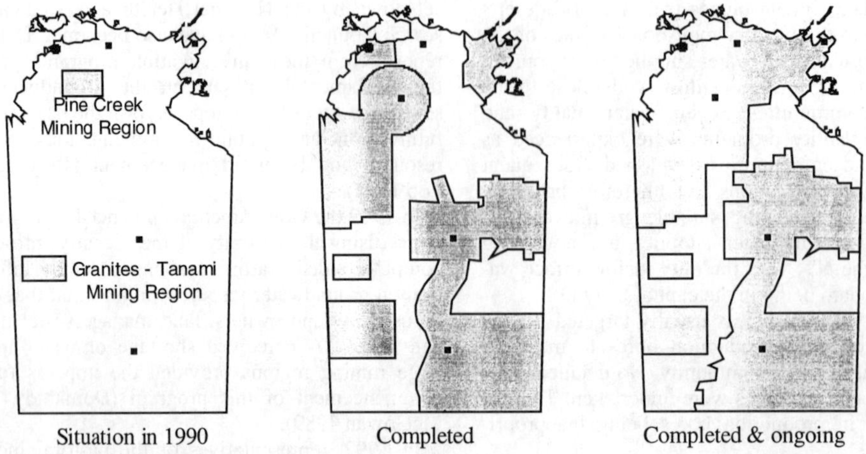

Figure 2. Evolution of the mapping program.

2 DATABASE DEVELOPMENT

The initial mapping programs in 1989 highlighted the need for an effective means of accessing the water resources data held in hard copy format. It was apparent that a structured, searchable database was required, as well as a means of assessing the vast amount of material that could not easily be entered into a structured database. The following section documents the path taken in moving from hard copy data storage to an online digital database.

In 1991 the Government identified that a requirement existed for providing online access to the data held on the groundwater resources of the NT. This comprises hard copy data of approximately 30,000 water bore data files, 1500 technical reports and an increasing range of map products. The preferred long-term solution was to provide online access via a spatial database. At the time however, suitable software and hardware was not available, and a decision was made to focus first on the development of a structured database, and the entry of groundwater data into it.

In the latter half of 1991, the Government commissioned a study that recommended the use of a structured database HYDSYS (Mann 1993). The following conclusions were reached:

1. HYDSYS software would meet current and future needs with respect to the management of water resources information.

2. The use of a server and networked personal computer was the preferred option.

3. Existing water resource data could be converted or input directly.

4. The software lent itself to seamless integration with GIS.

5. Support for the implementation of the strategy was available in the NT.

The Government has been using the software since 1992 for the input of groundwater data (eg. bore data such as location, strata, construction and water levels), water quality analysis results and hydrographic data (such as rainfall and streamflow records). This was done via a temporary "front-end" program, until the Natural Resources Division fully implemented the newly developed HYDSYS Groundwater Module in 1999.

This still left the problem of efficient access to and storage of the estimated 350,000 sheets of other hard copy data that could not be input into the structured database, such as technical reports. In 1998, digital image management software was purchased that could provide an online, indexed and searchable digital archive of scanned hard copy data. The platform for this archive is an Intel based Windows NT4.01 / Service Pack 3 server, with HPA InterView/Enterprise search and retrieval software (HPA 1998). This is used to manage and display the scanned digital information, and has also allowed digital back-up procedures to be implemented. The digital TIFF image files are stored and managed by an ORACLE database. The InterView/Enterprise software can be web-enabled, and a suitable web-browser program is being developed.

The conversion of hard copy data into digital format, its structured storage, and web-based access, has now provided the Division the opportunity to supply information in a range of ways.

3 ACCESS TO THE DATABASE

The Internet and e-mail have become major means by which professionals access and disseminate information. This type of communication has markedly changed the accessibility of up-to-date relevant information. This is particularly important when working in remote areas such as the Northern Territory.

Whilst the development of the Internet has resulted in a major shift in the way we work, the primary mechanisms to provide customers and decision-making managers with relevant data is still via maps, reports and brochures. To date the main advance has been in the way that these documents are produced, and the speed with which specific products can be developed and provided. Considerable development still has to occur before the Internet becomes the primary mechanism for data distribution.

At present, the only information on the groundwater resources of the NT available on the Internet is www.lpe.nt.gov.au/dlpe/advis/WATER/ground/DEFAULT.HTM. This educational site contains general comments on the availability, use and general hydrogeological characteristics of the groundwater resources. The use of this information requires the community to be well educated about hydrogeological issues. A community water resources awareness and participation program is being run as part of an existing national network (www.lpe.nt.gov.au/waterwatch). Programs are also tailored to the specific needs of the indigenous community through aboriginal Ranger training courses.

It is anticipated that by the end of this year, online access to detailed groundwater resources information will be available via the Internet. This will be the culmination of a 10-year program of work that began in 1991.

4 RECENT DEVELOPMENTS

In 1998 two initiatives provided further impetus and funding for the introduction of online access to the Government's digital data via a spatial database.

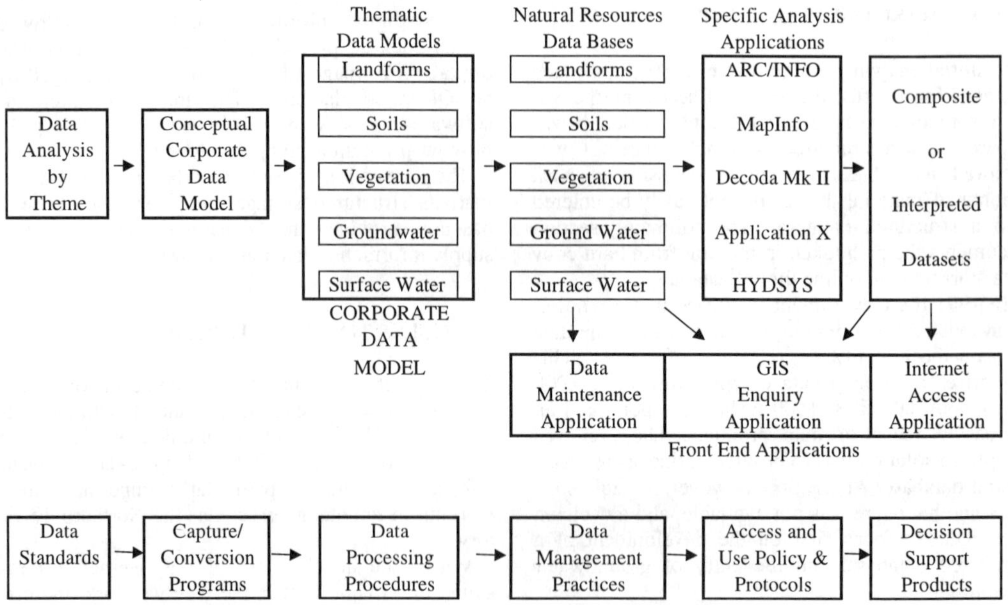

Figure 3. Model infrastructure for natural resources assessment and management.

4.1 *The National Land and Water Resources Audit*

The Commonwealth Government has identified that Australia's long-term economic prospects depend on the sustainable use of our natural resources, and conservation of our unique biological diversity. Funding of AUS$ 29.4 million has been made available nationally over the four years to June 2001, to undertake an audit of Australia's land and water resources, in order to provide a comprehensive appraisal of the national natural resource base (National Land and Water Resources Audit 1998).

The Audit aims to provide integrated, nationally compatible data sets, which are suitable for ongoing development and maintenance as a readily accessible national information system. One outcome will be a national water resource assessment that will show the extent of the surface water and groundwater resources, quality, supply capacity and use.

4.2 *Natural Resources Database (NRDB)*

In 1999 the NT Government commenced a project to develop a model infrastructure for natural resources assessment and management (refer Figure 3). The NRDB environment will comply with clearly defined design and development principles, and will be based on agreed protocols and agreed government and industry data collection and presentation standards (National Groundwater Committee 1999, Ryan & Cook 2000). It will allow effective access via Internet based software (subject to agreed security protocols) and common desktop GIS/CAD platforms such as ARC/INFO, ArcView, MapInfo and MicroStation. The project is sufficiently advanced to already allow preliminary online access to spatial data sets.

5 THE FUTURE

It is anticipated that by the end of this year, online access to information on the groundwater resources of the NT will be available via the Internet. The development of the NRDB is designed to enable Government, industry and the community to have appropriate online access to natural resources data and associated information held by the Department. This will provide a corporate environment for the effective management and use of natural resources information.

The data sources available will be:

1. A point source structured database, containing water resources data such as bore information and water levels.
2. An indexed image database, containing scanned images of technical reports and bore data records.
3. A groundwater educational web-site.
4. A selection of spatial datasets from the Natural Resources Database.

There is, however, still a considerable amount of work to do to complete this task. It is envisaged that the emphasis for the next three years will be:

1. Completion of regional water resources mapping, ensuring that relevant and appropriate products are produced.

2. Quality checking of existing spatial data.

3. Completion of data input into the Groundwater Module of the database.

4. Further development of educational Internet site.

5. Delivery of the National Land and Water Resources Audit database requirements, and implementation and modification of it as an ongoing water resource management tool.

6. Completion of the Natural Resources Database, which will provide the primary means of access to all data sets.

When this has been completed, fundamental information on the Northern Territory's groundwater resources will have shifted from the filing cabinets and bookshelves of the Government, to an online database accessible via the Internet. Professionals anywhere will be able to access this data. A mechanism will then be required to ensure that the results of studies using this data are also stored in the Government's online database, so that they are easily accessible to other researchers, the public and to those tasked with managing the groundwater resources of the Northern Territory.

REFERENCES

Domahidy, G. 1990. *Hydrogeology of the Granites Tanami Mining Region, Explanatory Notes and 1:250,000 Scale Map.* Water Resources Division, PAWA, Report 74/1990A.

Gallagher, R. 1998. *A new GIS extension for visualising water bore data and other tabular geological data.* Australian Geological Survey Organisation, Record 1998/12.

HPA InterView/Enterprise.*User Reference Manual for InterView/Enterprise Version 1.23.* Hermes Precisa Pty. Ltd. (www.hpa.com.au), 1998.

HYDSYS Version 7.27.4. hydsys@hydsys.com, (www.hydsys.com)

Jamieson, M. 1998. *Explanatory notes for the Groundwater Availability Maps of the Papunya – Yuendumu – Kintore Region Northern Territory, Wiluraratja Kapi (Western Water Study).* Natural Resources Division, DLPE, Report 20/1998D.

Jolly, P. & D. Yin Foo 1994. Investigation, development and management of groundwater resources for water supplies in remote coastal aboriginal communities in the Northern Territory, Australia. *Proceedings of Water Down Under 1994* : 423–428.

Mann, J. 1993. *Guide to HYDSYS Groundwater Data Entry and Viewing.* Water Resources Division, PAWA, Internal Report.

Matthews, I. 1997. *Hydrogeology of the Great Artesian Basin in the Northern Territory.* Natural Resources Division, DLPE, Report 52/1996A.

McGowan, R. J. 1989. *Hydrogeology of the Pine Creek Mining Region, Explanatory Notes and 1:250,000 Scale Map.* Water Resources Division, PAWA, Report 10/1989D.

National Groundwater Committee 1999. *The Australian National Groundwater Data Transfer Standard Release 1.0.* National Groundwater Committee Working Group on National Groundwater Data Standards, Canberra.

National Health and Medical Research Council 1996. *Australian Drinking Water Guidelines.* Agriculture and Resource Management Council of Australia and New Zealand.

National Land and Water Resources Audit*: Strategic Plan – 1998 to 2001.* Land and Water Resources Research Development Corporation, 1998.

Power and Water Authority 1996. *Levels of Service* (Internal Document).

Prowse, G., P. Jolly, D. Yin Foo & I. Matthews 1996. Cretaceous aquifers of the Northern Territory's Top End. Water Resources Division, PAWA, in *Extended Abstracts No. 43, Geological Society of Australia.*

Ryan, P. & R. Cook 2000. *Natural Resource Database (NRDB) Implementation Plan, Final Release V1.5.* Natural Resources Division, DLPE, Internal Report.

Sanders, R., L. Rajaratnam, & P. Jolly 1994. Water resources management at cattle station level in the wet-dry tropics of Northern Australia. *Proceedings of Water Down Under 1994* : 459–464.

Tickell, S. 1994. *Dryland salinity hazard map of the Northern Territory.* Water Resources Division, PAWA, Report 54/1994D.

Tickell, S. & L. Rajaratnam 1998. *Water resources of the Victoria River District.* Natural Resources Division, DLPE, Report 11/1998D.

Wischusen, J. 1998. *Western Water Study, Hydrogeology of the Yuendumu – Papunya – Kintore Region. Northern Territory.* Australian Geological Survey Organisation.

Zaar, U., G. Prowse & I. Matthews 1999. *Water resources of East Arnhem Land.* Natural Resources Division, DLPE, Report 2/1999D.

Groundwater: Past Achievements and Future Challenges, Sililo et al. (eds) © 2000 Balkema, Rotterdam, ISBN 90 5809 159 7

Benefits of incorporating remote sensing techniques as a methodological approach for improved borehole siting in fractured rock aquifers

N. Kellgren & P. Sander
Department of Geology, Chalmers University of Technology, Göteborg, Sweden

ABSTRACT: Satellite remote sensing data have been integrated with various groundwater indicative data in a GIS environment to delineate target areas for borehole siting. Selected target areas have been subject to detailed field investigations including geophysical techniques and exploratory drilling. The focus of the project has been to suggest a systematic and cost-effective approach to borehole siting for implementation in the groundwater development strategies of the Department of Water Affairs and Forestry of South Africa. The methodology is primarily aimed at rural community water supply in hard rock terrain. Drilling results show that success rates can be significantly improved by adopting a remote-sensing approach. Furthermore, a cost-benefit analysis clearly demonstrates that the accrued investment costs for the implementation of a remote sensing approach can be motivated already at minor improvements of borehole success rates and at relatively small scale groundwater exploration projects.

1 INTRODUCTION

1.1 *Background*

In groundwater exploration projects time and economy dictate the approach. For projects aimed at supplying larger communities, more resources can be spent on detailed investigations before drilling commences, while for single rural wells it is usually not feasible to apply costly investigation techniques. The common denominator regardless of project scale is the need for optimal use of existing information. An improved use of 'desktop data' such as remote sensing imagery, maps and borehole archives, can provide an efficient decision platform from which to select targets and exploration techniques. The integration of multiple data sets with various indications of groundwater availability decreases the uncertainty and leads to 'wiser' decisions. Many emergency drilling programmes undertaken in times of serious droughts have lacked a proper strategy for siting of boreholes, proper maintenance of equipment, and community participation (Calow, *et al.*, 1996). Sometimes selection of sites is based on the convenience of the site and land ownership rather than a proper understanding of the occurrence, movement, and recharge of groundwater. Consequently, large financial and manpower resources are spent on drilling unsuccessful boreholes.

1.2 *Objectives*

The research project has been carried out in close cooperation between the Department of Geology at Chalmers University of Technology, Sweden, and the Department of Water Affairs & Forestry (DWAF), South Africa. The primary objective has been to evaluate how remote sensing data can be integrated with other pertinent data in a GIS framework for decision support in groundwater development. The research project has specifically addressed groundwater development in hard rock environments under semi-arid conditions. This paper addresses the cost and time-effectiveness of the approach, relative to conventional practices for siting of boreholes.

2 GROUNDWATER EXPLORATION IN HARD ROCK AREAS

2.1 *Hydrogeological domains*

"Hard" rock includes metamorphic, igneous, and sedimentary formations with low primary porosity. Groundwater flow in hard rock is controlled by fractures mainly caused by deformation and weathering. Typically, the degree of deformation and weathering is locally highly variable and hydraulic conductivity and storativity may change significantly over a short distance. Success rates in borehole siting are commonly low and borehole yields are highly variable.

Fracture zones form the primary targets for groundwater exploitation in hard rock terrain. Fractures are caused by stress that commonly can be related to regional deformation, resulting in structural features such as joints and shear fractures, faults, folds, cleavage, foliation and lineation (Davis and Reynolds, 1996). However, fractures under tensile and shear conditions associated with brittle deformation in the uppermost layer of the crust are usually the most promising for groundwater development (Lloyd, 1999).Other important zones of enhanced fracturing are contacts, i.e. the zone between two lithological entities, and intrusive dykes. The emplacement of dykes is usually associated with fracturing, but the dykes themselves may, due to shrinkage during cooling, fracture more than the adjacent host rock (Lewis, 1987). Weathered zones are usually associated with the fractured zones and commonly targeted for borehole sites. However, the weathered profile is generally thin in semi-arid areas.

2.2 *Surface expressions*

With the exception of geophysical methods and the possible availability of detailed borehole data, groundwater exploration in hard rock terrain relies heavily on drawing the right conclusions from surface expressions of groundwater availability. Figure 1 illustrates the surface expressions of some typical groundwater targets in a semi-arid hard rock area.

Groundwater targets are usually selected from various groundwater indicators visible at surface, such as lineaments, phreatophytic vegetation, drainage patterns, etc. During the last three decades lineament mapping has been subject to much discussion and debate and there are as many approaches to the mapping of linear features and their subsequent interpretation to groundwater significance as there are authors of the subject. Despite this, mapping inferred structural features as lineaments is an important first step in identifying features for further investigations in hard rock areas. The key point is to be able to identify the lineaments with the highest groundwater potential, before commencing detailed and costly field investigations.

2.3 *Remote sensing and Geographical Information Systems*

Remote sensing data has been used extensively since the 1960's for the mapping of structural features in groundwater exploration programmes. Since the early 1980's various satellite sensor data have been used either alone or in combination with aerial photography. The development of user-friendly geographical information systems (GIS) in the last decade has greatly facilitated the analysis of and interpretation of remote sensing data, as well as integration with other supplementary data. However, methods for assessing the value of various input data in a GIS approach to groundwater exploration and the mutual inter-dependency of various input data, are still in their infancy.

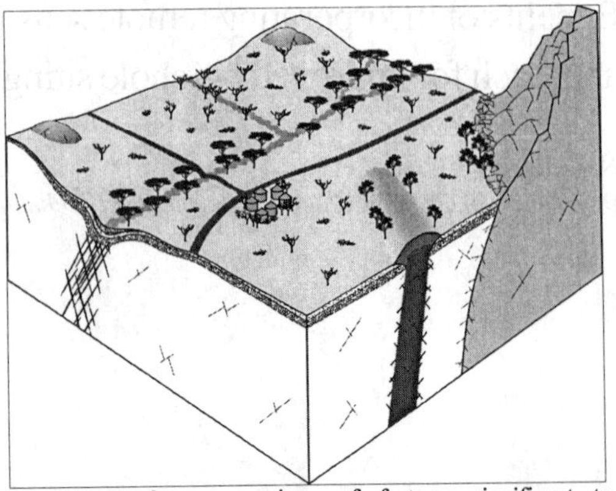

Figure 1. Surface expressions of features significant to groundwater exploration in hard rock terrain; fault zone, contact zone and dolerite dyke.

3 STUDY AREA

The study area is located in the northeastern corner of South Africa and covers approximately 1800 km^2. The area displays a large variation in topography, geology and groundwater conditions. The northern part of the study area is dominated by an older Precambrian, highly deformed, foliated and metamorphosed group of granitoid gneisses of the Limpopo Mobile Belt, known as the Beit Bridge Complex. The southern part is dominated by a shallow younger Precambrian sedimentary sequence of volcanic and arenaceous rocks of the Soutpansberg Group, while the centre is covered by Permian to Triassic argillaceous sediments and Jurassic basalts of the Karoo Group.

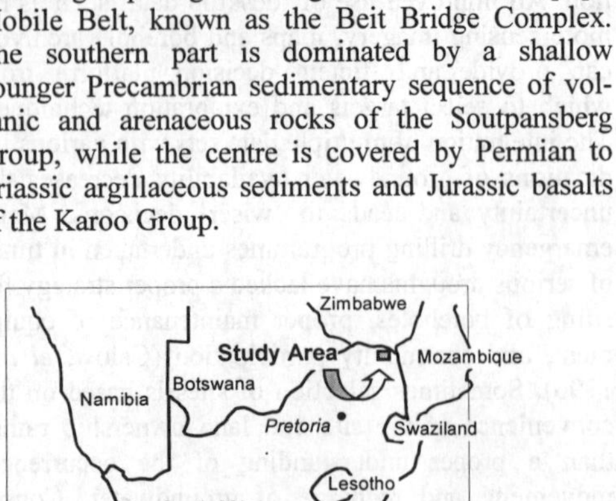

Figure 2. The project study area.

The entire study area can be regarded as hard rock terrain with no appreciable primary porosity. The exploitable groundwater resources are restricted to fractures, faults and zones of deep weathering. Previous drilling in the study area has given low yields, except in some parts of the Karoo Formation where a few high yielding boreholes have been sited on or adjacent to dolerite dykes.

4 METHODS

4.1 *Data acquisition*

Data has been collected from a number of different sources and can be divided into four different categories:

- *Groundwater target data*, primarily lineaments, which are limited in width and occurrence, and can be used as direct drilling targets, provided they can be identified in the field.
- *Complementary evidence data*, such as drainage data, boreholes and vegetation, which can add value to target data.
- *Explanatory evidence data*, which has regional coverage and provides additional information for groundwater exploration, such as lithology from geological maps, topography, and regional borehole statistics.
- *Supportive background data*, such as roads, communities, information on current water and land use, etc., which are not directly connected to the groundwater conditions.

Remote sensing data (primarily Landsat TM and SPOT) were used for the interpretation of lineaments and phreatophytic vegetation in an image processing system. The lineaments were categorised by the interpreter into three hydrogeological significance classes of fault/fracture zones and a fourth class representing inferred dolerite dykes. The hydrogeological significance class was assigned during interpretation and based on prominence, alignment with general structural trends, association with drainage and vegetation, etc.

Reproducibility tests, i.e. re-interpretation by the same and an additional interpreter, were used to decrease the subjectivity in the lineament interpretation. The interpretation resulted in approximately 800 lineaments, representing features of inferred hydrogeological significance. Extrapolation and interpolation during mapping and interpretation were kept to a minimum.

Through the use of a SAVI vegetation index inferred phreatophytic vegetation was extracted from the dry season Landsat TM image. This index is especially suited to the extraction of green vegetation in areas of low vegetation cover.

Various maps and databases were available in digital form and provided either directly or through intermediate processing the following data; topography (contour lines and DEM), drainage segments, lithology, geological structures and borehole information. Infrastructural data, such as roads and communities, were outdated on existing maps and were instead digitised from a recent SPOT panchromatic image. Existing boreholes were positioned using differential GPS techniques with an accuracy of +/- 5 meters or better. The borehole census resulted in 200 boreholes being identified in the field of which 40% could be tied to records available in the National Groundwater Data Bank (NGDB) of South Africa.

4.2 *Project investigations*

The ultimate goal for a successful approach in groundwater exploration is the reduction of exploration costs by providing (1) higher success rates, i.e. fewer dry boreholes and higher sustainable yields, (2) better use of existing data and (3) less extensive field investigations.

The aim of this project has been to develop and evaluate a methodological approach for achieving significantly reduced exploration costs in groundwater exploration. However, desk studies using satellite imagery supported by complementary data are not sufficient. Detailed geophysical field investigations at promising sites will remain the best tool for selecting the optimum borehole sites. The key to reduced exploration costs is, however, to limit their use only to the most promising areas.

The remote sensing interpretations were integrated with complementary data in the GIS. Promising lineaments in different geological settings were subsequently selected for detailed investigations using geophysical methods (EM and magnetics) followed by drilling and test pumping. The criteria used for selecting promising features were based on the lineament class and association with indicators favouring groundwater occurrence, such as phreatophytic vegetation, drainage segments, topography, etc.

The geophysical investigations were evaluated by DWAF independent of the lineaments mapped along the traverse lines. The geophysical sites were compared with the lineament mapping and drill sites were selected on the basis of correlation between geophysical anomalies and interpreted lineaments. In total 20 boreholes were drilled at twelve lineaments, representing different geological settings.

5 RESULTS

5.1 *Drilling*

The results from the exploratory drilling show a 100% success rate in intersecting structural features at the twelve investigated lineaments. However, not all of the boreholes proved to be significantly transmissive. As a means of evaluating the success rate of the approach, the results may be compared with regional borehole data from the NGDB. This data consists of 999 boreholes distributed within and around the study area, in a similar geological environment. Comparative statistics of project boreholes and NGDB boreholes are given in table 1 and figure 3. It should be noted that the true NGDB success rate is probably lower than shown, as many dry boreholes may not have been reported in the past. The median yield of boreholes drilled within this project is 1.2 l/s, compared with the NGDB of 0.5 l/s. The results demonstrate that success rates can be improved, and higher median yields can be obtained by adopting a remote sensing approach. The results are statistically significant to 97.5 % based on a Mann-Whitney test.

Table 1. Success rates in the NGDB and the remote sensing project.

	Success rate (boreholes ≥ 0.1 l/s)	Success rate (boreholes ≥ 2 l/s)
NGDB	68%	25%
Remote sensing project	85%	45%

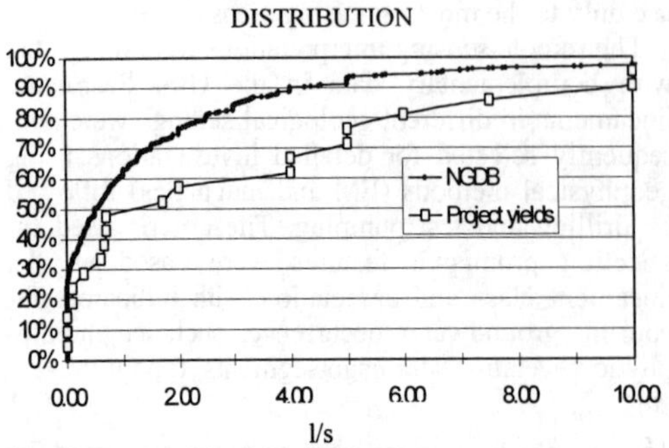

Figure 3. Cumulative distribution of borehole yields in the NGDB and the remote-sensing project.

5.2 *Cost and benefit of the approach*

The cost of establishing a borehole can be divided into two parts: siting and drilling. These costs are always involved, regardless of borehole success. The total cost will depend on the success rate, i.e. the number of boreholes needed to be drilled to obtain the required minimum yield.

The cost of siting includes equipment, personnel and transportation expenses. Total time spent for siting activities depends on the size of the area of investigation and the availability of information from previous work in the area. With an increasing number of boreholes, the siting cost per borehole can be expected to decrease. Siting costs also depends on whether or not an organisation has its own personnel and equipment, or if consultancy services are contracted for the work. Table 2 summaries typical costs associated with normal borehole siting procedures.

Table 2. Basic costs typically associated with borehole siting in South Africa. The example is given for three boreholes at the same location (modified from Sami & Murray, 1998).

Item	Cost*
Desk study 6 hrs @ R250	R1500
Community liaison 2 hrs @ R250	R500
Borehole siting 8 hrs @ R250	R2000
Field assistant 8 hrs @ R70	R560
Maps, aerial photos	R200
Geophysical equipment 1day @ R300/day	R300
Travel costs 400 km @ R3/km	R1200
Total	R6260
Siting cost per borehole	R2080

*) R1 approximately US$ 0.15 (1999)

The cost involved in drilling a borehole, whether successful or not, includes transportation, equipment, expendable supplies, drilling crew and drilling supervision. Expenses for installations are not in geology etc. Typical costs associated with borehole drilling are given in table 3.

Table 3. Basic costs (i.e. without installations) typically associated with drilling of a 100 m deep borehole in a hard rock formation (Sami & Murray, 1998).

Item	Cost
Set up at site	R500
Air drilling 216 mm, 15 m @ R 100/m	R1500
Air drilling 165 mm, 85 m @ R 70/m	R5950
Casing 165 mm, 15 m @ R 80/m	R1200
Development	R300
Total per borehole	R9450

The costs involved in a remote sensing approach include acquisition of satellite data, image processing, image interpretation, and GIS data compilation. Furthermore, costs for procurement of hardware and software need to be included. These can, however, be used in consecutive projects. The cost for a remote sensing approach is not dependent on the number of boreholes drilled, but will vary depending on the size of the study area, sophistication of software, availability of digital data, etc. An example of costs involved in a remote sensing approach is given in table 4.

Table 4. Costs associated with the implementation of a remote sensing-GIS approach in borehole siting. The costs are estimated for a hypothetical area covering 2000 km^2 with 40 villages.

Item	Cost
Satellite Data, 1 scene Landsat TM	R13000
Image interpretation and GIS data compilation, 300 hrs @ R250	R75000
Personal Computer *	R12000
Colour Printer*	R3000
Software: GIS, Image processing*	R30000
Total	R133 000

*Investment cost for equipment that can be used in consecutive projects

If more resources are spent on siting procedures higher success rates will most likely ensue. However, at a certain stage the costs associated with the use of more advanced site selection cannot justify the marginal improvement in success rate. The relationship between the two approaches for borehole siting can illustrated by the following formula, which shows in terms of cost-benefit how money can be saved through an improved success rate minus necessary investment costs (modified from van Dongen & Woodhouse, 1994).

$$CB = C_c - (C_{rs} + C_i) = N \times (C_s + C_d) \times \left(\frac{1}{R_c} - \frac{1}{R_{rs}}\right) - C_i$$

where:

CB = the cost-benefit of the approach

$C_c = \frac{N \times (C_s + C_d)}{R_c}$ = exploration cost, conventional approach

$C_{rs} = \frac{N \times (C_s + C_d)}{R_{rs}}$ = exploration cost, remote-sensing approach

R_c = the success rate using a conventional approach
R_{rs} = the success rate using a remote sensing approach
C_s = the cost of the site investigation (Table 2)
C_d = the cost for drilling (Table 3)
C_i = investment cost for a remote sensing approach (Table 4)
N = number of successful boreholes needed

The formula assumes that the costs associated with site investigations and drilling remain the same for both methods. However, the main idea of the approach is that time and money spent on site investigations will be less using a remote sensing approach.

Experience from this project and the basic costs as described above may be used in a hypothetical example, where 40 successful boreholes are needed in a 2000 km^2 study area. In order to equip a borehole with a motorised pump, a yield of 2 l/s can be considered as a minimum requirement. In the NGDB, 25% of the boreholes in the study area have a yield of 2 l/s or more, and out of these the median yield is 3.6 l/s.

The results achieved in the remote-sensing project give a success rate of 45% with a median yield of 5 l/s for boreholes exceeding 2 l/s. Consequently, assuming that conventional siting methods were used for the NGDB boreholes, 160 boreholes need to be drilled to achieve 40 successful boreholes with a minimum yield of 2 l/s. With the use of a remote-sensing approach 89 boreholes have to be drilled to achieve the same result. The following figures may be applied: $C_s = 2080$, $C_d = 9450$, $C_i = 133000$, $N = 40$, $R_c = 25\%$ (median yield 3.6 l/s), $R_{rs} = 45\%$ (median yield 5 l/s).

By inserting these figures into the cost-benefit formula, project savings from the use of a remote sensing approach, will be R687,000 (approximately US$ 100,000). Figure 4 illustrates project savings as a function of the number of successful boreholes, and figure 5 illustrates project savings as a function of an improved success rate. These figures demonstrate that the break-even point (i.e. CB = 0), at which a remote sensing approach is justified, can be

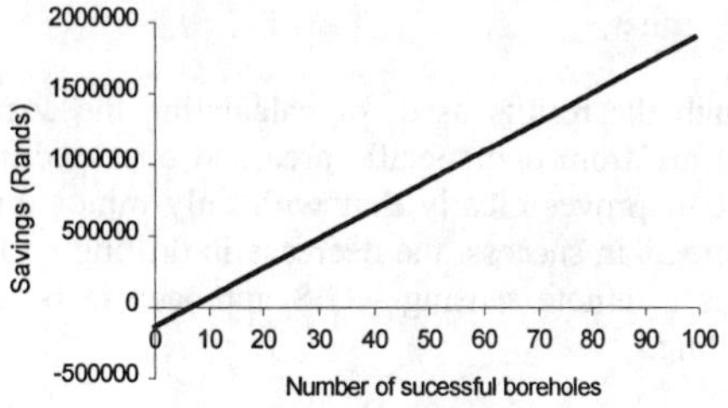

Figure 4. Savings as a function of number of successful boreholes (assuming R_c=25% and R_{rs}= 45%).

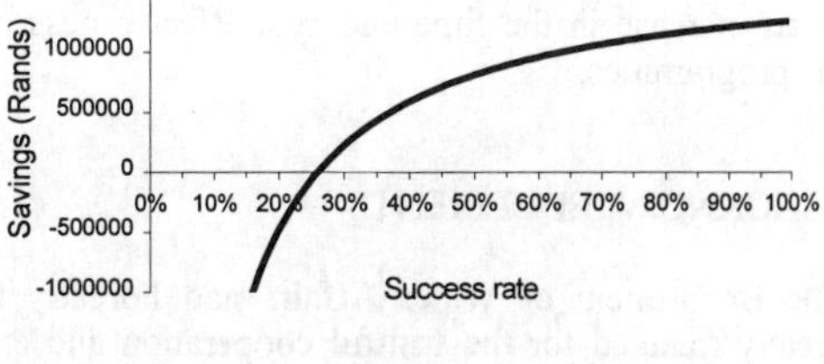

Figure 5. Savings as a function of improved success rates (assuming that 40 production boreholes are needed).

achieved with only 6 successful boreholes needed or when the success rate is improved from 25% to 27%.
Furthermore, in this hypothetical example, no attention has been given to the fact that the remote sensing approach has achieved an improvement of the average borehole yield. The total number of borehole required in a project may be less as higher yielding boreholes can supply more households.

6 DISCUSSION AND CONCLUSIONS

The field investigations performed within the project clearly verify the lineaments interpreted from the remote sensing data. The results indicate that success rates can be significantly improved by employing a remote sensing approach. Complementary evidence from independent data sources also clearly increases the confidence in interpreted features. The results also demonstrate the necessity of complementing lineament data with geophysical ground investigations.

At four of the twelve lineaments investigated, more than one borehole was drilled, typically with a significant variation in yields over a short distance. A dry borehole or a very low yield, might simply be the result of poor siting due to misinterpretation of the geophysical data. If only the most recently drilled borehole at each location is considered, then both the success rate and the median yield will increase further.

Although the results used for calculating the cost-benefit are from one specific area and one specific project, it proves clearly that with only minor improvements in success, the decrease in drilling costs justifies a remote sensing - GIS approach in borehole siting.

By emphasising the aspects of incorporating remote sensing data and combining these with other groundwater indicators as a systematic and integrated plan of action, will reduced uncertainties in groundwater exploration programmes. This leads to greater accuracy in the search for groundwater and to an increase in the time and cost-effectiveness of the programmes.

7 ACKNOWLEDGEMENTS

The Department of Water Affairs and Forestry is greatly thanked for the fruitful cooperation and extensive support with geophysical investigations, drilling and various borehole investigations. Special thanks go to the Directorate Geohydrology in Pretoria. The Swedish International Development Cooperation Agency (Sida) is thanked for providing funding for this study, through the SAREC programme.

REFERENCES

Calow, R.C., Macdonald, D.M.J and Orpen, W.R.G., 1996. *Groundwater Management in Drought Prone Areas of Africa: South African Inception Report*, British Geological Survey Technical Report WC/96/56, Second Draft

Davis, G.H., Reynolds, S.J.. 1996. *Structural geology of rocks and regions, second edition*, John Wiley & Sons Inc.

Lewis, M.A., 1987. *The analysis of borehole yields from basement aquifers, Proceedings of Groundwater Exploration and Development in Crystalline Basement Aquifers, June 1987*, Commonwealth Science Council, Vol. II, pp. 171–202

Lloyd, J.W. (editor), 1999. *Water resources of hard rock aquifers in arid and semi-arid zones*, Unesco, Studies and reports in hydrology 58

Sami, K., Murray, E.C., 1998. *Guidelines for the evaluation of water resources for rural development with an emphasis on groundwater*, Institute for Water Research Rhodes University, Water Research Commission Report No. 677/1/98

van Dongen, P., Woodhouse, M., 1994. *Finding groundwater A project manager's guide to techniques and how to use them*, UNDP-World Bank Water and Sanitation Program, Technical Report

Groundwater: Past Achievements and Future Challenges, Sililo et al. (eds) © 2000 Balkema, Rotterdam, ISBN 90 5809 159 7

The valuation of major aquifers of the Upper Silesia region

A. Kowalczyk, A. Różkowski & A.J. Witkowski
Department of Hydrogeology and Engineering Geology, University of Silesia, Sosnowiec, Poland

ABSTRACT: Potable groundwater in the Upper Silesia region occurs in the Cenozoic, Mesozoic and Paleozoic formations. Valuation of groundwater has been performed according to various hydrogeological conditions and by applying several quantitative and qualitative criteria. As the first stage, 18 major aquifers have been identified in the study area. They have then been valuated using a procedure which enables the identification of aquifers of fundamental importance for water supply and for management planning. This paper presents the hydrogeologic characteristics of these aquifers, the concept of their valuation and the results of this valuation.

1 INTRODUCTION

The Upper Silesia urban-industrial region (Southern Poland) constitutes one of the most industrialised areas in Europe. It is caused by a huge concentration of mineral deposits, including hard coal, zinc and lead ores and other raw materials. In total, all 56 mining areas, together with the hard coal mines, have a total area reaching 1 750 km^2 (Fig. 1). The population is about 3 908 000 inhabitants within an area of ca. 6650 km^2. Groundwater of the region is the second largest source of water. Its contribution is about 30% of the total water supply. Potable groundwater in this area occurs within the Quarternary, Tertiary, Cretaceous, Jurassic, Triassic, Carboniferous and Devonian formations.

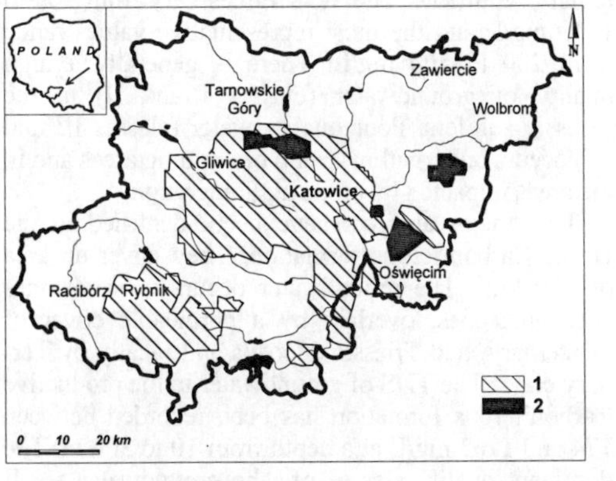

Figure 1. Locality map of study area showing the extent of mining areas in Upper Silesia. 1: coal mining, 2: Zn-Pb ore mining.

It is pumped by wells and dewatered by mines. Due to the variability of hydrogeological conditions, (quantity and quality of groundwater resources), possibility of future usage and for management planning reasons, the valuation of groundwater has been performed.

During the first stage, 18 major aquifers have been identified within the research area by applying generally accepted hydrological criteria (Różkowski et al. 1997). Next, the valuation of these aquifers has been conducted by applying several quantitative and qualitative criteria (Witkowski et al. 1998). This has enabled us to distinguish aquifers of fundamental importance for water supply and for management planning.

2 HYDROGEOLOGICAL UNITS

2.1 *Identification of major aquifers*

Potable groundwater in Upper Silesia occurs within the Quaternary, Tertiary, Cretaceous, Jurassic, Triassic, Carboniferous and Devonian formations. Within these formations, the major aquifers have been identified by their very advantageous hydrogeologic conditions. These were determined on the basis of the following quantitative and qualitative criteria proposed in the work edited by Kleczkowski (1990):

- transmissivity of the aquifer - over 10 m^{2}h^{-1}
- potential yield of individual wells - over 70 m^{3}h^{-1}
- total capacity of well fields - over 10 000 m^3 per day
- water quality - equivalent to I, II and III quality classes (according to classification proposed by the State Inspection of Environmental Protection), and

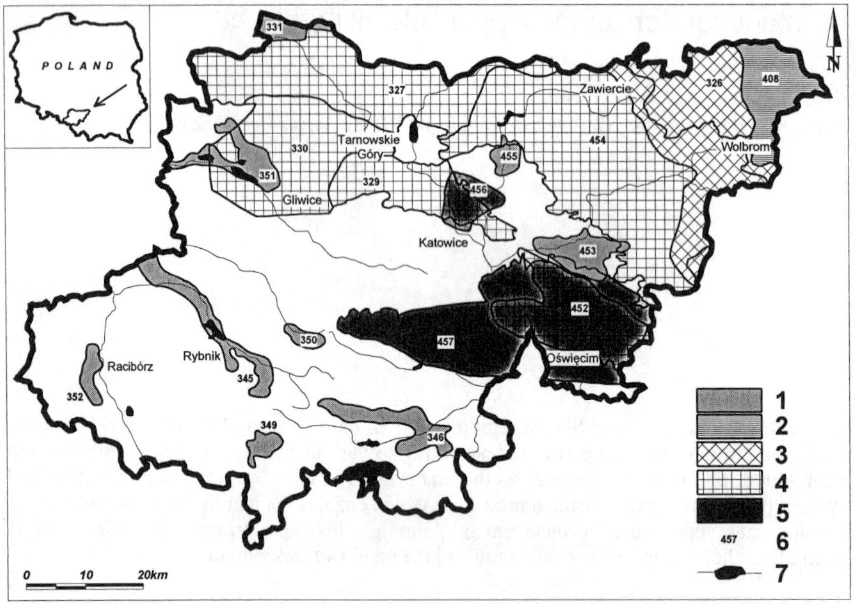

Figure 2. Map showing location of major aquifers in Upper Silesia within hydrogeological formation: 1-Quaternary, 2-Cretaceous, 3-Upper Jurassic, 4-Triassic, 5-Carboniferous, 6-N°_of major aquifers, 7-main rivers and lakes.

IV class – out of classification.

Some individual criteria have been applied considerably less strictly when compared with those cited for aquifers identified within the Quaternary and Carboniferous formations.

As a result 18 major aquifers have been identified with respect to the criteria cited above. They are presented on the map (Fig. 2). In some cases two to three aquifers occur in a single hydrogeologic profile. The total area of the aquifers is 4 228 km^2 and they occupy an area of 3 442 km^2 within studied region. The hydrogeologic characteristics of the major aquifers summarised in this section and in Table . 1, are based on the hydrogeologic monograph by Różkowski et al. (1997).

Within the Quaternary formation, nine major aquifers of porous character have been identified. They are situated in the areas of buried valleys and modern river valleys. In general they are unconfined aquifers, sometimes partly covered by a confined layer of clay. Their total area is 464 km^2. The TDS content ranges from 47 to 1374 mg/L. More than 80% of samples represent quality classes II and III in the four-class scale of water quality, because of urban and industrial impacts.

Seven major aquifers have been identified within the Mesozoic formations. The Upper Cretaceous major aquifer occupies an area of 144 km^2 in the study region. It is an unconfined to semi-confined aquifer, composed of fractured marls and limestones. Dominant quality classes are II and III, mainly due to higher concentrations of NO_3, Fe and Mn ions.

The fractured-karstic-porous Upper Jurassic aquifer occupies an area of 472 km^2. It an unconfined aquifer composed of limestones. It contains fresh water, for which the TDS varies from 200 to 500 mg/L and the dominant quality of water is class I.

Within the Triassic carbonate formation, five major aquifers have been identified. Their total surface area is 4100 km^2. They are composed of fractured-karstic-porous dolomites and limestones. This water-bearing complex is overlain partly by impermeable Keuper clays and by permeable Quaternary sediments. The TDS values vary from 103 to 1519 mg/L and the most representative values range from 200 to 500 mg/L. There is generally a high quality of groundwater (classes I and II) in the Triassic aquifers. Poor quality water (classes III and IV) occurs only within Zn-Pb ore mining areas and in the areas impacted by urban agglomeration.

Two major aquifers have been identified in the Upper Carboniferous formation. They cover an area of 770 km^2. The groundwater occurs in sandstones and mudstones, overlain by a permeable cover of Quaternary and Triassic deposits and locally by Tertiary clays. The TDS of groundwater in the productive Carboniferous formation has been recorded between 174 and 1763 mg/L at a depth from 10 to 300 m. The dominant quality classes of the groundwater are II and III, and occasionally class IV. In general, cumulative mine water is of poor quality.

Table 1. Hydrogeologic characteristics of major aquifers classified according to hydrogeologic formation (according to Różkowski et al. 1997, completed by authors).

Hydrogeologic formation	Number of main aquifers	Surface [km^2] a* / b*	Type of rock formation	Average thickness of aquifer [m]	Hydraulic conductivity [m/s]	Hydrodynamic conditions
Quaternary	9	464 / 464	porous	22	$4,16 \times 10^{-4}$	unconfined
Upper Cretaceous	1	144 / 144	fissured-porous	100	$0,9 \times 10^{-4}$	unconfined
Upper Jurassic	1	1207 / 472	karst-fractured-porous	250	$0,8 \times 10^{-4}$	unconfined
Triassic	5	4100 / 2378	karst-fractured-porous	110	$0,84 \times 10^{-4}$	confined-unconfined
Upper Carboniferous	2	770 / 770	fissured-porous	80	$0,2 \times 10^{-4}$	unconfined-confined

*a – total surface area; b – surface area within study region

2.2 *Groundwater dewatering and resources*

Groundwater dewatering by mining and wellfields has caused significant changes in rates of recharge and discharge, and ultimately in groundwater resources. This is of concern in the Triassic aquifers which are being dewatered by Zn-Pb ore mines (Kowalczyk et al. 1999), the Carboniferous aquifers dewatered by coal mines and two Quaternary aquifers intensively dewatered by sand-pits (Różkowski et al. 1997). Estimated groundwater resources of these aquifers vary from about 2 to 168 Mm^3/year (Table 2) (Różkowski et al. 1996) and the recharge ratio varies from 107 to 1521 m^3/d·km^2 (1,2 to 17,6 L/s·km^2). The highest recharge rate values are induced by intensive dewatering of groundwater in some mining areas. The estimated total value of groundwater resources is 564 Mm^3/year for all aquifers. About 50% of the resources are attributed to the Triassic aquifers.

The degree of development of the resources is variable (Table 2). For the valuation procedure, the following categories of estimated annual pumpage have been used: lower than 1 Mm^3/year, between 1 and 10 Mm^3, and more than 10 Mm^3. The total pumpage of groundwater from the major aquifers is about 385 Mm^3/year.

2.3 *Groundwater vulnerability*

Vulnerability assessment of groundwater is based on the travel time of water and pollutant from the surface to an aquifer. This time and its variability depends on the thickness of isolation cover overlying the aquifer or the depth to the unconfined aquifer (Paczyński 1997).

Table 2. Groundwater resources of major aquifers, their quality and vulnerability.

Nº	Nº of aquifer according to Kleczkowski (1990)	Stratigraphy	Surface [km^2]	Groundwater Resources [Mm^3/year]	Recharage rate [m^3/d· km^2]	Pumpage [Mm^3/year]	Ground-Water Quality Classes	Thickness of isolation cover or depth to aquifer [m]	Vulnerability
1	328	Q	87	8,19	258	<1	I,II	<50	high
2	351	Q	74	7,45	276	<1	III,IV,II	<50	high
3	455	Q	21	11,7	1521	11,67	III,II	<50	high
4	453	Q	67	28,29	1158	28,3	III,II	<50	high
5	352	Q	22	2,14	625	6,2	II,III,I	<50	high
6	345	Q	84	9,81	320	6,5	II,III,I	<50	high
7	349	Q	20	1,89	259	<1	III,II	<50	high
8	350	Q	15	2,25	411	2	II,III	<50	high
9	346	Q	74	8,6	320	<1	III,II	<50	high
10	408	Cr	144	11,35	216	1	II,III,I	<50	high
11	326	J	472	59,6	346	5	I,II,III	50-150	moderate/low
12	327	T	557	31,6	107	42	II,III,IV,I	50-250	moderate
13	330	T	411	45,1	298	29	I,II,III	50-150	moderate
14	329	T	130	26,4	553	26	II,III,IV,I	50-150	moderate
15	454	T	1033	168,36	449	150	I,II,III,IV	50-150	moderate
16	452	T	247	29,69	369	27	II,I,IV,III	50-150	moderate
17	456	C	60	15,5	708	15,5	II,I,IV	50-150	moderate
18	457	C	710	96,1	371	~31,0	II,I,III,IV	50-150	

Vulnerability assessment was performed according to the following main criteria, presented in Table 3:
-high vulnerability - travel time up to 25 years, weak isolation up to 50 m thick,
-moderate vulnerability - travel time 25-100 years, isolation cover 50-150 m thick,
-low vulnerability - travel time more than 100 years, isolation cover more than 150 m thick.

Results of the vulnerability assessment of major aquifers are presented in Table 5. The most vulnerable are the Quaternary aquifers, the Jurassic aquifer and the unconfined parts of the Triassic aquifers. This fact is confirmed by declining water quality in these aquifers.

3. GROUNDWATER VALUATION

There are serious differences between the various principles of groundwater valuation, aimed at identifying the groundwater systems which are most valuable to the society. Groundwater is valued mainly in economic terms (Custodio & Gurgui 1989, Winpenny 1996). In the past hydrogeologic assessment has been mainly limited to the evaluation of groundwater resources and quality of water. Recently the method of valuation of groundwater by assessment of its vulnerability to pollution as an important criterion have been developed (Albinet & Margat 1970, Vrba & Zaporozec 1994, Cost action 65 1995, Eaton & Zaporozec 1997; Vrba & Zaporozec 1999). However there are important differences between the principles of these valuation methodologies. For example, the valuation method proposed by Scharp et al. (1997) recognises such criteria as available quantities of groundwater, groundwater quality, present or planned use and sensitivity to changes in groundwater level. It does not take into account such important criterion as vulnerability of the groundwater.

3.1 *Concept and criteria of valuation*

The valuation presented in this paper was performed according to the assumptions and methodology proposed by Paczyński (1995, 1997) modified by the authors (Kowalczyk et al. 1998, Witkowski et al. 1998). The valuation procedure follows the basic criteria (Table 3) and supplementary criteria (Table 4). The basic criteria are water quality and vulnerability of a groundwater system.

As supplementary criteria the following have been included:
- the state of reserves of groundwater resources -"α",
- the importance of groundwater system in water supply expressed by the pumpage rate of water from an aquifer - "β"
- importance of the aquifer in regional water supply system -"γ"
- recharge rate of aquifer system expressed in m^3/day.km^2 - "δ". The higher value of the recharge rate the lower numerical index "δ".

The list of applied criteria and their numerical indices used for valuation of aquifers is shown in Tables 4 and 5.

Table 3. Basic valuation criteria (w)

Groundwater quality	Vulnerability as function of isolation cover or depth to aquifer		
	low > 150 m	moderate 150-50 m	high <50 m
very good	10	7	5
good	8	5	3
moderate	6	3	1
poor	0,5	0,5	0,5

The value of the groundwater, W, has been found as product of the basic valuation criteria index (w) compiled in Table 5, then multiplied by indices corresponding to the supplementary criteria, summarised in Table 5. The final result of valuation is described by equation (1):

$$W = w \cdot \alpha \cdot \beta \cdot \gamma \cdot \delta \quad (1)$$

Table 4. Supplementary criteria for valuation: α,β,γ,δ

Criteria:	Valuation parameter	Index
α - reserves of groundwater resources	< 25% of max	1,5
	25-50% of max	1,25
	>50% of max	1,0
β - pumpage	<1 Mm^3/year	1,0
	1-10 Mm^3/year	1,25
	>10 Mm^3/year	1,5
γ - importance of the aquifer in water supply	-the main or the only source of water	1,5
	-alternative source	1,0
δ - recharge rate of aquifer	100-300 m^3/day km^2	1,5
	300-500 m^3/day km^2	1,25
	>500 m^3/day km^2	1,0

3.2 *Results*

Results of valuation of aquifers are presented in Table 5. The aquifers have been classified into four classes of value on the basis of the valuation performed:
1. >15 points - extremely high (A)
2. 10 - 15 points - high (B)
3. 5 - 10 points - moderate (C)
4. < 5 points - low (D)

It has been found that 10 aquifers present low value (class D). Seven Quaternary aquifers and one of the Cretaceous, Triassic and Carboniferous fall into this category. Their groundwater is vulnerable to pollution, and it is already polluted to different degrees.

Four Triassic and one Jurassic aquifers show high value (classes A and B). The high rank of these aquifers is confirmed by the quantity and quality of their groundwater resources, and their importance for water supply.

Table 5. Results of valuation of major aquifers

N°	N° of aquifer according to Kleczkowski, (1990)	Stratigraphy	Criteria of valuation					Value of aquifer	
			w	α	β	γ	δ	points, W	class
1	328	Q	3	1,0	1,0	1,0	1,5	4,5	D
2	351	Q	0,5	1,0	1,0	1,0	1,5	0,75	D
3	455	Q	2	1,5	1,5	1,0	1,0	4,5	D
4	453	Q	2	1,5	1,5	1,0	1,0	4,5	D
5	352	Q	2	1,5	1,25	1,5	1,0	5,6	C
6	345	Q	2	1,25	1,0	1,0	1,5	4,7	D
7	349	Q	1	1,0	1,0	1,0	1,5	1,5	D
8	350	Q	2	1,5	1,25	1,5	1,25	7,0	C
9	346	Q	2	1,0	1,0	1,0	1,25	2,5	D
10	408	Cr	2	1,0	1,0	1,5	1,5	4,5	D
11	326	J	5	1,0	1,25	1,5	1,25	11,7	B
12	327	T	5	1,5	1,5	1,5	1,5	16,9	A
13	330	T	4	1,5	1,5	1,5	1,25	16,9	A
14	329	T	2	1,5	1,5	1,0	1,0	4,5	D
15	454	T	4	1,5	1,5	1,5	1,25	16,9	A
16	452	T	3	1,5	1,5	1,5	1,25	12,6	B
17	456	C	2	1,5	1,5	1,0	1,0	4,5	D
18	457	C	3	1,0	1,5	1,0	1,25	5,6	C

All analysed major aquifers are of importance for water supply in the region. However, the aquifers classified as belonging to the A and B classes are of the highest rank and their exploitation has to be limited to the supplying of drinking water. The special protection of water quality in these aquifers is also required.

4 CONCLUSIONS

For the Upper Silesia region, 18 major productive aquifers have been identified, based on certain hydrogeologic criteria. These aquifers are of varying importance for groundwater supply and for management planning.

With respect to our valuation assessment, the aquifers of high value, class A and B should be subject to special protection and groundwater exploitation should be strictly controlled.

The Triassic aquifers and the one Jurassic aquifer are the most important aquifers in terms of the usable water supply of the Upper Silesia region. Some of them have been too intensively and irrationally exploited by water intakes and they have been dewatered by mine workings. For these aquifers a new program of a rational exploitation and groundwater protection should be implemented. Particularly the proper localisation of wells, restriction of exploitation by industry and rational control of groundwater withdrawal is vital. Closure of some wells is also advisable. The protection areas of the aquifers under discussion, drawn on maps at a scale of 1:50,000 should be legally valid. In this new version of the protected areas, the latest state of hydrogeologic knowledge of aquifers, their resistance to anthropogenic pollution as well as areal distribution of national parks, landscape parks and areas of protected landscape, should be taken into consideration. All activities that represent a threat to the groundwater should be banned from these areas either because of the groundwater vulnerability or because of the value of the groundwater.

The valuation methodology presented in this paper gives priority to the quality of water and vulnerability of aquifers. These criteria seem to be the most important in the valuation of aquifers with very different hydrogeologic conditions. Additional criteria are dependent on regional conditions. For example, it is sometimes not possible to take into account an alternative source of ground- or surface-water.

Although unification of groundwater valuation methodologies is neither purposeful nor possible, valuation methods should still be developed, since they enable identification of the most valuable aquifers. They are also a valuable guide for planners and decision makers for improving rational water management and the protection of groundwater resources at regional level.

REFERENCES

Albinet, M.& J. Margat. 1970. *Cartographie de la vulnerabilite a la pollution des nappes d'eau souterraine*, Bull.BRGM, 2eme serie, section 3, nr 4 ,13-22.

Cost action 65, 1995. Karst groundwater protection. Final report. European Commission. Report EUR 16547 EN.

Custodio, E. & A. Gurgui (eds.) 1989. *Groundwater economics. Developments in water science.* Elsevier Science Publishers B.V.

Eaton, T. & A. Zaporożec 1997. Evaluation of groundwater vulnerability in an urbanizing area. In Chilton et. al. (eds.)

Groundwater in the Urban Environment: Problems, Processes and Management: 577-582. Rotterdam: Balkema.

Kleczkowski, A.S. (ed.) 1990. *The Map of the critical protection areas (CPA) of the major groundwater basins (MGWB) in Poland. 1:500000.* Kraków: Academy of Mining and Metallurgy .

Kowalczyk, A. et al. 1998. Groundwater valuation in the area of RZGW Katowice. *Unpublished report.* Sosnowiec ZBU Intergeo (In Polish)

Kowalczyk, A. et al. 1999. Groundwater flow system and water balance of the intensively exploited Chrzanów aquifer. Poland. *Proc. of the XXIX IAH Congress, Bratislava*: 785 – 790.

Paczyński , B. ed. 1995. *Hydrogeological Atlas of Poland 1 : 500000.* Warszawa: Polish Geol. Inst.

Paczyński, B. 1997. Valuation of groundwater: Future trends in vulnerability mapping. *Proc. of the XXVII IAH Congress. Nottingham.* UK. 21-27 Sept. 1997:641–645.

Różkowski A., A. Chmura. & A. Siemiński (eds.) 1997. Usable groundwater in the Upper Silesian Coal Basin and its margin. Warszawa*: Prace PIG* (In Polish*)*.

Różkowski, A., A. Kowalczyk & A. Witkowski 1996. Occurrence, resources and management of groundwater in Upper Odra and Upper Vistula catchments. (Katowice and Bielsko Biała region). *Przegląd Geologiczny.* 44(8) (In Polish).

Scharp, C, Alveteg T. & Johansson, P.-O. 1997. Assigning a groundwater protection value: Methodology development. In Chilton et. al. (eds.) *Groundwater in the Urban Environment: Problems, Processes and Management.* Rotterdam: Balkema.

Vrba, J. & A. Zaporożec A (eds). 1994. Guidebook on Mapping Groundwater Vulnerability, IAH, International Contributions to Hydrogeology, 16.

Vrba, J. & A. Zaporozec 1999. Uses and limitations of groundwater vulnerability maps in land use planning. *Proc.of the XXIX IAH Congress. Bratislava. SR.* 6-10 Sept. 1999: 115-118.

Winpenny, J.T. 1996. The value of water valuation. In Howsam & Crater (eds.) *Water policy: Allocation and management in practice.* London: E&FN Spon.

Witkowski, A. et al. 1998. The analysis of changes of groundwater quality of selected major aquifers in Katowice region in 1990 – 1996 for undertaken the protective measures. *Unpublished report.* Sosnowiec: University of Silesia (In Polish).

Groundwater: Past Achievements and Future Challenges, Sililo et al. (eds) © *2000 Balkema, Rotterdam, ISBN 90 5809 159 7*

Geologic factors influencing distribution of hardrock transmissivity

J. Krásný
Faculty of Science, Charles University, Prague, Czech Republic

ABSTRACT: In the hardrock environment of the Bohemian Massif (Czech Republic) inhomogeneity elements of at least three orders of magnitude were defined. Local fissure systems of different dip and strike (local inhomogeneities) might be seen at outcrops of hard rocks and are tapped by particular water wells or determined by well-logging techniques. Inhomogeneities of a medium scale (sub-regional inhomogeneities) usually follow river valleys and/or are represented by fractured zones characteristic by higher prevailing transmissivity. Regional scale inhomogeneities are evidently caused by different neotectonic activities. This hierarchic system of fracture inhomogeneities of different orders of magnitude is decisive for spatial distribution of transmissivity in hard rocks. Practical consequences for regional groundwater flow modelling, safe yield assessment, well siting and groundwater vulnerability studies are evident. Small differences in transmissivity caused by distinct lithologic composition of hard rocks are usually masked by rock fracturation, causing larger differences.

1 INTRODUCTION

Hydrogeologic environment is never homogeneous and isotropic. It is well known that results of hydrogeologic studies concerning transmissivity and/or hydraulic conductivity spatial distribution strongly depend on the extension of the study area ("size of a tested sample") and on its relation to the size of decisive inhomogeneity elements of the respective hydrogeologic environment. Rats (1967) and Rats & Chernyshov (1967) stated this in general terms for the first time. In a hydrogeologic environment characteristic by inhomogeneity elements of certain size (intergranular pores, small fissures etc.) hydraulic conductivity and transmissivity mean values become closer with increasing study area. Finally these values are practically the same, irrespective of the position of the tested area within this environment, as recently discussed e.g. by Schulze-Makuch & Cherkauer (1998) and Rovey II (1998). The extension of an area representing lowest limit above which practically no changes in mean transmissivity values occur is designated as a representative elementary volume (REV). This consideration is valid only if other, larger inhomogeneity elements do not occur in the respective hydrogeologic environment. If this is the case we have to consider one of the two possibilities:

1. To define a more extended area in order to reach a mean, spatially not changing transmissivity, representing such a new larger REV.

2. To consider the respective hydrogeologic environment as a complex system of two or more environments with different prevailing transmissivity.

The second approach is more recommendable as it reflects the real hydrogeologic conditions. It should be based, however, on detailed and time consuming hydrogeologic analysis of a study area.

Hydrogeologic studies performed during last years in the Czech Republic have brought important information on hardrock hydrogeology (Krásný 1996a). Statistical samples of transmissivity data were treated in different areas, according to different petrographic rock types and structural, geomorphologic and hydrogeologic position of water wells. The used methodology and classification system is described in Krásný (1993a).

In a local scale, any hardrock environment is characteristic by a chaotic permeability and transmissivity distribution. This is due to usually prevailing system of fractures and fissure zones (inhomogeneity elements) of different orientation that might be seen at outcrops of many hard rocks. In more extensive areas mean and prevailing transmissivity values (and those of hydraulic conductivity, too), as determined by results of aquifer tests, are usually very similar. These values can be considered as characterising the respective REV in a local scale. Occurrence of inhomogeneity elements of higher order may cause other – supplementary differences in transmissivity values. Studies on transmissivity distribution in hard rocks of the Bohemian Massif revealed at least two of such higher levels caused by larger inhomogeneity elements. Results of these studies that might be expressed by a hierarchic system of inhomogeneity elements of different order in hard rocks, i.e. local, medium scale (sub-regional) and regional scale inhomogeneities are summarised in this paper.

2 LOCAL SCALE INHOMOGENEITIES

In a local scale, irregular changes in permeability and transmissivity spatial distribution within the "near-surface aquifer" of hard rocks (i.e. its weathered and fissured zones) are common. They are evidenced by differences in yields and hydraulic parameters of near-by wells drilled in the same rocks. These differences may reach several (usually up to three but sometimes even up to four) orders of transmissivity magnitude (Krásný 1996a). Therefore the positive and negative anomalies (i.e. the extreme values of transmissivity) within the same environment (represented by a statistical sample) might differ considerably. The positive anomalies might indicate preferential paths of groundwater flow important for considerations on groundwater development possibilities (siting of water wells) and contamination transport and, on the other hand, remediation effectiveness. These differences are mainly attributed to distinct character and abundance of faults and joints in the middle (fissured) zone of hardrock environment (case *a* in Figure 1) and to changes in properties and thickness of the weathered zone in the uppermost part of a hydrogeologic environment.

3 SUB-REGIONAL (MEDIUM-SCALE) INHOMOGENEITIES

By comparing more extended areas differences in transmissivity and permeability typically decrease if larger inhomogeneity elements do not occur. Then regionally prevailing values („hydrogeologic background“) as well as extreme (maximum and minimum, i.e. anomalous) transmissivity values of various statistical samples covering distinct areas of the same hydrogeologic environment (indicated by distinct squares of case *b* in Figure 1) are usually very similar. Transmissivity background for many hardrock areas in the Czech part of the Bohemian Massif may be characterised by units of m^2/d up to slightly more than 10 m^2/d. These values correspond to transmissivity classes IV(-III) c,d after the classification of transmissivity magnitude and variation (Krásný 1993a).

Superposed upon the transmissivity background caused by a more or less regular fissuring (from a regional point of view - case *b* in Figure 1), significant regional differences in transmissivity magnitude have been proved due to the presence of inhomogeneity elements of a higher scale level. They belong usually to tectonically strongly affected zones and/or to belts of regional higher permeability along the river valleys.

First of them occur in areas where considerably higher transmissivity was determined in more water wells, thus indicating more extensive "anomalous" areas. Two examples from petrographically distinct hardrock environment in southern Bohemia with prevailing transmissivity between 40 and 90 m^2/d are mentioned by Krásný (1996b). These areas, apparently belonging to tectonically strongly affected zones, are represented by prevailing transmissivity almost one order of magnitude higher in comparison with surrounding areas that can be considered hydrogeologic background. Such positively anomalous zones might be of importance for groundwater development in hardrock environment. Anomalous and highly permeable deep-seated faults enable thermal and mineral water ascendant flux in some regions, too. Havlík & Krásný (1998) delimited both positive and negative anomalous areas in southern Bohemia.

Other significant sub-regional differences in transmissivity magnitude have resulted from comparison of statistical samples of transmissivity values from water wells sited in distinct geomorphologic and/or hydrogeologic position. More authors mention differences in prevailing permeability of rocks depending on distinct topographic or morphologic position (bottom of valleys, slopes etc. – LeGrand 1954, Krásný 1974, and Henriksen 1995).

In hard rocks of the Czech part of the Bohemian Massif prevailing transmissivity was estimated 2 to 4.5 times higher in zones of discharge (valleys) than in recharge zones (above all slopes and elevations). During the analysis, wells with possible or probable inflow from usually better permeable Quaternary cover were excluded from all sampled populations. In this way results reflect only permeability of underlying rocks, i.e. mostly their middle (fissured) zone. The differences in arithmetic means are also accompanied by differences in prevailing transmissivity values (hydrogeologic background). Results from various areas of the Bohemian Massif built by igneous and metamorphic rocks (granites, gneisses, mica schists, migmatites etc.), slightly metamorphosed Proterozoic deposits (mostly shales, graywackes, slates), Permocarboniferous deposits (mainly sandstones, arcoses and siltstones) and Cretaceous deposits (sandstones, calcareous and marly sandstones) are summarised in Table 1. These differences between transmissivity of elevations and depressions may reach one order of magnitude, or even more. They can be expressed by a ratio of respective arithmetic means: This ratio ranges from 1.6 to 38 in particular hydrogeologic environments as listed in Table 1.

Tectonic deformation may well be the main cause of these differences. It is generally assumed that the majority of valleys or depressions are tectonically predisposed. Nevertheless, in addition to a fracturing, hydrogeologic reasons can be admitted to have contributed to these differences. In recharge zones fine particles originating during weathering processes can be gradually transferred by the prevailing downward water percolation. As a result, a sealing of both interstitial pores and fractures occurs. On the

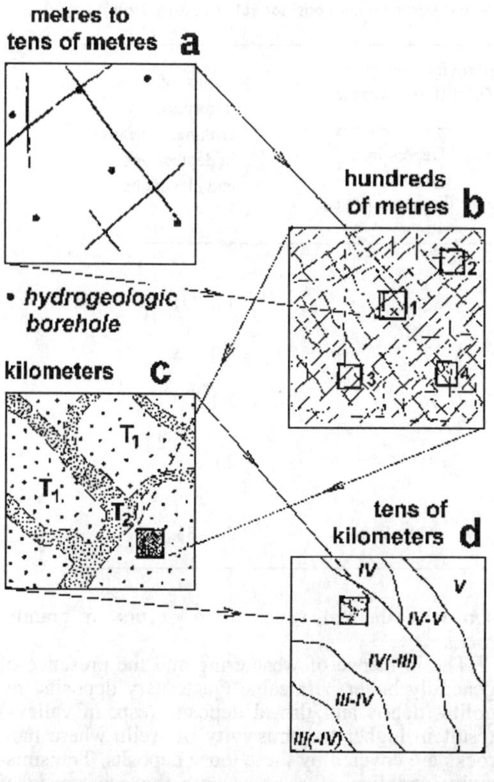

Figure 1. Relation of size of inhomogeneity elements to the extension of a study area. a – fissures and fissure zones in a local scale; b – (sub-) regional more or less regular fissuring representing hydrogeologic background; squares 1-4 represent different statistical samples characteristic of usually similar mean transmissivity magnitude and variation; c – sub-regional (medium-scale) inhomogeneities often following valleys with water courses: T_1 – lower prevailing transmissivity, T_2 – higher prevailing transmissivity; d – regional changes in transmissivity caused by different neo-tectonic activity; roman numbers express the class of transmissivity magnitude after Krásný (1993a)

other hand, in discharge zones fine particles can be removed from pores and fractures by issuing ground water. During long periods of geologic processes these recharge and discharge effects might change permeability of rocks noticeably (Krásný 1984). Considerable differences in prevailing permeability of rocks between groundwater discharge and recharge zones were indicated also in some nonindurated Tertiary and Quaternary deposits where influence of tectonic deformations can hardly be expected.

In all the environments studied until now mean transmissivity is (sometimes considerably) higher in “depressions” than in “elevations”. Nowhere opposite situation has been observed. It can be presupposed that the above-mentioned differences in permeability and transmissivity distribution are of general validity even though of distinct magnitude in different hydrogeologic environment. Irrespective of which are their causes this fact should be taken into account.

Therefore, hydrogeologic environment of hard rocks should not be considered as a regionally homogeneous but rather as a complex system where belts of regionally higher prevailing permeability, usually following the valleys and depressions, occur (case *c* in Figure 1). These belts can be perceived as manifestations of inhomogeneity elements of higher (sub-regional) order of magnitude superimposed over environment where smaller (local) inhomogeneity elements can be averaged.

4 REGIONAL SCALE INHOMOGENEITIES

During the hydrogeologic mapping program in the Czech Republic regional tendencies in prevailing transmissivity of hard rocks were discovered in extended areas of south Bohemia (Krásný et al. 1984). All this area consists mainly of gneisses and migmatites, partly with intercalations of quartzites and limestones. Granitic rocks intrude the sequence. In spite of geologic and petrographical similarities neotectonic activity had effected different parts of this area to a different degree. As stated by Kopecký (1989), since the Neogene-Palaeogene boundary tectonic activity has been intensified after a long-term "peneplain (base-levelling) regimen" lasting in the Bohemian Massif from the beginning of Mesozoic. During the interval Miocene-Recent a former low relief has been turned into a mountainous in some regions. Regional transmissivity distribution reflects the intensity of neotectonic activity: lower prevailing transmissivity belongs to areas with smaller neotectonic activity, higher to zones where neotectonic deformations have been more intensive (generally mountains) – Krásný (1996b), Havlík & Krásný (1998). As a result, these differences between prevailing transmissivity may somewhere reach more than one order of transmissivity magnitude. In Figure 1 (case *d*) regional differences are evident between the Šumava Mts. in the SW and the relatively flat environs of the South Bohemian basins in the NE. This may represent an important "shift" of the regional transmissivity background upon which local changes of transmissivity (positive and negative anomalies) have been superimposed (Havlík & Krásný 1998).

Similar gradual changes in regionally prevailing transmissivity of hard rocks caused apparently by different intensity of rock fracturing were reported also from Norway by Rohr-Torp (1994). The reasons of different degree of fracturing in both study areas are of course distinct. In the Bohemian Massif it is different intensity of neotectonic activities during Neogene and Quaternary, in Norway differences in isostatic uplift during ice-sheet melting after Quaternary Fennoscandian glaciation. In spite of different reasons the effect is very similar and can be well compared.

Table 1. Comparison of mean transmissivity in different geomorphologic (hydrogeologic) position (after Krásný 1998).

Hydrogeologic environment	Prevailing transmissivity in m^2/d (arithmetic mean of different samples)		Ratios of transmissivity arithmetic means in depressions and elevations
	Elevations (groundwater recharge zones)	Depressions (groundwater discharge zones)	
Igneous and metamorphic rocks (different parts of the Czech Republic)	1.2 - 2.8	4.5 – 13	1.6 – 11
Proterozoic deposits	0.95 - 4.2	17	4.0 – 18
Permocarboniferous deposits (Krkonoše Mts. Piedmont Basin)	11	26	2.4
Bohemian Cretaceous Basin (sandstones in its NE part)	4.7	95	20
Bohemian Cretaceous Basin (calcareous sandstones – Jizera region)	25	750 – 950	30 – 38

In both the mentioned regions the most and the least tectonically affected areas differ significantly in prevailing transmissivity. This might be of practical interest for groundwater supply studies and siting of waste deposits or deep repositories as well.

5 INFLUENCE OF LITHOLOGY

During the studies, prevailing transmissivity of different types of hard rocks has also been estimated. Areas built by crystalline limestones (marbles) are characteristic of higher prevailing transmissivity, usually half to one order of magnitude higher compared with other neighbouring crystalline rock types (Krásný 1996a). There are some indications that relatively higher permeability may be expected in areas formed by basic igneous and metamorphic rocks (Havlík & Krásný 1998). In some regions small differences between prevailing transmissivity of other rock types (granites, phyllites) have been found (Krásný 1993b). Except for the mentioned differences, however, influence of lithologic composition upon permeability and transmissivity cannot be considered significative. Possible small differences in transmissivity caused by distinct petrologic character of hard rocks are evidently masked by rock fracturation causing larger differences.

Based on correlation studies between the length of an open section of wells (or their depths) and transmissivity values Havlík & Krásný (1998) concluded that the decrease of permeability in granitic rocks and orthogneisses is evidently faster than in metasedimentary rocks in general. This might indicate an influence of a magmatic constituent, causing probably a different way of weathering and fracturing of different rocks. Different susceptibility of some hard rocks to deep fracturing might be the reason why thermal water often occurs in granitic rocks.

The influence of weathering and the presence of generally better permeable Quaternary deposits, regolith, debris and fluvial deposits (esp. in valleys) result in higher transmissivity of wells where hard rocks are covered by these loose deposits. Transmissivity variation of samples with the occurrence of Quaternary deposits is usually lower than that of samples of hard rocks without Quaternary cover. This indicates an equalizing effect of hydraulically more homogeneous Quaternary deposits.

6 CONCLUSIONS

In hard rocks of the Bohemian Massif a hierarchic system of inhomogeneity elements of different order of magnitude can be defined as follows:

1. Local fissures and fissured zones as tapped in particular boreholes or determined by well-logging techniques.
2. Inhomogeneities of a medium scale (subregional inhomogeneities) that usually follow river valleys and/or that occur at larger fractured zones typically expressed by higher prevailing transmissivity.
3. Regional scale inhomogeneities determined as a consequence of different neotectonic activities.

Prevailing transmissivity at all these scale levels might be expressed by mean or prevailing values within areas of such an extend that enables to consider the respective hydrogeologic environment more of less homogeneous. However a more detailed data analysis is recommendable to continue in revealing all the complexity of a hydrogeologic environment of hard rocks. A complicated hierarchic system of inhomogeneity elements of different size should be taken into account by implementing regional conceptual and numerical models, too. Practical consequences for regional groundwater flow modelling, safe yield assessment, well siting and studies of groundwater vulnerability are evident.

REFERENCES

Havlík, M. & Krásný, J. 1998. Transmissivity distribution in the southern part of the Bohemian Massif: Regional trends and local anomalies. In R.Annau, S. Bender & S. Wohnlich (eds.), *Hard Rock Hydrogeology of the Bohemian Massif, 3rd International Workshop Oct. 1998. Münchner Geol. Hefte*, B, 8: 11-18, München.

Henriksen, H. 1995. Relation between topography and well yield in boreholes in crystalline rocks, Sogn og Fjordane, Norway. *Ground Water*, 33, 4: 635-643. Dublin, Ohio.

Kopecký, A. 1989. Map of younger tectonic structures of the Bohemian Massif in 1:500,000 (In Czech). Ústř. Úst. Geol. Praha.

Krásný, J. 1974. Les différences de la transmissivité, statistiquement significatives, dans les zones de l'infiltration et du drainage. *Mém. Assoc. Intern. Hydrogeol.* 10, l. Communications: 204-211. Montpellier.

Krásný, J. 1984. Influence of hydrogeologic position of rocks on their permeability (In Czech). *Geol. Průzk.*, 26, 12: 342-345. Praha.

Krásný, J. 1993a. Classification of transmissivity magnitude and variation. *Ground Water,* 31, 2: 230-236. Dublin, Ohio.

Krásný, J. 1993b. Prevailing transmissivity of hard rocks in the Czech part of the Krkonoše and Jizerské hory Mts. In L. Poprawski & T. Bocheńska (eds.), *Proc. Symp. "Współczesne problemy hydrogeologii".* Polanica Zdrój: 79-86. Wroclaw.

Krásný, J. 1996a. State-of-the-art of hydrogeological investigations in hard rocks: The Czech Republic. In J. Krásný & J. Mls (eds.), *First Workshop on Hardrock hydrogeology of the Bohemian Massif* 3.-5.Oct.1994. *Acta Universitatis Carolinae Geologica*, 40, 2: 89-101. Praha.

Krásný, J. 1996b. Scale effect in transmissivity data distribution. In J. Krásný & J. Mls (eds.), *First Workshop on Hardrock hydrogeology of the Bohemian Massif* 3.-5.Oct.1994. *Acta Universitatis Carolinae Geologica*, 40, 2: 123-133. Praha.

Krásný, J. 1998. Groundwater discharge zones: sensitive areas of surface-water - groundwater interaction. In J. Van Brahana, Y. Eckstein, L.K. Ongley, R. Schneider & J.E. Moore (eds.) *Gambling with groundwater. Proceedings Joint meeting of the 28th IAH Congress and the Annual meeting of the AIH*: 111-116. Las Vegas.

Krásný J. et al. 1984. *Explanatory note to the Basic Hydrogeologic map of the ČSSR 1:200,000. Sheet 32 České Budějovice* (In Czech). Ústř. Úst. geol. Praha.

LeGrand, H.E., 1954. Geology and ground water in the Statesville area, North Carolina: North Carolina Dept. of Conservation and Development, Div. of Mineral Resources Bull. 68, 68 pp. In: Davis, S,N. & DeWiest, R.J.M., 1966. *Hydrogeology.* New York-London-Sydney: John Wiley and sons.

Rats, M.V. 1967. *Neodnorodnost gornykh porod i ikh fizicheskikh svojstv.* Moskva: Nauka.

Rats, M.V. & Chernyshov, S.N. 1967. Statistical aspect of the problem on the permeability of the jointy rocks. *Intern. Symposium on Hydrogeology of fractured rocks.* Dubrovnik.

Rohr-Torp, E. 1994. Present uplift rates and groundwater potential in Norwegian hard rocks. *Geological Survey of Norway, Bulletin* 426: 47-52.

Rovey II, CH. W. 1998. Digital simulation of the scale effect in hydraulic conductivity. *Hydrogeology Journal* 6: 216-225. Berlin: Springer.

Schulze-Makuch D. & Cherkauer D.S. 1998. Variations in hydraulic conductivity with scale of measurement during aquifer tests in heterogeneous, porous carbonate rocks. *Hydrogeology Journal,* 6: 204-215. Berlin. Springer.

Groundwater: Past Achievements and Future Challenges, Sililo et al. (eds) © 2000 Balkema, Rotterdam, ISBN 90 5809 159 7

A geophysical perspective in siting high yielding boreholes in the lower Karoo aquifer in western Botswana

H. Kumar & D. Sarma
Water Resources Consultants (Pty) Limited, Gaborone, Botswana

L. Carlsson
Department of Water Affairs, Gaborone, Botswana

ABSTRACT: Sandstones in the Karoo Supergroup contain some of the most promising aquifers within the country. The sandstones cover together approximately half of the country and are used both for regional and local water supplies. The sandstone aquifers are recognized as dual porosity aquifers where enhanced transmissivity zones are found in larger fracture zones. An integrated geophysical survey was conducted to locate sites for groundwater abstraction in the villages of Ncojane and Kule in the southwestern Botswana. The survey comprised HLEM and magnetic profiling combined with Vertical Electric Soundings at selected sites along the profiles. The boreholes drilled on sites selected form the geophysical background showed significantly higher yields (3-18 times) compared with boreholes drilled without any geophysical siting method applied.

1 INTRODUCTION

More than 90% of the water supply to villages and settlements in Botswana is based on groundwater. In a defined Rural Village Water Supply Programme, (RVWSP), Department of Water Affairs, DWA is annually conducting investigations for some 20 to 30 villages. Within the programme a groundwater investigation for the Ncojane and Kule villages was conducted on consultancy basis under the supervision of DWA with the primary aim to establish groundwater resources to meet a future water demand of 140 m^3/d.

The existing water supply to the villages is based on boreholes tapping the sandstone formation of the Ecca Group (Otshe Formation) of the Karoo Supergroup. The boreholes were not sited on any lineaments recognized from geophysical or other investigation and were yielding between 0.03 and 13.7 m^3/h.

It was considered that higher yields would be associated with fracturing and faulting in aquifer rock recognized as NW-SE, NE-SW and NNE-SSW aerial photo lineaments. An aerial photo interpretation study followed by ground geophysical survey was therefore carried out to locate production borehole sites to meet the objectives of the programme. It was also assumed that aquifers at depth could be higher yielding than the present one exploited.

Faults and fractured zones can be traced on the ground using various geophysical techniques. Ground magnetic and Horizontal Loop Electro-Magnetic (HLEM) profiling followed by Vertical Electrical Sounding (VES) was used in this project.

2 REGIONAL SETTING

2.1 *Location and Geological Framework*

Ncojane and Kule villages are located in the Ghanzi District in western Botswana (Figure 1). The climate is semi-arid with average annual rainfall of 265 mm (Bhalotra, 1987). Mean maximum and minimum surface temperature is 29.7°C and 12.9°C respectively. Evaporation and potential evapotranspiration are 2.12 m and 2.17 m respectively, far exceeding the annual precipitation. The villages are on an essentially flat area at an elevation of approximately 1250 mamsl covered by loose sand and occasional pans.

The geology of the area is described by Ramakate (1996), Smith (1984) and Carney et al. (1994). The study area is geologically situated on the northern part of the Gemsbok sub-basin of southwest Botswana containing rocks of the Karoo Supergroup west of longitude 22°E. The sub-basin is bounded in the north by the Precambrian Ghanzi Group rocks. It extends into South Africa and Namibia towards the south and west. Kalahari sediments cover most of the Karoo strata and information on the underlying Karoo rocks is gained from mineral and water exploration boreholes. The penetrated lithology below the Kalahari Beds consists of the Lebung Group, which has an upper aeolian unit – the Nakalatlou Sandstone Formation– followed by the arenaceous to argillaceous Dondong Formation. Below the Lebung is the

Kwelta (Kule) Mudstone Formation. This formation is underlain by the Ecca Group represented by arrenaceous sediments of deltaic environment with minor shale and coal horizons. The deeper sediments consist of argillaceous Kobe Formation underlain by glaciogenic sediments of the Dwyka Group. The Dwyka Group in this area was encountered at a depth of 349m in a coal exploration borehole W3 (Ellis, 1979). The applied methodologies did not penetrate to the Dwyka Group. The distribution of the Karoo in southwestern part of Botswana is shown in Figure 1 (Key, 1998).

2.2 *Groundwater Occurrence*

Data from existing boreholes in the area indicated minor water strikes in the Kalahari Beds and the presence of two aquifers within the Karoo Supergroup sediments. The shallow Karoo Supergroup aquifer within the Dondong Formation had water strikes in sandstone horizons at variable depths within 90 mbgl. Several village boreholes with low yields tapped this aquifer. A borehole that penetrated 3 m of sandstone below carbonaceous mudstone and coal indicated the presence of a second deeper Karoo Supergroup aquifer (Otshe Formation). The borehole was tested at a discharge rate of 17.2 m^3/h. The sandstone is described by Smith (1984) from coal exploration boreholes in the region as a deltaic prograde sequence with an uppermost coarsening upward sandstone horizon confined at the top by carbonaceous mudstone, coal and siltstone horizons.

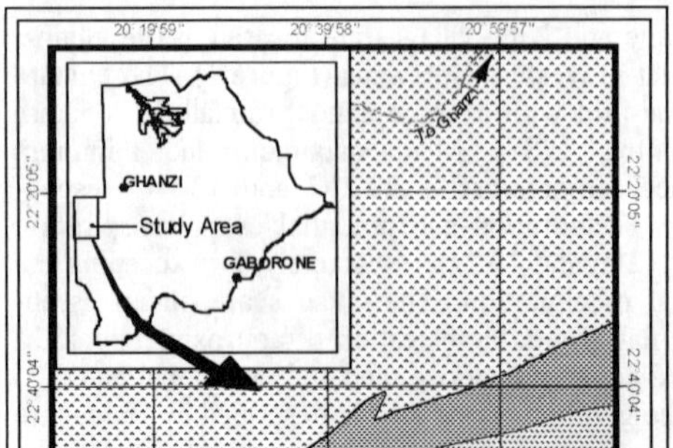

Figure 1: Location and Geology Map

The argillaceous Mokorobe Formation underlies the Otshe Formation. Smith (1984) suggested depths exceeding 150 m to the sandstone in the region.

3 GROUNDWATER DEVELOPMENT

3.1 *Approach*

As detailed geological and hydrogeological information in the exploration area was not available and surface exposures were lacking due to the Kalahari Beds cover it was necessary to employ geophysical methods to site the boreholes. The aim of the geophysical exploration was two-fold, (i) to determine the depths to and thickness of the upper and lower sandstones of the Karoo Supergroup, and (ii) precisely locate fractures and faulting within the Karoo Supergroup. It was known from past experience elsewhere in Botswana that high yields are associated with fractured or faulted Karoo Supergroup sandstone aquifers within the Lebung and Ecca Groups.

Information from several data sets were integrated including the national geological map of Botswana (1:1,000,000, Key, 1998), lineaments from an aeromagnetic data set, borehole information and 1: 50,000 scale aerial photographs to identify specific target areas for ground geophysical survey. Several sets of aerial photo lineaments were identified. The nature of these lineaments could not be verified due Kalahari Beds cover over the area.

Magnetic and HLEM profiling were considered to be fast and cost effective methods to delineate fault/fracture zones. VES method was used to estimate the thickness of overburden and depth to the aquifer.

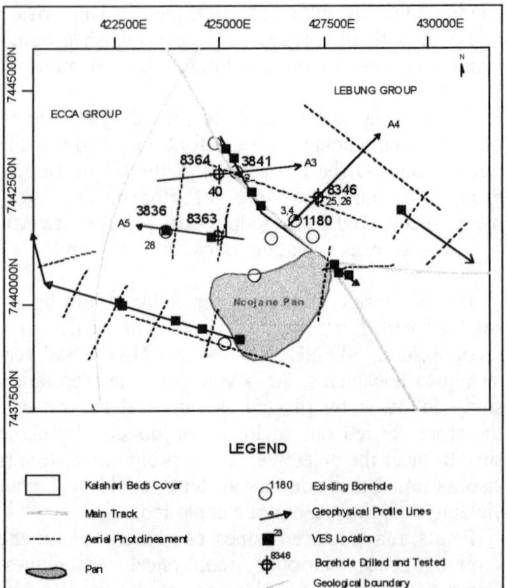

FIGURE 2: INTEGRATED BASE MAP

Three boreholes were drilled in the Ncojane area on geophysical anomalies coinciding with aerial photo lineaments. The boreholes tapped the uppermost sandstone horizon of the Otshe Formation underlying carbonaceous mudstone and coal. Two other boreholes were drilled in neighboring Kule area. The boreholes tap the same aquifer.

3.2 *Geophysical Methods Applied*

Magnetic and HLEM profiling followed by VES methods were employed in the study area. The methods complement each other and were used in order to overcome ambiguities inherent in the interpretation of each individual method.

A Proton Precession Magnetometer (Geotron Gee-5) was used at 20 m reading interval along each profile line followed by HLEM profiling.

HLEM profiling was carried out using Apex Max-Min II EM unit at 3555, 1777 and 888 Hz frequencies and 200 m coil separation (20m reading interval). The criteria of using 200 m coil separation were to achieve information within 200 m depth. However, the depth of penetration depends on the resistivity of different geological formations.

VES were conducted at selected locations identified from magnetic and HLEM profiling. The Schlumberger configuration was used with a maximum AB spacing of 800 m.

The survey comprised 29 profile lines of totally 60 km magnetic and HLEM profiling and 57 VES including calibration conducted at existing boreholes with known lithology. Magnetic and HLEM data was interpreted qualitatively and VES data interpretation was done using Resix-P computer software.

4 RESULTS AND DISCUSSION

4.1 *Depth Estimation*

In stratified sedimentary formations, the yield of a borehole depends on the thickness and hydraulic conductivity of water yielding layers penetrated. Prior to the current investigations, the bedrock in the study area was penetrated by boreholes reaching to a maximum depth of 163 m. Most boreholes were however drilled to much shallower depth. From investigations in the Karoo sandstone formations at other places it was known that layers of coal and shale are interbedded. Such layers would divide the formation in various aquifers having different yields. The most important issue was therefore to determine the depth for drilling to reach the most prominent sandstone aquifer.

A promising method to assess the depth and properties of the occurring layers is the VES. Based on the results from the application of VES and common knowledge of the geology it was concluded that successful boreholes must penetrate depths of more than 150 m or down to a sandstone layer at that depth identified by characteristic resistivity range.

Three boreholes were successfully drilled based on VES information tapping the lower Karoo, Ecca Group aquifer. The first borehole drilled, 8346, led to confirmation of the VES interpretation. The anticipated drilling depth, based on 30 Ω-m resistivity layer, was 180m (Figure 3). The borehole was drilled to 169 m depth with water strike at 145m bgl is sandstone below coal/ shale. The borehole was tested for 72 hours without any boundary effect and recommended for 48 m^3/h safe abstraction yield (Sarma and Carlsson, 1998) assuming no recharge for 5 years.

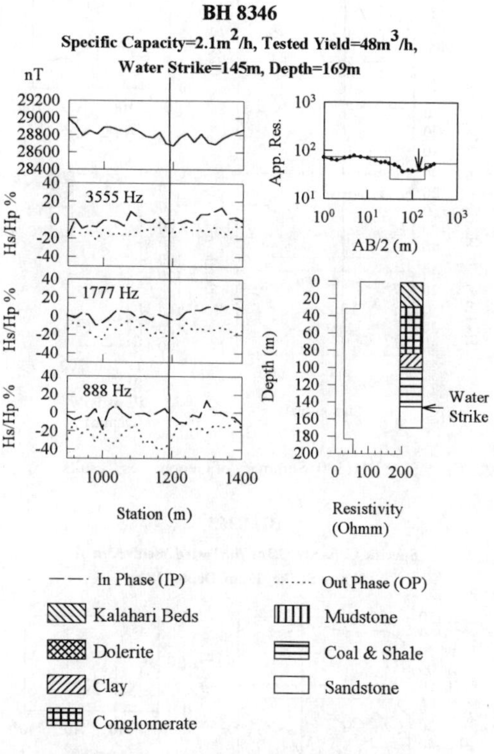

Figure 3: BH8346, Summary of Geophysics Results

An older borehole 1180, 73 m deep, located 800 m away was drilled only into the upper Karoo formation. Blow-out yield of the borehole was 13.7 m^3/h. Borehole 1180 was never tested, comparison of yields however indicated that new borehole has 4 times higher yield. Similar VES response to that in 8346 (Figure 4) suggested that borehole 1180 could have penetrated the deeper sandstone aquifer below coal/shale horizon had the borehole been drilled deeper. Water quality from borehole 8346 (TDS=450 mg/l) was potable and the results encouraged further exploration of this sandstone aquifer below coal/ shale horizon.

In the second borehole (8363), the depth of investigation from VES was only 110m (Figure 5). Thick Kalahari Beds sediments and resistive dolerite intrusive (sills) obstructed the VES from penetrating to the target depth. The decision to drill was however based on the drilling results of the first borehole. The borehole was drilled to 181m depth. Water was encountered at 172m, below the coal/shale and sandstone interface. The borehole was test pumped at 53 m^3/h for 72 hours with no boundary effect (Carlsson & Ntsatsi, 1998).

An older borehole 3836 (Figure 6), 131 m deep, located approximately 1 km away, had 5.1 m^3/h estimated yield. The borehole terminated within the coal/shale horizon. Borehole 3836 was never tested but the estimated blowout yield was approximately one tenth of the yield of borehole 8363.

The geological information obtained from the first two successful boreholes was utilised in re-interpreting VES data and selecting the third drilling

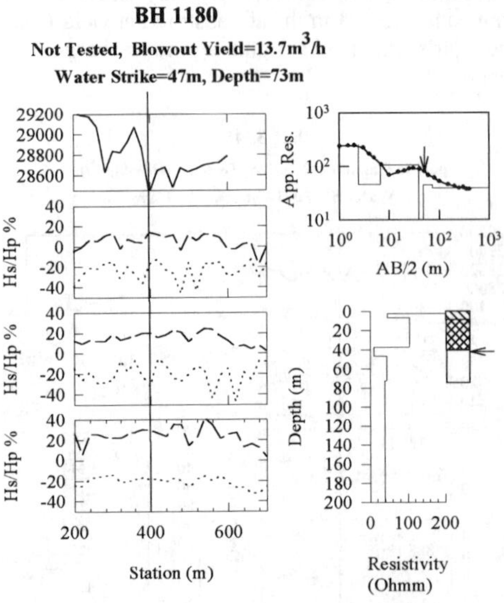

Figure 4: BH1180, Summary of Geophysics Results

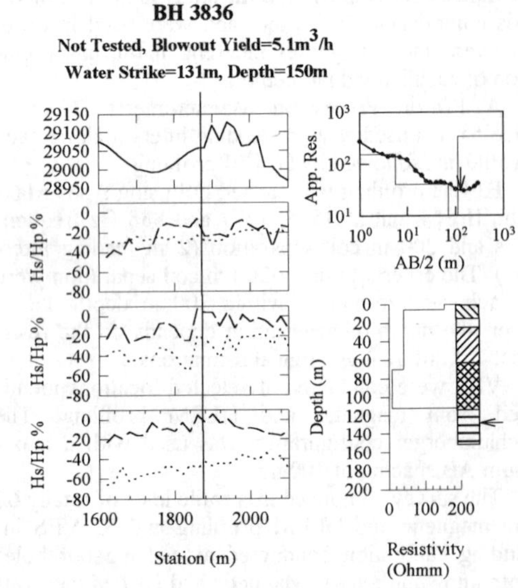

Figure 6: BH3836, Summary of Geophysics Results

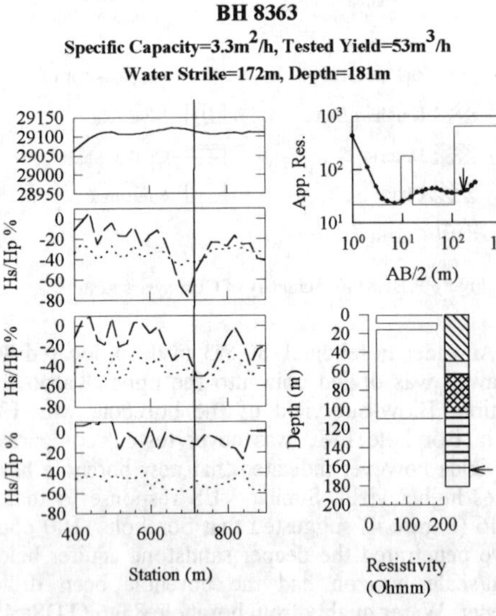

Figure 5: BH8363, Summary of Geophysics Results

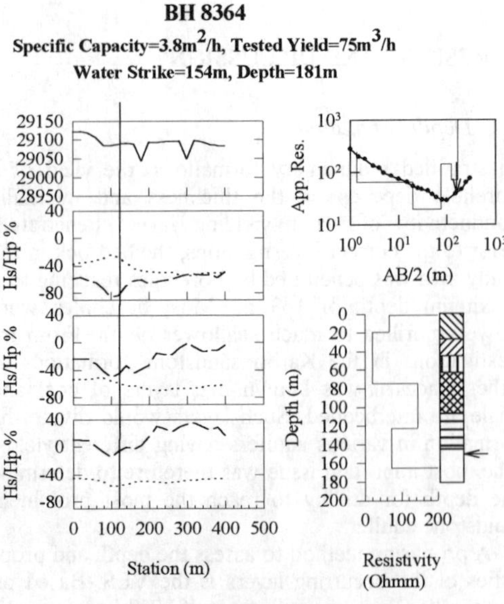

Figure 7: BH8364, Summary of Geophysics Results

BH 3841

Specific Capacity=1.1m2/h, Tested Yield=17.2m³/h

Water Strike=55, 152m, Depth=163m

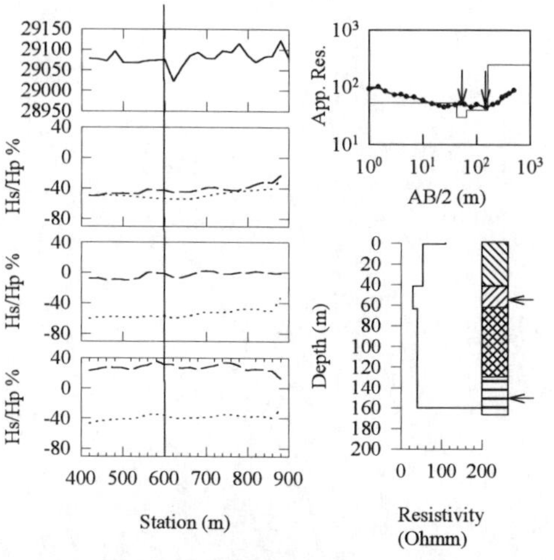

Figure 8: BH3841, Summary of Geophysics Results

site. The aquifer was expected between 123 and 168 m seen as a layer of 72 Ω-m resistivity from the VES interpretation (Figure 7). The water was struck in sandstone below the shale/coal at 154 m depth and the borehole was terminated at 181m depth. The borehole (8364) was tested for 66 hours with an abstraction yield of $45m^3/h$. No boundary effect was seen from interpretation of the test data (Sarma & Carlsson, 1998).

An older borehole 3841 (Figure 8), 163 m deep, was located approximately 500 m away, was tested at 17.2 m^3/h. A comparison of results indicate that borehole 3841 marginally penetrated the deeper sandstone.

The test pumping data interpretation showed that the older borehole gave approximately 38% of the yield of borehole 8363. The older borehole could have given better yield if it had penetrated deeper into the sandstone.

4.2 *Identification of Lineaments*

Enhanced hydraulic properties of a sandstone aquifer are generally encountered in connection with fracturing and faulting. Fracture zones of larger arial extent and dimensions can be identified using aerial photographs, remote sensing technique and/or geophysical methods. The project encompassed the use of magnetic and HLEM profiling on lineaments identified from aerial photographs. All project boreholes were sited on lineaments identified and further assessed by geophysical investigations. The results from test pumping of the boreholes showed differences in yield. This could be explained by the location of the borehole on lineaments representing fracturing of different intensity and extents. The boreholes with higher specific capacity were associated with significant negative IP anomaly at all HLEM frequencies applied.

Boreholes 8364 was sited at a fault/fracture zone identified by a considerable change in IP on all frequencies (Figure 7). This fault/fracture zone correlated with a NW trending lineament in the vicinity of the borehole (Figure 2). The specific capacity of the borehole from test pumping interpretation was 3.8 m^2/h.

Borehole 8363 was located on a fracture zone as interpreted from prominent IP and OP anomaly seen at 3555 and 1777 Hz frequencies (Figure 5). The fracture zone correlated with a NNE trending lineament (Figure 2). The specific capacity from test pumping data interpretation was 3.2 m^2/h.

Boreholes 8346 (Figure 3) was sited at a significant OP anomaly seen on 1777 and 888 Hz frequencies interpreted as weathered zones. IP anomaly however was not significant. Interpreted specific capacity from test pumping data was $2.1m^2/h$.

Borehole 3841 had a specific capacity of 1.1 m^2/h. HLEM signature at this borehole suggested no fracturing in the bedrock. Low specific capacity of the borehole could thus be due to two reasons; (i) no fracturing indicated by HLEM profiling data and, (ii) less penetration into the sandstone aquifer.

5 CONCLUSIONS

The use of geophysical survey in siting of production boreholes proved successful. A deeper and hitherto unexploited aquifer was identified. The drilling depth could be established in order to penetrate to this productive aquifer. Further, lineaments identified from aerial photographs could be detailed at ground level and boreholes located at precise sites.

The methodology was cost effective in the sense that no exploration drilling was required. Most of the existing boreholes had poor yields as they were either not located in fracture zones or they were not penetrating the potential aquifer (sandstone). The boreholes drilled at geophysical signatures close to existing boreholes had yields, 3 to 18 times, higher than the existing boreholes.

Typical HLEM anomalies at fracture zones were negative IP at all frequencies and small amplitude OP anomalies at lower frequencies. VES conducted at various locations provided reliable estimates of drilling depth.

6 ACKNOWLEDGEMENT

The authors wish to express their gratitude to the Permanent Secretary, Ministry of Minerals, Energy and Water Affairs for allowing them to publish this paper.

7 REFERENCES

Bhalotra, Y.P.R. 1987. Climate of Botswana, Part II Elements of Climate, 1. Rainfall, 2. Sunshine and Solar Radiation and Evaporation, 3. Temperature and Humidity of Air. Department of Meteorological Services, Botswana.

Carlsson, L & Ntsatsi, J. 1998: Village Water Supply in Botswana – Assessment of Recommended yield for Production Boreholes in Semi-Arid Environment. In *Abstract Volume of International Conference on The Role of a National Geological Survey in Sustainable Development, 23rd – 25th June 1998*: 11-15. Gaborone, Botswana.

Carney, J.N., Aldiss, D.T. & Lock, N.P. 1994. The Geology of Botswana. Department of Geological Survey, Lobatse, Botswana.

Ellis, C. J. R. 1979. Report on the exploration of Block 'W' Botswana. Shell Coal Botswana (Pty) Ltd.

Key, R. 1998. The 1998 edition of the national geological map of Botswana. Department of Geological Survey, Lobatse, Botswana.

Ramakate, V. 1996. Geological Report for the TGLP Ghanzi/Makunda Groundwater Survey. Department of Geological Survey, Lobatse, Botswana.

Sarma, D. & Carlsson, L. 1998. Lower Karoo Aquifer in Western Botswana – Fresh Groundwater Supply to Kule and Ncojane Villages. In *Abstract Volume of International Conference on The Role of a National Geological Survey in Sustainable Development, 23rd – 25th June 1998*: 29-33. Gaborone, Botswana.

Smith, R.A. 1984. The Lithostratigraphy of the Karoo Supergroup in Botswana. Department of Geological Survey, Lobatse, Botswana.

Groundwater: Past Achievements and Future Challenges, Sililo et al. (eds) © 2000 Balkema, Rotterdam, ISBN 90 5809 159 7

Groundwater evapotranspiration – Underestimated component of the groundwater balance in a semi-arid environment – Serowe case, Botswana

M.W. Lubczynski
ITC, International Institute for Aerospace Survey and Earth Science, Enschede, Netherlands

ABSTRACT: Steady state, quasi transient and true-transient modelling were used to study groundwater balance of the semi-arid Serowe aquifer in Botswana. Special attention was given to the spatial and temporal variations of groundwater evapotranspiration loss and its impact upon groundwater balance. As spatial variable groundwater evapotranspiration was determined through the integration of SEBAL algorithm in the steady state and quasi-transient modelling approach. A one-dimensional lumped model was also used to study the temporal impact of highly variable rainfall upon the effective recharge in true-transient modelling approach. Results of the various methods employed have been compared and evaluated.

1 INTRODUCTION

Recent developments in hydrology made it possible to determine the actual evapotranspiration spatially and temporally. To be solved remained the assessment of the contributions from the unsaturated and the saturated zone. This is not a simple matter and often requires tracer studies (Gieske 1992, Selaolo 1998). In this contribution various approaches for the groundwater budget have been used whereby all "in" and "out" components were considered. By comparing the results, the most realistic estimates were deduced.

2 DESCRIPTION OF THE SEROWE AREA AND NUMERICAL MODEL

The Serowe study area (Figure 1) is located in Botswana at the eastern fringe of the Kalahari Basin marked by the prominent 90-150m high escarpment. This escarpment splits the area into two contrasting parts: western and eastern. The higher western part, which slopes gently to the west, is covered by several tens of meters thick Kalahari Sand (Figure 2), where the topographical surface is dominated by the thornshrub "sandveld". In contrast, the eastern lower part slopes steeper, the Kalahari sand cover is much reduced (to a few meters maximum) or absent and savanna "hardveld" vegetation is denser and taller, especially close to the escarpment and along the drainage lines. It is noticeable that in the eastern area vegetation remains green even at the peak of the dry season in contrast to the western part, which is generally dry except for sparse, evergreen trees.

The long-term daily rainfall record since 1922, from Serowe meteorological station in the study area, indicates high temporal rainfall variation ranging from 200 to 1100mm/y. The rainy season occurs between October and March when ambient temperatures are at their highest value and rainfall occurs in the form of isolated, very high intensity storms (sometimes even above 100 mm/d). These infrequent storms constitute the principal source of groundwater recharge. In order to study hydrological fluxes, an Automatic Data Acquisition System (ADAS) has been installed in the eastern part of the area (Figure 1). The ADAS system was based on the Delta-t multi-channel logger providing the simultaneous operation of the 32 sensors monitoring meteorological, soil, water and groundwater variables at 0.5 and 1 hour resolution (Figure 3).

The hydrogeology of the Serowe study area has been extensively investigated, first by the Swedish Geological Survey (SGAB, 1988) and recently by Wellfield Consulting Services in association with ITC (WCS, 1999). A numerical groundwater model is described in WCS (1999) and summarised in Lubczynski & Mudzingwa (2000). The Serowe model is entirely located in the Karoo strata where the most important aquifer is the Ntane Sandstone layer, approximately 100m thick (Figure 2). This layer is underlain by the impermeable Mosolotsane Formation (mudstone, siltstone) and overlain by locally fractured Stormberg Basalt having variable thickness. Less or no recharge is assumed where thicker Stormberg Basalt overlays the Ntane Sand-

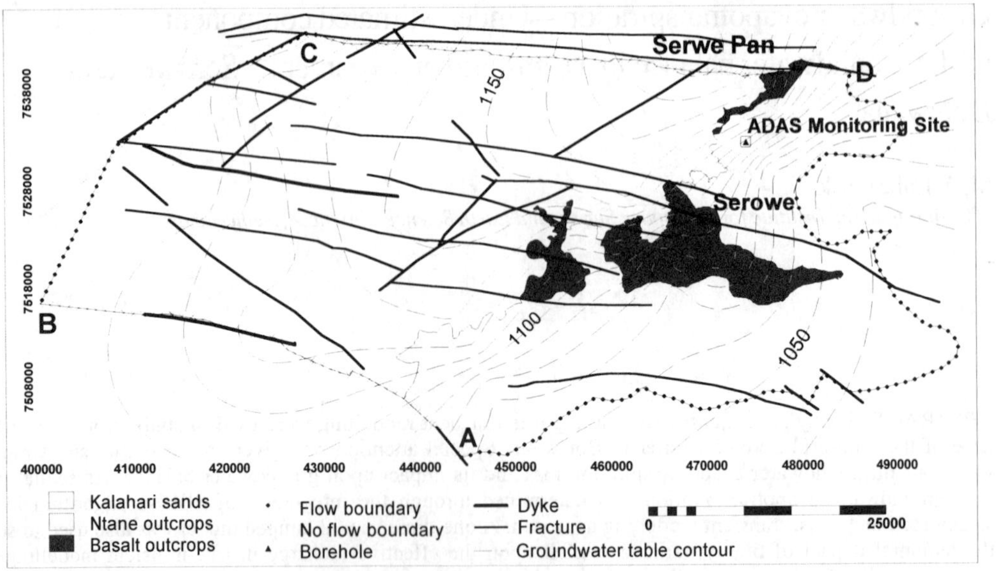

Figure 1. Location map.

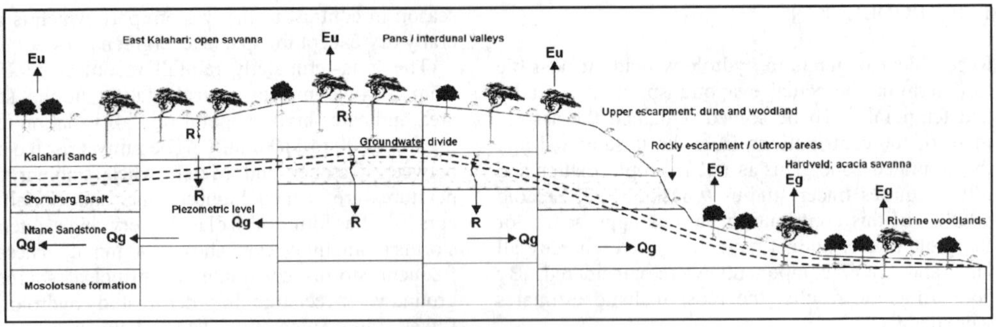

Figure 2. Conceptual hydrogeological cross section.

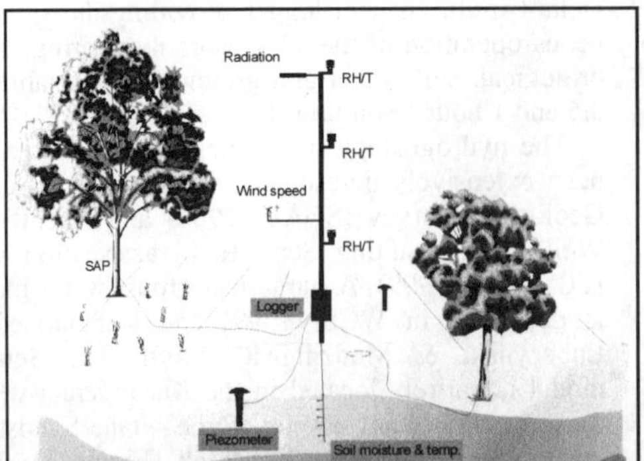

Figure 3. ADAS station setup.

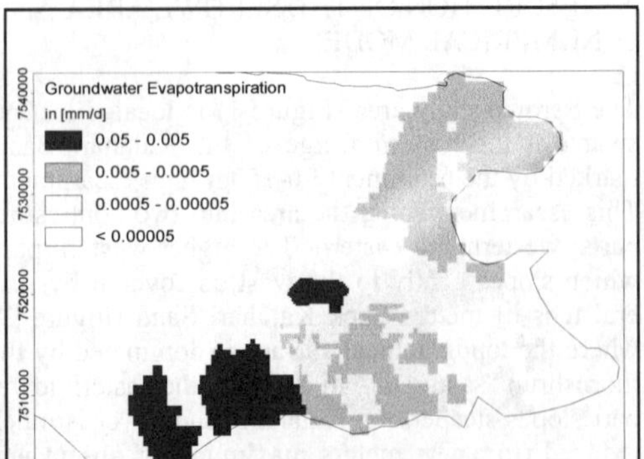

Figure 4. Average groundwater evapotranspiration.

stone layer. Structurally the area is characterised by a dominant WNW-ESE trending fault system (often intruded by dolerite dykes of 10-40m width), which breaks up the Karoo formation into a series of blocks like compartments. A structurally controlled groundwater flow west from the groundwater divide (central part of the study area) has a gentle gradient, whereas the one to the east has a steeper gradient (Figure 1). The model area is bounded: a) from the south by the regional extension of the impermeable dolerite dyke assumed as no-flow boundary A-B; b) from the west by the arbitrary head constant boundary B-C; c) from the north by the deep basaltic graben assumed as no-flow boundary (C-D); d) from the east and partly south by the eastern limit of the Ntane Sandstone aquifer where head dependent outflow conditions were assumed (D-A). For more information on the Serowe model setup see (Lubczynski & Mudzingwa 2000).

3 GROUNDWATER BALANCE

In a general sense the groundwater balance equation can be written in the following form:

$$Q_{Gin} + R = Q_{Gout} + Eg +/- \Delta S +/- Q_{ext} \quad (1)$$

where Q_{Gin} = groundwater inflow (lateral inflow in case of 2-D solution), R = groundwater recharge from rainfall, Q_{Gout} - groundwater outflow (lateral outflow in case of 2-D solution), Eg - groundwater evapotranspiration, ΔS - change of groundwater storage; Q_{ext} - external groundwater sink and sources (for example well abstraction).

Serowe study experience emphasised the usually underestimated importance of groundwater evapotranspiration (Eg) regarded as plant transpiration from the capillary and saturated zone when direct evaporation from groundwater can be neglected (sufficiently deep groundwater table). The difference between groundwater recharge (R) and groundwater evapotranspiration (Eg), was defined as effective recharge (Re). Groundwater evapotranspiration as such decreases the effective recharge, in extreme circumstances discarding the recharge completely or even taking a negative form emphasised by the groundwater table decline in time. This particular situation often occurs in the water scarce Southern Africa environment, where the world's deepest plant root systems are documented (Scott & Le Maitre 1998). Unfortunately, still very little is known about groundwater evapotranspiration rates.

Evapotranspiration (E) measured through the actual evapotranspiration (Ea) is defined as:

$$E = Es + Eu + Eg \quad (2)$$

where Es = surface evaporation, Eu = evapotranspiration from unsaturated zone, Eg = groundwater evapotranspiration.

So far, there is no direct method available to determine Eg. Considering equation 2 only E can be estimated non-spatially (Dingman 1994) and spatially (Bastiaanssen et al, 1998) as actual evapotranspiration (Ea). In Serowe case, in order to minimise the influence of Es and Eu upon Eg the dry season fluxes were calculated. Spatially this was provided using SEBAL algorithm (Timmermans & Meijerink 1999) in combination with MODFLOW numerical model and non-spatially using Bowen Ratio method, SAP flow measurements and 1-D groundwater modelling. All these solutions are discussed in the following groundwater balance approaches: steady-state, quasi-transient, true-transient and finally in the cross check evapotranspiration test.

3.1 *Steady state model approach*

In the steady state model approach it is assumed that the hydrogeological system is in equilibrium, so $\Delta S = 0$ and all the components of the groundwater balance are time independent including Q_{ext} which for natural systems is not considered. In such condition:

$$Q_{Gin} + R = Q_{Gout} + Eg \quad (3)$$

In circumstances where plant transpiration from groundwater is negligible, $Q_{Gout} = Q_{Gin} + R$. High Eg contribution however substantially reduces the groundwater outflow (Q_{Gout}). Such effect has been simulated in the 2-D Serowe numerical groundwater MODFLOW model where substantial Eg rates were expected in the eastern part of the model with regard to the wedging Ntane aquifer (figure 2). In the MODFLOW implementation of Eg (Lubczynski & Mudzingwa 2000), it is assumed that Eg decreases linearly with increasing depth of the groundwater table starting from "Maximum Eg Rate" when the groundwater table is at or above the "Elevation of the Eg Surface" and ending at the "Eg Extinction Depth" or below, at which the Eg = 0 (McDonald & Harbaugh 1988). In the Serowe model the "Extinction Depth" was "lumped" at 60m as maximal rooting depth, on the basis of the Botswana (WCS, 1999) and South Africa (Scott & Le Maitre 1998) experience. The "Elevation of the Eg Surface" was assumed at the topographical surface of the area. The "Maximum Eg Rate" was implemented in MODFLOW through the Ea fluxes from the SEBAL study (Timmermans & Meijerink 2000). The dry season evapotranspiration calculated by SEBAL consists of Eu and Eg with unknown contributions, where Eu with regard to MODFLOW solution represents non-extractable "noise". Therefore the dry season SEBAL fluxes were initially integrated in MODFLOW as "Maximum Eg Rates" and further calibrated in the model (Lubczynski & Mudzingwa 2000). The Eg

MODFLOW model result (Figure 4) shows that the application of the 60m "Extinction depth" filtered out the Eg from the western part of the model (irregardless of the Ea rates calculated by SEBAL) due to the deeper than 60m groundwater table. The MODFLOW calculated Eg fluxes in the eastern part of the area are substantially lower than those calculated by SEBAL (Timmermans & Meijerink 2000). The difference can be attributed to Eu but also to the SEBAL and MODFLOW solution errors.

Considering regional groundwater balance of the study area, in the western part of the model almost the entire amount of rain originated recharge flows out at the western boundary (3342 m^3/d) of the model. This represents 60% of the total R of the area. On the contrary, in the eastern part, 38% of the total R is released through Eg (2079 m^3/d) and only 2% (167 m^3/d) flows out through the Ntane Sandstone contact with the Mosolotsane Formation (Lubczynski & Mudzingwa 2000). Considering the eastern outflow as 100%, the groundwater evapotranspiration makes up 95% and the eastern boundary (A-D at Figure 1) outflow only 5%. The accuracy of the presented budget is constrained by the limited amount of data available for the numerical model calibration, particularly with regard to the model assumed recharge distribution, based mainly on the C^{14} data available through SGAB (1988).

3.2 *Quasi-transient model approach*

The quasi-transient numerical model solution (often referred to as transient) can be defined as a solution where model response is governed by man imposed, time dependent stress, and fixed as a time independent natural effective recharge. So far in groundwater management this particular approach is the most extensively used, because very rarely the sufficient R and Eg input data is available and moreover the prediction of rainfall is not possible.

In the quasi-transient groundwater models, R usually is fixed in time, while Eg can be simulated as linearly dependent on the well abstraction drawdown. In quasi-transient Serowe model such attempt has been made, however due to the complexity of the model, numerical instability and also limited time allocated for the numerical model calibration, this approach was abandoned and time independent Eg input values were applied.

Regarding the Serowe groundwater balance equation (1), the time variable abstraction imposes the time and stress dependent change of storage (ΔS) as well as the change of lateral fluxes (Q_{Gin} and Q_{Gout}) while R and Eg of course remain constant in time. For more information on quasi-transient model solution refer to Lubczynski & Mudzingwa (2000).

3.3 *True-transient model approach*

The true-transient modelling approach applies when all groundwater balance components are time dependent. Important R and Eg either can be determined externally using other methods or eventually they can be calibrated using groundwater table hydrographs.

In the Serowe case study, due to the lack of data only 1-D lump reservoir model called "EARTH" (Van der Lee & Gehrels 1990, 1997) was run as a true transient model approach for ADAS test site. As input data the EARTH model requires daily rainfall (taken from the Serowe Meteo Station) and daily potential evapotranspiration (PET) calculated using Penman-Monteith algorithm (Allen et al.1998) based on the daily temperature, solar radiation, relative humidity, and wind speed. The PET climatic data was available for the study area at the ADAS station for a period of half a year operation time only. The nearest location with long term record was Mahalapye Meteo Station. That data was used for extrapolation purpose, after regression with the ADAS data. The most important model parameter specific yield (0.04) was determined in the laboratory and confirmed by pumping tests. The calibration of the EARTH model was performed against the daily head values obtained from the groundwater logger called "Diver" installed in August 1997 in the piezometer at the ADAS station site (Figure 5).

Very interesting groundwater hydrograph indicates three different decline rates 4.5, 10.9 and 5.5 mm/d, which multiplied by the specific yield of 0.04 represent three different outflow rates 0.18, 0.44 and 0.22 mm/d respectively. These three values give also three different aquifer drainage characteristics with regard to three different stages of the groundwater table decline at one particular ADAS monitoring site.

The EARTH model outputs are the actual evapotranspiration (Ea) and the recharge (R). The Ea record (Figure 6) is calculated by EARTH model, based on a linear relationship with available soil moisture, so it represents only Es and Eu. The Ea values are substantially lower than PET values, particularly in the end of the dry season when Es = 0.

Considering saturated zone, the EARTH model assumes that groundwater table response is a func-

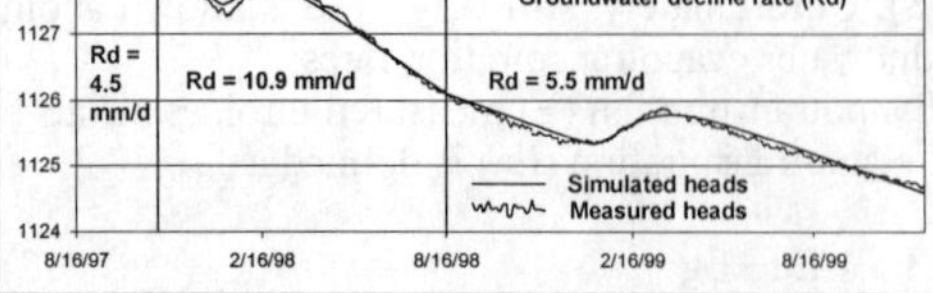

Figure 5. Calibrated versus measured daily heads.

tion of: a) deep percolation from unsaturated zone applied as recharge on the saturated zone; b) aquifer storage represented by the storage coefficient; c) reservoir outflow related to the combined effect of groundwater evapotranspiration and lateral flow, represented by the groundwater decline rate (Rd).

Considering particular contributions of the lateral flow and Eg in the drainage outflow, very low hydraulic conductivity (K) at ADAS site, acquired from the laboratory test, field pumping test and through the MODFLOW calibration (K<0.1m/d), suggests low importance of the lateral flux and predominant role of Eg. This assumption however is a matter of discussion, particularly with regard to the variable hydrograph decline rates (Figure 5) so further research needs to be devoted to it.

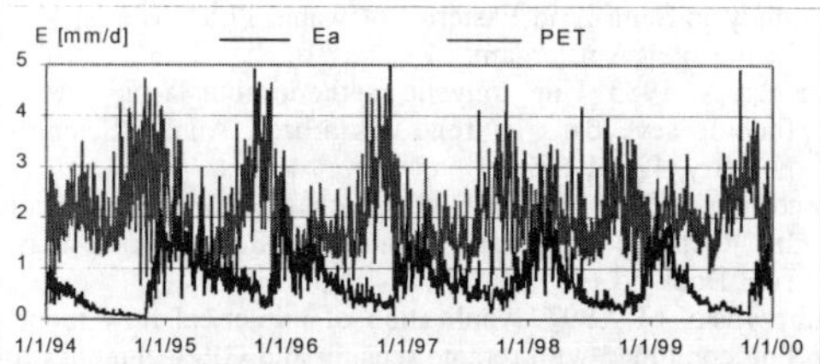

Figure 6. Potential and actual daily evapotranspiration

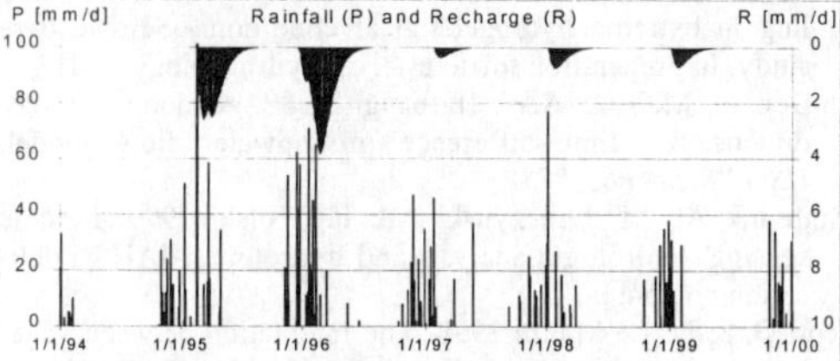

Figure 7. Daily rainfall and recharge.

Concerning EARTH calculation of recharge (R) in Serowe case (Figure 7), model simulates direct and relatively short time response reaction related to groundwater table movement, so the lateral flow has minor importance (discussed above). It is also expected that in the short and intensive recharge episodes, the unsaturated moisture remains at high level, so Eg can be assumed as negligible since Es and Eu are widely available.

3.4 *Cross-check of evapotranspiration estimates*

Evapotranspiration cross–check test was provided for the ADAS site on the pre-selected day of 14/8/98 because of the dry condition (Es = 0) and Landsat data availability. The SEBAL energy balance solution for the neighbourhood of the ADAS site indicated the E range equal to 1-2 mm/d and the Bowen Ratio Energy Balance (BREB), 0.5 mm/d, both described in Timmermans & Meijerink (2000). The Earth originated, groundwater evapotranspiration combined with lateral flow assumed as having minor importance, indicates 0.44 mm/d.

At the ADAS monitoring site, additional incidental data was obtained by SAP flow test using (TDP) Thermal Dissipation Probes (Granier 1985). The 2.5m high representative *Acacia* bush tree with a crown of 5m^2 indicated the SAP flow 6 l/d. To upscale this measurement, the EcoSurv vegetation map (EcoSurv 1998) was used. This map indicated approximately 40% canopy coverage at the neighbourhood of ADAS site. This compared with the SAP results indicated evapotranspiration of 0.48 mm/d, which seems to represent Eg because the upper 2.5m of the soil profile (measured by TDR instrument) was dry and the drillers reported similar dry soil condition down to the water table.

4 CONCLUSIONS

1. The Serowe groundwater balance study pointed to uncertainty in the solution of groundwater balance equation, mainly due to the problems with determination of the groundwater evapotranspiration component.

2. Steady state and quasi-transient groundwater balance approaches, contrary to the true-transient approach, oversimplify the balance Re=R-Eg. For example, the model cell R=4mm/y at steady state or quasi-transient condition, with assumption of Eg=0, gives the same balance as obtained with an average R=14mm/d and Eg=10mm/y, and so on, because in both cases Re=4mm/y. Considering the regional scale, the situation is more complex because higher values of R and Eg with the same entire balance of Re, result in the increase of the internal (lateral in 2-D) flow. In such a case higher transmissivities are needed to maintain the groundwater table at a certain level and may imply the calibration non-uniqueness problem discussed in Lubczynski & Mudzingwa (2000).

3. Serowe recharge is a highly temporally variable. It occurs randomly and may vary from 0 up to episodically even several tens of millimetres in the certain years with a sequence of heavy rainfall events.

4. The dry season evapotranspiration calculated on 14/8/98 for the ADAS test site indicates Es=0 and points at small Eu, so the measured Ea was equated to Eg. Eg estimated by Bowen Ratio method, SAP method and 1-D Earth modelling indicated convergent rates of about 0.4 - 0.5mm/d. The Ea values based on SEBAL indicated higher values 1-2mm/d. In fact the reason of that is not known, so further research need to be devoted to this particular aspect. The MODFLOW solution was not comparable with the test site results because R and Eg were simplified

in the MODFLOW model as time invariant. It is wise to note that the test results cannot be extrapolated over the entire area because of the specific site location (selected on purpose to reduce Eu) with a non-representative shallow groundwater table of about 10m depth below the surface.

5. The best identification of the hydrological regime at semi-arid environment is provided through the true-transient approach, which allows for temporal characterisation of the fluxes. This has been enlightened by the 1-D EARTH model solution, which unfortunately due to the various model simplifications can be treated only as a rough estimate.

6. So far, in countries like Botswana it's been assumed that the recharge values are in the range of a few mm/y. Shouldn't we interpret these figures as Re? If so it means that the total R could be much higher, being depleted by Eg, so temporally more of utilisable groundwater might be available in the systems than used to be thought. Supporting this concept for example is the common "too high" values in recharge determination, particularly in true transient calibration. Recent (beginning of 2000) extreme groundwater raises (up to even few meters!) as a result of the cyclonic, unusual high rainfalls, also support the concept of generally higher R values in Botswana.

7. The Serowe study, apart from some advances in understanding of groundwater balance in semi-arid areas confirmed a large demand, particularly with regard to the true-transient approach. It also emphasised the importance of the temporal hydrological data acquisition, including SAP flow, hydrochemistry and isotope data as well. This can be done best through the ADAS monitoring networks. Such data can then be implemented in the spatial recharge model (Lubczynski 1997, Meijerink et al. 1999) and evapotranspiration model (Timmermans & Meijerink 2000) both used in the semi-quantitative manner and finally interfaced with numerical groundwater model. This way designed and calibrated model provides the most reliable hydro(geo)logical system representation.

Acknowledgements

I thank Wellfield Consulting Services and DWA and GS of Botswana for cooperation and making Serowe data available. Prof A.M.J Meijerink and Dr. A. Gieske for critical readings and comments. W. Timmermans for the processed energy balance data and field contribution. Finally the ITC research funding scheme, particularly Dr E. Kosters for financial support.

REFERENCES

Allen R., L. Pereira, D. Raes & M. Smith 1998. Crop evapotranspiration. Guidelines for computing crop water requirements. FAO irrigation and drainage paper.

Bastiaansen, W. Menenti, R. Feddes & A. Holtslag 1998. A remote sensing surface energy balance algorithm for lands (SEBAL). Journal of Hydrology 212-213.

Beekman H., E. Selaolo & J. de Vries 1999. Groundwater Recharge and Resources Assessment in the Botswana Kalahari. Executive Summary GRES II. Geological Survey Botswana.

Dingman L. 1994. Physical hydrology. Macmillan Publishing Company, New York.

EcoSurv, 1998. Vegetation mapping and ground truthing for radar imagery. EcoSurv Report Gaborone, Botswana.

Gehrels H. 1999. Groundwater level fluctuations. Phd.Thesis, Vrije Universiteit Amsterdam.

Gieske A. 1992. Dynamics of groundwater recharge. A case study in Semi-Arid Eastern Botswana. Ph.D. Thesis, Vrije Universiteit Amsterdam.

Granier, A. 1985. Une nouvelle methode pour la measure du flux de seve dans le tronc des arbres. Annals Scientific Forestry, 42: 81-88.

Hendrickx J. & G. Walker. 1997. Recharge from precipitation. In "Recharge of Phreatic Aquifers in (Semi-) Arid Areas". IAH Book no.19.

Lubczynski, M. 1997. Application of numerical flow modelling combined with remote sensing and GIS techniques for the quantification of the regional groundwater resources in hard rock terrain's, IAHS Publication no. 241, ISSN 0144-7815.

Lubczynski, M. & B. Mudzingwa. 2000. Groundwater modelling in extreme hydrogeological conditions. Serowe case study. In preperation for Journal of Hydrogeology, IAH.

McDonald, M.G. & A.W. Harbaugh 1988. A modular three-dimensional finite-difference groundwater flow model. *USGS Report 83-875.*

Meijerink A., M. Lubczynski , & P. Wolski 1999. Remote sensing, hydrologic analysis and hydrotopes, IAHS Publication no 254.

Scott D. & D. Le Maitre 1998. The interaction between vegetation and groundwater: Research priorities for South Africa. Executive Summary. CSIR Report No. ENV/S-C 97161.

Selaolo E. 1998. Tracer studies and groundwater recharge assessment in the eastern fringe of the Botswana Kalahari. Ph.d. Thesis, Vrije Universiteit Amsterdam.

SGAB, 1988. Serowe groundwater resources evaluation. Swedish Geological Co report for Department Geological Survey, Botswana.

Timmermans W. & A. Meijerink, 1999. Remotely sensed actual evapotranspiration; implications for groundwater management.

Timmermans W., A. Meijerink & M. Lubczynski, 2000. Satellite derived actual evapotranspiration and groundwater modelling, Botswana.

Van der Lee J. & J. Gehrels 1990. Modelling aquifer recharge – Introduction to the Lumped parameter model EARTH. Free University of Amsterdam, The Netherlands.

Van der Lee J. & J.Gehrels 1997. Modelling of groundwater recharge for a fractured dolomite aquifer under semiarid conditions. . In "Recharge of Phreatic Aquifers in (Semi-) Arid Areas". IAH Book no.19.

WCS, 1999. Serowe Wellfield 2 Extension Project. Wellfield Consulting Services report for Department of Water Affairs, Gaborone, Botswana.

Groundwater: Past Achievements and Future Challenges, Sililo et al. (eds) © 2000 Balkema, Rotterdam, ISBN 90 5809 159 7

Location of a well field in a coastal alluvial aquifer of southern Sardinia, Italy, subjected to saline water intrusion

G. Luise, C. Sechi & P. Ciabatti
Hydrocontrol, Capoterra, Italy

ABSTRACT: This paper deals with a quantitative and qualitative study of the Capoterra plain alluvial aquifer (Southern Sardinia – Italy) with the aim of locating the more suitable area for public supply groundwater abstraction. Using data collected in the last years, in the paper is described the influence of the present groundwater abstraction rate on the aquifer equilibrium in the zone of dispersion between fresh and salt water. With the aim of defining the most suitable abstraction areas, a mathematical groundwater flow and transport model has been set up using the finite difference code MODFLOW and the particle tracking code MODPATH. The methodology proposed is based on the analysis of the aquifer hydraulic heads and hydrochemical data collected in these last years and the following setting up of the models for the location of the areas in which would be possible to abstract groundwater with the minimum drawdown. Using this methodology, the present paper presents a new groundwater resources management approach for the studied area, directed towards the containment of human activities impact on groundwater flow system and quantitative and qualitative depauperation of the resources.

1 INTRODUCTION

The Capoterra alluvial aquifer, located between the small town of the same name and the Cagliari pond (Southern Sardinia – Italy), has showed an increase of chloride pollution phenomenon in these last years. In the hydrological years 1994/95 and 1998/99, the groundwater qualitative characteristics deteriorated, mainly due to higher public supply groundwater abstraction rates. In fact, to the Tanca sa Turri municipal wellfield ($1{,}9 \times 10^6$ m^3/year) the new municipal wellfield of Su Bau Mannu was added, with an abstraction rate of $0{,}56 \times 10^6$ m^3/year. This fact caused a big change in the aquifer head distribution (Figure 1).

In the framework of this study, by means of a groundwater monitoring program, it has been possible to locate the area in which the salt water intrusion was more intense. At the same time, the setting up of a groundwater flow model allowed to locate a new position for the municipal wellfields in an area of very good hydraulic and qualitative properties of the aquifer and in which groundwater abstraction wouldn't cause drawdowns able to induce groundwater flow from the pond to the aquifer or upconing phenomenon. This was possible considering the studied area most up-to-date hydrogeological researchs results and the acquisition of new aquifer field data.

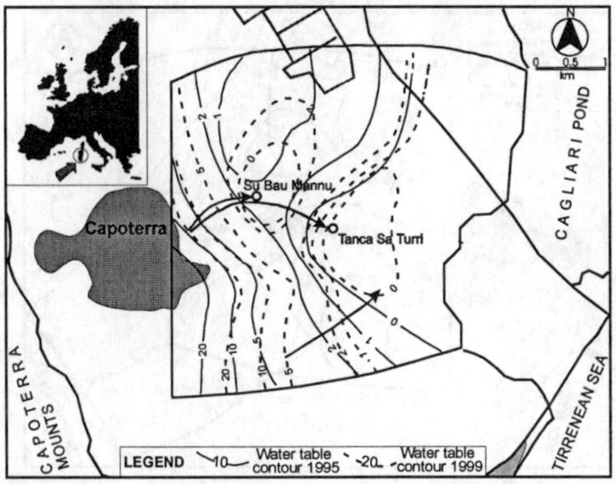

Figure 1. Water table countour map of the studied area

2 HYDROGEOLOGY

The studied area represents the southern part of the Capoterra plain. This plain is limited by the paleozoic shaly – granitic formation of the Capoterra Mountains to the west and the Cagliari pond to the east (total extent of about 11 km^2). The alluvial aquifer falls in a quaternary alluvial formation with medium-low to medium-high hydraulic conductivity (Ciabatti et al., 1997).

The aquifer thickness reaches the maximum values in the central-northern part of the studied area, getting thinner towards the Cagliari pond direction. The alluvial deposits outcrop all over the plain, representing a very important aquifer in which the paleozoic shaly - granitic fractured aquifer of the Capoterra Mountains descharges.

The presence, inside the alluvial deposits, of several clay lenses, causes the forming of different piezometric levels, which tend to connect in a single level because of the fact that the sandy - gravel levels join themselves both in vertical and horizontal direction. This geological conceptual model leads to the conclusion that in the area a single unconfined aquifer can be identified.

The aquifer water budget (Luise & Sechi, 2000) shows a last five years average recharge of the aquifer of $3{,}6 \times 10^6$ m^3/year. The abstractions were of $2{,}46 \times 10^6$ m^3/year from public municipal wellfields and $0{,}926 \times 10^6$ m^3/year from wells used for irrigation. Making a comparison between 1994/95 and 1998/99 hydrological years in and out fluxes of the aquifer, comes out that the aquifer presents a total negative water budget of $0{,}45 \times 10^6$ m^3/year in 1994/95 and $0{,}75 \times 10^6$ m^3/year in 1998/99. More in detail, during dry seasons of these two hydrological years, some wells of the Tanca sa Turri wellfield were closed for the high presence of chloride in the abstracted groundwater.

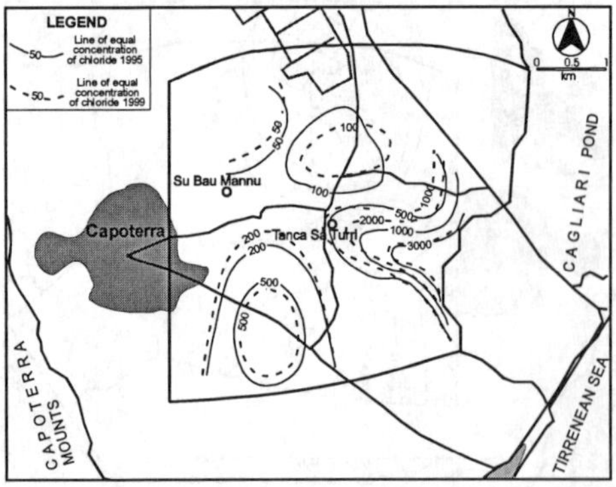

Figure 2. Chloride observed concentration

3 HYDROCHEMISTRY

During the period in which salt water intrusion phenomenon accentuated (November 1995 and November 1999), chemical-physical analysis of groundwater in 16 wells have been carried out at the Hydrocontrol Water Laboratories. Groundwater were classified as chlorinated – sulphate – alkaline (Luise & Sechi, 2000). Temperatures ranged between 12,4 and 19,5 °C, with a maximum value of 35,2 °C. This last value was measured in a well located in the southern sector of the studied area. This high temperature gives value to the hypothesis of thermal localized phenomenon originated by deep crystalline basement circulation groundwater, which laterally recharge the alluvial deposits aquifer of the plain. Nitrites and ammonia concentrations found are lower than the maximum values expected by the Italian legislation on drinking water (0,5 mg/l and 0,1 mg/l respectively). Also nitrates do not present concentration higher than the potability limit of 50 mg/l of the Italian legislation. Only in two wells a concentration of nitrites (0,7 mg/l) and nitrates (57 mg/l) are higher than the admissible limits.

The shape of isochlorine 3000 mg/l (Figure 2) shows that the salt water interface get into the aquifer from the Cagliari pond and extends in the direction of the Tanca sa Turri municipal wellfield. Here, chemical analysis pointed out chloride concentrations of 500 mg/l in 1995 and 1000 mg/l in 1999. The shape of the salt water interface in these two period remained the same, but the high chloride concentrations of 1999 can have been caused by the higher abstraction rate of the Su Bau Mannu municipal wellfield, that in the last two years raised of about $0{,}6 \times 10^6$ m^3/year.

The high measured chloride concentrations in the southern part of the studied area (500 and 200 mg/l isochlorine) can be traced back to the regional groundwater flow conceptual model, which provides a likely recharge of the alluvial quaternary deposits aquifer with water discharged by the fault system which is located between the plain and the paleozoic outcrops of the Capoterra Mountain (Cidu & Pala, 1987). These deep circulation groundwater mixes with the superficial aquifer one. Groundwater in this area shows costant chemical characteristics during the year. This fact shows a certain equilibrium in the in and out fluxes of this area of the alluvial aquifer.

In the north-western part of the studied area, chloride concentrations tend to be considerably lower, particularly during dry seasons of 1995 and 1999, not being influenced by salt water intrusion (42 mg/l in 1995 and 57 mg/l in 1999).

4 GROUND WATER MODELLING

An alluvial aquifer groundwater flow simulation model was set up using the MODFLOW code (McDonald & Harbaugh, 1988). The program PM – Processing Modflow for Windows (Version 4.1) (Chiang & Kinzelbach, 1991-97) was used to set up the input and output files needed to run and analize the results of each simulation. The studied area has been discretized using a single unconfined layer and a grid of 60 rows and 80 columns, which created a grid of 4800 square cells of 50 m of side (Figure 3). The grid is 4000 m long in the E - W direction and 3000 m long in the N - S direction (Luise & Sechi, 2000). Only 4308 cells of the grid are active, for a total extension of the modelled area of 10,77 km^2. 200 cells are constant head cells and represent the

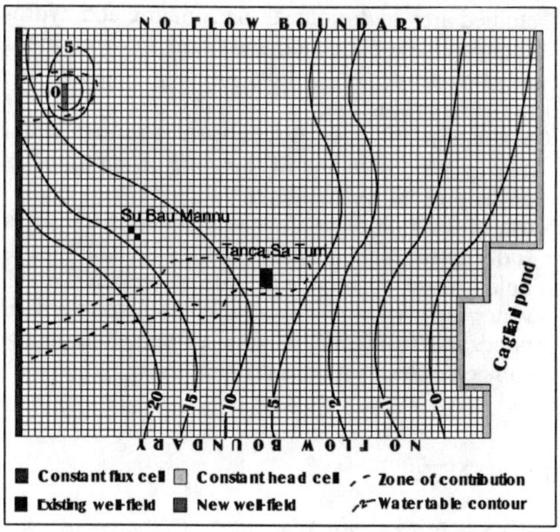

Figure 3. Model grid and specified head boundaries

coastal line between the aquifer and the Cagliari pond. The remaining 292 cells are inactive.

The calculated water balance of the aquifer (table 1) shows that the most part of its recharge is due to lateral fluxes entering into the model from the western boundary. This according to the conceptual model, which provides the main recharge of the alluvial deposits groundwater flow system from the paleozoic fractured aquifer of the Capoterra Mountains. Only a small part of the recharge is due to vertical infiltration of rainfall. By the costant head cells boundary, outflow is higher than inflow. Only in the area between the Tanca sa Turri municipal wellfield and the Cagliari pond, inflow from the Cagliari pond are higher than outflow, with a net recharge of the aquifer from the pond of 0,8 × 10^6 m^3/year.

Table 1. Steady state water budget calculation

	IN (m^3/years)	OUT (m^3/years)
Constant Head		100.000
Vertical Recharge	360.000	
Wells		3.386.000
Lateral Recharge	3.150.000	

Table 2. New well field water budget calculation

	IN (m^3/years)	OUT (m^3/years)
Constant Head		1.355.000
Vertical Recharge	360.000	
Wells		3.386.000
Lateral Recharge	4.391.000	

The zone of contribution of the Tanca sa Turri wellfield (Figure 3, Luise & Sechi, 2000) was delineated using the MODPATH code (Pollock, 1989). Looking at the shape of this zone, is possible to say that the main water contribution to the wellfield is from the Capoterra Mountains aquifer, but a little contribution is also from an area of the alluvial aquifer characterized of high chloride concentrations (Figure 2).

The next step in the study was to locate a new abstraction area whose position would have caused a slowing down of the saline water intrusion phenomenon. The location of the new abstraction area was based on the hydrochemical study of the area (chloride concentrations of 42-57 mg/l) and the hydrodinamic characteristics of the aquifer (transmissivity of 5 ÷ 0,5 × 10^{-2} m^2/sec). The best area for drilling new wells was then located in the north-western part of the studied area.

Using the codes MODFLOW and MODPATH, a scenario has been set up, which simulated the new municipal wellfield and its zone of contribution. In Figure 3 the new head distribution of the aquifer is showed. Making a comparison between the two head distributions of the dry years 1995 and 1999 with those calculated by the model (Figures 1 and 3), is possible to observe that the groundwater fluxes directions remain the same, with the pond that still is the final point of discharge of the alluvial aquifer.

With the new location of the municipal wellfield the following advantages could be achieved in terms of aquifer exploitation:

- the head distribution would be only slightly stressed;
- drawdowns in the coastal area would be very low and, for this reason, salt water intrusion phenomenon wouldn't be induced;
- the main groundwater flow directions would move to the new abstraction area.

In fact, the computed water balance for this new scenario shows that both the groundwater fluxes to the Cagliari pond and the aquifer recharge from the fractured paleozoic aquifer of the Capoterra Mountains are higher (table 2).

The calculated hydraulic gradients are generally very low in areas far from the new wellfield. Concerning the MODPATH code defined zone of contribution of the new wellfield, this is limited to the north-western part of the model with water that flows only from Capoterra Mountains fractured aquifer. No contribution is evident from the Cagliari pond.

5 DISCUSSION

The hydrogeological and hydrochemical study and the mathematical groundwater flow model set up for the alluvial aquifer gave the possibility to develop the following considerations.

For the hydrological years 1994/95 and 1998/99 the lateral recharge of the alluvial aquifer from the paleozoic fractured aquifer of the Capoterra Mountains have been calculated. Making a comparison between inflow and outflow of the alluvial aquifer was possible to define that in these two years the aquifer suffered a water deficit of respectively 0,45 and 0,75 $\times 10^6$ m^3/year.

During dry summer seasons of these years, the Tanca sa Turri municipal wellfield was not used for public water supply because of high chloride concentrations in the abstracted groundwater. All the wells located between this wellfield and the Cagliari pond showed the same pollution problem, causing the abandonment of agricultural practices in the area. Chemical – physical analysis of groundwater showed that aquifer do not present pollution due to organic compounds, fertilizers and pesticides used in agriculture. In fact, both nitrates, nitrites and ammonia concentrations are lower than the maximum values admitted by the Italian legislation. Only very high chloride concentrations were found in the aquifer area between the Tanca sa Turri wellfield and the Cagliari pond (4260 and 4899 mg/l). In the southern part of the studied area the high chloride concentrations (500 e 200 mg/l) are due to the upflow of deep and slow circulation groundwater. In the northern and north-western part of the studied area the chlorine concentrations are low (42 e 57 mg/l). No salt water intrusion or deep groundwater upflow can be assumed in this area.

The mathematical model, set up and calibrated in steady state condition using the MODFLOW code, was used to obtain a new head distribution of the alluvial aquifer in the case that the existing municipal wellfields of Tanca sa Turri and Bau Mannu were closed, and a new wellfield were drilled in the north –western part of the studied area (Figure 3).

The alluvial aquifer water budget, calculated using the groundwater flow model for a scenario in which only the new municipal wellfield were abstracting groundwater, clearly shows a decrease of salt water intrusion from the Cagliari pond. This scenario model calculated head distribution is much less stressed than the actual one, with very low hydraulic gradients in areas far from the wellfield.

MODPATH code simulation also gave the possibility to define the zone of contribution of the new wellfield. The shape this new zone clearly shows that no contribution from the pond is possible for the new proposed wellfield. The recharge of zone of contribution is, in fact, totally due to the upstream groundwater flow area.

6 CONCLUSIONS

The proposed methodology, based on the aquifer qualitative monitoring and flow simulation, has been used as an approach to locate a new position for the municipal wellfield, with particular care to the reduction of the saline water intrusion into the aquifer from the Cagliari pond. The study results allowed to conclude that the new position of the municipal wellfield could be identify by means of the following steps:

- studied area hydrological, piezometric and hydrochemical data collection and future integration for a better groundwater flow simulation model parameters evaluation;
- evaluation of the groundwater and salt water mixing processes and of the zone of contribution of the present wellfields;
- use of the model results and calculated transmissivity to simulate a new scenario with a new wellfield whose position was able to reduce the salt water intrusion into the aquifer from the Cagliari pond and to identify a zone of contribution not subjected to salt water intrusion.

REFERENCES

Anderson M.P., Woessner W.W. 1992. *Applied groundwater modeling*. San Diego, California: Academic Press.

Celico P. 1986. *Prospezioni idrogeologiche*. Napoli: Ed. Liguori.

Chiang W.H., Kinzelbach W. 1991-97. *Processing Modflow Version 4.10 Operation Manual*, Hamburg Heidelberg Germany: Scientific Software Group.

Ciabatti P., Liori G., Luise G. 1995. Setting up of a mathematical groundwater flow simulation model of the Capoterra plain alluvial aquifer (Southern Sardinia). First results. *2nd International meeting for young researchers in applied geology, Peveragno (Cuneo)11-13 October 1995:* 283-289.

Ciabatti P., Fontana M., Luise G., Sole C. 1997. Simulazione del flusso dell'acquifero della piana di Capoterra (Sardegna Meridionale). *L'Acqua Associazione Idrotecnica Italiana* 15/1997: 43-51.

Cidu R., Pala A. 1987. Acqusizione di nuovi dati idrogeochimici nel Campidano Meridionale. *XXth International Association of Hydrogeologists, Roma.*

Custodio E. 1983. *Hidrologia subterranea*. Barcelona: Ed. Omega.

Luise G., Sechi C., Spano E. 1998. Applicazione di una metodologia di perimetrazione, attraverso l'utilizzo dei codici di calcolo MODFLOW e MODPATH, per la definizione delle aree di salvaguardia relative a pozzi in acquiferi alluvionali. *3° Convegno Nazionale sulla Protezione e Gestione delle Acque Sotterranee per il III millennio. Supp. Spec. Quaderni di Geologia Applicata. Parma 13-15 October 1999:* vol. 4: 164-173.

Luise G., Sechi C. 2000. Studio idrogeologico e potenzialità del settore Sud dell'acquifero alluvionale della piana di Capoterra (Sardegna Meridionale). *Quaderni di Geologia Applicata* 6/1999: In press.

Mc Donald M.G., Harbaugh A.W. 1988. *Modflow, a modular three-dimensional finite-difference groundwater flow model.* Washington D.C. In Techniques of Water Resources sources Investigations of the United States Geological Survey, Book 6, Chapter A1,

Pollock D.W. 1988. Semianalytical computation of path lines for finite-difference models. *Ground Water* 26 (6): 743-750.

Pollock D.W. 1989. Documentation of computer programs to complete and display pathlines using results from the U.S.G.S modular three-dimensional finite-difference ground-water. *USGS Open File Report* 89-381, 81 p.

Groundwater: Past Achievements and Future Challenges, Sililo et al. (eds) © 2000 Balkema, Rotterdam, ISBN 90 5809 159 7

Hydrogeology of the coastal region of Jakarta, Indonesia

H. Maathuis
Saskatchewan Research Council, Saskatoon, Sask., Canada

W. Mak
IWACO, Maastricht, Netherlands

S. Adi
BPP Teknologi, Jakarta, Indonesia

ABSTRACT:
Rapid urbanization and industrial growth of Jakarta since the 1970s has led to mining of groundwater as surface water supplies are inadequate to meet the municipal and industrial demands. Groundwater withdrawal in the Jakarta area has resulted in changes of the recharge-discharge system. Water levels in deep wells have declined by up to 50 metres since the early 1900s. The geological setting is conducive to groundwater withdrawal-induced subsidence due to natural compaction and surface loading. Surface elevations have declined by up to 100 cm but can not be uniquely related to withdrawals. Prediction of groundwater withdrawal-induced subsidence is hampered by the complex setting, uncertainties in the distribution and withdrawals from wells, the lack of data on geotechnical properties for depths greater than 50 m, and the absence of subsidence measurements. The distribution and concentration of chlorides is due to the geological setting and history and not to classical seawater intrusion.

1 INTRODUCTION

As is the case for many other metropolitan areas in the world (Asian Development Bank 1994), in the Jakarta area, urbanization and industrialization and the large and rapidly increasing population have created significant problems with respect to reliable and sustainable supplies of potable drinking water. The potential impacts of inadequately controlled groundwater withdrawals on the environment in coastal zones of include: a) seawater intrusion, b) land subsidence, c) contamination of shallow and deep aquifers (Foster and Lawrence 1994). In this paper we address seawater intrusion and land subsidence.

The City of Jakarta (DKI Jakarta) is located along the northern coast of the island of Java, Indonesia in what is referred to as the Jakarta basin (Figure 1).

The climate of Java can be described as warm-tropical, with high rainfall and seasonal "monsoon" shifts of the winds due to atmospheric pressure changes on the Asian and Australian continents. The climate of the Jakarta basin is affected by factors such as geographic location, elevation and land cover. The elevation in the Jakarta basin increases inland to 200 m at Bogor and to 3000 m at the top of the volcanos south of Bogor. Average rainfall varies from 1500 mm/a near the coast to >4000 mm/a near Bogor. Class A pan evaporation is in the order of 2000 mm/a.

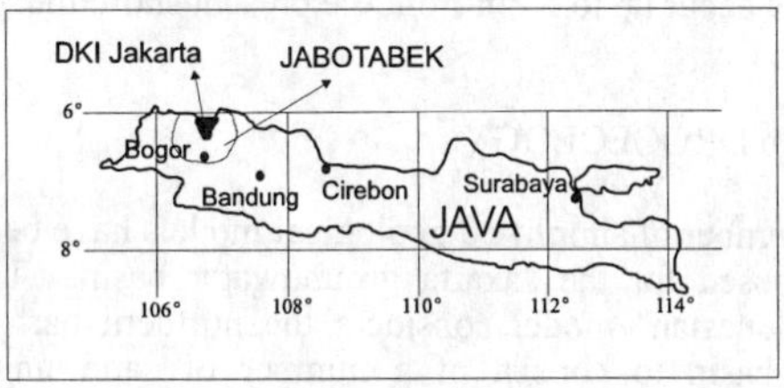

Figure 1. Location of DKI Jakarta and Jakarta basin

The rainy season commonly occurs between October and March, and the dry season between April and September.

The population of the DKI Jakarta has grown from about 3 million in the early 1960 to 9.3 million in 1996. The population in the Jakarta basin area is about 20 million.

In the mid 1990s the capacity of the water treatment plants was about 12.5 m^3/s and the actual production in the order of 11 m^3/s. Losses in the pipeline distribution system varied between 24% and 40%. It is estimated that the pipeline distribution system serves only 25% to 30% of the population and 45% of the industry and commerce needs. Therefore, there is a heavy demand for groundwater as a source of water supply. Shallow groundwater (less than 40 m deep) is mainly used for domestic supplies and withdrawals were crudely estimated in

1994 to be in the range 7 to 11 m^3/s. Industry and commerce mainly derive their water from deep wells. Recorded abstraction from registered deep wells was about 1 m^3/s in 1994, but actual withdrawals are estimated to be in the range 2 to 4 m^3/s.

2 GEOLOGY OF THE JAKARTA BASIN

During the Pleistocene, Java emerged as a result of both uplift from rising magma bodies and build-up of volcanos. Sedimentary basins beneath the present day northern coastal lowlands were formed and filled with eroded sediments from the emerging hinterland. The Jakarta basin has been filled with up to 200 to 250 m of Quaternary sediments and is underlain by Tertiary sediments. The depositional environment during the Quaternary in the Jakarta basin is a complex mixture of deltaic, lacustrine, lagoonal, swamp, terrestrial, flood plain and shallow marine environments. As a result, sand and clay deposits cannot be traced laterally over any significant distance (e.g. Soefner et al. 1986). Sand units in boreholes typically make up 20 to 25% of the total thickness, and clayey units between 75 and 80%. The thickness of individual sand units is commonly in the range 2 to 6 m (Soefner et al. 1986). The sand units are commonly comprised of silty sands and seldom consist of clean sands. Holocene marine silts and clays occur up to 5 km from the present shoreline.

3 HYDROGEOLOGY

A number of simplified geological models have been proposed for the Jakarta groundwater basin. The "Indonesian" model considers the northern part of the basin to consist of a number of sand units, separated by two major clay units. The same model was used by ILN (1987). Because of the absence of a stratigraphical framework and the fact that it is impossible to trace lithological units, the "German" model considers the fill of the basin as one single unit (Soefner et al. 1986). The "JWRMS" model used the same arguments, but differs from the "German" model in that a volcanic fan deposit in the upper part of the Quaternary sequence is shown as a separate unit (JWRMS 1994). In terms of hydrogeological units, the aquitard formed by the Holocene deposits and the aquifer formed by the Upper Pleistocene Volcanic Fan deposits are perhaps the only units which can be identified with some measure of confidence. It is not possible to identify individual aquifer and aquitard units within the underlying sequence of marine and non-marine Quaternary sediments. Consequently, this sequence forms an undifferentiated, and very complex, aquifer-aquitard system (Figure 2).

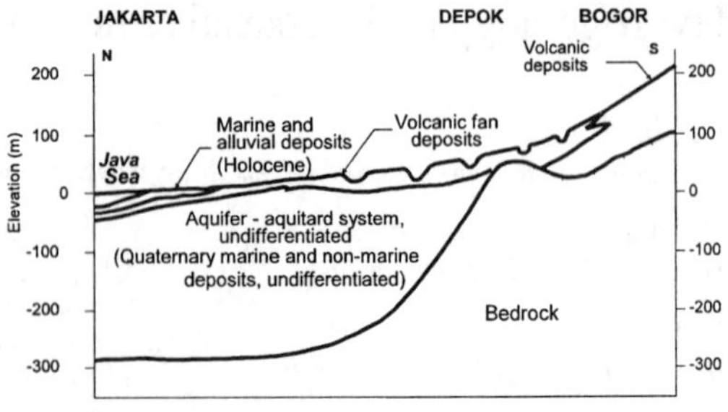

Figure 2. Schematic illustration of the geological and hydrogeological setting of the Jakarta basin.

Despite the fact that a classical differentiation between aquifers and aquitards is not possible, for practical reasons, it has become "standard" practice to subdivide the Quaternary sequence into hydrogeological "zones" or "horizons" (e.g. Soefner et al. 1986; JWRMS 1994a). The hydrogeological "zones" commonly used are: 0-40, 40-150, 150-200, and deeper than 250 m.

Few reliable data are available on the hydraulic properties. The transmissivity of the entire sequence is thought to decrease northward from 500 to 250 m^2/day (Soefner et al. 1986). Based on modelling JWRMS (1994b) suggested that the transmissivity might be less than this by a factor of 2. The sands have a hydraulic conductivity not more than a few metres per day, and the vertical hydraulic conductivity (K_v) of the silt and clay units is in the order of 8.6×10^{-5} to 4.3×10^{-4} m/day (Soefner et al. 1986; JWRMS 1994a; Maathuis et al. 1996).

4 IMPACTS OF GROUNDWATER WITHDRAWALS

Development of the groundwater resources started in the early 1970s, as a result of population and industrial growth. In the early 1900s, the water levels in the coastal zone were well above sea level and the area was a discharge zone. Recharge to the water table occurred over the entire region and discharge from shallow aquifers was toward rivers, canals, and by evapotranspiration. The present groundwater regime is schematically shown in Figure 3.

Due to the groundwater withdrawals from the deeper zones there is no longer upward discharge from the deeper zones in the coastal zone. Instead, strong downward vertical gradients have developed. These downward gradients, however, do little to increase recharge to the deeper zones because of the low hydraulic conductivity of the silt and clay units. Recharge to the water table still occurs across the entire region, but is affected by urbanization.

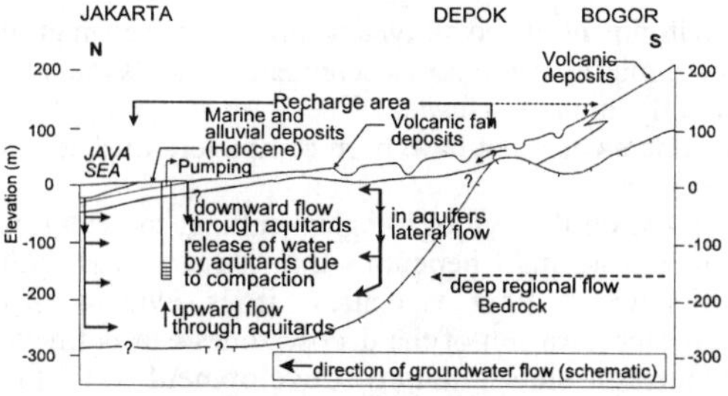

Figure 3. Schematic illustration of present ground- water flow regimes

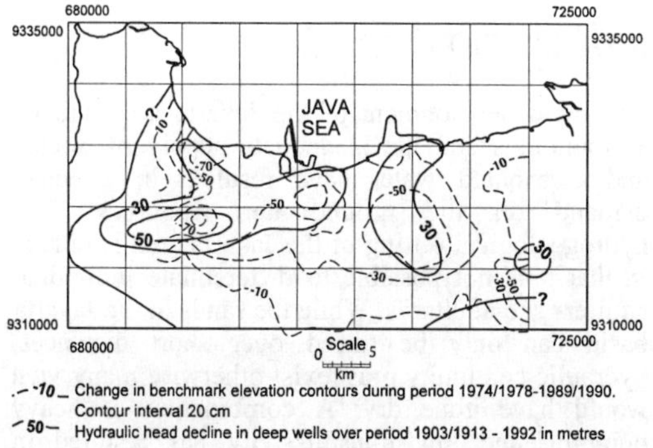

Figure 4. Benchmark elevation changes in relation to groundwater level decline

The groundwater withdrawals from the deeper zones have resulted in significant drawdowns. From the early 1900s until the early 1990s, water levels in the deep aquifers have dropped by up to 50 m (Figure 4).

In the period 1989 - 1995, water levels in the deep aquifers have dropped annually by 0.6 to 2.3 m. The continuously declining levels indicate that groundwater is being mined.

Figure 4 also shows changes in the elevations in benchmarks, over the period 1974/79 - 1989/1990.

As is shown in Figure 4, there is only a cursory relationship between benchmark changes and groundwater level decline. Quantitatively relating benchmark changes to withdrawals is difficult at best. The benchmark changes could be due to a number of causes, including: compaction of natural or man-made sediments (i.e. landfill materials), settlement due to load of construction, and subsidence due to groundwater withdrawals, or a combination of all.

Furthermore, there are significant uncertainties in the withdrawal data, and consequently in the spatial distribution of volumes withdrawn. Any quantitative analysis is also hampered by the absence of geotechnical data in the 50 to 250 m depth interval, and by the lack of actual subsidence measurements.

Considering the geological setting of the Jakarta basin and the experiences in similar basins elsewhere (e.g. Bangkok, Shanghai), subsidence along the coastal region of Jakarta is of concern. Subsidence can be counteracted by ceasing pumping and allowing water levels to recover, or by artificial recharge (*e.g.* Yong *et al.* 1994). A groundwater conservation plan has been in place since 1985 which prohibits construction of deep wells in the coastal zone. However, there has been little enforcement, as is evident from the continuing decline in water levels. Artificial recharge is excluded as there is an inadequate supply of surface water.

5 GROUNDWATER WITHDRAWALS AND SEAWATER INTRUSION

The term "seawater" intrusion is commonly used for migration of present day seawater into freshwater aquifers due to withdrawals in these aquifers, and implies that at some distance from the shore the aquifers are in direct hydraulic contact with the seawater. In any analysis of seawater intrusion, however, the geological history, the hydrogeological setting and the possible effects of eustatic sea level changes must be taken into account. It can not be assumed that the distribution of salt and fresh water in complex aquifer-aquitards system is in equilibrium with the present day sea level as a rising sealevel may result in entrapment of fresh water in underlying aquifers and aquitards (Kohut *et al.* 1977; Meisler *et al.* 1984).

The occurrence or non-occurrence of classical seawater intrusion in the costal region of Jakarta has been debated for decades. Purbo-Hadiwidjoyo (1960), Henanussa (1980), Puryanto (1982) and RIWRD (1985) suggested that the presence of seawater beneath the coastal zone was natural and related to the geological history. Soefner et al. (1986) concluded that along the coast seawater encroachment in the upper 100 m had reached 3 to 6 km landward by the mid 1980s and attributed small water quality changes in the deeper zones to slow landward-moving intrusion of seawater. Tjahjadi (1991) suggested that by 1990 seawater intrusion in the shallow aquifer had progressed to a distance of 10 to 15 km from the present shoreline. In deeper zones it would have spread to a distance of about 5 to 10 km. Rismanto and Mak (1993), and Maathuis et al. (1996) questioned the occurrence of seawater intrusion based on (1) the chloride distribution prior to major groundwater withdrawals, and (2) the hydrogeological setting and groundwater flow. The distribution of the chloride concentrations beneath Jakarta and the Java Sea in the early 1900s is shown in Figure 5.

Chloride concentrations beneath the Java Sea were much less than that in present day seawater (19,000

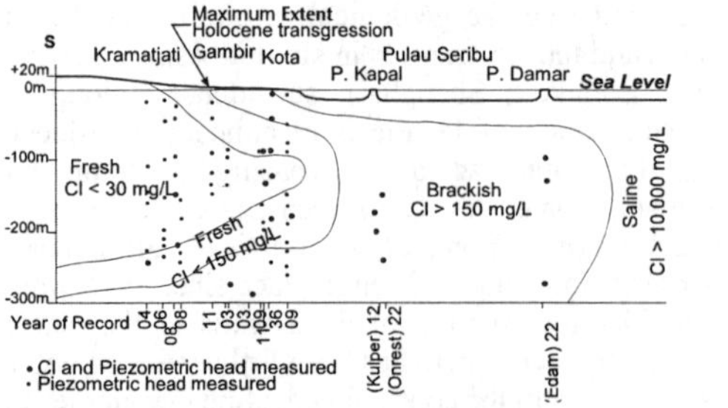

Figure 5. Distribution of chloride concentration beneath Jakarta and the Java sea in the early 1900s

mg/l). For classical seawater intrusion to occur, permeable zones would have to be in direct contact with the Java Sea. The Java Sea is a very shallow sea and therefore, it is unlikely that many sand layers will out crop at the sea bottom close to the present shore line.

The volcanic fan could extend beneath the shore line but would be covered by Holocene silts and clays.

Seawater would have to migrate vertically through these sediments and subsequently laterally landward. Any vertical advective flow through the clay and silt layers will be small because of the low hydraulic conductivity of these sediments (0.05 m/year for K_v= 8.6 x 10^{-5} m/d and gradient of unity). Assuming a horizontal hydraulic conductivity of less than 4 m/day for the volcanic fan deposits, a porosity of 0.3 and a hydraulic gradient of 0.001 (JWRMS 1994c), the flow velocity in the shallow aquifer zone would be in the order of 5 m/year. Consequently, recent seawater would have migrated a distance of less than 125 m in the 25 years since major development of the groundwater resources started. If shallow seawater intrusion had indeed progressed 6 km landward in 1985 (Soefner et al. 1986), or 10 km in 1990 (Tjahjadi 1991), the groundwater flow velocity would have to be in the order of 400 m/year. Furthermore, seawater intrusion several kilometres landward implies that wells near the coast would yield undiluted seawater. Shallow wells near the coastline may yield salty water, but not undiluted seawater.

The presence of salty water in the shallow aquifer in the coastal zone can be explained by considering the recent geological history. The sea level increased at 4500 years BP to about 5 m above the current level and retreated to the current level about 500 years ago (De Klerk 1983). During the encroachment of the sea the permeable volcanic fan deposits were likely invaded by seawater and there may have been some downward migration due to advection-diffusion. When the sea started to retreat, marine silts and clays were deposited which later were overlain by alluvial flood plain deposits. Because of the marine deposits and the near-shore depositional environment of the alluvial sediments, pore water in these marine and alluvial sediments was seawater or slightly diluted seawater. The pore water in these sediments may not have been completely flushed out yet.

Based on the hydrogeological setting, the chloride concentrations beneath the Java Sea and groundwater flow velocities, it is evident that seawater intrusion of the deeper zones can not have occurred since major development of the groundwater resources started in the early 1970s. This conclusions is also supported by the fact the conventional ^{14}C age of groundwater near the coast is greater than 25 ka (Geyh et al. 1986).

6 DISCUSSION

The rapid development of the Jakarta area and its surroundings and the inadequate supply of treated and distributed water have resulted in a heavy demand on the groundwater resources. The hydrogeological setting of the Jakarta area is unique in that it is not possible to differentiate individual aquifers and aquitards. While the sands in the Jakarta basin can only be traced over short distances, hydraulic continuity must exist otherwise many well would have gone dry. A combination of heavy pumping and low transmissivity has resulted in significant drawdowns in the coastal zone. Continuously declining water levels indicate that the groundwater resources are being mined. It has been demonstrated that the drawdowns have not resulted in classical seawater on a large scale intrusion. However, the geological setting and large drawdowns are conducive to groundwater withdrawal - induced subsidence. The latter can only be counteracted by significantly reducing withdrawals from the deeper zones in the coastal area. However, this will not be possible unless significant volumes of treated and piped water become available for domestic, industrial and commercial use. Furthermore, it would require discouraging the use of groundwater by means of enforcing strict regulations and by making treated water available at affordable cost.

REFERENCES

Asian Development Bank. 1994. *Managing water resources to meet megacity needs*. Proceedings of regional consultations. Asian Development Bank, Manilla, Philippines, 425 pp.

De Klerk, L.G. 1983. *Zeespiegels, riffen en kustvlakten in zuidwest Sulawesie, Indonesië; een morfogenetische bodemkundige studie*. Utrechtse Geografische Studies 27, Utrecht.

Foster, S.S.D. & A.R. Lawrence. 1994. *Groundwater quality in Asia: an overview of trends & concerns*. UNEP-

UNESCO-ICSU Scope 2nd Asian Regional Groundwater Workshop, IAH XXV Congress, Adelaide, Australia, November 21, 1994.

Geyh, M.A., M. Hobler & B. Söfner. 1986. *Jakarta groundwater study. Isotope investigations on groundwater samples from the confined aquifer system.* German Hydrogeological Advisory in Indonesia CTA-40. Directorate of Environmental Geology, Bandung, Federal Institute of Geosciences and Natural Resources, Hannover; Working Paper (HAG) 124, 19 pp.

Henannussa, P.E. 1980. *Pengaruh kedudukan muka hidrostatika terhadap penyusupan ke dalam Cekungan Artois Jakarta.* Riset Geologi dan Pertambangan, jilid 3, No. 1, LIPI, Bandung.

ILN 1987. *Cisadane River basin feasibility study. Groundwater.* Indec & Associates Ltd., Lavalin International Inc., & Nippon Koei Co. Ltd. Ministry of Public Works, Directorate General of Water Resources Development, Indonesia.

JWRMS 1994a. *Jabotabek Water Resources Management Study. Final Report, Volume 6, Annex 10. Groundwater resources.* Ministry of Public Works, Directorate of Water Resources Development, Indonesia, 75 pp.

JWRMS 1994b. *Jabotabek Water Resources Management Study. Final Report, Volume 6, Annex 11. Groundwater models.* Ministry of Public Works, Directorate of Water Resources Development, Indonesia, 44 pp.

JWRMS 1994c. *Jabotabek Water Resources Management Study. Final Report, Volume 7, Annex 12. Groundwater salinity.* Ministry of Public Works, Directorate of Water Resources Development, Indonesia, 55 pp.

Maathuis, H. et al. 1996. *Development of groundwater management strategies in the coastal region of Jakarta, Indonesia. Final report.* Saskatchewan Research Council, Publication R-1250-1-E-96, 62 pp.

Meisler, H., P.P. Leahy & L.L. Knobel. 1984. *Effect of eustatic sea-level changes on saltwater-freshwater in the northern Atlantic coastal plan.* U.S. Geological Survey Water-Supply Paper 2255, 27 pp.

Kohut, F.A., J.C. Hathaway, D.W. Folger, M.H. Bothner, E.H. Walker, D.F. Delaney, M.H. Frimpter, E.G.A. Weed, & E.C. Rhodehamel. 1977. *Fresh ground water stored in aquifers under the continental shelf: implications from a deep test, Nantucket Island, Massachusetts.* American Water Resources Association, Water Resources Bulletin, Vol. 13, No. 2, pp. 373-386.

Purbo-Hadiwidjoyo, M.M. 1960. *Text to accompany the tentative groundwater map of Java and Madura, Indonesia.* Geological Survey of Indonesia, Bandung.

Puryanto 1982. *Konservasi air tanah daerah Jakarta-Bogor-Tangerang-Bekasi.* DGTL, Bandung.

RIWRD 1985. *Survey penyelidikan wilayah basin air tanah Jakarta.* Bandung

Rismianto, G. & W. Mak. 1993. *Environmental aspects of groundwater abstraction in DKI Jakarta: changing views.* Proceedings 22nd Annual Convention of the Indonesian Association of Geologists, Bandung, December 6-9, 1993, Volume I, pp 327-345.

Soefner, B., M. Hobler & G. Schmidt. 1986. *Jakarta Groundwater study 1983-1985.* Final report. German Hydrogeological Advisory Group in Indonesia (CTA 40). Federal Institute of Geosciences and Natural Resources, Hannover. Directorate of Environmental Geology, Bandung, 71 pp.

Tjahjadi, B. 1991. *The impact of abstraction on groundwater quality and monitoring in the Jakarta region, Indonesia.* Directorate of Environmental Geology, Bandung.

Yong, R.N., E. Turcott & D. Gu. 1994. *Artificial recharge subsidence control, Bangkok.* Report for International Development Research Centre, Canada.

Groundwater: Past Achievements and Future Challenges, Sililo et al. (eds) © 2000 Balkema, Rotterdam, ISBN 90 5809 159 7

The effects of alien vegetation clearing on a coastal aquifer, Bushmans River Mouth, Eastern Cape

L.G.A. Maclear & J.C. Kotze
SRK Consulting, Cape Town, South Africa

ABSTRACT: The Department of Water Affairs and Forestry's Working for Water Programme is currently involved in country-wide clearing of invasive alien plants. This has resulted in increases of 25% to 120% in streamflow after clearing. To date, however, limited work has been carried out in South Africa to quantify the effect of alien vegetation removal on the water balance of an aquifer. A hydrogeological investigation of the Bushmans River Mouth Aquifer determined that the 1997 clearing programme of this area resulted in a 31 cm rise in water levels from measured differences in the average level and a 41 cm rise determined using the cumulative rainfall departure method. This equates to a 6% increase in aquifer storage and a 138% increase in effective recharge. It was noted that groundwater salinity worsened immediately after clearing, considered to be a result of remobilisation of vadose zone salts with a water level rise. This study demonstrates the effectiveness of a clearing programme on improving the water balance of an aquifer, thereby increasing the groundwater volume available for abstraction.

1 INTRODUCTION

The effects of changing vegetation cover on groundwater recharge have been well studied worldwide. In a water-scarce country such as South Africa, the introduction of water hungry alien vegetation species has resulted in a considerable cost to the environment and economy. This has prompted the need to improve our understanding of the changes to a catchment water balance in response to changes in vegetation type and density.

1.1 *Alien vegetation removal in South Africa*

The extent and rapid spread of invasive alien plants (IAPs) is a major cause for concern in South Africa, with a huge cost to the environment, specifically in 'lost' water. Woody alien invaders are estimated to use 3300 Mm^3/a of water in South Africa, with the large wattle (*Acacia spp*) accounting for 68% of the total water use of IAPs (Versveld et al. 1998).

In 1995 the Department of Water Affairs & Forestry (DWA&F) embarked on an ambitious project of eradication of IAPs from river catchments country wide under the leadership of Prof. Kader Asmal, then minister of the department. This project was named the Working for Water (WFW) Programme with two main aims: to rid catchments of thirsty IAPs and to create work opportunities for previously unemployed South Africans. To date, local studies have shown that IAP clearing has resulted in significant improvements in the water balance of some catchments, with streamflow increases varying from 25% (Prinsloo et al. 1996) to 120% (WFW 1998).

As the success of the WFW Programme became evident, a need was identified to quantify the water savings achieved, both within surface water and groundwater systems. This study represents one of only two case studies carried out to date on the effect of IAP removal on the groundwater balance of an aquifer. In the other study by Visser et al. (1999) an average water level rise of 20 cm in response to clearing of IAPs on the L'Agulhas Plains near Bredasdorp was noted.

1.2 *IAP removal at Bushmans River Mouth*

The Eastern Cape coastal towns of Bushmans River Mouth (BRM) and Kenton-on-Sea obtain all their water requirements from a shallow sand aquifer south of BRM, viz. the Bushmans River Mouth Aquifer. In the 1960's the area overlying the aquifer was vegetated with *Acacia cyclops* for dune stabilisation and reclamation purposes. The area was subsequently invaded by *Eucalyptus spp* and *Casuarina cunninghamiana* trees.

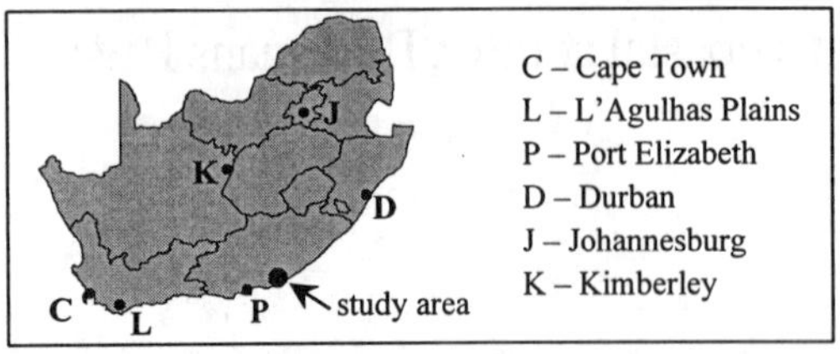

Figure 1. Study area location.

In early 1997 the WFW Programme cleared 20 ha of IAPs overlying the aquifer. This was done to reduce evapotranspirative losses from the BRM Aquifer and thereby increase the availability of groundwater for abstraction for domestic supply purposes.

This paper details the study approach, analytical methods, and results; and suggests a methodology for conducting similar studies.

2 STUDY AREA

2.1 *Physiography*

The study area (Figure 1) is located at the mouth of the Bushmans River in the Eastern Cape, 120 km east of Port Elizabeth. The topography is generally low lying with rounded hills. The coastal area is covered by a shifting dune belt that is fixed by vegetation inland. The area experiences rainfall throughout the year with a mean annual precipitation (MAP) of 660 mm and moderate average monthly temperatures ranging from 15° to 22° C. The portion of the aquifer supplying groundwater for domestic consumption is approximately 40 ha in extent.

2.2 *Hydrogeology*

The BRM Aquifer is a well-defined sandy coastal aquifer with shallow water levels for which a good database of water levels, hydrochemistry and groundwater abstraction figures exists. This study area was thus considered ideal for assessing the effect of IAP removal on a groundwater system.

The conceptual hydrogeology of the BRMA is shown in Figure 2. The aquifer is located within Tertiary to Recent deposits and comprises two hydrogeological units, viz. 1) an upper 6-8 m unconfined primary sand aquifer overlying 2) a deeper, semi-confined porous sandstone aquifer. The base of the semi-confined aquifer comprises a transmissive 1-2 m thick conglomerate unit of coarse semi-consolidated shelly gravel (Jolly 1983).

Poorly transmissive Bokkeveld Group shales form a bedrock aquitard separated from the overlying aquifer by a thin clay layer considered as an aquiclude for the study. The vegetated fixed dune immediately north of the aquifer acts as a groundwater divide, separating flow into a northern and southern component.

2.3 *Water supply and use*

BRM originally obtained all its water from rainwater conservancy tanks and two large diameter shallow wells. This supply has been upgraded over the years with well-points and boreholes being installed to presently supply 132 ML per year. In addition, three boreholes on the banks and within the tidal reach of the Bushmans River abstract a mixture of river and sea water to supply 122 ML of desalinated water for domestic use.

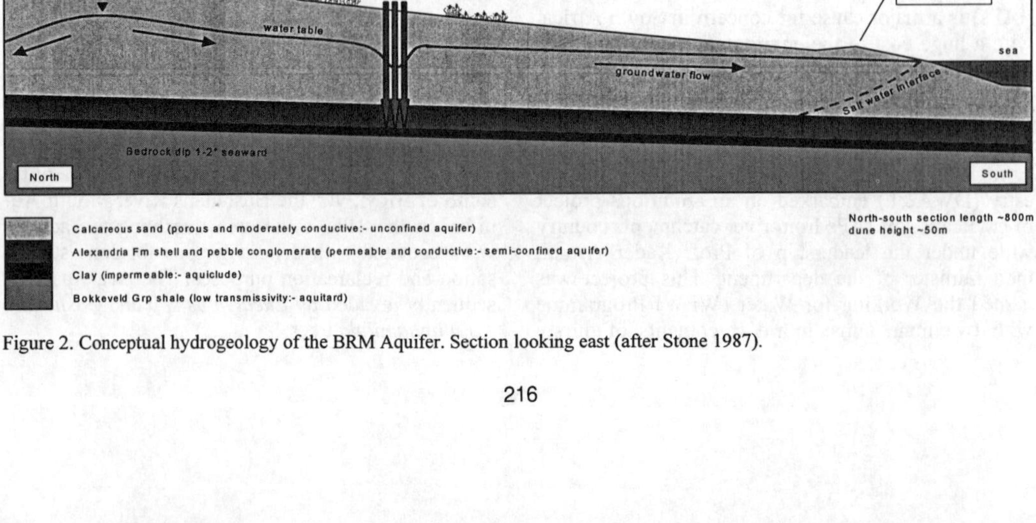

Figure 2. Conceptual hydrogeology of the BRM Aquifer. Section looking east (after Stone 1987).

The peak water use is in the December-January holiday season when the population of the town roughly doubles.

3 GROUNDWATER – VEGETATION INTERACTION

From an extensive literature survey the following are reported as the predominant effects of vegetation removal on an aquifer:

- A decrease in foliage interception and subsequent evaporation,
- A reduction in transpiration and therefore increased recharge,
- Suspension of groundwater abstraction from deep-rooted systems tapping an aquifer,
- Increased availability over time of disused trunk and root channels for preferential flow paths (pipe flow) for aquifer recharge.

Whilst the majority of a plant's roots are situated within the first 0.5 m of the soil profile, deep penetrating sinker roots are critical for a plant's survival, especially during times of drought. Alien plants grow rapidly and establish deep sinker roots whilst still young. *Acacia* sinker roots, in particular, can reach depths of up to 20 m enabling the plant to make maximum use of available soil and groundwater (Scott & Le Maitre 1998).

Within a coastal aquifer system with shallow water tables, rapid recharge and short groundwater residence times, there will thus be a significant degree of groundwater / vegetation interaction.

Although minimal information on the specific groundwater use characteristics of alien plants in South Africa was found to exist, the following hypotheses formed the basis of this study:

- Indigenous vegetation uses less water than alien plant species,
- Alien trees with deep sinker roots act as water 'pumps' during transpiration,
- Removal of alien vegetation reduces evapotranspiration, thereby increasing aquifer recharge,
- Water levels rise in response to an increase in recharge.

4 GROUNDWATER MODELLING

A good historic database of water levels, rainfall and groundwater abstraction was available for the study area. Analytical / statistical modelling methods were therefore used to quantify the impact of vegetation clearing on the BRM Aquifer, specifically with respect to a change in water levels.

4.1 *Cumulative rainfall departure (CRD) method*

The CRD method is a regression approach to establishing the relationship between rainfall and recharge, with the rationale that an equilibrium is established over time between the average annual precipitation and the hydrological response, i.e. runoff, recharge to and losses from the system. The relationship between the CRD and the groundwater level response is shown in the following formula:

$$\Delta h = a/S \times (Rf_i - k \times Rf_{av}) \qquad (1)$$

where Δh is the change in head, a is the fraction of rainfall that constitutes recharge, S is the aquifer storativity, Rf_i is the rainfall in period I, Rf_{av} the average rainfall over the monitored period, k is a constant where $k = 1$ represents a natural equilibrium system and $k > 1$ is a system being exploited, e.g. by pumping.

Bredenkamp et al. (1995) derived a regression approach that can be applied to obtain the CRD relationship for different time intervals where the water level response in an aquifer can be determined relative to the rainfall over preceding time intervals, e.g. 6 months or 3 years. This method is therefore a powerful tool to simulate water level fluctuations by incorporating short and long-term recharge events over time. This method was used to determine the response of the BRM Aquifer water levels to a change in vegetation type and density with clearing.

5 RESULTS

General direct observations of the trend in rainfall and water levels for the study area over the 12 year data period under consideration (September 1986 to September 1998) are the following (Figure 3):

- Rainfall remained relatively constant with a MAP of 660 mm and a monthly average of 51 mm,
- Water level rises in response to increased rainfall recharge, particularly during periods of rainfall two to three times the monthly average,
- With increases in groundwater abstraction there is a corresponding lowering of the water level, and *vice versa*,
- Towards the end of the study period, water levels rise whereas rainfall decreases, contrary to what would be expected in the absence of vegetation clearance. The average monthly rainfall following clearing is 43 mm versus 53 mm prior to clearing,
- There is a 31 cm rise in the average water level after IAP clearing (Mar 1997 to Sept 1998), compared with before clearing (Sept 1986 to Feb 1997).

The observed rise in water level after clearing was checked using the CRD method in order to confirm the hypothesis that IAP clearing over a coastal aquifer will result in increased effective recharge and therefore a rise in water level. The results of the analytical modelling are discussed below.

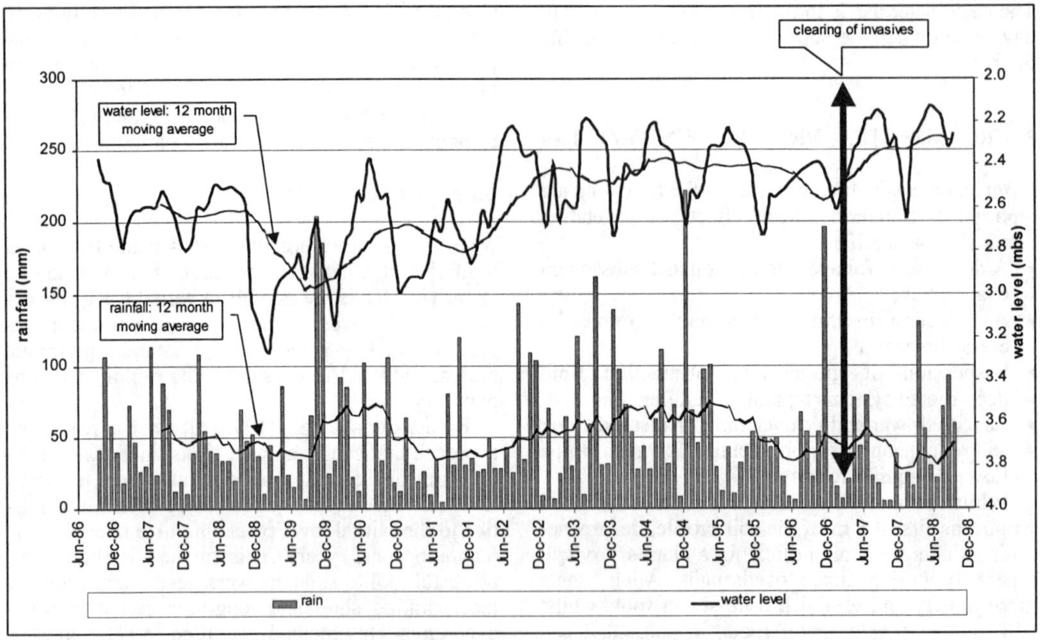

Figure 3. BRM Aquifer water level versus rainfall, 1986 to 1998.

5.1 *Cumulative rainfall departure method*

Based on analyses of rainfall records a number of CRD series were generated incorporating the effects of long and short term memory of rainfall events for the following time scenarios:

- Short-term memory (2, 3, 6, 9 and 12 months),
- Long-term memory (18, 24 and 36 months),
- One year memory combined with moving averages for 3, 6 and 9 months,
- Introduction of the effect of pumping by assigning k = 1.01 into equation 1.

The various CRD curves generated were compared with the actual water level fluctuation. The best correlation was achieved with a short (< 12 months) rainfall memory with poor correlations for long-term memory periods of 2 to 3 years.

Since the BRM Aquifer is unconfined with a shallow water table, effective recharge can be expected to be rapid occurring within a period of less that 1 year, thus validating the correlation obtained.

Once the memory period was defined, additional simulations were carried out using shorter memory periods and incorporating moving averages in the CRD series. The best results were obtained using a short-term memory of 6 months together with a long-term memory of 1 year (CRD 6,12 on Figure 4), with water levels generally following the trend of this CRD curve both before and after the period of vegetation removal. From Figure 4 it can be seen that there is a delayed response between the simulated (CRD curve) and the actual water level rise.

The effect of spatially and temporally variable pumping from the BRM Aquifer over the study period, however, is considered to have occasionally masked the degree of correlation between rainfall and water level. For examples, there are periods, such as the summers of 1991, 1992 and 1993, where a water level rise appears to correspond with a more negative CRD, contrary to what would be expected. Whilst it is apparent from Figure 3 that there is an immediate rise in the water level during periods of high rainfall – particularly for the 1989 floods (where 454 mm of rain fell in 3 months from Sept to Nov) – the CRD curve (simulated water level response) is offset. The degree of offset is dependent on the time period over which rainfall is averaged, where a rainfall averaged over long time periods (e.g. ≥ 1 year) will subdue the degree of rise in water level, shifting it forward and spreading the recharge from preceding rainfall events over a longer period.

Actual water level increases after vegetation removal are slightly higher than predicted by the CRD curve, attributed – from the available abstraction data – to out of season periods of lower pumping. In addition, the effects of vegetation clearing above the BRM Aquifer, specifically the recovery of water levels, are considered to be masked by groundwater abstraction as well as the irregular periods of abstraction from the aquifer. The difference between the two curves is therefore considered to represent the best approximation of the order of magnitude increase in water level that can be attributed to clearing of alien vegetation.

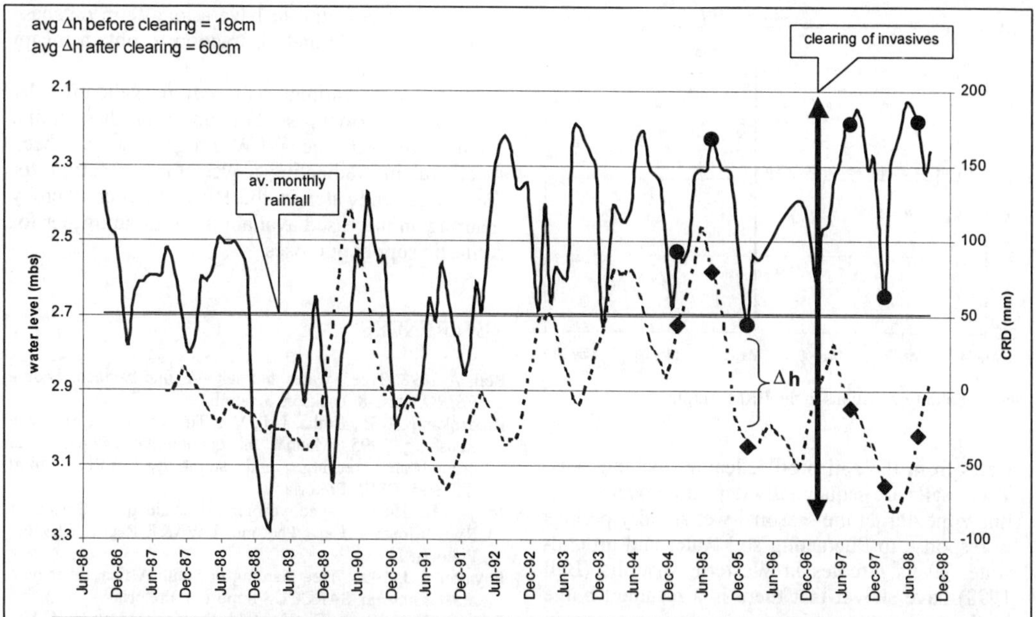

Figure 4. Moving average CRD versus water level, 1986 to 1998.

Thus, to enable the establishment of this order of magnitude increase in water level, the degree of difference (Δh) between the CRD and water level curves (Figure 4) before and after clearing was compared. In order to take the effect of pumping on water levels into account, specific periods representing the high and low pumping seasons (Dec – Jan and June – July, respectively) were chosen, marked with circles on the water level plot and diamonds on the CRD plot in Figure 4. The average difference in these two water levels (actual and simulated, respectively) before clearing is 0.19 m and 0.6 m after clearing. From this, a rise in water levels of 0.41 m on average is indicated, using the CRD method, as a direct result of the clearing. This figure compares well with the measured difference in average water level before and after clearing of 0.31 m.

This positive flux (V) in the water balance of the aquifer is calculated using the following formula:

$$V = A \times \Delta h \times \eta \qquad (2)$$

where A = the aquifer area, and η the porosity of the aquifer material.

Assuming an average saturated thickness for the BRM Aquifer of 6.5 m (from borehole log data) the total aquifer storage is calculated at 1.1 x 10^6 m^3. From equation 2, the 41 cm rise in water level equates to a volume increase of 67,700 m^3, i.e. a net increase of 6% in aquifer storage. Expressed differently, annual recharge to the aquifer is calculated at 49,000 m^3 and this groundwater volume flux increase therefore equates to 138% of the mean annual recharge to the aquifer.

It is understood that there is limited data for the period following clearing and that the observations made on the change in water balance apportioned to clearing in this study need to be confirmed with a follow-up study.

5.2 *Changes in aquifer salinity*

The BRM Aquifer is included in the DWA&F's national water quality monitoring network and a long-term database of hydrochemical analyses is therefore available for a period of 3 years before and 2 years after clearing. Figure 5 shows a marked and sustained increase in groundwater salinity immediately following the IAP clearing.

The salinity increased from an average of 256 mS/m before clearing to an average of 330 mS/m after clearing, i.e. an increase of 30%.

This increase in salinity is attributed to a combination of the following:

- Increase in evaporation rates during dry periods following clearing due to a reduction in vegetation cover, with a resultant increase in the salt load of the sandy soil above the water table which is leached into the aquifer during rainfall events,
- Vegetation root systems preferentially take up

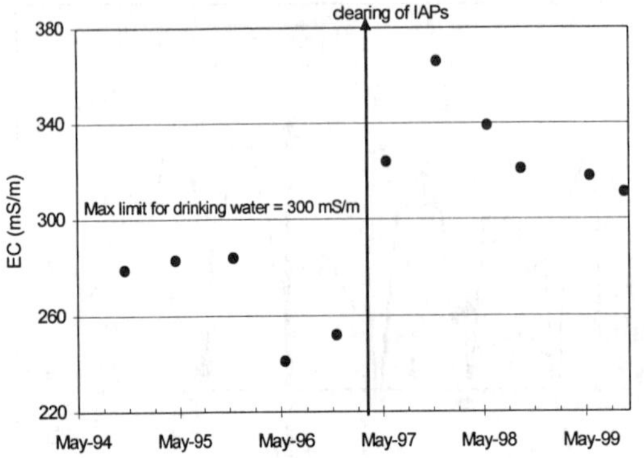

Figure 5. Salinity variations in the BRM Aquifer.

water from the soil profile, leaving behind salts. These salts are sequentially deposited over a wetting zone during the seasonal wet and dry periods in response to fluctuating soil water and groundwater levels. Studies in Western Australia (Bell 1998) have shown that there is a resultant bulge in the soil salt profiles (i.e. a discrete zone in which soil concentrations are higher) in the horizon in which roots have been most active. Remobilisation of salts, concentrated during the seasons over a wetting zone immediately above the pre-clearing water table, thus occurred with the rise in the BRM Aquifer water table,

- A large portion of the alien vegetation cleared consisted of Beef-Wood trees (*Casuarina cunninghamiana*). Since these trees have adapted to saline conditions and take up dissolved salts in their widespread root systems (Poynton 1995), it is postulated that these salts were released back into the soil profile from the root system remaining in the ground after clearing.

6 CONCLUSIONS

Based on the findings of this research project, the following conclusions can be made:

1. Clearing of IAPs above the BRM Aquifer caused a measured rise in water levels of 31 cm. Employing the CRD method, the difference in actual water levels and simulated water levels before and after clearing was determined to be 41 cm. This equates to a 6% increase in aquifer storage and a 138% increase in annual recharge, directly attributable to clearing of IAPs.
2. The salinity of the BRM Aquifer increased by 30%, considered to be mostly due to a remobilization of salts with a rise in the water table after clearing.
3. Direct analytical modelling, using the CRD method, is suggested as the preferred research method for future studies of this nature. In addition, it is essential to have a long-term database of water levels and abstraction to obtain meaningful results.

These observations will require validation by conducting follow-up studies. Based on these initial results, however, the WFW Programme has been successful in eradicating a wasteful competitor for shallow groundwater in the BRM Aquifer, thereby ensuring an increased availability of groundwater for domestic supply purposes.

REFERENCES

Bell, A 1998. Trees, water and salt – a fine balance. *Ecos – CSIRO*. 58: 2-8. Dickson: Australia.

Bredenkamp, D.B., Botha, L.J., Van Tonder, G.J. & H.J. Van Rensburg 1995. Manual on quantitative estimation of groundwater recharge and storativity. WRC Report TT73/95. CSIR: Pretoria.

Jolly, J.L. 1983. A geohydrological evaluation: Bushmans River Mouth – Cape Padrone. DWA&F Report Gh3262: Pretoria.

Poynton, R.J. 1995. Tree planting in South Africa, other genera, *Casuarina*. SARCCUS Report. 3. Pretoria.

Prinsloo, F.W., Scott, D., Versfeld, D. & G. Moses 1996. Hydrological monitoring of the Fynbos water project at Vergelegen and Knorhoek estates. CSIR report ENV/S C96010: Stellenbosch.

Scott, D.F. & D.C. Le Maitre 1998. The interaction between vegetation and groundwater: research priorities for South Africa. WRC Report 730/1/98: Pretoria.

Stone, A.W. 1987. Coastal aquifers: a fresh water resource. *Water sewage & effluent*. 31: 32-33.

Versveld, D.B., Le Maitre, D.C. & R.A. Chapman 1998. Alien invading plants and water resources in South Africa: a preliminary assessment. WRC report TT99/98. CSIR: Stellenbosch.

Visser, D., Norman, N. & W. Stadler 1999. Effects of invasive alien plant eradication on the groundwaters of the L'Agulhas plains, Western Cape Province. T & P report 980173b: Wynberg.

WFW 1998. The environmental impacts of invading alien plants in South Africa. Brochure produced for the Working for Water Programme. CSIR: Stellenbosch.

Groundwater: Past Achievements and Future Challenges, Sililo et al. (eds) © 2000 Balkema, Rotterdam, ISBN 90 5809 159 7

Hydrodynamic investigations in a recent volcanic aquifer, Reunion Island, using single well tracer test

J.S. Martial, J.L. Join & J.Coudray
Laboratoire des Sciences de la Terre, Université de la Réunion, Saint Denis, France

ABSTRACT: Hydrogeological investigations have been carried out on "Rivière des Remparts", south of Reunion Island, and their aim was to help the development of an area with a high potential of groundwater resources. Four boreholes already exist but inconsistent data collected from long term pumping tests compelled us to proceed to new investigations, but with the use of other methods. As fresh groundwater is poorly mineralized in Reunion Island, we performed tracer tests by means of chloride as tracer and conductivity cells for the monitoring. As natural gradient tracer tests between two wells failed, point dilution method was chosen for these tracing-tests The disappearance of the tracer after a brief injection was monitored by several conductimeters. Experimental data allow us to propose new hydrodynamical values ($1<K<3.10^{-2}$m/sec) for recent basaltic aquifer in this area of Massif de la Fournaise. Best-fit curves issued from experimental data are consistent with exponential curve of point dilution method theory. The flow of the aquifer was totally unknown. From the new hydrodynamical data established from tracer test, a numerical flow model has been built using ModFlow, and lends to estimate at 4,5m^3/sec the flow in the basaltic aquifer. We discuss the reliability of single well tracer test method, applied to young volcanic aquifers, in order to evaluate the velocities in such geological formations.

1 INTRODUCTION

The aim of this study is to understand the aquifer of Rivière des Remparts, which is located in the South of Reunion Island, Indian Ocean. (Fig. 1) The river starts at an altitude of 1800m and goes down 20 km toward the sea with a south direction. Its valley cuts the southern slopes of Piton de La Fournaise shield-volcano. The study concerns an area of 3.5km² (a transect of 3 km long), in the lower part of the valley, upwards the town of St-Joseph.

The area of hydrological watershed of Rivière des Remparts is about 54km²(Barcelo 1996). The average rainfall is over 3000 mm yearly. The discharge of Rivière des Remparts remains between 0.5 and 1.5m^3/sec. On the other hand, the flow measured on Rivière de l'Est, in the same volcano, is around 9m^3/sec in spite of a half-smaller watershed. Thus, an important part of the total discharge might be found as an underflow located in the recent lava filling the valley.

The upper underground is made of very recent basaltic lava flows. Understanding the hydrodynamic behavior of this kind of geological formation was the goal of the study.

Single well tracer tests have been executed to provide hydraulic parameters. Investigations were steered to estimate the flow of the basaltic aquifer. In this way, a model has been built.

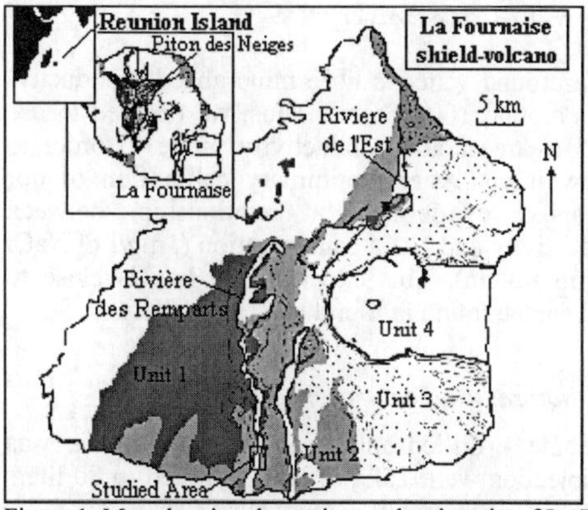

Figure 1. Map showing the various volcanic units of La Fournaise shield volcano and the location of the studied area.

2 HYDROGEOLOGICAL BACKGROUND

High cliffs (300m to 1100m), built by lava flows of Piton de la Fournaise first and second units, bound the valley. These materials are found at a depth of

88m under the river (Fig. 2). Over this low permeability weathered bedrock, the valley had been filled by 50m of alluvium. Very recent lava flows (40m) of La Fournaise fourth unit lie on the old alluvium formation.

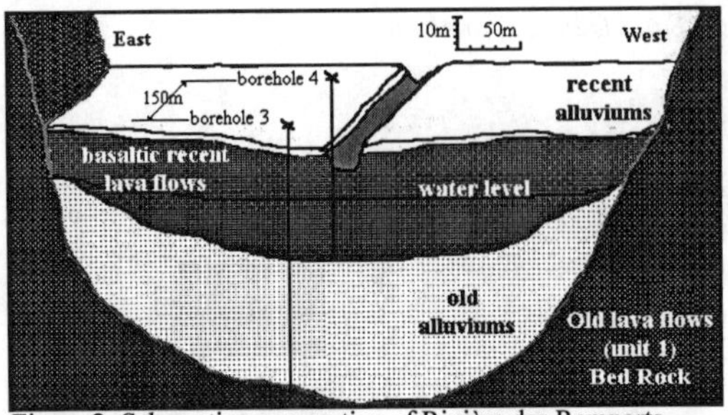

Figure 2. Schematic cross section of Rivière des Remparts.

Four boreholes have been drilled in the studied area. First, the alluvium aquifer was considered as the mainly productive one. The first drilled boreholes were slotted only in front of the alluvium. Last ones were slotted at the same time in front of basaltic aquifer and in front of the old alluviums.

Long-term pumping tests have given various values for hydraulic conductivity (K), ranging from $1.3e^{-3}$ $m.s^{-1}$ to $5.0e^{-2}$ $m.s^{-1}$ (Billard 1992). One should know that these hydraulic conductivities were poorly reliable, as draw down had shown abnormal curve.

3 MATERIAL & METHODS

Underground water is little mineralized (conductivity is around 100 µS/cm). Thus, we decided to use NaCl as tracer and conductivity probe in order to follow it up. After preliminary calibration of our sensors, we defined a relationship between conductivity and NaCl concentration (1mg/l of NaCl = 0.56 µS/cm). This experimental law is close to what can be found in literature (Hem 1982).

3.1 *Natural gradient tracer test*

Borehole 4, drilled only in the basaltic aquifer was the injection well. 25kg of NaCl, diluted in 80 liters of water was thrown into the well.

Downstream, borehole 3 was equipped at different depths with 3 sensor probes, which were linked to data loggers. The follow-up lasts 20 hours. The distance between both wells is of 150m.

3.2 *Single well tracer test*

This tracer test is based on point dilution method. (Banton & Banguoy 1997) Observing the tracer conductivity decrease after injection allows us to extract the dilution rate (*Vg*) (Drost et al. 1968). Thus, Darcy velocity can be estimated according to the simplified exponential decrease law:

$$C(t) = Co.e\left(-\frac{2.b.Vd}{\pi.r.h}.t\right) \quad (1)$$

$$Vd = K.i = \left[Ln(C(t)) - Ln(Co)\right]*\left[\frac{\pi.r.h}{-2.b.t}\right] \quad (2)$$

where :
Co = initial conductivity [time is initialized after injection]
b = aquifer thickness (m)
r = radius of the well (m)
h = height of water column in the well (m)
t = elapsed time after injection (s)
Vd = Darcy velocity = K.i , where
K = hydraulic conductivity ($m.s^{-1}$) and i = hydraulic gradient.

Single well tracer tests have been performed on borehole 4, which intercepts only the basaltic aquifer. (Fig. 3). Three sensor probes have been set at different altitudes. In this way, we were able to observe hydraulic behavior variations between each kind of layer of the volcanic aquifer.

We injected 10 kg of NaCl in the well, and the decrease of the conductivity was monitored. A pump was set down to the well, and ran in closed circuit, guaranteeing that the concentration of NaCl was kept homogeneous within the system (Novakowski et al. 1992).

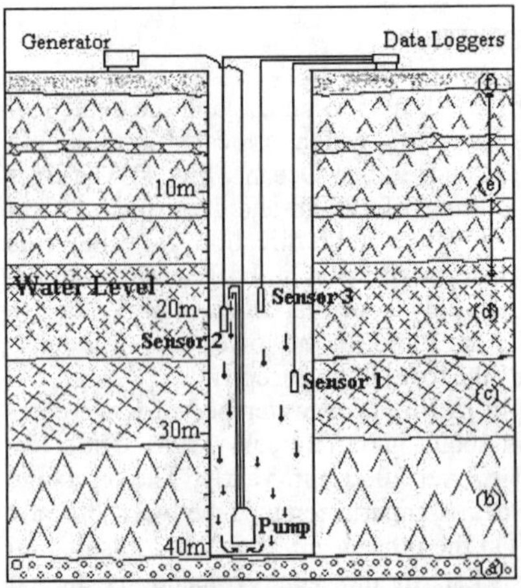

Figure 3. Experimental set - Single well tracer test.

Geological Description:
(f): young alluvium
(e): 3 last basaltic lava flows (5m thick) with associated thin scoriaceous inter-flow layer.
(d): thin basaltic layers & intercalated scoria
(c): very scoriaceous layer
(b): thick basement basaltic lava flow (10m)
(a):old alluvium

4 RESULTS OF TRACER TESTS

4.1 *Natural gradient tracer test*

25kg of NaCl injected 150m upstream in borehole 4, the conductivity remained constant at 70µS/cm into the borehole 3, during 20 hours of monitoring.

4.2 *Single well tracer test*

Result's charts are presented with time on x-axis and Ln[C(t)] on y-axis. By means of the slope of the line of best fit (linear regression), hydraulic conductivities are extracted from single well tracer tests. Depending on layers encountered in front of each sensor, hydraulic conductivities range from $9.0.10^{-3}$m/sec to $3.7.10^{-2}$m/sec. The latter has been obtained in front of the more scoriaceous layer, at 24m depth. Moreover, we observed that best R-squared values (R^2=0.94) have also been obtained there. We deduced that, greater the advective flow velocity was, better the homogenization ran. Therefore, the exponential decrease of conductivity was experimentally realized and complied with the requirements of the theory of single-well tracer test.

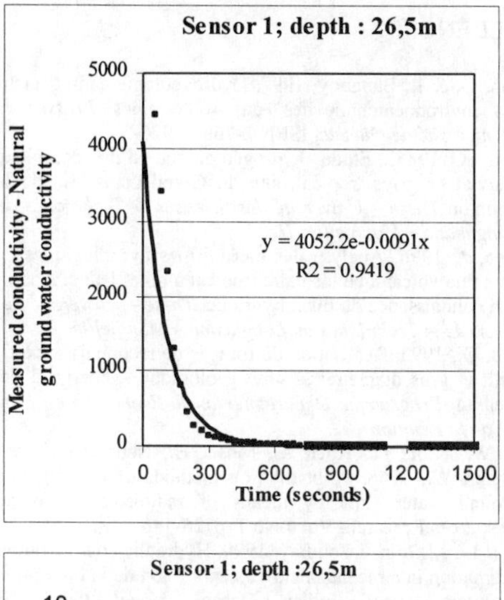

Figure 4. Typical example of result charts

The first chart shows experimental data [Cond(t)]. The natural groundwater conductivity was deduced from experimental points in order to obtain an exponential curve (1) that tends asymptotically to nil.

The second chart shows the linear decrease we get from equation (2), from which hydraulic conductivity, K, is extracted.

Concerning the very scoriaceous layer [Fig. 3(c)], the hydraulic conductivity is $3.7.10^{-2}$m/sec. In front of thin basaltic layers and intercalated scoria [Fig. 3(d)], K is $1.2.10^{-2}$m/sec.

5 MODELING

5.1 *Conceptual model and data input*

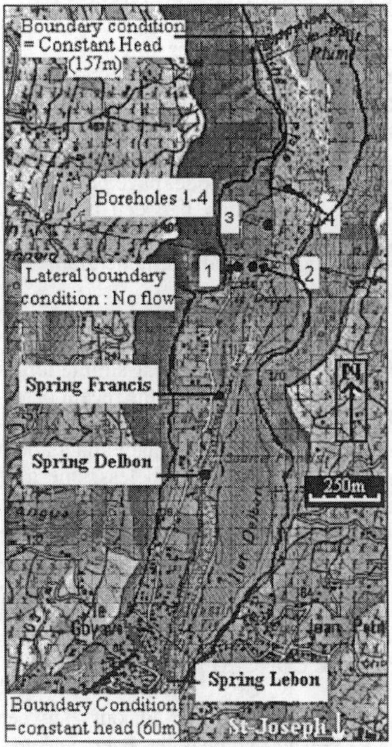

Figure 5. Model Map and boundary conditions

The scope of the modeling step was to simulate the underground flow in a recent basaltic aquifer, and to calculate the total flow of this aquifer.

The model has been built under Ground Modeling System (G.M.S.); flow calculation has been solved, in steady state conditions, using Modflow. The system modeled is a one-layer aquifer (the basaltic one). A grid of 8750 cells (square of 20m) wraps the studied area (width: 1150m; length: 3000m).

5.1.1 *Boundary conditions*

No-flow conditions have been set for lateral boundaries as Unit 1 lava flows are low-permeability strata in opposition to recent lava flow aquifer (10^{-6} m/sec against 10^{-2} m/sec) (Folio 1999). Moreover, flank landslides are common in Rivière des Remparts. Many contemporary events are evidences to compel no flow conditions as detrital materials, stemmed from massive landslides, cover the flanks.

Downstream boundary is given by Source Lebon elevation. This spring appears at an elevation of 60m. It induces a constant head boundary condition.

Upstream boundary conditions have been also set as constant head conditions. Piezometric data were not available straight below the boundary. Thus, we extrapolated the water level of borehole 4, which were 600m downstream. We preserve the hydraulic gradient of 4.5% existing between borehole 4 and borehole 3 until the northern boundary. (Fig. 5)

5.1.2 *Flow calculation parameters*

Constant and homogeneous hydraulic conductivity has been set at 1.3 10^{-2} m/sec for the whole area. This value is a means of all hydraulic conductivities, pondered by the thickness of the corresponding layer

Starting heads of the flow model have been calculated using linear interpolation between borehole's water levels, springs elevations, and additional fixed data, in accordance with general hydraulic gradient.

5.2 *Results*

The flow computed by the numerical simulation is 4.5m^3/sec. This value only concerns the recent basaltic aquifer. The old alluvium aquifer hasn't been inserted in the model as we lacked reliable data about hydrodynamical parameters. Nevertheless, one may note that, comparing to Rivière de l'Est, the calculated flow seems to be reasonable. Moreover, computed velocities are identical to the ones acquired from experimentation we carried out.

6 DISCUSSION & CONCLUSION

Natural gradient tracer test carried out between borehole 4 and 3 failed as no peak has been detected. Expected concentrations in the observation well have not been reached. The sensibility of our sensor probes does not allow us to know whether the tracer plume passed. Similar behaviors have already been shown in Reunion Island (Banton 1985). Further experimentations should bring suitable interpretation for the failure of this tracer test. On the other hand, single well tracing test appeared to be consistently reliable. Moreover, errors on construed values extracted from experimentations never exceed 5% ($\pm$0.15.10^{-2}m/sec). Nevertheless, one should take into account uncertainty related to borehole distortion (Drost et al. 1968).:

$$Vg = \alpha.Vd \qquad (3)$$

For our experimental setup, α factor should be between 1 and 2.

In further experimentations, thanks to packers, we will be able to bring more accurate data, which may solve hydrodynamical investigations in this kind of media. Therefore, point dilution method seems to be an interesting alternative for hydrogeological investigations to be performed in this kind of aquifer whereas pumping tests had shown their limits. As the implementation of this method is easy, it could be a systematic approach to define protection area.

The model we built has complemented the hydrodynamical characterization of an area. We simulated a 4.5m^3/sec flow for the recent basaltic aquifer. We could hardly achieve a better model as we lacked data concerning inflows and outflows. Therefore, the validity of the simulations can be argued. Nevertheless, these investigations bring sustainable data concerning this kind of aquifer, which was not well known. Thanks to a suitable survey, we will be able to build a more comprehensive model.

Acknowledgements: The funding for this study was provided by the Conseil Général of Réunion Island, as part of a large research program of La Fournaise-volcano hydrology.

REFERENCES

Banton, O. & L. Banguoy, 1997, Hydrogéologie - multisciences environnementale des eaux souterraines, *Presses de l'Université de Québec*, ISBN 2-7695-0926-5

Banton, O.,1985, Etude hydrogéologique d'un complexe alluvial en pays tropical, site du Grand-Etang, Ile de la Réunion *These – Université des Sciences & Techniques du Languedoc – Montpellier II*

Barcelo, A.,1996, Analyse des mécanismes hydrologiques en domaine volcanique insulaire tropical à relief jeune. Apport à la connaissance du bilan hydrique. *Thèse – Université des Sciences et Techniques du Languedoc – Montpellier II.*

Billard, G., 1992, Réalisation de forages de reconnaissance en petit et gros diamètres – suivi géologique – Forage d'Ilet Delbon, *Programme Départemental de Recherche en Eau, Ile de la Réunion.*

Drost W., Klotz D., Koch A., Moser H., Neumaier F. and Rauert W., 1968, Point dilution methods of investigating ground water flow by means of radioisotopes, Water Resources Research, Vol 4.No 1.,p125-146

Folio, J.L., J.L.Join, J.Coudray, 1999, Hydraulic conductivities estimation in a recent shield volcano: Piton de la Fournaise, Reunion Island, Indian Ocean. *Model Care 99, International conference on calibration and reliability in Groundwater Modelling. Coping with uncertainty,* Zürich, Switzerland, 20-23 Septempber 1999.

Hem, J.D., 1982, Conductance : a collective measure of dissolved ions, *Geological survey Menlo Park, C.A., Water Analysis*, Academic Press, Orlando FL, Volume I : inorganic species, Part I. Pag 137-161

Novakowski, K.S., P.A.Lapcevic, J.A.Voralek, R.Hoag, G.S.Bickerton, 1998. A note on a method for measuring the transport properties of a formation using a single well. *Water Resources. Research,* 34(5), 1351-1356.

Groundwater: Past Achievements and Future Challenges, Sililo et al. (eds) © *2000 Balkema, Rotterdam, ISBN 90 5809 159 7*

Conceptual model of groundwater flow in stratified mountainous terrain, Utah, USA

A.L. Mayo & T.H. Morris
Brigham Young University, Provo, Utah, USA

ABSTRACT: Thousands of springs issue from the 610+ m thick, gently dipping, bedrock sequence of clastic sediments in the 240 km long Wasatch Plateau and Book Cliffs, Utah, USA. Spring hydrographs correlate with seasonal and climatic variability and the waters have elevated ^{3}H contents and contain anthropogenic ^{14}C. In-mine (coal mines 300-600 m bgs) groundwater discharge rates decline rapidly and these waters have $\delta^{2}H$ and $\delta^{18}O$ compositions that are distinguishable from near surface groundwater. Almost all deep waters have no ^{3}H and ^{14}C have mean residence times ranging from 500 to 20,000 years. Discriminant analysis indicate that groundwater in each formation cannot be distinguished from each other on a regional basis. Locally the chemistries of most formation waters are statistically distinct from each other. Mass balance modeling of chemical evolution indicates that groundwaters in lower stratigraphic units have not chemically evolved from groundwater in overlying units. We propose a conceptual model involving both active and in-active flow regimes to describe the groundwater flow. The active zone extends about 75-150 m below ground surface and 150-300 m into cliff faces. The active zone has direct hydraulic communication with seasonal recharge. The in-active zone includes most of the interior portions of the Book Cliffs and Wasatch Plateau.

1 INTRODUCTION

Shallow and deep groundwater systems occur in the Book Cliffs and Wasatch Plateau, Utah, USA (Figure 1). The 6,500 km^2 mountainous region is underlain by a 610+ m thick sequence of gently dipping stratified bedrock. Economic coal deposits are mined near the base of the section. Coal mines provide underground access to groundwater systems 300-600 m below ground surface and up to 13 km horizontally into the mountain fronts.

The 2700-3200 m high Wasatch Plateau and Book Cliffs form a 240 km long, 16-40 km wide, arc shaped upland, which rises along imposing erosional escapements. Ancillary canyons and re-entrants, through which streams drain the interior uplands, cut the 460–915 m high, stair-stepped cliff faces. Seven early Tertiary and Late Cretaceous bedrock formations capable of transmitting groundwater occur in the region (Figure 2). Bedrock dips 2-5° away from the southern and eastern mountain fronts.

2 CLIMATE

Mean annual precipitation varies from less than 25 cm in the lower elevations to 65-90 cm in the higher elevations of the Book Cliffs and Wasatch Plateau, b respectively (WRCCR, 1999; USU Climate Center, 1998). Between December and March precipitation occurs mostly as snow and between June and September precipitation occurs mostly as rain (Figures 3a and 3b). The area includes portions of the Palmer Drought Severity Index (PDSI) Utah regions 4, 5 and 6 (NCDC, 1998; Figure 4a). The PDSI is a measure of relative wet and dry periods based on precipitation, soil moisture, potential evapotranspiration and other factors.

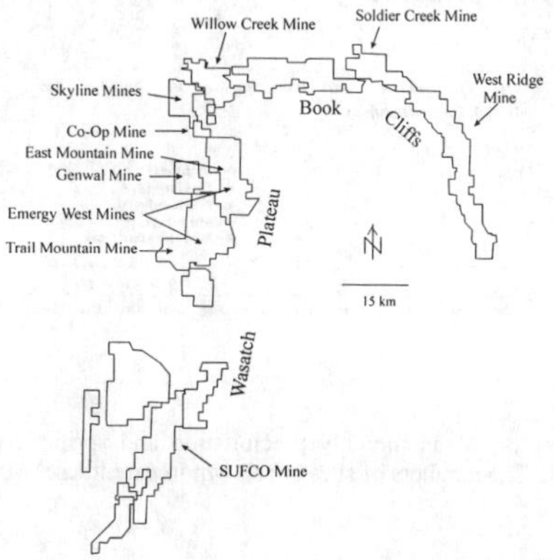

Figure 1 Location of coal mines in the Book Cliffs and Wasatch Plateau.

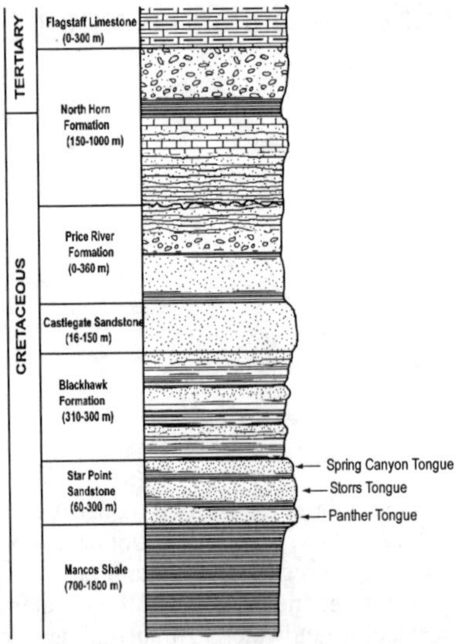

Figure 2 Geologic column.

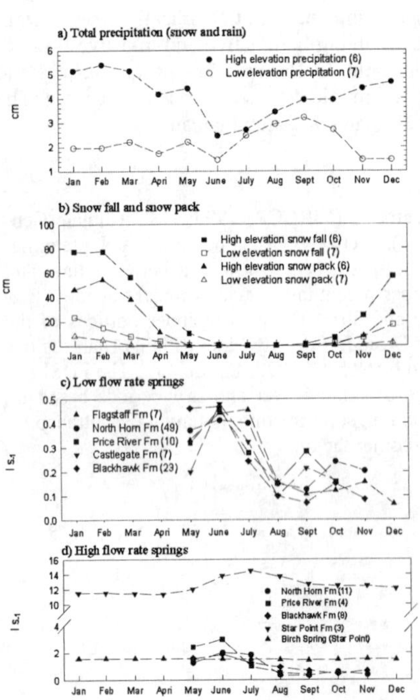

Figure 3 Mean monthly precipitation and spring discharge rates. The numbers of sites are shown in parentheses.

3 HYDROGRAPHS

Most springs have mean annual discharge rates less than 0.65 $l\ s^{-1}$ and are classified as having low discharge rates. Except for high discharge rate Star Point Sandstone springs, all springs have similar discharge characteristics and exhibit bimodal seasonal effects (Figure 3c and 3d). Discharge rates of all springs are also affected by climatic variability (Figures 4b and 4c). Except for the Star Point Sandstone, mean annual discharges of both low and high flow rate springs correspond well with wet and dry periods.

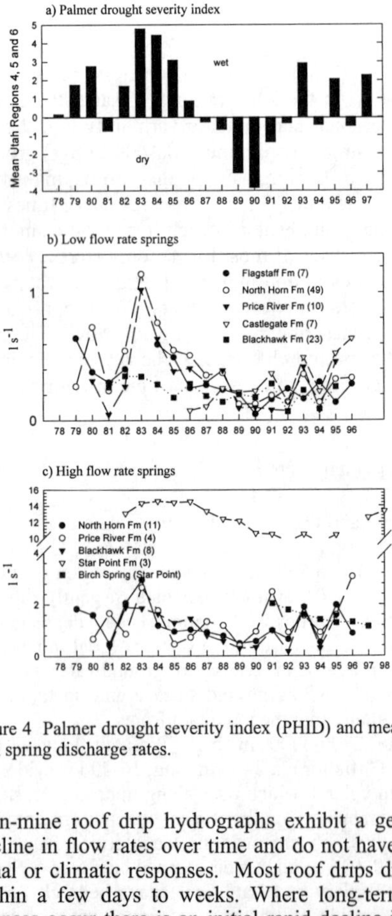

Figure 4 Palmer drought severity index (PHID) and mean annual spring discharge rates.

In-mine roof drip hydrographs exhibit a general decline in flow rates over time and do not have seasonal or climatic responses. Most roof drips dry up within a few days to weeks. Where long-term responses occur there is an initial rapid decline in inflow rate followed by a slower decline, which ultimately dry-up. The overall responses of total in-mine groundwater inflows are consistent with individual roof drip hydrographs. Groundwater discharge from the Skyline Mine illustrates this relationship. Between 1984 and 1997 there was general increase in groundwater inflows from roof drips as measured by the mine discharge (Figure 5a). Between 1984 and 1992 coal production increased and between 1992 and 1997 coal production remained relatively constant (Figure 5b). The fact that inflow rates per ton of coal declined rapidly as the total volume of coal mined increased (Figure 5c) means that groundwater inflow is not steady state and that the old mine workings rapidly dewater.

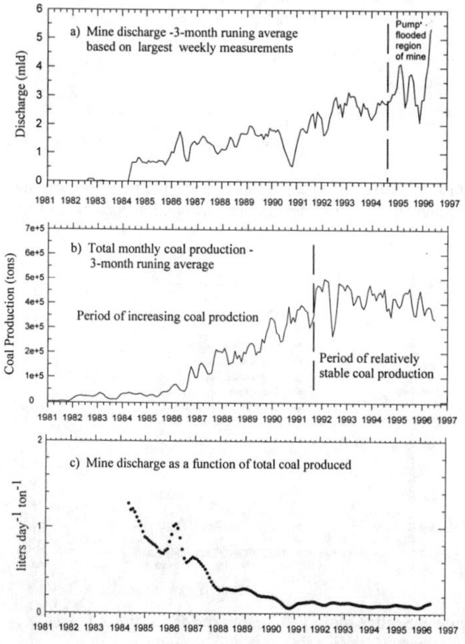

Figure 5 Skyline mine groundwater discharge and coal production.

4 SOLUTE COMPOSITIONS

Approximately 1930 chemical data sets from 164 springs and 50 in-mine locations were analyzed. Mean compositions are listed on Table 1. The data were analyzed by discriminat analyses. On a region wide basis Blackhawk Formation groundwater has the lowest probability of being correctly identified as Blackhawk Formation groundwater (31.75%), whereas Flagstaff Limestone groundwater has the highest probability (79.69%) of being correctly identified (Table 2). Except for the Flagstaff Limestone, groundwaters issuing from these formations are only somewhat chemically distinct. However, it is significant that in-mine Blackhawk groundwater has only 2.87 and 8.6% probabilities of being identified as Blackhawk spring and Star Point spring waters, respectively. When discriminant analysis was applied on a mine by mine basis, most formation groundwaters had a 70-90% probability of being correctly categorized and many had probabilities of 90-100%.

In an attempt to better understand the potential for vertical groundwater migration, the computer code NETPATH (Plummer et al., 1996) was used to evaluate plausible chemical evolutionary pathways at each mining area. Minerals and substances modeled included, calcite, dolomite, carbonic acid, gypsum, clay-ion exchange, and zeolite-ion exchange. Solute compositions modeled were based on the mean solute composition of each formation at each mine. It was not possible to model chemical evolution of Castlegate groundwater originating in the overlying Price River Formation. Only half of the Blackhawk-Star Point models were successful, and many Black-hawk-in mine models were either unsuccessful or required the unlikely condition of gypsum precipitation or reverse ion-exchange.

Table 1 Average solute composition of in-mine and groundwater samples according to bedrock lithology.

				mg l^{-1}			meq l^{-1}				
Formation	n	(°C)	pH	TDS	Ca^{2+}	Mg^{2+}	Na^{+}	K^{+}	HCO_3^-	SO_4^-	Cl^-
Flagstaff	8	6.8	7.6	265	4.3	1.0	0.2	0.05	4.8	0.6	0.16
Std.		2.1	0.4	41	1.2	0.4	0.3	0.09	0.6	1.1	0.17
North Horn	78	7.2	7.7	344	3.3	2.5	1.2	0.04	5.6	0.8	0.44
Std.		3.6	0.5	115	1.2	0.9	1.5	0.05	1.1	1.1	0.59
Price River	22	6.6	7.6	400	4.3	2.8	1.0	0.05	6.6	1.3	0.26
Std.		4.7	0.4	87	1.4	0.9	0.5	0.07	1.0	1.2	0.13
Castlegate	16	11.0	7.6	240	2.4	1.6	0.4	0.04	3.5	0.6	0.26
Std.		10.3	0.6	73	1.0	1.0	0.2	0.02	1.6	0.4	0.19
Blackhawk	35	6.7	7.3	255	3.1	1.4	0.3	0.03	4.0	0.6	0.22
Std.		3.2	0.8	100	1.0	1.2	0.4	0.04	1.4	0.6	0.30
In-mine	50	11.5	7.7	454	3.2	2.9	2.8	0.17	7.0	1.5	0.37
Std.		2.9	0.5	194	1.8	1.4	3.5	0.15	2.6	1.6	0.34
Star Point	5	17.0	7.4	546	5.0	3.8	1.1	0.06	6.5	2.5	0.86
Std.		8.4	0.6	150	1.4	1.3	0.7	0.04	0.6	1.2	0.68

Table 2 Regional discriminant analysis results. The upper line for each formation indicates the number of sites and the lower line indicates the percent probability that a groundwater will be correctly identified as discharging from that formation. For example, there is a 79.69 percent probability that Flagstaff Formation groundwater will be correctly identified. These differences are due to lithological conditions.

	Flagstaff	North Horn	Price River	Castlegate	Blackhawk	Star Point	In-mine	Total (%)
Flagstaff	51	2	2	1	8	0	0	64
	79.69	3.13	3.13	1.56	12.5	0	0	100
North Horn	61	341	67	45	61	59	39	673
	9.06	50.67	9.96	6.69	9.06	8.77	5.79	100
Price River	14	6	61	2	3	6	6	98
	14.29	6.12	62.24	2.04	3.06	6.12	6.12	100
Castlegate	1	14	4	32	3	0	0	54
	1.85	25.93	7.41	59.26	5.56	0	0	100
Blackhawk	123	82	45	145	194	9		611
	20.13	13.42	7.36	23.73	31.75	1.47	2.13	100
Star Point	4	6	25	0	0	42	1	78
	5.13	7.69	32.05	0	0	53.85	1.28	100
in-mine	4	26	70	33	8	24	114	279
	1.43	9.32	25.09	11.83	2.87	8.6	40.86	100
Total	258	477	274	258	277	140	173	1857
Percent	13.89	25.69	14.75	13.89	14.92	7.54	9.32	100
Error Rate	0.6825	0.4074	0.2031	0.4933	0.3776	0.4615	0.5914	0.4595
Priors	0.1429	0.1429	0.1429	0.1429	0.1429	0.1429	0.1429	

5 ISOTOPIC COMPOSITIONS

Stable isotopic compositions of δ^2H and $\delta^{18}O$ of 392 samples collected from 329 stream, spring, and in-mine locations were analyzed (Table 3). Isotopic compositions of waters in the arid Book Cliffs tend to be heavier than the isotopic compositions of waters in the more humid Wasatch Plateau (Figure 6). At most individual mine locations the average of δ^2H and $\delta^{18}O$ compositions of stream and spring waters are similar, suggesting similar histories (Figure

7). The compositions differ between mines reflecting the spatial variability in the isotopic composition of precipitation. This variability is due to both orographic affects and differences in storm tracking. Mean composition of most in-mine groundwaters are appreciably lighter than the corresponding near surface waters. These differences suggest that the in-mine groundwater is not part of the same groundwater system(s) that discharge at local springs. Unstable isotopic data presented below supports this idea.

Table 3 Average δ^2H and $\delta^{18}O$ compositions of surface, in-mine, and groundwaters.

	all samples			streams			springs			in-mine		
	n	δ^2H (°/oo)	$\delta^{18}O$ (°/oo)	n	δ^2H (°/oo)	$\delta^{18}O$ (°/oo)	n	δ^2H (°/oo)	$\delta^{18}O$ (°/oo)	n	δ^2H (°/oo)	$\delta^{18}O$ (°/oo)
Book Cliffs												
West Ridge Mine	26	-112.3	-14.8	9	-112.7	-14.8	17	-112.1	-14.7			
Soldier Creek Mine	15	-111.57	-15.6	4	-107.3	-14.8	3	-106.7	-15.1	8	-115.6	-16.1
Willow Creek Mine	30	-115.2	-15.2	13	-115.6	-14.9	8	-117.7	-15.4	9	-112.3	-15.3
Wasatch Plateau												
Skyline Mines	111	-126.9	-17.0	27	-124.2	-15.6	69	-126.6	-16.9	15	-134.1	-18.1
Co-Op Mine	60	-122.3	-16.2	9	-117.6	-15.4	40	-122.6	-16.2	11	-125.5	-17.0
Genwall Mine	84	-125.5	-16.5	20	-123.1	-16.1	26	-125.7	-16.6	15	-132.1	-17.6
Energy West Mines	21	-126.9	-16.7				11	-127.2	-16.6	10	-125.9	-16.6
SUCO Mine	76	-119.2	-16.0	24	-116.1	-15.6	36	-118.8	-15.9	16	-124.8	-16.8

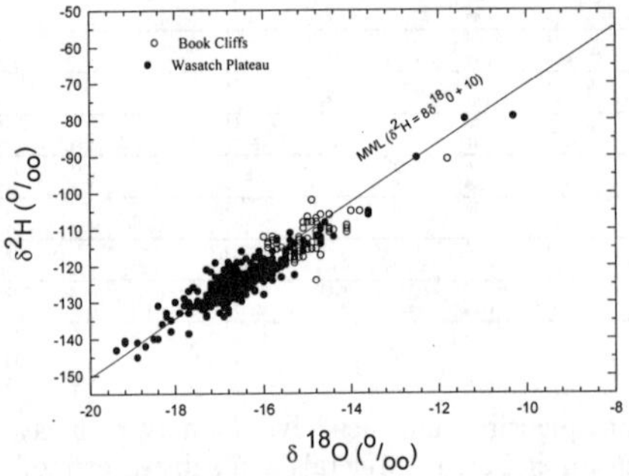

Figure 6 Stable isotopic compositions of groundwater in the Book Cliffs and Wasatch Plateau.

Unstable isotopic compositions, collected from 126 locations, indicate that spring and in-mine groundwaters have fundamentally different travel times. In-mine groundwater and fault related springs typically have zero or near zero ^3H contents and mean ^{14}C residence times ("ages") of 1,000 to 20,000 years (Figure 8). Almost all springs contain anthropogenic ^{14}C and have ^3H contents of 5 to 23 TU. A few samples contain components of both old groundwater and modern water.

6 DISCUSSION AND CONCLUSIONS

Two distinct types of groundwater systems have been identified in the Book Cliffs and Wasatch Plateau. Shallowly circulating, short flow path systems, which respond rapidly to seasonal and climatic variability, dominate the near surface environment regardless of bedrock lithology. Locally bedrock lithology controls solute compositions and geochemical evolution modeling indicates that the potential for interformational vertical migration of groundwater is limited.

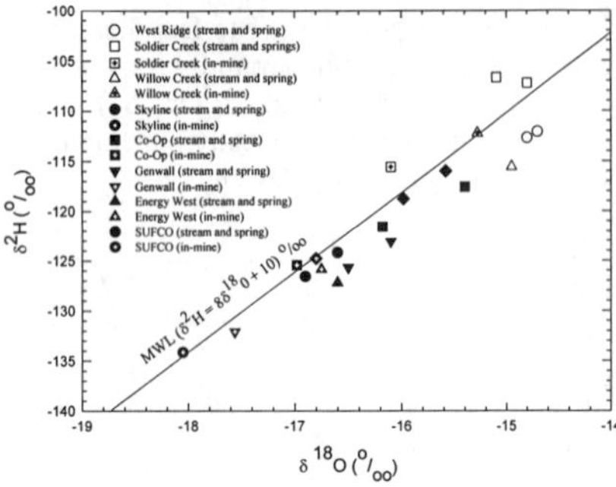

Figure 7 Mean stable isotopic compositions of surface, groundwater and in-mine samples grouped according to location.

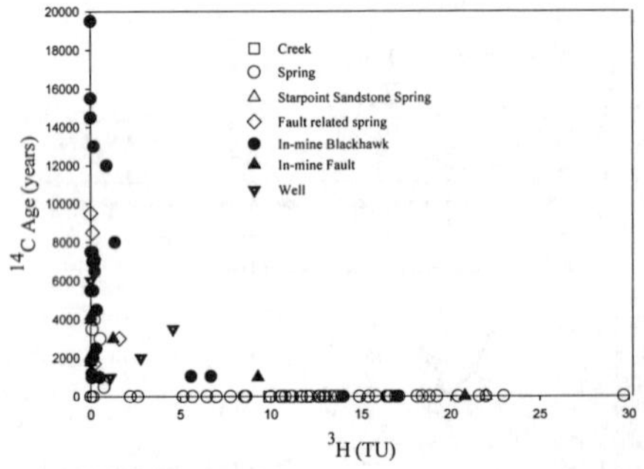

Figure 8 Scatter plot of ^{14}C mean residence times ("ages") vs. ^3H contents of surface, in-mine and groundwater.

Most in-mine groundwater discharges from sandstone channels in coal mine roofs, although occasionally upwelling from the underlying Star Point Sandstone occurs. In-mine hydrographs, δ^2H and $\delta^{18}O$ isotopic compositions, and ^{14}C groundwater "ages" demonstrate that deeply circulating groundwater systems have poor hydraulic communication with near surface groundwater. These deep groundwater systems dewater rapidly when encountered in the in-mine environment.

Groundwater flow can be described by a fairly simple conceptual model involving both active and in-active flow regimes (Figure 9). The active regime includes all near surface exposures, regardless of lithology. The near surface extends 75-150 m from the ground surface and 150-300 m into cliff faces. In the active region, fracturing, weathered rock, and matrix flow in fluvial sands control groundwater flow. Faults occasionally extend the active zone from the surface to deeper regions. The active zone is in direct hydraulic communication with seasonal recharge.

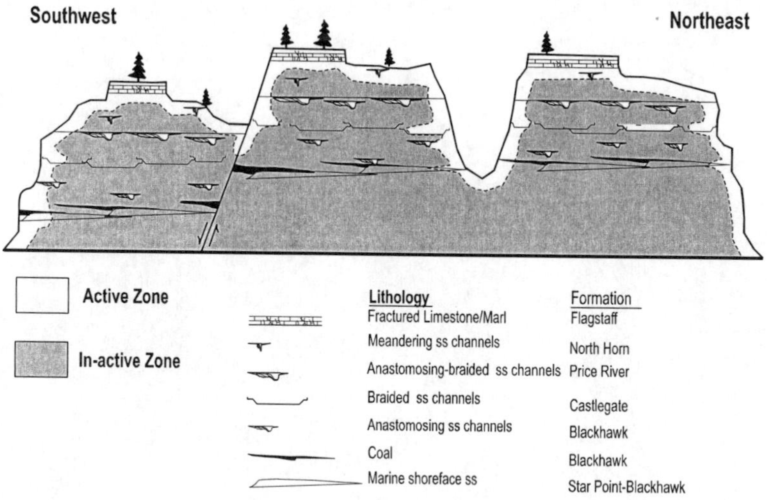

Figure 9 Conceptual model of active and in-active groundwater flow systems. The active and inactive regions are lithology independent. Diagram is not to scale. The active zone extends 75-150 m below land surface and 150-300 m into cliff faces. The active zone occurs in all bedrock formations that are exposed at the surface. The vertical distance from the top of the Flagstaff Limestone to the bottom of the Star Point Sandstone is about 600m.

The in-active zone includes most of the interior regions of the Book Cliffs and the Wasatch Plateau and most mine workings. Groundwater occurs in both vertically and horizontally isolated fluvial channel sands and isolated marine shoreface sands, which have limited hydraulic communication with the surface. When encountered in mine workings these channel sands are usually drained quickly. Coal seams are vertical hydraulic barriers.

REFERENCES

NCDC, 1998, National climatic Data Center online monthly climatic parameters: www.ncdc.noaa.gov/coop-precip.html.

Plummer, L.N., Prestmon, E.C., and Parkhurst, D.L., 1991, NETPATH - An interactive code (NETPATH) for modeling net geochemical reactions along a flow Path: U.S. Geol. Surv. Wat. Res. Invest. 91-4078, 94 p.

USU, 1998, Utah State University Climate Center online data for Emery, Huntington, Castledale, Salina and Scofield Dam weather station: www.climate.usu.edu/free/USA_UT.HTM.

WRCC, 1999, Western Region Climate Center, Desert Research Institute online precipitation data: www.wrcc.dri.edu/climsmut.html.

[illegible]

[illegible]

REFERENCES

[illegible]

Groundwater: Past Achievements and Future Challenges, Sililo et al. (eds) © 2000 Balkema, Rotterdam, ISBN 90 5809 159 7

Estimating the reliable output of wells

B.D.R. Misstear
Department of Civil, Structural and Environmental Engineering, Trinity College, Dublin, Ireland

S. Beeson
Mott MacDonald, UK

ABSTRACT: Estimating the reliable outputs of wells can be complicated because of the large number of variables that need to be taken into account. These variables relate to the characteristics of both the source and the aquifer. A methodology has been developed for estimating reliable outputs of water wells in the UK. This methodology includes a number of options depending on the complexity of the source and the amount of data available. The preferred option is based on a study of the operational history of the source, especially during periods of stress caused by drought. The methodology is compared for two contrasting aquifer types, the Chalk and the Sherwood Sandstone.

1 INTRODUCTION

The methodology described in this paper is based on a research project carried out by the authors for UK Water Industry Research. The aim was to develop a relatively simple, yet robust, methodology that could be applied to a wide range of groundwater source types, from single well sources in alluvial aquifers to large wellfields in complex fissure-flow bedrock aquifers, to karst springs.

Previous publications (Beeson *et al.* 1997, Misstear & Beeson 2000) have focused on the application of the methodology to the Cretaceous Chalk aquifer, the most important source of groundwater supplies in eastern and southeastern England. The present paper summarises the methodology and compares its application between a typical Chalk source and a typical Sherwood Sandstone source. The Permo-Triassic Sherwood Sandstone is the principal aquifer in central and northern Britain. It is a dual-porosity aquifer in which both intergranular and fissure permeabilities are frequently high (Downing 1993). In contrast, the Chalk is characterised by low intergranular permeability, owing to the very small pore sizes, and the main groundwater flow is through fissures; these are often highly transmissive in the upper horizons where they are enhanced by solution.

The output of a groundwater source depends on many variables. Some of these relate to the characteristics of the installation itself, including, for example, the number and type of wells, the presence or otherwise of regulatory controls on abstraction, the pump capacities and setting depths, the capacity of the raw water mains and the treatment plant. In addition to these source-related variables, the properties of the aquifer, including its transmissivity and storage, and the performance of the well or wells also have a major influence on the output of the source. It is therefore necessary to distinguish between the output that can be achieved with the existing setup of the source, and the output that could be achieved if the aquifer and well properties were the only limiting factors.

2 DEPLOYABLE OUTPUT AND POTENTIAL YIELD

Deployable output is defined as 'the output of a commissioned source or group of sources or of bulk supply as constrained by: licence (if applicable), water quality, environment, treatment, raw water mains and/or aqueducts, pumping plant and/or well/aquifer properties, transfer and/or output main - for specified conditions and demands' (Beeson *et al.* 1995). Potential yield, on the other hand, refers to 'the yield of a commissioned source or group of sources as constrained only by the well and/or aquifer properties for specified conditions and demands'.

The specified conditions and demands referred to in the definitions include normal and drought water level conditions and average and peak demand scenarios. For the purposes of the methodology, the drought condition is identified from long-term observation well records as the year in which groundwater levels were at their lowest. As such, the drought condition can be regarded as the condition of maximum aquifer stress. It was considered impractical to define drought conditions for individual groundwater sources on the basis of rainfall or recharge data. It must be emphasised that the methodology is intended for application to individual groundwater sources, and is not a tool for regional water resources assessment.

3 GENERAL APPROACH

The methodology was designed to be sufficiently flexible to accommodate wide variations in the quality of the available data as well as in the types of groundwater sources. It was also designed to be used by non-specialists as well as hydrogeologists.

The methodology comprises a so-called operational approach, which is divided into options A, B and C, and an analytical approach (option D). As its name suggests, the operational approach is based on the analysis of operational data whereas the analytical approach uses pumping test data. The options are summarised in the flow diagram in Figure 1.

3.1 *Application of the operational approach*

The choice of option depends on the extent and quality of the data available, with option A being the simplest and option C the most sophisticated. The procedure involves the completion of a standard assessment form and, for options B and C, a diagram summarising the performance of the source and the constraints on output.

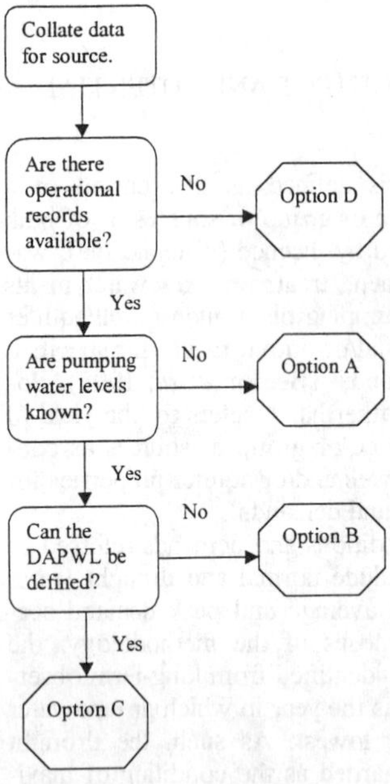

Figure 1. Choice of option.

An example of a summary diagram for a single well source is included in Figure 2. This shows the total output of the source (expressed in mega-litres per day, or Ml/d) plotted against pumping water levels (in metres above Ordnance Datum, or mAOD). Data from a step drawdown test are also included in the diagram. The non-pumping and operational water level data for the worst drought month on record (September 1992) have been used to fit a drought-bounding curve, the shape of which takes account of the shape of the step drawdown test plot. The slope of the drought-bounding curve increases at higher outputs. This is typical for a continuously pumped well in the Chalk aquifer, owing to the combined effects of well losses and the reduction in aquifer transmissivity that occurs as the upper flow horizons are dewatered.

The deployable output for both options B and C is 3.9 Ml/d. This is defined by the intersection of the drought-bounding curve with the pump cut-out level at -26 mAOD. If pumping water level data were not available, then the deployable output would be estimated by option A as 4 Ml/d, the annual abstraction licence quantity. (As can be seen from Figure 2, both of these estimates are less than the pump capacity of 4.5 Ml/d: owing to concerns about excessively low water levels, the pump at this source is throttled back to less than its full capacity).

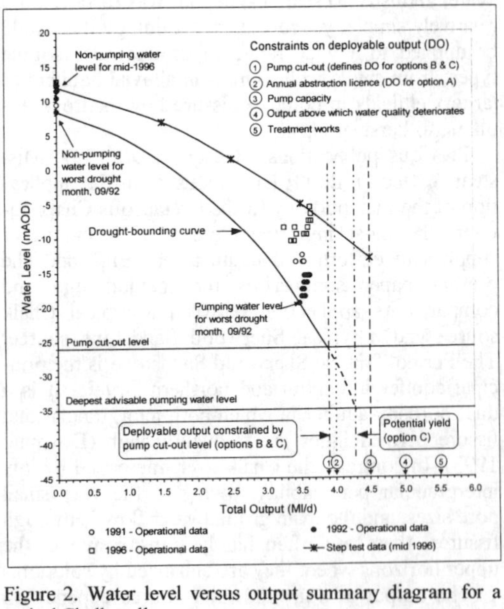

Figure 2. Water level versus output summary diagram for a typical Chalk well.

Option C also allows the determination of the potential yield by defining a deepest advisable pumping water level (DAPWL). This is the pumping water level below which there might be undesirable effects such as significant dewatering of the aquifer, sand pumping or saline water upconing. Borehole records, geophysical logs, test data and operational experience are useful for identifying the DAPWL. In the absence of such data, then the DAPWL is selected as the water level depth at which 30% or 50% of the aquifer remains saturated, the latter more conservative figure being used for a fissure-flow aquifer such as the Chalk. In Figure 2, the intersection of the drought-bounding curve and the DAPWL gives a potential yield of 4.3 Ml/d.

3.2 *Application of the operational and analytical approaches*

Figure 3 shows a water level versus output diagram for a Sherwood Sandstone source. This source comprises three operational wells that are used on a duty/standby/assist basis. The setup of the source permits only two of these to be pumped at the same time, and the reliable output is determined on this basis.

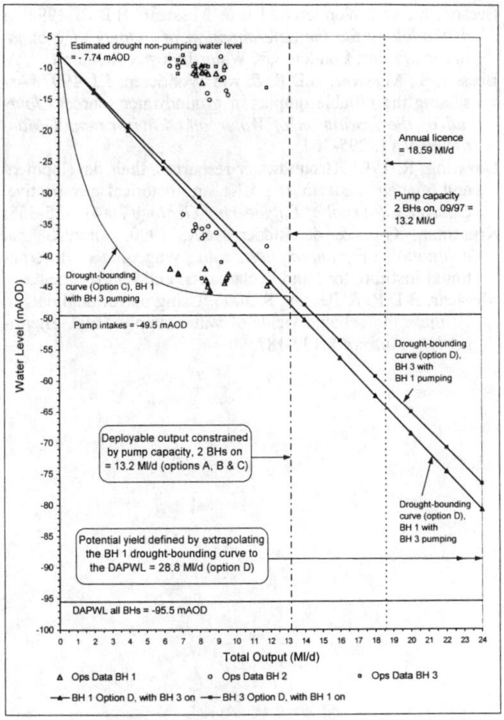

Figure 3. Water level versus output summary diagram for a Sherwood Sandstone multiple well source.

The summary diagram exhibits several important differences to the previous example. The boreholes are not pumped continuously, and this has the effect of shifting the pumping water level data points to the left compared to a continuously pumped source (because the x-axis gives the total output over a month, expressed as a daily average in Ml/d). The drought-bounding curve therefore assumes a concave rather than a convex shape. (Further information on fitting drought-bounding curves can be found in Misstear & Beeson 2000). In this example, nearby observation well data indicated that the worst drought on record occurred in 1993 (Fig. 4). In Figure 3, the operational water levels fall into two clusters: the shallow cluster at depths of between -7 mAOD and -15 mAOD represents non-pumping water levels, but generally for times when one of the other wells at the source is pumping, and the deeper cluster (between -33 mAOD and -45 mAOD) represents pumping water levels, again including interference effects.

The operational data are sufficient to be able to define a deployable output using option C of 13.2 Ml/d. This represents the pump capacity with two of the pumps running. However, the operational data do not permit an estimate of the potential yield. Based on the criterion that at least 30% of the aquifer should remain saturated, the DAPWL has been set at -95.5 mAOD, which is approximately 50 m lower than any of the pumping water levels shown on the figure. Therefore the analytical approach (option D) has been used to extrapolate a potential yield.

The analytical approach is described in Beeson *et al.* (1997) and is based on a) estimation of well losses from short step drawdown tests, b) calculation of additional drawdowns for extended periods of pumping using the well known Cooper-Jacob equation (see, for example, Kruseman and de Ridder 1990) and c) calculation of further drawdowns at each pumping well arising from interference effects from other wells. In the example in Figure 3, the predicted drought-bounding curves for two scenarios are shown, one for BH1 (with BH3 also pumping) and the second for BH3 (with BH1 pumping). That the predictions have some validity is shown by the fact that the predicted drought-bounding curve meets the operational drought-bounding curve at the relevant pump capacity. It can further be seen that BH1 has a slightly inferior predicted performance since it gives the greater pumping water levels, which is consistent with the operational data. This 'worst case' scenario is used to estimate the potential yield: extrapolating the option D drought-bounding curve for BH1 (with BH3 pumping) to the DAPWL gives a potential yield figure of 28.8 Ml/d. (Because the 'worst case' is used, further scenarios involving BH2 are not considered in this analysis as BH2 has a better performance than either BH1 or BH3).

The option D drought-bounding curves in this Sherwood Sandstone example are both straight lines. This is because the step test data for each of the wells did not indicate any well losses. This is typical of high-yielding wells in the Sherwood Sandstone in certain parts of England, where inflows occur throughout the open hole sections of the boreholes,

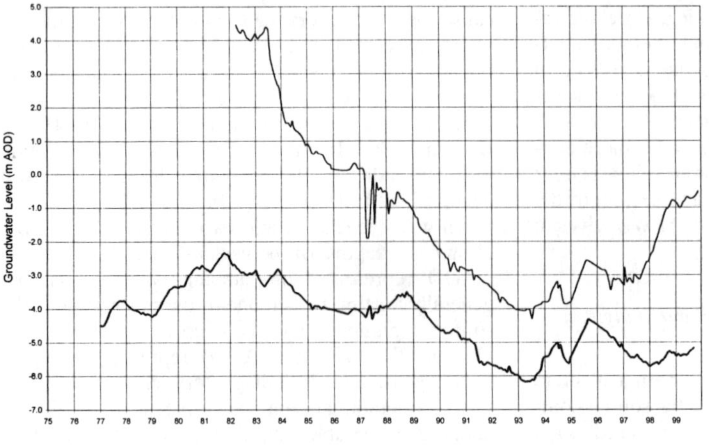

Figure 4. Hydrographs for two observation wells located in the vicinity of the Sherwood Sandstone source in Figure 3.

and contrasts with the convex slope of such curves in the Chalk, where the main flow horizons tend to be concentrated in the upper 30 m to 50 m. Nevertheless, some well losses would be expected at discharges higher than those at which the wells were tested. Also, aquifer transmissivity would be expected to decline where there are significant reductions in the water levels. Therefore, any extrapolation of the predicted performance of a source far beyond its operational performance history must be treated with caution.

4 CONCLUSIONS

1 The methodology for estimating deployable output and potential yield comprises several options, to account for variations in the quality and amount of data available at different groundwater sources.
2 The most detailed of the options that rely on operational data is option C, and this is the preferred method where sufficient data are available. The analytical approach (option D) can be useful in the absence of such data, but the predictions should be treated with caution.
3 The methodology can be applied to many different hydrogeological environments, including the two aquifer types contrasted in this paper.

ACKNOWLEDGEMENTS

The authors would like to thank the members the UK Water Industry Research steering group for the project, colleagues at Mott MacDonald, and the water companies that provided data to the original research and the examples used in this paper, including Cambridge Water Company and Yorkshire Water Services.

REFERENCES

Beeson, S., van Wonderen, J.J. & Misstear, B.D.R. 1995. *A methodology for the determination of outputs of groundwater sources*. London: UK Water Industry Research.

Beeson, S., Misstear, B.D.R. & van Wonderen, J.J. 1997. Assessing the reliable outputs of groundwater sources. *Journal of the Institution of Water and Environmental Management* 11: 295-304.

Downing, R. 1993. Groundwater resources, their development and management in the UK: an historical perspective. *Quarterly Journal of Engineering Geology* 26(4): 335-358.

Kruseman, G.P. & de Ridder, N.A. 1990. *Analysis and evaluation of pumping test data*. Wageningen: International Institute for Land Reclamation and Improvement.

Misstear, B.D.R. & Beeson, S. 2000. Using operational data to estimate the reliable yields of water-supply wells. *Hydrogeology Journal* 8:177-187.

Groundwater: Past Achievements and Future Challenges, Sililo et al. (eds) © 2000 Balkema, Rotterdam, ISBN 90 5809 159 7

The use of thermal imagery in ground water studies

K.L. Morton
KLM Consulting Services, Lanseria, South Africa

A.J. Franszen
Anglo American Technical Services, Johannesburg, South Africa

ABSTRACT: Airborne thermal imagery and its use in ground water location is not a new technique, however recent developments in Global Positioning Satellite (GPS) technology means that the ground water signatures on the image can be accurately pin pointed in the field. The low cost and simple technology makes it very attractive to development programs. Recent surveys in Southern Africa have been used to determine the position of dolomite aquifers, polluted ground water, seepage from slimes dams, geological structures, abandoned mines, underground fires and kimberlites. Examples of different applications are given in the text and as a projected presentation.

1 INTRODUCTION

Thermal images, taken from both satellite and airborne platforms have been used in geological investigations since the 1970s. Thermal data is useful for, amongst other applications, geological mapping and ground water detection. Thermal imagery is not new, thermal plots have been used since the 1970s but modern developments in Global Positioning Systems (GPS) mean that the image can now be georeferenced very accurately. Resolution of 2m on the ground is easily obtained and 6m has been obtained through low altitude flight scans.

A thermal image is a scan of the earth's temperature (emitted energy) just before dawn, and is scanned from an aircraft flown at about 2000 feet above ground level, preferably in winter when there is minimum soil moisture. Three variables, absolute temperature, total energy and wavelength contribute to the optimal conditions for collecting thermal data.

Typically swath widths of 1,4 km are used with 20% overlap.

The data is collected by airborne opto-mechanical radiometers. A typical detector images from 8 to 14 μm which coincides with the atmospheric window in that portion of the electromagnetic spectrum. The instantaneous field of view is typically 2 to 3 milliradians, which corresponds to a ground pixel of 2 to 3 m at an altitude of 1000m. This gives a far superior resolution than satellite imagery and also does not require the writing of complex algorithms for interpretation.

2 APPLICATION

A thermal image can reveal the structural geology of a site. The structural geology of an area can stand out like a sub surface "x –ray" of the site showing faults, fractures, bedding planes and many other features. Figure 1 illustrates a standard aerial photo of a tailings dam and surrounding area in the Free State, South Africa. Very little geological information can be interpreted. Figure 2 is the same area but is a thermal image. The geology is predominantly dolomite. Bedding planes, fractures and interconnecting fissures are clearly evident. The ground water seepage from the slimes dam in the upper left hand corner is exceptionally clear. The accurate geo-referencing of the image means that minimal geophysical investigation is required to locate high yield boreholes on the visible structures or to drill holes and dig trenches to intercept the seepage.

The accurate GPS referencing means that sites of

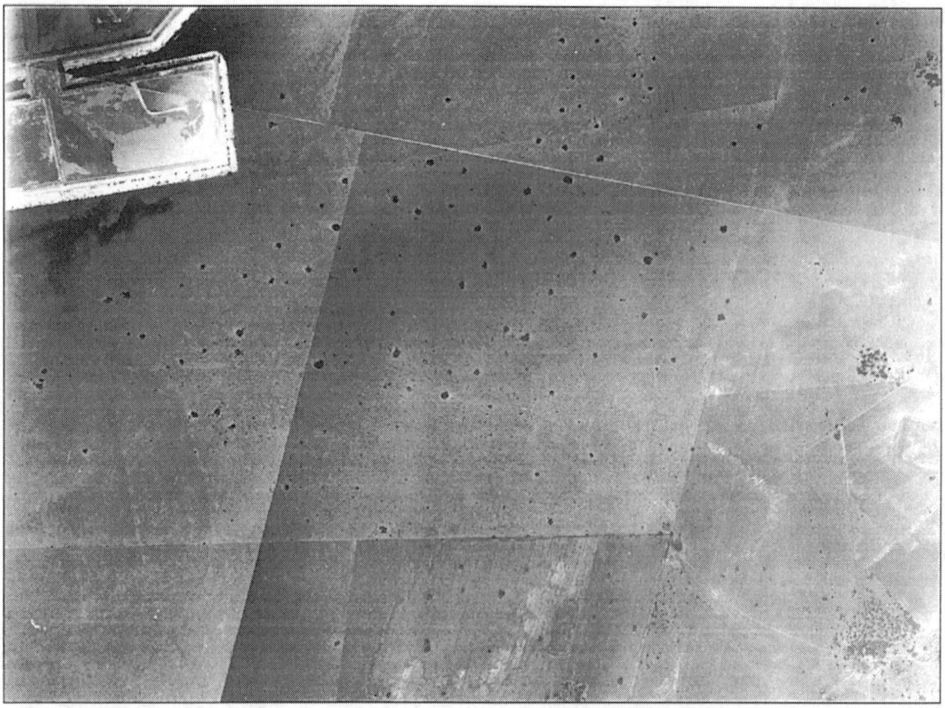

Figure 1. Aerial photograph of a mine dump and surrounding area in the Free State, South Africa

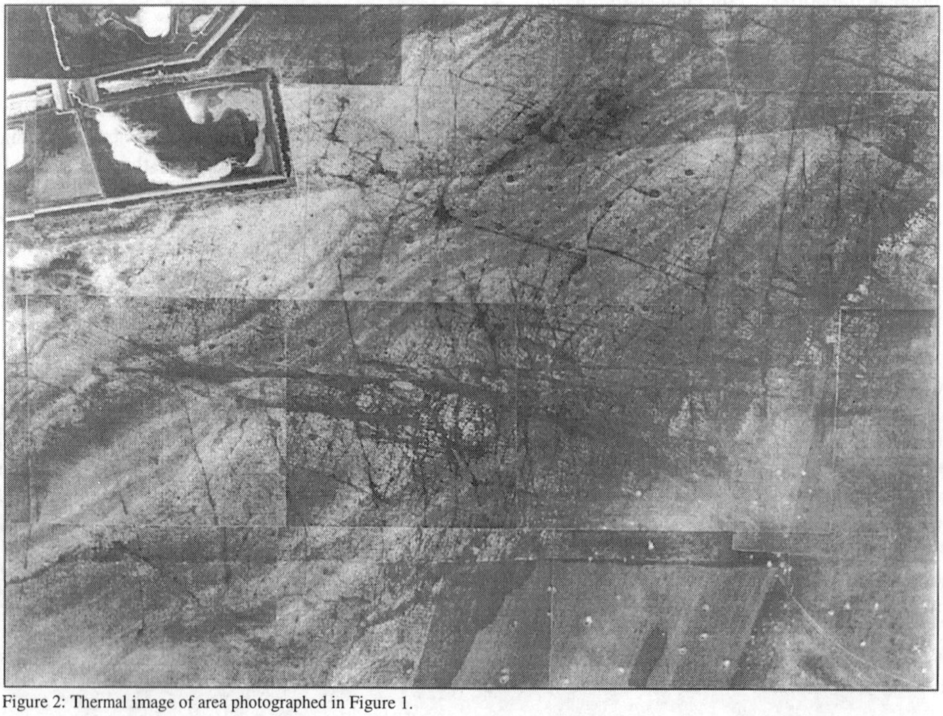

Figure 2: Thermal image of area photographed in Figure 1.

interest e.g. water filled dolomite cavities or seepage from dams can be located in the field with a sensetive portable GPS within one to two meters. If higher level of accuracy is required a tighter flight path and lower altitudes can be flown.

Figure 3 shows a slimes dam complex located near a karst system in a major dolomite plateau. The dark areas indicate cool ground water down to depths of about 60m in vertical structures within the dolomite. It also shows seepage from the overlying slimes dams. These seepage zones have been investigated using resistivity and target drilling. In all cases the resistivity profile and interpretation confirmed the predicted presence of underground water. A target for drilling a water supply boreholes was chosen by looking at the image on a high resolution computer screen. Drilling of the darkest areas in the major flow zone resulted in a borehole with a yield in excess of 69 000 l/hr.

The scan is in black and white with shades of grey showing temperatures to an accuracy of 0,1 degree Centigrade. This accuracy means that polluted water can be distinguished from unpolluted water. The exothermic reaction created by the pollutant e.g. sulphate, means that the polluted water is often warmer by a few degrees. Colour coding (false colour) of the temperature differences can be used to highlight temperature differences and emphasise specific targets such as polluted water or deep seated ground water.

Figure 4 shows a coal discard dump in Mpumalanga, South Africa. A retaining wall (berm) was built along its northern boundary to hold back any run-off from the dump. The image reveals that the berm does not contain the water. The run-off and seepage from the dump is warmer and can be seen to be moving under the berm and into the fresh water spruit which is cooler. This image has been used to tighten the collection of polluted water and to improve the siting of monitoring boreholes.

A major advantage of the technique is that it is very straightforward to interpret and is therefore

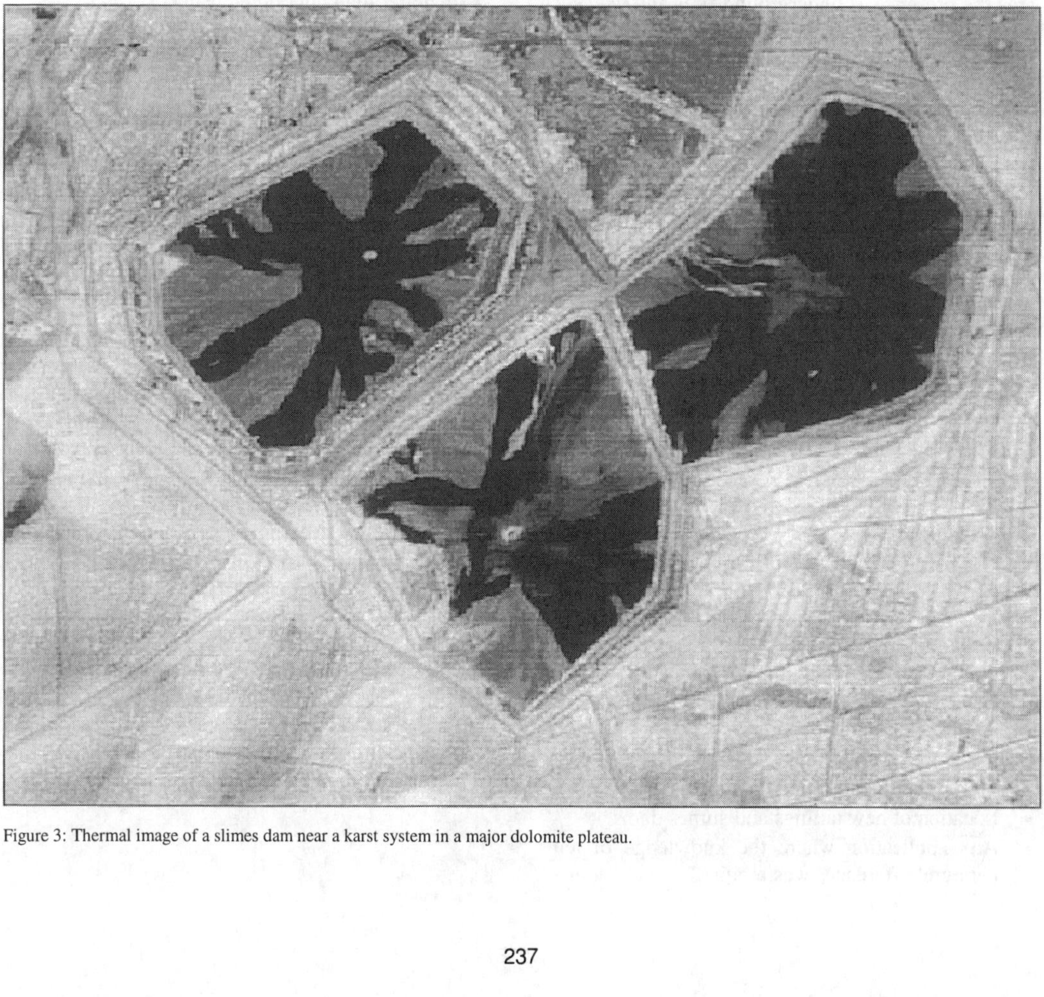

Figure 3: Thermal image of a slimes dam near a karst system in a major dolomite plateau.

Figure 4. Thermal image of coal discard dump.

ideal for managers without experience in remote sensing. Non technical people can grasp the concept very quickly. Thermal imagery has been used to demonstrate to community leaders in the Witbank area the presence of underground fires and why such areas are unsuitable for housing developments.

Cold is shown as black and represents cooler deep seated ground water, below surface seepage and underground cavities. Hot is white and shows areas of evaporation or warmer surface water bodies.

If required the data can have false colour applied to highlight temperatures with specific relevance, for example warmer, polluted water or moist clay rich mineral deposits.

During the 70's a company called Spectral Africa operated a thermal scanner over extensive areas of Southern Africa. In June 1998 and '99 a scanner was again used for earth science applications in South Africa. The purpose of the surveys was to illustrate the application of pre-dawn thermal data for geological mapping, to map sub surface fires in coal mines and to detect pollution plumes from mine dumps and slimes dams.

Thermal image surveys have been flown over different areas of South Africa for interpretation for use in:

- Ground water location
- Geological Mapping
- Dolomite cavity mapping
- Detection of seepage from tailings dams, rock dumps and impoundments
- Detection of pollution
- Plotting of underground coal mines
- Location of new tailings and slimes dams
- Any application where the knowledge of temperature differences was required.

The images are supplied on CD's in tif format and can be manipulated by non technical operators on a standard PC using a photoshop / paintshop type program to produce cut–outs of specific areas or enlargements of target areas. These can be printed on standard black and white printers.

In 1999 thermal scans were flown for the following South African mines: Impala platinum, Venetia, Finsch, The Oaks, Premier, Klipspringer, and an area of Secunda for SASOL. In previous years Anglogold Vaal River, West Wits and Free State operations were covered. Each image provided new uses for the thermal interpretation.

Some important discoveries were:

- Kimberlites – the clay rich kimberlite is damper and therefore cooler than surrounding areas.
- Sinkholes
- Seepage from slimes dams across mine boundaries
- Water bearing faults and fissures
- Ground water pollution from coal discard dumps (the exothermic reaction of the sulphate shows up hotter than unpolluted water)
- Underground abandoned coal mines
- Dolomite cavities
- Geological structures

Some examples of the use of the imagery are given in the slide presentation, the best viewing method is on a high resolution computer screen as some of the definition is lost in the print medium. The slides show that thermal imagery is a very powerful but simple and inexpensive tool for solving a range of problems. Typically an area of about 10 x 20km would cost about ZAR 150000 to fly, georeference, supply a hard copy mosaic of the area and cut to a Compact Disc. This is well within the budgets available for regional development and aid programs.

Airborne thermal remote sensing has proved an inexpensive tool for a whole range of applications with a much higher accuracy than satellite based data collection. Additional applications than those listed above are under investigation and further flights are planned for next winter season.

ACKNOWLEDGEMENTS

The authors would like to thank Anglo American Technical Services and the Manager of Finsch Mine for permission to publish this data.

REFERENCES

Franszen, A.J. 1998. Thermal remote sensing for earth science applications. Proceedings of the 1998 South African Symposium on communications and signal processing (COMSIG). Cape Town. Pp351-352. IEEE Catalogue no 98EX214.

Exploration, Research & Development Unit 1977 Better map ping through remote sensing. Coal Gold and Base minerals Vol. 25 No 9 September 1977.

Groundwater: Past Achievements and Future Challenges, Sililo et al. (eds) © 2000 Balkema, Rotterdam, ISBN 90 5809 159 7

Rational utilization of groundwater for water supply engineering

P. Némethy & D. Vlčková-Michalíková
Department of Hydrogeology, Faculty of Natural Sciences, Comenius University, Bratislava, Slovakia

ABSTRACT: Slovakia uses groundwater and surface water for supplying inhabitants with drinking water. This paper deals with groundwater. High groundwater withdrawal can cause discharge decreasing in near surface flow until the stream dried up. If we have wells and springs in the group water supply system, water resources can be utilized according to the climatic conditions. Springs have a high yield in the period of high precipitation so that greater amount of water can be taken, and the lower amount will be taken from wells and vice versa. The withdrawal will be regulated in a way, complying with the ecological criteria.

1 INTRODUCTION

Groundwater in Slovakia is mostly taken from springs and wells. Springs with the biggest yields can be found in Mesozoic karst and karst-fissure rocks. They are in many places of Slovakia, in our case are in Starohorský potok brook and Harmanecký potok brook catchments. Yields range from tens to hundreds of litres per second. In Mesozoic there is 25 % of exploitable amount of Slovakia`s groundwater. On alluvial deposits, 59 % of exploitable amount of groundwater is bound and it is therefore necessary to know the processes at water infiltration from river to aquifer (Ženišová 1999). In the areas, where there is no abundance of suitable water, it is generally considered the withdrawal of mine waters for drinking purposes (Hudáček 1997, Grmela 1999). In consequence of exploitation discontinuance, problems in some mines with endanger of surface and groundwater quality arise. Outflowing mine waters often have a high mineralization and heavy metal content which can cause accidental deterioration of water quality (Grmela & Jelínek 1999, Hudáček 1999). To solve protection of groundwater quality it can be advantageous to use mathematical modelling (Rapantová & Grmela 1998, Rapantová 1999), or treatment of water quality in situ (Škultétyová & Martoň 1999).

A big part of drinking water in the middle of Slovakia is delivered by Pohronský group water supply system. Springs in catchment of Starohorský potok brook and Harmanecký potok brook and water resource Podzámčok, where water is taken by wells, belong to its most important resources. Water withdrawal in unfavourable climatic conditions causes significant decrease of the brook discharge. To prevent this a result we propose to combine the utilization of springs and wells. In this manner it is possible to arrange high withdrawal of water while to keeping suggested ecological limits (Vlčková-Michalíková 1999, Kullman et al. 1994).

2 SPECIFICATION OF NATURAL CONDITIONS

Spring area lies on the south and southeast promontories of Veľká Fatra Mountains. There are five big springs in the Starohorský potok brook catchment and seven springs in the Harmanecký potok brook catchment up to evaluated stream-gauging station (Figure 1). Harmanecký potok brook flows from Zalámaná.

Precipitation in Starohorský potok brook catchment can be characterized by average precipitation amount for the years of 1982 - 1997 from the precipitation station Motyčky (Table 1). This station is near by the Štubne water resource.

Precipitation in Harmanecký potok brook catchment we can characterize by average precipitation amount for the years of 1982 - 1997 from precipitation station Dolný Harmanec (Table 1). The station is situated near a gauging station at Harmanecký potok brook. For the Podzámčok water resource, precipitation can be characterized by data from precipitation station in Sása located south of Podzámčok (Table 1).

The area of hydrological catchment of Starohorský potok brook up to the evaluated stream - gauging station is 62,61 km^2. Hydrogeological catchment is however larger and its area is still unknown.

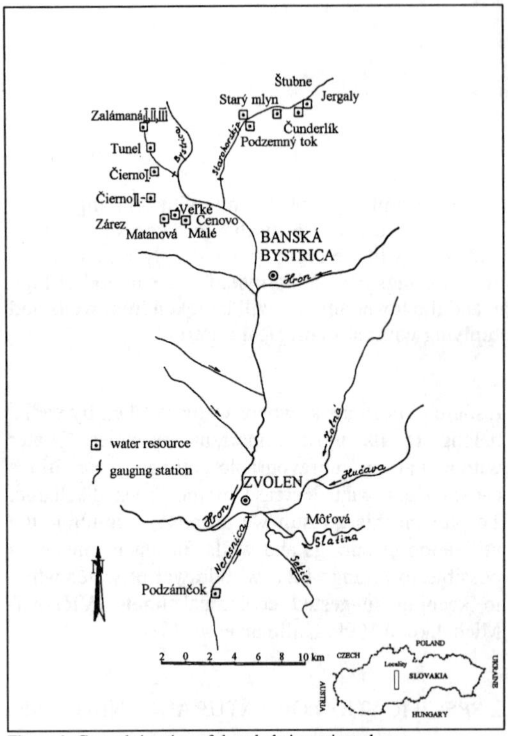

Figure 1. General situation of the whole investigated area

Table 1. Precipitation in stations [mm]

Location the Month	Motyčky	Dolný Harmanec	Sása
January	59	71	32
February	60	70	28
March	69	77	36
April	76	77	45
May	121	171	90
Jun	109	106	75
July	94	84	53
August	86	90	59
September	75	84	52
October	77	78	47
November	97	101	56
December	90	101	41
Year	1013	1056	614

Discharge measured in the stream - gauging station is influenced by both water withdrawal from springs and by operation of 2 power plants. The average discharges of Starohorský potok brook in the period of 1983 - 1997 are given in Table 2.

Similarly to Starohorský potok brook, Harmanecký potok brook has also an unknown hydrogeological catchment. The area of hydrological catchment up to the stream - gauging station is 23,01 km^2. The average influenced discharges of Harmanecký potok brook over the period of years 1979 - 1997 are shown in Table 2.

Table 2. Average influenced discharge in brooks [m^3.s^{-1}]

Location Month	Starohorský potok	Harmanecký potok	Neresnica
January	0.509	0.339	0.454
February	0.486	0.321	0.572
March	0.625	0.464	1.871
April	0.817	0.775	1.262
May	0.691	0.722	1.054
Jun	0.529	0.609	0.430
July	0.834	0.491	0.140
August	2.189	0.416	0.076
September	1.226	0.375	0.299
October	1.020	0.373	0.276
November	0.809	0.390	0.486
December	0.983	0.389	0.464
Year	0.893	0.472	0.615

Neresnica brook flows nearby water resource Podzámčok. Stream - gauging station is located several kilometres downstream. Average influenced discharges for years 1980 - 1997 are given in Table 2). Relevant discharges are observed for the period of water resource utilization.

Hydrogeological conditions in Starohorský potok brook and Harmanecký potok brook catchments are complicated.

Groundwater outflows on the surface or to the surface flows between formations or tectonic complexes of different hydrogeological properties. This area belongs to the most important water supply areas in mountain areas of Slovakia.

Water outflowing from springs in the catchment of Starohorský potok brook has its origin in Mesozoic limestones and dolomites of Krížna and Choč nappes.

On the base of hydrogeochemical properties it is assumed that the springs Jergaly, Štubne and Generál Čunderlík come from Krížna nappe, with the infiltration area of about 19,4 km^2. Other important springs like Starý Mlyn and Podzemný tok come from Choč nappe. In the catchment of Harmanecký potok brook there are 10 water supply important springs. Seven of them occur in the catchment of evaluated stream - gauging station. Groundwater comes from dolomites and limestones of Choč nappe and Krížna nappe creates impermeable underlying rock (Fľaková 1998).

Harmanecký potok brook is right-side tributary of the river Bystrica. In its catchment water is used from the springs Zalámaná I, II, III, Tunel, Čierno I, II, III, Zárez, Veľké and Malé Cenovo and Mátanová. The amount of outflowing water from the springs has to have greater infiltration area as it has its hydrological catchment. In the catchment of Starohorský potok brook are considering the same.

The water resource Podzámčok lies in the areas built of Neogene volcanic rocks. Effusive rocks, andesites and their pyroclastic rocks are most widespread. Rocks are reflected in fissure and porous permeability. Water is transported towards water resource by fissures from sizable distances. The water circuit has to be deep enough because water temperature is stable during the year, average temperature is 15°C.

3 POSSIBILITIES OF OPERATED WATER WITHDRAWAL

Within the Pohronský group water supply system it is possible to utilize water resource according to demands. It means that in some of them, withdrawal can be decreased or interrupted and needed amount of water can be obtained from other resources. Up to the present time practice in Slovakia did not take into consideration the ecological standpoint. Groundwater withdrawal can cause significant decrease of discharge in surface flow, what can unfavourably influence the fauna and flora. That is why there was an endeavour to propose such criteria for water withdrawal which will arrange ecological balance in the nature.

The average withdrawal amounts of water in $l.s^{-1}$ from Starohorský potok brook and Harmanecký potok brook catchments and from Podzámčok water resource for the period of years 1983 - 1997 are given in Table 3.

Table 3. Average withdrawal from watershed. [$l.s^{-1}$]

Location / Month	Starohorský potok	Harmanecký potok	Podzámčok*
January	345	438	139
February	341	445	152
March	302	442	139
April	402	520	146
May	374	536	139
Jun	350	537	148
July	332	508	146
August	315	495	138
September	309	487	142
October	324	473	137
November	334	468	139
December	288	454	135
Year	335	484	142

* Water resource

3.1 Estimation of global ecological limits

Works of Kullman et al. (1993), Patschová & Mihálik (1999) deal with global ecological limits with respect to amount of withdrawal. The global ecological limit $MQ_{(EKO)}$ is a value of natural discharge in surface flow in the vicinity of water resource which have to be preserved to ensure a biological life in the river. Global ecological limit for Starohorský potok brook was estimated by Kullman et al. (1994) from hydrological years 1984 - 1993. The average monthly withdrawal amount of groundwater from springs was added to average daily discharges in water - gauging station. The assumption that average monthly value of withdrawal amount equals average daily value withdrawals at a given month was accepted. By this way daily uninfluenced discharges were obtained, from which the cumulative frequency curve was constructed and according to formula: $MQ_{(EKO)} = (0.65 - 0.70).Q_{364}$ global ecological limit 188.5 - 203 $l.s^{-1}$ was calculated. Reduction coefficient (0.65 - 0.70) was divided on the basis of data of European Economic Commission, Austria, Poland, Hungary, Czech and Slovak republic. Q_{364} is discharge which is exceeded 364 days per year.

Global ecological limit for Harmanecký potok brook was estimated by the same methodology and its value was 335.4 - 361.2 $l.s^{-1}$ (Vlčková - Michalíková 1999).

3.2 Proposal of operated withdrawal by combined utilization of springs and wells

Mutual relationship between precipitation, withdrawal and discharges was found out and it was possible to estimate in which months it will be essential to decrease withdrawal from springs and increase withdrawal from wells. At Figure 2 it can be seen that from March to June withdrawals in the Starohorský potok brook catchment should be decreased. In the Harmanecký potok brook catchment, the influence of discharge is greater (Figure 3), because withdrawal is higher and concentrated in a smaller area. To keep the ecological limits withdrawal in the period of 8 months should be decreased. From Figures 4 - 6 it is possible to see mutual relationship between precipitation, withdrawals and groundwater level. The consequence of withdrawal decrease is increasing of the groundwater level. Tendency of water withdrawal decrease can be observed at the whole area of Slovakia. As a result of water meter installation in households and decreasing of economic activity we can observe significant decrease of water withdrawal since 1992. Such trend in investigated area is presented in Table 4.

According to the maximum depth of groundwater level, the time of possible water taking can be predicted. By using predicted methods of spring yield changes, overflows on surface flows and groundwater level depths can be predicted the manner of groundwater withdrawal for several months in advance (Fendeková 1999, Fendeková & Vous 1995, Némethyová 1999).

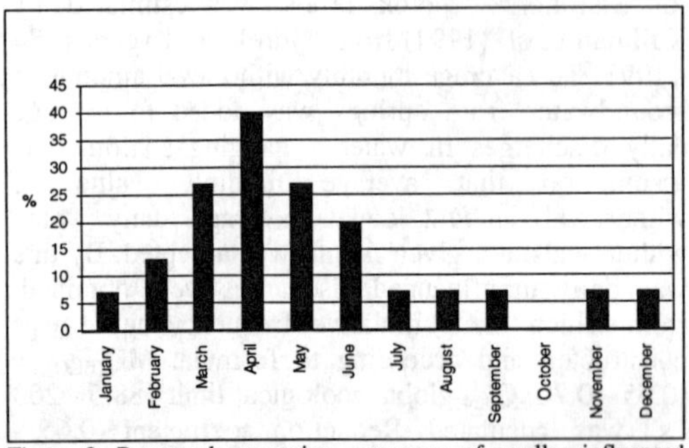

Figure 2. Percentual expression occurrence of smaller influenced discharges as estimated ecological limit 203 $l.s^{-1}$ for Starohorský potok brook in the period of years 1983 - 1997

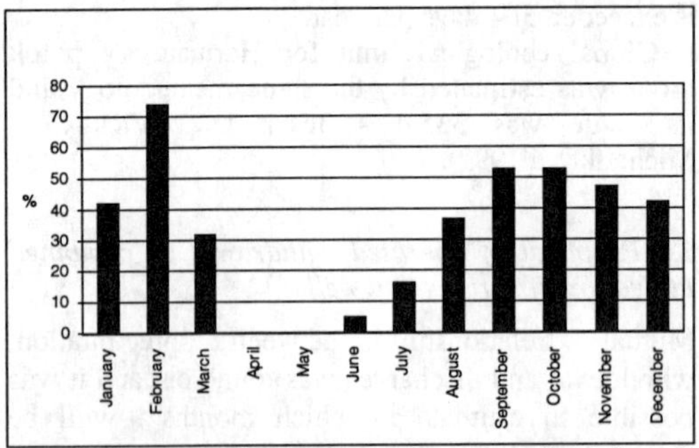

Figure 3. Percentual expression occurrence of smaller influenced discharge as estimated ecological limit 361,2 $l.s^{-1}$ for Harmanecký potok brook for the years 1979 - 1997

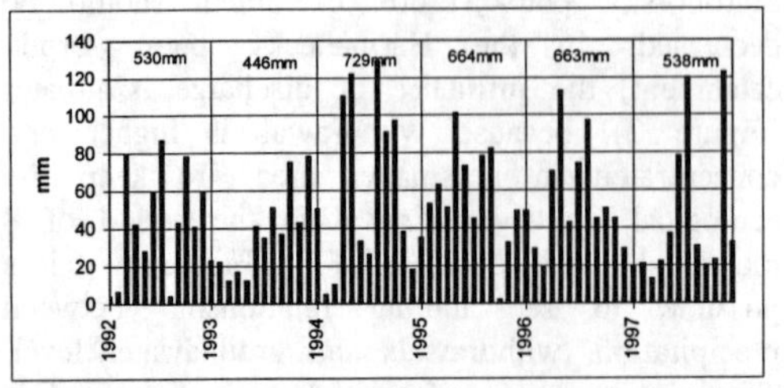

Figure 4. Average monthly precipitation amount at precipitation station Sása

4 CONCLUSION

At the example from middle of Slovakia we showed the possibility of combining the utilization of springs and wells for drinking water supply. By this means it is possible to obtain the needed amount of water by keeping the ecological discharges in surface flows and smaller operating costs. Water

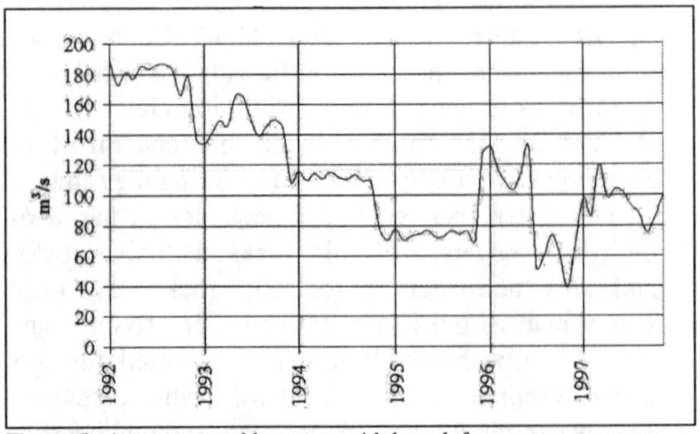

Figure 5. Average monthly water withdrawal from water resource Podzámčok in the period of 1992 - 1997 years

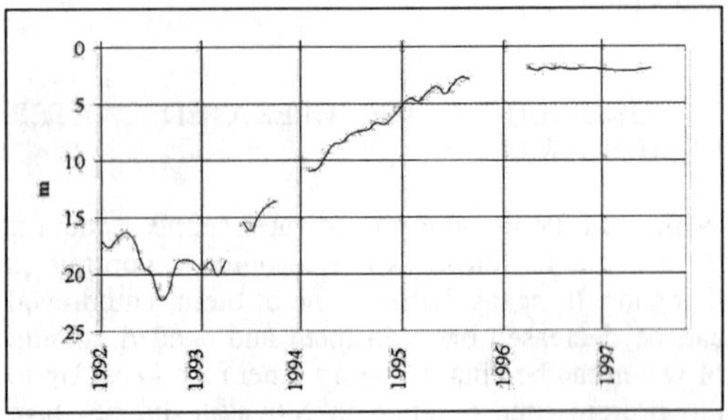

Figure 6. Average monthly depth of groundwater level in a well N - 3a in the period of 1992 - 1997 years

Table 4. Average withdrawal from watershed. [$l.s^{-1}$]

Location Year	Starohorský potok	Harmanecký potok	Podzámčok*
1992	450.5	483.0	179.9
1993	322.4	402.6	147.1
1994	429.1	509.0	109.0
1995	236.4	408.9	73.8
1996	210.5	371.8	95.9
1997	227.4	401.9	86.0

* Water resource

withdrawal from springs is substantialy cheaper because water pumping is not necessary. Base on evaluating withdrawal data and discharges in brooks, it was possible to find out the periods, when the increasing water amounts can be taken from springs and when it is essential to more utilize wells.

REFERENCES

Fendeková, M. 1997. Metódy prognózovania zmien výdatnosti prameov a ich aplikácie. *Podzemná voda*, III., . 2:46-55

Fendeková, M. & Vous, O. 1995. Vplyv horninového prostredia a odberov podzemných vôd na reim podzemného odtoku. *Geotechnické problémy ivotného prostredia; Zborník prác prvej slovenskej geotechnickej konferencie, Dom techniky ZS VTS s.r.o. Bratislava* :97-102

Faková, R. 1998. Formation and changes of groundwater chemical composition of the western carpatian carbonate systems. *Acta geologica Universitatis Comenianae Bratislava,* Nr. 53:5-25

Grmela, A. 1999. Mine water and its present utilization in the Czech republic. In M. Fendeková & M. Fendek (eds), *Hydrogeology and land use management; Proceedings of XXIX Congress of International association of hydrogeologists Bratislava 6. - 10. september*: 745-749

Grmela, A. & Jelínek, P. 1999. Mine waters and their changes related to mines closure in the Czech part of the upper silesian coal basin. In M. Fendeková &M. Fendek (eds), *Hydrogeology and land use management; Proceedings of XXIX Congress of International association of hydrogeologists. Bratislava 6. - 10. september*: 751-755

Hudáek, M. 1997. Zmeny chemizmu banských vôd po opustení banských diel. *Podzemná voda,* III.,. 2:112-117

Hudáek, M. 1999. Mining activity impact to hydrosphere of the Spišsko-gemerské rudohorie mountains, northern part (Western Carpathians). In M. Fendeková & M. Fendek (eds), *Hydrogeology and land use management; Proceedings of XXIX Congress of International association of hydrogeologists Bratislava 6. - 10. september*: 767-771

Kullman, E.S., Kullman, E.J. & Patschová, A 1993.. Generel ochrany a racionálneho vyuívania vôd. as C : *Ochrana vodného bohatstva. Ekologické limity pre vyuívanie vôd (vodných zdrojov) a pre hospodárenie a nakladanie s vodou z hadiska prírody a krajiny.*Manuskript Slovenský hydrometeorologický ústav Bratislava: 1-33

Kullman, E.S., Bujalka, P. & Kullman, E.J. 1994. Generel ochrany a racionálneho vyuívania vôd. as C : *Ochrana vodného bohatstva. Ekologické limity pre vyuívanie vôd (vodných zdrojov) a pre hospodárenie a nakladanie s vodou z hadiska prírody a krajiny. Stanovenie ekologických limitov vyuitenosti zdrojov podzemných vôd.* Manuscript Geovod Bujalka:56-93

Némethyová, S. 1999. Prognózovanie porušených asových radov modelom Sarima. *Podzemná voda,* V.,. 2:54-63

Patschová, A. & Mihálik, F. 1999. Hydrogeological limits of groundwater exploitation. .In M. Fendeková & M. Fendek (eds), *Hydrogeology and land use management; Proceedings of XXIX Congress of International association of hydrogeologists Bratislava 6. - 10. september*: 23-28

Rapantová, N. & Grmela, A. 1998. Utilization of analytical calculations of 1D transport of organic pollutants in groundwater for risk analysis in the area of Jan Šverma coke plant in Ostrava. *Zborník z odborného seminára Hydrogeochémia '98 IV.roník Bratislava 6. máj: 21-29*

Rapantová, N. 1999. Application of mathematical modeling for design of protection of drinking water withdrawal area Strata - Nov Vest. :In M. Fendeková & M. Fendek (eds), *Hydrogeology and land use management; Proceedings of XXIX Congress of International association of hydrogeologists Bratislava 6. - 10. september*: 227-231

Škultétyová, I. & Marto, J. 1999. In situ water treatment. : In M. Fendeková & M. Fendek (eds), *Hydrogeology and land use management; Proceedings of XXIX Congress of International association of hydrogeologists Bratislava 6. - 10. september*: 529-531

Vlková - Michalíková, D. 1999. Racionálne vyuitie podzemnej vody pre vodárenské úely. *Diplomová práca.* Katedra hydrogeológie PRIF UK Bratislava

enišová, Z. 1999. Chemical composition changes in alluvial deposits water during infiltration. In M. Fendeková & M. Fendek (eds), *Hydrogeology and land use management; Proceedings of XXIX Congress of International association of hydrogeologists Bratislava 6. - 10. september:* 587-592

Groundwater: Past Achievements and Future Challenges, Sililo et al. (eds) © 2000 Balkema, Rotterdam, ISBN 90 5809 159 7

Groundwater exploration in complex fractured rock

I. Neumann & K. Sami
Council for Geoscience, Department of Geohydrology, Pretoria, South Africa

ABSTRACT: The objective of groundwater exploration in fractured rock terrain is to ensure that targets offering the greatest probability of success in terms of borehole yield are identified. In complex structural environments exhibiting a large variability in borehole yields appropriate exploration techniques can lead to significant costs savings.

Research has been carried out in several regions in South Africa where historically low success rates have been achieved. Investigations involved structural mapping, photogeological and satellite image processing and filtering, a variety of surface geophysical methods and the drilling of exploration boreholes. Identified lineaments were assessed in terms of geological and structural processes to distinguish potential high yielding features from those, which have low groundwater potential. Drilling results and the cost-effectiveness of the approach are described.

1 INTRODUCTION

Large parts of South Africa suffer from rainfall deficit, either continuous in the semi-desert regions or with seasonal deficits, intensified by periodic droughts. As rainfall is the dominant control on recharge, potable water scarcity affects large parts of the country. Rural areas with their dispersed populations and with a poor economic base are especially affected by this water deficit. In rural communities groundwater is just one of different potential sources of potable water. However, groundwater is an attractive option due to its general availability, even in drought situations, and its relatively good quality. The dispersed nature of the rural population makes it necessary to extend groundwater development in South Africa into areas of particularly unfavourable hydrogeological conditions, like those found in the Natal Metamorphic Province.

The ca. 1000 Ma old Natal Metamorphic Province consists of igneous and metamorphic rocks of typically low productivity with little primary intergranular porosity or permeability. Thus, the occurrence and movement of groundwater is primarily controlled by the prevalence and orientation of secondary features, such as deep weathered zones, faults, fracture zones and intrusive dykes that enlarge storativity and transmissivity. Due to these unfavourable conditions 47% of all boreholes drilled into the rocks of the Natal Metamorphic Province are unsuccessful, thus large sums of money are presently being spend in the drilling of dry holes. As the area has a relatively dense settlement pattern and population growth rates are high, there is an urgent need to improve borehole success rates by developing a methodology for successful groundwater exploration in this complex terrain. At present the water abstraction through boreholes in the area under investigation equals approximately 1% of the areal recharge, hence there is obvious potential for improved water supply, if methods of increased abstraction which are both feasible and economic can be developed.

The objective of the research programme described below was to develop an exploration methodology that maximises the success rate of boreholes in the most cost effective and productive manner. The exploration technique followed a carefully considered plan of approach. To investigate the occurrence of groundwater in the rocks of the Natal Metamorphic Province and to identify the geological and structural controls that distinguish high yielding features from those, that have low potential, the following investigations were undertaken:

1. Determination of the geology and the structural set-up of the Natal Metamorphic Province on the basis of field mapping, aerial photo and LANDSAT imagery interpretation to identify

possible target features under tectonic extension in the area.

2. Geophysical ground surveys across observed lineaments and structural features to characterise the geophysical responses of potentially water bearing and poor target features.

3. Drilling of exploration boreholes to evaluate the ground water potential of the target features delineated during the research study.

4. Financial analysis to evaluate the extent to which additional exploration in the Natal Metamorphic province is financially viable.

2 SITE LOCALITY

The research area is situated in the Mapumulo District of KwaZulu-Natal, about 75km north of Durban, within the Natal Metamorphic Province (Figure 1). The area is essential rural with underdeveloped infrastructure. Because of poor historical success rates the area has been identified as a critical water deficit area by the Department of Water Affairs and Forestry (DWAF).

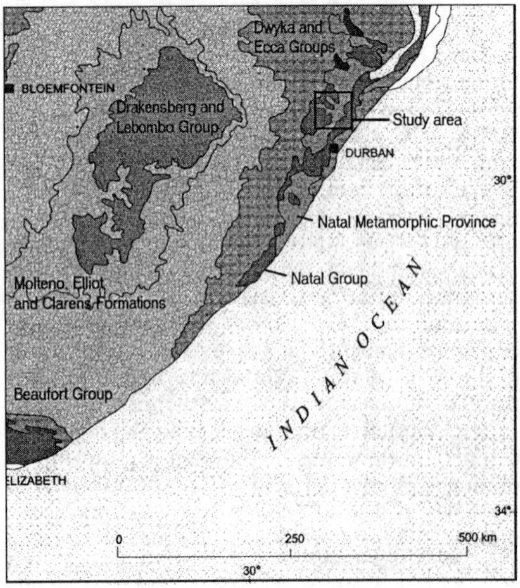

Figure 1. Site locality

3 RESULTS

3.1 Determination of the geology and the structural set-up of the Natal Metamorphic Province

The Natal Metamorphic Province is situated in the coastal belt of KwaZulu-Natal and outcrop is restricted to an erosional inlier that stretches N-S from the Tugela Valley to Port Edward. It consists of rocks varying in age from 1400 to 900Ma. According to Thomas (1989), the Natal Metamorphic province comprises three tectono-stratigraphic terranes separated by major southerly dipping thrust zones and intrusive granitoids which post-date thrusting. These are from the North to the South the Tugela Terrane, the Mzumbe Terrane and the Margate Terrane. The research area is situated within the Mzumbe Terrane. This terrane is made up of major ENE trending linear belts of highly deformed migmatitic rocks of the Mapumulo Group and belts of granitic intrusions.

3.1.1 Structure

An extensive and complex network of faults has been mapped in coastal Natal (Maud 1961). Faulting was related to the break-up of Gondwanaland and is associated with crustal extension normal to the continental margin (SE oriented tension) and/or strain along rotational couples resulting in strike slip movement parallel to the coastline along a transform fault known as the Agulhas Transform Fault. The regional pattern consists of series of fault arcs trending from SW-NE to N-S and swinging E-W along the coast (Figure 2). Theories about the origin of the arcuate fault system are still highly controversial.

A tectonic analysis of faulting in terms of the potential of faults and their related joint systems to be water bearing requires that the orientation of tension, compressional and shear couples be understood. On the Natal coast, this requires an understanding of tectonic processes at work on the Aghulas Transform Fault and the resulting onshore fault and fracture system:

Faulting in Natal can be explained by divergent wrenching on the Aghulas Fault, in which case the maximum extensional force would be directed perpendicular to normal faults, to the SE. In this case, ENE transtensile normal faults are expected to have the maximum SE oriented transtensional strain, while North trending faults are likely to be extensional normal faults. Northeast trending faults are likely to be synthetic and in shear (Neumann, Sami & de Kock, unpubl.).

Field structural mapping outlined the NNE and ENE trending faults with subordinate NE and NW trending faults as the most prominent and open features in the study area. This observation is backed up by remote sensing data. Additionally, it could be proven that ENE-WSW LANDSAT lineaments are of distinctively greater regional significance in the research area, whereas the N-S striking lineaments proved to be of only local importance.

From the above, it can be concluded that exploration needs to target the N-S and ENE-WSW lineaments, which are more likely to be tensional and open, with specific consideration of the ENE trending features as they are of regional significance in the area.

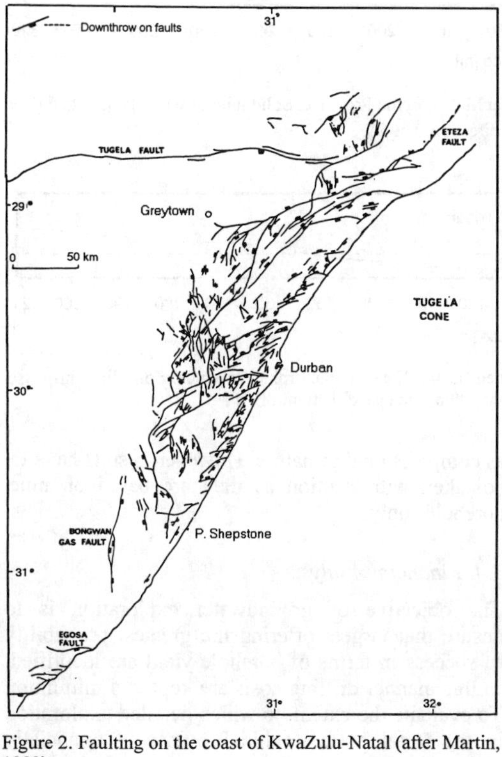

Figure 2. Faulting on the coast of KwaZulu-Natal (after Martin, 1983)

3.2 Ground Based Geophysics

To avoid that large sums of money are being spent on drilling "dry" holes the application of appropriate geophysical techniques is vital in the Natal metamorphic Province. An investigation into the exploration methods used by the consultants in the research area in the past revealed that the majority of the boreholes were sited with the GEONICS EM-34 in conjunction with the Magnetometer. The EM-34 penetrates to about 30m in the given lithology using a 20m coil separation. In retrospect, the hydrocensus showed that 50% of the water strikes in the area occur below 70m. Hence, the instrument does not always penetrate to the required depth under the given hydrogeological conditions and it is suspected that the high failure rate of boreholes may be attributed to the limited depth of penetration of the instrument together with an incomplete understanding of the controls on fractures aquifer occurrence.

Bearing this in mind, the geophysical methods used for the study area involved the Max-Min (HLEM) technique, magnetics and resistivity sounding. The Max-Min (HLEM) technique has a greater penetration than the EM-34 and is suitable for locating fractured aquifers (Botha et al. 1992) as long as the conductivity thickness product of the conductive zone increases with depth. Resistivity soundings were carried out to determine the depth of weathering or fracturing of the bedrock, whereas magnetics was applied to locate contact zones between different rock types, caused for instances by igneous intrusions such as dykes or sills.

The geophysical data showed that the Max-Min technique and the magnetic technique could be used in the field effectively to locate fault lines. In the case that a fault separated two different lithologies, the electromagnetic response was typical to that obtained over the edge of a horizontal conductor, where a positive response peak is located over the resistive host and a negative response peak is obtained over the conductor (Figure 3). If the investigated feature represented a fault within the same geological unit, the weathered fault zone was reflected by a valley discontinuity in the electromagnetic response: a negative peak is flanked by two positive shoulders. The centre of the negative anomaly was located over the top of the weathered fault zone.

Lineaments proved to be mainly associated with zones of more deeply weathered overburden. The electromagnetic responses obtained were mainly ramp discontinuities or trough-like anomalies, where the positive peak is obtained over the resistive host and the negative peak over the conductive overburden (Figure 4).

The magnetic data supported the electromagnetic data showing different magnetic field amplitudes for the host rock and the weathered overburden. Dykes or sill structures proved to be rare and of limited extend in the study area. With a maximum thickness of only a few meters, these structures were only detectable through magnetics and left no impact on the electromagnetic profile of the subsurface. Therefore igneous intrusions were not considered being of hydrological significance in the area.

3.3 Drilling Results

On the basis of the structural and geophysical data nine boreholes were drilled in the research area to investigate different geological and structural features. ODEX and air percussion drilling was used. The drilling sites were chosen to cover fault zones, lineaments in all major joint orientations, as delineated through the structural mapping and deeply weathered zones.

Results:

- Eight of the nine boreholes were successful with an average air lift yield of 2.55l/s (Table 1).
- From the geological logs it was evident that the

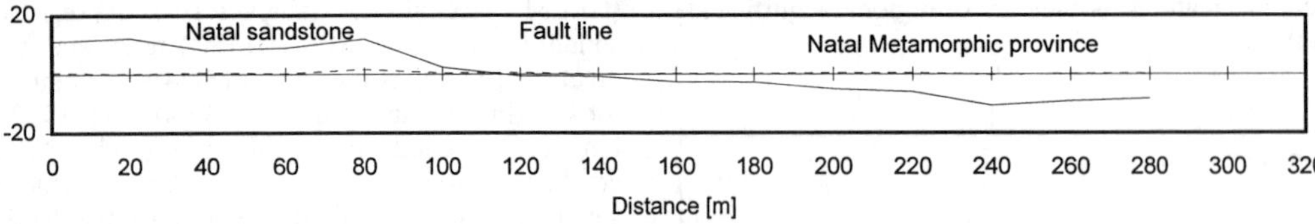

Figure 3. Max-Min electromagnetic profile over fault zone in the Natal Metamorphic Province. Solid line: in-phase [%], dotted line: out-of phase [%]; frequency: 220 Hz, coil separation: 100m, traverse direction: NE-SW.

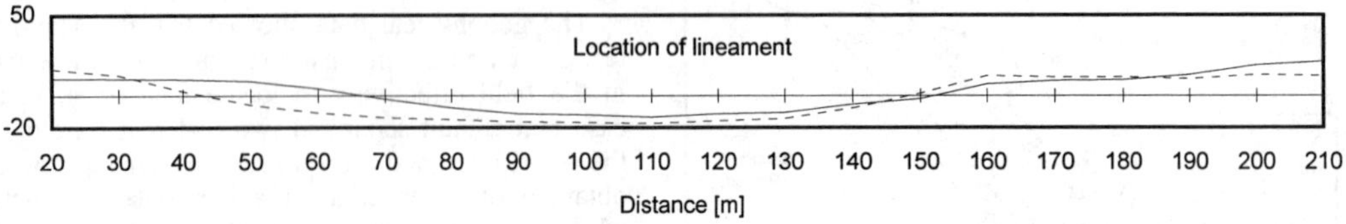

Figure 4. Max-Min electromagnetic profile over LANDSAT lineament in the Natal Metamorphic Province. Solid line: in-phase [%], dotted line: out-of phase [%]; frequency: 3520 Hz, coil separation: 100m, traverse direction: SE-NW.

Natal Metamorphic aquifers occur only within thick weathered overburden and within the fractured bedrock itself.

- High yields are generally derived from aquifers overlain by a thick saturated weathered overburden, or from large fault zones.
- The yield data set exhibited a strong correlation with lineament or fault length. The four highest yielding boreholes are associated entirely with regional scale structural features, whereas local features have a poorer groundwater potential.
- Lithological contacts did not appear to be water bearing neither did the contacts between dyke structures and the host rocks.
- A correlation between lineament or fault azimuth and yield shows that successful boreholes are associated with N-S, E-W and NE-SW striking structural features, orientations of extensional nature. The only unsuccessful borehole was drilled into a NW-SE striking fault, the orientation regarded as compressional in nature. However, results have to be taken with caution as they are based on nine boreholes only.

Table 1. Summary of drilling results

Borehole	Structural Feature	Yield [l/s]
G47151	NE-SW striking mylonite zone	0.9
G47152	NE-SW striking mylonite zone	4.0
G47153	N-S striking lineament, delineated from LANDSAT imagery and aerial photos	1.8
G47154	Prominent N-S striking fault zone	3.3
G47155	E-W striking lineament, delineated from LANDSAT imagery and aerial photos	1.0
G47156	E-W striking wide weathered zone	1.4
G47157	NW-SE striking fault zone	Dry
G47158	NE-SW striking lineament of regional scale	4.0
G47159	E-W striking fault zone	6.6

3.4 *Financial Analysis*

The objective of groundwater exploration is to ensure that targets offering the greatest probability of success in terms of borehole yield are identified. In this manner drilling costs are kept to a minimum. To evaluate the extent, to which detailed exploration in the Natal Metamorphic Province is financially viable in terms of a reduction in overall borehole establishment costs, a financial analysis was undertaken. Three scenarios were investigated. Historical drilling records obtained from the National Groundwater Database (NGDB) of South Africa (46% success rate), which are assumed to represent borehole siting without modern exploration techniques, were compared to data form the Critical Intervention Program of the middle 1990's. This drought relief program achieved a 63% success rate on the basis of the EM-34 method in conjunction with the magnetometer. This level of investigation can be considered as the current industry standard for rural water supply schemes. Both scenarios were compared with the extensive exploration method used in this project, including structural mapping, LANDSAT and aerial photo interpretation and geophysical exploration using the most appropriate method under the given hydrogeological conditions (
Figure 5). Costs for all three cases were extrapolated in order to determine the costs to establish ten yielding boreholes and to determine water production costs in terms of R/l/s. The estimated costs are based on the Council for Geoscience

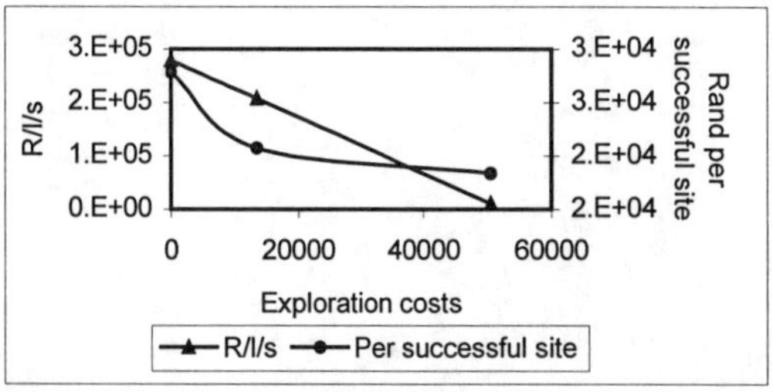

Figure 5. Exploration costs versus success rate

common consulting rates and man-hours expected during the investigation.

The investigations showed that with a larger budget for exploration a substantial cut in costs per successful borehole site or in terms of water production in R/l/s could be achieved. The improved exploration resulted in a 44% cost saving per successful borehole site in the research area. This analysis suggests that a more comprehensive exploration program could in the long-term result in significant cost savings, especially in large regional programs where the costs of regional geological exploration budgets are spread over many boreholes.

4 CONCLUSIONS

The research study showed that successful groundwater exploration for the Natal Metamorphic Province is possible. An evaluation of the structural geology and the hydrogeological conditions of the area together with a suitable geophysical method for the environment pushed the drilling success rate up from a historical 53% to 89%. Beside the improved success rates, the research did result in significantly improved yields. Whereas the historical median yield for the research area is 0.1 l/s, a median yield of 1.8l/s was achieved with the applied exploration method, which accordingly lowers water production costs considerably.

REFERENCES

Botha, W.J., F.E. Wiegmans, J.J. van der Walt & C.J.S. Fourie 1992. Evaluation of electromagnetic exploration techniques in groundwater exploration. *WRC Report* 212/1/92.

Martin, A.K. 1983. Fault pattern of coastal Natal – a result of strike-slip motion during Gondwanaland Break-up. University of Cape Town, Marine Geoscience Unit. *Technical Report* 14: 194-214.

Maud, R.R. 1961. A preliminary review of the structure of coastal Natal. *Trans. Geol. Soc. South Africa* 64: 247-256.

Neumann, I., K. Sami & G. de Kock (unpubl.). Groundwater development for rural water supply in complex and problematic terrain: am assessment of geological controls, geophysical exploration methods and the quatification of exploitation potential. *WRC Progress Report* No. K5/966/0/1.

Thomas, R.J. 1989. A tale of two tectonic terranes. *S. Afr. Jounral of Geol.* 92(4): 306-321.

Groundwater: Past Achievements and Future Challenges, Sililo et al. (eds) © 2000 Balkema, Rotterdam, ISBN 90 5809 159 7

Recharge-discharge relations of the Vipava karst aquifer, Slovenia

M. Petric
Karst Research Institute ZRC SAZU, Postojna, Slovenia

ABSTRACT: Due to highly variable recharge conditions and heterogeneous subsurface flow the hydrodynamic characteristics of karst systems are very complex. One of the research approaches used to define the characteristics of the groundwater flow and storage in karst aquifers is the comparison between the recharge and the discharge of the system. The input function – the recharge can be related to the output function – the discharge by a mathematical relationship called the transfer function that reflects the physical processes taking place within the aquifer.
This method was applied also in the study of a karst aquifer in the recharge area of the Vipava spring in south-western Slovenia. Special attention was focused on the estimation of the recharge function and its time distribution. Then the transfer functions were calculated based on the comparison between the daily values of the recharge components and the measured spring discharge. Combining the transfer functions and recharge components the discharges of the Vipava spring were simulated. The comparison with measured values indicated good correspondence.

1 INTRODUCTION

The karst aquifers are hydrodynamic systems with heterogeneous structure. Different methods are usually applied to define their functioning. The selection of the most appropriate method depends on different factors such as the supposed characteristics of the karst system, expected results, data availability, etc. Based on these criteria the linear system approach was chosen in a study of the karst aquifer in the recharge area of the Vipava spring. Using this method the input and output functions are compared, and no specific details of internal geometry are required. By such approach the conditions in the karst aquifers are not strictly met, but different researches showed that in deference to some limitations it can be applied also in karst. It describes regional scale transport in terms of the distribution of travel or residence times without requiring detailed knowledge of the internal structure in the aquifer (Knisel 1972, Ashton 1966, Dreiss 1989, Long & Derickson 1999, Wicks & Hoke 1999).
Only the input and output have physical meaning, and the relation between both signals is defined by empirical or mathematical functions which reflects the physical processes in the system (Chiew et al. 1993). In the hydrological studies they are called transfer functions or kernels. They may be useful in providing information about karst aquifers, even though they lump the effects of several complex processes (Dreiss 1982).
For lumped, time-invariant linear system the relation between the input and the response of the system can be given by the convolution integral:

$$O(t) = \int_{-\infty}^{\infty} h(t-\tau) \cdot I(\tau)\, d\tau \qquad (1.1)$$

where O(t) is the observed output of the system, I(t) is the input into the system, h(t-t) is the kernel function, t represents time and t is the variable of the integration.
For two isolated, finite series that are usually related, the convolution integral can be written in a discrete form. As the karst system is not exactly linear, and the input and output series can include errors, also this possible error is considered in the equation:

$$O(t) = \Delta t \cdot \sum_{k=0}^{j} h(j-k) \cdot I(k) + \varepsilon(j) \qquad (1.2)$$

where Δt is the time between data points, and ε(j) are the residuals of errors due to measurement, sampling interval, or model assumptions that are not strictly met.

2 RESEARCH APPROACH

The study of the recharge - discharge relation of the karst aquifer in the recharge area of the Vipava spring is also based on the described principles. The hydrological complexity of karst aquifers is primarily the consequence of changeable conditions of the recharge and heterogeneous characteristics of the groundwater flow. Therefore a special attention was focused on the evaluation of these processes. Different known methods for the estimation of the parameters that influence the infiltration (interception on vegetative surfaces and evapotranspiration, snow melting, water storage in soil) were used. Considering the principle of the water balance a model was set to calculate the recharge function. This general approach was further developed into the assumption that such accurate estimation of the recharge produces acceptable data which can be used as the input function of the studied karst system. In the next phase the time distribution of the recharge function in components was considered, too. Namely, the published results of several researches indicate that the epikarst zone as a control factor of such distribution has an important influence on the variable conditions of storage and flow in the karst aquifer. It directs a part of the infiltrated water quickly to the karst network. The rest is retained at the base of the epikarst and slowly feeds the low permeability zones of the aquifer. Therefore the time distribution of the recharge as the input function to slow and fast component was included into the model.

On the other side, the output function of the karst system is represented by the measured daily discharges of the Vipava spring.

The convolution relationship for the described recharge - discharge model can be written as (2.1):

$$Q(j) = \sum_{k=0}^{j} h_f(j-k) \cdot R_f(k) + \sum_{k=0}^{j} h_s(j-k) \cdot R_s(k) + \varepsilon(j)$$

where Q(j) is the spring discharge, $R_f(k)$ and $R_s(k)$ are fast and slow component of the recharge, $h_f(j\text{-}k)$ and $h_s(j\text{-}k)$ are kernels of fast and slow recharge, and $\varepsilon(j)$ are residual errors.

According to the upper equation each measured daily value of the discharge of the Vipava spring in the period from 25 August 1993 to 23 August 1995 can be defined as a function of the estimated values of the slow and fast recharge component and the corresponding kernels. In this way a system of linear equations was set in which the unknown quantities are the components of the transfer functions for the slow and fast recharge. To simplify the calculation the number of terms in each equation was defined according to the supposed duration of the transfer functions. These values were estimated by the comparison of the input and the output functions of the observed karst aquifer. The methods of autocorrelation, cross-correlation and spectral analysis were applied (Mangin 1984).

The least squares method was used to solve this system of linear equations. The result are the transfer functions for the fast and slow recharge components that give the best fit between the input and the output signals and the lowest residual errors. As a criterion for the model adequacy the accuracy of the simulation was set. This accuracy refers to the capability of the transfer function to reproduce the discharges from which it was derived. Therefore, to define the suitability of the model the correspondence between measured and calculated discharges was tested. And according to the structure and properties of the model the influence of the recharge function and its time distribution on the flow and storage in the karst aquifer was evaluated.

Using the described method it is not possible to exclude completely the possibilities of errors in the estimation of the input and output parameters of the system. Also the hydrodynamic conditions of the karst aquifer are not strictly met by the linear system approach. But in spite of this and assuming that the accuracy of the simulation is satisfactory I asses the evaluation of its applicability as sensible.

3 SITE DESCRIPTION

The Vipava karst spring is one of the most important water sources in the south-western Slovenia. In the period from 1961 to 1990 its discharge was in the range between 0,7 and 700 m^3/s, and the mean value was 6,8 m^3/s. The central part of its recharge area is represented by the Nanos and Hrušica karst plateaux. They are mostly built by Cretaceous limestones, only in the border areas also Jurassic limestones and dolomites and Upper Triassic dolomites can be found. Characteristic for the whole area is the thrust structure which is dissected by numerous neotectonic faults.

On three sides this karst background is bordered by impermeable Eocene flysch. More problematic due to specific characteristics of the karst aquifers (unknown paths of the underground water, bifurcation, changeable extent of the recharge areas at different hydrologic conditions, hidden groundwater flow in covered karst, etc.) is the position of the karst watershed in the Hrušica area. Therefore the position of the watershed in this north-eastern part of the studied aquifer was defined on the base of the known geological conditions, the results of the tracing tests, and the method of water balance in which the long-term mean values of precipitation and runoff in the recharge area were compared. The extent of the recharge area was thus estimated to around 150 km^2. This area is mostly karstic, only around 6 % covers the impermeable flysch border. The surface waters from this part sink at the contact with the karst aquifer and also drain towards the Vipava spring.

The carbonate complex in the central part is very thick and deep karst is developed in it. Intensive karstification is expressed by superficial karst features and numerous karst caves, mostly shafts. The main source of the recharge is the precipitation which sinks rapidly on the karstified surface and flows towards the Vipava spring. The precipitations are abundant and the average yearly amount is usually greater than 2000 mm. Also the snowfall is important. The first snow comes in November, and the snow cover lasts sometimes till May.

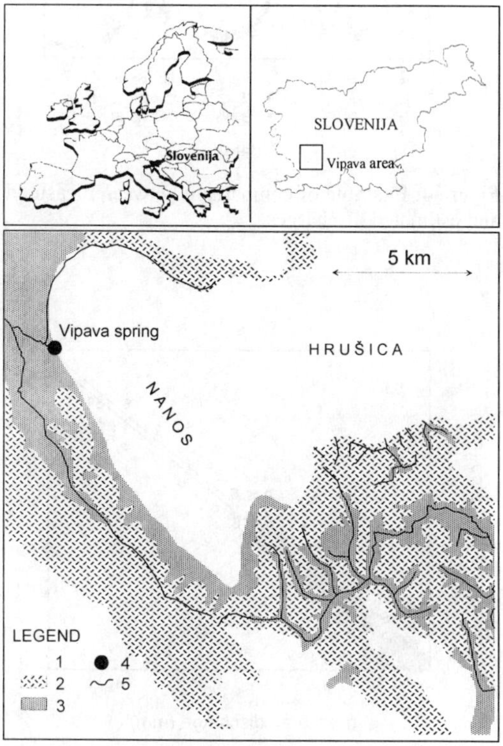

Figure 1. Map of the recharge area of the Vipava spring (1. karst aquifer, 2. impermeable flysch border, 3. porous aquifer, 4. karst spring, 5. surface streams).

4 RESULTS

The model for the estimation of the groundwater recharge was used first. In the observed period of two hydrological years the daily precipitation was regularly measured in three precipitation stations. Since around 4/5 of the recharge area is covered by wood the Rutter model was applied for the estimation of the interception loss. In essence the model calculates a running water balance of the canopy and the trunks (Rutter et al. 1971). Additionally, daily values of evapotranspiration were calculated according to the modified Penman equation (Shaw 1994). The snowpack may cover the recharge area from November to May, therefore the snow melting has an important influence on the amount of the recharge and its time distribution. To estimate these values the empirical equations of U.S. Army Corps of Engineers (1960) were used.

All the described parameters were included in the soil moisture balance. According to the soil hydrological characteristics the storage of water in the soil was estimated. Also the influence of the rapid recharge was tested. It occurs even when the soil moisture deficit exists, and describes the direct recharge of water through macropores and cracks. On the other hand the discharges of the sinking streams from the flysch surroundings were regularly measured, but this secondary recharge represents only around 6 % of the total recharge.

For the calibration of the described model the method of water balance was used. It is based on the assumption that the total amounts of the recharge and the discharge within individual hydrological years are approximately the same. Both values for the period of two hydrological years from 25 August 1993 to 23 August 1995 were compared and with application of some necessary corrections the above mentioned condition was fulfilled (Figure 2).

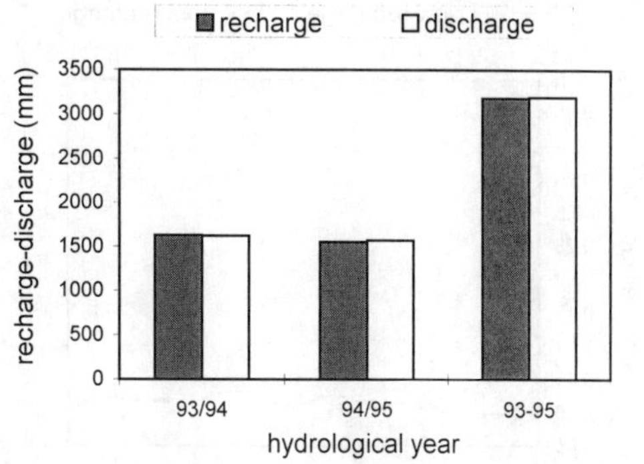

Figure 2: Comparison between recharge and discharge in the period of two hydrological years.

An important influence of the epikarst on the characteristics of the groundwater flow and the storage in karst was proved by several researches (Kiraly et al. 1995, Jeannin & Grasso 1995, Sauter 1992, Mohrlok & Sauter 1997). It directs a part of the infiltrated water quickly to the karst network. The rest is retained at the base of the epikarst and slowly feeds the low permeability zones of the aquifer. For the separation of both components the method of hydrological balance was used (Jeannin & Grasso 1995). The water balance over short periods with an estimation of recharge compared to rapid flow was defined. It gives also the information about the time variability of the distribution into slow and fast recharge component. The water balance is calculated within separate precipitation intervals which includes one flood or a succession of floods. In the observed period of two hydrological years the fast recharge dominates, but some intervals with a prevailing slow component were also found. The average values for the whole period are 62 % for the fast and 38 % for the slow component of the recharge. The estimated values are influenced also by the accuracy of the recharge estimation, the selected procedure of the hydrogram separation, and the chosen method for the evaluation of the fast and the slow component. Yet I assumed that they are suitable for the use in the further analysis. Next to this some caution is necessary and the influence of these parameters on the final results should be tested.

The estimated daily values of slow and fast recharge in the period of two hydrological years between 25 August 1993 and 23 August 1995 represent the input function, and the measured daily discharges of the Vipava spring the output function. Based on the comparison between them the corresponding transfer functions for both recharge components were defined by the method described in the chapter 2 (Figure 3).

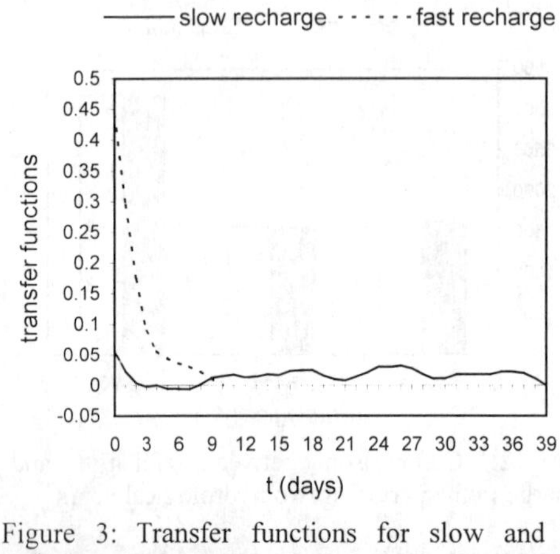

Figure 3: Transfer functions for slow and fast recharge components.

The form of both functions is in accordance to the assumption of the fast recharge, which infiltrates fast into the karst network and results in a fast increase of discharge, and slow recharge, which is retained in the aquifer for a longer time and flows slowly through the less permeable zones.

The convolution equation (2.1) was used to test the adequacy of this model recharge - discharge. The discharges of the Vipava spring were simulated on the base of the defined transfer functions and the estimated values of recharge (Figure 4).

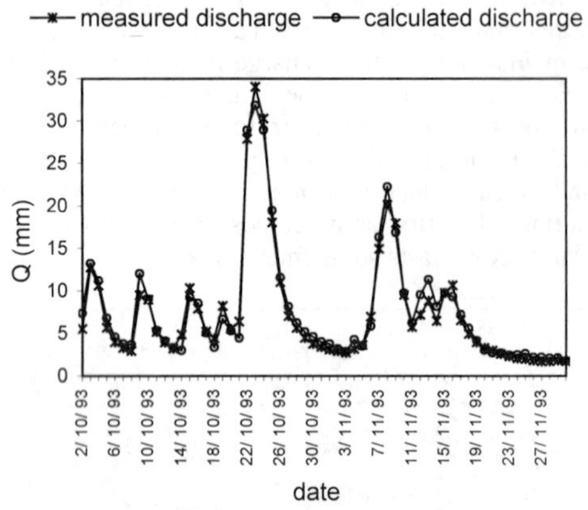

Figure 4: Example of comparison between measured and estimated discharges.

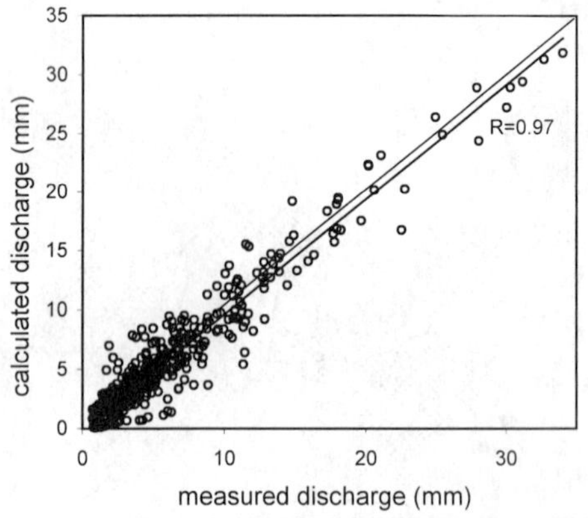

Figure 5: Correlation between measured and estimated values.

The accuracy of the simulation was tested, which refers to the capability of the transfer function to reproduce discharges from which it was derived. Different statistical methods were used to compare the measured and the estimated values of discharge (Figure 5). The coefficient of efficiency, which reflects the capability of the model to reproduce the measured values, is 0,9. According to the results of Chiew et al. (1993) such a model can be evaluated as perfectly acceptable. Still some deviations are present as a consequence of the assumption of the linearity of the system, or the errors in the estimation or measurement of the input or output data.

5 CONCLUSIONS

In spite of some limitations of the described method, which somewhat simplifies the actual conditions in the karst aquifers, certain conclusions about the characteristics of the studied karst system can be made. They are based on the satisfactory correspondence between the measured and the calculated discharges of the Vipava spring. An important influence of vegetation and the processes in the atmosphere, soil and epikarst on the amount and time distribution of the infiltration and through this also on the hydrodynamic characteristics of the aquifer was proved and evaluated by the introduction of the recharge function instead of the precipitation and its separation in the slow and the fast component. Still some discrepancies between the measured and the estimated values remain. Not just as a consequence of the limitations of the model, but also because of the errors in the estimation or measurement of its input and output data.

6 REFERENCES

Ashton, K. 1966. The analysis of flow data from karst drainage systems. *Transactions of the Cave Research Group.* 7: 163-203.

Chiew, F.H.S. & M.J. Stewardson & T.A. McMahon 1993. Comparison of six rainfall-runoff modelling approaches. *Journal of Hydrology.* 147: 1-36.

Dreiss, S. 1982. Linear kernels for karst aquifers. *Water Resources Research.* 18: 865-876.

Dreiss, S. 1989: Regional scale transport in a karst aquifer. *Water Resources Research.* 25: 117-134.

Jeannin, P.Y. & A.D. Grasso 1995. Estimation des infiltrations efficaces journalieres sur le bassin karstique de la Milandrine (Ajoie, JU, Suisse). *Bulletin d'Hydrogeologie.* 14: 83-93.

Kiraly, L. & P. Perrochet & Y. Rossier 1995: Effect of the epikarst on the hydrograph of karst springs: a numerical approach. *Bulletin d'Hydrogeologie.* 14: 199-220.

Knisel, W.G. 1972. Response of karst aquifers to recharge. *Hydrology Paper 60.* Fort Collins: Colorado State University.

Long, A.J. & R.G. Derickson 1999. Linear systems analysis in a karst aquifer. *Journal of Hydrology.* 219: 206-217.

Mangin, A. 1984: Pour une meilleure connaissance des systemes hydrologiques a partir des analyses correlatoire et spectrale. *Journal of Hydrology.* 67: 25-43.

Mohrlok, U. & M. Sauter 1997: Modelling groundwater flow in a karst terrane using discrete and double-continuum approaches-importance of spatial and temporal distribution of recharge. *Proceedings of the 12th International Congress of Speleology, Switzerland.* 2: 167-170.

Rutter, A.J. & K.A. Kershaw & P.C. Robins & A.J. Morton 1971. A predictive model of rainfall interception in forests. *Agric. Meteorol.* 9: 367-380.

Sauter, M. 1992: Quantification and forecasting of regional groundwater flow and transport in karst aquifer (Gallusquelle, Malm, SW Germany). *Tübingen geowissenschaftliche Arbeiten.* Tübingen

Shaw, E. 1994: *Hydrology in Practice.* London: Chapman & Hall.

U.S. Army Corps of Engineers 1960. Runoff from Snowmelt. *Engineering and Design Manuals,* EM 1110-2-1406.

Wicks, C.M. & J.A. Hoke 1999. Linear systems approach to modelling groundwater flow and solute transport through karstic basins. *Karst Modelling.* Charlottesville: Karst Waters Institute Special Publication.

Groundwater: Past Achievements and Future Challenges, Sililo et al. (eds) © 2000 Balkema, Rotterdam, ISBN 90 5809 159 7

Water table variation due to time-varying recharge and withdrawal

S.N. Rai & A. Manglik
National Geophysical Research Institute, Hyderabad, India

ABSTRACT: An analytical solution of a 2-D Boussinesq equation is developed to predict the spatial and temporal variations of the water table in response to time-varying recharge and withdrawal from any number of sources (recharging basins) and sinks (pumping wells). The rate of recharge (or withdrawal) is approximated by a series of linear elements of different slopes and lengths depending on the nature of variation of the recharge rate. By this approach any complex nature of recharge rate for any number of recharge operations can be approximated with more accuracy. The flow system is characterized by mixed boundary conditions. The solution is obtained by using extended finite Fourier cosine transform. Application of the solution in the prediction of water table variation and in the sensitivity analysis is illustrated with the help of an example problem.

1 INTRODUCTION

Recharging and pumping are the integral parts of any scheme of groundwater resources development and have significant influence on the dynamic behaviour of water table. Therefore, for the proper assessment, development, and management of groundwater resources, a prior knowledge of water table fluctuation in response to the proposed schemes of recharging and pumping is essential. Mathematical models play an important role in estimation of water table fluctuation. It helps in making the judicious selection of an appropriate scheme of recharging and pumping out of many proposed schemes to achieve the preset objectives of water resources management.

Many analytical models have been reported in the literature to predict the water table fluctuation in response to recharge from rectangular and circular basins. Warner et al. (1989) have presented a review of some of these models. Most of these models are based on the assumption of constant rate of recharge. However, the recharge rate largely depends on the infiltration rate which is known to vary with time due to many factors. Initially it decreases due to swelling and dispersion of the soil particles. This follows with increase due to release of the entrapped air and after reaching to a maximum value the infiltration rate again starts decreasing due to clogging of soil pores below the bottom of the basin. Recharge rate also follows more or less similar pattern of variation and when it falls below a prescribed low value the recharge operation is discontinued for some time. After drying and, if necessary, scrapping of the silted bottom of the basin the recharge rate is brought back almost to its initial value for the next cycle of recharge (Bear 1979, Detay 1995, Dickenson et al. 1995, Mousavi and Rezai 1999). Zomorodi (1991) has demonstrated that the solutions based on the assumption of constant recharge rate failed to predict the decline of water table which is due to decrease in the recharge rate. Therefore, it would be more appropriate to consider the recharge rate as time-varying to simulate actual field conditions. Some models have been developed to predict water table fluctuations from single basin (Rai et al. 1994, 1998, Rai and Singh 1995, 1996, Manglik et al 1997). Recently, Rai and Manglik (1999) presented a model to describe water table fluctuation due to time-varying recharge from four basins situated at equidistance from the respective coordinate axes.

In this work, an analytical model is presented to predict water table fluctuation in the presence of time-varying recharge and/ or withdrawal from any number of basins and wells.

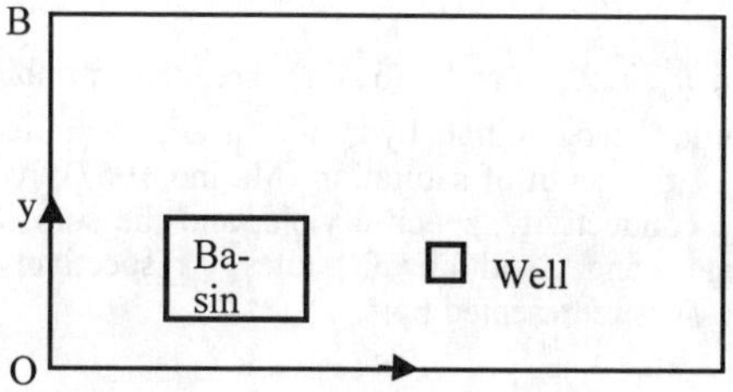

Figure 1. A plan view of recharge and withdrawal scheme.

2 MATHEMATICAL FORMULATION

Groundwater flow in an unconfined aquifer of length A and width B, as shown in Figure 1, is considered. Though only one recharge basin and one well are considered for illustration purposes but in principle recharge and withdrawal are considered from any given number of basins and wells. The wells are treated as discharge basins in rectangular form like recharge basins. The flow system is bounded by prescribed head boundary condition at $x=A$ and $y=B$ and zero flux condition along the coordinate axes. These two types of boundary conditions are commonly used in the modeling of groundwater flow problems. The prescribed head boundary condition is satisfied at sufficiently large distance where the static initial hydraulic head is not affected by recharging/ pumping. The zero flux boundary condition is satisfied in the presence of groundwater divide or impermeable boundary. Even in the absence of groundwater divide or impermeable boundary the zero flux boundary conditions is satisfied at sufficiently large distance where the recharging/ pumping does not alter the initial hydraulic head. To facilitate the mathematical treatment of the problem it is assumed that the aquifer is homogeneous and isotropic and initial water table is horizontal. Groundwater flow under Dupuit's assumption can be described by the following 2-D Boussinesq equation (Hantush 1967):

$$\frac{\partial^2 H}{\partial x^2}+\frac{\partial^2 H}{\partial y^2}+\frac{2}{K}P(x,y,t)=\frac{1}{a}\frac{\partial H}{\partial t} \tag{1}$$

The associated initial and boundary conditions are:

$$H(x,y,0)=0 \tag{2}$$

$$\frac{\partial H(0,y,t)}{\partial x}=\frac{\partial H(x,0,t)}{\partial y}=0 \tag{3}$$

$$H(A,y,t)=H(x,B,t)=0 \tag{4}$$

Here,

$H=h^2-h_0^2$, $a=K\bar{h}/S$,

and $h, h_0, \bar{h}, K, S$ and $P(x,y,t)$ are the variable hydraulic head, initial hydraulic head, weighted mean of the depth of saturation (Marino 1967), hydraulic conductivity, specific yield, and the sum of recharge and withdrawal rates, respectively. $P(x,y,t)$ is represented by:

$$P(x,y,t)=\begin{cases}\sum_{i=1}^{N}P_i(t) & \text{for } x_{i1}\le x\le x_{i2}, y_{i1}\le y\le y_{i2}\\ o & \text{elsewhere}\end{cases} \tag{5}$$

where N is the total number of basins, $P_i(t)$ is the time-varying recharge (or pumping) rate for the ith basin and $x_{i1}, x_{i2}, y_{i1}, y_{i2}$ are the coordinates of ith basin. $P_i(t)$ is positive for recharge rate and negative for the pumping rate and is approximated in the following form:

$$P_i(t)=\begin{cases}r_{ij}t+c_{ij} & t_j\ge t\ge t_{j+1},\ j=1,2,\dots,k-1\\ r_{ik}t+c_{ik} & t\ge t_k\end{cases} \tag{6}$$

where r_{ij} and c_{ij} are the slope and intercept of the jth linear element of the ith basin, and k is the number of elements. Advantage of this type of approximation is that any complex nature of variation of recharge (or pumping) rate for any number of recharge (or pumping) cycle can be approximated with more accuracy. Solution of the prediction problem lies in the solution of eq.(1) subject to the given initial and boundary conditions. The solution is obtained by using the extended finite Fourier cosine transform defined as:

$$C_e[H(x,y,t)]=\bar{H}(m,n,t)=\int_0^B\int_0^A H(x,y,t)\,.\cos\frac{(2m+1)\pi x}{2A}\cos\frac{(2n+1)\pi y}{2B}dxdy \tag{7}$$

where m and n are integers.

After applying this transformation and using conditions (3), (4), and (5), eq.(1) reduces to the following ordinary differential equation:

$$\frac{d\bar{H}}{dt}+\alpha\bar{H}=\sum_{i=1}^{N}g_i(m,n)P_i(t) \tag{8}$$

where

$$\alpha=\frac{a\pi^2}{4}\left[\frac{(2m+1)^2}{A^2}+\frac{(2n+1)^2}{B^2}\right] \tag{9}$$

and

$$g_i(m,n)=\frac{8aAB}{K\pi^2(2m+1)(2n+1)}\left[\sin\frac{(2m+1)\pi x_{i2}}{2A}-\sin\frac{(2m+1)\pi x_{i1}}{2A}\right]\cdot\left[\sin\frac{(2n+1)\pi y_{i2}}{2B}-\sin\frac{(2n+1)\pi y_{i1}}{2B}\right] \tag{10}$$

The solution of eq.(8) in the Fourier domain subject to the initial condition (2) is given by:

$$\overline{H}(m,n,t)=\sum_{i=1}^{N} g_i(m,n)\left[\sum_{j=1}^{k-1} R_{ij}+R_{ik}\right] \tag{11}$$

where

$$\begin{aligned}R_{ij}&=\frac{r_{ij}}{\alpha}\left[t_{j+1}\exp\{-\alpha(t-t_{j+1})\}-t_j\exp\{-\alpha(t-t_j)\}\right]\\&-\left(\frac{r_{ij}}{\alpha^2}-\frac{c_{ij}}{\alpha}\right)\left[\exp\{-\alpha(t-t_{j+1})\}-\exp\{-\alpha(t-t_j)\}\right]\end{aligned} \tag{12}$$

$$\begin{aligned}R_{ik}&=\frac{r_{ik}}{\alpha}\left[\mathrm{t}-t_k\exp\{-\alpha(t-t_k)\}\right]\\&-\left(\frac{r_{ik}}{\alpha^2}-\frac{c_{ik}}{\alpha}\right)\left[1-\exp\{-\alpha(t-t_k)\}\right]\end{aligned} \tag{13}$$

The space domain solution, $H(x,y,t)$ can be obtained by applying the following extended inverse finite Fourier cosine transform:

$$H(x,y,t)=\frac{4}{AB}\sum_{m=0}^{\infty}\sum_{n=0}^{\infty}\overline{H}(m,n,t)\cos\left(\frac{(2m+1)\pi x}{2A}\right)\cos\left(\frac{(2n+1)\pi y}{2B}\right) \tag{14}$$

After substituting the value of $\overline{H}(m,n,t)$ from eq.(11) into eq.(14), the desired solution for the variable water table height is obtained in the following form:

$$\begin{aligned}h^2(x,y,t)&=h_0^2+\frac{32a}{K\pi^2}\sum_{m=0}^{\infty}\sum_{n=0}^{\infty}\frac{1}{(2m+1)(2n+1)}\cos\left(\frac{(2m+1)\pi x}{2A}\right)\cos\left(\frac{(2n+1)\pi y}{2B}\right)\\&\sum_{i=1}^{N}\left[\sin\frac{(2\mathrm{m}+1)\pi x_{i2}}{2A}-\sin\frac{(2\mathrm{m}+1)\pi x_{i1}}{2A}\right]\left[\sin\frac{(2\mathrm{n}+1)\pi y_{i2}}{2B}-\sin\frac{(2\mathrm{n}+1)\pi y_{i1}}{2B}\right]\\&\left[\sum_{j=1}^{k-1}R_{ij}+R_{ik}\right]\end{aligned} \tag{15}$$

2.1 *Solution for constant rate of recharge*

As a special case of constant rate of recharge, P_0 from only one basin, i.e. i=1, the number of linear element k reduces to 1 with $r_{ij}=0$, $t_k=0$, and $c_k=P_0$. This reduces eq.(13) to the following form:

$$R_{11}=\frac{P_0}{\alpha}\left[1-\exp(-\alpha t)\right] \tag{16}$$

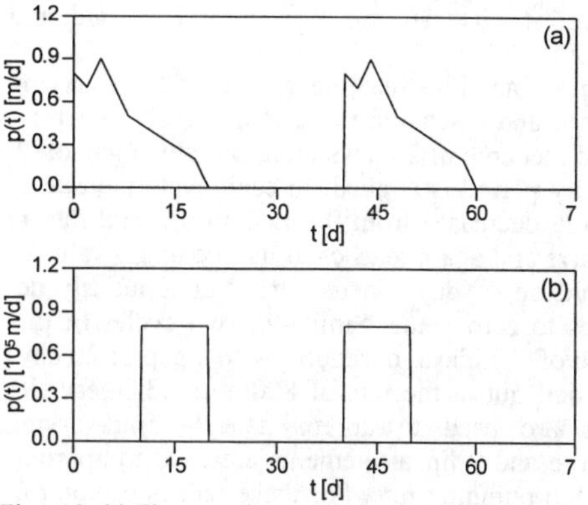

Figure 2. (a) Time-varying recharge rate and (b) pumping rate.

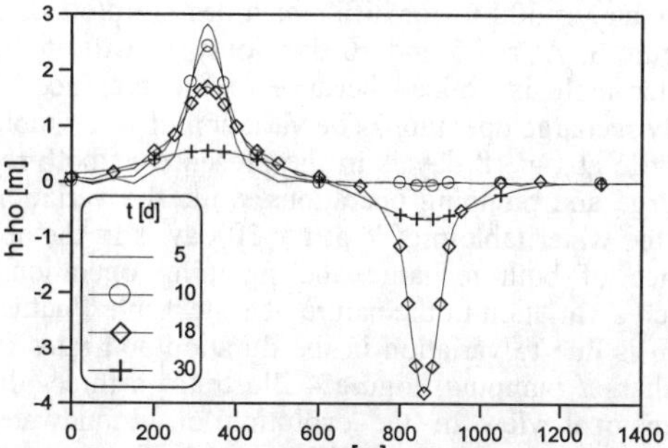

Figure 3. Water table profiles at t= 5, 10, 18, and 30 days.

and the solution (15) becomes:

$$\begin{aligned}h^2(x,y,t)&=h_0^2+\frac{32aP_0}{K\pi^2}\sum_{m=0}^{\infty}\sum_{n=0}^{\infty}\frac{1}{(2m+1)(2n+1)}\cos\left(\frac{(2m+1)\pi x}{2A}\right)\cos\left(\frac{(2n+1)\pi y}{2B}\right)\\&\left[\sin\frac{(2\mathrm{m}+1)\pi x_2}{2A}-\sin\frac{(2\mathrm{m}+1)\pi x_1}{2A}\right]\left[\sin\frac{(2\mathrm{n}+1)\pi y_2}{2B}-\sin\frac{(2\mathrm{n}+1)\pi y_1}{2B}\right]\\&\left[1-\exp\{-\alpha t\}\right]/\alpha\end{aligned} \tag{17}$$

Eq.(17) is identical to eq.(17) of Rai and Singh (1996).

3 NUMERICAL RESULTS AND DISCUSSION

An example of water table fluctuation due to recharging from one basin of 60 m x 40 m dimension centered at (330m, 320m) and pumping from one well of 10 cm x 10 cm dimension centered at (860m, 320m) is considered to demonstrate the application of eq.(15) in the prediction of water table fluctuation. Numerical values of other controlling parameters are: $A = B =$ 10 km, $h_0 =$ 25m, $K =$ 6m/d, and $S =$0.25.

Figure 2a,b illustrate the nature of time-varying recharge and discharge rates, respectively. Two cycles of recharge of 20 days duration with a gap of 20 days dry period is applied. In each cycle the rate of recharge decreases from 0.8 m/d to 0.7 m/d during two days and again reaches to its maximum value of 0.9 m/d on 4th day. Thereafter it continuously decreases to zero value. Similarly, two cycles of discharge of 10 days duration with a gap of 20 days is carried out at the rate of 800m^3/d. 13 linear elements are used to approximate the time-varying recharge and 9 linear elements are used to approximate the pumping rate. For these values of controlling parameters the water table profiles parallel to x-axis at y = 300 m for different times are plotted in Figure 3. At t = 5 and 10 days only growth of the water table is noticed because of the presence of only recharge operation. The variation of water table profile at t = 18 day is in the presence of both recharge and pumping operations while the variation of the water table profile at t = 30 day is in the absence of both recharge and pumping operations. Such a variation in the nature of water table fluctuation is due to variation in the duration and rates of recharge/ pumping. Figure 4 illustrates a three- dimensional view of the evolution of groundwater mound below recharge basin and cone of depression at the site of the pumping well at t = 18 day. The maximum growth and depression of the water table are 1.7 m and 3.8 m, respectively.

4 CONCLUSION

An analytical model (eq.15) is developed to predict the water table fluctuation in the presence of recharge and withdrawal from any numbers of basins and pumping wells. Numerical results reveal that the variations in the rates of recharge and pumping have significant effects on the nature of water table fluctuation. This model can be also used for testing the validity of numerical models and for sensitivity analysis of the effects of variation in the controlling parameters on the water table fluctuation. Such analysis is useful for selecting an appropriate scheme of recharging and pumping for sustainable development of groundwater resources. Accurate estimation of time-varying recharge rate is a major problem in groundwater resources management. If the time-history of water table variation at an observation well is known then the eq.(15) can be used for the estimation of varying recharge rate by using trial and error method such that the computed water table variation matches well with the observed one.

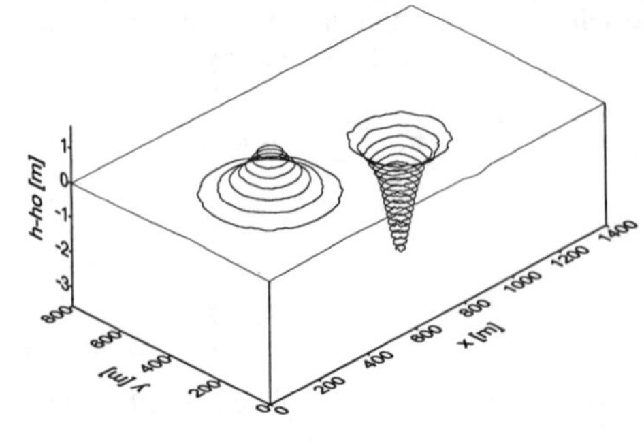

Figure 4. A 3-D view of growth and depression of water table.

REFERENCES

Bear, J. 1979. *Hydraulics of Groundwater*. New York: McGraw-Hill.

Detay, M. 1995. Rational groundwater reservoir management, the role of artificial recharge. In A.I. Johnson & R.D.G. Pyne (eds), *Artificial recharge of groundwater II*: 231-240. New York:ACSE.

Dickenson, J.M. & S.B. Bachman 1995. The optimization of spreading ground operations. In A.I. Johnson & R.D.G. Pyne (eds), *Artificial recharge of groundwater II*: 630-639. New York:ACSE.

Hantush, M.S. 1967. Growth and decay of ground water mounds in response to uniform percolation. *Water Resour. Res.* 3:227-234.

Manglik, A., S.N. Rai & R.N. Singh 1997. Response of an unconfined aquifer induced by time varying recharge from a rectangular basin. *Water Resour. Mgmt.* 11:185-196.

Marino, M.A. 1967. Hele-shaw model study of the growth and decay of ground water ridges. *J. Geophys. Res.* 72:1195-1205.

Mousavi, S.F. & V. Rezai 1999. Evaluation of scraping treatments to restore initial infiltration capacity of three artificial recharge projects in central Iran. *Hydrogeol. J.* 7:490-500.

Rai, S.N., A. Manglik & R.N. Singh 1994. Water table fluctuation in response to transient recharge from a rectangular basin. *Water Resour. Mgmt.* 8:1-10.

Rai, S.N. & R.N. Singh 1995. Two-dimensional modelling of water table fluctuation in response to localized transient recharge. *J. Hydrol.* 167:167-174.

Rai, S.N. & R.N. Singh 1996. On the prediction of groundwater mound formation due to transient recharge from a rectangular area. *Water Resour. Mgmt.* 10:189-198.

Rai, S.N., D.V. Ramana & R.N. Singh 1998. On the prediction of groundwater mound formation in response to transient recharge from circular basin. Water Resour. Mgmt. 12:271-284.

Rai, S.N. & A. Manglik 1999. Modelling of water table variation in response to time-varying recharge from multiple basins using the linearised Boussinesq equation. *J. Hydrol.* 220:141-148.

Warner, J.W., D. Molden, M. Chehata & D. K. Sunada 1989. Mathematical analysis of artificial recharge from basins. *Water Resour. Bull.* 25:401-411.

Zomorodi, K. 1991. Evaluation of the response of a water table to a variable recharge. *Hydrol. Sci.* J. 36:67-78.

Groundwater: Past Achievements and Future Challenges, Sililo et al. (eds) © *2000 Balkema, Rotterdam, ISBN 90 5809 159 7*

Hydrogeology of Nowergup Lake, Perth, Western Australia

Anwarul Qadeer Rathur
School of Applied Geology, Curtin University, Perth, W.A., Australia
Modesta Banda
Department of Water, Lilongwe, Malawi

ABSTRACT: Nowergup Lake is part of the wetlands system making up the Plio-Holocene interdunal depression of the Swan Coastal Plain. It is situated about 35 km north of Perth, is north-south trending and is a shallow flow-through lake. The lake area is approximately 60 ha and is 4 m deep with relatively steep sides. Nowergup Lake is surrounded by woodland, horticultural areas and farmland.Peat, mud and sand make up the sediments of Nowergup Lake and these unconformably rest on Tertiary-Quaternary sandy-calcareous Superficial Formation. This in turn rests unconformably on Cretaceous Leederville Formation. A total of 37 monitoring bores were drilled to varying depths (shallow, intermediate and deep) at distances up to 600m from the lake. Groundwater levels were monitored in some of these bores on a regular basis. Bathymetric map of the lake was supplied by Water Corporation of Western Australia which shows the lake to be oval in shape. Hydrogeologically Nowergup Lake is part of the regional groundwater mound in the Superficial Formation known as Gnangara Mound. Groundwater inflow is through the eastern margin and outflow occurs through the western margin of the lake. Water levels in the lake are affected by the drought-wet cycles in the Swan Coastal Plain and the lake has been artificially maintained by pumping deeper confined groundwater into the lake during hot, dry months during 1990s. Water balance studies of the lake show that out of an average annual input of 2.5 x 10^6 m^3 of water, 79.6% is from groundwater inflow and 9.88% from precipitation. These studies also show that evapotranspiration represents 25% and groundwater outflow represents 75% of the water loss from the lake.

1 INTRODUCTION

Nowergup Lake is part of the wetlands system making up the Plio-Holocene interdunal depression of the Swan Coastal Plain. It is situated 35 Km north of Perth and is a shallow flow-through lake. Periphery of the Lake is undisturbed and vegetation consists of swamp woodlands, banksia, jarrah, tuart. Nowergup Lake is recognised as an "A" class natural reserve and is of environmental significance. Hydrogeological and water balance studies of these wetlands can help in management and protection of these natural features. However, different competing land uses as well as groundwater abstraction may result in deterioration and endangering these wetlands.

1.1 *Physiography*

Nowergup Lake is one of the chain of lakes called Wanneroo Chain situated in northern Perth region(Fig. 1). Topography of the area is low and elevation not exceeding 100m AHD. It is 4m deep and covers an area of 60 hectares. Swamp woodland, banksia, jarrah and tuart form fringing vegetation of the lake. The lake is situated on the steepest gradient of the Gnangara Mound(70m above sea-level) groundwater flow system and has relatively steep sides. Bathymetric map of the lake has been supplied by the Water Authority of W. Australia (WAWA) (now Water Corporation), which indicates an oval shape for lake basin. (Martin 1988).

1.2 *Climate*

Climate of Nowergup Lake is mediteranean in nature with cool, wet winters and dry, warm summers. About 90% of annual rainfall occurs during winter and it exceeds Class A pan evaporation during May-August period. For Perth region av. ann. rainfall is 870 mm and pan evaporation is 1819 mm. Rainfall data used is taken from the Wanneroo weather station whereas evaporation data of Perth Airport station situated 40 km SE of Nowergup Lake, is used for this study.

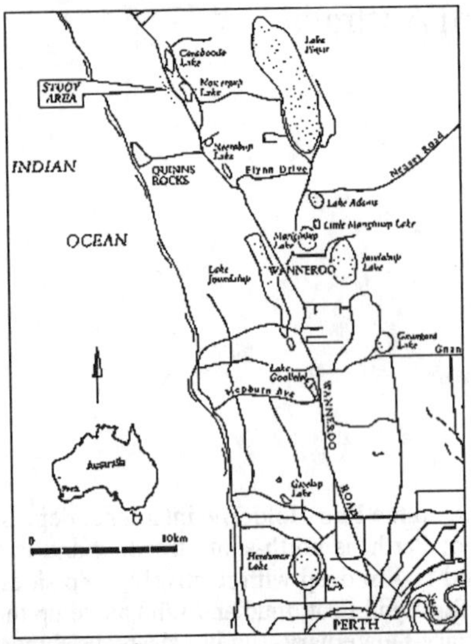

Figure 1. Location map.

2 METHODOLOGY

Geology and hydrogeology of Nowergup Lake has been monitored using 37 monitoring bores (shallow, intermediate and deep) around the lake at 23 different sites and ranged in depth shallow from 3 m below water table upto 60 m for deep bores.(Fig. 2). Various available logs of all monitoring bores were used to interpret the geology of the area. Information regarding the lake sediments were obtained from literature and Commonwealth Scientific and Industrial Research Organisation (CSIRO).

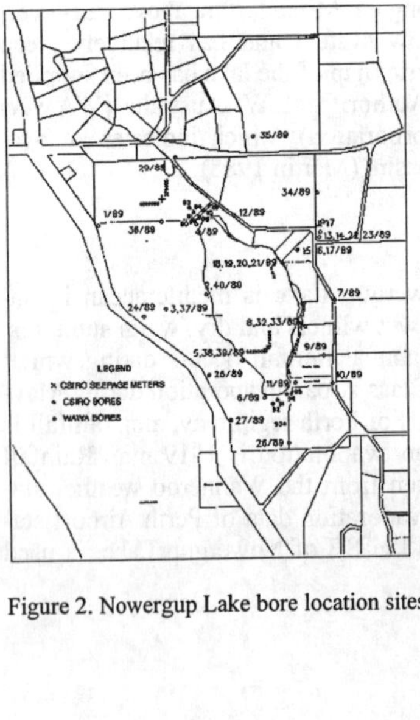

Figure 2. Nowergup Lake bore location sites (WAWA, 1992)

WAWA has measured the water levels in these bores since 1990. However, CSIRO also carried out some monitoring from 1989 to 1991.

2.1 *Geology*

The study area is part of the N-S trending Perth Basin on western margin of Australia.(Playford, et. al., 1976) It is a half-graben bounded on the east by the north- trending Darling Fault, which separates it from the Yilgran Block. Phanerozoic sediments in the Perth Basin exceeds 15000 m in thickness. Onshore central part of the Perth Basin is occupied by the Swan Coastal Plain which is a low-lying, gently undulating area made up of Holocene and Pleistocene coastal-dunes and shoreline deposits, with belts of alluvium in front of Darling Scarp and along the river valleys. Quaternary 'Superficial' Formations cover most of the Swan Coastal Plain and vary in thickness up to 100 m and unconformably overlies Cretaceous and Early Tertiary (Leederville Formation). It consists of sand, silt and clay, in varying proportions. In the east this formation is clayey whereas in the west sands laterally pass into limestone near the coast.(Smith and Allen 1987). 'Superficial' Formations consists in order of deposition; Ascot Formation, Yagoona Formation, Guildford Clay, Gnangara Sand, Bassendean Sand, Tamale Limestone, Beacher Sand and Safety Bay Sand. (Fig. 3)

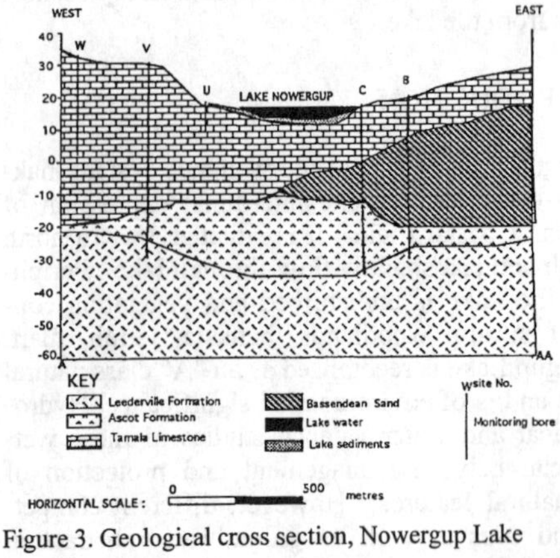

Figure 3. Geological cross section, Nowergup Lake

Nowergup Lake sediments described by Froend (1993) are relatively a function of distance from the shore i.e. closer to the shore, the sediments are mostly sand up to a depth of 20 cm, whereas, at some distance from the shore, the deposits are peat, dark and light grey sandy peat, black fibrous peat, and silt. These sediments are in contact with Bassendean Sand to the east and Tamala Limestone to the West.

2.2 *Hydrogeology*

The main aquifers of the Swan Coastal Plain are the unconfined 'Superficial' Formations aquifer, confined Leederville and Yarragadee aquifers.

2.2.1 *'Superficial' Formation Aquifer*

'Superficial' Formation forms a heterogeneous unconfined aquifer which varies both laterally and vertically. The aquifer has a saturated thickness of up to 60m and contains 10% to 35% by volume of groundwater.(Cargeeg, *et. al.*, 1987). The groundwater in the 'Superficial' Formations occurs in three major flow systems bounded by hydraulic boundaries formed by major rivers-- Gnangara Mound in the north, Jandakot Mound in the south, and Cloverdale flow system to the east. Hydrogeolgically, Nowergup Lake is part of the Gnangara Mound. The Gnangara Mound is 70m AHD and contains about $25.8 \times 10^{9}\ m^{3}$ water within 'Superficial' Formations aquifer and is in hydraulic connection with Nowergup Lake, inflow taking place through eastern margin and groundwater outflow occurring at the western margin of the lake.

2.3 *Water balance of the lake*

The following equation is used for the water budget of a lake:

$$P + Gw_i + R_a + S_i = E + GW_o + S_o + A \pm Ds \qquad (1)$$

where

P: Precipitation, GW_i: Groundwater inflow to the lake, R_a: Artifitial Recharge, S_i: Surface inflow to the lake, E: Evapotranspiration, GW_o: Groundwater outflow, S_o:Surface outflow, A: Abstraction and Ds: Change in storage

2.3.1 *Precipitation and evapotranspiration:*

Direct annual precipitation on the lake surface, is a product of lake area chosen as the average for 12 months, and annual rainfall. (P = R (mm) × lake area). Similarly, evapotranspiration from lake surface, is evaporation data, lake area, and pan coefficient factor. (E = Evaporation x lake area x Pan coefficient) (Table 1).

Table 1: Precipitation on and Evapotranspiration from the Nowergup Lake.

Year	Lake Area (m^2)	Rainfall (mm)	Precipitation on Lake (m^3)	Class A Pan Evaporation (mm)	Evapotranspiration from Lake (m^3)
90-91	334,400	583.2	195,022	2,036	544,671
91-92	334,400	1,044.5	349,281	1,946	520,647
92-93	334,400	801.4	267,988	2,004	536,217
93-94	334,400	735.4	245,917	2,428	649,512

2.3.2 *Groundwater inflow and outflow:*

(1) lake-aquifer interaction

Lakes with regard to their interaction with aquifer are divided into the following categories:

(a) recharge lakes

These are those lakes which recharge the aquifer and are situated high in the landscape.

(b) discharge lakes

These lakes abstract water from the aquifer and are located low in the landscape.

(c) flow-through lakes

These lakes receive water in their upstream and recharge aquifer in their downstream side and these are window to water table. Nield (1990) recognises six more configurations based on lake-groundwater systems.

2.3.3 *Nowergup lake hydrograph & water table*

This hydrograph shows that lake levels have fluctuated between 16.25m and 17.5m from 1973 to 1994 and it represents the normal seasonal variations(Fig: 4). As water table is in hydraulic continuity with the lake, hence the water table and the level in lake respond to periods of high rainfall or evapotranspiration. It appears that lake levels are lower than the water table in upstream side and higher in downstream side, thus indicating this lake to be a flow-through lake. This may be attributed to low transmissivities of the sediments in the capture zone and high transmissivities in the coastal limestones in the release zone.

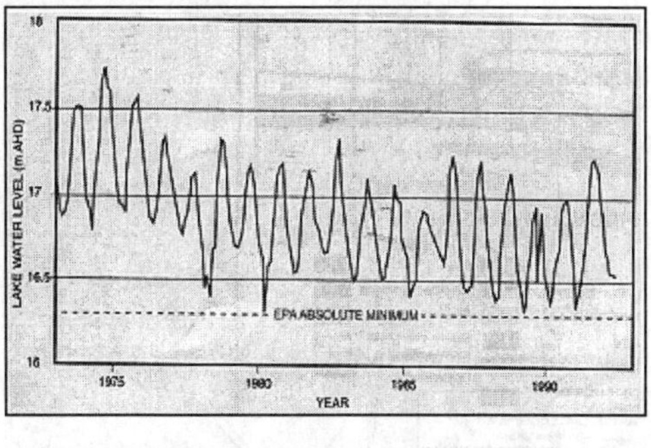

Figure 4 Hydrograph of water level for Nowergup Lake (WAWA, 1992)

Groundwater inflow and outflow were calculated using flow net analyses.The formula to calculate the groundwater inflow is :

$$Q = K.I.L.B$$

where: **Q**: Groundwater inflow **K**: Hydraulic conductivity of the aquifer
L: Width of flow section entering the lake
B:Thickness of aquifer contributing to the lake water and **I**: Hydraulic gradient of the aquifer in vicinity of upstream side of the lake

2.3.4 *Hydraulic conductivity (k)*

Based on available data, the hydraulic conductivity of the aquifer in the lake area varies from 8.2 m/d to 14.6 m/d which gives an average value of hydraulic conductivity of 11.4 m/d .

2.3.5 *Hydraulic gradient :*

Figures 5 and 6 show vertical head distributions and flow nets which are used for measuring hydraulic gradient of the aquifer. Davidson (1995) states that the elevation of water table in the upstream side of the flow through lakes is marginally higher than the downstream side resulting in a discharge of groundwater into the lake. Average hydraulic gradient for capture and release zones of the lake were calculated for the end of summer (March) and winter (October) each year (Table 2). These have been used to calculate groundwater inflows and outflows.

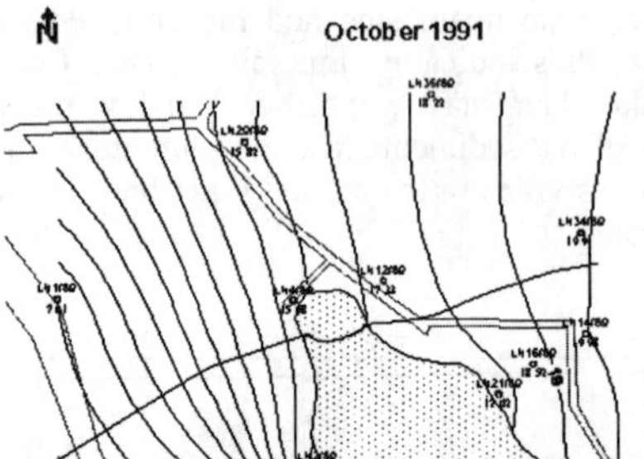

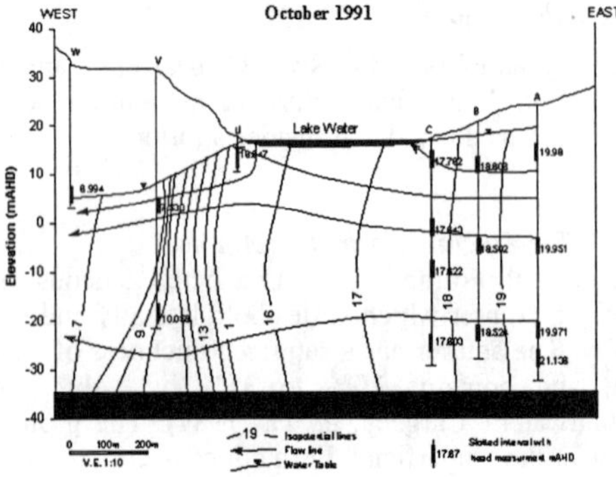

Figure 5 Flow nets and vertical head distribution of groundwater flow in October 1991 as example of four years water balance of Nowergup Lake

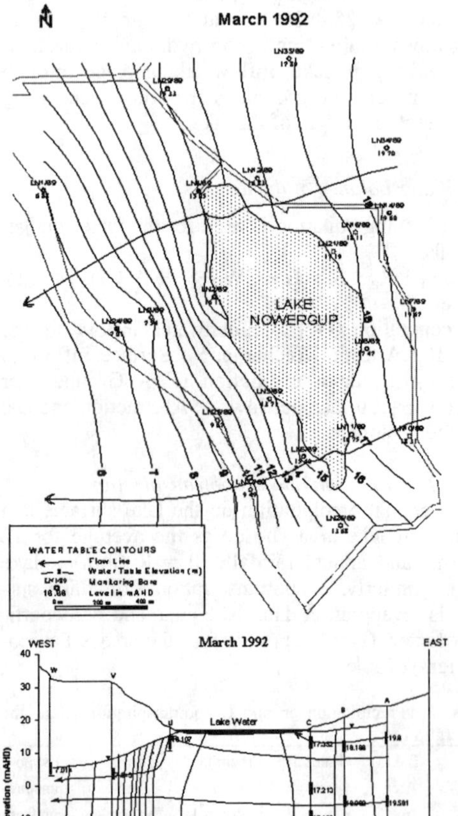

Figure 6 Flow nets and vertical head distribution of groundwater flow in March 1992 as example of four years water balance of Nowergup Lake.

Table 2: Hydraulic Gradient of up- and downstream sides of the Nowergup Lake.

Year	Upstream October	Upstream March	Downstream October	Downstream March
90-91	0.009	0.0088	0.0101	0.01
91-92	0.0095	0.0074	0.01	0.0099
92-93	0.009	0.0071	0.0105	0.01
93-94	0.009	0.007	0.01	0.009

2.3.6 *Width of flow section:*

Width of flow entering the lake is measured using the flow nets. (Fig: 5 & 6) The flow lines drawn perpendicular to the water table contours, enter into the lake and represent the width of the flow section. These have been used to calculate groundwater inflows and outflows (Table 3).

Table 3: The width of groundwater inflow and outflow in the Nowergup Lake area.

Year	October Inflow (m)	March Inflow (m)	October Outflow (m)	March Outflow (m)
90-91	1430	1450	1220	1230
91-92	1460	1510	1220	1230
92-93	1440	1490	1160	1170
93-94	1480	1500	1200	1250

2.3.7 *Thickness of aquifer contributing to the lake water (capture zone depth):*

Allen (1979) and Hall (1985), calculated the groundwater inflow to lakes Jandabup and Mariginiup, using the exaggerated scale of vertical head distribution, stated that only upper half of the aquifer contributes to lakes water. Cargeeg, et. al (1987) and Townley, et. al., (1993) rejected Allen's and Hall's approach and claimed that depending on the lake length and lake position within the groundwater flow system, the thickness of aquifer that feed the lake differs. They also suggested that for most of the Swan Coastal Plain Lakes, the capture zone depth is equal to the entire thickness of the aquifer.

2.3.8 *Daily and annual groundwater inflow and outflow:*

Daily groundwater inflow and outflow for October and March each year is calculated and averaged. This average value is multiplied by 365 to obtain an annual value. (Tables 4 & 5).

Table 4 Groundwater inflow in October, March, average daily and average annual.

Year	October (m^3/d)	March (m^3/d)	Av. Daily (m^3)	Av. annual ($x10^6m^3$)
90-91	5869	5819	5844	2.133
91-92	6325	5095	5710	2.084
92-93	5910	4824	5367	1.959
93-94	6074	4788	5431	1.982

Table 5 Groundwater outflow in October, March, average daily and average annual.

Year	October (m^3/d)	March (m^3/d)	Av. Daily (m^3)	Av. annual ($x10^6m^3$)
90-91	5619	5609	5614	2.049
91-92	5681	5563	5558	2.028
92-93	5568	5335	5452	1.990
93-94	5472	5130	5301	1.935

2.3.9 *Change in lake storage:*

The volume of water in the lake in April of one year and March of following years were obtained using Water Level-Volume Curve. The change in storage is negative if volume of water in March is less than the volume of water in April the previous year. (Table 6).

Table 6 Change in surface water storage.

	Lake depth (mAHD)	Volume of water stored (m^3)	Change in surface water storage (m^3)	Volume of water pumped into lake (KL)
April, 90	16.41	366,000		107,718
March,91	16.45	369,000	+3,000	
April, 91	16.49	392,000		252,349
March,92	16.63	427,000	+35,000	
April, 92	16.55	410.000		0
March,93	16.40	362,000	-48,000	
April, 93	16.42	367,500		161,420
March,94	16.41	366,000	-1,500	
April, 94	16.40	366,000		268,138

2.3.10 *Surface inflow:*

Environmental Protection Authority of W. Australia has set a low minimum lake levels in some lakes on the Swan Coastal Plain including Nowergup Lake. If the lake level drops below the recommended or absolute minimum level then lake levels are artificially maintained by pumping deeper confined aquifer waters into these lakes. Nowergup Lake is also artificially maintained. (see table 6).

Table 7 Annual water balance, not considering surface outflow for Nowergup Lake.

Year	Surface water inflow (Mm^3)	Precipitation (Mm^3)	Groundwater inflow (Mm^3)	vapotranspiration (Mm^3)	Groundwater outflow(Mm^3)	Changes surface Storage (Mm^3)	Inputs (Mm^3)	Outputs (Mm^3)
90-9	0.252	0.195	2.133	0.545	2.049	+0.003	2.580	2.597
91-9	0	0.349	2.084	0.521	2.028	+0.035	2.433	2.584
92-9	0.161	0.268	1.959	0.536	1.990	-0.048	2.388	2.478
93-9	0.268	0.246	1,982	0.649	1.935	-0.0015	2.496	2.582

2.3.11 *Uncertainties:*

Winter (1980) has comprehensively described the sources of uncertainty in water balance calculation of lakes. Some of these may be applicable to Nowergup Lake.

3 CONCLUSIONS:

Nowergup Lake is part of the wetlands system making up the Plio-Holocene interdunal depression of the Swan Coastal Plain. Nowergup Lake's sediments are mainly lacustrine which rest on 'Superficial' Formation. Groundwater inflow is through the eastern margin and outflow occurs through the western margin of the lake. Water levels in the lake are affected by the drought-wet cycles in the Swan Coastal Plain and the lake has been artificially maintained by pumping deeper confined groundwater into the lake during hot, dry months in the 1990s. Water balance studies of the lake show that out of an average annual input of 2.5 x 10^6 m^3 of water, 79.6% is from groundwater inflow and 9.88% from precipitation. These studies also show that evapotranspiration represents 25% and groundwater outflow represents 75% of the water loss from the lake.

4 ACKNOWLEDGEMENT:

Authors are indebted to the help and advice extended by Dr A. Davidson (Water & Rivers Commission) and Mr M.Martin (Water Corporation). Authors will also like to thank Water Authority of Western Australia (now Water Corp. and Water & Rivers Com.) for providing most of the data used in this study. Ms Banda is indebted to AusAID for financial support for this study.

REFERENCES:

Allen, A.D., (1979), "The hyrogeology of Lake Jandabup, Swan Coastal Plain, Western Australia. " W. Aust. Geol. Surv., Ann. Rept. for 1979, pp. 32-40.

Cargeeg,G.C., Boughton, G.N., Townley,L.R., Smith,G.R., Appleyard, S.J., & Smith, R.A., (1987), "Perth Urban Water Balance Study. vol 1-3" Water Authority of W. Aust. May 1987.

Davidson, W. A., (1995), Hydrogeology and groundwater resources of the Perth Region, Western Australia." West. Aust. Geol. Surv., Bulletin 142.

Froend,R.H., Farrell, R.C.C., Wilkins, C.F., Wilson, C.C., & McComb, A.J., (1993), "The effects of altered water regimes on wetlands plants." vol. 4 of a six volume report "Wetlands of the Swan Coastal Plain" prepared for the Australian Water Resources Advisory Council, Water Authority of W.A. & Envirnmental Protection Authority of W.A., pp. 144.

Hall, J., (1985), "The hydrogeology of Lake Mariginup, Perth, Western Australia." West. Aust. Geol. Surv., Prof. Papers, Rept. No. 14, pp. 1-13.

Martin, M. W., "Nowergup Lake proposed investigation." West. Aust. Geol. Surv. Hydrogeology Report No: 1988/33.

Nield, S.P., (1990), "The effect of lake lining and recharge on lake-aquifer interaction." M.Sc. Thesis, Department of Civil Engineering, Univ. of W. Aust. Unpublished, pp. 116.

Playford, P.E., Cockbain, A.E. & Low, G.H., (1976), " Geology of Perth Basin, Western Australia." West. Aust. Geol. Surv., Bull. 124, pp. 311.

Smith, R.A. & Allen, A.D., (1987), "The unconfined aquifer and effects of urbanisation, Perth, Western Australia" Proc. Internat. Conf. on Groundwater Systems Under Stress, Brisbane, Australia, pp. 577-586.

Townley, L. R., Turner, J. V., Barr, A. D., Trefrey, M. G., Wright, K. D., Gailitis, V., Harris, C. J. & Johnston, C. D., (1993), "Interaction Between Lakes, Wetlands and Unconfined Aquifers." vol. 3 of a six volume report "Wetlands of the Swan Coastal Plain" prepared for the Australian Water Resources Advisory Council, Water Authority of W.A. & Envirnmental Protection Authority of W.A., pp. 144.

Water Authority of Western Australia (WAWA), (1992), "The Artificial Maintenance of Nowergup Lake – A case study". Water Authority of Western Australia, Perth.

Winter, T.C., (1980), "Uncertainties in estimating the water balance of lakes." Water Res. Bull., Vol. 17, pp.82-115.

Groundwater: Past Achievements and Future Challenges, Sililo et al. (eds) © 2000 Balkema, Rotterdam, ISBN 90 5809 159 7

Groundwater in hard rocks, Oppland county, Norway

E. Rohr-Torp
Geological Survey of Norway, Oslo, Norway

ABSTRACT: The Oppland county of c. 25 000 km^2 is situated in Southern Norway. The bedrock is composed of gneisses and mafic rocks of Precambrian age along with metasedimentary rocks, sandstones, quartzites, conglomerates, shales and limestones of late Precambrian age, and shales, limestones and phyllites of Cambrian-Silurian age. Regional metamorphic grades range from medium to high. Based on fieldwork, existing geological maps, Landsat images and collection of information from c. 4 500 drilled wells, an interpretation of the water yield and water quality within the different rock units was carried out. The investigation has demonstrated that differences exist between the different lithological units with regard to capacities and water quality in drilled wells.

1 INTRODUCTION

The Oppland county has been chosen as a test area for the development of methods for hard rock hydrogeological mapping in Norway. The work started with the collection of data from c. 4 500 drilled wells. The information was gathered both from drilling companies and by inquiries in the field. Based on the well data, field work and existing geological maps and publications, an interpretation of the water yield and water quality from the different lithologies in the county was carried out. Of great importance for the work has been:, Englund & Myhrstad (1980), Siedlecka, Nystuen, Englund & Hossack (1987), Nystuen, Kirkhusmo & Naterstad (1993) and Lutro & Tveten (1996).

Most drilled wells in Norway supply a single house, some few houses or a small farm. A well, 80–100 m deep, with a yield of 100 l/h (0,02-0,03 l/s), has sufficient capacity to supply a household with water.

2 RESULTS

Below is given a short description of the different lithological units and their water yield and water quality in drilled wells in Oppland. A simplified bedrock map of Oppland is given in Figure 1.

2.1 *Schist, phyllite and greenschist*

The schists, phyllites and greenschists of Cambro-Silurian age, occupy relatively large areas in Oppland (Figure 1). These are soft rocks, and the overburden closes deeper fractures, so a typical yield from drilled wells is less than 300 l/h (0,08 l/s), and dry wells are not uncommon. Locally, the schists and phyllites are carboniferous. In such areas, the water quality often is unsatisfactory. Common problems are high contents of hydrogen sulphide, radon iron and manganese.

Hydrofracturing of low-yielding wells in these rocks is usually successful. A common yield after hydrofracturing is between 500 l/h (0,14 l/s) and 1000 l/h (0,28 l/s) in a drilled well.

2.2 *Limestone*

Limestones occur in the southern part of Oppland (Figure 1). The age is partly late Precambrian, and partly Cambro-Silurian. The limestones are predominantly good aquifers, partly caused by karstic dissolution. which creates transmissive groundwater conduits. The problem in Oppland is that the late Precambrian limestone (the Biri limestone) is interbedded with black shale, and an Ordovician limestone (the Orthoceras limestone) which is only c. 40 m thick, is underlain by a black,

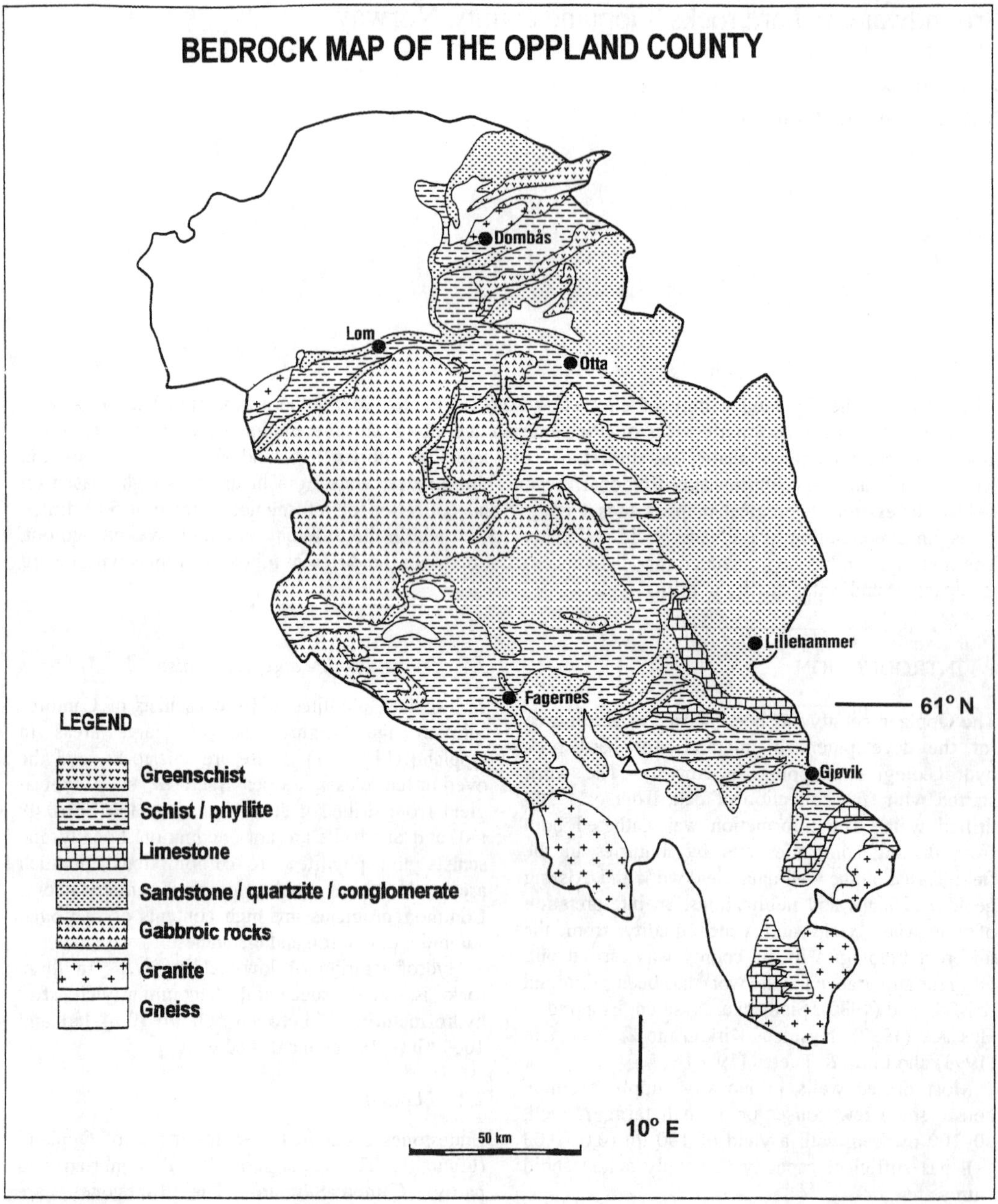

Figure 1. Simplified bedrock map of the Oppland county

carboniferous shale. Typical yields in drilled wells in these limestones are 50 l/h (0,014 l/s) to 1000 l/h (0,28 l/s). Furthermore, most wells will intersect black shales, which influences the groundwater quality. Common quality problems are: Hydrogen sulphide, radon, iron and manganese in addition to high hardness. Especially in the Biri limestone, hydrogeologists often advise against well drilling because of the probable water quality problems.

A Silurian limestone, the Mjøsa limestone, is a pure limestone of more than 100 m thickness. This is a good aquifer. Drilled wells usually have yields ranging from 1000 l/h (0,28 l/s) to 5000l/h (1,4 l/s). Except for hardness, the water quality is good in this limestone.

2.3 *Metasandstone and conglomerate*

The late Precambrian metasediments occur in the eastern and central parts of Oppland (Figure 1). Metasandstone and conglomerate are competent rocks, which are capable of keeping water-bearing fractures open to large depths (more than 150 m, which is the approximate limit for well depths in Norway). Common yields in drilled wells are between 1000 l/h (0,3 l/s) and 3000 l/h (0,85 l/s). The water quality is usually good.

2.4 *Quartzite*

The late Precambrian quartzite appear in the same areas as the metasandstone and conglomerate.Where the quartzites are tectonically disturbed, their brittle deformation makes them among the best aquifers in Oppland. Common yields in drilled wells are from 2000 l/h (0,6 l/s) to 3500 l/h (1 l/s). However, in areas where the quartzites are little deformed, with more or less flat lying schistosity, they are poor aquifers. Common yields in drilled wells in such areas, are less than 300 l/h (0,08l/s). In areas where rust deposits frequently occur on schistosity and fracture planes, high contents of iron and also manganese may occur in the groundwater. In most cases, however, the groundwater quality is good in the quartzites.

2.5 *Metasandstone interbedded with shale*

The lower part of the late Precambrian metasedimentary sequence is developed as massive sandstone beds alternating with dark shale. This so-called Brøttum formation is found in the same areas as the metasandstone and conglomerate (2.3). The sandstone beds are usually heavily fractured, while the soft shale is almost unfractured. The combination of fractured contact zones between the sandstone beds and the shale and the fractured sandstone beds makes the Brøttum formation an average aquifer. Drilled wells will usually yield between 500 l/h (0,14 l/s) and 2000 l/h (0,6 l/s). Because of the interbedded dark shales, the groundwater quality can be unsatisfactory with high content of hydrogen sulphide, in some cases also of iron and manganese.

2.6 *Gabbroic rocks*

These allochthonous Precambrian mafic rocks (Figure 1), make up a high mountain area, named Jotunheimen. The Jotunheimen area is scarcely populated, so only a few drilled wells are known in the gabbroic rocks. From the registered well yields, however, the gabbroic rocks seem to be comparable aquifers to the Brøttum formation. High contents of iron and manganese are not uncommon.

ROCK TYPE	YIELD, l/h (0 1000 2000 3000 4000 5000)	WATER QALITY
Schist, phyllite, greenschist	Hydrofractured XXXXXXX	Carboniferous rocks often high in H_2S, Rn, Fe and Mn
Limestone	Biri & Orthoceras … Mjøsa limestone XXXXXXXXXXXXXXXXXXXXXXXXXXXXXXXXXXXXXXXXXXXXXX	Biri & Orthoceras high in H_2S, Rn, Fe and Mn Mjøsa limestone good. Hardness high in all
Metasandstone and conglomerate	XXXXXXXXXXXXXXXXXXX	Good
Quartzite	Undeformed … Deformed XXXXXXXXXXXXXXXXXXXXXXXXXXXXX	High contents of Fe and Mn occur locally
Metasandstone inter-bedded with shale	XXXXXXXXXXXXX	High contents of H_2S, Fe and Mn rather common
Gabbroic rocks	XXXXXXXXXXX	High contents of Fe and Mn not uncommon
Gneiss and granite	Granite … Gneiss XXXXXXXXXXXXXXXXX	Mostly good

Figure 2. Groundwater yield and quality in Oppland

2.7 *Gneisses and granites*

In the north-western and southern part of Oppland, the bedrocks are composed of gneisses and granites (Figure 1). The gneisses and granites are competent rocks, having hydrogeological propertiessimiliar to the metasandstone and conglomerate. These rocks are usually moderately fractured, and solid parts do occur. Drilled wells will usually have medium-good yields, between 500 l/h (0,14 l/s) and 2 500 l/h (0,7 l/s). Most often, the water quality is good, but high contents of iron and manganese may locally occur.

3 SUMMARY AND CONCLUSION

Figure 2 shows a synopsis of what has been said above. The figure shows that most drilled wells in Oppland have yields between c. 500 l/h (0,14 l/s) and c. 3000 l/t (0,85 l/s). Combined with a storage reservoir, this is sufficient to supply 35 and 200 persons respectively. The results of the project serve as good guide lines for future well drillings in the Oppland county.

REFERENCES

Englund, J.O. & J.A. Myhrstad 1980. Groundwater Chemistry of some Selected Areas in South eastern Norway. *Nordic Hydrology 11, 33-54.*

Lutro, O. & E. Tveten 1996. Geologisk kart over Norge, berggrunnskart Årdal M 1:250 000. *Norges geologiske undersøkelse.*

Nystuen, J. P., L. A. Kirkhusmo & J. Naterstad 1993. International Association of Hydrogeologists XXIVth Congress. Excursion No 1. Geology and hydrogeology in South Central Norway. *Norges geologiske undersøkelse.*

Siedlecka, A., J. P. Nystuen, J. O. Englund & J. Hossack 1987. Geologisk kart over Norge, berggrunnskart Lillehammer M 1:250 000. *Norges geologiske undersøkelse.*

Groundwater: Past Achievements and Future Challenges, Sililo et al. (eds) © 2000 Balkema, Rotterdam, ISBN 90 5809 159 7

The MRS technique for groundwater resources evaluation – Test results from selected sites in Southern Africa

J. Roy & M.W. Lubczynski
International Institute for Aerospace Survey and Earth Science,, Enschede, Netherlands

ABSTRACT: ITC, together with ten collaborating agencies from seven countries is involved in a research project on the MRS (Magnetic Resonance Sounding) technology since 1996, when IRIS Instruments introduced the NUMIS instrument. Thanks to regional collaboration, the MRS tests using the NUMIS instrument, were made in Botswana, Namibia and South Africa. The fieldtests were carried-out under a combination of low earth's magnetic field and high ambient noise conditions. The test sites included different aquifers from the point of view of lithology, depth and effective porosity • thickness product. The MRS technique, some of its operational constraints and practical field test results are presented and critically discussed. Observations are made on instrumentation performance, on the best time to execute an MRS survey and on the linkage between MRS determined quantities and hydrogeological parameters.

1 INTRODUCTION

The use of the NMR (Nuclear Magnetic Resonance) technology for non-invasive assessment of groundwater resources from the surface has been patented by Varian (1962). The first field tests were made by Semenov (1987) in Russia using the Hydroscope. This MRS instrument was an improvement over Varian's concept as it included a sounding capability. More recently, IRIS Instruments (1996) in France introduced a new instrument called NUMIS, which made the MRS technique commercially available in the Western world.

MRS Evaluation work - As soon as the NUMIS instrument became available, a project to evaluate the MRS technology was set-up at ITC (International Institute for Aerospace Survey and Earth Sciences). The objective of this project was to determine the capabilities and limitations of the method and eventually propose improvements to the MRS technique and the NUMIS instrumentation with regard to its technical performance and suitability for groundwater resources evaluation. In that respect, a wide range of survey targets and operational conditions were defined and are gradually being tested. So far, tests have been performed over various aquifers including alluvial, sandstone, carbonates and granites with variable degree of weathering and fracturing, over water bearing layers of different depths, thickness, water contents and structural complexity (when more complex than 1D). The tests were made under different conditions relative to the earth's magnetic field magnitude, homogeneity and dip, to the noise level from cultural and natural sources and to the distortions due to conductive formations or to magnetic permeability effects etc.

2 MRS PRACTICAL IMPLEMENTATION

Several descriptions of the MRS technique have been published in the last ten years (Schirov et al 1991), (Goldman et al 1994), (Lieblich et al 1994), (Legchenko & Beauce 1999), (Roy & Lubczynski 1999, 2000) etc. These descriptions will not be repeated, however, practical considerations related to the MRS concept and its application will be listed.

2.1 *MRS application background*

The MRS technique can be considered as an adaptation of the proton precession magnetometer (Telford et al, 1990), where the magnetometer sensor is replaced by a large loop. In the MRS implementation, the large loop is energised with an alternating current (AC) tuned to the local precession frequency f of the in-situ groundwater's hydrogen nuclei ($^1H^+$). This frequency is found by running a short magnetometer survey prior to the MRS survey [f (in Hz) = $\gamma \cdot B$, i.e. $\approx 0.0426 \cdot B$ where B is the average value of the earth's magnetic field in nT]. A resonance excitation is used instead of the DC excitation used for magnetometers because of the large subsurface volumes involved in the MRS technique and because the excitation mode easily allows a sounding mode (i.e. with a small amount of excitation, only the $^1H^+$ located near the loop will be excited enough to contribute to the signal while with a larger excitation, the $^1H^+$ at a larger depth will contribute to the measured NMR signal). In this context, the sounding parameter is the excitation moment (Q) i.e. the product of the current in the loop times the pulse duration, which is also called pulse parameter. By tuning the instrument at a frequency $f = \gamma \cdot B$ (this frequency is site specific as it depends on the value of the earth's field B at the location), only the groundwater's $^1H^+$ nuclei are excited; moreover only the signal at that particular frequency are detected thus insuring a good selectivity for groundwater. In a subsequent step, the same loop is used as a receiver coil and the NMR signal from the groundwater's precessing $^1H^+$ is detected and processed. The recorded NMR signal is analysed in terms of its initial amplitude, its decay rate, its precession frequency and its phase referenced to the excitation pulse. The initial amplitude after inversion provides the subsurface free-water content, which below the water table is related to the effective porosity (Φ_{eff}) and/or the specific yield in layers as a function of depth (Legchenko & Shushakov 1998). The only water detected by the MRS technique is the free-water; i.e. it does not take into account clay-bound and micropore-bound water. The decay rate is used as an indicator of the layer's pore size and for a given lithology it is combined with Φ_{eff} to yield a hydraulic conductivity estimate (Kenyon 1997). The last two parameters, precession frequency and the phase are used for quality control. The phase may also be used as an additional constraint in the inversion with electrically conductive formations.

2.2 *MRS field application*

The actual MRS field implementation is quite different from a magnetometer and is rather similar to a single loop TDEM set-up. A large square loop (e.g. L = 75 to 150 m where L is the length of one side of the square) is connected to an MRS instrument (in our case the NUMIS instrument) with a total system mass of about 300-450 kg depending on the loop size. The MRS instrument has a double function: loop excitation and NMR signal detection/processing. The MRS survey data acquisition and storage is controlled from a PC-type portable computer.

An MRS survey needs time for field installation and actual data acquisition. Typical combined set-up and uninstall time may vary from 2 hours up to 6 hours depending on the terrain difficulties and on the number of modifications made to the loop being a function of the local noise conditions. The data acquisition itself takes approximately ½ hour in cases of good signal/noise ratio and up to 6-7 hours in conditions of high ambient noise.

2.3 *MRS environmental constraints*

A useful MRS data set can only be acquired when the NMR signal from the groundwater's precessing $^1H^+$ is large enough to be extracted in the presence of the ambient noise. The quality and reliability of MRS data set is a direct function of its signal to noise (S/N) ratio.

The magnitude of the NMR signal is controlled by the shape and size of the MRS loop, by the location and quantity of groundwater, i.e. depth, thickness (b) and Φ_{eff} of the water bearing layers, and by a series of environmental factors such as the electric conductivity and thickness of the subsurface formations and the earth's magnetic field homogeneity, its magnitude and its dip (Legchenko et al 1997).

Considering the ambient noise, there is presently no simple way of predicting its magnitude at a planned MRS survey location. In practice the noise is measured at each location through the MRS survey loop after its field installation (or optionally through a more portable auxiliary loop). The two main sources of noise observed so far while performing MRS surveys are: (1) man-made electric installations (power line, electric machinery in general, powered underground cables, pumps, etc.), which generate harmonics up to the $^1H^+$ precession frequency f (over land areas this is normally in the range of 1 to 2.7 kHz) and (2) natural sources, which are dominated by lightning and may thus have a diurnal and seasonal time component associated.

It is useful to consider an illustrative set of 'favourable MRS survey conditions', to know, under which conditions, reliable MRS data sets can be acquired easily. Such a set of conditions is listed next: (1) water quantity > 0.5 m ($\Phi_{eff} \cdot b$), (2) shallow aquifer e.g. depth to top < $L/2$ and < $\delta/2$ where δ is the skin depth at f i.e. low electric conductivity down to aquifer's base, (3) coarse pore size (e.g. coarse sand, or gravel), (4) large (e.g. > 45000 nT) and homogeneous earth's field, negligible magnetic susceptibility contrast, (5) primary Φ_{eff} and 1D stratification parallel to loop plane, (6) low ambient noise (cultural & natural) to allow S/N >> 1. Two items should be noted: (a) except for very shallow aquifers i.e. depth to top < $L/6$, the dip of the earth's field is in fact not a limiting constraint, (b) with current and improved MRS technology, valid MRS data sets can be acquired under worst conditions than those listed in (1) to (6).

3 WHY SOUTHERN AFRICA?

While getting involved with MRS technique evaluation work, it was soon realized that not all sites in the world were equally easy to survey and sites of varying difficulty must be studied. Thus the evaluation work included easy sites (large MRS signal, unobstructed terrain access, low ambient noise) useful to collect data sets with high reliability, average sites representative of 'usual' conditions and difficult sites (low water content, deep-lying water bearing layers, difficulties with terrain access, high ambient noise level). The idea of such a variety of sites was to test the reliability of the MRS performance in various conditions and if necessary to adapt the technology to unfavorable cases. An advantage of such adaptation is that sites, which used to be called difficult, may change their status to normal or even easy ones. In this regard, Southern Africa constitutes an interesting and difficult area for MRS field tests because of the following reasons: (1) many aquifers of interest in Southern Africa (like Karoo aquifers) have for example a lower Φ_{eff} and are located at deeper levels than representative economic aquifers in Europe; (2) Southern Africa is characterized by a low value of the earth's magnetic field (B < 30,000 nT) in most of Botswana, Namibia and South Africa (See Figure 1). Since the MRS signal is proportional to B^2, this means that a given aquifer (Φ_{eff} , thickness b & depth) will produce an MRS signal at least 5 times smaller in Cape Town, SA (~ 26,300 nT) than in Melbourne, Australia (~ 60,500 nT) despite the fact that these two cities are roughly at the same latitude;(3) at least during some time of the year and some time of the day, the ambient noise from natural sources is likely to be considerably higher than in Europe.

The combination of low Φ_{eff}, deep water bearing layers, low earth magnetic field B and high ambient noise, at least during some periods of the year makes MRS surveying in Southern Africa a challenge with regard to S/N ratio. While the reduction of water content and the increase in depth can simply be modeled with lower signal amplitude, the investigations on the nature of the ambient noise and the designs of improved strategy to cope with interfering noise of man-made and natural origins must be based on real data sets.

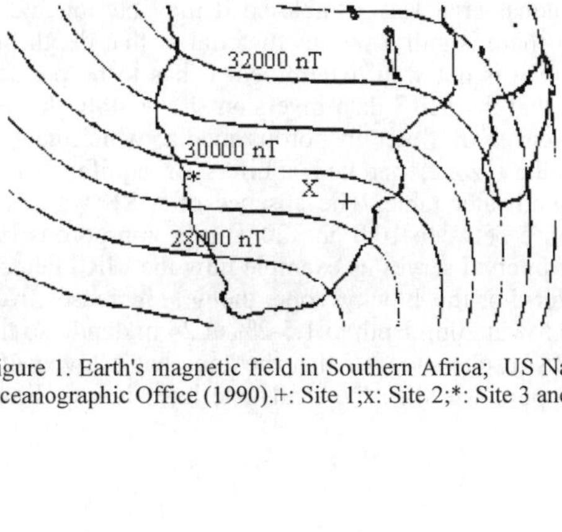

Figure 1. Earth's magnetic field in Southern Africa; US Naval Oceanographic Office (1990).+: Site 1;x: Site 2;*: Site 3 and 4.

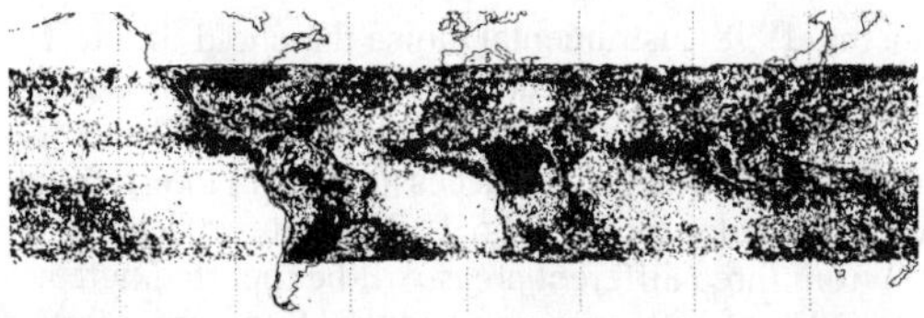

Figure 2. Worldwide 35°N to 35°S latitude strip, two-years average lightning frequency as observed by LIS space-borne sensor from 97/12 to 99/12, darker, high lightning frequency areas are observed in particular in SE U.S., SE South America and Central and Southern Africa, NASA (2000)

The main source of natural noise in the MRS signal frequency band is lightning. Figure 2 illustrates the worldwide distribution of lightning where Central and Southern Africa are among the areas with high natural noise levels. Additional studies on the space and time distribution of noise have been published by Chrissan & Fraser-Smith (1996, 1997) as monthly and hourly averages. It suggests substantial seasonal and diurnal changes of noise especially in high noise areas. This has implications with respect to MRS applications conditions. Considering the current MRS technological development, when MRS surveys are executed during the period of the year with the highest noise level (i.e. highest lighting occurrence frequency), in environment like Southern Africa, they must be done at a particular time of the day (e.g. in our case late night to early morning). MRS surveys are best executed during the lowest noise period of the year; long term monitoring will be used to determine these periods.

Several organizations from Botswana, Namibia and South Africa were involved in the collaboration on the MRS tests in their countries. The sites were identified following local recommendations and field tests of the MRS technique were carried-out in November-December 1998. The data set reported here was acquired, processed and inverted using instrumentation and software supplied by IRIS Instruments as part of their NUMIS system.

4 TEST RESULTS

The MRS test sites were classified in 3 categories: (1) MRS survey executed and reliable data acquired, (2) MRS survey executed with S/N $\leq$ 1, (3) MRS survey not executed due to noise level too high to perform the survey.

Considering the sites unsuitable for MRS (3), both natural and man-made sources of noise were observed. In the case of natural noise, the proximity of lightning storms was the cause while in the case of man-made noise, high noise levels were observed in wellfields even with disconnected power lines.

Considering sites with low S/N ratio (2), the observed MRS signal levels were lower or comparable to the 1998 instrumental noise threshold of 5 to 10 nV. The results from these sites, although processed in a normal way, can only yield an upper limit for the water content. The successful sites (1) are those with S/N ratio higher than 1. Four sites of category (1) from three different areas will be reported here as examples of MRS survey results in Southern Africa. During the field tests, all the data sets were acquired with the MRS loop located around existing boreholes with the exception of two sites selected with regard to other priorities. The hydrogeological information (parameters), from these boreholes is currently available for only part of these sites. In this report the sites are identified only by rough localization; the specific identification of the boreholes is excluded until the identification clearances are clearly granted.

4.1 *Site 1*

Site 1 is located in South Africa, approximately 100 km North of Pretoria. The area was cultivated with arable soil (~ 1 m thickness) overlying a clay rich material of about 5-8 meter thickness. Below that sequence up to the depth of 24m a basalt layer is found underlain by sandstone. The lowest 4m section of basalt is fractured with some cemented joints. Together with the sandstone it creates a confined aquifer with static groundwater table at borehole depth of 16.8m. The MRS data set (Figure 3a) was acquired with a square loop of dimension L = 115 m. The local value of the earth's magnetic field B was ~ 28900 nT. An MRS data set is acquired by measuring the NMR response (initial amplitude E_0 in nV on

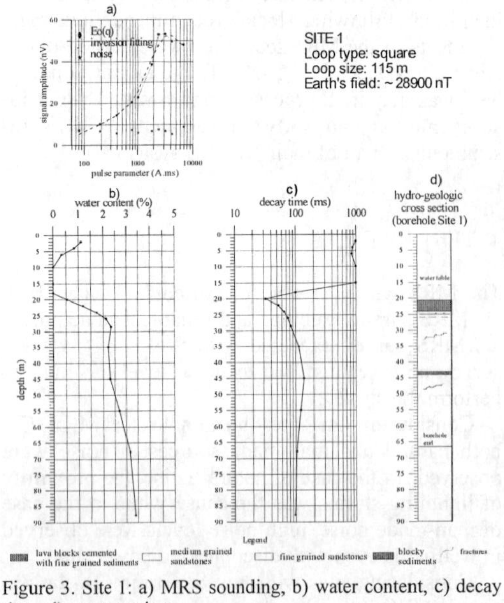

Figure 3. Site 1: a) MRS sounding, b) water content, c) decay time, d) cross-section

the Y-axis) as a function of a series of different excitation moments (Q in A·ms on the X-axis). The 'initial amplitude' E_0 refers to the amplitude of the MRS signal at the instant when the excitation pulse is terminated. In Figure 3a, the dots joined by a continuous line are the observed data points, the dotted line is the best fit inverted model as described by Figure 3b, while the stars shows the noise level (measured before or in absence of excitation pulse). In this implementation, the rate of MRS signal decay is characterized by fitting an exponential model curve. This curve quantifies the decay rate as a time constant identified as T_2^* in NMR terminology and as 'decay time' in Figures 3a, 3c. A numerical inversion of both the initial amplitude and the decay time is done as a function of the excitation moment Q. The inversion takes into account the environmental conditions (i.e. loop size & shape, earth's magnetic field magnitude and dip, geoelectrical layering: layers' resistivity and thickness). The inversion yields two hydrogeologicaly important parameters: the water content as a function of depth (Figure 3b) and the decay time as a function of depth (Figure 3c).
Below the water table, the water content parameter represents the volumetric percentage of free extractable water in each of the modeled layers referred to directly as Φ_{eff} or specific yield (Figure 3b).

The decay time parameter is related to the average pore size of the layer, empirically related to the hydraulic conductivity of the media (Figure 3c) for sandstones and carbonates (Kenyon, 1997). It has to be pointed out that any hydraulic conductivity determination is highly sensitive to the scale of the measurement (e.g. Worthington 1997). Therefore it needs adjustments/calibration between the different types of determinations, from core samples to pump tests, to MRS surveys and to watershed scales (e.g. petrophysical determinations are handling volumes < 1 liter, while MRS field measurements are sampling up to 10^6 m^3).

Figure 3b shows a small quantity of water in a shallow interval i.e. < 1 % over the first 5 m, next a dry interval down to about 20 m, and then a porous interval containing ~ 2 to 3 % water. The information corresponding to depth larger than ~ 0.7 L may be considered less reliable so it may be concluded that there is still a porous material at that depth but its Φ_{eff} is not well determined. It has to be pointed out that the MRS data inversion shows only the occurrences of water in porous media, which can be capillary zone, perched aquifers or aquifer below groundwater table. The absence of MRS water response between 16.8 and 20 m in a non-porous basalt interval serves as example how the MRS detects water. For the 'blocky' zone, the Φ_{eff} increases from 0-0.5% at 20m depth to 1.5-2% at 24 m depth, so the MRS results suggests that the zone has a lower Φ_{eff} and pore-size than the sandstone layer below. From

24 m downward, the MRS-derived Φ_{eff} is 2 to 3.5 % and agrees with the typical Φ_{eff} of fresh sandstone in the investigated area. Thus, the MRS derived information fits well the hydrogeological data as confirmed with a video record made inside the borehole located within the MRS survey loop location.

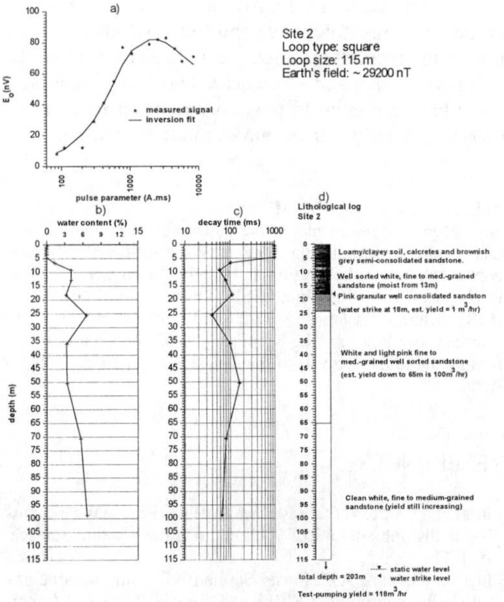

Figure 4. Site 2: a) MRS sounding, b) water content, c) decay time, d) cross-section

4.2 *Site 2*

Site 2 is located in Botswana near the place where the Tropic of Capricorn crosses the road Gaborone-Mahalapye. The site is located within the Karoo formation where the main Ntane sandstone aquifer is highly fractured and has extremely productive hydrogeological condition - well abstraction > 100 m^3/h. Figure 4a documents the MRS test made. If compared with Figure 3, the MRS response has a higher peak (>80 nV for Site 2 vs <60 nV for Site 1) and the peak is located at a lower value of the excitation moment ($Q \approx 2500$ vs $Q \approx 5000$ for Site 1). An examination of Figures 4b-4d shows that the groundwater distribution pattern as a function of depth is also different than in Figures 3b-3d. There is less water in the first 5 m but the water table is shallower (observed 11.5 at Site 2 vs 16.8 at Site 1). In fact from 9 m down to the depth of 70 m, the water content is always larger than 3 % and reaches a peak of 6.5 % at 25m depth. Except for the two lower determinations at 9 and 25 m, the decay time fluctuates between 80 and 160 ms in the interval of 6 to 70 m. Considering the first 70 m, the $\Phi_{eff} \bullet b$ product where b is the porous media thickness [i.e. the integral of the free water content (equivalent to Φ_{eff} below the water table) along the depth dimension] for this site is 2.5 m, which is the highest value for all the sites tested during this MRS tests series in Southern Africa. In this case the main water contribution is from 9 m down to the arbitrary evaluation limit of 70 m, which translate into an average Φ_{eff} over the interval of 2.5 / (9-70m) i.e. 4%. The hydrogeological borehole data indicates that the aquifer is unconfined with 8 m of Kalahari sand (black organic soil, clay, loamy sand) and underlain by well-sorted and highly fractured Ntane Sandstone. The groundwater table stabilizes at depth of 11.5 m. The hydraulic conductivity and transmissivity are 10.5 m/day and 1000 m^2/day respectively and were obtained through the pumping test. It has to be pointed out, that any NMR estimator of hydraulic conductivity would yield a value much lower than 10.5 m/day for this site because of scale effect. The NMR estimators available so far are at the core sample scale and no 'scale-up' factor is yet available for MRS. Also the MRS inversion uses a 1D model while this fractured site is probably best modeled as a 3D zone.

4.3 *Site 3 & Site 4*

Site 3 and Site 4 are both located in the Omaruru Delta in Namibia. In each of these two sites the MRS loop was placed around a borehole location. Unfortunately the hydrogeological data from the two boreholes have not yet been made available to the MRS Evaluation Project. The two sites are located at a distance of a few kilometers from each other, within the same alluvium aquifer, but at different

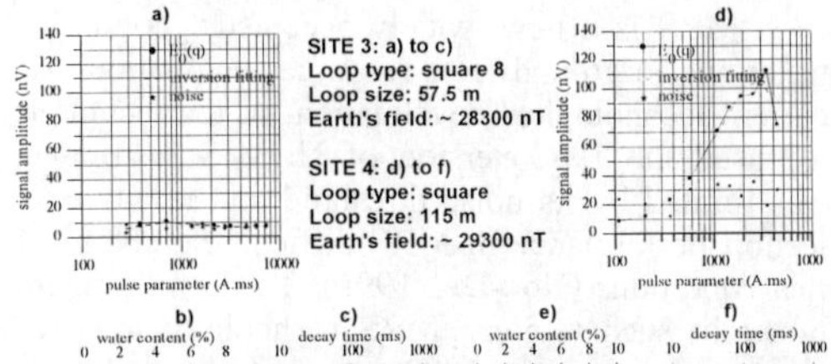

Figure 5: Site 3: a) MRS sounding, b) water content, c) decay time; Site 4: d) MRS sounding, e) water content, f) decay time

earth field value (B difference ~ 1000 nT) however at the scale of the MRS loop, each site has an homogeneous B. This could mean a significant hydrogeological difference between the two locations. A comparison between Site 3 and Site 4 MRS data sets (Figures 5a & 5d) shows that the data from Site 3 is barely above noise level while the data from Site 4 has an MRS signal peak approximately ten times higher than Site 3 peak. Site 3 has only one significant water bearing layer centered around 10 m depth (see Figure 5b), with a $\Phi_{eff} \cdot b$ product of about 0.2 m, spread over the 5-20 m depth interval, and a very short decay time of 30 ms, which represents a material with very small pores (e.g. clayey fine sand). The deeper profile information (1% Φ_{eff} with 550 to 700 ms decay time for depth > 45 m), considering the low signal level and the square-8 loop dimension (L = 57.5 m), is highly uncertain. Site 4, contrary to Site 3, indicates two water-bearing layers (Figure 5e,f). The shallow one, similar to the shallow layer at Site 3, is centered at around 12 m with a 0.1 m $\Phi_{eff} \cdot b$ product spread over the 8-14 m depth interval. The deeper layer is centered at depth of 37 m and has a $\Phi_{eff} \cdot b$ product of about 2.2 m spread over the 19 to 75 m interval (i.e. an average Φ_{eff} of nearly 4% over that interval). The deeper water bearing layer has a $\Phi_{eff} \cdot b$ product, which is more than 10 times larger than the $\Phi_{eff} \cdot b$ product of the shallow water bearing layer observed at Site 3 and at Site 4. Moreover the longer decay time of about 100 ms indicates coarser pore-size (e.g. fine to medium sands) indicating better hydraulic conductivity.

5 CONCLUSIONS

1. MRS is a new, widely accepted, geophysical technique in groundwater resources evaluation. The current technological development of the technique, we used the 1998 version of NUMIS instrument, was limited by its noise threshold and its noise rejection performance in MRS-difficult Southern Africa conditions (Nov-Dec 1998). To be fully operational in such regions, MRS technology improvements (relative to the 1998 status) are needed in the areas of noise rejection capability, higher sensitivity to detect deep and low Φ_{eff} aquifers etc.

2. At least for some time of the year, unfavorable natural noise conditions were observed in Southern Africa, which made MRS survey impractical at some of the planned test sites and restricted the use of the instrument to a certain time of the day (late night to early morning). The sites with high enough water content at shallow depth, allowed the recording of valid MRS data sets (i.e. with acceptable S/N ratios) and resulted in inverted parameters compatible with hydrogeological informations.

3. Concerning the MRS output it is postulated (not verified yet) that the inverted free water content (%) over the investigated volume of the aquifer below the water table can be directly used in hydrogeology as Φ_{eff} and/or specific yield for this volume. The second parameter called decay rate indicates relation with hydraulic conductivity but the nature of this relation requires more research.

4. After adequate improvements, the MRS technique will constitute a fast and low cost tool suitable for groundwater resources evaluation at sites in Southern Africa and elsewhere. The MRS technique has a high potential to play an important role in numerical groundwater resources modelling tasks.

Acknowledgments - The MRS fieldtests in Southern Africa have been possible thanks to the collaboration and support of many individuals and organizations. We want to acknowledge with thanks the collaboration and support from CSIR, DWA-Botswana, DWA-Namibia, GSD-Botswana, GSN-Namibia, IRIS Instruments-France, ITC and WCS-Botswana. E. Martinez del Pino and L. Kgotlhang have contributed to the figures and to the analysis of the data sets acquired in Southern Africa.

REFERENCES

Chrissan, D.A., & A.C. Fraser-Smith 1996. Seasonal variations of globally measured ELF/VLF radio noise. Radio Science, 31(5): 1141-1152.

Chrissan, D.A., & A.C. Fraser-Smith 1997. Diurnal variations of globally measured ELF/VLF radio noise. Technical Report D177-2, Space, Telecommunications and Radioscience Laboratory, Stanford, CA, Stanford Univ.

Goldman M., B. Rabinovich, M. Rabinovich, D. Gilad, I. Gev, & M. Schirov 1994. Applications of the integrated NMR-TDEM method in groundwater exploration in Israel. J. of Applied Geophysics 31: 27-52.

IRIS Instruments 1996. NUMIS data sheet.

Kenyon, W.E. 1997. Petrophysical principles of applications of NMR logging. The Log Analyst, March-April: 21-43.

Legchenko, A., A. Beauce, A. Guillen, P. Valla, & J. Bernard 1997. Natural variations in the magnetic resonance signal used in PMR groundwater prospecting from the surface. EEGS-ES Journal 2: 173-190.

Legchenko, A.V. & O.A. Shushakov 1998. Inversion of surface NMR data. Geophysics 63: 75-84.

Legchenko, A. & A. Beauce 1999. Surface-PMR: State of the art. SNMR Workshop, Berlin, 99/03/22-24.

Lieblich, D.A., A. Legchenko, F.P. Haeni & A. Portselan 1994. Surface nuclear magnetic experiments to detect subsurface water at Haddam Meadows, Connecticut. Proceedings SAGEEP-1994: 717-736, Boston, EEGS.

NASA, 2000, LIS and OTD web site: http://www.thunder.msfc.nasa.gov.

Roy, J. & M. Lubczynski 1999. The SGW-NMR technique - a user's perspective. SNMR Workshop, Berlin, 99/03/22-24.

Roy, J. & M. Lubczynski 2000. MRS: Introduction to a new geophysical technique and its current status. Geophysics'2000, La Havana, 00/03/21-24.

Schirov, M., A. Legchenko & G. Creer 1991. A new direct non invasive groundwater detection technology for Australia. Exploration Geophysics 22: 333-338.

Semenov, A.G. 1987. NMR Hydroscope for water prospecting.

Proc. Seminar on Geotomography: 66-67, Hyderabad, Indian Geophysical Union.
Telford, W.M., L.P. Geldart & R.E. Sheriff 1990. Applied Geophysics, Second Edition. Cambridge University Press.
U.S. Naval Oceanographic Office, 1990, Earth's magnetic field: Total Intensity (F, nT) WMM-90.
Varian, R.H. 1962. Ground Liquid Prospecting Method and Apparatus. US Patent 3,019,383.
Worthington, P.F. 1997. Petrophysical estimation of permeability as a function of scale. Developments in Petrophysics, Spec. Pub. No. 122: 159-168, London, The Geological Society.

Groundwater: Past Achievements and Future Challenges, Sililo et al. (eds) © 2000 Balkema, Rotterdam, ISBN 90 5809 159 7

Providing ground water supply and assessing borehole yield in a disaster relief context

F. Ruden
Geological Survey of Norway, (presently: Council for Geoscience, Pretoria, South Africa

K. Sami
Council for Geoscience, Pretoria, South Africa

ABSTRACT: During the influx of refugees into Tanzania from neighbouring Rwanda and Burundi in 1994 large settlements of up to 350 000 persons had to be supplied with water immediately. The 65 000 refugees at the Kitali Hills camp, situated along the Ruboroga intermittent river, was one of the smaller such settlements. No perennial surface water source existed in the vicinity and water was initially tankered in from various sources. Initial drilling into quartzites, phyllites and granites yielded no water. The Ruboroga alluvium is saturated but with too high silt content for direct water abstraction. Tapping the underlying confined fractured quartzites and granites proved to be productive and a number of wells provided the camp with water for 2 years. Due to lowering of the piezometric surface, gravity drainage or leakage from the alluvium may account for the sustainability of supply.

1 INTRODUCTION

Applied hydrogeology in a disaster relief context does not traditionally allow for much detailed study. Experiences from the 1995 run-up of the water supply for the Kitali Refugee Camp of Ngara District, Tanzania shows that under stressful circumstances, where lives are at stake there is a direct correlation between available water supplies and infant mortality, the main tasks for the responsible hydrogeologist are: a) to locate adequate sources of water, b) to predict output in terms of water, and c) to inform the camp administrators of these water sources, if any. When working with crystalline aquifers, limited resources and under time constraints these tasks are difficult, at best. (Banks et al, 1993).

By the spring of 1995, with the steady influx of refugees crossing over to Tanzania from Burundi, the population of the *United Nations High Commissioner for Refugees (UNHCR)* administrated Kitali Hills Refugee Center had swollen from nil to approximately 65 000 persons in less than two weeks, and continued to increase. Surface water is relatively scarce in this area and water supply was initially based on a fleet of tankers, trucking water in from a number of sources. Even though rains continued through April and May, the sources dwindled while roads were deteriorating, not least due to the water traffic. The commercial road transport to and from neighbouring Burundi virtually ground to a halt by May. A series of water related crises emerged, including local sources of water being depleted before the onset of the dry season, creating an understandable concern on part of the local landowners.

The target daily minimum *per capita* water requirement, as defined by the UNHCR is 20 litres, i.e. for the Kitali population this translated into 1.300 m^3/d, or 15 L/s. During the road transportation phase, the *per capita* figure often dropped under the critical 5L level. With no permanent surface water sources within reach, alternatives to the costly tankering business were pressing. A piped water scheme was discussed throughout the life of the Kitali Camp, but as a new drilling rig & accessories reached the area by April 1995 the ground water option was given a chance. The equipment, mounted on 6x6 tactical vehicles coped well during the incessant rains and undertook a series of parallel activities including exploration drilling, well construction and soundings, without the need for the traditional time consuming preparation of road access to sites. Water from test pumping was conserved and distributed within the camp.

2 THE APPROACH

The Kitali Hills refugee center was established in the wake of the larger refugee centers further North, where a relatively successful water supply had been achieved by drilling some 30 wells, and with minimal efforts as to the actual siting. The same strategy was initially envisaged for the much smaller Kitali

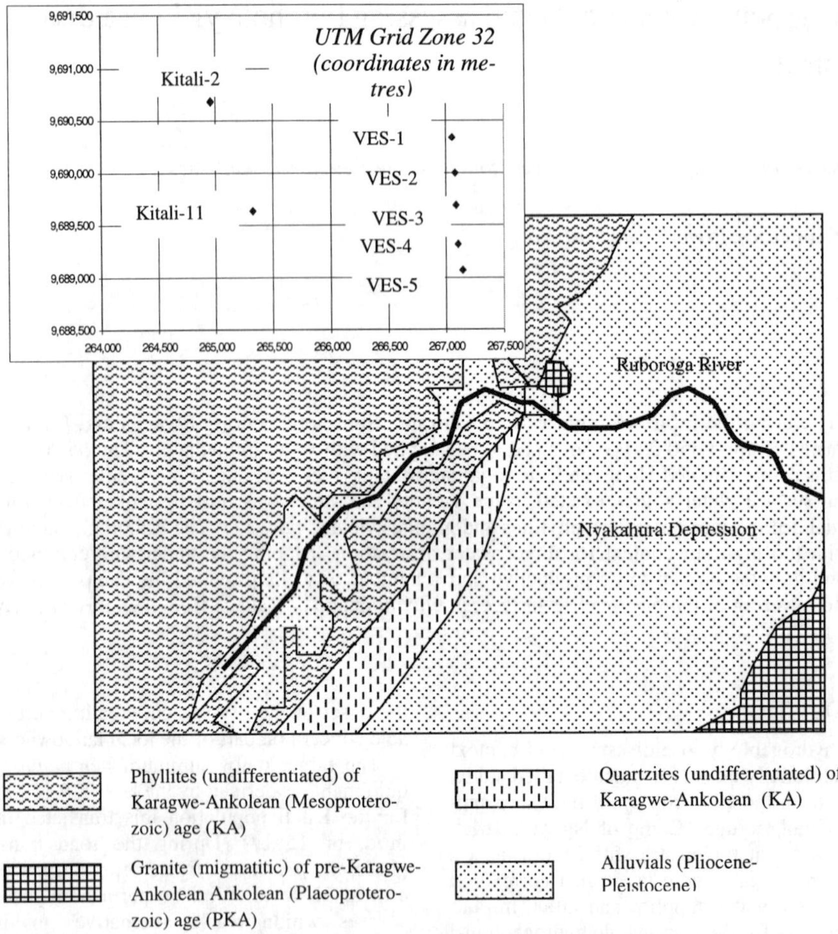

Figure 1. Simplified geological sketch of the Ruboroga River basin, and site map of Kitali Hills Refugee Center (inset) showing VES locations and pumped well Kitali-11. The area is within NW Tanzania, close to the Burundian border.

Hills, but the consequences of a radically different geological setting may not have been fully appreciated at the time.

Initial reconnaissance surveys in the Kitali Hills area indicated that the target of 15 L/s from bedrock aquifers could be problematic, and in order to increase the success rate a multidisciplinary approach was adopted. This included a hydrocensus, review of historical & geological data, infiltration tests, drilling and test pumping. Mapping of local bedrock geology revealed little water potential due to low fracture densities and lack of primary permeability. Shifting towards a geomorphological approach provided slightly more insight. A subsequent VES survey and classification of hydrogeological units provided geometrical and qualitative clues towards the presence of a possible bedrock aquifer. A conceptual aquifer model was the main tool of reference for communication with the increasingly concerned camp managers.

As drilling and testing proceeded, new geological and hydraulic information both corroborated and refuted the early conceptual model. However, as more and more water became available, camp managers grew less interested in how water was found; the concern was now justifiably shifted towards the question of aquifer sustainability.

3 THE SETTING

The siting of refugee centers can be determined by political rather than hydrogeological criteria. Formations present within the area (Figure 1) included a massive outcropping dome of migmatitic granite of *Pre-Karagwe-Ankolean (PKA)* Plaeoproterozoic age outcrops of dense phyllites of *Lower series, Karagwe Ankolean (KA)*, and a quartzite ridge *(KA)* of Mesoproterozoic age. The alluvial plain of the Ruboroga intermittent river were of unknown hydraulic properties. With no observed features of interest in a hydrogeological context, the prospects for groundwater development were bleak. Drilling into the outcropping phyllites and granites on either sides of the river encountered no water bearing features.

Saprolitic aquifers are an important source of water in rural Africa. Only rarely can these aquifers be counted on for large-scale supply purposes, such as required in the Kitali Hill case.

Kitali Hills is situated at the mouth of the Ruboroga River basin, along the western flank of the Nyakahura depression (Figure 1). The latter is filled with clastics. Both within and around this depression, including the Kitali Hills area, are conspicious remnants of a "low level surface at 4650ft" (1550m) assigned by McKinlay (1967) to be part of the African, possibly a later, erosional surface. Numerous 'mesa' remnants of this erosional surface continue along both flanks of the Ruboroga sub-basin, in part protected by lateritic sheets (Sivarajasingham, 1962), in places with complete saprolitic profiles preserved and exposed. Much of these ancient saprolitic aquifers now rest well above the regional ground water table, but the infiltration potential of these prominent and exposed features were considered significant.

A series of samples from the Ruboroga alluvial basin revealed sediments consisting of poorly to moderately sorted quartzitic sand, set in a matrix of silt. The silt content could be attributed to the combination of old age of the clastics and their predominantly phyllitic provenance. This meant adverse conditions for drilling, and precluded direct abstraction of water by means of traditional screens and gravel pack. On the other hand, the saturated and highly porous sediments hosts a significant volume of water; a volume in the order of $10^7 m^3$ is estimated from assumed basin geometry.

With no water encountered in the adjoining granites, phyllites and quartzites, with the apparent absence of water in the adjoining elevated saprolites, and the technical impossibility of abstracting water from the saturated alluvials, the prospects of a quick solution to the Kitali Hills water supply were bleak. The focus was consequently shifted towards the crystalline aquifers underlying the alluvials.

A number of wells were drilled through the alluvials; cased throughout and sealed into the bedrock. Percussion drilling into the granitic and quartzitic bedrock soon bedrock revealed a number of water bearing fractures. The individual contribution from each of these fractures, i.e. the 'blow yield' was soon masked by factors including considerable and unsteady water influx, the use of various types of flushing additives, variations in fracture and hole geometry and varying effective air pressure gradient over the drilling tool due to water back pressure. With the possible exception of the uppermost aquifer section, the inherent unreliability of the 'blow yield' concept under circumstances as described above rendered water contribution estimates of the lower water bearing fractures unreliable as indicators of well performance. To provide clues towards a sustainable yield, a rigorous hydraulic testing of each well was required. Due to restricted well bore diameter and the urgent need to install permanent pumps with no provision for monitoring devices, this was often not possible.

The catchment area of the Ruboroga sub-basin is approximately 100 km^2; draining via the Nyakahura Depression and Lake Tanganyika, and eventually to the distant Atlantic Ocean. The Ruboroga is an intermittent river with no dry season surface flow, possibly due to the phyllitic and impermeable nature of the catchment area above the Kitali Hills.

4 VERTICAL ELELCTRIC SOUNDING (VES)

The only geophysical means available to the project on short notice was an antiquated but still useable electric sounding equipment. Using Schlumberger configuration and with maximum AB/2 = 200m, 12 vertical electric soundings were made across the Ruboroga plain, with electrode arrays aligned with the direction of the river. Using the well Kitali-11 to calibrate the VES soundings a classification of apparent resistivity (cut off) values was established (Table 1).

Table 1. Cut-off values of apparent formation resistivity

Formation	From:(Ω m)	To:(Ωm)
Topsoil	100	40
Alluvials	40	500
Fractured bed-rock	500	2500
Massive bed-rock	2500	>> 2500

Assuming that the calibration and classification of VES-data against the drilled well Kitali-11 is sufficiently representative in the area, the sounding profile (Figure 2) indicates a skewed alluvial basin underlain by an irregularly shaped fractured aquifer basin.

5 SUSTAINABILITY

The question of sustainability of the aquifers was a salient point throughout the history of the Kitali Camp. As a basis for such assessments, as drilling progressed and VES data became available, the two

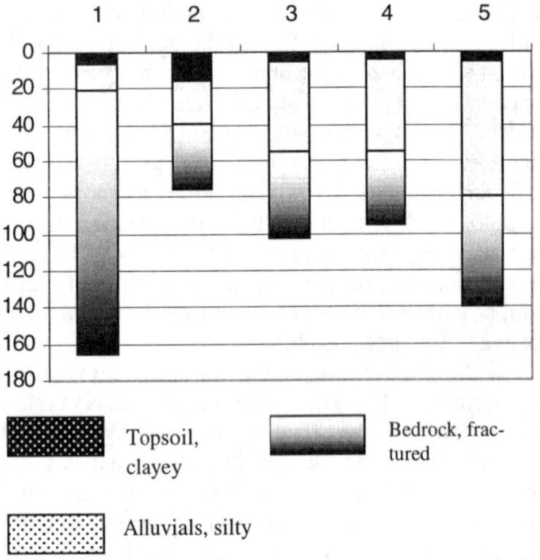

Figure 2. Apparent resistivity profile (metres below surface) across the Ruboroga. Location of VES profiles given in fig.1.

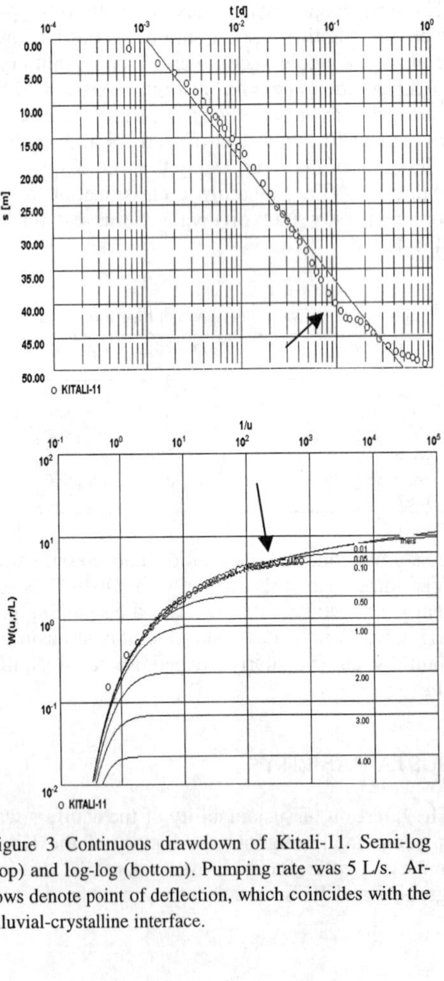

Figure 3 Continuous drawdown of Kitali-11. Semi-log (top) and log-log (bottom). Pumping rate was 5 L/s. Arrows denote point of deflection, which coincides with the alluvial-crystalline interface.

main working hypotheses were a) the presence of considerable storage of water in the alluvials, released by leakage, and b) the presence of hydraulic connection between the alluvials and the fractures of the underlying bedrock.

The presence of the considerable volume of water stored in the alluvials provided the strongest argument against failure of the ground water based water supply system, and this argument was considered adequate by camp managers at the time. For the hydrogeologist, providing a convincing explanation and evidence rather than a conceptual model represented the next challenge. At this point in time, as new wells came on line, water was being pumped at an increasing rate, to everyone's satisfaction. How long the aquifer would be able to maintain this rate of pumping was not known. For the worried camp managers this eventually became the crucial point. Until the emergence of long term test pumping data, the conceptual model was still referred to, but the premises remained untested.

6 THE AQUIFER MODEL

According to the conceptual hydrogeological model, the pumping wells tap the underlying fractured bedrock aquifer. Having high T-values but low storativity, the bedrock aquifer would then effectively transmit the pressure drop during pumping over a large area of the alluvial-saprolitic interface. The fractured bedrock aquifer thus represents a means of abstracting water from the saturated, highly porous, troublesome but also less transmissive alluvials, by vertical leakage. The exposed saprolites on either flank of the basin were assumed important recharge areas, supplying the Ruboroga aquifer system. The model describes the aquifer architecture, but does not give any indication of sustainability of supply.

7 TEST PUMPING

A series of wells were eventually drilled into the bedrock aquifers under the Ruboroga Basin, producing a total of approximately 650 m^3 per day for nearly two years. Due to pump installations and restricted annular space, monitoring of water levels within the pumped wells was usually not possible. However, with new tools arriving on site the well Kitali-11 could be drilled with a sufficiently large diameter to accommodate a monitoring device. The well was pump tested at 5 L/s for a duration of 1200 min. The semi-log curve (Figure 3, top) indicated an apparent recharge boundary and the initiation of possible leakage at a drawdown of 42m (indicated by an arrow). The base of the alluvium and casing correspond to approximately 25 m

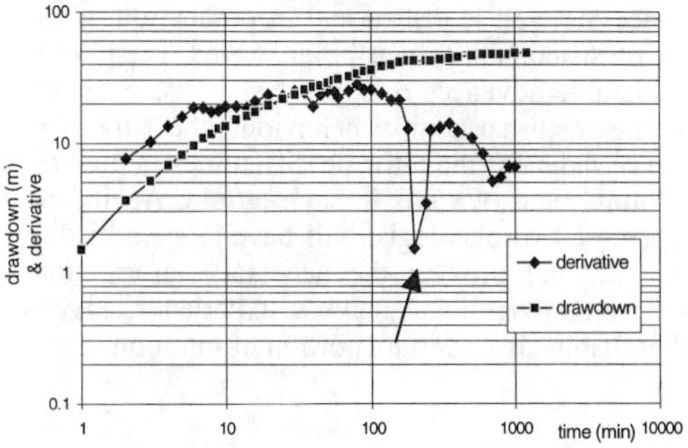

Figure 4 Log-Log plot of drawdown and first derivative of drawdown, Kitali-11. Arrow indicates effects of main water bearing fracture..

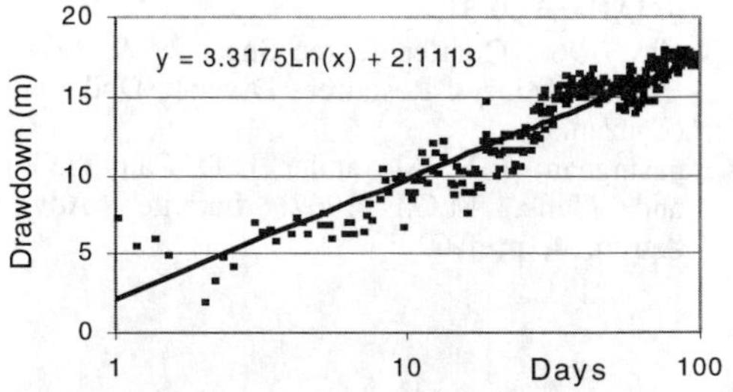

Figure 5. Drawdown in a monitoring well, located 10m E of production well Kitali-2. Discharge=7.5 L/s.

of drawdown. Weathered quartzite extends to 35 m, after which water inflows are encountered in quartzitic bedrock.

The log-log curve (Figure 3, bottom) indicates a leaky (Hantush) aquifer with a r/L value of around 0.05-0.10. Other wells in the area revealed a similar behaviour, with characteristic leakage deflections at a point close to the bedrock interface. As all productive wells also encountered their principal water bearing fractures close to this interface, drawdown effects originating from dewatering of the main water strike and effects from leakage across the alluvial/bedrock interface may coincide.

A main water bearing fracture shows up as a dip on the derivative of drawdown curve (arrow, Figure 4). The following 4-5 points could indicate dewatering effects once the water level drops below the fracture intercept, i.e. corresponding to a transition from confined to unconfined conditions. This effect could subsequently be masked by leakage, as indicated by the negative gradient of the derivative curve. Both leakage and dewatering effects could be at work here, and only a longer test pumping period could have indicated a development in the direction of a Boulton delayed yield curve, or leakage. All wells drilled in the Ruboroga basin show the same tendency: a) reduced rate of drawdown immediately below the first water strike, b) change of drawdown gradient invariably coinciding with the water level passing under the top of the bedrock. However, the recovery data from most wells show a zero t/t' intercept value of around 2, which does not substantiate the leakage explanation.

Long-term water level observations from Kitali-2 illustrate the deviation of the well response from a characteristic 'leaky' response. (Figure 5). The well was pumped at 7.5 L/s for 90 days. The water level dipped under the strike after a few hours, and the rate of (semilog) drawdown remained essentially unchanged throughout this period.

Significant throughflow of water within the fractured aquifer could explain the perseverence of the aquifer. However, the relatively low T-values of the wells ranging between 1.5-3.5 m^2/d, the obsereved hydraulic gradient of less than 0.005 and an average basin width of some 800m, can not provide sufficient ground water throughflow in the fractured bedrock to adequately explain the availability of water and the reduced rate of drawdown under the strike.

The transition from confined to unconfined conditions seems to offer the most plausible explanation for the observed aquifer behaviour and vertical leakage affects are believed to be less important.

8 A REVISED AQUIFER MODEL

Figure 6 illustrates the deduced conceptual model of the Ruboroga Basin alluvial and fractured crystalline aquifers. The aquifers are recharged from the adjoining saprolites, which discharge into both the alluvial and crystalline aquifers. Seasonal floods recharge the alluvials by vertical infiltration of the floodplain. The well (in center) is fully cased into the bedrock and draws water from the crystalline fractured aquifer, which in turn receives water by means of vertical gravity drainage of the overlying alluvials.

9 DISCUSSION AND CONCLUSION

The Ruboroga aquifer provided around $7x10^5 m^3$ of water during the period 1995-1997 while showing little signs of declining yields or substantial drops in piezometric head. Long-term aquifer drawdown caused by pumping exhibited a conventional Cooper-Jacob type drawdown, with no evidence of low permeability boundary conditions. Vertical leakage, i.e. a release of water from the alluvial aquifer as a response of pressure drop in the underlying fractured aquifer was one of the early conceptual pillars of the Kitali ground water supply project. This mechanism is believed to only partially account for the 'leaky

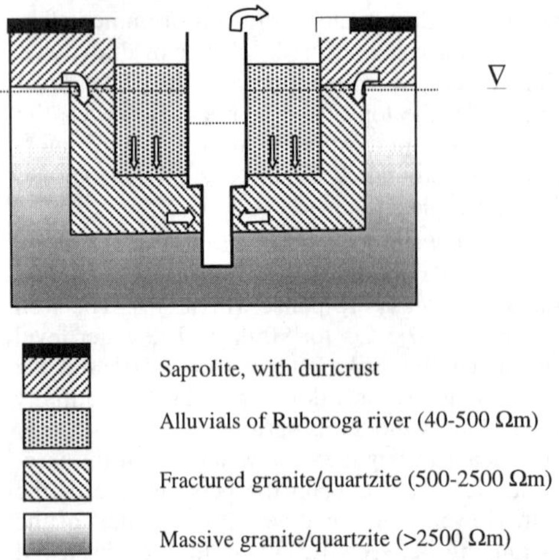

Figure 6. A revised aquifer model. The well (centre) is cased throughout the alluvials, and draws water from the crystalline fractured aquifer only, which in turn is recharged by vertical leakage from alluvials and influx from adjoining saprolites. Note water level (V).

type response of the drawdown curve. Contributions from gravity release of water in the quartzite aquifer due to lowering of water levels under the fracture strikes; i.e. transition to unconfined conditions are significant, as indicated by the long term drawdown. The relative contributions from leakage vs. drainage of the quartzite aquifer has not been resolved. Neither can the aquifer's sustainability be adequately explained in terms of water available as throughflow within the underlying fractured aquifer. However, the influx of water from adjoining saprolites could represent an important recharge mechanism for both the alluvial and fractured crystalline aquifers.

The Kitali Hills camp was eventually closed down in 1997 following repatriation. In light of the sheer volume of the system and the recharge mechanisms, it is speculated that there would have been adequate amounts of water at an abstraction rate of 15 L/s for several years longer.

The role of a hydrogeologist working under relief situations should ideally follow a careful approach such as recommended by Banks et al (1993). However, an overly conservative approach may often lead to overly pessimistic estimates due to limited data. In situations where water is a matter of life and death, calculated risks on part of the hydrogeologist may be justified. This involves necessary shortcuts such as a 'drill first and ask afterwards' mentality that would not be acceptable under more normal circumstances. A traditional ground water survey takes considerable time, and doe not provide critically needed water. The Drilling of a few wells takes about the same amount of time, but provides answers more quickly.

Once the well is drilled and in production, questions of sustainability can be answered. In spite of considerable advances over the last years in understanding fractured aquifer behaviour, the Kitali experience indicates the intrinsic difficulties in predicting future yield of wells in such terrain. Again, the responsible hydrogeologist will have to stick his/her neck out, and provide the necessary information, based on conscientious analyses, experience, advise, and probably also a certain portion of intuition.

10 REFERENCES

Banks, D., Rohr-Torp,E., Skarphagen,H., 1993: Groundwater resources in hard rock; experiences from the Hvaler study, southeastern Norway. In: Memoires of the XXIV Congress of IAH, pp.39-51.

McKinlay, 1967: Quarter degree Sheet 29 & 29W 'Ngara'. Mineral Resources Divison, Dodoma (Tanzania)

Sivarajasingham, L.T., Alexander, L.T., Cady, J.G. and Cline, M.G., 1962: Laterite. Adv. Agron.14, pp.1-60.

Groundwater: Past Achievements and Future Challenges, Sililo et al. (eds) © 2000 Balkema, Rotterdam, ISBN 90 5809 159 7

Change of subsurface temperature caused by climatic change in Japan

Y. Sakura – *Department of Earth Sciences, Chiba University, Japan*

Y. Uchida – *Geological Survey of Japan, Tsukuba, Japan*

M. Taniguchi – *Department of Earth Sciences, Nara University of Education, Japan*

I. Kayane – *Graduate School of Literature, Aichi University, Toyohashi, Japan*

M.P. Anderson – *Department of Geology and Geophysics, University of Wisconsin, Madison, Wis., USA*

ABSTRACT: Generally, the temperature profile in a borehole is influenced by both groundwater flow and past climatic change. Geophysicists tend to remove such effects from a temperature profile to make corrections in terrestrial heat flow values. It is also well known among hydrogeologists that the distribution of subsurface temperature is affected by groundwater flow, and that temperature data can be used to estimate the direction and velocity of that flow. To date, there are few studies simultaneously concerned with both effects. Borehole temperature profiles and the distribution of temperature inversions observed in the Nobi Plain around Nagoya, Japan are presented as an example, showing the effects of both groundwater flow and surface warming caused by global warming and urbanization.

1 INTRODUCTION

To obtain accurate measurements of subsurface temperature, thermal equilibrium between water in a borehole and the surrounding subsurface temperature is necessary. Therefore, it is required that water in the borehole does not move by convection or artificial disturbance. Temperature profiles obtained under these conditions yield continuous vertical information whereas groundwater heads and solute concentrations yield discrete information controlled by the position of the well screen. Temperature profiles are one-dimensional sequential data arrays so that spatial distributed temperature profiles provide three-dimensional subsurface information.

Generally, the normal geothermal gradient of the subsurface thermal regime is affected by groundwater flow and changes in surface temperature. Domenico & Palciauskas (1973) presented a theoretical analysis of temperature field distortion by topographically driven groundwater flow as shown in Figure 1. Their results show that groundwater temperature increases with depth under the influence of both heat conduction and advection. At the same altitude, subsurface temperature in the recharge area is lower when groundwater flow is included than under no flow conditions. In the discharge area, temperature is higher with groundwater flow than under no flow conditions. Thereafter, Figures 1c and 1d provide a conceptual and theoretical framework for

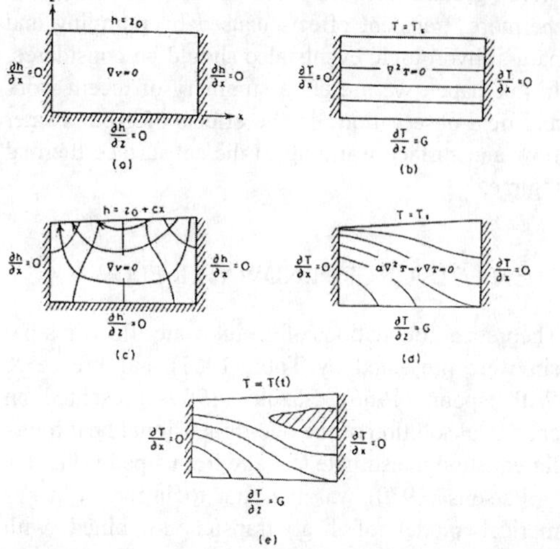

Figure 1. Subsurface thermal regime influenced by a groundwater flow system and surface warming, modified from Domenico and Palciauskas(1973). (a) static groundwater, (b) thermal regime under condition of (a), (c) simple regional groundwater flow system, (d) thermal regime under condition of (c) and (e) thermal regime caused by the effects of groundwater flow and surface temperature warming (Hatching shows the zone of temperature inversion which shows anomalous low temperature). Where, h, v, T, T_s, G, x, y and t are groundwater hydraulic head, groundwater flow velocity, subsurface temperature, constant surface temperature, constant temperature gradient, x-y coordinate and time, respectively.

analyzing subsurface temperature distribution in a groundwater basin.

The effects of past climatic changes on subsurface temperatures have been recognized for a long time (Lane, 1923). The effect of the Pleistocene ice-age, which lasted for several hundred thousand years, is usually assumed to extend to a few thousand meters below land surface, while the effect of climatic changes in the last thousand years extend to 500 m (Jessop, 1990). The effects of seasonal fluctuations extend to a few tens of meters and the effect of daily changes to a few meters. The most recent major climatic change, i.e., the rapid warming for the last 100 years, has caused a temperature inversion in many parts of North America, Europe and Australia at a depth of 50 to 100 m, depending on the thermal properties of the rock (Cermak, 1971). In this paper, the depth below which the effects of seasonal temperature change are negligible is called the isothermal layer.

In analyzing temperature profiles to depths of a few hundred meters, it is necessary to take into consideration two effects, namely topographically driven groundwater flow and surface warming. Furthermore, transient effects caused by pumping and natural hydrologic events also should be considered. In this paper, we present a summary of recent work and field observations on the effects of groundwater flow and surface warming in the subsurface thermal regime.

2 EFFECTS OF GROUNDWATER FLOW

Theoretical foundations of groundwater flow in a basin were presented by Toth (1963) and Freeze & Witherspoon (1966). Suzuki (1960) presented an analytical solution of the one-dimensional heat transfer equation to estimate leakage from a paddy field.

Parsons (1970) was the first to introduce a numerical model of heat transfer combined with groundwater flow and used it to simulate the groundwater flow and thermal regime in northern Ontario, Canada. Domenico & Palciauskas (1973) presented the theoretical relationship between heat transfer and regional groundwater flow using analytical techniques assuming steady-state two-dimensional flow. The International Union of Geodesy and Geophysics Symposium " Hydrogeological Regimes and Their Subsurface Thermal Effects" was held in August 1987 at the General Assembly in Vancouver, British Columbia (Beck et al., 1987). Papers from this symposium showed that subsurface temperatures within a depth of several thousand meters were affected by deep groundwater circulation. Sakura (1993) showed the existence of regional groundwater flow caused a significant disturbance of thermal regimes and produced a zone of high temperature in the central part of the Yonezawa Basin, Northeast Japan.

Recently, Person et al. (1996) presented a review on basin-scale hydrogeologic modeling of coupled groundwater flow, heat transfer and chemical mass transport to provide the impetus for multidisciplinary basin studies between hydrologists and geologists. As groundwater flow is also causally related to tectonic activity, the relationship between groundwater flow and subsurface temperatures have been studied in a variety of Earth science disciplines.

3 CLIMATE CHANGE AND SUBSURFACE TEMPERATURE

The effect of past climate changes on the geothermal gradient was recognized a long time ago and first described by Lane (1923). He found that the rate of increase in temperature with depth in Lake Superior mines in the shallow layer was smaller than in the deep layer and pointed out that this is due to an increase in surface temperature about 10,000 years ago

Surface temperature changes are transferred downward slowly, and are recorded in a borehole temperature profile. Such detailed temperature records provide important palaeoclimatological information. Cermak (1971) made the inverse application of the theory used for climatic correction to terrestrial heat flow and showed that borehole temperatures measured in the Kapuskasing area, northeastern Ontario, Canada clearly provided evidence for the Little Climatic Optimum" (A.D 900 - 1300), Little Ice Age"(A.D. 1500-1800) and recent warming from A.D.1850.

From meteorological data, global temperature changes show recent warming trends of about 0.5 globally over the last century (Hansen & Lebedeff, 1987). Ellsaesser et al. (1986) reviewed methods for detecting climatic change from observational data including air temperature, sea surface temperature, tropospheric temperature, tree ring thickness, temperature from isotopic fractionation, temperature profiles in a borehole and snow and ice cover data. These data suggested significant cooling in the late 1700' with recovery to natural early levels rather than global warming. Problems in interpreting temperature records are caused by the effects of significant urban warming, station change and changes in measurements.

The most recent major climatic change, the rapid warming during the last 100 years, has caused a temperature inversion that is manifested as a low or inverse temperature gradient in the subsurface shallow layer in many parts of the world. Hence, a temperature profile in a borehole is influenced both by surface warming and by groundwater flow.

4 EFFECTS OF BOTH GROUNDWATER FLOW AND SURFACE WARMING

In the Nobi Plain, declines in hydraulic head accompanied by land subsidence have been severe during the last 30 years owing to pumping of groundwater. More than 50 observation wells ranging from 50 to 400 m below the land surface have been installed to monitor groundwater levels and land subsidence. The Nobi Plain is about 1300 km^2 in area. To the west it is adjacent to an alluvial fan at the Yoro Mountains (Figure 2). River terraces near the Owari hills lie to the east, and Paleozoic basement rock crops out in mountains to the north. The Ise Bay borders the plain to the South. The Nobi Plain is a sedimentary basin tilted westward and is composed of sand, gravel, sandy-clay and clay deposited during the Pliocene and Pleistocene (Figure 3). The main aquifers are three layers of gravel composed of riverbed gravel deposited during glacial advances. Confining layers are composed of sandy-clay and clay deposited during inter-glacial stages.

Temperature profiles were measured in 1993 and 1994 in 41 observation wells as shown in Figure 2. A thermister attached to a 300 m cable was used to measure temperature at two meter intervals with a precision of 0.01 in each well. Temperature profiles are classified into three types according to the temperature gradient (Figure 2 and 3). Figure 2 shows that discharge type locates in the southern part of the study area, near the Ise Bay, recharge type locates in the northern part and intermediate type locates in the middle of those parts. Nine borehole temperature profiles along line A-A' in Figure 2 are shown in Figure 3 as representative examples (Uchida & Sakura, 1999).

In these temperature profiles, the temperature inversions, which show the minimum temperature in the profile below the isothermal layer (about 30 m below the surface), are recognized as of the recharge type.

From the distribution of hydraulic head in the Nobi Plain, groundwater recharge occurs at the foot of the mountains and hills peripheral to the plain. Groundwater flows to the central part of the plain and discharges to Ise Bay. Subsurface temperatures and gradients in the recharge area are lower than in

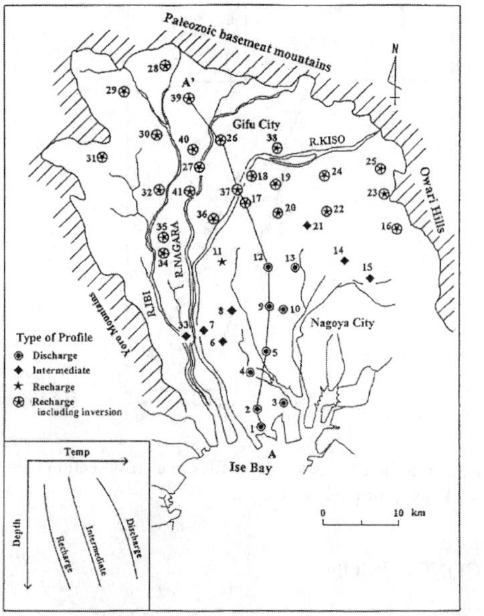

Figure 2. Location of study area and observation wells.

the discharge area and there is a zone of high temperature in the central portion of the plain. Furthermore, almost all of the recharge type profiles in Gifu Prefecture to the west and north of Kiso River show temperature inversions and most of the recharge type and some of the intermediate type profiles in Aichi Prefecture to the east and south of Kiso River also show temperature inversions.

Inversions are observed only outside the discharge area where subsurface temperature is influenced more by the warmed surface temperature than by regional groundwater flow.

In Gifu, average annual air temperature rose from 14 to 15.5 for the last 60 years. Temperature profile in Gifu is shown as well26 in Figure.3. This inversion is due to the global and urban warming, and groundwater flow. Figure 1e shows the conceptual model of complex effects.

Taniguchi et al. (1999) presented an analytical method of groundwater flow estimated from the inversion of a borehole temperature profile in Tokyo and suggested that its inversion plays a role of tracer to understand the groundwater flow system.

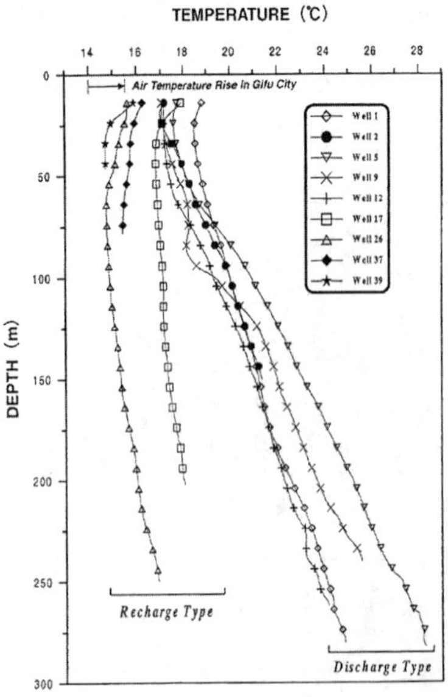

Figure 3. Borehole temperature profiles in a cross section along line A-A' as shown in Figure 2.

5 CONCLUSION

There are some studies dealing individually with the effect of groundwater flow and surface temperature change on the subsurface thermal regime, but there are few studies considering both effects. Temperature profiles and the distribution of temperature inversions in the Nobi Plain, Japan were presented as examples of profiles showing both effects. The conclusions are summarized as follows:

(1) It is generally recognized that the subsurface thermal field within several thousand meters of the surface is affected by deep groundwater circulation; temperatures and thermal gradients in the shallow layers are relatively high in the structural center of the basins, grabens and coastal plains, and are low in the mountain area. Temperature profiles in both the Nobi Plain and the Tokyo area follow this pattern.

(2) In the Nobi Plain, there are many temperature profiles outside the discharge area that show temperature inversions. Inversions appear in the temperature profile as a near surface minimum temperature below the isothermal layer. Inversions are observed where subsurface temperature is influenced more by the warmed surface temperature than by regional groundwater flow. Hence, inversions are rarely observed in the discharge areas.

(3) The subsurface thermal regime is influenced not only by groundwater flow, but also by surface climatic change. Surface warming is enhanced in urban areas owing to the effects of urbanization.

REFERENCES

Beck, A. E., G. Garven, & L. Stegena 1987. *Hydrogeological Regimes and Their Subsurface Thermal Effects*. Geophysical Monograph 47, IUGG 2, AGU.

Cermak, V. 1971. Underground temperature and inferred climatic temperature of the past millennium; *Palaeogeography, Palaeoclimatology, Palaeoecology*; 10: 1-19.

Domenico, P. A. and V. V. Palciauskas 1973. Theoretical analysis of forced convective heat transfer in regional groundwater flow. *Geol. Soc. Amer. Bull.*, 84: 3803-3814.

Ellsaesser, H. W., M. C. MacCracken, J. J. Walton, and S. L. Grotch 1986. Global climatic trends as revealed by the recorded data; *Reviews of Geophysics*; 24: 745-792.

Freeze, R. A. and P. A. Witherspoon 1966. Theoretical analysis of regional groundwater flow: 1. Analytical and numerical solutions to the mathematical model. *Water Resour. Res.*, 2: 641-656.

Hansen, J., and S. Lebedeff 1987. Global trends of measured surface air temperature; *J. Geophys. Res.*, 92: 13345-13372.

Jessop, A. M. 1990. *Thermal Geophysics*. Development in Solid Earth Geophysics, 17, Elsevier.

Lane, A. C. 1923. Geotherms of Lake Superior Copper Country; *Bulletin of the Geological Society of America*. 34: 703-720.

Parsons, M. L. 1970. Groundwater thermal regime in a glacial complex. *Water Resour. Res.*, 6: 1701-1720.

Person, M. A., Raffensperger, J. P., Ge, S. and G. Garven 1996. Basin-scale hydrogeologic modelling. *Reviews of Geophysics*, 34: 61-87.

Sakura, Y. 1993. Groundwater flow estimated from temperatures in the Yonezawa basin, northeast Japan. *Int. Assoc. Hydrol. Publ.* 215: 161-170.

Suzuki, S. 1960. Percolation measurements based on heat flow through soil with special reference to paddy fields. *J. Geophys. Res.*. 65: 2883-2885.

Taniguchi, M., J. Shimada, T. Tanaka, I. Kayane, Y. Sakura, Y. Shimano, S. Dapaah-Siakwan, and S. Kawashima 1999. Distributions of temperature-depth profiles due to surface climate change and subsurface water flow: 1. An effect of linear Increase in surface temperature caused by global warming and urbanization in the Tokyo metropolitan area, Japan. *Water Resour. Res.*, 35: 1507-1517.

Toth, J. 1963. A theoretical analyssis of groundwater flow in small drainage basins. *J. Geophys. Res.*, 68: 4798-4812.

Uchida, Y., and Y. Sakura 1999. Subsurface temperature profiles in the Nobi Plain, central Japan. *Bull. Geol. Surv. Japan*, 50: 635-659.

Groundwater: Past Achievements and Future Challenges, Sililo et al. (eds) © 2000 Balkema, Rotterdam, ISBN 90 5809 159 7

Scientific support for sustainable groundwater management: A modelling study in Botswana using environmental tracers

B. Schaffner, Y.Y. Chen, T. Kelepile & W. Kinzelbach
Institute of Hydromechanics and Water Resources Management, ETH Zürich, Switzerland
L. Carlsson & I. Mannathoko
Department of Water Affairs, Gaborone, Botswana

ABSTRACT: The single most important quantity required as a basis for sustainable aquifer management is the recharge rate. This figure is especially difficult to determine in arid and semi-arid environments. For the Palapye aquifer in Botswana a model was constructed which integrates all available knowledge in order to arrive at an estimate of recharge and an understanding of the functioning of the aquifer. The fractured nature of the aquifer and an ephemeral river required special modelling solutions. Recharge was initially characterized using environmental isotope (δ^2H and $\delta^{18}O$) as well as chloride measurements. The simulated chlorofluorocarbon (CFC) distribution in the aquifer satisfactorily reproduced the measurements and confirmed the model concept. Three recharge variants were distinguished to account for non-uniqueness of calibration. The scenarios were the basis for the determination of sustainable pumping rates. The modelling process showed that isotope, CFC and chloride measurements together with geophysical survey data can provide the essential information for the understanding of this type of fractured aquifers.

1 INTRODUCTION

The village of Palapye is located in the Central District of Botswana, about 265 km north of Gaborone on the Francistown Road. The Palapye wellfield aquifer is situated to the southeast of the village. It is currently the main water supply for the village.

The climate is semi-arid with cold, dry winters and hot summers. The precipitation is characterized by local, high intensity and short duration events that occur mostly between November and April.

The aquifer is situated on a gently northerly sloping topography. It is bordered to the north by the Lotsane River and to the south by the Tswapong Hills. The Lotsane and the Tswapong formations form the confining and water-bearing layers of the aquifer, respectively (Figs. 1, 2). The Lotsane formation comprises strata consisting mainly of shales with some siltstones and occasional quartzite beds. It overlies the northern half of the wellfield area and outcrop exposure is very poor. The beds weather to clayey soils of low permeability causing high surface run-off towards the Lotsane River. The main aquifer is the Tswapong formation, which is fractured and consists mainly of sandstones. Its groundwater potential is highly variable. This is evidenced from the varying yields of the existing boreholes, which strongly depend on the connections to the system of large fractures. The Tswapong formation forms the topographic relief of the Tswapong Hills. Sandy soils, gravels and boulders cover part of the outcrop area. They create highly permeable soils limiting surface run-off and facilitating groundwater recharge. Superimposed on this structure are numerous faults and intrusions of dolerite and gabbro. Some of these structural features act as barriers to flow, while others act as conduits and have good groundwater potential.

The aquifer is unconfined in the southern part of the wellfield on the slopes of the Tswapong Hills where the Tswapong formation is either exposed or near the surface. The aquifer becomes confined towards the northern part of the wellfield due to the overlying Lotsane formation (DWA 1997; Neumann-Redlin 1984; GCS, in press).

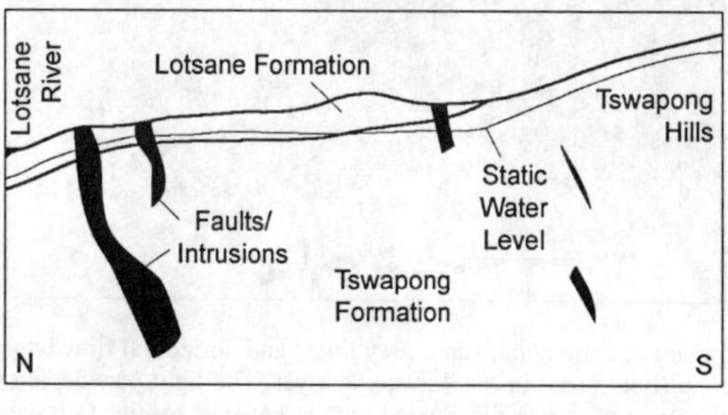

Figure 1: Schematic geological cross section. The Lotsane formation is up to 100 m deep. The N-S distance is about 12 km on the average.

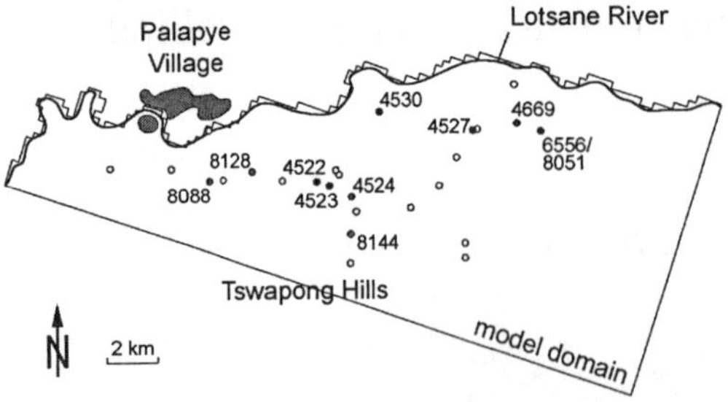

Figure 2: The model domain. Borehole numbers of the boreholes sampled for chlorofluorocarbon (CFC) are given. Production boreholes are marked by black circles, observation boreholes sampled for CFC by grey circles, and additional observation boreholes and wells by open circles. All boreholes were sampled for chloride, δ^2H and $\delta^{18}O$.

2 MODELLING

The aquifer system was modelled using the package *Processing Modflow 5.0*. A two-layer model (i.e. the Lotsane and Tswapong formations) was constructed using geophysical data and drilling logs as well as geological maps. The model domain is shown in Figure 2. A two-layer version was preferred to a probably sufficient one-layer model of the Tswapong formation in order to be able to mimick exchange mechanisms in a more physical way. Measurements of stable isotopes, chloride and chlorofluorocarbons helped further characterize the aquifer.

2.1 *Modelling a fractured aquifer*

Groundwater flow in the aquifers is presumed to be dominated by the flow in the system of large fractures. For modelling, these fractures were conceptualized (Fig. 3):

- Cells along the major lineaments were treated as zones with distinct hydraulic characteristics and were assigned common conductivity values higher than those of the background cells.
- Structural features acting as barriers to flow (such as intrusive dikes) were modelled by assigning «horizontal-flow barriers» (i.e. low permeability barriers) to the edges of the cells along those features.

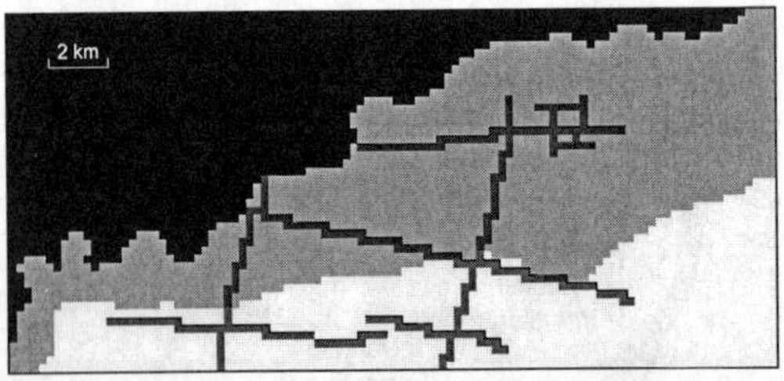

Figure 3: Fault zones (dark grey lines) and horizontal flow barriers (bold lines) in the Tswapong layer. The light part depicts the outcrop, while the shaded part is covered by the Lotsane layer.

2.2 *Modelling an ephemeral river*

The Lotsane River is ephemeral and carries water only for 3 months of the year. During the rainy season it acts as a drainage while in the dry season groundwater is lost from the riverbed by evapotranspiration. This situation was modelled by establishing an automatically adapting dual boundary consisting of a drain and an evaporation boundary:

1. The drain boundary removes water from the model when the groundwater head is above the drain elevation. The drain elevation of each river cell was set 10 m below the top elevation of the cell, taking into account the incision of the riverbed. The drain conductance was chosen high enough to remove any excess water from the model domain.

2. The evapotranspiration (ET) boundary removes water from model cells when the water table is at or above the ET extinction depth. The rate of removal decreases linearly from the full rate at the ET surface to zero at the ET extinction depth. As for the drain, the elevation of the ET surface was chosen at 10 m below ground elevation. Extinction depth was chosen as 3 m. The maximum ET rate was estimated from climatic data.

In this way, the behaviour of the ephemeral river is captured: When the water table is above the drain elevation, water is removed both by drainage and maximum ET. For a water table below the drain elevation, water is removed by a linearly decreasing ET rate. At a head of 3 m below the drain elevation, the ET rate drops to zero and from that level on no more water is removed (Fig. 4).

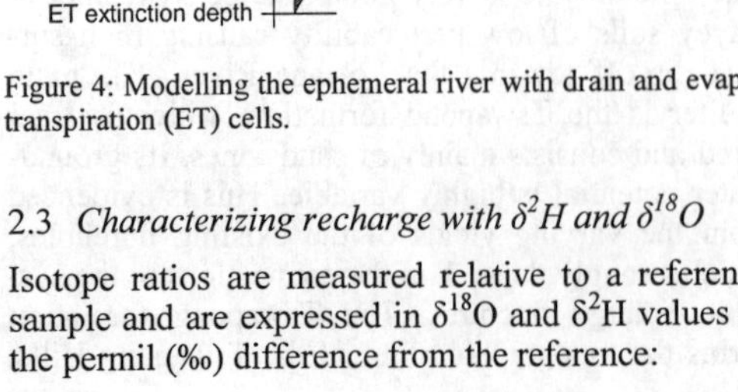

Figure 4: Modelling the ephemeral river with drain and evapotranspiration (ET) cells.

2.3 *Characterizing recharge with δ^2H and $\delta^{18}O$*

Isotope ratios are measured relative to a reference sample and are expressed in $\delta^{18}O$ and δ^2H values as the permil (‰) difference from the reference:

$$\delta^{18}O_{sample} = \left(\frac{\left(^{18}O/^{16}O\right)_{sample}}{\left(^{18}O/^{16}O\right)_{reference}} - 1 \right) \cdot 1000‰ \text{ VSMOW} \quad (1)$$

$$\delta^{2}H_{sample} = \left(\frac{\left(^{2}H/^{1}H\right)_{sample}}{\left(^{2}H/^{1}H\right)_{reference}} - 1 \right) \cdot 1000‰ \text{ VSMOW} \quad (2)$$

where VSMOW is the reference used, in this case Vienna Standard Mean Ocean Water (Clark & Fritz 1997).

Stable isotopes in water are affected by meteorological processes that provide a characteristic fingerprint of their origin. Craig (1961) found that $\delta^{18}O$ and $\delta^{2}H$ in rain water correlate on a global scale according to the «global meteoric water line». A refined version suggested by the International Atomic Energy Agency (IAEA) is given by:

$$\delta^{2}H = 8.17(\pm 0.07)\delta^{18}O + 11.27(\pm 0.65)‰ \text{ VSMOW} \quad (3)$$

Water becomes progressively enriched in both ^{18}O and ^{2}H during evaporation. On a $\delta^{18}O$ versus $\delta^{2}H$ diagram, the result is a deviation from the meteoric water line along with a lower slope, which depends largely on the relative humidity (Clark & Fritz 1997).

In Figure 5, $\delta^{18}O$ is displayed against $\delta^{2}H$ for samples taken in the wellfield. Two types of water can be distinguished:

1. Boreholes and wells reaching into the Tswapong formation all lie on the meteoric water line, which signifies that the groundwater in the Tswapong layer is subject to little or no evaporation before reaching the water table, despite the usually strong evaporation in semi-arid regions. This can be explained by the fractures which are partially exposed or lie below a rather thin unsaturated zone, and – during strong precipitation events – permit the fast movement of mobile water to the water table with very limited mixing with the isotopically enriched water found in the upper parts of the soil profile. The measurements all lie very close together, which means that the groundwater was recharged at a similar altitude and temperature, i.e. in the outcrop of the Tswapong formation.

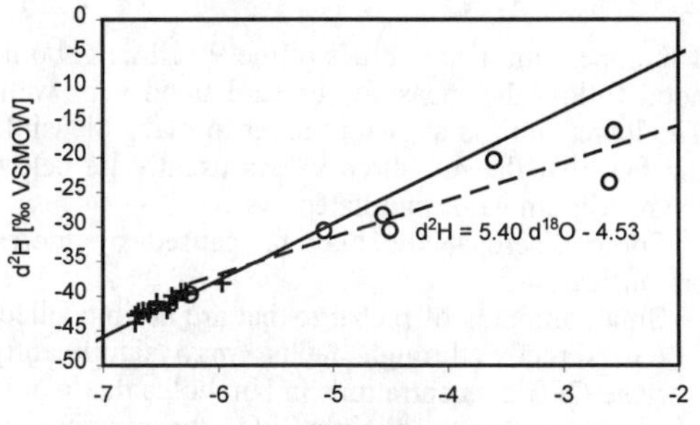

Figure 5: $\delta^{18}O$ vs. $\delta^{2}H$ measurement results. Boreholes in the Tswapong formation are marked by crosses; boreholes and wells in the Lotsane formation by circles. The solid line depicts the global meteoric water line, the dashed line depicts the local water line of the Lotsane formation.

2. Values from wells and boreholes of the Lotsane formation deviate from the meteoric water line. This could be due to evaporation, possibly through slow infiltration in the shaly and little pervious Lotsane formation or through direct evaporation from the large diameter wells. As this water is of no interest to the recharge of the main aquifer, these values are not further discussed here.

2.4 *Estimating recharge with chloride*

Recharge was estimated by measuring the chloride content in groundwater samples from the Tswapong outcrop. According to Wood and Sanford (1995), a groundwater mass-balance technique can be applied if the only source of groundwater in the aquifer is from precipitation, which is true in this case. Chloride concentrations in the aquifer reflect the degree to which the chloride in rainwater has been concentrated by evaporation. Thus, knowing the amount of annual precipitation and the chloride concentration in groundwater and precipitation, it is possible to calculate the recharge flux. The steady-state chloride mass balance across the water table can be given as:

$$q\text{Cl}_{gw} = P\text{Cl}_{wap} \quad (4)$$

where q is the groundwater recharge flux, Cl_{gw} is the chloride concentration in groundwater, P is the precipitation and Cl_{wap} is the weighted average chloride concentration in precipitation (including wet and dry deposition). As the water of the aquifer is not affected by evaporation, the question arises how the chloride method can be used for recharge estimation here. As recharge occurs only in large rain events, salt deposited in the soil layer during small and completely evaporating rain events will be carried down to the aquifer subsequently by large rains. The salinity of the groundwater accounts for the evaporated water by averaging and can still be used for recharge estimation from total precipitation.

The surface run-off, r, from the outcrop of the Tswapong formation directly to the Lotsane River during heavy rains may remove some chloride from the recharge area, which will not show in the groundwater:

$$q\text{Cl}_{gw} = P\text{Cl}_{wap} - r\text{Cl}_{ro} \quad (5)$$

where Cl_{ro} is the chloride concentration in the surface run-off.

For Palapye, recharge was calculated with Equation 4 to lie between 8.3 mm/a and 29.8 mm/a, which is 2% to 10% of the mean annual rainfall, based on an annual chloride deposition of around 430 mg/(m^2a) (Gieske, 1992). These values are an upper estimate for recharge, because the direct surface run-off to the Lotsane River was neglected (Equation 5). They are in line with the estimate of 5% of the mean annual rainfall by Neumann-Redlin (1984) and Smoor (1988).

2.5 *CFC measurements confirm the model concept*

Chlorofluorocarbons (CFCs) are man-made organic compounds. There are no known natural sources of CFCs. Chlorofluorocarbons CFC-11 ($CFCl_3$) and CFC-12 (CF_2Cl_2) have relatively long residence times in the atmosphere (44 and 180 years, respectively), and so their atmospheric concentrations are uniform over large areas. Thus atmospheric input to the groundwater can be known with a high degree of precision, even for remote sites. CFC concentrations in the atmosphere have steadily increased over the past 50 years. CFC-11 and CFC-12 are considered to be essentially stable in groundwater under aerobic conditions (Cook & Solomon 1997).

Apparent CFC ages in the groundwater can be obtained by converting measured CFC concentrations in groundwater to equivalent air concentrations using known solubility relationships and the recharge temperature. Corrections for excess air are made if necessary. However, this method of comparison implies the assumption of plug flow. Here another strategy was followed by modelling the transport of CFC directly in a distributed model put on top of the flow model.

CFC input to the aquifer was simulated by using the CFC concentration curve of the Southern Hemisphere (Busenberg 1999). This curve was discretized into 29 2-year periods from 1941 to 1999 and assigned to the model as a recharge concentration series.

Water samples were taken in oxidized copper tubes which were closed gas tight by squeezing them shut with a tightening screw mechanism. The analysis was performed in the lab in Switzerland.

Effective porosity values in the faults and in the background of the two layers were adjusted to obtain the best result for the modelled CFC distribution (Fig. 6). In Figure 7, observed CFC concentrations are compared with the minimum and maximum calculated concentrations found within a 9-cell-array around the cell of each sampled borehole. A 9-cell array was chosen for comparison because the CFC concentration distribution has very large gradients along faults that are not modelled precisely due to coarse model discretization. The samples were ordered according to measured concentration value.

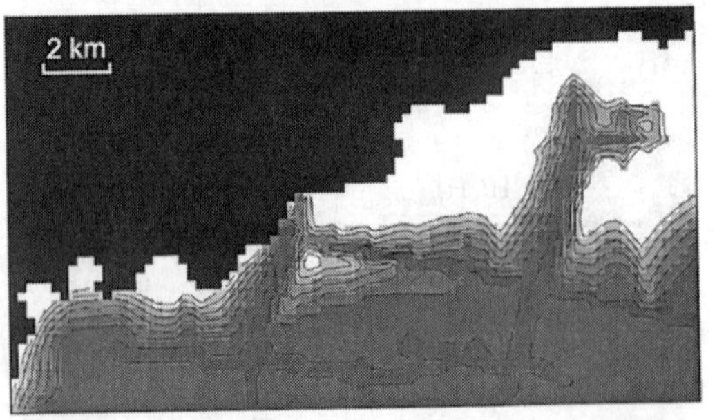

Figure 6: Simulated CFC-12 distribution in the Tswapong layer. The contour interval is logarithmic, the ratio of two neighboring contours being 2.17.

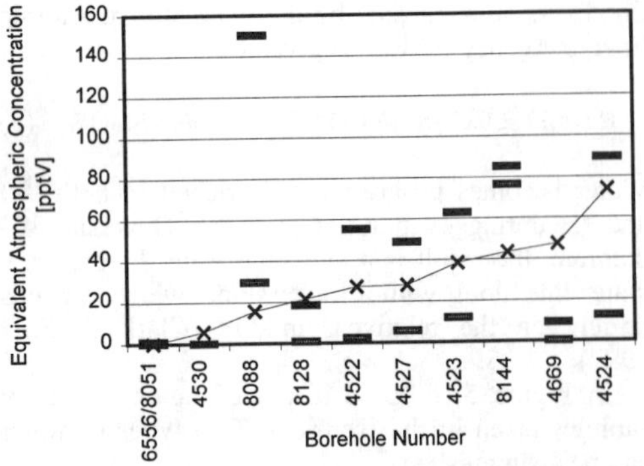

Figure 7: Observed CFC-12 concentrations (crosses) compared to modelled concentrations. The bars show the maximum and minimum modelled concentrations within a 9-cell array around each borehole. Present day atmospheric CFC-12 concentration in the Southern Hemisphere is around 500 pptV. Compare Figure 2 for borehole numbers.

The upper simulated values of the 9 cell neighbourhood follow the measured upward trend very well. The lower limit is shown in order to make plausible the fact that the measured values usually lie below the maximum value simulated.

Some discrepancies may be caused by model simplifications:

- Small amounts of recharge that are not modelled, e.g. directly through faults, may significantly raise CFC concentrations in boreholes that would otherwise show negligible CFC concentrations.
- Fault zone cells were assigned common conductivity and porosity values, even though conductivity and porosity may greatly vary in faults.

Further error possibilities arise from the measurement and analysis procedure, which is confirmed by differing values where double samples were taken.

While pumped boreholes were preferred for sampling, three observation boreholes were also used (4524, 8088 and 8144). These could not be completely purged during sampling. Thus, some CFC contamination via contact with the atmosphere might have occurred. Errors could also have arisen during the sampling procedure, where direct air contact of the sample has to be prevented.

3 SUSTAINABLE WELLFIELD MANAGEMENT

3.1 *Recharge variants*

The calibration shows that there is a minimum recharge below which no fit of the observed heads is possible. At larger recharge values a good fit between observed and measured head values is achieved. The fit, however, is not unique and allows a range of recharge rates. Taking this into consideration, it was decided to run three recharge variants for the management scenarios:

1. The upper recharge variant adopts the highest estimate for the recharge flux calculated from chloride concentration in the groundwater (32 mm/a).
2. The lower recharge variant is run with the lowest recharge parameter with which model calibration still delivers acceptable results (19 mm/a).
3. The best estimate recharge variant adopts the mean of the two recharge fluxes (25 mm/a) .

3.2 *Sustainability and dynamic water levels*

The «natural» limit to water abstraction is, of course, the total input to the aquifer, i.e. the total annual recharge, which may be calculated for each recharge variant.

However, the maximum possible annual abstraction is much lower than the annual recharge for two reasons: First, not all of the recharge can be pumped at the boreholes. Some groundwater will inevitably reach drains or evaporate (e.g. near the river). Second, sustainable drawdowns in boreholes are limited by the depth of the main water strike in this kind of fractured aquifer. If the dynamic water level descends below this limit, the main water strike is quickly dewatered. This could lead to its collapse and/or the possibility of iron or manganese clogging due to oxygenation of the strike itself (GCS, in press). For safety, sustainable water levels are chosen at 5 m above the main water strikes. By positioning the pump intakes at that depth, borehole depletion below the main water strike is automatically avoided.

The water levels for the production boreholes calculated in the model are valid for a borehole radius that can be computed with the following formula (Prickett & Lonnquist 1971):

$$r_{sim} = \frac{a}{4.81} \qquad (6)$$

where a is the grid distance (250 m) and r_{sim} [m] is the radius of the borehole simulated by the model. To calculate the effective dynamic water level of the borehole with the real radius r_{bh}, a corrective term Δh has to be subtracted from the calculated water level (Equation 7, Prickett & Lonnquist 1971). All of the boreholes are situated in or next to faults that act as flow barriers. Therefore the cones of depression develop only towards one side. For this reason, Δh is subtracted twice (Equation 8):

$$\Delta h = 0.3665\left(\frac{q}{T}\right)\log\left(\frac{a}{4.81 r_{bh}}\right) \qquad (7)$$

$$h_{eff} = h_{calc} - 2\Delta h \qquad (8)$$

where h_{eff} [mamsl] is the effective dynamic water head, h_{calc} [mamsl] is the calculated water head, q [m^3/s] is the abstraction rate, T [m^2/s] is the aquifer transmissivity in the vicinity of the borehole, and r_{bh} [m] is the radius of the borehole.

3.3 *Prediction scenarios*

In order to facilitate sustainable wellfield management, several scenarios were generated. As an example, Figure 8 shows the results of a scenario assuming continuous use of the wellfield with different pumping rates for production borehole 4669. According to the recharge variant, the maximum sustainable rate lies between 120% and 200% of the 1998 abstraction rate for this borehole.

As the aquifer will in the future probably be used as a backup water supply, it was also of interest to study how long the aquifer takes to recover, and how long and how much the aquifer can be overpumped.

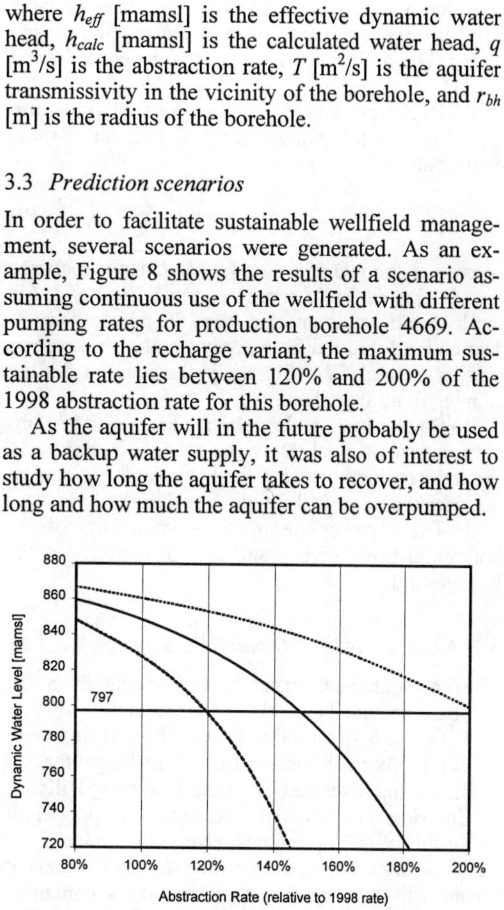

Figure 8: Dynamic water levels in borehole 4669 for continuous pumping rates (relative to 1998). Upper (dotted), best estimate (solid) and lower (dashed) recharge variants are displayed. The line at 797 m depicts the sustainable water level.

In summary, it can be said that the present total pumping rate is still well below the average value of recharge. However, no substantial increase in pumping rate can be justified.

4 CONCLUSIONS

4.1 *Groundwater Modelling*

The model satisfactorily reproduces measurements both qualitatively and quantitatively. The high degree of qualitative reproduction indicates that the model reflects the major characteristics of the wellfield. The functioning of the wellfield seems to be understood.

The quantitative correspondence of observed and calculated data does not necessarily mean that the parameters were correctly chosen. There are further possible parameter sets that reproduce measurements just as precisely. This uncertainty was taken into consideration by choosing three recharge variants with different parameter sets in the prediction scenarios. The main causes of non-uniqueness are listed below.

1. Recharge and hydraulic conductivities are linked by the following simplified relation of mass conservation:

$$rA_1 = kA_2 I \tag{9}$$

where r is the recharge flux [m/s], A_1 is the recharge area [m^2], k is the hydraulic conductivity [m/s], I is the hydraulic gradient, and A_2 is the aquifer's cross-section [m^2]. For a different, say smaller value of r, a smaller value of k will satisfy the relation for the same hydraulic gradient.

2. Background and fault conductivities are inversely proportional. Only the product of the background conductivity and fault conductivity can be accurately determined for given recharge.

3. The same applies to the conductivity of the horizontal-flow barriers and the conductivity of the background.

4.2 *Chloride, δ^2H, $\delta^{18}O$ and CFC data*

The environmental tracer measurements from the wellfield provided essential information:

- $\delta^{18}O$ and δ^2H sampling provided proof that there is no other relevant source of recharge for the Tswapong layer than from the Tswapong Hills.
- Chloride data allowed estimating an upper limit for the range of possible recharge flux values.

CFCs indicated by their presence that there is recent recharge to the aquifer as they can only be contained in groundwater that is younger than 50 years. The decisive samples come from deep pumped boreholes where contamination through atmospheric contact can be excluded. CFC measurements were used to calibrate porosity. They prove by their presence in the production boreholes that fault zones are preferential flow paths that conduct most of the available flow. The alternative hypothesis of a homogenous flow through the bulk rock would not allow CFC to reach the boreholes in many years to come.

CFC sampling was initially intended to serve in the determination of recharge. However, it was found that CFC are not suitable to estimate recharge in a fractured aquifer because the porosity (which is needed to convert the pore velocity to a recharge flux) may not be determined with the necessary precision. Nevertheless, CFC sampling allowed qualitative verification of the modelling concept and calibration of porosities by transport modelling.

ACKNOWLEDGEMENTS

The authors wish to express their gratitude to the Permanent Secretary, Ministry of Minerals, Energy and Water Affairs for allowing them to publish this paper. We would also like to thank Markus Hofer at the Swiss Federal Institute for Environmental Science and Technology (EAWAG) for his kind assistance with CFC sampling and analysis as well as Stefano Bernasconi at the Geological Institute of ETH for his valuable support with the δ^2H measurements. GCS is thanked for help with the field work and financial support by the Alliance for Global Sustainability, ETH, is gratefully acknowledged.

REFERENCES

Busenberg, E. 1999. Written communication.

Clark, I.D. & P. Fritz 1997. *Environmental Isotopes in Hydrogeology.* Lewis Publishers, New York.

Cook, P.G. & Solomon, D.K. 1997. *Recent Advances in Dating Young Groundwater: Chlorofluorocarbons, $^3H/^3He$ and ^{85}Kr.* Journal of Hydrology 191 (1997): 245–265.

Craig, H. 1961. *Isotopic variation in meteoric waters.* Science, 133: 1702–1703.

DWA 1997. *Palapye Water Supply Supplementation Programme.* Wellfield Consulting Services, Botswana.

GCS, in press. *The Palapye Wellfield Monitoring Report.* Geotechnical Consulting Services, Botswana.

Gieske, A. 1992. *Dynamics of Groundwater Recharge in Semi-Arid Eastern Botswana.* Ph. D. thesis, Free University Amsterdam.

Neumann-Redlin, C. 1984. *The Palapye Groundwater Exploration Project.* Department of Geological Survey, Lobatse, Botswana.

Prickett, T.A. & C.G. Lonnquist 1971. *Selected Digital Computer Techniques for Groundwater Resource Evaluation. Illinois State Water Survey.* Bulletin 55, State of Illinois.

Smoor, P.B. 1988. *Recharge to Groundwater at Palapye Well Field.* Department of Geological Survey, Botswana.

Wood, W.W & W.E. Sanford 1995. *Chemical and Isotopic Methods for Quantifying Ground-Water Recharge in a Regional, Semiarid Environment.* Ground Water, 33, (1995).

Groundwater: Past Achievements and Future Challenges, Sililo et al. (eds) © 2000 Balkema, Rotterdam, ISBN 90 5809 159 7

Theory and procedure of thermal fluid logging in discontinuous rock mass

Martin Schreck, Kameichiro Nakagawa & Hiroshi Suenaga
Geology Department, Abiko Research Laboratory, Central Research Institute of Electric Power Industry, Japan

ABSTRACT: The known-data of parameters describing fluid-flow in discontinuous rock mass are still unsatisfactory for the needs of input data for fracture flow models. Despite the sharp rise in the mathematical performance of fracture models with the introduction of ever faster computers, the quality of the results still leaves much to be desired. New methods for parameter identification in discontinuous rock must be developed to improve the quality of input data for fracture-flow-models. This work introduces a method for determining the influx direction, influx rate and permeability of discontinuous rock using thermal fluid logging by applying **D**istributed **I**ntrinsic **F**iber **O**ptic temperature measurement. After heating a borehole by warm water injection, the recovery phase was monitored using the DIFO-3D-Monitoring system. One fiber cable was attached to four sides of an injection tube in two loops. Differences in temperature data between each side of the injection tubing at the same depth level were linked to influx direction. A numerical heat-exchanger-model was then applied to compute the influx rate and permeability from the decrease in temperature versus depth and time during the recovery phase. The data derived are compared to results obtained from standard measurements such as RQD measurement and a constant-head injection test.

1 Introduction

Chang (1) introduced the fluid logging method for determining the location of water influx zones. In his study, the change in salinity was monitored and this change versus time enabled the determination of groundwater-conducting fractures. Later, the same principle was used, but instead of the change in salinity inside the borehole, the change in temperature was measured using fiber-optic temperature measurements (2). In each case, fluid logging was only applied to determine the depth of influx zones.

Both fluid logging methods have proved reliable in various field tests. However, the amount and accuracy of field data describing a discontinuous rock mass is still not sufficient for the needs of input data for fracture flow models. Although the mathematical performance of fracture models is swiftly increasing with the arrival of faster computers, the quality of the results is still unsatisfactory. Therefore, new methods for parameter identification in discontinuous rock are required.

Measuring the temperature in a borehole at the same depth but at different positions along the circumference enables the influx direction to be computed. Furthermore, the decrease in temperature versus time and depth can be used to compute the influx rate and permeability. However, this task necessitates a small temperature measurement system (with negligible hydraulic head loss) which is highly accurate and enables continuous monitoring over depth in a borehole in discontinuous rock.

We present the mathematical routines for the computation of flow rates in a borehole using temperature measurement, the experimental setup and the interpretation algorithm of data from such an experiment. To evaluate the reliability of computation, the results are compared to results derived using standard measurement techniques.

2 Theory

2.1 Mathematical model

Determination of flow direction

For a water-filled borehole in a rock formation, the temperature (T) in K is given by:

$$T = f(x, y, z)\,. \qquad (1)$$

Heat transfer here is governed by three separate mechanisms: (1) heat conduction ($\dot{Q}_{Formation}$) in the formation matrix, (2) heat migration by the fluid phase $\ddot{Q}_{FluidPhase}$) in $\frac{W}{m^2}$, and (3) heat exchange between the two phases, which depends on the temperature difference. For very short heating periods, (1) and (3) become very small. Thus we can write:

$$\frac{Q_x}{Q_y} = \frac{\ddot{Q}_x}{\ddot{Q}_y} \tag{2}$$

According to *Bernoulli's law* the balance of heat migration due to fluid flow at the section of the borehole for short periods of time can be written as:

$$\ddot{Q} = \lambda\frac{\partial T}{\partial x} + \lambda\frac{\partial T}{\partial y} + \lambda\frac{\partial T}{\partial z} = 0 \tag{3}$$

With a lack of vertical flow ($\frac{\partial T}{\partial z} = 0$) and for $\partial x = \partial y = \partial d_a$ we can write Eq. 2 as:

$$\frac{Q_x}{Q_y} = \frac{\Delta T_x}{\Delta T_y} \tag{4}$$

Thus the rate of flow and its direction can be estimated using the tracer temperature if the water is injected for a short period of time from the borehole into the formation, and the flow back into the borehole is monitored. For $x \perp y$ the direction of fluid flow to a borehole section in discontinuous rock can be written as:

$$\tan\xi = \frac{\Delta T_x}{\Delta T_y} \tag{5}$$

Determination of influx

The temperature variation along a borehole profile during fluid logging arises because of heat transfer due to inflow into the borehole; time-dependent convection within the borehole; heat transfer between the formation and the annulus of the borehole; heat transfer between the injection tubing and the annulus of the borehole.

The heat transport inside the borehole takes place in either direction between formation and borehole fluid as well as between the fluid in the injection line and the fluid in the annulus. This results in a heat exchanger (3; 4). The heat losses are given by the heat exchange between annulus and formation and are equal to the heat efficiency. As the heat efficiency is expected to be constant, the case in question is a quasi-stationary case (5). The function $H_f(t)$ defines the heat transfer coefficient from formation to annulus in $\frac{W}{mK}$ (5)(P, R: Bessel's functions):

$$H_f(t) = \frac{\lambda_F}{G_0(t)} \,. \tag{6}$$

The heat and mass transport processes of this system can be described in a balance equation as follows:

$$\begin{aligned} 0 = {} & \rho_w C_w Q\frac{\partial T}{\partial z} + \pi({r_B}^2 - {r_I}^2)\rho_w C_w\frac{\partial T}{\partial t} \\ & - H_f(T_B - T_F) - C_I(T_B - T_I)\,. \end{aligned} \tag{7}$$

λ heat migration $\frac{W}{mK}$

ρ density $\frac{kg}{m^3}$

r_I radius of injection tubing m

r_B radius of borehole m

C_I coefficient of heat transition from injection tubing into the borehole $\frac{W}{mK}$

C_w specific heat capacity of injected fluid $\frac{J}{kgK}$

The equation is solved for Q in $\frac{m^3}{s}$ and we obtain:

$$Q(z,t) = -\frac{\pi(r_B^2 - r_I^2)\frac{\partial T}{\partial t} - H_f\frac{T_B - T_F}{\rho_w C_w} - C_I\frac{T_B - T_F}{\rho_w C_w}}{\frac{\partial T}{\partial z}}\,. \tag{8}$$

The heat exchange for quasi-stationary heat transfer between the formation and the annulus of the borehole was first described by (5):

$$T_B - T_F = \frac{Q}{\lambda_F}G_o(t)\,, \tag{9}$$

where:

$$G_0(t) = \frac{8}{\pi^3}\Big[\frac{\rho C}{\rho_w C_w}\Big]^2 \int_0^\infty \frac{1 - \exp(-\kappa t\frac{u^2}{r_B^2})}{u^3\Delta u}du \quad \text{, with} \tag{10}$$

$$\Delta u = P^2 + R^2 \quad \text{and} \tag{11}$$

$$P = uJ_0(u) - J_1(u)\frac{2}{\kappa\rho_w C_w} + \frac{Ku^2}{r_B H_R}, \tag{12}$$

$$R = uY_0(u) - Y_1(u)\frac{2}{\kappa\rho_w C_w} + \frac{Ku^2}{r_B H_R}. \tag{13}$$

For a sufficiently exact computation of the integral in Eq. 10 within the limits $(0, \infty)$, the Shannon Theorem (6) has to be taken into account, which defines the highest discrete frequency by:

$$u_{max} = \frac{\pi t_{max}}{N}. \tag{14}$$

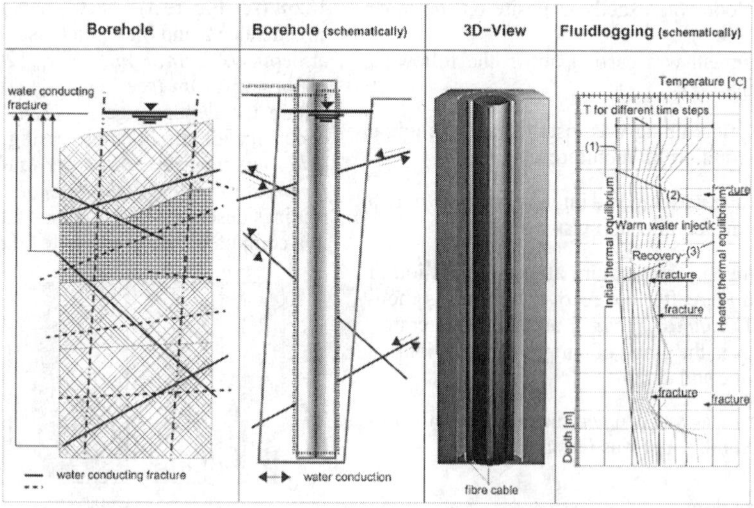

Figure 1: Experimental setup of the temperature monitoring system.

The heat exchange between the injection line and the borehole can be computed in a similar manner by using the equation (7):

$$C_I = \frac{1}{\frac{1}{\alpha_i d_i} + \frac{1}{2 \cdot \pi \cdot \lambda_I} \ln \frac{d_a}{d_i} + \frac{1}{\alpha_a d_a}}. \quad (15)$$

κ thermal diffusivity $\frac{m^2}{s}$

α coefficient of heat transfer $\frac{W}{m^2 K}$

The numerical part is treated by standard techniques: first-order-explicit time steps and central differences for the spatial derivatives.

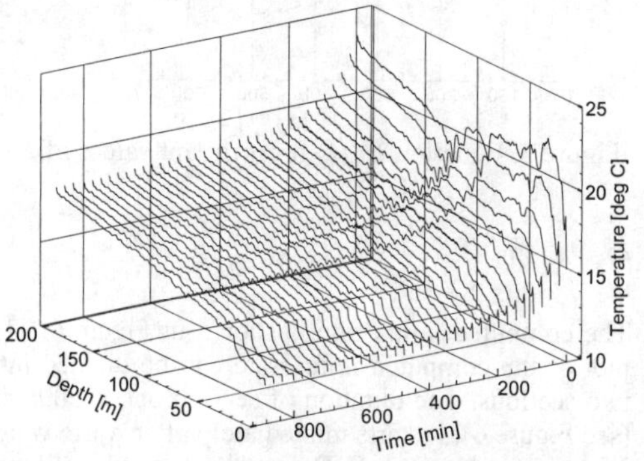

Figure 2: Temperature profile versus time.

3 Application

3.1 Experimental setup

The experimental setup is shown in Figure 1. A $201.3m$ deep borehole with an inner diameter of $076m$ was drilled in crystalline rock. The subsurface consisted mainly of Gneiss. The borehole was drilled with a vertical inclination of 1^0C. Due to the instability of the annulus, the upper $77m$ of the borehole was fully cased, while no casing was installed along the following $131.3m$. Various preliminary investigations were undertaken, such as measurement of RQD-value (rock quality designation (8)), hydraulic conductivity and borehole TV. These preliminary investigations revealed that the main fracture-zones are located at depths of $100m$, $120m$ and $190m$.

A PVC injection tube with an external diameter of $0.43m$ was installed inside the borehole for the experiment. The tube consists of 101 tube sections (each $2m$ long) connected by thread. The injected water flows down this tubing to the bottom of the borehole. Using this system, a constant flow of hot or cold water from the bottom of the borehole can be created as shown in Figure 1. The injected water flows downwards in the injection tubing, out of the tubing bottom, and then moves upwards through the annulus between the tube and the well.

A single fiber cable was attached in two loops to the outside of the injection tubing. This kind of fiber cable setup minimizes the loss of laser light intensity due to fiber cable connections, improving the quality of measurement. The two loops were perpendicular to each other. Each loop was attached such that one half loop was on one side of the injection tubing while the second half loop was exactly opposite on the other side (see Figure 1).

The experiment was carried out in the following steps (see Figure 1):

(1) The injection tubing was installed and the initial thermal equilibrium monitored.

(2) Warm water was injected into the borehole until a new thermal equilibrium was reached.

(3) The change in temperature after the cessation of injection (hereafter the recovery phase) is monitored. The *influx-fractures* are shown as crater-like drops in the graph of temperature distribution versus time and depth.

The experiment was deemed finished when the initial thermal equilibrium was reestablished.

Temperature data

The measurement data is shown in Figure 2. Temperature measurement was carried out at a spatial resolution of $0.5m$. The accuracy of the fiber-optic measurement system is $0.1K$-systematic error. The temperature resolution of the fiber-optic measurement system is $\theta = 0.05K$. Since only temperature differences are interpreted for fluid logging, the accuracy in this case is $0.05K$. The temperature was continuously monitored at a time interval of $260sec$.

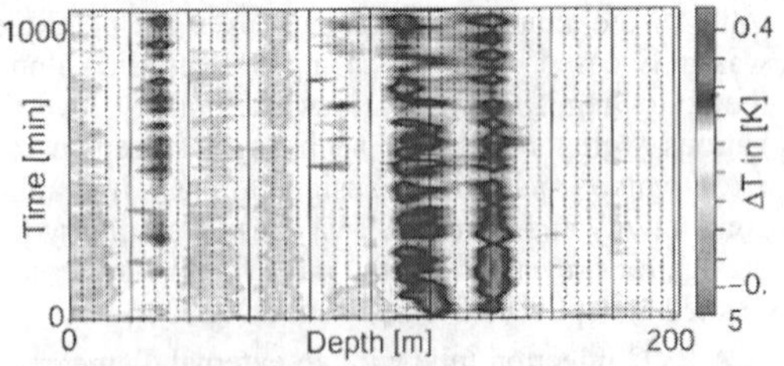

Figure 3: Temperature differences between sides 1 and 4 during the recovery phase after warm water injection.

At first, the average temperature distribution (over time) before any water injection (henceforth referred to as the initial temperature distribution) during thermal equilibrium was determined. Then thermal differences between the opposite sides of the borehole not exceeding the accuracy of the temperature gauge were set to 0.

The depth of inflow zones and their flow direction were recorded in order to gain an initial impression of the inflow zones along the borehole profile. The temperature differences between sides 1 and 4 show such temperature effects at depths of $30m$, $70m$, $90m$, $100m$, $110m$, $120m$, $140m$, $150m$, $162m$, and $182m$ (see Figure 3). The temperature differences between sides 2 and 3 exhibit these temperature effects at depths of $20m$, $80m$, $100m$, $120m$, $150m$, $162m$, $180m$ and $190m$ (see Figure 4). Temperature effects within the depth range $0 - 77m$ are effects which occur outside the borehole casing. *The data of each side were superposed in order to derive the centered temperature profile of the borehole.* The mathematical routines described by Equations 6 to 15 were applied in a computer code to compute $Q(z,t)$.

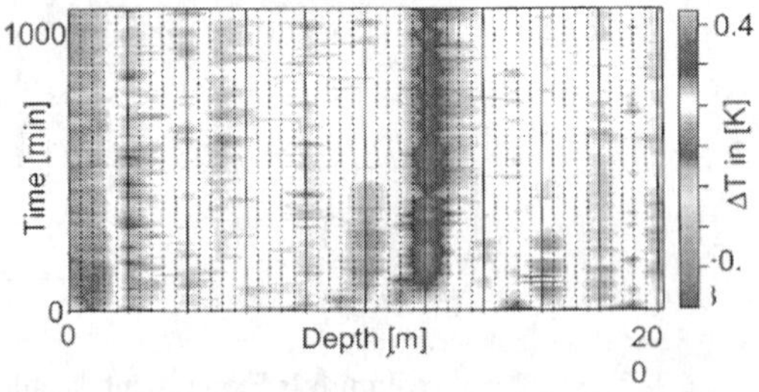

Figure 4: Temperature differences between sides 2 and 3 during the recovery phase after warm water injection.

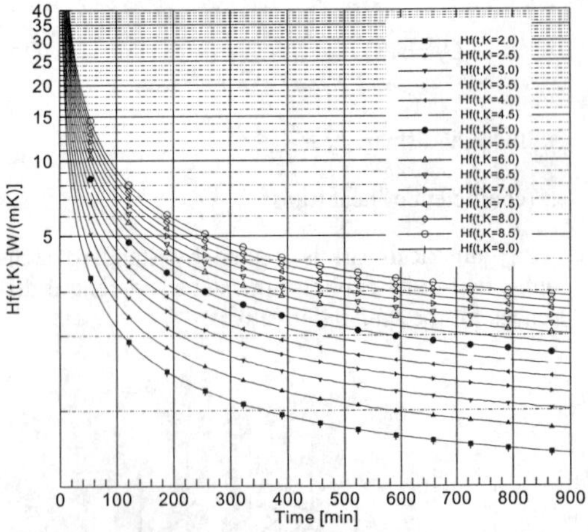

Figure 5: Function $H_f(t)$ for different values of λ_F.

4 Results

4.1 Influx

The computed influx rate is shown in Figure 6. The plot of the computed influx rate can be divided into two sections. The duration of section one is $200min$ (see Figure 6). It starts immediately after warm water injection was turned off. During this time the well was pumped to facilitate the hydraulic gradient for influx into the borehole. Within the first $200min$ during the recovery phase, the influx into the borehole decreases

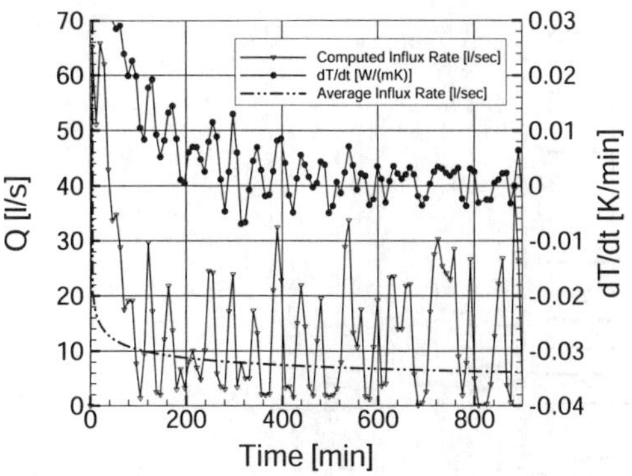

Figure 6: Computed influx rate.

from $65l/s$ to $9l/s$. Section 2 shows the final state, during which the influx decreases to $8l/s$, as shown in Figure 6. The plot of the influx rate versus time shows a frequency 6) stemming from the term $\frac{\partial T}{\partial t}$ of Eq. 7. The reason for this frequency is not yet certain. One explanation could be that the environment to which the DIFO measurement system was exposed was too harsh, causing error measurement. Another might be that with the experiment site being located in a mine, the control of the surrounding drainage system influenced the experiment site. Hence the frequency of the temperature plot was attributable to non-steady geohydraulic conditions. Establishing the source of the frequency is to be the task of future thermal fluid logging tests. An influx rate of $8l/s$ is used to compute the hydraulic conductivity versus depth.

4.2 Hydraulic conductivity

An average hydraulic conductivity estimate (over the whole profile) of $1.7 \times 10^{-7} \frac{m}{s}$ was obtained from the thermal fluid logging experiment.

Conductivity was also determined by means of constant-head injection tests (9) (see also Figure 8). The average hydraulic conductivity was determined to be $1.8 \times 10^{-7} \frac{m}{s}$ with this method. The finest vertical resolution for this measurement method was $3m$.

The spatial resolution of the computed hydraulic conductivity along the profil was very high. Influx zones can be located within $1m$. Comparing these results to those of standard measurement systems, the data correspond to both the RQD measurement and the conductivity. The same zones of higher conductivity were determined using both standard and thermal approaches.

Temperature differences between opposite sides of the injection tube with a magnitude of $2 \times \theta = \triangle T = (-0.1K \leqq 0K \leqq 0.1K)$ were neglected and set to $0K$. The influx direction was determined from the temperature data as shown in Figure 7.

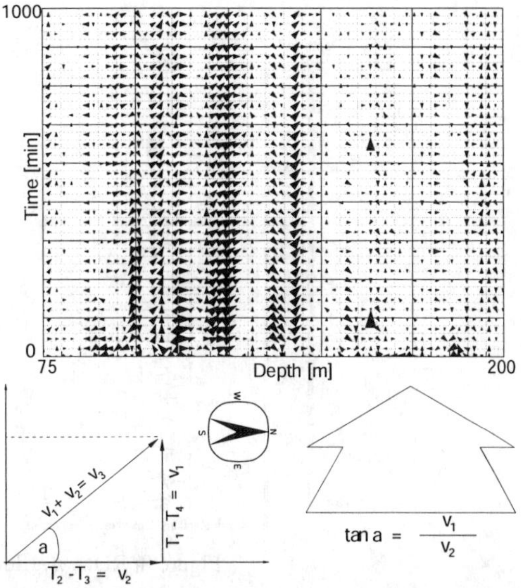

Figure 7: Computed influx direction versus depth and time.

Table 1 contains a summary of the results shown in Figures 8 and 7. In total, 17 levels with an influx of $Q \gtrapprox 0.03l/s$ were found. These levels supply 17.5% of the total influx of $8l/s$. The remaining 82.5% results from influx zones with a $\dot{Q}$ smaller than $Q = 0.03l/s$ and therefore not considered for the localization of influx zones and the determination of influx direction.

5 Conclusion

DIFO measurement was used to determine quantitative and qualitative rock mass parameters. The fiber-optic approach allows data to be acquired with high accuracy and spatial resolution. The fact that the sensor is very small and a profile can be determined at a given instant in time without hydrodynamic impact opens up new possibilities for identifying hydraulic parameters. The arrangement of the fiber-optic cable into two loops on the injection tubing produced data which were accurate and temporally discrete enough to describe radial temperature processes inside the borehole. If the proposed interpretation algorithm is applied together with the proposed mathematical model, the temperature processes monitored can be linked to the magnitude and direction of influx, and to hydraulic conductivity. The results of this experiment provide the better vertical resolution of data than standard measurement techniques. Standard measuring systems deliver data with a vertical spatial resolution of data points of $2-3m$. When using DIFO measurement, the depth of inflow fractures can be lo-

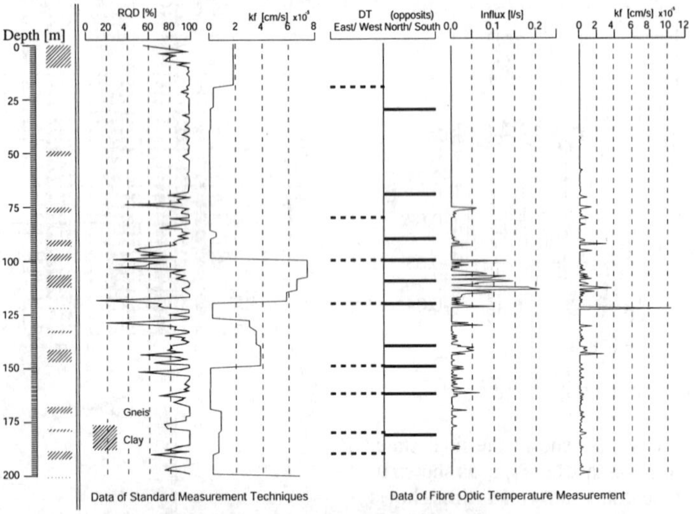

Figure 8: RQD, k_f, flow direction and influx rate.

calized within $1m$ if the temperature is measured with a spatial resolution of $0.5m$. The difference between the magnitude of the hydraulic conductivity derived from DIFO measurement and from standard measurement obtained during this experiment was 5%.

Table 1: Influx Direction of the Influx Zones with 17.5% Share of the Total Influx Rate.

No.	Z [m]	$\dot{Q}$ [$\frac{l}{s}$]	Direct.	Direct. Zone
1	76	0.06	SSE-NNW	A
2	93	0.05	S-N	B
3	101	0.13	W-E	C
4	103	0.04	SSE-NNW	A
5	107	0.09	SSE-NNW	A
6	109	0.13	E-W	D
7	110	0.15	SSE-NNW	A
8	116	0.21	S-N	B
9	117	0.10	SSE-NNW	A
10	123	0.10	SSE-NNW	A
11	130	0.04	E-W	D
12	131	0.07	E-W	D
13	142	0.05	SSE-NNW	A
14	143	0.05	SSE-NNW	A
15	145	0.03	SSE-NNW	A
16	163	0.07	S-N	B
17	172	0.03	S-N	B
		Total: 1.40 (17.5%)		

6 Acknowledgements

The authors would like to thank DIA Corporation Japan amd GESO GmbH Jena for their extensive collaboration.

REFERENCES

C.-F. Chang; P. Hufschmied; F. Hale. Determination of fracture parameters with borehole fluid conductivity logging method. *Water Resources Research*, pages 519–524, 1990.

E. Hurtig; S. Großwig; M. Jobmann; K. Kühn; P. Marschall. Fibre-optic temperature measurement in shallow boreholes: Experimental application for fluid logging. *Geothermics*, 23(4):355–364, 1994.

J. Dornstädter. Das Temperaturangleichverfahren (TAV) - Ein Verfahren zur in-situ Bestimmung Thermischer Gesteinseigenschaften. Master's thesis, Karlsruhe University, 1987.

J. H. Blackwell. A Transient-Flow Method for Determining of Thermal Constants of Insulting Materials in Bulk. *Journal of Applied Physics*, pages 137–144, 1954.

H. S. Carslaw; J.C. Jaeger. *Conduction of Heat in Solids*. University Press Oxford, 1946, 1956.

W. Heinz. *Schwingungslehre Für Ingenieure - Theorie, Simulation, Anwendung*. BI Wissenschaftsverlag Mannheim/ Wien/Zürich, 1989.

H. Landolt; R. Börnstein. *Numerical Data and Functional Relationships in Science and Technology : New Series*. Springer-Verlag, 1992.

Cheng-HawLee; Ian Farmer. *Fluid Flow in Discontinuous Rocks*. CHAPMANN & HALL, 1993.

K. Nakagawa; T. Shidahara; Y. Edo; A. Nozaki. Field studies on geohydrological aspects of deep sedimentary rock for compressed air storage. Technical report, Central Research Institute of Electric Power Industry, 1998.

Investigation of water resources in arid climate using radar

D.Schulze-Makuch
Department of Geological Sciences, University of Texas, El Paso, Tex., USA
P.Haddad
Department of Environmental Science and Engineering, University of Texas, El Paso, Tex., USA

ABSTRACT: A water exploration investigation was conducted in an arid basin in West Texas using ground penetrating radar (GPR) and airborn synthetic aperture radar (AirSAR). The shallow, perched aquifer was found to consist of localized syncline structures that are able to retain water. The depth to the water table of the basin-fill aquifer is at about 60 m below ground surface based on both GPR results and hydraulic head measurements. AirSAR was not able to confirm the obtained GPR results because of the small-scale nature of the localized syncline structures. It remains unclear whether the perched water table would have been detectable if it was continuous. However, AirSAR was sensitive enough to detect changes in surface soil properties that resulted from irrigation practices more than 10 years ago.

1 INTRODUCTION

Previous studies were conducted to evaluate the future requirements of the city of El Paso regarding the availability of drinking water sources (e.g. New Mexico-Texas Water Commission, 1998). The studies showed that the city's current water resources would not meet the city requirements with its increasing population demand for drinking water. The studies also showed that the major water shortage would be within the next 20 years. Thus, new resources are needed to eliminate the predicted water shortage.

The area selected for water exploration was the northern portion of the Ryan Flat sub basin of the Salt Basin near Valentine, Texas (Figure 1). It is located approximately 200 km southwest of the city of El Paso. The Valentine Salt Basin extends between 30.25° N and 31° N latitude at the northern most end of the Marfa Basin. The area is characterized by its subtropical arid climate, the summers are hot and winters are mild.

Parts of that basin were a major area for agricultural activities between 1979 and 1990 for a variety of grain and vegetable crops. The GPR was conducted in a part of the basin where a perched aquifer was assumed to exist, based on shallow depths to water recorded in some wells and historical data (Figure 2).

The Ryan flat subbasin is described by a series of isolated, northwest trending mountain ranges separated by large structural basins (grabens) shaped by late Tertiary and Quaternary normal faulting.

These basins are packed with alluvium and volcanic flows from nearby mountains and usually have closed drainage systems. The Ryan Flat subbasin is underlain by basin-fill alluvium and surrounded on the northwest by the Rubio Dome, an outcrop of volcanic rock; on the southeast by the drainage divide at the southern end of the Salt Basin; to the northeast by the Davis mountains, (Tertiary volcanic field of alkalic igneous rocks); and on the west by Sierra Vieja, (a sequence of Tertiary flow pyroclastic rocks) (Bartolino, 1997).

The Ryan Flat sub basin consists of the basin-fill aquifer with a water table at a depth of about 60 m and a deep basalt aquifer. Also, there is evidence from field observations for a shallow perched aquifer a few meters below the ground surface.

The main objectives of this study were (1) to determine whether the perched aquifer continues throughout the study area, (2) to determine the depth to water in the perched aquifer and in the basin-fill aquifer, (3) to determine whether the perched aquifer is an exploitable water resource. A further objective was to compare the GPR (Ground Penetrating Radar) results to AirSAR (aircraft synthetic aperture radar) results and to investigate whether AirSAR results could confirm GPR results. Remote sensing detection of perched water tables were attempted as early as 1978 (Estes and Simonett, 1978); while GPR is used as a common tool for water exploration (e.g. Blindow et al., 1987; Overmeeren, 1994).

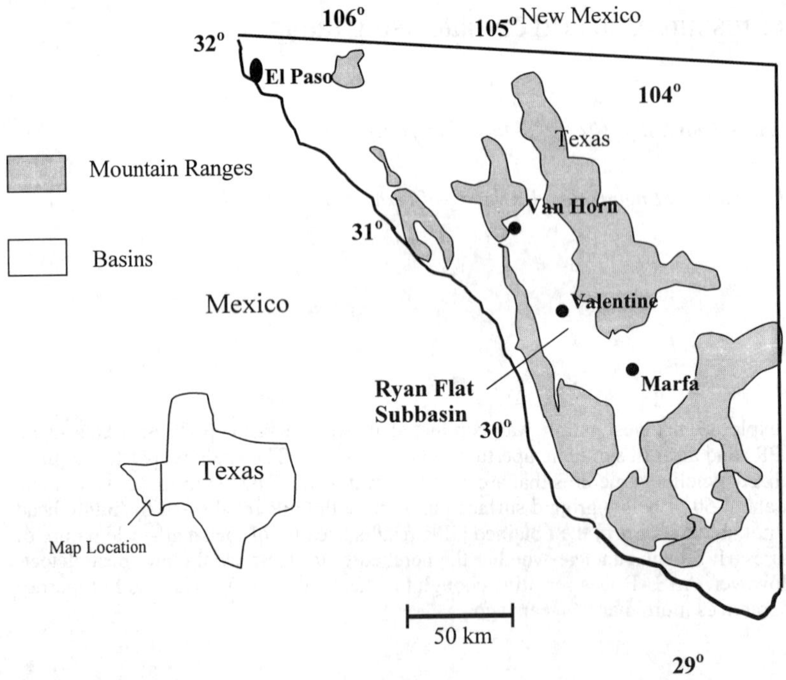

Figure 1. Location of the study area

2 METHODS

The site investigation was conducted by using Ground Penetrating Radar pulse EKKO 1000 with a dominant frequency of 110 MHz, the antenna separation was 1.00 m, the survey mode was reflection, time window was estimated to be 500 nanoseconds (ns). Thus, in normal geological material the electromagnetic velocity is about 0.10 m/ns. Two survey methods were used at the site, a step-wise survey and a continuous survey method. The step-wise method had a step-size of 0.25 m. It was obtained in areas were the continuous survey method showed major changes of the underground structure and where a better resolution of the subsurface variability and continuity was needed. The continuous survey velocity was about 1.0 km/hr. However, the velocity of the towing car varied to some degree depending on the roughness of the surface terrain. The data collection was based on two main survey lines, one located S-N along the extraction wells, and the other survey E-W roughly perpendicular to the first line.

Synthetic aperture radar (SAR) was flown over the study area from an aircraft. AirSAR is a radar system in which fine azimuth resolution is achieved by storing and processing data on the Doppler shift of multiple return pulses in such a way as to give the effect of a much longer antenna (Sabins, 1996). The l, p, and c-bands of the AirSAR images were analyzed and compared to the GPR results.

3 RESULTS

Based on the analysis and evaluation of the GPR data, there is no continuous perched aquifer present in the study area as previously thought. However, several syncline structures of about 15 m in diameter were detected (Figure 3).

These structures appear to hold at least seasonally water based on changes in vegetation pattern. We do not know what these structures are, but they might be collapsed caliche structures that are impermeable enough to retain water. The GPR was able to detect the water table of the basin-fill aquifer at a depth of about 60 m at locations where syncline structures were not present. Our interpretation is that GPR is not able to detect the basin-fill water table in areas where the syncline structures are filled with water, because (1) the water contained in the structures diffuses the signal strength too much, and (2) the structure itself may be a too strong of a reflector to allow enough signal strength to penetrate to the basin-fill water table.

The syncline structures detected by GPR correlate to shallow water observed by auguring in some of the locations in the study area, as did the vegetation pattern. Thus, there is direct evidence that shallow

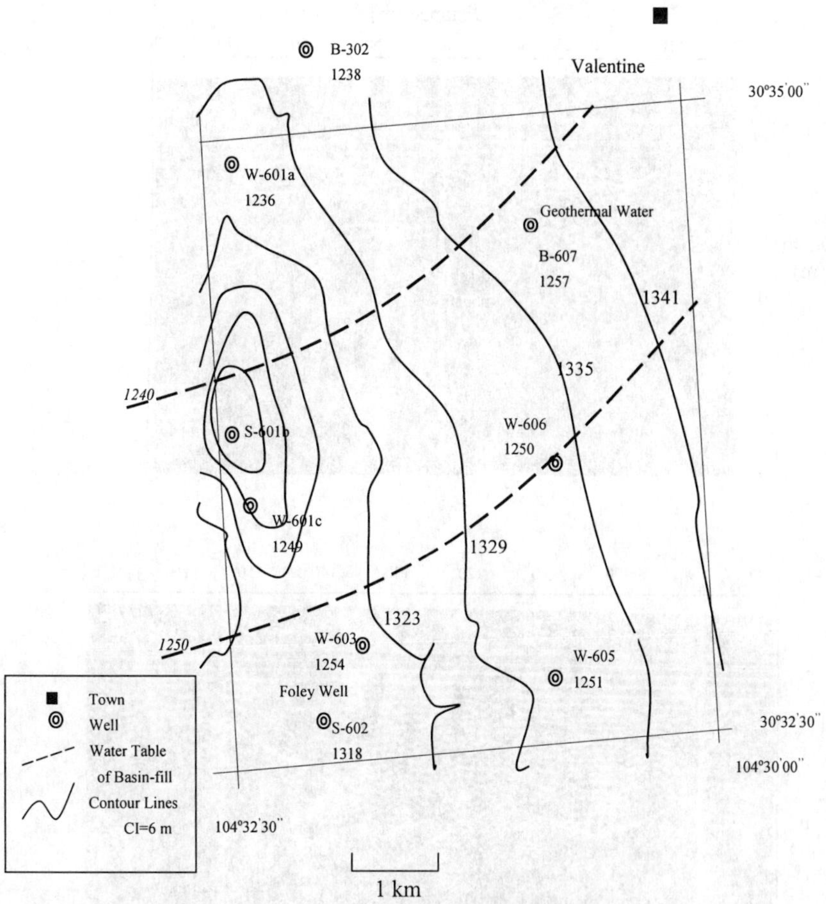

Notes: Contour lines and water table elevations are in meters above mean sea level.
S-wells are screened in the shallow aquifer, W-wells are screened in the basin-fill aquifer, and B-wells are screened in the deep basalt aquifer.

Figure 2. Schematic map of the Valentine Basin showing the water table and well locations.

water is retained and accumulated by the syncline structures. No continuous perched aquifer is present in the study area. Thus, the so-called perched aquifer is not a suitable ground-water resource because of its limited quantity and limited spatial extent. The depth to water table in the basin-fill aquifer is about 60 m below ground surface (bgs) based on both hydraulic head measurement of available wells and GPR. Ground water in the basin-fill aquifer flows to the NNW at a gradient of about 0.005 (Figure 2). The potentiometer surface of the deep basalt aquifer was not detected with GPR. Based on hydraulic head measurement the depth to water in this confined aquifer is about 50 m below ground surface (bgs), data were insufficient to construct a potentiometeric surface map of the deep aquifer.

We were not able to successfully confirm the GPR results with AirSAR. One problem is the shallow penetration depth into the soil of the smaller wavelength AirSAR. The other problem was that no continuous shallow perched aquifer was present that might have been detectable. Instead, the perched aquifer proved to be discontinuous and of small-scale nature. The airborne radar resolution is not high enough to resolve the highly localized syncline structures of the discontinuous perched aquifer. Thus, it is unclear whether AirSAR and GPR results are compatible for the exploration of ground water. However, the AirSAR results surprisingly revealed crop irrigation circles even though there has not been active irrigation in the area for the least 10 years. That shows that irrigation practices change

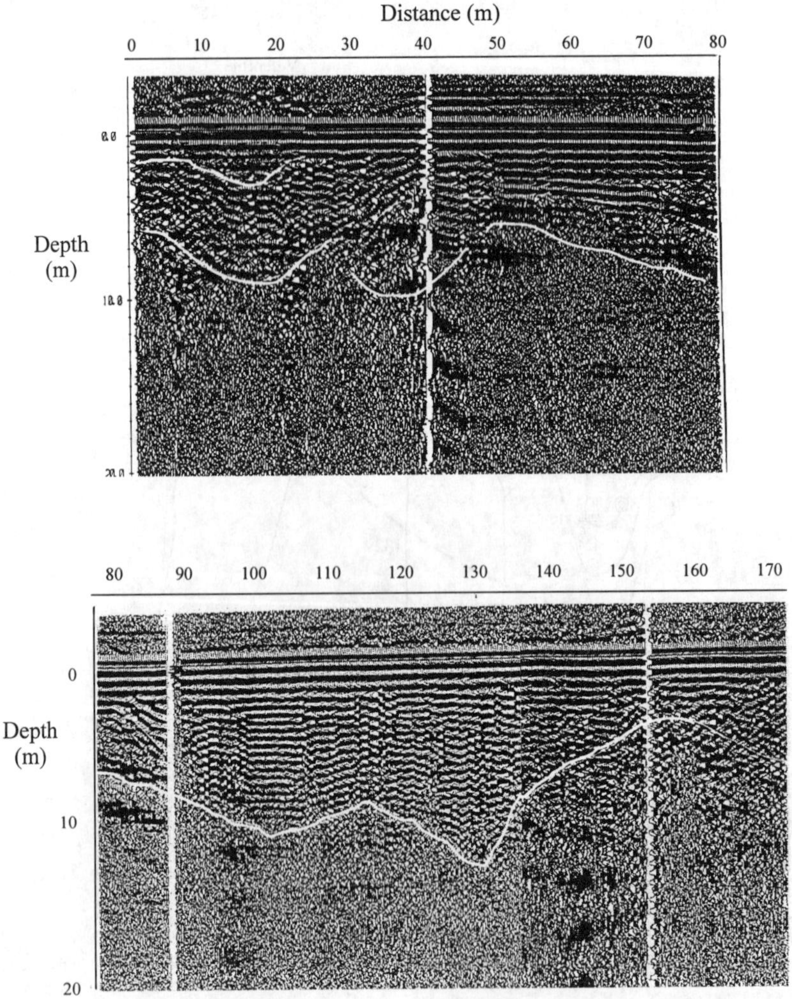

Figure 3. The syncline structures as shown by GPR. Reflectance boundaries are enhanced with white marker.

the surface soil properties significantly enough to be detected by airborne radar long after the irrigation practices have been terminated.

4 REFERENCES

Bartolino, J. 1997. Chlorofluorocarbon and tritium age deter mination of ground-water recharge in the Ryan Flat Subbasin, Trans-Pecos, Texas. USGS Water-Resources Investigation Report 96-4245. 29 p.

Blindow, N., Ergenzinger, P., Pahls, H., Scholz, H. and Thyssen, F. 1987. Continuous profiling of subsurface structures and groundwater surface by EMR methods in southern Egypt. Berliner geowiss. Abh. 75.2: 575-628.

Estes, J.E. and Simonett, D.S. 1978. Remote sensing detection of perched water tables. California Water Resources Center. Contribution No. 175.

New Mexico-Texas Water Commission. 1998. NEPA Alterna tives Preliminary Screening Project. August 5, 1998-workshop on El Paso-Las Cruces Regional Sustainable Water Project.

Overmeeren, R.A. 1994. High speed georadar data acquisition for groundwater exploration in the Netherlands. Proceedings of the 5th International Conference on Ground Penetrating Radar., 3: 1057-1073, Ontario, Canada.

Sabins, F.F. 1996. Remote sensing: orinciples and interpreta tion. New York: Freeman and Company, 494 p.

Groundwater: Past Achievements and Future Challenges, Sililo et al. (eds) © 2000 Balkema, Rotterdam, ISBN 90 5809 159 7

The role of bypass- and matrix flow in the unsaturated zone for groundwater protection

K.-P.Seiler, S.v.Loewenstern & S.Schneider
GSF, Institute of Hydrology, Neuherberg, Germany

ABSTRACT: In the unsaturated zone bypass-flow may compete with matrix-flow. Deuterium tracing proves that bypass-flow exceeds flow velocities of 0.5 m/d and matrix-flow ranges at 0.7 m/year (Loess), 1.2 m/year (sands) and 2 m/year (gravels). In humid climates and unconsolidated rocks bypass-flow penetrates to depths of less than 1m (Loess) and more than 3m (gravels) below ground before it finally incorporates either into matrix-flow or generates interflow in hilly areas. Bypass-flow depends upon geologic, biologic, pedologic and anthropogenic factors. Bypass- and connected interflow enhances vertical and lateral export of agrochemicals and particles like humic substances from the effective root zone. Thus interflow diminishes the import of agrochemicals to groundwater in hilly areas; contrary, in flat lands bypass-flow causes direct impacts to groundwater. Bypass-flow seems to be strongest in wet and dry soils and more dominant in coarse than fine grained-sediments; in most lacustrine and marine sediments bypass-flow is scarce.

1 INTRODUCTION

The unsaturated zone has a regulatory function for discharges like (Fig.1)

- Evapotranspiration,
- Overland flow,
- Interflow and
- Groundwater recharge.

All discharges except evapotranspiration transport pollutants either dissolved or carry it particle favoured and may achieve adverse impacts on ground and surface waters. The main pools of contaminants are waste disposals and the effective root zone of cultivated soils.

Porous media often exhibit a variety of natural heterogeneities, such as shrinking cracks, macro pores, interaggregate pores and destruction features from humans, animals and plants. All these structures may affect water and solute movement at the macroscopic level by creating widely different flow-velocities, often referred to as slow matrix (Hillel, 1971; Feddes et al., 1988) and fast preferential-flow (Scotter, 1978; Beven & German, 1982; Germann, 1990; van Genuchten, 1994). These phenomena have been studied in structured soils by laboratory and field experiments (Beven & Germann, 1982, Harum et al., 1997). Fast-flow leads to an apparent no equilibrium situation with respect to the pressure head or the solute concentration (Wang, 1991) and severely limits our ability to define initial boundary conditions of flow and reliable forecast of contaminant transport in unsaturated media.

Bypass-flow interacts with matrix-flow by means of diffusion and incorporation into the matrix or produces interflow in hilly terrains. However, it is not well understood

- How much of the infiltrating water consists of bypass-flow,
- What is its penetration depth as related to the type of sediments, the weathering history, the bioturbation, shrinking and swelling processes and the intensity of the infiltration process itself and
- How its interaction with matrix-flow is governed by the whole array of capillarities or by surface tension pre-existing before the respective infiltration event.

To answer some of these questions tracer experiments have been executed and have been compared with results of discharge analysis.

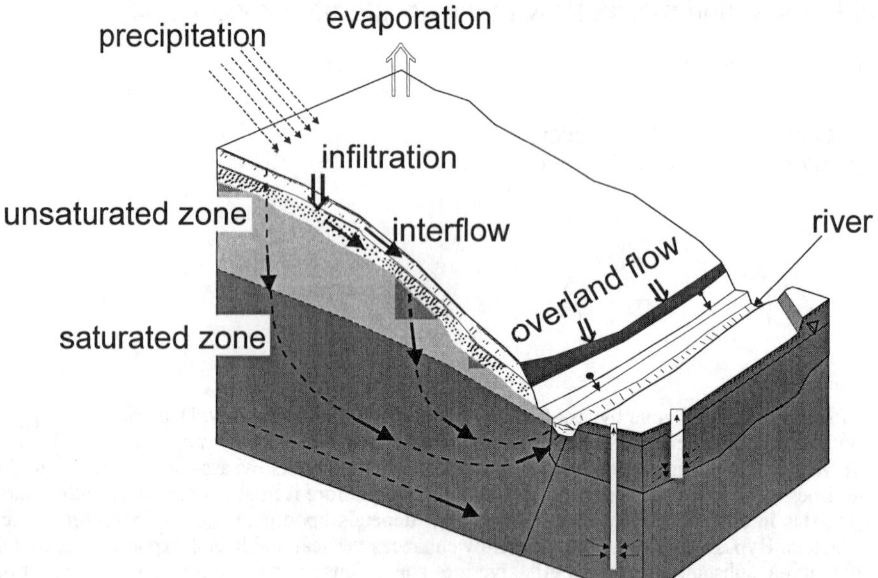

Figure 1. Block diagram of a landscape with the four most important discharge components

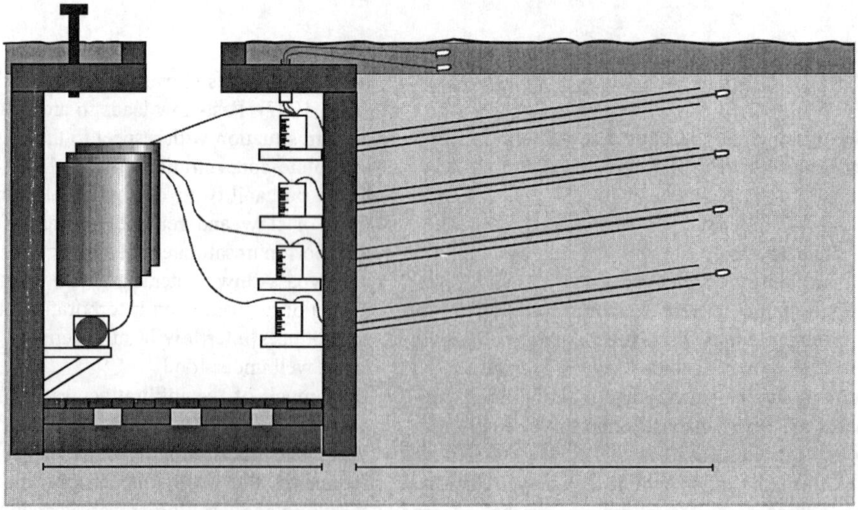

Figure 2. Design of a natural lysimeter with suction cups 3 m away from the shaft

2 DETERMINATION OF BYPASS-FLOW

In field experiments melting snow covers have been traced with Deuterium (^2H) on areas of 50 to 100 m^2. The ^2H-tracer is non-reactive, does not undergo significant isotope fractionation in winter and does not change ion balance of soil water. Deuterium break-through has been observed in natural lysimeter (Fig. 2) at 10 cm, 20 cm, 50 cm, 90 cm, 130 cm and 180 cm below the traced surface using suction cups.

Tracing the melting snow cover produces a Dirac signal along a hydraulic potential plane. As far as flow was homogeneous the Dirac signal produced Gauss-distributions of concentrations at defined time steps all over the profile; this, however, has never been observed in 8 performed experiments (Fig. 3). Obviously flow in the unsaturated zone is inhomogeneous; analyses of break-through-curves confirms

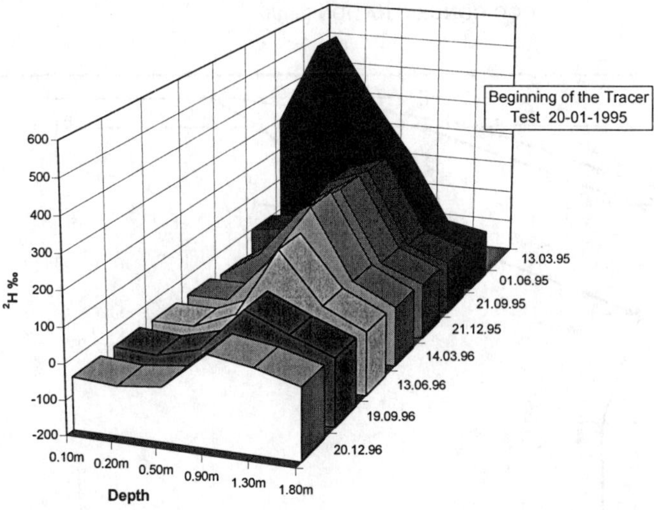

Figure 3. Concentration distribution of Deuterium all over the profile at given time intervals

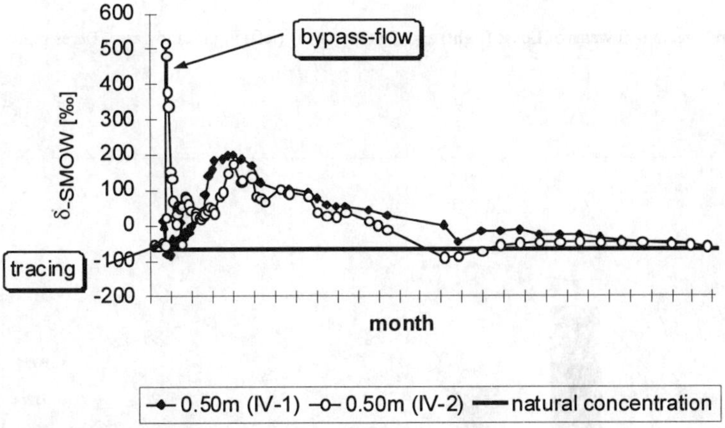

Figure 4. Break through curves of deuterium at the same depth documents matrix and fast bypass-flow in Losses. Line at –60‰ represents mean annual ^{2}H concentration in precipitation

that flow is dominated by slow matrix-flow, superimposed at the beginning of the tracer experiments by positive and thereafter by negative peaks in the break-through-curve (Fig. 4) from bypass-flow. The observation of Deuterium break-through at the same (Fig.4) and at indifferent levels documents also this inhomogeneous flow.

From eight Deuterium experiments it gets obvious that matrix-flow in Loess ranges at 0.7 m/year, in Tertiary sands at 1.2 m/year and in Quaternary gravels at 2 m/year. Infiltration ranges at 150 mm/year in the studied Loess, at 200 mm/year in the Tertiary sands and at 300 mm/year in the Quaternary gravels. Contrary to matrix-flow bypass-flow ranges in all studied sediments similarly in between 0.5 m/day and 2 m/day and has been observed up to a depths of 1 m in Loess, 1.5 m in Tertiary sands and more than 3 m in Quaternary gravels; these flow velocities equal widely to the flow velocity of overland-flow (v.Loewerstern, 1997).

Traditional calculations of seepage velocities based on simultaneous tensiometer- and TDR-observations did not show any significant differentiation between slow and rapid flow. Through ^{2}H experiments this differentiation is possible, however require that injected tracer concentrations exceed

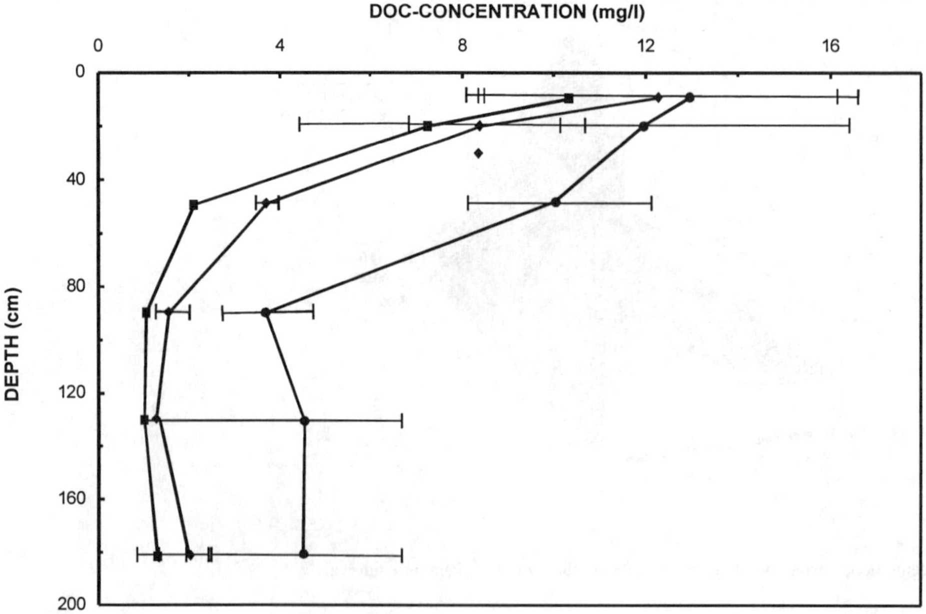

Figure 5. DOC-concentrations in soil water of Loess (right) and Tertiary sands (left) in the study area. Bares represent the width of annual variations.

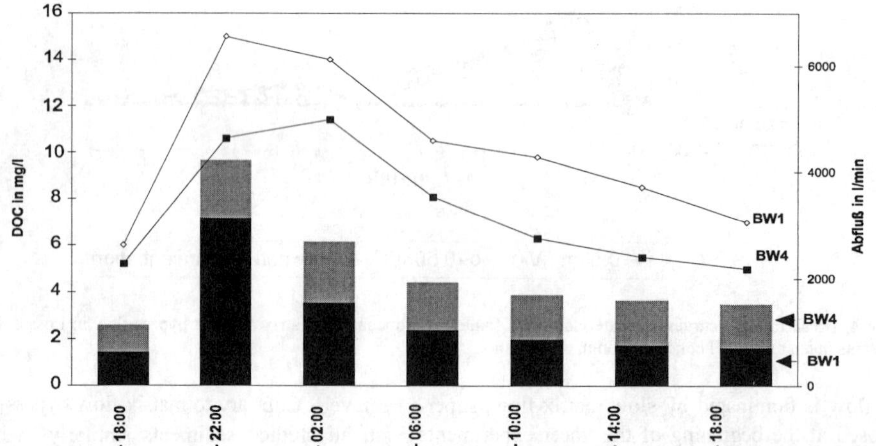

Figure 6. Discharge (columns) and DOC-concentrations in a river of the study area during a discharge event. BW1=100 % forested, BW2=60 % forested and 40 % arable land

several orders of magnitude the mean of natural annual ^{2}H-concentrations (Fig. 4); bypass-flow contributes at indivudual events only few to infiltration and thus is hardly to recognized at too low tracer concentrations.

3 BYPASS-FLOW AND INTERFLOW

Seiler & Baker (1985) have shown, that permeability changes along interfaces may favour lateral flow of seepage (principle of capillary barriers) according to the inclination of the interface. Such interfaces transform fast-seepage flow into fast interflow in hilly terrains. Most of these interfaces are of anthropogenic, rock dilatation, rooting or in high latitudes of permafrost origin and thus parallel morphology. v. Loewenstern (1998) demonstrated on the catchment scale of the study area that discharge (200 mm/year) produce 7 % of overland flow, 14 % of interflow and 79 % of groundwater recharge (150 to 160 mm/year).

The annual mean of 14 % of interflow varies in between 0 and 60 % of infiltration; it depends on initial hydraulic conditions and the intensity, duration and quantity of rain events. Only precipitation exceeding 2 to 4 mm/day produce overland and interflow (v. Loewenstern, 1998) and interflow is strongest under very wet and very dry conditions and less pronounced in between.

4 COTRANSPORT OF BYPASS AND INTERFLOW

In hilly terrains, it is often observed that discharge events increase the concentrations and freights of agrochemicals in rivers; but it is not well known, which is the contribution of solute or particle favoured transport.

Typical distributions of DOC (organic particles <0.45μm) in seepage waters of the study area are shown in Fig. 5. Only within the effective root zone of arable lands DOC exceeds 2 to 4mg/L; in groundwater it amounts maximum to 2mg/L. Rain events increase the DOC-concentrations proportional to surface discharge; this increase reflects only DOC-concentrations typically occurring in the effective root zone and is stronger in the forest than in the arable land. Either the DOC production in forests is higher or the effective root zone of arable lands is more leached.

Also highly soluble chlorides, nitrates and non-soluble pesticides get exported through and proportional to discharge events (v. Loewenstern, 1998; Schülein, 1998). This happens either soluted (chlorides) or fixed on DOC (pesticides). It has been observed that chloride experiences a more intensive export than DOC particles; the nitrate export is situated in between.

Interflow export produce unproductive losses of agrochemicals and chock impacts to surface waters and thus contributes to groundwater protection in areas with existing interflow.

5 CONCLUSIONS

In the study area only mass transport studies on soil and small catchment scale contribute in combination directly to distinguish quantitatively between bypass and matrix-flow; the classical hydraulic observations (tension, water contents and its evaluation by estimating hydraulic functions) in unsaturated soils does not as well.

The velocities of bypass-flow are in the same order of magnitude as overland flow and produce in hilly terrains interflow along inclined interfaces of discontinuities of hydraulic conductivities paralleling morphology. This interflow has a significant export potential for water, nutrients and applied agrochemicals either soluted or particle bound; in this way it reduces groundwater recharge and favours to some extend groundwater protection.

Bypass and interflow are significantly connected to human, animal and plant activities in the effective root zone. Therefore hydrograph analysis demonstrate, that mean residence times of interflow are shorter in the vegetation period than during winter and early spring season.

REFERENCES

Beven, K., German, P. 1982. Macropores and water flow in soils. *Water Resour. Res.* 18: 1311-1325.

Feddes, R.A., Kadat, P.van Bakel, P.J.T Bronswyk, J.J.B. Halbertzsma, J. 1988. Modeling Soil Water Dynamics in the Unsaturated Zone. - State of the Art. *J. Hydrol.* 100: 69-111.

Genuchten van, M. Th. 1994. New Issues and Challenges in Soil Physics Research. *15th World Congress of Soil Science, Vol. 1: Inaugural and State of the Art Conferences*: 5-27, Acapulco, Mexiko..

Germann, P. F. 1990. Preferential flow and the generation of runoff. 1. Boundary layer flow theory. *Water Resources Research* 26: 3055-3063.

Harum, T., Fank, J., Stichler, W. 1997. Tracer investigation in the unsaturated zone under different cultivation types in a mountainous catchment area. *Tracer Hydrology 97*: 153-160, Balkema.

Hillel, D. 1971. Soil and Water. - *Acad. Press*, New York, London.

Loewenstern v., S. 1998. Separierung und Bewertung von Abflusskomponenten für den Stoffaustrag aus Einzugsgebieten mit Tertiärsedimenten (Scheyern, Oberbayern). *GSF-Ber.* 7/98: 142 p., Neuherberg.

Schülein, J. 1998. Untersuchungen zum lateralen Transport von Herbiziden und Herbizidmetaboliten in einem Agrarökosystem. *FAM-Ber.* 24: 145 p., Neuherberg.

Scotter, D. R. 1978. Preferential solute movement through larger soil voids I. Some computations using simple theory. *Austr. J. Soil. Res.* 16: 257-267.

Seiler, K.-P., Baker, D. 1985. Der Einfluss der Schichtung auf die Sickerwasserbewegung bei punkt- und linienförmiger Infiltration. *Z. dt. geol. Ges.* 136: 659-672.

Wang, J.S.Y. 1991. Flow and transport in fractured rocks. *Rev. Geophys.* 29: 254-262.

Groundwater: Past Achievements and Future Challenges, Sililo et al. (eds) © 2000 Balkema, Rotterdam, ISBN 90 5809 159 7

Groundwater recharge deduced from He isotopes in the Botswana Kalahari

E.T. Selaolo
Geological Survey Department, Lobatse, Botswana

D.R. Hilton
Geosciences Research Division, Scripps Institute of Oceanography, La Jolla, Calif., USA

H.E. Beekman
German Technical Co-operation, Harare, Zimbabwe

ABSTRACT: The *long-term average areal recharge* was evaluated for the Mmamabula aquifer in the eastern fringe of the Kalahari Basin using helium isotopes and abundances. Helium concentrations (^4He) and isotopic $^3He/^4He$ ratios were found to vary considerably both in the Mmamabula aquifer and along the Meratswe Valley. In the recharge areas, ^4He contents range from 50 to 350 nccSTP/gH$_2$O whereas $^3He/^4He$ ratios lie between 0.2 and 0.8R_A. The highest ^4He concentration (~ 50 μccSTP/gH$_2$O) and the lowest $^3He/^4He$ ratios (~ 0.014R_A) are found in the confined parts of the aquifer. The *long-term average areal recharge rate is ~ 2.2 to 3.1 mmyr^{-1}* using the ^4He method. This range of recharge values is within the same order of magnitude as those obtained from the chloride mass balance, groundwater flow and hydrochemical-isotope modeling.

1 INTRODUCTION

Noble gases (He, Ne, Ar, Kr and Xe) dissolved in groundwaters are increasingly being used in hydrogeological studies, particularly in characterising aquifers and investigating the genesis of groundwaters by means of $^3He/^4He$ ratios and ^4He concentrations (Torgersen & Ivey, 1985; Heaton, 1984). In particular, $^3He/^4He$ ratios and ^4He abundances have been used to delineate recharge from discharge areas (Stute et al., 1992) and in dating groundwaters with ^4He for time scales that are too old for ^{14}C dating (i.e. > 50 000 years: Andrews et al., 1991). The Ne, Ar, Kr and Xe abundances can preserve temperature conditions at the time of recharge and hence are useful in palaeoclimatology (Mazor, 1972; Herzberg & Mazor, 1979; Andrews & Lee, 1979). These developments in the application of noble gases rely on the fact that they occur in low abundances in nature and most importantly, they are chemically inert and thus do not take part in water-rock interactions.

This paper focuses on helium isotopes (^3He and ^4He) which were determined in groundwaters of the Mmamabula aquifer and along the Meratswe Valley in the Letlhakeng-Botlhapatlou area of the eastern Kalahari. Detailed geological and hydrogeological background can be found in Beekman & Selaolo (1997) and Selaolo (1998).

2 THEORETICAL BACKGROUND

Helium isotopes (^3He and ^4He) are produced by radiogenic processes in terrestrial rocks (Mamyrin & Tolstikhin, 1984). ^3He is generated in the earth's crust by the neutron action of ^6Li as a result of (n, α) reactions on light nuclei (with natural tritium as an intermediary daughter product) while ^4He is produced during the decay of U and Th and their daughter isotopes. Tritium (^3H) can also be directly subjected to β -decay thereby producing tritiogenic ^3He, particularly where groundwaters show significant amounts of bomb ^3H penetration. The effect of tritiogenic ^3He is that it increases the $^3He/^4He$ ratio in the course of time.

On the other hand, the rate of accumulation of helium (^4He) in groundwaters will mainly vary as a function of time and distance along flow paths. The basic assumption governing this process is that ^4He is generated uniformly throughout the aquifer and that there are no additional sources of helium other than that produced in-situ. However, complications arise when the helium produced in the aquifer is supplemented by additional sources such as that from the underlying crustal rocks and/or the mantle.

In stable cratonic regions, such as southern Africa, crustal degassing is considered to be the most likely mechanism by which helium could be transported from greater depths into aquifer systems (Torgersen & Clarke, 1985; Torgersen & Ivey, 1985; Bottomley et al., 1984).

2.1 Helium isotope nomenclature

The $^3He/^4He_{sample}$ ratio (R) measured in a sample is usually normalised to the air $^3He/^4He$ ratio where $^3He/^4He_{air}(R_a)$ = 1.4 x 10^{-6} (Ozima & Podosek, 1983). For air, therefore, the ratio R/R_a = 1. The measured ratios in an aquifer vary between $0<R/R_a<1$. For mantle-derived He, much higher ratios of $8<R/R_a<35$ are observed (Mamyrin & Tolstikhin, 1984). The $^3He/^4He$ ratios in groundwater generally decrease from values characteristic of air in the recharge areas with predominantly younger waters, to much lower values in older waters down-gradient in the confined parts of aquifers. Similarly, the ^4He concentrations in the recharge areas are generally close to or slightly in excess of the equilibrium solubility value (~ 4.41 x 10^{-8} ccSTPg^{-1} H_2O at 25 ^oC). In the confined parts of aquifers, ^4He concentrations are often several orders of magnitude higher than the equilibrium values as a result of in-situ production and possible additional sources of ^4He entering the aquifer (Torgersen & Ivey, 1985; Heaton, 1984).

The subsurface production of neon by nuclear processes is usually considered insignificant. However, it is used as a monitor of air-derived helium to enable correction of the measured $^3He/^4He$ ratios and ^4He concentrations as discussed below.

2.2 Radiogenic $^3He/^4He$ and ^4He in aquifers

The measured $^3He/^4He$ ratios and ^4He concentrations in groundwater represent a two-component mixing of *atmospheric helium* and *radiogenic helium*, the latter being produced either in the aquifer or transported into the aquifer from elsewhere.

Measured Ne concentrations may be used to filter out He of atmospheric origin by making use of the (He/Ne) ratio and their differing solubilities in water. For example, $(He/Ne)_{air}$ = 0.2882 (Weiss, 1971) and for pure water at 25 ^oC, $\beta Ne/\beta He$ = 1.165 (where β = Bunsen solubility coefficient). In this way, the air-derived ^4He component can be subtracted from the measured $(^4He)m$ to give the corrected helium concentration $(^4He)c$ as follows:

$$He_C = \frac{He_m(X-1)}{X} \qquad (1)$$

where He_C refers to the corrected ^4He concentrations (ccSTPg$^{-1}H_2O$); He_m are the measured concentrations (ccSTPg$^{-1}H_2O$) in groundwater and X = $[(He/Ne)_m/(He/Ne)_{air}]$ * βNe/ βHe.

Similarly, the air-corrected $^3He/^4He$ isotope ratio is given by:

$$(R/R_a)_C = \frac{[(R/R_a)_m * X]-1}{(X-1)} \qquad (2)$$

Corrected ratios $(R/R_a)_C$ in aquifers which accumulate radiogenic helium are usually less than 0.1.

2.3 ^4He groundwater dating

Radiogenic helium has been observed to increase in the direction of groundwater flow in predominantly porous media aquifers (Torgersen & Ivey, 1985; Torgersen & Clarke, 1985 and Heaton, 1984). Considering the in-situ production of ^4He only, the solution rate (He_S) in an aquifer is given by Andrews & Lee (1979):

$$He_S=\rho\phi^{-1}\{1.19 \times 10^{-13}[U]+ 2.88\times10^{-14}\,[Th]\} \qquad (3)$$

where He_S is the ^4He solution rate (ccSTPg$^{-1}H_2O$yr^{-1}); [U] and [Th] are the uranium and thorium concentrations in the aquifer rock (ppm); ρ is the bulk density of the aquifer rock (gcm^{-3}) and ϕ is the fractional effective porosity of the aquifer rock. From Equation 3, the apparent groundwater ages can be calculated as follows:

$$T_r = \frac{He_C}{He_S} \qquad (4)$$

where T_r is the apparent age of groundwater (yrs) along a flow path. From Equation 4, *groundwater travel times* can be estimated if He_C is replaced by He_{c_1}-He_{c_2} where the subscripts c_1 and c_2 denote respectively, concentrations of first water strikes in the up and down-gradient sections of the confined aquifer. Groundwater travel times by use of Equation 4 are often several orders of magnitude greater than the ^{14}C derived travel times (Bottomley et al., 1984; Heaton, 1984). Difficulties arise if parameters used in Equation 3 are not uniformly distributed throughout the aquifer and if additional sources of ^4He from underlying crustal rocks contribute significantly to the ^4He concentrations. In such situations it is thus necessary to estimate crustal fluxes from other sources to obtain more reliable results and to enable a more realistic comparison with other dating methods.

2.4 Estimation of crustal He fluxes

Stute et al. (1992) provide an approach by which helium fluxes from the underlying crustal rocks may be estimated, which enables realistic relative ages to be obtained.

The corrected apparent groundwater age is then estimated from:

$$T_{Corr} = \frac{He_C}{\frac{Jo}{\phi Z_O p} + He_S} \tag{5}$$

where T_{corr} is the apparent age (yrs) taking into account additional fluxes; and the term Jo ($ccSTPcm^{-2}yr^{-1}$) is the He flux entering the bottom of the aquifer; φ is the effective porosity of the aquifer; Z_O is the depth (m) at which this flux enters the aquifer and p is the density of water (~ 1 gcm^{-3}). The term $J_O/(Z_Op)$ can be equated to a *solution rate factor* which accounts for additional sources of ^4He diffusing into the aquifer.

2.5 Calculation of recharge fluxes using the ^4He method

The long-term average areal recharge rate R_T ($mmyr^{-1}$) can be calculated from the balance equation (Beekman et al., 1996):

$$R_T = 1000\frac{(V_u A_R H_u) + \phi A_T H_S}{(A_R T_T)} \tag{6}$$

where V is the average volumetric moisture content in the recharge area; A (m^2) is the surface area; H (m) is the average layer thickness; φ is the average effective porosity of the aquifer; T (yrs) is the total average residence time. Subscripts represent u: unsaturated zone, R: recharge, T: total and s: saturated zone. T_T (yrs) is the sum of the average transit times in the unsaturated zone (T_u), the unconfined (T_{unc}) and the confined (T_c) parts of the aquifer (i.e. $T_T = T_u + T_{unc} + T_c$, yrs). The basic assumptions are steady-state groundwater flow conditions and that V_u and A_R remain constant.

3 RESULTS

The results of helium isotopes and their abundances in groundwaters from the Mmamabula and Ntane aquifers and along the Meratswe Valley are discussed, in the light of aquifer characterization and groundwater dating. These results clearly show that noble gases are useful in studying regional trends, as the case is, in the Mmamabula aquifer. Furthermore, groundwater travel times were calculated which ultimately are used to derive the long-term average areal recharge rate by the ^4He method.

3.1 Aquifer characterisation and regional trends

Regional trends were examined from the Mmamabula aquifer and along the Meratswe Valley using helium isotopes dissolved in groundwater. In both areas, the recharge areas are characterised by the highest $^3He/^4He$ ratios (which approach $1R_A$) and then decrease to radiogenic values under confined conditions. The measured $^3He/^4He$ $(R/R_a)_m$ and the corrected $^3He/^4He$ $(R/R_a)_c$ ratios are presented in Figure 1 (a) and (b) respectively. Figure 1 (a) also shows how the measured isotopic composition of helium corresponds to younger groundwaters in the recharge areas having generally high ratios ($0.2 < R/R_a < 0.8$ whereas older waters in the confined regions have low radiogenic ratios of $< 0.1R_a$.

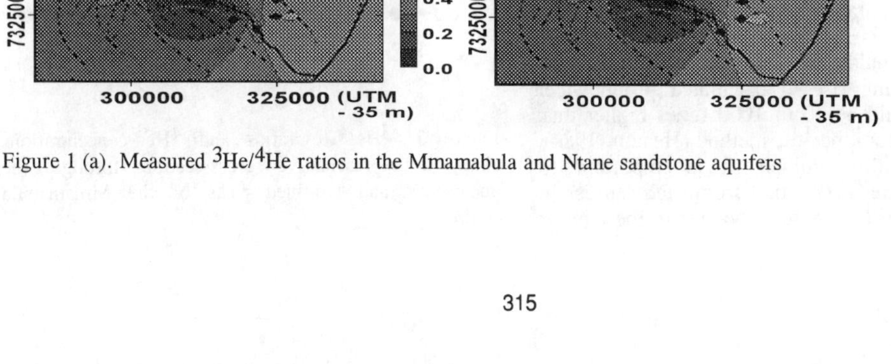

Figure 1 (a). Measured $^3He/^4He$ ratios in the Mmamabula and Ntane sandstone aquifers

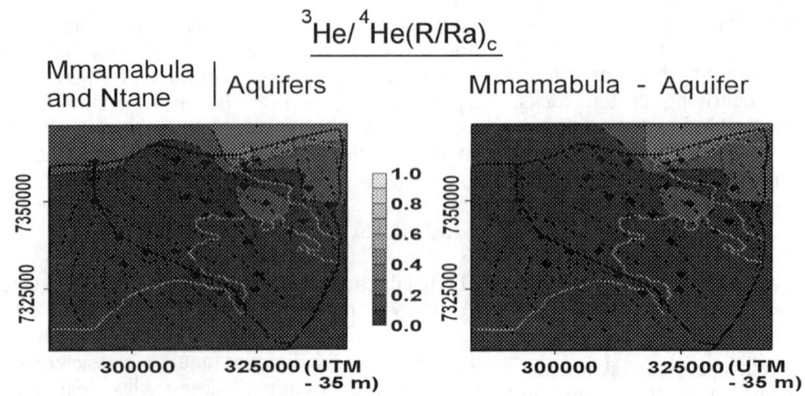

Figure 1 (b). $^3He/^4He$ ratios in the Mmamabula and Ntane sandstone aquifers corrected for the atmospheric component

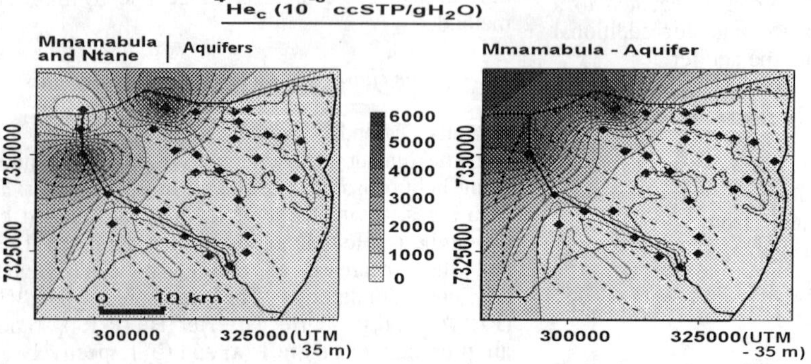

Figure 2. ^4He concentrations in the Mmamabula and Ntane sandstone aquifers corrected for the atmospheric component

The corrected ^4He concentrations are shown in Figure 2. Lower concentrations are found in the recharge area, whereas much higher values, in excess of 50 μccSTPg^{-1}H$_2$O, characterize the confined regions. The *aging effect* of the groundwaters is apparent from this figure in spite of helium migration from deeper sources.

3.2 Groundwater relative ages

Groundwaters gain radiogenic ^4He as a result of the decay of uranium and thorium and their daughter products in the aquifer rocks. Additional ^4He may accumulate by diffusion from underlying formations giving rise to calculated groundwater ages which could be 10 to 1000 times higher than those derived from the ^{14}C method (Heaton, 1984). To overcome this problem it is important to estimate He fluxes originating from such sources in order to correct the measured values in the aquifer (Stute et al., 1992).

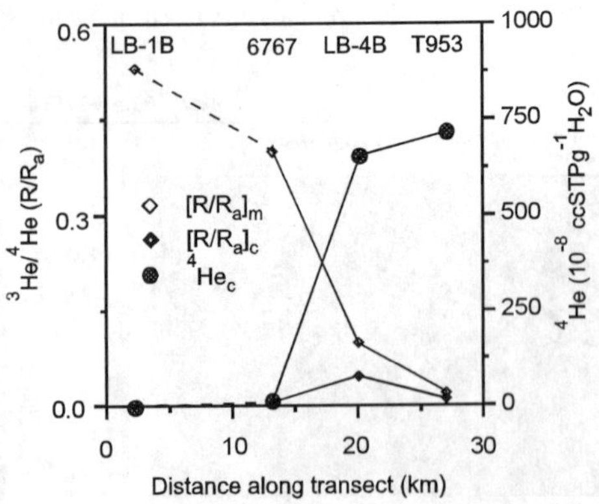

Figure 3. $^3He/^4He$ ratios and ^4He concentrations along a representative cross-section through the unconfined and confined parts of the Mmamabula aquifer.

In the study area, ^4He concentrations in the aquifer were observed to increase with distance along flow paths from the recharge area to the confined sections of the aquifer. Figure 3 above shows that, along a representative cross-section through the unconfined and confined parts of the Mmamabula aquifer, the measured and corrected $^3He/^4He$ ratios decrease in the direction of groundwater flow.

The groundwater relative ages for the Mmamabula Formation were estimated from the corrected helium (^4He_c) values and the in-situ ^4He solution rate (He_s) based on the following parameters: weighted averages for U = 5 ppm; Th = 12 ppm; weighted average effective porosity (based on cored borehole LB-4C) ϕ_1 = 0.05 and overall effective porosity of the aquifer ϕ_2 = 0.1. Using ϕ_2 = 0.1 and ϕ_1 = 0.05 in Equation 3 gives ^4He solution rates ranging from about 2.493 x 10^{-11} to 4.985 x 10^{-11} $ccSTPg^{-1}H_2Oyr^{-1}$ respectively.

Groundwater relative ages were calculated from both the ^4He and ^{14}C methods. Results indicate that the rate of accumulation of ^4He in the study area may be accounted for by two processes. The young groundwaters (< 10 Ka) in the recharge areas appear to be dominated by the in-situ production of ^4He from the alpha decay of U and Th radioelements in the aquifer rocks. However, in the confined part of the aquifer, the ^4He flux entering the bottom of the aquifer is dominant. For groundwater residence times generally >> 10 Ka, the estimated ^4He crustal flux of between 0.66 to 1.33 x 10^{-6} $ccSTPcm^{-2}yr^{-1}$ reduces the erroneously high ages in the confined area by at least a factor of 10^2. Accounting for the effects of crustal fluxes in the recharge zones gives ages which are < 100 years. These relatively young ages, are considered unlikely as ^{14}C apparent ages reveal turnover times of more than 1 000 years; suggesting that the crustal flux component is negligibly small under unconfined conditions due to dilution from mixed waters and atmospheric influences.

3.3 Long term average recharge rate calculated from the ^4He method

Assuming in-situ production of ^4He in the recharge area, the average apparent groundwater residence time, from well mixed samples, based on weighted effective porosity ϕ_1 = 0.05 and overall effective porosity ϕ_2 = 0.10 for the Mmamabula aquifer, ranges from ~ 1200 to 2400 yrs.

Groundwaters in the confined part of the aquifer are strongly influenced by the ^4He flux from the basement rocks. If one considers a representative cross-section in the direction of groundwater flow, starting at Bh 6767 to T953, then a transit time ranging from ~ 5140 to 5230 yrs is estimated for the confined part of the aquifer. The transit time through the unsaturated zone is estimated at ~ 800 yrs based on the chloride mass balance method.

The long-term average areal recharge rate R_T ($mmyr^{-1}$) can be calculated from the balance Equation 6. A summary of the parameters used in the calculations is as follows: V_u = 0.1; A_R = 902 km^2; H_u = 50 m; ϕ_1 = 0.05 ϕ_2 = 0.10; A_T = 1878 km^2; H_s = 100 m; T_{T1} = T_u + T_{unc1} + T_{c1} = 800 + 1206 + 5140 = 7146 ~ 7150 yrs and T_{T2} = T_u + T_{unc2} + T_{c2} = 800 + 2412 + 5230 = 8442 ~ 8440 yrs.

From the above, the *long-term average areal recharge rate* using the ^4He method falls in the range 2.2 to 3.1 $mmyr^{-1}$.

4 DISCUSSION AND CONCLUSIONS

Regional trends and the *long-term average areal recharge* were evaluated using helium dissolved in groundwaters within the Mmamabula and Ntane aquifers and along the Meratswe Valley.

Results indicate that the recharge areas are dominated by air-derived gases which are indicative of younger groundwaters, whereas higher concentrations of radiogenic helium reveal much older waters in the confined regions. The most sensitive parameter is the effective porosity and should be determined as accurately as possible for both the aquifer and the underlying crustal rocks.

In the study area, groundwater residence times derived from ^4He are similar to those obtained by the ^{14}C method for ages < 10 Ka. However, for ages >> 10 Ka, ^4He ages are consistently older than ^{14}C ages, which is indicative of an additional source of radiogenic helium, mixing with an older component of groundwater, or both. We suggest that the excess helium is primarily from the crystalline basement rocks of the Waterberg Formations. It is particularly pronounced in the confined regions of the Mmamabula aquifer.

Using the ^4He method, the *long-term average areal recharge rate is in the order of ~ 2.2 to 3.1 $mmyr^{-1}$*. This range of recharge rates falls within the same order of magnitude as estimates of 4 to 10 $mmyr^{-1}$, 5.7 $mmyr^{-1}$ and 1 $mmyr^{-1}$ obtained respectively by the chloride mass balance, groundwater flow and hydrochemical/isotope modeling. Continuing research aims at integrating the noble gas isotopes with other hydrochemical and hydraulic parameters in order to obtain refined recharge estimates for the Mmamabula aquifer and the general Kalahari environment.

ACKNOWLEDGEMENTS

This paper is based on the results of a long-term co

-operation research programme on groundwater recharge and water resources evaluation in Botswana (GRES) carried out by the Botswana Department of Geological Survey (DGS), the University of Botswana (UB- Geology Department) and the Vrije Universiteit Amsterdam, The Netherlands (VUA - Centre for Development Co-operation Services and Faculty of Earth Sciences).

We would like to thank Prof.dr. J.J. de Vries (VUA) who initiated the GRES research programme and guided it through the various stages. The Director of DGS (Mr. T.P. Machacha) and the Vice - Chancellor of UB (Prof.dr. T. Tlou) gave full support to the programme. Logistical support and laboratory analysis were provided by DGS, VUA and the Centre for Isotope Research in Groningen, The Netherlands.

We thank the Botswana government (Ministry of Minerals, Energy and Water Affairs) and the Netherlands government (Directorate General for International Co-operation of the Ministry of Foreign Affairs) for their financial support to the GRES programme.

REFERENCES

Andrews J.N. & Lee D.J., 1979. Inert gases in groundwater from the Bunter Sandstone of England as indicators of age and palaeoclimatic trends. *J. Hydrol.*, 41: 233-252.

Andrews, J.N., Drimme, R.J., Loosli, H.H. & Hendry, M.J., 1991. Dissolved gases in the Milk River aquifer, Alberta, Canada. *Applied Geochemistry,* 6: 393-403.

Beekman, H.E. & Selaolo, E.T., 1997. *Recharge in the eastern Botswana Kalahari - Hydrochemical, isotope and noble gases tracer study in the Letlhakeng-Botlhapatlou area. GRES II project: Technical report,* Department of Geological Survey, Lobatse; University of Botswana and the Free University Amsterdam, The Netherlands.

Beekman, H.E., Gieske, A & Selaolo, E.T., 1996. GRES - groundwater recharge studies in Botswana (1987-1996). *Botswana Journal of Earth Sciences, Volume III:* 1-17.

Bottomley, D.J., Ross, J.D. & Clarke, W.B., 1984. Helium and neon isotope geochemistry of some groundwaters from the Canadian Precambrian Shield. *Geochimica et Cosmochimica Acta 48:* 1973-1985.

Heaton, T.H.E., 1984. Rates and sources of 4He accumulation in groundwater. *Hydro. Sci. J.* 29: 29-47.

Herzberg, O. & Mazor, E., 1979. Hydrological applications of noble gases and temperature measurements in underground water systems: Examples from Israel. *J. Hydrol.*, 41: 217-231.

Mamyrin, B.A & Tolstikhin, L.N., 1984. *Helium isotopes in nature. Developments in Geochemistry 3.* Elsevier Science Publishers B.V. Amsterdam, pp. 273.

Mazor, E., 1972. Paleotemperature and other hydrological parameters deduced from noble gases dissolved in groundwaters; Jordan Rift Valley, Israel. *Geochimica et Cosmochimica Acta 36*: 1321-1336.

Ozima, M. & Podosek, F., 1983. *Noble gas geochemistry. Cambridge University Press,* Cambridge, pp: 367.

Selaolo, E.T., 1998. *Tracer studies and groundwater recharge assessment in the eastern fringe of the Botswana Kalahari - The Letlhakeng-Botlhapatlou area. Ph.D. Thesis, Free University Amsterdam,* pp. 229.

Stute, M., Sonntag, C., Deak, J. & Shlosser, P., 1992. Helium in deep circulating groundwater in the Great Hungarian Plain: Flow dynamics and crustal and mantle helium fluxes. *Geochimica et Cosmochimica Acta 56*: 2051-2067.

Torgersen, T. & Clarke, W.B., 1985. Helium accumulation in groundwater, I: An evaluation of sources and the continental flux of crustal ^{4}He in the Great Artesian Basin, Australia. *Geochimica et Cosmochimica Acta* 49: 1211-1218.

Torgersen, T. & Ivey, G.N., 1985. Helium accumulation in groundwater, II: A model for the accumulation of the crustal ^{4}He degassing flux. *Geochimica et Cosmochimica Acta 49*: 2445-2452.

Weiss, R.F., 1971. Solubility of helium and neon in water and sea-water. *J. Chem. Eng. Data 16*: 235-241.

Groundwater: Past Achievements and Future Challenges, Sililo et al. (eds) © 2000 Balkema, Rotterdam, ISBN 90 5809 159 7

Predicting fractured rock hydrogeological parameters from field and laboratory data

J.M.Sharp, Jr, T.Halihan, M.M.Uliana, G.P.Tsoflias, M.T.Landrum & R.Marrett
Department of Geological Sciences, University of Texas, Austin, Tex., USA

ABSTRACT: Standard hydrogeologic methods are difficult to apply in fractured and karstic aquifers because permeabilities vary over orders of magnitude; anisotropy is significant; flow rates increase with scale; and contaminants and tracers appear at unexpected locations. Studies demonstrate that outcrop and laboratory data can predict fractured aquifer properties. Mapping of fracture orientations and densities delineate regional fracture domains and fracture control of regional flow. Permeability scale effects are quantified as a combination of matrix, fracture, and dissolution features. Geostatistical analyses of pumping tests and outcrop data show that wells exploit fractures and conduits control regional flow in karst aquifers. Ground-penetrating radar can monitor drawdown in single fractures, resolve vugs, and predict anisotropy of carbonate dissolution zones. Field and laboratory data on fracture skin and matrix properties are input into mathematical models to predict contaminant migration.

1 INTRODUCTION

Characterization of fractured rock and karstic aquifers provides many examples of the difficulty in applying standard hydrogeologic techniques. Permeability can vary by orders of magnitude in an aquifer, flow directions are unpredictable, groundwater velocities increase with the scale of measurement, and contaminants appear more quickly than expected at unexpected locations. Recent studies at The University of Texas at Austin have shown that information can be obtained from outcrop studies, ground-penetrating radar (GPR), and laboratory measurements that permit better-constrained hydrogeologic interpretations.

Fractures create permeability anisotropy and generally act as high permeability features. However, the quantitative influence of fracture orientation on the fluid flow direction is hard to predict, and problems occur when attempting to characterize the anisotropy from fractures. Bias errors inherent in sampling influence the ability to assess the influence of the fractures on flow anisotropy. In addition, other factors influence permeability, including connectivity, fracture aperture variations, roughness and channeling (Sharp 1993). Outcrop data from the Edwards aquifer of central Texas, from Trans-Pecos Texas, USA, indicate that flow directions and permeability trends are predictable and that the permeability scale effect observed in karst aquifers can be quantified as a combination of carbonate matrix with fractures and dissolution features. GPR data from a dolomite well site in Door County, Wisconsin, USA, illustrate the ability to monitor drawdown in single fractures, to resolve vugs in selected carbonate beds, and to predict hydraulic anisotropy of dissolution-enlarged bedding planes. The significant effect of fracture skins on transport is investigated with a combination of laboratory analysis and mathematical modeling. Laboratory and field mapping procedures allow *a priori* predictions of aquifer parameters and flow systems at various scales.

2 METHODS

Below we discuss regional flow system delineation from mapping fractures, scale permeability effects from measuring fracture apertures and spacing, GPR evaluation of shallow fractures, and laboratory/modeling analyses of the effects of fracture skins on transport.

2.1 *Regional flow delineation from fracturing mapping*

Fracture trace maps and aerial photo analysis can be used to characterize fracture systems and investigate the influence of fracture systems on regional flow. Fracture-mapping studies can then be integrated with modeling and with groundwater geochemical and isotopic data to delineate groundwater flow patterns, infer general groundwater residence times, and identify recharge areas. An example of fracture mapping analysis associated with a regional scale groundwater study is currently underway at The University of Texas at Austin (Sharp and others, 1999; Uliana and Sharp, 1999). Fracture orientations and fracture densities (defined as the total length of fracture per unit area [L^{-1}]) in Trans-Pecos Texas were analyzed using aerial photos and geologic maps from previous studies. Fracture traces were identified on the aerial photographs, traced, and compared to mapped faults from previous field studies. Rose diagrams for each 25-km^2 area of the map were plotted based on fracture orientations determined from the aerial photos. These rose diagrams were length weighted; that is, the rose diagrams were plotted based on 0.1-km length segments of fracture, rather than using each total fracture length as a single fracture. The rose diagrams graphically represent the changes in fracture orientations throughout the model area (Fig. 1). Rose diagrams with larger petals represent greater fracture density.

Most of the rose diagrams indicate that the fractures tend to follow a N10W to N30W structural grain. However, the rose diagrams near the center of the map, in southeast Culberson County, follow a ~N70W trend that changes to a more northwesterly trend that parallels the border of Reeves and Jeff Davis counties. This fracture trend is also associated with the largest fracture densities. This zone of high fracture densities and distinct fracture orientations appears to correspond to regional groundwater flow paths identified by Uliana and Sharp (1999).

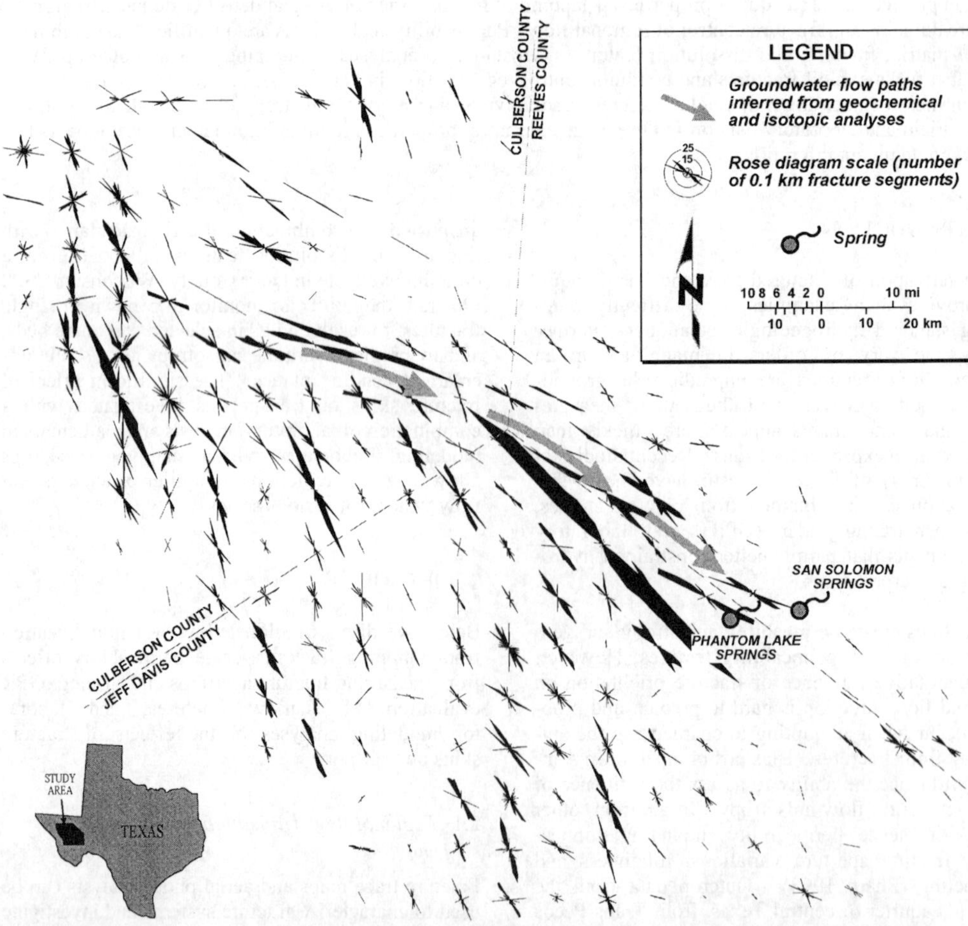

Figure 1. Length-weighted rose diagrams of study area in Trans-Pecos Texas. The previously inferred flow system to the springs is confirmed by the surface mapping of fractures. The shapes of the rose diagrams correspond to fracture domains of differing permeabilities. Rose diagrams with larger petals represent greater fracture density.

They used groundwater chemistry to determine rock-water interactions and mixing, $\delta^{18}O$ data to infer general groundwater residence times and identify regional and local recharge, and $^{87}Sr/^{86}Sr$ data to trace the regional flow path to San Solomon and Phantom Lake springs. Fracture analysis results, coupled with the geochemical and isotopic results, suggest that major springs in Trans-Pecos Texas are partially recharged by a regional groundwater flow system that is controlled by fracture trends. The fracture orientation and density data, and fracture domains determined from those data, can be used as model input to generate computer groundwater models of the regional system.

These findings are similar to those inferred by Mayer and Sharp (1998) for the Diablo Plateau in Texas and New Mexico and by Zahm et al (1998) for the Edwards aquifer. Mayer and Sharp used air photo analysis to identify fracture domains and develop a finite-element model of the groundwater system north of Dell City, Texas. They also integrated their fracture analysis with geochemical data to verify groundwater flow paths. Zahm et al mapped fractures at outcrop and regional scales and used groundwater gradients and fracture anisotropy to calculate groundwater flow paths. Their calculated flow paths were verified with artificial trace tests. In all of these studies, the researchers have observed regional fracture trends that connect major recharge and discharge areas over great distances. They have also found that quantitative fracture data can constrain two-dimensional permeability distributions in carbonate aquifers and define fracture anisotropy domains.

2.2 *Permeability scale effect from outcrop data*

Estimating permeability in karst aquifers is not simple. Permeability varies by many orders of magnitude depending on the scale, location, and direction of measurement. Permeability is also the most statistically sensitive parameter in laminar and turbulent groundwater flow equations. Kiraly (1975) first described the scale effect in which permeabilities continually increased from the small- to the regional-scale. He hypothesized that the increase from small scale to well scale was caused by fractures and that the largest permeabilities on the regional scale were caused by karstic conduits.

Halihan et al. (2000) tested the Kiraly hypothesis using a permeability databases (Hovorka et al. 1998; Mace & Hovorka 2000) for the Edwards aquifer of central Texas and found that Kiraly's hypothesis applied.

Conduits. Conduit permeabilities are difficult to estimate because most conduits have variable diameters, and have limited lengths. Permeability calculations indicate that on a regional scale, a meter-scale diameter conduit would dominate the flow field (Halihan et al. 2000), but the permeabilities are very high and not often observed because of either pumping test limits or incorrect modeling assumptions in regional models. However, some work suggests that conduits should follow a lognormal or power law distribution so that, when present, a continuous single conduit will dominate the flow in an aquifer.

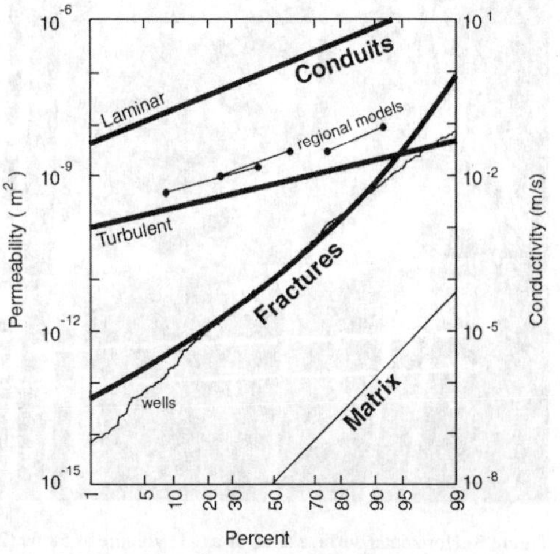

Figure 2. Permeability scale effect in the Edwards aquifer (modified from Halihan et al. 1999. Depicted are data for matrix (core plugs), wells (pumping tests) (Hovorka et al. 1998), and regional scale models (Klemt et al. 1979; Maclay & Land 1988; Thorkildson & McElhaney 1992). Combination matrix with 50-fracture Monte Carlo model (**Fractures**) using matrix and fracture permeability distributions matches hydraulic conductivities from over 1000 pumping tests. The 2-D turbulent conduit model (**Conduits - Turbulent**) based on outcrop data matches regional conductivities calculated from numerical simulations.)

Fractures. The distribution of apertures among a population of genetically related fractures typically follows a power-law distribution over at least several orders of magnitude (Marrett et al. 1999). When a power-law distribution is assumed for calculating permeability using the cubic law, generally a single fracture or a small number of fractures will dominate (Marrett 1996). Although many fractures may be present, the distribution of apertures indicates that only a few will control the flow in an aquifer. Because the size of the largest fracture increases with the size of the spatial domain considered, scale-dependent fracture permeability should be expected.

Matrix. Published matrix permeabilities tend to range from $5x10^{-17}$ to $5x10^{-13}$ m^2 for limestone and dolomite, but values extend from below 10^{-17} m^2 where they become difficult to measure up to 10^{-11}

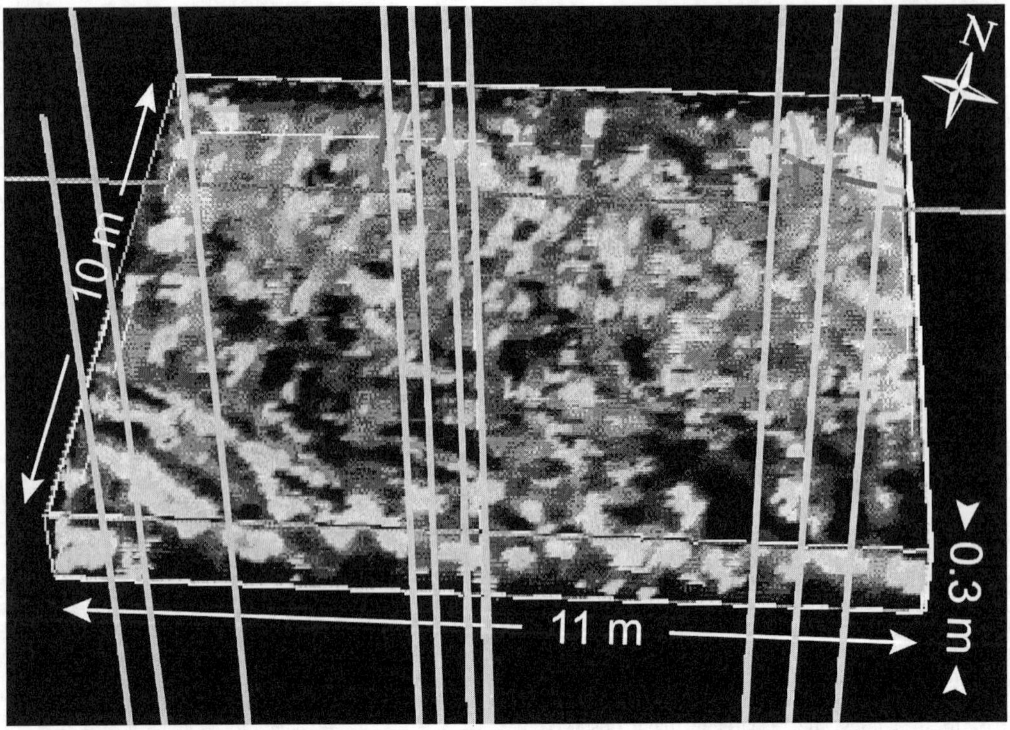

Figure 3. Horizontal surface slice of a 3-D volume at 58 ns (2.75m) through a dissolution zone characterized by poorly connected vugs. Light areas are negative amplitudes indicating increased water content. Wellbores are shown intersecting the horizon.

m^2. These latter may include small voids or fractures (Hovorka et al. 1998) (Fig. 2). Matrix permeability is commonly lognormally distributed (Halihan et al. 1999).

Monte Carlo simulations of Edwards matrix and fracture permeabilities estimated from the laboratory and outcrop data, respectively, provide a good match for the pumping test data collected for the aquifer. Conduit permeability data were also estimated from outcrop exposures (Halihan et al. 2000), assuming turbulent Darcy-Weisbach flow (Halihan et al. 1998). These calculations provide a good match for the regional permeabilities needed to calibrate numerical models.

2.3 *Fracture delineation with ground-penetrating radar (GPR)*

GPR surveying is a suitable technique for high-resolution investigation of fractured carbonate aquifers because of the high dielectric contrast between water and rock formations. The resolution of high frequency electromagnetic surveying makes possible the detection of groundwater flow paths. Proper survey design and processing (Tsoflias 1999) can provide accurate 3-D representations of the subsurface and can determine some hydrogeological properties based on GPR signal attributes (velocity and amplitude).

At the Wisconsin site, horizontal fractures and diagenetic zones were identified along with small-scale features (e.g., dissolution enlarged vugs and cross bedding). Cyclic electromagnetic wave velocity variations correlated with cyclic alternating open and restricted marine carbonate facies with varying rock matrix volumetric water content. Amplitude variation along flow conduits suggested centimeter scale spatial aperture variation, with significant effects on aquifer flow properties. Figure 3 (Tsoflias & Sharp 1998) depicts GPR 3-D imaging of a vuggy zone from which the vug connectivity patterns can be inferred. Multi-polarization GPR surveying also identified vertical zones of dielectric constant anisotropy (e.g., fractures). Field observations, 3-D numerical modeling and analytic solutions demonstrated that EM wave high-angle incidence transmission through a fracture causes a phase shift dependent on the polarization of the wavefield. When the incident electric field is oriented perpendicular to the plane of the fracture, the transmitted electric field undergoes a greater phase-shift than presence of a phase lead between orthogonal pairs of polarization data (Tsoflias 1999).

2.4 *Fracture delineation with ground-penetrating radar (GPR)*

GPR surveying is a suitable technique for high-resolution investigation of fractured carbonate aquifers because of the high dielectric contrast between water and rock formations. The resolution of high frequency electromagnetic surveying makes possible the detection of groundwater flow paths. Proper survey design and processing (Tsoflias 1999) can provide accurate 3-D representations of the subsurface and can determine some hydrogeological properties based on GPR signal attributes (velocity and amplitude).

At the Wisconsin site, horizontal fractures and diagenetic zones were identified along with small-scale features (e.g., dissolution enlarged vugs and cross bedding). Cyclic electromagnetic wave velocity variations correlated with cyclic alternating open and restricted marine carbonate facies with varying rock matrix volumetric water content. Amplitude variation along flow conduits suggested centimeter scale spatial aperture variation, with significant effects on aquifer flow properties. Figure 3 (Tsoflias & Sharp 1998) depicts GPR 3-D imaging of a vuggy zone from which the vug connectivity patterns can be inferred. Multi-polarization GPR surveying also identified vertical zones of dielectric constant anisotropy (e.g., fractures). Field observations, 3-D numerical modeling and analytic solutions demonstrated that EM wave high-angle incidence transmission through a fracture causes a phase shift dependent on the polarization of the wavefield. When the incident electric field is oriented perpendicular to the plane of the fracture, the transmitted electric field undergoes a greater phase-shift than when the incident electric field is oriented parallel to the plane of the fracture. Thus, the location and azimuth of a fracture can be determined based on the presence of a phase lead between orthogonal pairs of polarization data (Tsoflias 1999).

2.5 *Analysis of fracture skin/matrix properties and extrapolation from laboratory to modeling*

Fracture skins consist of alteration products, precipitated and/or dissolved minerals in the rock matrix adjacent to the fracture, mineral coatings on the fracture surface, debris such as soil or crushed rock, and organic material. In the lab, we have measured porosity, diffusivity, and permeability of fracture skins when the incident electric field is oriented parallel to the plane of the fracture. Thus, the location and azimuth of a fracture can be determined based on the and the rock matrix. The fracture skins may have higher or lower porosity, effective diffusivity, permeability, and sorption distribution coefficient than the rock matrix. We utilize closed-form analytical solutions, based upon Laplace transforms, (Robinson et al. 1998) for steady-state and transient flow and transport of a decaying contaminant in a set of finite, parallel fractures with or without fracture skins. These analyses, based upon skin data obtained from laboratory analyses (Fu et al. 1994; Kreisel & Sharp 1996) indicate that fracture skins have a significant effect on contaminant transport and the fracture skin can either accelerate or retard contaminant transport in fractures relative to no-skin conditions. The most important properties of the skin are its thickness and its relative porosity, diffusion coefficient, and retardation coefficient relative to the matrix. There are maximum and minimum skin thicknesses beyond which changing the thickness has little effect.

3 CONCLUSIONS

Results indicate that *a priori* analysis of regional fracture systems by fracture-mapping techniques can improve models of groundwater flow and transport. Commonly, geological controls are only considered *post priori* when needed to calibrate a model. Analysis of conduit, fracture, and matrix permeabilities in a fractured karstic aquifer using outcrop data may allow prediction of well yields (controlled by fractures) and regional flow systems (controlled by conduits). The scale effect is also defined by the combination of these permeabilities. GPR surveying can monitor real-time response of a fractured aquifer to a pumping test and provide detailed information about the structural architecture and fluid flow properties of fractured rocks. GPR waveforms identified saturated-unsaturated transitions in fractures. Finally, laboratory analyses of matrix and fracture skin properties may be utilized to make predictions of transport in fractured rocks using mathematical models.

ACKNOWLEDGEMENTS

We acknowledge the Owen-Coates Fund of the Geology Foundation for support in manuscript preparation. We appreciate the insights of our colleagues Chris Zahm, Sarah Lindsay Tsoflias, Neville Robinson, Kitty Milliken, Jim Mayer, Carla Matherne, Andrew Love, Glenn Ledder, Ilan Kreisel, Fu Li, Peter Cook, and Don Blankenship.

REFERENCES

Fu, L., K. L. Milliken, & J. M. Sharp, Jr. 1994. Porosity and permeability variations in the fractured and liesegang-banded Breathitt Sandstone (Middle Pennsylvanian), eastern Kentucky: Diagenetic controls and implications for modeling dual porosity systems. *Journal of Hydrology*. 154: 351-381.

Halihan. T., C. M. Wicks & J. F. Engeln 1998. Physical response of a karst drainage basin to flood pulses: Example of the Devil's Icebox cave system (Missouri, USA). *Journal of Hydrology*. 204:24-36.

Halihan, T., J. M. Sharp, Jr., & R. E. Mace 1999. Interpreting

flow using permeability at multiple scales. In A. R. Palmer, M. V. Palmer & I. D. Sasowsky (eds.), *Karst Modeling* Special Pub. no. 5: 82-96. Charlottesville, VA: Karst Waters Inst.
Halihan, T., R. E. Mace & J. M. Sharp, Jr. 2000. Flow in the San Antonio segment of the Edwards aquifer: matrix, fractures, or conduits. In I. D. Sasowsky & C. W. Wicks (eds.), *Groundwater Flow and Contaminant Transport in Carbonate Aquifers*, 129-146. Rotterdam: Balkema.
Hovorka, S. D., R. E. Mace & E. W. Collins 1998. Permeability structure of the Edwards aquifer, south Texas - Implications for aquifer management. *Texas Bur. Econ. Geol. Rept. Inv. No. 250*. Austin, TX: Univ. Texas.
Kiraly, L. 1975. Rapport sur l'etat actuel des connaissances dans le domaine des caracteres physiques des roche karstiques. In A. Burger & L. Dubetret (eds.), *Hydrogeology of Karstic Terrains,* Series B, no. 3: 53-67. Paris: Int. Assoc. Hydrogeologists.
Klemt, W. B, T. R. Knowles, G. R. Elder & T. W. Sieh 1979. Ground-water resources and model applications for the Edwards (Balcones fault zone) aquifer in the San Antonio region, Texas. *Texas Water Dev. Board Rept. No 239*.
Kreisel, I., & J. M. Sharp, Jr., 1996. Fracture skins in the Brushy Canyon Formation. In W. D. DeMis & A. G. Coles (eds.) *The Brushy Canyon Play in Outcrop and Subsurface: Concept and Examples*: Pub. 96-38, p. 147-152. Midland, TX: Permian Basin Sec. SEPM.
Mace, R. E. & S. D. Hovorka 2000. Estimating porosity and permeability in a karstic aquifer using core plugs, well tests, and outcrop measurements. In I. D. Sasowsky & C. W. Wicks (eds.), *Groundwater Flow and Contaminant Transport in Carbonate Aquifers*, 93-111. Rotterdam: Balkema.
Maclay, R.W. & L.F. Land 1988. Simulation of flow in the Edwards aquifer, San Antonio region, Texas. *U.S. Geol. Survey Water-Supply Ppr. 2336-A*.
Marrett, R. 1996. Aggregate properties of fracture populations. *Jour. Structural Geol.* 18: 169-178.
Marrett, R., O. J. Ortega & C. M. Kelsey 1999. Extent of power-law scaling for natural fractures in rock. *Geology* 27: 799-802.
Mayer, J. M. & J. M. Sharp Jr. 1998. Fracture control of regional ground-water flow in a carbonate aquifer in a semi-arid region. *Geol. Soc. America Bull.* 110: 269-283.
Robinson, N. I., J. M. Sharp, Jr. & I. Kreisel 1998. Contaminant transport in sets of parallel finite fractures with fracture skins. *Jour. Contaminant Hydrology*. 31: 83-109.
Sharp, J. M., Jr. 1993, Fractured aquifers/reservoirs: Approaches, problems, and opportunities. In D. Banks & S. Banks (eds.), *Hydrogeology of Hard Rocks*, Memoires, 24th Congress, 24(1): 22-38. Oslo, Norway: Int. Assoc. Hydrogeologists.
Sharp, J. M. Jr., Uliana, M. M., and Boghici, R., 1999. Fracture controls on regional groundwater flow in a semiarid environment and implications for long-term maintenance of spring flows: Water 99 Joint Congress, Inst. Engineers, Australia, Brisbane, v. 2, p. 1212-1217.
Thorkildsen, D. & P. D. McElhaney 1992. Model refinement and applications for the Edwards (Balcones fault zone) aquifer in the San Antonio region, Texas. *Texas Water Dev. Board Rept. No 340*.
Tsoflias, G. P. 1999. Hydrologic Characterization of Fractured Carbonate Aquifers Employing Ground-Penetrating Radar: *unpub. Ph.D. dissertation, Univ. Texas, Austin, TX.*
Tsoflias, G. P., & J. M. Sharp, Jr., 1998. Three-dimensional hydrogeologic characterization of fractured carbonate aquifers using ground-penetrating radar. *Trans., Gulf Coast Assoc. Geol. Socs.*, 48: 439-447.
Uliana, M. M. & J. M. Sharp, Jr. 1999. Tracing regional flow paths to major springs in Trans-Pecos Texas using naturally occurring geochemical and isotopic species. *Geol. Soc. America, Abs. with Programs (Ann. Mtg.),* 31: A-396.
Zahm, C. K., R. Marrett, R. & J. M. Sharp, Jr. 1998. Prediction of regional groundwater flow and permeability using fractured outcrop exposures, Edwards aquifer, central Texas: In J. V. Brahana, Y. Eckstein, L. K. Ongley, R. Schneider & J. E. Moore (eds.) *Gambling with Ground Water – Physical, Chemical, and Biological Aspects of Aquifer-Stream Relations*, 28th Congress, 697-705. Las Vegas, NV: Int. Assoc. Hydrogeologists.

Groundwater: Past Achievements and Future Challenges, Sililo et al. (eds) © 2000 Balkema, Rotterdam, ISBN 90 5809 159 7

Estimation of groundwater recharge to the Kanye dolomite aquifer in Botswana

M. Simonic
Hydromedia Solutions (Pty) Limited, Pretoria, South Africa

A. Adams & L. Carlsson
Department of Water Affairs, Gaborone, Botswana

C. Marobela
Geoflux (Pty) Limited, Gaborone, Botswana

ABSTRACT: The groundwater potential of the Kanye area (present abstraction about 1.5 million m^3/a) has been investigated for nearly two decades. Following the extensive drilling program a need for a reassessment of original recharge estimates resulted in application of Saturated Volume Fluctuation and Cumulative Rainfall Departures methods. Recent recharge estimates have been shown to be lower than previously established (based on 50 borehole hydrographs). Depending on the net storativity and selected scenario the results vary between 8 and 32 mm/a. Storativity is also smaller ~ 30-50% of the original estimate of 0.01. While for previously accepted storativity values the computed recharge coincides with earlier estimates (~20 mm/a), for the revised net storativity (0.0029-0.0056) the derived recharge values are significantly lower (10-11 mm/a). The paper examines the applied methods and discusses variability of recharge estimates as a result of methodology and as a spatial and temporal phenomenon.

1 INTRODUCTION

In the semi-arid to arid climate prevailing in Botswana, groundwater replenishment through rainfall recharge is the most important parameter in the assessment of the sustainability of groundwater resources. Around the village of Kanye in southern Botswana a potential groundwater resource abstracting 1.5 million m^3/a has been assessed with the goal to abstract water under sustainable or mining conditions for a period of 20 years. The dolomite aquifer has been investigated for almost twenty years. As a result a sizeable amount of monitoring data was established. A number of recharge assessment techniques were applied at various stages of investigations. Most recently, the relative wealth of groundwater level data led to the use of two techniques, namely the Cumulative Rainfall Departure (CRD) and Saturated Volume Fluctuation (SVF) methods. These methods complement the existing range of recharge estimates derived for the Kanye area.

The history of recharge estimation in the Kanye area includes first comprehensive studies by BRGM (1988), during 1984-87. According to the BRGM study recharge was estimated at 20 mm/a. A semi-global hydrologic model was used to arrive at this recharge estimate. Under GRES study, Gehrels & van der Lee (1990) applied a number of different techniques, but focused on a lumped parameter model (EARTH) to quantify recharge for specific boreholes. Using their model results combined with stochastically generated rainfall records they derived a long-term average recharge at 26 mm/a. By making use of the chloride method the recharge estimate according to their calculation is 23 mm/a.

The paper is focused on the application of CRD and SVF techniques, the various scenarios under which they were applied and the results in comparison with previous recharge estimates.

2 HYDROGEOLOGY

Kanye is situated in the southeastern Botswana, on the edge of the lower Transvaal Supergroup formations. These formations occupy an area of over 300 km^2 extending south southwestward and more to the south further down. Except for the mild hills around the main wellfield, the area is flat, at an average altitude of 1300 m a.m.s.l. The region is complex in terms of the soil cover, but can be effectively categorized into two: the sandveld to the north and hardveld to the south. Semi-arid conditions prevail, with average rainfall of 568 mm/a and sporadic high amounts up to 700 mm.

There are three main aquifer units, the Waterberg sandstone in the northernmost part (Selokolela wellfield), the chert breccia around the currently used wellfield (Kgwakgwe) and the dolomite formations

(Ramonnedi Wellfield) of the Taupone Dolomite Group, which consists of five formations, classified by its chert contents. The chert-free formations are built by massive, dark dolomite, forming depressed landscape. The chert-rich formations developed mild surface elevations with karst features. In the southern sections the area is covered by calcrete. Around the Kgwakgwe wellfield chert breccia occurs and forms the hilly landscape. The confining units are built of arenacous deposits of the Waterberg Group (north); magmatic intrusions (west - Segwagwa and Masoke Complex); and sandstone and intrusive members of the Lobatse Volcanic Group (east). Dolerite sills and dykes strike through the main dolomite body and are subject to intensive weathering close to the surface. Ferruginous crusts have developed over the bedrock containing sufficient iron.

The area is highly faulted, fractured and karstified. The major tectonic features are the Taupone Fault stretching from NE to SW through the central section, Kgwakgwe fault trending N-S, Nneneke fault trending E-W, and Gamorotswana fault trending NE-SW.

Karstification developed to a maximum depth of 100 m. Under the calcrete cover and in the area with intensive dolerite network continuous karst is poorly developed or missing. Dolomitic outcrops, especially in chert-rich sections, are well karstified creating pathways for rainfall recharge.

Transmissivities obtained from the pumping tests vary from 0.1 m^2/d to 2700 m^2/d. The areal distribution of transmissivity values vary over short distances indicative of the aquifer's hydrogeological anisotropy and heterogeneity.

3 METHODS

3.1 *Saturated volume fluctuation method (SVF)*

The SVF method transforms groundwater level fluctuations above a given datum plane to equivalent amounts of water. A simplified (saturated) water balance can be expressed as follows:

$$S\frac{\Delta V}{\Delta t} = Q_i - Q_o + R - Q_a \qquad (1)$$

Where Q_I *and* Q_o = lateral inflows and outflows [mm], respectively during time increment Δt; Q_a = net abstraction or discharge from the aquifer [mm]; R = recharge [mm], S = storativity and ΔV is a change in saturated volume of the aquifer.

When the lateral flow can be neglected, the net volume change is dependent only on recharge and the abstraction. The recharge is equal to abstraction when there is no change in groundwater level status and the storativity has not changed - this is the basis of equal volume method. By correlating the equal volume rainfall and abstraction the recharge as a portion of rainfall can be calculated.

The SVF technique is a lumped parameter method – the hydrological status of the aquifer is integrated into a composite integrated hydrograph derived from water level records in the studied area. The integration of water levels is achieved over a Thiessen polygon network. The usual assumptions include (1) the base of aquifer to be impermeable and (2) the evapotranspiration losses are included in the recharge term.

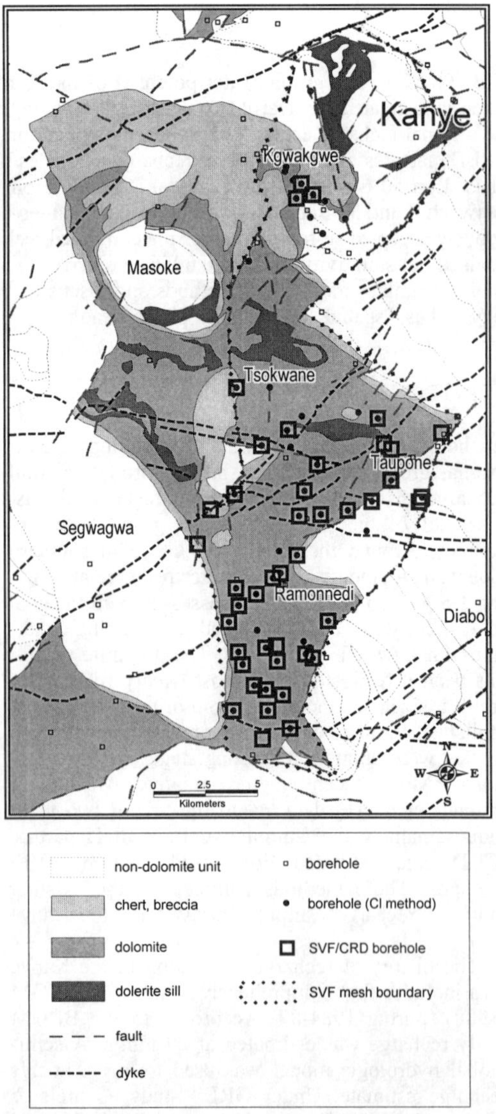

Figure 1. Situation map of the studied area

When the changes over a time period are zero the storativity term is eliminated. The zero change conditions usually apply to different periods, implying recharge/abstraction conditions for various monthly rainfall values. The average recharge can thus be calculated using recharge (from equal volume periods) and rainfall relationship as follows:

$$R = kR_a + (Q_i - Q_o) \quad (2)$$

Where R_a = mean rainfall for the evaluated period.

3.2 *Cumulative rainfall departure method (CRD)*

The cumulative rainfall departures are defined as follows:

$$\hat{R}_i = R_i - k\overline{R} + \hat{R}_{i-1} \quad (3)$$

Where $\hat{R}_i$ is cumulative rainfall departure (CRD) for month i [mm]; R_i is rainfall in month i [mm]; $\overline{R}$ is average monthly rainfall for the entire rainfall record [mm]; k is constant representing pumping or injection (equals 1 for no pumping) [dimensionless].

Most often the acceptable correlation exists between the CRD and the groundwater level, expressed as a proportionality constant a:

$$h_i = a\hat{R}_i + H_w \quad (4)$$

where H_w = average depth of the groundwater level below surface; or:

$$\Delta h_i = a\left(R_i - k\overline{R}\right) \quad (5)$$

Constant a represents a lumped coefficient of recharge and storativity, R_p/S. The objective is to find a fixed proportion of rainfall for every month. The correlation between the CRD and the groundwater levels can be improved by introducing the concept of lag effects:

$$\hat{R}_i = \frac{1}{m}\sum_{j=1-(m-1)}^{i} R_j - k\frac{1}{n}\sum_{j=1-(n-1)}^{i} R_j + \hat{R}_{i-1} \quad (6)$$

where m denotes a time lag required to observe the effect of a recharge event on the water table (short memory) [month]; and n is a redistributed effect of recharge over long term, usually reflecting subtle climatic or hydrological cyclicity (long-term memory) [month]. Considering the change in saturated volume approximated to CRD we obtain:

$$V_i \propto a\left(\frac{1}{m}\sum_{j=i-(m-1)}^{i} R_j - k\frac{1}{n}\sum_{j=1-(n-1)}^{i} R_j + \hat{R}_{i-1}\right) \quad (7)$$

The relationship between CRD and water levels is sought using iteration/optimization techniques. By changing parameters such as m and n and/or k (if abstraction is important) iteratively, the best fit can be found. The optimization routine can be carried out for individual boreholes (to assess a point value of recharge) or for a domain (global recharge estimate).

4 RESULTS AND DISCUSSION

The approach was to apply the SVF and CRD to the domains of various sizes assuming that the hydraulic stress applied to these domains is representative of recharge. The variable sizing was chosen because the extent and geometry of the aquifer system is uncertain.

4.1 *Domain selection*

The selection of domains was based on (1) the shape and extent of the potentiometric surface (Fig. 2) and cone of depression; (2) lithological boundaries (Fig. 1); and (3) structural boundaries (faults, dykes; Fig. 1). The "entire" domain (Scenario 1) thus includes the whole dolomite/chert aquifer (160 km^2), affected by abstraction (Fig. 1). The abstraction in the northern section may however not affect groundwater levels in the southern section. The central part of the aquifer is dissected by numerous dykes, which may affect the hydraulic response to abstraction in Kgwakgwe. Therefore the "northern" domain (Scenario 2) stretches from the central part (north of Taupone) to the north (118 km^2). The immediate area of the Kgwakgwe wellfield, covering the apparent cone of depression is evaluated in Scenario 3. Unfortunately, the borehole cover in both the northern and Kgwakgwe domains is poor (Fig. 1), which may affect the accuracy of the recharge calculation.

4.2 *Recharge and storativity (SVF and CRD)*

Both recharge and storativity was calculated for all three scenarios using SVF (equal volume periods). Using the CRD method recharge was calculated for the scenarios as well as for the individual boreholes.

The goodness of the obtained fit decreases with the number of observed boreholes, especially for Scenario 3. The variations of the groundwater potentiometric surface and associated hydraulic stresses cannot thus be adequately expressed. In addition, the evaluated period for the Scenario 3 was shorter than for the other two scenarios, because of insufficient data. The Scenario 3 results are therefore considered less reliable.

During the evaluated period (1988-1996) there was a general decline of water levels in the studied area (e.g. Fig. 3). The decline became steeper after 1992. The overall decline represents about 3 m, so the rate of recession between 1992 and 1995 is just below 1m/a. The groundwater recharge mound is present in the central-to-NW part of the area (around Tsokwane). The groundwater level there is at 1280 m a.m.s.l. and more. The lowest water level is in the southern portion of the dolomite segment, at 1200 m a.m.s.l. This constitutes a maximum groundwater gradient of approximately 80 m through the studied area (4 m per km, or 0.004).

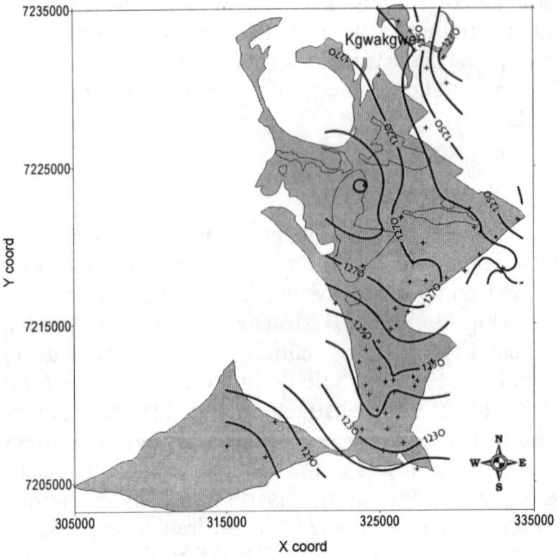

Figure 2. Groundwater levels in January 1996

By studying groundwater levels no major changes in the shape of the flow field with time could be deduced. During the low rainfall periods the wellfield draws water from larger area than during recharge periods. This is visible on annual basis, suggesting that recharge is not limited to major rainfall events and that it is laterally redistributed.

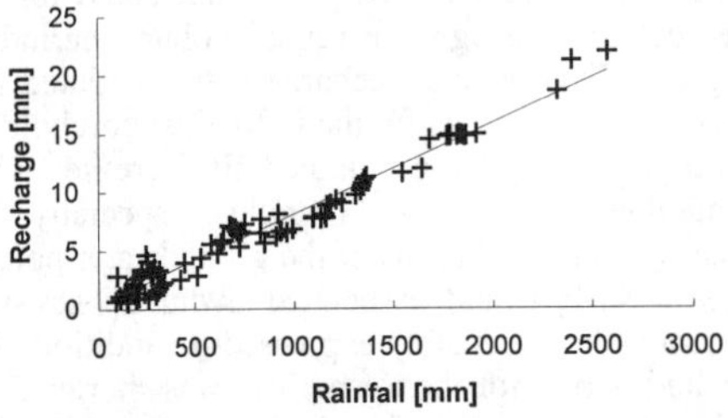

Figure 3. An example of the SVF (equal volume) application for Scenario 1. Recharge volumes calculated for equal volumes are depicted on rainfall/recharge graph. The regression lines describes the average annual recharge for the evaluated period.

The SVF results suggest that the recharge is lower than was previously determined. This depends on accuracy of abstraction rates and potential losses from the aquifer system. For this evaluation the losses from the system were taken equal to inflows.

The estimates for different domains are not meant to represent recharge in different segments of the aquifer. They are attributed to domains of different estimated size because it is difficult to accurately define the capture area of the Kwagkwe pumping field. The capture area of the wellfield may go beyond the evaluated dolomite/chert aquifer, but the existing network of monitoring boreholes cannot document the extent of this effect.

The SVF results also bring lower than previously-determined storativity values (2-3 times, Table 1). According to the SVF results the Kgwakgwe wellfield may be pumped at or above its capacity.

Table 1. Recharge and storativity estimates derived from the SVF method

Scenario/Domain	Recharge	Recharge	Storativity
	[mm/a]	[mil m^3/a]	
(1) Entire	4-5	0.66-0.77	0.0029
(2) Northern	8-10	0.64-0.76	0.0056
(3) Wellfield (Kgwakgwe)	63-74	1.21-1.41	0.0036

The proportionality constant a, derived from the CRD application, lumps recharge and storativity together. Reliable estimates of storativity are therefore required in order to arrive at actual recharge figures. In this evaluation two storativity values were used: 0.01 as the best previous global estimate and estimates derived from the SVF method for different domain scenarios.

Figure 4 shows an example of the fitted model to the domain's integrated hydrograph (Scenario 1):

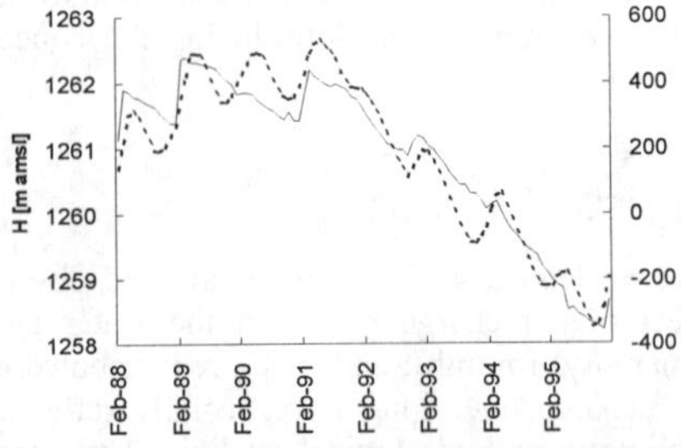

Figure 4. An example of the CRD application (Scenario 1). The graph shows fit between the CRD and water level integrated over the entire domain (correlation coefficient = 0.94).

Table 2, (CRD results) reveals that the calculated values represent a wide range. The values derived from storativity of 0.01 (17-31 mm) correlate with values that were determined by previous studies. The CRD method would therefore confirm previous estimates. Storativity = 0.02 is considered too high, producing very high recharge figures.

Delay lag parameters represented as short- and long-term memory (*m, n*) were determined by fitting the CRD curve to respective hydrographs. The average short-term memory is 3-4 months for all domains except for Kgwakgwe (12). The long-term memory is 77-91 months for large domains and 47 months for the Kgwakgwe wellfield scenario.

Table 2. Recharge estimates derived from the CRD method (global)

Scenario/Domain	Recharge [mm/a]	Correlation	Storativity
(1) Entire	6-8	0.94	0.0029 (SVF)
	22-26		0.01 (estimate)
(2) Northern	10-11	0.93	0.0056 (SVF)
	17-20		0.01 (estimate)
(3) Wellfield	10-11	0.74	0.0036 (SVF)
	27-32		0.01 (estimate)

For individual boreholes the recharge calculation revealed considerable differences, which is to be expected. Rather than discussing recharge variations computed for 50 boreholes we focus on whether earlier recharge estimates differ from the CRD-derived values. Gehrels & van der Lee (1990) calibrated the recharge and storativity for 17 boreholes in the Ramonnedi area. It was thus possible to compare the results related to 14 boreholes from their report and our CRD implementation.

Using nonparametric statistical tests such as Wilcoxon paired test and Spearman ranks correlation coefficient it is possible to quantify the global difference between two sets of results. A large p value greater than the chosen significance value (e.g., 0.01) is usually interpreted to mean that there is no significant difference between two sets. In our case it also means that there is no evidence to conclude that the average recharge from both techniques is significantly different. The comparison between the GRES and CRD using Wilcoxon matched pair test and Spearman ranks correlation coefficient is as follows:

The calculated p-levels are in general higher than 0.05, or 0.01 except for the comparison between "long" and CRD S(GRES) scenario estimates (Table 3), suggesting statistically significant difference between GRES "short" and CRD estimates. The comparison between CRD results (with GRES estimates of storativity) and GRES results shows a weak correlation (r_s=0.56-0.63), which is marginally statistically significant ($0.01<p<0.05$).

Table 3. Statistical comparison between EARTH (long and short term estimates) and CRD (for storativity at 0.01 or GRES value) results for the set of 14 matched individual boreholes

EARTH scenario	CRD scenario	p(W)	r_s	p(Sp)
short	S=0.01	0.70	-0.14	0.62
long	S=0.01	0.41	-0.26	0.37
short	S(GRES)	0.34	0.56	0.04
long	S(GRES)	0.002	0.63	0.02

Note: p(W) = p-level for Wilcoxon matched pair test; r_s = Spearman ranks correlation coefficient; p(Sp) = p-level of Spearman ranks correlation; S(GRES) = storativity estimate from GRES study (Gehrels & van der Lee, 1990)

4.3 *Recharge derived from the chloride method*

The principles of the chloride method are well known (e.g. Wood & Sanford, 1995) and were previously applied to the project area (e.g. BRGM, 1988; Gehrels & van der Lee, 1990). Chloride is assumed to be conservative and does not volatilize, react with or get adsorbed on the aquifer matrix. It is assumed that steady-state conditions exist between average rainfall and its chloride concentration. It is further assumed that the unconfined aquifer is receiving recharge only by direct rainfall on the surface and subsequent recharge through unsaturated zone. The chloride concentration is considered to be in temporal and spatial steady state with respect to rainfall, experiences no net sorption or plant uptake and there is no dissolution or precipitation of chloride minerals. For these conditions the steady-state mass balance across the water table can be given as follows:

$$Rc_{gw} = R_a c_r \tag{8}$$

Where R = recharge flux [mm/a], c_{gw} = chloride concentration in groundwater [mg/l], R_a = annual rainfall [mm/a] and c_r = weighted average concentration of chloride (wet and dry) in rain water [mg/l]. The input concentration was taken from Gehrels & van der Lee (1990) being 0.6 mg/l, which includes the dry deposition component.

The results, compared to previous calculations, are in Table 4. The GRES data were concentrated in the southern section of the dolomite aquifer. The data used for this paper come from the boreholes located within the SVF- and CRD-defined aquifer boundaries. The present coverage should therefore be more representative for the entire studied area and should allow for comparison between water-level based methods and the chloride balance method. Considering the median values to be the best estimates, the currently derived recharge value for the entire aquifer do not differ from the original GRES estimate.

The GRES averaging methodology apparently differs from the usual approach. The average groundwater chloride concentration is used in Eq. 7. In the GRES study recharge figures were calculated for each borehole on individual basis and the resulting recharge figures were averaged. The results of such averaging are therefore different from the usual approach. In Table 4 the GRES, 1990 values are based on chloride values from 22 boreholes using

GRES methodology. Using the usual approach the values were recalculated and denoted as GRES, 1990, corr. The values were also recalculated to reflect the updated mean annual precipitation (568 mm). After these adjustments GRES data show different recharge figures, and, after updating for MAR = 568 mm/a, even lower values than our study. The GRES study evaluated only a part of the domain and is based on 22 values as opposed to the current study, which is based on 71 values (or 67 after removing outliers).

Table 4. Recharge estimates derived from the chloride balance method (mean annual precipitation=568 mm; rainfall chloride concentration=0.6 mg/l).

Reference	mean [mm]	median [mm]	geomean [mm]	harmean [mm]
BRGM, 1988	20			
GRES, 1990	22	21	19	16
GRES, 1990, corr.	18[#]	22[#]	20[#]	25[#]
GRES, 1990, corr.	19	16	17	14
Current	12(18*)	21	21	30

Note: geomean = geometric mean; harmean = harmonic mean; [#] mean annual precipitation = 586 mm; * outliers removed after 95 percentile

On a spatial basis, the southern domain is characterized by an average chloride recharge, i.e. between 10 to 20 mm. The southernmost part however has higher recharge values, which are about twice as high as the average chloride recharge.

High chloride recharge values are associated with eastern to northeastern perimeters of the dolomite aquifer. The groundwater chloride concentrations are smaller there suggesting more active mode of recharge and possibly the influence of fresh water in non-dolomite area. These values correlate with the SVF estimates for the Kgwakgwe wellfield area, but not with those belonging to the northern domain.

The comparison between water balance (CRD and SVF) and chloride balance methods shows that chloride recharge estimates are close to SVF and CRD results for storativity = 0.01. For SVF-derived storativity, the recharge estimates are comparatively smaller. There are two main reasons: (1) either the aquifer delineation was inaccurate; or (2) the average computed chloride value does represent the actual groundwater average value. These two hypotheses will have to be pursued further by (1) extending the monitoring cover around the wellfield to delineate the extent of the depression created by pumping; and (2) subject the average chloride determination to a similar process of integration as was applied to groundwater levels.

5 CONCLUSIONS

The recharge to the dolomite/chert aquifer near Kanye, SE Botswana was reassessed using new methods and data. The application of CRD and SVF methods results in more conservative recharge estimates than previously reported. It was concluded that when using storativity = 0.01 the CRD results are similar to original estimates in the range of 17-32 mm/a. Using SVF-derived storativity (0.0029-0.0056, i.e. 2-3 times less), the recharge is comparatively smaller, 5-10 mm/a (SVF) and 6-11 mm/a (CRD). SVF-derived values for Scenario 3 reflect the fact that the calculation was based on area that probably does not cover the entire capture zone. The results are potentially charged with error(s), because the monitoring records around the wellfield are insufficient. A larger sample of monitoring records is thus necessary to confirm CRD and SVF results.

Using the consistent methodology, the chloride balance recharge estimates from GRES study have been shown to be smaller, from 21 to 16 mm/a (median). Our results based on 71 boreholes show consistency with previous estimates ~ 21 mm/a. Combining chloride and CRD figures it seems that storativity = 0.01 would be acceptable. It is however suggested that the application of chloride method be extended from mere arithmetic calculation of central value to more appropriate weighed value based on spatial distribution of sampling points.

REFERENCES

Bredenkamp, D.B., Botha, L.J., van Tonder, G.J., Janse van Rensburg, H. 1995: Manual on quantitative estimation of groundwater recharge and storativity. *Water Research Commission project T3/95, Pretoria*

Bureau de Recherches Geologiques et Minieres 1988: Kanye dolomite groundwater basin, hydrogeological investigation, resource assessment & development project – phase II. *Technical report, Orleans*

Gehrels, J.C. & van der Lee, J. 1990: Rainfall & recharge. A critical evaluation of the atmosphere-soil-groundwater relationship in Kanye, semi-arid Botswana. *GRES report, Free University of Amsterdam*

Simonic, M. 1998: Groundwater resource evaluation project: Kanye, Ramonnedi and Moshaneng area - recharge study. *Technical report for Geoflux (Pty) Ltd. Hydromedia Solutions (Pty) Ltd, Pretoria*

Wood, W.W. & Sanford, W.E. 1995: Chemical and isotopic methods for quantifying ground-water recharge in a regional, semiarid environment. *Ground Water* 33(3): 458-468

Groundwater: Past Achievements and Future Challenges, Sililo et al. (eds) © 2000 Balkema, Rotterdam, ISBN 90 5809 159 7

Natural groundwater recharge assessment: Role of preferential flow in semi-arid fractured aquifers

B.S. Sukhija, D.V. Reddy, P. Nagabhushanam & S. Hussain
National Geophysical Research Institute, Hyderabad, India

ABSTRACT: Understanding groundwater recharge processes and quantitative assessment of recharge is vital in semi-arid areas for proper groundwater management. The piston flow model of soil moisture movement has been widely used for different geological formations in India over the past three decades for estimating natural groundwater recharge using different tracers. The groundwater recharge values thus estimated constitute only minimum values as the preferential flow component of recharge was not accounted for. Though piston flow modeling of groundwater movement is valid for alluvial tracts, we observed that preferential flow is equally important for consolidated hard and fractured rocks. The present study used environmental chloride as a tracer in estimating the preferential flow component of recharge in a weathered/fractured granitic area where recharge measurements using injected tritium are available. The comparative analysis of results shows that in total recharge, the preferential flow component is either equal to or higher than the piston flow component. This calls for revision of earlier recharge estimates wherever the preferential flow is significant.

1 INTRODUCTION

In order to protect the groundwater resources from environmental problems like continuously declining groundwater levels, increasing pumping costs, degradation of quality of groundwater, seawater ingression etc., it is vital to assess the groundwater recharge and understand recharge processes. Groundwater recharge evaluation assumes great significance in countries having an arid character and monsoon climate and areas covered with hard rocks. Several conventional methods used for the above purpose require data like specific yield, surface runoff and evapotranspiration, and generally such data are neither available nor reliable. Recently, isotopic and geochemical tracer methods have been developed for evaluation of groundwater recharge, which are considered to be more reliable than the conventional methods (Gee & Hillel 1988, Lerner et al. 1990). Munnich and his co-workers (Zimmermann et al. 1967) made pioneering contributions to the development of tracer methods. From their experiments using environmental and injected tritium as tracers they postulated that in the case of homogenous soil, without many cracks and fissures, the bulk of water movement from unsaturated zone to saturated zone takes place in layered form. Thus recharge waters of different years of precipitation are stacked one above the other, as younger water does not overtake the older water, leading to development of the concept of the piston flow movement of water. This concept provided a convenient method of evaluation of recharge using the thermonuclear tritium peak of 1963-64 in precipitation which could be easily identified in the unsaturated zone. Following this concept many workers (Datta et al. 1973, Sukhija & Shah 1976, Verhagen et al. 1979, Athavale & Rangarajan 1988, Sukhija et al. 1996b) have used environmental and injected tritium to estimate groundwater recharge in areas situated in semi-arid and arid environments. Allison & Hughes (1978) developed the environmental chloride profile method, by studying the conjuctive use of environmental tritium and chloride for recharge estimates; and Edmunds & Walton (1980) demonstrated that the environmental chloride profile method can be effectively utilized where the assumptions of no additional sources and sinks of chloride in the profile are valid. Sharma & Hughes (1985) compared recharge rates by the chloride profile method and by chloride concentration in groundwater in relation to that of precipitation and thus obtained evidence of preferential flow. Recently, Wood et al. (1997) estimated the macropore recharge to be about 60 - 80% of the total recharge in a semi-arid area, Taxas, New Mexico. Another study by Davidson et al. (1998) in unsaturated tuff in Arizona, using detailed hydrochemical and isotopic analyses on pore water extracted from the core samples collected to a depth 157 m, shows evidence of a substantial quantity of recent recharge through fractures. The purpose of this paper is to bring forth the importance of preferential flow in recharge assessment, especially through weathered and fractured hard rocks in comparison to unconsolidated rocks, where the water moves primarily through the matrix.

2 RECHARGE ESTIMATES IN INDIA

Aquifers in India occur in three major hydrogeologic groups based on the water storage and yielding capacities. These groups are:

1 Unconsolidated sediments
2 Semi-consolidated sediments and
3 Consolidated rocks (Fig. 1).

Recharge estimates in India are obtained mainly using tracer techniques of environmental and injected tritium and natural chloride (Sukhija et al. 1996a).

2.1 *Unconsolidated sediments*

Environmental tritium profiles were used for the estimation of natural recharge in the alluvial tracts of Gujarat situated in the western part of India (Fig.1, no.2). Two methods of environmental tritium tracers i.e., total tritium and peak tritium, are used to evaluate recharge at six sites in sandy loam and clayey soils. As indicated by the typical tritium profile (Fig. 2a), measurements repeated over a two year period provided reproducible recharge results (Table 1, 5-7% of local rainfall, 11% in the recharge area and 3.0% at sites close to Rann off Kutch, near the Thar Desert). Recharge rates calculated using a groundwater diffusion flow model for the Mehsana aquifer are consistent with the environmental tritium method (Sukhija & Shah 1976). Datta et al. (1974) also indicated that groundwater recharge is 5-7% of average annual rainfall in the Sabarmati Basin part of the alluvial tracts of Gujarat.

Table 1. Average annual recharge for the alluvial tracts of Gujarat, as estimated by tritium methods (Sukhija & Shah 1976).

Site	Total ^3H method [mm/yr]	Peak ^3H method [mm/yr]	Annual recharge [mm]	% Local rainfall
Ahmedabad	46	66	56	7.0
Balol	25	28	26	4.3
Kosamba	50	63	56	4.5
Sankeshwar	13	17	15	3.0
Taranga	>27	58	56	10.9
Varahi	15	16	15	3.3

Injected tritium experiments (Fig. 2b) carried out in western Uttar Pradesh, the largest alluvial tracts in the country, and in Punjab and Haryana (Goel et al. 1975; Fig. 1, no. 3-5) indicated a recharge rate of 17 to 22 % of rainfall.

2.2 *Semi-consolidated sediments*

Extensive recharge studies were carried out in the Cuddalore sandstones of Tertiary age located in Pondicherry (160 km south of Chennai) and Neyveli (200 km south-southwest of Chennai) covering about 2000 km^2. Table 2 shows the annual average recharge estimated by the chloride (Fig. 2c) and injected tritium methods, which varies from 20-28% in Pondicherry, and 12 to 13% in Neyveli. In order to assess the preferential flow, if any, chloride concentration in rainfall as well as groundwater is made use of as discussed later.

Table 2. Average annual recharge for semi-consolidated formations, estimated by chloride and injected tritium methods.

Parameter	Pondicherry aquifer#		Neyveli aquifer##	
	Chloride method	Tritium method	Tritium method	
			1985-86	1986-87
No of sites	3	3	54	26
Injection time	-	6/1984	7/1985	7/1986
Sampling time	-	3/1985	2/1986	2/1987
Rainfall [mm/yr]	1200*	1080**	1367**	746**
Recharge range [mm]	160-310	180-300	0-550	0-220
MAR [mm/yr]	220	260	161	112
% of rainfall	19	28	11.8	13.4

#Sukhija et al. (1988), ##Rangarajan et al.(1989), *Average of 35 years (1950-85), **Rainfall during the study period, MAR = mean annual recharge.

2.3 *Consolidated formations (fractured rocks)*

Detailed injected tritium studies in basalts (Fig 1 no. 8-10, Athavale and Rangarajan 1988), assumed a piston flow model for tritium profiles, but the profiles do indicate some variation due to dispersion. Recharge estimates for these regions ranged between 7 and 12% of annual rainfall.

A large part of southern Peninsular India is under lain by granitic and gneissic complexes. Typically, these rocks have a thick soil cover that overlies weathered, semi-weathered, fissured and jointed rock mass. The areas have been investigated under various basins and sub basins (Fig. 1 nos. 13-17). Some of the recent experiments for the study of preferential flow are discussed below.

2.3.1 *Experiments in Aurepally Water Shed*

The Aurepally water shed located about 70 km from Hyderabad, Andhra Pradesh, India, typically represents the granite gneissic area where groundwater occurs in weathered and fractured zone typically at about 8-10 meters depth. Recharge measurements were carried out by Rangarajan et al., (1987) by tritium injection at 15 sites at a depth of 80 cm before monsoon (June 1984 & 1985) and tritium profiles

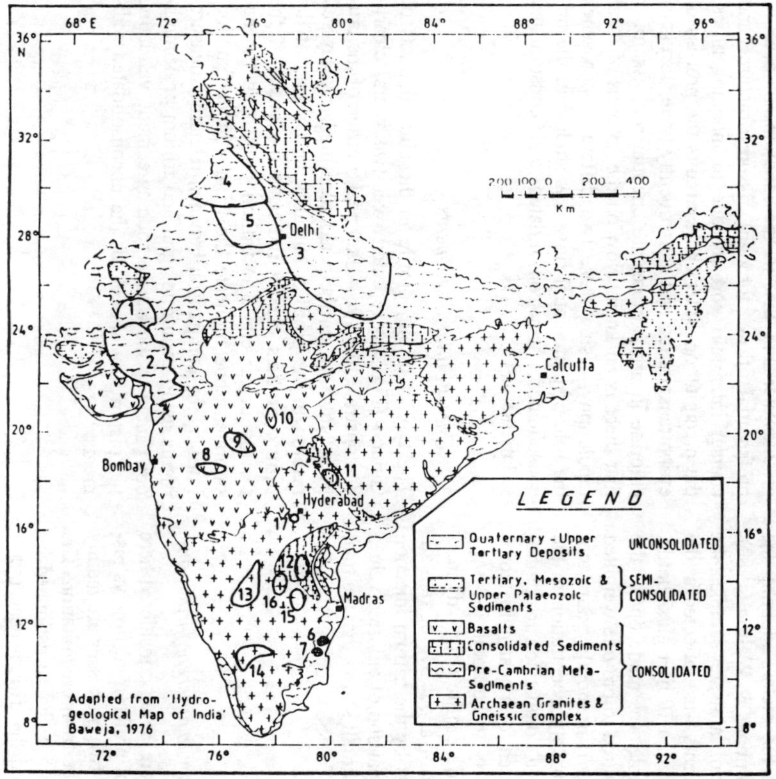

Figure 1. Hydrogeology of India, showing areas selected for recharge studies. 1) Aeolian deposits of western Rajastan; 2-5) Allivial tracts of Gujarat, westarn Uttar Pradesh, Punjab, Haryana; 6-7) Semi-consolidated sediments of Pondicherry and Neyveli; 8-10) Kukadi-Godavari-Purna and Jam basins in basalts; 11-12) Lower Maner and Kunderu basins in consolidated sediments; and 13-17) Vedavati, Noyil-Vattamalai Karai-Ponnai, Chitravati, Marvanka and Auropally basins in Archaean Gneissic complexes.

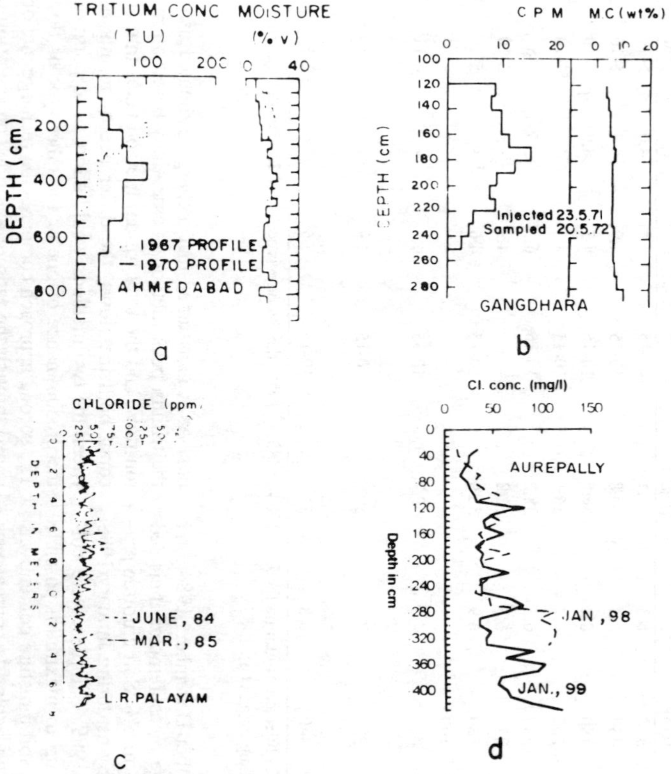

Figure 2. a) Typical environmental-tritium profiles in alluvial tracts of Gujarat (Sukhija & Shah 1976), b) Typical injected tritium profile in alluvial tracts of Uttar Pradesh (Datta et al., 1973), c) Typical environmental chloride profiles in sandstone, Pondicherry (Sukhija et al. 1988), d) Typical environmental chloride profiles in granitic terrain, Aurepally.

Table 3. Recharge through piston flow and preferred path flow in hard rock area (Aurepally water shed).

Site	Year	Cl profile depth [cm]	Moisture [wt %]	Av. Cl in profile [mg/l]	Recharge (% rainfall)			
					Total	Piston flow		Preferred path flow
						(1)	(2)	
Konapur	1998	330	8.3	52.1	10-15	4.8	4.0*	5-10
	1999	430	6.3	50.8	10-15	4.9	4.0*	5-10
Hanumanaya-ka tanda	1998	180	8.2	57.8	10-15	4.3	4.0*	5-10
	1999	200	4.4	56.1	10-15	4.5	4.0*	5-10
Akuthotapally	1998	230	6.5	38.2	10-15	6.5	1.5**	3-8
	1999	270	6.3	51.3	-	4.9	-	5-10
Akuthotapally (a)	1998	180	6.9	53.9	10-15	4.6	-	5-10
	1999	100	-	-	-	-	-	-
Akuthotapally (b)	1998	85	4.6	59.7	10-15	4.2	-	6-11
	1999	-	-	-	-	-	-	-
Aurepally	1998	90	7.6	50.8	10-15	4.9	-	5-10
	1999	-	-	-	-	-	-	-
Mudwin	1999	340	6.8	31.5	10-15	7.5*	7.6	2-7

(1) Chloride profile method; (2) Injected tritium method, (* 1984 & '85, ** 1985; the average is 5.6% for 15 sites during 1984 and 3% for 12 sites during 1985; Rangarajan et al. 1987).

were obtained in November-December 1984 and 1985. Two sets of recharge measurements indicated the recharge to be about 3 to 6%. Soil chloride profile studies were initiated by us during January 1998 and repeated in January 1999. Samples were collected at 10 cm interval, up to maximum possible depth using a hand auger. Soil moisture content was measured gravimetrically and chloride content was measured using a chloride ion meter, adopting the elutriation technique of Edmunds & Walton (1980). For chloride measurements of rainwater samples, the rainwater for individual events was collected in the raingauge and transferred into a 10 liter glass bottle. A monthly composite sample was made out of the rainwater, which was measured to provide weighted chloride concentrations (mg/l) in monthly rainwater samples. The average chloride concentration in the profiles is about 50 mg/l, in comparison to rainfall chloride, which was measured for a nearby meteorological station at Kalwakurty to be about 2-3 mg/l for 1997. A typical chloride profile (Fig. 2d) indicates piston flow recharge of about 5% of rainfall, which agrees well with that by the tritium injection method. However the comparison of chloride values in groundwater to that in rainfall shows that the total recharge is about 10-15% (Table 3).

3 DISCUSSION

3.1 *Role of preferential flow in different formations*

Basically two approaches are being considered here to study the phenomenon of preferential flow vs piston (matrix) flow, as these two processes are mainly responsible for total recharge from precipitation. The first approach is to compare recharge estimated by the total tritium method and the tritium peak methods. As it is assumed that recharge estimated using the tritium peak method is essentially based on piston flow, and the total tritium method takes into account both the flows, the difference of results between the two methods would allow us to point out the predominance of one or the other type of flow. The second approach compares the recharge rate obtained through the ratio of chloride concentration of groundwater and that in precipitation. In case of preferential flow through fissures/fracture and coarser intergrain porosity due to inhomogenities, the precipitation water short circuits the process of evapotranspiration and consequently lower levels of chloride in groundwater are encountered, assuming that there is no lateral flow on the grounds of local topography. Using these two different approaches the role of preferential flow was studied in the unconsolidated, semi-consolidated and consolidated sediments.

3.2 *Unconsolidated sediments*

The typical tritium profile in the alluvial tracts of Gujarat (Fig. 2a) repeated over two years, clearly demonstrates the 1963-1964 peak in each of the profiles. Not only are the peaks identified in a layered fashion in the profiles, but also the tail end of the tritium profiles clearly indicates the onset of the thermonuclear era. The tritium profiles in the unsaturated zone replicate the tritium input function. Further, the comparison between recharge measurements using tritium integral and tritium peak methods (Table 1) show quite good agreement, validating the piston flow model for the unconsolidated sediments.

3.3 *Semi consolidated sediment*

The Cuddalore sandstone aquifer was studied at Pondicherry and Neyveli. The recharge results in the Pondicherry, using chloride profiles and injected tritium method, are in agreement with the 30% (Table 2). The injected tritium method shows a high recharge value (28%), which is understandable as the value corresponds to 1985, a wet year, whereas the chloride profile method provides long term piston flow recharge (19%). The ratio of chloride in groundwater (25 mg/l) and that in precipitation (8 mg/l) indicates total recharge to be 30%. The detailed work carried by Rangarajan et al. (1989) in the Neyveli Cuddalore sandstone aquifer indicates a piston flow recharge of 12 to 13%, and the total recharge using groundwater and rainfall chloride ratio is 16.6% (Sukhija et al. 1996b). Thus there is an indication of by pass flow of about 30 to 50% in addition to the piston flow recharge in the semi-consolidated sediments.

3.4 *Consolidated formation*

However the difference between the piston flow model recharge and total recharge are at a maximum in the consolidated formation in granites and gneisses. Table 3 shows the estimation of groundwater recharge using environmental chloride (piston flow as well as total recharge) and the injected tritium methods. For this region, using the injected tritium method, piston flow recharge is estimated as 5.6% for 1984 and only 3% of total rainfall for the year 1985 (Rangarajan et al. 1987). Using the environmental chloride method, piston flow recharge is estimated at about 5% of the mean annual rainfall. Total recharge is estimated to be 10 - 15%, using the groundwater and rainfall chloride concentration ratio. The results show a significant contribution (equal to the piston flow component) to groundwater through preferential flow. Thus the recharge values estimated using the piston flow method are under estimated and represent a minimum value, while the preferential movement of water through fractures/fissures and other inhomogeneties is considerable.

4 CONCLUSIONS

Study of recharge processes in three different types of formations indicate that preferential flow is a considerable component especially in the case of semi-consolidated and fractured aquifers. Thus there is need to take into account this component for realistic assessment of groundwater recharge rates.

Acknowledgements: The authors express their gratitude to Dr. Harsh K. Gupta, Director, NGRI for his keen interest and permission to publish the paper. We are also grateful to IAEA for funding the project through research contract no. 10724/RO.

REFERENCES

Allison, G.B. & M.W. Hughes 1978. The use of environmental chloride and tritium to estimate total recharge to an unconfined aquifer. *Aust. J. Soil Res.* 16: 181-195.

Athavale, R.N. & R. Rangarajan 1988. Natural recharge measurements in the hard rock regions semi-arid India using tritium injection - a review. In Simmers, I ed., *Estimation of natural recharge of groundwater*: North Atlantic Treaty Organisation, Advanced Science Institute Series 222: 175-194.

Baweja, B.K. 1976. *Hydrogeological map of India*: Central Ground Water Board, India, 20p.

Datta, P.S., P.S. Goel, Rama, S.P. Sangal 1973. Groundwater recharge in western Uttar Pradesh. *Proc. Ind. Acad. Sci. LXXVII*, Sec. A.I, 1-12.

Datta, P.S., B. Desai, & S.K. Gupta 1974. *Hydrological investigations in Sabarmati basin: Groundwater recharge estimation using tritium tagging method: Ahmedabad, India*, Physical Research Laboratory, Report no. HYD-77-05: 30 p.

Davidson, G.R., R.L. Bassett, E.L. Hardin & D.L. Thompson 1998. Geochemical evidence of preferential flow of water through fractures in unsaturated tuff, Apache Leap, Arizona. *Applied Geochemistry* 15: 185-195.

Edmunds, W.M. & N.R.G. Walton 1980. A geochemical and isotopic approach to recharge evaluation in semi-arid zone: past and present. *Proc. Symp. Arid Zone Hydrol., Invest. Isot. Tech.* IAEA, Vienna, 47-68.

Gee, G.W. & D. Hillel 1988. Groundwater recharge in arid regions, review and critique of estimation methods. *Hydrological Processes* 2: 255-266.

Goel, P.S., P.S. Datta, Rama, S.P. Sangal, Hans Kumar, Prakash Bahadur, R.K. Sabherwal & B.S. Tanwar 1975. Tritium tracer studies on groundwater recharge in the alluvial deposits of Indo-Gangetic plains of Western U.P., Punjab and Haryana, In Athavale, R.N. and Srivastava, V.B., eds., *Approaches and methodologies for development of groundwater resources*: Indo-German workshop held at NGRI, Hyderabad, India, 309-322.

Lerner, D.N., A.S. Issar & I. Simmers (Eds) 1990. *Groundwater recharge: A guide to understanding and estimating natural recharge.* IAH, 8, Heinz Heise, Hannover, 345 p.

Rangarajan, R., G.K. Hodlur, T.G. Rao & R.N. Athavale 1987. *Measurement of natural recharge to phreatic aquifers of two alfisol watersheds in semi-arid tropical India.* Tech. Rep. of ICRISAT, 34 p.

Rangarajan, R., S.D. Deshmukh, D. Muralidharan & T.G. Rao 1989. *Recharge estimation in Neyveli groundwater basin by tritium tagging method.* NGRI tech. Rep. no. NGRI-89-Environ-65, 63 p.

Sharma, M.L. & M.W. Hughes 1985. Groundwater recharge estimation using chloride, deuterium and oxygen-18 profiles in the deep coastal sands of western Australia. *J. Hydrol.* 81: 93-109.

Sukhija, B.S. & C.R. Shah 1976. Conformity of groundwater recharge rate by tritium method and mathematical modelling. *J. Hydrol.* 30: 167-178.

Sukhija, B.S., D.V. Reddy, P. Nagabhushanam, R. Chand 1988. Validity of the environmental chloride method for recharge evaluation of coastal aquifers, India. *J. Hydrol.* 99: 349- 366.

Sukhija, B.S., P. Nagabhushanam & D. V. Reddy 1996a. Groundwater recharge in semi-arid regions of India: an over view of results obtained using tracers. *Hydrogeology Journal*, 4/3: 50-71.

Sukhija, B.S., D.V. Reddy, P. Nagabhushanam, Syed Hussain, V.Y. Giri & D.J. Patil 1996b. Environmental and injected tracers methodology to estimate direct precipitation recharge to a confined aquifer. *J. Hydrol.* 174: 77-97.

Wood, W.W., K.A. Rainwater & D.B. Thompson 1997. Quantifying macropore recharge: examples from a semi-arid area. *Groundwater* 35: 1097-1106.

Verhagan B. Th, P.E. Smith, I. McGeorge & Dziembowski 1979. Tritium profiles in Kalahari sands as a measure of rainwater recharge. *Proc. of Isotope Hydrology*, held in Neuherberg, June 1978, IAEA, Vienna, 733-749.

Zimmerman, U., K.O. Munnich and W. Roether 1967. *Downward movement of soil moisture traced by means of hydrogen isotopes.* Geophys. Monogr., Am. Geophys. Union, 11, 28-36.

Groundwater: Past Achievements and Future Challenges, Sililo et al. (eds) © 2000 Balkema, Rotterdam, ISBN 90 5809 159 7

Geology and resistivity characteristics of the Stampriet Artesian Basin

G. Symons – *Poseidon Geophysics Namibia (Pty) Limited, Windhoek, Namibia*

R. McG. Miller – *Windhoek, Namibia*

Y. Hata & Y. Yamasaki – *Pacific Consultants International, Tokyo, Japan*

K. Sugawara – *Mitsubishi Materials Natural Resources Development Corporation, Tokyo, Japan*

U. Noell – *Department of Water Affairs, Windhoek, Namibia*

ABSTRACT: Kalahari Sequence sediments, forming an upper aquifer, overlie the Karoo Sequence, which contains two artesian to subartesian sandstone horizons sandwiched between shales, the upper Auob sandstone, generally over 100 m thick, and the lower thinner Nossob sandstone. The Karoo rocks overlie an older basement. Pre-Kalahari erosion cut deep into the Auob sandstone in the south-eastern part of the basin where the Kalahari aquifer becomes highly saline. A Time Domain Electromagnetic survey comprising over 200 soundings has been used to determine the resistivity characteristics of the basin. These soundings detected predictable resistivity responses from the Kalahari and Karoo Sequences, the saline Salt Block, and Damaran basement. An unexpected feature of this survey was the discovery of a Resistive Zone dividing the main fresh water aquifer area from the Salt Block. The position of this Resistive Zone correlates with the position of magnetic features such as the Gochas Magnetic Body, a late stage intrusion, and north-easterly trending basement features, possibly indicating faulting. These features may have influenced the overlying Karoo and Kalahari Sequences creating barrier to the flow of aquifer water from the north and west, and contributing to the formation of the Salt Block.

1 INTRODUCTION

The Stampriet Artesian Basin of south-eastern Namibia extends from approximately 22°S to 26°S and from 18°E to the Botswana border. This study was carried out in order to obtain a better understanding of the three main aquifers in the basin, the Tertiary Kalahari aquifer and the underlying, lower Permian artesian to subartesian Auob and Nossob aquifers, prior to drilling, testing and the development of management plans for the aquifers.

2 GEOLOGY

The stratigraphy of the basin is given in Figure 1. The Carboniferous to Jurrassic Karoo Sequence of the Stampriet Artesian Basin rests unconformably on Neoproterozoic to Cambrian sandstones and shales of the Kamtsas Formation and Nama Group.

The upper Carboniferous Dwyka Group at the base of the Karoo Sequence consists of a basal tillite overlain by several deglaciation units containing scattered faceted and striated dropstones (Heath, 1972, Bangert et al., in press). The deepest test wells will stop in the uppermost part of the Dwyka Group.

The stratigraphically lowest aquifer, the Nossob Formation, lies directly on the Dwyka Group with a slight disconformity. Generally between 20 and 30 m thick, it reaches a local maximum of 55 m and gradually decreases in thickness southwards. It consists of two to three upward-coarsening cycles of grey shale and porous, white siltstone and medium- to coarse-grained sandstone. Grain size decreases southwards. Springs occur at the base of the formation where it outcrops in the Weissrand escarpment south of Mariental.

The overlying Mukarob Formation is an upward-coarsening succession consisting of a relatively thick, basal, dark grey to black, carboniferous shale. A very thin, black, intermittent, marker limestone and local gypsum layers occur at the top of the shale. The shale passes upwards rather abruptly into a sequence of thin, grey, highly micaceous and highly bioturbated, fining-upward siltstone-shale cycles which in turn pass upwards into similarly micaceous and disturbed fine-grained sandstone-siltstone-shale cycles. Upwards these coarser grained cycles interfinger with and are eventually overlain by fine- to medium-grained, light brown sandstones which are less micaceous and are not bioturbated. Well developed low-angle cross bedding occurs throughout these upper sandstones.

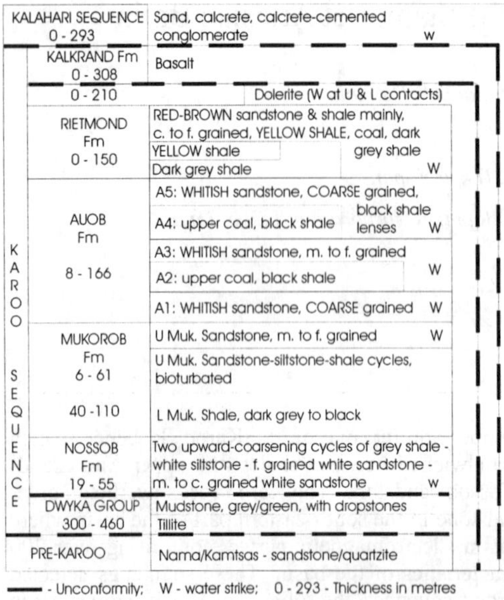

Figure 1. Stratigraphy of the Stampriet Basin (compiled from Heath, 1972 and Kingsley, 1985)

In the western part of the basin, outcrops of this upper succession of siltstone and sandstone have mistakenly been referred to as the Auob Formation. However, they form part of the upward-shallowing Mukarob succession (Kingsley, 1985). Springs occur near the base of the sandstone succession. The sandstones are porous and form the basal part of the Auob aquifer.

The overlying formation, the Auob Formation, is the main artesian aquifer in most parts of the basin. On the basis of over 70 core boreholes drilled into the basin during coal exploration, Kingsley (1985) was able to divide the Auob Formation into five distinct units, three sandstone units separated by two dark grey to black, carbonaceous, marker shale units with associated coal. These are designated A1 to A5 in Table 1. In the poorly exposed outcrop area due east of Mariental, in the north-east near the Botswana border and in places in the south, the shale units are absent but they occur throughout most of the rest of the basin. A1 is a coarse-grained, indistinctly cross bedded, highly porous, feldspathic, channel sandstone which is the only unit that outcrops. Generally white to very light grey or cream coloured when fresh, in outcrop it is light pink coloured east of Mariental and light cream east of Asab. In the latter locality only the basal 4 m are exposed and rest as channel-fill units with erosive bases on the Mukarob sandstones. Reaching 38 m in thickness in the north, this sandstone is only one meter thick in places in the southern part of the basin. The A2 shale/coal unit, 1 to 38 m thick, occurs throughout most of the basin but is locally absent due either to non-deposition or to erosion by the overlying sandstone. Where A2 is thick, it contains many interbedded sandstone and siltstone layers. The A3 unit ranges from 5 to 63 m in thickness but is usually about 50 m thick. Although very variable along strike, A3 consists mainly of fining-upward cycles of varying thickness of white to very light grey, medium- to fine-grained, crevasse splay sandstone and siltstone with grey shale tops in places. It also contains lenses of grey to black shale, a thin intermediate coal horizon, and lenses of white, coarse-grained sandstone which can be as much as 20 m thick. The A2 shale/coal unit varies from 0.25 to 40 m in thickness and is similar to the A4 unit in variability. In places it contains lenses of coarse-grained sandstone. A5 is a very light grey to white channel sandstone between 22 and 79 m thick and is similar to A1. In places, basal A5 channel erosion has removed the A4 shales and coals and cut deep channels into the A3 sandstone to produce continuous sections of coarse-grained sandstone up to 79 m thick. Cementation by calcite is very limited in the Auob Formation and where present is concentrated in the upper few cm of psammitic layers. A few of the siltstones and fine-grained sandstones in the formation are argillaceous and have a low porosity but all other sandstones and siltstones are well sorted and highly porous.

The lower Permian Rietmond Formation lies conformably on the Auob Formation. It consists of an upper sandstone/siltstone/shale succession underlain by shales, which where present, form an effective seal to the Auob artesian sandstones. The basal shale is dark grey to black, reaches 70 m in thickness and is indistinguishable from the dark Auob shales. It is overlain in places by a yellow to beige shale, which is distinctive for the Rietmond Formation. The upper Rietmond sandstones are fine to coarse grained and together with the siltstones are generally rather micaceous and brown to reddish in colour although grey and white layers are also present. The interbedded shales are red, brown, grey, black, white and yellow in colour. Thin coal seams are also present. Where the basal Rietmond shale is absent or thin, it can be extremely difficult to distinguish the Rietmond sandstones from the Auob sandstones (Martin, 1958) and one has to rely on rock colour and the level at which artesian or subartesian water was struck to determine the boundary.

A time gap of 100 million years separates the Rietmond Formation from the Jurassic Kalkrand basalts, which occur north of Mariental and unconformably overstep all underlying Karoo units to eventually lie directly on pre-Karoo basement west of the basin margin. The basalts appear to fill a deep pre-Kalkrand depression as they increase rapidly in thickness westwards. A second occurrence of basalt is located beneath relatively thin Kalahari cover in and north-eastern parts of the basin, the Karoo and underlying successions were extensively intruded by dolerite sills up to 210 m thick during basalt volcanism. These sills intrude mainly the upper part of the Karoo sedimentary sequence. Pre-Kalahari erosion removed most of the rocks overlying the sills.

The Kalahari Sequence increases in thickness from the margins to the center of the basin where it reaches a maximum of 293 m. It fills a deeply eroded pre-Kalahari valley that extends in a south-westerly direction from the Botswana border just north of 24°S to 25°S and then bends to a south-easterly direction to exit the Namibian border just south of 26°S. This valley probably formed a tributary to the Molopo River. Numerous deep tributaries feed into this valley and the pre-Kalahari topography was clearly very rugged with many steep valley sides where erosion cut though the dolerite sills. The valley has cut right through the Auob Formation in its deepest parts which has major implications for groundwater in the region and for modelling of the aquifers.

3 GROUNDWATER

Although groundwater is encountered in the Rietmond sandstones and sporadically in the Dwyka Group, the Kalahari Sequence and the artesian to subartesian Auob and Nossob sandstones are the main aquifers. Water quality deteriorates and TDS increases in a south-easterly direction into the Salt Block in all aquifers but particularly in the Kalahari Sequence (Tredoux, 1981; Tredoux *et al.*, 1979). The Salt Block is underlain by the deep pre-Kalahari valley. Narrow strips of near-surface fresh water follow the courses of the Nossob and Auob Rivers where they flow across the Salt Block. Deep water boreholes have been drilled in the basin to obtain better quality water and/or greater yields.

4 GEOPHYSICS

Time Domain Electromagnetic (TDEM) soundings and physical property logging of newly drilled boreholes comprise the geophysical component of the current study. The aim of the TDEM soundings is to provided information on the resistivity characteristics of the Kalahari and Karoo Sequences, especially the water bearing aquifers such as the Auob Formation and to a lesser extent the deeper Nossob Formation. The logging portion of the study is underway at the time of writing.

4.1 *TDEM Calibration and Methodology*

An initial calibration survey was carried out in the south-eastern portion of the survey area at the Vreda borehole (see Figure 2). A sounding using a 300 m loop at a 4 Hz duty cycle is shown in Figure 3, together with modelling including an equivalence analysis. The Kalahari is made up of a resistive (250 Ohm.m) upper unit and a less resistive (40 Ohm.m) lower unit. A dolerite sill is resolved with a resistivity of 200 Ohm.m. The entire Auob Formation is resolved as one layer with a resistivity of 60 Ohm.m. The lower boundary to the Auob Formation falls within the range of equivalent solutions. A thick low resistivity (10 Ohm.m) modeled layer lumps the stratigraphy of the Mukarob, Nossob and a relatively thick Dwyka Formation together. A smooth layered model of the same data, which is not presented, gives essentially the same results. Low resistivities (<250 Ohm.m) appear to characterize the Karoo stratigraphy. On the basis of this sounding and the expectation that no boreholes would be drilled deeper than 300m, it was decided that the 300m loop size at a 4Hz duty cycle was adequate for the entire survey. It is noted that an investigation of the Dwyka Formation was not one of the objectives of this project.

The main TDEM survey consisted of 200 soundings arranged along eight traverses (Figure 2). An average of 8 km was maintained between stations along the traverses. The equipment used consisted of a Zonge ZMG 7.5 generator, a Zonge GGT 10 transmitter, a Zonge GDP 32 receiver and a Zonge TEM/3 antenna. A current of 17 A into a single turn 300 by 300 m transmitter loop was used. The receiver coil measuring the Z or vertical component of the secondary magnetic field was located at the center of the transmitter loop. Three repeat readings were made at a 4 Hz duty cycle using a 64-cycle stack, resulting in the measurement of a time decay extending from 59 μs to 48 ms. A GPS coordinate of the receiver position at each station, accurate to 50 m was measured. The height for each sounding position above AMSL was digitized from 1: 50 000 topographical sheets.

4.2 *Results and Interpretation*

The data was processed and modelled using the program TEMIX XL by Interpex Limited. All soundings were inverted using a 4-layer model, together with an equivalence analysis as well as a 19-layer smooth model inversion.

The smooth modelling provides a product that appears to have more resolution in low resistivity areas and is used for the production of resistivity sections. The smooth modelled resistivity section, together with a geological section as well as an interpretation of the smooth model resistivity section for Travers H is presented in Figure 4. The geological section is constructed from an analysis of boreholes in the vicinity of each sounding location.

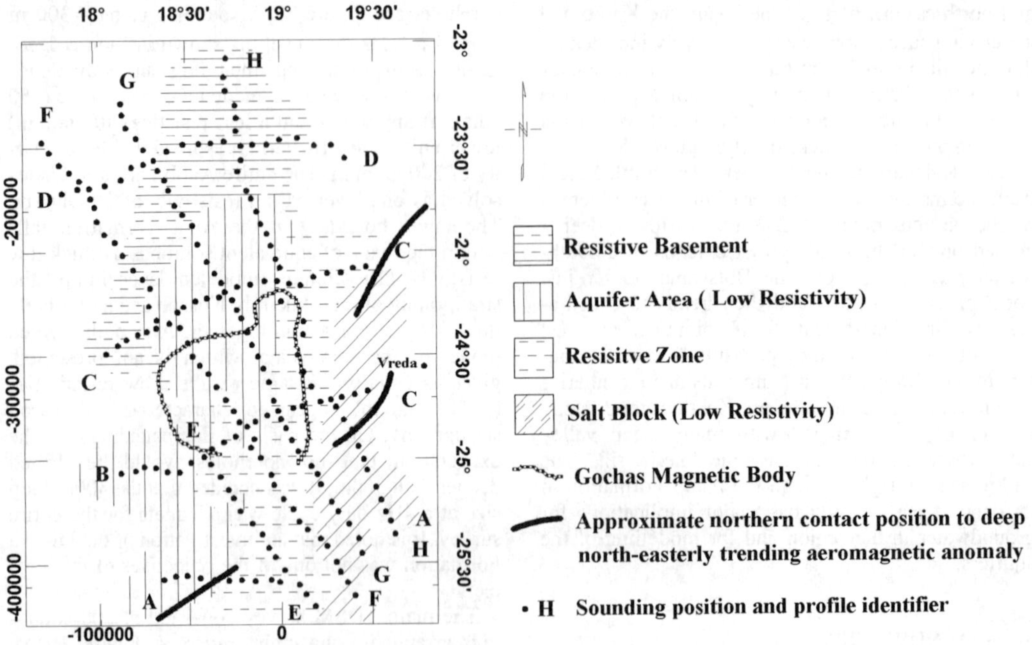

Figure 2. Interpretation of smooth layered model at 900 m AMSL together with features derived from aeromagnetic data

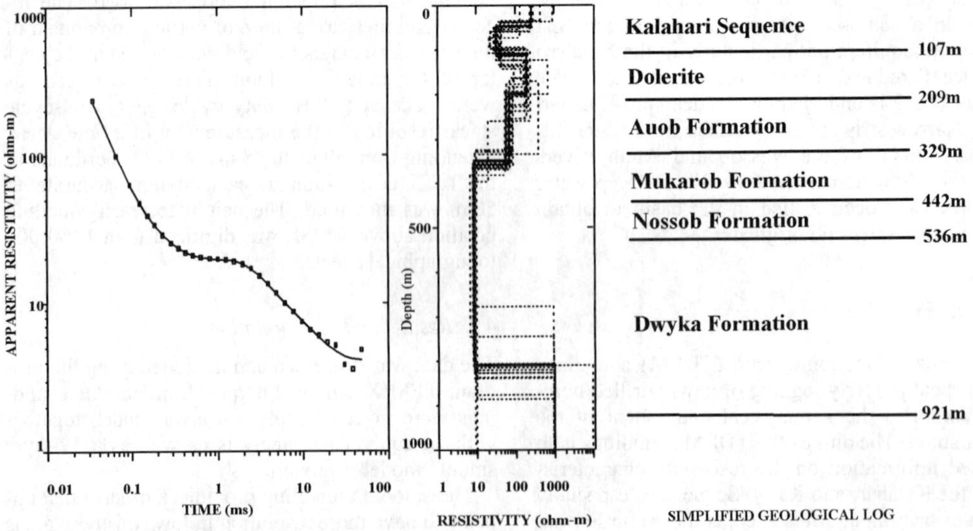

Figure 3. Layered and equivalence sounding interpretation at Vreda borehole

Smooth Model Resistivity Section

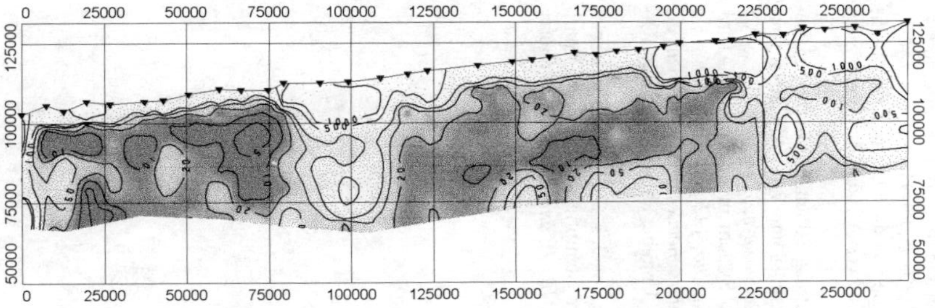

Geophysical Interpretation Section

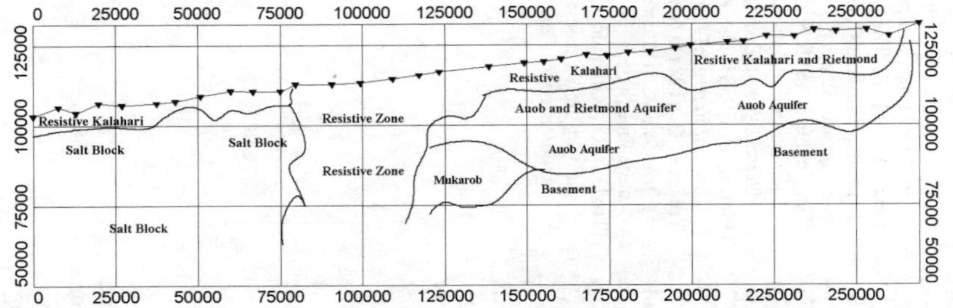

Geological Section

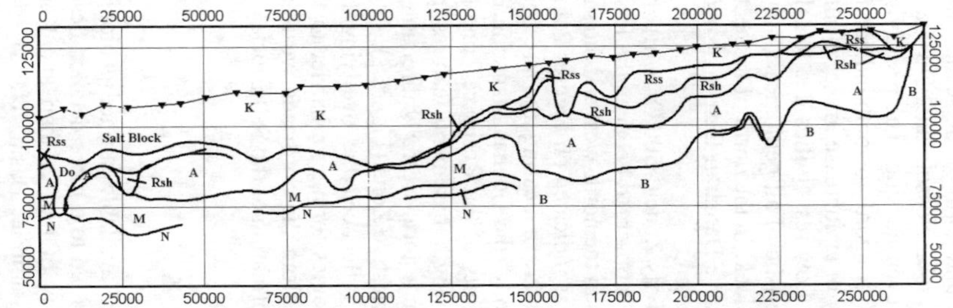

K-Kalahari | Do-Dolerite | Rss-Rietmond sandstone | Rsh-Rietmond shale | A-Auob and Mukarob sandstone | M-Mukarob shale

N-Nossob sandstone | B-Basement | ▼ Sounding position

Figure 4. Smooth Model Resistivity Section, Geophysical Interpretation Section and Geological Section for Profile H

It is noted that the resistivity response of lithological units in this geological environment is governed by porosity, water saturation, clay content and water resistivity as discussed by Bussian (1983) and de Witte (1957).

The geological section shows Kalahari thickening from north to south with deep channels cutting into the underlying Karoo stratigraphy in the south. The resistivity of the Kalahari is over 100 Ohm.m where unsaturated, dropping below 100 Ohm.m were saturated and even reaching resistivities below 10 Ohm.m in the south within the Salt Block. The Kalahari and Rietmond respond as one equivalent unit of intermediate resistivity (100-200 Ohm.m) where the Rietmond is present. The Auob has a low resistivity (< 150 Ohm.m), forming a distinct resistivity low in the centre and north of the section where the underlying Mukarob, Nossob and Dwyka Formations are thin or even missing. The resistivity of the Auob decreases from north to south indicating an increase in porosity or an increase in the salinity of the contained water. Low resistivities due to saline water are a characteristic of all geological units present within the Salt Block. In the north of the section, the presence of resistive Damaran Basement underlying the Karoo is detected.

A high Resistivity Zone (>500 Ohm.m) cross cutting stratigraphy is observed between the fresh to saline aquifer areas in the north-west and the salt block to the south-east as indicated on Figures 2 and 4. An aeromagnetic feature indicating the presence of a large intrusive body (the Gochas Magnetic Body) is found in the vicinity of the Resistive Zone together with long wavelength basement features possibly faults, trending in a north-westerly direction, almost parallel to the Resistive Zone (Figure 2). These features may have influenced the overlying Kalahari and Karoo lithologies forming a barrier to the flow of aquifer water from the north and west and contributing to the formation of the Salt Block.

5 CONCLUSIONS

A reevaluation of available surface geological and borehole data has led to a more detailed description of the stratigraphy of the Kalahari and Karoo Sequences in the Stampriet Basin. TDEM soundings have characterized the resistivity response of the Kalahari Sequence, the Rietmond and Auob Formations of the Karoo Sequences, the Damaran basement as well as the saline water containing Salt Block. The TDEM survey has discovered the presence of a resistive zone which correlates with magnetic features. This Resistive Zone may have had an influence on the formation of the Salt Block. Results from ongoing drilling and physical property logging are awaited.

6 ACKNOWLEDGEMENTS

This study was funded by Japan International Cooperation Agency (JICA) for the Ministry of Agriculture, Water and Rural Development in Namibia. The authors would like to thank JICA for their kind support and permission to present this paper.

REFERENCES

Bangert, B., Stollhofen, H., Geiger, M. and Lorenz, V. In press. Fossil record and high-resolution tephrostratigraphy of Carboniferous glaciomarine sandstones, Dwyka Group, southern Namibia. *Comms geol. Surv. Namibia, Henno Martin Commemorative Volume,* 12.

Bussian, A.E. 1983. Electrical conductance of a porous medium. *Geophysics* 48(9): 1258-1268.

de Witte, A.J. 1957. Saturation and porosity from electric logs in shaley sands. *Oil and Gas J.* 55: 89-93.

Heath, D.C. 1972. Die geologie van die Sisteem Karoo in die gebied Mariental-Asab, Suidwes-Afrika. *Mem. geol. Surv. S. Afr.* 61: 35 pp.

Kingsley, C.S. 1985. Sedimentological analysis of the Ecca sequence in the Kalahari Basin, South West Africa/Namibia. *Unpubl. Rep. CDM Mineral Surveys*, 35 pp.

Martin, H. 1958? The Karoo System in South West Africa. *Unpubl.Rep. geol Surv. Namibia:* 122 pp.

Tredoux, G. (1981). *'n Geohidrochemiese ondersoek van die Artesiese Kom Stampriet.* Unpubl. PhD thesis, Univ. Orange Free State.

Tredoux, G., Kirchner, J., and Vogel, J. C. (1979). Geohidrologie van die Artesiese Kom Stampriet. *Unnumbered confidential report; Parts IA, IB, II and III, CSIR*, Pretoria.

Groundwater: Past Achievements and Future Challenges, Sililo et al. (eds) © 2000 Balkema, Rotterdam, ISBN 90 5809 159 7

Groundwater recharge estimation in the Pingtung Plain, Taiwan

Cheh-Shyh Ting
Department of Civil Engineering, National Pingtung University of Science and Technology, Taiwan

ABSTRACT:Groundwater forms an essential part of the total water resources for Pingtung Plain. Knowledge of the processes and amount of groundwater recharge is therefore of fundamental importance for establishing an adequate water resource development plan for the area. The methods that have been applied are Soil Moisture Budgeting, Chloride Mass Balance and Simulation Model Methods. From the three methods, the recharge rate is 25% to 31% for 2740 mm of rainfall.

1 INTRODUCTION

Groundwater forms an essential part of the total water resources for the Pingtung Plain (Figure 1). Knowledge of the processes and amount of groundwater recharge is therefore of fundamental importance for establishing an adequate water resource development plan for the area. Quantifying the current rate of groundwater recharge is a basic prerequisite for efficient groundwater resource management in these regions, where such resources are often the key to economic development. Unfortunately, of all the factors in the evaluation of groundwater resources, the rate of aquifer replenishment is one of the most difficult to determine.

A recharge area can be defined as that portion of the drainage basin in which the net saturated flow of groundwater is directed away from the water table. Groundwater recharge can be defined as the entry into the saturated zone of water made available at the water table suface, together with the associated flow away from the water table within the satured zone. Two principal types of recharge can be recognized: i.e. direct and indirect recharge. Direct recharge is defined as water added to the groundwater reservoir in excess of soil moisture deficits and evapotranspiration, by direct vertical percolation of precipitation through the unsaturated zone. Indirect recharge results from percolation to the water table following runoff and accumulation in joints, as ponding in low-lying areas, or through the beds of surface watercourses.

In 1961 the Provincial Groundwater Development Bureau made first estimates for the amount of recharge in the Pingtung Plain by subtracting average annual runoff and actual evapotranspiration from the total rainfall, resulting in 903 Mm^3 water (i.e. 746 mm/year) (Ting et al. 1998). This amount, being the net resource to the groundwater body, is equivalent to 35% of the total mean annual rainfall.

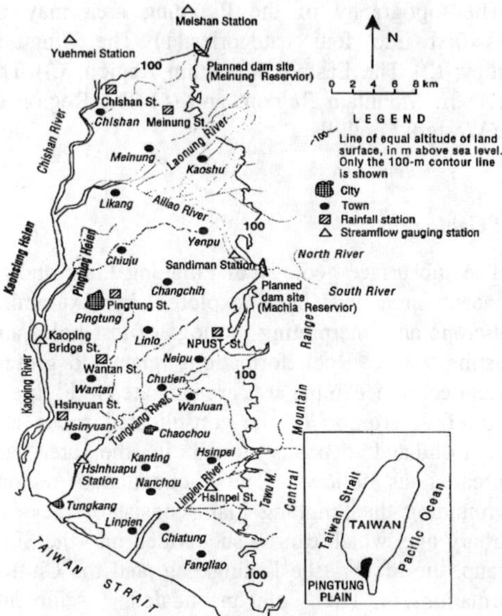

Figure 1. Location map of Pingtung Plain.

Groundwater recharge estimations are often subject to considerable uncertainty and large errors: aquifer overdevelopment can occur if active groundwater recharge is overestimated. For optimum resource modelling and management, quantitative information on recharge is critical. There is a need for a better understanding and more detailed research on recharge mechanisms. It is also essential to use more than one technique for result verification; there three methods that have been determined by this study's objectives, i.e. soil water budgeting, chloride mass balance and numerical model methods. These are briefly described in the following sections.

2 PHYSICAL FEATURES OF THE PINGTUNG PLAIN

2.1 *Geomorphology*

The Pingtung Plain is located in the southwestern part of Taiwan. It includes most of Pingtung Hsien and some of Kaohsiung Hsien, and covers an area of 1210 km^2. It is bounded in the north by low hills, by foothills lying along the right bank of the Kaoping River in the west, by the Central Mountain Range in the east and by the Taiwan Strait to the south. It is rectangular in shape, 22km wide in an east-west direction and 55 km long in the north-south direction. Being surrounded by hills in the north, west and east, this plain may be considered in a broad sense as a valley.

The topography of the Pingtung area may be classified into four categories:(1) The Pingtung Valley; (2) The Eastern Mountain Region; (3) The Northern Mountain Region; and (4) The Region of the Western Foothills.

2.2 *Geology*

The subsurface geology of Pingtung Plain and its adjacent area has been explored by examining out-crops and interpreting the logs of test holes and existing wells. Rock formations related to ground water occurrence in the area are indicated in Figure 2. A brief description of the consolidated rocks and unconsolidated deposits found in the area are presented as follows: The mountainous regions surrounding the Pingtung Plain consist of rocks of Tertiary age, which can be subdivided into the Suao Group, the Mucha, the Kutingkeng, and the Chiting Formations. The unconsolidated sediments underlying Pingtung Plain are of Quaternary age and from the main aquifer. With respect to their character, they may be divided into the Linkou conglomerate, an older alluvium and recent alluvium which may be divided into three categories: a zone of coarse gravel, a zone of alternating gravel and sand, and a zone of fine sediments.

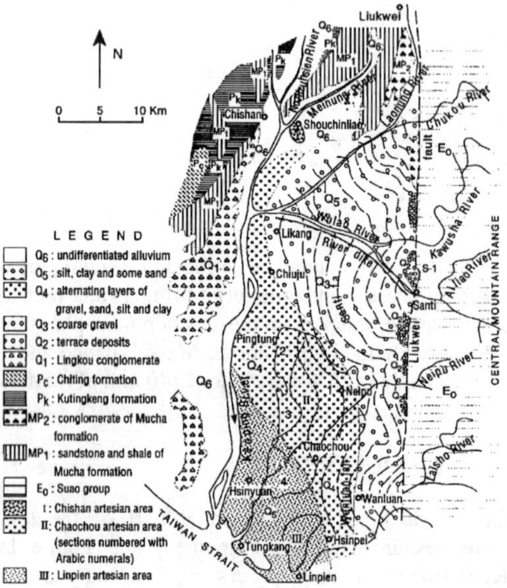

Figure 2. Geological map of the Pingtung Plain Region.

2.3 *Climate*

The climate of Pingtung Plain is sub-tropical, in accordance with its position just south of the tropic of Cancer. Rainfall is alternately affected by typhoons in summer and monsoons in winter. During the summer, typhoons and thunderstorms occur regularly, producing most of the rainfall, while during the winter the monsoons cause a dry season. The rainy season lasts from May to October. Precipitation in the remaining six months amounts to only about 10 percent of the total; there is usually less than 1 mm of rain per month in the dry season.

Total precipitation for the month with highest rainfall at Pingtung Meteo station was 927 mm in August 1988; the lowest monthly precipitation was zero in November 1989 at the same location. The average annual precipitation was 2118 mm (period of record 1981-1990). Rainfall characteristics are therefore the main factors controlling water resource availability; precipitation distribution is alternately affected by typhoons in the summer wet season (May to October) and monsoons in the dry winter season (November to April). Spatial and monthly precipitation distributions are clear in the plain.

Average annual temperature in Pingtung is 24°C, with the highest monthly mean value in July (28°C) and the lowest in January (19°C).

Open water (pan) evaporation observations have been made at six hydrometric stations over a period of 10 years (1981-1990). Average annual pan evaporation is about 1200 mm.

3 ESTIMATION TECHNIQUES FOR THE PRESENT RESEARCH

3.1 *Recharge Estimation Using Soil Moisture Budgeting Method*

The method was originally developed to estimate soil moisture content and actual evapotranspiration, recharge estimates being only a by-product. The water budget method can be applied to any hydrological system. In the present case it is applied to the upper unsaturated zone; more precisely, the zone above the level (below the roots) where during dry periods a zero flux plane develops. Below this zone only downward percolation is assumed; groundwater is deep enough in the area to prevent capillary rise to the root zone. Another assumption is that fluxes are mainly vertical. Soil-water-plant-atmosphere interactions may be more complex but are not considered in detail within this study. For each period of time, for example days, the soil water budget can be assessed according to:

$$R = P - R_o - ET_{act} - S \quad (1)$$

Where R = recharge (mm/day)
P = precipitation (mm/day)
R_o = runoff (mm/day) + surface water evaporation
ET_{act} = actual evapotranspiration (mm/day)
S = storage change (mm/day)

3.2 *Chloride Mass Balance Method*

Natural tracers such as the chloride ion (Cl^-) can be used for estimation of groundwater recharge through the unsaturated zone. The chloride ion is used in chemical recharge studies because of its conservative nature. The chloride mass balance method developed by Eriksson & Khunakasem (1969) compares total chloride deposition at the surface with chloride concentrations in the soil moisture beneath the active root zone. Assuming chloride is a conservative ion and rainwater is the only source of Cl, conservation of mass leads to a relation between precipitation and recharge:

$$P_{eff} \times Cl_p = R \times Cl_{gw} \quad (2)$$

Where P_{eff} = effective precipitation in mm (yearly average)
Cl_p = chloride concentration of precipitation in ppm (yearly average)
R = recharge in mm/yr
Cl_{gw} = constant chloride concentration of groundwater in ppm

Effective precipitation is used in these equation because in South Taiwan most precipitation falls during the summer typhoon season, when extremely high rainfall intensities occur. A large part of this rainfall runs off as overland flow. When using Cl_{gw} total recharge is estimated. Using Cl_{gw} is more complicated as more knowledge is required about the groundwater flow path upstream and other sources of chloride along the path.

Sharma and Hughes (1985) have introduced a model which includes by-pass flow. Assuming this to be sufficiently rapid to be not available for root uptake, recharge can hence be described as:

$$R_{tot} \times Cl_{gw} = R_d \times Cl_{sm} + R_{bp} \times Cl_{bp} \quad (3)$$

Where R_{tot} = total recharge (mm/year)
R_d = recharge by diffuse flow (mm/year)
R_{bp} = recharge by localized flow along preferred paths (mm/year)
Cl_{bp} = concentration of the preferred flow (ppm), taken as the average chloride concentration in precipitation ($Cl_{bp} = Cl_p$)

In Pingtung Plain irrigation may also contribute to the recharge. For this situation equation 2 is therefore modified to:

$$P_{eff} \times Cl_p + I_{irr} \times Cl_{irr} = R \times Cl_{gw} \quad (4)$$

Where I_{irr} = irrigation

3.3 *Simulation Model Method*

The Areal Recharge in MODFLOW (McDonald and Harbaugh, 1988) is designed to simulate spatial distributed recharge to the groundwater flow system. Most commonly, areal recharge is from direct precipitation infiltration; i.e. $R=\alpha \times P$, where R is areal recharge, α is an infiltration coefficient and P is precipitation. Because of the differences in soil properties, geological characteristics, vegetation, rainfall intensity and phreatic water table, etc, different (time varying) infiltration coefficient values are required for each of the different sub-regions. Taking into account the spatial variability of precipitation and availability of rain gauge stations, 22 precipitation sub-areas are defined. Monthly total precipitation was derived. A program was developed to calculate monthly recharge for each model cell from precipitation and the relevant calibrated infiltration coefficients.

4 DISCUSSION OF THE ESTIMATION RESULTS

4.1 *Water Balance Calculations*

Detailed soil moisture budget calculation is performed on a 10-day basis for the two dominant agricultural types in the Pingtung Plain; i.e. double rice and sugarcane. In addition, water balances are calculated for upland crops and fruit trees on a 10-day time step.

The computer programme CROPWAT (Smith, 1989) from the Food and Agricultural Organization of the United Nations also allowed an approximate, yet less time-consuming determination of effective precipitation and irrigation requirements, leading to preliminary recharge estimates for the years 1988 to 1994. The calculations were based on monthly meteorological data. The programme is used here for water budget calculations for the years 1988 to 1994 according to the soil water budgeting method. Crops included mango, sugarcane, vegetables and wetland rice. The input parameters are: evapotranspiration and effective precipitation, crop water requirement, irrigation requirement. The results are listed table 1.

Table 1. Regional recharge estimates for Pingtung Plain (1988).

Land use	Average recharge (mm)	Area (ha)	Recharge total (Mm^3)	Irrigation requirement (mm)	Irrigation total (Mm^3)
Paddy rice	1078	12431	134.0	1033	128.4
Sugarcane	744	7871	58.6	775	61.0
Upland crops	870	14623	127.2	311	45.5
Fruit trees	676	39371	266.1	339	133.5
Fallow	553	40329	223.0	-	-
Urban	-	4436	-	-	-
Aquaculture	-	1937	-	-	-
Total	-	121000	830.5 Mm^3	-	368.4 Mm^3

It should be noted here that in all these calculations potential recharge is estimated from numerous assumptions and generalizations, disregarding any actual recharge-discharge mechanisms in the Pingtung Plain. It is therefore interesting to compare these calculated values with results from other methods such as the chloride mass balance, which can provide more insight to the actual recharge.

4.2 *Chloride Mass Balance Calculations*

Four investigation sites (Figure 3) have been chosen on a north-east to southwest running section from the Ailiao River to Pingtung and reflect the transition from the coarse sediments of the alluvial fan near the mountains to the finer deposits along the Kaoping River. The section runs parallel to the direction of the regional groundwater flow. Three of the four boreholes were drilled with the cable tool method during the dry season on 13 and 14 October 1994. For the depths of 0-1, 1-2, 2-3, and over 4 m, the intervals are 0.1, 0.20, 0.25, and 0.50 m respectively.

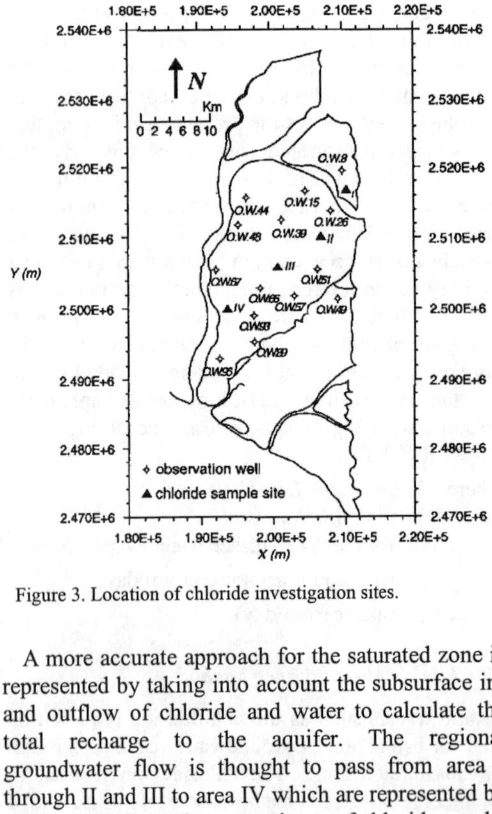

Figure 3. Location of chloride investigation sites.

A more accurate approach for the saturated zone is represented by taking into account the subsurface in- and outflow of chloride and water to calculate the total recharge to the aquifer. The regional groundwater flow is thought to pass from area I through II and III to area IV which are represented by sites I to IV. In each area an input of chloride can be considered from precipitation on that area supplemented with an input from the area situated upstream. This model calculates the chloride balance according to equation 3 and 4. The results are listed table 2.

The calculation of recharge using this method confirms the low and even negative recharge for areas III and IV, respectively. The potential recharge can be high for the upper part of the profiles but the actual recharge in those areas appear to be negligible. Area IV might even be considered as a discharge area in this respect.

Table 2. Chloride mass balance.

Site	Cl_{sm} (ppm)	Cl_{gw} (ppm)	Cl_p (ppm)	Cl_{irr} (ppm)	Diffuse recharge (mm)	Total recharge (mm)	P (mm/year)	I_{rr} (mm/year)	R_{bp}/R_d (%)
1	2.9	2.9	0.9		676 (31%)	676 (31%)	2177		
$I_{irr}+I_p$		2.9	0.9	3.6	1057 (48%)		2177	307	
II	8.3	6.3	0.95		240 (11%)	316 (15%)	2096		9.8
III	22.8	18.5	1.52		149 (7%)	184 (8%)	2237		2.4
IV	3.8	33.4	2.19		1279 (58%)		2219		

4.3 *Simulation Model Calculations*

The conceptual hydrological model for Pingtung Plain, with the various water fluxes, is depicted in Figure 4. These fluxes are quantified as elements of the overall water balance in Table 3, which displays the measured components and those calculated by MODFLOW. The upper right corner lists the "unknown" components and paramenters; the bottom section gives the various water balance equations. Equation (1) (Table 3) is in fact the result from MODFLOW and hence has no unknowns. There are therefore nine equations with 12 unknown components and paramenters, which means that three values need to be estimated. The most "independent" variables are the rainfall-runoff coefficient (a), the irrigation runoff coefficient (e), and overall (average) irrigation efficiency (b). The water balance has been solved by assuming values of 0.55, 0.50 and 0.40 for a, b and e respectively.

A runoff coefficients for the entire catchments are estimated to be about 75%, but these values are not representative of the Pingtung Plain; the catchments include large areas of the northern and eastern mountains, which are characterized by steep valleys, high rainfall intensities and impermeable soils. Runoff coefficients for the Pingtung Plain must therefore lie in the range 40-60%. An estimate of 40% for the irrigation runoff coefficient, e, may seem high, as runoff on irrigated fields is likely to be negligible. However, about 60% of abstractions are for the fish farms, where water is constantly being replenished. The estimate of 0.50 for overall irrigation efficiency (b) also agrees with the estimates 4 for various crops, viz:80% for fruit trees, 60% for sugarcane and 40% paddy rice. This last value is based on estimated 5 mm/day field losses. It should be noted that approximately 10% conveyance losses must also be deducted.

Given these assumptions, the total estimated evapotranspiration (ETtot=ETgw+ETrai+ETirr) is 1046 mm/year, this indicates a pan evaporation coefficient (d) of 0.77, which again seems appropriate. It is interesting to note that recharge from irrigation (231 mm/year) is a substantial proportion of the total recharge, as it amounts to more than half that from rainfall (448 mm/year). It should be noted that these figures vary spatially and also during the year. In the lower part of the valley (area III) there is probably no recharge at all, while in the upland areas recharge from irrigation may even exceed that from rainfall. Although values for the "unknown" parameters a, b, and e may vary within certain limits, the general conclusions are considered realistic.

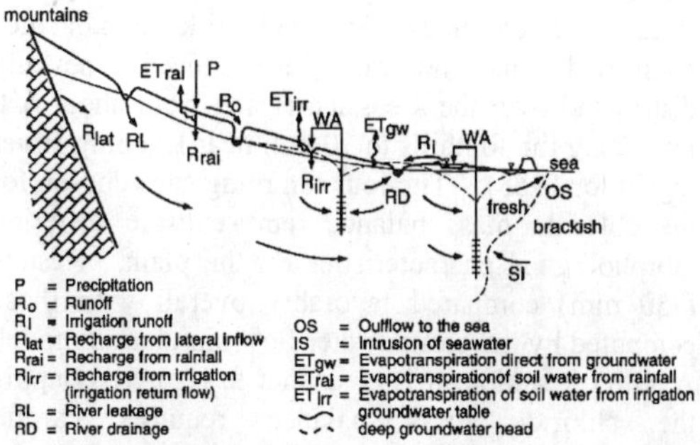

Figure 4. Schematic conceptual model of ground-water circulation.

Table 3. Summary water balance for Pingtung Plain.

MODFLOW components (mm/yr) and coefficients			Unknown components (mm/yr) and coefficients		
Total recharge	Rtot	730	Runoff	Ro	1507
River leakage	RL	50	Irrigation runoff	Ri	308
Seawater intrusion	SI	30	Recharge by lateral inflow	Rlat	51
River drainage	RD	25	Recharge from rain	Rrai	448
Outflow to sea	OS	25	Recharge from irrigation	Rirr	231
Evap.from	ETgw	30	Evapotranspiration total	ETtot	1046
groundwater		-40	Evap. rain water from soil	ETrai	785
Storage change	DS	0.07	Evap. Irrig. Water from soil	ETirr	231
Rlat/Rtot	c		Runoff coefficient	A	0.55
			Irrigation efficiency	B	0.50
			Pan evaporation factor	D	0.77
			Irrig. Runoff coefficient	e	0.40

Measured components (mm/year)		
Rainfall	P	2740
Well abstraction	WA	770
Potential pan evap.	EP	1360

Equations		
Saturated zone (MODFLOW)	(1)	Rtot+ RL + SI = RD + OS + ETgr + WA + DS
Unsaturated zone	(2)	P- R = ETrai + Rrai
Evapotranspiration	(3)	ETtot = ETgw + Etrai + Etirr
Recharge	(4)	ETtot = d ×EP
Runoff	(5)	ETirr = b× (WA- Ri)
	(6)	Rtot = Rlat + Rrai + Rirr
	(7)	Rlat = c× Rtot
	(8)	Rirr = (1-b) (WA- Ri)
	(9)	Ro = a×P
	(10)	Ri = e× WA

5 CONCLUSIONS AND RECOMMENDATIONS

The following general conclusions and recommendations are drawn from this study viz: soil water budgeting a chloride mass balance, and simulation model. These methods are applied for the first time on the Pingtung Plain. Knowledge of the various recharge processes and their relative contributions are of fundamental importance for establishing a realistic water resources development plan for the area. Estimated recharge values (for 1988) using the water balance method vary from 687 mm to 870 mm, with rainfall totaling 2740 mm. However, since the soil water budgets were derived from only limited meteorological data, additional information on rainfall spatial variability is required. Recharge determination from a chloride mass balance confirmed that the rates are indeed unevenly distributed over the area, decreasing from about 650 mm along the foothills to 10 mm near Kaoping River (again for 1988). The four sampling sites chosen for the chloride mass balance represent the differing morphological characteristics on the plain. Results (730 mm) compared favorably overall with those computed by the groundwater flow simulation model. A note of caution: it is clear that to rigorously apply the chloride mass balance requires careful determination of the various irrigation practices and monitoring of chloride input from irrigation water, pesticides and fertilizers, as well as that from wet and dry atmospheric deposition.

Development of a hydrogeological conceptual model is the first phase of numerical groundwater flow simulation; first step is presented. The development and optimization for Pingtung Plain are encouraging, with the final simulations affording much better insight to the hydraulic behavior of the aquifer system.

REFERENCES

Eriksson, E. & Khunakasem, V. 1969. Chloride concentrations in groundwater, recharge rate and rate of deposition of chloride in the Israel coastal plain. *J. Hydrol.* , 7:178-197.

McDonald, M. G. and Harbaugh, A.W. 1988. *A Modular three-dimensional finite-difference groundwater flow model.* U.S. Geological Survey open file report, 83-875, Book 6.

Sharma, M.L. and Hughes, M.W. 1985, Groundwater Recharge Estimation using Chloride, Deuterium and Oxygen-18 profile in the deep coastal sands of Western Austrial. *J. Hydrol.*, 81: 93-109.

Smith, M. 1989. Manual for Cropwat (Version 5.5). *Food and Agricultural Organization of the United Nations.* Rome.

Ting, C.S., Zhou, Y., De Vries, J.J. and Simmers, I. 1998. Development of a Preliminary Ground Water Flow Model for Water Resources Management in the Pingtung Plain, Taiwan. *Ground Water.* 36(1):20-36

Groundwater: Past Achievements and Future Challenges, Sililo et al. (eds) © 2000 Balkema, Rotterdam, ISBN 90 5809 159 7

Analysing the geometry of South African fractured rock aquifers

I.van der Voort & G.J.van Tonder
Institute for Groundwater Studies, University of the Orange Free State, Bloemfontein, South Africa

ABSTRACT: Fractured rock aquifers underlie most of South Africa and, as 60% of South Africa experiences semi-arid to arid conditions, these aquifers are an important water resource. Previous investigations of Karoo fractured rock aquifers have led to the conclusion that the geometry of the fractured rock aquifer plays an important role in the physical behavior of the aquifer (Botha *et al*, 1998). The aim of this paper is to briefly discuss various methods of determining the geometry of fractured rock systems in South Africa. The methods range from fractal based analyses of pumping tests to using three dimensional numerical models. The focus of the paper will be a case study highlighting the advantages and disadvantages of the methods.

1 INTRODUCTION

Water resources in South Africa are already being stressed and the country is slowly becoming a water-scarce country (Braune, 1996). This presents a challenge to all water resource managers to ensure the basic water needs of every South African are met. As fractured rock aquifers underlie more than 80% of South Africa, emphasis must be placed on these systems.

Most fracture rock systems consist of rock matrices intersected by discrete discontinuities such as fractures, joints and shears. Open fractures can form conduits for the flow of water and contaminants through an otherwise relatively impermeable rock matrix. Major factors affecting groundwater flow through fractured rock include fracture density, orientation, effective aperture width and the nature of the rock matrix.

2 DETERMINING THE GEOMETRY OF FRACTURED ROCK AQUIFERS

The fracture geometry of an aquifer plays an important role in the physical behavior of groundwater systems. It is therefore of utmost importance that every effort should be made to depict these aspects as accurately as possible.

One of the major difficulties in determining the geometries of fractured rock aquifers is the limited efficiency of the methods that can be used for this purpose. Indeed, there are only two classes of methods available for this purpose – field methods and modeling. Some of these methods will briefly be discussed in the following section.

2.1 *Field methods*

Knowledge of the geology of the area in which an aquifer is found provides one with a concept of the fracture geometry of that aquifer.

Geophysical techniques are useful in the structural mapping of an area, particularly to locate features such as faults, dykes and fractured zones.

A relatively easy, inexpensive way to analyze the geometry of fractured aquifers is with the aid of a pumping test. However, an assumption in most modern analyses is that the aquifer is confined. As will be shown later, this assumption is usually invalid for the shallow fractured rock aquifers found in South Africa. However, a worthwhile method to consider when analyzing a pumping test was developed by Barker and Bangoy. Barker (1998) generalized the Theis equation by including a term called the flow dimension (n), making it applicable in an arbitrary fractured confined aquifer. The flow dimension is also known as the fractal dimension. A dimension of larger than 1 implies fracture flow conditions. If $n = 1$ then the flow towards the borehole is linear, if $n = 2$ then the flow towards the borehole is radial.

Bangoy *et al* (1992, 1994) went a step further and derived equations for characterizing an arbitrary fractured aquifer system ($n<2$).

Van Wyk *et al* (1999) obtained interesting results from tracer tests performed in Karoo formations. He also discusses how tracers can be used as a tool in determining the geometry of fractures.

2.2 *Modeling*

A model is a representation of an object or situation. One such geostatistical model known as the Bayesian Markov Geostatistical Model (BayMar) was developed by Lars Rosen (1995). Given a reasonable distribution of known points, the method can be applied if the fracture geometry is not too complicated.

A fractal is a geometrical figure in which an identical pattern repeats itself in an ever diminishing scale (Lauwerier, 1991). Many fractal models have been used to represent the geometry of fractures. The Levy-Lee is an example of such a fractal model (Black, 1993).

Numerical flow and transport models can be used to generate fracture geometries. Examples of such models are Modflow and FracMan.

3 THE FRACTURED KAROO SEQUENCE

The so-called *Karoo Sequence* of geological formations underlies approximately 50% of the country. The Karoo Sequence consists mainly of sandstones, mudstones, shales and siltstones. The isostatic uplift of Karoo sediments, together with the intrusion of Drakensberg lavas and dolerites, has caused intensive fracturing, particularly in the sandstone layers which are less elastic than the rest of the Sequence.

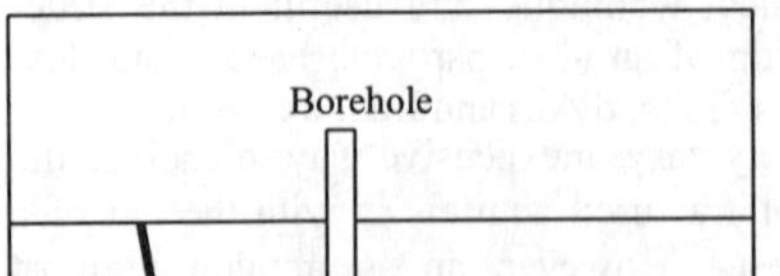

Figure 1. A typical conceptual model for a Karoo fractured rock aquifer.

Figure 1 is a typical conceptual model for a shallow fractured Karoo aquifer. There is normally at least one bedding plane fracture, which is usually semi-confined. Vertical fractures intersect the bedding plane fractures. The dotted line indicates the water level of the bedding plane fracture and the matrix, and the piezometric level of the vertical fracture. When performing a constant rate pumping test on the borehole in Figure 1, the following behavior is evident: initially (after well bore storage) the flow towards the borehole will be linear, suggesting that all the water is coming from the bedding plane fracture. The flow then becomes bi-linear as water is flowing to the borehole from both the fracture and the rock matrix. Finally the water level in the borehole (depending on the abstraction rate) will either remain constant (above the level of the bedding plane fracture) and the flow to the borehole will be pseudo-radial or the water level will drop (below the level of the bedding plane fracture) indicating that the transmissivity of the matrix is very small.

This characteristic drop in water level is shown in Figure 2. A Karoo borehole was pumped at 4 L/s. When the fractures were dewatered, most of the flow was coming from the matrix.

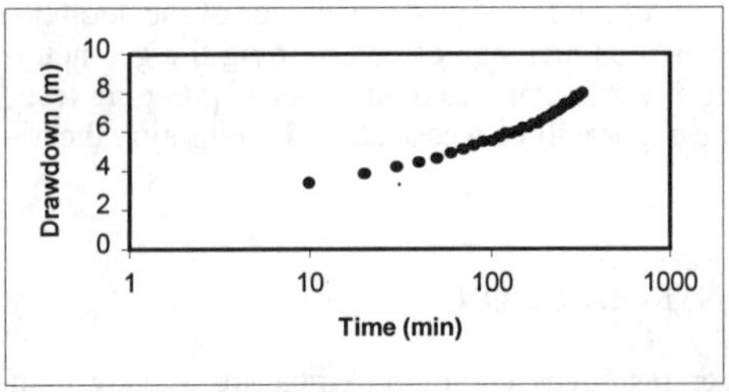

Figure 2. Pumping test results from a borehole in the Karoo Sequence. The borehole was pumped at a constant rate of 4L/s.

The characteristic flow geometries (Figure 3) can be derived from a log-log plot of the drawdown versus time graph and the derivative of the drawdown versus time plot.

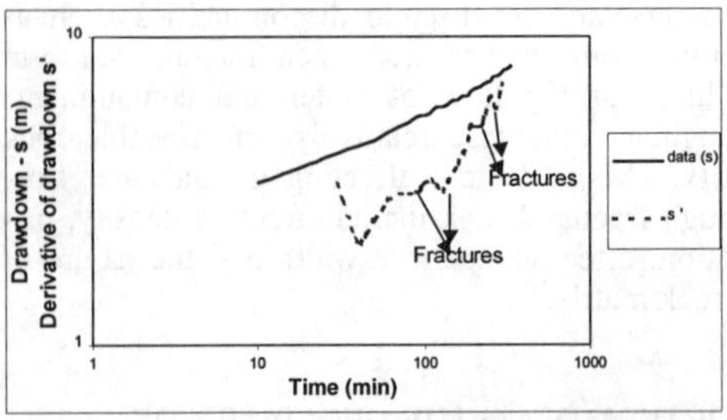

Figure 3. Graphs showing characteristic flow geometries.

Most analytical methods available today for example Moench (1984), Cinco and Samaniego (1981 a&b) and Gringarten (1974) could not be used to analyze this data set as the methods do not include boundary effects. The steep increase in drawdown towards the end of a pumping test indicates the boundary of a fracture has been reached and most of the flow to the borehole is from the matrix. The Barker method, although for confined aquifers, is the only analytical method that takes into account the steep increase in drawdown. It must be noted that this method is not ideal as most of the fractured rock aquifers are semi-confined. However currently Barker method presents the best results when analyzing these Karoo fractured aquifer systems.

A case study will now illustrate how by integrating a few of the methods discussed in section 2, an accurate geometry of a fracture can be determined.

4 A CASE STUDY – THE CAMPUS TEST SITE

4.1 *General*

The Campus test site is located on the grounds of the University of the Free State, South Africa, and covers an area of approximately 180x192 m^2. The aquifer is intersected by 30 percussion boreholes (Figure 4).

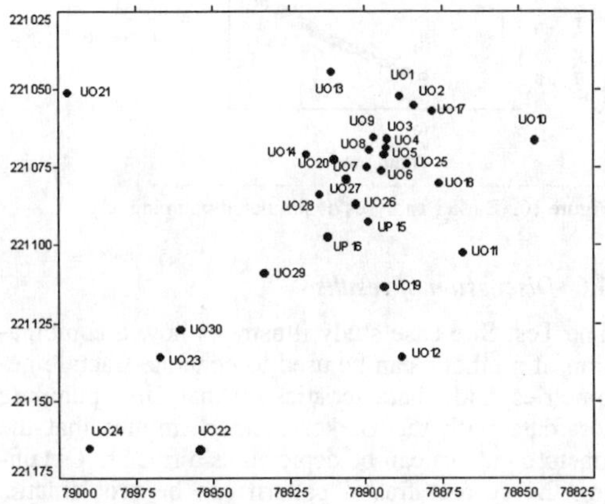

Figure 4. The Campus Test Site

4.2 *Geology*

The thickness of the aquifer on site is approximately 50m. The aquifer is situated in the Karoo Sequence and the geology consists of sandstone, mudstone and shale deposited under fluvial conditions.

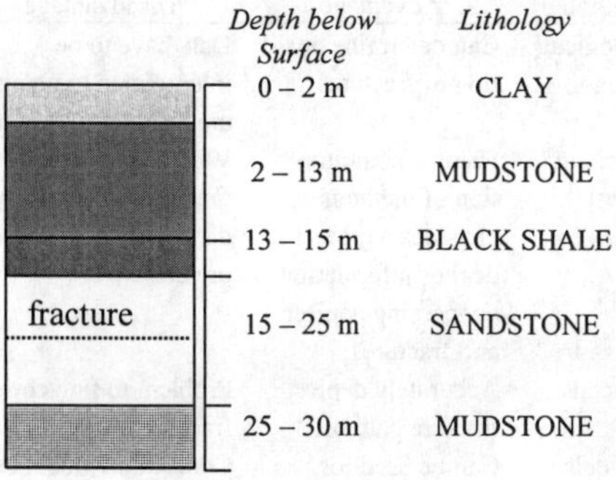

Figure 5. Simplified borehole log of UO26

Core samples indicate parallel horizontal fractures, the most significant of which is at a depth of 21 m. In more weathered sections of the aquifer, diagonal fractures intersect the bedding plane fractures. The sandstone containing the most horizontal fractures also forms the main water carrying formation (Figure 5).

4.3 *Pumping test*

On 3/3/2000 a pumping test was performed on borehole UO26, while observing the drawdown in boreholes UO28 and UO29. The borehole was pumped at a constant rate of 0.7 L/s for a day. Figure 6 illustrates the drawdown curves.

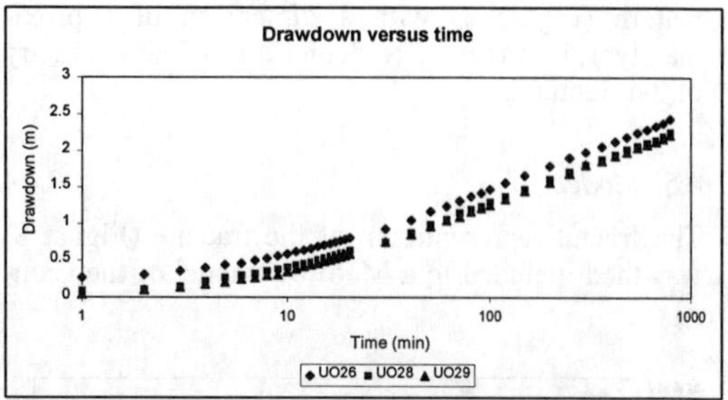

Figure 6. Drawdown curves for constant rate pumping test.

Table 1. Results of pumping test .

Borehole name	Distance from UO26 (m)	Transmissivity (m^2/d)	Storativity
UO26	-	10.6	-
UO28	14	10.5	4.5e-4
UO29	34	8.8	9e-4

Table 1 is a summary of the values obtained for the transmissivity and storage when the data was analyzed using the Cooper-Jacob method.

Figure 7 is a graph showing the flow characteristics of borehole UO26.

The data was then analyzed using the Barker method and the results are listed in Table 2. The

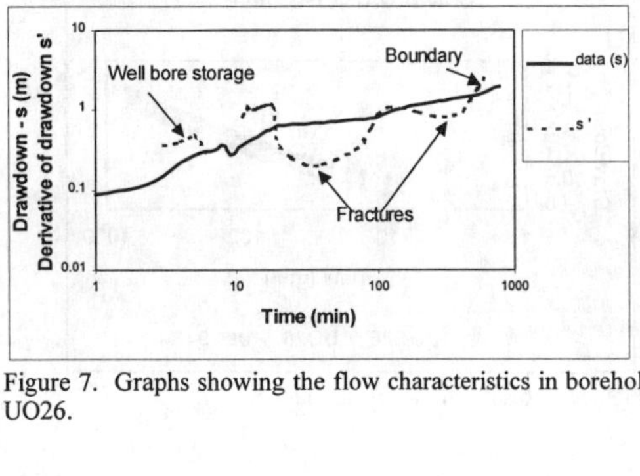

Figure 7. Graphs showing the flow characteristics in borehole UO26.

fractal dimension of approximately 1.6 implies that there is bi-linear flow from both the fracture and the matrix to the boreholes.

Table 2. Barker analysis of the pumping test data.

Borehole name	Fractal dimension (n)
UO26	1.39
UO28	1.69
UO29	1.64

4.4 *Fractal patterns*

By superimposing a horizontal fractal pattern over the aquifer and correlating the pattern with known positions of the fracture, it is possible to indicate the geometry of the fracture. Such a stochastic fractal pattern (Figure 8) with a dimension of approximately 1.5 was used to depict a possible geometry of the fracture.

4.5 *Models*

The fractal representation of the fracture (Figure 8) was then included in a Modflow model of the Campus Site. The values in Table 1 were as input parameters for the matrix. The fracture was assigned a transmissivity value of 720 m^2/d. The model was then used to simulate the pumping test discussed in section 4.3. The results of the pumping test as shown in Figure 9.

These results of the simulated pumping test were then once again analyzed using the Barker method (Figure 10). The fractal dimensions obtained are summarized in Table 3.

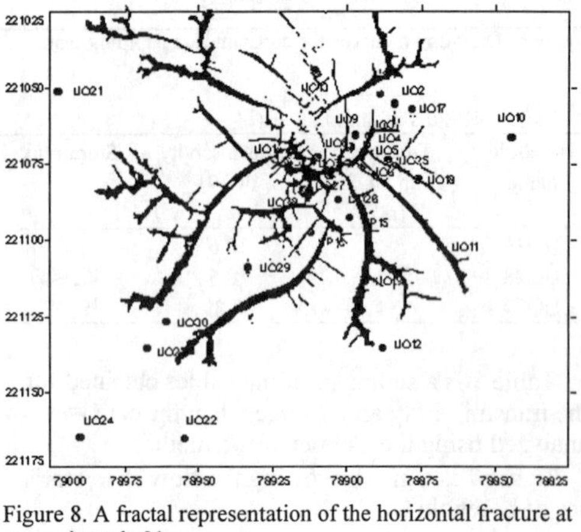

Figure 8. A fractal representation of the horizontal fracture at approximately 21 m.

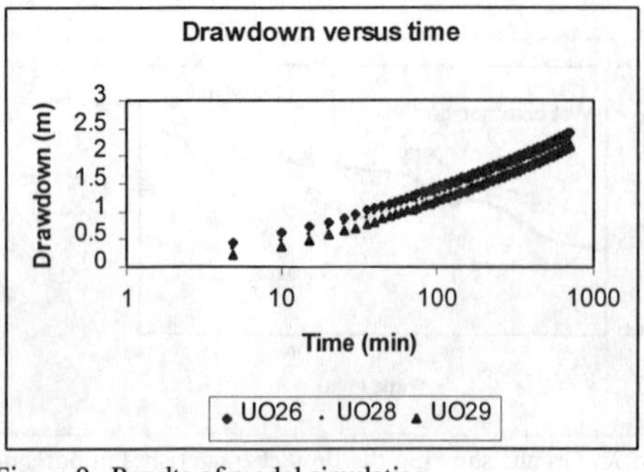

Figure 9. Results of model simulation.

Table 3. Barker analysis of simulated pumping test data.

Borehole name	Fractal dimension (n)
UO26	1.45
UO28	1.64
UO29	1.42

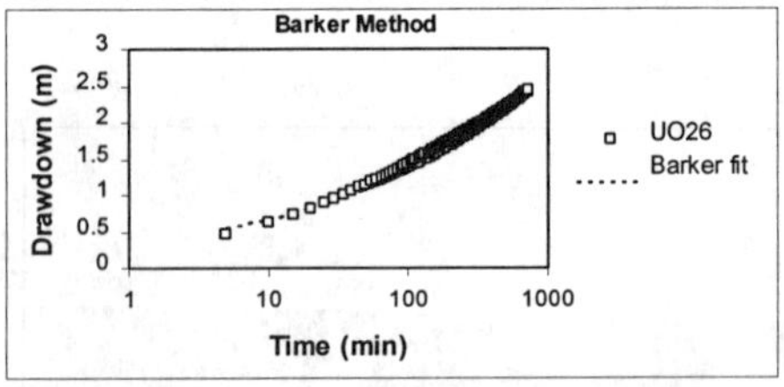

Figure 10. Barker analyses of simulated pumping test

4.6 *Discussion of results*

The Test Site case study illustrates how a combination of methods can be used to generate fracture geometries and characteristics. Analyzing pumping test data with the Barker method implies that the fracture pattern can be depicted as a fractal. Obtaining the correct fractal pattern can be problematic. Therefore a flow model (which included the fractal-fracture) was used to simulate the flow conditions. The results of the simulated and observed data compare favorably. Therefore it seems as though the

Table 4. Summary of the advantages and disadvantages of the methods used in the case study.

Method	Advantage	Disadvantage
Geological logs	Can determine position of fractures	Data have to be interpolated to obtain areal extent of fracture
Pumping test	Using a combination of methods provides a great deal of information concerning aquifer and fractures	When considering methods individually, data can be misinterpreted.
Fractals	Accurately depict fracture patterns	Problem to find correct fractal pattern
Models	Can be used for management purposes	Conceptual model can be incorrect

fractal pattern used depicts fractures and bi-linear flow patterns present on the Site accurately.

The advantages and disadvantages of the methods used in the case study are summarized in Table 4.

5 DISCUSSION

South Africa is becoming a water-scarce country and it is therefore important to manage and protect resources, especially fractured aquifers. Defining the geometry of these fractures is an important component in the management and protection process. This paper has highlighted a few methods that can be used in this process. More accurate results can be obtained when using a combination of methods.

For example standard pumping test methods can only give one an indication of the average transmissivity and storage values. By combining these methods with the flow characteristic and Barker methods the following information can be obtained:

- Depths of fractures
- Possible boundaries
- Fractal dimension which will indicate whether flow to the borehole is linear, bilinear or radial.

Models have been classified as the worst remediation idea of the millennium (Nyer, 1999). This is not because of the assumptions made in the mathematics behind models, but because users have incorrect conceptual models. However, by combining the information obtained from geological logs with the information obtained from the various pumping test methods, an accurate conceptual model for an area can be created. This in turn can be used to create an accurate flow model, which after validation can be used for management purposes. The accurate flow model can then be used as a sound basis for a mass transport model. The movement of pollution plumes will then be traced more accurately.

ACKNOWLEDGEMENTS

The research discussed in this paper emanates from a project funded by the Water Research Commission entitled: "Decision Tool for establishing a Strategy for protecting groundwater resources: Data requirements, Assessment and Pollution Risk". The financing of the project by the Water Research Commission is gratefully acknowledged.

The authors would also like to thank the students at the Institute for Groundwater Studies for collecting the data discussed in this paper.

REFERENCES

Bangoy, L.M. and Bidaux, P, Drogue, C., Plégat, R. and Pistre, S. 1992. A new method of characterizing fissured media by pumping tests with observation wells. *Journal of Hydrology,* 138, 77 – 88.

Bangoy, L.M. and Drogie, C. 1994. Analysis of intermittent pumping tests in fissured fractal aquifers: theory and applications. *Journal of Hydrology*, 158, 47 – 59.

Barker, J.A. 1988. A Generalized Radial Flow Model for Hydraulic Tests in Fractured Rock. Water Resources Research, 24, 1796 – 1804.

Black, J.H. 1993. Hydrogeology of Fractured rocks – A question of uncertainty about geometry. In: Memoires of the XXXIVth congress of IAH, Oslo.

Botha, JF , Verwey, JP, Van der Voort, I, Vivier, JJP, Buys, J, Colliston, WP and Loock, JC 1998. Karoo Aquifers: their geology, geometry and physical properties. WRC report no. 487/1/98. Water Research Commission, Pretoria.

Braune, E. 1996. Sustainable Water Resource Management – A groundwater perspective. Department of Water Affairs and Forestry, Private Bag X313, Pretoria.

Cinco-Ley, H. & Samaniego, V. 1981a. Transient Pressure Analysis for Fractured Wells. SPE 7490, Richardson, Texas: Society of Petroleum Engineers, American Institute of Mining, Metallurgical and Petroleum Engineers.

Cinco-Ley, H. & Samaniego, V. 1981b. Transient Pressure Analysis: Finite Fracture versus Damaged Fracture case. SPE 10179, Richardson, Texas: Society of Petroleum Engineers, American Institute of Mining, Metallurgical and Petroleum Engineers.

Gringarten, A.C., Ramey Jr., H.J., & Raghavan, R. 1974. Unsteady-state pressure distribution created by a well with a single infinite-conductivity vertical fracture. SPE 4051, In Proceedings of SPE-AIME 47th Annual Fall Meeting, San Antonio, Texas, Oct. 8-11, 1972. Richardson, Texas: Society of Petroleum Engineers, American Institute of Mining, Metallurgical and Petroleum Engineers.

Moench, A.F. 1984. Double-Porosity Models for Fissured Groundwater Reservoir with Fracture Skin. Water Resources Research, Vol. 20, No 7: 831-846.

Nyer, E.K. 1999. Treatment Technology: Looking Back. GWMR. Fall edition: 54 – 56.

Lauwerier. H. 1991. *Fractals: Endlessly Repeated Geometrical Figures*. Princeton University Press. Princeton, New Jersey.

Paillet, F.L. 1993. Application of borehole geophysics in characterising the hydrogeology of fractured igneous and metamorphic rocks. In: Memoires of the XXXIVth congress of IAH, Oslo.

Rosen, L. 1995. Estimation of hydrogeological properties in vulnerability and risk assessment. Dissertation at the Department of Geology, Publ. A79, ISSN 1104-9839, Goetenborg.

Van Wyk, B., Xu, Y., De Lange, F., Van Tonder, G. and Chiang, W.H. 1999. Draft Final Report: *Utilization of Tracer Experiments for the development of rural water supply management strategies for secondary aquifers*. Institute for Groundwater Studies, University of the Free State, Bloemfontein.

Groundwater: Past Achievements and Future Challenges, Sililo et al. (eds) © 2000 Balkema, Rotterdam, ISBN 90 5809 159 7

A tracer test in a fractured aquifer that gave a groundwater flow rate of kilometers per day

A.E.Van Wyk
Department of Water Affairs, Windhoek, Namibia

E.C.Murray
Council for Scientific and Industrial Research, Stellenbosch, South Africa

G.J.van Tonder
Institute for Groundwater Studies, University of the Orange Free state, Bloemfontein, South Africa

ABSTRACT: Tracer and dilution tests were done during a borehole injection test in a fractured quartzite aquifer in Windhoek, Namibia. The injection tests formed the basis of a study into artificial recharge to the Windhoek aquifer. During one of the tests, at an injection rate of 17.1 L/s, a tracer was added to the injection water and monitored in a borehole 786 m away. The tracer took 216 minutes to reach the observation borehole, giving a flow velocity of 3.6 m/min. The hydraulic gradient, as a result of injection was approximately 0.01. This paper describes the geology of the Windhoek aquifer, the tracer test and a dilution test on the observation borehole.

1 INTRODUCTION

Tracer and dilution tests were conducted during a borehole injection test in order to enhance our understanding of groundwater flow mechanisms in a quartzitic aquifer. Abstraction from this fractured aquifer in excess of natural recharge has led to a decline in groundwater levels. Artificial recharge is being tested with the aim of rapidly restoring the water levels and using the full storage capacity of the aquifer. Pilot borehole injection tests have been completed and the results appear favourable for implementing a full-scale artificial recharge scheme.

During one of the borehole injection tests, dilution tests in observation boreholes were conducted to locate the zones of increased groundwater flow. A tracer test was then conducted between the injection borehole and a borehole that proved to have a high through-flow. The aim of the test was to establish the rate at which the injected water would move in the dominant fractures.

The results demonstrate that groundwater can flow at high velocities in fractured aquifers..

2 THE WINDHOEK AQUIFER

The Windhoek aquifer consists east-west striking quartzites and schists that dip 15° - 30° to the north (Fig 1). High yielding production boreholes are located on the breccias and fracture zones associated with the numerous N-S, NW-SE and NE-SW striking faults (Murray & Hughes 2000).

The quartzites, being brittle, are highly fractured as a result of folding and faulting and have developed secondary porosity and permeability. The schists on the other hand are ductile and have poorly developed secondary permeability (Carr Barbour & Associates 1999). Both the schists and the quartzites are thought to have no primary porosity (Colvin 1994). The natural groundwater flow direction within the study area is from the mountains in the south to the city in the north.

3 THE INJECTION BOREHOLE 9/8A AND THE OBSERVATION BOREHOLE 9/9

The injection borehole, Bh 9/8A was drilled into a permeable fault zone (Fig 1). The borehole penetrates laminated, fractured, sericitic quartzites from 5 m to 138 m; graphitic schist from 138 m to 165 m; and intercalated fractured, sericitic quartzites and graphitic schists from 165 m to it's completion depth of 202 m. Water strikes were recorded at 100 to 108 m; 115 to 117 m and 168 to 173 m below surface; the main strike being at 173 m.

Unfortunately, for injection purposes, the borehole was cased from surface to 112.9 m with plain steel casing (273 mm OD), and thereafter, to it's completion depth at 202 m, with slotted steel casing. The injection pipe was installed to 150 m below surface. Water entering the borehole at this point could flow into the aquifer at the following depths:

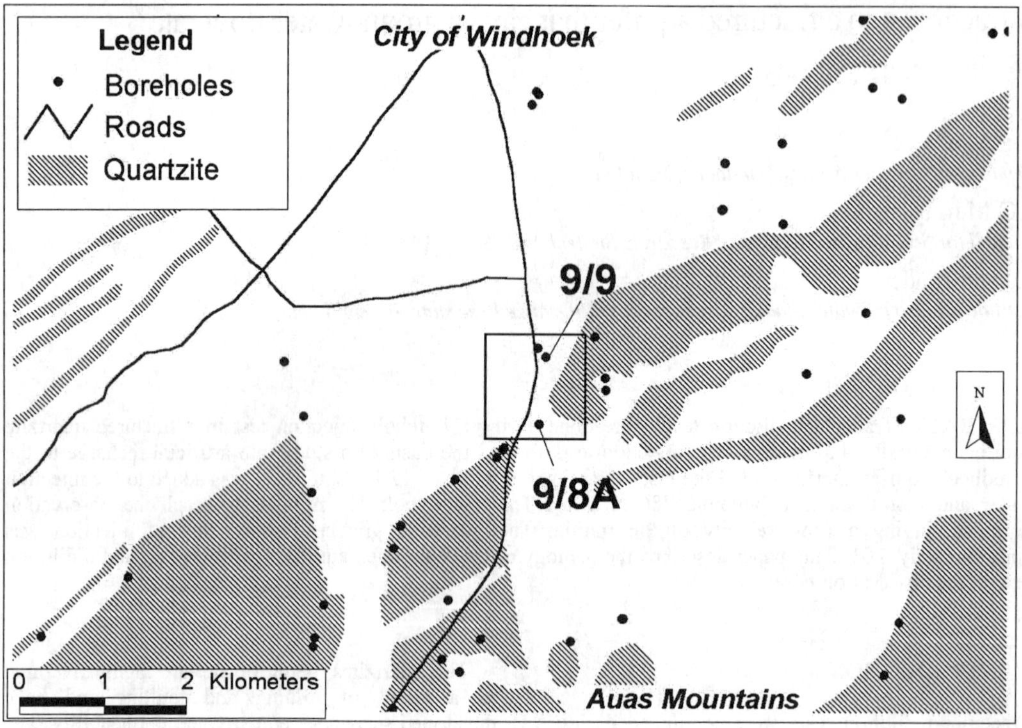

Figure 1. Outcropping quartzites and borehole locations.

1. The water strike zones around 115 to 117 m and 168 to 173 m;
2. It could rise up the annulus to enter the water strike zone around 100 to 108 m; and
3. It could rise up the annulus to enter permeable areas within the unsaturated zone above the water level, which was 84 mbgl at the time.

Borehole 9/9, located 786 m from the injection borehole was used as the observation borehole for the injection test and the tracer test. It is not known whether this borehole is located on the same fault as borehole 9/8A, however evidence from geological mapping and surface geophysics tends to suggest that they do penetrate the same fault zone.

This 214 m deep borehole was drilled in 1959. The only recorded water strike corresponded to the water level at that time, which was 58 m below surface. The dilution tests (discussed below) suggest that the main hydraulically conductive zone lies between 80 m – 130.

A recently drilled observation borehole next to borehole 9/9 confirms that the lithology is similar to that at borehole 9/8A. Weathered, fractured and fresh sericitic quartzites dominate up to a depth of 90 m, followed by biotite-rich schists up to the borehole's completion depth of 120 m. Water strikes were recorded at 80 m, 99 m and 102 m with the main water strike at 102 m.

4 WATER LEVEL RESPONSES IN THE BOREHOLE 9/9 TO THE INJECTION AT BOREHOLE 9/8A

The injection test ran for 195 days, at an average of 17.1 L/s. The water used was good quality drinking water from the town supply system. Prior to commencement of the injection test, this part of the aquifer had been heavily pumped and was characterised by a hydraulic cone of depression. Long-term water level data indicates that it would take more than five years without significant pumping for the water level to recover, and that full recovery to the original, pre-pumping water levels, would take much longer.

The water level response in borehole 9/9 to injection is shown in Fig 2. On the semi-log plot, the slope stops increasing with time after 60 days. This indicates that barrier boundary conditions have not been reached, and that the injected water is supplying water to the vast network of micro-fractures that make up this double-porosity aquifer system.

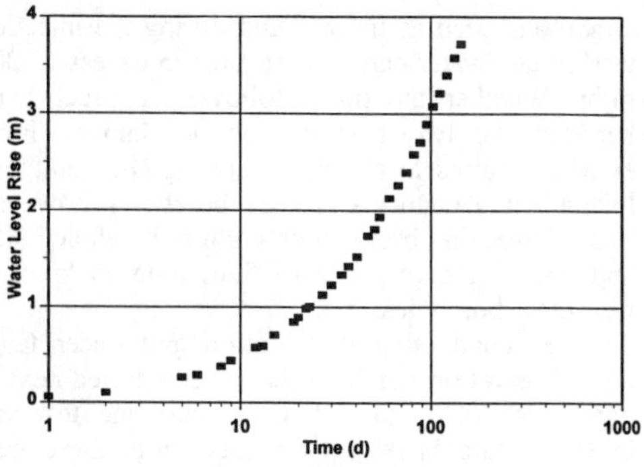

Figure 2. Water level response to injection.

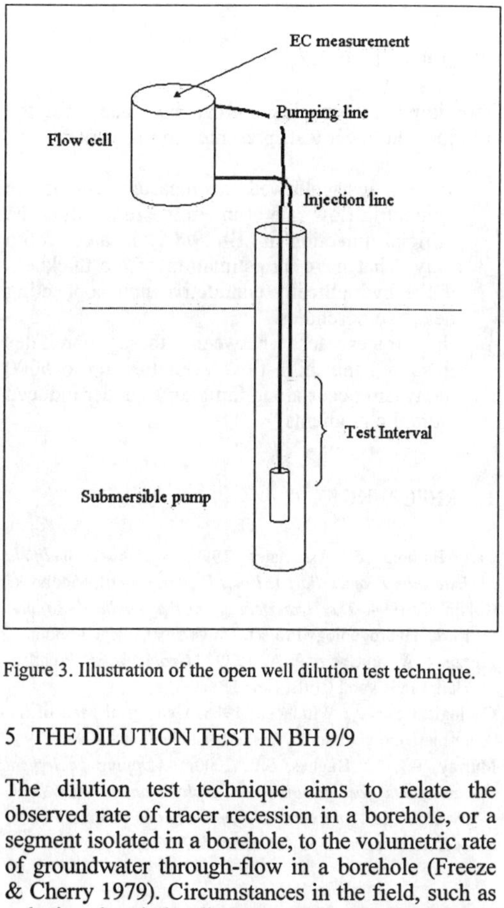

Figure 3. Illustration of the open well dilution test technique.

5 THE DILUTION TEST IN BH 9/9

The dilution test technique aims to relate the observed rate of tracer recession in a borehole, or a segment isolated in a borehole, to the volumetric rate of groundwater through-flow in a borehole (Freeze & Cherry 1979). Circumstances in the field, such as reducing borehole diameters with depth, lack of knowledge of the exact positions of water strikes and on the lithology intersected by borehole 9/9 did not favour the traditional use of straddle packers to conduct the dilution tests. Instead the dilution test methodology was adapted to an open well dilution test technique (Fig 3).

A test interval, defined by the position of the pump and the injection line (Fig 3), was used instead of the conventional packer chamber. Detection of the NaCl tracer was accomplished by pumping the water to a flow-through cell located at the surface. The water was fed back into the borehole under gravity flow via the injection line. Achieving a uniform distribution of tracer is a prerequisite for correct interpretation, but with the configuration used (Fig 3), this was not possible. The tracer was therefore introduced at once, as a spike, to the injection line, and this produced a tracer plume that circulated through the system. Fig 4 shows the resulting field curves (the example used here is for the section 120 m – 216 m).

The rate at which the tracer was removed from the plume by groundwater through-flow was measured in the field. The volumetric groundwater flow rate was then obtained by modelling the field data (Fig 4).

6 RESULTS OF THE DILUTION TESTS

The dilution test showed that the main zone at Bh 9/9 that responded to injection in Bh 9/8A was the section from 80 to 130 m below the ground surface, and that the total volumetric flow rate was in the order of 3 – 6 l/sec.

The Darcy velocity (q) that corresponds to the volumetric flow rate measured for each test section was computed using equation 1.

$$q=Q/A \tag{1}$$

where Q = volumetric flow rate obtained from the dilution test, and A = borehole area tested during the dilution test.

The volumetric flow rate and the area used, were obtained from each tests section (80 – 100 m; 90 – 110 m; etc. The results are listed in table 1.

7 DISCUSSION OF THE DILUTION TEST

Even though water was being injected into borehole 9/8A at 150 mbgl, or 23 m above the main water

Table 1. Volumetric flow rate and corresponding Darcy velocities in Bh 9/9 as estimated with the point dilution tests and using a borehole distortion factor of 1.

Section	Borehole radius	Sample area	Volumetric flow rate Q	Darcy velocity q
mbgl	m	m^2	m^3/day	m/day
80 – 100	0.17	10.68	177.98	16.6
90 – 110	0.13	8.17	138.24	16.9
110 – 130	0.0905	5.69	172.8	30.3
120 – 214	0.0905	26.7	62.2	2.32

strike, the dilution tests demonstrated that groundwater flow occurs over an extensive part of borehole 9/9 and that most of the flow took place in the upper 80-130m zone.

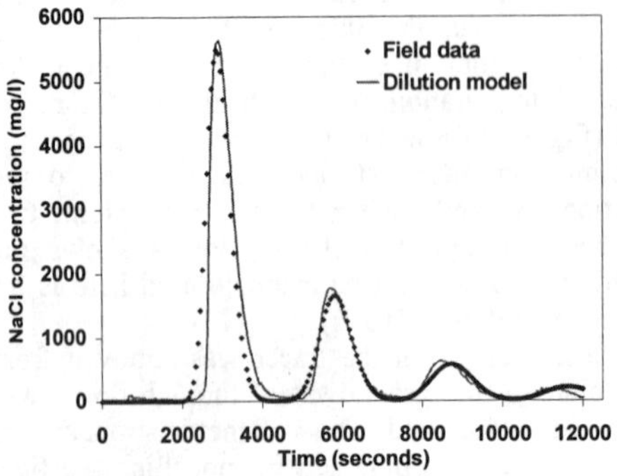

Figure 4. Dilution test data and modelled data for borehole section 120 m – 216 m.

8 THE TRACER TEST

The tracer test was conducted after 187 days of injection into borehole 9/8A. The hydraulic gradient between Bh 9/8A and Bh 9/9 is estimated to be 0.01. This value is based on the estimated water level in the aquifer at Bh 9/8A during injection, and takes the efficiency of the borehole into account.

Fluorescein (alternative names Uranine, Sodium fluorescein), a green organic dye was selected due to it's low adsorption on sediment surfaces if the organic content is low. Four kilograms of tracer was injected at a rate of 17 l/sec at 150 mbgl.

The tracer took 216 minutes to reach the observation well, Bh 9/9, giving a flow velocity of 3.6 m/min (Fig 5).

9 DISCUSSION ON THE TRACER TEST

The tracer test demonstrated that in a fractured quartzitic aquifer, under specific hydraulic conditions such as those found during the injection test, groundwater can flow at rates in excess of 200 m/hr. Whether the tracer followed a direct route between the two boreholes is not known. Field evidence points to a steeply dipping N-S fault that hydraulically connects the two boreholes, however, this is based on observations south of borehole 9/8A, and not on the presence of fault induced breccias near these boreholes.

Since conducting the dilution and tracer tests, new observation boreholes have been drilled next to boreholes 9/8A and 9/9. Once pumping test and injection data has been obtained from these new boreholes, it should be possible to determine the true hydraulic properties of the fault zones within the Windhoek quartzites.

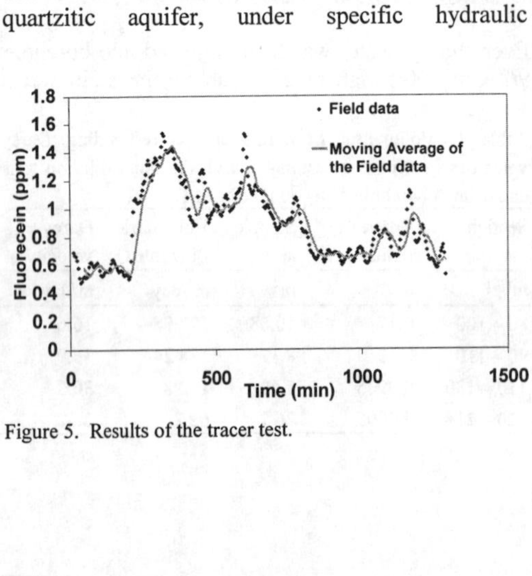

Figure 5. Results of the tracer test.

10 CONCLUSIONS

Two important conclusions can be made from the dilution and tracer tests presented in this paper:

1. Dilution tests allowed the measurement of the volumetric flow rate Q in Bh 9/9 responding the artificial injection in Bh 9/8A, located 786m away. That gave an estimation of the thickness of the hydraulically conductive unit connecting these two boreholes.
2. The tracer test between these boreholes illustrated that high flow velocities, up to 5000 m/day can occur along fault zones under induced hydraulic gradients.

11 REFERENCES

Carr Barbour & Associates 1999. *Windhoek Wellfield Extension Project. Part 1: Final Report*. City of Windhoek.

Colvin, C. 1994. *The Hydrogeology of the Windhoek Aquifer*. M.Sc. Hydrogeology Project. University College London.

Freeze, R.A. & Cherry, J.A. 1979. *Groundwater*. Prentice-Hall, Englewood Cliffs, New Jersey.

Geological Survey, Windhoek. 1988. Geological Map of the Damara Orogen. South West Africa/Namibia - 1998.

Murray, E.C. & Hughes, S.P. 2000. *Mapping geological lineaments for groundwater exploration in Windhoek, Namibia. 28th International Symposium on Remote Sensing of Environment*. Cape Town, South Africa.

Groundwater: Past Achievements and Future Challenges, Sililo et al. (eds) © 2000 Balkema, Rotterdam, ISBN 90 5809 159 7

Deep flowing groundwater in the Table Mountain Group quartzite

John M.C.Weaver
Groundwater Group, Environmentek, Council for Scientific and Industrial Research, Stellenbosch, South Africa

A.Siep Talma
Quaternary Dating Research Unit, Environmentek, Council for Scientific and Industrial Research, Pretoria, South Africa

ABSTRACT: The concept of a dynamic and deep-flowing (up to 3000 m deep), aquifer system in the Table Mountain Group quartzite has been the subject of debate in South Africa. Most of our fractured aquifers are shallow, with groundwater rarely occurring more than 100 – 120 m below surface. Research on groundwater temperatures, hydrochemistry and isotopes from boreholes and hot springs confirms that deep flow occurs. Aquifer recharge calculations using chloride and $\delta^{18}O$ show potential recharge of 50%. The implication for water resources is vast. If this is a reality then the water in storage, within a radius of 200 kilometres of Cape Town, could be as much as 66 000x10^6 m^3, with annual recharge of 2 600x10^6 m^3. Cape Town consumes 500 million cubic metres annually and is facing a water shortage in the near future. The TMG has great potential as an untapped resource for city water-supply.

1 INTRODUCTION

Cape Town, the surrounding metropolitan area and the Western Cape has been experiencing rapid growth since the democratization of South Africa. This trend is expected to continue for the future. A limiting factor is water-supply and unlike Gauteng, the importing of water is not an economically viable option. This is because the inland areas are semi-desert. The closest possible water source is the Orange River 700 kilometres to the north and this is already a stressed system.

In 1989 a study called "Western Cape System Analysis" (Ninham Shand 1994) examined all the possible sources of water for this area. Twenty-four possible methods of increasing the water-supply were identified, ranging from building the Skuifraam Dam through to importation of water by tanker and included two groundwater resources. None of these 24 options considered groundwater from the quartzites of the Table Mountain Group (TMG). This is despite the fact that it is the second largest aquifer/geological unit in South Africa and is extensively and successfully exploited by hundreds of farmers.

This rather incredible situation is a reflection of the absence of research and investigation work that has been directed towards the TMG aquifer. As an example, in 1995 the Water Research Commission commissioned a state-of-knowledge report (Bredenkamp et al. 1995) on groundwater recharge. There are thirty investigations listed, not one of which is in the TMG, yet the "dolomite aquifer", the largest South African aquifer, has nine listed investigations.

2 GEOLOGY AND HYDROGEOLGY OF THE TMG

The Lower Palaeozoic Cape Supergroup was deposited in the Cape Trough, an aborted rift sedimentary basin of which the original extent is poorly known (Tankard et al. 1982). The Cape Supergroup comprises three conformable series. The Table Mountain Group at the base and the Witteberg Group at the top are mainly quartzose sandstones. The Bokkeveld Group in the middle is distinguished from the overlying and underlying units by its essentially argillaceous character as well as by the generally "dirtier" and more feldspathic nature of the sandstones that do occur in this unit (Geological Survey 1980)

The hydrological characteristics of these three groups differ. The differences are ascribed to the varying amounts of argillaceous material, which in turn has had an effect on the degree of open fracture formation during the Cape Orogeny and later periods of deformation. This Permo-Triassic Cape Orogeny started about 278 ±2Ma and lasted until about 230 ±3Ma (Halbich 1983). Table 1 shows the various units, their thicknesses in the type area, lithology and general hydrogeological character.

Table 1. Lithostratigraphy of the Cape Supergroup and general hydrogeological characteristics of the various units. (adapted from Geological Survey (1980))

Age (Ma)	Group	Subgroup	Formation	Thickness (m)	Lithology	General hydrogeological characteristics
300						
		LAKE MENTZ	Waaipoort	50	Shale	
CARBONIFEROUS			Floriskraal	60	Shale, Sandstone	
360			Kweekvlei	130	Shale	Perceived to have a low borehole yield potential. Little hydraulic research has been done on this unit. The preferred drilling target is the Witpoort Formation.
	WITTEBERG		Witpoort	310	Sandstone	
		WELTEVREDE	Swartruggens	450	Shale, Sandstone	
			Blinkberg	80	Sandstone	
			Wagen Drift	70	Shale, Sandstone	
		BIDOUW	Karoopoort	50	Shale	Generally a poor groundwater resource, as groundwater usually has a high salinity. The source of salinity has not been researched, but is probably related to low permeability and the marine deposition environment of the shales.
			Osberg	55	Sandstone	
			Klipbokkop	170	Shale	
DEVONIAN			Wuppertal	65	Sandstone	
			Waboomberg	200	Shale	
	BOKKEVELD	CERES	Boplaas	30	Sandstone	This unit is quite heavily exploited due to its easy accessibility. The shales produce very good farming soil. Groundwater is obtained from the quartzite formations. High yielding boreholes over 10 L/s are quite common. Salinities vary and where low salinities are obtained it is assumed this is due to the flushing effect of good quality water from the TMG, which invariably is at higher elevations and thus has a hydraulic head.
			Tra-Tra	85	Shale	
			HexRiver	100	Sandstone	
			Voorstehoek	115	Shale	
			Gamka	135	Sandstone	
			Gydo	160	Shale	
		NARDOUW	Rietvlei			The relatively high percentage of argillaceous material has limited the development of open fractures. Boreholes usually yield some water, but rarely are yields more than 10L/s obtained. The weathering pattern produces a very rough topography (Skurweberg) which gives this aquifer a high recharge potential.
408			Skurweberg	500	Sandstone	
SILURIAN			Goudini			
438			Cederberg	120	Shale	Aquiclude. Some fractures may yield water, but generally avoided.
			Pakhuis	40	Sandstone	Minor.
	TABLE MOUNTAIN		Peninsula	1550	Sandstone	The prime TMG target for obtaining groundwater. The very clean nature of the sandstone has resulted in large open fractures, capable of sustaining boreholes with high yields. Yields of more than 10 L/s are common. The main restriction to exploitation is drill-rig accessibility into the mountains. Rough topography combined with high mountains causing increased orographic rainfall results in a high recharge potential.
ORDOVICIAN			Graafwater	440	Sandstone, Shale	Minor. The thickness varies, and only along the west limb are these units more than a few 10's of metres thick.
500			Piekenierskloof	800	Conglomerate, Sandstone	

3 GROUNDWATER TEMPERATURE AND DEPTH OF FLOW

The heat flow figures for Southern Africa show reasonably low values ranging from 33 to 66 mW.m^{-2} (de Beer 1976) No regions of abnormally high heat flow have been found and there are no geothermal regions. The KaapVaal Craton shows the lowest values with the minimum being recorded in the Witwatersrand Goldfields, which is why such great depths can be mined quite comfortably. The Cape Fold Belt has not been investigated. However 4 sites close to the boundary, range from 55 to 66 mW.m^{-2}

Kent (1969) calculated, thermal gradients ranging from 36 to 70 m/°C and averaging 51 m/°C for boreholes in the Karroo sequences. Diamond & Harris (submitted), calculated temperature gradients of 47 and 55 m/°C for two deep boreholes just to the north of the TMG, Thus a reasonable average figure to use would be 50 m/°C.

Weaver (2000) reviewed the population of boreholes recorded on the DWAF groundwater database for which temperatures were recorded. For areas with a reasonable number of boreholes an

assessment was made as to which were in the recharge zone (higher elevation), usually a "window" of Peninsula Formation, and which were in the discharge zone. Three areas were identified as having a sufficient number of boreholes to give representative results, Piketberg, Kammanassieberg and Rooiberg (Fig. 1).

In the Piketberg area, 120 km north of Cape Town, 8 boreholes in the recharge zone have water temperatures from 17.4°C to 20.8°C, (average 18.9°C). In the discharge zone 20 to 30 km to the north-west, temperatures range from 20.9 to 23.5°C, (average 22.2°C). This is an increase of 3.3°C and implies a minimum depth of flow of 165 m, (using the 50m/°C average).

In the Oudtshoorn area variable temperatures were recorded. The Cango Caves air temperature is 17°C (Talma & Vogel 1992) and recorded groundwater temperatures from the Kammanassieberg average 16.2°C. The Cango Caves are at a lower elevation than the mountainous recharge areas, so 16.2°C is accepted as the recharge temperature. For the 11 boreholes in the discharge zone, temperatures range from 20.4°C to 24.5°C, (average 23.4°C). This is an increase of 7.2°C and implies a minimum depth of flow of 360 m.

The Calitzdorp wellfield is 80 km away on the western edge of the synclinal Oudtshoorn basin. The

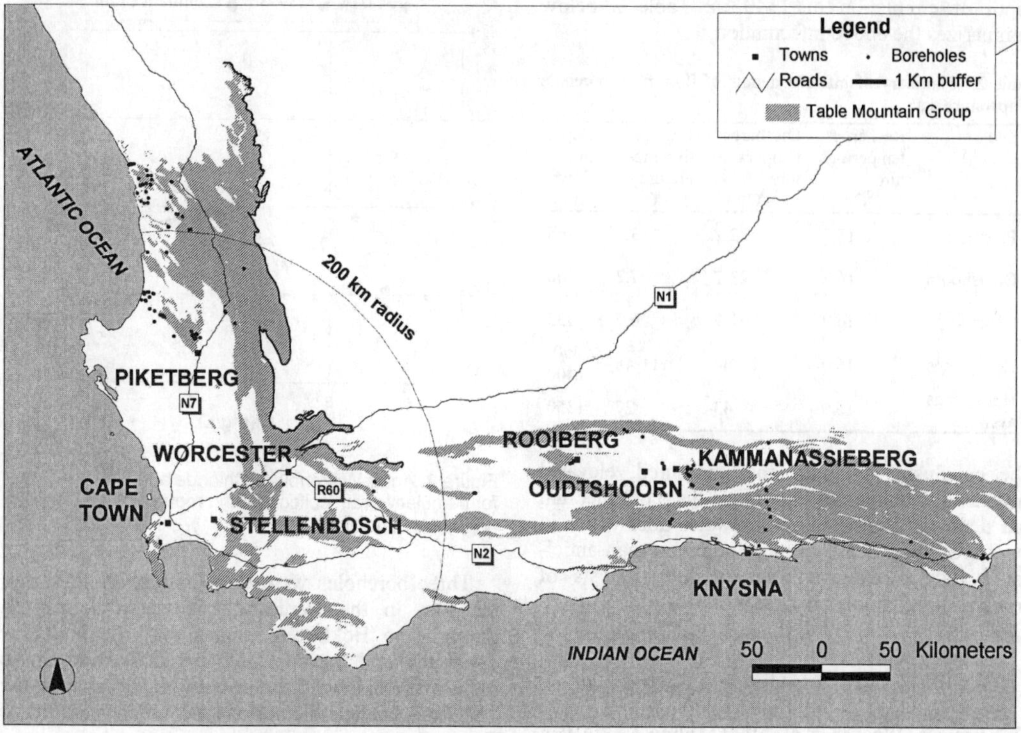

Figure 1. Areal distribution of TMG quartzites and borehole positions with recorded water temperatures.

recharge zone for this wellfield is the Rooiberg and is to the west of the wellfield. The average temperature of the wellfield boreholes is 22.9°C. Using 16.2°C as the recharge zone groundwater temperature, the increase of temperature is 6.7°C, which implies a minimum depth of flow of 335 m.

These are minimum figures for depth of flow. If one allows 2°C of warming for the recharge borehole ie an average borehole depth of 100 m then these three examples show that water can circulate between 265 and 460 m in the TMG. Also allowing for some cooling to take place, then circulation deeper than 500 m can be accepted quite readily.

There are at least 11 hot (>26°C) springs associated with the TMG. The water temperatures range from 27°C to 64°C, the latter being the hottest in South Africa. The chemical characteristic of all these springs are similar to that of shallow groundwater in the TMG quartzites. The exception is that SiO_2 is mostly the dominant constituent (Kent 1969), which can be ascribed to increased silica solubility at higher temperatures.

The temperatures of the hot springs indicate depths of circulation of 500 to 2300 m. The average temperature is 43°C, which implies, disregarding loss of heat on the return to surface, an average depth of circulation of 1340 m. Table 2 below summarizes the above information.

Table 2. Summary of inferred depths of flow from borehole temperature data.

Site	Recharge temperature °C	Discharge temperature °C	Change of temperature °C	Depth of flow m
Piketberg	18.9	22.2	3.3	165
Oudtshoorn	16.2	23.7	7.2	360
Calitzdorp	16.2	22.9	6.7	335
Hot springs	16.0	27-64	11-48	550-2400
Hot springs average	16.0	43	27	1350

The conclusion is made that measured, elevated groundwater temperatures confirm that both medium and deep circulation of groundwater does occur in the TMG. It is possibly more common than anticipated and an extensive survey and analysis of groundwater temperature data will be useful.

4 STUDIES IN THE AGTER-WITZENBERG

This project (Weaver et al. 1999, Talma & Weaver, this volume) looked at using hydrochemical and isotope time-series to gain insights into recharge and resource evaluation in the TMG. The valley is a syncline of TMG, with the central part of the valley comprising Bokkeveld Group shales. These shales conformably overlie the TMG. The valley sides are inward dipping at 30° or more. No major faulting occurs, but fracturing due to the folding is present. A transect of boreholes from the top of the access pass into the valley allowed for groundwater sampling at various levels.

Chloride and $\delta^{18}O$ measured in five cumulative rain samplers was compared to groundwater from a borehole in the mountain, recharge zone. The data implies a recharge potential of up to 50% using a chloride mass balance. Figure 2 shows this data, with the rain having about 5 mg/L of chloride and the groundwater about 10 mg/L. The slight shift of $\delta^{18}O$ from rain to groundwater shows the effect of evaporation.

Heavy pumping of groundwater during summer for fruit production causes groundwater levels to drop 30 m in the valley boreholes, and by up to 80 m in the farmers wellfields. During winter complete water level recovery occurs. This seasonal fluctuation of water levels enhances flow in the aquifer and also seasonal variations of both hydrochemistry and isotope content.

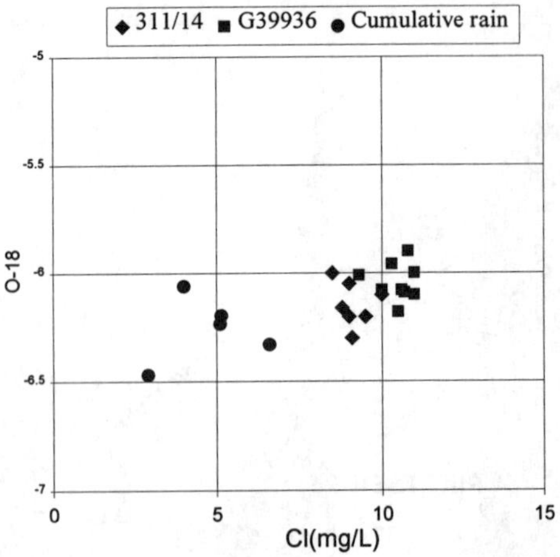

Figure 2. Agter-Witzenburg. Chloride and $\delta^{18}O$ for cumulative rain collecters and mountain boreholes G39936 and 311/14.

Three boreholes were drilled in the valley floor, the first in the Nardouw quartzites, the second through the Bokkeveld shales into the Nardouw quartzites, and the third into the Bokkeveld shales only. The Bokkeveld shales conformably overlie the TMG quartzites and comprise the middle zone of the Cape Supergroup. The quartzite-only borehole, which is closest to the mountain and thus to the recharge zone, shows young groundwater (^{14}C > 100 pmc), with some influence of recharge from a stream, and has a water temperature of 17°C at 100 m depth. The third borehole, furthest from the recharge zone at 800 m from the first borehole has older groundwater (^{14}C = 60 pmc), and has a water temperature of 17.5°C at 150 m depth. However the middle borehole which is only 200 m from the first borehole, has the oldest groundwater (^{14}C = 40 pmc), has $\delta^{18}O$ closest to the mountain recharge signal, and has a water temperature of 19°C at 160 m. This middle borehole therefore intersects deep flowing groundwater, evidenced by the elevated temperature, and, because of the longer flow-path, the older water.

5 RECHARGE POTENTIAL OF TMG

The TMG is an orthoquartzite with insignificant shale bands with the exception of the Cedarberg Formation, which is a quartzitic shale band. This results in a sandy soil with no clay horizons. This soil layer, especially in the mountainous areas, is also quite thin, usually less than 5 m.

The parent rock is coarse-grained, being a lithified beach deposit, and although it does not have any primary porosity, when it weathers the resulting soil is also coarse. The rock is often extensively exposed and invariably jointing is prominent. All of these features are favourable to a high recharge potential.

Further enhancing this recharge potential is the Mediterranean climate of relatively low intensity rainfall occurring in winter, ie less post-rain evaporation. Also the quartzites tend to form mountains, which increases the rainfall due to orographic effects.

At Agter-Witzenberg, Weaver et al. (1999) used chloride enrichment of rainwater to estimate aquifer recharge of 50% (Fig. 2). A regional TMG study to the north of Agter-Witzenberg in the Citrusdal area has established a recharge rate of 43% for the mountainous area between Citrusdal and Agter-Witzenberg (Hay & Hartnady, CAGE Project, report in press). These figures are way beyond the experience of Southern African hydrogeologists who are more comfortable with 5% to 15% recharge. Under normal conditions much of this water merely daylights lower down the mountain as ephemeral springs. However if a wellfield management system of heavy exploitation during summer and resting during winter is applied, then this resource can be used much more effectively. This is essentially what the Western Cape farmers are doing when heavily exploiting their boreholes for summer crop production, although they are probably quite unconscious that they are applying sophisticated groundwater management techniques.

6 WATER SUPPLY POTENTIAL OF THE TMG

Potentially, the TMG offers a wonderful opportunity for providing an integrated water supply source for the Cape Metropole. Our peak water demand is in mid-summer, which is caused mainly by increased garden-use plus the normal increased domestic water use of summer. For example the Weaver household uses ~120m^3/month in Dec/Jan, while in the winter months of June/July this drops to less than 10m^3/month.

Figure 1 shows the TMG outcrop in a 200 kilometre radius around Cape Town. The area of TMG outcrop was measured using GIS methods and, allowing a 1km buffer for the area of TMG quartzites dipping under the Bokkeveld, the exploitable area is 13200 km^2. Taking a storage of 10^{-2} and an average aquifer thickness of 500m, the water in storage is 66 000 million m^3. Using a 33% recharge rate and 600 mm rain per annum, the potential recharge is 2600 million cubic metres. The current annual consumption of water by Cape Town is 500 million cubic metres, so it is fairly obvious that the TMG has the potential to satisfy future increase in demand for Cape Town.

Two factors which could alter these figures are the storage and recharge rate. Hay & Hartnady (pers comm) support a storage of 10^{-1} while Kotze (pers comm.), working in the Klein Karroo supports 10^{-2} to $5x10^{-2}$. Murray & Tredoux (2000), working in the alternating quartzite and shale horizons of the Windhoek aquifer, have calculated a storage of $3x10^{-3}$. Thus 10^{-2} would be a reasonable figure to use. The recharge rate of 33% of 600 mm used in these calculations is a conservative estimate. In the high mountains 40-50% recharge of 1500 to 2000 mm of rain is expected, but this is balanced by lower recharge rates and lower rainfall in the lower-lying valleys.

The proposed method of use would be to establish deep (over 500 m) boreholes, which are capable of high pumping yields. These would be "overpumped" according to current wellfield management techniques, during summer, and water-levels allowed to drop up to 200 m. In winter pumping would cease allowing full recovery of the water-levels. This is what the Agter-Witzenberg farmers are unwittingly doing at present. The thinking behind the method is that firstly, the fractured zones have very high hydraulic conductivities, and thus water can physically move hundreds of metres a day. van Wyk et al. (this volume) have measured such rates in the quartzite aquifer at Windhoek. Secondly, by lowering the water-table, a large "hole" is created that can then easily be filled by the 50% potential recharge, which at present does not actually recharge the aquifer, because it is already full.

7 RECOMMENDED FUTURE RESEARCH

A regional detailed hydrocensus of TMG boreholes and groundwater temperatures is recommended. This survey could include down-hole temperature measurements. In addition ^{14}C measurements could be carried out as a supporting measurement of relative water ages. Deep flowing groundwater would have elevated temperatures and would be older water due to the long flow-path

Recharge rates for the TMG and the Cape Supergroup as a whole need much more research. The recharge of 50% is derived from 1 year of data from 3 boreholes in Agter-Witzenberg is inadequate for regional water-resource calculation. A programme of research using various methods of recharge estimation is needed.

The groundwater pumping by farmers in the Agter-Witzenberg valley has resulted in an ideal test-site for studying the summer "overpumping" method of aquifer management. A project which examines the effects of overpumping on the water-balance should be developed and established in the Agter-Witzenberg.

"Overpumping" in summer could have some impact on summer baseflow. The South African water law requires that the first portion of the water resource be set aside for the ecological reserve. Conceptual models, methods and protocols of implementation need to be developed and tested.

8 ACKNOWLEDGEMENTS

A big thankyou to Rooseda Pieters of DWAF, Bellville, for her application in matching databases, so as to enable temperature to be assigned to a borehole location. To Julian Conrad and Andiswa Silinga of CSIR, for taking this data and preparing it in GIS map format.

The Water Research Commission is thanked for funding the Agter Witzenberg study and the CSIR for funding the TMG groundwater temperature study.

9 REFERENCES

Bredenkamp, D.B., L.J. Botha, G.J. van Tonder and H.J. van Rensburg 1995. Manual on quantitative estimation of groundwater recharge and aquifer storativity. *Report TT 73/95, Water Research Commission, Pretoria.*

De Beer JH, 1976, Magnetometer array studies in South Arica. Unpublished PhD Thesis. University of Alberta, Edmonton, Canada.

Diamond, R.E. and C. Harris. Submitted. Oxygen and hydrogen isotope geochemistry of thermal springs of the Western Cape, South Africa: recharge at high altitude? *Draft unpublished paper* (submitted).

Geological Survey 1980. Stratigraphy of South Africa. Handbook 8, Dept of Mineral and Energy Affairs, South Africa

Halbich, I.W. 1983. A Tectogenesis of the Cape Fold Belt *in* Geodynamics of the Cape Fold Belt, Special Publication No 12 of the Geological Society of South Africa.

Kent, L.F. 1949. The thermal waters of the Union of South Africa and South West Africa. *Trans. Geol. Soc. S. Afr.*52:231-264.

Kent, L.F. 1969. The thermal waters in the Republic of South Africa. *23 Intl. Geol. Congress*, 18:143-164.

Murray E.C., and G. Tredoux, 2000. Windhoek artificial recharge study, Phase 1; Pilot borehole injection tests. Progress Report, Project K5/967, Water Research Commission.

Ninham Shand 1994. Western Cape System Analysis, *Consulting Report for DWAF and City of Cape Town.*

Talma, A.S. and J.C. Vogel 1992. Late quaternary paleotemperatures derived from a speleothem from Cango Caves, Cape Province, South Africa. *Quat Res 37:203-213, 1992.*

Talma, A.S. and J.M.C. Weaver. This volume. CFC tracing of groundwater in fractured rock aquifers aided by 14C and tritium to identify water mixing. *IAH2000 Conference Proceedings, Cape Town.*

Tankard, A.J., M.P.A. Jackson, K.A. Erikson, D.K. Hobday, D.R. Hunter, and W.E.L. Minter, 1982. Crustal Evolution of Southern Africa - 3.8 Billion Years of Earth History, Springer-Verlag Inc., New York, 523pp.

Van Wyk, B., E.C. Murray and G.J. van Tonder. This volume. Tracer test which demonstrate groundwater flow rates of kilometres per day in fractured aquifers. *IAH2000 Conference Proceedings, Cape Town.*

Weaver, J.M.C, A.S. Talma and L. Cavé 1999. Geochemistry and isotopes for resource evaluation in the fractured rock aquifers of the Table Mountain Group. *Report 481/1/99 Water Research Commission, Pretoria*

Weaver, J.M.C. 2000. Borehole water temperature and deep groundwater. *Internal Report, CSIR, Stellenbosch.*

2 Groundwater modelling

Topic co-editors:

Gerrit van Tonder
University of the Orange Free State, South Africa

Wen-Hsing Chiang
University of the Orange Free State, South Africa

Ingrid van der Voort
University of the Orange Free State, South Africa

Groundwater: Past Achievements and Future Challenges, Sililo et al. (eds) © 2000 Balkema, Rotterdam, ISBN 90 5809 159 7

Comparison of genetic algorithms with other methods for the ambient groundwater monitoring network planning

L.-C. Chang & H.-C. Su
Institute of Civil Engineering, National Chiao Tung University, Hsinchu, Taiwan
C.-L. Tien
Water Conservancy Agency, Ministry of Economic Affairs, Taiwan

ABSTRACT: This work concentrates on the ambient groundwater monitoring network design. This kind of planning largely concerns itself with constructing a flexible and efficient monitoring network. This study presents a novel procedure that combines genetic algorithms (GAs) with geostatistical theory. The proposed method is compared to other methods that also integrate geostatistical theory with other optimization schemes, including the sequential design method (SDM), branch and bound method (BBM) and non-linear programming method (NPM). These methods are implemented and applied to a simplified field case. The findings indicate that the SDM technique provides a computationally efficient solution for a preliminary study. On the other hand, the multiple choices given by the GAs provide decision-makers with flexibility to consider factors that geostatistics can not.

1. INTRODUCTION

Hydrogeological engineers cannot provide proper and objective analyses without sufficient background information on the regional groundwater physical characteristics. To resolve this problem, a sufficient number of monitoring wells should be constructed. Estimating the number of samples and obtaining them are always problematic owing to the high cost of hydrogeological investigations. The objective of this study is to develop a feasible methodology for the design of a monitoring network for a region with a limited number of monitoring wells. This work investigates four geostatistics based methods for determining the feasible additional monitoring locations. The geostatistical estimation variance, co-kriging variance, is of primary importance here. The idea of kriging variance was initially applied to determine the optimal locations of sampling sites for the groundwater monitoring network design (Sophocleous 1982, Carrera et al. 1984). Four technologies are incorporated with the co-kriging interpolation method in this study. The first technology is based on variance reduction reduction (Rouhani 1985) which is termed herein as the sequential design method (SDM). The other three operation research algorithms, which are separately incorporated with co-kriging, are genetic algorithms (GAs), branch and bound method (BBM), and non-linear programming method (NPM) (Lin 1995).

All of these algorithms are applied to a simplified field case in the Pingtung Plain in the Southern part of Taiwan. Moreover, the computational reliability of these algorithms and the related application issues are discussed. In this work, the distribution of the ambient monitoring network is sparse. Five additional monitoring locations are specified as the goal in the first stage. If the candidate locations are specified with a high degree of certainty then SDM and NPM schemes can be selected. Otherwise GAs are a better choice.

2. CO-KRIGING METHOD

Co-kriging (Matheron 1971, Myers 1982) is a type of kriging that interpolates the value using the data from interrelated variables. The estimator is expressed as a linear combination of all the available data:

$$Z_j^*(X_0) = \sum_{i=1}^{M} \sum_{k=1}^{N} \lambda_{ij}^k Z_i(X_k) \tag{1}$$

where $Z_j^*(X_0)$ are the variables to be estimated at location X_0. λ_{ij}^k denotes the co-kriging coefficients which present the weight to the variable Z_j using the observation variable Z_i at location X_k. M represents the total number of variables and N is the total number of observation points for variable Z_i. By applying the unbiased condition and minimum variance criteria, the co-kriging system can be derived (Myers 1982). The co-kriging estimation variance is

$$\sigma^2 = Tr\left(\sum_{j=1}^{N} \bar{\gamma}_{x_j 0}\Gamma_j\right) + Tr\left(\bar{\mu} - \bar{\gamma}_{00}\right),$$

$$subject\ to\ \sum_{j=1}^{N} \Gamma_j = I \qquad (2)$$

where Tr denotes the trace of the matrix, σ^2 represents the co-kriging estimation variance, $\gamma_{x_j o}$ is the cross variogram of the variables (Myers 1982), and $\bar{\mu}$ denotes Langrange multipliers matrix. By assuming that multiple points, are estimated, the total estimation variance can be written as follows:

$$\sigma_T^2 = \sum_{\beta=1}^{Q}\left[Tr\left(\sum_{j=1}^{N} \bar{\gamma}_{x_j 0}\Gamma_j\right) + Tr\left(\bar{\mu} - \bar{\gamma}_{00}\right)\right] \qquad (3)$$

The variance reduction caused by adding L new monitoring wells into the network can be expressed as:

$$VR_{*0} = V_0(N) - V_0(N+L) \qquad (4)$$

3. FRAMEWORK OF OPTIMAL ALGORITHMS

3.1. Genetic algorithms (GAs)

1. Express the objective function as the total co-kriging variance reduction maximum.
2. Randomly generate 100 chromosomes (network designs) initially and the number of chromosomes remains constant for all of the generations.
3. Evaluate the fitness for all of the chromosomes: In GAs, the number of wells for each chromosome may be greater or less than the required number of well numbers. Therefore, a penalty factor is here defined while the proposed number of wells is greater than the required well numbers: $F = \Delta V \times \left(\frac{1}{c}\right)^2$ (Goldberg 1989), where c equals the number of wells for the chromosome when they are greater than L (eq.(3)). On the other hand, c equals 1 when the number of wells is less than the specified number. ΔV represents the total variance reduction for the (chromosome) network design.
4. Verify the terminating conditions: Herein, we apply the simple criteria to stop the calculation after fifty generations.
5. Select (reproduction) the chromosomes: The roulette wheel method (Goldberg 1989) is applied to select the chromosomes for crossover in this study. This operation simulates the natural selection. The greater the number of high fitness chromosomes, the greater the survival probability.
6. Perform crossover of the chromosomes: Randomly select two chromosomes from the pool and, then, perform the crossover to generate new offspring. In this study, one-point crossover is utilized.
7. Conduct mutation of the chromosomes: The mutation restores the lost or unexplored genetic material into the population to prevent the GAs from prematurely converging to a local minimum. A mutation probability allows random mutations to be made to individual genes. 0.1 is adopted as the mutation probability. (Lis 1995, Chen 1998) After completing this procedure, a new population of chromosomes is generated. The algorithm then goes back to step 2.

3.2. Sequential Design Method (SDM)

1. Define the objective function as the total co-kriging variance reduction maximum.
2. Prepare three sets of data points: the first set evaluates the total estimation variance. The second set is the available wells and the third set is the candidate locations.
3. Select a point X_* from the candidate set and evaluate the new total estimation variance. Repeat the same procedure for all of the candidate points. The well location that causes the new network to have minimum total variance is the optimal well location.
4. Move the above optimal point out of the candidate set and into the observation data set.
5. Repeat procedures 3 and 4 until the required number of additional monitoring wells is selected.

3.3. Branch and Bound Method (BBM)

BBM is a simultaneous search as Ben-Jemaa (1994) proposed. This algorithm is formed as a sequential procedure in which a well is conceptually deleted (dropped) at a time while moving along the tree branches. On the other hand, each node in the tree represents a possible network design. Figure 1 illus-

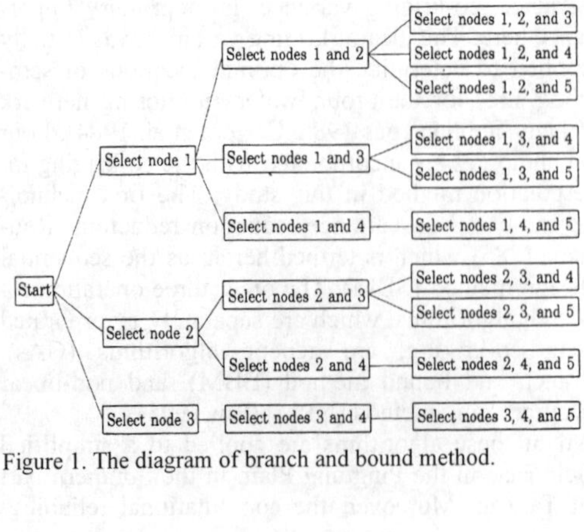

Figure 1. The diagram of branch and bound method.

trates a search tree of this algorithm in which three wells are selected from five candidates. Based on the evaluated estimation variance at each node, the search either proceeds forward along the same branch to delete a new well or leaves the current branch and moves one step back to another branch. In principle, the BBM method can determine the global optimal solution for the network design problem. Nevertheless, computational load for this algorithm is too heavy for a practical problem.

3.4. *Non-linear Programming Method (NPM)*

1. Define the objective function as the co-kriging variance reduction maximum.
2. Choose an initial solution: Herein, we adopt SDM solution as the initial well locations.
3. Evaluate the gradient d^k (via central-difference approximation) of the objective function on the well location design.
4. Select a search step S^k using a one-dimensional search method and, then, update the well locations using $X^{k+1} = X^k + S^k d^k$ and re-evaluate the value of the total estimation variance.
5. Define an acceptable stopping criterion ε. Repeat procedures 2 and 3 till the ratio of the variance reduction is less than or equal to ε.

4. APPLICATION AND COMPARISONS

4.1. *Study Area*

This study applies the above algorithms to a simplified groundwater monitoring network design problem in the Pingtung Plain, one of nine groundwater regions in Taiwan. Located in the Southern part of Taiwan, Pingtung Plain is about 1210km^2. The topography may be classified into four categories: (1) The Pingtung Valley; (2) The Eastern Mountain Region; (3) The Northern Mountain Region; (4) The Region of the Western Foothills. (Ting 1998) The groundwater table and hydraulic conductivity are the parameters in this groundwater monitoring network plan. The groundwater table is the parameter of primary concern, while the hydraulic conductivity is an auxiliary parameter. Three sets of sites must be defined before the optimal schemes are employed: (1) a set of existing data sites: fifteen installed observation wells. (2) a set of candidate sites: forty candidate sites are chosen as the locations for constructing new monitoring wells. (3) a set of basis nodes: Thirty-eight nodes are uniformly defined in our study area to calculate the total co-kriging variance. Table 1 and Figure 2 presents the distribution of the above three sets of sites for the Pingtung Plain.

Table 1. Solutions of the different methods.

Method	SDM		NPM		GAs	
Candidate(km)	X	Y	X	Y	X	Y
Site1	219.9	2532.3	212.9	2532.9	212.9	2532.3
Site2	207.6	2479.2	207.0	2532.9	207.6	2532.3
Site3	203.3	2533.4	200.9	2532.9	203.3	2533.4
Site4	182.9	2490.9	182.9	2491.0	182.9	2490.9
Site5	206.3	2515.1	206.9	2515.0	206.3	2515.1
Initial TEV	59.3571		59.3571		59.3571	
TEV*, after the above 5 sites added	48.8479		46.4772		48.8479	

* TEV: total estimation variance; [m^2]

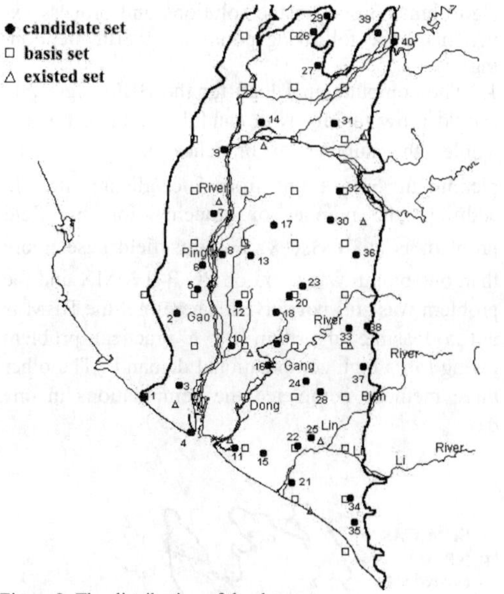

Figure 2. The distribution of the data sets.

4.2. *Geostatistical Structure Analysis*

By analyzing the data for the Pingtung plain using the GeoEAS, which was developed by the EPA, the semivariogram and cross-semivariogram for the ground-water table and hydraulic conductivity can be summarized as follows:

1. Semivariogram for the water table (W):

$$\gamma_W = 1.7\left[1 - e^{\frac{-3h}{30}}\right]$$, where h is the distance between any two points

2. Semivariogram of the natural of hydraulic conductivity log (ln(K)):

$$\gamma_{\ln(K)} = 0.7\left[1 - e^{\frac{-3h}{20}}\right]$$

3. Cross-semivariogram of water table and natural log of hydraulic conductivity was taken by the following formula (Myers 1982):

$$\gamma_{W\ln(K)}(h) = \frac{1}{2}\left[\gamma^{+}_{W\ln(K)}(h) - \gamma_W(h) - \gamma_{\ln(K)}(h)\right]$$
$$= \frac{1}{2}\left[0.85(1 - e^{\frac{-3h}{25}}) - 0.7(1 - e^{\frac{-3h}{20}}) - 1.7(1 - e^{\frac{-3h}{30}})\right]$$

4.3. Models Application

The objective is to add five additional wells to the existing monitoring network. The final network will contain twenty monitoring wells. Figure 3 demonstrates the optimal locations for all of the different algorithms. Based on the solutions and process experiences, the following points are worth mentioning:

1. The computational load for the BBM algorithm should grow rapidly with candidate points. For example: the number of branches is $C_3^7 = 35$ while picking up 3 locations from 7 candidate sites. In addition, the number of branches for this field problem is $C_5^{40} = 658008$. For this field case, more than one month was spent on PC P-II MMX and the problem was still not solved. Therefore, the BBM is not a feasible algorithm for a practical problem owing to its high computational demand. The other three methods completed the computations in one day.

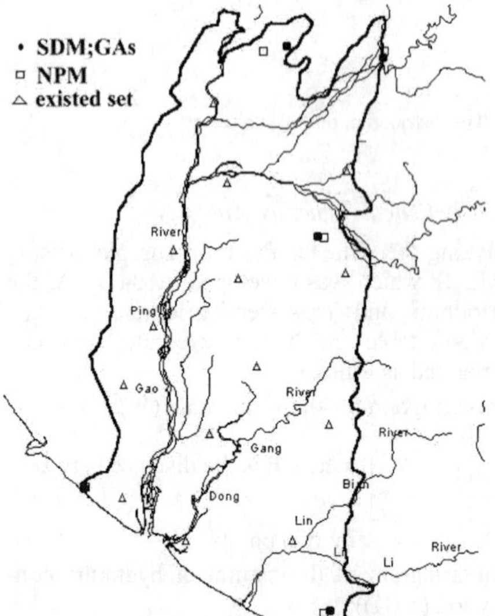

Figure 3. The distribution of solutions from each optimal algorithm.

2. As mentioned earlier, GAs can provide users with multiple solutions simultaneously. For our problem, the optimal fitness for each generation is almost the same after twenty-three generations. This finding implies that the chromosomes will have a fitness value near the optimal fitness value, 10.509, after twenty-three generations. After collecting all of the chromosomes with a fitness value near the optimal, there are fifty-one chromosomes. In addition, each chromosome represents a network design, implying that there are fifty-one near optimal network designs for this problem. Table 2 lists the solutions as well. For all candidate sites, points 40, 1, 34, 32, 27, and 26 are selected with high frequency. Points 2, 3, 4, 6, 7, 8, 9, 10, 11, 14, 15, 19, 20, 21 and 23 are never selected.

Table 2. The total co-kriging variance for the GAs solutions.

Solution (Selected Sites*)	TEV	Solution (Selected Sites*)	TEV
1,26,32,34,40	48.845	25,27,32,35,40	50.605
1,27,32,34,40	48.920	1,26,27,32,40	50.056
1,22,26,34,40	49.029	1,22,30,34,40	50.112
1,16,26,34,40	49.689	1,25,27,32,40	50.491
1,24,32,34,40	49.407	1,28,31,32,40	50.468
1,24,26,34,40	49.050	1,27,34,39,40	50.281
1,26,31,34,40	49.341	1,30,32,34,40	50.236
1,27,30,34,40	49.644	1,25,32,34,40	50.250
1,24,26,32,40	49.574	1,24,34,39,40	50.616
1,13,27,34,40	49.579	1,22,26,38,40	50.700
1,22,27,34,40	49.098	1,26,32,37,40	50.443
1,24,27,35,40	49.435	16,22,27,34,40	50.299
1,24,27,34,40	49.119	1,29,32,36,40	50.599
1,22,32,34,40	49.385	18,22,26,34,40	50.549
1,24,27,32,40	49.648	1,31,33,34,40	50.654
1,25,27,34,40	49.963	1,22,27,36,40	50.113
1,26,27,34,40	49.530	1,27,33,36,40	51.123
1,29,32,34,40	49.579	1,25,33,34,40	51.332
1,26,33,34,40	49.929	1,29,32,38,40	51.249
22,26,32,34,40	49.352	1,30,32,36,40	51.438
1,27,33,34,40	49.998	1,13,27,33,34	51.955
1,26,32,36,40	49.868	1,30,32,38,40	51.920
1,27,32,35,40	49.235	1,17,33,34,40	51.698
1,31,32,34,40	49.646	5,25,27,34,40	51.000
12,26,32,34,40	49.618	1,22,33,36,40	51.594
1,24,31,34,40	49.775		

3. The fitness of GAs contains a penalty factor to restrict the number of added wells. Therefore, the total estimation variance for their represented network designs may not be the same value when the chromosomes have the same fitness value. Hence, the total variance must be re-calculated to evaluate the efficiency of the network design represented by each chromosome. The actual values of the total estimation variance for the previous fifty-one different chromosomes are re-evaluated using co-kriging as summarized in Table 2. According to this table, the difference between the minimum total variance and

the maximum is 3.108, which is insignificant for hydrogeological estimation in practice. This finding confirms that all fifty-one solutions can be accepted as the monitoring network design. Our results also suggest that the network design can be flexible. In fact, the geostatistics can not consider factors such as the geological condition, land availability and other administrative constraints. The multiple choices given by the GAs can provide decision-makers with flexibility in considering these factors. On the other hand, the single solution provided by the other methods places unnecessary restrictions on the decision-makers.

5. DISCUSSION AND CONCLUSION

This study presents four geostatistics based algorithms (GAs, SDM, BBM, and NPM) to design an ambient groundwater monitoring network. The proposed algorithms are applied to a simplified field case. Based on the results and experiences in this study, we conclude the following:

1. The BBM solution is a global optimal, indicating that the SDM and GAs can find the optimal solution as well. Nevertheless, for a moderate size field case problem, the BBM method fails to solve the problem within a reasonable computational time, implying that it may be inappropriate for practical applications.
2. The decision variables from the NPM method are continuous in space. Nevertheless, for a practical regional network project, a future study should determine whether or not the available sites for constructing monitoring wells should be implemented before the network is developed. Therefore, the continuous nature of the NPM method may not be beneficial for practical applications. Results in this study indicate that, when the density of the candidate sites is high, the solution from the NPM method is close to solutions from other methods.
3. The monitoring wells should be well maintained when located at schools and national regions in Taiwan. Though the candidate sites are distributed in a mesh, the most troublesome points (such as: private land, river, etc.) can be eliminated before the optimal algorithms are implemented. As we know, geostatistics cannot consider special geological situations or external factors. For the four algorithms that we use, the SDM, NPM and BBM methods can only offer a single solution. GAs can provide multiple near optimal solutions. The multiple choices given by the GAs can provide decision-makers with the flexibility to consider other real factors (such as: the geological situations) which are not easily defined in the monitoring design algorithms. Additionally, the single solution provided by other methods might place unnecessary restrictions on decision-makers.
4. The solutions from the SDM method are close to the solutions from other three methods despite the fact that this algorithm is not an optimization scheme in general. Therefore, while considering the efficiency and reliability of the proposed algorithms, the SDM method is the most practical solution for a network design problem if the flexibility provided by multiple solutions is not a concern.

REFERENCES

Ben-Jemaa et al. 1994. Multivariate Geostatistical Design of Ground-Water Monitoring Networks. *J. of Water Resource Planning and Management* 20(4): 505-522.

Carrera, J. ,d Usunoff, E. and Szidarovszky, F. 1984. A Method for Optimal Observation Network Design for Groundwater Management. *J. of Hydrology*. 73: 147-163.

Chen, Huag-chia 1998. *Application of Genetic Algorithms to the Groundwater Monitoring Network Planning. Master Dissertation. Institute of Civil Eng.*, National Chiao Tung University, Taiwan, R.O.C.

Goldberg, D. E. 1989. *Genetic Algorithms in Search, Optimization, and Machine Learning*, Addison-Wesley.

Lin, Y. P. 1995. *Multiple-Point Variance Analysis for Groundwater Monitoring Network Design. In Fulfillment for the Requirements for the Degree Doctor of Philosophy in Civil Engineering.*

Lis, Joanna 1995. Genetic Algorithm with the Dynamic Probability of Mutation in the Classification Problem. *Pattern Recognition Letters*. 16: 1311-1320.

Matheron G. 1971. *The Theory of Regionalized Variables and its Applications, Paris School of Mines*, Cah. Cent. Morphologic Math.. Fontainnebleau. France.

Myers, D.E. 1982. Matrix Formulation of Co-Kriging. Math. Geol. 14(3): 249-257.

Rouhani, S 1985. Variance Reduction Analysis. *Water Resources Research* 21(6): 837-846.

Ting, Cheh-Shyh et al. 1998. Development of a Preliminary Ground Water Flow Model for Water Resources Management in the Pingtung Plain, Taiwan. *Ground Water* 36(1): 20-36.

Groundwater: Past Achievements and Future Challenges, Sililo et al. (eds) © 2000 Balkema, Rotterdam, ISBN 90 5809 159 7

Construction of groundwater model for Pusan area, South Korea

Byong-Wook Cho, Bong-Joo Lee, Ig-Hwan Sung, Byung-Dae Lee & Chang-Oh Choo
Korea Institute of Geology, Mining and Materials, Taejeon, Korea

ABSTRACT: An attempt was made to construct a groundwater model of the urbanized Pusan area with coverage of 523.5 km^2. Two layers for the model were considered; Layer 1 is alluvium and weathered bedrock and layer 2 is slightly fractured bedrock. Average thickness and hydraulic conductivity of each layer is 20 m, 150 m and 1 x 10^{-4} m/s, 5 x 10^{-9} m/s, respectively. The main inflow is leakage from drinking water network and sewage water network and rainfalls infiltration. The main outflow is overflow by sewage water network and pumping from boreholes. Calculation at final simulation shows that recharge from surface to layer 1 is 655,260 m^3/d. Among them, 67.7 % are retaken by the sewage network, 26.4 % are pumped by boreholes and 5.7 % enters the Nakdong river.

1 INTRODUCTION

Urban development with rapid economic growth and increase of the population can change the spatial distribution of groundwater recharge (Appleyard, 1995). Urbanization may also either increase or decrease groundwater recharge (Lernar et al., 1990). Recently, there has been increasing concern as to the quantity and quality of groundwater resources in Korea.

The study area, Pusan city, is highly subjected to groundwater contamination due to its unfavorable geographical features; it is located in the estuaries of the Nakdong river, most of the area is composed of highlands, and the large population resides in the small downhill and flat area. Heavy pumpage and deterioration of groundwater are currently found to be significant compared to other major cities, resulting in shortage of water resources and contamination of groundwater. Depth to water table which were measured at 489 wells in April 1999 averages 12.96 m, showing that its value is two times deeper than those of other cities.

The aim of this paper is construction of groundwater model of the urbanized Pusan area to provide guideline for the management of groundwater resources.

2 HYDROGEOLOGY

2.1 Topography and geology

The study area is located in the southeastern part of the Korean peninsula (Figure 1). The area is characterized by the low relief in the western part and relatively steep relief in the eastern part. Topographic elevation ranges from 0 to 801 m. The main river, Nakdong, flows from north to south entering the South Sea. Geology can be divided into four groups; Cretaceous sedimentary rocks, volcanics, granites and alluvium (Chang et al., 1983). Except for delta area, most of area is composed of crystalline rocks and thin alluvium. The thickness of alluvium is usually less than 20 m.

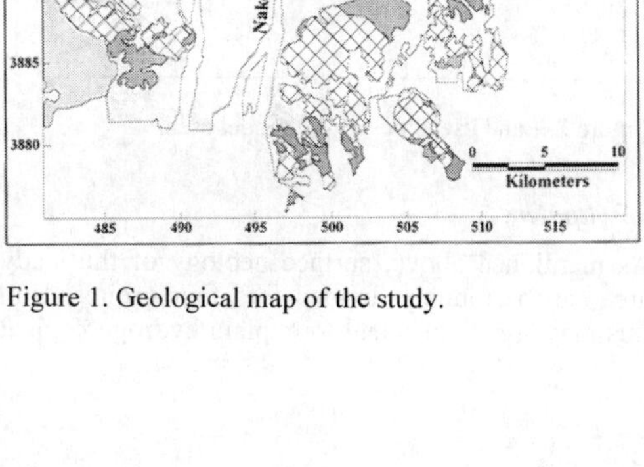

Figure 1. Geological map of the study.

2.2 Hydraulic conductivity and storativity

At thirty boreholes, hydraulic tests like pumping test and lugeon test were performed on the bedrock. The hydraulic conductivity (K) is low to very low, ranging from 8.1 x 10^{-10} m/s to 7.7 x 10^{-5} m/s, most of which range from 1 x 10^{-7} m/s to 5 x 10^{-6} m/s. Storage coefficient (S) varies from 1.7 x 10^{-5} to 7.2 x 10^{-4} (KIGAM, 1999).

2.3 Potentiometric surface map

Potentiometric surface levels are highly correlated with ground elevation due to high hills in the Pusan area. Therefore, DTW map was drawn instead of potentiometric surface map. Contour map for the depth to water table was drawn using 489 boreholes measured in April 1999. Some interpolation was made at hilly areas where few borehole data were available.

3. CONCEPTUAL MODEL

3.1 Land use

The boundary of the model area is limited to the administrative boundaries of Pusan city. Although some hilly parts may have no aquifers, they were included in the model area because of their potential capabilities of natural recharge. Using a land use map drawn by Pusan City Hall, land use zones were roughly divided into four classes, taking infiltration into account (Figure 2, Table 1).

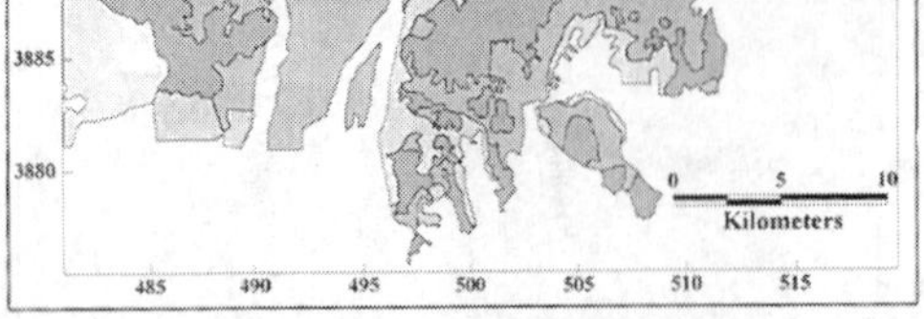

Figure 2. Land use zones of the model area.

3.2 Aquifers

As mentioned above, surface geology of the study area can be roughly divided into four groups. But, this maybe not sufficient to explain hydrogeological characteristic of the study area. To delineate thickness distribution of layer 1 (alluvium, weathered rocks), 341 borehole logs were collected and analyzed. Figure 3 shows an interpolated thickness distribution map. To avoid interpolation errors resulting from the extensive hilly areas where few boreholes are available, weathered bedrock was considered to have average 10 m thickness along outlines of the hills.

Table 1. Coverage of land use zones.

Zone No.	Name	Area (km^2)
1	Urban	123.6
2	Industrial	43.4
3	Agricultural	117.1
4	Forest	216.6
	(River)	(22.8)
Total		523.5

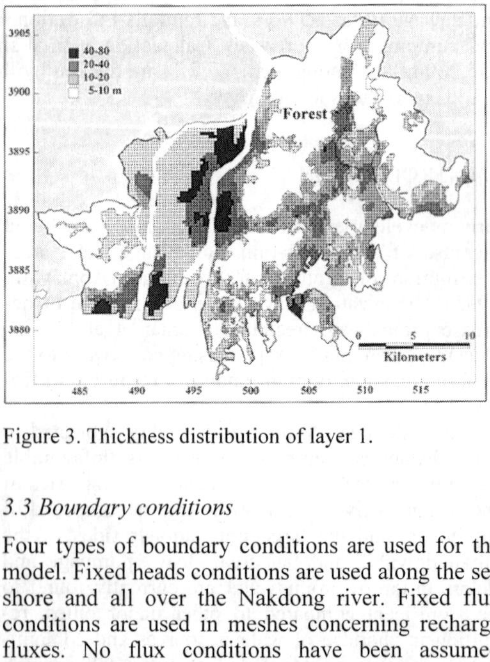

Figure 3. Thickness distribution of layer 1.

3.3 Boundary conditions

Four types of boundary conditions are used for the model. Fixed heads conditions are used along the sea shore and all over the Nakdong river. Fixed flux conditions are used in meshes concerning recharge fluxes. No flux conditions have been assumed throughout the Pusan model boundaries, except sea shore. Overflow conditions are defined to the area where leakage occurs from drinking water and sewage water networks.

3.4 Inflows

Inflows to the model area result mainly from rainfalls and leakage from sewage and drinking water networks. The average yearly rainfalls over the past 20 years (from 1979 to 1998) in the study area is 1499 mm/y (Bureau of Meteorology, 1999). The natural recharge from rainfalls is assumed to be efficient only in the forest and agricultural areas, but not in the urban and industrial areas (Lerner, 1990).

Recharge to the aquifers in the urban and industrial zones is in a close relationship with the leakage from both drinking water and sewage water networks. Production and leakage rates of both network is listed in Table 2.

The known leakage rate of drinking water network is 17.1 % in 1997 and 16.5 % in 1998 (average leakage 250,000 m^3/d), respectively. Although the leakage rate of sewage water network is not known, 10 % leakage was estimated (160,000 m^3/d). Therefore, 410,000 m^3/d is assumed to leak into aquifers and riverbeds. This yields concerns about mainly urban area (123.6 km^2) and industrial area (43.4 km^2). This inflow is equivalent to uniform infiltration of 897 mm/y and seems to very excessive. For this leakage, it is reasonable to think that an average "leakage density" can be applied to all along each network branch, which means that the leakage flow in an area is proportional to the total network length running through it. Using MapInfo the two return flows have been spread all over the model area.

Table 2. Production rates (m^3/d) of drinking water and sewage water in Pusan city.

Year	Drinking water Network	Sewage water Network
1997	1,506,056(17.1)	1,538,000
1998	1,434,734(16.5)	1,601,000
1999	-	1,665,000

( ) leakage rate

3.5 Outflows

Besides the natural outflow towards the sea and the rivers, the main groundwater outflow in the Pusan area is pumping from boreholes and dewatering from subway stations.

The registered borehole number in the study area in 1997 is 9, 908 and the accessed total abstraction rate is about 60,000,000 m^3/year. Among 9,908 boreholes, only 1,843 boreholes have UTM coordinates. So, it is more proper to count the borehole number per village, and to total up the global discharge per village. When dividing the global discharge of a village by the village area (km^2), we can have the groundwater abstraction density per village ($m^3/y/km^2$).

Figures 4 shows the spatial distribution of this parameter. In fact, as boreholes are mainly located in the flat areas, abstraction density was calculated using the flat area of each village instead of the total village area.

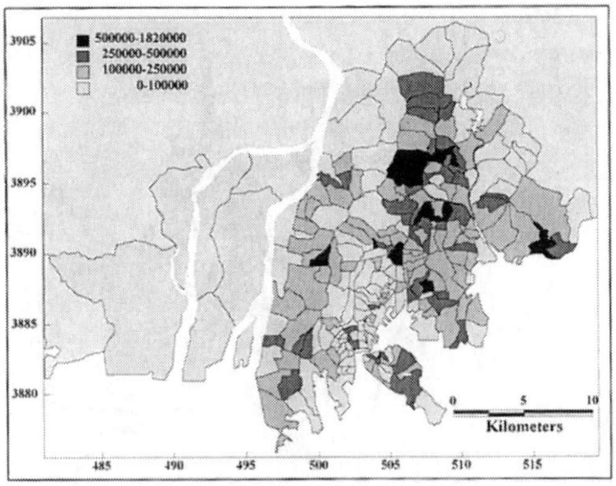

Figure 4. Pumping rate density ($m^3/y/km^2$) of each village.

There is two subway lines in the Pusan area. Line 1 have 37 stations dewatering groundwater about 3,300 m^3/d, then diverted to the sewage water network. Line 2 is now under construction and only half of the stations is operating. The number of stations dewatering rate is 24 and 5,300 m^3/d, respectively.

4. MODEL CONSTRUCTION

4.1 Model grid and layer

Regular mesh of 250 m was selected considering density and distribution of the available data. Two layers were considered for the model. Layer 1 is alluvium deposit and weathered bedrock. Thickness of layer 1 has been interpolated from 341 boreholes logs. In the hilly areas where the bedrock is not covered with alluviums, its weathered thickness is assumed to be 10 m. In the Nakdong delta area where few boreholes data exist, the aquifer thickness is assessed to be 20 m. Layer 2 is fractured bedrock. Usually, aquifer capabilities of such rocks are relatively poor, but not negligible. Thickness of layer 2 was assumed to be 150 m.

4.2 Infiltration and abstraction

Infiltration rates of rainfalls were allocated to meshes of agriculture and forest zones. Leakage from drinking water supply network and sewage network to each mesh was computed using MapInfo software. The sum of leakage from both networks is illustrated in Figure 5.

Abstraction rates of each mesh is the sum of subway stations and boreholes. Abstraction rate from boreholes was computed with the hypothesis that boreholes are mainly drilled in flat areas with homo-

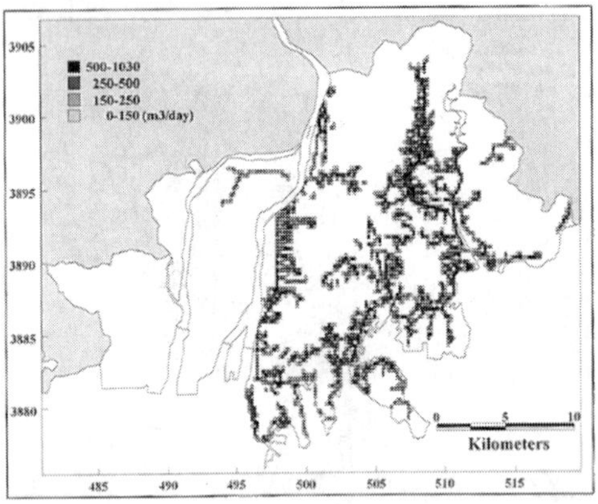

Figure 5. Infiltration rates by drinking water network and sewage water network in the flat area.

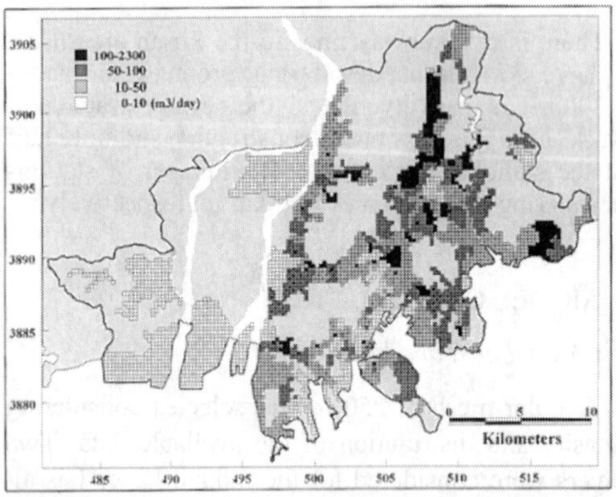

Figure 6. Abstraction rates by boreholes and subway stations in the flat area.

geneous density in each village. The sum of the two abstraction rates in each mesh of the model is shown in Figure 6.

4.3 Steady state condition calibration

After several simulations, best results have been achieved with an uniform hydraulic conductivity of 1×10^{-4} m/s in the whole upper layer and 5×10^{-9} m/s in the lower layer. At this steady state calibration phase, the simulated water table fits globally well with the potentiometric surface maps interpolated from depth to water table measurements performed in April and in September 1999.

The steady state water budget calculated by the model is shown in Table 3. One can see that among the global 665,260 m^3/d which is recharging to layer 1, two-thirds of the recharge is immediately retaken by the sewage water network, without real flowing as groundwater, 26 % are pumped by boreholes and subway stations, 6 % reach the South Sea, and only the small rate (0.2 %) of the amount underflows to layer 2. The small rate from layer 1 to layer 2 is ascribed to the low hydraulic conductivity of layer 2 and all layer 2 is covered by layer 1.

It must be emphasized that all these figures are highly dependent on the inflows resulted from leakage in both networks. Because leakage rate of sewage water network was only estimated and the pumping by unregistered production wells was not considered to the model, the water budgets shown in Table 3 must be considered as a guideline. Nevertheless, it appears that the sewage water network acts as the main outflow.

Table 3. Water budgets after calibration.

Layers	Flows	Flow rates (m^3/d)
Layer 1	Recharge from rainfalls	196,200
	Recharge from drinking water network	250,000
	Recharge from sewage water	160,000
	Recharge from Nakdong river	49,060
	Pumping from boreholes	164,408
	Pumping from subway station	8,600
	Outflow to South sea	37,480
	Overflow to sewage water network	443,300
Layer 2	Transfer to Layer1	1,472
	Pumping from boreholes	1,472

5. CONCLUSION

An attempt was made to preserve the Pusan groundwater resources. In consideration of the complexities of the urban hydrogeology and the available data, it is not easy to construct the Pusan groundwater model.

Two aquifers were assumed for the model. Layer 1 is alluvium and weathered zone and layer 2 is fractured bed rocks. Main inflows to the model area is leakage from drinking water and sewage water networks, and rainfalls infiltration. Main outflows from the model resulted from sewage water network and pumping in production wells. Steady state water budgets at final simulation show that among total inflows to layer 1, 67 % is overflowed by sewage water network and 26 % is pumped by boreholes and subway stations. Therefore, both drinking water and sewage water networks play an important role in the Pusan groundwater system.

Considering the hydrogeological data applied to this model, it is not sufficient yet to delineate the urbanized Pusan groundwater system. Nevertheless, this model can be used as a reference model for the future groundwater management.

REFERENCES

Bureau of Meteorology 1999. Meteorology report (1979~1998), Published Bureau of Meteorology report.

Chang T.W. et al. 1983. Geological map of Pusan, Gadeog. Published KIGAM Report:22.

Korea Institute of Geology, Mining and Materials (KIGAM) 1999. Study on protection and reclamation for the groundwater resources in Pusan area. Published KIGAM Report:541

Lerner, D.N. 1990. Recharge due to urbanization. Hydrology J. 8:201-214.

Appleyard S.1995. The impact of urban development on recharge and groundwater quality in a coastal aquifer near Perth, western Austrialia. Hydrology J. 3(2):65-75.

Groundwater: Past Achievements and Future Challenges, Sililo et al. (eds) © *2000 Balkema, Rotterdam, ISBN 90 5809 159 7*

Hydraulic fracturing of water wells in Botswana

V.Devalpally & M.S.Nidamarty
Geotechnical Consulting Services (Pty) Limited, Gaborone, Botswana

L.Carlsson
Department of Water Affairs, Gaborone, Botswana

ABSTRACT: In order to obtain 1.5 m^3/h minimum yield of a production borehole for village water supply, hydraulic fracturing was applied on 21 existing low yielding boreholes. About 90% of the boreholes subjected to hydraulic fracturing showed an improvement. The improvement ratio factor based on the recommended production yield prior to and after hydraulic fracturing was found to be the most comprehensive factor to assess the success of hydraulic fracturing. The improvement ratio was around 2 for boreholes fully hydraulic fractured using sections of 6-9 m straddle packer configuration. Where partial sections of the borehole were hydraulic fractured, the improvement ratio was less than 1.5. The hydraulic fracturing proved to be a technically and economically viable technique for improving borehole yields with 35-40 % saving compared to drilling of a new borehole.

1 INTRODUCTION

Until the end of 1999, about 9,200 governmental boreholes were registered in Botswana. The boreholes, mainly drilled through Department of Water Affairs (DWA), are for various purposes. About 40% of these are declared blank and therefore left unused. Boreholes with yield less than 1 – 1.5 m^3/h or of poor water quality (TDS>2,000 mg/l) are usually categorized as blank boreholes. Original blank boreholes can however be commissioned provided their yield is increased. One method to increase the yield could be hydraulic fracturing if found cost effective compared to siting and drilling of new boreholes.

Hydraulic Fracturing (HF) is a borehole development process involving injection of water under high pressure into selected zones of a borehole thereby increasing the size and extent of existing fractures, initiation of new fractures and cleaning of fractures.

During 1998 DWA carried out a project to establish HF as one of the applicable borehole rehabilitation methods and to assess cost effectiveness of the method (DWA, 1998). In total 21 boreholes falling in three different geological environments in Botswana namely the Gaborone granites, Waterberg group of metamorphic rocks and the North Eastern gneisses were subjected to HF.

2 HYDRAULIC FRACTURING PROCESS

2.1 *Equipment*

The HF equipment used in the project comprised of:

1) HF pump: A triplex plunger pump with a capacity of injecting 200 litres of water per minute at 280 bar pressure. It was powered by Deutz BF61913 (6 cylinder – turbo charged). The same engine also powered the HF rig.
2) Booster pump: A high pressure centrifugal booster pump to boost the total system pressure thereby increasing the efficiency of the HF pump.
3) Hydraulic pump: A positive displacement pump driven by a 4.5 HP petrol engine for inflating the packers. It had a pressure capacity of 560 bar.
4) Packers: Two types of packers, inflatable and compressible, used in single or straddle configurations.

2.2 *Hydraulic fracturing performance*

The project was executed in 4 stages. Stage I included selection of boreholes for hydraulic fracturing, identification of fracture zones by geophysical logging methods, and assessing hydrogeological and aquifer characteristics by test pumping. Stage II involved hydraulic fracturing of the selected boreholes. Stage III involved post test pumping works and provided estimates on well and aquifer characteristics for assessing improvement due to hydraulic fracturing. Evaluation of results formed the last phase to establish technical and economic viability of hydraulic fracturing in Botswana.

During the first stage, selected boreholes were geophysically logged using acoustic scanner, caliper and focused electrical logging. The boreholes were further tested by step drawdown (SDT) and constant rate (CRT, 24 h) test pumping.

In Stage II the HF process applied on all boreholes comprised: (i) cleaning and stabilization of the borehole, (ii) water injection/pressurizing the system, (iii) back flushing, & (iv) return flow of water.

HF procedures adopted in the present study fall in two categories referred to as partially and fully HF boreholes. In the first category, only selected fracture zones were hydraulic fractured while in the second, all the fracture zones in a borehole were hydraulic fractured. Table 1 summarizes the categories applied versus the three areas investigated. Fourteen boreholes fall in the first category where the hydraulic fracturing was carried out in 3-6 m straddle interval, 5-10 minutes of treatment time per zone, and injection of 5-10 m^3 of water. The remaining 7 boreholes fall in the second category where the whole boreholes were hydraulic fractured with 6-9 m straddle interval, 20-25 minutes of treatment time per zone, and 30-50 m^3 of water injection.

Table 1. No of boreholes hydraulic fractured.

	Area 1	Area 2	Area 3
No of boreholes	12	4	5
Partially HF	11	3	
Fully HF	1	1	5
No of sections	2 - 5	3 – 8	4 – 7
Bedrock	Granite	Metamorphic rocks	Gneisses

3 EVALUATION OF HYDRAULIC FRACTURING

3.1 *Hydraulic parameters*

One of the prerequisites to evaluate the effectiveness of hydraulic fracturing was to possess data on borehole and aquifer characteristics prior to and after the HF process. The evaluation was carried on the concept of borehole drawdown (s_w) behavior expressed as:

$$s_w = Q(W(u,\beta,m) + 2\zeta)(4\pi T)^{-1} + CQ^2$$

where Q = pumping rate; $W(u,\beta,m)$ well-function for double porosity aquifer; ζ = skin factor; T = transmissivity; C = factor for non-linear loss.

The C-factor was obtained from the step drawdown test whereas evaluation of skin factor assumed that storativity values were known. In the evaluation, the storativity values were taken from literature studies or from other tests performed in the areas. As the effects of hydraulic fracturing are expected to extend only to limited distances into the aquifer, it was reasonably assumed that storativity will not change due to hydraulic fracturing.

The CRT data were evaluated using similar procedure as in the rural village water supply programme (Carlsson and Ntsatsi, 1998).

In order to assess the effectiveness of the applied hydraulic fracturing, the hydraulic parameters obtained from pre and post test pumping were compared to ascertain the most representative parameters which could be used in the over all evaluation of the success of hydraulic fracturing.

The general assumption in the evaluation was that the process of injecting water under high pressure primarily affected the immediate surroundings of the borehole. Thus the parameters expressing linear and non-linear well-losses would be affected. Impact on transmissivity would be limited with the pressure and treatment time applied. Main effects related to transmissivity changes will be expressed by changes in linear well-losses (skin). Also changes in storativity values were assumed to be limited. It was however not possible in all interpretations of post testing to assume the same storativities as in the pre testing. Late storativity had to be changed in order to obtain a proper interpretation.

Parameters such as specific capacity, nonlinear well loss coefficient, skin, transmissivity and recommended yield are discussed hereunder.

3.2 *Specific capacity*

The Specific Capacity (SPC) obtained from the pre and post step drawdown tests was calculated from the first steps. Figure 1 summarises the results.

The increase in SPC value was high in case of fully hydraulic fractured boreholes compared to the partially hydraulic fractured boreholes.

3.3 *Non linear well losses*

The non linear well losses were assumed to be proportional to the square of the pumping rate (Jacob, 1947). The factor ‘C” is referred to as the non linear well loss coefficient is calculated from the SDT prior to and after the HF. The relation between the pre and post C-factor is defined as the Improvement Ratio for C-values (IR_C) and summarized by Figure 2.

Increase in C means an increase in well loss component resulting in reduced yields. An IR_C higher than one will thus show a reduction of the C-factor due to HF.

The median and average values of improvement ratio are shown in Table 2. A considerable improvement in reduction of well losses was obtained in boreholes injected with large volumes of water, long treatment time and fully HF.

The probability distribution of the C-values prior to and after HF were used to establish the 50% probability values and a corresponding statistically improvement ratio, also presented in Table 2.

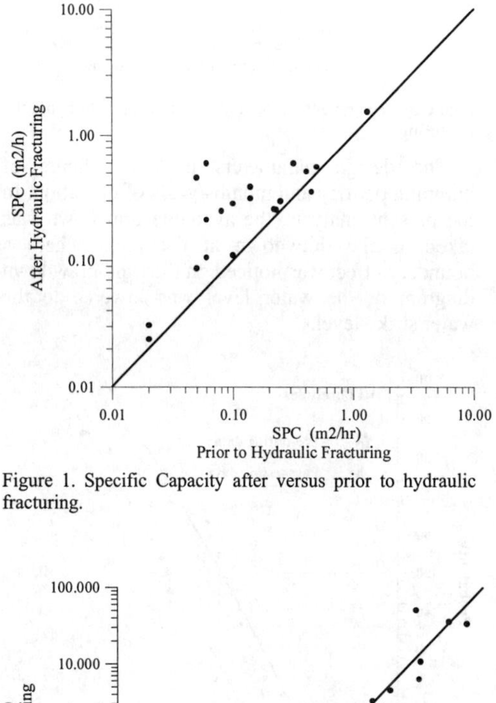

Figure 1. Specific Capacity after versus prior to hydraulic fracturing.

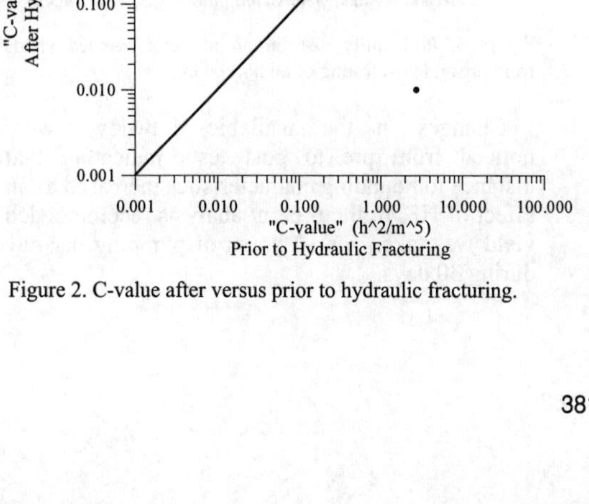

Figure 2. C-value after versus prior to hydraulic fracturing.

Table 2. Median and average value of improvement ratio (IR_C) regarding C-value.

Population	No	Median	Average	Prob.
All boreholes	21	1.8	25	4.4
Partial HF	14	1.6	10	2.7
Fully HF	7	7.7	58	10.6

The improvement ratio obtained from the probability distributions reflects the fact that high improvement ratios were observed in the case of fully hydraulic fractured boreholes.

3.4 *Skin*

Skin relates to the zone within the immediate vicinity of a borehole influenced by drilling, cleaning and development activities. The zone is characterized by a hydraulic conductivity and by an assumed steady state flow during pumping. The head loss due to the skin effect or skin is a linear well loss proportional to the pumping rate. The relationship is expressed through a skin factor, ζ. The skin factor can be positive or negative. A positive factor indicates that a zone of decreased hydraulic conductivity has developed around the borehole. A negative skin factor on the other hand indicates a good communication between the aquifer and borehole through a zone of increased hydraulic conductivity.

Hydraulic fracturing affected the skin factor values. Decreased values in the skin reflect improvement as a result of hydraulic fracturing hence a parameter to consider for evaluation. Figure 3 shows ζ (the skin factor) prior to HF versus ζ after HF. As a general conclusion the skin factor was improved in about 50% of the hydraulic fractured boreholes. No difference was found between partially and fully HF boreholes.

3.5 *Total well loss*

The total well loss, referred as a total skin factor (ζ_{tot}), is expressed as the sum of skin and non-linear losses expressed as:

$$\zeta_{tot} = \zeta + 2\pi T C Q$$

A reduction in the total well loss, taken as an improvement of the borehole, was calculated for a pumping rate of one m^3/h (Figure 4).

3.6 *Transmissivity*

The yield of a borehole is directly related to Transmissivity (T) and as such an increase in T due to HF will directly result in a yield improvement. The transmissivities from pre and post hydraulic fracturing evaluated from Constant Rate Tests (CRT) showed marginal increase in only 30% of the boreholes. A very high increase (>300 times) recorded in one borehole could be due to its favorable structural location.

3.7 *Recommended yields*

Recommended yield is the yield at which a borehole can be pumped on a long-term sustainable basis without deterioration in quantity or quality of water. It was assessed using hydraulic parameters from test pumping (transmissivity, storativities, skin, C-factor, etc) and available drawdown from borehole logging (Carlsson & Ntsatsi, 1998).

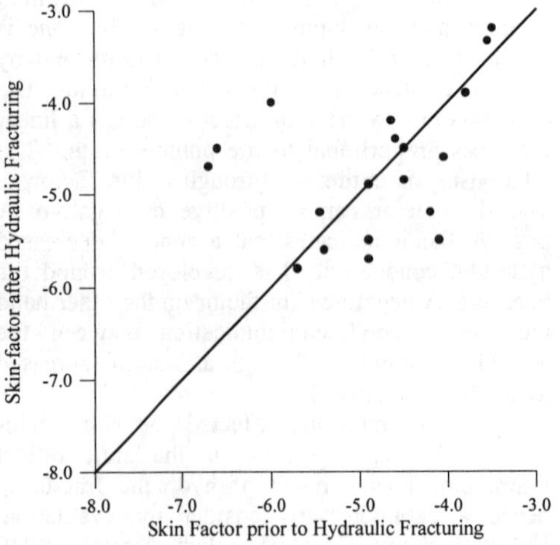

Figure 3. Skin factor after versus prior to hydraulic fracturing.

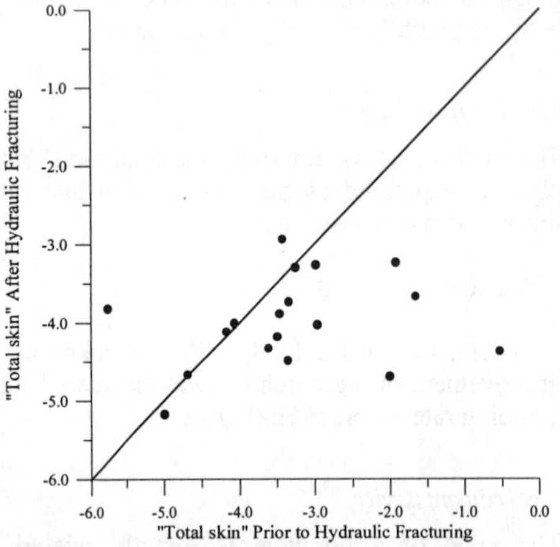

Figure 4. Total well loss expressed as a total skin for abstraction rate of 1 m^3/h after versus prior to hydraulic fracturing.

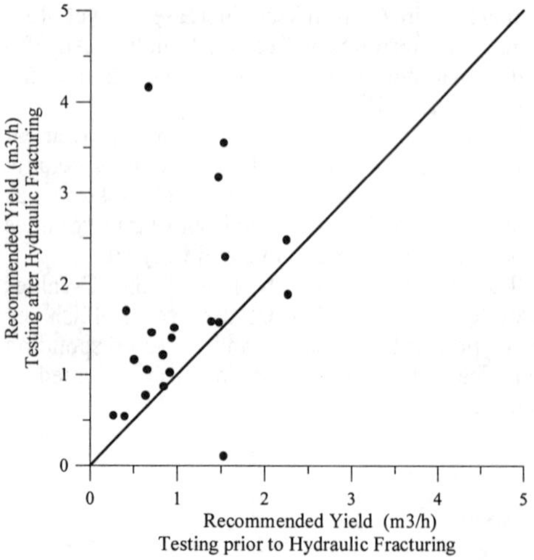

Figure 5. Recommended yield after versus prior to hydraulic fracturing.

The design parameters used are hours of pumping per day and months/years of operation. In the present analysis, the available drawdown was taken as the drawdown at the point where a boundary effect was noticed on the time-drawdown diagram or the water level was lowered to the water strike level.

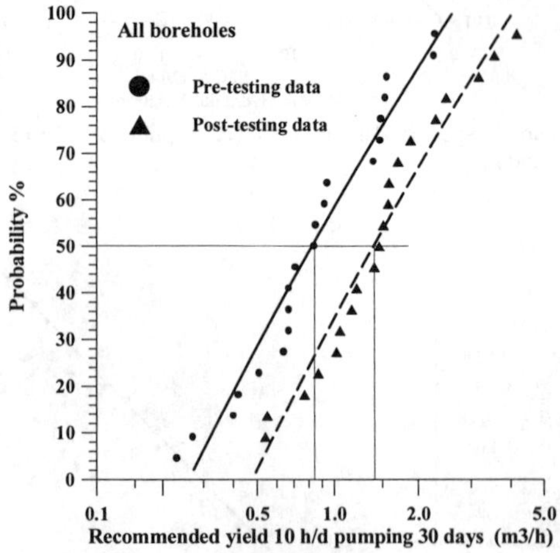

Figure 6. Probability distribution of recommended yields from pre and post testing of all boreholes.

Changes in the available drawdown were noticed from pre to post tests indicating that distance to bounding characteristics increased as an effect of HF. In the present analysis recommended yield was taken for 10 hours of pumping per day during 30 days.

The pre- versus post-recommended yields of the boreholes are illustrated in Figure 5.

It was recognized that the recommended yield calculations included all parameters, which could be effected by HF. An improvement ratio (IR_{rec}) defined as the ratio between the recommended yield after and prior to HF was established to show the success of the fracturing process. Table 3 summarizes the findings.

The distribution of the recommended yields prior to and after HF is presented in probability diagrams (Figure 6, 7 and 8). The relation between the 50% probability values in the diagrams produced a statistically improvement ratio, also presented in Table 3.

The recommended yields were further statistically analyzed and an improvement ratio versus recommended yield prior to hydraulic fracturing was obtained (Figure 9).

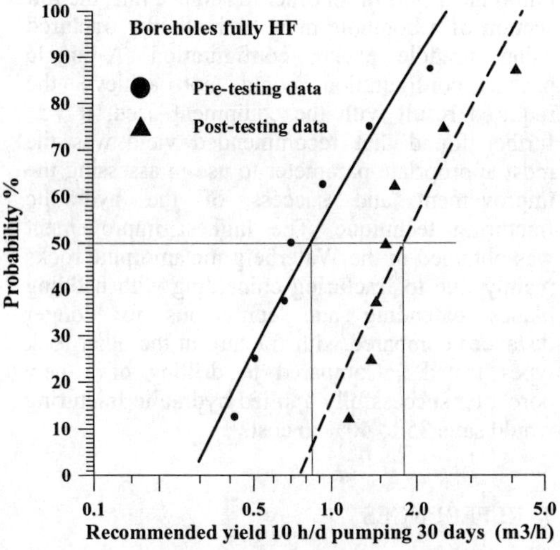

Figure 7. Probability distribution of recommended yields from pre and post testing of completely hydarulic fractured boreholes.

Table 3. Median, average and 50% probability values of improvement ratio (IR_{rec}) regarding Recommended yields.

Population	No	Median	Average	Prob.
All boreholes	21	1.5	1.8	1.9
Partial HF	14	1.4	1.4	1.7
Fully HF	7	2.1	2.6	2.1

4 EVALUATION OF HYDRAULIC FRACTURING PERFORMANCE PARAMETERS

From the evaluation of hydrogeological parameters, it was revealed that improvement ratio of recommended yields was a representative parameter for HF results. IR_{rec} was examined against the injected volume of water and treatment time.

4.1 *Injection of water for hydraulic fracturing*

Table 4 indicates the results of hydraulic fracturing on the basis of volumes of injected water.

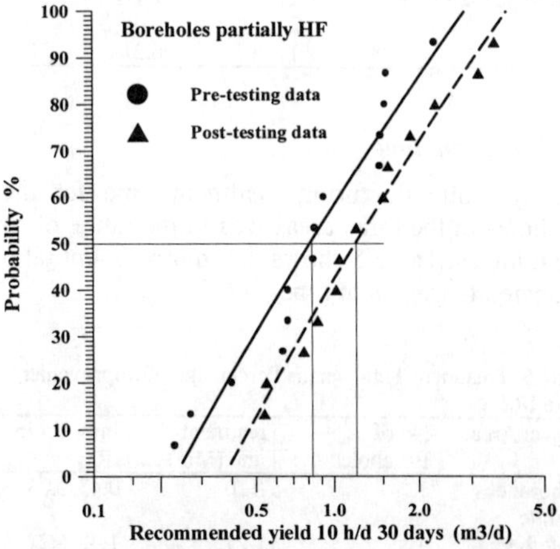

Figure 8. Probability distribution of recommended yields from pre and post testing of partially hydarulic fractured boreholes.

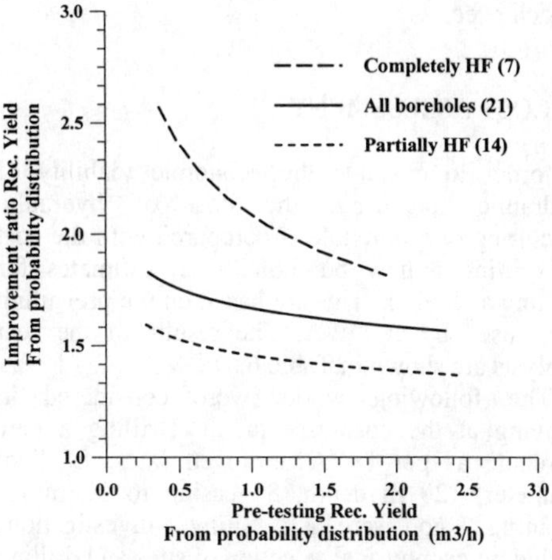

Figure 9. Improvement ratio IR_{rec} versus recommended yield from pre-testing obtained from probability distributions.

The table indicates that high IR_{rec} was achieved by injecting large volumes of water during hydraulic fracturing process.

Table 4 Comparison of Injected Water versus Improvement Ratio IR_{rec}.

Vol. of Injected Water (m^3/borehole)	No of Boreholes	Impr. Ratio IR_{rec}
Less than 5	1	1.02
5-10	2	<1.0
	5	1.0-1.5
	2	1.5-2.0
	2	2.0-2.32
10-15	2	0.43-1.20
15-20	2	1.49-4.05
More than 20	2	1.1-1.14
	2	2.0-2.27
	1	6.21

4.2 *Treatment time*

The hydraulic fracturing treatment time for the boreholes in the three areas was in the range of 5-130 minutes. Table 5 shows the improvement ratio compared to treatment time.

Table 5. Treatment Time versus Percentage of Improvement Ratio IR_{rec}

Project Area	No of Boreholes	Treatment Time (Min)	Impr.Ratio IR_{rec}
Gaborone Granite	12	5-120	0.43-2.06
Waterberg	4	35-215	1.48-6.21
North East Gneisses	5	70-130	1.10-4.05

The results showed that by increasing the treatment period, a higher improvement ratio could be achieved.

5 COST ASSESSMENT

In order to evaluate the economic viability of hydraulic fracturing, the cost of hydraulic fracturing of a borehole is compared with the cost of drilling a new borehole. Cost estimates for drilling and other items are based on the prevailing rates used by at DWA. The results of the cost analysis are shown in Table 6.

The following works were considered in arriving at the cost assessment. Drilling a new borehole as per DWA approved rates, 8" final diameter, 120 m depth, 8" casing to 20 m for grouting, no screens. Siting investigations including geophysical selection of sites and drilling supervision. Test Pumping including SDT and CRT. Hydraulic fracturing of a borehole comprising the total penetrated aquifer, logging, post test pumping, water sample analysis.

It was concluded that a cost saving of approximately 35-40% could be achieved through HF compared to drilling of new borehole.

Table 6. Cost (Botswana Pula 1998) for drilling of a new borehole and for hydraulic fracturing of existing borehole.

Item	New Borehole	Hydraulic Fracturing
Drilling of 120 m borehole	55,000	
Siting of borehole	35,000	
Test pumping	25,000	
Total for new borehole	115,000	
Hydraulic fracturing		40,000
Logging, test pumping, water sampling		30,000
Total for HF of a borehole		70,000

6 CONCLUSIONS

Hydraulic fracturing as a method to improve the yield of a production borehole is usually applied on very low yielding boreholes for single households, small farms etc. For the supply of settlements and villages in Botswana the required rate for a borehole to be considered for production is minimum 1.5 m^3/h. In order to achieve this, the full section of a borehole must be hydraulic fractured using straddle packer configuration. A single packer configuration could not achieve the required result with the equipment used. It was further found that recommended yield was the most appropriate parameter to use in assessing the improvement and success of the hydraulic fracturing technique. The highest improvement was obtained in the Waterberg metamorphic rocks mainly due to fracturing coinciding with bedding planes extending and continuous for longer distances compared with fracture in the other rock types tested. Compared to drilling of a new borehole, successfully applied hydraulic fracturing would save 35 to 40% in cost.

7 REFERENCES

Carlsson, L & Ntsatsi, J. 1998: Village Water Supply in Botswana – Assessment of Recommended yield for Production Boreholes in Semi-Arid Environment. In *Abstract Volume of International Conference on The Role of a National Geological Survey in Sustainable Development, 23rd – 25th June 1998*: 11-15. Gaborone, Botswana.

Department of Water Affairs, DWA 1998. Hydraulic Fracturing. *Final report to DWA prepared by Geotechnical Consulting Services*. Gaborone, Botswana. December 1998.

Jacob, C.E. 1947. Drawdown test to determine effective radius of artesian aquifer, *Trans. Of American Soc. Of Civil Engineers*. Vol 112, paper 2321, pp 1047-1064.

Groundwater: Past Achievements and Future Challenges, Sililo et al. (eds) © 2000 Balkema, Rotterdam, ISBN 90 5809 159 7

Complementarity of electromagnetic prospecting and groundwater modelling to study shield volcano hydrogeology: Piton de la Fournaise volcano case study, Reunion Island

J.L. Folio, J.L. Join, B. Robineau & J. Coudray
Laboratoire des Sciences de la Terre, Université de la Réunion, Saint Denis, France

M. Ritz
IRD, Département Géophysique Appliquée, Dakar, Senegal

ABSTRACT: A geoelectrical survey on Piton de la Fournaise volcano provided a water table mapping. Until geoelectrical model interpretations have been validated though borehole observations, the problem of the hydrodynamical reliability of such result remains. In order to test this reliability, a groundwater flow profile model was developped using the geometrical information provided by geophysical model interpretations. Despite the basic assumptions made to define the conceptual model, the main result of this work is the clear correspondence between the water table map resulting from geoelectrical model interpretations and the one computed at Piton de la Fournaise.

1 INTRODUCTION

Shield volcanoes are considered to be built by fast accumulation of basaltic lava flows. Layers of pyroclastics and volcanic detrital material are intercalated with flows. On all scales, radial and lateral heterogeneity may increase as a result of vertical or horizontal intrusions (sills and dykes) as well as of topographic inversions. Thus, these volcanoes are characterized by a complex, poorly understood, hydrogeological behaviour. Generally, determination of the location of water saturated zones is very difficult. Exploring aquifers through drilling is neither easy nor realistic because the water table is usually very deep (several hundred meters) and as a consequence, boreholes are very expensive. The use of indirect methods of investigation is therefore necessary. In the case of Piton de la Fournaise volcano (Reunion Island, Indian Ocean), extensive geophysical surveys (AMT, TDEM, CSAMT) were conducted because of their ability to overcome these economical and technical limitations. They have provided reliable information about geological structures, existence of impervious layers and in some cases water table maps. Though, the problem of the hydrodynamical reliability of such result remains until it has been validated through borehole observations. The scope of this paper is to test this reliability with the help of a groundwater model. We first present a geoelectrical survey composed of nine controlled source audiomagnetotelluric (CSAMT) soundings across Piton de la Fournaise volcano (Fig. 1). A groundwater flow profile model was developped using the geometrical informations provided by electrical model interpretations. Then, a comparative analysis of the water table map provided by the two methods has been performed.

2 GEOPHYSICAL INVESTIGATION

2.1 *CSAMT method*

Controlled-source audiomagnetotellurics is a frequency-domain electromagnetic (EM) sounding technique which uses artificial and natural E.M. fields to measure resistivity variations in the ground. In our study area, CSAMT measurements were recorded using the Stratagem tensor system of Geometrics.

2.2 *CSAMT data and inferred resistivity structure*

Nine CSAMT soundings have been performed across Piton de la Fournaise volcano. Data analyses show that the subsurface structure can be regionally approximated by a 1D isotropic layered-earth model. The geological section revealed by the rampart above Rivière de l'Est spring gives information which can be use to evaluate the geophysical interpretation. Comparison of the geoelectric layers determined from the CSAMT data and the lithologic log of the rampart, shows a very general agreement. Interpretation results are illustrated in models with four layers :

- A resistive surface layer (2000 to 10000 Ωm) which is typical for dry basaltic rocks and which

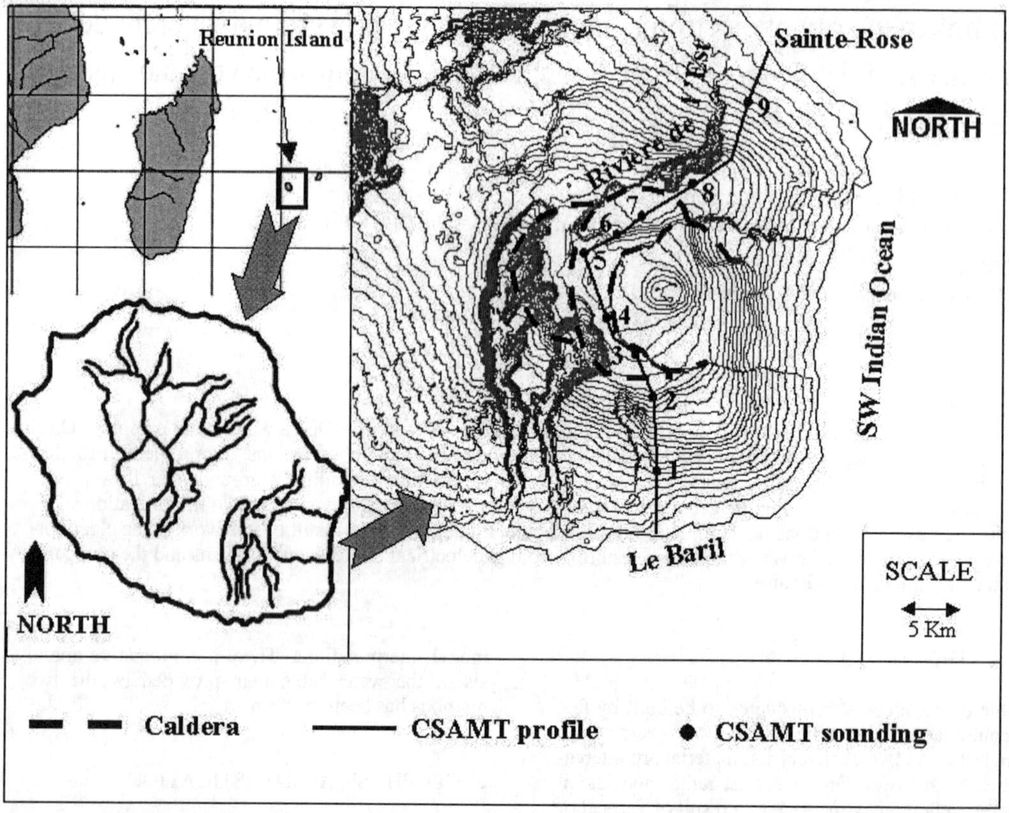

Figure 1 :Topographic map of Piton de la Fournaise massif showing the location of main volcano-tectonic discontinuities and CSAMT soundings. Contour lines are every 100 m.

is identified as the recent unsaturated volcanic series.

- A moderate resistivities layer (100 to 900 Ωm) interpreted in term of water content beneath the water table in the first layer.
- A conductive layer (<100Ωm) corresponding to the top of the old volcanic series.

A very conductive basement (<10 Ωm) which has not been observed on the geological section.

This last layer (substratum) has been already described on the Piton de la Fournaise volcano. The lithology of this conductive basement has been discussed by several authors (Courteaud et al. 1997, Lénat et al. 1999).

As a result, the geoelectrical cross section, presented in Fig. 2, reveals a pattern of smoothly varying structures.

3 GROUNDWATER MODELLING

3.1 *Selection of a groundwater flow model*

In the study area, the elevation varies from 0 to 2300 m ASL. Piton de la Fournaise volcano exhibits a mean surface slope of 15 %. Water table obtained by interpolation of geophysical results is roughly parallel to the surface topography. We assume that the water head distribution is strongly vertically oriented. Therefore, a three dimensional (3D) groundwater flow model is needed to compute flows and heads in Piton de la Fournaise volcano. The modelling has been performed using Femwater (Yeh et al. 1992), a 3D finite elements, saturated and unsaturated, density driven, flow and transport model. This version of Femwater is a special version that is supported by the graphical interface GMS (Owen et al. 1996). The flow circulation is solved for an unconfined aquifer in steady-state condition.

3.2 *Geometry of the conceptual model and hydraulic conductivity distribution*

The geometry of the conceptual model has been determined by taking into account i) the geometry of the volcanic series interpolated from geophysical interpretations, ii) the hydraulic conductivity distribution throughout the volcano. Statistical

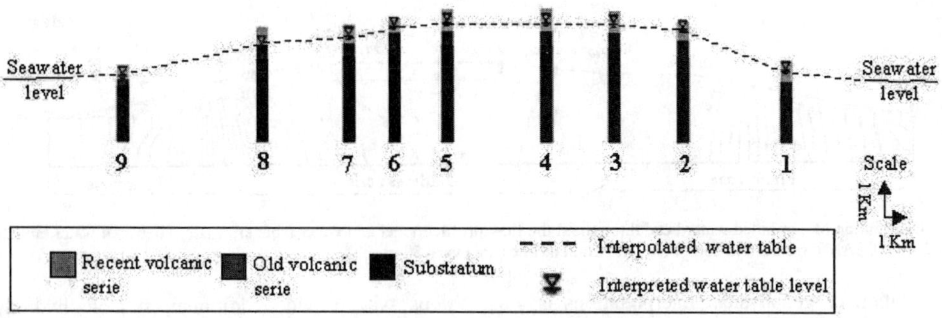

Figure 2 : One dimensional model cross-section built from individual models obtained at each CSAMT sounding

analyses of transmissivity data have been performed from pumping tests and hydraulic conductivities were measured with the help of an easy handling, compressed air permeameter (Folio et al. 1999). They have shown that the hydraulic conductivity parameter is roughly homogeneous at the scale of volcanic series. Thus, we assume that the volcanic series can be considered as hydrogeological layers that can be used in a modelling process.

On the basis of these assumptions, the three geophysical layers have been chosen as hydrogeological layers of our conceptual model. An hydraulic conductivity of 10^{-4}, 10^{-6} and 10^{-7} m s^{-1} is assigned to the recent volcanic serie, the old volcanic serie and the substratum, respectively. Note that the small hydraulic conductivity value of the substratum is arbitrary since this layer has never been observed on outcrop or in borehole. Nevertheless, several authors (Courteaud et al. 1997, Lénat et al. 1999) assume that this substratum constitutes a hydrogeological limit for groundwater.

3.3 *Boundary conditions*

In Reunion island, the climate is intertropical. Rainfall, evapotranspiration, run-off and infiltration are strongly heterogeneous in space and time. Previous studies (Barcelo 1996) have lead to the estimation of each term of the hydrologic balance. These data have been used to compute annual mean infiltration distribution over the entire massif of Piton de la Fournaise volcano. Due to the existence of heavy rainfall episodes and highly permeable subsurface volcanic formations, infiltration is of major importance. In our study area, this parameter varies from 1000 mm $year^{-1}$ to no more than 10000 mm $year^{-1}$.

In order to simplify the resolution of the flow problem, our model does not consider seawater intrusions. Nevertheless, as seawater intrusions basically act as an impermeable flow barrier, exit faces of our model are limited to the first two mesh elements. Exit faces have therefore a thickness of about 100 m. We assume that this value is consistent with regard to low hydraulic gradient usually observed in the coastal area (from 0.2 to 3.2 ‰) and Ghyben-Herzberg assumption. Dirichlet boundary conditions are assigned to nodes belonging to these exit faces.

4 RESULTS AND DISCUSSION

For each CSAMT sounding, the water table error (Er_{wt}) between water table level provided by geophysical interpretations and computed ones was estimated using the following relationship :

$$Er_{wt} = \frac{\left(H_{csamt} - H_{computed}\right)}{L_{aquifer}} \qquad (1)$$

Where H_{csamt} = water table level interpreted from geophysical results, $H_{computed}$ = water table computed by mean of Femwater, $L_{aquifer}$ = thickness of the aquifer layers (recent series and old series).

A mean water table error of 21% (standard deviation is 13 %) was computed with the help of all Er_{wt} values (Fig. 3). This value constitutes a reasonable performance since our model has not been calibrated. Despite the basic assumptions made to define our conceptual model, the main result of this work is the clear correspondence between the water table map, provided by geoelectrical model interpretations, and the computed one at Piton de la Fournaise scale (Fig. 3). Water table errors (Er_{wt}) can be sorted in two categories depending of their spatial localization : the flank areas and the caldera areas :

In the flank areas, water table errors are small with the mean and standard deviation of Er_{wt} both being 11%. The computed water head distribution is consistent with the water table provided by geophysical model interpretations except near sounding 9 (Fig. 1 and 3). The good correspondence between the two water table maps globally shows that the geometry of each volcanic series restituted

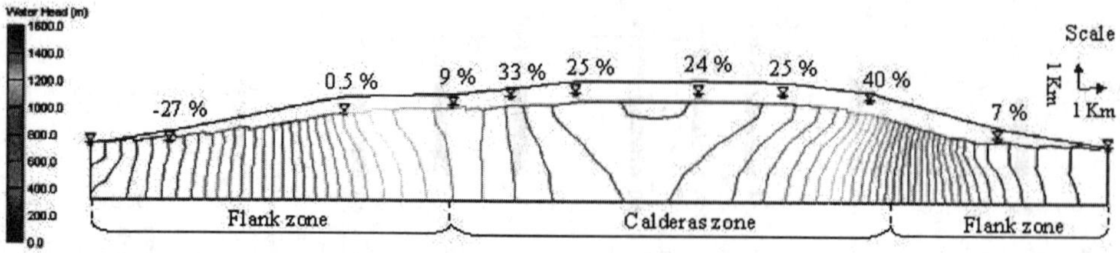

Figure 3 : Distribution of computed water head throughout the flow profile model and water table error (Er_{wt}) between water table level interpreted from CSAMT results and computed one. Contour lines are every 50 m.

by interpolation of geophysical interpretations is compatible with the assumed geological model in flank area. One possible reason for the existence of a strong negative error for sounding 9 is that our conceptual model does not include the existence of a very permeable upper layer in the coastal area (Pumping tests and well specific discharges point to transmissivity values of more than 10^{-2} m^2 s^{-1} for depths of about 50 m). This layer, not detected by geophysical survey seems to have no effect in the case of sounding 1 (Fig. 1 and 3) located near Le Baril coastal area.

In the caldera areas, water table errors are higher than in flanks area with the mean and standard deviation of Er_{wt} being 29% and 7%, respectively. Computed water table levels always underestimate the geophysical ones. Possible explanations for these large errors are the hydraulic conductivity anisotropies at volcanic series scale but particularly the existence of calderas. Our conceptual model does not consider all the parameters mentioned above. Previous interpretations of the hydrogeological behaviour of these calderas have been discussed by several authors (Coudray et al. 1990, Violette et al. 1997). In our study area, two or three calderas may disrupt the geological structure (Fig. 1). Earlier geological observations (Bachelery and Lénat. 1993) have shown that each volcanic series begins with summit calderic collapses into which lava flows have set. Like tectonic faults, these calderas change the distribution of hydraulic conductivity throughout summit regions of Piton de la Fournaise. The average net slip of these calderas is not known. Our geophysical results do not reveal the existence of layer displacements probably because of the low density of CSAMT soundings along the profile.

5 CONCLUSION

As a result, this rough modelling confirms the hydrodynamic reliability of our geophysical profile interpretations. This coupled approach provides a valuable example of the necessity to use different methods in order to better understand the hydrogeological behaviour of heterogeneous and complex volcanic environments. At the scale of Piton de la Fournaise volcano, new geophysical prospections will provide additional, suitable hydrodynamical information for a regional, three dimensional model calibration.

ACKNOWLEDGEMENTS

The funding for this study was provided by the Conseil Général of Réunion Island, as part of a extensive research program of the hydrology of La Fournaise volcano.

REFERENCES

Bachelery P. & J. F. Lénat 1993. Le Piton de la Fournaise. *Mém. Soc. géol. France*, n° 163, 221-229.

Barcelo A. 1996. Analyse des mecanismes hydrologiques en domaine volcanique insulaire tropical à relief jeune. Apports à la connaissance du bilan hydrique. Massif du Piton de la Fournaise (Ile de la Réunion), *Phd thesis*, USTL Montpellier II.

Coudray J. , P. Mairine, E. Nicolini and J. P. Clerc 1990. Approche hydrogéologique du Piton de la Fournaise. Le volcanisme à la Réunion. *Monographie, J. F. Lénat Ed., CRV Clermont-Ferrand* : 307-356.

Courteaud M. , M. Ritz, B. Robineau, J.L. Join and J. Coudray 1997. New geological and hydrogeological implications of the resistivity distribution inferred from audiomagnetotellurics over La Fournaise young shield volcano (Reunion Island). *Journal of Hydrology* : 203, 93-100.

Folio J. L. ; J. L. Join and J. Coudray 1999. Hydraulic conductivities estimation in a recent shield-volcano : Piton de la Fournaise, Réunion Island, Indian Ocean. *Model Care 99, International Conference on Calibration and Reliability in Groundwater Modelling. Coping with uncertainty*, Zürich, Switzerland, 20-23 September 1999.

Lénat J. F. , D. Fitterman, D. B. Jackson and P. Labazuy 1999. Geoelectrical structure of the central zone of Piton de la Fournaise volcano (Réunion), *Bulletin of Volcanology* : in press.

Owen S. J. , N. L. Jones, J. P. Holland 1996. A comprehensive modeling environment for the simulation of groundwater flow and transport. *Engineering with Computers* (1996) : 12 : 235-242.

Violette S. , E. Ledoux, P. Goblet and J. P. Carbonnel 1997. Hydrologic and thermal modeling of an active volcano : the Piton de la Fournaise, Réunion. *Journal of hydrology* : 191(1997), 37-63.

Yeh G. T. , S. S. Hansen, B. Lester, R. Strobl and J. Scarbrough 1992. 3DFEMWATER/3DLEWASTE : Numerical codes for delineating wellhead protection areas in agricultural regions based on the assimilative capacity criterion. *US Environment Protection Agency*.

Groundwater: Past Achievements and Future Challenges, Sililo et al. (eds) © 2000 Balkema, Rotterdam, ISBN 90 5809 159 7

Groundwater conditions in the Upper Shire Valley, Malawi: A modelling synthesis

S.L. Horrocks & W.G. Burgess
University College London, UK

P. Mkandawire
Ministry of Water Development, Lilongwe, Malawi

ABSTRACT: Aquifers in sediments of the Upper Shire Valley of southern Malawi have recently been developed for rural water supply over an area of 1000 km^2. There is concern about sustainability of the supplies, and about vulnerability of the aquifer to drought. A modelling-based study to address these issues incorporates results from the "Kalembo Groundwater Project". Diffuse recharge, which includes river-bed seepage, is about 8% of rainfall. Groundwater flow from the Basement Complex bounding the valley is also significant. Baseflow to the Shire River, and in the dry-season evapotranspiration from marshes adjacent to the River Shire dominate the discharge. As with crystalline basement aquifers, this sedimentary aquifer is spatially extensive but hydraulically discontinuous. Current abstraction is insignificant in a regional sense, but unsuitable aquifer characteristics result in unsustainable yields and poor water quality in places. The model demonstrates the impact of drought, and may be used as a guide to monitoring.

1 INTRODUCTION

The topography and geological structure of Malawi are dominated by the southern part of the Rift Valley, where Lake Malawi and the smaller, shallow Lake Malombe, are situated. The lakes drain southwards to the Shire River, and the Upper Shire Valley extends southwards from Lake Malombe (Figure 1). The valley, approximately 35 km wide, is sharply defined between the eastern and the steeper western Rift Valley escarpments, and is infilled with Quaternary lacustrine and alluvial sediments (Geological Survey of Malawi 1996). Clay-rich "black cotton soils" cover much of the area. Deep dessication cracks form in these soils during the dry season, but disappear during the wet-season, resulting in poor drainage. Ground surface slopes gently from approximately 600 m aOD on the valley margins to approximately 470 m aOD at its centre. The plateau areas beyond the escarpments lie at 900 to 1300 m aOD, and are underlain by weathered Palaeozoic/Pre-Cambrian basement.

The Shire River is 750 m wide on average, with a substantial, though variable, flow – discharge at Liwonde is 420 m^3/s (1960 – 1964 average). Numerous ephemeral streams drain from the plateau areas down the escarpment towards the Shire River. These are dry for most of the year and show rapid response to rainfall.

Prior to 1995, provision of protected water supplies in the Upper Shire Valley was limited, and many communities relied on shallow hand dug wells. Water demand has remained relatively small for many years, due to the modest standard of living. In Malawi, normal water use is estimated to be <10 l/person/day, increasing to 10 – 25 l/person/day where there is a reliable source within a few metres of the home (as previously estimated by Smith-Carrington and Chilton 1983).

A programme of borehole drilling and hand-pump installation in the Kalembo and Liwonde Traditional Areas of the Machinga District, within the Upper Shire Valley, between 1995 and 1998 (the "Kalembo Groundwater Project", KGP) aimed to provide rural communities with safe drinking water supplies (Hankin 1998). The projected population of the study area for the year 2000 is 121,000. Design targets of 240 people per borehole, with a maximum walk of 500 m to collect water, were set. 388 boreholes were drilled during the project, of which 68 were dry and abandoned. There are now 380 operational boreholes within the KGP area, lined with 110 mm diameter PVC casing and screen, and fitted with Afridev hand-pumps. Typically, yields are little more than 0.2 l/s, and average daily abstraction is about 5 m^3 per borehole. It is estimated that 75% of the population of the KGP area is served to the standard of the design criteria (Hankin 1998).

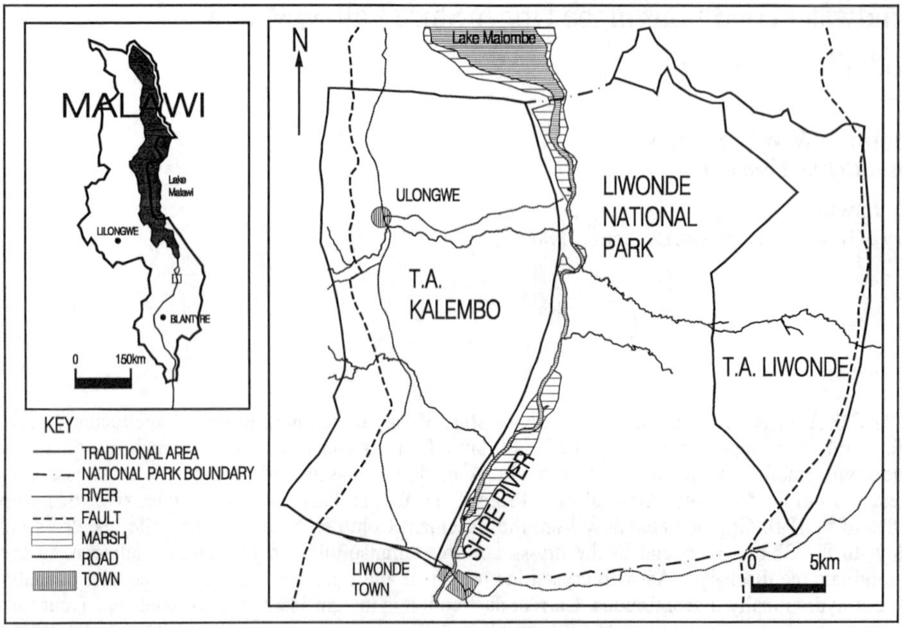

Figure 1. Location Map

Rainfall records indicate a number of droughts during the 1980s and 1990s, with annual rainfall being less than half the long-term average during this time. Following the failure of the 1991-1992 monsoons, Malawi suffered its worst drought in over 40 years. The impact of this drought on groundwater supplies was profound but difficult to evaluate due to lack of data. Health statistics record outbreaks of diarrhoea, cholera and shigella dysentery. Large numbers of boreholes failed, but to what extent this was due to falling groundwater levels rather than maintenance problems (exacerbated by increased use) is unclear. Although drought is a recurring event in this part of Africa, institutional planning for this situation is very limited (Calow et al. 1998). An integration of the large amount of data collected during the KGP was prompted by concern over the sustainability of the groundwater abstraction, the vulnerability of the aquifer to drought and long-term interest in the possible extension of water supply provision in the region.

Data available to this study included basic geological logs, water levels measured at the time of drilling, major ion analyses and pumping-test data for the 380 boreholes constructed during the KGP. The standard approach of the KGP to aquifer testing included a constant discharge test of 4 to 8 hours duration, with drawdown and recovery monitored in the pumping borehole. Meteorological data for stations within and close to the study area were available, as well as river flow data for the Shire River immediately downstream of the study area, and for one ephemeral tributary within the study area. The quality and consistency of the data are variable. Borehole design was often not suitable for the results of the pumping tests to be valid for analysis of aquifer permeability, for example. Previous knowledge of the aquifer system (Msonthi and Chivunga 1987) was largely generalized for the whole of the Rift Valley, and based on very limited hydrogeological data. Nevertheless, an astute assessment of groundwater resources for the whole of Malawi, produced by Smith-Carrington and Chilton (1983), furnished valuable background information. In this paper we present an integration of the large amount of data gathered during the KGP, and consider the implications for future monitoring, the vulnerability of groundwater supplies to drought, and the potential for further development.

2 THE AQUIFER SYSTEM

Boreholes exploit groundwater in the Quaternary sediments of the Upper Shire Valley, which were deposited under environments similar to those of today. The sedimentary sequence thickens markedly by over 100 m between the valley margins and the centre, but the overall geometry of the basin is poorly known, as few boreholes penetrate the full thickness of sediments. Periodic expansion of Lakes Malawi and Malombe has resulted in extensive layers of silt and clay, and evaporative episodes have led to localized salt pans in the central areas.

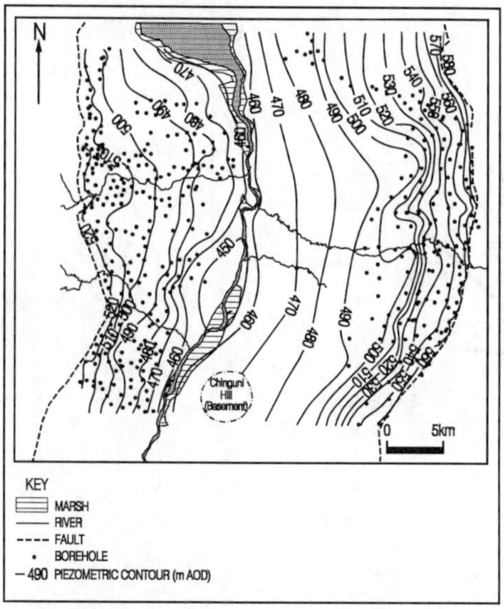

Figure 2. Observed Piezometry

Tributaries flowing from the plateau areas have deposited coarser sediments – fine sands to gravels - along the gentle slopes of the valley floor. Ribbons of interconnected channel sands so deposited with a west-east orientation are thought to comprise the aquifers of the Shire Valley alluvial sequence.

Pumping-tests indicate that the sediments act as a leaky system of aquifers and intervening aquitards. Connectivity between aquifer units is variable. Combining analysis of the pumping-tests with grain size determination and lithological observation suggests an overall hydraulic conductivity in the region 10 to 35 m/day, higher in a west-east than north-south direction due to the orientation of the channel sand deposits.

Piezometric contours (Figure 2) indicate groundwater flow from the margins of the valley towards the Shire River and Lake Malombe. Hydrochemical trends of generally increasing salinity towards the centre of the valley are consistent with this. Locally, extremely saline Na^+-HCO_3^- type groundwater (up to 8140 μS/cm) occurs close to the Shire River. At approximately 10% of boreholes sampled by the KGP, groundwater has electrical conductivity (E.C.) of over 3000 μS/cm. We postulate that evaporite (trona) deposits occur in the central region of the valley.

Available hydrological and hydrogeological data have been interpreted to estimate components of the water balance and hydraulic characteristics of the aquifer system (Table 1).

Throughflow from the basement rock complex adjacent to the valley is thought likely to augment recharge from rainfall infiltration. Infiltration is also likely from the beds of the ephemeral streams which drain from the escarpments and cross the valley floor, although there are no data by which to quantify this. Enhanced levels of dissolved silica in some places suggest there is a geothermal component to the flow system, but by analogy with the Kenya Rift Valley (Allen et al. 1989) the geothermal contribution is thought to be very small. An illustration of the aquifer system and a conceptual description used as a basis for the groundwater flow model are given in Figure 3.

3 A SIMPLE SINGLE-LAYER MODEL OF GROUNDWATER FLOW

A simple model of groundwater flow in the Shire River alluvial aquifer has been developed using the code FLOWPATH (Franz and Guiguer, 1991) to enable a preliminary evaluation of the available data

Table 1. Water balance & model parameters for the Upper Shire Valley aquifer system (see text for explanation)

Component / variable	Estimated value	Value in calibrated model
Direct recharge (mm/year)	rainfall infiltration: 8 – 61 mm	66 mm; 2 mm in extreme east (see river-bed seepage below)
Heads driving basement throughflow	520 m (west) 570 m (east	as estimated
River-bed seepage	unknown	incorporated in direct recharge
Measured baseflow to Shire River	488 – 3760 m^3/day per km length	4000 – 8000 m^3/day per km stretch of river (also see evaporative losses)
River Shire riparian evaporative losses	5140 m^3/day per km length	incorporated as discharge with baseflow to River
Hydraulic conductivity, K (m/day)	10 – 35 (alluvial sands) 1.0 – 2.8 (basement)	K_x 20 m/day; 1.2 m/day in extreme east K_x/K_y : 2
Alluvial aquifer thickness (total sand thickness)	0 – 50 m	20 m

Rainfall recharge is based on river hydrograph analysis of the Lower Shire Valley by Smith-Carrington and Chilton (1983)

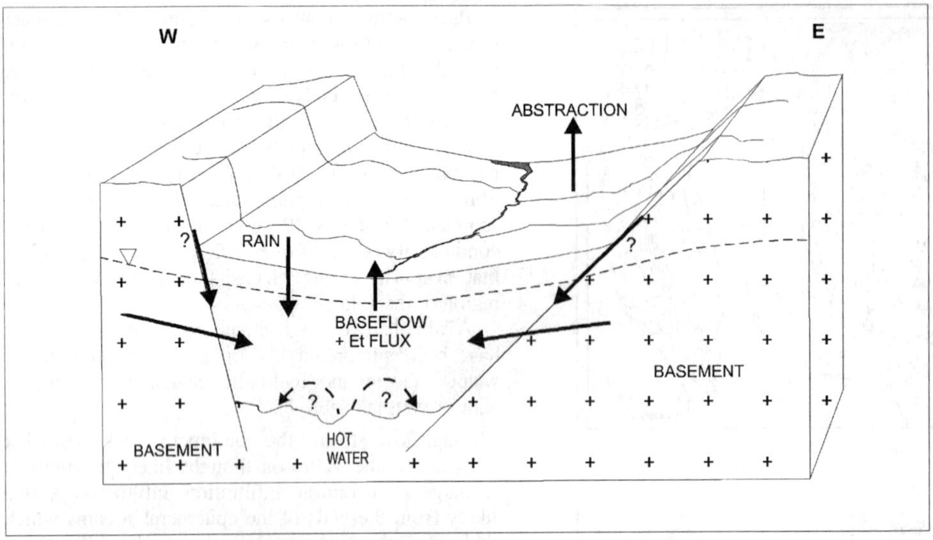

Figure 3a. The Upper Shire Valley alluvial aquifer system

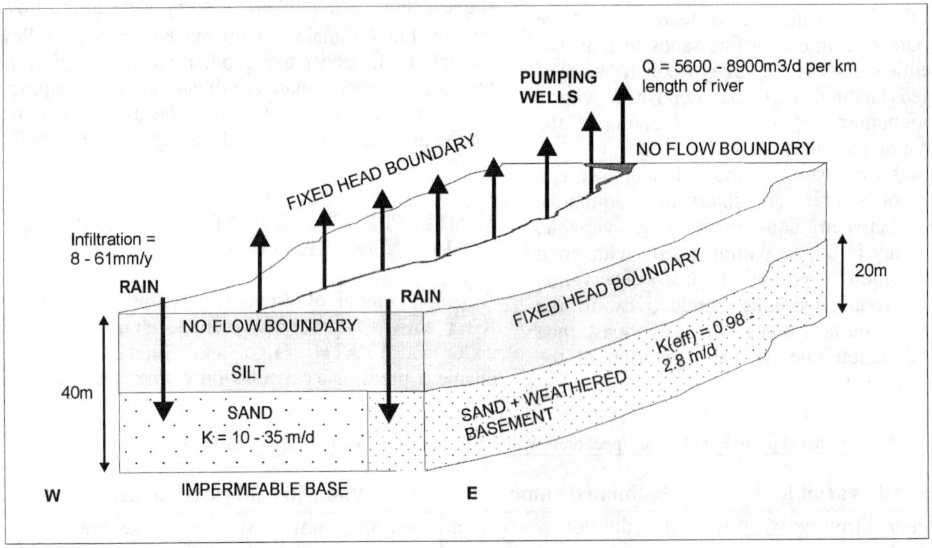

Figure 3b. A simplified conceptual model of the aquifer system

and to guide management of the groundwater resource (Horrocks 1998). The geometry of the aquifer is unknown in detail and minimal transient data are available, therefore a single layer steady-state model has been developed. Water-level observations demonstrate confined conditions and refer to the upper 40 m of the sedimentary sequence, in which coarser sand layers make up approximately 50% of the total. Therefore the aquifer was represented as a semi-confined layer of 20 m thickness. The quality of the available data do not warrant a more detailed representation of the aquifer geometry. An orthogonal grid of 500 m spacing was used, over a model domain of ca. 1000 km^2. The northern and southern limits of the modelled region approximate to groundwater flow lines and are represented as no-flow boundaries. The western and eastern limits of the valley are represented as fixed head boundaries to allow recharge of the Shire Valley aquifer from the adjacent Basement Complex.

Values for the water balance and model parameters are listed in Table 1. The availability of data limited

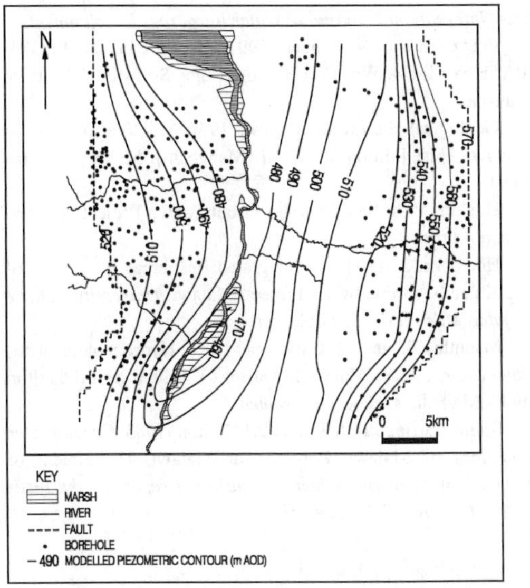

Figure 4. Modelled Piezometry

the model representation of diffuse recharge to a combination of rainfall infiltration and river-bed seepage. Groundwater discharge as baseflow to the River Shire and evaporative losses from the marshes adjacent to the river were jointly represented as a series of abstractions along the line of the river. Estimates of both discharge routes were based on assessment of available data, but are poorly constrained and introduce considerable uncertainty. Hydraulic anisotropy of the sedimentary aquifer was incorporated by assigning a K_x / K_y ratio of 2 over most of the region. This is a pragmatic approach to acknowledge an expectation of anisotropy in permeability of the channel sands. A calibrated K_x value of 20 m/day represents an aquifer transmissivity of 400 m^2/day, in broad agreement with the majority of pumping test results. In the easternmost part of the region K_x was assigned a lower value, 1.2 m/day, to account for the considerably lower transmissivities where the basement is shallow. 1.2 m/d is close to the overall permeability for a geological sequence comprising 50% sand and 50% basement, determined by the harmonic mean of 0.5 m/d and 20 m/d.

The model was calibrated against piezometric data from 40 boreholes evenly distributed over the region, taking as an additional constraint a minimum groundwater level below the Shire River of 440 m aOD, to justify continuous evaporative losses represented by abstraction nodes along the river. Calibration was by adjustment of recharge throughout the domain, and by reduction of recharge and transmissivity in the easternmost regions where basement is shallow or exposed. Simulated groundwater levels in the calibrated model (Figure 4) were within 10 m of calibration targets in 72% of boreholes (Horrocks 1998), a standard that must be considered in the context of significant inaccuracies in recorded water levels, borehole locations and published topography. A water balance error of –3.2% is acceptable, and due in part to dry-season groundwater levels being used as calibration targets, but rainfall recharge and groundwater discharge to the river being treated as annual average values.

There are two areas of poor calibration, in the central-west and south-east of the region. In the former, marshlands may be more extensive in reality than on published maps, and greater evaporative losses would improve the calibrations in this area. In the latter area of poor calibration, numerous small hills of exposed basement occur which are not incorporated into the model in detail.

4 RESULTS AND IMPLICATIONS OF MODELLING

The model represents a pragmatic approximation of a complicated aquifer system. Complex geometry of the aquifer is simplified, vertical flows and localized indications of geothermal activity are ignored, and the model has many limitations imposed by the scarcity of available data and the poorly constrained mechanisms of recharge and discharge. Nevertheless, results lend support to our conceptualization of the aquifer system. Observed groundwater conditions are represented in the model with reasonable accuracy over most of the area.

Total groundwater abstraction is only 1% of total recharge, of which 65% is diffuse recharge and 35% is lateral inflow from the adjacent regions of crystalline basement. The modelled groundwater levels are most sensitive to diffuse recharge. The mechanisms and timing of rainfall infiltration are poorly known in this area of black cotton soils. Yet diffuse recharge is only 8% of annual rainfall in the calibrated model, not unreasonable in this environment. We postulate that the first rains of the wet season result in significant recharge before the soils swell and restrict further drainage. Regionally, the current rates of abstraction are sustainable, and there is scope for extending groundwater-based rural water supply. In some areas, improved borehole design and installation would allow higher yields, but the aquifer characteristics are highly variable and in places are unsuitable. Siting high-yielding boreholes remains a challenge, and vulnerability to drought is a major concern.

Drought conditions were simulated in the model by a combination of (a) halving the direct recharge, to simulate the effects of reduced rainfall and river-bed seepage (see Section 1 above), (b) reducing the fixed head conditions along the escarpment boundaries by 10 m, to simulate the effect that reduced rainfall (Smith-Carrington and Chilton 1983) would have on the water table in the basement rocks with a drainable storage of 0.1, and (c) maintaining a fixed head condition of 450 maOD along the River Shire and around Lake Malombe, to simulate a fully stressed aquifer state, based on an estimate of the maximum depth of influence of evapotranspiration in fine-grained alluvial sediments. Two drought simulations were run, one combining modifications (a), (b), (c) and one solely (b) and (c) while maintaining direct recharge as in the calibrated model. The latter case was to allow for the possibility that reduced rainfall under drought conditions contributes to recharge via the deep dessication cracks observed in the black cotton soils of the region. Groundwater levels fall by between 11 m and 14 m on average in response to these changes. Larger declines occur in the east, where depth to basement is shallower and the valley floor slopes more steeply than elsewhere. These simulations of drought conditions using a steady-state model are indicative only. However they demonstrate that the impact of a serious and prolonged drought could be that pumps would fail in around 7 to 22 % of boreholes and 4 % of boreholes might become dry.

The Rift Valley alluvial aquifers have been thought to be less vulnerable to drought than the shallow aquifers developed on the crystalline basement, but our assessment indicates that the impact of drought may nevertheless be extremely serious. The model highlights the importance of a better understanding of the mechanism of rainfall recharge, the scale of inflow from the escarpment, and the extent of evaporative losses from the central marshes. Reliable transient data of groundwater levels are required, particularly in the regions more vulnerable to the effects of drought.

ACKNOWLEDGEMENTS

The Kalembo Groundwater Project was funded by the German Bank for Reconstruction (KFW), and implemented by Gitec-Consult between 1995 and 1998. We are grateful to the Malawi Ministry of Water Development and to Gitec-Consult for making available the data they collected as part of the Project, and to Phil Hankin of Gitec-Consult for his helpful advice. Sarah Horrocks acknowledges provision of an Advanced Course Studentship from the UK Natural Environment Research Council.

REFERENCES

Allen, D., W.G.Darling & W.G.Burgess. 1989. Geothermics and hydrogeology of the southern part of the Kenya Rift Valley with emphasis on the Magadi-Nakuru area. *British Geological Survey, Research Report SD/89/1.*

Calow, R., N.Robins, A.MacDonald, D.MacDonald, B.Gibbs, W.Orpen, P.Mtembezeka, A.Andrews & S.Appiah. 1998. Groundwater management in drought-prone areas of Africa. *International Journal of Water Resources Development.*

Franz T. and N.Guiguer. 1991. User's Manual, FLOWPATH version 4. Waterloo Hydrogeologic Software, Waterloo, Canada.

Geological Survey of Malawi. 1996. Geological Atlas of Malawi, First Edition, Sheet III (Mangochi). Malawi Government.

Hankin, P. 1998. Kalembo Groundwater Project. *Unpublished report.*

Horrocks, S. 1998. An Integrated Resources Asessment of the Kalembo Groundwater Project, Malawi. *University College London MSc thesis (Unpublished).*

Msonthi, B. and C.Chivunga. 1987. Hydrogeological reconnaissance map: Mangochi (Sheet 6). Edited by P.J.Chilton and S.Mainala. *Malawi Government.*

Smith-Carrington, A.K. and P.J.Chilton. 1983. Groundwater resources of Malawi. Republic of Malawi, Department of Lands, Valuation and Water. *Unpublished report by the Institute of Geological Sciences, U.K.*

Groundwater: Past Achievements and Future Challenges, Sililo et al. (eds) © 2000 Balkema, Rotterdam, ISBN 90 5809 159 7

The interaction between coastal lakes and the surrounding aquifer

B.E. Kelbe & T. Germishuyse
Department of Hydrology, University of Zululand, KwaDlangezwa, South Africa

ABSTRACT: The freshwater coastal lakes in the Richards Bay area of South Africa are the main water resources for domestic, mining and industrial use in this area. Utilization of these resources has reached threshold levels that require management intervention through frequent restrictions. The increasing demand on these water resources necessitates further study of the dynamics of the system to develop a greater understanding of the lake systems and their dependence on groundwater. Maximum rates of interaction were simulated by assuming no sedimentary restriction through a constant head boundary where flow dynamics were controlled by the hydraulic characteristics of the aquifer. The inclusion of sedimentary control was conducted through the application of the lake module. The groundwater model was calibrated against borehole observations while the lake-groundwater model was calibrated, for present conditions with known abstraction, against observed lake levels. The model simulations of flow conditions for Lake Mzingazi, including municipal and industrial use, have been used to understand the relative importance of geophysical features of the system. The model simulations have also been used to conduct a frequency analysis and derive the water balance components of the lake to determine the relative importance of the local primary aquifers in sustaining the lake during periods of flooding and drought.

1 INTRODUCTION

Water resources in most areas of South Africa are very limited. Along the east coast of Southern Africa, there is heavy reliance on the coastal lake systems for freshwater supply. These natural lake systems have been over exploited and many restrictions have been placed on their abstraction limits in recent times. Very little is known about the dynamics of these systems which are believed to be extensions of the local aquifer. Exploratory studies of several lakes in the region have been undertaken by Kelbe & Germishuyse (1999). This study used numerical models to estimate the system dynamics of Lake Mzingazi, one of the principle water supply lakes in the Richards Bay area.

The study area covers part of the primary aquifer that stretches along the east coast of South Africa (Figure 1). It is part of an extensive low lying plain that stretches along the coast of Mozambique. The flat coastal plains consist mainly of highly permeable sands that encourage rapid recharge to the shallow aquifer. Exposed wetlands and lakes are abundant in the region, many of which are remnants of estuarine systems that have become closed during recent geological periods forming freshwater lake systems. These freshwater lakes, because they are generally considered to be surface extensions of the primary aquifer, are likely to exhibit varying flow through dynamics (Townley et al. 1993).

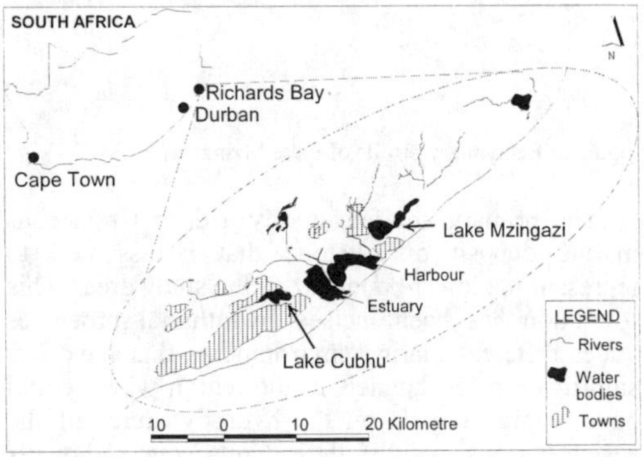

Figure 1. Location of the study area

The South African National Water Act (1998) requires specification of a reserve on all water resources to sustain their basic social and aquatic ecological function. For the coastal lake systems, their natural water level fluctuations are related to their ecological function. However, a recent study to determine the Lake Water Level Requirements (LWR) of coastal lakes, highlighted the complexities in determining the dynamics of the systems and the importance of the interaction between these lakes and the surrounding aquifer (Kelbe & Germishuyse 2000).

A numerical study of Lake Mzingazi is presented to examine the concepts and assumptions readily employed in studies of dynamic lake system in the region where there is very limited information. The study used MODFLOW, a finite difference model described by McDonald & Harbough (1983), with several additional modules simulating internal boundary conditions that describe the interaction between the surface and groundwater components.

2 STUDY AREA

Lake Mzingazi is a natural freshwater lake system with a raised weir at the outlet. The lake comprises two compartments that have been isolated during extreme drought conditions (Figure 2). The deepest compartment extends 14 m below mean sea level (msl) and is protected from saline intrusion by maintaining a freshwater wetland between the lake and the saltwater estuary (Rawlins et al. 1977).

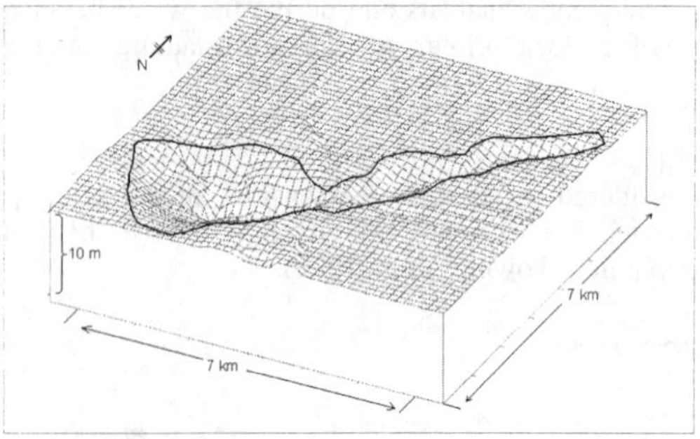

Figure 2. Bathymetry profile of Lake Mzingazi

The primary aquifers overly a deep Cretaceous marine deposit of siltstone that is assumed to represent the lower boundary of the study area. This formation has been incised by alluvial processes under different marine environments that have left significant paleochannels in the region which could have a large impact on the hydrodynamics of the system. Consequently, these Cretaceous sediments have been mapped as accurately as possible by Germishuyse (1999) using all available borehole data and interpolated surfaces from previous geological and geophysical studies. The upper surface elevation for this layer is shown in Figure 3 and indicates a deep paleochannel running through the deepest part of Lake Mzingazi.

Overlying the Cretaceous siltstone is an intermittent layer of coarse sandy calcareous conglomerate of the Miocene age. The spatial extent of these deposits was estimated from an inadequate borehole network by Germishuyse (1999). The location of the borehole network and the spatial extent of these formations are shown in Figure 4. The deposits are intermittent layers that generally occur on the higher regions of the Cretacous sitlstone formation.

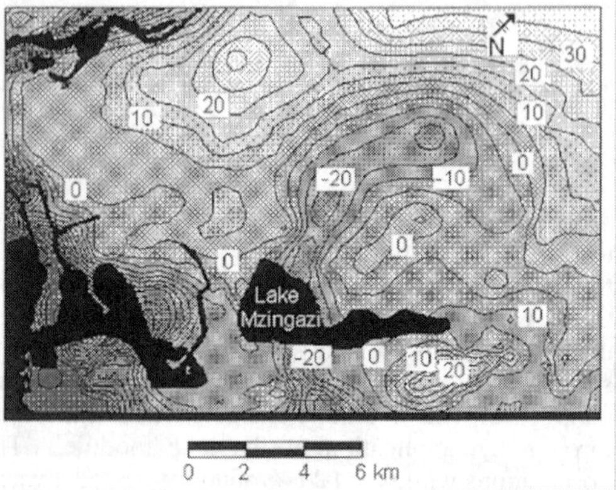

Figure 3. Base elevation of the primary aquifer

The Cretaceous and Miocene deposits are overlain by more extensive layers of clayey sands of marine origin and more recent aeolian cover sands that are highly permeable and promote rapid recharge. Efforts to map these layers has been inhibited by the lack of suitable borehole information.

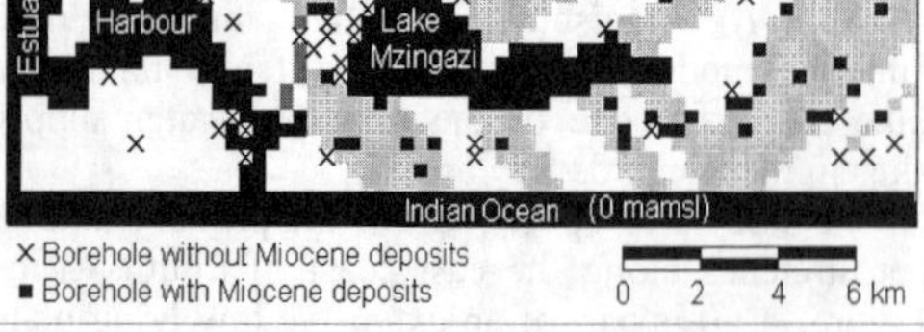

Figure 4. Intermittent layers of calcareous conglomerates of the Miocene age (grey areas)

3 NUMERICAL MODEL

The numerical method chosen for this study was MODFLOW. This is a three dimensional finite difference model that is constrained by specified boundary conditions.

A lake package developed by Council (1997) was incorporated to simulate a head dependent boundary that responds to surface recharge and groundwater seepage through lake bed sediments. This package simulates fluctuating lake levels and computes a lake water budget by balancing the changing volume with multiple inflow and outflow processes. The degree of lake-groundwater interaction is controlled by the hydraulic gradient and the lake bed sediment conductance which had to be estimated from calibration studies.

4 CONCEPTUAL MODEL

All the stratigraphic layers of the aquifers have high sand content and they are generally less than 30m in thickness. For a model grid spacing of 333m x 333m, it was assumed that the regional groundwater flow dynamics could be simulated as a single unconfined aquifer, above the Cretaceous (Kelbe & Germishuyse 1999). Further studies have shown that a multi layer model is needed to account for the effect of paleochannels and the more permeable conglomerate deposits in the immediate vicinity of the lake.

The study area is constrained by the Indian Ocean, the Mhlathuze Estuary (both at 0 mamsl), Lake Nsezi (6 mamsl) and the Nseleni river (6-15 mamsl) which were configured as constant head boundaries on three sides of the model domain (Figure 4). The other side of the model domain was assumed to be a no flow boundary.

Vertical flow boundaries were configured to simulate the effect of land use through evaporation from the shallow water table. The rooting depth that regulates evapotranspiration from the capillary zone (for vegetation in hydraulic contact with the water table) was varied according to the land use. The vertical recharge rate was assumed to be uniform over the entire spatial domain but was allowed to vary over time. The recharge and evapotranspiration rates were estimated using a crude rainfall-runoff model. This model accounted for surface evaporation, interception, and runoff, but allowed for additional evapotranspiration when plant roots were in hydraulic contact with the water table (Kelbe & Germishuyse 1999).

5 MEASUREMENTS

The water level elevation for Lake Mzingazi as well as the discharge rate through a sharp crested rectangular weir, have been monitored since 1975. No measurements of stream flow into the lake have been recorded and rainfall is only monitored several kilometers to the northeast of the lake.

There were very few boreholes in the model domain where the water table elevation was documented and where the hydraulic characteristics of the aquifers could be determined or estimated. Consequently, all the hydraulic parameters were estimated from studies in other parts of the region and only a few boreholes were available for calibration studies.

6 CALIBRATION OF THE AQUIFER MODEL

The lack of surface and groundwater observations and measurements in the region restricted the initial application of MODFLOW to a single layer model because the relative depth of the aquifer was much less than the grid spacing. A uniform hydraulic conductivity of 10^2m/day was used for the one layer model. This value was obtained by fine adjusting of the steady state model to achieve the best fit between the observed and simulated water table elevation for the few dispersed borehole measurements (Germishuyse 1999). These water table measurements were assumed to represent normal climatic conditions but may have been obtained under different climatic conditions.

The aquifer model was then calibrated against three observation boreholes in the immediate vicinity of the lake. The transient flow model parameters were adjusted to produce a similar fluctuation to the observations shown in Figure 5.

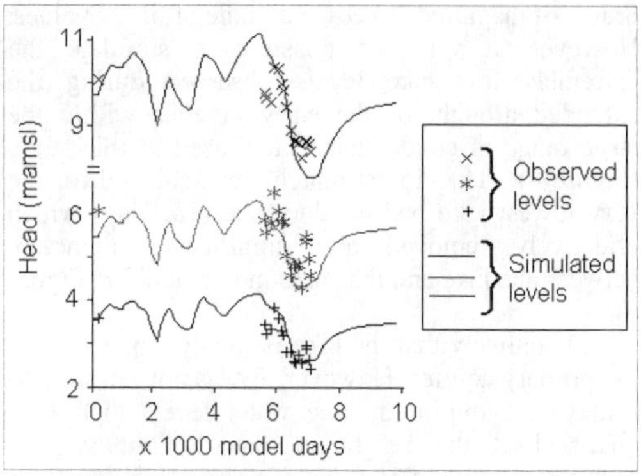

Figure 5. Comparison of observed water table elevation and simulated levels for three boreholes near the lake.

7 LAKE MODEL CALIBRATION

The lake package of Council (1997) was configured for the one layer groundwater model where the surface groundwater interaction would be controlled by the leakage rate through lake bed sediments. This lake leakance parameter (or conductance) defines the level of interaction between the aquifer and the lake.

For very large conductances the lake would act like an unconstrained boundary which would be controlled by the aquifer properties and hydraulic gradients. For very low conductances, the lake behaves like an isolated system dependent only on the surface recharge and discharge conditions.

The area of each cell was known but the thickness and the hydraulic conductivity of the lake bed sediments are unknown. Calibration studies were used to derive a best estimate of these parameter values using observed lake levels (Figure 6) during extreme hydrological events.

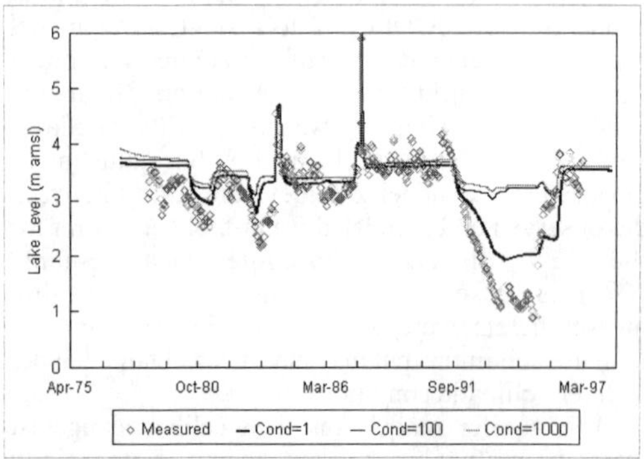

Figure 6. Calibration runs with different lake bed conductances (m^2/day) and measured lake levels

The lake bed conductance was varied over three orders of magnitude to cover a range of likely values. However, it was not possible to simulate the extremely low lake levels observed during the extended drought of the early nineties within the large range of conductance rates used in this study (Figure 6). The closest match was achieved for the very lowest lake bed conductance (1m^2/day) which effectively removed any significant interaction between the lake and the surrounding aquifer (Figure 6).

It is unlikely that the lake is totally isolated from the primary aquifer. However, it was not possible to achieve the minimum lake water levels that were observed at the height of the drought with a conductance rate of 1m^2/day. To derive a better fit to the observation a range of unknown water loss factors (between 2 000 m^3/day and 10 000 m^3/day) were introduced into the calibrations in order to achieve a more reasonable level of agreement between simulated and observed lake levels (Figure 7).

The arbitrary loss factor needed to produce the agreement in Figure 7 is approximately the same order of magnitude as the abstraction by local authorities. While there is still some uncertainty in the total water abstraction rate and the use of water by riparian users, this unaccountable water loss factor is considered to be uniform and too large to be attributed to these users. Consequently, other factors needed to be identified as probably sinks within the system.

It is expected that the surface-groundwater interaction is significant and needs to be simulated with a conductance value of at least 1000 m^2/day. To achieve better comparisons of lake levels, it would be necessary to induce greater losses from the system through subterranean systems, hence the need for additional hydraulic layers..

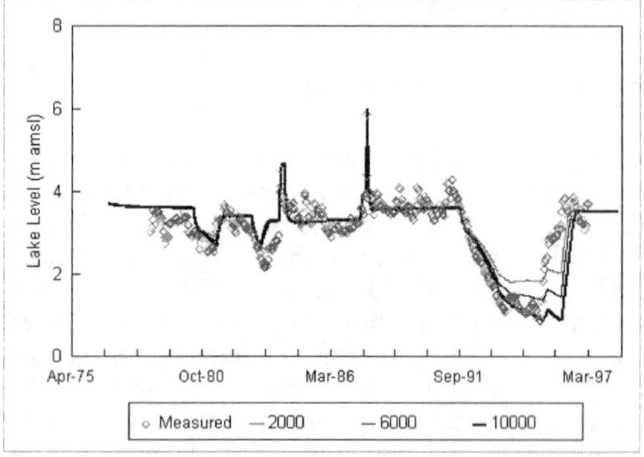

Figure 7. Calibration runs using additional losses (m^3/day)

8 TWO LAYER MODEL

An additional layer was configured for the aquifer model to represent the Miocene conglomerate deposits which are known to have a higher hydraulic conductivity value (25m^2/day). This layer is linked to the lake and it is speculated that it acts as a conduit which drains water from the lake directly into the Indian Ocean. For this model the lake bed conductance was increased by an order of magnitude to 10 m^2/day (within the original range indicated in Figure 6). For this simulation, no unknown loss factor was necessary to achieve the comparative

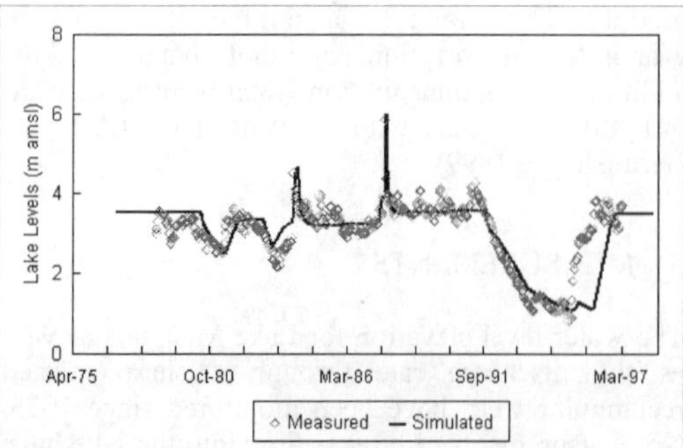

Figure 8. Simulated lake levels using the two layer model

agreement with lake levels shown in Figure 8 during the extreme drought period from 1993 - 1995.

9 MAXIMUM SEEPAGE RATES

The seepage rates derived from the above studies indicate the best estimate using available information. To estimate how these rates compare to maximum, unconstrained conditions, the conductance parameter was increased by an order of magnitude to 100 m^3/day. The lake was also simulated as a constant head boundary where there is no restrictive layer between the aquifer and the lake. The constant head values were derived from observed lake levels which were adjusted for average conditions in each stress period to account for major fluctuations during extreme events. The three simulated series are shown in Figure 9.

There is very little difference between the constant head simulation and the lake model with a highly conductive sedimentary layer (100 m^2/day) for the period up to the floods of 1987. The seepage rates for the restrictive sedimentary layer model (conductance value of 10 m^2/day) during the same period, are approximately one quarter of the maximum rates.

There is reverse flow into the aquifer during the extreme rainfall event (1987) when the lake responded more rapidly to surface runoff.

There is a huge difference between the unconstrained lake model (conductance value of 100 m^2/day) and the constant head model for the period following the 1987 floods. The lake levels for the constant head model were derived directly from observed conditions. However, the unconstrained lake model simulated lake levels that were considerably higher than the observed values (*cf* Figure 6), which has caused this large discrepancy between seepage rates shown in Figure 9.

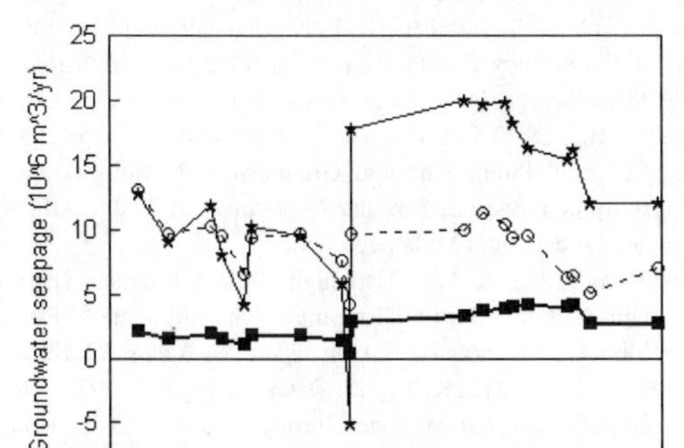

Figure 9. Simulated seepage from different models

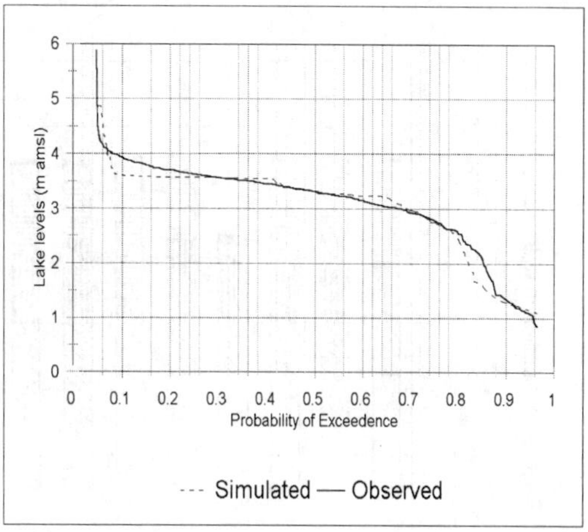

Figure 10. Depth-duration curve for observed and simulated lake levels

The seepage rates of the constant head and the restrictive sedimentary model during the height of the drought converged to almost the same value implying that the system was no longer constrained to the same extent by the sedimentary layer.

10 FREQUENCY ANALYSIS

There is close agreement shown in Figure 10 between the observed lake level frequencies and the simulated levels for the two layer model with a restrictive lake bed sedimentary layer (10 m^2/day). The main discrepancies are due to the selection of stress periods that filter out some of the high rainfall events and intermediate drought conditions.

11 WATER BALANCE

The two layer model simulations have been used to investigate the water balance components for the lake during each stress period. Figure 11 shows the relative contributions of each water balance component for each stress period. Positive values contribute to inflow while negative values represent outflows. For the minimum level of interaction simulated using a lake bed conductance of 10 m^2/day, the relative contribution to the water balance from the groundwater is important during the severe drought period from 1993.

12 CONCLUSIONS

Freshwater lakes in the highly populated and industrialized east coast region of South Africa are

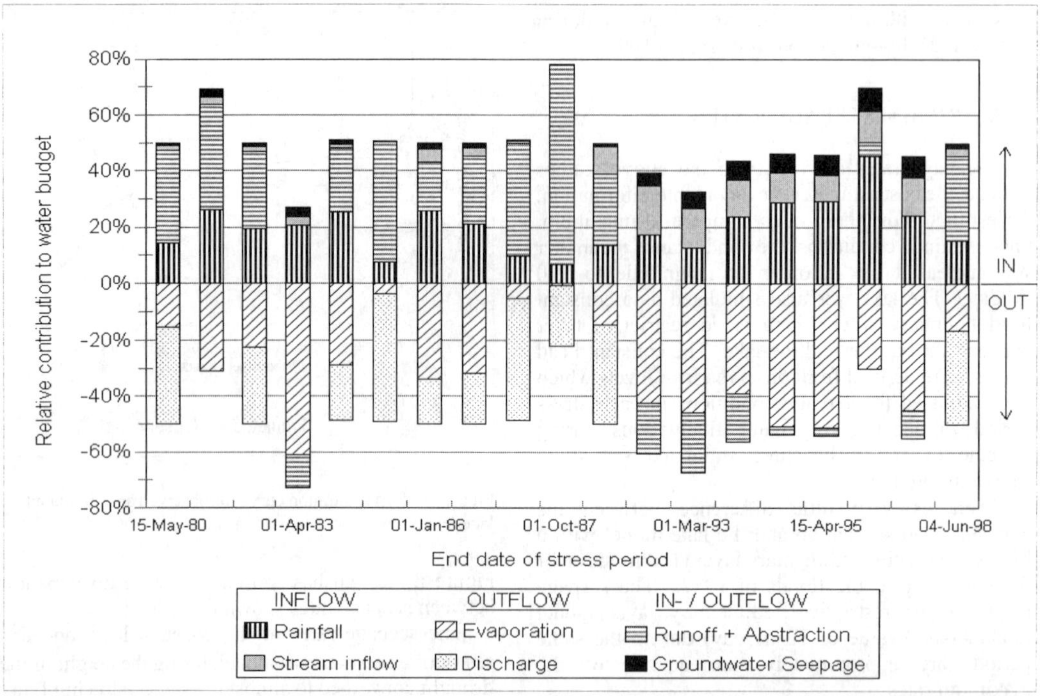

Figure 11. Relative contributions (%) of the lake water balance components using the two layer model with lake bed sediments.

critical water resources that are being over exploited. Their reliance on groundwater for part of their assured yield was investigated using numerical methods. The inadequate geological and hydrological data require an investigative approach of methods used in a study of this nature.

The regional aquifer model provided an adequate framework for investigating the interaction with the lake. However, it was necessary to include additional layers to achieve the desired calibration of the lake water balance.

This numerical study of surface-groundwater interaction has provided a first estimate of the hydrodynamics of the coastal lake system. The results indicate the importance of understanding the methodologies employed in numerical studies. They also highlight the important role of lake bed sediments in determining the flow dynamics of these systems.

ACKNOWLEDGMENTS

We are grateful to many organizations for the financial and material support. The Water Research Commission funded this research under project K5-720. Uthungulu Regional Council together with many of the main companies in Richards Bay provided hydrogeological information.

REFERENCES

Council, G.W. 1997. Simulating Lake-Groundwater Interaction with MODFLOW. *Proceedings of the 1997 Georgia Water Resource s Conference, held 20 - 22 March, at The University of Georgia.* Editor: K.J. Hatcher.

Germishuyse, T. 1999. Geohydrology of the Richards Bay Area. *MSc dissertation submitted to the science faculty, University of Zululand*

Kelbe, B.E. & T.Germishuyse 1999. Geohydrological Studies of the Primary Coastal Aquifer in Zululand. *Report to the Water Research Commission (K5/720).* Under review.

Kelbe, B.E. & T.Germishuyse 2000. Mhlathuze Operating Rule and Future Phasing: Groundwater hydrology, joint study commissioned by the Department of Water Affairs and Forestry and Mhlathuze Water.

McDonald, M.G. & A.W. Harbaugh 1983. A Modular Three-Dimensional Finite-Difference Ground-Water Flow Model, *US Geological Survey, Open-File Report 83-875.*

Rawlins B.K., B.E.Kelbe & T.Germishuyse 1997. The Potential for Saline Water Intrusion to A Coastal Lake Under Drought Conditions, Investigations at Richards Bay, Zululand. *Proceeding of the Eighth SANCIAHS Symposium, Pretoria*

Townley, L.R., J.V.Turner, A.D.Barr, M.G.Trefry, K.D.Wright, V.Gailitis, C.J.Harris, & C.D. Johnston 1993. *Wetlands of the Swan Coastal Plain. Vol 3.* Interaction between Lakes, Wetlands and Unconfined Aquifers. Water Authority of Western Australia.

Groundwater: Past Achievements and Future Challenges, Sililo et al. (eds) © 2000 Balkema, Rotterdam, ISBN 90 5809 159 7

Assessing uncertainty in sustainable yield

N. P. Merrick
National Centre for Groundwater Management, University of Technology, Sydney, N.S.W., Australia

ABSTRACT: The sustainable yield of an aquifer system is usually based on an estimate of long-term average annual recharge, which is generally difficult to quantify. Numerical groundwater models can provide a quantitative basis for estimating recharge, but long-term averages are sensitive to the length of the averaging period and the start date for the averaging. By calculating all possible averages for a chosen minimum averaging period, and using random sequencing of the periods to avoid bias, the average recharge estimates can be ranked and assigned probabilities. The resulting cumulative distribution function allows managers to quantify the risk in setting a value for sustainable yield, and provides users with a measure of confidence in the surety of groundwater supply. The main limitations of this approach are the appropriateness of the conceptual model, the robustness of the numerical model, and the representativeness of the simulation time period. Nevertheless, the method is an advance on current practice.

1 INTRODUCTION

The sustainable yield of an aquifer system is generally difficult to quantify. At the present time there is increasing pressure on groundwater managers to refine estimates and definitions of the sustainable yield of an aquifer, and to provide an indication of the likely uncertainty in the estimate. This issue is particularly important in stressed aquifer systems where competing users insist on justification for the quantum of water subject to distribution as well as the policy which governs water distribution. Too often, an estimate of sustainable yield is stated as a hard and fast value with no indication of its uncertainty. This paper presents a methodology for assessing the uncertainty in sustainable yield. It requires as a basis a reliable groundwater flow model of the aquifer under investigation.

In Australia at present, significant water reforms are being implemented which are underpinned by the concept of sustainability. The Agriculture and Resource Management Council of Australia and New Zealand (ARMCANZ 1996) designed a national framework for improved groundwater management, articulated in a set of 12 key reforms. Amongst other issues, the reforms advocate adoption of "the principles of ecologically sustainable development", recognition of a need for "an agreed nationally consistent definition and approach to sustainable groundwater yield", and "identification of the sustainable yield", with due provision for environmental water requirements. Agencies were advised to "develop strategies to reduce abstractions to sustainable levels" where the resource is currently over-allocated.

In New South Wales, the working definition of sustainable yield is (Kalaitzis et al. 1999): *"Sustainable Yield is that proportion of the long term average annual recharge which can be extracted each year without causing unacceptable impacts on groundwater users or the environment"*. In the absence of firm data on the needs of the environment, the default proportion has been set at 70%.

There are many stressed and over-allocated aquifers in New South Wales. At the present time, strategies are being articulated and implemented for reducing groundwater usage to sustainable levels. Any reduction in farmer allocations is understandably contentious, but there are likely to be very real social and economic consequences from an overhaul of regulations. There is increasing pressure on groundwater managers to quantify and refine estimates and definitions of the sustainable yield of an aquifer. Competing users insist on justification not only for the policy which governs water distribution but also for the volume of water subject to distribution.

This paper offers a contribution to the definition of "long term average annual recharge", from which sustainable yield usually is derived. Even with a well calibrated numerical model, the definition is far from straightforward. The approach taken recommends replacement of a single deterministic value by a cumulative distribution function, or probability distribution, which acknowledges and quantifies the uncertainty in the definition. In this way, both managers and groundwater users will be conscious of the degree of risk in matching usage to sustainability.

The approach introduced here assumes that the aquifer under consideration has a robust groundwater flow model. In this paper, the existing model of the lower Namoi Valley in northern New South Wales will be used for purposes of illustration (Merrick 1999). The model consists of three layers, 30 rows and 50 columns, with uniform cell size 2500 m x 2500 m. Calibration has been achieved over the period September 1980 to August 1994 with a stress period of one month. The simulation period includes a significant flood event in 1984. The model was developed with the PMWIN graphical user interface (Chiang & Kinzelbach 1996, 2000) in a Windows 95 environment with MODFLOW finite difference software (McDonald & Harbaugh 1985).

2 SOFTWARE

The most effective way of estimating the average recharge over a long period of time is by interrogating the water budget details of a robust numerical groundwater model which replicates faithfully the main recharge processes in a natural system. A model code such as MODFLOW (McDonald and Harbaugh, 1985) provides optionally a water budget at the end of every time step within a stress period. For use in sustainability estimates, a cumulative budget over a long period of time is required. The standard MODFLOW code provides both cumulative budgets (since the start of the simulation) and instantaneous rates for the whole model area, but not for individual layers and not for subdivisions of the model area. At times, groundwater managers might need to define sustainable yields for zones smaller than the coverage of a model, or for various combinations of zones.

Apart from the standard MODFLOW code, modelling environments based on graphical user interfaces to MODFLOW provide a facility for examining detailed water budgets; for example PMWIN (Chiang & Kinzelbach 1996) and Visual MODFLOW (Guiger & Franz 1997). As a rule, they do not report the cumulative budgets which are required for long-term estimates. Instead, they provide instantaneous budgets for user-defined zones for individual layers and for aggregated layers. Utility programs extract budget information from MODFLOW's binary BUDGET.DAT output file for zones defined using the graphical interface. However, there is no provision for composite zones.

The ZONEBUDGET program written by Harbaugh (1990) for extracting budget information from the BUDGET.DAT file comes closest to providing the information needed. Zones are defined by the user interactively or by editing an input file. As a zone number applied to a surface subregion does not extend to layers lying directly beneath (as it does in PMWIN), zone numbers must be defined for separate layers. ZONEBUDGET allows the user to define composite zones, which may be used to aggregate layers which are vertically aligned, or aggregate all zones across an individual layer, or aggregate all zones to give a global water balance. The program reports instantaneous budgets for individual zones and composite zones but does not report cumulative budgets.

New software (called ZBUDGET) has been written by the author to give cumulative water budgets and long-term averages for individual zones and composite zones. The program, based on the ZONEBUDGET code, produces output files of three types: (1) Individual zone cumulative budgets at the end of nominated stress periods; (2) Composite zone cumulative budgets at the end of nominated stress periods; (3) Time series files for each individual zone and for each composite zone, for nominated stress periods, reported in ML/d units (1 ML = 1,000 m^3) for budget components and GL/a units (that is Mm^3/a; 1 GL = 1,000,000 m^3) for total recharge, total discharge and change in storage.

In this study, "total recharge" is defined as the sum of all recharge components including lateral inflows.

3 AVERAGING PERIODS

A simple average of the net recharge over the simulation period of a model is not appropriate, as estimates of long-term average total recharge vary according to the length of the averaging period and the start date for the averaging.

For the lower Namoi Valley case study, which has a simulation period of 14 years from September 1980 through to August 1994, higher recharge rates are observed as the start date approaches 1984 because a major flood event at that time is distributed over fewer years (Fig. 1). Recharge rates reduce with time as the effect of the 1984 flood event is spread out. If the start date is later than 1984, recharge rates are lower but stable.

Figure 1 shows how the long-term average total recharge varies with time, as averaging is done over a progressively longer time period. For example, the values at August 1994 are the averages over 14 years when simulation starts at September 1980; over 13 years when simulation starts at September 1981; etc. The values at August 1987 are averages over seven years when simulation starts at September 1980; over six years when simulation starts at September 1981; etc. The value at September 1981 is the "av-

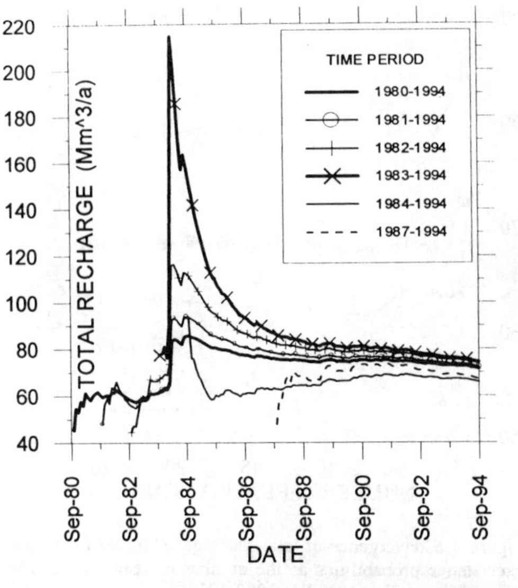

Figure 1. Estimates of total recharge averaged over different time periods (lower Namoi Valley 1980-1994).

erage" over one year (that is, the actual monthly volumes accumulated for one year). Values between September 1980 and September 1981, where the time period is less than one year, are extrapolations from a partial year to a full year. Hence, a value within 12 months of the start date can be unduly affected by a strong event within one month. This explains the high peak for 1983-1994 data, due to a flood in 1984.

For different simulation time periods, estimates of total recharge (based on at least one full year of record) could range from 60 to 160 Mm^3/a during the 1980-94 period, but a value of 65 to 80 Mm^3/a is more likely. It is clear that there is considerable uncertainty in an estimate of sustainable yield which is based on a proportion of total recharge.

To better quantify the uncertainty inherent in estimates of sustainable yield, one could determine the average total recharge for all combinations of contiguous stress periods from an agreed minimum averaging period out to the maximum possible averaging period. Let N be the maximum number of stress periods in a simulation, and let n be the minimum number of stress periods in an averaging period.

Then the number of possible averaging periods of different lengths is:

$$M = N - n + 1 \quad (1)$$

and the total number of combinations of contiguous averaging periods is:

$$S = \frac{1}{2} M (M + 1) \quad (2)$$

For the lower Namoi Valley case study, for a simulation period of 168 months, there are 157 possible averaging periods and 12,403 combinations of contiguous averaging periods for a minimum averaging period of 1 year (12 months). For 5 years (60 months) minimum, there are 109 possible averaging periods and 5,995 combinations.

The many total recharge values $(k=1,2,...,S)$ can be ranked and each assigned a probability $(P=k/S)$. The resulting cumulative distribution functions (cdf), shown in Figure 2, suggest that the median $(P=0.5)$ total recharge is about 72 Mm^3/a for averaging down to one year, and about 75 Mm^3/a for averaging down to five years. There is 90% exceedance probability $(P_e=1-P)$ that total recharge exceeds 64 Mm^3/a (1 year minimum) or 68 Mm^3/a (5 years minimum), and 10% probability that total recharge exceeds 85 Mm^3/a (1 year minimum) or 80 Mm^3/a (5 years minimum). A shorter minimum averaging period allows more extreme values to occur at low and high probabilities.

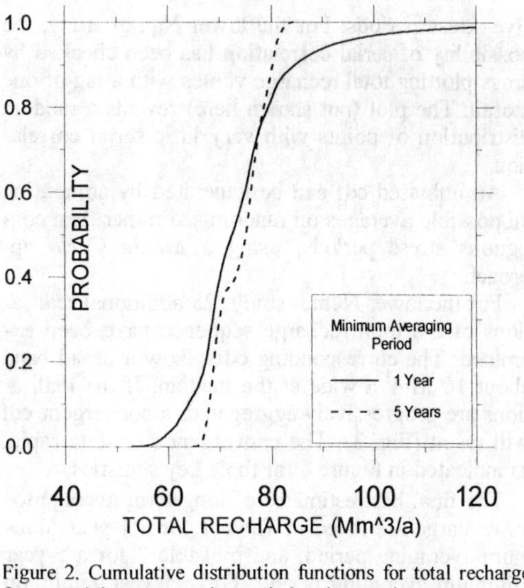

Figure 2. Cumulative distribution functions for total recharge for two minimum averaging periods using contiguous stress periods (lower Namoi Valley 1980-1994).

4 CUMULATIVE DISTRIBUTION FUNCTIONS

Use of all possible combinations of contiguous averages will be biased in favour of the stress periods near the centre of the simulation period, as they will participate in more averages than do periods near the start or end of the simulation. The longer the minimum averaging period, the more biased will be the cdf.

There might also be bias arising from serial correlation between total recharge estimates in succes-

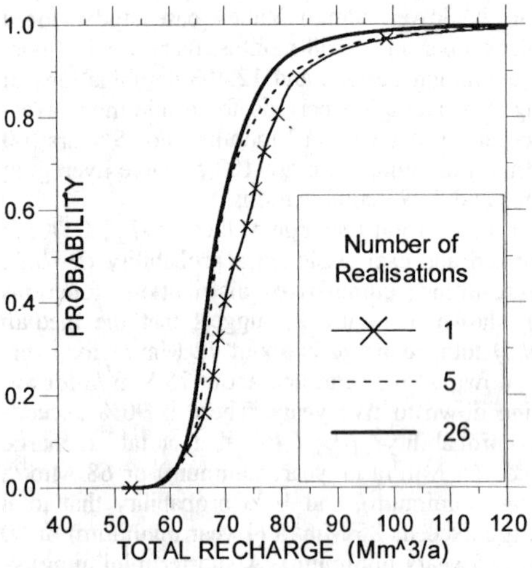

Figure 3. Cumulative distribution functions for total recharge as the number of realisations is increased (lower Namoi Valley 1980-1994).

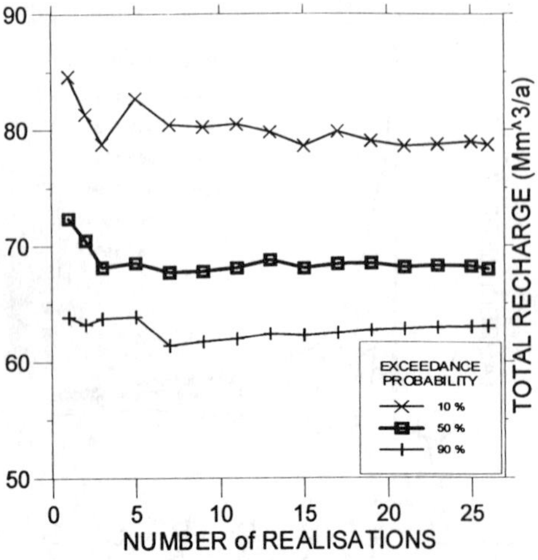

Figure 4. Convergence of total recharge estimates for various exceedance probabilities as the number of realisations is increased (lower Namoi Valley 1980-1994).

sive stress periods. For the lower Namoi study, the possibility of serial correlation has been checked by cross-plotting total recharge values with a lag of one month. The plot (not shown here) reveals a random distribution of points with very little serial correlation.

An unbiased cdf can be generated by computing all possible averages on randomised rather than contiguous stress periods, using a *Monte Carlo* approach.

For the lower Namoi study, 25 additional realisations of the total recharge sequence have been examined. The corresponding cdfs show a broad band about 10 Mm^3/a wide at the median. If the realisations are progressively aggregated, a convergent cdf will result (Fig. 3). The convergence is quite rapid, as indicated in Figure 4 for three key statistics.

The final best estimates of long-term average total recharge are listed in Table 1 for a 1-year minimum averaging period, and in Table 2 for a 5-year minimum averaging period. For a 1-year minimum averaging period there is 80% probability that total recharge during the 1980-1994 period averaged between 63 and 79 Mm^3/a, with the expected value being 68 Mm^3/a. For a 5-year minimum averaging period there is 80% probability that total recharge during the 1980-1994 period averaged between 64 and 81 Mm^3/a, with the expected value being 72 Mm^3/a.

Table 1. Best estimates for total recharge for the lower Namoi Valley for 1-year minimum averaging period

Total recharge (Mm^3/a)	EXCEEDANCE PROBABILITY 90%	50%	10%
Contiguous months	64	72	85
Randomised months	63	68	79

Table 2. Best estimates for total recharge for the lower Namoi Valley for 5-year minimum averaging period

Total recharge (Mm^3/a)	EXCEEDANCE PROBABILITY 90%	50%	10%
Contiguous months	68	75	80
Randomised months	64	72	81

5 CONCLUSION

Dependable estimates of sustainable yield require aggregation and averaging over long periods of time of total recharge values. The most effective method for obtaining dynamic and cumulative recharge estimates is numerical modelling of groundwater flow. However, long-term averages are sensitive to the length of the averaging period and the start date for the averaging. To counter this uncertainty, all possible combinations of averages should be considered for a specified minimum averaging period. Multiple realisations of the stress sequence are warranted in order to remove the bias introduced when averages are based on contiguous months. Aggregated realisations lead to a convergent cumulative distribution function, from which total recharge estimates can be derived with assigned exceedance probabilities.

Although the methodology is robust, the final estimates are limited by the integrity of the numerical groundwater model, its underlying conceptualisation, how representative the simulation period is of a full climatic cycle, and by an arbitrary choice of the minimum averaging period. The latter should be consistent with the adopted management period for the aquifer. The parameters in a groundwater model are non-unique, even when calibration appears to be good. This adds another level of uncertainty to the determination of sustainable yield which is not addressed by the method presented here. In spite of these limitations, the method is an advance on current practice which provides a single deterministic estimate with no hint of uncertainty. The cumulative distribution function and the derived total recharge estimates will require updating as the numerical model for a particular aquifer is improved, as the range of non-uniqueness in model parameters is narrowed, and the model is extended to longer simulation periods.

This probabilistic approach allows managers to quantify the risk in setting a value for sustainable yield .based on total recharge. It also provides groundwater users with a measure of confidence in the security of supply. The approach may be applied to an entire aquifer system, or to any number of smaller groundwater management zones covered by the model extent.

ACKNOWLEDGMENT

The author acknowledges the support of the Department of Land and Water Conservation (New South Wales, Australia) in undertaking this study. However, the opinions, methodology and estimates put forward in this paper should not be construed as having any official government endorsement.

REFERENCES

ARMCANZ. 1996. Allocation and use of groundwater: A national framework for improved groundwater management in Australia. Commonwealth of Australia, Occasional Paper Number 2, 9p.

Chiang, W-H. & Kinzelbach, W. 1996. Processing Modflow for Windows: A simulation system for modelling groundwater flow and pollution. Users manual.

Chiang, W-H. & Kinzelbach, W. 2000. 3D Groundwater Modeling with PMWIN - A Simulation System for Modeling Groundwater Flow and Pollution. Springer: Heidelberg, Berlin, New York, 356p. (in press)

Guiger, N. & Franz, T. 1997. User's manual for Visual MODFLOW. Waterloo Hydrogeologic Inc.

Harbaugh, A.W. 1990. A computer program for calculating subregional water budgets using results from the U.S. Geological Survey modular three-dimensional finite-difference ground-water flow model. US Geological Survey, Reston, Virginia, Open-File Report 90-392, 46p.

Kalaitzis, P., Brownbill, R. & Jamieson, M. 1999. Environmental provisions in determining sustainable yield for groundwater management plans in the lower Namoi Valley, NSW. Proceedings, Murray Darling Basin Groundwater Workshop, Griffith, 55-60.

McDonald, M.G. & Harbaugh, A.W. 1985. A modular three-dimensional finite-difference ground-water flow model. US Geological Survey, Reston, Virginia, 528p.

Merrick, N.P. 1999. Lower Namoi groundwater budgets: Simulation 1980-1994. Insearch Limited Report for Department of Land and Water Conservation (Tamworth), Project No. C99/44/008, April 1999, 72p. (unpublished).

Groundwater: Past Achievements and Future Challenges, Sililo et al. (eds) © 2000 Balkema, Rotterdam, ISBN 90 5809 159 7

Three-dimensional numerical models of groundwater flow and solute transport in fractured granites: Impact of tunnel construction in the groundwater system at the Äspö island, Sweden

J. Molinero & J. Samper
E.T.S. Ingenieros de Caminos, Universidad de la Coruña, Spain

R. Juanes
Department of Civil and Environmental Engineering, University of California, Berkeley, Calif., USA

ABSTRACT: The Äspö Hard Rock Laboratory (HRL) is an underground facility operated by the Swedish Nuclear Waste Company, SKB. The Äspö tunnel was constructed between October 1990 and June 1995, reaching a maximum depth of 450 m. In order to simulate the impact of the tunnel construction on the hydrogeology of the Äspö island, a 3-D transient groundwater flow and multi-component solute transport numerical model was carried out. The model accounts for 20 major fracture zones and extends over 6 years. The tunnel front advancement was simulated dynamically by means of internal time-varying boundary conditions. Model results were compared to measured data at 45 control points. The calibrated flow model provides an excellent fit of both water inflows into the tunnel and drawdowns at selected boreholes. The transport model, which was constructed with no calibration, is in a very good agreement with measured concentrations of conservative species breaking through the tunnel and boreholes. Observed discrepancies for reactive species such as calcium, sulfate, bicarbonate and sodium are attributed to hydrochemical processes.

1 INTRODUCTION

The Äspö HRL is an underground laboratory operated by the Swedish Nuclear Waste Company, SKB, which provides a multitude of data to improve the knowledge of the crystalline bedrock. This underground facility constitutes a unique opportunity for testing models of groundwater flow and solute transport. The Äspö island is situated in the south-east part of Sweden, 400 km south of Stockholm. It is a granitic coastal island in the Baltic Sea, separated from the mainland and other islands by shallow estuaries (Figure 1). The rocks of the area are predominantly granitoids belonging to the TranScandinavian Igneous Belt. These granitoids have a range of mineralogical composition between true granites to granodioritic and dioritic compositions (Kornfalt & Wikman, 1988). The Äspö HRL facilities consist on a 3600 m length tunnel with a long ramp coming from the Simpevarp peninsula and an spiral part under the Äspö island (Figure 1). The tunnel goes down more than 450 m depth.

The large amount of research work performed during the pre-investigation and construction phases (1986-1995) at Äspö has provided a strong conceptual model for the hydrogeology of the area (Rhén et al., 1997). Under natural conditions the water table ranges between 0 and 4 m.a.s.l.. The amount of water that infiltrates into the rock has been estimated to be between 5 and 35 mm/year (Rhén et al., 1997). Around 20 major fracture zones were identified at the Äspö site which control recharge, discharge and groundwater flow through the island (Smellie et al., 1995). There is a clear increase of salinity with depth. The fresh water lens below Äspö has a thickness of 100-200 m under natural conditions. Below this fresh-water lens salinity increases, reaching a value of 20 g/L at a depth of 800 m. The salinity of the surrounding Baltic estuaries is approximately equal to 6 g/L.

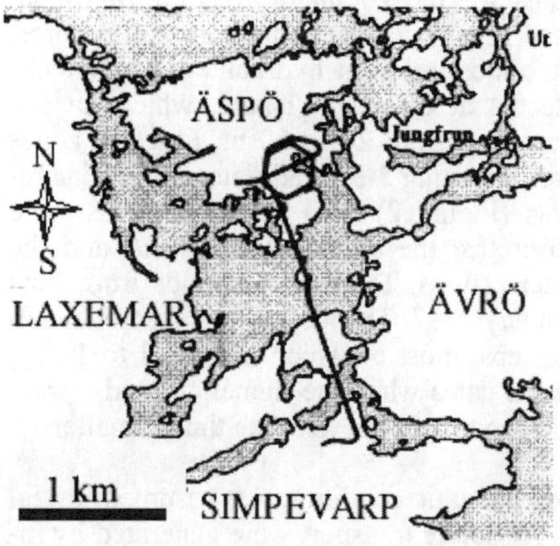

Figure 1. Location of the Äspö site. Plan view of the Äspö HRL tunnel showing the access ramp and the spiral part under the Äspö island.

2 NUMERICAL MODEL

In order to simulate the impact of the tunnel construction on the hydrogeology of the Äspö site, a numerical model was carried out which accounts for groundwater flow and solute transport, including advection, hydrodynamic dispersion and molecular diffusion. According to Hautojärvi et al. (1994) density effects in the groundwater flow are negligible once the system is altered by the tunnel construction, so these effects were neglected.

The numerical model was solved with TRANMEF-3D (Juanes et al., 2000), a domestic numerical code, developed and available at the Hydrogeology Group of the University of A Coruña, for solving multi-dimensional groundwater flow, multi-component weakly-reactive solute transport and heat transfer using the Finite Element Method. TRANMEF-3D handles up to 6 different types of elements that can be combined arbitrarily in any simulation problem. Moreover, these elements do not need to have the same dimensions. They can be either 1-D, 2-D or 3-D. This capability allows one to simulate, for example, 1-D features (wells, tunnels...) and 2-D features (such as fractures) within a general 3-D porous matrix. Juanes & Samper (2000 a,b) demonstrate that by representing fracture networks with elements of lower dimension than the matrix domain, one can obtain more accurate and efficient numerical solutions of groundwater flow and solute transport. Contrary to most available codes, TRANMEF-3D allows the user to handle time-varying boundary conditions. Such capability was used in order to simulate dynamically the process of tunnel advancement by using 29 stages for the discretization of the tunnel construction.

The model domain consists on a 3-D block of 2 x 2 km on the upper surface and 1 km depth (Figure 2), and accounts for the 20 major fracture zones detected at the Äspö site. These fracture zones correspond to the most relevant hydraulic conductor domains detected at the Äspö Island, which have a thickness ranging from 20 to 50 m. Fracture zones were discretized using 2-D quadrilateral and triangular elements (Figure 2). 1-D linear elements were used to discretize the tunnel, the elevator and the fracture intersections. The model extends from June 1991 to January 1997. This period was discretized in 2275 time steps, most of which are equal to 1 day, except for the dates when the tunnel crossed a fracture zone, when time steps were ten times smaller.

An initial hydrostatic condition was assumed. Initial conditions for solute transport were generated by interpolation of available data during undisturbed conditions. A Dirichlet condition was imposed at lateral boundaries as well as at the Baltic Sea. The bottom boundary was assumed impervious. A recharge of 5 mm/year was assigned on the emerged lands of the upper boundary.

3 GROUNDWATER FLOW MODELING RESULTS

Groundwater flow model was intensively calibrated by comparing computed results with measured inflows into the tunnel at 14 control points and groundwater heads at packed intervals of boreholes at 15 control points. The final model leads to excellent fit at all selected control points. Figure 3 illustrates the consistency between computed and measured heads and water inflows.

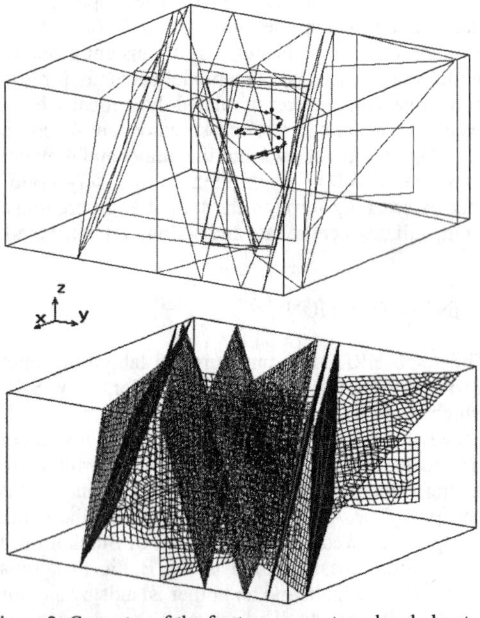

Figure 2. Geometry of the fracture zones, tunnel and elevator shaft (top) and finite elements mesh used for the fracture zones (bottom).

The absolute value of the differences between computed and measured values has a mean value of 3.1 m for groundwater heads, and 45.6 L/min for tunnel inflows. Figure 4 shows the evolution of the total amount of water flowing into the tunnel.

It should be noticed that the adopted numerical approach does not account for the contribution of the rock blocks between fractures. One should expect that computed flow rates should be smaller than measured values. The model reproduces accurately the measured trends in inflows because they are controlled mainly by the intersection of the tunnel and the fracture zones. The largest discrepancy is observed at the transient stage between days 800 and 1200. Later, these discrepancies decrease once steady state is reached (Figure 4). The role of the rock blocks between fracture zones can be evaluated from the comparison of computed and measured flow rates into the tunnel. It can be stated that granite blocks between fracture zones contribute approximately 20% of the total amount of water flowing through the tunnel walls during the transient stage. This contribution, however, decreases gradually to less than 5% at steady state (Figure 4). This analysis indicates that the granite blocks at the Äspö site show a noticeable storage capacity and a much lower ability for water transmission. The water storage capacity in a fractured rock mainly depends on fracture density while the transmissivity depends on connectivity of the fracture network

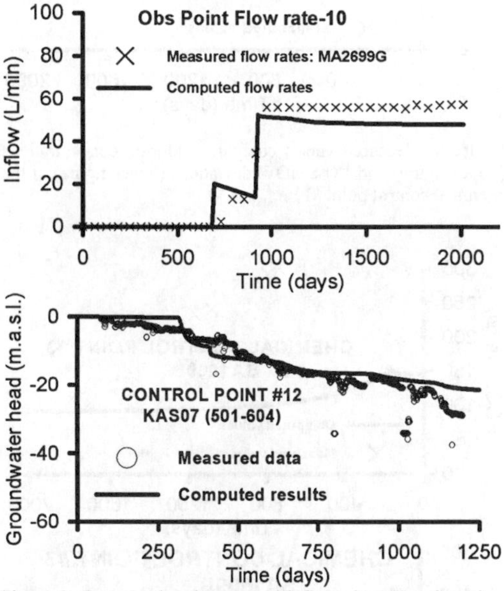

Figure 3. Computed and measured inflows in a tunnel section (top) and groundwater heads at borehole KAS07 (bottom).

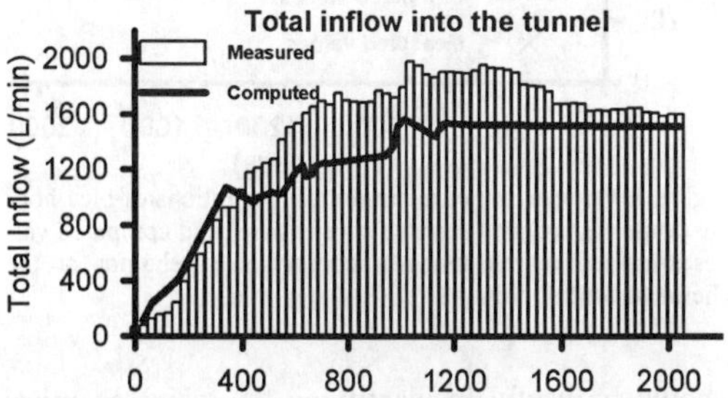

Figure 4. Computed and measured cumulative water inflow into the tunnel.

It is worth noting that calibrated values of hydraulic conductivity and specific storativity lie in most cases within the range of variability of field values. Figure 5 shows a comparison between field estimations and calibrated values of hydraulic conductivity for each fracture zone. Only at 2 of the 20 fracture zones (zones 5 and 15) the calibrated conductivities are slightly lower than measured values (Figure 5).

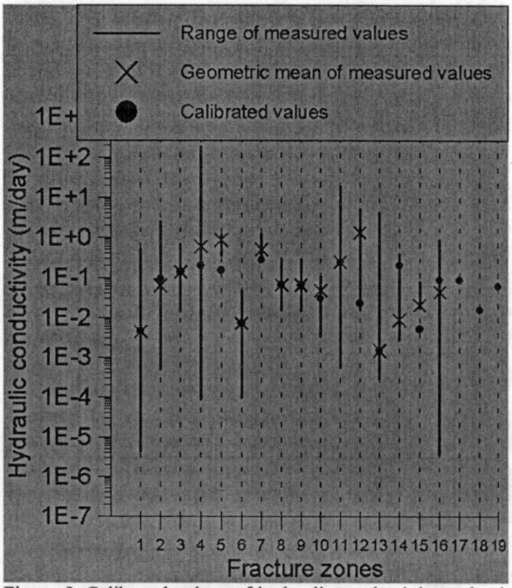

Figure 5. Calibrated values of hydraulic conductivity and values obtained by hydrogeologic characterization. Initial guesses for the numerical model were the geometric mean values of field estimations.

4 SOLUTE TRANSPORT MODELING RESULTS

The construction of the tunnel modifies drastically the groundwater system at the Äspö site. Groundwater chemical composition is also affected by tunnel construction. Groundwater chemical evolution reflects the coupled effect of physical and chemical processes. In order to understand and characterize mass transfer processes, attention was focussed on the behavior of conservative species such as chloride and ^{18}O.

Contrary to the groundwater flow numerical model, no calibration was performed on the transport model. Transport parameters (kinematic porosity and dispersivities) were derived from available tracer tests (Rhén et al., 1997). These authors propose several empirical relationships for transport parameters at the Äspö site, which were used to derive the values of kinematic porosity and dispersivity at those conductive features with no field data. Proposed empirical relationships include:

$$\phi_e = 34.87\, K^{0.753} \quad (1)$$

$$\alpha_L = 0.053\, s^{1.210} \quad (2)$$

where ϕ_e is kinematic porosity, K is hydraulic conductivity (m/s), α_L is longitudinal dispersivity (m) and s is spatial scale (m). The spatial scale represents the distance between the injection and sampling points.

Initial concentration fields were generated by using a geostatistical interpolator (kriging), based on chemical data of 32 water samples collected prior to tunnel construction. Initial chloride concentrations range from almost 0 (meteoric water) on the land surface to 20 g/L (deeper saline water) following a more or less stratified pattern. The initial distribution of $\delta^{18}O$ shows a different pattern with a clear maximum (heaviest waters) located at a depth of around 250 m. This maximum of $\delta^{18}O$ has been interpreted as the signature of old recharge under colder climatic conditions, probably dominated by glacial-melt waters (Smellie et al., 1995).

More than 15 points were selected to compare the results of the numerical model to measured concentrations for selected species. Most of the control points correspond to short boreholes drilled from the tunnel walls, but others were boreholes drilled from the island surface, far away from the tunnel location. A very good agreement was found between model computations and measured values except for 2 points. Figure 6 shows a comparison between measured values and computed results for chloride concentrations and ^{18}O activities at one of the short boreholes drilled from the tunnel. This figure illustrates the ability of the model to reproduce the impact of the tunnel construction on the hydrochemical evolution of the Äspö site.

A decrease with time of both measured and computed $\delta^{18}O$ was observed at all the control points. Heavy waters correspond to relict (fossil) waters probably infiltrated during the last glaciation. The replacement of these heavier waters (initially located at an intermediate depth) by lighter waters coming from the upper and bottom boundaries is responsible for the observed and computed behavior of $\delta^{18}O$. Chloride data show a general dilution pattern in points located between the land surface and 350 m depth due to a predominant arrival of fresh water from the emerged lands. Breakthrough curves at points located deeper than 350 m show the influence of upconing of deep saline waters towards the tunnel location.

Solute transport modeling was also performed for the major dissolved species by ignoring chemical reactions. Since the numerical model reproduces the behavior of truly conservative species, computed concentrations for chemically-reactive species

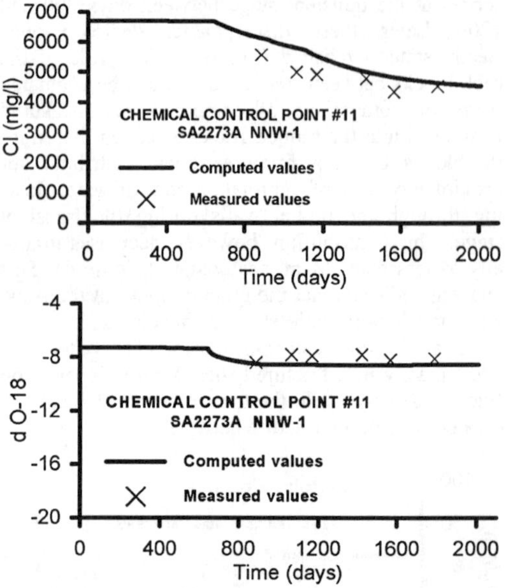

Figure 6. Measured versus computed chloride concentrations (upper figure) and ^{18}O SMOW deviations (lower figure) at the chemical control point #11.

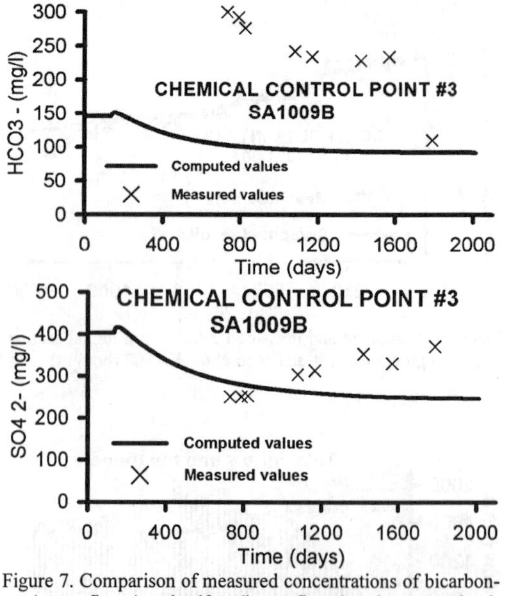

Figure 7. Comparison of measured concentrations of bicarbonate (upper figure) and sulfate (lower figure) and computed values obtained by assuming a conservative behavior at the chemical control point #3.

should be useful in identifying the major chemical processes. Computed and measured concentrations of bicarbonate, sulfate, sodium and calcium show significant differences (Figure 7), which are caused by chemical processes such as cation exchange, calcite dissolution and organic matter degradation. These processes control the evolution of sodium, calcium and bicarbonate. An additional source of sulfate is required in order to explain measured data.

5 CONCLUSIONS

A numerical model of transient groundwater flow and solute transport of the Äspö site has been performed. The numerical model accounts explicitly for a dynamic treatment of the tunnel construction process by means of an inner time-varying boundary condition. The model matches simultaneously the observed water inflows in the tunnel, groundwater drawdowns at boreholes and breakthrough curves of conservative species. The role of granite blocks between fracture zones has been evaluated to be approximately 20% of the total amount of water flowing through the tunnel walls during the transient stage, decreasing to less than 5% at steady state. Granite blocks at the Äspö site show a noticeable storage capacity and a much lower ability for water transmission. A non-conservative behavior has been detected mainly for calcium, sodium, bicarbonate and sulfate for which deviations between measured and computed concentrations allow evaluating the role of chemical processes. Cation exchange, calcite dissolution and organic matter degradation have been proposed as the main chemical processes controlling the evolution of sodium, calcium and alkalinity. An additional source of sulfate is required in order to explain measured data.

ACKNOWLEDGMENTS

This work has been carried out within the framework of Research Projects funded by ENRESA (FEBEX Project-703231 and Validation Project-703334) as well as from the Spanish Scientific & Technological Commision (HID98-0282). The FEBEX Project was partially funded by the European Community. Access to the Äspö database was provided within the framework of the collaborative agreement between ENRESA and SKB. Authors want to thank SKB and all the people involved in the Äspö Task #5 for their help and fruitful comments and suggestions.

REFERENCES

Juanes, R.; Samper, J. & Molinero, J. 2000. A generalized approach to modeling fractures in the finite element method. ECCOMAS'2000. European Congress on Computational Methods in Applied Sciences and Engineering. Barcelona, 11-14 September 2000. In press.

Juanes, R. & Samper, J. (2000 a). A general and efficient formulation of fractures and boundary conditions in the F.E.M.: I. Theoretical aspects. (In Spanish with abstract in English). Revista Internacional de Métodos Numéricos para Cálculo y Diseño en Ingeniería. (Accepted-In Press).

Juanes, R. & Samper, J. (2000 b). A general and efficient formulation of fractures and boundary conditions in the F.E.M.: II. Synthetic cases. (In Spanish with abstract in English). Revista Internacional de Métodos Numéricos para Cálculo y Diseño en Ingeniería. (Accepted-In Press).

Kornfalt, K.A. & Wikman, H. 1988. The rocks of the Äspö island. Description of the detailed maps of solid rocks including maps of 3 uncovered trenches. SKB PR 25-88-12.

Hautojärvi, A.; Koskinen, L. & Löfman, J. 1994. Hydraulic modeling of the large scale redox experiment at Äspö HRL. In: Hydraulic modeling and tracer tests on the redox experiment in the Äspö HRL. SKB PR 25-94-37.

Rhén, I.; Gustafson, G.; Stanfors, R & Wikberg, P. 1997. Models based on site characterization 1986-1995. SKB Technical Report 97-06

Smellie, J.; Laaksoharju, M. & Wikberg, P. 1995. Äspö, SE Sweden: a natural groundwater flow model derived from hydrochemical observations. J. Hydrol. 172, 147-169.

Groundwater: Past Achievements and Future Challenges, Sililo et al. (eds) © 2000 Balkema, Rotterdam, ISBN 90 5809 159 7

Robust groundwater management through second-order cone programming (SOCP)

J.M.Ndambuki, C.B.M.Stroet & E.J.M.Veling
Delft University of Technology, Netherlands

T.Terlaky
McMaster University, Canada

ABSTRACT:Although the problem of groundwater flow simulation and optimization has largely been treated deterministically, the problem is intrinsically stochastic. This is because of the uncertainties that inherently exist in specifying the input parameters to the flow model. As a result, the recent past has witnessed a shift from the traditional deterministic optimization approach to the more recent stochastic optimization one. The most popular way of treating stochastic groundwater optimization problems has been through the Monte Carlo approach. This approach aims at getting a distribution of either the optimal solutions or the potentiometric heads after which the optimization can be carried out. A shortcoming of this approach is the number of forward problems which have to be solved to give a reliable distribution. This means more computer time and hence high expenses. In this paper, we present an approach which is more efficient in terms of computer time. Furthermore, the solutions obtained through this approach are robust, meaning that they are relatively insensitive to slight data perturbations. In our approach, the stochastic groundwater flow simulation and optimization problem is transformed into a Second-Order Conic Programming (SOCP) problem. We demonstrate the application of this methodology through a hypothetical example.

1 INTRODUCTION

Robust programming is an optimization process in which uncertainty resulting from data perturbations is explicitly taken into consideration and the resulting optimal solution is insensitive to data perturbations. Consider a linear program of the form:

$$\begin{array}{ll} \text{minimize} & c^T x, \\ \text{subject to} & a_i^T x \leq b_i,\ i = 1, \cdots, m. \end{array} \tag{1}$$

where $c \in R^n$, $x \in R^n$, $a_i \in R^n$ and $b_i \in R$. In such a problem, the parameters a_i, b_i and c may not be known with precision. What may be known is a domain Ξ (uncertainty set) containing the actual unknown values. A realistic solution to this problem requires that all actual constraints are satisfied by the solution x whatever the actual values realized by the parameters from the uncertainty set.

For exposition, assume that b_i and c are deterministic while a_i are uncertain parameters lying in given ellipsoids, Ξ_i, (uncertainty sets) defined as:

$$a_i \in \Xi_i = \{\overline{a}_i + Q_i u \mid \|u\| \leq 1\}. \tag{2}$$

where Q_i is an $n \times \xi$ matrix of perturbations, $u \in R^\xi$, $\overline{a}_i$ are the nominal values and restriction $\|u\| \leq 1$ enforce the perturbations to vary in ellipsoids. Since in the worst case scenario a robust solution requires satisfaction of all constraints for all possible outcome of the uncertain parameters, a robust formulation of (1) can be written as:

$$\begin{array}{ll} \min & c^T x, \\ \text{s.t.} & a_i^T x \leq b_i,\ \forall a_i \in \Xi_i,\ i = 1, \cdots, m. \end{array} \tag{3}$$

Formulation (3) is now a "deterministic" problem having infinitely many constraints. A solution to this problem is robust feasible if and only if for every $i = 1, \cdots, m$ we have

$$\left[\overline{a}_i^T x + (Q_i u_i)^T x - b_i\right] \leq 0, \forall u_i : u_i^T u_i \leq 1 \tag{4}$$

which is equivalent to:

$$\max_{u_i : u_i^T u_i \leq 1} \left[\overline{a}_i^T x + (Q_i u_i)^T x - b_i\right] \leq 0, \tag{5}$$

i.e.

$$\overline{a}_i^T x - b_i + \max_{u_i : u_i^T u_i \leq 1} (Q_i u_i)^T x \leq 0, \tag{6}$$

or

$$\overline{a}_i^T x + \|Q_i x\| \leq b_i. \tag{7}$$

Constraint (7) is a second-order cone constraint (also known as "Lorentz cone" or "Ice Cream cone" constraints). A solution to formulation (3) is robust feasible if and only if it satisfies the system of inequalities given in (7). Formulation (3) can, therefore, be written as a Second-Order Cone Program (SOCP) frequently referred to as Conic Quadratic Programming Problem as follows:

$$\begin{array}{ll} \min & c^T x, \\ \text{s.t.} & \overline{a}_i^T x + \|Q_i x\| \leq b_i,\ i = 1, \cdots, m. \end{array} \tag{8}$$

The norm terms in formulation (8) act as 'regularization terms' or 'penalizing terms', discouraging large values of decision variables, x, in directions of high magnitudes of uncertainty. Solving program (8) results to a robust solution which remains feasible for all possible data perturbations from the given ellipsoidal uncertainty set.

SOCP problems can be solved efficiently using interior-point methods (Lobo et al. 1998, Andersen & Andersen 1999, Boyd et al. 1994). A few applications of SOCP problems have been reported in the literature. They include Antenna array weight design, Filter design, Grasping force optimization, Portfolio optimization, Truss design, and Equilibrium of system with piecewise-linear springs design (Lobo et al. 1998, Boyd et al. 1994, Ben-Tal & Nemirovski 1998, Boyd et al. 1998, El Gaoui & Lebret 1997). From the literature, the reported applications are basically in the areas of Electrical engineering, Mechanical engineering, Economics, and Structural engineering. To the best of our knowledge, this methodology has not been applied in the area of Geohydrology.

2 ROBUST GROUNDWATER MANAGEMENT

To demonstrate the power of this novel methodology, we are going to solve a groundwater management problem where our objective is to minimize the operational cost of a groundwater aquifer subjected to some constraints. The constraints define the feasible domain of the optimization problem. We start by formulating the problem, and then follow with an application. Our optimization problem is defined as follows:

$$\underset{x}{\text{minimize}} \left[Z = \sum_{j=1}^{N_w} c_j x_j \right], \tag{9}$$

subject to:

$$\overline{a}_i^T x + \|Q_i x\| \leq b_i,\ i = 1, \cdots, N_c, \tag{10}$$

$$\sum_{j=1}^{N_w} x_j \geq W_d, \tag{11}$$

$$0 \leq x_j \leq U_j,\ j = 1, \cdots, N_w, \tag{12}$$

where $x = (x_1, \cdots, x_j, \cdots, x_{N_w})^T$, $\overline{a}_i = (\overline{a}_{i,1}, \cdots, \overline{a}_{i,j}, \cdots, \overline{a}_{i,N_w})^T$ and the other parameters and variables are defined as follows:

- N_w is the number of pumping wells;
- N_c is the number of control points;
- c_j is the aggregated daily cost of pumping and transportation in monetary units (MU) per unit volume at location j;
- x_j is the pumping rate in cell j;
- $a_{i,j}$ is the response at cell i due to pumping in cell j;
- b_i is the constraining value at control point i;
- W_d is total water demand; and
- U_j is the maximum pumping rate allowed in cell j.

2.1 *Application to an example*

We use a hypothetical example to demonstrate the applicability of the methodology outlined above. The problem we seek to solve is the one we have already formulated (problem 9-12). This problem has a total of 49 decision variables (our decision variables are concerned with the determination of location of boreholes and their corresponding pumping rates), 225 second-order cone constraints (coefficients of these constraints are uncertain, hence the constraints are stochastic), and 99 linear constraints (including nonnegativity constraints) which are assumed to be deterministic.

Our goal is to supply domestic water to a distribution center situated in the middle of a single confined square aquifer of thickness 35m. The aquifer is an island of dimensions 30km and with parameters as shown in Figures 1 and 2. The potential candidate well locations are shown in Figure 3.

3 DISCUSSION OF RESULTS

Two scenarios which include a deterministic model where only the nominal values were considered (i.e., the perturbation matrices are all zero); and a robust model where both the nominal values and the perturbations were taken into account were analysed.

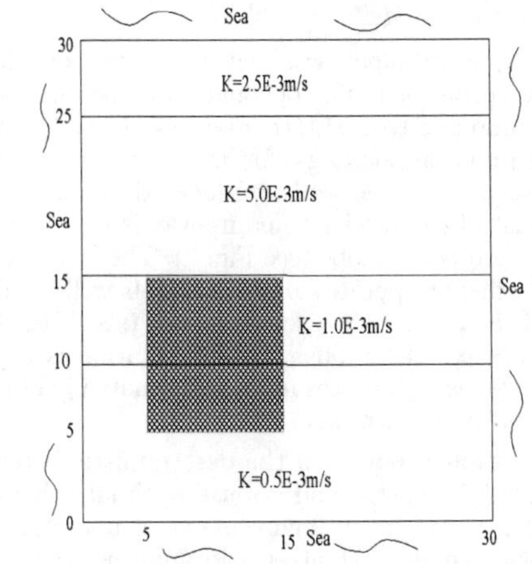

Figure 1: Ecological protection zone and conductivity zones

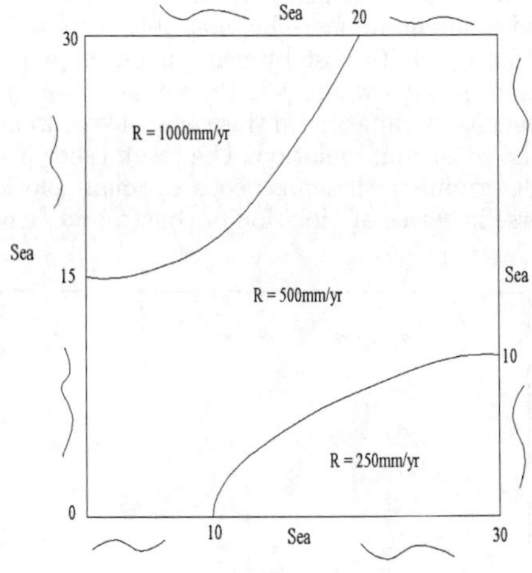

Figure 2: Recharge zones

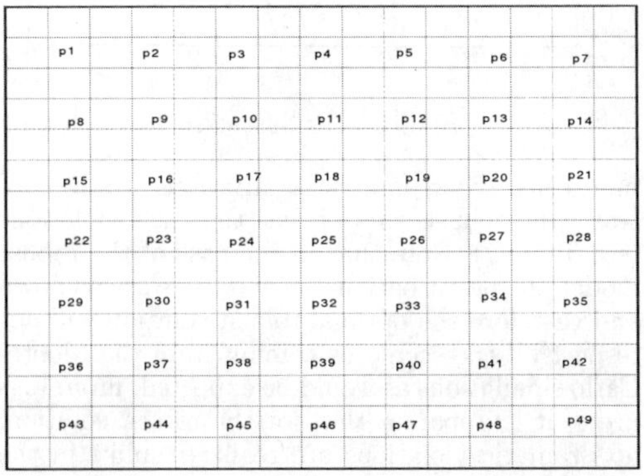

Figure 3: Location of potential pumping wells

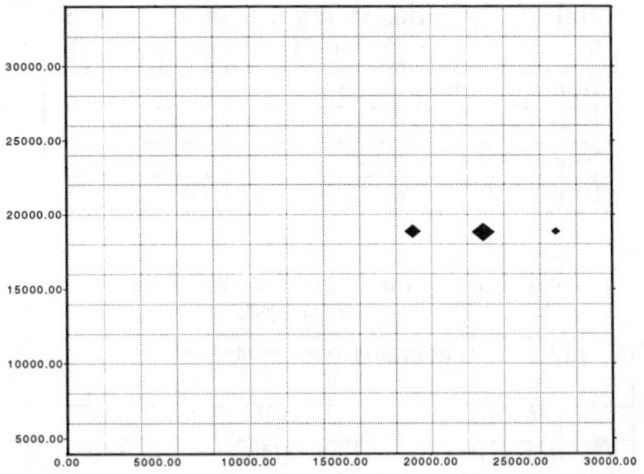

Figure 4: Homogeneous case

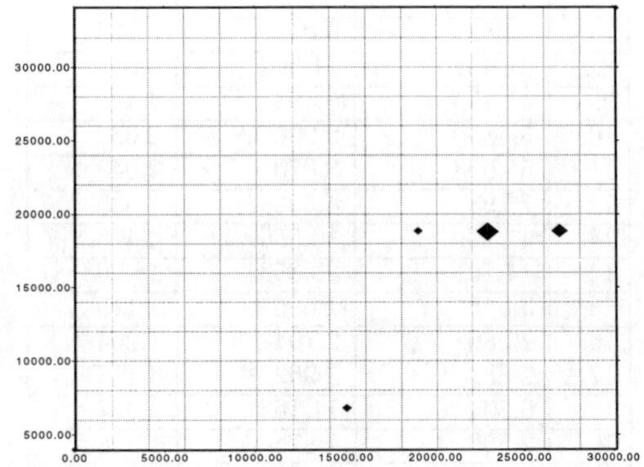

Figure 5: Heterogeneous case

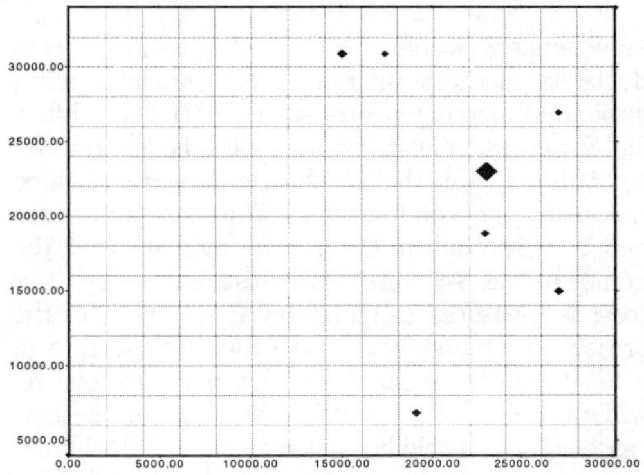

Figure 6: Stochastic case

3.1 *Deterministic model*

This analysis was carried out so as to make comparison of the optimal strategies between a fully

Table 1. Scenarios vs number of wells in use

Scenario	Cost (MU)	Active wells
Homogeneous	1.38	3
Heterogeneous	1.70	4
Stochastic	1.95	7
Robust	2.07	16

Table 2. Penalty costs (MU) for various models

Real. No.	Heterogeneous	Stochastic	Robust
1.	51.732	5.088	3.954
2.	1.810	16.004	9.390
3.	8.790	3.341	1.283
4.	10.354	4.483	4.795
5.	44.333	9.547	8.095
6.	23.184	2.263	2.079
7.	4.288	0.046	0.133
8.	27.431	11.504	6.016
9.	41.255	6.494	4.470
10.	5.485	0.000	0.032
11.	55.603	3.070	2.048
12.	7.793	4.556	1.535
13.	4.717	10.566	4.473
14.	14.101	23.353	16.968
15.	3.545	1.060	1.119
16.	20.802	2.194	1.264
17.	2.883	3.962	3.117
18.	6.504	11.162	9.927
19.	2.559	1.875	1.828
20.	11.794	3.253	1.884
	$\sum = 349.0$	$\sum = 124.3$	$\sum = 84.4$

Real. No. = Realization number

homogeneous scenario (where the hydraulic conductivity of the aquifer does not vary spatially), a banded heterogeneous scenario (where different zones assume different hydraulic conductivity values) and the stochastic scenario (where the hydraulic conductivity values are uncertain and vary spatially). The results indicate that the homogeneous scenario under-estimates the cost (cost is measured in monetary units, MU) of the project compared with the other two scenarios. The stochastic scenario is the most expensive, while the cost of the optimal strategy of the heterogeneous scenario lies between that of the homogeneous and stochastic ones (see Table 1). Figures 4-6 clearly show the locations of the wells for the three optimal scenarios. The optimal scenario corresponding to the stochastic case operates more wells (7) which are more spread out than the other two scenarios (compare Figs 4-6). This tends to suggest that the higher the heterogeneity, the higher the number of active wells and consequently the higher the operational costs.

3.2 *Robust model*

The same problem was solved when considering perturbations. The optimal objective function value amounts to $2.07MU$. Figure 7 shows the optimal pumping strategy for this scheme. As expected, most of the wells are located in the north and east of the model domain away from the ecological protection zone (see Fig. 7). The robust optimal scheme operates more well (16 wells) than any of the previously discussed cases (see Table 1). As expected, the robust optimal solution is more expensive as it operates more wells than any of the other optimal solutions.

To establish how well the deterministic heterogeneous, stochastic and robust optimal schemes would perform in an environment of uncertainty, a post-optimality Monte Carlo analysis was performed using 20 hydraulic conductivity realizations. Violations of the constraints were calculated for each optimal scheme as the hydraulic conductivity field (and hence the responses $a_{i,j}$) assumed different values. The violations were translated into penalty cost by using a linear penalty function: $\rho = qv$ where ρ is the penalty cost, q is the penalty parameter (in this case set to 0.2) and v is the constraint violation. The results show that the deterministic heterogeneous optimal solution is worse in terms of violation of constraints, hence

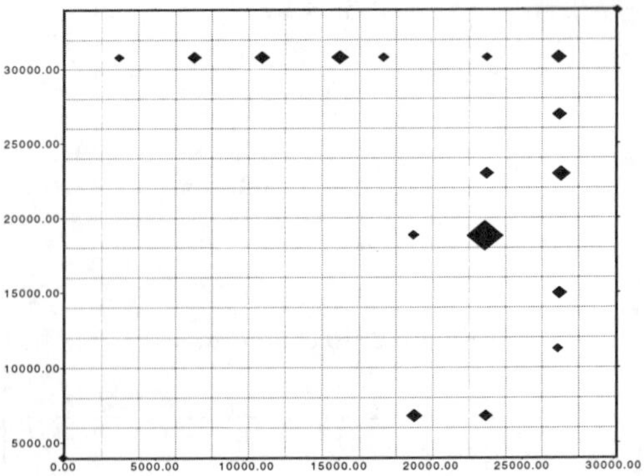

Figure 7: Robust case

more sensitive to parameter variations. To the contrary, the robust solution has the least violations (see Table 2). Note that in the case of the robust model, the perturbation matrices were scaled down and therefore the optimal robust solution will not respect all the scenarios coming from the Monte-Carlo simulations as would be expected. From Figure 8, it is apparent that for the robust solution, no dramatic violations are realized unlike in the other solutions. Further, it is interesting to note that for this robust solution, the objective func-

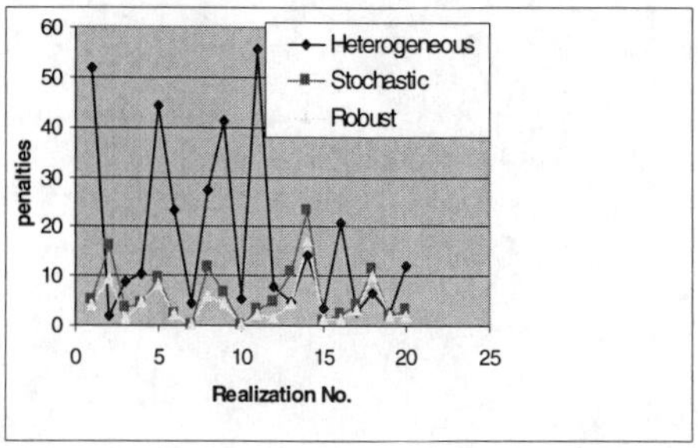

Figure 8: Comparison of solution robustness

tion value of $2.07MU$ is not much different from that of the stochastic case of $1.95MU$ (an increase of about 6%). However, the robustness of the robust solution has substantially increased by about 32% from that of the stochastic one.

4 CONCLUSIONS

We have shown that when confronted with uncertain linear programming problems, such problems can be cast as Second-Order Conic Programs (SOCPs). Further, we have solved a hypothetical example using this technique. Results from the example indicate that the solutions are indeed more robust in comparison with solutions obtained when the parameters are assumed to be known without error. Moreover, the price to pay for a more robust solution (in terms of percentage increase in objective function value) is relatively smaller compared to the gains (in terms of percentage increase in robustness). This means that by approaching the uncertain problem through SOCP, one is assured of robust solutions which are unlikely to be adversely affected by relatively small perturbations within the problem parameters. By recognizing that every system is fraught with uncertainty, it is the desire of every DM to be presented with such robust solutions.

REFERENCES

Andersen, E. and K. Andersen (1999). Exploiting parallel hardware to solve optimization problems. *SIAM News 32*(4).

Ben-Tal, A. and A. Nemirovski (1998). *Convex Optimization in Engineering.* Technion-Israel Institute of Technology.

Boyd, S., C. Crusius, and A. Hansson (1998). Control applications of nonlinear convex programming. *Journal of Process Control 8*(5-6), 313–324.

Boyd, S., L. Vandenberghe, and M. Grant (1994). Efficient convex optimization for engineering design. In *Proceedings IFAC Symposium on Robust Control Design*, Rio de-Janeiro, Brazil.

Lobo, M., L. Vandenberghe, S. Boyd, and H. Lebret (1998). Applications of second-order cone programming. *Linear Algebra and its Applications 284*, 193–228.

Groundwater: Past Achievements and Future Challenges, Sililo et al. (eds) © 2000 Balkema, Rotterdam, ISBN 90 5809 159 7

Hydrogeological investigations in the Otavi Mountain Land

G. Schmidt & D. Ploethner
Federal Institute for Geosciences and Natural Resources, Hannover, Germany

ABSTRACT: The hydrogeological and meteorological conditions in the Otavi Mountain Land are more favourable than in other parts of Namibia and have led to high expectations from groundwater exploration. A numerical model simulates the groundwater system of the Grootfontein karst aquifer at a regional scale. It includes the effects of the drought on the groundwater regime and shows that the hydraulic system has been draining predominantly naturally. The situation might improve only if there is a long period of seasons with above-average rainfall that compensates for the deficits of the past years of drought. Considering the present conditions in the Otavi Mountain Land, the groundwater resources could be used in the short-term for an emergency water supply for the central part of the country. Preference should be given to groundwater from abandoned mines.

1 INTRODUCTION

1.1 *General Situation*

Namibia is the driest country of sub-Saharan Africa. The surface water resources of the country are almost completely utilised, with large dams providing the regional drinking and industrial water supply and small dams on farms the local water supply. However, these dams cannot meet the increasing water demand in future. Omatako Dam, Von Bach Dam and Swakoppoort Dam in the central part of Namibia are used to store the runoff of ephemeral rivers and play a major role in the water supply. They are connected with the Grootfontein-Omatako Canal (part of the Eastern National Water Carrier, ENWC), which conveys groundwater abstracted predominantly from mines in the Otavi Mountain Land to rural areas in the east and to the central areas, including Windhoek. The long-term mean annual precipitation (1926 to 1992) of about 550 mm in the Otavi Mountain Land is considerably higher than in other parts of the country. Owing to this favourable rainfall and the good quality of the soil, cattle and maize are important products of parts of these mountains and their foreland. For nearly two decades, however, precipitation has remained mostly below the long-term mean.

1.2 *Hydrogeology*

The Grootfontein karst aquifer is made up of three synclines: (I) Otavi-Valley-Uitkomst, (II) Grootfontein-Berg Aukas and (III) Harasib-Olifantsfontein; synclinorium IV is part of the Tsumeb karst aquifer (Figure 1). The fractured and locally karstified dolomite of these aquifers has, in absolute terms, a low hydraulic conductivity. It is considerably higher, however, than that of the metasediments of the Nosib Group and the granite of the Grootfontein basement in the foreland of the Otavi Mts. The expectation of a high groundwater potential is based on the high influx of water into the abandoned Berg Aukas, Tsumeb, Abenab and Abenab West mines and the still pumped Kombat mine (mining ceased in April 1998) and the large Otavifontein, Strydfontein and Olifantsfontein springs, as well as the earlier Grootfontein and Rietfontein springs. The cavities formed by underground mining represent easily accessible groundwater.

The hydrogeological structure differs in different parts of the study area. The fractured dolomite in the synclines is karstified near the surface over wide areas and locally covered with soil. Deep karstification and paleo-karst (e.g., Tsumeb and Berg Aukas) occur only locally. The metasediments of the underlying Nosib Group and the basement crop out in the anticlines. They have relatively good soils, which are being farmed. The groundwater recharge

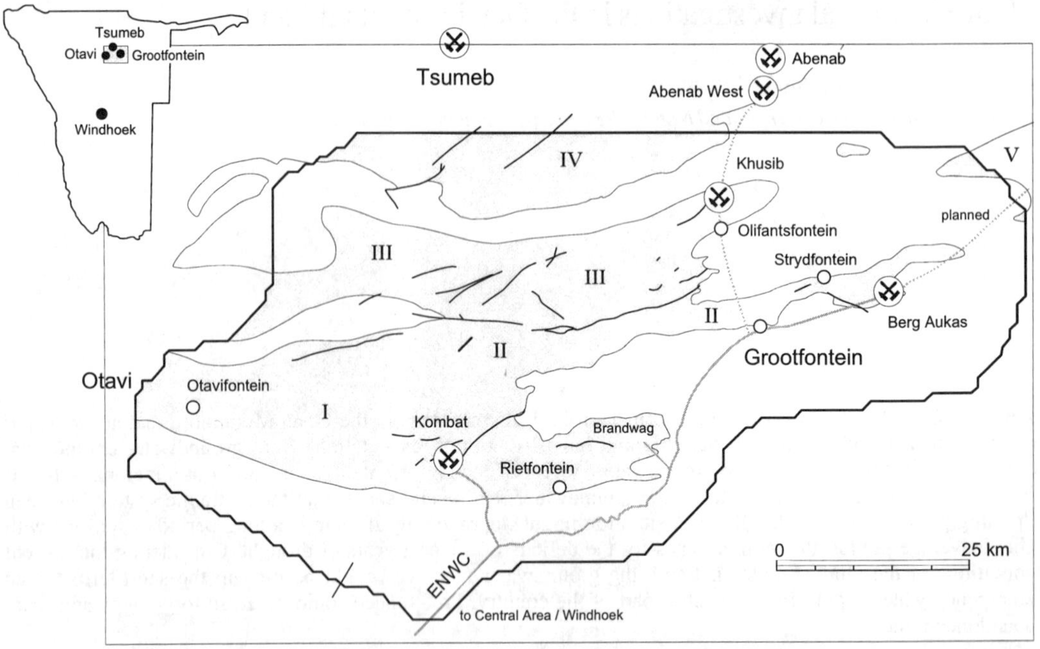

Figure 1. The study area in the Otavi Mountain Land is in northern Namibia and covers 4900 km².

has been below normal for almost twenty years, with a lowering of the water table in the Otavi Mts. of about 2 m per year. This trend is also observed in the foreland, but less strongly.

The project area covers most of the southern part of the semi-arid Otavi Mountain Land, rising up to 500 m above the surrounding plain. The aquifer system is composed of fractured, locally karstified dolomites with a storativity of up to 2 % and a transmissivity of about 100 m^2/d. Both aquifer parameters characterize a fractured rather than a karstified aquifer, therefore, the terms "Grootfontein karst aquifer" and "Tsumeb karst aquifer" are considered to be misleading since they indicate a higher groundwater potential than actually exists. Nevertheless, the dolomite aquifer system of the Otavi Mountain Land is Namibia's most important groundwater resource.

1.3 *Conceptual Model*

The water levels in the lakes (e.g., Otjikoto and Guinas), in karst caves (e.g., Dragon's Breath Cave), and in the shafts of abandoned mines (Berg Aukas and Abenab) vary in the same manner as the groundwater table. The hydraulic system does not react abruptly to rainfall, and discharge from the Olifantsfontein, Strydfontein and Otavifontein springs can be viewed as nearly constant over the year. Analysis of the hydrogeological situation as a whole and the observed groundwater levels leads to the conclusion that the groundwater flow system may be viewed macroscopically as more or less homogeneous (Figure 2).

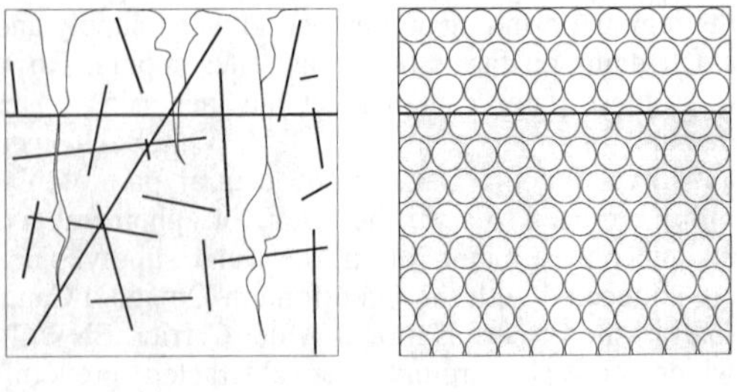

Figure 2. The natural aquifer system in fractured rock (locally karstified) is modelled by a hydraulically equivalent pore water aquifer.

2 GROUNDWATER MODEL

The hydrogeological units consisting of a sequence of aquifers and aquitards are spatial (usually synclines) and are cut by fault zones. Furthermore, the depressions in the groundwater table resulting from the pumping of mine water, as well as the vertical groundwater movement caused by the relief require a three-dimensional model approach. The generalized section through a model syncline (Figure 3) illustrates the basic hydraulic situation and shows the flow components in the model. The groundwater in the dolomite is recharged from precipitation, QINF. The aquifer discharges laterally into the neighbouring rock, QFIX, locally discharges from springs or seepages QEXF, or is extracted, QABS, e.g., during dewatering of mines or for the water supply (e.g., at Grootfontein).

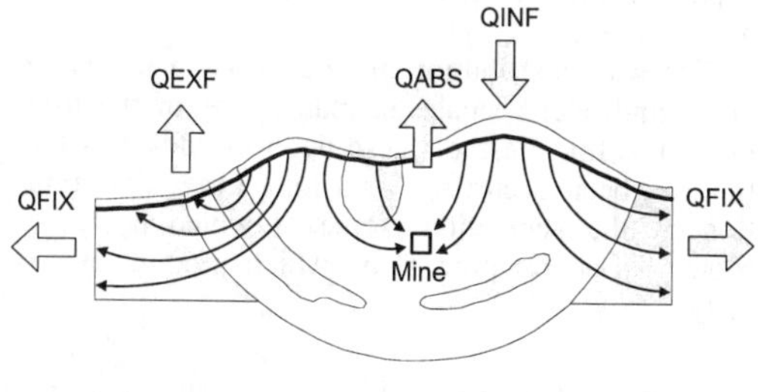

Figure 3. Generalized model section with flow components.

The aquifer system is divided into 14 hydrogeological units on the basis of lithology and hydraulic significance. Each unit is either an aquifer (e.g., dolomite), an aquitard (e.g., schist), or an aquiclude (e.g., granite or gneiss) and the hydrogeological units are assigned values for horizontal and vertical hydraulic conductivity, for storativity, and for the potential recharge rate. The model is discretized horizontally into squares 1 km to a side. The top of the Grootfontein granite/gneiss complex is taken as the bottom of the model. In those areas where this unit crops out, e.g., at Grootfontein, it is included in the flow system as an aquitard.

The partial differential equation for groundwater movement at constant temperature and density of the pore water is solved iteratively, using the preconditioned conjugated gradient method. At the margins of the modelled area, the groundwater levels have been lowered only slightly. The constant levels assumed at the boundaries of the model are based on long-term observation. In the southwest the boundary follows a flow line.

The simulation is for the period between 1979 and 1997 and is based on an initial, calculated groundwater-table distribution, which represents the hydraulic situation at the end of 1978. This initial situation is characterized by good replenishment during at least three years of above average rainfall.

3 MODEL RESULTS

The simulation model may be taken as a reasonable mass balance model for simulating the hydraulic conditions in the Otavi Mountain Land at a regional scale with respect to variable recharge and extraction. The groundwater in the model conforms to the natural flow from the topographic highs in the central part of the mountains, which is the recharge area, to the lower lying foreland. The calculated lowering of the groundwater table reflects, in essence, the long-term natural drainage of the system.

3.1 *Model Calibration*

Comparison of calculated and observed groundwater levels generally shows good agreement. This is also true for recorded hydrographs (available since 1985) and for the long-term recovery of the groundwater table at a local scale after mining at Berg Aukas ceased in 1978. The considerable lowering of the groundwater table since 1978 has occurred mainly in the dolomite aquifer. In comparison, the lowering in the foreland is slight, but not negligible.

3.1.1 *Hydraulic Parameters*

The range of values for storage coefficients and hydraulic conductivity, which was previously derived from aquifer test data, was confirmed by the model calibration. The aquifers have storage coefficients on the order of 1 % (with the exception of Grootfontein in syncline II with up to 2 %). The aquitards and aquicludes (basement with a very low hydraulic conductivity) are assigned values of 0.4 – 0.5 %. The hydraulic conductivity values cover the range from 0.0035 to 3.5 m/d, with the highest values in the Grootfontein area of syncline II (Figure 4).

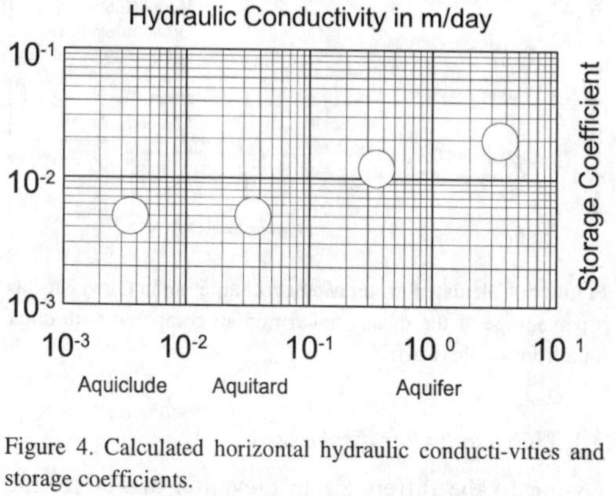

Figure 4. Calculated horizontal hydraulic conducti-vities and storage coefficients.

3.1.2 *Groundwater recharge*

The mean annual rainfall for the study area as a whole, based on data recorded between 1926 and 1992, is about 550 mm. The distribution of the annual rainfall differs and is very uneven in the area. Isotope analyses (Vogel & Van Urk 1975) show that most of the groundwater recharge is from the heavy rains in the summer season. The assumption is made for the model that the rainfall distribution is homogeneous throughout the modelled area and the effective recharge period within the rainy season totals one month.

The regional mean rainfall for 1979 – 1997 yields a mean regional groundwater recharge between 7 and 28 mm for the dolomite aquifer area. The model calculations show that up to 28 mm (i.e., about 5 %) of the long-term annual mean precipitation in the Otavi Mts. reaches the groundwater table during an above-average rainy season. These recharge conditions occurred during the 1975/76 to 1978/79 seasons and the 1988/89 und 1993/94 seasons. The simulations of all of the other years since 1978/79 (i.e., 17 years of lower than average rainfall) show that the regional recharge is about 7 mm (i.e., 1.3 % of the long-term mean). The simulation of 19 years since the last time the aquifer was full shows a mean annual groundwater recharge of about 9 mm (1.7 % of the long-term annual mean rainfall in the Otavi Mts., Figure 5).

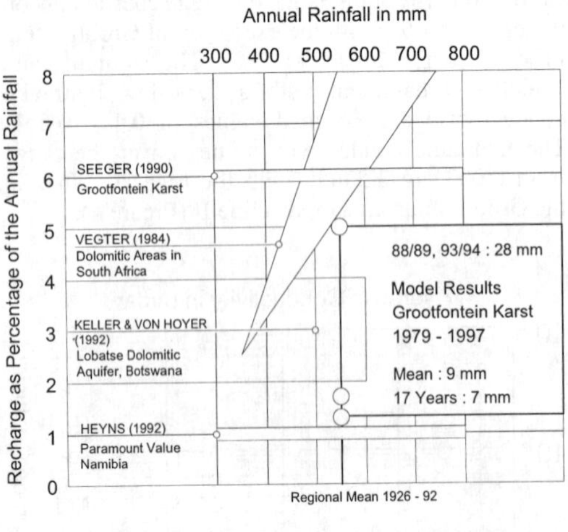

Figure 5. Calculated groundwater recharge in the Otavi Mts. as a percentage of the mean annual rainfall compared with other data from southern Africa.

3.1.3 *Groundwater discharge*

Owing to the difference in elevation of 500 m, the relatively flat areas of the foreland south of the mountains and the plain to the southeast represent the local base level for the groundwater recharged in the mountains, draining further to the Omatako and Omambonde ephemeral rivers. Differences between the hydraulic conductivity in the mountains and the foreland amplify the effect of this difference in elevation. The near-surface groundwater discharges from springs or evaporates from seeps. These areas are simulated in the model, for example, the springs at Khusib, Olifantsfontein, and Mariabronn and at Brandwag in the eastern part of syncline I, where there is more vegetation than in the surrounding area. The effects of the drought can be seen in the groundwater discharge of the springs and seepage areas, which in the model calculations dropped from about 36 million m^3/a in 1979 to about 13 million m^3/a in 1997.

The wide distribution of calcrete is an indication of groundwater seepage in many parts of the foreland. The thickness of these deposits indicates that this has been occurring for a long time. Transpiration by the vegetation (some of which is deeply rooted) is also a process by which groundwater is lost.

3.1.4 *Groundwater budget*

Due to the unfavourable climatic conditions since 1978, the groundwater budget in the Otavi Mts. is characterized by the natural depletion STOR, which is equal to the sum of the discharge at the surface QEXF, of the outflow across the boundaries of the model (at a constant groundwater level) QFIX and of the amount of abstracted groundwater QABS. This sum is reduced by the recharge QINF from precipitation (Figure 6).

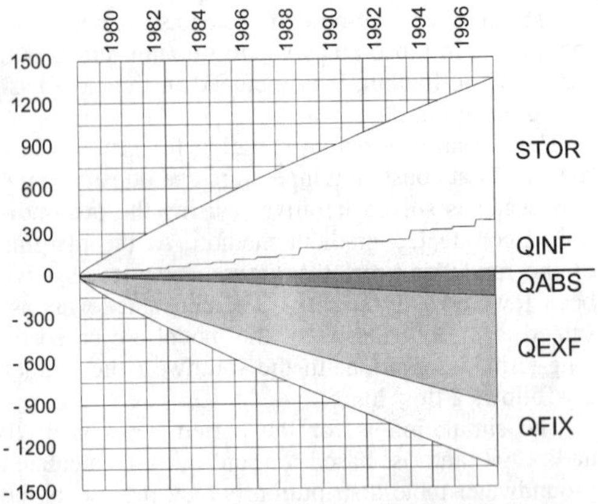

Figure 6. Calculated cumulative groundwater quantities from 1978 to 1997 in million m³.

The outflow across the system boundaries decreases with decreasing hydraulic gradient. When the system was nearly full in 1978 the total outflow from the modelled area amounted to 34.2 million m³/a.

After years of little recharge, it amounted to only 22.4 million m³/a (Figure 7).

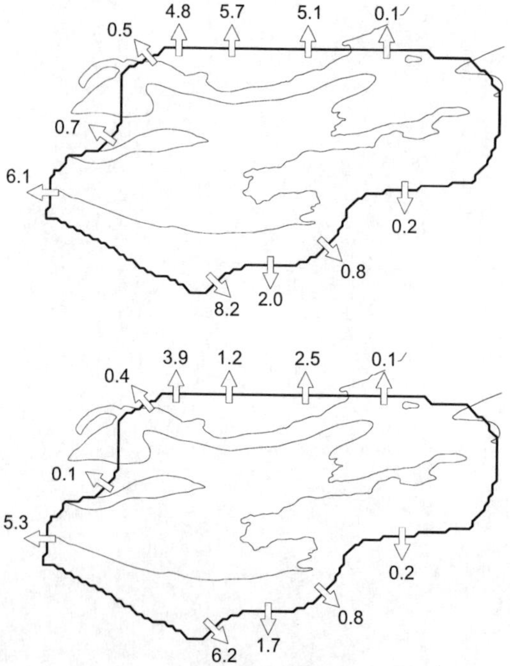

Figure 7. Calculated groundwater flow across the model boundaries in 1978 (top) and in 1997 (bottom) in million m³/a.

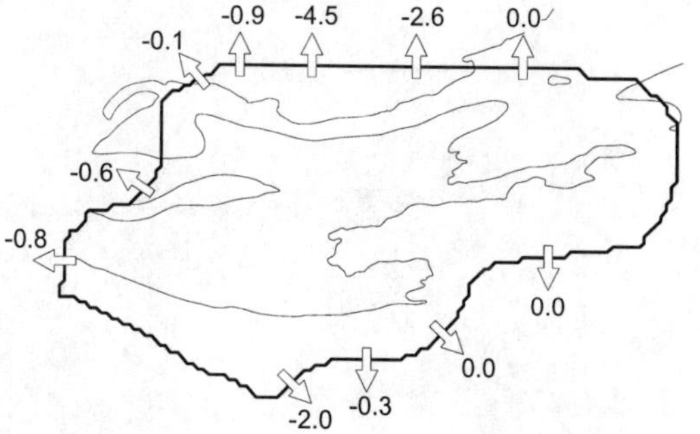

Figure 8. Calculated differences in groundwater flow across the model boundaries from 1978 to 1997 in million m³/a.

The largest reduction was at the northern boundary with a decrease from 16.2 million m³ in 1978 to 8.1 million m³ in 1997. Outflow in the south decreased from 11.2 to 8.9 million m³, in the west from 6.8 to 5.4 million m³. Outflow in the east was a negligible, unvarying 0.3 million m³. As shown by calculations for selected sections of the model boundary, outflow varied somewhat, with increases correlating with the seasonal rains. The differences in boundary flow are illustrated in Figure 8.

4 CONCLUSIONS

The groundwater system has been draining naturally most of the time since 1978. During this time the potential groundwater resources of the Otavi Mts. decreased by a third. The results indicate that the determination of a sustainable yield and the conventional development of the dolomite aquifer system based on relatively shallow wells may not be appropriate. However, the cavities formed by underground mining represent easily accessible groundwater. Groundwater can be abstracted from Kombat and Berg Aukas mines and conveyed through the Eastern National Water Carrier (ENWC) to the capital of Windhoek to overcome water supply shortages during prolonged droughts. The abandoned mines Abenab, Abenab West and Tsumeb may be linked with the ENWC in future as well.

5 ACKNOWLEDGEMENTS

The authors want to extend their sincere thanks to all parties involved during the various stages of this bilateral German-Namibian Project. The contributions of the Namibian Ministry for Agriculture, Water and Rural Development (MAWRD), the Department of Water Affairs (DWA) and the Namibian Water Corporation (Namwater) are much appreciated. On the German side, the Ministry of Economic Cooperation and Development (BMZ) created the overall framework, granted the funds and commissioned the Federal Institute for Geosciences and Natural Resources (BGR) to conduct this technical cooperation project.

REFERENCES

Heyns, P. 1992. Namibia's water resources: A new approach needed.– *Namibia Review*, 1, 17-29.

Keller, S. & von Hoyer, M. 1992. Erkundung und Bewirtschaftung von präkambrischen Karst-aquiferen im südlichen Afrika.– *Zt. dt. geol. Ges.*, 143, 277-290, Hannover.

Ploethner, D. et al. 1997. German-Namibian ground-water exploration project, Vol. D-III. Hydrogeology and isotope hydrology of the Otavi Mountain Land and its surroundings. BGR file No. 10793/97, BGR arch. no. 116049, DWA file no. 12/1/2/16/1; Windhoek, Hannover.

Ploethner, D. & Schmidt, G. 1999. German-Nami-bian groundwater exploration project. Summary Report.

Groundwater studies in parts of western and northern Namibia, 1992-1998. BGR file No. 6022/99, BGR arch. no. 0119249, Windhoek, Hannover.

Schmidt, G. 1996. German-Namibian groundwater exploration project. Otavi Mountain Land: Data base evaluation and conceptual model KARST_01 and KARST_02 (Vol. E-II). Groundwater modelling KARST_01 (part 1), KARST_02 (part 2) (Vol. E-III). BGR file No. 11267/95, BGR arch. no. 113752; Windhoek, Hannover.

Schmidt, G. & Ploethner, D. 1999. Grundwasserstudie zum Otavi-Bergland, Namibia.– *Z. an-gew. Geol.*, 43(3), 114-123, Hannover.

Seeger, G. 1990. An evaluation of the groundwater resources of the Grootfontein karst area. Unpubl. DWA report, no. 12/5/G2, 187 pp, Windhoek.

Vegter, J.R. 1984. Dolomitic water supplies with special reference to southern and western Transvaal. Workshop on groundwater drilling in the Republic of South Africa, Johannesburg.

Vegter, J.R. 1993. Concepts for mapping South Africa's hydrogeology. Africa needs groundwater – a ground water convention 6–8 Sept. 1993, conv. pap., Vol. II: 11 pp; Johannesburg.

Vegter, J.R. & Foster, M.B.J. 1992. The Hydrogeology of dolomitic formations in the southern and western Transvaal. In: Paloc, H. & Back, W. (eds.): *Hydrogeology of selected karst regions. International contributions to Hydrogeology*, 13: 355–376; Hannover (International Association of Hydrogeologists).

Vogel, J.C. & Van Urk, H. 1975. Isotopic composition of groundwater in semi-arid regions of southern Africa.– *Journal of Hydrology*, 25, 23-36.

Groundwater: Past Achievements and Future Challenges, Sililo et al. (eds) © *2000 Balkema, Rotterdam, ISBN 90 5809 159 7*

On variable-density groundwater flow in heterogeneous porous media

C.T. Simmons
Flinders University of South Australia, Adelaide, S.A., Australia

J.M. Sharp Jr
University of Texas, Austin, Tex., USA

ABSTRACT: In certain hydrogeological situations, fluid density variations occur because of changes in the solute concentration and temperature of the groundwater. These include seawater intrusion, high-level radioactive waste disposal, groundwater contamination, and design of saline water disposal basins, such as in the Murray-Darling Basin of Australia. When the density of the invading fluid is greater than that of ambient groundwater, density-driven free convection leads to rapid and extensive solute transport. The critical role that heterogeneity plays in the onset, growth, and/or decay of convection is examined by numerical experiments. Different styles of heterogeneity are considered and range from continuously "trending" heterogeneity to discretely fractured geologic media. The results indicate that onset of instabilities and their subsequent growth and decay are intimately related to the structure and variance of the permeability field. Resultant plume patterns, migration rates, and pathways are not presently amenable to prediction.

1 INTRODUCTION

Many important problems in groundwater and of environmental concern involve transport issues. These include groundwater contamination, saline water intrusion, radioactive waste disposal, geothermal energy development, groundwater-surface water interaction, and storage of fluids, energy, and materials.

Field and laboratory studies both demonstrate that fluid density gradients caused by variations in concentration and temperature play a significant role in the transport of solutes in groundwater systems. Density-driven free convective flow may manifest itself in the form of lobe-shaped instabilities or fingers. The instability mechanisms have been studied for nearly a century and are called, variously, Benard, Rayleigh, solute, or thermal instability. The study of instability probably began when Benard (1900) observed hexagonal patterns in an unstably stratified horizontal fluid layer heated from below. Rayleigh (1916) analyzed an idealized model of instability and determined the characteristics of the fluid layer that caused the instability to occur. The dimensionless Rayleigh Number (*Ra*) is commonly used to predict and describe the occurrence of instability. *Ra* combines key properties of the porous medium and the fluids, including the permeability. For *Ra* greater than some critical value, fluid system becomes unstable and free convection occurs.

Free convection enhances hydrodynamic mixing. It is significant for three reasons: (1) the total quantity of solute involved in the convective mixing process is typically greater than for diffusional transport; (2) the time scales associated with the mixing process are much less; and (3) the dimensions of the mixing zone are typically larger and the solutes spread over much greater distances.

There are many situations and hydrogeological processes in which variable-density groundwater flow and solute transport are important. Some of these are compiled in Table 1.

Table 1. Some hydrogeologic situations in which free convection is significant (e.g., Combarnous & Bories, 1975; Duffy, 1991; Hassanizadeh & Leijnse, 1988; Land, 1991; Oostrom et al., 1992; Ritsema et al, 2000; Schincariol & Schwartz, 1997; Simmons et al., 1999; Voss & Souza, 1987).

Sea water intrusion
Infiltration of leachates from waste disposal sites
DNAPL flow and transport
Infiltration through hydrophobic soils
Flow through salt formations in high level disposal sites
Dolomitization by seepage reflux
Hydrocarbon migration and entrapment mechanisms
Diagenesis of sedimentary basin sediments
Processes beneath playas and playa lakes
Heat and fluid flow in geothermal systems
Design of saline (and irrigation) water disposal basins

Below we briefly discuss a case history of density-driven free convection beneath saline water disposal basins in the Murray-Darling Basin of Australia; outline the significance of the heterogeneous conditions in free convection; and present the results of numerical simulations in assessing the effects of heterogeneous permeability distributions.

2 SALINE WATER DISPOSAL BASINS

The Murray-Darling Basin of Australia is the agricultural heart of Australia, and like many arid and semi-arid regions, irrigation is an integral to agricultural production. The clearance of natural vegetation combined with irrigation practices in the Murray-Darling Basin has led to salinity problems. Rising groundwater levels and increased soil and groundwater salinity must be mitigated by drainage and the disposal of the large volumes of saline effluent is a critical management issue. The most common approach is the diversion and storage of saline water at the land surface in saline disposal basins. Concerns have been raised about the possible impacts of these disposal basins on underlying groundwater systems and, subsequently, on surrounding low-lying land and river systems. The continued use of the disposal basins as wastewater evaporation sites therefore requires a detailed understanding of groundwater dynamics.

Evaporative concentration of saline effluent can lead to situations where dense brine overlies less dense groundwater. Simmons et al. (1999) have shown that under certain conditions, the dense basin brines become hydrodynamically unstable causing them to mix with groundwater over distances several orders of magnitude greater than calculated due to diffusion alone. This process is significantly exacerbating salinization of underlying aquifers. Understanding the factors that enhance or reduce saline water recharge into the underlying aquifers is critical. Simmons et al. showed that heterogeneity in permeability is one of the important, and most poorly understood factors. In many cases we cannot predict if density-driven free convection will occur. The conceptual model (Fig. 2) used in the numerical analysis represents a saline disposal basin.

3 THE EFFECTS OF HETEROGENEITY IN THE PERMEABILITY FIELD

Almost all quantitative hydrogeological studies on free convection have assumed homogeneous and isotropic conditions, and those that haven't have yielded conflicting conclusions. Schincariol and Schwartz (1990) examined variable-density groundwater flow in homogeneous layered and lenticular media. They noted that density differences as low as 0.8kg/m^3 were sufficient to produce instabilities in a lenticular medium, but that the combination of convective dispersion and nonuniform flow caused by heterogeneities created large dispersion that dissipated instabilities. Schincariol et al. (1997) conducted numerical simulations incorporating heterogeneity in the permeability field and found that the mean, variance, correlation length scales of the permeability were critical in the onset and subsequent growth or decay of instabilities. However, relevant stability criteria for describing these processes in heterogeneous porous media are still unclear.

In a numerical study of density-dependent solute transport in discretely-fractured geologic media, Shikaze et al. (1998) found that fractures with aperture values as small as 25 μm significantly enhance contaminant migration in the saturated zone relative to the case where the fractures are absent. They concluded that where there is a high degree of nonuniformity in fracture apertures and spacings that the migration rates, directions, and patterns of dense plumes might not be amenable to prediction. Shikaze et al. contradict the findings of Schincariol and coworkers that heterogeneities tend to increase dispersion and dissipate instabilities. We surmise that both findings are correct and that delineation of the style and structure of the heterogeneous permeability field explains the differences in behavior.

4 MATHEMATICAL MODELING

In order to examine variable-density flow and transport processes in heterogeneous geologic media we utilized the variable-density groundwater flow and solute transport code SUTRA (Voss 1984). The style of heterogeneity considered varies from a "random" heterogeneous porous media to a more regular periodic style of continuous heterogeneous porous media, from which we infer results for a discretely fractured system (Fig. 1). Such periodic variations are the result of both fracture clustering or sediment depositional processes. We utilized both stochastic (random) and ordered/periodic permeability fields to simulate a variety of natural systems. We also examine the applicability of Rayleigh number for predicting onset conditions for density-driven convection in heterogeneous porous media.

4.1 *The SUTRA model*

SUTRA simulates density-dependent groundwater flow and solute transport in 2 dimensions using the governing equations of conservation of fluid mass and conservation of either solute mass or thermal energy. It computes flow velocities using Darcy's Law. SUTRA employs a hybrid finite-element mesh discretization and integrated finite-difference time-stepping method to solve the governing equations.

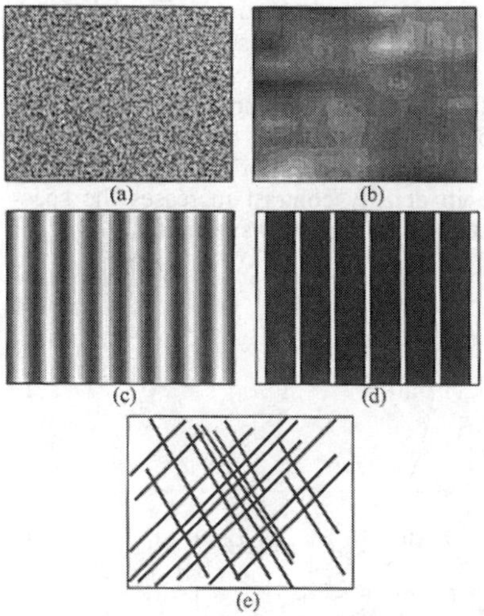

Figure 1. Different styles of geologic media: (a) homogeneous porous media, (b) heterogeneity represented by a stochastic random spatial function, (c) continuous periodic sinusoidal heterogeneity, (d) fractured geologic media represented by regular vertical fracturing and (e) fractured geologic media with non-uniform fracture aperture, separation distances and orientation.

4.2 *Conceptual model*

A conceptual model of a Murray-Darling saline water disposal basin is shown in Figure 2. The disposal basin (or any other dense fluid contaminant source) with concentration C_U *overlies an aquifer with an initial concentration* C_L*. The basin is on the surface at a position described mathematically by* $x_L<x<x_R$

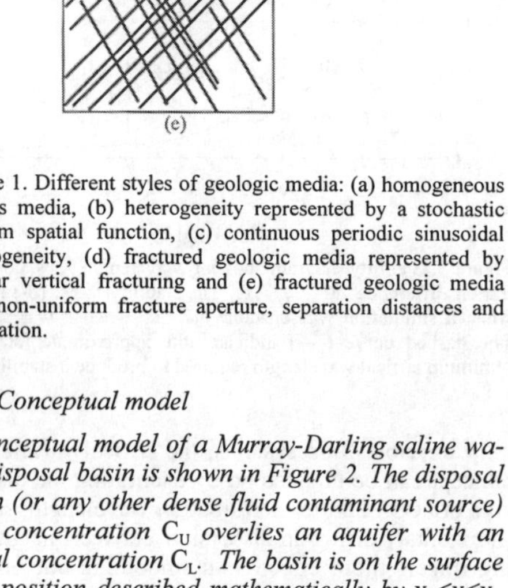

Figure 2. Simulation geometry, boundary and initial conditions for the simulation of a dense contaminant plume source.

where x is the horizontal Cartesian coordinate relative to the left hand boundary. The flow and trans port boundary conditions are no flow except for the dense fluid source. Initially, pressure throughout is hydrostatic with an initial concentration in the aquifer given by C_L.

4.3 *Spatial and temporal discretization*

The cross section was discretized in a uniform mesh of 2761 nodes and 2500 square elements, by choosing 250 finite elements horizontally (Δx=1m) and 10 finite elements vertically (Δy=1m). Transient simulations were run with 100 time steps to cover a simulation time of about 15 years. Time step sizes increased from one-month to yearly steps during the simulations.

4.4 *Hydraulic conductivity distributions*

Numerous studies suggest that permeability (k) distributions can often be approximated as log normal. If k is log-normally distributed, we can define a new parameter $Y=log(k)$ that is normally distributed (i.e., Y: $N[\mu_y, \sigma_y]$ where μ_y is the mean and σ_y is the standard deviation). A normal distribution of Y is converted to a log-normal distribution using the exponential transform $k=e^{2.3Y}$.

For permeability distributions represented by a random spatial function, the spatial dependence between neighboring values of the random variable is defined in terms of a stochastic model that specifies the spatial autocorrelation or alternatively its variogram throughout the system. We use the exponential variogram with effective range 3τ, defined as the distance at which the variogram value is 95% of its plateau or sill. For $h>3\tau$, permeabilities are uncorrelated. In a vertical slice, it is necessary to specify a correlation anisotropy factor $anis=\tau_H/\tau_V$ to reflect geologic structures, such as bedding, microfractures, or clast orientation.

To evaluate the effect of heterogeneity structure and, in particular, the wavelength of fracture clusters or sedimentary structures, the permeability value $k_{x,y}$ at any x and for all y was assigned using:

$$k_{x,y}(x,y) = k_{AV} 10^{\psi(\sin\frac{2\pi x}{\lambda})} \tag{1}$$

Heterogeneous permeability distributions are defined in terms of an average k_{AV}, an amplitude ψ (representing the order of magnitude variation around the average value), and a wavelength λ.

4.5 *Model parameters*

The input parameters, based upon Murray-Darling Basin studies (Simmons et al., 1999; Simmons and Narayan, 1997), are given in Table 2. These were held constant throughout all simulations unless otherwise specified. Parameters such as k_{AV}, C_U, λ and ψ were varied in specific simulations. For the stochastic permeability distributions μ_y, σ_y, τ_H and $anis=\tau_H/\tau_V$ were varied between simulations.

Table 2. Model parameters used in sinusoidal and stochastic permeability studies based on Murray-Darling Basin studies (e.g., Simmons, et al., 1999; Simmons and Narayan, 1997).

Model Parameter	Value
Aquifer concentration (C_L)	0 mgl^{-1}
Fluid dynamic viscosity	10^{-3} $kgm^{-1}s^{-1}$
Coefficient fluid density change	700 kgm^{-3}
Water compressibility	4.5 x 10^{-10} Pa^{-1}
Soil compressibility	$1x10^{-8}$ Pa^{-1}
Porosity	0.35
Intrinsic permeability (k)	k(x,y) m^2
Longitudinal dispersivity	1 m
Transverse dispersivity	1 m
Molecular diffusivity	10^{-9} m^2s^{-1}

4.6 *Analysis of model results*

Below we discuss how periodic vertical permeability distributions control free convection and then evaluate the effects of stochastic permeability distributions.

Periodic hydraulic conductivity distributions. These simulations examined the effect of fracture clusters or depositional fabric periodicity (varying sinusoidal wavelength) on stability. A set of numerical simulations were run with $k_{AV}=10^{-13}$ m^2, $\psi=1$ and with C_U= 5000, 10000, 20000, 30000, 40000, 50000 and 100000 mgl^{-1} and with permeability distribution wavelength varied from λ=2, 5, 50 and 100m. Lobe-shaped protrusions or "fingers" that sink into the underlying aquifer under gravitational influence manifest instability. Output results show that not all configurations are unstable. By defining a stability index where simulations at a given time (e.g., 12 years) were classed as either stable (SI = stability index = 0) or unstable (SI=1), the dependence of stability on both concentration and wavelength is evident.

The two instability criteria are: (i) the amplitude of the initial perturbation grows with time as evidenced by lobe shaped protrusions or fingers of the concentration and (ii) zones of convective motion (upward and downward fluid movement) evident in fluid velocity vector plots. For small concentration differences, progressively larger wavelengths are required to propagate instabilities (Fig. 3). As the concentration difference increases, the spectrum of unstable wavelengths increases and the minimum critical wavelength required to perturb the system is reduced. In flow through natural porous media, interfacial disturbances are continuously generated due to the heterogeneities of the medium (Moissis and Wheeler 1990). Therefore, perturbations to flow occur over many spatial scales. Slight differences in pore geometry to larger heterogeneities on the scale of the dense plume source are sufficient to generate instabilities. The maximum wavelength should be similar to the dimensions of the basin (Marle 1981). The results of this analysis agree with List (1965) in that: (i) horizontal two-dimensional flows in saturated porous media with a denser fluid overlying a less dense fluid are always potentially unstable, although instability may not occur when the perturbations are below the necessary critical wavelength; (ii) longer wavelengths are more unstable; and (iii) an increase in density contrast increases the spectrum of unstable perturbation wavelengths.

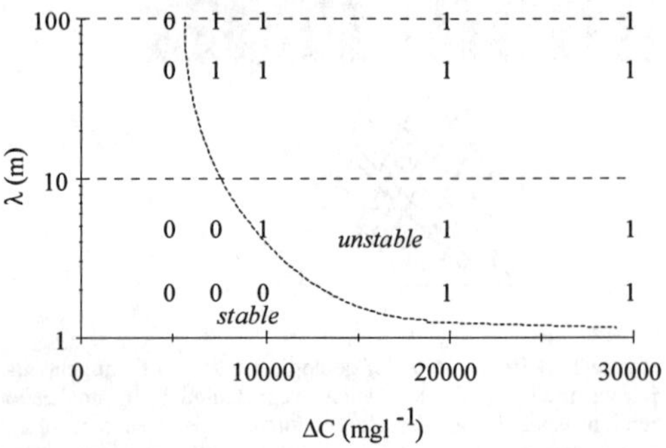

Figure 3. Stability diagram showing wavelength λ vs. concentration difference $\Delta C=C_U-C_L$. The stability index (*SI*) is "1" when a simulation was unstable and "0" when it was stable. The dashed curve (----) indicates the approximate locus of minimum critical wavelength required to produce instabilities.

In miscible fluid systems, the critical wavelength concept can be understood by examining the dispersive effects. Shorter perturbation wavelengths make (horizontal) diffusion/dispersion between neighboring fingers more effective at smoothing out fingers before they can grow. In fact, the existence of a mixing zone (in which the viscosity and density of the fluid vary progressively from those of the displacing to those of the displaced fluid) diminishes the principle cause of the instability - differences in fluid properties (Marle 1981).

Simulations tested the effect of average permeability of a porous layer on the stability of a dense plume. As expected, reducing the average permeability of the field stabilized flows. More interesting was the effect of permeability distribution amplitude. A set of simulations was performed with four orders of magnitude permeability variation (i.e., ψ=2 as opposed to ψ=1), $k_{AV}=10^{-13}$ m^2, and C_U=5,000 mgl^{-1}. Results for all wavelengths tested in the range λ=2m to λ=100m now showed instabilities. Other simulations confirmed that the permeability amplitude alone could alter system stability, even with average permeability k_{AV} and basin salinity C_U held constant. This is because increasing the amplitude of a permeability distribution creates: zones of both: (i) higher permeability where instabilities are more likely to occur and are easier to propagate and (ii) lower permeability that hinder lateral dispersion between fingers. The limiting case is a set of periodic and uniform high conductivity fractures in a low-permeability matrix. This is a highly unstable configuration.

It is clear that changing the amplitude of the permeability distribution alone changes the stability state of the system. Such changes are not reflected in a *Ra* based upon an average hydraulic conductivity. Therefore, *Ra* is not be a useful predictor of instability in heterogeneous media; it does not account for the spatial variability and structure of the permeability field, both of which are critical in the onset, growth, and/or decay of instabilities.

Stochastic hydraulic conductivity distributions. Ten random permeability fields were generated for use in the model. Each of the fields was defined by $Y: N[\mu_y = -6.0, \sigma_y = 0.5]$. The structural constants used were τ_H=300m and *anis*=200. The resulting field had an average hydraulic conductivity of ~ 0.1 m day^{-1} or an average intrinsic permeability $k_{AV} \sim 10^{-13}$ m^2. The permeability distributions, plotted as *-log(k)* are given for four stochastic fields in Figure 4a. The concentration in the disposal basin (C_U) was 50,000 mgl^{-1}. Figure 4b shows the computed concentration profiles at a simulation time of 12 years for these permeability distributions.

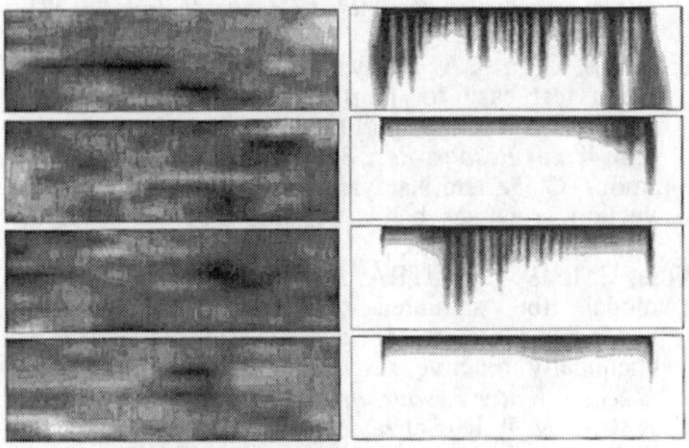

Figure 4. (a) Permeability distributions, plotted as *-log(k)* for four stochastic fields defined by $Y: N[\mu_y = -6.0, \sigma_y = 0.5]$. The structural constants used were τ_H=300 and *anis*=200. Average permeability $k_{AV} \sim 10^{-13}$ m^2. Lighter regions correspond to high permeability regions and darker regions to lower permeability regions, (b) Corresponding cross-sectional showing concentrations after 12 years.

For these *statistically equivalent permeabilities* in the aquifer, entirely different behaviors occur depending upon the spatial location of the low and high permeability zones. The spectrum ranges from entirely erratic and unstable behavior to almost completely stable behavior. The stability of the plumes is locally determined as a function of the permeability field in the vicinity of the plume boundary layer. Lower permeability regions dissipate instabilities and higher permeability regions promote their growth. A *Ra* based upon mean permeability cannot predict the stable or unstable states.

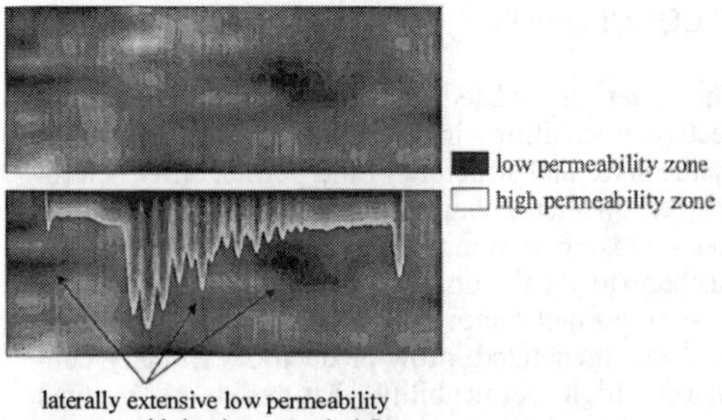

Figure 5. Conceptual model of instability in heterogeneous porous media. Long, vertically connected high permeability conduits enhance instabilities and laterally extensive low permeability conduits tend to dissipate them.

To test the effect of varying the standard deviation (σ) of the permeability, a set of ten random permeability fields were then generated and defined by $Y: N[\mu_y = -6.0 , \sigma_y = 1.0]$. Increasing σ increases the spread of permeability. Larger scale mixing enhances dispersion within the system and the resultant number of unstable cases is lower as σ increases. Increasing σ also produces local zones of lower permeability that stabilize the flow by providing a barrier to the necessary vertical flows associated with convective instabilities (Fig. 5).

This finding contrasts with that of the periodic permeability distributions where increased amplitude or variance in the permeability distribution promotes instability. This fundamental difference is related to the structure or style of the heterogeneous field. The periodic (modeled as sinusoidal) distributions are highly ordered with longer continuous conduits; vertically propagating fingers do not encounter low permeability barriers. The larger variance in the periodic case reduces lateral dispersion, but the stochastic distributions are more disordered. If a finger begins to propagate in a sandy conduit, it is likely to encounter a lower permeability. At the same time, dispersive mixing is enhanced due to the irregular heterogeneous structure and continually dissipates the buoyancy force driving the instability.

The effect of permeability structure or correlation length scale was examined in a set of numerical experiments with varying structural constants and with $Y: N[\mu_y = -6.0, \sigma_y = 0.5]$. The horizontal continuity values were τ_H =1000m, 300m, 100m and 10m. The anisotropy factor was varied in each case to keep the vertical continuity constant at τ_V =1.5m. The results showed that increased horizontal correlation length and low vertical-correlation length scales promote stability. This is because higher correlation length scales produce more laterally extensive low permeability zones that impede or dampen the vertical development of instabilities.

5 CONCLUSIONS

The onset and subsequent growth or decay of convective instabilities in hydrogeology are controlled by the style and structure of the permeability heterogeneity. Heterogeneity is important for two main and opposing reasons - it serves as the triggering mechanism for the onset of instabilities and it is the most important factor controlling whether instabilities, once generated, grow or decay. Vertically continuous high permeability features (e.g., vertical fractures, conduits, sand dikes) enhance growth conditions where intermediate low-permeability zones provide resistance to horizontal dispersive mixing. In lenticular or layered structures, increased variance and increased horizontal correlation create laterally extensive low permeability barriers to the vertical flows necessary to maintain convection. More realistic non-uniform distributions for fracture apertures, separation distances and orientations should be the subject of further study (Fig. 1).

The onset conditions, growth rates, shapes, and resultant plume directions are not presently amenable to prediction. In many cases, the Rayleigh Number cannot predict the occurrence of instabilities in heterogeneous porous media. The relevant stability criteria are unclear and unanswered questions arise include: (i) what is the level of permeability characterization required to predict transport processes accurately? (ii) Can we uncover the requisite details and, if not, what simplifications are permissible and yet allow accurate depictions of density-dependent phenomena? We contend that it is difficult to predict the fate of a dense contaminant plume in the field. However, given the significance of variable-density flow in many hydrogeological processes, evaluation of these phenomena is needed.

ACKNOWLEDGEMENTS

We acknowledge the United States Department of Energy, Geosciences Research Program, for partial research support under grant DE-FG03-97ER14772. Author CTS wishes to acknowledge partial support provided by an Australian Research Council grant.

REFERENCES

Combarnous, M. & S. A. Bories, 1975. Hydrothermal convection in saturated porous media. *Advances in Hydroscience*, 10: 231-307.

Duffy, C. J. 1991. Haline convection of groundwater in the mountain and playa-lake deserts of the Great Basin: Int. Assoc. Hydrogeologists, *Hydrogeology of Low Permeability Environments, Selected Papers*, 2: 69-81.

Hassanizadeh, S. M. & T. Leijnse 1988. On the modeling of brine transport in porous media, *Water Resour. Res.*, 24(2): 321-330.

Land, L. S. 1991. Evidence for vertical movement of fluids, Gulf Coast sedimentary basin*: Geophysical Research Ltrs.*, 18: 919-922.

List, E.J. 1965. The stability and mixing of a density-stratified horizontal flow in a saturated porous medium, Rep. KH-R-11, Calif. Inst. of Technol., Pasadena.

Marle, C.M. 1981. *Multiphase Flow in Porous Media*, Gulf, Houston, Tex. 257 pp.

Moissis, D.E. & M.F. Wheeler 1990. Effect of the structure of the porous medium on unstable miscible displacement, *Dynamics of Fluids in Hierarchical Porous Media*, Academic, San Diego, CA.: 243-271.

Oostrom, M., J.S. Hayworth, J.H. Dane & O. Güven 1992. Behavior of dense aqueous leachate plumes in homogeneous porous media, *Water Resour Res.*, 28(8): 2123-2134.

Rayleigh, Lord (J.W. Strutt) 1916. On convection currents in a horizontal layer of fluid when the higher temperature is on the under side, *Philos. Mag., Ser.* 6, 32: 529-546.

Ritsema, C.J., J.L. Nieber, T.S. Steenhuis, L.W. Dekker & J. C. van Dam 2000. Principles and modeling approaches of preferential flow in hydrophobic soils, *EOS*, 81(19), S195.

Schincariol, R.A. & F.W. Schwartz 1990. An experimental investigation of variable density flow and mixing in homogeneous and heterogeneous media, *Water Resour. Res.*, 26(10): 2317-2329.

Schincariol, R.A., F.W. Schwartz & C.A. Mendoza 1997. Instabilities in variable density flows: Stability analyses for homogeneous and heterogeneous media, *Water Resour. Res.*, 33(1): 31-41.

Shikaze, S.G., E.A. Sudicky, & F.W. Schwartz 1998. Density-dependent solute transport in discretely-fractured geologic media: is prediction possible?, *J. Contam. Hydrol.*, 34: 273-291.

Simmons, C.T., K.A. Narayan, & R.A. Wooding 1999. On a test case for density-dependent groundwater flow and solute transport models: The salt lake problem, *Water Resour. Res.*, 35(12): 3607-3620.

Simmons, C. T., and Narayan, K. A., 1997. Mixed convection processes below a saline disposal basin, *J. Hydrol.*, 194: 263-285.

Voss, C. I. 1984. SUTRA: A finite-element simulation model for saturated-unsaturated fluid density-dependent groundwater flow with energy transport or chemically reactive single-species solute transport, *U.S.G.S. Water Resour. Invest.* Rep. 84-4369, 409 p.

Voss, C.I., & W.R. Souza 1987. Variable density flow and solute transport simulation of regional aquifers containing a narrow freshwater-saltwater transition zone, *Water Resour. Res.*, 23(10): 1851-1866.

Groundwater: Past Achievements and Future Challenges, Sililo et al. (eds) © 2000 Balkema, Rotterdam, ISBN 90 5809 159 7

Groundwater investigation and modelling of effluent irrigation: A case study

J.J.G.Vermaak & D.G.Walker
Wates, Meiring and Barnard (Pty) Limited, South Africa

J.C.F.Barnes
E. Martinelli and Associates, South Africa

ABSTRACT: The paper documents a case study in which effluent is discharged by irrigation within a valley area. It has been identified that groundwater flow is dominated by fracture systems within a weathered dolomitic aquifer and also by high permeability alluvial paleo-channels overlying the dolomitic aquifer. Groundwater discharges mainly at three springs located downstream of the irrigation area. A 3D-groundwater model was constructed and was calibrated against time-series borehole and spring flow data. The model accurately predicted that the river does not act as a geohydrological boundary. The model was used to predict groundwater and surface water impacts and to assess various remediation strategies. The model predicted a significant improvement in groundwater quality within nine years by lowering the contamination load at source. However, during this period, contamination loads towards the river will increase as the pollution plume moves through the system and discharges at the springs.

1. INTRODUCTION

In the study area, paper mill effluent irrigation for about 3 decades has impacted upon a river in the Mpumalanga province, South Africa. Concerns have been raised that downstream receiving water would be of unacceptable quality to some of the users, particularly tobacco farmers.

Numerous investigations have revealed that a pollution plume has developed underneath the irrigation area and is moving down-gradient along the valley and towards the river. The investigations revealed several seepage fronts alongside the river. In 1998, dewatering wells were installed to limit seepage to the river at one of the seep zones.

Coinciding with the investigations, it was observed that decreasing water qualities are emanating from three springs (eyes), situated downstream of the irrigation area, two of which are situated opposite of the river (Figure 1). It became apparent that the contaminant plume moves along preferential pathways from where it mainly discharges in the river at the springs. By 1999, the combined spring flow of 460ℓ/s resulted in a salt load exceeding 22.7 tons per day, affecting stream water quality.

An investigation was initiated as part of the Integrated Water Management Plan with the purpose to understand the physical controls on the groundwater system and to evaluate a number of remedial strategies with the aid of a calibrated numerical model.

The success of the first dewatering scheme created the expectation that one or more additional areas of groundwater abstraction could be identified similar to the first. Groundwater abstraction was considered to be a viable and, possibly, better alternative than collecting the spring water at their points of discharge. However, there was a need to understand the geohydrological regime in sufficient detail to be able to predict the location of preferential flow paths and to target these for limited drilling and testing to provide an acceptable confidence limit of the project viability.

2. SITE DESCRIPTION AND GEOHYDROLOGY

2.1 Geohydrological Setting

The unique and complex combination of geology and geomorphology of the region has given rise to a singular geohydrological setting. Stratigraphy and structure control the surface drainage and erosion. Thus, the key to understanding the flow of the groundwater from the irrigation fields to the river lies in the correct interpretation of the geology and structure and, in turn, the interaction with the younger geomorphological processes.

Studies highlighted the presence of thick alluvial and colluvial soils covering dolomite bedrock, which has been intruded by one or more post-Transvaal intrusive bodies. A combination of the downstream Black Reef Quartzite outcrop and the deeply weathered and karstic dolomite up stream of it resulted in the development of a mature river system with a significant amount of groundwater, some of which emerges into the river in the form of springs.

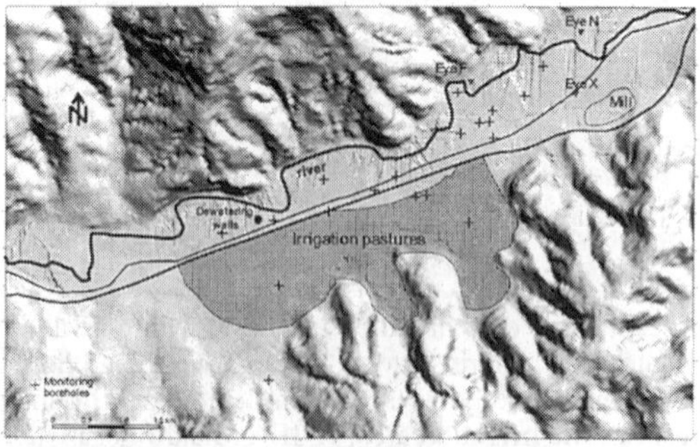

Figure 1. Layout of the investigation area.

Previous work recognised the presence of the springs and their obvious correlation with the dolomite. However, they were apparently unable to satisfactorily explain the mechanisms of solute transport from the irrigation fields to the springs, some of which are on the opposite flanks and at elevations above the river.

2.1.1 Air Photo Interpretation

There are three dominant joint sets indicated by observed photo-lineaments. The first is a northwest - southeast trending set that appears to be concordant with the bedding of the sediments in the valley and is interpreted as representing major bedding planes. The second joint set is a northeast – southwest trend of which at least one fault was inferred. The third joint set trends almost east – west and approximately follows the valley.

The dolerite intrusives are more difficult to discern on the air photos, however, the high ground east of the irrigation pastures has preserved one of the known dolerite sills above the valley floor. The second sill is covered by thick alluvial sediments.

2.1.2 Geomorphology

The palaeo-valley floor is much wider than the current flood plain. As the valley floor was filled with sediment, the river meandered back and forth across the valley. There is evidence to suggest, that there is at least one major channel south of and parallel to the current river course.

A process of scarp and valley side retreat following the deep incision led to a period of deposition, caused by a reduction in the hydrodynamic flow regime as the steep slopes receded and sedimentary deposition ensued. Local variations in the flow regime, storm events, underlying lithology and geomorphology led to the recent sedimentary sequence observed today, i.e. an irregular palaeo valley floor filled with coarse sediments, covered by finer sediments, with lenses/bars of gravel and sand. Talus, colluvial and hillwash soils mix with the alluvial sediments in the valley flanks as well.

The presence of the Black Reef Quartzite downstream of the springs played a significant role in the development of the essentially mature river reach now observed up stream of the mill for many kilometres. The erosion resistant quartzites form a lower barrier or "dam" behind which continued weathering and erosion of the dolomites resulted in the relatively wide and flat flood plain with a mature river reach observed today.

2.1.3 Geophysical Surveys

The geophysical survey was undertaken to supplement past works and to ascertain whether groundwater flow from the irrigated areas to the three springs is structurally controlled. The survey concentrated on the running of EM lines using a Frequency Domain Electro Magnetometer with Vertical Electrical Soundings (VES) to support the EM technique. The VES were run using the Schlumberger array, which allowed a quantitative interpretation of thickness of a certain resistivity.

The EM revealed a conductive (clay) layer of indeterminate thickness blanketing a possible aquifer zone. Specific feature identification was difficult from the EM, but it does support the general hypothesis.

The VES defined a number of geo-electric zones, including confirmation of a deep structural feature, potentially a good aquifer.

2.1.4 Drilling & Yield Testing

The boreholes drilled intersected large volumes of groundwater and blowing yields of 10 to 50ℓ/s was encountered.

The maximum discharge that the boreholes could be tested at was 20l/s, due to hole size dictated by the ODEX drilling method and the limitation of pump that could be installed for the test. Because the aquifers intersected are highly permeable and transmissivities range from $20m^2/d$ to $8000m^2/d$, only limited drawdowns were recorded in the test boreholes.

2.1.5 Aquifers

The aquifers intersected can be classified according to the water bearing formation or contact in which they occur:

- Aquifers within the sedimentary sequence and dolerite contact: Groundwater is associated with the gravel and boulder horizons, which produce modest blowing yields (<5ℓ/s).
- Dolerite Aquifers: Locally the dolerite may form an aquifer. A water strike with a blowing yield of

7ℓ/s was encountered in one borehole at the base of the completely weathered dolerite.

- Contact Dolerite/Dolomite Aquifers: In each of the boreholes where dolerite intrusions were intersected, a water-bearing horizon at the dolerite/dolomite contact was encountered. Blowing yields in excess of 5ℓ/s were recorded. This contact aquifer is widespread and is probably a direct source of recharge into the underlying dolomite aquifers.
- Dolomite Aquifer: All exploratory boreholes intersected aquifers within the dolomite and chert. Blowing yields in excess of 20ℓ/s and as high as 50ℓ/s were recorded.

The dolomite constitutes a major regional aquifer underlying the valley. The widespread weathering of dolomite and chert, especially along fault zones, have produced highly permeable conditions. Since the dolomite aquifer is in direct hydraulic continuity with the overlying aquifer it is vulnerable to contamination from the irrigation of the effluent.

2.2 *Conceptual Groundwater Model*

The three major springs in the valley are Eye F, Eye N and Eye X. The data collected indicates that Eye F is associated with the inferred fault. Eye X is associated with the palaeo drainage feature(s) south of the present day river course. Eye N has an association with a combination of structure, geomorphology and its close proximity to the Black Reef outcrop, causing an up-welling of groundwater at surface. The implication is that one would expect that:

- Eye F, being associated with a regional feature would have the highest flow. Conversely, being a regional feature, the highest degree of mixing with regional groundwater flow is also expected, with the lowest influence on quality by the pasture irrigation;
- Eye X, being a shallow localised feature, controlled by alluvial sedimentation is expected to have a much lower flow and being in direct hydraulic continuity with the irrigation pastures to have the poorest water quality;
- Eye N, is similarly expected to have a lower flow than Eye F, but a water quality between that of the other two.

This is exactly the trend observed by the monitoring data collected over time. The springs have been monitored since approximately 1984 and show the expected relative trends.

The apparent anachronism of the presence of strong-flowing springs on the opposite bank of the river implies that there is an extensive but confined aquifer system beneath the riverbed. The evidence from river quality profiling suggests that the river has chiefly been impacted upon at three distinct locations, i.e. the outflow areas of the three springs into the main river channel. However, there also has to be a mechanism that allows water to rise through the confining layer(s) to surface and in close proximity to the river.

The hypothesis that explains this is presented below:

- The exploratory fieldwork conducted conclusively demonstrates the presence of a multilayered aquifer system with multiple water strikes associated with different lithologies.
- The pump testing demonstrated that the aquifer system or components thereof is regional in extent with high transmissivity and storage.
- The underlying geology is conducive to the development of a strong karst aquifer, i.e. karstic dolomite, including chert rich formations, intruded by dolerite.
- The Black Reef Formation is present as the lower confining layer (base) of the aquifer system. The quartzite outcrop represents the surface expression of this layer and eastern boundary to the aquifer system.
- The geomorphological processes are also significant. The southern flank of the river is characterised by relatively gentle slopes where one finds a thick succession of well-sorted sediments and deep residual soils. The steep almost cliff-like northern slopes imply a rapid lateral facies change, leaving opportunity for the development of springs such as found at Eye F.
- The irrigation pastures are elevated above the river and are between 30 and 70m above the Black Reef outcrop, which allows the presence of the springs at surface.

The conceptual groundwater model is schematically represented in Figure 2, which is a section along the river valley. It illustrates the different aquifer systems and its relationship with the springs.

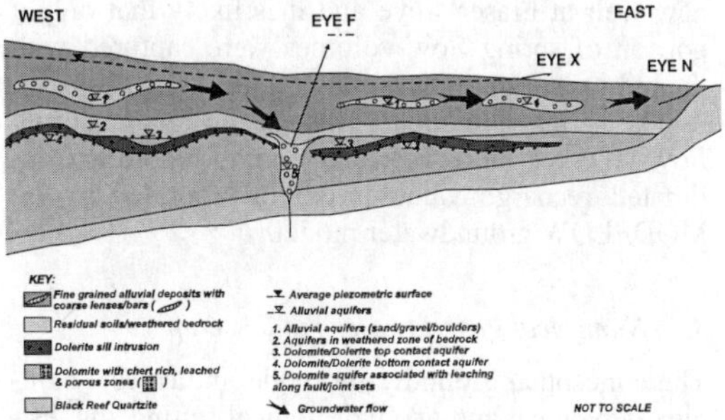

Figure 2. Conceptual Groundwater Model, Schematic Section along River.

3. GROUNDWATER MODELLING APPROACH

3.1 *Spring flow analysis*

The linear relationship between rainfall recharge and the flow rates at springs have long been recognised

(Bredenkamp et al. 1992; Bredenkamp et al. 1995, Kok, 1992). It was recognised that if a similar relationship could be found at the investigation area, since the geohydrological investigation revealed that a significant portion of groundwater manifest at the three springs in the area (Xu, 2000). The spring flow analysis approach was applied to the study area using the REME software (Van Tonder & Xu, 2000). Irrigation was incorporated in the model as an "injection well" and spring flow volumes were then related to rainfall.

The results of the spring flow analysis are shown in Figure 3.

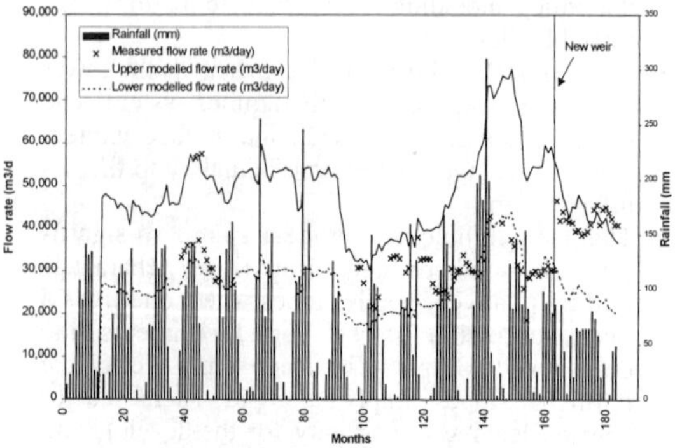

Figure 3. Results of the spring flow analysis.

The results indicate a dynamic groundwater system. Response times after which the eyes react to a particular rainfall event was calculated to be approximately 10 – 12 months. A recharge rate of 25% annual rainfall plus irrigation was determined for the irrigation pastures.

The "jump" in spring flow rates measured from January 1998 coincided with the construction of a new weir at Fraser's eye and it is likely that only a portion of spring flow volumes were captured with the old weir.

The spring flow model was used to predict spring flow volumes for a particular rainfall record and estimated recharge values were incorporated in the MODFLOW groundwater model.

3.2 Numerical groundwater model setup

The conceptual groundwater model indicates a complex geological and geohydrological setting suggesting that a 3D numerical groundwater model could represent the site. The MODFLOW96 code was used to simulate the groundwater situation using the "Visual Modflow Ver 2.8.1" software. The MT3D1.5 and MT3DMS (Multi species) codes were used as solute transport packages.

In order to simulate groundwater flow in the karst aquifers, the Equivalent Porous Medium (EPM) approach was followed. The EPM approach assumes that the fractured material can be treated as a continuum and that a Representative Elementary Volume (REV) of material with effective hydraulic parameters can be defined.

The model covered an area of 56.4 km^2, incorporating the irrigation pastures and extending up to the quartzite outcrop, which was defined as a hydraulic barrier. The grid design comprised irregular sized grids of 167 columns, 132 rows and 5 layers, necessitated by the complex geology.

Both the dolomitic and palaeo-valley aquifers were incorporated in the model. The valley sides were excluded from the model area (i.e. no flow conditions). A preferential recharge zone was provided along the foot of the valley sides representing recharge from runoff of the valley sides along the high-permeability talus zone. A constant head boundary was provided along the upstream boundary of the model window representing groundwater inflows from up-stream catchments. Outflow along the eyes was calculated by the zone budget package, assigned to the respective eyes, represented by constant head boundaries.

An initial groundwater contour map was produced using the SURFER software and applying the kriging method to average water levels as measured in fourteen boreholes in the area.

The source of contamination for the solute transport model was provided by a recharge boundary with a concentration equal to that of the irrigation effluent. The contaminants simulated included chloride, sulphate and sodium. Chloride was treated as a conservative anion and no chemical interactions were simulated. Sorption processes were simulated for sulphate and sodium species using the Linear Isotherm option. No adsorption processes were simulated for the dolomitic aquifer.

3.3 Calibration of the groundwater model

Calibration of the groundwater model implies a demonstration that the groundwater model is capable of producing field-measured heads and flows (flow model) and concentrations (solute transport model).

The flow model was calibrated against groundwater levels of seven boreholes within the model area (transient conditions). The Normalised Root Mean Square (NRMS) value was measured at 4.70% and are well within the 10% norm.

The flow model was also calibrated using spring flows of the three springs under steady state conditions (using average rainfall, irrigation and spring flow rates). The results are summarised in Table 1.

The lower modelled spring flow at Eye F can possibly be attributed to groundwater inflows originating from the valley side, which is not considered by the model.

After the flow model was calibrated, the solute transport model was executed. The solute transport

Table 1. Comparison between measured and modelled spring flows at the respective springs.

	Average measured spring flow (m^3/day)	Average modelled spring flow (m^3/day)
Eye X	8 820	7 408
Eye F	18 673	13 338
Eye N	6 814	6 624

model served as verification of the flow model. The solute transport model was calibrated against time-series chloride concentrations as measured in the monitoring boreholes as well as what was measured at the three springs. Chloride was chosen for calibration purposes because of its conservative characteristics. The NRMS of 12% is acceptable for solute transport models especially considering the time period of the model domain.

After calibration of the solute transport model for the chloride ion, the sulphate and sodium ions were calibrated incorporating adsorption processes.

Good calibration were accomplished at all of the springs and the boreholes, suggesting that the groundwater model is an accurate representation of the actual groundwater system and can be used for prediction purposes. The measured and modelled chloride concentrations at the respective springs are shown in Figure 4.

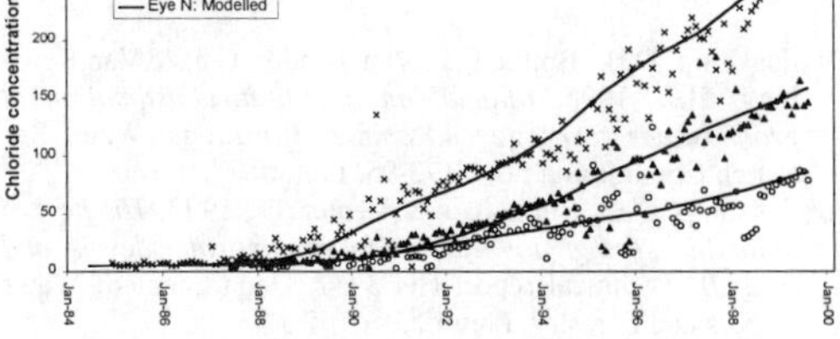

Figure 4. Comparison between measured and modelled chloride concentrations.

3.4 *Sensitivity analysis*

A sensitivity analyses was conducted in order to identify and quantify the uncertainty of the flow model caused by the uncertainty of aquifer parameters and indeed the geometry of the system. The sensitivity analyses comprised two aspects namely:

i. Uncertainty of the geometry of the systems, i.e. position of important fracture zones in the dolomitic aquifer, position and geometry of the dolerite sill and dyke layers intruding the dolomitic aquifer, the position and extent of the paleochannel and the position of the Black Reef Quartzite.
ii. Uncertainty of aquifer parameters i.e. representative transmissivity and storativity data for the dolomitic and alluvial aquifers.

The results of the sensitivity analyses indicate that the flow model is highly sensitive to changes in the hydraulic conductivity of the alluvial and dolomitic aquifers. This was expected, since groundwater flow is dominated by the hydraulic conductivity of these highly permeable units. The results also indicate that the model is insensitive to changes in the geometry of low-permeability units, which verified the validity of the EPM approach assumed for the model set-up.

4. GROUNDWATER MODEL PREDICTIONS

The groundwater model was used to quantify the response of the groundwater system to future events by introducing stresses into the calibrated groundwater system. The purpose of these predictions was to assess the affect of the introduced stresses on salt load discharges at the three springs and at the seepage front along the river.

A number of remedial scenarios whose aim was to attain short-term and long-term compliance were modelled. Two scenarios identified for predictive analysis are discussed namely:

i. Scenario 1: No rehabilitation strategies are implemented in the investigation area. A constant chloride load of is applied to the irrigation pastures.
ii. Scenario 2: The irrigation effluent has been subjected to remedial treatment, resulting in a decreased chloride concentration at source.

4.1 *Scenario 1: No rehabilitation strategies are implemented*

For this scenario, no changes were made to the groundwater model. The groundwater model predicts that continuous irrigation at rates similar to the current situation will result in an expanding contamination plume developing within the upper and dolomitic aquifers. Consequently, the salt load at the three eyes and at the river seepage front will increase with time. The expected contaminant plume after 30 years of irrigation are indicated in Figure 4.

The salt loads, emanating from the springs, will be a function of the spring flow rate and this will depend on the preceding rainfall patterns. The groundwater model indicates that total chloride, sulphate and sodium loads emanating from the eyes will continue to increase up the seventeenth year followed by a more gradual increase in chloride load.

4.2 *Scenario 2: Lowering of contaminant load at source*

A treatment plant has been implemented at the paper

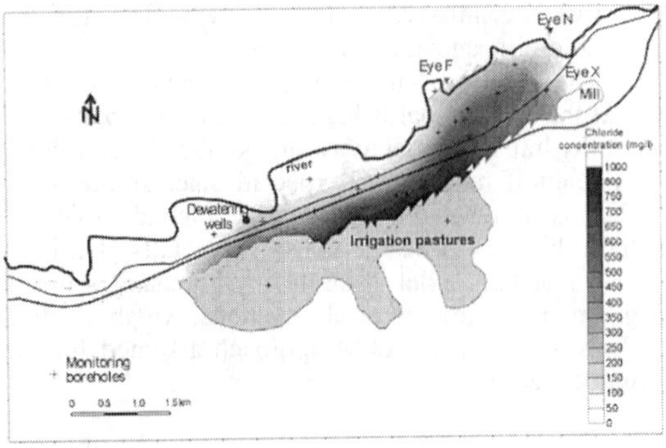

Figure 5. Expected pollution plume 30 years after commencement of irrigation with no treatment of effluent.

mill which purpose is to reduce the chloride content of the irrigation effluent. In order to simulate the effect of this treatment facility, the groundwater model was modified to account for the lower concentrations. The expected combined chloride loads emanating from the eyes are indicated in Figure 6.

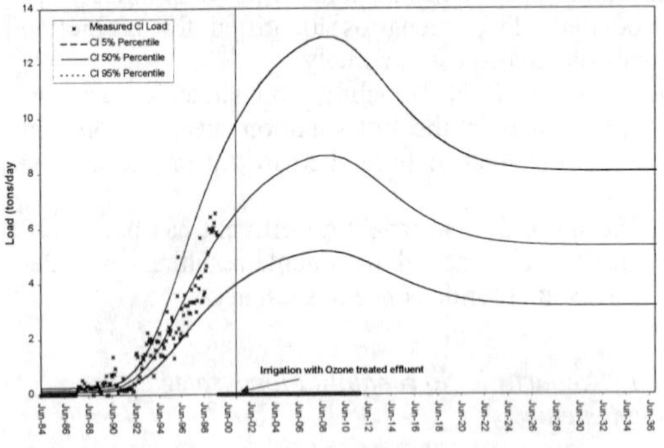

Figure 6. Predicted chloride load after treatment of effluent.

The results indicate an initial increase in chloride content after the introduction of treated effluent. The chloride concentration peaks approximately 9 years after introduction of treated effluent where after a decrease in chloride concentrations are expected. The chloride concentration will reach equilibrium approximately 15 years after introduction of treated effluent.

5. CONCLUSION

The geohydrological situation is unique in that approximately 84% of groundwater flows towards three springs in the area, while relatively little water seep directly towards the river. The geohydrological situation resembles a large sub-surface dam where groundwater is confined to the area upstream of the Black Reef Formation (resembling a dam wall) and water leaves the system mainly by means of springs (resembling spillways).

The groundwater model has shown that the quality of water emanating from the springs will deteriorate steadily with time if no remedial measured are introduced.

It is concluded that capturing of contaminated groundwater could be achieved by either establishing scavenging well field(s) located in the valley or by simply collecting contaminated water from the springs.

The abstraction wells would have to be close to the river to have the desired short-term effect, i.e. a reduction in salt load within a matter of months. However, the volume of water that would require abstraction would be similar to the spring flows with similar concentrations and it would be therefore more cost effective to abstract water directly at the springs.

The optimum positions of the wells are some distance from the riverfront, where relatively smaller volumes of water (compared to spring flow) could be abstracted at higher concentration. However, the modelling solutions indicate that the existing salt plume will not be affected in the short-term. The groundwater model has further shown that contaminated water emanating from the springs should be managed for a period of approximately 15 years before long-term measures will take effect.

6. REFERENCES

Bredenkamp, D.B., Botha, L.J., Van Tonder, G.J. & Van Rensburg, H.J., 1995. *Manual on quantitative estimation of groundwater recharge and aquifer storativity*. Water Research Commission no. TT73/95, Pretoria

Bredenkamp, D.B., Botha, L.J. & Venter, D., 1992. *The flow of dolomitic springs in relation to groundwater levels and rainfall*. Technical report HH 3757, Department of Water Affairs and Forestry, Pretoria.

Kok, T.S., 1992. Recharge springs in South Africa. Technical report GH 3748. Department of Water Affairs and Forestry, Pretoria.

Van Tonder, G.J. and Xu, Y., 2000. Recharge Estimation Model in Excel. Software developed for DWAF under supervision of E. van Wyk.

Xu, Y. 2000. Pers comm.

Groundwater: Past Achievements and Future Challenges, Sililo et al. (eds) © *2000 Balkema, Rotterdam, ISBN 90 5809 159 7*

Optimised calibration of regional groundwater model through inverse modeling

Y.S.Yang, A.A.Cronin & R.M.Kalin
School of Civil Engineering, Queen's University, Belfast, UK

ABSTRACT: Regional groundwater modeling is always constrained by lack of data and lack of understanding of system uncertainty. Inverse modeling of groundwater flow provides both an efficient automated parameter calibration and an additional uncertainty assessment of physically based model. The inverse groundwater modeling was used to identify model parameters and understand uncertainties associated with the regional hydrogeology and groundwater model in Belfast, Northern Ireland. The automated calibration of the model parameters, boundary condition and regional recharge was carried out using non-linear least squared optimisation. The regional groundwater flow model was constrained by uncertainty analysis during the optimal calibration. The sensitivity and confidence intervals obtained from the inverse modeling indicate that the major uncertainties were insufficient head observations and effect of the complex topography in the region. The approach is proved valuable for better understanding of the regional hydrogeology.

1 INTRODUCTION

Numerical groundwater models have provided valuable tools to understand regional hydrogeology and simulate groundwater flow system at spatial and temporal scales. However, regional numerical modeling is always constrained by lack of field data, lack of understanding of conceptual complexes and uncertainty of the systems. Inverse groundwater modeling can provide automated hydrogeological parameter calibrations (Finsterle & Najita 1998, Sun 1994, Wagner 1992). It facilitates assessment of prediction reliability because the results yield not only parameter estimates, heads or concentrations simulated for aquifer stresses of interest, but also confidence intervals for the estimated parameters, heads or concentrations (Poeter & McKenna 1995). The automated calibrations are able to effectively model the flow system through uncertainty assessment and provide functioning insights of the physical system (Poeter & Hill 1997). The challenge for the calibration approach is to develop a useful representation of an aquifer system from necessarily limited data and to reduce uncertainty associated with the modeling.

The automated calibration is optimised and leaves time for modeller to re-evaluate the conceptual model. The effective parameter automation and evaluation of conceptual alternatives can improve the system calibration and increase the confidence of model prediction in regional groundwater modeling.

Optimised calibration of a regional groundwater flow model was carried out through inverse modeling of hydrogeologic parameters on regional aquifers of Belfast, Northern Ireland. The methodology was a 3-step process: to build a regional groundwater flow model using ModFlow, to calibrate the model parameters using inverse automation, and to assess the uncertainty associated with the system modeling. The hydrogeological properties, model boundaries and recharge of the regional aquifer were automated in parallel based on initial values. This proves that inverse modeling is beneficial and efficient in flow validation, uncertainty estimation and better understanding of regional aquifer systems.

2 REGIONAL SURVEY

The study area is located around the city of Belfast, Northern Ireland with much of the province's population and industries. It can be divided into two subregions, the Lagan Valley to the southwest and the Newtownards area in the east (Fig. 1). Average rainfall is 900 mm per annum and evapotranspiration 400 mm.

The Lagan Valley includes some of the richest agricultural land in the province of Ulster (Manning et al. 1970). The valley is flanked by large hills to the west and north with the scarp of the Antrim Plateau reaching 480 m OD. The Lagan is a major river in the area, then the Ravernet a small one joining to

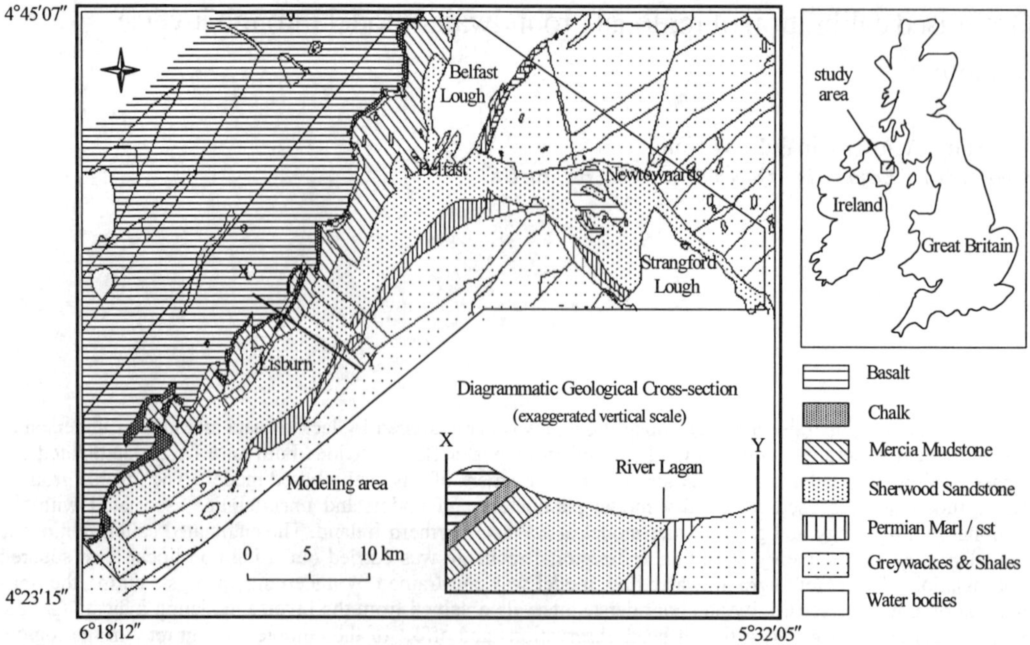

Figure 1. Map showing the location of the modelling area and the conceptual model of the study (GSNI 1997).

the Lagan near Lisburn. The eastern area between Belfast and the towns of Newtownards is fairly flat except for the imposing Scrabo Hill. The Enler River flows eastern into the Strangford Lough.

The Lagan Valley exhibits a typical Caledonian NE-SW trend and glacier U-shape. The main geology is summarised in Figure 1. The Permo-Triassic strata dip shallowly at about 14° to the NW. Dykes are the Tertiary intrusions, which runs through all the solid rock aquifers. Over 300 dykes are known in the area, the majorities are not wider than 3.1 m in size (Manning et al. 1970). The geology is further complicated by the presence of numerous faults running parallel and at right angles to the axis of the Lagan valley.

The Triassic Sherwood Sandstone is Northern Ireland's most important aquifer, but other formations such as the sand and gravel drift is also important water supply source locally. The groundwater of the sandstone flows towards the centre of the Lagan Valley from the both sides then to the sea (Fig.1). Flowing artesian conditions have only been encountered in a few locations (Bennett 1976).

Limited recharge to aquifer from precipitation occurs along the valleys, where the drift of the relatively high ground is absent or thin (Robins & Shearer 1994). The Sherwood Sandstone receives significant recharge from the overlying formations (Chalk, Mercia Mudstone and Basalt) along the northwest side of the Lagan valley (Bennett 1976). Recharge through the till may mainly occur along preferential pathways (Robins 1996).

3 REGIONAL GROUNDWATER MODEL

This regional modeling area covers all the Sherwood Sandstone aquifer in the Lagan Valley and Newtownards area. It is within a rectangular area of W. Long. 5°32′05″ to 6°18′12″ and Lat. 54°23′15″ to 54°45′07″. The model system is identical with surface water catchments in the area.

3.1 *Conceptual model*

The Sherwood Sandstone was the main objective aquifer of the modeling exercise. However the drift was also modelled to reveal the role of the Quaternary and its relationship to solid geology even it is extremely variable in thickness and extent in the region (Robins & Shearer 1994). Three layers were conceptualised: the alluvium, sand and gravel, and subglacial till. Six layers of the solid geology: the Ordovician and Silurian greywackes, Permian Marl, Triassic Sherwood Sandstone, Mercia Mudstone and Sandstone, Chalk and Tertiary Basalt, were taken in the conceptual model (Fig. 1). In the light of aperture and fracture development two layers of solid geology were conceptualised to represent the permeability-depth profile of the sandstone.

To model the regional groundwater flow the modeling area included all the solid strata. For the Quaternaries, the alluvium (layer 1) and the sand and gravel (layer 2) are only few meters thick. The till (layer 3) is fairly complete with thickness up to 35 m. According to the effective depth of the aquifer, thickness of 60-120 metre and 50-80 metre were used respectively for the Sandstone layers in most of the modeling area (Bennett 1976).

3.2 *Numerical modeling*

A regional numerical groundwater model was built using ModFlow (Harbaugh & McDonald 1996) based on the conceptual model. Regular finite difference grids of 500 metre were used.

Extensive data was gathered to run the model and to better understand the regional hydrogeology and geochemistry (Cronin et al. 2000). The data of meteorology, hydrology, borehole monitoring and pump test was analysed and integrated into the model. Constant heads of 0 m were placed at Belfast and Strangford Loughs. For effective recharge incorporating evapotranspiration and runoff, an initial figure of 30 mm was taken for the centre of the valleys and urban areas (Hartley 1935). A higher values of 150-300 mm were used on areas where the till cover is thin or preferential flows exist. The rivers of the Lagan, Ravernet and Enler were assigned in the model. The value of riverbed conductance varied from 2000 to 3400 m^2/d. The aquifer property values were taken from various pumping tests results and previous reports carried out on the study area (Kalin & Roberts 1997, Robins 1996). Geological structures and intrusion dykes were taken into account when assigning permeability values.

To understand catchment parameters of the regional system, seasonal fluctuations of the water head were analysed. The ranges were low, 1-2 m, at most sites except some bores near boundaries of the Sandstone. Therefore, a steady-state flow model was used for the simulation so that the system can be calibrated under the uncertainty of data scarcity in an average sense. Groundwater head distribution was obtained from the steady-state model to delineate the regional groundwater characteristics.

Because of the quite steep gradients in the modeling area, the non-uniform thickness and the large difference in thickness between the Quaternary and the solid layers, cells of the model in upper layers dried out quickly. Drying out of the thin layers of the Quaternary thus would induce more dry cells nearby (Harbaugh & McDonald 1996). On the other hand, the regional scale of the modeling made the finer 3-D grid spaces unpractical in the model operations. Considering the availability of observation data in the meantime, the borehole observations in the layers of the Sandstone were used as targets in the system validation by the inverse modeling.

4 FLOW SYSTEM VALIDATION

The regional groundwater flow system was calibrated through inverse numerical modeling. A model-independent parameter estimator – PEST 98 by Watermark Computing (1998) was employed. It uses Gauss-Marquardt-Levenberg non-linear least square algorithm.

The approach was a 3-step process: to build a regional groundwater flow model (forward), to calibrate the model parameters using inverse automation and to assess the uncertainty associated with the system modeling. Because of the complexity of geology and topography and therefore wide spatial variations of parameters in this area, it is indeed a complex modeling problem and a challenge of the system calibration. Initial set-up and run of the regional groundwater model of the area have revealed some issues to the automated system calibration: a) calibration of hydraulic conductivities of the regional strata, b) lateral recharge into the aquifer and relationships between Quaternary and bedrocks, c) spatial pattern of the recharge load to the regional system, and d) the uncertainty of the modeling study.

A reasonable initial data set of the adjustable parameters is useful for the convergence of inversion process and the overall iterations of optimisation. Initial parameter values were estimated based on the previous investigations. The observation data was taken from the means of 1971-1975 at 18 observation targets and 4 supplementary measurements in the Newtownards area. The parameters were optimally adjusted until the fits between model-generated and borehole groundwater heads were reduced to a minimum in the weighted least squares sense.

4.1 *Parameter calibration*

Hydrogeological parameters of aquifer are important characteristics to describe a groundwater flow system. They were calibrated through the inverse modelling based on the forward model. The initial values of the hydraulic conductivities K_x, K_y were estimated based on the previous study (Kalin & Roberts 1997, Robins 1996). PEST 98 for Windows (WPEST) was run with a Pentium 233 machine running Windows 95. As the time required to run each groundwater flow model was about 20-30 minutes, use of WPEST for optimisation was 2 to 3 days.

System uncertainty may exist due to limited observation data and the different measurements. To regularize the parameter estimation and reduce the uncertainty, less weights were assigned for the 4 observations in the Newtownards area in the inversion parameter optimisation (Doherty 1998). The optimal objective function, the sum of squared weighted residuals of groundwater heads, was achieved after 5 inversion iteration runs. The sensitivity of each parameter (hydraulic conductivity) with respect to all the weighted observations during 5 optimisation iterations was shown in Figure 2 (the order and description of the parameters are shown in Table 1). This illustrates a general contribution of the parameters to the objective function in the optimisation process. The inversion of the regional flow model was sensitive to the parameter values of the sand and gravel, till and the Sherwood Sandstone in the middle part of the Lagan Valley, and then to the basalt and greywackes. Zonation of hydraulic conductivity was determined upon the basis of the geology, land use, topography, and hydrology.

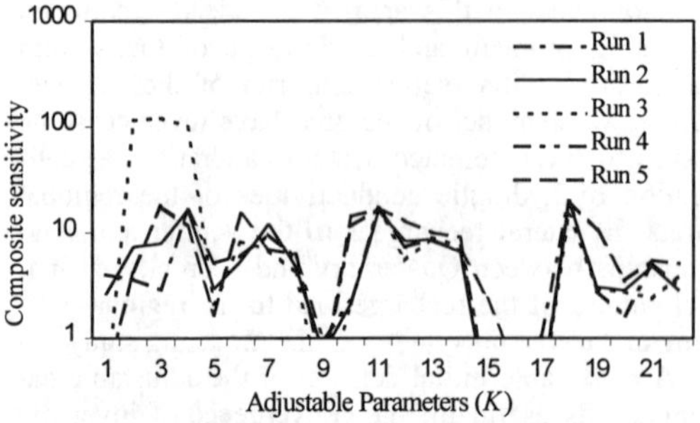

Figure 2. Model sensitivity to the *K* in the inversion.

4.2 *Boundary uncertainty assessment*

To identify the uncertainty of the boundary condition of the groundwater flow model at spatial and quantitative scales several separate inverse scenarios have been carried out. The inflows or outflows were identified along the boundaries between the greywackes and the modelled aquifers in the eastern and northeastern parts, and between the Basalt and the aquifers in the west and southwest.

Wells have been assigned to represent the lateral flows along the boundaries between the Sherwood Sandstone and the greywackes in the eastern and northeastern modeling areas. Uncertainty caused by unknown lateral inflows or outflows, spatial locations along the boundaries has been identified by the inverse modeling.

The overall runs of the inversion showed that the model was only sensitive to the inflows in the lower layers along the greywackes boundaries, not to that in the quaternary drift. This means the fracture groundwater from the greywackes interacts laterally with the Sandstone but little water flows to the drift. This demonstrates that the low permeable till prevents groundwater from moving into the Quaternary aquifers. Figure 3 is a typical diagrammatic cross-section presenting the boundary geology in the northeastern area.

The inverse model was not sensitive to the western and northwestern boundaries of the modeling area, which suggests that the lateral inflow recharge along these boundaries is insignificant to Sandstone aquifer. This is identical to the regional hydrogeoloy. The Mercia Mudstone (known also as the Keuper Marl) comformably overlying the Sandstone forms an aquiclude in the regional sense and prevents groundwater from penetrating it. The basalt is also a stratum with low permeability/porosity. The groundwater then comes out as forms of springs and flows to the Sandstone as a form of surface recharge. This was validated by the characterisation of spatial recharge in the following paragraphs. The Sandstone aquifer discharge in the southwest boundary as lateral outflow at 578 m^3/d (0.083 m^3/d per metre).

The lateral recharge of the Sandstone aquifer from the greywackes in the northwestern Newtownards area was optimised by the inverse modeling as 10550 m^3/d (2.11 m^3/d per metre). The composite sensitivity of the observations to the lateral recharge is relatively low because a) most observations are far from this boundary; b) lateral recharge through the Basalts and Mudstone to the Sandstone is little in the west region; c) the Newtownards area and the Lagan Valley are two separate groundwater systems with a groundwater divide.

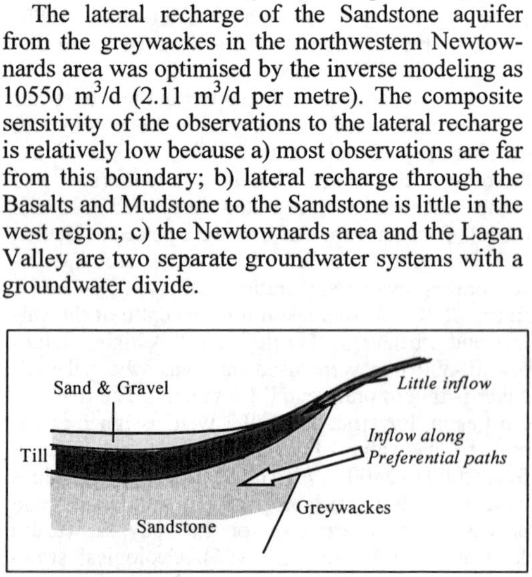

Figure 3. Diagrammatic presentation of the northeast boundary.

4.3 *Recharge characterisation*

The spatial distribution of the recharge was characterised after the parameter and boundary calibrations. 11 recharge zones were designed with consideration of the land use, geology and topography. The

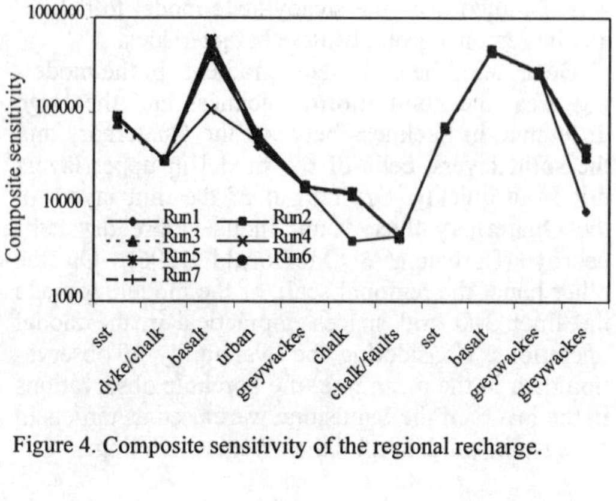

Figure 4. Composite sensitivity of the regional recharge.

upper/lower bounds of the variables in the inversion were properly set to reach logical values. The model was very sensitive to the recharge as indicated in Figure 4. The composite sensitivity is correlated to areas of the recharge zones.

The calibrated recharge is 420 mm/year around the dyke in the Newtownards area and along the Chalk in the west of Belfast where springs are located. It is only 10 mm/year in the southern area. The value on some cells where the faults cross the boundary between the Chalk and Sandstone reaches as high as 970 mm/year. This demonstrates the most recharge to the Sandstone happens along some preferential flow paths. Little recharge for the impermeable areas such as the basalt, the greywackes and the urban development. Actually most of the effective recharge comes out in the springs at the Basalt/Chalk escarpment to the west of the Lagan Valley.

4.4 *Discussion and uncertainty analysis*

The hydraulic conductivity, boundaries and recharge were calibrated in parallel to assess their individual effects on the system validation.

The contour of the modeled heads, measured points and their location of the Sandstone aquifer are shown in Figure 5. It shows overall good match even with the uncertainty arisen during the regional model validation. The lateral inflow/outflow estimated by the inversion may contain uncertainty due to the lack of observations.

The estimated values for the hydraulic conductivity are listed in Table 1. The values of the Quaternaries are greater than the initial estimations especially for the sand and gravel deposit, but they fall into normal range (Freeze & Cherry 1979). This may also indicate an exaggeration phenomenon of the model to the Quaternaries in spatial scale. The K values of the solid geological strata are all located within reasonable range and identical to the initial estimations. The much smaller value of the Mercia Mudstone suggests a very low velocity and long travel time of groundwater recharge through the escarpment and hills in the western part of Belfast.

For the Sherwood Sandstone aquifer, the spatial distribution of the hydraulic conductivity K is characterised by a general decrease, without surprise, from shallow to deep in the vertical direction. Horizontally, K is relatively greater in north and middle (around Lisburn) of the region than that in the rest of the area. This features the high anisotropy of the Sandstone aquifer as a result of the geological structures, as numerous faults run parallel around the north of the Lagan Valley (Kalin et al. 1997). The significant fault zone, reactivated a number of times in geological history, may well provide preferential flow paths for groundwater flow. The secondary fractures associated with the faults increased the anisotropy in Lisburn and Belfast areas where much drilling has shown the variability of well yields over short distances (Bennett, pers. Comm. 1998).

Table 1. Estimated parameter values of K (m/d).

No[a]	Geological layer	K	No[a]	Layer	K
1	Alluvium	7.71	12	sst (up 3)[b]	0.06
2	Sand/Gravel	49.82	13	sst (up 4)[b]	3.8
3	Till	4.32	14	sst (up 5)[b]	0.03
4	Basalt	10.24	15	sst (low 1)[b]	0.33
5	Chalk	35.44	16	sst (low 2)[b]	0.06
6	Greywackes	0.016	17	sst (low 3)[b]	0.06
7	Permian sst	0.22	18	sst (low 4)[b]	0.06
8	Permian Marl	0.16	19	sst (low 5)[b]	0.02
9	Mercia Mudstone	2.33	20	sst (low 6)[b]	0.32
10	sst (up 1)[b]	4.55	21	sst (low 7)[b]	0.02
11	sst (up 2)[b]	0.34	22	sst (low 8)[b]	0.81

[a]The numbers of the adjustable parameter.

[b]sst –Sandstone, upper/lower layers with zonation number.

The confidence intervals of the parameters from the inversion indicate relative correlation and sensitivity between parameters. For example, the parameters of the Quaternaries, Chalk, and Mercia Mudstone, with very wide (ill-posed ranges) confidence limits, are well correlated with much uncertainty; whilst the Basalt, Permian Marl and Sandstone, with comparatively small range, are well correlated with more certainty. The parameters of the Sherwood Sandstone have relatively small confidence intervals, however the hydraulic conductivity in north and east zonations, where there is no or less observation data, contains more uncertainty. The wide confidence interval may also demonstrate too few observations and relatively too many parameters in the automation process. The objective function of the parameter automation may sensitive to the individuals of the highly correlated parameters, appears insensitive to all of them if they were varied in concert (Doherty 1998).

Considering the overall automated calibration to the hydrogeologic properties, boundary condition and recharge, the results were identical and provided a better understanding of the conceptual model. The hydraulic conductivity of the Chalk was exaggerated but its effect on this regional system may be larger than it supposed to be. The inverse model was not sensitive to the lateral flow, which was validated by more recharge occurrence along the west Sandstone boundary. Little recharge was calibrated in the areas of the greywackes, but the groundwater replenishes to the Sandstone through the lateral inflows. This is validated by the production of relatively old water in the area (Bennett 1976).

5 CONCLUSIONS

Inverse modeling was carried out to calibrate a regional flow model by incorporating the Quaternary and solid bedrock aquifers in Belfast, Northern Ire-

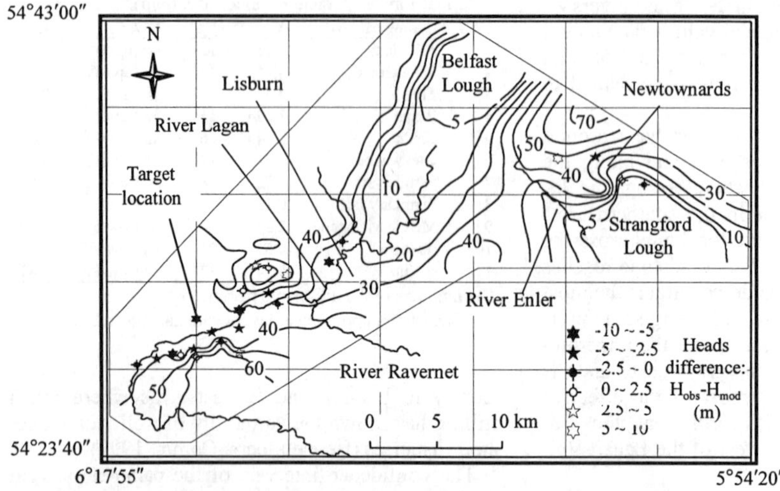

Figure 5. Map showing the regional groundwater contour and the calibration results (in metre).

land under the conditions of limited data availability with system uncertainty. The automated calibrations to the hydrogeologic properties, boundaries and recharge were done using a non-linear squared algorithm PEST 98. It is proved to be useful in model automation and better understanding of the regional hydrogeology.

The inverse groundwater modeling revealed uncertainty including insufficient observations at the regional modeling scale, effects of the complex topography, and the importance of the parameter zonations. The ranges of the hydraulic conductivity K were estimated as 49 m/d for the sand and gravel and 0.02-4.5 m/d for the Sherwood sandstone in this area. The lateral inflow was estimated as 2 m^3/d per metre in the northeast area. The Recharge was mainly from the areas where preferential paths exist.

The inverse model was sensitive to the K values of sand and gravel, Sherwood Sandstone in the middle part of the Lagan Valley, and then to the Basalt and greywackes. The resulting wide confidence intervals demonstrate too few observations for the parameter estimation in the area and may indicate the effect of dry cells on the upper Quaternary layers. Soft data in the term of head measurements may be introduced with appropriate weighting to their heads in the future study so as to reduce the modeling uncertainty of the inversion.

REFERENCES

Bennett, J.R.P. 1976. *Hydrological Study of the Lagan Valley*, Geological Survey of Northern Ireland Open Report No. 57. Northern Ireland.

Cronin, A., Elliot T., Yang, Y. & Kalin R.M, 2000. Geochemical modelling and Isotope studies in the Sherwood Sandstone Aquifer, Northern Ireland. Tram'2000, Belgium.

Doherty, J. 1998. Some issues in using PEST, Workshop at the MODFLOW'98 International Conference, Golden Colorado, USA.

Finsterle, S. & Najita J. 1998. Robust estimation of hydrogeologic model parameters. *Water Resources Research*, 34 (11): 2939-2947.

Freeze R & Cherry J 1979. *Groundwater*. Prentice-Hall, NJ.

GSNI 1997. Solid Geology Map (1:250 000) Solid Edition. Geological Survey Northern Ireland.

Harbaugh A.W. & McDonald M.G. 1996. User's Documentation for MODFLOW-96, an update to US Geological Survey Modular Finite-Difference Groundwater Flow Model. USGS Open-file Report 96-485.

Hartley, J.J. 1935. The Underground Water resources of Northern Ireland, Talk to Institution of Civil Engineers, N. Ireland.

Kalin, R. M, Roberts, C. 1997. Groundwater Resources in the Lagan Valley Sandstone Aquifer, Journal of Chartered Institution of Water and Environmental Management - Vol 11, April 1997.

Manning P.I, Robbie, J.A., Wilson, H.E. 1970. Geology of Belfast and the Lagan Valley Memoir - Geological Survey of Northern Ireland.

Poeter, E.P. & Hill M.C. 1997. Inverse models: a necessary next step in groundwater modeling, *Groundwater* (35): 250-260.

Poeter, E.P. & McKenna, S.A. 1995. Reducing uncertainty associated with groundwater flow and transport predictions. *Ground Water* 33(6):899-904.

Robins, N.S., Shearer T.R. 1994. The Permian and Triassic Aquifers at Comber, from Determination of recharge capture areas for aquifers at Bellagh and Comber, British Geological Survey Report WD/94/39.

Robins, N.S. 1996. Hydrogeology in Northern Ireland - British Geological Survey (HMSO), ISBN 0 11 8845241, 60pp.

Sun, N-Z, 1994. *Inverse Problems in Groundwater Modeling*. Kluwer Academic Publishers, Dordrecht, the Netherlands.

Wagner B.J. 1992. Simultaneous parameter estimation and contaminant source characterization for coupled groundwater flow and contaminant transport modeling. *Journal of Hydrology* 135: 275-303.

Watermark Computing 1998. PEST, Model-Independent Parameter Estimation. User's Manual.

3 Chemical and isotopic characterization of groundwater

Topic co-editors:

Siep Talma
Council for Scientific and Industrial Research, South Africa

Balt Verhagen
University of Witwatersrand, South Africa

Lisa Cavé
Council for Scientific and Industrial Research, South Africa

Gideon Tredoux
Council for Scientific and Industrial Research, South Africa

Chris Harris
University of Cape Town, South Africa

John Tellam
University of Birmingham, UK

Groundwater: Past Achievements and Future Challenges, Sililo et al. (eds) © 2000 Balkema, Rotterdam, ISBN 90 5809 159 7

Groundwater salinization in parts of southern Ghana

T.E.K.Armah
Department of Geology, University of Ghana, Accra, Ghana

ABSTRACT: Groundwater developments in the crystalline terrain underlying parts of the Central Region of Ghana have often been found to be hampered, among other problems, by high conductivity water with values reaching more than 5,000 µS/cm. This deteriorating groundwater quality has consequently forced a number of the inhabitants at the affected areas to revert to their old and unhygienic surface water sources.

A research study has been initiated to investigate the causes of groundwater salinity in these areas. Stable isotope analyses indicate that there is no relation between the groundwater system and possible seawater intrusion. Preliminary physico-chemical analyses indicate that the groundwater quality varies even within the individual rock units, therefore suggesting different causes for the salinization.

1 INTRODUCTION

In certain areas of Ghana the most appropriate and widely used source of drinking water is groundwater. However in some of these areas groundwater developments have been found to be hampered among others problems, by the water quality. This quality problem, which includes high levels of fluoride, iron, nitrates, manganese and high conductivity reaching excess of 5,000 µS/cm, is particularly prevalent in the Greater Accra Plains, Volta, Upper West and the Central Regions,

With all the efforts that have gone into developing the groundwater resources in these areas, the principal issue that has not been addressed is the sources of salinization.

In order to address the issue involving the groundwater salinization, a research study is currently being conducted at the Department of Geology, University of Ghana under the sponsorship of DANIDA-ENRECA to identify the sources of the salinization. This involves the determination of the physico-chemical parameters as well as the chemical processes characterizing the aquifer systems.

2 STUDY AREA

The area under study is located in the southern part of the Central Region of Ghana, figure 1. It is underlain by gently rolling terrain and mostly drained by ephemeral rivers. The climate is a tropical type and characterized by double rainy seasons with a mean annual rainfall reaching about 1200mm.

3 GEOLOGICAL SETTING

The terrain is characterized by precambrain crystalline basement rocks consisting of the Birimian metavolcanics and metasediments as well as granitoid intrusives. The metavolcanics classifield as the Upper Birimian (UB), comprise of metamorphosed and pyroclastic lava which are volcaniclastic and marine origin sediments and have been altered largely to chloritized and epidotised rocks. Effects of dynamothermal metamorphism on some of these rocks have resulted in the formation of hornblende schists and amphibolites. The metasediments, forming the Lower Birimian (LB), consist of alternating phyllites, greywakes, schistose, slates and argilla-

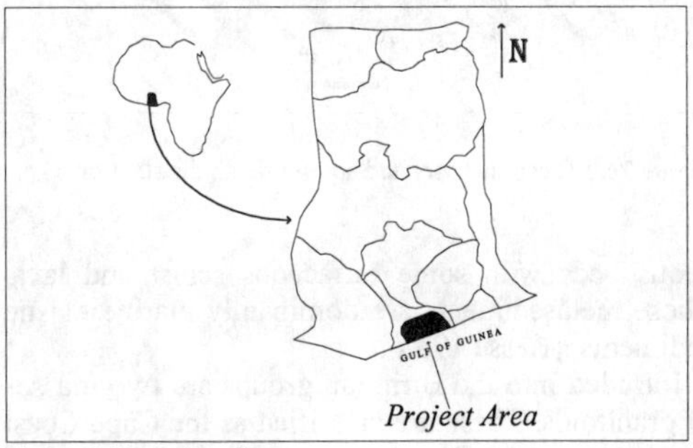

Figure 1. Study location.

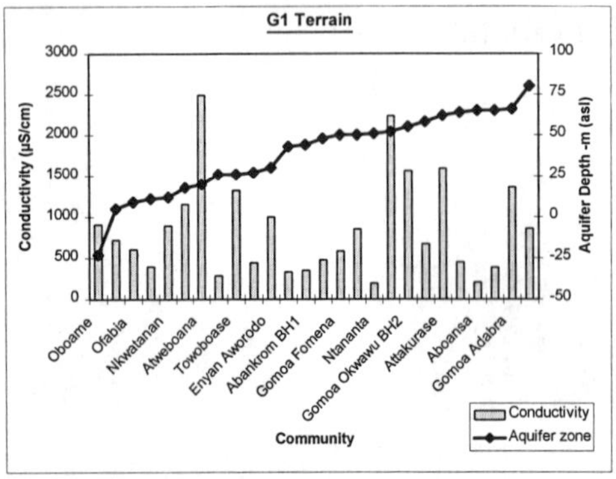

Figure 2a. G1 conductivity and aquifer depth distribution.

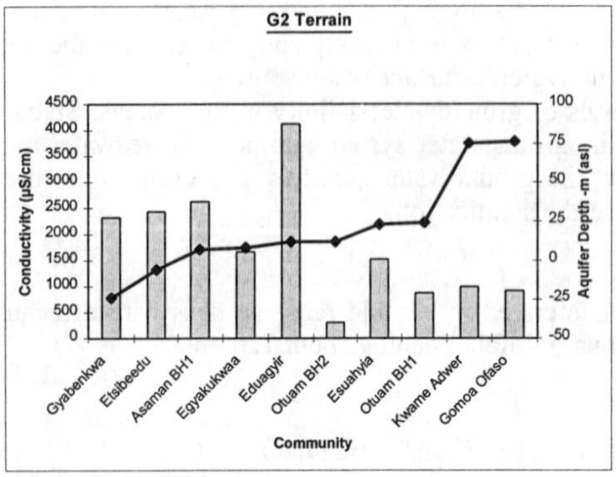

Figure 2b. G2 conductivity and aquifer depth distribution.

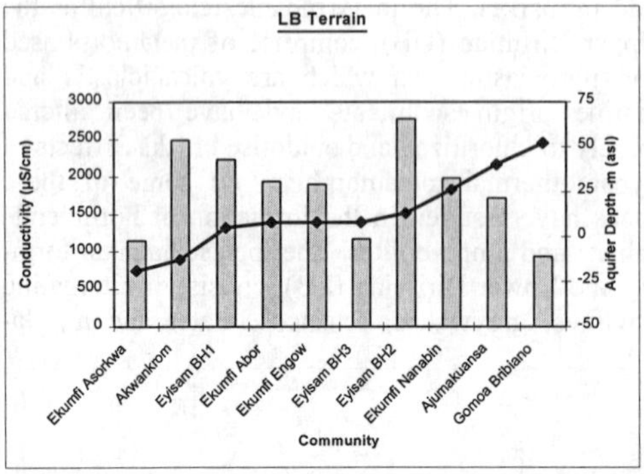

Figure 2c. LB conductivity and aquifer depth distribution.

ceous beds with some tuffaceous schist and lava. These metasediments are dominantly marine clastic sediments, (Kesse 1985).

Intruded into the Birimian groups are two masses of granitoids. These are classified as the Cape Coast type (G1) and the Dixcove type (G2) complexes.

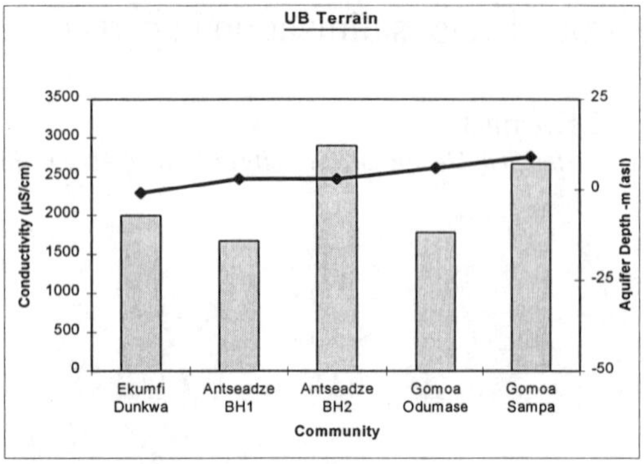

Figure 2d. UB conductivity and aquifer depth distribution.

The G1 granites complex are foliated and often migmatic potach-rich muscovite-biotite granite and granodiorite, porphyroblastic biotite gneiss, aplites and pegmatites. These granitoids, characterised by enclaves of schists and gneisses, are generally associated with the Birimian metasediments.

The G2 complex, younger than the G1 type, consists of soda-rich hornblende-biotite granite or granodiorite, which grades into quartz diorite and hornblend diorite. Compared to the G1 type, this complex has lower SiO_2 and Al_2O_3 but slightly higher CaO contents and higher Na_2O/K_2O ratio (Kesse 1985).

4 HYDROGEOLOGY

Potential aquifers in this terrain are not restricted to any particular rock type. There are two main types as described by Wright (1992). The first is the high porosity and low permeability weathered basement type and the second relates to the fractured basement having low porosity and high permeability characteristics. Both vary in depth from the near surface, 15m to about 100m with a mean depth of 39m. Yields range between 0.1 l/s and 10 l/s. The variations in yields possibly suggest that the various structural features such as faults, fractures and shear zones in the underlying crystalline terrain, which act as ingress and storage facilities for the groundwater systems in this area may have little or no hydraulic interconnections.

Recharge takes place by direct infiltration of rainfall and to a lesser extent from influent streams and rivers.

Generally, the quality of the groundwater is good for multipurpose use, except the presence of low pH (3.5 – 6.0) waters, high levels of iron, manganese and fluoride as well as high total dissolved solids (TDS) in the range of 1,500 – 2,000 mg/L (Ghana's Water Resources 1998).

In a number of cases there have been reports on occurrence of saline water in this area. Though the average conductivity level is within the standard acceptable limits, values in excess of 5000 μS/cm have been recorded. Incidentally some of these high conductivity areas are remote from the sea and are underlain by crystalline rocks which are inherently impermeable.

5 FIELD SAMPLING

Ninety-two samples from boreholes, hand-dug wells and ponds in the different geological in the study area were collected for physico-chemical and stable isotopic analyses. At each borehole site the well was pumped for about 5 minutes to obtain fresh water for sampling. Parameters such as conductivity, pH and oxidation potential as well as temperature were measured immediately at the various points of collection. Each water samples was further analyzed for their total dissolved solids (TDS) and stable, naturally occurring isotopes of oxygen (oxygen-18) and hydrogen (deuterium) in the laboratory.

6 RESULTS

In this paper results of preliminary investigation is presented. The samples were analyzed against the various geological terrains.

6.1 *Water quality*

The conductivity distribution in the area indicates that there are probably four types of groundwater systems classified as:

Fresh Water	< 500 μS/cm
Marginal Water	500 – 1500 μS/cm
Brackish Water	1500 – 5000 μS/cm
Saline Water	> 5000 μS/cm

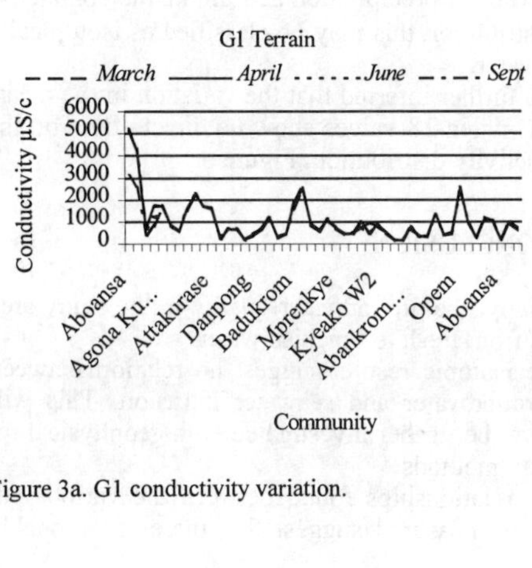

Figure 3a. G1 conductivity variation.

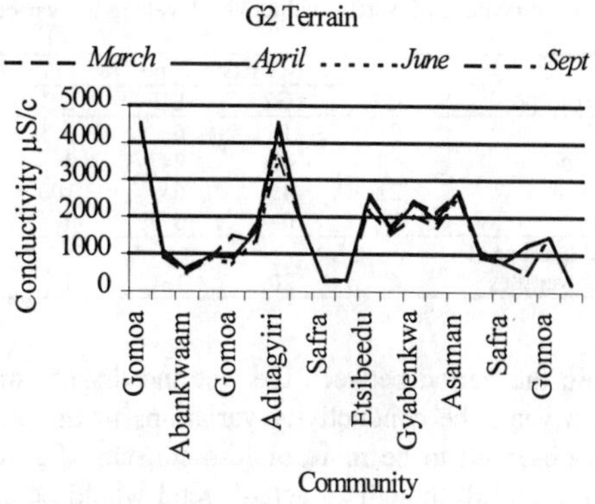

Figure 3b. G2 conductivity variation.

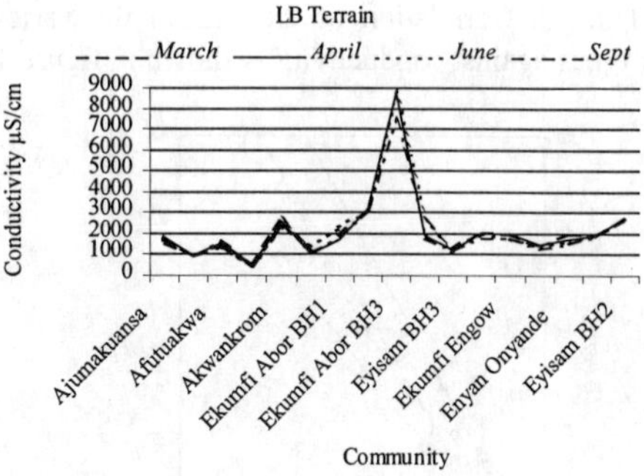

Figure 3c. LB conductivity variation.

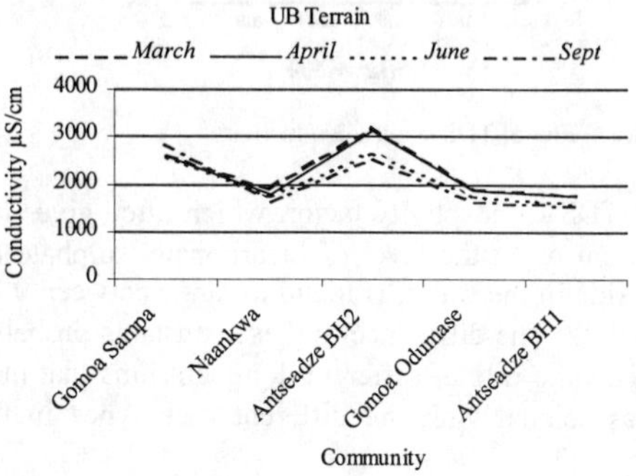

Figure 3d. UB conductivity variation

Their percentage occurrences in the various rock types are shown in table 1. It is inferred that the Birimian metavolcanics and metasediments terrains have fair to poor quality waters.

The conductivity variations as against the borehole depths do not show any relationship within any particular rock type, figures 2a, b, c and d.

Table 1. Percentage of water-conductivity levels in the various rocks.

	Rock Type			
Water Type	G1	G2	LB	UB
Fresh	31	15	6	0
Marginal	48	40	24	0
Brackish	21	45	65	100
Saline	0	0	5	0
Total number of water samples	48	20	18	6

Within the period between the wet and dry seasons over a year, the conductivity variations in the area were observed to be more or less constant, figures 3a, b, c and d, though an actual trend would be established after a long term observation.

The TDS determined from residues of the evaporated samples range between 65 mg/L to a high of 6200 mg/L. Correlation of these from the various rock types against conductivity is shown in figure 4.

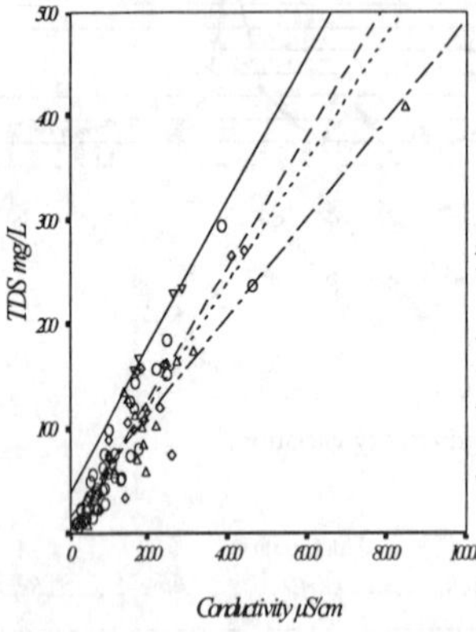

Figure 4. Plot of TDS against conductivity.

The TDS-Conductivity factor, which often gives indication as to the level of bicarbonate, sulphate or chloride in the water, is found to range between 0.86 and 0.99. The difference in these constants probably reflects the role of different dominant ions that may be associated with the different rock types in the area.

pH distributions in the groundwater system range from 4 to 9. Seventy-seven percent of these are in the range of 5.0 – 7.0 and 22% have values higher than 7.0. The distributions in the various rocks are given in table 2.

Table 2. pH distribution in water from the various rocks.

Rock type	G1	G2	LB	UB
pH range	4 - 9	5 - 8	5.5- 7.5	5.5 – 7.5

6.2 *Stable isotope*

Forty-two samples were analyzed for the levels of chloride ions as well as stable isotopes of oxygen-18 and deuterium. Seawater sample was also analyzed to determine the level of stable isotope present and how this relates with the groundwater system in the study area. A plot of δ^2H against $\delta^{18}O$ of the analyzed samples is shown in Figure 5.

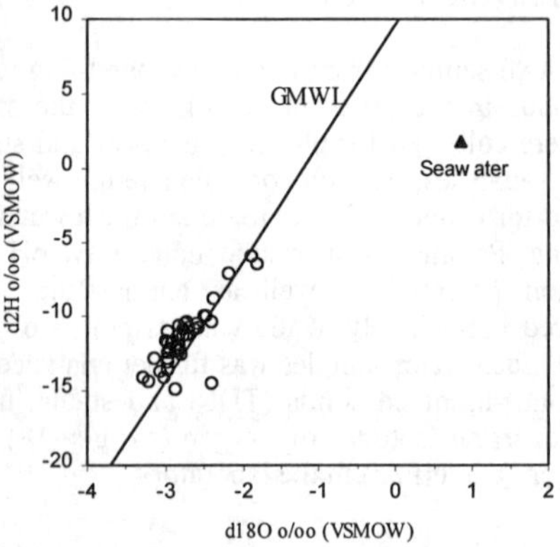

Figure 5. Stable isotopes plot.

The deviation of the stable isotope content of the seawater sample from the groundwater samples indicates that there is probably no relation between the groundwaters in the area and seawater intrusion. Apart from the seawater sample, results of the other sources can be found to cluster around the Global Meteoric Water Line (GMWL), given by the equation

$$\delta^2H = 8.13\,\delta^{18}O + 10.3 \qquad (1)$$

after Craig (1961).

The groundwater in the area might therefore be derived from precipitation and unlike the conductivity distribution this may be classified as isotopically freshwater.

It is further inferred that the variation in the measured Oxygen-18 values show no direct effect on the conductivity distribution, Figure 6.

7. CONCLUSION

Generally, the groundwater quality in the study area varies from fresh to brackish water.

The isotopic results suggest no relation between the groundwater and seawater intrusion. This will however be further investigated with geophysical resistivity methods

The relationships found between the various parameters measured suggest that there are probably many scenarios for the salinization in the area. Hydrochemical analyses are to be conducted to determine the type of groundwater present.

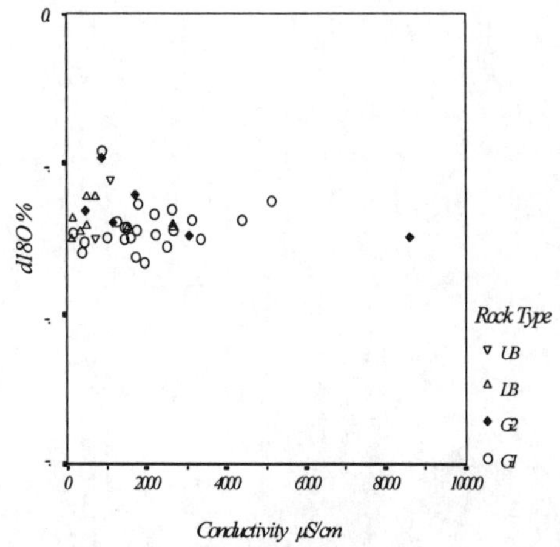

Figure 6. Plot of $\delta^{18}O$ against conductivity.

REFERENCES

Craig, H. 1961. Isotope variations in meteoric water. *Science* 133:1702 - 1703.

Ghana's Water Resources Management Challenges and Opportunities, 1998. 16-17.

Kesse, G.O. 1985. *The mineral and rock resources of Ghana.* 12-27.

Groundwater: Past Achievements and Future Challenges, Sililo et al. (eds) © *2000 Balkema, Rotterdam, ISBN 90 5809 159 7*

A study on the mixing phenomena between surface water and groundwater induced pumping using environmental isotope tracers

Daeseok Bae, Chunsoo Kim, Yongkwon Koh & Kyungsu Kim
Korea Atomic Energy Research Institute, Taejeon, Korea

ABSTRACT: One of the important issues for the environmental impact assessment for groundwater abstraction by the bottled water industries is groundwater contamination during the operation. For the characterization of a groundwater reservoir mixed with surface water, the ratios of deuterium (^2H) and oxygen-18 (^{18}O) were monitored during the rainy season in Korea as well as a chemical analysis. The testing wells of one deep well and two shallow observation wells were selected for this study, which are located in the area of igneous rocks. Water sampling was carried out for all water including rain, stream water and groundwaters from shallow and deep well everyday under the condition of continuous pumping in a deep well.

There were no indications observed in the mixing behavior between the aforementioned waters from the chemical components. The values of δD and $\delta^{18}O$ in groundwater and stream water, however, show good relations. The variation pattern of the $\delta^{18}O$ values of stream water and shallow groundwater with time is almost identical, whereas the pattern of deep groundwater is relatively stable. However, the typical peak values indicate that rain also contributes to deep groundwater under the continuous pumping condition when the rainy season starts. Environmental isotopes should be used as good tracers under limited geologic conditions.

1 INTRODUCTION

One of the many questions, which should be answered in an environmental impact assessment for the development of the bottled water industry, is the contamination of deep groundwater due to mixing with surface water. The flow condition of surface water to a groundwater system is generally studied by numerical flow modeling based on piezometric contour maps of the aquifer and the groundwater level. It would, however, be difficult to build confidence for flow paths, in the case of large variations in the hydrogeologic properties such as the hydraulic conductivities and gradients. It is also difficult to solve whether any hydraulic connection would be present or not.

As constituents of the water molecule, ^{18}O or ^2H and ^3H are regarded to be conservative tracers in the water cycle and useful tools in hydrological research (Mazor, 1991). Since the isotope ^{18}O is considered as a conservative tracer, the seasonal input variations of its concentration in precipitation can be transformed into the output variation in aquifers mainly by mixing of water particles of different residence times along the flow paths. Many case studies have been reported using isotopic approaches that provide a tool for identifying the actual mass transport of water from a surface water system to a groundwater system (Carlin, et al. 1974, Vogel et al. 1975, Dincer et al. 1978, Stichler & Moser 1979, Maloszewski et al. 1992, Vitvar & Balderer, 1997). In this study, environmental isotope tracers were used to evaluate the contribution of surface water to a groundwater reservoir during the rainy season.

2 STUDY AREA DESCRIPTION

2.1 *Geological Setting*

The vicinity of the study area is characterized as an alluvial mountain valley in steep mountainous terrains at the upstream area of the Nakdong river basin, which is located in the area of Cretaceous igneous rock bodies. They are comprised of andesite, biotite granite, alkali granite, felsite porphyry and granite porphyry (Figure 1). The stream flows to the northward direction. This direction is supposed to be an inferred fault zone.

2.2 *Hydrological Condition*

The climate of the study area is represented as a monsoon with an annual precipitation of about 1,260mm. However, the majority of rain falls in the area from mid-June to early August. The drainage also indicates high runoff receding within 4-5days after a precipitation event.

Chemical analyses and in-situ measurements (pH, EC, alkalinity, Temp.) of the representative water samples were carried out during this study.

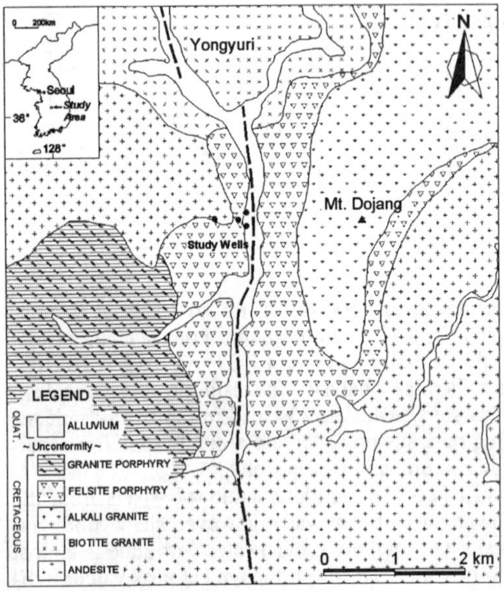

Figure 1. Geologic map and sampling sites.

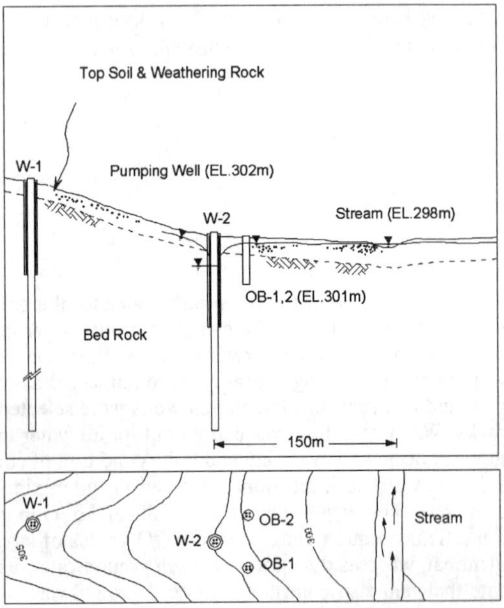

Figure 2. Well locations.

3 METHODS EMPLOYED

3.1 *Water sampling*

All water samples, such as rain, stream water and groundwater from shallow and deep wells, were taken almost daily under the condition of continuous pumping from a deep well from May to September in 1996 and 1997.

The study wells consist of one deep well (W-2) with a depth of 200m and two shallow observation wells of about 30m (Figure 2). The well (W-2) located 150m from the nearest stream was grouted up to 70m from surface to protect any intrusion of surface water. The pumping rate was about 100m^3 per day. All rainwater was sampled excluding a small amount of rainfall.

3.2 *Analysis in laboratory*

Isotopic analyses for all water samples were carried out at KAERI (Korea Atomic Energy Research Institute). Oxygen-18 and deuterium in the waters were analyzed using the CO_2 gas equilibration technique with VG SIRA II and the zinc reduction method with an Optima mass spectrometer. The precisions were found to be within ±0.15‰ for ^{18}O and ±1.5‰ for ^2H. The samples for tritium were also measured by a liquid scintillation counter (Packard 2770TR/SL) for 500 minutes with a precision of 0.5 TU in common after electrolytic enrichment.

4 RESULTS

4.1 *Chemical characteristics in the waters*

The major physicochemical data of stream water and groundwater of the study area are listed in Table 1. The chemical composition of all the water belongs to Ca-HCO_3 type. The groundwater from the pumping well (W-2) is slightly enriched in Ca, Na, SiO_2 and HCO_3 compared to shallow groundwater (OB-1 and 2), indicating that the groundwater is relatively evolved by a water-granite interaction. It can be considered that the groundwater has a longer residence time. However, their mixing phenomena could not be evaluated quantitatively because the chemical composition of pumping well showed no significant difference from the shallow wells.

Table 1. Representative physicochemical properties of stream water and groundwater.

Samples No.	pH	EC (μS/cm)	Na	K	Ca (mg/L)	Mg	SiO_2	HCO_3	SO_4	Cl
Stream	6.69	50.0	2.6	1.1	4.9	1.0	8.8	13.4	4.2	3.5
OB-1	7.19	62.5	3.5	1.2	6.2	1.2	9.6	23.2	5.0	3.7
OB-2	7.05	63.7	3.6	1.2	6.4	1.2	10.3	22.5	5.0	3.7
W-2	7.59	95.9	7.4	0.8	9.5	1.2	27.8	41.7	6.3	3.5

4.2 *Isotopic compositions in the water*

The $\delta^{18}O$ versus δD diagram (Figure 3) shows that all water is plotted close to the world wide meteoric water line (Craig 1961). The $\delta^{18}O$ and δD values of water are distributed from –17.4 to –3.4‰ and –128.9 to –10.2‰, respectively, showing a wide range. The average values ($\delta^{18}O$ and δD) for rainwater are –9.5‰ and –63.4‰, for surface water –9.6‰ and –65.0‰, for shallow groundwater –9.4‰ and –66.7‰, and for the pumping well –9.4‰ and –65.8‰. These isotopic values of groundwater from the pumping well do not show any typical altitude effect (0.19‰/100m for $\delta^{18}O$ in Korea, Kim and Nakai 1988). This indicates that the groundwater of the pumping well (200m below ground surface) is not recharged from the high altitude area compared to shallow groundwater or that the groundwater is highly mixed with shallow groundwater and/or surface water. However, the tritium content of groundwater (5-10 TU) is similar to that of shallow groundwater, whereas that of groundwater from another well (W-1), which is located near the pumping well with a similar depth (about 300m below ground surface), is almost tritium free. Considering continuous pumping of W-2, the mixing process in W-2 has been accelerated, whereas no mixing process is indicated in W-1 by pumping.

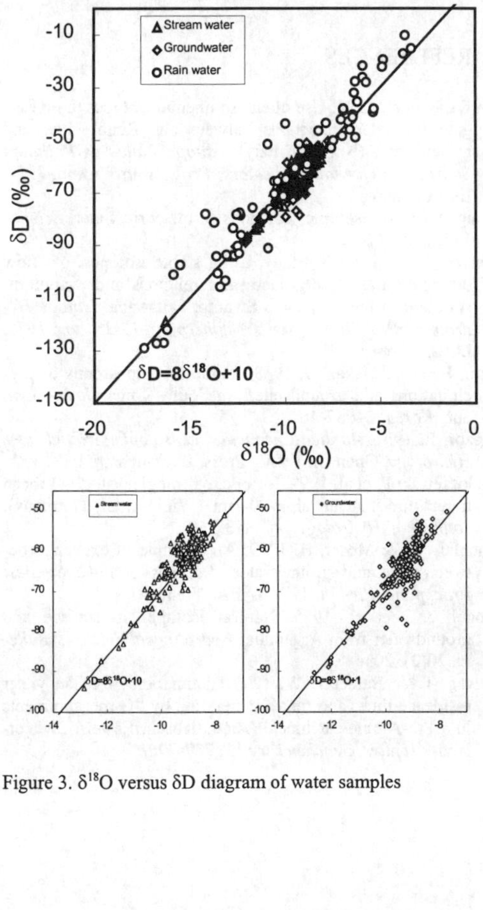

Figure 3. $\delta^{18}O$ versus δD diagram of water samples

4.3 *Isotopic variations in the waters*

During the summer period of 1996 and 1997, stream water, shallow groundwater and deep groundwater from pumping wells were sampled almost daily and the isotopic values were measured, though every rain event was not sampled. The amount of precipitation and the $\delta^{18}O$ value of each water are plotted in Figure 4, including the $\delta^{18}O$ values of some rain events.

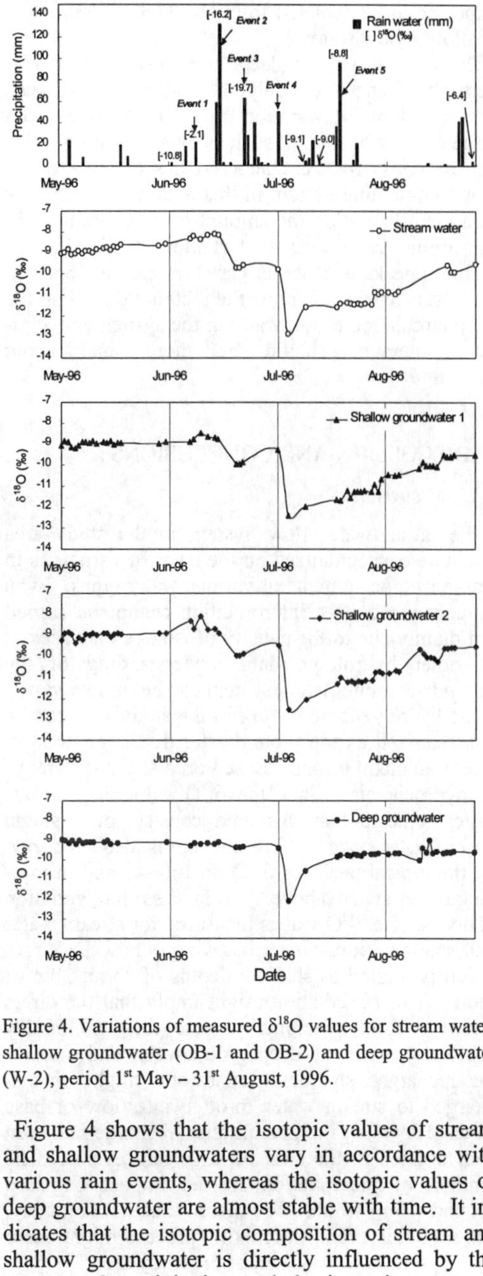

Figure 4. Variations of measured $\delta^{18}O$ values for stream water, shallow groundwater (OB-1 and OB-2) and deep groundwater (W-2), period 1st May – 31st August, 1996.

Figure 4 shows that the isotopic values of stream and shallow groundwaters vary in accordance with various rain events, whereas the isotopic values of deep groundwater are almost stable with time. It indicates that the isotopic composition of stream and shallow groundwater is directly influenced by the amount of precipitation and the isotopic compositions of rainwater without a time delay. From May to August, the rainouts in the study area are divided into five events based on the amount of precipitation. The $\delta^{18}O$ values of stream water and shallow groundwater were changed by rain event 1 (-2.1‰) and rain event 2 (-16.2‰), whereas the deep groundwater did not recognize any change. Although the $\delta^{18}O$ value was unfortunately not measured on rain event 4, the isotopic value of event 4 is expected to be depleted from the data of stream and shallow groundwater.

The $\delta^{18}O$ values of deep groundwater were discernibly changed on July 4th and in mid-August. These patterns of variation might be attributed to a large amount of rainfall such as events 2 and 5. This can be considered as characteristics of the groundwater circulating system in this area, which depends to a great extent on the amount of precipitation. In the groundwater system of shallow depth, the discernible peaks of isotopic variation can be observed on a rain event of more than 20mm/day. For the deep circulation of rainwater in the system a precipitation amount should be more than about 100mm/day.

5 DISCUSSION AND CONCLUSIONS

5.1 *Discussion*

The groundwater flow system in the study area could be conceptualized on the basis of variations in stable isotope contents of various water types. Even though quantitative interpretation cannot be carried out mainly due to the paucity of isotope data as well as other hydrologic data, understanding of the groundwater circulating system can be improved.

The hydrogeologic setting in the study area can be characterized as the groundwater discharge zone in steep mountain terrain. Based on a slight difference in hydrochemical data, the $\delta^{18}O$ values of various water indicate that this area consists of a single groundwater system. This concept is also supported by the typical peaks of $\delta^{18}O$ on July 4th and in mid-August as stated above. From a similar variation pattern of the $\delta^{18}O$ values measured for stream water and shallow groundwater, active groundwater circulation is limited to shallow depths of about 30m or more. This observation might imply that the direct recharge from precipitation less than 60mm/day could not reach the deep groundwater and most of the recharged shallow groundwater might be discharged to stream water through interflow or base flow. However, the precipitation recharge has been well defined in the case of large rain events. The amount of precipitation for an impact on deep groundwater would be more than 100mm/day. Through this event, the mixing process might occur in the groundwater system of the study area.

On the other hand, according to the hydrogeological investigation of the deep well, the highly fractured zone at a depth from 130 to 150m was identified. Therefore, it could not be ruled out that the fracture zone crossing the deep well, having high hydraulic conductivity, was not connected to the fractured reservoir containing the stream water zone. The pumping of the deep well might also result in the inflow of stream water. The wells are pumped with 135 and 149 m^3 per day, which were recorded as the typical peaks on July 4th and July 9th, respectively, which is identical to the dates showing the depleted isotopic values of deep groundwater. A question is still remains as to the uncertainty why stream water can contribute to the deep well within a day. This issue will be solved from the on-going isotopic study in the area.

5.2 *Conclusions*

This study was conducted to characterize the mixing process, the whether a deep groundwater system would be affected by stream water or not. This study is limited to provide decisive answers on these issues. However, environmental isotopes in summer rains could be used as a good tracer to improve understanding of the groundwater circulating system in the case of limited geologic conditions.

6 REFERENCES

Carlin, F. et al. 1974. Use of environmental isotopes to investigate the interconnections between the Reno River and groundwater (Northern Italy), *Isotope Ratios as Pollutant Source an Behaviour Indicators; Proc. symp. Vienna, 1974*, IAEA, Vienna .

Craig, H. 1961. Isotopic variations in meteoric water, *Science* 133: 1702-1703.

Dincer, T. et al. 1978. Study, using stable isotopes, of flow distribution, surface-groundwater relations and evapotranspiration in the Okavango Swamp, Botswana. *Isotope Hydrology; Proc. Intern. symp. Neuherberg, 19-23 June 1978*, IAEA, Vienna.

Kim, K. H. & Nakai, N. 1988. Isotopic compositions of precipitations and groundwaters in South Korea. *Jour. Geol. Soc. Korea* 24: 37-46.

Mazor, E. 1991. *Applied Chemical and Isotopic Groundwater Hydrology*, Open University Press, Buckingham, UK.

Maloszewski P. et al. 1992. Isotope hydrological study of mean transit times in an alpine basin (Wimbachtal, Germany). *Journal of Hydrology* 140: 343-360.

Stichler, W. & Moser, H. 1979. An example of exchange between lake and groundwater, *Isotopes in Lake Studies; Proc. panel Vienna, 1977*. IAEA: Vienna.

Vogel, J.C. et al. 1975. Natural isotopes in surface and groundwater from Argentina, *Hydrological Sciences Bulletin* 20(2): 203.

Vitvar, T. & Balderer, W. 1997. Estimation of mean water residence times and runoff generation by ^{18}O measurements in a Pre-Alpine catchment (Rietholzbach, Eastern Switzerland). *Applied Geochemistry* 12: 787-796.

Groundwater: Past Achievements and Future Challenges, Sililo et al. (eds) © 2000 Balkema, Rotterdam, ISBN 90 5809 159 7

Groundwater sampling during the mud-rotary drilling process

A. Bittner
Bittner Water Consult, Windhoek, Namibia

G. Christelis
Department of Water Affairs, Windhoek, Namibia

ABSTRACT: The EDL-technology aims at extracting liquids from an unsleeved / unstabilised borehole without having to lift or remove the drill string. The technology enables hydrogeologists to sample water once or several times at whatever stage of the mud-rotary drilling process, and to analyse key parameters of the aquifer before completion of the well construction. A test borehole was drilled at Eenyama village where previous drilling only yielded boreholes with brackish water from the main deep Kalahari aquifer. Clear water samples were taken from three different depths. The sample taken at 153 m depth showed a conductivity higher than 5000 μS/cm, while the sample taken at 240 m depth showed a conductivity of 1130 μS/cm. A filter was installed at the depth of the freshwater aquifer while the brackish aquifer was sealed with cement. The borehole was subsequently recommended for installation in order to supply safe drinking water to the village.

1 INTRODUCTION

As part of the annual rural water supply drilling programme in the Ohangwena, Omusati and Oshikoto regions a test borehole was drilled at Eenyama, approximately 12 km southeast of Eehnana in north-central Namibia (Fig. 1). The aim was to identify and develop a potential freshwater source below the regional brackwater aquifer within the unconsolidated sediments of the Owambo Kalahari basin.

The EDL (Exploration Drilling for Liquids)-technology was developed in Europe and applied in the mineral water industry as well as in the field of groundwater pollution control (Leßmöllmann 1996). The Department of Water Affairs, Namibia, was the first institution to apply the EDL-technique outside of Europe.

2 REGIONAL HYDROGEOLOGY

In the study area the groundwater mainly occurs within unconsolidated to semi-consolidated sediments of the Kalahari Sequence, which can reach a thickness of more than 400 m in the north-central part of the Owambo basin (Momper 1982).

A confined freshwater aquifer, which is replenished by lateral throughflow from the recharge area in southern Angola, underlies the larger part of the eastern Ohangwena and northern Oshikoto regions (Geyh 1997).

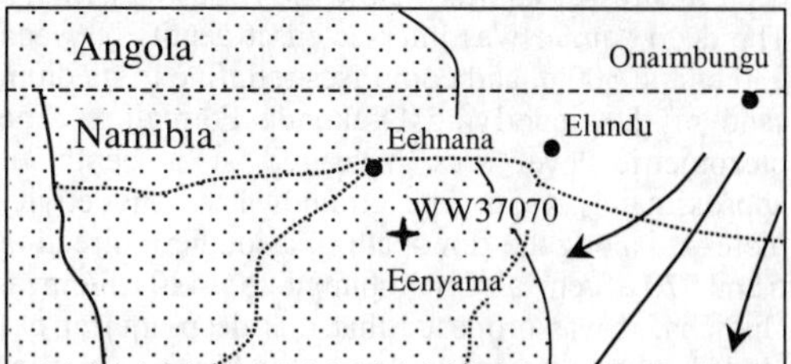

Figure 1. Location of test borehole WW37070 at Eenyama.

This Main Deep Aquifer (MDA) is found between 70 m and 150 m depth within the alluvial Andoni Formation. West of the line Eehnana – Okankolo the MDA becomes increasingly saline.

Eenyama village is situated in an area where the MDA is known to be brackish or saline. The existing boreholes were drilled to depths between 80 m and 150 m. The nearest borehole yielding freshwater fit for human consumption is located some 30 km northeast at Elundu.

Between 1996 and 1998 some 40 boreholes were drilled to depths of up to 180 m into the freshwater MDA between Elundu and Okongo in the eastern part of the Ohangwena Region.

The confined aquifer was intersected at varying depths between 70 m and 150 m, and comprises green, medium to coarse-grained sand of the Andoni Formation. The freshwater MDA is the only source of safe drinking water in the eastern Ohangwena Region.

As part of the Ohangwena Exploration Drilling Project, a 200 m, 300 m and 400 m deep borehole group was drilled along the dormant Odila River at Onaimbungu, near the Angolan border, and at Ondjaba, 40 km northeast of Okankolo, respectively (Fig. 1). One of the findings was the existence of a deep freshwater aquifer below the freshwater MDA. The deep aquifer was intersected at depths between 230 m and 300 m and comprises red, fine to medium sand of the underlying Olukonda Formation. The piezometric level was measured at a depth of approximately 46 m. The groundwater conductivity increased along the flow-path in a southern direction from 570 µS/cm at Onaimbungu to 1420 µS/cm at Ondjaba. It was proposed that this deep aquifer extends further west and underlies the brackish aquifer in the Eehnana area (Bittner, unpubl.). The water quality was expected to be similar to that of the Ondjaba borehole.

3 THE EDL-TECHNOLOGY

The EDL-technology comprises a procedure as well as equipment aimed at extracting liquids from an unsleeved / unstabilised or partially sleeved / stabilised borehole without having to lift or remove the drill string.

The equipment is thereby utilised as a temporary well. An extraction tool with filter sections is placed on top of the drill bit. After reaching the desired depth the drilling is halted and filling material is poured into the annulus or pumped through the drill string until the extraction tool is covered (Fig. 2).

The removal of the drill head is required to lower a submersible pump or other extraction devices within the drill string. Thereafter, the purging of the drill string can commence.

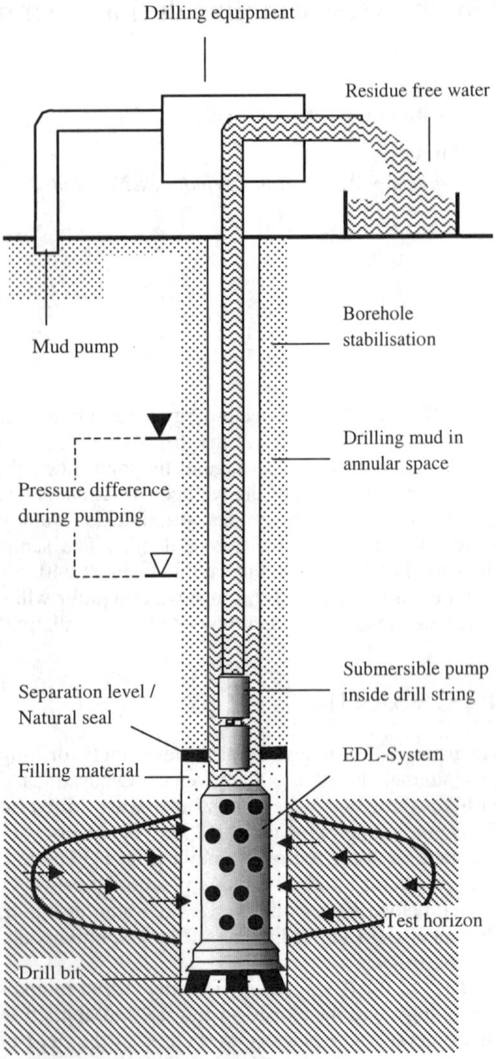

Figure 2. Principle of the EDL-technology.

The pressure difference, caused by the purging of the drill string, forces the stabilising drilling mud in the annular space to flow towards the extraction tool. The fine particles within the drilling mud settle on top of the filling material, resulting in the development of a natural seal, which prevents the drilling mud from entering the extraction tool through the test horizon.

Hence only water from the desired test horizon is sampled and the unsleeved borehole remains stable, due to the static drilling mud. The pumping of water must be continued until the sample is free of drill residue. Depending on the geology, depth and rate of flow, this process can take between 2 and 6 hours.

To enable a final decision on the continuation of the drilling, the sample could then be tested in situ for key variables like conductivity – temperature – pH – nitrate – etc., determined by the project aims.

The obvious advantage this method offers, is the option of making an informed decision regarding the continuation or discontinuation of the drilling process and the well construction phase based on detailed in situ knowledge of the aquifer. It also allows the early determination of the positioning of filtration equipment.

The EDL-technology enables the sampling of different aquifer horizons or multi-layered aquifer systems without the necessity to drill a large number of exploration boreholes. This is important for the exploration of thermal and mineral waters as well as for the detection of groundwater contamination and other hydrogeological investigations, where exact water quality information is required.

4 TEST BOREHOLE WW37070 AT EENYAMA

Eenyama was chosen as a test site for the application of the EDL-technology because it was high on the priority list of the Rural Water Supply Directorate. The village has a new school but no safe drinking water source. Water for human and livestock consumption is taken from pans and shallow hand-dug wells, which are water filled only during the wet season.

After preparation and organisation of all personnel and equipment necessary on site, the actual drilling commenced on 29 April 1998. The EDL-technology was tested at 51 m, 153 m and 240 m depth. Water samples could only be taken from 153 m and 240 m depth, the horizon at 51 m depth was dry.

It was not possible to retrieve a clear sample from 153 m depth within 3 hours pumping, most probably because there was only little water within the basal sandstone of the Andoni Formation. The conductivity of the sample was measured as 3470 μS/cm, while the conductivity of the original drilling mud was measured as 1210 μS/cm. Taking in consideration that the sample was a mix of groundwater and drilling mud the conductivity of the groundwater is estimated to be greater than 5000 μS/cm.

The drilling was again stopped at 240 m depth when a layer of red sand was intercepted. After 2 hours of purging and pumping the water became clear and a sample was taken. The conductivity was measured as 1130 μS/cm. The mud-rotary drilling was then continued until the bottom of the sand aquifer was reached at 259 m depth (Fig. 3).

The duration of one EDL-test depends on the depth of the aquifer, which also effects the settling time of the fill material and the aquifer material, which will determine the developing time.

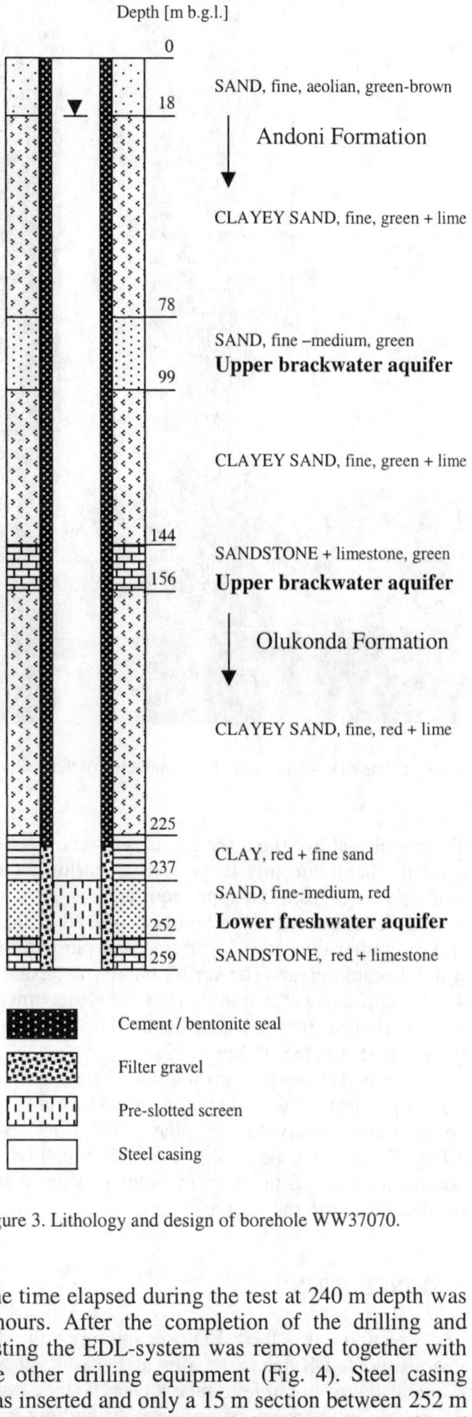

Figure 3. Lithology and design of borehole WW37070.

The time elapsed during the test at 240 m depth was 5 hours. After the completion of the drilling and testing the EDL-system was removed together with the other drilling equipment (Fig. 4). Steel casing was inserted and only a 15 m section between 252 m and 237 m depth was installed with 1 mm pre-slotted screens. Filter gravel was poured into the annulus to fill 28 m from the bottom of the borehole to a depth of 231 m, 6 m above the screen.

Figure 4. The EDL-equipment after completion of the drilling.

The annulus above the filter gravel was sealed with a cement / bentonite mix to prevent an inflow of the brackish water from the upper aquifer.

The development of the borehole proved difficult, and only after two days of surging and pumping the water became clear. The airlift yield was tested at 4.5 l/s and the water tasted fresh. The piezometric level was measured at 18 m depth. The final completion date was the 05 May 1998.

The abstraction rate during the 3-day constant discharge test was calculated at 5 l/s and the conductivity analysed in the laboratory was 1030 μS/cm. The borehole was subsequently recommended for installation in order to supply safe drinking water to the village.

5 CONCLUSIONS

The testing of the EDL-technology in the unconsolidated Kalahari aquifer at Eenyama was a success. It was possible to take a clear water sample during the mud-rotary drilling process without prior completion of the borehole.

A freshwater aquifer was intersected at a depth between 237 m and 252 m. A thick layer of clayey sediments separates the aquifer from the upper brackwater aquifer, which is located at a depth between 78 m and 156 m.

The lithology of the upper aquifer comprises green, medium to coarse-grained quartz sand of the Andoni Formation, while the lithology of the lower freshwater aquifer consists of red, fine to medium-grained quartz sand of the Olukonda Formation.

The duration of the EDL-test depends on the aquifer material and the depth of the test zone. The duration of the drilling, the EDL-testing and completion of the 259 m deep borehole at Eenyama lasted approximately six days. The yield of the freshwater aquifer was tested at 5 l/s.

In the past drilling requests from the Eehnana - Okankolo area were rejected due to the high salinity of the known groundwater resources. The application of the EDL-technology made it possible to develop a deep freshwater aquifer, at reasonable cost, in an area where drinking water would have had to be supplied by more expensive means such as tanker service or pipelines.

The EDL-technology is a cost effective alternative for the sampling of individual aquifer horizons or multi-layered aquifer systems. Other possible applications for the EDL-technology are in the fields of thermal and mineral water exploration as well as in the detection and quantification of groundwater pollution.

REFERENCES

Bittner, A. unpublished. Results of the 1997/98 annual drilling in the 4-O regions. *Department of Water Affairs Memorandum*, File No. 12/12/1. Windhoek.

Geyh, M.H. 1997. Isotope hydrological study in eastern Owambo, Etosha Pan, Otavi Mountain Land and central Omatako catchment including Waterberg Plateau. *BGR Report*, File No. 6532/97. Hannover.

Leßmöllmann, M. 1996. Numerische Modellierung der hydraulischen und hydrogeologischen Bedingungen bei tiefenspezifischer Grundwasserentnahme aus Vertikalbrunnen. *Ruhr-Universität Bochum*. 45p.

Momper, J.A. 1982. The Etosha basin reexamined. *Oil & Gas Journal*. 5:262-287.

Groundwater: Past Achievements and Future Challenges, Sililo et al. (eds) © 2000 Balkema, Rotterdam, ISBN 90 5809 159 7

Application of environmental isotope techniques to hydrological and pollution problems in the urban environment

M.J. Butler & B.Th. Verhagen
Schonland Research Centre, University of the Witwatersrand, Johannesburg, South Africa

M. Levin
Africon Engineering International, Pretoria, South Africa

ABSTRACT: Environmental isotope hydrology has been very successful in solving a variety of hydrological, pollution and leakage problems in urban areas in South Africa. Environmental tritium was used in studies to evaluate the potential of monitoring boreholes as early detectors of pollution. The discovery of anomalously high levels of tritium in landfill leachate led to the realisation that tritium has the potential to be used as a tracer of leachate into the environment.

The isotope ratios $^2H/^1H$ and $^{18}O/^{16}O$ were successfully applied in Bellville (Cape Town) to trace pollution to sewerage sludge ponds and not a landfill as was generally accepted.

Both tritium and stable isotopes were employed to trace leakage in urban water reticulation systems. Leakage sites in Pretoria were pin-pointed, the extent of the leakage estimated, and valuable information obtained on the effect of physiography on small-scale ground water systems.

The vulnerability to pollution of dolomite aquifers was clearly demonstrated in the Midrand (Johannesburg) area, where both radioactive and stable isotopes were employed to prove incipient pollution and allow for source apportionment of pollutants.

1 INTRODUCTION

The study of urban hydrology increases in importance as cities develop and expand. In South Africa, ground water is widely extracted for municipal supply and by private standowners. With increasing fresh water scarcity, management and protection of this resource requires the development of novel approaches to the study of ground water in the urban environment.

Environmental isotope hydrology has become an established field of interdisciplinary applied science. It provides geohydrological information which may be difficult or even impossible to obtain by other known techniques. During recharge, the chemistry of infiltrating water may change radically in the vadose zone. The isotopes of hydrogen and oxygen on the other hand are conservative tracers.

2 ISOTOPE TECHNIQUES

The radioactive isotopes of tritium (half-life 12.43 a) and radiocarbon (half-life 5730 a) are produced by cosmic ray interaction with the atmosphere. Oxidised in the form of water and carbon dioxide, these isotopes enter ground water through the infiltration of rain. In the early sixties, thermonuclear tests increased atmospheric levels, producing a "bomb" peak which could easily be traced through hydrological systems. Although most thermonuclear tritium has now decayed away, the near natural levels continue to constitute a unique and simple tracer of recent ground water given sensitive detection methods. The isotope ratios of hydrogen ($^2H/^1H$) and oxygen ($^{18}O/^{16}O$), undergo small changes due to evaporation and condensation of water, expressed as relative differences (δ ‰) w.r.t. a standard, often diagnostic of the pathway water has followed in the environment.

In southern Africa, environmental isotope techniques have been applied to geohydrological problems for the past 30 years and have revolutionised aspects of geohydrological thinking. More recently, their potential is being exploited in the fields of urban hydrology, ground water pollution and waste disposal impact

3 LEAK DETECTION AND URBAN HYDROLOGY

The distinctive δD - $\delta^{18}O$ composition which characterised Rand Water (RW) up to 1996 has been used to distinguish mains water from naturally recharged ground water (Butler & Verhagen 1997).

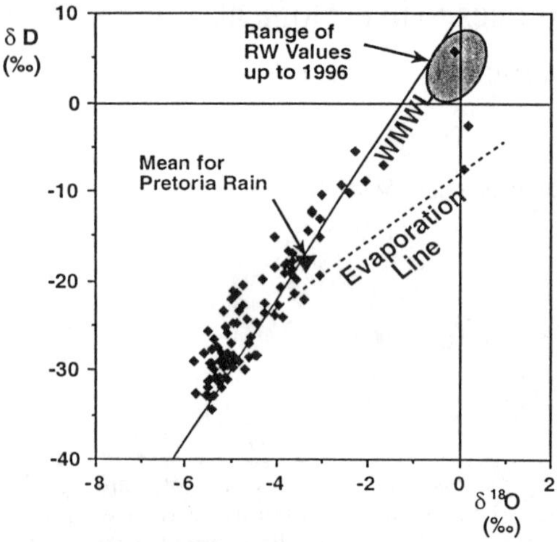

Figure 1. δD - δ^{18}O plot for ground water in the northern suburbs of Pretoria, showing the WMWL, the long-term weighted mean for Pretoria rain and the range of values measured in RW mains (up to 1996).

Figure 1 shows the spread of ground water values plotted on a δD - δ^{18}O diagram, the long-term mean for Pretoria rain and the range for RW mains. Also shown is the world meteoric water line (WMWL). RW mains contains tritium at about 4 TU. Numerous private boreholes in the northern suburbs of Pretoria were used as sampling points to detect mains leakages into the ground.

In townships in the north of Pretoria (Figure 2) there are many reported water mains leakages. Ground water levels are assumed to follow the gentle surface slope, gradient 0.008. Values of δD and EC scatter considerably without geographic trends except for rapidly decreasing values on the much steeper hillslope in the NE.

On a similar steep slope to the west in Mountain View, a pronounced correlation between EC and δ^{18}O was found (Figure 3) ascribed to rain recharge selection and active down-slope drainage. In most of the area shown in Figure 2 there is practically no such correlation (Figure 4). As in Mountain View, alkalinity correlates significantly with EC, reflecting normal soil processes during infiltration. This physiographic control on isotopic and chemical values shows that there is very little ground water flow, except for the hillslopes. This is important information, in terms of the localisation of leakage water and the potential spread of pollution.

The percentage mains water p_m is given by

$$p_m = \frac{\delta - \delta_g}{\delta_m - \delta_g} \times 100\%$$

where δ = measured value, δ_g = background for the area, δ_m = value in mains water.

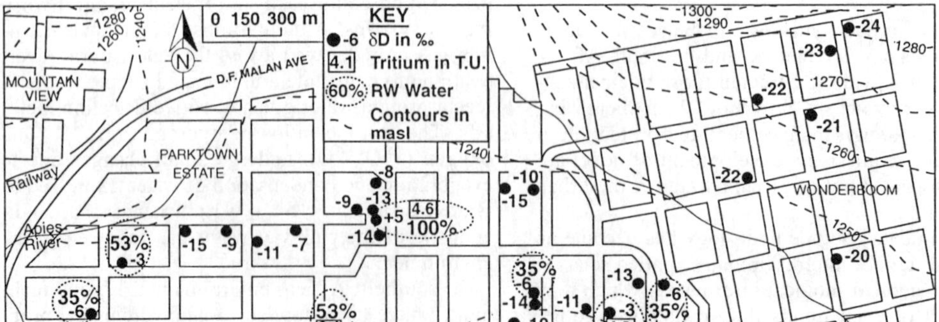

Figure 2. Map of suburbs in northern Pretoria showing topographic contours, δD values, tritium values and calculated percent mains water.

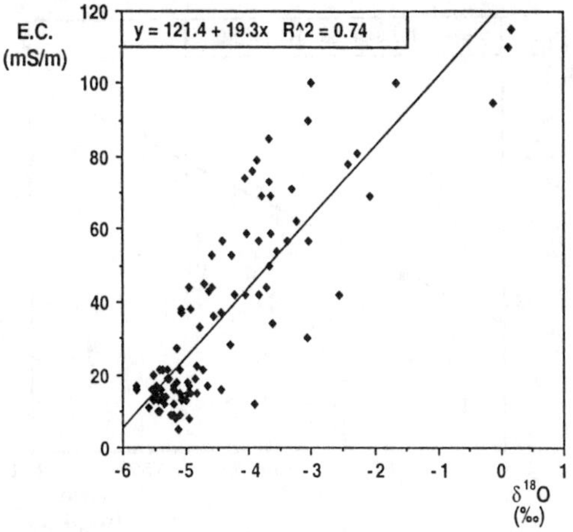

Figure 3. Plot of E.C. - $\delta^{18}O$ for Mountain View ground water.

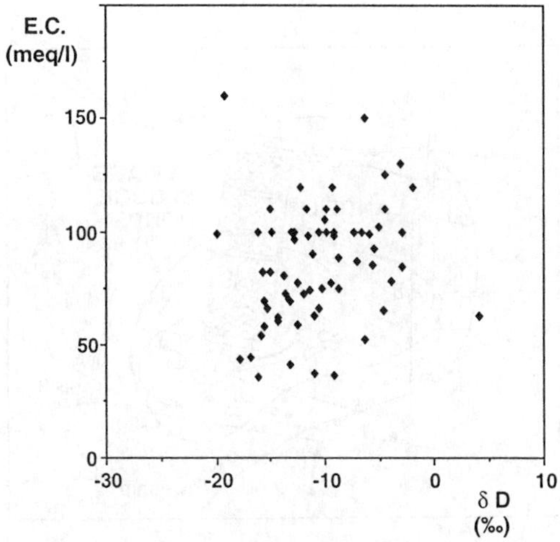

Figure 4. Plot of E.C. - δD for ground water in suburbs shown in Figure 2.

The values of p_m are shown for individual sampling points in Figure 2.

Many sampling points show between 30% and 56% of mains water, confirming the suspected numerous mains leaks. Tritium at several TU shows active recharge in the townships. The estimate of 100% mains for one borehole is supported by tritium at 4.6 TU.

4 GROUND WATER POLLUTION IN A DOLOMITE AQUIFER

The area of study at Clayville is a mixed urban, industrial and agricultural environment. The principal aquifer is extensively karstified Proterozoic dolomite, compartmentalised by dykes. Major springs and several high-yielding boreholes provide 30% of the water supply of the capital Pretoria and some of its satellite towns. There is concern at the presence of landfills and moderate to severe pollution in some surface water.

The dominant ground water chemical type is Ca, Mg-HCO_3. The overall mineralisation is fairly low, with the range of ions which in this study are regarded as proxy tracers of pollution (in mg l^{-1}): $SO_4^{2-} < 70$; $Cl^- < 40$; $NO_3^- < 25$.

Many tritium values > 1 TU in borehole and spring samples indicate significant rain recharge. Recharge to the karstic dolomite can occur along a variety of pathways, from almost direct inflow to soil zones of tens of metres in thickness. There is a significant correlation of $\delta^{18}O$ with ^3H (Figure 5). This was interpreted (Verhagen et al. 1994; Verhagen & Butler 1998) as mixtures of ground water: the deeper, longer residence time, component (low ^3H and more negative $\delta^{18}O$) with local recharge (higher ^3H and a more positive $\delta^{18}O$). The $\delta^{18}O$ and δD generally lie on the WMWL (Figure 1) but compositions span a considerable range, interpreted as natural rainfall infiltration, and ingress of waste water and leakages derived from the RW mains (see previous section).

Figure 6a shows plots of the concentrations of pollution indicators SO_4^{2-} and NO_3^- against tritium values. Higher concentrations are clearly associated with higher values of tritium. They are therefore actively being transported from the surface. In Figure 6b the concentration of these ions is plotted against $\delta^{18}O$. Higher SO_4^{2-} corresponds with less negative $\delta^{18}O$ values; NO_3^- with more negative values, closer to the weighted mean for rain. Sulphate therefore tends to enter ground water through domestic

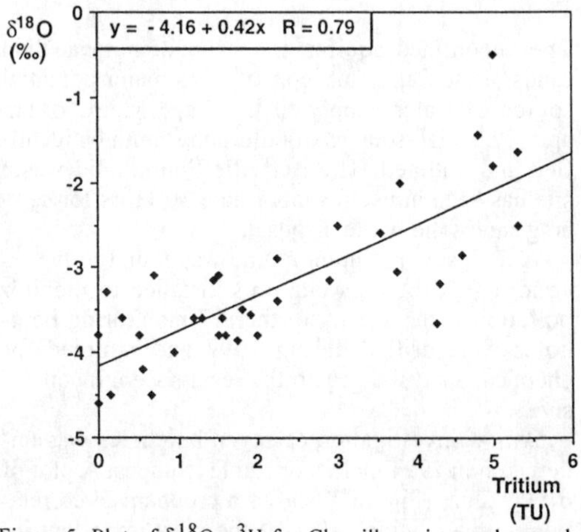

Figure 5. Plot of $\delta^{18}O$ - ^3H for Clayville springs and ground water.

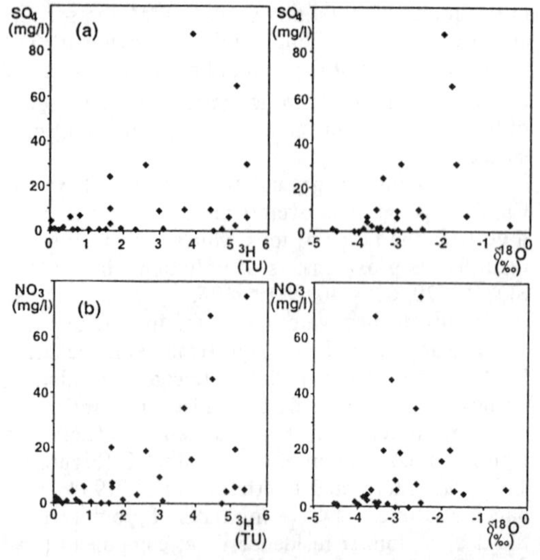

Figure 6a and b. Ground water in the Clayville area: (a) Plots of SO_4 against ^3H and $\delta^{18}O$; and (b) plots of NO_3 against ^3H and $\delta^{18}O$.

and industrial effluent, derived from mains water. Nitrate is introduced largely through rain water and ground water-derived irrigation, rather than through urban and industrial activity.

These ionic concentrations would not have been ascribed to pollution without the isotopic association, giving rise to the important concept of "incipient pollution" which draws attention to the need for timeous measures to limit the impact and mitigate the sources.

5 STABLE ISOTOPES IDENTIFY A POLLUTION SOURCE

The unconfined aquifer of the Quaternary aeolian sands of the Cape Flats constitutes a major potential source of water supply to the Cape Metropolitan area. Potential sources of pollution should be identified and outlined. The Bellville South solid waste site has been in use for more than 30 years for sewerage and solid waste disposal.

A definable pollution plume was found to lie beyond a field of 12 wellpoints installed to monitor pollution at the site. A further 17 monitoring boreholes were drilled during 1992 and sampled for chemical analysis. Again the results were inconclusive.

A new investigation of seven boreholes was undertaken in 1994 including stable isotopes. A plot of $\delta^{18}O$ values (Figure 7) shows a pronounced correlation with ammonia, taken as the most important indicator pollutant. The $\delta^{18}O$ and ammonia contours (Figure 8) show the highest enrichment in ground

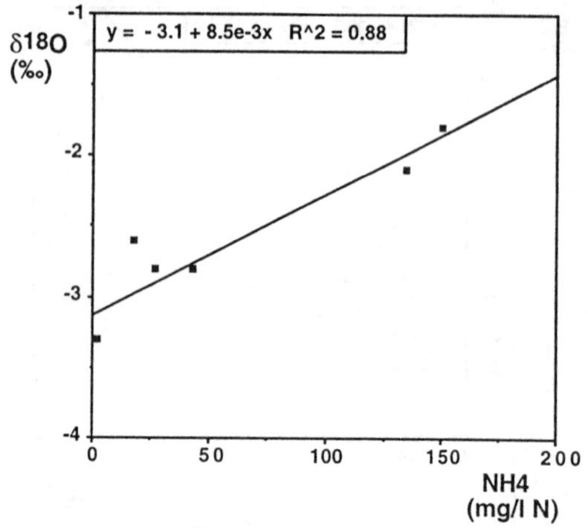

Figure 7. Plot of $\delta^{18}O$ against ammonia concentration for the Bellville South solid waste disposal site.

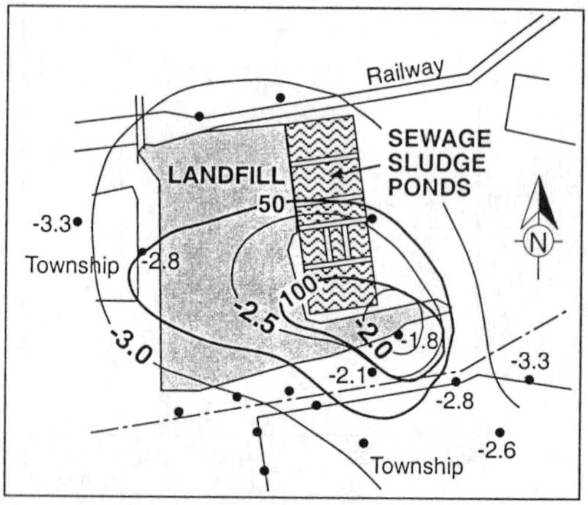

Figure 8. Map of the Bellville South solid waste disposal site showing $\delta^{18}O$ (light lines) and ammonia (heavy lines) contours.

water below the sewage sludge ponds, the enriched oxygen-18 being evidence of evaporative enrichment in the ponds.

Lining of the sewage sludge dams was therefore recommended as well as a program to monitor the dispersion of the pollution plume.

6 TRITIUM IN SOUTH AFRICAN LANDFILLS

Tritium concentrations higher than environmental levels were first discovered in South Africa in association with landfill site leachate in 1995 (Verhagen et al.1998) and numerous others of various types from domestic to toxic have since been identified (Figure 9)

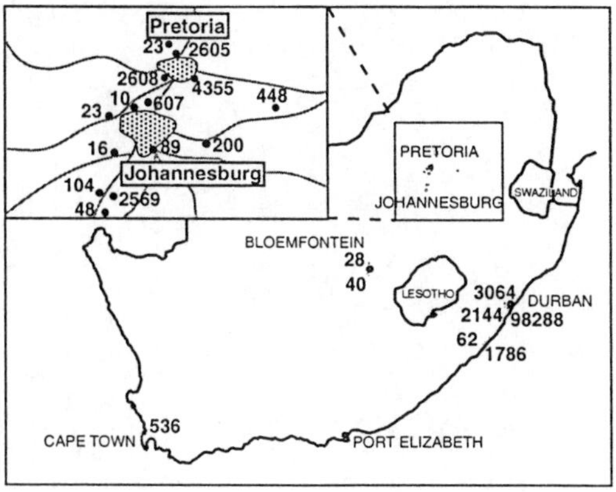

Figure 9. Location of landfill sites in South Africa with associated elevated tritium values.

Tritium exchanges readily between tritium-containing material and moisture moving through a landfill. In the oxidised state it becomes a conservative tracer of water. Southern hemisphere tritium values in rain have returned to near natural levels (≤ 5 TU). This renders very sensitive the tracing of leachate containing waste tritium in the southern African environment. In this study tritium was measured either directly in leachate, or in ground water directly associated with the landfill. Levels were found to range from a few tens of TU to nearly 10^5 TU, approaching the limit for drinking water.

The most probable sources of landfill tritium are luminous dials of watches and clocks containing up to 280 Mbq, and gaseous tritium light sources and signs containing up to 1 TBq of tritium. Even a single damaged object in the landfill, releasing tritium as gas or in a compound, will readily find its way into moisture.

6.1 *A failed landfill*

The highest tritium level observed to date (~10^5 TU) was in the integrated impoundment of leachate at a landfill emplaced in a small, steep valley. When the landfill failed after heavy rain leachate appeared at different points from the landfill mass along the toe. The considerable differences in tritium level observed at these points (405 TU to 12828 TU) suggest that the source of the tritium at this site is highly localised.

In an environmental impact assessment low-level tritium measurements on water samples from monitoring boreholes around the site showed values approaching the limit of detection (0.2 TU). It could be concluded that the ground water contained substantially less than 1 % of leachate. This confirms the prediction that local recharge to the ground water would be minimal on account of the argillaceous nature of the terrain.

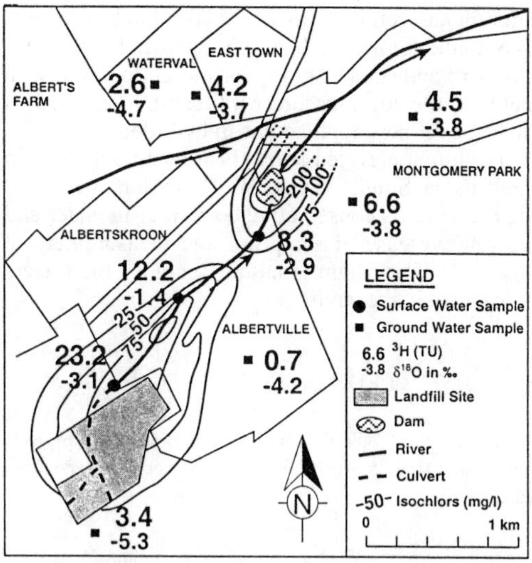

Figure 10. Map of Johannesburg landfill site and surroundings showing surface drainage, borehole and surface water sampling positions, ^3H and $\delta^{18}O$ values and previously inferred isochlors for shallow ground water.

6.2 *A closed landfill*

Tritium levels up to 25 TU were observed in a small stream issuing from a culvert passing underneath a Johannesburg landfill, surprisingly as it was officially closed since 1975. A survey of the area downstream from the landfill revealed levels of tritium above environmental in shallow ground water, earlier been found to be contaminated (Figure 10). Bank infiltration from the stream was therefore shown to be a more likely mechanism for the spreading of ground water contamination than the sub-surface transport previously proposed. Changes of tritium and oxygen-18 levels along the flowing stream revealed a leaking sewer, which previously went unnoticed.

7 CONCLUSIONS

Studies in quite disparate southern African environments show that environmental isotope techniques are often indispensable in understanding geohydrological processes, in uniquely identifying and quantifying recharge, and elucidating transport of pollutants.

Environmental tritium in particular remains a useful and unique tool for studying shallower, more dynamic ground water even though southern African rainfall has returned to pre-thermonuclear levels. The methodology of "isotopic snapshots" with radioactive and stable isotopes is straightforward, with state of the art measurement techniques, and is finding ever new applications in hydrology.

Tritium at elevated levels is almost ubiquitous in landfills in South Africa. With ambient values at their lowest, its conservative behaviour in water and the relative ease of sampling and measurement at levels > 5 TU tritium is unique amongst the tracers of leachate in the environment.

8 ACKNOWLEDGEMENTS

Support by the Water Research Commission under projects K5/311 and K5/628 which initiated these studies, is gratefully acknowledged.

REFERENCES

Butler, M.J. & Verhagen B. Th. 1997. Environmental isotopic tracing of water in the urban environment of Pretoria, South Africa. *Groundwater in the Urban Environment: Problems, Processes and Management*, 101-106. Chilton et al. (eds). Balkema, Rotterdam.

Verhagen B.Th., Levin, M. & Walton, D.G.1994. Diffuse and point source groundwater pollution studies with environmental isotopes. *Future Groundwater Resources at Risk.* 363-378. IAHS Pub. No. 222. IAHS Press,UK.

Verhagen B. Th., Levin, M., Walton, D.G. & Butler, M.J. 1998. Environmental isotope, hydrogeological and hydrogeochemical studies of ground water pollution associated with waste disposal. Final Report to the Water Research Commission: Project #K5/311.

Verhagen B. Th. & Butler, M.J. 1998. Environmental isotope studies of urban and waste disposal impact on groundwater resources in South Africa. *Isotope Techniques in the Study of Environmental Change*, 411. IAEA, Vienna.

Groundwater: Past Achievements and Future Challenges, Sililo et al. (eds) © 2000 Balkema, Rotterdam, ISBN 90 5809 159 7

Geochemical modelling and laboratory experiments to test the feasibility of artificial recharge

L.C.Cavé
Division of Water, Environment and Forestry Technology, CSIR, Stellenbosch, South Africa

ABSTRACT: Chemical reactions may influence the success of artificial recharge and subsurface water storage schemes. Primary concerns are deterioration of water quality and clogging effects. Geochemical evaluation was required for a feasibility study involving a planned borehole injection scheme in the semi-arid Western Karoo, South Africa. A breccia pipe has been identified as a potential underground repository for emergency water supplies. Geochemical modelling was used to simulate pH-buffering effects and solute activity changes across a mixing front during injection of various recharge sources. Saturation indices for carbonate and ferric iron minerals were then used to predict whether the formation of precipitates would be thermodynamically favoured. Modelling was supplemented by laboratory leaching tests to identify readily soluble components in the aquifer materials, and by water-rock equilibration batch tests, which assessed the capacity for aquifer rocks to sorb and immobilise dissolved species from a range of recharge waters.

1 INTRODUCTION

Artificial recharge (AR) is becoming increasingly popular worldwide as a practical, cost effective and environmentally acceptable water management tool (Harpaz 1971, Brouwer et al. 1990, Pyne 1995, Brouwer 1996). Hydrogeological, geochemical and socioeconomic factors need to be considered carefully before implementing an AR scheme to avoid unnecessary expense or loss of water resources.

This paper will focus on the geochemical aspects of AR by borehole injection into fractured hard rock. A case study from Calvinia, South Africa, demonstrates how a geochemical approach was used to anticipate potential problems and how these could be minimised through effective management.

1.1 *The Calvinia study site*

AR has been proposed as an innovative strategy for managing the water supplies of Calvinia, near the western edge of the semi-arid Karoo, South Africa (Murray et al. 1998). Acquiring water for the 10,000 residents of the town is a difficult task considering the low total rainfall (MAP <250 mm), erratic rainfall patterns and high evaporation for the region. Surface runoff is scarce and unpredictable and water from the town dam frequently has to be supplemented by groundwater, which remains a vital resource for the town and surrounding farms. Groundwater exploration around Calvinia, however, tends to meet with moderate success, both in terms of quantity and quality.

The geology of the area consists of Permian age mudstones, shales and siltstones. These have been inundated and baked by numerous sheet-like dolerite intrusions during the Jurassic. Several isolated pipes of brecciated sediments occur in the region, associated with the intrusion of the dolerites into the Karoo basin sediments. Groundwater yields from the argillaceous rocks are generally low (<1 L/s). Weathered shales and alluvial deposits along the valley floors may yield stronger flows, but the water in this environment is mostly brackish. Fractured dolerite-shale contact zones and breccia pipes, which allow significant and relatively rapid recharge, are favoured targets for exploration.

1.2 *The borehole injection scheme*

The difficulty of acquiring additional water sources has led Calvinia to seek alternate means of conserving their available water. A highly porous and permeable breccia pipe (Fig.1) is planned to act as an underground reservoir, reducing evaporative losses and providing storage for occasional surplus water, which can later be used for emergency supply. The layout of the scheme is shown in Figure 2.

Two boreholes (I1 and I2) are equipped with injection pipes to recharge surplus groundwater from boreholes in the town wellfield (B1, B2 and B3) or surface runoff from the town dam into the breccia at

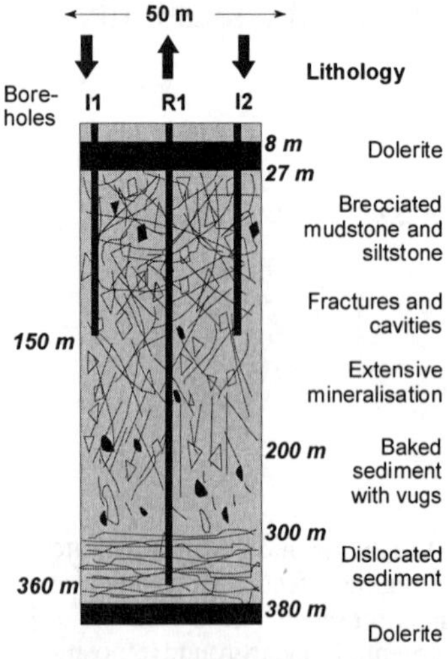

Figure 1. Schematic cross section through the pipe.

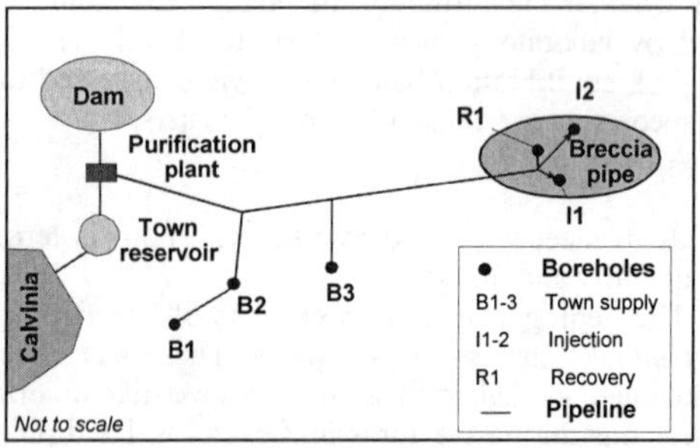

Figure 2. Layout of the injection scheme.

a rate of approximately 10 L/s. The stored water will later be abstracted via a deep recovery borehole (R1) in the centre of the breccia pipe.

The subsurface reservoir is a roughly cylindrical body of brecciated, low-grade metaclastites containing open vugs up to a few cm in diameter. Porosity measurements and pump tests indicated a storage capacity of about 80,000 m^3 (Murray et al. 1998).

A maximum blow yield of 80 L/s was initially measured when drilling into the breccia pipe, promising a valuable water resource for Calvinia. The groundwater, however, has not been suitable for production because water level recovery after pumping is extremely slow and, although the water is of low salinity, its chemical composition is undesirable for domestic consumption without advanced treatment or blending. The highly brecciated portion of the pipe is bounded above and below by two later dolerite sills and the permeability of the pipe itself exceeds that of the surrounding shales to such an extent that it may be viewed as an isolated groundwater compartment. While this may limit the sustainability of natural abstraction, it makes the breccia pipe ideal for low-cost, underground storage.

1.3 *Geochemical factors affecting AR*

Introducing recharge water with different chemical properties from that of the *in situ* groundwaters may shift chemical equilibria in the recharge and native waters and affect their chemical interaction with aquifer solids. Changes in the physical and chemical properties of the system, such as temperature, pressure, gas concentrations, pH and redox potential may also enhance or inhibit reactions, or cause one reaction to be favoured over another (Pyne 1995).

Two major issues that arise from chemical reactions are changes in the permeability of the system and changes in the quality of the recovered water. Clogging of borehole screens or the aquifer itself (from mechanical blockage by suspended solids, biological activity, gas entrainment or mineral precipitation) is a major problem affecting borehole injection. Injection of turbid water, mobilisation of clay particles, nucleation of precipitates and dissolution of cementing materials could cause clogging by suspended solids. Typical precipitation problems involve carbonates or Fe or Mn oxyhydroxides, but any sparingly soluble phase may precipitate under favourable conditions.

Changes in water quality can be beneficial, as in the case of soil-aquifer treatment of wastewater or dilution of unwanted groundwater constituents to acceptable levels using high quality recharge water. AR could, however, mobilise contaminants such as Rn, As, F, Fe, Mn, NH_4 and organic compounds from natural sources in the aquifer, degrading the recovered water. Residence time is an important factor in determining the extent of accumulation.

2 GEOCHEMISTRY OF WATER SOURCES AND THE UNDERGROUND RESERVOIR

Chemical analysis of samples from the various water sources and mineralogical and bulk chemical analysis of rock samples from a drilling core were conducted. This allowed the various chemical systems to be delineated, so that predictions could be made regarding the fate of chemical species during mixing and water-rock interaction.

2.1 *Composition of water sources*

Both the town dam and some of the higher yielding municipal boreholes have occasional water surpluses, which would be available for storage. Chemical characteristics of the recharge sources are compared with those of the native groundwater in

Table 1. Major ions were analysed by ion chromatography, molybdate-reactive silica by colorimetry, total iron by flame atomic absorption spectrophotometry and arsenic by inductively coupled plasma-mass spectroscopy.

Table 1. Chemical composition of water sources in the Calvinia borehole injection scheme (data in mg/L unless other units specified).

	Potential recharge sources				Breccia
Parameter	Dam	B1	B2	B3	R1
Field pH	7.1	8.1	7.8	7.7	9.8
EC (mS/m)	19	435	571	850	71
T.Alk (meq/L)	1.20	5.21	7.54	8.00	3.06
Na	13	590	645	1270	170
K	0.8	5.8	3.9	5.3	1.7
Ca	17	220	340	260	5.3
Mg	12	210	340	600	2.6
NH_4	<0.1	<0.1	<0.1	<0.1	1.3
Cl	16	1570	1970	2585	120
SO_4	20	150	480	1200	54
NO_3	0.4	<0.1	<0.1	<0.1	<0.1
F	0.1	1.1	0.4	0.8	12
Si	3.7	5.0	7.7	7.0	8.6
Total Fe	0.08	0.07	0.13	0.10	0.36
As (μg/L)	0.8	12	6.2	7.7	325

The dam stores runoff water from a nearby mountain catchment. This water is of very high quality after treatment, having low salinity and near-neutral pH. The dominant major ions in solution are Ca^{2+}, Mg^{2+} and HCO_3^-. The water is oxygenated and may cause oxidation of reduced species in the breccia pipe.

The municipal boreholes abstract groundwater from shallow alluvial and weathered shale aquifers. These three waters are of similar composition, being hard, brackish and more reducing than the surface water. They are generally blended with the dam water when supplies are critically low, but undiluted they are not desirable for human consumption. The dominant major ions are Na^+, Mg^{2+} and Cl^-.

The natural breccia pipe water (R1) is of the Na-Cl/HCO_3 type and has an unusually alkaline pH and low salinity. Species such as NH_4^+ and H_2S in solution indicate a reducing environment. High concentrations of F^- and As would pose a health risk for long-term consumers of the water and could become problematic if similar levels were to accumulate in water stored for emergency supply. One of the considerations of the borehole injection scheme is the possibility of using recharge waters to lower the pH and dilute F^- and As concentrations to acceptable levels within the breccia pipe.

2.2 *Aquifer chemistry and mineralogy*

The mudstones and shales that comprise the upper portion of the breccia pipe have been subject to hydrothermal activity which has caused some alteration of the original detrital minerals and precipitated new phases, especially along the cavities and fractures. Rock mineralogy was determined by optical microscopy and X-ray powder diffraction analysis. The bulk argillaceous material comprises quartz, feldspar and clay minerals of the illite and chlorite groups. The hydrothermal minerals include chlorite, calcite, metal sulphides (especially pyrrhotite) and a fluoride-rich species of apophyllite. Reaction with these minerals may have a strong influence on the groundwater composition in the breccia pipe, particularly since they are concentrated along zones of high permeability.

Bulk elemental composition was analysed by X-ray fluorescence spectrometry. Table 2 gives mean chemical concentrations based on the analysis of 14 core samples from the top 300 m of the breccia.

Table 2. Bulk composition of breccia pipe rocks.

Major oxides	mean wt %	s.d.*	Trace elements	mean mg/kg	s.d.*
SiO_2	60	3.4	F	360	320
Al_2O_3	18	1.7	As	11	6
Fe_2O_3	7.3	1.6	Ba	780	210
MgO	2.4	1.4	Mn	840	220
CaO	1.8	2.5	Cu	16	22
Na_2O	2.4	0.9	Pb	22	22
K_2O	3.7	1.1	Zn	53	25

*s.d. = standard deviation, n =14.

Base cations in the rock are dominated by K, Mg and Na, but Ca concentrations are generally low. High standard deviations for Ca, F and the trace metal concentrations reflect localisation of minerals containing these elements (such as calcite, apophyllite and metal sulphides) in the breccia pipe. Accumulation of F^- and As in the natural groundwater relates to the alkaline environment and redox conditions, rather than an abundance of these elements in the aquifer.

3 EVALUATION OF GEOCHEMISTRY

Quantitative geochemical data from water and rock analyses and from simple batch experiments conducted in the laboratory were used to evaluate the potential impact of geochemical factors on the AR scheme. Computerised geochemical modelling was used to anticipate the effects of mixing water types, potential mineral precipitation problems and the influence of the native groundwater on recovered water quality.

3.1 *Mixing effects*

Mixing of different water types may drive several chemical reactions, leading to a final water composition that is different from a simple conservative mixture of the components in the two end-members. Redox and pH buffering effects are likely to be observed as the species in solution are redistributed among redox states and protonated or deprotonated to achieve a new equilibrium.

The program PHREEQC (Parkhurst 1995) was used to investigate the effects of mixing on the aqueous geochemistry of the injection system. Thermodynamic constants from the WATEQ4F database for natural waters were used for the calculations (Ball & Nordstrom 1991). In the mixing simulations, each of the recharge water sources was added to the breccia pipe water at mixing ratio increments of 10% and the equilibrium solution composition calculated, to investigate the range of solution compositions that could be found across a mixing front. Initially, only the effects in the aqueous solution were investigated as no solid phases were specified in the model. Redox conditions were approximated using estimated pe values of +3 for the breccia pipe and +2 for the recharge boreholes, based on average Eh readings for similar boreholes logged in the field. The default pe of +4 in PHREEQC was used for the surface water source. Simulations used field-measured values for pH and temperature. Predicted pH mixing curves are shown in Figure 3.

Buffering by the carbonate system, in particular, but also by other proton-producing or consuming reactions, including redox reactions, will control the pH of the mixed waters. The prevalence of deprotonated species in the alkaline breccia pipe water will favour reactions that consume H^+ and shift the pH to higher values than those from a conservative mixing process. The effect is likely to be most pronounced for the dam water, which has the lowest alkalinity and lacks the buffering capacity imparted by the carbonate species in the other sources.

Mineral saturation indices (SI) for a range of sparingly soluble solids were calculated during the mixing simulations to evaluate the potential for mineral precipitation. Saturation indices indicate the thermodynamic tendency for a mineral to dissolve (SI<0) or precipitate (SI>0), but do not account for the reaction rates of solid-solution phase changes.

Mixed solutions (Fig. 4) show a greater degree of supersaturation with respect to calcite, ferrihydrite and fluorite than for the pure end-member solutions. The risk of calcite precipitation is greater if groundwater rather than surface water is used for AR. This was confirmed by a simple test tube demonstration where approximately equal proportions of R1 and B1, R1 and B2 or R1 and B3 produced a white precipitate from the original clear solutions. No precipitate formed from Dam water and R1. Model sensitivity analysis showed that the calculation of carbonate mineral SI's was very sensitive to pH above pH 7. SI's for calcite, dolomite, aragonite and magnesite are increased significantly by raising the simulation pH from 8 to 10. Predicting the clogging potential of carbonate minerals rests on the ability to anticipate pH changes in the AR scheme.

Saturation indices of ferric hydroxide minerals decrease slightly with increasing pH in the pH 7 to 10 range. The solubility of minerals such as goethite and ferrihydrite is extremely sensitive to redox conditions. Saturation indices are below zero when negative pe values are used, because of the increased tendency for dissolved iron to occur as the more soluble Fe^{2+} form under reducing conditions. With-

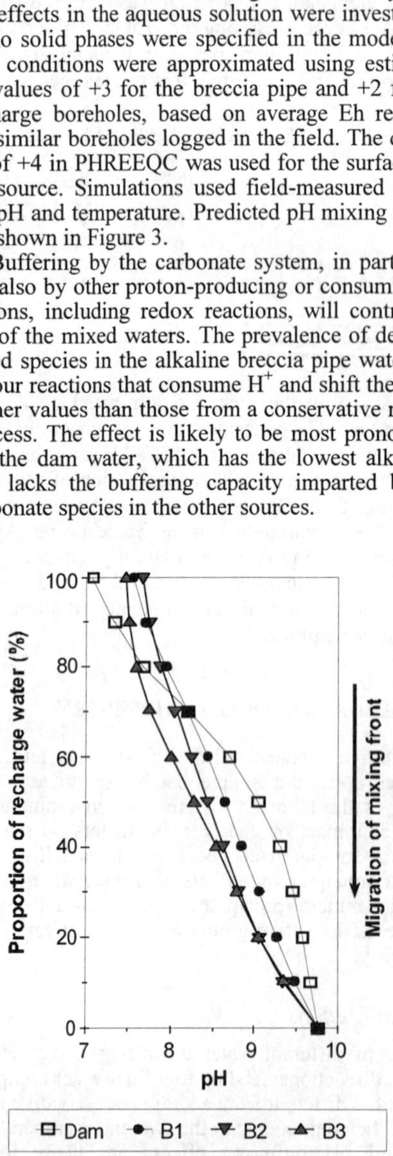

Figure 3. Simulated pH values for mixing of potential recharge sources with native breccia pipe water.

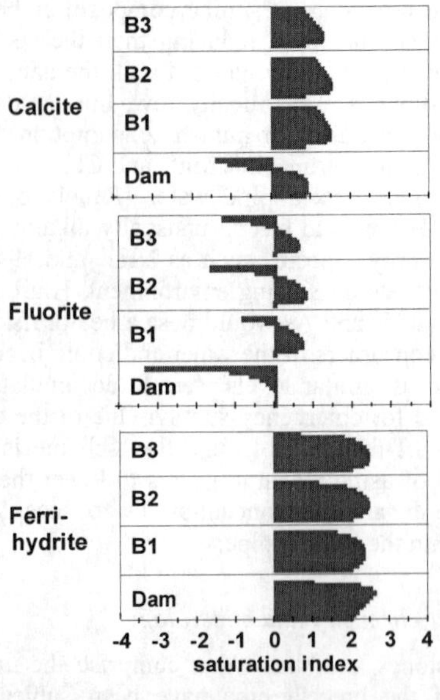

Figure 4. Saturation indices for calcite, fluorite and ferrihydrite calculated for water mixtures. Each simulation starts at 10% recharge water mixed with breccia water at the bottom of the graph and increases by 10% increments to 100% recharge water (all breccia water replaced) at the top.

out accurate knowledge of solution pe and *in situ* determination of both Fe^{2+} and Fe^{3+} concentrations, rather than total dissolved Fe, it is difficult to have any confidence that Fe oxides will precipitate, as is suggested by the saturation indices.

The potential severity of clogging by precipitation was assessed by rerunning the mixing simulations, this time allowing solid calcite, ferrihydrite and fluorite to precipitate. PHREEQC was used to calculate the mass of solid that would need to be precipitated per litre of water for the SI to reach zero (Fig. 5).

Even though ferrihydrite SI's are higher than those of the carbonate minerals, low total dissolved Fe concentrations mean that the mass of Fe(III) oxide that could precipitate is negligible. Carbonate mineral clogging would appear to pose a far greater risk to the injection scheme in terms of the volume of material involved, especially when using the borehole recharge sources. Depending on the relative rates of precipitation, calcite, dolomite or aragonite precipitation could also remove calcium from solution, thereby inhibiting the natural removal of dissolved fluoride and improvement of groundwater quality by fluorite precipitation.

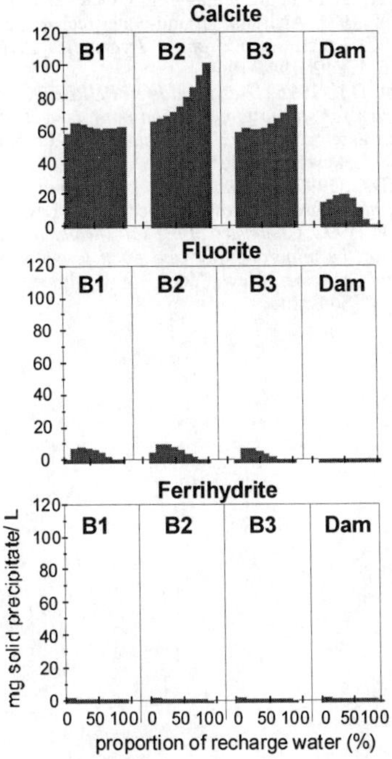

Figure 5. Calculated total mass of mineral solids that could precipitate from mixtures of breccia water with various recharge waters.

3.2 *Water-rock interactions*

The previous geochemical modelling calculations have relied on the assumption that the subsurface reservoir is chemically inert and does not react with the mixing aqueous solutions. This is unlikely in the case of the breccia pipe, which contains reactive clay mineral surfaces and secondary mineral phases.

Rock samples from the breccia pipe were crushed and reacted in laboratory batch-type experiments to investigate the potential impact of water-rock interactions on the quality of water recovered from the breccia pipe after storage. Bulk samples of three different particle size fractions (<63 μm, 63 μm - 2 mm and 2-10 mm) were leached with deionised water. A 333 g/L slurry of each sample was placed horizontally on a reciprocating shaker in a polyethylene container and shaken at 1 cycle per second for 500 hours to determine the products of mineral dissolution under atmospheric conditions (Table 3). An additional sample containing a concentrate of secondary hydrothermal minerals (2-10 mm fraction) was treated in the same manner (sample M1).

Table 3. Results of rock leaching experiments.

Parameter mg/L	Size fraction <63 μm	63μm - 2 mm	2 - 10 mm	M1
pH	8.5	8.6	8.3	7.4
Na	53	50	58	44
K	50	27	29	40
NH_4	7.5	2.4	3.6	5.7
Mg	7.0	7.4	3.2	18
Ca	16	19	8.9	46
F	0.8	1.0	1.2	5.4
Cl	9.4	5.1	6.4	4.5
NO_3	0.7	0.6	0.7	<0.1
SO_4	35	50	90	270
T.Alk (meq/L)	4.10	2.45	2.45	0.82

The rock samples contain significant amounts of soluble Na^+, K^+ and alkalinity and were also found to release NH_4^+, SO_4^{2-} and F^- into solution, all of which could be added to low salinity recharge waters. K^+, NH_4^+, F^- and SO_4^{2-} were readily leached from hydrothermal clay minerals, apophyllite and metal sulphides upon oxidation. Sulphide mineral oxidation and addition of atmospheric CO_2 caused the pH to decrease with time. Alkalinity is generated by most weathering reactions (Stumm 1992) and increased with increasing solid surface area in these experiments.

A 350-hour batch equilibration test was also undertaken, reacting the surface and groundwater recharge sources from Calvinia with a fine fraction (<63 μm) of bulk rock material from the breccia pipe, following the same procedure as the water-rock leaching tests. Reaction of the low salinity dam water produced similar reactions to those found when leaching the rock with deionised water. A small, but significant quantity of As was also released to the dam water. The pH rose immediately to 9.3 on contact with the rock and then gradually fell to 8.0 over the course of the experiment.

High solute concentrations in the groundwater sources dominated over any solutes added by the rock and buffered the pH at lower values than for the dam sample. Na^+, Ca^{2+} and Mg^{2+} concentrations decreased slightly in the brackish groundwater, by cation exchange (Na^+) and carbonate precipitation (Ca^{2+} and Mg^{2+}), but in all cases the removal efficiency was negligible and could not be considered a significant improvement in water quality. The proportion of NH_4^+ and K^+ in the groundwaters increased substantially after reaction with the finely divided rock. The artificial powdering of the rock material and limitations of working under laboratory conditions cannot be ignored when applying information from the batch tests to the borehole injection scheme.

4 IMPLICATIONS FOR AR

The municipal groundwater sources are not recommended for borehole injection, because of their already inferior quality and the high risk of carbonate precipitation in the aquifer and scaling of pipelines and other infrastructure. In comparison, the risk of clogging by iron precipitation is low.

The treated dam water is a suitable recharge source if high mixing proportions can be maintained by dewatering most of the natural groundwater from the pipe before recharge. Care will need to be taken to minimise air entrainment and cascading in the injection boreholes. pH is a controlling variable for many of the potential reactions during recharge and should be measured regularly.

Dissolved oxygen in the surface water may oxidise metal sulphides in the breccia pipe causing increased dissolved sulphate, but the rocks contain sufficient alkalinity to neutralise any acid generated by this process and the high pH should immobilise heavy metals which might be released. Monitoring of F^-, SO_4^{2-}, K^+, NH_4^+ and As concentrations are recommended to protect the stored water quality.

The usefulness of a modelling exercise is limited by uncertainties in input data and the modeller's conceptual understanding of the hydrochemical system. Laboratory experiments are also restricted by the inability to artificially reproduce all possible controlling variables of the natural environment. Yet geochemical modelling and equilibration tests can be a powerful tool for investigating hypothesised chemical reactions. Forward modelling can be used to predict water composition, mineral stabilities and mass transfers in response to changing physical and chemical conditions during AR. Supplemented by simple batch experiments examining potential water-rock interactions, these inexpensive methods allow preliminary assessments to be made of the feasibility of recharge activities. Verification of these predictions will now be attempted during pilot-scale field tests during the next phase of the project.

ACKNOWLEDGEMENTS

This work was funded by the CSIR, South Africa and was undertaken under the supervision of Professors Martin Fey and James Willis at the University of Cape Town. Ricky Murray and Dr. Gideon Tredoux of the CSIR initiated the borehole injection scheme at Calvinia in collaboration with Calvinia Municipality. Dr. Luc Chevellier of the Council for Geosciences assisted with geological descriptions of the breccia pipe and Stewart Whyte with the editing of this paper.

REFERENCES

Ball, J.W. & D.K. Nordstrom 1991. *User's manual for WATEQ4F, with revised thermodynamic data base and test cases for calculating speciation of major, trace and redox elements in natural waters.* USGS Open-File Rep. 91-183.

Brouwer, H. 1996. Issues in artificial recharge. *Wat. Sci. Tech.* 33(10/11): 381-390.

Brouwer, H., R.D.G. Pyne & J.A.Goodrich 1990. Recharging ground water. *Civil Engineering* June 1990: 63-66.

Murray, E.C., G. Tredoux & K. Murphy 1998. Developments in artificial recharge in Southern Africa. In Brahana et al. (eds.). *Gambling with groundwater - physical, chemical and biological aspects of aquifer-stream relations.* Proc. XXVIII IAH Congress, Las Vegas, October 1998: 345-351.

Harpaz, Y. 1971. Artificial ground-water recharge by means of wells in Israel. *Journal of the Hydraulics Division.* Proc. ASCE, HY12, 1947-1964.

Parkhurst, D.L. 1995. *User's guide to PHREEQC- a computer program for speciation, reaction path, advective-transport, and inverse geochemical calculations.* WRI report 95-4227, USGS, Lakewood, Colorado.

Pyne, R.D.G. 1995. *Groundwater recharge and wells. A guide to Aquifer Storage Recovery.* Boca Raton: Lewis.

Stumm, W. 1992. *Chemistry of the solid-water interface. Processes at the mineral-water and particle-water interface in natural systems.* New York: Wiley-Interscience, John Wiley & Sons, Inc.

Groundwater: Past Achievements and Future Challenges, Sililo et al. (eds) © *2000 Balkema, Rotterdam, ISBN 90 5809 159 7*

Geochemical processes affecting the composition of mineral waters in the South Portuguese Zone, Portugal

A.Chambel
Departamento de Geociências, Universidade de Évora, Portugal

C.Almeida
Departamento de Geologia, Facultade Ciências, Universidade de Lisboa, Portugal

ABSTRACT: The Iberian Pyrite Belt is an important sector of the southernmost geostructural domain of the Iberian Peninsula Palaeozoic Massif, the South Portuguese Zone (SPZ). It is composed of metamorphic rocks of volcano-sedimentary origin, overlain by metasediments of the Mértola Formation, both formed in a submarine environment. In this sector, several highly mineralised sulphide waters occur, having all a Na-Cl matrix. In order to evaluate the geochemical processes affecting the composition of these waters, forward and inverse geochemical modelling were used. The results show that the final composition could be the result of water/rock interaction of meteoric water with the following phases: calcite, dolomite, pyrite, goethite, NaCl, CO_2 gas and cationic exchange. Minerals like calcite, dolomite, pyrite and NaCl are dissolving in an open system, as indicated by the large amounts of O_2 consumed during the evolution, accompanied by the precipitation of goethite.

1 INTRODUCTION

The South Portuguese Zone (SPZ) is the southernmost geostructural domain of the Iberian Massif. The presence of a Pyrite Belt (Iberian Pyrite Belt), of extraordinary economic importance for Portugal and Spain, and associated lithologies is related to several occurrences of highly mineralised sulphide waters (Calado & Chambel 1999).

The aim of this paper is to evaluate the geochemical processes controlling the composition of these waters, using forward and inverse geochemical models and considering the final composition as the result of the interaction between meteoric water and rocks.

2 GEOLOGICAL SETTING

The Iberian Massif is an important geological unit of the Iberian Peninsula and is sub-divided into several geostructural domains. The southernmost of these domains is the SPZ, characterised by the presence of three main units that are the result of the collision of the "middle European" and "North European/South Iberian" continents (Bard et al. 1980) accompanied by the closing of a Palaeozoic ocean just north of the SPZ (Fig. 1):

- the Pulo do Lobo Sub-Zone, in the northern part
- the Pyrite Belt
- the Baixo Alentejo Flysch Group, in the southern part

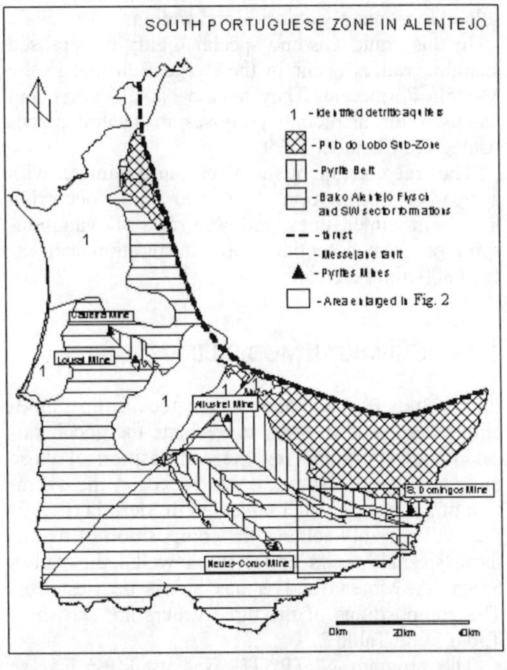

Figure 1. Tectono-stratigraphic unit of the South Portuguese Zone, in South Portugal, and its sub-divisions. The rectangle represents the area enlarged in Figure 2.

The Pulo do Lobo Sub-Zone is the oldest of all the units of the SPZ, consisting of phylites, quartzites and some layers of acid and basic volcanics. The Pyrite Belt consists of original sedimentary and volcanic rocks containing massive sulphides. It is composed of alternating decimetric to metric layers of acid and basic volcanic rocks, shales, graywackes, quartzowackes, siltstones, pellites, quartzites, sandstones, rare conglomerates and limestones and other types of rocks. South of the Pyrite Belt there are the most recent sediments of the SPZ, a turbiditic sequence corresponding to the Baixo Alentejo Flysch Group. The northernmost sub-division of this last Group is the Mértola Formation, followed by the Mira Formation and Brejeira Formation (Oliveira 1988, Silva 1989).

The collision was oblique and created an important left shear zone that forced, by transtension, the opening of grabens and the bimodal volcanism of the Pyrite Belt (Fig. 1).

3 HYDROGEOLOGY

The hard rock aquifers of this area, represented by metamorphic rocks, are scarce in water resources. The median of the well yields is less than 1 L/s and the quality is poor, with high concentrations of dissolved ions, mainly chloride and sodium.

In this context, some special highly mineralised sulphide waters occur in the Pyrite Belt and in the Mértola Formation. They have a Na-Cl matrix and are used for medicinal purposes in several places (Calado & Chambel 1999).

The area corresponds to a semi-arid climate, with a yearly precipitation near 500 mm, occurring mainly in winter times, and high rates of evapotranspiration. The potential evapotranspiration can exceed 800 mm per year.

4 GEOCHEMICAL MODELLING

In this study, forward and inverse geochemical modelling were used in order to evaluate the geochemical processes influencing the composition of those mineral waters. The study was based on the evolution of a meteoric water collected in Mértola (Fig. 2) that will possibly achieve the composition of two of these special waters collected in wells: the Águas Santas da Morena and Águas Santas de Besteiros. The compositions of the three waters are shown in Table 1 and Table 2.

The program NETPATH was used for inverse modelling (Plummer et al. 1991), and for both final waters the following constraining phases were assumed:

CO_2 gas + Pyrite + Calcite + Dolomite + NaCl + Goethite

The possibility of a concentration process by evapotranspiration was also considered since this may be likely in the upper part of the soil where high temperatures are reached in this semi-arid climate.

Among the possible models resulting from the modelling exercise, the results show that evapotranspiration is not needed to explain the composition after meteoric water interaction with the phases considered. In the two final waters the results were identical (Table 3), both involving the oxidation of pyrite, dissolution of calcite, dolomite and NaCl and precipitation of goethite.

The water of Água Santa da Morena seems to be more evolved than that of Águas Santas de Besteiros. The great amounts of NaCl needed for the dissolution can be justified by its presence in the meta-sedimentary rocks, resulting from the accumulation

Table 1. Analysis of rainwater collected in Mertola.

	Unit	Rain water
Electric conductivity	µS/cm	18
pH		7
Total hardness	mg/l	14
Cl^-	mg/l	1.59
NO_3^-	mg/l	0.62
SO_4^{2-}	mg/l	1.19
Ca^{2+}	mg/l	1.33
Mg^{2+}	mg/l	0.21
K^+	mg/l	0.69
Na^+	mg/l	1.31
SiO_2	mg/l	0.3
Fe	µg/l	20.35
Al	µg/l	26.5

Table 2. Analysis of the groundwater of Águas Santas da Morena and Águas Santas de Besteiros.

	Unit	A S Morena	A S Best.
Temp.	ºC	21.7	21.3
PH		7.45	7.19
Eh	MV	-172	-185
EC	µS/cm	2160	1595
Tot. alcal.	ml/l	72.3	66.6
SiO_2	mg/l	16.4	18.9
F^-	mg/l	1.6	1.7
Cl^-	mg/l	464	320
HS^-	mg/l	0.2	0.4
SO_4^{2-}	mg/l	139	66
HCO_3^-	mg/l	441	406
NO_3^-	mg/l	<0.05	0.24
Li^+	mg/l	0.29	0.21
Na^+	mg/l	390	261
K^+	mg/l	4.1	2.2
Mg^{2+}	mg/l	35.8	30.5
Ca^{2+}	mg/l	64.8	60.2
NH_4^+	mg/l	0.26	<0.06
Br^-	µg/l	1750	1292
I^-	µg/l	34	17
B	µg/l	310	164
As	µg/l	24	28
Mn	µg/l	83	38
Fe	µg/l	70	13

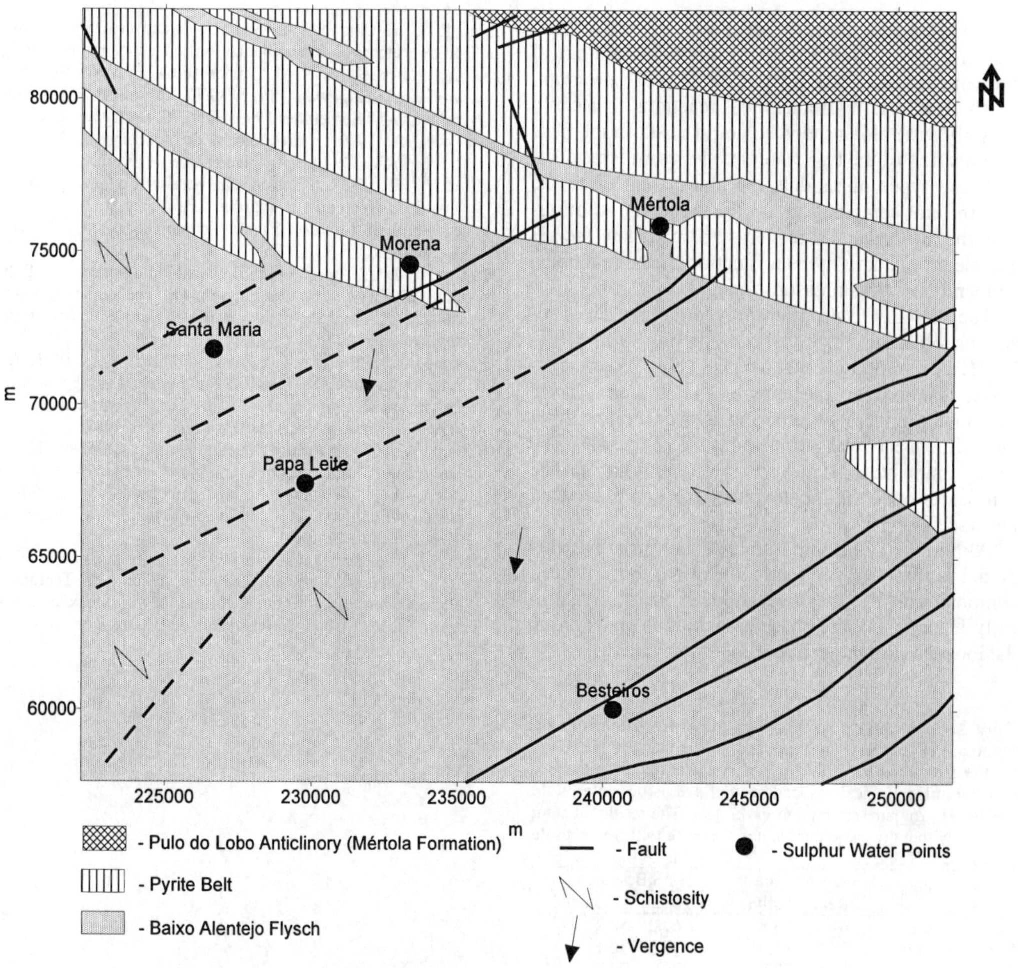

Figure 2. Map of the studied area, with the geological units and the water points identified.

of brines in Palaeozoic times during orogenesis.

The great amount of oxygen consumed in pyrite oxidation (each mole of pyrite needs 3.5 moles of oxygen) shows that this reaction occurs in an open system.

The products of pyrite FeS_2 oxidation, are iron and sulphate. The NETPATH interpretation shows that all the iron resulting from pyrite dissolution is used to form goethite:

$Fe^{3+} + 2H_2O = FeOOH$ (goethite) $+ 3H^+$

To test the thermodynamic consistency of the reactions obtained by the NETPATH program, the geochemical modeling program PHREEQE (Parkhurst et al. 1980) was used. This is a model that can execute a great number of geochemical calculations and allows evaluation of how the water composition changes in response to reactions (mineral dissolution, gas phase equilibria, etc.) or to temperature changes, for example.

Table 3. Mass transfer values (α) for each mineral, calculated for the evolution of the rain water to reach the composition of two waters collected in Águas Santas da Morena and Águas Santas de Besteiros. The results were obtained with the program NETPATH (Plummer et al. 1991). Positive values represent dissolution and negative values precipitation (in mmol/kg water.

	A S Morena	A S Besteiros
CO_2	2.716	3.619
Pyrite	0.718	0.338
Calcite	2.055	1.382
Dolomite	1.466	1.247
Ionic exchange	1.935	1.159
NaCl	13.063	8.992
Goethite	-0.718	-0.338

In the present case, the goal was to verify if the meteoric water interacting with the mineralogical component, indicated by the NETPATH program, and with all the hydrochemical reactions that could possibly arise along the way, could result in the final composition of the two waters under consideration.

These types of models are able to calculate the equilibrium composition of the aqueous solutions with the minerals, add reagents and calculate the pH and electronic activity as a function of the reactions, temperature or gas pressure changes.

The PHREEQE program proved that the set of reactions obtained by inverse modelling with NETPATH is thermodynamically plausible. So, the proposed reactions are acceptable and it seems likely that all the mineral weathering is occurring without the influence of evapotranspiration (Table 4). The final composition achieved is very similar to the composition of the final waters for some selected ions (see Table 3).

Although it is not possible to confirm that this model is the only one that can explain the observed compositions, the set of reactions is thermodynamically feasible and the phase associations plausible in the local environment where the waters occur.

Table 4. Estimated values, based on the hydrochemical model program PHREEQE (Parkhurst et al. 1980), using as initial water the rain water of Table 1 and considering its interaction with the mineralogical associations defined as possible by the NEPATH program, referred to in the text. The results of some ions show that the estimated values are close to the real results presented in Table 2.

	Units	A S Morena		A S Besteiros	
		Real	Estimated	Real	Estimated
Ca^{2+}	mg/l	64.8	56.8	60.2	58.2
Mg^{2+}	mg/l	35.8	31.3	30.5	32.1
Na^{+}	mg/l	390	377	261	244
Cl^{-}	mg/l	464	462	320	321
SO_4^{-}	mg/l	139	115	66	54.5
HCO_3^{-}	mg/l	441	386	406	365

5 CONCLUSIONS

Inverse and forward geochemical modelling shows that the special type of composition represented by the Morena and Besteiros mineral waters can be explained by the dissolution of calcite and dolomite which provides a source of carbon, calcium and magnesium; pyrite oxidation, which provides a source for sulphur and enhances carbonate dissolution by proton release, ionic exchange and halite dissolution. I

REFERENCES

Bard, J. Burg, J. Matte, P. & Ribeiro, A. 1980. La Chaîne Hercynienne d'Europe Occidentale en termes de tectonique des plaques: *Mem. Bur. Rech. Geol. Min. Fr.* 6(26): 233-246.

Calado, C. & Chambel, A. 1999. Un unexpected mineral sulphide water type in the Iberian Pyrite Belt (South Portugal), In M. Fendeková & M. Fendek (eds), *Hydrogeology and land use management* - Proc. XXIX Congress of IAH, Bratislava, 6-10 September 1999: 671-676. Bratislava: IAH.

Chambel, A. 1999. Hidrogeologia do concelho de Mértola. *Doctoral Thesis*. Évora: Universidade de Évora.

Oliveira, J.T. 1988. Contribuição para o Conhecimento da Evolução Tectono-Estratigráfica da Zona Sul Portuguesa em Portugal. *Doctoral Thesis*. Lisboa: Serviços. Geológicos de Portugal.

Parkhurst, D.L. Thorstenson, D.C. & Plummer, L.N. 1980. PHREEQE - A computer program for geochemical calculations: *U.S. Geological Survey Water-Resources Investigations Report*: 80-96

Plummer, L.N. Prestemon, E.C. & Parkhurst, D.L. 1991. An interactive code (NETPATH) for modeling net geochemical reactions along a flow path: *U.S. Geological Survey Water-Resources Investigations Report*: 91-4087

Ribeiro, A. 1979. Le cadre structural et géotectonique; Essay de reconstitution paléogéographique par cycles orogéniques. Le cycle hercynien. In A. Ribeiro et al. (eds) *Introduction à la géologie générale du Portugal*: 8-45. Serviços Geológicos de Portugal.

Silva, J.B. 1989 Estrutura de uma Geotransversal da Faixa Piritosa: Zona do Vale do Guadiana. Estudo da Tectónica Pelicular em Regime de Deformação Não Coaxial. *Doctoral Thesis*. Lisboa: Universidade de Lisboa.

Groundwater: Past Achievements and Future Challenges, Sililo et al. (eds) © 2000 Balkema, Rotterdam, ISBN 90 5809 159 7

Dissolved radon concentrations in fractured rock aquifers

P.G.Cook & J.C.Dighton
CSIRO Land and Water, Glen Osmond, S.A., Australia

ABSTRACT: Dissolved radon concentrations are used to estimate groundwater flow rates and fracture aperture sizes. Groundwater radon concentrations have been measured both before and after purging nested piezometers, and radon emanation from aquifer matrix materials has been measured on samples obtained during drilling. The paper shows how the ratio of radon concentration before purging to that after purging can be used to quantify the groundwater flow rate. The ratio of radon concentration in groundwater after purging, to radon emanation rate is related to the fracture aperture size. These results show that both groundwater flow rate and fracture aperture decrease with depth in fractured shales in the Clare Valley, South Australia.

1 INTRODUCTION

Radon, with a half-life of 3.8 days, is produced from radioactive decay of uranium-series isotopes. Radon concentrations in groundwater should be controlled by the rate at which radon is released from the aquifer solid phase, by the aquifer porosity, and by the location of radon-bearing minerals relative to the pore spaces (Lawrence et al. 1991, Torgersen et al. 1992). For these reasons, several authors have suggested that radon concentrations in fractured rock aquifers could be used to determine fracture apertures and aquifer transmissivities (Lawrence et al. 1991, Torgersen et al. 1992, Folger et al. 1996). They argue that large fracture apertures are associated with low rock surface area to water volume ratios, and so the emanated radon becomes diluted. Thus, other factors being equal, large fracture apertures should correlate with high aquifer transmissivities, and lower radon concentrations. Only relatively few studies have sought to directly measure radon emanation from aquifer materials (e.g., Andrews et al. 1986).

This paper presents measurements of radon concentrations from fractured shale aquifers in the Clare Valley, South Australia, together with measurements of radon emanation rates from aquifer materials, and uses this information to estimate groundwater flow rates and fracture apertures.

2 THEORY

2.1 *Groundwater flow rate*

Cook et al. (1999) and Hamada (1999) showed that the ratio of radon concentrations measured before purging to those measured after purging could be used to estimate the horizontal flow rate. If the flow rate through the well is very low, then the water in the well will be stagnant, and so most of its radon will have decayed to background. Significant radon concentrations should be present in well water only if there is sufficient flow into the well to replenish the radon faster than it can decay. One might therefore expect radon concentrations in unpumped wells to increase as the groundwater flow velocity increases. At some stage, a point will be reached where the flow through the well is sufficiently high, that radon decay within the bore becomes negligible. Beyond this stage, there would be no further increase in radon concentration with increasing flow velocity, and the radon concentration within the well would be equal to the concentration within the aquifer. If the water within the screened interval of the piezometer is perfectly mixed, then the radon concentration within the well can be written:

$$\frac{\partial}{\partial t}(c\,V) = Qc_0 - Qc - \lambda\, c\, V \qquad (1)$$

where c and c_0 are the radon concentrations in the well and in the aquifer respectively, Q is the volumetric flow rate through the well [L^3 T^{-1}], V is the well volume [L^3] and λ is the decay constant for radon [0.18 day^{-1}]. The inflow of radon is Qc_0 and the outflow is Qc. Under steady state conditions ($\partial c/\partial t=\partial V/\partial t=0$), substituting $V = \pi r^2 h$, where r is the well radius and h is the height of the screened or open interval, this can be written as:

$$q = \frac{c}{c_0 - c}\frac{\lambda \pi r}{2} \qquad (2)$$

where q is the flow rate into the well in units of m yr^{-1} ($Q=2rhq$). Solutions to Equation 2 for well radii of 0.05, 0.1 and 0.15 m are depicted in Figure 1. They suggest that variations in radon concentrations in unpurged wells should be sensitive to groundwater flow rate, provided that the latter is less than about 40 m yr^{-1}.

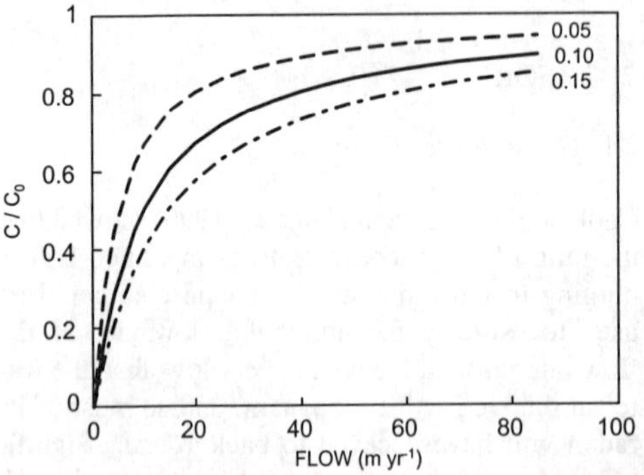

Figure 1. Theoretical relationship between radon concentration in an unpurged borehole, c, relative to that in the aquifer, c_0, and groundwater flow rate for bore radii of 0.05, 0.1 and 0.15 m (Equation 2).

2.2 *Fracture apertures*

In porous media of uniform mineralogy (where radium is evenly distributed), the radon emanation rate, E (Bq kg^{-1}), and the radon concentration within the aquifer matrix, c_m (Bq L^{-1}), will be related by:

$$E = \frac{c_m \varepsilon}{\rho_s (1-\varepsilon)} \qquad (3)$$

where ρ_s is the density of the solid phase (kg cm^{-3}) and ε is the porosity (cm^3 cm^{-3}). However, in a fractured rock system, radon concentrations within the fracture will be less than those within the matrix, because the emanated radon is diluted. Suppose that the rock matrix is homogeneous with respect to both porosity and mineralogy, and that it is transected by a series of planar, parallel fractures of aperture, $2b$ (Fig. 2). Suppose also that the water within the fracture is perfectly mixed, such that there is no variation in radon concentration across the width of the fracture. The radon concentration within the rock matrix can be expressed:

$$D\frac{\partial^2 c}{\partial x^2} + \gamma - \lambda c = 0 \qquad 0 \le x \le L \qquad (4)$$

with boundary conditions:

$$\frac{\partial c}{\partial x} = 0 \qquad x = L$$

$$c = \frac{\partial c}{\partial x}\frac{D}{b\lambda} \qquad x = 0 \qquad (5)$$

where D is the effective diffusion coefficient within the aquifer matrix, γ is the radon production rate, $2b$ is the fracture aperture, and $2L$ is the fracture spacing.

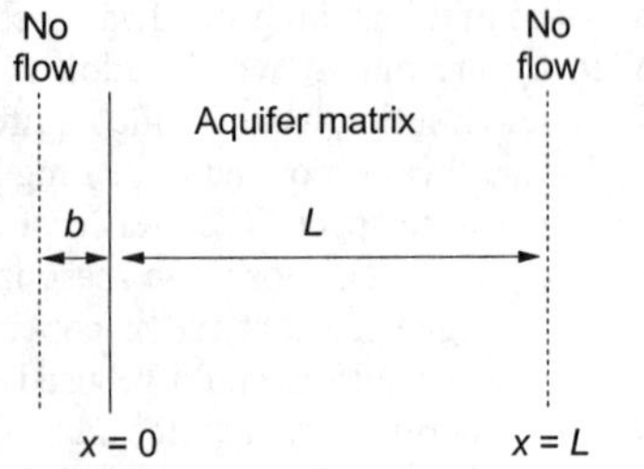

Figure 2. Schematic representation of an aquifer matrix, transected by planar, parallel fractures.

The solution to Equations 4-5 is:

$$c = \frac{\gamma}{\lambda}\left[1 - A\exp\left(-x\sqrt{\lambda / D}\right) + B\exp\left(x\sqrt{\lambda / D}\right)\right] \qquad (6)$$

where:

$$A^{-1} = 1 + \frac{D\sqrt{\lambda / D}}{b\lambda} + \left(1 - \frac{D\sqrt{\lambda / D}}{b\lambda}\right)\exp\left(-2L\sqrt{\lambda / D}\right) \qquad (7)$$

$$B = -A\exp\left(-2L\sqrt{\lambda / D}\right) \qquad (8)$$

For $x = 0$ (concentrations within the fracture), and large L, this reduces to:

$$\frac{c\lambda}{\gamma} = \frac{1}{1 + b\sqrt{\lambda / D}} \qquad (9)$$

Figure 3 shows the concentration within the fracture ($x = 0$), as a function of the fracture half-aperture, b, for different values of the matrix diffusion coefficient, D. Provided that L is greater than a few millimetres, then this has no effect on the concentrations. The plot shows that concentrations within the fracture decrease with increasing fracture aperture, and also with decreasing diffusion coefficient.

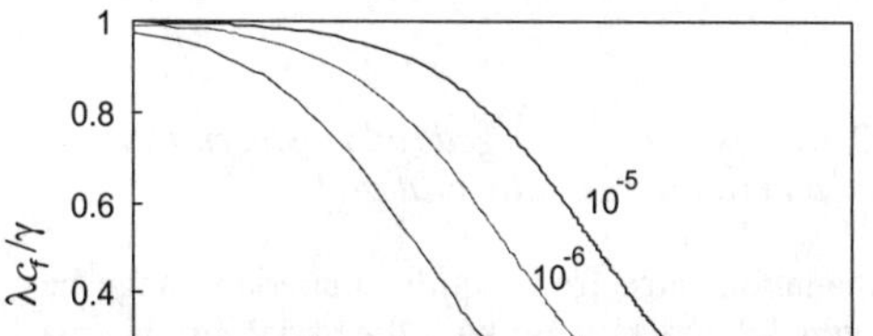

Figure 3. Theoretical relationship between radon production rate, γ, decay coefficient, λ, concentration of radon in fractures, c_f, fracture half-aperture, b, and matrix diffusion coefficient, D. Curves indicate diffusion coefficients of 10^{-5}, 10^{-6} and 10^{-7} m^2 yr^{-1}.

3 METHODS

The Clare Valley, located approximately 100 km north of Adelaide, South Australia, consists of low-grade metamorphosed, folded and faulted sedimentary rocks of Proterozoic age (Morton et al. 1998). The major lithologies are sandstone, shale, quartzite and dolomite. In this paper, data are presented from a nest of piezometers within the Mintaro Shale. A 200 mm diameter bore was drilled to 100 m depth, and a 250 mm bore to 26.4 m depth. (The two bores are spaced only 5 m apart.) A nest of six 50 mm PVC piezometers was installed in the 250 mm bore, and three 50 mm piezometers in the 200 mm bore. Screen lengths range from 0.5 to 5 m, being longer in the deeper piezometers. Gravel packs were used around the screens, which were separated using cement plugs. The nest of nine piezometers allows groundwater samples to be collected from depths between 8.5 and 95 m below the land surface.

Mean annual rainfall in the study area is approximately 650 mm yr^{-1}, with approximately 450 mm falling in the six months between May and October. The position of the water table varies seasonally, from approximately 4 m depth in October - November to 7 m in July - August.

Fracture properties have not been directly measured within the Mintaro Shale, but have been determined in an outcrop of the Saddleworth Dolomite, which is of similar age to the Mintaro Shale, and part of the same structural feature (Hill River Syncline). The two units also have similar hydraulic conductivity distributions. The outcrop exposure is approximately 70 m long and 6 m deep, and located approximately 10 km north of the bore installations. Outcrop data indicates 5 sets of fractures are present, four steeply dipping sets and one gently dipping set. For this site, on the west limb of the Hill River Syncline, the aquifers are on end, with the majority of fractures and bedding planes oriented vertically. The bedding comprises one vertically dipping fracture set that strikes 151°. A second set of vertical fractures strikes 59°. A set of conjugate shear fractures has the same trend as the second set of vertical fractures with dips of approximately 60°. The only shallow dipping fracture set has a strike of 345° and a dip of 27°. Fracture spacing is approximately log-normally distributed, with a mean fracture spacing of approximately 0.16 m. Fracture apertures are also approximately log-normally distributed, with a median aperture of approximately 100 μm, and the 90th percentile of 2 mm (Halihan, unpubl.).

During drilling, core samples were taken at selected intervals for laboratory determination of porosity, matrix hydraulic conductivity and diffusion coefficient. On a core sample from 8.7 m depth, the primary porosity was measured to be approximately 2.2% by helium porosimetry. Subsequent measurements on rock chips using mercury porosimetry gave porosities between 1 and 5%. The hydraulic conductivity of the core sample from 8.7 m was 6×10^{-13} m s^{-1}. A single measurement of the diffusion coefficient was made on a core sample from 26 m depth, and determined to be approximately 2×10^{-5} m^2 yr^{-1}.

Measurements of radon emanated from aquifer materials were made on rock chips collected during drilling of the bores. Approximately 40 g of oven-dried chips were sealed in 60 ml brass containers, with 20 ml of mineral oil scintillant. The balance of the volume (~ 20 ml) was filled with distilled water. After six weeks, the mineral oil was sampled and its radon concentration was measured in the laboratory by liquid scintillation counting (Herczeg et al. 1994). By using a series of radium standards, the efficiency of this process

Table 1. Field and laboratory measurements of radon concentration in groundwater and radon emanated from aquifer materials, and parameters calculated from these measurements.

	Radon Concentration in Groundwater		Emanation Rate	Calculated Parameters	
Screen Depth (m)	c (Bq L^{-1})	c0 (Bq L^{-1})	E (Bq kg^{-1})	c/c_0	$\lambda c_0/\gamma$
5.5-8.5	2.9	10	8.4	0.29	.023
9.5-11.35	6.3	10	5.4	0.63	.036
12.3-13.8	2.8	15.7	8.9	0.18	.034
14.5-16.2	4.0	33	8.3	0.12	.078
17.5-20.6	2.5	40		0.06	
24.0-26.0	2.1	45	4.2	0.05	.209
33.0-38.5	1.5	88	11.6	0.02	.148
73.0-78.0	0.3	88	4.5	0.00	.381
94.2-99.2	0.5	88		0.01	

(percentage of emanated radon that is captured in the scintillant) was determined to be approximately 61%. Radon emanation rates from the rock samples are expressed as Bq per kg of dry rock.

Initially, measurements of radon concentrations in groundwater were obtained from within the screened intervals of piezometers before the wells were purged, using a specially-designed bailer (Cook et al. 1999). After the bailer is filled and withdrawn from the well, a sample of water is transferred to a low diffusion vial containing mineral oil scintillant. Subsequently, three well volumes were purged from each piezometer, after which the wells were resampled. Radon concentrations are expressed as Bq per litre of groundwater.

4 RESULTS

4.1 *Comparison of purged and unpurged concentrations*

Radon concentrations measured before and after purging are shown in Table 1 and Figure 4. Before purging, the highest radon concentration (6.3 Bq L^{-1}) occurred in the piezometer screen centred at 11 m depth. Background concentrations ($\leq$ 0.5 Bq L^{-1}) were measured below 40 m. Radon concentrations for purged wells increase from 10 Bq L^{-1} above 12 m depth, to 88 Bq L^{-1} below 30 m, and are always greater than those measured before purging. The ratio of radon concentration measured before purging to that measured after purging ranges between zero and 0.63. Horizontal flow velocities have been calculated from these ratios using Equation 2. (Corrections have been made to the equation to account for the volume occupied by the gravel pack in the piezometer nests, and for the volume occupied by PVC pipe. It is assumed that the gravel pack does not produce any radon.) The ratio c/c_0 is greater than 0.1 above 17 m depth, suggesting flow velocities greater than 0.5 m yr^{-1}. The maximum value of c/c_0 is 0.63 at 11 m depth, and this corresponds to a flow rate of 9.7 m yr^{-1}. Below 30 m, c/c_0 is less than 0.03, and flow rates approach the detection limit of the method.

4.2 *Comparison of purged radon concentrations and radon emanation rates*

Emanation rates from aquifer materials range between 4.2 and 11.6 Bq kg^{-1}, the variability in emanation rate being much less than that of the purged radon concentration, c_0. Assuming a value for matrix porosity of $\varepsilon = 0.05$, and a solid phase density of ρ_s = 2.7 g cm^{-3}, then radon concentrations within the aquifer matrix would be expected to range between 212 and 595 Bq L^{-1} (Eq. 3). However, values measured in purged piezometers are much less than this, suggesting dilution of emanating radon within fractures, or that radon emanation adjacent to the fractures is much lower than that measured in the emanation tests. If we assume that it is due to the former, then we can calculate fracture apertures from Equation 9.

The radon production rate, γ (Bq L^{-1} yr^{-1}), is related to the measured emanation rate, E (Bq kg^{-1}) by:

$$\gamma = \frac{E(1-\varepsilon)\rho_s\lambda}{\varepsilon} \tag{10}$$

Values of $c_0\lambda/\gamma$ for the field site are given in Table 1, and increase with depth, from 0.02 to 0.38. Assuming a diffusion coefficient of 10^{-5} m^2 yr^{-1}, this is consistent with fracture half-apertures that decrease with depth, from 18 mm at 5 – 10 m depth, to 625 μm at 75 m depth. For the fracture apertures between 5 and 10 m depth determined using radon concentrations to be equivalent to those measured on the outcrop to the north ($2b$ = 2 mm), would require a diffusion coefficient of the order of 10^{-7} m^2 yr^{-1}. This same value of diffusion coefficient would then suggest a fracture aperture at 75 m depth of approximately $2b$ = 125 μm.

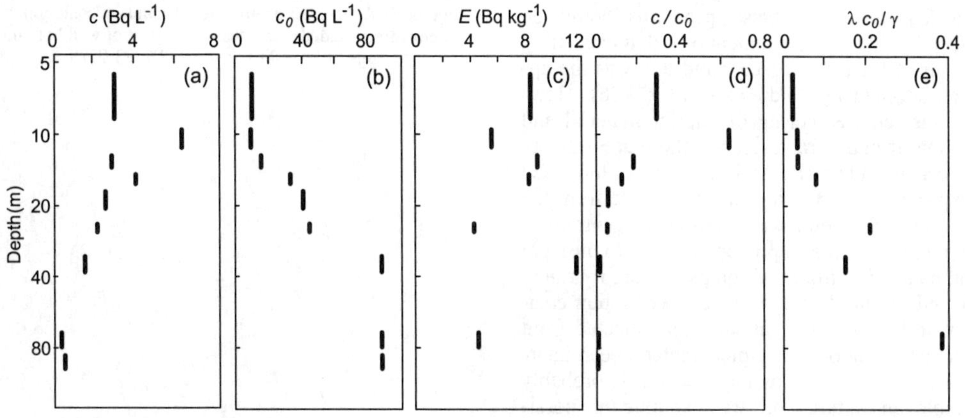

Figure 4. Radon concentrations and emanation rates from Mintaro Shale, Clare Valley. (a) radon concentrations in unpurged piezometers; (b) radon concentrations in purged piezometers; (c) radon emanated from aquifer materials; (d) ratios of unpurged to purged radon concentrations; (e) normalised ratios of purged radon concentrations to radon emanation rates.

5 DISCUSSION AND CONCLUSIONS

In this paper, the ratios of radon concentrations in unpurged boreholes to those in purged boreholes are used to estimate groundwater flow rates through the bores. Similar approaches have been used by Nelson et al. (1983), Hammond et al. (1988), Cook et al. (1999) and Hamada (1999). At this site, groundwater flow rates are found to decrease exponentially with depth in a fractured shale aquifer. Of course, the flow rates represent an instantaneous measurement, and considerable variation in unpurged radon concentrations would be expected throughout the year, due to variations in head gradients and hence flow rates (Cook et al. 1999). Furthermore, because the well constitutes a local zone of high hydraulic conductivity, convergence of groundwater flowlines near the well will mean that the flow rate within the well, per unit area, will be greater than the flow rate within the aquifer (Drost et al. 1968, Zlotnik & Logan 1996). The magnitude of this effect will vary with the bore geometry (bore and well screen radius), and with the hydraulic conductivity of the aquifer and the gravel pack (if present). The effect can be quantified for particular situations, and usually ranges between two and four times. For both of these reasons, the calculated flow rates are best considered as relative, rather than absolute flows.

Measurements of fracture apertures have been made by comparing radon emanation rates with radon concentrations from purged piezometers. To our knowledge, this is the first time that groundwater radon concentrations have been compared with radon emanation rates from aquifer material from the same location (i.e. from rock chips obtained from the screened interval of the piezometer being sampled). Andrews et al. (1986) estimated fracture apertures within granites by comparing emanation rates measured on outcrop and core samples with mean radon concentrations from groundwaters within the Carnmenellis Granite and Stripa Granite. Fracture apertures were estimated to be between 80 and 250μm and 30 and 250μm respectively, but spatial distributions of emanation rates and groundwater concentrations were not considered. Our results suggest a decrease in fracture aperture with depth, which is consistent with the decreasing flow rates. However, the magnitudes of the aperture values obtained are much greater than appear likely from outcrop measurements, or from hydraulic testing (not included here). There are two possible explanations for this. The first is that the measured value of the diffusion coefficient is much higher than the actual value for the undisturbed rock material. Obtaining the cores, and then cutting them to size for the diffusion measurement, may have resulted in microfractures, thus giving an artificially high value. Secondly, the distances from the radium sites to the pore spaces may be greater in the intact rock than in the rock chips, and hence our estimates of radon concentrations within the aquifer matrix may be overestimates.

There are a number of other assumptions inherent in using radon emanation rates to estimate fracture apertures. The first is that radon emanation rates from rock chips are representative of those from aquifer materials. We carried out some preliminary tests on radon emanation rate versus particle size, which suggests no obvious increase in emanation rate with decreasing size. However, it is also possible that the emanation rate from aquifer matrix materials is not representative of the emanation rate adjacent to the fractures. This has yet to be tested. Some of these problems could be avoided by direct measurement of radon emanating from intact fracture surfaces, and this was the approach adopted by Andrews et al. (1986). However, this requires coring of aquifer material and sampling of intact fractures, and these are unlikely to be easily achievable. Andrews et al. (1986) used quarry samples and core samples remote from the sites at which groundwater radon concentrations were measured. Our approach has been to measure radon emanation from rock chips that are routinely collected during drilling. This allowed radon emanation to be measured on samples collected from the precise location of the piezometer screen interval. A combination of two approaches is probably desirable, and future work will examine this possibility.

REFERENCES

Andrews, J.N., N.Hussain, A.S.Batchelor, & K.Kwakwa 1986. ^{222}Rn solution by the circulating fluids in a "hot dry rock" geothermal reservoir. *Appl. Geochem.* 1(6):647-657.

Cook, P.G., A.J.Love & J.C.Dighton 1999. Inferring ground water flow in fractured rock from dissolved radon. *Ground Water* 37(4):606-610.

Drost, W., D.Klotz, A.Koch, H.Moser, F.Neumaier & W.Rauert 1968. Point dilution methods of investigating ground water flow by means of radioisotopes. *Water Resour. Res.* 4:125-146.

Folger, P.F., E.Poeter, R.B.Wanty, D.Frishman & W.Day 1996. Controls on ^{222}Rn variations in a fractured crystalline rock aquifer evaluated using aquifer tests and geophysical logging. *Ground Water* 34(2):250-261.

Hamada, H. 1999. Estimation of groundwater flow rate using the decay of ^{222}Rn in a well. *J. Environ. Radioact.* 47:1-13.

Hammond, D.E., B.W.Leslie & T.-L.Ku 1988. ^{222}Rn concentrations in deep formation waters and the geohydrology of the Cajon Pass borehole. *Geophys. Resear. Lett.* 15(9):1045-1048.

Herczeg, A.L., J.C.Dighton M.L.Easterbrook & E.Salomons 1994. Measurement of radon-222 and radium-226 in ground water by liquid scintillation counting. In Radon and Radon Progeny Measurements in Australia. Symposium, Canberra, 18 February 1994. pp79-82.

Lawrence, E., E.Poeter & R.Wanty 1991. Geohydrologic, geochemical, and geologic controls on the occurrence of radon in groundwater near Conifer, Colorado, USA. *J. Hydrol.* 127:367-386.

Morton, D., A.J.Love, D.Clarke, R.Martin, P.G.Cook & K.McEwan 1998. Clare Valley Groundwater Resources. Progress Report 1. Hydrogeology, Drilling and Groundwater Monitoring. Primary Industries and Resources, South Australia. Report Book 98/00015.

Nelson, P.H., R.Rachiele & A.Smith 1983. Transport of radon in flowing boreholes at Stripa, Sweden. *J. Geophys. Res.* 88:2395-2405.

Torgersen, T., J.Benoit & D.Mackie 1992. Lithological control of groundwater ^{222}Rn concentrations in fractured rock media. Isotopes of Noble Gases as Tracers in Environmental Studies. IAEA, 1992. p.263-283.

Zlotnik, V.A. & J.D.Logan 1996. Boundary conditions for convergent radial tracer tests and effect of well bore mixing volume. *Water Resour. Res.* 32:2323-2328.

Groundwater: Past Achievements and Future Challenges, Sililo et al. (eds) © 2000 Balkema, Rotterdam, ISBN 90 5809 159 7

High nutrient and chloride loads to surface waters in polder areas due to groundwater seepage

P.G.B.de Louw & J.Griffioen
TNO Netherlands Institute of Applied Geoscience, National Geological Survey, Delft, Netherlands

G.A.P.H.van den Eertwegh
Rijnland Water Board, Netherlands

ABSTRACT: Both hydrological and hydrochemical processes will affect the contribution of groundwater seepage to the nutrient and chloride loads to the surface water system in polder areas in the coastal lowlands of the Netherlands. The interaction of these processes is studied in a deep polder called 'De Noordplas'. High loads due to groundwater seepage lead to eutrophication (PO_4 and NH_4) and salinisation of the local and regional surface waters, adversely affecting the aquatic ecosystems and agricultural practices. The seepage water, originating from a confined sandy aquifer, reaches the surface water in ditches and drain tubes via a low-permeable confining layer of Holocene clay and peat. Buried creek channels and so-called 'hot spots' occur in this confining layer, leading to preferential seepage zones within the polder area. The area of these zones accounts for about 10% of the total polder area but contributes for about 50-90% to the solute loads by groundwater seepage. Raising the surface water levels along these preferential seepage zones turns out to be an effective measure to reduce the chloride and nutrient loads by groundwater seepage and to improve local and regional water quality.

1 INTRODUCTION

High nutrient and chloride loads to surface waters in deep polder areas in the Netherlands are not only induced by human activities such as agriculture and sewage, but also by (upward) groundwater seepage. The surface level elevation of deep polders and the local water management practices cause deep groundwater to exfiltrate within the polder. The seepage water is often brackish or saline, nutrient-rich (PO_4 and NH_4), and anoxic. Consequently, the local surface water quality is poor. Also the regional surface water quality is negatively influenced by the polder excess water, pumped out of the polder system. This adversely affects the aquatic ecosystem and agriculture production. Groundwater seepage rates can be reduced by raising the surface water level in the polder. Little is known about the interaction between hydrological and hydrogeochemical processes involved.

A research project has been set up in a deep polder area in the western part of the Netherlands called 'Polder de Noordplas', hereafter called Noordplas (research) area. The project aims at a better understanding of the interrelationships among the various processes involved, both hydrological and physical-chemical. A monitoring program on ground and surface water quantities and composition, field surveys, field experiments, and modeling activities have been conducted. This paper describes the most important results.

2 GEOHYDROLOGY OF DEEP POLDERS

2.1 *History of polders*

The Noordplas research area is located in the western part of the Netherlands and part of the so-called reclaimed drained lakes district. Having been a coastal lagoon during the Holocene epoch, peat accumulated, forming a landscape with small rivers, levees, and raised bogs. The resulting shallow subsurface is characterized by a complex pattern of peat, loamy marine, and fluviatile deposits. As the need for fuel increased during the middle ages, peat was cut and used. A network of large channels, smaller laterals, and parallel ditches was constructed for drainage and peat transportation purposes and lakes emerged. The reclamation of the lake area started in the 17th century, resulting in deep polders. Their surface elevation is now 4-6 m below mean sea level (NAP).

2.2 *Seepage in polder areas*

The thickness of the confining layer of Holocene peat and clay deposits ranges between 3 and 6 m (also called the Holocene cover layer). A sandy Pleistocene aquifer lies below this confining layer. The hydraulic head in this first aquifer is on average 1-2 m above the surface level of the polder area. Therefore, deep groundwater flows from the aquifer into the phreatic groundwater system of the polder (Figure 1). The seepage rates are about 0.5-3 mm/d. The horizontal groundwater velocities in the aquifer are about some tens of meters per year.

Fossil brackish and saline groundwater flows from the aquifer to the phreatic system of the polder. Besides the high concentrations of chloride, this groundwater also consists of high concentrations of PO_4 (tot-P: 2-6 mg/l) and NH_4 (NH_4-N: 30-40 mg/l).

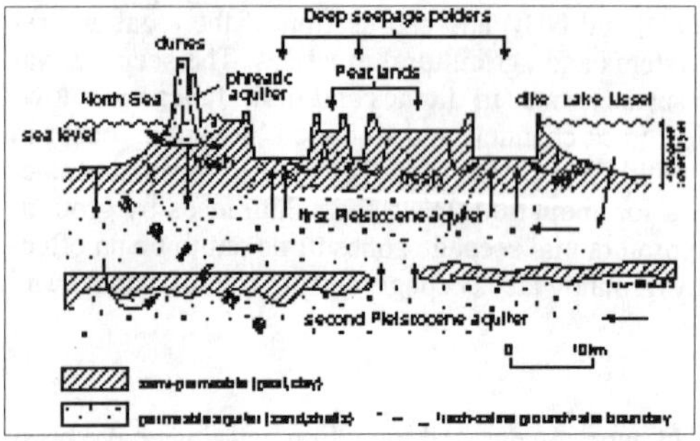

Figure 1. The geohydrological situation in the western part of the Netherlands and the position of the deep seepage polders.

2.3 *Water management in polder areas*

In the western part of the Netherlands, the surface water system is subdivided into two main systems. The regional surface system, the so-called 'boezem', consists of large channels and lakes. It is below mean sea level but elevated several meters above the second system, the polder surface water systems. Both systems have more or less fixed and controlled water levels. Winter surface water levels usually are about 0.1-0.3 m below summer water levels.

The water management practice is such that in the winter months, the excess water of the polder areas is removed by pumping the water into the regional system. The latter system is again provided with pumps that transport the regional-scale excess water (in)directly to the North Sea. During the summer period, water is provided by the regional system to the polder areas for water level control, for irrigation purposes, and to flush the polder channel system. In general, the flow direction in winter is from polder to boezem, in summer from boezem to polder areas.

The excess water of deep polders is a mixture of precipitation excess and groundwater seepage. Depending on the weather conditions and season of the year, the seepage water composition is of large influence on the surface water quantity and quality at both polder and regional scale.

3 GROUNDWATER SEEPAGE

3.1 *Different types of seepage*

The seepage rate of deep groundwater not only depends on the head difference but also on the resistance of the confining Holocene layer. The average flow resistance of this layer is about 3000 d. The peat layer at the bottom of the confining layer mainly determines the vertical flow resistance. This peat layer has been removed at some locations by erosion processes, and sandy creek deposits occur in the lower part of the confining layer. The top of these buried sandy creek deposits lies 3-4 m below soil surface. Consequently, the resistance of the confining layer is strongly reduced in areas where these creek deposits occur. In the Noordplas area, the buried sandy creek deposits were mapped in detail (Figure 2).

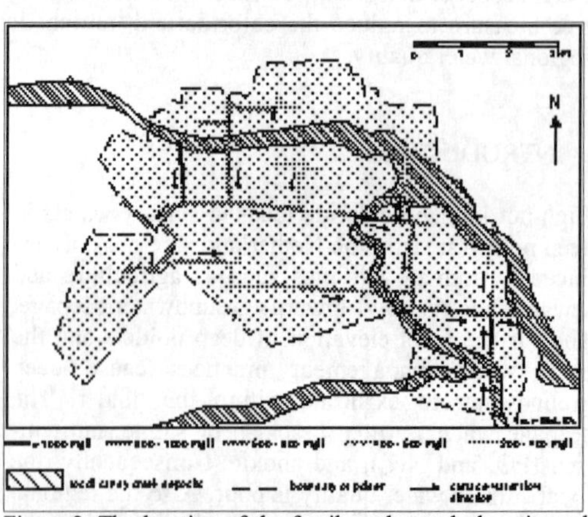

Figure 2. The location of the fossil sandy creek deposits and the chloride concentration of the surface water in the summer of 1999 at the Noordplas research area.

So-called seepage hot spots, to be seen as small-scale subsurface cones of sand connected with the sandy aquifer, are observed at several locations in ditches and channels of the Noordplas area. These hot spots are visually indicated in the field situation by little sand volcanoes, and also by their red-brownish color due to the extensive oxidation of Fe(II) dissolved in exfiltrating groundwater. The location of these hot spots correspond spatially for 50% with the buried creek channels. High chloride concentrations (800-3000 mg/l) are observed in surface waters at these point-source seepage locations.

Obdam and Cleveringa (2000) described that the presence of methane in the shallow subsurface plays an important role in the development of the hot spots. Methane is released during methanogenic degradation of organic matter within the redox-reactive Holocene confining layer and the Pleistocene sediments. Methane degasses from the groundwater at high degradation rates and accumulates below badly permeable layers, i.e. the Holocene confining layer. When enough pressure is built up, it preferably flows through inhomogeneities within the confining layer like the buried creek channels. Methane gas strongly decreases the grain pressure and grains may become suspended. The gas or water permeability is then strongly increased and associated seepage rates become very high. This mechanism, in combination with the occurrence of the sandy creek deposits in the Holocene cover layer, is held responsible for the development of the seepage hot spots in the Noordplas area.

3.2 *Chloride routing in surface water*

A chloride-routing was conducted in the main water courses of Polder de Noordplas in the summer of 1999 in order to locate seepage sources. Water samples were collected from the surface water every 250 m and the chloride concentration was analyzed (Figure 2). The data show that the chloride concentration varies from 200 to over 1000 mg/l. The high Cl-concentrations are mostly associated with the occurrence of the buried sandy creek deposits and hot spots, where seepage water preferentially enters the surface water. The seepage rate is also higher at the location of the sandy creek deposits. The contribution of these locations to salinisation of surface water is therefore high.

3.3 *Characterizing the confining layer*

From the geological mapping and the chloride routing, the Holocene confining layer was split up in three different geohydrological types (Table 1). Seepage catchers were installed to measure the resistance of the confining layer and the seepage rates for these three types. The highest rates were measured at the hot spots. At one hot spot, an extremely high seepage rate of 4000 mm/d was measured. At regions where sandy creek deposits are present in the Holocene confining layer, the rates are about 6 times higher than outside these areas.

Table 1. Subdivision of the Holocene cover layer

Type of confining layer	Average resistance in (d)	Average seepage rate in (mm/d)
Sandy creek deposits not present	3000	0.5
Sandy creek deposits present	500	3.0
Seepage hot spots	0-50	20-4000

3.4 *Reduction of seepage*

A measure to reduce the chloride load and also the nutrients load by seepage is to increase the surface water level in the polder area. As a result from the water level rise, the head difference decreases, and subsequently does the seepage rate. A regional groundwater model was applied using the MODFLOW code (McDonald and Harbaugh, 1988) in order to calculate the effect of such a measure (Raat, 2000). Figure 3 shows the effect of a water level increase of 0.5 m in all ditches and watercourses on the seepage rate. The largest reduction was calculated at the locations where buried creek channels are present in the Holocene confining layer (0.5-1.5 mm/d). The reduction was only 0.05 mm/d outside these locations. The average reduction of the seepage rate for the total polder area was from 1.15 to 0.98 mm/d.

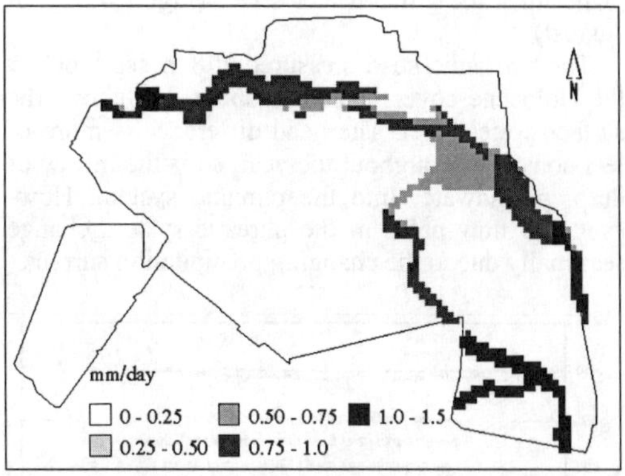

Figure 3. Reduction of the seepage rate due to a water level increase of 0.5 m of all the ditches and watercourses in the Noordplas research area, calculated with MODFLOW.

The regional-scale groundwater flow modeling results showed that the largest reduction of seepage rates are found in and around buried creek channels. Therefore, water flow and salt transport around a buried creek channel were modeled in detail (Van de Scheur, 2000) using the HYDRUS_2D code (Simunek *et al.*, 1999). The modeling was done for a 2D vertical transect with a length of 100 m and a depth of 30 m, containing a 30 m long buried creek channel of 2.5-3 m thickness.

In the reference model, subsurface drains were installed at 1.5 m depth with a 10 m spacing (present situation). In the scenario model, an identical spacing is used, but a 0.5 m raised tile drains level is implemented at the 30 m long buried creek channel. The reference model shows that chloride concentration levels along the buried creek channel are higher and more close to the root zone, compared to chloride fronts in other parts of the confining layer left and right of the buried channel. High salt concentration levels at or near the root zone might cause crop production problems. Calculation results for the scenario show that on an annual basis, the overall seepage rate is reduced by 15%, as well is the subsurface drains discharge. The evapotranspiration rate is equal to the reference situation. If all subsurface drains would be raised by 0.5 m, a 30% seepage rate reduction would be achieved. Locally raising the subsurface drains level at buried creek channels has the largest effect on the reduction of the seepage rate.

3.5 *Temporal variability of seepage*

Figure 4 shows that during the winter period, phreatic groundwater levels are about 0.1-0.8 m higher than the surface water levels. In these periods, drainage of groundwater by subsurface drains, ditches and channels occurs. During summer, infiltration can occur in case the surface water levels are higher then the groundwater levels (Figure 4; e.g. in August).

The hydraulic head measured at 8 m depth below the Holocene cover layer lies about 2 m above the surface water level. The head difference is more or less constant throughout the year, so is the inflow of deep groundwater into the phreatic system. However, the flow paths in the phreatic system change seasonally due to the changing precipitation surplus.

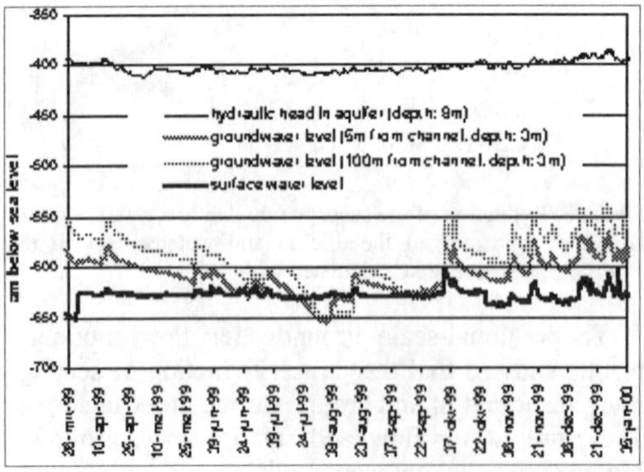

Figure 4. Time series of the hydraulic head in the confined aquifer, the phreatic groundwater level in the Holocene cover layer and the surface water level, measured in the Noordplas area

The different groundwater flow paths were simulated with the water flow and solute transport model, HYDRUS_2D. Simulation results show that in the winter period, the largest part of the deep groundwater flows directly to the ditches. A minor part is discharged by subsurface drain tubes installed at the parcels between ditches. During summer, actual evapotranspiration is larger than precipitation, and subsequently groundwater levels fall. In this season, a minor part of the groundwater flows to ditches and the subsurface drains run dry. Deep groundwater is not drained, but supplies the soil water storage and facilitates agricultural crop production by plant transpiration.

4 HYDROCHEMICAL ASPECTS OF SEEPAGE

4.1 *Groundwater quality*

The surface water and groundwater composition was determined at 14 locations during April and May 1999. Samples were taken from the surface water, from the groundwater just below the channel bottom, from groundwater at a depth of 5 m below soil surface, and from groundwater present in the sandy aquifer. The spatial variation is characterized together with variations with depth. The summarized data are presented in Table 2. Note that the surface water at a particular location is a mixture of local-scale exfiltrating pore water and of water, originating from upstream sources.

Table 2. Average and standard deviation (between brackets) of dissolved solute concentrations (mg/l) in surface water and groundwater for 14 locations in Noordplas research area, during April and May 1999.

Parameter	Surface water	Groundwater		
		channel bottom	5m depth	sandy aquifer
Cl	390 (275)	350 (470)	170 (170)	230 (225)
SO_4	240 (135)	365 (420)	210 (270)	15 (30)
Fe	5.2 (5.1)	5.1 (4.1)	8.5 (13.4)	11.0 (11.7)
NH_4	3.7 (3.0)	17 (12.7)	33 (11.9)	35 (11.0)
total-P	0.4 (0.4)	2.9 (2.7)	4.5 (4.0)	2.9 (2.1)

The dissolved solute concentrations vary largely in space as can be deduced from the large standard deviations. The average surface water concentrations of chloride and sulfate approach the channel bottom concentrations. The surface water composition seems to be influenced largely by the seepage water composition for chloride and sulfate, but not as far as ammonium and phosphate are concerned. Note the differences between surface water concentrations and groundwater just below the channel bottom for ammonium and total-phosphate and the low value of sulfate concentration in the sandy aquifer. Ammonium is possibly transformed into nitrate and/or is adsorbed at or near the channel bottom. Phosphorus is likely to precipitate or to adsorb to the channel bottom soil matrix.

As stated in section 3, buried creek channels occur within the lower part of the confining layer. Six measurement locations are located within areas with buried creek channels. In these areas, the seepage rate is high and deep groundwater is likely to determine the composition of channel bottom groundwater and surface water. The groundwater contains more chloride at the areas with the buried creek channels than outside these areas. The average Cl-concentration is 393 (± 246) mg/l for these areas and 95 (± 65) mg/l for the other areas. SO4, Fe, NH_4, and total-P on the contrary, are a factor of 0.3 to 0.6 smaller within the areas then outside. Salinisation of surface water is, therefore, majorly determined by seepage of more saline groundwater via the buried creek channels.

4.2 *Phosphorus chemistry during outflow of anoxic groundwater*

High phosphate concentrations occur in the deep groundwater flowing into the polder. The phosphate loads to the surface water are strongly controlled by processes during oxidation of the outflowing anoxic groundwater, especially immobilization of PO_4 by coprecipitation with Fe[III]-oxyhydroxides. Griffioen (1994) showed that the Fe:PO_4 ratio plays an important role in determining the immobilization of PO_4 during Fe-oxyhydroxide precipitation. Complete immobilization of PO_4 is impossible when the molar Fe:PO_4 ratio is smaller than 1, because the precipitation stoichiometry is 1 to 1 at maximum PO_4 uptake.

The molar Fe:PO_4 ratio varies from 0.4 to 40 for groundwater originating from the first aquifer, except for one sample which has an unusually low Fe-concentration of 0.2 mg/l. No difference was observed for locations within the areas with buried creek channels and outside. The average of the log-value of the molar ratio is 0.5 (± 0.7). The general conditions for immobilization of PO_4 are thus favorable, but throughput of PO_4 can be expected during exfiltration of deep groundwater at several locations within the polder studied.

5 CONTRIBUTION OF SEEPAGE TO EUTROPHICATION AND SALINISATION OF SURFACE WATER

Discharge-dependent sampling of the surface water has been performed at 4 measurement locations since June 1999, using automated equipment. Figures 5 and 6 show the relationship between chloride concentrations and of the so-called sumparameter nitrite+nitrate (NO2+NO3) and ammonium (NH4) respectively, for one sub-catchment within the Noordplas area. Comparable graphs can be shown for the other 3 measurement sites. Seepage water is known to be brackish. Seepage rates are more or less constant throughout the year. The precipitation surplus varies with time and therefore determines the variation of the chloride concentration in the surface water. High chloride concentrations (seepage water influence mainly) coincide with low nitrate+nitrite concentrations (Figure 5). The nitrate loads originate mainly from agricultural subsurface drainage of the precipitation excess. On the contrary, ammonium shows a positive correlation with the chloride concentration (Figure 6). Therefore, the source of the ammonium loads can only originate from groundwater seepage. Solute concentration measurements can in this way be used for environmental tracer purposes to find solute sources.

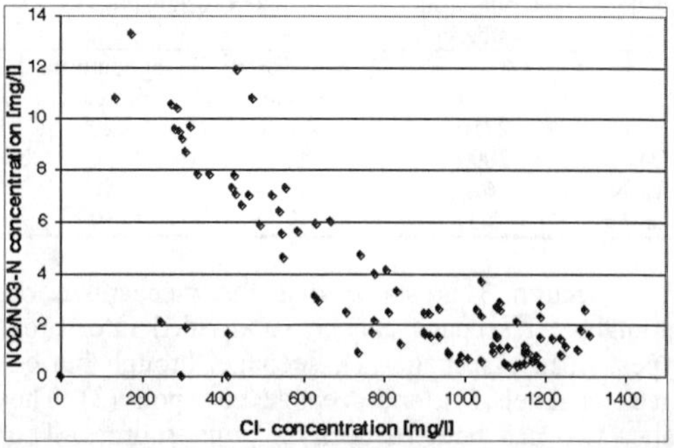

Figure 5. Cl- vs. (NO_2+NO_3)-N concentrations in the surface water for the period June 1999 – June 2000. Measurements for sub-catchment '25C' in Noordplas research area.

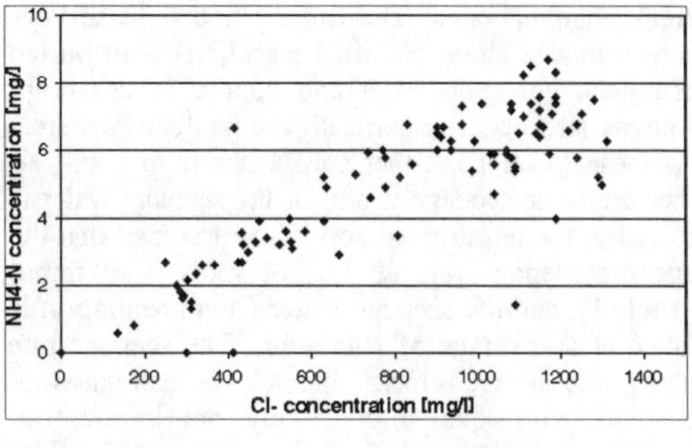

Figure 6. Cl- vs. NH_4-N concentrations in the surface water for the period June 1999 – June 2000. Measurements for sub-catchment '25C' in Noordplas research area.

Based on the one-year time series of concentration measurements and long-term average drainage and seepage rates, solute loads from the polder area to the regional surface water system can be calculated (Table 3). As can be seen from Table 3, it depends on the solute source and solute load to be reduced whether one should focus on e.g. water management measures or on e.g. agriculture practices to improve the surface water quality. As an example, NO_3-N and SO_4 in the surface waters mainly originate from agricultural drainage during the winter season when the precipitation excess exceeds the seepage rate. On the other hand, the influence of

NH_4-N present in seepage water is dominant during summer. It is noted that total-N consists of about 50% of NO_3-N and 50% NH_4-N. The SO_4 loads are, besides the total-N and total-P loads, of importance to the eutrophication problem as well due to its secondary effects on the mobilization of phosphorus from ditch and channel bottom sediments.

Table 3. Estimated annual solute loads to the surface waters of the Noordplas area and the relative contribution of the sources groundwater seepage and agricultural drainage.

Solute	Annual load [kg/ha/y]	Source, origin [%]	
		groundwater seepage	agricultural drainage
Cl	2750	80	20
SO4	2000	10	90
total-N	65	50	50
total-P	2.5	70	30

In section 3 is stated that the mechanism of groundwater seepage can be subdivided into three types: diffuse seepage (1), seepage through buried creek channels (2), and seepage hot spots (3). The latter two can both be seen as point sources. The solute loads that originate from groundwater seepage can now be subdivided into diffuse sources and point sources. This calculation was done for chloride.

The pattern in Figure 2 shows that at about 10% of the polder surface water area sections with buried creek channels exist. The routing and field surveys shows that at about 5% of the sections with buried channels, hot spots exist and appear. These point sources also occur underneath the land surface area. At these locations, either subsurface drains will either drain the seepage water, or the seepage water is stored in the unsaturated zone. It is assumed that the average seepage rate at the hot spots is 50 mm/d (Table 1), and the seepage water Cl-concentration is taken at an average of 1000 mg/l. The seepage rate along the buried creek channels is estimated at 3 mm/d, with equal high Cl-concentrations as observed at the hot spots. For other parts in the polder area (diffuse seepage), the seepage rate is estimated at 0.5 mm/d with a Cl-concentration equal to 300 mg/l. Using these numbers, it is calculated that from the total water volume and chloride load originating from groundwater seepage, point sources take care of about 50% of the water volume and 80-90% of the chloride load. The chloride loads from point-source seepage thus form a major factor compared to diffuse sources of groundwater seepage. The same results can be found for NH4-N and total-P. Water management measures to reduce the chloride and nutrient loads from seepage should preferably be focused on the regions where buried channels and hot spots are present in the Holocene confining layer.

6 CONCLUSIONS

Seepage of deep groundwater accounts for 50% of the water volume drained, but for 80-90% of the chloride load to the surface water in the Noordplas research area. As far as other solute loads are concerned, the percentages are 50% (total-N), 70% (total-P), and 10% (SO_4) of the total loads. The other part of the solute loads originate from agricultural activities in the area. Sulfate is of importance to the eutrophication problem as well due to its secondary effects on the mobilization of phosphorus from ditch and channel bottom sediments.

Seepage in deep polders is not a spatially uniform process but happens locally where permeable sandy creek deposits happen within the lower part of the Holocene cover layer. Besides this, seepage rates turned out to be high at so-called hot spots. These two sources contribute for about 50% of the total seepage water volume whereas they only exist in 10% of the total polder area. The chloride loads and nutrient loads by groundwater seepage happen for about 50-90% at these point sources.

Increasing the surface water level has turned out to be a effective measure to reduce the different solute loads to the surface water by groundwater seepage. The largest reductions can be expected around the preferential seepage zones.

REFERENCES

Griffioen, J. 1994. Uptake of phosphate by iron hydroxide during seepage in relation to development of groundwater composition in coastal areas. Environ. Sci. Technol. (28), pp. 675-681.

McDonald, M.G., and Harbaugh, A.W. 1988. A modular three-dimensional finite-difference groundwater flow model. U.S. Geological Survey Techniques of Water-resources Investigations Report. Book 6, Chap A1.

Obdam, A & P. Cleveringa 2000. Gas in the shallow subsoil. Netherlands Institute of Applied Geoscience – Geological Survey (in concept).

Raat, K.J. 2000. Regional groundwater flow model of Polder de Noordplas. The calculated effects of different water management practices. Netherlands Institute of Applied Geoscience – Geological Survey. TNO-report 00-126-B.

Simunek, J., M. Sejna, and M.Th. van Genuchten 1999. Hydrus_2D / Meshgen_2D. Simulating water flow and solute transport in two-dimensional variably saturated media. U.S. Salinity Lab., California, 227 p

Van de Scheur, W. 2000. How Holland turns brackish. Survey of spatial and temporal variation of groundwater seepage in Polder de Noordplas area. MSc. thesis Wageningen University, Dept. of Water Resources.

Groundwater: Past Achievements and Future Challenges, Sililo et al. (eds) © 2000 Balkema, Rotterdam, ISBN 90 5809 159 7

Isotope based assessment of groundwater renewal: An international study of the unsaturated zone

W.M.Edmunds
British Geological Survey, Wallingford, UK

B.Th.Verhagen
Schonland Research Centre, University of the Witwatersrand, Johannesburg, South Africa

ABSTRACT: Geochemical and environmental isotope techniques in unsaturated zone moisture profiles are becoming established and reliable tools in assessing direct - or diffuse - recharge rates in arid to semi-arid regions. A cooperative Research Programme organised by the International Atomic Energy Agency brought together 13 research institutes worldwide in a study of the unsaturated zone in a wide variety of environments and materials. The results represent perhaps the most extensive survey of the unsaturated zone ever undertaken. The paper presents and discusses profiles from a few selected sites and presents the main conclusions drawn from the information. Where chloride concentrations approach a steady state and can be averaged, the mean concentration is inversely proportional to recharge. Stable isotopes may reflect evaporation and recharge history, but cannot be used for quantitative recharge calculations. Thermonuclear tritium "peaks" under conditions approaching pure piston flow present clear time markers, whilst more diffuse tritium may place constraints on the interpretation of other methods. Bypass flow has been shown to contribute significantly to recharge in various environments.

1 INTRODUCTION

Geochemical techniques using chloride and tritium contained in unsaturated zone moisture profiles are becoming established as a reliable tool for measurement of direct - or diffuse - recharge rates in semi-arid regions (Allison et al. 1994). The thermonuclear tritium 'bomb peak' in the early 60's has been widely used as a time marker in temperate zones and less commonly in arid zones (eg Edmunds and Walton 1980; Aranyossy and Gaye 1992; Gaye and Edmunds 1996). Limitations to its use are cost, the radioactive decay (half life 12.43 years) since the thermonuclear moratorium, and the loss of tritium to the vapour phase during evapotranspiration. Nevertheless the position and shape of the tritium peak in certain areas provides convincing evidence of 'piston displacement' as well as an estimate of recharge rate.

In contrast to tritium, chloride from atmospheric deposition is conserved in the soil zone and concentrated by evapotranspiration and may be used for estimating recharge. Whilst the timing of the tritium spike is well known, the use of chloride relies on knowing the depositional flux and hence details of present and past distributions in rainfall (Edmunds and Gaye (1994), Allison et al. (1994)). Chloride together with stable isotope data may also provide a record of past recharge and help reconstruct the antecedent climatic conditions. The unsaturated zone therefore may act as an additional climatic archive in continental areas where records are scarce. Such records may be found in Africa covering the century or millennial scale with typically decadal resolution (Cook et al. 1992).

A Cooperative Research Programme, organised in the late nineties by the International Atomic Energy Agency, brought together 13 research institutes in as many countries, mostly in Africa but also in the Middle East, Asia, America and Europe (see list below). An impressive body of data has been accumulated during the nearly three years of project activities, representing perhaps the most extensive survey of the unsaturated zone in arid to semi-arid regions ever undertaken.

A number of benchmark sites is now available, with detailed information on physiography, lithology, rainfall, and a variety of determinands on the unsaturated zone. Profiles in a range of unsaturated media, mostly hand-cored, were obtained to depths of 40 metres or more. Typical parameters measured on the cores were moisture content, chloride (occasionally nitrate), oxygen-18 and deuterium as well as environmental tritium, which gives a measure of transport rate. Much of the data has been analysed in terms of groundwater recharge. In some cases, complimentary information was obtained on the underlying saturated zone, the ultimate resource the assessment of which formed the target of the project.

This paper presents a summary of much of the work and discusses a few typical profiles from selected sites in a variety of environments and presents the main conclusions.

2 METHODS AND THEORY

The chloride balance method has now been successfully used in a range of environments to determine recharge. Details of the method and a review of studies in various parts of the world have appeared elsewhere (Allison et al. 1994; Edmunds et al. 1988).

Samples of moist sand were obtained by augering or other dry drilling techniques at regular intervals through the unsaturated zone. Moisture contents are measured gravimetrically through drying and chloride is determined on samples obtained either by centrifugation (Kinniburgh and Miles 1983) or by elutriation and high precision colorimetry in some instances.

Isotopic measurements are typically performed on water vacuum distilled from soil samples. Tritium can be measured directly, giving a measurement error typically of ± 5 TU. For lower tritium levels, larger water samples from greater lumped depth intervals are pre-enriched, reducing the error to ± 0.5 TU. In some cases δ^2H was measured by direct reaction with the moist soil sample (e.g. Darling and Talbot 1989). In a method developed during this CRP, $\delta^{18}O$ can be measured by direct equilibration with the soil sample (Butler and Verhagen 2000)

Rainfall (total average deposition) inputs must be known. Errors associated with rainfall measurement and the spatial variability of rainfall are likely to constitute the largest uncertainties in recharge estimation using chloride (Beekman et al. 1996).

3 REGIONAL EXAMPLES

3.1 *Senegal*

An example of an integrated study from Senegal using Cl, stable isotopes and tritium is shown in Figure 1, based on results given in Gaye and Edmunds (1996) and Aranyossy and Gaye (1992). The tritium peak defines clearly the 1963 moisture layer and piston displacement. A recharge rate of 26mm a^{-1} is indicated. The Cl profile shows oscillations due to variable climatic conditions. A mean concentration of Cl (23.6 mg l^{-1}) gives a recharge rate of 34.4 mm a^{-1}, in close agreement with the tritium, using a rainfall Cl value of 2.8 mg l^{-1} for an average rainfall 290mm a^{-1}. The most enriched stable isotope values correspond to high Cl and periods of drought, the Cl peak around 6-10m reflecting the major Sahelian drought commencing in the late 1960's. Further profiles from the Louga site have been published in the references given above and two further profiles were obtained in the present work which will be published elsewhere.

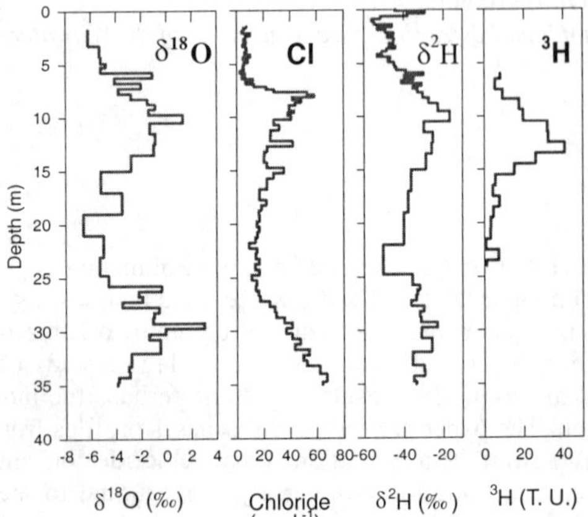

Figure 1. Profiles in a deep core from northern Senegal, reflecting the Sahelian drought period and showing the 1963 'bomb' tritium peak.

3.2 *Nigeria*

Two profiles were obtained in NE Nigeria to 15m. One of these is from a site 40km north of Maiduguri lying on stabilised dune sands where the mean annual rainfall is approximately 450 mm. The profile

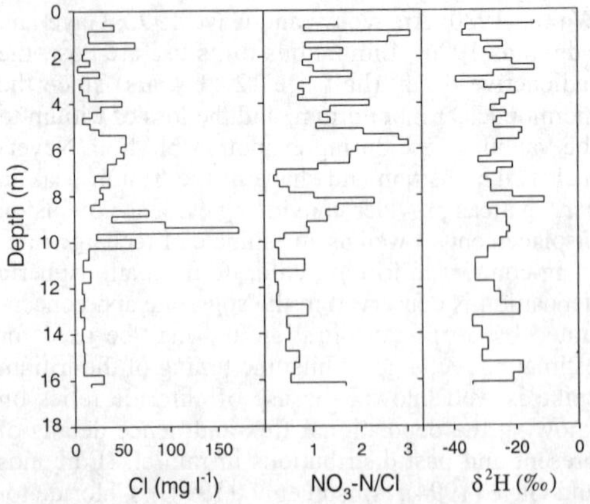

Figure 2. Profiles from NE Nigeria showing values of Cl, the ratio of NO_3-N/Cl and deuterium.

has a mean chloride concentration of 29.5 mg l^{-1}, and its relative uniformity down the profile (Figure 2) demonstrates the homogenous displacement of an atmospheric input value. With 1.7 mg l^{-1} Cl in rainfall a long term average recharge of 22 mm a^{-1} is found spanning 21 years. The δ^2H is relatively uniform consistent with active recharge and ranges from -14 ‰ to -39 ‰ with an average of -26 ‰. High nitrate concentrations, well above atmospheric, must represent fixation by leguminous vegetation as described elsewhere in the Sahel (Edmunds and Gaye 1997).

3.3 *Jordan*

Two profiles were drilled in Cretaceous sandstone by percussion methods near Jarash in northern Jordan down to depths of 21m on which chloride, nitrate and deuterium were measured. Chloride mass balance in this area with a mean rainfall of 500 mm a^{-1} indicated recharge values of 37 mm a^{-1}. Four shallow holes (up to 10.5m) were also drilled by hand auger near Al Quwayrah in southern Jordan where the mean annual rainfall is only 65mm a^{-1}. Cl concentrations in the unsaturated zone of between 10000 and 25000 mg l^{-1} clearly indicate that negligible recharge is occurring in this region. Nitrate concentrations are also enriched under the aerobic conditions in these profiles with variations following those of chloride. Deuterium enrichment and depletion patterns in the Jarash profiles tend to follow higher and lower chloride values, which are interpreted as reflecting periods of lower and higher precipitation.

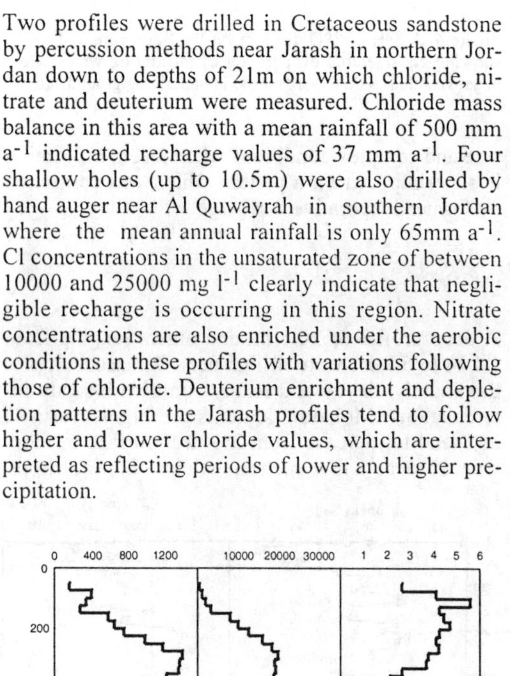

Figure 3. Profiles in Jordan showing denitrification and increasing chloride with depth, indicating zero recharge.

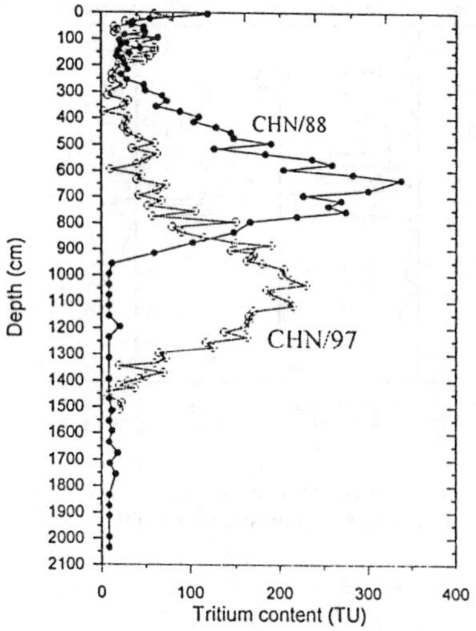

Figure 4. A 'Bomb' tritium peak measured in a drilled core in Loess in Inner Mongolia in 1997 compared with the peak found in a dug pit in 1988.

3.4 *China*

The profiles from China (Inner Mongolia) are of special interest in that the unsaturated zone medium is a Loess deposit. Loess has special properties which make it a near-ideal medium: it is fine-grained and has considerable mechanical competence. The area studied is semi-arid, with seasonal rainfall averaging 360 mm a^{-1}. Samples were collected at intervals of 10 cm and 20 cm from two deep cores drilled one year apart to depths more than 15 metres. Soil moisture was extracted by vacuum distillation and tritium and δ^{18}O measured. Chloride was determined through elutriation. As tritium values range up to more than 200 TU, they could be determined by direct counting, allowing for impressive depth resolution. Profiles were compared with results from test pits sunk in the middle eighties (Figure 3). Using the position of the tritium peak and the average moisture contents of the profiles, infiltration rates of 47 mm a^{-1} and 68 mm a^{-1} respectively were obtained. For the deeper (steady-state) section of the profiles, recharge rates of 85 mm a^{-1} and 288 mm a^{-1} were obtained using chloride balance. Whereas the concentrations drop to some 10 TU or less at the bottom of the profiles, the underlying phreatic ground water has values in the range of 30 - 40 TU, suggesting mechanisms other than intergranular flow in recharge of the ground water.

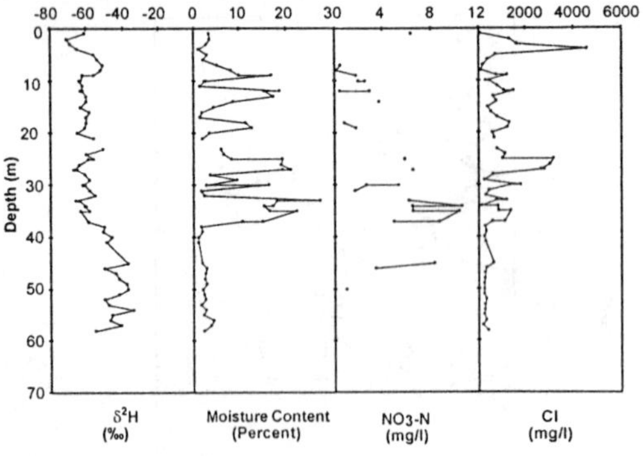

Figure 5. Parameters measured on deep core in Mexico, showing the effects of variations in grain size with depth.

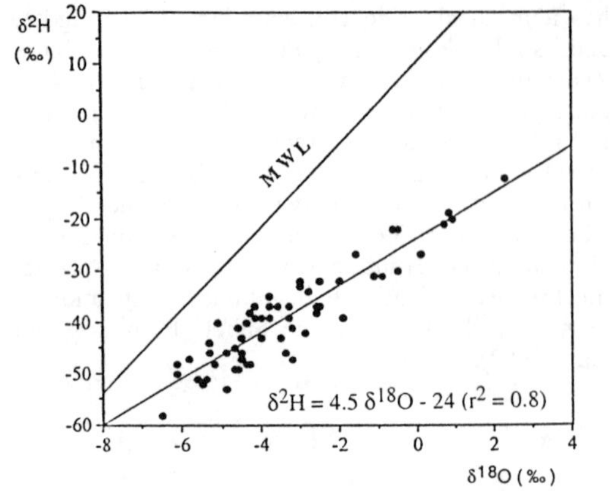

Figure 6. A δ^2H - $\delta^{18}O$ diagram for the Nefta core in Tunisia, showing considerable evaporation of infiltrating water. More negative values tend towards those in underlying groundwater.

3.5 *Mexico*

Three unsaturated profiles were obtained from Mesilla Bolson and Guaymas in northern Mexico. A 58 m borehole located near Cd Juarez, Chihuahua state and drilled by dry percussion was sampled at metre intervals (Figure 5). The other two boreholes with depths 5.50 m and 8.50 m in the southwestern portion of Sonora state were sampled every 0.25 m by dry drilling (Figure 5). Samples were analysed for moisture content, chloride, deuterium and nitrate (NO_3-N). Average precipitation at Mesilla Bolson is 230 mm a^{-1} with 1 mg l^{-1} chloride. With Cl content in the unsaturated profile of 977 mg l^{-1} below the zero flux plane Cl mass balance gives a long term recharge of about 0.24 mm a^{-1} with a probable time scale for the profile of 11,000 years. At Guaymas the recharge range is 0.11-0.16 mm a^{-1}, for average rainfall of 320 mm a^{-1}, Cl content of 3.4 mg l^{-1} and mean profile Cl of 6,900 and 9,470 mg l^{-1} below the zero flux plane, respectively.

3.6 *Tunisia*

In southern Tunisia a profile of 13.5 metres was drilled in the Corbeille Nefta, a depression in an arid environment with average rainfall less than 100 mm a^{-1}. The core was found to consist of a succession of sandy strata with silty intercalations. Moisture contents oscillates between 2% and 13% down the profile, with peaks corresponding to silty layers. Chloride and $\delta^{18}O$ values similarly show wide ranges of fluctuation of some 5000 mg l^{-1} and 7‰ respectively, tending at depth towards values observed in the underlying groundwater, rest level at -28 m at the time of sampling but close to surface in 1950. δ^2H - $\delta^{18}O$ values for soil moisture show strong evaporation (Figure 6). The profiles show the difficulty in interpretation with large variations in grain size and complex conditions when recent infiltration is superimposed on a pattern which was controlled by former evaporative discharge.

3.7 *Syria*

A deep profile of 20.6 m, and others ranging from 1.5 m to 5.9 m, were drilled near Damascus in predominantly clayey and loamy strata with sandy intercalations. Moisture contents are highly variable and tend to increase from 5% near surface to more than 25% at a few metres, whilst chloride drops

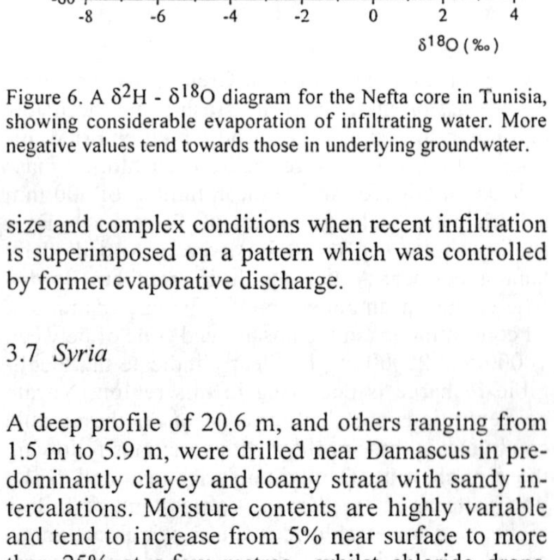

Figure 7. Depth profiles of deuterium and chloride in a core drilled in clayey material near Damascus. Note evaporation inversion in δ^2H near surface.

from several thousands to a few hundreds of mg l^{-1} and NO_3 from over 100 to a few mg l^{-1} (presumably through denitrification). At more than 12 m in the deep profile, chloride values drop sharply (Figure 7) suggesting increased recharge rates in the past. The unsaturated zone at present effectively protects the underlying aquifer from surface salinity and pollution. Deuterium profiles clearly show the inversion near the surface (Figure 7), evidence for evaporative losses of 5-6 mm a^{-1}, using the model of Allison and Barnes (1983).

3.8 *South Africa*

The study area, the only one in the southern Hemisphere, in the southern extent of the Kalahari thirstland, mean summer rainfall 300 mm a^{-1} is covered by some 50m of sand. The underlying gravel aquifer at about 100 m is confined by clay but isotopic and chemical evidence suggests that ephemeral saturation may occur above the clay (Verhagen et al. 1978). Three hand auger cores to 15 m, 18 m and 22 m were sampled at 1 m intervals. Shallower soundings were sampled every 0.1 m depth. Moisture contents is variable near surface and increases with depth from 1 to 3 weight %. Variable chloride values range up to a few hundreds mg l^{-1} (Figure 8). Electrical conductance often increases more rapidly than Cl values, suggesting another ion, probably HCO_3 may be a significant solute. The lowest Cl values give a maximum recharge rate of 4.9 mm a^{-1}. Saturated zone chloride concentrations, one order of magnitude less than the lowest values in the profiles, suggest bypass flow as the principal recharge mechanism. Tritium concentrations in two of the deepest profiles lie in the range 3 - 10 TU with no clear trends nor suggestions of a 'bomb' peak, giving maximum recharge rates of some 13 mm a^{-1}. Lower values of deuterium excess indicate evaporation, but no inversion (Allison and Barnes 1983) is seen, even at a depth resolution of 0.1 m. At depth, all profiles show a zone of stability with lower $\delta^{18}O$ values and higher deuterium excess. The upward trend in $\delta^{18}O$ below 18 m suggests mixture with material dislodged from higher in the profile.

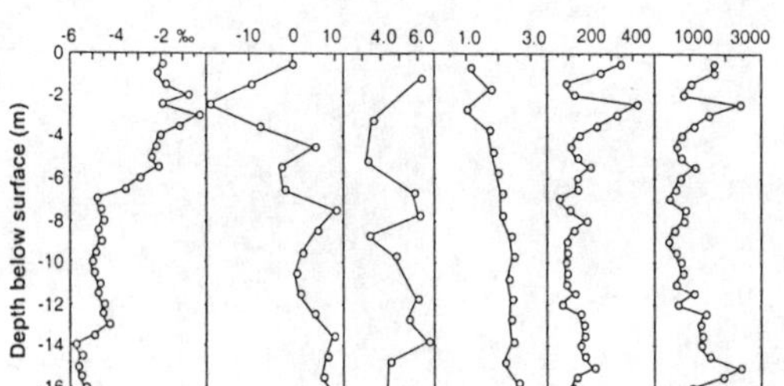

Figure 8 Various parameters measured on a sand core from the southern Kalahari. Note variations in deuterium excess.

4 CONCLUSIONS

Several profiles obtained during the present CRP and reported in the country papers provide evidence that where chloride concentrations approach steady state and can be averaged in the unsaturated zone, mean chloride in the profiles is inversely proportional to recharge in unsaturated zone profiles. Stable isotope values may reflect evaporative enrichment of Cl, contain information on the recharge history, but cannot be used quantitatively to calculate rates of recharge. Nitrate was found to act as a conservative tracer in the profiles, similar to chloride. Under conditions approaching true piston flow, bomb tritium 'peaks' give a clear time marker. More diffuse tritium can act as a constraint to the interpretation of other parameters. The results of several studies suggest that bypass flow is a significant contributor to recharge in various environments. Clear guidelines for future investigations can be derived from this study and the data obtained is to be published as an archive by the IAEA.

5 ACKNOWLEDGMENTS

We wish to acknowledge all the participants below who have allowed profiles from the various countries to be used in this paper and made individual contributions. The project was managed and supported by the IAEA. This paper is published with the permission of the Director, British Geological Survey, NERC.

REFERENCES

Allison, G.B., Gee, G.W. & Tyler, S.W. 1994. Vadose-zone techniques for estimating groundwater recharge in arid and semi-arid regions. Soil Sci.Soc.Am.J.58: 6-14.

Allison, G.B.& Hughes, M.W. 1978. The use of environmental chloride and tritium to estimate total local recharge to an unconfined aquifer. Aust.J.Soil Res.16:181-195.

Aranyossy, J.F. & Gaye, C.B. 1992. La recherche du pic du tritium thermonucléaire en zone non-saturée profonde sous climat semi-aride pour la mesure de recharge des nappes: première application au Sahel. C.R.Acad.Sci.Paris 315 (ser.II):637-643.

Beekman, H.E., Gieske, A. & Selaolo, E.T. 1996. GRES: Groundwater recharge studies in Botswana 1987-1996. Botswana Jnl of Earth Sciences 3: 1 - 17.

Butler, M.J. and Verhagen, B.Th. 2000. Isotope studies of a thick unsaturated zone in a semi-arid area of southern Africa. In: IAEA TecDoc (in preparation).

Cook, P.G., Edmunds, W.M.& Gaye, C.B. 1992. Estimating palaeorecharge and palaeoclimate from unsaturated zone profiles. Water Resour.Res.28:2721-2731.

Darling, W.G. and Talbot, J.C. 1989. Extraction methods for the isotopic analysis of porewaters Technical report WD/89/66. BGS, Nottinghamshire.

Edmunds, W.M., Darling, W.G. & Kinniburgh, D.G. 1988. Solute profile techniques for recharge estimation in semi-arid and arid terrain. pp 139-158. In: Estimation of Natural Groundwater Recharge. NATO ASI Series. Reidel, Dordrecht.

Edmunds, W.M. & Gaye, C.B. 1994. Estimating the spatial variability of groundwater recharge in the Sahel using chloride. J.Hydrol.156:47-59.

Edmunds, W.M. & Walton, N.R.G. 1980. A geochemical and isotopic approach to recharge evaluation in semi-arid zones, past and present. pp 47-68. In "Arid Zone Hydrology, investigations with isotope techniques". IAEA, Vienna.

Edmunds, W.M.& Gaye, C.B. 1997. High nitrate baseline concentrations in groundwaters from the Sahel. Journal of Environmental Quality, 26, 1231-1239.

Gaye, C.B. & Edmunds, W.M. 1996. Intercomparison between physical, geochemical and isotopic methods for estimating groundwater recharge in northwestern Senegal. Environmental Geology, 27, 246-251.

Kinniburgh, D.G. & Miles, D.L. 1983. Extraction and chemical analysis of interstitial water from soils and rocks. Environ.Sci.Technol.17:362-368.

Verhagen, B.Th., Smith, P.E., McGeorge, I. & Dziembowski, Z.M. Tritium profiles in Kalahari sands as a measure of rain-water recharge. Isotope Hydrology 1978 2: 733-751. IAEA, Vienna.

Wood, W.W. & Sanford, W.E. 1995. Chemical and isotopic methods for quantifying groundwater recharge in a regional semi-arid environment. Ground Water 33:458-468.

PARTICIPANTS

Participants in the IAEA Coordinated Research Programme: 'Isotope-Based Assessment of Groundwater Renewal and Related Anthropogenic Effects in Water Scarce Regions'

B Abou Zakhem,
Syrian Atomic Energy Commission, Syrian Arab Republic
I Alsagaby,
King Abdulaziz City for Science & Technology, Saudi Arabia
H Amro and S Kilani,
Water Quality Department, Jordan
M J Butler and B Th Verhagen,
University of the Witwatersrand, South Africa
W M Edmunds,
British Geological Survey, UK
M Fahmy Hussein,
Cairo University, Egypt
I B Goni and G B Ibrahim,
University of Maiduguri, Nigeria
C B Gaye and A Faye,
Cheikh Anda Diop University, Senegal
C Gutierrez,
Instituto Mexicano de Technologia del Agua, Mexico
A Leis,
Institut fuer Hydrogeologie und Geochemie, Austria
Ruifen Lin and Keqin Wei,
Guangzhou Institute of Geochemistry, China
S Navada,
Bhabha Atomic Research Centre, India
E Ramirez,
Universidad Mayor de San Andrès, Bolivia
K Zouari,
Ecole National Ingenieurs, Tunisia

Groundwater: Past Achievements and Future Challenges, Sililo et al. (eds) © 2000 Balkema, Rotterdam, ISBN 90 5809 159 7

Evolution of high-pH Na-HCO_3 groundwaters in anorthosites: Silicate weathering or cation exchange?

B. Frengstad
Department of Geology and Mineral Resources Engineering, Norwegian University of Science and Technology, Trondheim & Geochemistry and Hydrogeology Section, Geological Survey of Norway, Trondheim, Norway

D. Banks
Holymoor Consultancy, Chesterfield, UK

ABSTRACT: A significant number of groundwaters are found to be of Na-HCO_3 type, with low Ca and high pH, in a range of crystalline lithologies in Norway. This water-type is also found in the Ca-rich Egersund anorthosites which have a relatively simple mineralogy. Thus, this lithology was chosen to investigate whether silicate weathering or cation exchange may provide an adequate mechanism for the generation of alkaline Na-HCO_3 waters. The modelling code PHREEQC was employed to simulate simplified anorthosite weathering systems and to assess the influence of feldspar composition and CO_2 conditions on the resultant water. Although the ion-exchange model is widely accepted, silicate weathering, or a combined model, are alternative possibilities to explain the generation of such water compositions. The three models yield similar net results. However, there is a significant difference in the rate of CO_2 consumption and it appears that a closed CO_2-system is necessary to achieve the analysed water chemistry by anorthosite weathering.

1 INTRODUCTION

In many aquifers throughout the world it is common to observe a hydrogeochemical evolution pathway culminating in soft, sodium bicarbonate waters of high pH and low calcium concentrations. Calcium removal typically permits the solution of high concentrations of fluoride (if present in the aquifer matrix), which can have adverse health impacts when consumed in excess. Such waters appear to be unrelated to specific aquifer lithologies and are reported from for example, the British Lincolnshire Limestone (Edmunds 1981), a gabbro-anorthosite pluton in Ontario (Bottomley et al. 1990), Dutch Quaternary coastal sediments (Appelo & Postma 1996) and the Coal Measures of Eastern England (Banks 1997).

The explanation offered for these waters is often cation exchange (Appelo & Postma 1996, Bottomley et al. 1990, Edmunds 1981) following calcite dissolution, according to the following equations (where X is a cation exchange site):

$$CaCO_3 + H_2O + CO_2 = Ca^{2+} + 2HCO_3^- \quad (1)$$

$$Ca^{2+} + 2HCO_3^- + Na_2\text{-}X = 2Na^+ + 2HCO_3^- + Ca\text{-}X \quad (2)$$

Ca is thus removed from the system and the water remains undersaturated with respect to calcite. Calcite dissolution may continue until dissolved CO_2 has been consumed, at which stage the pH of the solution may increase significantly:

$$CaCO_3 + H_2O = Ca^{2+} + HCO_3^- + OH^- \quad (3)$$

Banks (1997) noted the coincidence of reducing conditions with Na-HCO_3 waters in the English Coal Measures and noted that sulphate reduction would tend to have a similar effect, introducing alkalinity to the system and driving calcite to saturation, thus removing Ca from the system.

$$SO_4^{2-} + 2CH_2O = H_2S + 2HCO_3^- \quad (4)$$

Na would continue to be introduced by, for example, feldspar weathering (i.e. equation 5).

During a survey in 1996 and 1997, the Norwegian Radiation Protection Authority (NRPA) and the Geological Survey of Norway (NGU) carried out a nation-wide survey of the groundwater quality in relatively shallow (typically < 100 m) Norwegian boreholes in Precambrian and Palaeozoic crystalline bedrocks. The analytical results from 1604 samples have been reported by Banks et al. (1998), and Frengstad et al. (2000a).

It was initially expected that many parameters would exhibit a clear correlation with lithology. In fact, only a relatively small number of parameters (including, e.g. radon, uranium and fluoride) exhibited such a lithological dependence. Many parameters, including pH and the major ions, were found to be controlled by more universal thermodynamic equilibria (e.g. the calcium carbonate system) and kinetic factors (Banks et al. 1998).

A significant number of waters were found to be of Na-HCO_3 type, with low Ca and high pH, in a range of lithologies. Such waters were also, to our initial surprise, found in Ca-rich mafic rocks and anorthosites, such as those of the Egersund complex (Duchesne & Michot 1987, Fig. 1).

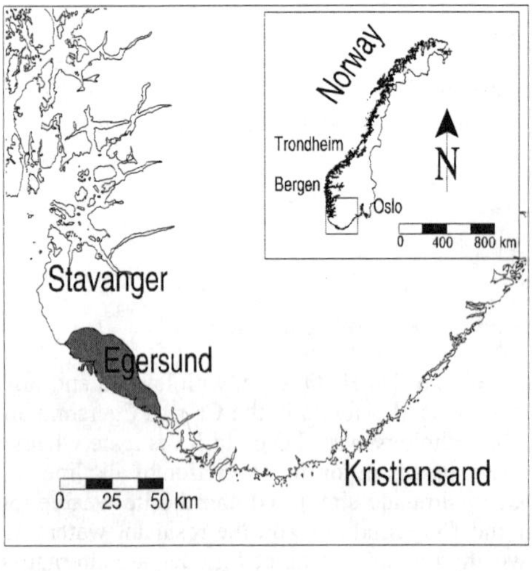

Figure 1. Map of southern Norway showing Egersund. The area of anorthositic rocks is outlined.

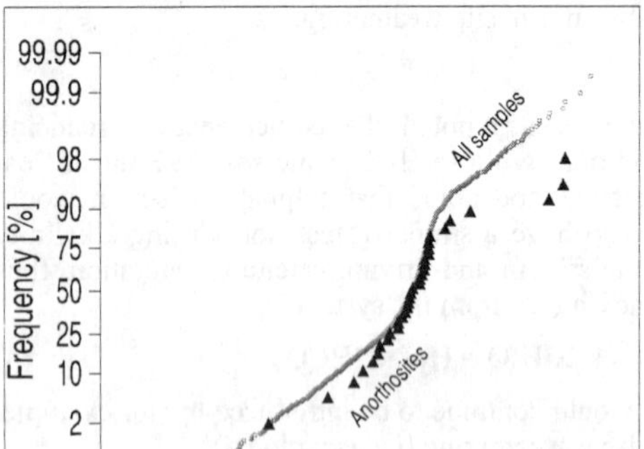

Figure 2. Cumulative frequency diagram showing the distribution of pH values for all 1604 analysed crystalline bedrock groundwaters from the Norwegian national survey (grey squares) and for groundwaters from Precambrian anorthosites only (black triangles - lithological group 93 of Banks et al. 1998). Note the probability scale on the y-axis.

Figure 2 shows cumulative frequency distribution curves for pH in all 1604 analysed crystalline bedrock groundwaters, together with a similar curve for Precambrian anorthosite groundwaters. Banks et al. (1998) interpreted the sharp kink at pH = 8.0 - 8.2 as indicative of a calcite precipitation buffering mechanism. Figure 3 shows the relationship of alkalinity, Ca and lithogenic Na (Na corrected for marine input on the basis of Cl concentrations) to pH. Note the low Ca and high Na concentrations of the highest pH waters, an observation confirmed in the plot of the Na/Ca ratio versus pH. The Na/Ca ratio undergoes a sharp increase at around pH 8.2, suggesting removal of Ca from the dissolved phase, either by ion-exchange or calcite precipitation.

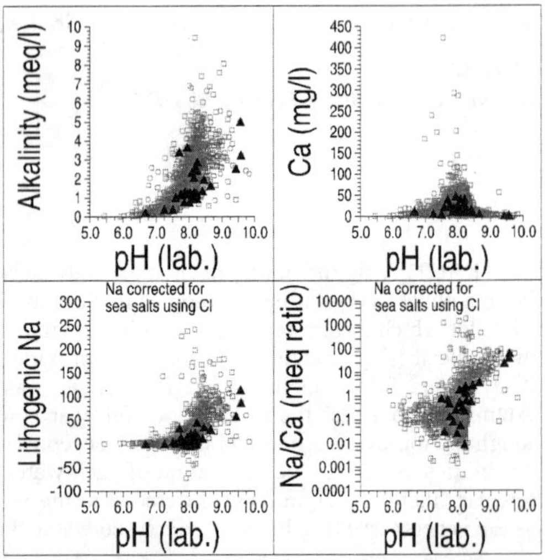

Figure 3. x-y diagrams showing the relationship of (a) alkalinity, (b) Ca, (c) lithogenic Na, (d) lithogenic Na:Ca meq ratio to pH for all 1604 analysed crystalline bedrock groundwaters from the Norwegian national survey (grey squares) and for groundwaters from Precambrian anorthosites only (black triangles). Lithogenic Na is Na corrected for input of marine salts and is calculated by the formula Na-(Cl*10600/19000).

2 EGERSUND ANORTHOSITE AQUIFER

Egersund is situated in the county of Rogaland, southeast of Stavanger on the southwestern coast of Norway (Fig.1). The Precambrian (950-900 Ma BP) Egersund anorthositic province is part of the Rogaland igneous complex which comprises three large anorthosite massifs, two smaller leuconorite bodies, a layered leuconorite lopolith and three granitoid plutons. The Egersund-Ogna anorthosite massif has a monotonous rock composition of coarse-grained, homogeneous, 1-3 cm plagioclase crystals (An_{40-50}), with sub-ophitic aggregates and megacrysts of plagioclase and high-Al orthopyroxene megacrysts (Ashwal 1993). They exhibit very little Quaternary cover, are relatively fresh at outcrop (having been recently glacially scoured) and contain little calcite.

As the Egersund anorthosite complex contained a wide range of differing hydrogeochemical waters (see Fig. 3), including high-pH Na-HCO_3, 9 of the wells sampled in the National Survey were revisited in May 1997 by NGU hydrochemists and sampled more rigorously. Sampling and analytical methods are documented in detail by Frengstad et al. (2000b).

Low Eh measurements and the presence of H_2S were not typically noted in the field, suggesting that significantly reducing conditions are not common in these relatively shallow groundwaters. Selected analytical results are shown in Table 1, from which the same trends may be identified as seen in Figure 3.

Table 1. Selected parameters for the 9 samples collected from boreholes in the Egersund anorthosites in 1997, sorted according to increasing pH. Alk = alkalinity (titrated to pH 4.3); given as meq/l. Element concentrations are given as mg/l, except for Sr, Al, B, and Li, given as µg/l.

	1	2	3	4	5	6	7	8	9
pH field	5.40	5.59	6.43	7.16	7.90	8.00	8.20	9.72	9.93
Alk	0.5	0.2	0.3	3.2	2.2	3.0	1.0	3.0	3.4
Na	12.3	20.5	27.9	52.9	40.7	28.0	13.1	83.2	103
Ca	12.3	7.86	6.64	37.7	40.9	21.2	18.1	2.49	1.50
Mg	3.16	4.02	2.49	7.18	16.3	19.1	3.52	0.97	0.33
K	0.97	2.16	0.25	7.53	3.11	3.55	0.57	0.63	<0.5
Cl	24.5	34.2	41.5	53.5	18.1	16.9	22.2	30.9	30.5
SO_4	9.55	13.9	11.8	15.8	33.9	27.1	7.02	12.1	18.4
F	<0.05	<0.05	<0.05	0.06	0.06	0.05	0.05	0.07	0.13
Si	4.19	2.06	4.26	5.01	3.41	14.4	3.68	4.16	5.61
Sr *	63.2	57.1	42.6	84.7	287	179	78	23.5	12.9
Al *	49.4	64.8	18.6	9.1	24.8	5.49	5.83	12.1	23.3
B *	13.8	12.3	11.0	25.5	25.7	17.0	8.48	68.3	449
Li *	0.06	0.06	0.24	0.74	5.58	3.40	1.03	1.02	0.45

* values in µg/l

From Figures 2 and 3, anorthosite groundwaters may be characterised as follows:
- a wide range of pH values, from around 6 to 10.
- a bias towards pH values of ~ 8.0-8.2, possibly suggesting a pH buffering system.
- low Ca and high Na concentrations in waters above pH ~ 8.2.
- alkalinity concentrations of up to around 6 meq/l in the highest pH waters.
- Ca concentrations of up to 70 mg/l, typically at pH ~ 8.
- lithogenic Na concentrations of some 30-50 mg/l up to pH ~ 8.2, with maximum concentrations of some 100 mg/l at the highest pH values.

Additionally, Table 1 suggests that:
- high-pH anorthosite waters have elevated B and F.
- Al concentrations are (as one might expect, from the solubility of the amphoteric Al hydroxide) elevated in the low and high pH waters.
- Sr, Mg and Li exhibit low concentrations in the high-pH, Ca-poor waters.

3 OBJECTIVES

Due to their mineralogical simplicity, this paper will consider the Egersund anorthosites as a case study to elucidate possible mechanisms for the evolution of Na-rich, Ca-poor alkaline groundwaters in silicate rocks. It is proposed that silicate weathering pathways alone may provide an adequate explanation. In simplified form, such a model can be described by the following equations, starting with the weathering of labradorite feldspar by CO_2-charged rainwater:

$$2NaCaAl_3Si_5O_{16} + 6CO_2 + 9H_2O = 2Ca^{2+} + 2Na^+ + 6HCO_3^- + 4SiO_2 + 3Al_2Si_2O_5(OH)_4 \quad (5)$$

Na, Ca and bicarbonate alkalinity are released, and pH is elevated. It is important to note that an excess of alkalinity over Ca is released (in terms of milliequivalents). When the ion concentrations and pH are sufficiently high, calcite saturation occurs and calcite is precipitated (releasing protons and thus providing a pH buffer):

$$Ca^{2+} + HCO_3^- = CaCO_3 + H^+ \quad (6)$$

The presence of orthopyroxene $(Mg,Fe)SiO_3$ also provides another non-calcium balanced source of alkalinity which serves to drive the removal of Ca. Thus the remaining solution is dominated by Na (from feldspar weathering) and HCO_3.

In a closed system, where CO_2 has been consumed, the pH is driven rapidly up and Ca can begin to accumulate in the water once more:

$$2NaCaAl_3Si_5O_{16} + 9H_2O = 2Ca^{2+} + 2Na^+ + 6OH^- + 4SiO_2 + 3Al_2Si_2O_5(OH)_4 \quad (7)$$

4 ALTERNATIVE MODELS

The opposing models for generation of $Na\text{-}HCO_3$ waters can be summarised as single equations.

4.1 Model 1

The calcite dissolution / ion-exchange model (equations 1 and 2) is summarised as:

$$CaCO_3 + H_2O + CO_2 + Na_2\text{-}X = Ca\text{-}X + 2Na^+ + 2HCO_3^- \quad (8)$$

where X indicates the exchanger.

4.2 Model 2

The silicate weathering model (equations 5 and 6) is summarised, for labradorite/andesine feldspar, as:

$$4NaCaAl_3Si_5O_{16} + 8CO_2 + 14H_2O = 4CaCO_3 + 4Na^+ + 4HCO_3^- + 8SiO_2 + 6Al_2Si_2O_5(OH)_4 \quad (9a)$$

In a closed system (equation 7), the net reaction will tend towards complete consumption of CO_2 as outlined in equations 9b-9e. The ratio of Ca and Na in the final solution will depend on the amount of weathering which has taken place in the absence of CO_2:

$$6NaCaAl_3Si_5O_{16} + 8CO_2 + 21H_2O = 6CaCO_3 + 6Na^+ + 2HCO_3^- + 4OH^- + 12SiO_2 + 9Al_2Si_2O_5(OH)_4 \quad (9b)$$

$$8NaCaAl_3Si_5O_{16} + 8CO_2 + 28H_2O = 8CaCO_3 + 8Na^+ + 8OH^- + 16SiO_2 + 12Al_2Si_2O_5(OH)_4 \quad (9c)$$

$$12NaCaAl_3Si_5O_{16} + 8CO_2 + 46H_2O = 8CaCO_3 + 4Ca^{2+} + 12Na^+ + 20OH^- + 24SiO_2 + 18Al_2Si_2O_5(OH)_4 \quad (9d)$$

$$20NaCaAl_3Si_5O_{16} + 8CO_2 + 82H_2O = 8CaCO_3 + 12Ca^{2+} + 20Na^+ + 44OH^- + 40SiO_2 + 30Al_2Si_2O_5(OH)_4 \quad (9e)$$

Note that equations 9a-9e have been normalised to a fixed amount of available CO_2 (8 moles). The system progresses from a circum-neutral Na-HCO_3 water (9a) in an open CO_2 system (or a closed system prior to CO_2 consumption) to an alkaline Na-HCO_3 water (9b) in a closed system. Theoretically, the closed system could further develop (9c, 9d) to highly alkaline Ca-dominated waters (9e). This is not observed in our groundwater samples, but has been noted in deep groundwaters of the Canadian Shield (Bottomley et al. 1990).

4.3 Model 3

Finally, a third model may be proposed, where silicate weathering by CO_2 (equation 5) is followed by ion exchange (equation 2):

$$2NaCaAl_3Si_5O_{16} + 6CO_2 + 9H_2O + 2Na_2\text{-}X = 2Ca\text{-}X + 6Na^+ + 6HCO_3^- + 4SiO_2 + 3Al_2Si_2O_5(OH)_4 \quad (10)$$

5 PHREEQC MODELLING

Weathering of a monomineralic silicate system, namely an anorthosite comprising only plagioclase feldspar, was simulated using the code PHREEQC (Parkhurst 1995). The input file essentially specified that a given amount of plagioclase (1.5×10^{-3} moles) was dissolved in CO_2-charged pure water in 30 equal steps. The phases calcite, chalcedony, kaolinite and gibbsite were permitted to precipitate when they became oversaturated. Saturation with respect to feldspar phases was not considered to be a limiting factor, due to kinetic factors and the uncertainty around equilibrium conditions for feldspar dissolution/precipitation (Strömberg & Banwart 1999). Other specifications include the initial concentration of CO_2, whether the system was open or closed with respect to CO_2, and the temperature (7°C, typical for southern Norway). The modelling procedure was run for a closed system with $pCO_2 = 10^{-1.5}$ atm. using five idealised plagioclase compositions, namely:

- Albite, $Na_2Al_2Si_6O_{16}$, simulated by adding Na, Ca, Al, and SiO_2 in the ratios 2:0:2:6
- Oligoclase, $Na_{1.5}Ca_{0.5}Al_{2.5}Si_{5.5}O_{16}$, (Na, Ca, Al, and SiO_2 in the ratios 1.5:0.5:2.5:5.5)
- Labradorite, $NaCaAl_3Si_5O_{16}$, (Na, Ca, Al, and SiO_2 in the ratios 1:1:3:5)
- Bytownite, $Na_{0.5}Ca_{1.5}Al_{3.5}Si_{4.5}O_{16}$, (Na, Ca, Al, and SiO_2 in the ratios 0.5:1.5:3.5:4.5)
- Anorthite, $Ca_2Al_4Si_4O_{16}$, (Na, Ca, Al, and SiO_2 in the ratios 0:2:4:4)

In all cases, the feldspar formula have been normalised such that the stoichiometry of the defined reactions is analogous for all five model runs. The results of the labradorite modelling are shown graphically in Figure 4.

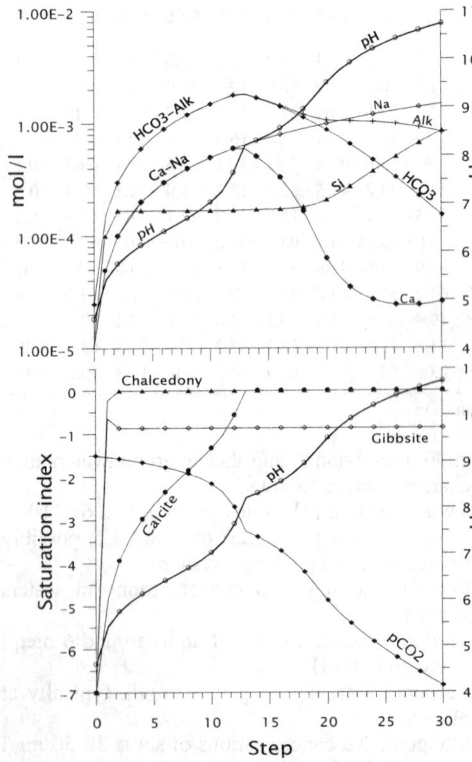

Figure 4. Modelled evolution of groundwater chemistry, using PHREEQC, by dissolving $1.5*10^{-3}$ moles labradorite in water at 7°C and initial $pCO_2 = 10^{-1.5}$ atm. (closed CO_2 system). Precipitation of calcite, kaolinite, gibbsite and chalcedony are permitted if oversaturated. Upper diagram shows the evolution of total element concentrations, and lower diagram shows saturation indices. Note that alkalinity is defined as the milliequivalent sum of HCO_3^-, CO_3^{2-} and OH^-.

Additionally, the weathering of labradorite was simulated in: i) open systems of $pCO_2 = 10^{-3.5}$ (atmospheric pressure), $10^{-2.5}$ and $10^{-1.5}$ atm. (soil gas in soils with high biological activity). ii) closed systems of initial $pCO_2 = 10^{-3.5}$ and $10^{-2.5}$ atm. The results from atmospheric pCO_2 are shown in Figure 5.

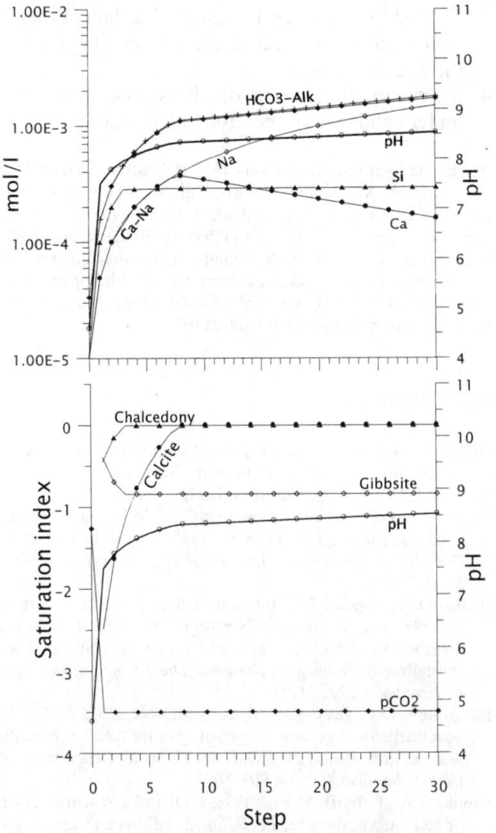

Figure 5. Modelled evolution of groundwater chemistry, using PHREEQC, by dissolving $1.5*10^{-3}$ moles of labradorite in water at 7°C in an open system with $pCO_2 = 10^{-3.5}$ atm. Precipitation of calcite, kaolinite, gibbsite and chalcedony are permitted if oversaturated. Upper diagram shows the evolution of total element concentrations, and lower diagram shows saturation indices. See comments on alkalinity in caption for Figure 4.

In closed systems three phases of groundwater evolution are predicted to occur:

- Phase I - accumulation of Na, Ca, and alkalinity rise in solution. Rapid increase of pH. Corresponds to equation 5.
- Phase II - removal of Ca and alkalinity decline due to calcite precipitation; continued accumulation of Na, buffering effect on pH. Equation 6.
- Phase III - reaccumulation of Ca in system, increase in alkalinity, and in pH (8.5 – 11).

The composition of feldspar is important for determining the evolution of water chemistry in a closed system. In particular, at which stage Ca is removed by calcite precipitation and the quantities of Na and Ca released during Phases I and II. In the case of labradorite, with $pCO_2 = 10^{-1.5}$ atm., about 35 mg/l Na is released during Phases I and II, with only some 24 mg/l Ca released during Phase I. The weathering of oligoclase results in more Na and less Ca being released, while the weathering of bytownite has the reverse effect. In the case of bytownite, although Ca accumulates faster in the water in Phase I, it is also removed faster during Phase II, and the beginning of Phase III may be observed in the final stages of the modelling run. In all cases, the pH buffer lies between 8.0 and 8.5.

Modelling of open systems indicates that initial partial pressure of CO_2 has some effect on the evolution of groundwater systems, particularly in the amounts of Na and Ca released during the consumption of CO_2. A rather low pCO_2 of around 10^{-3} atm is required to attain a pH value of ~ 8.2 (characteristic of many groundwaters from Norwegian crystalline bedrock). Such open silicate weathering systems can result in Ca removal by calcite precipitation (Fig. 5) and Na-HCO_3 rich waters, but do not typically achieve the highest pH values observed in actual waters from Norwegian anorthosites.

The modelled open system with $pCO_2 = 10^{-1.5}$ atm. attains pH values of around 7, and can accumulate greater amounts of dissolved Ca (~ 100 mg/l), Na (~ 60 mg/l) and alkalinity (~ 7 meq/l), before calcite saturation is achieved.

6 COMPARISON WITH EMPIRICAL DATA

The data in Table 1 suggest that Mg and Sr are also removed from the high-pH, Ca-poor waters, which may be explained either by co-precipitation with Ca as a carbonate phase, or by cation exchange. However, Li is also depleted in the high-pH waters, which is not readily explained by carbonate precipitation, but may indicate ion exchange. In the larger national data-set (n=1604), Mg and Sr exhibit a pH-related trend similar to that of Ca, while Li shows no clear tendency to lower concentrations in high-pH waters (Frengstad et al., in prep). The evidence from Li must thus be regarded as inconclusive.

The apparent pH buffer at 8.2 (Fig. 2) can be explained either by a closed-CO_2 silicate weathering system or by an open system at a low pCO_2 of some 10^{-3} atm., followed by cation exchange (Fig. 5).

The actual data (Fig. 3 and Table 1) indicate the accumulation of some 30-50 mg/l lithogenic Na in groundwaters in anorthosite rocks up to, and across the hydrochemical transition (Phases I and II). Modelling suggests that a pCO_2 of around 10^{-2} - $10^{-1.5}$ atm. is required to generate these quantities by silicate weathering in a closed system. Such concentrations could be generated using an open system model, even with a relatively low pCO_2 of $10^{-3.5}$ atm.

The observed alkalinities of up to 6 meq/l and Ca concentrations of up to 70 mg/l in anorthosite groundwaters cannot be explained by the closed CO_2 system models alone (Figs 4, 5). Some component of open system CO_2 weathering is required to generate such concentrations (Fig. 5).

It must be remembered that the anorthosite groundwater samples are not derived from a groundwater evolution pathway in a single homogeneous hydrological system. In reality:

- the anorthosite massif is *geochemically* heterogeneous, ranging from mono-mineralic plagioclase rocks (of the type simulated using PHREEQC) to more complex multi-mineralic assemblages including mafic and accessory phases;
- the boreholes possibly draw water from more than one fracture system. Such systems are likely to be *hydraulically* and *hydrogeochemically* heterogeneous;
- groundwater evolution may have been subject to water-rock interaction under both open (e.g. in superficial Quaternary deposits with significant calcite content) or closed (deep fracture) conditions; and
- the sampled waters may have been recharged in vegetation-poor (highland anorthosite outcrop), low pCO_2 conditions, or lowland soil-covered areas with high pCO_2 conditions.

7 CONCLUSIONS

The existing evidence does not clearly rule out either cation exchange or pure silicate weathering as a mechanism for generation of high-pH, $Na\text{-}HCO_3$ waters in crystalline silicate rocks (especially anorthosites). The hydrogeochemical evidence presented in this paper suggests two plausible scenarios:

1 a pure silicate weathering pathway under closed CO_2 conditions, but with a high initial pCO_2 of maybe $10^{-1.5}$ atm. (required to generate the observed quantities of dissolved ions). This scenario is favoured for its simplicity. Nevertheless, contemporary surveys suggest that, in southern Norway, the partial pressure of CO_2 in soil gas is unlikely to exceed 10^{-2} atm. in the anorthosite recharge waters (Brook et al. 1983).

2 an open silicate (and carbonate) weathering system with a more realistic pCO_2 of some 10^{-3} atm. is adequate to generate $Na\text{-}HCO_3$ waters up to pH ~ 8.2 (where a calcite precipitation buffer controls pH). Such a model also generates quantities of Ca, HCO_3 and Na more closely resembling actually observed concentrations. However, to generate pH values in excess of ~ 8.5 it appears necessary either to invoke cation exchange to remove Ca and thus negate the calcite precipitation buffer, or subsequent system closure.

Four lines of evidence will be pursued in future work, in order to identify the most appropriate model:

1 measurements of *in situ* pCO_2.

2 the identification of recent secondary carbonates or the identification of mineral phases which might act as cation exchangers.

3 ^{13}C and ^{14}C analyses of dissolved carbon phases in water and of any carbonate phases identified in the aquifer system.

4 laboratory simulations of plagioclase weathering under conditions of controlled CO_2 access.

Acknowledgement: U. Siewers (BGR) kindly provided ICP-MS analyses. We would like to thank H. Skarphagen, A.K. Skrede, J.R. Krog, A. Dagestad and J. Cramer (NGU) for support and assistance. Y. Dutova (TPU), B. Heidenstrøm (UiO), B. Aarrestad (NTNU) and P. Shand (BGS) provided stimulating discussions. The Sedatives® were responsible for an excellent soundtrack. EU's "Leonardo da Vinci" program provided funds for Bjørn Frengstad to work at BGS.

REFERENCES

Appelo, C.A.J. & Postma, D. 1996. *Geochemistry, groundwater and pollution 3rd ed.* Rotterdam: Balkema.

Ashwal, L.D. 1993. Anorthosites. Berlin: Springer-Verlag.

Banks, D. 1997. Hydrogeochemistry of Millstone Grit and Coal Measures groundwaters, south Yorkshire and north Derbyshire, UK. *Quarterly Journal of Engineering Geology* 30: 237-256.

Banks, D., Frengstad, B., Midtgård, A.K., Krog, J.R. & Strand, T. 1998. The chemistry of Norwegian groundwaters: I. The distribution of radon, major and minor elements in 1604 crystalline bedrock groundwaters. *The Science of the Total Environment* 222: 71-91.

Bottomley, D.J., Gascoyne, M. & Kamineni, D.C. 1990. The geochemistry, age and origin of groundwater in a mafic pluton, East Bull Lake, Ontario, Canada. *Geochemica et Cosmochemica Acta* 54: 993-1008.

Brook, G.A., Folkoff, M.E. & Box, E.O. 1983. A world model of soil carbon dioxide. *Earth Surface Processes and Landforms* 8: 79-88.

Duchesne & Michot 1987. The Rogaland intrusive masses: Introduction. In: C. Maijer & P. Padget (eds), The geology of southernmost Norway: an excursion guide. *Norges Geologiske Undersøkelse Special Publication* 1: 48-50.

Edmunds, W.M. 1981. Hydrogeochemical investigations. In Lloyd, J.W. (ed.), *Case Studies in Groundwater Resources Evaluation*. Oxford Science Publications: 87-112.

Frengstad, B. Skrede, A.K.M., Banks, D., Krog, J.R. & Siewers, U. 2000a. The chemistry of Norwegian groundwaters: III. The distribution of trace elements in 476 crystalline bedrock groundwaters, as analysed by ICP-MS techniques. *The Science of the Total Environment* 246: 21-40.

Frengstad, B., Banks, D., Skarphagen, H. & Siewers, U. 2000b. Hydrogeokjemien i filtrerte og ufiltrerte prøver fra 30 bergborede brønner i granitt, gneis og anortositt. [The *hydrogeochemistry of filtered and unfiltered samples from 30 drilled wells in granite, gneiss and anorthosite – in Norwegian]. Norges Geologiske Undersøkelse Report* 2000.051.

Frengstad, B., Banks, D. & Siewers, U. (in prep). The Chemistry of Norwegian Groundwaters: IV. The pH-Dependence of Element Concentrations in Crystalline Bedrock Groundwaters. Submitted to the Science of the Total Environment.

Parkhurst, D. 1995. User's guide to PHREEQC – A computer program for speciation, reactive path, advective transport and inverse geochemical calculations. US Geological Survey, Lakewood, Colorado.

Strömberg, B. & Banwart, S.A. 1999. Weathering kinetics of waste rock from the Aitik copper mine, Sweden; scale dependent rate factors and pH controls in large column experiments. *Journal of Contaminant Hydrology* 39: 59-89.

Groundwater: Past Achievements and Future Challenges, Sililo et al. (eds) © 2000 Balkema, Rotterdam, ISBN 90 5809 159 7

Regional scale solute transport modelling of a Permo-Triassic sandstone aquifer

B.V. Furlong, J.H. Tellam & R. Mackay
School of Earth Sciences, University of Birmingham, UK

J.A. Ingram
Environment Agency, Warrington, UK

A.W. Herbert
Environmental Simulations International, Shrewsbury, UK

ABSTRACT: A previous study completed upon the Lower Mersey Basin Permo-Triassic Sandstone Aquifer of England gathered a substantial amount of hydrochemical data. Analysis of these data suggested that groundwater flow rates and directions within the aquifer system could be qualitatively deduced from the regional hydrochemical distribution. This study describes an attempt to develop that concept into a quantitative description of the regional hydrochemistry, linking the aquifer flow history to the observed present day hydrochemical distribution. To achieve this goal a three dimensional regional groundwater flow model has been developed. The results of the model show that the flow system is consistent with the gross features of the hydrochemical distributions, such as saline water ingress distances, modern recharge water distributions, and the occurrence of old saline groundwater (with the longest residence times). Work continues with particle tracking, reactive transport flow path modelling and explicit representation of each sampled well.

1 INTRODUCTION

Aquifers often display regional scale patterns in groundwater chemistry. These patterns are products of the flow system, and although they are often taken into account qualitatively when developing conceptual models, they are rarely interpreted quantitatively. The object of completing a quantitative interpretation i.e. historically fitting a flow model not just against aquifer heads and flows but also against hydrochemical distributions, are: 1) more reliance can be placed on the conceptual model as it has been tested against a wider range of data; 2) the developed model can be used for water quality predictions; 3) the regional scale solute transport parameters obtained may be useful as first estimates for similar aquifers where regional hydrochemical patterns are not so well developed.

The aim of the work described here is to develop a regional groundwater flow model, testing it against groundwater head, groundwater flow and hydrochemical distribution data. The aquifer described is the Lower Mersey Basin Permo-Traissic Sandstone aquifer in NW England (Fig. 1). It has been the subject of previous investigation and there exists a regional groundwater flow model and qualitative interpretation of the regional hydrochemistry (University of Birmingham 1981, Tellam 1994).

2 HYDROGEOLOGICAL BACKGROUND

The Lower Mersey Basin consists of two main aquifer units: the Permian Collyhurst Sandstone Formation and the Triassic Sherwood Sandstone Group. The sandstones dip south at 5° and are up to 500m thick. The base of the aquifer is limited either by the underlying Manchester Marl or Bold Formation which separates the sandstones in the east of the area or by a Carboniferous sequence dominated by mudstones (Fig. 1). Evaporite derived saline groundwater is present in the lower parts of the sandstone. Below the Mersey and in central and eastern areas the saline interface is less than 50m from the ground surface (Tellam et al. 1986). The aquifer is overlain by the Mercia Mudstone Group in the south and by highly heterogeneous, vertically variable, Quaternary deposits across most of the area (Fig. 1).

The aquifer is bounded to the south by the Mersey Estuary and Manchester Ship Canal, to the north by the Carboniferous, to the east by a groundwater divide and to the west by a fault where transmissivity is reduced. Regional modelling by the University of Birmingham (1981) and Environmental Simulations Limited (1998) indicate that the upper-most portion of the aquifer contributes the majority of the transmissivity, reflecting the greater frequency of fractures at shallow depths in the sandstone (University of Birmingham 1981, Campbell 1986).

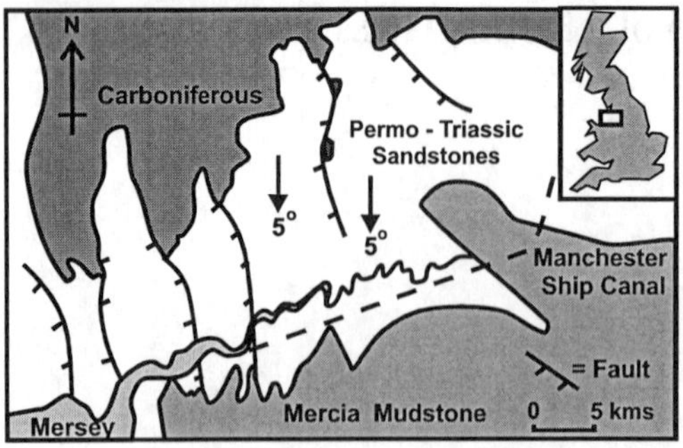

Figure 1. Schematic geological map. Quaternary deposits not shown.

Accurate measurements of specific yield are hindered by the lack of piezometers (Rushton & Howard 1982). As a result the specific yields are often substantially less than the measured porosity (Allen et al. 1999). Modelling indicates a regional specific yield of 0.14 and a confined storage coefficient of 0.007 (University of Birmingham 1981).

An extensive sampling program completed as part of the University of Birmingham (1981) study provides a "snap shot" of the hydrochemistry throughout the aquifer. The study used all accessible industrial, public and private boreholes across the region. The hydrochemistry of the groundwater was examined in detail and reported by Tellam (1983), the work culminated in a regional pattern of hydrochemistry based upon a number of water types (Tellam 1994) (Fig. 2, Fig. 3). The hydrochemical distribution correlates strongly with the broad patterns of flow in the aquifer. Recently recharged water (Type 1, Fig. 3) can be found in known areas of recharge, in contrast to older fresh groundwater (Type 4 and 5) and saline groundwater (Type S1) located in areas of the aquifer with low hydraulic gradients. The effect of modern abstraction is reflected by saline intrusion (Type S2) and recently induced recharge through the Quaternary deposits (Type 2).

3 PROJECT APPROACH

The observed hydrochemistry at a borehole results from sampling and historical processes. The hydrochemistry at any point in the aquifer is the product of two historical factors: the time of recharge (in relation to anthropogenic pollution) and subsequent flow and hydrochemical evolution to the point of sampling, dependent upon the period of residence of the groundwater and interaction with the aquifer.

The effect of sampling is then imposed upon the existing hydrochemical distribution. A bulk sample obtained from a borehole, as opposed to a piezometer, will mix groundwater within the hole removing any stratification present and altering the hydro-

Type	Description	Characteristics	Predominant Cause
1	Recent Recharge Water	Calcite unsaturated	Low amount of calcite encountered
		High NO_3, SO_4	Fertiliser pollution
2	Mix of Type 4 and Type 1	Calcite saturated	Type 4 + recharge through Quaternary
		Low SO_4, NO_3	
3	Intermediate Isolated	Range of compositions	Possible isolation due to faulting.
		Intermediate Cl, SO_4, NO_3	
4	Old Confined Groundwater Under Quaternay Deposits	Calcite saturated	Calcite dissoltion
		Low Cl, SO_4, NO_3	Groundwaters are pre-anthropogenic pollution
5	Ion exchanged Type 4 Groundwater	High Na/Cl ratio	Type 4 waters with saline influence
		Low NO_3, SO_4	
S1	Saline Groundwater	Saline chemsitry	Evaporite derived saline groundwater
S2	Saline Groundwater	Saline chemsitry	Saline water ingress from Estuary and Canal

Figure 2. Hydrochemical water types (after Tellam 1994).

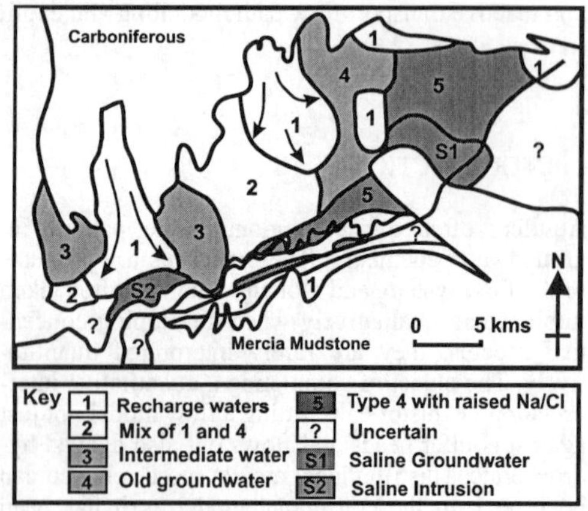

Figure 3. Regional Hydrochemical patterns observed within the aquifer (after Tellam 1994).

chemistry (Stagg et al. 1998). These factors need to be considered before comparison of predicted hydrochemistry with sampled groundwater takes place.

The issues concerning historical and sampling processes are to be addressed by using 3D models of groundwater flow and solute transport rates to provide quantitative information regarding the likely mix of groundwater types in a given borehole, the timing of recharge and the residence time of contributing groundwater. Particle tracking will be used to identify transport rates through the aquifer to the point of sampling. Where hydrochemical evolution of groundwater through the aquifer is important a separate 1D reactive flow code will be used to model changes in hydrochemistry along the flow line predicted by the 3D model.

The comparison of model and field data will lead to an iterative updating as the increased understanding of the system gained from the modelling work may alter the interpretation of the sampled hydrochemistry (Fig. 4). The potential end point of this work is a transport model of the aquifer system that has been developed with consideration for the regional hydrochemistry.

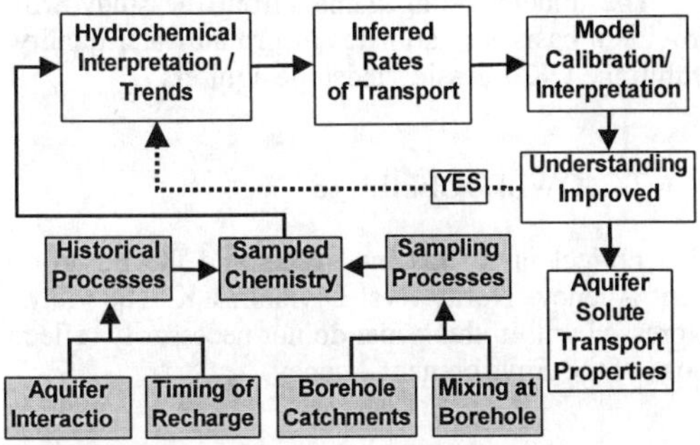

Figure 4. Flow diagram of the project approach.

4 GROUNDWATER MODELLING

A groundwater flow model of the aquifer system is under development using PARADIGM a 3D finite difference code (Environment Agency 1998). Before the effects of abstraction a large proportion of water in the system was lost either to spring flow or by recharge rejection. The ability to discharge water from the top surface of the model when groundwater heads were above the ground surface was of primary importance in selecting PARADIGM.

The modelling period stretched from before significant abstraction to the point of hydrochemical sampling at 1980, a duration of 133 years. The model area was divided into a uniform 250m grid giving a high resolution to the computed groundwater head distribution. A three layer model was developed, enabling the Quaternary deposits to be represented as a single unit, layer 1. The sandstone aquifer was divided into 2 layers reflecting the 200m zone of high conductivity in the upper most part of the aquifer (University of Birmingham 1981).

The University of Birmingham (1981) model provided the data sets for the recharge, abstraction and characterisation of the drift cover. A steady-state model incorporated only recharge and a constant head boundary to represent the Mersey Estuary. The transient model included 84 abstracting cells and river boundaries to represent flows from the Manchester Ship Canal, Carboniferous sandstone units and flow across the western model boundary. Hydrograph data for the modelling period consisted of 16 different observation wells across the region. The data provide good spatial coverage given the time at which measurement was made (pre-1980).

5 LINKING FLOW AND HYDROCHEMISTRY

The early results of the regional groundwater flow model confirm the link between the aquifer flow system and the pattern of hydrochemistry. Flow modelling shows that dewatering of the aquifer has resulted in the Mersey Estuary and the Manchester Ship Canal changing from groundwater sinks (steady-state model, Fig. 5) to sources (1980 model, Fig. 6) resulting in saline intrusion (Fig. 3). By assuming simple advective transport the approximate position of the saline front can be estimated from the simulated head data. Intrusion from the estuary is in the order of 2km reflecting the high connectivity to the sandstone aquifer, as opposed to the Manchester Ship Canal, where intrusion is estimated at less than 1 km. The estimated positions of the intrusion front compare well with field data (Fig. 3).

Dewatering has increased the unconfined area of the aquifer across the region. This process has been particularly significant in the central region where groundwater heads fall dramatically during the last 30 modelled years to approximately 30m below the steady-state groundwater levels. This has lowered the water table below the base of the Quaternary deposits resulting in recharge through the drift. This region coincides with the occurrence of Type 2 groundwater characterised by fertiliser pollution due to recent recharge.

The simulated groundwater heads in the eastern region show little temporal variation, as they are distal to the main areas of abstraction. The steady-state model predicts surface discharge and, although groundwater heads have declined with time, the aquifer is still confined with some surface discharge persisting. This accounts for the lack of recently recharged water in the east of the region (Fig. 3).

In the 50 years since significant fertiliser application the model predicts lateral transport away from recharge areas to be in the order of 1km. However, vertical transport of recent recharge through thin drift that surrounds the recharge area limits the use of this technique for tracing the progression of Type 1 groundwater through the aquifer.

Type 5 groundwater has an ion exchange signature indicating that the aquifer previously contained saline groundwater and that it has not yet achieved equilibrium with fresh groundwater (Fig. 3). Early work using the 1D reactive transport code PHREEQM (Nienhuis et al. 1994) suggest that approximately 20 pore volumes of fresh groundwater flushing are required before equilibrium is attained. The areas of Type 5 groundwater correlate strongly with the areas of surface discharge (where the longest residence times are predicted) throughout the modelling period. The model also predicts very low hydraulic gradients in the Type 5 areas reducing the rates of flushing through the aquifer (Fig. 5- 6).

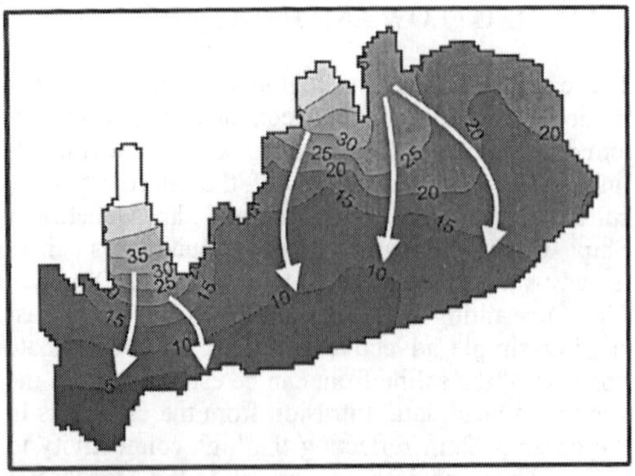

Figure 5. Modelled steady-state groundwater head contours (meters above sea level).

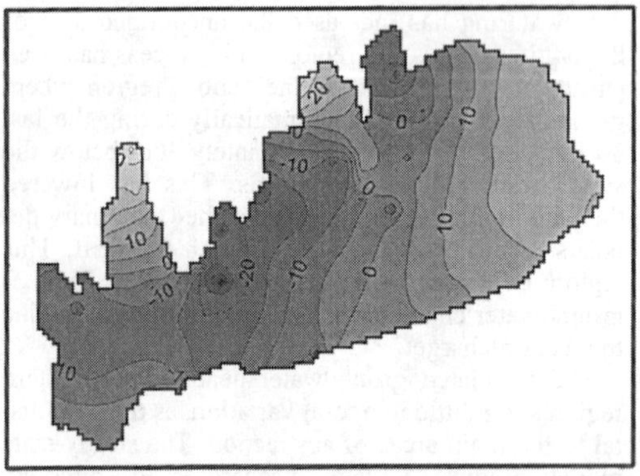

Figure 6. Modelled groundwater head contours for 1980 (meters above sea level).

The examples above are the initial stages of a quantitative description of the hydrochemistry. It is clear from this work that the hydrochemical patterns have identifiable processes that may be examined in detail through the use of computer models. Developing the flow model to allow particle tracking is the next step, facilitating a more comprehensive quantitative description of the aquifer system.

6 DISCUSSION

The hydrochemistry of the Lower Mersey Basin has resulted from a modern transient flow system following a prolonged period of flow under pseudo-steady-state conditions. A complete understanding of a "snapshot" of the hydrochemistry can only be achieved, if the pre and post anthropogenic history of the aquifer is modelled. Flow modelling of the aquifer has demonstrated that solute transport rates in the system can be tied to the hydrochemistry and provides the basis for quantitative analysis.

The next phase of the investigation is to carry out particle tracking analysis. From this work the timing of recharge, residence time of groundwater, borehole catchments and evolution of groundwater to the point of sampling will be understood, providing a quantitative interpretation of individual hydrochemical samples and the regional trends they form.

The understanding gained from the study will provide a basis for estimates of groundwater quality within the UK Triassic sandstone aquifers.

7 ACKNOWLEDGEMENTS

This project is funded by NERC and the Environment Agency, North West Division UK. The views expressed within this paper do not necessarily reflect those of the Environment Agency.

REFERENCES

Allen, D.J., Brewerton, L.J., Coleby, L.M., Gibbs, B.R., Lewis, M.A., MacDonald, A.M., Wagstaff, S.J., & A.T. Williams. 1999. The physical properties of major aquifers in England and Wales. *British Geological Survey Technical Report* WD/97/34. 312. Environment Agency R&D Publication 8.

Campbell, J.E. 1986. West Cheshire saline groundwater investigation (Phase 3). Unpublished Report.

Environment Agency. 1998. PARADIGM - A groundwater flow and transport code for the hydrogeological modelling of the faulted rocks of the Sellafield region.

Environmental Simulations Limited. 1998. Improvements to groundwater protection zone network: Landside model. Appendix 10 and 11. *Report Reference* 6061R2D1.

Nienhuis, P.R., Appelo, C.A.J. & A. Willemsen. 1994. PHREEQM Version: 24 July 1987. Updated Jan 10 1994. Free–University, Amsterdam, Netherlands.

Rushton, K.R., & K.W.F. Howard. 1982. The unreliablity of open observation boreholes in unconfined aquifer pumping tests. *Groundwater.* 20 (5) :546-550

Stagg, K. A., Tellam, J.H., Barrett, M.H. & D.N. Lerner. 1998. Hydrochemical variations with depth in a major UK aquifer: the fractured, high permeability Triassic sandstone. In Brahana, J.V., Eckstein, Y., Ongley, L.K., Schneider, R. & J.E. Moore. (ed.) *Gambling with groundwater. Proceedings volume of the joint conference of the I.A.H. and A.I.H.* Las Vegas. 1 :53-58.

Tellam, J.H. 1983. Hydrochemical Studies of the Lower Mersey Basin Permo-Triassic Sandstone Aquifer System. Unpublished. *Ph.D.* Thesis. University of Birmingham. 350.

Tellam, J.H. 1994. The groundwater chemistry of the Lower Mersey Basin Permo-Triassic Sandstone Aquifer system, U.K: 1980 and pre-industrialisation-urbanisation. *J. Hydrology.* 161 :287-325.

Tellam, J.H., Lloyd, J.W. & M. Walters. 1986. The morphology of a saline groundwater body: its investigation, description and possible explanation. *J. Hydrology.* 83 :1 –21.

University of Birmingham. 1981. Saline Groundwater Investigation Phase 1 – Lower Mersey Basin. Unpublished. Final Summary Report to the North West Water Authority. 81.

Groundwater: Past Achievements and Future Challenges, Sililo et al. (eds) © 2000 Balkema, Rotterdam, ISBN 90 5809 159 7

Arsenic in groundwaters of the southeast of Tucumán Province, Argentina

M.G.García – *Centro de Investigaciones y Transferencia en Química Aplicada, Universidad Nacional de Tucumán, Argentina*

D.S.Fernández – *Servicio Geológico Minero Argentino, Delegación Tucumán (SEGEMAR), Argentina*

M.del Valle Hidalgo – *Centro de Investigaciones y Transferencia en Química Aplicada, Universidad Nacional de Tucumán, Argentina*

M.A.Blesa – *Universidad de Actividad Química, Comisión Nacional de Energía Atómica, Buenos Aires, Argentina*

ABSTRACT: The presence of high and intermediate concentrations of arsenic in groundwaters of the southeastern plain of Tucumán, NW Argentina, have been repeatedly reported. High As contents (>50 μg/L) in the phreatic waters are associated with the leaching of volcanic ashes distributed in the loess that covers the eastern part of the plain. Intermediate levels of As (10-50 μg/L) in the confined groundwaters of the plain could be correlated with the thermal anomalies produced by a basaltic layer located 7 km deep in the south of the province.

1 INTRODUCTION

The Salí River Hydrogeological basin is located in the southeastern part of the Tucumán Province, NW Argentina (Fig. 1). The area, of about 4300 km^2, consists of an alluvial plain, which contains one of the most important groundwater reservoirs in the country.

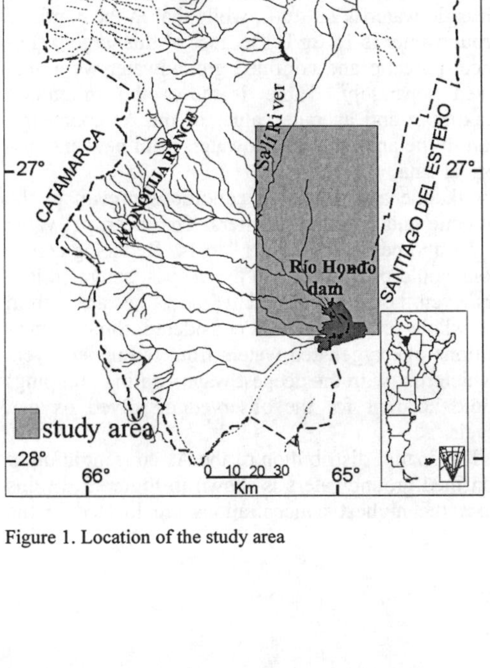

Figure 1. Location of the study area

Along the western and southern borders of the basin, the Aconquija range is developed. Metamorphic rocks and granitic intrusions form the nuclei of these mountains; cretaceous and tertiary sediments outcrop in the pedemontane area. The eastern border of the study area is determined by the state limit with the province of Santiago del Estero, while the northern limit corresponds to a deep anticline that acts as a groundwater dividing line (Tineo et al. 1998).

Weather is hot, with an extremely wet summer. The mean annual temperature is 19°C, and rainfall varies between 600 to 1000 mm/year, decreasing to the east.

The most important economic activity of the area is agriculture, and the chief crop is sugar cane. Grains are farmed along the eastern border, while citrus and vegetables represent the chief crops in the pedemontane zone.

Because of the serious organic pollution of the Salí River waters and its tributaries (Chaile et al. 1999, Perondi et al. 1999; Hidalgo et al. 1998), water supply is carried out almost exclusively by groundwater exploitation. The depth of the boreholes varies widely, from 80 to more than 400 metres below surface (mbs). Domestic wells, with depths ranging from 6 to 18 metres are common, especially in the west, where the water table is closer to the surface.

Since many decades ago, inhabitants of the zone have been affected with skin diseases, which were associated with the presence of high arsenic levels in drinking water (Bergoglio 1963, Biagini et al. 1974, Biagini & Salvador 1983, Hopenhayn-Rich et al. 1996, Concha et al. 1998). In groundwaters of the east of Tucumán and the Chacopampean Plain, arsenic occurs in concentrations exceeding both the provisional guideline concentration for drinking water set by the World Health Organisation (10 µg/L; WHO 1993) and the Argentinean standard limit (50 µg/L, CAA 1994), (SIPROSA 1982/1986, Nicolli et al. 1989, Merino 1999, Smedley et al. 1998, Nicolli et al. 1997), but there is still uncertainty about the sources and the spatial distribution of this element in the provincial territory.

The objective of this work is to describe the spatial distribution of arsenic concentrations in the phreatic and confined groundwater of the Salí River Hydrogeological basin and to discuss its probable sources.

2 HYDROGEOLOGICAL SETTING

The Salí River Hydrogeological basin is a structural depression, which was filled with quaternary and tertiary sediments reaching a thickness of about 3000 metres (Mon & Vergara 1987, YPF profiles 1996). The upper 30 m have been differentiated as Tucumán Formation (Bonaparte & Bobovniqov 1974), which covers the eastern part of the alluvial plain. The deposits consist of partially reworked yellow brown loess, with a relatively low content of calcium carbonate and quartz and a widespread presence of volcanic glass (Sayago 1995). They alternate with some fluvial sandy channels. The western and central parts of the basin are occupied by the Salí River flood plain deposits, which consist of sands and conglomerates. Basal quaternary sediments are constituted by highly permeable gravels and conglomerates that alternate with layers of silts and clays. The thickness of these deposits range from 70 to more than 200 m, according with their position in the basin (Tineo et al. 1998). Sediment grain size decreases to the east and south, where silts and clays predominate.

Rainfall infiltrates in the pedemontane area (the recharge area of the multi-layered aquifer), and flows in the direction of the regional slope: NW–SE.

In the southern part of the basin, geothermal groundwater is exploited with temperatures that oscillate between 24 and 51°C; the higher values were found near the Lamadrid village, where the deepest wells are located.

3 EXPERIMENTAL

Both new and historical data were considered for this study. The collection of new data was carried out during 1998 and included the sampling of 28 deep and 4 domestic wells. Historical data was supplied by SIPROSA and included depth and arsenic concentrations, measured in more than 40 deep boreholes and in more than 15 domestic wells during 1982 and 1986, as part of a regional project. No other chemical information was available. Historical As data were determined by the silver diethyldithiocarbamate method (APHA et al. 1989) and the analytical accuracy was not specified.

Water samples collected for this study were filtered using 0.45 µm membrane filters and acidified to pH 2 with pure HNO_3. Total As concentrations were determined by graphite furnace AAS (Hitachi Z-5000), following the methodology proposed by APHA et al. (1989). The estimated error is about ±5%.

Dissolved oxygen and pH data were used to calculate the redox conditions of the phreatic and confined aquifers. This, as well as the assessment of dissolved As speciation, was performed using the computer speciation program Mineql+ version 3.0 (Schecher & McAvoy 1994)

Confined and phreatic groundwater sampling points were georeferenced with the GIS MapInfo 4.0, and the As spatial distribution maps were obtained using Vertical Mapper version 1.5 (Northwood 1996). The method of interpolation selected was the Inverse Distance Weighting Interpolation which calculates a value for each grid node by examining surrounding data points that lie within a user defined search radius. The node value is calculated by averaging the weighted sum of all the points.

4 RESULTS

Both new and historical arsenic levels measured in the phreatic water are much higher than those determined in confined groundwater. The total mean As content (considering historical and new data) in the phreatic water is 83 µg/L, while that in the confined groundwater is 15 µg/L. Because of this large difference, phreatic and confined groundwater were analysed separately. Table 1 shows the minimum, maximum and average values of the As concentration in the analysed groundwater, from new and historical data.

Alkaline and oxidising conditions prevail in the phreatic and confined aquifers. The mean pH value is 7.8 and pe ranges from 11 to 14. Reducing conditions could prevail only in the deepest saturated layers, even though the calculated pe values are high in the whole set of boreholes considered. Small contributions of oxygenated waters from the upper layers (which mix with the deepest waters during pumping) could account for the observed dissolved oxygen levels.

The spatial distribution of the As concentration in confined groundwaters is shown in Figure 2. In this case, the highest concentrations are located in the south, near the cities of Graneros and Lamadrid. This zone is typically hydrothermal, with temperatures ranging from 30° to more than 50°C (Mon & Vergara 1987, Vergara et al. 1998). Lower concentrations (<10 μg/L) are found in the west, all along the pedemontane area. In the northeastern part of the basin, the concentration of arsenic increases, reaching values that range from 20 to 25 μg/L.

Table 1. As concentration in μg/L.

		Minimum	Maximum	Average
New data 1998	Confined	1	43	12
	Phreatic	8	67	50
Historical Data 1982/1986	Confined	4	45	17
	Phreatic	8	256	97

Figure 3 shows the spatial distribution of the As content in the phreatic water. Despite the scarcity of data points, it can be seen that the higher contents, with values over than 250 μg/l, are found in the northeastern part of the study area. Towards the west and southwest, the contents seem to decrease to less than 10 μg/L, but more sampling points are needed to confirm the trend.

5 GROUNDWATER GEOCHEMISTRY

The spatial distribution and the range of concentrations of As are different in the phreatic and confined aquifers. The mean concentration of As in the phreatic aquifer is almost five times higher than the mean concentration determined in the confined aquifers (see Table 1).

5.1 *Occurrence of arsenic in phreatic groundwater*

Nicolli et al. (1989, 1997) have reported high values of As concentration in the phreatic waters of the Pampean Plain, and on the basis of geochemical analysis of the sediments and waters, they concluded that the main source of As is the volcanic glass that is present in the loess, which covers the entire plain.

The origin of the high As contents in the phreatic groundwater of the east of Tucumán is probably the same, but with some differences. The mean As content in the plain of Tucumán is noticeably lower, almost half of the mean As concentration determined in the phreatic waters of the Pampean Plain. The lower content of volcanic glass registered in the loess of Tucumán (Sayago 1995) could account for this difference. According to geochemical data supplied by Camino (1988), the composition of the loess in Tucumán corresponds to that of an andesite, while the composition of the loess in the Pampean Plain is equivalent to that of a dacite (Nicolli et al. 1989).

The recharge of the phreatic aquifer in the plain is produced by the infiltration of rain, and the water-table level is strongly dependent on the precipitation – evaporation balance (Figueroa et al. 1996). During infiltration, water dissolves the high-As volcanic glass and releases As, which finally accumulates in the phreatic aquifer. Towards the east, the higher concentrations registered are probably a consequence of the combination of two factors. On the one hand, the finer sediment grain size and the low slope result in a longer residence time of the water in the aquifer, which enhances the weathering of the volcanic glass; and on the other, the drier and warmer conditions promote evaporative concentration of As in the water.

5.2 *Occurrence of arsenic in deep groundwater*

On the basis of chemical relationships and lithology, confined groundwater has been classified into two groups: that located in predominantly loess-like deposits (in the east) and that developed in alluvial sands and gravels (in the west and south).

Groundwater of the first group shows a chemical behaviour similar to that described by Smedley et al. (1998) in the phreatic water of the Pampean Plain. As contents are associated with waters containing dominant alkali metals in the cation composition, and there is a positive correlation between As and HCO_3^- concentrations. Despite the experimental data scatter, clear trends are observed. The dissolution of As-rich volcanic glass distributed in the loess releases As, HCO_3^- and alkaline ions to the groundwater over the first 30 m depth, but the mixing with deeper low-As content groundwater, could produce the observed scatter of points.

Even though the majority of the sampled wells exhibit thermal anomalies, only those in the second group span an adequate range of depths and temperatures (measured at the well heads) to permits the demonstration of the existence of a positive correlation between these parameters and arsenic concentrations (Figures 4 and 5). The relationship with depth and temperature suggests that the As content could be related to the presence of a conducting layer located 7 km deep under Lamadrid village. According to Baldis et al. (1983), this layer is constituted of partially molten basalt at a temperature near 900°C. Several papers have associated the presence of naturally occurring arsenic in groundwaters with volcanic rocks of basalt composition (Hinkle & Polette 1999; Nickson et al. 2000) and with the upflow of geothermal water (Ball et al 1998, Wilkie and Hearing, 1998).

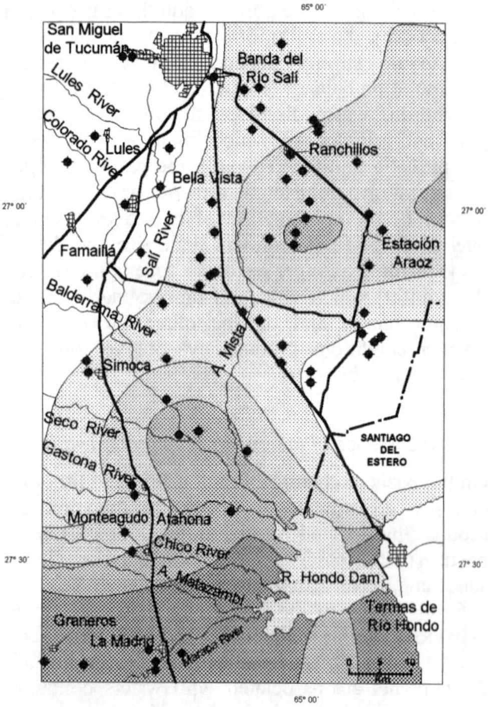

Figure 2

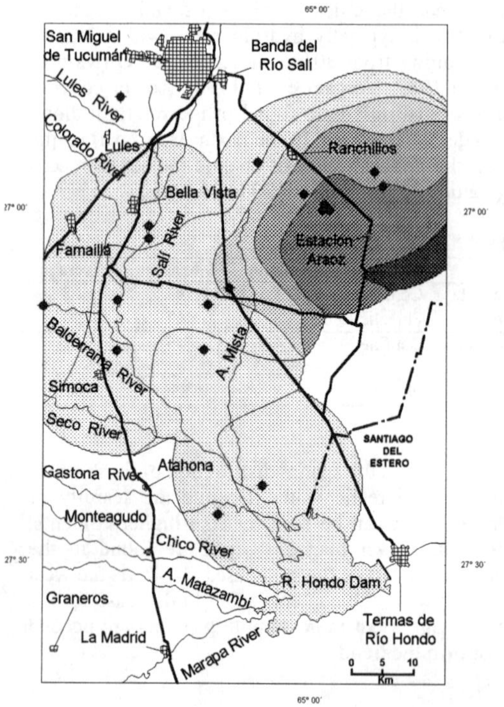

Figure 3

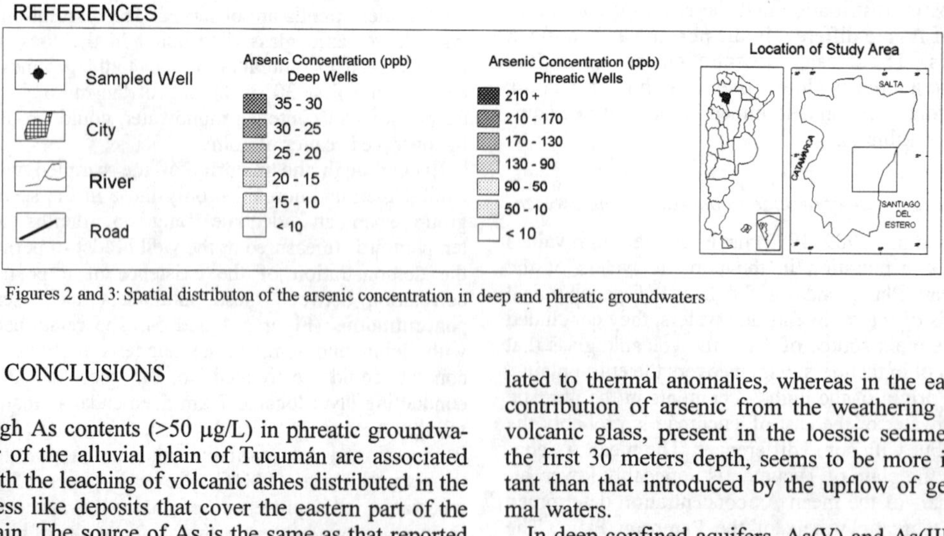

Figures 2 and 3: Spatial distributon of the arsenic concentration in deep and phreatic groundwaters

6 CONCLUSIONS

High As contents (>50 μg/L) in phreatic groundwater of the alluvial plain of Tucumán are associated with the leaching of volcanic ashes distributed in the loess like deposits that cover the eastern part of the plain. The source of As is the same as that reported in the phreatic water of the Pampean plain, in the centre of Argentina.

Intermediate levels of As (10-50 μg/L) have been found in confined groundwater of the study area. Two sources of As can be distinguished in this case, associated with the lithology. The As content in the confined aquifers of the SW seems to be more related to thermal anomalies, whereas in the east, the contribution of arsenic from the weathering of the volcanic glass, present in the loessic sediments of the first 30 meters depth, seems to be more important than that introduced by the upflow of geothermal waters.

In deep confined aquifers, As(V) and As(III) species could be present. Under the measured pH conditions, negatively charged oxyanions $H_2AsO_4^-$ and $HAsO_4^{2-}$ are the dominant species, but some small contribution from the uncharged H_3AsO_3 specie is expected. Arsenate adsorbs strongly on Fe-oxide minerals in middle-alkaline or acidic pH conditions, whereas the affinity between arsenite and Fe-oxides

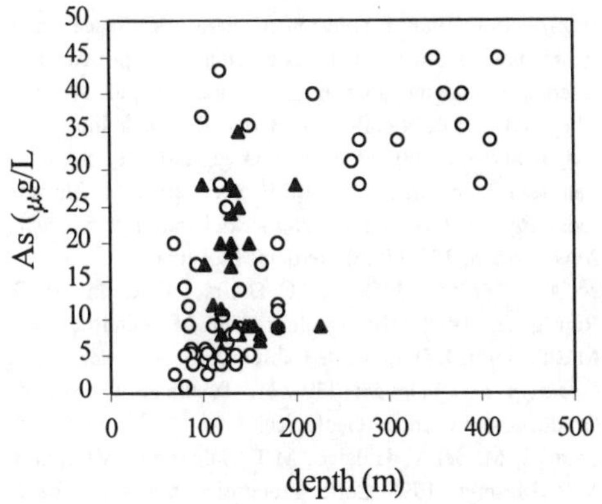

Figure 4. Relationship between As and depth in confined groundwaters. (▲) corresponds to aquifers developed in loessic deposits, (O) aquifers developed in alluvial sands and gravels. Points correspond to new and historical data.

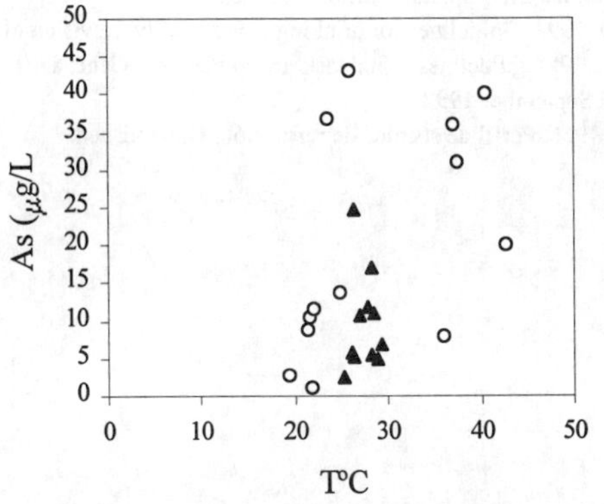

Figure 5. Relationship between As and temperature in confined aquifers. (▲) corresponds to aquifers developed in loessic deposits, (O) aquifers developed in alluvial sands and gravels. Points correspond to new data.

is rather weak (Dzombak & Morel 1990). The maximum adsoption of arsenate species on montmorillonite and kaolinite occurs at pH 5 (Griffin & Shimp 1978) resulting in a negligible influence over the removal of arsenic at the pH range of the studied waters.

In the upper confined groundwater of the east and the phreatic water, As (V) species are dominant. The affinity between arsenate and Fe-oxide surfaces is adequately high to remove As as the water infiltrates through the rich Fe-oxide sediments. The removal is efficient in lowering the As contents in the deepest groundwater which, at this depth, is no longer in contact with the volcanic glass. However, the loss of As is probably partially compensated by the contribution of geothermal waters.

The As concentrations observed in phreatic waters must be governed by the balance between the rates of dissolution of the high-arsenic volcanic glass and the removal by adsoption or precipitation.

Further sampling points, as well as geochemical studies of the sediments, are needed in order of better know the origin of arsenic in this part of the Chacopampean plain.

ACKNOLEDGEMENTS

The authors thank H. Nicolli for his support and Lisa Cave for her helpful review and comments, which significantly improved this paper. This work was supported by CIUNT (project 26/G104), CNEA (program 5), CONICET (PIP 4196/96) and ANPCYT (PICT 00806/97). Miguel Angel Blesa is a member of CONICET. María Gabriela García acknowledges a Doctoral fellowship from CONICET.

REFERENCES

APHA, AWWA & WPCF. 1989. Standard Methods for the Examination of Water and Wastewater. 17th Edition. USA.

Baldis, B., J. Demicheli; J. Febrer; H. Fournier; E. García; J.C. Gasco; M. Mamaní & M.C. Pomposiello. 1983. Magnetotelluric results along a 1200 km long deep profile with an important geothermal area and its north-west end in the provinces of Tucumán and Santiago del Estero in Argentina. *Acta Geodact. Geophys et Montanist. Hung. 18*: 489-499.

Ball, J.W.; D.K. Nordstrom; E.A. Jenne & D.V. Vivit. 1998. Chemical Analyses of hot springs, pools, geysers and surface waters from Yellowstone National Park, Wyoming and vicinity, *1974-1975 U.S.Geol.Surv.Open-File Rep. 98-182*: 45.

Bergoglio, R.M. 1963. Mortalidad por cáncer y tumores malignos en zonas de aguas arsenicales en la provincia de Córdoba. *V Congr. Iber.Amer.Dermat.*: 1111-1118.

Biagini, R.E. & M. Salvador. 1983. Arsenicismo crónico en la provincia de Salta. *Arch. Argent. Dermat,34* (4): 193-197.

Biagini, R.E.; G.C. Quiroga & V. Elías. 1974. Hidroarsenicismo crónico en Urutaú (Prov. Santiago del Estero). *Arch. Argent. Dermat, 24*: 8-11.

Bonaparte, J.F. & J. Bobovniqov. 1974. Algunos fósiles pleistocenos de la provincia de Tucumán y su significado bioestratigráfico. *Acta Geol. Lilloana 12* (11): 171-183.

Camino, M. 1988. Estratigrafía y evolución paleoambiental durante el cuaternario del valle de La Sala. MSc thesis (unpublished). Fac. de Cs. Naturales. Universidad Nacional de Tucumán.

CAA. 1994. Art.1 Res. MS y AS N°494. Ley 18284. Dec. Reglamentario 2126. Anexo I y II. Buenos Aires: Marzocchi.

Chaile, A; N.C. Romero; ,M.J. Amoroso; M. Hidalgo & M.C. Apella 1999. Plaguicidas clorados en el Río Salí, Tucumán–Argentina *Congreso Boliviano de Limnología y Recursos Acuáticos* (Abstract): 45. Cochabamba, Bolivia.

Concha G., B Nermell & M Vahter. 1998. Metabolism of inorganic arsenic in children with chronic high arsenic exposure in northern Argentina. *Environmental Health Perspectives, 16* (6).

Dzombak, D.A. & F.M.M Morel. 1990. Surface complexation modelling. Hydrous ferric oxide. USA: John Wiley & Sons.

Figueroa, L.R., L.F Medina & A.M. Pietroboni. 1996. Variaciones del nivel freático en la llanura deprimida de Tucumán. *Serie Monográfica, 3*. Tucumán: *INTA - CRTS*.

Griffin, R.A. & N.F. Shimp. 1978. Attenuation of pollutants in municipal landfill leachate by clay minerals. *EPA 600/2-78-157*.

Hidalgo, M. del V.; M.C. Apella & M.A. Blesa. 1998 Surface Water Surveillance in the Southern Region of Tucumán. *Proc. School of Environmental Sciences and Technology* (abstact). II: 499. Buenos Aires.

Hinkle, S.R. & D.J. Polette. 1999 Arsenic in Ground Water of the Willamette Basin, Oregon. *Water Resources Investigations Report 98*-4205. USGS.

Hopenhayn-Rich, C.; L. Biggs; A. Fuchs; R. Bergoglio; E.E. Tello; H. Nicolli & A. Smith. 1996. Bladder cancer mortality associated with arsenic in drinking water in Argentina. *Epidemiology*, 7: 117-124.

Merino, M.H. 1999. Arsénico en aguas subterráneas. Un problema generalizado en la Llanura Chacopampeana. *Gerencia Ambiental*: 528-569.

Mon, R. & G. Vergara. 1987. The geothermal area of the eastern border of the Andes of North Argentina at Tucumán Province. *Bulletin of the International Association of Engineering Geology. 35*: 87-92. París.

Nickson, R.T.; J.M. McArthur; P. Ravenscroft; W.G. Burgess & K.M.Ahmed. 2000. Mechanism of arsenic release to groundwater, Bangladesh and West Bengal. *Applied Geochemistry 15*: 403-413. Elsevier.

Nicolli, H.B.; J.M. Suriano; M.A. Gómez Peral; L.H. Ferpozzi & O.A. Baleani. 1989. Groundwater contamination with arsenic and other trace elements in an area of the pampa, Province of Córdoba, Argentina. *Environ Geol Water Sci 14 (1)*: 3-16.

Nicolli, H.B. & M.H. Merino. 1997. Aguas contaminadas, geoquímica ambiental y salud. *Protección del Medio Ambiente (F.A.A), 1* (1): 68-82. Argentina.

Northwood Geoscience Ltd. 1996. Vertical Mapper For MapInfo. Contour Modeling and Display Software. Version 1.5. Canada

Perondi, M.E; M.B. Vece; M.C Galindo; M. Hidalgo & M.C. Apella. 1999. Materia orgánica y calidad de aguas superficiales, Tucumán, Argentina. *Congreso Boliviano de Limnología y recursos acuáticos* (abstract): 69. Cochabamba, Bolivia.

Pierce, M.L. & C.B. Moore. 1982. Adsorption of arsenite and arsenate on amorphous iron hydroxide. *Water Res.*, 16: 1247-1253.

Sayago, J.M. 1995. The argentine neotropical loess: An overview. *Quaternary Science Reviews*. 14: 755-766.

Schecher, W.D. & D.C. McAvoy 1994. MINEQL+: A chemical equilibrium program for personal computers. *User's manual. Version 3.0. Environmental Research Software*, Mallowell, Maine.

SIPROSA 1982/1986. Dirección General de Saneamiento Ambiental de Tucumán. Investigación de la presencia de arsénico en el agua para consumo humano (unpublished).

Smedley, P.L.; H.B. Nicolli; A.J. Barros & J.O. Tullio. 1998. Origin and mobility of arsenic in groundwater from the Pampean Plain, Argentina. In E.B. Arenhart & J.R. Hulston (eds*), Proc. 9th Symp. on Water - Rock Interaction, Taupo, New Zeeland*: 275-278. Rotterdam: Balkema.

Tineo, A.; C. Falcón; J. García; C. D'Urso; G. Galindo & G. Rodríguez. 1998. Hidrogeología. In M. Gianfrancisco; M.E Puchulu; J. Durango de Cabrera & G. Aceñolaza (eds), *Geología de Tucumán*: 259-274. Tucumán: Colegio de Graduados en Ciencias Geológicas.

Vergara, G, M. del V. Hidalgo; M.T. Balegno de Vergara & M.P. Masmut. 1998 Corte geotérmico transversal de la Provincia de Tucumán. In M. Gianfrancisco; M.E Puchulu; J. Durango de Cabrera & G. Aceñolaza (eds), *Geología de Tucumán*: 161-178. Tucumán: Colegio de Graduados en Ciencias Geológicas.

Wilkie, J.A. & J.G. Hearing. 1998. Rapid oxidation of geothermal arsenic (III) in steamwaters of the eastern Sierra Nevada. *Environ. Sci. Technol.* 32: 657-662.

WHO. 1993. Guidelines for drinking water quality. Revision of the 1984 guidelines. Final task group meeting, Geneva, 21-25 September 1992.

YPF. 1966. Perfil de sísmica de refracción. Unpublished.

Groundwater: Past Achievements and Future Challenges, Sililo et al. (eds) © 2000 Balkema, Rotterdam, ISBN 90 5809 159 7

Hydrochemical characteristics of the Faria basin, northeastern West Bank

M.Ghanem & B.Merkel
Hydrogeological Department, Geological Institute, TU Bergakademie Freiberg, Germany

ABSTRACT: A hydrochemical study is conducted in the Faria basin / NE of the West Bank to define water types and to determine hydrochemical parameters of the aquifer system. More than 150 water samples were analysed, which covered all springs and wells. The physical and chemical properties were determined. The major cations Ca^{2+}, Mg^{2+}, Na^{+} and K^{+} and the major anions HCO_3^-, Cl^-, SO_4^{2-} and NO_3^- were analysed. The minor constituents of SiO_2, F^-, Br^- and PO_4^{3-} and the trace elements Cu, Cd, Fe, Mn, Cr, Zn and Pb were determined. Indicators for water quality were determined indicating that the confined aquifer is not polluted.

1 INTRODUCTION

There is no previous water chemistry study of the project area, however some data were available from *Rofe and Raffety* (1965) for the springs in the study area as part of the hydrogeological study of the West Bank. There is a lack of trace elements analyses in the area of study. This study aims at determining the hydrochemical characteristics of the aquifers system.

2 THE STUDY AREA

The Faria drainage basin lies in the northeastern part of the West Bank (Fig. 1). The area of study lies within the following coordinates:

Latitude : national 160000 - 195000 m N
(international 32° 2′ - 32° 12′ N)
Longitude: national 175000 - 200000 m E
(international 35° 12′ - 35° 35′E)

The elevation varies from 704 m above sea level in the western parts to 302 m below sea level in the Jordan Rift Valley. The Faria basin has an area of 320 square km. The climate of Faria is classified as arid in the eastern part; while it is semi-arid in the western part. Two major aquifers are located within this basin; phreatic and confined aquifers. The phreatic aquifer consists of Eocene, Pleistocene and Neogene sub-aquifers and the confined aquifer consists of the lower and upper Cenomanian sub-aquifers.

3 WATER SAMPLING, FIELD MEASUREMENTS AND METHODS OF LABORATORY ANALYSES

Two water sampling campaigns were carried out in the Faria Basin during the years 1996 and 1997 to study the hydrochemical parameters of the aquifer system. More than 150 analyses were included in the two rounds of sampling; before and after recharge. For the same sites, 20 100mL samples were collected for trace element analyses. These samples were filtered 450 nm and treated with Super Pure HNO_3 in the field to prevent any possible reactions and to prepare them for the determination of heavy metals. The water sampling campaign covered all springs as well as major groundwater wells and surface water (wadi) in the area. For all collected samples, the following parameters were measured during fieldwork at the site itself: temperature in °C, electrical conductivity (EC) in mS/cm, dissolved oxygen (DO) in mg/L, pH - value and redox potential (Eh) in mV. Alkalinity and acidity tests were done for selected samples. The water samples were analysed at the Water and Soil Environmental Research Unit (WSERU) of the Department of Chemistry at Bethlehem University in the West Bank. The following parameters were measured: Alkalinity and total dissolved solids (TDS), major cations in mg/L: Ca^{2+}, Mg^{2+}, Na^{1+} and K^{1+}, major anions in mg/L: HCO_3^-, Cl^-, SO_4^{2-} and NO_3^-, minor constituents: Silica (SiO_2), F^-, Br^- and $PO4^{3-}$ and trace inorganic constituents in µg/L: Fe, Mn, Pb, Zn, Cd and Cu.

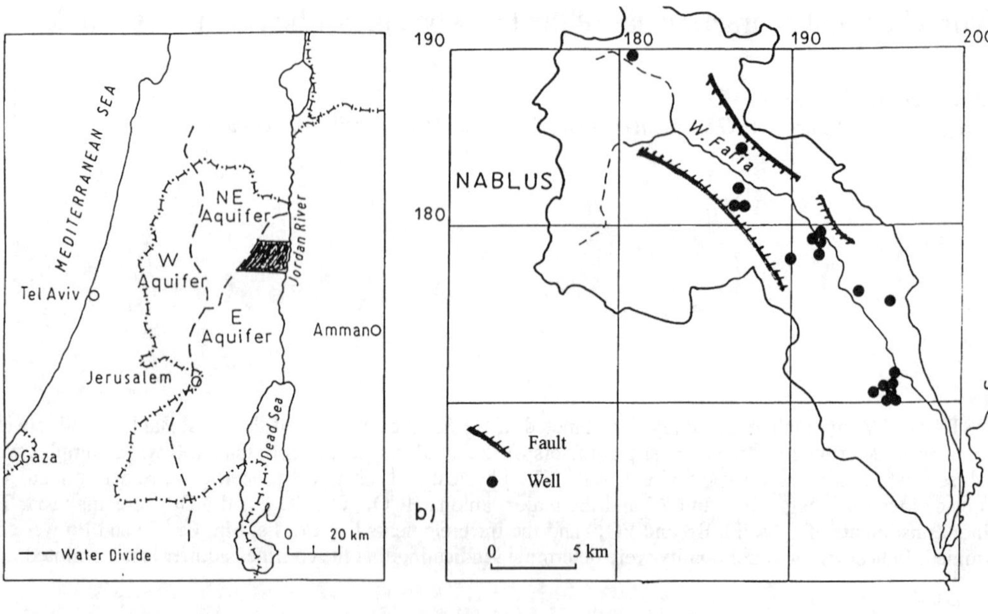

Fig.1: The location of the study area, A: The major groundwater basins of the West Bank
B: The Faria Basin with the sampling sites

4 INTERPRETATION OF FIELD MEASUREMENTS

Electrical conductivity: The EC of the Eocene sub-aquifer ranges between 752 µS and 725 µS/cm in the western parts and it ranges between 2843 to 3386 µS/cm in the eastern parts. The area can be subdivided into two main EC regions: Western phreatic and confined aquifers have low EC, and the eastern phreatic aquifer has very high EC.

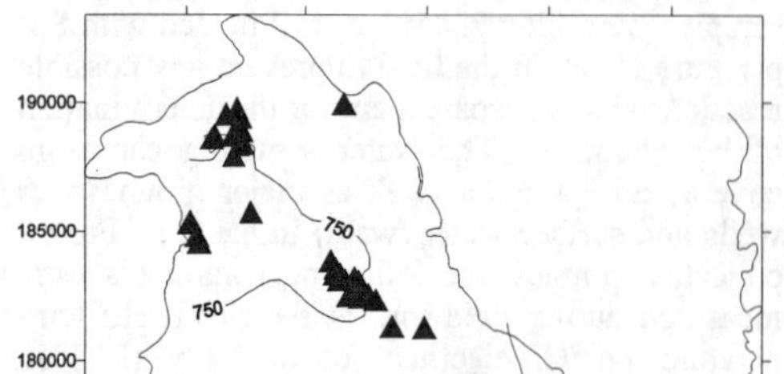

Fig. 2: The contour map of the EC (µS/cm) in the Faria basin.

According to this, the eastern phreatic aquifers are unusable for both drinking and agricultural purposes (Fig. 2).

pH value: Most of the groundwater in the study area has pH values ranging from 7.4 to 8.2. Slightly alkaline conditions generally occur in limestone and dolomite terrain.

Temperature: The average water temperature of the whole aquifer system is 23.3°C and its average for the phreatic and confined aquifers are 23.2°C and

23.6°C, respectively. The observed temperatures are close to temperatures expected from the local geothermal gradient.

Dissolved oxygen: The average dissolved oxygen is measured at the field to be 5.2 mg/L for the phreatic aquifer and 2.7 mg/L for the confined aquifer. A trend of decreasing DO is shown towards the east within the phreatic sub-aquifers (5.7 to 3.7 mg/L).

Redox potential (Eh): The mean Eh is measured to be 553 mV for the Eocene western sub-aquifer, 534 mV for the Neogene sub-aquifer, 570 mV for the Pleistocene sub-aquifer, and 601 mV for the Eocene eastern sub-aquifer. The Eh of the confined aquifer is measured to be 609 mV. The smallest value is seen in the springs of the study area.

5 TOTAL DISSOLVED SOLIDS (TDS)

TDS of the phreatic aquifer increases rapidly towards the east. The Eocene western sub-aquifer shows the lowest value of TDS (402 mg/L) and reaches 1551.5 mg/L in its eastern parts. Pleistocene sub-aquifer shows a mean value of 1608.4 mg/L. The confined sub-aquifers have a TDS ranging between 636.8 and 608.6 mg/L for their western and eastern parts, respectively. According to TDS, the area of study could be divided into two categories: fresh and brackish water. Fresh water lies within the phreatic western sub-aquifer (Eocene western and Neogene) and the confined aquifer. Brackish water lies within the phreatic eastern sub-aquifers. As water travels down the flowpath, it increases the total dissolved solids from 15 meq/L in the Faria area of the western part to the Jordan Rift of the eastern part to 67.1 meq/L. The average mean of major ions of all sub-aquifers of the Phreatic aquifer in the Faria basin shows an increasing trend towards the east and southeast - towards the Jordan Rift valley (Fig.3).

6 MAJOR IONS

Calcium and Magnesium (Ca^{2+} and Mg^{2+}): Calcium is derived from calcite dissolution and therefore balanced by HCO_3^-. Ca^{2+} has a mean of 128 mg/L in the study area. A big range of Ca^{2+} is recorded in phreatic sub-aquifers (479.6 mg/L) with a mean of 134.7 mg/L, compared with the range of the sub-confined aquifers (218 mg/L) that have a mean of 109 mg/L. This amount of Ca^{2+} defines the aquifers as carbonatic in nature. Magnesium shows a mean of 35.2 mg/L for the whole system and 38.8 and 28.9 mg/L for the phreatic and confined aquifers, respectively. Towards the Jordan Rift, an increase in Mg^{2+} is observed. This increase in Mg^{2+} can be attributed to water-rock interactions with the aquifer rocks which change in composition from limestones to dolomites eastwards.

Sodium and potassium (Na^+ and K^+): The mean of Na^+ in the phreatic aquifer is 33 mg/L. A decreasing trend of Na^+ is shown towards the east in the confined aquifer. The mean of Na^+ for the whole system is 74.4 mg/L. Potassium shows a mean of 3.4 mg/L for the confined aquifer and 6.1 mg/L for the phreatic aquifer. A small trend of increasing K^+ is shown in the Cenomanian aquifers towards eastern sub-aquifers.

Bicarbonate (HCO_3^-) m- and p-Value (HCO_3^- and CO_3^-, respectively): The amount of HCO_3^- in both phreatic and confined aquifers (301 and 316 mg/L), reflects the carbonate nature of the aquifer systems in the area. The p-value is measured to be 2.5 mmol for the phreatic aquifer and shows an increasing trend towards the east, reaching a value of 5.4 mmol /L. P-value is measured to be 2.2 mmol /L for Cenomanian sub-aquifers. The m-value is found to be 5.2 mmol /L for the phreatic aquifer.

Sulfate (SO_4^{2-}): The phreatic aquifer as well as the confined aquifer show a range of SO_4^{2-} between 18.4 to 36.4 mg/L; the latter shows smaller amounts of SO_4^{2-} than those of the former. The increasing trend of SO_4^{2-} towards the east is recorded.

Chloride (Cl^-): Chloride content was found to be increased in the east and southeast provinces (60 to 900 mg/ L) (Fig. 4). The high amount of Cl^- in the eastern phreatic sub-aquifers is due to chlorine-bearing rocks in the alluvium as well as Lisan formation, which contains a high amount of NaCl. The water of the eastern phreatic sub-aquifers is not potable for drinking and is used for agriculture.

Nitrate NO_3^-: The NO_3^- has a mean of 43.5 mg/L for the phreatic aquifer and 21.8 mg/L for the confined aquifer. The eastern phreatic aquifer shows less NO_3^- (22.2 to 31 mg/L) (Fig. 5). The pollution of the Eocene western sub-aquifers is due to leakage from septic tanks and cesspools in the area. The pollution of the Neogene sub-aquifer is due to

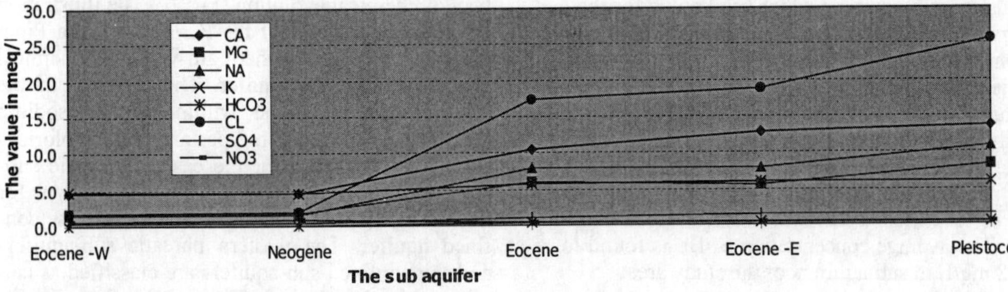

Fig. 3: the average mean of major ions in the W-E and NW-SE profile direction.

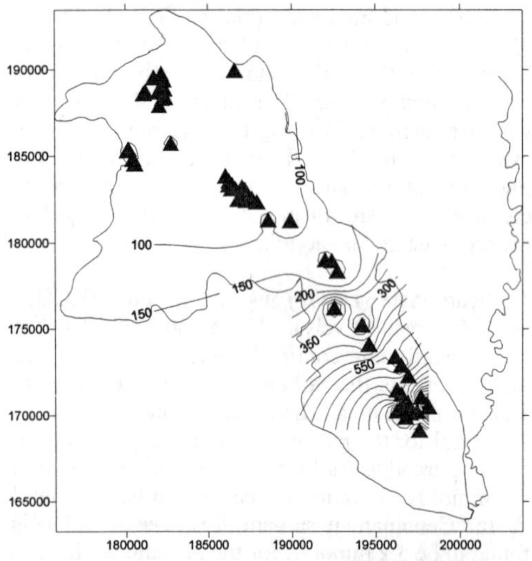

Fig. 4 : The contour map of the Cl (mg/L) in the Faria basin.

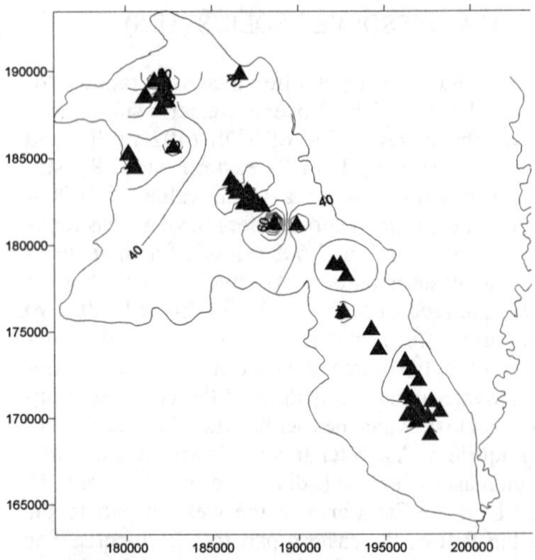

Fig. 5: The contour map of the NO_3 (mg/L) in the Faria basin.

the wastewater from the Nablus municipality. Agricultural fertilizers are also cause an increase in the amounts of NO_3^- in the groundwater in both the Eocene and Neogene sub-aquifers. The confined aquifer is not yet polluted with NO_3^- due to its confined nature and greater mean residence time.

Silica SiO_2 **:** An increase in SiO_2 towards the east in the confined aquifer is observed. It shows a mean of 38.7 mg/L for the western Cenomanian and 29 mg/L for the eastern Cenomanian sub-aquifers.

7 MINOR IONS AND TRACE ELEMENTS

Iron (Fe): Iron values are higher in the phreatic sub-aquifers, which is due to the ferrygenous lithology of the aquifer. The average concentration of Fe is found to be 22.1 and 17.8 µg/L in the phreatic and confined aquifers, respectively.

Manganese (Mn^{2+}): Manganese is enriched in the eastern sub-aquifers, which are known for their manganese mineralization. The mean of Mn^{2+} for the phreatic aquifer is found to be 1.4 and 1.8 µg/L for the confined aquifer.

Fluoride (F^-) and Bromide (Br^-): Fluoride is mainly enriched in waters of the southeastern parts of the Faria basin reflecting the close association with phosphorites and the effects of desert dust. The F^- is enriched in southeastern and eastern parts. The average concentration of Br^- is found to be 0.8 mg/L in sub-aquifers of the study area.

Phosphate (PO_4^{3-}): Phosphate reaches up to 0.5 mg/L for the whole aquifer system with an average of 0.2 mg/L.

Cadmium (Cd): The phreatic and the confined sub-aquifers show an average concentration of Cd 4.5 µg/L and 9.4 µg/L, respectively.
Copper (Cu): Copper shows an average of 4 µg/L for the whole aquifer system.
Lead (Pb): Lead is found of 3.8 µg/L for the phreatic aquifer and 2.7 µg/L for the confined aquifer.
Zinc (Zn): The aquifer system of the whole area has an average Zn concentration of 12.6 µg/L.

8 GRAPHICAL PRESENTATION OF RESULTS

Groundwater for Windows (*GWW*, 1994) and Hydrowin software were used to represent graphs of water chemistry data. Piper diagrams were used for classifying the groundwater and defining a pathway of chemical evolution (Fig. 6). The illustration of all water samples of the Faria basin in a Piper diagram indicate that there are very few chemical changes within the aquifer. The majority of the groundwater are found within calcium sodium cation types. Two anion types are found: chloride-sulphate-bicarbonate for the eastern phreatic sub-aquifers and bicarbonate-chloride-sulphate type for the western phreatic sub-aquifer as well as the confined aquifer. The western phreatic sub-aquifers and the confined sub-aquifers are classified as calcium - bicarbonate type. The eastern phreatic sub-aquifers are classified as calcium - chloride type

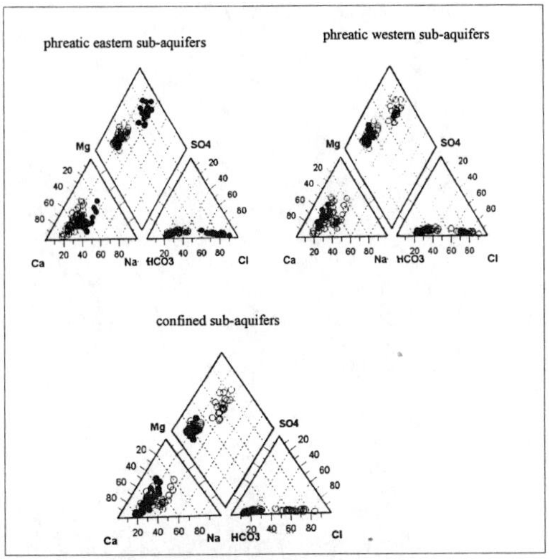

Fig. 6: Piper diagram illustrates chemical analyses of water samples of the Faria basin.

and some of them as calcium-magnesium-chloride type.

9 WATER TYPES

The phreatic western sub-aquifers and the confined aquifer show an increasing trend of Ca^{2+}> Mg^{2+} > Na^{+}+ K^{+}. An increasing trend of Ca^{2+}> Na^{+} + K^{+} > Mg^{2+} is shown for the eastern phreatic sub-aquifers. Three anions increasing trend orders are observed: HCO_3>Cl>NO_3>SO_4^{2-} in the phreatic western sub-aquifers, HCO_3>Cl>SO_4^{2-}>NO_3^- in the confined aquifer and Cl^->HCO_3> SO_4^{2-}>NO_3^- in the phreatic eastern sub-aquifers. The increasing trend of NO_3^- in reference to SO_4^{2-} in the phreatic western sub-aquifers is due to pollution caused by cesspools and septic tanks, as well as the wastewater, that are draining into the Wadi Faria.

The phreatic western sub-aquifers and the confined aquifer are composed of carbonate rocks (limestone and in some places dolomitic limestone) with low salinity (< 800 mg/L). Otherwise, the phreatic eastern sub-aquifers consist of alluvium and lisan formations (gypsum-marl-lisan) with high salinity (> 1000 mg/L). Thus, the composition of rocks and soils has a direct influence on water quality. The chemical composition of the confined aquifer is homogeneous in comparison with the phreatic aquifer. The ratio Mg/Ca is between 0.3 and 0.8, within the same ration of rainfall, which indicates aquifers of limestone and dolomitic nature.

10 HYDROGEOCHEMICAL MODEL

In this study, the computer program PhreeqC (*Parkhurst et al.*, 1980) was used to calculate the equilibrium speciation and saturation indices. Generally, the western phreatic sub-aquifers and the confined sub-aquifers show the same distribution species and slightly different from those of the phreatic eastern sub-aquifers. The saturation indices of all water samples were determined using the PhreeqC programme.

The following phases are found to be in the trend of equilibrium with the solution: Calcite, Dolomite, Aragonite, $Fe(OH)_3$, Chalcedony, Quartz and SiO_2. The phases Calcite ($CaCO_3$), Dolomite ($CaMgCO_3$) and Aragonite ($CaCO_3$) represent the limestone-dolomitic origin of major sub-aquifers in the area.

11 CONCLUSIONS

A hydrochemical study is conducted in the Faria basin indicates that the TDS is increasing towards the east. The physical and chemical properties (pH, dissolved oxygen, major ions, trace elements) were determined. The hydrogeochemical model was used to calculate distribution of species and saturation indices. The water has a calcium - bicarbonate type in the western parts and sodium-chloride type in the eastern parts. The ion concentrations of water show an increasing trend towards the east. The chemical composition of the confined aquifer is homogeneous in comparison with the phreatic aquifer. The confined aquifer is not polluted.

REFERENCES

GWW. 1994. Groundwater for Windows (manual and software): United Nation, New York.

Paarkhurst, D. L., Thorsternson, D.C. and Plummer, L. N. 1980. Phreeqe - a computer program for geochemical calculations. U.S: geological Survey, Water Resources Invest., 80-96, NTIS Tech. Report. PB81-167801, Springfield, 210pp.

Rofe and Rafferty Consulting ENG. 1965. West Bank hydrology: Nablus district water resources survey, geological and hydrological report, 120pp.

Groundwater: Past Achievements and Future Challenges, Sililo et al. (eds) © 2000 Balkema, Rotterdam, ISBN 90 5809 159 7

Modelling mixing of waters in a layered basaltic aquifer in a tropical climate at Rivière de l'Est, Reunion Island

J.L. Hoareau & E. Nicolini
Laboratoire des Sciences de la Terre, Université de la Réunion, Saint-Denis, France

S. Violette
Laboratoire de Géologie Appliquée, Université Pierre et Marie Curie, Paris, France

ABSTRACT: The drainage gallery of Rivière de l'Est on Reunion Island (Indian Ocean), allows the investigation of the deep groundwaters from two environments of different permeabilities and lithologies. From the modelling study of the mixing between the two identified main kinds of waters, the functioning of the different drain pipes in a constructed drainage gallery can be explained. The study of the saturation state of these waters (mainly with respect to clay and secondary minerals) provides valuable information about the chemical evolution of the system. Consequently, we suggest a hydrochemical and hydrogeological model for such a volcanic domain.

1 INTRODUCTION

1.1 *Study area*

Rivière de l'Est is one of the three main perennial rivers of the Piton de la Fournaise basaltic massif on Reunion Island, Indian Ocean. It is located in the North of the massif and its orientation is South-North (Fig. 1).

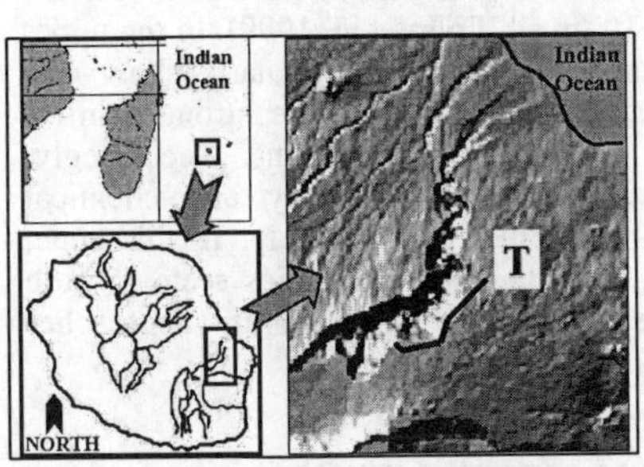

Figure 1. Location of La Rivière de l'Est and the EDF (Electricité de France) tunnel (T) on the massif of Piton de la Fournaise. Reunion island, Indian Ocean.

It drains a 27 km² topographic catchment and has a high precipitation rate during the wet season (from November to April), with an average annual rainfall ranging from 6500 to 12000 mm (Barcelo 1996). Thus, it has a permanent torrential flow with a low water level flow of 3.5 $m^3 s^{-1}$.

A drainage gallery consists of 40 drains, each 50 metres long, at an angle of 45° from the horizontal, and strained on the last 15 metres. It was drilled through the right side of the river in order to delay the collapse of a rock flack which could endanger the actual hydroelectric equipment. Thirty one of these drains are located on the left external side of the gallery, meaning at the river side (ES : External Side) whereas the other nine drains are situated on the right internal side (IS : Internal Side) (Fig. 2).

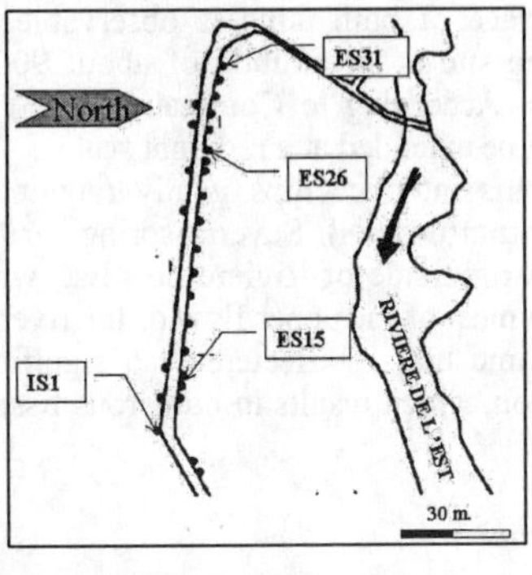

Figure 2. Map of the drains layout and their situation in relation to the river.

This drainage gallery allows to investigate the deep groundwater of the Rivière de l'Est massif.

1.2 *Hypothesis*

The basaltic shield volcanoes consist of formations of highly variable porosities and fracturation stages. This study incorporates the important decompression fractures of the edge of the massif adjacent to gallery.

The drains we have investigated are more or less related among each other geological subsets with regard to their physical properties.

We hereby propose to establish a hydrogeological relation between the drains and the considered geological setting by means of hydrodynamical and hydrochemical data.

2 HYDROCHEMICAL AND HYDROGEOLOGICAL CHARACTERIZATION OF THE DRAINS

The drainage gallery is situated at the bottom of a 1000 metres high cliff, at an altitude of approximately 900 metres. The gallery was constructed across the cliff, about 70 metres from the river bed, parallel to its edge.

2.1 *Geological environments*

One geological section of the cliff reveals two superposed lithological units.

The lower part (where the tunnel was drilled) comprises cotectic basalts which are extremely compacted and brecciated. Their interstitial porosity is higher than their fracture porosity.

This unit is overlaid by a pile of aphyric basalts and olivine basalts which are extremely jointed and fractured (from 1000 to 1900 metres high).

The interface of both units is observable upstream of the site at an altitude of about 900 to 1000 metres. According to Courteaud (1996) the interface can be extended at a regional scale.

Further, this interface dips gently (about 15° downstream (northwards). Several springs are located on the right side of Rivière de l'Est, which contribute to most of the upper flow of the river.

The volcanic unit is affected by a significant decompression, which results in numerous fissures of the cliff.

2.2 *The drains*

The drain pipes of the gallery can be divided into two groups (i) IS drains and (ii) ES drains. The former are characterized by a stable flow, electric conductivity and hydrochemistry. The latter vary with the season.

During the rainy season the ES drains constitute a homogeneous group with a low conductivity (from 50 to 150 µS cm^{-1}) and high flow directly related with rainfalls.

During the dry season :

- The flow of the upstream ES drains (ES31 and ES30) decreases considerably, with no major change in the conductivity (Fig. 3).
- The ES29 to ES18 and ES12 to ES7 drains are almost completely dry.
- The flow in the ES17 to ES13 drains (intermediate drains) drops considerably after which their hydrochemical facies resemble the IS facies (reaching about 400 µS cm^{-1}) (Fig. 3).

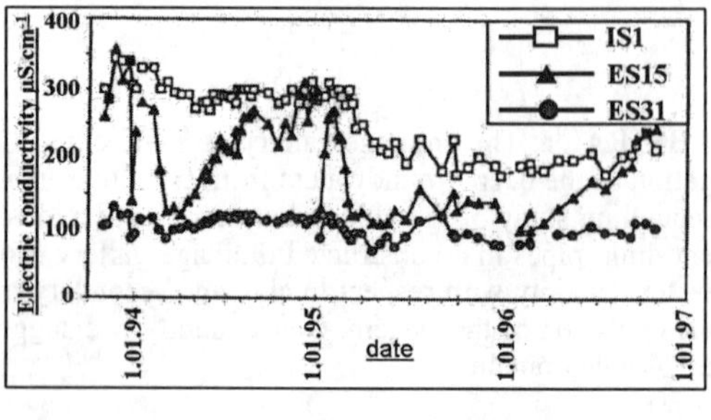

Figure 3. Evolution of ES15 conductivity compared which ES31 and IS1.

2.3 *Hydrogeological and hydrochemical environments*

In a permanent flow situation, the lower part of the aquifer consists of highly mineralized waters which are stable in time. Their facies are mainly calcium bicarbonate and magnesium bicarbonate. The interstitial porosity seems to be the dominant hydrodynamic factor in this environment (permeability of 10^{-6}m s^{-1}, Folio et al. 1999). In the upper portion, the water bodies particularly show medium bicarbonate facies and are strongly influenced by the seasonal conditions. Accordingly, their flow, electric conductivity and chemical composition fluctuate considerably. In this zone, fracturing and interstitial porosities seem to both contribute to the flow (permeability approaches 10^{-4} m s^{-1}).

3 MODELLING THE MIXING

The typology of the drains allows to define two groups being extremely characteristic of two specific reservoirs. The behavior of the drains seems to be explained by the proportionally varying mixture of these two reservoirs.

Thus we have used the mixing function of the PHREEQC modelling code provided with Aquachem software (Waterloo Hydrologic®) in order to simulate the mixing process. We have not considered changes in the saturation state that may occur during the mixing process.

A thermodynamical model (Kindis modelling code, Madé 1991) was then used to calculate the saturation indexes.

3.1 *Description of the mineralization sources*

We can consider three possible sources of mineralization :

- Meteoric waters, slightly mineralized (10 to 20 μS cm^{-1}) mainly consisting of sodium chloride facies ([Na] = 1.2 mg l^{-1}).
- Waters in relation with fractured formations (ES31 type) are medium bicarbonate facies (mean conductivity up to 110 μS cm^{-1})
- Waters in contact with lower formations (IS1 type) are calcium bicarbonate and magnesium bicarbonate facies, with high conductivity (up to 400 μS cm^{-1}).

In the following paragraph, we consider the drains which completely dry up (ES 26 type), as well as the intermediate drains (ES15 type).

3.1.1 *The drains of the ES26 type*

These drains are extremely responsive to rainfall. Their conductivity ranges between 45 μS cm^{-1} at the beginning of the solicitation and 80 μS cm^{-1} just before the drying up. The mineralization of these drains fluctuates between those of the ES31 type waters and those of rainwater.

We demonstrate that a mixing of these two end-members (35% and 65 % respectively) leads to a water type which is similar to the ES26 type at the beginning of its outflow increase (after the storm). Likewise, at the end of the response, the proportions are 53% and 47%, respectively (Fig. 5).

3.1.2 *The drains of ES15 type*

According to Figure 3, we logically considered the drains of the ES15 type as a mixture of the IS1 and ES31.

However, the chemical composition of such a mixed body is not usable since the amount of Na is too low (in the modeled water) in relation to the equivalent natural water. Even though, both components of the mixing barely react to rainfall, whereas we noticed that the ES15 type waters were highly responsive to rainfall (within three days). As a result, we consider a third component in the mixture, meteoric water.

Thus, ES15 can be simulated by mixing 15% rain, 35% ES31 and about 50% IS1 during low water level periods (Fig. 4). During the wet season, these figures are 80%, 12% and 8%, respectively.

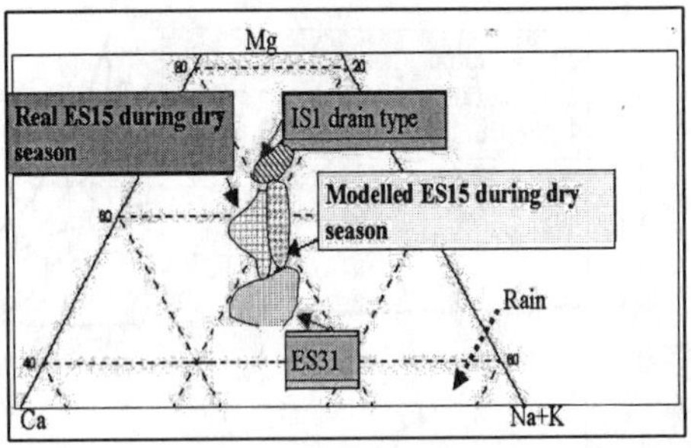

Figure 4. Representation of the chemical variations of the waters from selected drains. Notice the concordance between observed and modeled ES waters during the dry season.

We therefore propose the following interpretation of the relation between the different mineralization sources (Fig. 5).

ES26 results from the more or less significant dilution of ES31 by rain water depending on the season. On the other hand, ES15 represents a mixture of seasonally dependant and therefore variable proportions of IS1, ES31 and meteoric waters.

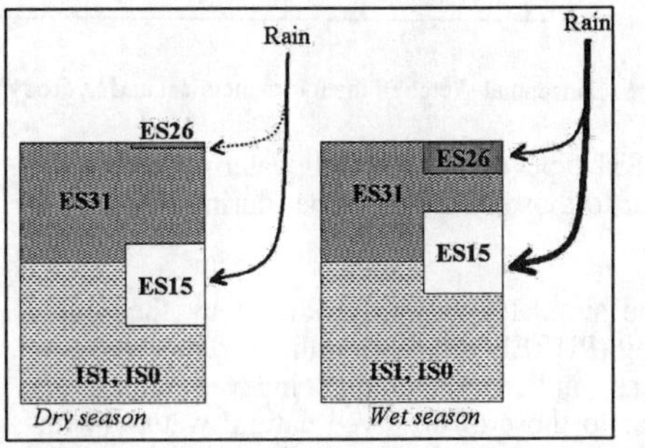

Figure 5. interpretative sketch of the contributions leading the ES15 and ES26 type waters.

4 SATURATION INDEXES AND THEIR HYDROCHEMICAL IMPLICATIONS

4.1 *Saturation indexes (SI)*

Kindis (Madé, 1991) is a thermodynamic dissolution model used in static configuration to calculate the equilibrium state of minerals in solution. Its application to the water types of the selected drains shows a state of undersaturation with respect to primary minerals as well as secondary minerals like clays.

First, calcite is more saturated in the IS drains (SI = -0.1 to -0.3) than in the ES drains (-1 to -2). Secondly, intermediate drains (ES15) show saturation indexes which are either similar to those of

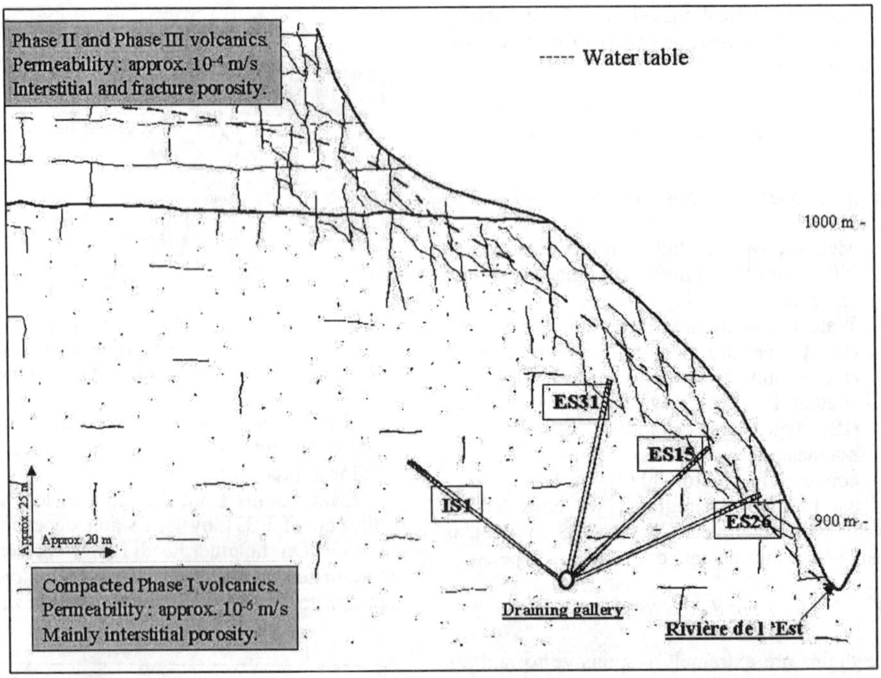

Figure 6. Conceptual sketch of the hydrochemical and hydrodynamical functioning of the draining gallery of Rivière de l'Est.

the ES31 type (during the high water period) or similar to those of the IS1 type (during the dry period).

The simulated waters produced by theoretical mixing (PHREEQC), during the dry and the rainy seasons, indicate saturation indexes which are similar to those of observed natural waters (without secondary precipitation). In conclusion, the study of the saturation indexes confirms the hypothesis of mixing in the drainage gallery.

4.2 *Clay minerals*

The clay minerals (illite, montmorillonite, kaolinite for example) are widely undersaturated in natural waters. In all the previous physical investigations no clay phase precipitation in the study area could be found.

Recent studies of static models (Hoareau et al. 2000) have demonstrated that, during the dissolution of basaltic rocks, clays control the total amount of silica in waters. During the simulation of basaltic rock alteration, we are able to either allow the precipitation of clays or to prevent it. In the first case, the calculated amount of dissolved silica averages 25 to 35 mg l^{-1}. In the second case, the amount reaches up to 55 to 65 mg l^{-i} at the same step of the alteration process.

Since the dissolved silica content in our natural waters reaches 60 mg l^{-1}, the hypothesis of a lack of neogenic complex clay phases in the aquifer can be confirmed.

5 CONCLUSION

We hereby propose the following model of circulation (Fig. 6) and identify four groups of drains, each one being representative of one type of circulation :

1. The IS1 type is typical of a deep environment, in which low interstitial porosity predominates (basalts of Phase I).
2. The ES31 type drain is characterized by less mineralized waters from the upper part of the aquifer in which fracturing of basaltic Phase II and Phase III layers, and interstitial porosity of scoria inter-layers, both contribute to flow.
3. The ES15 type is also situated in the lower part of the aquifer. These waters result from the seasonally mixing of both previous types of groundwater and of rainwater which rapidly penetrates the drain through decompression fractures.
4. The drains of the ES26 type which are situated in the unsaturated environment. These waters circulate in the decompression frac-

tures connected to the fractures of the upper zone.

This study of a drainage gallery considers two types of circulations simultaneously, providing better knowledge of the complex hydrodynamic pattern in a volcanic environment characterized by varying permeabilities.

REFERENCES

Barcelo A.1996. Analyse des mécanismes hydrologiques en domaine volcanique insulaire tropical à relief jeune. Apports à la connaissance du bilan hydrique. Massif du Piton de la Fournaise (Ile de la Réunion). *Unpubl. Doctoral dissertation.* Montpellier II Univ..

Courteaud M. 1996. Etude des structures géologiques et hydrogéologiques du massif de la Fournaise par la méthode audiomagnétotellurique. *Unpubl. Doctoral dissertation.* Reunion Island Univ.

Folio JL, Join JL, Coudray J. Hydraulic conductivities estimation in a recent shield volcano : Piton de la Fournaise, Reunion island, Indian Ocean. *Model Care 99, International Conference on Calibration and Reliability in Groundwater Modelling.* Zürich, Switzerland

Hoareau JL, Nicolini E, Fritz B. 2000. Modélisation des altérations des roches basaltiques. Essais de comparaisons avec les travaux expérimentaux. *Special SGF Session.* Paris.

Madé B.1991.Modélisation thermodynamique et cinétique des réactions géochimiques dans les interactions eau-roche. *Unpubl. Doctoral dissertation.* Louis Pasteur Univ. Strasbourg, France.

Groundwater: Past Achievements and Future Challenges, Sililo et al. (eds) © 2000 Balkema, Rotterdam, ISBN 90 5809 159 7

Sustainable management of groundwaters in dryland salinity affected Spring Creek catchment, New South Wales, Australia

J. Jankowski & S. Shekarforoush
UNSW Groundwater Centre, School of Geology, University of New South Wales, Sydney, N.S.W., Australia

ABSTRACT: Environmental degradation and salinisation processes in the Southern Tableland of NSW are associated with groundwater discharge from fractured bedrock aquifers causing loss of agricultural productivity. The driving mechanism behind rising groundwater is clearing of native vegetation in the beginning of the last century. Discharging groundwaters from the confined fractured aquifers interact with the overlying colluvium causing erosion, gullying and soil degradation. The upward movement of saline groundwater in discharge zones on foot-slopes and valley floors, together with the mobilisation of soluble salts stored in clay colluvium, are shown to be mechanisms for the development of dryland salinity. Water-rock interaction and hydrogeochemical processes such as reverse ion-exchange in fractured aquifers and ion-exchange reactions in colluvium clays are the dominant chemical processes. The stable isotopes of deuterium and oxygen-18 provide no evidence for evaporative concentration in these shallow groundwaters. Sustainable management of these groundwaters is achieved through the understanding of hydrogeochemical processes in the mixing zone where deep discharging groundwaters are reacting with colluvium.

1 INTRODUCTION

Large areas of the agricultural land of the Australian continent are suffering from environmental degradation caused by dryland salinity. The Southern Tablelands of New South Wales are severely affected by dryland salinity. Clearing of native vegetation 80 years ago changed hydrological properties of the aquifer. Now more rainwater is recharging the aquifer than can be intercepted in the soil, root and unsaturated zones. Because of the change in water budget in the catchment, the position of the water table in the recharge area is hence well above the valley floor and groundwaters from the deep confined fracture aquifer are flowing under artesian pressures.

This discharge is producing a saturated colluvium and resulting in the development of seepages and outflow of salt into surface waterways (Jankowski & Acworth 1998). The presence of colluvium in the form of debris flows is causing the bedrock aquifer to be semi-confined or confined. As the debris flow deposits contain wind-blown deposits including soluble salts, their release is causing salinisation and soil degradation (Acworth et al. 1997, Acworth & Jankowski In press, Jankowski & Acworth 1993, 1997). The shallow groundwater system saturated by the discharging deep groundwater represents a mixing zone between discharging groundwaters and waters present and already chemically modified in colluvium.

The Yass area in Southern Tableland developed classical examples of dryland salinity. The major focus on the recently described dryland salinisation processes have been directed towards hydrogeochemical processes and reactions in fractured groundwater systems and in mixing zones (Jankowski 1993, 1997, Jankowski et al 1998a, b, Shekarforoush 1999), and origin of salt in colluvium (Acworth 1999, Acworth et al. 1997, Broughton 1992). Evans (1994, 1998) investigated salt balance and groundwater dynamics at the Begalia site.

The studies of stable isotopes deuterium and oxygen-18 for the Williams Creek catchment in the Yass area (Turner et al. 1991) and for the Wattle Retreat catchment in the Cootamundra region (Waite & Turner 1989) have found that evaporative concentration is not causing the development of dryland salinity. Turner et al (1991) and Waite & Turner (1989) suggest that the predominant salt concentration mechanism is the process of ion exclusion during the uptake of water by pre-cleared vegetation. The studies in the Yass area have shown that the remediation action and methods to prevent further developments of salinity must be changed to account for salts that are stored in the colluvium and are not associated with evaporation from discharging groundwater.

2 ENVIRONMENTAL SETTING

The study area is located about 15 km north-northeast of Yass. The investigated site known as the Begalia catchment covers approximately 230 ha of the Spring Creek catchment. The catchment lies in the headwaters of the Lachlan River, a tributary of the Murray River. The southern part of the catchment is covered by undulating hills with slopes of about 6-7 degrees. The northern part has more gentle topography with slopes of 2-3 degrees where the salinity scalds are exposed. The elevation varies from 730 m (AHD) in the highest point to 620 m (AHD) at the outlet of the catchment.

The site is located within the Lachlan Fold Belt of eastern Australia and is a part of the Cowra-Yass synclinorial zone. The bedrock contains mainly dacites and rhyodacites of the Middle Silurian Hawkins Volcanics. The volcanic unit mostly of blue-grey dacitic crystal tuff contains quartz, feldspars and biotite, with disseminated pyrite present in fractures and quartz veins. Minor calcite, dolomite, haematite and goethite, and traces of gypsum and halite were found in fresh and weathered rocks. The fresh rock is well fractured and foliated with the dominant north-south direction paralleling the major fault line. Intense weathering during the Tertiary has produced deep weathering profiles to variable depths reaching up to 26 m in some areas. Weathering has altered biotite to chlorite, and feldspar to sericite. Clay comprises up to 30% of fresh rock (Lertsirivorakul 1990), and XRD analyses indicate that the dacite is largely altered to montmorillonite, illite, chlorite and kaolinite. The bedrock is overlain by a Cainozoic unconsolidated clay-rich gravity flow depositional unit. Broughton (1992) divided the Cainozoic sediments into three units: (1) a lower complexly graded sequence of gravel to coarse sand-size sediments; (2) a compacted clay-rich gravity flow deposit, and: (3) an upper clay-rich matrix supported clast deposit. The lower clay unit is characterised by blue-grey veins along weakly developed polygonal cracks formed by desiccation processes. The blue-grey colour is related to the upward leakage of the groundwater from the underlying fractured bedrock. The clay-rich material which incorporates the matrix supported bedrock clasts, is considered to be derived from wind-blown parna and remobilised as debris flows (Acworth et al. 1997). In the northern part of the catchment on the clay colluvium two types of soils are present: (1) skeletal soils on higher slopes, and; (2) duplex soils on lower slopes (Evans 1995).

The meteorological data taken from the closest station in Yass indicates that the mean annual rainfall is 639 mm and is evenly distributed throughout the year. The annual evaporation ranges between 1220 and 1575 mm with an average of 1372 mm. The period October-March accounts for 75% of the annual evaporation.

The Begalia site was cleared around 1910 and up to 1944 and only traces of salinity outbreaks are seen on aerial photos. By 1953 the site was severely affected by salinity and by 1963 there had been significant increase in salinity sites with general direction to the north. In 1962 the major outbreaks were unsuccessfully treated with contour banks. After 1973 some regeneration of the salt affected area has occurred as a result of stock exclusion. For the Begalia site a considerable lag time of 30 years has occurred between clearing and the outbreaks of salinity.

3 HYDROGEOLOGY

Groundwater system in the Begalia site is present in both unconsolidated sediments in colluvium as a shallow aquifer system, and fractured bedrock aquifer as a deep system. The fractured groundwater flow is developed in fractures, veins and joints and flow is controlled by catchment topography. Groundwater recharge occurs on the hills where bedrock outcrops, and flows via fractures to the valley floors, where groundwater discharges under flowing artesian pressures. The fractured aquifer represents a local aquifer system with the recharge zone on the neighboring hills. The fractured aquifer is unconfined except where the lower slopes and valley floors are overlain by colluvium, and aquifer behaviour changes for semi-confined to confined. Most groundwater supplies are from fractures at depths of 20 to 45 m (Lawson 1989). The bore yields vary between 0.1 and 2.5 L/s. The shallow aquifer is present in colluvium when the debris-flow deposits are permanently saturated by the discharging groundwater from the fractured aquifer. Piezometric surfaces in piezometers installed in the scalded area generally are at or above the ground surface for much of the year and show little response to recharge events. The lack of response is due to low hydraulic conductivity of the clay material. As piezometric surfaces are above the ground surface this prevents any recharge of rain water into the colluvium. The scalded area is waterlogged by seeping groundwater from the shallow system in colluvium. Measurements of hydraulic conductivity in these sediments indicate that values are highly variable from 10 mm/day to 150 mm/day (Evans 1995).

4 HYDROGEOCHEMISTRY

Sixteen deep bores ranging from 11 to 102 m in depth, and 45 shallow bores and piezometers have been intermittently sampled over the last few years from the Begalia site and environs. All shallow piezometers are located in scalded areas. Selected chemical analyses of shallow and deep groundwaters are shown in Table 1. Outside the waterlogged and scalded areas shallow groundwater is fresh (samples

Table 1. Selected chemical analyses of deep and shallow groundwaters from the Begalia catchment.

Bore/Piezo No	Depth m	EC μS/cm	TDS mg/L	pH	Na	K	Ca	Mg	Cl mg/L	SO_4	HCO_3	SiO_2
B36760	102	746	611	6.76	78	1	23	43	62	13	357	21
B36757	52	1965	1441	6.68	114	2	137	99	347	45	668	28
B36758	62	2010	1410	6.61	122	2	116	104	383	39	608	29
B36756/3	92	1954	1337	6.72	71	2	172	82	363	25	584	26
P24	2.15	78	72	5.75	7	1	3	3	5	2	34	16
P25	2.12	339	192	5.44	38	3	2	10	81	3	3	20
P48	2.75	1336	697	5.91	135	5	4	71	379	5	75	23
SB1	8.15	3080	1991	7.67	240	2	50	226	775	39	632	29
P27	2.29	4620	2957	7.27	744	2	26	126	1039	71	925	24
P3	2.42	9960	5893	7.03	1517	6	28	315	3001	7	995	19

B – Bore; P – Piezometer; SB – Shallow Bore

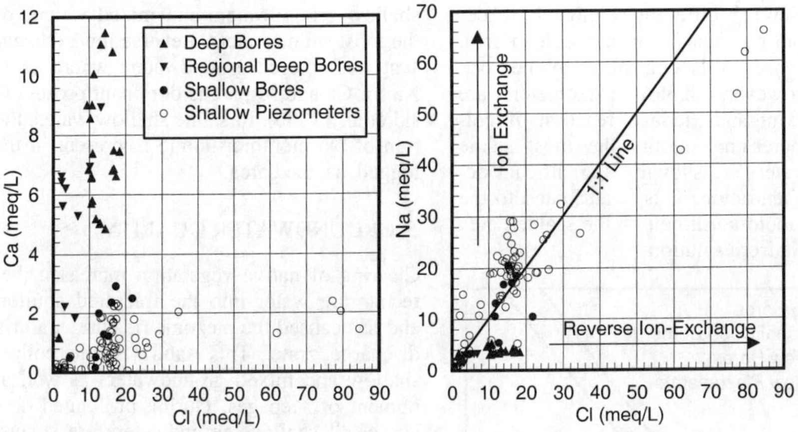

Figure 1. Relationship between Ca versus Cl (A) and Na versus Cl (B) for Begalia groundwaters.

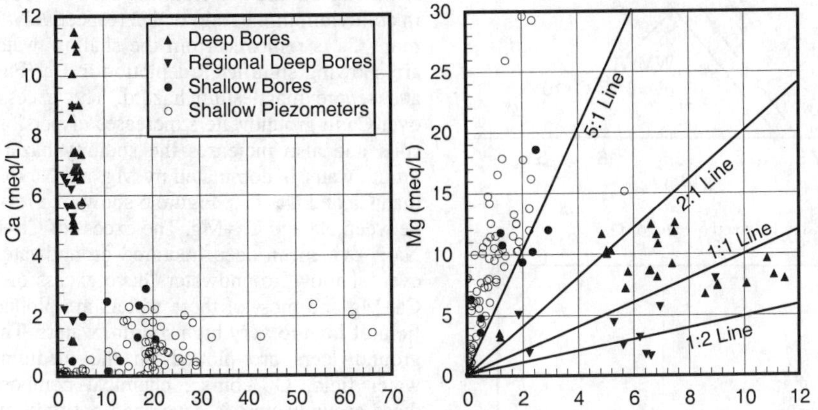

Figure 2. Relationship between Ca versus Na (A) and Mg versus Ca (B) for Begalia groundwaters.

P24 and P25), however, in the scalded area salinity is extremely high (sample P3). The shallow, fresh ground waters are slightly acidic with Na and Cl as dominant ions. The most saline shallow waters have nearly neutral pH and dominance of Na and Cl. The chemical composition of deep groundwaters in the fractured aquifer originates from the water-rock interaction. The dominant process is reverse ion exchange which elevates the Cl:Na ratio well above 3 (Jankowski et al. 1998b). High concentrations of

HCO_3 and low of SO_4 are attributed to redox reactions in the fractured aquifer. Figures 1A and 1B clearly show the difference between deep and shallow groundwaters with respect to chemical composition and hydrogeochemical reactions. The fractured rock groundwaters are dominated by Ca and Mg, however, shallow groundwaters are dominated by Na (Figure 2A & 2B). Also the shallow groundwaters have a high ratio of Mg:Ca which is above 5:1, whereas in deep groundwaters the ratio is below 2:1. Mixed groundwaters are showing a chemical composition, which develops during slow movement of deep groundwaters through the low permeability colluvium. These waters are located on Figures 1, 2 and 4 between both populations. The location of single points representing different shallow bore depends on the depth and location in the scalded area. Shallow bore chemistry is controlled by bedrock groundwaters, however, shallow piezometers are controlled by mixing and chemical reactions in colluvium. With increasing salinity the most saline shallow groundwaters are showing a significant decrease in Na concentration. This is attributed to the formation of Na-montmorillonite in the scalded area and removal of Na from solution.

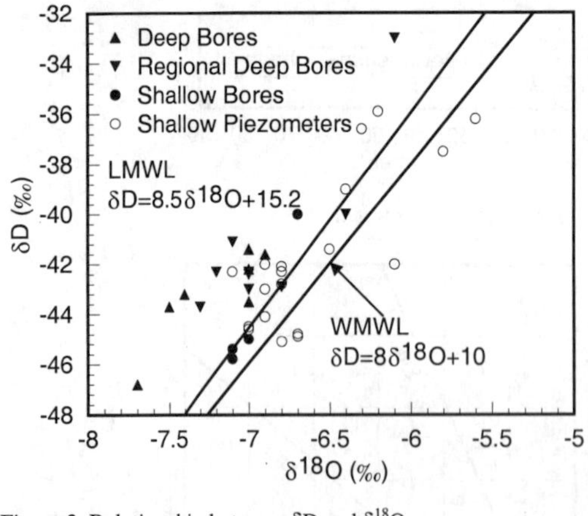

Figure 3. Relationship between δD and $\delta^{18}O$.

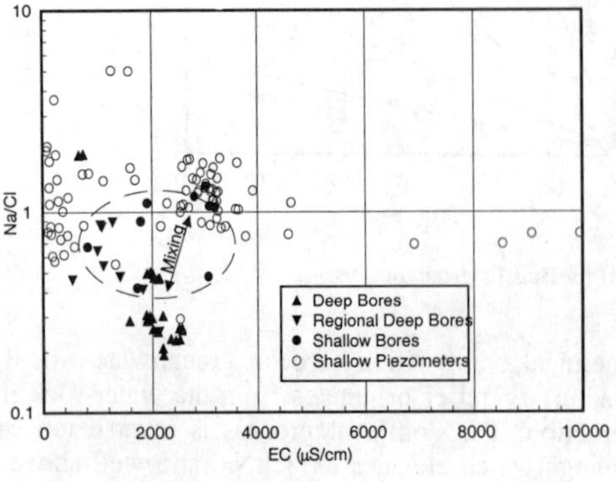

Figure 4. Relationship between Na/Cl versus EC.

Stable isotope plots of D and ^{18}O (Figure 3) show that most groundwaters are located to the left of the Local Meteoric Water Line (LMWL) showing water-rock interaction signature. Only two samples from shallow piezometers are showing evaporative enrichment. These two samples contain the highest concentration of Cl and evaporation from very shallow groundwater is possible particularly during prolonged drought. The isotopic data from the Begalia site show that evaporation is not the major process responsible for the chloride increase in shallow groundwaters. This is consistent with other dryland salinity studies in the Southern Tablelands (Jankowski et al. 1998c, Turner et al. 1991). Figure 5 shows that ion exchange is the dominant process in shallow groundwaters and mixed waters, except in the most saline waters. Reverse ion exchange is present in fractured groundwaters where exchange of Na for Ca and Mg is the dominant process (Jankowski et al. 1998b). In saline shallow waters the formation of Na-montmorillonite is present in the waterlogged, scalded area.

5 GROUNDWATER QUALITY

Clearing of native vegetation increased the flux of recharging water into the fractured aquifer system and has caused the piezometric surface to rise in the discharge zone. This saturates the colluvium by shallow and mixed groundwaters as well as development of seepages. During prolonged dry periods nearly all shallow groundwaters are supersaturated with respect to calcite and dolomite. This process is removing Ca and Mg from the shallow groundwater in colluvium in the ratio of 1:2 respectively. Because more Ca is removed from the shallow waters, they are showing significant depletion in Ca (Figure 2B) and severe magnesium hazard. The excess of Mg over Ca in groundwaters increases dispersion of clay soils and also increases the sodicity hazard where groundwater is dominated by Mg and Na ions rather than Ca and Na ions. Figure 6 shows the relationship between Na and Ca+Mg. The excess of Ca+Mg over Na is evident in deep fractured groundwaters, however, shallow groundwaters have excess of Na over Ca+Mg and most of these waters are plotted on the field of high to very high sodium waters. The mixed groundwaters are plotted on the medium-sodium water field. Of course, chemical composition of these groundwaters is developed not only by simple mixing but also by colluvium-water interaction. The mobilisation of salts in colluvium and ion exchange reactions are elevating the Na concentration. The high Na concentration in shallow groundwaters is causing significant sodium hazard. Most of the shallow groundwaters are plotted on the medium to very high sodium hazard, however, all deep groundwaters are located on the low sodium hazard field. The mixed groundwaters are located on the medium sodium hazard field. The groundwaters that originate from the mixing between deep and shallow systems are plotted on the medium sodium hazard field. The chemical composition of mixed groundwaters is showing that the origin of these waters is clearly associated with the slow movement of groundwater from the fractured bedrock aquifer and shallow groundwaters present in colluvium. The interaction of mixed groundwaters with sediments further modifies their chemical composition.

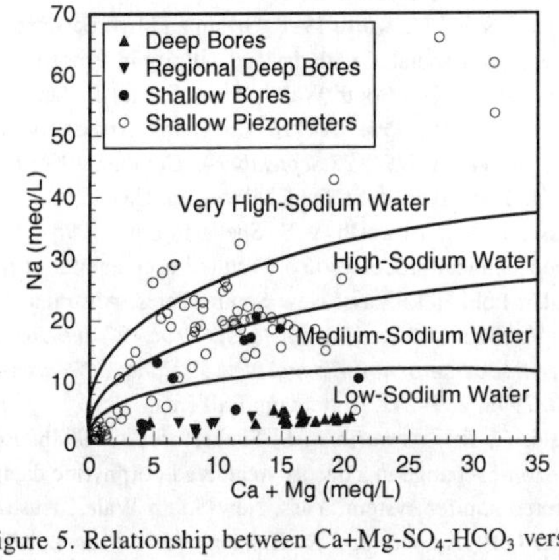

Figure 5. Relationship between Ca+Mg-SO_4-HCO_3 versus Na-Cl.

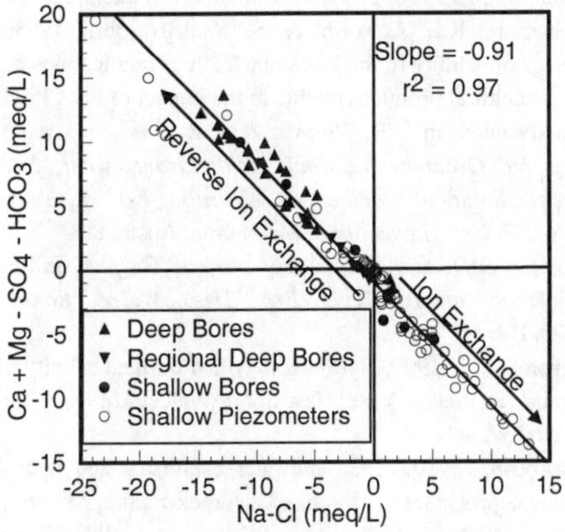

Figure 6. Relationship between Na versus Ca+Mg.

The scalded area is waterlogged and clay soils are saturated by the sodium rich groundwaters. The sodium-saturated soil clays are highly dispersive and very easily eroded causing removal of top-soil. Un der these conditions development of gullying is a rather rapid process. The processes of erosion, dispersion, destabilisation of soils and development of gullying are increased when the clays are in disequilibrium with discharging deep groundwaters from the fractured aquifer. As this clay material was deposited by aeolian accession and reworked by debris-flow processes during the Pleistocene Epoch the deep groundwaters are in chemical disequilibrium with clays and chemical processes of ion exchange and mineral precipitation will be relatively quick. Because of this, the process of degradation always starts when the deep groundwater reaches the ground surface in discharge zone and the clay become saturated. Similar processes of this same origin are present in other small catchments in Southern Tablelands of NSW (Acworth et al. 1997, Jankowski & Acworth, 1997). The high levels of sodium, magnesium, chloride and salinity in seeping groundwaters are toxic for native plants and trees.

6 SUSTAINABLE MANAGEMENT

Sustainable management of this environment is an important task to prevent further increase of salinity as well as to understand the discharge zone and to undertake action in other sites before the total degradation process is well advanced and agricultural land is lost. The land value in the Southern Tablelands is not as high as in lowland areas and remediation action on a big scale is not necessary, however, the flux of saline waters from seepages is adversely affecting waterways, as more salts are transported by run-off from salinity sites. The main problem with the land degradation is the challenge of preventing the development of salinity. From several studies in the Southern Tablelands it was found that the problem can be significantly reduced if the hydrological system and groundwater chemistry in deep and shallow systems is better understood. From these studies it is possible to differentiate clays from weathering (deposited in situ) and from aeolian origin which are causing salinity problems. This can be an important task for further understanding of hydrological and hydrogeochemical processes leading to salinisation. As it was found in the Southern Tablelands (Acworth et al. 1997; Jankowski & Acworth, 1997) dryland salinity processes are not developed in the absence of clay even when groundwater from the bedrock aquifer is discharging on foot-slopes and valley floors. In the absence of clay the weathered zone or soils are not causing chemical disequilibrium as is occurring in the areas with clay soils. The upward discharging groundwater from the fractured bedrock aquifer is causing destabilisation of clays and ion exchange brings more Na into clays causing sodicity and finally dispersion and disintegration of clay particles. Chemical processes in clays through sediment-water interaction decrease hydraulic conductivity in colluvium and as an effect of this drainage is reduced and surface runoff and erosion of the top soil are increased.

The most important sustainable management strategies to prevent the upland degradation in areas of dryland salinity are: (1) assessment of the origin of soil clays and colluvium; (2) isolate or minimise contact of deep groundwater on foot-slopes and valley floors from aeolian clays by depressurisation of fractured rock aquifer, and; (3) develop rehabilitation strategies for degraded soils which include planting of trees with long roots system which are tolerant for toxic elements in water such as Mg, Na, Cl and HCO_3.

The mobilisation of salts in colluvium clays of aeolian origin and their saturation by slowly moving discharging groundwaters will also increase salinity of shallow groundwaters by evaporation leading to much more rapid chemical processes and finally land degradation.

7 CONCLUSIONS

Detailed hydrogeological, hydrogeochemical and geophysical studies in the Southern Tableland of Australia showed that the origin of dryland salinity is clearly associated with the presence of airborne salts in Cainozoic colluvium (Acworth et al. 1997, Acworth and Jankowski in press, Jankowski and Acworth, 1997). The interaction between groundwater and sediments modify chemical composition of waters, when ion exchange is a dominant process responsible for the significant increase of sodium concentration. The sodium concentration can be also be associated with the ion exclusion during uptake by pre cleared vegetation Turner et. al (1991). The excess of sodium in colluvium makes it more erodible, causing soil dispersion and later development of gullying. The conventional form of remediation such, as plantation of trees is not the best option for this area as trees are dying because salt is still present in colluvium.

REFERENCES

Acworth, R.I. 1999. Investigation of dryland salinity using the electrical image method. *Aust. J. Soil Sci.* 37: 623-36.

Acworth, R.I., A. Broughton, C. Nicoll & J. Jankowski 1997. The role of debris-flow deposits in the development of dryland salinity in the Yass River catchment, New South Wales, Australia. *Hydrogeology J.* 5(1): 22–36.

Acworth, R.I. & J. Jankowski In press. Salt source for dryland salinity – evidence from an upland catchment on the Southern Tablelands of New South Wales. *Australian J. Soil Research.*

Broughton, A.K. 1992. The effect of parna-rich debris flow deposits on upland salinity occurrences in the Southern Tablelands, New South Wales. *MAppSc Thesis*, UNSW.

Evans, W.R. 1994. Regional salt balance and implications for dryland salinity management. *Proc. 25th Cong. Int. Assoc. Hydrogeol. Adelaide, Australia, 21-25 November, 1994*: 349-353. Vol. 2, Inst. Eng. Aust., Barton, ACT.

Evans, W.R. 1995. Begalia catchment dryland salinity study. *Tech. Rep. Dep. Water Resour.* TS.95.005.

Evans, W.R. 1998. What does Borowa tell us? Salt stores and groundwater dynamics in a dryland salinity environment. In T.R. Weaver & C.R. Lawrence (eds), *Proc. Int. Groundwater Conf. 1998: Groundwater: Sustainable Solutions, Melbourne, Australia, 8-13 February 1998*: 267-274. University of Melbourne, Australia.

Jankowski, J. & R.I. Acworth 1993. The hydrogeochemistry of groundwater in fractured bedrock aquifers beneath dryland salinity occurrences at Yass, NSW. *AGSO J. Aust. Geol. Geophys.* 14(2&3): 279-285.

Jankowski, J. & R.I. Acworth 1997. Impact of debris-flow deposits on hydrogeochemical processes and the development of dryland salinity in the Yass River catchment, New South Wales, Australia. *Hydrogeology J.* 5(4): 71-88.

Jankowski, J & R.I. Acworth 1998. Origin and hydrochemistry of seepages associated with dryland salinity in the southern Tablelands of New South Wales, Australia. In: J. Van Brahana et al. (eds), *Proc. XXVIII Cong. Int. Assoc. Hydrogeol., Las Vegas, USA, 27 September–2 October, 1998*: 67–74. Am. Inst. Hydrol., St. Paul, Minnesota, USA.

Jankowski, J., R.I. Acworth & S. Shekarforoush 1998a. Hydrogeochemical processes in a fractured rock aquifer of the Lachlan Fold Belt: Yass, New South Wales, Australia. In: G.B. Arehart & J.R. Hulston (eds), *Proc. 9th Int. Symp. Water-Rock Interaction, Taupo, New Zealand, 30 March–3 April, 1998*: 239–242. Rotterdam: Balkema.

Jankowski, J., R.I. Acworth & S. Shekarforoush 1998b. Reverse ion-exchange in a deeply weathered porphyritic dacite fractured aquifer system, Yass, New South Wales, Australia. In G.B. Arehart & J.R. Hulston (eds), *Proc. 9th Int. Symp. Water-Rock Interaction, Taupo, New Zealand, 30 March–3 April 1998*: 243–246. Rotterdam: Balkema.

Jankowski, J., R.I. Acworth & S. Shekarforoush 1998c. Sources of salinity in the Kyeamba Valley: Implications for the agricultural productivity due to the impact of Na-Cl rich groundwaters. In T.R. Weaver & C.R. Lawrence (eds), *Proc. Int. Groundwater Conf. 1998: Groundwater: Sustainable Solutions, Melbourne, Australia, 8-13 February 1998*: 287-292. University of Melbourne, Australia.

Lawson, S. 1989. Dryland salinity project: Begalia drilling completion report. *Tech. Rep. Dep. Water. Resour.* TS.88.018.

Lertsirivorakul, R. 1990. Hydrogeology and dryland salinity of Begalia catchment, Yass, New South Wales. *MAppSc Thesis*, UNSW.

Shekarforoush, S. 1999. Groundwater chemistry and hydrochemical processes in fractured bedrock aquifers beneath dryland salinity occurrences in SE Australia. *PhD Thesis*, UNSW.

Turner, J.V., T.D. Waite & J.M. Bradd 1991. The conjunctive use of isotopic techniques to elucidate solute concentration and flow processes in dryland salinized catchments. *Proc. Int. Hydrol. Water Resour. Symp. Perth, Australia, 2-4 October 1991*: 118-125. Vol.1. Inst. Eng. Aust. Barton, ACT.

Waite, T.D. & J.V. Turner 1989. Solute concentration and flow processes in the Cootamundra region of New South Wales: Application of hydrogeochemical and isotope techniques. *Australian Nuclear Science and Technology Organisation Report* ANSTO/C97.

Groundwater: Past Achievements and Future Challenges, Sililo et al. (eds) © 2000 Balkema, Rotterdam, ISBN 90 5809 159 7

Biofouling in boreholes of the Klein Karoo Rural Water Supply Scheme, Oudtshoorn, South Africa, and the effect on well scheme management

J.L.Jolly
Groundwater Consulting Services, Cape Town, South Africa

ABSTRACT: Development of production borehole supplies for the Klein Karoo Rural Water Supply Scheme between Calitzdorp and Dysselsdorp in South Africa, is relatively well understood. However, iron problems in the production boreholes have complicated the abstraction. Iron related biofouling has resulted in clogging of pumps and borehole screens, further resulting in a decrease in the borehole efficiencies and a decrease in borehole yields. As a result boreholes have had to be rehabilitated at considerable cost and disruption to the scheme management. To minimize clogging problems, changes have been made to the boreholes, pumps and to abstraction conditions. Boreholes are pumped at lower rates, but for longer, while pump positions have been raised to levels above the first water strike. Monitoring has also been introduced to assess biological activity and borehole efficiencies, thus allowing better control of rehabilitation that might be necessary.

1 INTRODUCTION

The Klein Karoo Rural Water Supply Scheme provides approximately 1 million cubic metres of groundwater to the rural populations between Dysselsdorp and Calitzdorp in the Klein Karoo area of South Africa on an annual basis (Fig.1). In many holes biofouling has effected borehole use.

Biofouling has been studied for many years. The clogging process can be extremely rapid, reducing the borehole yields to below their recommended capacity within months of installing (Hackett 1987); (Cullimore 1993). Boreholes can however be remediated, although effective remediation requires knowledge on the causes of clogging (Smith 1995).

Groundwater Consulting Services (GCS) and the CSIR Groundwater Group (GWG) undertook in a joint venture investigation into the iron problems experienced in boreholes in the Klein Karoo Rural Water Supply Scheme. The work was undertaken for the South African Department of Water Affairs and Forestry. The investigation is split into three distinct phases. Phase 1 consisted of a desk study accumulation of all the available information related to iron precipitation and iron bacterial problems with specific reference to management measures that can be introduced to reduce the problems. Phase 2 covered the evaluation of historical data related to iron problems and the accumulation of new data on the iron chemistry and bacterial status of the groundwater in all the production boreholes on the scheme. Samples were collected from 15 boreholes. Detailed chemical and microbiological analyses were undertaken to evaluate the microbial populations present and the conditions influencing their growth as well as the chemical/biological precipitation of iron. The final phase included the formulation of management procedures or treatment systems to reduce iron problems.

2 BACKGROUND

The boreholes abstract water from the Table Mountain Group (TMG) Aquifer, a quartzitic sandstone, secondary aquifer system, with two abstraction areas, namely Calitzdorp wellfield in the west and five wellfields adjacent to Dysselsdorp in the east (Fig.1). The boreholes are capable of high yields (5 - 35 l/sec), although the low rainfall of the area (350 - 450 mm per annum) results in low recharge rates and lower sustainable borehole yields of 2 - 10 l/sec.

Historically, boreholes were pumped for variable periods of time (between 8 - 24 hours) based on the demand requirements and at rates far above the sustainable rates. During peak demand periods, holes were often over abstracted, sometimes for 2-3 months at a time.

The fieldwork and data evaluation undertaken on the Klein Karoo boreholes was aimed at identifying the cause of the borehole clogging and investigating options for the management of clogging in the future.

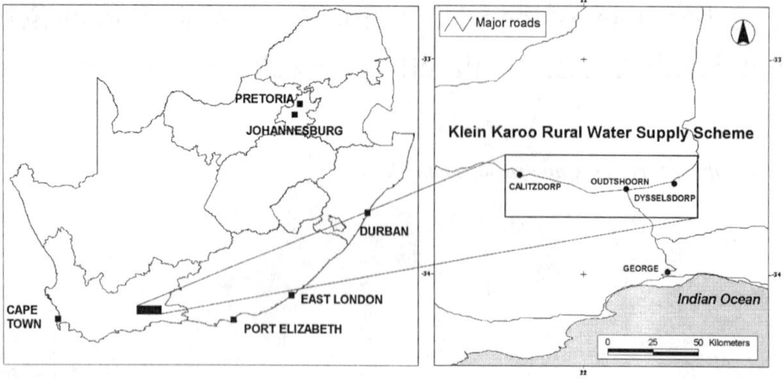

Figure 1 : Location of the Study area

3 DATA INTERPRETATION AND FIELD RESULTS

3.1 *Borehole yields*

The majority of the boreholes in the Table Mountain Sandstones in the Klein Karoo are being pumped at less than 20% of the maximum achievable pumping rates. This is a result of the low rainfall and limited recharge which results in the borehole's sustainable yield being far lower than the maximum yield achievable. As a result, historical clogging of screens has had limited impact on the ability of most of the boreholes to maintain the low pumping rates set, although borehole efficiencies have dropped.

3.2 *Iron Concentrations*

Iron is usually measured in a groundwater sample as total iron (both dissolved and solid state) - this is acceptable in a clean, clear groundwater sample because the solid state is limited. However, if the groundwater is turbid, there is usually a large contribution to total iron from the solid state and the dissolved iron component will be considerably less. It is important, from a comparison perspective, to know if the iron recorded is total or dissolved iron.

Assessment of historical changes in iron levels did not show any conclusive results of an overall increase in the iron levels - in fact the opposite appears to have occurred. Iron concentrations have dropped since 1992 in all but 3 of the holes. This may be as a result of oxidation of the ferrous iron in solution, causing precipitation of ferric oxyhydroxides and a lowering of the dissolved iron concentrations. Increases in iron concentration should only take place under circumstances where reduction, due to acid generation, results in the disillusion of any iron scale developed.

Iron levels fluctuate wildly in many of the boreholes (2-8 mg/l being a typical variability). These erratic changes are as a result of the sloughing of the biofilm and signifies that bio-accumulation is in an advanced phase.

3.3 *Odour/Turbidity*

A number of boreholes have a turbidity problem and the groundwater has a typical red to orange tinge associated with iron bacteria. This discoloration is normally worst during the initial few minutes when pumping starts, improving with time, indicating bacterial growth in and directly adjacent to the hole. Odours commonly occur, including a musty earthy odour, or a sulphurous rotten egg smell, or an earthy musty smell.

3.4 *Specific Capacity*

Specific capacity is a measure of the "productivity" of a borehole, defined by the yield per unit of drawdown after a given time has elapsed. Comparison of specific capacity in a borehole is only meaningful at a constant pumping rate. Variations in pumping rates in the Klein Karoo boreholes has result in changes in specific capacity, thus making it exceptionally difficult to evaluate whether clogging is causing any decrease in the specific capacity in a borehole. An assessment of changes in specific capacity was made by evaluating specific capacity calculated from step tests undertaken during 1995 and further tests undertaken during 1997/98. In 50% of the holes tested the specific capacities dropped - the drop being between 10 - 30%.

3.5 *Groundwater Chemistry*

The water is typically of a sodium chloride nature, with electrical conductivities ranging between 20 mS/m in the east and 70 mS/m. at Calitzdorp.

The field pH measured shows the water to be acidic, with the Calitzdorp boreholes being the most neutral (pH variations between 6.42 and 6.2). The remainder of the boreholes have a pH range between 6.3 and 4.8. Two borehole (in the Bokkraal wellfield at Dyssesldorp) have a pH between 4.2 and 3.4 - these pH conditions are not ideal for the growth of bacteria, being too low. Under most of the prevailing pH conditions bacterial growth would be retarded and only specific bacteria would thrive, the conditions suiting *Thiobacillus*.

The Eh was not measured directly in the field but was calculated from the dissolved oxygen (DO) measurements undertaken in the field. Dissolved oxygen levels in groundwater are commonly related to the amount of organic material in the soils through which rainfall must infiltrate before entering the groundwater. Where thick soils with high humic concentrations exist DO oxygen levels are commonly low while fractured rock environments with thin soil covers will have higher DO levels. In areas where biological activity in boreholes takes place it is likely that this biological action will reduce the dissolved oxygen content in the groundwater - this explains the low DO levels in the Calitzdorp wellfield (high bacterial activity) and the high DO levels in the Vermaaks river wellfield (limited biological activity). The lower limit of dissolved oxygen necessary for the existence of most aerobic bacteria is considered to be about 0.05 mg/l. Based on the dissolved oxygen levels in the Klein Karoo most of the bacteria existing should be aerobic bacteria .

The Eh / pH relationship (Fig.2) indicates an oxidised environment.

3.6 *Heterotrophic plate counts*

The majority of the boreholes have exceptionally high heterotrophic plate counts (HPC) with the exception of the Vermaaks river boreholes which showed low levels.

Plate counts do not adequately take into account anaerobic bacteria. As a result the total bacterial populations cannot be accurately gauged. It is interesting to note that the HPC in the Vermaaks river boreholes decreases with distances into the valley (and thus also with greater depth to groundwater level). The low HPC for one of the boreholes in the Bokkraal wellfield (which has a know iron clogging problem) is surprising and is probably related to the anaerobic nature of the bacteria in the borehole. The HPC for the boreholes varied between 25000 - 72000 in the worst holes, to less than 500 in irons with limited clogging problems.

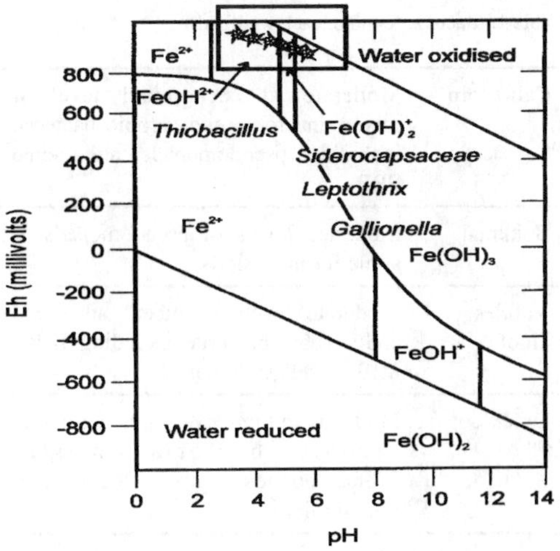

Figure 2. Eh/pH environments found in the boreholes (indicated by stars in block).

3.7 *BARTS Tests*

The BARTS biodetectors are test kits developed to determine the aggressivity and forms of biofouling taking place in water (Cullimore 1993). The test vial is so designed that bacteria in the water added to the test vial are provided with a number of different environmental niches, based on the formation of a redox gradient in the vial. Different growth mediums and nutrients are placed in different vials, allowing growth of different microorganisms.

The iron reducing bacteria (IRB) and sulphate reducing bacteria (SRB) test vial were used during this investigation. The BARTS tests (the reaction time, the reactions, the aggressivity and the bacteria found in each hole) indicated that pseudomonads are the most common bacteria found. The detailed bacterial breakdown per wellfield is shown in Table 1.

Variability in the results from repeat tests were put down to the effect of sloughing of the biofilm - high sloughing would result in high aggressivity readings. Sulphur oxidizing bacteria may be dominant under the prevailing low pH conditions existing in the Bokkraal wellfield. The difference for the Vermaaks river boreholes between the BARTS test results (showing high aggressivity) and the heterotrophic plate counts (showing limited bacteria) was ascribed to the BARTS tests culturing more aggressive and diverse bacteria than the agar spread plate.

The BARTS test indicate low growth (background levels) of sulphate reducing bacteria in most of the holes. There is however a concern with the SRB levels measured - SRB often grow under

Table 1. Bacteria found in each wellfield

Calitzdorp	Moderate to background levels of mixed anaerobic and aerobic bacteria, including pseudomonads and some SRBs.
Bokkraal	Moderate levels of pseudomonads & slime forming SRBs
Varkies-kloof	Moderate levels of mixed anaerobic and aerobic bacteria, including IRBs, SRBs and Pseudomonads
Droëkloof	Moderate to high levels of anaerobic & aerobic bacteria, mostly Pseudomonads, plus IRB's and Desulfovibrio
Vorsorg	Moderate to high levels of mixed anaerobic, enteric, SRB and Pseudomonad bacteria plus slime formers capable of functioning aerobically
Vermaaks river	Moderate to high levels of mixed anaerobic and aerobic bacteria, dominated by pseudomonads and Desulfovibrio

the biofilm in the anaerobic zone. Unless considerable sloughing of the biofilm takes place, the SRB are not released into the water and the BARTS tests could well be under-estimated of SRB growth. Moderate to high levels of iron related bacteria exist, with non-IRB pseudomonads being dominant.

3.8 *Scale Chemistry*

Four samples of precipitate, as well as analysis of the groundwater chemistry drawn from the boreholes in question, were submitted to the Department of Geological Sciences at the University of Cape Town for analysis. Analysis included examination for elemental composition undertaken by Energy Dispersive Spectroscopy, (EDS), morphology and particle size determination by Scanning Electron Microscopy (SEM), phase identification performed using X-Ray Diffraction (XRD), and qualitative tests using hydrogen peroxide (H_2O_2) & hydrochloric acid (HCl) to test for the presence of manganese oxides or calcite respectively.

The effervescence tests showed a reaction to the addition of hydrogen peroxide on all the samples, suggesting than manganese oxides and/or organic material exist. There was however, no effervescence observed after the addition of HCl, indicating that no carbonates exist. In view of the absence of Mn and the presence of carbon in some samples (seen from the EDS analysis), the effervescence observed upon H_2O_2 addition was probably derived from organic carbon and not manganese dioxide.

The major elemental composition of the precipitates are Fe, C and O - it is noticeable that there was no Mn. The mineral is mostly in the form of a poorly crystalline ferrihydrite, although an alteration to goethite has taken place in some holes, suggesting a more advanced stage of precipitation.

The saturation indices provide an indication of the potential of a specific mineral to precipitate out from solution. The results showed a strong potential for the precipitation of Goethite.

4 SUMMARY

The fieldwork showed elevated plate counts, moderately aggressive BARTS reactions and conditions conducive to chemical precipitation of ferrihydrites and goethite. Oxgenation of the water, due to water cascading down the hole from upper water strikes, has resulted in highly oxidized conditions where pseudomonads are growing strongly. All these conditions are conducive to borehole clogging.

5 MANAGEMENT ISSUES

To reduce the rate of clogging, some changes to the operation of the boreholes were introduced.

5.1 *Reduce oxygenation of the groundwater*

Oxygenation of the water is directly related to pumping at high rates with too large a drawdown, thus allowing water to cascade down the sides of the borehole. Oxygenation is decreased by limiting the drawdown in the borehole, achieved by reducing pumping rates and extending pumping times to achieve the same daily abstraction volume. This was implemented.

5.2 *Monitoring*

Rehabilitation of a boreholes should be based on sound monitoring data which would indicate a necessity to undertake expensive rehabilitation. The aim of the monitoring is to assess the bacterial and chemical "health" of the hole, the effect of clogging and the need for borehole rehabilitation. A hierarchical order of monitoring was implemented and is shown in Table 2.

Table 2. Monitoring measures introduced.

Heterotrophic Plate Count	Undertaken on the utilized production boreholes, on a monthly basis.
Dissolved Oxygen	Measured in the field (monthly) and converted to Eh so that the redox potential of the water can be assessed. This is important in evaluating the chemical precipitation potential and potential changes from oxidizing to reducing conditions.
Specific Capacity	A set monthly procedure implemented to ensure comparable specific capacity measurements
BARTS tests	Undertaken on those holes where the specific capacity is dropping, utilizing tests for IRB, SRB & slime forming bacteria (SLYM). Four samples collected, a week apart, to average out sloughing effects.
Ground-water chemistry	Full chemical analyses undertaken on a quarterly basis, with monthly monitoring of iron levels
Video camera	Only holes showing a decrease in specific capacity and aggressive bacteria will be viewed by camera.

5.3 *Regular Bacterial disinfection*

Six monthly disinfection of the wells is planned, utilizing liquid sodium hypochlorite. The aim of the disinfection is not to restore lost capacity, but merely to retard bacterial growth.

5.4 *Scale removal*

Holes showing a decline in specific capacity and where clogging is confirmed by video camera inspection, will have to be rehabilitated. Rehabilitation should be aimed at breaking down and removing the scale. The exact nature of the rehabilitation treatment depends on the groundwater chemistry and the bacteria growing in the hole.

5.5 *Replacement of screens*

It appears that stainless steel screens clog at a faster rate that PVC screens. Any new holes drilled, or existing holes re-screened, should be equipped with PVC screens. If the formation is stable, screens / casing should not be used at all, since the screens only promote turbulent flow and decrease the borehole efficiency. Screens only need be emplaced to protect the pump and should therefore not be placed below anticipated pump depths.

5.6 *Correct equipping of boreholes*

In future new holes drilled in the TMG aquifer (especially in the Nardouw Formation rocks) or in other aquifers where biofouling is known to be a problem, should be drilled and equipped in such a way as to seal off initial shallow water strikes, thus making certain that no cascading takes place in the hole. This requires a change in current drilling practises, or the courage of the geohydrologist to seal off initial water strikes !!

6 REFERENCES

Cullimore, D.R. 1993. *Practical manual of ground-water microbiology*. CRC Press, Florida, p 412.

Hackett, G. 1987. A review of chemical treatment strategies for iron bacteria in wells. *Water Well Journal*, February, pp 37-42.

Smith, S.A.1995. *Monitoring and Remediation Wells*, Lewis Publishers, Boca Raton.

Groundwater: Past Achievements and Future Challenges, Sililo et al. (eds) © 2000 Balkema, Rotterdam, ISBN 90 5809 159 7

Isotopic characterization of groundwater and surface water in the Keta Basin, Ghana

N.O. Jørgensen
Geological Institute, Copenhagen, Denmark

B.K. Banoeng-Yakubo
Geology Department, Legon, Ghana

ABSTRACT: This paper presents the application of environmental isotopes ($^{18}O/^{16}O$, D/H and $^{87}Sr/^{86}Sr$) to groundwater and surface water studies with emphasis on saline groundwater in aquifers in the Keta Basin, Ghana. The isotopic compositions of oxygen and hydrogen of deep groundwater reveal small variations and plot close to the global meteoric water line. Shallow groundwater and surface water reveal considerably larger variations in the isotopic compositions and show pronounced evaporitic enrichment. The $^{87}Sr/^{86}Sr$ ratios of the groundwater and surface water primarily reflect the geology of the formations in the aquifers and the recharge areas. However, groundwater from a number of deep wells and dug wells in near-coastal aquifers is characterized by relatively high Cl contents. A significant marine influence in these aquifers is indicated by well-defined mixing lines for both strontium isotopes, hydrogen and oxygen isotopes with the isotopic composition of modern seawater as one end-member.

1 INTRODUCTION

The Keta Basin is one of the fault-controlled sedimentary basins in West Africa along the southeastern coast of Ghana and is dominated by the Volta River estuary (Akapati 1978) (Figure 1). The sedimentary sequence in the Keta Basin constitutes one of the most profilic aquifer systems in Ghana (Nerquaye-Tetteh 1993, Bannerman 1994, Frempong & Kortatsi 1994). Groundwater plays a vital role in the domestic water supply and in farming economies and, furthermore, supplies the fairly large communities living along the coastal strip between the outlet of Volta River and the Ghanaian-Togo border.

However, a major problem in drinking water quality and domestic water supply management in the Keta District is high salinity of groundwater in dug wells as well as from deep bore holes (Gill 1969, Bannerman 1994). A large number of prospected well sites have not been carried out due to salinity problems in the aquifers and existing water wells and dug wells have been abandoned due to increasing salinity over time. The extensive occurrences of groundwater with relatively high chloride concentrations are therefore particularly noteworthy and may warrant further consideration.

The objective of this study is to analyse four major aquifers in the Keta Basin by means of oxygen, hydrogen and strontium isotopes to establish a multi-isotopic hydrogeochemical investigation of the groundwater and surface water and to elucidate the source of salinization of deep and shallow groundwater.

2 HYDROGEOLOGICAL SETTING

Four major aquifers in the Keta Basin can be distinguished according to the geography and the geological setting (see Figure 1): 1) Weathered Dahomeyan gneisses in the northeast rim of the basin. They outcrop around Akatsi and are locally overlain by Recent to Neogene continental sedimentary deposits of sands and gravels; 2) Neogene continental deposits of unconsolidated to semi-consolidated limonitic argillaceous sands in the northeast and central part of the basin; 3) Cretaceous-Eocene marine limestones and sandstone beds exploited in the central and southeastern seaward part of the basin and 4) Quaternary deposits including Holocene to Pleistocene coastal marine sands and gravels in the Volta River estuary and Keta Lagoon area.

3 ANALYTICAL TECHNIQUES

Water sampling took place during two periods, first in the rainy season October 1996 and second in February-March 1998, towards the end of the dry season. In all, groundwater samples were collected

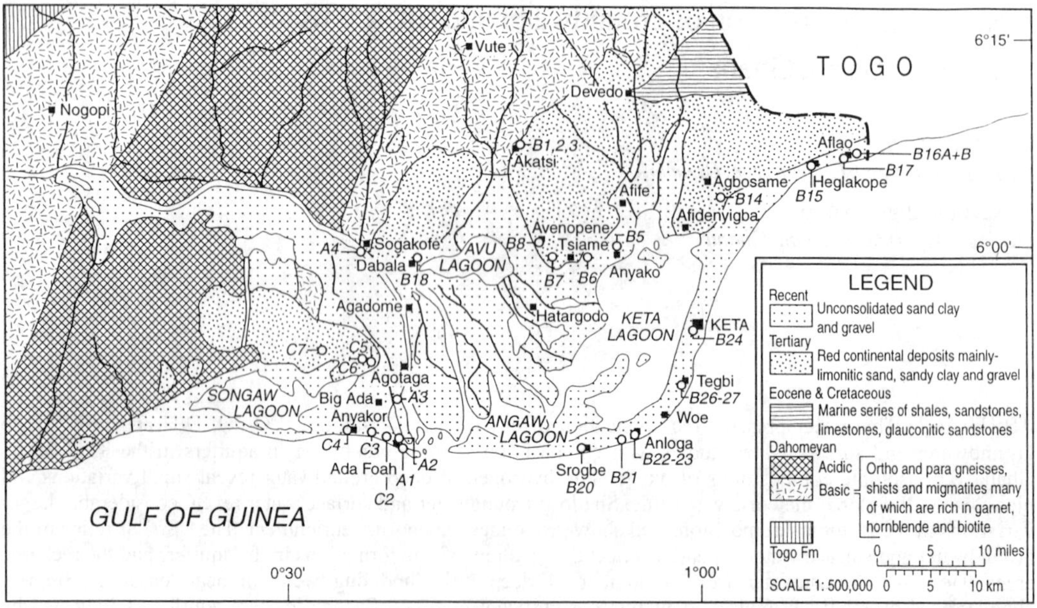

Figure 1. Geological map of the Keta Basin, Ghana showing the locations of wells and surface water sampling sites investigated.

twice from 16 boreholes and 18 dug wells. Surface water includes freshwater from the Volta River, brackish water from the Keta Lagoon and seawater. On-site analyses comprised temperature and electrical conductance (EC) corrected to 25^{0}C. Samples were collected in acid-washed polyethylene bottles and filtered (0.45 μm) for subsequent laboratory analyses. The Cl contents of water samples were determined by a digital titrator using the mercuric nitrate method and the Sr analyses were carried out with atomic absorption spectrophotometry.

The oxygen stable-isotope analyses of water samples were obtained on a VG Sira 10 mass spectrometer with automatic inlet. The water samples for stable hydrogen isotope analyses were reduced to H_2 by use of zinc in quartz ampoules at 700°C. The analyses were carried out in a Finnigan MAT 250 triple collector mass spectrometer. The results for both isotopes are expressed in per mille deviation from the VSMOW standard using the standard δ notation. Reproducibility of $\delta^{18}O$ is better than 0.1‰ and about 1‰ for δ^2H. Freeze-dried 50 ml water samples were were analysed for $^{87}Sr/^{86}Sr$ ratios. The dry matter was dissolved in 6N HCl and evaporated to dryness. Subsequently, Sr was separated from a 2.5N HCl solution by conventional ion exchange. Approximately 1μg $SrCl_2$ was mounted on tantalum filaments with phosphoric acid. The Sr isotopic composition was measured on an 8-collector Fisons Sector 54-30 mass spectrometer. Data were collected by multidynamic analysis, and exponential mass-fractionation was assumed in the applied correction to a $^{86}Sr/^{88}Sr$ ratio of 0.1194. Within-run precision was typically ±0.000005 (1 s.d.) on $^{87}Sr/^{86}Sr$ and always better than ±0.000007. In the course of data acquisition, the average of measured values of the international standard material NBS987 was $^{87}Sr/^{86}Sr$ = 0.710245 ±0.000011 (1 s.d.), n = 43.

4 RESULTS

4.1 *Surface water*

The surface water studied includes samples of freshwater from the Volta River, brackish water from the Volta Estuary and from the Keta Lagoon. Seawater was collected close to the seashore west of Ada Foah and off Keta along the coastal strip (see Figure 1). The chloride contents of brackish water and freshwater vary from c. 13,000 mg/L Cl in the seaward part of the Keta Lagoon to 16 mg/L Cl in the Volta River at Sogakofe. Seawater shows $\delta^{18}O$ and δ^2H values very close to zero for both isotopes as reported for normal oceanic seawater (Figure 2). Surface water collected at the delta outlet close to Ada Foah and from different sites in the estuary and the Keta Lagoon reveal a strong marine influence on the isotopic compositions with δ^2H and $\delta^{18}O$ values slightly depleted but close to zero. Surface water from the Volta River collected further from the sea reveals isotopic compositions closer to but below the global meteoric water line (Craig 1961) and are significantly enriched in comparison to the

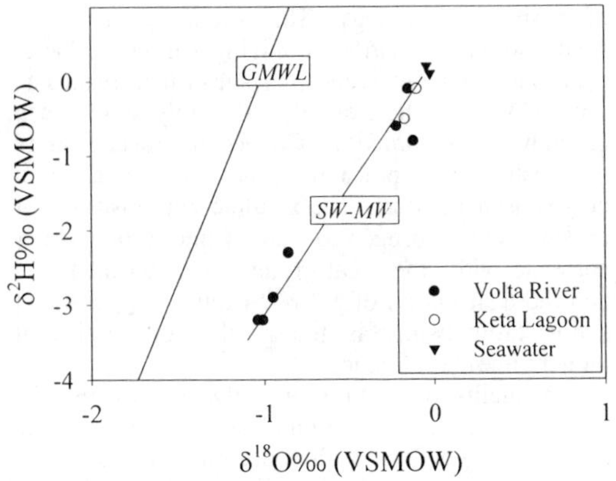

Figure 2. $\delta^{18}O$ vs δ^2H diagram for surface water in the Keta Basin, Ghana. The global meteoric water line (*GMWL*) and the seawater-meteoric water mixing line (*SW-MW*) are shown for reference.

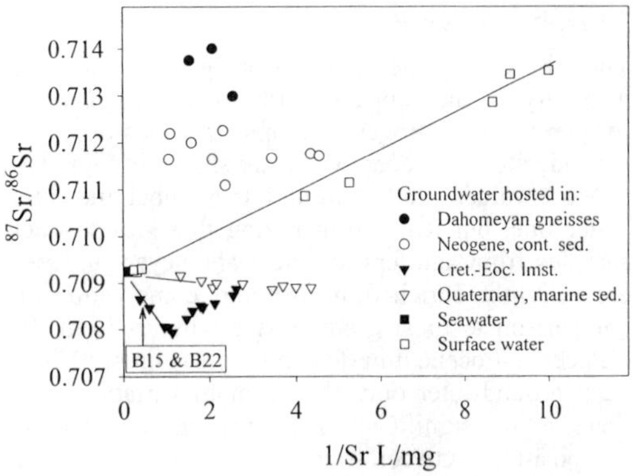

Figure 3. $^{87}Sr/^{86}Sr$ ratios vs 1/Sr diagram for groundwater and surface water from the Keta Basin, Ghana, and seawater from the Gulf of Guinea. The seawater-meteoric water mixing lines are shown for reference.

groundwater. The Sr isotopic compositions of the surface water vary from relatively high isotopic ratios of 0.7130 for the Volta River water to approximately 0.7112 for the seaward parts of the estuary and the Keta Lagoon and to 0.7092 for the seawater (Figure 3). The $^{87}Sr/^{86}Sr$ ratio at 0.7092 corresponds to the isotopic ratio normally reported for oceanic seawater (Faure 1986). The Sr isotopic ratios and the reciprocal Sr concentrations of the surface water samples reveal a mixing line (r^2=0.99) with seawater as one end-member and meteoric water highly influenced by Sr released from Proterozoic crystalline basement as the other end-member.

4.2 *Shallow groundwater*

Shallow groundwater may reveal relatively high chloride contents in the range up to 500 mg/L Cl. The oxygen and hydrogen isotopic compositions of shallow groundwater are shown in Figure 4. The diagram shows a wide range of values scattering around the global meteoric water line. However, a significant number of samples are enriched with respect to both isotope in comparison to the isotopic composition of normal groundwater, which is indicative of evaporation and/or influence of modern seawater. The variation in the isotopic composition of shallow groundwater is significantly greater than in the case of the deep groundwater (Figures 5 and 6). Furthermore, groundwater samples collected in the dry season of 1998 are generally slightly enriched in $\delta^{18}O$ compared to water samples collected in the the rainy season of 1996. Shallow groundwater from dug wells in aquifers consisting of Neogene continental deposits of sandy clay, sands and gravels shows relatively high Sr isotope ratios in the range 0.7111 - 0.7126 (Figure 3). There is a considerable scatter in the values and no correlation exists between the isotope ratios and the reciprocal Sr concentrations in the groundwater and the Sr isotope ratios show no evidence of modern marine influence. Shallow groundwaters from dug wells in Quaternary unconsolidated coastal marine sands and gravels in the seaward part of the Keta Lagoon areas show Sr isotope ratios in the range 0.7089 - 0.7091. Furthermore, the Sr isotope ratios show a significant correlation (r^2=0.58) with the 1/Sr contents indicating a well defined mixing line with the isotopic composition of seawater (0.7092) and relatively high Sr contents as one end-member. The other end-member is most likely dissolved Sr in low concentrations from Holocene-Pleistocene biogenic carbonates or reworked Neogene skeletal material.

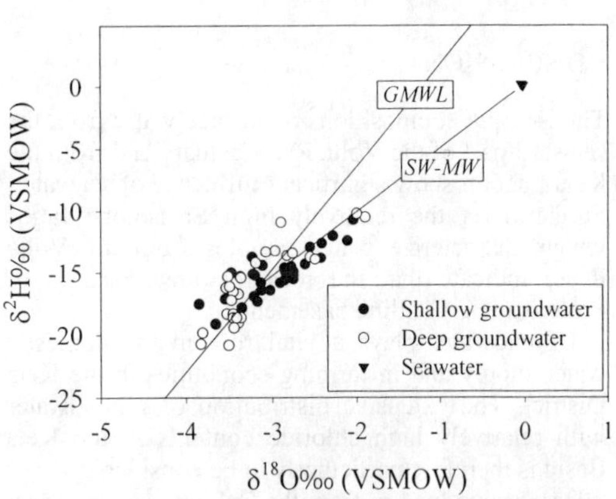

Figure 4. $\delta^{18}O$ vs δ^2H diagram for groundwaters and seawater in the Keta Basin, Ghana. The global meteoric water line (*GMWL*) and the seawater-meteoric water mixing line (*SW-MW*) are shown for reference.

4.3 *Deep groundwater*

The chloride contents of deep groundwater are generally in the range of 100 to 200 mg/L. The oxygen and hydrogen isotopic compositions of groundwater from deep wells are shown in Figure 4.

Most values scatter around the global meteoric water line, but it is worth noting that groundwater samples from aquifers in the Dahomeyan gneisses are slightly enriched in isotopic compositions in comparison to most groundwater samples from the Cetaceous-Eocene limestone aquifers (Figure 6). The deep groundwater only shows small variations and there is no significant difference in the isotopic compositions between water samples collected in the dry season 1998 and the rainy season 1996 (Figure 6). However, two boreholes in the Cretaceous-Eocene limestone aquifers, B15 and B22, yield saline groundwater with chloride contents of 1500 mg/L Cl and 7500 mg/L, respectively. Groundwater from these two boreholes have significantly high δ^2H and $\delta^{18}O$ values (Figure 6).

The Sr isotopic compositions of deep groundwater are shown in Figure 3. The groundwater from the fractured and weathered Dahomeyan gneisses has higher $^{87}Sr/^{86}Sr$ ratios than other groundwater sampled and is characterized by relatively high Sr isotopic ratios in the range of 0.7130 - 0.7140.

The groundwater samples from the Cretaceous-Eocene aquifers show Sr isotopic compositions that reflects the stratigraphic age of the marine carbonate in the formations (Burke et al. 1982, Jørgensen & Holm 1994 and 1995). However, the two wells, B15 and B22, which yield saline groundwater show isotopic compositions that fit well with the local seawater - groundwater mixing line for both strontium isotopes and oxygen and hydrogen isotopes with the isotopic compositions of modern seawater as one end-member (Figures 3 and 4).

5 DISCUSSION

The isotopic compositions of surface water from the seaward part of the Volta River estuary and from the Keta Lagoon show significant influence of seawater. Furthermore, the relatively high Sr isotope ratios which characterize water samples from the Volta River indicate that the recharge area consists of Proterozoic crystalline basement.

Groundwater plays a vital role in the domestic water supply and in farming economies in the Keta District. The extensive distribution of groundwater with relatively high chloride contents in the Keta Basin is therefore particularly to be considered.

Deep groundwaters from the Dahomeyan basement and Cretaceous-Eocene aquifers are considered to yield potable water of relatively high quality when applying careful well construction and consideration of abstraction strategy. This groundwater shows small isotopic variations implying similar recharge conditions and relatively long subsurface residence time. However, the examples of highly saline deep groundwater recorded in Cretaceous-Eocene strata demonstrate that precautions have to be taken in near-coastal aquifers. The isotopic compositions of oxygen and hydrogen of the saline groundwater correlate with the local groundwater-seawater line and the distribution of $^{87}Sr/^{86}Sr$ ratios suggests that saline groundwater is the result of significant modern marine influence.

The quality of shallow groundwater is cause for greater concern. This groundwater shows a wide range of Cl contents with concentrations up to 500 mg/l, which is far above the limit for a satisfactory potable water supply. Significant evidence of modern marine influence is found in shallow groundwater from aquifers in Holocene-Pleistocene marine deposits in the seaward part of the Keta Basin. Seawater intrusion is indicated by the $^{87}Sr/^{86}Sr$ ratios vs 1/Sr diagram (Figure 3) which show well-defined mixing lines with end members with relatively high Sr contents and isotopic compositions close to modern seawater.

In contrast, the isotopic compositions of shallow groundwater in continental Neogene aquifers do not give any evidence of modern marine influence although the recorded Cl contens may be even higher than in groundwaters collected from near-coastal Quaternary marine deposits. However, the oxygen and hydrogen isotopic compositions suggest that evaporation plays an important but secondary role in Cl distribution. The Sr isotopes reveal no significant correlation between $^{87}Sr/^{86}Sr$ ratios and the reciprocal Sr concentrations which indicates that the dissolved Sr may have multiple sources due to weathering processes of a variety of rock materials of different origin. Thus, It is believed that the relatively high Cl concentrations found in these aquifers are the result of a combination of evaporation and weathering processes.

6 CONCLUSIONS

The hydrogen, oxygen and strontium isotopic compositions allow to identify three distinct groups of surface water and groundwaters in the Keta Basin:

1) Surface water from the Volta River and the Keta Lagoon reveals isotopic compositons which indicate modern marine influence and show well-defined mixing lines with seawater as one end-member and freshwater from the Volta River as the other endcharacterized by strong evaporitic enrichment and relatively high $^{87}Sr/^{86}Sr$ ratios due to prevailing recharge from Proterozoic crystalline terraines.

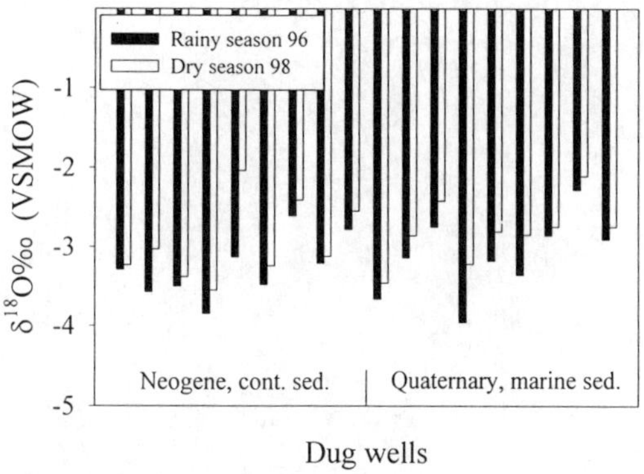

Figure 5. $\delta^{18}O$ distribution in shallow groundwater obtained from dug wells in the Keta Basin, Ghana, in the rainy season of 1996 and the dry season of 1998.

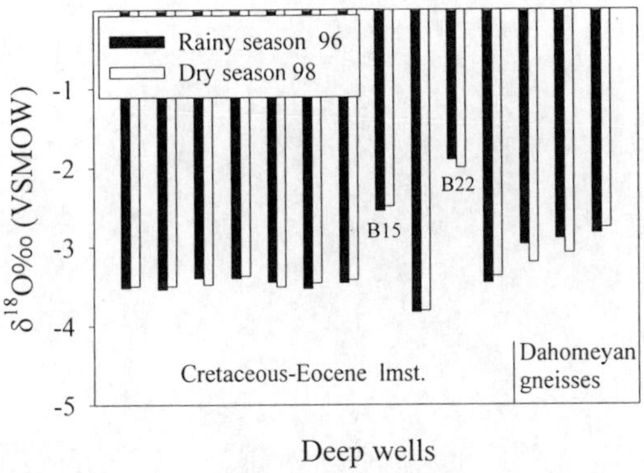

Figure 6. $\delta^{18}O$ distribution in deep groundwater collected from deep wells in the Keta Basin, Ghana, in the rainy season of 1996 and the dry season of 1998. B15 and B22 are wells influenced by seawater intrusion.

2) Groundwaters from aquifers in the Dahomeyan basement and the Neogene continental deposits are characterized by relatively high $^{87}Sr/^{86}Sr$ ratios which reflect the weathering effects and the mineralogical compositions of the reservoir rocks, i.e. Proterozoic crystalline rocks and their derivative weathering products. These groundwaters may have relatively high Cl contents but show no indication of marine influence.

3) Groundwater collected from the Quaternary and Cretacoues-Eocene limestone aquifers in general reveal isotopic compositions which reflect the stratigraphic age and water-rock interactions in the respective reservoir rocks. However, groundwater from a number of deep wells and dug wells in near-coastal aquifers are characterized by relatively high Cl contents and significant modern marine influence isindicated by well-defined mixing lines for both strontium isotopes and hydrogen and oxygen isotopes with the isotopic compositions of modern seawater as one end-member.

The interpretations derived from environmental isotopes of shallow and deep groundwaters and surface water from the Keta Basin demonstrate that the multi-isotopic approach provides a useful supplement to traditional hydrochemical analyses indentifying the origin and sources of saline groundwaters and in understanding the qualitative contribution of seawater intrusion, evaporation and water-rock interactions on the quality of groundwater.

REFERENCES

Akapati, B.N.1978. Geological structure and evolution of the Keta Basin, Ghana, West Africa. *Geological Society of America Bulletin.* 89:124-132.

Bannerman, R.R. 1994. Appraisal of the limestone aquifer of the Keta Basin, Ghana. In . Soveri & T. Suokko (eds), *Future Groundwater Resources at Risk, IAHS Publication No. 222*: 315-321.

Burke, W.H., R.E. Denison, E.A. Hetherington, R.B. Koepnik, H.F. & J. Otto 19982. Variation of seawater $^{87}Sr/^{86}Sr$ throughout the Phanerozoic time. *Geology* 10:516-519.

Craig, H. 1961. Isotopic variations in meteoric water. *Science* 133:1702-1703.

Faure, G. 1986. *Stable isotope Geochemistry*, 2nd edn., Springer Verlag, Berlin.

Frempong, D.G. & B.K. Kortatsi 1994. Some problems associated with the development of groundwater resources in Ghana. In T. Suokko & J. Soveri (eds) *Future Groundwater Resources at Risk, IAHS Publication No. 222:* 389-398.

Gill, H.E. 1969. A ground-water reconnaissance of the replubic of Ghana, with a description of geohydrologic provinces. *Geological Survey Water-Supply Paper 1757-K.*

Jørgensen, N.O. & P.M. Holm 1994. Isotope studies ($^{18}O/^{16}O$, D/H and $^{87}Sr/^{86}Sr$) of saline groundwater in Denmark. In J. Soveri & T. Suokko (eds), *Future Groundwater Resources at Risk, IAHS Publication No. 222:* 231-238.

Jørgensen, N.O. & P.M. Holm 1995. Strontium-isotope studies of chloride-contaminated groundwater, Denmark. *Hydrogeology Journal* 3:52-57.

Nerquaye-Tetteh, B. 1993. Water resources of the Keta Basin. In *Proc. 19th WEDOCCont*:102-108.

Groundwater: Past Achievements and Future Challenges, Sililo et al. (eds) © 2000 Balkema, Rotterdam, ISBN 90 5809 159 7

An aquifer model based on chemistry, istopes and lineament mapping: Little Karoo, South Africa

J.C. Kotze
SRK Consulting, Cape Town, South Africa

B.Th. Verhagen & M.J. Butler
Schonland Research Centre, University of the Witwatersrand, Johannesburg, South Africa

ABSTRACT: A major ground water abstraction scheme, developed in the fractured Table Mountain Group sandstone aquifer of the Little Karoo, is falling short of designed capacity. A multidisciplinary study, encompassing structural (including satellite imagery), hydrogeological, hydrochemical and environmental isotope methods, aims at arriving at a conceptual model with which to guide development and management strategies. Stable isotopes show that thermal springs are fed by regional faults, which are recharged in catchments at higher altitude. The study centers on the fault-controlled Vermaaks River valley of the Kammanassie Mountains, where high-yielding boreholes feed into a regional water supply. Water balance considerations using abstraction figures give a recharge estimate for the catchment of 5 x 10^5 m^3 a^{-1}. Radiocarbon values can be interpreted in terms of ground water mean residence times, which with porosity estimates give an independent assessment of recharge. Only the highest of the range of 1% to 5% for porosity, re-interpreted on structural considerations, brings these figures into correspondence. It is concluded that the storage of the catchment as a whole is substantially higher than previously accepted.

1 INTRODUCTION

1.1 *Background*

The Table Mountain Group Sandstone (TMG) is a major lithological unit in the Western and South-Western Cape (Figure 1). Regional scale tectonic movements during the Cape Fold Belt (CFB) orogenesis led to the loss of primary porosity through cementation and metamorphism. Extensional faulting (due to over-folding) during and after the CBF orogenesis created fracture sets, which not only control the hydraulic properties of the aquifers, but also the geomorphology of the landscape. These fracture sets ranges from local, often shallow fractures, mostly confined to the mountain valleys, to regional fault and fracture systems connecting several surface water catchments. Good examples of the above are found in the Little Karoo study area, situated in the Western Cape Province of South Africa, between the towns of Calitzdorp and Uniondale (Figure 1).

The assumption that ground water could be found only in shallow fracture systems, locally recharged, and the semi-arid nature of the region, rainfall 200 to 400 mm/a, gave rise to a hydrogeological conceptual model, based on the water balance calculations for a limited area, postulating low recharge and storage.

High-yielding boreholes have been developed in narrow mountain valleys (i.e. the Vermaaks River Valley, situated in the Kammanassie Mountains, indicated by No.3, Figure 1), which carry ephemeral and perennial drainage from the mountain catchments. These valleys are fault controlled and are often intersected by further faults, resulting in the convergence of local, and what could be regarded as regional, fracture systems.

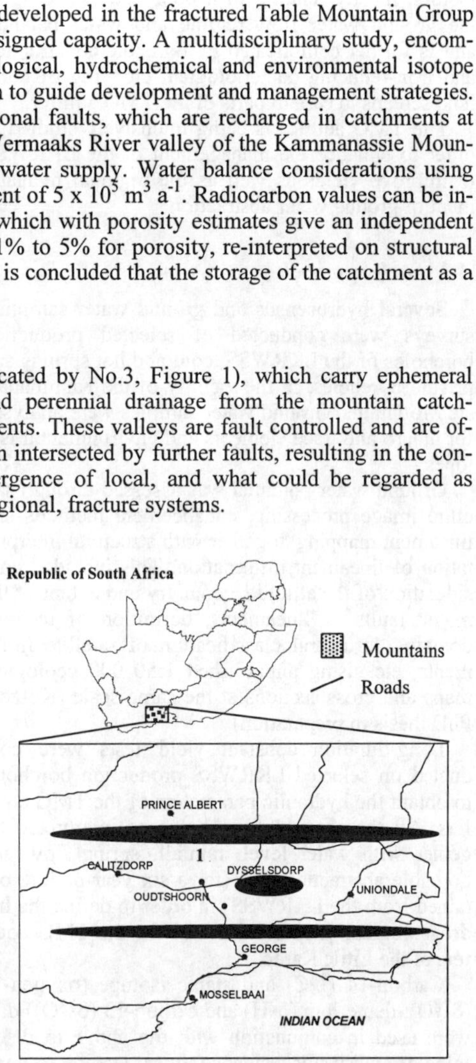

Figure 1. Schematic locality map showing TMG aquifer.

The Little Karoo Rural Water Supply Scheme (LKRWSS) is designed to supply up to 4.7 x 10^6 m^3/a of purified water. It is fed by an Eastern Sector (Kammanassie Mountains near Dysselsdorp: 13 boreholes of which 5 constitutes the Vermaaks River Wellfield), and a Western Sector (Calitzdorp: 5 boreholes), all but one of the boreholes drilled into TMG aquifers. Some 400 km of pipeline delivers the abstracted ground water to two purification plants (at Dysselsdorp and Calitzdorp, respectively), and to end-users.

Current ground water abstraction from the LKRWSS is approximately 1.0 x 10^6 m^3/a, which is considerably less than the original designed capacity. This difference can be attributed to a number of factors, mostly relating to a poor understanding of the ground water flow regime and water balance, and is further complicated by borehole construction and iron-reducing bacteriological clogging of borehole screens in certain parts of the TMG aquifer.

The TMG aquifer is being intensively studied in order to enhance the management of the LKRWSS to improve yields as well as the regional management of ground water abstraction.

1.2 *Study approach*

Several hydrocensus and ground water sampling surveys were conducted of selected production boreholes of the LKRWSS, cold and hot springs and private boreholes, in the vicinity of the Kammanassie Mountains. Ground water samples were analysed for macro and trace elements and environmental isotopes.

Ground water potential was assessed through satellite image processing, enhancement methods and lineament mapping, together with structural interpretation of lineament information. This included consideration of the attitude, geometry and extent of the major faults or lineaments, definition of tectonic domains, structural classification of satellite lineaments, etc. using unpublished 1:50 000 geological maps and cross sections at the same scale (Kotze – PhD thesis in preparation).

Long-duration constant yield tests were conducted on selected LKRWSS production boreholes to obtain the hydraulic parameters of the TMG aquifers. All the above information was interpreted together with water level, rainfall, spring flow and borehole abstraction data (for a six year period, obtained from the LKRWSS) in order to define the hydrogeological conceptual model for the TMG aquifers of the Little Karoo.

Carbon-14 (^{14}C) and stable isotope (oxygen-18 ($\delta^{18}O$); deuterium (δ^2H) and carbon-13 ($\delta^{13}O$)) data were used in conjunction with the above to determine ground water residence times, leading to aquifer storage estimates and identification of recharge areas, respectively.

2 HYDROGEOLOGY

2.1 *Hydrostratigraphic units*

The following three major hydrogeological units together constitute the TMG super-aquifer:

- Nardouw Aquifer: Baviaanskloof, Kouga and Tchando Formations;
- Cedarberg Aquitard (C/S): the Cedarberg shale layer;
- Peninsula Aquifer: the Peninsula Formation quartzites.

The Nardouw Aquifer has a relatively high shale and feldspar content, is prone to ductile deformation, and generally has low permeabilities. The Peninsula Aquifer is more quartzitic with associated brittle deformation, and has generally higher permeabilities.

Numerous springs originate on the contact between Peninsula and the C/S shale, which constitutes an effective boundary to ground water flow.

So-called aquizones (Al-Aswald and Al-Basaam, 1997) have been identified in the Kammanassie Mountains. These are inter-connected fracture networks, which can host high yielding, good quality ground water when connected with the recharge area, irrespective of the lithology of the host aquifer. They often cut across more than one hydrogeological unit, such as locally connecting the recharge area (Peninsula Aquifer) and the Nardouw Aquifer via a network of highly interconnected and permeable fractures.

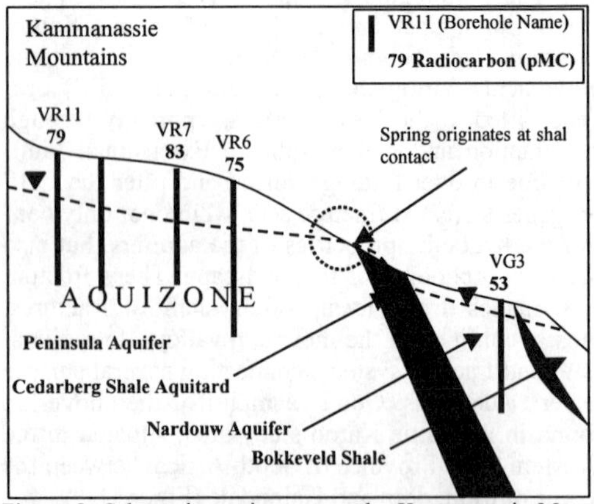

Figure 2. Schematic section across the Kammanassie Mountains, showing the Vermaaks River well field.

Generally the Nardouw Aquifer outcrops at lower altitudes, with less direct rainfall recharge (annual precipitation 300 mm/a), relative to the Peninsula aquifer outcropping at higher altitudes and associated higher rainfall (> 500 mm/a).

In the absence of aquizones, the Nardouw aquifer generally hosts lower yielding boreholes, with poorer ground water quality and iron reducing bacteriological clogging of borehole screens is common. On the contrary boreholes drilled in the Nardouw aquifer inside the aquizone, are high yielding with good ground water quality, i.e. similar to that of the Peninsula aquifer. In these conditions iron-reducing bacteriological clogging of boreholes screens are limited.

Figure 2 shows a schematic east-west cross-section along the Vermaaks River Valley (No. 3, Figure 1), incised into the north-western flank of the Kammanassie Mountain anticline. Four boreholes were drilled in an aquizone of the Peninsula Formation and one in the Nardouw Formation, respectively above and below the shale band, an effective aquitard at this location, as a spring rises at its intersection with the valley.

2.2 *Hydrochemistry and environmental isotopes*

Hydrochemical and environmental isotope data are summarised in Table 1.

Ground water in the Peninsula Formation of the Table Mountain Group sandstone is characterised by its low mineralisation (TDS < 300 mg/l), low pH (< 6) and very low TAL (<< 0.5 meq/l). The low pH values in this formation show that the total dissolved inorganic carbon (TDIC) in this mainly quartzitic formation is present mainly in the form of free CO_2, whilst the narrow range of $\delta^{13}C$ values (-19 to -22 ‰) is evidence of minimal carbonate dissolution. Somewhat higher values of $\delta^{13}C$ are found away from the fracture bound steep mountain valleys, in ground water held in local fracture systems, and in the much older thermal spring water. This allows the radiocarbon values measured in quite disparate sections of the regional system to be compared without necessitating isotopic dilution corrections.

To obtain samples adequate for conventional radiocarbon analysis, it was necessary to extract carbonate in the field from some 300 L or more of this low TDIC water. This was made possible by using the polyethylene bag method of precipitation, which allows as much as 150 L of water to be treated in each bag at the wellhead (Verhagen et al. in preparation).

^{14}C values for wells above the shale band in the Vermaaks River Valley (Figure 2) range 75-84 pMC (percent modern carbon), with practically no measurable tritium. The well below the shale band in the Nardouw gives a much lower value of 53 pMC. Apart from the similar $\delta^{13}C$, the $\delta^{18}O$ and δD values are almost identical to those above the shale band. The sudden "ageing" suggests that the band acts as an effective aquitard, inhibiting subsurface drainage further downstream at this site. Surface drainage of mobile ground water from above the clay band could also re-infiltrate to form less mobile ground water below it. The results further suggest that the water pumped from the upper valley, in spite of considerable - if steady - drawdown, is largely derived from the Kammanassie Mountain catchment rather than from the regional fracture system.

A similar situation is encountered for wells in smaller valleys to the south and west of the Kammanassie Mountains complex. Here too the mobility of the ground water, as evidenced by the ^{14}C values, is the only distinguishing feature, the hydrochemical and stable isotope signal being similar to the extent of suggesting genetic similarity with ground water in the main valley. A correlation was found between ^{14}C-values and fracture systems connecting the water-bearing structure with the recharge areas in the mountains. Highly permeable aquizones cross-cutting the C/S aquitard produce higher pMC values in the Nardouw aquifers (70 – 85 pMC), versus values ranging from 40 – 70 pMC in the Nardouw aquifers in the absence of these fractures (e.g. Vermaaks River).

The thermal springs, regarded as expressions of deeply circulating regional ground water, and rising in low-lying terrain, have $\delta^{18}O$ and δD values similar to those observed for the mountain catchments (Figure 3). Where ground water has clear local associations away from mountain recharge sources, sometimes actively recharged in the intermontane regions (e.g. DP18, situated in the valley between the Swartberg and Kammanassie Mountains – No 1 and 3, respectively on Figure 1), the stable isotope values show significantly more positive values.

These differences are ascribed to the altitude effect (Verhagen et al. 1991). As mentioned earlier,

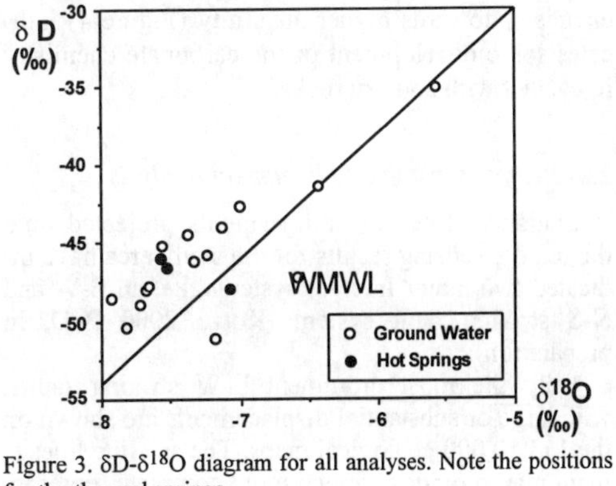

Figure 3. δD-$\delta^{18}O$ diagram for all analyses. Note the positions for the thermal springs.

Table 1. Summary of hydrochemistry and environmental isotope data for the Little Karoo study area.

NAME	AQUIFER	EC (mS/m)	TDS (mg/l)	pH	Na (mg/l)	Mg (mg/l)	Ca (mg/l)	Cl (mg/l)	NO3 (mg/l)	SO4 (mg/l)	TAL (Ca) (mg/l)	Si (mg/l)	K (mg/l)	$\delta^{18}O$ (‰)	δ D (‰)	C-14 (pMC)	Tritium (T.U.)
Toorwater Spa	Hot spring	16.5	135	7	14	4	11	22	0.04	4	60	15.1	7.9	-7.08	-48	25.6	
Calitzdorp Spa	Hot spring	21.5	133	7.3	17	4	9	34	0.04	9	42	17.7	9.4	-7.53	-46.6	13	
DL13	NAR-DOUWWEST	49.5	313	7.2	66	8	12	115	0.05	24	59	5.3	16.2	-7	-42.7	21	
KG1	NAR-DOUWWEST	39.9	198	6.68	38.1	8.1	12	78.6	0.005	11.2	33.2	5.17	9.59	-6.26	-27.7	51.2	0.2±0.1
DL17	NAR-DOUWWEST	50.9	266	6.69	47.9	11.9	15.7	98.9	0.001	24.6	42.7	5.7	14.43	-6.63	-32.2	21.3	0.3±0.1
DL16	NAR-DOUWWEST	48.6	278	6.93	57.3	8.6	13.2	94.2	0.006	22.5	54.8	5.75	15.18	-6.63	-32.4	13.1	0.2±0.2
DP29	NARDOUW	20.8	107	6.4	25	4	5	36	0.04	14	15	4.9	4.2	-7.72	-50.8	41.2	0.1
DP28	NARDOUW	18	100	5.9	22	3	3	39	0.04	18	9	5.4	3.6	-7.93	-48.6	62.3	0.4
DP15	NARDOUW	30.2	124	3.7	21	3	3	32	0.04	60	4	5	3.7	-7.66	-47.8	52	
DP12	NARDOUW	22.2	126	6.4	32	4	5	41	0.04	16	19	5	4.3	-6.58	-47.3		0.3
DG110	NARDOUW	22	102	5.54	24.6	3.9	4	47.5	0.093	10.2	8.1	5	1.31	-7.7	-49.5	41	0.0±0.2
VG3	NARDOUW	15.5	78	5	23	3	2	40	0.93	5	4	4.4	0.7	-7.68	-43.3	78.1	1.3±0.2
VR11	PENINSULA	7.5	35	6.69	10.9	1	1.3	14.6	0	2.1	3.4	4.26	0.37	-7.06	-35.5	67.4	0.0±0.1
VR6	PENINSULA	11.5	52	6	14	2	2	21	0.26	4	9	4	1.2	-7.68	-48.1	79.4	0.4
VR11	PENINSULA	9.5	44	5.9	11	2	1	19	0.11	4	7	4.4	1.6	-7.18	-51.1	74.6	0.3
VR7	PENINSULA	9	43	5.5	9	2	2	17	0.14	1	6	4.6	0.3	-7.27	-35.7	79.3	0.4±0.1
WN3	NARDOUW	25.1	143	5.76	30.1	4.6	8.7	50.8	0.088	31.6	8.5	7.69	6.46	-7.65	-53.2	84.4	
WK4	NARDOUW	26	121	5.47	31.3	4.4	3.5	58	0.513	13.6	5.4	5.02	1.08	-7.23	-50.9	70.4	
WN2	NARDOUW	25.9	167	6.2	35	5	9	55	0.08	40	12	7.5	8.5	-7.82	-50.4	72	0.3
SL2	NARDOUW	83.7	473	6.23	54.9	19.9	57.4	126.7	0.007	158.6	34.1	13.5	13.18	-7.03	-49.1		0.3
YR2	NARDOUW	18.1	97	5.6	19.4	3.1	7	31.5	0.083	20.1	8.7	7.32	4.7	-6.93	-47.4		
WK3	AQUIZONE	17.4	104	5.84	23.8	3.4	2.6	39.8	0.249	24.1	6.1	5.05	1.61	-7.21	-50.6	69.2	
RF8	AQUIZONE	10.4	60	5.33	14.8	1.6	1.7	19.9	0.205	15.7	3.9	4.7	0.62	-7.23	-40.8	82.2	
RF2	AQUIZONE	11.9	54	5.7	17	2	2	25	0.17	4	6	4.3	0.5	-7.13	-44	75.4	0.1
LD2	AQUIZONE	13.9	67	5.84	18.6	2.2	2.4	27.5	0.186	3.2	10.1	4.72	0.35	-7.42	-50.3	75	0.0±0.1
DR2	NARDOUW	32.3	217	6.9	25.9	4.8	26.2	43.5	0.011	45.8	49.3	9.56	9.71	-6.98	-46.5		

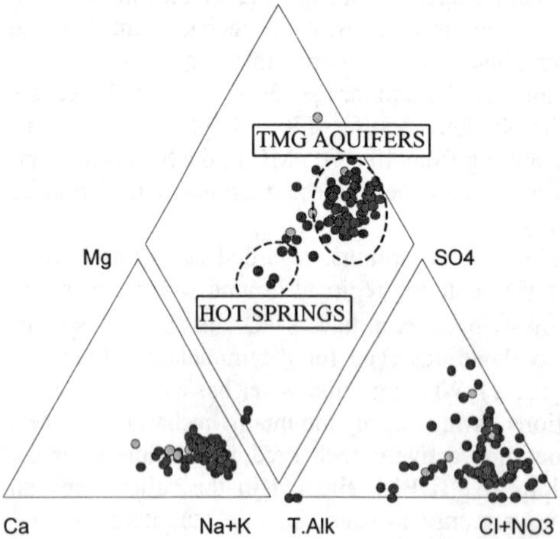

Figure 4 Piper diagram of hydrochemistry of TMG aquifer. Note the shift for the thermal springs (C/S, T/W and WBG).

thermal springs have slightly more positive $\delta^{13}C$ values, which along with the slight shift in hydrochemistry towards higher alkalinity (Figure 4) indicates some development of the carbonate chemistry in contact with country rocks.

2.3 *Remote sensing and lineament analysis*

Statistical filtering of lineaments projected onto the remote sensing results for the study area have indicated two major fracture systems, i.e. an E-W and N-S striking fault system (Kotze 2000 PhD, in preparation).

Only the most prominent E-W striking faults, with large or substantial displacements are shown on the 1 : 250 000 geological maps. The satellite lineament map provides a much more complete representation of the predominant fracture sets (more fractures and longer extent), which are: E-W, WNW and ENE trending, the first two being the most dominant structures.

The N-S trending system consists of shorter, more discontinuous fractures, generally a dense network of joints often not showing any displacements on geological maps or satellite images. These fractures or joints can be very penetrative and form prominent morphotectonic features. The two main fracture orientations, i.e. NNW and NNE trending, form a conjugate set, with a third, less prominent N-S trending fracture orientation.

3 HYDROGEOLOGICAL CONCEPTUAL MODEL FOR TMG AQUIFERS

The following hydrogeological conceptual model is therefore generally confirmed. Recharge occurs principally in the highly fractured mountainous local catchments. These partially produce local subsurface drainage and surface runoff in mountain streams, and partially feed the major regional fractures, which are manifested in the low-yielding thermal springs, situated near fracture zones in low-lying intermontane areas. The balance of this transport is probably lost through diffuse upward discharge from these zones over wide areas.

Ground water quality and yield of TMG aquifers varies according to the lithological unit intercepted, i.e. Nardouw or Peninsula Aquifer as well as the presence of highly permeable aquizones, cross-cutting the C/S aquitard. These aquizones correlate with remote sensing and lineament mapping. These lead to high yielding, good quality boreholes in the Nardouw Aquifer inside the aquizone, whereas elsewhere yields are lower and quality poorer.

Similar chemical and environmental isotope patterns were found in the Agter-Witzenberg TMG groundwater (Weaver et al., 1999), in accord with the conceptual model developed here.

As yet, the characteristics of the regional fractures are not well understood. The wide separation and low yield of the thermal springs suggest limited deep fracture flow.

Increased exploitation from the highly fractured catchment valleys therefore appears to be the most economical solution to increasing water supply at present. This will clearly have to be at the expense of the baseflow of mountain streams and a drastic local lowering of rest levels. Eventually, the transport of ground water in regional fractures will be reversed, producing resistance levels in the yield-drawdown relationship. In the light of the present information, this would not, in the first instance, appear to have major water quality consequences.

4 CALCULATION OF STORAGE

Figure 5 shows the rainfall figures for the Vermaaks River Wellfield and Figure 6 the rest water levels versus abstraction in VR7 (the highest yielding borehole drilled into the Peninsula Formation, Vermaaks River Valley) over a six year period ending in March 1999. The wellfield was pumped at a combined average rate of 5 x 10^5 m^3 a^{-1} over six years with very moderate drawdown. For the purpose of this calculation it is assumed conservatively that this withdrawal equals the annual recharge rate.

Tritium in samples from production boreholes lies at or below the limit of detection (0.2±0.2 TU). CFC data from the Vermaaks River boreholes (Weaver and Talma 1999) are consistent with tritium. It can therefore be assumed to a first approximation that the pumped water as yet does not contain significant amounts of thermonuclear radiocarbon.

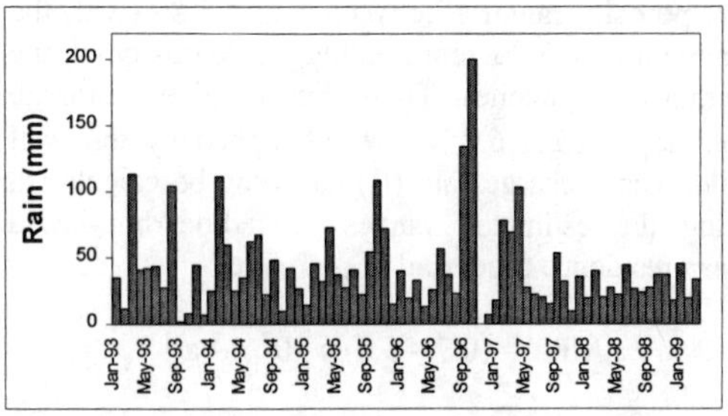

Figure 5 Monthly rainfall figures - Vermaaks River Wellfield.

Assuming that the system as a whole can be regarded as phreatic, the mean residence time (MRT) of the ground water (Verhagen et al. 1991; Gieske 1995) is given by:

$$MRT = 8270 \{(A_0/A) -1\}$$

where A_0 is the initial (recharge) value and A the measured value of radiocarbon. In the absence of carbonate dissolution (see Section 2.2) A_0 is taken conservatively to lie in the range of 90-95 % of atmospheric. This gives values for MRT:

Borehole	Measured ^{14}C A (pMC)	Initial (Est.) ^{14}C A_0 (pMC)	MRT (years)
VR 6	79.4	90 - 95	1100 -1600
VR 7	83.0	90 - 95	700 - 1200
VR 11	74.6	90 - 95	1700 - 2300
		Mean	1200 –1700

Hälbich and Greeff (1995) studied the structure of

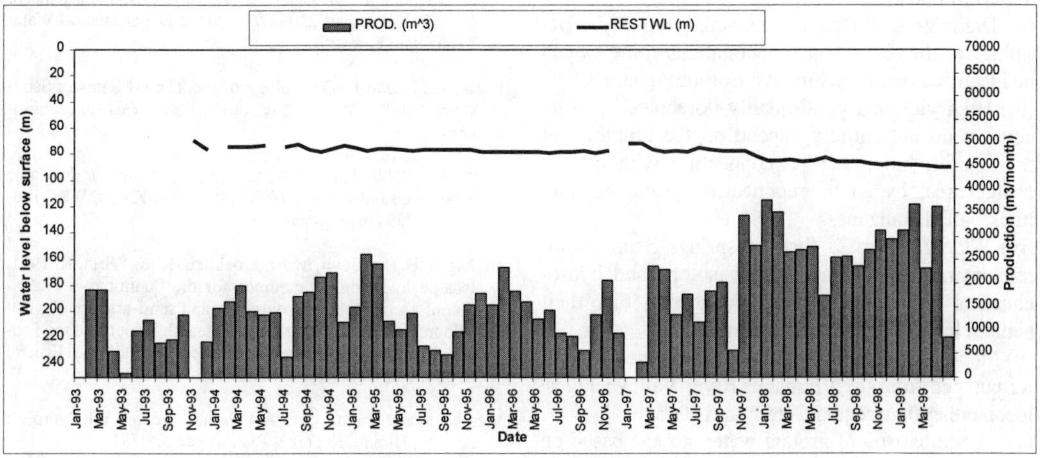

Figure 6 Water levels and abstraction figures for VR7 – production borehole in the Vermaaks River Wellfield

the Kammanassie Mountains and arrived at a figure of 1.2×10^{10} m^3 for the volume of rock which could be tapped by the well field and estimated a total fracture porosity ranging between 1 and 5 % (with the maximum of 5 % representing the porosity of the Vermaaks aquizone). These figures give a storage volume of $1.2 - 6.0 \times 10^8$ m^3 tapped by the well field. The recharge rate (R) can now be calculated, using the estimated ranges of radiocarbon-based mean residence times and of porosities:

$$R = V / MRT = (0.7 - 5.0) \times 10^5 \ m^3 \ a^{-1}$$

The withdrawal rate at balance, or effective recharge rate R_E :

$$R_E = 5 \times 10^5 \ m^3 \ a^{-1}$$

corresponding to the upper limit of estimated R.

As substantial amounts of ground water are likely to drain out of the storage over and above the abstraction from the well field, R_E, and therefore the discrepancy between the figures, could be greater than these figures suggest. The MRT values based on radiocarbon values are unlikely to be significantly smaller than the range calculated above.

It is therefore concluded that the porosity of the entire catchment feeding the well field is in the same order of the estimate from structural considerations for aquizones, and therefore considerably higher, on average, than previously accepted.

5 CONCLUSIONS

(i) The stable isotope signature of ground water recharged in the mountain catchments differs significantly from local recharge in intermontane areas due to the altitude effect.
(ii) Drainage in different directions around the periphery of the Kammanassie Mountains complex can be identified by this distinctive isotope signature.
(iii) High yielding, good quality boreholes in TMG aquifers do not entirely depend on the lithology of the aquifer, but rather its connection with the recharge area, by an interconnected system of fractures, called aquizones.
(iv) Widely separated thermal springs rising in intermontane regions have stable isotope and hydrochemical signatures strongly suggesting that their sources lie in mountain catchments.
(v) Modest mean residence times determined for water pumped from the Vermaaks River Valley suggest local, rather than regional structural storage.
(vi) A comparison of ground water storage based on water balance with structural analysis and ^{14}C data suggests that fracture porosity values of the TMG catchment may approximate those observed in the aquizones, implying high storage.
(vii) The results from the Vermaaks River Valley are sufficiently encouraging for other mountain catchments to be investigated for ground water exploitation. There is considerable ground water exploitation potential from and storage in, the fractured TMG, provided suitable conductive structures can be identified.
(viii) The updated information on the Vermaaks River Valley is of great significance to the development of the conceptual model of local and regional fracture flow and the emerging quantitative data and boundary conditions are invaluable as input data for numerical modelling.

6 ACKNOWLEDGEMENTS

The authors wish to thank the Water Research Commission, under contract K5/729, the Department of Water Affairs and Forestry and the International Atomic Energy Agency, under contract 8983/RO/RBF for financial support.

REFERENCES

Al-Aswald and Al-Bassam, 1997. Proposed hydrostratigraphical classification and nomenclature: application to the Palaeozoic in Saudi Arabia. J. Afr. Earth Sci., 24 (4) 497-510.

Gieske, A. 1995. Hydrodynamics of ^{14}C analysis in unconfined aquifers. Proc Conf. *Ground Water Recharge and Rural Water Supply*, 26-28 Sept. 1995, Midrand, South Africa

Hälbich, I. W. and Greeff, G. J. 1995. Final report on a structural analysis of the west plunging nose of the Kammanassie anticline. University of Stellenbosch

Kotze, J. C. and Rosewarne, P. N. 1996 Little Karoo scheme: Hydrogeological overview. Steffen Robertson and Kirsten, Cape Town. Report 230827/1,2 to the Department of Water Affairs and Forestry.

Kotze, J.C., 2000 Hydrogeology of the Table Mountain Sandstone Aquifer, in the Little Karoo, PhD thesis *(in preparation)*.

Kotze, J.C. 2000. Towards a management model for the Table Mountain sandstone aquifers in the Little Karoo, WRC Project K5/729 (*in preparation*).

Verhagen, B.Th., Geyh, M.A., Froehlich, K. & Wirth, K. 1991. Isotope hydrological methods for the quantitative evaluation of ground water resources in and semi-arid areas. - Development of a methodology. Research Reports of the Federal Ministry for Economic Cooperation of the Federal Republic of Germany. BMZ, Bonn.

Weaver J. and Talma, S. 1999 Unpub. Report WRC Project K5/481; Unpub. Report WRC Project K5/731.

Groundwater: Past Achievements and Future Challenges, Sililo et al. (eds) © 2000 Balkema, Rotterdam, ISBN 90 5809 159 7

Assessment of fluoride concentration in groundwater, Saldungaray, Argentina

E.E.Kruse & J.E.Ainchil
Facultad de Ciencias Astronómicas y Geofísicas, UNLP, La Plata, Argentina

ABSTRACT: Spatial variations of fluoride concentration in groundwater in the town of Saldungaray (southern Buenos Aires Province) affect water quality for human supply and decrease the aquifer reserves. The study region is a piedmont area, situated between a hilly area to the west, and the fluvial valley of Sauce Grande River to the east. Two geological units can be identified: Paleozoic bedrock and Cenozoic clastic sediments. The sediments, known as Pampean sediments, consist of sandy silt with a variable amount of calcium carbonate. They reach their greatest thickness of 60 m near the river. Groundwater flow coincides with topography, but has a lower gradient. Fresh water, with low salinity content (dissolved solids 400 to 800 mg/l), is exploited from the clastic unit. The fluoride concentration of this water varies between 0.2 and 5 mg/l. Spatial variations of fluoride content are analysed in relation to the groundwater flow and hydrogeological characteristics.

1 INTRODUCTION

Fluoride concentrations in drinking water are important for public health. Fluoride contributes to dental health and to the maintenance of appropriate bone density. The fluoride concentration in most natural water that has a dissolved solids concentration of less than 1000 mg/l, is less than 1 mg/l (Hem 1985). The factors that influence the fluoride concentrations in natural water are geological and hydrogeological, geochemical and anthropogenic (Gosselin et al. 1999).

Fluoride concentrations of more than 2 mg/l have been reported in low salinity water of the "Pampas" region in the provinces of Buenos Aires, La Pampa, Córdoba, and Santa Fe in Argentina (Castro et al. 1998). There are sites in this huge region where fluoride concentrations can reach up to 20 mg/l. This limits the use of the water for human consumption. Furthermore, high fluoride concentration often occurs near the towns where groundwater is the sole water source, as is the case of Saldungaray (Fig. 1).

In this paper we analyse the hydrogeological features of the region and examine the relationship between spatial variations of the fluoride concentration and the behavior of groundwater.

2 SITE DESCRIPTION

The study area includes a piedmont region and the fluvial valley of the Sauce Grande River. Piedmont slopes are 1%, and occur between 210 and 250 m.a.s.l.

The climate is humid temperate. The mean annual rainfall is about 850 mm/year and the mean temperature is 16°C (60°F). The monthly rainfall distribution is variable with higher rainfall in summer, leading to a maximum in March (110 mm).

Figure 1. Location of the study area.

Empirical formulas have been used to estimate that actual evapotranspiration represents about 75% of rainfall. The monthly water budget shows water excess in winter and spring. The possibilities of groundwater recharge are higher during these periods, despite these being times of lower rainfall (Kruse et al. 1995).

3 MATERIALS AND METHODS

A field survey including well surveys, vertical electrical soundings, geological and geomorphologic reconnaissance was conducted. These tasks allowed us to characterize groundwater. The survey area was approximately 3500 hectares, including the urban area of Saldungaray and its surroundings.

Aerial photographs on a scale of 1:20,000 were used to identify geologic and geomorphologic characteristics that might influence hydrogeological behavior.

The well surveys included water level measurements and sampling of 40 wells, which represented an average of approximately one well per 90 hectares. The aim of the survey was to determine the hydrodynamic and hydrochemical conditions of the groundwater.

Measurements of vertical electrical soundings were made at 18 stations using the Schlumberger configuration. The objective was to provide specific information about the thickness and depths of the aquifer.

Two exploration wells were drilled for the collection of lithological samples and to verify and interpret the vertical electrical sounding data. These wells were also used to test pump the aquifer and to take water samples.

4 HYDROGEOLOGICAL SETTING

In the study area we can identify two major geological units forming the bedrock and aquifer system. Palaeozoic bedrock outcrops in the hills to the west of the study region. Quartzitic rocks are dominant in this unit. A system of fractures, joints and stratification planes provide a low secondary permeability.

The aquifer system is a clastic unit of Cenozoic age with primary permeability. Groundwater is exploited from this unit. It consists of a basal conglomerate and a sequence, known locally as Pampean sediments, which is composed predominantly of eolian clay to sand-silt, or loess, with variable amounts of calcium carbonate. Very fine sand and silt are the predominant particle sizes in this formation. The sediments contain calcic plagioclase, alkali feldspar, volcanic rock fragments, vitroclasts and unstable heavy minerals, such as hornblende and augite. Trace quantities of micas, apatite, zircon and epidote are also found. The Pampean sediments are derived primarily from Mesozoic to Pleistocene age Andean volcanoes (Teruggi et al. 1957). These sediments are characterized by a relatively high content of vitroclasts, in some cases more than 50% of the mineralogical composition (Riggi et al. 1986).

The percentage of vitroclast material allows us to distinguish three different sediment horizons. The upper horizon extends to a depth of approximately 10 metres and the vitroclast content varies between 0 and 10%. The middle horizon is defined as being between 10 and 25 m depth with vitroclast distribution between 15 and 55% and the lower horizon contains variable values between 0 and 15%.

The sediment thickness was found to be significant in the study of fluoride concentrations. The variation in thickness can be observed in Figure 2. Over 60 m of sediment are found to the west of the city. There is also an increase in sediment thickness toward the river, where they reach over 60 or even 80 m.

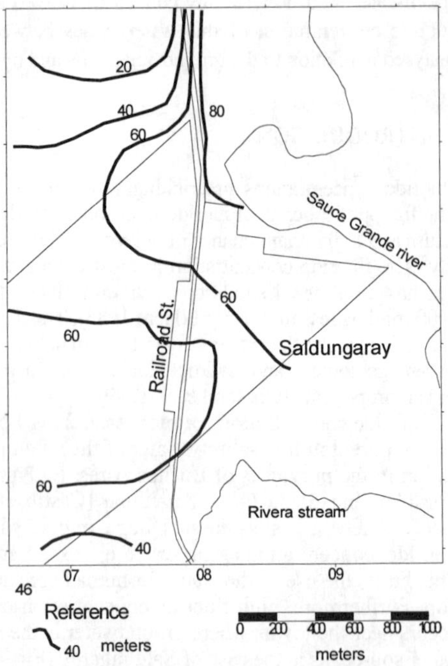

Figure 2. Thickness of the Pampean sediment unit.

5 GROUNDWATER

5.1 *Dynamic aspects*

The characteristics of groundwater movement and its behaviour within the primary aquifer unit can be established from the water table morphology. The

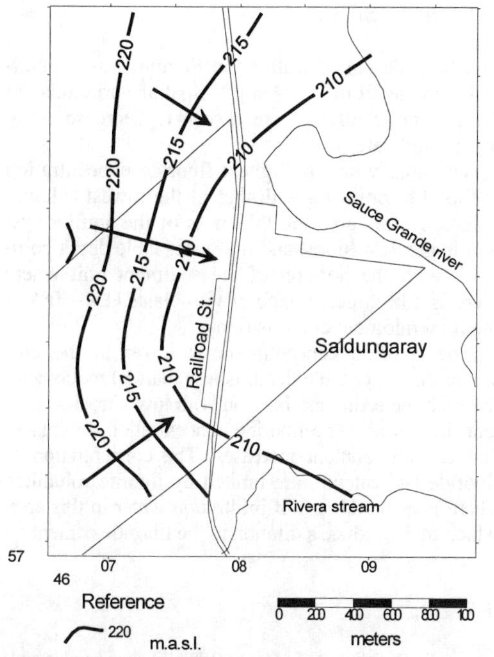

Figure 3. Groundwater flow

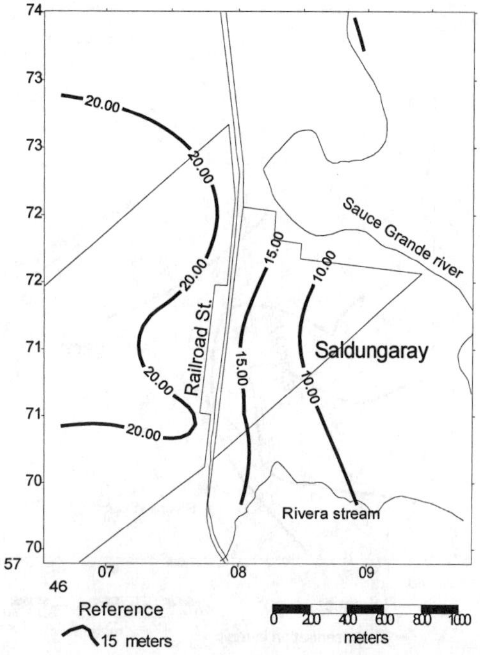

Figure 4. Depth to water table

general direction of groundwater flow coincides with the topography, though it has a lower gradient (Fig. 3).

The natural process of groundwater recharge to the aquifer is by the infiltration of excess rainfall. Regional hydraulic gradients are about 0.4 % with values increasing in the headwater region, to the west of the city, and decreasing toward the river.

The groundwater discharge area is the Sauce Grande River. The water table is more than 20 m deep to the west of the city and 18 m deep near the railroad station. To the east, near the river, the water level increases to 10 m below surface (Fig. 4).

5.2 *Hydrochemical aspects*

The groundwater has low salinity (dissolved solids 400 to 800 mg/l). The water is calcium bicarbonate type. The lowest values of bicarbonate (200-250 mg/l) and calcium (20 to 40 mg/l) occur to the west of the city and their content increases with the flow direction. The chloride concentration varies from 40 to 150 mg/l and its content increases with the flow direction. Sulphate varies between 10 and 20 mg/l. The major ion concentrations do not impair the quality of the water supply for drinking water purposes. The same applies to heavy metals. High values of nitrate (40 mg/l) in some isolated points are associated with contamination due to human activities.

6 FLUORIDE CONCENTRATIONS

The fluoride concentration varies from 0.2 mg/l (in only one well) to 5 mg/l. The values typically range between 2 and 4 mg/l for over 90% of the data. The fluoride concentration is a limiting factor to using the water for human supply.

The fluoride concentration shows a tendency to decrease with the direction of the groundwater flow (Fig. 5). Values ranging from 1 to 2 mg/l appear in the groundwater discharge area (Sauce Grande River).

The areas with the highest fluoride concentration are found to coincide with those with the lowest calcium and bicarbonate values. The highest fluoride values (above 3 mg/l) correspond with the increase in sediment thickness (60 m) to the west of the city. This area also has the deepest water table, at a depth greater than 20 m. Although this area includes rural activities, there is no use of phosphorated fertilizers.

The relationship between spatial variations of fluoride concentrations and groundwater flow and hydrogeological characteristics shows that the increase in fluoride content is associated with natural processes. Fluoride can be related to dissolution of vitroclasts derived from volcanic ash distributed in the sediment unit.

In the area with the highest values of fluoride concentration, the water table (20 m) is located in the

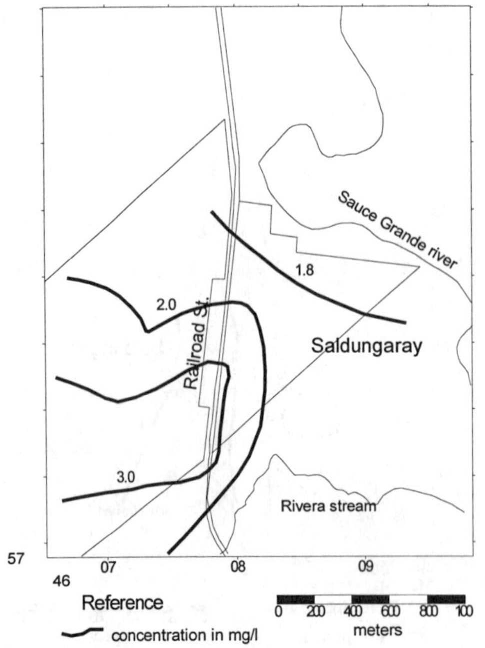

Figure 5. Fluoride concentration in mg/l

middle horizon of the mineralogical zones of the sediment unit, in which there is the greatest percentage of volcanic ash. The fluoride concentration is lower in the area where the water table depth is less than 10 m, coinciding with the upper horizon, which has a low percentage of vitroclasts.

The calcium content increases and the fluoride concentration decreases with the direction of the groundwater flow. The precipitation of fluorite may constitute the main sink for fluoride in aquifer (Pekdeger et al. 1992). The concentration of fluoride and calcium is influenced by the equilibrium solubility of calcite and fluorite (Nordstrom & Jenne 1977). When calcium concentrations are low (to the west of the city), fluoride from vitroclasts may dissolve, but once the water has a higher calcium content (in the east) it becomes supersaturated with respect to fluorite and the mineral precipitates, removing fluoride from the solution. In this case, since the bicarbonate concentration has also increased, the higher calcium appears to have been added by dissolution of calcium carbonate from the Pampean sediments along the flow path. There is also significant infiltration in the region, which increases to the east of the city, where the topographic slopes are less steep. In this area, addition of infiltration water may cause some dilution of the fluoride content.

7 CONCLUSIONS

The fluoride concentrations in Saldungaray groundwater range from 2 – 4 mg/l. Spatial variations in fluoride concentration are shown to decrease along the groundwater flow path.

The area with the highest fluoride concentration is found to coincide with that of the lowest calcium values. In this area the thickness of the aquifer system is greatest (60m) and the water table depth coincides with the horizon of the sediment unit where there is a high percentage of vitroclasts (15 –50% of the mineralogical composition).

The fluoride concentration is lower in the area where the water table depth is less than 10 m, coinciding with the sediment horizon with low vitroclast content. In this area the fluoride concentration diminishes and calcium content increases. The concentration of fluoride and calcium are limited by fluorite solubility. There is also addition of infiltration water in this area, which may produce a dilution in the fluoride content.

REFERENCES

Castro, E., C. Schulz & E. Marino 1998. El agua potable en la Provincia de La Pampa. Consecuencias por presencia del flúor y arsénico. *Proceeding II Reunión Nacional de Geología Ambiental y Ordenación del Territorio: 42 –53.*

Gosselin, D., J. Headrick, F. Harvey, R. Tremblay & K. McFarland 1999. Fluoride in Nebraska groundwater. *Ground Water Monitoring and Remediation, 19*: 87-95.

Hem, J.D. 1985. *Study and interpretation of the chemical characteristics of natural waters.* U.S. Geological Survey, Water Supply Paper 1473, Washington, D.C.

Kruse, E., J. Ainchil & M. Giusso 1995. Abastecimiento de agua potable a la localidad de Saldungaray. *Universidad Nacional de La Plata.* Departamento de Geofísica Aplicada: 1-24.

Nordstrom, D.K. & E.A. Jenne 1977. Fluorite solubility equilibria in selected geothermal water. *Geochimica et Cosmochimica Acta* 41: 175 – 188.

Pekdeger, A., N. Ozgur & H. Schneider 1992. Hydrogeochemistry of fluorine in shallow aqueous system of the Golcuk area, SW Turkey. In Kharaka & Maest (eds) *Water – Rock Interaction*: 321 – 324.

Riggi, J.C., F. Fidalgo, O. Martinez & N. Porro 1986. Geología de los sedimentos pampeanos. *Asociación Geológica Argentina. XLI (3-4)*: 316-333.

Teruggi, M., M.C. Etchichury & J. Remiro 1957. Estudio sedimentológico de las barrancas de la zona de Mar del Plata – Miramar. *Museo Argentino de Ciencias Naturales Bernardino Rivadavia. Revista, IV, 2*, Geología: 167 – 250.

González Bonorino, F. 1955. Mineralogía de las fracciones arcilla y limo del Pampeano en el área de la ciudad de Buenos Aires y su significado estratigráfico y sedimentológico. *Asociación Geológica Argentina, Revista, XX (1):* 57 – 150.

Groundwater: Past Achievements and Future Challenges, Sililo et al. (eds) © 2000 Balkema, Rotterdam, ISBN 90 5809 159 7

Hydrogeology, groundwater resources and salinification processes in the lower Okavango Delta, northwestern Botswana

F. Linn & M. Masie
Water Resources Consultants, Gaborone, Botswana

A.J. Rana
Vincent Uhl Associates, Washingtons Crossing, Va., USA

ABSTRACT: The Okavango Delta is a large alluvial fan complex in the northwest corner of Botswana. Both surface and groundwater in this area are important to wildlife, flora and human settlements. The study covered an area of over 14,000 km^2. The primary control on groundwater development potential is water quality, with much of the area characterised by brackish/saline aquifers. This largely reflects the high rate of potential evapotranspiration. As a consequence, freshwater aquifers are strongly dependent on recharge from surface water during flooding. The transition zone in shallow aquifers from fresh to brackish and saline conditions is abrupt. The transition is characterised by increasing levels of dissolved solids and migration of saline water from underlying aquifers. Lateral inflow from surrounding saline aquifers does not appear to be a significant. The impacts of evapotranspiration indicate the importance of regular surface water recharge to sustainable development of groundwater resources.

1 INTRODUCTION

The Okavango Delta is located in northwestern Botswana on the fringe of the main Kalahari Basin, forming large perennial and seasonally flooded wetlands over an area of greater than 20,000 km^2. Due to low rainfall (450 mm/a) and a long dry season (April-October) in Botswana, the Okavango Delta is an extremely important water source for both humans and wildlife. In the distal reaches of the Delta (where much of the human habitation is located) surface flow is seasonal (2 to 4 months/year) and settlements, villages and towns are dependent on groundwater. The main town in this area, Maun (pop. 42,000), is supplied almost completely by groundwater. However, groundwater over much of the area is not potable due to salinity.

The Okavango Delta (Figure 1) is actually a very low gradient alluvial fan, fed by the Okavango River with headwaters in the Angolan interior (Stanistreet and McCarthy, 1993). The total complex has an area of over 65,000 km^2 (Thomas & Shaw 1991). Within the Delta permanently flooded swamps occupy approximately 6,000 km^2, while an additional 6,000 to 8,000 km^2 is annually flooded forming a region of seasonal swamps.

The Okavango Delta is a unique geohydrologic environment due to its seasonal flow regime, low rainfall and extremely high rates of evapotranspiration. The outflow from the Delta represents approximately 2% of total inflow (abstraction is negligible). As groundwater outflow is minimal, the resultant figure of evapotranspiration across the Delta is nearly 98% (Gieske, 1996). The high rate of evapotranspiration is primarily responsible for the brackish to saline groundwater conditions in the Delta, outside of river valleys and flood plains. The aim of this paper is examination of the mechanisms of salinisation of aquifers due to evapotranspiration as well as abstraction in light of extensive new data on the hydrological environment of the delta.

2 BACKGROUND

The Okavango Delta forms an important surface water and groundwater resource for communities in

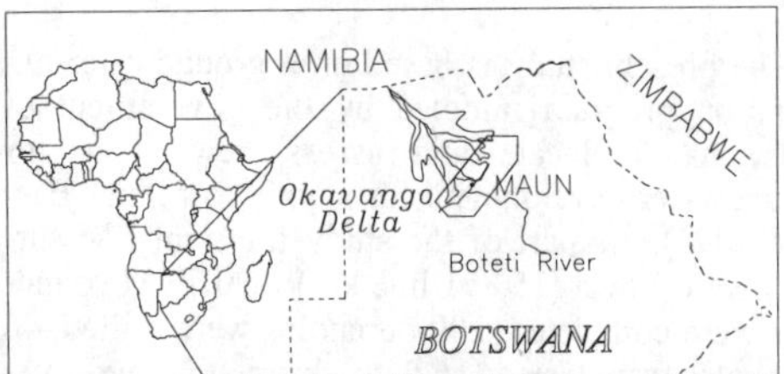

Figure 1. Location Map

this region. In particular, the urban center of Maun (Fig. 2), with a rapidly growing population and tourist industry, relies on groundwater extracted from a wellfield developed along the Shashe River, a nearby distributary.

The Shashe Wellfield began operation 1986 with shallow boreholes (<20m) completed in an unconfined aquifer (BRGM, 1986). However, the Shashe River has not flowed since1990 and without surface water recharge, water levels in the shallow aquifer have declined to an extent that it is no longer a viable resource. As a result of increasing demand and decreasing yields, additional boreholes were completed in an underlying semi-confined aquifer between 1992 and 1995, which has since become the primary source for the town supply. Under the present level of abstraction, portions of the semi-confined aquifer are beginning to show increases in TDS (Eastend, 1997).

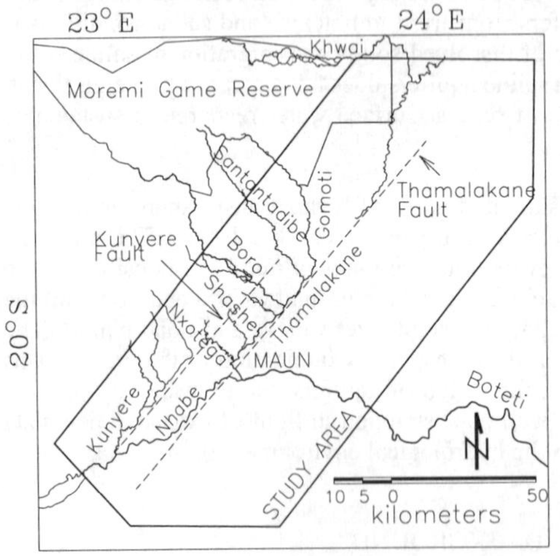

Figure 2. Location of the Study Area.

3 METHODOLOGY

The present study is based on a groundwater exploration program initiated by the government of Botswana to locate and assess new areas for groundwater development for supply to Maun (Eastend 1997). As part of the study transient EM surveys were flown (15,871 line km), 690 TEM soundings were completed, 109 boreholes were drilled, 46 boreholes were tested (16 with observation piezometres, 2 in clusters), 31 split spoon samples were sieved, 140 groundwater samples were analysed for inorganic composition and stable isotopes. Geophysical techniques (transient EM) were found to be extremely effective in delineating the lateral and vertical extent of the freshwater zone. The airborne EM survey data closely defined the lateral and vertical extent of fresh groundwater and with TEM soundings along profile lines providing definition in terms of hydrostratigraphic layers (based on resistivity).

4 GEOHYDROLOGIC SETTING

The study area extends along the length of the distal portion of the Okavango Delta adjacent to its terminus along the Thamalakane fault, covering an area of 14,000 km^2 as indicated in Figure 2. The river channels present are only seasonally flooded and there are only limited areas of permanent swamp. Evapotranspiration in the delta environs is four times as high as rainfall at 1,860 mm/yr (Wilson & Dincer 1976). An important effect of the high evapotranspiration is that much of the shallow groundwater in the study area is brackish to saline, particularly away from river channels.

Outflow in the distal delta has been highly variable during the period of historical record. In general, outflow from most of the rivers was consistent during much of the 1970's and early 1980's, while the overall flow out of the Delta decreased considerably as from the late 1980's. In recent years all but the Boro/Thamalakane, Santantadibe and Khwai Rivers have largely ceased to flow.

4.1 *Geology*

The whole of the Okavango Delta is underlain by a mixed assemblage of unconsolidated sediments and minor calcrete and silcrete of the Kalahari Beds. The thickness of this sedimentary blanket ranges from approximately 100 m east of the Thamalakane fault to 500 m northwest of the Kunyere fault. The lithologies present in the Kalahari Beds underlying the delta area are primarily fine sand, silty and clayey sand and clay, reflecting deposition in alluvial, aeolian and lacustrine environments (McCarthy et al. 1991; Thomas & Shaw, 1991).

4.2 *Hydrogeology*

Prior to the present study, only limited information existed on the hydrogeology of the lower Okavango Delta. Aside from a few recent boreholes drilled in villages surrounding the delta, existing borehole drilling and testing data was only available primarily for the Shashe River Wellfield and a limited portion of the Thamalakane River near Maun (Figure 2). On the other hand, considerable research has been carried out on the sedimentology (McCarthy & Metcalfe, 1990; McCarthy et al. 1991; Stanistreet et al. 1993) and surface water hydrology of the Delta (SMEC 1987), including outflow modeling (Dincer et al. 1987; SMEC 1987; Gieske 1996).

Data from existing and project boreholes indicate that there is a relatively consistent hydrostratigraphy across the area. Aquifer systems in most of the area are composed of a shallow unconfined aquifer of 5 to 25 metres thickness with one or more underlying semi-confined aquifers. Variably developed clay, sandy clay and silty sand separate the underlying semiconfined aquifers layers which form leaky aquitards. The thickness of the freshwater zone (<1,500 mg/l TDS) varies from 25 to 120 metres.

Water quality changes rapidly both laterally and vertically from fresh to brackish and saline conditions. Laterally this transition occurs over 5 to 20 metres while vertically it commonly occurs over less than 5 metres. The vertical transition zone is not always defined by any lithologic change.

The distribution of fresh groundwater is strongly controlled by the frequency and extent of the annual flooding and to a much lesser extent by the lithologic variation. The geometry of the freshwater zone clearly follows the river valleys and areas of seasonal swamp. The dominance of surface water recharge to the freshwater aquifers is also evident in $\delta\ ^{18}O$ analyses, with groundwater plotting closely on the surface water evaporation line (Figure 3).

Grain size of the aquifer units is extremely uniform. Grain size is in the range of 1.9 to 2.3ϕ and very well sorted. Transmissivities range from 10 to 80 m^2/d. Vertical hydraulic conductivity values for aquitards are mostly 10^{-3} to 10^{-4} m/d, with some up to 10^{-2} m/d. Test pumping in semiconfined aquifers consistently reflect the leaky nature of the system. Groundwater levels are shallow, 1-10 metres in river valleys and 10 to 15 metres in interfluve areas. Regional groundwater flow is radially away from the delta area while local groundwater flow systems are generally downstream and away from river valleys.

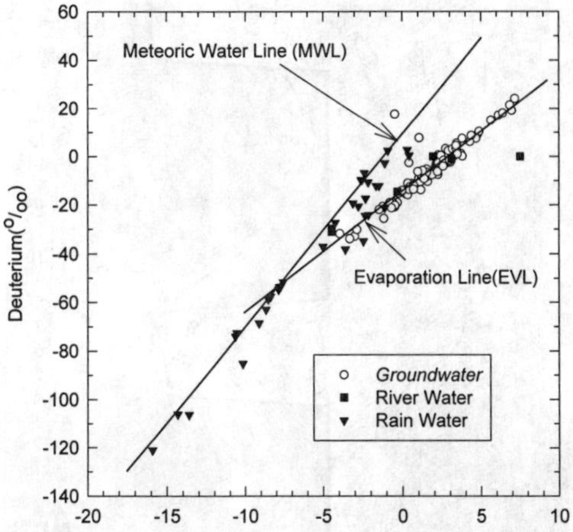

Figure 3. Deuterium-Oxygen 18 relationship: groundwater, river water and rain water.

5 WATER QUALITY AND SALINISATION PROCESSES IN AQUIFERS

The primary limiting factor on groundwater development in the distal Okavango Delta is salinity. Significant freshwater aquifers are developed only along river valleys and areas of seasonal/perennial flooding. Some of the river valleys (Gomoti, Kunyere, upper Boro Rivers) have very broad and deep freshwater zones up to 10 kilometres wide and extending to more than 100 metres depth. Typically, however, river channels in the distal delta have freshwater zones on the order of 2 kilometres wide and 40 to 60 metres deep. A typical cross section (across the Shashe River Wellfield) is shown in Figure 4.

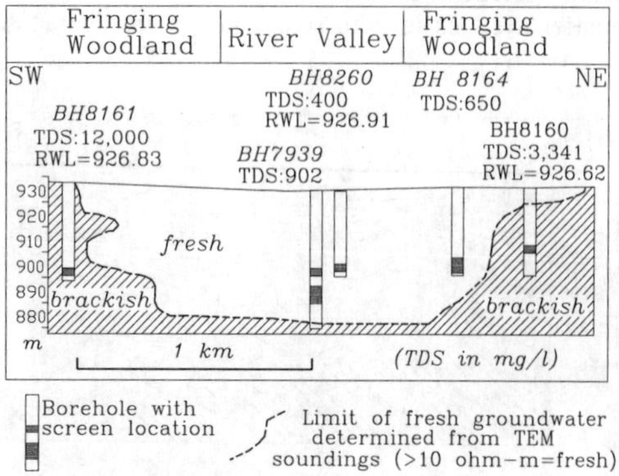

Figure 4. Geo-electric cross section, Shashe River.

In seven of the river valleys, borehole/piezometer clusters were installed. In undeveloped rivers under natural conditions, vertical gradients are consistently downward from shallow to deeper aquifers from fresh toward brackish/saline aquifers (Figure 5). In the one natural system where observation boreholes were located 1 to 2 kilometres outside of the river channel (Lower Thamalakane), water levels away were from 8 to 13 metres deeper than in the river valley. Lateral gradients in the Shashe Wellfield are much lower in the immediate vicinity of the river valley and may reflect the effects of abstraction (Figure 4).

In the absence of flood recharge, the vertical component of the gradient from fresh to brackish water can be reversed. An example from a natural river system based on geophysical evidence is illustrated in Figure 6. In this area, the transition zone is not marked by any lithological change but occurs within a sequence of fine to medium sand. Of particular interest is the distortion of the transition zone beneath a small (<100m across) vegetated island within the channel.

If it assumed that during regular flooding, the whole of the channel area had a consistent thickness of fresh groundwater, then in the approximately 6 years since this portion of the river last flooded, the fresh water lens beneath the vegetated area within the channel has thinned considerable. This is interpreted as a result of upward migration from the underlying brackish water zone in response to transpiration from trees and scrub. Additionally, the TDS within the freshwater aquifer (10 to 30 metres) has increased in relation to surrounding portions of the aquifer (1300 mg/l, borehole 8301, 486 mg/l, borehole 8354).

This process is also apparent on a larger scale when comparing two adjacent rivers with very different flood histories, the Nxotega and Shashe Rivers (Figure 2). Both rivers have similar meandering geomorphologies with valley margins defined by faults (BRGM, 1986; Eastend, 1997). Freshwater aquifer systems in both river valleys are characterised by a shallow unconfined aquifer and an underlying semi-confined aquifer. The primary difference between the rivers is their recharge history, the Nxotega River has not flooded in more than 25 years, while the Shashe River last flooded in 1989. As a result, the channel area of the Nxotega River is increasing vegetated with shrubs and trees.

Comparing the airborne EM signature for the rivers in the shallow zone (channel 3, approximately to 30 metres depth) both river valleys are well defined as broad regions of low conductivity (< 4 S) indicative of fresh groundwater (Figure 7).

However, there is a marked difference in the EM signature of the two valleys at greater depth, shown in Figure 8 (channel 8, approximately 50-60 metres

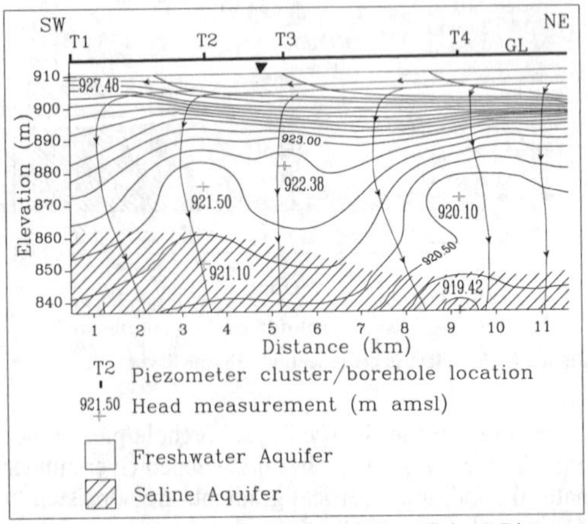

Figure 5. Conceptual vertical flow net, Thamalakane River.

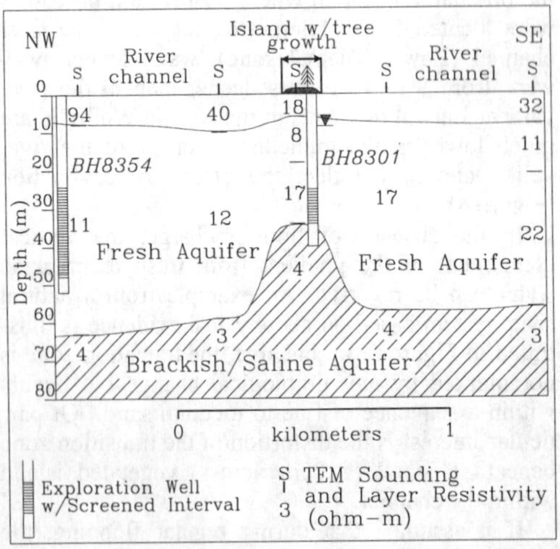

Figure 6. Geo-electric cross section, Upper Shashe River.

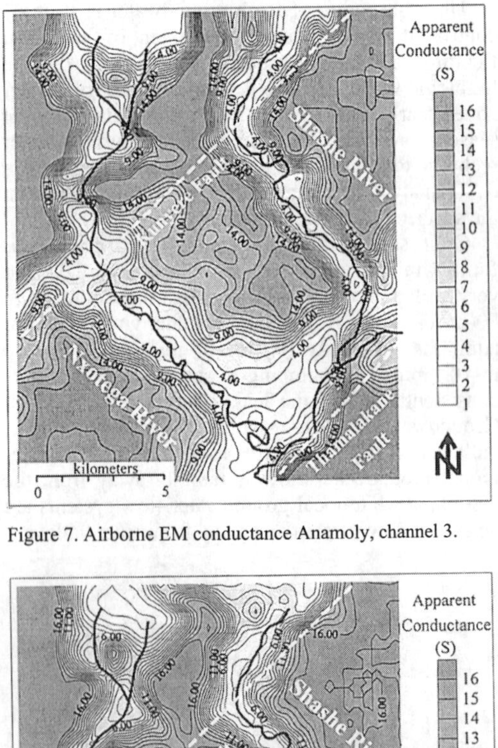

Figure 7. Airborne EM conductance Anamoly, channel 3.

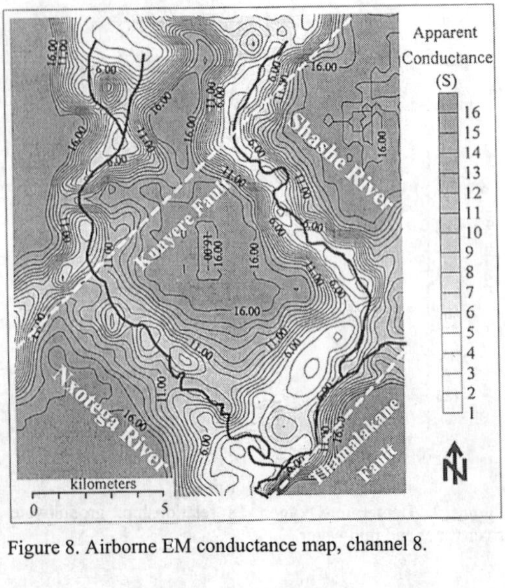

Figure 8. Airborne EM conductance map, channel 8.

depth). The Shashe River continues to show low conductivity representing the lower semi-confined aquifer (fresh). However, in this same depth range, the Nxotega River valley is characterised by conductivities greater than 6 Siemens, which has been shown to represent brackish groundwater. The airborne EM data are confirmed by drilling results in terms of the thickness of the freshwater lens (37 m in Nxotega, 60m in Shashe) and quality, with higher TDS in the freshwater zone (2,437 mg/l in Nxotega, 700 -1200 mg/l in Shashe).

The comparison of the Nxotega and Shashe River valleys also indicates that in the absence of flood recharge, freshwater aquifers in a natural system tend to become saline in their deepest portions first. The mechanism is suggested to be upward migration from underlying saline aquifers. This is induced by reduction in heads in the fresh aquifers relative to underlying saline aquifers due to removal of fresh groundwater through evapotranspiration and/or abstraction. Even with the effects of abstraction and evapotranspiration, the Shashe valley continues to have a thicker freshwater lens than the Nxotega valley due to a more recent flooding event. This suggested salinisation process contrasts with previously hypothesised mechanisms of possible downward flow of saline water formed in the near surface in island environments (McCarthy et al., 1991, Gieske, 1995). Similar to the Shashe Wellfield, the shallow unconfined aquifer in the Nxotega valley is largely dewatered. Although the vertical extent of the freshwater zone has decreased significantly in the Nxotega River valley, the lateral extent has not changed between shallow and deep (Figures 7 and 8). This suggests that lateral flow from the surrounding saline zone is not significant.

6 IMPLICATIONS FOR GROUNDWATER DEVELOPMENT

A large quantity of data has been collected in the Shashe River Wellfield since its development in 1986 (BRGM 1986; DWA 1993; Eastend 1997). As such it makes a good case study of wellfield abstraction during periods of both annual flooding and multiple years of no flooding. As described previously, the shallow unconfined aquifer was largely abandoned in 1992 when production from the underlying semi-confined aquifer commenced. The Shashe River has experienced no flooding during production from the lower aquifer to date. Annual abstraction is on the order of 1 million m^3 with the bulk of abstraction concentrated in the southern 4 kilometer stretch of the river valley. Similar to the findings in other river valleys, the vertical transition from fresh to brackish/saline groundwater is rapid, occurring within 1 to 5 metres. Although significant water level declines have not been noted in the wellfield,

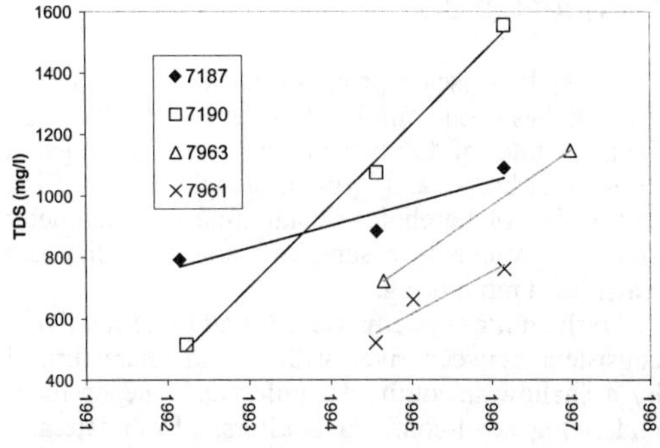

Figure 9. TDS increase in production boreholes, Shashe Wellfield.

various degrees of water quality deterioration has been clearly evident (Figure 9).

Similar to natural systems under the influence of evapotranspiration, the change in water quality in the Shashe Wellfield is attributed to migration of poor quality water upward from underlying aquifers. Abstraction from the semi-confined aquifer induces leakage from both overlying and underlying semi-confining beds. This is illustrated in a vertical flow net constructed in the southern wellfield in the vicinity of two production boreholes where piezometer clusters were installed (Figure 10).

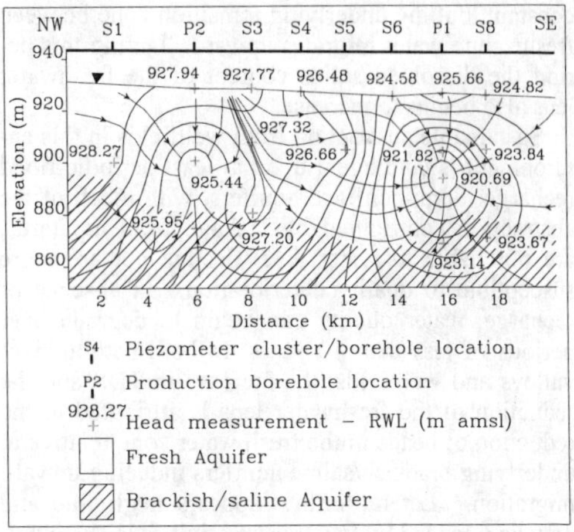

Figure 10. Conceptual vertical flow net in the vicinity of production boreholes, Shashe Wellfield.

Unlike the natural conditions in a river system receiving regular flood recharge (Figure 5), the head gradients between the fresh semiconfined aquifer and underlying brackish/saline aquifer are locally reversed, inducing upconing of poor quality water. Lateral inflow from does not appear to be as significant based on geophysical data, although the lateral head relationships are not well constrained.

7 CONCLUSIONS

The first large scale groundwater exploration programme has been completed in the lower Okavango Delta. A total of 11 river valleys were investigated through airborne and ground geophysics, drilling and testing of boreholes, installation of peizometer clusters, groundwater sampling and analysis, and water level monitoring.

Fresh aquifer systems were found to be relatively consistent between river valleys and characterised by a shallow unconfined aquifer and one or more underlying semi-confined aquifers. Both laterally and vertically the transition from fresh to brackish/saline aquifers is abrupt. The source of recharge to freshwater aquifers is annual flooding of the rivers, with rainfall recharge being insignificant in relation to evapotranspiration. The primary difference between river systems is the thickness of the freshwater lens, which varies from as little as 30 metres to more than 120 metres. This variation is attributed primarily to the frequency and lateral extent of surface water flooding.

Geophysical data and the comparison of river valleys with different flooding histories suggest that the freshwater lens underlying river valleys is temporal and begins to significantly decrease in response to evapotranspiration in periods of less than 20 years without flood recharge. In these cases, the lateral boundaries of the freshwater lens remain relatively constant, but the underlying transition zone between fresh/saline water migrates upward. During this period the dissolved solids content of the freshwater lens also begins to increase.

Sustainable abstraction from wellfields in this environment is contingent upon at least periodic flood recharge. Semi-confined aquifers, with much of the abstraction contributed by leakage from aquitards, are not liable to water level decline but are more susceptible to quality deterioration. In absence of recharge, water quality can begin to degrade over periods of less than 10 years. In both natural river valleys and wellfields, the increase in TDS and the reduction of the freshwater lens is attributed to the reduction of heads in the fresh water zone relative to underlying brackish/saline aquifers inducing upward migration. Lateral inflow appears negligible and may be impeded by low transmissivities in surrounding saline aquifers, but additional study is required to test this hypothesis.

Sustainable development and management of groundwater resources will require additional study particularly focusing on more data collection in the brackish/saline aquifers. Due to groundwater development objectives, the existing data set is highly biased toward only the fresh groundwater regions. An understanding the hydraulic/hydrochemical relationship between saline and fresh aquifers will be critical to understanding and developing groundwater in this region.

REFERENCES

Bureau De Recherches Geologiques Et Minieres (BRGM), 1986: Maun Water Supply Hydrogeological Survey. Final Report to Botswana Department of Water Affairs. February, 1986.

Department of Water Affairs, 1993: Data Report on Shashe Maun Pilot Borehole Construction Project, September 1992 - March 1993, Gaborone, Botswana.

Dincer, T., Child, S. and Khupe, B.B.J., 1987: A simple mathematical model of a complex hydrological system; Okavango Swamp, Botswana, J. Hydrol., 93, 41-65.

Eastend Investments, 1997: Maun Groundwater Development Project, Phase I: Exploration and Resource Assessment. Final Report to Botswana Department of Water Affairs, October, 1997.

Gieske, A., 1995: Vegetation driven groundwater recharge below the Okavango Delta (Botswana) as a solute sink mechanism - an indicative model. Conference on Groundwater Recharge and Rural Water Supply, Johannesburg, South Africa, 26-28 September 1995, pp 119-124.

Gieske, A., 1996: Modeling of outflow from the Jao/Boro flow system - Okavango Delta, Botswana. Journal of Hydrology, 193, 214 – 239.

McCarthy, T.S. and Metcalfe, J., 1990: Chemical sedimentation in the semi-arid environment of the Okavango Delta, Botswana: Chemical Geology, vol. 89, pp. 157 - 178.

McCarthy, T.S., Stanistreet, I. G., and Cairncross, B., 1991: The sedimentary dynamics of active fluvial channels on the Okavango Fan, Botswana: Sedimentology, vol. 38, pp. 471 - 487.

McCarthy, T.S., McIver, J. R. and Verhagen, B. Th., 1991: Groundwater evolution, chemical sedimentation and carbonate brine formation on an island in the Okavango Delta swamp, Botswana: Applied Geochemistry, vol. 6, 577-595.

McCarthy, T.S. and Ellery, W.N., 1994: The effect of Vegetation on Soil and Groundwater Chemistry and Hydrology of Islands in the Seasonal Swamp of the Okavango Fan, Botswana – J. Hydrol. 154: 169-193.

Snowy Mountains Engineering Corporation (SMEC), 1987: Southern Okavango Integrated Water Development. Phase 1. Final Report, Technical Study to Botswana Department of Water Affairs.

Stanistreet, I. G. and McCarthy, T. S., 1993: The Okanvango Fan and the classification of fan systems: Sedimentary Geology, vol. 85, pp. 115 - 133.

Thomas, D.S.G. and Shaw, P.A., 1991: The Kalahari Environment. Cambridge University Press, 284p.

Wilson, B.H. and Dincer, T., 1976: An introduction to the hydrology and hydrography of the Okavango Delta, Symposium on the Okavango Delta, pp. 33-48 (Botswana Society, Gaborone.

Groundwater: Past Achievements and Future Challenges, Sililo et al. (eds) © 2000 Balkema, Rotterdam, ISBN 90 5809 159 7

Chemical character of groundwater in a basalt aquifer, North Queensland, Australia

K.L.Locsey & M.E.Cox
Queensland University of Technology, Brisbane, Qld, Australia

ABSTRACT: Due to the increasing demands on the groundwater resources on the Atherton tableland, North Queensland over the last 15 years, a number of investigations are being undertaken to determine the extent and sustainability of the resource. The groundwater supplies are predominantly sourced from fresh and weathered basalt, varying from highly vesicular to massive in character, with fracture zones also present. The chemical character of groundwater contained within the basalt can be used to define trends in groundwater movement, based on its chemical evolution. The groundwater chemically evolves from water with no dominant ions to slightly HCO_3^--rich water in areas where direct rainfall infiltration predominates, to a Ca, Mg – HCO_3 type water along flow gradient and at discharge zones. Trends in major ion concentrations and relationships between dissolved constituents, indicate that reactions with clinopyroxene, plagioclase feldspars and olivine are the principle controls on the chemical character of these groundwaters.

1 INTRODUCTION

The basaltic volcanic pile of the Atherton tableland is the major source of groundwater supply for irrigation and other agricultural use. Some groundwater is also sourced from underlying fractured metamorphosed sediments and granites. Groundwater use in the region has increased substantially over the last 15 years, due to a change from low to high water use crops. In the Atherton Shire, centred on the townships of Atherton, Tolga and Kairi (Fig. 1), groundwater use was estimated at 3000 ML per annum in 1985 (Leach, unpubl.), and is currently estimated at 12,000 ML per annum (Barclay & Pearce, unpubl.).

The increasing demands on groundwater prompted a number of investigations in the region, to accurately define the groundwater resource, so that a management plan could be developed to ensure sustainable future use. A component of this investigation is a detailed study of the hydrochemistry of the groundwater to better define its form of occurrence and movement.

2 SITE DESCRIPTION

The Atherton tableland is situated approximately 45km southwest of Cairns, a regional city in North Queensland (Fig. 1), and extends from Walkamin (10 km northwest of Tolga) in the north, to Millaa

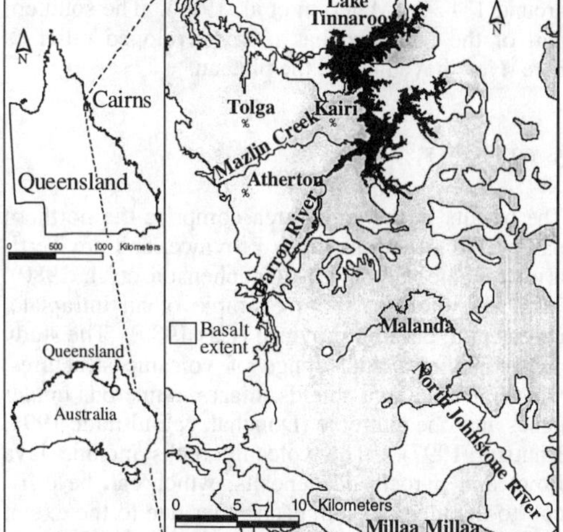

Figure 1. Location of Atherton tableland, North Queensland, showing the extent of basalt flows.

Millaa in the south, and covers an area of approximately 20 km (E-W) by 35 km (N-S).

The Atherton tableland is a plateau, which rises to 600 – 1000 m above sea level and has a sub-tropical monsoonal climate, with hot humid summers and mild dry winters. Mean annual rainfall is seasonal and decreases from southeast to northwest, ranging from approximately 2700 mm at Millaa Millaa to 1300 mm at Tolga. Most rain falls in the summer months (December to March), usually during high intensity thunderstorms. The six months of May to October are relatively dry receiving less than 25% of the annual rainfall (Laffan 1988). On an average monthly basis, rainfall exceeds evaporation only from January to March, indicating that rainfall during these months is a significant component of aquifer recharge (Leach, unpubl.).

The tableland supports a growing agricultural industry of crops such as sugar cane and tea-tree and other horticultural activities, as well as cattle grazing and dairy in the southern section. Some forested areas have been listed as World Heritage Wet Tropical Rainforest Reserves. Two major catchments drain the study area, with the Barron River flowing to the north and the North Johnstone River to the southeast (Fig. 1). The Barron River and Mazlin Creek flow into Lake Tinnaroo, which is a major water resource in the region, supplying water to the irrigation areas to the north, as well as the Atherton township.

The morphology of the region is highly variable. The edges of the tableland have steep slopes in the east and south (> 35%), and to the west is the steep Great Dividing Range of eastern Australia. On the plateau itself are a variety of basaltic volcanic features, and slopes of most areas in the north are around 1 – 3% (Malcolm et al. 1999). The southern part of the tableland has more developed relief as here rivers have incised the plateau.

3 GEOLOGY

The basalts in the study area comprise the northern part of the Atherton Basalt Province, and are Tertiary to Quaternary in age (Stephenson et al. 1980). This volcanic area is an example of an intraplate, continental, basaltic province (Cas 1989). The study area contains a wide range of volcanic structures, which include lava shields, maars, numerous cinder cones and one diatreme (Donchak & Bultitude 1994, Knutson 1997). The volcanic rocks include lava flows and pyroclastic deposits, which can be difficult to distinguish from each other due to the extent of weathering (Bultitude & Robertson 1997, Buck, unpubl.). Highly weathered, clay rich palaeosols are common between basalt flows, as observed in drill cuttings, as well as in stream outcrop in the north of the tableland.

A recent investigation by Sheldrick (unpubl.) identified four basaltic rock types in the study area, not including pyroclastic material, which is crystal-rich ash. The basalt mineralogy is predominantly olivine, plagioclase (most likely albite and anorthite) and clinopyroxene (augite).

The volcanic rocks of the Atherton Basalt Province erupted over a variety of rocks of Devonian to Early Permian age. These basement rocks include metamorphosed sediments of the Devonian Hodgkinson Formation, Late Carboniferous to Early Permian granitic and volcanic rocks, and Permian granitic rocks. Quartz-rich and micaceous sediments ranging in texture from silt to pebble, commonly occur at the base of the basalt, and are interpreted to be granitic and metamorphic weathering products.

Assessments of drainage patterns, basement rock topography and locations of volcanic features (Buck, unpubl.), as well as an analysis of airborne magnetic data (Moloney, unpubl.), identified linear NW-SE and NE-SW trending structural zones in the basement rocks. Those studies indicate that the location of volcanic centres in the area is structurally controlled. Post-volcanic re-activation of some of these structures, appears to have some influence on surface morphology and may have locally produced zones of increased porosity.

4 HYDROGEOLOGY

The basaltic aquifer can be regarded as a generally unconfined, layered system. Within this, numerous basalt flows sourced from a number of vents, are separated by palaeo-weathering surfaces and minor alluvial gravels of palaeo-drainage channels. Layers of massive basalt and clay-rich weathered zones act as local aquitards, and some local perched aquifers are also present.

Within this current study 19 new monitoring bores were drilled and cuttings logged. The basalt in the study area varies from massive to highly vesicular. Vesicle size ranges from < 2 mm to > 1 cm. The degree of vesicle interconnection also varies considerably, and weathering products are commonly observed, such as clay and zeolite infilling of vesicles. Permeability is shown to be largely due to weathered zones between basalt flows, highly vesicular flows and a variety of fracturing.

The thickness of the basalt pile is variable, and is greatest in the area bounded by the townships of Atherton, Tolga and Kairi. Basalt in this area ranges from 50 m to greater than 100 m in thickness. The depth of weathering can be significant, with up to 30 m of the basaltic material weathered to clays. Leach (unpubl.) notes that good irrigation supplies in the region are most often obtained from highly weathered vesicular basalt layers at depths from 35 to 75 m; yields range from 5 to 50 L/s. Standing water levels in the study area are generally within 15 to 40 m from the surface, and seasonal variations in water level can be up to 20 m.

Recharge by direct rainfall infiltration occurs throughout the tableland. Recharge rates may be expected to reflect the rainfall distribution pattern, with rates decreasing from SE to NW. Preliminary recharge estimates by Cook et al. (unpubl.) using a chloride mass balance, indicate that the rate of recharge decreases towards the NW, from approximately 600 mm/yr near Malanda, to 200 mm/yr in the Atherton, Tolga, Kairi area.

The general groundwater flow direction within the northern section of the basaltic area is towards the NNW, and generally reflects the surface topography. There are numerous locations where groundwater discharges as springs along the edges of lava terraces, and as baseflow within incised streams. Recent work (Barclay & Pearce, unpubl., Cook et al., unpubl.) confirms that the Barron and North Johnstone Rivers, Mazlin Creek, and other creeks receive significant volumes of groundwater as baseflow.

Transmissivity calculations from pump tests by Leach (unpubl.) are highly variable and range from 17 to 3500 m^2/day. Leach (unpubl.) notes that values in excess of 500 m^2/day are associated with highly weathered vesicular basalt or scoria, and that values below 40 m^2/day are associated with jointed basalt or unweathered vesicular basalt.

Groundwater is also obtained from the upper profile of granitic and metamorphic basement rocks, predominantly in the SE of the study area. Due to low yields (generally < 5 L/s) these supplies are generally used for stock and domestic purposes only.

5 GROUNDWATER CHEMISTRY

The chemical composition of groundwater from the Atherton tableland has been obtained for more than 800 samples from the Queensland Department of Natural Resources (QDNR). Fifty percent of these samples were collected during 1998 and 1999, and another 25% collected from 1990 – 1997.

As part of this study 90 samples were collected from bores intercepting basalt, in October 1998 and May and October 1999. Samples of water for cation analyses were collected in polyethylene bottles aged in nitric acid, and samples for anion analyses were collected in non-acidified bottles. The samples were stored at approximately 4°C until analysis. The samples were analysed in the Queensland University of Technology geochemical laboratory. Cations and sulfate were analysed using inductively coupled plasma atomic emission spectroscopy (ICP-AES). The ICP was calibrated against appropriate standards and against a blank prepared with nitric acid. Chloride and sulfate (the latter for comparison with ICP results) were measured by high performance ionic liquid chromatography. The alkalinity as HCO_3^- and CO_3^{2-} was determined by titration with 0.009 N Standard HCl following procedures described by Greenberg et al. (1992). All reported values had an ionic balance within 5% of error.

Groundwater from the Atherton tableland (based on the chemical data from 800 QDNR samples) is of low salinity with total dissolved solids generally ranging from 30 – 250 mg/L and electrical conductivity (EC) approximately 40 – 350 μS/cm at 25°C. Groundwater pH can range from 5 to 11, although the majority of samples (> 80%) are within 6 – 8 pH units. The major dissolved constituents are HCO_3^- and H_4SiO_4. On average dissolved silica (H_4SiO_4) comprises 35% of the total dissolved load.

Although these groundwaters have low total dissolved solids (TDS), trends in some dissolved ion concentrations and ionic ratios show a relationship with groundwater flow patterns, and indicate a degree of hydrochemical evolution. Based on the distribution of the chemical composition data, the basalt-derived samples can be sub-divided into three categories. The data for the following categories and discussion are based on the 90 basalt-derived groundwater samples collected between October 1998 and October 1999.

5.1 *Category One*

The average EC and pH of groundwater in this category is 99 μS/cm and 6.0 units, respectively. The concentrations of Na^+ and Ca^{2+} are high relative to Mg^{2+} and K^+, and HCO_3^- is more than three times the concentration of Cl^-. The average H_4SiO_4 concentration is 53 mg/L. Concentrations of SO_4^{2-} are negligible. These groundwaters range from no dominant chemical type to slightly HCO_3^- enriched. Ratios of HCO_3^- / Cl^- and Mg^{2+} / Cl^-, calculated by equivalent weight, are on average 2.73 and 0.56, respectively.

The chemical compositions of five representative samples in category one are shown in Table 1, as well as average values (based on the 90 basalt-derived groundwater samples).

Groundwater in this category is predominantly derived from direct infiltration of rainfall. Chemical analyses undertaken as part of this study on 58 monthly rainfall samples collected from five rainfall stations across the tableland seem to support this interpretation. The rainfall samples were provided by the CSIRO Land and Water. Chemical analyses for these rainfall samples, collected from November 1998 to November 1999, show Mg^{2+} / Cl^- ratios calculated by equivalent weight ranging from < 0.01 to 0.51. These values are similar to the Mg^{2+} / Cl^- ratios for category one groundwaters.

Table 1. Representative chemical analyses for category one groundwaters (mg/L).

Bore	Na	K	Ca	Mg	HCO_3	Cl	H_4SiO_4
55	9.6	2.2	6.5	4.5	45	10	69
56	11.5	1.4	6.5	5.1	58	8.2	77
63	7.6	2.2	5.3	4.1	36	11.5	67
66	12	1.2	5.3	3.4	26	16.5	46
103	7.8	1.6	2.8	3	11	8.3	32
Av.	**8.9**	**1.7**	**6.5**	**3.6**	**37.3**	**9.8**	**53**

5.2 *Category Two*

The average EC and pH of groundwaters in category two are 204 μS/cm and 7.3 units, respectively. The average concentration of Ca^{2+} in these groundwaters is only slightly greater than Na^+ and Mg^{2+} concentrations. Concentrations of K^+ are greater than in category one, with an average of 2.5 mg/L, although the percentage of K^+ of total cations in solution remains < 10% for these groundwaters. The dominant anion is HCO_3^-, at greater than six times the concentration of Cl^-. Some CO_3^{2-} is also present in low concentrations (average = 0.3 mg/L). Concentrations of SO_4^{2-} remain negligible and H_4SiO_4 concentrations are greater than in category one, at an average of 76 mg/L. Representative chemical analyses for groundwater in this category are shown in Table 2, as well as average values for the larger data set.

Groundwaters in category two are Ca, Mg – HCO_3 type waters. Average HCO_3^- / Cl^- and Mg^{2+} / Cl^- ratios calculated by equivalent weight, are greater than in category one, at 3.97 and 0.99, respectively. These groundwaters represent a mixed water type, which while likely comprising a component of direct rainfall, essentially represent an early stage of chemical evolution as a result of contact with basaltic minerals.

5.3 *Category Three*

Groundwaters in this category are also Ca, Mg – HCO_3 type waters, although more enriched in HCO_3^- than category two groundwaters. The concentrations of all ions have increased, with Ca^{2+}, Na^+ and Mg^{2+} average concentrations of 18.7, 16.7 and 14.3 mg/L, respectively. Concentrations of HCO_3^- are also greater, at an average of 158 mg/L, and CO_3^{2-} concentrations have also increased slightly. The average pH of these groundwaters is similar to the previous category at 7.4 units, and the average groundwater EC is 279 μS/cm. Ratios of HCO_3^- / Cl^- and Mg^{2+} / Cl^- calculated by equivalent weight, have also increased substantially over the previous categories to 10.73 and 2.38, respectively. Representative chemical analyses for groundwater in this category are shown in Table 3, as well as average values for the larger data set.

These groundwaters are interpreted as being mixed water comprising direct rainfall, in addition to being influenced by groundwater migration (residence) and irrigation return flow in agricultural areas.

Table 2. Representative chemical analyses for category two groundwaters (mg/L).

Bore	Na	K	Ca	Mg	HCO_3	Cl	H_4SiO_4
114	10.5	2.3	10.5	7.6	79	14.5	61
115	11	2.4	13	9.8	94	12.5	78
133	10	2.5	12.5	9.8	90	12.5	88
Av.	**10.5**	**2.5**	**14.4**	**9.2**	**93**	**13.75**	**76**

Table 3. Representative chemical analyses for category three groundwaters (mg/L).

Bore	Na	K	Ca	Mg	HCO_3	Cl	H_4SiO_4
92753	14.5	2.3	11	15	125	9	96
109108	14	2.7	20.5	9.8	135	7.1	67
105	41	7.8	40.5	33.5	385	15	38
118	13	2.2	18	17	160	10	72
131	10.5	2.1	14	8	100	5.1	96
Av.	**16.7**	**3.2**	**18.7**	**14.3**	**158**	**8.3**	**69**

6 DISCUSSION

The increasing concentrations of HCO_3^-, H_4SiO_4 and cations from category one to three, and the relationship between cations (e.g. Mg^{2+}) and HCO_3^- (Fig. 2), indicate that groundwater chemical composition is predominantly influenced by basalt mineralogy.

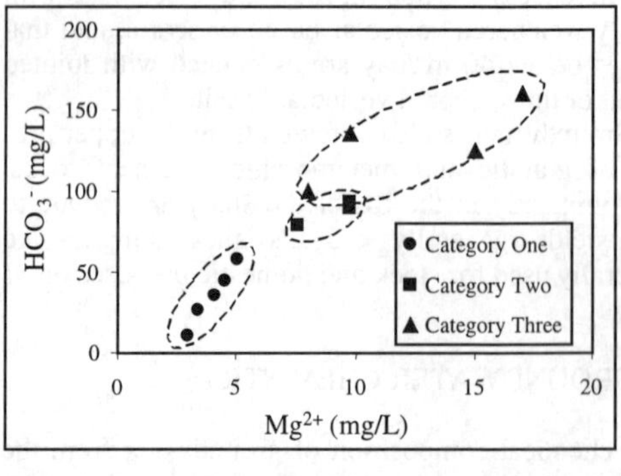

Figure 2. Relationship between Mg^{2+} and HCO_3^- concentrations in basalt-derived groundwater.

The predominant mineralogy of the Atherton basalts is clinopyroxene (augite), plagioclase feldspars (probably albite and anorthite) and olivine. Groundwater composition is influenced by the reactions with these minerals. Concentrations of ions in solution are likely to be a function of contact time and surface area per unit volume of water. Where soil development and plant growth are rapid, as is the case in the sub-tropical monsoonal climate of the Atherton tableland, substantial CO_2, and hence, hydrogen ions are available to circulating water. The extent to which reactions with the minerals occurs, is related to the availability of H^+, as well as the factors noted above (Hem 1985).

The average Mg^{2+} / HCO_3^-, Na^+ / HCO_3^- and Ca^{2+} / HCO_3^- ratios calculated on an equivalent moles of charge basis, are 0.44, 0.45 and 0.46, respectively. The sum of these ratios is 1.35. This is slightly higher than the expected release of Mg^{2+}, Na^+ and Ca^{2+} on a 1:1 equivalent moles of charge ratio relative to consumption of H^+ (i.e. increase in HCO_3^-), due to dissolution of olivine, albite and anorthite. The dissolution of these minerals is shown in equations 1 - 3, with the albite and anorthite weathering reactions written with kaolinite as the end product.

The alteration of albite to montmorillonite (equation 5) consumes no acid, but releases Na^+ to solution. A degree of weathering of albite to montmorillonite would account for an excess of cations over HCO_3^-. This is supported by the range of H_4SiO_4 / Na^+ molar ratios in these groundwaters as discussed below.

olivine (forsterite)

$$Mg_2SiO_4 + 4CO_2 + 4H_2O = 2Mg^{2+} + H_4SiO_4 + 4HCO_3^- \quad (1)$$

albite

$$2NaAlSi_3O_8 + 2CO_2 + 11H_2O = Al_2Si_2O_5(OH)_4 + 2Na^+ + 4H_4SiO_4 + 2HCO_3^- \quad (2)$$

anorthite

$$CaAl_2Si_2O_8 + 2CO_2 + 3H_2O = Al_2Si_2O_5(OH)_4 + Ca^{2+} + 2HCO_3^- \quad (3)$$

The average molar ratios of Mg^{2+} / H_4SiO_4 in these groundwaters are 0.27, 0.48 and 1.01 for categories one to three, respectively. The theoretical values of Mg^{2+} / H_4SiO_4 for the equations of dissolution of the clinopyroxene augite, and olivine (considering only the Mg^{2+} end-member, forsterite in equation 1) are 0.7 and 2.0, respectively. The weathering of augite to produce kaolinite and cations is shown in equation 4.

augite

$$[Ca_{1.15}MgAl_{0.3}Si_{1.7}]O_6 + 4.3CO_2 + 5.25H_2O = 0.15Al_2Si_2O_5(OH)_4 + 1.15Ca^{2+} + Mg^{2+} + 1.4H_4SiO_4 + 4.3HCO_3^- \quad (4)$$

The groundwater data indicate that dissolution of these minerals contributes to the concentrations of Mg^{2+} and H_4SiO_4. The increasing concentrations of Mg^{2+} and H_4SiO_4 from category one to three indicate that residence time is reflected in the groundwater chemical composition. The variation in Mg^{2+} / H_4SiO_4 molar ratios between the categories indicates that there may be a shift in the proportions of minerals dissolving.

The theoretical H_4SiO_4 / HCO_3^- molar ratios for the dissolution of olivine, augite to kaolinite, and albite to kaolinite are 0.25, 0.32 and 2.0, respectively, as shown in equations 1, 4 and 2. Molar ratios of H_4SiO_4 / HCO_3^- for the Atherton basalt groundwaters range from 0.12 to 1.85, with an average of 0.62. The H_4SiO_4 / HCO_3^- ratios obtained for the Atherton basalt groundwater indicate that olivine, augite and albite weathering contribute to the concentrations of H_4SiO_4 and HCO_3^-. Olivine and augite would be expected to weather preferentially to the sodic plagioclase, albite, as defined by the Goldich (1938) weathering sequence for silicate minerals.

The theoretical H_4SiO_4 / Na^+ molar ratios for albite dissolution to montmorillonite, kaolinite and gibbsite are 0.5, 2.0 and 3.0, respectively. The weathering of albite to montmorillonite and gibbsite is shown in equations 5 and 6, respectively. The weathering of albite to montmorillonite assumes the presence of Mg^{2+} leached from pyroxenes.

albite

$$3NaAlSi_3O_8 + Mg^{2+} + 4H_2O = 2Na_{0.5}Al_{1.5}Mg_{0.5}Si_4O_{10}(OH)_2 + 2Na^+ + H_4SiO_4 \quad (5)$$

$$NaAlSi_3O_8 + CO_2 + 8H_2O = Al(OH)_3 + Na^+ + 3H_4SiO_4 + HCO_3^- \quad (6)$$

Molar ratios of H_4SiO_4 / Na^+ range from 0.19 to 3.60, with an average of 1.72, which indicates that albite dissolution is likely to be occurring. This is supported by preliminary analysis of rock mineralogy using X-ray diffraction (XRD), which shows a substantial presence of montmorillonite and kaolinite as well as other weathering products such as zeolites, in addition to the predominant primary minerals of augite, anorthite and albite.

Groundwater obtained from a bore near the township of Millaa Millaa showed a H_4SiO_4 / Na^+ molar ratio of 3.60, indicating possible weathering of albite to gibbsite. Some gibbsite (< 5%) has been identified in basalt derived soils in the southern part of the study area, particularly around Millaa Millaa and southeast of Malanda (Laffan, 1988). Gibbsite occurs in the kaolinite dominated Maalan soils, which are well-drained soils on steeply sloping (> 32%) hills (Malcolm et al. 1999). In this area annual rainfall is 2500 – 3000 mm and intense leaching conditions persist. This favours the formation of gibbsite under conditions of vigorous weathering.

The chemical character of the Atherton basalt groundwater is variable, both laterally and vertically, over small distances. This is due to the high degree of hydraulic variability as discussed in section 4. Where subtle changes in chemical composition are used to categorise groundwater evolution and movement, it is difficult to spatially illustrate these categories on a regional scale. If the northern part of the Atherton tableland is considered (Fig. 1), some generalised spatial interpretations are possible. In the area between Atherton and Tolga, and around and slightly west of Kairi, groundwater is classified as within category one. Category two groundwater is present in a central N – S zone between Atherton, Tolga and Kairi, in the area of thickest basalt. To the NW of Tolga, groundwater is defined as category three water. Category three groundwaters were also identified near known spring locations towards the Barron River, east of the Atherton township. These interpretations correlate well with groundwater flow patterns based on standing water levels.

7 CONCLUSIONS

Trends in major ion concentrations and ratios of Na^+, Ca^{2+}, Mg^{2+}, HCO_3^- and H_4SiO_4 concentrations, can be useful indicators of evolution of the chemical character of groundwater in basaltic aquifers. The relationships between these ions have enabled the classification of the Atherton basalt groundwaters into three general categories. This chemical grouping reflects groundwater movement from recharge to discharge zones, despite the low concentrations of TDS. These groundwaters evolve chemically from a water of no dominant chemical type to slight HCO_3^- enrichment in recharge areas, to increased Ca, Mg – HCO_3 enrichment along flow gradient towards discharge zones.

The predominant primary mineral phases present in the Atherton basalts are clinopyroxene (augite), plagioclase feldspars (most likely albite and anorthite) and minor olivine. Kaolinite and montmorillonite are the dominant clay minerals. Reactions with these minerals and release of ions to solution are the principle controls on the chemical character of the groundwater. Interpretations of groundwater chemical composition with respect to rock mineralogy are particularly helpful in characterising not only the chemical character of the groundwater, but in defining the chemical evolution. This latter aspect can be used to determine the nature of groundwater occurrence and movement.

Further work with this data should include the application of more rigorous statistical methods, such as principal component analysis, as well as mass balance calculations to determine the proportions of the various minerals dissolving at different stages along the flow path.

ACKNOWLEDGMENTS

The authors thank S. Price (QUT School of Natural Resource Sciences, geochemistry laboratory) for assistance with chemical analyses, and M. Preda and J. Harbison (QUT) for assistance with XRD analyses, reviewing this manuscript and many helpful discussions. We also thank B. Pearce (Queensland Department of Natural Resources, Brisbane) for provision of additional data, J. Bean and G. Herbert (QDNR, Mareeba) for field assistance, and P. Cook and K. McEwan (CSIRO) for provision of rainfall samples.

This work was conducted within a research project funded by the Land and Water R & D Corporation and the Queensland Department of Natural Resources.

REFERENCES

Barclay, D. & B. Pearce 1999. A preliminary review of available groundwater reserves in the Atherton tableland basalts. Brisbane: Queensland Department of Natural Resources (unpubl.).

Donchak, P. & R. Bultitude 1994. Geology of the Atherton 1:250 000 Sheet Area. Brisbane: Queensland Geological Record 1994/5.

Buck, L. 1999. Physical features of volcanism and their relationship to groundwater, Atherton Basalt Province, North Queensland. BAppSc (Hons) thesis. Brisbane: Queensland University of Technology (unpubl.).

Bultitude, R. & A. Robertson 1997. Atherton Basalt Province. In R. Bultitude & P. Garrad (eds), *Geology of the Innisfail Region*: 117-144. Brisbane: Queensland Geological Record 1997/2.

Cas, R. 1989. Physical Volcanology. In R. Johnson (ed.), *Intraplate volcanism in eastern Australia and New Zealand*: 55-87. Sydney: Cambridge University Press.

Cook, P.G., A.L. Herczeg & K.L. McEwan 1999. Groundwater recharge and contribution to stream baseflow, Atherton tablelands, Queensland. Adelaide: CSIRO Land and Water (unpubl.).

Goldich, S.S. 1938. A study in rock-weathering. *Journal of Geology* 46: 17-58.

Greenberg, A.E., L.S. Clesceri & A.D. Easton 1992. *Standard methods for the examination of water and waste water.* Washington, D.C: APHA-AWWA-WEF.

Hem, J.D. 1985. *Study and interpretation of the chemical characteristics of natural water.* Geological Survey Water-Supply Paper 2254. Washington: United States Government Printing Office.

Knutson, J. 1997. Volcanics. In J. Bain & J. Draper (eds). *North Queensland Geology.* Canberra: Australian Geological Survey Organisation.

Laffan, M.D. 1988. Soils and land use on the Atherton tableland, North Queensland. Soils and Land Use Series 61. Canberra: CSIRO Australia.

Leach, L.M. 1986. Groundwater resources of the Atherton basalt, Atherton Shire. Mareeba: Queensland Water Resources Commission (unpubl.).

Malcolm, D.T., B.K.A. Nagel, I. Sinclair & I.J. Heiner 1999. Soils and agricultural land suitability of the Atherton tablelands, North Queensland. Land Resources Bulletin DNRQ980091. Brisbane: Queensland Department of Natural Resources.

Moloney, M. 1999. The magnetic method applied within a Quaternary volcanic plateau – Atherton, North Queensland. BAppSc (Hons) thesis. Brisbane: Queensland University of Technology (unpubl.).

Sheldrick, M. 1999. Stratigraphic interpretation of a lava field using petrogenetic modelling. BAppSc (Hons) thesis. Brisbane: Queensland University of Technology (unpubl.).

Stephenson, P., T. Griffin & F. Sutherland 1980. Cainozoic volcanism in northeastern Australia. In R. Henderson & P. Stephenson (eds), *The geology and geophysics of northeastern Australia.* Brisbane: Geological Society of Australia.

Groundwater: Past Achievements and Future Challenges, Sililo et al. (eds) © 2000 Balkema, Rotterdam, ISBN 90 5809 159 7

Boron isotope geochemistry as a tool to delineate natural groundwater processes

J.D. Mather & N.C. Porteous
Department of Geology, Royal Holloway University of London, Egham, UK

ABSTRACT: Boron isotopic ratios, together with total concentrations of boron and major cations and anions, have been measured in two UK sandstone aquifers. In the marine Lower Greensand, increasing boron concentrations in unconfined groundwaters are associated with decreasing $\delta^{11}B$ and increasing SO_4. Boron isotopic ratios suggest that boron is mobilised through the incongruent dissolution of glauconite, which is enhanced by pyrite oxidation. In confined groundwaters boron concentrations increase through mixing with connate waters. In the non-marine Hastings Beds, boron concentrations are high, up to 316 µg/l, and appear to be associated with ferroan carbonates. The research demonstrates that boron isotopic ratios can be used to identify the source of boron in groundwaters.

1 INTRODUCTION

Boron is a relatively rare element and has a crustal abundance of around 9 µg/g. It occurs naturally mainly in the form of borates associated with volcanic activity, in evaporites and in tourmaline which is found in pegmatites, aplites and granites. It is used in the manufacture of glass and enamels, in metallurgy and in herbicides and fertilizers. It is also a major constituent of washing powders, since it hydrolises H_2O and acts as a bleaching agent in hot water.

Boron has two stable isotopes with atomic masses of 10 and 11. Isotopic variations are expressed relative to a standard as $\delta^{11}B$, where:

$$\delta^{11}B = \frac{^{11}B/^{10}B \text{ sample}}{^{11}B/^{10}B \text{ standard}} - 1 \times 1000‰$$

The standard used is NIST SRM 951 which has a $^{11}B/^{10}B$ ratio of 4.04367.

The large difference in atomic mass leads to significant isotopic fractionation in geological and environmental systems. Boron is present in aqueous solutions as the $B(OH)^{4-}$ anion and as undissociated boric acid, $B(OH)_3$. At a pH of 7, around 95% of the boron present is in the form of the $B(OH)_3$ species. ^{11}B is preferentially incorporated into the $B(OH)_3$ species in solution whereas ^{10}B is incorporated into $B(OH)^{4-}$ in the solid phase (Kakihana, et al, 1977). Thus one might expect most natural waters to have an enriched signature compared to rocks and minerals. This is largely correct with a mean seawater signature of around+40‰ compared with a range of around -8‰ to + 16‰ in magmatic and metamorphic rocks (Figure 1).

This paper considers the groundwater chemistry of two sandstone aquifers with rather different mineralogies. Both contain groundwaters with boron concentrations of over 100µg/l. The geochemical evolution of boron in these two aquifers is followed from precipitation chemistry through to groundwaters in both the unconfined and confined zones. Data on major element geochemistry, boron concentrations

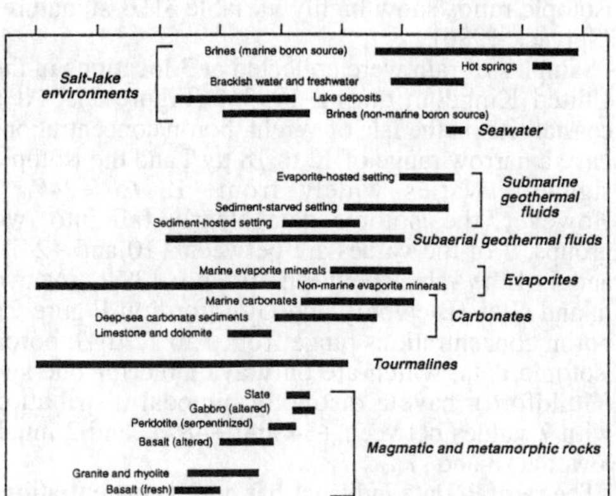

Figure 1. Variations in $\delta^{11}B$‰ from a range of environments (after Barth 1993)

and boron isotopic ratios are used to establish the processes involved in the evolution of boron geochemistry.

2 METHODS

Baron isotopic ratios have been measured using inductively-coupled plasma mass spectrometry (ICP-MS). Little sample preparation is required and experiments using the NIST SRM 951 standard have demonstrated that precisions of between 0.1 and 0.2% RSD (where RSD equals standard deviation/mean x 100%) can be attained at boron concentrations of around 200μg/l. A preconcentration step has been introduced to increase boron concentrations to this level. The methods used are fully described by Porteous et al (1995). Total boron was measured using inductively-coupled plasma atomic emission spectrometry (ICP-AES) and other parameters were determined by standard hydrogeochemical methods (Mather 1997).

3 BORON IN PRECIPITATION

Most groundwaters in the UK are derived ultimately from atmospheric deposition, rain and snowmelt although deep groundwaters may reflect contributions from additional sources such as fluid inclusions or connate waters. In general, the composition of rain is controlled by the dissolution of atmospheric gases, particularly carbon dioxide and by the washout of airborne particulates derived from the sea, the land and from pollution. Boron is present in the atmosphere in both gaseous and particulate forms in an approximate ratio of 90:10 (Fogg and Duce 1985). In precipitation boron concentrations range from about 1 to 95 μg/l and the limited data on boron isotopic ratios show highly variable $\delta^{11}B$ signatures (Spivack 1986).

Samples of rain were collected at 3 locations in the United Kingdom during 1993/4 (Figure 2). At a coastal site on the Isle of Wight, boron concentrations have a narrow range of 12 to 18 μg/l and the isotopic signature varies widely from -13 to +24‰. However, the isotopic data clearly fall into two groups, 7 of the values are between +10 and +24‰ and 4 of the values between -10 and -13‰. At two inland sites (Blewbury and Guildford on Figure 2), boron concentrations range from 1 to 12 μg/l; boron isotopic data, which are only available for one site (Guildford), have a distinctly bimodal distribution with 7 values between +34 and +48‰ and 2 much lower at -8 and -1‰.

The rainfall data indicate that boron concentrations in British rainfall are likely to range up to 20 μg/l in coastal regions with somewhat lower concentrations inland. Isotopic ratios can be highly variable and unpredictable with a range of at least -13 to +48‰. There are insufficient data to indicate if there is a relatively consistent annual weighted mean as with some other isotopes.

4 BORON IN GROUNDWATERS

In southern Britain most groundwater recharge takes place in the winter months between October and April. Measurements and calculations by the Water Resources Board (1972) have indicated that the total dissolved solids in precipitation could be concentrated by a factor of around 3 by evapotranspiration during the initial stages of groundwater recharge. With boron concentrations in rainfall of up to 20 μg/l, this means that boron concentrations of up to 60 μg/l in groundwaters could be attributed to precipitation alone. The isotopic fractionation which might take place during evapotranspiration is unknown but, on the basis of the maximum values of $\delta^{11}B$ measured in the outcrop waters it is estimated that recharge waters in southern Britain are characterised by $\delta^{11}B$ values near +20 ±5‰.

Groundwater samples have been collected from 2 aquifers in the southern part of Britain and analysed for major species as well as boron and its isotopes. The aquifers selected are (Figure 2);

1. Cretaceous Lower Greensand of the London Basin (Marine Sandstones).
2. Cretaceous Hastings beds of the Weald (Non-marine Sandstones).

4.1 *The Marine Lower Greensand of the London Basin*

The Lower Greensand is exposed on the northern and southern limbs of the London Basin which plunges towards the east. It is a predominantly arenaceous formation deposited in a shallow marine near-shore environment. It is not continuous in the central part of the London Basin but in the western part around Slough the sands form a confined aquifer. To the south two distinct aquifers can be distinguished, the Folkstone Beds and the Hythe Beds separated by the Sandgate Beds which form an aquiclude. To the north and beneath Slough, the sands are undifferentiated but are lithologically similar to the Folkestone Beds in the south.

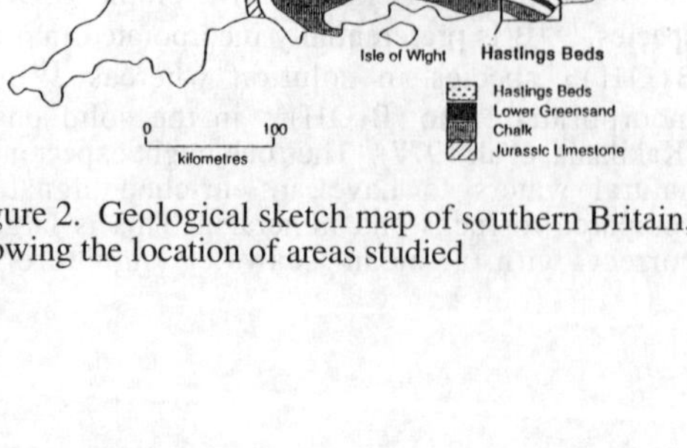

Figure 2. Geological sketch map of southern Britain, showing the location of areas studied

Samples for hydrochemical analysis have been taken from both the northern and southern outcrops, from the confined area at Slough and from the confined aquifer immediately to the north of the southern outcrop. In general terms the unconfined groundwaters are typical recharge waters of Ca/HCO_3 type whereas the confined groundwaters are cation exchanged and dominated by Na and HCO_3. There are no distinct differences between the boron chemistry of the Hythe and Folkestone Beds and they are dealt with together.

The ranges in boron concentrations and isotopic ratios for the various aquifer areas are as follows. In the unconfined southern outcrop (samples 1-9), boron concentrations range from 14 to 25 µg/l and $\delta^{11}B$ from 0 to +25‰. In the unconfined northern outcrop (samples 10-15), the ranges are boron 28 to 53 µg/l and $\delta^{11}B$ -14 to +10‰. In the confined areas, boron concentrations range from 115 to 145 µg/l in the south (samples 16-18) and 75 to 100 µg/l around Slough (samples 20-23) and isotopic ratios vary from -6 to +17‰ in the south to + 12 to + 18‰ around Slough.

A plot of $\delta^{11}B$ against B (Figure 3) shows that the samples from different areas plot in separate fields and in fact it is possible to identify the source of an unknown sample from its boron chemistry.

If the distribution of isotopic ratios in the southern outcrop area (Area 1, Figure 3) is examined in more detail it is clear that groundwaters with the lowest isotopic ratios (between 0 and +4‰) occur in the southern part of the sampled outcrop and groundwaters with the highest ratios (+14 to +25‰) occur in the northern part. These are shown as Areas 1A and 1B in Figure 3. If the groundwaters within Area 1B are ignored, there is a general correlation within the southern and northern outcrop areas with the isotopic signature becoming more depleted with increasing boron concentrations (Figure 3).

There are no direct correlations between boron or boron isotopic ratios and any of the major dissolved species in the groundwaters except for SO_4. The plot of boron against SO_4 (Figure 4) clearly defines the same 4 areas as were identified in Figure 3. Plots of

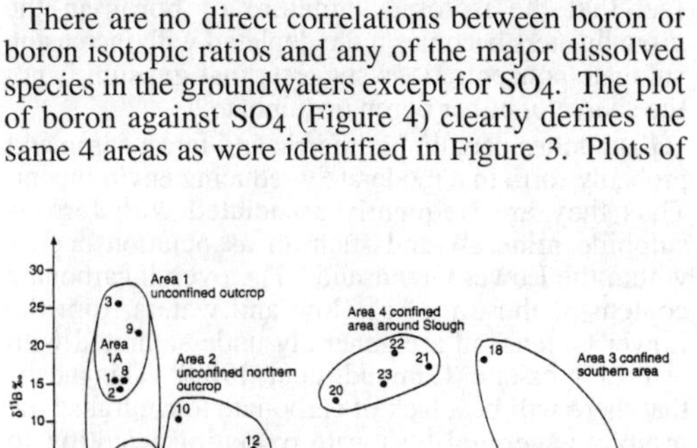

Figure 3. Plot of B against $\delta^{11}B$ for Lower Greensand samples

total boron and $\delta^{11}B$ against SO_4 in the outcrop waters (Areas 1 and 2) demonstrate that boron increases as SO_4 increases, and if the groundwaters within Area 1B are again ignored, there is a good correlation between increasing SO_4 and an increasingly depleted boron isotopic signature (Figure 5).

The confined Lower Greensand groundwaters (Areas 3 and 4) are ion exchanged waters in which Ca has been replaced by Na. They also contain higher Cl concentrations than the outcrop waters. If the Ca:Na ratio is taken as a measure of the degree of cation exchange, boron concentrations increase and $\delta^{11}B$ decreases with increasing exchange of Na for Ca.

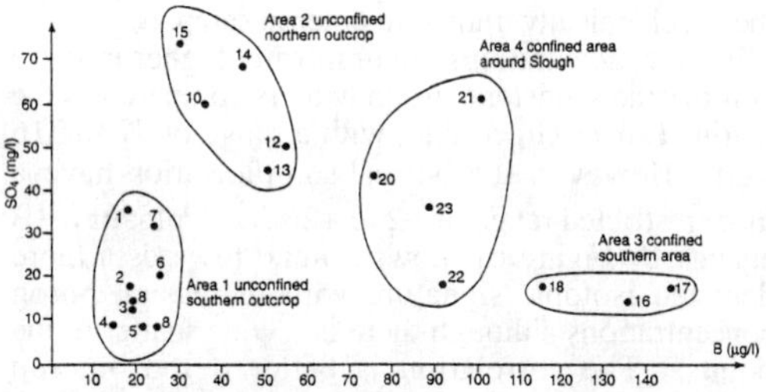

Figure 4. Plot of B against SO_4 for Lower Greensand samples

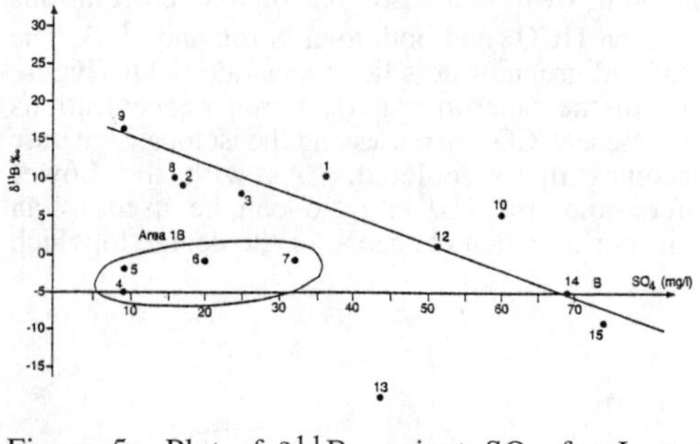

Figure 5. Plot of $\delta^{11}B$ against SO_4 for Lower Greensand samples

4.2 *The Non-marine Hastings Beds*

The Hastings Beds occupy the core of a large anticline to the south of the London Basin (Figure 2) and are part of a braidplain deposit laid down in a period of intense deposition of channel sands and overbank flood deposits separated by periods of little stream activity and the development of muddy swamps. The principal aquifer is the Ashdown Beds which are locally confined by the Wadhurst Clay. They are locally important for public water supply and consist of fine-grained, silty sandstones and siltstones with subordinate shales and mudstones.

Groundwater samples were collected from 9 water supply boreholes. They are generally rich in iron and

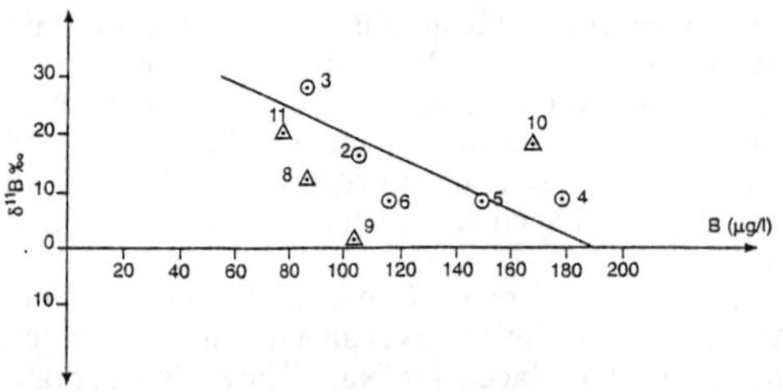

Figure 6. Plot of B against $\delta^{11}B$ for Hastings Beds samples

have a low pH. Those samples designated as confined generally have a higher pH, electrical conductivity and alkalinity and are more evolved than the geochemically immature outcrop samples.

Boron concentrations are uniformly higher in these non-marine sandstone groundwaters compared to the marine Lower Greensand, with a range of 77 to 316 µg/l. However, the boron isotopic ratios have a more restricted range of +2 to +26‰. A plot of $\delta^{11}B$ against B (Figure 6) shows a trend towards a more depleted isotopic signature with increasing boron concentrations although there is a wide scatter to the points. The correlation is better if the outcrop samples only are considered.

In this aquifer there is no correlation between boron and sulphate but if the outcrop samples are examined on their own there are reasonable correlations between HCO_3 and both total boron and $\delta^{11}B$. The confined groundwaters lie in separate fields (Figure 7). In the outcrop waters, boron concentrations increase as HCO_3 increases and the isotopic signature becomes more depleted. As with the Lower Greensand, the Ca/Na ratio can be used as an indicator of cation exchange or the degree to which

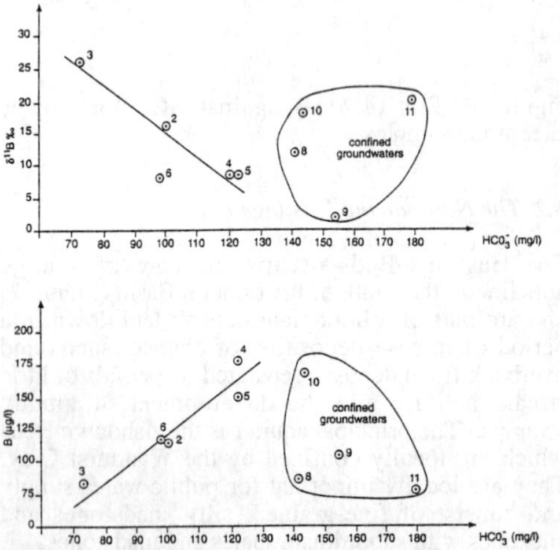

Figure 7. Plots of HCO_3 against $\delta^{11}B$ and B for Hastings Beds samples.

water-rock interaction has taken place. Although there are few data points, similar trends to those observed in the Lower Greensand are seen, with an increase in boron concentrations and isotopic depletion with increasing exchange of Na for Ca.

5 DISCUSSION

5.1 *Lower Greensand*

In outcrop waters it has been demonstrated that, in the marine aquifer, increases in boron concentrations are associated with decreasing $\delta^{11}B$ and an increase in sulphate concentrations. Ishikawa and Nakamura (1993) suggest that boron isotopic variation in modern sediments is controlled by the simple mixing of four constituents, detritus of continental origin, marine smectite, biogenic carbonates and biogenic silica. Because of its arenaceous character and its formation in a near-shore environment, the Lower Greensand is likely to be dominated by continental detritus which is characterized by low $\delta^{11}B$ in the range of -13 to -8‰ (Ishakawa and Nakamura 1993). The boron concentration in continental detritus is largely controlled by the concentration of illite. The Lower Greensand is rich in glauconite, which is a green illite-related mineral and which gives the rock its name. It is therefore possible that the slow incongruent dissolution of glauconite in these groundwaters releases boron and systematically lowers the groundwater $\delta^{11}B$ signature as boron concentrations increase.

The concentration of sulphate in these groundwaters is likely to be controlled by the oxidation of pyrite and/or the dissolution of gypsum. Boron occurs in concentrations of up to 500 ppm in marine anhydrite and gypsum (Garrett 1998). However, if boron came from such a source it would have isotopic ratios characteristic of marine borates i.e. over +30‰ . The fact that the isotopic signature of boron in the groundwaters becomes more depleted with increasing sulphate concentrations suggests that gypsum is not the source of either boron or sulphate.

Glauconites contain both ferric and ferrous iron and probably form in a moderately reducing environment. Thus they are frequently associated with ferrous sulphide minerals and such an association occurs within the Lower Greensand. The overall carbonate content of these rocks is low and waters from the Lower Greensand are generally undersaturated with respect to calcite (Edmunds et al. 1989). This means that there will be a lack of carbonate to neutralise the protons generated by pyrite oxidation resulting in reactive groundwaters and enhanced dissolution of glauconite. This could explain the observed correlations between increasing boron, decreasing $\delta^{11}B$ values and increasing sulphate in Lower Greensand groundwaters. The samples from area 1B in the southern outcrop do not fit the above scenario as $\delta^{11}B$ values remain constant while both boron and sulphate concentrations increase. Further work is required to understand the reasons for this.

If a boron isotopic signature of -13 to -8‰ is characteristic of the detritus from which glauconite is derived, as boron concentrations increase, the $\delta^{11}B$ signature should stabilise at around this value. It would therefore be reasonable to anticipate such signatures from the confined zone. This is not the case and most samples from the confined zone have $\delta^{11}B$ values of +12 to +18‰. The elevated chloride concentrations in these confined groundwaters suggest that some mixing with old formation waters might be taking place. Information on the $\delta^{11}B$ signatures of formation waters is sparse. Land and Macpherson (1992) report values ranging from +12 to +20‰, with a mean of +17‰ for a Cenozoic reservoir in Texas. This is close to the +12 to +18‰ range of samples from the Lower Greensand and it is suggested that the isotopic signatures in the confined zone could be the result of mixing with connate waters. Such mixing could increase total boron concentrations and result in higher $\delta^{11}B$ signatures within the confined zone.

In summary, within the Lower Greensand aquifer the data suggest that boron is originally derived from rainfall. Interaction with glauconite results in increased boron concentrations and a depleted isotopic signature of around -10‰. Within the confined zone mixing with connate waters occurs, doubling boron concentrations and increasing the $\delta^{11}B$ signature to around +12 to +18‰.

5.2 *Hastings Beds*

In the non-marine Hastings Beds, increases in boron concentrations within outcrop waters are associated with depleted boron isotopic signatures but with an increase in bicarbonate concentrations rather than an increase in sulphate. In the Lower Greensand, $\delta^{11}B$ seems to stabilise in the outcrop samples at a value of -10 to -15‰ whereas in the Hastings Beds this value is higher at around +8‰. In addition boron concentrations in outcrop samples are much higher than in the Lower Greensand, 85 to 179 mg/l compared to 14 to 53 mg/l. These observations imply that different processes are involved.

The argillaceous rocks contain illite, kaolinite and mixed layer minerals and Allen (1948) records detrital glauconite as microcrystalline aggregates and as thin films on quartz grains. Limestones are rare and are virtually confined to the clay facies. Ferroan calcite cements are rare in the major arenaceous formations such as the Ashdown Sands, but common in porous silt and sand bodies within the clay formations (Allen, 1989). Iron concentrations are high in these groundwaters and Edmunds et al (1989) demonstrated that the majority of samples were above saturation with respect to $Fe(OH)_3$ and in a few cases saturation was approached with respect to siderite.

The correlation with bicarbonate and the suggested equilibrium value of around +8‰ for $\delta^{11}B$ both suggest that boron is associated with the ferroan carbonate in these rocks. A $\delta^{11}B$ value of +8‰ lies within published values for carbonate rocks (Barth 1993; Garrett 1998) and Vengosh et al. (1991) obtained values of +1.5 to 8.4‰ for Cretaceous carbonate rocks in Israel.

As with the Lower Greensand samples it might have been anticipated that the equilibrium value of +8‰ which appears to have been reached in the unconfined zone would have persisted into the confined zone. Three of the four confined zone samples have $\delta^{11}B$ values between +12 and +20% (which are within the isotopic range of connate waters) but the fourth sample with a value of +2‰ suggests extended water rock interaction with neritic material. The hydrogeology of the Hastings Beds is known to be complex and it is probable that a variety of processes have contributed to boron concentrations in the confined aquifers, including mixing and further water/rock interaction.

6 CONCLUSIONS

The data on boron geochemistry can be interpreted in terms of water/rock interaction and mixing within both the marine Lower Greensand and the non-marine Hastings Beds. In both aquifers rainfall interacts with organic matter and weathering products within the soil zone prior to groundwater recharge. A $\delta^{11}B$ signature of +20 ±5‰ is suggested as a reasonable estimate for this recharge water. Boron concentrations may approach 60 µg/l after concentration of the rainfall by evapotranspiration, however, concentrations are likely to be lower than this at inland sites.

In the unconfined Lower Greensand aquifer, incongruent dissolution of glauconite results in the mobilisation of boron, an increase in groundwater boron concentrations and a reduction in the $\delta^{11}B$ value relative to the recharge waters. A correlation with sulphate suggests that dissolution of glauconite is related to the oxidation of pyrite. In the confined zone, mixing with old formation waters is postulated. This further increases boron concentrations and results in an enrichment in the isotopic signature.

Boron concentrations in the Hastings Beds are much higher than in the Lower Greensand and the $\delta^{11}B$ values stabilize at a level of around +8‰ rather than -10‰. A correlation with bicarbonate in the unconfined groundwaters suggests that boron is associated with the dissolution of ferroan carbonates. In the confined groundwaters some mixing with connate waters also may occur but the hydrogeology is complex and a variety of processes may be operating.

7. REFERENCES

Allen, P. 1948. Wealden petrology: the Top Ashdown Pebble Bed and the Top Ashdown Sandstone. *Quarterly Journal, Geological Society London,* 104, 257-321.

Allen, P. 1989. Wealden research - ways ahead. *Proc. Geologists Association*, 100, 529-64.
Barth, S. 1993. Boron isotope variations in nature: a synthesis. *Geologische Rundschau,* 82, 640-51.
Edmunds, W.M., Cook. J.M., Kinniburgh, D.G., Miles, D.L. and Trafford, J.M. 1989. Trace elements in British groundwaters. *British Geological Survey Report* SD/89/3, B.G.S. Keyworth.
Fogg, T.R. and Duce, R.A. 1985. Boron in the troposphere: distribution and fluxes. *Journal Geophysical Research,* 90-2, 3781-96.
Garrett, D.E.1998. Borates. *Handbook of deposits, processing, properties, and use.* Academic Press, San Diego.
Ishikawa, T. and Nakamura, E. 1993. Boron isotope systematics of marine sediments. *Earth Planetary Science Letters,* 117, 567-80.
Kakihana, H., Kotaka, M., Satoh, S., Nomura, M. and Okamoto, M. 1977. Fundemantal studies on the ion exchange separation of boron isotopes. *Bull, Chemical Society, Japan,* 50, 158-63.
Land, L.S. and Macpherson, G.L. 1992. Origin of saline formation waters, Cenozoic Section, Gulf of Mexico Sedimentary Basin. *American Association of Petroleum Geologists, Bulletin,* 76, 1344-62.
Mather, J.D. 1997. Collection and analysis of groundwater samples. In: O.M. Saether and P. de Caritat (eds). *Geochemical processes, weathering and groundwater recharge in catchments.* A.A. Balkema, Rotterdam. 167-184.
Porteous, N.C., Walsh, J.N. and Jarvis, K.E. 1995. Measurement of boron isotope ratios in groundwater studies. *Analyst,* 120, 1397-1400.
Spivack, A.J., Boron isotope geochemistry. PhD dissertation. MIT/WHOI, Joint Program in Oceanography.
Vengosh, A., Kolodny, Y., Starinsky, A., Chivas, A.R. and McCulloch, M. 1991b. Co-precipitation and isotopic fractionation of boron in modern biogenic carbonates. *Geochim. et Cosmochim. Acta,* 55, 2901-10.
Water Resources Board, 1972. *The hydrogeology of the London Basin.* Water Resources Board, Reading.

Groundwater: Past Achievements and Future Challenges, Sililo et al. (eds) © 2000 Balkema, Rotterdam, ISBN 90 5809 159 7

Groundwater quality and sustainability in an alluvial aquifer, Australia

W. McLean & J. Jankowski
UNSW Groundwater Centre, School of Geology, University of New South Wales, Sydney, N.S.W., Australia

N. Lavitt
Centre for Water and Waste Technology, University of New South Wales, Sydney, N.S.W., Australia

ABSTRACT: The heterogeneous alluvial aquifer systems of the Lower Namoi River catchment, constitute some of the most valuable groundwater resources in New South Wales, Australia, supporting the state's cotton industry and other highly productive agricultural activities. For the last thirty years, irrigation demands have placed considerable stress on the aquifers, with groundwater abstraction exceeding long term projected recharge. Significant water level declines, reversal of vertical and lateral hydraulic gradients and localised land subsidence have resulted from heavy abstraction. Additionally, depressurisation of the two most productive aquifers, the Gunnedah and Cubbaroo Formations, has resulted in the leakage of poorer quality, saline shallow groundwaters from the overlying Narrabri Formation and Na–HCO_3 artesian groundwaters from the underlying bedrock of the Great Artesian Basin. Artesian recharge has increased the Na concentration and subsequently the Sodium Adsorption Ratio of alluvial groundwaters and reduced their potential for irrigation.

1 INTRODUCTION

The sustainability of the heterogeneous aquifer systems of the Lower Namoi River catchment has been threatened by the heavy exploitation of groundwater resources for irrigation over the last thirty years. With the expansion of the New South Wales cotton industry, centred in the catchment, and groundwater management strategies of the early 1980s which allowed abstraction of groundwater in excess of recharge, the groundwater resources of these productive aquifers have been depleted (Kalaitzis et al. 1999).

Groundwater recharge studies carried out by Williams et al. (1987) based on hydrochemical and isotopic studies by Calf (1978) and groundwater modelling (Merrick 1989, 1999) have shown that groundwater extraction greater than recharge rates has led to declining water levels and long term depletion in certain localised areas. Land subsidence due to aquifer compaction has also occurred in the areas of heavy abstraction of groundwater (Ross & Jeffery 1991).

A study of hydrological and hydrogeochemical data taken from New South Wales Department of Land and Water Conservation (DLWC) monitoring boreholes has shown that water level depletion, aquifer depressurisation and reversal of hydraulic gradients has resulted in the decrease of water quality for productive aquifers. This is believed to be due to leakage of saline groundwater from overlying shallow aquifers and increased leakage of Na–HCO_3 type groundwaters from the underlying Great Artesian Basin (GAB).

2 GEOLOGY

The Lower Namoi River catchment covers 5100 km^2 of semi–arid country in Northern New South Wales, Australia (Figure 1a). The sediments of the alluvial aquifers were deposited by the Namoi River over a period of 15 million years and form highly heterogeneous alluvium that overly the pre-Tertiary bedrock surface of the GAB. The consolidated sediments of the bedrock consist of Jurassic–Cretaceous age sandstones interbedded with shales and coaly siltstones.

Three multiple aquifer systems in the Namoi catchment have been classified on the basis of age and compositional differences (Figure 1b) (Gates 1980, Williams 1993). The Cubbaroo Formation, comprised predominantly of quartz sand, gravel and wood fragments supported by a clay matrix, is restricted to the main Namoi palaeochannel that runs north-northwest of the present river course. This palaeochannel is up to 130 m deep and is restricted by a buried bedrock high in the north of the catchment. The Gunnedah Formation overlies this formation and is an extensive fluviolacustrine unit of Pliocene to early Pleistocene age. The sand and gravel deposits in the Gunnedah and Cubbaroo Formations provide the most productive aquifers for irrigation.

The Pleistocene and Holocene age Narrabri Formation, which is between 10 and 40 m thick, overlies the Gunnedah Formation and is comprised predominantly of brown clay with some sand and gravel.

(a)

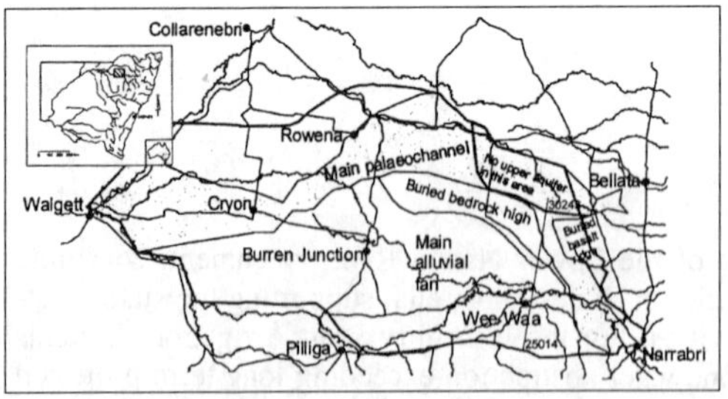

(b)

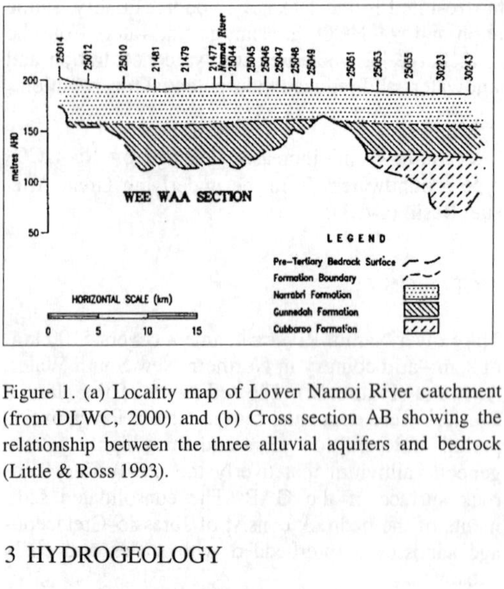

Figure 1. (a) Locality map of Lower Namoi River catchment (from DLWC, 2000) and (b) Cross section AB showing the relationship between the three alluvial aquifers and bedrock (Little & Ross 1993).

3 HYDROGEOLOGY

Groundwater flow in all aquifers is generally in a northwesterly direction away from the Namoi River towards a major palaeochannel at the centre of the catchment. Groundwater recharge occurs in the east of the catchment, with the major recharge sources being; the Namoi River and its tributaries and diffuse recharge during major flood events. Other contributions to recharge come from diffuse rainfall infiltration, inflow from the neighbouring Upper Namoi River catchment and leakage from the GAB.

In order to predict vertical potential head components, the difference between the potential head data for the alluvial aquifers and bedrock boreholes was calculated, based on 5 yearly weighted averages. The potential for interconnection between aquifers was assessed on the response of borehole hydrographs from the three alluvial aquifers and bedrock, the geology taken from drilling logs and the presence of aquitards between aquifers and the calculated potential vertical head. Examination of vertical potential head components and geology has shown that a degree of vertical hydraulic connection exists between the aquifers. In the west of the catchment, there is an upward flux from the Cubbaroo to Gunnedah Formation. In the southeast of the catchment there is upward leakage from the GAB and in certain areas in the east there is downwards leakage from the Narrabri Formation. Heavy localised groundwater abstraction has resulted in seasonal and, in some instances, long term reversal of vertical hydraulic gradient and flow conditions between aquifers.

Hydrographs were constructed from DLWC monitoring data from 1968 to 1999. Hydrographs indicate that net water level declines of greater than 30 m have occurred north of a bedrock high that constricts groundwater flow in the north of the catchment. Linear regression analysis of average yearly weighted water level data was performed to determine annual rate of change. Yearly water level declines up to 1.5 m/yr occurred in the most heavily exploited aquifers.

4 GROUNDWATER CHEMISTRY

New hydrochemical data collected during 1999 for the Lower Namoi River catchment provide new insight about groundwater chemistry, quality and hydrogeochemical processes. Groundwaters in the east (recharge zone) and the west of the catchment (discharge zone) have distinctly different water chemistries indicating the gradual chemical evolution of waters along the flowpath of a large alluvial system.

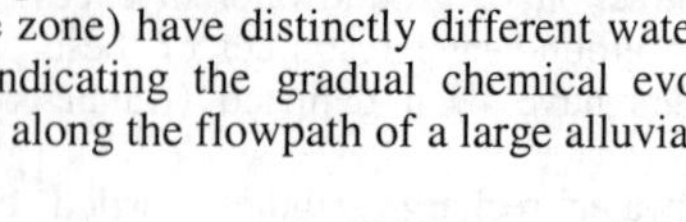

Figure 2. Piper diagram for Gunnedah Formation groundwaters.

The chemical composition of groundwaters from the Gunnedah Formation can be seen on the Piper diagram (Figure 2). Groundwaters from the east range from Na dominant to Na–Ca–Mg dominant, with HCO_3 as the dominant anion. This change in cation distribution produces a trend line in the left-hand side of the diamond diagram. In the west of the catchment, river recharge is limited and groundwaters display a strong right hand trend on the diamond diagram. Groundwaters are still Na-rich but have evolved to a Cl or HCO_3–Cl mixed dominance. In the west SO_4 concentrations increase, and SO_4 comprises up to 16% of the equivalent weight percentage load.

Table 1 identifies the representative chemical compositions of the different groundwater types in the catchment. Saline groundwaters occur in the western part of the catchment and fresh groundwaters of either Na–HCO_3 or Na–Ca-Mg–HCO_3 dominance in the recharge zone. In the west, groundwater in the deepest aquifer, the Cubbaroo Formation, has a chemical composition dominated by Na-Cl-HCO_3 and is classified as saline with EC values ranging from 1200 to 2200 μS/cm. Salinity of the groundwater from this palaeochannel increases along the flowpath with groundwater in the east having EC values <1000 μS/cm. The pH also increases along the flowpath from slightly acidic to slightly alkaline. The GAB groundwaters have pH values ranging from 7.5 to 8.1 and EC values ranging from about 1000 to 1300 μS/cm. They are dominated by two ions; sodium and bicarbonate

Table 1. Representative chemical analyses of groundwaters from selected boreholes, Lower Namoi Valley.

Bore-hole No	Geology	Depth m	EC μS/cm	TDS mg/L	pH	Ca	Mg	Na mg/L	HCO_3	SO_4	Cl
36314/1	NF	33.0	29700	21020	7.11	663	772	6423	530	2934	9698
36387/1	GF	45.7	22000	14424	7.09	364	493	4594	358	1397	7218
30099/3	GF	64.0	301	264	6.44	10	5	57	172	9	11
30281/2	GF	45.4	265	217	6.30	22	11	20	143	7	14
36398/2	CF	115.8	1237	898	7.19	8	4	270	336	63	217
36387/3	CF	100.5	1461	1029	7.51	10	7	320	378	34	280
30142/1	B	172.2	1110	1054	7.95	3	1	314	686	ND	50
30309/3	B	84.7	1113	1046	7.82	3	1	302	690	ND	50

NF – Narrabri Formation; GF – Gunnedah Formation; CF – Cubbaroo Formation; B – Bedrock ND – Not Detected

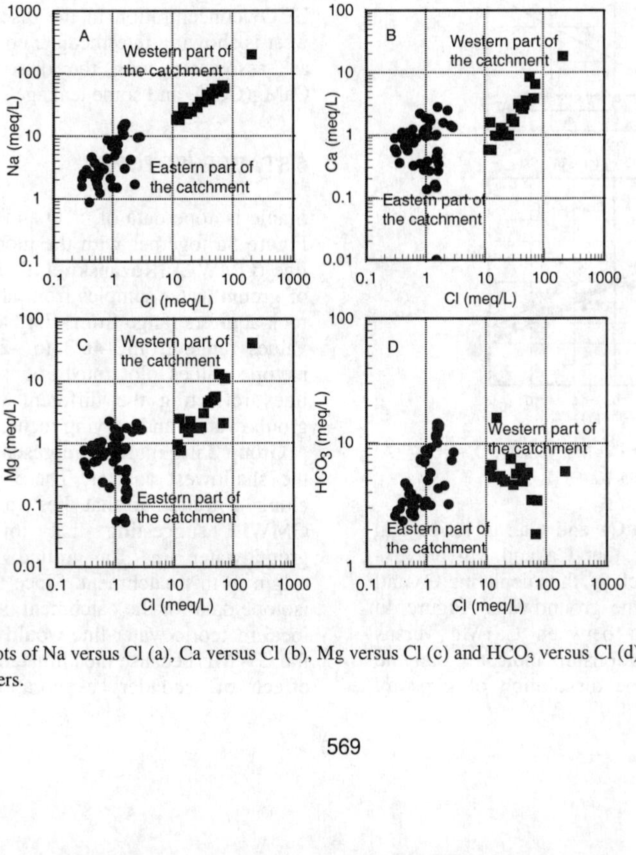

Figure 3. Bivariate plots of Na versus Cl (a), Ca versus Cl (b), Mg versus Cl (c) and HCO_3 versus Cl (d) for the Gunnedah Formation groundwaters.

(Habermehl 1980). In the recharge zone groundwaters in the three aquifers are generally dominated by Na and HCO_3, with Cl dominance increasing in the Narrabri Formation. Electrical Conductivity (EC) generally decreases with depth and pH increases.

In the western part of the catchment Na is the dominant cation in all aquifers and Cl is a dominant anion. However, in the deeper formations, HCO_3 increases, producing mixed HCO_3–Cl waters. Electrical conductivity and pH also increase with depth. Generally groundwater EC increases along the flowlines from the east to west, away from the recharge area. Values in the Gunnedah Formation range from 280 μS/cm to 22,000 μS/cm in the west.

The two distinct groundwater types are again apparent by looking at bivariate plots of the major ions versus Cl (Figure 3). Groundwater from the west exhibits a simple linear trend, with all major cations increasing with chloride concentration. In the east, there is no simple linear trend because of numerous hydrogeochemical processes adding and removing ions from the system and the mixing of different waters (river water, GAB and Upper Namoi River catchment groundwaters).

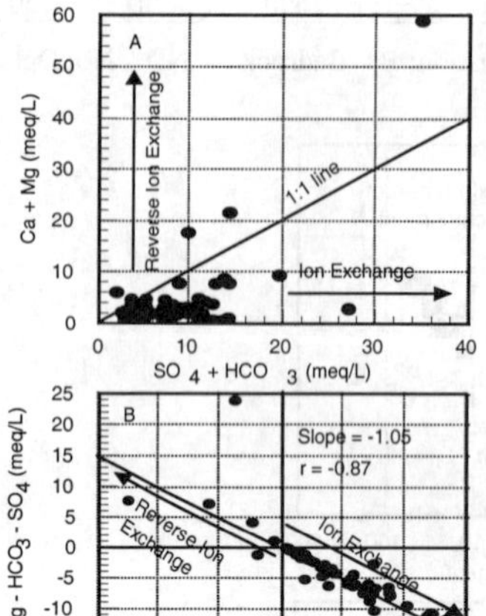

Figure 4. Bivariate plots of Ca+Mg versus SO_4+HCO_3 (A) and Ca+Mg-SO_4-HCO_3 versus Na-Cl (B).

In the east Na:Mg, Na:Ca and Ca:Mg ratios and bivariate plots indicate that Ca and Mg are exchanging for Na in the clays, thus depleting Ca and Mg concentrations in the groundwater. Figure 4a shows the relationship between Ca+Mg versus SO_4+HCO_3. A 1:1 relationship indicates that the dominant process is the dissolution of gypsum, calcite and dolomite. Groundwaters fall below the 1:1 dissolution line indicating that ion exchange is occurring, with Ca and Mg being depleted with respect to SO_4 and HCO_3. Figure 4b shows the relationship between Ca+Mg–SO_4–HCO_3 versus Na-Cl. Waters plotting close to the zero value on the x-axis are not influenced by ion exchange. The addition of Ca+Mg-SO_4-HCO_3 will also yield close to zero values if the dissolution of calcite, dolomite and gypsum is congruent and ion exchange does not occur. Waters undergoing ion exchange will plot along a line with a slope of –1 (Jankowski et al. 1998). Groundwaters from the Gunnedah Formation plot on a line with a slope of –1.05, showing that ion exchange is occurring. Ion exchange is also occurring in the discharge zone, however, Mg:Ca ratios are greater than in the east, indicating that dissolution of $CaCO_3$ and $CaMg(CO_3)_2$ is continuously supplying Ca and Mg to the system and ion exchange of Ca for Na is occurring simultaneously.

A bivariate plot of HCO_3 versus Cl (Figure 3d) delineates three chemically distinct groups of groundwater. In the east there are two groups, one with low HCO_3 concentration (northeast) and one with higher HCO_3 concentrations (southeast). The high HCO_3 groundwaters in the southeast show upward vertical leakage of Na–HCO_3 rich groundwater from the GAB. Additionally, the reduction of SO_4, the flux of CO_2 from recharging waters and calcite dissolution in the soil zone are contributing to the increase of HCO_3 to the system. HCO_3 concentration in the discharge zone to the west is showing intermediate concentrations which are associated with the dissolution of $CaCO_3$, $CaMg(CO_3)_2$ and some leakage from GAB.

5 STABLE ISOTOPES

Stable isotope data of $\delta^{18}O$ and $\delta^{2}H$ are plotted in Figure 5a together with the global meteoric water line (GMWL) (Rozanski et al. 1993). $\delta^{18}O$ values of groundwater samples from all alluvial and bedrock aquifers range from –7.42 to –3.28‰ and $\delta^{2}H$ values range from –46.3 to –22.5‰. The stable isotope values plot into groups and along trendlines reflecting the different chemical types of groundwaters and varying recharge conditions.

Group 2 in Figure 5a represents groundwaters of the shallowest aquifer. These groundwaters plot along a trendline that lies to the right of the GMWL, suggesting that this represents the groundwater line for groundwaters of meteoric origin in this catchment. Since there is no rainfall isotope data for the catchment, it is assumed that a local meteoric water line would fall to the right of the GWML because the climate is semiarid and the effects of secondary evaporation in a lower humidity climate would enrich $\delta^{18}O$ values compared to the global precipitation average. Stable isotope data indicates that group 2 groundwaters receive river recharge and this is supported by spatial data that locates these groundwaters in the northeast of the catchment near the Namoi River. Several boreholes in the Gunnedah Formation also plot on this line indicating that in some parts of the catchment, good hydraulic connection exists between the shallow and middle aquifers.

Groups 3 and 4 on Figure 5a represent Na-HCO_3 rich groundwaters of the Narrabri and Gunnedah Formations respectively occurring in the southeast part of the catchment. These groundwaters are depleted in $\delta^{18}O$ compared to the GMWL but show no depletion in δ^2H. This depletion may be due to recharge under wetter climatic conditions but more likely is the result of mixing between groundwaters of alluvial and bedrock aquifers and resulting changes in the carbonate system. Group 6 represents GAB groundwaters and it can be seen from Figure 5a that these bedrock groundwaters have a similar isotopic signature to groups 3 and 4.

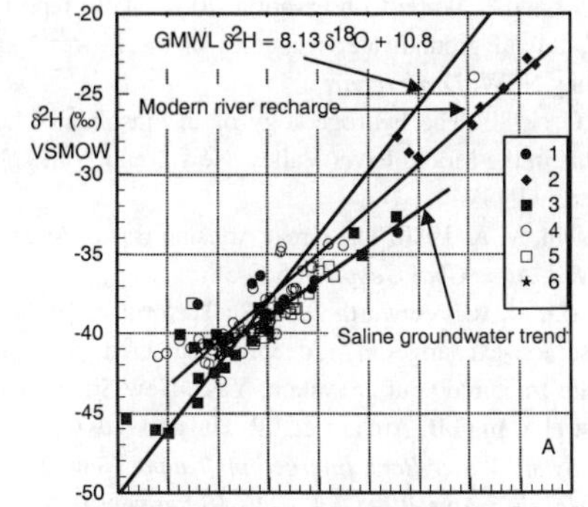

Figure 5. $\delta^{18}O$ versus δ^2H (a) $\delta^{18}O$ versus chloride (b).Group 1 – Narrabri Formation (saline groundwaters), 2 – Narrabri Formation groundwaters receiving river recharge, 3 – Narrabri Formation (Na-HCO_3 groundwaters), 4 – Gunnedah Formation (eastern catchment), 5 – Gunnedah Formation (western catchment) and 6 – bedrock.

Group 1 represents saline groundwaters from the shallow aquifer and group 5 comprises the saline groundwaters of the Gunnedah Formation occurring in the western part of the catchment. Both groups are enriched in $\delta^{18}O$ compared to the GWML. Since it is unlikely that evaporation processes would affect groundwater at depths greater than a few metres, this enrichment suggests that these groundwaters could possibly be palaeowaters recharged under drier climatic conditions or conditions of lower rainfall intensity, with recharge water being enriched in $\delta^{18}O$ by evaporation prior to or during infiltration. The groundwater groups identified by stable isotope data in Figure 5a can also be seen in the plot of $\delta^{18}O$ against chloride concentration (Figure 5b). This diagram shows that the oxygen isotopic composition is independent for Na-HCO_3 groundwaters and groundwaters receiving river recharge in the northeast of the catchment. Group 5 groundwaters have similar $\delta^{18}O$ values with increasing salinity. Group 1 groundwaters appear to become isotopically heavier with increasing chloride concentration.

6 GROUNDWATER QUALITY

Since the groundwater of the Gunnedah Formation is used extensively for irrigation, the quality of irrigation water is important to the sustainability for water and soil resources in the catchment. Groundwater quality in the catchment is assessed by a number of criteria including Electrical Conductivity (EC), sodium adsorption ratio (SAR) and residual sodium carbonate (RSC).

The SAR is an index reflecting the hazard of the sodium ion to plant and soil health and the RSC is an index to determine the HCO_3 hazard. High HCO_3 levels in groundwater can stunt plant growth and lead to calcite precipitation, decreasing soil permeability, lowering infiltration capacity and increasing erosion. The RSC hazard is calculated by:

$$RSC = [(HCO_3^- + CO_3^{2-}) - (Ca^{2+} + Mg^{2+})]$$

where concentration units are in meq/L.

Table 2. Water quality criteria for irrigation.

Salinity Hazard EC µS/cm		Sodium Hazard SAR		RSC Meq/L
C1 <250	L	S1 <10	L	<1.25 Suitable
C2 250-750	M	S2 10-18	M	1.25-2.5 Marginal
C3 750-2250	H	S3 18-26	H	>2.5 Not suitable
C4 >2250	VH	S4 >26	VH	

L – Low; M – Medium; H – High; VH – Very High

The classification of water quality for irrigation according to these criteria is summarised in Table

2. Groundwaters in the west of the catchment are unsuitable for irrigation, having a high to very high salinity, with EC up to 29,700 μS/cm. SAR values are also greater than the threshold value of 8, ranging from 8 to 37. Groundwaters in the west are of Na–Cl, Na–HCO_3–Cl or Na–Cl–HCO_3 type. Sodium-chloride groundwaters are suitable for irrigation with respect to RSC values but due to higher HCO_3 concentration the latter chemical types have unsuitable RSC values. The increase of HCO_3 concentration in the Gunnedah Formation in the western part of the catchment is likely to be due to mixing with Na-Cl-HCO_3 type groundwaters from the underlying Cubbaroo Formation and also from the dissolution of $CaCO_3$ as indicated by saturation indices. Groundwaters from the Gunnedah Formation in the east are generally fresh, with EC ranging from 310 to 1370 μS/cm. Na–HCO_3 groundwaters in the southeast have a high to very high SAR. Na concentrations are high in this area as a result of mixing with Na-HCO_3 rich groundwater from the GAB and ion exchange reactions. These groundwaters also have high bicarbonate hazard with RSC ranging up to 14 meq/l.

Comparison between water quality indicators calculated for recent and historical groundwater data has shown increases in SAR and RSC values in deeper groundwaters from the southeast of the catchment. Given the potential vertical hydraulic gradient, this could be attributed to increased vertical leakage of the GAB Na-HCO_3 groundwaters due to depressurisation of the middle aquifer. Increases in EC occurred in some deeper boreholes, indicating mixing between Gunnedah Formation groundwaters and more saline shallow groundwaters.

7 CONCLUSIONS

Hydrogeochemical data has identified several types of groundwaters including Na-HCO_3 groundwaters in the southeast of the catchment, Na-Ca-Mg-HCO_3 groundwaters near the Namoi River and Na-Cl groundwaters in the west. Stable isotope data has identified mixing between groundwater types and different recharge sources for the different groundwater types. The sustainability of both the quality and quantity of these groundwater resources have been threatened by irrigation practices over the last 30 years. Hydrologic data reveals steady declines in groundwater levels over this period and increases in RSC and SAR values indicate a decline in groundwater quality in certain parts of the catchment.

Ongoing work as part of the principal author's research aims to use stable and radiocarbon isotopic data, trace element data, coupled with mechanistic hydrogeochemical modelling to quantiry the degree of mixing between groundwater for the different aquifers. The data will also be used to ascertain the age of groundwater and flow directions allowing recharge and sustainability calculations.

8 ACKNOWLEDGEMENTS

The authors thank the New South Wales Department of Land and Water Conservation for the grant provided for the hydrogeochemical investigation in the Lower Namoi River catchment. We thank Phil Kalaitzis and Michael Jamieson from the DLWC, Tamworth for technical support and Dorothy Yu from the School of Geography, UNSW for laboratory work.

REFERENCES

Calf, G.E. 1978. An investigation of recharge to the Namoi Valley aquifers using environmental isotopes. *Aust. J. Soil Res.* 16: 197-207.

Dept. of Land & Water Conservation. 2000. Status report for the alluvial groundwater resources of the Lower Namoi Valley NSW. *Draft report*

Gates, G. 1980. The hydrogeology of unconsolidated sediments in the Mooki River Valley, New South Wales. *MSc Thesis*, UNSW.

Habermehl, M.A. 1980. The Great Artesian Basin, Australia. *BMR J. Aust. Geol Geophys.* 5: 9-38.

Jankowski, J., R.I. Acworth & S. Shekarforoush 1998. Reverse ion-exchange in a deeply weathered porphyritic dacite fractured aquifer system, Yass, New South Wales, Australia. In G.B. Arehart & J.R. Hulston (eds), *Proc. 9th Int. Symp. Water-Rock Interaction, Taupo, New Zealand, 30 March–3 April 1998*: 243–246. Rotterdam: Balkema.

Kalaitzis, P., R. Brownbill & M. Jamieson 1999. Environmental provisions in determining sustainable yield for groundwater management plans in the Lower Namoi Valley, NSW. *Proc. Murray Darling Basin Groundwater Workshop, Griffith, NSW, 14-16 September 1999*: 55-60.

Little, W. & J.B. Ross 1993. Review of groundwater use and water level behaviour in the Lower Namoi Valley. *Tech. Rep. Dept. Water Resour.* TS93.078

Merrick, N.P. 1989. Lower Namoi Valley Groundwater Model. Report by Heritage Computing for DWR. *Tech. Rep. Dept. Water Resour.* CR2163.

Merrick, N.P. 1999. Lower Namoi groundwater budgets: simulation 1980 – 1994. *Tech. Rep. Insearch Limited*, National Centre for Groundwater Management, University of Technology, Sydney.

Ross, J.B. & L. Jeffery 1991. Ground subsidence and borehole collapse associated with groundwater withdrawals – Namoi Valley, NSW. *Tech. Rep. Dept. Water Resour.* TS91.007.

Rozanski, K.L. Araguas-Araguas & R. Gonfiantini 1993. Isotopic patterns in modern global precipitation, conti-

nental indicators of climate. *American Geophysical Union Monograph.*
Williams, R.M. 1993. The Cainozoic geology, hydrogeology and hydrochemistry of the unconsolidated sediments associated with the Namoi River in the Lower Namoi Valley, NSW. *Tech. Rep. Centre Nat. Resour.* CNR97.093.
Williams, R.M., N.P. Merrick & J.B. Ross 1987. Natural and induced recharge in the Lower Namoi Valley, New South Wales. In: M.L. Sharma (ed.), *Proc. Groundwater Resour. Symp., Perth*, 239-252.

Groundwater: Past Achievements and Future Challenges, Sililo et al. (eds) © 2000 Balkema, Rotterdam, ISBN 90 5809 159 7

Environmental isotopes as indicators for groundwater recharge conditions below glaciated areas in crystalline rocks of the central Swiss Alps

U.S.Ofterdinger, W.Balderer & S.Loew
Swiss Federal Institute of Technology, Zürich, Switzerland

T.Vitvar
State University of New York, N.Y., USA

ABSTRACT: Groundwater recharge conditions in a high alpine catchment are a crucial factor for understanding the groundwater flow regime. Stable isotopes (^{2}H, ^{18}O) and tritium analysis of groundwater samples support the finding from hydrological modeling in a high alpine research area, that a substantial contribution of glacial meltwater and snowmelt enter the groundwater flow system. This influence is documented in anomalous low temperatures and depleted δ^{18}O values in groundwater samples taken in a gallery passing beneath glaciated areas. Low tritium concentrations in these waters show the imprint of meltwater from old submodern glacial ice. Lumped-parameter modeling, applied to the δ^{18}O signal yields mean residence times of $\sim$2–4.5 years for the encountered waters.

1 INTRODUCTION

For understanding the groundwater flow regime in a partly glaciated high alpine catchment, knowledge of the groundwater infiltration conditions as the upper boundary condition is a key issue. To characterize these conditions for a research area in the western Gotthard-Massif a hydrological model was applied (PREVAH-ETH) to asses the temporal and spatial distribution of infiltration rates (Vitvar & Gurtz 1999). One finding from this model are high infiltration rates in the glaciated areas during the ablation periods. Even though only a small portion of the total catchment is glaciated ($\sim$14.6%) significant effects on the groundwater flow regime are anticipated, especially regarding the flow to an unlined tunnel, which partly passes beneath the research area and its glaciated areas.

This Bedretto-tunnel, an unlined northwest-southeast trending segment to the Furka-basetunnel, runs through the Hercynian, fractured and locally cataclastic faulted Rotondo-Granite. Topographic elevations in the research area range from 1500–3200 m a.s.l. Groundwater samples were obtained from several inflows along the profile of the Bedretto-tunnel for a period of one year. The sampling points are mainly associated with steeply northwest dipping cataclastic fault zones or zones of dense fracturation where inflows occur. The observation point T9 is situated within a major fault zone, striking approximately at 75° (Fig.1).

The question addressed in the following is whether the influence of the glaciated areas on the groundwater infiltration conditions is also documented in the temperature and isotope signal of groundwater samples from the gallery beneath the study area.

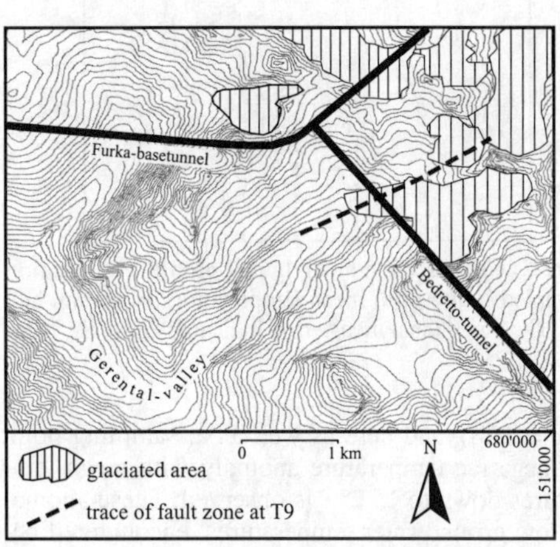

Figure 1: Enlarged plan view of research area with indicated glacial areas and tunnel as well as fault trace.

2 TEMPERATURE PROFILE

Figure 2a shows the mean temperature of the sampled inflows along the tunnel profile. At most of the sampling points the temperatures lie in the range of 19–20°C. The temperature measurements at the individual sampling points show only little variability over time from their mean value (standard deviation $0.04 \leq \sigma \leq 0.28$). Variability is the highest at location

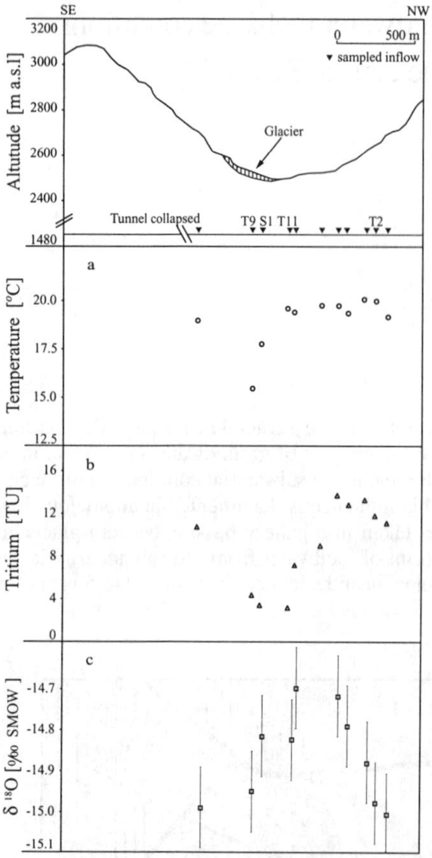

Figure 2: Cross-section along tunnel profile with location of sampled inflows, (a) mean temperature, (b) tritium and (c) $\delta^{18}O$ values (analytical error indicated) at these locations.

T9 ($\sigma = 0.28$), and here as well as at sampling point S1 a negative temperature anomaly with mean temperatures down to 15.2°C is observed. These anomalous low groundwater temperatures, encountered below the glacier, indicate the contribution of cold waters to groundwater recharge.

3 TRITIUM

The mean tritium concentrations observed at the sampling points along the tunnel profile (Fig. 2b) show values of 10.5–13.6 TU in the northwestern end of the tunnel, followed to the southeast by a transitory zone with tritium values of 7.2–8.9 TU,. Values of 3.1–4.4 TU are observed towards the southeastern end of the profile below the glacier. These latter values can be qualitatively attributed to a mixing of a submodern (precipitation prior to 1952) component with recent recharge.

Two possible explanations might elucidate the origin of the submodern component.

The sampling sites showing these low tritium-concentrations (T9, T11 and S1) are situated below the valley bottom and might be influenced by upwelling older groundwater.

Another explanation with regard to the influence of the glaciated areas above the observation points is also plausible. Meltwater from the glaciated areas is derived from a source, which accumulates winter precipitation with seasonal low tritium contents. However, the observed tritium concentrations in recent winter precipitation reach their minimum at ~ 6 TU (observations during the 1990's at the stations Guttannen, Grimsel and Meiringen). Therefore, groundwater infiltration from this accumulated recent winter precipitation can only explain the observed data with lower values down to ~ 3 TU, if the residence times of these waters are assumed to exceed at least one half-life cycle of tritium (12.43 years).

Another variant of this last explanation is regarded to be the most plausible scenario. Assuming that not only recent accumulated winter precipitation with seasonal low tritium concentrations contribute to the groundwater recharge but also meltwater from old submodern glacial ice, a source of groundwater recharge is present, which can explain the measured low tritium contents.

The intermediate values of 4–10 TU would then represent the decreasing influence of the glacial component towards the northwest, while the values of ~ 10–13 TU are even closer to the long-term mean annual tritium concentration in recent precipitation with values from 15.5–16.2 TU (measurements during the 1990's at the above stations).

The influence of groundwater recharge from the snowcover during the melting season is regarded as an explanation for the remaining difference between the tritium concentration in groundwater samples and mean annual precipitation. This depleted source shifts the tritium content in the groundwater away from the mean annual value in precipitation towards more negative values.

However, as long as no certainty on the residence times of the groundwater and therefore on the significance of the natural decay of the tritium signal in the flow system is achieved, the above qualitative interpretation of the measurements is not compelling and influences of groundwater with higher residence times of deeper flow paths can not be fully ruled out on the basis of the tritium data alone.

4 STABLE ISOTOPES FROM PRECIPITATION AND GROUNDWATER SAMPLES

Data on the $\delta^{18}O$ and $\delta^{2}H$ content of precipitation (in ‰ SMOW) were obtained from three meteorological stations in the Haslital (Guttannen, Meiringen and Grimsel; 1970–1999) and from five additional stations in the vicinity of the research area (Gütsch, Andermatt, Oberwald, Binn, Robiei; 1998–1999). From

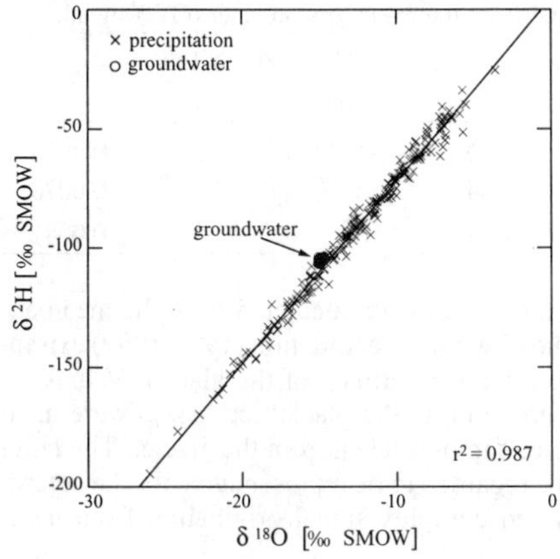

Figure 3: Local meteoric water line based on δ^2H/δ^{18}O-data from meteorological stations together with δ^2H/δ^{18}O-data from groundwater samples

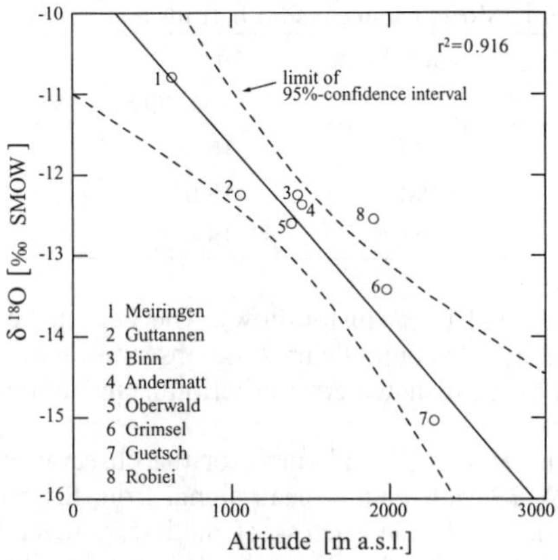

Figure 4: Relationship between altitude and mean annual δ^{18}O content in precipitation at the investigated meteorological stations

this data, the local meteoric water line (LMWL) for the study area was derived following

$$\delta^2 H = (7.65 \pm 0.05) \times \delta^{18}O + (4.7 \pm 0.77) \quad (1)$$

The alignment of the groundwater samples with the LMWL (Fig. 3) underlines their meteoric origin.

From the same data set, the δ^{18}O-altitude (z) relationship was calculated as

$$\delta^{18}O = (-0.0023 \pm 0.0003) \times z - (9.4 \pm 0.5). \quad (2)$$

For equation 2, the value from the Ticinese station Robiei was excluded from the regression analysis, because its value lies outside the 95%-confidence interval (Fig. 4).

This probably reflects the different origin of the air masses at this station coming preferentially from the south. The mean annual deuterium excess for 1998 at this station of 13.05 compared to 5.25–11.47 at the other stations also reflects the different source region. The above calculated δ^{18}O-altitude correlation is in good agreement with published data (Pearson et al. 1991), giving a correlation of

$$\delta^{18}O = (-0.0021 \pm 0.0006) \times z - (10.6 \pm 0.7). \quad (3)$$

The δ^{18}O values encountered in the tunnel show, if at all, only damped seasonal variations. Figure 2c shows the mean δ^{18}O values measured at the different sampling points along the tunnel profile. Noticeable are the more negative values at the sampling points T9, T11 and S1 below the glaciated areas. These values do not show the otherwise observed trend following the δ^{18}O-recharge altitude dependency, where potential recharge zones at higher elevations are generally characterized by depleted δ^{18}O values.

The δ^{18}O anomaly at T9, T11 and S1 can not be explained by the δ^{18}O signal in precipitation alone. Rather, the influence of a δ^{18}O-depleted component, most probably originating from accumulated depleted winter precipitation in the glaciated areas is reflected by this anomaly.

5 LUMPED-PARAMETER MODEL AND MEAN RESIDENCE TIMES

To analyze the potential contribution of glacial meltwater to groundwater recharge and the mean residence time of the encountered waters, the observed δ^{18}O-values were introduced in a lumped-parameter model. The δ^{18}O input function (1970–1999) for precipitation was derived from the stations Guttannen, Meiringen and Grimsel (1970–1999) as well as from additionally analyzed data from the stations Gütsch, Andermatt, Oberwald, Binn and Robiei (1998-99). The extrapolation to the mean recharge altitude of the massif above the tunnel was carried out following equation 2. Finally, weighing of the input function by the mean monthly groundwater recharge rates, derived from the hydrological model (Vitvar & Gurtz 1999) was performed. The applied code (Maloszewski & Zuber 1996) solves the direct problem, calculating the theoretical output concentration from a given input concentration (input function) and for a chosen transit time distribution (parameterized by the mean residence time) under constant flow assumption. The inverse problem is solved by a trial-and-error procedure with the possibility to choose between six model-types for the transit time distribution (exponential, piston-flow, linear, dispersion, exponential-

Table 1: Model parameters for best fits to δ^{18}O data from T9, T2 and from the mean groundwater (GW) signal.

Data	Model type	$c\beta$ [‰ δ^{18}O SMOW]	β	η	Mean residence time [months]	σ
T9	EPM	-16.2	0.58	1.08	21	0.0082
T2	EPM	-18.6	0.34	1.04	52	0.0070
mean GW	EPM	-18.6	0.26	1.04	50	0.0081

piston-flow and linear-piston-flow). The best fit between the modeled output curve and observation data is judged by a calculated error criterion (standard deviation σ).

The modeling was performed for the observation points T9, T2 and for the mean signal from the remaining points. In the process of modeling, several of the above models have been tested, but best fits were only achieved in all cases applying the exponential model combined with a piston flow component (EPM). The applied ratios of total volume to the volume with the exponential distribution of transit times, η, were hereby 1.03–1.08.

The necessity to apply a combined model in order to achieve a good fit as in this case is most probably linked to the nature of the granitic rocks. These provide, due to zones of dense fracturation and cataclastic faults as well as more intact massive rock, different flow systems with different transit time distributions. As the majority of inflows to the tunnel, especially T9, are associated with cataclastic fault zones or zones of dense fracturation, it is not surprising that a combined model leads to the best fits.

Furthermore, another finding during the modeling is that in all cases, the theoretical output curve calculated from the precipitation input function alone showed enriched δ^{18}O values compared to the observed data (Fig. 5a). This means that the input function derived from precipitation data alone can not explain the observed δ^{18}O-depleted data. An additional depleted component needs to be added to the input function. For simplification, the applied code only allows adding a component with constant concentration ($c\beta$). Only after adding such a depleted component, a best fit could be obtained for the observed data (Fig. 5b). Table 1 gives an overview of the applied model parameters together with the modeled mean residence times for the best fits.

For the case of T9, below the glacier, this δ^{18}O concentration was assumed to originate from glacial meltwater. This glacial meltwater itself is a combination of snow- and icemelt. At the onset of the ablation period the contribution from snowmelt dominates while at later times, melting of glacial ice becomes increasingly important (Malard et al. 1999). As each of these two components can be regarded as dominant for an equally long time during the total ablation period, the overall δ^{18}O concentration of glacial meltwater ($c\beta$) was derived from the arithmetic mean of $c\beta_{snow}$ and $c\beta_{ice}$. The δ^{18}O concentration of the snowmelt ($c\beta_{snow}$) was calculated from the mean annual value of winter precipitation (1970-1999) extrapolated to the mean altitude of the glacier. Values for the concentration in the glacial ice ($c\beta_{ice}$) were measured in ice samples taken from the glacier. The latter can not be regarded to be representative for the heterogeneous and complex signal originating from melt-

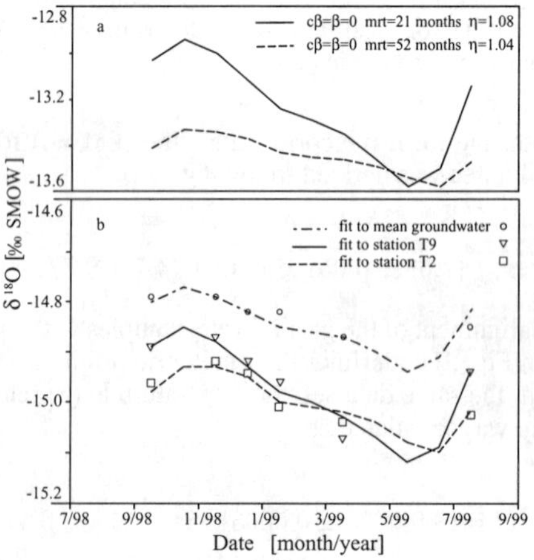

Figure 5: a) Modeled output functions from precipitation input without depleted meltwater component (mrt=mean residence time). b) Observed δ^{18}O data (from T9, T2 and mean groundwater) and best fit output functions for model parameters of Table 1.

ing of the glacial ice (Stichler & Schotterer 2000). It should rather be considered as an estimate for the potential contribution of glacial icemelt to the overall glacial melt component. The calculated estimate for the $\delta^{18}O$ concentration in the glacial meltwater amounts to $c\beta = -16.2‰$. When such an additional depleted component is introduced in the model it can be shown, that a substantial fraction ($\beta > 50\%$) of the total water flow needs to be attributed to this component in order to achieve a good fit to the observed data (see Table 1).

This large contribution of the additional depleted component reflects the significance of groundwater recharge by glacial meltwater in this part of the study area.

For the modeling of T2 and the mean groundwater signal such an additional depleted component was also needed to achieve good fits between the modeled output and the observed data. Here, the δ^{18}O-signal of this component was derived only from the extrapolated mean winter precipitation ($c\beta$=-18.6‰), because no glaciated areas exist along the slope above the sampling points. The depleted recharge component is mainly derived from snow and firn. The models show that here, less contribution from the depleted additional source is necessary to achieve a good fit (see Table 1).

6 INTERPRETATION AND DISCUSSION

The hydrological model of the research area (Vitvar & Gurtz 1999) shows significant groundwater infiltration rates below the glaciated areas during the ablation period. Observations in the Bedretto-tunnel, partly passing beneath these areas show an apparent negative anomaly in groundwater temperatures at sampling points below the glacier. This negative anomaly is most probably caused by infiltration of cold glacial meltwater. Especially at T9, this effect is enhanced by the orientation of the major fault zone associated with this tunnel inflow, which projected trace is covered to a high percentage by glaciated area in this region.

The measured tritium concentrations along the tunnel profile by themselves are not necessarily conclusive for determining the influence of the glaciated meltwater. While the authors regard the influence of glacial infiltration, also indicated by the observed temperature anomaly, to be the most plausible explanation for the low tritium contents in groundwater samples taken below the glaciated area, the possibility of encountering older groundwaters can not be excluded on the basis of the tritium observations alone.

The δ^{18}O values of the analyzed samples from the Bedretto-tunnel are depleted compared to the mean values in precipitation. This depletion is coherent along the tunnel profile and is the result of the influence from accumulated δ^{18}O-depleted winter precipitation - be it dominantly from the recent snowcover as for the cases at the northwestern end of the tunnel (resulting in depleted δ^{18}O + tritium concentrations closer to recent precipitation) or from recent winter precipitation together with submodern glacial ice as for the cases at the southeastern end of the tunnel below the glacier (resulting in depleted δ^{18}O + low tritium concentrations).

Additionally, the calculated mean residence times show residence times of $\sim$2 years for samples below the glacier and up to $\sim$4.5 years for samples in the core of the massif. Keeping in mind that in applying the exponential transit time distribution as part of the combined model, theoretically an infinitely long transit time is possible, the calibrated mean residence times indicate, that upwelling submodern groundwater is if at all less significant within the system.

The calibrated mean residence times for T2 of $\sim$4.5 years is in agreement with the qualitative interpretation of the tritium content at this northwestern end of the tunnel (10–13 TU). For continental regions Clark & Fritz (1997) attribute tritium contents in this observed range to modern groundwater (recharge <5–10 years).

7 CONCLUSIONS

The temperature profile along the Bedretto-tunnel indicates the influence of infiltrating cold glacial meltwaters. The tritium concentrations observed together with analysis of δ^{18}O support this assumption. Lumped-parameter modeling of the δ^{18}O measurements from groundwater samples show that a substantial contribution of glacial meltwater and snowmelt enters the flow system as predicted by the hydrological model of the study area. The calibrated residence times lie in the range of $\sim$2–5 years, depending on the position along the tunnel profile.

Acknowledgements

Samples from the meteorological stations at Gütsch, Andermatt, Oberwald, Binn and Robiei were kindly taken by the station supervisors. Data from the stations Guttannen, Meiringen and Grimsel where obtained from the Schweizerische Meteorologische Anstalt-SMA, Zürich and extracted from the GNIP database provided by the IAEA, Vienna. Laboratory analysis of δ^{18}O, δ^{2}H and tritium were carried out at the Institut für Hydrologie at the Forschungszentrum für Umwelt und Gesundheit-GSF, Neuherberg. Geograhical data was provided by the Landestopographie, Bern. Access to the tunnel and logistic help was kindly provided by the Furka-Oberalp-Bahn.

REFERENCES

Clark, I. D. & Fritz, P. (1997). *Environmental Isotopes in Hydrogeology*. Boca Raton: Lewis Publishers.

Malard, F., Tockner, K., & Ward, J. (1999). Shifting dominance of subcatchment water sources and flow paths in a glacial floodplain, Val Roseg, Switzerland. *Arctic, Antarctic and Alpine Research 31 (2)*, 135–150.

Maloszewski, P. & Zuber, A. (1996). Manual on mathematical models in isotope hydrology. *Technical Report TECDOC-910*, IAEA.

Pearson, F., Balderer, W., Loosli, H., Lehmann, B., Matter, A., Peters, T., Schmassmann, H., & Gautschi, A. (1991). *Applied isotope hydrogeology-a case study in Northern Switzerland.* Technical Report 88-01, NAGRA.

Stichler, W. & Schotterer, U. (2000). From accumulation to discharge: Modification of stable isotopes during glacial and post-glacial processes. *Hydrological Processes* in press.

Vitvar, T. & Gurtz, J. (1999). *Spatially distributed hydrological modeling in the Rotondo area, western Gothard-Massif.* Technical Report 3465/16, ETH-Zürich, Engineering Geology, ETH-Hönggerberg.

Groundwater: Past Achievements and Future Challenges, Sililo et al. (eds) © 2000 Balkema, Rotterdam, ISBN 90 5809 159 7

Predicting ion exchange parameters for Triassic Sandstone aquifers

K.E.Parker & J.H.Tellam
Department of Earth Sciences, University of Birmingham, UK

M.I.Cliff
Formerly: M.J.Carter Associates, Atherstone, UK

ABSTRACT: In an attempt to quantify the ion exchange behaviour of UK Triassic Sandstone, the following issues are being investigated. Issue 1: rules for allowing upscaling from laboratory results to fieldscale; 2: determination of appropriate ion exchange relationships; 3: estimation of sandstone exchange properties from mineralogical composition data. Issue 1 has been examined using numerical experimentation: it was found that linear upscaling is only approximately valid when exchange heterogeneity is limited. Issues 2 and 3 are being examined by laboratory experimentation; it has been found that the exchange behaviour of the clay minerals in the sandstone is best described by power functions, and that the properties vary linearly with mixing fractions in mixtures of clays. It therefore seems likely that the exchange properties of rocks can be estimated from knowledge of their mineral compositions.

1 INTRODUCTION

Ion exchange reactions are common in groundwater systems. They are important not only because they directly affect the migration of the exchanging species (e.g. NH_4^+), but also because they can disturb solubility equilibria (e.g. calcite, fluorite) initiating changes in a range of other species. It would be useful to be able to predict ion exchange behaviour quantitatively, and there has been much research effort devoted to this. Research has usually been concentrated in the soil science context (e.g. Bolt, 1967; McBride, 1994; and Evangelou, 1998), but also to some extent in the groundwater context (e.g. Valocchi et al., 1981; Ceazan et al., 1989; and Bjerg & Christensen, 1993).

The work presented here concerns three basic issues.

Issue 1: Rules for upscaling from laboratory experimental data to field systems. Water entering a borehole will be a mixture of waters which have reached it via different flow paths. If the exchange properties of the rock associated with each pathway are different, it is unclear how the rock properties should be upscaled to allow prediction of the chemistry of the mixed sample.

Issue 2: Ion exchange "laws". There are a number of suggested approaches for quantifying exchange behaviour and it is unclear which are the most appropriate to use.

Issue 3: Estimating ion exchanger properties. Because the determination of ion exchanger properties is normally very labour intensive, it would be very useful to develop a method that would allow estimation of exchange parameters given detailed rock properties.

The work detailed here has been tackling these issues in the context of the UK Triassic Sandstone, which was found by El-Ghonemy (1997) to show non-ideal behaviour. Issue 1 (upscaling) has been tackled using numerical experimentation. Issues 2 (ion exchanging "laws") and 3 (estimation of exchanger properties) has been tackled by laboratory determination of the ion exchange behaviour of each of the exchanger phases within the sandstones. So far the major cation exchange behaviour of the clay minerals present in the sandstone has been explored: the effect of the iron oxyhydroxides has not yet been considered.

2 THEORY

A general binary exchange reaction can be represented, for example, as:

$$CaX_2 + 2K^+ \leftrightarrow 2KX + Ca^{2+} \quad (1)$$

where X is the exchanger.

The exchanger is often thought of as having a fixed ion exchange capacity, though this is usually not strictly correct.

The thermodynamic exchange equilibrium constant is given by:

$$K_{ex} = \frac{(KX)^2(Ca^{2+})}{(K^+)^2(CaX_2)} \quad (2)$$

where ( ) represent activities.

There are various ways of approximating the concentration of MX; for example according to the Vanselow (1932) and Gaines and Thomas (1953) conventions, equation (2) is rewritten in concentration terms respectively as:

$$K_v = \frac{(\overline{N}_K)^2(Ca^{2+})}{(K^+)^2(\overline{N}_{Ca})} \quad (3)$$

$$K_{GT} = \frac{(\overline{E}_K)^2(Ca^{2+})}{(K^+)^2(\overline{E}_{Ca})} \quad (4)$$

where $\overline{N}$ and $\overline{E}$ are mole and equivalent fractions respectively of K^+ and Ca^{2+} in the adsorbed phase. A third common convention is the Gapon (1933) convention

The equilibrium constant (K_{ex}) is related to K_v and K_{GT} by:

$$K_v = K_{ex}\left(f_{Ca} / f_K{}^2\right) \text{ and} \quad (5)$$

$$K_{GT} = K_{ex}\left(g_{Ca} / g_K{}^2\right) \quad (6)$$

where f and g are the activity coefficients of adsorbed cations. As f and g vary with the composition of the exchanger sites, the "selectivity coefficients" K_v and K_{GT} also vary with exchange site composition. If the values of the selectivity coefficients are known as functions of exchange site composition (i.e. $\overline{N}, \overline{E}$), then for the Vanselow convention:

$$\ln f_{Ca} = \overline{E}_K \ln K_v - \int_0^{\overline{E}_K} \ln K_v d\overline{E}_K \quad (7)$$

$$2\ln f_K = -(1-\overline{E}_K)\ln K_v + \int_{\overline{E}_K}^{1} \ln K_v d\overline{E}_K \quad (8)$$

where $\overline{E}$ denotes the equivalent fraction of K on the exchanger phase (Argersinger et al., 1950).

From (7) and (8)

$$\ln K_{ex} = \int_0^1 \ln K_v d\overline{E}_K \quad (9)$$

This relationship can be used with laboratory determined values for K_v and $\overline{E}_K$ to determine K_{ex} (e.g. Evangelou & Phillips, 1989; Sparks, 1995).

Experimentation has indicated that power functions sometimes describe ion exchange reactions rather well (Rothmund & Kornfeld 1918; 1919; Langmuir, 1981; Bond, 1995):

$$\frac{(K)^2}{(Ca)} = K_{K/Ca}\left(\frac{N_K{}^2}{(N_{Ca})}\right)^n \quad (10)$$

Assuming a solid solution model for sorbed ion activities, it can be shown that a power function relationship might be expected. In this case

$$n = 1 - \frac{B}{2RT} \quad (11)$$

where B is a constant independent of composition, R is the universal gas constant, and T is absolute temperature.

3 ISSUE 1: UPSCALING

The 1D reactive transport code PHREEQM (Appelo & Postma, 1993) has been used to model the process of sampling. The aquifer was represented by a series of parallel non-interacting flow tubes, each of which was associated with perfect Gaines - Thomas or Gapon exchangers, each exchanger having different properties (selectivity coefficients and cation exchange capacity (CEC)). At the end of the flow tube, the modelled concentrations in the discharging water from each flow tube are averaged to represent the effect of mixing during sampling. This "actual" breakthrough curve (BTC) can be compared with BTCs predicted using, for example, (weighted) averages of exchange properties of each of the flow tubes. The results presented here relate to simple two flow-tube systems involving binary exchange with Gaines - Thomas and Gapon exchangers. Parameter values were chosen to be representative of likely values in the Triassic Sandstones (El-Ghonemy, 1997). CEC ranged from 1-2.75 meq/l with a standard value of 1.2meq/l. Gaines - Thomas selectivity coefficients values for heterovalent exchange ranged from 0.35-0.5 and for homovalent from 0.2-0.5 (with respect to Na^+).

As the theory dictates, the models showed that the CEC values for the whole profile is simply the sum of the exchange capacities of the flow tubes. However, again as expected, the "macro system" selectivity coefficient values are not so easily estimated. Figure 1 shows results from numerical experiments on a two layer, binary exchange (K/Mg) system. The abscissa is the ratio of the selectivity coefficients in the two layers: the CEC in each layer was the same (1.2 meq/l). The ordinate (labelled "poorness of fit") is the root mean square of the deviation of the "actual" BTC concentrations from the BTC expected using averaged selectivity coefficient values. Figure 1 shows that the poorness of fit rapidly increases as the selectivity coefficients for the two layers diverge. Similar results were obtained for a wide range of K_{GT} values, and for Gapon exchangers.

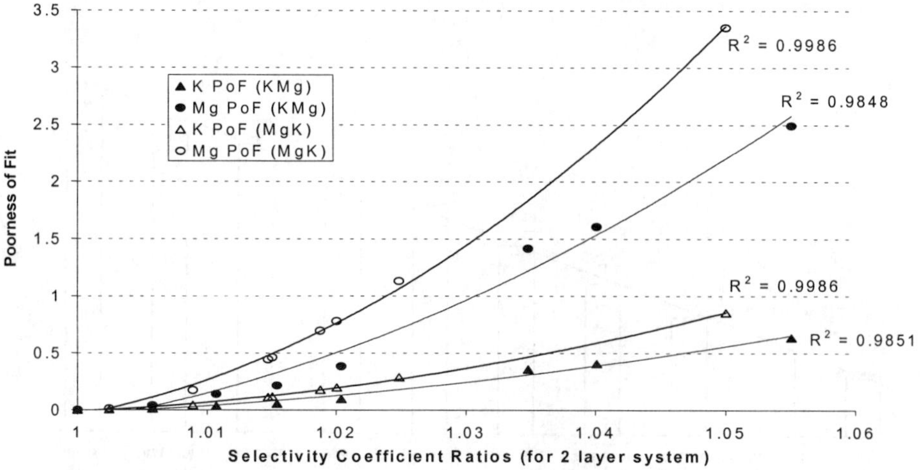

Figure 1. Poorness of fit curves for the PHREEQM modelled binary systems of K/Mg and Mg/K

Attempts were made to find K_{GT} values for a single exchanger that would reproduce the "mixed" model results. However, only in a few cases was it possible to significantly reduce the poorness of fit index.

More work remains to be done, but the results to date indicate that:

1) If all the aquifer units are Gaines - Thomas (or Gapon) exchangers, then the system as a whole behaves to a good approximation as a Gaines - Thomas (or Gapon) exchanger with 'macroscopic' selectivity coefficients equal to the average of the properties for each unit, but only if:
   (i) the exchange properties of each unit are similar (extent as yet unquantified) and
   (ii) only narrow ranges of exchange site occupancies are considered (extent as yet unquantified).
2) If the properties of the units are dissimilar, an equivalent Gaines - Thomas (or Gapon) exchanger cannot adequately represent the system as a whole, whatever the values of the parameters chosen.
3) If all the aquifer units are Gaines - Thomas (or Gapon) exchangers with fairly similar properties and dissimilar ranges of site occupancy, then the system as a whole can be described approximately as a Gaines – Thomas (or Gapon) exchanger but with properties that are not necessarily weighted averages of the individual units.

Preliminary calculations indicate that the upscaling effects can make a significant difference to predicted water chemistries. However, this needs to be confirmed in the context of the heterogeneity displayed by individual beds within the UK Triassic Sandstone.

4 ISSUES 2 AND 3: ION EXCHANGE "LAWS" AND ESTIMATION OF EXCHANGER PROPERTIES

4.1 *Approach*

The exchanger phases in the UK Triassic Sandstones are clay minerals and Fe-oxyhydroxides. El-Ghonemy (1997) found that the power-function expression provided a good description of the ion exchange properties of the UK Triassic Sandstone. In order to determine if this might be the case (Issue 2), the major cation exchange behaviour of individual and mixed clay mineral assemblages are being determined. The possibility of being able to estimate exchange properties, given information on rock mineral composition, will also be investigated from these results. Eventually the exchange behaviour of the oxyhydroxides will also be examined.

Kaolinite (KGa-1b), montmorillonite (SCa-3) and illite (1Mt-2) have been purchased from the clay minerals repository Missouri, USA. Separating the clay phases from the rock proved impracticable. CEC was determined using a 1M NH_4Cl flush, with account being taken of any carbonate dissolution.

Selectivity coefficients were obtained by the method of Jenson & Babcock (1973) at a constant ionic strength of 0.05. Chloride salts were used in all cases. No pH control was imposed.

4.2 *CEC Results*

The CEC values obtained were typical of values quoted in the literature (Van-Olphen & Fripiat, 1979). As expected CEC values for clay mixtures varied linearly with mixing proportions (Figure 2).

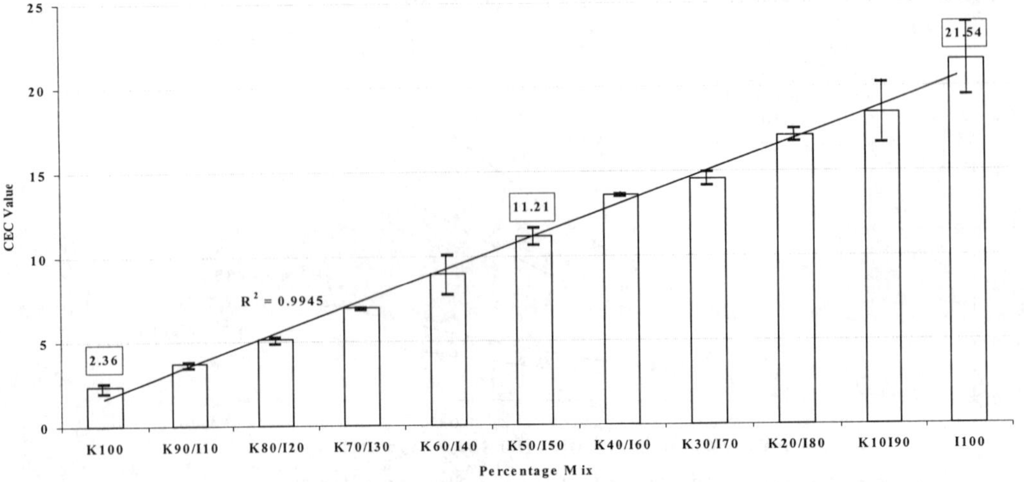

Figure 2. CEC values for Kaolinite / Illite mixtures (100%-0%) Error bars are shown for 1 standard deviation

4.3 *Selectivity Coefficient Results*

Experiments conducted on kaolinite/illite and montmorillonite/illite mixtures for the K/Na, K/Mg and Mg/Ca binary systems produced the isotherms presented in Figure 3. In most cases the variation in isotherm shape is continuous and simple between clay end members. It is clear that the selectivity coefficients vary with exchange site occupancy.

Log A/B v Log AX/BX graphs are displayed for 2 systems in Figure 4. The figure shows that a power-function expression can describe the exchange behaviour very well. Calculations of n (slope of line) and K_{ex} (intercept) provide the necessary data for calculating B values (Eq(11))(Table 1).

The exponent n from Eq (11) varies linearly with clay proportion mixing with r^2 values ranging from 0.91 – 0.98 for the K/Mg and Mg/Ca systems for kaolinite / illite and montmorillonite /illite mixtures. B, a constant independent of exchange site composition, denotes the energy of interaction between the two ions at the exchanger surface. Greater values are seen in the heterovalent systems when compared with the homovalent, and also in those with an increasing illite composition. The K/Na system however has an almost constant n and B value. K_{ex} values ranged from 1.5-3.2, 11-37.2 and 1-3 for K/Na, K/Mg and Mg/Ca exchange on kaolinite/illite respectively and from 1.6-4 and 0.01-0.4 for K/Mg and Mg/Ca exchange on montmorillonite/illite. These findings are currently being evaluated.

Sorbed phase activity coefficients, as calculated using Equations (7) and (8), are shown in Figure 5. The data plotted are for kaolinite/montmorillonite mixtures for Mg/Ca exchange, but the shape of the curves is typical of other systems. The step at about 40% mole fractions has also been found by Jensen & Babcock (1973) on Yolo Loam, and Deist & Talibudeen (1967) involving Ca/K exchange on Harwell series soil.

Table 1. Table of n and B values calculated using Eq 10 and 11

	Kaolinite / Illite			Mont / Illite	
	n values			n values	
%Clay Mix	K/Na	K/Mg	Mg/Ca	K/Mg	Mg/Ca
0.00	1.24	1.71	1.47	1.71	1.34
25.00	1.21	1.68	1.41	1.37	1.28
50.00	1.24	1.59	1.29	1.18	1.18
75.00	1.25	1.45	1.23	1.09	1.14
100.00	1.23	1.27	1.07	1.01	1.03
	B values			B values	
0.00	-0.47	-1.03	-0.75	-1.03	-0.59
25.00	-0.44	-0.99	-0.67	-0.62	-0.52
50.00	-0.47	-0.89	-0.53	-0.40	-0.40
75.00	-0.48	-0.72	-0.45	-0.29	-0.35
100.00	-0.46	-0.50	-0.26	-0.19	-0.22

4.4 *Discussion*

Major cation exchange on kaolinite, illite and montmorillonite, and on mixtures of these clays, can be described well by a power-function relationship. The parameters of the power function appear to vary sys tematically with mixing proportion. Hence, if for a rock the proportion of clay minerals present is known, it may be possible to estimate the rock's bulk exchange parameters. Although issues concerning pH dependence and more complex mixtures of ions and clays have yet to be investigated, it is encouraging that the style of exchange seen in the UK Triassic Sandstone is similar to that seen in pure samples of clay minerals that it contains.

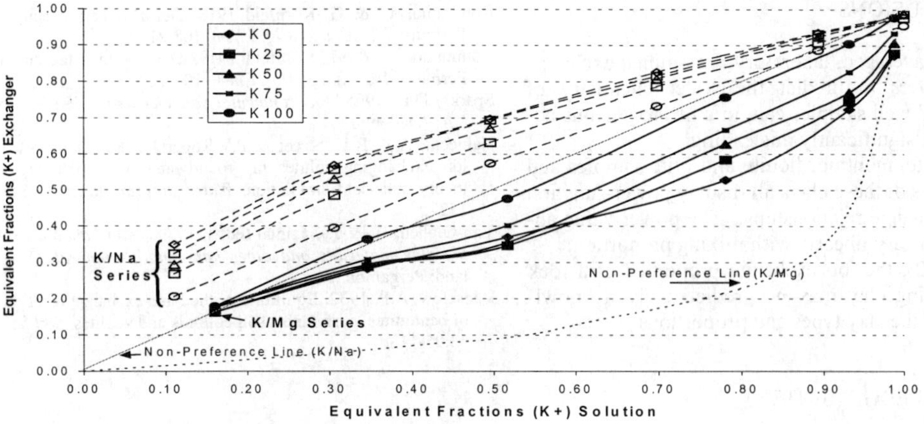

Figure 3. Isotherms for K/Na and K/Mg on kaolinite/illite mixtures (K0 - K100 representing kaolinite/illite mixtures from kaolinite 0% to 100%)

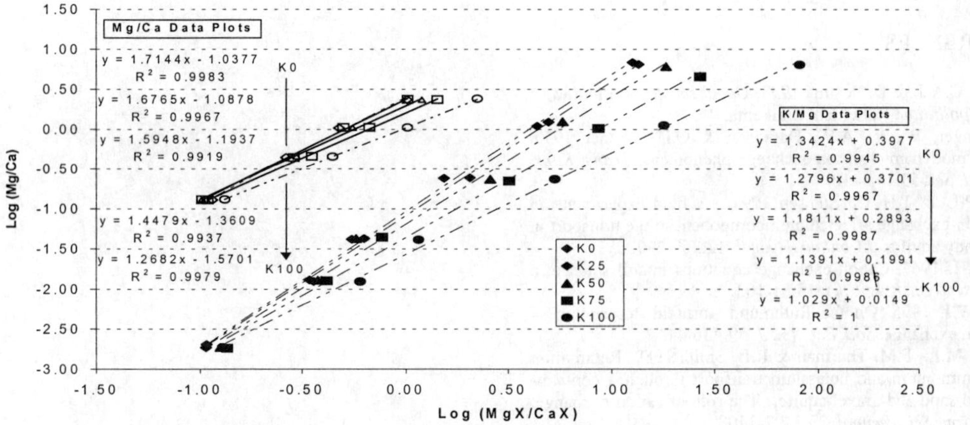

Figure 4. Power function plots for Mg/Ca and K/Mg exchange on kaolinite / illite mixtures

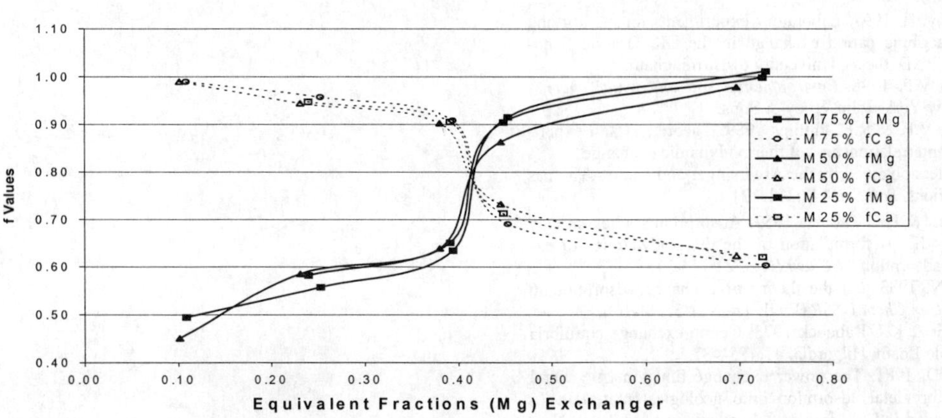

Figure 5. *f* value plots for Mg/Ca exchange on montmorillonite / illite mixtures (75 – 0%)

5 CONCLUSIONS

1. Use of averaged, laboratory-determined exchange parameters to predict the water chemistry of pumped well samples can, in a heterogeneous aquifer be significantly misleading.
2. Kaolinite, montmorillonite and illite samples and mixtures of these clays all display power-function ion exchange relationships. The power value appears to vary linearly with mixing proportion.
3. From (2), the ion exchange properties of a rock containing clay may be predicted given knowledge of the clay types and proportions.

ACKNOWLEDGEMENTS

The research reported in this paper was supported by funding from the National Environmental Research Council (Diagnostic Program) and M.J Carter Associates UK. (Contact Dr C.Pointer)

REFERENCES

Appelo, C.A.J. & D. Postma 1993. *Geochemistry, groundwater and pollution*. Rotterdam: Balkema.

Argersinger, W.J.Jr., A.W. Davidson & O.D. Bonner 1950. Thermodynamics of ion exchange phenomena. *Trans. Kans. Acad. Sci.* 53:404-410.

Bjerg, P.L. & T.H. Christensen 1993. A field experiment on cation-exchange affected multicomponent solute transport in a sandy aquifer. *J.Contam. Hydrol.* 12:269-290.

Bolt, G.H. 1967. Cation exchange equations in soil science: a review. *Neth.J. Agr. Sci.* 15:81-103.

Bond, W.J. 1995. On the Ruthmund-Kornfeld description of cation exchange. *Soil. Sci. Am. J.* 59:436-443.

Ceazan, M.L., E.M. Thurman & R.L. Smith 1989. Retardation of ammonium and potassium transport through a contaminated sand and gravel aquifer: The role of cation exchange. *Environ. Sci. Technol.* 23:1402-1408.

Deist, J. & O. Talibudeen 1967. Ion exchange in soils from the pairs K-Ca, K-Rb and K-Na. *Journal of Soil Science.* 18:125-137.

El-Ghonemy, H. 1997. Laboratory experiments for quantifying and describing cation exchange in the UK Triassic Sandstones. Ph.D. thesis, University of Birmingham.

Evangelou, V.P. 1998. *Environmental soil and water chemistry.*, New York: John Wiley & Sons.

Evangelou, V.P. & R.E. Phillips 1989. Theoretical and experimental interrelationships of thermodynamic exchange parameters obtained by the Argersinger and Gaines-Thomas conventions. *Soil Sci.* 148:311-321.

Gaines, G.L. & H.C Thomas 1953. Adsorption studies on clay minerals II. A formulation of the thermodynamics of exchange adsorption. *J.Chem.Phys.* 21:714-718.

Gapon, Y.N. 1933. On the theory of exchange adsorption in soils. *J.Gen.Chem USSR* (Engl. Transl.). 3:144-160.

Jensen, H.E. & K.L Babcock 1973. Cation exchange equilibria on a Yolo Loam. Hilgardia. 41:475-487.

Langmuir, D. 1981. The power exchange function: A general model for metal adsorption onto geological materials. In D.H. Tewari (ed.), *Adsorption from aqueous solutions*. New York: Plenum Press.

McBride, M.B. 1984. *Environmental chemistry of soils*. Oxford: University Press.

Rothmund, V. & G. Kornfeld 1918. Der Basenaustausch im Permutit. *I.Z. Anarg. Allg. Chem.* 108:215-225.

Rothmund, V. & G. Kornfeld 1919. Der Basenaustausch im Permutit. *I.Z. Anarg. Allg. Chem.* 103:129-163.

Sparks, D.L. 1995. *Environmental soil chemistry*. San Diego, CA: Academic Press.

Valocchi, A.J., R.L. Street & P.V.Roberts 1981. Transport of ion exchanging solutes in groundwater: Chromatographic theory and field simulation. *Water. Resour. Res.* 17:1517-1527.

Van-Olphen, H. & J.J. Fripiat (eds.), 1979. *Data handbook for the clay materials and other non-metallic minerals*. England: Pergamon.

Vanselow, A.P. 1932. Equilibria of the base-exchange reactions of bentonites, permutites, soil colloids and zeolites. *Soil Sci.*. 33:95-113.

Groundwater: Past Achievements and Future Challenges, Sililo et al. (eds) © 2000 Balkema, Rotterdam, ISBN 90 5809 159 7

Geochemical evolution of groundwater in metamorphic rocks of Trás-Os-Montes, Portugal

M. Rosário Pereira
Departamento de Geologia, Universidade de Trás-os-Montes e Alto Douro, Vila Real, Portugal

C. Almeida
Departamento de Geologia, Centro de Geologia, Facultad de Ciências, Universidade de Lisboa, Portugal

ABSTRACT: Inverse and forward geochemical modelling were used in order to identify the water-rock interaction processes that control the geochemical evolution of groundwater occurring in metamorphic rocks of Trás-os-Montes (Portugal). For such purpose, physical and chemical data from 30 water samples, collected in springs and wells in Paleozoic schists, quartzites and metavolcanics from hercynian shield in NW of Iberian Meseta (NE Portugal). The results obtained suggest that calcium bicarbonate composition, characteristic of most of the samples, can be explained by plagioclase, biotite or chlorite and carbonate dissolution (the last one only present in rock cement) in an open system, with precipitation of clay minerals (usually kaolinite). A small group of samples shows a significant deviation from the general pattern, with enrichment in total mineralization, especially in DIC, SO_4^{2-} and Na^+. Isotopic data point to a significant contribution of a carbon source depleted of ^{14}C and enriched in ^{13}C. Modelling results suggest that the composition of this group could be explained by pyrite oxidation and by plagioclase and carbonate dissolution.

1 INTRODUCTION

This paper describes some of the results of a modelling study (Pereira 1999) to identify geochemical reactions controlling water chemistry in metamorphic rocks of hercynian shield in NW of the Iberian Peninsula (NE Portugal).

Inverse and forward geochemical modelling were used in order to identify the most probable reactions that could explain the final composition of groundwaters, based on the knowledge of mineral composition of rocks, chemical data for water samples and speciation calculations.

2 GEOHYDROLOGIC SETTING

The geology of the area is rather homogeneous and is dominated by metasedimentary sequences (phyllites, greywackes, and quartzites) and metavolcanics. In the northern part of the area some hercynian granites are present.

The region is separated from the coast by the presence of two mountainous ridges, give a continental character to the climate. The mean annual precipitation ranges from 500 mm/year to 1300 mm/year and the mean annual temperature between 10 °C and 14 °C.

The wells located in metamorphic rocks are characterised by higher yields (1.5 L/s in average) than the ones located in granites (0.2 L/s in average) (Pereira & Almeida 1994).

The chemical features of waters from metamorphic rocks were characterised on the basis of physical and chemical data from 30 water samples collected in springs and wells (Fig. 1).

Hydrochemical facies vary between sodium chloride, in a few springs, to calcium bicarbonate, in most of the well samples (Fig. 2). The electrical conductivity ranges from 30μS/cm to 1000μS/cm with a median of 250μS/cm and the pH between 5.5 and 9.3 with median of 7.

In order to evaluate the equilibrium state of the water in relation to the minerals present in rocks, speciation calculations were made using WATEQ (Plummer et al. 1992).

The saturation indices obtained indicate oversaturation with respect to quartz and chalcedony and frequently to goethite and epidote.

A few cases of oversaturation with respect to calcite, dolomite, talc, and tremolite were also observed.

3 MASS BALANCE REACTION MODELS

Inverse modelling calculations provide a way to evaluate the most probable chemical reactions between water and minerals. The mass balance calculations were made using chemical data of the recharge

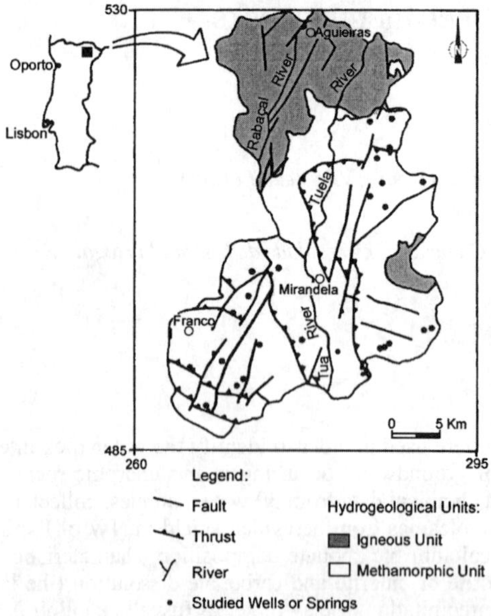

Figure 1. Map of study area with location of sampling points. (Hydrogeological Units according to Pereira 1999).

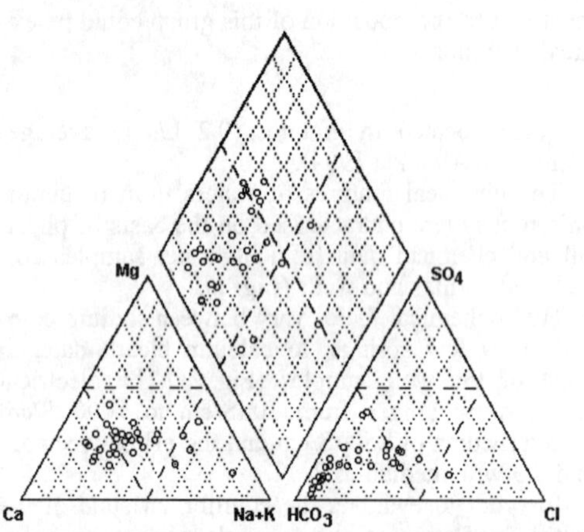

Figure 2. Piper diagram for waters from metamorphic rocks of NE of Portugal.

water and downgradient wells, or between two wells situated in the same flowpath NETPATH (Plummer et al. 1992). The plausible phases were selected according to the mineralogical composition of the rocks and the results of the speciation model.

In order to verify the kinetic and thermodynamic consistency of results forward geochemical modelling was conducted using PHREEQE (Parkhurst et al. 1980) and the sequence of reactions needed to justify the mass balance obtained by the selected inverse model were chosen.

4 GEOCHEMICAL INVERSE MODELS

4.1 *Sodium chloride waters*

To simulate the evolution from rainwater to sodium chloride spring water we used chemical data for two samples: rainwater and a spring water from Franco village (Table 1).

Table 1. Chemical variation between rainwater and spring water of Franco village in Trás-os-Montes (Portugal).

Mineral element	Rain (mmol/L)	Spring (Franco) (mmol/L)	Variation (mmol/L)
C	0.110	0.744	0.634
Al^{3+}	-	-	-
Mg^{2+}	0.002	0.039	0.037
Na^{+}	0.015	0.113	0.098
K^{+}	0.004	0.008	0.004
Cl^{-}	0.010	0.124	0.114
Si^{4+}	0	0.121	0.121

For the inverse modelling NETPATH program options were used to select mineral species, phases, evaporation and ion exchange according to the mineralogical composition of rocks and anticipated geochemical processes (Table 2).

Table 2. Chemical species, mineral phases and parameters selected to simulate the evolution between rainwater and NaCl spring water of Franco village in Trás-os-Montes (Portugal).

Species	Phases	Parameters
C, Cl^-, Na^+, Mg^{2+}, K^+, Al^{3+}, Si^{4+}	Plagioclase (An_{38}), Biotite, Chlorite, CO_2(g), Illite, Kaolinite, Silica, NaCl	Ion exchange (Ca/Na) Evaporation

From several models obtained the following one is the most probable.

Minerals	Rain→Spring (Franco) (mmol/L)
Plagioclase (An_{38})	+0.127*
Biotite	+0.012
Kaolinite	-0.036
Illite	-0.051
NaCl	+0.053
Exchange Ca^{2+}/Na^+	-0.064
Concentration factor	7.188

* + Dissolution of minerals
– Precipitation of minerals.

This model corresponds to plagioclase, biotite and sodium chloride dissolution and precipitation of kaolinite and illite. Some cation exchange between Ca^{2+} and Na^+ is admitted.

The concentration factor corresponds to an infiltration rate of about 14%, which agrees with other methods of infiltration calculation (Pereira 1999).

4.2 *Calcium bicarbonate waters*

The geochemical evolution from rainwater to calcium bicarbonate water was modelled using four well samples.

The chemical composition of the well waters is presented in Table 3.

Table 3. Chemical composition of four calcium bicarbonate waters from wells located in metamorphic rocks of Trás-os-Montes (Portugal).

Mineral element	Passos (mmol/L)	Orelhão (mmol/L)	Pereira (mmol/L)	Carvalhal (mmol/L)
C	1.490	1.875	2.551	2.939
Ca^{2+}	0.145	0.389	0.891	0.789
Al^{3+}	-	-	-	-
Mg^{2+}	0.111	0.243	0.276	0.309
Na^{+}	0.213	0.526	0.535	0.801
K^{+}	0.023	0.038	0.033	0.046
Cl^{-}	0.104	0.127	0.172	0.200
Si^{4+}	0.226	0.443	0.534	0.511

Chemical species, phases and selected parameters for NETPATH modelling are shown in Table 4. These are the same as in the last example, except for the inclusion of calcite, present in rock cement.

Table 4. Chemical species, mineral phases and parameters selected to simulate the evolution between rainwater and calcium bicarbonate waters from wells located in metamorphic rocks of Trás-os-Montes (Portugal).

Species	Phases	Parameters
C, Cl^-, Na^+, Ca^{2+} Mg^{2+}, K^+, Al^{3+}, Si^{4+}	Plagioclase (An_{38}), Biotite, Chlorite, Calcite, CO_2(g), Illite, Kaolinite, Silica, NaCl	Ion exchange (Ca/Na) Evaporation

Although the concentration of aluminium was unknown this element was included in inverse models. This is justified because aluminium is usually conserved in the weathering reactions and so the changes in the solutions are in general negligible when compared with mass tranfer among phases.

The four samples, Passos, Orelhão, Pereira and Carvalhal, have different degrees of mineralization (80, 182, 253 and 252 µS/cm respectively) but have undergone similar geochemical evolution. For that reason the same species, phases and parameters were used to model the evolution between rainwater and each of the samples.

The selected models obtained using NETPATH show that the most plausible reactions include:

- dissolution of CO_2, NaCl, Calcite, Plagioclase (An_{38}) and Biotite;
- precipitation of Kaolinite and Illite;
- cation exchange, between Ca^{2+} and Na.

The concentration factor corresponds to an infiltration rate varying between 15 % and 5% of annual precipitation.

5 FORWARD MODELS

The forward models were based on the inverse models obtained by using NETPATH and were performed using the PHREEQE program. They included:

1. Evaporation, with a concentration factor of 13 times
2. Dissolution of 0.922 mmol of plagioclase (An_{38}) and 0.212 mmol of biotite maintaning the equilibrium with kaolinite and goethite and allowing the dissolution of calcite (up the saturation index of -1.2). The CO_2 partial pressure considered was $10^{-2.1}$ atm. This reaction was added in six steps: 8%-20%-40%-60%-80%-100%.

Table 5 gives the results of these models for sodium, carbon and pH.

Table 5. Results of forward models proposed for the evolution from rainwater to groundwater in metamorphic rocks of Trás-os-Montes.

% of reaction added	Sodium (mmol)	Carbon (mmol)	pH
8%	0.242	1.563	6.98
20%	0.311	1.631	7.00
40%	0.426	1.760	7.04
60%	0.541	1.908	7.08
80%	0.656	2.077	7.13
100%	0.711	2.265	7.17

Figure 3 presents the carbon and sodium values for the water samples and the evolution line for the proposed geochemical models.

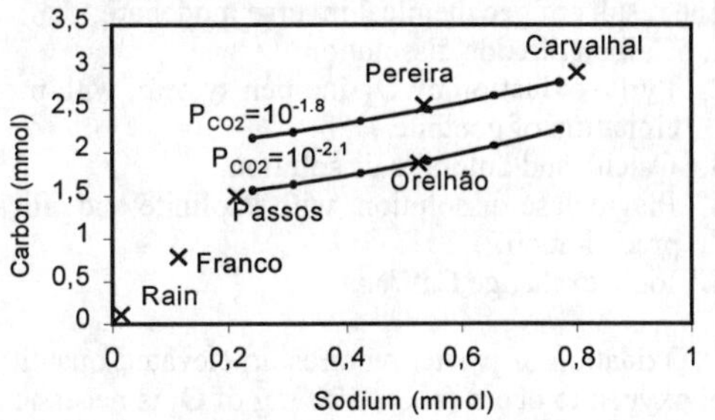

Figure 3. Geochemical evolution of groundwater associated with metamorphic rocks in Trás-os-Montes.
Trend lines represent the modelled water compositions and x symbols observed compositions.

Sodium chloride water from Franco spring has low values of sodium and carbon, since it is near the beginning of the evolution path. Orelhão groundwater results from evolution of Passos water.

In the case of Pereira and Carvalhal, the evolution process seems to be related to different infiltration conditions.

A soil with a higher content of organic matter could explain the more elevated carbon present in the water. Based on this assumption, the same model was run using a partial pressure of CO_2 of $10^{-1.8}$ atm.

The respective evolution line, present in Figure 3, shows that Carvalhal is the evolution of Pereira groundwater.

6 SULPHATE ENRICHED WATER SAMPLES

A few water samples are characterised by much higher mineralization (about 1000 μS/cm) due to special enrichment in DIC, SO_4^{2-} and Na^+ (Pereira & Almeida 1996).

Hydrochemically they are of sodium bicarbonate type and are all associated with sulphide mineralization present in quartz veins in the schists.

The principal chemical components of one selected sample (Freixeda) are compared with rainwater composition in Table 6.

Table 6. Chemical variation between rainwater and sodium bicarbonate groundwater of Freixeda in Trás-os-Montes (Portugal).

Element	Rain (mmol/L)	Freixeda well (mmol/L)	Variation (mmol/L)
C	0.110	5.90	5.79
Ca^{2+}	0.008	0.5	0.492
Al^{3+}	-	-	-
Mg^{2+}	0.002	0.68	0.678
Na^+	0.015	8.97	8.955
S^{6+}	0.01	2.2	2.19
Fe^{3+}	0.001	0.002	0.001
Si^{4+}	0	0.31	0.31

The results of geochemical inverse model are:

1. Carbon dioxide dissolution.
2. Pyrite oxidation by O_2, in open system, with precipitation of goethite.
3. Calcite and dolomite dissolution.
4. Plagioclase dissolution, with kaolinite and silica precipitation.
5. Ionic exchange Ca^{2+}/Na^+.

Oxidation of pyrite, requires an elevated quantity of oxygen to occur (about 120 mg of O_2 is necessary to oxidise 1 mmol of pyrite) which means that it should happen in an open system.

This reaction is responsible for the high content in sulphate and, as a proton donor, promotes the dissolution of silicates (with precipitation of kaolinite and silica) and carbonates.

All these reactions led to a mineralization that is very distinct from the normal compositions of the area.

To explain the elevated concentration of carbon in this water, isotopic analyses of ^{13}C and ^{14}C were conducted (Pereira & Almeida 1997). A value of 6.75 TU indicates modern water.

Carbon isotopes indicates a low ^{14}C content ($a^{14}C = 37.88$ pmc) and a source of heavy carbon ($\delta^{13}C = -8.49$ ‰). These values indicate source of older carbon without ^{14}C and enriched in ^{13}C, such as by dissolution of carbonate minerals.

The geochemical simulations give thermodynamically consistent results which compare well with real values, including isotopic data.

7 HYDOCHEMICAL FACIES VERSUS GEOCHEMICAL EVOLUTION

Based on geochemical modelling results and observed hydrochemical facies, different evolution stages were defined for groundwaters related to metamorphic rocks in NW Portugal:

A. SODIUM CHLORIDE WATERS:
Have a low degree of evolution and a shallow circulation.
Dissolution of very small quantities of plagioclase and biotite with precipitation of illite and kaolinite is consistent with the water chemistry.

B. CALCIUM BICARBONATE WATERS:
Correspond to a higher degree of geochemical evolution and a deeper circulation.
Hydrolysis of silicates and dissolution of carbonates could explain the composition and the higher mineralization.

C. SODIUM BICARBONATE WATERS:
These waters have a very high conductivity (1000 μS/cm) and are associated with oxidation of sulphide minerals in schists where some carbonate minerals are present in the rock cement.

8 CONCLUSIONS

The most frequent rock-water interactions in metamorphic rocks of Trás-os-Montes (Portugal) are the dissolution of plagioclase and biotite or chlorite, which induce the precipitation of clay minerals, usually kaolinite.

In some lithologies, carbonate minerals (calcite) are present as cementing material and, in these cases, the mineralization of water samples is higher and attributed to the easy dissolution of carbonates.

When sulphide minerals are present, associated with mineralised quartz veins, the oxidation of these minerals gives a high concentration of dissolved sulphate and the H^+ ions released during oxidation promote the dissolution of silicate minerals and carbonates.Under such hydrogeologic conditions, the waters have a high concentration of inorganic carbon, SO_4^{2-} and Na^+.

REFERENCES

Parkhurst et al. 1980. PHREEQE - a computer program for geochemical calculations. USGS *Water Resources Investigations Report*, 80-96.

Pereira, M. R. 1999. Hidrogeologia das rochas fracturadas da Terra Quente Transmontana. Tese de Doutoramento. Univ. de Trás-os-Montes e Alto Douro. 299p.

Pereira, M. R. & Almeida, C. 1994. Captação de Água Subterrânea em Rochas Cristalinas: Factores que Influenciam a Produtividade. Comunicações do II Congresso Nacional da Água. **II**: 95-106.

Pereira, M. R. & Almeida, C. 1996. Fácies Hidroquímicas influenciadas pela oxidação de sulfuretos. Exemplo da Freixeda (Mirandela). Actas do 3° Congresso da Água, Vol. III: 187:194.

Pereira, M. R. & Almeida, C. 1997. Evidências químicas e isotópicas da dissolução de carbonatos induzida pela oxidação de sulfuretos em rochas metamórficas. Secction Séptima Hidrogeoquímica das Actas do I Congresso Ibérico de Geoquímica e VII Congresso de Geoquímica de Espanha, Sória: 434-441.

Plummer et al. 1992. NETPATH: An interactive code for interpreting NET geochemical reactions from chemical and isotopic data along a flow path. In Kharaka & Maest (eds) Water Rock Interaction: 239-242. Rotterdam: Balkema.

Groundwater: Past Achievements and Future Challenges, Sililo et al. (eds) © 2000 Balkema, Rotterdam, ISBN 90 5809 159 7

Rapid water quality changes in a thick clay sequence

B.A.Robinson & J.H.Tellam
School of Earth Sciences, University of Birmingham, UK

J.B.Joseph
RfS, Reading, UK

A.W.Herbert
Environmental Simulations International, Shrewsbury, UK

ABSTRACT: A number of landfill sites in south-east Britain, are located on the Jurassic Oxford Clay which provides natural containment over a thin (approximately 3-4m) low permeability formation, the Kellaways Sand. Monitoring piezometers in the Kellaways Sand show surprisingly rapid water quality changes of up to an order of magnitude within a month time period. Four main hypotheses have been proposed for these variations: 1) Chemical heterogeneity 2) Fracture flow 3) Sampling method 4) Drilling damage. Exploration of the nature of variations in groundwater chemistry with time, porewater hydrochemistry analyses and computer modeling studies have suggested that the variations are probably the product of oxygen ingress into the system due to (i) dewatering of the Kellaways Sand during sampling and (ii) flow through drilling induced fractures. Research is still in progress.

1 INTRODUCTION

A number of landfill sites in the Bedford to Buckingham area of south-east Britain, are located on the Jurassic Oxford Clay. The Oxford Clay provides natural containment over a thin (approximately 3-4m) low permeability formation, the Kellaways Sand, whose groundwater flux is very low. Monitoring piezometers in the Kellaways Sand could be expected to indicate relatively stable solute distributions. This, however, is not the case and concentration variations in the order of 100% are very common, especially in Na, Ca, Cl and SO_4.

The aim of the investigation described here is to develop a conceptual model to explain the observed groundwater solute distributions in the Oxford Clay/Kellaways Sand sequence in the Bedford/Buckingham sector of the Oxford Clay outcrop. A detailed study of the nature of groundwater quality variation with time for piezometer samples is outlined in the context of a study of porewater hydrochemistry profiles, physical flow parameters and computer modeling studies.

2 GEOLOGICAL SETTING

The research area is located in the Bedford to Buckingham sector of the Oxford Clay outcrop in the centre of a band of Jurassic sediments which stretches from north-east to south-west of Britain. The general geology (as shown in Table 1) is composed of a sequence of clays, sandy clays and limestones dipping at a few degrees towards the south-east (described in more detail by Callomon (1968) and Barron et al. (1992)).

Table 1. General geological succession in the study area.

Lithology	Description	Thickness
Oxford Clay (Peterborough Member)	Variably silty mudstone	>25m
Kellaways Sand	Hetrogeneous sandy, silty clay	3m
Kellaways Clay	Fissile mudstone	2m
Cornbrash	Hard micritic limestone	5m
Blisworth Clay	Calcareous muds and silty clays	3m
Blisworth Limestone	Detrital limestone	12m

The oldest rock exposed locally is the Middle Jurassic detrital limestone of the Blisworth Formation. This is overlain by up to 3m of the calcareous muds and silty clays of the Blisworth Clay and 1.5m of hard micritic shelly limestone of the Cornbrash. Next comes the Kellaways Formation, comprising 2m of the Kellaways Clay, a fissile mudstone, and the heterogeneous sandy, silty clay of the Kellaways Sand which averages 3m in thickness. Above the Kellaways Sand is the Oxford Clay Formation. The base of the Oxford Clay consists of a lower bituminous, variably silty mudstone unit, the Peterborough Member, which is up to 25m thick. This is overlain in places by the more homogeneous and plastic Stewartby Member and glacial deposits comprising valley gravels and reworked clays.

3 VARIATION IN CHEMISTRY

3.1 *Piezometer design*

A large number of piezometers have been set up around landfill sites in the Oxford Clay to monitor groundwater quality within the Kellaways Sand. Historical monitoring data are available for up to 10 years from piezometers in the Kellaways Sand in the sudy area. The piezometers are all completed with a HDPE cased section down through the Oxford Clay to the top of the Kellaways Sand and are gravel packed and installed with slotted casing (Fig. 1). Most of the piezometers have dedicated pumps located either within or just above the screened section.

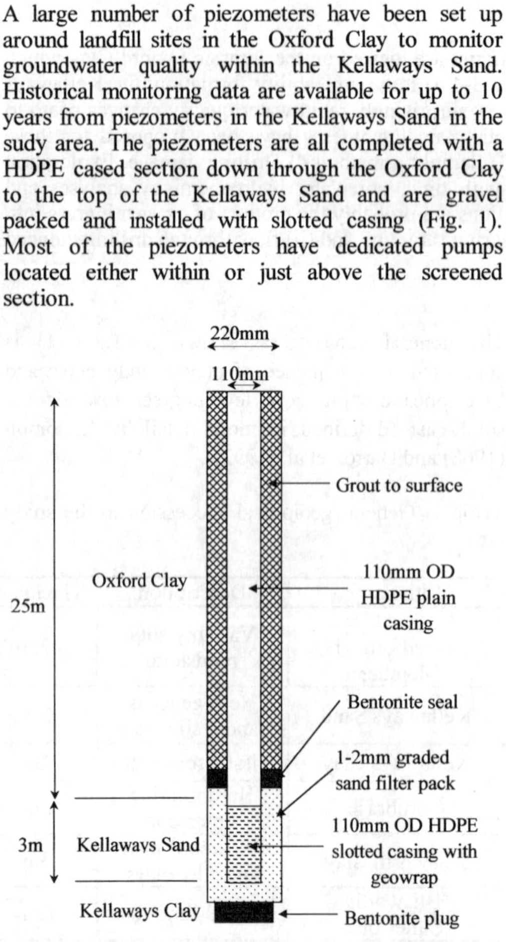

Figure 1. Piezometer completion details with typical formation thicknesses.

3.2 *Permeability data and flow rates*

The average linear groundwater velocity in the Kellaways Sand is about 0.09 ma^{-1} (Joseph et al. 1998). Reported hydraulic conductivity (K) values measured in the laboratory range from 10^{-9} to 10^{-11} ms^{-1} for the Oxford Clay and 10^{-6} to 10^{-9} ms^{-1} for the Kellaways Sand (Halliday 1997, Klinck 1992, Klinck & Noy 1992). Field K values are on average at least one order of magnitude higher than laboratory results (Halliday 1997) suggesting that fractures or sedimentary structures may be important. Smearing may have biased the results, however, and the lab/field differential could, in fact, be greater than an order of magnitude.

3.3 *Sampling protocol*

The standard sampling procedure involves pumping the piezometer, using the dedicated pump installed just above the screened section, until it runs dry. The piezometer is then allowed to recover and a sample taken from the pump level. If it is not possible to pump the piezometer dry, pumping is continued for 30 minutes before a sample is collected.

3.4 *Data*

The data indicate rapid and large variations in groundwater quality on a time scale shorter than a month, coupled with apparent increased concentrations around the landfills. The ions most affected are chloride, sulphate, sodium and calcium (Fig. 2).

Spatial evaluation of the data shows no clearly distinguishable plumes, and comparison of recent sample analyses and those taken at the turn of the previous century indicates that there has been no significant change in water quality over the period (Woodward & Thompson 1909). It is therefore suggested that there is no indication of leachate migration.

This is confirmed by low levels of BOD in the groundwater and studies of stable isotope ratios using chloride (Loomis et al. 1998).

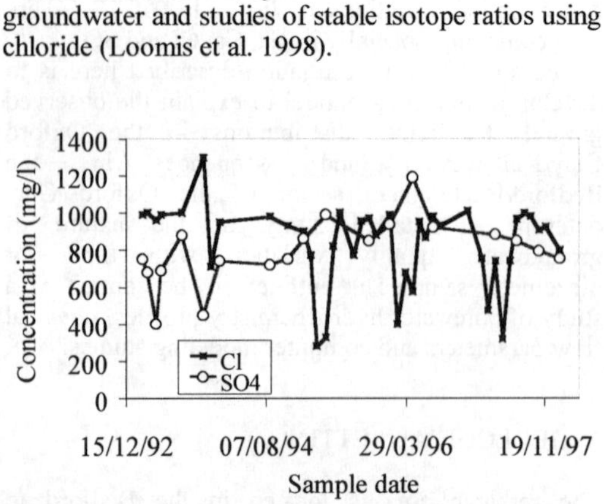

Figure 2. Chloride and sulphate borehole sampling profiles taken over a six year period for a typical Kellaways Sand monitoring borehole, Brogborough, Marston Vale, UK.

4 HYPOTHESES

Four main potential explanations for the observed water chemistry changes have been explored to date.

- Is the observed chemical variability a product of drawing water from different sections of the formation and therefore, related to small-scale chemical heterogeneity in the system (Section 5)?
- Is fracture bypass flow responsible for changes in chemistry as it could allow faster than expected solute transit times through the system (Section 6)?
- Are variations in sampling depth, method and duration responsible for the observed chemical variations (Section 7)?
- Is drilling-induced fracturing responsible for the observed chemical variability by facilitating oxidation reactions (Section 8)?

5 CHEMICAL HETEROGENEITY

5.1 *Porewater analyses*

In an effort to provide information on matrix water chemistry and also to establish the feasibility of the chemical oxidation hypothesis (as developed from the observed chemical variability data) an investigation of porewater hydrochemistry has been undertaken.

Large quantities of existing Oxford Clay and Kellaways Sand core, collected from sites across the area and a limited number of purposely obtained fresh cores were used in this investigation. The fresh cores were sealed into core liners at ground level immediately on retrieval. The samples analysed were all taken from the centre of core sections (usually one metre in length) to eliminate, as far as possible, the effects of water or air ingress into the core during drilling or storage.

Conventional methods of porewater extraction using high speed centrifuges and vacuum equipment have been shown to be inappropriate for use on the Oxford Clay (Halliday 1997). Instead a procedure was developed to measure the porewater solutes indirectly. The rock is first crushed until it will pass a 100μm sieve. 25g of this clay is then added to 50ml of distilled water and shaken continuously for twelve hours. The mixture is then centrifuged, filtered and analysed. A further set of experiments was conducted on the fresh core in a nitrogen atmosphere to establish the effect of pyrite oxidation processes on the porewater chemistry. The experiments indicate that Cl concentrations are unaffected by degree of oxidation of core prior to porewater extraction.

5.2 *Porewater data*

Current chloride concentration results, corrected to take original moisture content into account, are in the range of 200-2000mg.l^{-1}. This is similar to the concentrations seen in piezometers in the Bedford to Buckinghamshire sector of the Oxford Clay outcrop. The results indicate an increase in chloride with depth in the Oxford Clay with a sharp decrease at the top of the Kellaways Sand whilst sulphate levels drop steadily after an initial high near the surface and then suddenly increase in the Kellaways Sand (Fig. 3).

The consistency of these profiles from cores from several boreholes suggests that chemical heterogeneity on a micro scale does not exist within the Oxford Clay. Macro scale variations through the sequence, however, may well be very important. These data possibly suggest the existence of a $CaSO_4$ rich water in the Callow and Kellaways Sand and a NaCl rich water in the main body of the Oxford Clay (see Section 7.2).

The chloride data indicate that the Oxford Clay has undergone extensive flushing over the last 150Ma, whilst the sulphate levels are probably the product of relatively recent oxidation reactions

The original marine Jurassic connate waters are thought to have been diluted from a chloride concentration in the region of 20,000mg.l^{-1}. This view is supported by previous porewater studies conducted by Ross et al. (1989). Chloride concentrations in the Oxford Clay Formation remain relatively high, however. Even if it is assumed that flushing did not

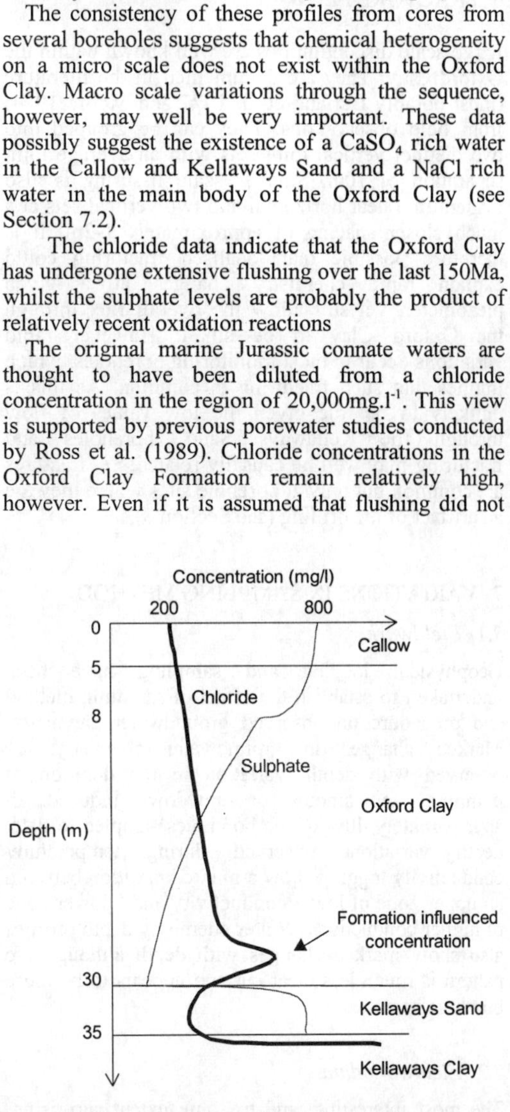

Figure 3. Average porewater chloride concentration depth profiles for three boreholes at Calvert and L'Field Landfill Sites in the Marston Vale, England.

start until exposure of the Oxford Clay at the end of the last glaciation (ca 10,000 years ago) this still implies a flow rate of only $7x10^{-4}$ $m.a^{-1}$. This fits well with the vertical flow estimate of $2x10^{-4}m.a^{-1}$ used by Klinck and Noy (1992) in their groundwater flow model for the area. This flow rate could explain the chloride profile observed on the basis of flushing by infiltrating rainwater over a period of around 5000 years.

6 FRACTURE FLOW

Large-scale discontinuities are well known within the Oxford Clay. They are documented in the literature (most notably Burland et al. 1977 and Moore 1974) from field observations. They can be grouped into two distinct vertical joint sets with an average joint separation of 10m. Smaller scale fissuring is also evident in a near horizontal and two vertical sets at a much closer spacing of approximately 1-2m. It is therefore possible that localised fracturing could explain rapid chemistry changes at a given piezometer, yet still allow the overall flux through the Oxford Clay to be small. However rapid variations occur in most monitoring boreholes, which implies that such fracturing is common. As this is unlikely to be true given the low values of flow through the Kellaways Sand, boreholes and fracturing may well be causally related, i.e. fractures are common but only at borehole sites where they are an artifact of the drilling (see Section 8).

7 VARIATIONS IN SAMPLING METHOD

7.1 *Fieldwork*

Geophysical logging and sampling have been undertaken to establish the effect of sampling method and procedure on observed groundwater chemistry. Marked changes in groundwater chemistry are observed with depth, variation in the duration of pumping, and amount of drawdown induced. In approximately 40% of the boreholes sampled to date, depth variations observed during temperature/ conductivity logging show a marked partition between an upper zone of lower conductivity and a lower zone of higher conductivity. Water chemistry depth profiles also show marked changes with depth although the pattern is much less clear cut and appears to be more borehole specific.

7.2 *Fieldwork data*

The most interesting, and to some extent surprising, changes in chemistry occur between samples taken over a very short time span of tens of minutes with an average pumping rate of $1l.s^{-1}$ (Fig. 4).

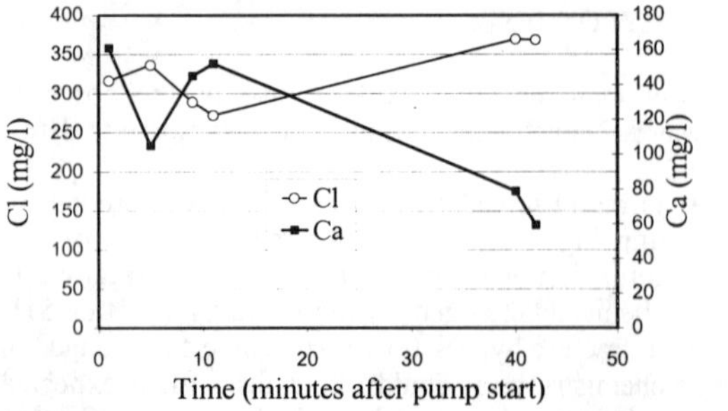

Figure 4. Change in groundwater chemistry with varying pumping time in a Kellaways Sand monitoring borehole in the Marston Vale, UK.

These data lead to some important observations:

- Large variations occur in Na, Ca, Cl and SO_4 concentrations, but there are also small changes in other species including NO_3 and NH_4.
- The observed changes in chemistry are insufficient to account for the magnitude of the change seen in the monthly sampling data.
- NO_3 is present in approximately 10% of the boreholes sampled to date suggesting some rapid water movement from the near surface.
- Carbonate saturation indices are greater than zero, which may suggest active degassing or mixing.
- Mass balance calculations suggest that the redox system is far from equilibrium, which may also suggest recent mixing.
- Na and Cl concentrations vary inversely with Ca and SO_4, possibly due to low NaCl oxidised water entering the system (this could also be related to pyrite oxidation and anion exchange (Loomis et al. 1998) or the presence of two waters within the system.
- The Na/Cl ratio increases with decreasing Cl, possibly due to greater Na exchange as more fresh water flushes through the system.

It is possible that oxidation reactions may explain the changes in SO_4 concentrations. Oxygen could enter the system if the water within the borehole is drawn down below the screened section or when localised drilling induced fractures are dewatered during sampling. In the latter case a connection could be made from the Callow (the weathered top five metres of the Oxford Clay) through the Oxford Clay and into the Kellaways Sand allowing a pathway for oxidation products from the Callow to be transported to the monitoring well screen. The borehole purging volumes are likely to be large in comparison to fracture storage, so the transfer could well be rapid.

Coupling together the Na, Ca, Cl and SO_4 variations observed during fieldwork studies and the porewater profiles referred to in Section 5.2 and Figure 3, it is also possible that observed chemical changes could be due to the presence and mixing of two different waters within the system. These would consist of a $CaSO_4$ rich water in the Callow and Kellaways Sand and a NaCl rich water in the main body of the Oxford Clay.

7.3 *Modelling*

A series of 2D and 3D computer modelling studies using NAMMU (Hartley & Jackson, 1994), a finite element code, are in hand. The models are designed to clarify issues such as local drainage within the geological succession when a borehole is pumped.

At present this work suggests that to cause an appreciable change in the piezometric surface in the Oxford Clay, it is first necessary to dewater a section of Kellaways Sand and let water from the Oxford Clay seep slowly down into it. It would appear that during sampling an unsaturated wedge develops in the Kellaways Sand. This allows passage of entrained oxygen into the Kellaways Sand. This would result in pyrite oxidation and might be a contributing factor in leading to rapid changes in the concentration of sulphate and other oxygen affected species during sampling.

8 DRILLING DAMAGE

There is some evidence for drilling-induced fracturing with the system. Boreholes are drilled in pairs throughout the area to monitor water quality separately in the Kellaways Sand and Blisworth Limestone. In at least 5% of cases there is evidence of connection between boreholes.

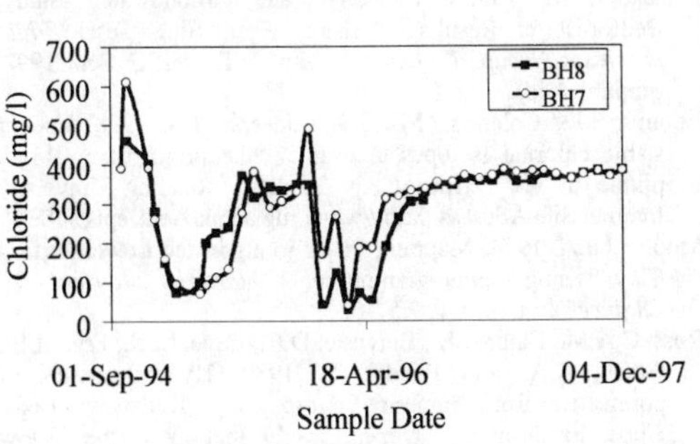

Figure 5. Chloride profiles for a borehole pair at Bletchley, Milton Keynes, UK.

Figure 5 shows the evidence of hydraulic connection between a Blisworth Limestone borehole (BH7) and a Kellaways Sand monitoring hole (BH8) situated 8m apart at the Bletchley site near Milton Keynes.

Most sites show no evidence of connection between the boreholes in each pair because they have been drilled at some distance purposely to avoid the problem.

9 DISCUSSION

Drawing together the various, rather disparate, data discussed above the following points can be made concerning the Oxford Clay/Kellaways Sand system in south-east England:

- Pronounced variations, typically of up to 100%, in borehole water chemical concentrations are observed in data from monthly sampling.
- Major ions affected are Cl, SO_4, Na and Ca. Ca and SO_4 concentrations are inversely related to Na and Cl, which may indicate periodic influx of low concentration oxidised water.
- Spatial distribution, chemical and stable isotope data show that the variations in chemistry are not related to landfill leachate migration.
- Porewater chemistry studies, calculation of times for diffusive equilibrium between fractures and matrix water chemistry, and hydraulic property data indicate that the observed erratic behaviour must be connected with the presence of the boreholes rather than a phenomenon of natural groundwater behaviour in the Oxford Clay.

10 CONCLUSION

10.1 *Hypothesis one: Chemical heterogeneity*

Porewater studies suggest that small scale chemical heterogeneity within the Oxford Clay is unlikely, so the observed chemical variations are not related to different flow rates or paths through this formation. Macro scale heterogeneity between the Oxford Clay and the Kellaways Sand, however, may well be a contributory factor. This hypothesis rests on the fact that two different waters may exist within the system; a $CaSO_4$ rich water within the Callow and Kellaways Sand and a NaCl rich water within the main body of the Oxford Clay. Varying mixtures of these constituent waters could possibly, therefore, explain the observed chemical changes within the piezometers. This is suggested by porewater studies and sampling in the field but still requires further clarification.

10.2 *Hypothesis two: Natural fracture flow*

Abstraction during sampling could cause rapid movement of porewater from near surface through natural sparse and narrow aperture fractures. This is numerically possible. This, however, is thought to be unlikely because whilst discontinuities undoubtedly exist within the system it is difficult to reconcile the general low permeability of the sequence with the fact that chemical variability is observed at most boreholes.

10.3 *Hypothesis three: Sampling dewatering of the Kellaways Sand*

Abstraction during drilling could dewater the Kellaways Sand partially, inducing an influx of oxygen. This is suggested by numerical modeling. The variations encountered during sampling fieldwork, however, suggest that variations attributable to this hypothesis alone are insufficient to account for the observed changes. This hypothesis would also require some further explanation for the observed chloride variations.

10.4 *Hypothesis four: Drilling induced fracture flow*

Abstraction during sampling could cause movement of fresh, initially oxygenated, water from near surface through drilling induced fractures around each borehole. These would link the Kellaways Sand with the shallow weathered zone at the top of the Oxford Clay. This is suggested by evidence of connection between closely (<8m) spaced piezometers drilled to different depths.

It should be borne in mind that the proposed mechanisms are not necessarily mutually exclusive. At present it is proposed that the variation is a product of sampling dewatering (hypothesis 3) and drilling induced fracturing (hypothesis 4) and is possibly heavily influenced by the existence of two waters within the system (hypothesis 1), although this still needs further clarification.

11 FURTHER WORK

Work is still in progress to confirm the conclusions presented here and to try to determine which of the proposed mechanisms are dominant. At present this is focused largely on the fieldwork aspects of the inquiry and the numerical modeling of these data. The results of this work will have important implications for the design of sampling and drilling protocols, and possibly for the understanding of flow in apparently "impermeable" formations.

ACKNOWLEDGEMENTS

Shanks Group (previously Shanks and McEwan) have provided important background data, samples and site access, without which this work could not have been done. The authors acknowledge with gratitude all of those provisions, together with substantial assistance in funding and permission to publish this paper. The views expressed in the paper are the authors' own and do not necessarily represent either the views or the policy of the Shanks Group. The project, originally funded by Shanks and McEwan and the Natural Environmental Research Council is now funded by Research for Sustainability.

REFERENCES

Barron, A.J.M., Horton, A. and Sen, M.A. 1992. Geology of the Marston Vale. *BGS Technical Report* WA/92/23C.

Burland, J.B., Longworth, T.I. and Moore, J.F.A. 1977. A study of ground movement and progressive failure caused by a deep excavation in Oxford Clay. *Geotechnique*. 27(4):557-591.

Callomon, J.H. 1968. The Kellaways Beds and the Oxford Clay. 264-290. In: Sylvester Bradley, P.C. and Ford, T.D. (eds). 1968. The Geology of the East Midlands. *Leicester University Press*.

Halliday, D. 1997. The hydrogeology of the beds underlying the Oxford Clay and the impact of excavation and landfilling. PhD thesis, Royal Holloway, University of London (unpublished).

Hartley, L.J. and Jackson, C.P. 1994. NAMMU (Release 6.2) User Guide. AEA Technology report, Hydrogeology Department, Harwell Laboratory, Oxfordshire, UK.

Joseph, J.B., Halliday, D. and Mather, J.D. 1998. Balanced assessment; relating groundwater protection to the real level of risk. In: Groundwater – Physical, Chemical and Biological Aspects of Aquifer-Stream Relations. 207-211. Ed. Geological Society London.

Klinck, B.A. and Noy, D.J. 1992. A Groundwater Flow Model for the Marston Vale. *BGS Technical Report*. WE/92/21C. Oct 1992. (unpublished).

Klinck, B.A. 1992. Marston Vale Groundwater Study, Bedfordshire: Results of rising Head Slug Tests. *Fluid processes Group. Technical Report*. WE/92/19C. Sept 1992. (unpublished).

Loomis, J.L., Coleman, M.L., and Joseph, J.B. 1998. Use of stable chlorine isotopes to evaluate the origin of a Cl-rich plume in the Oxford Clay, England. Reading University Internet Site Address: http://www.rdg.ac.uk/AcaDepts/sl/Sed

Moore, J.F.A. 1974. Mapping major joints in the Lower Oxford Clay using photogrammetry. *Quarterly Journal of Engineering Geology*. 7:57-67.

Ross, C.A.M., Bath, A.H., Entwisle, D.C., Cave, M.R., Fry, M.B., Green, K.A. and Reeder, S. 1989. Hydrochemistry of porewaters from Jurassic Oxford Clay, Kellaways Beds, Upper Estuarine and Upper Lias Formations at the Elstow site, Bedfordshire. *BGS Technical Report*. WE/89/28. May 1989. (unpublished).

Woodward, H.B. and Thompson, B. 1909. The Water Supply of Bedfordshire and Northamptonshire, from underground sources: with records of sinkings and borings. *Mem. Geol. Surv. Eng. Wales. HMSO*.

Groundwater: Past Achievements and Future Challenges, Sililo et al. (eds) © *2000 Balkema, Rotterdam, ISBN 90 5809 159 7*

Example of O- and H-isotope use to identify surface water pollution in groundwater

I.C.Saayman
Council for Scientific and Industrial Research, Environmentek, Stellenbosch, South Africa

S.Adams
Earth Science Department, University of the Western Cape, Bellville, South Africa

C.Harris
University of Cape Town, Rondebosch, South Africa

ABSTRACT: Varying levels of pollution were observed in water samples taken from observation boreholes that surround a waste disposal site and a waste water treatment works. In an attempt to establish which of the sources was responsible for groundwater pollution, water samples were taken and analyzed for the isotopes ^{18}O and deuterium. The O- and H- isotopes were used successfully to define the level of influence recharge from surface pollution sources had on observation boreholes. From these results conclusions could be made on the relative impact of one of the two possible sources of pollution. This allows for decisions on the management of the facility to focus on elimination of the surface water body as a source of pollution.

1 INTRODUCTION

The Cape Flats phreatic aquifer represents an important potential source of water for the Cape Town Metropolitan area. Unfortunately urban development has taken place over many parts of the aquifer. Waste disposal sites in particular have the potential to introduce pollutants into the environment. The literature reports on a number of instances of groundwater pollution within the Cape Flats aquifer (e.g. Tredoux 1984, Weaver & Tworeck 1988.). In 1998 a study was initiated to establish whether leaching from a waste disposal site was causing pollution of groundwater. The results of the study were however complicated by the presence of other potential sources of pollution (Saayman 1999).

The Cape Flats aquifer consists of Cenozoic deposits underlain by essentially impervious Malmesbury Shales and Cape Granite. Sedimentation of these deposits initially occurred in a shallow marine environment, subsequently progressing to intermediate beach and windblown deposition and finally to Aeolian and marsh (peat) conditions. A feature of the sediments is the presence of shelly material. The deposits cover an area of approximately 630 km^2 and extends in a northerly direction along the West Coast. The sand body is generally stratified horizontally, with several lithostratigraphic units. The sand layer in the study area has an average thickness of about 25 meters. Transmissivity values for the aquifer range between 50 and 650 m^2/day. (Wright & Conrad 1995). Transmisivity in the area of the study site vary between 50 and 100 m^2/day (Parsons, pers. comm.). The observation boreholes that surround the site penetrate aquifer to a depth of 15 meters. One borehole (Bh 17) penetrate to a depth of 30 meters.

Particularly high concentrations of potassium, sulphate and orthophosphate, with elevated concentrations of other ions and heavy metals (magnesium, chloride, COD, electrical conductivity, nickel and lead), indicate that groundwater pollution had taken place (Saayman 1999). Although the initial study focused on the possibility of pollution emanating from the nearby waste disposal site, interpretation of the results required the influence of a nearby wastewater treatment works on groundwater quality to also be considered.

2 THE STUDY SITE

The study site has served as a place of refuse and sewage disposal since the 1930's. Presently only domestic waste and non-toxic refuse from light industries are accepted at the waste disposal site. Present management at the site is guided by the standards outlined by the Department of Water Affairs and Forestry (DWAF 1994) on the disposal of domestic waste. A number of observation boreholes surround the site, six of which were used during the isotope sampling.

Located to the north and east of the waste disposal site is a waste water treatment works. The waste water treatment works have been operational

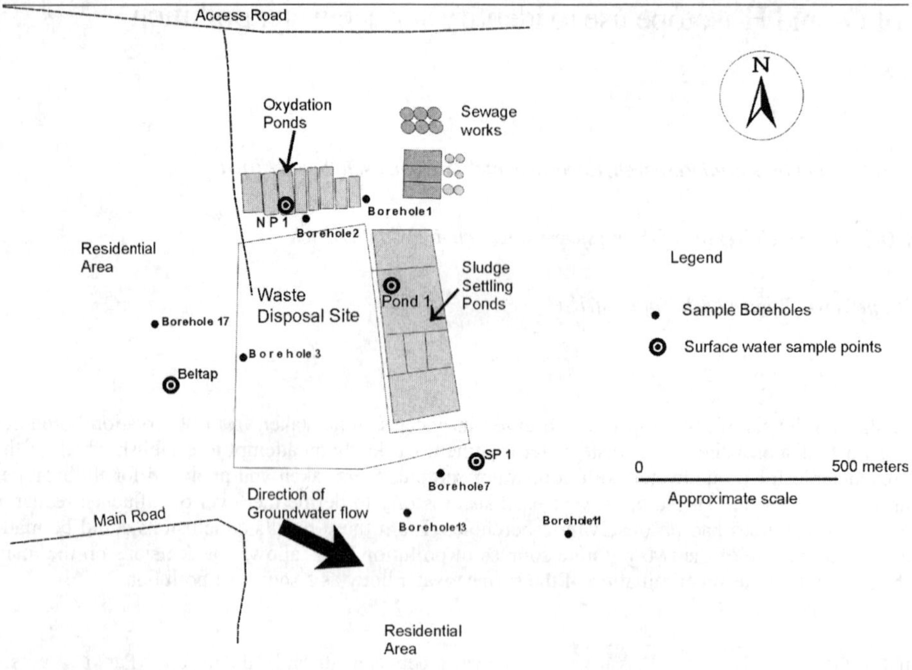

Figure 1. Layout of study site, with the location of the observation boreholes and the direction of groundwater flow.

since 1951. Waste water from residential and industrial premises in the surrounding area is treated at the site. The treatment process includes aeration and aerobic and anaerobic digestion, with separation of the resultant sludge and water. Water is channelled to a system of oxidation dams, while the sludge is channelled to sludge settling ponds (Fig. 1).

The study site is located on the Cape Flats aquifer at an altitude of about 60 meters above mean sea level. The surface has a gentle slope towards the south-east. To the south, immediately adjacent to the site, remnants of the Cape Flats dune landscape remain. Water accumulates in topographic lows during the rainy season. To the east of the site a large wetland is found.

3 THEORY AND METHOD

Both the waste disposal site and waste water treatment works are recognized as potential sources of pollution. Both are unlined and suspected of recharging the underlying aquifer. It was hoped that the O- and H-isotope analyses would yield an indication of which boreholes within the study area are affected by recharge from the sludge ponds. This would give an indication of the impact each source of pollution has on the environment and of contaminants introduced into the environment by each.

Cape Town receives most of its water supply from large reservoirs, resulting in the reticulated supply having a typical evaporated isotopic signatures. Harris, et al. (1999) found that on average a shift of 0.6 ‰ occurs in O^{18} content between treated and untreated water during summer months. This enrichment in the heavier O isotope content is almost certainly the result of evaporation. This water is ultimately disposed in waste water treatment works. During the waste water treatment cycle the water is further exposed to evaporation, resulting in further enrichment of heavy isotopes. Water recharged to the Cape Flats aquifer from waste water treatment works thus carries a distinctive signature of evaporation. The isotopic composition of mains water have been found to show a difference of up to 9 ‰ H^2 and 1.6 ‰ O^{18} from that of groundwater (Harris, et al., 1999).

During the study, samples were taken from six boreholes, from a domestic water supply tap (Beltap), from one of the oxidation ponds north of the landfill (NP 1), from the sludge settling ponds east of the landfill (Pond 1), and from a surface pond south of the landfill (SP 1).The position of

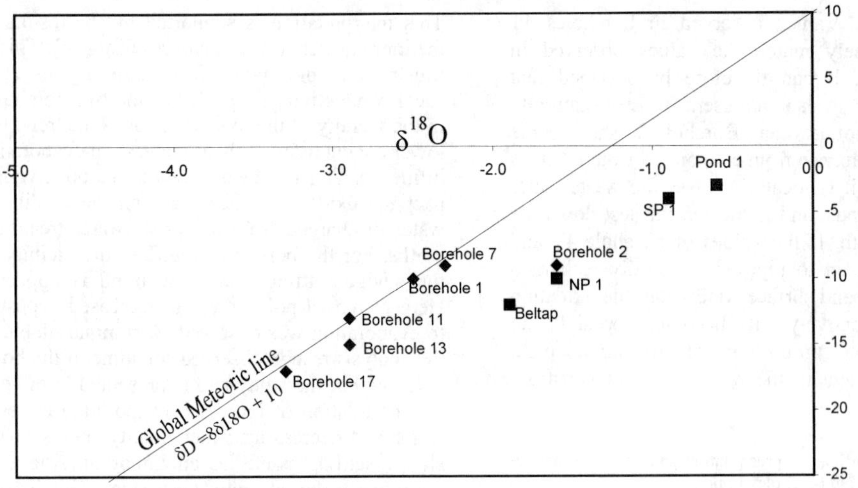

Figure 2. Results of isotope analysis plotted with the Global meteoric line.

the sampling points is shown in Figure 1. The samples were collected in 500 ml glass bottles, and were sent to the isotope laboratories at the University of Cape Town for analysis. Here the CO_2 equilibrium method was used in determining the oxygen isotope ratios, while a variant of the Zn closed tube reduction method was used for H.

4 RESULTS

The results of the analysis (Table 1) are plotted in Figure 2. The observed $\delta^{18}O$ value for borehole 17 is similar to that observed in precipitation on the central Cape Flats. The observed long term average $\delta^{18}O$ values for Cape Town International airport is $\delta^{18}O = -2.7$ ‰, with the average for 1996 being $\delta^{18}O = -3.35$ ‰ (IAEA, 1997). This corresponds well with the values observed by Walton, et al. (1993) for Pretoria. The observed weighted long-term average for Pretoria is $\delta^{18}O = -3.7$ ‰. The isotopic content of water introduced to the Cape Town reticulation system is $\delta^{18}O = -2.0$ ‰ (Harris, et al. 1999). This is very similar to that observed in the sample taken from the local reticulation system (Beltap - $\delta^{18}O = -1.9$ ‰).

When plotting the results, trends in the degree of exposure of the water to evaporation are observed for both groundwater and surface water samples:

Groundwater:

Increased exposure to evaporation 11 →

→ Borehole 1 → Borehole 7 → Borehole 2

Table 1. Results of the isotope analysis (as per mil).

Sample ID	$\delta^{18}O$	$\delta^{2}D$	Water sampled
Borehole 1	-2.5	-10	Groundwater
Borehole 2	-1.6	-9	Groundwater
Borehole 7	-2.3	-9	Groundwater
Borehole 11	-2.9	-13	Groundwater
Borehole 13	-2.9	-15	Groundwater
Borehole 17	-3.3	-17	Groundwater
Beltap	-1.9	-12	Reticulation system
SP 1	-0.9	-4	Surface water body
NP 1	-1.6	-10	Waste water oxidation pond
Pond 1	-0.6	-3	Sludge settling pond

Surface water

Increased exposure to evaporation →

Beltap → NP 1 → SP1 → Pond1

From the plotted results it can be seen that the surface water samples show the clearest signature of evaporation. The groundwater samples appear more a mixture of precipitation and evaporation exposed water.

Based on historical precipitation data (IAEA, 1997) and the location of borehole 17 up-gradient of sources of surface water recharge (Fig. 1) it is proposed that the sample taken from borehole 17 represents mostly precipitation recharge of groundwater. Values observed in boreholes 11 and 13 closely match the values observed in borehole 17. It can therefore be assumed that these samples also represent mostly naturally recharged groundwater. Borehole 2 show some degree of influence from evaporated water. This is expected as it is located next to the waste water oxidation ponds, and is located furthest down the treatment path. If the values of borehole 17 and NP 1 are taken to represent naturally recharged groundwater and surface water from the oxidation ponds respectively, it becomes possible to calculate the proportion of oxidation pond recharge present in the water sample taken from borehole 2.

Table 2. Proportions of precipitation and surface water recharge observed in boreholes 1 and 2.

	Precipitation Recharge	Oxidation pond recharge
Borehole 1	81%	19%
Borehole 2	33%	66%

A similar procedure was followed in assessing the proportion of surface water recharge to boreholes 7, 11 and 13 (Table 3). Precipitation recharge is taken to be represented by the $\delta^{18}O$ ratio of borehole 17, while surface recharge that would influence these boreholes are represented by the $\delta^{18}O$ ratio of Pond 1. The results presented in tables 2 and 3 shows that boreholes 2 and 7 are most strongly influenced by recharge from the water bodies associated with the waste water works.

Table 3. Proportions of precipitation and surface water recharge observed in boreholes 7, 11 and 13.

	Precipitation Recharge	Sludge Ponds Recharge
Borehole 7	59%	41%
Borehole 11	74%	26%
Borehole 13	74%	26%

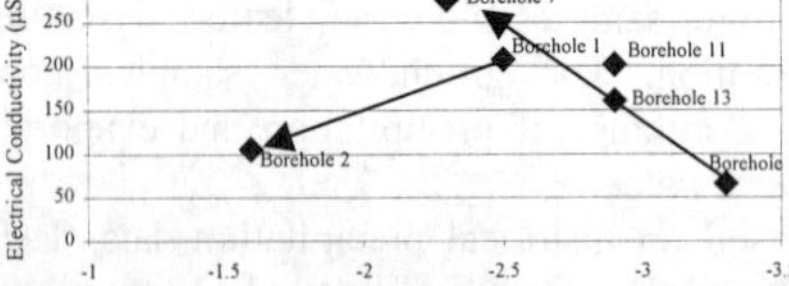

Figure 3. The observed correlation between $\delta^{18}O$ values and electrical conductivity in the observation boreholes.

This interpretation is supported by the results of the inorganic chemistry analyses (Table 4). This is illustrated by plotting the $\delta^{18}O$ values against electrical conductivity (Fig. 3). For the two boreholes in the vicinity of the oxidation ponds the trend is a lower conductivity with an increase in evaporative influence. This can be ascribed to the positive impact the oxidation process has on the quality of water discharged from the waste water treatment works. For the boreholes found in the vicinity of the sludge settling ponds the trend is opposite. Here increased pollution with increased exposure to evaporation was observed. A summary table of the analysis results observed for some of the boreholes within the study site is presented in table 4. The correlation of increased evaporation exposed water and decreasing water quality points to the sludge settling ponds as an important source of pollution to the groundwater system. This has serious implications for the management of the two facilities, as it points to pollution emanating from the sludge ponds and the oxidation ponds.

Table 4. Analytical results for some of the boreholes within the study site.

	Bh 02	Bh 07	Bh 11	Bh 17
pH	7.8	7.4	7.5	7.8
Conductivity @ 25°C mS/m	106	276	202	66
Chloride Cl mg/L	105	125	215	24
Sulphate SO4 mg/L	34	<1	118	52
Alkalinity CaCO3 mg/L	335	1314	616	245
Sodium Na mg/L	92	125	160	36
Potassium K mg/L	6	55	10	26
Calcium Ca mg/L	80	156	167	51
Magnesium Mg mg/L	11	34	34	10
Zinc Zn µg/L	38	52	134	42
Cadmium Cd µg/L	<5	6	5	<5
Nickel Ni µg/L	33	33	<25	<25
Lead Pb µg/L	70	106	74	<50

5 CONCLUSIONS AND DISCUSSION

Calculating the proportional contribution of evaporation exposed recharge to groundwater showed boreholes 2 and 7 to be the most strongly influenced by recharge from surface water bodies. The location of these boreholes, and the known direction of groundwater flow points to water bodies associated with the waste water treatment works as the source of recharge. Borehole 7 shows the highest levels of pollution within the study site. A trend was observed in the proportion of evaporated water within samples and its level of pollution. This indicates that the sludge settling ponds are making a considerable contribution to the high level of pollution observed in borehole 7.

Similarly the oxidation ponds are contributing to the pollutants observed in borehole 2. Based on these result it was recommended that use of the sludge settling ponds and the oxidation ponds be re-evaluated.

Results of this study made it possible to determine the level of influence of recharge from surface water bodies to groundwater. This allowed for a more complete interpretation of water quality of boreholes that surround the waste disposal site and waste water treatment works.

Waste disposal sites and waste water treatment works are often found in close proximity to each other. This practice often makes it difficult to establish which of these are impacting on groundwater quality. Through the use of O- and H- isotopes it is possible ascribe the influence of waste water recharge on groundwater samples.

It is foreseen that such output would assist in the management of sources of pollution, and in the rehabilitation of polluted groundwater. Waste disposal sites and waste water treatment works introduce different pollutants into the environment, sometimes requiring different measures in rehabilitation and treatment. In applying the strategy followed during this study to similar sites, it may be possible to develop a more complete understanding of processes that impact on water quality in such environments.

ACKNOWLEDGEMENTS

This study was made possible through funding by the Swedish International Co-operation Agency (STINT) and the South African Foundation for Research Development (now NRF).

Special thanks to the researchers at the University of the Western Cape, especially Mr. Rian Titus. Thanks to Mr. John Weaver of the Council for Scientific and Industrial Research for his input and encouragement, and to Dr. Chris Harris and his staff at the University of Cape Town who conducted the isotope analysis. I am also grateful to the Cape Metropolitan Council for allowing me access to their facilities and for allowing me to conduct this study.

REFERENCES

Department of Water Affairs and Forestry 1994. Minimum Requirements for the Establishment of a Waste Disposal Site, DWAF, Pretoria.

Harris, C. B. M. Oom, and R. E. Diamond. 1999. A preliminary investigation of the oxygen and hydrogen isotope hydrology of the greater Cape Town area and an assessment of the potential for using stable isotopes as tracers, Water SA, Vol. 25, No. 1, pp 1-10.

IAEA station 6881600 >Malan' (Cape Town) South Africa, Global Network of Isotopes in Precipitation (GNIP), http://www.iaea.or.at/programs/ri/gnip/gnipmain.htm .

Parsons R (pers.comm)

Tredox, G. 1984. The Groundwater Pollution Hazard in the Cape Flats, Wat.Pollut. Control 1984, pp. 473 B 483.

Saayman, I. C. 1999. Case study on the chemical characteristics of a pollution plume and a determination of its direction of movement at the Bellville Waste Site, Cape Town, South Africa, Unpublished M.Sc. Thesis, Royal Institute of Technology (KTH), Stockholm.

Walton, D. C. B. T. Verhagen, and D. M. Taussig. 1993. The practical applications of environmental isotopes in pollution studies. In: Procs. Africa Needs Ground Water – An International Ground Water Convention, '93.

Weaver, J. M. C. and W. C. Tworeck. 1988. Hydrogeological Investigation of leachate Attenuation at the Bellville South Solid Waste Disposal Site, Report to the Foundation for Research Development, CSIR.

Wright, A., and J. Conrad. 1995. The Cape Flats Aquifer, Current Status, Groundwater Programme Watertek, CSIR, Report No. 11/95.

Groundwater: Past Achievements and Future Challenges, Sililo et al. (eds) © 2000 Balkema, Rotterdam, ISBN 90 5809 159 7

Hydrogeology of the Ballimore region, central New South Wales, Australia

S.Schofield
Oil Search Limited, Sydney, N.S.W., Australia

J.Jankowski
Groundwater Centre, School of Geology, University of New South Wales, Sydney, N.S.W., Australia

ABSTRACT: The integrated use of geophysical, geological, hydrochemical, and hydrogeological data has allowed the development of a plausible conceptual model for groundwater flow in the Ballimore region. A realistic model for this under-explored system could not be derived solely by the use of hydrogeological data. Interpretation of the available datasets indicates that two groundwater systems are active: a regional, and a local system. These are separated by a regionally extensive aquiclude. The local groundwater system is subdivided into the deep, intermediate and shallow cells. The intermediate cell is a leaky aquitard that acts as a mixing zone between the deep and shallow cells.

1 INTRODUCTION

This paper describes the hydrogeology of the Ballimore region, which until recently, was poorly understood. The Ballimore region is located approximately 400km northwest of Sydney in central New South Wales. Elevations vary from 740m in the east to 240m in the west. The region is dissected by two drainage systems the Talbragar River and the Macquarie River, both flow westward (Figure 1). Habermehl (1980) interprets this region to provide recharge to the Great Artesian Basin (GAB).

Schofield (1998) collated and interpreted the available geological, geophysical, hydrochemical and hydrogeological data from published, unpublished and original work (Table 1). Waterbore locations and elevations are not surveyed when drilled: topographic maps were used for borehole levelling. Recent geological mapping helps constrain the surface distribution of geological units (Colquhoun et al. 1997). The regional geophysical datasets include gridded aeromagnetic and station based gravity data. Borehole geophysical logs were available for the mineral exploration boreholes. Only lithological logs are available for water boreholes. Samples from the one core available (Soda-1) were taken, and thinsections were prepared with dyed resin to allow visual porosity analysis. These thinsections were described, and porosity estimates made. The available hydrochemical data was collated, and both duplicate and additional samples were collected and analysed by the authors.

The integration of these data sets has allowed the authors to resolve much of the complexity of this interesting system.

The stratigraphy of the area can be divided into basement rocks and cover rocks (Figures 1, 2). Basement rocks of the Lachlan Fold Belt are of Palaeozoic age, and crop-out extensively in the south of the study area. These rocks are folded, faulted, fractured and regionally metamorphosed to greenschist-grade. They are dominantly of marine origin, with lithologies including metasedimentary, volcanic, and carbonate units.

The cover rocks are faulted and fractured, and are Permian to Jurassic sediments deposited in the Gunnedah and Surat Basins. These sediments thicken to the north, off the faulted edge of the basement. The Middle Permian and younger

Table 1. Database.

Data type	Number
Topographic maps (1:100000)	4
Geological map	5
Regional geophysical datasets	2
Mineral exploration boreholes	32
Water boreholes	126
Cored boreholes	1
Old hydrochemical data	33
New hydrochemical data	71
Water level measurements	56

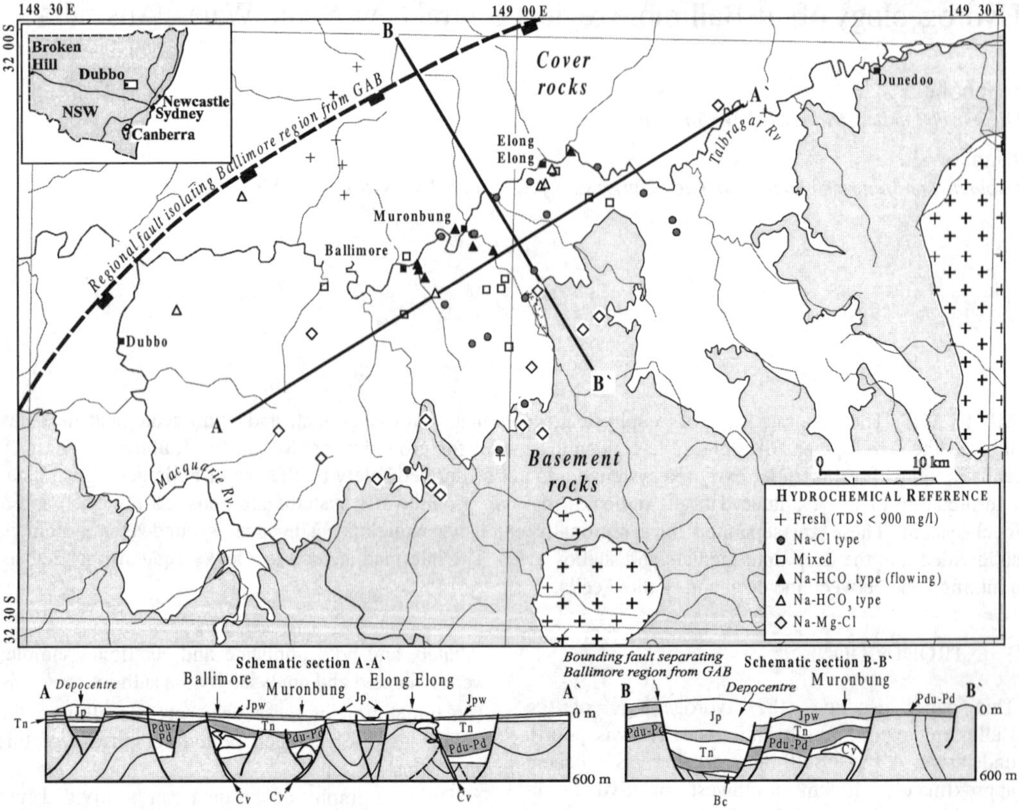

Figure 1. Location map of the Ballimore region.

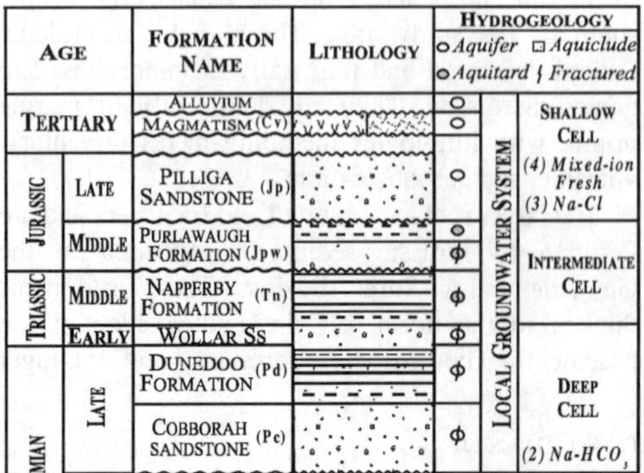

Figure 2. Aquifer summaries.

sequences were deposited under terrestrial conditions. Syndepositional faulting controlled accommodation space and therefore deposition of the cover rocks. The Pilliga Sandstone (Figure 2) is a key unit, it thickens into localised, fault controlled, rectangular, depocentres. These are recognised on the gridded gravity image and are confirmed by drilling.

Miocene magmatism led to the emplacement of intrusive bodies within the cover rocks; in places these break through to the surface as volcanoes. These are mapped in outcrop and intrusions are evident on the regional magnetic image. Most of the sequence is overlain by Quaternary alluvium.

Four end-member water types are recognised: (1) Na-Mg-Cl; (2) $Na-HCO_3$; (3) Na-Cl; and, (4) fresh mixed-ion groundwaters (Table 2).

The groundwaters of end-member 2 are unusual as they are effervescent. These $Na-HCO_3$-rich groundwaters have flowed from boreholes under artesian pressure for over a century. They have very low calcium, magnesium, sulphate and chloride concentrations (Table 3b). They are slightly acidic, with Na and HCO_3 comprising more than 95% of the ions in solution (Figure 3). The ratio of $Na:HCO_3$ is close to unity from low to high salinity, indicating that ion exchange is not a controlling hydrogeochemical process. These groundwaters carry up to 1g/l of CO_2, and have calculated P_{CO2} concentrations that are up to 1000 times atmospheric levels (Schofield & Jankowski 1998: Table 3a). Even though these waters are strongly reducing, alkalinity (Alk) to dissolved inorganic carbon (DIC) data indicate that the CO_2 in these waters does not result from redox processes (Figure 4: see Deffeyes 1965). Modelling of isotopic data shows that this CO_2 gas is of a mixed mantle-magmatic origin with a primary $\delta^{13}C$ of –5.9‰ (Schofield 1998).

Table 2. End-member water types.

Type	TDS (mg/l)	Description
(1) Na-Mg-Cl	3,300-7,700	Deep regional system
(2) Na-HCO$_3$	2,000-8,000	Local system: deep cell
(3) Na-Cl	900-7,600	Local system: shallow cell
(4) Fresh mixed-ion	< 900	Local system: depocentres in shallow circulation cell

Geophysical and geological data indicate that magmatism in the area led the emplacement of intrusions and the emergence of Na-rich volcanics to the north and south of the study area. Potassium-argon dating of these rocks indicates that they erupted 14-12Ma ago, and therefore are of Miocene age (Wellman & McDougall 1974). This magmatism provided a sodium source (syenitic intrusions and trachytic flows), and allowed the influx of mantle gasses. As the magmas forced their way to the surface, contact metamorphism is thought to have introduced magmatic CO_2 to the system.

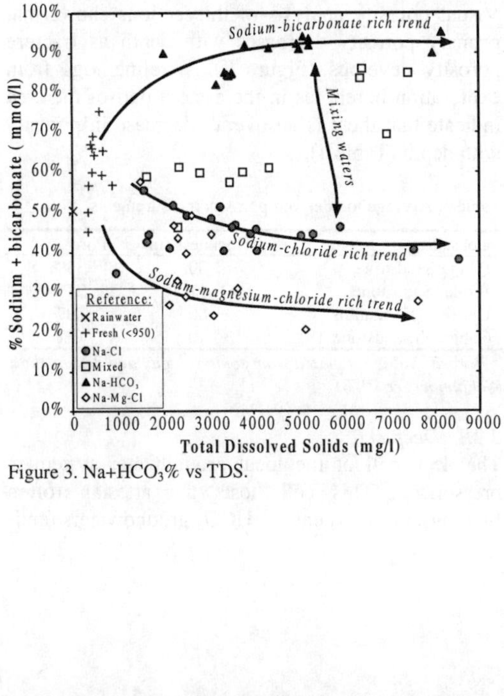

Figure 3. Na+HCO$_3$% vs TDS.

The reservoir for this CO_2 gas is poorly understood: it may be present in structural and stratigraphic traps, or crustal traps (see Chivas et al. 1987, Giggenbach et al. 1991, Bailey 1994).

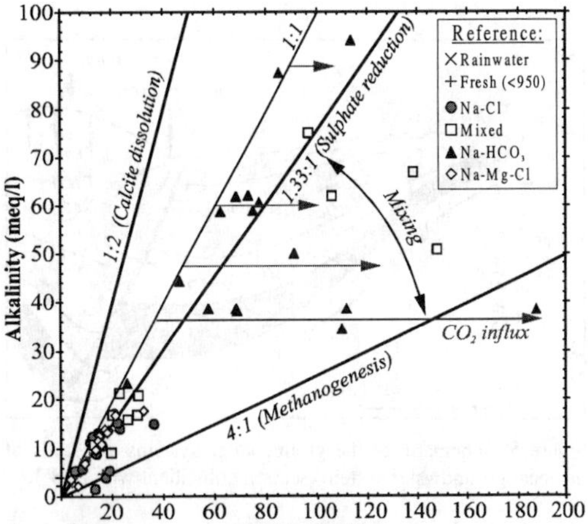

Figure 4. Alk vs. DIC.

Table 3a. Physical and calculated parameters.

Water type	Temp °C	TDS mg/l	PH	CO_2 mg/l	Log P_{CO2} Atm
Rainfall	-	20	5.6	-	0.03
Na-Mg-Cl	19.9	3,151	6.8	145	0.11
Na-HCO$_3$	21.7	4,531	6.3	1,088	0.85
Mixed	20.8	4,356	6.6	533	0.70
Na-Cl	20.7	3,720	6.7	164	0.16
Fresh	21.1	535	6.6	103	0.07

Table 3b. Chemical compositions (mmol/l).

Water type	Na	Mg	Ca	SO$_4$	HCO$_3$	Cl
Rainfall	0.1	0.0	0.1	0.0	0.2	0.1
Na-Mg-Cl	18.3	12.2	4.7	1.1	11.2	39.9
Na-HCO$_3$	44.1	2.3	2.3	0.1	52.1	3.7
Mixed	48.5	4.1	2.2	1.4	32.5	25.0
Na-Cl	39.3	7.8	3.5	2.4	10.2	47.5
Fresh	3.1	1.0	0.9	0.1	4.6	2.1

While investigating the origin of these Na-HCO$_3$-rich groundwaters the authors developed a model that describes the evolution of each of the water types shown in Table 3a and b.

2 GROUNDWATER FLOW

Two groundwater systems are recognised: the regional, and the local groundwater system. The regional groundwater system is confined to the basement rocks by an Early Permian aquiclude: the Dubbo unit (Figure 5). The local groundwater system is restricted to the cover rocks. This system is subdivided into the deep, intermediate and shallow groundwater cells. The intermediate cell is a mixing zone where the pressurised Na-HCO_3-rich artesian waters of the deep cell rise and mix with the sinking mixed-ion fresh and Na-Cl waters of the shallow cell.

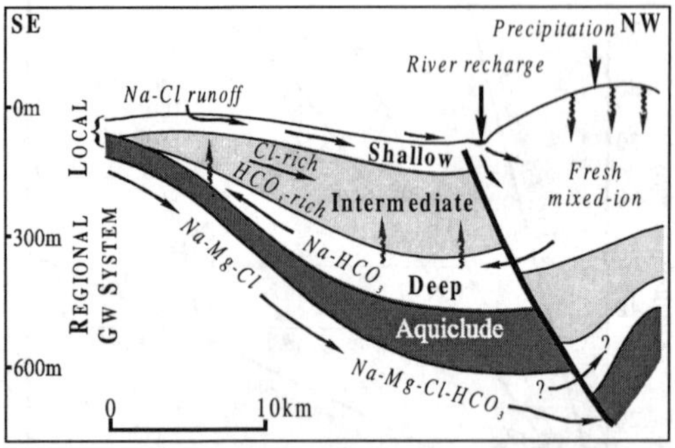

Figure 5. Schematic of the groundwater systems and cells of the local groundwater system (section coincident with B-B`).

The boreholes used for the construction of each potentiometric head map were selected using the chemical composition of the sampled groundwater (Table 2). This proved to be the only means of generating geologically reasonable groundwater flow maps. This is because flow in the deeper aquifers is fracture controlled, and syndepositional faulting has led to variations in the thickness and continuity of units.

2.1 *Regional Groundwater System*

The regional groundwater system is characterised by Na-Mg-Cl groundwaters. Rainfall and percolating surface waters recharge this fractured aquifer in the south of the region where the basement rocks crop out. As these waters infiltrate the Palaeozoic rocks they dissolve halite and weather chlorite. This increases salinity and results in the evolution end-member type 1.

Groundwater of this composition is recognised in boreholes that draw water from the cover rocks, but these are limited to the fringe of the Permo-Jurassic basins.

The regional structural grain and well-developed foliation trends north-northwest: this controls groundwater flow. Piezometric head data for this water type indicate that the water flow to the northwest (Figure 6). As these groundwaters move along the flowpath they become more saline. In the same direction the aquifer changes from unconfined to confined. A regionally extensive Early Permian claystone acts as an aquiclude, isolating the regional groundwater system (Figure 2).

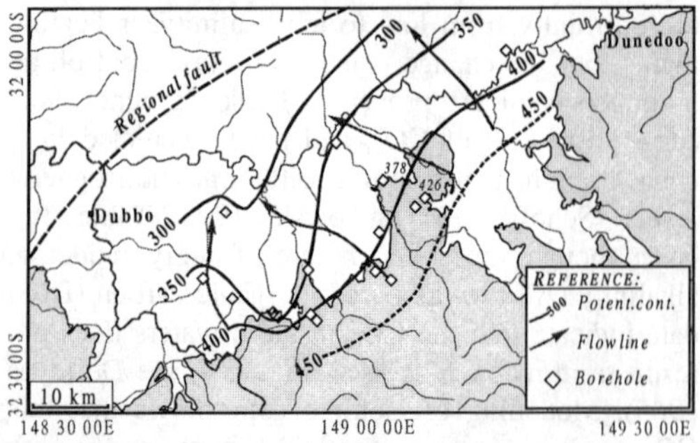

Figure 6. Potentiometric head map for regional groundwater system.

These waters occasionally have elevated CO_2 gas concentrations, but do not effervesce. This CO_2 is generally isotopically lighter than the CO_2 gas in the local groundwater system. This is attributed to the mixing of heavy $\delta^{13}C$ (inorganic) and a light $\delta^{13}C$ (organic-biogenic) carbon source. The heavy source is either: (1) the dissolution of carbonate minerals; or (2) remnant magmatic-mantle CO_2 gas that has migrated beneath the aquiclude that separates the regional system from the deep cell of the local groundwater. The light $\delta^{13}C$ source is likely to be CO_2 gas from root respiration or redox processes.

2.2 *Local Groundwater System*

The local groundwater system is isolated from the chloride-rich groundwaters of the regional system by the Dubbo unit. Hydrochemical data show that groundwater flow within the local groundwater system can be separated into three cells.

Groundwater flow in the local groundwater system is variably fracture and porosity controlled. Visual porosity analysis of thinsections shows that primary porosity decreases with depth as fracture porosity develops (Figure 7). Wireline logs from exploration boreholes in the eastern part of the area indicate that there is an overall decrease in porosity with depth (Table 4).

Table 4. Average log derived porosity measurements.

Formation	Thickness (m)	Porosity *
Pilliga Sandstone	30	33%
Wollar Sandstone	11	21%
Dunedoo Formation	60	20%
Cobborah sandstone	40	3%

** Porosity value estimated from sonic velocity and density data (Schlumberger 1978)*

2.2.1 *Deep cell*

The deep cell of the local groundwater system is pressurised. This cell hosts the artesian (often flowing), effervescent Na-HCO_3 groundwaters (end-member 2). The cell extends from the top of the Dubbo unit to the base of the argillaceous Triassic sediments (Napperby Formation). Figure 7 and Table 4 suggest that fracturing controls groundwater flow in this cell. This inference is confirmed by the observation: the aquifer interval in adjacent boreholes which produce chemically identical Na-HCO_3-rich groundwaters draw water from different stratigraphic and lithological units (including shale, coal and sandstone).

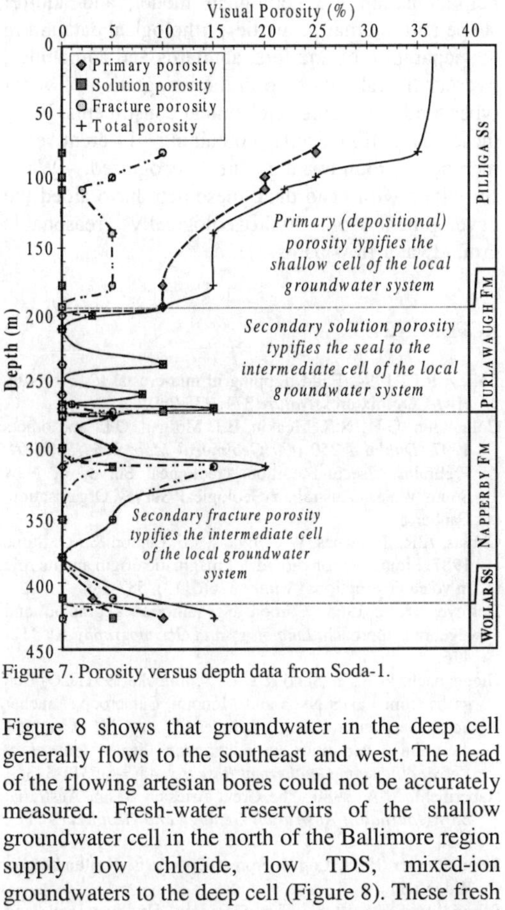

Figure 7. Porosity versus depth data from Soda-1.

Figure 8 shows that groundwater in the deep cell generally flows to the southeast and west. The head of the flowing artesian bores could not be accurately measured. Fresh-water reservoirs of the shallow groundwater cell in the north of the Ballimore region supply low chloride, low TDS, mixed-ion groundwaters to the deep cell (Figure 8). These fresh water reservoirs have higher potentiometric head than the Na-HCO_3-rich waters. As these fresh mixed-ion groundwaters move south and west they pass through CO_2 gas reservoirs, and become saturated with CO_2, their pH drops. Acidic groundwaters are aggressive, and when they intersect Miocene intrusions they rapidly weather Na-rich feldspar (nepheline is the dominant mineral in the Miocene magmatic rocks). This produces the observed Na-HCO_3 ratio in mmol/l of 1:1 (Table 3b). Gas lift then drives these groundwaters upward along fractures until they reach the basal seal of the deep cell. Observation shows that Na-HCO_3-rich waters are encountered whenever boreholes intersect fractures within the deep groundwater cell, near the intrusive bodies recognised on the aeromagnetic data.

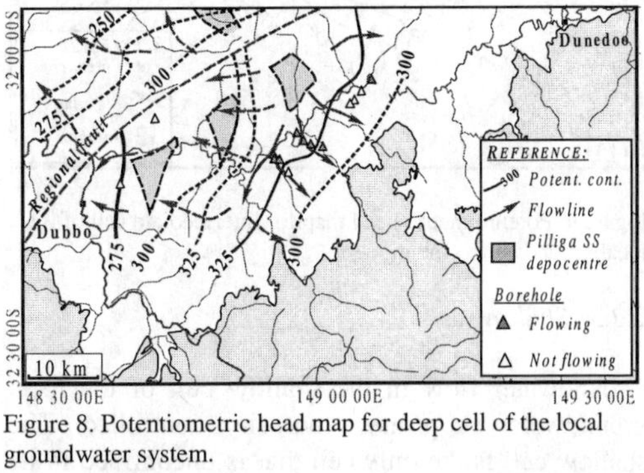

Figure 8. Potentiometric head map for deep cell of the local groundwater system.

In combination with geological data, Schofield (1998) uses chemical and isotopic comparisons to show that the Na-HCO_3-rich groundwaters of the deep cell do not recharge the GAB.

2.2.2 *Intermediate cells*

The intermediate cell of the local groundwater system is a leaky aquitard. This cell extends from the base of the Napperby Formation to the top of the Jurassic argillaceous sediments (Purlawaugh Formation: see Figure 2). These rocks maintain the pressure within the deep cell by not being an effective aquifer. Flow is thought to occur through a mixture of primary and secondary porosity (Figure 7). Figure 9 shows that the intermediate cell generally flows to the northwest, but relative flow rates are thought to be low due to the low porosity (and hence inferred permeability) of the strata.

This cell is recognised by mixed water types (Table 3a, b): between upwelling gassy Na-HCO_3 waters; laterally migrating fresh mixed-ion groundwaters; and downward percolating Na-Cl waters. The samples taken from this cell have chemistries that indicate mixing between the deep and shallow cells. These groundwaters are commonly of Na-HCO_3-Cl and Na-Cl-HCO_3 type. These groundwaters often have CO_2 gas concentrations that are higher than those of the regional groundwater system, but lower than those of the deep cell of the local groundwater system. Isotopic data confirms that these groundwaters have heavier carbon isotopic signatures than are found in the regional groundwater system, but lighter than are measured in the Na-HCO_3 waters.

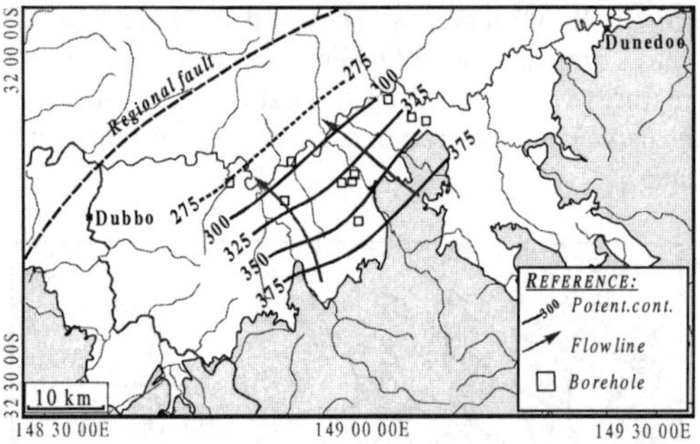

Figure 9. Potentiometric head map for intermediate cell of the local groundwater system.

2.2.3 *Shallow cell*

Groundwater flow in the shallow cell of the local groundwater is primary porosity controlled. The shallow cell is the only cell that is unconfined in its entirety. The piezometric head map for this unit (Figure 10) is constrained better than those of the deeper cells or regional groundwater system as it is more frequently drilled. Generally water flow westward away from the elevated areas.

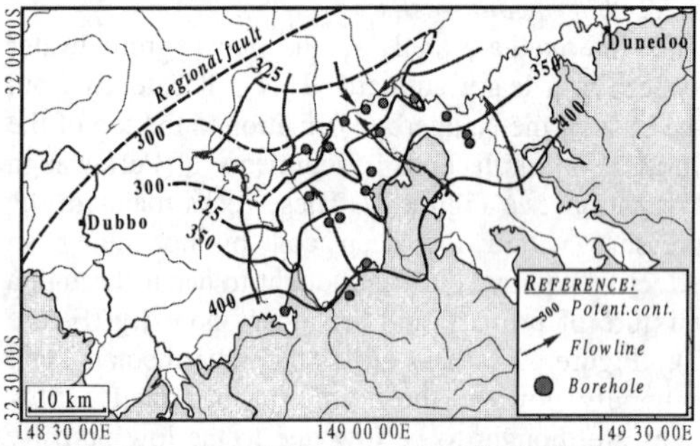

Figure 10. Potentiometric head map for shallow cell of the local groundwater system.

The shallow cell is characterised by the presence of two groundwater end-members: (1) end-member 3, Na-Cl waters in the argillaceous Quaternary alluvium; and, (2) end-member 4, fresh mixed-ion waters in the porous, clean, quartzose, Pilliga Sandstone. This unit is cliff-forming and crops out in the elevated areas in the north and east. Groundwaters from this unit are fresh (<900mg/l) with a low chloride concentration and have a mixed-ion composition.

The Quaternary alluvial rocks vary from being sandy to argillaceous. Boreholes that draw water from sands within these units typically contain Na-Cl type waters. These waters originate from dryland salinisation processes where marine basement rocks crop out. The resulting saline surface waters enter regional drainage channels. These drain north and west onto the cover rocks. The Na-Cl waters in the shallow cell accumulate largely through riverbed recharge

3 CONCLUSIONS

A conceptual model for the hydrogeology of the Ballimore region has been devised through the integrated use of all available data. Geophysical data help constrain the geological model, and aquifer shape and internal properties; lithological data aid in the separation of aquifers, aquitards and aquicludes; hydrochemical data help in the recognition of water types, and the source of chemical constituents. From these data flow cells containing different end-member groundwaters are recognised. When combined with head data, these data have aided the development of hydrogeologically reasonable groundwater flow paths.

REFERENCES

Bailey, R.C. 1994. Fluid trapping in mid-crustal reservoirs by H_2O-CO_2 mixtures. *Nature* 371: 238-240.

Colquhoun, G.P., N.S. Meakin, E.J. Morgan, O.L. and others 1997. *Dubbo 1:250 000 Geological Map Sheet SI/55-04.* Preliminary second edition, Geological Survey of New South Wales, Australian Geological Survey Organisation, Canberra.

Chivas, A.R., I. Barnes, W.C. Evans, J.E. Lupton & J.O. Stone 1987. Liquid carbon dioxide of magmatic origin and its role in volcanic eruptions. *Nature* 326(6113): 587-589.

Deffeyes, K.S. 1965. Carbonate equilibria: a graphic and algebraic approach. *Limnology and Oceanography* 10: 412-426.

Giggenbach, W.F., Y. Sano & H.U. Schmincke 1991. CO_2-rich gases from Lakes Nyos and Monoun, Cameroon; Laacher See, Germany; Dieng, Indonesia, and Mt Gambier, Australia - variations on a common theme. *Journal of Volcanology and Geothermal Research* 45(3-4): 311-323.

Habermehl, M.A. 1980. The Great Artesian Basin, Australia. *BMR Journal of Australian Geology and Geophysics* 5(1): 9-38.

Schlumberger 1978. Log interpretation charts, Schlumberger, Ridgefield.

Schofield, S.K. 1998. The Geology, Hydrogeology and Hydrogeochemistry of the Ballimore region, Central New South Wales. *PhD Thesis*, University of New South Wales.

Schofield S. & J. Jankowski 1998. The origin of sodium-bicarbonate groundwaters in a fractured aquifer experiencing magmatic carbon dioxide degassing, the Ballimore region, central New South Wales, Australia. In: G.B. Arehart & J.R. Hulston (eds), *Proc. 9th Int. Symp. Water-Rock Interaction, Taupo, New Zealand, 30 March–3 April, 1998*: 271-274. Rotterdam: Balkema.

Wellman, P. & I. McDougall 1974. Potassium-Argon ages on the Cainozoic volcanic rocks of New South Wales. *Journal of the Geological Society of Australia* 21(3): 247-272.

Groundwater: Past Achievements and Future Challenges, Sililo et al. (eds) © 2000 Balkema, Rotterdam, ISBN 90 5809 159 7

Salinisation of a coastal aquifer in Mauritania: Salt wedge or bedrock alteration

B.M.Semega
Faculté des Sciences et Techniques, Université de Nouakchott, Mauritania

M.Lacroix, H.Q.Wang & J.P.Dupont
Laboratoire de Géologie, Université de Rouen, Mont Saint-Aignan, France

ABSTRACT : The Trarza aquifer, located in the Senegalese-Mauritanian sedimentary basin, has a problem of salt-water intrusion. This paper discusses the hydrochemical evolution of groundwater in this multi-layered aquifer. To do so, we use logs of chemical analyses and precipitation records from the Idini pumping field (60 km east of Nouakchott).
The hydrochemical facies of the water can be divided into three major groups : 1) chloride-sulfated, 2) sulfate-chlorinated, and 3) chloride-bicarbonated. These facies are also sodo-calcic, which implies a mixed continental-marine origin. The evolution of sodium and chlorides does not reflect an influence of brackish water from the salty marine fringe. The comparison of relationships between ionic species in groundwater and diluted seawater confirms the important role of the rock-water interactions in the the mineralisation. The 1978-1986 sahelian drought did not provoke a really significant increase of the mineralisation rate in the Idini pumping field. The only real augmentation is located at the northern boreholes. It is the result of the increasing hydraulic gradient which provokes an improvement of the leakage from the water which dissolves the evaporitic minerals of the surrounding sebkhra.

1 INTRODUCTION

In coastal areas, the appearance of a salty intrusion frequently limits the exploitation of groundwater (Lacroix et al., 1982). In the Trarza plain, groundwater is the only source of fresh water, so its salinisation is a matter of grave concern. There are two potential sources of contamination for this groundwater : the surrounding sebkhras or the atlantic ocean.

This paper considers the origin of the regional water mineralisation by studying a large water supply plant.

2 GEOLOGICAL AND HYDROGEOLOGICAL SETTINGS

The Senegalese-Mauritanian depresssion is filled by continental sediments of Paleosozoic to Cenozoic age (fig. 1). It is a shallow syncline bounded by the Atlantic Ocean in the west, precambrian basements rocks in the north (Dorsale Reguibat) and by cambrian shales in the east (Mauritanide Ranges). Tectonic movements associated with the opening of the Atlantic during the Mesozoic have caused major faults zones or trenches in the eastern boundaries. The depression was filled by continental sediments from Middle Jurassic to Middle Eocene (Continental Intercalaire formation), with some later episodes (Upper Cretaceous to Middle Eocene, i.e. Continental Terminal formation) of lagoonal and coastal/marine sedimentation.

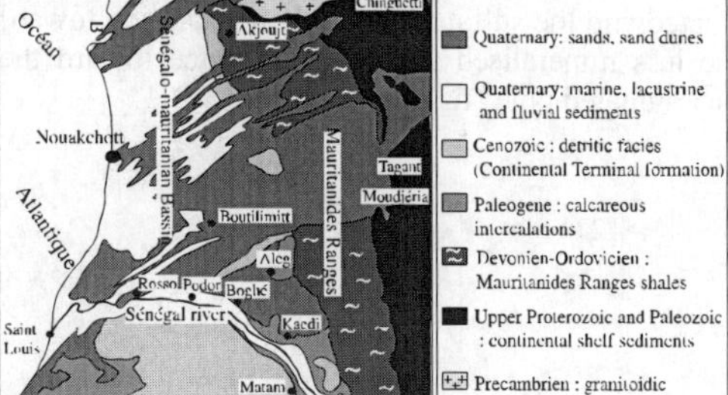

Figure 1. Senegalo-mauritanian basin.

The latter are composed of clays, sandstones with calcareous intercalations containing some gypsum, halite and other evaporitic minerals. Quaternary deposits consist of sandy loam alluvial deposits, sand/clay dunes, lacustrine deposits and sebkhras. The aquifer of the Continental Terminal is multilayered with confined and unconfined regions (Puyoo et Schwartz, 1987). The salinity depends of the depth, 1 g.l^{-1} between 10 to 40 m, 0.15 to 0.4 g.l^{-1} between 60 to 90m and 4 g.l^{-1} between 150 to 170 m. It is globally drained towards the north-east with a gradient from .1 to .05%. The fresh water recharge to this aquifer generally occurs through filtration of precipitation in the east and through drainage of the Senegal river alluvial aquifer. There is any outlet for the groundwater, and the Continental Terminal aquifer can be qualified of "endoreic" (fig. 2) (Archambault, 1987 ; Aranyossy & N'Diaye, 1993).

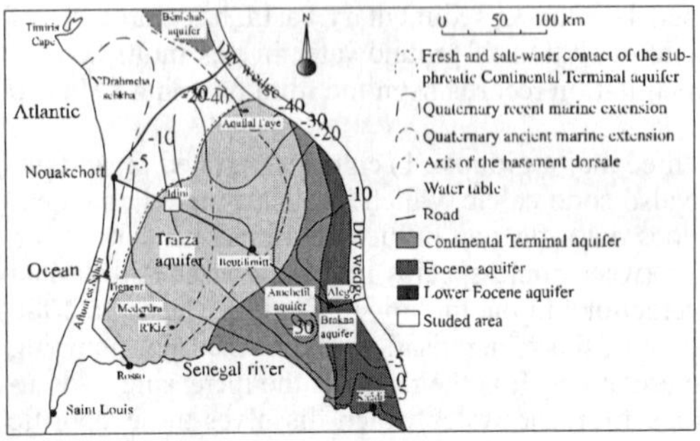

Figure 2. Continental Terminal aquifer of the Trarza.

3 METHODS AND RESULTS

The sampling was done in the Idini wells site, tapping the aquifer of the Continental Terminal and the Quaternary. The hydrochemical and isotopic study of the water were carried out between 1978 and 1993. The Piper diagram (Piper 1944 ; Derec & Louvrier 1973) shows that the more mineralised water of Idini have a median position between the chlorinated and the sulfated poles, their evolution toward the less mineralised implies a tendance toward the bicarbonated pole (fig. 3).

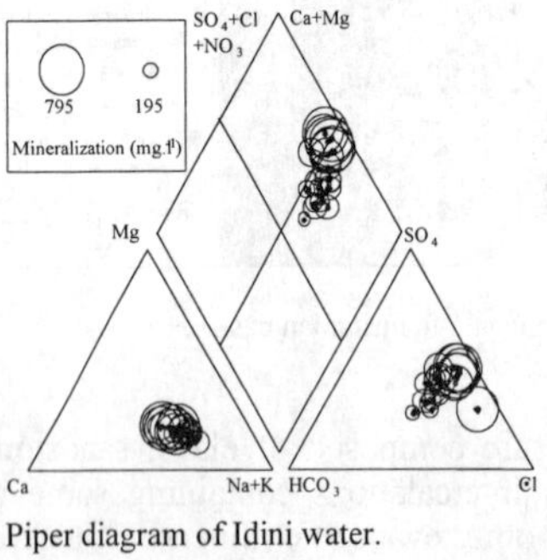

Figure 3. Piper diagram of Idini water.

Square diagrams (Semega, 1995) display three facies: (C1) chlorido-bicarbonated caracterised by some mineralisation above 500 mg.l^{-1}, (C2) sulfato-chlorinated caracterised by a mineralisation between 250 to 500 mg.l^{-1}, and (C3) chlorido-sulfated caracterised by a mineralisation under 250 mg.l^{-1}. A gradient of mineralisation exists from the north-west to the south-east, which follows the succession of the three facies (fig. 4).

Between 1978 and 1993, the ions concentrations (fig. 5) have not evolved significantly except for the salt wedge closest wells which had a slight increase of their salinity.

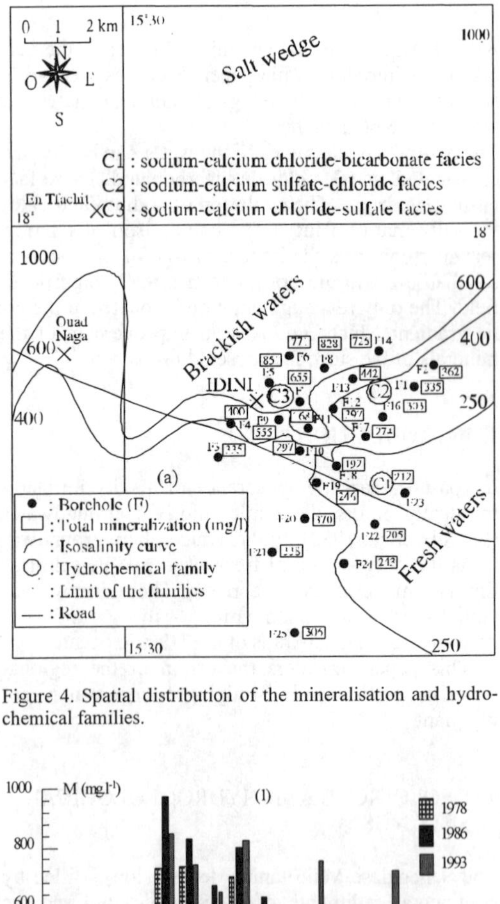

Figure 4. Spatial distribution of the mineralisation and hydrochemical families.

Figure 5. Temporal variation of the Total Mineralisation.

4 INTERPRETATION AND DISCUSSION

The study of saturation rates for the principal minerals leads to determine the influence of dissolution on the surrounding rock. The Idini waters are under saturated for gypsum and anhydrite (-2.6<IS<-1.2), calcite, aragonite, dolomite and magnesite (-1.9<IS<-.2) and strongly undersaturated for halite(IS<-6.5). The equilibrium in relation to calcium minerals indicates an important supply of this ion through dissolution of evaporitic rocks (Fontes et *al.*, 1991 ; Andrews et *al.*, 1994). The ratio of chloride concentrations with other ion types compared with the linear dilution of sea water gives an indication on the origin of the mineralisation, continental or marine. The Idini results (fig. 6) show a particularly heavy load for Ca^{2+} and SO_4^{2-}, and in some slighter proportions for Mg^{2+}, Na^+.

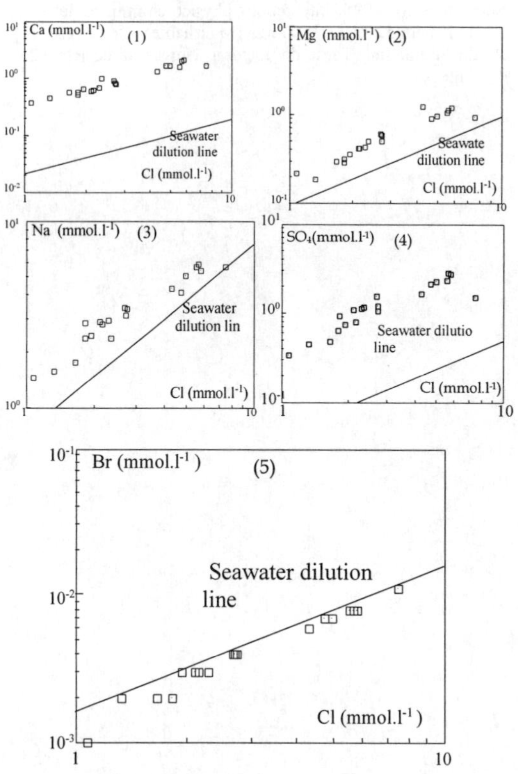

Figure 6. Ionic concentrations (mmol.l^{-1}) versus chloride concentrations of the waters of Idini.
Ca^{++} = vs Cl^- (1), Mg^{++} = vs Cl^- (2), Na^+ = vs Cl^- (3), SO_4^{--} = vs Cl^- (4), Br^- = vs Cl^- (5)

This means that the water dissolved some gypsum in high quantities during his underground transit, but also some magnesium minerals. The sodium highest concentrations appear for ratios Na^+/Cl^- close from the marine ones. This should lead to a marine influence on the brackish limit, as the ratio Br^-/Cl^- seems to prove. The incertitude is on the datation of the acquisition of these ions. The brome minerals associated with the halite keep the sea water Br^+/Cl^- ratio in the evaporites (Andreasen & Fleck, 1997).

The isotopic analysis of the waters from Trarza aquifer reported on the figure 7-1 shows that the meteoric water line (Craig, 1961) is cut on the same point that the water line draw with some evaporated water from the Sénégal river underflow (Diagana, 1984).

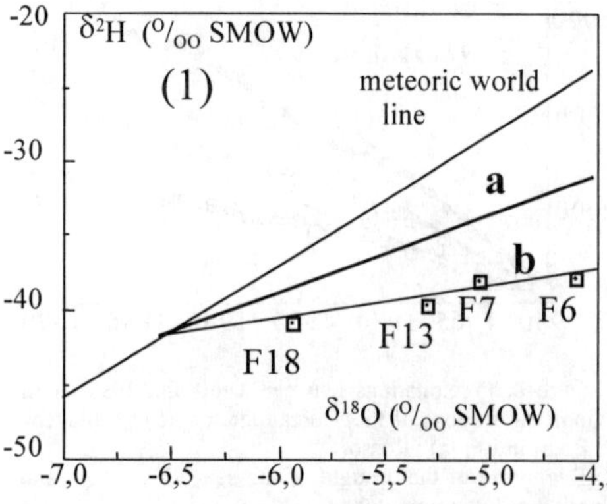

Line "**a**" is the evaporation line of Senegal river underflow (Diagana, 1984), "**b**" is the Idini groundwater

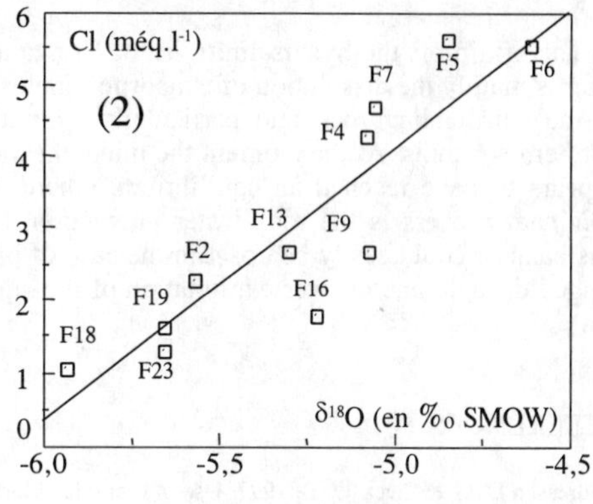

Figure 7-1 and 7-2. relations between Cl^- and ^{18}O (1) and ^{2}H as a function of ^{18}O (2) of the Idini waters.

The only difference is the rate of evaporation. The correlation between the Cl^- and the $_{18}O$ (fig 7-2) show that there is a relation between the evaporation and the chlorine concentration.

So the climatic variations should have their importance. An analysis of cumulated precipitations between 1960 and 1991 shows three distinct rainfall periods (fig. 8). These periods (rainy, dry, and intermediate) correspond to the evolutionary stages in the mineralisation. The dry period of 1978-1986 has been the highest ionic charge increase in the Idini pumping site. The causes were the increase in salt diffusion from the sabkhas and the increase of the hydraulic gradient towards the continent, accelerating the penetration of the salt wedge.

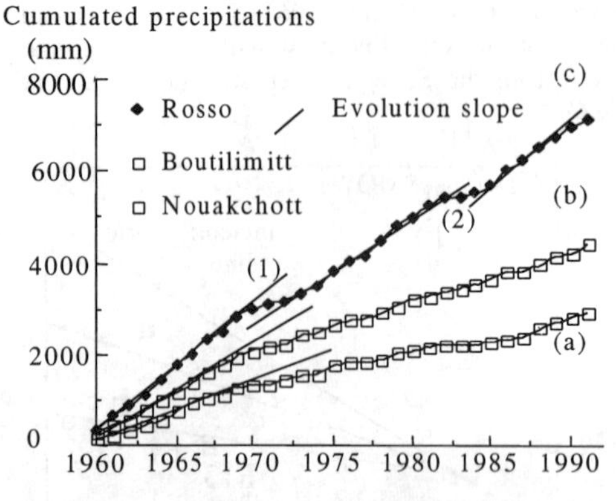

Figure 8. Precipitations between 1960 and 1991 of three pluviometric stations of the Trarza region : (a) : Nouakchott ; (b) : Boutilimitt ; (c) : Rosso.
(1) : Beginning of the drought of the seventies ; (2) : end of the period of intense drought

5 CONCLUSION

The origin of the hypersalinity in the Trarza aquifer is mainly the dissolution of evaporites enclosed in the surrounding rock and particularly from the northern sebkhras. At the moment the mineralisation appears to have reached an equilibrium, where the dominant process is the rock-water interaction, but this balance could easily be upset in the case of prolonged drought and/or over-exploitation of the aquifer.

REFERENCES

Andreasen D. C. & Fleck W. B. 1997. Use of bromide:chloride ratios to differentiate potential sources of chloride in shallow, unconfined aquifer affected by brakish-water intrusion. Hydrogeology Journal, v. 5, n° 2, pp. 17-26.

Andrews J.-N., Fontes J.-C., Aranyossy J.-F., Dodo A., Edmunds W.-M., Joseph A. & Travi Y. 1994. The evolution of alkaline groudwaters in the continental intercalaire aquifer of the Irhazer plain, Niger. Water. Ress. Research. Vol. 30, n°1, pp. 45-61.

Aranyossy J.-F. & N'Diaye B. 1993. Etude et modélisation de la formation des dépressions piézométriques. Revue des Sciences de l'eau, 6; pp. 81-96.

Archambault J. 1987. Réflexions sur l'alimentation et l'évaporation des nappes phréatiques en Afrique subsaharienne. Hydrogéologie (2). III. 2, 69-78.

Craig H. 1961. Isotopic variation in meteoric waters. Science. 133, 1702-1703.

Derec F. & Louvier M. 1973. Représentation graphique des analyses d'eau - le programme Piper. Orléans, France. Bull. BRGM (2) III, n°2, pp. 107-112.

Diagana A. 1994. Etudes hydrogéologiques dans la vallée du fleuve Sénégal de Podor à Bakel : relations eaux de surface et eaux souterraines. Université Dakar, Sénégal. 130p.

Fontes J.-C., Andrews J.-N., Edmunds W.-M., Guerre A. & Travi Y. 1991. Paleorecharge by the Niger River (Mali) deduced from groundwater geochemistry. Water. Ress. Research. Vol. 27, n°2, pp. 199-214.

Lacroix G., Custodio E., Ganoulis J. & Potie L. 1982. Utilisation rationnelle des ressources en eau dans les régions côtières. Bull. BRGM. (2). III. 2, 185-196.

Piper M. 1944. A graphic procedure in the geochimical interpretation of water analyses. Transactions American geophysical Union, Papers Hydrology, pp. 914-928.

Puyoo S. & Schwartz J. 1987. Alimentation en eau potable de Nouakchott : études pour le renforcement et l'extension du champ captant d'Idini. Rapport BRGM 87, MRT 040, Ministère de l'Hydraulique et de l'Energie – Sonelec; Nouakchott, Mauritanie.

Séméga B. M. 1995. Interactions physico-chimiques des eaux de la nappe côtière du Trarza (Mauritanie) à Idini et le long du littoral sud. Thèse de doctorat, Université de Nice, 232 pages.

Groundwater: Past Achievements and Future Challenges, Sililo et al. (eds) © 2000 Balkema, Rotterdam, ISBN 90 5809 159 7

Arsenic and acidity in a groundwater system affected by acid sulfate soils, Stuarts Point, Australia

J.V.S. Smith & J. Jankowski
UNSW Groundwater Centre, School of Geology, University of New South Wales, Sydney, N.S.W., Australia

J. Sammut
School of Geography, University of New South Wales, Sydney, N.S.W., Australia

ABSTRACT: Elevated arsenic concentrations were found in the Stuarts Point coastal sands aquifer, Northern New South Wales, Australia. The aquifer contains acid sulfate soils, which during exposure to atmospheric oxygen or fresh groundwaters rich in oxygen, generate large amounts of proton and mineral acidity. The dissolution of FeOOH in acidic and reduced environments as well as the oxidation of arsenopyrite, mobilise arsenic present in the aquifer. The Yarrahapinni Wetlands and acid sulfate soils present in the aquifer produce large amounts of organic acids and organic matter causing the groundwater system to be strongly reduced. Acid neutralising reactions involving $CaCO_3$ shell material dispersed in the sedimentary phases of the aquifer elevate pH to circumneutral level. The aquifer is under stress as demands on groundwater resources for reticulated domestic water supply and irrigation have risen in recent years.

1 INTRODUCTION

Increasing demands on groundwater resources for irrigation and reticulated domestic consumption have raised questions over the long-term sustainability of the Stuarts Point coastal sands aquifer. Water–pollution has also threatened the sustainability of groundwater resources in the area. Recent investigations have identified problems arising from increased use of fertilizers, pesticides and saltwater intrusion. In addition, elevated arsenic concentrations, sand mining operations and acidification from disturbance of acid sulfate soils (ASS), have degraded groundwater quality. This region also supports several fragile groundwater dependent ecosystems whose long-term survival will be threatened if impacts to groundwater quality and sustainability continue to go unknown and unaddressed.

The unconsolidated sediments along the eastern and northern coasts of Australia contain significant concentrations of ASS (White et al. 1996). Pyrite (FeS_2) is the primary mineral in ASS that, through oxidation, acidifies the soil and groundwater. The oxidation of FeS_2 produces sulfuric acid through the oxidation of sulfide to sulfate. This process may be triggered by natural lowering of the watertable, water abstraction, drainage or excavation of pyritic sediments which enable oxygen to enter the soil profile. The oxidation of pyrite mobilises metals present in ASS and in the aquifer matrix. Metals such as Fe^{2+}, Fe^{3+}, Al^{3+} and Mn^{2+} produce mineral acidity because they hydrolyse in water and produce protons. Mineral acidity forms an important part of the total acidity in this aquifer. Some of the ground- and surface-water acidity is neutralised by the dissolution of carbonates (Dent, 1986).

Elevated concentrations of arsenic in the groundwater system have already been identified with the highest concentrations occurring in town water supply bores in the Stuarts Point region. Concentrations of 14 to 80 μg/L have been monitored in the last few years (Piscopo 1996). Elevated concentrations of arsenic in drinking water are considered a hazard to human health. The World Health Organisation set the guideline concentration of arsenic at <10 μg/L (WHO 1994). In Australia, the National Health and Medical Research Council has recommended a threshold of 7 μg/L (NHMRC 1996). The occurrence of arsenic in drinking waters sourced from groundwater systems is a problem which occurs in many places around the world (Nickson et al. 2000, Peters et al. 1999, Smedley et al. 1998, Saha & Chakrabarti. 1995, Welch et al. 1988, Bottomley 1984).

Arsenic solubility is extensively influenced by redox and pH conditions and hence varies considerably within soil profiles. It has been shown that As has the potential to accumulate in ASS (Dudas 1986, Gustafsson & Tin 1994). This paper presents initial information from hydrogeochemical studies in the Stuarts Point aquifer and is a part of a larger research program in this area.

2 GEOLOGY

The unconsolidated sediments of the Stuarts Point aquifer can be divided into 3 geological units:

(1) Pleistocene barrier and Holocene marine sand deposits consisting primarily of siliceous sands, deposited by wind and wave action;
(2) Poorly drained alluvial swamp deposits overlying estuarine clayey sands; and
(3) Holocene estuarine clay deposits, which overlay the Pleistocene deposits and consist primarily of grey silty sandy clays.

The Holocene clay deposits are classified as having a high probability of potential acid sulfate soils (Eddie 2000) (Figure 1).

The unconsolidated sediments overlie the Late Carboniferous to Early Permian lithic sandstones, mudstones, pebbly sandstone and minor conglomerates of the Kempsey beds (Leitch 1972). Lying to the north, adjacent to the Kempsey beds and forming the upper reaches of the catchment is the lithologically similar Early Permian Dee Pee beds comprised primarily of slaty siltstones, lithic sandstones and minor diamictite (Gilligan et al. 1992, Leitch 1972).

The Yarrahapinni Adamellite intrudes into the Dee Pee beds in the region. This intrusion consists of a series of massive discordant coastal granitoids of Late Triassic age (Thomson 1975) composed of leucocratic hornblende-biotite monzogranite, biotite monzogranite, granite and alaskite (Gilligan et. al. 1992). The contact aureoles associated with the Yarrahapinni granitoids in places span up to 2 km wide and comprise foliated sandstones and siltstones, phyllites and sandy slate (Masman 1973).

Considerable mineralisation occurs within the Yarrahapinni Adamellite and the surrounding contact aureoles. Mineralisation occurs as outward zoning deposits grading from silver-lead zones to silver-arsenic zones (Gilligan et al. 1992). Within these zones facture-controlled siliceous sulfide veins are present and have been shown to contain considerable arsenopyrite (FeAsS) with some mineralized deposits producing assays results of up to 9.4% arsenic of the mineralized mass (Kenny 1937, Suppel & Hobbs 1977).

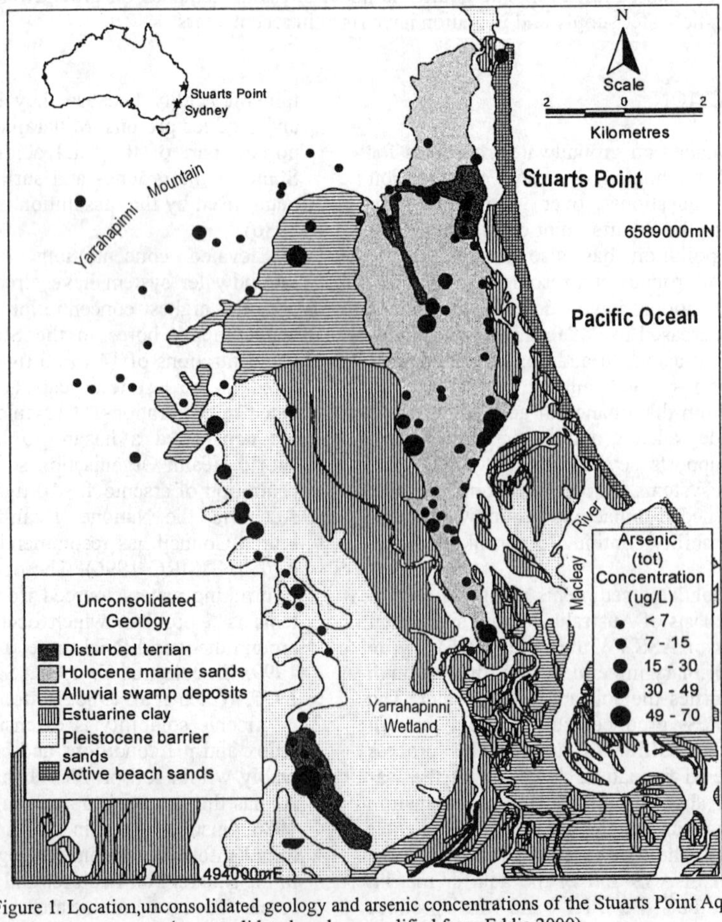

Figure 1. Location,unconsolidated geology and arsenic concentrations of the Stuarts Point Aquifer (unconsolidated geology modified from Eddie 2000)

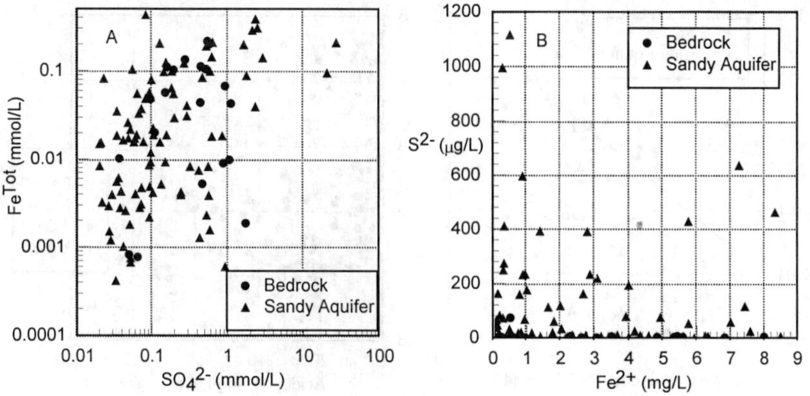

Figure 2. Relationship between Fe^{Tot} versus SO_4 (A) and S^{2-} versus Fe^{2+}(B).

3 HYDROGEOLOGY

Due to increasing urban and horticultural development in the region, demands for groundwater resources for irrigation and human consumption, have placed the region's groundwater systems under considerable stress. Low water-holding capacity and high permeability of the barrier sands makes retention of irrigated waters difficult. Consequently, irrigation demands are high and large quantities of groundwater are abstracted. The agricultural region's close proximity to the coast, south of Stuarts Point (Figure 1) also raises questions as to the potential impacts on water quality from saltwater intrusion. Areas within the aquifer are already embargoed from further water extraction due to the sensitivity of the system.

The Yarrahapinni–Stuarts Point region consists of two dominant groundwater systems. The first of these is the heterogeneous Stuarts Point coastal sands aquifer comprising unconsolidated Pleistocene barrier deposits and Holocene estuarine clays. The second system contains the consolidated Permian sandstone, Triassic coastal granitoids and associated metasediments, forming the Yarrahapinni fractured rock aquifer.

Flow within the system generally follows the topographic gradient from the dominant recharge areas in the NNW of the catchment, to discharge areas within the estuaries in the ESE. Considerable recharge occurs from direct infiltration through the barrier sands as well as from wetlands in the ESE corner of the catchment (Figure 1). Indurated soils within the Pleistocene barrier deposits, resulting from natural leaching of decaying organic material and subsequent cementation or impregnation of barrier sands, act as temporary perched aquifers during times of extend precipitation (Hail 1969). These layers influence localised flow patterns to a limited extent, but are not continuous and/or consolidated to an extent were they would exhibit the properties of a confining layer.

4 WATER CHEMISTRY

The chemical composition of groundwaters in the fractured aquifer system west from Stuarts Point aquifer is closely related to water rock-interaction. Recharge groundwaters are oxidised (dissolved oxygen values (DO) up to 7 mg/L), acidic (pH 4-5) with very low content of dissolved ions (electrical conductivity (EC) values around 100 μS/cm). Oxidation of metal sulfides and weathering of aluminosilicates in the fractured system rapidly elevates EC values to more than 3000 μS/cm, DO values decrease and pH is around 4. Sodium and chloride are dominant ions with appreciable SO_4 concentrations and low content of Ca, Mg and HCO_3. Generation of H^+ during oxidation of FeS_2 causes mobilisation of metals such as Mn, Zn, Fe and Al. The concentration of As reaches 17 μg/L in the fractured aquifer system.

The Stuarts Point aquifer is recharged by the flux of water from the underlying bedrock aquifer and by diffuse recharge from precipitation. The shallow system in the sandy aquifer contains very fresh (EC values 30-100 μS/cm), oxidised and strongly acidic (pH values 3-4) groundwaters due to oxidation of FeS_2 present in ASS. The major ions are Cl, Na and SO_4. HCO_3 concentrations are variable and the content of Ca and Mg is low. Ion exchange reactions remove Ca and Mg from solution and carbonates dissolve as a result. Oxidation of FeS_2 produces elevated iron and sulfate concentrations (Figure 2a).

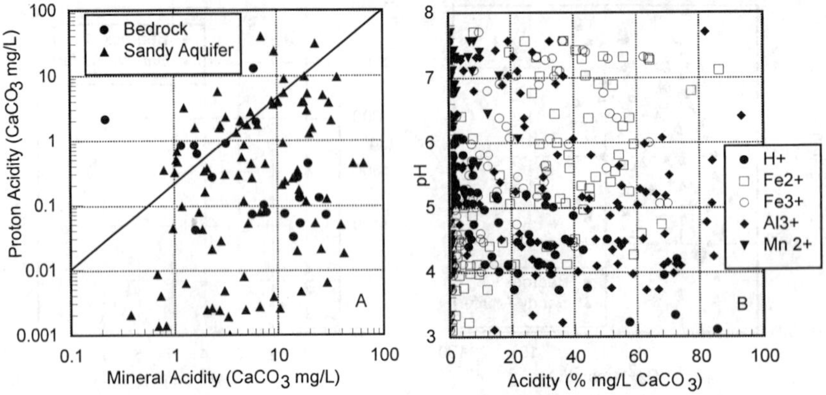

Figure 3. Relationship between proton acidity versus mineral acidity (A) and pH versus acidity (B).

5 ACIDITY AND METAL MOBILISATION

The oxidation of ASS generates significant amounts of H^+ acidity and mineral acidity from the concentration of metals undergoing hydrolysis and dissociated in water. The pH decreases to 3.1-3.5 and the concentration of metals in this acidic groundwater reach 5.5 mg/L Al^{3+}, 8.5 mg/L Fe^{2+}, 17 mg/L Fe^{3+} and 1450 µg/L Mn^{2+}. Other elements occurring in high concentrations include S^{2-} (1.1 mg/L), Ba and Ni. The oxidation of FeS_2 in ASS in shallow parts of the sandy aquifer causes generation of H^+ according to the reaction:

$$FeS_2+15/4O_2+7/2H_2O \rightarrow Fe(OH)_3+2SO_4^{2-}+4H^+ \quad (1)$$

The presence of both Fe^{2+} and SO_4^{2-} in the groundwater indicates that the oxidation of pyrite must be incomplete. This can be represented by the reaction:

$$FeS_2+7/2O_2+H_2O \rightarrow Fe^{2+}+SO_4^{2-}+2H^+ \quad (2)$$

The generation of H^+ from reactions 1 and 2 maintains a low pH and high concentrations of Fe^{3+}, which is more stable in acidic environments. Some Fe^{2+} is removed by the oxidation to Fe^{3+} (reaction 1) and precipitates as $Fe(OH)_3$. The Fe^{2+} would also be removed from solution by S^{2-} in reduced groundwater and precipitated as FeS, as some groundwaters are depleted in DO and have redox values of –200 to –230 mV (Figure 2b).

Proton acidity is calculated from pH and the mineral acidity is determined from the milliequivalents of metals involved in the hydrolysis reactions. Both proton and mineral acidities are presented as $CaCO_3$ equivalent in mg/L. Equations used for the calculation of mineral acidity are:

$$Fe^{2+}+1/4O_2+1\tfrac{1}{2}H_2O \rightarrow FeOOH_{(s)}+2H^+ \quad (3)$$
$$Fe^{3+}+2H_2O \rightarrow FeOOH_{(s)}+3H^+ \quad (4)$$
$$Al^{3+}+3H_2O \rightarrow Al(OH)_3+3H^+ \quad (5)$$
$$Mn^{2+}+1/4O_2+1\tfrac{1}{2}H_2O \rightarrow MnOOH+2H^+ \quad (6)$$
$$\text{Mineral acidity} = 2(Fe^{2+}) +3(Fe^{3+}) + 3(Al^{3+}) +2(Mn^{2+}) \quad (7)$$

where ( ) denote molar concentrations

From the calculation it can be seen that mineral acidity is the dominant acidity and only 20% of water samples show excess of proton acidity over mineral acidity (Figure 3a). Proton acidity is up to 40 mg/L acid equivalents of $CaCO_3$ and mineral acidity is up to 65 mg/L acid equivalents of $CaCO_3$.

Proton acidity and mineral acidity with respect to individual metals are shown in Figure 3b. At circumneutral pH values the mineral acidity originates from the hydrolysis/dissolution of Fe^{2+} and Fe^{3+} and for pH values between 4 and 6 mineral acidity originates from the hydrolysis of Al^{3+}. Proton acidity is dominant at low pH i.e. values below a pH of 5.

The coastal aquifer systems in eastern Australia contain dispersed shell material sometimes in significant proportions. Acidic groundwaters flowing through sandy material in the presence of shell results in the dissolution of the associated carbonates, these reactions result in alkalinity levels being elevated and accordingly a decrease in acidity. The process of acid neutralisation by the dissolution of shell material releases Ca, Mg, Sr and HCO_3 into groundwater. This process can be seen from the correlation exhibited between cations and bicarbonate (Figure 4). Hence, the dissolution of carbonates elevates pH in the sandy aquifer to the observed level of 7 and above.

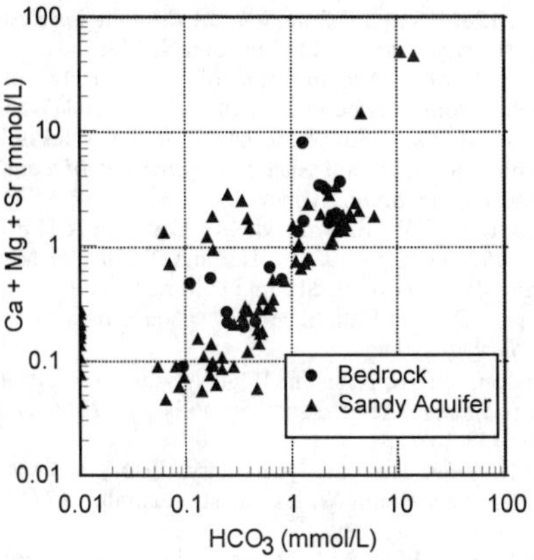

Figure 4. Relationship between Ca + Mg + Sr versus HCO_3.

6 ARSENIC OCCURRENCES WITHIN THE AQUIFER SYSTEM

The presence of As in the Yarrahapinni fractured rock groundwater system is putatively linked to the presence of FeAsS, FeS_2, pyrrhotite ($Fe_{1-x}S$) and chalcopyrite ($CuFeS_2$) minerals associated with the Yarrahapinni Adamellite silver-arsenic zones. Identified FeAsS enriched mineralised fractures run parallel to regional lineaments and comprise northerly strikes. The orientation of mineralised zones complements flow systems and observed As concentration in bores within the fractured system.

The discharging groundwater from the bedrock brings As as a dissolved phase into the sandy aquifer. Arsenic concentrations in the sandy aquifer vary considerably across the groundwater system. The absence of correlations between As and SO_4^{2-}, As and S^{2-} and As and HCO_3 potentially indicates the presence of a complex series of controlling factors acting upon As mobility within the unconsolidated sediments.

The highest As concentrations are related to low values of DO (Figure 5a) and a slightly reduced environment from 0 to –200 mV (Figure 5b). Some waters having higher concentrations of DO probably contain As(V) as the dominant arsenic species. The concept that arsenic is derived from mixed sources is also supported by the poor correlations between As and Fe^{Tot} (Figure 5c), Fe^{2+} and Fe^{3+}. If

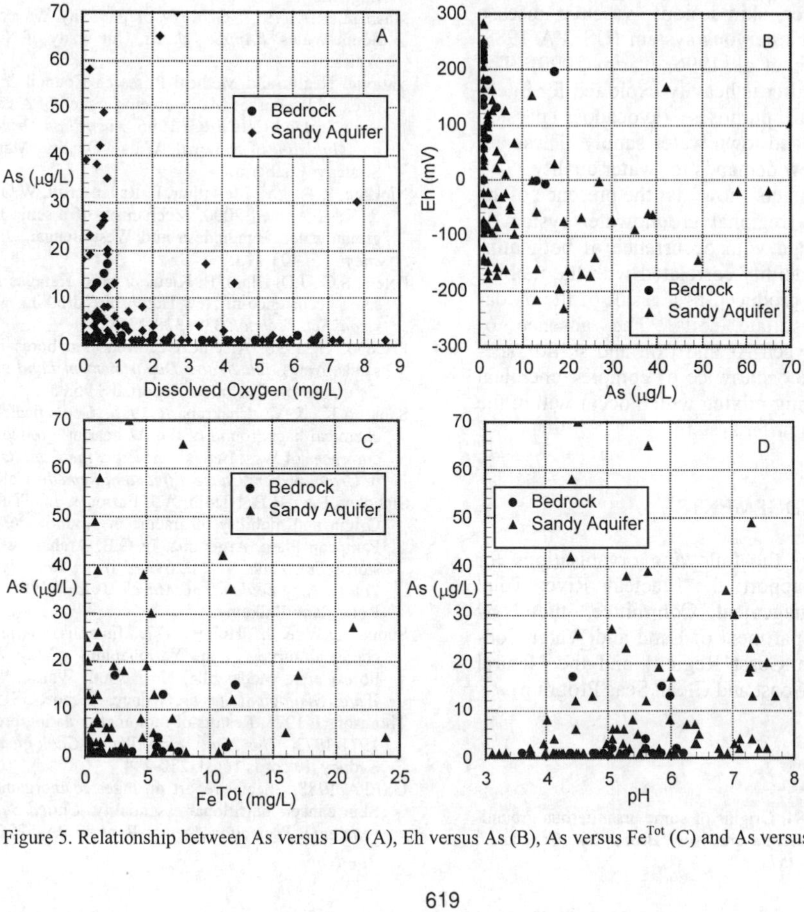

Figure 5. Relationship between As versus DO (A), Eh versus As (B), As versus Fe^{Tot} (C) and As versus pH (D).

a good correlation is present between ferric iron and As the origin of arsenic can be associated with the dissolution of FeOOH. The mixed origin of As is also seen in Figure 5d where the highest concentrations of As are related to two pH ranges of 4.5-5.5 and 7-8.

The As associated with acidic groundwaters is probably from the oxidation of FeS_2 present in ASS and As present in circumneutral groundwaters with dissolution of FeOOH and the potential release of sorbed As(V) associated with ASS clay material.

Research is currently underway to associate elevated As concentration within the various unconsolidated geological units and delineate the complex series of pH/Eh based processes occurring within the sandy aquifer system.

7 CONCLUSIONS

Arsenic concentrations reach 70 µg/L and 30% of wells sampled showed concentration above 7 µg/L which is the threshold for drinking water in Australia. High arsenic concentrations are considered a hazard for human health, having the potential to cause skin cancer, skin lesions, vascular disease and effects on the nervous system (USEPA 1988, Hopemhayn-Rich et al. 1996, Karim 2000). The Stuarts Point aquifer is heavily exploited for intensive agricultural purposes (avocados, passion fruits, potatoes) and town water supply. These activities placed new demands for water quality.

Initial assessments show that the presence of arsenic within the regional groundwater system is strongly associated with occurrence of both mineral and proton acidity correlated with the mobilisation of metals produced as a result of the oxidation of acid sulfate soils. The absence of correlations between As and iron and sulfide species suggests the occurrence of complex chemical processes including mixing which occur within the unconsolidated aquifer system.

8 ACKNOWLEDGEMENTS

We acknowledge the following organisations for their research support: the Macleay River Total Catchment Management Committee, the New South Wales Department of Land and Water Conservation (North Coast Region) and the Natural Heritage Trust (Coast and Clean Seas Program).

REFERENCES

Bottomley, D.J. 1984. Origins of some arseniferous groundwater in Nova Scotia and New Brunswick, Canada. *J. Hydrol.* 69: 223-257.

Dent, D. 1986. *Acid Sulphate Soils: A Baseline for Research and Development.* ILRI Publication No. 39.

Dudas, M.J. 1986. Accumulation of native arsenic in acid sulphate soils in Alberta. *Can. J. Soil Sci.* 67: 317-331.

Eddie, M.W. 2000. Soil Landscapes of the Macksville & Nambucca 1:100 000 Sheets, *Department of Land and Water Conservation*, Sydney.

Gilligan, L.B., J. W. Brownlow, R.G. Cameron & H.F. Henley 1992. Dorrigo – Coffs Harbour 1:250 000 Metallogenic Map SH/56-10, SH/56-11: Metallogenic Study and Mineral Deposit Data Sheets. *New South Wales Geological Survey*, Sydney.

Gustafsson, J. P. & N.T. Tin 1994. Arsenic and selenium in some Vietnamese acid sulphate soils. *Sci. Total. Environ.* 151: 153-158.

Hail, J.R. 1969. The Late Quaternary History of three areas on the New South Wales Coast, Australia. *PhD.Thesis*, University of Sydney.

Hopemhayn-Rich, C., M.L. Biggs, A. Fuchs, R.E. Bergoglio, H.B. Nicolli & A.H. Smith 1996. Bladder cancer mortality associated with arsenic in drinking water in Argentina. *Epidem.* 7: 117-124.

Karim, M. 2000. Arsenic in groundwater and health problems in Bangladesh. *Wat. Res.* 34(1): 304-310.

Kenny, E.J. 1937. A geological reconnaissance of portions of the north coast of New South Wales. *New South Wales Geological Survey*, Sydney – Report GS1937/054.

Leitch, E.C. 1972. The Geological development of the Bellinger – Macleay Region – a study of the tectonics of the New England Fold Belt. *PhD. Thesis*, University of New England.

Masman, K.A. 1973. Geology of the Way Way area New South Wales. *BAppSc. Thesis*, University of New South Wales.

National Health and Medical Research Council & Agriculture and Resources Management Council of Australia and New Zealand (NHMRC) 1996. *Australian Drinking Water Guidelines*. National Water Quality Management Strategy, Canberra.

Nickson, R.T., J.M. McArthur, P. Ravenscroft, W.G. Burgess & K.M. Ahmed 2000. Mechanism of arsenic release to groundwater, Bangladesh and West Bengal. *Appl. Geochem.* 15: 403-413.

Peters, S.C., J.D. Blum, B. Klaue & M.R. Karagas 1999. Arsenic occurrence in New Hampshire drinking water. *Environ. Sci. Technol.* 33: 1328-1333.

Piscopo, G. 1996. Arsenic Occurrence in bores in coastal sands. *New South Wales Department of Land and Water Conservation.* Technical Report, TS 96.004.

Saha, A.K. & C. Chakrabarti 1995. Geological and geochemical background of the As bearing groundwater occurrences of West Bengal. In: *Proc. Int. Conf. On Arsenic in Groundwater: Cause, Effect and Remedy*, Calcutta.

Smedley, P.L., H.B. Nicolli, A.J. Barros & J.O. Tullio 1998. Origin and mobility of arsenic in groundwater from the Pampean Plain, Argentina. In G.B. Arehart & J.R. Hulston (eds), *Proc. 9th Int. Symp. Water-Rock Interaction, Taupo, New Zealand, 30 March–3 April 1998*: 275–278. Rotterdam: Balkema.

Suppel, D.W. & J.J. Hobbs 1977. Mineralisation in and geochemical survey of the Yarrahapinni – Way Way State Forest area, Macksville, New South Wales. *New South Wales Geological Survey,* Sydney – Report GS1977/124.

Thomson, J. 1975. Results of radiometric dating programme, 1971-1973. *The New South Wales Geological Survey*, Sydney, Records 16(3): 230-244.

USEPA 1988. Special report on ingested inorganic arsenic. Skin cancer; nutritional essentiality. *United States Environmental Protection Agency*, Report EPA-625/3-87/013.

Welch, A.H., M.S. Lico & J.L. Huges 1988 Arsenic in groundwater of the Western United States. *Ground Water* 26(3): 333-344.
White, I., M.D. Melville, J. Sammut, B.P. Wilson & G.M. Bowman 1996. Downstream impact from acid sulfate soils. In: Hunter, H.M., A.G. Eyles & G.E. Rayment (eds), *Downstream Effects of Land Use*, Department of Natural Resources, Queensland, Australia, 165-171.
World Health Organisation 1994. Recommendations, second ed. In: Guidelines for drinking water quality, Vol. 1. *World Health Organisation*, Geneva.

Groundwater: Past Achievements and Future Challenges, Sililo et al. (eds) © 2000 Balkema, Rotterdam, ISBN 90 5809 159 7

Geohydrology and isotope hydrology of the Lomagundi dolomite, Zimbabwe

S.Sunguro – *Ministry of Energy and Water Resources and Development, Groundwater Branch, Harare, Zimbabwe*

K.Froehlich – *International Atomic Energy Agency, Vienna, Austria*

B.Th.Verhagen – *Schonland Research Centre, University of the Witwatersrand, Johannesburg, South Africa*

W.Wagner – *Federal Research Institute for Geosciences and Mineral Resources, Hannover, Germany*

ABSTRACT: The karstified Lomagundi dolomite in NW Zimbabwe is being heavily exploited for irrigation agriculture. Concerns about over-exploitation and pollution of the valuable resource prompted geohydrological, hydrochemical and isotope hydrological studies since 1981 by various groups. The groundwater is of a Ca,Mg-HCO_3 type at relatively low mineralisation. Time series of tritium values indicate mean residence times of 30 to 100 years, qualitatively confirmed by ^{14}C measurements. Assuming values of porosity, recharge rates of some 50 mm a^{-1} are estimated. Although lower than results from water balance calculations, they are taken as conservative figures for resource evaluation. Stable isotope data show recharge only during major rainfall events and significant contributions from irrigation return flow. This emphasises long-term vulnerability, seen also in the correlation between nitrate and tritium.

1 INTRODUCTION

The Lomagundi dolomite is an intensively karstified aquifer in NW Zimbabwe (Figure 1) with moderate to high productivity. It is exploited mainly for irrigation, which allows a second crop to be grown in the dry winter season (May-October). By the late 80's, commercial farmers in the Chinhoyi - Mhangura area had drilled some 140 wells with a total groundwater abstraction of 19 M m^3 per year for irrigation of around 3500 ha of wheat, coffee, citrus, tobacco and maize.

The study area lies north-west of Chinhoyi (Figure 2), covering about 1,500 km^2. The long-term annual rainfall in the region, averaging 820 mm per year (Harare) is restricted to the summer season (November to April). Estimated potential evaporation is 1700 mm a^{-1}.

Lower than average rainfall in the early 80's provoked concerns about a possibly continued decline of water levels. Various geohydrological studies were conducted in cooperation with the Zimbabwe Department of Water Development since that time, focusing on the assessment of the hydrologic balance of the aquifer, based mainly on water level observations and surveys of groundwater extraction.

Isotope hydrological studies were undertaken to complement the appropriate assessment of groundwater recharge rates. The first study was conducted in 1981 by the University of Witwatersrand (Wits), South Africa. Follow-up studies were carried out by the Federal Institute for Geosciences and Natural Resources (BGR), Germany (1983-1992) and an International Atomic Energy Agency (IAEA) - sponsored technical co-operation project (1988-1991).

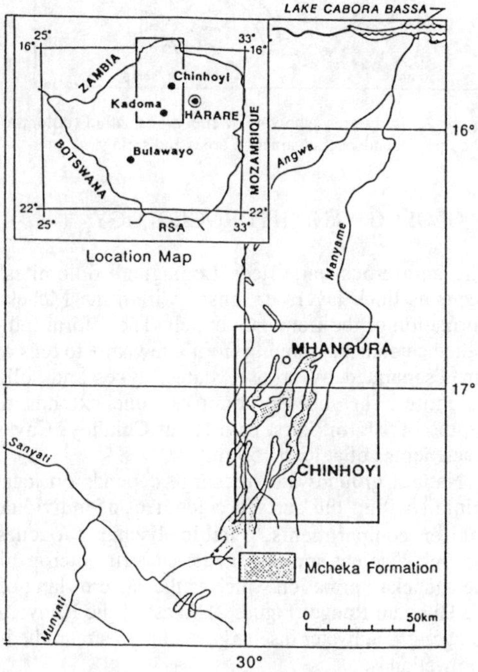

Figure 1. Location diagram of study area, showing regional outcrops of the Mcheka Formation.

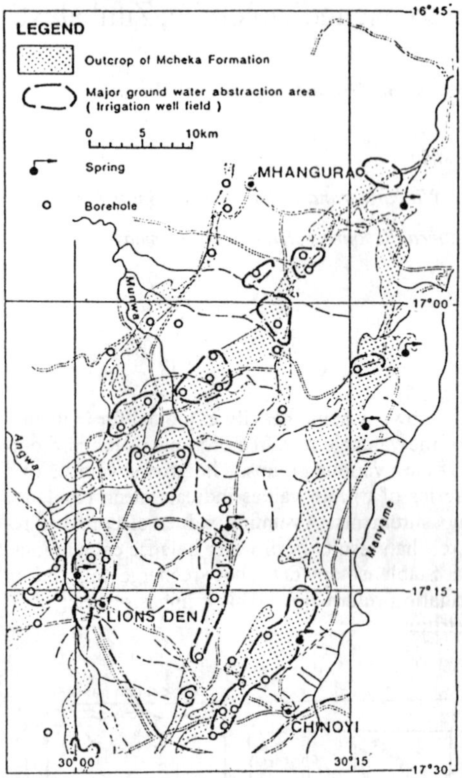

Figure 2. Study area, showing Mcheka Formation (dolomite) outcrops, groundwater abstraction areas and springs.

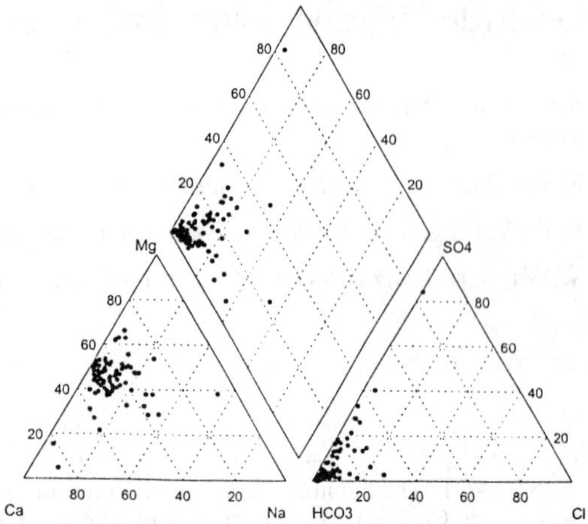

Figure 3. Piper diagram of Lomagundi dolomite groundwater. Deviations from the Ca,Mg-HCO_3 type are due to local sulphate mineralisation and incipient pollution.

2 GEOLOGY AND HYDROGEOLOGY

The aquiferous karstified "Lomagundi dolomites" occur as thick layers or lenses within the Mcheka Formation of the Lomagundi Belt. These form individual catchments ranging from a few km^2 to tens of km^2, separated by arkose, slates, dykes and solid dolomite. The most productive zone extends to depths of 100 m; karst cavities at Chinhoyi Caves reach depths of at least 150 m.

Natural groundwater recharge depends on local rainfall within the surface catchment of individual aquifer compartments. Visible discharge occurs through discrete springs on the eastern outcrop of the Mcheka Formation which is the eastern flank of the Hunyani Range (Figure 2). West of the Hunyani Range, groundwater discharge occurs in stream beds and wetlands.

The exploitation of the Lomagundi dolomite aquifer is characterised by rapid rest level response to both rain recharge and intensive abstraction, strongly seasonal demand, and less than optimal well field layout determined by demand and access to electric power.

3 HYDROCHEMISTRY

The dolomite groundwater is characterized by low to moderate mineralisation (300-800 mg l^{-1} total dissolved solids). Predominant ions are HCO_3 and Ca+Mg at more than 80 meq% (Figure 3). HCO_3 is low in spring water of the Hunyani Range, but higher in irrigation areas (up to 450 mg l^{-1}). Sulphate concentrations are low but reach 130 mg l^{-1} and higher in areas with sulphide mineralization. Nitrate concentrations are increasing and are a cause for concern. The WHO recommended maximum for potable water is 45 mg l^{-1} and values close to 40 mg l^{-1} have been measured in the dolomite groundwater.

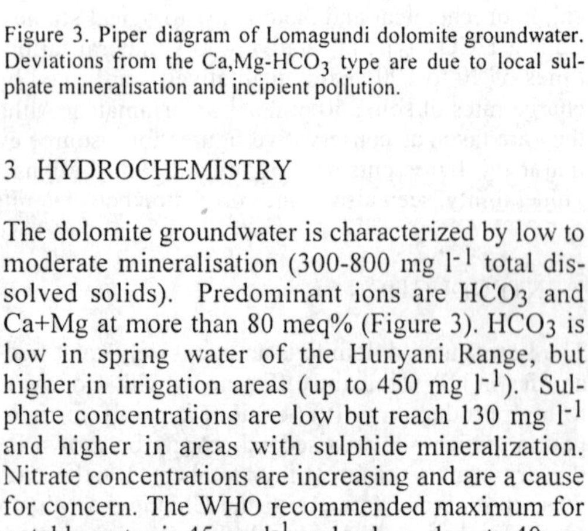

Figure 4. Nitrate concentrations plotted against tritium suggest irrigation return flow as the source.

Values above about 10 mg l^{-1} are all associated with tritium above 1.5 TU (section 5.3) which suggests strongly that the nitrate is derived from fertilisers in irrigation return flow (Figure 4). Other contaminants such as pesticides are probably reaching the saturated zone as well.

4 GROUNDWATER BALANCE ESTIMATES FROM HYDROGEOLOGICAL DATA

Pumping tests conducted by BGR gave transmissivity values of 100 to 1000 m^2 per day. Hydraulic data is however inadequate to assess reliably the groundwater flow pattern.

Recharge was estimated using the water balance equation :

$$R - D_{nat} - D_{art} - S^* \, dh/dt^* \, A = 0 \qquad (1)$$

where R [m^3/s] = groundwater recharge; D_{nat} [$m^3\ s^{-1}$] = natural discharge; D_{art} [$m^3\ s^{-1}$] = groundwater abstraction; S = storage coefficient; A [m^2] size of aquifer outcrop; dh/dt [$m\ s^{-1}$] = rate of change of water level. Subsurface inflow and outflow were neglected.

On the basis of eqn. 1, the groundwater balance was established for several catchments of the study area and the average recharge rate was estimated. The values range between 95 and 178 mm a^{-1}. At Lusaka (Zambia), an average annual recharge of around 180 mm has been assessed. Using a rainfall/recharge relationship for dolomite aquifers in South Africa (Bredenkamp et al. 1995) gives a recharge of 152 mm for Chinhoyi, mean annual rainfall 820 mm. Assuming a mean recharge rate of 120 mm a^{-1} gives a total recharge of 45 million $m^3\ a^{-1}$ over a dolomite outcrop of about 38O km^2. Though this volume is nearly double the estimated abstraction and natural groundwater discharge (19 M $m^3\ a^{-1}$ and 5.4 M $m^3\ a^{-1}$ resp.) for the same area in 1986/87, groundwater over-exploitation may occur locally.

The total storage in the aquifer, estimated thickness of 80 m and porosity 0.03 is 2400 mm water column, giving a seasonal groundwater turnover of 5 % of stored volume. The very approximate nature of these groundwater balance estimates is evident, prompting environmental isotope studies to independently assess recharge rates.

5 ISOTOPE HYDROLOGY

5.1 *Environmental isotopes in precipitation and groundwater*

As no isotopic data are available for rainfall in the project area, deuterium, oxygen- 18 and tritium compositions are inferred from Harare, the nearest IAEA/WMO GNIP (Global Network for Isotopes in Precipitation) station with data at intervals from 1960 to 1983 (IAEA, 1992).

Groundwater samples were taken from springs and numerous pumping wells from 1981 to 1990, variously by Wits, BGR and the IAEA in collaboration with the Groundwater Branch, Harare. Isotopic analyses were conducted by the Environmental Isotope Laboratory, Wits, and the Isotope Hydrology Laboratory, IAEA. Chemical analyses were performed by BGR and IAEA.

5.2 *Stable isotopes deuterium and oxygen-18*

Monthly mean deuterium and oxygen-18 data of Harare precipitation observed between 1960 and 1976, plotted in Figure 5 fit the Local Meteoric Water Line (LMWL): $\delta D = 7.2\ \delta^{18}O + 10$ (‰), the slope of which is slightly lower than the World Meteoric Water Line (WMWL): $\delta D = 8\ \delta^{18}O + 10$ (‰), also shown in Fig.5. The deuterium excess, d, is defined as: $d = \delta D - 8\delta^{18}O$

Plotting the $\delta^{18}O$ values versus rainfall, an ellipsoidal curve ("hysteresis") is obtained (Figure 6) which reflects the rainfall amount effect on stable isotopes (Dansgaard, 1964). The hysteresis phenomenon seems to be due to the movement of the Inter-Tropical Convergence Zone (I.T.C.Z.), and the phase shift between air and sea surface temperature at the oceanic source region of the precipitation (Froehlich et al., in preparation).

The deuterium vs. oxygen-18 plot of mean monthly precipitation values and of spring samples (Figure 7) indicates that the spring water tends to be replenished by monthly rain (December - March; > 50 mm) which has $\delta^{18}O$ lower than about -6‰. Spring values of $\delta^{18}O$ = -2.4‰ and -4.4‰ lying off the MWL indicate evaporative isotopic enrichment from the spring pools sampled.

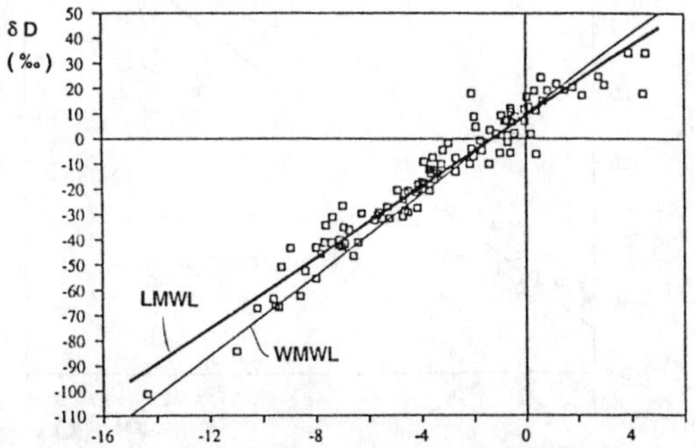

Figure 5. δD - $\delta^{18}O$ diagram for Harare rainfall stable isotope values (IAEA 1992). Also shown: LMWL and WMWL (see text).

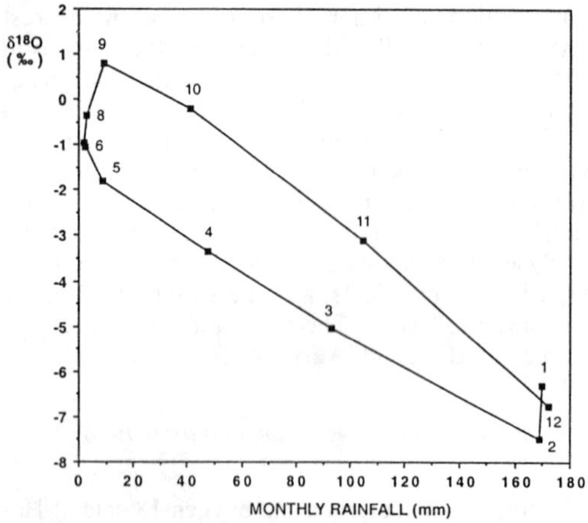

Figure 6. $\delta^{18}O$ values versus mean monthly rainfall at Harare (IAEA 1992) showing "hysteresis" effect. Labels on points indicate months (Jan - Dec).

Pumping wells are slightly to significantly enriched in heavy isotopes (Figure 8; labels: S = spring, I = irrigation well). Long-term heavy pumping may enhance recharge from rainfall of lower intensity whilst isotopically enriched values, some with lower deuterium excess (d), show irrigation return of sprinkle irrigation, isotopically enriched through evaporation.

5.3 *Tritium and radiocarbon*

The rainfall tritium input function (Figure 9) was reconstructed from partial records for Harare rain, supplemented by data from Pretoria and Ottawa GNIP stations (IAEA 1992).

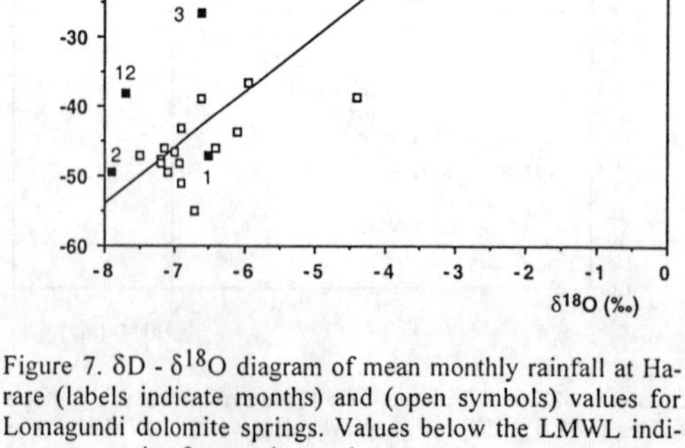

Figure 7. δD - $\delta^{18}O$ diagram of mean monthly rainfall at Harare (labels indicate months) and (open symbols) values for Lomagundi dolomite springs. Values below the LMWL indicate evaporation from spring pools.

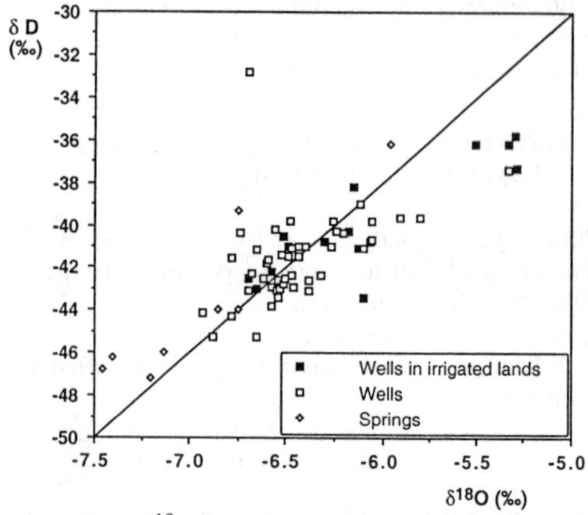

Figure 8. δD - $\delta^{18}O$ diagram for pumped groundwater and for springs during three sampling campaigns. S: springs I: wells situated in irrigated lands.

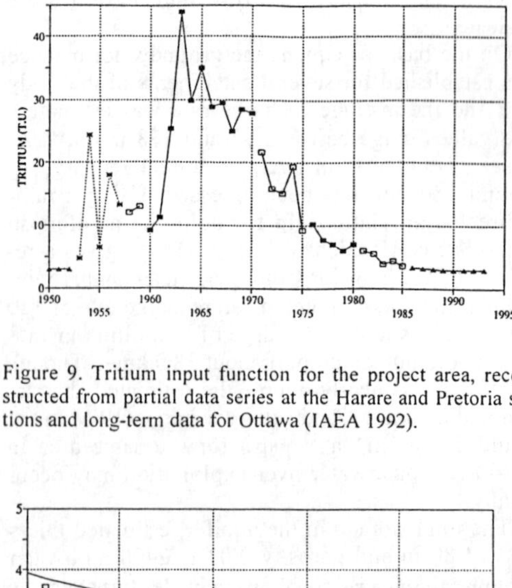

Figure 9. Tritium input function for the project area, reconstructed from partial data series at the Harare and Pretoria stations and long-term data for Ottawa (IAEA 1992).

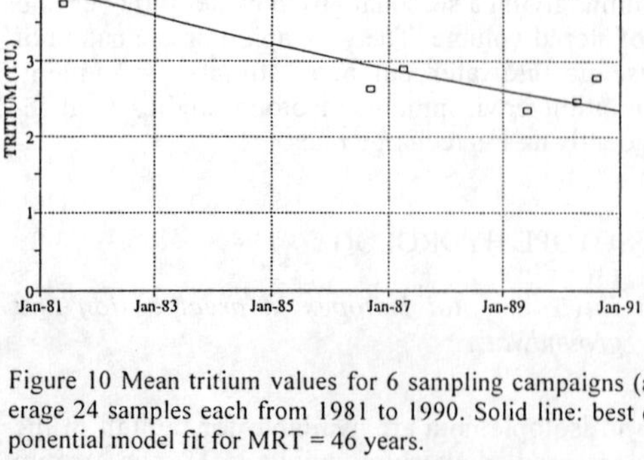

Figure 10 Mean tritium values for 6 sampling campaigns (average 24 samples each from 1981 to 1990. Solid line: best exponential model fit for MRT = 46 years.

The tritium concentration of groundwater samples taken during the six sampling campaigns between 1981 and 1990 (averaging 24 sampling points) shows a spatial and temporal variability from 0 to more than 6 T.U. Mean tritium concentrations calculated for the different sampling periods tend steadily to decrease from about 4 TU to 2.5 TU (Figure 10).

The solid curve represents the theoretical tritium output function derived from the equation:

$$C(t) = \int_0^{\infty} C_0(t-\tau)e^{-\lambda t} f(\tau)d\tau \qquad (2)$$

where $C(t)$, C_o (t -τ) = output, input function, respectively; τ = groundwater residence time; λ = decay constant for tritium (0.558 a^{-1}); f(τ) = residence time distribution function which defines the fraction of water $\alpha(\tau_o)$ which entered the system during a given period prior to the observation time. If this period is called τ_o and the fraction of 'young water' (Gellermann et al. 1990):

$$\alpha(\tau_o) = \int_0^{t0} f(\tau)d\tau \qquad (3)$$

(3)

The Lomagundi dolomite aquifer system can be considered as unconfined and the residence time distribution to a good approximation follows an exponential function (Verhagen et al. 1991):

$$f(\tau) = \frac{e^{-\tau/\bar{\tau}}}{\bar{\tau}} \qquad (4)$$

$\bar{\tau}$ is the mean residence time, which for unconfined aquifers with porosity (p), aquifer thickness (H), and recharge rate (R) is given by:

$$\bar{\tau} = \frac{p.H}{R} \qquad (5)$$

The curve in Figure 10 represents the best fit obtained for a mean residence time of 46 years. Individual sampling sites give mean residence times from about 30 years to more than 100 years. Most of the ^{14}C (half-life 5730 years) values observed (Figure 12) lie above 80 pMC and nearly half above 100 pMC. These indicate various contributions of bomb-^{14}C to the TDIC of the groundwater.

The solid line in Figure 11 shows the theoretical locus of expected ^3H and ^{14}C values for different MRT in a phreatic aquifer. A number of points lie significantly below this line. These are interpreted as the result of isotope dilution of ^{14}C due to continued hydrochemical interaction with the aquifer following recharge (Verhagen et al. 1991).

Considering eqs. (2) and (3), the fraction of water infiltrated during the past 30 years ($\tau_o(30)$) can be derived from the mean residence times estimated. This fraction of "young" water ranges from about 25% to 60% with an average of 45% (mean residence time = 50 years). Groundwater is therefore vulnerable to pollution, especially in areas under heavy irrigation. This is also seen in the correlation of nitrate concentrations with tritium (section 3).
The mean recharge rate can be estimated using eqn. 5. The porosity of the dolomite is estimated between 3 and 10 %. Adopting a mean saturated thickness of 40 m penetrated by all wells and a mean residence time of 50 years, we obtain a recharge rate R ranging from 24 mm/yr (p = 3%) to 80 mm/yr (p = 10%). This range is generally lower than the one found on the basis of water balance considerations. The simplified water balance seems to over-estimate the natural discharge by neglecting water table fluctuations and the influence of sub-surface discharge. However, stable isotope data suggest, if as yet only qualitatively, that irrigation return contributes significantly to regional recharge. The two results can be considered to be complementary, i.e. the isotopic approach defines the *minimum* recharge rate to be considered in the water balance of Lomagundi Dolomite groundwater resources.

The discordance in the estimated recharge rates calls for the need to establish more accurate aquifer parameters, especially the critical value of porosity. Rain recharge is subject to climatologic variations and locally to hydrogeologic conditions (heterogeneity of the aquifer system) as is revealed by the wide ranging tritium concentrations.

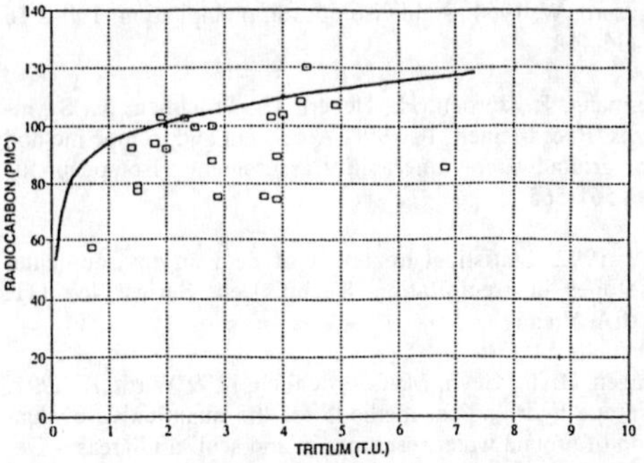

Figure 11 Radiocarbon values plotted against tritium for a number of sampling sites. The solid curve is the theoretical locus of values in a phreatic aquifer for different values of MRT.

6 CONCLUSIONS

(i) Isotope techniques have established that, next to precipitation, irrigation returns are an important source of groundwater recharge in various locations of the area of study

(ii) A conservative estimate of the natural replenishment of the groundwater resources in the study area can be made on the basis of isotope data under prevailing climatic conditions.

(iii) Both stable and radioactive isotope techniques have established considerable aquifer vulnerability to pollution and allow minor trends in the hydrochemistry to be interpreted as an early warning of incipient pollution

(iv) In the light of the important conclusions regarding the management and protection of this valuable resource, there is an urgent need to continue and expand these studies, which were conducted nearly a decade and more ago.

7 ACKNOWLEDGEMENTS

The authors are indebted to Dr P Wurzel, formerly of the Department of Water Development and the University of Zimbabwe, who initiated this study. They acknowledge the financial and material support of Department of Water Development, Harare, the University of the Witwatersrand, Johannesburg, the Federal Institute for Geoscience, Hanover and the International Atomic Energy Agency, Vienna. Michael Butler is thanked for his considerable efforts in preparing the final version of this paper.

REFERENCES

Bredenkamp, D.B., Botha, L.J., van Tonder, G.J. & van Rensburg, H.J. 1995 Manual on the quantitative estimation of groundwater recharge and aquifer storativity. Water Research Commission, Pretoria

Dansgaard, W. 1964. Stable isotopes in precipitation. Tellus 16 : 436-468

Gellermann, R., Jordan, H., Hebert, D., Froehlich, K., Szymczak, P. & Meinert, N. 1990. A concept and isotope method for groundwater vulnerability assessment. Isotopenpraxis 26: 561-565.

IAEA. 1992. Statistical treatment of data on environmental isotopes in precipitation. Techn. Rep. Series No. 331. IAEA, Vienna

Verhagen, B.Th., Geyh, M.A., Froehlich, K. & Wirth, K. 1991. Isotope hydrological methods for the quantitative evaluation of ground water resources in and semi-arid areas. - Development of a methodology. Research Reports of the Federal Ministry for Economic Cooperation of the Federal Republic of Germany. Bonn.

Groundwater: Past Achievements and Future Challenges, Sililo et al. (eds) © 2000 Balkema, Rotterdam, ISBN 90 5809 159 7

Karst aquifer characterization using physico-chemical spring data, Gran Sasso, Central Italy

M.Tallini & D.Ranalli
Dipartimento Ingegneria Strutture Acque Terreno, Università de l'Aquila, Italy

M.Petitta
Dipartimento Scienze della Terra, Università di Roma 'La Sapienza', Roma, Italy

S.Tersigni
Dipartimento Scienze Ambientali, L'Aquila University, Italy

ABSTRACT: During the hydrogeological study of the Gran Sasso karst aquifer, typical of the Mediterranean area, new insights were gained on regional groundwater circulation through statistical analysis of physico-chemical data sets. The Gran Sasso carbonate aquifer, laterally confined by impermeable sediments, has a high recharge, due to tectonic pattern, climate and karst features. The springs of the Gran Sasso may be divided into 6 groups. These springs, which drain the massif, have an average discharge of about 18 m^3/s. The construction of two highway tunnels obviously interfered with and involved many changes in this hydrogeological setting. The processing of temperature and resistivity/conductivity data sets showed significant differences between the springs. These findings suggest that the conceptual model of the aquifer, developed on the basis of hydrogeological and discharge data, should be redefined and that new chemical studies should be planned.

1 INTRODUCTION

The characterization of groundwater in time and space demands an in-depth knowledge of its flowpath, of residence time inside the original aquifers, as well as of the recharge areas of springs. Common methods are based on the use of tracers (Meus & Bakalowicz 1997, Zuppi 1993), which may be natural or artificial.

In addition to these methods, useful insights into groundwater circulation may be derived from the study of its temperature and electrical conductivity in time and space, based on the hydrological regime of their springs or on the pattern of piezometric surfaces (Freixes et al. 1998, Maréchal et al. 1999, Tulipano 1998). As is known, the temperature of groundwater depends on the heat capacity of the aquifer, on its residence time inside it and on the temperature of rain waters. Electrical conductivity, instead, is conditional upon water-rock interaction, residence time and conductivity of the original water. These considerations apply to isolated aquifers, without mixing with foreign groundwater.

The paper discusses the findings from the study of the temperature and conductivity patterns of the groundwater contained in a fractured carbonate massif, which is typical of the Mediterranean area.

2 HYDROGEOLOGICAL SETTING

The above-mentioned methodology was applied to the Gran Sasso massif (surface area: about 800 km^2; altitude from 2,912 to 300 m a.s.l.). The massif holds a regional aquifer, whose permeability is due to fracturing and karst and whose recharge (net recharge: 700 mm/yr) is ensured by precipitation (average: 1,100 mm/yr). The aquifer supplies local springs of perched aquifers, and especially the large basal springs which are located at the margins of the Gran Sasso (total discharge: 18 m^3/s, Fig. 1).

The carbonate structure of the Gran Sasso consists of a Meso-Cenozoic slope to basin sequence (grain- and mud-supported limestone, marl and marly limestone; at the bottom dolomite), with an overthrust in its northeastern part and multiple monoclines, downthrown by extensional faults, in its southern part (Vezzani & Ghisetti 1998).

The boundaries of the regional aquifer are well defined by stratigraphic and tectonic contacts with terrigenous (sandstone, clay and marl, Upper Miocene) or clastic continental deposits (gravel, sand, clay and breccia, Quaternary). The permeability of these deposits is much lower than the one at the core of the Gran Sasso (Boni et al. 1986). Hydrogeologically, these deposits act as aquicludes or aquitards, as they cause the drainage of karst regional groundwater through basal springs. The Capo Pescara springs (discharge = 7 m^3/s) are also drained by Sirente Mt. Aquifer (Fig. 1).

From 1968 to 1980, two parallel highway tunnels were drilled through the Gran Sasso at an elevation of 970 m a.s.l. (total length 10 km). The construction of the tunnels interfered with the hydrogeological setting of the Gran Sasso. During the drilling, many hydrogeological studies were carried out (Monjoie 1975), about discharge, chemical and physico-chemical characteristics of the main springs. As the tunnels intercepted the regional aquifer of the Gran Sasso, they have been draining sizeable amounts of groundwater from it.

At present, the groundwater drained from the tunnels (drinking use) is estimated to amount to 1 m^3/s on the northern side of the Gran Sasso and 0.4 m^3/s on the southern one. This drainage has been decreasing the discharge, especially from the springs of the northern side of the mountain (Group 1), but also from the southern springs (Groups 3 and 5).

Subsequently, the regional aquifer gradually adjusted to the new situation and the drainage groundwater can now be considered as coming from two new "springs", defined as constant-head points.

Schematically, the Gran Sasso springs can be divided into 6 main groups (Fig. 1), with different groundwater circulation. The northern side of the Gran Sasso accommodates multiple springs of groundwater draining the regional aquifer. These springs, which flow at elevations of over 1,000 and have a variable discharge, come from the most elevated sectors of the Gran Sasso (Ruzzo springs: Group 1). The southern side of the Gran Sasso, instead, holds several springs with a moderate discharge. These local springs are connected to perched aquifers (Raiale springs; Group 2). The springs with the highest discharge drain groundwater from the southern boundary of the basal regional aquifer. The springs of the Vera River (Group 3), which have a very steady discharge, are located in the area of L'Aquila. Other basal springs (L'Aquila Plain springs: Group 4a) lie a few kilometers to the east of L'Aquila. These springs are fed by the northwestern sector of the aquifer and receive significant contributions from the L'Aquila-Scoppito Plain multi-layer aquifer (Group 4b). The Tirino and Popoli springs (Group 5) drain the aquifer at the lowest elevations; their groundwater represents the final drainage of the basal aquifer (average discharge about 15 m^3/s). The groundwater from the regional aquifer that is drained in the Gran Sasso tunnels (discharge about 1.4 m^3/s) belong to Group 6.

3 GEOCHEMICAL SPRING FEATURES

Analyses of major elements (Ca, Mg, Na, K, HCO_3, SO_4, Cl) and of SiO_2, were selected from 27 springs (ANAS 1991, Stigliano et al. 1999). For each spring, the samples varied both in number and sampling periods (1968-1999). From the geochemical point of view, the groundwater is of alkaline-earth bicarbonate type, with a particularly high calcium ion content (Fig. 2). The analysis of the main chemical components of the water samples does not allow to discriminate among the different springs.

Based on typical Ca/Mg ratios (Fig. 3), the springs fed by the regional basal aquifer (Groups 1-3-4a-5-6) can be distinguished from those connected with perched aquifers (Group 2) and from springs characterized by mixing with multi-layer aquifer (Group 4b). The groundwater chemical evolution depends on its interaction with carbonate lithologies, namely on their residence time within the karst aquifer and on the mineralogical characteristics of the rocks.

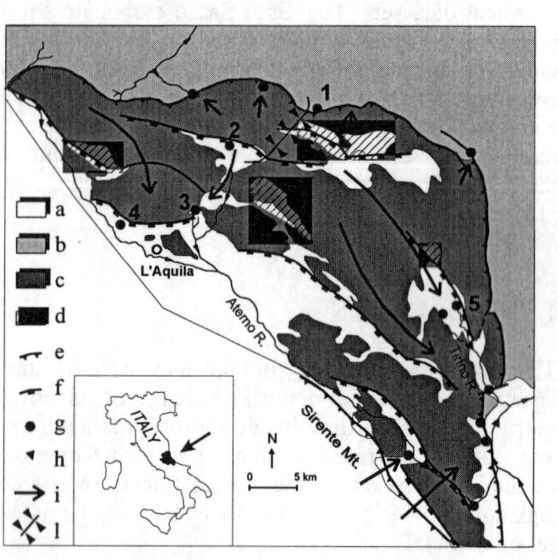

Figure 1. Hydrogeological setting of Gran Sasso massif. a) quaternary clastic continental deposits (aquitard); b) Upper Miocene terrigenous foredeep deposits (aquiclude); c) Meso-cenozoic carbonate deposits (aquifer); d) Upper Triassic dolomite (low permeability sequence); e) overthrust; f) normal fault; g) main spring; h) linear spring; i) groundwater flowpath; l) tunnel drainage. 1) Ruzzo spring group; 2) Raiale spring group; 3) Vera spring group; 4) L'Aquila Plain spring group; 5) Tirino spring group; 6) Gran Sasso tunnel drainage.

A preliminary study of the groundwater, via multivariate statistical analysis and computation of chemical equilibria in water solutions (Kharaka et al. 1988), led to identify the following main groups:

- Short residence time groundwater which do not reach equilibrium with calcite (Group 1);
- Long residence time spring waters, in equilibrium with calcite (Group 4a);
- Spring waters mixing with the groundwater from the multi-layer aquifer which lie in the clastic deposits of the L'Aquila Plain (Group 4b);
- Groundwater from the basal aquifer drained near

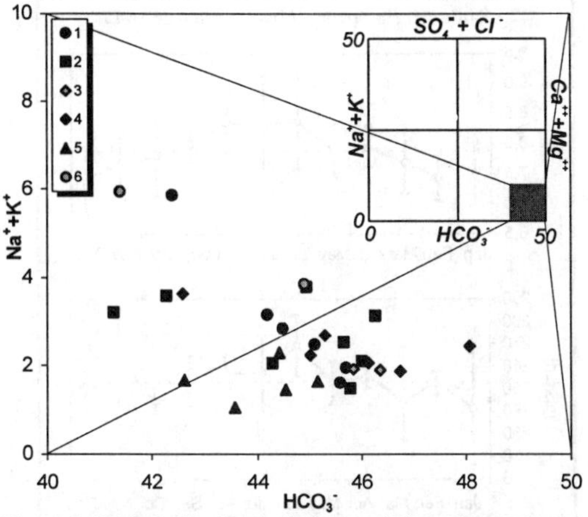

Figure 2. Main hydrochemical characteristics of Gran Sasso groundwater, plotted in Chebotarev diagram. 1) Ruzzo spring group; 2) Raiale spring group; 3) Vera spring group; 4) L'Aquila plain spring group; 5) Tirino spring group; 6) Gran Sasso tunnel drainage.

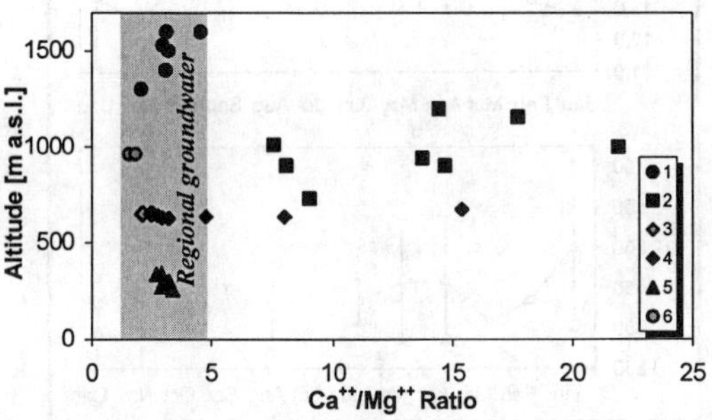

Figure 3. Altitude versus Ca^{++}/Mg^{++} ratio diagram. 1) Ruzzo spring group; 2) Raiale spring group; 3) Vera spring group; 4) L'Aquila plain spring group; 5) Tirino spring group; 6) Gran Sasso tunnel drainage. Grey area shows regional groundwater.

its recharge area, not saturated in terms of carbonate-calcium ion equilibria (Groups 3 and 6);

- Spring waters which have a long residence time inside their karst aquifer, are calcite-saturated and receive the contributions of short residence time recharge water (Group 5).

4 PHYSICO-CHEMICAL DATA DISCUSSION

Temperature and conductivity data derive from a monthly monitoring of groundwater (28 springs) in the Gran Sasso aquifer (1970s) and in the L'Aquila Plain (1990s). Most of the data (about 3,600 temperature and conductivity values) were kindly made available by ANAS (Italian National Highway Company). The data were processed with statistical software programs, which yielded the main physico-chemical characteristics of the groundwater. The results confirmed the proposed division into six groups of springs and provided additional data on groundwater circulation inside the Gran Sasso. Figure 4 displays the monthly temperature and conductivity values for one spring in each group.

The S.Nicola I spring (Group 1) shows a significant seasonal variability of temperature and, to a lesser extent, conductivity values which have a recurrent cyclical pattern: in Spring, snow thawing causes the recharge of the karst aquifer at high elevation and thus the spring. The other springs of Group 1 benefit of the input of groundwater with shorter and less deep flowpath.

The parameters of the S.Maria spring (Group 2) display less significant seasonal variations. The group includes springs connected with local aquifers. The average temperature and conductivity values are not very high and very similar to those of Group 1.

The temperature and conductivity values of the Tempera spring (Group 3) are very constant over time. The spring is thus supposed to be directly fed by the basal aquifer, as it shows no mixing with shallow groundwater and a poor water-rock interaction with respect to the Gran Sasso core aquifer (Group 6).

The Boschetto spring (Group 4) represents the overflow of the basal groundwater in the L'Aquila Plain (Group 4a). This groundwater reflects mixing with the multi-layer aquifer (Group 4b). In this case too, seasonal variations are clear, and absolute values are higher than those of Group 3, because the spring is more distant from the regional aquifer core.

The Capodacqua groundwater (Group 5) represents the lower drainage of the regional aquifer (about 300 m a.s.l.). The discharge is considerable and constant over time. The physico-chemical characteristics of this groundwater are very stable but with much higher values than those of Group 3.

The groundwater in the tunnels (lowest values, Group 6) represents the core contribution of the regional aquifer, where it is directly recharged by rain waters (vertical pathway in the vadose zone).

To summarize the relations between the 6 groups of springs on the basis of their physico-chemical parameters, two elevation-temperature (Fig. 5) and elevation-conductivity (Fig. 6) diagrams were drawn.

Figure 5 shows that the groundwater of Group 6 represents the extreme end-member from which almost all other spring waters evolve, taking different flowpath inside the aquifer. In fact, the temperature values (Fig. 5) have a clear evolutionary trend, which departs from the groundwater of the tunnels (Group 6), reaches the springs at the lowest elevations (Group 5), and passes through the basal springs of Group 3.

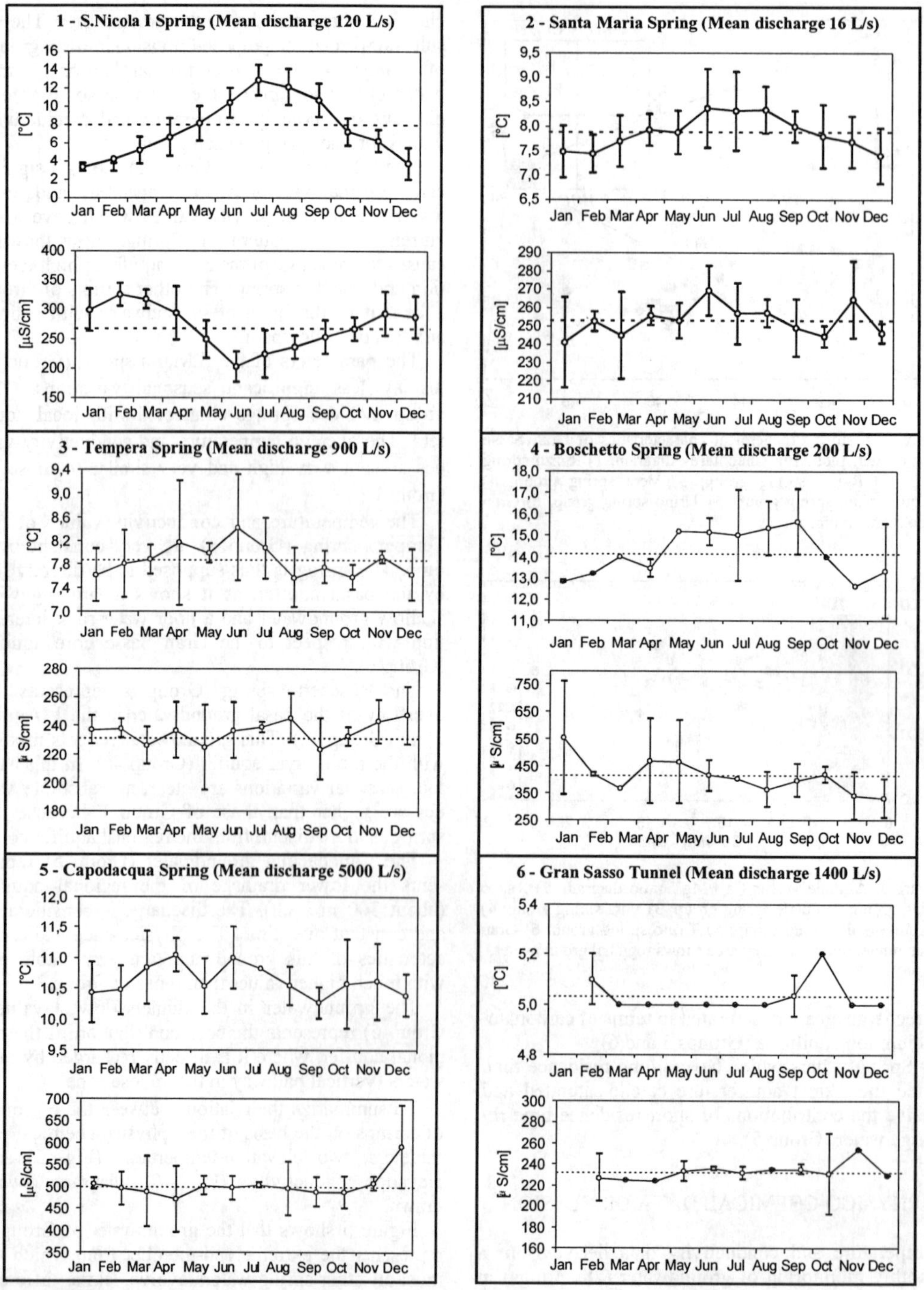

Figure 4. Temperature and conductivity regime of a representative spring for each Gran Sasso spring group. 1) Ruzzo spring group; 2) Raiale spring group; 3) Vera spring group; 4) L'Aquila plain spring group; 5) Tirino spring group; 6) Gran Sasso tunnel drainage. In each diagram the vertical lines represent standard deviation and the horizontal dotted line represents mean value.

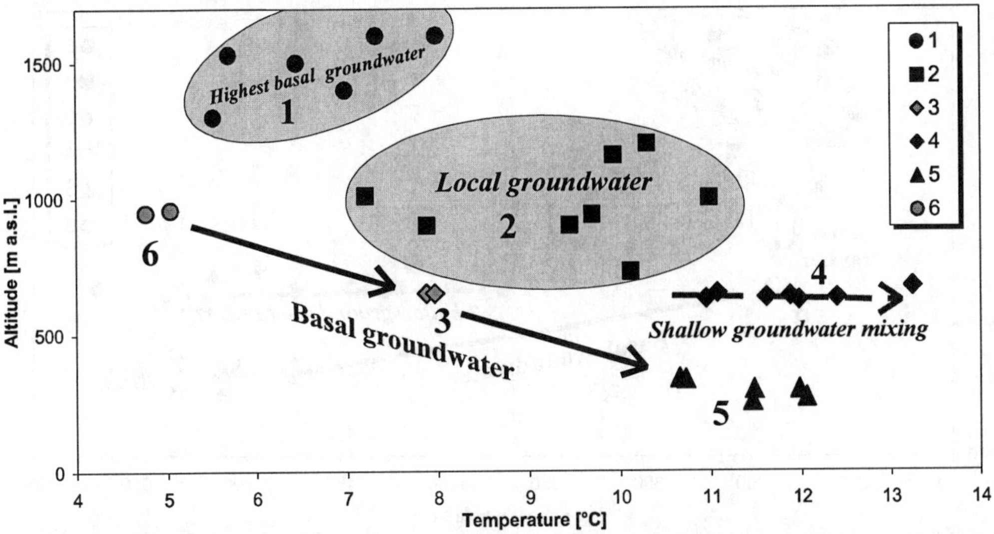

Figure 5. Altitude versus average temperature diagram of Gran Sasso groundwater. Black arrows show flowpath trend of basal groundwater. Grey dotted arrow shows mixing between Gran Sasso basal groundwater and L'Aquila Plain groundwater. Numbers refer to the single spring group (Fig. 1).

This trend infers that the longer the distance covered by the basal groundwater, the higher its temperature. The same trend identifies mixing of the spring water of Group 4a with the groundwater connected with multi-layer aquifer (Group 4b). The springs on the northern side of the Gran Sasso (Group 1) have temperature values similar to those at aquifer core. These springs are recharged from the most elevated sectors of the basal aquifer and individual differences are to be ascribed to local hydrostructural features. The Group 2 springs, with low discharge, show a more significant scatter of temperature values, suggesting that recharge is affected by local specificities.

The main difference between Figure 5 and Figure 6 lies in the constancy of conductivity values when passing from Group 6 to Group 3. This finding implies a reduced water-rock interaction, which may be justified by fast drainage through karst conduits. This assumption is validated by the relationship between the discharge of the Vera springs (Group 3) and the drainage in the tunnels. The springs of Group 5 have constant conductivity, with values that are practically twice those of Group 6. This trend is reflective of a significant water-rock interaction, which was not evidenced by typical chemical ratios (Fig. 3). The groundwater of Group 4 confirms the scatter of values which is due to mixing with shallow groundwater having a higher salinity. This mixing results from water-rock interaction processes which are different from karstic ones (clay minerals, silicates), but also from anthropic contamination of the multi-layer aquifer of the L'Aquila Plain.

For Group 1 springs, the values are fairly homogeneous. Values lower than those at the aquifer core (Group 6), are indicative of recharge from the most elevated areas of the regional basal aquifer.

The values of Group 2, instead, are extremely scattered, just as temperature values (Fig. 5). The assumption that the springs of this group are in part recharged by local shallow groundwater (rapid circulation, limited recharge areas) is further substantiated by some of its conductivity values, that are lower than those at the core of the regional aquifer.

5 FINAL REMARKS

By applying the above methodology to the hydrogeological study of the Gran Sasso aquifer, the following results were achieved (Figs 5-6):

- identification of groundwater flowpath;
- discrimination of groundwater flowing along karst conduits from groundwater flowing through fissured aquifer (Groups 3-6);
- discrimination of springs recharged by the basal aquifer (Groups 3-6) from those connected with perched aquifers (Group 2);
- groundwater mixing from the regional aquifer (Group 4a) with shallow groundwater (Group 4b);
- preliminary data on the different rates of groundwater transfer from the core to the boundaries of the hydrogeological system (Groups 6-3-5).

This methodology also offers the advantage of a straightforward and cheap process of data collection. Therefore, it is suitable for the preliminary stages of

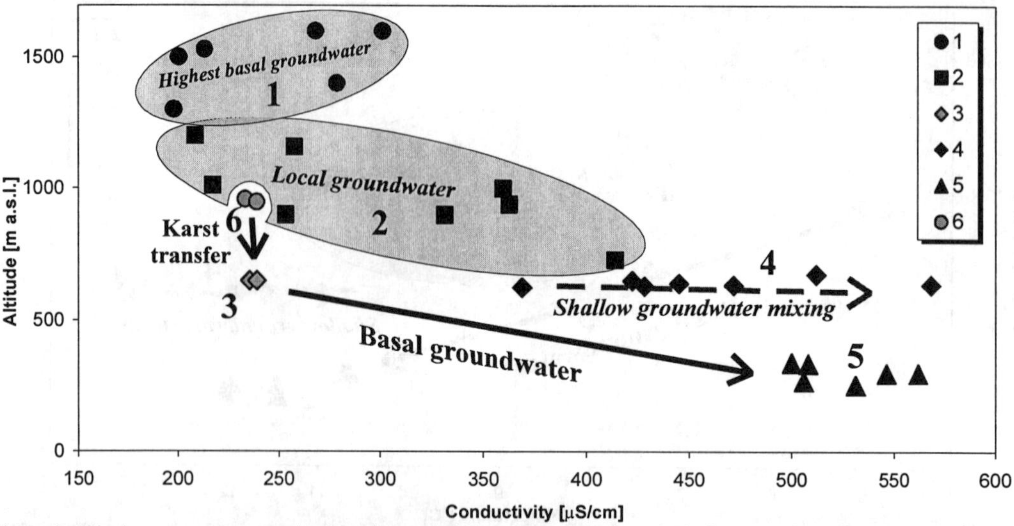

Figure 6. Altitude versus average conductivity diagram of Gran Sasso groundwater. Black arrows show flowpath trend of basal groundwater. The vertical arrow between Groups 6 and 3 shows a presumed karst groundwater transfer. Grey dotted arrow shows mixing between Gran Sasso groundwater and L'Aquila Plain groundwater. Numbers refer to the single spring group (Fig. 1).

groundwater chemistry studies. After these stages, more comprehensive studies may be initiated, minimizing the number of sampling points and consequently costs.

REFERENCES

A.N.A.S., 1991. Completamento ed adeguamento delle strutture del laboratorio di Fisica Nucleare del Gran Sasso. Studio di impatto ambientale, 4-5-6, Unpublished report.

Boni, C., P. Bono & G. Capelli 1986. Schema Idrogeologico dell'Italia Centrale. *Mem. Soc. Geol. It.* 36: 991-1012.

Freixes A., M. Monterde & J. Ramoneda 1998. Spring water temperature as a characteristic feature of karst aquifer behaviour: Examples from the Central Pyrenees (NE Spain). *Hydrogéologie* 1: 31-40.

Kharaka Y.K., W.D. Gunter, P.K. Aggarwal, E.H. Perkins & J.D. DeBraal 1988. Solmineq88: a computer program for geochemical modelling of water-rock interactions. *U.S. Geol. Survey, Water-Resources Inv. Report* 88-4227.

Maréchal J.C., P. Perrochet & L. Tacher 1999. Long-term simulations of thermal and hydraulic characteristics in a mountain massif: The Mont Blanc case study, French and Italian Alps. *Hydrogeology Journal* 7: 341-354.

Meus P. & M. Bakalowicz 1997. Les tracages artificiels, outils de reconnaissance et d'etude des aquifères karstiques. *Hydrogéologie 3: 43-50.*

Monjoie A. 1975. Hydrogéologie du Massif du Gran Sasso (Apennin Central). *Publ.Fac.Sc.Appl., Univ. Liège* 53:1-60.

Stigliano P., Altieri A., Barberi R., De Caterini G., Miccadei E., Petitta M. & Salvati R., 1999. Studi per un monitoraggio idrologico del massiccio del Gran Sasso (Abruzzo). *Studi Geologici Camerti,* vol.spec. 99/1.

Tulipano L. 1988. Temperature logs interpretation for the identification of preferential flow pathways in the coastal carbonatic and karstic aquifer of the Salento Peninsula (Southern Italy). In: *Karst Hydrogeology and Karst Environment Protection.* IAHS Publ. 176: 956-961.

Vezzani, L. & F. Ghisetti 1998. Carta geologica dell'Abruzzo (1/100.000). S.EL.CA., Firenze

Zuppi G.M. 1993. Impiego dei traccianti per la caratterizzazione dell'acquifero e la valutazione delle aree di salvaguardia. *IGEA* 2: 127-131.

Thanks go to "Consorzio Ricerca Gran Sasso" for the contribution to this research and to ANAS (L'Aquila Department) for giving us part of the data.

Groundwater: Past Achievements and Future Challenges, Sililo et al. (eds) © 2000 Balkema, Rotterdam, ISBN 90 5809 159 7

CFC tracing of groundwater in fractured rock aided with ^{14}C and ^{3}H to identify water mixing

A.S.Talma
Quaternary Dating Research Unit, Environmentek, Council for Scientific and Industrial Research, Pretoria, South Africa

J.M.C.Weaver
Groundwater Group, Environmentek, Council for Scientific and Industrial Research, Stellenbosch, South Africa

L.N.Plummer & E.Busenberg
Water Resources Branch, US Geological Survey, Reston, Va., USA

ABSTRACT: Interpretation of CFC analyses in groundwater samples from fractured rock aquifers is complicated due to mixing of water within the aquifer and during borehole sampling. Recharge dates derived directly from atmospheric CFC master curves are meaningless, since samples contain water from a range of flow paths in different proportions. Additional tracers, with appropriate input functions need to be used to quantify these effects. Groundwater from the fractured Table Mountain Group produced a range of inconsistent ^{14}C ages, tritium ages and CFC recharge dates when interpreted as unmixed, piston-flow water samples. Plots of the tracer concentrations against each other were compared to that of tracer input curves. They demonstrate mixing between a young and an old component that is consistent with the flow patterns established earlier. This approach offers possibilities to use CFC analyses in mixed-water aquifers containing significant amounts of water recharged during the last few decades.

1 INTRODUCTION

Analysis of CFC (chlorofluorocarbons or freons) in groundwater has become a useful tracer technique to identify recharge events of the last few decades (Plummer & Busenberg 1999). The technique has been successfully applied in primary aquifers where flow patterns are regular and water samples represent single years of recharge (Dunkle et al. 1993, Cook & Solomon 1997). In fractured rock aquifers, which form 90% of South Africa's water bearing formations, these conditions usually do not exist. The water encountered at a certain point usually consists of water parcels that have moved through different fractures and water samples therefore represent mixtures of water that has recharged at different times. In many boreholes in which investigations are conducted, the boreholes also tap more than one water strike and water pumped from such boreholes represents a mixture of water from different strikes intercepted by the borehole. The aim of the geohydrologist is to nevertheless describe the recharge history in some way or other.

Of the three CFC gases used for groundwater tracing, CFC-11 and CFC-12 have very similar input functions (Figure 1) and thus complement each other.

CFC-113 entered the market later and its input function is sufficiently different to emphasise later recharge (Talma et al. 1998). In cases where there is a significant component of much older water, ^{14}C and tritium ^{3}H can be used as additional tracers.

This paper shows an application where these three tracers are used in combination to indicate mixing of water with a large range of recharge ages.

2 STUDY AREA: AGTER-WITZENBERG VALLEY

The Agter-Witzenberg valley is a syncline of Table Mountain Group sandstone overlying Bokkeveld shales, 140 km north of Cape Town. The formation is very steeply dipping and highly fractured enabling rapid recharge. Rainfall in the area is mainly in winter, averaging 890 mm/a, when springs develop in the foot of the mountain and water recharges the valley aquifer while the rest drains away. During summer large quantities of water are pumped by local farmers that cause groundwater levels in the valley to be seasonally lowered by tens of meters.

A study was undertaken between 1993 and 1997 during which six boreholes along a line from the mountain peak towards the valley were monitored for a variety of chemical and isotopic parameters (Weaver et al. 1999). Borehole details are (Figure 2):

- G36: high up in the mountain, 22m depth, single water strike (21m);
- G14: existing borehole in the valley, at the foot of the mountain, no details available;

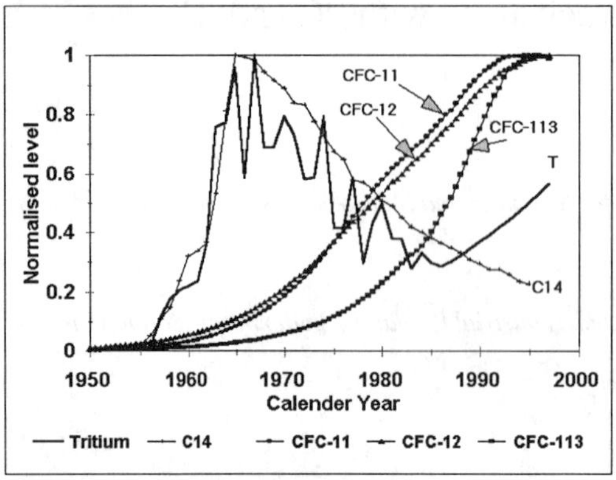

Figure 1. Time series of the relative variations of environmental tracers in groundwater applicable to the W Cape Province, South Africa. Data have been normalized between the estimated value at 1950 AD and the maximum value attained later.

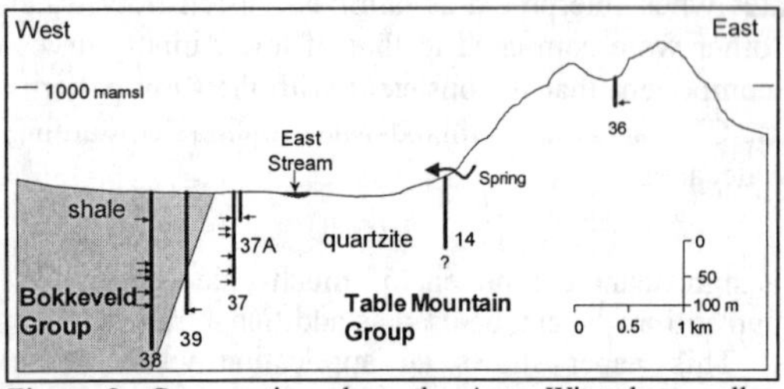

Figure 2. Cross-section along the Agter-Witzenberg valley profile showing topography, presumed geology, borehole positions and water strikes.

- G37A: 21m depth, in TMG, single water strike (19m).
- G37: 109m depth, in TMG, multiple water strikes;
- G38, 146m depth, in shales with multiple water strikes;
- G39: 171m depth borehole through the shales (cased off) into underlying TMG sandstone, several water strikes at the sandstone/shale contact (169-171m)

This study showed the presence of bomb ^{14}C, some tritium, low salinity and low pH in the water of the TMG, but low ^{14}C and low, but detectable (>0.3 TU) tritium, with higher salinity and higher pH in the boreholes in the valley (G38 and G39) (Table 1). These latter boreholes showed seasonal changes in isotope content and chemistry in parallel with the water levels (Weaver et al. 1999). The data indicate that recharge occurs in the mountains as well as in the valley. The boreholes in the valley also showed a strong inflow of deeper, and older, water (Weaver & Talma 2000). Sampling for CFCs indicated (piston-flow) recharge 'dates' ranging from 1950 to 1990 AD that were not concordant within themselves nor with 'ages' estimated from tritium and ^{14}C. In a parallel study in a Karoo aquifer with only very irregular recharge, it was shown that discrepancies between CFC-12 and CFC-113 can be used to separate recharge of the 1980s from pre-1960 recharge (Talma et al. 1998). In the present case the mixing relations were satisfactorily resolved by using CFC with ^{14}C and tritium.

3 TRACERS AND THEIR INPUT FUNCTIONS

Atmospheric CFC levels have increased since 1950, flattened off in the 1990's and are in slow decline for the last few years (Figure 1). The solubility of CFCs in water as a function of temperature and the partial pressure (due to the recharge altitude), enables one to calculate the CFC level in water that is in contact with the atmosphere. Analysis of CFCs in the water and comparison with the atmospheric input curve then yields a model recharge date, which is the date when the water last had contact with the atmosphere under the assumption that it is a single parcel of

Table 1. ^{14}C, tritium and CFC from the Agter-Witzenberg valley (from Weaver & Talma, 1999). CFC data are expressed as fractions of the maximum likely values, in pmf (percent modern freon). For details see text. The old water fraction was calculated from the position of the sample data on the mixing lines of Figure 3 averaged over the useful parameters.

Bh	^{14}C pmc (±0.5)	Tritium TU (±0.3)	CFC-11 pmf	CFC-12 pmf	CFC-113 pmf	Old water fraction
G36	111-115	2.8	17-37	22-38	25-49	0
G14	102-104	2.3	46-55	69-72	51-80	27%
G37	102-106	1.6	18-22	23-32	11-16	40%
G37A	100-104	1.8	18-22	26-29	14-24	38%
G38	54-72	1.2	6-7	12-13	-	72%
G39	29-53	0.4	0-6	1-10	0	90%

water, i.e. piston-flow (Busenberg & Plummer 1992). In South Africa the groundwater table is generally deep (>10 m) and the recharge temperature can be taken as the mean annual air (=soil) temperature.

^{14}C levels of CO_2 in the atmosphere during the last 50 years are dominated by the input from the nuclear weapon tests between 1954 and 1961 which reached the southern hemisphere a few years later (JC Vogel pers. comm.). The original (pre-bomb) level of 95 pmc increased to a maximum of 170 pmc from 1954 to 1963 and has since gradually reduced to about 110 pmc. ^{14}C levels in groundwater less than 95 pmc can be due to dilution by dissolution old carbonate rocks and/or decay of the pre-bomb levels which can indicate ages as much as 40 000 years.

For the Agter-Witzenberg study the ^{14}C input has been taken from the ^{14}C data series for Pretoria air. The pmc values have been reduced by 5% to account for a small amount of dilution by old carbonate. This is justified by the ^{13}C values of all of this groundwater being between -23 and -18 ‰ PDB (Weaver et al. 1999). Rainfall tritium for this study has been estimated as the mean value of the Cape Town (marine) and Pretoria (inland) data (IAEA 1999) and adjusted for decay to 1997.

The pre-bomb tritium level in South African rainfall was probably about 3-5 TU (Vogel et al. 1974). The maximum bomb-tritium levels were never higher than 70 TU (IAEA 1992, 1999). The bomb-tritium peak in groundwater has now decayed to about 10 TU and is still about double the present (2000AD) rainfall value (Figure 1).

Atmospheric CFC input curves for the Southern Hemisphere were constructed for this project. The post-1978 CFC air curves are based on the atmospheric measurements of at Cape Grim, (Tasmania, Australia) (P. Fraser http://cdiac.esd.ornl.gov/ftp/ale_gage_agage/gage/complete/tasmania/). The pre-1978 CFC air curves were reconstructed from the AFEAS production and release data (http://www.afeas.org/) using the method outlined by Plummer and Busenberg (1999).

The relative variation of each of these tracers during the last 50 years, normalized to the minimum and maximum value that occurred during this period is shown in Figure 1 as applicable to the Western Cape interior. The main features are:

- Steady rise of the CFCs, flattening off in the 1990s;
- Rapid rise and slow fall of the ^{14}C levels until the present low levels;
- A similar rise of tritium, but with the present level settling midway between the two extremes.

These curves each have their own characteristics and form the boundary conditions for a mixing model for groundwater. They emphasize the usefulness of CFC data to date water recharged since the 1970s while ^{14}C and tritium are more suited to trace somewhat older water.

4 RESULTS AND DISCUSSION

CFC samples from the boreholes in the study area were collected at various times between 1995 and 1997. Samples were obtained using a sampler of the USGS design (Busenberg & Plummer 1992) directly connected to the sampling pump line. CFC concentrations were converted from pg/kg via the solubility in water, local air temperature (9°C) altitude (980m) to the partial pressure of CFC in air in equilibrium with the water. The partial pressure was divided by the maximum atmospheric CFC concentration (in ppt, between 1990 and 1997) to yield a fraction ranging, in this case, from 0 to 0.8. We think that this parameter will be more useful than concentrations or partial pressures when CFC is used for the interpretation of mixing processes where one deals with straight lines on mixing diagrams and not recharge dates derived from a non-linear curve. Normalizing concentrations also enables one to place the various tracers on a common footing (Figure 1). By analogy with ^{14}C, we express this fraction in percent and name it 'percent modern freon' (pmf).

Plotting of the input functions of the different tracers against each other shows the likely relations between different water samples that are possible in piston-flow situations (Figure 3 a-f). Mixing of two water types is indicated by the two parameters plotting on straight lines located in the 'space' between these curves. The end members of the mixing lines are then located on the tracer plots.

The ^{14}C-T diagram shows all the samples other than G38 and G37 to be of post-bomb age since they cluster along the average relation line (Figure 3a). Using CFC-11 and CFC-12 (Figure 3 c,d,e&f) shows that G36 is of post-bomb age and that the intermediates (G14, G37 and G37A) are mixtures of the old component, represented by G39, and a younger component with average date of 1975AD. Tritium data (Figure 3d&f) support the mixing better than ^{14}C (Figure 3 c&e) does. CFC-113 resolves the younger components somewhat better (Figure 3b).

3c,d,e&f) that the borehole samples lie on a line extrapolating to a recharge date of 1975. The line with CFC-113 (Figure 3b) extrapolates to 1985. It is unlikely that the young component would be recharge of only this date. It is more likely to be a mixture of a number of recharge years well integrated within the mountain, but there are no detailed samples from the mountain to resolve that mixing pattern.

The water samples from the valley can therefore be described as mixtures between a young

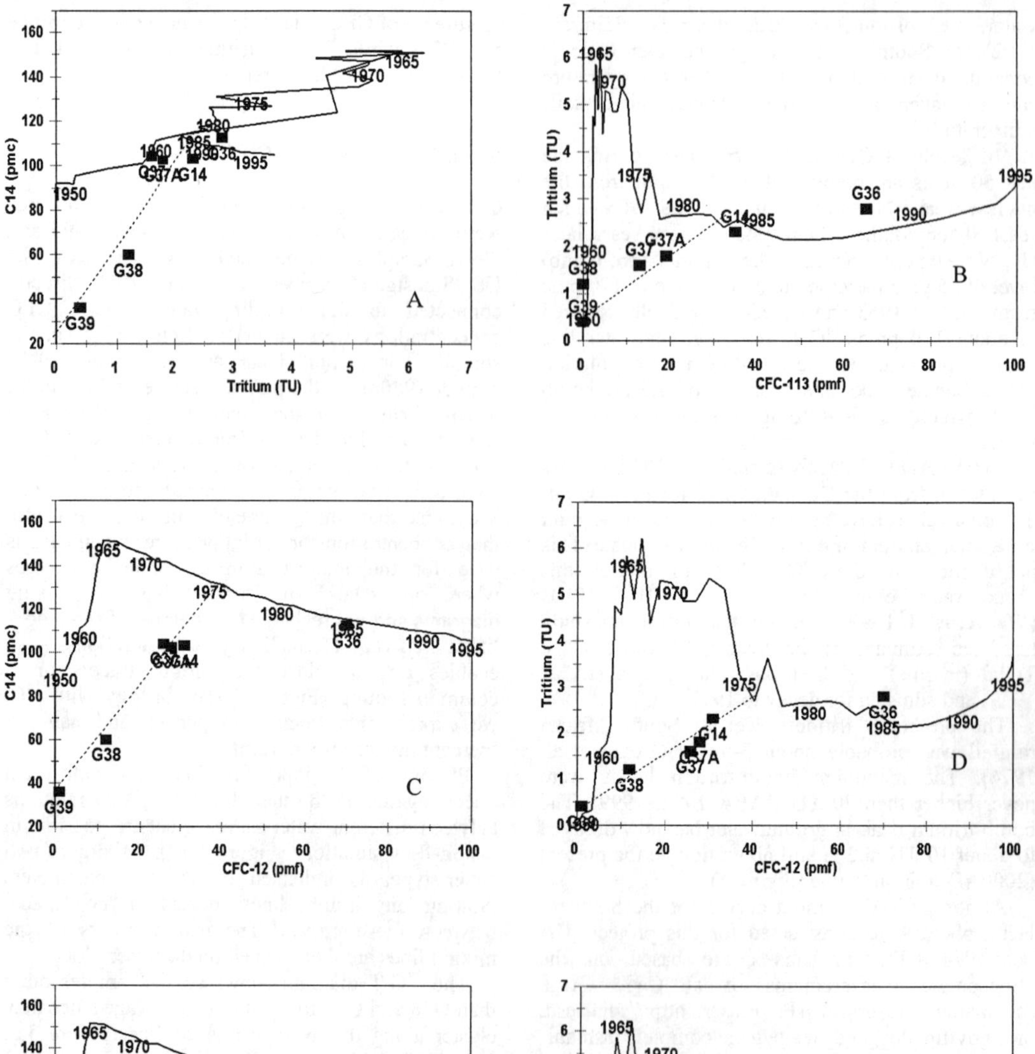

Figure 3: Plots of tracer concentrations derived from input curves of Figure 1 together with the analytical data obtained from the ground water samples. Borehole numbers (G) and recharge dates on the input curves are indicated. The plots are: (a) ^{14}C vs tritium, (b) Tritium vs CFC-113, (c) ^{14}C vs CFC-12, (d) Tritium vs CFC-12, (e) ^{14}C vs CFC-11, (f) Tritium vs CFC-11.

component (with ^{14}C 135 pmc; tritium 3 TU; CFC-11 35pmf; CFC-12 45pmf, CFC-113 30pmf) on the one hand and an old component with ^{14}C of 30 pmc (age 9000 years) and none of the other tracers present. Mixing ratios between these two components can be calculated for each borehole sample based on the position of the sample points on the mixing lines (Table 1).

In more detail the separation of the borehole water as different ratios can be formulated as follows:

- G36 (mountain) is located along the input curves and could be interpreted as representing a single recharge package. Given that the input curves are all nearly straight lines in this age range, it is more likely that G36 represents a mixture of recharge water over a decade or so. At a water strike depth of 16 meters below the ground water level, this seems more reasonable.
- G39 is a single water strike from the TMG underlying the shales. It represents the oldest water end member and evidence for deep groundwater in the TMG (Weaver & Talma 2000). The lowest ^{14}C levels are found in summer when water levels are at their lowest due to local abstraction (Weaver et al. 1999) and indicate residence times in the order of 9000 years for the old component.
- G38 resembles G39 and essentially taps water from the same depth. It contains a quantity of younger water, probably due to the presence of superficial water strikes that have not been cased off in this particular borehole (Figure 2).
- G14 at the foot of the mountain (with unknown number of water strikes) is likely to be a mixture of young recharge waters in the date range of 1970 to 1980, with a small amount of old water.
- G37 (deep) and G37A (shallow) represent real mixtures between the end members with the shallow borehole having marginally more young water (Figure 3b).

5 CONCLUSION

Mixing processes in groundwater can best be described by mixing lines with tracer concentrations plotted on linear scales. By using a normalized concentration (such as pmf) it is possible to view the three CFC gases collectively. In the present study site it was possible to use CFCs for the young (post-1970 AD) component of the aquifer and ^{14}C for the older component. Tritium confirmed the mixture lines and enabled refinements to be made.

6 ACKNOWLEDGEMENTS

CFC analyses were done at the USGS, Reston VA, USA. Tritium analyses were done at the University of the Witwatersrand, Johannesburg, and ^{14}C at Quadru, CSIR, in Pretoria. Financial support from the Water Research Commission and the CSIR is acknowledged.

7 REFERENCES

Busenberg, E. & L.N.Plummer 1992. Use of chlorofluorocarbons as hydrologic tracers and age-dating tools: The alluvium and terrace system of Central Oklahoma. *Water Resources Research* 28: 2257-2283.

Cook, P.G. & D.K.Solomon 1997. Recent advances in dating young groundwater: chlorofluorocarbons, $^{3}H/^{3}He$, and ^{85}Kr. *J Hydrol* 191:245-265.

Dunkle, S.A., L.N.Plummer, E.Busenberg, P.J.Philips, J.M.Denver, P.A.Hamilton, R.L.Michel & T.B.Coplen 1993. Chlorofluorocarbons as dating tools and hydrologic tracers in shallow ground water of the Delmarva Peninsula, Atlantic coastal plain, United States. *Water Resources Research* 29: 3837-3860.

IAEA 1992. *Statistical Treatment of Data on Environmental Isotopes in Precipitation.* Technical Report 331, Vienna: International Atomic Energy Agency.

IAEA 1999. GNIP: IAEA/WMO Global Network of Isotopes in Precipitation, IAEA, Vienna *http://www.iaea.or.at/programs/ri/gnip/gnipmain.htm*.

Plummer, L.N. & E.Busenberg 1999. Chlorofluorocarbons. In: P.G.Cook & A.Herczeg (eds), *Environmental Tracers in Subsurface Hydrology*, Ch 15. Kluwer, Dordrecht.

Talma, A.S., J.M.C.Weaver & G.J. van Tonder 1998. Dating recent ground-water recharge: a case study using CFC gases in the Free State Karoo. In: *Geocongress 1998* : 279-284. Pretoria: Council for Geosciences.

Vogel, J.C., L. R. Thilo & M.C. van Dijken 1974. Determination of groundwater recharge with tritium. *J. Hydrol.* 23:131-140.

Weaver, J.M.C. & A.S.Talma 1999. *Field studies of Chlorofluorocarbons (CFC's) as a Groundwater Dating Tool in Fractured Rock Aquifers.* Pretoria: Water Research Commission, Report 731/1/99.

Weaver, J.M.C., A.S.Talma & L.Cave 1999. *Geochemistry and Isotopes for Resource Evaluation in the Fractured Rock Aquifer of the Table Mountain Group.* Pretoria: Water Research Commission, Report 481/1/99.

Weaver, J.M.C. & A.S.Talma 2000. Deep flowing groundwater in the Table Mountain Group quartzite. This volume.

Groundwater: Past Achievements and Future Challenges, Sililo et al. (eds) © 2000 Balkema, Rotterdam, ISBN 90 5809 159 7

Characteristics of groundwater quality in the United Arab Emirates

C.Tang
Graduate School of Science and Technology, Chiba University, Japan

S.Shindo & Y.Sakura
Faculty of Science, Chiba University, Japan

I.Machida
Center for Environmental Remote Sensing, Chiba University, Japan

ABSTRACT: Groundwater is the only water resource for many countries in arid and semi-arid regions. In order to get a scientific base for groundwater assessment in the United Arab Emirates (UAE), the authors carried out a 4 years groundwater monitoring program from 1994 to 1997. Water samples were taken from 150 wells throughout UAE. Major ions and isotopes in water were analyzed. Chemical analyses allow groundwater to be grouped in five groups: mountain regions, fans and wadis, sand dune, inland sabkha and coastal sabkha. It was found that chemical components and isotopes of groundwater in mountain regions are very similar to that in fan and wadis, which means that the wadis and fans are very important recharge areas. Authors also compared the chemical variations in groundwater before and after a heavy rain event that caused big floods in wadis. It was found that water quality changed very greatly in groundwater of mountain region, fans and wadis, and little in other regions. Finally, groundwater quality was assessed from the point of view of both hydrogeology and hydrochemistry.

1 INTRODUCTION

Judicious management of groundwater water resources in arid areas is the key to solve the substantial development problems in these areas. In fact, recharge is frequently the most important factor in evaluating regional aquifer systems in arid and semiarid environments, and unfortunately it often is also the most difficult to quantify. Allison (1988) gave a thorough review of the chemical and isotopic techniques available to evaluate recharge in arid environments. Radiocarbon dating of latchstring deposits in the southern Arabian subcontinent suggests a paleoclimate with two wet or pluvial periods, one approximately 21,000 to 30,000 year BP and the other approximately 6,000 to 9,000 year BP (McClure 1976). Dating of travertine in northern Oman indicates wet periods of 19,000 to 33,000 year BP and 6,500 to 12,500 year BP (Clark and Fontes 1990). During the older wet period annual rainfall in the Liwa Crescent area, UAE, was inferred as much as 200 50 mm/year or approximately five times the present rate (Wood and Imes 1995). Wood and Sanford (1995) showed the combined use of geochemical and isotopic methods to quantify groundwater recharge in a regional semiarid environment.

The traditional physical approach for determining recharge flux is based on direct measurement of critical properties such as infiltration capacity, evapotranspiration, and water table at numerous points in the system. In the present study, groundwater chemical composition in the United Arab Emirates (UAE) has been monitored from 1994 to 1997. In these periods, water samples have been taken from 150 wells throughout the whole country. Major ions and isotopes in water have been analyzed. In this paper, it is proposed that recharge regions can be identified by checking carefully the variations of chemical compositions in groundwater before and after heavy rain events, which can provide independent constraints on these physically based techniques and also provide insight unattainable by physical methods.

2 HYDROGEOLOGICAL SETTINGS

Figure 1 shows the map of the United Arab Emirates. There are no permanent streams in the UAE. According to meteorological observations in Ai Alin from 1971 to 1997, the average temperature, annual rainfall, potential evaporation, and humidity were 27°C, 132mm, 3975 mm and 56.3%, respectively. The rainy seasons in UAE extend primarily from January to March and from July to August. Statistically, about 66% and 20% of annual rainfall occurs during those two periods, respectively. Climatologically, evaporation is much higher than rain in arid regions, it seems that rain will not

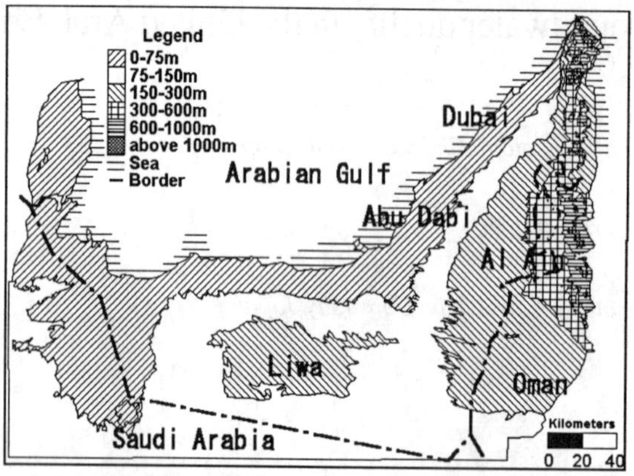

Figure 1. Map of United Arab Emirates

recharge groundwater. Although rainfall is usually negligible, heavy rains in UAE are available for every six to ten years. At those times, flooding occurs in wadis and it is expected that groundwater can be recharged.

Statistical meteorology data in UAE show that about 72% of the annual rainfall is lost by evapotranspiration. Half of the remaining precipitation runs directly into the sea, while the other half infiltrates to groundwater. The amount of fresh water consumed exceeds one billion cubic meters per year and 70% is not renewed. Consequently, some wells have dried up and many are salinized.

Prior to the 1960s the amount of water consumed using traditional methods such as manual wells were balanced by natural renewal. More modern methods, using mechanical pumps to draw out large amounts of groundwater have resulted in receding quantity and quality of groundwater. In 1985 there were 70 rotary drills in use, yielding up to 4000 wells annually. In addition, there are many ordinary percussion drills belonging to private drilling establishments, the official department of water and electricity and the army drilling unit.

More than 85% of the groundwater is naturally salinized due to the high solubility of the salt and limestone layers. The fresh water reservoirs in the gravel plains and the Quaternary alluviums (particularly around the Al Ain area) have started to deteriorate because of vertical movement of deep waters from saline reservoirs. The coastal plains have intrusions of seawater as a result of intensive exploitation of inland and fresh groundwater.

Irrigated land expanded in area from 13000 hectares in 1973 to 43800 hectares in 1990. Traditional irrigation methods have remained prevalent. Studies have shown that by using modern irrigation techniques such as the sprinkler trickle or the use of closed pipes and channels, savings up to 350 million m^3/year are achievable. Surface water flowing into the sea could be stored and the government has established a network of dams in suitable areas to aid this. Desalinated sea and groundwater is an important source of water in the UAE, especially for cities and isolated villages. The amount of desalinated water exceeded 200 million m^3 in 1989. Work started in the late 1970s to establish purification and treatment plants for sewage water. The resulting outflow water (62 million m^3 in 1989) could be used for irrigation and the fixation of dunes around cities and near major highways. Desertification studies have resulted in the identification of certain plant species, which can be irrigated by salty water. They have also succeeded in developing new irrigation techniques, which prevent the formation of a salty layer on the surface of the soil.

3 METHODS

Four field campaigns were carried out between 1993 and 1997. During the first preliminary survey, groundwater samples were taken in parts of regions in UAE in September 1993 and November 1994. In the first period, some wells controlled by the UAE government were checked, and in the second period, wells around the Al Ain area were studied. The third time we sampled more than 150 wells all over the UAE in February 1996. In March of that year, there was the biggest rain event since 1970 that recorded more than 200mm in UAE. In order to study where and how rainwater recharges the groundwater in UAE, we carried out the fourth field survey in February 1997. During our field surveys we used a GPS camera to record both position and took photographs of the sampling points to build into our hydrological database. Based on this, we could take samples from the same points every time, and check changes of water chemistry during the study period. Water table, water temperature, EC as well as pH of water have been measured on site. Water samples were analyzed in the laboratory at Chiba University. Anions such as NO_3^-, Cl^- and SO_4^{2-} were measured by ion chromatography and cations such as Na^+, K^+, Ca^{2+} and Mg^{2+} by ICP. HCO_3^- was measured by titration. Stable isotope ^{18}O and D were measured by mass spectrometer.

4 RESULTS

4.1 *Regional characteristics of groundwater*

The northern mountains are distinguished by an abundance of carbonate rocks. They contain major fresh water that is fed by direct rainfall and overflows at the surface in a number of karstic springs. It also feeds the marine Khor Khuwer spring (40 l/sec), where it infiltrates to the adjacent ground layers. This unit is distinguished by its high productivity.

The southern mountains, composed of ophiolite, granite, gabbros and a variety of volcanic and metamorphic rocks, are distinguished by their compactness and impermeability. Groundwater is scarce but fissures and fractures that accompany faults allow for the formation of groundwater reservoirs. Some of this water, in addition to the water of the alluviums, flows over into the Gulf of Oman.

The gravel plains of the Western Bahada, composed of gravel, fine alluviums and conglomerates of rocks, range in thickness from 200m in the north to 600m in the Al Ain region and are considered the most important UAE groundwater reservoirs. Modern canals and wells have replaced ancient ones. They yield a very high productivity in the plains adjacent to the Carbonate mountains and moderate productivity in the regions adjacent to metramorphic and ultra-basic rocks. Productivity is associated with the quantity of rock conglomeration; permeability decreases with the increasing compactness.

The gravel plains of the Eastern Bahada are composed of conglomerates and are distinctive for their high compactness and low permeability. Groundwater is only found in the fractured regions, or at the contact of the conglomerates with ultrabasic rocks. Wells drilled in the eastern region have yielded different results; some were dry, but most yielded moderate productivity (2-10m^3/hr). The quaternary detritus permeable alluviums, situated on the eastern coast below the 15m contour line, (which conforms with the ancient coastline) have good hydraulic characteristics and have traditionally formed the borders of agricultural areas that were irrigated by shallow wells. The large number of wells and excessive consumption of fresh water has caused a serious infiltration of seawater to these alluviums.

The sandy desert plains in the western region are underlain by Quaternary alluviums with a moderate re-availability of acceptable salinity. This has helped the continuation of oases (e.g. Mahder Liwa). Layers under these alluviums have saline and carbonate facies, which make the groundwater very salty. Reconnaissance wells in Liwa with depths exceeding a thousand meters have yielded an abundant water supply, which is useless, due to its salinity.

The low coastal plains around the Arabian Gulf have good productivity but the prospects of renewing and supplying fresh water through rainfall are extremely scarce in this arid milieu. Sea intrusion has salinated most of the water and its salinity is particularly severe in the salt domes of Jabalaz Zannah and the UAE's islands.

4.2 *Regional characteristics of groundwater chemistry*

In arid regions, recharge is much less than that in other regions since the evaporation potential is much higher than precipitation. There are no rivers in UAE, and chemical transport in the water cycle only occurs by groundwater. Cl^- is the dominating ion in the study area. In fact, chemical types of groundwater in UAE are almost Na+K-Cl from recharge area to discharge area. Paying attention to the Ca/Mg mole ratio, the Ca/Mg value is larger than 1 in Liwa region and less than 1 in other regions of UAE. Regional characteristics of groundwater EC in UAE is shown in Figure 2 based on the results of field work in February 1996.

Temperature of groundwater decreased from recharge area to discharge area. The highest temperature was found in the north part of UAE. In the area near the borderline between Oman and UAE, groundwater temperatures were higher than 30°C. However, inter-sanddunes in Liwa and mountain regions in Oman, groundwater temperatures were below 30°C. The pH values of groundwater were found at about 7.0 in the recharge area and decreased from recharge area to coast. The pH values of groundwater in the limestone area were greater than 8.0.

Figure 2 shows the distribution of EC in groundwater. EC was higher than 4000 $\mu S/cm$ in all regions except the recharge area. As usual, EC values increase from recharge area to discharge area. In the Oman mountain area which is the recharge areas for groundwater in the UAE, the EC in groundwater was found as low as 120 $\mu S/cm$.

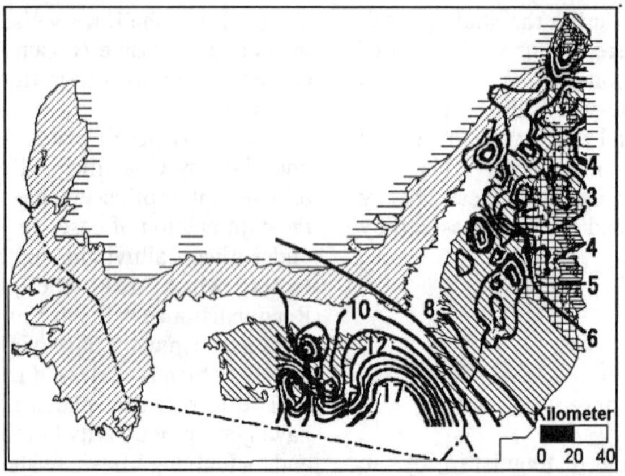

Figure 2. Distribution of EC in groundwater of UAE (1996) (Units: X10^4 $\mu S/cm$)

However, EC values of groundwater in Liwa region were either lower than 5000 $\mu S/cm$ for deep wells or higher than 10000 $\mu S/cm$ for shallow wells that were affected by evaporation. In the Oman mountain and oases such as Al Ain, the Cl^- concentration in groundwater was less than 500 mg/l. After groundwater flows down, Cl^- concentration increased to above 2000 mg/l. As groundwater flows to discharge area, the water table becomes shallower and shallower. Salt ponds or salinized soil can be found along the highway from Abu Dabi to Tarif. Analyzed results show that the Cl^- concentration was as high as 222.0 g/l in salt pond water. However, there was small recharge area in Liwa where the Cl^- concentration in groundwater was above 4000mg/l.

High concentrations of nitrate were found in the groundwater of old oases, such as Al Ain and Liwa regions or near highways along which there are many farms for producing tomatoes. Nitrate concentrations were as high as 200mg/l along the highway and 300mg/l in Liwa region. Therefore, nitrate concentration distribution in UAE was coined with landuse for agriculture. At the same time, many chemical fertilizers and pesticides are used for agriculture.

4.3 *Chemical Variations in groundwater between 1996 and 1997*

In March 1996, there was 200mm rainfall in regions along Oman Mountains, the heaviest rain UAE since 1970. During the rain period, floodwater flowed down wadis and formed ponds in the lowlands. Some of those ponds existed for more than two months after the floods. It is believed that recharge for groundwater should occur along flood courses as well as ponds. Up to now there was little research concerning the recharge of heavy rain to groundwater in UAE. We compared the changes of chemical compositions in groundwater before and after the heavy rain based on the analysis data of the same well in February 1996 and February 1997. The following formula was used to compare the change rates of water quality C_j.

$$C_j\,(\%)=(C_{j1997}-C_{j1996})/C_{j1997}\times 100$$

where C_{j1997} and C_{j1996} are item j such as temperature, EC and other elements in 1997 and 1996, respectively.

As expected, concentrations of chemicals in rainfall were very low compared to groundwater. If there were rapid infiltrations of rainwater to recharge groundwater, water chemical concentrations would change. From the analyzed data, chemical changes increased in recharge area. Little change can be found in discharge area. As a result, chemical changes in groundwater were coined with recharge area.

Figure 3 shows the change rate of EC from 1996 to 1997. Changes of EC in the recharge area were large, and EC of groundwater decreased 30% in Al Ain. However, EC of groundwater in the discharge area increased. It was believed that fresh water infiltrating into groundwater caused the EC value decreased in recharge area on one hand, and continuing pumping in discharge area made EC increased on the other hand. EC in Liwa area changed little, since no heavy rain occurred during our working period from 1996 to 1997. Change of Cl^- shows the same pattern as EC.

Except for the northern part of UAE, groundwater temperature in many areas decreased. For example, in Al Ain it decreased by 20%. The areas where groundwater temperature decreased follow well with the distribution of heavy rain occurring during March 1996. It means that groundwater was recharged from the Oman mountain as well as flood in wadis. At the same time, there was little change of groundwater temperature in the discharge areas.

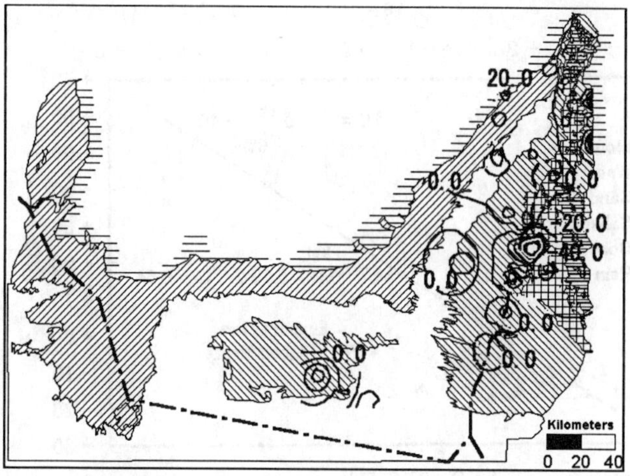

Figure 3 Distribution of EC change rate (%) in groundwater of UAE

5 DISCUSSIONS

As is well known, groundwater quality is affected greatly by the groundwater flow system, geomorphology and geology. The local hydrogeomorphology can be divided into five areas: Oman mountain area, wadis, sand dunes, inland sabkhas and coastal sabkhas. Based on this classification, groundwater qualities can be looked at in more detail.

The Piper diagram (Figure 4) shows the groundwater pattern in 1996. First all, Cl^- ion was the dominant anion, and SO_4^{2-}, NO_3^- and HCO_3^- showed regional characteristics. In the recharge area, the concentration of SO_4^{2-} was small, but both NO_3^- and HCO_3^- vary considerably. It is considered that water infiltrated in mountain and wadis. However, in discharge area such as sabkhas and coastal regions, HCO_3^- and NO_3^- changed little and SO_4^{2-} changed. As a result, groundwater types can be classified as follow.

Mountain area:	Na+K-Cl
Wadis:	Na+K-Cl
Sand dunes:	Na+Ca-Cl+SO4
Inland sabkhas:	Na+K-Cl+SO4
Coastal sabkhas:	Na+Mg-Cl

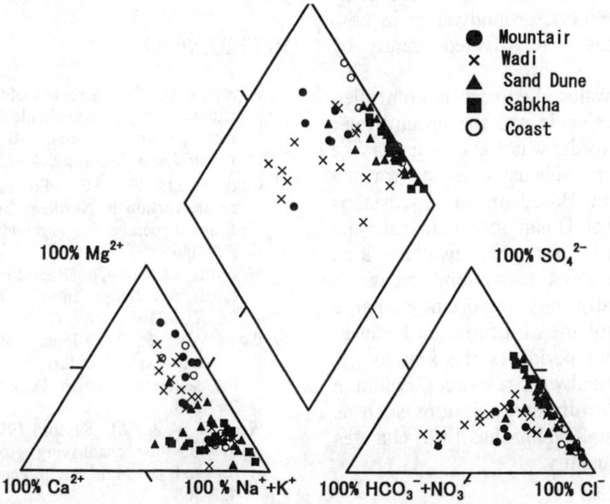

Figure 4. Piper diagram of UAE groundwater water

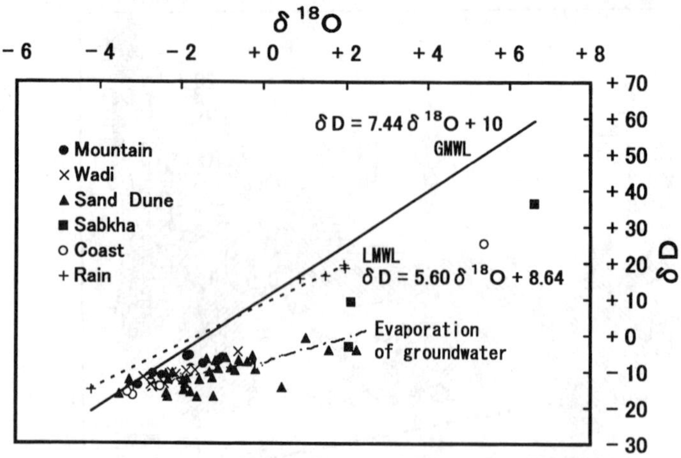

Figure 5 Relation between $\delta^{18}O$ and δD by considering the hydrogeomorphology

Figure 5 shows the relation between $\delta^{18}O$ and δD in UAE based on our hydrogeomorphological classification. From the figure, it is found that $\delta^{18}O$ and δD in water increase when water moves from mountain area (recharge) to the coastal region (discharge). The groundwater in wadis is a little heavier than that in mountain area. As a result, global meteoric water line (GMWL) and local meteoric water line (LMWL) of stable isotopes can be described as the equations:

GMWL: $\delta D = 7.44\delta^{18}O + 10$
LMWL: $\delta D = 5.60\delta^{18}O + 8.64$

Comparing with GMWL of stable isotopes, it can be found that LMWL in mountain area is very close to GMWL, but LMWL in other areas were away from the GMWL. Especially, groundwaters in both sand dunes and sabkhas were affected greatly by evaporation.

As is well known, water chemistry is controlled by water balance, water cycle and the groundwater flow system. In other words, water chemistry is like a mirror to show a comprehensive image for the water cycle in the basin. Based on the discussions above, it was found that Oman mountains are the permanent recharge areas for groundwater which form the basic groundwater flow system and wadis are temporary but locally very important recharge areas. As a result, the volume of infiltration water in wadis during heavy rain period is the key to get exact assessment of groundwater resources available in UAE. Groundwater in discharge regions such as inland sabkha and coastal sabkha has little changes in both its quality and quantity.

6 CONCLUSIONS

UAE groundwater has been used to illustrate the application of geochemical methods to the evolution of recharge in an arid area. Groundwater quality patterns in UAE were affected greatly by hydrogeomorphology. Groundwater quality and isotopes in mountain areas show the same characteristics in wadis, which means that rainwater recharges to groundwater directly in wadis during flood period. Based on the results in 1996 and 1997, it can be expected that groundwater would be recharged when heavy rain occurred and flood in wadis, which hints that wadis are very important recharge areas in arid regions, even if they only experience seasonal floods every six to ten years.

REFERENCES

Allison, G.B. 1988. A review of the physical, chemical and isotopic techniques available for estimating groundwater recharge. In I. Simmers(ed.), *Estimation of Natural Groundwater Recharge:* 49-72. D. Reidel Pub.

Clark, I.D. & J.C. Fontes 1990. Palaeoclimatic reconstruction in Northern Oman based on carbonates from hyperalkaline groundwater. *Quat. Res.* 33: 320-336.

McClure, H.A. 1976. Radiocarbon chronology of late Quaternary lakes in the Arabian Desert. *Nature*, 263:755-756.

Wood,W.W. & J.L. Imes 1995. How wet is wet? Precipitation constraints on late Quaternary climate in the southern Arabian Peninsula, *Jour. Hydrol.* 164: 263-268.

Wood,W.W. & W.L. Sanford 1995. Chemical and isotopic methods for quantifying groundwater recharge in a regional, semi-arid environment. *Ground Water*. 33(3): 458-468.

Groundwater: Past Achievements and Future Challenges, Sililo et al. (eds) © 2000 Balkema, Rotterdam, ISBN 90 5809 159 7

Calibration of a groundwater flow model for water management in Hamburg, Germany, using groundwater age measurements and hydrochemical data

R.Taugs & L.Moosmann
Geologisches Landesamt, Hamburg, Germany

ABSTRACT: A 3D-groundwater flow model for aiding optimized management of the deeper groundwater systems in the city of Hamburg was calibrated using groundwater age measurements and hydrochemical data. Extensive pumping over several decades has resulted in mixing of groundwater of different ages complicating the model calibration and validation. It could be shown that numerical simulation methods in combination with environmental tracers and groundwater age data allow a more objective estimate of the uncertainties in the estimated rates and paths of water movement.

1 INTRODUCTION

Municipal drinking water supply in the city of Hamburg (Germany) depends on groundwater resources from Quaternary and Tertiary porous sediments reaching depths of 500 m below surface. Groundwater systems in Hamburg are in hydraulic interconnection as a result of the presence of buried channel systems of Elsterian age deeply incised into the pre-Pleistocene strata.

Due to these interconnections a continous recharge of the deeper aquifers takes place from the upper aquifers. But not only water of good quality is transported to the aquifers, man-made contamination can also find its way into the groundwater system. Beyond that the public water supply in Hamburg is locally affected by increasing salinization resulting from extensive pumping up to the late seventies.

At present about 140,000,000 m^3/a of groundwater of drinking water quality are required for the public water supply in Hamburg. In addition another 26,000,000 m^3/a of high-quality groundwater is needed in trade and industry.

This demand for unaffected groundwater of excellent quality requires effective management and a sustainable development of groundwater resources. Therefore a 3D-groundwater flow model was built as a tool to aid optimisation of groundwater management in Hamburg.

Part of this project was to investigate how far model calibration and validation could be improved by groundwater age determination (^{14}C ages) in combination with hydrochemical data. ^{14}C ages span up to 40,000 years in the Hamburg region. Mixing of groundwater of different ages and the effects of hydrochemical reactions caused by extensive pumping over several decades had to be considered.

2 HYDROGEOLOGY

The groundwater resources in Hamburg are divided in six upper and deeper aquifers of Quaternary and Tertiary age, separated by layers of silt, clay and till sequences. The upper aquifers comprise Holocene fluvial sands and sand and gravel accumulations of the Weichselian glaciation in the ice-marginal valley of the river Elbe, Saalian and Elsterian sediments and the Pliocene kaolin sands. Most important for municipal drinking water supply are the lower aquifers in Hamburg. Sediments of the Upper Tertiary, such as the Middle Miocene Upper Lignite Sands and the Lower Miocene Lower Lignite Sands, together more than 200 m thick, contain most of the deeper groundwater resources. These Tertiary aquifers reach down to 500 m below surface. During the Elsterian glaciation channels exceeding a depth of more than 400 m were cut into the pre-Quaternary strata. The channels were refilled with glaciofluvial sand and gravel, later with silt and clay from a large ice-dammed lake (Lauenburg Clay). The coarse grained fillings in the lower parts of the channels form another important lower aquifer.

The lower aquifers are in most areas separated by layers of low permeability such as the Middle Miocene Hamburg Clay and the Upper Miocene Mica Clay. As a result of three large glaciations the Pleistocene strata covering the pre-Quaternary bedrock is very heterogeneous. Hydraulic interconnections between the upper aquifers are quite common whereas the Upper Miocene Mica Clay reaching a thickness of locally more than 100 m protects the lower aquifers in most parts of Hamburg against influences from the surface. In the Elsterian buried channels the Mica Clay is mostly eroded, but in these cases the Elsterian Lauenburg Clay separates upper and lower aquifers. However, locally these natural barriers between upper and lower aquifers are absent, so that in the Hamburg region all aquifers are interconnected and form a complex hydraulic system.

Problems of the municipal drinking water supply not only result from man-made groundwater contamination in the upper aquifers but arise as well from salt domes causing a salinization of the lower aquifers. Mobilized Rotliegendes and Zechstein salt formed the Hamburg the salt domes of Langenfelde, Quickborn and Reitbrook. The gypsum caprock of the Langenfelde salt dome had moved upwards to about 30 m below surface. Due to the solution of gypsum at the flanks and on the tops of the salt dome parts of the lower aquifers are highly mineralized by sulphate and chloride. In other places the salinization of the lower aquifers may be caused by ascending salt water from deeper salt structures or underlying older formations.

Significant chloridic mineralization in the Lower Lignite Sands is found in more than one third of the area of Hamburg. Also parts of the Upper Lignite Sands and of the Elsterian buried channels show chloridic and sulphatic mineralization. The salinization of such considerable parts of the lower aquifers cannot be explained only by natural solution and transport processes. There are many arguments that extensive pumping from the lower aquifers over several decades mobilized the salinated water and initiated a slow displacement of the salt water/fresh water boundary. Another consequence of heavy pumping from the Lower Lignite Sands was a lowering of the water table by more than 20 m. Due to the obvious effects of overuse of groundwater resources serious efforts were made to improve the situation. So far they have been quite successful. But one important consequence of that critical situation was that careful management practices had to be imployed to better manage the groundwater resources in the lower aquifers.

3 GROUNDWATER FLOW MODEL

An effective and sustainable management of groundwater resources in Hamburg requires the ability to predict flow and transport of solutes especially in the lower aquifers and their response to changes in natural or human-induced stresses. Already in the seventies a 2D-computer simulation model had been developed to analyze the groundwater situation in Hamburg, at that time on the base of a rectangular grid with a spacing of 2 km. By using this very unwieldy tool a quantitative approach to describe the groundwater systems was executed for the first time. In 1995 the Geological Survey of Hamburg started to build and refine a numerical 3D-groundwater flow model for the lower aquifers, initially representing only steady-state conditions. The finite-difference simulation system used for modelling groundwater flow and transport processes is Processing Modflow for Windows (PMWIN 5.0).

The area of investigation encompasses about 3,500 km^2 and was chosen on the basis of natural hydrologic and geologic boundaries of the Tertiary Lignite Sands. The active model grid covers an area of about 1,900 km^2. To minimize any impact of conceptual errors associated with boundary conditions the boundaries of the investigated area have been placed as far as possible away from the area of interest.

After formulating a conceptual model the relevant geological and hydrological information has been assigned to the active model grid with cell dimensions of 250 by 250 m.

When building the model grid, in general it must be kept in mind that every model is only an approximation of the very complex real system. It is used to simplify reality in a manner that captures or represents the essential features and processes relative to the problem at hand. So an appropriate level of model complexity has to be selected. In this case it means that a more finely discretized model would not yield greater accuracy because a finer resolution is not matched by the available database.

Model validation in general is attempted by comparing calculations with field measurements. In most cases matching observed potentiometric surfaces are used as a basis for recalibration and validation. But in this case model validation can be improved additionally by using hydrochemical data and measurements of radiometric water ages.

4 GROUNDWATER MONITORING

4.1 *Groundwater levels*

Recording of potentiometric levels in Hamburg started at the beginning of 20th century as a consequence of increasing groundwater exploitation. From these early measurements it can be determined that originally the deeper aquifers were mainly confined before industrialization caused excessive withdrawal of groundwater from the lower aquifers. Originally, groundwater drainage was directed from outside Hamburg towards the Elbe valley. Groundwater movement was only influenced or locally hampered by the N-S aligned salt structure of Quickborn-Langenfelde.

Groundwater recharge of the lower aquifers takes place mainly to the south and the northeast in the neighbouring federal states of Schleswig-Holstein and Niedersachsen.

Despite the changing hydraulic conditions of the groundwater systems in the growing urban region, groundwater flow is still directed from all around towards the low-lying central part of the city. Despite increased groundwater withdrawal from the lower aquifers up to the seventies, recent groundwater movement is only slightly deflected towards the main production wells.

4.2 *Hydrochemical data*

Since 1988, the lower aquifers have been monitored. The aim of this programme is the characterization and recurrent control of the groundwater quality. Every two years groundwater samples are taken from 79 monitoring wells screened in the Lower and Upper Lignite Sands and in the Elsterian buried channels. Analyses of groundwater samples include major cations and anions as well as trace elements and organic compounds.

In general the groundwater system of the lower aquifers is under reducing conditions. Dissolved oxygen is mainly zero. Concentrations of dissolved iron and manganese are low, trace elements in most cases are not detectable. In areas influenced by salinization the electrical conductivity is about 2000 µS/cm and higher. Maximum salinity measured in Hamburg is about 100 g/l. Whereas fresh groundwaters are predominantly of the Ca-HCO_3 type, distinctly sodium-dominated fresh waters are found close to the salt water contact. Mixing of groundwater types can be observed in regions of main groundwater withdrawal.

Groundwater coming from the main recharge areas show very distinct chemical characteristics. These variations in hydrochemistry reflect mainly differences in the depositional environment resulting in differing liquid/solid phase interactions in the system. So groundwater from the southern recharge area typically has an electrical conductivity of about 250 µS/cm, a Na : Ca ratio of 6 or 7, a very low mineralization with sulphate and chloride concentrations of about 10 mg/l and bicarbonate contents of 100 mg/l. Groundwater coming from the northeast shows an electrical conductivity of about 400 µS/cm, bicarbonate contents of more than 200 mg/l, whereas sulphate is not detectable. Na : Ca ratio and chloride concentrations are comparable to groundwater samples coming from the south. Both groundwater types do not vary in quality with depth.

On the basis of these hydrochemical characteristics, the distribution of both groundwater types can be mapped. The observed hydrochemical relations in combination with measured changes in salinization forced by pumping are used as "environmental tracers" in the groundwater flow model.

4.3 *Stable Isotope data*

During the last 3 years 66 groundwater samples have been taken from monitoring wells screened in the lower aquifers and analyzed for various environmental isotopes including ^{14}C, ^{13}C and tritium. Another 33 samples have been analyzed for the stable isotopes ^{18}O and ^{2}H.

Using the stable isotopes ^{18}O and ^{2}H the palaeoclimatic conditions during groundwater recharge can be reconstructed. δ^2H varies in most samples between -49 and -60 ‰, $\delta^{18}O$ between -7.5 and -8.5 ‰. In more depleted water samples which are mostly highly mineralized, δ^2H decreases to about -67 ‰ and $\delta^{18}O$ to about -9.5 ‰. Stable isotope measurements show that mineralized water and fresh water can be distinguished very clearly, so this may be suggesting on recharge periods with differing climatic conditions.

Also a correlation can be registered between depletion of stable isotopes and low amounts of ^{14}C. This correlation may be due to the presence of "palaeo-waters" found preferentially in areas with very low groundwater flow velocities. The salt structure of Langenfelde in particular forms such a flow barrier. (Loehnert et al. 1992). By using the results of stable isotope analyses more "stagnant systems" can be clearly differentiated from "dynamic systems" with low mineralization and higher groundwater flow velocities.

4.4 *Radiometric water age data*

^{14}C data were used to determine groundwater ages. Conventional water ages (half-life after Libby/NBS-standard) based on ^{14}C concentrations (as percent modern carbon) are 6,000 to 32,000 years in the Upper Lignite Sands, whereas water samples from the Lower Lignite Sands are 7,000 to 40,000 years old. Water samples taken from the deep buried Elsterian channels show ^{14}C based conventional water ages of 2,200 to 35,000 years.

A number of ^{14}C age correction methods (Ingerson & Pearson 1964, Tamers 1975, Fontes & Garnier 1979) were employed. The initial ^{14}C activity estimated by the Tamers method (1975) is about 76 %, and the method of Ingerson & Pearson (1964) gives 53%. Using the method of Fontes & Garnier (1979), an initial ^{14}C activity of 52 % (with variations from 26 % to 76 %) has been estimated. Taking a mean initial ^{14}C activity of 53 % the model ages estimated on this base are about 5,100 years younger than conventional water ages. Differences of initial ^{14}C activities estimated in samples taken from one aquifer are less than 2 %, i. e. negligible.

5 VALIDATION OF THE MODEL

Groundwater age data from the lower aquifers have been compared with predictions from the steady-state groundwater flow model. Along an idealized pathline aligned from the southern recharge area to the Elbe valley, groundwater flow velocities have been estimated. Calculating using the corrected water ages, groundwater flow velocity in the Upper Lignite Sands is about 1 m/a and in the older Lower Lignite Sands about 1.5 m/a. These ^{14}C based values have been compared with the simulated flow velocities from the groundwater flow model which has been calibrated using the present hydraulic conditions. The steady-state model gives much higher flow velocities which do not match very well with the ^{14}C values: in the Upper Lignite Sands 80 m/a were simulated, and in the Lower Lignite Sands 50 m/a.

The resulting discrepancies can have various reasons:

1. One complication may be that dissolution of carbonate minerals or the oxidation of organic matter may add "dead" ^{14}C to the water, giving an erroneous water age.

2. Another complication may be mixing of relatively young and "dead" water. In general ^{14}C measurements can be interpreted as ages only where mixing is insignificant.

3. A large portion of the observed discrepancies may be due to the non-representative conductivity (resulting from aquifer tests) in the simulation model.

4. All observations from groundwater monitoring point to the drastically changing hydraulic conditions caused by increasing groundwater withdrawals since the beginning of the 20th century. The extensive pumping led to an increased groundwater gradient towards the Elbe valley whereas the palaeoflow velocity in the Hamburg region must have been much lower.

From these considerations it becomes evident that a steady-state model covering the whole time period would be unsuitable for simulation. So it was decided to build a transient model to come to a better fitting simulation of groundwater flow.

6 CONCLUSIONS

It was shown in the project that environmental tracers and numerical simulation methods in combination are effective tools that complement each other and provide a means to estimate the flow rate and path of water moving through a groundwater system. Moreover, environmental tracers and numerical simulation methods provide a certain feedback that allows a more objective estimate of the uncertainties in the estimated rates and paths of water movement to be made. Together both enabled the development of an understanding of the system which is now requiring in some cases additional data collection and/or refinement of the simulation.

Difficulties with this large-scale groundwater modelling study arose mainly from the great degree of heterogeneity in the aquifer materials that could not be discretized into the model, from errors in estimating the magnitude and distribution of conductivity and from the various possible carbonate reactions that may have occurred along the groundwater flowpaths, which cannot be reconstructed. Even given this, it can be considered that ^{14}C based ages can be used to improve models of groundwater flow.

The steady-state model of the lower aquifers in Hamburg is now being extended to consider changes in flow and boundary conditions in time. Also the linking up of the lower aquifers with the more disturbed and inhomogeneous upper Quaternary aquifers has to be done. This will be an important further step towards a much better understanding of the whole groundwater system in Hamburg.

REFERENCES

Fontes, J.-C. & J.M. Garnier 1979. Determination of the initial ^{14}C activity to the total dissolved carbon – A review of the existing models and a new approach. *Water Resour. Res.* 15: 399-413.

Ingerson, E. & F.J. Pearson 1964. Estimation of age and rate of motion of groundwater by the ^{14}C-method. In Y. Miyake & T. Koyama (eds), *Recent researches in the fields of atmospheric, hydrosphere, and nuclear geochemistry, Sugawara Festival Volume*: 263-283. Tokyo: Maruzen Co., Ltd.

Loehnert, E.P., W. Bauhus & C. Sonntag 1992. Hydrogeochemical processes in the Hamburg deep aquifer system. In G. Matthess, F. Frimmel, P. Hirsch, H.D. Schulz & H.-E. Usdowski (eds), *Progress in hydrochemistry. Organics-carbonate systems-silicate systems-microbiology-models*: 258-269. Berlin: Springer.

Tamers, M.A. 1975. Validity of radiocarbon dates on groundwater. *Geophys. Surv.* 2: 217-239.

Groundwater: Past Achievements and Future Challenges, Sililo et al. (eds) © 2000 Balkema, Rotterdam, ISBN 90 5809 159 7

Depth variations in aquifer hydrochemistry using a low-cost, multilevel piezometer

R.G.Taylor & M.H.Barrett
Robens Centre for Public and Environmental Health, University of Surrey, Guildford, UK
O.P.Baines
Department of Civil and Environmental Engineering, University of Bradford, UK
S.A.Trowsdale, D.N.Lerner & S.F.Thornton
Department of Civil and Structural Engineering, University of Sheffield, UK

ABSTRACT: Studies of depth variations in aquifer hydrochemistry are constrained by the expense of collecting depth-specific groundwater samples. To overcome this restriction, a low-cost, 'bundled' (multilevel) piezometer has been adapted for a study of the depth to which recent recharge and contaminants have penetrated the Triassic sandstone aquifer in urban areas of the United Kingdom (UK). The design features 10 sampling intervals and can be installed manually in open boreholes. Field testing of the design in an existing borehole in the unconfined sandstone aquifer underlying the city of Nottingham (UK) demonstrates the efficacy of the design in fissured bedrock as significant hydrochemical variations are observed over a few metres in depth. The availability and low cost of materials (US$30/metre) as well as the ease of construction and installation permit widespread application.

1 INTRODUCTION

Contamination of groundwater in urban areas is well documented but the vertical profile of contamination within these aquifers often remains poorly resolved. In the Triassic sandstone aquifer of central England, contamination of groundwater by chlorinated solvents and transition metals (e.g. chromium) has been indicated (e.g. Jackson and Lloyd, 1983; Rivett et al. 1990; Ford and Tellam, 1994; Rivers et al. 1996) but these data derive from samples collected in existing industrial boreholes that are open through much of the vertical extent of the aquifer (i.e. depth-integrated samples). There is consequently very limited information about the depth of penetration of urban contaminants, vertical hydraulic gradients and the existence of usable groundwater at depth in the sandstone aquifer. This latter uncertainty is of particular concern in central England where urban abstraction may be necessary to offset rising water tables resulting from a decline in industrial abstraction (Greswell et al. 1994) and leakage from pressurised water mains (Lerner, 1986; Yang et al. 1999). Depending upon the quality of pumped groundwater, urban abstraction may also serve to reduce the problem of overpumping groundwater in rural areas.

Investigations of aquifer hydrochemistry with depth are, however, restricted by the considerable expense of collecting depth-specific groundwater samples. To overcome this restriction and permit regular depth-specific sampling at six locations in two cities, Birmingham and Nottingham, an inexpensive multilevel piezometer with 10 depth-specific, sampling intervals has been employed. A trial installation to a depth of 51 metres below ground level (mbgl) in the unconfined Triassic sandstone aquifer underlying the city of Nottingham (Figure 1) has been conducted in an existing open borehole. Hydrochemical variations with depth recorded using this new multilevel piezometer, are reported here.

2 LOW-COST (MULTILEVEL) PIEZOMETER

The 'bundled' (multilevel) piezometer is easily manufactured and consists of a series of small diameter, flexible high-density polyethylene (HDPE) tubes of varying lengths that is attached to a rigid, polyvinyl chloride (PVC) support casing (Figure 2). The number and diameter of individual tubes (piezometers) which can be attached to the central PVC casing, depends upon the diameter of the open borehole. The multilevel design was adapted from the bundled piezometer design of Cherry et al. (1983). Thornton et al. (in preparation) have recently field tested a 'bundled' piezometer that is similar to the design employed in this study.

The mutilevel piezometer was installed in a borehole with a diameter of 165 mm. A 48 mm - outside diameter (OD) PVC support casing was used with ten, 16 mm (OD) HDPE tubes. Tubes were bundled to the support casing using nylon cable ties. Slots of 4 mm were drilled at the base of each tube

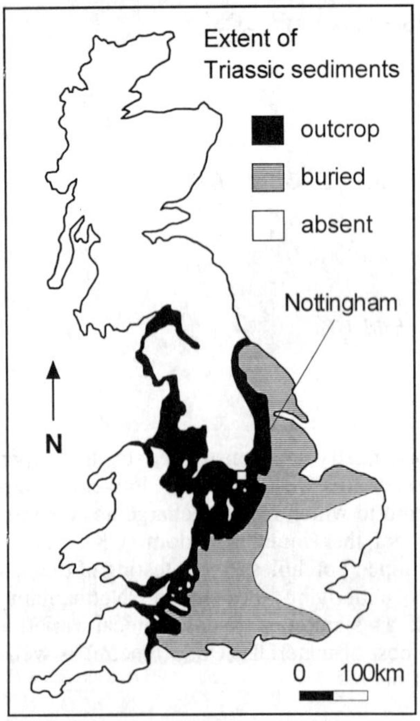

Figure 1. Distribution of Triassic sediments in Britain (adapted from Anderton et al. 1987) and location of study site.

over a 100 mm interval (Figure 2). The slotted section was sheathed in a stainless-steel mesh (0.01 mm spacing) to act as a filter to silica-sand fill (particle size: 0.125-0.250 mm). Installation was conducted manually with a hand-operated winch and quadpod. In the United Kingdom, material costs are in the order of USD$30 per metre. Purging and sampling was performed manually using an inertial Waterra™ handpump. Key advantages of the 'bundled' design over the traditional 'nested' piezometer and the recently developed, continuous multichannel tubing design (Einarson and Cherry, in press) are the capacity to install it in existing, narrow diameter (150-200 mm) boreholes, and the large number of sampling intervals (10-11) that permit greater resolution in aquifer hydrochemistry with depth.

3 STUDY SITE

The unconfined aquifer underlying the city of Nottingham comprises sandstone with mudstone bands. Sediments were eroded from a dessicated, upland landscape during the Triassic (Anderton et al., 1987; Charlsley et al., 1990). Lithological variatons with depth from sandstone to mudstone are considered to reflect changes in deposition by fast-flowing rivers to intermittent rivers and ephemeral lakes caused by increased aridity. Triassic sediments are underlain by marl that effectively forms the base of the aquifer. The aquifer is highly layered (Rushton and Salmon, 1993). Fissures and coarse-grained horizons give rise to high-flow zones whereas mudstone bands form low-conductivity layers. The borehole is located in north-central Nottingham and was drilled by air-flush rotary methods to a depth of 52 mbgl on March 24, 1997. The well casing extends to a depth of 9 mbgl to prevent the collapse of poorly consolidated materials into the borehole. Further details of the borehole are given in Stagg et al. (1998).

Downhole geophysical logging of the borehole by Stagg et al. (1998) identified low-conductivity horizons (mudstone bands) as well as inflow and outflow zones that are considered to result from fissures (Figure 3). Sampling intervals on the

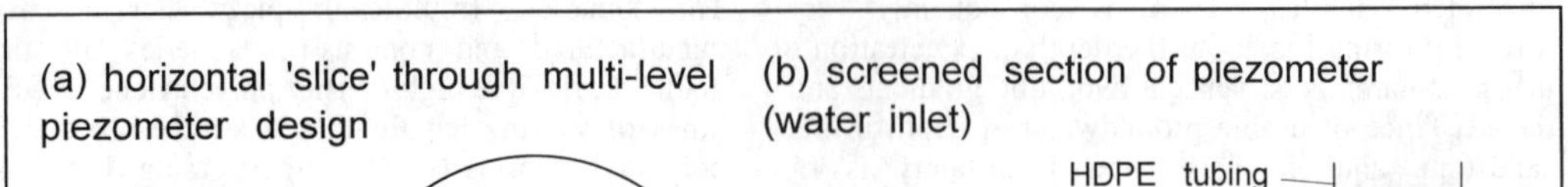

Figure 2. 'Bundled' multilevel piezometer design (adapted from Cherry et al. 1983)

multilevel piezometer were selected on the basis of these data and are located at 12, 16, 19, 21, 25, 28, 31, 35, 40, and 46 mbgl (Figure 3). Installation of the multilevel piezometer occurred on February 17, 1999. Purging and sampling of each piezometer followed on April 22, 1999. In addition to wellhead physico-chemical measurements, samples were analysed for major ions, transition metals and volatile organic compounds.

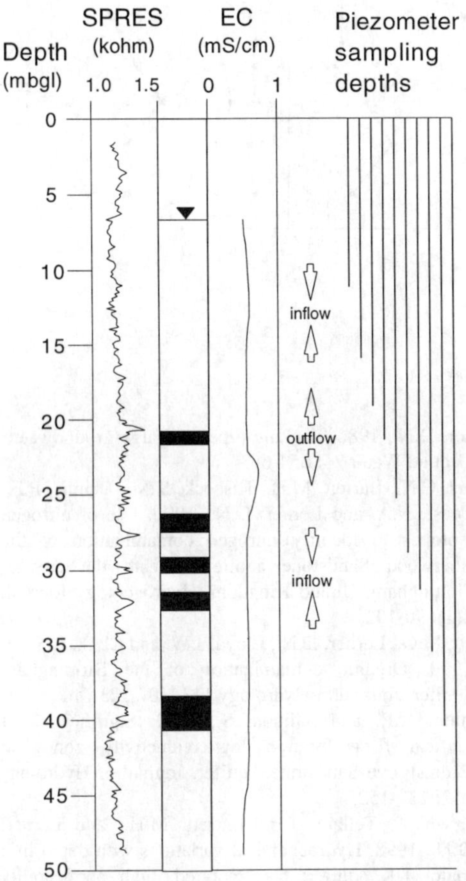

Figure 3. Geophysical logs (single-point resistivity, conductivity) with lithological (mudstone bands) and flow interpretations of Stagg et al. (1998). Sampling depths of multilevel piezometer are also shown.

4 RESULTS AND DISCUSSION

Inorganic hydrochemical data (pH, electrical conductivity, major ions) derived from sampling the installed 'bundled' piezometer are plotted with depth in Figure 4. The pH decreases with depth from 6.5 at 12 mbgl to 7.7 at a depth of 46 mbgl, near the base of the borehole. Depth variations in electrical conductivity (EC) of groundwater range from 0.30 mS/cm to 1.2 mS/cm with a peak value occurring at 21 mbgl that is recognised in the geophysical log (Figure 3) and considered an inflow zone of more highly conductive (i.e. more saline) groundwater. Increases in the concentration of major-ion species and calcium (Ca^{+2}) and chloride (Cl^{1}), in particular, correspond with the EC peak at 21 mbgl (Figure 4). Significantly, the EC and concentration of all major ions, except the bicarbonate anion, are lower below 21 mbgl.

Inorganic data demonstrate that significant depth variations in hydrochemistry exist in the Triassic sandstone aquifer. Clearly, depth-specific sampling is required not only to indicate the depth to which contaminants have penetrated the aquifer at multiple locations but also to investigate the processes and pathways by which contaminants may be transported and attenuated. Installation of the adapted 'bundled' piezometer design will occur at 4 more locations in Birmingham and Nottingham this year and form the basis of a detailed investigation of the depth and extent of penetration of contaminants into sandstone aquifer underlying long-term, industrial cities. Data derived from the 'bundled-nested' piezometers will also be used to identify whether high-quality groundwater may be found at depth in these urban areas.

5 CONCLUDING DISCUSSION

Significant hydrochemical variations, observed over depths of a few metres, demonstrate that the adapted 'bundled' (multilevel) piezometer is capable of sampling depth-specific groundwater samples in the Triassic fissured-sandstone aquifer of central England. The common availability of the piezometer materials and low cost as well as the ease of construction and manual installation all serve to permit widespread application of this multilevel piezometer design to other aquifer environments.

6 ACKNOWLEDGMENTS

Funding for this study was provided by the Natural Environment Research Council and Environment Agency, United Kingdom. Analysis of major ions was conducted by Paul Dewsbury of the Groundwater Protection and Restoration Group, University of Sheffield.

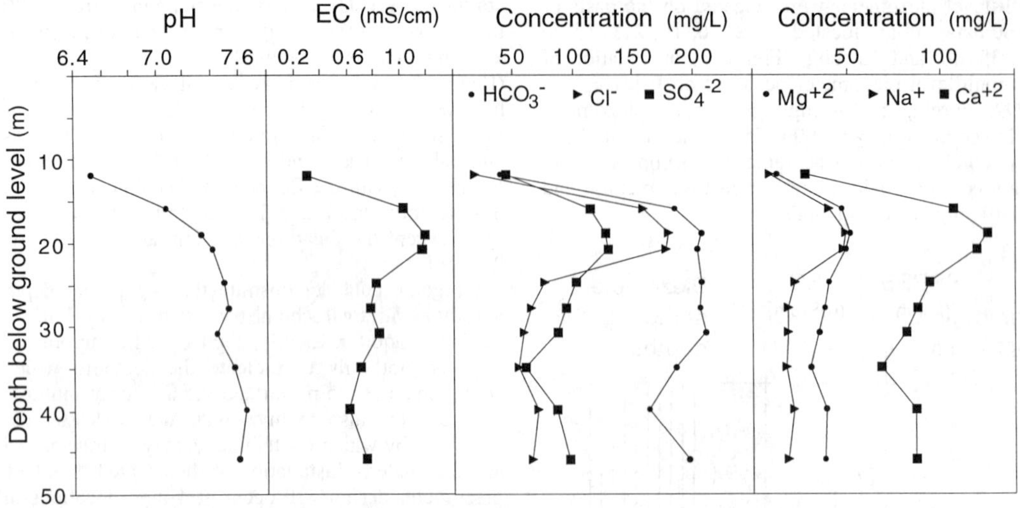

Figure 4. Depth variations in inorganic hydrochemistry derived from sampling the 'bundled' piezometer installed in the Triassic sandstone aquifer of north-central Nottingham.

7 REFERENCES

Anderton, R., Bridges, P.H., Leeder, M.R. and Sellwood, B.W., 1987. A dynamic stratigraphy of the British Isles. London: George Allen.

Charlsley, T.J., Rathbone, P.A. and Lowe, D.J., 1990. Nottingham: A geological background for planning and development. British Geological Survey Technical Report WA/90/1, p. 82.

Cherry, J.A., Gillham, R.W., Anderson, E.G., Johnson. P.E., 1983. Migration of contaminants in groundwater at a landfill: A case study. 2. Groundwater monitoring devices. Journal of Hydrology 63: 31-49.

Einarson, M.D. and Cherry, J.A., 2000. A new low-cost multi-level groundwater monitoring system for unconsolidated sediments and bedrock aquifers. Ground Water Monitoring and Remediation (in press)

Ford, M. and Tellam, J.H., 1994. Source, type and extent of inorganic contamination within the Birmingham urban aquifer system. Journal of Hydrology 156: 101-135.

Greswell, R., Lloyd, J.W., Lerner, D.N. and Knipe, C.V., 1994. Rising groundwater in the Birmingham area. In: E. Wilkinson (ed), Groundwater problems in urban areas: 330-341. London: Thomas Telford.

Jackson, D. and Lloyd, J.W., 1983. Groundwater chemistry of the Birmingham Triassic sandstone aquifer and its relation to structure. Quarterly Journal of engineering Geology (London) 16: 135-142.

Lerner, D.N., 1986. Leaking pipes recharge groundwater. Ground Water 24: 654-663.

Rivers, C.N., Barrett, M.H., Hiscock, K.M., Dennis, P.F., Feast, N.A. and Lerner, D.N., 1996. Use of nitrogen isotopes to identify nitrogen contamination of the Sherwood Sandstone aquifer beneath the city of Nottingham, United Kingdom. Hydrogeology Journal, 4(1): 90-102.

Rivett, M.O., Lerner, D.N., Lloyd, J.W. and Clark, L., 1990. Organic contamination of the Birmingham aquifer. Journal of Hydrology 113: 307-323.

Rushton, K.R. and Salmon, S., 1993. Significance of vertical flow through low-conductivity zones in Bromsgrove Sandstone Aquifer. Journal of Hydrology 152: 131-152.

Stagg, K.A., Tellam, J.H., Barrett, M.H., and Lerner, D.N., 1998. Hydrochemical variations with depth in a major UK aquifer: the fractured, high permeability Triassic sandstone. In: Brahana et al. (eds), Gambling with Groundwater-Physical, Chemical and Biological Aspects of Aquifer-Stream Relations: 53-58.

Thornton, S.F., Lerner, D.N. and Baines, O.P. Vertical profiling of natural attenuation in a sandstone aquifer using a dedicated multilevel groundwater sampling system. (in preparation)

Yang, Y., Lerner, D.N., Barrett, M.H. and Tellam, J., 1999. Quantification of groundwater recharge in the city of Nottingham, UK. Environmental Geology 38(3): 183-198.

Groundwater: Past Achievements and Future Challenges, Sililo et al. (eds) © *2000 Balkema, Rotterdam, ISBN 90 5809 159 7*

Groundwater quality trends related to aquitard salt storage at selected sites in the Lower Murrumbidgee alluvium, Australia

W. Timms & R. I. Acworth
Groundwater Centre, School of Civil and Environmental Engineering Water Research Laboratory, University of New South Wales, Manly Vale, N.S.W., Australia

J. Jankowski
Groundwater Centre, School of Geology, University of New South Wales, Sydney, N.S.W., Australia

S. Lawson
Department of Land and Water Conservation, Leeton, N.S.W., Australia

ABSTRACT: The risk of declining groundwater quality due to induced leaching of salts is a major concern in the over-stressed Lower Murrumbidgee alluvial groundwater system, Australia. Significant groundwater quality decline has occurred at some sites, with TDS increasing by up to 133 mg/L/yr in the shallow Shepparton aquifer, though no TDS increases were observed in the deep Calivil aquifer. Similar hydrochemical changes were observed to occur in the Shepparton aquifer, over 15 years, and to a lesser degree during the 1998-99 irrigation season. Preliminary mass balance modelling suggests that between 10 and 38% of groundwater volume was due to leakage of aquitard pore waters. Mixing was associated with NaCl dissolution, and reverse ion-exchange. The potential for further leaching is substantial, given the 4.25 kg/m^2 of salt within the Upper Shepparton aquitard, which accounted for 81% of total clay salt storage. Increased hydraulic gradient due to pumping was directly related to increased salinity in the Shepparton aquifer.

1 INTRODUCTION

The role of aquitards in controlling the rate of infiltration and discharge, as a source and sink for leachable chemicals, and in determining evolution of groundwater chemistry, is well recognised (Back 1986, Rudolf et al. 1991). In the central Murray-Darling Basin, the Mid-Tertiary Geera Clay has been identified as a major source and sink for salts (Jones et al. 1994). Little is known about Late Quaternary aquitards which overlie important irrigation aquifers of the Lower Murrumbidgee alluvium.

The potential for salt leaching from overlying aquitards has contributed to the highest risk classification, in terms of groundwater quality decline, of any inland groundwater system in New South Wales. Total groundwater usage in this over-stressed system is 241,000 ML/yr which is within the range of sustainable yield estimates (Lawson & Webb 1998). The allocation of additional entitlements has been embargoed since 1997. The system may be subject to further stresses in the future if all of the 500,000 ML/yr already allocated is utilized, or if pumping schemes to lower watertables within the Coleambally Irrigation Area (CIA) are expanded.

The role of deep groundwater pumping in mitigating local watertable rise has been limited by generally low aquitard hydraulic permeability and concern over induced leakage of poor quality water.

Previous studies have concluded that the risk of increasing salinity in the Calivil Formation from downwards leakage is minimal due to large dilution effects, with salinity expected to reach a maximum within 120 years (Williams 1987). This study predicted a maximum salinity of about 3000 μS/cm for the deep Calivil Formation. Based on average regional groundwater salinity, van der Lelij (1989) observed that higher salinity groundwater had dispersed downwards by 5 m between 1966 and 1980. Electrical Conductivity (EC) of pumped water from the Calivil aquifer during a 468 day pump test near Coleambally increased from 510 to 670 μS/cm. The associated shallow groundwater level decline due to leakage was calculated to be 0.2m (Lawson & Webb 1998).

This paper examines recent trends in groundwater quality together with new hydrochemical and aquitard hydrochemical data at selected sites in the Lower Murrumbidgee alluvium.

2 HYDROGEOLOGY

The Lower Murrumbidgee alluvial fan is comprised of over 200 m of unconsolidated clay, sand and gravel. The basal unit is the Mid-Tertiary Renmark Group which consists of medium-grained quartzose sands. This is overlain by coarse quartzose sand and fine gravel with lenses of kaolinitic clay of the Late Tertiary Calivil Formation. These high yielding Tertiary aquifers constitute the major groundwater resource in the area.

Table 1 Representative chemical composition of groundwater and surface waters (February 1999)

System*	Bore	pH	EC uS/cm	TDS mg/L	DO	Na	K	Ca	Mg	HCO_3	Cl	SO_4
Shep CL	12985	7.1	9290	6712.2	0.7	1275.6	12.4	433.8	385.8	376.9	2781	1363.0
	12986	7.7	11490	7330.3	1.5	1853.3	14.1	548.6	562.0	439.7	3475	2236.4
Shep SD	36372-1	6.8	4120	2558.7	7.8	407.9	5.2	216.5	176.3	164.1	1085.7	418.3
	30489-1	7.4	448	294.9	0.3	68.2	1.3	18.3	11.8	133.2	66.6	13.5
Shep MX	30349	6.5	524	493.9	-	36.1	1.1	33.7	26.2	198.0	56.1	19.7
Cal CL	B5	7	245	158.5	2.2	41.5	0.9	6.8	4.9	132.3	15.9	9.1
Cal SD	36372-2	7.1	294	187.9	0.1	39.9	1.3	11.7	9.6	139.0	26.5	9.7
	30348	6.7	424	243.8	2.1	57.7	9.0	11.6	12.1	151.1	55.5	4.0
Cal GR	30489-3	7.3	252	218.9	0.8	26.8	2.1	12.0	9.6	126.3	30.5	2.8
	36774-2	6.5	683	510.8	0.9	94.4	3.7	22.1	19.4	133.6	157.3	6.4
River	@Euroley	7.3	92	60.1	6.2	6.3	1.3	6.7	3.4	46.5	9.1	4.0
Additional analyses for NETPATH modelling												
Shep SD	$B1_{Oct-98}$	5.9	822	459.1	3.5	139.0	2.1	23.5	15.3	169.0	108	117.0
	$B1_{May-99}$	6.8	972	570.6	5.3	130.1	1.8	35.2	23.9	150.2	161.3	131.3
	$36372\text{-}1_{Sep-86}$	7.1	305	148.0	-	34.9	1.6	9.2	8.9	124.5	19.1	11.1
Shep CL	$Shep\text{-}CL_{av}$	-	1722	729.1	-	262.5	7.1	36.9	28.9	259.5	233	187.5

* Shep = Shepparton formation, Cal = Calivil formation, CL = clay, SD = sand, GR = gravel, MX= mixed clay/silt/sand/gravel

Variegated clay, silt and polymitic sand of the Late Pliocene-Pleistocene Shepparton Formation conformably overlie the main aquifers. Clay was deposited between shoestring/lensoid sands in a fluvio-lacustrine environment. Low permeability clay often limits irrigation drainage and diffuse recharge, and as a source of salts may contribute both to salinity in the root zone as watertables rise, and saline contamination of deep groundwater.

3 SALINITY TRENDS

Salinity trends in the Shepparton aquifer at selected sites over the past 20–25 years, were variable (Fig. 2). No corresponding trends were evident in the Calivil aquifer. Decreasing and stable EC values at the Gundaline and Tubbo sites respectively, contrasted with increasing EC values in shallow aquifers over the past 5 years at sites within and to the north of the CIA.

The most rapid rates of increase occurred between 1986 and 1999 at North Coleambally (36372-1, 47 m depth) where EC increased from <500 to 4120 μS/cm. This was equivalent to an average increase of total dissolved salts (TDS) of 133 mg/L/yr. There is no low permeability aquitard below the Shepparton aquifer at this site to protect the Calivil aquifer. At South Darlington Point (30342, 50.7 m depth), EC values rose two fold between 1980 and 1999, from <600 to 1260 μS/cm.

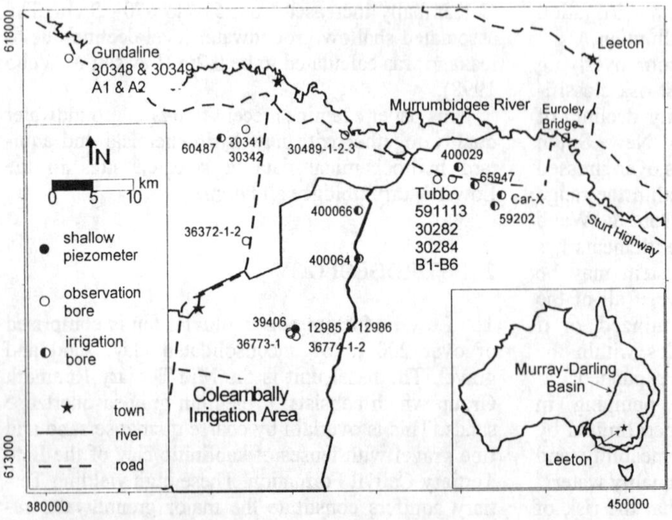

Figure 1 Location map of the Lower Murrumbidgee area

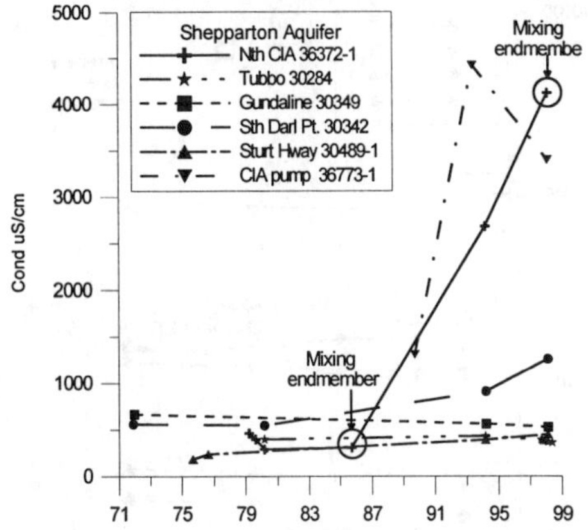

Figure 2 Salinity trends between 1971 and 1999, Shepparton Aquifer

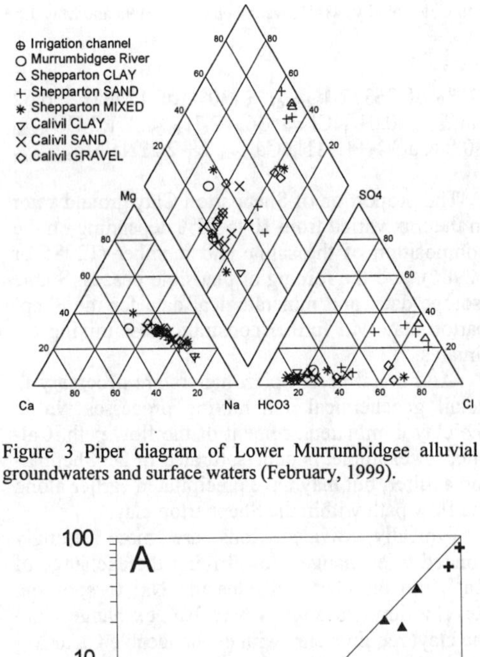

Figure 3 Piper diagram of Lower Murrumbidgee alluvial groundwaters and surface waters (February, 1999).

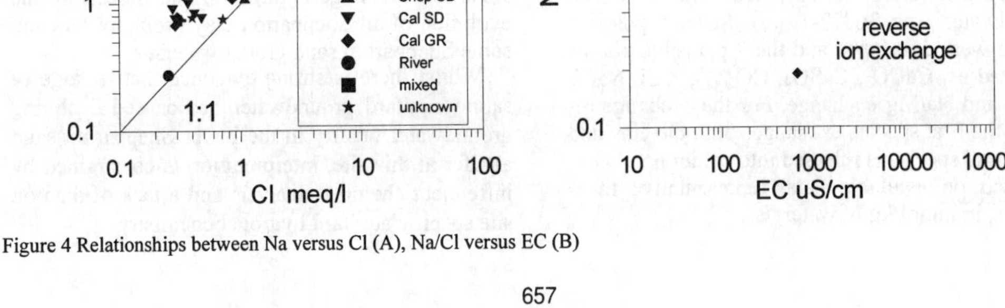

Figure 4 Relationships between Na versus Cl (A), Na/Cl versus EC (B)

4 GROUNDWATER CHEMISTRY

EC values for 34 samples ranged between 92 and 11490 μS/cm, with an average of 1135 μS/cm. Groundwater at a depth of less than 5 meters within clay aquitards of the CIA exceeded 10,000 μS/cm. Three samples from the Shepparton sand were over 3000 μS/cm, and were of a similar groundwater type to Shepparton clay groundwater (Fig. 3).

Groundwater quality was generally good with nitrate and phosphorous levels below 3 mg/L. Dissolved oxygen (DO) concentrations ranged from 0.1 to 9.2 mg/L with a mean of 3.2 mg/L, and groundwaters were near neutral pH.

Groundwaters were dominanted by Na-Mg-Ca-HCO_3-Cl and Na-Mg-HCO_3 type, contrasting with Murrumbidgee River water of Ca-Mg-Na-HCO_3-Cl type. Median sodium adsorption ratio (SAR) and exchangeable sodium ratio (ESR) values were 1.8 and 1 respectively and the mean magnesium hazard (MH) value was 55.7. Anhydrite, gypsum, dolomite and calcite were undersaturated in all samples except 12985, 12986, 36774-1 and 36773-1 which were saturated with respect to calcite, with saturation index (SI) exceeding zero. All groundwaters were in equilibrium with kaolinite.

Relationships between Na versus Cl and Na/Cl versus EC indicate that groundwater chemistry is significantly influenced by dissolution of NaCl; ion exchange in the Shepparton and Calivil aquifers, and by reverse ion exchange in the Shepparton Clay, (Fig. 4).

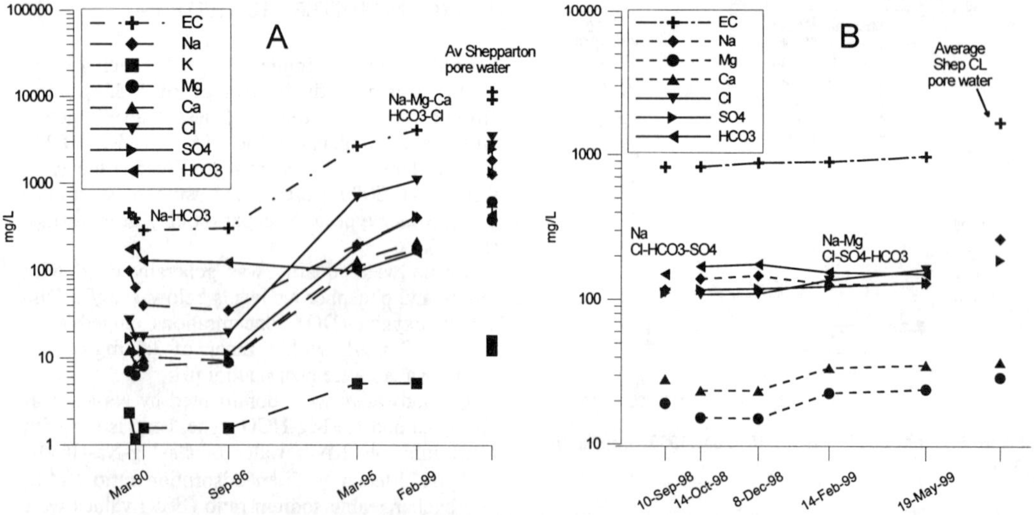

Figure 5 Hydrochemical variation A) Between 1980 and 1999 at North Coleambally B) Between September 1998 and May 1999 at Tubbo

5 GROUNDWATER LEAKAGE AND MIXING AT NORTH COLEAMBALLY

Detailed analysis of groundwater leakage and mixing at the North Coleambally site (36372) was possible on the basis of a number of chemical analyses between 1980 and 1999. At this site, located to the north of the irrigation area, rising groundwater levels between 1979 and 1996 in both the Shepparton and Calivil coincided with increasing seasonal drawdown. By 1997, drawdown in the Calivil aquifer was so great that seasonal recovery was minimal and groundwater levels in the Shepparton aquifer had declined. The downwards hydraulic gradient between the Shepparton and Calivil aquifer in February, 1999 was 0.052.

Over this period of time, groundwater evolved from Na-HCO_3 to Na-Mg-Ca-HCO_3-Cl type (Fig. 5A), and MH for irrigation increased from 48.8 to 57.3, which is above recommended levels, while the SAR remained low, decreasing from 5.6 to 5.0.

The computer program NETPATH (Plummer 1994), was used to account for the differences in mass balance between a computed and observed groundwater composition. The two component mix consisted of 36372-$1_{Sep\text{-}86}$ and a saline Shepparton clay groundwater (12985), and the observed groundwater was 36372-$1_{Feb\text{-}99}$. Several possible models were developed and the 7 plausible phases consisted of: $CaCO_3$, $CaSO_4$, $CO_{2(gas)}$, KCl, NaCl, Na/Ca and Na/Mg exchange. For the exchange reaction the first species exchanges onto the clay and the second species is released into solution.

Based on available data, representative mass transfer, in mmol/kg of water, is:

$$(70\% \text{ of } 36372\text{-}1_{Sep\text{-}86}) + (30\% \text{ of } 12985) + 6.62\ NaCl + 0.01\ KCl \Rightarrow 36372\text{-}1_{Feb\text{-}99} + 1.2\ CO_{2(gas)} + 0.88CaCO_3 + 1.11Na/Ca_{exchange} + 2.22Na/Mg_{exchange}$$

The proportion of Shepparton Clay groundwater in the mix varied from 10 to 38% depending on the composition of the saline end member (12985 or 12986), and the forcing of plausible phases. Stable isotope data, and mineralogical data for the Shepparton clay will further constrain these mixing estimates.

Additional nested piezometers are necessary to detail geochemical and mixing processes within the clay dominated segment of the flow path. Calcite, for example, is undersaturated in the Shepparton aquifer, but may have precipitated earlier along the flow path within the Shepparton clay.

Typically, divalent ions are more strongly bonded to exchange sites, driving the exchange of Ca^{2+} into the clay and releasing Na^+ to solution. Reverse ion-exchange, where Na^+ exchanges into the clay, requires clay with exchangeable Ca or Mg and a flowing water to have a relatively high concentration of Na to drive a reaction as the system equilibrates (Jankowski et al. 1998). Relationships between dissolved species (Fig. 4) indicate that reverse ion exchange is an important process in the evolution of all Shepparton clay groundwaters and some Shepparton sand groundwaters.

Whilst there is strong evidence that leakage of saline aquitard groundwater has caused declining groundwater quality in the lower Shepparton sand aquifer at this site, interpretation is constrained by infrequent chemical analysis and a lack of data on site specific aquitard hydrogeochemistry.

6 GROUNDWATER AND AQUITARD HYDROCHEMISTRY AT TUBBO

A close relationship between hydraulic gradient, pumping rates, groundwater chemistry and aquitard hydrogeochemistry was observed at the Tubbo nested piezometer site, during the 1998 -1999 irrigation season. Detailed monitoring was possible at this site where nested piezometers between 22.5 and 94.5 m deep were installed next to an irrigation bore pumping at 280 L/sec in a multi-level aquifer-aquitard system.

Between September 1998 and May 1999, 2500 ML was abstracted from the Calivil aquifer, associated with a drawdown of 0.3 m in the Shepparton sand aquifer. Groundwater levels did not respond to rainfall events. The drawdown was similar to that observed near Coleambally suggesting that despite the lateral discontinuity of clay lenses, site based leakage estimates may be indicative of leakage on a regional scale.

The downwards hydraulic gradient within the Shepparton aquifer reached a maximum of 0.36 in February 1999. However, the maximum hydraulic gradient averaged between the Shepparton and Calivil aquifer was just 0.076. Applying a hydraulic gradient averaged between the 2 aquifers may therefore result in a gross underestimation of leakage at the aquitard-aquifer interface of the Shepparton formation.

Variation in chemical composition in the Shepparton sand aquifer at Tubbo (B1 – 22.5m depth) between September 1998 and May 1999 was similar, but of a lesser magnitude than that observed at North Coleambally (36372-1) (Fig. 5). Groundwater salinity in B1 increased from 818 to 972 μS/cm, an increase of 154 uS/cm, similar to the average rate of rise at North Coleambally of 360 μS/cm/yr. Despite the likely differences in hydraulic gradient and the quantity of salt available to leach at each site, similar mixing and hydrogeochemical processes may occur.

Selected physical and chemical parameters of the Tubbo clay core are found in Table 2. An average porosity of 0.42 was calculated based on moisture content from wet and dry density data for subsections of the Tubbo clay core. Total dissolved salts, calculated from chemical data of 1:5 clay dilutions (Fig. 6), were used to approximate salt storage. Total subsurface salt storage contained in the total aquitard thickness (35 m) was about 5.2 kg/m^2. Of this salt store, the shallow Shepparton aquitard accounted for 4.3 kg/m^2 (81%), the middle Shepparton aquitard for 0. 74 kg/m^2 and the deep Calivil aquitard for 0.23 kg/m^2. Assuming lateral homogeneity and continuity, about 43 tonnes/ha of salts are stored in the uppermost 15 m of the clay profile.

Table 2 Aquitard salt storage estimates for Tubbo site

Aquitard	Thick m	N	ECav* μS/cm	TDSav mg/L	ϕ**	Salt-TOT Kg/m^2
U. Shep	15	14	1668	729	0.42	4.25
L. Shep	14.5	10	165	88	0.42	0.74
Cal	5.5	9	135	72	0.44	0.23

* corrected by $EC_{1:5}$ dilution factor
** ϕ calculated from dry & wet density & moisture content

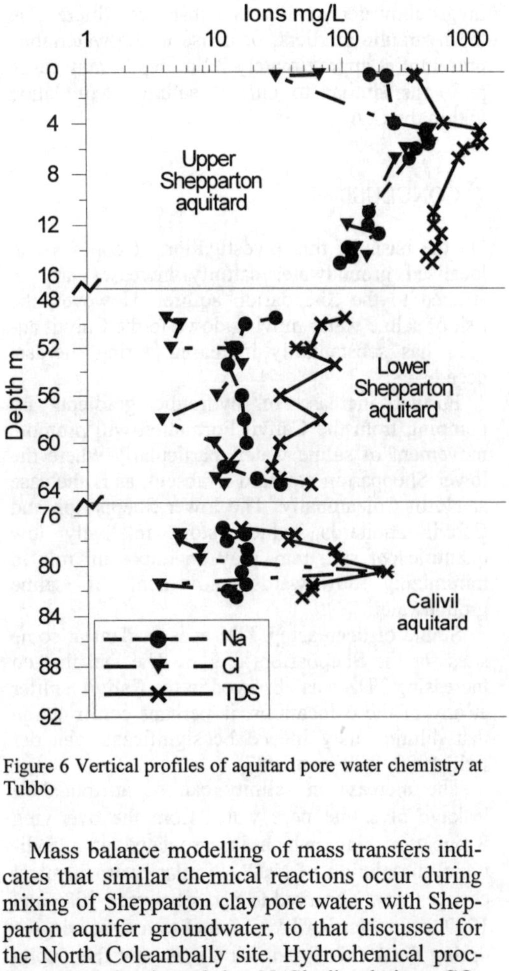

Figure 6 Vertical profiles of aquitard pore water chemistry at Tubbo

Mass balance modelling of mass transfers indicates that similar chemical reactions occur during mixing of Shepparton clay pore waters with Shepparton aquifer groundwater, to that discussed for the North Coleambally site. Hydrochemical processes are dominated by NaCl dissolution, CO_2 outgassing and reverse ion exchange. The composition of $B1_{May-99}$ is developed by a mixing of about 20% of Shepparton clay pore water with 80% of $B1_{Oct-98}$. Based on available data, representative mass balance is:

(80% of $B1_{Oct-98}$) + (20% of ShepCL) + 0.38 NaCl ⇒ $B1_{May-99}$ + 5.4 $CO_{2\,(gas)}$ + 0.4 $CaCO_3$ + 0.7 Na/Ca-exchange + 0.24 Na/Mg-exchange + 0.04 K/Na-exchange

The proportion of Shepparton clay groundwater in the mix varied from 14 to 28 % depending on the forcing of certain phases. Stable isotope data and aquitard mineralogical analysis will allow a more precise estimate to be made.

During mixing of groundwater the net TDS increase in the Shepparton aquitard was 112 mg/L which accounts for a small proportion of the total salt storage in the overlying Shepparton clay. The risk of additional leaching and associated decline in groundwater quality is substantial. Increasing the hydraulic gradient, or a rise in the watertable, presently at approximately 20 m depth, may cause problems similar to current salinity degradation within the CIA.

7 CONCLUSIONS

On the basis of this investigation, it appears that localised groundwater salinity increases are restricted to the Shepparton aquifer. However, the risk of saline water moving down to the Calivil aquifer has substantially increased during the last decade.

Further increase of hydraulic gradients by pumping from the Calivil Formation will promote movement of saline water, particularly where the lower Shepparton aquitard is absent, as is the case at North Coleambally. The lower Shepparton and Calivil aquitards, which store relatively low quantities of salt, may play an important role in minimizing downwards movement of saline groundwater.

Stable or decreasing TDS was evident at some sites for the Shepparton aquifer. The fact that no increasing TDS was observed in the Calivil aquifer at any of the 6 locations, is perhaps confirmation that dilution may indeed be significant (van der Lelij 1989, Williams 1987).

The increase in salinity can be attributed to leakage of saline pore water from the overlying Shepparton clay, which is associated with additional dissolution of NaCl. Similar hydrochemical changes were observed to occur between 1986 and 1999 at the North CIA site, and to a lesser degree during the 1998-99 irrigation season at the Tubbo site. Mass balance modelling suggests that between 10 and 38% of groundwater in the Shepparton aquifer has leaked from the overlying clay aquitard.

Variation in Shepparton clay thickness, permeability and salt storage, and differences in hydraulic gradients induced by pumping, may help explain the differences in medium term salinity changes on a regional scale.

Further work is in progress to constrain mixing models using stable isotopes and aquitard mineralogy. Site specific hydrodynamic and solute modelling is required to quantify the extent of Shepparton aquifer mixing with both the Shepparton aquitard, and Calivil aquifer.

Sustainable groundwater management in this area must incorporate both quality and quantity considerations. These results show that increased hydraulic gradient due to pumping is directly linked to increased salinity in the lower Shepparton aquifer. An improved understanding of variations of the lateral extent, thickness and salt balance of the Shepparton aquitard may help in identifying areas where groundwater is at risk from saline leakage.

ACKNOWLEDGEMENTS

This project was possible thanks to funding agencies, Land and Water Research and Development Corporation, the Natural Heritage Trust, the Department of Land and Water Conservation and the UNSW School of Civil and Environmental Engineering. The cooperation of local landholders, particularly the owner and manager of Tubbo Station is also acknowledged. Technical contributions by M. Groskops, R. Beck, J. De Martin and field assistance of G. Dacey, T. Friedenoff and B. Timms were greatly appreciated.

REFERENCES

Back, W. 1986. Role of aquitards in hydrogeochemical systems: a synopsis. *Applied Geochemistry*, 1:427-437.

Jankowski, J., S. Shekarforoush & R.I. Acworth 1998. Reverse ion-exchange in a deeply weathered porphyritic dacite fractured aquifer system, Yass, New South Wales, Australia. In *Arehard G.B. & Hulston R. (eds) Proceedings of the 9th International Symposium on Water-Rock Interaction, Taupo, New Zealand, 30 March – 3 April Rotterdam: Balkema, 243-246.*

Jones, B.F., J.S. Hanor & W.R. Evans 1994. Sources of dissolved salts in the central Murray Basin, Australia. *Chemical Geology*, 111:135-154.

Lawson, S & E. Webb 1998. *Review of groundwater use and groundwater level behaviour in the Lower Murrumbidgee Valley*. Department of Land and Water Conservation, Murrumbidgee Region, Technical Report No. 98/05.

Plummer, L.N. 1994. An interactive code (NETPATH) for modelling NET geochemical reacions along a flow PATH, Version 2.0. *USGS Water Res. Invest. Rep. 94-4169.*

Rudolf D.L., J.A. Cherry & R.N. Farvolden 1991. Groundwater flow and solute transport in fractured lacustrine clay near Mexico City. *Water Resources Research*, 27(9):2187-2201.

Van der Lelij, A. 1989. Groundwater accessions in the Coleambally Irrigation Area of N.S.W. In *Sharma (Ed) Groundwater Recharge, Proceedings of the symposium on groundwater recharge, Mandurah, 1989.* Rotterdam: Balkema.

Williams, R.M. 1987. The Murray Basin aquifer system with some reference to developing salinity problems. In *Australian Water Resources Council Conference Series, Proceedings of the International Conference on Groundwater Systems Under Stress, Brisbane, 1986.*

Groundwater: Past Achievements and Future Challenges, Sililo et al. (eds) © 2000 Balkema, Rotterdam, ISBN 90 5809 159 7

Groundwater flow in fractured rocks of central Namaqualand, South Africa

R.A.Titus & S.Adams
Department of Earth Sciences, University of the Western Cape, Bellville, South Africa

K.C.Pietersen
Water Research Commission, Pretoria, South Africa

G.Tredoux
Council for Scientific and Industrial Research, Stellenbosch, South Africa

ABSTRACT: Delineating groundwater flow in central Namaqualand is difficult due to the irregular distribution of the boreholes. Most boreholes in the Namaqualand region are located within structurally controlled valleys. In a few instances boreholes target fractures at higher elevations. Springs, large diameter wells and boreholes have been sampled for major element chemical analysis to assist with the identification of groundwater flow patterns. Joint orientations were measured to support the interpretations. Variations in groundwater chemistry were observed spatially as well as vertically within individual boreholes. Boreholes located close together within a structurally controlled valley can yield groundwaters with varying quality while a significant and fairly marked change in electrical conductivity is observed with depth. Shallow circulating and relatively young groundwater systems were recognised together with older, deeper and slower circulating groundwater systems. The groundwater chemistry for these flow systems is a result of varying flow-path length (residence time) and resultant water-rock interaction reactions.

1. INTRODUCTION

The Namaqualand region in the Northern Cape, South Africa, is one of the driest regions of the country. The rural communities are entirely dependent on groundwater for their existence. Only large towns like Springbok, Steinkopf, Nababeep and Okiep as well as towns supported by mines (e.g. Kleinsee) receive water by pipeline from the Orange River. It is not economically viable for most communities to import water by pipeline over large distances.

Delineating groundwater-flow patterns is difficult to construct due to paucity of boreholes, borehole position and spatial distribution of the boreholes. Fritz et al. (1990) mentioned similar constraints for groundwater flow patterns in desert terrains. Furthermore, almost no data was collected during the drilling of most boreholes. Insufficient test pumping results are available and groundwater chemistry data were analysed based on guidelines for domestic water supply. The groundwater is naturally of a poor and variable quality in most of the area. Variations in groundwater chemistry were observed spatially as well as vertically within boreholes. The geochemical data seems to support a stratification model with shallow circulating and relatively young groundwater systems overlying older, deeper and slower circulating groundwater systems.

The conceptual model for flow is based on: (1) The identification of spatial and vertical variations in groundwater chemistry; (2) The numerous lineaments, at all scales, visible on a satellite (i.e. Landsat) image. These observations are supported by a wealth of data on Proterozoic deformation phases; (3)The presence, to varying intensity, of horizontal joints (i.e. tension relief joints) in all lithologies; and (4) The structural control on both major and minor drainage systems.

2. DESCRIPTION OF STUDY AREA

2.1 Physical setting

The study area includes the Buffels River Catchment (F30) and parts of catchment F40 (Fig. 1). The area is characterised by three major zones: (1) coastal lowland, (2) intermediate escarpment zone and (3) Bushmanland plateau. The intermediate escarpment zone is characterised by the Springbok high (ca. 1300masl) and Kamieskroon high (ca. 1600masl) towards the southeast of the the Buffels River Catchment (BRC), and the Buffels River low (characterised by local relief within the escarpment).

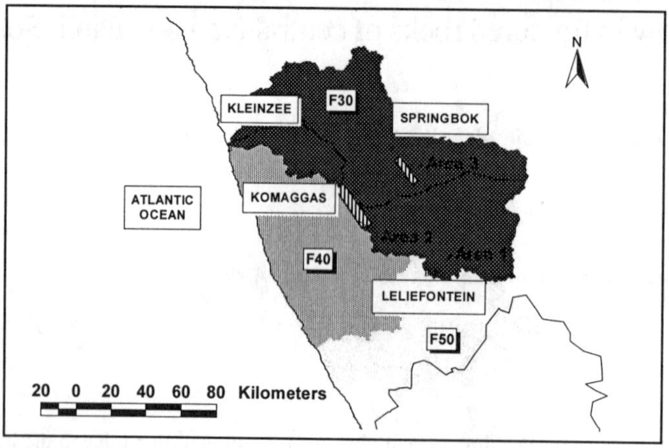

Figure 1. Locality map of study area. (The shaded area represents the BRC)

Recent unconsolidated sediments cover the coastal lowland of catchment F40.

2.2 Geological framework

An Early to Late Proterozoic (or Lower to Upper Precambrian) episode (approximately 2500 Ma to 570 Ma) of intense crustal instability can be divided on a chronologic and geographic basis into: (a) an Early to Middle Proterozoic tectonism resulting in the Namaqua and Natal Provinces, and (b) a long chain of Late Proterozoic (i.e. Namibian erathem) geosynclines, termed the Pan African Geosynclines, along the present southwestern coast and Namibian interior, including the Gariep Complex (Tankard et al. 1982; Visser et al. 1989). The Proterozoic Namaqua mobile belt, including the Richtersveld Subprovince, was subjected to the 1000 – 1200 Ma Namaquan Orogeny (Blignault et al. 1983 and Van der Merwe, 1995) characterised by thrusting, shearing and folding. The margins of the Namaqua Province are largely obscured by younger cover rocks of the Gariep, Nama and Karoo sequences, as well as with Cenozoic surficial sediments to the east and west.

Various researchers have mentioned unmapped and isolated occurrences of calcareous rocks in the pre-Bushmanland region (Moore 1989; Albat 1984 and Joubert 1971). Joubert (1971) reported that unmapped, small and isolated crystalline rocks may occur throughout the study area.

2.3 Hydrological framework

Most of the BRC receives winter rainfall while the eastern parts experience summer rainfall. The area has an arid to semi-arid climate. The average annual rainfall varies from 120mm in the west (F40) to 380mm in the east (F30). The average rainfall for Springbok and Kommagas are 210mm and 120mm, respectively. The Buffels River has its main source in the Kamiesberg Mountains. It is fed by the western Bushmanland peneplane (900-1100m a.m.s.l.) at the eastern border and the Springbok and Kamiesberg Mountains with maximum elevations of respectively 1300 and 1600m above mean sea level.

Groundwater is abstracted from the fractured and weathered crystalline rocks, with minor extractions from the alluvium of the Buffels River. The depth of weathering extends to between 54m and 60m below surface for both the metasediments and the gneissic rocks. Differential weathering of rock occurs with depth together with localised weathering on fracture systems. Toens et al. (1991) dated, using carbon-14 and tritium, some groundwater north of Springbok and found ages between 4500 and 9900 years. The tritium analyses indicated that no recent recharge (<50 years) occurs in the area. This may not be the case in the higher areas of the Buffels River Catchment.

3. HYDROCHEMISTRY

3.1 Sampling and Analyses

A total of 60 groundwater samples were collected. Four samples had ionic balance errors above 10% but less than 15%, and were included in the data set. Alkalinity, electrical conductivity (EC), pH, temperature and redox potential (Eh) were measured and recorded at the time of sampling. Alkalinity was measured using a Hach® field titration kit. The EC, pH, temperature and Eh were measured using portable Orion® and Cyberscan® EC and pH meters. Water was collected mainly from wind pumps. Samples were filtered in the field with a hand-held syringe using 0.45 μm membrane filters and stored in new polyethylene bottles. All sampling bottles were rinsed with the filtered sample. Major and minor ion analyses were performed using a Varian Vista ICP Atomic Emission Spectrometer. Nitrate and fluoride were analyzed using the cadmium reduction column method for nitrate and a specific ion electrode for fluoride.

3.2 Results and Discussion

Kamiesberg Mountain Ranges (Area 1).

Two groups of groundwaters can be distinguished on a bicarbonate (HCO_3^-) vs. pH diagram (Fig. 2). There is a rise in the pH and bicarbonate concentrations with increasing salinity. This could be due to the weathering of aluminosilicate minerals. The groundwaters within Group 1 can be regarded as relatively young, reflecting the weathering processes associated with a dilute and chemically aggressive infiltrating and recharging water. These groundwaters, collected from springs and a well, may be representative of shallow circulation with a relatively quick through flow. The weathering processes of aluminosilicate minerals associated with the infiltration of chemically aggressive waters are also described by Kay (1995) as well as Freeze and Cherry (1979).

Group 2 comprises of groundwater sampled at windpumps, a borehole and wells. The wells may have a topographic location that allows groundwater with this distinctive pH and HCO_3 values to be present in the wells. Groundwaters of Group 2 may be indicative of relatively deeper circulation at the onset of reducing conditions.

Alternatively, the rise in pH and bicarbonate concentrations may also be attributed to carbonate dissolution and to redox reactions, in particular sulfate reduction reactions. Such processes were also described by Domenico and Schwartz (1990) as well as Freeze and Cherry (1979).

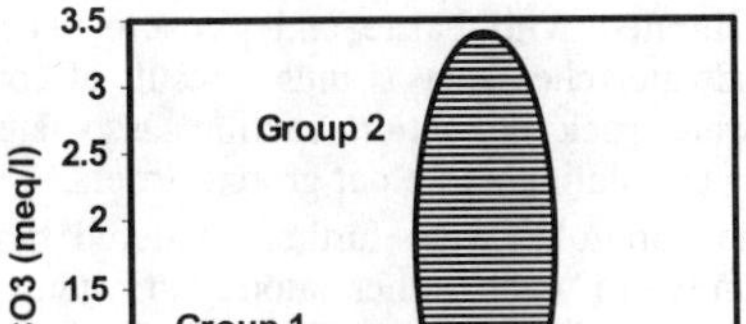

Figure 2. A bicarbonate vs. pH diagram for groundwaters on the Kamiesberg Mountain range.

Group 1 groundwaters are classed as sodium-bicarbonate type waters, based on the major cation and anion in a sample, and sodium-chloride (Na-Cl) type waters. These groundwaters have Na-Cl ratios (Fig. 3) above the ratio for sea water, as proposed by Goldberg (1963), indicating an increase in the sodium content with respect to chloride. A (Ca+Mg):HCO_3 ratio of 0.5 is representative of groundwaters associated with the weathering of carbonate minerals as well as amphiboles and pyroxenes. Sami (1992) used this ratio, relative to increasing salinity, to discuss the origin of dissolved Ca and Mg concentrations. These groundwaters have (Ca+Mg):HCO_3 ratios that are well above the ratio of 0.5, and which increase with increasing salinity.

The rest of the groundwaters show both Na enrichment and depletion when compared to the Na:Cl ratios for oceanic water. The Na/Cl vs. Cl relation shows increasingly lower Na:Cl ratios while the (Ca+Mg)/HCO_3 vs. Cl relation indicate higher Ca+Mg:HCO_3 ratios, with increasing salinity, for these groundwaters.

Wildeperdehoek ranges (Area 2).

This range forms a surface drainage boundary between the surface drainage catchments F30 (i.e. the Buffels River catchment) and F40. Again two groups of groundwater can be recognised, which in turn are further subdivided into Groups 1a and 1b, and Groups 2a, 2b and 2c (Fig. 4). The Group 1 groundwaters were collected from boreholes and springs in both the metamorphosed, intrusive rocks of the Little Namaqualand Suite and the interlaminated schists and quartzites of the Khurisberg Subgroup (Okiep Group).

Mineral saturation indices (SI), calculated using the WATEQF chemical speciation model of Plummer et al. (1994), indicate that the groundwaters of groups 1a and 1b are predominantly oversaturated with regard to silica. The groundwaters within Group 1a can be regarded as relatively young with limited water-rock interaction representing shallow circulating groundwaters (Fig.4). These groundwaters compare, in terms of salinity, with Group 1 groundwaters of the Kamiesberg mountain ranges. Group 1b groundwaters, collected from deeper boreholes located within the Khurisberg metasediments and the intrusive Little Namaqualand Suite, may representative of relatively longer flowpaths (i.e. residence time) compared to group 1a.

The saturation indices indicate, with increasing salinity, a complex hydrochemical pattern. Group 2a groundwaters are oversaturated with silica, plagioclase feldspar (albite), fluorite and barite and undersaturated with carbonate minerals and gypsum. These groundwaters occur predominantly in the intrusive rocks of the Little Namaqualand Suite.

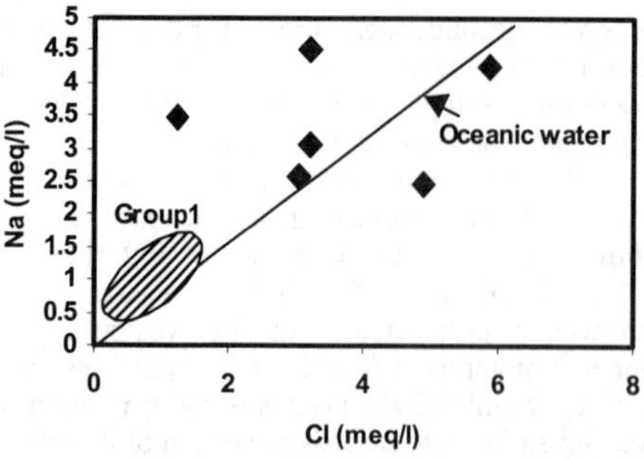

Figure 3. A Na vs.Cl relation compared to similar ratios for oceanic water (Kamiesberg Mountain range). Diamond symbols represent sample points.

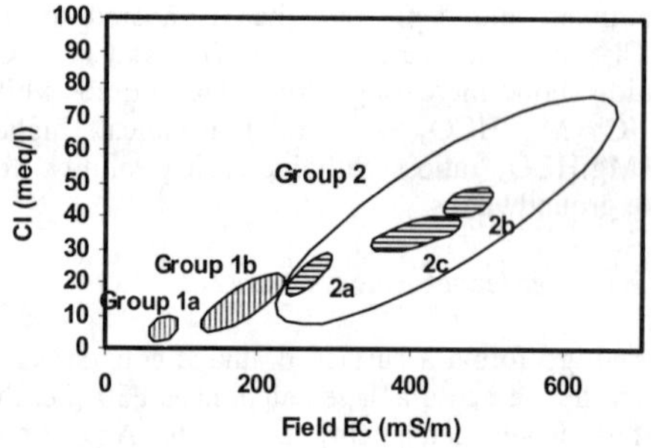

Figure 4. A Cl vs. EC (field) relation for groundwaters of the Wildeperdehoek ranges.

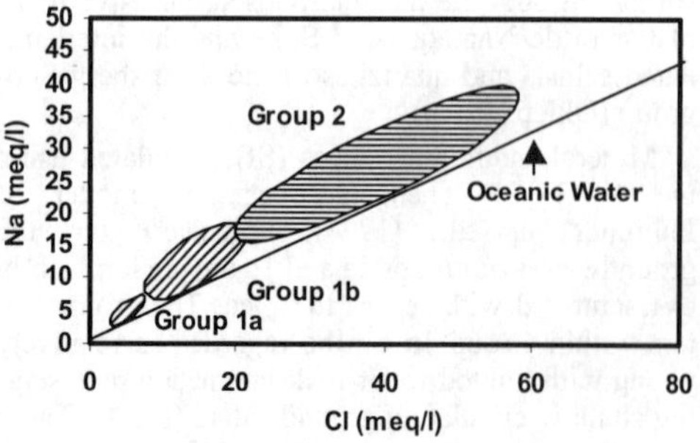

Figure 5. Na vs. Cl relation compared to the sea water ratio (Wildeperdehoek range).

Group 2b groundwaters are similar except that these groundwaters are undersaturated with regard to fluorite. The Group 2b groundwaters occur predominantly in the gneissic rocks of the Garies Subgroup (Okiep Group). Group 2c groundwaters are oversaturated with regard to carbonate minerals, fluorite and barite, saturated with silica and generally undersaturated with plagioclase feldspars (albite and anorthite). This variability may be the result of the isolated and patchy occurrence of carbonate-bearing host rocks. The Group 2 groundwaters may be indicative of older waters associated with deeper and slower circulation.

The Na^+ vs. Cl^- relation (Fig. 5) further refines the Group 1 groundwater. These groundwaters have Na:Cl ratios above the ratio for sea water again indicating an increase in the sodium content with respect to chloride. The groundwaters have (Ca+Mg):HCO_3 ratios that are well above the ratio of 0.5 and increasingly so with increasing salinity.

The Group 1b groundwaters show slightly higher average ratios of Na:Cl and (Ca+ Mg):HCO_3, at a higher salinity, compared to similar ratios for Group 1a. The Group 2 groundwaters have increasingly lower average Na:Cl ratios and much higher (Ca+Mg):HCO_3 ratios with increasing salinity.

NNW fault southeast of Springbok (Area 3).

A major NNW striking fault is located southeast of the town of Springbok. In the central Buffels River Catchment (BRC) groundwater is commonly abstracted from boreholes located within major fault controlled valleys. Groundwaters were collected from boreholes located in the gneisses of the intrusive Little Namaqualand Suite as well as the gneissic rocks of the Garies Subgroup. Again, the hydrochemistry indicates two groundwater types.

Saturation indices (SI) calculations indicate that the groundwaters of Group 1 are oversaturated with respect to silica, saturated to undersaturated with fluorite and generally undersaturated with regard to carbonates and plagioclase feldspar (albite). These groundwaters can thus be regarded as shallow circulating, relatively younger groundwaters having undergone limited water-rock interaction (Fig.6).

Group 2 groundwaters are oversaturated with silica, plagioclase feldspars (albite and anorthite) and fluorite. The Group 2a groundwaters are predominantly undersaturated with carbonate minerals while the Group 2b groundwaters are generally saturated to oversaturated with carbonate minerals. Groundwaters of Group 2 can also be undersaturated to oversaturated with barite and gypsum. The latter groundwater chemistry is thus a result of considerable water-rock interaction reactions associated with deeper circulating and older groundwaters.

The boreholes were further clustered based on proximity to each other along the fault zone. Groundwater from cluster 2 (i.e. 2a, 2b, 2c and 2d) is characterised by varying groundwater chemistry, hence the inclusion in both Groups 1 and 2b. Similar trends are observed for groundwaters of the other clusters, in particular clusters 1 and 3. A rose diagram (Fig.7) developed from lineaments for this area, illustrate the multitude of vertical to subvertical fractures orientated obliquely or at right angles to the NNW striking fault zone. A stereo-graphic plot for joint planes (Fig. 8) indicates the varying orientations for the sub-vertical to vertical joints. In addition, the stereo-graphic plot also indicates the presence of horisontal joint planes measured in all lithologies in the immediate vicinity of Springbok.

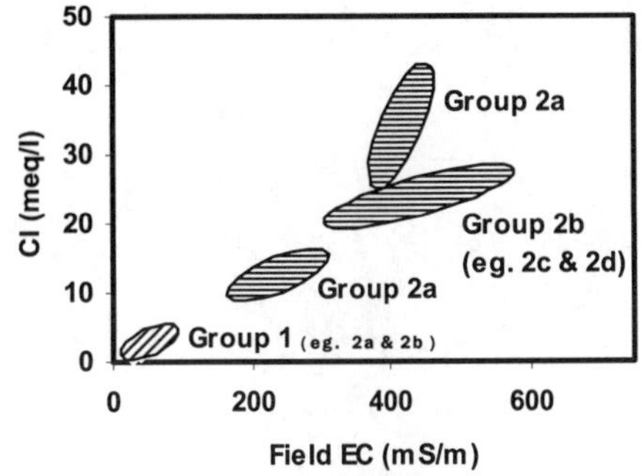

Figure 6. A Cl vs. EC (field) relation for groundwaters on a major NNW fault southeast of Springbok.

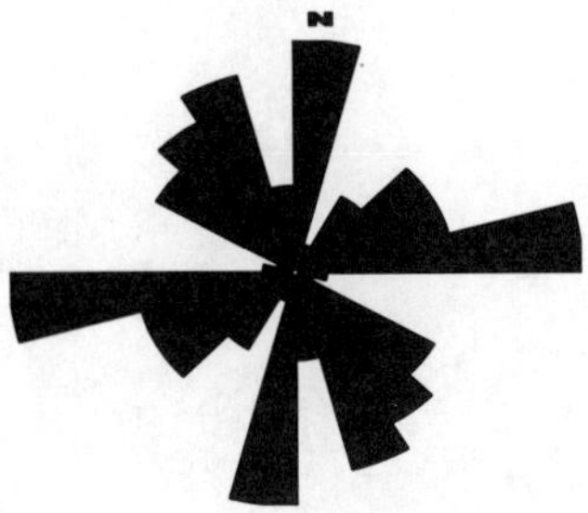

Figure 7. Rose diagram depicting from lineaments orientated obliquely or at right angles to the major NNW fault.

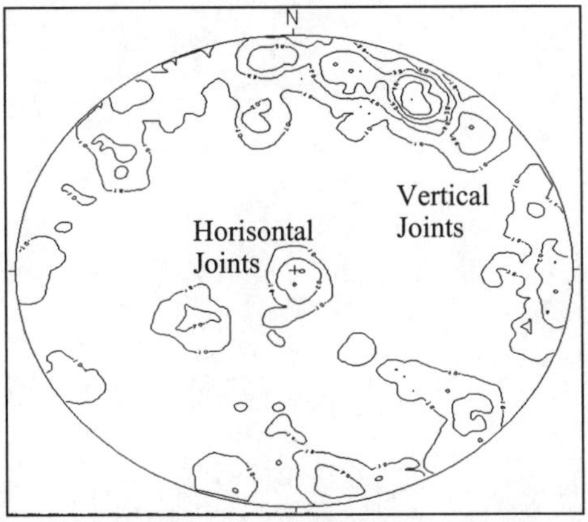

Figure 8. A stereo-graphic plot of both vertical to subvertical joints (along circumference) as well as horisontal joints (in the center) for the Little Namaqualand Suite.

The latter probably represents tension relief joints that formed as a result of erosional unloading. The subvertical to vertical joints planes may act as preferred pathways for infiltration while the horisontal joint planes predominantly act as conduits for horisontal flow.

4. SUMMARY

Similar flow regimes were observed for all the areas, including the NNW orientated fault. Shallow circulating and relatively young groundwater systems are recognised together with older, deeper and slower circulating groundwater systems.

The data indicates that;

1. The groundwater chemistry in these fractured crystalline rocks seems to be dependant on the position of the sampling point within either shallow, fast circulating (i.e. active) or deeper, slower circulating flow systems. Johnston (1982) came to a similar conclusion to explain the variable groundwater chemistry in crystalline rocks.
2. The deeper circulating, higher salinity groundwaters probably experience insignificant recharge under present climatic conditions.
3. The fracture patterns in most of the lithologies are well developed with relatively active groundwater flow systems. The broad range of orientations for these fracture systems may indicate that most of the fractures are connected with complex groundwater flow paths. Kay (1985) also concluded that well developed fracture networks with active groundwater flow systems occur granitic rocks.

The general movement of groundwater for these semi-confined aquifer systems wil be a function of head differences associated with topography. Groundwater flow from topographic highs occurs along major fractures as a function of the naturally high vertical hydraulic gradients in these regions. Groundwaters circulating deeper and slower are associated with a reduction of the hydraulic gradients and hydraulic conductivity with depth.

ACKNOWLEDGEMENTS

The authors thank the Water Research Commission of South Africa for funding this project. The authors also wish to thank all involved in the analyses and logistical components of the project.

REFERENCES

Albat, H.M. 1984. The Proterozoic granulite facies terrane around Kliprand, Namaqualand Metamorphic Complex. Bulletin 33. Precambrian Research Unit, UCT.

Blignaut, H.J., Van Aswegen, G.J., Van der Merwe, S.W. and Colliston, W.P. 1983. The Namaqualand Geotraverse and environs: part of the Proterozoic Namaqua mobile belt. *Spec. Publ. Geol. Soc. S. Afr. 10:1-29.*

Domenico, P.A., and Schwartz, F.W. 1990. Physical and Chemical Hydrogeology. John Wiley and Sons.

Freeze, R. and Cherry, J.A. 1979. Groundwater. Prentice-Hall Inc, New Jersey.

Fritz, S.J. et al. 1990. Elucidating Ground-water Flow paths in a Desert Terrane by Geochemical Methods. *Ground Water,. Vol. 28. No. 4. Pp 551–558.*

Goldberg, E.D. 1963. Chemistry in the oceans as a chemical system. In Composition of sea water, comaparative and descriptive oceanography. Wiley-Interscience, New York.

Johnston, L.M. 1982. A review of the geochemistry of groundwater in crystalline rocks. Rept. Atomic Energy of Canada Ltd. TR-154. 99 pp.

Joubert, P. 1971. The regional tectonism of the gneisses of part of Namaqualand. Bulletin 10. Precambrian Research Unit, UCT, Cape Town.

Kay, R.L.F. 1985. Review of applications of hydrogeochemistry to groundwater development in tropical basement and regolith. Rep. British Geological Survey WD/OS/85/4. 51 pp.

Kay, R.L.F. 1987. Groundwater exploration and development in crystalline aquifers. Technical Paper 273. Volume II, Sessions 6 to 15, pp. 69-92. Commonwealth Science Council.

Moore, J.M.,1989. A comparative study of metamorphosed supracrustal rocks from the Western Namaqualand Metamorphic Complex. Bulletin 37. Precambrian Research Unit, UCT, Cape Town.

Plummer, L.N., Prestemon, E.C. and Parkhurst, D.L. 1994. An interactive code (NETPATH) for modelling net geochemical reactions along a flow path. Water-Resources Investigations rteport 94-4169, USGS.

Tankard, A. J., Jackson, M. P. , Eriksson, K. A., Hobday, D. K., Hunter, D. R and Minter, W. E. L. 1982. Crustal Evolution of South Africa: 3.8 Billion Years of Earth History. Springer-Verlag, New York.

Toens, P.D., Levin, M. and Verhagen B.Th. 1991. Results of pumping tests and environmental isotope analysis at Steinkopf. Toens Progress Report no. 40. Cape Town.

Sami, K. 1992. Recharge mechanisms and geochemical processes in a semi-arid sedimentary basin, Eastern Cape, South Africa. Elsevier Science Publishers; pp.26-30.

Van der Merwe, S.W. 1995. The relationship between thrusting, vertical shear and open folds in the western part of the Namaqua mobile belt. *S.Afr., J.Geol. 98:68-77.*

Visser, D. J. L. et al. 1989. The geology of the Republics of South Africa, Transkei, Bophuthatswana, Venda and Ciskei and the Kingdoms of Lesotho and Swaziland. Dept. Of Mineral and Energy Affairs, Pretoria.

Groundwater: Past Achievements and Future Challenges, Sililo et al. (eds) © 2000 Balkema, Rotterdam, ISBN 90 5809 159 7

In-situ measurement of physico-chemical parameters down a borehole as a tool for resource evaluation

G.Tredoux, L.C.Cavé & J.F.P.Engelbrecht
Council for Scientific and Industrial Research, Stellenbosch, South Africa

ABSTRACT: Parameters such as electrical conductivity, pH, Eh, dissolved oxygen, temperature and other chemical constituents can be logged with confidence up to depths of 200 metres below the water table. This technology provides the ability to determine *in situ* physical and chemical properties of the groundwater in the subsurface. This is essential because temperature and gas partial pressures, which are controlling variables in chemical equilibria, are significantly different in the subsurface compared to those at the surface. Thus logging allows for a closer approximation of the chemical conditions driving the actual water-rock equilibria in the aquifer than measurements taken at the wellhead or in a laboratory. Important applications of this technology are in pollution studies for the identification of contamination and determining the optimum groundwater sampling depth. It also provides a basis for the evaluation and understanding of the groundwater resource.

1 INTRODUCTION

Chemical equilibria in groundwater establish themselves at a specific temperature and pressure. Changes in temperature and pressure will cause a shift in the concentrations of reactants and products as a new equilibrium condition is established. This affects a number of physico-chemical parameters in groundwater, including pH, Eh and dissolved gases. It is commonly accepted that such parameters have to be measured at the wellhead. However, even though such measurements approach the true hydrochemistry more closely, they still do not allow determination of the actual groundwater composition in the aquifer.

In the past, the measurement of the oxidation-reduction (or redox) potential (Eh) has been difficult and attempts to obtain subsurface readings had to be very innovative (e.g. Mannheim 1961). In other cases readings were taken from flowing boreholes in order to get an indication of subsurface conditions (Germanov et al. 1959).

As a consequence of large improvements in sensor technology, much more reliable measurements are presently possible down a borehole, at much greater depths and temperatures, than were possible even at the surface two decades ago. It is, therefore, not surprising that down the hole physico-chemical logging is practised much more widely and involves measurement of many more parameters than before. As with geophysical techniques, it is necessary to start in areas where the geology is known. Extrapolation to adjacent areas without geological logs then becomes possible.

More than two decades ago, down the hole temperature logging was proposed as a means of establishing the temperature gradients which could then be used to calculate the rate of natural and artificial groundwater recharge (Boyle & Saleem 1979). Current technology has made this method far more accessible, even for non-specialised groundwater projects.

This paper considers applications of down the hole logging for monitoring pollution and the identification of redox barriers, where denitrification, sulphate reduction, or other chemical processes occur. Logging in dynamic pumping situations and the effects of leaky borehole casings are also considered.

2 pH, EH AND DISSOLVED GASES

Solute chemistry and mineral stability relationships in groundwater systems may be described in terms of the controlling variables Eh and pH (Stumm & Morgan 1970). The pH reflects the acid-base status while the Eh of natural water depends on the combination of oxidation-reduction couples present in the water.

Baas-Becking et al. (1960) have tabulated measurements of the pH and Eh in an extensive range of natural environments and found that these are within definite limits. The upper and lower limits for aqueous environments are defined by the pH-Eh relationships for redox reactions involving oxygen, water and hydrogen.

Dissolved gases such as carbon dioxide and oxygen are an important component of many chemical equilibria in groundwater. They are often closely linked to the pH and Eh of the system by their participation in acid-base reactions (e.g. CO_2 in the carbonic acid system) or redox reactions (e.g. O_2 as an oxidising agent for organic matter in the aquifer). The concentrations of these gases in solution are dependant on their partial pressures as described by Henry's law. Subsurface temperature conditions also influence the concentration of dissolved gases and indeed that of all dissolved species involved in rapid chemical equilibria.

In general, several oxidation-reduction couples will be present in the water which together will determine the Eh. However, in the subsurface environment, geological conditions may cause certain couples to dominate. This is the type of situation which can provide for useful *in-situ* Eh measurements, e.g. the strongly reducing environment of carbonaceous shale, which may overlie or underlie strata with totally different characteristics.

Apart from the reactions providing the upper and lower limits for redox potentials in the aqueous medium, many other redox reactions are strongly pH dependent. This provides the basis for a pH-Eh relationship which is often noticed during borehole logging. Generally, where such redox couples dominate, an increase in pH will be accompanied by a decrease in Eh. Evangelou (1998) provides a number of examples of such reactions.

Borehole clogging is often related to the presence of iron and manganese in the groundwater or the host rock. As pH-Eh relationships are directly responsible for iron and manganese solubility control, problems of borehole clogging involving these elements and their compounds cannot be resolved without detailed *in-situ* pH and Eh measurements. This should be accompanied by well head pressure filtration, Fe-species determination, and preservation of samples.

3 FIELD MEASUREMENTS

A Hydrolab MiniSonde 4a with a depth range of 200 m and equipped with sensors for electrical conductivity, temperature, pH, Eh, and dissolved oxygen was used for the logging of open boreholes in various parts of South Africa and Namibia. The sensors were calibrated as set out in the manual and overall it was found that the sensors were relatively stable with little recalibration required.

The logger was initially held just below the water surface for at least five minutes to allow the sensors to equilibrate with the groundwater before taking measurements. The probe was then lowered slowly down the borehole, allowing sufficient time for at least two readings to be recorded at each depth increment. Logging was ceased within 1 to 5 metres of the bottom of the borehole to avoid lowering the probe into muddy sediments which could foul or damage the sensors. Readings were also recorded while withdrawing the probe, using greater intervals between successive measurements. Readings taken at the same depth, both while lowering and raising the logger, were found to vary by 10% or less. The repeatability of the measurements was also confirmed at boreholes where logs were repeated after a brief time interval. In most applications, it was found that relative changes in the parameters, rather than the absolute measurements gave the most valuable information.

Figure 1 shows the physico-chemical log for an uncased borehole in the southern area of the Springbok Flats. This borehole penetrates 24 m of basalt before entering sandstone. Both formations are of Karoo age.

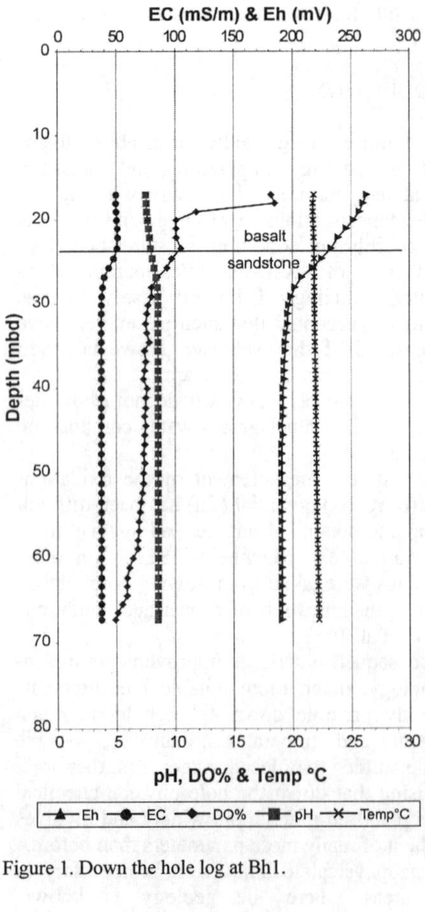

Figure 1. Down the hole log at Bh1.

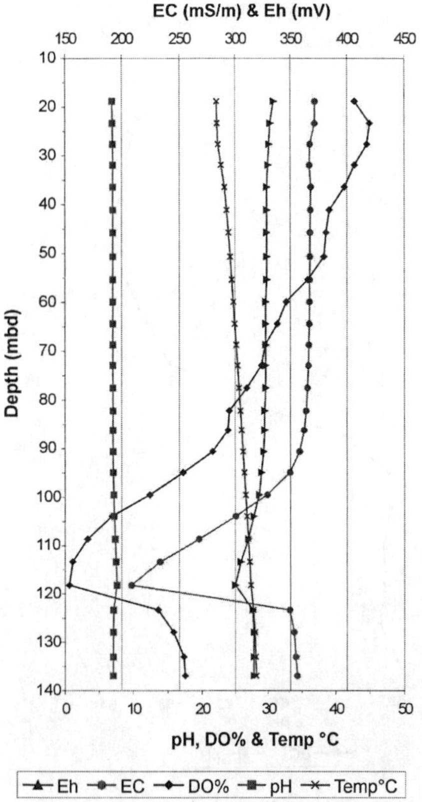

Figure 2. Bh2 log while pumping (October 1999).

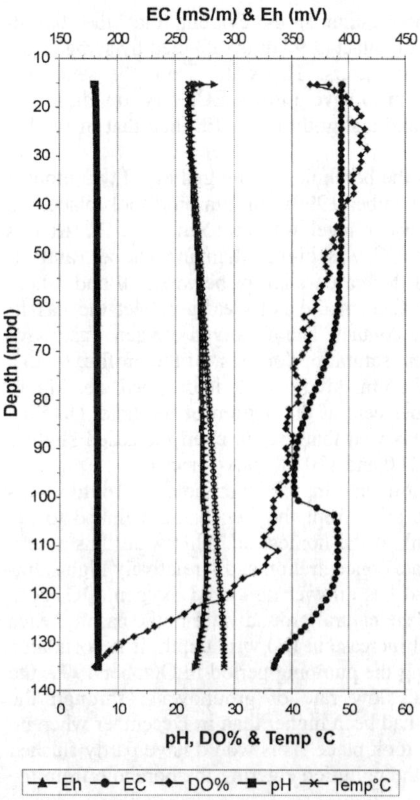

Figure 3. Bh2 when not pumping (December 1999).

The borehole was inspected with a down the hole video camera and the exact depth of the contact established. Most of the parameters change near 24 m, i.e. in the vicinity of the contact between the basalt and the sandstone. The salinity (EC), dissolved oxygen, and the oxidation-reduction potential (Eh) all decrease, while the pH shows an increase. Thus a relationship between water quality and the geological formations is clearly evident.

In Figure 2, five parameters are shown for the hydrochemical log of another borehole in the Springbok Flats (designated Bh 2 for the purpose of this paper). It was logged as part of a nitrate investigation. This borehole, which is 137 m deep, penetrates the basalt and enters the underlying sandstone at 97 m. The borehole is situated approximately 30 m from a homestead and chemical and microbiological analyses have shown it to be polluted. When it was logged in October 1999, it was newly drilled and test pumping was in progress at another new borehole more than a kilometre away.

At the start of logging the water level in Bh 2 was at 18.5 m, but it dropped to 21.5 m during the time the logging was carried out. Five hours later, it had dropped another 8 m as a result of the pumping, despite the intervening distance of more than one kilometre and the fact that the aquifer is not confined. Furthermore, water was running slowly into the borehole from a depth of approximately 16 m. Under these conditions, horizontal flow through the borehole was suspected as well as a degree of vertical flow.

The EC of Bh 2 was approximately 370 mS/m from the water level down to 90 m below surface. It then decreased to reach a minimum of 210 mS/m at 118 m, after which it returned to the original level. The dissolved oxygen declined steadily from just below the water level to also reach a minimum of virtually zero at a depth of 118 m, with a slight increase at greater depth. Eh showed a corresponding minimum at the same depth. The trends for the other parameters cannot be seen at this scale, but pH followed an inverse trend compared to the Eh. The temperature followed a steady increase with depth except for a slight deviation between the water level and 38 m below surface.

From the results it is evident that the basalt-sandstone contact at 97 m is reflected by some of the logged parameters. The water from the sandstone would seem to have a lower EC, very low dissolved oxygen, and a slightly lower Eh than that of the basalt.

When the borehole was logged again two months later (December 1999), no pumping took place and the rest water level was at 15 m (Fig. 3). At this stage the EC was higher than in October, and although it decreased slightly between 70 and 103 m below surface, it did not clearly reflect the basalt-sandstone contact. The dissolved oxygen was above 40 per cent saturation for most of the profile, up to a depth of 95 m, after which it dropped steadily to about 5 per cent at the bottom of the hole. Detailed analysis showed that the Eh even increased slightly between 100 and 120 m below surface.

Pollution entering the borehole at 16 m is expected to arise from the french drain linked to the septic tank at the homestead. This water has a very high nitrate concentration and a relatively high salinity. It also has a lower dissolved oxygen (DO) content than the natural groundwater in the basalt, hence the initial increase in DO with depth. It is postulated that during the pumping period in October 1999, the horizontal flow rate of groundwater through the borehole had been higher than in December when no pumping took place. This would have partly flushed some of the pollution entering the borehole from the top.

The deviation in the upper part of the temperature-depth curves could relate to the "artificial recharge" from the french drain. This could be analysed using the approach of Boyle & Saleem (1979) in order to determine the (total) recharge rate.

For the purposes of the nitrate investigation, it was evident that the basalt generally contained sufficiently high levels of dissolved oxygen to prevent bacterial denitrification. This was confirmed at other boreholes. However, when the groundwater reaches the underlying sandstone, oxygen levels are depleted and denitrification commences, provided sufficient organic substrate is available.

In a separate investigation, boreholes at a waste disposal site were logged in order to assess the possibility of groundwater stratification to ensure a successful monitoring programme. The EC and pH logs for a monitoring borehole, Bh 3, are shown in Figure 4. It is evident that the groundwater quality is adversely affected below a depth of 35 m. For proper monitoring, groundwater sampling should ideally take place at a depth of 42 m. Pumping at a low rate is indicated.

Comparing the data for several study areas, most boreholes showed an inverse relationship between pH and Eh. Some examples are shown in Figure 5. Most of these boreholes were located in basalt overlying sandstone such as in the Springbok Flats.

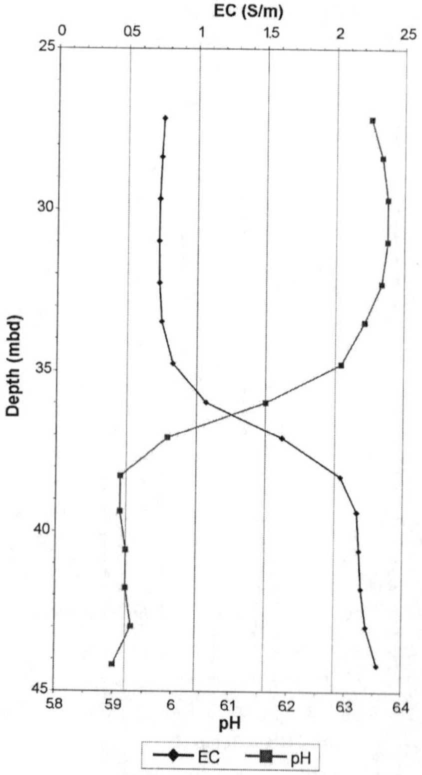

Figure 4. Polluted Bh3 at waste disposal site.

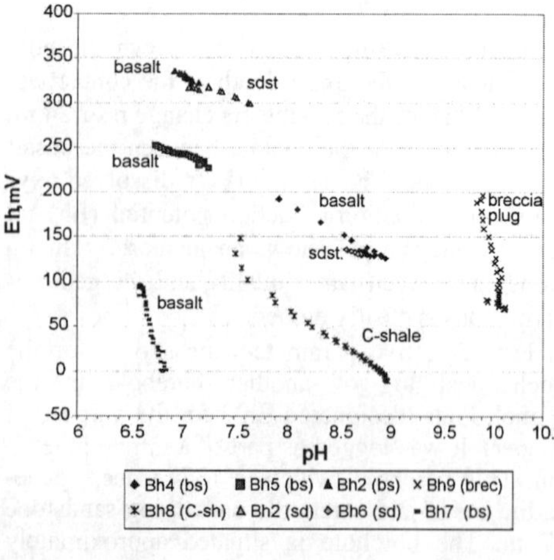

Figure 5. Redox potentials vs. pH at various boreholes. C-sh = carbonaceous shale, sdst = sandstone, bs = basalt, brec = breccia plug.

Each of these boreholes had very different pH and Eh values but also a distinctive pH-Eh relationship.

The borehole in carbonaceous shale, for example, showed a high pH and low Eh, while groundwater in a nearby breccia plug had an even higher pH value, but was not as strongly reducing as in the shale.

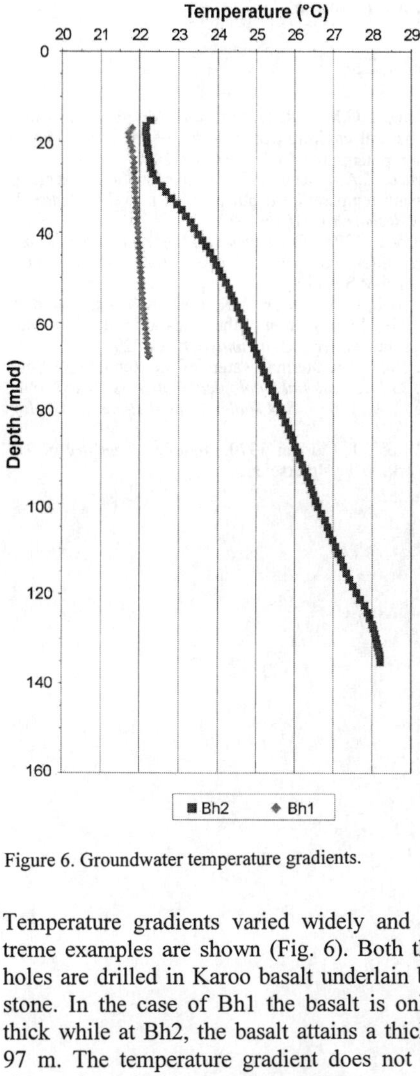

Figure 6. Groundwater temperature gradients.

Temperature gradients varied widely and two extreme examples are shown (Fig. 6). Both the boreholes are drilled in Karoo basalt underlain by sandstone. In the case of Bh1 the basalt is only 24 m thick while at Bh2, the basalt attains a thickness of 97 m. The temperature gradient does not seem to change from the basalt to the sandstone.

The gradient at Bh1 is not unusual for this geological environment. Kent (1969) calculated a range of thermal gradients from 1.4 to 2.7 ºC/100 m for boreholes in Karoo rocks. The temperature gradient at Bh2, however, is close to 5 ºC/100 m.

The area around borehole Bh2 consists of maize fields on so-called black cotton soils, while borehole Bh1 is surrounded by grassland used for grazing cattle. It is postulated that differences in recharge rates due to tilling of the soil, and heat flux, due to the bare black soils, could at least partially explain the different gradients.

In another study in an arid, granite-gneissic terrain, hydrochemical logging showed that the groundwater below a depth of 60 to 80 m was significantly more saline than nearer to surface. This is of great significance as correct management of the abstraction will yield a substantially lower salinity water for village supply. Without the down the hole logging of these boreholes, only the "average" salinity value obtained during pumping would have been known.

4 DISCUSSION

The equipment used for the fieldwork in this paper could operate at depths of up to 200 m below the water level, but units are now available for reaching depths of 300 m. The calibration of the sensors for field measurements is carried out at the surface. Obviously, it is virtually impossible to simulate conditions down the borehole unless one has appropriate research facilities with the necessary pressure vessels. So in general, one has to accept the specifications of the manufacturers. However, this may not be a too serious problem for groundwater resource development where it is often more important to identify changes or trends rather than the absolute values. It is also recognized that the down the hole measurements provide the best data for use in models for calculating geochemical speciation and mass transport, because they give the closest approximation of true aquifer values. The advantages of such data, compared with parameters measured at the wellhead or, even worse, in the laboratory, are considered to outweigh any inaccuracies in the pressure compensation built into the equipment.

Groundwater movement in the subsurface is largely controlled by the physical characteristics of the host rock, i.e. the pores or fractures conducting the fluid. In primary aquifers, the layered nature of the geological formations often causes stratification of the water quality. Sampling disturbs this stratification and leads to incorrect conclusions regarding the groundwater chemistry. This is all the more important where an uncased borehole penetrates two or more overlying aquifers. In the case of confined aquifers, the borehole provides a potential conduit for the exchange of water. Logging the physico-chemical parameters enables the identification of such phenomena as well as other conditions, such as leaky borehole casings. Once the detailed hydrochemical information is available, it can be confirmed by sampling at the identified depths, or by inspection with a down the hole camera, e.g. for confirming leaks.

One aspect of physico-chemical down the hole logging that deserves attention, is the availability of open boreholes. It was found that newly drilled boreholes, with as little casing or well screens as possible, were ideal for logging. Thus an attempt needs to be made to log the borehole immediately after drilling. Most of the older boreholes that were suitable for logging, since they were not equipped, were in a bad condition because they had been abandoned for some or other reason.

5 CONCLUSIONS

Determining a series of parameters such as pH, redox potential, temperature, dissolved oxygen, electrical conductivity and total alkalinity at the wellhead is necessary for getting to grips with the hydrochemistry of a particular groundwater. These parameters change when the sample is removed from the borehole and, therefore, have to be measured immediately at the wellhead. This is the standard approach in groundwater investigations. Physico-chemical logging of the borehole goes one step further as it measures physical and hydrochemical parameters *in situ*. Down the hole logging is essential for investigating the groundwater chemical equilibria in the aquifer. For establishing the clogging potential of a borehole, the information on the *in situ* chemical equilibria in the aquifer is indispensable.

Down the hole logging provides a means of identifying fractures. Alternatively, when the depths to each of the fractures are known, the quality of water derived from each of these can be determined. Leaky borehole casings have been identified by irregular temperature-depth profiles and salinity-depth profiles.

Groundwater sampling disturbs the stratification in the aquifer, especially while boreholes are pumped. However, pumping is essential for determining the water quality to be expected from a production borehole. Thus, although pumping is preferred as a sampling method for routine operations, for detailed studies a different approach may be indicated. For instance, in pollution studies it is essential to determine the appropriate depth for sampling by logging a profile of the physico-chemical measurements down the borehole under static conditions.

Longer term deployment of logging equipment in a borehole for recording time related trends is of importance in groundwater management studies, e.g. when applying artificial recharge. Placement of the logger at a strategic depth in a borehole for continuous logging allows realtime measurement of changing parameters.

It is concluded that down the hole logging can add great value to groundwater investigations by providing additional information not possible from wellhead measurements alone. This will allow for better understanding of the aquifer geochemical system and consequently better management of the groundwater resource.

REFERENCES

Baas-Becking, L.G.M., I.R. Kaplan & D. Moore 1960. Limits of the natural environment in terms of pH and oxidation-reduction potentials. *J. Geol.* 68:243-284.

Boyle, J.M. & Z.A. Saleem 1979. Determination of recharge rates using temperature-depth profiles in wells. *Water Resources Research* 15:1616-1622.

Evangelou, V.P. 1998. *Environmental Soil and Water Chemistry: Principles and Applications* p. 246 & 254. New York: John Wiley & Sons Inc.

Germanov, A.I., G.A.Volkov, A.K. Lisitsin & V.S. Serebrennikov 1959. Investigation of the oxidation-reduction potential of ground waters. *Geokhimiya* 3:322-329.

Kent, L.F. 1969. The thermal waters of the Republic of South Africa. *23 International Geological Congress* 18:143-164.

Mannheim, F.T. 1961. *Stockholm Contributions in Geology* 8:27.

Stumm, W. & J.J. Morgan 1970. *Aquatic Chemistry* p. 319. New York: Wiley-Interscience.

Groundwater: Past Achievements and Future Challenges, Sililo et al. (eds) © 2000 Balkema, Rotterdam, ISBN 90 5809 159 7

Environmental isotopes assist in ground water sustainability assessment of the Taaibosch fault zone, Northern Province, South Africa

B.Th.Verhagen & M.J.Butler
Schonland Research Centre, University of the Witwatersrand, Johannesburg, South Africa

M.Levin
AFRICON Engineering International, Pretoria, South Africa

E.van Wyk
Directorate of Geohydrology, DWAF, Pretoria, South Africa

ABSTRACT: Large investments in rural water supply have been made in South Africa, but very limited information is available on which sound management can be based. A pilot study of the Taaibosch fault area in Northern Province integrates environmental isotope hydrology techniques into a multidisciplinary investigation, and forms part of a regional project of the International Atomic Energy Agency. High yielding boreholes have been developed along the Taaibosch fault. An earlier conceptual model proposes ground water flow towards the fault from an extensive plain and drainage along the strike of the fault. Initial isotopic and hydrochemical investigations place limits on aspects of this model. Stable isotopes and hydrochemistry suggest that the flow pattern is not as simple as the model proposes. Isotope-based assessment of diffuse recharge to the plain and the fault zone question the earlier estimates of sustainable average yield for the fault zone.

1 INTRODUCTION

A major government drive of ground water development is underway in South Africa's Northern Province to supply the needs of 3.6 million people. Unmanaged rural development poses threats of pollution and over-exploitation to ground water resources.

The application of environmental isotopes is seen as a major step towards characterizing and quantifying ground water in the Taaibosch fault zone community water supply scheme (Figure 1). The project was selected as South Africa's contribution to a regional initiative by the International Atomic Energy Agency to stimulate the use of environmental isotopes in groundwater resource development.

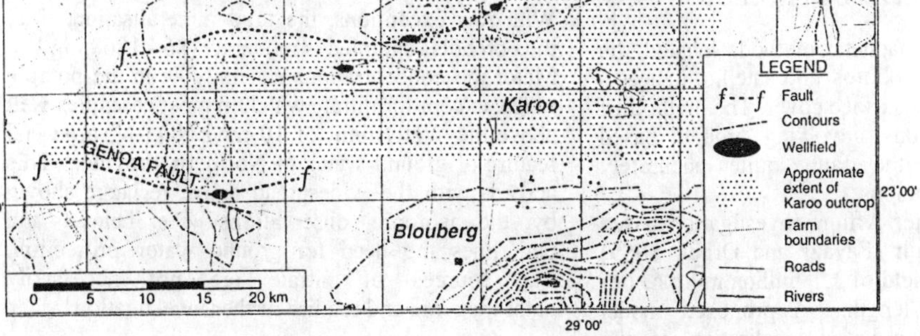

Figure 1. Map of Taaibosch study area showing borehole locations and schematic geology

2 REGIONAL GEOLOGY OF STUDY AREA

Rock formations in the northern portion of the Bochum District were exposed to several phases of metamorphism and tectonic movement under the collective name of the Limpopo Mobile Belt.

The Taaibosch regional fault (Figure 1) separates the metamorphic rocks from Karoo sediments and basalt to the south. Approximately 20 km further south the Genoa fault separates the graben system of younger Karoo sediments faulted into the basement, from the Soutpansberg Group, quartzitic rocks with extrusive basalts of the Blouberg.

In the east, the Taaibosch fault lies within the Karoo, is undisturbed, but with intersecting dykes; in the west, secondary faults run parallel to the main fault within Karoo and against gneiss, with no discernible dykes but possible transverse faulting. Extensive calcretes cover part of the fault towards the SW.

3 HYDROGEOLOGY

The depth of weathering in the fractured Karoo basalt, the principal aquifer, depends on the lithological character but in general does not exceed 40m. The depth of fresh rock fracturing is also limited. Supply wells are thus drilled no deeper than 60-80m. Depth to water level varies between 10m and 20m. The rest level gradient should mimick the surface gradient of about 0.004 in a northerly and westerly direction towards the Mogalokwena River (Figure 1).

Boreholes drilled into the basalt rarely yield in excess of 1.0 l s^{-1}. Yields from Karoo sandstone are generally poor. The exploration target is the regional intensively fractured Taaibosch fault zone with blow-off yields of 20 l s^{-1}, and the Genoa fault. The supply to 26 villages is planned through discrete smaller well fields with reticulation systems, away from villages and cattle. This would allow for ground water protection, monitoring and management, public ownership of the resource and for domestic and enterprise development beyond mere subsistence levels.

4 THE TAAIBOSCH FAULT PROJECT

Magnetic and electro-magnetic methods located the fault - identified on air photos and satellite images - where it is obscured by surface cover. The method of surface radon emissions, involving ground based detectors, pinpoints the fault quite accurately (Figure 2) in many locations.

A Department of Water Affairs investigation based on the Taaibosch Fault (Fayazi and Orpen 1989) estimated an average yield of 1.1 million m^3 p.a. per one km of the fault length. A proposed model (Figure 3) represents a broad idealisation of the natural flow-rate towards the Taaibosch Fault.

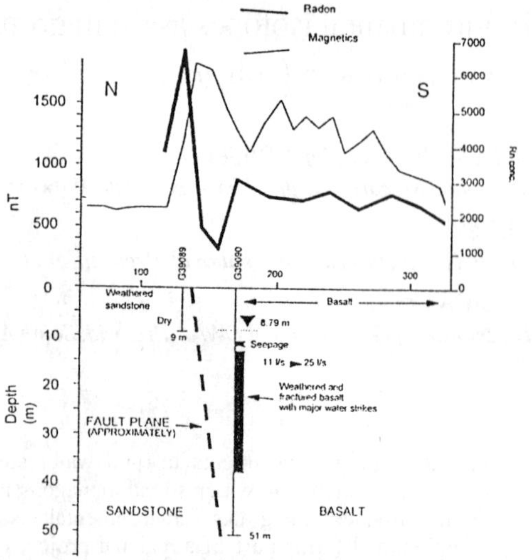

Figure 2. Electromagnetic and radon traverses, showing the precise location of the fault position by the radon method.

The proposed Taaibosch Fault catchment area is a well-vegetated, featureless plain extending 25 km southwards towards the water divide at the Blouberg The mean annual rainfall of 400 mm p.a. could be lost by evapotranspiration and run-off. Only above average rainfall (such as the seasons of 1998 - 1999 and especially 1999 - 2000, with more than 1000 mm) may produce significant recharge.

The model proposes that the fault may act as a sink for ground water drainage out of the area, enhanced by exploitation of production boreholes in the fault zone. Flow should occur mainly along the deeper, fractured fresh basalt zone with vertical leakage from overlying weathered rock. The permeability of this deeper zone should be an order of magnitude less than the Taaibosch Fault.

This paper reports work in progress addressing some of the uncertainties surrounding the behaviour and resource of groundwater in the area.

5 HYDROCHEMISTRY

With few exceptions, the total mineralisation of all the boreholes sampled is moderate (86 - 1868 mg l^{-1}). Major ion concentrations generally lie in the potable range, except for nitrate which in most cases lies well above the WHO limit of 10 mg/l NO_3-N, a general feature of ground water in Northern Province. This accords with the widespread active recharge shown by the isotopic observations (see below) and underlines the need for ground water protection. The sources of nitrate are not necessarily anthropogenic and require further investigation.

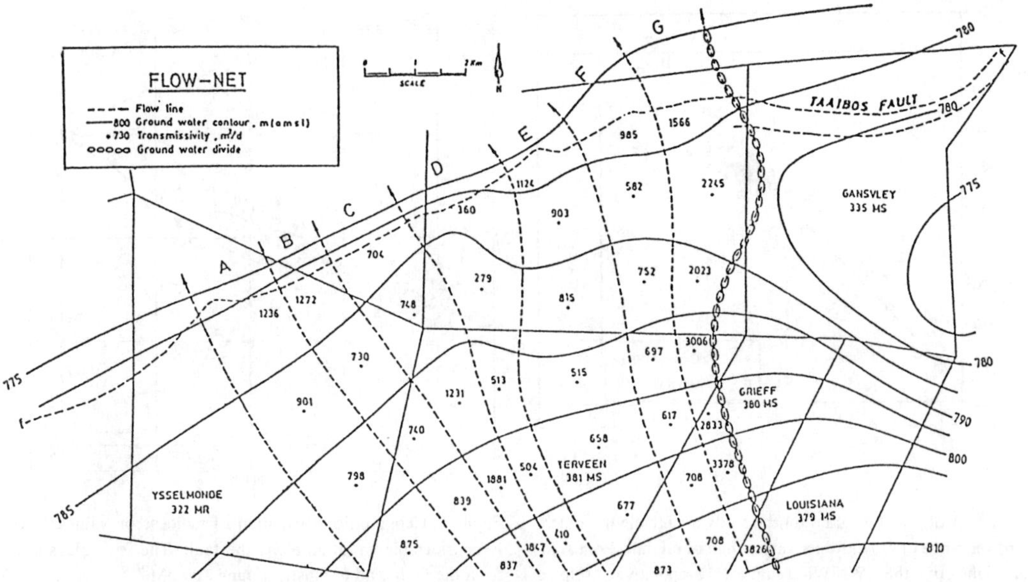

Figure 3. Model of flow towards the fault showing rest level contours (m.a.m.s.l.) and inferred flow lines (from: Fayazi & Orpen 1989)

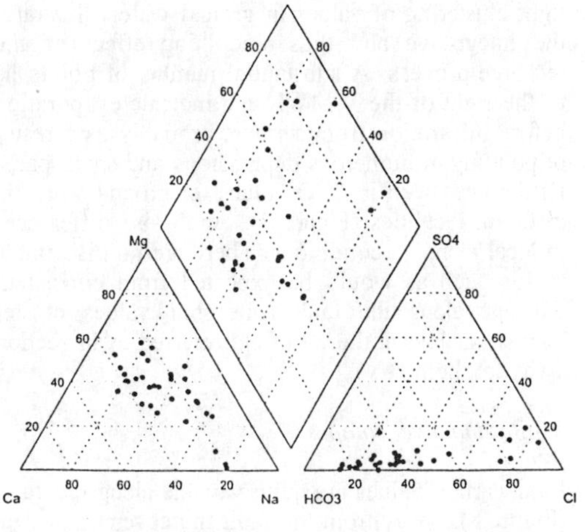

Figure 4. Piper diagram of groundwater hydrochemistry at different sampling sites.

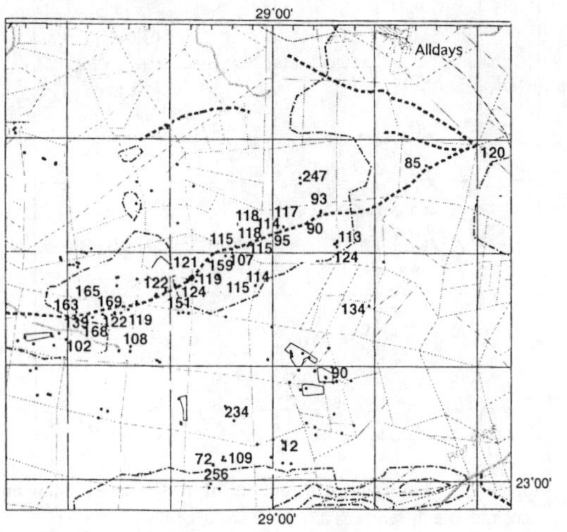

Figure 5. Geographic distribution of EC in groundwater

The chemical data is plotted on the Piper diagram in Figure 4. The majority of the points cluster towards the Ca,Mg-HCO_3 dominant quadrant of the diamond field, in agreement with the active recharge shown by the isotope data. The geographic distribution of EC values is shown in Figure 5. Mineralisation along the fault strike varies by almost a factor of 2 with no discernible trends, indicating the importance of local recharge conditions. In contrast, one of the wellfields shows very uniform chemical parameters as well as stable isotope values.

6 ENVIRONMENTAL ISOTOPES

6.1 *Concepts*

The isotope "snapshot" of the area presented here gives unique information in the absence of intensive or long-term geohydrological observations.

For an idealised uniformly recharged phreatic aquifer the vertical distribution of a radioactive isotope, and the derived mean residence time, is independent of position (Verhagen et al. 1991). Ground water flow

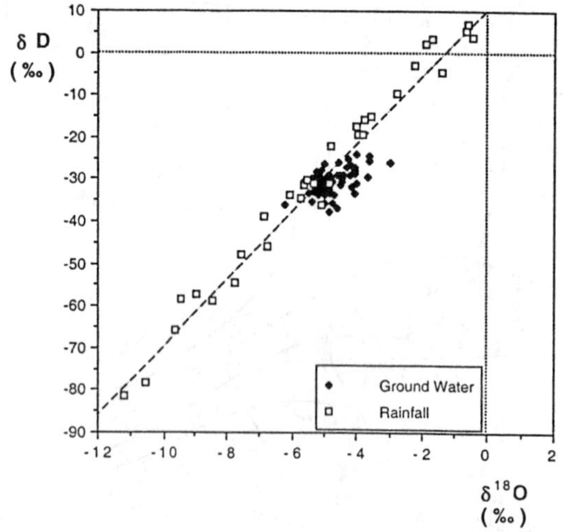

Figure 6. A δD-δ^{18}O diagram for the study area. Note the much narrower range of groundwater values than of rainfall. Points to the right of the WMWL suggest evaporative isotope enrichment

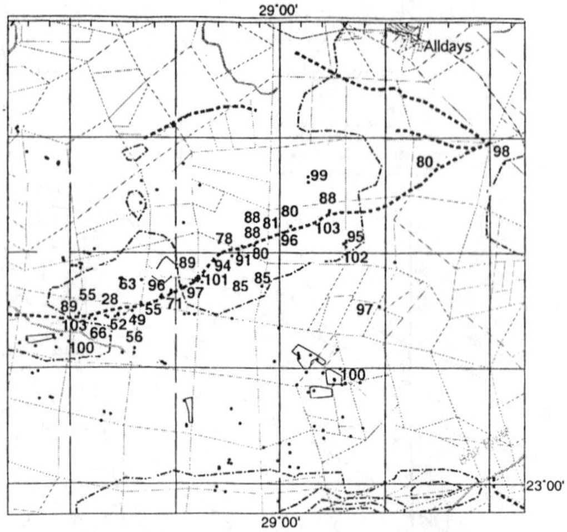

Figure 8. Geographic distribution of radiocarbon values. There is considerable variation along the fault. The few values away from the fault tend to cluster around 100 pMC.

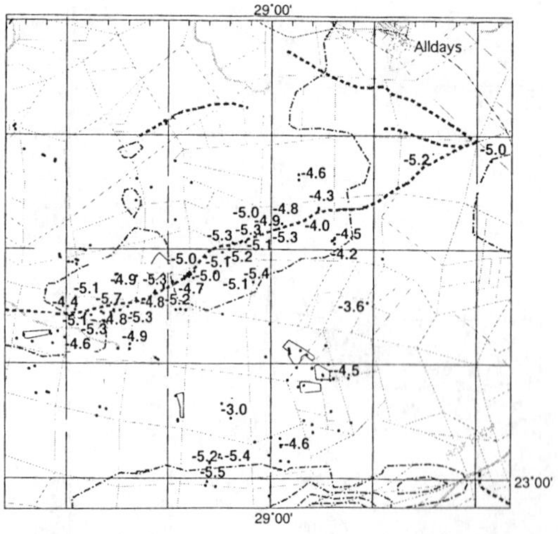

Figure 7. Geographic distribution of δ^{18}O values, showing considerable differences along the strike of the fault.

therefore can be assessed only by mass balance in terms of isotope-based recharge values, the continuity of stable isotope "fingerprints", and of hydrochemistry.

6.2 *Stable oxygen and hydrogen isotopes*

Figure 6 shows a δD-δ^{18}O diagram for individual major rainfall events at three rainfall collection stations in the 1998/99 season, with the world meteoric water line (WMWL). The enormous range (δ^{18}O ~ 10 ‰) of values in individual rainfall events/episodes could lead to significant differences in groundwater values (see below). The relatively tight clustering of values in ground water illustrates the integrative and thus smoothing effect of the recharge process. A substantial number of points lie to the right of the WMWL and indicate evaporation before infiltration from surface, probably as a result of ponding in numerous depressions and small pans. Differences seen in δ^{18}O values for ground water in different localities (Figure 7), emphasise differences in local recharge conditions. There are no discernible trends such as would be expected from consistent drainage along the fault zone, but values cluster locally, such as at the well field referred to in section 5 (hydrochemistry).

6.3 *Radioactive isotopes*

Radiocarbon values are quite variable along the fault (Figure 8). Away from the fault the coverage of data points is as yet inadequate. The radiocarbon frequency histogram (Figure 9) shows a predominant cluster around 95 pMC with a minor cluster around 65 pMC, representing mainly samples from the SW part of the fault. Tritium, a conservative tracer of water, can act as a constraint in estimating the initial or recharge value of radiocarbon. Many radiocarbon values around 100 pMC in groundwater, usually ascribed to post -1963 (thermonuclear) recharge, are accompanied by tritium values at the limit of detection (Figure 10). Deep tree roots could transport radiocarbon to the saturated zone more rapidly than tritium, which labels the infiltrating rain water. Whichever the mechanism, the data suggests an initial or recharge value for radiocarbon around 95% atmospheric.

The exponential mixing model integrates the known input functions for either isotope with time to the groundwater system, and the mean concentration at any time A(t):

$$A(t) = \int_0^\infty A_o(t-\tau)e^{-\lambda\tau}\frac{e^{-\tau\sqrt{\tau}}}{\bar{\tau}}d\tau$$

where A_o is the value at recharge, λ is the decay constant and τ is the mean residence time. The exponential transit time distribution function in this equation has been shown (Verhagen et al. 1991) to be an acceptable approximation for a phreatic aquifer. The inverse problem finds the most likely MRT for an observed value A(t) in ground water. When water is pumped from a well, it may be assumed that it represents a mixture from the entire saturated profile penetrated by the well. For a phreatic aquifer, recharge is then given by (Verhagen et al. 1991)

$$R = \frac{p.H}{\bar{\tau}}$$

where the product of the mean porosity (p) and the saturated zone thickness (H) gives the equivalent water column.

As in many ground water investigations, aquifer porosity, a critical parameter in resource estimation, is poorly known. A high priority in this ongoing study is the assessment of realistic porosity values. Cores are to be taken in future drilling exercises on which porosity values can be measured directly. Down-the-hole observations and radon measurements can generalise spot measurements. In view of the nature of the aquifers, such assessments will of necessity at best remain limited in their accuracy.

6.4 *Estimating MRT*

The solid line shown in Figure 10 is the locus of values (Verhagen et al. 1991) for different mean residence times (indicated) according to the exponential mixing model, using tritium and ^{14}C input functions for southern Africa. The scatter of points is ascribed to hydrogeological heterogeneity and mixtures of deeper, lower radiocarbon water with shallower, higher tritium water.

Points cluster around MRT $\approx$ 500 a. Significant deviations from the exponential model are for samples in the SW section of the fault zone on the farms Fontaine de Champ and Eldorado (Figures 1 and 8). Slumping along the fault zone developed complex juxtapositions of lower and higher permeability rocks with possibly confined and perched conditions.

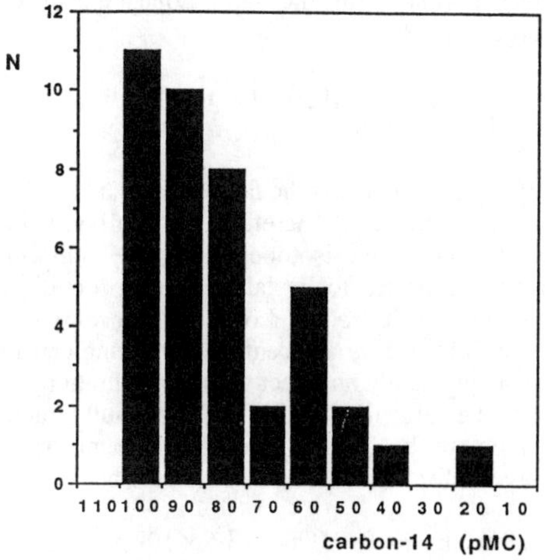

Figure 9. Frequency histogram of radiocarbon values, showing a bi-modal distribution

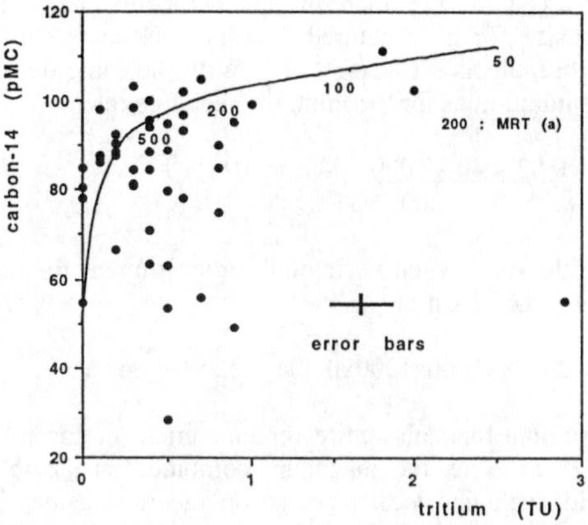

Figure 10. Plot of radiocarbon against tritium values. The solid line is a plot of predicted exponential mixing pairs of the two variables for different groundwater mean residence times

Drilling records often record a single significant water strike. In down-the-hole video observations, a wider spread of smaller pores, joints and fractures is observed, suggesting more uniform water-bearing and transporting properties. Prawn-like animals seen in video images of several boreholes suggest the existence of macropores with regional interconnectivity.

6.5 *Estimating recharge*

A most probable median MRT for the bulk of the data (the fault zone) shown in Figure 9 is 500 a.

Taking the mean height of saturated water column as 40 m and a (probably high) estimate of 0.08 for the fracture porosity for the fault zone, the median recharge value R:

$$R = \frac{p.H}{T} = \frac{0.08 \times 40 \times 1000}{500} = 6.4 \text{ mm a}^{-1}$$

The regional model sees the fault zone as a "sink" for groundwater drainage. Therefore, radiocarbon values, chemistry and stable isotopes along the fault could have been expected to be fairly uniform or at least exhibit consistent trends. Local differences suggest that the fault and its adjacent fracture zone can rely for recharge mainly on direct vertical infiltration.
Taking the effective width of the fault fracture catchment to be 2 km, the total amount of rain recharge R_t for 28 km of fault line would be:

$$R_t = 28 \times 1000 \times 2000 \times 0.0064 = 3.6 \times 10^5 \text{ m}^3 \text{ a}^{-1}$$

The few available radiocarbon values significantly away from the fault appear to be marginally higher than along the fault itself. A mean residence time of 200 a (Figure 9) is therefore taken for this area. The porosity for the fractured country rock away from major faults is set at p = 0.02. With the same depth of saturation as for the fault, the recharge rate:

$$R_t = \frac{0.02 \times 40 \times 1000}{200} = 4 \text{ mm a}^{-1}$$

and the total potential rainfall replenishment for the area (20 x 28 km):

$$R = 20 \times 28 \times 1000 \times 1000 \times 0.004 = 2.2 \times 10^6 \text{ m}^3 \text{ a}^{-1}$$

Assuming that this entire replenishment drains into the fault zone, the maximum combined sustainable yield, including local recharge on the fault zone, is 2.6×10^6 m^3 a^{-1}, or one order of magnitude less than the earlier average yield estimates (Fayazi and Orpen 1989) of 10^6 m^3 a^{-1}per kilometre of fault length. Although major uncertainties are inherent in both these estimates, it is clear that average yield estimates should be approached with caution in terms of their long-term sustainability and further development strategies in the area.

7 CONCLUSIONS

(i) Radiocarbon values in the range 88-103 pMC indicate widespread active rain recharge in the area.
(ii) Measurable tritium up to several TU confirms the radiocarbon data and constrains model-based mean residence times to few hundreds of years.
(iii) Variable radiocarbon, oxygen-18 and hydrochemistry along the fault zone, possibly enhanced by localised groundwater abstraction, suggest that, in addition to possible regional drainage, significant localised recharge is occurring to the fault zone itself.
(iv) Recharge rates of 4 mm a^{-1} and 6.4 mm a^{-1} respectively are found for the plain and the Taaibosch fault zone using rough estimates of saturated zone thickness and fracture porosity.
(v) Isotope-based estimates of total rain recharge to the entire area to be harvested by well fields along the fault amount to some 10% of estimates arrived at during an earlier study. This discrepancy should sound a word of caution in terms of the development of a regional groundwater supply.

8 ACKNOWLEDGEMENTS

This study is being supported by the International Atomic Energy Agency. The authors acknowledge the assistance of members of staff of the Department of Water Affairs and Forestry and of the University of the Witwatersrand. Residents of the Bochum study area are thanked for their hospitality and logistical support.

REFERENCES

Fayazi, M. & Orpen, W.R.G. 1989. Development of a water supply for Alldays from groundwater resources associated with the Taaibos fault. Techn. Rep. GH 3664, Department of Water Affairs and Forestry, Pretoria.

IAEA. 1992. Statistical treatment of data on environmental isotopes in precipitation. Techn. Rep. Series No. 331. IAEA, Vienna

Verhagen, B.Th., Geyh, M.A., Froehlich, K. & Wirth, K. 1991. Isotope hydrological methods for the quantitative evaluation of ground water resources in and semi-arid areas. - Development of a methodology. Research Reports of the Federal Ministry for Economic Cooperation of the Federal Republic of Germany. Bonn.

Groundwater: Past Achievements and Future Challenges, Sililo et al. (eds) © 2000 Balkema, Rotterdam, ISBN 90 5809 159 7

3D groundwater modelling of isotope data at Äspö, Sweden

B.Wallin
Geokema AB, Lidingö, Sweden

K.Innings
Saltsö-Boo, Sweden

Abstract: 3-D modeling of isotope data ($\delta^{18}O$, $\delta^{2}H$ and ^{3}H, $\delta^{87}Sr$) and chloride concentration (Cl^{-}) in the groundwater at the island of Äspö, Sweden has been performed by use of the computer program Voxel Analyst. This work is a part of the development of a method performed by Geokema AB for studying groundwater data in a 3-D perspective. Voxel Analyst is a general-purpose data visualization and analysis tool where a 3-D volume dataset is created by interpolation of input data. The model can be displayed as volumes, surfaces or cutting planes etc. The volume model is built from arrays of 3-D pixels called voxels. A 3-D model of the isotope signatures and chloride concentration has been generated from data collected from boreholes . The input data consist of X, Y and Z co-ordinates and concentration values for the included variables. Background values are added to the model. All data are imported to Voxel Analyst as ASCII file format. In this study, Voxel Analyst was found to be an helpful tool for studies of groundwater system at Äspö. This 3-D perspective method can be a powerful modeling tool in future management plans for sustainable development of water resources.

1 INTRODUCTION

It is decided that spent nuclear fuel from Swedish nuclear power plants will be stored in deep repositories in the crystalline basement. The Swedish Nuclear Fuel and Waste Management Company (SKB) has the responsibility to develop and construct the final disposal of this fuel. SKB has chosen the Äspö island in south-eastern Sweden as a site for a study of final nuclear fuel disposal in Precambrian bedrock. An underground laboratory, Äspö Hard Rock Laboratory (ÄHRL) situated 450 meters below sea level is constructed. The basement rock comprises granites and greenstones. These rocks have low permeability and offer good possibilities for a safe repository but the transport mechanisms in the rock are still insufficiently investigated. The transport of different groundwater elements is carried out by groundwater movements in the fracture system of the rock. One important part of the research work at Äspö is thus the studies of the groundwater conditions at the site. Different groundwater constituents and data of stable ($\delta^{18}O$, $\delta^{2}H$ and ^{3}H) and radiogenic isotopes ($\delta^{87}Sr$) are used in order to

characterize the origin of the groundwater and the mixing of waters with different origins. To facilitate the interpretation of the results, a 3-D

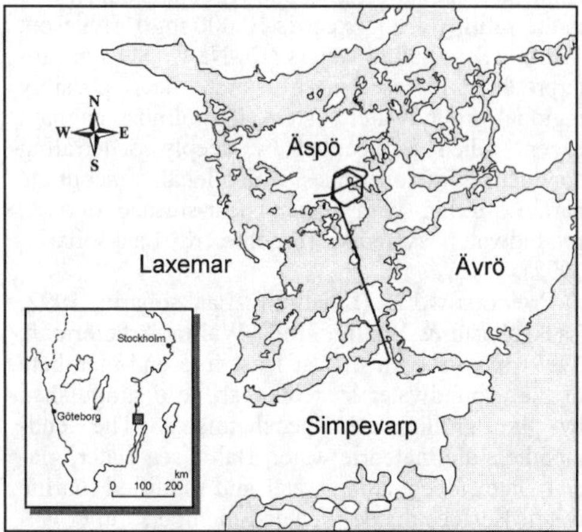

Figure 1. Location of the Äspö area, Sweden

model can be constructed to visualize the variation of geochemical data in the groundwater.

2 GEOLOGY AND GROUNDWATER GEOCHEMISTRY

The Äspö area is part of the Precambrian basement in SE Sweden where Småland-Värmland granite intrusions dominate (Kornfelt & Wikman, 1987). These intrusions are 1700 to 1800 million years old (Johansson, 1988) and regarded as late- to postorogenic in relation to the Svecofennian orogeny. Småland granite, with associated greenstones and fine-grained granite/aplite varieties, dominates. Within the Äspö area, all rock units, except the anorogenic granites, show a pronounced elongation in approximately ENE-WSW. This direction is also parallel with the general foliation in the area. The deformation is rather weak, but in some zones it can be somewhat stronger. To the west the foliation usually turns into WNW-ESE. (Wikman & Kornfelt, 1995). Highly fractured zones (5-10 m wide), alternating with more normally fractured rock, complicate the shear structure which also has undergone substantial weathering, seen as oxidation of magnetite to hematite and Fe-oxyhydroxides.

The groundwater flow at Äspö is controlled by the major tectonic fractures and discontinuities. The general flow trend is from south to north, at least in the upper 500 m or so, facilitated by the NNW-trending vertical to subvertical fractures. The major E-W trending shear zone is a strong recharge feature (Smellie & Laaksoharju, 1992). The hydrochemical situation at Äspö is characterized by fresh water (HCO_3-Na-Ca) in the upper 40 to 70 m of the bedrock. The fresh water zone is followed by aquifers with saline waters with increasing chlorine content versus depth. At depths below 800 m the salinity often exceeds 10 000 mg/L (Nilsson, 1989). These saline waters (Ca-Na-Cl-SO_4) are interpreted as representing a separate, deep, possibly regional groundwater system. It is almost stagnant except when intercepted by deeply penetrating conducting fracture zones, which locally encourage partial mixing with upper, near-surface derived groundwater systems. (Smellie & Laaksoharju, 1992).

Previous works (Smellie & Laaksoharju, 1992; Laaksoharju & Wallin, 1997, Wallin & Peterman, 1999) have shown that at least five end-members of the groundwater systems, can be distinguished by the groundwater geochemistry. The end-members are meteoric water, Baltic sea water, glacial water, deep saline water and modified marine water. Besides the defined end-members, other water types may be distinguished, such as relict sea water, possibly originated from Litorina sea water.

3 ISOTOPES

The utility in applying the multiple isotope approach to the study of groundwater results from the fact that the different isotope signatures represent different processes. Studies of the groundwater chemistry at Äspö suggest a very complicated history. However, the isotope data demonstrate that it is possible to postulate mixing of different groundwater members to explain the changes in the groundwater flow pattern. Isotope records of the groundwater at Äspö are sufficient to provide information for defining the end-members in the groundwater system and the paleohydrological history of different waters.

The stable isotopes of oxygen ($\delta^{18}O$) and deuterium (δ^2H) are used to distinguish water originating from different sources (each source has a distinct isotopic signature through isotopic fractionation) in the assumption that they remain unchanged from recharge to groundwater flow conditions. The radiogenic tritium isotope (^3H) is used as a radioactive tracer. The characteristics of the five end-members distinguished in the groundwater systems at Äspö (Smellie & Laaksoharju, 1992; Laaksoharju & Wallin, 1997; Wallin & Peterman, 1994) can roughly be described by the isotopic signatures. The meteoric recharge water at Äspö displays $\delta^{18}O$ values around −10 ‰ and δ^2H values of −77 ‰ representing the annual mean value for the site. The Baltic sea water surrounding the island of Äspö has uniform $\delta^{18}O$ values of −6.4 ‰ and δ^2H values of −52.3 ‰. The glacial water recognized in the area has significantly lower $\delta^{18}O$ values, reaching −15 ‰. The deep saline water below 600 m depth has a uniform $\delta^{18}O$ value around −13 ‰. The fifth end-member, modified marine water, is characterized as a Baltic sea water which has undergone chemical changes. It roughly shows $\delta^{18}O$ values similar to the Baltic sea water. The δ^2H and $\delta^{18}O$ values are referenced to SMOW (Standard Mean Ocean Water) and expressed in per mil (‰) deviations (δ values) from the appropriate standard. $\delta^{87}Sr$ are related to the value for modern sea water using USGS standard EN-1 which is a *Tridacna* shell collected from Enewetok Lagoon in the western Pacific Ocean.

The tritium isotope (^3H) values vary between the different water types, typically showing different ages. The higher values measured at shallow depth, may be an input of tritium high waters generated during the nuclear test period in the early sixties. In the following 3-D modeling, the isotope signatures of the groundwater at Äspö are used to predict different hydrological changes that may have been induced by the excavation of the tunnel at Äspö.

4 METHODS

Stable ($\delta^{18}O$ and δ^2H) and radiogenic (^3H) isotopes and chloride concentration have been used to model the groundwater conditions at Äspö before and after the construction of the tunnel. All data are taken from the SKB database GEOTAB. The $\delta^{18}O$ and $\delta^{2}H$ values are referenced to SMOW (Standard Mean Ocean Water). The ^{3}H is expressed in tritium units (TU). Isotopic values for the stable isotopes are expressed in per mil (‰) deviations (δ values) from the appropriate standard where δ is defined as:

$$\delta = [\ (R_{sample}/R_{standard}) - 1\ ] \times 1000$$

and R is either $^{2}H/^{1}H$ or $^{18}O/^{16}O$.

21 samples from the drilled bore holes KAS02, KAS03, KAS04 and KAS06 taken 1988-1989 and 11 samples from the drilled bore holes HAS02, HAS03, HAS05, HAS06, HAS07 and HAS13 taken 1987 are used to describe the stable groundwater conditions at Äspö before the tunnel

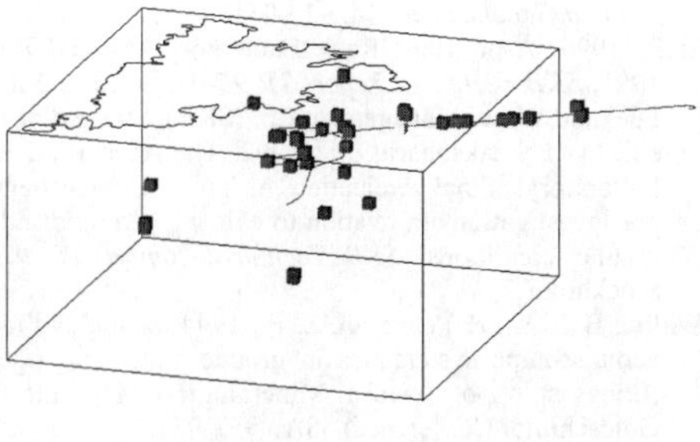

Figure 2. Source points for the model in side view.

construction (Figure 2). In April 1994 the access tunnel reached a depth of 420 m, and the main drilling work was finished (Hammarström & Olsson, 1996). Data from 52 sample points in the SA- and KA-bore holes in the tunnel, taken from February to June 1994, are used to describe the groundwater conditions after the tunnel construction. Data from the bore holes KAS03, KAS05, KAS07, KAS08, KAS09 and KAS14 taken at the same time complete the input values. Totally, 60 points are used in the model (Fig. 2). The sampled tunnel bore holes were of two types: probing holes (SA-) and core drilled bore holes (KA-) (Nilsson, 1995). The sampling and quality control of the HRL tunnel groundwater sampling has been evaluated in detail and described by Nilsson (1995).

In the 3-D modeling the volume model is built from arrays of 3-D pixels called voxels. Various 3-D models of the isotope signatures and chloride concentration has been generated. The input data consist of X, Y and Z co-ordinates and concentration values for the included variables. Background values are added to the model. All data are imported to Voxel Analyst as ASCII file format. The size of the model box is limited to a length and width of 1500 m and a depth of 915 m after the source point distribution. The whole tunnel could not be included in the model, due to lack of data.

5 RESULTS AND DISCUSSION

The clearly stratified system at Äspö displays continuous gradual changes with depth as well as a wide variation in the stable isotope signatures. This is easily visualized in the 3-D modeling exercise by Voxel Analyst. The shallow groundwater is isotopically ($\delta^{18}O$ = −10 ‰) similar to the meteoric water in the area, and hence the rainwater is the important recharge water at Äspö. Previous works (Banwart et al., 1995a; Banwart et al., 1995b) have shown that the shallow recharge (in the upper 70 m) mainly occur through fractures, however the assumption of having a homogeneous media in this model, does not allow verification of these features.

At greater depths (200-300 m), the dissolved load increases as indicated by the increase in chlorinity. Local changes in the isotopic composition reflect isolated patches of the different end-members involved in the system, such as glacial water, older marine intrusions as well as modified Baltic sea water (Wallin & Peterman, 1999; Peterman & Wallin, 1999).

The deep saline water, here characterized by a homogeneous $\delta^{18}O$ value around −13 ‰ and a $\delta^{2}H$ value of −98 ‰, reflects a stagnant water body (Fig 3). The interpretation of the stable isotope data in the transitional areas between the deep saline and the intermediate groundwater has to be accompanied by the interpretation of the chlorinity. This is mainly due to the fact that end-members in the system with different evolution processes or ages may attain similar isotopic signatures.

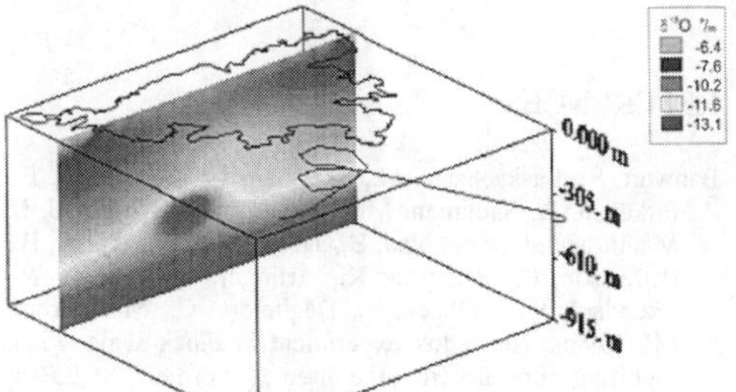

Figure 3. Distribution of $\delta^{18}O$ in the groundwater under the northern part of Äspö after the tunnel construction.

The tritium isotope (^{3}H) has been used as a radioactive tracer. It gives as a first approximation a rough picture of the age relation between the different waters involved in the system. The shallow groundwater below the island of Äspö has a ^3H value of 25 TU. In contrast, the Baltic sea water in the intermediate vicinity has elevated values of 39 TU. Due to the short life time of ^3H, relatively high values in the shallow groundwater indicate a rapid recharge of meteoric water. Alternatively, it may reflect mixing with ^3H enriched water originating from the nuclear test period in the early sixties.

The compositionally gradual system at Äspö appears to reflect a dynamic system recorded in the scatter in the isotopic compositions. Although the chloride concentration as well as most dissolved ions increase systematically with depth, their variability reflects the general inefficiency of advective mixing for developing smooth hydrochemical gradients. However, the mixing which occur between the different end-members involved in the system can not be traced by this 3-D modeling, but the different water types can be distinguished in the model. The hydrodynamic differences and isotopic variations with depth at Äspö seen in this model may ultimately be related to large scale differences in fracture-controlled conductivity in the rock masses. This is clearly demonstrated in the cutting planes of the major fracture systems (EW-1, NE-1 and NE-3) at Äspö.

ACKNOWLEDGEMENTS

The study was financed by SKB (Swedish Nuclear Fuel and Waste Management Company). Peter Wikberg encouraged the work throughout. The authors wish to thank Karl-Göran Nederfeldt, Katinka Klingberg and Anna Säfvestad who kindly helped with the ground water sampling and Ann-Christin Nilsson who supported us with some data from the database.

REFERENCES

Banwart, S., Laaksoharju, M., Skårman, C., Gustafsson, E., Pitkänen, P., Snellman, M., Landström, O., Aggeryd, I., Mathiasson, L., Sundblad, B., Tullborg, E.-L., Wallin, B., Pettersson, C., Pedersen, K., Arlinger, J., Jahromi, N., Ekendahl, S., Hallbeck, L., Degueldre, C., Malmström, M., 1995a. The redox experiment in block scale -Final reporting of results from the three year project. *SKB Progress Report, PR 25-95-06*, Swedish Nuclear Fuel and Waste Management Co. (SKB), Stockholm.

Banwart, S., Gustafsson, E., Laaksoharju, M., Pedersen, K., Tullborg, E.-L., Wallin, B., Wikberg, P., Snellman, M., Pitkänen, P., Leino-Forsman, H., 1995b. The Äspö redox investigations in block scale -Project summary and implications for repository performance assessment. *SKB Technical Report TR 96-26*, Swedish Nuclear Fuel and Waste Management Co. (SKB), Stockholm.

Hammarström, M. & Olsson, O., 1996. Äspö Hard Rock Laboratory -10 years of research. Swedish Nuclear Fuel and Waste Management Co. (SKB), Stockholm.

Johansson, Å., 1988. The age and geotectonic setting of the Småland-Värmland granite-porphyry belt. *Geol. Fören. Stockholm Förh., 110*, 105-110.

Kornfelt, K-A. & Wikman, H., 1987. Description of the map of solid rocks around Simpevarp. *SKB Progress Report, PR 25-87-02*, Swedish Nuclear Fuel and Waste Management Co. (SKB), Stockholm.

Laaksoharju, M. & Wallin, B., 1997. Integration of hydrochemical and hydrogeological models of Äspö. *Proceedings of the second Äspö International Geochemistry Workshop*, June 6-7, 1995. In manuscript.

Nilsson, A-C., 1989. Chemical characterization of deep groundwater on Äspö 1989. *SKB Progress Report PR 25-89-14*, Swedish Nuclear Fuel and Waste Management Co. (SKB), Stockholm.

Nilsson, A-C., 1995. Compilation of groundwater chemistry data from Äspö 1990-1994. *SKB Progress Report PR 25-95-02*, Swedish Nuclear Fuel and Waste Management Co. (SKB), Stockholm.

Peterman, Z., & Wallin, B., 1999. Synopsis of strontium isotope variations in groundwater at Äspö, southern Sweden. *Applied Geochemistry* 14: 939-951.

SKB, 1996. Äspö Hard Rock Laboratory -Annual Report 1995. *SKB Technical Report TR 96-06*, Swedish Nuclear Fuel and Waste Management Co. (SKB), Stockholm.

Smellie, J. & Laaksoharju, M., 1992. The Äspö Hard Rock Laboratory: Final evaluation of the hydrogeochemical pre-investigations in relation to existing geologic and hydraulic conditions. *SKB Technical Report TR 92-31*, Stockholm.

Wallin, B G. K., & Peterman, Z. E., 1994. Stable and radiogenic isotope systematics in ground water and fracture fillings at Äspö, Sweden. Mineralogical Magazine (VM Goldschmidt Conference) 58A: 953-954.

Wallin, B. & Peterman, Z., 1999. Calcite fracture fillings as indicators of paleohydrology at Laxemar at the Äspö Hard Rock Laboratory, southern Sweden. *Applied Geochemistry* 14: 953-962.

Wikberg, P., Gustafsson, G., Rhén, I. and Stanfors, R., 1991. Äspö Hard Rock Laboratory. Evaluation and conceptual modeling based on the pre-investigations 1986-1990. *SKB Technical Report TR 91-22*, Swedish Nuclear Fuel and Waste Management Co. (SKB), Stockholm.

Wikman, H. & Kornfelt, K-A., 1995. Updating of a lithological model of the bedrock of the Äspö area. *SKB Progress Report, PR 25-95-04*, Swedish Nuclear Fuel and Waste Management Co. (SKB), Stockholm.

Groundwater: Past Achievements and Future Challenges, Sililo et al. (eds) © 2000 Balkema, Rotterdam, ISBN 90 5809 159 7

Pilliga Sandstone hydrogeochemistry, Great Artesian Basin, Australia

C.A.Wolfgang & I.White
Centre for Resource and Environmental Studies (CRES), and The Water Research Foundation (WRF), The Australian National University, Canberra, A.C.T., Australia

ABSTRACT: Demands on both surface and groundwater resources in the Coonamble Embayment portion of the Great Artesian Basin (GAB) continue to increase. The freshwater aquifers of the Coonamble Embayment, while being of importance locally are also managed in the context of the GAB and so present complex issues in natural resource management. The Pilliga Sandstone is the only freshwater artesian aquifer in the GAB and has been used as a water source since the late 1800's. However, its characteristics in the Coonamble Embayment are yet to be fully described. A geologic model constructed from well logs and interpolated using the ANUDEM package reveals the complex geometry within which the aquifer exists. Existing groundwater chemistry data have been analysed and a discrete database representative of the Pilliga Sandstone aquifer has been produced. ANUDEM was then used to model the spatial distribution of major ions. These reveal recharge zones and mixing with other waters.

1 INTRODUCTION

Water from the Pilliga Sandstone aquifer within the Coonamble Embayment of Australia's Great Artesian Basin (GAB) is used by urban centres, irrigators, and graziers. The Macquarie, Bogan, and Castlereagh catchments comprise the Coonamble Embayment. The area under irrigated cotton production within the Macquarie catchment is estimated to have increased by 400% from 1981 to 1992 and now dominates the farming region (Kingsford & Thomas 1995). Demands on both surface and groundwater resources in the Macquarie catchment continue to increase.

Groundwater resources in the Embayment are contained in unconsolidated sediments, consolidated sandstones, fractured rocks, and basalts. Mesozoic-age Pilliga Sandstone, Keelindi beds, Drildool beds, and Rolling Downs Group contain groundwater of varying quality and quantity. Of these, the Pilliga Sandstone is the only artesian aquifer and with significant fresh waters. This research focuses on the properties of the Pilliga Sandstone aquifer.

Though the Pilliga Sandstone aquifer has been used as a water source since the late 1800's the aquifer has never been fully characterised in terms of its geology, hydrology, and geochemistry although a recent summary of the GAB presents a suggested model (Radke et al., in press).

1.1 *Geologic History and Hydraulic Head*

The western portion of the Coonamble Embayment is bordered on the south and west by the Palaeozoic Lachlan Fold Belt (LFB) (Watkins & Meakin 1996). The LFB is exposed in an arc around the GAB sequence in the lower Macquarie catchment, with its western extent of outcrop passing through Dubbo and then curving north, following the western edge of the Bogan River. It also forms the basement to the GAB as it dips northward, unconformably underlying the GAB in the lower Macquarie catchment.

The Pilliga Sandstone is influenced by the basement structure, and has regions of greater thickness in scattered 'potholes'. These are in Palaeozoic depressions and may act as areas of stagnant groundwater. To the west and south there is a comparatively shallow Palaeozoic basement and Pilliga Sandstone is absent. It is expected that GAB groundwater is less in this region. In the eastern part of the Coonamble Embayment, the Pilliga Sandstone appears to have been deposited in an alluvial fan environment. The Pilliga Sandstone is thickest (300m) in the east of the Macquarie catchment and shows as the core of a fan-like projection splayed outwards from the Pilliga Sandstone outcrop near the Warrumbungle Mts. This is called here the Yungundie Fan.

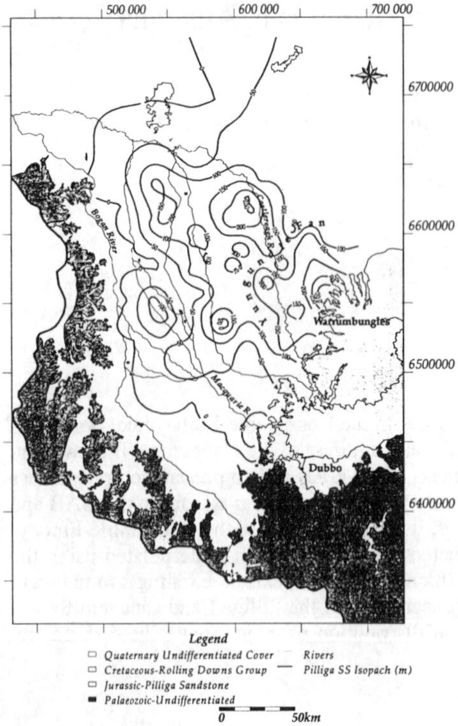

Figure 1. Pilliga Sandstone Isopach (m) and Coonamble Embayment Surface Geology.

The structural top of the Pilliga Sandstone is deepest in the northeast (-600m Australian Height Datum, AHD). It is uplifted in the east of the Embayment (Warrumbungle Mts). The elevations of the Pilliga Sandstone confirm that the Sandstone was deposited in the lows of the Palaeozoic basement.

Temperature and neutron logs offer further information on the Pilliga Sandstone aquifer. Twenty temperature logs and 13 neutron logs were used spanning an area of approximately 50km in a broad NE-SW line in the middle of the Embayment (Habermehl et al. 1997). Temperature logs indicate that generally deeper waters are warmer and that there are no major hotspots in the Embayment. Neutron logs show that high moisture areas are present in all geologic units, particularly in the Pilliga Sandstone and along boundaries between the geologic units. Low moisture areas (possibly aquitards/aquifuges) are present in all units particularly in the Cainozoic-age strata and the Rolling Downs Group.

Hydraulic head data also define the flow direction in the western Embayment. The general flow direction is from the southeast to the northwest. Most recent head data available in the mid-1980's range from 160m AHD near the Warrumbungle Mts. to 130m AHD in the northwest.

2 AQUIFER CHEMISTRY ANALYSIS

Examination of the several hundred bore water chemistry samples in the New South Wales (NSW) Department of Land and Water Conservation (DLWC) database for 1890-1980, revealed questionable and inconsistent sampling spatially and temporally. Only analyses which had a charge balance of ±5% meq/L of total dissolved ions are accepted (Lloyd & Heathcote 1985). The most comprehensive sampling available with good distribution was in the 1970's.

Knowledge of the bore perforation geometry is essential in characterising groundwater samples (Mazor 1997). Screen positions and details are included in the DLWC databases. From these it was noticed that screening was often placed beyond the first or primary source of water described in the drillers' logs. This raises the possibility of non-Pilliga Sandstone water entering the screen. A selection was made of bores that reached at least to the structural top of the Pilliga Sandstone and whose screen positions must not be beyond the structural top of the confining layer above the Pilliga Sandstone.

The resulting spatial distribution of the database of 78 bores was reasonable, with bore locations including the areas of spatial variability of the Pilliga Sandstone and covering an area progressing away from recharge locations in the east to the Pilliga Sandstone areal limit in the west.

The geochemical analysis of the Pilliga Sandstone groundwater included: (1) Piper (Piper 1944) and Schoeller (Schoeller 1954) plots of major ions from AquaChem (Calmbach & Waterloo Hydrologic 1997); and (2) spatial analysis with ANUDEM (Hutchinson 1997) of the areal distribution of major ions and groundwater temperature.

The Piper diagram for the Pilliga Sandstone in Figure 2 shows a vector running outwards from the Na apex, and a group of bores clustered in the HCO_3 apex. This is consistent with the observation that GAB waters are Na-HCO_3 waters (Habermehl 1983). The bores furthest from the Na apex are closest to the recharge area, just to the west of the Warrumbungle Mts. The bores further away from the recharge area are closer to the Piper diagram Na apex. This indicates progressive exchange with clays as water moves across the Embayment or upward or downward leakage of marine-based formation waters (Herczeg et al 1991).

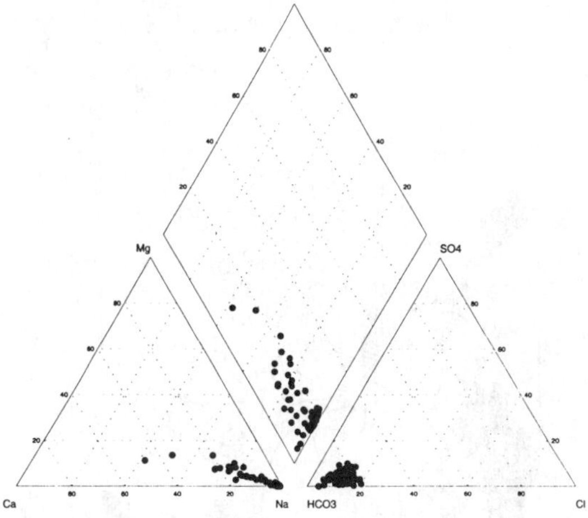

Figure 2. Piper Diagram for Pilliga Sandstone Aquifer.

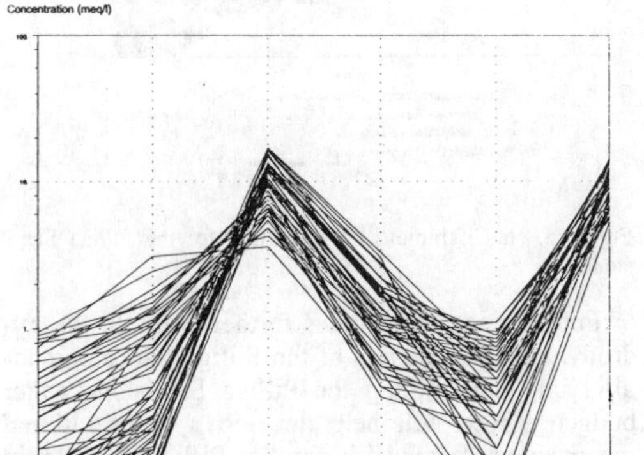

Figure 3. Schoeller Plot for Pilliga Sandstone Aquifer.

Bicarbonate may be generated either by the reaction of water with $CaCO_3$ or by microbial reduction processes. The current hypothesis is that most HCO_3 in the GAB is due to reduction processes (Radke et al., in press). Bores have relatively low chloride and low sulphate.

The Schoeller Diagram is another way to graphically represent the chemical data (Schoeller 1954). From Figure 3 it is clear that Pilliga Sandstone waters vary compositionally but that the waters are generally Na-HCO_3 dominant. The ions Ca, Mg, and SO_4 show considerable variation. The different patterns may reflect the evolutionary stages of the water in the Pilliga Sandstone or mixed waters.

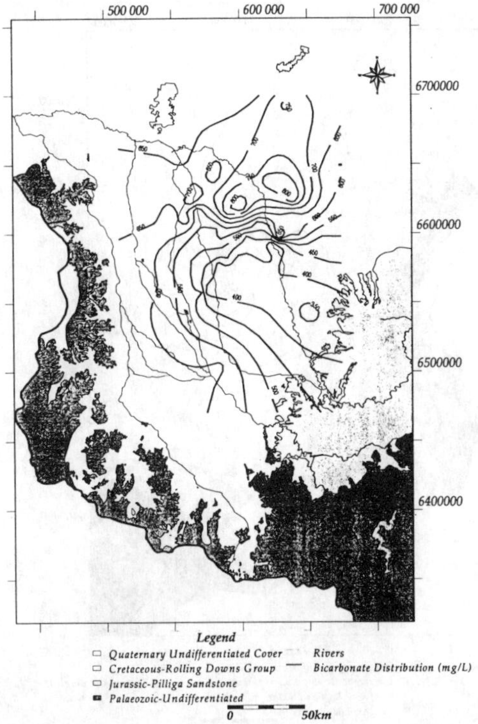

Figure 4. Pilliga Sandstone HCO_3 Distribution (mg/L).

The analysis of how major ions and compounds are distributed in the aquifer helps to interpret the aquifer mechanics and reveals flow paths and potential recharge areas. The distribution of HCO_3 shows higher values (>700mg/L) towards the north and northwestern portion of the Embayment, while lower values (<400mg/L) are near the Warrumbungle Mts.

HCO_3 concentrations in the GAB normally increase down the hydraulic gradient and with groundwater flow and residence and as such are an excellent indicator of flow direction and preferential permeability and possible mixing (Herczeg et al. 1991). Flow lines, if drawn on Figure 4 would run from the southeast towards the northwest. This is consistent with hydraulic head analysis and support the Warrumbungle Mts. as a recharge area.

The cation distributions all show an area of rapid change occurring along a portion of the Castlereagh River, as illustrated by Ca distribution (Figure 5). Values change from 50mg/L to 25mg/L within 25km. This may indicate mixing or mineralogical changes in the Pilliga Sandstone and is being investigated further.

The ionic ratio distribution for Na:Cl (in mmol/L) shows a general increase in values moving towards the west and northwest from the Warrumbungle Mts. (Figure 6). The general trend is similar to the HCO_3.

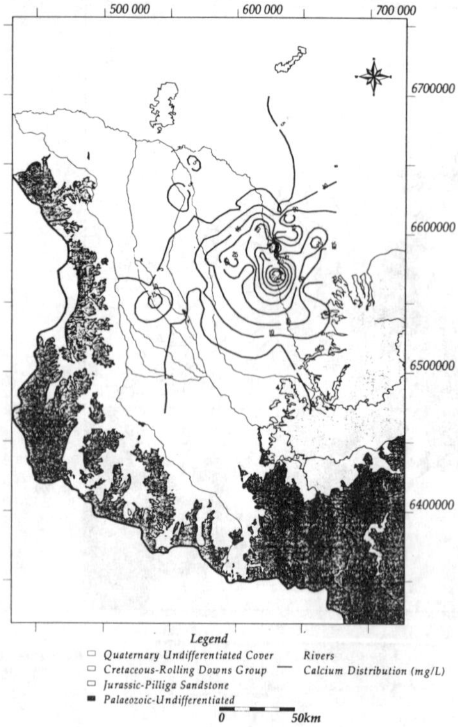

Figure 5. Pilliga Sandstone Calcium Distribution (mg/L).

Figure 6. Na:Cl (mmol/L) Distribution for the Pilliga Sandstone.

High values of 19 in the mid-west portion are anomalies and may indicate mixing or stagnant areas.

3 CONCLUSION AND FUTURE WORK

In keeping with findings elsewhere in the GAB the Pilliga Sandstone waters in the western Coonamble Embayment are essentially Na-HCO_3 dominant. The Piper diagram of the Pilliga Sandstone water reveals trends in the data with Ca and Mg dominance changing to Na dominance as bore sample locations move away from the recharge area.

HCO_3(mg/L) and Na:Cl(mmol/L) distributions help confirm the hydraulic head data by showing overall movement from the southeast towards the northwest. It may also show pools of relatively stagnant water and preferred routes of flow. Stagnant areas show relatively higher Na:Cl, higher Na, higher Cl, higher Total Dissolved Ions and smaller concentrations of Ca and Mg. These may be controlled by the fan depositional history of the Pilliga Sandstone. Ca along with other cations identified possible mixing or mineralogical changes for the Pilliga Sandstone along the Castlereagh River.

Future research includes further defining the hydrodynamic parameters of the Pilliga Sandstone aquifer and determining the Pilliga Sandstone water budget. These will help develop a more rational management framework for the Pilliga Aquifer in the Coonamble Embayment. It is clear that any framework should recognise the possible interaction of groundwater in the Pilliga Aquifer with other aquifers and surface waters.

ACKNOWLEDGMENTS

We acknowledge gratefully the financial support of DLWC New South Wales and contribution of data and allowing access to databases. The Land and Water Division of the Bureau of Rural Science, particularly Dr. Richard Cresswell, Dr. Jim Ferguson (ret.), and Dr. Bruce Radke were helpful in discussions of the broader hydrogeochemical model in the GAB.

REFERENCES

Calmbach, L. & Waterloo Hydrogeologic 1997. AquaChem for Windows 95/NT, version 3.6.

Habermehl, M.A. 1983. Hydrogeology and hydrochemistry of the GAB, Australia. In *Australian Water Resources Council Conference Series, Papers of the International Conference on Groundwater and Man, Sydney, 5-9 December, 1983*. 8(3):83-89. Canberra: Australian Government Publishing Service.

Habermehl, M.A., Tucker, A.G., Sedgemen, A.L. & Revell, N.G. 1997. Wire-line logged waterbores in the Great Artesian Basin-digital data of logs and waterbore data. Australian Geological Survey Organisation Bulletin No. 240. Canberra: Australian Geological Survey Organisation.

Herczeg, A.L., Torgersen, T., Chivas, A.R. & Habermehl, M.A. 1991. Geochemistry of groundwaters from the Great Artesian Basin, Australia. *Journal of Hydrology*. 126:225-245.

Hutchinson, M.F. 1997. ANUDEM Version 4.6. Canberra: CRES, The Australian National University.

Kingsford, R.T. & Thomas, R.F. 1995. The Macquarie Marshes in arid Australia and their waterbirds: A 50-year history of decline. *Environmental Management* 19:867-878.

Lloyd, J.W. & Heathcote, J.A. 1985. *Natural inorganic hydrochemistry in relation to groundwater*, Oxford: Clarendon Press.

Mazor, E. 1997. *Chemical and isotopic groundwater hydrology*, 2nd.ed., USA: Marcel Dekker, Inc.

Piper, A.M. 1944. A graphic procedure in the geochemical interpretation of water analysis. *American Geophysical Union Transcripts*. 25:914-923.

Radke, B., Ferguson, J., Cresswell, R.G., Ransley, T.R. & Habermehl, M.A. in press. The hydrochemistry and implied hydrodynamics of the Cadna-owie-Hooray Aquifer, Great Artesian Basin, Australia. Canberra: Bureau of Rural Sciences Bulletin.

Schoeller, H. 1954. Arid zone hydrology-recent developments. Paris: UNESCO.

Watkins, J.J. & Meakin, N.S. 1996. Explanatory notes Nyngan and Walgett 1:250,000 Geological Sheets SH/55-15 & SH/55-11. Sydney: Geological Survey of NSW, Department of Mineral Resources.

4 Groundwater pollution vulnerability and remediation

Topic co-editors:

John Chilton
British Geological Survey, UK

Steve Appleyard
Water and Rivers Commission, Australia

Mike Barrett
University of Surrey, UK

Julian Conrad
Council for Scientific and Industrial Research, South Africa

Matthias Eiswirth
Department of Applied Geology, Karlsruhe, Germany

Groundwater: Past Achievements and Future Challenges, Sililo et al. (eds) © 2000 Balkema, Rotterdam, ISBN 90 5809 159 7

Sources of faecal contamination of shallow groundwaters in Kampala

M. H. Barrett, K. Johal, G. Howard & S. Pedley
Robens Centre for Public and Environmental Health, University of Surrey, Guildford, UK
M. Nalubega
Department of Civil Engineering, Makerere University, Kampala, Uganda

ABSTRACT: An important element in groundwater protection and management is the identification of pollution sources. This study demonstrates that protected spring sources in Kampala are highly susceptible to contamination, with impacts on discharge quality observable within hours of rainfall events. Landuse in Kampala is complex, and land uses providing potential for both human and animal faecal contamination are observed. This study has demonstrated that sorbitol fermenting bifidobacteria and *Rhodococcus coprophilus* are useful indicators of human and animal faecal contamination in shallow groundwaters, and that both human and animal faecal sources are contaminating Kampala's groundwater in both peri-urban and high-density population areas.

1 INTRODUCTION

Land use in Kampala (as in many other rapidly urbanising areas of sub-Saharan Africa) is complex. Areas of high and low-density populations are juxtaposed and small-scale industrial and agricultural development is often incorporated within residential areas. A variety of metasedimentary rocks (e.g. quartzites and phyllites) underlie Kampala. Differential weathering has resulted in a pronounced topography with thin weathered mantles of limited extent (Taylor and Howard, 1998) containing shallow (2-20m) groundwater flow systems that discharge to valley springs. These are fed by a combination of baseflow and seasonally derived interflow. Many of these springs are used, untreated, by lower income communities with limited access to higher service levels of piped water supply.

Faecal contamination of shallow groundwater in Kampala is widespread (Barrett et al. 1999a, Barrett et al. 1999b, Howard et al. 1999). However, in view of the complex land-use, the specific sources have remained unresolved. Protection of catchments for groundwater sources (predominantly springs in Kampala) is a complex issue under such circumstances and will involve analysis of both hazards and pathways. An important element in this process is the identification of pollution sources. Where funds are limited, and multiple potentially contaminating activities are present, it is important to identify which activities impact groundwater quality in order that effective (and not unnecessary) remedial actions are taken, intercepting the source-receptor pathway. This study specifically seeks to establish methods for distinguishing between human and animal faecal contamination of shallow groundwater systems.

Mara & Oragui (1985) identified bacteriological techniques for distinguishing between human and animal faecal pollution as being a valuable tool in water pollution control programmes as they facilitate the tracing of faecal pollution of drinking water supplies and assist the assessment of source protection, especially in developing countries. Their field studies took place in Nigeria and Zimbabwe, and were largely restricted to faecally contaminated surface waters. They demonstrated that sorbitol fermenting bifidobacteria (unique to human faeces) and *Rhodococcus coprophilus* (unique to animal faeces) could be used with confidence to distinguish between human and animal faecal pollution of surface waters. However, no faecally contaminated boreholes were sampled, leaving the method untried in groundwaters, and doubts remained as to whether transport and survival characteristics would prevent their use in groundwater systems, particularly as sorbitol fermenting bifidobacteria are known to have limited survival outside the human intestine (Mara & Oragui, 1983). It is important to state that non-sorbitol fermenting strains of bifidobacteria have been isolated from both humans and animals. Mara & Oragui experimented with other species, finding *Streptococcus bovis* to be unsuitable, being isolated from both human and animal faeces. Earlier work (Wheater et al., 1979) investigated the possibility of using *Escherichia coli* to faecal streptococci ratios as identification tools (the ratios being different in fresh faeces from different animals and humans). However, a serious limitation is the fact that the two species have different transport and survival characteristics, meaning that ratios will vary with time after a pollution event.

Grabow et al. (1995) investigated the incidence of somatic coliphages, male-specific coliphages and *Bacteroides fragilis* HSP40 phages in human and animal excreta. *B. fragilis* HSP40 phages were found to be highly specific to human faeces. Somatic and male-specific coliphages were found to be excreted in substantial numbers by both humans and animals. Jagals & Grabow (1996) further investigated the use of sorbitol fermenting bifidobacteria as a means of identifying sources of faecal contamination. They confirmed its specificity to human origin, however, they also determined that the species is less robust in the environment than faecal coliforms due to its anaerobic nature (a time-span of 3 hours-6 days was estimated), and consequently is likely to indicate recent pollution of human origin. Again, the work examined surface waters and no analysis of groundwaters was undertaken. For this study, intensive sampling of springs in Kampala was undertaken to establish whether sorbitol fermenting bifidobacteria and *Rhodococcus coprophilus* are useful indicators of human and animal faecal contamination in shallow groundwaters and to indicate the relative importance of these two faecal sources to urban groundwater contamination.

2 FIELD SAMPLING AND ANALYSIS

Sampling of protected springs was undertaken in Kampala on three successive days; 28-30/09/99. The wet season was in progress, and rainfall occurred on every day of sampling. On the night of 28-29/09/99, a significantly heavier rainfall event took place. The sampling sites are listed in Table 1. Brief landuse surveys were carried out at each sampling site. It should be noted that a number of sites contain more than one risk element. The sites were chosen to reflect areas of high-density and low-density population (peri-urban). Analysis of the samples for faecal coliforms, faecal streptococci, coliphage and sorbitol fermenting bifidobacteria and *Rhodococcus coprophilus* was undertaken.

Analyses for faecal coliforms, faecal streptococci, coliphage, and sorbitol fermenting bifidobacteria were carried out at the Department of Civil Engineering, Makerere University, commencing on the day of sampling. Analysis for *R. coprophilus* was carried out at the Robens Centre, U.K. Unfortunately, only samples from 30/09/99 were available. The analytical methods used for *Rhodococcus coprophilus* and sorbitol fermenting bifidobacteria were those of Mara & Oragui (1985). Standard methods were employed for faecal coliforms, faecal streptococci and coliphage.

Table 1. Sampling Sites.

Site	Population density*	Pit latrines within 20m?	Surface waste dump within 50m?	Scattered waste within 50m?	Animal husbandry within 100m?
1. Auntie Claire	L	N	N	N	N
2. Gayaza Road	H	N	N	Y	N
3. Kalerwe	H	N	N	N	N
4. Kazungu	H	N	Y	Y	Y
5. Kikandwa	L	Y	Y	N	Y
6. Kyebando	L	N	Y	N	N
7. Luzige	H	Y	N	Y	N
8. Musoke2	H	N	Y	Y	N
9. Nakatanza	H	N	Y	Y	N
10. Nsooba Parents	L	N	N	N	N
11. Sentema Road	L	N	N	N	N
12. Slaughter House	H	N	N	Y	Y
13. Swimming Pool	L	N	N	N	N
14. Wakaliga	L	N	Y	N	N
15. Washing Bay	H	N	Y	Y	N

N.B. Slaughter House (central Kampala) is in a low-density population area (704/km^2). However, due to the direct and indirect activities associated with the slaughter house, there are generally a large number of people within the immediate catchment area during the day. The site is best described as an animal industry complex.

*Population density is, subjectively, taken as being that in the immediate spring catchment area. The population density around sites 3, 4, 8, 9 and 15 exceeds 30,000 per km^2. The low-density sites are generally <5,000 per km^2.

Table 2. Results from samples taken in Kampala

Site	Faecal Coliforms (cfu/100ml)			Faecal Streptococci (cfu/100ml)			Coliphage (pfu/ml)			*R. coprophilus* (cfu/100ml)			Sorbitol Fermenting Bifidobacteria (cfu/100ml)		
	28/9	29/9	30/9	28/9	29/9	30/9	28/9	29/9	30/9	28/9	29/9	30/9	28/9	29/9	30/9
Auntie Claire	10	87	24	6	70	14	0	0	0	-	-	0	1	0	0
Gayaza Road Mkt	18	Tntc	1636	3	Tntc	520	0	7	0	-	-	0	-	Tntc	3
Kalerwe	119	3360	927	59	388	168	0	0	0	-	-	0	0	120	7
Kazungu	400	970	373	93	68	64	0	0	0	-	-	0	1	0	8
Kikandwa	194	5500	873	57	1210	107	0	6	0	-	-	0	8	450	22
Kyebando	17	Tntc	445	2	Tntc	23	0	3	0	-	-	0	1	0	3
Luzige	210	2170	1409	4	87	49	0	1	1	-	-	1000	0	12	5
Musoke2	2360	3100	1182	177	880	290	0	0	0	-	-	0	0	11	14
Nakatanza	920	2320	2036	82	1230	590	0	1	1	-	-	0	7	9	0
Nsooba Parents	22	308	636	1	228	216	0	0	0	-	-	500	1	0	0
Sentema Road	177	4900	2464	38	1920	391	0	2	0	-	-	0	7	72	7
Slaughter House	1760	9300	2209	1010	2480	718	3	6	3	-	-	1000	0	0	0
Swimming Pool	5	43	43	0	56	19	0	0	1	-	-	500	0	0	0
Wakaliga	120	640	670	5	59	55	0	-	0	-	-	1000	0	3	0
Washing Bay	17	36	21	114	224	92	0	0	0	-	-	0	7	0	0

Tntc: too numerous to count

3 RESULTS

The results are given in Table 2. There is a clear 'pulse reaction' of spring water quality within 12 hours of the major rainfall event (28-29/09/99), as demonstrated in Figures 1 (faecal coliforms) and 2 (sorbitol fermenting bifidobacteria). Increased flow rates were observed, but not measured. There is a significant deterioration in microbiological groundwater quality within 12 hours of the rainfall event and significant 'recovery' within 36 hours (although bacterial counts remain slightly higher than pre-rainfall levels). It is also clear (from Figure 1) that every site sampled shows a degree of faecal contamination (the use of enterotubes established the dominant faecal coliform type to be *E. coli*).

There are some important implications from these results. The first is that any monitoring of protected springs in these localised groundwater flow systems in weathered crystalline aquifers must be interpreted in the context of rainfall events. It is known that there are seasonal variations in groundwater quality in such systems (e.g. Barrett et al., 1999a). This study demonstrates that major variations occur even on a daily (and possibly hourly) basis. Secondly it is clear that the role of interflow (locally recharged waters moving laterally through unsaturated, heterogeneous materials) and/or surface runoff infiltrating the hardcore behind protected springs (where drainage ditches, backfill or retaining walls are absent or damaged) contributes to spring water quality in such systems. Figure 3 shows the typical design of a protected spring.

The dramatic fluctuations are unlikely to result only from baseflow contributions in such a short time span (vertical hydraulic conductivity being very low in these clay rich systems). As all the springs sampled produce discharge year-round, it is assumed that they also receive some permanent baseflow contribution. Sorbitol fermenting bifidobacteria were detected in a significant number of samples, and therefore successfully indicated human faecal contamination in this localised, shallow weathered crystalline groundwater system. Indeed only 2 sites (Slaughter House and Swimming Pool) gave negative results over the entire 3 day sampling period. *R. coprophilus* was found to be present at fewer sites, although, where present, concentrations were high (comparable to faecal coliform levels). The results obtained are likely to be an underestimation of true concentrations, as the samples were in transit for several days between sampling and analysis in the UK. The limited data produced for this species none the less allow several sites to be identified as being contaminated by animal faeces.

4 DISCUSSION

This study has utilised bacterial indicators (sorbitol fermenting bifidobacteria and *Rhodococcus coprophilus*) to identify, and distinguish, human and animal sources of faecally contaminated groundwaters.

R. Coprophilus was detected at 5 sites, clearly indicating that animal faecal contamination of groundwaters may occur in urban areas:

1) Slaughter House: This is the most heavily microbiologically contaminated site of those surveyed, and is centrally located in Kampala (and used by staff at the site in food preparation). The faecal origin appears to be dominantly, if not entirely, animal. This is one of only 2 sites where sorbitol fermenting bifidobacteria were not detected. Whilst it is known that this species may not survive well in shallow groundwaters, this site (as all others) shows a rapid quality response to rainfall, and as

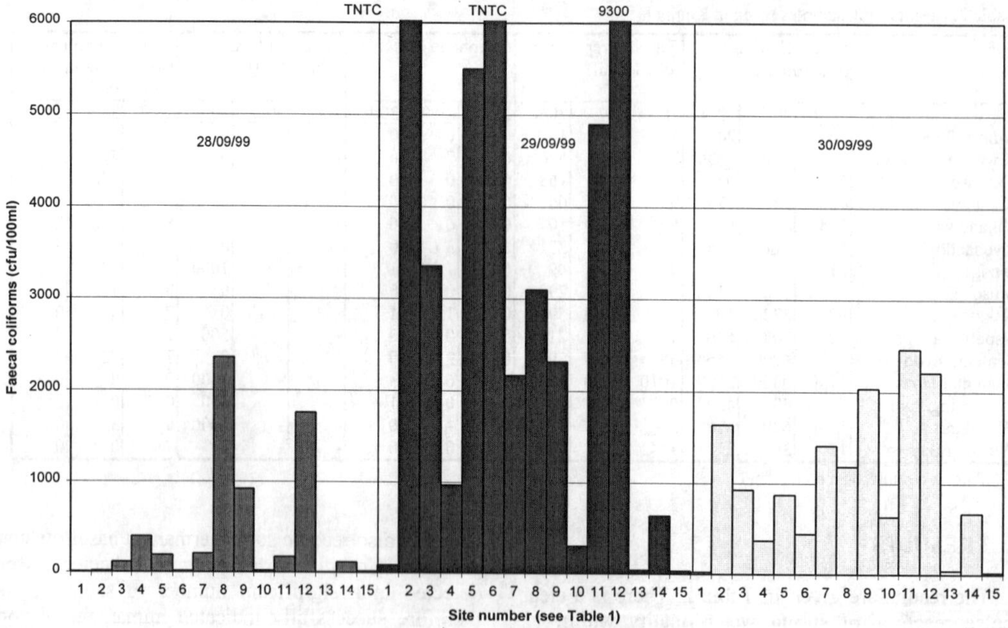

Figure 1. Temporal variations in faecal coliform counts at sampling sites

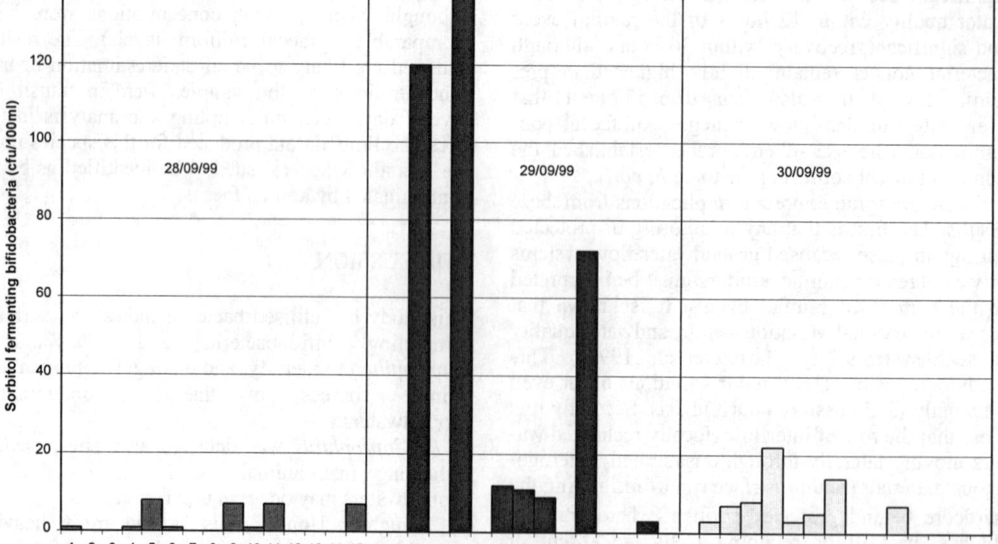

Figure 2. Temporal variations in sorbitol fermenting bifidobacteria counts at sampling sites

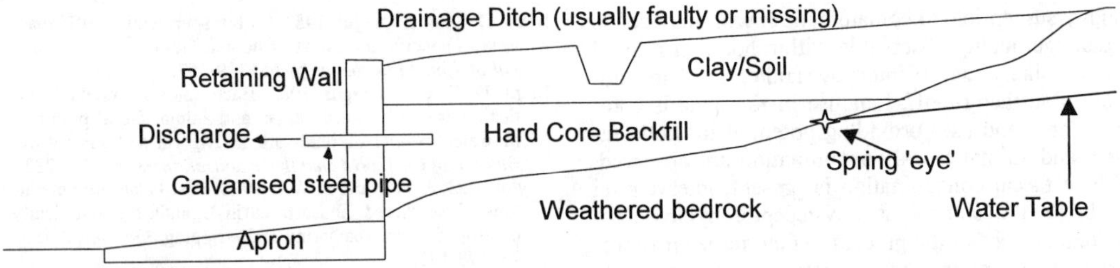

Figure 3. Protected spring design

sorbitol fermenting bifidobacteria are present at almost all other sites in very similar hydrogeological settings, the absence here is taken to indicate the absence of significant human faecal loading. The animal source is clear, with large quantities of livestock frequently being present awaiting slaughter.

2) Nsooba Parents: This is a peri-urban site with low population density and no obvious contaminant sources in the immediate catchment area. However, ongoing risk assessment (Howard et al. 1999; Howard & Luyima, 1999) demonstrates that the poor state of spring protection (eroded backfill, deteriorating masonry, absence of adequate drainage and fences) exerts a major control on spring water quality. This site demonstrates poor protection, and the likely contaminant source is ingress of surface waters contaminated by animal faeces (only one detection for sorbitol fermenting bifidobacteria was made over the 3 days, suggesting a low human faecal loading).

3) Swimming pool: This is also a peri-urban site, but again risk inspections show faulty protection works. There are no other obvious pollution sources. At this site (like Slaughter House) no sorbitol fermenting bifidobacteria were detected suggesting contamination to be dominantly, or exclusively, of animal origin.

4) Wakaliga: Waste dumps were seen to be present in the vicinity, and up-slope, of the spring. The presence of sorbitol fermenting bifidobacteria indicates human as well as animal contamination. It is possible the waste contains discarded human faeces, and is likely also to be a focus of scavenging animals.

5) Luzige: By contrast, this is a high-density population area. Pit latrines are present, and may contribute to the human contamination observed. The animal faecal contamination may again result from the presence of scattered waste in the vicinity of the spring, a likely site for animal scavenging.

Sorbitol fermenting bifidobacteria were detected at all sites but 2, reflecting widespread loading of human sewage to groundwater in Kampala. Clearly this species is a useful indicator of human faecal contamination of the Kampala groundwater system. Interestingly, only 2 sites were adjacent to pit-latrines, and both these sites are also at risk from waste dumps and/or scattered waste. In the absence of adjacent pit-latrines at many of these sites, with the rapid emergence of human faecal material at the springs after rainfall events, and knowing the poor state of the protection works at many of the springs, it seems likely that the major risk of bacterial contamination of a human source is scattered waste and waste dumps, the transport route being either localised infiltration of runoff where protection is poor, and/or interflow.

Further work is required during the dry season to establish how robust the tested indicators are. Ongoing monitoring of the springs indicates faecal contamination to persist, although even in high density areas, dry season quality is generally better than in the wet season (Howard & Luyima, 1999b).

Laboratory studies of the survival of *Rhodococcus coprophilus* and sorbitol fermenting bifidobacteria in Kampala groundwater are underway. Investigations into transport controls (attenuation) are also required. In summary, it can be seen that animal faecal contamination of groundwater is present at many locations in Kampala, but that human faecal contamination is ubiquitous. Major improvements of bacterial quality of protected spring water could be achieved through simple improvement of sanitary protection measures to prevent inundation with contaminated surface water, although improvements in design are also required (Howard et al., 1999). Results from monitoring the ongoing rehabilitation of springs in Kampala indicate significant improvements in water quality (Howard et al., in prep).

5 CONCLUSIONS

The shallow groundwater system in the weathered crystalline bedrock of Kampala is extensively contaminated by faecal material. The current study and previous surveys demonstrate microbiological contamination to be most severe during the rainy (recharge) season. The data presented above demonstrate that protected spring sources in Kampala are highly susceptible to contamination, with impacts on discharge quality observable within hours of rainfall events, likely due to interflow and/or local infiltration of surface runoff. Landuse in Kampala is complex, and land uses providing potential for both human and animal faecal contamination are observed. Where faecal contamination is present, intervention to break the source-pathway-receptor cycle is required. As part of the process to determine an appropriate intervention, the source(s) of contamination should be identified. This study has demonstrated that sorbitol fermenting bifidobacteria and *Rhodococcus coprophilus* are useful indicators of human and animal faecal contamination in shallow groundwaters, and that both human and animal faecal sources are contaminating Kampala's groundwater in both peri-urban and high-density population areas.

ACKNOWLEDGEMENTS

The research was funded by DFID (UK). The views expressed are those of the authors and do not necessarily represent the official views of DFID. The authors are grateful for the comments and advice of the reviewers.

REFERENCES

Barrett, M. H., Nalubega, M., Howard, G., Taylor, R. G., & S. Pedley 1999a. The impact of on-site sanitation on urban groundwater quality in Uganda. In: Fendekova and Fendek (eds), International Association of Hydrogeologists: 335-340.

Barrett, M. H., Nalubega, M., & S. Pedley 1999b. On-site sanitation and urban aquifer systems in Uganda. *Waterlines* 17 (4): 10-13.

Grabow, W. O. K., Neubrech, T. E., Holtzhausen, C. S., & J. Jofre 1995. *Bacteriodes Fragilis* and *Escherichia Coli* Bacteriophages: Excretion by Humans and Animals. *Water Science and Technology* 31 (5-6): 223-230.

Howard, G. 1999. On-site sanitation and groundwater: the art of balancing unknown risks? *Waterlines* 17 (4): 2-5.

Howard, G. & P. G. Luyima 1999a. Urban water supply surveillance in Uganda. In J. Pickford (ed), Integrated Development for Water Supply and Sanitation, Proceedings of the 25th WEDC Conference, Addis Ababa, 1999: 290-293.

Howard, G & P. G. Luyima 1999b. Report on water supply surveillance in ten selected urban areas of Uganda. Ministry of Health, Uganda.

Howard, G., Bartram, J. K., & P. G. Luyima 1999. Small water supplies in Urban Areas of Developing Countries. in Cotruvo, J.A. Craun, G.F. and Hearne, N. (eds) Providing safe drinking water in small system: technology, operations and economics. Lewis Publishers, Washington, DC. USA: 83-93.

Howard, G., Nalubega, M. & R. Mutambazi (in prep) Rehabilitation of protected springs in Kampala. Paper to be presented at the 26th WEDC Conference, Dhaka, 2000.

Jagals, P. & W. O. K. Grabow 1996. An evaluation of sorbitol-fermenting bifidobacteria as specific indicators of human faecal pollution of environmental water. *Water SA* 22 (3), 235-238.

Mara, D. D. & J. Oragui. 1983. Sorbitol-fermenting bifidobacteria as specific indicators of human faecal polution. *Journal of Applied Bacteriology* 55: 349-357.

Mara, D. D. & J. Oragui. 1985. Bacteriological methods for distinguishing between human and animal faecal pollution of water: results of fieldwork in Nigeria and Zimbabwe. *Bulletin of the World Health Organisation* 63 (4): 773-783.

Taylor, R.G. & Howard, K.W.F. 1998. Post-Palaeozoic evolution of weathered landsurfaces in Uganda by tectonically controlled deep weathering and stripping. *Geomorphology* 25. 173-192.

Wheater, D. W. F., Mara, D. D., & J. Oragui 1979. Indicator systems to distinguish sewage from stormwater run-off and human from animal faecal material. In: A. James and L. Evison (eds), John Wiley, Chichester: 21-27,

Groundwater: Past Achievements and Future Challenges, Sililo et al. (eds) © 2000 Balkema, Rotterdam, ISBN 90 5809 159 7

Using K_d-Test data in the estimation of contaminant attenuation and risk assessment

P.H. Beck
ERM Contaminated Site Solutions, Crows Nest, N.S.W., Australia

J. Jankowski
UNSW Groundwater Centre, School of Geology, University of New South Wales, Sydney, N.S.W., Australia

ABSTRACT: K_d-Tests are an economic means of assessing the interaction between soils and solutes. This type of data is often essential in estimating natural attenuation of contaminants and environmental risk assessment. Consultants in Australia commonly undertake assessment of natural attenuation and environmental risk and, depending on the results, decide if remediation of the site is required. Most of these studies do not use K_d-Tests to assist in the assessment process. This paper outlines how K_d-Test data can be used in assessment of natural attenuation and environmental risk. K_d-Tests were used to study the interaction between heavy metals and three soil types commonly encountered in the Botany Sands aquifer. Heavy metals used in the study included cadmium, chromium, copper, lead, nickel and zinc. Solution pH decreased notably during the K_d-Tests due to sorption process. Retardation factors and soil sorption capacities for each soil were calculated, based on the K_d-Test results. This type of data is invaluable in the assessment of an aquifers attenuation properties, contaminant transport, environmental, property and public health risk assessment. Retardation factors are required to estimate the transport velocity of contaminants. Sorption capacities are required in the assessment of the attenuation capacity of the aquifer soils and the assessment of the contamination risk to the soil, both on the site and on adjoining properties. Changes in the chemical environment due to contaminant interaction with soils could affect the mobility of contaminants or deteriorate structures on the site or adjoining properties and therefore will need to be considered.

1 INTRODUCTION

1.1 *Background*

Natural attenuation studies, fate and transport modelling and risk assessment, as a means of dealing with and managing contaminated sites, in Australia, has increased in popularity over recent times. Usually these assessments involve soil and groundwater quality data but do not include a study of the interaction between the soil and groundwater solutes. K_d-Tests are an economical laboratory based means of assessing the interaction between soils and groundwater solutes.

Knowledge of the soil solute interaction is often critical in assessment of natural attenuation of contaminants and environmental risk assessment (Fuller et al. 1996).

Roy et al. (1991) outlined K_d-Test theory, described the major processes and developed a laboratory based methodology for undertaking the tests. The methodology outlined in Roy et al. (1991) was used for this paper, with the methodology and results discussed in pervious publications by the authors (Jankowski & Knight 1991, Beck & Jankowski 1998, 1999). K_d-Tests are synonymous with Batch Sorption Tests, which are used by soil scientists.

K_d-Tests are reported in scientific publications, but most of these focus on specific sorption sites. Sorption site types studied included iron oxyhydroxides (Benjamin & Leckie 1981b), carbonates (Dudley et al. 1988, Madrid & Diaz-Barrientos 1992), silicon and aluminium oxides (Koeppenkastrop & De Carlo 1993; Schulthess & Huang 1990) and hydrous manganese oxides (Loganathan et al. 1977; Zasoski & Burau 1988).

1.2 *Purpose of K_d-Test*

K_d-Tests reported in the literature have only limited applicability to soil and water from different environments. The soils at the Eastlakes Experimental (ELE) site are likely to contain a mixture of all or most of the sorption sites described above. As a result a series of K_d-Tests using the three most common soils, found in the Botany Sands Aquifer, Sydney, NSW were undertaken to study the specific groundwater environmental conditions at the ELE site (Beck & Jankowski 1998).

2 SUMMARY OF RESULTS AND DISCUSSION

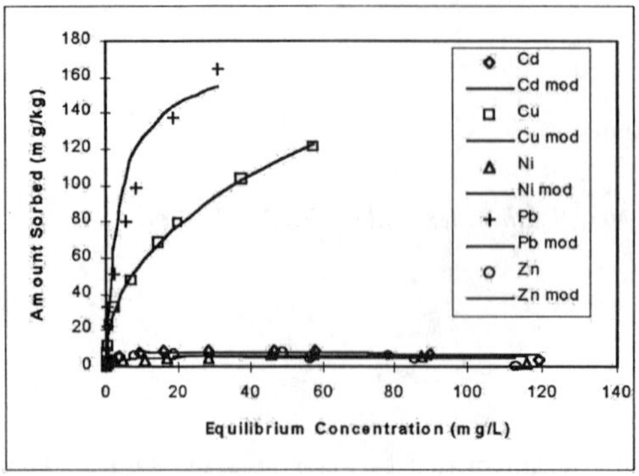

Figure 1a. Sorption plot for Soil A combined element K_d-Test (mod = Model Isotherm) (Beck & Jankowski 1998).

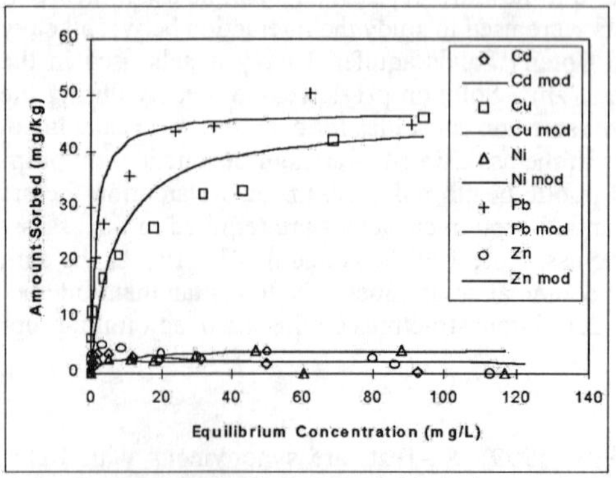

Figure 1b. Sorption plot for Soil B combined element K_d-Test (mod = Model Isotherm) (Beck & Jankowski 1998).

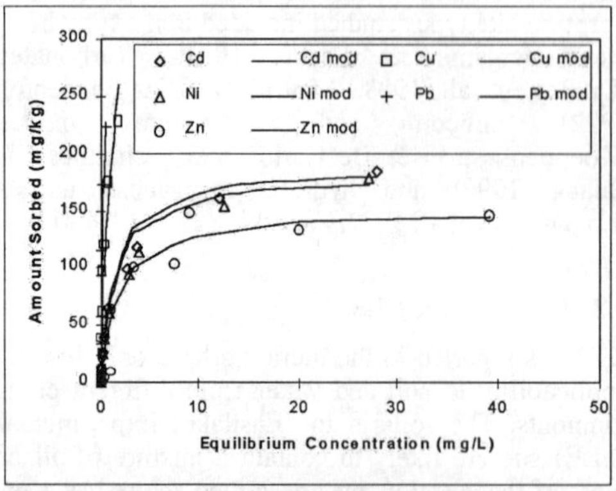

Figure 1c. Sorption plot for Soil C combined element K_d- Test (mod = Model Isotherm) (Beck & Jankowski 1998).

Figures 1a-1c summarise the results of the combined element K_d-Tests. Sorption preference of Soil A was Pb>Cu>Cd~Zn~Ni (Fig. 1a), when all five heavy metals were in solution. Soil B had sorption preference of Pb>Cu>Zn>Ni>Cd (Fig. 1b). Soil C showed the highest sorption of the three test soils, with sorption preference for Pb>Cu>Cd>Ni>Zn (Fig. 1c)

3 DISCUSSION

3.1 *Discussion of chemical processes*

Competitive sorption processes between heavy metals caused an overall reduction in the amount of metal solutes removed from solution by sorption.

An increase in the concentrations of major cations as well as a decrease in solution pH were also observed with increasing metal input concentrations.

Sorption of heavy metal ions onto iron, aluminium or silicon oxyhydroxide surface sites can cause the release of hydrogen ions into solution (Benjamin & Leckie 1981b). Benjamin and Leckie (1981) proposed the following mechanisms to explain the decrease in solution pH:

$$Me^{2+} + \underline{SOH_x} = SOMe + xH^+ \quad (1)$$

$$Me^{2+} + H_2O + \underline{SOH_y} = SOMeOH + (y+1)H^+ \quad (2)$$

where

$\underline{SOH_x}\ \underline{SOH_y}$ = oxy-hydroxide sorption surface

These reaction mechanisms would result in the release of hydrogen ions into solution as metal ions sorb, resulting in the decrease of the pH.

Decrease of the pH causes mineral dissolution, which results in the observed increase in major cation (Ca, Mg, Na, K) concentration with increased metal ion input.

Dissolution was considered to be more likely than ion exchange, as sorption of mono valent ions is preferred at lower pH. Therefore a decrease in the rate of Ca concentration increase (ie non linear trend with decreasing pH) is likely with decreasing pH (Appelo & Postma 1994). Shell fragments, observed during petrographic analysis (Evans 1993) are the likely source of the calcium.

3.2 *Modelling of chemical processes*

As equilibrium conditions existed at the end of the K_d-Test mixing period, thermodynamic equilibrium modelling of the chemical processes postulated in section 3.1 was feasible. MINTEQ-A2 (Allison et al. 1991) was used to simulate the sorption and mineral dissolution processes described in section 3.1.

As sorption processes occur at a significantly faster rate than mineral dissolution (Appelo & Postma 1994), the model was set up as follows:

a) The measured pH and solute concentrations of the standard solutions were entered into the model and the solute speciation calculated.

b) Using the modelled solute species in solution, formula (1) was modified with the correct molar prefixes for each of the five elements used in the K_d-Tests.

c) The modified sorption reaction equations were then added to the model to simulate metal removal from solution and hydrogen ion release.

d) Finally small amounts of calcium carbonate were dissolved stepwise by the model to simulate mineral dissolution.

Step a). model results indicated that the heavy metals speciated into a variety of mono and divalent complexes at the starting pH of 5. The metal speciation calculated by the model is summarised in Table 1.

Table 1. Metal speciation at injection pH.

			mmol/l
Cadmium			
44%	as	Cd^{2+}	0.208
45.10%	as	$CdCl^{+}$	0.213
2.70%	as	CdCl2	0.013
8%	as	$CdBr^{+}$	0.038
Copper			
97.60%	as	Cu^{2+}	0.867
2.20%	as	CuCl	0.019
Nickel			
96.80%	as	Ni^{2+}	0.904
2.60%	as	$NiCl^{+}$	0.024
Lead			
68.50%	as	Pb^{2+}	0.169
26.20%	as	$PbCl^{+}$	0.065
4.50%	as	$PbBr^{+}$	0.011
Zinc			
97.60%	as	Zn^{2+}	0.767
2.30%	as	$ZnCl^{+}$	0.018

Step b). The miliequivalants of charges removed and released by metal sorption were then estimated. Equation (1) was then modified to allow for metal speciation and entered into the model.

Step c). The modified sorption reaction equations are presented below. (Beck & Jankowski 1998)

$$1.411SOH + Cd^{2+} = S_{1.411}O_{1.411}\text{-}Cd + 1.411H^{+} \quad (3)$$
$$1.974SOH + Cu^{2+} = S_{1.974}O_{1.974}\text{-}Cu + 1.974H^{+} \quad (4)$$
$$1.962SOH + Ni^{2+} = S_{1.962}O_{1.962}\text{-}Ni + 1.962H^{+} \quad (5)$$
$$1.677SOH + Pb^{2+} = S_{1.677}O_{1.677}\text{-}Pb + 1.677H^{+} \quad (6)$$
$$1.975SOH + Zn^{2+} = S_{1.975}O_{1.975}\text{-}Zn + 1.975H^{+} \quad (7)$$

Once the sorption reactions were entered, the model was run and the equilibrium pH calculated. The results indicated that pH values as low as 2 were feasible. Mineral dissolution, particularly of calcium carbonate, is the likely cause of the higher pH values measured during the K_d-Tests.

Step d). Small amounts of calcium carbonate were allowed to dissolve during each model run and the pH calculated. As no calcium ions were included in the solutions, calcium was considered as an indicator of the amount of calcium carbonate dissolved.

Thus two checks on the model performance were available. One was the estimated pH compared to the measured pH and the second was the predicted calcium concentrations compared to the actual measured concentration. The model achieved good agreement between the measured and estimated pH and calcium concentrations.

3.3 *Analysis of results*

Sorption isotherms, using the linear, Freundlich and Langmuir model, were fitted to the K_d-Test results. Once the model with the best fit to the data was established a retardation factor and, where possible, soil sorption capacity were calculated. Model fitting and parameter estimation was carried out using the method outlined by Roy et al. (1991). Each of the three models was fitted to the data and the one with the highest R^2 value was used to calculate the retardation factor and, where applicable, the soil sorption capacity (Bowman et al. 1984). Table 2 summarises the results of the data analysis, while the models fitted to the data are presented on Figures 1A-1C.

Table 2. Summary of K_d-Test Analysis (Beck & Jankowski 1998).

Metal	Isotherm	Rfactor	Sor Cap
Cd Soil A	Langmuir	1.000	7.04
Cd Soil B	Langmuir	1.000	1.87
Cd Soil C	Langmuir	1.002	131.58
Cu Soil A	Freundlich	1.112	NA
Cu Soil B	Langmuir	1.000	45.45
Cu Soil C	Freundlich	2.506	NA
Ni Soil A	Langmuir	1.000	6.40
Ni Soil B	Langmuir	1.000	0.75
Ni Soil C	Langmuir	1.003	131.58
Pb Soil A	Langmuir	1.001	172.41
Pb Soil B	Langmuir	1.001	46.73
Pb Soil C	Langmuir	1.027	303.03
Zn Soil A	Langmuir	1.000	4.52
Zn Soil B	Langmuir	1.000	1.89
Zn Soil C	Langmuir	1.002	151.52

Notes:	Rfactor	Retardation factor
	Sor Cap	Sorption Capacity
	NA	Not Applicable

Retardation factors calculated for all test data show that Ni and Zn in solution would travel at or near the groundwater flow velocity when travelling through any of the three soils used in the tests. Cd, Cu and Pb showed a tendency for retardation in Soils A and particularly Soil C, while little or no retardation would occur in Soil B.

The Langmuir isotherm applied to the majority of elements and allowed calculation of the soil sorption capacities. Soil sorption capacity results are summarised on Table 2. Competitive sorption processes resulted in a reduction of soil sorption capacity of up to 98% for some heavy metal solutes (Beck and Jankowski 1999). In the combined element K_d-Tests soil sorption capacities could be calculated for all metal elements, with the exception of copper in the Soil A and C K_d-Tests.

3.4 *Application of K_d-Test results*

K_d-Test results were then used to assess the potential for the soils to become contaminated by heavy metals.

The K_d-Tests results indicated that the sorption capacities of the three soils are relatively low. As a result accumulation of unacceptably high metal concentration in the soils over time is unlikely. Under current Australian guidelines only the sorption capacity of Soil C for lead would exceed applicable environmental or health guidelines. This is one aspect of contaminated site assessments that is often not addressed, as generally no data is available. K_d-Tests provide this type of data, which can then be used to assess contamination risk to soils.

Conversely, transport distances of metal contaminants in groundwater are likely to be enhanced , particularly for high concentration, high mass sources, as the soils would only have a limited attenuation capacity and retardation factors are generally low. Once the soils reach their sorption capacity their ability to attenuate contaminants decreases dramatically and hence contaminant transport would become conservative (Harmon et al. 1992).

From a risk perspective, the soils would have a low risk of becoming contaminated to unacceptably high levels but would allow greater transport distance therefore increasing the risk of down-gradient environmental and health impacts.

4 CONCLUSION

K_d-Tests provide valuable information for natural attenuation studies, fate and transport modelling and risk assessments without the expense of field trials. In particular K_d-Tests allow accurate prediction of sorption effects, which are particularly important in the transport behaviour of heavy metals.

A series of K_d-Tests using Cd, Cu, Ni, Pb and Zn metal ions in solution were undertaken to gain a better understanding of potential metal transport behaviour. The sorption behaviour of the five metals onto three different soil types was studied. Test conditions were designed to simulate the groundwater environment commonly encountered in the Botany Sands aquifer.

The K_d-Tests showed relatively little retardation and attenuation is likely to occur when all of the metals are present in solution. Competitive sorption processes would significantly enhance the mobility of heavy metal species in solution, particularly those of Cd, Ni and Zn.

Major cation and anion concentrations in solution were also measured. By measuring the concentrations of the major cations (Ca, K, Mg and Na), at the end the K_d-Tests, other chemical processes were indicated. The pH of the solution was influenced by metal sorption, which released hydrogen ions, lowering the pH and causing mineral dissolution, particularly of calcium carbonate, which raised solution pH. Modelling of these processes using the thermodynamic equilibrium model MINTEQ-A2, showed that metal species sorption would decrease solution pH. A reduction in the pH of the solution would reduce the tendency of metals to sorb (Appelo & Postma 1994), therefore enhancing metal mobility.

Traditionally K_d-Tests have been used to calculate retardation factors and soil sorption capacities (where applicable) for use in contaminant transport models. The results of the K_d-Test described in this paper demonstrate the validity of the method when used in multi solute and sorption site environments.

Based on the K_d-Test results, the sandy soils at the site would preferentially sorb copper and lead from a solution containing dissolved Cd, Cu, Ni, Pb and Zn solute species. This preferential removal of Cu and Pb would significantly enhance the mobility of Cd, Ni and Zn, due to a decrease in available sorption sites.. Similar behaviour of heavy metal solutes could be excepted in other aeolian sand deposits in the Botany Basin.

In the absence of the K_d-Test data, prediction of metal transport behaviour would have been difficult, as insufficient relevant published data was available. This would therefore not allow accurate modelling of the metal transport behaviour and any assessment of environmental of health risk would be subject to potentially significant errors.

REFERENCES

Allison J D, Brown D S, and Novo-Gradac K J 1991. *MINTEQA2/PRODEFA2, A geochemical assessment model for environmental systems: Version 3.0 user's manual*: US EPA EPA/600/3-91/021.

Appelo C A J, and Postma D 1994. *Geochemistry, groundwater and pollution*: Rotterdam, AA Balkema, 536 p.

Atanassova I D 1995. Adsorption and Desorption of Cu at High Equilibrium Concentrations by Soil and Clay Samples from Bulgaria*: Environmental Pollution,* Vol. 87, p. 17 - 21.

Beck P, and Jankowsky J 1998. Adsorption tests of heavy metals on three sand soils from the Botany aquifer,

Sydney, Australia. Implications for estimation of contaminate transport, *International Groundwater Conference "Groundwater Sustainable Solutions", The University of Melbourne, 8-13 February 1998, Melbourne, Australia.*

Beck P, and Jankowsky J 1999. Laboratory based sorption studies to estimate the attenuation of contaminants in saturated sand aquifers, *8th Australian New Zealand Conference on Geomechanics, 15 - 17 February 1999, Hobart, Australia.*

Benjamin M M, and Leckie J O 1981a. Conceptual Model for Metal - Ligand - Surface Interactions during Adsorption: *Environmental Science & Technology,* Vol. 15, No. 9, p. 1051 - 1056.

Benjamin M M, and Leckie J O 1981b. Multiple Site Adsorption of Cd Cu Zn and Pb on Amorphous Iron Oxyhydroxide: *Journal of Colloid and Interface Science,* Vol. 79, No. 1, p. 209 - 219.

Bowman R S, Urquhart N S, and O'Connor G A 1984. Statistical Evaluation of Sorption Isotherm Data: *Soil Science*, Vol. 137, No. 5, p. 360 - 369.

Dudley L M, McLean J E, Sims R C, and Jurinak J J 1988. Sorption of Copper and Cadmium from the Water - Soluble Fraction of an Acid Mine Waste by two Calcareous Soils: *Soil Science*, Vol. 145, No. 3, p. 207 - 214.

Evans D J 1993. *A Physical and Hydrochemical Characterization of a Sand Aquifer in Sydney, Australia: Sydney*, UNSW, Masters, 242 p.

Fuller C C, Davis J A, Coston J A, and Dixon E 1996. Characterization of metal adsorption variability in a sand and gravel aquifer, Cape Cod, Massachusetts, U.S.A.: Journal of Contaminant Hydrology, v. 22, p. 165 - 187.

Harmon T C, Semprini L, and Roberts P V 1992. Simulating Solute Transport using Laboratory - Based Sorption Parameters: Journal of *Environmental Engineering*, Vol. 118, No. 5, p. 666 - 689.

Jankowski J & MJ Knight 1991. *Astralobe Park Landfill Leachate and its Impact on the Botany Wetlands*: Unisearch Limited R9309 (unpubl.).

Koeppenkastrop D, and De Carlo E H, 1993. Uptake of Rare Earth Elements from Solution bt Metal Oxides: *Environmental Science & Technology*, Vol. 27, p. 1796 - 1802.

Loganathan P, Burau R G, and Fuerstenau D W 1977. Influence of pH on the Sorption of Co Zn and Ca by Hydrous Manganese Oxide: *Soil Science Society of America Journal*, Vol. 41, p. 57 - 62.

Madrid L, and Diaz-Barrientos E 1992. Influence of carbonate on the reaction of heavy metals in soils: *Journal of Soil Science*, Vol. 43, p. 709 - 721.

Roy W R, Krapac I G, Chou S F J, and Griffin R A 1991. *Batch-Type Procedures for Estimating Soil Adsorption of Chemicals*: Illinois State Geological Survey EPA/530-SW-87-006-F.

Schulthess C P, and Huang C P 1990. Adsorption of Heavy Metals by Silicon and Aluminum Oxide Surfaces on Clay Minerals: *Soil Science Society of America Journal,* Vol. 54, p. 679 - 688.

Zasoski R J, and Burau R G 1988. Sorption and Sorptive Interaction of Cadmium and Zinc on Hydrous Manganese Oxide: *Soil Science Society of America Journal*, Vol. 52, p. 81 - 87.

Groundwater: Past Achievements and Future Challenges, Sililo et al. (eds) © 2000 Balkema, Rotterdam, ISBN 90 5809 159 7

Simulation of mass transfer kinetics for chemically enhanced removal of residual trichloroethene

T.B. Boving – *Department of Geosciences, University of Rhode Island, Kingston, R.I., USA*
Z. Zhang – *Department of Soil, Water and Environmental Science, University of Arizona, Tucson, Ariz., USA*
W. Ji – *Arcadis Geraghty and Miller Incorporated, Melville, N.Y., USA*
M.L. Brusseau – *Department of Soil, Water and Environmental Science, and Department of Hydrology and Water Resources, University of Arizona, Tucson, Ariz., USA*

ABSTRACT: This paper presents the results of a series of chemically enhanced flushing experiments using an anionic surfactant (SDS) and a cyclodextrin-based complexing agent (HPCD) to remove residual phase TCE from porous media. Compared to a water flushing experiment, both agents enhanced the dissolution of TCE by approximately 10 times when present at concentrations of 2.5% SDS and 20% HPCD, respectively. A mathematical model was used to simulate mass transfer between the TCE and the aqueous phase. The calculated mass transfer rate coefficients were compared to the values obtained for the water-TCE flushing experiment. It was found that the mass transfer rate coefficients for both SDS and HPCE were similar to that for the water system

1 INTRODUCTION

Chlorinated organic compounds and complex mixtures of these compounds have been identified as a common cause of groundwater contamination. The generally poor performance of conventional groundwater pump-and-treat schemes has led to research and application of in-situ chemically enhanced flushing methods. Chemically enhanced-flushing technologies are based on flushing contaminated porous media with chemical agents to increase contaminant solubility and concomitantly the mass removal rate, and thus reduce the time and cost of remediation.

In-situ treatment technologies are particularly useful for the treatment of dense nonaqueous phase liquid (DNAPL) source zones. Chemical treatment of contaminated zones may be attractive when (1) large amounts of contaminant mass have to be removed from the source zone, (2) alternative methods (e.g. bioremediation) are incompatible or will not function effectively with respect to rate or extent of treatment, or (3) access to the contaminated soil and groundwater is difficult due to restricting surface structures or uses.

The dissolution of immiscible liquids by solubilization-enhancing agents, such as surfactants and complexing agents, is governed by mass transfer processes. The potential significance of mass transfer rate limitations on the dissolution of NAPLs has been extensively investigated (e.g. Miller et al.

1990; Powers et al. 1991 and 1992; Abriola et al. 1993; Geller and Hunt 1993; Illangasekare et al. 1995).

Mass transfer limitations between NAPL and the aqueous phase may affect the efficiency of ground water remediation (Hunt et al. 1988). For example, the potential importance of rate-limited NAPL dissolution was demonstrated by Powers et al. (1991), who conducted a theoretical assessment of NAPL dissolution rates and showed that the concentration of a contaminant in the flushing solution decreases as the volume of the NAPL in the source zone is reduced. The objective of this paper is to evaluate the mass transfer parameters that describe the dissolution kinetics of residual level trichloroethene (TCE) in the presence of an anionic surfactant (SDS) and a complexing agent, hydroxypropyl-β-cyclodextrin (HPCD).

2 SIMULATION OF MASS TRANSFER KINETICS

Several models have been developed to simulate NAPL dissolution (e.g., Miller et al. 1990; Imhoff et al. 1993; Powers et al. 1994; Ji and Brusseau 1998). Generally, a linear driving force expression is assumed appropriate to represent the interphase mass transfer (Eqn.2.1):

$$\frac{\partial M_n}{\partial t} = -k_f a_0 \left(C_S - C_w\right) \qquad (2.1)$$

where M_n is the mass concentration of NAPL residual (ML^{-3}), k_f is the mass-transfer coefficient ($L\,T^{-1}$), a_o is the specific surface area across which the mass flux occurs (L^{-1}), C_s is the equilibruim aqueous phase concentration of the NAPL at the interface (M L^{-3}), and C_w is the bulk aqueous phase concentration (M L^{-3}). Because it is not currently possible to measure exactly the concentration at the surface C_S is usually assumed to be equal to the equilibruim aqueous phase solubility of the NAPL. If the specific surface area cannot be determined independently, a_o and k_f are usually lumped together into a mass transfer rate coefficient, k_L (T^{-1}). This parameter depends on properties of the porous medium, aqueous phase velocity, the molecular diffusion coefficient of the NAPL compound, and the NAPL distribution in the porous matrix (Miller et al. 1990; Powers et al. 1992; Imhoff et al. 1993; Powers et al. 1994b).

If it is assumed that the magnitude of the enhanced driving force is dependent on the enhancement factor, E (Eqn. 2.2):

$$\frac{\partial M_n}{\partial t} = -k_f a_o (C_S - C_w) = -k_L \left(\frac{C_S^m}{E} - \frac{C_c^m}{E} \right)$$

$$\text{For complexing agents}: \quad E = 1 + K_{pt} C_A^{aq} \tag{2.2}$$

For surfactants:

$$E = 1 + K_{pt}(C_A^{aq} - CMC) \quad for \ C_A^{aq} > CMC$$

$$E = 1 \quad for \ C_A^{aq} \leq CMC$$

where C_S^m is the apparent solubility of contaminant in the bulk solution (M L^{-1}), C_c^m is the total concentration of the contaminant in the bulk solution (M L^{-1}), C_A^{aq} is the dissolved concentration of the flushing agent (M L^{-3}), CMC is the critical micelle concentration (M L^{-3}), and K_{pt} is the partition coefficient ($M^{-1}L^3$). Ji and Brusseau (1998) developed a mathematical model used for the simulation of chemically enhanced dissolution of NAPL. It accounts for enhanced solubilization, rate-limited dissolution of immiscible-phase liquid, enhanced sorption of contaminant due to sorption of the agent (if applicable), and nonlinear, rate-limited sorption of both the agent and the contaminant. The simulated flushing process includes two parts: the transport of the contaminant under the influence of a chemical flushing agent (Eqn.2.3) and the transport of the flushing agent (Eqn 2.4).

$$\frac{\partial C_c^m}{\partial t} + \frac{\rho}{\theta}\frac{\partial S_C^1}{\partial t} + \frac{\rho}{\theta}\frac{\partial S_C^2}{\partial t} + \frac{\rho}{\theta}\frac{\partial S_{adm}^2}{\partial t} + \frac{1}{\theta}\frac{\partial M_n}{\partial t} = D\frac{\partial^2 C_c^m}{\partial x^2} - v\frac{\partial C_c^m}{\partial x} \tag{2.3}$$

$$\frac{\partial C_A^{aq}}{\partial t} + \frac{\rho}{\theta}\frac{\partial S_A^1}{\partial t} + \frac{\rho}{\theta}\frac{\partial S_A^2}{\partial t} = D\frac{\partial^2 C_A^{aq}}{\partial x^2} - v\frac{\partial C_A^{aq}}{\partial x} \tag{2.4}$$

where S^1 and S^2 are the instantaneous and rate-limited sorbed concentration of the (A) agent and (C) contaminant, respectively (M M^{-1}), S_{adm} is the enhanced sorption concentration (M M^{-1}), D is the dispersion coefficient (L^2 T^{-1}), and v is the interstitial velocity. The two-domain approach is used for modeling rate-limited versus instantaneous sorption. Depending on the sorption characteristics of the flushing agent, nonlinear sorption can be simulated by either a Freundlich or Langmuir isotherm.

Mass transfer rate coefficients, k_L, are calculated based on the power law correlation for the Sherwood number, S_h,(Powers et al. 1994b):

$$S_h = \beta_0 R_e^{\beta_1} \delta^{\beta_2} U_i^{\beta_3} \left(\frac{\theta_n}{\theta_{n_0}} \right)^{\beta_4} \tag{2.5}$$

where U_i is the uniformity index (d_{60}/d_{10}), and δ is a normalized grain size ($\delta = d_{50}/d_M$). The parameter d_M refers to "medium" sand (d_M = 500 µm), as defined by the United States Department of Agriculture (Powers et al., 1994b). In order to express the NAPL content of the system as a function of the porous medium, NAPL saturation, S_n, is combined with porosity, n, of the systems to yield the NAPL volumetric fraction, θ_n. The initial volumetric fraction of NAPL in the system is θ_{no}. The dimensionless Reynolds number, R_e, is defined as:

$$R_e = \frac{v \rho_w d_{50}}{\mu_w} \tag{2.6}$$

In Eqn. 2.5, the coefficients β_1, (= 0.598 ± 0.073), β_2 (= 0.673 (± 0.156), and β_3 = 0.369 (± 0.119) were determined by Powers et al. 1992 (for 0.012 < Re < 0.2). The mass transfer rate coefficients, k_L, is then calculated from equation 2.7:

$$k_L = \frac{S_h D_L}{d_{50}^2} \tag{2.7}$$

where D_L is the free-liquid diffusion coefficient (L^2T^{-1}) (for TCE: 8.30x10^{-6} cm^2s^{-1}). The model was used for this work to evaluate the mass-transfer rate coefficient for the dissolution of TCE by an anionic surfactant and a cyclodextrin complexing agent. Sorption was not simulated because the cyclodextrin agent does not sorb (Brusseau et al. 1994) and sorption of the surfactant was considered negligible given the very low organic carbon and clay content of the porous medium employed. The model was fitted to the experimental data using a nonlinear optimization routine to obtain values for β_0.

3 MATERIALS AND METHODS

Analytical-grade trichloroethene (>99% purity) and SDS (dodecyl sulfate, sodium salt; 70% purity) were purchased from Aldrich Chemical Co., Inc. Hydroxypropyl-b-cyclodextrin (HPCD; 90% purity) was supplied by Cerestar USA Inc. Aqueous solutions of 2.5 wt% SDS and 40 wt% HPCD, respectively, were prepared. At these concentrations, both agents were found to enhance the dissolution of TCE by approximately 10 times. The solutions were flushed through a glass column (5 cm length, 2.5 cm id) filled with a sand and TCE at residual saturation. The flushing experiments were conducted following the procedures described in Boving and Brusseau (2000). A TCE flushing experiment with water was conducted for comparison. Table 1 summarizes the experimental conditions.

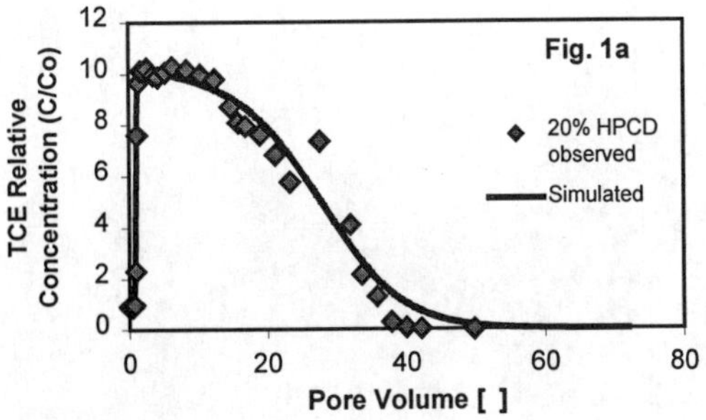

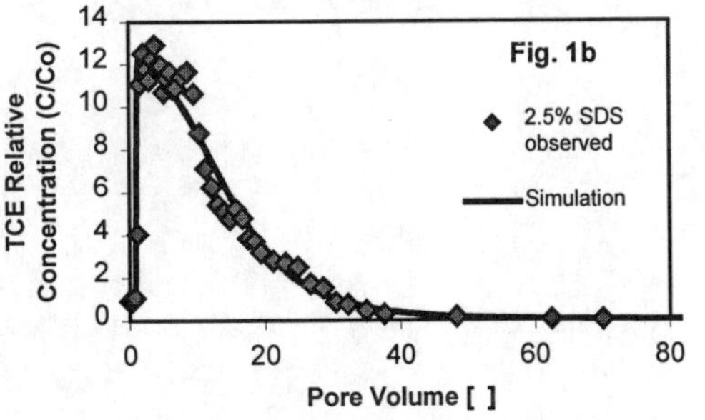

Figure 1. Results of chemically enhanced flushing of residual-phase TCE with (1a) 20% HPCD and (1b) 2.5% SDS solution.

4 RESULTS AND DISCUSSION

Figure 1 shows the results of the 2.5% SDS and 40% HPCD flushing experiment and of the model simulations. The results of the water/TCE flushing experiment are not shown for scale reasons.

Observed data and the results of the simulation are in good agreement. The corresponding Sherwood numbers and mass transfer rate coefficients are summarized in Table 2.

Table 1. Experimental Conditions

Property	**20% HPCD**	**2.5% SDS**	**Water**
Initial TCE Saturation (%) S_n (%)	19.9	14.0	11.2
Pore Volume (cm^3)	10.0	9.9	10
Average linear flow Velocity (cm/hr)	24.0	31.41	21.1
Partition Coefficient, K_{pt} (L g^{-1})	0.0465	0.541	0.0
Peclet number, P ()	68	92	68

Table 2. Results of optimized simulations

Property	**20% HPCD**	**2.5% SDS**	**Water**
Optimized β_0	111 (±25.7)	101 (±17.8)	11.5 (±4.7)
Sherwood number, S_h *()*	9.2	9.8	1.0
Enhanced driving force mass transfer rate coefficient k_L/E (1/hr)	21.1(±4.9)	20.5(±3.6)	23.6(±9.6)
Dimensionless initial mass transfer rate coefficient $(\omega_c)_0=k_L L/v$	44.3	36.2	5.6
Reynolds number, R_e ()	0.021	0.028	0.019

The calculated enhanced driving force mass transfer rate coefficients for SDS (20.5±4.9) and for HPCD (21.1±3.6) are similar to that of water (23.6±9.6).

This indicates that the mass transfer rate coefficient in the presence of SDS and HPCD is not significantly different compared to water flushing. The increased contaminant removal rate by chemical agents is mainly contributed by the enhanced contaminant solubilization, rather than by the change of mass transfer rate coefficient.

The dimensionless initial mass transfer rate coefficient $((\omega_c)_0)$ for SDS (36.2) and HPCD (44.3) was significantly large relative to the transport rate, indicating initial equilibrium dissolution conditions. With increasing mass of TCE removed and Corresponding decreasing interfacial area, the value of (ω_c) decreased to 0.3 (SDS) and 0.8 (HPCD) after 50 pore volumes. This decrease indicates that the TCE dissolution become rate limited at later times. Similar results were obtained by Ji and Brusseau (1998).

5 CONCLUSIONS

The results of this study demonstrate that the optimized transport model can be used to simulate experimental data obtained from flushing a TCE contaminated porous medium with an anionic surfactant and a cyclodextrin based complexing agent. The calculated mass transfer rate coefficients for the enhanced driving force model were comparable for both agents and similar compared to water flushing. Also, it was shown that the TCE dissolution becomes rate limited after initial equilibrium dissolution conditions. These observations are in good agreement with previous studies.

The calculated mass transfer rate coefficients provide the means to quantitatively evaluate the chemically enhanced remediation of organic contaminants by complexing agent or surfactants. This is beneficial because knowledge about the mass transfer rate coefficient allows for a more accurate estimation of cleanup times and therefore remediation costs.

REFERENCES

Abriola, L.M., T.J. Dekker & K.D. Pennell 1993. Surfactant enhanced solubilization of residual dodecane in soil columns 2.Mathematical modeling. *Envrion. Sci. Technol.*, 27(12): 2341-2351.

Boving, T.B., Wang, X., M.L. Brusseau 1999: Cyclodextrin-enhanced solubilization and removal of residual chlorinated solvents from porous media. *Environ. Sci. Technol.*, 33(5): 764-770

Boving,T.B. & M.L. Brusseau 2000. Solubilization and removal of residual trichloroethene from porous media: comparison of several solubilization agents. *Journal of Contaminant Hydrology,* 42(1): 51-67.

Brusseau, M.L, Wang, X. & Q. Hu 1994. Enhanced transport of low-polarity organic compounds through soil by cyclodextrin. *Environ.Sci.Technol,* 28: 952-956.

Geller, J.T. & J.R. Hunt 1993. Mass transfer from nonaqueous phase organic liquids in water-saturated porous media. *Water Resour.Res.*, 29(4): 833-845.

Hunt, J.T., N. Sitar, & K.S. and K.S. Udell 1988. Nonaqueous -phase liquid transport and cleanup. *Water Resour.Res.*, 24(8): 1247-1258.

Illangasekare, T.H., J.L. Ramsey, K.H. Jensen & M.B. Butts 1995. Experimental study of movement and distribution of dense organic contaminants in heterogenous aquifer, *J.Contam.Hydrol.*, 20(1-2): 1-25.

Imhoff, P.T., P.R. Jaffe, P.R. & G.F. Pinder 1993. An experimental study of complete dissolution of a nonaqueous phase liquid in saturated porous media. *Water Resour.Res.*, 30: 307-320.

Ji, W. & M.L. Brusseau 1998. A general mathematical model for chemical-enhanced flushing of soil contaminated by organic compounds. *Water Res.Research*, 34(7): 1635-1648.

Miller, C.T., M.M. Poirier-McNeil., & S.A. Mayer, S.A. 1990. Dissolution of trapped nonaqueous phase liquids: Mass transfer characteristics. *Water Resour.Res.*, 26(11): 2783-2796.

Powers, S.E., C.O. Loureiro, S.M. Abriola, & W.J. Weber 1991. Theoretical study of the significance of nonequilibrium dissolution on nonaqueous phase liquids in subsurface systems. *Water Resour. Res.*, 27(4): 463-467.

Powers, S.E., Loureiro, C.O., Abriola, L.M., & W.J. Weber 1992. Theoretical study of the significance of nonequilibrium dissolution of nonaqueous phase liquids in subsurface systems. *Water Resour.Res.*, 27: 463-477.

Powers, S.E., L.M. Abriola, & J.S. Dunkin 1994a. Phenomenological models for transient NAPL-water mass-transfer processes. *J. Contam. Hydrol.*, 16: 1-33.

Powers, S.E., L.M. Abriola, & W.J. Weber 1994b. An experimental investigation of nonaqueous phase liquid dissolution in saturated subsurface systems: Transient mass transfer rates. *Water Resour.Res.*, 30(2): 321-332.

Groundwater: Past Achievements and Future Challenges, Sililo et al. (eds) © 2000 Balkema, Rotterdam, ISBN 90 5809 159 7

Field relationships and models of arsenic in aquifers of southern Bangladesh

W.G. Burgess, M. Burren, M.O. Cuthbert, S.E. Mather, J. Perrin & M.K. Hasan – *University College London, UK*

K.M. Ahmed – *University of Dhaka, Bangladesh*

P. Ravenscroft – *Mott MacDonald, Dhaka, Bangladesh*

M. Rahman – *Bangladesh Water Development Board, Bangladesh*

ABSTRACT: Arsenic pollutes groundwater throughout southern Bangladesh, yet arsenic concentrations are highly variable spatially and with depth. Detailed patterns and profiles of arsenic in groundwater and aquifer sediments have been established at two locations. Arsenic is concentrated at discrete depths in the sediments. Spatial patterns of arsenic in groundwater are consistent with a sedimentological control. Hydrochemical relationships support the hypothesis of arsenic release to groundwater by reductive dissolution of iron oxyhydroxides, but favourable reducing conditions are pervasive and do not control arsenic concentrations in detail. Groundwater flow exerts a strong influence, and time is vital factor. Highest arsenic concentrations, and the greatest range and spatial variability, are found in the discharge of shallow tube-wells. Arsenic concentrations are lower and more uniform in groundwater from deeper tube-wells. Numerical models of arsenic in the discharge of shallow hand-pumped tube-wells raise important issues concerning the sustainability of safe water supplies from the alluvial aquifers.

1 INTRODUCTION

The scale and extent of the groundwater arsenic crisis in Bangladesh have been defined through a recent series of regional and local surveys (summarized by the British Geological Survey & Mott MacDonald 1999). The arsenic contamination occurs across all of southern Bangladesh. Approximately 50% of wells pump groundwater with arsenic above the WHO Guideline Value of 0.01 mg/l, and 35% of wells are above the Bangladesh Standard of 0.05 mg/l (BGS & MM 1999). In certain parts of the south-central and eastern districts more than 90% of wells shallower than 150 m have arsenic above 0.05 mg/l. At many tube-wells the arsenic concentration exceeds 0.3 mg/l and there are a number of instances where arsenic concentration exceeds 1.0 mg/l.

The arsenic is associated with fine-grained iron oxyhydroxides in the widespread Holocene alluvial sediments of the Ganges/Meghna/Brahmaputra floodplain (Perrin 1998). These sediments contain aquifers which have been developed as sources of water for domestic and agricultural use since the early 1980s. They currently provide drinking water for 95% of the population of the region, and are also widely used for irrigation. Chronic arsenicosis had become recognised extensively in southern Bangladesh by 1995, and more than 20 million people are now thought to be at risk of arsenic poisoning on account of drinking contaminated water pumped from the Holocene aquifers. The older, Pleistocene, aquifers of the Madhupur and Barind Tracts of central and western Bangladesh are free of arsenic; the residents of Dhaka, who are supplied from the Dupi Tila aquifer of the Madhupur Tract, are therefore not affected (Hasan et al. 1999).

Studies of the hydrochemical context of arsenic occurrence (Nickson et al. 1998) have suggested that arsenic is released from the sedimentary iron oxyhydroxides under chemically reducing conditions, in the absence of oxygen and the presence of organic matter. These conditions are prevalent in the Holocene aquifer, yet arsenic concentration in groundwater is highly variable both spatially and with depth. There are many examples of grossly contaminated tube-wells pumping groundwater with arsenic at concentrations close to 1.0 mg/l, separated by only a few tens of metres from tube-wells pumping groundwater with arsenic at concentrations less than 0.01 mg/l (eg. Safiullah 1997). Knowledge of the detail of this fine-scale variability is a pre-requisite for understanding the processes controlling the distribution of arsenic in the groundwater, and hence for developing approaches to mitigation.

Groundwater provides a bacteriologically safe source of water for drinking and irrigation for many millions of people in Bangladesh. If it is not to be abandoned, the processes governing arsenic occurrence must be established at the scale and point of use so that action can be taken to mitigate the impact. Key questions are therefore: How is the arsenic source distributed within the alluvial sediments? What is the pattern of its movement to tube-wells? How can tube-well placement, design and pumping regime be managed to minimise the arsenic concentrations in tube-well discharges?

In seeking answers to these questions, we review here recent results and implications of the London-Dhaka Arsenic in Groundwater Programme (www.ucl.ac.uk/GeolSci/people/burgess/) from two project areas (Fig.1): detailed studies of the spatial and depth distribution of arsenic at Meherpur in western Bangladesh (Burren 1998; Perrin 1998) and Chaumohani in eastern Bangladesh (Mather 1999), and development of numerical models of arsenic concentration at representative hand-pumped tube-wells (Cuthbert 1999) which reproduce the essential features of the observed variability to a depth of 60 m.

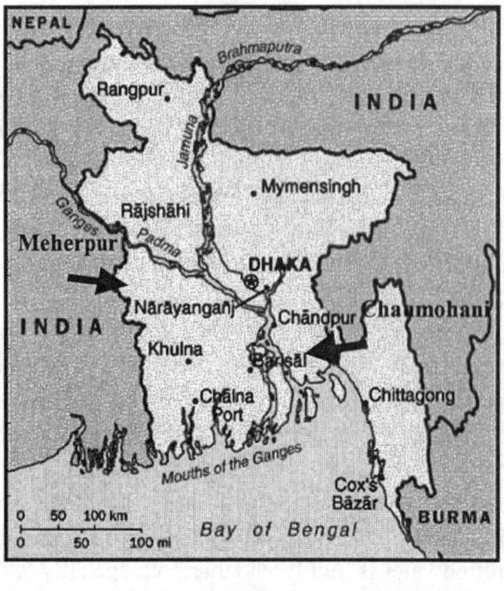

Figure 1. Location of study areas

2 THE SPATIAL DISTRIBUTION OF ARSENIC IN GROUNDWATER

In the most regionally-extensive and quantitative survey of arsenic concentration in Bangladesh groundwater to date (BGS & MM 1999), 2022 tube-wells were sampled at a spatial frequency of one tube-well per 37 km^2. The results demonstrate the extent of groundwater contamination by arsenic across southern Bangladesh, but it should be noted that 65% of the tube-wells sampled had arsenic concentration below 0.05 mg/l and 49% were below 0.01 mg/l. The variability was comparable in three "special study areas", each approximately 400 km^2 in area, in west, south-central and south-eastern Bangladesh, sampled at a spatial frequency of one tube-well per 7 km^2 (BGS & MM 1999).

Descriptions of the spatial distribution of arsenic in groundwater, at a scale two orders of magnitude more detailed than in the "special study areas" of BGS & MM, are illustrated over 15 km^2 at Meherpur (Fig.2) and over 10 km^2 at Chaumohani (Fig.3). In both cases the spatial frequency of sampling was approximately 10 tube-wells per km^2 (Burren 1998; Mather 1999).

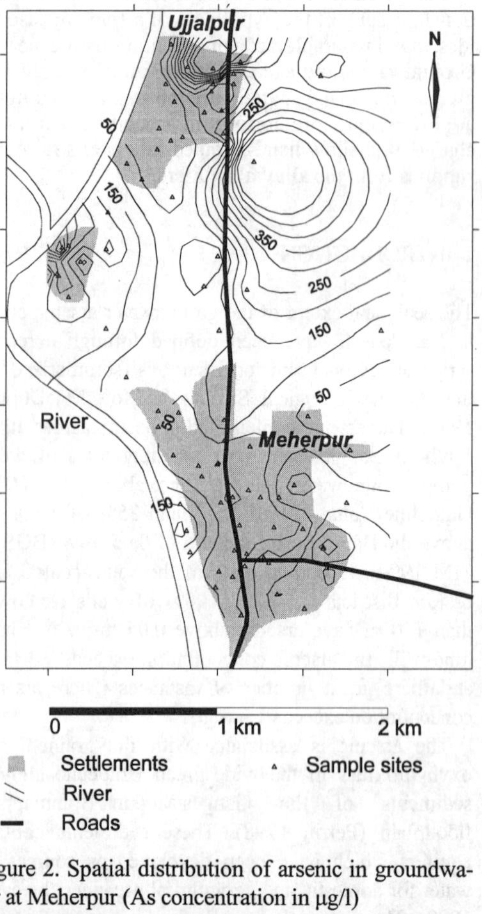

Figure 2. Spatial distribution of arsenic in groundwater at Meherpur (As concentration in μg/l)

At Meherpur, 55% of the tube-wells sampled exceeded the 0.05 mg/l national standard for arsenic. However, this detailed survey reveals a linear belt 500 m wide, linked to the present river system, within which arsenic concentration is less than 0.05 mg/l, suggesting a sedimentological control on the distribution of the arsenic source. Within this "low arsenic belt" the average arsenic concentration is 0.015 mg/l; beyond it the average is 0.198 mg/l (maximum 0.890 mg/l). The spatial variability of arsenic concentration is considerable, 0 to 0.25 mg/l over just 105 m on the margin of the low-arsenic belt. A sedimentological control is confirmed by the inverse correlation between arsenic and grain size in the sediments, and strong correlation between arsenic, iron and aluminium (Perrin 1998). Arsenic is concentrated in the fine-grained, clay-rich abandoned meander and over-bank deposits, relative to the coarser sediments of the braided river channels. The "low arsenic belt" may arise from the existence of an abandoned braided-river channel at shallow depth, choked with relatively arsenic-free coarser sediment.

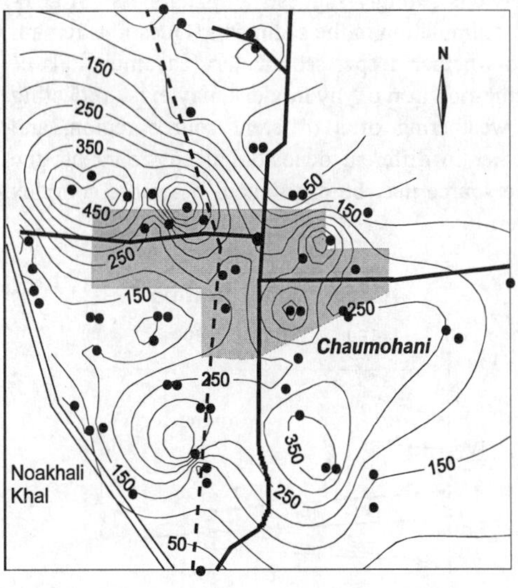

Figure 3. Spatial distribution of arsenic in groundwater at Chaumohani (As concentration in µg/l)

At Chaumohani in the south-east of Bangladesh, high arsenic concentration in the groundwater is more extensive and the spatial variability is less marked. Arsenic occurrence is restricted to the upper 30 m of sediments; 87% of the tube-wells shallower than 30 m exceed 0.05 mg/l. Arsenic concentrations above 0.25 mg/l occur in an arcuate zone up to 1000 m wide which traverses the town. The arcuate pattern of elevated arsenic may reflect the existence of a back-filled meander at this location, where the depositional environment is of lower energy than the high-energy braided-river environment of further north and west.

The sedimentological associations at Chaumohani are less distinct than at Meherpur, but these tentative interpretations demonstrate the possibility of linking spatial patterns of arsenic concentration in groundwater to sedimentological features that may have geomorphological manifestation, and hence the importance of detailed spatial surveys.

3 THE DEPTH DISTRIBUTION OF ARSENIC IN THE AQUIFER

Determination of the depth distribution of arsenic is fundamental to a strategic response to the crisis. Early presumptions that arsenic entered the groundwater on oxidation of detrital arseno-pyrite, following water level decline as a consequence of excessive pumping, had the implication that arsenic concentration in groundwater should decrease with depth. When the mechanism of arsenic release to groundwater was shown to be the reduction of arsenic-rich iron oxyhydroxides (Nickson et al. 1998), an important corollary was that arsenic concentration might increase with depth as reducing conditions became more pronounced. Limited investigations at three locations in south-central Bangladesh did suggest an increase of arsenic concentration in groundwater with depth (Nickson 1997), but this was an error based on too limited a set of data in a system that has subsequently been shown to have considerable variability. The depth distribution of arsenic in groundwater should be determined by the pattern of accumulation of the arsenic-rich oxyhydroxides in the aquifer sediments, which remains largely unknown, modified by groundwater flow.

The depth distribution of arsenic in groundwater at Chaumohani is illustrated in Fig.4, from a survey of 65 tube-wells of 9 to 340 m depth (Mather 1999), and that for Meherpur is illustrated in Fig.5, from a survey of 80 tube-wells of 15 to 220 m depth (Burren 1998). The related depth distribution of arsenic in porewater and sediments at Meherpur, determined from core material collected at 3 m intervals from a single 60 m deep borehole (Perrin 1998) are illustrated in Fig.6.

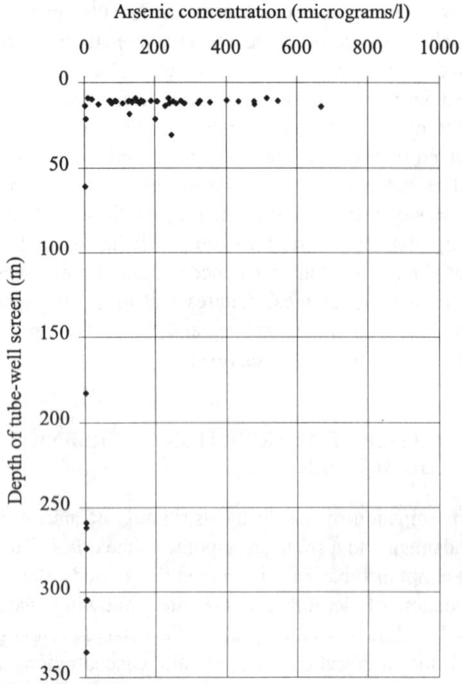

Figure 4. The depth distribution of arsenic in groundwater at Chaumohani

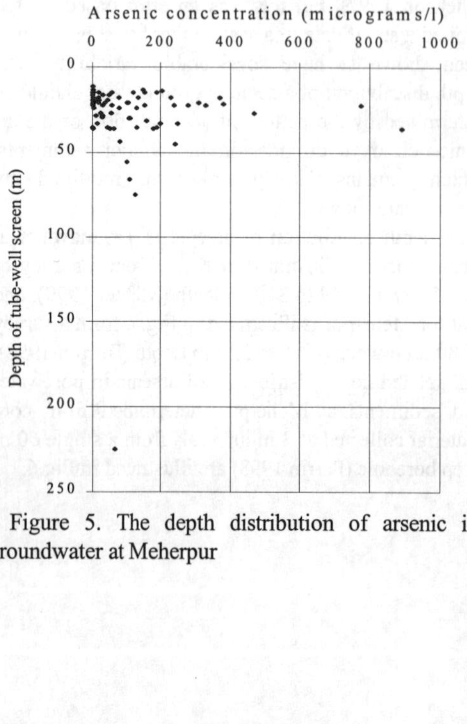

Figure 5. The depth distribution of arsenic in groundwater at Meherpur

These depth profiles demonstrate that:

1) the highest arsenic concentrations are in groundwater pumped from relatively shallow depths (15 to 45 m at Meherpur; 9 to 30 m at Chaumohani), which also exhibit the greatest range of arsenic concentration (0 to 0.89 mg/l at Meherpur; 0 to 0.67 mg/l at Chaumohani).

2) groundwater pumped from greater depths has a considerably lower and more limited range of arsenic concentration (*ca.* 0.1 mg/l at Meherpur; < 0.005 mg/l at Chaumohani).

3) the highest porewater concentrations at Meherpur, over a restricted depth interval 18 to 21 m, are related to a marked but modest accumulation of arsenic in the alluvial sediments.

The Meherpur and Chaumohani observations suggest that the sedimentary source of arsenic is unimodal, occurring over a single, restricted depth interval in both cases. Furthermore, the spatial variability of arsenic concentration in the discharge of shallow tube-wells strongly suggests a "patchiness" of arsenic accumulation in the sediments. Questions remain as to whether deeper sedimentary accumulations of arsenic-rich iron oxyhydroxides may exist, reflecting the weathering of a different source region, and whether in different depositional environments the same source may be distributed in a more complex pattern.

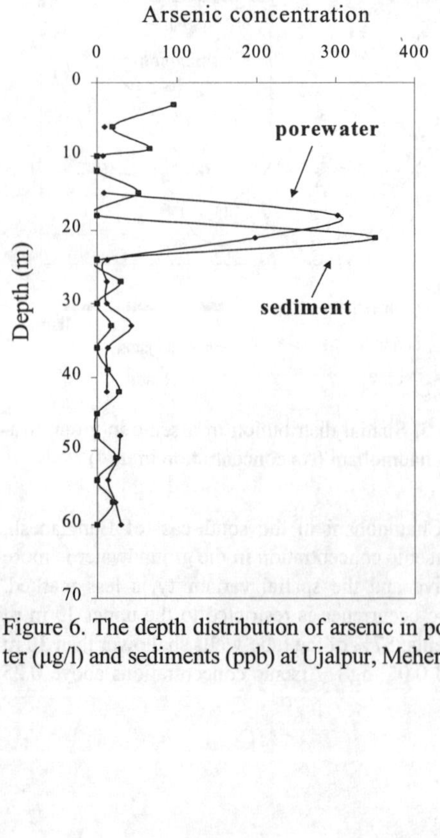

Figure 6. The depth distribution of arsenic in porewater (μg/l) and sediments (ppb) at Ujalpur, Meherpur

4 CONCEPTUAL AND NUMERICAL MODELS OF ARSENIC IN PUMPED GROUNDWATER

We propose that the spatial variability and depth distribution of arsenic in the groundwater reflect the patterns of accumulation of arsenic-rich iron oxyhydroxides in the alluvial sediments, in relation to the pattern of groundwater flow to individual tube-wells whose catchments vary in shape and size according to their yield, depth and the hydraulic characteristics of the sediments (Burgess et al. 2000).

Manually-pumped tube-wells used for individual domestic supplies draw water from a radius of only a few metres in this aquifer, which is renowned for being hydraulically "leaky". The groundwater has an arsenic concentration governed by the degree to which the tube-well catchment coincides with a sedimentary accumulation of arsenic. Complete over-lap results in the maximum arsenic concentration in the tube-well discharge, and lack of over-lap may result in an arsenic-free catchment and the absence of arsenic in the tube-well discharge. Thus adjacent tube-wells of comparable depth may pump groundwater with markedly different arsenic content. Deeper tube-wells, generally pumping at higher rates for irrigation or for distributed public water supply, have larger catchments, typically 100 m or more in radius, and are consequently more likely to incorporate a region of arsenic accumulation in the sediments but with much greater dilution. This results in a lower, more uniform, arsenic concentration in the groundwater pumped.

Numerical models have been developed to test the validity of this conceptual model (Cuthbert 1999; Cuthbert et al. 2000). Groundwater flow and arsenic transport to manually-pumped tube-wells in a variety of settings in relation to the arsenic source have been simulated, using the codes MODFLOW and MT3D with reference to hydraulic and geometrical parameters and boundary conditions described for Meherpur.

The range of arsenic concentrations observed spatially at Meherpur is contained within the range of model results obtained for tube-wells with a variety of depths and positions in relation to the arsenic source (Cuthbert 1999). The pattern of the observed depth distribution of arsenic concentration in groundwater at Meherpur has also been reproduced, and the issue of time-dependency of arsenic concentration in shallow tube-well discharges has been explored (Fig.7).

5 HOW MAY GROUNDWATER ARSENIC CONCENTRATION CHANGE WITH TIME?

Recognition of the occurrence of arsenic in Bangladesh groundwater is too recent for changes in arsenic concentration to have been monitored. Proxy indications, considering arsenic concentration for a range of tube-well ages at Meherpur, suggest a trend of increasing arsenic concentration with time (Burgess et al. 2000), at least for up to 15 years after tube-well installation. Similar indications are apparent from the extensive regional database of BGS/MMI (1999). Apart from these proxy indications, the evolution of arsenic concentration with time at individual tube-wells is unknown. Modelling (Cuthbert 1999; also Fig.7) allows an evaluation of the likely trends.

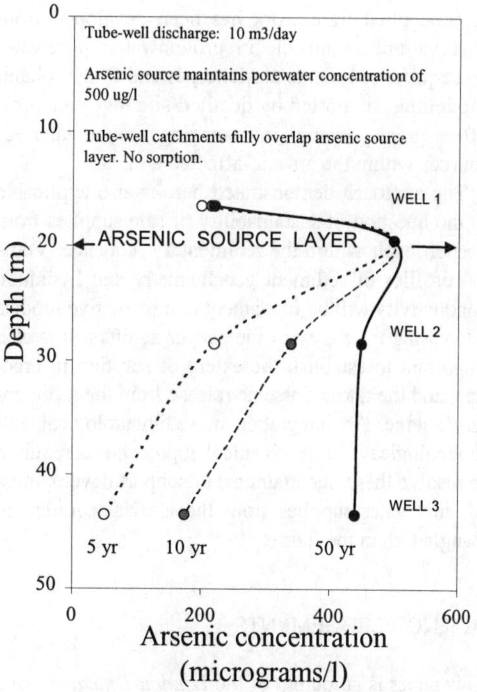

Figure 7. Modelled depth distribution of arsenic in shallow tube-well discharges, with time

Our results suggest that:

1) the observed arsenic distributions in groundwater represent a transient situation, particularly for those tube-wells further (vertically) from the sedimentary source of arsenic. Many tube-wells, though initially free of arsenic, will eventually become contaminated; it is important to establish regular monitoring.

2) the presence of deep, high-yielding tube-wells enhances the mobility of arsenic to manually-pumped tube-wells drawing water from below the sedimentary accumulation of arsenic.

3) manually-pumped tube-wells drawing water from more than 10 m below the level of arsenic accumulation in the sediments can expect to pump groundwater relatively free of arsenic for some years after installation. Thereafter the maximum arsenic concentration may not be reached for many years, even decades. The timing of arsenic breakthrough and build-up is sensitive to sorption, which is largely unquantified for the particular conditions of the alluvial aquifers of Bangladesh.

SUMMARY

A conceptual framework has been developed from observations of arsenic in groundwater, porewater and aquifer sediments at Meherpur and Chaumohani. Modelling, supported by detailed site investigations, offers great potential in managing groundwater resources within the arsenic-affected area.

The approach demonstrated here is also applicable to the question of sustainability of safe supplies from deeper levels within the sedimentary sequence. Vertical profiles of sediment geochemistry and hydraulic conductivity will be fundamental to predictive models addressing the safety of the deeper aquifers. It is also important to establish the extent of sorption in practice, and the rates of arsenic release from the sedimentary source. An integration of sedimentological, hydrogeological and geochemical approaches is required to resolve these uncertainties, to support development of safe water supplies from the alluvial aquifers of Bangladesh in the future.

ACKNOWLEDGEMENTS

This paper is an output of the *London-Dhaka Arsenic in Groundwater Programme*. Financial support from NERC and logistical support from DPHE, BWDB and Mott-MacDonald in Dhaka is gratefully acknowledged. Chemical analyses were carried out by the Robens Institute for Public & Environmental Health at Surrey University, at the Environmental Mineralogy lab. of the Natural History Museum in London, the NERC ICP-AES facility at Royal Holloway College, London, and at the Wolfson Geochemistry Lab. at UCL; grateful thanks to Andrew Taylor, Chris Stanley, Vic Din, Nikki Paige and Tony Osborn.

REFERENCES

British Geological Survey and Mott MacDonald Ltd. 1999. Groundwater Studies for Arsenic Contamination in Bangladesh. Phase 1: Rapid Investigation Phase. *UK DFID Report to Govt. of Bangladesh*. 5 volumes.

Burgess, W.G., M.Burren, M.O.Cuthbert, S.E.Mather, J.Perrin, K.M.U.Ahmed & L.Connell 2000. Is groundwater development from arsenic-contaminated aquifers of southern Bangladesh sustainable? Abstracts Volume, *Geoscience 2000*, Manchester, UK (April, 2000), p34.

Burren, M. 1998. Small scale variability of arsenic in groundwater in the district of Meherpur, western Bangladesh. MSc thesis, University College London (unpublished).

Cuthbert, M.O. 1999. Modelling the transport of arsenic to hand pumped tube-wells in the Holocene alluvial aquifer of Bangladesh. MSc thesis, University College London (unpublished).

Cuthbert, M.O., W.G.Burgess, L.Connell & K.M.U.Ahmed. 2000. Preliminary numerical models of arsenic variability in groundwater of southern Bangladesh. Abstracts Volume, *Geoscience 2000*, Manchester, UK (April, 2000). p37.

Hasan, M.K., W.G.Burgess & J.Dottridge. 1999. The vulnerability of the Dupi Tila aquifer of Dhaka, Bangladesh. In J.Bryan Ellis (ed.) *Impacts of Urban Growth on Surface Water and Groundwater Quality*. IAHS Publication No.259, pp.91-98.

Mather, S.E. 1999. The vertical and spatial variability of arsenic in the groundwater of Chaumohani, south-east Bangladesh. MSc thesis, University College London (unpublished).

Nickson, R. 1997. Origin and distribution of arsenic in groundwater, central Bangladesh. MSc thesis, University College London (unpublished).

Nickson, R., J.M.McArthur, W.G.Burgess, K.M.U.Ahmed, P.Ravenscroft & M.Rahman. 1998. Arsenic poisoning of groundwater in Bangladesh. *Nature*, 395, p.338.

Nickson, R.T., J.M.McArthur, P.Ravenscroft, W.G.Burgess & K.M.U.Ahmed. 2000. Mechanism of arsenic release to groundwater, Bangladesh and West Bengal. *Applied Geochemistry*, 15, 403-413.

Perrin, J. 1998. Arsenic in groundwater at Meherpur, Bangladesh: a vertical porewater profile and rock/water interactions. MSc thesis, University College London (unpublished).

Safiullah, S. 1997. Report on monitoring and mitigation of arsenic in the groundwater of Faridpur municipality. *CIDA Arsenic Project Report*. 89pp.

Groundwater: Past Achievements and Future Challenges, Sililo et al. (eds) © 2000 Balkema, Rotterdam, ISBN 90 5809 159 7

The closure of existing waste piles in terms of the South African Department of Water Affairs and Forestry Waste Management Series

I.S.Cameron-Clarke & J.A.Palmer
SRK Consulting, Johannesburg, South Africa

ABSTRACT: Chemical, mining and metal processing industries are frequently faced with the problem of how best to rehabilitate existing piles of potentially hazardous wastes in an environmentally responsible and appropriate manner without creating new problems and without incurring excessive costs. This paper presents an overview of the South African Department of Water Affairs and Forestry Management Series relating to the minimum requirements for waste disposal, and describes the retrospective design and flexibility required for developing closure plans for existing waste piles. Specific examples include small residue stockpiles from a heavy metal refining operation, a complex disposal area used historically for the disposal of slimes containing chlorinated organic chemicals and metals, and a large chemical complex containing a large number of piles of hazardous and non-hazardous wastes.

1 INTRODUCTION

The South African Department of Water Affairs and Forestry (DWAF) Waste Management Series relating to the minimum requirements for waste disposal provide guidelines for:

1. Handling, Classification and Disposal of Hazardous Wastes [1].
2. Waste Disposal by Landfill [2].
3. Monitoring of Water Quality at Waste Facilities

The guidelines, generally referred to as the Minimum Requirements, aim to present the lowest acceptable standards for the disposal and management of wastes, and although focusing mainly on proposed new facilities, recognise that existing waste piles may not be able to comply in the strict sense as set out, and that a Best Practicable Environmental Option (BPEO) approach may need to be adopted.

The main objectives of the Minimum Requirements are to:

1. Prevent water pollution and ensure sustained fitness for use of the country's water resources.
2. Attain and maintain minimum waste management standards.
3. Provide a systematic and nationally uniform approach to waste disposal.
4. Gain international acceptability.

In many instances, ground water contamination is one of the most significant environmental impacts resulting from existing waste piles, and is also often the most difficult to remedy. Managing, rehabilitating and closing existing waste piles in terms of the Minimum Requirements is often difficult and even if practically achievable may be extremely costly. Whilst, it is accepted that, Industry being the generators of the waste needs to act responsibly to mitigate against adverse environmental impacts, and the Department of Water Affairs and Forestry being the custodian of South Africa's water resources needs to adopt a strong stance in the application of the Minimum Requirements, in the rehabilitation and closure of existing waste piles, innovation and flexibility will be required if practical and cost effective solutions are to be found.

The Minimum Requirements also include certain procedural aspects regarding the involvement of interested and affected parties. These are recognised, but are not addressed in this paper, which attempts to focus more on technical and practical aspects to be dealt with in the rehabilitation and closure of existing waste piles in the spirit of the Minimum Requirements.

2 APPLICATION OF THE MINIMUM REQUIREMENTS

The Minimum Requirements identify procedures for permitting and closure of the following three waste site categories:

1. Operating Landfills (Non-permitted or Concept permitted).
2. New Landfills.
3. Closed Landfills (Non-permitted or Concept permitted).

Conceptual and technical designs provided in Section 8 of the Minimum Requirements for Waste Disposal by Landfill, are aimed at the construction of New Landfills, but are nevertheless indicated as being applicable to the rehabilitation and closure of existing waste sites which cannot meet the minimum requirements.

In practical terms, for the rehabilitation and closure of existing waste piles, the application of the Minimum Requirements is therefore retrospective design aimed at attempting to achieve, as far a practically and financially possible, the objectives of the Minimum Requirements, with similar or alternative technologies, viz. the option that provides the most benefit or least damage to the environment as a whole at acceptable cost in the long and short term (the BPEO approach).

The essence of the BPEO approach is the sustainability of the rehabilitation and management measures, and a responsibility to establish a system, which will not leave future generations with a bad legacy. It is suggested that, the Minimum Requirements, by implication, should make provision for technologies which could result in future improvements to the affected environment.

3 MINIMUM REQUIREMENTS FOR DESIGN

The design guidelines provided by the Minimum Requirements aim to:

1. Mitigate against any potential adverse environmental impacts identified in the site investigation and from the EIA.
2. Prevent leachate pollution of adjacent ground and surface water.
3. Provide an environmentally and aesthetically acceptable cover.

The recommended design procedure is a two-stage approach involving a conceptual design, followed by the detailed technical design. The conceptual design, which addresses the principals of the intended design without detailed specifications, is submitted as part of a permit application. In cases where a liner may be required, a technical specification for the liner will be required with the permit application.

For non-permitted or concept permitted existing sites, the permit application is again retrospective and the design has to be related to an assessment of the actual risks associated with adverse environmental impacts derived from the waste pile, and a practically achievable end-use for the site. From a ground water perspective it is a required that the local aquifer system should be classified in terms of its yield potential and significance [(2)]. In the case of existing waste piles where adverse impacts are usually "fait accompli", the design philosophy should be to reduce existing and future impacts to a minimum in a sustainable and cost effective manner.

3.1 Conceptual Design

The conceptual design for a new facility is required, as a prerequisite, to include information about the site classification, airspace, and the site layout, as well as a Preliminary Closure Plan (PCP) based on the end-use vision for the site. For existing waste sites the conceptual design will in reality represent a PCP based on the results of investigations aimed at classifying the nature of the waste in the pile, the identification of impacts and mechanisms of contaminant movement from the waste in the various media of concern, and an assessment of health and/or other environmental risks associated with the impacts.

It is proposed that the following main elements should be the minimum requirements for the development of the PCP (conceptual design) for existing waste piles, which require retrospective permitting:

1. Type and nature of the waste.
2. Size (volume and areal extent) of waste pile.
3. The location of the waste pile with respect to potential receptor environments.
4. A hydrogeological report defining the nature and extent of existing ground water contamination derived from the waste pile.
5. A soils report defining the physical and chemical characteristics of soils around the waste pile.
6. The quantification of existing adverse impacts, particularly on surface water bodies, ground water and soil and air in the vicinity of the waste pile.

7. A screening level risk assessment.
8. End-use vision.

In the case of hazardous wastes the PCP should make provision for containment of the waste. For new sites the implication is that the waste pile and any derived leachate must be completely separated from the surrounding formations by means of a liner and leachate collection system. For existing waste piles this may be a very difficult or impossible objective to achieve, and a containment strategy aimed at minimising future contamination and remediating the source and the areas external to the source may be more realistic and more sustainable in the longer term.

3.2 *Technical (Closure) Design*

The Technical Design is supposed to quantify all necessary aspects of the Conceptual Design and must be approved by the DWAF. For existing waste piles the technical design represents the Closure Plan. The design should adhere to the principals set out in the Minimum Requirements for a new facility and should include the following main elements:

1. Design of up-slope surface water interception and clean/dirty water separation systems.
2. Design of the system for separation of the waste pile and leachate from the ground water.
3. Design of a lining system to reduce leachate flow to a stipulated minimum.
4. Design of a leachate detection and leachate collection system.
5. Design of a leachate treatment system.
6. Design of a final cover or capping.
7. Design of a gas management system.
8. Design of a monitoring system.
9. Design of a maintenance system.

For existing waste piles some of the specified elements may not be feasible and alternative methods, based on agreed and substantiated objectives, may be required. From a ground water perspective, the separation of the leachate from the ground water and the placing of a physical liner beneath and existing waste pile without upliftment of the waste will be, in some instances, practically unachievable.

The sustainability of a Closure Design in the long term is also a prerequisite of the Minimum Requirements. The optimal approach is the removal of the waste, either at source (i.e. minimise waste generation, recycling) or, in the case of an existing waste pile, upliftment to an appropriately designed facility, or reuse. This may not necessarily be a practical option however, and gradual reduction of the hazardous nature of the source or the leachate (contaminated ground water) over time may be an acceptable alternative which could be included in the Closure Design.

The rehabilitation and closure of existing waste piles can be complex and costly. The effectiveness of selected options may not be absolutely predictable at design stage, and the Closure Design should therefore be based on an overall Conceptual Design consisting of individual elements with sufficient flexibility to permit modifications as rehabilitation progresses.

4 EXAMPLES

Four existing situations are used to illustrate the sorts of rehabilitation and closure designs being adopted or proposed for existing waste piles in order to meet the general objectives of the Minimum Requirements. These include:

1. Small stockpiles of copper/cobalt and cadmium residues.
2. A 20-hectare complex of slimes dams and sludge ponds containing a variety of chlorinated organic chemicals and heavy metals.
3. An 800-hectare chemical-manufacturing complex containing numerous waste piles.

4.1 *Small Stockpiles*

Small stockpiles (tonnes) of copper/cobalt and cadmium residues had historically been stored on unprotected surfaces without cover. As part of the environmental upgrading of the operation it was recognised that the stockpiles constituted piles of hazardous wastes (H: H) which in terms of the Minimum Requirements should be contained in fully lined and covered facilities.

The stockpiles were located within the area of the metal refinery plant and future land use would be restricted to plant extensions as appropriate. There remained a possibility that the residues may be processed in the future to extract some of the metal content. A hydrogeological investigation of the area indicated that notwithstanding some evidence of ground water contamination from the stockpiles, and over-riding source of contamination was a large tailings dam complex adjacent to the plant. Soil investigations indicated some soil contamination around the stockpiles. There was no risk to users of ground water or surface water from the stockpiles.

There were no adverse windborne impacts.

The Closure Plan for the stockpiles in terms of the Minimum Requirements included the following main elements:

1. Construction of concrete bunkers with leach detection and leachate collection facilities adjacent to the original stockpiles.
2. Upliftment and placement of the residues in the bunkers.
3. Covering of the residues with HDP liners to prevent rain water infiltration.
4. Monitoring of the ground water immediately downgrade of the area.
5. Shaping and capping of the footprint areas.
6. Run-off control from and onto the area.
7. Sampling and rehabilitation of the footprint areas as development plans for the area evolved.
8. Provision for earlier rehabilitation if monitoring indicated an unacceptable environmental risk.

4.2 Slimes Dam Complex

Chlorinated organic chemicals and other wastes had historically been deposited in slimes dams and sludge ponds located on the most elevated areas of a large chemical-manufacturing complex. Wastes contained in drums from other sites had also been imported to the area. General characterisation studies indicated off-site ground water contamination, which could potentially place residents of adjacent areas at risk. Further investigations revealed that the off-site contamination was largely being driven by the hydraulic head created by the slimes dams. Studies on the dams themselves and ground water monitoring showed that degradation of the chlorinated organics is occurring and that some form of remediation appears to be taking place in conjunction with the development of a natural plant habitat on the surfaces of parts of the area.

After identification of the problem immediate intervention measures taken included:

1. The provision of an alternative water supply to affected users.
2. The establishment of hydraulic control between the source and the affected area off-site.
3. A health risk assessment in the affected area.
4. Communication to the authorities, the public and affected parties.

Following the decision to de-commission the area the area was declared as a hazardous waste site and permitting and rehabilitation are required. The end-use vision is for a natural environment with restricted access. From a practical perspective the letter of the Minimum Requirements cannot be met and a Closure Plan containing the following elements has been proposed:

1. Hydraulic containment using a system of de-watering wells located within and around the slimes dams.
2. Head reduction by induced de-watering with the wells within the dams.
3. Storm water diversion around the area.
4. Conventional capping of selected lower lying areas known to contain buried wastes and heavy metals.
5. Vegetative capping of the slimes dams.
6. Run-off control with minimal dirty water generation.
7. Treatment of leachate and dirty water run-off.
8. Monitoring and provision for revision of the plan as appropriate.
9. Natural or enhanced bio-remediation of the ground water.

The proposed closure plan is aimed at meeting the overall objectives of the Minimum Requirements in terms of sustainability and looking to the legacy to be retained by future generations. The details of closure will only emerge as rehabilitation progresses so flexibility in the design is essential.

4.3 Industrial Complex

The area of concern in this instance is approximately 800 hectares in extent and contains upwards of 40 existing piles of wastes varying from builder's rubble to hazardous materials. The general philosophy being proposed in this instance is to address each waste pile in terms of specific adverse impacts relative to other contamination on the site. The actual rehabilitation and closure of each waste pile will aim to address the Minimum Requirements but the emphasis will be placed on those piles, which pose the greatest environmental risk.

5 CONCLUSIONS

The paper has attempted to illustrate that the rehabilitation and closure of existing waste piles in the context of the DWAF Minimum Requirements is a retrospective design procedure which needs to provide for sufficient flexibility to include modifications to the design as the effectiveness of rehabilitation measures are assessed. The regulatory authorities need to appreciate these aspects and also be open to alternative methods which may in the longer term prove to be more sustainable and environmentally more acceptable than those prescribed in the Minimum Requirements.

6 REFERENCES

1) Department of Water Affairs and Forestry. Minimum Requirements for the Handling, Classification and Disposal of Hazardous Waste. 2nd Edition. 1998.

2) Department of Water Affairs and Forestry. Minimum Requirements for Waste Disposal by Landfill. 2nd Edition. 1998.

Groundwater: Past Achievements and Future Challenges, Sililo et al. (eds) © 2000 Balkema, Rotterdam, ISBN 90 5809 159 7

Effect of rapidly urbanizing environment on the Thiaroye, Senegal, unconfined sandy aquifer

S.Cissé, S.Faye, C.B.Gaye & A.Faye
Geology Department, University C.A.Diop, Dakar, Senegal

S.Wohnlich
Institute of General and Applied Geology, University of Munich, Germany

ABSTRACT: Nitrate is the most widespread contaminant in the Thiaroye groundwater system. The elevated concentration of nitrate (up to 540 mg/l NO_3) detected in parts of the aquifer system is perfectly correlated with the densely populated settlements in the Dakar suburb area. The study was carried out using a geographical information system (ARC/INFO). Thematic maps based both on surface features and reservoir characteristics are combined to identify in the aquifer domain various and gradual vulnerability zones.

1 INTRODUCTION

Dakar, the capital city of Senegal, has grown dramatically in terms of population (2 millions people) during the last three decades. This rapid demographic growth has caused an anarchic settlement in the city suburb with improper urban sanitation system. This latter is believed to be the main pollution source of the local groundwater system. Part of the region drinking water supply is provided by the Thiaroye aquifer which participates up to 12,000 m^3/d (MH/PNUD, 1994) to the modern network distribution. However, the aquifer system is facing an excessive quality deterioration because of the nitrate contamination in the vicinity of the densely populated suburb areas. The elevated nitrate concentrations (up to 540mg/l NO_3) (Collins & Salem, 1989; Tandia, 1997; Cissé, 2000) measured in the pumping boreholes could jeopardized the future of these groundwater resources. The emphasis of this paper is on the nitrate contamination of the aquifer and its link to human settlement in the region. The objective is to evaluate the anthropogenic inputs to this aquifer in relation to the urban development and land uses, and to assess the vulnerability with regard to nitrogen contamination.

2 STUDY AREA

The Thiaroye unconfined aquifer belongs hydrogeologically to the Senegalese superficial aquifer system. It is located along the Atlantic Ocean facade and encompasses about 300 km^2 from Dakar to Kayar in the northeast of Senegal (Fig.1). The aquifer reservoir is composed of an unconsolidated quaternary sands. The sediments consist largely of fine and medium grained sands with porosity value of about 20% and hydraulic conductivity values between 0.06 to $11x10^{-4}$ m/s (OMS, 1972). The reservoir overlies a tertiary marly basement, the elevation of which varying from +10 m in the south-eastern part (where it outcrops) to -95 m northwards at Kayar. The aquifer thickness varies correlatively in the same direction from 5 to 70 m. The ground surface relief varies between 2 and 35 m above mean sea level (Hébrard, 1966).

3 METHODOLOGY

A recently developed methodology by Klima (1996) has been used for this study. The essence of this methodology is similar to the GIS technology; it consists of compiling basic maps (land use, soil type, nitrate distribution, water table depth, piezometric maps). The set of maps, laid one over another is utilized to define vulnerability areas within the aquifer system. In developing basic maps, we re-evaluate surface and reservoir features according to their degree of impact on the water quality. We have used for this reason two terms: potential polluting zones for land use features and vulnerable zones for reservoir features.

The land use map was established using visual interpretation of a SPOT (XS+P) scene (December 23th, 1995). The map shows various surface features which are grouped according to the study objective into potential highly, moderately and lower polluting zones:

- A suburb area characterized by a densely populated (4,000 inhabitants/km^2) agglomeration. The

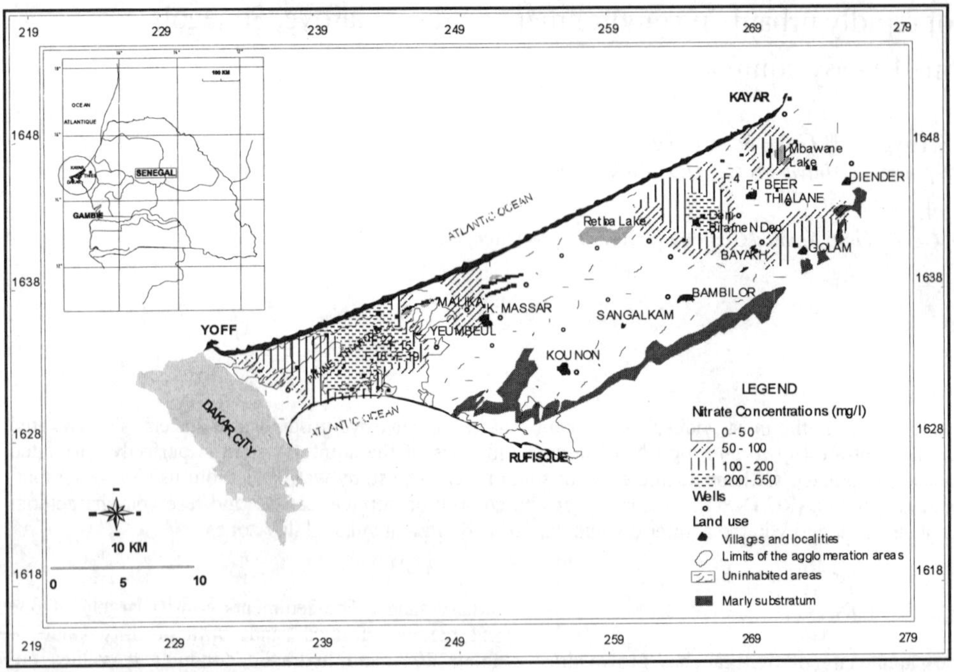

Figure 1. Location of the Cap-Vert peninsula in the western part of Senegal and nitrate concentration distribution correlated with land use

population is estimated today at more than 1 million inhabitants. This area includes settlements in Thiaroye, Pikine, Malika, Yeumbeul, Keur Massar localities. These localities are known by their inexistant sanitation system network. Most of the houses are built with an improper septic tank which allows seepage from its bottom to the water table. These areas are classified as a potential highly polluting zones.

- Rural habitations in villages and small localities from Keur Massar to Diender are also present. They are characterized by a less important anthropogenic pressure on the water reservoir because of their relatively low population density (600 inhabitants per km^2). These areas are classified as a potential moderately polluting zones.
- In addition, the land use map individualises uninhabited zones in the sand dunes, agricultural areas and the "Niayes" (local term used to designate coastal depression zones with superficial groundwater levels and where agricultural practices are developed), and also a street network. These areas where practically no impact is detected are designated as potential lowly polluting zones.

Soil typology was mapped according to an early study (USAID/RSI, 1986). Three type of soils are identified:

- The tropical ferruginous soils located in the suburb areas and in the north-eastern part of the study area including Mbayakh, Beer Thialane, Golam and Diender localities. Clay contents in this type of soil are less than 3%.
  The coastal zone from Yoff to Kayar is characterized by fine grained sand.
  The homogeneous soil texture and fine grains size explain the relatively fast infiltration rate in these areas and thus aquifer in these localities is defined as a highly vulnerable zone.
- The soils of the "Niayes" zone are riche in organic matters and the clay contents is estimated at 12%. The recent soil formations in Keur Massar, Sangalkam, Bambilor present more clay contents (20%).
  Soil characters in these areas define a moderately vulnerable zone.
- The southern-eastern part of the aquifer system where the tertiary marly basement of the reservoir outcrops is covered by a clayey sand with clay contents up to 30%. This area is classified as a lowly vulnerability zone.

The hydraulic conductivity distribution in the aquifer domain has been obtained from a final calibration data of a hydrodynamic model (Cissé, 2000) using PM3 (W. Chiang & W. Kinzelbach, 1993) and Modflow code (McDonald & Harbaugh, 1988). High Values ranging between 5 to $10x10^{-4}$ m/s are located in the Thiaroye pumping area, the coastal zone from Malika to Retba lake and in the north-eastern part of the study area. The water budget obtained after calibration of the model shows an important rainwater infiltration (36% of the total recharge) in the suburb which is classified as a highly vulnerable zones.

Area from Keur Massar to Diender is characterized by hydraulic conductivity ranges from 1 to $6x10^{-4}$ m/s, it corresponds to a moderately vulnerable zones.

In the south-eastern part of the system from Kounon to Bambilor, the results of the calibration shows low hydraulic conductivity values, ranging from $0.11x10^{-5}$ to $1x10^{-4}$ m/s, which defined a lowly vulnerable zones.

The water table depth distribution (Cissé, 2000) shows that the phreatic aquifer is shallow (less than 6 m) which define high vulnerable zones. The depth between 8 and 12 m characterized a moderately vulnerable zones. Maximum depths (up to 16 m) located in the Beer-Thialane and Diender areas define a lowly vulnerable zones.

The piezometric distribution shows flows from south-eastern to northern part of the aquifer with a low hydraulic gradient (0.06%).

Chemical data of the aquifer and spatial distribution of the nitrate contamination were obtained through field work during June-July and October 1997. In-situ measurements (pH, conductivity, dissolved oxygen and alkalinity) were made after pumping the monitoring wells. Samples analysis for major ions were performed in the laboratory of the Institute of General and Applied Geology in Munich. Major ions (Cl^-, Na^+, Ca^{2+}, Mg^{2+}, K^+, SO_4^{2-}, NO_3^-) were measured using ion chromatography. Nitrite and Ammonium were measured using the Hach Spectrophotometer DR/2000.
Bacteriological analysis concerning the faecal coliforms were carried out (ESP/Dakar) in few samples.

A Geographic Information System (PC ARC/INFO version 3.4D ESRI, 1990) was used to analyze and store the data listed above for mapping the vulnerability with regard to nitrogen contamination on the Thiaroye aquifer system.

4 RESULTS

4.1 *Influence of land use*

The inhabited districts as mentioned above are grouped into potential highly and moderately polluting zones. The agglomeration area (Fig.1) classified as a potential highly polluting zone is characterised by houses with septic tanks which allow seepage to the aquifer system. The second zone is the rural habitations in villages and small localities. The relatively low population density in this zone comparing to the suburb areas is likely to induce less pollution from septic tanks.
The uninhabited zones which are designated as a potential lower polluting zone include sand dunes, agricultural areas and the "Niayes" (Fig.1). In these areas, a lower impact is detected on the groundwater quality.

The Mbeubeuss landfill which is used for the city solid waste disposal system is located in the NE of Dakar. The landfill seems to be another source of pollution for the aquifer. Unfortunately, there is no observation wells in the vicinity of the landfill to monitor the impact.

4.2 *Correlation between groundwater quality and land occupation*

Statistical analysis of the hydrochemical data (Table 1) grouped according the land use / occupation show high nitrates concentrations (arithmetic mean 310 mg l^{-1}) in the agglomeration zone, which decrease by 1/3 in rural areas. In uninhabited zones, nitrates values are very low (arithmetic mean less than 20 mg l^{-1}) and groundwater quality in these areas is found to meet the drinking water standard (less than the WHO standard of 50 mg l^{-1}). Sulfates and chlorides are not correlated with nitrates, however high values of sulfates are observed in B and C, and high values of chlorides in A, B and C.
Specific correlation observed between the human settlement with poor sanitation infrastructures and the nitrates concentrations (Fig.1) indicates an anthropogenic input by seepage from septic tanks to the groundwater.
The bacteriological analysis shows in the suburb values of faecal coliforms between 19 to 160 (UFC/ml).
Statistical distribution of the nitrate concentrations, the pH values and the dissolved oxygen (Fig.2) in relation to land occupation give evidence of aerobic conditions in the agglomeration (Suburb) zones. In fact, in this area, dissolved oxygen values are high because of the shallow water table depth (less than 6m). Combination of the low pH values and the open system nature of the groundwater should explain the absence of bacterial reduction process in the area.

Table 1. Some variables of groundwater chemistry in Thiaroye (arithmetic means, mgl^{-1}).

Land use	Cl^-	SO_4^{2-}	NO_3^-	Ca^{2+}	Na^+	NH_4^+	n
Groundwater chemistry in inhabited zones							
A	182	45	310.5	68	125	1.6	10
C	196	235	105.1	98	117.6	0.53	10
Groundwater chemistry in uninhabited zones							
B	143	226	15.1	44.4	112	0.30	6
D	34.2	8.02	6.51	32	25	0.24	9
E	74.3	96.4	19	47	47	0.43	15

A – Suburb C - Villages B - Sand dunes D - Niayes
B - Sand dune E - Agricultural areas n - Number of samples

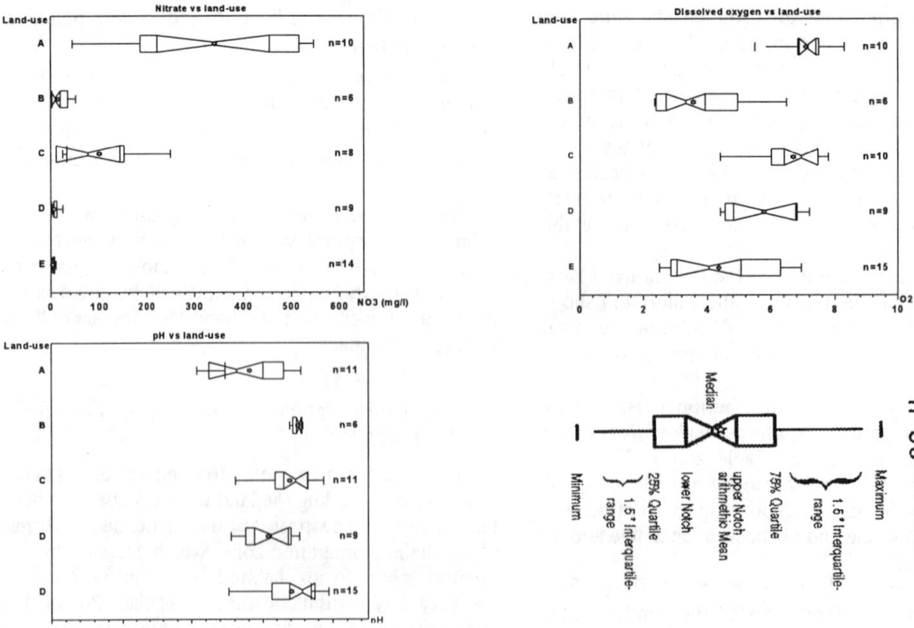

Figure 2. Statistical analysis of some chemical parameters in relation to land use

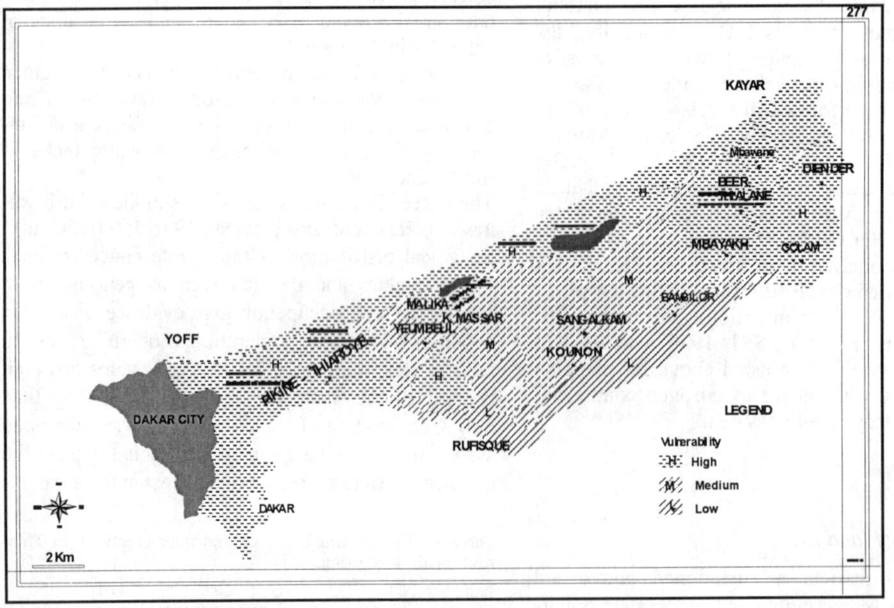

Figure 3. Groundwater vulnerability zones

4.3 *The aquifer vulnerability*

Aquifer vulnerability was assessed using GIS technology. It consists of overlaying basics maps of land use, reservoir characters, soils types, water table depth and piezometric. These parameters are re-evaluated qualitatively into three categories according to their possible impact on the groundwater quality. Results show three different vulnerability zones (Fig.3).

- A high vulnerability zone which includes the agglomeration areas and specific localities at Diender, Beer Thialane and the coastal zone from Malika to Kayar. High permeability values and thin unsaturated zone thickness contribute to the high vulnerability of these areas.
- A medium vulnerability zone which characterizes rural areas with thick unsaturated zone. In these areas, the sediments facies type show higher clay contents and thus give a better aquifer protection to contamination from the ground surface. However, high nitrate concentrations observed at villages and small localities are due to local contamination around the dug wells in these areas.
- A lower vulnerability areas located in the southeast of the aquifer. The lower permeability values associated to an uplift of the tertiary marly basement of the aquifer give a real protective cover overlaying the system.

5 CONCLUSIONS

Results from this study provide evidence that the high nitrate concentrations in the Thiaroye aquifer are directly linked to population growth in the suburban area rather than agricultural activities (fertilizers). This pollution (up to 540 mg/l NO_3^-) maintained by a poor waste water management in these areas places these resources in danger. In fact, improper urban sanitation, reservoir characteristics and depth of the water table are the main factors defining the gradual vulnerability with regard to nitrate contamination. As far as deterioration in the water quality is concerned, it is important to take urgent and local actions by controlling settlement and by introducing zone of protection around areas of intensive pumping. This has to be done before building a proper urban sanitation and installing a working systems for disposal of solid wastes.

ACKNOWLEDGEMENT

This study was financially supported by a grant of the German Academic Exchange Service (DAAD).

REFERENCES

Collin, J.J. & Salem, G. 1989. Pollution des eaux souterraines par les nitrates dans les banlieues non assainies des pays en développement. Le cas de Pikine (Sénégal). *Note tech. no. 089/27 BRGM SGN / 3E* 34p.

Chiang W.H.& Kinzelbach W. 1993. *Processing modflow Version 3.0.* Hamburg: Heidelberg, 48p.

Cissé, S. 2000. Nappe libre des sables quaternaires Thiaroye/Beer Thialane: Etude de la contamination par les nitrates sur la base d'un système d'information géographique (PC ARC/INFO). *Thes. 3°cycle. Dakar Univ., Sénégal* 234pp.

ESRI. 1990. *Understanding GIS. The ARC/INFO Method.* New York: Environmental Systems Research Institute INC. 550p.

Hébrad, L. 1966. Les formations tertiaires et quaternaires de l'isthme de la presqu'île du Cap-Vert (Feuille Thiaroye, Sénégal) *Rpp. Lab. Géol. Fac. Sc. Dakar n°.16. 75p.*

Klima, A.A. 1996. Methodology for mapping shallow groundwater quality in urban areas: A case study from Lithuania. *Environ. Geol*: 27, p320-323.

McDonald, M.G., Harbaugh, A.W. 1988. *A modular three - dimensional finite difference groundwater flow model.* Washington D.C: Technique of water Resources Investigations of the U.S. Geological survey, Scientific Software. Book 6, 548p.

Ministère de l'hydraulique - PNUD. 1994. Bilan-Diagnostic des ressources en eau du Sénégal. *Rapport MH/PNUD/ DASG-SEN/87/006 345p.*

OMS, 1972. Etude sur modèle mathématique du système aquifère de la presqu'île du Cap-Vert. *Géoh.336. Rpp. Préliminaire n°1. 18p.*

Tandia, A.A. 1997. Origine, process and migration of nitrates compounds in the Dakar region aquifer (Senegal): First results of the contribution of the isotopes techniques. *Rpp. Cont. Rech. AIEA. 1997* no. 7279/R1/RB 37p.

USAID/RSI. 1986. Cartographie et Télédétection des ressources de la république du Sénégal. Etude de la géol. De l'hydrog. de la vég. Et des potentiels d'utilisation des sols. *Doc. SDSURSI- 86-01, 653p.*

Groundwater: Past Achievements and Future Challenges, Sililo et al. (eds) © *2000 Balkema, Rotterdam, ISBN 90 5809 159 7*

Pollution vulnerability assessment in karstic aquifers: A case study of the Matese Mountains, Southern Italy

Alfonso Corniello & Daniela Ducci
Dipartimento di Ingegneria Geotecnica, Sezione di Geologia Applicata, Università di Napoli, Italy

ABSTRACT: In order to construct aquifer vulnerability maps at medium scale (e.g. 1:50.000-1:100.000), two parametric methods, SINTACS and EPIK, have been applied in a karstic area, located in Southern Italy, and compared using qualitative and statistical approaches. For both methods the final vulnerability maps are markedly affected by the lithology. The EPIK method, easily applicable, requires sometimes subjective decisions based on the hydrogeologic knowledge of the examined areas; the SINTACS method allows a more detailed definition of the vulnerability, but requires greater effort to obtain the input data.

1 INTRODUCTION

In Central-Southern Italy groundwater represents the main source of drinking water. Due to the increase in human economic activities (e.g. industrial, agricultural, cattle-breeding, etc.), which constitute real or potential sources of groundwater contamination, the demand for evaluation of aquifer vulnerability to pollution is also increasing. The construction of aquifer pollution maps can be considered as a first-step in implementing groundwater management programs.

To assess pollution vulnerability several methods and simulation models have been developed since the seventies. In the last ten years the advent of GIS, permitting the inventory, archival, retrieval and display of spatial data and the link to numerical rating systems, resulted in the widespread use of the parametric methods, based on the hydrogeologic setting: e.g. DRASTIC, largely used in the USA, and SINTACS (Civita 1994; Civita & De Maio 1997), an Italian method.

In previous papers the point count system vulnerability methods have been successfully applied to alluvial/porous aquifers located in Central and Southern Italy, by the authors (Corniello et al. 1995; 1998). In this hydrogeological context a comparison between three property-based index models for vulnerability assessment (including DRASTIC and SINTACS) demonstrated that they were all considered suitable (Corniello et al. 1997). In contrast, DRASTIC and SINTACS give markedly different results where fissured and karstified rocks crop out; SINTACS, which offers a specific set of weights, appears much more suitable (Alvino et al. 1998).

Therefore, the above mentioned SINTACS method and the EPIK method (Doerfliger et al. 1997, Tripet et al. 1997), which was developed for karstic aquifers, were selected to assess pollution vulnerability in the Matese Mts. The thematic maps and the vulnerability maps are drawn up using a GIS.

It is hoped that this paper contributes to the definition of aquifer contamination vulnerability in karstic and fissured environments, comparing the two vulnerability maps using both qualitative and statistical approaches.

2 HYDROGEOLOGICAL SETTING

The western sector of the Matese Mts. encompasses approximately 170 km^2 and is located between the Campania and the Molise regions (Southern Italy). The area where carbonate rocks (limestone and dolomite) crop out is approximately 130 km^2, with a mean elevation of 950 m a.s.l. and large sectors are characterised by several forms of karst origin, which influence the groundwater flow. The groundwater discharges via springs, located at the foot of the mountains, about 2 m^3/s.

In order to evaluate the amount of groundwater resource the hydrogeological budget was performed in Corniello et al. 1999: on the basis of

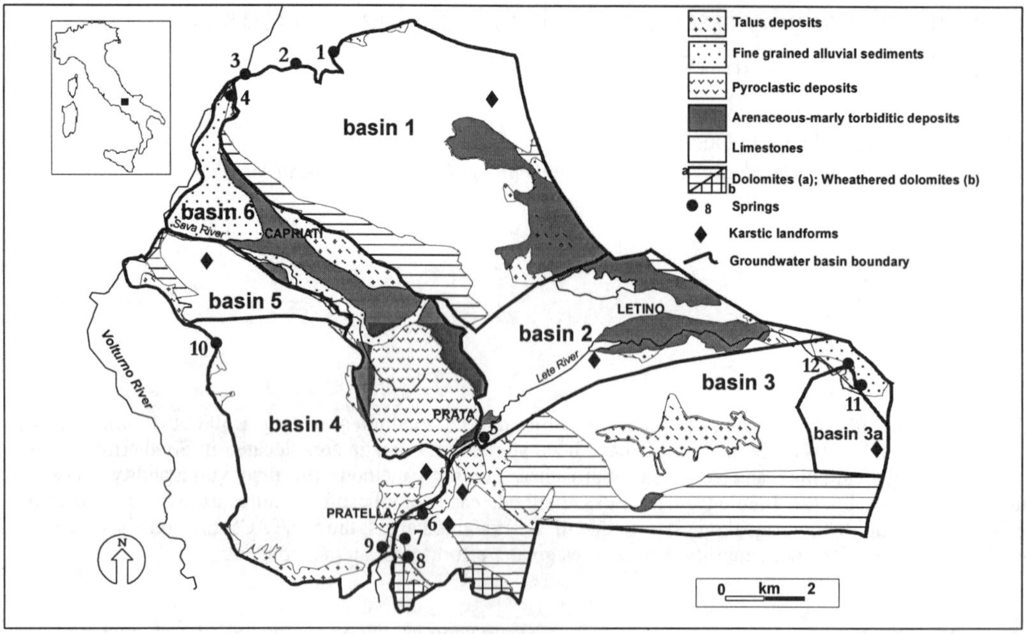

Figure 1. Hydrogeological scheme of the Matese Mts. area and partition into groundwater basins.

geological, structural and hydrochemical data the whole carbonate area has been divided into five main hydrogeological basins (fig. 1). In some basins the discharge from the springs corresponds to the recharge; in other cases, there are basins where the groundwater body also feeds the rivers. In detail, the groundwater basin 1 feeds the springs 1 to 4 (0.5 m^3/s) and the Volturno River; basin 2 feeds the spring 5 (0.3 m^3/s); the groundwater of the basin "Valle Agricola" (0.8 m^3/s) moves deep down (basin 3) towards springs 6 to 8 and the Lete River and with a shallow flow (basin 3a) towards the springs 11 and 12; in basin 4 groundwater flows towards the springs 9 and 10 (0.4 m^3/s) and in the basin 5 exclusively towards the Volturno River.

The groundwater basin 6 consists of impervious arenaceous-marly torbiditic deposits locally covered in the NW sector by a thickness of fine-grained alluvial sediments. In the SE sector about 20 m of these arenaceous-marly sediments overlie the carbonatic bedrock and are extensively covered by pyroclastic deposits (~ 20 m).

3 POLLUTION VULNERABILITY IN KARST AQUIFERS

3.1. *The SINTACS method*

SINTACS method (Civita 1994; Civita & De Maio 1997), originally derived from DRASTIC, in the latest release 4, retains only the structure of DRASTIC. It evaluates the vertical vulnerability using the same seven parameters: depth to groundwater (S), recharge action (I), attenuation potential of the vadose zone (N), attenuation potential of the soil (T), hydrogeologic characteristics of the aquifer (A), hydraulic conductivity (C), topographic slope (S). Each mapped factor is classified into ratings (from 1 to 10) which have an impact on pollution potential. Weight multipliers are then used for each factor to balance and enhance their importance. The final vulnerability index (I_v) is a weighted sum of the seven factors (Table 1).

The weight classes used by SINTACS depend on the hydrogeological features of each area. It is possible to use, in the same map, different weight classes in different sectors:

a. relatively uncontaminated scenario (absence of non-point sources of contamination);
b. urbanised and/or industrial zones or highly agro-chemical supported agriculture areas;
c. areas in which surface waters and groundwater are closely linked because of the drainage;
d. karstic areas with rapid connection between the aquifer and the recharge water (the vadose zone plays a negligible role and the net recharge is often equal to the precipitation rate);
e. sectors dominated by an elevated fracture index and limited or absent karst features.

In the study area the classes adopted are the following: a. in basin 6; d. in the sectors where the epikarst features are evident and e. where carbonatic rocks are less pervious fissured rocks.

The range variation of each parameter and the rating assigned are shown in Table 1. The final Vulnerability Index (Figs. 2, 4) ranges between 75 and 212 in carbonate area and between 63 and 120 in basin 6. In sub-basin 3a, where the groundwater flow is shallow, the superficial vulnerability has also been evaluated using the class of weight e. The parameters S and I are different from the main aquifer. In this zone the recharge of the main underlying aquifer has been reduced by the discharge rate of the springs fed by sub-basin 3a. The vulnerability of this sub-basin is prevalently high, with a small sector with very high vulnerability (Fig. 2).

3.2. *The EPIK method*

The EPIK multi-attribute method (Doerfliger et al. 1997, Tripet et al. 1997), developed by the Centre of Hydrogeology of Neuchatel University, is specific for karst aquifers. The groundwater flowing in these aquifers is highly heterogeneous in space (presence of open joints, sink-holes, etc.) and variable in time (see regime of karst springs).

Table 1. Simplified layout of the SINTACS method for contamination vulnerability assessment. Vulnerability classes established in Civita & De Maio 1997.

SINTACS Parameter	Range of the values Comments	Predominant Rating (P)
S Depth to water	The values range between few meters close to the springs and more than 1000 m in the eastern sector (500 in the sub-basin 3a).	1
I Recharge action	The values, achieved in the hydrogeologic budget (CORNIELLO et al. 1999), range between 150 and 900 mm/year.	4 sub-basin 3a: 3
N Effect of the vadose zone	From the graphs in CIVITA & DE MAIO 1997.	8-10
T Effect of the soil media	Partially from data furnished by "Autorità di Bacino del F. Volturno, Liri, Garigliano".	9
A Characteristic of the aquifer	From the graphs in CIVITA & DE MAIO 1997.	8-9
C Hydraulic conductivity	The values range between $1*10^{-7}$ and $1*10^{-2}$ m/s	4 and 9
S Topographic slope	From the digital elevation model using the GIS ILWIS. The values range generally between 0 (basin 6) and 30%, but often there are >50%;	1
Iv (vulnerability index)= $\Sigma\ [P_{(1,7)} * W_{(1,n)}]$ where $W_{(1,n)}$ is the weight in each class		
VULNERABILITY DEGREE: Very high (Iv ≥187); High (186≥Iv ≥141); Moderate (140≥Iv ≥106)I_V >25); Low (Iv ≤105)		

Table 2. Simplified layout of the EPIK method for contamination vulnerability assessment. I=Index; R=Rating. Vulnerability classes established in Doerfliger et al. 1997.

EPIK FACTOR Weighting Coefficient	I	R	Description	% in the area
E Epikarst	E_1	1	Epikarst connected to the karstic network	3
	E_2	3	Epikarst connected to a fissured zone	17
α=3	E_3	4	Epikarst absent	80
P Protective Cover	P_1	1	No protective cover or cover without notable effect	14
	P_2	2	Cover with fairly good protective effect	36
	P_3	3	Cover with good protective effect	42
β=1	P_4	4	Low-permeability cover with good protective effect	7
I Infiltration Conditions	I_1	1	Point infiltration	5
	I_2	2	Intermediate case	73
γ=3	I_3	3	Diffuse infiltration	22
K Charts Network	K_1	1	Presence of a well-developed karstic network	15
	K_2	2	Presence of a poorly-developed karstic network	1
δ=2	K_3	3	Spring in a porous media recharged by a karst aquifer or a fissured, non karst aquifer	84
F (protection factor)= $\alpha E_i + \beta P_j + \gamma I_k + \delta K_l$				
VULNERABILITY DEGREE: Very high (F≤19); High (20<F<25); Moderate (F>25); Low (Presence of P_4)				

The authors emphasize that the residence time of groundwater is not directly related to the distance between the recharge area and the catchment area (spring or well), such as SINTACS in the weight class "e". The aptitude of an aquifer to be contaminated by human activities depends on four factors: the epikarst (E), the type and thickness of the cover (P), the recharge conditions (I), and the development of the karst network (K). In a similar way to SINTACS each parameter is classified into ratings (from 1 to 4), and characterised by a weighting coefficient (from 1 to 3) indicated with a Greek letter. Based on the values of the protection factor F, which is the inverse of the vulnerability, four vulnerability classes are established. A simplified layout of this method is given in Table 2.

In the study area (Fig. 3) the F value ranges between 9 and 30. The method has been applied to the basins 1÷5 and comparative analysis with SINTACS map has been performed for these basins.

The whole basin 6 with the EPIK method would belong to the low vulnerability class because of the presence of P4.

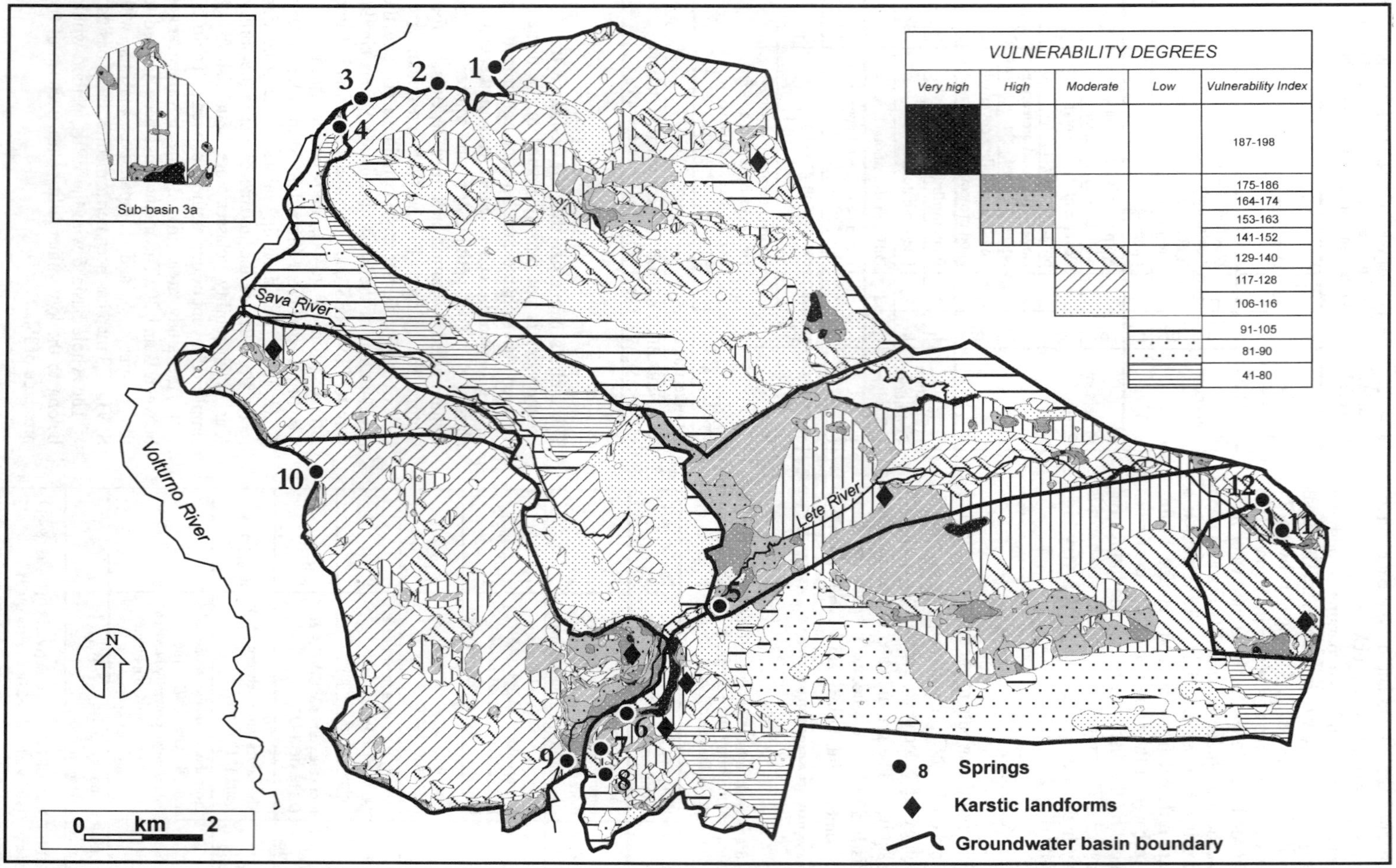

Figure 2. Aquifer contamination vulnerability map of the Matese Mts. computed using SINTACS method. In the frame thevulnerability of the shallow aquifer of the sub-basin 3a.

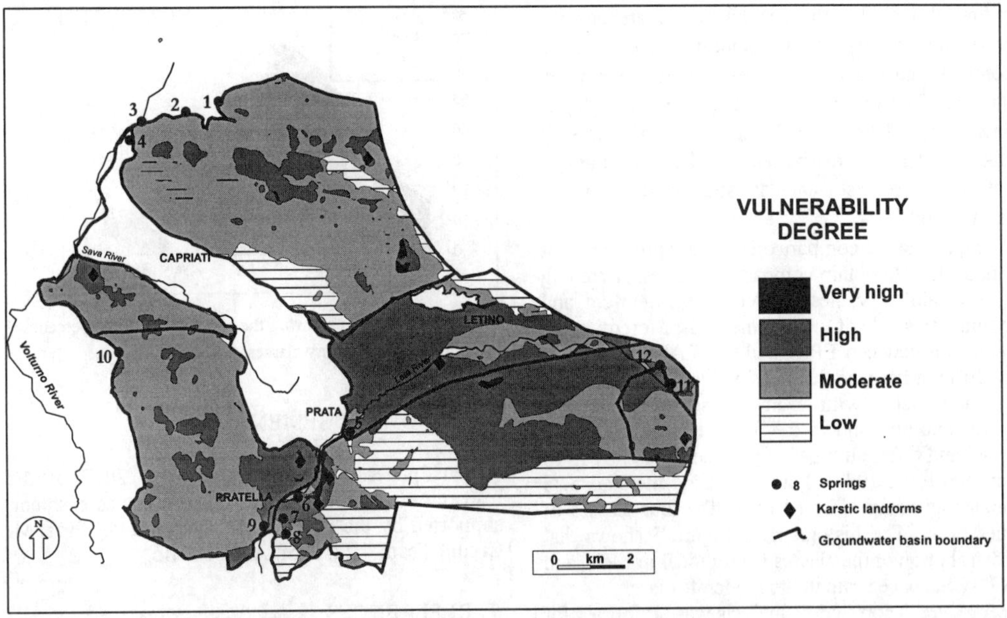

Figure 3. Simplified aquifer contamination vulnerability map of the Matese Mts. computed using SINTACS method.

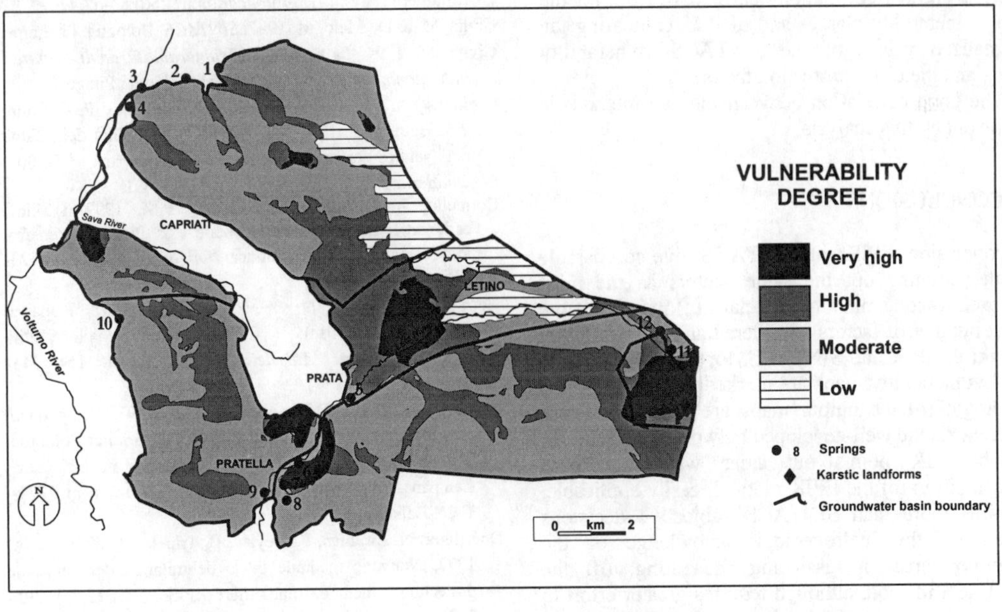

Figure 4. Aquifer contamination vulnerability map of the Matese Mts. computed using EPIK method.

4 COMPARISON BETWEEN THE TWO VULNERABILITY MAPS

The vulnerability maps (Figs 2-4) show that the prevalent vulnerability degree is moderate for both methods (more than 50% of the carbonatic area). Areas with high vulnerability are pointed out by SINTACS and EPIK in the sectors where karstified limestones crop out. The main differences between the two maps are in the low class: this class on the map is wider in SINTACS than in EPIK because in the first it includes the sectors where arenaceous-marly torbiditic deposits or dolomites occur, in the second only the outcropping area of arenaceous-marly sediments. The differences in extent of "very high" vulnerability class depend on the distribution of surface karst landforms because of the low values of the factors "epikarst" and "karstic network" in the EPIK method.

The quantitative comparison includes procedures of normalization to obtain comparable values, stretching the vulnerability values between the minimum and maximum possible of the two methods. Moreover, the comparison between EPIK and SINTACS method is very difficult because the EPIK values are in terms of Protection Factor (with an inverse scale compared to the vulnerability) and in the two methods a different number and different types of parameters are involved (Corniello et al. 1997). For this reason the statistical comparison was performed using the four classes of vulnerability. The histogram of Figure 5 shows that the distribution of the classes is unimodal in EPIK and SINTACS, concentrate in the moderate class.

Moreover, a correlation analysis was performed for each pixel (20 m x 20 m) using the GIS ILWIS, which allows spatial statistics, to compare the maps obtained using the two methods. The results of the correlation analysis show a correlation coefficient of 0.7 for the four vulnerability classes and of 0.25 comparing the normalized values of the SINTACS vulnerability index and the EPIK protection factor.

The good correlation between the two methods is borne out by this analysis.

5 CONCLUSIONS

In conclusion, EPIK and SINTACS give comparable results pointing out the same sectors as the most exposed to contaminants. In detail: (a) in spite of the great number of factors considered and their different weight in EPIK and SINTACS, for both methods the final vulnerability maps are markedly affected by the lithology; (b) the output maps are indeed different where a karstic well-developed network is present, for which EPIK points out areas with very high vulnerability; (c) the EPIK method, easily applicable, requires more than SINTACS subjective decisions based on the hydrogeologic knowledge of the examined areas in assigning the rating; (d) the SINTACS method, although requires greater effort to obtain the input data, allows a more detailed definition of the vulnerability; the precise delimitation of the most vulnerable areas is very important where there are legal restriction imposed on land use.

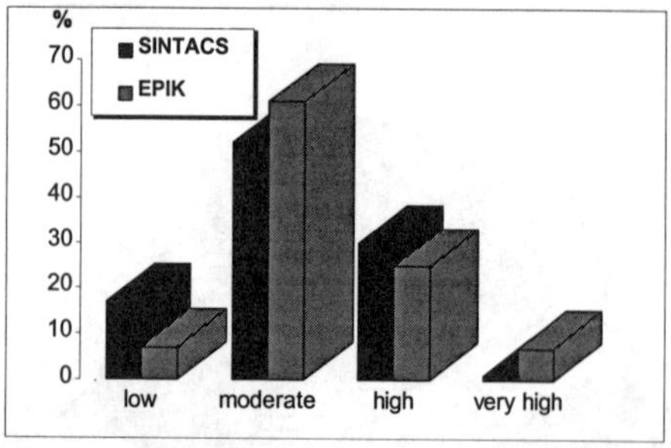

Figure 5. Histogram showing the comparison of the percentage areas in the vulnerability classes.

6 ACKNOWLEDGEMENTS

This paper is the publication N. 2077 of the research project on aquifer vulnerability assessment, supported by the CNR-GNDCI, Theme 4, (Prof. M. Civita) Team 4.22: Prof. A. Corniello.

7 REFERENCES

Alvino, R. Corniello, A. & Ducci, D. 1998. Aquifer Pollution Vulnerability in the Roccamonfina Volcano Area (Southern Italy). *IAHS Publication* n° 250: 63-67, NL.

Civita, M. & De Maio, M. 1997. *SINTACS*. Bologna: Pitagora.

Civita, M. 1994. *Le carte della vulnerabilità degli acquiferi all'inquinamento. Teoria e pratica*. Bologna: Pitagora.

Corniello, A. de Riso, R. Ducci, D. Napolitano, P. Guarino, P.& Bellucci F. 1995. Carta della vulnerabilità delle falde del settore orientale della Piana Campana 1/50.000. *Quaderni di Geologia Applicata*: 311-324, I.

Corniello, A. Ducci, D. & Guarino, P.M. 1999. I rilievi carbonatici del Matese occidentale e la Piana di Venafro: idrogeologia ed idrogeochimica. *Boll. Soc. Geol. It.*118: 523-535.

Corniello, A. Ducci, D. & Monti, G.M. 1998. Piana di Venafro (Molise): Carta della vulnerabilità all'inquinamento. *"Atti dei Convegni Lincei"* 154: 311-315.

Corniello, A. Ducci, D. & Napolitano, P. 1997. Comparison between Parametric Methods to Evaluate Aquifer Pollution Vulnerability Using a GIS. An example in the Piana Campana. *Engineering Geology and the Environment*: 1721-1726, Balkema.

Doerfliger, N. Zwahlen, F. Meylan, B. Tripet, J.P. Wildberg, A. 1997. Nouvelle méthode de la delimitation des zones de protection - méthode multicritère EPIK. *GWA* (5,77): 295-302.

Tripet, J.P. Doerfliger, N. Zwahlen, F. 1997. Vulnerability mapping in karst areas and its uses in Switzerland. *Hydrogéologie*, 3: 51-57.

Groundwater: Past Achievements and Future Challenges, Sililo et al. (eds) © *2000 Balkema, Rotterdam, ISBN 90 5809 159 7*

Natural redox processes and bioremediation in the Gardermoen glaciofluvial aquifer, Norway

A. Dagestad
Geological Survey of Norway, Trondheim, Norway

ABSTRACT: The natural redox processes and the reducing potential of uncontaminated sediments have been investigated in the saturated zone of the glaciofluvial aquifer at the Oslo International Airport, Gardermoen. The study was aimed at determining the potential of redox sensitive minerals in the sediments in reducing the effects of oxidizing agents injected into the aquifer for stimulating biodegradation of organic contaminants. Groundwater samples were collected from multi-level samplers in the upper 5 m of the aquifer. The sampling revealed an abrupt shift in the redox condition between 7.5 and 8.5 m below surface, dividing the aquifer into an upper 3 m oxic groundwater zone above a deeper anoxic groundwater. This change in groundwater composition, together with a corresponding increase in the concentration of pyrite in the sediments with depth, show that the redox shift in the aquifer is due to oxidation of pyrite. Air sparging tests carried out in the anoxic zone of the aquifer also showed that pyrite oxidation impaired the attempt to increase the oxygen in this zone. Even after several weeks of air injection, the increase in oxygen concentration within the sparging zone was insignificant, indicating that pyrite oxidation buffers the redox conditions and severely impairs the possibility to stimulate aerobic biodegradation in the deeper anoxic part of the Gardermoen aquifer.

1 INTRODUCTION

Leakage of hydrocarbons and other organic compounds comprise a threat to ground water quality in many aquifers, and remedial actions have to be carried out to protect groundwater resources. Due to the high costs involved in traditional protection or restoration of contaminated aquifers (physical or hydraulic containment, pump and treat etc.) in-situ bioremediation has gained more interest (Alexander 1994, Christensen et al. 2000). In-situ bioremediation involves the native microorganisms' capability to degrade organic contaminants either intrinsically or artificially promoted. To promote in-situ biological oxidation of organic contaminants, oxidizing agents like O_2, H_2O_2 or NO_3^- may be introduced to the contaminated part of the aquifer.

However, the range of the biodegradation is highly dependent on the natural redox condition in the aquifer, as well as on the biodegradability of the organic contaminants and soil temperature. The natural redox conditions in the aquifer are not often taken in to account when designing remedial actions using bioremediation, and the efforts may fail to reach the remediation standard set by the environmental authorities.

A new international airport was constructed on the Gardermoen glaciofluvial delta during the late nineteen-nineties. The storage and use of large amounts of different hydrocarbons and deicing chemicals at the airport pose a potential major threat to the large groundwater reservoir in the delta deposits. As part of a protection program for the aquifer, investigations were carried out to evaluate the potential of using in-situ bioremediation techniques to restore the aquifer in case of accidental release of organic contaminants to the groundwater. An important part of these investigations was to reveal the natural redox condition in the aquifer and what effect it may have on the in-situ bioremediation potential. To ensure that reliable measurements of the aquifer's redox potential could be made, the investigations were carried out in a pristine area of the aquifer.

2 HYDROGEOLOGY

The investigations were carried out at the Moreppen research site, located adjacent to Oslo international airport at Gardermoen 60 km northeast of Oslo (Fig.1). The Gardermoen aquifer is formed by one of the largest glaciofluvial deltas in Norway (ca. 70 km^2), deposited ca. 9500 years BP during the last deglaciation of Fennoscandia (Tuttle 1997, Longva

& Thoresen 1989). The hydrostratigraphy of the delta aquifer consists of three major units, which mainly correspond to the delta's topset, forset and the bottomset stratigraphic units. The coarse sandy forset unit is the most important for groundwater flow in the aquifer, since the upper topset unit is unsaturated and the bottomset unit consists predominantly of glaciomarine and marine silt and clay with low hydraulic conductivity.

The aquifer is phreatic and entirely recharged from precipitation. The general groundwater drainage pattern is radial in the investigated area, with groundwater flow towards several lakes in the northeast, and a number of spring ravines discharging at the edge of the coarser delta deposits in the southwest (Fig. 1).

At the Moreppen research site the delta top is relatively flat-lying at an average elevation of 205 m above sea level. It is underlain by 2-3 m of gravely sandy horizontal topset layers which overly 12-14 m of coarse to medium sandy forset layers. The transition zone between the permeable forset unit ($k \geq 10^{-3}$-10^{-4} m/s) and the low-permeablity bottomset unit ($k \geq 10^{-6}$-10^{-9}) is 15-18 m below surface, with sporadic interfingering of coarser layers into deeper silty deposits.

The depth to the groundwater table at the research site fluctuates between 4 - 5.5 m due to seasonal variation in the precipitation and infiltration, with

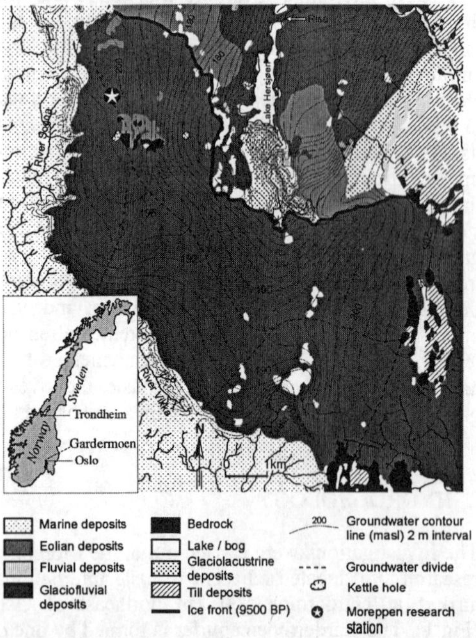

Figure 1. Quaternary map of the Gardermoen glaciofluvial delta showing groundwater contours and the location of the research station at Moreppen.

the highest groundwater level in the springtime (May-June) after the snow melting period. The average annual infiltration is approximately 400 mm/m^2 (Jørgensen & Østmo 1990). The site is close to the groundwater divide and the natural average hydraulic gradient is 1.7×10^{-3}. Tracer tests carried out in the vicinity of the research site under natural gradient conditions recorded a groundwater flow velocity of about 0.03 m/d towards the northeast (Rudolp-Lund 1997).

3 INSTRUMENTATION

The research site at Moreppen was instrumented with one pumping well (well 2) and one injection well (well 1), which made it possible to carry out experiments under a forced hydraulic gradient (Fig. 2). In addition, several observation wells were installed at the site, together with 9 air-sparging wells evenly distributed along the midline between the two pumping wells. Around the central air-sparging well, In5, six multi-level samplers (MLS) were established with a horizontal distance from In5 of between 0.5 to 6 m (Fig. 3).

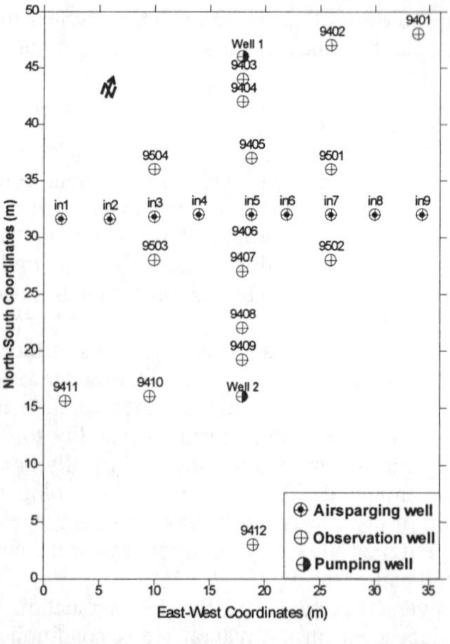

Figure 2. Well installations at the Moreppen research station.

The multi-level samplers were equipped with microscreens (L= 4 cm, Ø= 0.5 cm) in the upper 5 m of the aquifer; with one meter vertical interval between the screens. Each filter-screen was connected to the surface by separate Teflon tubes. To avoid vertical gasflow along the multi-level samplers during airsparging performance, or mixing of water from various levels during water sampling, a bentonite seal was placed between each screen-level. A multiport peristaltic pump was used for water sampling which allowed simultaneous sampling of all levels. This pumping, in combination with the hydraulic separation of the micro-screens, made it possible to collect depth-specific water samples from the upper 5 m of the aquifer.

4 HYDROGEOCHEMISTRY

Several weeks after the multi-level samplers were installed, and prior to the air-sparging experiments, groundwater samples were taken from the multi-level samplers to examine the natural hydrogeochemical conditions in the upper 5 m of the aquifer at the research site. Measurements of electric conductivity (EC), pH and dissolved oxygen were carried out on site, whilst the concentration of major elements, anions and alkalinity were determined in the laboratory using ICP-AES, IC and standard titration, respectively. The most characteristic features of the natural groundwater composition are shown in Figure 4. The investigation revealed a major change in the groundwater chemistry between 7.5 m and 8.5 m below surface, dividing the aquifer into an upper 3 m oxic-groundwater zone above deeper low oxic groundwater. This major shift in the redox condition in the aquifer coincides with an increase in the EC, mainly due to an increase in the concentration of Ca^{2+}, SO_4^{2-} and HCO_3^-. The pH of the groundwater increases two orders of magnitude across the redox change. The concentration of dissolved iron was below detection limit (<0.01 mg/L) in all samples.

Despite detectable dissolved oxygen (DO) in the groundwater below 8.5 m depth, it is assumed that the deeper part of the aquifer is anoxic, and measured DO is due to diffusion of oxygen through the Teflon tubes during sampling. Other experiments using hydrostatic pumps and low diffusion tubes for sampling in wells with filter screens in the anoxic zone support this assumption.

During the construction of groundwater well 2, 1 m mixed sediment samples were taken between 1 m and 14 m depth below surface. Using XRF analysis of bulk and fractionated samples (125-250 µm), the variation in concentrations of calcium and sulfur in the sediments versus depth were obtained (Fig. 5). The pyrite concentration in the sediment samples was calculated from the sulfur analysis, assuming that all sulfur is present as pyrite. The observed chemical changes in the aquifer between 7.5 and 8.5 m below surface suggest that pyrite

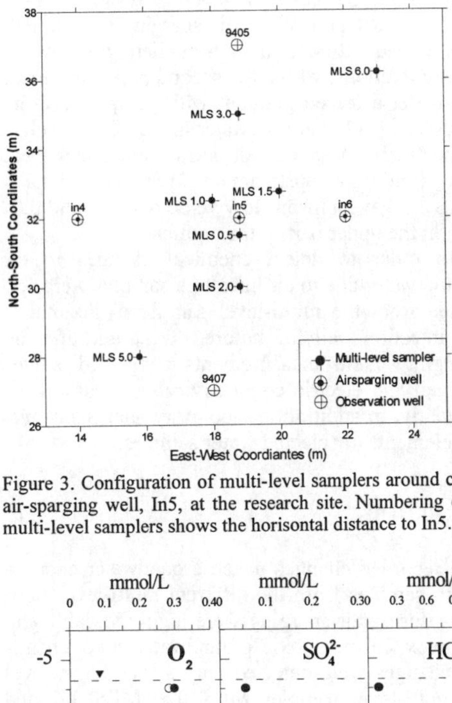

Figure 3. Configuration of multi-level samplers around central air-sparging well, In5, at the research site. Numbering of the multi-level samplers shows the horisontal distance to In5.

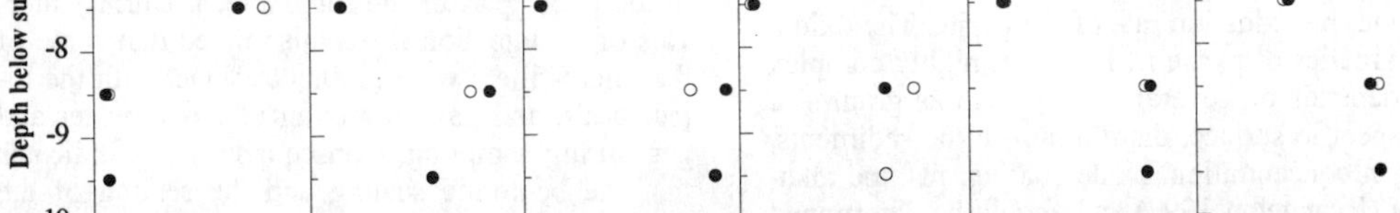

Figure 4. Chemical composition of groundwater versus depth at the research site. Groundwater samples collected from multi-level sampler MLS 0.5 (●) and MLS 1.5 (O). For location of the multi-level samplers see figure 3.

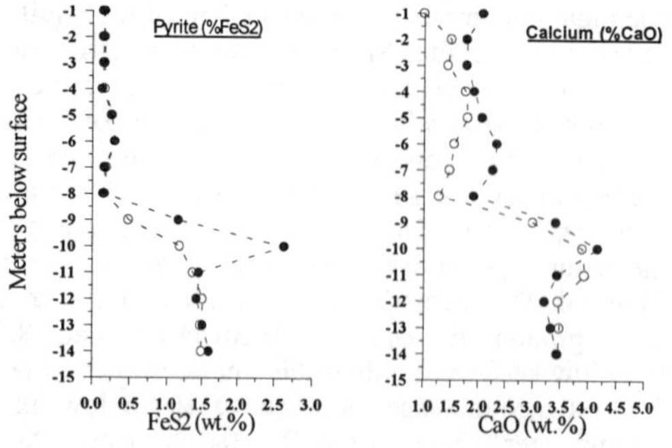

Figure 5. Pyrite and calcium versus depth in sediment samples at the research site (● bulk sample, ○ fractionated sample (125-250 μm)).

oxidation and calcite dissolution are the controlling processes.

Furthermore, the abrupt shift in the redox condition within such a limited vertical section of the aquifer indicates that the oxidation of pyrite proceeds relatively fast, consuming the dissolved oxygen transported down from the oxic groundwater zone. The pyrite concentration in the anoxic zone shows that the aquifer has a significant natural reduction potential that strongly influences the mobility and accessibility of oxygen in the deeper part of the aquifer. Consequently, the natural redox condition in the anoxic part of the aquifer will inhibit intrinsic aerobe biodegradation of organic contaminants, and may also impair the possibility of promoting aerobe biodegradation by injection of electron acceptors. This is likely due to consumption of most of the introduced electron acceptors by pyrite oxidation, rather than the biological oxidation of organic contaminants. Both the degradation rate of the organic contaminants and oxidation kinetics of pyrite in the sediments will thus effect the efficiency of remediation processes. If the oxidation rate of pyrite is slow compared to the aerobe degradation rate of the targeted contaminants, the oxygen concentration and the biodegradation rate may increase with simultaneous pyrite oxidation. However, if the pyrite oxidation proceeds faster, the possibility of stimulating biodegradation seems to be limited. Hence, it is important, prior to making a decision regarding remedial actions, to determine the oxidation rate of the pyrite. The oxidation kinetics of pyrite is, however, highly complex and depends on several factors such as grain size and specific surface, distribution in the sediments, oxygen concentration, oxide coating, pH and alkalinity (Evangelou 1997) and should be determined for each specific site.

5 AIR SPARGING EXPERIMENTS

To estimate the oxidation rate of pyrite in the aquifer, and the extent to which pyrite oxidation may influence the possibility to stimulate biodegradation, air sparging experiments were carried out at the research site. The main objective of the experiments was to investigate if it was possible, by air injection over a limited period of time, to increase the oxygen concentration in the anoxic lower part of the aquifer with simultaneously ongoing pyrite oxidation.

The air sparging experiments were carried out over a period of three weeks. Only the central air sparging well, In5 (Fig. 2), with a filter-screen between 9 to 10 m below surface, was used during the experiments. The air was injected in pulses (10 min on/ 5 min off) with an average flow rate of 4.3 L/s. The first part of the air sparging experiments was carried out under natural gradient groundwater flow conditions, whilst the second part was carried out under a forced gradient with pumping and injection of 1.6 L/s groundwater in well 2 and well 1, respectively. A tracer test carried out under these flow conditions upstream of multi-level sampler MLS 1.5 gave a mean flow velocity of around 0.04 m/h in the upper part of the aquifer .

In order to detect chemical changes in the groundwater due to air injection, samples were collected from the multi-level samplers proximal to the injection well In5 before, during and after air sparging. Field measurements of pH, dissolved oxygen and electric conductivity were carried out regularly, in addition to laboratory analysis of major elements of selected water samples.

6 RESULTS

The recorded changes in the groundwater chemistry in samples from the different multi-level samplers during air sparging were highly variable and complex. In the anoxic groundwater zone, chemical changes were detected only at the 8.5 m level in multi-level sampler MLS 0.5, MLS 1.5 and MLS 2.0 (MLS 1.0 malfunctioned). No changes were recorded below this level, indicating that the airflow out of the injection well was concentrated in the upper part of the filter-screen. During intervals of air injection it was also noted that some of the micro-filters were in direct contact with the injected air, and gave a mixing of groundwater and gas during sampling. Consequently, this indicates that the bentonite sealing and the sealing of the Teflon tubes prior to sampling could not prevent the formation of preferential flow paths and increased airflow along the multilevel samplers in the vicinity of the air sparging well.

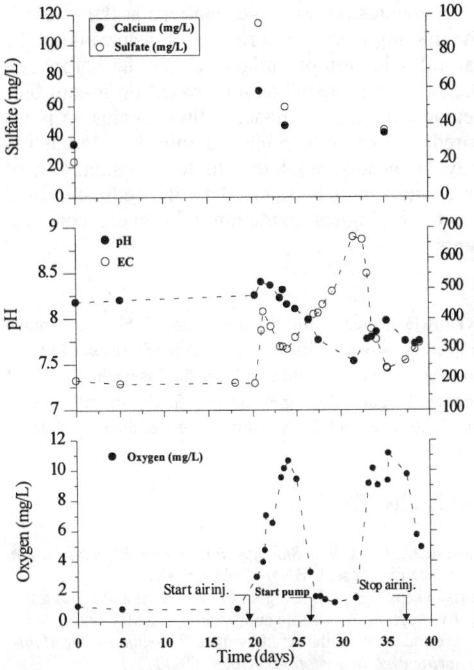

Figure 6. Chemical changes in groundwater from multi-level sampler MLS 0.5 at filter level 8.5 meter below surface prior to, during and after air injection in air sparging well In5.

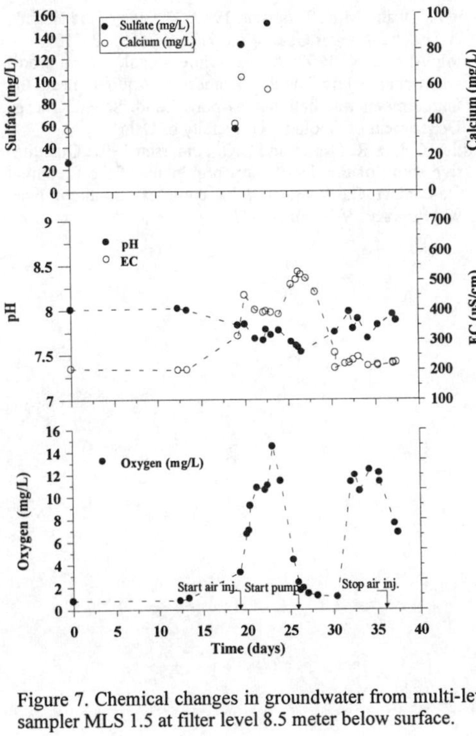

Figure 7. Chemical changes in groundwater from multi-level sampler MLS 1.5 at filter level 8.5 meter below surface.

Figure 6 shows the recorded changes in field parameters (pH, EC, DO) and sulfate and calcium concentrations in samples from MLS 0.5. The monitoring revealed a quick response to the onset of the air injection with a significant increase in dissolved oxygen followed by an increase in conductivity. The recorded changes in conductivity were mainly due to an increase in the sulfate and calcium concentration, which indicated an extensive ongoing pyrite oxidation buffered by calcite dissolution, shown by a simultaneous moderate drop in pH (Fig. 6). The oxygen concentration increased to close to saturation within a few days, followed by a major drop two days before the start of the groundwater pumping. This incident corresponded in time with a gradual disappearance of the gas intrusion in the sampled level, probably due to an increased air flow closer to the injection well. The start of the pumping and injection of groundwater did not change the low oxygen concentration at this sample level, but resulted in a major increase in the conductivity caused by horizontal groundwater flow from the area upstream of the multi-level sampler around injection well In5. After several days of groundwater circulation and low oxygen concentration, gas was suddenly found to be flowing in the same sampling levels that previously experienced gas intrusion (Fig. 6; day 33). This incident was in time concurrent with a restart of air injection after a 17 hour stop due to a malfunction of the air compressor. The gas intrution resulted in a renewed increase in the oxygen concentration which lasted until the air injection ceased.

Similar changes in the oxygen concentration during the air injection experiments were also recorded at the 8.5 m level in multi-level sampler MLS 1.5 (Fig. 7) and MLS 2.0 (not shown).

The changes in conductivity in MLS 1.5 differ somewhat from those recorded in MLS 0.5. This is probably due to the location of MLS 1.5 upstream from air injection well In5. The similarity in the hydrogeochemical changes observed for the multi-level samplers reflects the air-channel distribution in the aquifer during air injection, and the extensive pyrite oxidation along the air-channels at these sampled levels. The recurrence of gas in some sampled levels, after a collapse in the horizontal spreading of air channels, shows that the distribution of air channels may be reestablished after a longer cessation of the air injection.

7 DISCUSSION AND CONCLUSION

The observed hydrogeochemical changes during the air sparging experiments showed that the oxygen supply to the anoxic groundwater zone resulted in extensive pyrite oxidation. Due to the observed preferential airflow along some of the multi-level samplers, the oxygenation of the groundwater and aquifer sediments in the vicinity of the samplers was probably much larger than the general oxygenation of the aquifer close to the air sparging well, In5. Hence, the recorded chemical changes in the groundwater samples collected from the multi-level samplers do not reflect the general chemical changes in the aquifer proximal to the air injection well, but may display the extreme oxidation conditions and subsequent chemical changes in groundwater close to air channels.

Nevertheless, the recorded fluctuations in the concentration of dissolved oxygen indicate that the oxygenated volume of the aquifer is limited to the proximity of air channels. Further, the disappearance of gas intrusion at sample level 8.5 m in MLS 0.5, followed by an immediate drop in the oxygen concentration and a simultaneous increase in conductivity (Fig. 6), show clearly that the oxidation of pyrite proceeded rapidly. Another observation, which supports this explanation, is the low oxygen concentration, accompanied with high EC, recorded at the same sample port during the first days of groundwater circulation. This indicates that oxygen transferred to the groundwater close to the injection well is rapidly consumed by pyrite oxidation, resulting in an anoxic groundwater when it reaches the extraction points less than 0.5 m downstream.

The reoccurrence of gas intrusion in some of the sampled points in the anoxic zone did not indicate any significant reduction over time in the oxidation rate of pyrite. The reduction in conductivity compared to the previous period with gas intrusion, does not indicate a decrease in oxidation rate, but reflects the shorter residence time of groundwater in contact with air channels during groundwater circulation. Furthermore, the cessation of air injection also resulted in a rapid drop in oxygen concentration followed by an increase in conductivity in groundwater sampled from 8.5 m in MLS 0.5, MLS 1.5 and MLS 2.0.

All together the air sparging experiments show that the air injection over many days is not sufficient to significantly reduce the pyrite oxidation rate even in a high oxidizing environment close to air channels. Consequently, the possibility to increase the oxygen concentration over a larger volume than the vicinity of air channels seems to be limited by the highly reactive pyrite in the aquifer sediments.

Based on these observations it is expected that the potential to promote aerobe biodegradation of organic contaminants, by injection of oxidizing agents, is limited in the anoxic zone of the aquifer due to the highly reactive pyrite in the sediments.

In conclusion, the investigations at the research site highlight the importance of determining the natural reduction potential of an aquifer before any decision on remedial actions based on in-situ biodegradation can be made. If this conclusion is ignored, efforts to stimulate aerobe biodegradation may be unsuccessful due to the consumption of oxidizing agents by reduced aquifer sediments, and not by biological oxidation of organic contaminants.

Acknowledgements: This study was part of the "Gardermoen Project, the environment of the subsurface", financed by the Norwegian Aviation Authority and the Research Council of Norway. I thank Jan Cramer and Paul Shand for critical reading of the paper and for the improvement of the English text.

REFERENCES

Alexander, M. 1994. *Biodegradation and Bioremediation.* Academic Press. ISBN 0-12-040860-X.

Christensen, T.H., P.L. Berg and P. Kjeldsen 2000. Natural Attenuation: A feasible approach to remediation of ground water pollution at landfills ?. *Groundwater Monitoring and Remediation*, Winter: 69-77.

Evangelou, V.P. 1995. *Pyrite oxidation and its control.* CRC Press. ISBN 0-8493-4732-7.

Jørgensen P. and S.R. Østmo 1990. Hydrogeology in the Romerike Area, southern Norway. *NGU Bulletin* 418: 19-26.

Longva, O. and M.K. Thoresen, 1989: The age of the Hauerseter delta. *Norges Geologiske Tidsskrift* 69: 131-134.

Rudolph-Lund, K. 1997: A three-dimensional, natural gradient tracer study in the Romerike Aquifer near the Gardermoen International Airport. Cand. Scient. Thesis, Department of Geology, University of Oslo.

Tuttle, K. J., S.R. Østmo and B.G. Andersen 1997: Quantitative study of the distributary braidplain of the Preboreal ice-contact Gardermoen delta complex, Southern Norway. *Boreas.* Vol. 26: 141-156.

Groundwater: Past Achievements and Future Challenges, Sililo et al. (eds) © 2000 Balkema, Rotterdam, ISBN 90 5809 159 7

Proving natural attenuation with modeling at a site contaminated by ammonium and phenolics

Ruth M. Davison & David N. Lerner
Groundwater Protection and Restoration Group, University of Sheffield, UK

ABSTRACT: A former coal carbonisation plant has contaminated the underlying aquifer with ammonium and organic compounds. The source term, from analysis of wash water from a similar plant, is estimated to contain 7700 mg/l of phenol, 12000 mg/l of ammonium and 7000 mg/l of chloride. The site has been extensively characterised with 39 boreholes and multilevel samplers giving a total of 120 monitoring points. Concentrations of ammonium up to 400 mg/l have been observed however no phenol has been found within the field site. The low retardation of phenol relative to ammonium could mean that the phenol has simply left the field site. However source term reconstruction, an electron balance, field data and modelling studies have indicated that the phenol mass could have degraded within the aquifer. Stochastic biodegradation and solute transport modelling were used to find the most likely position of the phenol if natural attenuation were not complete. One new borehole has been drilled on the periphery of the predicted plume to validate the modelling and the analyses show virtually no phenol but high concentrations of chloride (>1000 mg/l).

1 INTRODUCTION

Environmental legislation in the UK is currently undergoing dramatic changes. The local councils, from April 2000 onwards, will be responsible for identifying all contaminated land in their area and ensuring that it is remediated. Sites that are identified as particularly problematic will be classified as 'special sites' and the Environment Agency will be the enforcing authority in this case. With the new legislation has developed an increased interest in natural attenuation, as more sites are required to be remediated.

A contaminated site near Mansfield in Nottinghamshire, UK is the focus of an Environment Agency led study to use risk based techniques to predict the future of contamination rather than immediately implementing costly engineered approaches. It is hoped that this type of approach will be more widely applied throughout the UK in the future.

The aim of this study is to prove whether natural attenuation will remove phenolic compounds from the aquifer before they pose any risk to potential targets.

2 THE CONCEPTUAL MODEL

A former coal carbonisation plant in Mansfield, UK has contaminated the underlying Triassic Sandstone aquifer with ammonium liquor. The main components of the liquor are ammonium and phenolics with other inorganic and organic compounds present in lower concentrations. The ammonium liquor entered the subsurface through two main areas, firstly the start of an unlined drainage ditch next to the plant itself and secondly at a storage lagoon, 500 m to the East (Fig. 1).

Two plumes of ammonium can be observed downstream from both infiltration zones and have been retained around the field site. The ammonium is posing limited risk to a local factory borehole and surface water. The phenol contamination is not however observed around the study area as it is mobile relative to ammonium resulting in the chromatographic separation of the two components. It is known that an organic plume has passed through the aquifer as much Total Inorganic Carbon (TIC) exists on the site with ferrous iron and positive analyses for phenol were found at the farm borehole in 1986 (Fig. 1). The general groundwater flow direction is towards the north-east and flow velocity in the contaminated upper layer of the aquifer is 96 myr^{-1}.

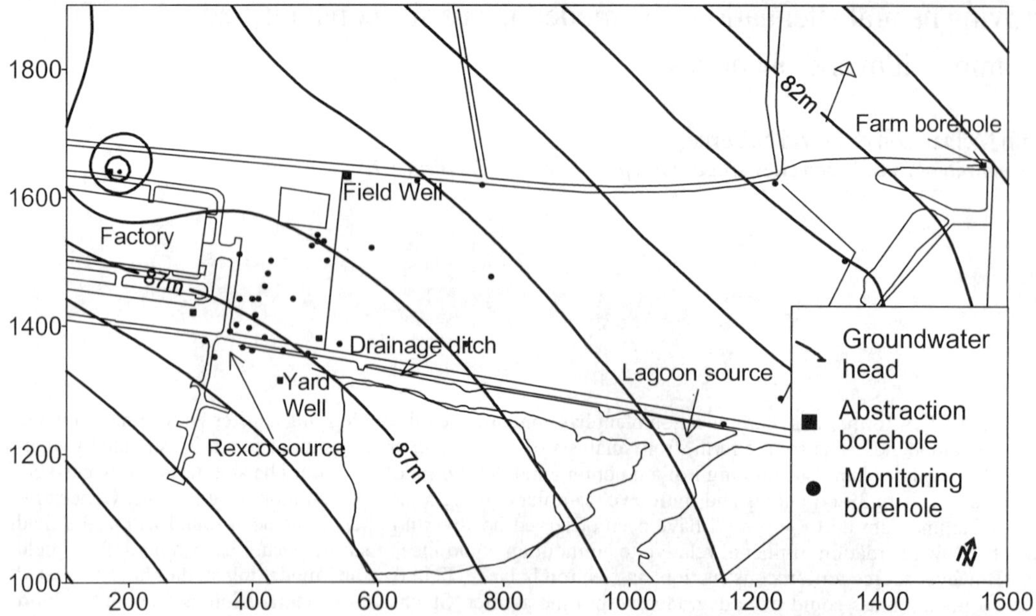

Figure 1: The field site, with modelled groundwater heads for 1996.

As the phenol is more mobile than the ammonium it has the potential to progress towards a public water supply borehole four kilometers to the N.E.E.. Previous studies including laboratory microcosms (Broholm & Arvin 2000), electron balancing (Thornton et al. 1998) and biodegradation modelling (Davison & Lerner 2000) have indicated a high potential for degradation of phenolics at the site. Therefore it is possible that the phenol plume has entirely degraded in the aquifer and poses no threat to any targets.

To assess the validity of this hypothesis a mix of stochastic biodegradation modelling and fieldwork were used.

3 FLOW MODEL

The flow conditions around the field site have been complicated over the past 12 years as changing industry has led to changing abstraction patterns. These altering abstractions have caused four major flow direction changes over the site with a maximum change of 90 degrees. To simulate the flow at the site two levels of modelling were completed; regional and local scale. The regional model covered the entire unconfined width of the Triassic Sandstone in Nottinghamshire, UK. The regional model consisted of a steady state and transient version and was well calibrated to both groundwater heads and river groundwater interactions. A local scale model, also with steady state and transient versions, was telescoped from the regional to maintain the regional flow properties but to facilitate a finer grid resolution. This flow model just covered the area of suspected contamination and was used as a base for the solute transport and biodegradation modelling. The local scale base model was well calibrated to heads across the field site with an RMS error of just 0.3 m (Fig. 2). The flow model may be very well calibrated to the field data but uncertainty is still inherent in the results. To take some account of this uncertainty three further flow models where created that were also well calibrated to the field data but with slightly changed input parameters. The details of these flow models are contained in Table 1 and their calibrations shown in Figure 2.

Table 1. The four base flow models

Flow model	Change	RMS error
Flow calibration 1	Original	0.3 m
Flow calibration 2	Field well pumping rate decreased to half original	0.5 m
Flow calibration 3	Recharge decreased by 5%	1.1 m
Flow calibration 4	Hydraulic conductivity increased by 10%	0.5 m

To assess the impact that the small changes in flow model imposed on the solute transport simulations a standard solute transport simulation was completed in each flow model.

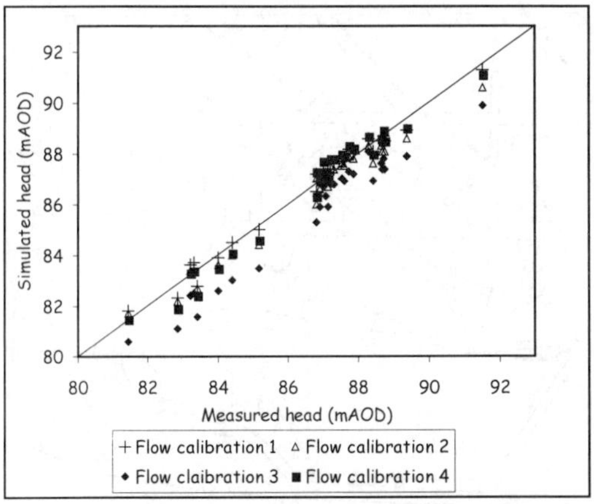

Figure 2: The calibration of the 4 alternative flow models

The positions of the plumes were found to be quite dramatically different with plume centers shifted by 200 m perpendicular to the flow direction (Fig. 3). This simple exercise shows how a single flow model could cause an incorrect assessment of a plume's location and could lead to misleading conclusions. Ideally hydraulic parameters will be randomly selected and many alternative flow models used to make the analysis. In this example we changed parameter values but only accepted flow models well calibrated to the field data.

4 THE STOCHASTIC BIODEGRADATION MODELLING

The solute transport and biodegradation modelling of phenol, ammonium, oxygen and nitrate were completed with an adapted version of RT3D (Clement et al. 97, 98, pers. comm. Whittaker). Instantaneous degradation of phenol was assumed and degradation was only limited by the availability of oxygen and nitrate. Initial concentrations of oxygen and nitrate were set at background field levels, 6 mgl^{-1} and 50 mgl^{-1} respectively. Additional inputs of the electron acceptors were placed around the southern and western model boundaries to represent the oxygenated water passing into the model. The biodegradation simulations were completed in the steady state (1948-1984) and the transient flow model (1984-2005). The fine grid, transient conditions and many solutes resulted in a slow running model with an average simulation time of 36 hours. A detailed description of the model parameters can be found in Davison and Lerner (2000).

The solute transport parameters were also uncertain and variable and the model was shown to be particularly sensitive to source flux, retardation and dispersion (Davison and Lerner 1998). Predetermined ranges were estimated for the uncertain parameters from field data. Parameter sets were obtained by randomly sampling from the predetermined ranges and randomly selecting one of the 4 flow models. Due to the time constraints only 26 simulations were completed. The overlain plume centers from the models are shown in Figure 4.

5 THE FIELD DATA

The field data collection phase is currently underway. One borehole has been drilled on the expected periphery of the plume and contains low concentrations of phenolics and high concentrations of chloride (Table 2).

Table 2. The concentrations of phenolics and chloride in Borehole 1

Depth (m)	Total phenolics (μg/l)	Chloride (mg/l)
25	1.418	1440
50	5.949	1450
80	1.948	1540

Three further boreholes are planned over the next three months and Borehole 1 will be packer tested to identify the depth of the main phenol contamination. Each of the three new boreholes will contain multilevel samplers with 10 levels to enable a three dimensional view of the plume to be ascertained and a more accurate estimate of the current mass. It is suspected that we will find some phenol contamination but the concentrations will be lower than the modelling suggests due to the contribution of iron as an electron acceptor to the biodegradation processes. The new field data will then be used to condition a further set of stochastic biodegrdation models to predict the likelihood that natural attenuation will remove the phenol contamination before it impacts the public water supply borehole.

6 CONCLUSIONS

1. The stochastic biodegradation modelling has identified a potentially contaminated zone that has been used to direct the field data collection.

2. Well calibrated groundwater flow models can give very imprecise estimates of plume location.

3. The current field data fits with the groundwater flow modelling.

4. The modeling suggests that a plume of phenol still exists within the aquifer although it is position is yet to be confirmed from field data. The mass of phenol currently within the aquifer is likely to be less than modeled due to the contribution of iron on the aquifer solids as an electron acceptor. Following the fieldwork phase a stochastic mass balance will be completed at the site to assess the current status of natural attenuation and the risk to local boreholes.

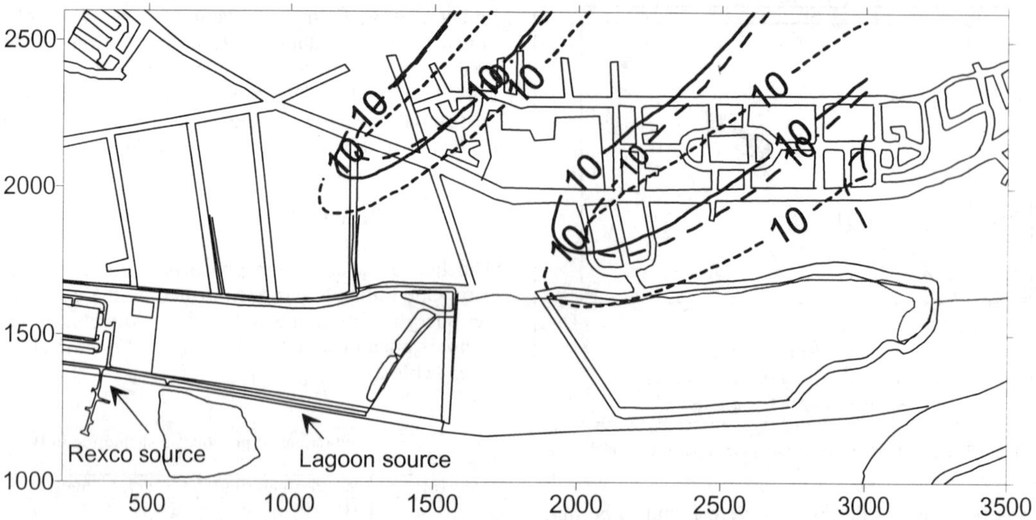

Figure 3: The effect of flow model on current plume position

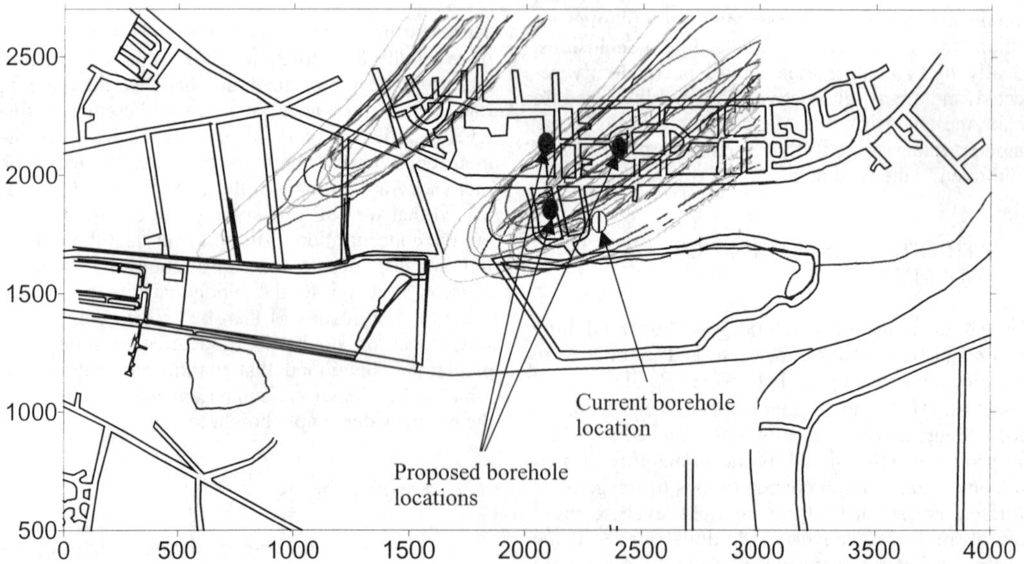

Figure 4: Superimposed plumes from the two sources.

REFERENCES

Broholm, M.M., and Arvin, E., 2000. Biodegradation of phenols in a sandstone aquifer under aerobic conditions and mixed nitrate and iron reducing conditions. Accepted for publication in J. Contam. Hydrol.

Clement, T.P. 1997. A modular computer model for simulating reactive multi-species transport in three dimensional groundwater systems. Pacific Northwest National Laboratory, PNNL-SA-11720. Richland, Washington:PNNL.

Clement, T.P., Y. Sun, B.S. Hooker, and J.N. Petersen. 1998. Modelling multispecies reactive transport in groundwater. *Ground Water Monitoring and Remediation* 18 (2): 79-92.

Davison, R.M., and D.N.Lerner. 2000. Evaluating natural attenuation of groundwater pollution from a coal carbonisation plant: Developing a local scale model using Modflow, MODTMR and MT3D. Accepted for publication in *Water and Environmental Management Journal*.

Davison, R.M., and D.N.Lerner. 1998. An uncertainty based risk assessment for ammonium contamination emanating from a coal carbonisation plant. IAHS proceedings, GQ98, Tubingen, Germany. IAHS publication number 250: 41-47.

Thornton, S.F., R.M. Davison, D.N. Lerner, and S.A. Banwart. 1998. Electron balances in field studies of intrinsic bioremediation. IAHS proceedings, GQ98, Groundwater quality, remediation and protection, Tubingen, Germany. IAHS publication number 250: 273-282.

Groundwater: Past Achievements and Future Challenges, Sililo et al. (eds) © 2000 Balkema, Rotterdam, ISBN 90 5809 159 7

Delineation of borehole protection zones in fractured rock aquifers

S.S.de Lange & G.J.van Tonder
Institute for Groundwater Studies, University of the Orange Free State, Bloemfontein, South Africa

ABSTRACT: A new approach to delineate protection zones around boreholes in fractured rock aquifers is suggested because of some shortcomings in previous approaches and standards. Based on the German approach by suggesting three protection zones, a simple program, BPZONE, was developed by making use of a spreadsheet. Criteria used in the decision making process is population, fracture extent, travel times and hazardous substances.

1 INTRODUCTION

Results from many tracer tests indicate that the minimum distance between a pitlatrine and borehole as proposed by Xu and Braune (1995) will not be adequate in many practical cases to protect the borehole from being polluted. The idea of three protection zones, like in Germany (Kinzelbach *et al.*, 1991), for boreholes at on-site sanitation areas is proposed.

The U.S. Environmental Protection Agency (EPA) was also involved in delineating wellhead protection areas (WHPA). In order to achieve the overall goal of delineating the zone of influence (ZOI) or the zone of contribution (ZOC) of a well in an unconfined fractured-rock aquifer, the methods used were (EPA, 1991):

1. Arbitrary fixed radius.
2. Calculated fixed radius.
3. Vulnerability mapping.
4. Flow-system mapping:
   -with time of travel (TOT) calculations.
   -with analytical equations.
5. Residence time approach.
6. Numerical flow/transport modeling.

After the WHPA is determined, the following steps are proposed by the EPA to protect the groundwater resource (Xu, 1998):

1. Identifying potential sources of contamination in the WHPA.
2. Establishing management approaches to protect the groundwater in the WHPA.
3. Developing a contingency plan for pollution events.
4. Instituting programs for public education and participation.

2 PROTECTION ZONE I: FENCING

For protection zone I (i.e. the immediate fenced area around the borehole), it is proposed that the distance of the fence around the borehole must be at least 5 m. For a borehole that is supplying water to less than say 20 persons, a well-constructed sanitary seal is regarded as enough. Quality monitoring is, however, very important.

3 PROTECTION ZONE II: BACTERIAL AND NITRATE POLLUTION

A second protection zone around the borehole is proposed. The idea with this zone is to protect the drinking water from microbial (bacteria and viruses) and nitrate pollution. Many case studies have shown that bacteria usually die within 30 days after being introduced into the soil. For the delineation of this zone, table 6 in the report by Xu and Braune (1995) will be adequate in some cases. They proposed an absolute minimum distance of 50 m between a pitlatrine and a borehole. However, in many cases in fractured aquifers, this will not be adequate as shown in a following section.

In the case of the delineation of protection zone II, the areal extent of the fracture is very important, and a fracture could be viewed as an extended borehole with a very high T-value and a small S-value. It is thus very important to estimate the areal extent of the fracture. Once bacteria or an element reaches the fracture, its movement will be very rapid towards the abstraction borehole and one of the major mechanisms that will play a role in bacterial die-off time, is the distance from the ground surface to the water level. If a vertical fracture intersects a pitlatrine, the movement of the pollutant to the water level could be very rapid, even in the case of a very deep water level. This illustrates the difficulty to estimate travel times from the surface to the water level in a fractured aquifer. For this reason it is proposed that the whole domain above the fracture be regarded as protection zone II and must be protected. In the following section a method to estimate the size of this capture zone is discussed.

3.1 *Estimation of fracture extent*

3.1.1 *General*

A very good idea of the areal size of the fracture could be obtained by making use of early pumping test data. Considering early data of a pumping tests, it was found that the ratio s/Q (i.e. drawdown in abstraction borehole/ abstraction in L/s) after 1 minute of abstracting water from a borehole, gives a good first approximation of the extent of the fracture and the following generalisation applies:

1. If $s/Q < 0.5$ very good fracture extent
2. If $s/Q < 1$ but > 0.5 good fracture extent
3. If $s/Q > 1$ limited fracture extent

3.1.2 *Estimating the T-value of the fracture*

The following equation (Equation 1-adapted Logan equation) will give an estimate of the T-value of the fracture:

$$T = 10^{(Log\frac{104Q}{s} + Log15Q)/2} \quad (1)$$

where T = transmissivity in m^2/d; Q = abstraction rate in L/s; s = drawdown in m after 1 minute.

3.1.3 *Estimating the S value of the fracture*

For the estimation of the S-value of the fracture, two other parameters must first be determined:

First the theoretical specific storage (S_s) must be determined. The S_s is calculated by applying Equation 2 (Kruseman and De Ridder, 1994).

Secondly the thickness of the fractured zone (D) must be determined (Equation 3).

3.1.3.1 Estimation of S_s.

$$S_s = \rho g\,(\alpha + n\beta) \quad (2)$$

The following values for n and α (obtained from inverse modeling and tracer tests) are proposed:

$\rho g = 9804\ N/m^3$ (specific weight of water)
$n = 0.13$ (porosity)
$\alpha = 5.56 \times 10^{-9}\ m^2/N$ (compressibility of the rock)
$\beta = 4.74 \times 10^{-10}\ m^2/N$ (compressibility of water)

After the application of Equation 2 a proposed value for S_s is 5.6×10^{-5}.

3.1.3.2 Estimating the thickness of the fractured zone (D), (de Lange, 1999).

To estimate S, the thickness of the fracture zone is required and the following equation (obtained from experience from the tracer tests) can be used:

Thickness D(m) of fracture zone = $(0.2*Q/s)*0.14$ (3)

Where s = drawdown (m) in borehole after 1 minute of abstraction at a rate of Q (L/s) from borehole.

After the specific storage and thickness of the fracture zone have been estimated the S-value for the fracture can be estimated by the following equation:

$$S = S_s \times D \quad (4)$$

with S_s = Specific Storage; D = Thickness of fracture zone.

3.1.3.3 Estimating the fracture half length

By applying the Equtions 5 – 9, the half-length of the fracture (x_f in the case of vertical fractures) or the radius of the fracture extent (r_f for horizontal fractures) can be estimated.

a. Vertical fracture (Gringarten and Ramey, 1974)

$$x_f = \frac{Q\sqrt{t}}{2s_w\sqrt{\pi T_f S_f}} \quad (5)$$

where x_f = half length (m); Q=abstraction rate in m^3/d; s_w = drawdown in m after time t (minutes); T_f=T-value of fracture; S_f=S-value of fracture]

b. Horizontal fracture: early storage (Gringarten and Ramey, 1974)

$$r_f = \sqrt{\frac{Qt}{\pi s_w S_f}} \quad (6)$$

c. Vertical or horizontal fracture (Proposed equation - assumed bi-linear flow at early time in fracture).

BPZONE

Estimation of:
(i) Protection Zones I, II and III.
(ii) Maximum NO_3 as N.
(iii) Areal extent of Fracture.
(iv) Travel Times.

Borehole Name:UO5

(I) Protection Zone I : Fencing around borehole and sanitary seal

Dependent on number of persons using on-site sanitation

(II) Protection Zone II : Protection against microbial and nitrate pollution

(a) Maximum NO3 (as N) estimation for different number of people

Concentration of urine as N (mg/l) =	13000
Percentage N reaching water level =	50
Volume urine/person/day [L] =	1
Recharge (mm/a) =	15
Sustainable abstraction rate [l/s] =	0.43
Number of persons using on-site sanitation =	0
in an area with circle around BH (m) =	537

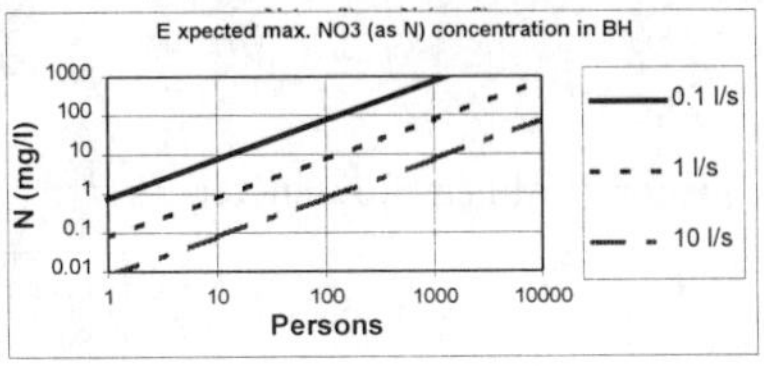

Expected max. N (mg/l) in borehole = 0.000

Risk for Infants = No risk

(b) Areal extent of fracture (vertical or horizontal fracture)

Use early time and drawdown during pumping test

	t(min)	s(m)	Q (l/s)		
Enter other values for t,s,Q if required	1.50	0.177	1.26		
t,s,Q used =	1.5	0.177	1.26		
Est. of D(m) =	0.199	(D(m)= thickness of fracture zone)		Est. of S_f=	9.87E-06
D(m) =	0.199	T_f=	118	118	

Gringarten		
$(x_f^2S_f)$=	0.2659	
x_f (m)=	164	x_f = half length of vertical fracture (use early linear flow)
r_f (m)=	165	r_f = radius of horizontal fracture (use early storage)
r_f (m)=	150	Use early bi-linear flow in fracture
r_f (m)=	148	Combination of linear/bilinear flow in fracture
x_f (m)=	218	Adapted Bohmer equation

Average half length/radius of fracture =	169
with standard deviation (m) =	29

(c) Estimation of travel time from surface to water strike: for bacterial die-off time

K-values of unconsolidated and consolidated material

Unconsolidated material

clay content (%) = 3

K (m/d) = 12.197

Consolidated material Vertical K-value (if unknown)

Use K (m/d) = 0.01 for first estimate

(usually between 0,001 and 0,1 m/d)

(Note: if more than one layer exists, use the harmonic mean of the K-values of the layers)

	Thickness	K (m/d)	Porosity	time (d)
(a) unconsolidated material	2	12.197	0.2	0.033
(b) consolidated material to water strike	19	0.005	0.1	380.000
(c) Depth to water level (m) =	6			
Travel time to water strike (d) =	380.03			
Travel time to water level (d) =	80.03			
Risk of microbial pollution in BH =	Low			

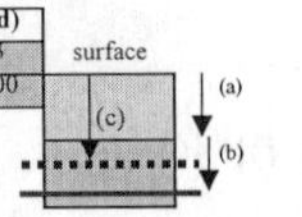

(III) Protection Zone III: Estimation of BH catchment area for hazardous pollution

Do hazardous elements exist? (0/1) =	0	(0 = No; 1 = Yes)
recharge (mm/a) = enter	15	
Catchment area (m^2) =	904032	
i.e. an area around BH with radius=	537	
Protection zone III necesarry =	No	

Final Results

Protection Zone I:	No fencing around borehole required : only sanitary seal	
Protection Zone II:	Protect an area around BH with radius (m) =	85
Protection Zone III:	No Protection Zone III required	

Figure 1. Sample output of program BPZONE for borehole UO5 on the Campus Site.

$$r_f = \frac{Q(t)^{0.25}}{4s_w(T_f S_f)^{0.25}} \quad (7)$$

d. Vertical or horizontal fracture (Proposed equation - assumed a combination of linear/bi-linear flow in fracture at early time).

$$r_f = \frac{Q(t)^{0.333}}{4s_w(T_f S_f)^{0.333}} \quad (8)$$

e. Adapted Bohmer equation (bi-linear flow in vertical dyke/fault)

$$x_f = \frac{Q(t)^{0.25}}{2.74s_w(T_f S_f)^{0.25}} \quad (9)$$

4 PROTECTION ZONE III

If persistent hazardous non-degradable elements are present, the whole catchment area of the borehole must be protected. Because of the use of the word "hazardous", some consideration must be given to its implications. Therefore the importance of risk assessment must be considered.

5 PROGRAM *BPZONE*

5.1 *Explanation of program* BPZONE

Figure 1 is an example of the output of program BPZONE for borehole UO5 on the Campus Test Site of the University of the Free State.

5.1.1 Protection Zone I

If more than 20 persons are using pitlatrines, it is proposed that a fence of at least 5 m must be constructed around the borehole. Otherwise just a sanitary seal is required.

5.1.2 Protection Zone II

The maximum expected NO3 (as N) is estimated by using the following information:

1. Concentration of urine = 13000 mg/l (can be changed)
2. Percentage N leaching = 50% (can be changed).
3. Volume urine/person/day = 1 L (can be changed).
4. Average annual recharge in mm.
5. Abstraction rate of the borehole (i.e. sustainable yield) in L/s.
6. Number of persons using on-site sanitation in the catchment area of the borehole (the catchment area is estimated from the abstraction rate and recharge).

   Program BPZONE then estimates the maximum expected N in boreholes and the risk to infants according to table 1

Table 1: Nitrate value intervals and the risk (to infants) associated with each interval.

Parameter	no risk	low risk	high risk	very high risk
N (mg/l)	<5	5-10	10-20	>20

The next step is to estimate the extent of the vertical or horizontal fracture by using early pumping test data (i.e. the drawdown after 1 minute). To estimate the extent, the program uses the five equations presented earlier (Equations 5 – 9). The mean of the estimated fracture extent together with the standard deviation is given.

Step 3 is the estimation of the travel time from the surface to the water level and the position of the water strike. For the estimation of travel times a number of parameters is required, i.e.

i) The saturated vertical K-value of the unconsolidated material. If the K value is not known, it was decided to try and estimate it by means of the clay content of the unconsolidated material. Data were gathered of soils with known values for clay content and K-values. A graph was drawn of clay content vs. K-value and the equation of the exponential trend line fitted with the data was incorporated in the program BPZONE.

$$K_u = 30\, e^{-0.3*cl} \quad (10)$$

Where cl = clay content (%)

ii) Vertical K-value of the consolidated material. If unknown it is suggested that a value of 0.01 be used (usually between 0.001 and 0.1 for South African rocks).

iii) Distance from surface to water level.

iv) Thickness of the unconsolidated and consoli dated material and their kinematic porosities.

The total travel time to the water strike is estimated from (assuming a gradient of 1):

$$t = \frac{n_{eu} D_u}{K_u} + \frac{n_{ec} D_c}{K_c} \quad (11)$$

where t=travel time (d); n=kinematic porosity; D=Thickness (mm) and K=hydraulic conductivity (m/d) . Subscripts u and c denote unsaturated and saturated zone repectively.

Risk of microbial pollution is assigned by using the following criteria:

Table 2: Travel times and the risk of microbial pollution assigned to each.

Parameter	no risk	low risk	high risk	very high risk
Travel time (days)	>100	30-100	3-30	<3

The program then estimates protection zone II by using the following criteria:

Table 3: Criteria involving risk and proposed distances for protection zone II.

Risk	Protection zone II
High or very high risk of microbial pollution.	2 times fracture half-length.
Low risk for microbial pollution.	1 times fracture half-length.
No risk for microbial pollution.	0.5 times fracture half-length.
High risk for N for infants.	2 times fracture half-length.
Low risk for N for infants.	1 times fracture half-length.
No risk for N for infants.	0.5 times fracture length.

5.1.3 Protection zone III

If hazardous chemical elements are present the program estimates the catchment area of the borehole by using the recharge and the abstraction rate (Equation 12).

$$AREA = Q / Recharge \qquad (12)$$

6 JUSTIFICATION WITH *MODFLOW* GENERATED EXAMPLES

The well-known *MODFLOW 3D* finite difference program (*PMWIN*, Chiang and Kinzelbach, in press) was used to generate three typical case studies in fractured-rock aquifers. The model constructed has 10 layers with a thickness of 1 m each, except for the bottom layer (i.e. the fracture zone) which was assigned a thickness of 0,2 m (i.e. the typical situation on the Campus Aquifer). The following parameters were assigned to the different layers (Table 6.7):

Table 4: Parameter values assigned for the generated Modflow model

Layer	S_s	K_h(m/d)	K_v(m/d)
1-9	5e-5	0.1	0.005
10	5e-5	3600	.005

Where S_s = specific storativity [1/m]; K_h = horizontal K-value; K_v = vertical K-value

This gives a T-value of the matrix = 1 m^2/d and T-value of the fracture = 720 m^2/d as on the Campus of the University of the Free State Test Site.

A fracture zone was assigned to the bottom layer (i.e. layer 10) and three fracture radii were used, i.e. 60m, 100, and 200m (i.e. the case of a horizontal bedding plane fracture). The model was run for 300 minutes for each of the set ups and the drawdown after 1 minute (with an abstraction rate of 1.25 l/s) was supplied to the *BPZONE* program and the results are shown in Table 5.

Table 5: Comparison between *MODFLOW* and *BPZONE* results (mean and standard deviation of the 5 equations) for the estimation of the radius of the fracture

PROGRAM	r_f(m)	r_f(m)	r_f(m)
MODFLOW (value assigned)	60	100	200
BPZONE (mean of 5 equations)	76 ± 15	113 ± 19	175 ± 40

7 CONCLUSIONS AND RECOMMENDATIONS

1. Due to the heterogeneity of South African conditions concerning the groundwater environment, this document can be seen as a continuation of the work that has already been done on delineating protection zones. The information in this paper must be regarded as such and not as the ultimate solution to assigning protection zones. There is room for many more approaches as well as studies on this topic.

2. In this paper the focus is on nitrates and microbes due to their association with pitlatrines and septic tanks. The same approach can be applied to more parameters depending on what type of study is being done.

3. The application of risk assessment and management is of great importance. However, great care must be taken and the person(s) responsible should be adept in the process and be able to evaluate a given situation with care before any major decisions are reached.

4. The program BPZONE must be used with great care and the answers obtained must be viewed within the existing conditions of a given situation or problem.

REFERENCES

Chiang, W.H. and Kinzelbach, W. (in press). Three dimensional groundwater flow and transport modeling. Springer, Berlin, Heidelberg, New York.

De Lange, S.S. (1999). Environmental impact of point pollution sources. Unpublished M.Sc.thesis, University of the Free State, Bloemfontein, South Africa.

EPA, (1991). Delineation of wellhead protection areas in fractured rocks. Wisconsin geological and natural history survey. Ground-water protection division, Office of groundwater and drinking water, U.S. environmental protection agency, Washington, DC 20460.

Gringarten, A. C. and Ramey, H. J. (1974). Unsteady-state pressure distributions created by a well with a single horizontal fracture, partial penetration, or restricted entry. *Society of Petroleum Engineers Journal.* **14** (5), 413-426.

Kinzelbach, W., Marburger, M., Chiang, W. (1991). Determination of groundwater catchment areas in two and three spatial dimensions. *Journal of Hydrology*, 134 (1992): 221-246.

Kruseman, G.P. and De Ridder, N.A. (1994). Analysis and evaluation of pumping test data. ILRI publication 47, Wageningen, The Netherlands.

Xu, Y. (1998). Delineation of borehole protection areas in fractured aquifers. Unpublished Ph.D thesis, University of the Free State, Bloemfontein, South Africa.

Xu, Y. and Braune, E (1995). A Guideline for Groundwater Protection for the Community Water Supply and Sanitation Programme. Department of Water Affairs and Forestry, Pretoria, South Africa.

Groundwater: Past Achievements and Future Challenges, Sililo et al. (eds) © 2000 Balkema, Rotterdam, ISBN 90 5809 159 7

Degradation behaviour of halogenated aliphatics influenced by concurrent chromate reduction in subsurface Fe^0-reactor systems

M. Fruth
Faculty of Agriculture and Environmental Protection, University of Applied Science Weihenstephan, Triesdorf, Germany

M. Schneider
Institute of Geology, Geophysics and Geoinformatics, Free University of Berlin, Germany

ABSTRACT: Halogenated aliphatics and redox-sensitive inorganic species like hexavalent chromium, Cr(VI), can be transformed by abiotic reduction processes. Installing zero-valent iron in a permeable reactor system in the flow path of contaminated groundwater, leads to adsorption of the aliphatics (e.g. TCE, DCE, cDCE, VC) at the iron surface and to reductive dehalogenation. If groundwater is contaminated by halogenated aliphatics and chromate, the process of reductive dehalogenation is significantly inhibited by the concurrent process of chromate reduction. Column experiments were conducted to examine these processes. It could be shown that the precipitation of the reaction products on the metal surface leads to passivity of the reactive material and decreasing reduction rates.

1 INTRODUCTION

Halogenated aliphatics like trichloroethylene (TCE), cis-, trans-, 1,1-dichloroethylene (DCE) and vinyl chloride (VC), as well as the inorganic contaminant chromate (Cr(VI)) are not degraded or very slowly degraded in groundwater. Therefore it is necessary to eliminate those toxic contaminants or to reduce their further dissemination through appropriate remedial actions. Nowadays passive hydraulic groundwater remediation methods are intensively discussed for this purpose. By this approach subsurface in situ reactor systems are installed in the groundwater flow path in order to remove dissolved contaminants. In redox-sensitive reactor systems elemental iron can be used to remove chromate and chlorinated solvents by reduction processes.

The main degradation process for halogenated aliphatics is reductive dehalogenation under abiotic conditions by applying zero-valent iron as reducing agent. The first tests of reductive dehalogenation of halogenated organic compounds by zero-valent iron were patented by Sweeny & Fischer (1972) (in: Gillham & O'Hannesin 1994). These tests were followed by various investigations of different materials which can be used as reductive agents for abiotic dehalogenation (Bachmann et al. 1991; Gillham & O'Hannesin 1994; Matheson & Tratnyek 1994). It could be shown that degradation rates are of magnitudes higher than those of natural systems which led to further developments of this technology. More detailed tests have shown that degradation rates of chlorinated solvents can be influenced by the presence of CO_2^{2-}, HCO_3^{2-} and PO_4^{2-} due to their inhibiting effects (reduction of reactivity and permeability) on the iron surface. Plumes of contaminated groundwater usually contain more than one contaminant, so the question arises as to whether there are further negative influences on the degradation rate of halogenated aliphatics.

One typical scenario is the occurrence of organic compounds and high concentrations of chromate which can be reduced to Cr(III). Column tests are conducted to investigate the influence of chromate reduction on the degradation of TCE, DCE and VC by applying zero-valent iron. The main objective was to determine how long a specific material can be used in a reactor system for removing dissolved contaminants in a descrete environment.

2 CHEMICAL BACKGROUND

The process of reductive dehalogenation involves reactions at the Fe^0 surface. This means a direct electron transfer from the iron to the adsorbed chlorinated species. The dechlorination of TCE occurs in three steps and leads to the metabolites DCE (1,1-DCE, 1,2-DCE, t-DCE) and VC, which is finally reduced to ethene.

$$3Fe^0 + 3H_2O + C_2HCl_3 \Leftrightarrow 3Fe^{2+} + C_2H_4 + 3OH^- + 3Cl^- \quad (1)$$

In a Fe^0-H_2O system corrosion occurs as shown in equation (2).

$2Fe^0 + 2H_2O \Leftrightarrow H_2 + 2Fe^{2+} + 2OH^-$ (2)

In the presence of hexavalent chromium as an additional contaminant in groundwater the main reaction is described in equation (3).

$Fe^0 + CrO_4^{2-} + 4H_2O \Leftrightarrow FeCr(OH)_6 + 2OH^-$ (3)

Degradation of TCE, iron corrosion and reduction of chromate result in increasing pH values. This pro??cess causes the precipitation of slightly soluble mineral phases like magnetite (Fe_3O_4) and Fe(III)-Cr(III)-hydroxide phases, leading to passivation of the metal surface and therefore to reduced degradation rates.

3 COLUMN EXPERIMENTS

3.1 Apparatus and materials

For the simulation of the degradation of TCE, several columns (two 25 cm in length and one 100 cm in length) were packed with iron grains (diameter 1.0-1.6 mm, specific weight 7600 kg/m³, 90% Fe^0). A peristaltic pump was used to generate a constant flow rate of the solutions from the column base upward.

3.2 Sampling and analytical methods

Samples of the TCE/Cr(VI)-solution were collected from sampling ports located along the column. Concentrations of halogenated aliphatics were determined by using chromatographic methods. Chromate was determined by ion chromatography.

4 RESULTS

The inhibition velocity (in cm/d) of the chromate reduction was determined on the base of concentration profiles of Cr(VI) in the test column. With the inhibition velocity the delay of Cr(VI) reduction respectively TCE degradation due to passivation of iron surface can be expressed. Increasing Cr(VI) input concentrations (52 mg/l and 156 mg/l) resulted in an increase of inhibition velocity from 0.61 cm/d to 1.5 cm/d. Under consideration of the installed mass of zero-valent iron in the test columns and their specific surface, the inhibition rate of chromate reduction in m^2 of passivated iron surface per mg Cr(VI) can be computed. On average 0.34 m^2 surface is passivated per mg Cr(VI) input.

The column tests were conducted in 4 phases with stepwise increases of Cr(VI) input concentration (0mM; 0,19mM; 1mM; 3mM). For each step, 6 concentration profiles of chromate, TCE and the metabolites DCE (1,1-DCE; 1,2-DCE) and vinyl chloride were detected. The computation of the flow path when 50% of TCE is degraded (s_{50}) gives the inhibition velocity in cm/d, characterising the influence of chromate on TCE degradation. Additionally, the time for degrading of 50% of the initial concentration of TCE (t_{50}) can be computed (see Fig. 1).

Figure 2 shows the TCE degradation front along the flow path dependent on the particular Cr(VI) input concentration. In the test phase with input concentrations of 10 mg/L (0.19 mM) and 52 mg/L (1mM) Cr(VI), TCE degradation is slightly influenced. When the initial concentration is increased to 156 mg/L (3 mM), the TCE degradation front is shifted to the upper part of the test column. With increasing Cr(VI) input concentrations, an accelerated inhibition velocity can be observed.

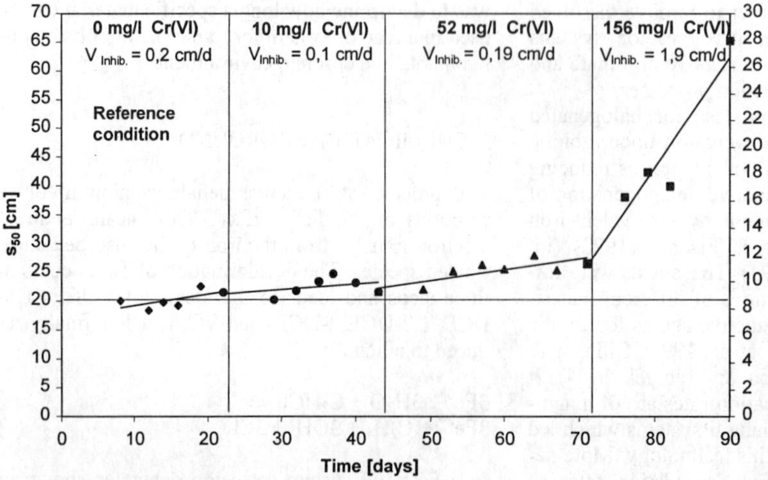

Figure 1. s_{50} and t_{50} data versus time and inhibition velocities of each test phase.

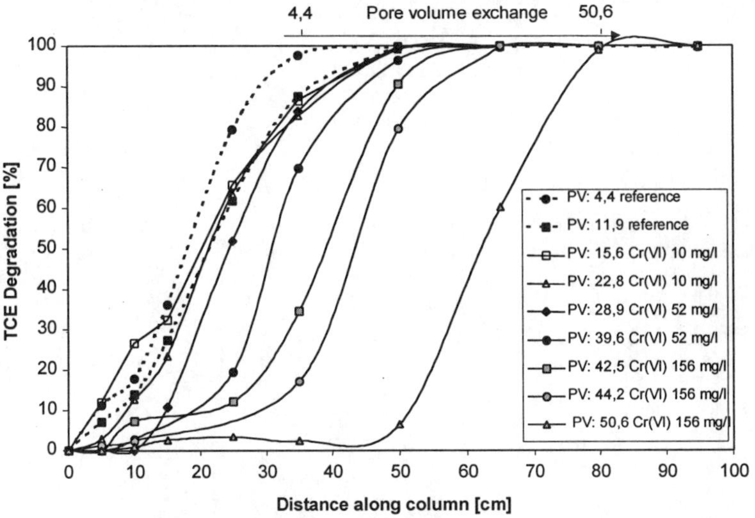

Figure 2. Movement of TCE degradation front along the column. PV = Number of exchanged pore volumes.

The structure of the precipitates on the iron grains was investigated by X-ray diffraction methods. The analyses have shown that magnetite (Fe_3O_4), chromite ($FeCr_2O_4$) and greenalite ($Fe_3Si_2O_5(OH)_4$) can be considered as the main compounds of the precipitates. TEM (transmission electron microscopy) analysis in comparison with investigations of Pratt et al. (1997) indicated the presence of geothite (α-FeOOH) and escolaite (Cr_2O_3).

Column tests conducted at the Institute of Hydraulic Engineering at the University of Stuttgart indicate that the reaction of carbonate (CO_3^{2-}), phosphate (PO_4^{2-}) und sulfide (S^{2-}) with Fe^{2+}, leads to the formation of solid phases on the iron surface. Whereas the coating of phosphate compounds decreases reactivity, the formation of iron sulfides causes an increase in chemical activity (Schlicker et al. 1998).

5 CONCLUSIONS

The column tests confirmed that the degradation of halogenated aliphatics is influenced by concurrent chromate reduction at the iron surface. With passing time and increasing chromate input concentration, the TCE degradation was significantly inhibited.

In the absence of chromate, the TCE degradation rate depends on the sorption capacity of the iron surface and on the transport velocity of TCE molecules to the metal surface. The mass transfer rate is particularly reduced if coatings form on the iron grains, passivating the metal surface. In the Fe^0-H_2O system too, the precipitation of sparingly soluble Fe(III)-compounds can be observed as indicated in the stability diagram for iron. It was clearly shown that, in the test column, chromate reduction has priority over TCE degradation. So it can be stated that chromate acts as an inhibiting factor in the system, causing TCE degradation to be delayed. Due to precipitation of reaction products, the available surface area is continuously reduced, inhibiting the rate of reduction. More detailed analysis is necessary to identify the structure of the precipitates.

6 REFERENCES

Bachmann, T., E. Heitz & I. Vermes 1991. Abbau umweltbelastender Kohlenwasserstoffe durch Metalle. In *Dechema – Monographien* Band 124, Weinheim (VCH Verlagsgesellschaft mbH): pp 221-239.

Gillham, R.W. & S.F. O`Hannesin 1994. Enhanced Degradation of Halogenated Aliphates by Zero-Valent Iron. In *Ground Water*, Vol. 32 No. 6: pp 958-967.

Matheson, L.J. & P.G. Tratnyek 1994. Reduktive Dehalogenation of Chlorinated Methanes by Iron Metall. In *Environ. Sci. Technol.* Vol. 28 No.12: pp 2045-2053.

Pratt, A.R., D.W.Blowes & C.J. Ptacek 1997. Products of Chromate Reduction on Proposed Subsurface Remediation Material. In *Environ. Sci. Technol.* Vol. 31: pp 2492-2498.

Schlicker, O., M. Ebert, H. Schad, W. Wüst & A. Dahmke 1998. Geochemische Modellierungen zur Abschätzung der Grundwassergüte im Abstrom von Fe^0-Reaktionswänden. In *Terra Tech Sanierungspraxis* Band 1: pp 43-46.

Groundwater: Past Achievements and Future Challenges, Sililo et al. (eds) © 2000 Balkema, Rotterdam, ISBN 90 5809 159 7

Bacterial contamination in Norwegian groundwater wells in bedrock

S.Gaut & G.Storrø
Geochemistry and Hydrogeology Section, Geological Survey of Norway, Trondheim, Norway

B.Brattli
Department of Geology and Mineral Resources Engineering, Norwegian University of Science and Technology, Trondheim, Norway

ABSTRACT: This paper describes the extent of bacterial contamination in waterworks in Norway that use groundwater derived from bedrock. Bacteriological analyses were obtained on water from 195 waterworks where most of the facilities supports more than 100 persons or 20 households. 26% of the waterworks have no bacterial contamination whereas 74% either have too high heterotrophic plate count (HPC) or are contaminated by coliforms. Bacterial contamination related to poor well design and protection, may be indicated, among others, by seasonal changes in the bacterial content and high content of Fe and Mn. Physiochemical data indicate that wells without bacterial contamination have low alkalinity and low conductivity. Waterworks along the coast appear to have a greater chance to get bacterial contamination. This may be related to lack of surficial deposits to protect the aquifers from surface water contamination.

1 INTRODUCTION

This paper describes the extent of bacterial contamination in waterworks in Norway that use groundwater derived from bedrock. It presents the first part of a PhD thesis where the aim is to look at the vulnerability of groundwater wells in bedrock concerning bacterial contamination. The factors that contribute to the bacteriological quality of the water are examined and a method for protection around groundwater wells in bedrock will be formulated.

Many small waterworks and private houses in Norway use groundwater derived from bedrock. Each year between 3000 and 4000 new wells are constructed and 3.4-6.8 % (Morland 1997) of the population is supplied with groundwater. The majority of the bedrock wells are located in rural areas where municipal waterworks can not offer an alternative water supply (Morland 1996). In the past few years the focus has been on both the physiochemical and bacteriological quality of groundwater from these wells. Information from the National Institute of Public Health and various departments of the Norwegian Food Control Authority indicate bacterial contamination in many of these groundwater wells and the National Institute of Public Health is considering to require disinfection of all drinking water from these waterworks.

This study examines how many of these waterworks use water where the heterotrophic plate count (HPC) or the number of coliforms (Total coliforms (TC) and fecal coliforms (FC)) exceed the national standards of drinking water quality (Sosial- og helsedepartementet 1995). In addition, the bacteriological data and the location of the waterworks will be examined in an effort to identify what causes the bacterial contamination.

2 DATA COLLECTION

The first part of the study included collecting bacteriological data from waterworks. A list of waterworks that use groundwater derived form bedrock were put together from databases mainly at the National Institute of Public Health and the Geological Survey of Norway (NGU). This list was presented to the different departments of the Norwegian Food Control Authority in 1998 and they contributed with bacteriological analyses of water from 195 waterworks in Norway, which support primarily more than 100 persons or 20 households with groundwater derived from bedrock. The rest of the data relate to groundwater wells supplying water to cafés, schools and public institutions. In the request for bacteriological analyses data from the years 1996 and 1997 were asked for. Data from these years are received from the departments of the Norwegian Food Control Authority that replayed quickly. However, collection of data continued until the end of 1999 and for some of the waterworks bacteriological analyses from 1998 and 1999 are received instead of or in addition to data from 1996 and 1997. The most common bacteriological analyses and the Norwegian standards for the bacteriological quality of drinking water are shown in Table 1. The drinking water is sometimes also analysed for sulfite reducing clostridium and fecal streptococcus.

Table 1. Type of bacteriological analysis and Norwegian standards for drinking water quality.

Type of analysis	Guidance level	Maximum allowable concentration
HPC* (22°C) /ml	100	-
HPC (37°C) /ml	10	-
TC** (37°C) /100ml	0	0
FC*** (44°C) /100ml	0	0

* HPC = Heterotrophic plate counts
** TC = Total coliforms
*** FC = Fecal coliforms

In according to the Norwegian standards for drinking water quality, water samples for bacteriological analyses must be taken at least once a month. This is only done by 50% of the waterworks participating in this study. The other waterworks have sample frequencies of between 1 and 9 samples per year, with the majority taking at least 4 samples a year.

The results of the bacteriological analyses were examined and classified in one of five groups:

1 No problems with bacteria.
2 Often problems with too high HPC
3 Seldom problems with too high HPC
4 Often problems with too high number of TC and/or FC.
5 Seldom problems with too high number of TC and/or FC.

Each group were then given a symbol and a colour to help locate the bacteriological quality when each waterwork were marked on a map. This map was then used to examine any obvious connection between geographical site and bacteriological quality. The map was also used to pick out approximately 40 waterworks, which will be more closely examined by fieldwork during the summer/autumn 2000.

3 DISCUSSION OF DATA

The bacteriological data show that 1/4 of the waterworks have no problem with bacterial contamination whereas 3/4 do have problems with the bacteriological quality of the drinking water (Table 2). About half of the waterworks with bacterial contamination only have a problem with the HPC. The second half has problems with coliforms alone or both coliforms and HPC. More than 60% of this latter group have often a number of coliforms that exceed the national standards for drinking water quality.

Table 2. Bacterial contamination in the 195 groundwater wells in bedrock being a part of this study, expressed in percentages and fraction.

	Bacterial contamination		
	None	HPC	Coliforms*
Group	1	2 and 3	4 and 5
Per cent (%)	26	34	40
Fraction	1/4	3/4	

* Some wells also have too high HPC

One problem with the bacteriological data is that the samples from several of the waterworks are not collected directly from the supply well but are tap water from a household or a public institution. It means that the pipeline will affect the water quality and the sample will not be entirely representative for the groundwater from the well. Because of this the authors expect some of the problems with too high HPC to be caused by the pipeline and not by the groundwater itself. Cases were the groundwater is treated with UV, chlorination or other methods and the analyses of tap water show a high HPC (>100) lend weight to this assumption. To be able to describe the extent of this problem further study will be done to properly divide between those samples taken from the well and those that are tap water.

Poor design or protection of the groundwater well may cause bacterial contamination. A groundwater well in Norway is supposed to be enclosed by a fence with a radius of at least 5 m. In addition the well should be inside a well house and the top of the borehole closed with a lid. When this requirement is not met, different kind of activities by humans or animals close to the well can cause bacterial contamination of the groundwater. This is especially important if surface runoff is allowed to drain directly into the well.

Most groundwater recharge in Norway occurs during spring and autumn. A higher occurrence of bacteria, especially coliforms, in the wells during these seasons may indicate poor well design or protection. When examining seasonal changes, data were only used from the waterworks that followed the regulations. These data show that for some groundwater wells the content of bacteria in the drinking water increases during April and May and again in September and October.

Some of the waterworks had their groundwater analysed in connection with a joint project between the Norwegian Radiation Protection Authority (NRPA) and NGU in 1996 (Banks et al. 1998). This project looked into the problems of radon in drinking water taken from groundwater wells in bedrock. In addition to radon, water samples were also analysed for physio-chemical parameters. Comparing the content of different elements with the bacteriological data only indicate poor correlations. However, wells contaminated with bacteria often show high contents of Fe and Mn. The data also indicate that wells without bacterial contamination have low alkalinity. A study of groundwater from bedrock wells in North-Trøndelag (Hanssen 1998) also indicates such a correlation as well as a correlation between low bacterial content and low conductivity.

The correlation between bacteriological quality of the water and geographical site is also examined. A trend is emerging where waterworks situated close to the coast or on islands have more often problems with bacteria, whereas waterworks in the mountains or those that are situated in a higher area than the surroundings have less bacterial contamination. This difference might have several reasons. On the smallest islands there are fewer places to drill a groundwater well than on the mainland and the well will more easily be influenced by human activities. In addition, along the coast of Norway the surficial deposits are only very thin (<1.0 m and discontinuous) or non existing (Thoresen 1990). This means that the surficial deposits will have little potential to filter out bacteria from the surface water that infiltrates into the fracture aquifer. There is less human activity in the mountainous areas and together with more extensive surficial deposits further from the coast, this will better protect the groundwater aquifer and well against contamination.

4 POSSIBLE SOURCES OF BACTERIA

Infiltration of contaminated water has been mentioned as a possible cause for bacteria in the groundwater wells. One of the sources of this contamination might be leakage from manure pits or from the spreading of the manure on nearby fields. In addition human activities or animals may contaminate wells without proper well protection such as a well house or a fence

Leakage to the groundwater from septic tanks and sewage systems might be a source for bacterial contamination but this is not expected to be a common problem in connection with the waterworks. Contamination of the groundwater from septic tanks, sewage systems and outdoor toilets is probably more often a problem in connection with private wells.

As mentioned analysed water samples are often tap water and will be influenced by the condition of the pipeline. Sources of bacteria can be leakage through cracks in the pipeline or from bacterial coating inside the pipeline.

Other sources of bacteriological contamination can be landfills or lagoons. These are however sparsely distributed and the authors do not think this is a big problem.

5 CASE STUDY

The Sørlandet waterwork is situated at the island Værøy in Lofoten (Figure 1). It is one of the biggest waterworks in Norway utilising groundwater from bedrock (Hilmo and Anda 1999, Hilmo and Gaut 2000). Because of capacity- and water quality problems the local government asked NGU to examine the capacity and the quality of each groundwater well. Several of the wells contain too high HPC and at least in one of the wells the groundwater contains coliforms. Fieldwork with inspection of the wells was done in 1998. The well field, situated in a flat area surrounded by steep mountainsides, consisted in 1998 of 18 drilled wells with a depth of 27-80 m. In the well field surficial deposits occur with thicknesses up to 10 m. These sediments consist of shoreline sand, gravel deposits and bog and talus along the mountainsides. The groundwater level is high and in wet periods the water rises to ground level. Inspection of the wells showed that Sørlandet waterwork is a good example on how poor design and protection of the groundwater wells can lead to bacterial contamination. The oldest wells had fence and concrete well-protection cover but neither were tight. The other wells had only a lid on top of the well casing. For several of the unprotected wells the top of the casing were cut close to or almost below ground level, and this was also the case inside the well-protection (Fig. 2). Thus, surface water can flow into the well and this is probably the reason for the bacterial contamination of the groundwater.

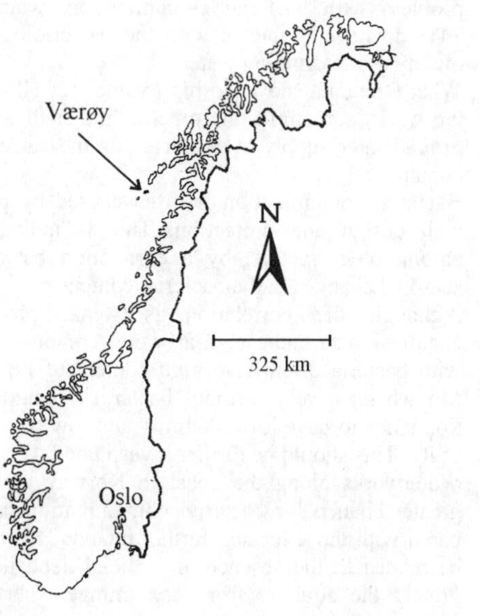

Figure 1 Map of Norway showing the position of Værøy.

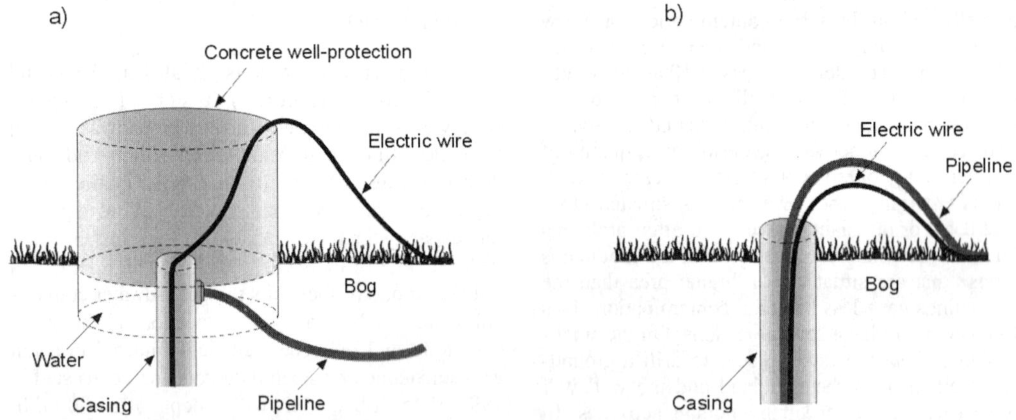

Figure 2. The figure shows examples of poor well protection. a) Well with concrete well-protection where water flows into the well. b) Well without any protection.

6 FURTHER WORK

During the summer/autumn of 2000, approximately 40 waterworks will be examined. During fieldwork pollution sources will be located and examination of the well design, fractures and soil cover will be carried out to identify the causes of bacterial contamination. Steps to solve existing problems will be evaluated.

7 CONCLUSIONS

1 Examination of the collected bacteriological data shows that 26% of the waterworks do not have a problem with bacterial contamination, whereas 74% do have problems with the bacteriological quality of the drinking water.
2 Where the data show too high values for HPC in tap water, sometimes the pipeline rather than the groundwater supply is suspected as the cause of contamination.
3 Bacterial contamination is often caused by poor well design and protection. This is indicated, among other factors, by a correlation between seasonal changes in the bacterial content.
4 Although the correlation is weak, physio-chemical data indicate that wells contaminated with bacteria often have high content of Fe and Mn whereas wells without bacterial contamination tends to have low alkalinity and low conductivity. This should be further investigated.
5 Waterworks along the coast in Norway have a greater chance for bacterial contamination compared with those located further inland. This may be related to the absence of surficial deposits to protect the aquifers from contaminated surface water and/or human activity in the vicinity of the groundwater wells. This is a major issue that will be further examined during fieldwork.

REFERENCES

Banks, D., Frengstad, B., Midtgård, Aa. K., Krog, J.R. & Strand, T. 1998. The chemistry of Norwegian groundwaters. I. The distribution of radon, major and minor elements in 1604 crystalline bedrock groundwaters. *The Science of the Total Environment* 222: 71-91.

Hanssen, G. 1998. *Bakterier og grunnvann.* Prosjektarbeid. Department of Geology and Mineral Resources Engineering, Norwegian University of Science and Technology, Trondheim, Norway

Hilmo, B.O. & Anda, S. 1999. Sørlandet grunnvannsanlegg på Værøy. Kapasitet og vannkvalitet. *Kommunalteknikk* 6/99: 10-14

Hilmo, B.O. & Gaut, S. 2000. A water-supply system using groundwater from bedrock wells at Værøy, Lofoten. Oral presentation 24th Nordic Geological Winter Meeting, 6-9 January 2000, Trondheim, Norway. Abstract in *Geonytt* 1/2000: 83-84

Morland, G. 1996 Bruk av grunnvann i Norge [Use of groundwater in Norway]. *Norges geologiske undersøkelse report 96.082*

Morland, G. 1997. Petrology, lithology, bedrock structures, glaciation and sea level: Important factors for groundwater yield and composition of Norwegian bedrock boreholes? *Geological Survey of Norway report 97.122*

Thoresen, M.K. 1990. Kvartærgeologisk kart over Norge. Tema: Jordarter. M 1:1 mill. *Geological Survey of Norway.*

Sosial- og helsedepartementet 1995. Forskrift om vannforsyning og drikkevann m.m. [Directive on water supply and drinking water etc. - in Norwegian]. No. 68. I-9/95.

Groundwater: Past Achievements and Future Challenges, Sililo et al. (eds) © 2000 Balkema, Rotterdam, ISBN 90 5809 159 7

Soil pore manipulations to increase bioaccessible pore volume

J.H.Görres, M.A.Stolt & J.A.Amador
Department of Natural Resources Science, University of Rhode Island, Kingston, R.I., USA

C.Schulthess
Department of Plant Science, University of Connecticut, Storrs, Conn., USA

P.W.Johnson
Electron Microscope Facility, College of the Environment and Life Science, University of Rhode Island, Kingston, R.I., USA

ABSTRACT: Nanopores residing in iron oxide coatings of soil grains may sequester and retain organic pollutants, making them less available for bioremediation. We investigated nanopore (diameters < 0.1 μm) and bioaccessible pore volume (diameters > 1 μm) in four soils before and after iron oxide removal. The specific nanopore volume of the soils varied from 12% to 38% of total specific pore volume. The removal of iron oxide coatings altered pore structure, increasing the bioaccessible pore space of three out of four soils. The manipulation of pore volume by reduction and removal of iron oxides may, thus, affect the bioremediation potential of soils. However, redox reactions may also have undesirable effects on the fate of pollutants and their use to enhance the efficacy of remediation needs to be evaluated.

1 INTRODUCTION

The sequestration and retention of organic chemicals affect their fate in the environment. Sorption to organic matter is regarded as the overriding factor in the retention of organic compounds in soils and sediments (Chiou et al. 1979). However, physical entrapment and sorption in micropores in the nanometer range, so called nanopores, may also affect the retention of organics. Nanopore volumes of natural soils, as part of a multilinear regression model, contributed to the sequestration of atrazine into soil (Alexander, pers. comm.). Nanopores have been hypothesized to be important in providing sorption sites for organic pollutants (Steinberg et al. 1987, Pignatello 1989).

The nanopore structure of iron oxides coating mineral grains may provide sites for the adsorption of and voids for the entrapment of organic pollutants. Sorption of organic pollutants is correlated to Fe or Al oxides (Werkheiser & Anderson 1996, Ukrainczyk & Ajwa 1996). In soils where pollutants may bypass the A-horizon in preferential flow (e.g. Jarvis et al. 1991), sorption to iron oxides in subsurface horizons may provide an alternative retention mechanism. Compounds sorbed to nanopore surfaces are immobilized and removed from spaces accessible to microorganisms, thus protecting them from biodegradation (Alexander, pers. comm.). However, they may be remobilized when soils are exposed to redox fluctuations, resulting in the formation or dissolution of iron oxide coatings. Whether the effects of nanopores (diameters < 0.1 μm) on sequestration of organic compounds in soils is significant depends on how much nanopore space is created by the iron oxides. The removal of iron oxides may influence pore structure at different spatial scales. Pore structure changes may occur at the grain scale, where iron oxides act as cementing agents and interstitial fillers. Other changes may occur at the scale of the iron oxide crystals whose internal pore structure is thought to contribute to nanopore volume. Both changes can impact the accessibility of a pollutant to microorganisms and, thus, alter the potential for bioremediation.

We altered the pore structure of soils with different iron contents by exposing them to a reduction agent. We also examined the pore structure of natural aggregates of goethite crystals. Goethite is a common iron oxide mineral in soils and its crystal structure is well understood. Our objective was to verify that nanopores existed within goethite crystal aggregates, to quantify the pore volume distribution in different soil pore size intervals and to assess the bioaccessibility of pore spaces before and after treatment. Pores with diameters > 1 μm are regarded as bioaccessible. Our hypotheses were that:

- there is a smaller fraction of biologically accessible pore space in the untreated than in the treated materials.
- nanopore volumes are reduced when iron oxide coatings are removed from grain surfaces.

2 MATERIALS AND METHODS

2.1 *Soils and soil characterization*

We obtained B-horizon material from the Hiwassee, Cecil, Christiana and Bridgehampton soil series. The colours of these materials suggested varying degrees of iron content. We determined the extractable iron concentration of the soils from dithionite citrate bicarbonate (DCB) extracts using standard atomic absorption spectrometry. Particle size distributions were determined using sieves for the sand fraction and the pipette method for the silt and clay fractions.

2.2 *Treatments*

Iron oxide coatings were removed by shaking the soil samples (treated samples) in a solution of DCB for 16 hours (Holmgren 1967). They were washed three times with a 10% NaCl solution and rinsed with distilled water. The samples were dried in a forced-air oven at 105 °C.

2.3 *Mercury Intrusion Porosimetry (MIP)*

Pore size distributions of soil samples were measured using mercury intrusion porosimetry (Lawrence 1978). The method allows the determination of pore volume distributions for pores with diameters of 3 nm to 300 μm from small samples. MIP does not distort soil pore structure even at high pressures (Lawrence 1978).

In MIP, pressure is applied step-wise and the intrusion volume is measured for each pressure step. The Washburn equation relates pressure, P_n, to the hydraulic pore diameter, D_n, of the pores:

$$D_n = -\frac{\sigma \cos(\gamma)}{P_n} \quad (1)$$

where γ is the contact angle of mercury and σ is the surface tension. The volume of mercury that intrudes into the sample at P_n gives the pore volume accessible through pore necks in the pressure interval (P_{n-1}, P_n).

We measured pore size distributions for the treated and untreated samples from > 100 μm to 0.003 μm using a Micromeritics 9420 Poresizer (Micromeritics Ltd., Norcross, Georgia, USA). For each soil series and treatment, three replicate samples, each weighing between 0.40 and 0.84 g, were analyzed. The samples were placed in powder penetrometers with measurable pore volumes of up to 1.16 cm^3. They were evacuated in three stages to a vacuum of 50 μm Hg. The analysis was performed in the pressure range from 10 kPa to 200 MPa.

From these data we computed the volumes of pores with hydraulic pore diameters >10 μm, 1 - 10 μm, 0.1-1 μm, 0.01-0.1 μm and 0.003 - 0.01 μm. We calculated the pore volume that was bioaccessible as the volume of pores with neck diameters greater than 1 μm.

2.4 *Scanning Electron Microscopy (SEM)*

Grains from each sample were mounted on adhesive tape and then splatter-coated with a palladium-gold film. Grain surfaces were inspected using a JEOL 1200 EX STEM in scanning electron microscope mode (JEOL Ltd., Tokyo, Japan). To inspect the internal pore space of single goethite grains, thin-sections of goethite grains were cut after embedding in resin. They were examined with the JEOL 1200 EX STEM in transmission electron microscope mode.

3 RESULTS

3.1 *Soil characteristics*

The iron content of the soils varied from 12 g Fe/kg soil for the Bridgehampton sandy loam to >60 g Fe /kg soil for the Hiwassee clay soil (Table 1). The clay soils, Hiwassee and Cecil, comprised 40% - 50% clay size material, while the clay fraction in the sand sample, Bridgehampton, only accounted for 8% of its mass.

Table 1. Soil properties for the B-horizon materials.

Soil Series/ Horizon	Texture	Fe - content g Fe/kg soil
Bridgehampton/ Bw	sandy loam	12
Hiwassee/ Bt2	clay	61
Christiana/ Bt2	silty clay	52
Cecil/ Bt2	clay	36

Soil colours prior to treatment were red to red-ochre. After treatment with DCB the soils had a white to gray colour indicating the successful removal of surface iron oxides.

3.2 *Visual inspection of electron micrographs*

Goethite grains contained pore spaces that were up to 500 nm long and as narrow as 40 nm (Figure 1A).

These pores permeated the goethite grains providing potential, tortuous diffusion paths into the internal interstices. The surfaces of soil grains coated with iron-oxides appeared to be covered by small granules of circular cross-sections. Pore openings in between granules were between 10 and 100 nm in width (Figure 1 B). Grains in treated soils had less granular surface texture than the DCB treated soils (Figure 1 C).

3.3 *Pore Size Distributions*

Pore size distributions varied significantly between soils. ANOVA showed that there were between-soil differences in the pre-treatment soil pore volumes in all pore size intervals except for in the interval from 0.003 - 0.01 μm. Means separation tests further showed that the fewest significant differences between pore volume in the five intervals were found for the Cecil and the Hiwassee clay soils. The Bridgehampton sandy loam differed significantly in pore volume from any of the other soils in 2 or 3 pore size intervals, mainly in the large pore diameter ranges.

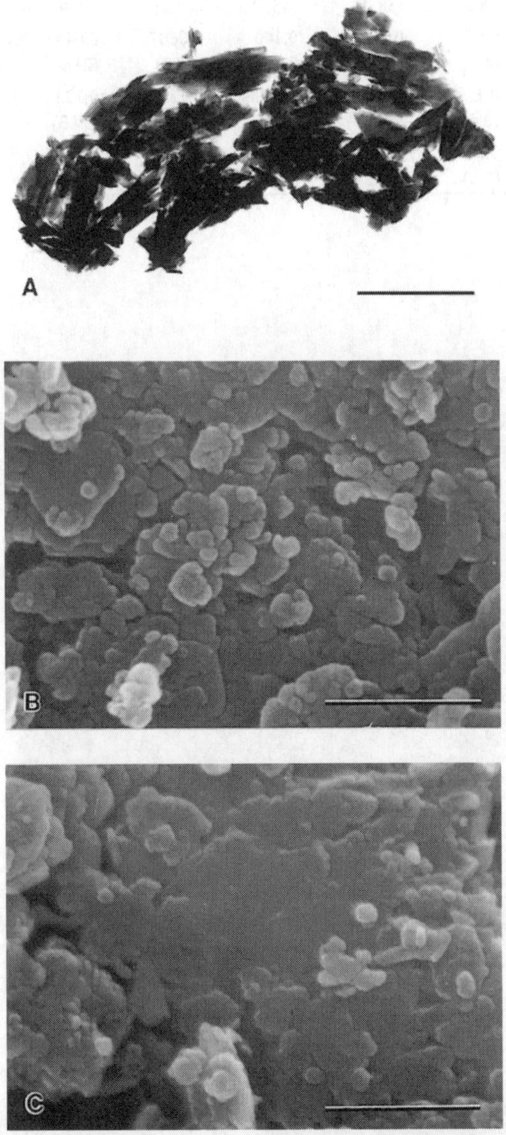

Figure 1. (A) Cross-section of a goethite crystal showing the internal nanopore space. (B) Nanopores at the surface of the Cecil clay soil before treatment. (C) Nanopores at the surface of the Cecil clay soil after treatment with DCB.

Pore volumes of three soils (Cecil, Christiana and Bridgehampton) were significantly altered in several pore size intervals by the DCB treatment (Table 2). Most of the significant changes were in bio-accessible pore size intervals (>10 μm, 1 - 10 μm) but some occured in the pore diameter interval from 0.1 - 1 μm. Increases in large pore size volumes were observed in particular in the > 10 μm and 1 - 10 μm intervals. For the sandy loam soil, there was a decrease in the 1 -10 μm interval. We consistently observed decreases in nanopore volumes with DCB treatment. However, these decreases were only statistically significant for the Cecil soil. For the Christiana and Hiwassee soil, there was little decrease in pore volume with treatment in the 0.003-0.01 μm range.

Pore volume not bioaccessible ranged from 38% for the Bridgehampton to 53% for the Hiwassee soil before treatment and from 15% (Bridgehampton) to 55% (Hiwassee) after treatment. Significant increases in bioaccessible pore volume (volume in pores greater than 1 μm) were observed for Cecil ($P < 0.036$), Christiana ($P < 0.022$) and Bridgehampton ($P < 0.048$) but not for the Hiwassee soil ($P < 0.45$).

4 DISCUSSION

Treatment of soils with a reducing agent alters the pore structure of soils, increasing, for three of the four soils investigated, the volume fraction of bio-accessible pores. The changes in pore structure and associated release of pollutants caused by the reduction of iron oxides may be helpful in increasing bioaccessibility of pollutants. However, current practices aiming at enhancing bioremediation, for example, the addition of oxygen to anoxic soils and ground water, may have antagonistic effects because it may lead to the crystallization of iron oxides which in turn may occlude pollutants and render them inaccessible.

Reduction-oxidation cycles occurring naturally in areas with fluctuating water tables or alterations in water regime may release and retain pollutants through the formation and dissolution of iron-oxide coatings. More research is needed to assess the interactions between nanopore structure and bioremediation.

Table 2. Pore volumes for five pore diameter ranges for B-horizon soil materials in DCB treated and untreated soil. Significant differences and significance levels between treatments are indicated by asterisks.

Pore Diameter [um]	Average specific pore volume in pore size interval [cm ^{3} /mg soil] (Coefficient of variation)							
	Bridgehampton		Cecil		Christiana		Hiwassee	
	untreated	treated	untreated	treated	untreated	treated	untreated	treated
>10	0.170	0.338*	0.259	0.279	0.105	0.177**	0.225	0.171
	(3.3)	(10.5)	(5.0)	(3.8)	(19.7)	(8.9)	(11.7)	(48.8)
1 - 10	0.190	0.116*	0.0712	0.0973*	0.189	0.334	0.0337	0.0390
	(2.6)	(12.4)	(7.7)	(7.8)	(58.0)	(45.7)	(17.2)	(29.9)
0.1 - 1	0.0814	0.0180***	0.0738	0.0504*	0.136	0.100	0.0607	0.0483
	(9.1)	(31.6)	(13.9)	(4.1)	(14.6)	(65.0)	(17.0)	(21.9)
0.01 - 0.1	0.0623	0.0241	0.187	0.130	0.0424	0.0308	0.155	0.133
	(68.4)	(87.2)	(19.2)	(0.5)	(51.4)	(87.3)	(45.1)	(57.3)
0.003-0.001	0.0780	0.0403	0.0277	0.0182*	0.0091	0.0116	0.0769	0.0796
	(66.4)	(68.1)	(1.3)	(5.5)	(62.2)	(63.5)	(103)	(94.0)

* $P < 0.1$; ** $P < 0.05$; *** $P<0.01$

5 CONCLUSIONS

Reduction of iron oxides in soils can significantly alter the pore structure of soils. In particular, the removal of iron oxides may increase bioaccessible pores and, thus, the potential for bioremediation.

REFERENCES

Alexander, M. 1999. Personal communication, Cornell University.

Chiou, C.T., L.J. Peters & V.H. Freed. 1979. A physical concept of soil water equilibria for nonionic organic compounds. *Science* 206:831-832.

Holmgren, G.G.S. 1967. A rapid citrate-dithionite extractable iron procedure. *Soil Sci. Soc. Am. Proc.* 31:210-211.

Jarvis, N.J., R.S. Bengström & L. Stenström. 1991. A model to predict pesticide transport in macroporous field soils. In Preferential Flow, Proceedings of the National Symposium, Chicago, Dec. 1991., St. Joseph, Michigan: ASAE.

Lawrence, G.P. 1978. Stability of soil pores during mercury intrusion porosimetry. *J. Soil Sci.* 29:299-304.

Pignatello, J.J. 1989. Sorption dynamics of organic compounds in soils and sediments. In Sawney and Brown (ed.), *Reactions and movement of organic chemicals in soils* SSSA Spec. Publ. No. 22, Madison, Wisconsin: SSSA-ASA.

Steinberg, S.M., J.J. Pignatello & B.L. Sawney. 1987. Persistence of 1,2-dibromoethane in soils: entrapment in interparticle micropores. *Envir. Sci. Tech.* 21:1201-2108.

Ukrainczyk, L. and H.A. Ajwa.1996. Primisulfuron sorption on minerals and soil. *Soil Sci. Soc. Am J* 60:460-467.

Werkheiser, W.O., and S.J. Anderson. 1996. Effect of soil properties and surfactant on primisulfuron sorption. *J. Envir. Qual.* 25: 809-814.

Groundwater: Past Achievements and Future Challenges, Sililo et al. (eds) © 2000 Balkema, Rotterdam, ISBN 90 5809 159 7

Using urban aquifers: Assessing water quality issues

R. Greswell, A. Thomas, K. A. Shepherd, R. I. Sauer, R. G. R. Mitchener, J. M. Harris, R. Mackay, J. H. Tellam & M. O. Rivett
School of Earth Sciences, University of Birmingham, UK

ABSTRACT: Many cities in the UK and the rest of the world are founded upon aquifers that have contributed to their growth and prosperity. Unfortunately, this has often led to pollution of these urban aquifers and in many cases the perceived loss of this important groundwater resource. Birmingham is the UK's second largest conurbation and has been a major centre of industry for two centuries. The groundwater quality in the underlying Triassic sandstone has been regarded as poor due to continual contamination from inorganic and more recently, organic pollutants. However, examination of the distribution of contamination shows that it is mainly confined to industrial areas and that water quality at depth is good. It is thought possible that zones exist within the aquifer that may be developed without inducing inflow of neighbouring contaminants. The aim of the work described here is to assess the hydrogeological issues associated with recovering good quality water from urban aquifers using the Birmingham aquifer as the main example. Using a GIS and a contaminated land database the distribution of pollutants is being defined and will be used to estimate temporal and spatial contaminant fluxes to the aquifer. A 3D flow and transport model of the aquifer will then be used to predict variation through time of groundwater quality. At the borehole scale, a test site has been set up to examine local flow mechanisms. This is being used to develop a borehole catchment scale numerical model that will be used to assess the effect of different pumping regimes and well designs on pumped water quality.

1 INTRODUCTION

Demand for water continues to increase in the UK, as it does in the rest of the world. As many urban aquifers have a history of industrial development and pollution it is important to be able to assess their current and future value as a resource. This may be considered at two scales: regional questions are often faced by regulators and water companies who need to understand area-wide and long-term water quality issues and; local scale problems are faced by use of a single wells when attempting to design wells and pumping regimes. The question being asked at both scales is 'can water of usable quality (whether for domestic or industrial supply) be sustainably abstracted from the urban aquifer'?

The aim of the work reported here is to identify the contributory processes governing the observed water quality patterns at the regional and local scale and to determine how this knowledge can be employed to utilise the aquifer. The three main investigation tools are:

- *A pollutant recharge flux model*
- *A regional solute transport model*
- *A borehole scale solute transport model*

The overall structure of the project, and how the tools link is shown in Fig. 1. Central to the investigation is a GIS-based pollutant recharge flux model. Pollutant sources are diverse. They may be point, multi-point, area-wide or diffuse in nature. A GIS based approach allows the areal distribution of based approach allows the areal distribution of such source types to be considered and their transport behaviour through the unsaturated zone to be described based upon arealy variable geological and geochemical properties. Contaminant flux is also dependent on recharge and so the distribution of recharge must also be accounted for in the model. Contaminant fluxes calculated by the GIS model will be input to a three-dimensional regional groundwater flow and transport model that will consider the effects of geochemical and hydraulic variations below the water table. Ultimately, this will allow investigation of regional time-variant strategic water quality issues to be made. The borehole scale model will investigate flows and transport around a single pumping well and is being tested against data obtained from a purpose designed field test site. Subsequent application of the model will explore the sensitivity of pumped water quality to hydraulic and geochemical properties and to a range of borehole designs and pumping regimes. Following interpretation, aspects of the behaviour of the system at a local scale may be applied to the regional scale model.

The models are being developed using data from the Birmingham Aquifer. This paper describes the progress and results from the first stages of the project.

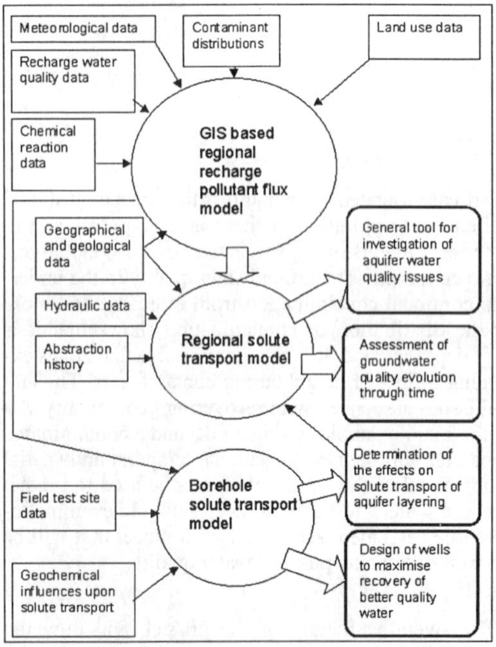

Figure 1. Project structure.

2 THE BIRMINGHAM AQUIFER

2.1 *Use*

The City of Birmingham is founded upon a Permo-Triassic sandstone aquifer that has been utilised as a water supply for at least 300 years. From 1860 onward, a series of high capacity wells were drilled to augment supply to the population. Following completion of reservoirs in Wales, public supplies were imported into the city. However, industrial demand continued to rise and peaked in the 1960's. Since then the total abstraction from the aquifer has rapidly declined to less than that in 1900. Birmingham was once dubbed *'The City of a thousand trades'* a name derived from the huge range of products that it manufactured. Predominant among these were the metal industries that produced everything from jewellery to heavy machine tools. The legacy of this has been a long history of pollution of both inorganic and more recently, industrial organic contaminants such as chlorinated hydrocarbons.

2.2 *Groundwater Quality*

It has been assumed that the groundwater is contaminated. Indeed, some public supply wells had become polluted and were abandoned before 1943 (Butler and Lee, 1943). However, several factors suggest that this potentially valuable resource is currently under-exploited.

- Studies of the water quality indicate that on average the quality is good and deeper water is excellent (Ford and Tellam, 1994)
- Contamination levels will be reduced by the large dilution capacity of the aquifer and by significant amounts of mains leakage (Greswell, 1992)
- Contamination and groundwater flows are often localised - pollution sources are frequently localised and the permeability of the sandstone is lower vertically than horizontally
- The aquifer has been shown to have a significant attenuation capacity (Ford and Tellam, 1994; Tellam, 1995)

The conclusions that may be inferred are: poor quality groundwater is localised to areas of past and present industry; reasonable quality water should exist beneath residential parts of the city; and by using a controlled abstraction regime it may be possible to produce a sustainable supply from the Birmingham aquifer.

2.3 *Geology and hydrogeology*

The unconfined Permo-Triassic sandstones overlie low permeability carboniferous mudstone sandstone sequences that outcrop along the western boundary of the unconfined aquifer (Fig. 2). To the east, the Birmingham Fault acts as a partial barrier to groundwater flows: east of the fault the downthrown sandstones are confined beneath the low permeability Triassic Mercia Mudstone group. To the north a groundwater divide defines the limit of the study area. The majority of the unconfined aquifer is covered by a complex distribution of glacially derived Pleistocene drift consisting of sands and clays of up to 20m in thickness. The Permo-Traissic sandstones have varied lithologies characterised by varying grain size and degrees of cementation. Interbedded laterally extensive mudstones are common. The sandstones are typically red in colour resulting from oxy-hydroxide coatings of the grains. Organic carbon concentrations are very low. Throughout the sequence horizontal fracturing has been recorded during geophysical and CCTV logging. The regional groundwater flows are toward the River Tame which crosses the middle of the unconfined aquifer, flowing from west to east. Limited easterly groundwater flow takes place across the Birmingham fault, but it has low permeability and significant head gradients have developed during periods of high groundwater abstraction. Pumping tests show the permeability of the sandstone to be approximately 2.0 m/d but laboratory measurements of rock sampled from a site at the University exhibit a wide range of values from 5×10^{-3} to 20 m/d and porosities that range from 14% to 30%.

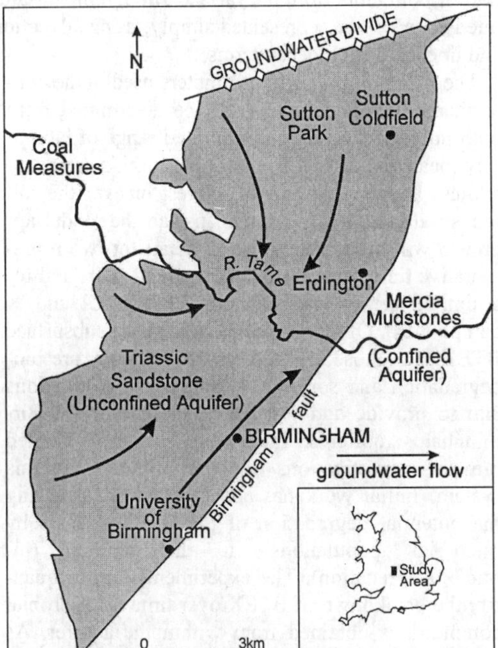

Figure 2. Simplified geological and hydrogeological map showing the Birmingham Aquifer.

Changes in groundwater head distribution have been marked over the last 150 years with significant regional drawdown resulting from abstraction. Since 1960 groundwater levels have been rebounding leading to concern about possible effects on foundation stability and lateral mobilisation of contaminants in the unsaturated zone as inundation occurs. (Knipe et al, 1993).

Clays, grain coatings and matrix cements have the most important influence upon the geochemical properties of the sandstone in relation to contaminant transport. Laboratory measurements (including clay-rich sandstones) of cation exchange capacities (cec) range from 1 to 20 meq/100g. Acidic recharge waters (principally resulting from industrial pollution) have developed a calcite depleted zone in the upper part of the aquifer. The effect of this has been to reduce the acid buffering capacity of this zone and to increase the mobility of some contaminants compared with the deeper, carbonate-bearing rock (Tellam, 1995).

3 THE REGIONAL APPROACH

3.1 *The Pollutant Recharge Flux Model*

In order to calculate the distributed regional contaminant flux, both recharge and contaminant distributions must be determined. By combining these data with a simple vertical transport model to simulate the effects of the unsaturated zone, the mass of a specified pollutant entering the aquifer can be estimated.

3.1.1 *Recharge Calculation*

Using ArcView GIS, a prototype runoff-recharge model has been developed to estimate the potential recharge (direct recharge) from urban land use. The potential recharge calculation procedure for each part of the urban aquifer is dependent on the urban land use and the infiltration properties of the underlying soils and drift. The model will initially examine the time period from 1980 to 2000. The input data for this model are:

- Land use map
- Hydrologic soil group map (based on superficial and bedrock geology)
- Meteorological data (potential evaporation, actual evaporation, soil moisture deficit and total rainfall)

The recharge model has three parts. The first estimates the aerial rainfall by producing a precipitation grid from the rain gauge station measurement data through the inverse distance weighted interpolation method. The second part estimates the excess rainfall or runoff taking place from each part of the land use units using the USDA Soil Conservation Curve Number Method (USDA, 1986) The third part calculates the 'potential' recharge from each land use unit as follows. Using the input parameter data, the program generates a land use grid, a soil hydrologic group grid, a curve number grid, a rainfall grid, and a runoff grid. Subtracting the runoff and evaporation values from the original precipitation grid gives the 'potential' recharge associated with each land use. In the vegetated areas, the calculation procedure takes into account the soil moisture deficit derived actual evapotranspiration. The calculated recharge value is a 'potential' recharge value as no account is taken of interflow processes at depth. Figure 3 shows an example of calculated potential recharge for a hypothetical rainfall event of 10 cm. Note the differences between the potential recharge in the largely open ground of Sutton Park in comparison to the heavily developed city centre of Birmingham itself.

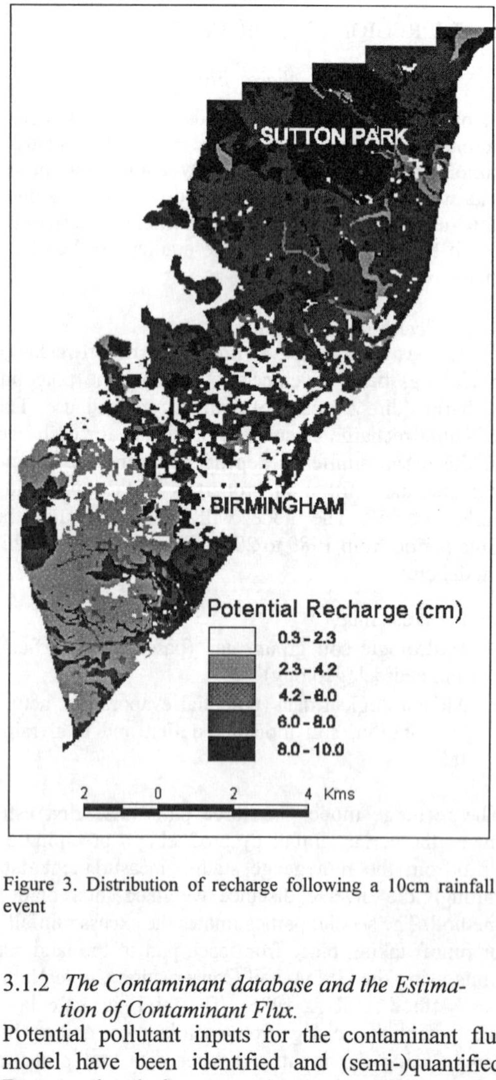

Figure 3. Distribution of recharge following a 10cm rainfall event

3.1.2 *The Contaminant database and the Estimation of Contaminant Flux.*

Potential pollutant inputs for the contaminant flux model have been identified and (semi-)quantified. Data on chemical use associated with different urban land uses were obtained from sources such as municipal and regulatory bodies, utility companies and geotechnical consultancies. The land uses considered include industrial, non-industrial and residential built up areas, plus parkland and roads. Data collected to date includes: location of petrol stations and various industrial facilities, road salt loading data and mains water quality. Work is also being carried out to determine surface run-off in residential and non-industrial built up areas. Levels of pollutants associated with urban fill, or 'made ground' have been investigated through the analysis of soil chemistry data taken from 123 sites in the study area. Data were limited to those chemical species detailed in the ICRCL guidelines (ICRCL, 1987). This work has highlighted a number of interesting trends such as the variation of contaminant levels between made ground and superficial deposits, with the former having significantly greater concentrations of inorganic contaminants. The transport of pollutants from the surface through the unsaturated zone will be represented simply, using advection and linear sorption/decay terms.

The transport property parameters used in the contaminant flux calculations will be determined from both published values and a limited range of laboratory measurements. It is unfeasible for this project to address all contaminants of interest or over the full industrial history of the area. Instead the model approach will be demonstrated in detail for two representative time periods over the last 150 years and for a limited set of determinands (BTEX, Cl and N compounds). Of the processes controlling subsurface BTEX fate, those selected for further study are biodegradation and sorption. Laboratory investigations aim to provide additional data specific to the Birmingham Aquifer as there have been very limited previous investigations of these pollutants in this system. Initial work has concentrated on assessing the potential degradation of the BTEX species by microbial populations in the aquifer (biotic/biodegradation). The experiments aim to quantify the breakdown of BTEX over time by microbial communities obtained from within the aquifer. As the aquifer matrix has very low levels of organic material experiments will also examine sorption properties. Following evaluation of the data, a range of possible transport parameter values will be applied in the GIS transport model. Finally, these distributed fluxes will form the primary input to the regional flow model.

3.2 *The Regional Solute Transport Model*

A three dimensional model of the system is being developed based on an existing single layer 'MODFLOW' representation of the system which was developed to study the effects of rising groundwater (Greswell, 1992). A multi-layer model is necessary to adequately represent the various lithologies within the sandstone as well as the calcite depleted upper layer and hence the effects of anisotropy on flow and transport. Distributed contaminant flux data from the Pollutant Recharge Flux model will be input to this model and allow the three dimensional regional distribution of selected contaminants to be studied over time. The completed model will allow a wide range of groundwater issues to be examined including the sensitivity of water quality to aquifer and land-use development, and the importance of water table elevation relative to the base of the calcite depleted zone of the aquifer.

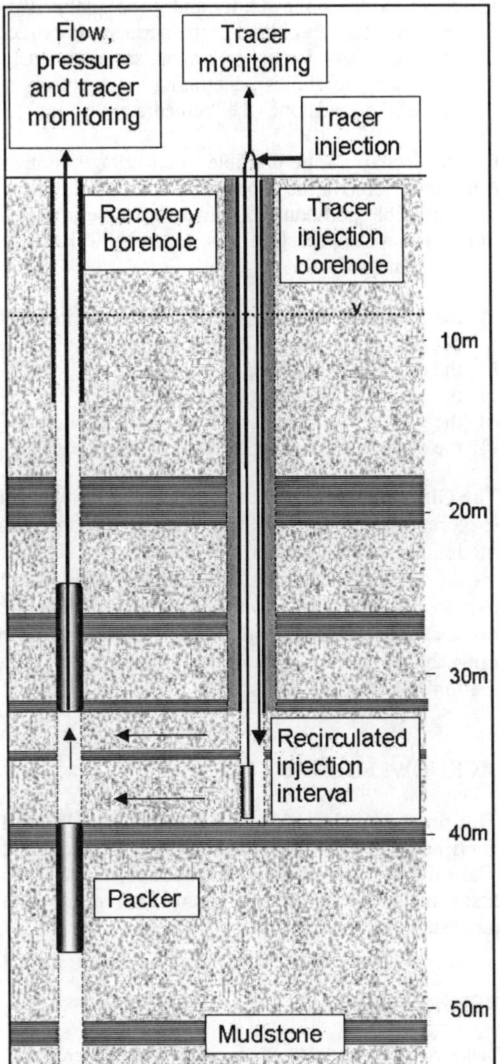

Figure 4. Horizontal tracer test between boreholes in test site.

4 THE SITE SPECIFIC SCALE

4.1 *4.1 Approach*

In a layered aquifer such as that considered here, anisotropy can have significant influence upon flow and transport. The presence of interbedded mudstones produces a system where the vertical permeability is significantly lower than the horizontal permeability. Where contamination is largely confined to the near surface it should be possible to design a well such that a combination of casing design and pumping regime will minimise the downward migration of pollutants to an extent that extraction of usable quality water can be maximised. The aim of this part of the project is to address the problem of *site-specific* characterisation and well design in order to achieve this goal. Methodologies derived from this exercise should be applicable to future well designs in similar systems and any wider implications of layering may be used to supplement the development of the regional model. To do this a new borehole array has been installed at Birmingham University. This complements an existing research site that contains an observation well, piezometer, tensiometer and resistivity arrays. Four additional boreholes penetrate the formation and, based on geophysical and CCTV logs, were drilled to allow the investigation of the effects of different casing depths.

4.2 *Investigation*

A double packer system was constructed and used in the observation well (uncased below 18m) to isolate specific sandstone horizons and allow both hydraulic characterisation and horizontal tracer tests to be conducted. Pumping tests were carried out on each borehole, and in order to prevent bypass occurring through the observation well, a 60m double-skinned polythene packer was inflated in the observation well. A pump and combined double packer system was installed in the observation well allowing cross-hole horizontal tracer tests to be performed from each of the other wells using the rhodamine WT and fluorescein. A fluorimeter monitored the concentration at the discharge well. A pumped recirculation loop within the injection well ensured continuous mixing within the injection interval and, by taking discrete samples, allowed the concentration of tracer in the well to be monitored. A schematic cross section of part of the array and a tracer test is shown in Fig 4.

Initial interpretation of the hydraulic testing shows the permeability of the sandstone to be similar to the mean intergranular permeability of core samples measured in the laboratory at around 2.3 m/d but a full analysis will only be possible after the numerical model has been more fully developed. The breakthrough curve from the tracer test between boreholes 1 and 5 is shown in Fig. 5. The result is important as it shows the transport properties of the rock to be significantly different from those anticipated assuming intergranular flows alone. The curve shows four peaks that are thought to result from transport through a series of at least four fractures. The indication is that, for this interval at least, the matrix has an insignificant role in the short-term transport of solutes. The breakthroughs from the other test intervals show similar, though much less marked features. The results suggest that the test site exhibits significant heterogeneity even

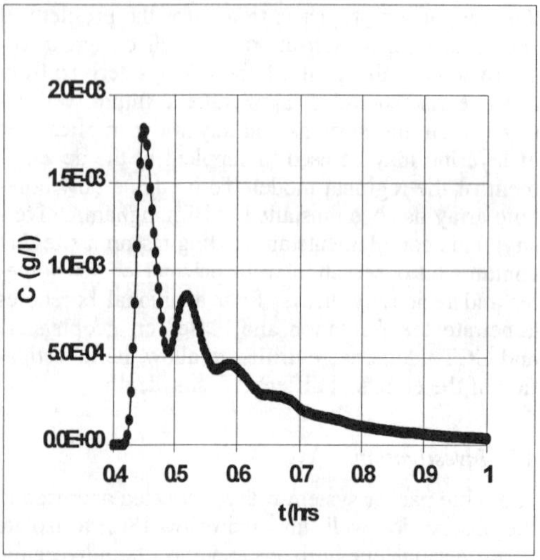

Figure 5. Rhodamine WT Tracer breakthrough from horizontal test between boreholes 1 and 5.

over vertical distances of a few meters and the rate of lateral transport of contaminants will be varied and depth dependant. Vertical flow and transport properties will be investigated during a series of experiments scheduled for the summer of 2000. Extensive analysis of core material from the site indicates that correlations between geochemical and hydraulic properties of the sandstones are weak.

5 USES OF THE MODELS

If urban aquifers are not to be abandoned it is necessary to learn how to manage the resource. It must be accepted that considerable pollution will probably have occurred in the past and that, even given more awareness of pollution issues, there will be continued pollution in the future. Shallow aquifers with distributed pollutant sources may well be too polluted to do anything other than treat groundwater after pumping. Deep aquifers offer more possibilities, especially if the pollution sources are concentrated in localised areas. There are two main options:

- extract good quality water, leaving poor quality water in the aquifer (the "mining option")
- either extract poor quality and good quality water separately and blend at the surface, or extract poor and good quality water simultaneously from deep boreholes penetrating a large proportion of the sandstone (the "remediation option").

Clearly it may not be possible to encourage enough dilution to remove the problem of some of the more toxic possible pollutants. To investigate the implications of these approaches, it is necessary to have an understanding of:

a) the distribution of poor quality water and its relationship with land and chemical use;
b) the effects of hydraulic layering;
c) the effects of geochemical layering;
d) the effects of pumping on lateral migration; and
e) the effects of pumping on vertical migration.

The GIS pollutant flux recharge model will be used to explore (a), and coupled with the 3D regional flow model, issues (b) to (e) on a regional scale. The GIS model coupled to the borehole model will be used to explore issues (b) to (e) on the local scale. It will be necessary to examine both scales in order to determine the relative viability of the "mining" and "remediation" options.

ACKNOWLEDGEMENTS

This project has been funded by the NERC, UK with contributions from Birmingham City Council and The Environment Agency. The authors also wish to thank the members of the Steering Committee who have supplied data and other resources.

REFERENCES

Butler, A.J. & Lee, J. 1943. Water supply from underground sources of the Birmingham and Gloucester district. *Wartime Pamphlet No. 32. Geological Survey of Great Britain.*

Ford, M. & Tellam, J.H. 1994. Source, type and extent of inor ganic contamination within the Birmingham urban aquifer system. UK. *Journal of Hydrogeology* 156, 101-135

Greswell, R., 1992. The modelling of groundwater rise in the Birmingham area. *Unpublished MSc Thesis.* University of Birmingham. UK

ICRCL (Inter-departmental Committee on the Redevelopment

of Contaminated Land) 1987. *ICRCL 59/83: Guidance on the assessment and redevelopment of contaminated land (2nd ed.).*
C. Knipe, J.W. Lloyd, Lerner, D. N., R.Greswell. 1993. Rising groundwater levels in Birmingham, and their engineering implications. *CIRIA Special Pub. 92.* London. 114pp.
Tellam, J.H., 1995. Urban groundwater pollution in the Bir mingham Triassic Sandstone aquifer. *Procxeedings of IBC conference on groundwater pollution.* London.
USDA (United States Department of Agriculture), 1986. Urban Hydrology for small watersheds. *Technical review 53.*

Groundwater: Past Achievements and Future Challenges, Sililo et al. (eds) © 2000 Balkema, Rotterdam, ISBN 90 5809 159 7

Colloid release due to degradation of organic contaminants at a former gas plant – Gaisburg/Germany

T. Hofmann
Applied Geology, Mainz University, Germany

T. Schäfer
Institute for Nuclear Waste Management, FZ Karlsruhe, Germany

ABSTRACT: The transport of various contaminants such as heavy metals, radionuclides, and polar organic substances can be enhanced by colloidal transport. The efficiency of this transport mechanism is linked to the colloid concentration, colloid stability and contaminant-colloid interaction. Research was focused on colloid concentrations and colloid stability. The high colloid concentrations at the former gas plant discussed in this paper are linked to redox driven dissolution of sediment coatings, whereas the redox environment is controlled by degradation of organic contaminants. Colloid concentration would be several orders of magnitude lower without this process due to the high ionic strength of the groundwater.

1 INTRODUCTION

Contamination of aquifers by organic contaminants released from gas plants is widespread and of public concern. It is known that a organic contamination can induce a redox shift from oxic to anoxic conditions due to the higher input of organic carbon as electron donator and thus alter the sediment chemical properties (Barcelona & Holm 1991, Loveley et al. 1994, Schwarzenbach et al. 1993). The main objective for remediation strategies is to define the major pathways of contaminant transport and degradation.

Most approaches to describe and predict transport phenomena treat groundwater as a two phase system: a mobile aqueous phase and an immobile solid phase. Contaminants partition between this two systems and those which have a strong tendency to sorb readily to the solid phase are considered to be virtually immobile or retarded (Fig. 1).

However, colloidal and suspended particles also may be mobile in groundwater environments. The composition of these colloids is expected to be chemically similar to that of the surface of the aquifer material. Thus, they can also sorb organic and inorganic contaminants (McCarthy & Zachara 1989, Ouyang et al. 1996, Hofmann & Schöttler 1998).

Many nonpolar organic and biogenic substances, e.g. polycyclic aromatic hydrocarbons, geosmine, toluene and m-xylene have a strong tendency to adsorb to the solid phase. Inorganic trace substances, like heavy metals and radionuclides can show the same tendency (Backhus & Geschwend 1990, McCarthy & Zachara 1989, Stumm 1987) or build up particles themselves. Predictions on contaminant transport may fail if colloid facilitated transport is not taken into consideration (Fig. 1).

Taking this into consideration we can constitute the following key questions evaluating the possibility of colloid bound transport:

I. Are colloids abundant in the studied environment?
II. Are the colloids mobile or do they suffer filtration?
III. Are the colloids stable regarding groundwater hydrochemistry?
IV. Is the colloid concentration high enough to provoke a significant influence on transport patterns?
V. Do the colloids interact with the contaminants?

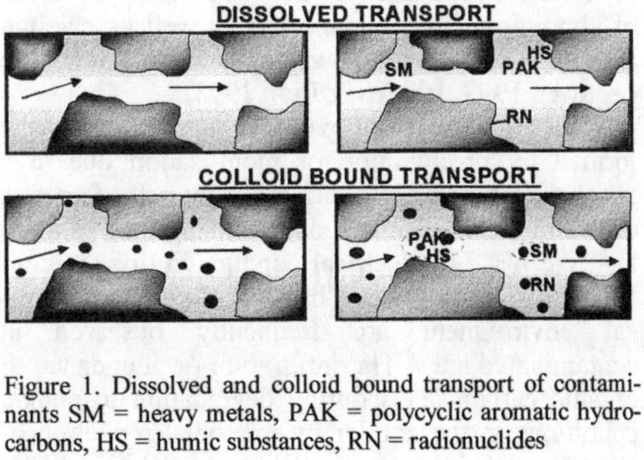

Figure 1. Dissolved and colloid bound transport of contaminants SM = heavy metals, PAK = polycyclic aromatic hydrocarbons, HS = humic substances, RN = radionuclides

The aim of this paper is to study the first 4 key questions at a former gas plant in Stuttgart/Germany (Gaisburg test site). The interaction of colloids and contaminants can be deduced if colloid mineralogy and contaminant sorption behaviour is known. At this test site a strong hydrochemical gradient, i.e. a redox gradient, favours the release of colloids by dissolution of carbonate and ferric coatings.

2 ORIGIN AND TRANSPORT OF PARTICLES

2.1 *Formation and origin of particles*

In this paper we refer to colloids/particles as potential mobile solids including macromolecular (or "dissolved") components of organic carbon. They can cover a broad size range from a few nm up to 10-20 µm. Thus, common filtration techniques in groundwater quality monitoring, i.e. filtration through 0.45 µm filters, may exaggerate the concentration of dissolved elements and underestimate the total element content.

Mobile solids can be colloidal stable or suspended. The colloidal stability and transport behaviour of a colloid is a complex function of the surface charge, density, surface chemistry, water chemistry and flow velocity. These conditions change along the flow path in the aquifer. Thus, for practical reasons colloids and suspended particles are not differentiated in this study.

Many different colloids can be expected in natural subsurface aquatic environments. Apart from the transport of colloids from the the vadose zone, the aquifer itself can consist to a large part of small particles < 20 µm.

These particles were deposited during sedimentation of the aquifer or formed by post-depositional weathering of thermodynamically unstable primary minerals. The composition of these particles depends on the sedimentation conditions, the chemistry of the pore water solution and the mineral content of the original geologic material. Dominating inorganic particles are aluminosilicates, oxides and hydroxides of aluminium, iron, manganese as well as calcium and magnesium carbonates and silica (Buffle & Leeuwen 1993, Jenkins & Lion 1993).

A second group of particles in the aquifer is formed by precipitation or mobilisation due to a geochemical gradient along the flow path. Particles can form or be mobilised due to changes of parameters like pH, Eh, electrical conductivity or the content of dissolved gases. Changes in the hydrochemical environment are frequently observed at contaminated sites. The infiltration or abundance of organic carbon leads during degradation to anoxic conditions in the aquifer by microbial reduction of O_2, NO_3^-, NO_2^-, MnO_2, Fe(III)-oxide, SO_4^{2-} and CO_2 (Appelo & Potsma 1999). Iron and Manganese coatings of the aquifer material are dissolved during this process and liberate formerly fixed particles and elements in these coatings. Nevertheless, the total content of iron and manganese, released by dissolution of these cements, is limited by the solubility of carbonates and sulfides. Due to CO_2 production by microbiological degradation of contaminants carbonates can also be dissoluted and thus release particles (McCarthy and Zachara 1998).

Biogenic substances are from a third group of microparticles. Organic macromolecules and humic substances can be released from the aquifer material or brought in by the infiltrated surface water. Bacteria attached to the matrix may become unattached or can be infiltrated. They may release extracellular matter or cell fragments. Another rare but very important and conceivably dangerous group are spores and viruses. Most of these organic substances and micro-organisms have a strong tendency to adsorb to the solid phase and therefore also to inorganic particles or build up aggregates with them. This effect can give rise to serious problems during the disinfection of drinking water.

Research on colloids in subsurface environments has shown that without any exception mobile solids are present in all aquifers (Buffle and Leeuwen 1993). Thus, the first of our key questions must not be considered further.

2.2 *Transport and filtration of particles*

Suspended or colloidal particles can be mobile in the aquifer if they are not susceptible to filtration and stable with respect to agglomeration and dissolution. Filtration effects can be either straining filtration or physico-chemical collection. While the maximum size of mobile particles is limited by straining filtration and pore velocity,. the concentration and size distribution of particles in the nanometer size range are controlled by physico-chemical filtration. A minimum filtration rate can be observed at a size range close to 1 µm, where straining filtration is less effective in most pore aquifers and physico-chemical filtration does not yet reach its maximum (Bedbur 1993).

The collection of particles on the aquifer material strongly depends on the surface charge. An opposite surface charge of particle and aquifer matrix enhances collection. Differently charged particles favour coagulation and consequently sedimentation. This effect is even more complex regarding clay minerals, because the edges and faces may show opposite charge. A charge reversal of particles can occur due to adsorption processes of polyelectrolytic substances. The adsorption of humic substances to positive charged particles like ferrihydrite can cause a charge reversal and facilitate transport.

In previous work we have shown that colloids are stable in hydrochemical environments with a low

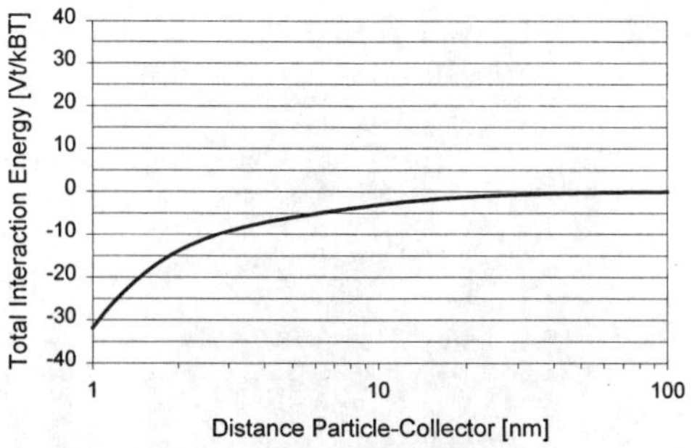

Figure 2. Calculation of interaction energy between colloids/particles and collector (sediment). Ionic strength 30 mmol/l, zeta potential of particles/sediments 30 mV, Hamaker constant $1 * 10^{-20}$ J, temperature 10 °C, particle size 0,2 µm.

ionic strength (Hofmann 1998, Hofmann & Schöttler 1998). Filtration processes are less effective with increasing mean pore size diameter.

One approach in understanding forces between two particles or a particle and a collector, i.e. the aquifer, is the electrostatic model developed independently by two groups of researchers (Derjaguin & Landau, Verwey & Overbeek) the so called DLVO-theory (Stumm & Morgan 1996). In this model electrostatic repulsive forces are balanced by attractive van der Waals forces. The repulsion and thus the stability of a colloidal system is therefore mainly a function of the surface charge of the particle and the ionic strength of the solution. A repulsive interaction (by definition positive) between particles promotes a colloidal stable system, whereas attractive interactions (by definition negative) favour coagulation and filtration.

In Figure 2 the interaction energy between two particles or a particle and a collector is calculated following the approach of Ruckenstein & Prieve (1973) and Spielman & Cukor (1973). Using the measured ionic strength, particle size, temperature, and zeta potential found at the test site, only attractive interactions can be modelled. Thus, due to the high ionic strength, mobile solids are colloidaly destabilised. This means, that most of the contacts between a colloid/particle and a collector will result in attachment. The deposition of colloids/particles is taking place under favourable conditions (Ryan & Elimelech 1996).

Altogether we can assume that due to the hydrochemical conditions colloids are due to filtration and are not expected to be mobile at Gaisburg test site. We will see later that this statement is in contrast with field observations.

3 TEST SITE DISCRIPTION

The investigation site is located in the Neckar valley/Stuttgart in southern Germany (test site Gaisburg). Gas production started in 1875 and was linked to pollution of the aquifer with organic contaminants, e.g. PAH, BTEX, heavy metals (Ebert et al. 2000, Weber et al. 2000, Zamfirescu & Grathwohl 1998). Contamination occurred mainly from spills and leaks in tanks. The distribution of BTEX is shown in Figure 3.

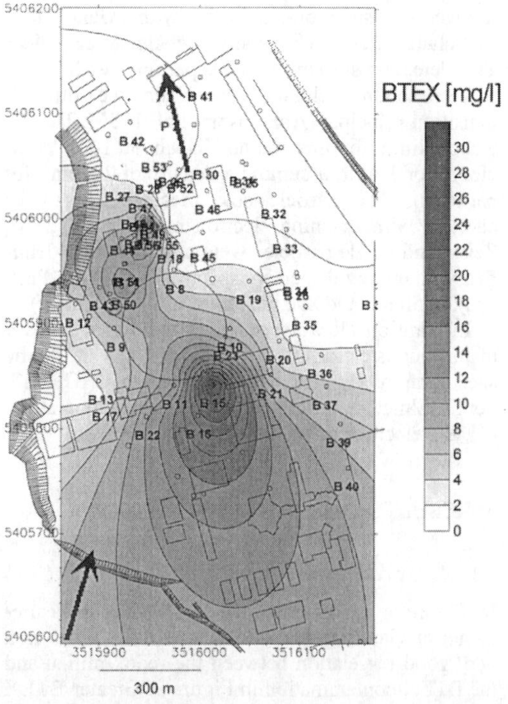

Figure 3. BTEX concentration in mg/, test site Gaisburg, Stuttgart/Germany. Arrows indicate groundwater flow direction. Map is orientated north. Data with kindly permission from K. Weber, Karlsruhe.

The aquifer consists of quarternary medium-grained carbonate gravel with sand and silt lenses. The aquifer thickness is about 3 - 4 m, underlain by Upper Triassic rocks. Hydraulic conductivity is about 130 m/d, groundwater flow velocity about 2 - 3 m/d. The hydraulic gardient is from south/south-west to north/north-east, bending in the northern part to north/north-west (Fig. 3).

Weber et al. (2000) reported east-west striking faults in the northern part of the test site, partially disrupting the Triassic rocks and thus allowing higher mineralised water to rise.

4 MATERIALS AND METHODS

All samples were taken with a peristaltic pump respecting low-flow minimal-drawdown guidelines (EPA 1995). Flow was regulated to 100 ml/min.

Particle concentration and size range above 0.5

μm were measured on-site in-line with a Galai CIS-1 by Schäfer, Mainz University (Schäfer et al. 1998). This detection system measures particles with the laser extinction technique based on the time-of-transition principle (Aharonson et al. 1986). The detection limit for this method is about 1000 particles/ml or 1 μg/l (assuming a density of 2 g/cm^3 for particles). The particle size measurements were checked with scanning electron microscopy (SEM, Zeiss) and single particles were analysed and identified with energy dispersive spectroscopy EDX, Phillips) on 50 nm and 200 nm Isopore filters (Costar).

In addition all water samples were analysed for major anions/cations with IC (WATERS) and flame spectography (Eppendorf), heavy metals (ICP-MS, Perkin Elmer) and field parameters, i.e. temperature, pH, E_H, electric conductivity, and O_2 (WTW).

5 RESULTS

5.1 *Redox conditions*

In Figure 4 redox potentials from field measurements at Gaisburg test site are plotted. There is a very good correlation between the redox milieu and the BTEX contamination in Figure 3. Greater BTEX degradation rates are observed under aerobic conditions (Edwards & Gribic-Galic 1994, Lovley & Longergan 1990). In anaerobic systems Fe(III) from the solid phase are regarded as the most important electron acceptors and thus will determine the oxidation capacity of an aquifer (Heron et al. 1994).

Where BTEX values are high at Gaisburg site, redox values are as low as - 200 mV. This values correspond to the reduction of SO_4^{2-} to HS^-. Analysis of water samples in this area reveal a significant decrease in SO_4^{2-} and HS^- was detected. HS^- concentrations were lower than total SO_4^{2-} decrease. A dissolution of Manganese and Iron Oxid-Hydroxides can be presumed at these low E_H values. Thus, Fe^{2+} and Mn^{2+} will be released to groundwater. At the same time geochemical modeling shows a supersaturation of Pyrite/Monosulfide, in such way the low HS^- concentrations can be easily explained be precipitation of dissolved iron as iron sulfides (as ascertained by Ebert et al. 2000).

Downstream of the contamination increasing E_H values in the range of 0 - 100 mV indicate an equilibrium of Fe(II)/Fe(III). At Gaisburg in this part of the test site dissolved iron can be detected and no more sulfide is traceable. Along the groundwater flow path, subsequent Mn^{2+} and O_2 can be detected (Where dissolved Fe^{2+} is decreasing again).

High BTEX concentrations were linked to increased CO_2 concentrations in the soil.

The discussed redox environment could lead to several mechanism for particle formation/release: by precipitation of iron sulfides, by precipitaion of dissolved iron downstream, by dissolution of Fe(III) coatings and simultaneous release of previously fixed particles and finally dissolution of carbonate cement due to the higher CO_2 concentrations by microbial degradation.

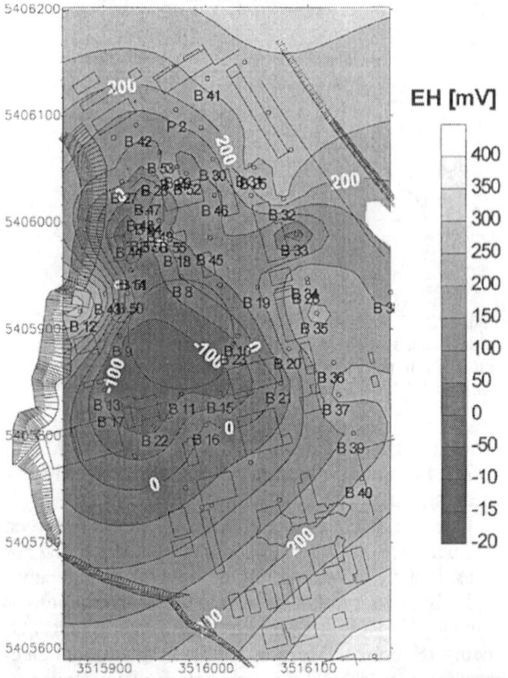

Figure 4. EH values at test site Gaisburg. Map is orientated north

5.2 *Particle concentration*

The particle concentration measured with laser-shading (CIS-1) at Gaisburg site are plotted in Figure 5. Particle concentration vary from detection limit (1 μg/l) in the south-eastern and northern path to a maximum of 500 μg/l downstream of the contamination in the western part of the test site.

The particle size (not plotted) was in the lowest range of the detection range between 0.5 and 1.5 μm. SEM probes showed particles down to 20 nm. Two distinct spots of very high particle concentrations can be detected in the north-eastern and north-western part of the test site. Beside the generally increased particle concentration this two spots can not be directly explained by contamination linked particle formation. Possibly mixing between rising groundwater from deeper layers and groundwater from the shallow aquifer at east-west striking faults can be an explanation which has to be proofed.

SEM/EDX analysis identified mainly alumosili-

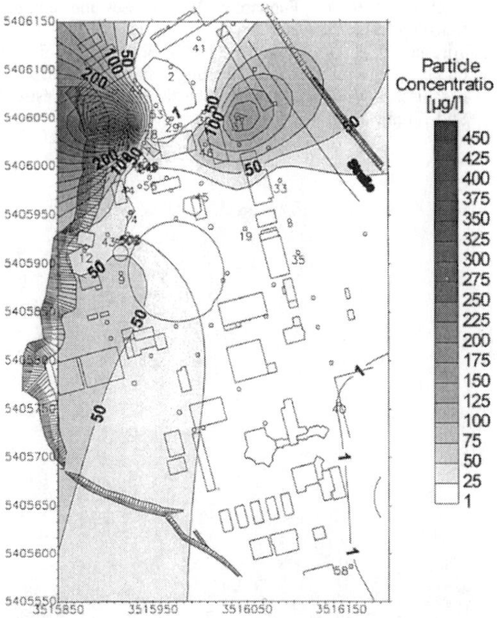

Figure 5. Particle concentration measured with CIS-1 (laser shading) at Gaisburg site. Map is orientated north.

cates, quartz, carbonate and iron/sulfur particles. While the first three groups of particles are from the matrix of the aquifer and can be mobilized by dissolution of cements or hydraulic perturbations, iron/sulfur particles are expected to be formed by precipitation of FeS/FeS_2. At Gaisburg test site the alumosilicates, silicates and carbonates can be regarded stable at the prevalent groundwater chemistry.

6 DISCUSSION

Mobile particles, which are abundant at all natural aquatic environments, can enhance contaminant transport. The key questions for a significant particle bound transport is the mobility, stability, concentration, and interaction with the contaminant itself.

Besides mobility and stability the particle concentration is of major importance for transport phenomena. At the studied test site Gaisburg/Germany the measured particle concentration inside the contaminant plume were significantly increased in respect to uncontaminated groundwater upstream and downstream of the contaminant plume. This values can be explained by a geochemical gradient downstream of the BTEX contamination. Particles are release and/or generated due to the dissolution of the aquifer matrix. Mixing between shallow and ascending deeper groundwater types must also be

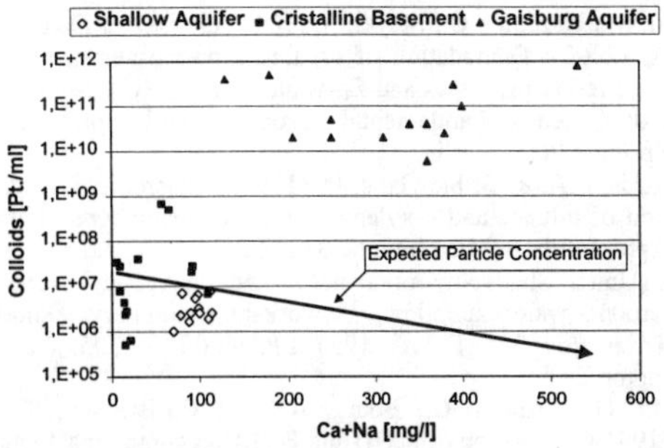

Figure 6. Colloid concentration from three different test sites. Data crystalline basement aquifer from Degueldre 1994, data uncontaminated shallow aquifer from Hofmann 1998. Colloid concentration is expected to decrease with increasing Ca + Na concentration due to their ability as coagulant. Calcium has the strongest coagulation potential and similar influence on humic material.

taken into consideration for extraordinary high particle concentrations.

The relevance of a geochemical gradient to particle formation can be deduced from Figure 6. Particle concentrations at three different test sites are plotted against the sodium and calcium concentrations of the groundwater. Where sodium and calcium is high, particle concentration is expected to be low due to electrostatic destabilization of colloids. Colloids will be attached to the aquifer matrix under this conditions (Buffle & Leeuven 1993, Degueldre 1994, Ryan and Elimelech 1996). Thus, at Gaisburg site a very low particle concentration is expected without a geochemical gradient (following the arrow in Figure 6). However, increased concentrations downstream of the contamination indicates that not ionic strength but another process must govern particle release. For Gaisburg site this is a redox gradient.

REFERENCES

Aharonson et al. (1986). Galai CIS-1. A novel approach to aerosol particle size analysis. J. Aerosol Sci., 17/3: 530-536.

Backhus D. A. & P. M. Gschwend (1990): Fluorescent polycyclic aromatic hydrocarbons as probes for studying the impact of colloids on pollutant transport in groundwater. *Environ. Sci. Technol.* 24: 1214-1223.

Barcelona, M.J. & Holm, T.R. (1991): Oxidation-reduction capacities of aquifer solids. Environ. Sci. Technol. 25:1565-1572.

Bedbur E. (1993): *Laboruntersuchungen zum Einfluß sedimentologischer und hydraulischer Parameter auf die Filterwirkung gleichförmiger Sande.* Dissertation, Christian-Albrechts-Universität, Kiel.

Buffle, J. & Van Leeuwen, H.P. (1993.), Environmental particles, Vol. 2, 247-316, Lewis Publishers, London.

Degueldre, C. (1994): Colloid properties in groundwater from crystalline formations. Technical Report 92-05, NAGRA, PSI, Würelingen, Villingen.

Ebert, M., Hümmer, O., Mayer, M., Schliecker, O. & Dahmke, A, (2000): Degradation of organic groundwater contaminats: redox processes and E_H-values. In: SCHÜRING et al. (eds.): Redox. Fundamentals, processes and applications. Springer 2000, Berlin.
Edwards, E.A. & Gribic-Galic, D. (1994): Anaerobic degradation of toluene and o-xylene by methanogenic consortium. Appl. Environ. Microbiol. 60:313-322.
EPA (United States Environmental Protection Agency) (1995): Ground water sampling – a workshop summary. Dallas, Texas Nov. 30 – Dez. 2, 1993, EPA/600/R-94/205, Washington DC.
Heron, G., Crouzet, C., Bourg, A.C. & Christensen, T.H. (1994): Speciation of Fe(II) and Fe(III) in contaminated aquifer sediments using chemical extraction techniques. Environ. Sci. Technol. 28:153-158.
Hofmann T. & Schöttler U. (1998b): Microparticle facilitated transport of contaminants during artificial groundwater recharge. In Peters J. et al., Artificial recharge of Groundwater, S. 205-210, Balkema, Rotterdam.
Hofmann T. (1998): Kolloidale und suspendierte Partikel. Herkunft, Transport und Relevanz von mobilen Festphasen im Hinblick auf die künstliche Grundwasseranreicherung. Ph.D. Thesis, Veröffentlichungen des Institutes für Wasserforschung GmbH Dortmund und der Dortmunder Energie- und Wasserversorgung GmbH, Dortmunder Beiträge zur Wasserforschung Nr. 56, Dortmund.
Jenkins M. B. & L. W. Lion (1993): Mobile bacteria and transport of polynuclear aromatic hydrocarbons in porous media. *Appl. Environ. Microbiol.* 59: 3306-3313.
Loveley, D.R., Chapelle, F.H. & Woodward, J.C. (1994): Use of dissolved H_2 concentrations to determine distribution of microbial reducible ferric iron in aquatic sediments. Environ. Sci. Technol. 28:1205-1210.
Lovely, D.R. & Lonergan, D.J. (1990): Anaerobic degradation of toluene, phenol and p-cresol by dissimilatory iron-reducing organisms. Appl. Environ. Microbiol. 56:1858-1864.
McCarthy J. F. & J. M. Zachara (1989): Subsurface transport of contaminants. *Environ. Sci. Technol.* 23: 496-502.
Ouyang, Y., Shinde, D., Mansell, R.S. & Harris, W. (1996): Colloid-enhanced transport of chemicals in subsurface environments: a review. Environ. Sci. Technol. 26:189-204.
Ruckenstein, E. & Prieve, D.C. (1973): Rate of deposition of Brownian particles under action of London and double-layer forces. *J. Chem. Soc. Faraday Trans.* 2, 69(10):1522-1536.
Ryan, J.N. & Elimelech, M. (1996): Colloid mobilization and transport in groundwater. *Colloid Surf. A.* 107:1-56.
Schäfer, T., Knoke, H. & Schenk, D. (1998): Transport of particulate matter in groundwater depending on natural and anthropogenic hydraulic variations. In BRAHANA, J.V. et al. (Hrsg.), Gambling with groundwater – pysical, chemical, and biological aspects of aquifer-stream relations, Americ. Inst. of Hydrol., St. Paul, Minnesota, Las Vegas, Nevada, 541-547.
Schwarzenbach, R.P., Gschwend, M. & Imboden, D.M. (1993): Environmental Organic Chemistry. Wiley-Interscience.
Spielmann, L.A. & Cukor, P.M. (1973): Deposition of non-Brownian particles under colloidal forces. *J. Coll. Interface Sci.* 43:51-61.
Stumm W. (1987): *Aquatic Surface Chemistry*. John Wiley, New York.
Stumm, W. & Morgan, J.J. (1996): Aquatic chemistry. Chemical equilibria and rates in natural waters. 3. Aufl., Wiley-Interscience, New York.
Weber, K., Brandsch, N., Reichert, B., Eiswirth, M., Hötzl, H., Hümmer, O. & Dahmke, A. (2000): Redox zones in the plume of a previously operationg gas plant. In: SCHÜRING et al. (eds.): Redox. Fundamentals, processes and applications. Springer 2000, Berlin.
Zamfirescu, D. & Grathwohl, P. (1998): Schadstoff-emission durch Desorption und Lösung standortspezifischer organischer Verbindungen im Schadenszentrum, Testfeld Süd. Grundwasser 4/3:167-182.

Groundwater: Past Achievements and Future Challenges, Sililo et al. (eds) © 2000 Balkema, Rotterdam, ISBN 90 5809 159 7

Transport of suspended solids from karst to alluvial aquifer: Column/field experiment

N. Massei, M. Lacroix, H.Q. Wang & J.P. Dupont
Laboratoire de Géologie, University of Rouen, Mont Saint-Aignan, France

ABSTRACT: We study the influence of the karst-alluvium interface on the transport of suspended sediment. Column results and field measurements were confronted to identify the transport mode from a karstic basin to a borehole located in the alluvial plain, cased through the alluvium and open to the underlying chalk.

1 INTRODUCTION

Suspended sediment (SS) plays a major role in the contamination of aquifers, because of the tendancy of hydrophobic contaminants (Mahler et al., 1999) and bacteria (Pommepuy et al., 1992; Burton et al. in (Palmateer et al., 1993) to sorb on particulate matter. In the Haute-Normandie, where virtually 100% of potable water is from karst, SS has important implications for public health (gastro-intestinal epidemics), as has been previously demonstrated (Beaudeau et al., 1999). There are a few wells constructed in the alluvial plain of the Seine River and intercepting the chalk aquifer below; these wells, although cased through the alluvium and open to the chalk, are not affected by turbidity and have had few problems with bacterial pollution. Recharge from the overlying alluvial water table aquifer has been shown in most cases. Here we examine the influence of the karst-alluvium interface on the transport of SS both by column experiments and measurements made *in situ*.

2 COLUMN EXPERIMENTS

A solution of fluorescein and clay particles in suspension (median diameter 25 μm, measured with a Coulter Multisizer) was injected instantaneously into a column (6.6 cm interior diameter) filled with silica particles with an average diameter of 3 mm (Fig. 1). The flow velocity was varied from 0.05 to 0.87 l/min. We compared the transport of the dissolved and particulate tracers. The experimental curves described allow the determination of the recovery rates of the two tracers, while the hydrodispersive parameters of the medium (effective porosity, dispersivity, coefficient of deposition) can be calculated by the linear graphic method (Wang, 1987). The results show that the average velocity of the particles is greater than that of the dissolved tracer. The lag between the two tracers increases inversely with flow velocity; the difference between the two was 3 seconds for a flow velocity of 0.55 l/min, and 20 seconds for a flow velocity of 0.07 l/min (Figure 2). The recovery rate, despite the relatively high velocities of the experiments, varied from 10 to 50%, without ever exceeding the higher value; the recovery rate of the particles increased with the Darcian velocity, as well as the trapping efficiency (trapping coefficient of the medium K_d), following a power law (Figure 3). This behavioral law (also demonstrated by Kretzschmar et al., 1997) explains the existence of an upper threshold for the recovery. Particle size analysis demonstrated that, for the most part, only the smallest particles underwent breakthrough: at maximum breakthrough, the mode was 7 μm; in general,

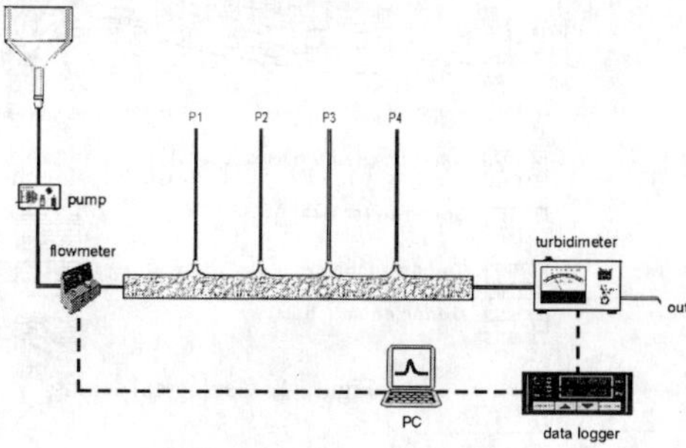

Figure 1. Experimental setup

particles with a diameter greater than 11 μm were removed from flow.

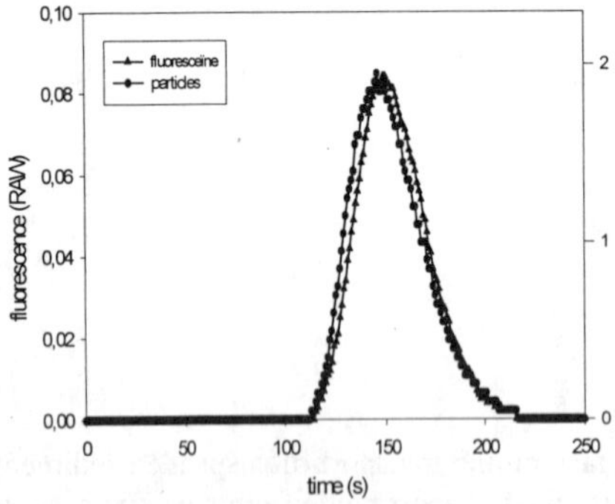

Figure 2. Fluorescine and suspended particle breakthough curves

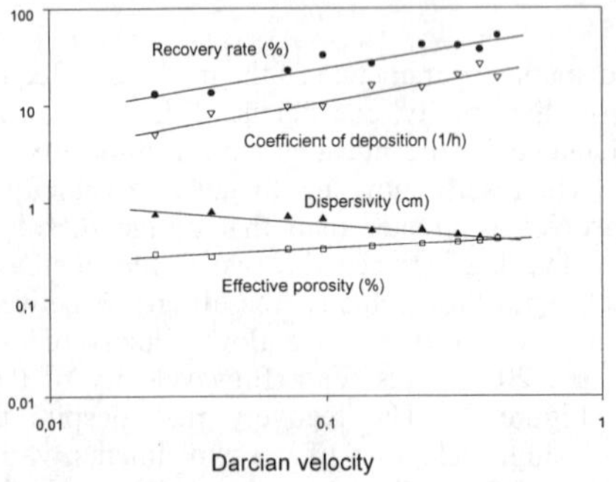

Figure 3. Calculated hydrodispersive parameters

3 FIELD STUDY AND *IN SITU* MEASUREMENTS

3.1 *Geologic setting*

The karst system of Norville is located in the Coniacien chalk (Figure 4), on the northwest edge of the Paris Basin. Hannetôt Spring, located at the interface of the Seine River alluvium and the karstic chalk, drains the chalk aquifer. A portion of its recharge comes from a sinking stream (Bébec Creek). This perennial stream has a watershed of about 8 km^2, whose landuse consists of agriculture and grazing. After heavy rainfall, Bébec Creek carries a high load of SS, resulting in discharge with high turbidity from Hannetôt Spring. A well was drilled through the alluvium and into the chalk 150 m from the spring. Water from the well has very low turbidity and good biological and chemical quality. These three sites — the spring, the sinkhole, and the well — were used for *in situ* measurements in response to rainfall.

3.2 *Materials and methods*

The sinkhole and the spring were equipped with data probes (YSI 6820) for measurement of turbidity, conductivity, and temperature at 15 minute intervals. The data recorder was connected to an autosampler (ISCO 3700). The discharge was measured with a Doppler flowmeter (ISCO 4150). The well was equipped with a similar autosampler. The particles collected were examined under a scanning electron microscope (SEM) after being counted by a Coulter Multisizer.

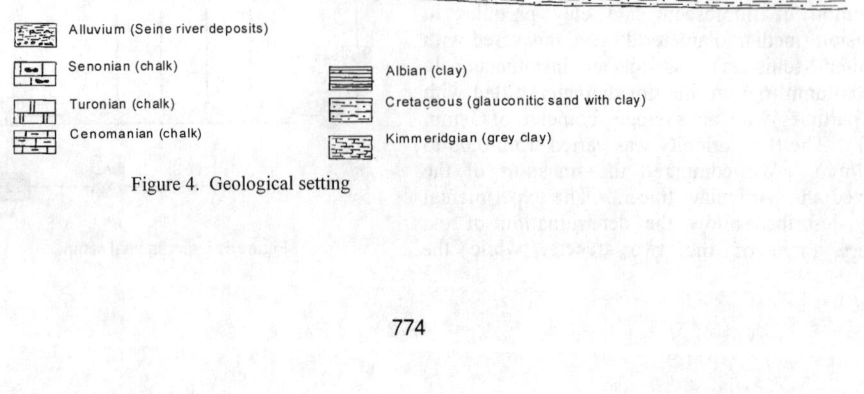

Figure 4. Geological setting

3.3 *Results*

3.3.1 *Synthetic tracer*

Several traces have been carried out from the sinkhole to the well. All flow velocities were calculated based on the arrival time of the tracer. The fluorescein breakthrough at the spring gives an apparent transport velocity of 75 to 125 m/h (transport time of 10 to 24 hours), and a high recovery rate (90-95%). At the well, the fluorescent signal is dampened, with a relation between the maximum concentration at the spring and the well Cw_{max} / Cs_{max} of 14% under low-water conditions (Figure 5), and 3% under high-water conditions. The lag between the arrival of the fluorescein at the well and at the spring varies from 6 to 10 hours, giving an apparent flow velocity between the spring and the well of 22 m/h for the tracing carried out in May 1997.

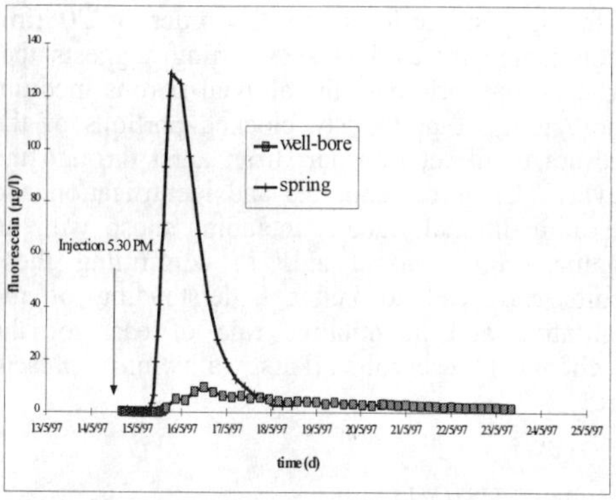

Figure 5. Fluorescein breakthrough curves at spring and well (May 97 dyeing campaign).

3.3.2 *Natural tracer*

The turbidities shown in Figure 8 were recorded after rainfalls of 23.2 and 21.4 mm on the 24 and 25 of December, 1999, respectively. An increase in the discharge of the spring during rainfall indicates increased drainage of the chalk aquifer related to changes in the water table level, but uncorrelated to recharge via the sinkhole (Fig. 6). In contrast, the turbidity at the spring (Fig. 7) is directly correlated to increased turbidity of the sinkhole recharge, with a dilution of about 65% and a time lag of the same order of magnitude as those seen during the fluorescein tracings. At the well, the time lag in relation to the spring was between 4.5 and 12.5 hours (the uncertainty is because of an 8-hour sampling interval at the well), with a turbidity 50 times more dilute.

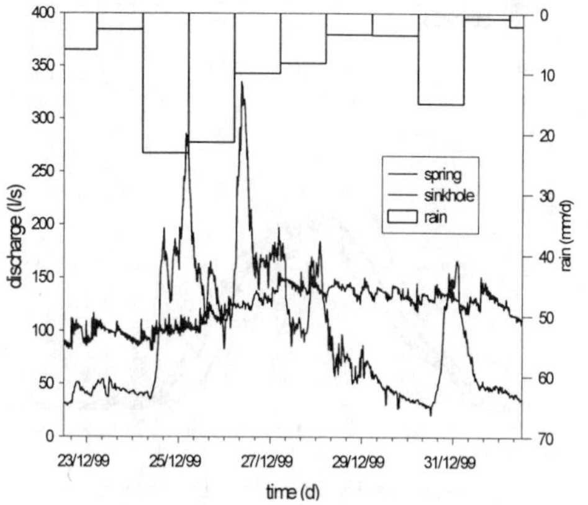

Figure 6. Flow response at sinkhole and spring related to two consecutive storm periods.

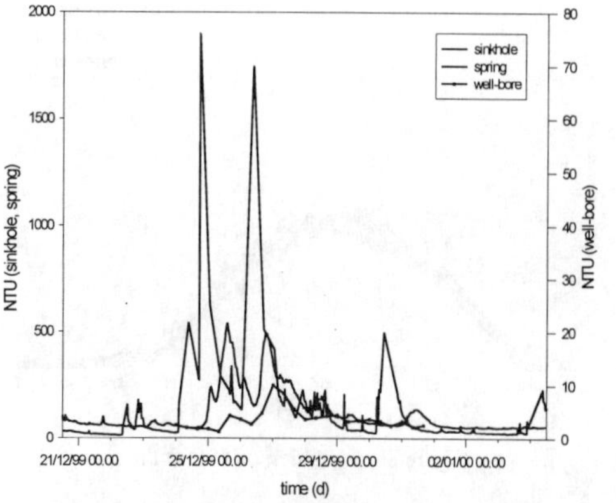

Figure 7. Induced turbidity at sinkhole, spring and well

Comparison of the particle-size distributions (PSD) at the two sites during an episode of high turbidity affords better understanding of the heterogeneities of the medium affecting the transport of the sediment. Figures 8a-c show the results of sampling during the second turbidity episode (second peak). This analysis shows a similar evolution of the particle-size distribution at the sinkhole and the spring, with the proportion of large particles increasing with time, suggesting a direct connection between the two points. In contrast, the evolution of particle-size distribution seen at the well is very different: the sample collected on Dec. 26 at 8 a.m., corresponding to the beginning of breakthrough, contains a relatively high proportion of large particles. Size-sorting with time is evident and reflects the influence of the porous medium on particle transport.

Examination of the particles under SEM shows a morphologic similarity between the particles recharging at the sinkhole and those discharging

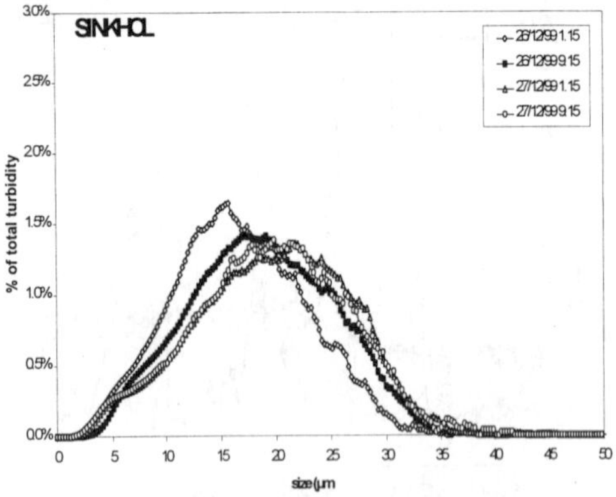

Figure 8a. Evolution of PSD at the sinkhole

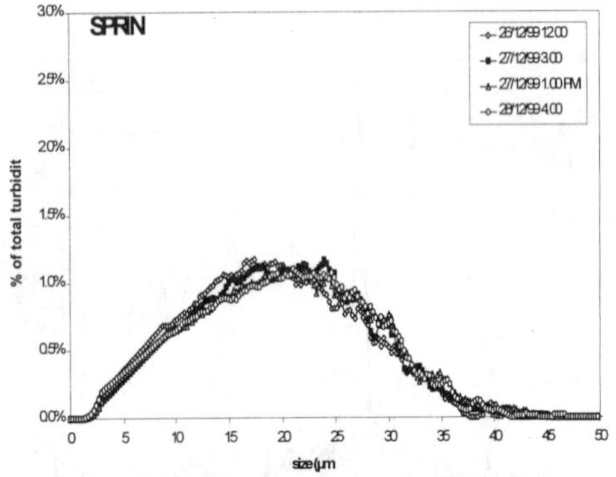

Figure 8b. Evolution of PSD at the spring

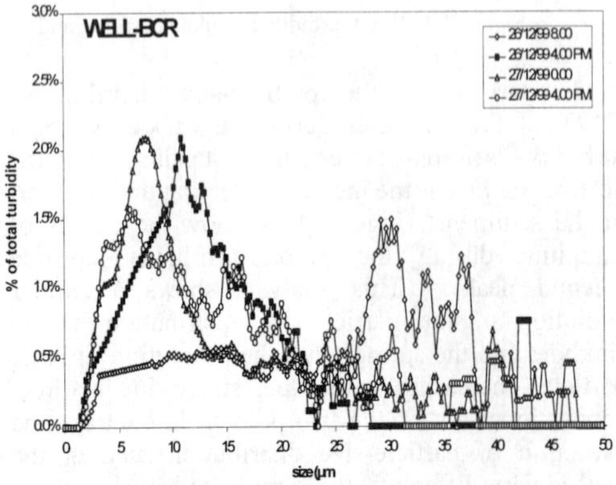

Figure 8c. Evolution of PSD at the well-bore

from the spring, while the particles collected at the well have characteristics distinct from those at the other sites. This difference is particularly obvious for the particulates of biological origin (a large number of diatom, thecamoeba, etc.).

4 DISCUSSION

The high discharge and the virtual absence of turbidity clearly indicate the influence of the overlying alluvium on the recharge of the well at Norville. The acidification carried out during well construction demonstrated the relation between the screened interval and the alluvium. The transport velocity and the duration of the breakthrough of the tracer imply passage through a very permeable medium. The interface between the chalk and the alluvium is karstified throughout the region, and overlain by a layer of gravel (Lefebvre et al., 1993). This porous medium acts as a horizontal filter for upward flow from the confined chalk aquifer. For a gravelly material with similar characteristics, the experiments and simulations show that the recovery rate remains small (always less than 50%) even for the highest velocities. The fate of the trapped particles is uncertain, but includes the possibility that they will at some point clog the permeability. At the onset of the high water, the arrival at the well of the largest particles (on the order of 20 µm) accompanied by evident size sorting suggests that particles may originate the alluvial porous medium (unclogging of previously blocked portions of the medium or direct transport from karst through the gravel). Chemical analyses and identification and use of additional tracers, including those with an organic origin, would aide in confirming these results and lead to better understanding of the qualitative and quantitative role of each of the geochemical reservoirs (karst, alluvium) present here.

5 CONCLUSION

The study of particle transport in porous media, by column experiments, allows the observation of a systematic blocking of the medium by the largest size fraction, and, for the fine particles which are not trapped, a transport velocity greater than that of the dissolved phase. The lag between these two increases with decreasing velocity (increase in the effect of dispersion). The release of the trapped particles under the affect of a pressure transfer may occur. Because of the strong relation between suspended sediment and bacteria, these events may result in an increased risk for public health. The application of particle-size analysis as a predictor of susceptibility to bacterial contamination for public water supply wells may thus be of great value.

REFERENCES

Beaudeau, P. et al., 1999. A time series study of anti-diarrheal drug sales and tap-water quality. International Journal of Environmental Health Research, 9: 293-311.

Kretzschmar, R. et al., 1997. Experimental determination of colloid deposition rates and collision efficiencies in natural porous media. Water Resources Research, 33(5): 1129-1137.

Lefebvre, D., de La Querière, P. and Leboulanger, T., 1993. Nappes captives et sous-écoulements: le rôle des alluvions en Basse Seine. Quaternaire, 4(2-3): 103-110.

Malher, B.J., Lynch, F.L. and Bennett , P.C., 1999. Mobile sediment in an urbanizing karst aquifer: implications for contaminant transport. Environmental geology (sous presse).

Palmateer, G.A., MacLean, D.E., Kutas, W.L. and Meissner, S.M., 1993. Suspended particulate/bacterial interaction in agricultural drains. S.S. RAO: 1-40.

Pommepuy, M., Guillaud, J.F., Derrien, A., Le Guyader, F. and Cormier, F., 1992. Enteric bacteria survival factors. Wat. Sci. Tech., 25(12): 93-103.

Wang, H.Q., Crampon, N., Huberson, S., Garnier, J.M., 1987. A linear graphical method for determining hydrodispersive characteristics in tracer experiments with instantaneous injection. Journal of Hydrology, 95(143-154).

Groundwater: Past Achievements and Future Challenges, Sililo et al. (eds) © 2000 Balkema, Rotterdam, ISBN 90 5809 159 7

Characterization and remediation of a low permeability formation contaminated with halogenated solvents

B.A. Memon, L.D. George & M.M. Azmeh
P.E. LaMoreaux and Associates, Tuscaloosa, Ala., USA

ABSTRACT: Ground water which occurs in diversified geologic environments has been impacted by contamination hazards such as uncontrolled storage and/or disposal of waste materials, particularly used solvents and chemicals; accidental spills; and leachate from waste management facilities. As a result of environmental awareness, regulatory standards and development of remedial technologies, cost effective efforts are made to clean up contaminated groundwater and restore water quality to acceptable limits within reasonable time frames. Ground water and soils of the regolith at an operating manufacturing facility contain halogenated solvents and degradation products at concentrations above regulatory limits. Remediation of ground water has been successfully implemented based on a thorough understanding and characterization of the geologic and hydrogeologic setting, the nature of contaminants, and configuration of the plume at the facility. Hydraulic conductivity of the heterogeneous, low permeable residuum was beneficiated by applying pneumatic fracturing technology. The ground water is being remediated by pump-and-treat methodology. Multi-phase extraction techniques will be employed to enhance remedial action. Solvents with low solubility coefficients may be recovered using pump and treat methodology to remediate low permeability environments. Such recovery is slow but cost effective. More rapid recovery can be achieved by application of Multi-Purpose Extraction Technology but at a higher cost and capital expenditure.

1 INTRODUCTION

Releases of solvents occurred several years ago at a site in the Tennessee Valley District of the Highland Rim Physiographic Province of southeastern United States resulting in a concentrated solvent plume in the low permeable residuum overlying carbonate bedrock.

The release of halogenated solvents, namely 1,1,1-Trichloroethane (TCA), was the source of contamination in the underlying soils and ground water at the site. A field investigation was conducted to evaluate the hydrogeologic setting and characterize the impact of the released solvents on the ground water. The investigation was also directed to develop a conceptual hydrogeologic model of the site, to evaluate the distribution and fate of the solvents, and to establish remedial measures.

The objective was to determine the most efficient, cost effective methodology to remediate the ground water and soils, and to implement such methodology.

2 HYDROGEOLOGIC SETTING AND OCCURRENCE OF CONTAMINANTS

The conceptual hydrogeologic model for the site was based on review of published information and interpretation of the drilling data from soil borings, HydroPunch^{TM} borehole data, drilling and installation of wells (Fig. 1), in-situ testing, and water-level measurements.

The site is underlain by the Tuscumbia Limestone, residuum from weathering of the Tuscumbia Limestone, and construction fill material that was derived from the residuum. The top of bedrock at the site is a fairly planar surface that slopes gently to the southeast, although the Tuscumbia Limestone exhibits karst features in the region. The residuum at the site is approximately 16-24 meters in thickness. It consists primarily of clayey sand/ sandy clay matrix with thin clay-rich layers separating the residuum into two main saturated zones, termed the upper zone and lower zone. The upper zone is from land surface to a maximum depth of 12 meters. This zone includes

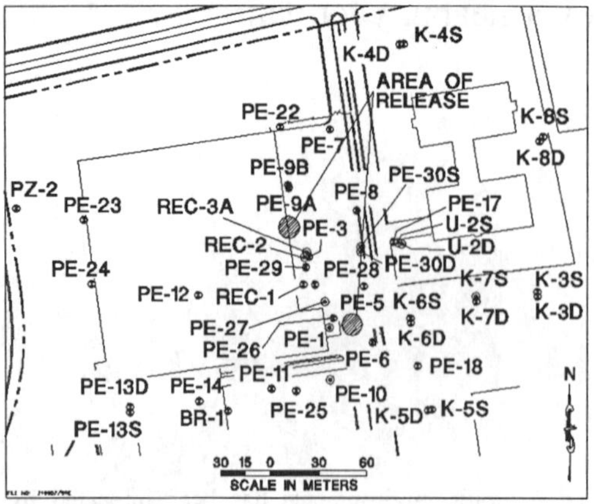

Figure 1. Site Map

some discontinuous thin clay–rich layers that further divide the upper zone. The contaminants in soil and ground water are predominantly present in this upper zone. The hydraulic conductivity of the upper zone ranges from 4.1 x 10^{-4} to 7.5 x 10^{-7} cm/sec.

The lower zone is from 12 to 24 meters below land surface. This zone is more permeable than the upper zone although the lithologic composition is very similar, except for particle size. Hydraulic conductivity and particle size increase with depth. The hydraulic conductivity in this zone ranges from 7.5 x 10^{-3} to 1.4 x 10^{-4} cm/sec.

The occurrence and movement of ground water within the residuum is complex. The residuum is not completely saturated. Dry or partially saturated intervals were encountered below saturated zones during continuous split-spoon sampling. Because grain size within the residuum generally increases with depth the hydraulic conductivity increases downward toward bedrock. The clay layers vary in thickness and serve as semi-confining units. The water levels and water-level responses to natural recharge or pumping (or lack of) in nested and adjacent wells indicate vertical hydraulic separation. The residuum is not a single, simple hydrologic unit.

The majority of the wells installed are screened either at or in the rubble zone (in proximity to the bedrock interface), or at a depth of approximately 9-12 meters below land surface (the upper zone). Water levels in nested wells indicate a downward vertical gradient. The head differentials vary from about a meter to approximately 9 meters. Water-level data have been used to prepare maps illustrating the configuration of the water surface and direction of ground-water movement. The direction of ground-water movement in the lower residuum at the site is toward the east-southeast, consistent with regional ground-water movement. As expected water levels in the wells tapping the upper zone vary significantly, because of complex lithologic makeup, as a result of different recharge rates due to varying hydraulic conductivities, and type of cover at land surface (asphalt, concrete, or grass). Manmade features, such as pavement, ditches, culverts, underground utilities, drains, water lines, and sewer lines, etc., divert or influence infiltration and movement of water. As such, the direction of ground-water movement within the upper zone is complex.

Review of water-level data and interpretation of the water-surface maps over time demonstrate that ground-water levels fluctuate seasonally. Water levels were highest in March and lowest in September in both the upper and lower zones. The magnitude of fluctuation monitored over the hydrologic cycle is about 6 meters.

Water samples were collected from monitoring wells and analyzed on a routine basis. Volatile Organic Compounds (VOCs) comprised of TCA and degradation products, 1,1-Dichloroethylene (DCE), and 1,1-Dichloroethane, were the predominant constituents in ground water where VOCs were detected. In general, TCA and DCE were detected at the highest concentrations in the immediate vicinity of the release area. The highest concentrations were generally restricted in extent, laterally and vertically, and decline significantly in concentration with distrance from the release area. Total VOC concentrations ranged from 950,000 parts per billion (ppb) to 4,700 ppb within the plume in the upper zone in March 1995. These VOCs were mainly composed of 1,1,1-Trichloroethane (TCA) and 1,1-Dichloroethene (DCE). Within the plume, concentrations of TCA ranged from 660,000 ppb to 1,300 ppb (Fig. 2) and concentrations of DCE ranged from 120,000 ppb to 5,500 ppb. The solubility of TCA and DCE is 1,300 and 3,350 mg/L, respectively.

3 SITE ASSESSMENT AND REMEDIAL ALTERNATIVES

Site assessment was performed to characterize subsurface geologic conditions and hydro-geologic setting for development of a conceptual hydrogeologic model and understanding of movement of contaminants. The model is critical for selection and design of ground-water remedial technologies to remediate heterogeneous residuum of relatively lower hydraulic conductivity.

Due to low hydraulic conductivity, the plume of contamination is confined to the upper zone. Only traces of contaminants have been detected in this lower zone except at a location where the clay-rich separating layer formed a natural depression allowing the heavier-than-water solvents to settle and migrate. This hot spot at depth has been abated by remedial action, namely pump and treat technology.

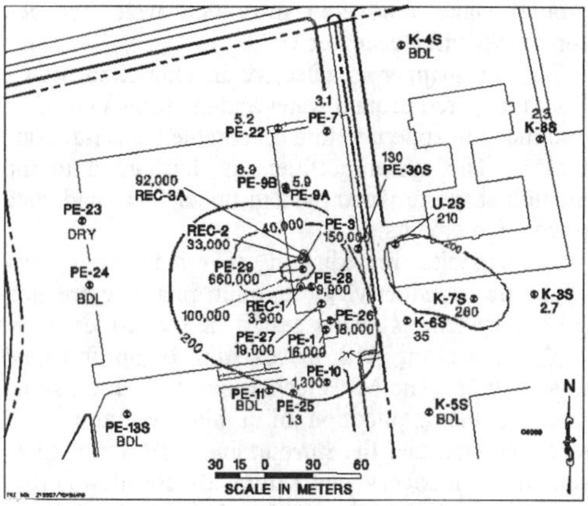

Figure 2. TCA Upper Zone March 1995

Prior to initiation of remediation, the nature of the problem was assessed and remedial alternatives were evaluated based on environmental impacts, risk assessment, optimal solution, cost-effectiveness analysis, cost/benefit analysis, and sensitivity analysis. Methodologies considered for remediation of the residuum at the site included: surface caps, excavation and disposal of contaminated soils, thermal treatment, physical treatment of soil, physical vertical barriers, vapor extraction, and pump and treat with physical treatment of ground water (air stripping). Based on these analyses and consideration of the various possible remediation methods, "Pump and Treat" technology was selected to address the hot spots at the site. To enhance hydraulic conductivity, and to expedite and more effectively remediate, soil fracturing was performed prior to initiation of pump and treat remediation.

Pilot tests were performed to determine recovery rates, areal distribution of monitoring network, and determine geographic location of recovery wells to control the plume both vertically and laterally and also create hydraulic barrier to off-site migration of contaminants. Based on the results of the pilot tests and monitoring, the recovery wells were established in the plume and near center of the monitoring network as shown in Figures 1 and 3. The automatic monitoring and control of the system is performed through a telemetry system. Pumping created a depressed zone of influence within the contaminated area creating a hydraulic barrier preventing further migration of contaminated ground waters.

4 PNEUMATIC FRACTURING

Hydraulic conductivity of the upper zone of the residuum was enhanced using pneumatic fracturing to facilitate the movement of water and recovery of contaminants. Pneumatic fracturing technology was applied in the area of the release. Four wells were drilled for this purpose. In-situ hydraulic conductivity in the shallow residuum determined prior to fracturing was 2.00 x 10^{-6} to 7.5 x 10^{-7} cm/sec. Fracturing was performed by injecting air under pressure through discrete 70 centimeters intervals in each of the fracture wells. Packers attached to the injection tool defined the discrete intervals. Fracture propagation was limited to horizontal direction and controlled by design of injection tool and screened interval. During each injection, a transducer installed on the injection tool was used to monitor down-hole pressure within the fracture zone.

Fracture initiation pressure (pressure at which the formation yields and fracture propagation is initiated) and fracture maintenance pressure (pressure required to maintain fracture as it propagates outward from the fracture point) were recorded. Initiation pressures ranged from 260 to 340 pounds per square inch (psi) and maintenance pressures ranged from 46.9 to 68.8 psi. A tiltmeter heave monitoring system detected ground-surface heave as much as 7.62 meters away from the injection well in all directions. The maximum vertical heave recorded was 7.6 mm. The extent and orientation of fracturing shown by the tiltmeters correlated well with ground-water level data collected during injections. Hydraulic conductivity of the upper zone was increased by approximately an order of magnitude (i.e. improved from 10^{-7} and 10^{-6} to 10^{-6} and 10^{-5} cm/sec).

5 PUMP AND TREAT REMEDIATION SYSTEM

An underground piping system connecting recovery and selected monitoring wells (Figure 3) was installed to protect discharge lines, air lines, and monitoring equipment cables from heavy traffic and freezing. A total of nine pumps were installed to recover ground water at a total rate of approximately 0.25 l/sec. Pressure transducers were installed to continuously monitor water levels in all pumping wells and selected monitoring wells. A remediation center was installed housing: (a) four-tray air stripper (*White Water Shallow Tray*), (b) air compressor to supply air to the pumps, (c) 7950 liters batch tank receiving discharge water pumped from the wells.

Ground water is pumped from nine recovery wells, and discharged through a common discharge line to the batch tank. Level controls were installed in the batch tank to activate or deactivate the air stripper. A digital flowmeter and a pH analyzer/recorder were installed to monitor quantity and quality of effluent.

In addition, a second tank 9465 liters capacity was

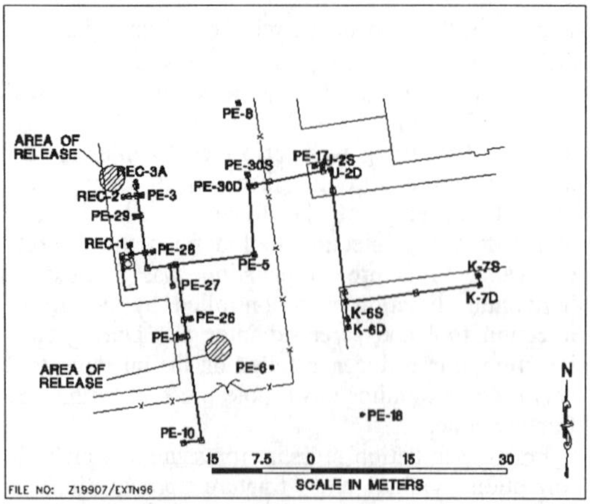

Figure 3. Remediation System Plan.

installed outside the treatment center to provide emergency storage capacity.
The two tanks are connected to provide for overflow from the inside batch tank to the outside tank.

A telemetry system was designed and installed to remotely monitor and control the remediation system on a continuous basis

6 RESULTS

The remedial system in operation has been successful in reducing the concentration and size of the contaminant plume and preventing off-site migration with the hydraulic barrier created by pumping. The areal extent of the TCA plume has been reduced from its original dimension, Figure 2, to a smaller and contained plume as shown in Figure 4. Data demonstrates the decline in concentrations of contaminants at the recovery wells, as shown in Figure 5 which includes the cumulative volume of

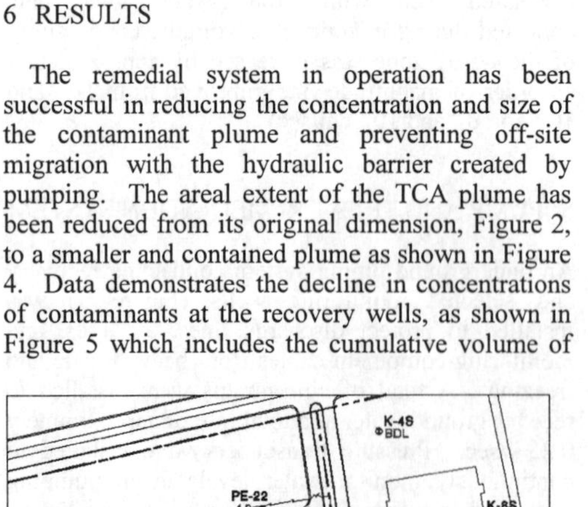

Figure 4. TCA Upper Residuum March 1999.

ground water recovered from that well, between June 1995 and December 1999.

The air stripper is effective in extracting VOCs from recovered ground water and emitting VOCs into the atmosphere within acceptable concentration limits. The system effluent is discharged to the municipal waste water system under local and state permit agreements.

To expedite remediation of residuum of low hydraulic conductivity, a multi-phase extraction (MPE) system is being tested and considered for application along with the existing pump and treat remediation. The MPE technology is a process that involves the application of a high vacuum on a recovery well and the surrounding soil to maximize soil vapor recovery and saturated-zone dewatering as well as enhanse the recovery of dissolved phase VOCs in ground water. Saturated-zone dewatering also results in the extraction of soil vapor from previously saturated soil.
It is expected that both pump and treat and MPE will reduce the remainder of contaminants in a timely and cost effective manner.

Solvents with low solubility coefficients are recovered from low permeability environments by using pump and treat methodology to recover ground water containing these solvents then treating it by air stripping. Such remediation is slow but cost effective. As more efficient equipment became available, more rapid recovery can be achieved by application of MPE applying high vacuum to contaminated soils to extract ground water and air containing volatalized solvents.

Ground-water remediation and aquifer restoration is expensive and time-consuming and is a multi-disciplinary effort. It is important that all regional and site-specific data related to geologic settings, hydrogeologic characterization, physical and chemical characteristics of the pollutant, source of

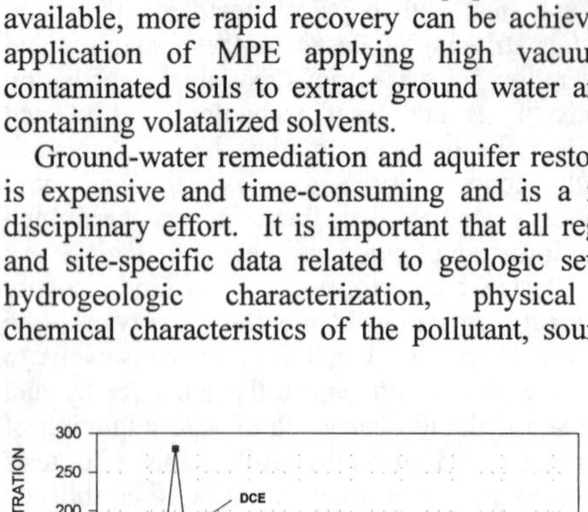

Figure 5. Ground water recovery and TCA/DCE concentrations, REC-2

pollution, transport path ways, applicable technologies, engineering design and construction, regulatory requirements, health effects, costs and benefits are evaluated and understood prior to initiating a remediation program.

This case history represents a challenging scenario for ground-water remediation because of the hydrogeologic setting and the nature of the contaminants. It is anticipated that completion of corrective action will occur by meeting water-quality standards to be delivered by risk-based approaches upon the state's adoption of such a program.

REFERENCES

Bouwer, H. 1989. The Bouwer and Rice Slug Test-An Update. *Ground Water* 27(3):304-09.

Bouwer, H. & H.C. Rice 1976. A slug test for determining hydraulic conductivity of unconfined aquifers with completely or partially penetrating wells. *Water Resources Research* 12:423-28.

Fetter, C.W. 1994. *Applied hydrogeology. Third Edition*. New York: Macmillan.

LaMoreaux, P.E. & Associates, Inc. 1990-2000. Various consultants reports of on-site investigations, corrective action plans, and effectiveness monitoring.

Pankow, J.F. & J.A. Cherry. 1996, *Dense Chlorinated Solvents and other DNAPLs in Groundwater*. Waterloo Press.

Schuring, J.R., P.C. Chan, C.D. Liskowitz, C.D. Fitzgerald & U. Frank. *Pneumatic fracturing of low permeability formations*.

Todd, D.K. 1980. *Groundwater hydrology. Second Edition*. New York: John Wiley & Sons.

U.S. Environmental Protection Agency 1988, *Functional guidelines for organics*. EPA/540/I-89/002.

U.S. Environmental Protection Agency 1993, *Accutech pneumatic fracturing extraction and hot gas injection, Phase I*. EPA/540/AR-93/509.

U.S. Environmental Protection Agency 1993. Pneumatic fracturing increases VOC extraction rate. In *Tech Trends*, EPA/542/N-93/010.

U.S. Environmental Protection Agency 1996. *Environmental investigations, standard operating procedures and quality assurance manual*. U.S. Environmental Protection Agency, Region IV, Athens, Georgia.

Groundwater: Past Achievements and Future Challenges, Sililo et al. (eds) © 2000 Balkema, Rotterdam, ISBN 90 5809 159 7

Hydrogeological characterization of fractured rock for Sweden's nuclear waste storage

M. Morosini
Swedish Nuclear Fuel and Waste Management Company (SKB), Figeholm, Sweden

I. Rhén
SWECO, Sweden

ABSTRACT: During the past 25 years Sweden has conducted research on the feasibility of storing spent nuclear fuel in an underground repository. A concept of engineered barriers surrounded by natural rock barriers was proposed as the main alternative for storage. The barrier concept is a necessary requirement for containment and delay of potential radioactive leakage. The process of finding a suitable site has now reached the stage where a generic site characterization programme has been set up. Once a potential site have been identified a site specific programme will be formulated. Efforts are focused to produce all necessary programme for the site characterization by the end of the year 2001. The characterization programme is based on a hydrogeological concept where the rock is divided into different hydraulically conducting features (e g fracture zones) hosted in rock possessing a background transmissivity. The paper presents possible strategies, methods and concepts to be employed within the hydrogeological site characterization program.

1 INTRODUCTION

The Swedish Nuclear Fuel and Waste Management Company (SKB), with whom is vested the task of developing an acceptable system for storing spent nuclear waste in Sweden, proposed a technical solution, called KBS-3, SKBF/KBS (1983). The KBS-3 concept is based on depositing the spent nuclear fuel at a depth of somewhere between 400m and 700m in the granitic bedrock. The nuclear fuel will be surrounded by engineered barriers: steel cladding, copper canister , bentonite clay and by the natural barrier of the rock (Figure 1).

This multi-barrier concept of containing a potential leak from the nuclear waste depends ultimately on the capability of the host rock to retain hazardous solutes and to transport the solutes from canister to biosphere. Although the complete waste storage system comprises a multitude of issues, we will only give an overview of the Swedish approach to hydrogeological site characterization and focus on strategies and concepts. The complete geoscientific site characterization program is structured into different subject-specific disciplines which in addition to hydrogeology consist of geology, rock mechanics, thermal issues, hydrogeochemistry, solute transport and biosphere interaction. During the execution of the investigations, integration of work between the disciplines is important for the reliability of the geoscientific models.

2 OBJECTIVES

The overriding objectives with the site characterization is to identify a suitable volume of rock which may host a repository. The suitability is defined with respect to functional issues:

- constructability of the repository
- radiological safety
- long-term performance of the repository

From these functional issues a number of hydrogeological issues have to be addressed, of which the most central are:

a) where can hydrogeological problems be expected during construction and what inflow can be expected?
b) is there sufficient groundwater to adequately saturate the bentonite?
c) is the advective transport time from canister to-biosphere sufficiently low?
d) is the dilution of the radioactive dose from canister to biosphere sufficiently high?

Sorption processes are of course very important for the transport time, but are not discussed here.

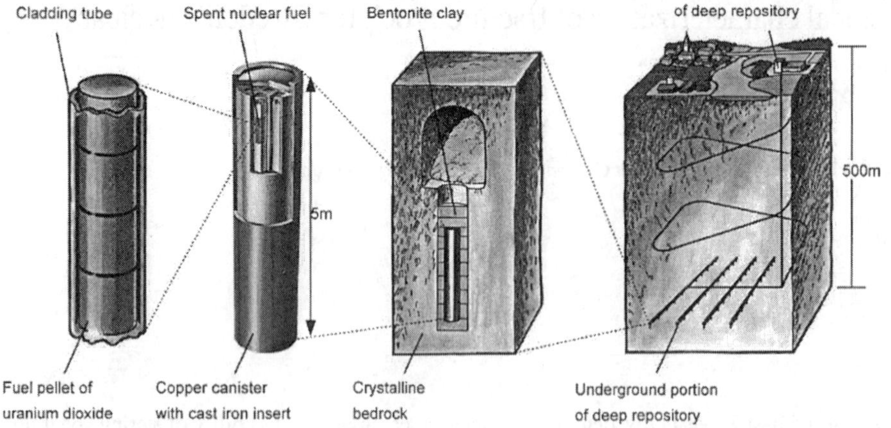

Figure 1. The nuclear waste storage concept of KBS-3.

3 SITE CHARACTERIZATION PROGRAMME

The approach to characterization is first to gain a regional hydrogeological understanding of processes in the area. The second stage is to perform more detailed investigations to a smaller area which have the potential of accommodating the repository. The characterization should supply sufficient data to allow the construction of a hydrogeological model.

The hydrogeological model should then be integrated with other subject-specific models (geology, chemistry, hydrogeology, etc) in order to deduce one comprehensive geoscientific model for the site. During the site investigation, these models will be continuously developed in parallel, as the data is collected from the field.

Note that an initial conceptual model will be provided through the geological characterization programme by non-intrusive characterization methods such as remote sensing, airborne and ground geophysics, and seismic. In the following we discuss the hydrogeological model only.

3.1 *Conceptual model*

Our approach to address the above issues (a-d) is based on a conceptual hydrogeological model where we distinguish between the soil aquifer and the rock mass aquifer. The rock mass is subdivided into homogeneous blocks separated by fracture zones. This concept is shown in Figure 2.

The fracture zones will be classified according to their size and transmissivity, they will be termed hydraulic conductor domains (HCD). The rock volume between the fracture zones will be classified according to its transmissivity, it will be termed hydraulic rock domain (HRD). The data which will form the base for the classification into HCD and HRD is the geological model (structural and lithological) in conjunction with data from the borehole flowlogging and hydraulic testing. The requirement for a volume of rock to be classified within a certain domain is to

- be continuous geometrically, and
- have a transmissivity within a certain range

There may be several volumes of rock, separated in space, each being geometrically continuous with the same transmissivity.

In Figure 2, the local major fracture zone has been classified as HCD1 and the local minor fracture zone as HCD2 because they have different transmissivities. In case the transmissivity would be the same they would both be given the same classification, e g HCD1.

Hence, the classification into HCD and HRD is a convenient way of discretizing the rock based on hydrogeologically relevant criteria.

Based on experience from field investigations throughout Sweden, e g Ahlbom et al (1992), and in particular from work conducted at the Äspö Hard Rock Laboratory (Rhén et al, 1997), the fracturing of the rock may be categorized into four different classes according to their size:

Regional fracture zone have an extent of over 10 km and a width may be more than 100m, these will be established deterministically, with some uncertainties though;

Major local fracture zones have a length of 1-10km and a width of 5-100m and will also be established on deterministical grounds, with some uncertainties though;

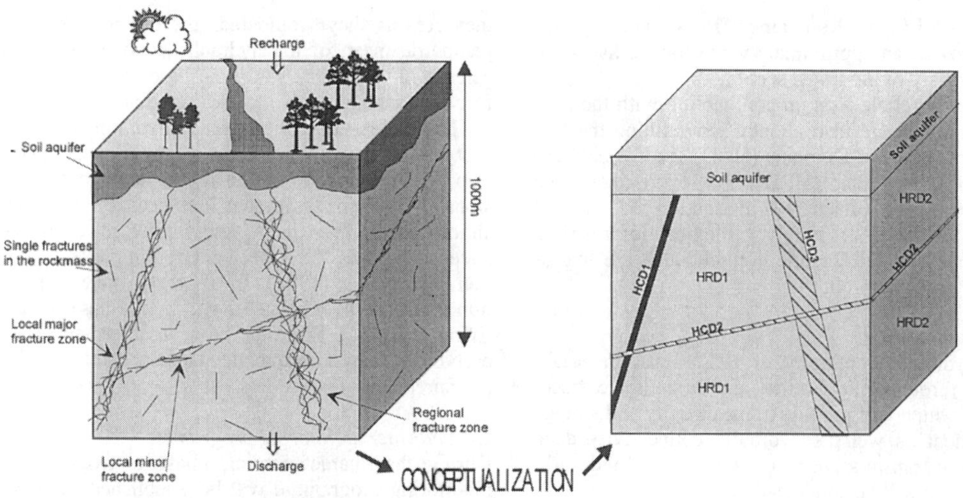

Figure 2. Conceptual hydrogeological model

Minor local fracture zones have a length of 10m-1km and a width of 0.1-5m. This type of zone will predominantly be established statistically although some may be established deterministically;

Single fractures in the rock mass that are less than 10m in length and less than 0.1m wide will only be established statistically.

Near the coast the aquifer is furthermore assumed to have saline water in the deeper parts.

3.2 *Parameters*

A thorough review has been completed where all parameters of importance during geoscientific site investigation were identified and their relative importance assessed (Andersson et al 1998). That review form the base on which the field investigation to determine hydrogeological parameters has been made.

During the site characterization the following data will be collected:

- Meteorological and hydrological data such as run-off, water-divide and spring discharge. Estimation of hydraulic conductivity, specific storage coefficient and aquifer thickness of the soil aquifer.
- Transmissivity and specific storage coefficient of deterministically modeled fracture zones.
- Transmissivity distribution, porosity and specific storage coefficient of stochastically modeled rock mass.
- Spatial distribution of salinity and temperature of the groundwater
- Boundary conditions will be established with respect to pressure and salinity distribution, to groundwater flow and to recharge/discharge areas.

3.3 *Characterization methods*

The hydrogeological site characterization programme sets out to obtain field data on the parameters mentioned above in order to enable the construction of a quantitative model of the rock volume of interest. Below is outlined what measurements may be made during the characterization.

Drilling

Characterization will be performed during the borehole drilling phase with respect to:

- drilling fluid pressure as well as in-going and out-coming flow
- water inflow to the borehole at different depths
- hydraulic head at different depths by monitoring overnight recovery of head
- transient hydraulic testing through the drilling rod with a custom designed testing and water sampling unit (wire-line technique)

Upon drill completion water samples from selected borehole intervals will be collected for chemical characterization.

Borehole logging

A suite of geophysical borehole logs of which temperature and fluid resistivity will partly provide the fluid density along the borehole.

Two complementing flowlogs will be conducted with different flowrange. Major inflows will be identified with the high range flowlog (2-700 L/min). It will be followed by the low range logging

(0.0001–5 L/min) (Rouhiainen, 1998). This log will also provide an approximate value for the hydraulic conductivity of the tested section.

These borehole logs, in conjunction with the geological model are instrumental for defining the Hydraulic Conductor Domains (HCD) and the Hydraulic Rock Domains (HRD). As opposed to the fracture zones (discussed in section 3.1 above) which are based on purely geological information, the HCD and HRD definition includes the hydrological aspect as well.

Hydraulic testing

The hydraulic continuity of HCD must be established through interference and large scale tracer testing which will provide transmissivity and storage coefficient as well as hydraulic connectivity data. For these transmissive features transient fluid withdrawal test will be conducted.

Hydraulic characterization of HRD will be done through hydraulic injection testing. Since the HRD are defined on statistical grounds it is necessary to provide a sufficient number of tests in order to make up the statistics. Part of the borehole which penetrate the HRD will be subjected to double packer injection testing with section lengths of possibly 5m.

All hydraulic tests will be interpreted with theories developed for continuous porous media (Raghavan, 1993). This approach is based on experiences from testing fractured rock in Sweden where in many cases the hydraulic response show similarities to the continuous porous media response. Where difficulties or inconsistencies are encountered in the interpretation the generalized general flow approach proposed by Barker (1988) may be utilized in order to give alternative ideas on the flow regime.

Natural flow measurements

A number of flow measurements under undisturbed conditions will also be made. This is of particular value in order to:

- appreciate flow velocities and subsequently transport time in the neighbourhood of the repository
- provide calibration measures for the numerical transport modeling
- provide an approximate value for the hydraulic gradient at repository depth

SKB have developed a probe to measure natural groundwater flow through a borehole section through the dilution technique.

Statistical measures

With regards to the defined HRD the approach is to derive a frequency distribution of the hydraulic conductivity (K) from tests of equally long test sections. It is essential to exclude the HCD in the statistical measures as they would bias the distribution. The geometric mean of the hydraulic conductivity (K_g) calculated from this statistical base will be carried forward to the groundwater flow modeling.

The K_g measure is the effective hydraulic conductivity (K_e) for two dimensional steady flow in an isotropic porous medium. The K_e for a three dimensional heterogeneous system has been found to be slightly larger than the K_g and various relations between K_g and K_e have been derived in the scientific literature, Gelhar (1993). In any case, whatever relation is ultimately chosen K_g will be the base for arriving at the K_e. There are however alternatives for deriving K_e based on fracture statistics and hydraulic measurements.

Groundwater monitoring programme

During the characterization phase a groundwater monitoring programme will be established supplying real-time and online data on groundwater levels and on hydraulic head from packed off sections along the boreholes. It will also allow the sampling of water from sections and introduction of tracers. This system will provide data on long-term fluctuations in hydraulic head and be utilized for the hydraulic interference testing. It will also provide baseline data for the environmental impact assessment and possibly monitor the pressure head after the repository has been closed.

The gradual integration of different type of data and parameters will enable the production of new conceptual models which will be utilized for the numerical modeling of the groundwater flow.

3.4 *Numerical modeling*

The descriptive, conceptual model will be the base for all numerical modeling. Parameters established through the characterization programme are carried forward, with due upscaling if necessary, to the quantitative groundwater flow modeling exercise. Modeling will be performed for two volumes: a large regional volume providing sound and stable boundary conditions to a smaller site volume extending about one km beyond the area of the repository.

There are several approaches to quantitative groundwater flow modeling in fractured rock. During the course of the site investigations both the stochastic continuum and the discrete fracture network formulation will be utilized in parallel.

The modeling will be implemented from the very beginning of the investigations. Different evolving versions of the conceptual model are implemented in the numerical model as more data become available in order to

- test the effects of different hypothesis
- provide guidance on field data collection
- assess the sufficiency of the collected data

The hypothesis testing will include alternative interpretations of the geological structural model.

4 PROGRAMME STATUS

Feasibility studies are presently being conducted at six localities in Sweden. All of them will be completed by the end of 2000, at which time SKB will select at least two sites to recommend for a detailed site investigation. These sites have to be approved by the concerned municipalities before SKB may proceed. Presently, work is in progress in writing programme for

- Geoscience
- Environment & society
- Function & safety
- Layout & constructability

In the near future the generic geoscientific programme will be completed while the work on site specific programme, methodology manuals and activity plans will commence. Efforts are being made to have a complete programme available by the end of 2001, at which time the application for conducting the site investigations should be submitted to the regulators.

Upon completion of the site investigations, one of the sites will be chosen for a detailed investigation which will take place during the construction phase of the repository.

5 SUMMARY

The hydrogeological site investigation programme in fractured rock will aim at identifying and characterizing the fracture zones and the rock mass between the zones. The fracture zones, termed hydraulic conductor domains (HCD), will be characterized deterministically when they are large and statistically if they are smaller. The rock mass between the fracture zones, termed hydraulic rock domain (HRD), will be characterized statistically

The data which will form the base for the classification into HCD and HRD is the geological model (structural and lithological) in conjunction with data from the borehole flowlogging and the hydraulic testing.

The HCD and HRD will form the base when producing a conceptual model. This will be implemented in several and gradually evolving numerical groundwater flow models utilising both the stochastic continuum and the discrete fracture approach. An important component of the numerical flow modeling will be to include alternative interpretations of the structural geological model.

An overall geoscientific model will be produced by integrating the hydrogeological results from the characterization with results from the other subject-specific disciplines: geology, rock mechanics, thermal issues, hydrogeochemistry, solute transport and biosphere interaction. During the execution of the investigations, integration of work between the disciplines will be important for the reliability of the geoscientific models.

A programme for the geoscientific investigation of a nuclear waste repository site is presently (March 2000) in preparation and will be officially published in due time.

NOTE

Statements and opinions expressed in this paper are those of the authors and not necessarily those of SKB.

REFERENCES

Ahlbom, K., Andersson, J-E., Andersson, P., Ittner, T., Ljungren, C., Tirén, S. 1992. Klipperås study site. Scope of activities and main results. SKB TR 92-22. Skensk kärnbränslehantering AB, Stockholm, Sweden.

Andersson, J., Almén, K-E., Ericsson, L O., Fredriksson, A., Karlsson, F., Stanfors, R., Ström, A. 1998. Parameters of importance to determine during geoscientific site investigations. SKB Technical Report 98-02, Svensk kärnbränslehantering AB, Stockholm, Sweden.

Barker, J. 1988. A generalised radial flow model for hydraulic tests in fractured rock. Water Resources Research, v.24, no.10, p1796-1804.

Gelhar, L. 1993. Stochastic subsurface hydrology. Prentice Hall, Englewood Cliffs, New Jersey, USA.

Raghavan, R. 1993. Well test analysis. P T R Prentice Hall, Englewood Cliffs, New Jersey, USA.

Rhén, I. (eds), Gustafson, G., Stanfors R., Wikberg, P. 1997. Äspö HRL Geoscientific evaluation 1997/5- Models based on site characterization 1986-1995. SKB Technical Report TR-97-06, Svensk kärnbränslehantering AB, Stockholm, Sweden.

Rouhiainen, P. 1998. A case history of the difference flow measurement at the Hästholmen site in Loviisa. in "Characterization and evaluation of sites for deep geological disposal of radioactive waste in fractured rock", Proceedings from the 3rd Äspö International Seminar in Oskarshamn, June 10-12, 1998. SKB Technical Report TR-98-10, Svensk kärnbränslehantering AB, Stockholm, Sweden.

SKBF/KBS 1983. Final storage of spent nuclear fuel- KBS-3. Parts I-IV. Svensk kärnbränslehantering AB, Stockholm, Sweden.

Groundwater: Past Achievements and Future Challenges, Sililo et al. (eds) © 2000 Balkema, Rotterdam, ISBN 90 5809 159 7

Practical and responsible approaches to soil and groundwater contamination management at a South African metal plating factory

R. D. Morris
Morris Environmental Groundwater Alliances (MEGA), Cape Town, South Africa

H. G. Allerton III
TRW Automotive, Washington, Mich., USA

J. P. W. van der Westhuysen
VdW&H Consulting Engineers, Stellenbosch, South Africa

ABSTRACT: Inadequate housekeeping and lack of employee environmental awareness at metal plating facilities will result in the release of toxic chemicals to the environment. Chemicals of interest may include the more common metals of zinc, chromium, nickel, copper and cadmium, for which cyanide is used as a complexing agent to keep them in solution. Metal cleaning and preparation prior to plating involves degreasing and, traditionally solvents comprising mainly chlorinated aliphatic hydrocarbons, have been used. In 1995, TRW Automotive from the USA bought a South African metal plating operation in the Atlantis Industrial area, north of Cape Town. In accordance with TRW's Company environmental policy to comply with local regulations to protect public health and the environment at its manufacturing facilities, TRW commissioned an Environmental Site Assessment to identify potential pre-existing environmental impacts at the facility prior to acquisition. Contamination of soil and groundwater was found to be a problem and, following discussions with the regulatory authorities, a contamination management programme was initiated. This involved the excavation and safe disposal of impacted soil and the installation of a groundwater pump and treat programme. Of relevance within the South African context is the fact that this was most likely the first such remedial programme in the country using these technologies within an agreed scope of work where remedial targets were agreed with the authorities. The paper describes the nature and extent of the problem, the hydrogeological setting, the corrective action plan and the results obtained. The demonstrative and proactive approach adopted, coupled with the involvement of the regulatory authorities is of relevance.

1 INTRODUCTION

The raw materials used and wastes generated by the metal plating industry are harmful to the environment and have potential to result in severe contamination if not handled in a responsible manner. Many of these products are toxic, persistent within the environment, and a potential threat to human health. Consequently, regulating authorities have prescribed low maximum contaminant levels (MCL's) and stringent clean-up requirements. Experience indicates that the cost of cleaning up a contaminated site is normally far more than preventing contamination in the first place. It is thus industry's responsibility and sound business practice to ensure that contamination does not take place.

An understanding of the types, sources and impacts of contaminants arising from the metal plating industry provides an incentive to management to eliminate contamination. The consultant tasked with carrying out detailed site assessment and remediation would also require an understanding of the fate and mobility of the contaminants, when associated with the hydrogeological characteristics of the area.

This paper summarizes the criteria that should be considered when undertaking site assessment and remediation for subsurface contamination related to the metal plating industry. A case study example involving the management of contaminated soil and groundwater at a metal plating facility in the Atlantis Industrial area, north of Cape Town, South Africa, is discussed. Atlantis acquires its domestic water supply from a nearby primary aquifer. Of importance to this case study is the manner in which the regulatory authorities were involved and the achievement of the remedial objectives and closure of the programme.

2 CHEMICALS OF INTEREST

The chemicals of interest related to the metal plating industry can be divided into two categories, namely:

1. Pretreatment metal degreasing agents: which include solvents, deburring solutions, alkaline soapy

solutions such as caustic soda or potash, and pickling solutions (acids) usually H_2SO_4 and HCl.

2. Process compounds: comprising the metals listed in the abstract and alkali cyanide solutions.

The most commonly used degreasing agents are solvents, mainly trichloroethene (TCE), 1,1,1-trichloroethane (TCA) and tetrachloroethene (PCE or PER). The physical, chemical and biological properties of the chlorinated solvents make these chemicals particularly serious soil and groundwater contaminants. These properties are summarized (Pankow and Cherry 1995) as follows:

1. High volatility: once in the saturated zone high volatility is of little assistance in removing the solvent under natural conditions.

2. High density: typically 1.2 to 1.7 g/cm^3. Therefore, they may sink through the aquifer.

3. Low viscosity: allows rapid downward movement. Mobility increases with increasing density/viscosity ratios.

4. Low interfacial tension: solvents can easily penetrate small fractures and pore spaces.

5. Low absolute solubility: solvents may migrate as a dense non-aqueous phase liquid (DNAPL) and pool on less permeable layers.

6. High relative solubility: they can dissolve at concentrations considered harmful to human health.

7. Low soil partitioning: solvents do not bind easily to soil and rock and therefore there is low retardation by sorption processes.

8. Low degradability: either by biological or abiotic-chemical reactions, therefore subsurface lifetimes are long.

Some characteristics of the most commonly plated metals and other chemicals used in this industry are summarized as follows:

1. Chromium: dissolved chromium exists as cations in which the oxidation states are Cr^{6+} and Cr^{3+}.

2. Nickel and cadmium: exist in the +2 oxidation state and are subject to precipitation by hydroxide at higher pH or by sulphide at lower pH. The rinse waters are the major avenue by which Ni and Cd are released to the environment (Sawyer and McCarty 1978).

3. Zinc: occurs in one oxidation state (+2) and is more soluble than Ni.

4. Cyanide: is contained within the rinse waters from metal plating operations. The CN- ion acts as a complexing agent to lower the concentration of free metal ion so that the cation does not plate too rapidly.

3 TYPICAL SOURCE AREAS

The typical source areas of contamination at a metal plating factory include the following:

1. Improper storage of raw materials.

2. Spillage during the plating and process operations themselves.

3. Improper handling, treatment and storage of the effluent, wastes and sludges; and

4. When wastes are released to the environment through disposal in unsuitable landfills, leaking storage tanks and uncontrolled contaminated stormwater runoff.

Experience at a number of metal plating facilities has shown that poor housekeeping practices are the major cause of the release of contaminants to the subsurface.

4 BACKGROUND TO THE CASE STUDY

TRW is a world leader in the design and manufacture of automotive components and aerospace systems integration. Consequently, TRW has instituted corporate policies to protect its employees health and safety and the environment. TRW's policy on protection of the environment specifically requires:

- TRW's employees to understand the potential environmental impacts associated with their job duties and to use appropriate preventative practices.
- TRW management will establish appropriate preventative practices for all manufacturing operations and implement environmental management systems.
- TRW business will seek to limit or eliminate the use of hazardous materials in its products and processes.
- The Company will establish worldwide standards to minimize potential risks of environmental concerns at TRW facilities.

TRW has implemented this policy at the TRW Occupant Restraints South Africa (TORSA) facility. The facility was acquired in late 1995 to provide seat belt assemblies to TRW Global automotive customers in South Africa, Europe and the Pacific Rim. TRW's investment in the TORSA facility has included integration of the aforementioned environmental policies. Successful projects completed to date include installation of process air emission control equipment, upgrade of the plating process waste water treatment system, construction of a secure chemical storage area, elimination of use of chlorinated solvents, and remediation of pre-existing soil and groundwater contamination.

In establishing a remedial action plan, the team actively presented the findings and proposed remedy so that target clean-up values could be established with the regulatory authorities. Given that there are no South African remedial standards, reference was made to US EPA and other international remediation guidelines to establish a target appropriate to the local environmental and hydrogeological setting.

5 DESCRIPTION OF THE CASE STUDY SITE

The TRW South African factory is located in the Atlantis Industrial area, approximately 50 km north of Cape Town (Figure 1), midway to Saldanha Bay along the west coast road. Atlantis has approximately 65 000 inhabitants and the water supply for both industrial and domestic purposes is obtained from a wellfield located about 5 km to the west of the town, closer to the coast (Figure 1).

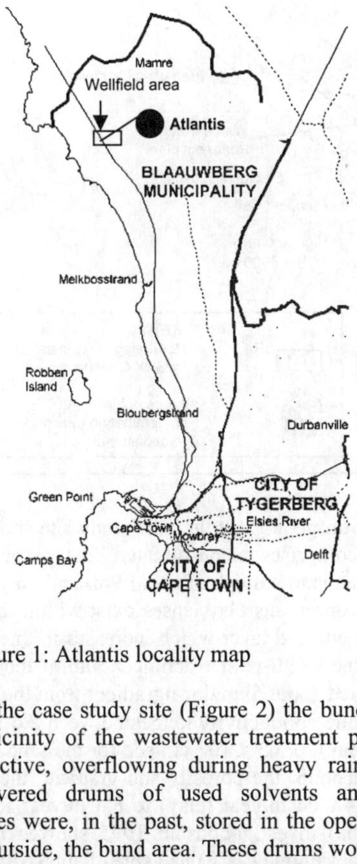

Figure 1: Atlantis locality map

At the case study site (Figure 2) the bund wall in the vicinity of the wastewater treatment plant was ineffective, overflowing during heavy rain events. Uncovered drums of used solvents and waste sludges were, in the past, stored in the open within, and outside, the bund area. These drums would overflow during rain events impacting soil with subsequent leaching to shallow (±3 mbgl) groundwater.

6 FATE AND MOBILITY CONSIDERATIONS

The fate and mobility of chemicals associated with the metal plating industry depends upon their chemical speciation in the soil and groundwater. Consequently, chemicals will not necessarily migrate within an aquifer at the same velocity as the groundwater flow. Different solutes may react through precipitation/dissolution, adsorption, and/or partitioning (Palmer and Johnson 1989).

Factors which influence the fate and mobility of metal plating contaminants are summarized as follows:

1. Cadmium: the fate of cadmium in water depends on its chemical speciation, which is determined by the pH and hardness of the water and the presence of ligands and co-existing metal cations. The lower the pH, the higher the solubility (Galvin 1996).

2. Chromium: Cr^{3+} ions form complexes with chlorides, fluorides, sulphates, ammonium, cyanides, sulphocyanides, oxylates and citrates. Chromates are complexed by H_2O_2. Chromates of the alkali metals are soluble. The majority of the chromates of the other metals are of very low solubility (Pourbaix 1966). Cr^{6+} salts are more soluble than the Cr^{3+} salts. The lower the pH the higher the solubility.

3. Nickel: Nickel salts are reasonably soluble in water, except the ferri- and ferrocyanides and sulphide. Nickel occurs as complexes and the following are known, hydrochloric acid (very stable), amines (fairly stable), oxalic thiocyanide, metaphosphoric and cyanide (Pourbaix 1966).

4. Zinc: occurs mainly in water as Zn^{2+}, $Zn(OH)^+$ and $ZnCl^-_3$ and forms the following complexes, amine, cyanide, thiocyanide, oxalic, hydrochloric, ethene diamine, pyridine, aniline and hydrazine.

5. Trichloroethene: when in the dissolved phase may be affected by hydrolysis to yield alcohol and/or an alkene (Johnson et al. 1989).

From the above, the complexation of contaminants from the metal plating industry plays an important role in determining their fate and mobility in the subsurface. In a complexation reaction, a metal ion reacts with an anion which functions as a ligand. The metal and the ligand bind together to form a new soluble species called a complex. Transition metals are the most important metals involved in complexation; alkaline earth metals only form weak complexes, while alkali metals essentially do not form complexes at all (Johnson et. al. 1989). The order of complexing strength of metals is as follows:

$$Fe(III) > Hg > Cu > Pb > Ni >$$
$$Zn > Cd > Fe(II) > Mn > Ca > Mg$$

The implication of the above is that complexes

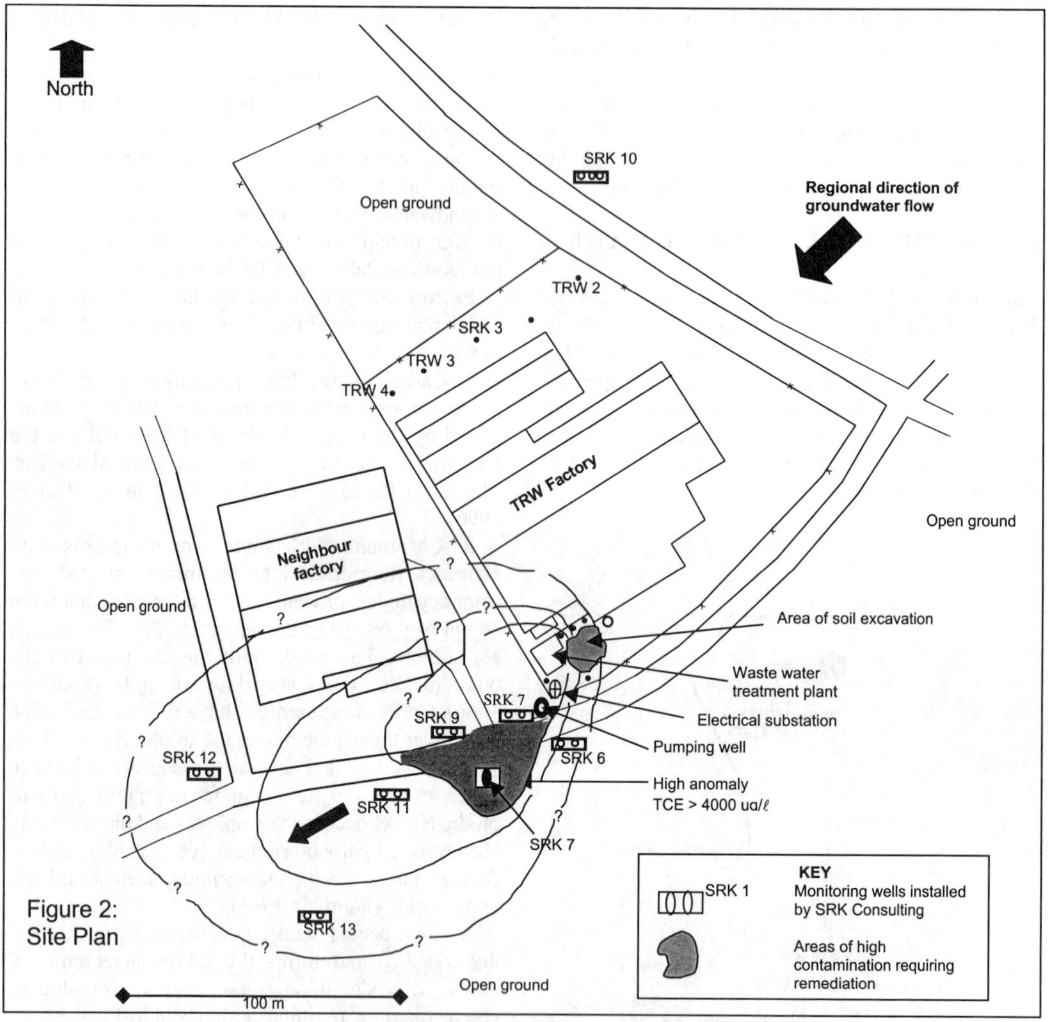

Figure 2:
Site Plan

decrease the concentration of free ions in solution and thus less metal may adsorb onto aquifer material or precipitate; ie the metal is more soluble because it is bound up in the soluble complex. Complexed metal may be less toxic to aquifer microbes (Johnson et al. 1989).

Careful consideration of sampling procedures is required in aquifers characterized by conduit flow where metals can migrate in groundwater as colloidal particals via a phenomenon known as 'facilitated transport'. The US EPA (1992) recommend that groundwater for metals analysis should not be filtered. Colloidal metals will not pass through a 0.45 micron filter and, although present in the groundwater under natural conditions, they would be removed, leading to inaccurate measurement of their concentration.

The hydrogeology at the case study site shown in Figure 2 comprises unconsolidated, fine grained calcareous sands to a depth of about 9m, underlayen by clay. Discontinuous clay lenses exist within the unconsolidated sand layer which increases in thickness towards the south-west reaching >50m at the wellfield located about 5km downgradient from the Site.

Hydraulic conductivity ranges between 2.6 x 10^{-6} cm/sec (clay) to 5.3 x 10^{-4} cm/sec for the sand. Taking into account the porosity and gradient, the flow velocity is 0.04 m/year (clay) to 8.4 m/year for the sand. Initial investigations in 1995 showed that a contaminant plume (TCE) had spread approximately 60m off-site (Figure 2). This coincides with the higher flow velocity and duration of factory operation (8 years). Environmental hazards and risks and remediation issues are discussed later.

7 HAZARDS AND RISKS IN TERMS OF EXPOSURE AND IMPACTS

Incorporation of a hazard and risk assessment component into the site assessment programme facilitates the decision making process with regard to the urgency and scheduling of the remediation process, and any negotiations with regulatory authorities regarding acceptable clean-up levels. The latter is of particular relevance in those countries which do not have clean-up standards for soil and groundwater, as is the case in South Africa. A typical risk and hazard assessment process should address four issues, namely: 1. Chemicals of interest; 2. Source areas, viz extent, quantity; 3. Pathways (routes and exposure mechanisms); 4. Targets or impacts.

The types of data that would be considered in such an assessment are summarized below and are taken from the TORSA study.

1. Chemicals of interest: these are best subdivided into categories which comply with the source areas, eg soil and groundwater contaminants. Contaminants in the soil may not be (or become) contaminants of concern in the groundwater due to their low solubility, low leaching potential and/or complexation with other chemicals, which may render them less toxic. The chemicals of interest at the case study site were TCE, arsenic and chromium.

2. Critical source areas: typically, five possible source areas of contamination were identified at the case study site and these were prioritized in terms of their potential to form a longer term source. They were: (i) directly from continuous releases, ie spillage; (ii) latent contamination which still exists in the soil matrix; (iii) contamination in the soil vapour phase; (iv) contaminants which are dissolved within the groundwater; and (v) contaminants which are either in the soil and/or groundwater and exist as either free phase product, or dissolved and sorbed forms.

3. Pathways: these are the contamination migration routes or transport mechanisms, which include both natural (eg seepage, plumes and dust), and unnatural (eg transport to waste sites) movement.

4. Targets (impacts/effects/health): are related to the points of exposure and include humans, plants, animals and ecological systems as a whole. The effects tend to be overshadowed by contaminant toxicity and human health implications.

8 ASSESSMENT FINDINGS AT THE CASE STUDY SITE

Multilevel soil samples were taken from fourteen positions in the vicinity of the waste water treatment works and analysed for $Cr_{(total)}$, Zn, CN and Total Petroleum Hydrocarbons (TPH – modified EPA 418.1 method using GC-FID).

CN was found to be the most prevalent soil contaminant at elevated concentrations in the range from 15 to 210 mg/kg. Depths of contamination varied and it was possible to map-out the depth of excavation required to remove the contaminated soil.

The primary chemical of interest in groundwater was TCE with concentrations in the range of 6 μg/ℓ to 4 089 μg/ℓ. Seasonal fluctuations in concentrations were experienced between winter (high rainfall) and summer (low rainfall). Sampling to determine the fluctuations was undertaken in September (Spring = end of winter rainfall) and April (autumn). Generally, TCE concentrations in groundwater were lower at the end of the winter rainfall season. This is attributed to the recharge and shallower piezometric surface.

Groundwater quality was also monitored during an extended pumping test carried out on the proposed abstraction well. It was found that TCE concentrations more than doubled in some of the monitoring wells located between the source area and abstraction well, eg well STI 1 at the start of the test had a TCE concentration of 333 μg/ℓ and at the end of the test the concentration was 884 μg/ℓ.

For the case study and using the US EPA (Region IX) Preliminary Remediation Goal data for TCE, arsenic and chromium, the combined risk of getting cancer from drinking the groundwater was calculated at 7.8 in 1000. However, this risk should be considered in more practical terms taking into account the many assumptions built into the EPA calculations such as, duration of exposure, water intake, etc. Also, for the case study, groundwater abstraction does not occur from the contaminated plume area (Figure 2).

9 CONTAMINATION MANAGEMENT AT THE CASE STUDY SITE

Contamination management at the case study site involved the following:

1. Source control: managing the release sources, ie spillage, operating procedures, constructing proper bunding for storage of chemicals, managing waste storage areas, changing certain chemical types used in the factory, eg TCE, etc.
2. Excavation and disposal of contaminated soil, which essentially comprised a secondary source of continued groundwater contamination;
3. Groundwater pump and treat: for both hydraulic control and remediation of the high concentration plume area.

Excavation and disposal of contaminated soil

Excavation of contaminated soil was required in places to a depth of 1.2m below ground level. Approximately 90 m^3 of soil was removed, placed in waste hoppers for temporary storage before permission was obtained for safe disposal at a nearby landfill site.

The soil excavation and disposal programme was well documented and photographed with a closure report being issued to the regulatory authorities who approved the programme.

Groundwater pump and treat

The groundwater pump and treat system comprised a sieve tray type airstripper followed by activated carbon adsorption of TCE. What made this project special in South Africa is that remediation equipment is not readily available in the country. Accordingly, the air-stripper and off-gas treatment system had to be designed from first principles, followed by an evaluation of the equipment performance and generation of design parameters on a pilot scale, before the final design was completed. Using laboratory facilities a close-loop air-stripper was developed, which was subsequently patented.

The system that was installed was small in comparison and designed to treat 0.5 ℓ/s of contaminated groundwater with an inlet concentration of 4 000 ppb to a treated concentration of 5 ppb. The stripper was a two-sieve tray unit followed by two activated carbon columns. About 2 000 m^3 of groundwater was treated during a two year period before the system was decommissioned after achieving the cleanup level agreed to.

It should be noted that although the primary objective of the system was to treat the TCE contaminated groundwater, this was driven through the need to induce hydraulic control over the plume area with elevated (> 4 000 μg/ℓ) TCE concentrations, which was leaving the Site (Figure 2).

10 CONCLUSION

Subsurface contamination arising from inadequate employee environmental awareness and housekeeping practices in the metal plating industry results in an intricate mix of toxic contaminants which may form complexes with other ligands thus altering their solubility, and consequently, their fate and mobility within the subsurface.

Assessment of contamination arising from metal plating operations requires a multi-disciplinary team to provide a good understanding of the interrelationships between the hydrogeological and geochemical processes. Only then is it possible to ascertain the impact of the contaminant suite on the subsurface environment and any other secondary impacts such as human health. Whereafter, following a risk based approach, clean-up levels can be set and the necessary remediation programme put in place to achieve these levels.

At the case study site the remediation programme has entailed excavation and safe disposal of heavily contaminated soils and installation of a groundwater pump and treat system, mainly to provide hydraulic control over the plume thus allowing time for intrinsic remediation. Improved housekeeping was also instituted as a priority. Over a 30 month period TCE concentrations in groundwater have reduced from >4000 μg/ℓ to < 480 μg/ℓ in wells located furthest off-site. TCE in groundwater from wells in the vicinity of the source area has been remediated to below detection limits (1 μg/ℓ). It has been shown that once hydraulic control and the pump and treat operation was instituted over the plume that the higher TCE concentration groundwater continued to migrate forming a breakaway zone from the source area. Natural attenuation has played an important part in reducing the concentration of TCE in the off-site plume, however, at a slower rate than in the plume area under influence of the pump and treat radius of influence (±25m).

It is apparent that contamination from the metal plating industry is potentially very intricate, and may involve a wide range of chemicals. Consequently, "educate to prevent, is far better than remediate with money ill spent".

REFERENCES

Galvin, R.M. 1996. Occurrence of metals in waters: An overview. *Water SA, Vol. 22, No. 1*, January 1996.

Johnson, R.L. et al. 1989. Subsurface Chemical Processes. *In EPA 625: 4-89 019 document, Seminar Publication: Transport and Fate of Contaminants in the Subsurface.*

Palmer, C.D. and Johnson, R.L. 1989. Physical Processes Controlling the Transport of Contaminants in the Aqueous Phase. *In EPA 625: 4-89 019, Seminar Publication: Transport and Fate of Contaminants in the Subsurface.*

Pankow, J.F. and Cherry, J.A. 1995. *Dense Chlorinated Solvents and other DNAPLS in Groundwater*. Waterloo Press, Ontario, Canada.

Pourbaix, M. 1966. *Atlas of Electochemical Equilibria*. Pergamon Press, Oxford, England.

Sawyer, C.N. and McCarty, P.L. 1978. *Chemistry for Environmental Engineers*. Third Edition, McGraw Hill.

US EPA, 1992. RCRA Ground-water monitoring: draft technical guidance, EPA 530-R-93-001, November 1992.

Groundwater: Past Achievements and Future Challenges, Sililo et al. (eds) © 2000 Balkema, Rotterdam, ISBN 90 5809 159 7

Risk-based decision analysis for the selection of remediation strategy at a landfill

J. Norrman
Department of Environmental Technology, Swedish Geotechnical Institute, Linköping & Department of Geology, Chalmers University of Technology, Göteborg, Sweden

ABSTRACT: A risk-based decision framework is described for the purpose of selecting the most cost-efficient remediation strategy at a municipal landfill in Estonia. The evaluation of three different remediation alternatives is based on a risk-cost-benefit decision analysis, where the alternative that maximizes the economical outcome for the decision-maker is regarded to be the most favorable. An à priori descriptive conceptual model forms the basis for risk identification. The efficiency of an alternative influences the risk reduction, which is compared with the cost of the alternative. In this context, the risk is defined as a probabilistic cost and possible shortcomings of this definition are discussed in the paper. The analyzed example is based on limited existing information. However, the conceptual model can be updated by additional investigations.

1 INTRODUCTION

The Aardlapalu landfill, some 10 km south of Tartu City in Estonia, was established in 1971 and is still in use. It is currently supposed to act as a contaminant source. New regulations regarding waste disposal in Estonia have brought attention to the question of older landfills posing a possible threat to human health and the environment and current policies on waste disposal are under reconstruction. The number of landfills will be reduced and the deposition of waste concentrated to a few, carefully planned and constructed sites.

Regarding the future at Aardlapalu, there are three possibilities: 1) to close and remediate the site, 2) to close and remediate the existing landfill and construct a new landfill at a site following the new regulations, or 3) to continue using the existing landfill. This study considers closing the present landfill and the selection of remediation alternatives at the site.

This paper presents a method on how to structure and analyse limited data using an à priori risk based decision analysis. The aim is to use the framework to select a remediation strategy at the site. Additional data is needed for a final choice, but the study can be used as a preliminary evaluation and as a basis for further investigations. The paper presents a decision model and an event tree to describe the possibility of different outcomes. Costs for three different remediation alternatives are estimated with associated uncertainties. Probabilities of different outcomes are subjectively estimated. Objective functions for the different alternative strategies are calculated by Monte Carlo simulations.

2 WASTE MANAGEMENT

All landfills can be considered to be potential contamination sources. The impact on groundwater from a leaking landfill differs between different hydrogeological settings. New regulations have been worked out by the European Union (Directive 99/31/EEC) on waste management and on the construction of new landfills. The main aim is to optimise waste management economically and environmentally. The technical demands on the construction of new landfills are generally very high, depending on what kind of waste is to be disposed (hazardous, non-hazardous or inert waste).

The question of older landfills is somewhat intricate. Records of the waste are not always kept, making the estimations of the content rather difficult. Analyses of the leachate give a picture of the present situation and possibly an indication of whether there is non-hazardous or hazardous waste present in the landfill, but it does not predict the composition of future leachate.

When remediation of a former landfill is to be evaluated, a risk-based analysis of the problem could serve as a useful tool. There are many uncertain variables and the costs involved are generally high. There is a demand to be able to handle decisions under uncertainty cost-efficiently. The key for doing this is to a large extent to identify uncertainties of importance and to structure information consequently.

3 RISK-BASED DECISION ANALYSIS

Generally we can express risk as a combined effect of the probability of a harmful event to occur and the magnitude of the consequence. It is possible to differentiate between qualitative, semi-quantitative, and quantitative measures of risk. In addition, it is possible to formulate the quantitative measures stochastically. The different measures of risk are being used for different purposes, from qualitative prioritization and screening to quantitative risk assessment and decision analysis. For the purpose of this study the risk is being viewed as a quantitative measure, defined as

$$R = P_f \times C_f \qquad (3.1)$$

P_f is the probability of failure (failure to fulfil a defined objective) and C_f the cost of the consequence of this failure. The definition is in accordance with that given by Freeze et al. (1990). A hydrogeological simulation model with input of both geological uncertainty and parameter uncertainty in combination with an engineering reliability model calculates the probability of failure (P_f). The transport model can be of varying degrees of complexity depending on the problem at hand. In this case, information is so scarce that a very basic decision model was used.

The risk or the probabilistic cost is fitted into a risk-cost-benefit analysis to identify the most cost-effective remediation alternative among a number of defined alternatives. The decision model formulated by Freeze et al. (1990) consists of maximization of an objective function (Φ). The objective function (Φ_j) for each alternative (j=1...N) is defined as the net present value of the expected stream of benefits (B), costs (C), and risks (R) taken over a certain time horizon (T), and discounted at a certain rate (i):

$$\Phi_j = \sum_{t=0}^{T} \frac{1}{(1+i)^t} \left[B_j(t) - C_j(t) - R_j(t) \right] \qquad (3.2)$$

4 SITE DESCRIPTION

4.1 *Geology and hydrogeology*

Aardlapalu landfill is situated in the watershed of the Porijõgi River, which is a tributory to River Emajõgi being the largest river in Southern Estonia. The Southeast Estonian moraine plain south of Tartu has an elevation of 30-60 m above sea level (a.s.l.). The relief is undulated and valleys dissect the landscape (Mander et al. 1996a). The valleys were formed by streams during the Pleistocene, which cut their way through the bedrock consisting of red Devonian sandstone. The landfill is situated in the Aardla polder (31-35 m.a.s.l), a drained peat land in one of the valleys.

Quaternary deposits in the valley have a thickness up to 40 m while on the adjacent heights, the glacial deposits (loamy sand-till) are normally between 1-5 meters in thickness. The stratigraphy of the valley is complex, as it has been remolded by glaciers of the last glaciation (Mander et al. 1996b). Investigations by Kobras (1998) reveal some information of the uppermost layers at the site (Fig. 1). The lower sand layer have no data of thickness. It is overlaid by varved clay and another sand layer. The top layer consists of peat.

Groundwater flows from the sandstone, recharging the layers of sand and gravel in the valley. The groundwater in the valley is artesian in some places with a main flow direction to the northwest. The groundwater table at the landfill slopes gently from east to west.

4.2 *Landfill*

The present landfill construction is simple. Waste is deposited directly on the ground and night soil is disposed in an open pond. A ditch surrounds the area to collect leachate (Figs 1-2). Drillings indicate that the bottom of the landfill is in direct contact with the top sand layer. The groundwater table marked in Figure 1 is measured in the sand layer and indicates that leachate is possibly produced not only by rainfall, but also by groundwater inflow.

The content of the landfill is in principal unknown. The landfill has received municipal waste but possibly also industrial waste since the start of operation. Geophysical investigations indicate a zone of low resistivity some two hundreds meters to the west of the northwest corner of the landfill area (Aaltonen & Olofsson 1997). Chemical analyzes of groundwater samples in VPK-3 and 4 indicate high

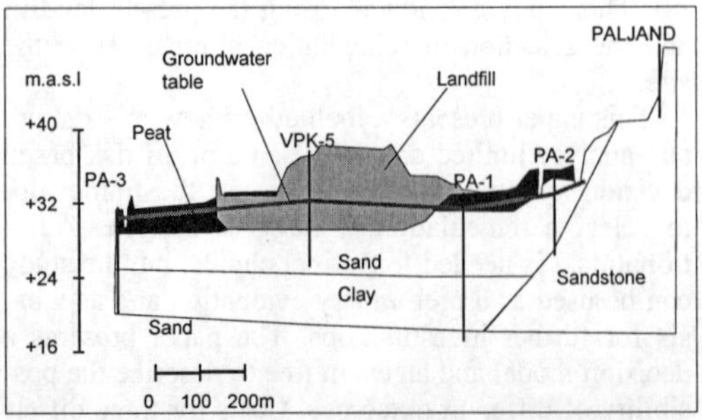

Figure 1. A profile view of the landfill. The upper parts of the stratigraphy in the ancient river valley is shown (after Kobras 1998).

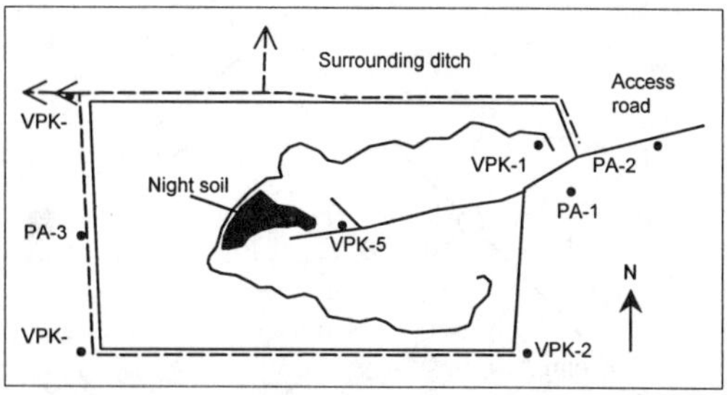

Figure 2. The landfill area with a size of 28.4 ha with marke boreholes. The dashed line symbolizes the main ditch (afte Kobras 1998).

conductivity compared with unaffected samples upstream of the landfill (Kobras 1998).

5 DESCRIPTIVE CONCEPTUAL MODEL

The following assumptions and simplifications are made to describe the situation: 1) If leachate is produced it will reach the groundwater; 2) the transport of substances in the groundwater is independent of how the leachate is produced; and 3) leachate that is drained within the landfill area is considered to be under control and is not included in the analysis. The analysis considers the transport of substances with groundwater that is not drained within the area. The fourth assumption is that there exists flow lines from the bottom of the landfill that will reach the regional groundwater pattern. Hence, processes like dispersion, adsorption and degradation will control the amount of substances leaving the area.

For the purpose of this study the failure criteria is defined as *any hazardous contaminant leaving the landfill area with the groundwater at concentrations exceeding the level at which the surroundings will be negatively effected.* Normally, a certain level of contaminant release is allowed from the area provided it will not have any negative effects on the surroundings.

There are not enough investigations conducted to be able to exclude a transport of contaminants with the groundwater in the top sand layer. The second sand layer may be protected by the clay layer and by a possible artesian pressure. No chemical data of the groundwater status in the second sand layer is available.

6 CONSEQUENCES

The consequences of a failure depend on how many people are living in the surroundings and how they are using the area. The consequences also depend on the sensitivity of the ecological system, the water living organisms, birdlife, and wildlife in the area.

A division in use values and non-use values can be done in order to describe consequences. Use values can be agricultural production, industrial production or drinking water production to mention some examples. Non-use values are the values of the presence of a clean area, a healthy population in the area and a healthy ecological systems and a flourishing flora and fauna. The difference between the two kinds is not always clear.

Consequences that have a direct economical impact on the owner/operator of the landfill are the costs for new remediation of the landfill and possible penalties because of contaminant release.

7 REMEDIATION ALTERNATIVES

Three alternative actions are considered.

1. Leaving the landfill with no actions taken.
2. To cover the top of the landfill, which will prevent leachate production due to rainfall. The success of coverage depends on stability and settlements in the landfill.
3. Total isolation of the landfill including covering the top combined with vertical cutoff screens and design to prevent both rainfall and groundwater from generating leachate.

8 COSTS AND BENEFITS

Costs are presented below. The currency used is €, EURO (in April 2000, 1 € was approximately US$ 0.95). The analysis considers closing the landfill, which is assumed to generate no income, thus no benefits are estimated.

8.1 *Costs of consequences*

Failure is defined as any hazardous contaminant leaving the landfill area. The consequences if the remediation alternative fails are described in general in Section 6. To estimate the cost of these consequences is difficult, especially the non-use values. Economists have developed a number of methods to estimate non-market values in monetary terms. Use values can often be related to an existing market and estimated by its relation to this (travel cost method, hedonic pricing method, dose-response functions etc). Non-use values however are often estimated by direct methods such as the contingent valuation method. For further information on valuation methods the reader is referred to National Research Council (NRC) 1997, Brent 1996 or Hanley & Spash 1995. For the purpose of this paper the costs of consequences will be used as a variable to be able to find the value of C_f where there is a shift of the most beneficial alternative.

8.2 *Investment costs*

All investment costs and areas are estimated with an interval, defined as a uniform distribution in the Monte Carlo simulation. Table 1 summarizes the remediation costs (C_j). Cost estimations are made on the basis of discussion with experienced people and from Lundgren (1995). Since the remediation alternatives are so generally formulated the construction costs are very uncertain. Variations depend to a large extent on local conditions and material supply. The areas considered are shown in Table 2.

Table 1. Investment costs for remediation alternatives.

C_j	Description	Cost [€/m^2]	Interval
C_1	No action	0	-
C_2	Coverage	17	12-22
C_3	Isolation:		
	Coverage	17	12-22
	Vertical screen	132	84-180

Table 2. Area of the landfill.

	Area [m^2]	Interval
Horisontal	120,000	100,000-140,000
Vertical	10,000	8000-12,000

9 DECISION-ANALYSIS

9.1 *Event tree*

The decision analysis can be structured with the help of an event tree, Figure 3. An event tree illustrates possible outcomes of a number of events. The events considered here are A) production of leachate and B) spread of contaminants with unacceptable concentrations with the groundwater out from the area. According to the assumptions in Section 5, A and B are independent events. The general objective function is formulated as:

$$\Phi_{alt\,j} = P[A_j] \times P[B] \times C_f + C_j \qquad (9.1)$$

where $j = 1, 2, 3$. The most beneficial alternative will minimize the objective function. The risk and costs are discounted with a zero discount rate. In environmental economics concerning natural resources, low discount rates are often used for intergenerational equity. High positive discount rates make the net present value of future investments low and thus less worthwhile to today's generation. Landfills are features that will last for thousands of years, and it was therefore assumed reasonable to use a zero discount rate in this study. For further discussions on the selection of discount rates in environmental economics, see NRC 1997, Brent 1996, Hanley & Spash 1995.

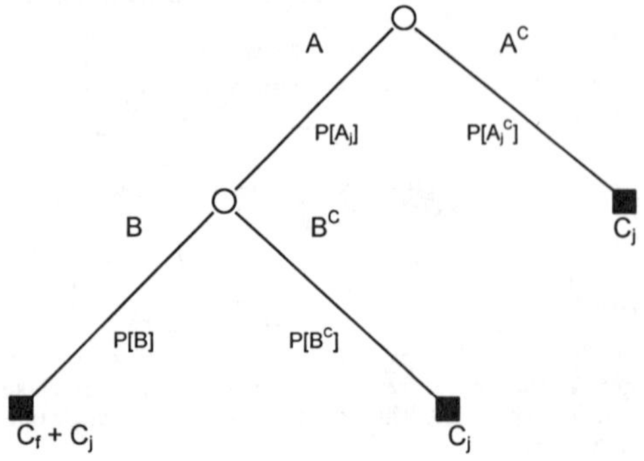

Figure 3. Event tree. The tree shows two events A) production of leachate and B) spread of contaminants with unacceptable concentrations with the groundwater out from the area. A^C and B^C represents the complementary events. P[*A*] etc represents the probabilities for the different outcomes.

9.2 *Probabilities*

Probabilities are subjectively estimated for the purpose of this à priori decision analysis. Table 3 summarizes the different probabilities. Equation 9.2 (Rosén & Gustafson 1995) was used to estimate representative uncertainty intervals expressed as +/- the variance:

$$\text{Var}(p_x) = p_x \times (1 - p_x) \qquad (9.2)$$

In the Monte Carlo simulation, the intervals are given with triangular distribution.

Table 3. Estimated probabilities.

	$P[A_j]$	Interval	$P[A_j^C] = 1 - P[A_j]$
$j = 1$	1.00	-	0.00
$j = 2$	0.50	0.25 - 0.75	0.50
$j = 3$	0.01	0.0001 - 0.0199	0.99
	$P[B]$		$P[B^C] = 1 - P[B]$
$j = 1, 2, 3$	0.80	0.64 - 0.96	0.20

9.3 *Monte Carlo simulation*

The analysis was performed in a spreadsheet with Monte Carlo simulations. Number of trials was selected to 10 000.

10 RESULTS

Alternative 2 is not beneficial using the costs and probabilities assumed here. Table 4 shows the most beneficial choice as a function of C_f. For lower values of C_f, alternative 1 is most beneficial and for higher values of C_f, alternative 3 is preferred. For the 50th percentiles this breakeven point is 4.2 million € and for the 95th percentiles at 5.6 million €. Table 5 presents the total investment costs that will not be exceeded for 50 % certainty and 95 % certainty respectively. The cost of alternative 1 is 0 € and for alternative 3 the cost will not exceed 4.2 million € with 95 % certainty. The probability of failure, P_f is presented in a similar way in Table 6. P_f of alternative 1 will not exceed 0.91 by 95 % certainty. P_f of alternative 3 will not exceed 0.014 by 95 % certainty.

Table 4. Choice of alternative given as a function of C_f at 50th and 95th percentiles.

50th percentile	If $C_f < 4.2 \times 10^6$ [€]	If $C_f > 4.2 \times 10^6$ [€]
Choose alternative	1	3
95th percentile	If $C_f < 5.6 \times 10^6$ [€]	If $C_f > 5.6 \times 10^6$ [€]
Choose alternative	1	3

Table 5. Total investment costs [€] for alternatives 1-3 given as 50th and 95th percentiles.

C [EURO]	50th percentile	95th percentile
1	0	0
2	2,032,885	2,730,259
3	3,335,904	4,230,619

Table 6. Probability of failure for alternatives 1-3 given as 50th and 95th percentiles.

P_f	50th percentile	95th percentile
1	0.80	0.91
2	0.40	0.55
3	0.008	0.014

11 DISCUSSION AND RECOMMENDATIONS

The most common criticism of risk-cost-benefit analysis is the formulation of the consequences in monetary terms. Wladis et al. (1999) prepared eight different consequence costs by both direct and indirect methods. Freeze et al. (1990) uses the term *acceptable risk* to describe limitations on selection of remediation alternative because of large uncertainties in the outcome. In the case of no action the level of uncertainty is very high (Table 6) and may well serve as a limitation on the selection of remediation strategy. For all decisions of this character a qualitative discussion needs to be added to the analysis.

Assumption 2 in Section 5 is a simplification of the situation. The transport of contaminants may be independent of how the leachate is produced. However, the concentration of the transported contaminants is not. In spite of this the assumption was used for the purpose of this study.

By viewing the problem as a system where geology (the natural setting) and technology interact uncertainties of relevance for the whole system can be identified. To complete a refined decision analysis at Aardlapalu landfill a number of additional data are needed.

1. Data worth analysis should be performed in order to find a cost-efficient investigation strategy to reduce uncertainties.

2. To be able to refine the definition of failure as a concentration that is allowed to be released, human health and ecological risk assessments are needed. Characterization of the contaminant source is needed. A transport model on total amount of contaminants transported off the site should be developed and receptors identified.

3. A safety assessment of the technical design is needed in order to be able to quantify the uncertainty in the technical performance. This includes long term effects on the performance of the technical construction, changes in natural conditions and stability of the landfill under decomposition of the waste material.

4. Design of remediation alternatives. Vertical screens are possibly only needed upstream of the landfill with a drainage ditch downstream. There are several different possibilities depending on the local material availability to design different coverage. Each remediation alternative thus has a possible number of variations of design.

5. Costs of consequences. Analyze of the consequences and level of the accompanying costs must be done in order to have a reality-based estimation of this parameter.

12 CONCLUSIONS

From this analysis alternative 1, which is no action at all, or alternative 3, which is total isolation of the landfill, should be chosen, depending on how large the consequence costs are. To be able to choose a remediation alternative, consequence costs need to be estimated.

The probability of failure for alternative 1 is so high that it may not be acceptable to choose that alternative whatever the consequence costs.

Alternative 3 is the most expensive alternative and it can therefore be worthwhile to refine the failure criteria and to reduce uncertainties in contaminant transport estimations. Further investigations should be done accompanied by a data worth analysis.

The design of the three remediation alternatives can be improved regarding design and material available in the surrounding to obtain more realistic cost estimates and reduce uncertainties in technical performance.

The case study presented is characterized by scarce and uncertain data and is done on a preliminary basis. The main advantage is the consequent structuring of data in an early stage of a project and the support of a holistic view of the problem at hand. Data needed for a complete study is both of technical and hydrogeological character as well as analyzing any impacts of the landfill on the surrounding.

13 ACKNOWLEDGEMENTS

This study was made possible through the financial support of the Swedish Institute. The help provided by the personnel of the Environmental Protection Institute, Kobras and at the Environmental Department of the local government in Tartu has been invaluable to this study. Many thanks to all the persons who contributed with fruitful discussions and constructive comments.

REFERENCES

Aaltonen, J. & B. Olofsson 1997. Geophysical Measurements at Swedish and Estonian Landfills - Case studies from four sites. *Proceedings 3rd Meeting, environmental & engineering geophysics, Aarhus, Denmark, 8-11 September 1997.*

Brent R.J. 1996. *Applied cost-benefit analysis.* Cheltenham, UK: Edward Elgar.

Freeze, R.A., Massmann, J., Smith, L., Sperling, T. & B. James 1990. Hydrogeological Decision Analysis: 1. A Framework. *Ground Water.* 28(5): 738-766.

Hanley, N. & C.L. Spash 1995. *Cost-benefit analysis and the environment.* Aldershot, UK: Edward Elgar.

Kobras 1998. *Tartu Linna Aardlapalu Prügila Prügilavee Kogumis- ja Puhastussüsteem.* (In Estonian). Tartu.

Lundgren, T. 1995. *Sluttäckning av avfallsupplag. Krav, material, utförande och kontroll.* (In Swedish). Swedish Environmental Protection Agency, Report 4474. Stockholm: Naturvårdsverket.

Mander, Ü., Tamm, V., Kuusemets, V. & R. Karjus 1996a. Long-term study on nutrient losses in the Porijõgi River catchment. In K. Alekand (ed.), *Investigations of nutrient cycling in agricultural watersheds*: 10-25. Transactions of the Estonian Agricultural University, 186. Tartu.

Mander, Ü., Kuusemets, V. & K. Treier 1996b. Variation in groundwater quality in the Porijõgi River catchment. In K. Alekand (ed.), *Investigations of nutrient cycling in agricultural watersheds*: 26-38. Transactions of the Estonian Agricultural University, 186. Tartu.

National Research Council 1997. *Valuing Ground Water. Economic Concepts and Approaches.* Washington DC: National Academy Press.

Rosén, L. & G. Gustafson 1995. A Bayesian Markov geostatistical model for canister positioning in a nuclear waste repository. *Proceedings, IAH Solutions '95, Edmonton, Canada, June 4-10, 1995.*

Wladis, D., Rosén, L. & H. Kros 1999. Risk-Based Decision Analysis of Atmospheric Emission Alternatives to Reduce Ground Water Degradation on the European Scale. *Ground Water*. 37(6): 818-826.

Groundwater: Past Achievements and Future Challenges, Sililo et al. (eds) © 2000 Balkema, Rotterdam, ISBN 90 5809 159 7

Groundwater pollution, landuse and environmental impacts on Lusaka aquifer

I.A. Nyambe
Geology Department, School of Mines, University of Zambia, Lusaka, Zambia

C. Maseka
Water and Sanitation Program, East and Southern Africa, World Bank Resident Mission, Lusaka, Zambia

ABSTRACT: Lusaka, the capital city of Zambia is built in a fractured rock environment of karstic dolomitic marble aquifer, which is sensitive to recharge fluctuations and easily susceptible to pollution. The increase in population over the last 30 years from below 500,000 to about 1.2 million people in 1998 has meant an increase in the rate of pollution from landfills, rubbish pits, septic tanks and pit latrines. Unfortunately, half of the city's urban water supply including industry is extracted from groundwater. The water quality in some areas of the aquifer shows high electrical conductivity values in the range 1040-1400 μS/cm suggesting poor water quality, coupled with high nitrate-nitrogen levels (15-40 mg/l), which are well above World Health Organisation (WHO) limit of 10 mg/l. Human activity in the recharge zone has modified the recharge mechanism resulting in reduced stream discharges in the Chalimbana catchment and well yields over the aquifer.

1 INTRODUCTION

The city of Lusaka (Fig. 1), capital of the Republic of Zambia, lies on a plateau with an elevation of about 1300 metres above sea level, with a population of about 1.5 million people. It forms one of the fast growing urban cities in sub-Saharan Africa. It is underlain by a dolomitic karstic aquifer, which serves as an important source of water for the city of Lusaka. The aquifer supplies 70,000 m^3/d of water representing 42% of the city's urban water supply. The rest of the city's water supply is obtained from Kafue River via a 60 kilometre long pipeline south of the city. Between 1950 and 1965, 40,000 m^3/d of water was pumped from this aquifer for the urban water supply. From 1978 the pumpage from the aquifer was increased to 110,000 m^3/d and by 1996 the pumpage was reduced to 80,000 m^3/d due collapsing of boreholes.

Groundwater abstraction for the city's urban supply and other purposes such as farming and industries have been going on without any control. There is also an indiscriminate drilling of wells and groundwater extraction by various private users. Human activities accompanying urbanization are going on without any regard to the aquifer's protection. The Lusaka South Forest reserve, considered the recharge zone for this aquifer, was degazetted around 1985, and this coupled with poor land use practices such as the cutting down of trees in the forest reserve may lead to reduced groundwater recharge in the long run. The removal of vegetation canopy exposes soils to erosion, increases runoff and evaporation from the soil. This results in percolation of solutions rich in iron (Nyambe & Sichingabula 1995). These solutions concentrate around soil particles forming lateritic nodules and concretions, which locally have led to reduction in percolation of water to the aquifer.

Most of the city's industries lie on the karstic aquifer and future development plans continue to be put directly on the aquifer. The aquifer also suffers from over pumping, indiscriminate drilling and pollution from industrial effluents and domestic pollution. The future of Lusaka's groundwater resources is indeed at risk.

This paper therefore addresses a number of issues including assessment of groundwater potential for the aquifer, evaluation of groundwater pollution resulting from industrial and human activity directly associated with a sensitive aquifer. Landuse practices and environmental impacts around the aquifer recharge zone on the groundwater resources of the Lusaka area are also highlighted.

2 GEOLOGY, DRAINAGE AND HYDROGEOLOGY

The Lusaka landscape is controlled by the geology with the countryside underlain by schists and quartzites, hilly and intersected by many valleys. The city lies on a plateau at an elevation of about 1300m reducing to just below 1100m to the north eastern part (Fig. 1). Schists and quartzites underlie the northeastern part, and it is hilly and intersected by valleys. The study area comprises Pre-Cambrian rocks of the Katanga Supergroup namely, schists, quartzites, limestones, and dolomites belonging to the Chunga, Cheta and Lusaka formations. Overlying the Katangan rocks are Quaternary to Recent alluvial and colluvial deposits.

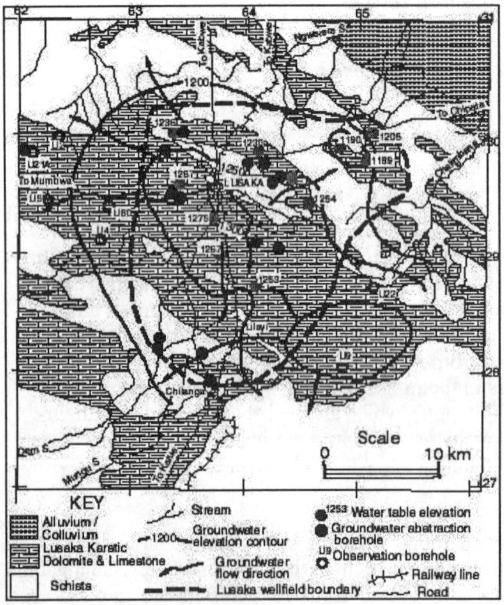

Figure 1. Geology and hydrogeological map of Lusaka. Note the location of Lusaka City area. Modified from Maseka & Nyambe 1998.

The Chunga Formation consists of a sequence of coarse and fine-grained schists, quartzites and minor calcareous horizons. These rocks show a general north-west to south-east trend and occur in the upper part of the study area.

The Cheta Formation consists of a sequence of northwest to southeast trending calcareous dolomites to dolomitic limestones and fine-grained schists occurring mostly in the southwest, the central upper parts of the study area. The dolomites are fine-grained, white, pink or grey. The limestones are coarser and grey to bluish in colour. Argillaceous limestones occur around the city.

The Lusaka Dolomite outcrop stretches for about 70 kilometres along strike form Mwembeshi in the northwest to the Lusaka South Forest Reserve. The vast outcrop consists of blue-grey and white limestones and dolomitic limestones, lithologically similar to those of the Cheta Formation. Dolomite is less abundant and easily identified on weathered surfaces by its closely spaced network of joints. On a large scale the distribution of dolomite limestone appears to be stratigraphically controlled, since horizons for banded limestone can be traced for considerable distances. Exploration drilling of up to 120 m in 1977 for the Lusaka water supply did not intersect the base of the Lusaka dolomite (FIGNR 1978).

The large body of the Lusaka Dolomite is thought to be a down - faulted synclinorial block with internal minor synclines and anticlines co-axial to the major trend. Lambert (1963) and FIGNR (1978) describe occurrences of breccia in the dolomites and limestones. Fault breccias consists of angular and sub-angular fragments of dolomites and limestone enclosed in massive calcite. Collapse braccias are often rounded fragments enclosed in compact reddish calcite-siderite sand or mud. The breccias are elongated along the strike of the formation and are thought to represent the ruble-fillings of solution cavities of a paleokarst system. Lambert (1963) suggests that the solution activities of the paleokarst took place along tectonic structures like thrust planes. This would explain the alignment of the breccias along the strike of the formation.

The alluvial and colluvial deposits, which include sand, gravels and clays, are found in the northeastern part of the study area.

Surface drainage in this area is well developed, with streams beginning at the contact between limestone/dolomite and schists locally (Fig. 1). The central part of the study is mostly underlain by the dolomites and limestones, which are generally flat, with surface drainage being notably absent. However, subsurface drainage is well developed in the highly permeable karstic dolomites and limestones. In this area, streams emerge where the sub-surface drainage meets a hydraulic barrier such as an impermeable schist, and flow on the schist terrain and disappear underground where they intersect another dolomite and limestone area, only to reappear further down stream when they again meet a schist terrain.

Rainwater drains into fissures and sinkholes in areas where the rocks outcrop and infiltrates through the laterite cover in areas with thick overburden and hence joining the underground water. The Lusaka Dolomite plateau forms the watershed among the Chunga River flowing westwards to join the Mwembeshi River, a tributary of the Kafue, the Ngwerere and Chalimbana rivers flowing northeastwards to join the Chongwe River, a tributary of the Zambezi River and the smaller streams heading south and flowing directly into the Kafue (Fig. 1).

There are three distinct seasons in Zambia; a dry, cool season or winter from mid April to August, a hot season from September to October, and a rainy season from November to April. The average rainfall is about 820mm. Temperatures rarely exceed 20°c and often fall to around 8°c during night or on cloudy days. The mean minimum and maximum are 17° and 32° C. The hydrogeology of the Lusaka area has been evaluated by previous workers and their findings indicate the following:

(a) The Lusaka Dolomite aquifer, with an average aquifer thickness of 18m and average specific capacity of 50.2m^2/d, has the best groundwater potential to support the city water supply and for large scale exploitation (Yachiyo Engineering 1995).
(b) The availability of groundwater in non-karstified carbonates such as the Cheta Formation carbonates is minor (Lambert 1963).
(c) The other lithologies such as schists and quartzites have poor groundwater potential (poor aquifers) and hence cannot be used for large-scale water supply (Lambert 1963, FIGNR 1978, Jones 1971).
(d) monthly climate data analysis showed that the annual recharge to the aquifer was 162-200mm, depending on vegetation cover, representing 20-40% of the total rainfall (FIGNR 1978).

The annual rainfall account for 30-45 % to the recharge of the aquifer in other Karstic areas (Hadwen 1972, Wimpey Laboratories Limited 1983) such as the Kabwe carbonates (Jones 1971) about 140km north of Lusaka.

The hydrogeology of the Lusaka area arrived at from this study is given Figure 1. It shows groundwater circulation patterns. Groundwater circulation and subsequently karstification is controlled by the water base level given by less permeable schists surrounding the carbonate rocks. The main flow in the Lusaka Dolomite is directed towards northwest (FIGNR 1978, this study). Though the Lusaka Dolomite is considered to be the main aquifer, the Cheta Carbonates above are increasingly becoming an important source of water. Natural discharges occur in certain areas along the schist-dolomite contact through small intermittent springs and seepages.

Alluvial sands, gravels and clays are dominantly fine-grained and of low permeability. These deposits make moderate uniform aquifers suitable for small-scale supplies.

3 GROUNDWATER RESOURCE ANALYSIS

The climate in Zambia is characterized by the annual alternation between the wet and the dry season, with the wet season lasting for about 6 months, from November to April at an average of about 820 mm of rainfall received. For Lusaka, the average is about 714 mm. Groundwater static levels from 11 observation wells show an annual drop in water levels of 6.2 m. An example of groundwater hydrographs is given in Figure 2 for observation wells U-7 and U-9. The annual variation in water levels in the production wells is 10.6 m. The average dynamic water level within Lusaka Urban in July 1997, three months into the dry season was found to be 12.2 m from the 10.6 m datum. There are, however, areas of concern with dynamic water levels far below the average.

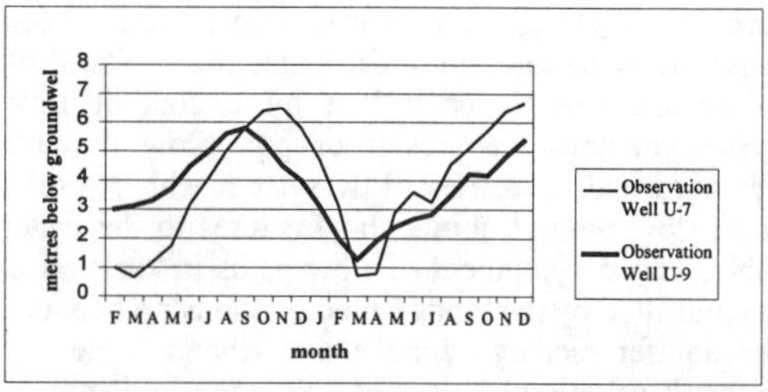

Figure 2. Groundwater hydrograph for Observation Wells U-7 and U-9 Feb 1996 - Dec 1997.

3.1 *Groundwater abstraction versus water demand*

The Lusaka Urban water supply has at least 50 production wells. An average of about 40 wells are used daily to pump about 90,000m^3/d. This figure is only 47% of the total 190,000 m^3/d of surface and ground water utilised for the urban water supply. Out of the 40 production wells about 80% of the wells are pumped continuously for 24 hours per day. The population of Lusaka Urban for 1998 as projected from the National Water Master Plan Study stands at 1,122,500 million people. Using 180 litres/capita/day, this gives a domestic water demand of 202,050m^3/d. The industrial water demand for 1998 as projected from the same study is 86,580 m^3/d. The total water demand (domestic and industry) for Lusaka Urban is therefore 288,630m^3/d. Compared with the total water pumped per day of 190,000 m^3/d, the demand outstrips the supply.

3.2 *Groundwater potential*

The city of Lusaka is underlain by fractured schists and dolomites/ limestones. In fractured rocks ground-water occurs in weathered zones, fractures, joints, faults and solution opening. It follows that the aquifers produced in these fractured rocks are heterogeneous and anisotropic, often isolated from one another.

The Chunga and Cheta formation schists form one hydrogeological unit, where occurrence of ground-water is depended on tectonic fracturing, quartz veining and the extent of chemical weathering. The schists are characterized by low yields ranging from 9.6 m^3/d – 432m^3/d. Boreholes are drilled up to about 60m depth depending on the extent of the weathering zones. Groundwater occurs under confined and semi-confined conditions. Specific capacities range from 0.24 – 240m^2/d (FIGNR 1978). The schists have low infiltration rates and high runoff. They are therefore poor aquifers with localized ground-water occurrences.

The Lusaka Karstic Dolomite aquifer is generally tectonically deformed and consists of partly metamorphosed rocks. Ground-water flow occurs through fractures and joints, which have been enlarged by dissolution of carbonate rocks. The flow of groundwater is controlled by a conduit flow system occurring as result of pipe flow in cave passages and by diffuse flow system which results from flow through joints and fractures to the water table. The development of flow paths in karst areas is controlled by the topographic relationship between the aquifer recharge zone and discharge zones. A network of enlarged fissures following joint planes and cavitation in the form of sheet fissures occur at various depths cutting the limestone and dolomite aquifer. The zone of greatest circulation and karstification is between 0 and 25m. Solution cavities also occur at 65-80m depth and 125-150m with laterite infilling being encountered. This suggests a vertical link with the higher aquifer base. The epikarst zone is at 25m and the general aquifer base at 150m.

Lambert (1963) observed that where collapse breccias occur in soluble bands of dolomite there is often extensive cavitation, which acts as a reservoir with free interconnections with the figure network. Such occurrences are often favourable sites for high yield boreholes.

The karstic dolomite and limestone aquifer has very high permeabilities and porosities varying with depth. Average borehole yields are between 360 - 720m^3/d. Exploration drilling for the Lusaka water supply encountered borehole yields of 3024 - 8640 m^3/d. Specific capacities vary between 0.72 – 2400m^2/d. Based on a flow net analysis, Jones (1972) obtained a transmissivity of 372m^2/d for the Kabwe dolomites that are similar to those of the Lusaka area. Evaluation of the unconfined storage based on annual variations in the groundwater table gave a storage of 3%. This is comparable with Jones' (1971) and Wimpey Laboratories Limited (1983) findings of 3% for Kabwe dolomites and Mpongwe dolomites respectively. Maseka (1994) found storage values of 3 to 5%, which were based on daily rainfall to a one dimensional catchment water balance model.

The main ground-water flow direction over the aquifer is from the south-east to the north-west, branching to the north-east and south-west (Fig. 1). The highest potentiometric levels are around 1300m over the Lusaka Forest Reserve and this appears to be the principle recharge area. Ground-water flow in this part of the aquifer is confined and the water table follows topography. The discharge areas are located in low lying swamps or dambos around Kalingalinga to University of Zambia, Natural Resources Development College to Avondale, Chamba Valley to Ngwerere and Chunga to Barlaston Park. The discharge zones are associated with the dolomite-limestone/schist contact which is characterized by perennial springs.

3.3 *Groundwater recharge*

Changes in the levels of the hydrograph during the year reflect gross changes in the flow of water to the aquifer in the area around the observation point providing a way of assessing aquifer response to recharge, and to losses by seepage and evapotranspiration. Ground-water recharge, may therefore be calculated from the balance between input from rainfall and output in the form of losses to runoff, seepage and evapotranspiration. The present well-field for Lusaka Urban water supply stretches over an area of 16 km by 12 km giving a total area of 192 km^2. Using an annual variation in ground-water level of 6.2m and a storage coefficient of 0.03, the change in ground-water storage is therefore 97,841 m^3/d. This change is comparable to 90,000 m^3/d of groundwater pumped from the Lusaka Urban supply well-field (Fig. 1). From an annual variation of 6.2m observed in the Lusaka aquifer and from an annual rainfall of 714 mm, a recharge of 186mm of rainfall was obtained. This represents about 27% of the annual rainfall. This gives an annual storage of 35,712,000 m^3/year in Lusaka Urban water supply wellfield.

Urbanisation in Lusaka led the degazetting of the Lusaka South Forest no. 26 in 1985 and was sub-divided into smaller farms (small holdings). Human settlements in the area led to cutting down of trees, cultivation and construction activities significantly affecting the recharge zone to the aquifer. This has resulted in reduced stream discharges over the Chalimbana catchment and reduced well yields over the aquifer due to increased density of impermeable surfaces.

4 GROUNDWATER QUALITY AND POLLUTION

For most ground-waters, electrical conductivity (EC) is a good measure of the relative mineralisation with waters containing dissolved inorganic salts recording high conductivity values and may not be suitable for use. EC measurements taken from 33 boreholes indicated values between 324-820 μS/cm, except for 4 areas where values between 1040 -1400 μS/cm were obtained suggesting poor quality water. Although these values may not be indicative of polluted water, the water resource is increasingly becoming threatened as a result of human activity such as solid waste disposal, septic tanks, pit-latrines in the informal settlements. Leachate from these and liquid waste will sooner or later find its way to the aquifer. Three major sources of pollution include urbanisation, industrial activity and agriculture.

4.1 *Impact of urbanization on ground water*

The city suffers from an inadequate mains sewerage system. Only about 32% of the city is served by a sewer line, about 43% serviced by septic tanks and 25% pit latrines. Ironically, the areas serviced by pit latrines, though smaller in size, have higher population densities than those served by septic tanks and sewer lines. Pollutants from these services easily get into the groundwater. Worse still, is the fact that Lusaka has many shallow private boreholes providing drinking water, which are inadequately protected from pollutants. Some of these show poor microbiological quality, for example, nitrate-nitrogen levels of between 15 and 18 mg/l recorded in Chainda and Chunga boreholes have levels higher than the World Health Organisation (WHO) limit of 10 mg/l. The source of the nitrate is considered to be decomposing organic matter associated to pit latrines and septic tanks.

About 400,000 tonnes/year of domestic and industrial garbbage is generated and less than 20% of this is collected by the City Council. Uncontrolled tipping is common within the boundaries of Lusaka Urban and no records are kept of the nature of wastes disposed even at the controlled disposal sites. Rain water leaches chemicals from the gabbage and these find their way into the aquifer resulting in serious groundwater pollution. A good example is the Libala dump site which is located in the groundwater recharge zone.

4.2. *Pollution from industrial activity*

The heavy industrial area in Lusaka, covering a wide range of industries such meat processing, vegetable oil, paper, breweries, metals, various materials and engineering workshops, is located directly on the karstic aquifer. Service industries such as metal workshops, dry cleaners, photo processing and printers use considerable quantities of potentially toxic contaminants and these are poorly disposed off along trenches and drains along road sides.

Chemical spills and leaking underground tanks also pose a danger of ground-water contamination There is little or no information on the life-span of the underground fuel tanks situated on this aquifer. It is highly probable that many of them have slow leakages.

Quarrying and mining activity for building industry (crushed stones) and marble have trebled in the recent years due to increase in population and the need for housing. Most of these quarries, upon intersecting the water table are abandoned and used as dumps for refuse. These waters eventually find their way into boreholes and are a source of pollution. Of great concern are the relatively small pits that have been dug close to a number of roads in all over Lusaka in order to get dolomite boulders which are subsequently crushed manually into smaller aggregates to produce crushed stones.

4.3 *Pollution from agriculture*

A number of commercial farms are located on the outskirts of Lusaka Urban, and are directly on the aquifer. The intensive use of nitrogen-rich fertilizers and of pesticides is a problem to water resources. A high level of nitrate-nitrogen and in some areas, pesticides in groundwater is attributed to intensive agriculture. However, the extent of the pollution from fertilizers and pesticides is not accurately known in the area.

5 DISCUSSION AND CONCLUSION

Groundwater levels observed have been used as a basis for evaluating groundwater recharge. Annual variations in water levels averaged 6.2m in observation wells and 10.2 in production wells. The groundwater recharge to the aquifer is estimated at 186mm. This is a direct result of an annual rainfall of 714mm. This gives annual groundwater storage of 35,712,000 m^3/year in the Lusaka Urban water supply well field. This is comparable to 32,850,000 m^3/year abstracted for the urban water supply. However, if private boreholes are considered the groundwater storage will indeed be lowered.

Areas within the well field around Show grounds, Chelstone and Shaft 5 show signs of overpumping as observed from the dynamic water levels. Analyses of water samples indicate high conductivity levels in range of 1000-1400 μs/cm in some suburbs, coupled with high nitrate-nitrogen levels (15-40 mg/l). Human activity in the Forest Reserve considered to be the principle recharge zone has modified the recharge mechanism resulting in reduced stream discharges in the Chalimbana catchment and reduced well yields over the karstic aquifer.

Urbanization and the heavy demand of water that accompanied it have brought increased pressure on the water resource in the Lusaka area. Coupled with low service levels for urban water in Lusaka and non-existent in some places, it has resulted in indiscriminate drilling of boreholes in the city and caused a dramatic number of private boreholes. Whereas only 52 boreholes are used for the urban water supply, all located in an area of about 20km x 20km, the aquifer which stretches for over 45km has between 3,000 to 4,000 private boreholes.

The dramatic increase in population meant an increase in the rate of pollution from landfills, rubbish pits, septic tanks and pit latrines amounting to an increase in costs of water treatment and public health through efforts to control water borne diseases arising from polluted ground water. Because groundwater pollution is irreversible, controlling abstraction through enforced regulation, pollution through monitoring and an expansion of the Lusaka Urban Water Supply groundwater resource base are highly recommended. The Department of Water Affairs should make proposals to amend the Water Act to include groundwater controls, which is currently still regarded as private. This in effect would reduce the indiscriminate drilling of boreholes in Lusaka and the country. Recharge areas should be zoned and activities within these areas restricted.

REFERENCES

FIGNR 1978. Groundwater and management studies for Lusaka Water Supply, Hannover.

Hadwen, P. 1972. The groundwater resources of the Greater Ndola area. *Groundwater resources paper 1.* Department of Water Affairs, Lusaka.

Jones, M. J. 1971. The old pump station Lusaka - a Hydrogeological study. *Department of Water Affairs Hydrogeological Report No. L24.*

Lambert H. H. J. 1963. The geology and groundwater resources of the Lusaka area, Part 2, Report no. 16, Geological Survey, Lusaka.

Maseka, C. 1994. A groundwater resource evaluation of fractured rock aquifers Lusaka, UNSW Unpublished M. App. Sc. Thesis.

Maseka, C. & I A. Nyambe. 1998. Lusaka's groundwater resources at risk: Final Report. Unpublished Environmental Capacity Enhancement Project, Lusaka. 56p.

Nyambe, I. A & H. M. Sichingabula. 1995. Sand and gravel as basic constituents and their market share in the building and construction industry around Lusaka. 1995. International Conference on Industrial Minerals: Investment Opportunities in Southern Africa. Conference Proceedings, Council for Geoscience, South Africa. Pp. 205-216.

Simpson, J. G., R. A. Drysdall & J. H. H. Lambert 1963. Geology and groundwater resources of Lusaka area. *Northern Rhodesia geological Survey Report No 16*

Tague, M. 1965. The groundwater reserves of the city of Lusaka, Zambia. *Hydrogeological Report No. 50.*

Yachiyo Engineering Company Limited 1995. The study on national water resources master plan in Zambia.

Wimpey Laboratories Limited 1983. Review of the groundwater development potential – Mpongwe Development Project.

Groundwater: Past Achievements and Future Challenges, Sililo et al. (eds) © *2000 Balkema, Rotterdam, ISBN 90 5809 159 7*

Modelling and field trials for the development of a rehabilitation strategy for an existing hazardous waste site

J.A. Palmer & I.S.Cameron-Clarke
SRK Consulting, Johannesburg, South Africa

A.J.Oliver
Umbogintwini Industrial Complex, South Africa

ABSTRACT: Seepage from a waste disposal site, on the Umbogintwini Industrial Complex, comprising five unlined slimes dams, a depository of buried drums and two mercury ponds, has resulted in contamination of the underlying ground water. The major chemicals of concern are chlorinated hydrocarbons, which originate from the slimes dams and the drum area. The increased head beneath the slimes dams has modified the natural flow conditions, resulting in movement of the contaminated ground water towards springs located in an informal settlement. A number of management measures have been implemented to address the problem. This paper describes the modelling and field testing and the systems that have been implemented as part of the overall rehabilitation and closure strategy for the area.

1 INTRODUCTION

1.1 *Site description*

The Umbogintwini Industrial Complex, located on the south east coast of Kwazulu Natal in South Africa has been used for the manufacture of various chemicals during the past 60 years. Liquid and solid wastes from the operations have been placed in a number of disposal sites. The largest of these, comprising a group of unlined slimes dams, a depository of buried drums containing tars and chlorinated hydrocarbons (CHC's) and old and new mercury ponds, is referred to as the North West Waste Disposal Area.

The site boundary is located a couple of hundred metres to the north west of the waste site. A small ridge, which forms a watershed, runs parallel to this boundary. An informal settlement, called KwaMakutha, is located on the other side of the ridge.

1.2 *Site characterisation*

The site is covered by a clayey sand, which varies in thickness from five to twenty metres, beneath the North West Waste Disposal Area. Its permeability ranges from 10^{-6} to 10^{-8} m/s. It is underlain by an unconfined to semi-unconfined weathered/fractured aquifer with a permeability of between 10^{-5} to 10^{-7} m/s. Both aquifers are generally low yielding, although the weathered/fractured aquifer is higher yielding. The yields are also higher near dams and in areas of increased seepage, caused by site activities.

During a general site characterisation study a number of monitoring piezometers were installed to sample the ground water. The ground water beneath and around the North West Waste Disposal Area and along the site boundary was found to be contaminated by a number of inorganics (sulphate, chloride, ammonia) and by CHC's. Ground water sampled from a piezometer installed adjacent to the drum site indicated total CHC concentrations of 100 000 ppb at a depth of 30m, indicating the presence of a DNAPL (Dense Non Aqueous Phase Liquid) beneath the drum site. Mercury was only detected in two piezometers, north west of the dams, where it was present at concentrations of less than 3 ppb.

The development of a preliminary ground water flow and transport model indicated that contaminated ground water could be emerging as springs in the adjacent informal settlement, KwaMakutha. Field inspections confirmed the presence of springs, which were being used foe domestic purposes. Sampling of the springs showed high salt contents and the presence of CHC's, particularly vinyl chloride.

The initial historical review of site activities indicated that the drum disposal site was the only source of CHC contamination. However, further ground water monitoring and more detailed modelling of the ground water flow in the area showed that the drum site could not be the source of contamination for the springs. Additional piezometers installed between the slimes dams and the north western boundary recorded the presence of CHC's in the ground water. The ground water flow was such that these CHC's could not originate from the drum site. This was supported by the nature of the CHC's. The drum site was characterised by the CHC 1,2-dichloroethane while the springs and the piezometers around the dams indicated the presence of tetrachloroethene and its breakdown products trichloroethene, cis-1,2 dichloroethene and vinyl chloride (Weigand, 1998). After further discussions with site personnel, the effluent dams were confirmed as a source of CHC contamination, which had been deposited on the dams since 1953.

Remodelling of the CHC contaminant migration showed that the hydraulic head created by the slimes dams was resulting in migration of ground water in all directions, away from the dams. The ground water flow directions are indicated in Figure 1. The flow from the dams towards KwaMakutha has created a CHC plume, which has reached the springs and which, is advancing at a rate of about 23 m/yr.

A management programme for providing alternative water supplies to the community, initiating the construction of drains in KwaMakutha to cut off the springs, intercepting the plume migration towards KwaMakutha, cessation of effluent disposal on the slimes dams and rehabilitation of the waste site was implemented. The hydraulic control of the plume migration from both the slimes dams and the remainder of the North West Waste Site will be discussed further.

2 MANAGEMENT MEASURES FOR HYDRAULIC CONTROL OF PLUMES

2.1 *Installation of an hydraulic barrier*

Extended test pumping of the aquifer along the north western boundary of the slimes dams was undertaken to establish the feasibility of using an abstraction point (or points) to create an hydraulic barrier between the slimes dams and KwaMakutha. Rest water levels in five boreholes, located at varying distances from the test pumping borehole, and the flow rate at the closest spring, were monitored. The results of the test indicated an asymmetrical cone of drawdown, extending into KwaMakutha, but not reaching the spring.

Modelling of the effect of abstraction points along the north western boundary indicated that three abstraction points would reduce the contamination levels reaching the springs. The total CHC concentrations reaching the springs in 2018 would be reduced from about 3700 ppb (without a barrier) to approximately 1500 ppb (with a barrier). The positions of the abstraction points are indicated on Figure 1. In addition, flow directions in the vicinity of the abstraction points would be reversed back towards the dam. However, that part of the contaminant plume, which has passed the watershed, will continue migrating towards the springs.

Three abstraction points were installed and pumping commenced at the end of 1998. Rest water levels in piezometers along the north western boundary and flow rates in KwaMakutha, as well as water quality have been monitored to test the effectiveness of the barrier. Figure 2 presents the decrease in seepage rates and the drop in rest water level in BH-M2. Initial data indicates that the pumping has decreased seepage into KwaMakutha and lowered the rest water levels in piezometers 300 m away from the abstraction points, but this needs to be monitored over a longer period of time. Monitoring of water quality is inconclusive at this stage.

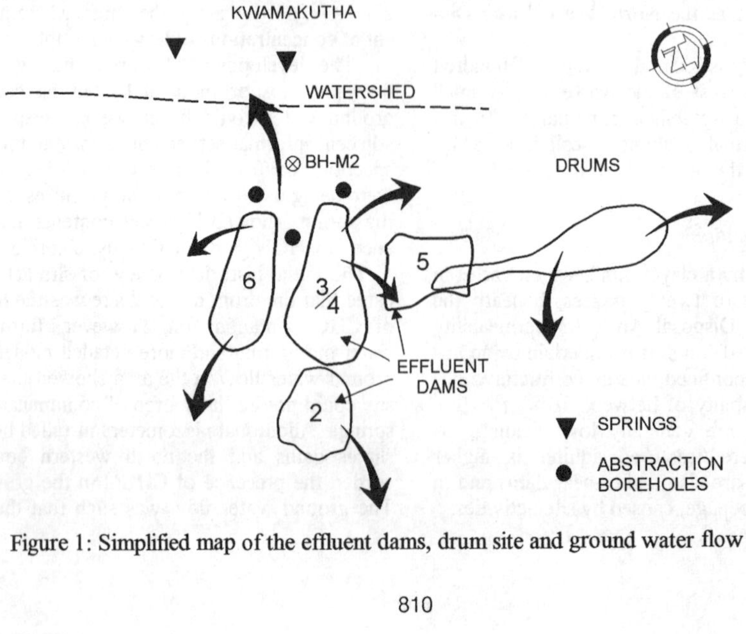

Figure 1: Simplified map of the effluent dams, drum site and ground water flow directions

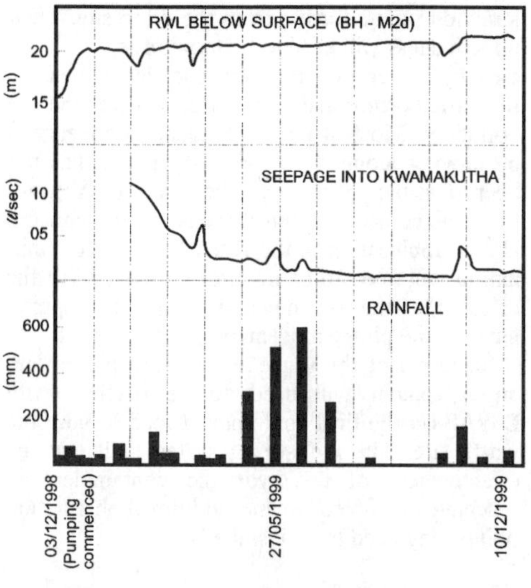

Figure 2: Monitoring results of hydraulic abstraction

2.2 *Dam dewatering*

It appears that, prior to the presence of the slimes dams, the ground water flow directions were predominantly in a south easterly direction, from KwaMakutha, through the present day North West Waste Site and towards the sea. This is supported by the positions of springs observed on old aerial photographs and by flow directions elsewhere in the area.

The hydraulic head in the dams has created an hydraulic gradient between the dams and KwaMakutha, has changed the natural ground water flow directions, and provides the driving force for the plume migration in all directions away from the dams. Modelling of the hydraulic barrier indicated that the barrier would not be completely effective in cutting off the plume migration towards KwaMakutha unless the hydraulic head created by the slimes dams was reduced and ground water flow reverted to pre-dam directions.

Modelling of the effects of dam dewatering (and therefore removal of the hydraulic head) predicted a plume concentration of 410 ppb total CHC's reaching the springs in 2018, compared to a concentration of 1490 ppb using an hydraulic barrier only. In addition, removal of the hydraulic head created by the dams would affect the ground water movement away from the drum site and mercury dams. It was therefore recognised that, as part of the management of the contaminated plume from both the slimes dams and drum site, the slimes dams would need to be dewatered.

Undisturbed samples of vertical profiles through each of the dams were taken, logged and sent for laboratory testing. The logging of the samples indicated that the slimes were comprised of a wet silt with horizontal layering. A strong organic odour was noted in all the samples. The slimes were underlain by either natural soils or by calcine, produced during historical operations on site, above the weathered/fractured aquifer. A black clay layer of varying thickness underlies the calcine.

The laboratory testing indicated low permeabilities within the slimes, 10^{-9} m/s. Permeabilities within the underlying calcine and soils were higher (10^{-7} m/s). Moisture contents within the slimes indicated super-saturated conditions with values ranging from just over 100% to greater than 200%. A three dimensional model of the slimes and the underlying aquifers, using parameters obtained from the laboratory testing and from field testing undertaken during the initial characterisation phase, was set up, using the finite difference package MODFLOW, to test the effectiveness of various dewatering options.

Various methods of dam dewatering were investigated. These included natural drainage, drainage using underdrains and dewatering of the underlying aquifer using abstraction points, thereby inducing increased drainage from the slimes. It was found that the factor which has the most limiting influence on the dewatering rate is the low permeability of the slimes. This permeability is expected to decrease as the slimes become less saturated and they consolidate, but this decrease has not been quantified.

Practical constraints precluded underdrains. The topography around the dams was unsuitable for installation of the underdrains at a sufficiently low elevation and the low permeability of the slimes resulted in a vary small area of influence adjacent to the underdrains. In order to effectively dewater the slimes it would be necessary to install the underdrains at intervals of about 10m.

The various rates of dewatering, calculated from the modelling for the remaining methods, are presented in Figure 3. Natural drainage, as would be expected, shows the slowest drainage rate. It was estimated that, using natural drainage, it would take about thirty years to dewater the dams. The installation of abstraction points beneath the dams, causing the development of an unsaturated zone between the base of the dams and the underlying weathered fractured aquifer and thereby enhancing the rate of vertical seepage from the base of the dams, indicated the highest rates of dewatering. This option would also result in a reversal of ground water flow directions, back towards the dams. However, due to the current super-saturated nature of the dams, it appeared that installation of the abstraction points through the dams would not be feasible.

It was therefore decided to dewater the dams using a system of abstraction points around the dams. The effectiveness of this system would be monitored using a system of piezometers to measure water levels within, beneath and around the dams. It is recognised that additional abstraction points may need to be installed if the monitoring system indicates that the dewatering is not occurring at a sufficiently fast rate.

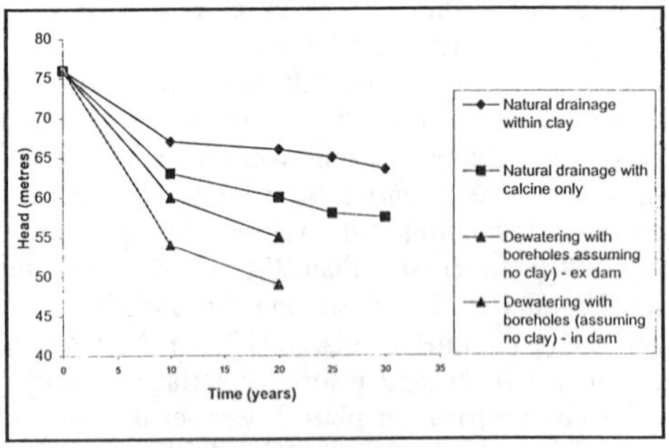

Figure 3: Graph showing different dewatering rates

2.3 *Closure of the North West Waste Site*

In addition to managing the contaminated ground water plume originating from the North West Waste Site, the waste site needs to obtain a closure permit. This needs to be done in terms of the South African Department of Water Affairs and Forestry (DWAF) requirements for the closure of waste sites. Various issues, such as surface water runoff, capture of all shallow seepage, selection of a suitable capping material, hydraulic containment of the area and disposal of all seepage, dirty water runoff and abstracted ground water. Only the hydraulic containment of the area is discussed in this paper.

There are a number of flow paths, indicated from the modelling, which are moving away from the North West Waste Site. Contaminant migration is taking place along all these flow paths, mainly from the slimes dams and the drum area. Hydraulic containment of the waste site would need to intercept all these paths. The major flow paths are:

1. Movement north west from the slimes dams towards KwaMakutha. This has been addressed by the installation of the three abstraction points between the dams and KwaMakutha.
2. Ground water flow out from the dams towards the rest of the Umbogintwini site. The dewatering of the dams should reverse this flow.
3. Flow moving in an easterly direction, from the drum site, towards the eastern portion of the Umogintwini Complex. Test pumping of the aquifer along this flow path indicated that if an abstraction point is installed too close to the drum site, the DNAPL beneath the drum site may be mobilised. The present understanding of DNAPLS is that they should be left untouched, particularly if their distribution is not fully understood. It was therefore decided to install two abstraction points further from the drum site. Their effectiveness will need to be monitored.
4. Ground water flowing from the dams, beneath the buried drums and northwards towards the site boundary. Most of this contamination appears to be moving along a fracture, although it has not been possible to intercept the fracture. Abstraction was considered, but there is again a concern about mobilisation of the DNAPL. This contamination will need to be monitored to establish the effects that the dam dewatering and the capping have on the plume movement.

In terms of the waste site closure permitting, an agreement will need to be reached with DWAF on the final contaminant load leaving the waste site. The monitoring system will test the effectiveness of the hydraulic containment in reaching this objective and additional abstraction points may need to be installed.

3 CONCLUSIONS

1. Initial site characterisation indicated that contamination was moving off-site from the Umbogintwini Complex in a north westerly direction towards KwaMakutha
2. This plume movement was confirmed by sampling in the field
3. Modelling of the plume migration indicated the slimes dams as a source of the KwaMakutha contamination
4. The hydraulic head created by the slimes dams was recognised as providing the major driving force to the contaminant movement.
5. Management measures were implemented, including an alternative water supply, a human health risk assessment and the installation of an hydraulic barrier.
6. Initial monitoring indicates a reduction of seepage into KwaMakutha since commisioning of the hydraulic barrier.
7. Modelling of various methods for dewatering of the dams indicated that dewatering of the aquifer beneath the dams would be the most effective
8. Hydraulic containment is being implemented around the rest of the waste site, but the positions of the abstraction points are influenced by a decision taken not to disturb the DNAPL present.

4 REFERENCES

– Weigand, M.A. 1998. Natural attenuation of PCE and TCE: A case study. In G Wickramanayake and R Hinchee (ed.), *Natural attenuation: Chlorinated and recalcitrant compounds C1-3:* 287-291. Columbus: Battelle Press

Groundwater: Past Achievements and Future Challenges, Sililo et al. (eds) © 2000 Balkema, Rotterdam, ISBN 90 5809 159 7

Enteric virus detection in groundwater using a glass wool trap

K.L.Powell, M.H.Barrett & S.Pedley
Robens Centre for Public and Environmental Health, University of Surrey, Guildford, UK

J.H.Tellam, K.A.Stagg, R.B.Greswell & M.O.Rivett
School of Earth Sciences, University of Birmingham, UK

ABSTRACT: The transmission of waterborne disease through contaminated groundwater is a recognised problem, particularly in developing countries where untreated groundwater supplies are often used for human consumption. The precise behaviour and survival of viruses in groundwater is still relatively poorly understood. Routine microbiological monitoring of groundwater does not often involve viral analysis due to the inconvenience, expense and insensitivity of current standard methods. Glass wool is a cheap and effective material for adsorbing enteric viruses. The application of glass wool in a 'trap' design for monitoring viral contamination of groundwater in the field is described. Use of the trap for enterovirus concentration from groundwaters in Nottingham and Birmingham has proved, in this instance, to be a superior detection method than that of the current standard filtration method using cellulose-nitrate membranes.

1 INTRODUCTION

Snow (1854) first demonstrated the association of disease outbreaks with the consumption of untreated water. There are now several hundred known enteric viruses (any virus transmitted by the faecal-oral route) that are potentially important agents of waterborne disease. The viruses most frequently associated with waterborne outbreaks are hepatitis, Norwalk, coxsackie, rotavirus and the echoviruses (Huffman & Rose 1999). Rotavirus is the major cause of viral gastroenteritis throughout the world causing over one million cases of severe diarrhoea in the 1-4 age group in the USA annually (Ho et al. 1988).

The most commonly studied virus group in natural waters has been the *Picornaviridae* family, which includes poliovirus, coxsackie virus A and B, echovirus and several enterovirus (types 68 to 72). Hepatitis A virus (HAV) (enterovirus type 72) has been proven responsible for waterborne epidemics traced to sewage contamination of water (Rao & Melnick 1986). Other viruses for which there is direct evidence of involvement in waterborne outbreaks are adenovirus and Norwalk (Keswick & Gerba 1980).

Inadequate treatment of groundwater used for human consumption continues to be an important cause of waterborne disease outbreaks. Contamination of groundwater with viruses is of great concern due to the resistant nature of the viral structure and their colloidal size (20nm) which reduces filtration in aquifer materials (unlike the larger bacteria). Studies in the US have found viruses in 20-30% of the groundwater where faecal coliforms were not predictive of viral contamination (Huffman & Rose 1999). Factors affecting virus survival in groundwater are complex and poorly understood. Soil texture, organics, humic acids, pH, ionic strength, permeability, virus type and climatic conditions have all been reported to influence survival of viruses in the subsurface (Keswick & Gerba 1980).

Monitoring of groundwater for faecal contamination involves the use of standard indicators (faecal coliforms and faecal streptococci). Well-established, cheap and simple methods for the detection of bacteriological indicators mean that virus analysis is often over-looked. However, there are numerous reports of the failure of these indicators to effectively determine the presence/absence of pathogenic viruses in groundwater (Keswick & Gerba 1980, Slade 1985).

There are few reports on virus isolation from groundwater due to the lack of sensitive detection methods. The levels of human enteric viruses in contaminated water samples are often very low and must be concentrated from the water before a successful, cost-effective assay can be performed. The detection of potentially low numbers of virus particles therefore involves a preliminary concentration stage. Directed adsorption has generally become the preferred option for recovering viruses from large volume water samples (Hurst 1997), and forms the basis of the standard method for enteric virus detection in water (APHA 1998). Other methods (ultrafiltration, flocculation, affinity chromatography) have seen restricted application because they cannot be used with highly turbid water or water containing high concentrations of organics, and none can easily be used in the field.

Due to the unique surface properties of viruses, they readily adsorb to a number of substances, including alumina gel, starch, minerals, fabrics and various resins. Wallis & Melnick (1967) reported the first application of virus adsorbing filters when they described a technique for concentrating viruses from sewage. Currently the most widely used method is the microporous filter (cellulose-nitrate membrane) adsorption-elution technique, which has seen the greatest use in the detection of viruses in water samples. Several papers have reported on the use of glass powder (Sarrette et al. 1977, Schwartzbrod & Lucena-Gutierrez 1978) and glass wool (Vilaginès et al. 1993, 1997) for recovering enteroviruses from water through passive adsorption. The recovery of poliovirus from 10L and 50L of tap water using glass powder was with only limited success (45% and 55% respectively) (Sarrette et al. 1977, Schwartzbrod & Lucena-Gutierrez 1978). Oiled sodocalcic glass wool has been used successfully to monitor sewage contaminated surface waters in France (Vilaginès et al. 1993, 1997), and glass wool filters have been used to monitor drinking water supplies from surface waters in South Africa (Grabow et al. 1999). Little research has been carried out into the potential use of glass wool for the successful detection of enteric viruses from groundwater. This paper describes the use of oiled sodocalcic glass wool, and the development of a 'trap' for on site monitoring of groundwaters for the presence human enteric viruses.

2 GROUNDWATER MONITORING USING GLASS WOOL

Oiled sodocalcic glass wool (Rantigny 725, Isover St. Gobain) was tested for its ability to concentrate enteric viruses from groundwater samples. The glass wool was first treated to improve virus adsorption using a modified washing method to that of Vilaginès et al. (1993). The wool was washed in 1M HCl, rinsed with distilled water and then washed with 1M NaOH to ensure any adsorbed materials were effectively removed. It was then thoroughly rinsed in distilled water until the pH of the washing solution returned to 7. Finally the wool was packed into polypropylene tubing (67mm length, 16mm ID) and securely fitting polyethylene end caps were glued in place. The end caps enabled the tubes to be easily attached to a pump via polyethylene tubing (Fig. 1). Following sample filtration, traps were either transported directly to the laboratory for elution (Tellam et al. 1999), or were eluted in the field following the method of Powell et al. (2000). Elution was carried out with 0.05M glycine with 3% beef extract (pH 9.5), and the pH immediately adjusted to 7. Eluates collected in the field were placed in cold storage and immediately transported back to the laboratory. Depending on eluate volume, samples were either assayed directly or reconcentrated using organic flocculation (Katzenelson et al. 1976). Standard filtration analysis of groundwater samples was also undertaken on various occasions. Five litre samples of groundwater were collected from selected boreholes and transferred back to the laboratory for filtration following the method of APHA (1998). Enteroviruses were enumerated by a plaque assay method incorporating an agarose overlay. Rotavirus were enumerated by an immunocytochemical technique involving an enzyme-substrate reaction, the product of which stains the infected cells. Eluates were inoculated onto Buffalo Green Monkey kidney cells (BGM) for enteroviruses, or Foetal Rhesus Monkey kidney cells (MA-104) for rotavirus. For both methods, plaques were counted and expressed as PFU/L (Plaque Forming Units/L).

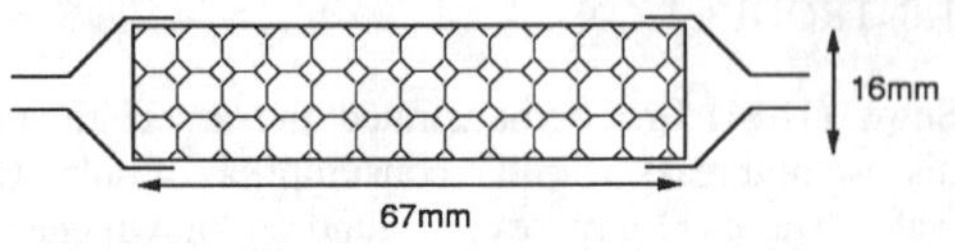

Figure 1. Schematic of glass wool trap.

3 FIELD SAMPLING AND ANALYSIS

Field sampling was undertaken in Nottingham and Birmingham between March 1999 and March 2000. The boreholes sampled may be divided into 3 groups. The first (B, E, F, D) were shallow piezometers (approximately 10m deep) installed into river terrace sands and gravels in the river Trent flood plain in Nottingham. These sediments overly the Triassic sandstone, one of the UK's major aquifers, and the groundwater in the terrace system is assumed to be in hydraulic continuity with the underlying sandstone. The groundwater in the river terrace system is thought to largely comprise recent urban recharge and is known to be impacted by leaking sewers (Barrett et al. 1999). The next group were industrial boreholes (10, 17, 8, 2) and are located within the Triassic sandstone of Nottingham. These are deep boreholes, generally penetrating the entire thickness of the aquifer (30-100m). The final group of boreholes (U2, B1-55K) were also deep boreholes located in the Triassic sandstone aquifer beneath Birmingham. A total of 21 samples were

Table 1. Detection of enterovirus and rotavirus in groundwater from Nottingham and Birmingham using glass wool traps and standard filtration.

Date	Site	Sample volume (L)	Glass wool trap		Standard filtration (5L)	
			Enterovirus (PFU/L)	Rotavirus (PFU/L)	Enterovirus (PFU/L)	Rotavirus (PFU/L)
26/03/99	F	594	<1	<1	-	-
21/04/99	B	5	30	450	-	-
21/04/99	D	5	65	<1	-	-
05/05/99	D	25	1	<1	<1	<1
19/05/99	8	25	2	<1	<1	<1
19/05/99	17	25	1	<1	2	<1
19/05/99	B	5	2	<1	<1	<1
28/05/99	10	100	1	<1	-	-
11/06/99	17	75	<1	<1	<1	<1
18/06/99	10	100	<1	170	<1	20
18/06/99	U2	510	<1	<1	-	-
18/06/99	B1-55K	100	<1	8	-	-
31/08/99	D	8.5	<1	<1	<1	<1
31/08/99	B	5	<1	<1	-	-
29/02/00	B	37	<1	-	-	-
29/02/00	D	29	4	-	-	-
29/02/00	E	23	<1	-	-	-
29/02/00	F	27	4	-	-	-
09/03/00	2	25	<1	-	-	-
09/03/00	8	12	9	-	-	-
09/03/00	17	88	2	-	-	-

taken and concentrated using the glass wool traps. Seven of these samples were analysed for enteric viruses by both glass wool trap and standard filtration methods (APHA 1998) for comparison. The self-contained glass wool 'traps' were used in the field attached directly to the rising main. In field tests volumes up to 594L were passed through the traps (the constraints being flow rate and time availability). In instances where direct use of the rising main was not possible, groundwater samples were collected (>10L) and transported back to the laboratory where the samples were passed through a glass wool trap using a peristaltic pump. A 0.45µm prefilter (Waterra Ltd, UK), was used with groundwater samples which were visibly turbid to prevent clogging the trap with sediments. Where prefilters were used, these were also eluted in the laboratory to recover any virus that may be adsorbed to retained sediments or to the prefilter itself. Previous work by the authors (Tellam et al. 1999) had determined the glass wool trap to work well at the range of pH values found in the Nottingham and Birmingham groundwater systems. Therefore no initial pH adjustment of the groundwater samples was required.

4 RESULTS OF FIELD SAMPLING

Enteric viruses were detected in over 60% of the groundwater samples analysed (Table 1). Enteroviruses were detected in 52% of groundwater samples and rotavirus were detected in 21% of samples tested for rotavirus. Groundwater from piezometer B contained both entero- and rotavirus. Enteroviruses were recovered by standard filtration from one sample only (borehole 17), where enterovirus had also been recovered by a glass wool trap. Rotaviruses were recovered by standard filtration from one sample only (borehole 10), where rotavirus had also been recovered by a glass wool trap. On no occasion did a glass wool trap fail to recover viruses from groundwater samples, which were subsequently detected by the standard filtration method. The likely source of the viral contamination is groundwater recharge from leaking sewers, previously demonstrated to impact shallow groundwater in Nottingham by Barrett et al. (1999).

5 CONCLUSIONS

The analysis of groundwater for contamination with viral pollutants is not routinely undertaken in water quality monitoring schemes. The cost of specialised equipment, lack of specially trained personnel, and the necessity to carry large volumes of groundwater back to a laboratory have made viral analysis expensive and time consuming. The use of glass wool has proved to be an effective material for the concentration of enteric viruses from surface water (Vilaginès et al. 1993) and groundwater (Tellam et al. 1999). With the development of a 'trap' and field protocol (Powell et al. 2000) allowing the sampling of large volumes of groundwater whilst in the field, glass wool is demonstrated to be a viable means of monitoring viral pollution of groundwater.

The glass wool traps have proven to be cost effective and simple to use, and will enable large surveys of viral groundwater pollution to be undertaken. The cost of filtration equipment that is required for the standard method (APHA 1998) is expensive, normally costing over £1200. New membranes and prefilters are required for each analysis (£4) and this, together with the necessity for pre-treatment of each water sample, means that the method becomes time consuming and expensive. Glass wool traps are relatively inexpensive; there is no initial outlay for expensive equipment, each trap including glass wool costing only approximately £1. Additionally no sample pre-treatment is required; water can be passed directly through the traps in the field.

A limitation of the trap is the need for a prefilter in sediment rich groundwaters to prevent the trap from clogging. Any viruses adsorbed to particulate material will be retained within the prefilter and not within the glass wool trap. Elution of potential viruses from the prefilters must therefore be undertaken. There are well-established methods for the recovery of solid-associated viruses from filters (APHA 1998), and these may be employed for the recovery of such viruses.

REFERENCES

APHA, AWWA, WEF (1998). *Standard methods for the examination of water and wastewater 20th edition*. L.S. Clesceri, A.E. Greenberg, A.D. Eaton (eds).Washington, DC: APHA.

Barrett, M.H., Hiscock, K.M., Pedley, S., Lerner, D.N., Tellam, J.H., French, M.J. 1999. Marker species for identifying urban groundwater recharge sources: a review and case study in Nottingham, UK. *Water Research*. 33(14): 3083-3097.

Grabow, W.O.K., Taylor, M.B., Clay, C.G., de Villiers, J.C. 1999. Molecular detection of viruses in drinking - water implications for safety and disinfection. In *The safety of water disinfection: balancing chemical and microbial risks*. Second Int. Conference, November 15-17, International Life Sciences Institute, Florida USA.

Ho, M.S., Glass, R.I., Pinshy, P.F., Anderson, L.L. 1988. Rotavirus as a cause of diarrheal morbidity and mortality in the United States. *Journal of Infectious Disease*. 158: 1112-1116.

Huffman, D, Rose, J. 1998. Emerging waterborne pathogens. *Water Quality International*, December 1998: 14-18.

Hurst, C.J. 1997. Detection of viruses in environmental waters, sewage, and sewage sludges. In C.J. Hurst, G.R. Knudsen, M.J. McInerney, L.D. Stetzenbach, M.V. Walter (eds), *Manual of Environmental Microbiology*: 168-175 Washington, D.C: ASM Press.

Katzenelson, E., Fattal, B., Hostovesky, T. 1976. Organic flocculation: an efficient second-step concentration method for the detection of viruses in tap water. *Applied and Environmental Microbiology* 32(4): 638-639.

Keswick, B.H., Gerba, C.P. 1980. Viruses in groundwater. *Environmental Science and Technology Features*: 1290-1297

Powell, K.L., Barrett, M.H., Pedley, S., Tellam, J.H., Stagg, K.A., Greswell, R., Rivett, M.O. 2000. Optimisation of virus detection in groundwater. National Groundwater and Contaminated Land Centre Report NC/99/40, Environment Agency, UK.

Rao, V.C., Melnick, J.L. 1986. Environmental virology. *Aspects of Microbiology* 13. Van Nostrand Reinhold (UK) Co. Ltd.

Sarrette, B.A., Danglot, C.D., Vilaginès, R. 1977. A new and simple method for recuperation of enteroviruses from water. *Water Research* 11: 355-358.

Schwartzbrod, L., Lucena-Gutierrez, F. 1978. Enterovirus concentration from water by adsorption to glass powder: proposal of a simplified apparatus. *Microbia* 4: 55-68.

Slade, J.S. 1985. Viruses and bacteria in a chalk well. *Water Science and Technology* 17: 111-125.

Snow, J. 1854. Snow on Cholera. In a reprint of two papers together with a biographical memoir by B.W. Richardson and an introduction by W.H. Frost, 1936. The Commonwealth Fund; London, Humphrey Milford, Oxford University Press.

Tellam, J., Stagg, K., Barrett, M., Pedley, S., Powell, K. 1999. Enteroviruses in groundwater. Final report to the EPSRC (UK) on feasibility project GR/M33952.

Vilaginès, Ph., Sarrette, B., Husson, G., Vilaginès, R. 1993. Glass wool for virus concentration at ambient water pH level. *Water Science and Technology* 27(3-4): 299-306.

Vilaginès, Ph., Sarrette, B., Champsaur, H., Hugues, B., Dubrou, S., Joret, J.C., Laveran, H., Lesne, J., Paquin, J.L., Delattre, J.M., Oger, C., Alame, J., Grateloup, I., Perrollet, H., Serceau, R., Sinègre, F., Vilaginès, R. 1997. Round robin investigation of glass wool method for poliovirus recovery drinking water and seawater. *Water Science and Technology* 35 (11-12): 445-449.

Wallis, C., Melnick, J.L. 1967. Concentration of viruses from sewage by adsorption on Millipore membranes. *Bulletin World Health Organisation* 36: 219-225.

Groundwater: Past Achievements and Future Challenges, Sililo et al. (eds) © 2000 Balkema, Rotterdam, ISBN 90 5809 159 7

Natural resistance to agricultural contamination of a limestone aquifer

S.C. Roberts
Department of Geological Sciences, University College London, UK (Presently: Komex H2O Science, California, USA)

J.M. McArthur
Department of Geological Sciences, University College London, UK

ABSTRACT: The Lincolnshire Limestone, a regionally important aquifer in the UK, is in hydraulic continuity with rivers draining intensively farmed arable land. To assess the impact of river chemistry on groundwater quality, nutrient concentrations were monitored in surface and ground waters during a two-year period. Fertiliser applications and heavy rain resulted in winter concentrations of nitrate in rivers of up to 145 mg l^{-1}. Concentrations of dissolved organic carbon (DOC) in rivers peaked in low-flow summer conditions (2.0-13.0 mg l^{-1}), due to algal contributions and treatment works effluents and were lowest in winter (1.9-8.5 mg l^{-1}) due to dilution. Concentrations of DOC in recharge were sufficient to reduce <40% of recharging nitrate. Microbial consumption of DOC in the aquifer rapidly reduced DOC concentrations to 0.6-1.0 mg l^{-1} in shallow confined groundwater. The Lincolnshire Limestone has been naturally resisting fertiliser contamination, but rising concentrations of nitrate indicate that the aquifer is unable to cope with increasing loads.

1 INTRODUCTION

Agricultural contamination of surface waters, due to overuse of fertilisers, is a world-wide problem. Where such waters are in hydraulic continuity with aquifers, groundwater is threatened. Such is the case with the Lincolnshire Limestone, shown in Figure 1, part of a regionally important aquifer in the United Kingdom that supplies water for public and agricultural use. The aquifer has become threatened by escalating concentrations of nitrate, due to over-application of fertilizers and the abstraction of large volumes of water (Edmunds et al. 1973, Edmunds & Walton 1983, Smith-Carrington et al. 1983). In the unconfined zone, groundwater nitrate concentrations often exceed 50 mg l^{-1} (the European Union maximum admissible concentration for drinking water). Concentrations will increase if nitrate moves through the aquifer as a pollution front, even if the amount of nitrate leached from soils remains stable.

Much of the nitrate invading the south Lincolnshire Limestone aquifer recharges from the beds of the East and West Glen Rivers and numerous swallow holes, which provide up to 50 % of local recharge (Rushton et al. 1982). In the subsurface, contaminants can flow rapidly through the fissured limestone (10 to >3000 m d^{-1}; Booker 1977), towards points of abstraction. Some public supply boreholes in the unconfined and shallow confined zones have been abandoned and less contaminated sources are blended with waters of lower nitrate concentrations to maintain consent standards. However, nitrate concentrations decrease rapidly down the flow path and have remained absent from the deeper confined zones, even where tritium is present (Downing et al. 1977). These observations are evidence of denitrification within the aquifer. Denitrifying bacteria have been found in concentrations up to three million per litre, primarily in association with fissures (Lawrence & Foster 1985). These bacteria metabolise organic matter (OM), including DOC. Whether DOC in recharge represents a significant part of the redox budget in aquifers was not known. This study, therefore, monitored for two years the chemical composition of the Glen rivers, recharge entering swallow holes, and the groundwaters, with the aim of assessing how such recharge affects groundwater quality.

1.1 *Geological & hydrogeological setting*

The Middle Jurassic Lincolnshire Limestone crops out in the west of the study area (Fig. 1) where it attains an elevation of 120 m as upland downs. A generalised cross section of the aquifer is shown in Figure 2. The Lincolnshire Limestone attains a maximum thickness of about 40 m. It is a shallow water carbonate unit with individual carbonate beds generally >90% $CaCO_3$ and coloured grey to black by fine-grained disseminated pyrite when unweathered, and buff, grey, white or pink when weathered. The limestone dips at <1° eastwards where it passes

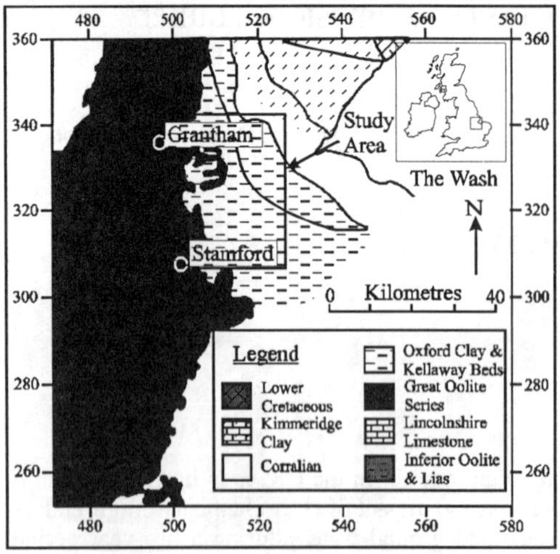

Figure 1. The regional location of the study area.

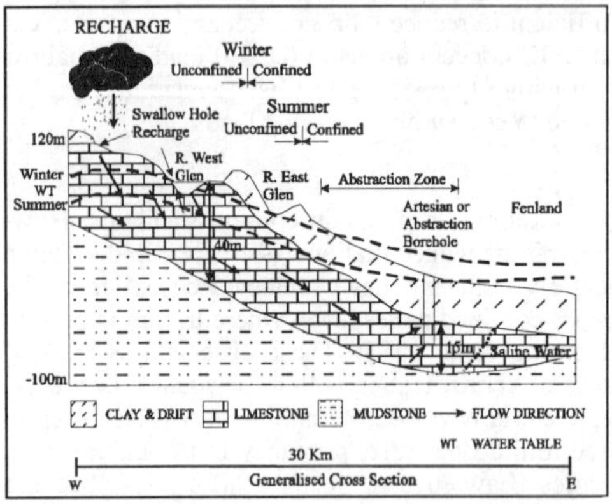

Figure 2. Generalised cross section of the Lincolnshire Limestone aquifer (after Rushton et al. 1982).

beneath confining beds of the Upper Estuarine Series and overlying strata, extending beyond the UK coastline before pinching out in the subsurface (Emery & Dickson 1991). In the deep confined zone, older saline connate water (>25000 years, Downing et al. 1977) is present (Fig. 2). Water in the aquifer moves predominantly *via* fissure flow, with flow velocities from 0.3 to >3000 m d^{-1}, transmissivity >1500 m^2 d^{-1} and storage coefficients from 5 x 10^{-3} to 5 x 10^{-5}. Water levels fluctuate greatly (>12 m) between summer and winter, moving the point of confinement east and west on a seasonal basis. The West Glen river flows over Lincolnshire Limestone for much of its length, whilst the East Glen river flows mostly over drift but cuts through in small inliers to the underlying limestone in several places. River quality reflects land runoff, flow-thro and small inputs from sewage treatment works (STWs) serving individual houses or villages.

2 MATERIALS AND METHODS

2.1 *Sampling*

Surface and groundwater samples were monitored for a wide range of chemical parameters, of which only DOC, nitrate, sulphate, dissolved oxygen (DO) and flow (surface waters) are reported in this paper. Surface waters were sampled from rivers, swallow holes, land drains, highway drains and sewage outfalls. For a 14 month period, ten sites on the Glen rivers were sampled at intervals of six weeks. Samples of swallow hole recharge, highway drainage and land drainage were collected during or shortly after storm events. Additional samples of riverine water were taken from four sites during exceptional rainfall in the winter of 1998.

Groundwaters were collected from springs and boreholes in the unconfined and confined zones of the aquifer. In order to assess seasonality in groundwater chemistry, several boreholes were sampled repetitively, including boreholes along four lines of profile in the direction of groundwater flow. Owing to the fractured nature of the aquifer, boreholes that affect regional flow patterns (those pumping large volumes of water) were preferentially targeted. Boreholes on any flow line were sampled within a 4-day period. To investigate the chemical composition of the aquifer material, samples were collected from quarries and were analysed for total carbon, organic carbon and total sulphur (i.e. pyrite).

2.2 *Analytical methods*

Surface water samples collected for DOC, nitrate and sulphate were filtered using cellulose nitrate filter papers (3 μm and 0.45 μm in series), preflushed with sample. Boreholes were purged before sample collection. Being clear on collection, groundwater samples were not filtered. Samples for DOC analysis were collected in triplicate in 40 ml Shimadzu© glass vials that were pre-cleaned with concentrated chromic acid and stored full of biocide (5% $Na_2S_2O_5$) to preserve sterility. On collection, sample vials were capped with aluminium foil, acidified to pH 2.75 with 1.6 N Aristar® grade sulphuric acid and sparged with N_2 (15 minutes) to remove free CO_2, a substrate for autotrophic bacteria. This procedure preserved DOC concentrations for up to one month at room temperature. Blanks of ultrapure water were prepared at all sites. Samples for nitrate and sulphate were collected in plastic bottles. Analysis of DOC was done using high-temperature catalytic oxidation in a Shimadzu 5000 TOC Analyser (Benner & Hedges 1993, Hedges 1993). Analysis for nitrate and sulphate was done by DIONEX 2000i ion chromatograph (IC). Aquifer materials were hand-ground and analysed for total carbon, organic carbon and total sulphur, using a LECO C/S 125 Analyser.

3 RESULTS

3.1 *Surface waters*

Seasonal flow variations for the Glen rivers are displayed in Figure 3. Flow volume changes along the Glen's length as water recharges to or discharges from the underlying aquifer, with which the rivers are in hydraulic continuity. The majority of flow in the East Glen and upper West Glen occurred during the winter and spring months of 1996/97 and 1997/98, including regional flooding in April 1998. Exceptional storm events yielded brief river-length flow at other times. Two perennial tributaries augmented the lower reaches of the West Glen river.

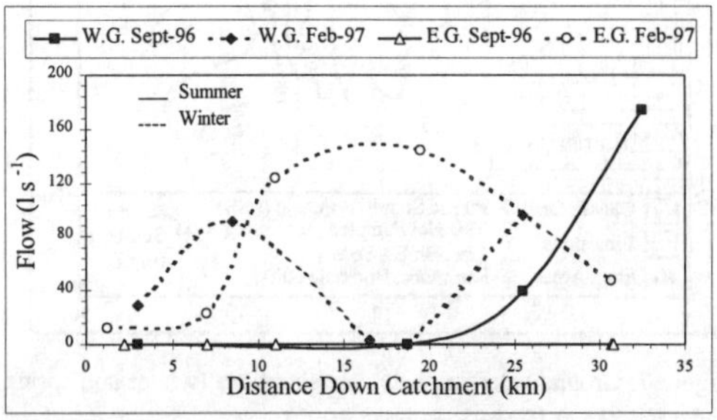

Figure 3. Seasonal flow variations in the East and West Glen rivers.

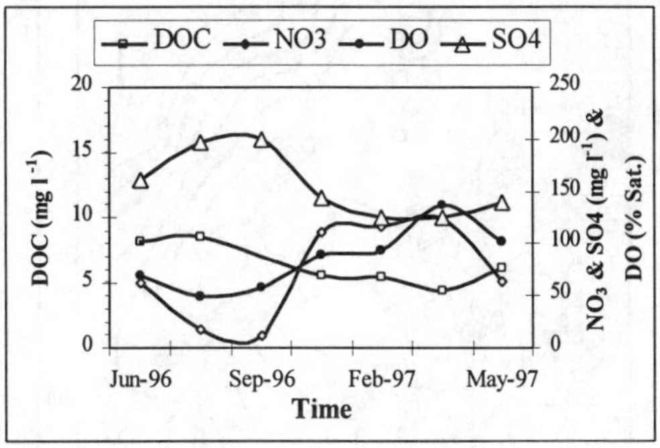

Figure 4. Seasonal variations in hydrochemistry in the Glen rivers

Seasonal changes in riverine chemistry are shown in Figure 4. From a summer high (9.1 mg l^{-1}), DOC concentrations decreased through the autumn, reaching a minimum in winter (4.7 mg l^{-1}) before increasing again in spring. Sulphate also peaked in summer (199 mg l^{-1}) and was at a minimum in winter (125 mg l^{-1}). Concentrations of nitrate showed the converse pattern, peaking in winter (124 mg l^{-1}) and showing minimum concentrations in summer and autumn (12 mg l^{-1}). Dissolved oxygen was lowest in summer (50 % saturated) and peaked in winter (140 % saturated). During regional flooding DOC concentrations increased to as much as 9.54 mg l^{-1} (Glen's mean 5.56 mg l^{-1}, n=280), whereas nitrate was diluted to 22.9 mg l^{-1}.

The major influences on chemical trends along the Glen Rivers were the location of STWs, inflowing tributaries and localised land use practices, as reviewed in more detail in Roberts & McArthur (1998) and Roberts (1999). Swallow holes located in the Glens received recharge of similar chemistry. However, those isolated in fields received only storm runoff from surrounding farmland. Therefore, recharge from these sources was influenced by the chemical characteristics of the soil in that area. For example, storm water entering one swallow hole contained 90 mg l^{-1} of nitrate and 7.0 mg l^{-1} of DOC. Water sampled from land drains had low concentrations of DOC (2.0-3.0 mg l^{-1}) and high concentrations of nitrate (180-200 mg l^{-1}). Effluent from STWs, located in the upper reaches of the East and West Glens, contained a wide range of nitrate (7.1-134 mg l^{-1}) and DOC (4.9-18 mg l^{-1}) concentrations. Flow from the treatment works ranged between 0.1 and 0.6 l s^{-1}, being relatively constant throughout the year.

3.2 *Groundwater and aquifer material*

Groundwater nitrate concentrations are shown in Figure 5, contoured using SURFER®, as were all maps displayed. Nitrate concentrations were highest in the unconfined zone of the aquifer (50-109 mg l^{-1}). Concentrations decreased eastwards, down the flow gradient, and nitrate was undetectable in the deep confined zone of the aquifer. Nitrate penetration into the confined zone was greatest in the central, shallow confined part of the area. This coincided with the region most densely covered with abstraction boreholes, particularly public supply boreholes. A pronounced seasonality for nitrate concentrations occurs in the unconfined and some shallow confined boreholes (Smith-Carrington et al. 1983), although these fluctuations were less pronounced than in surface waters. Groundwater DO concentrations are shown in Figure 6. Dissolved oxygen was rapidly consumed on recharge, unconfined groundwaters typically containing 70-80 % DO, which decreased to below detectable limits in the confined zone. Dissolved oxygen is absent from a zone where nitrate concentrations remain high (*cf.* Fig. 5 with Fig. 6). Groundwater sulphate concentrations in the study area are shown in Figure 7. In outcropping areas of the Lincolnshire Limestone, sulphate concentrations were ~90-100 mg l^{-1}. Concentrations increased in the shallow confined zone to 152 mg l^{-1}. A decline in sulphate concentrations occurred in the deeper confined zones, where hydrogen sulphide can be detected.

Concentrations of DOC in groundwater are shown in Figure 8 (mean 1.14 mg l^{-1}, n=357). In unconfined and shallow confined areas DOC concentrations

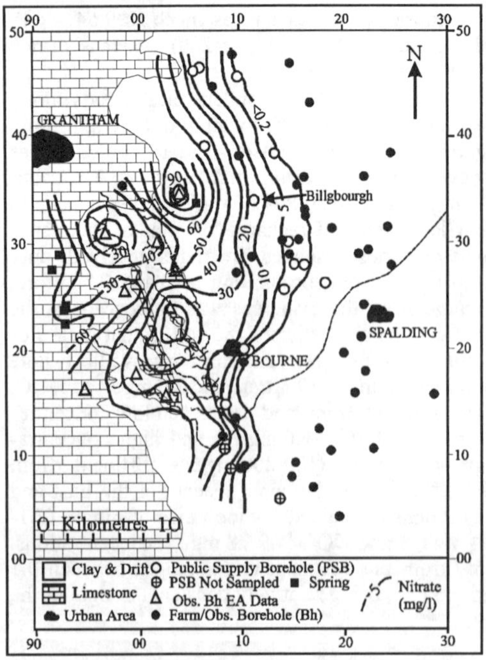

Figure 5. Groundwater nitrate concentrations (winter and spring data collected in 1997/98; Roberts, 1999).

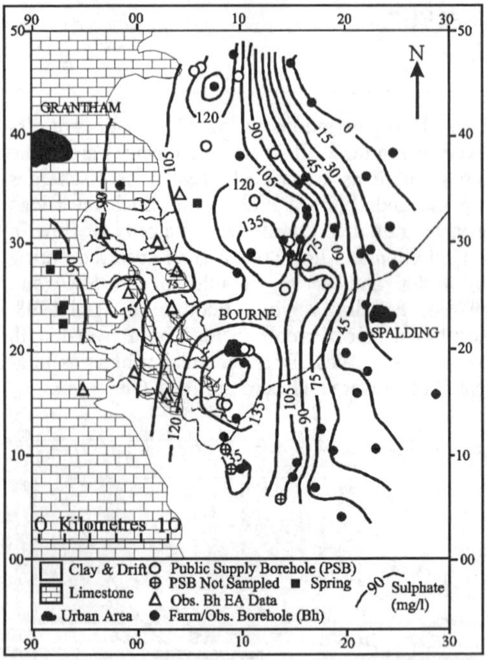

Figure 7. Groundwater sulphate concentrations (winter and spring data collected in 1997/98; Roberts, 1999).

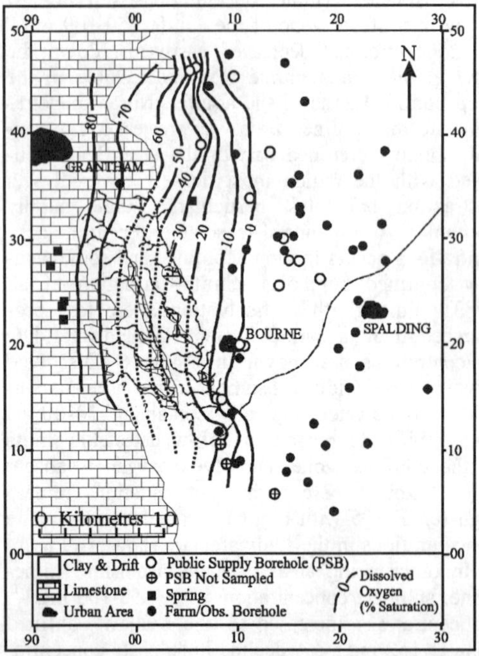

Figure 6. Groundwater DO saturation (winter and spring data collected in 1997/98; Roberts, 1999).

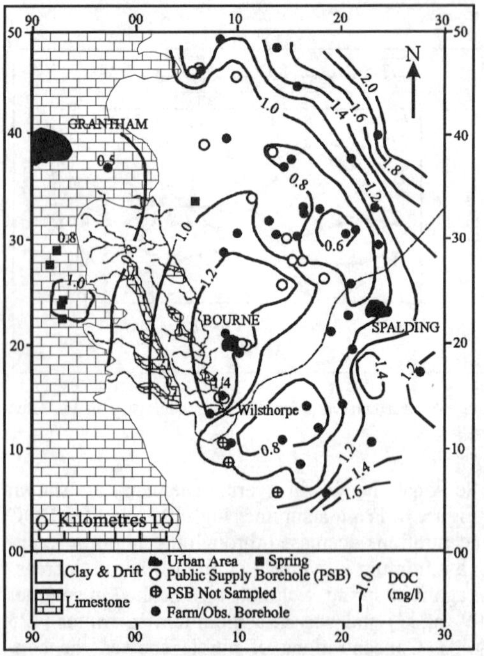

Figure 8. Groundwater DOC concentrations (winter and spring data collected in 1997/98; Roberts, 1999).

were generally below 1.2 mg l^{-1}. This regional uniformity did not however extend to the deep confined zone, where concentrations rose to 2.0 mg l^{-1}. Other exceptions included the public supply boreholes at Bourne and Wilsthorpe in the lower Glen catchment (Fig. 8), a spring and borehole where rapid flow from surface water to point of discharge was apparent. In the unconfined zone, concentrations of DOC appeared to display seasonality, but no trends have been established.

Figure 9 profiles the amount of solid organic carbon and total sulphur (i.e. pyrite) across an interface between weathered (oxidised) and unweathered (reduced) aquifer rock. Amounts of organic carbon differ little between unweathered and weathered materials (0.08 to 0.09 % Org. C, n=125) but the total sulfur concentration (i.e. pyrite), differed greatly (<0.04 to 0.58 % S, n=126).

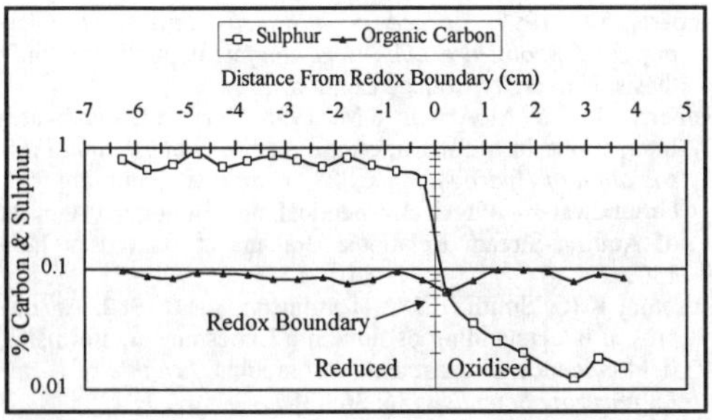

Figure 9. Organic carbon and sulphur analysis on a redox block.

4 DISCUSSION

Over abstraction of the Lincolnshire Limestone aquifer has resulted in an easterly migration of the point of confinement. As a consequence the Glen Rivers have become more influent, making the aquifer more vulnerable to surface water contamination from treatment works effluents and fertilisers. Also, rapid recharge mechanisms allow pollutants to enter the aquifer with little or no attenuation.

In winter, fertiliser applications, combined with high rainfall, led to high riverine nitrate concentrations (40-145 mg l^{-1}). Much of the fertiliser was lost *via* overland flow under storm conditions or as seepage through land drains. During flooding, the flux of nitrate in the Glen rivers was >195 times larger than at any other time during the sampling period. In summer, elevated concentrations of nitrate were attributed to sewage and agrochemical effluents, which were of low flow and therefore not significant to the overall flux of nitrate entering the subsurface. In the shallow confined aquifer, nitrate concentrations have been increasing during the last twenty years, particularly at public supply wells in the shallow confined zone (e.g. Bourne from 1.6 to 8.3 mg l^{-1}, Billingbourgh from 6.6 to 17.0 mg l^{-1}, Fig. 5). However, nitrate was not found to have advanced through the aquifer, unlike sulphate, which has displayed an easterly migration (Roberts 1999). The lack of penetration of nitrate is evidence that denitrification is occurring at a higher rate than sulphate reduction.

Concentrations of DOC in the Glen Rivers varied with flow, which, combined with other factors such as STW outfall volume and seasonal algal growth, produced the variations seen in this study. The extent of anthropogenic inputs and autochthonous production of DOC meant that despite leaching of DOC from soil during winter recharge, summer DOC concentrations (2.0-14.6 mg l^{-1}) were higher than those of winter (1.9-8.5 mg l^{-1}). Only exceptional flooding raised winter DOC concentrations to summer levels. The concentrations of DOC in water from land drains and springs were low, (0.8-3.0 mg l^{-1}) indicating that little DOC is lost from soil through-flow. Recharge entering the aquifer *via* swallow holes (DOC up to 15 mg l^{-1}), whether in the river bed or elsewhere, bypasses the soil zone. The DOC in this recharge must be rapidly consumed on entering the aquifer as groundwater concentrations of DOC in outcrop, and the shallow confined zone were low (0.8-1.0 mg l^{-1}). This residual DOC may consist of unreactive humic substances. In the deep confined zone, DOC concentrations increased to 2.0 mg l^{-1}, possibly as DOC leaches from shales and clays in confining beds, as the limestone decreases in thickness (Krumholz 1999).

The activity of microbial faunas can significantly alter influent water quality, particularly by reducing nitrate and sulphate. Populations of both denitrifying and sulphate reducing bacteria have been isolated from the aquifer (Downing et al. 1977). The activity of these bacteria, if heterotrophic, is controlled by the presence of organic carbon as a food source. Both oxygen consumption and nitrate reduction will utilise DOC with a stoichiometry of 1 mg l^{-1} DOC consuming 2.7 mg l^{-1} DO ($1C \equiv 1O_2$) and 4.1 mg l^{-1} of nitrate ($5C \equiv 4NO_3$). Dissolved oxygen may also be consumed by the oxidation of pyrite in the aquifer rocks, but were this so for all recharging DO, and all DOC was used to reduce nitrate, no more than 40% of the influent nitrate (<20 mg l^{-1}) could be removed by this mechanism. Since nitrate concentrations in the aquifer decrease from ~70 mg l^{-1} to undetectable, other reduction mechanisms must affect nitrate. A comparison of weathered and unweathered aquifer material showed that they contained similar amounts of organic carbon but that weathering removes most of the pyrite (Fig. 9). Quarry exposures show that this oxidation takes place underground as water moves through the aquifer. Since pyrite can reduce nitrate (Hiscock, et al. 1991, Ottley et al. 1997), and the reaction is catalysed by autotrophic bacteria, we postulate that most nitrate removal from the aquifer occurs *via* reduction by pyrite. In addition, nitrate may be reduced by inorganic reduction by iron (II) (Ottley et al. 1997), although a detailed mineralogical investigation of the aquifer material would be required to identify the by-products of likely reaction pathways.

5 CONCLUSIONS

The Lincolnshire Limestone aquifer is naturally resisting agricultural contamination, notably high influxes of nitrate. This is achieved by a combination of heterotrophic and autotrophic nitrate reduction processes. Heterotrophic bacterial denitrification is limited by the availability of organic carbon. Dissolved organic carbon entering the aquifer from surface waters, even as rapid recharge, is quickly consumed. Aquifer material from the unconfined zone showed that solid organic matter contributes little to the aquifer's reducing capacity. Autotrophic reduction of nitrate is likely to be sustained by pyrite oxidation. If denitrification mechanisms are sufficiently robust, then increases in concentrations could be less than previously predicted. However, nitrate levels are rising in the unconfined and shallow confined zones. This suggests that the natural protection afforded by the aquifer may not be able to cope with rising nitrate loads. Sulphate concentrations are increasing down the aquifer as a pollution front, although retarded by sulphate reduction. It is, therefore, concluded that the aquifer is potentially threatened by agricultural contamination. Regulatory control over the amount and timing of fertilizer applications could help to reduce this threat. Parts of the Lincolnshire Limestone and surrounding areas have been designated as "Nitrate Vulnerable Zones" since December 1998 (EA 1998). The Glen Valleys water quality would benefit from inclusion in this program.

REFERENCES

Benner, R. & Hedges, J.I. 1993. A test of the accuracy of freshwater DOC measurements by high-temperature catalytic oxidation and UV-promoted persulphate oxidation. *Marine Chemistry*. 41: 161-165.

Booker, I.R. 1977. *The Significance of Swallow Holes in the Hydrology of the Southern Lincolnshire Limestone*. Progress report 1976-1978. Unpublished. University of East Anglia: 107.

Downing, R.A., Smith, D.B., Pearson, F.H., Monkhouse, R.A. & Olet, R.L. 1977. The age of groundwater in the Lincolnshire Limestone, England, and its relevance to the flow mechanism. *Journal of Hydrology*. 33: 201-216.

EA, 1998, *Nitrate Vulnerable Zones*, Environment Agency, Public information leaflet, AN-10/98-1.5K-C_BEBR.

Edmunds, W.M. 1973. Trace element variations across an oxidation-reduction barrier in a limestone aquifer, *Proceedings of Hydrogeochemistry and Biogeochemistry*. Tokyo 1970. Vol. 1: 500-526.

Edmunds, W.M. & Walton, N.R.G. 1983. The Lincolnshire Limestone-Hydrogeochemical evolution over a ten-year period. *Journal of Hydrology*. 61: 201-211.

Emery, D. & Dickson, J.A.D. 1991. The subsurface correlation of the Lincolnshire Limestone formation in Lincolnshire. *Proceedings of the Geological Association*. 102: 109-122.

Hedges, J. 1993. Measurement of dissolved organic carbon and nitrogen in natural waters: Workshop report. *Marine Chemistry*. 41: 5-10.

Hiscock, K.M., Lloyd, J.W. and Lerner, D.N., 1991, Review of natural and artificial denitrification of groundwater, *Water Resources*, 25, No. 9, p1099-1111.

Krumholz, L.R. 1999. Microbial communities in the deep subsurface. *Hydrogeology Journal*. 8: 4-10.

Lawrence, A.R. & Foster, S.S.D. 1986. Denitrification in a limestone aquifer in relation to the security of low-nitrate groundwater supplies. *Journal of the Institute of Water Engineering Science*: 40. 159-172.

Ottley, C.J., Davison, W. & Edmunds, W.M. 1997. Chemical catalysis of nitrate by iron (II). *Geochemical et Cosmochimica Acta*. 61. No. 9: 1819-1828.

Roberts, S.C. 1999. *Surface/Groundwater Interactions affecting the Lincolnshire Limestone aquifer*. unpublished Ph.D thesis. University College London: 299.

Roberts, S.C. & McArthur, J.M. 1998. Surface/Groundwater Interactions In a UK Limestone Aquifer. *International Association of Hydrogeologists 28th Congress*. Gambling with Groundwater – Physical, Chemical, and Biological Aspects of Aquifer-Stream Relations. Brahana et al.. (eds.): 125-130.

Rushton, K.R., Smith, E.J. & Tomlinson, L.M. 1982. An improved understanding of flow in a Limestone aquifer using field evidence and mathematical models. *Institute of Water Engineering Scientists*. 36: 369-387.

Smith-Carington, A.K., Bridge, L.R., Robertson, A.S. & Foster, S.S.D. 1983. *The nitrate pollution problem in groundwater supplies from Jurassic limestones in Central Lincolnshire*. Report. Institute of Geological Sciences. NERC. 83/3: 22.

Groundwater: Past Achievements and Future Challenges, Sililo et al. (eds) © 2000 Balkema, Rotterdam, ISBN 90 5809 159 7

Passive in situ treatment of Cr(VI)-contaminated groundwater by zero-valent iron: Results of a field study

M. Schneider
Institute of Geology, Geophysics and Geoinformatics, Free University of Berlin, Germany

ABSTRACT: Subsurface in situ reactor systems are a viable alternative to conventional pump-and-treat systems, especially for heterogeneous aquifers and if the source of contamination cannot be entirely removed, so that the remedial system has to be operated over a period of time. In this case, in situ reactors are more efficient and more economical than pump-and-treat. At a contaminated site in northern Bavaria (Germany), a drainage system was installed to collect groundwater contaminated by chromate (Cr(VI)) and hydrophobic organic compounds. The groundwater plume contained about 5 mg/L chromate with maximum values of 80 mg/L and an average concentration of 2 mg/L halogenated aliphatics. In accordance with the regulations established by the competent authorities, a conventional pump-and-treat system was installed as a final site remedy. In the frame of a research project, part of the drainage system was connected with a reactor shaft containing zero-valent iron. Due to reduction of Cr(VI) to Cr(III), coupled with the oxidation of zero-valent iron to Fe(II) and Fe(III) and the precipitation of Fe(III)-Cr(III)-(oxy)hydroxide phases, the chromate concentration can be reduced below the target value of 0.2 mg/L. Experiments with iron in various forms show that the rate of Cr(VI) removal using steel wool is greater than that for iron granules. The degradation of the organic compounds in the reactor system was insufficient, so the remedial target for these pollutants was not reached.

1 INTRODUCTION

Within the scope of a research project (supported by the Ministry of Economics, project no. 826/97), a concept was developed for a pilot-scale in situ remedial of groundwater contaminated with Cr(VI) and halogenated aliphatics (tetrachloroethylene, PCE; trichloroethylene, TCE and cis 1,2-dichloroethylene, cDCE). The site selected was a galvanization plant near Coburg in northern Bavaria. Due to difficult hydrogeological conditions and very low groundwater flow velocities, a drainage system connected with a reactor shaft was installed for collecting the contaminated groundwater. In order to remove the dissolved contaminants, zero-valent iron was used as the reactive material in the reactor shaft. Zero-valent iron is capable of transforming toxic Cr(VI) to slightly soluble Cr(III) compounds and of degrading halogenated aliphatics to ethene and other non-toxic compounds. The relevant oxidation, reduction and precipitation reactions are described by Gould (1982), Powell et al. (1994), Gillham & O'Hannesin (1994) and Matheson & Tratnyek (1994). The interaction between chromate and halogenated aliphatics are discussed in Fruth & Schneider (this volume).

The installed unit for in situ treatment is part of the full-scale remediation system at the contaminated site which is operated as a conventional pump-and-treat system and approved by the regulatory agency. In situ reactors installed in the flow path of contaminated groundwater are the subject of intense discussions as a low-cost and efficient remediation technology compared to energy- and maintenance-intensive pump-and-treat systems (McMurty & Elton 1985, Starr & Cherry 1994, Teutsch et al. 1997, Schad & Teutsch 1998).

2 CHARACTERIZATION OF THE FIELD SITE

The galvanization plant had been in use over 30 years. While the plant was in operation, chromium wastes and chlorinated solvents contaminated the soil and groundwater. The plume of contaminated groundwater shows a mean concentration of 2 mg/L halogenated aliphatics and 5 mg/L chromate (maximum value: 80 mg/L). A disused waste pit was identified as the source of contamination. The estimated total amount of pollutants in the plume is in the order of 100-200 kg for halogenated aliphatics and 10-20 kg for chromium compounds. For Cr(VI), the detected maximum concentration exceeds the target value for remediation by a factor of 400.

The base of the shallow sandstone aquifer (Upper Triassic) is 5-6 m below ground level (Figure 1). Groundwater flow velocity is relatively low, the hydraulic conductivity lies between 10^{-5} and 10^{-6} m/s. The field site is located in an area of a groundwater divide, so flow directions are extremely variable.

3 REMEDIATION CONCEPT

Because of the small aquifer thickness and low groundwater discharge, a subsurface drainage system was installed approximately 5 m below ground surface. It consists of a side drain which is connected through a reactor shaft to a main drain. The mixed water from both drains finally enters a main shaft and is pumped into an on-site treatment unit where the contaminants are removed by conventional methods (ion exchange and activated carbon). The remediated groundwater from the on-site unit is discharged into a nearby stream. Figure 2 shows the installations at the field site and the position of the contaminant plume. This concept allows the operation of the in situ reactor system under vary conditions, even if the quality of the water discharged from the reactor is not in accordance with the regulations. In a full-scale passive remediation system, the remediated water would be infiltrated directly into the aquifer.

The construction of the reactor shaft is presented in Figure 3. The shaft has a diameter of 1.5 m and is subdivided into 4 compartments in order to optimize the flow path of the contaminated water through the permeable reactive material which has a volume of approximately 1 m^3. The hydraulic gradient in the reactor is 0.0033.

Iron in various forms (grains, chips and wool) was investigated in laboratory batch and column

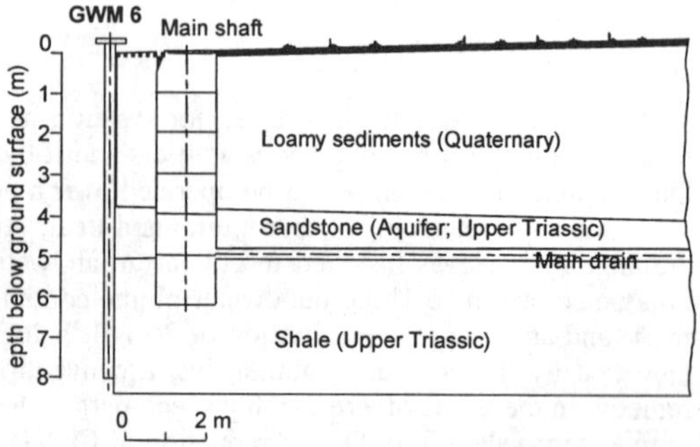

Figure 1. Geological profile (GWM=Groundwater monitoring well)

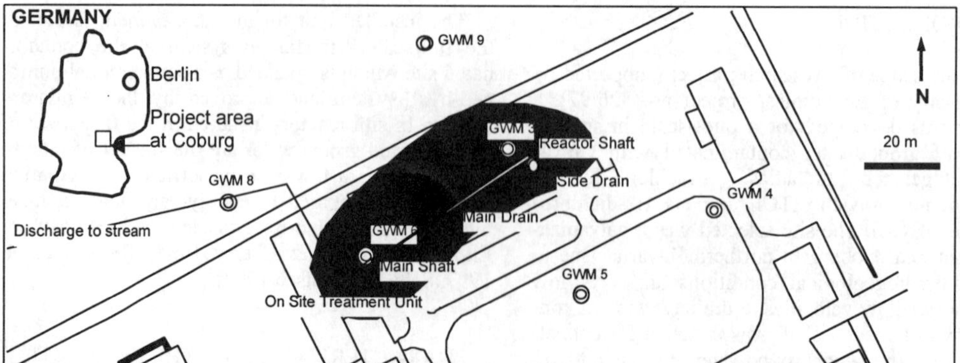

Figure 2. Installations at the field site and position of the contaminant plume (GWM = Groundwater monitoring well).

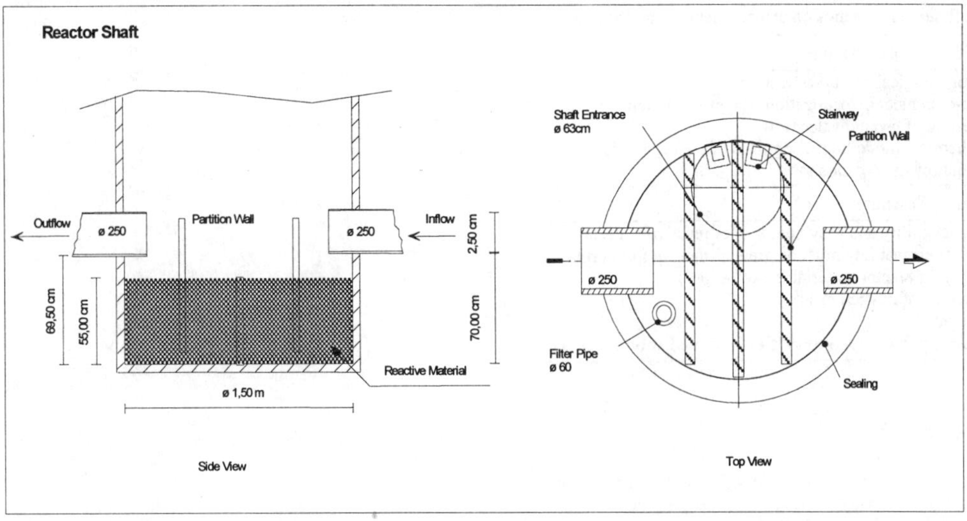

Figure 3. Construction of the reactor shaft.

tests. Due to the large specific surface, steel wool achieved the best results. For this reason, steel wool was choosen as the reactor material and mixed with calcite to keep the pH-value within a narrow range. The content of Fe^0 in the steel wool is 98%.

4 RESULTS AND DISCUSSION

For a field test, 45 kg of steel wool mixed with calcite in 64 single packs was inserted in the reactor. Over a period of 32 days, the mean rate of flow through the reactor was 30 l per hour. Water samples were taken at the inflow, outflow and from the various compartments of the reactor system. The mean inflow concentration was 2.1 mg/L for Cr(VI) and 0.9 mg/L for halogenated aliphatics. During the test period, the Cr(VI) concentrations in the effluent samples has a mean value of 0.03 mg/L and were below the target value of 0.2 mg/L at all time (Figure 4). Regarding the halogenated organic compounds, the remediation target (0.04 mg/L in groundwater) was not reached, in most cases the outflow concentration is in the same order as the inflow concentration. It is assumed that passivation of the surface of the steel wool - due to concurrent chromate reduction - and insufficient residence time of the contaminated water in the reactor led to the unsatisfactory degradation performance. The pH-values were stabilized between 6.7 and 7.3.

Using steel wool as the reactive material, it was shown that a stable remediation performance was achieved throughout the test phase, removing chromate from groundwater even when inflow concentrations and water quantities fluctuated greatly. Reduction of permeability within the reactive material due to precipitation of mineral phases was not observed. More investigations are in progress to develop adequate systems for mixed contaminations comprising anorganic and organic pollutants. The installation of a passive in situ remediation system and its optimal operation require a detailed hydrogeological site investigation. Modelling groundwater flow is advisable in order to optimize the position and dimensions of the subsurface structure. Essential demands on the project planning and structure are summarized in Table 1.

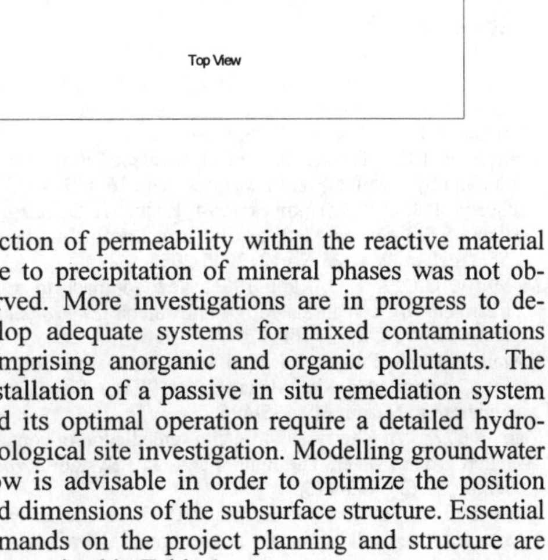

Figure 4. Temporal change in Cr(VI) concentration (Inflow and outflow).

Table 1. Essential demands on project planning and structure.

Demands on Project Planning:
– Hydrogeological site investigation – Three-dimensional localization of contamination – Modelling of groundwater flow – Geochemical modelling: Prognosis of influence of reactor on groundwater compostion
Demands on Structure:
– Exchangeability and easy handling of reactive material – Reactor permeability must be greater than aquifer permab. – Stability (if reactor material is exchanged) – Avoidance of oxygen influx – Longevity – Monitoring facilities: Supervision of remediation process

5 REFERENCES

Gillham, R.W. & S.F. O'Hannesin 1994. Enhanced Degradation of Halogenated Aliphates by Zero-Valent Iron. In *Ground Water*, Vol. 32 No. 6: 958-967.

Gould, J.P. 1982. The Kinetics of Hexavalent Chromium Reduction by Metallic Iron. *Water Res.*, vol. 16: 871-877.

Matheson, L.J. & P.G. Tratnyek 1994. Reductive Dehalogenation of Chlorinated Methanes by Iron Metall. In *Environ. Sci. Technol.* Vol. 28 No.12: 2045-2053.

McMurty, D.C. & R.O. Elton 1985. New approach to in-situ treatment of contaminated groundwaters. *Environmental Progress*. V. 4, no. 3: 168-170.

Powell, R.M., R. Puls & C.J. Paul 1994. Innovative Solutions for Contaminated Site Management. Water Environment Federation, Miami, FL: 485-496.

Starr, E.C. & J.A. Cherry 1994. In situ remediation of contaminated ground water: The Funnel-and-Gate System. *Ground Water*, 32 (3): 465-476.

Teutsch, G., J. Tolksdorff & H. Schad 1997. The design of in-situ reactive wall systems - A combined hydraulical-geochemical-economical simulation study. *Journal of Land & Reclamation*, 10.

Schad, H. & G. Teutsch 1998. Reaktive Wände – Aktueller Stand der Praxisanwendung. Geotechnik, 21, no. 2: 73-82.

Groundwater: Past Achievements and Future Challenges, Sililo et al. (eds) © 2000 Balkema, Rotterdam, ISBN 90 5809 159 7

Risk assessment, modeling groundwater contamination, capping and optimizing hazardous waste landfills from metal smelters

J.E.Sorlie & K.Rudolph-Lund
Norwegian Geotechnical Institute, Oslo, Norway

ABSTRACT: At the industrial site of Mo i Rana in northern Norway three major smelting plants are located on a huge alluvial delta. The waste was usually buried in the deep ravines, more or less uncontrolled, later the hazardous wastes were placed in controlled landfills with sealings and leachate control. The electric arc furnace dust landfill was terminated 5 years ago due to too high concentrations of heavy metals in the leachate. The ferrochromium producer requested to take over the dust pit for depositing his own scrubber sludge. However, the authorities demanded a thorough environmental assessment of the landfills with respect to possible unfavorable leaching effects and worst case scenarios such as failing bottom drainage in the ravine valley, failing leachate pipelines and ruptured sealings. Investigations for installing more monitoring and exploratory wells were performed as well as laboratory investigations for waste characterization and leaching properties. A comprehensive 3-D model was developed to simulate various effects such as failing drainage and remedial measures.

1 INTRODUCTION

In connection with the restructuring of a norwegian steel mill (1950 to 1988) in Mo i Rana, the production was altered to electric arc furnace iron production (Fundia/Rautaruukki) in 1990 for iron waste, collected from the whole country. This smelter production produces the electric arc furnace dust (EAF dust), the well-known hazardous waste, with high content of heavy metals and some polycyclic aromatic hydrocarbons (PAH), see table 1. The dust was placed as a controlled landfill in a 25 m deep ravine.

This industrial site, see Figure 1, is a huge, flat alluvial delta approximately 50 m above sea level, consisting of fine sands, which become more silty and clayey with depth. The groundwater table is approximately 15 to 20 m below the surface and bedrock is at 40 to 60 m depth.

There are two more smelters at this site producing ferrochromium (Elkem) and ferrosilisium (Rana Metal). The ferrochromium scrubber sludge, is also a hazardous waste high in PAH but low in heavy metals, see table 1. The sludge has been deposited in the ravine in a separate landfill next to the EAF dust landfill. These two landfills are a part of a plan to reclaim the terrain of the ravine. In the bottom of the ravine there is a draining culvert to maintain a groundwater table below the landfills. The landfills are sealed at the bottom and on the sides, but open at the top. The leachates from the landfills are collected in a closed drainage system with monitoring wells. Today the leachate drains directly to the sea. During the earlier steel smelting activities, the wastes, mostly slag and granulate, were generally filled in an uncontrolled manner into this long ravine both upstream and downstream of these two new landfills. Considerable amounts of groundwater contamination are transported to this drainage system and flows directly to the sea via a small creek. In periods the EAF dust landfill produced too high concentrations of heavy metals. The steel plant (Fundia) was therefor requested by the environmental authorities to terminate and prepare plans to seal off the landfill and evaluate the environmental impact. Simultaneously, Elkem proposed to utilize the remaining pit for depositing their ferrochromium sludge.

1.1 *Scope of work*

A solution that would serve both Fundia and Elkem was deemed to be ideal. The owners were requested to investigate the following:

1. A plan for sealing/capping of the EAF dust landfill that at the same time serves as a bottom lining for a new sludge landfill.
2. The plan must show that the final solution would not increase the leaching, preferably reducing the leaching from the EAF dust landfill.

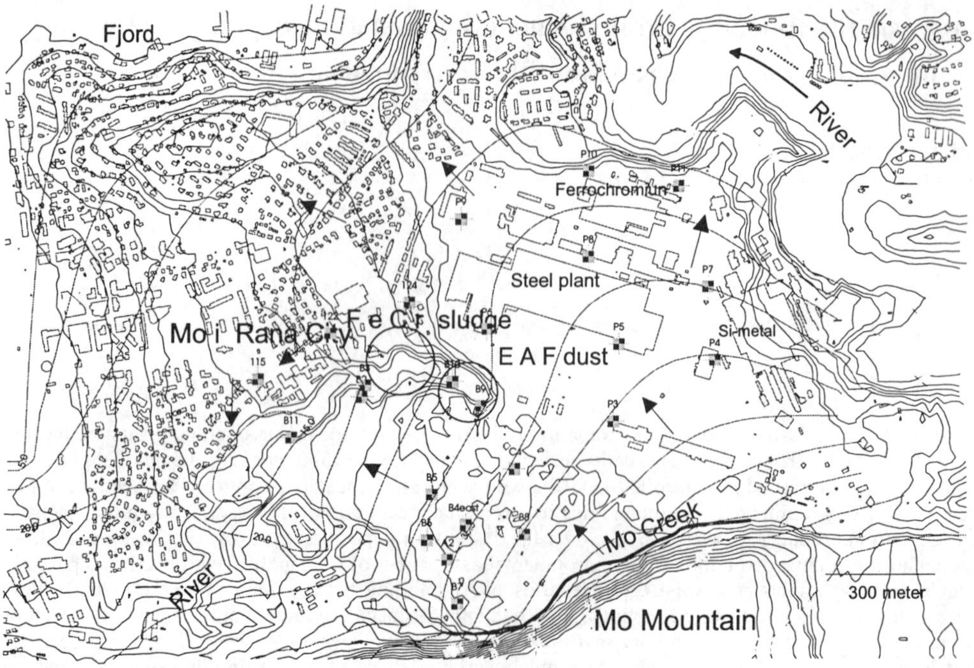

Figure 1. Site plan with monitoring wells, the landfills and equipotential lines

3. An overall environmental risk assessment of the selected solution with regard to worst case scenarios, for example the bottom drainage stops functioning, the membrane around the EAF dust is ruptured or the groundwater rises.

1.2 *Recommended investigations*

From previous work, there was considerable information from groundwater monitoring wells, contaminant transport and monitoring of leachate from both landfills.

To establish the documentation needed for this study the following investigations were performed:

1. A thorough characterization of the EAF dust waste and the ferrochromium sludge and what factors that govern the leaching of the heavy metals.
2. Investigate if leachate from the ferrochromium sludge would have any adverse effects on the leaching of the EAF dust waste.
3. Investigate various sealing methods and, if necessary, investigate the materials.
4. Investigate the actual contamination status in the groundwater downstream the landfills.
5. Establish a groundwater model where it would be possible to simulate the various scenarios.

Table 1. Content of waste and leachate

	EAF dust mg/kg	EAF dust-leachate*	Sludge mg/kg	Sludge-leachate*
Ca	2.62	271	0.072	41
CD	0.054	18	0.001	
Cr	0.21	6.2	1.91	0.048
Cu	0.24	141	0.015	6
Fe	19.9	<10	0.53	173
K	2.06	13.100	0.40	437
Mn	1.39	<2	0.39	<2
Na	2.9	13.600	1.29	313
Ni	0.25		0.009	0.061
Pb	3.52	56.2	0.18	0.061
Si	1.14	2.23	9.8	1.86
Zn	27	4.3	4.65	0.011
PAH	70	22	200-2.550	205
pH		13.2		11.2
Cond.		6529		288

* Real leachate from the landfills, μg/l

2 RESULTS OF THE INVESTIGATIONS

2.1 *Characterization and leaching of the wastes*

The two wastes are quite different as shown from table 1. The EAF dust contains high concentrations of most heavy metals whereas the real leachate shows high concentrations of Cd, Cu, Pb, Zn and is very alkaline with pH equal to 13.2. The ferrochromium sludge contains polycyclic aromatic hydrocarbons (PAH), only small amounts of heavy metals and a pH of 11.3.

The leaching tests were performed with both deionized water and leachate from the sludge. The results showed that the leaching increased with increasing liquid to solid ratio (L/S from 1:5 to 1:25) regardless of liquid, with more heavy metals being leached by deionized water than by the sludge leachate. The pH remained constant throughout the batch testing. Under very unfavorably conditions over 100 years the L/S ratio would be maximum 1:7 and the type of liquid would have little effect.

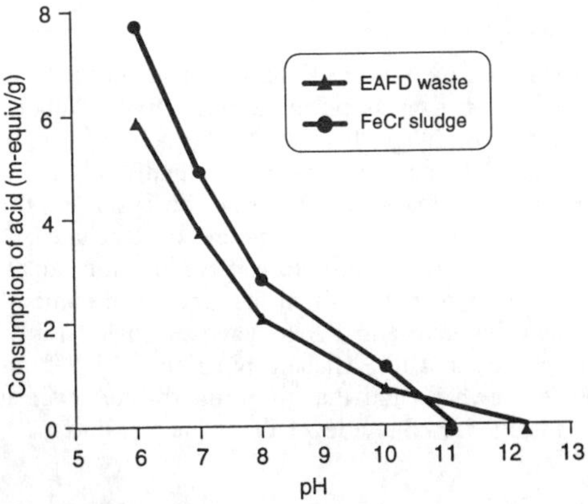

Figure 2. The titration curves

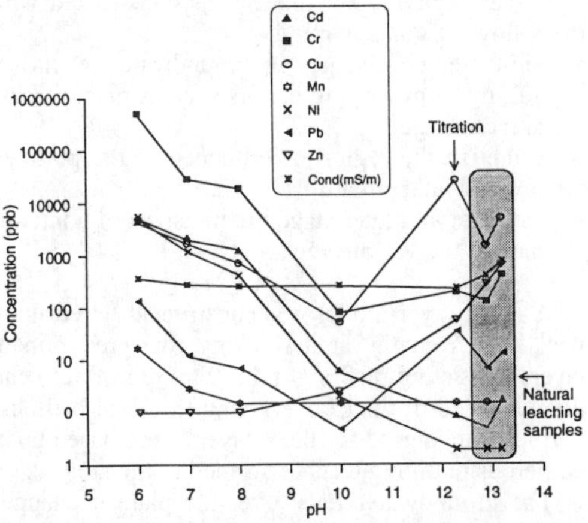

Figure 3. Leaching from EAF dust versus the effect of pH

The result of the titration tests for both wastes show in Figure 2 that the buffer capacity of both wastes is very large. The titration curve of the EAF dust waste, assuming acid precipitation (pH = 4) and unfavorable infiltration conditions, show that it will take 4000 years to reduce the pH from 12.3 to 10. The pH curves for the various heavy metals are shown in Figure 3. The leaching of Pb, Zn and Cu decreases with a decreasing pH between 10 and 11 and then increases again with decreasing pH. For Cd and Mn the leaching increases continuously with increasing pH. Cr leaching is unchanged with decreasing pH to 10 and then increases rapidly. No Cr^{+6} was detected (<0.01 mg/l).

The titration curve for the sludge showed that the leaching of heavy metals increased by reducing the pH to 10 and from then on was either unchanged or increasing.

2.2 *The sealing method*

Several conventional membrane type solutions were considered and found to be technically feasible but expensive. It was then decided to investigate the use of the rejected low-grade microsilica from Rana Metal at the site as a possible capping material. This material, being free of charge, has many unique properties and could possibly serve as both a seal and an adsorbent. The microsilica is a waste form the Si and FeSi-metal production. It contains about 90 % amorphous silica, 1 to 4 % organic carbons and 5 % various oxides. The content of heavy metals is insignificant, but leaches very small quantities of As and PAH.

Sorption tests, mixing microsilica and leachate in batch tests, showed that even most of the heavy metals and the PAHs from the EAF dust leachate would be adsorbed. As there was no previous experience using microsilica as a sealing material various geotechnical parameters had to be investigated. The microsilica particles are less than 0.1 μm and are agglomerated into about 1 mm round particles. With a plastic limit of 55 % and liquid limit and 62 %, pilot compaction tests showed that it had to be placed in layers of 0.25 m and compacted dry. The permeability coefficient, k was found to be less than 5 $x10^{-8}$ cm/sec. Due to the amorphous structure of the silica, the samples became never fully saturated during the testing and it appeared that the k value became less as the saturation increased. The ultimate k was calculated to reach $1x10^{-10}$ cm/sec. Figure 4 shows how the 5,000 m^2 area sealing and the adsorption layer was constructed. Figure 6 shows the completed sealing of the EAF dust deposit ready for the FeCr sludge.

The remaining 15 m of the pit would be filled up with sludge in 15 years where it will be exposed to infiltration water and then permanently sealed off. Calculations indicate that the leaching from the EAF dust landfill will decrease from the current 700 l/hr to 0.5 l/hr with a corresponding reduction of the contaminant flux by a factor of 1400.

The sealing was completed in 1998 and the rate of leaching has already started to reduce.

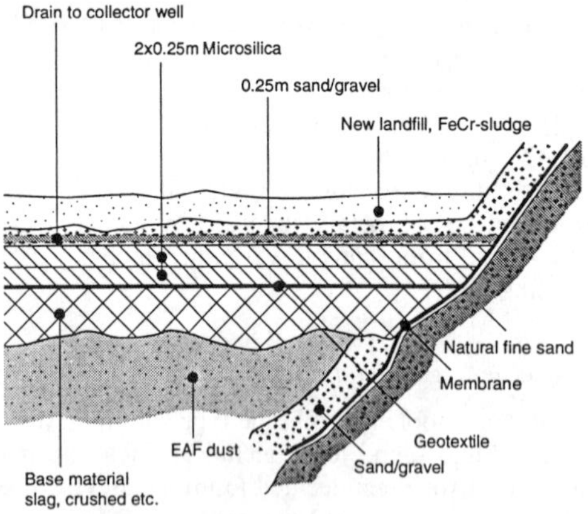

Figure 4, Sealing/capping of the dust landfill and bottom lining for the sludge.

2.3 *Uncontrolled leaching into the groundwater.*

It was necessary to investigate if the EAF dust landfill contaminated the groundwater by uncontrolled leaching. Supplementary monitoring wells were placed just upstream of the landfill and downstream of the landfill. The results showing that there were hardly any differences, were not unexpected. The heavy metals and the pH were more or less the same and the PAH was considerably higher upstream. The ravine upstreams of the EAF dust landfill, with several side ravines are already filled with industrial waste from the previous smelting activities. As the groundwater is already contaminated it will be difficult to monitor possible ruptures/holes in the sealing. It is especially the former granulate, with its high content of calcium, that maintains the high pH throughout the former ravine valley.

2.4 *Groundwater table*

The groundwater table at the site has been monitored regularly at 21 observation wells since 1995. At a few places the groundwater fluctuation have been recorded since 1990.

In the area of the landfills, the groundwater table is estimated to have risen totally about 5 m since 1990 and 2 m since 1995. The last two years of observations show that the groundwater table has stabilized. This indicates that the bottom drainage is damaged somewhere and is not functioning as it should. Underneath the two landfills the groundwater table is raised just to the bottom of the landfills.

2.5 *Bottom drainage*

In 1990 a bottom drainage, an open draining culvert (400 mm cement pipes), was placed at the bot tom. This should have secured a permanent groundwater table as it was before the reclaiming of the valley started.

The bottom drainage is built with a sand and gravel layer around the culvert. Outside the sand a geotextile is placed to avoid damage and mixing of the filter material with the ordinary fill. Experience from other places show that this type of geotextile is too fine-textured and it acts like a filter retaining the suspended particles. Over some years it may become impervious.

2.6 *Leachate pipeline*

The leachate pipeline (320 mm cement pipes) from the two landfills was placed about 2 to 4 m above the bottom drainage. It is a closed pipeline for only the generated leachate from the two landfills.

Both flux measurements and analysis of the leachate showed that the amount of leachate increases about 10 times before it reaches the outlet. The analysis also shows that the amount of some of the heavy metals and PAH increases considerably. This proves that the leachate pipeline is also damaged somewhere and that the groundwater level at this place is raised to a level above the pipeline.

2.7 *Groundwater modeling*

A Visual MODFLOW model was constructed with the following goals in mind:

- simulate the changes in groundwater levels resulting from varying factors such as precipitation and drainage
- establish the region of influence of the polluted groundwater stream
- estimate at a later stage the transport of contaminant to the various recipients

A multi layer model was constructed based upon well logs from the area showing two predominant layers consisting of first sand (20 to 30 m thick) and then clay (20 m thick) overlaying the local bedrock. Permeabilities used for these layers were based upon numerous pumping tests performed in the area.

The groundwater level over the plateau changes throughout the year, but is approximately 20 masl. The direction of the natural groundwater flow is towards the west - northwest.

Calibration was based on groundwater level measurements recorded during the construction of the landfills. Calibration was performed on the conductivity of all three layers.

Various simulations performed with varying drainage in the ravine valley and the Mo creek showed a number of interesting results:

1 As long as the drainage is functioning in the ravine valley, the majority of the groundwater coming from the polluted sediments on the pla-

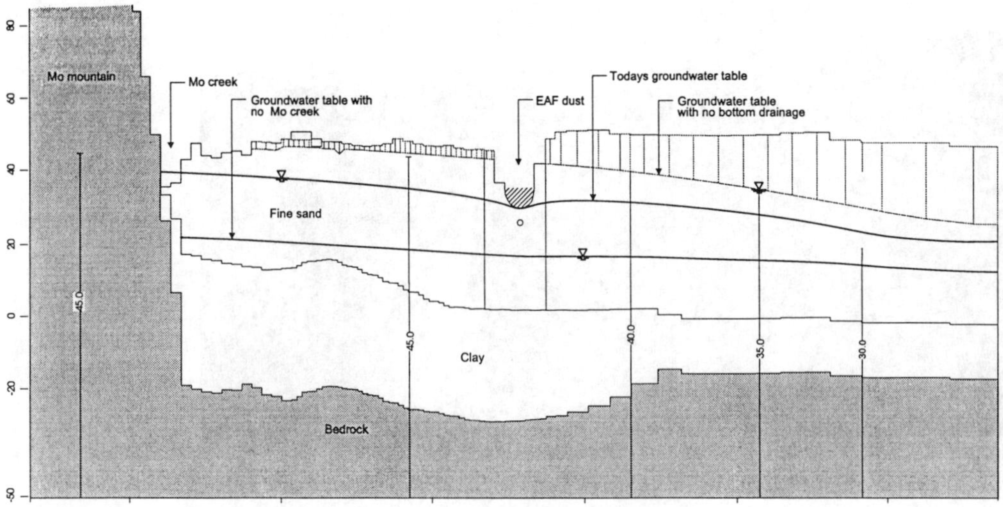

Figure 5. Cross section through the plateau and the EAF dust landfill.

teau are captured by the drainage system, see Figure 2 and 5.

2 If the drainage system in the ravine valley does not function, the groundwater table will rise approximately 10 m and change the main direction of the flow away from the valley towards the town of Mo i Rana, see Figure 5.

3 The Mo Creek along the foot of Mo Mountain is the main source of recharge. If this water was harvested to reduce recharge, the water table under the plateau would be lowered by approximately 25 m, see Figure 5.

3 DISCUSSION OF RESULTS

3.1 *Not functioning bottom drainage and raising groundwater table*

The groundwater modeling shows that the combination of no bottom drainage and more or less impervious fill material and waste in the valleys, raises the groundwater table approximately 10 m. This will also change the main direction of the groundwater flow away from the valley and more towards the town. In general one would avoid a situation with rising groundwater level as this will create a new and unpredictable situation with regard to leaching of contaminants, stability problems, settlements etc. However, today all the contaminants are trasported directly to the sea. A drain would force the contaminants into a long lasting transportation pathway that would retard contaminants considerably and reduce the current pollution to the sea.

With increasing gradient and pressure towards the bottom drainage, internal erosion may occur and could cause considerable damage such as subsidence to the fill areas which supports roads, drainage, houses etc.

During the modeling it became clear how the Mo creek controlled the groundwater table throughout the plateau. The effect of infiltration (1000 mm annual rainfall) is insignificant due to the large areas of buildings and asphalt pavements where the intercepting rainfall is piped off the site. The effect of removing the Mo creek is dramatic as the lowered groundwater table is shown on figure 5. To "remove" the creek along the foot of the mountain is not costly and it has become a very interesting option as possibly the most effective remedial measure at this site. It would lower the groundwater table below all fills and wastes and the contaminant transport would be insignificant. Another possible solution is a partial removal such that a certain groundwater flow is maintained towards some of the recipients.

3.2 *Not functioning leachate pipeline*

If the landfill is not drained due to a clogged pipeline, theoretically the EAF dust fill will be fully saturated, unstable and not operational. However it is known that the membrane may have several damages that may have caused ruptures and the pore water inside the fill is expected to freely drain through natural fine sand. The surrounding sand has a hydraulic conductivity about 100 times higher than the dust. There is also a draining sand layer between the membrane and the dust. The higher gradient inside the fill will drain excess pore water from the landfill. The leaking leachate may eventually drain into the bottom drainage or follow the natural groundwater flow. If poor drainage should cause operational dif-

Figure 6. The sealed EAF dust deposit

ficulties it may be possible to pump the leachate from the collector well and/or cover up the surface area against rainfall.

3.3 *Ruptured membrane*

In addition to what is mentioned in Section 3.2, it is important to keep in mind that the groundwater flowing past the EAF dust fill is already contaminated. Some leakage of EAF dust leachate into the groundwater will have insignificant effect. The two monitoring wells at the landfill and the well for leachate control have been monitored closely for two years prior to completing the sealing of the fill. This database will make it possible to follow the effect of the sealing and the new landfill of sludge on the top.

4 CONCLUSIONS

A 3-D groundwater model over such a large area is an invaluable tool when completed. The main conclusions based on the modeling work and investigations are:

- If leachate from the FeCr sludge should penetrate the microsilica layer this will not increase the leaching from the EAF dust. Generally the trend is such that if the pH is reduced some, this will decrease the leaching of heavy metals.
- No infiltration/leaking is capable of reducing the pH because of the extreme buffer capacity in the dust.
- If the bottom drainage fails, the groundwater table near the EAF dust fill will raise a maximum of 10 meters. The higher groundwater flow will bypass the drainage that currently leads the contaminated groundwater directly to the sea. The contaminant transport will follow new and longer pathways with increased remediation of the contaminated groundwater on its way to the fjord.
- Possible uncontrolled situations of internal erosion and subsidence may occur due to increased groundwater levels. It is therefore recommended to avoid a high groundwater table.
- If the leachate pipeline fails, this is not expected to have any serious effect.
- If the membrane ruptures, this will not have any measurable effect on the contaminated groundwater. It may be necessary to change the filling operation to reduce infiltration.
- A possible solution to nearly eliminate the flow of contamination from this site is to remove the Mo creek fully or partially. This will lower the groundwater table below all fills and deposits, and reduce the importance of the bottom drainage.

REFERENCES

Sorlie, J.E. 1997. The EAF dust landfill, Mo i Rana. Sealing and future use. NGI report 974081-1.

Sorlie J.E. 1999. The EAF dust landfill, Mo i Rana. Risk assessment and environmental consequences. NGI report 984029.

Kjeldsen, P.&T.H. Christensen 1990. Leaching tests to evaluate pollution of combustion residues from an iron recycling industry. Waste Management and Research, vol 8, no 4.

Groundwater: Past Achievements and Future Challenges, Sililo et al. (eds) © 2000 Balkema, Rotterdam, ISBN 90 5809 159 7

In situ measurement of oleoporic fracture aperture by DNAPL surrogate injection

A. Steele & D. N. Lerner
Groundwater Protection and Restoration Group, University of Sheffield, UK

ABSTRACT: Predicting the migration pathways and depths of penetration of dense non-aqueous phase liquids (DNAPLs) is a formidable challenge, especially in the UK where the major aquifers commonly possess significant degrees of fracturing. One of the most challenging field parameters to record is fracture aperture, due to inherent difficulties in measuring small aperture fractures. A new field technique has been developed to measure oleoporic fracture aperture *in situ* by the injection of a harmless NAPL (sunflower oil) into a fracture isolated by a double packer array beneath the water table. Regulatory permission has been obtained for NAPL injection into a fractured Permian sandstone aquifer in Scotland. Here we present the experimental design, with single and dual phase injection experiment results. The technique provides a new dataset of *in situ* fracture aperture for a sandstone aquifer, including detailed evidence for constrictions and enlargements of fracture aperture at increasing radial distance from the borehole.

1 INTRODUCTION

Accidental spills and other releases of organic chemicals such as chlorinated solvents, PCBs and coal tars have resulted in widespread groundwater contamination in many parts of the world. These chemicals often possess physical properties resulting in their subsurface migration for extended periods as dense non-aqueous phase liquids (DNAPLs). The delineation and recovery of DNAPLs is proving to be a significant challenge to the scientific community. The migration of DNAPLs is controlled by capillary as well as gravity and viscous forces, and consequently DNAPLs flow relatively independently of the prevailing groundwater gradient, preferring to move along the path of least capillary resistance. Non-wetting DNAPLs will always preferentially enter and flow in the largest available space, and therefore in strata having significant dual permeability such as fractured sandstones, migration will preferentially be in fractures, as fracture apertures are typically greater than the largest pore throats. The complex geometry and connections of fracture networks prevent complete characterisation by conventional penetrative site investigation techniques. Real rock fractures have rough walls and variable apertures (Yeo et al. 1998), and consequently the properties of individual fractures that often dominate DNAPL migration is poorly understood.

Measurement of fracture apertures is however extremely important in solving contaminant transport and DNAPL penetration depth problems for fractured domains (Murphy & Thomson 1993 ; Wealthall & Lerner 2000). Numerical simulation of NAPL migration in fracture systems is complex and constrained by the lack of real data. Whilst fracture orientations, intensities and lengths are relatively easily recorded in the field by detailed measurement, the apertures of these fractures which control migration are very difficult to measure due to their small size and high variability. Previous modeling studies (Sudicky & Slough 1998) have shown that the spreading of DNAPLs within fracture networks may be affected by characteristics of the fracture network such as spatial variations in fracture apertures and zones of closure. It has also been shown that aperture variability may control the distribution of DNAPL and water within a fracture plane, and hence the ability of the fracture to transmit these fluids (Zhou 1995).

Attempts have been made to measure apertures in several ways. Back calculating the bulk aperture from hydraulic testing using the cubic law is a relatively easy method, but can overestimate the aperture applicable to DNAPL flow (Jorgensen et al. 1998). Estimates of fracture aperture have also been performed by injecting NAPLs into fractures and back calculating fracture apertures from entry pressure required to initiate flow (Reitsma & Kueper 1994; Jorgensen et al. 1998; Hinsby et al. 1996). Reitsma & Kueper (1994) used this technique in the laboratory to estimate aperture distributions, but concluded that the effect of isolated zones introduced errors. Variation in fracture aperture has been examined by the use of enhanced imaging techniques including high resolution measurement of fracture apertures in sandstone using a computer aided tomography (CAT) scanner (Keller 1998). Detailed physical measurement of aperture variation in a natural sandstone fracture has also been performed by measurement of a silicon rubber cast of the fracture (Yeo et al. 1998). There is a scarcity of information available in the literature on fracture aperture in fractured sandstone, especially *in situ*, because of the inherent difficulties in obtaining data.

The concept of oleoporic aperture describes the smallest aperture constriction where NAPL first forms a fully connected pathway across a variable aperture fracture plane, conceptually corresponding to the entry pressure of a fracture. Oleoporic aperture obtained from field experiments should provide a better approximation of the fracture aperture critical to the formation of a continuous NAPL pathway than the conventional graphical interpretation of entry pressure (Steele & Lerner *in subm.*) Our field results suggest that a single characteristic injection pressure does occur, and this results in the injection of a capacious volume of NAPL into a fracture over long injection periods. Oleoporic aperture can be derived from this injection pressure.

In this paper we present details of a developmental field technique where a DNAPL surrogate is injected into a fractured sandstone aquifer. The novelty of this technique is that is creates real two-phase flow conditions in the field, and allows fracture aperture to be determined using reverse numerical modeling.

2 APPROACH

The aim of this work is to reduce the uncertainty in field determination of fracture aperture *in situ*. We have adopted two techniques to intensively characterise a single borehole. The first technique is conventional hydraulic testing using straddle packer apparatus, and the second is a new technique using NAPL injection into isolated fractures in the borehole.

A research site has been established on the western edge of the Dumfries basin in south-west Scotland (Fig. 1). The basin is of Permian age, and consists of over 1000 m of sandstones and breccias, the latter being more persistent in the west (Robins & Buckley 1988). The breccia was derived from a hardened muddy flood gravel or debris flow and has little inter-granular permeability (Ball 1997). The Dumfries basin is a key water resource in Scotland, and produces over 50% of the local public water supply (Ball 1997).

A 145 mm borehole has been cored to 30m below ground through breccia and occasional sandstones. The borehole was geophysically logged using the acoustic televiewer (ATV) technique which produced a very detailed image of the borehole wall, angle and orientation of fractures and exact depths of lithological changes. The logs gave good correlation with the core. The major benefit of ATV logs in this work was the exact determination of fracture depths, enabling accurate placement of testing equipment over identified fractures and fracture zones. The borehole was developed by over-pumping prior to testing.

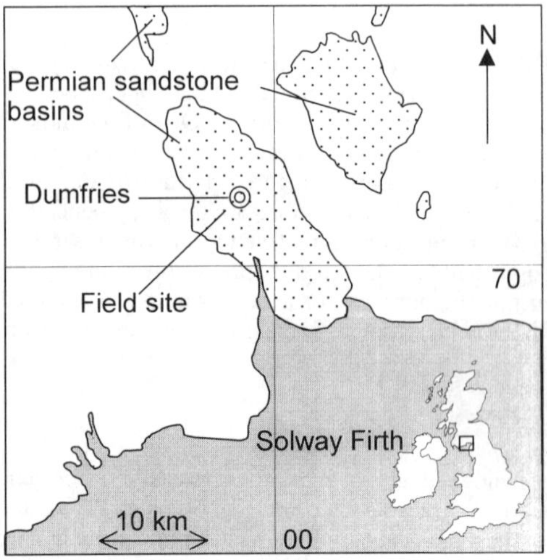

Figure 1. Location of the field site

3 HYDRAULIC TESTING

Conventional hydraulic testing was performed over 15 fracture zones using straddle packer apparatus (Fig. 2a) with a test zone length of 0.73 m. The equipment was lowered down the borehole and the packers inflated over the test zone from a surface nitrogen supply. The borehole was then allowed to equilibrate. Pressure (head) was monitored both in the test zone and above the test zone using electric pressure transducers connected to a digital pressure indicator and a data logger. Pressure (head) was monitored above the test zone in order to detect head changes above the packers during testing, indicating short circuiting around the packers. Slug tests were performed over all the fractures in the borehole by pouring a slug of water into the test zone creating a near instantaneous rise in head. Multiple tests were performed in several zones to check repeatability of results. A typical hydraulic response is shown in Figure 3. This shows the response of test zone 5, lo-

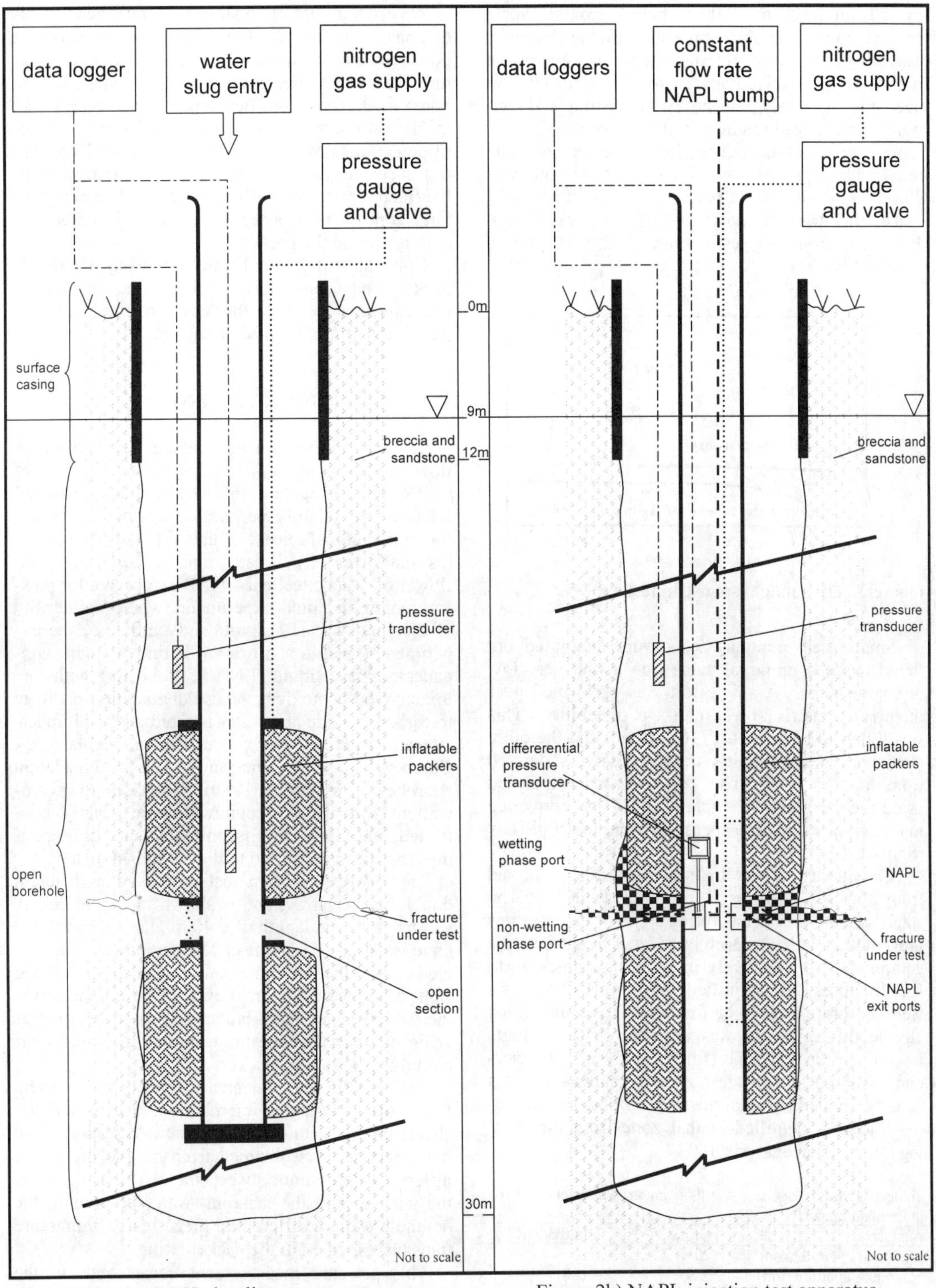

Figure 2a) Hydraulic test apparatus

Figure 2b) NAPL injection test apparatus

cated from 16.27 to 17.00 m below ground over a fracture zone containing three identifiable fractures. The above test-zone pressure transducer records a rise in water level compared to the static column over approximately one hour, suggesting a downward vertical head gradient in this borehole. The response shows rapid equilibration of the test zone due to its high transmissivity. Two slug tests both produced rapid responses back to hydrostatic conditions. Bulk transmissivity for this zone was 1.57 x 10^{-5} m/sec from slug test one and 1.20 x 10^{-5} m/sec from test two.

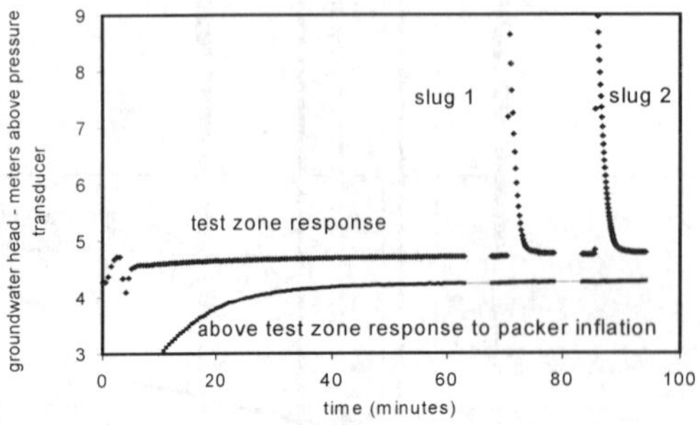

Figure 3. Hydraulic response in test zone 5

Small scale pumping tests were attempted but found to be of minimal value due to the very low discharge rates. A single test was performed over the breccia matrix to test matrix permeability. This was found to be very low (10^{-9} m/sec) and therefore almost all the borehole's transmissivity is derived from fracture flow. The slug tests were analysed using the method of Hvorslev (1951) and converted into equivalent fracture apertures using the cubic law (Snow 1968).

A summary of the hydraulic apertures derived from slug test analysis is given in Table 1. The small spacing distance between some of the fractures prevented isolation of each individual fracture by the equipment. Consequently in some test zones up to four fractures were identified. Two types of analysis were therefore used. The first analysis assumed that all the transmissivity was attributable to a single fracture in the test zone ("gross analysis"): the second assumed that the test zone transmissivity was equally divided between the total number of individual fractures identified in that zone from the ATV log ("individual analysis").

Table 1 Hydraulic aperture results from straddle packer testing in Permain breccia

Analysis	Gross	Individual
Minimum aperture (μm)	20	20
Maximum aperture (μm)	259	180
Mean aperture (μm)	82	74
Number of zones	15	30

A very high (0.27) downward vertical head gradient has been identified on this site, suggesting a poorly connected fracture network with small aperture fractures. The low sustainable yield of the borehole derived from development pumping (< 35 m^3/day) supports this hypothesis. However, a small response was observed above some of the 18 packer test zones during slug testing, showing that some of the fractures are vertically connected. The results of three zones were ignored due to excessive short circuiting around the packers.

Conventional hydraulic testing using a straddle packer setup has established a baseline set of fracture apertures to which the results of the NAPL injection experiment can be compared

4 NAPL INJECTION TESTING

The second experiment generates real two-phase flow conditions in an *in situ* fracture by injection of a DNAPL surrogate into a hydraulically isolated fracture. Regulatory permission has been obtained for an intentional release of up to 50L of triglyceride (as sunflower oil) directly into the aquifer. Sunflower oil is an excellent DNAPL surrogate for pressure controlled field experiments where the density of the fluid will not significantly influence the experimental results. It is a harmless light non-aqueous phase liquid (LNAPL) foodstuff with low toxicity and ecotoxicity whose degradation pathway to carbon dioxide and water is rapid and well understood. Triglyceride is a hydrocarbon regulated by European Community legislation (European Community 1980). Releases of hydrocarbons directly to groundwater are strictly controlled, and this is the first time that such an experiment has been permitted to the authors knowledge.

The NAPL injection test employed a different double packer array, and is shown in Figure 2b. A shorter test zone length of 0.40 m allowed more discrete isolation of fractures. The equipment was lowered down the borehole and the packers inflated from a surface nitrogen supply. The borehole was then allowed to equilibrate. Pressure (head) was again monitored above the test zone to detect short circuiting around the packers.

The NAPL was injected between the packers from surface using a micro-flowrate constant discharge pump with a high pressure capacity. The main data set was obtained from monitoring of the differential pressure between the NAPL isolated by the packers and the standing water column in the borehole using a differential piezo-resistive pressure transducer located in the packer string.

This pressure measurement corresponds to the capillary pressure, being the pressure difference between the non-wetting phase NAPL and the wetting phase groundwater at the same point. The packers were placed such that there was always a continuous NAPL pathway from the pressure monitoring point to the leading edge of the NAPL front in the fracture. Short term changes in the absolute pressure of groundwater due to, say, rainfall or abstractions can affect these measurements. However, this technique of using the standing water column as representative of the wetting phase pressure offers some limited compensation for this. The data were recorded using a surface laptop PC with data logging software.

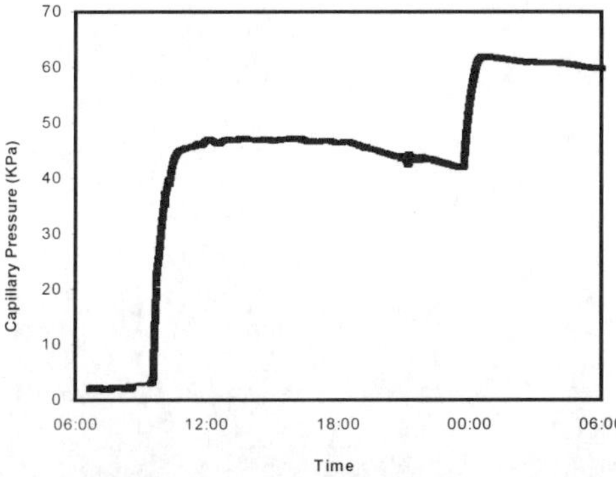

Figure 4. NAPL injection test 1 pressure curve

The results of two experiments are given here. NAPL injection test 1 was performed on a saturated fracture with a gross hydraulic aperture of 46 μm. The capillary pressure verses time curve (Fig. 4.) clearly shows a series of pressure steps which we interpret to be aperture constrictions at increasing radial distance from the point of injection. The later stages of the test are characterised by a relatively constant pressure that enables NAPL to migrate through any aperture constriction in the fracture under test. Steele & Lerner (*in subm.*) describe the concept of oleoporic aperture to explain this phenomenon, whereby the oleoporic aperture describes the smallest aperture constriction where NAPL first forms a fully connected pathway across a variable aperture fracture plane.

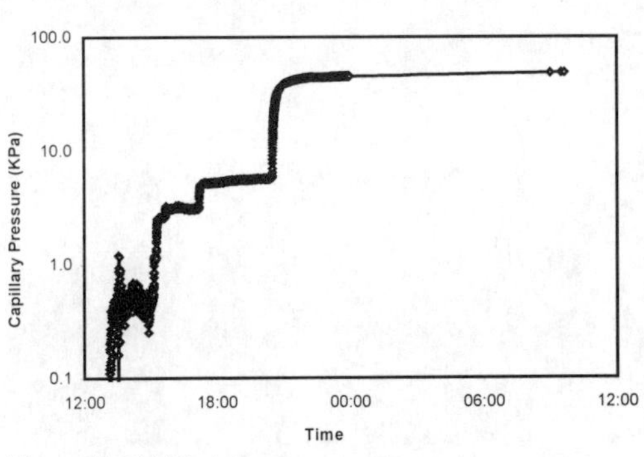

Figure 5. NAPL injection test 2 pressure curve

NAPL injection test 2 was performed on a saturated fracture with a gross hydraulic aperture of 27 μm. The capillary pressure verses time curve (Fig. 5.) shows a large initial pressure step followed by a gradual fall in pressure and a late stage pressure increase. This pattern is interpreted as the NAPL migrating through a near borehole aperture constriction into a large aperture area or void. Once this void is filled, a late stage pressure increase is required to initiate flow through a more distant aperture constriction.

Further work will focus on analysing the data using a numerical two-phase flow model. Matching of experimental capillary pressure curves against time to simulated curves will allow generation of a dataset of fracture apertures and oleoporic apertures from *in situ* two-phase flow tests.

5 DISCUSSION

Hydraulic apertures are small in this borehole and range from 20 μm to 180 μm. The developmental NAPL injection test presented here enables much more detailed resolution of variations in fracture aperture than is possible using hydraulic testing, including clear evidence for significant aperture constrictions and enlargements *in situ*. Work is underway to compare the results of the hydraulic testing with those of the NAPL injection experiment, specifically to determine if hydraulic apertures are a representative approximation of true fracture apertures derived from the NAPL injection testing. This would be significant for field investigations as hydraulic testing is a quick and cost effective method of determining a fracture aperture value, where as NAPL injection testing is more lengthy, detailed, expensive and not appropriate for all sites. Reverse numerical modeling of the NAPL injection tests is underway to determine the oleoporic aperture for each test, the controlling aperture for DNAPL migration in fractures. Detailed site characterisation such as this NAPL injection experiment will give increased confidence in the prediction of DNAPL penetration depths and mass distribution in the subsurface, and better populate the increasingly sophisticated modeling codes now available to simulate DNAPL transport in fractured rocks.

6 ACKNOWLEDGEMENTS

Funding for this work came from the UK Engineering and Physical Sciences Research Council, the UK Environment Agency and ICI Chemicals and Polymers Ltd. DuPont (UK) Ltd and the British Geological Survey are thanked for their support.

REFERENCES

Ball, D. F. 1997.The Dumfries Basin Aquifer: Hydrogeology In , Fox, I. (ed), *Upland Groundwater;* Meeting of the UK Groundwater Forum, Edinburgh, 23 October 1997. UK Groundwater Forum.

European Community 1980. EC Directive on the Protection of Groundwater Against Pollution Caused by Certain Dangerous Substances (80/68/EEC).

Hinsby, K., McKay, L. D., Jorgensen, P., Lenczewski, M. and Gerba, C. P. 1996. Fracture Aperture Measurements and Migration of Solutes, Viruses, and Immiscible Creosote in a Column of Clay-Rich Till. *Ground Water,* 34, 1065-1075.

Hvorslev, M. J. 1951. Time lag and soil permeability in ground-water observations. *Waterways Experiment Station report, US Army Corps of Engineers,* 36, 50p.

Jorgensen, P. R., Broholm, K., Sonnenborg, T. O. and Arvin, E. 1998. DNAPL transport through macroporous, clayey till columns. *Ground Water,* 36, 651-660.

Keller, A. 1998. High resolution, non-destructive measurement and characterization of fracture apertures. *International Journal of Rock Mechanics and Mining Sciences,* 35, 1037-1050.

Murphy, J. R. and Thomson, N. R. 1993. 2-Phase Flow in a Variable Aperture Fracture. *Water Resources Research,* 29, 3453-3476.

Reitsma, S. and Kueper, B. H. 1994. Laboratory Measurement of Capillary-Pressure Saturation Relationships in a Rock Fracture. *Water Resources Research,* 30, 865-878.

Robins, N. S. and Buckley, D. K. 1988. Characteristics of the Permian and Triassic Aquifers of Southwest Scotland. *Quarterly Journal of Engineering Geology,* 21, 329-335.

Snow, D. T. 1968. Rock Fracture Spacings, openings and porosities. *Journal of Soil Mechanics, Foundations Division, Proceedings of the American Society of Civil Engineers,* 94, 73-91.

Steele, A. and Lerner, D. N. *in subm.* Oleoporic aperture: a better approximation of the controlling aperture for DNAPL migration in fractures. *Submitted to the Journal of Contaminant Hydrology.*

Sudicky, E. A. and Slough, K. J. 1998. DNAPL migration in fractured porous rock media: parameter sensitivity, prediction uncertainty and implications for remediation (eds), *GQ98: Groundwater Quality: Remediation and Protection* 250 International Association of Hydrological Sciences, Tubingen, Germany, 157-165.

Wealthall, G. P. & Lerner, D.N. 2000. A volume-based approach to predicting the fate of DNAPLs in fractured aquifers. In *Groundwater: Past achievements and future challenges*; Proc. Int. Assn. Hydrogeologists XXX Congress, Capetown, 26 November - 1 December 2000. Rotterdam: Balkema.

Yeo, I. W., DeFreitas, M. H. and Zimmerman, R. W. 1998. Effect of shear displacement on the aperture and permeability of a rock fracture. *International Journal of Rock Mechanics and Mining Sciences,* 35, 1051-1070.

Zhou, Z. 1995.Migration behaviour of dense non-aqueous phase liquids in water-saturated fractured rock, PhD thesis. Imperial College of Science and Technology, London.

Groundwater: Past Achievements and Future Challenges, Sililo et al. (eds) © 2000 Balkema, Rotterdam, ISBN 90 5809 159 7

Experiences with an in situ groundwater remediation of VOCs in Hamburg, Germany

J.Taugs
Taugs Geo-Consult, Bendestorf, Germany

ABSTRACT: On a former industrial site in Hamburg, northern Germany, groundwater contamination with trichlorethylene, tetrachlorethylene, 1,2-cis-dichlorethylene and vinyl chloride as main pollutants has to be treated. Special hydrochemical conditions make a conventional "pump-and-treat" hydraulic remediation highly uneconomic. So it was decided to use the C-sparge^{TM} process, a combination of insitu air stripping with encapsulated ozone, to oxidize the contaminants. After a pilot test, groundwater remediation started in January 1999. Experiences with the C-sparge^{TM} process during the pilot test and actual results of the groundwater remediation are presented.

1 INTRODUCTION

In urban regions with a long industrial history, groundwater contaminated with dissolved volatile organic chemicals (VOCs) is quite common. Conventionally, the treatment of polluted groundwater is done by extracting the contaminated water, treating it on site above ground, and re-injecting or discharging the clean water. The extracted pollutants must be disposed of separately. These technologies known as "pump-and-treat" require considerable investment over a long time (in some cases 30 years or longer), and may not clean up the source of the contamination. Generally this is not very satisfactory, from neither ecological nor economical points of view. Consequently, there is considerable interest and effort being expended on alternative, innovative treatment technologies for contaminated groundwater.

Investigations on a former industrial site in Hamburg, northern Germany, showed a groundwater pollution with chlorinated volatile organic chemicals (VOCs). The contamination of the aquifer, consisting of Quaternary fine to coarse grained sand and gravel, covers an area of more than 40,000 m². Main contaminants are trichlorethylene (TCE), tetrachlorethylene (PCE), 1,2-cis-dichlorethylene (DCE) and vinyl chloride (VC). Maximum concentration of VOCs measured before starting the groundwater remediation was about 2,100 µg/l. DNAPLs were not found.

Due to special hydro-chemical conditions in the contaminated area, a conventional "pump-and-treat" hydraulic remediation could be carried out only with an uneconomically high input of material and energy. So it was decided to use an innovative in situ techno-logy for groundwater remediation. The chosen C-sparge^{TM} process consists of a combination of in situ air stripping with encapsulated ozone to oxidize the contaminants. After a pilot test where the C-sparge^{TM} process was tested successfully in the contaminated area in Hamburg, groundwater remediation started in January 1999. Experiences with the C-sparge^{TM} process during the pilot test and actual results of the VOC elimination are presented.

2 CASE STUDY

In the northwestern part of Hamburg a house building project started on a former industrial site for manufacturing of rubber products. Investigations of the subsoil, soil air and groundwater showed contamination with VOCs. Groundwater remediation was demanded by the responsible water authority and arranged by contract with the owner of the site.

The detected groundwater contamination covers an area of more than 40,000 m². The Quaternary aquifer consists of Saalian meltwater sands and gravels covered by near surface anthropogenic fill material followed by a glacial till layer. In monitoring wells piezometric heads of 10 to 12 m below surface were measured. The thickness of the saturated zone ranges between 12 and 15 m. In the area of investigation groundwater normally is unconfined except those places where locally a thin interposed silt layer is present. In the course of pumping tests k_f-values of 2.5 x 10^{-4} m/s to 4 x 10^{-4} m/s have been found. Groundwater flow velocity is about 25 m/a. In the polluted area groundwater flow has an easterly to northeasterly direction.

In order to explore the extent of groundwater pollution a net of 20 monitoring wells has been installed. The main contaminants detected are the VOCs TCE, PCE, DCE and VC. Other organic compounds, as mononucleous aromatic hydrocarbons (BTEX), polycyclic aromatic hydrocarbons (PAH) and mineral oil hydrocarbons were detected intermittently and in slight traces, so they are of secondary importance in the contaminated area.

Field investigations suggested the existence of various pollutant sources on the former industrial site. In the western area of the site, a former storage house for TCE was located. Therefore TCE at a concentration of about 220 µg/l is found in groundwater at this place. In the eastern area of the site however in the former production plant a degreasing facility was operated where PCE was used. The maximum concentrations of PCE in groundwater are around 200 to 220 µg/l as well. The decomposition product DCE was found with the highest single compound concentration in all of 1200 µg/l. VC was measured at various monitoring wells showing concentrations from 1.9 µg/l up to 45 µg/l. Indications of pollutant entries in phase (DNAPL) have not been found during the investigations. The measured concentration of VOCs exceed the official limits in Hamburg, which are for VOCs in total 10 µg/l, for VC as carcinogenic single compound 1 µg/l.

Besides the organic pollutants, inorganic main and trace substances have been analyzed in order to get more information about the general groundwater quality relevant to the planned remediation. Groundwater in this region is characterized by reducing conditions. Beyond that groundwater shows a high portion of hardening agents. The determined total hardness is between 33.8 and 84.7 °dH. Corresponding to these results are sulphate concentrations between 390 mg/l and 1100 mg/l. In addition to these extraordinary high sulphate concentrations also high concentrations of chlorides and iron may have disturbing consequences for water treatment. Chloride values between 36 mg/l and 190 mg/l as well as iron values between 5.7 mg/l and 94 mg/l had to be taken into account in the remediation process technology design.

3 THE C-SPARGE™ PROCESS

The reported special hydro-chemical conditions in the contaminated area forced a decision against a conventional "pump-and-treat" hydraulic remediation procedure. So a technology was chosen allowing an in situ elimination of VOCs in groundwater, the C-Sparge™ process.

Via special C-Sparge™ wells an ozone-air mixture is blown into the aquifer. By means of installed in well pumps a hydraulic circulation is generated around the well. Micro bubbles of the ozone-air mixture are injected in the circulating stream. The magnitude of bubbles is dimensioned that way, that they can infuse the strata without fracturing it. Magnitude of the bubbles as well as their flow velocity are controlled by so called Sparge units in the C-Sparge™ wells and by injection pressure.

During the C-Sparge™ process at first the VOCs dissolved by in situ air stripping are extracted from aqueous solution into the gas phase of the bubbles because of their relative high vapor pressure (Kerfoot 1997). The extracted VOCs react with the encapsulated ozone in a gas/gas reaction following the Criegee mechanism as a primary ozonide. This ozonide is very unstable and forms a carboxyl oxide which reacts with water (hydrolysis) after exiting the bubble giving the reaction end products Cl^-, H_2O and CO_2.

This process focuses ozone reaction selectively to air strippable compounds which invade the bubbles. That way the strong oxidation power of ozone is restricted to a great extent to the volatile compounds. That means, if the encapsulated ozone concentration is maintained at a low multiplier of the strippable VOCs, in contrast to the usual oxidative waste water treatment, no additional ozone is available for side reactions with other dissolved organic compounds with a low Henry's number.

Other oxidizing substances in groundwater (e.g. iron) can be transformed as well. The generated (hydr-)oxides are precipitated over the entire strata lying in the range of the hydraulic circulation, so that relevant changes of hydraulic parameters in the aquifer are not produced as proved by pumping tests performed before and at the end of several C-sparge™ tests. The yielded iron hydroxides for instance preferentially first deposit in the hydraulically ineffective "dead end" pores (Groth et al. 1997).

4 PILOT TEST

To test the principal applicability of the C-Sparge™ process under the specific local conditions a pilot test has been executed by Mateboer Milieutechniek B.V., Netherlands, lasting 18 days in November 1997.

The test comprised the installation of a C-Sparge™ well (diameter: 273 mm) consisting of an in ground sparge point, an in well sparge point, a packer and a fluid pump, and the installation of six monitoring wells (diameters: 50 mm) at a distance of about 10 to 15 m around the C-Sparge™ well.

The double screened C-Sparge™ well was installed to 27 m below surface. The integrated pump had to generate the hydraulic circulation in the aquifer. The ozone-air mixture was blown out of the lower screen with a pressure of 1.8 bar during the test, whereas the re circulating water should enter the C-Sparge™ well through the upper screen.

Three monitoring wells were installed to 12 m, 15 m and 16.5 m below surface with special screen tubes in the lowest 1.5 m. There KVA model 40 GeoFlo® instruments were installed to record water movement during the test. The highly sensitive GeoFlo® instruments use the heat pulse principle. The correct initiation and the range of the hydraulic circulation should be determined with this equipment.

Three further monitoring wells were installed to 22.5 m below surface screened over the total saturated zone for taking water samples during the test period.

About 10 g/h ozone were brought into the aquifer. During the 18 days pilot test period in total 4.32 kg ozone were injected.

Control units (ozone- and oxygen generator, compressor, control system) were stored in a container near the test field.

Measurements of VOCs, pH value, conductivity, redox potential, dissolved oxygen, iron, temperature, pressure changes, piezometric heads, groundwater flow were carried out before, during and after the pilot test. Moreover samples of soil air were taken from the unsaturated zone to measure a possible discharge of pollutants and the ozone content of the soil air. Also a pumping test was performed before and at the end of the pilot test to determine the influence of precipitated oxidized compounds, here especially iron, on the hydraulic parameters of the aquifer.

Already after a few hours of test time, changes in oxygen content could be observed in the C-Sparge™ well and in two monitoring wells. The oxygen contents in all monitoring wells after one week corresponded to an about 25 m wide and 30 m long zone of influence in the aquifer with oxygen contents of more than 0.5 mg/l.

The absence of micro bubbles in the upper screen area of the C-Sparge™ well and in those monitoring wells screened in the upper region of the aquifer during the test period were indications for an evidently incompletely developed hydraulic circulation in the test field. Assuming a normal circulation development micro bubbles would have been transported with the circulating water flow in the upper area of the aquifer and to the upper screen of the C-Sparge™ well.

VOC concentrations first increased in the C-Sparge™-well during the test period from initially 395 µg/l to more than 1200 µg/l. Until test end VOC content decreased to values distinctly below the initial concentration values. In one of the monitoring wells screened in the lower part of the aquifer the VOC concentration decreased from 520 µg/l to 341 µg/l during the test period whereas concentrations in the upper parts of the aquifer increased.

Phenomena relating to vertical permeability could be attributed to a small silt layer with low hydraulic conductivity disturbing the expected development of water circulation, which led to the following effects during the pilot test:

1. Obviously only a small amount of bubbles could reach the upper screen of the C-Sparge™ well. Consequently a weak circulation and a very poor removal activity in the upper screen area resulted.

2. Due to the silt layer the bubbles blown out of the lower screen of the C-Sparge™ well could only intrude the strata of the lower part of the aquifer. So removal activity could be observed only in the monitoring wells screened in the lower parts of the aquifer.

3. Because no complete circulation was achieved, increasing VOC concentrations due to the general mobilization of pollutants caused by pumping without noticeable elimination were observed in the upper parts of the aquifer.

Though the pilot test was performed for only 18 days and the generated water circulation was not optimal a certain elimination of VOCs was achieved. The observed removal rate of VOC was about 1600 µg/l of VOC during the test period.

The elimination of dissolved volatiles with the C-Sparge™ process is an oxidative decomposition reaction of volatile compounds in groundwater forced by the injection of the ozone-air mixture. It is assumed that the process of removal of dissolved volatiles is similar to detoxification in human bodies. Elimination and detoxification processes correspond to first order reactions where the rate of decrease in concentration of the toxic substance is directly proportional to the concentration of the substance. The direct relation between rate of elimination and the concentration of the dissolved volatile compounds can be expressed in the following differential equation:

$$dc / dt = -bc \qquad (1)$$

where dc = change in concentration of dissolved volatile, dt = change in time, b = fraction of volatile substance that is removed from groundwater in one unit of time, c = concentration of volatile compound.

To determine the actual amount of substance eliminated per unit of time, the initial concentration was compared to later concentrations at increasing time intervals from start of operation. The amount of material eliminated is obtained from

$$Bc\,(dc / dt) = fc \qquad (2)$$

where Bc = volume of groundwater block containing the chlorinated substance, c = concentration of volatile substance in groundwater (in µg/l or ppb), f = a constant.

The solution of the differential equation is given by:

$$c = c_0\, e^{-bt} \qquad (3)$$

where e = Euler number, c_0 = initial concentration of volatile organic compound in groundwater (µg/l).

The exponential term e^{-bt} expresses the phenomenon of the "exponential decay". Using the results of a graph representing the solution of the equation the yielded removal rate could be roughly estimated under the observed local circumstances during the pilot test.

Beside VOCs other oxidizing substances like dissolvable Fe(II) or Mn(II) react as well with the injected ozone-air mixture and are precipitated as Fe hydroxide and Mn oxide in the range of influence of the circulating stream. During the pilot test near to the C-Sparge™ well about 5.1 mg/kg Fe hydroxide had been precipitated, around the monitoring wells about 0.9 mg/kg Fe hydroxide. The amount of manganese precipitated could be neglected (on average about 0.05 mg/kg Mn oxide). Principally the reactions of other oxidizing substances beside the VOCs do not disturb the remediation progress because the reactions with VOCs take place in the gas phase as a gas/gas reaction and not in water phase like the oxidation of the other substances. Therefore, also in presence of Fe(II) ions the removal of VOCs may proceed.

5 GROUNDWATER REMEDIATION

After evaluating the pilot test data and under consideration of the quite homogeneous strata in the contaminated area a successful application of the C-Sparge™ process could be expected in Hamburg. So it was decided to start groundwater remediation with this in situ technology. As it was the first time the C-Sparge™ process should be applied in Germany it lasted about half a year to get the required allowance from the responsible water authority. Mainly discussed problems were the injection of ozone into the groundwater, the oxidation and precipitation of other substances in the groundwater and possibly resulting changes of hydraulic properties of the aquifer.

After detailed planning the groundwater remediation was designed to be carried out with 26 C-Sparge™ wells and 27 monitoring wells to cover the whole contaminated area of about 40,000 m². In addition 5 monitoring wells screened over the whole saturated zone had to be installed requested by the water authority. The operation of the C-Sparge™ wells will be controlled by three control units. The wells and the control units were installed till end of 1998. Groundwater treatment executed by Mateboer Milieutechniek B.V., Netherlands, started in January 1999. A remediation time of one year and a half was projected. It is expected that a full scale treatment will bring the observed VOC concentrations to a level between acceptable commercial and drinking water quality.

To fulfill the conditions of the given allowance a variety of physical and chemical measurements, e.g. groundwater head, groundwater flow, temperature, pH, Fe, dissolved oxygen, VOCs including PCE, TCE, DCE and VC, have to be performed every six weeks including all C-sparge™ and monitoring wells. Though the monitoring program is very expensive, it is accepted as necessary to control the complex in situ reaction processes during C-sparging under full scale treatment conditions.

After starting the C-sparge™ process in January 1999 at first a certain increase of VOC concentrations could be observed in several wells, obviously caused by mobilization processes forced by the initiated water circulation around the C-sparge™ wells. But until summer 1999, after six months of treatment, a significant and continuous decrease of VOC concentrations of about 50 to 60 % could be seen in most of the C-sparge™ monitoring wells.

Unfortunately this positive trend was stopped. The VOC concentrations increased rapidly. In some cases the initial concentrations were observed again. Looking for explanations for this unexpected development it was found that after more than six months of treatment the generation of ozone was disturbed massively. There are several hints that a sufficient ozone supply could not be realized because of technical problems, so that most probably only air was injected into the aquifer instead of the required ozone-air mixture. Some of the installed sparge-units were however clogged with precipitated Fe-hydroxide. Beyond these technical complications also other problems occurring during the groundwater treatment had to be analyzed. Another reason for rising VOC concentrations may be locally high amounts of fine grained charcoal particles in the saturated zone. The charcoal, which is quite common in this contaminated Quaternary strata reacts with the injected ozone-air mixture so that more ozone (than injected into the aquifer) may be needed to force the desired VOC removal.

At the beginning of the year 2000 the situation seems to have stabilized. VOC concentrations decrease again. But it is obvious that the projected treatment time of one and a half years cannot be met.

REFERENCES

Groth, P. Czekalla, C. Dannöhl, R. Kölle, W. Liebfeld, R. Meyerhoff, R. Olthoff, R. Rott, U. & Wiegleb, K. 1997. Unterirdische Enteisenung und Entmanganung – aktualisierter Statusbericht. Gwf Wasser Abwasser. 138 (4): 182-187.

Kerfoot, W.B. 1997. Extremely rapid PCE removal from groundwater with a dual gas microporous treatment system. In P.T. Kostecki, E.J. Calabrese & M. Bonazountas (eds), *Contaminated soils, Vol. 2.*: 275-284. Amherst (MA): ASP Publishers.

Groundwater: Past Achievements and Future Challenges, Sililo et al. (eds) © 2000 Balkema, Rotterdam, ISBN 90 5809 159 7

A volume-based approach to predicting the fate of DNAPLs in fractured aquifers

G.P. Wealthall & D.N. Lerner
Groundwater Protection and Restoration Group, Department of Civil and Structural Engineering, University of Sheffield, UK

ABSTRACT: Many of our aquifers are polluted by dense non-aqueous phase liquids (DNAPLs). Their complex distribution in the subsurface is governed both by the fluid properties and the geological structure. Rapid migration is promoted by the presence of fractures, which are ubiquitous in the majority of bedrock aquifers. However, heterogeneity, inherent at a range of scales in all fractured rocks, suggests that accurate prediction of the fate of DNAPLs will be limited by our abilities to measure, interpret and represent key fracture parameters. This paper presents a framework for estimating DNAPL penetration depths, and reports new fracture data from a field research site. A novel approach is introduced for predicting DNAPL penetration depths, based on the estimation of potential flow path volumes from detailed fracture network characterisation and discrete fracture network modelling. The development of a 3D invasion percolation model, which will interface with the stochastic fracture network simulator, is described.

1 A VOLUME-BASED CONCEPT

The connected fracture volume (porosity) in fractured porous media (dual porosity) is typically two to three orders of magnitude lower than the porosity of the matrix of equivalent porous media. The respective porosities are a measure of the potential storage capacity of the two systems for DNAPLs. However, the resulting distribution of the non-aqueous phase liquid will be markedly different for each system. In the fracture-dominated system the release volume has the tendency for lateral spreading and significant vertical penetration.

This study introduces a new approach to estimating the depth of penetration of DNAPLs in fractured aquifers. This approach calculates penetration depth by determining the specific volume of DNAPL required to traverse a given rock mass (m^3 DNAPL per m^3 rock) - or the specific connectivity. Whether a DNAPL will cross a given volume of rock therefore becomes a function of the size and type of the spill, and the available (connected) fracture volume.

2 FRACTURE NETWORK CHARACTERISATION

A field research facility has been established to the west of Dumfries Scotland, UK. The site was selected based on *a priori* knowledge of the geology underlying the site, and the proximity to a disused railway cutting, which provides a 300-400 m exposure of bedrock.

The research site is situated on the western margins of the Dumfries Basin and is underlain by Permian and Triassic sediments, which rest unconformably on Lower Palaeozoic strata (Robins & Buckley 1988). Faulting and subsidence associated with the western boundary fault has resulted in the formation of wedge-shaped breccias (Doweel Breccia Formation) which thin to the east. These sediments form the focus of this study.

The Dumfries aquifer is a highly productive source of potable water serving private, public and industrial users. A large proportion of the flow is derived from secondary permeability associated with bedding-parallel and high-angled fractures. Elongate drawdown cones have been observed locally (Robins & Buckley, *op cit.*) and result from anisotropic secondary permeability, which is aligned subparallel to the adjacent NW-SE trending western boundary fault.

The Doweel Breccia Formation typically has low intergranular permeability and high secondary (fracture) permeability These hydraulic characteristics are similar to the Chalk aquifer of southern England, where the ratio of laboratory (matrix) to field (fractures and matrix) permeability is greater than 1:1000 (Price 1996).

Initial characterisation of the fracture network involved fracture mapping at outcrop, and drilling and hydraulic testing of a fully cored borehole to 30 m depth. The aim of this work was to acquire fracture properties for the fracture network modelling study, and to derive equivalent parameters for the invasion percolation model.

2.1 *Fracture mapping*

The successful interpretation of fracture data is an iterative process that progresses by data collection and synthesis, the development of a conceptual model, and construction of a numerical model (Jones et al. 1999). Description of the fracture parameters for fracture network analysis is based on the separation of data into discrete sets, which are described by their statistical properties and by approximation to theoretical distributions.

Quantitative field measurements of fractures were obtained using 1D scanline surveys, 2D tracemaps (Priest & Hudson 1981; Gillespie et al. 1993), and aperture profiling (Renshaw 1995). In addition to measuring fracture orientation and intensity, the data were recorded using the criteria for fracture network modelling recommended by Jones et al. (1999).

Initial interpretation of the fracture properties has focused on the statistical analysis of the field data, and fracture set identification based mostly on orientation data. This has led to the development of a conceptual model of the fracture system, as a precursor to construction of a numerical model.

The formation exposed in the disused railway cutting represents an analog for fracture distribution in the same formation at shallow depths under the field research site. The cutting is located approximately 1 km to the south west of the research site.

The fracture distributions mapped in the Doweel cutting (Figure 1) are characterised by 4 fractures sets (1 bedding-parallel, 3 high-angled); bedding-parallel fractures (typically >10 m in length); high-angled contained fractures and multi-layer fractures (length uncertain, but typically >8 m); mean block-size of 3.3 x 0.7 m; sub-millimetre fracture aperture: and 16% fractures are infilled.

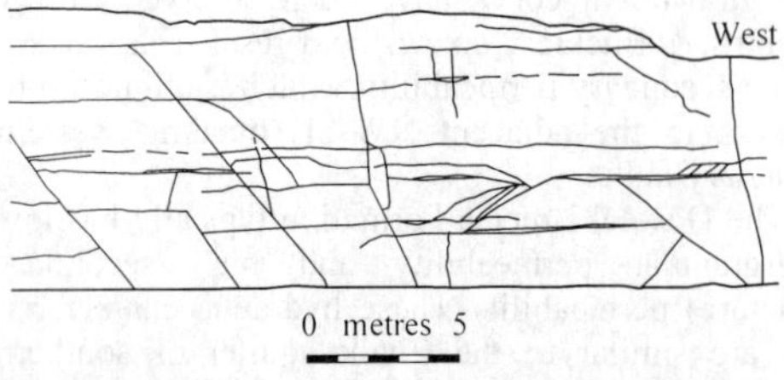

Figure 1. Tracemap of fracture distributions in the Doweel cutting. The figure illustrates the elongate (lozenge-shaped) block structure, and censoring of through-going high-angled fractures.

Further statistics for fracture spacing and fracture aperture data are presented in Table 1.

Table 1. Fracture spacing (m) and fracture aperture (mm) statistics from the Doweel cutting

	Population	Mean	Standard deviation	Minimum	Maximum
Spacing (ha)	40	3.29	2.70	0.15	10.85
Spacing (bp)	12	0.67	0.31	0.14	1.20
Aperture (ha)	25	0.41	0.21	0.00	0.80
Aperture (bp)	73	0.76	0.79	0.00	5.00

ha High-angled fractures, bp Bedding-parallel fractures

The cutting is 7-9m in height, was blasted during construction, and has undergone weathering and stress unloading due to erosion and the retreat of ice sheets. It is suggested that measurements of fracture orientation and spacing are a reasonable reflection of the *in situ* distributions of the fracture network, due to pseudo-3D mapping of fractures on both sides of the cutting. However, the values derived from the aperture profiles will be biased towards larger fracture apertures, as a result of the reduction of overburden pressure, and should be interpreted with caution.

A fully cored borehole of 130 mm nominal diameter (PWF) was drilled to 30.8 m depth. The borehole is located approximately 1 km north-east of the Doweel cutting. The lithological succession in the borehole comprises 25.1 m of fractured breccias and sandstones, overlain by 5.7 m of glacial drift. The upper 8.15 m of the sandstone-breccia sequence is weathered, this led to wash-out of the borehole and precluded hydraulic testing above c. 14 m depth.

Fracture distribution has been determined from a combination of core and down-hole Acoustic TeleViewer (ATV) logs. This approach has proved to be essential for discriminating drilling-induced core breaks from *in situ* fractures in a heterogeneous matrix. Fracture intensity ranges from 0 to 8 fractures m^{-1}. Mean fracture spacing is 0.54 m.

Discrete fracture network modelling is currently underway using the stochastic fracture network simulator FracMan (Dershowitz et al. 1998).

2.2 *Hydraulic testing*

Hydraulic testing was undertaken to discriminate hydraulically active components of the fracture network, and to derive cubic law equivalent fractures apertures (here termed "equivalent aperture"), for use in the fracture network and invasion percolation modelling. Additional studies are being undertaken at the field site (Steele & Lerner 2000) to compare apertures derived by injection of a DNAPL surrogate with those obtained from the hydraulic testing.

A double-packer system (Price et al. 1982) was used to characterise the hydraulic properties of the rock mass to a depth of 30 m. The test-zone (inter-packer) spacing was set to 0.73 m. Test zone locations were determined from the combined core-log and ATV fracture picks.

The borehole was developed using a submersible pump at a flow rate of 0.4 L s^{-1} (ca. 35 m^3 d^{-1}). Falling head slug tests were used to determine hydraulic conductivity and the results were analysed by the method of Hvorslev (1951). The measured matrix hydraulic conductivity was 8×10^{-10} m s^{-1} and lower, therefore test zone transmissivity was assumed to result primarily from the fracture system.

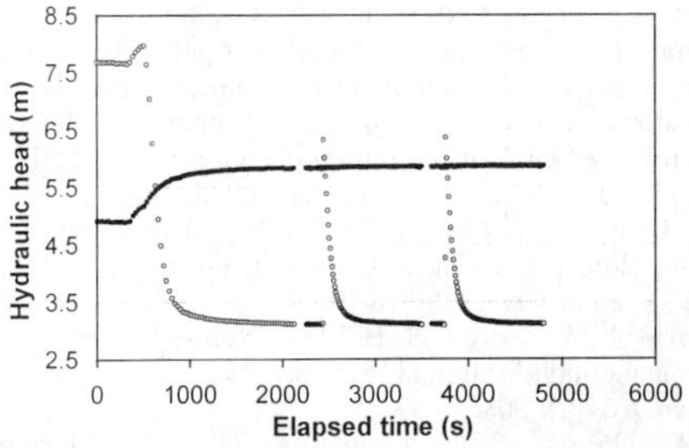

Figure 2. Plot showing test zone (open circle symbol) and above test zone (solid triangle symbol) response following packer inflation (400 seconds) and two falling head slug tests ($H_0 \approx 2500$ seconds and 4000 seconds). The test zone isolates a single fracture at 23.98 m depth with a transmissivity of 9.5×10^{-6} m^2 s^{-1}, and an equivalent aperture of 248 μm. The test zone response records a low head zone, relative to the open borehole head, whereas the above test zone head increases, indicating a high downward gradient.

Two estimates of transmissivity were resolved. Firstly a 'test zone' transmissivity was determined from the contribution of all of the fractures in the test zone, and secondly an 'individual' transmissivity was calculated, assuming an equal contribution from all fractures in the test zone, by dividing 'test zone' transmissivity by the number of fractures in the test zone. Equivalent aperture values were determined from both 'test zone' and 'individual' transmissivities. The aperture values are given in Table 2. A spreadsheet-based analysis package, which autofits the recovery data, was developed to derive transmissivity and equivalent apertures.

Table 2. Fracture aperture (μm) derived from 'test zone' and 'individual' transmissivities

	Population	Mean	Standard deviation	Minimum	Maximum
'Test zone'	26	74	82	10	248
'Individual'	26	65	81	6	248

The low sample population precludes fitting to a theoretical distribution. Further hydraulic testing, planned for subsequent field studies, will increase the sample population and increase confidence in the data.

3 3D INVASION PERCOLATION MODEL DEVELOPMENT

A fully 3D macroscopic-scale invasion-percolation model, which utilises the output from the fracture network simulator, is being developed. The model will test the hypothesis of DNAPL connectivity, by predicting penetration depths based on spill volume and available fracture porosity. The model incorporates some of the physics of two-phase flow, but assumes that the capillary forces are dominant and that viscous forces are negligible.

Invasion is initiated from a non-wetting NAPL discrete volume source with a given density and interfacial tension (IFT). The invasion front progresses by successively invading connected fractures where the capillary pressure (P_c) of the invading fluid exceeds the threshold (entry) pressure (P_e) of the given fracture (Kueper & McWhorter 1991). The capillary pressure (1) and entry pressure (2) are defined by the following expressions:

$$P_c = (\rho_{nw} - \rho_w)gh \tag{1}$$

where ρ_{nw} is the non-wetting phase density; ρ_w is the wetting phase density; g is the gravitational constant; and h is the height of the pool of DNAPL

$$P_e = \frac{2\sigma\cos\theta}{e} \tag{2}$$

where σ = interfacial tension between the DNAPL and water; θ = contact angle on the fracture walls; and e = fracture aperture.

At present the fractures are represented by homogenous 2D planes, based on a parallel-plate model assuming constant aperture, with an inter-model aperture distribution which is a function of the statistical input parameters of the fracture network simulator.

The non-wetting phase residual saturation (S_{NWr}) is determined as a function of dip (Longino and Kueper 1999) for individual pathway fractures. End-members for residual saturation are: dip 0 ° $S_{NWr} \approx$ 21 %, and dip 90° $S_{NWr} \approx 3.2$ %. The derived bulk retention capacity (a function of S_{NWr}) is then used to estimate DNAPL penetration depths by comparison with given spill volumes.

A number of modelling scenarios are envisaged based on spill volume, spill area, and variable fluid properties. Future developments aim to include accessibility of the resident fluid through the fracture walls to the matrix, applying aperture distributions to individual fractures, and a visual model output.

4 DISCUSSION

This study will develop a methodology for estimating DNAPL penetration depths in fractured aquifers, based on parameters which can be measured at most sites which have been impacted by DNAPLs. The approach to estimating penetration depths, using the volume-based DNAPL connectivity concept, gives an indication of the likely impact of DNAPL spills, and will aid risk assessment.

Modelling the migration of limited volume DNAPL spills in fractured systems requires detailed representation of the fracture system. Clearly, this can not be achieved through the use of REVs or equivalent porous media approaches. Therefore, a stochastic fracture network model is employed to provide discrete representation of the characteristics of heterogeneity in the fractured rock mass.

The invasion percolation model is based on the output from the 3D fracture network simulator. Description of two-phase flow in the model is subject to the limiting assumptions of both models. Therefore, future work will consider the following issues:

1. Increasing confidence in calculating fracture volumes from fracture network models. This includes identification of the critical scales and types of measurements needed to constrain realistic representations of fracture systems, and determining how unknown microscopic variations in aperture geometry can be represented at the macroscopic scale.
2. The significance of matrix entry in the early invasion of a DNAPL in a fractured porous aquifer. particularly in coarse-grained sandstones with high intergranular permeability and large pore throat diameters
3. Determining if spills are small enough and adequately documented to constrain the source term input to the invasion-percolation model.
4. Assessing whether problem owners and their consultants can acquire the required data cheaply enough for practical site investigation.

ACKNOWLEDGEMENTS

The authors gratefully acknowledge the Engineering and Physical Sciences Research Council, the Environment Agency, and ICI Chemicals and Polymers for funding this study. Golder Associates Limited have provided their fracture network code (FracMan) and continuing advice. We are also grateful to DuPont, UK, Limited for access and support at the field research site.

REFERENCES

Dershowitz, W., Lee, G., Geier, J., Foxford, T., LaPointe, P. & Thomas, A. 1998. *FracMan. Interactive discrete feature data analysis, geometric modelling, and exploration simulation.* User documentation, version 2.6. Golder Associates Inc. Seattle, Washington.

Gillespie, P., Howard, C. B., Walsh, J. J. & Waterson, J. 1993. Measurement and Characterisation of Spatial Distributions of Fractures. *Tectonophysics* 226: 113-142.

Hvorslev, M. J. 1951. Time lag and soil permeability in ground-water observations. Waterways Experimental Station, Corps of Engineers, United States Army. Bulletin Number 36: 50 pp.

Jones, M. A., Pringle, A. B., Fulton, I. M. & O'Neill, S. 1999. Discrete Fracture Network Modelling Applied to Groundwater Resource Exploitation in Southwest Ireland. In: McCaffrey, K. J. W., Lonergan, L. & Wilkinson, J. J. (eds). *Fractures, Fluid Flow and Mineralization.* Geological Society, London, Special Publications 155: 88-103.

Kueper, B. H. & McWhorter, D. B. 1991. The Behavior of Dense, Nonaqueous Phase Liquids in Fractured Clay and Rock. *Ground Water* 29; 716-728.

Longino, B. L. & Kueper, B. H. 1999. Nonwetting phase retention and mobilisation in rock fractures. *Water Resources Research* 37(7): 2085-2093.

Price, M. 1996. *Introducing groundwater 2nd edition.* Chapman & Hall, 2-6 Boundary Row, London. SE1 8HN, UK.

Price, M., Morris, B & Robertson, A. 1982. A study of intergranular and fissure permeability in Chalk and Permian aquifers, using double-packer injection testing. *Journal of Hydrology* 54: 401-423.

Priest, S. D. & Hudson, J. A., 1981. Estimation of Discontinuity Spacing and Trace Length Using Scan-Line Surveys. *International Journal of Rock Mechanics, Mining Sciences and Geomechanical Abstracts* 18: 183-197.

Renshaw, C. E. 1995. On the Relationship Between Mechanical and Hydraulic Apertures in Rough-Walled Fractures. *Journal of Geophysical Research* 100(B12): 24,629-24,636.

Robins, N. S. & Buckley, D. K. 1988. Characteristics of the Permian and Triassic aquifers of Southwest Scotland. *Quarterly Journal of Engineering Geology* 21: 329-335.

Steele, A & Lerner, D. N. 2000. *In situ* measurement of oleoporic fracture aperture by DNAPL surrogate injection. Presented at the International Association of Hydrogeologists XXXth Congress - Groundwater: Past Achievements and Future Challenges, Cape Town, 26 November to 1 December 2000.

Groundwater: Past Achievements and Future Challenges, Sililo et al. (eds) © *2000 Balkema, Rotterdam, ISBN 90 5809 159 7*

Changes of quantity and quality of groundwater in the Triassic carbonate formation in Upper Silesia, Poland

A.J.Witkowski & A.Kowalczyk
Department of Hydrogeology and Engineering Geology, University of Silesia, Sosnowiec, Poland

ABSTRACT: Karst-fractured Triassic aquifers are the most important and valuable source of potable water for Upper Silesia. These aquifers have been drained by Zn-Pb ore mines and numerous wells. Intensive long-term drainage has caused lowering of the groundwater table up to 260 m, and creation of extensive regional cones of depression covering an area up to about 500 km^2. Overexploitation of aquifers has caused significant changes of hydrogeochemical conditions resulted in deterioration of groundwater quality. Currently observed groundwater quality in the Triassic aquifers has been creating by negative impact of various pollution sources situated on the ground surface and changes of geochemical conditions in the unsaturated zone, related mainly to oxidation of sulphides. This impact is mainly reflected in increase of SO_4 (up to 6000 mg/L) and NNO_3 (up to 20 mg/L) concentration in groundwater. Starting of gradual flooding-of mines has worsened this situation.

1. INTRODUCTION

Karst-fractured Triassic aquifers are the most important and valuable source of potable water for Upper Silesia. Groundwater of these aquifers is intensively exploited by numerous intakes. Unfortunately because of the occurrence of zinc and lead ores in the Triassic dolomites and limestones these aquifers have been subjected to additional intensive mining drainage.

Additional disadvantages to these aquifers are related to their location in Upper Silesia which is the most urbanised and industrialised part of Poland. This location has caused an extremely high negative human impact, which together with generally high vulnerability to pollution of recharge areas of the Triassic aquifers, has resulted in significant risk of groundwater contamination (Witkowski & Rubin, 1999).

In this paper the problems mentioned above, are shown using selected areas of major well fields and Zn-Pb ore mines. The analysis of changes of water chemistry is based mainly on results of the regional groundwater quality monitoring which has been performed in this area from 1993 (Witkowski 1997).

2. HYDROGEOLOGICAL SETTING

The investigated area is situated in the southern part of Poland and belongs to the Silesian-Cracow Monocline built up of the Triassic and Jurassic formations which discordantly overly the folded and faulted Paleozoic basement.

This area is the only region of Zn-Pb ore exploitation in Poland. Ores occur in the Middle Triassic dolomites and limestones.

Two water-bearing horizons can be differentiated in the profile of the Triassic carbonate formations (limestones and dolomites): the Muschelkalk and the Roetian horizons. They are frequently considered jointly as one aquifer known as the complex of the Triassic carbonate series (Różkowski et al. 1990). The hydraulic structure of fractured and karstified Triassic rocks consists of three types of void: pores, fissures and caverns. Limestones represent the fissured-cavernous type of aquifer whereas dolomites represent the porous-fissured-cavernous type (Motyka 1998). Fissures and karstic channels are favourable pathways of groundwater flow while the pore space is the main water reservoir.

Five major Triassic aquifers (MA) have been determined in the considered region: Lubliniec-Myszków (MA 327), Bytom (MA 329), Gliwice (MA 330), Chrzanów (MA 452) and Olkusz-Zawiercie (Ma 454) (Różkowski ed.1990) (Fig.1). Total area of these aquifers is over 3 600 km^2. They are recharged directly in outcrops areas or indirectly through the permeable Quaternary and Jurassic sediments. The Triassic aquifers are partly covered by isolating clayey Rhaetian-Keuper sediments, and locally Tertiary formation (Fig.1).

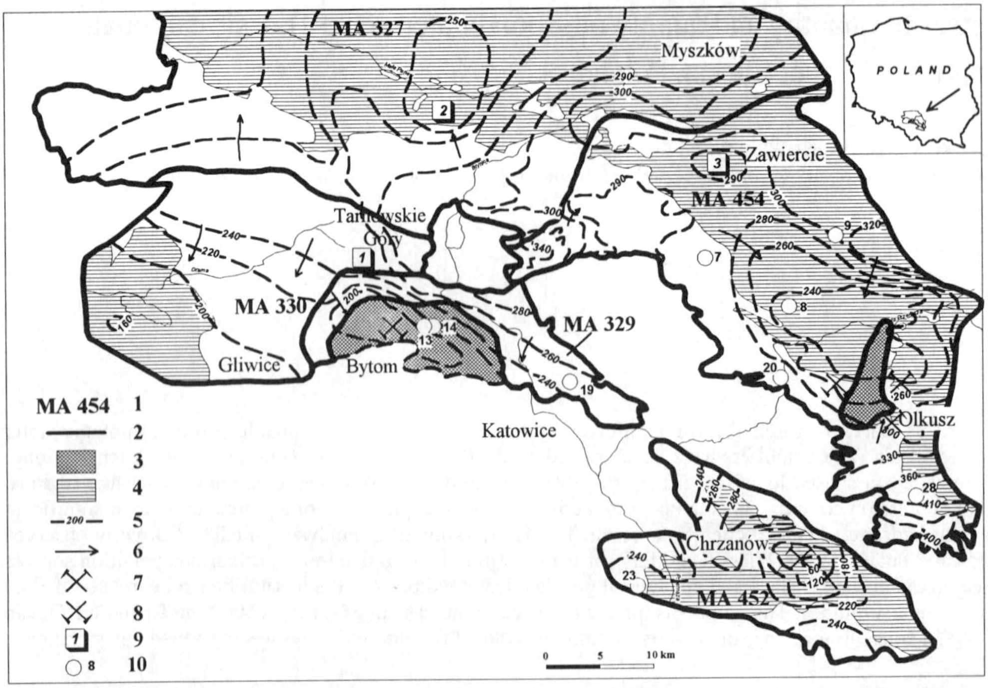

Figure 1. Hydrogeological sketch of the Triassic carbonate aquifers.
1- number of the major aquifer, 2- border of the major aquifer, 3- areas excluded from the range of major aquifers, 4- extent of isolating clayey Rhaetian-Keuper and Tertiary sediments, 5- groundwater contour lines (in metres above the sea level), 6- groundwater flow direction, Zn-Pb ore mines: 7-active, 8- abandoned, 9- major well fields (1- Staszic, 2- Bibiela, 3-Łazy Błędowskie), 10- regional groundwater quality monitoring sites.

These carbonate Triassic sediments are the most important and resourceful aquifers being the major sources of potable water for the Upper Silesian agglomeration. The thickness of this water- bearing formation varies from 20 to 160 m. The average yield of water pumped out from aquifers by the wells is 110 m^3/h, at drawdowns reaching a dozen or so metres (Różkowski et al. 1997). The dynamic resources of these aquifers have been established at the rate of 1,1 Mm^3/d.

3. GROUNDWATER ABSTRACTION

The Triassic aquifers are intensively exploited by numerous single wells and well fields. Additionally the Bytom, Olkusz-Zawiercie and Chrzanów major aquifers are subjected to concentrated intensive groundwater abstraction by mining activity (directly by Zn-Pb ore mines and quarries, and indirectly by coal mines). Mining drainage in some regions (the Olkusz and the Tarnowskie Góry towns) has lasted for more than 400 years.

Maximum groundwater withdrawal from the Triassic aquifers reaching about 400 Mm^3/y (Różkowski et al. 1990) was noticed in the mid eighties. This long lasting intensive drainage has led to the significant depletion of groundwater resources and general overexploitation of the Triassic aquifers (Kropka et al. 1992, Witkowski & Kowalczyk 1995). A gradual decrease of groundwater withdrawal has been observed since 1990. Recently total groundwater withdrawal from these aquifers has changed from about 323 Mm^3/y in the beginning of the nineties (Kowalczyk & Różkowski 1997) to about 282 Mm^3/y in 1999.

Groundwater abstraction by wells from the analysed Triassic aquifers has decreased by about 13% in the considered period. The Łazy Błędowskie well field is the best example of this trend. Groundwater abstraction by this intake has been reduceded from about 30 Mm^3/y in the mid eighties to about 6 Mm^3/y in 1999 (Fig.2).

Over the same period groundwater abstraction by mines has increased from about 173 Mm^3/y to about 215 Mm^3/y. It has been caused by very high precipitation recorded in 1997 and 1998.

Particularly significant increase of water inflow

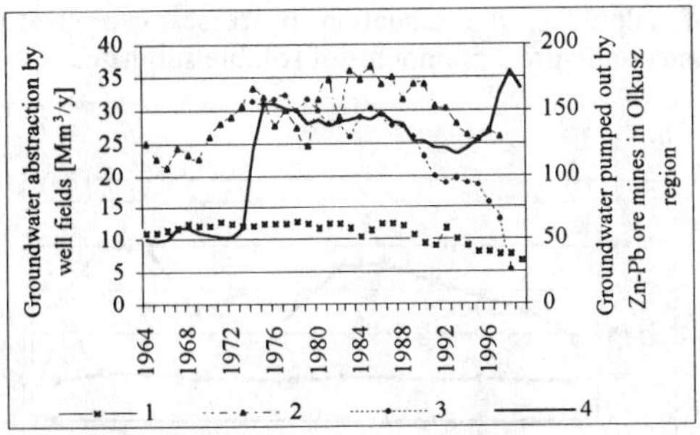

Figure 2. Selected examples of groundwater abstraction from the Triassic carbonate aquifers
1- the Staszic intake, 2- the Bibiela well field, 3- the Łazy Błędowskie well field, 4–withdrawal from Zn-Pb ore mines in the Olkusz region (mines: Bolesław, Olkusz & Pomorzany)

to the mines situated in the Olkusz region has been observed. Groundwater abstraction from these mines has increased from over 120 Mm^3/y to about 180 Mm^3/y in the period 1991-1998 (Fig. 2).

4. CURRENT HYDRODYNAMIC CONDITIONS

Very large long-term withdrawals from the Triassic aquifers have caused the groundwater table to decline about 30 –70 m, in wells (Fig. 3), and 100 - 260 m, in mining areas (Fig.1). Vast deep regional cones of depression have been created. For example the area of such cones caused by mining activity performed by Zn-Pb ore mines in the Olkusz area is about 500 km^2 and over 170 km^2 in the Chrzanów region. Long lasting intensive drainage has led to profound transformation of hydrodynamic conditions resulting in changes of flow directions, increase of hydraulic gradients, and changes of hydrogeochemical conditions on the regional scale.

In the natural conditions rivers were the local drainage basis for the aquifers. Currently there are

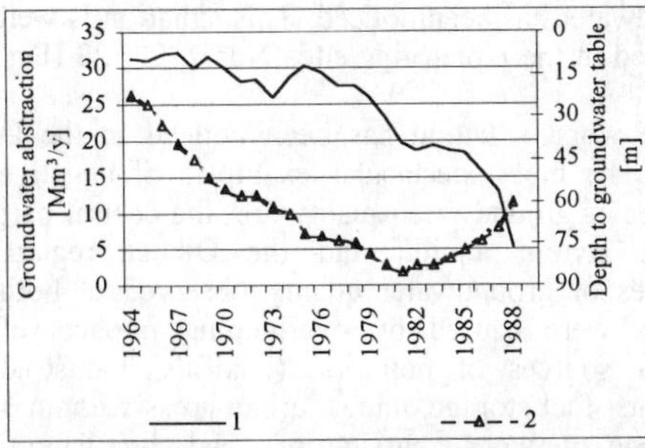

Figure 3. Changes of groundwater abstraction and depth to water table in the Łazy Błędowskie well field.
1- groundwater abstraction, 2- depth to groundwater table.

centres of intensive man-made drainage (active and abandoned but still drained Zn-Pb ores mines and major well fields) (Fig.1).

Since the mid-1990s a decrease in groundwater withdrawal and simultaneous increase in precipitation have resulted in a general rise of the water table, over 20 m in some parts of the Triassic aquifers (Fig.3).

Further significant changes in hydrodynamic conditions are connected with the commencement of gradual closure of Zn-Pb ores mines. This relates mainly to the Chrzanów (closure of the Trzebionka mine) and Olkusz-Zawiercie (closure of the Bolesław and the Olkusz mines) aquifers.

5. GROUNDWATER QUALITY

The chemical composition of the groundwater in the Triassic aquifers'- is variable. TDS varies from about 100 mg/L to more than 8500 mg/L (Motyka & Witkowski 1999). The SO_4 content varies from 4 to about 6000 mg/L, Cl varies from 1 to about 1200 mg/L, and NNO_3 varies from 0 to 19.7 mg/L. Locally an increased content of some other pollution indicators (Pb, Cd, Zn, Bo, Ba, lignosulfonates) has been observed.

The quality of this groundwater depends not only on the differences in local hydrogeological conditions but also on the character of the land use. Considering these criteria the analysed aquifers could be divided into four major parts:

A) areas covered by thick very low permeability Upper Triassic or Tertiary sediments - not disturbed by intensive drainage

B) areas covered by thick very low permeability Upper Triassic or Tertiary sediments – disturbed by intensive drainage (by mines and large well fields)

C) uncovered areas not subjected to the intensive drainage and negative impact of local contamination sources

D) uncovered areas subjected to intensive drainage and negative impact of local contamination sources.

The very important problem of groundwater quality of the Triassic aquifers is related to both still active and abandoned Zn-Pb ore mines. A contemporary exploitation takes place in the Olkusz and Chrzanów regions (Fig.1). The central part of the Bytom aquifer is still being intensively drained by two water galleries despite the closure of Zn-Pb mines. So in these areas, problems with groundwater quality are focused. The three areas of mining drainage mentioned previously, represent different part of aquifers. The Chrzanów region represents the B part, the Bytom one represents the D part and the Olkusz region represents some kind of mixture of the D and B parts.

Stable chemical composition and a very good groundwater quality have been observeded in the A part. Groundwater in this area is represented by monitoring well No 9 (MS 9) (Figs 4-6).

a.

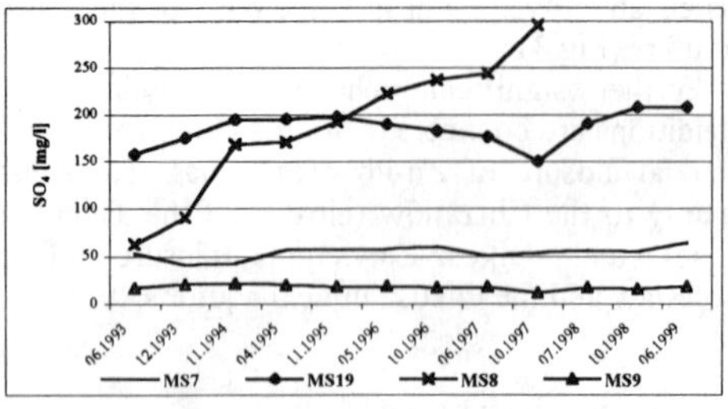

b.

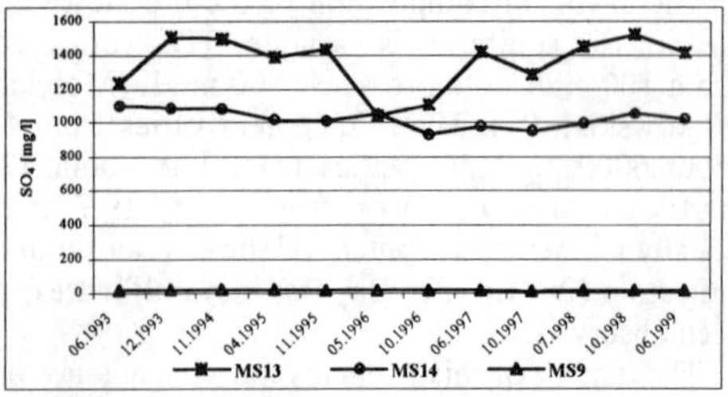

Figure 4. Changes of sulphate concentrations in groundwater in selected monitoring sites.

In the B areas groundwater has variable chemical composition and quality. Groundwater is significantly changed in the Chrzanów area affected by mining activity of the "Trzebionka" mine. Increase of sulphates (up to more than 200 mg/L) and periodically of Zn and Pb have been observed.

Despite the intensive abstraction of groundwater its quality is good and stable in some areas. This situation has been observed in the "Bibiela" well field (Witkowski et al. 1998). Gradual groundwater quality deterioration reflected by the significant increase of sulphates content has been noticed in the others. The "Łazy Błędowskie" well field, represented by monitoring well No 8 (MS 8), is a good example of such a situation. Increase of sulphate content from 63 mg/L (1993) to about 300 mg/L (1997) has been observed in this well (Fig. 4a). The observed increase of groundwater contamination caused gradual closure of several wells of this well field.

The reason for such differences in reaction to intensive drainage is probably the occurrence of various amounts of sulphides in the unsaturated zone. In the areas characterised by a higher content of sulphides, the oxidation processes can create potentially greater amounts of soluble sulphates.

a.

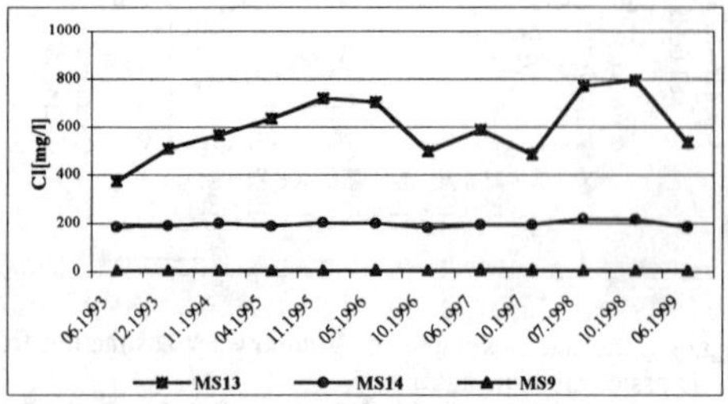

b.

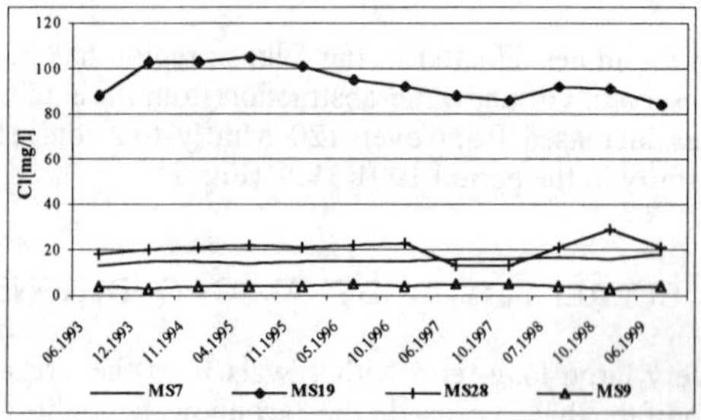

Figure 5. Changes of chloride concentrations in groundwater in selected monitoring sites.

Groundwater in the C parts of the aquifers has a rather stable chemical composition and still good quality. These areas are represented by the monitoring site No 7 (MS 7). Groundwater monitored by this site has low and stable concentrations of such pollution indicators as sulphates and chlorides (Fig.4a & 5a). Unfortunately, in the areas of intensive agriculture and in the urban areas without a sewerage network the significant increase of nitrate-nitrogen content in groundwater has been noticed. This situation is well reflected in the monitoring sites Nos 7, 23, 28 (Fig. 6).

The worst situation has been noticed in the D areas. The most spectacular examples of negative changes of groundwater quality are: the central part of the Bytom aquifer and the Olkusz region. Changes of groundwater quality observed in both regions were caused by overlapping impacts of various sources of pollution (landfills, industrial effluents, fuel storage, mines, urban areas without a sewerage network, agriculture) and by natural geochemical processes related mainly to the transformation of the deepened unsaturated zone.

The central part of the Bytom aquifer has been

monitored by two monitoring sites (MS 13 and MS 14) situated in both galleries (Fig.1). The quality of monitored water is very low and concentrations of many pollution indicators (sulphate, chloride, ammonium-nitrogen, cadmium, zinc) exceed standards for potable water (Kropka & Witkowski 1994). Observed concentrations of sulphates (Fig. 4b) and chlorides (Fig. 5b) are about 20-times higher than average concentrations in the analysed Triassic aquifers. Even in the peripheral parts of the aquifer groundwater quality is low due to higher contents of sulphates and chlorides. These areas are represented by the monitoring site No 19 (Figs.4a & 5a).

Similar negative groundwater quality changes have been noticed in areas of mining activity in the Olkusz region where there are two still active Zn-Pb ore mines. Very high concentration of SO_4, reaching 6000 mg/L has been observed locally. The concentration of sulphates in groundwater of the Triassic aquifer in the Olkusz region is presented in Figure 7. (according to Motyka & Witkowski 1999). Groundwater in this region is additionally polluted by lignosulfonates.

Two areas have been proposed to be excluded from the range of major aquifers because of serious groundwater contamination (Fig.1) (Kropka 1996, Różkowski et al. 1997).

Unfortunately due to the current gradual recovery of the primary pressure in some parts of the Triassic aquifers further negative changes of groundwater quality are observed. This situation is visible in the area of the Łazy Błędowskie well field and in the areas intensively drained by Zn-Pb ore mines, in which some mines are being closed. Gradual flooding of the “Bolesław” mine, in the Olkusz region, and the “Trzebionka” mine, in the Chrzanów region, has caused deterioration of groundwater quality mainly by significantly increasing the sulphate content.

There are more examples of local significant groundwater contamination of analysed Triassic aquifers in areas not subjected to current mining activity. The best known example is the Tarnowskie

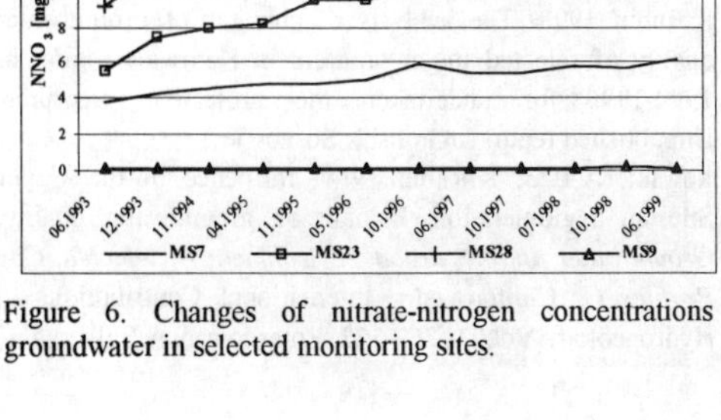

Figure 6. Changes of nitrate-nitrogen concentrations in groundwater in selected monitoring sites.

Góry region where approximately 1,7 Mtons of dangerous wastes from a chemical factory have been dumped over an area of 0,3 km^2 directly on the ground without any protection. This has caused contamination of the Triassic aquifer reflected in higher content of Cl (up to 259 mg/L), SO_4 (up to 309 mg/L), Pb (up to 0,109 mg/L), B (up to 77 mg/L), Sr (0,96 mg/L) (Rubin et al. 1999).

6. CONCLUSIONS

A long-term intensive drainage by mines and well fields has resulted in overexploitation of the Triassic aquifers reflected by significant lowering of the groundwater table and the geochemical transformation of the created thicker unsaturated zone.

The decreased abstraction from the Triassic aquifers, caused by the recent decline of heavy industry and economically constrained smaller communal use combined with increased precipitation, has resulted in a general rise of the water table. Slow recovery of primary pressure is currently observed in mining areas where some mines are being gradually flooded.

All these changes have caused considerable increase of sulphate content in groundwater in parts of the aquifers where Zn-Pb ores occur, particularly in areas where sulphide minerals have dominated.

The observed gradual deterioration of groundwater quality of valuable Triassic aquifers is

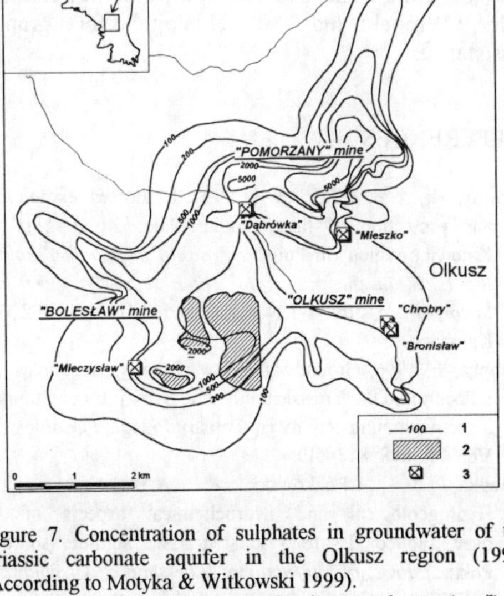

Figure 7. Concentration of sulphates in groundwater of the Triassic carbonate aquifer in the Olkusz region (1997) (According to Motyka & Witkowski 1999).
1- contour of sulphates concentration in groundwater (mg/l), 2- sources of pollution (landfills), 3- shafts.

the result of the overlapping impact of factors mentioned and the changeable, intensive drainage and the negative impact of a great number of pollution sources situated on the ground surface.

There is an alarming increase of nitrate content in the analysed groundwater observed in uncovered parts of the Triassic aquifers.

7. THE EXPECTED FUTURE FATE

Predicting the future sulphate content in groundwater is difficult due to the following factors: unstable hydrodynamic conditions (progressive recovery of the groundwater pressure), high variability of sulphides content in the Triassic rocks and the short period of observations. Investigations dealing with this problem have started only recently. The future content of SO_4 in groundwater depends on the water recovery. Unfortunately the final decision concerning partial or total flooding of Zn-Pb ore mines has not been taken yet. Starting in 1997, gradual flooding of the lowest parts of the "Trzebionka" mine has led to temporary increase of SO_4 concentration in groundwater from about 200 mg/L to 400 mg/L (Szuwarzyński et al.1999). Later on, a gradual decrease has been observed. Concentration of SO_4 has decreased to about 250 mg/L after a dozen of months. We can expect the same scenario of changes in other areas.

ACKNOWLEDGEMENTS

The authors are grateful to the State Committee for Scientific Research for providing financial support (project No 9 T12B 011 16). We would like to thank Mr J.Wróbel and Mr G.Wojtal for graphic assistance.

REFERENCES

Kowalczyk, A. & A.Różkowski 1997. Resources, exploitation and protection of the Triassic Major Aquifers in the Katowice region (in Polish). *Proc. of the Confer. "Role of the GPW in the system of water supply – today and tomorrow"*. Ustroń 5-7 November 1997: 103-116. PZIiTS: Katowice.

Kropka, J., 1996. Groundwater flow pathways, resources and utilisation in the Triassic aquifer of Bytom in conditions of an active mining activity (in Polish). *Przegląd Geologiczny*. Vol. 44, no 8: 845-849.

Kropka, J., A.Różkowski & A.Witkowski 1992. Hydrogeological and hydrochemical aspects of the overexploited fissure - karst triassic aquifer (southern Poland). *Proc. of International Workshop on Groundwater and Environment*.:14 –20. Beijing: Seismological Press.

Kropka, J. & A.Witkowski 1994. Groundwater quality degradation in the area of abandoned Zn-Pb ore mines (Bytom syncline-Southern Poland), *Proc. of 5th International Mine Water Congress*. Nottingham. Vol.2: 799-809.

Motyka, J., 1998. A conceptual model of hydraulic networks in carbonate rocks, illustrated by examples from Poland. *Hydrogeology Journal* 6: 469-482.

Motyka J. & A.J.Witkowski 1999. Sulphates in groundwater of the karst-fractured Triassic aquifers in areas of intensive mining drainage (the Olkusz and the Bytom regions). *Mine, Water & Environment* (R.F.Rubio, ed.). Vol.I. IMWA International Congress: 189-195. Sevilla.

Różkowski, A. (ed) 1990. *Fissure-karst groundwater basins of the Cracow-Silesian monocline and problems of their protection*. Monograph CPBP 04.10. Vol 57 (in Polish). Warszawa: Wydawnictwo SGGW-AR.

Różkowski, A., A.Kowalczyk, J.Kropka, K.Rubin & A.Witkowski 1996. Hydrogeology of Triassic carbonate complex of Silesia-Cracow monocline. *Proc.of the Internat. Conference on Karst-Fractured Aquifers – Vulnerability and Sustainability*. Prace Naukowe Uniwersytetu Śląskiego No 1563: 205-221. Katowice: Wydawnictwo Uniwersytetu Śląskiego.

Różkowski, A., A.Chmura & A.Siemiński (ed.) 1997. *Usable groundwater in the Upper Silesian Coal Basins and its margin* (in Polish). Prace PIG CLIX. Warsaw: PIG.

Rubin, K, H.Rubin & A.Kowalczyk 1999. Endangering of groudwater quality of the Triassic aquifer in the area of chemical factory in Tarnowskie Góry (in Polish). *Współczesne problemy hydrogeologii*. Vol. IX. Warsaw: PIG.

Szuwarzyński, M., J.Motyka,A.Kowalczyk 1999. Groundwater contamination as a potential result of closing down of the Trzebionka mine, southern Poland. Proc. of the 7th IMWA Congress. Katowice (in press).

Witkowski, A.J. 1997. *Groundwater quality monitoring in the area of Katowice Regional Water Management* Council – Report on investigations carried in 1993-1996 (in Polish). Seria: Nauki o Ziemi. Katowice.

Witkowski A. & A.Kowalczyk 1995. Impact of ground-water overexploitation on Triassic aquifer in Zawiercie region (Souhtern Poland), *Water Resources at Risk*. Selected papers presented at the conference "Water Resources at Risk". American Institute of Hydrology: SI-124-132. Denver.

Witkowski, A.J, A.Różkowski, J.Motyka, A.Kowalczyk, K.Rubin 1998. The analysis of changes of groundwater quality of selected major aquifers in Katowice region in 1990-1996 for undertaken the protective measures. Unpublished report (in Polish). Sosnowiec.

Witkowski, A.J. & K.Rubin 1999. Influence of the Upper Silesian agglomeration, Poland, on groundwater quality. *Groundwater in the urban environment - Selected City Profiles* (J. Chilton, ed.), International Contributions to Hydrogeology.Vol.21: 267-272. Roterdam:A.A.Balkema.

Groundwater: Past Achievements and Future Challenges, Sililo et al. (eds) © 2000 Balkema, Rotterdam, ISBN 90 5809 159 7

Effects of soil environment changes on the behavior of VOC's in unsaturated zone

M.Yoshimura & C.Tang
Graduate School of Science and Technology, Chiba University, Japan

ABSTRACT: Behavior of Volatile Organic Compounds (VOC's) in an unsaturated-saturated capillary zone plays an important role in the understanding of pollutant mechanisms. Most of the previous experimental studies had problems of scale, temperature and size of porous media. Lysimeter experiments were then conducted to study the behavior of VOC's in unsaturated zone by simulated rain. TCE and Bromide were used in the experiments to trace respectively the behavior of VOC's and water. By comparing the vertical profile of bromide and the one of TCE, it was found that liquid's TCE does not always move straight downward to the water table. It rather remains at the top layer, while TCE gas moves faster than water. This is considered to be the result of downward pushing of water with dissolved TCE and TCE gas by percolating water. Furthermore, heterogeneous soil property, one of the factors causing the distribution of VOC's is also discussed.

1 INTRODUCTION

Groundwater is an important resource also in Japan and, accounts for approximately 23 % of life water and 34 % of industrial water. In the results of groundwater quality nation-wide investigation proceeded by the Environment agency government of Japan, it was found that Trichloroethylene and Tetrachloroethylene are detected in more than 30 % of the objective wells. In the meantime, 1,750 mm of annual precipitation of Japanese Islands exceed largely world major city average of 970 mm. Japanese major plains and those of their peripheries are covered with late Neogene and Quaternary formations which is composing the major aquifers. For these reasons, groundwater table in Japan is relatively shallow and chemicals that enter the subsurface reach groundwater table sooner or later. Groundwater contamination tends then to spread widely.

In general, and though the aqueous solubility of organic liquids contamination is a low in the subsurface, it is still large enough to degrade water quality. The mobility of liquid's VOC's depends on spilled volume in site and its physico-chemical properties as well as hydraulic properties of porous media. In the unsaturated zone, entrapped organic liquids can volatilize and be transported by diffusion or convection in the vapor phase or they can solubilize into infiltrating surface precipitation (e.g.,Schwille 1988). Several models have been developed to describe multiphase flow of organic contaminants in the subsurface with exchange of organic mass between fluids phases. Most of these utilize the assumption of "local equilibrium" to describe the extent of partitioning between VOC's and the aqueous phase. The local equilibrium assumption implies that if the concentration of contaminant in one phase is known, its concentration in other phase at the same spatial location can be described by equilibrium partitioning relationships (Abriola & Pinder 1985). On the contrary, Lam et al (1983) indicate that the rate of mass transfer of surfactants from the aqueous phase to residual oil blobs is frequently not sufficiently fast to reach equilibrium conditions. It is clear then that the local equilibrium assumption has yet to be validated for a reasonable range of conditions, which are typically encountered in contaminated subsurface settings. If nonequilibrium conditions are to be considered, it could significantly impact our understanding of the dynamics of polluted sites and associated remedial techniques (Powers 1991).

Most experiments have been studied to quantify the dissolution of VOC's rates with homogeneous media and a uniform distribution of residual VOC's (Miller et al. 1990). Both experimental observations and model evaluations indicate that very localized variations in the VOC's saturation, and consequently the relative permeability, contribute to preferential flow paths and the establishment of dissolution fingers in columns packed homogeneously with uniform sand (Powers et al. 1994; Imhoff et al. 1994). With the assumption that the local equilibrium assumption is appropriate, it was showed that observed lower concentrations could arise from dilution effects, even for homogeneous sands and relatively uniform residual VOC's saturations (Soerens et al. 1998). Several studies to predict NAPL dissolution in heterogeneous media described by random distributions of the hydraulic conductivity fields have been conducted (Mayer & Miller 1996).

The behavior of VOC's in the unsaturated-saturated capillary zone plays an important role in the understanding of pollutant mechanisms. However, most of the previous experimental studies of its behavior have problems as scale, temperature and the size of porous media. It is therefore necessary to conduct middle scale experiments between laboratory and the field scale to simulate a behavior similar to actual phenomenon. For that reason, lysimeter experiments were in unsaturated zone by simulated rain. TCE and Bromide were used in the experiments to trace respectively the behavior of VOC's and water. Furthermore, heterogeneous soil property, one of the factors in causing the distribution of VOC's is also considered.

2 LYSIMETER EXPERIMENTS

Infiltration Experiments were conducted with a lysimeter of sand volume 120cm×120cm×210cm as illustrated in Figure 1. Tensiometers, soil water samplers, gas samplers and thermometer were set at 10 depths. A neutron soil moisture meter and time domain reflect meter (TDR) was applied to measure soil water content. Those locations are shown in Table 1. Soil suctions and effluent water from the bottom of the lysimeter were measured automatically and recorded in a data-logger. Hydraulic potentials were calculated by setting the lysimeter surface at zero cm.

Soil properties such as porosity, soil particle distribution and hydraulic conductivity were obtained every 10 cm depth in the lysimster.

Trichloroethylene (TCE) and sodium bromide (NaBr) were used to trace the behavior of VOC's and water. Water was sprinkled from the top of the lysimeter with an intensity of 25 mm/h. Water volume was 33.3 mm in the first eight experiments, then 66.6 mm and 15.6 mm respectively in the last two experiments. At the beginning of the first experiment, 2,420ml pure TCE and bromide-dated water were put into the lysimeter at 10 cm depth. Soil water and gas in the unsaturated zone were sampled during and after rain events. Soil water was extracted by n-Hexane and analyzed by gas chromatography on Shimadzu GC-14B equipped with a flame ionization detector (FID). Soil gas was analyzed by gas chromatography on Nihondenshi GC-311 equipped with a photon ionization detector (PID). Variations of bromide concentration in soil waters were analyzed with an ion chromatography on Shodex.

3 NUMERICAL SIMULATIONS

To investigate how does the heterogeneity of soil properties affect the behavior of TCE, numerical simulations were carried out by finite element code of multiphase flow and multi component transport (MOFAT). Soil properties used in the calculation was showed in table 2. Seven layers were set in Case 1 and one layer was set in Case 2, respectively. These data for case 1 was obtained from laboratory experiments and data for case 2 by averaging those of case 1. And, a constant flux for water was imposed at the upper boundary condition.

4 RESULTS AND DISCUSSION

4.1 Relationship between the movements of water and TCE

Figure 2 shows the time-variation of soil water content during experiments where the water table was set at the depth of 170 cm including ten rain events.

Table 1. Sensors used in the experiments

Sensors	Setup depth cm
Tension meter, Soil water sampler Gas sampler	10,32,55,77,100,122 ,145,167,190,212
TDR	0-15,15-30,30-60,60-90,90-120
Neutron soil moisture meter	Every 10

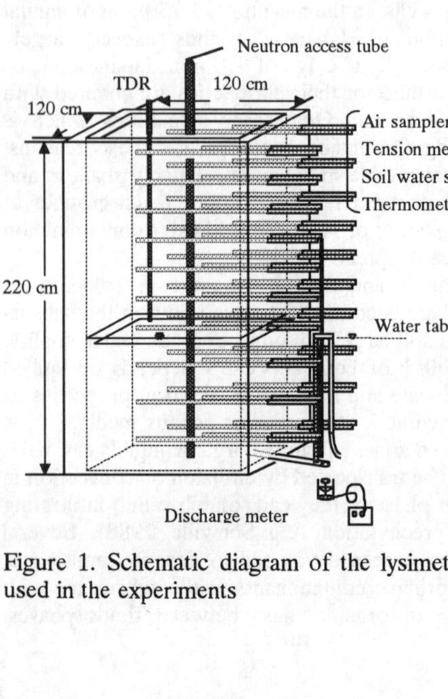

Figure 1. Schematic diagram of the lysimeter used in the experiments

Table. 2 Comparison of soil parameters used in the simulations

Depths	Hydraulic conductivity		Porosity	
	Case 1	Case 2	Case 1	Case 2
Number of layers	7	1	7	1
cm	cm/sec	cm/sec	-	-
0-22	3.14×10^{-3}		0.48	
22-44	3.14×10^{-3}		0.44	
44-66	3.14×10^{-3}		0.42	
66-88	3.14×10^{-3}	3.14×10^{-3}	0.40	0.41
88-110	3.14×10^{-3}		0.42	
110-132	3.14×10^{-3}		0.37	
132-170	3.14×10^{-3}		0.37	

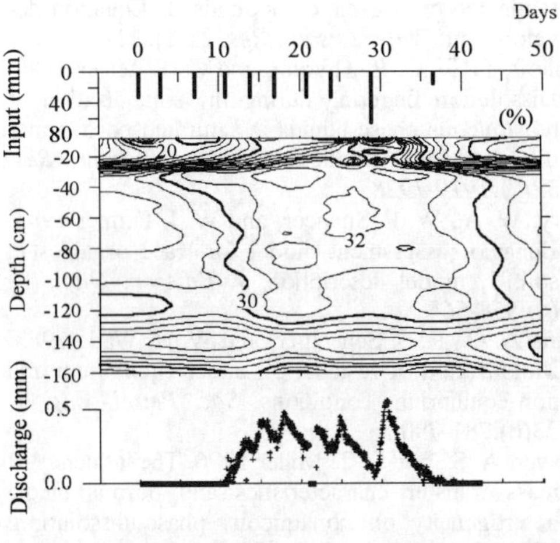

Figure 2. Time-variation of water input, content and discharge during the experiments

After supplying with rainwater, soil water content from 30 cm to 90 cm depth remained the same value in spite of increasing at the surface. However, by 3rd and 4th sprinkles, water content at the 70 cm and 100 cm increased. On the other hand, Discharge began around 4th sprinkles. It seems to vary with the interval of input.

The variations of hydraulic potential at the beginning of experiments were almost the same at every depth shown in Figure 3. It means that no vertical flow movement occurs shown. During the first rain period, hydraulic potentials increased at depths of 10 cm and 32 cm. After the 4th rain period, patterns of hydraulic potentials responding to rain were the same at each depth above 77 cm. It means the percolation reached to a steady state. While, hydraulic potentials at depths of 122 cm and 167 cm responded to rain less since the discharge would occur as soon as the infiltration front reached the top of a capillary zone.

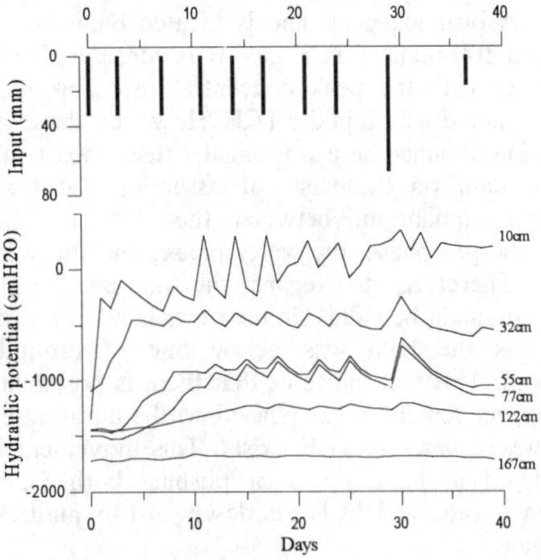

Figure 3. Variations of hydraulic potential during the experiments

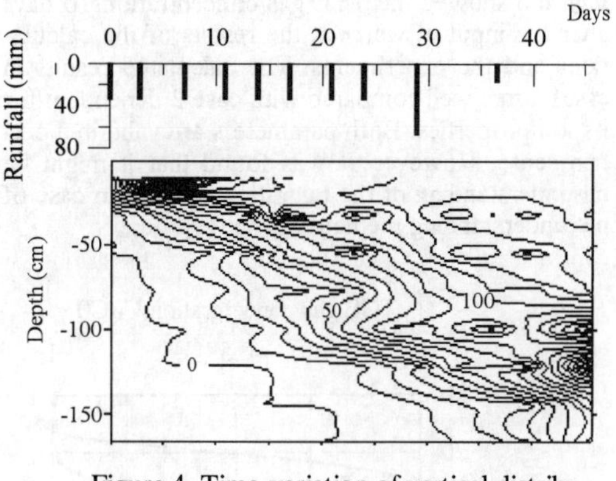

Figure 4. Time-variation of vertical distribution for TCE gas concentration (ppm)

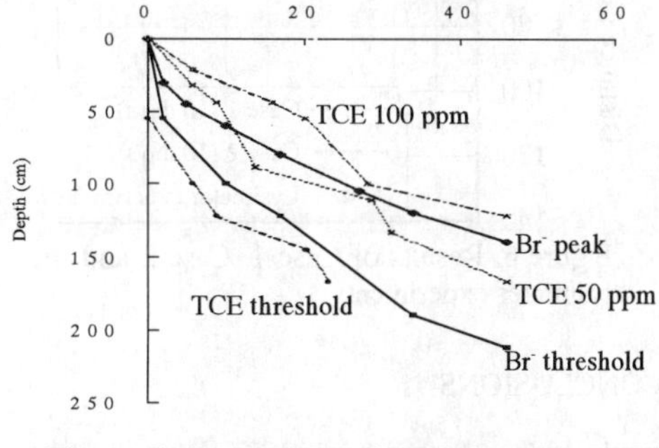

Figure 5. Comparison of the infiltration speed between bromide and TCE gas

Figure 5 illustrates the velocity of infiltration for the bromide labeled water and TCE gas shown in Figure 4. In general, the peak concentration of bromide shows the movement of water during experiments. A bromide peak line is located between 50 ppm and 100 ppm of TCE gas. It is considered that the water with the peak concentration of bromide was contacted with liquid's TCE. However, the concentration obtained here was smaller than one of calculated data on the basis of assuming the local chemical equilibrium between the liquid's TCE phase, the gas phase, the water phase, and the solid phase. Therefore, to regard the non-equilibrium condition might be valid. On the other hand, a line of TCE gas threshold was below one of bromide threshold. This fact indicates that there is possibility of quick movement in gas phase into the unsaturated zone where gaseous TCE exists. This movement is considered to be a result of pushing both TCE-dissolved-water and TCE gas downward by infiltrating water.

4.2 Effect of the soil properties

Figure 6 showed the TCE gas concentration 10 days after the input of water at the results of the calculations and the experiments. The calculated results in case1 agree well compared with case 2 dependent for its soil properties. Both parameters are valid in a certain sense. However, it was found that it might be misunderstanding of the migration of TCE in case of not understanding the heterogeneity.

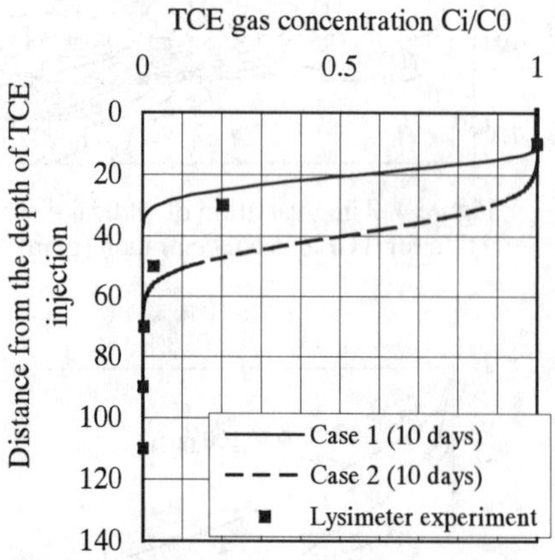

Figure 6. Results of Case 1, Case 2 and the lysimeter experiments

5 CONCLUSIONS

From the lysimeter experiments:

1. Liquid TCE does not always move straight downward immediately.
2. TCE gas moves faster than water in unsaturated zone where TCE migration in the gas phase is important.

From the numerical simulations by MOFAT:

3. It might be misunderstanding of TCE migration in case of not understanding the heterogeneity.

ACKNOWLEDGMENTS

The authors gratefully acknowledge Professor, Emeritus Shindo, S. of Chiba University, Professor, Sakura, Y. of Chiba University and Professor Hirata, T. of Wakayama University for their discussions on these results.

REFERENCES

Abriola, L. M., and G. F. Pinder 1985. A multiphase approach to the modeling of porous media contaminated by organic compounds, 1, Equation development. *Water Resour. Res.* 21(1): 11-18

Imhoff, P. T., G. P. Thyrum, and C. T. Miller 1994. Dissolution fingering during the solubilization of nonaqueous phase liquids in saturated porous media, 1, Model predictions, *Water Resour. Res. 32(7):1919-1928*

Jury, W. A., W. F. Spencer, and W. J. Farmer 1983. Behavior assessment model for trace organics in soil, 1, model description. *J. Environ. Qual.* 12 (4):558-564

Lam, A. C., R. S. Schechter, and W. H. Wade 1983. Mobilization of residual oil under equilibrium and non-equilibrium conditions. *Soc. Petrel. Eng. J.* 23(6):781-790

Mayer, A. S., and C. T. Miller 1996. The influence of mass transfer characteristics and porous media heterogeneity on nonaqueous phase dissolution. *Water Resour Res.* 32(6):1551-1568

Miller, C. T., M. M. Poirer-McNeill, and A. S. Mayer 1990. Dissolution of trapped nonaqueous phase liquids: Mass transfer characteristics. *Water Resour. Res.* 26(11):2783-2796

Powers, S. E. 1991. Theoretical study of the significance of nonequilibrium dissolution of nonaqueous phase liquids in subsurface systems. *Water Resour. Res.* 27(4):463-477

Powers, S. E., L. M. Abriola, and W. J. Weber Jr. 1994. An experimental investigation of NAPL dissolution in saturated subsurface systems: Transient mass transfer rates, *Water Resour. Res.* 30(2):321-332

Schwille, F. 1967. Petroleum contamination of the subsoil-A hydrological problem, *in The Joint Problems of the Oil and Water Industries. Proceedings of a Symposium.* edited by P. Hepple. pp.23-54. Institute of Petroleum. London

Soerens, T. S., D. A. Sabatini, and J. H. Harwell 1998. Effects of flow bypassing and nonuniform NAPL distribution on the mass transfer characteristics of NAPL dissolution. *Water Resour. Res.* 34(7):1657-1673

5 Integrated water resource management

Topic co-editors:

Peter Dillon
CSIRO, Australia

Eberhard Braune
Department of Water Affairs, South Africa

Christine Colvin
Council for Scientific and Industrial research, South Africa

Steve Appleyard
Water and Rivers Commission, Australia

Groundwater: Past Achievements and Future Challenges, Sililo et al. (eds) © 2000 Balkema, Rotterdam, ISBN 90 5809 159 7

Groundwater status of the City of Stockholm

M.Aastrup, J.Anderberg & B.Thunholm
Geological Survey of Sweden, Uppsala, Sweden

ABSTRACT: The groundwater status in the City of Stockholm was investigated by the Geological Survey of Sweden on commission by the Environment and Health Protection Administration and the Swedish Environmental Protection Agency. The investigation comprised a hydrogeological survey, the design of a quality sampling network and an evaluation of the groundwater contamination situation in Stockholm. The survey produced geographical databases on groundwater resources and vulnerability. The groundwater quality investigation showed significantly elevated concentrations of heavy metals such as Cu and Hg, but also locally high contents of nitrogen, phosphorus, bacteria, biocides and petroleum products. In four selected watershed areas within the City of Stockholm groundwater flow as well as transport of heavy metals, phosphorus and nitrogen were simulated using the model codes MODFLOW and MT3D. Modeling results indicated a relatively small contribution of heavy metals to surface waters by groundwater. The transport of contaminants through artificial drainages is generally more extensive.

1 INTRODUCTION

For more than 700 years the soils and waters of Stockholm have been polluted by emissions from households, industry, waste deposits, traffic, accidental spills etc. It has been assumed that the groundwater is polluted, but to what extent has been unknown.

Groundwater is considered a valuable natural resource in rural as well as in urban areas. Even though the City of Stockholm today relies on surface water for its drinking water supply, the groundwater is however essential as it discharges into lakes, watercourses, springs and wetlands of importance to the ecosystem.

The objectives of the investigation have been:

- to map the groundwater resources
- to get a picture of the groundwater quality
- to estimate the contribution of groundwater transported heavy metals, phosphorus and nitrogen compounds to the major surface water bodies (Lake Mälaren and The Baltic Sea).

The investigation has been carried out by the Geological Survey of Sweden on commission by the Environment and Health Protection Administration in Stockholm (EHPAS) and the Swedish Environmental Protection Agency (SEPA) (EHPAS, 1997; Aastrup & Thunholm, 1999; Aastrup & Thunholm, 2000). Partly the work is a contribution to the SEPA/EHPAS program on "Metals in urban areas" studying the metal fluxes from technosphere to biosphere.

2 MATERIALS AND METHODS

2.1 *Hydrogeological survey*

The objective of the hydrogeological survey was to present the EHPAS with digital groundwater information as a foundation for the design of a groundwater quality monitoring network and the evaluation of the groundwater contamination situation in Stockholm. The survey was based entirely on existing information in archives and databases. All database handling, analysis and map production were performed in a GIS environment using MapInfo and ArcInfo software.

The results of the survey were presented as groundwater resources and vulnerability maps, both to the scale of 1:20 000. The resources map shows the main aquifers and features of the groundwater system. The vulnerability map shows the sensitivity of the groundwater to the infiltration and spread of contaminants to the subsurface environment.

The geological framework was based on engineering geology maps to the scale of 1:10 000. In addition, hydrogeological data, including close to

1000 well point records on groundwater levels and stratigraphical data, were collected.

2.2 *Groundwater quality investigation*

Areas for groundwater quality sampling were selected according to special criteria, which include different types of existing or previous polluting activities. Within the chosen areas, existing groundwater level observation tubes were used for sampling. Ten areas were selected at random, to get a fairly representative picture of the groundwater quality in Stockholm. The observation tubes are made of polythene. Sampling has been carried out at 75 sites. Only the groundwater of glacial tills and glaciofluvial deposits was sampled.

The investigation of groundwater quality comprised major ions, heavy metals, nitrogen, phosphorus, bacteria and organic pollutants. The sampling was conducted according to a clean procedure adopted in the Swedish National Groundwater Monitoring Program (Nordic Council of Ministers, 1989).

2.3 *Groundwater modelling*

The aim of the modelling was to estimate the contribution by groundwater to the leakage of metals into the main surface water bodies, Lake Mälaren and the Baltic Sea. The calculation was made on a yearly basis by multiplying the estimated concentrations of the various compounds in groundwater in the discharge areas by the amount of discharging groundwater.

The modelling was conducted in two steps, the first step being modelling of groundwater flow with the 3D finite-difference groundwater flow model MODFLOW (McDonald & Harbaugh, 1988). The second step, transport modelling, aims at getting a better understanding of the metal concentration distribution within the watersheds, the time perspective of the transport and an estimate of the metal concentration in the zone of discharge into the lakes. The transport model used is referred to as MT3D, which simulates advection, dispersion and chemical reactions of contaminants in groundwater flow systems in either two or three dimensions (Zheng, 1990). The Freundlich isotherm has been used in the retardation factor of the MT3D code, using literature values for the distribution coefficient and Freundlich exponent, transformed to fit the assumed prevailing soil properties of the saturated zone in Stockholm.

Simulation of groundwater flow and transport of heavy metals was conducted in three and of phosphorus and nitrogen in four catchments in central Stockholm (Figure 1). The total area of the four catchments is 22 km^2, which is about 12% of the total area of the City of Stockholm.

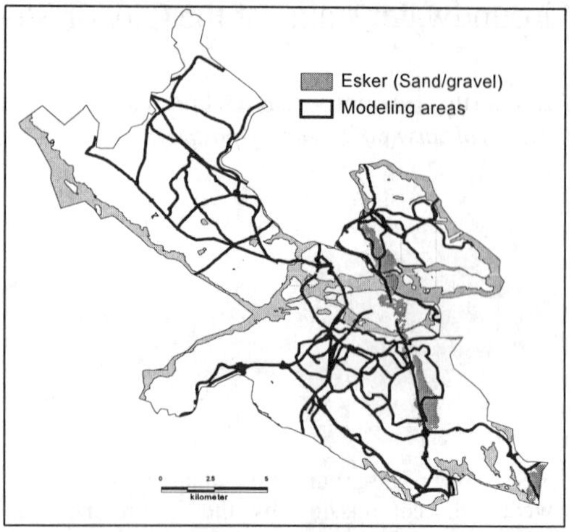

Figure 1. Locations of modelled catchment areas.

For the modelling, the areas were divided into two layers. The depth of layer 1 was approximated to 4 m, which is an average of the available data on depths of the clay covering the coarser friction material, mainly till deposits, in the valleys. Layer 2 represents the friction material and has no specified bottom, as this layer has hydraulically been characterised by the transmissivity.

Based on the geological maps the areas were divided into geological units. The units were given estimated hydraulic properties as conductivities or transmissivities, leakage factors, total and effective porosities.

The morphology of the surface of layer 2 was adjusted according to a detailed digital elevation model of the Stockholm area. The artificial fillings that cover parts of the primary geological units are not considered in the set-up, but are considered in the calculation of groundwater recharge.

The model was calibrated against known groundwater levels. This was done by adjusting the values of groundwater recharge.

3 RESULTS

3.1 *Hydrogeological characteristics*

The only significant groundwater resource in Stockholm is the Stockholm esker consisting of glaciofluvial sand and gravel deposits (Figure 1). The parts of the esker not covered by clay layers are important recharge areas to a number of small, local, aquifers with open or confined conditions. The Stockholm region is generally characterised by pronounced valleys with glacial clay deposits covered by post-glacial clay and peat in the lowlands. The glacial till deposits and crystalline bedrock underlying the clays contains groundwater, often in amounts enough to provide a couple of households with drinking water. Although these resources are unimportant for the public water supply of Stockholm, they are important as transporting agents for pollutants towards discharge areas and finally to Lake Mälaren and the Baltic Sea.

As for all major urban areas there are significant human impacts on the groundwater in Stockholm. Groundwater recharge is greatly inhibited by land sealing and surficial drainage systems. Tunnels and other subsurface constructions affect groundwater levels and flow directions. Dual or triple groundwater levels are common, e.g. in the superficial fillings covering impermeable clays, in the underlying glacial deposits and in the fractured bedrock aquifer.

3.2 *Groundwater quality*

The Stockholm groundwater generally has a high electrical conductivity (66 $mS.m^{-1}$), pH (7.5) and TOC (13.2 $mg.L^{-1}$) compared to median values in Sweden (16 $m.Sm^{-1}$, 6.4 and 3.4 $mg.L^{-1}$ respectively). Especially calcium (111 $mg.L^{-1}$), alkalinity (360 $mg.L^{-1}$ as HCO_3), sodium and chloride (62 $mg.L^{-1}$) concentrations are remarkably high. The concentrations are 6 - 10 times higher than the country average and about double the concentrations for the region, which, for Swedish standards, has a very high total hardness and alkalinity due to high calcareous content in the soils. De-icing of roads with sodium chloride evidently causes the high chloride concentrations, even though marine clays can contribute somewhat to the elevated levels. Low nitrate concentrations in relation to ammonium concentrations indicate a low redox potential in the groundwater under the clays.

The most frequent pollutants of the groundwater are heavy metals. Practically all sites have elevated concentrations compared to groundwater in forested areas. Most severe are extremely high mercury and generally high copper concentrations. Bacteria indicating fecal and other pollution were found at more than half of the sites, while biocides and petroleum hydrocarbons were found at almost half of the sites. Strongly elevated nitrogen and phosphorous concentrations were found along a passage in the centre of the city indicating leakage of the sewage system. PAHs were found in almost a quarter of the sites, while PCBs were not detected at all. Tables 1 and 2 presents some statistics on concentrations of the compounds that were of interest for transport estimations.

Table 1. Median values of heavy metal concentrations in Stockholm groundwater compared to groundwater in forested areas. Metal concentrations are given in $\mu g.L^{-1}$ with the exception for Hg, which are given in $ng.L^{-1}$. TOC is given in $mg.L^{-1}$. *Based on few values.

Variable	Median		CV_{Sthlm} %	$Median_S/Median_F$
	Sthlm	Forest		
As	0.54	0.12	163	4.5
Cd	0.052	0.027	156	1.9
Co	1.27	0.21	273	6.1
Cr	0.79	0.22	115	3.6
Cu	8.63	0.84	122	10.3
Hg	15.6	2.0*	435	7.8
Ni	7.03	2.19	122	3.2
Pb	0.56	0.18	109	3.2
Zn	30.7	17.0	125	1.8
PH	7.50	6.40	6	
TOC	9.75	3.4	87	

Table 2. Median and mean concentrations ($\mu g.L^{-1}$) of nitrogen and phosphorus compounds in Stockholm groundwater compared to groundwater in forest ecosystems

Variable	Median		Mean		CV_{Sthlm}
	Sthlm	Forest	Sthlm	Forest	
NH_4-N	84	23	921	56	549
NO_3-N	25	57	303	220	224
Tot-N	1 305	-	2 666	-	185
Tot-P	98	-	434	-	259

3.3 *Modelling of groundwater flow*

The simulation with MODFLOW resulted in a detailed picture of flow directions and estimates on the quantity of groundwater leaving the four catchments directly to the main surface water bodies.

The yearly total outflow of groundwater was 560 000 m^3, which equals 26 mm.

3.4 *Groundwater transport modelling*

The most challenging issue was to estimate the concentrations of the different heavy metals, nitrogen and phosphorous discharging into the major surface water bodies. There were very few sampling sites per area and even fewer were located close to the discharge areas. Only one was actually situated at the border of discharge and could thus be used directly for the estimation. In the remaining discharge areas, the concentrations predicted by the model simulations have been used.

Sampling sites with elevated concentrations in groundwater have served as sources, i.e. the initial concentrations of the simulations. Concentrations generated by atmospheric deposition have been added to the surface of layer 1. The model has been allowed to simulate concentration propagation for different time spans up to 109 years. Simulations giving the highest concentrations in the areas of discharge have been used for the transport calculations. These concentrations are still very low and close to the detection limits and have served as input for the calculation of a lower limit of the groundwater transported metal contribution. The use of median concentrations (Table 1) probably gives a more realistic estimation of the transport. Medians have been used here as the main alternative. To calculate a plausible upper limit of the transport, the 90th percentile values were used. The extreme outliers of the investigation have thus been excluded. According to the simulation of concentration propagation, these have not had an immediate influence on the concentrations in the discharge areas. Extrapolating the results from one representative catchment, the annual total contribution of heavy metals, nitrogen and phosphorous transported by groundwater has been estimated (Table 3).

Table 3. Annual contribution of some heavy metals and nutrients to Lake Mälaren and The Baltic Sea and the areal loss in the City of Stockholm.

Determinand	Annual contribution ($kg.y^{-1}$)	Areal loss ($g.ha^{-1}.y^{-1}$)
As	2.2	
Cd	0.2	0.036
Co	4.9	
Cr	3.1	
Cu	34	
Hg	0.06	0.011
Ni	28	
Pb	2.2	0.40
Zn	118	
Tot-N	4 712	250
Tot-P	425	22

4 DISCUSSION AND CONCLUSIONS

The groundwater in Stockholm has elevated concentrations of all heavy metals investigated. Most conspicuous are the locally extremely high Hg concentrations and the generally high concentrations of Cu. The model simulation however, shows that the extremely high Hg concentrations have not yet been present in groundwater discharged to the surface waters, provided the source has not been there for more than 109 years.

The results on heavy metal concentrations suggest that, apart from point and diffuse polluting sources, the chemical conditions of Stockholm groundwater play a major role for the concentration range. Hence Zn and Cd are the least elevated heavy metals in the groundwater of high pH values. The conditions are more favourable for Hg and Cu, which to a great extent are transported in association with organic matter.

Extremely high nitrogen and phosphorus concentrations were located along a passage in the centre of the town indicating leakage of the sewage system.

The direct contribution of heavy metals and nutrients to the main surface water bodies by groundwater is merely one third of the total amount transported with groundwater. The remainder flows into streams and small lakes, seeps into tunnels or is drained by the drainage system.

The transportation of the metals Hg, Cd and Pb with groundwater is much smaller in Stockholm than it is with shallow groundwater in a forested catchment area in the middle of Sweden. In the forested catchment area the fluxes of Hg, Cd and Pb were 0.034, 0.063 and 6.27 $g.ha^{-1}.y^{-1}$ (Aastrup et al., 1991 and 1995) compared to the total fluxes in Stockholm, which have been estimated to 0.011, 0.036 and 0.40 $g.ha^{-1}.y^{-1}$ respectively. The reason is the small groundwater recharge in the urban area due to land sealing and artificial drainage. This implies that the major part of the urban effective precipitation take other routes towards the major surface water bodies. The extremely elevated nitrogen and phosphorous concentrations were located within one the modelled catchments. The areal loss from that catchment was in the same order as from a strongly cultivated catchment, but from Stockholm as a whole the contribution by groundwater transported nutrients to the eutrophication of the Baltic is of minor importance.

REFERENCES

Aastrup, M., Johnson, J., Bringmark, E., Bringmark, L. & Iverfeldt, Å. 1991, Water, Air, and Soil Pollut. 56:1, 55-67.

Aastrup, M., Iverfeldt, Å., Bringmark, L., Kvarnäs, H., Thunholm, B. & Hultberg, H. 1995, Water, Air, and Soil Pollut. 85:2, 755-760.

Astrup, M. & Thunholm, B. 1999. Nitrogen and Phosphorus transport by groundwater to Mälaren, Saltsjön and Brunnsviken within the city of Stockholm - a modeling study, 32 pp (in Swedish).

Anderberg, J. 1997. Groundwater resources and vulnerability mapping of the City of Stockholm. In: Ottoson, L. (Ed) Proceedings 18th ICA/ACI International Cartographic Conference 97, Stockholm Sweden, 23 – 27 June 1997, vol 4, 1997 - 1984.

Environment and Health Protection Administration in Stockholm (EHPAS) 1997. Groundwater in Stockholm - Resource, Vulnerability, Quality. Report, 103 pp (in Swedish).

McDonald, M. G. & Harbaugh, A. W. 1988. A modular three-dimensional finite-difference groundwater flow model: US Geological Survey Techniques of Water Resources Investigations, Book 6, Chapter A1, 586 pp.

Nordic Council of Ministers 1989. Methods for Integrated Monitoring in the Nordic Countries, Miljörapport 1989:11.

Zheng, C. 1990. MT3D, A Modular Three-Dimensional Transport Model for Simulation of Advection, Dispersion and Chemical Reactions of Contaminants in Groundwater Systems. Prepared for United States Environmental Protection Agency and Robert S. Kerr Environmental Research Laboratory, Ada, Oklahoma 74820.

Groundwater: Past Achievements and Future Challenges, Sililo et al. (eds) © 2000 Balkema, Rotterdam, ISBN 90 5809 159 7

A review of impacts of forestry on groundwater and implications for forest management

A.R.Allen & D.V.Chapman
Department of Geology, University College Cork, Ireland

ABSTRACT: Forestry affects groundwater by diminishing groundwater recharge to as little as a tenth of that under grass or heathland, largely due to the greater uptake of soil water by trees and the increased water-holding capacity of forest soils. Furthermore, afforestation can lead to a significant reduction in surface runoff, by as much as 20%, mainly due to interception of rainfall by the forest canopy. Reduced streamflow is compensated by enhanced loss of groundwater to effluent streams. Groundwater quality may be affected by enhanced acidification and nitrification processes under forests, due mainly to a significant increase in the organic content of forest soils. The reduced recharge rate under forests means that any deterioration in groundwater quality resulting from afforestation may take a decade or more to become apparent. Implications for forest management are discussed.

1 INTRODUCTION

Many countries have invested heavily in forestry over the past century, as a means to exploit land unfit or only marginally fit for agriculture, or to restore some of the forests which have been decimated over the centuries for industry, fuelwood, or other agricultural activities. Most reafforestation has been for commercial purposes, using fast growing, commercially exploitable species rather than native species often more suited to the ambient climatic conditions.

Ireland is an example of a country in the process of rapidly increasing its forestry holdings. Over the centuries, Ireland has had its landscape denuded of forests to the extent that, despite the fact that it has a climate particularly suited to the growth of trees, it has the lowest proportion of land under forestry (8%) than any other European country (av. 33%; Stanners & Bordeau, 1995). In order to change Ireland's position from a net importer of timber and other forest products to a net exporter, the Irish Government has recently embarked on a scheme to more than double the proportion of land under forestry to 17% over the next 35 years. It plans to achieve this through conversion of large areas of mainly arable and grazing land to forestry, concentrating mainly on coniferous woodlands, and in particular on one fast growing species, *Picea sitchensis* (Sitka Spruce). The scheme will undoubtedly stimulate rural economies and create jobs, but has failed to recognize potential negative environmental impacts of such a rapid major land use change.

In general, major land use changes tend to be accompanied by significant environmental impacts, the most striking modern example being the rapid degradation and loss of productivity of tropical soils associated with the destruction of the world's tropical forests. Conversion of large areas of land, often previously arable or grassland, to forestry will also have major environmental impacts. Many of the impacts are visually apparent, such as changes in landscape patterns and in terrestrial and aquatic wildlife habitats. However, other impacts may be effectively "invisible" and only become recognized when an important human resource is affected. Groundwater falls into the latter category, suffering from the "out of sight, out of mind" syndrome. Growing international concern for the quality and quantity of fresh water resources, has led to considerable research into the various impacts on water bodies. This paper considers the impacts of forestry on groundwater and implications for forestry management.

Two potential impacts of forestry on groundwater are :-

- Diminished Recharge – reduction in the quantity of rainwater percolating down to the water table.
- Decline in Quality – the chemistry of soil water percolating downwards, due to the unique biogeochemical processes operating in forest soils.

Most research on forestry impacts on hydrology has been concerned with surface water, because the impacts of forestry on groundwater are far more complex and difficult to measure and, therefore, to quantify.

2 RECHARGE

The main impact of afforestation on water resources is without exception and regardless of the type of forest cover, a reduction in water yield (the proportion of total rainfall reaching the ground surface to undergo infiltration or surface runoff) (Bosch & Hewlett 1982, McCulloch & Robinson 1993). Conversely, clearfelling, i.e. reduction in forest cover, increases water yield.

Reduction in water yield by forests is due to interception of rainfall by the forest canopy (Horton 1919), and increased evaporation, leading to a loss of rainwater reaching the ground surface. This results in a reduction in surface runoff (estimated to be of the order of 20% for upland areas in the UK) (Calder & Newson 1980) and erosion, and ultimately in streamflow. Both negative and positive impacts accrue from this relationship. On the positive side, flooding is partially alleviated and flood crests are reduced (Hoyt & Langbein 1955). Negative impacts include the drying up of streams and a significant overall reduction in water yield (Bosch & Hewlett 1982), which is particularly serious in situations where forestry has been established in the catchments of rivers harnessed for hydroelectric power or used for public water supplies (Calder & Newson 1980). Furthermore, reduced summer low flows (Robinson et al. 1991), attributed to higher summer transpiration and evaporation, as well as decreased groundwater recharge in winter arising from increased winter evaporation, have significant consequences for agriculture.

A significant proportion of the intercepted rainwater undergoes evaporation directly back to the atmosphere. This is known as interception loss (Stewart 1968) and for coniferous forests in the UK and Ireland it is roughly twice that of adjacent grassland areas (Calder & Newson 1979). Interception loss depends on the climate (primarily rainfall and wind regimes and evaporative demand during rainfall) and the structure of the vegetation which determines its water holding capacity and its roughness (Hall et al. 1996). Interception losses expressed as a fraction of annual precipitation for UK uplands are about 30–40%, their values decreasing with increasing rainfall (Harding et al. 1992). Interception is greater for coniferous and eucalyptus forest cover, than for deciduous species. A very crude estimate of the change in annual water yield per 10% change in forest cover is ~ 40 mm for conifers and ~ 25 mm for deciduous species (Bosch & Hewlett 1982), but these are global figures which do not take into account local climatic conditions, particularly mean annual precipitation (MAP).

Forests also affect the water budget by transpiration loss. This is the transfer of water from the soil through the tree roots, trunk and leaves. In the UK, the water use of coniferous plantations ranges from about 600 mm per year in the east, where transpiration loss is the main component, to about 1200 mm per year in the upland areas of the west, where interception loss dominates (Hall et al. 1996). Less data on the water use of deciduous trees is available, but mean annual transpiration losses of ~330 ± 35 mm measured by Roberts (1983) for a number of forest sites in NW Europe indicate remarkably similar transpiration totals for both coniferous and deciduous forests. This is indicative of a strong physiological control on transpiration.

The most significant effect of forest cover on the surface/groundwater balance is an increased proportion of the total rainfall, which infiltrates down through the forest canopy and the underlying forest litter into the soil, compared to that for grasslands or arable land (McCulloch & Robinson 1993). This suggests that recharge should be enhanced by forest cover. However, the rooting depths of trees are generally greater than for short crops, and the water requirements of trees are significantly greater than that of grass or crops. The combined result is a greater removal of water from the soil by tree roots and from deeper levels in the soil profile, outweighing the benefits of higher infiltration (McCulloch & Robinson 1993). Calder et al. (1992) found that the annual water use of trees without access to groundwater can be as much as twice that of an annual agricultural crop. This will reduce the moisture content of the soil, leading to a decreased proportion of soil water in the unsaturated zone percolating down to the water table.

Studies in the UK have confirmed this. Cooper (1980) found that percolation to the water table under coniferous forest was 44% less than under grass in a clearing, whilst Moss & Edmunds (1989) determined that the rate of recharge of the water table under forest is at least three times lower than under heathland (~330 mm/a), and may be as little as about 20 mm/a. This indicates that the residence time for water in the unsaturated zone under forests is significantly greater than that under heathlands, possibly as much as a decade or two (Moss & Edmunds 1992). A reduction in rate of recharge of this order means that any deterioration in groundwater quality, associated with a change in land use to forestry, will take many years to be reflected in the quality of groundwater pumped from neighbouring water wells (Hall et al. 1996). This increased water-holding capacity is due mainly to the higher organic content of forest soils (Virzo De Santo 1992).

A further impact of forests on drainage and recharge is their effect on the location of the zero flux plane. During dry periods (mainly summer), removal of soil water, by direct surface evaporation and by plant roots, gives rise to a soil moisture deficit, leading to the establishment of a zero flux plane (ZFP), which divides upwards groundwater movement in the upper part of the profile from downwards draining fluxes below (Richards et al. 1956). The soil moisture deficit is significantly greater under forests than grassland, so the depth of the ZFP, which varies with the magnitude of the soil moisture deficit, is generally greater under trees than under grass (Cooper, 1980). Where the water table is relatively close to the surface, the ZFP may underlie the water table, so groundwater from the saturated zone may be moved upwards by capillary action to higher levels in the soil profile in order to reduce the soil moisture deficit. Ultimately, such groundwater may be lost by a combination of plant removal and evaporation leading to a reduction in the level of the water table (a negative recharge). Thus, despite enhanced infiltration of rainwater into the soil under forest cover, there appears to be a reduction in recharge as a proportion of total precipitation, and during summer dry spells there may even be a negative recharge (discharge).

One further impact of forestry on groundwater resources, is the overall balance between surface flow and groundwater. As a response to the reduction in streamflow within a catchment, brought about by forestry, the proportion of groundwater feeding effluent streams in the catchment may be enhanced. As a result, discharge of groundwater will tend to increase, leading to an overall lowering of the water table in the catchment.

3 QUALITY

The quálity of both surface water and groundwater is a function of a number of factors (Hall et al. 1996):-

- Chemistry of atmospheric inputs, e.g. rainwater.
- Weathering of minerals in soils and rocks.
- Inputs derived from diffuse agricultural sources (fertilisers or enhanced mineralisation).
- Anthropogenic inputs from point sources, e.g. factories, or more generally from urban areas.
- Amount of rainfall and evaporation.

In general, surface waters are more susceptible to a deterioration in water quality than groundwater. Soil and geological formations can mediate groundwater contamination in the unsaturated zone before pollutants reach the water table. Even in the saturated zone, exchange processes and reactions between solid mineral grains and the groundwater can bring about attenuation of contaminants in the groundwater. The two most critical aspects of water quality related to forest cover are 'acidification' and 'nitrification'.

3.1 Acidification

Trees scavenge atmospheric pollutants about 2–3 times more efficiently than short crops (Brechtel 1992), because of their greater aerodynamic roughness and larger leaf area (Harding et al. 1992). Because conifers are aerodynamically rougher, they are likely to be more efficient scavengers than broadleaf species. The role of forests in acidifying upland streams is now primarily attributed to the enhanced scavenging of acid pollutants, i.e. deposition on the tree canopy of industrial emissions of acidic oxides, principally SO_2 and NO_x derived from fossil fuel combustion, non-ferrous metal smelting, fertiliser and sulphuric acid manufacture and other industrial processes, together with ammonia emissions from agricultural sources (Meybeck et al. 1989). This occurs both by wet deposition of cloud and mist droplets, which contain significantly higher ionic concentrations than larger rain droplets, and by dry deposition through impaction of particles and from reactive gases (Fowler et al. 1989). These pollutants are then washed down to the ground surface by stemflow and leafdrip processes, where they either infiltrate into the ground or are washed into surface streams. Marked reductions in salmonid fishstocks, stream invertebrate diversity and wildlife dependent on rivers for their food supply are attributed to acidification of upland streams.

There are additional effects associated with forest development, that also contribute to acidification, such as loss of base cations from the soil to the biomass, increased supply of highly acidic organic matter (leaf litter) to the soil, and changes in hydrological flow pathways. An increased supply of acidic leaf litter influences strongly the acidification of water infiltrating into the ground and percolating down through the unsaturated zone. Increased concentrations of H^+, SO_4^{2-} and NO_3^-, scavenged from the atmosphere and released by the forest litter and humus of the surface layers, gives rise to highly acidic infiltrating water (pH < 5). This water interacts with the soil or bedrock as it percolates downwards, leading to gradual loss of essential nutrients, such as the base cations, Ca^{2+} and Mg^{2+}, from these media, which in turn may lead to forest decline (Becquer et al. 1992). Highly siliceous bedrock and overburden is especially sensitive to acidification.

The interaction between groundwater and the media through which it passes, leads to weathering and ion exchange processes in the soil and bedrock, giving rise to a complementary mediating effect on the groundwater that reduces its acidity. The magnitude of this neutralising effect will be dependent on the mineral composition of the rock or soil, particularly the carbonate and clay contents and the quantity of organic matter present as well as on the residence time of water passing through. However, dissolution of other minerals, e.g. aluminosilicate minerals such as feldspar, may, in the absence of carbonates or high clay contents, also have a significant neutralising effect (Moss & Edmunds 1992). Carbonate bedrock, bedrock with a carbonate cement, or overburden with abundant carbonate float (such as might be deposited by a glacier which had sampled a carbonate terrain) have the greatest buffering capacity and would normally completely neutralise highly acid groundwater. Intermediate to mafic volcanic and plutonic igneous rocks, mafic, pelitic and calcsilicate metamorphic rocks and argillaceous sediments have a moderate buffering capacity, whilst acid igneous rocks, quartzofeldspathic metamorphic rocks and arenaceous sediments have a low buffering capacity. Clay-rich overburden could be expected to have a moderate to high buffering capacity, depending on the clay mineral types present.

Organic processes involved in mediating groundwater composition in the unsaturated zone usually increase the acidity of the percolating groundwater, and include nitrification, base cation uptake by vegetation, production of organic acids in decaying vegetation and oxidation of reduced forms of sulphur (Reuss et al. 1987). Organic-rich soils underlying forests buffer interstitial water at a lower pH than heathland soils (Moss & Edmunds 1992), which in turn are more acid than grassland soils. Furthermore, the rate of assimilation of strong acid anions, e.g. SO_4^{2-}, is higher in forest soils.

The inorganic processes of groundwater mediation include dissolution of carbonate by the acidified groundwater, to produce Ca^{2+} and less frequently Mg^{2+}, rapid cation exchange reactions, principally with clay minerals, and slow mineral weathering. In carbonate-free rocks and soils, the buffering capacity will be dependent on the rate of mineral weathering reactions in relation to water flux, and the magnitude of the acid neutralising capacity of the mineral reactions (Moss & Edmunds 1992). Rates of mineral reactions are highly dependent on the pH of interstitial water, for example the rate of alkali feldspar dissolution is greater at lower pH. However, rates of mineral dissolution above the water table will also be critically dependent on the flux of water draining downwards, which in forest soils is significantly lower than in grassland soils. So mineral dissolution in forest soils is likely to be considerably slower than in grassland soils.

Weathering of parent aluminosilicate minerals, e.g. alkali feldspar, produce a variety of secondary clay minerals, such as kaolinite, montmorillonite and gibbsite (Meybeck et al. 1989). Where groundwater has a pH < 4, gibbsite and other aluminium hydroxides may undergo further dissolution, releasing toxic labile monomeric Al^{3+} (Driscoll 1985). Increasing acidity may also mobilise other trace metals, e.g. Cd, Mn, Fe, As and Hg from soils (Meybeck et al. 1989). Mobile anions, e.g. SO_4^{2-}, arising from atmospheric deposition, may leach positively-charged ions to maintain electroneutrality (Seip 1980). In well-buffered unsaturated zones, SO_4^{2-} will be leached down to the saturated zone accompanied by the base cations Ca^{2+} and Mg^{2+}, whereas in poorly-buffered unsaturated zones, H^+ and Al^{3+} will be leached, resulting ultimately in decreases in alkalinity and pH (Meybeck et al. 1989). As indicated above, forest soils may be able to absorb a large proportion of the deposited SO_4^{2-}, which may temporarily inhibit groundwater acidification. Where forest soils become saturated with nitrate, NO_3^- may contribute significantly to acidification of groundwater (Henriksen & Brakke 1988).

3.2 Nitrification

Nitrogen inputs, principally in the form of nitrate, can affect groundwater quality and its suitability for human and livestock consumption. Atmospheric nitrogen may be scavenged by forests as NO_x, mainly derived from fossil fuel combustion, fertiliser manufacture and other industrial processes, and NH_4 emissions from agricultural sources, such as animal manures. Nitrogen may also be derived from soil organic matter and fertiliser application (Hall et al. 1996), the latter being the dominant source of groundwater nitrate contamination in Europe and North America (van der Voet et al. 1996). Both the increased efficiency of pollutant scavenging by forests and the significantly enhanced supply of organic matter to the soil, can lead to 'nitrogen saturation' of forest soils (Gundersen & Rasmussen 1990), i.e. nitrogen in excess of the demands of tree growth, particularly in mature and slowly-growing forests (Hall et al. 1996). Once a forest becomes N-saturated, the excess nitrate is leached by percolating groundwater down to the saturated zone (Johnson 1992).

The major soluble forms of nitrogen in water are nitrate and ammonia. High levels of organic matter in forest soils promote ammonification, provided decomposition of the organic matter keeps pace with its accumulation. In the event that it doesn't, such as in young or fast growing forests (Nihlgård 1997), excess nitrogen, will be converted to nitrate. Nitrification gives rise to a net acidification effect if the nitrification rate in the soil (an H^+ producing process) exceeds the rate of uptake of nitrate by roots and soil micro-organisms (an H^+ consuming process) (Bredemeier & Marmann 1992). Leaching of nitrogen from forest soils after nitrification, coincides with increased acidification and accumulation of nitrate and toxic aluminium concentrations in the groundwater (Tietema et al. 1992).

Nitrification can be reversed, and thereby acidification of the soil inhibited, by denitrification, an anaerobic respiratory process, which results in the return of nitrogen, principally in the form of nitrous oxide, a greenhouse gas, to the atmosphere (Virzo De Santo 1992). Which of these competing physiological processes predominates in forest soils, is dependent on a variety of factors, including soil composition and texture, soil moisture content, soil water pH, quantity of organic matter in the soil surface layers, climate, particularly rainfall and temperature, microbial population, and availability of carbon and oxygen (Virzo De Santo 1992). Seasonal variation in climatic conditions and the growth activity of trees can lead to switches from nitrification to denitrification and vice versa (Alfani *et al.* 1983), as can short-term weather variations such as rainfall (Binstock 1984). In addition, clearfelling can also affect the balance of nitrification/denitrification processes, because this leads to a two- to three-fold increase in N-mineralisation (Khanna et al. 1992). Where slash is left on the ground after clearfelling, nitrification is enhanced significantly, whereas removal of surface organic debris during site preparation for replanting favours denitrification processes (Virzo De Santo 1992). Fertiliser addition and liming to improve soils and promote tree growth also promotes N-mineralisation and nitrification (Khanna et al. 1992), but the latter, by increasing the base cation supply to the soil, also reduces acidification (Rapp 1992). Fires may also affect the nitrogen balance in forest soils temporarily, bringing about a two- to five-fold increase in soil mineral N-levels (N-mineralisation) as a consequence of heating in the surface (0-5cm) soil (Keith & Raison 1992). This could also lead to enhanced leaching of nitrate by groundwater.

3 DISCUSSION AND CONCLUSIONS

The relationships between forestry and groundwater outlined above, have consequences for forestry development and management. Furthermore, the strong correlation between the extent of forest cover in a catchment and flooding (Hoyt & Langbein 1955), suggests that strategic afforestation of upland source areas of flood prone rivers could significantly reduce the frequency of flooding, and also moderate flood crests. Where major river systems traverse international boundaries, cooperation between governments in the planning development and management of forests is essential. Likewise, harvesting of forests needs to be coordinated, because wholescale clearfelling over a short time period of large areas of forest cover within individual catchments can have disastrous consequences in terms of flood events, even hundreds of kilometres away in the case of major river systems. Ireland, a country with major winter flooding problems, could benefit from strategic forestry planning. Similarly, strategic forestry planning could potentially alleviate some of the catastrophic flooding that has affected countries occupying the lower reaches of some of the world's major rivers in recent years. International cooperation in regional forestry management could have significant potential benefits for flood-prone regions of the world.

However, there are other potential impacts of afforestation on water balance. In particular the significantly reduced recharge of groundwater under forests could, in arid and semiarid regions and where rainfall is unevenly distributed throughout the year, have serious impacts on groundwater supplies for drinking and irrigation. Furthermore, certain forest species, e.g. eucalypts, have a particularly heavy demand on groundwater, and agriculture adjacent to large eucalyptus plantations may be adversely affected by depletion of local groundwater resources.

Acidification and nitrification processes associated with forests can result in deterioration of groundwater quality under forest cover. However, this may not be detected for perhaps a decade or so after planting, due to the tenfold reduction in recharge flux, by which time these processes may be difficult to reverse. Furthermore, introduced species may not be so well adapted to the climatic regime of a region as native species, but this may only become apparent in second or third generation growth due to forest decline brought about by the development of an imbalance in nutrient cycling (Nihlgård 1997). Nutrient cycling in forest soils is a delicately balanced process, disturbed by climatic fluctuations, groundwater deterioration, absence of diversity of forest species, and forests under stress. Nutrients lost from the cycle are not readily replaced under forests, as the diminution in recharge flux will severely restrict the process of mineral weathering in the soil, a major mechanism of nutrient replenishment. Monospecies plantations lead to loss of biodiversity and landscape degradation, and run the risk of catastrophic financial loss and major environmental degradation if the single species becomes afflicted by a species-specific disease or pest infestation. Despite its disease-resistant character, Ireland's reliance on Sitka Spruce (60% of all new plantings) could ultimately be very short sighted.

In the light of the above, it is essential that national governments exercise control over development and management of forests, in order to ensure that sound ecological and environmental principles are maintained. Many private forestry operators are concerned only with exploitation of forests as a resource, and even the most enlightened companies in the private sector are commercial enterprises, driven by financial motivation. In the event of a conflict of interest, environmental principles are likely to become subordinated to economic considerations.

REFERENCES

Alfani, A., Fioretto, A., Virzo De Santo, A., & Russo, G. 1983. Denitrification potential of beech soils as influenced by the seasonal cycle. *Pedobiologia* 25: 149-156.

Becquer, T., Herbillon, A., Merlet, D., Boudot, J.P., & Rouiller, J. 1992..Importance of the nitrogen and sulfurcycles in the proton budget in a declining fir forest and its relation to aluminium toxicity. In: A. Teller, P. Mathy & J.N.R. Jeffers (eds.) *Responses of forest ecosystems to environmental changes*: 679-680. New York: Elselvier Applied Science.

Binstock, D.A. 1984. Potential denitrification in an acid forest soil; dependence on wetting and drying. *Soil Biol. Biochem.* 16: 287-288

Bosch, J.M. & Hewlett, J.D. 1982. A review of catchment experiments to determine the effect of vegetation changes on water yield and evapotranspiration. *J. Hydrol.* 55: 3-23.

Brechtel, H.M. 1992. Impact of acid deposition caused by air pollution in Central Europe. In: A. Teller, P. Mathy & J.N.R. Jeffers (eds.) *Responses of forest ecosystems to environmental changes*: 594-595. New York: Elselvier Applied Science.

Bredemeier, M. & Marmann, P. 1992. Acidification pulses due to nitrification in forest soils. In: A. Teller, P. Mathy & J.N.R. Jeffers (eds.) *Responses of forest ecosystems to environmental changes*: 732-734. New York: Elselvier Applied Science.

Calder, I.R. & Newson, M.D. 1979. Land use and upland water resources in Britain – a strategic look. *Water Resource Bull.* 16: 1628-1639.

Calder, I.R. & Newson, M.D. 1980. The effects of afforestation on water resources in Scotland. In: *Land assessment in Scotland. Proc Symp Roy Geog Soc Edinburgh* 51-62

Calder, I.R., Swaminath, M.H., Kariyappa, G.S., Srinivasala, N.V., Srinivasala Murthy, K.V. & Mumtaz, J. 1992. Deuterium tracing for the estimation of transpiration from trees. *J. Hydrol.* 130: 37-47.

Cooper, J.D. 1980. Measurement of moisture fluxes in unsaturated soil in Thetford Forest. *Inst. of Hydrology Rept.* No 66: 97pp.

Driscoll, C.T. 1985. Aluminium in acidic surface waters: chemistry, transport and effects. *Environ. Health Perspec.* 63: 93-104.

Fowler, D., Cape, J.N. & Unsworth, M.H. 1989. Deposition of atmospheric pollutants in forests. *Phil. Trans. Roy. Soc. London* 324: 247-265.

Gundersen, P. & Rasmussen, L. 1990. Nitrification in forest soils: effects from nitrogen deposition on soil acidification and aluminium release. *Revs. Env. Contam. Toxicol.* 113: 1-45.

Hall, R.L., Allen, S.J., Rosier, P.T.W., Smith, D.M., Hodnett, M.G., Roberts, J.M., Hopkins, R., Davies, H.N,, Kinniburgh, D.G., & Gooddy, D.C. 1996. Hydrological effects of short rotation energy coppice. *ETSU Rept. B/W5/00275*: 204pp.

Harding, R.J., Neal, C. & Whitehead, P.G. 1992. Hydrological effects of plantation forestry in north-western Europe. In: A. Teller, P. Mathy & J.N.R. Jeffers (eds.) *Responses of forest ecosystems to environmental changes*: 445-455. New York: Elselvier Applied Science.

Henriksen, A. & Brakke, D.F. 1988. Increasing contributions of nitrogen to the acidity of surface waters in Norway. *Water Air Soil Pollut.* 14: 183-201.

Horton, R. E. 1919. *Monthly Weather Review* 47: 603-623

Hoyt, W.G. & Langbein, W.B. 1955. *Floods*: Princeton: Princeton University Press.

Johnson, D.W. 1992 Nitrogen retention in forest soils. *J. Env. Qual.* 21: 1-12

Khanna, P,K., Meiwes, K,J. & Bauhus, J. 1992. Nitrogen dynamics in some beech forests as measure of resilience to changes due to management practices and anthropogenic impacts. In: A. Teller, P. Mathy & J.N.R. Jeffers (eds.) *Responses of forest ecosystems to environmental changes*: 664-665. New York: Elselvier Applied Science.

Keith, H. & Raison, R.J. 1992. Effects of prescribed fire on nitrogen cycling and tree growth in an Australian eucalypt forest. In: A. Teller, P. Mathy & J.N.R. Jeffers (eds.) *Responses of forest ecosystems to environmental changes*: 889-891. New York: Elselvier Applied Science.

McCulloch, J.S.G. & Robinson, M. 1993. History of forest hydrology. *J. Hydrol.* 150: 189-216

Meybeck, M., Chapman, D.V. & Helmer, R. (eds.) 1989. *Global freshwater quality – A first assessment.* Blackwell Reference

Moss, P.D. & Edmunds, W.M. 1989. Interstitial water-rock interaction in the unsaturated zone of a Permo-Triassic sandstone aquifer. In: Miles, (ed.) *Water-Rock Interactions.* Rotterdam: Balkema.

Moss, P.D. & Edmunds, W.M. 1992. Processes controlling acid attenuation in the unsaturated zone of a Triassic sandstone aquifer (U. K.), in the absence of carbonate. *Applied Geochemistry* 7: 573-583.

Nihlgård, B.J. 1997. Forest decline and environmental stress. In: D. Brune, D.V. Chapman, M.D. Gwynne & J.M. Pacyna (Eds.) *The global environment. Science, technology and management*: 422-440. Weinheim: VCH.

Rapp, C. 1992. Effects of liming and N-fertilization on soil chemistry, biomass and nutrient content of fine roots in a mature beech stand in the Solling area (Germany). In: A. Teller, P. Mathy & J.N.R. Jeffers (eds.) *Responses of forest ecosystems to environmental changes*: 829-830. New York: Elselvier Applied Science.

Reuss, J.O., Cosby, B.J. & Wright, R.F. 1987. Chemical processes governing soil and water acidification. *Nature* 329: 27-32.

Richards, L.A., Gardner, W.R., Ogata, G. 1956. Physical processes determining water loss from soil. *Soil Sci. Soc. Amer. Proc.* 20: 310-314.

Roberts, J.M. 1983. Forest transpiration: a conservative hydrological process. *J. Hydrol.* 66: 133-141.

Robinson, M., Gannon, B. & Schuch, M. 1991. A comparison of the hydrology of moorland under natural conditions, agricultural use and forestry. *Hydrol. Sci. J.* 36: 565-577.

Seip, H. M. 1980. Acidification of freshwater - sources and mechanisms. In: D. Drabløs & A. Tollan (eds.) Ecological impact of acid precipitation. Proceedings of an International Conference, Sandefjord, Norway, 11-14 March, 1980: 358-366. Oslo

Stanners, D. & Bourdeau. P. (eds.) 1995. *Europe's environment: The Dobris assessment.* Copenhagen: European Environment Agency.

Stewart, J.B. 1968 Evaporation from forests. *Institute of Hydrology Report.* No 3

Tietema, A., Riemer, L. & Verstraten, J.M. 1992. Nitrification of acid forest soils in the Netherlands; vertical distribution, regulating factors and implications for soil acidification. In: A. Teller, P. Mathy & J.N.R. Jeffers (eds.) *Responses of forest ecosystems to environmental changes*: 763-764. New York: Elselvier Applied Science.

van der Voet, E., Kleijn, R.K. & Udo de Haes, H.K. 1996. Nitrogen pollution in the European Union – origins and proposed solutions. *Environ. Conserv.* 23: 120-132

Virzo De Santo, A. 1992 Denitrification in forest soils. In: A. Teller, P. Mathy & J.N.R. Jeffers (eds.) *Responses of forest ecosystems to environmental changes*: 208-222. New York: Elselvier Applied Science.

Groundwater: Past Achievements and Future Challenges, Sililo et al. (eds) © 2000 Balkema, Rotterdam, ISBN 90 5809 159 7

Current and future challenges for managing groundwater quality in Perth, Western Australia

S.J.Appleyard
Water and Rivers Commission, Perth, W.A., Australia

ABSTRACT: Perth is situated on large groundwater resources which provide 273 million m^3/year, or 70% of the region's water requirements. Water for public supply is abstracted from both confined and unconfined aquifers, and a regional unconfined aquifer is also used for garden watering from more than 130 000 mostly unlicensed domestic bores. The sandy soils and generally shallow water table make groundwater in the region especially vulnerable to contamination from surface landuse. The management of groundwater quality in Perth is currently focussed on restricting landuse and regulating the storage and handling of chemicals in so called Underground Water Pollution Control Areas (UWPCAs) where groundwater is pumped for public water supply. However, the management of groundwater quality in urban areas of Perth will become increasingly important because of potential health impacts on private groundwater users, and because of environmental impacts caused by groundwater discharge to wetlands, waterways and the coastal marine environment in Perth.

1 INTRODUCTION

Perth is the only large urban centre within Western Australia (Figure 1). It has a population of only 1.3 million people, but this comprises about 80% of the population of Western Australia. Perth is also the only major city in Australia that lies on a major resource of fresh groundwater. Groundwater forms an important component of the city's water supply, providing 70% of all water used, and 40% of the municipal water supply for Perth. Groundwater occurs both in confined and unconfined aquifers, but the unconfined aquifer that underlies the region is particularly vulnerable to contamination from landuse because of the sandy nature of soils and the generally shallow watertable. Perth is also a rapidly growing city and there are competing demands between the need to preserve undeveloped land to protect groundwater quality, and the need to build 400 000 houses and urban infrastructure over the next 25 years to house an additional million people.

However, despite the development pressure, the protection of groundwater for potable supply has been secured under Western Australian planning legislation. This paper discusses past achievements in identifying, securing and managing groundwater quality for potable use in Perth, and also looks at the more difficult challenge of managing groundwater quality in urban areas. In these areas there are potential health risks from the unregulated use of groundwater from domestic bores, and environmental effects where contaminated groundwater discharges to wetlands, estuaries and the coastal marine environment.

2 HYDROGEOLOGICAL SETTING

Perth is situated on the Swan Coastal Plain, an alluvial and aeolian plain largely formed of Quaternary sand and sand/limestone dunes that are as much as 110 m thick. The plain lacks surface drainage due to the high permeability of the sediments, and the only fresh surface water in the region is in wetlands in interdunal depressions which are surface expressions of the watertable. The Swan Coastal Plain is underlain by a 12 km thick sequence of sediments of Permian to Cretaceous age belonging to the Perth Basin. The sediments are separated from crystalline Archaean rocks forming the Darling Range by the Darling Fault (Figure 2).

There are four major aquifers in the Perth area (Davidson 1995): an unconfined aquifer comprised of mainly Quaternary sediments, and three confined aquifers comprised of Jurassic and Cretaceous sediments. The unconfined aquifer is up to 70 m thick, and in some localities groundwater in the confined aquifers is fresh to a depth of at least 1000 m. Groundwater in the unconfined aquifer is recharged

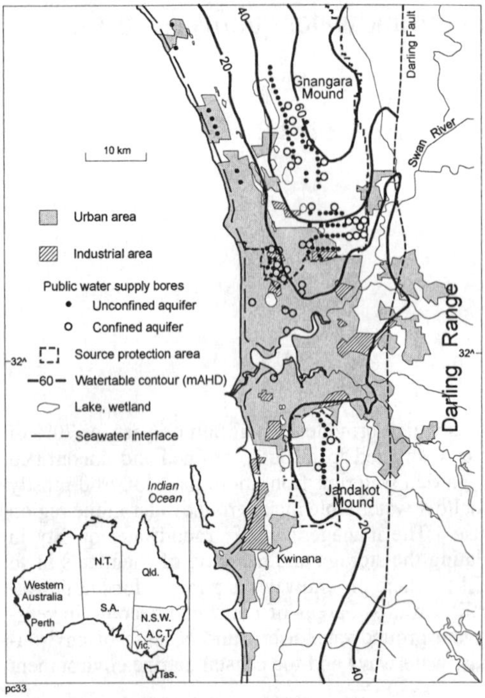

Figure 1. Location map of the Perth Region showing the extent of the urban area and groundwater features.

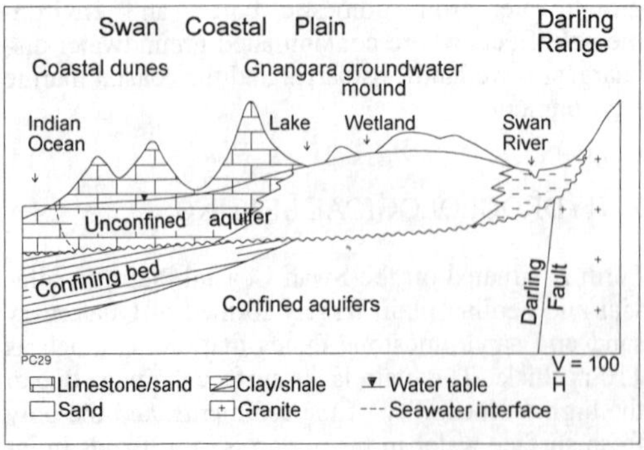

Figure 2. East-west hydrogeological section across the Gnangara groundwater mound.

by direct rainfall recharge on the coastal plain, and the deeper aquifers are recharged by leakage from the unconfined aquifer.

The unconfined aquifer is the most intensely used aquifer in Perth, and provides a total of 200 million m^3/year of water out of a total groundwater usage of 273 million m^3/year (Appleyard et al 1999). Groundwater in this aquifer occurs in two distinct groundwater flow systems to the north and south of Perth: the Gnangara and Jandakot Mounds (Fig. 1). Groundwater flows radially from the crests of the groundwater mounds towards major urban areas and the coast.

3 PAST ACHIEVEMENTS – PROTECTING GROUNDWATER FOR MUNICIPAL SUPPLY IN PERTH

The Swan Coastal Plain near Perth may have been inhabited by aboriginal peoples for more than 60 000 years, and it is likely that groundwater has been abstracted from dug soaks in the region for much of this period. The aboriginal custodians of the land in the region, the Noongyar people, had some understanding of the importance of groundwater in the region. They recognised the underground connection between wetlands and the Swan River through creation stories involving the Wagyls or rainbow serpents, which travelled underground from the Swan River to create the wetlands to the north and south of the river.

Immediately following European settlement in 1829, the water supply for the new Swan River Colony was from shallow wells. However, by the1880s it was recognised that there was a significant health risk due to contamination from cesspits. The development of new drilling technology following exploration in the Great Artesian Basin, led to the development of artesian bores for water supply for the city. The confined groundwater resource continued to be the major component of the water supply until 1940, after which surface water reservoirs were built in the Darling Range, 30 to 50 km to the east.

In the 1970s, a radical change in Perth's water supply took place when undeveloped sources of unconfined groundwater on the Gnangara Mound were utilised for public water supply (Allen 1997) conjunctively with surface water resources. At this time it was also recognised that there were limits on available surface water resources in the region, and that Perth's future water requirements would be met by groundwater. As a consequence, exploratory drilling programs defining both confined and unconfined resources began, and various borefields for public water supply were progressively commissioned on the Gnangara and Jandakot Mounds.

The new groundwater source areas were located on undeveloped land, and consequently groundwater was of a high quality and free from contamination. Water source protection areas (so called Underground Water Pollution Control Areas or UWPCAs) were proclaimed under appropriate legislation which allowed regulatory agencies (currently the Water and Rivers Commission) to have some control over landuse in these areas to protect groundwater quality. In addition, wellhead protection zones were defined to protect individual production bores.

Urban development in Perth continued to extend

in long corridors around the groundwater protection areas, but there was increasing pressure to develop land within the protection areas. The issue culminated in a parliamentary enquiry in the early 1990s into the competing demands for urban development and groundwater protection. As part of the enquiry, a group of parliamentarians visited a large number of cities throughout the world where groundwater was an important part of the city's water supply. The enquiry found that Perth was in a very fortunate position to have uncontaminated groundwater so close to the city centre. The enquiry reaffirmed the need to preserve the current water source protection areas on the Gnangara and Jandakot Mounds (Western Australia, Legislative Assembly 1994; Carew-Hopkins 1996) and identified poor planning decisions as the main cause of the progressive encroachment of urban development onto the groundwater mounds. Some of the major outcomes of the enquiry were:

- a greater role for water agencies in land planning and development;
- agreement between all government agencies that the protection of groundwater quality will be given primacy over all other issues in Priority 1 groundwater protection areas;
- strengthening of water resource protection policies in the Ministry for Planning and in planning legislation.

The linking of urban planning to protection of groundwater resources has helped ensure that Perth will continue to have access to good quality groundwater supplies well into the future. The success of hydrogeologists, engineers, environmental scientists and water resource managers in communicating with politicians, combined with the public demand for an unpolluted water supply, has raised the profile of the issue. As a result, government currently has the will to protect groundwater for public supply, and resist the pressure for inappropriate urban development.

4 CURRENT AND FUTURE CHALLENGES – MANAGING THE EFFECTS OF CONTAMINATION IN URBAN AREAS

Perth is very much a garden city. Most people live in detached houses with large gardens of lawn and exotic shrubs on 500 to 1000 m^2 blocks. The gardens require watering for about six months of the year due to the dry Mediterranean-type climate, and are responsible for up to 80% of domestic water use. The ready availability of groundwater has contributed to the garden culture, and there are currently more than 130 000, mostly unlicensed private bores which are used for garden irrigation. The gardens are major sources of groundwater contamination from excessive fertiliser use. The widespread use of garden bores (on average, one household in four has a bore) has meant that there are potential health risks for householders who may be exposed to contamination from contaminants including solvents, pesticides and petroleum hydrocarbons from up to 2000 potential contaminated sites in the city (Appleyard and Manning 1997; Appleyard at al. 1999). The use of garden bores is deeply entrenched in Perth's culture, and has become an alternative water distribution system. It would be politically very difficult to license and regulate their use, and they have a positive benefit of reducing the usage of high quality treated water for garden use.

The challenges for groundwater management in urban Perth arise because the issues cannot controlled by "command and control" legislative processes, but require the whole community to become better informed about the potential environmental or health impacts of their lifestyle choices and to voluntarily make changes where necessary. Water resource management agencies in this context are required to be information providers, educators and facilitators of change, rather than prescribers of community actions.

This is particularly the case with fertiliser use in Perth, which is a major cause of algal blooms in waterways and wetlands in the area. Currently, about 1600 tonnes of nitrogen and 480 tonnes of phosphorus is applied annually to lawn areas in Perth (Sharma et al. 1996). Although much of the phosphorus is bound up in soil profiles, up to 80% of the applied nitrogen may leach to the watertable (Sharma et al. 1996). About 160 tonnes of nitrogen is discharged by groundwater to the Swan River each year, and up to 10 tonnes of nitrogen for each kilometre of coastline is discharging annually into the marine environment (Appleyard and Powell, 1998). Although the community in Perth have a good understanding of the importance of groundwater for water supply, there is a poor understanding of the interaction between groundwater and surface water bodies and of the potential the environmental effects of excessive fertiliser use.

To redress this, the Water and Rivers Commission has included a groundwater component into its "Ribbons of Blue" school and community water resource education program. Groundwater education resources so far implemented have included materials for school curricula, videos, and "hands on" workshops where students can install and monitor bores next to wetlands, and physically measure groundwater discharge to demonstrate the link between groundwater and surface water bodies. The Commission also has run media campaigns promoting the wise use of water and fertiliser, and has also promoted the benefits of using local native plants in gardens which do not need fertiliser (Appleyard and Powell, 1998).

A number of community action groups are also becoming interested in the issue of measuring and managing groundwater discharge to wetlands and waterways. Several groups are working in partnership with local and State government agencies, some with the assistance of Commonwealth (Federal) government funding. These may be the beginnings of an Integrated Catchment Management (ICM) approach to dealing with groundwater-related environmental issues in Perth. This approach is being successfully used in rural areas to manage environmental problems caused by land salinisation and the eutrophication of waterways, and should also help manage environmental problems in urban areas.

The public need to have access to high quality information about the distribution of groundwater contamination to make informed choices about where groundwater use is appropriate, particularly in Perth where groundwater contamination plumes may extend up to 2 km from large contamination sources. The Water and Rivers Commission has published a "Groundwater Atlas" showing groundwater features and the distribution of contaminated sites in Perth, and this is widely used in the community. As part of its commitment to proposed contaminated sites legislation, the Commission is also developing a groundwater contamination database (Nelson et al. 1999) to capture and quality assure contamination data. It is intended that the database will be accessible by the public with the enactment of the new legislation.

5 IMPLICATIONS FOR WATER RESOURCE MANAGEMENT IN OTHER URBAN AREAS

This case study has illustrated the important role that the community have in helping protect groundwater quality in urban areas. Without community support, it is unlikely that there would have been the political will to maintain the current level of groundwater protection for public water supply in Perth. It also illustrates the changing role that water resource management agencies are having in urban areas. It is no longer possible for these agencies to work in isolation, but they need to work closely in partnership with planning agencies and specific industry and community groups to achieve management objectives. It is also important that the wider community are aware that groundwater is not only important for water supply, but is an important environmental agent that can affect the health of waterways, wetlands and the coastal environment. Communities in urban areas dependent on groundwater need to be well informed about groundwater management issues to ensure that the political process allows sound management policies to be enacted. To ensure this happens, water resource management agencies need to become effective communicators, and providers of community-based education programs and specific groundwater resource information.

REFERENCES

Allen, A.D. 1997. Groundwater: the strategic resource. A geological perspective of groundwater occurrence and importance in Western Australia. *Western Australia Geological Survey, Report 50.*

Appleyard, S.J. & P. Manning 1997. Managing the impact of contamination on domestic groundwater use in an urban setting. *Proc. "Wai Whenua", 24th Hydrology and Water Resources Symposium, Auckland, New Zealand*: 241-245.

Appleyard, S.J. & R.J. Powell 1998. Preserving urban bushland and growing local plants to protect water resources. *Proc. "Managing our bushland" conference, Perth, October 1998:* 72-75.

Appleyard, S.J. W.A. Davidson & D.P. Commander 1999. The effects of urban development on the utilisation of groundwater resources in Perth, Western Australia, *in* Chilton, J. (ed.), *Groundwater in the Urban Environment: Selected City Profiles. Volume 2 City Case Studies*, pp 97-102. Balkema publishers, Rotterdam.

Carew-Hopkins, D. 1996. Findings of the WA Parliamentary Select Committee on Metropolitan Development and Groundwater Supplies. *Proc. Groundwater and Land-use Planning Conference, Fremantle, September 199:* 216-227. Centre for Groundwater Studies Publication.

Davidson, W.A., 1995. Hydrogeology and groundwater resources of the Perth Region, Western Australia. *Western Australia Geological Survey, Bulletin 142.*

Nelson, S.W., P. Wallace-Bell & X. Yu 1999. SiteLEGACI: A systematic approach to groundwater contamination investigations in Western Australia. *Proc. Contaminated Site Remediation: Challenges Posed by Urban and Industrial Contaminants, Fremantle, March 1999*: pp 251-258.

Sharma, M.L. D.E. Herne J.D. Byrne & P.G. Kin 1996. Nutrient discharge beneath urban lawns to a sandy coastal aquifer, Perth, Western Australia. *Hydrogeology J.* 4(1): 103-117.

Western Australia Legislative Assembly 1994. Select Committee on Metropolitan Development and Groundwater Supplies. *State Print, Perth, 286p.*

Groundwater: Past Achievements and Future Challenges, Sililo et al. (eds) © 2000 Balkema, Rotterdam, ISBN 90 5809 159 7

Groundwater resources management in Africa with the aid of databases and GIS

I.F. Blecher
VSA Consulting Group (Pty) Limited, Silverton, South Africa

ABSTRACT: The management of groundwater, and other water resources in general, is the single most important aspect of groundwater related exploration and development projects. In order to manage a resource successfully it is essential that a comprehensive relational and spatial database is compiled which can provide the user with information and analysis tools for decision making purposes. Users of such a database can range from policy makers, authorities and managers in regional and local structures to hydrogeological consultants, engineers and researchers. The data stored in such a database should not only include basic site information and hydrogeological data on a specific site but should also include information on exploration techniques, project management, site management recommendations and operation and maintenance. The integration with the capabilities of a Geographic Information System (GIS) should visualize the data within the environment of other spatial data like topography and geology to provide the user with a complete picture of the resource to be managed.

1 INTRODUCTION

When it comes to exploiting the resources of our earth we must ask ourselves the question if we are managing these resources properly. In particular we have to be concerned about the question of managing our groundwater resources properly, as groundwater plays and will play in future a very important role in urban and rural water supply, not only in Africa, but all over the world. If we believe that we manage our groundwater resources properly, what tools do we use?

The management of groundwater, and other water resources in general, is the single most important aspect of groundwater related exploration, development and management projects. No rural water supply project, for example, can be carried out efficiently without a comprehensive relational and spatial database which can provide the user and groundwater manager with information and analysis tools for decision making purposes. Role players in such a project and therefore potential users of such a database range from policy makers, authorities and managers in regional and local structures to hydrogeological consultants, engineers and researchers.

2 HISTORY

2.1 Existing databases

Currently only a few groundwater information system exists in South Africa or elsewhere which can handle the enormous variety of data accumulated by earth scientists and engineers for rural, urban and industrial water supply projects, environmental investigations, mine pit de-watering exercises and modelling studies.

Existing water related databases include paper based "databases", which are often unreadable, get lost or gather dust in filing cabinets, mainframe based surface, groundwater and water quality databases, mainly residing at government institutions and therefore not accessible to outside users, and various PC based groundwater and chemistry databases and applications. The PC based applications include Groundwater for Windows, a UN developed package with strong graphics capabilities, but is very unstable, has a format which is not compatible with other common database formats and is not easily customizable for additional tables and entry fields for later use with the graphing and mapping functions. The PC based South African software products HydroCom, MuniWater and WISH (Windows

Interpretation System for Hydrogeologists) for many years provided and still provide limited groundwater database capabilities, all with some sort of reporting, charting and mapping facilities, but no GIS integration. Due to the unavailability of suitable groundwater database packages many users have developed their own dBASE, Paradox, Access and spreadsheet databases which often have only weak querying and reporting facilities and rudimentary graphing capabilities.

Because of the lack of suitable groundwater database applications a groundwater consultancy in South Africa has developed Aquabase, a PC based database/GIS software which caters for almost every facet of data accumulated on surface and groundwater projects. Database formats that can be used for the storage and access of data include Dbase, Oracle, Informix, MS Access, MS SQLServer, Interbase and others allowing enterprise wide distribution and use of data.

2.2 Status of databases/GIS's in sub-Saharan Africa

The only sub-Saharan countries operating some sort of organized water related databases are South Africa, Botswana and Namibia. There are, however, fundamental deficiencies in terms of entry fields, output and integration with GIS's, mainly due to isolation of the countries, lack of know-how and funds. In recent years these three countries have made effort and some progress in trying to adapt and customize existing databases or have conducted database systems analyses to determine requirements for a new water related database. Other sub-Saharan countries that have some sort of water related database, although rather un-organized, include Zimbabwe, Kenya, Uganda and Ghana. Malawi only has a rather organized "paper database". All other countries do not have any form of water database or GIS.

Most sub-Saharan countries have just started to think about the need for a water database due to recent droughts and poor management of existing water resources. Some made various attempts to set up some form of database/GIS, but in most cases failed due to the lack of funds and know-how. Unfortunately millions of dollars from donor organizations have been wasted because of poor after-sales service by IT and other consultants. This and other factors pose a major problem to IT consultants and donors for future water database projects. In many cases expensive hardware and software have been purchased and either have never been set up or have been installed and never used.

Often countries want to move from nothing to highly sophisticated database management systems (Client/Server), which are costly and require constant maintenance to avoid failure. Such systems have been developed for use in first world countries and are very often not suitable for second and third world countries due to the lack of infrastructure, unavailability of reliable maintenance and support companies and, most importantly, do not provide fields and tables specifically for third world country or African hydrogeological conditions and parameters. These database systems therefore have to be adapted to accommodate specifically rural water supply and hard rock aquifer parameters and the design of these systems requires good knowledge of these.

Unfortunately such database systems are often financed by foreign aid money and the donor countries often require that IT and other consultants from these countries be involved to a large degree in the design and implementation of the systems resulting in very costly solutions.

2.3 Status of databases in South Africa, Botswana and Malawi

In order to shed some light on existing groundwater database/GIS systems and data management structures research has been conducted for three countries, namely South Africa, Botswana and Malawi.

2.3.1 Groundwater data management in South Africa

The groundwater management environment in South Africa has changed dramatically during the last six years due to the increasing priority of community water supply projects and the changes in the legal status of groundwater. These have resulted in the need for a new GIS based groundwater information system which would replace the current National Groundwater Database (NGDB)(Orpen 1996).

The present NGDB, maintained by the Department of Water Affairs and Forestry, Directorate: Geohydrology is a mainframe based groundwater database devoid of any analysis, interpretation or mapping facilities. It has been maintained and adapted during recent years to accommodate additional fields and entry forms. Additionally various business rules have been implemented to create a database system with high integrity but which has also resulted in the creation of default values for various parameters with no value to the user. Sites that can be stored are strictly groundwater related (boreholes, wells, springs etc.) and no provision is made for the storage of sampling points from surface water or rainwater harvesting stations, which have become an important source of water for rural communities. Neither geophysical exploration data nor project information data like site management recommendations and expenditure can be stored. The NGDB also does not provide for the storage of groundwater related chemistry data, which is stored and maintained by the Department's Directorate: Hydrology on a newly implemented Water Management System (WMS). In order to bring both data sets together a tedious site linking process has to be conducted.

The biggest drawback of the present NGDB is the fact that it is not directly accessible to users outside Government. Data from the NGDB can only be obtained through export files and have to be edited and updated by either the HydroCom, MuniWater or Aquabase software programs in order to be compatible with the upload process back to the NGDB.

By the end of 2000 the Directorate: Geohydrology hopes to replace the current mainframe based NGDB with the "Open-NGDB" system running under the Informix database platform, allowing access from the regional offices which will be responsible for the maintenance of the regional data (Appelcryn 1996). Eventually this system will be known as the National Groundwater Archive (NGA).

In order to access the groundwater data for use in Arc/Info and ArcView, the Department's preferred GIS packages, a consultancy from the Netherlands will adapt their existing Regional Geohydrological Information System (REGIS), an ArcView/Oracle application, to accommodate and interpret the data on the NGDB by the end of 2000. Until such time when the REGIS system will be able to work with the Informix based NGDB data directly a data replication process will enable the use of the NGDB data in the GIS applications.

The present NGDB contains some 206 000 boreholes/wells/springs and related hydrogeological information, most of the records being obtained from simple borehole record sheets filled in by the state drilling personnel. Unfortunately most of these records do not have accurate coordinate information and can therefore mainly be used for statistical purposes. The greatest volume of adequately accurate data lies with private consultancy and engineering firms and in many cases does not reach the NGDB due to the lack of interest in a common groundwater database or the building of "private data warehouses" for own use and unwillingness to share.

2.3.2 Groundwater data management in Botswana

All groundwater data in Botswana is managed by two Government institutions, the Geological Survey and the Department of Water Affairs. While the Department of Water Affairs handles the registration of boreholes the Geological Survey updates the so-called Borehole Archive, which is presently running on a simple Dbase database system. This system has only limited querying and reporting facilities and is not integrated with any mapping or GIS system.

As all boreholes drilled in Botswana since the late 1920's had to be registered and one would assume that the Borehole Archive contains most of the boreholes drilled. The archive is, however, totally outdated and only contains some 12 500 boreholes. It is estimated that between 3500 and 4500 boreholes still reside in "paper databases" due to the lack of funds and personnel capacity.

The Department of Water Affairs additionally runs a MS SQLServer based database software called Wellmon, which is intended for the monitoring of wellfields in Botswana. This application, however, has only limited querying and reporting functionality and does not provide GIS integration. At present the Geohydrology Division of the Department is also in the initial phase of developing a new interface and database to handle the geohydrological data.

Both Government institutions operate chemical laboratories for the analysis of groundwater and have used arcane chemical data storage methods, which were not compatible with the Borehole Archive. Both laboratories are presently in the process of switching to the Aquabase software to handle both Borehole Archive and chemistry data. The laboratory of the Department of Water Affairs has stored in access of 22 000 chemistry analyses which will be handled together with the Borehole Archive data on MS SQL Server.

2.3.3 Groundwater data management in Malawi

Malawi does not have any electronic database system but rather keeps records of boreholes in paper files at the Ministry of Water Development. A groundwater data management system is non-existent. Aggravating the situation is the fact that presently all groundwater supply projects are handled and managed by drilling contractors which in turn subcontract the hydrogeological consultant for the siting. There is no requirement by the Ministry to obtain borehole information on the projects and therefore presently most information either gets lost is stored on the drilling contractor's or consultants databases in various formats.

There are an estimated 16 000 boreholes in Malawi. However, only 9 700 of these holes are currently in production. It is alleged that the Government plans to drill another 15 000 to 20 000 in the coming years to meet demand (Siebert 2000). No attempts have been made to assess the impact of the existing and planned boreholes on the groundwater balance. Matters are exacerbated by "unregistered" boreholes drilled by the numerous NGOs operating in Malawi, without the knowledge of the Ministry.

Some of the NGOs and donor organisations have only recently realised the urgent need for a groundwater management system in form of a database with GIS to get some control over the existing and new boreholes, at least on their own financed projects.

3 DATABASE MANAGEMENT

3.1 General requirements

Data entry and management requires a sound knowledge of the information to be stored. The database manager should therefore be a hydrogeologist or have a good background of hydrogeology and other related fields. The database operators (data entry) should have at least some basic knowledge of the hydrogeological data in order to identify valuable from invaluable data. Additionally data entry and database managers should know the database structure so that data is stored in the correct positions in the database. Database managers and operators should form a well-trained team in order to handle database tasks like data entry, verification, auditing and reporting as efficiently as possible.

Ideally a central database, which is managed by a database management team, should be set up and should be updated on a regular basis with data from regional databases, which in turn are maintained and managed by users and regional database managers, and various other databases from outside users like groundwater and other consultants and institutions.

The data stored in the databases described above should not only include basic site information and hydrogeological data on a specific site but should also include information on exploration techniques, exploration project management, site management recommendations and operation and maintenance. The integration with the capabilities of a Geographic Information System (GIS) should visualize the data within the environment of other spatial data like topography, environmental data, geology and remote sensing data to provide the user with a complete picture of the resource to be managed and allow a spatial analysis and subsequently decision making.

3.2 Database/GIS structure

A database with integrated GIS consists of three major components which play an important role in the success of such a system: i.) The relational database itself for storing the alpha-numeric data like site information, project information, geophysical, operations and environmental information, with other words all the information that can be connected/related to a feature on a map; ii.) The spatial database for storing spatial information like topographic and cadastral information, geology/hydrogeology and satellite and air photo interpretations and images; iii.) The user interface for accessing the stored data for viewing and editing, analysis and interpretation, graphing and reporting and subsequently presentation.

4 CONCLUSION

A database with integrated GIS is one important tool in groundwater resources development and management. It aids the groundwater manager, who has the task of ensuring sustainable and clean sources of water for urban and rural water supply, in the planning and decision making processes, not only in exploration and development projects, but also the follow-up projects like site management, operation and maintenance. Africa urgently needs this tool to effectively manage her groundwater resources which play an increasingly important role in water supply projects. A database with GIS does not have to be costly and highly sophisticated to fulfil its purpose. Such basic systems are readily available and with some customisation suitable for water resources management, not only for the African continent and other third world countries, but all countries where groundwater has to be protected and managed for use by future generations.

REFERENCES

Appelcryn, D.T 1996. Development of a strategy for a Groundwater Information System for South Africa. Notes from Workshop 1 and 2.

Orpen, W.R.G. 1996. Business case for Project: National groundwater Information System. Project Documentation, Release Nr. 2.0, Final.

Siebert, C. 2000. Water perspective Malawi. Development Bank of Southern Africa, Internal Report.

Groundwater: Past Achievements and Future Challenges, Sililo et al. (eds) © 2000 Balkema, Rotterdam, ISBN 90 5809 159 7

Groundwater management below sea-level coastal areas in the Po Delta, NE Italy

M. Bondesan, A. Gargini & A. Giovannini
Department of Geological and Paleontological Sciences, University of Ferrara, Italy

ABSTRACT: This paper reviews groundwater management in low alluvial plains along the coastal area of the Padania plain (Northern Italy) in the delta region of the Po River. In particular, the focus is on the problem of artificial maintenance and salinization of shallow, fresh phreatic groundwater of limited thickness exploited for both agricultural and ecological purposes.

1 WATER MANAGEMENT ISSUES IN LOW-LANDS

Coastal plain low-lands below sea level are real frontier and amphibious realms where well defined borders do not exist, both between sea and land as well as within the hydrological cycle between fresh surface-water, sea-water and groundwater. Such areas are a great challenge to scientists, technicians and decision-makers involved in water resource management. In order to prevent environmental hazards and make territorial planning more efficient and cost-effective an integrated use of corrective hydrological measures for surface and groundwater is required. This becomes even more important since these lands only exist as a result of anthropic reclamation carried out in order to develop an economy that heavily depends on a well co-ordinated use of the water supply (agriculture, seaside resorts, fisheries, etc.). Moreover, many environmental sites such as wetlands, dunes and coastal forests rely on a delicate and continuously unstable equilibrium of geomorphological, hydrological and hydrogeological processes where anthropogenic activities may have a negative effect on the equilibrium (Fig. 1).

Control of the fluvial network, land reclamation and groundwater pumping are the main environmental hazards to cause pressure over the coastal low-lands; although the former are part and parcel of the existence of emerged low-lands if they are not carried out on a sound hydrological basis and well co-ordinated management there may well be an increase in environmental risks. The undesired effects of bad management could include land subsidence (induced either by groundwater pumping or reclamation of more compressible lands like peat and clay), salinization of coastal groundwater, and artificial and/or cementation of coastal dune ridges.

The strong relationships between the various physical components of the low-lands increase the chance of negative feedback effects. The destruction and urbanisation of coastal dunes decreases the direct recharge of groundwater while increasing saline encroachment in correlation with the rise of the interface's elevation and the reduction in groundwater dilution. The increasing rate of land subsidence can either augment the pumping rate required for reclamation (which in its turn supports groundwater salinization) or it can increase the level difference between the water-table and surface water reservoirs (terminal sections of rivers, fishery coastal lakes connected to the sea, canals) thus raising saline water intrusion. The resulting increase in groundwater salinity creates a greater need for more fresh water for agricultural purposes.

The hydrogeological system is also affected by the geomorphological problems that are brought about as a result of the increased hydraulic risks and the effects of coastal erosion induced by the load decrease in the rivers (Bondesan 1989). The problem may be considered as part of the global rise in the sea-level to which anthropic action may have contributed (Bondesan et al. 1995).

The coastal plain of the Po River Delta along the shore of the Northern Adriatic Sea (Italy) is an example of the aforementioned problems.

In this paper prominence is given to the specific hydrogeological features of this region, considered as representative of other similar areas in the same condition, and to the efforts which have been made to address the problems of management.

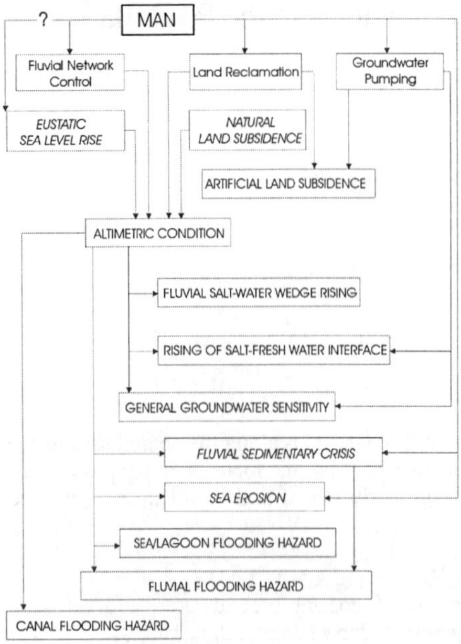

Figure 1. Flow chart of present issues of low-lands. The natural or mainly natural issues are in italic.

2 HYDROGEOLOGY OF A DELTAIC COASTAL PLAIN

The altimetric features of a typical deltaic low-land are shown in Figure 2. Most of the area is below sea-level; the strong control of recent (Holocene) geologic and geomorphological evolution upon the altimetry is reflected by the pattern of above sea-level terrains in the elongated stripes of varying thickness either sub-perpendicular to the shore line (outcropping paleo-river beds) or sub-parallel (beach-ridges and paleo-dunes as successive markers of the prograding deltaic processes). The grey areas (Fig. 2) denote emerged morphologic depressions, since the second half of the nineteenth century, as a result of land reclamation; the coast is characterised by the presence of dispersed lagoons and lakes, either fresh or brackish. From Figure 2 it may be seen just how many factors have affected phreatic groundwater recharge: direct recharge from precipitation (particularly from below more permeable paleo-dunes), lateral recharge by hanging channels of the delta, and by irrigation channels, exchanges of fresh, brackish and salt waters alongside the coast.

The geomorphological evolution of the area, shown in Figure 3, is the outcome of the holocenic progradation of continental (coastal plain), littoral and deltaic deposits over off-shore muds (Fig. 4). The local hydrogeological structure is, therefore, characterised by a mono-stratum phreatic aquifer (Phreatic Sandy Aquifer Complex), formed mainly by medium-fine grained sands with an average thickness of 10-15 m, underlain by a low permeability layer of fine-grained sediments of continental origin towards the land (alluvial plain silts and clays) and of littoral origin towards the sea (open-sea muds). Further down in the hydrogeologic succession there is the first confined aquifer unit (corresponding to the upper part of Aquifer A of the Po Plain according to the Regione Emilia Romagna & Eni/Agip 1998), formed by alluvial plain sands of Holocenic progradation, the top of which is at about a depth of 30 metres and which has an almost complete hydraulic confinement compared to the phreatic aquifer above (Bondesan at al. 1999a).

Figure 5 presents the morphology of the water-table along with the specific electrical conductivity of phreatic groundwater relative to a sector of the southern part of the delta area of the Po River (Bondesan et al. 1999b). The water-table, which is mostly brackish, is permanently below sea-level; lateral recharge from the main surface-water areas such as the Adriatic Sea through the Sacca di Goro lagoon, Po di Goro and Po di Volano Rivers and the effects of water-scooping machine pumping are evident. The only areas with fresh groundwater are controlled by irrigation canals that drain water out of the area from the main Po River.

As a result of this hydrogeological system, a sort of "pillow" of fresh or brackish groundwater, supplied either by direct recharge or by hydraulic loading of lateral recharge and irrigation, "floats" over salty groundwater inside the Sandy Phreatic Aquifer Complex. The elevation of the interface, shown in Figure 6, as evidenced by a geoelectrical survey carried out in the southern portion of the area, varies from a few metres to over 15 metres below sea-level (Abu Zeid et al. 1999). Minimum thickness is located near the head boundaries along the shoreline and near the coastal salty lakes ("Valli di Comacchio"), whereas maximum thickness is found where hydraulic loading from irrigation is high and the draining effect of pumping plants is low.

3 HYDROLOGICAL-BUDGET ASSESSMENT

The definition of a hydrologic budget in such a situation represents a fundamental tool in groundwater resource management (Bondesan et al. 1999b); Figure 7 reports the different components of a hydrogeologic budget for the area represented in Figure5. The only parameter found to be different from the others is lateral recharge from river and sea.

In such an "artificial" area the hydrologic budget output depends mainly on the amount of surface water being pumped from the canal network and irrigation can be considered an important input com-

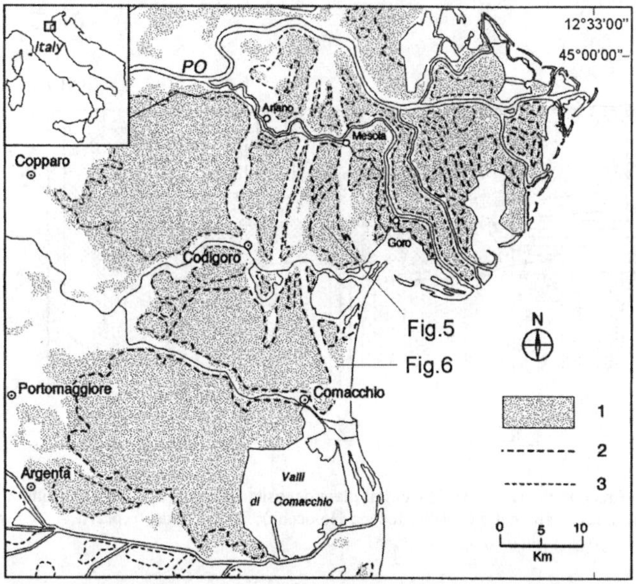

Figure 2. Altimetrical map of the Po delta region. 1) Areas below sea level; 2) boundary of reclaimed saltwater marshes; 3) boundary of reclaimed freshwater marshes.

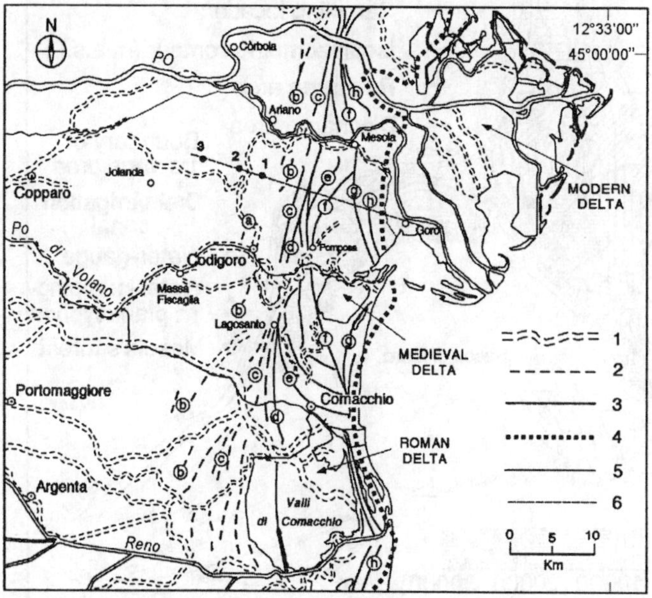

Figure 3. Geomorphological map of the Po delta region. 1) Main paleo-riverbed; 2) buried or half-buried beach-ridges; 3) outcropping beach-ridges; 4) coastline of the XVI century; 5) present day coastline; 6) geological section. Age of littoral structures: a) Atlantic/Atlantic-Sub-boreal transition; b) Sub-boreal; c) Bronze Age; d) Iron Age; e) Roman Age; f) High Middle Ages; g) X century; h) XIV century.

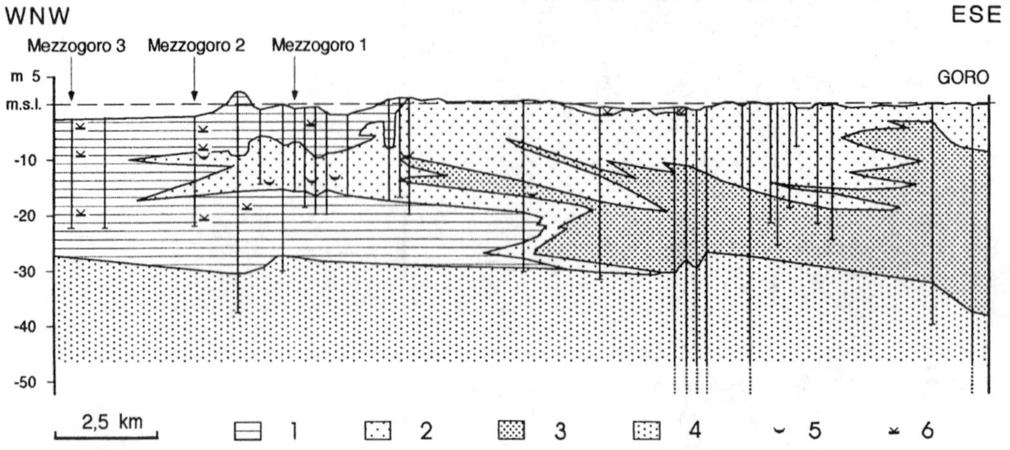

Figure 4. Geological section of the Po delta region (trace in the figure 3). 1) Coastal plain deposits (predominant clays); 2) littoral deposits (predominant sands); 3) off-shore muds; 4) sands of alluvial plain (Pleistocene-Holocene); 5) shell beds; 6) peats.

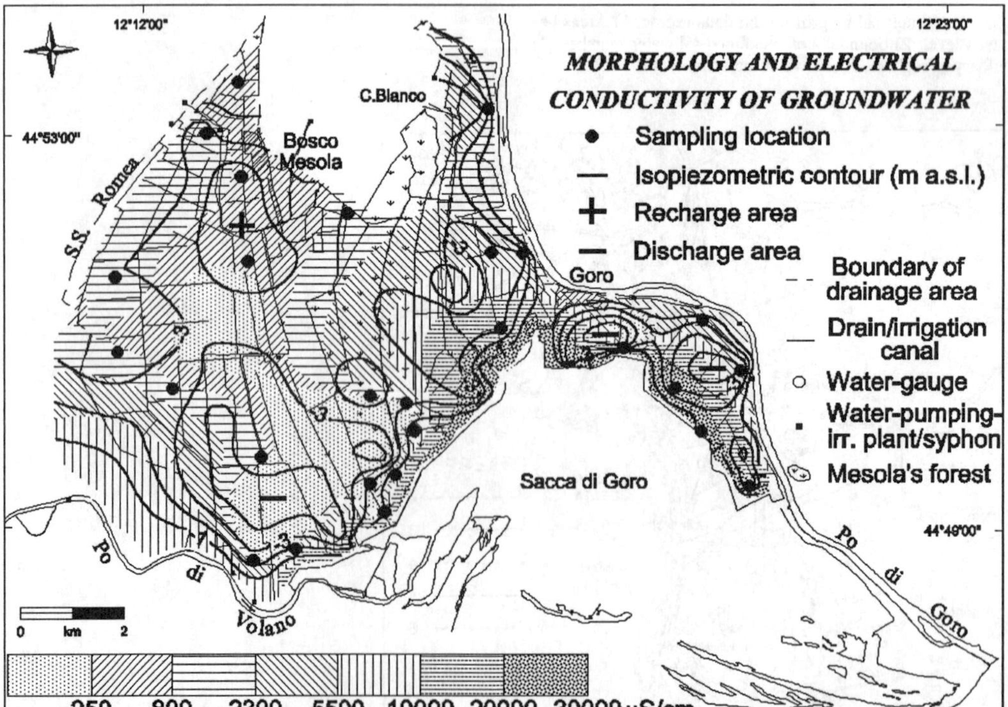

Figure 5. Piezometric morphology and electrical conductivity of phreatic acquifer in the *Mesolano* reclamation district (survey of October 1997). The piezometric contour interval is equal to 1 m.

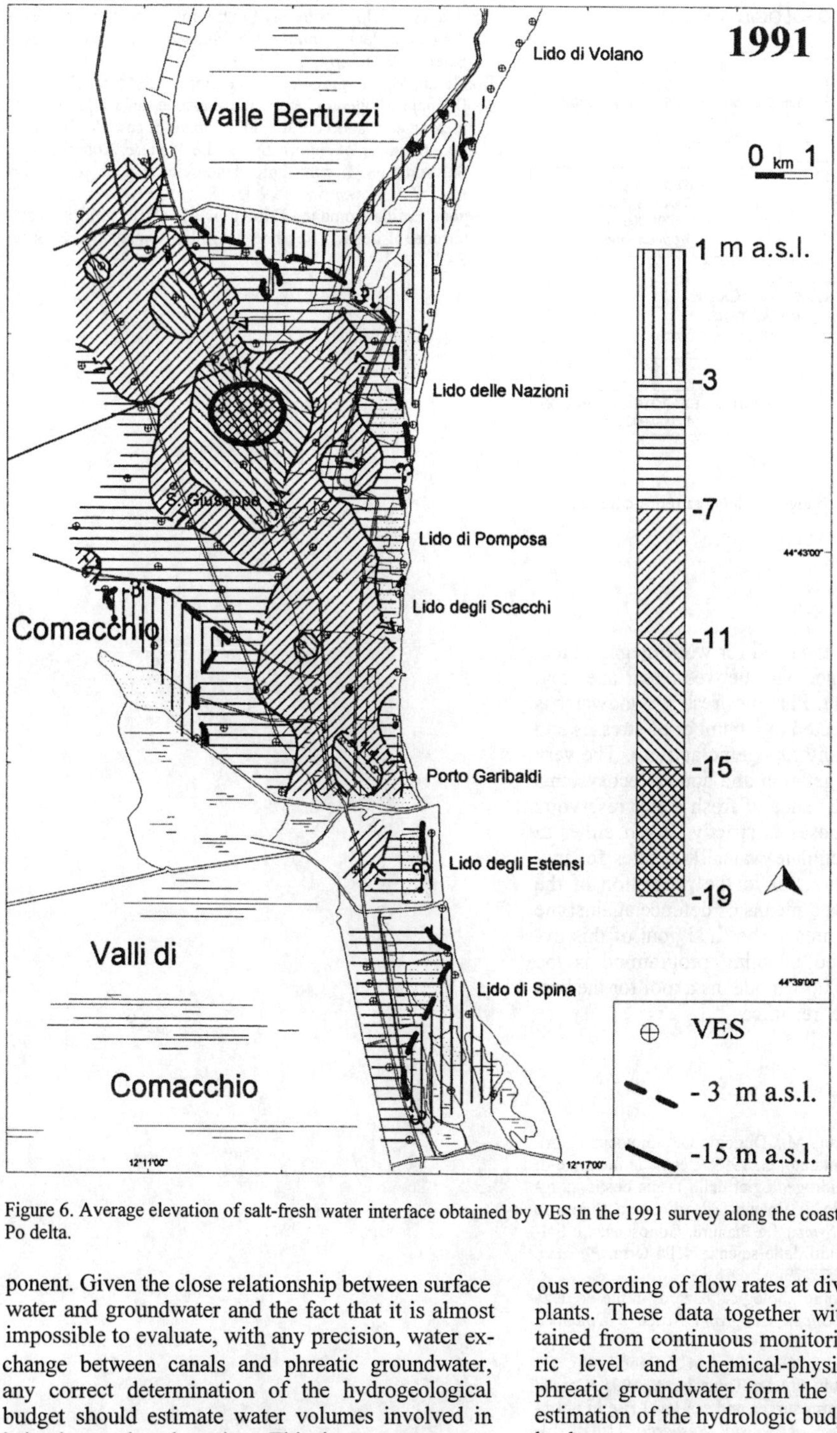

Figure 6. Average elevation of salt-fresh water interface obtained by VES in the 1991 survey along the coastal strip to the south of Po delta.

ponent. Given the close relationship between surface water and groundwater and the fact that it is almost impossible to evaluate, with any precision, water exchange between canals and phreatic groundwater, any correct determination of the hydrogeological budget should estimate water volumes involved in irrigation and reclamation. This is no easy matter given the complexity of the network irrigation and reclamation canals, as well as the lack of a continuous recording of flow rates at diversion and pumping plants. These data together with the findings obtained from continuous monitoring of the piezometric level and chemical-physical parameters of phreatic groundwater form the basis of an optimal estimation of the hydrologic budget of a coastal lowland area.

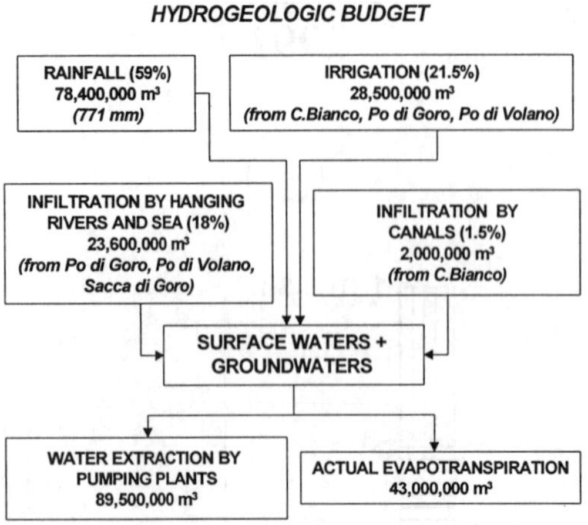

Figure 7. Hydrogeologic budget of the area represented in Figure 5.

4 CONCLUSIONS

In coastal low-lands the need for well integrated and co-ordinated management between surface and ground water is great. Phreatic fresh groundwater is a thin "pillow" subjected to a number of stresses and very susceptible to anthropogenic activity. The very survival of many vegetation and aquatic ecosystems as well as the maintenance of fresh water reservoirs for agricultural purposes is strictly linked either to man's capacity to regulate water flow rates for irrigation and reclamation, or to the protection of the coastal dunes both as a means of defence against the sea and as recharge areas. The final goal of this hydrogeologically based planning programme is the implementation of a flow model as a tool for the best management of water resources.

REFERENCES

Abu Zeid, N., Bondesan, M., Dugoni, G., Giovannini, A., Mantovani, E. & Santarato, G. 1999. Tecniche integrate di studio dei problemi idrogeologici della fascia costiera: applicazione all'area Po di Volano-Lido di Spina (Ferrara). *Abstracts of Intern. Symp.*, Le Pianure. Conoscenza e Salvaguardia. Il contributo delle scienze della terra. *Ferrara, 8-11 November 1999*:77-79.

Bondesan, M. 1989. Geomorphological hazards in the Po delta and adjacent areas. *Suppl. Geografia Fisica e Dinamica Quaternaria* II:25-33.

Bondesan, M., Castiglioni, G.B., Elmi, C., Gabbianelli, G., Marocco, R., Pirazzoli, P.A. & Tomasin, A. 1995. Coastal areas at risk from storm surges and sea-level rise in north-eastern Italy. *Journal of Coastal Research* 11(4):1354-1379.

Bondesan, M., Calderoni, G., Cattani, L., Ferrari, M., Furini, A.L., Serandrei Barbero, R. & Stefani, M. 1999a. Nuovi dati stratigrafici, paleoambientali e di cronologia radiometrica sul ciclo trasgressivo-regressivo olocenico nell'area deltizia padana. *Annali dell'Università di Ferrara,* Sez. Scienze della Terra, 8 (1):1-34.

Bondesan, M., Gargini, A., Giovannini, A. & Masè, G. 1999b. Bilancio idrogeologico e salinizzazione della falda freatica in zone al di sotto del livello del mare: il caso del Delta del Po. *Abstracts of Intern. Symp.,* Le Pianure. Conoscenza e Salvaguardia. Il contributo delle scienze della terra. *Ferrara, 8-11 November 1999*:53-55.

Regione Emilia-Romagna-Eni-Agip 1998. Riserve idriche sotterranee della Regione Emilia-Romagna. Firenze: S.EL.CA:1-120.

Groundwater: Past Achievements and Future Challenges, Sililo et al. (eds) © 2000 Balkema, Rotterdam, ISBN 90 5809 159 7

US regulation of animal feeding operations: Need for a watershed approach

T.J.Centner & M.E.Wetzstein
Department of Agricultural and Applied Economics, University of Georgia, Athens, Ga., USA

J.W.Worley
Department of Biological and Agricultural Engineering, University of Georgia, Athens, Ga., USA

ABSTRACT: Animal feeding operations (AFOs) and other agricultural activities are conspicuously contributing to the impairment of water quality in the United States. With the marked expansion of AFOs and problems created by large numbers of animals in a confined area, US federal and state governments have enacted laws and regulations to address the problems associated with animal waste. Overall, the regulations are unduly dependent on animal units at a single location and show a lack of meaningful criteria that relate excessive nutrient discharges to water quality needs of individual watersheds. Rather than adopting blanket regulatory proscriptions, regulators should look for alternative signalling and screening devices to address site-specific problems.

1 INTRODUCTION

Agriculture has been identified as a major source of pollutants for various water bodies (EPA 1998), with estimations suggesting that animal feeding operations account for about 9% of impaired rivers and streams by distance (GAO 1995). The US Department of Agriculture estimates there are 450,000 animal feeding operations (AFOs) in the United States (USDA & EPA 1999). Of this number, nearly 20,000 have large numbers of animals and may be classified as confined animal feeding operations, known as CAFOs (EPA 1999).

Particularly troublesome is the plentiful supply of manure from AFOs that may provide too many nutrients for an area. Excessive nutrients in water bodies have led to eutrophication of rivers, lakes, and coastal waters.

A study that analysed the nutrients from CAFOs in each county of 48 US states found a potential for excess nitrogen and phosphorus (Lander et al. 1998). The simulation model forecast that 35 counties may have too much nitrogen and 107 counties may have too much phosphorus from CAFOs for available crop acreage (Lander et al. 1998). While the simulations suggesting excess nitrogen and phosphorus may be limited in their ability to accurately predict situations at the individual field level, the fact that most manure is applied to fields close to where it is produced presages potential nutrient problems (Letson & Gollehon 1999).

Due to the large quantities of manure produced at CAFOs and the expense involved in transporting it to distant locations, it may be expected that CAFOs will be linked to some nutrient contamination problems. Crop land receiving manure applications year after year may have too much nitrogen or phosphorus for the crop. Concentrations of animals at CAFOs increase the probability that excess nutrients may find their way into the ground and groundwater.

Animal manure has been linked to the impairment of water quality due to excessive concentrations of phosphorus and nitrogen. Research has shown that too much manure on crop land may exacerbate the amounts being transported into ground and surface waters (Bosch & Napit 1992). Because of this contamination problem, nutrient management of animal manure has become a major public and legislative issue (Huang 1997).

Commencing with the Clean Water Act in 1972 (33 USCA § 1251 et seq., 1999), the federal government has taken steps to control water pollution and improve the quality of the nation's waters. Federal law classifies pollutants as discharges from either point sources or nonpoint sources. Outside of the US, nonpoint source pollution is generally referred to as diffuse pollution.

The definition of point source pollution in the US Code of Federal Regulations (CFR) includes CAFOs. Thus, those AFOs that meet the regulatory definition of a CAFO are subject to the permitting requirements for point sources and need to comply with a federal permitting system (USDA & EPA 1999). Data suggest that only about 2,000 AFOs have been issued permits pursuant to federal regulations (USDA & EPA 1999).

The Clean Water Act establishes national thresholds with a goal of attaining water quality that protects fish and wildlife and provides for recreation. State governments adopt water quality standards for waters within the state, and may adopt water quality standards more stringent than those required by federal law. In response to pollution issues presented by AFOs, states have moved to adopt new legislation and regulations governing animal wastes. Therefore, the federal Clean Water Act establishes a mandatory minimum set of requirements that impact AFOs, but states are dictating more exacting requirements responding to public demands for reductions of pollutants.

A conspicuous omission from the federal and state responses to pollution problems presented by AFOs is the absence of appropriate consideration of the needs for individual watershed systems. With respect to the impairment of water quality by excess nutrients, some systems can absorb more nutrients than others. Thereby, a single set of governmental regulations for diverse watershed systems fails to account for the ability of some systems to absorb nutrients without appreciable denigration of water quality. In addition, current regulations do not consider the cumulative effect of a number of AFOs in a watershed. Several small operations may be more harmful than one large operation.

2 FEDERAL GUIDANCE PERTAINING TO ANIMAL WASTES

Although the US government is not directly regulating diffuse pollution from AFOs that are not subject to the point source pollution regulations, it is looking at this pollution issue. On March 9, 1999, the US Department of Agriculture and the Environmental Protection Agency adopted a *Unified Strategy* aimed at minimizing water pollution from AFOs and the land application of manure (USDA & EPA, 1999).

The federal *Unified Strategy* announces a set of guidance recommendations concerning mandatory and voluntary programs regarding AFOs. It prioritizes CAFOs having significant manure production and unacceptable conditions for permitting and enforcement actions.

A perusal of the guiding principles of this *Unified Strategy* identifies priorities and areas of possible governmental involvement. First, the government seeks to minimize water quality and public health impacts from AFOs, with a focus on AFOs that represent the greatest risks to the environment and public health. Second, the government seeks to make appropriate use of diverse tools including voluntary, regulatory, and incentive-based approaches that complement the long-term sustainability of livestock production in the United States. Next, through regulations, the government proposes to establish a national goal and environmental performance expectation for all AFOs.

To provide incentives for the adoption of practices to reduce pollution, the *Unified Strategy* delineates financial assistance opportunities to support AFOs in meeting the established national performance expectation.

The *Unified Strategy* addresses a need to minimize water pollution from all confinement facilities and the land application of manure. This is to occur through a national performance expectation for the implementation of economically feasible Comprehensive Nutrient Management Plans (CNMPs). A CNMP specifies actions or priorities that will be followed to meet clearly defined nutrient management goals at an agricultural operation. CNMPs will be mandatory for about 5% of AFOs and voluntary for other AFOs.

The development and implementation of a CNMP are the responsibilities of the AFO operator. The Natural Resources Conservation Service Field Office Technical Guide is the primary technical reference for the development of CNMPs (USDA 1990). This guide contains technical information about utilization and conservation of soil, water, air, plant, and animal resources, and is localized to consider the particular characteristics of the geographic area for which it is prepared.

3 DIFFUSE POLLUTION FROM ANIMAL WASTES

Because a defined permit system exists for point source pollution under US law, diffuse pollution has become the focus for further efforts to reduce pollutants in many water bodies. It is generally felt that US federal and state legislative bodies are unlikely to subject diffuse water pollution to explicit controls, such as the permit system employed for point source pollution. Therefore, regulatory efforts may be expected under current authority. Four avenues exist for further mandatory regulations.

3.1 *Reinterpretation of CAFOs*

The US federal regulations delineating which AFOs are subject to the federal permitting requirements are ambiguous due to a proviso. The regulations provide that AFOs with more than 1,000 animal units are deemed CAFOs under federal law and that AFOs with 301—1,000 animal units will be considered CAFOs if certain discharge conditions are met (CFR 1999).

Following these explicit provisions, a proviso delineates an exception: "Provided, however, that no animal feeding operation is a concentrated animal feeding operation as defined above if such animal feeding operation discharges only in the event of a 25 year, 24-hour storm event" (CFR 1999). The issue is whether this proviso applies to CAFOs with more than 1,000 animal units.

In its 1999 "*Draft Guidance Manual and Example NPDES Permit for CAFOs*," the EPA suggested that the presence of more than 1,000 animal units at an AFO creates a rebutable presumption that there is a reasonable likelihood for a discharge in the future (EPA 1999). Under this interpretation, the EPA enumerates a policy that will protect water resources against potential discharges from large AFOs. Thereby, the policy adopts the concept that the number of animal units forms a signal that mandatory governmental regulation is needed to protect water quality. The EPA is expected to adopt a final policy on this issue in the Spring of 2000.

3.2 *Case-by-case designation*

Current federal regulations delineate authority for the EPA to analyse an AFO on a case-by-case basis to determine whether it should be designated as a CAFO. Under the case-by-case designation regulations, AFOs do not have to meet the criteria delineated by Appendix B and may be designated as CAFOs despite not having a discharge. Rather, an appropriate official "may designate any animal feeding operation as a concentrated animal feeding operation upon determining that it is a significant contributor of pollution to the waters of the United States" (CFR 1999). This regulation provides the flexibility to employ further information to determine whether an AFO needs to be regulated to protect water quality.

3.3 *State regulations*

Section 319 of the Clean Water Act delineates federal requirements for diffuse source management programs to attain or maintain applicable water quality standards for navigable waters. States identify and describe state and local programs for controlling pollution emanating from diffuse sources. Pursuant to section 319 directives, states have adopted voluntary programs to encourage the adoption of practices to reduce water contamination.

Due to the environmental problems associated with animal manure, states are enacting specific regulatory management measures for AFOs. Similar to the federal government, most state governments employ the number of animal units to decide when mandatory rules apply. Three aspects of the composition of state regulation are significant.

1. Many of the state AFO regulations require larger operations to adopt manure management plans to help protect water quality, enhance crop performance, and reduce conflicts with others. A manure management plan generally involves testing a representative manure sample to determine the nitrogen, phosphorus, and potassium available through manure. The manure testing results can then be coordinated with nutrient soil testing results to determine the nutrient needs of each field for a specific crop.

2. State AFO legislation and regulations often contain various requirements concerning the licensing and certification of persons for designing facilities, preparing nutrient management plans, and operating livestock facilities. The primary objective of the regulations is to oversee the skills of persons dealing with operations and facilities dealing with nutrients.

3. Odours from AFOs have been a major issue addressed with individual state legislation or regulations. One strategy is to enact rules concerning lagoon construction, timing of application of manure, and the employment of odour control management guidelines. A second strategy is to incorporate setback rules for AFOs. Setbacks prescribe minimum distances that AFO facilities must be sited from other buildings and land uses.

3.4 *Total Maximum Daily Loads*

US law delineates provisions for states to identify waters whose quality is below prescribed ambient water quality standards and to calculate the total maximum daily loads (TMDLs) of pollutants that waters may receive and still meet those standards (33 USCA § 1313(d), 1999). From state lists of impaired waters, priorities can be established for responding to pollutants, including the allocation of pollution reductions.

While the TMDL requirements have existed for more than 25 years, they have not been employed extensively to confront pollution problems (Wenig 1998). Due to several citizen suits against governments and proposed changes to federal TMDL regulations, this may change. Moreover, some experts feel that the TMDL program may offer the best mechanism to address diffuse pollution through a watershed approach (Wenig 1998).

The federal TMDLs are important because they may be employed to reduce future pollution. TMDLs may result in stricter federal permits for existing facilities. Existing facilities may not be permitted to expand their discharges of pollutants. New facilities may be prohibited from obtaining federal discharge permits or may be forced to obtain an offset.

The EPA is considering changes to its TMDL rules (FR 1999). One of the changes would include listing water bodies pursuant to section 303(d) of the Clean Water Act impaired by diffuse pollutants. The EPA advances the argument that "there is no express exclusion of nonpoint source impacted water bodies from the statute's requirements." Thus, the EPA envisions TMDL requirements being used to address diffuse pollution.

Others argue that Congress never intended the TMDL program to cover diffuse pollution (Houck 1998). The absence of coverage of diffuse pollution in section 303(d) and the coverage of diffuse pollution by section 319 discloses that the TMDL program was designed as a tool for establishing effective water quality-based effluent limitations for point source pollution.

The EPA's proposed regulations raise the possibility that the TMDL program may lead to regulations that would impact the disposal of animal wastes. Agricultural and forestry groups are expected to contest regulations that interpret TMDLs to cover diffuse pollution.

4 WATERSHED MANAGEMENT

For nearly a decade, the US EPA has been advocating a watershed protection approach as part of its efforts to improve water quality (EPA 1996). The generalized benefits from watershed management form the basis for the EPA's approach. Environmental, financial, social, and administrative considerations can be brought together to enhance local and regional economic viability in ways that are environmentally sound. Three guiding principles are delineated.

1. The employment of partnerships of those people most affected by management decisions to implement activities that consider multiple criteria.

2. The use of a geographic area, generally a watershed, for considering activities.

3. The application of scientific data, tools, and techniques in an iterative decision making process.

Individual states are the main vehicle for implementing these approaches. States have flexibility to develop and implement their own management approaches. No special federal legislation exists that specifically furthers the watershed protection approach, and most states also lack specific laws and regulations that oversee this strategy.

4.1 *Watershed management and AFO provisions*

A majority of the federal and state regulations for AFOs have been based on number of animal units. Although the individualized state responses provide more flexibility for considering multiple criteria for regulating AFOs, they generally are based on animal units and employ uniform setback requirements. Such criteria fail to account for the ability of some systems to handle problems associated with animal wastes.

Scientific knowledge pertaining to nutrients, soils, topography, rainfall, vegetation, management practices, and modelling programs has shown that diverse systems operate quite differently when exposed to nutrient inputs (Adams et al. 1995; Letson et al. 1998). Management practices have been found to be basic in minimizing nutrient pollution (Edwards et al. 1996). The "one size fits all" approach declines to take into account site specific features that are important in reducing discomforts associated with animal wastes.

Total numbers of animals in a given area of a watershed system may be more important than the number of animals at a single production location. Therefore, alternative signalling measures to provide information that identify which AFOs are likely to cause unacceptable denigration of water quality should be sought. Some other mechanism may offer a means to differentiate AFOs so that only those with prominent environmental loadings would be subjected to governmental regulation.

Alternatively, the government might be able to collect additional information about AFOs to screen out operations that are not major sources of pollution. Through the use of this new information, regulations could be revised so that qualifying AFOs could be exempted from the mandatory regulatory controls.

The development of alternative signalling and screening mechanisms will involve costs. Final adoption of a mechanism will depend on achieving lower costs or greater benefits than exist under the current regulatory system. However, the incorporation of new mechanisms into governmental regulations may allow livestock operations that are not meaningfully impairing water quality in a watershed system to be excepted from some of the existing costly regulatory controls.

5 CONCLUSION

The US federal regulations and some of the state requirements regarding AFOs are not related to scientific information and watershed management strategies. Federal and state regulations based on number of animal units fail to address the issue of whether the nutrients from these operations are impairing water quality. Research has identified soils, topography, rainfall, vegetation, land area, and management practices as criteria that might be employed to devise more effective regulations for potential pollutants generated by AFOs.

Rather than adopting blanket regulations that entail considerable expenses for livestock operators and require a bevy of officials to oversee compliance, regulators should look at alternative signalling and screening devices. As the cost of obtaining scientific information declines, and the precision of the information generated increases, new regulations can be adopted to differentiate those AFOs creating environmental problems. New regulatory provisions providing for the reduction of nutrients may also maximize environmental benefits per dollar spent. Such a regulatory response may provide a superior solution to eliminating pollution from AFOs.

REFERENCES

Adams, R., S.M. Dunn, R. Lunn, R. Mackay & J.R. O'Callaghan 1996. Assessing the performance of the NELUP hydrological models for river basin planning. *J.Env. Planning & Manage.* 38:53-76.

Bosch, J. & K. Napit 1992. Economics of transporting poultry litter to achieve more effective use as fertilizer. *J. Soil & Water Conserv.* 47:342-346.

CFR 1999. *Code of Federal Regulations*, title 40, part 122.

Edwards, D.R., T.C. Daniel, H.D. Scott, J.F. Murdoch, M.J. Habiger & H.M. Burks 1996. Stream quality impacts of best management practices in a northwestern Arkansas basin. *Water Resources Bulletin* 32:499-509.

EPA 1999. *Draft guidance manual and example NPDES permit for concentrated animal feeding operations review draft.* Washington: Environmental Protection Agency, http://www.epa.gov/owm/afo.htm.

EPA 1998. *The quality of our nation's water: 1996.* Washington: Environmental Protection Agency.

EPA 1996. *Watershed approach framework.* Washington: Environmental Protection Agency, http://www.epa.gov/owowwtr1/watershed/framework.html.

FR 1999. Proposed revisions to the water quality planning and management regulation: Proposed rule. *Federal Register* 64:46011-46055.

GAO 1995. *Animal agriculture: Information on waste management and water quality issues.* Washington: Government Accounting Office, GAO/RCED-95-200BR.

Huang, W. 1997. Nutrient management. In *Agricultural Resources and Environmental Indicators, 1996-97*, 97-114. Washington: US Department of Agriculture, Economic Research Service, Natural Resources and Environment Division, Agricultural Handbook No. 712, http://www.econ.ag.gov/epubs/pdf/ah712/.

Houck, O.A. 1998. TMDLs III: A new framework for the Clean Air Act's ambient standards program. *Env. Law Reptr.* 28:10415-10443.

Lander, C., D. Moffitt & K. Alt 1998. Nutrients available from livestock manure relative to crop growth requirements. Washington: US Department of Agriculture, Natural Resources Conservation Service, http://www.nhq.nrcs.usda.gov/land/pubs/nlweb.html.

Letson, D. & N. Gollehon 1999. Confined animal production and the manure problem. *Choices* 3d Quarter:18-24

Letson, D., N. Gollehon, V. Breneman, C. Kascak & C. Mose 1998. Confined animal production and groundwater protection. *Rev. Agric. Econ.* 20:348-364.

USCA 1999. *US Code Annotated.* Title 33.

USDA 1990. *Field office technical guide, national handbook of conservation practices.* Washington: US Department of Agriculture.

USDA & EPA 1999. *Unified national strategy for animal feeding operations*. Washington: US Department of Agriculture and Environmental Protection Agency. http://www.epa.gov/owm/finafost.htm.

Wenig, M.M. 1998. How "total" are "total maximum daily loads"? - legal issues regarding the scope of watershed-based pollution control under the Clean Water Act. *Tulane Env. L.J.* 12:87-187.

Groundwater: Past Achievements and Future Challenges, Sililo et al. (eds) © 2000 Balkema, Rotterdam, ISBN 90 5809 159 7

Experiences from implementation of artificial recharge projects in India

D.K.Chadha
Central Ground Water Board, New Delhi, India

ABSTRACT: Accelerated development of ground water for various uses in India has led to continuous lowering of water levels. About 10% of the country's area is affected by problem of over-exploitation. To overcome this problem, a number of artificial recharge projects have been implemented under various hydrogeological environs, which have shown encouraging results. A preliminary study has indicated that it is possible to create additional storage potential of 214 billion cubic meter (bcm) by way of artificial recharge by utilizing surplus surface run-off during monsoon period, out of which 160 bcm will be utilizable. By the implementation of artificial recharge measures, local areas have been benefited a lot, which have registered rise in water levels and improved water quality. The outcome of artificial recharge projects implemented in India under different hydrogeological conditions have been elaborated.

1 INTRODUCTION

Over the recent years, the concept of artificial recharge has gained importance in India in order to meet ever-growing demands of water for various uses particularly irrigation. The main objective of artificial recharge has been to reduce, stop or even reverse the declining trends of ground water levels. Advantages of storing water underground are (i) cheaper cost of recharge operations than that of surface reservoirs and (ii) ground water is not directly exposed to pollution and evaporation. Besides diverting storm water underground for augmenting ground water storage, artificial recharge operations may also reduce flood hazards, conserve energy by reducing pumping lift and improve quality of ground water by way of dilution particularly in areas affected by problems of salinity ingress, arsenic and fluoride.

2 NEED FOR ARTIFICIAL RECHARGE

In the post-independence era (after 1947), the pressing need of self-reliance spurred the country's efforts for socio-economic revival and all round development in agriculture and industrial sectors for its ever-growing population. India, which was an importer of food grains till 1960s, has emerged as a marginal exporter in recent years with increase in production of food grains from around 50 million tonnes in1950-51 to 192 million tonnes in 1997-98. Increase in irrigation facilities has been the major contributor in food grain production. The irrigation potential has increased from merely 195 thousand sq km in 1950-51 to 900 thousand sq km in 1997-98. The irrigation potential from ground water has increased from merely 65 thousand sq km in 1950 –51 to 450 thousand sq km in 1997-98.

The development of ground water resources in areas with scarce surface water resources, particularly in hard rock and coastal areas has brought problems like decline in ground water levels, failure of wells/ tubewells, shortage in supplies, saline water intrusion, land subsidence, etc. About 249 areas fall under "Over-Exploited" category (withdrawal >100% of the annual recharge), and 179 areas under "Dark" category (withdrawal >85% but <100% of the annual recharge) in various states like Andhra Pradesh, Bihar, Gujarat, Haryana, Karnataka, Madhya Pradesh, Maharashtra, Punjab, Rajasthan, Tamil Nadu and Uttar Pradesh. Such areas necessitate deepening of existing abstraction structures thereby leading to increased pumping lifts and pumping costs. Such areas also face problem of increase in concentration of solute thereby resulting in quality deterioration. In coastal areas, over-exploitation of ground water has led to sea water ingress.

Fig.1 Map of India showing areas where artificial recharge studies have been carried out.

Urgent steps are, therefore, called for in these areas to augment the ground water resources through various suitable recharge measures.

Based on the preliminary studies, a National Perspective Plan has been prepared for recharge to ground water by utilizing surplus monsoon run-off. The ground water storage potential in the country is summarized in Table 1. The guiding principles are: (a) availability of non-committed surplus monsoon run-off in space and time, and (b) identification of suitable hydrogeological environment and sites for creating sub-surface reservoir through cost effective recharge techniques.

Table 1. Ground water storage potential.

Available water for recharge	872 bcm
Water required to create sub-surface storage up to 3 m below ground level (mbgl)	591 bcm
Feasible ground water storage	214 bcm
Utilizable ground water storage Potential	160 bcm

To mitigate the problem of water scarcity in over-exploited areas, a number of artificial recharge projects have been implemented in India. The experiences from these projects are briefly discussed in the succeeding paragraphs.

3 EXPERIENCES FROM ARTIFICIAL RECHARGE PROJECTS

To prove efficiency of various types of recharge structures as well as that of aquifers, artificial recharge studies have been carried out across the country in different hydrogeological set-ups. The location of sites, where artificial recharge studies have been conducted, is shown in Figure 1.

3.1 *Artificial recharge in Gujarat*

A pilot project to study the technical feasibility and economic viability of artificial recharge to the depleted aquifers in Mehsana area, and for the control of salinity in coastal Saurashtra, Gujarat was taken up during 1980-84. An area of 5100 sq. km. in parts of Mehsana, Sabarkantha and Banaskantha districts was selected for artificial recharge to depleted aquifers. In coastal Saurashtra, 900 sq. km. area was selected for artificial recharge to control sea water ingress.

3.1.1 *Artificial recharge of alluvial aquifers in Mehsana area*

In Mehsana, alluvial plain slopes gently towards south-west. Much of the northern and north-eastern part of this district is uncultivated because of the rolling and sandy nature of the landscape. Rivers Sabarmati, Saraswati, Banas and Rupen traverse the plains in west-southwesterly direction and disappear in the little Rann of Kutch. In the south and south-western Mehsana district and western Ahmedabad district, there is no surface drainage except very minor internal inter-dunal drainage.

The water levels of period 1959-82 showed progressive decline in the north-eastern, eastern and central adjoining areas of Mehsana. The decline in water levels was of the order of 1 to 2.5 m/ year, the total decline being 10 to 18 m. The decline in the eastern section is due to over-development of ground water, where natural recharge is somewhat hampered due to highly silty nature of the soil near the surface and presence of kankar bed between 6 to 8 m of depth.

On the basis of studies, the narrow strip on the north and northeastern part was found suitable for artificial recharge. The area in central Mehsana was not able to reap the benefits of recharge due to slow movement of ground water (17 to 21 m/year) in the recharge zone. It was considered that the best way to provide immediate recharge to these areas was through injection wells or shafts. Surplus water available in the saturated flood plain deposits of Saraswati river and excess water from the canal system of Dharoi Dam, could be utilised for this purpose. The excess surface water available during flash floods of Rupen and Pushpavati river system in Central Mehsana area could also be diverted underground through injection shafts.

The area around Kamliwara in Central Mehsana was selected for injection experiment, where sufficient water was available from Saraswati river during monsoon period till the onset of summer season. Experiment was carried out with the help of a 15 m deep and 35 cm diameter source well in the river bed for supplying silt-free water to 125 m deep injection well. Recharge was done at the rate of 18 lps and 12 lps for a period of 100 minutes each. A total of 173 m^3 water was injected. The recharge mound dissipated in about 45 hours. Figure 2 shows design of injection well in Kamliwara, Mehsana.

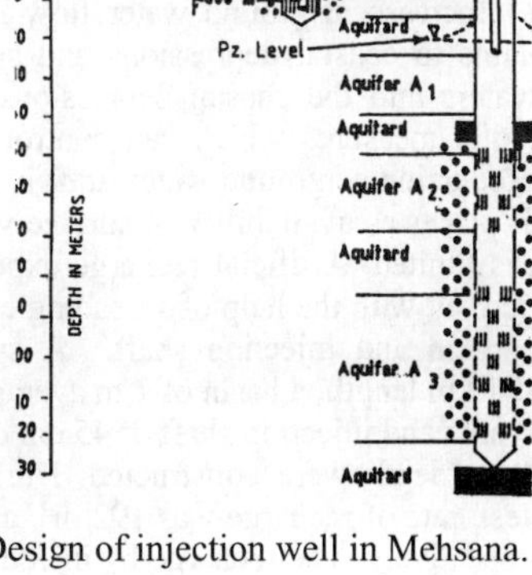

Fig.2 Design of injection well in Mehsana.

Injection experiment was also carried out by injecting water from the source well in Saraswati river bed to the injection well at a rate of 225 m^3/day for 250 days. There was a build up of 5 m^3 in the injection well and 0.6 to 1.0 m in wells 150 m away. These injection experiments clearly indicate the feasibility of artificial recharge of ground water through injection wells.

In Mehsana area, artificial recharge experiments through spreading of water were also conducted. A spreading channel of 3.6 m width and 400 m length was constructed. Silt free canal water was fed into this channel for 46 days. The recharge rate was 260 m^3/day. The recorded build up was up to 2 m in 50 days, 15 m from the channel and 20 cm at a distance of 200 m.

3.1.2 *Studies on control of salinity in coastal Saurashtra, Gujarat*

The coastal tract of Saurashtra is about 15 km in width and can be divided into three distinct land features, namely the coastal plain, coastal depressions and coastal ridges. The coastal ridges run parallel to the sea coast at a distance of 100 to 200 m from the coast with minor gaps on account of erosion action of streams and rivers. The sea water during high tides constantly invades the coastal plains and the ridges. The area is drained by a number of NE-SW flowing ephemeral rivers and streams. Geological formations in the area comprise of alluvium and sand dunes, Deccan Traps, Gaj beds and Miliolite limestones. The two main geological formations namely the Gaj and Miliolite formations, constitute potential aquifers. The occurrence of ground water in Miliolite limestone differs from that in yellow fossiliferous Gaj limestone because the former contains solution cavities and the latter is less cavernous and possesses alternating clay bands. The clay bands give rise to confining conditions and also to quality deterioration due to limited movement of ground water.

Two areas were selected for detailed studies on artificial recharge – (i) Chorwad area (spreading method) (ii) Maliya area (spreading and injection methods). Increase in ground water flow from the coastal plains to coastal depressions and spreading of tidal waters into the coastal depressions are the only remedial measures which can control salinity ingress. Reduction of ground water draft is not possible at all places as availability of surface water for irrigation is limited. Artificial recharge experiments were carried out with the help of spreading channels, spreading basin and injection shaft. A spreading channel of 1 km length, a basin of 1 m average depth and 80 m^2 area and injection shaft of 45 cm diameter and 16.50 m depth were constructed. During preliminary test, rate of recharge was 192 m^3 and 2600 m^3 /day respectively. The recharge by injection shaft was found to be cheaper and required less maintenance.

The deterioration of water quality in coastal plains is likely to decrease by adopting measures like spreading channels, vertical shafts and check dams on streams and rivers in the elevated lands, 10 to 15 km from the coast.

3.2 *Artificial recharge studies in Ghaggar river basin,Haryana*

Artificial recharge studies were carried out in the Ghaggar river basin at three sites.

3.2.1 *Kandi area*

Kandi is a narrow belt bordering the Siwalik hills in the north-eastern part of Haryana state. The aquifer consists of boulders, pebbles, gravel and sand with inter-calations of clay and silt. Infiltration studies at various points revealed that most part of the river and stream beds in the area are suitable for recharge through ponds and spreading basins provided that the beds are improved a little by way of removing the covering clay layers.

3.2.2 *Left Bank of Ghaggar river, Kurukshetra*

Experiment on induced recharge was performed at Tatiana site situated on the left bank of the Ghaggar river in Kurukshetra district, Haryana. The purpose of the experiment was to assess feasibility of recharging the aquifers by induced recharge method. A test well and four observation wells tapping an unconfined aquifer of fine to medium sand with some kankar and gravel down to depth of 24.5 mbgl were constructed. Two observation wells were constructed parallel to the river (20 m from the test well) while two observation wells were constructed perpendicular to the river (50 m from the test well). From the analysis of a pumping test, the transmissivity, permeability and specific yield were estimated as 400 m^2/day, 22.2 m/day and 0.18 respectively.

Resistance of the river bed was found to be 94 m which indicates high degree of clogging. It was estimated that if a well field along the river at a distance of 100 m is developed with a spacing of 200 m between two wells and if each well is pumped at a rate of 1728 m^3/day, about 97 % of the well discharge will be contributed by the river to the aquifer by means of induced recharge and it takes 94 days for the river water to reach the aquifer after commencement of pumping. It was further observed that if clogging from the river bed is removed by scrapping the surface clay-silt layer from the river bed and resistance of the river bed is nil, well discharge increases by 35 % with the same drawdown and the river water takes only 40 days to reach the aquifer. The induced recharge method was not effective during the monsoon when the river bed was very much clogged with clay and silt deposits settled on it.

3.2.3 *Narwana Branch area, Dabkheri*

A site in Narwana Branch Canal area, Dabkheri in Kurukshetra district, Haryana was selected for artificial recharge studies by the injection well method as water levels in the area were declining due to overexploitation of ground water. Two test wells and four observation wells were constructed. The aquifer predominantly consisted of sand and kankar with intercalations of clay, silt and kankar. From the analysis of a pumping test conducted on the test well, the transmissivity, permeability and storage coefficient values were determined as 2424 m^2/ day, 33.4 m/ day and 0.0185 respectively.

The injection experiment started by pumping water in the well at the rate of 40 lps. After 44.5 hours of injection, the rate was increased to 57.7 lps by putting additional water by siphon method and continued for 5 hours after which the injection was again continued at 40 lps upto 16 days. The injection rate was again reduced to 22 lps and the injection was stopped after 17 days.

From this injection experiment it was concluded that through a properly constructed well, recharge even at the rate of 57.7 lps is possible for a period of 15 days.

3.3 *Injection well at Ahmedabad City*

Excessive pumping from a large number of tubewells in Ahmedabad city had resulted in decline of water levels at the rate of 2 to 3 m/ year. The annual pumpage of ground water in the city area was of the order of 200 mcm/ year against a safe yield of about 80 mcm/ year. Artificial recharge by injection method was attempted in the city in 1974, in which clean, treated and chlorinated water was injected at the rate of 160 m^3 / hour for 72 hours. The results established feasibility of injecting nearly 4500 m^3/ day of water.

An alternative approach was undertaken in 1977 near Hansol, on the outskirts of Ahmedabad city. In this experiment, water was transferred by siphon (primed by a vacuum pump) from source well to injection well. The initial head difference, in the aquifers of the source well and the injection well, which was 7.06 m was used as the source of energy for transfer of water. Initially, the rate of injection was 15 lps which subsequently got stabilized at 9.8 lps. The efficiency of the injection well was estimated as 58% and it was noticed that if the efficiency of the injection well increases, the recharge rates also go high. The experiment established the feasibility of injection by siphon method and need for improving the well design to improve the injection efficiency.

3.4 *Sub-surface dykes in Kerala*

In Kerala, three sub-surface dykes were constructed. The first attempt was made by a farmer in 1963. The water was extracted from the aquifer through a drainage system along the upstream side of the dyke to a series of collector wells feeding into a pumping well. A sub-surface dyke was constructed in 1979 in the state seed farm, Ananganadi using plastered brick wall and tarfelt sheets. The length of the dyke was 160 m and the maximum height was 5 m. Two wells, interconnected with each other were constructed and one well was pumped.

Another sub-surface dyke was constructed in 1988 in Kerala Agricultural University. Plastered brick wall of thickness 25 to 60 cm was constructed. A low-density polythene sheet was put on the upstream side of the dyke. The length of the dyke was 80m and the maximum height 8m. A large diameter (4.25 m) well was constructed on the upstream side of the dyke and seven sets of piezometers were constructed for water level monitoring. During summer months, the water is used to irrigate additional 370 coconut trees, 60 coconut seedlings, 4000 m^2 of vegetable garden and for distillation of oil from lemon grass.

3.5 *Maharashtra*

In Maharashtra, studies were confined to Orange growing tracts of Amravati district and Banana growing areas of Jalgaon district where water levels had been declining at an alarming rate. In both the areas, detailed integrated studies were carried out in respect of hydrometeorology, hydrology, hydrogeology and geophysics. Based on these studies, sites for 6 percolation tanks, 10 cement plugs (Check Dams), one injection well, 2 recharge shafts were identified and conversion of 2 village tanks into percolation tanks was also taken up. The live storage capacity of percolation tanks varies from 15 to 220 thousand cubic meters (tcm) and that of cement plugs/ check dams varies from 2.10 to 7.42 tcm. Brief findings of the studies are outlined below.

3.5.1 *Amravati District*

The efficiency of percolation tanks constructed varies from 78 to 91 %. The seepage loss ranges from nil to 8 %, while evaporation loss accounts up to 8%. Around 445 tcm surplus runoff could be conserved in three percolation tanks and around 380 tcm is the additional recharge to ground water benefiting an area of about 3 sq km. The rise in ground water level is even upto 10 m in the command area. The benefited area of individual percolation tank varies from 0.5 sq km to 1sq km. Correlation of water level in percoaltion tank and observation wells at one of the sites in Amravati is shown in Figure 3.

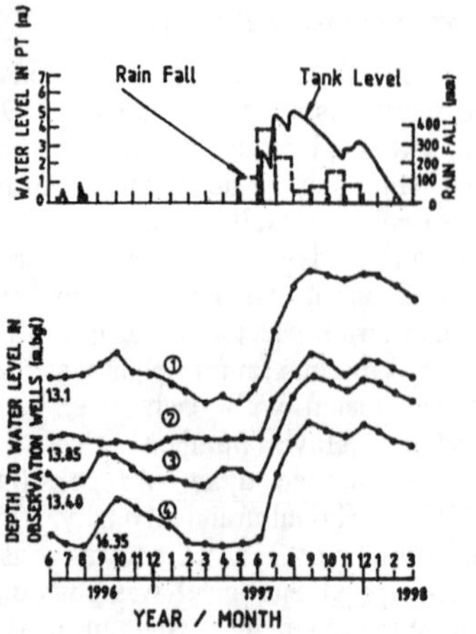

Fig.3 Correlation of water levels in percolation tanks and water wells in Amravati.

The efficiency of 10 cement plugs constructed under the project varies from 81.1 to 97.5 % while the evaporation losses range up to 15 %. Higher evaporation losses were observed in cement plugs, which stored water for longer period. The area benefited from cement plugs varies from 20,000 to 150,000 m^2. The rise of water level in such areas ranges from less than a m to over 3 m. In general, each cement plug is filled 3-4 times during the rainy season depending on the incidence of rainfall and at times capacity utilization is up to 300 %. Around 50,000 m^3 of surface water could be conserved in 10 cement plugs benefiting an area of about 1sq km.

3.5.2 *Jalgaon District*

The efficiency of percolation tanks constructed in the area varies from 95 to 97 %. The evaporation losses vary from 3 to 5 %. Around 425 tcm of surplus runoff was conserved in three percolation tanks and 80 tcm in two village tanks converted into percolation tanks and additional 100 tcm through nalla diversion in existing tank at Nagjhira. Thus 605 tcm water was conserved of which around 580 tcm (96%) was recharged to ground water reservoir benefiting more than 10 sq km of land. It was also observed that existing percolation tank at Nagadevi never filled to its full capacity. Thus water from the adjoining catchment was diverted by constructing a channel and for the first time in 1996, the tank was filled up to 75% of its capacity. This additional filling led to rise in water levels in dugwells upto a distance of 5 km. The farmers have started cultivating sugarcane and banana in place of conventional crops.

Experiment was also carried out for recharging the ground water through injection tubewell. Silt free Hatnur canal water was used for recharge. As a precautionary measure on-line sand filter was used utilizing siphon technique. About 1.5 tcm of canal water was recharged in the year 1997 and the rise in water level upto 0.5 m was observed at a distance of 50 m.

Recharge experiment was also carried out in an old disused dug well at Giradgaon. The surplus water from Hatnur lift irrigation system was used for recharging the ground water through the dug well. Around 6.58 tcm water was recharged in 354 hrs. spread over 35 days at the maximum rate of 40,000 lph. The rise in water level was observed over an area of 20,000 m^2. This effort has opened a new vista and it was established that thousands of existing dug wells in such hydrogeological environment could be used as recharge structures. The farmers of adjoining area have started replicating this technique. In Bamnod lift irrigation scheme, around 50 wells are being used as recharge well utilizing surplus water from the scheme. Efforts are being made by the State Government and local people to make 100 such wells as recharge wells in the command of the irrigation scheme.

3.6 *Karnataka*

In Karnataka, studies were taken up in Gauribidanur and Mulbagal taluks of Kolar district. The study area is underlain by granite gneiss. Two existing minor irrigation tanks were converted into percolation tanks, two micro-watersheds were treated and 7 recharge wells and one roof top rainwater harvesting system were constructed. Treatment of Basavapura micro-watershed in Gauribidanur taluk has shown very encouraging results. The piezometer at Basavapura recorded rise in water levels of the order of 5 - 10 m. An irrigation dug well located close to the micro-watershed has received water after treatment and the farmer has started irrigating about 110 m^2 of land drawing water from this dug well. The farmers have reported an improvement in the yield of irrigation borewells in Basavapura village. Short duration injection tests conducted in the gravity recharge well fields at Hussainpura and Belchikkanahalli have indicated good intake rate of about 4 lps proving the possibility of creating additional storage using the excess water.

Recharge well fields were prepared within the nala bed or adjoining canals by constructing point recharge structures which involved construction of a borewell for recharge and designing of filter bed around the recharge borewell. The surface water either from a tank or from the nala was allowed to flow over the filter bed and the silt free water from the filter bed was recharged through the borewell tapping fractured aquifer.

3.7 *Delhi*

The National Capital Territory of Delhi has been the fastest growing city in India. Decline in water levels by 2 to 8 m. has been observed in the most parts of the city during past decade. Jawahar Lal Nehru University (JNU) - Indian Institute of Technology (IIT) - Sanjay Van area was selected for artificial recharge. These two adjoining areas represent two different hydrogeological set ups - JNU Campus is underlain by quartzites and Sanjay Van - IIT Complex by quartzites and alluvium. Three check dams in JNU, one check dam and one roof top rainwater harvesting system at IIT Complex were constructed. Sixteen piezometers were also constructed in this area for assessment of benefited zone. The efficiency of the check dams was found to be 95 – 98% of reservoir capacity excluding evapo-transpiration losses. Area benefited from check dams ranges from 90,000 to 300,000 m^2. Through two injection wells, 468 m^3 of roof top rainwater was recharged, which resulted in rise of 0.5 m in water levels.

3.8 *Chandigarh*

A roof top rain water harvesting system has been constructed in CSIO Complex in Chandigarh. Total area of roof top is 3550 m^2 and total rainfall during 1998 was 585.9 mm for 47 days, which generated a runoff of 2037 m^3. Out of this, 1628 m^3 water was recharged through recharge well. Sudden rise in water levels was observerd in the piezometers after each rainfall, which lasted for a couple of hours.

4 CONCLUSION

In the present scenario of water scarcity and increasing water demands, artificial recharge is the need of the hour. The experiences from artificial recharge projects carried out in India have indicated feasibility and economic viability of various types of recharge structures. Encouraged by the results of these studies, artificial recharge schemes on a large scale have been launched in the country. Emphasis is being laid on creating awareness among the users on the need for conservation and augmentation of ground water resources and involve them in implementation of the recharge measures..

Groundwater: Past Achievements and Future Challenges, Sililo et al. (eds) © 2000 Balkema, Rotterdam, ISBN 90 5809 159 7

Assessment of saline groundwater resources in the goldfields of Western Australia

D.P.Commander & S.L.Johnson
Water and Rivers Commission, Perth, W.A., Australia

ABSTRACT: Saline and hypersaline groundwater resources in Tertiary palaeochannels are being depleted for gold and nickel ore processing. Assessment is aimed at determining the groundwater storage, and management relies on periodic reassessments of storage depletion. Consideration of storage alone is conservative, as considerable leakage into the palaeochannel aquifer has been demonstrated. Regional assessments based on broad-scale investigations have shown that groundwater resources are sufficient to last several decades at current rates of abstraction.

1 INTRODUCTION

Saline or hypersaline groundwater from Eocene palaeochannel aquifers is the main source of water for gold and nickel ore processing in the Eastern and Northern Goldfields of Western Australia (Fig. 1). This resource has mainly been developed since the mid 1980s when saline water carbon-in-pulp processing came into use. The region produces 150 tons of gold annually and contains two of the world's five largest nickel mines, one of which is also the world's largest cobalt producer.

The hypersaline groundwater is being mined from storage, therefore the lifetime of the resource in relation to the available ore is important. As early as 1987 alternatives were proposed, such as piping sea water from the coast 330 km to the south of Kalgoorlie. A regional exploration program showed that there were considerable resources in the palaeochannel aquifers and that alternative sources would not be required for at least several decades. Even though there are significant reagent costs for the hypersaline groundwater, when compared with sea water or freshwater, groundwater is cheaper than upgrading the present 660 km potable pipeline from Perth to Kalgoorlie, or a new sea water pipeline.

Following the exploration program around Kalgoorlie (Commander et al. 1992) a similar program of widely spaced transects and 1:250 000 hydrogeological mapping was completed in the Northern Goldfields in 1999, providing a regional hydrogeological context for assessing the requirements of large proposed nickel producers (Johnson et al. 1999). A reassessment of groundwater abstraction from the palaeochannels (Ion 1998) was included in a comprehensive study of aspects of water supply and demand for the Kalgoorlie-Boulder WaterLink study (Water Corporation 1999).

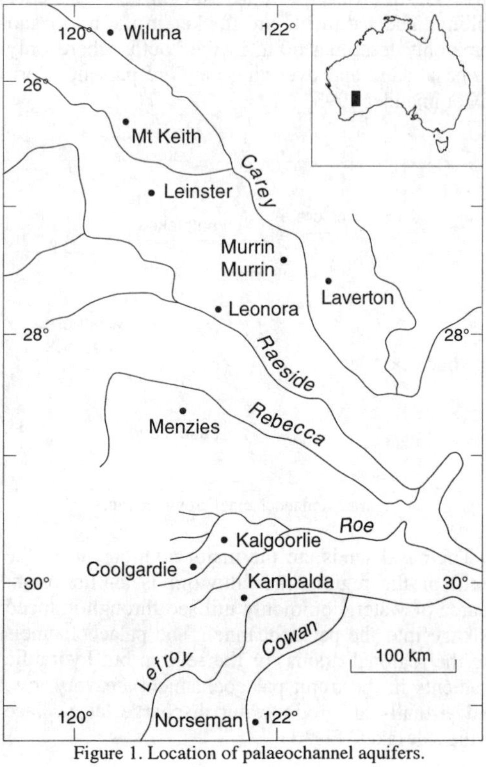

Figure 1. Location of palaeochannel aquifers.

The region is semi-arid and receives about 300 mm of rainfall with summer and winter peaks. Infrequent tropical cyclones may result in half the annual average and cause stream flow in ephemeral drainage lines. Evaporation of about 3000 mm/a exceeds rainfall in all months.

The topography is subdued, with a relief of 200 m, characterised by linear ridges of Archaean greenstone belts (volcanic and metasedimentary rocks) with intervening sandplain over granite batholiths. A dissected laterite represents an early Tertiary land surface which has undergone little subsequent modification. Drainage is internal, with flat-bottomed palaeodrainages now occupied by discontinuous chains of salt lakes and gypsiferous dunes (Morgan 1993). The region is vegetated with eucalypt woodland in the south and mulga (acacia) shrub steppe in the north.

2 HYDROGEOLOGY

The valleys are filled with as much as 130 m of sediments consisting of a basal Eocene sand overlain by a plastic clay, in turn overlain by alluvial material in which calcrete has been developed (Fig. 2). The basal sands range up to about 40 m thick and are up to 1 km in width. These concealed palaeochannels can be located using gravity and electromagnetic surveys, followed by rotary air-blast or air-core drilling. The sediments are thicker in the north and commonly less than 60 m in the south, where only the basal sand and overlying clay are present (Kern & Commander 1993).

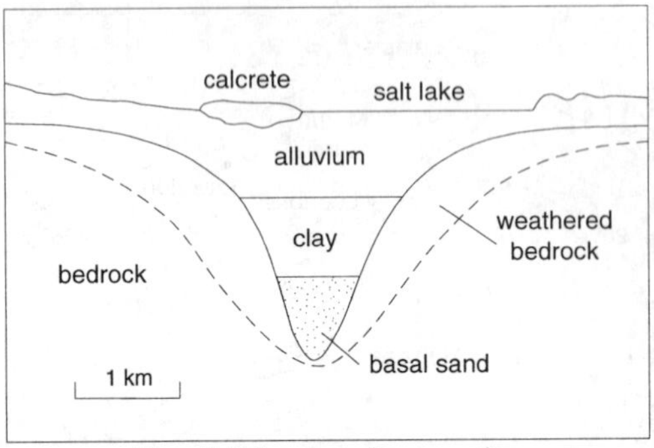

Figure 2. Palaeochannel cross-section.

The basal sands are the major aquifer, but in the north of the region, the alluvium is an important source of water, commonly utilised through induced leakage into the palaeochannel. The palaeochannels are the regional drains for the region, but hydraulic gradients in the trunk palaeochannels are very low, and virtually all groundwater discharge takes place to the salt lakes (Fig. 3).

Fresh groundwater is only present in the uppermost parts of the tributaries in the north of the region. There is commonly a sharp interface between brackish to saline groundwater in the tributaries and hypersaline groundwater in the trunk palaeochannels. Groundwater salinity exceeds 200 g/l in the vicinity of salt lakes and varies along the trunk palaeochannels with ingress of less saline groundwater from buried tributaries.

Groundwater recharge constitutes a very small proportion of rainfall, with losses related to evaporation, utilisation by vegetation, and runoff into the playa lakes. Recharge is only likely to occur during periods of heavy rainfall, and in the upper reaches of the tributary palaeochannels, where low salinity groundwater is present (Johnson, in prep).

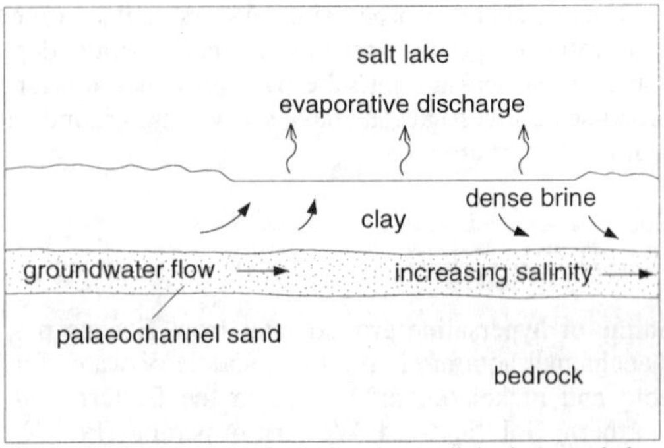

Figure 3. Diagrammatic section to show salinity increase below salt lake.

Isotopic studies using Carbon-14 and Chlorine-36 imply residence times of hundreds of thousands of years to account for the stored salt which is derived from meteoric accession (Turner et al. 1994, Commander et al. 1994). These studies indicated recharge contributing to groundwater outflow is only in the order of 0.03% of rainfall on the catchment.

Groundwater storage has been estimated for the Kalgoorlie Region (Roe Palaeodrainage) and for the North Eastern Goldfields (Carey and Raeside Palaeodrainages) from the cross sectional, area based on widely spaced drilling transects, and by applying a specific yield of 0.2 (Commander et al. 1992, Johnson et al. 1999).

3 GROUNDWATER ABSTRACTION

The palaeochannel sand is the major aquifer in the region, yielding up to 1000 kL/d from bores which are commonly placed at one kilometre spacings (Fig. 4). Annual abstraction from the palaeochannel aquifers (Table 1) is estimated to be around 20 GL/a in the Kalgoorlie region (Ion, 1998) and 22 GL/a in the Northern Goldfields (Johnson et al. 1999). Most gold mines use less than 1 GL/a, whereas the nickel operations use up to 16 GL/a each. The nickel mines of Mt Keith, Leinster and Murrin Murrin in the Northern Goldfields have a combined future requirement of about 35-40 GL/a. At present the Murrin Murrin process requires a groundwater salinity of less than 8 000 mg/l.

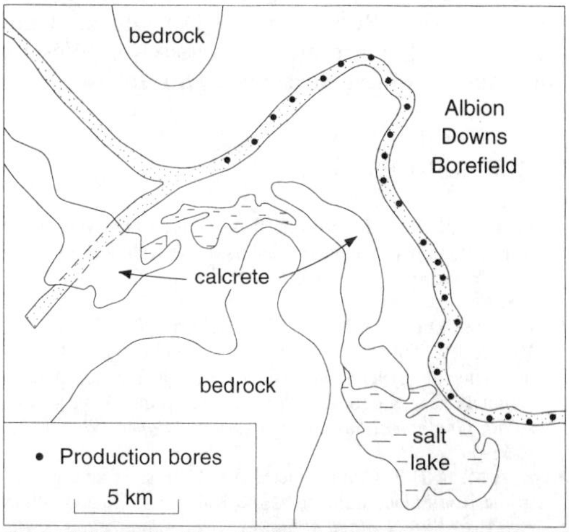

Figure 4. Example of borefield in palaeochannel (after Johnson 1999).

Groundwater is also produced from a few fractured rock borefields and from the bedrock in the course of mine dewatering. In the north of the region, groundwater is also drawn from the alluvial and calcrete aquifers.

4 GROUNDWATER MANAGEMENT

Groundwater abstraction in Western Australia is regulated by the Water and Rivers Commission according to the Goldfields Groundwater Area Management Plan (Water Authority 1994). This plan allows for the mining of brackish (>2000 mg/L) and saline groundwater, in contrast to other areas of Western Australia where groundwater is allocated on a sustainable basis. Mining is considered beneficial use for these brackish and saline groundwater resources. For the purpose of groundwater management the region is divided into sub-areas coinciding with the palaeodrainages. This is to enable cumulative abstraction to be periodically estimated against the remaining storage.

Mining companies are required to carry out exploration to prove the groundwater storage before an abstraction licence is granted. An annual monitoring report must be submitted with abstraction, water level and water quality data, with reviews of borefield performance carried out at intervals of three to five years. These reports allow a periodic review of groundwater storage on a regional basis.

An important consideration in the assessment of groundwater storage is the transition from confined to unconfined conditions. During the period of abstraction when the aquifer is confined, the groundwater storage in the palaeochannel aquifer is virtually unchanged, and groundwater is produced by leakage from the overlying alluvium and adjacent weathered bedrock. When the palaeochannel aquifer becomes unconfined, the groundwater storage in the palaeochannel itself is being depleted (Fig. 5).

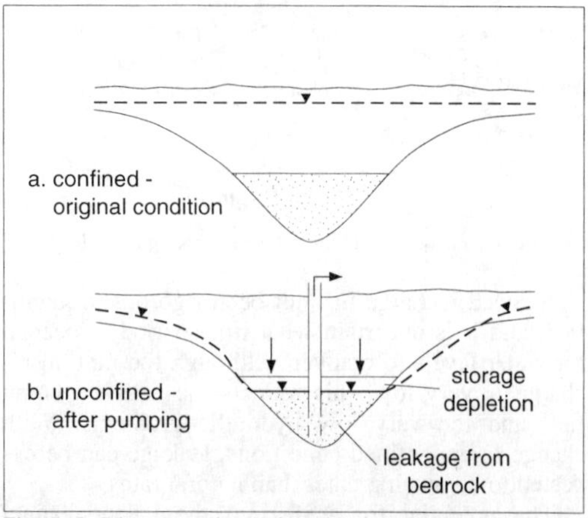

Figure 5. Diagrammatic section showing transition to unconfined state.

The initial study of the Kalgoorlie Region (Commander et al. 1992), estimated 630 GL in storage in the Roe Palaeochannel. A reassessment in 1998 (Ion 1998) revealed that of the cumulative abstraction of 100 GL, estimated storage depletion was 37 GL, leaving 587 GL in storage, representing several decades of use at current rates. In the Northern Goldfields, a storage of 3400 GL has been estimated by Johnson et al. (1999) for the Carey and Raeside Palaeochannels (Table 1).

Table 1. Groundwater storage and abstraction from palaeochannel aquifers

Palaeochannel	Groundwater storage GL	Cumulative abstraction GL	Annual abstraction GL/a
Carey*	2040	-	22
Raeside*	1320	-	0
Roe**	586	100	20
Lefroy**	-	8	2

* Johnson et al. 1999. ** Ion 1998

A key to monitoring the transition from confined to unconfined conditions (Ion 1998, Fig. 6) is the availability of water level data from monitoring bores. While operators are required to submit annual monitoring reports, most water levels are from production bores, and few borefields are monitored with adequate observation bores. The review by Ion

(1998) identified problems with incomplete returns and a general lack of monitoring bores.

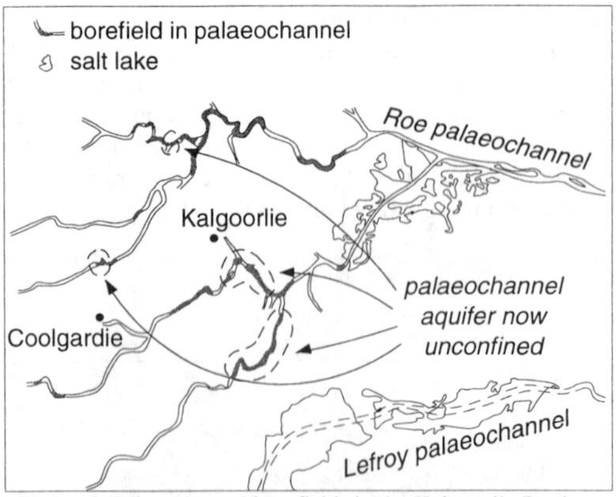

Figure 6. Palaeochannel borefields in the Kalgoorlie Region.

Induced recharge has not been rigorously quantified, and it is uncertain what time period is needed for water levels to recover. Although the natural recharge is very low, this reflects the aquifers being full, and the very low hydraulic gradients. With change to unconfined conditions, leakage can be expected to be much greater than natural rates.

The potential for ASR into the palaeochannel sand has been evaluated in studies by CSIRO and the mining industry. The intermittent ponding of fresh water in the salt lakes provides a sources of water which can be potentially injected, especially where the aquifer has become unconfined. Also mine dewatering injection has been tested, however, both trials have encountered problems with chemical precipitation.

The compilation of hydrogeological maps (e.g. Johnson 1999) enables the spatial relationships of borefields to be easily determined, assisting both management and further exploration.

5 CONCLUSIONS

The Eastern and Northern Goldfields of Western Australia have a unique hydrogeological environment with a buried integrated palaeodrainage system containing brackish to hypersaline groundwater, which is being progressively mined to support nickel and gold ore processing.

Management of the groundwater resource is based on available groundwater storage over the life of the proposed mining projects. While natural recharge is negligible, induced recharge from bedrock and overlying alluvium will occur as the aquifers are stressed, so that allocation of storage only is a conservative position. Very large nickel operations will need careful management over the next few decades. As groundwater abstraction increases, so will the importance of monitoring and regular regional reviews of abstraction.

6 ACKNOWLEDGEMENTS

Part of this work has been undertaken through the Centre of Water Research in the Department of Environmental Engineering, University of Western Australia as a M.Eng.Sc degree by S.L Johnson.

7 REFERENCES

Commander, D.P., A.M. Kern & R.A. Smith 1992. Hydrogeology of the Tertiary palaeochannels in the Kalgoorlie Region (Roe Palaeodrainage). *Western Australia, Geological Survey, Record 1991/10.*

Commander, D.P., L.K. Fifield, P.M. Thorpe, R.F. Davie, J.R. Bird & J.V. Turner 1994. Chlorine-36 and carbon-14 measurements on hypersaline groundwater in Tertiary palaeochannels near Kalgoorlie, Western Australia. *Western Australia, Geological Survey, Professional Papers, Report 37, p. 53-60.*

Kern, A.M. & D.P. Commander 1993. Cainozoic stratigraphy in the Roe Palaeodrainage of the Kalgoorlie Region. *Western Australia, Geological Survey, Professional Papers, Report 34, p. 1-11.*

Johnson, S.L. 1999. Sir Samuel, W.A. *Water and Rivers Commission, 1:250 000 Hydrogeological Series.*

Johnson, S.L. in prep. Hydrogeology of the Northern Goldfields, Western Australia. *University of Western Australia, M.Eng.Sc Thesis.*

Johnson, S.L., D.P. Commander & C.A. O'Boy 1999. Groundwater resources of the Northern Goldfields. *Western Australia, Water and Rivers Commission, Hydrogeological Record HG 2.*

Morgan, K.H., 1993. Development, sedimentation and economic potential of palaeoriver systems of the Yilgarn Craton of Western Australia. *Sedimentary Geology, v.85, p.637-656.*

Ion, N. 1998. Assessment of hypersaline groundwater use in the Kalgoorlie-Norseman region. *Western Australia, Water and Rivers Commission, Hydrogeology Report HR 120 (unpublished).*

Turner, J.V., L.R.Townley, M.R.Rosen & N. Milligan 1994. Groundwater recharge to palaeochannel aquifers in the Eastern Goldfields of Western Australia. *International Association of Hydrogeologists XXV Congress Proceedings, Adelaide, November 1994, p.511-516.*

Water Authority of Western Australia 1994. Goldfields Groundwater Area Management Plan. *Western Australia, Water Authority, Groundwater and Environment Branch, Report WG157.*

Water Corporation 1999. Kalgoorlie-Boulder WaterLink, A study of future water needs and sources in the Goldfields Region. *Western Australia, Water Corporation.*

Groundwater: Past Achievements and Future Challenges, Sililo et al. (eds) © 2000 Balkema, Rotterdam, ISBN 90 5809 159 7

Classification of groundwater resources under the South African National Water Act

J.Conrad & I.van der Voort
Groundwater Group, Environmentek, Council for Scientific and Industrial research, Stellenbosch, South Africa

ABSTRACT: The South African government has introduced a new National Water Act, radically changing the thinking around the management and protection of water resources. In order to assist with the management and protection of these resources, a classification system has been introduced. The purpose of a classification system is to provide a set of nationally consistent rules to guide decision making about water resources. A national classification system should allow transparency, accountability and long-term goal setting to be incorporated into water resources management. Water resources which need to be improved can be identified, and the necessary control measures implemented to meet the requirements associated with the assigned classification. This paper reviews the steps needed to comprehensively classify the current conditions of a groundwater resource. A case study is then used to illustrate the results and uncertainties of such a classification.

1 INTRODUCTION

In 1994, the South African government introduced the Reconstruction and Development Programme (RDP) to improve the quality of life for all South Africans. Water supply is an important component of the RDP since one of its objectives is to ensure all households have a clean, safe water supply of at least 25 litres per capita per day within 200 metres distance, and an adequate, safe sanitation facility per site. In addition, the government revised legislation governing water. Classification is necessary to assess the current status of a resource by considering the degree of modification and consequently the degree of risk of irreversible damage to the resource.

2 THE NEW NATIONAL WATER ACT

During the last few years the thinking around water management and protection in South Africa has changed radically. A prime example of these changes is the National Water Act (Act No 36 of 1998) which introduces many new concepts including *resource directed measures* (RDM), the *Reserve* and *resource quality objectives* (RQOs). The Reserve considers both the quantity and quality of water required to satisfy basic human needs and to protect the functioning of aquatic ecosystems. RQOs for a resource are a numerical or descriptive statement of the conditions that should be met in terms of resource quantity and quality, in order to ensure the resource is protected (MacKay 1998).

The purpose of a classification system is to provide a set of nationally consistent rules to guide decision making about water resources. A national classification system should allow transparency, accountability and long-term goal setting to be incorporated into water resources management. Water resources that need to be improved can be identified, and the necessary control measures can be implemented to meet the requirements associated with the assigned class (MacKay 1998).

3 A COMPREHENSIVE CLASSIFICATION SYSTEM FOR GROUNDWATER RESOURCES

Under a national protection-based classification system, water resources can be grouped into classes representing different levels of protection. The risk that can be accepted in each class is related to the level of protection required for that class. For water resources that are especially important, sensitive, or of high value, little or no risk would be acceptable, and they would be assigned a high protection class. In addition, certain activities or impacts would be regulated or controlled to a certain degree in each class (MacKay 1998).

A comprehensive classification of water resources indicates a relatively high confidence in these classes supported by extensive field investigations.

4 THE CLASSIFICATION METHOD

The process of classifying a resource is part of a broader methodology, which is prescribed for RDM determinations. The RDM methodology can be summarized in the following seven steps:

1. Delineate geographical boundaries of the significant water resource
2. Delineate geohydrological and eco-geohydrological units
3. Determine reference conditions for each unit
4. Determine current status and classify each unit
5. Set future management classes
6. Determine the Reserve and RQOs
7. Implement monitor each unit.

Once the study has been initiated the geographical boundaries need to be established. At present the quaternary catchment boundary is used as the defining boundary. The quaternary catchment is the basic areal drainage unit used in surface water resources (Midgley et al. 1994), and is used as the broad boundary for which groundwater resources are investigated.

A quaternary catchment is then subdivided into geohydrological units. These units must be relatively homogeneous with respect to groundwater and, therefore, require their own Reserve specifications and protection criteria.

The geology of a region is an important factor to consider when studying the groundwater in that region, as it plays a major role in the structure and behaviour of an aquifer. The major geological boundaries present in the quaternary catchments are therefore delineated. All the geological structures such as faults, dykes and lineaments are taken into account. If the same geological formation intersects two or more quaternary catchments, it must be flagged so as to indicate that the resource directed measures for the geohydrological unit will be interdependent on resources directed measures in adjacent catchments. Two other factors that need to be taken into consideration when delineating geohydrological boundaries are: the vulnerability of the aquifer and the yield of an aquifer.

As aquatic ecosystems play an important role in the National Water Act (Act No 36 of 1998), the geohydrological units can be subdivided into eco-geohydrological units.

The eco-geohydrological units delineate regions in which groundwater plays a role in the sustainability of aquatic ecosystems.

The next step is to determine the reference conditions for a geohydrological unit. Reference conditions describe the natural pristine or 'unimpacted' conditions of a geohydrological unit. They present a base-line from which the current classes can be calculated. Groundwater reference conditions include: water quantity and water quality.

Having determined groundwater reference conditions, the current conditions can now be assessed. The current conditions of the unit take in account the degree of modification from reference conditions as well as an assessment of current and predicted water uses, land uses and socio-economic conditions. The importance and vulnerability of a water resource is used to guide or influence the decision on the level of protection required.

Classification of a water resource, and the subsequent derivation and setting of objectives, should ideally be carried out in a formal process of negotiation and consensus seeking among all stakeholders. Stakeholder groups that should be represented in this process include water user sectors, industrial sectors, agricultural sectors, public sectors, special interest groups, local and regional government, as well as other government departments responsible for resource development and for resource protection.

Following on from the above process is the setting of the Reserve and resource quality objectives for each resource unit. Thereafter a monitoring programme must be established to improve the confidence of future RDM determinations and to monitor the response of the resources to the Reserve and RQOs.

5 CASE STUDY – THE PIENAAR'S RIVER CATCHMENT

5.1 *Site Description*

The Pienaars River catchment, shown in Figure 1, is located north of Pretoria. The six quaternary catchments associated with the Pienaars River cover an area of approximately 4800 km^2. The Pienaars River initially flows in a northerly direction, however approximately midway through the study area it flows in a westerly direction to the Klipvoor Dam.

5.2 *Data Collation*

There is usually limited hydrogeological data available for study areas, however it is important to obtain as much data as possible concerning the hydrogeology of the area.

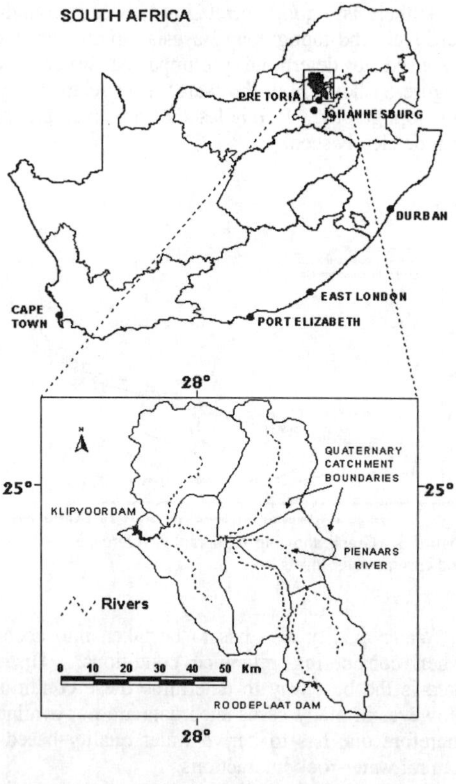

Figure 1. The Pienaar's River catchment and contributing river catchments.

5.3 *Determining the current class*

The case study will focuses on only one of the six quaternary catchments (Figure 2). This quaternary catchment is then divided into geohydrological units taking into account geology, aquifer yield and vulnerability (Figure 3). The geological formations present in the study area include fractured Karoo sedimentary rocks (sandstone and shale) and igneous rocks (granite). Assessment of borehole yields is important in delineating geohydrological units and ideally for a comprehensive study a hydrocensus would be necessary to augment and verify NGDB abstraction values. The vulnerability of study area was determined from a national groundwater vulnerability study carried out by Parsons and Conrad (1998), which was based on the DRASTIC approach (Aller et al 1987).

A summary of all data collected and the data sources used for this study is outlined in Table 1.

Once the geohydrological units have been delineated eco-geohydrological units can be determined. For the whole of South Africa ecoregion boundaries have been delineated at a very broad scale (Level I). Attributes such as physiography, climate, rainfall, geology and potential natural vegetation were evaluated and 18 Level I ecoregions were identified (Kleynhans et al 1998). Larger scale Level II typing is still being carried out, and has not yet been completed for the entire country. The same characteristics as for Level I typing are used, but in more detail. Level II typing produces sub-catchment scale eco-

Table 1. A summary of all the data and sources used in the study.

Data	Source
Maps	
Quaternary catchments	Water Resources '90 (Midgley et al, 1994)
Geology	Council for Geoscience
Topography	Surveys and Mapping
Borehole positions	National groundwater database (NGDB)
	VSA consultants
Land cover /land use	ARC/CSIR
Groundwater	Vegter (1995)
Literature	
Reports	Parsons (1999), Hobbs (1996)
Physical data	
Drilling logs	VSA consultants /NGDB/ Atomic Energy Corporation
Water levels of ~400 boreholes	VSA consultants /NGDB
Abstraction rates	NGDB
Number of people dependent on groundwater	Parsons (1999)
Groundwater chemistry	National groundwater chemistry database
Rainfall	Water Resources '90 (Midgley et al, 1994)
Surface water information	Water Resources '90 (Midgley et al, 1994)
Ecology	Institute for Water Quality Studies

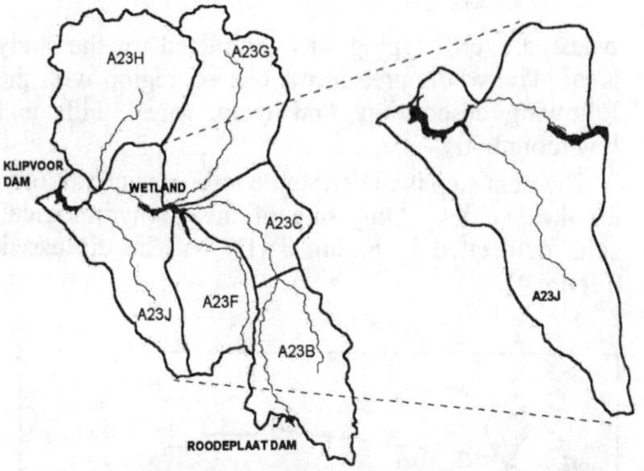

Figure 2. Quaternary catchment to be dicussed in the case study.

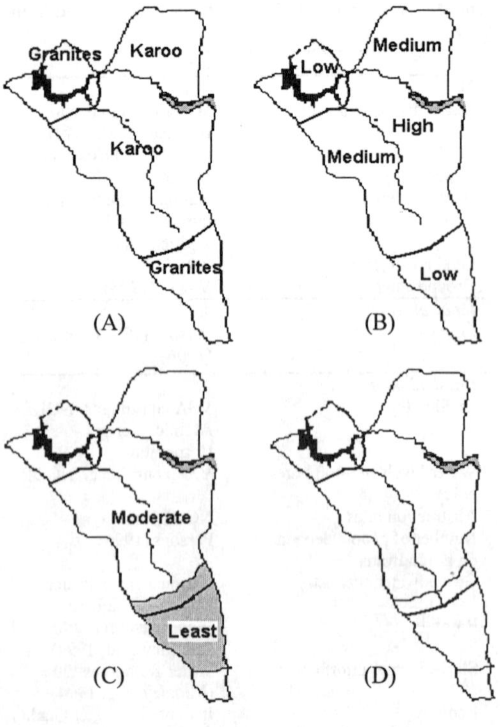

Figure 3. (A) The geology, (B) the yield of the aquifers, (C) the vulnerability of the groundwater to contamination and (D) the final geohydrological units.

types. Level II typing was completed for the study area. The whole area forms one ecoregion with the following description: "bushveld, forest, hills and low mountains".

The next step is to determine reference conditions for these units. Only one of the geohydrological units delineated in Figure 3 (D) will be discussed (Figure 4).

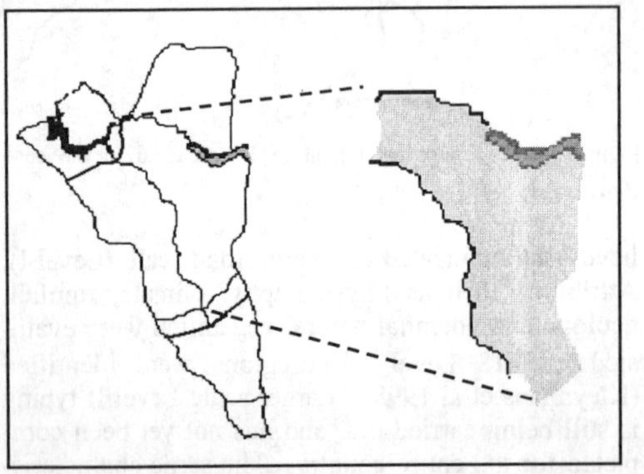

Figure 4. Geohydrological unit to be discussed in case study.

If there is a good correlation between groundwater levels and topography Bayesian interpolation can be used to determine pre-impacted water levels. Figure 5 shows how the water level relationship to the topography. Boreholes that are over pumping can be clearly seen.

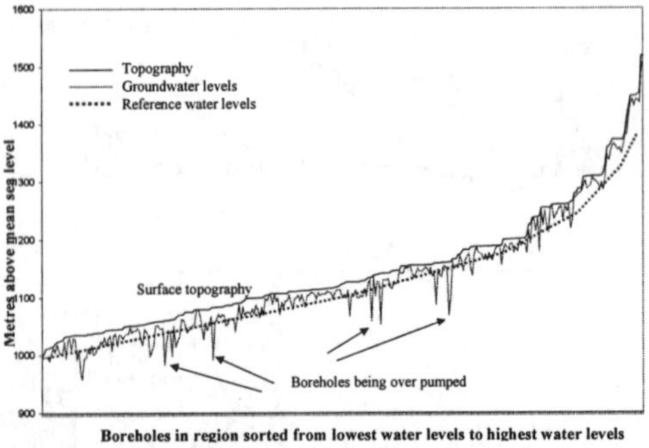

Figure 5. Graph showing topography, reference water levels and current water levels.

Water quality also has to be taken into account when considering reference conditions. Historic data is the best way to determine these conditions. However in many cases such data are not available, therefore one has to derive water quality based on natural water-rock interactions.

There are elevated electrical conductivity and nitrate values in the unit under discussion, which is most likely associated with the agricultural activity. Unfortunately there is no further information available for water quality. The current water quality in the area compares favourably with typical values associated with geology.

It is important to note that it is not always possible to determine the reference conditions for a unit. Sometime it will be necessary to 'reset' the reference conditions.

When comparing the reference conditions and current conditions present in the unit it can be concluded that the water levels and water quality has basically remained constant except for localized changes and can therefore, according to Table 2, be rated as a category B.

Other aspects to be taken into account when classifying the unit are:

- Protection of the resource and
- Use of the resource.

Table 3 is used to define the classification system describing protection and use.

From Figure 3 one can see the unit is moderately vulnerable and has a medium yield. Therefore it can be classed as a C for protection.

Within the geohydrological unit there are 40 000

Table 2. A broad classification for resources (DWAF, 1998).

Class	Definition
A	Unmodified or approximates natural conditions
B	Largely natural conditions with only a few localised modifications; no negative effects apparent
C	Moderately modified; moderate changes are apparent
D	Largely modified; a widespread loss of natural functioning
E	Seriously modified; the losses of natural functioning are extensive
F	Critically modified; modifications have reached a critical level and the region has been modified completely with almost complete loss of natural functioning

Table 3. Protection and use classes.

Class	Definition
A	Allow no change
B	Allow slight risk of change
C	Allow moderate risk of change
D	Allow high risk of change

people dependent on groundwater for basic human needs which amounts to $0.4e^{-6}$ m^3/a. The amount of water needed to maintained ecological sustainability is $1.4e^{-6}$ m^3/a. When calculating the water balance it can be determined that the amount of groundwater that can be allocated is $5\ e^{-6}$ m^3/a. Therefore basic human needs and ecological requirements are only 5% and 20% respectively of the total amount of water to be allocated. These percentages are low and therefore the unit can be classified as a C for use as well.

6 DISCUSSION AND CONCLUSION

This paper has focused on classifying geohydrological units. Presently only surface water drainage and geology are taken into consideration when defining geohydrological units. However there are many other aspects that might need to be taken into consideration, such as; aquifer parameters, recharge zones, the kind of ecosystems associated with the systems, and the nature and extent of dependency as well as their vulnerability and significance (Hatton 2000).

It is important to note, however, that the processes discussed in this paper are in their initial stages of development and the Pienaars River is the first case study area on which these procedures were tested. Field work still has to be conducted to achieve a higher confidence in the results.

7 ACKNOWLEDGEMENTS

The research discussed in this paper emanates from a project funded by the Water Research Commission entitled: "The determination of the Groundwater Reserve: Delineation, Reference Conditions and Classification". The financing of the project by the Water Research Commission is acknowledged gratefully.

8 REFERENCES

Aller, L., Bennet, T., Lehr, J., & Petty, R. 1987. DRASTIC – a standardized system for evaluating groundwater pollution potential using hydrogeologic settings. *United States Environmental Protection Agency Report EPA/600/2-87/035*

Department of Water Affairs and Forestry. 1998. Part3: Development of the methodology for determination of the ecological reserve. Pretoria, South Africa.

Hatton, T. 2000. Review of the Water Research Commission Programme: The Comprehensive Determination of Water Resource Reserve: Groundwater Component. Unpublished. Report submitted to the WRC.

Hobbs, P.J. 1996. JICA study-situation report on geology, soils and hydrogeology of the Maglies Water Extended Supply Area: Report prepared for ENV Consulting Engineers by VSA GeoConsultants, Pretoria.

Kleynhans, C., Silberbauer, M. & Kemper, N. 1998. Preliminary ecoregion I classification for South Africa, unpubl. Internal report, Institute of Water Quality Studies, Department of Water Affairs and Forestry.

MacKay, H.M. 1998. Towards a classification system for water resources in South Africa. Institute for Water Quality Studies, Department of Water Affairs and Forestry, Pretoria.

Midgley, D., Pitman, W., & Middleton, B. 1994 *Surface Water Resources of South Africa 1990.* Water Research Commission, Pretoria.

National Water Act 1998. Government Gazette, 19182. Act 38, 26 August 1998. Pretoria, South Africa.

Parson, R. & Conrad, J. 1998. Explanatory notes for the Aquifer Classification Map of South Africa. *WRC report No. KV 116/98.* Water Research Commission.

Parsons, R. 1999. Intermediate reserve method pilot test: Pienaars River between Roodeplat Dam and Klipvoor Dam. P.O. Box 2606, Somerset West, 7129.

Vegter, J.R. 1995. An explanation of a set of National Groundwater Maps. *WRC report no. TT74/95.* Water Research Commission.

Groundwater: Past Achievements and Future Challenges, Sililo et al. (eds) © 2000 Balkema, Rotterdam, ISBN 90 5809 159 7

The potential of riverbank filtration for drinking water supplies when aquifers are brackish

P.J. Dillon
Centre for Groundwater Studies and CSIRO Land and Water, Glen Osmond, S.A., Australia

W. Rhebergen
Faculty of Earth Sciences, Vrije Universiteit, Amsterdam, Netherlands

M. Miller & H. Fallowfield
Department of Environmental Health, Faculty of Health Sciences, Flinders University of South Australia, Adelaide, S.A., Australia

ABSTRACT: The River Murray in South Eastern Australia contains turbid water, and on occasions, toxic cyanobacteria (blue green algae). For numerous small communities that obtain water supplies directly from the river, bank filtration appears to be an attractive form of natural water treatment. However the ambient groundwater of unconfined aquifers skirting some reaches of the river is saline. This paper reports a modelling study to examine the compatibility of two constraints on the quality of water recovered from bank filtration schemes; that (1) removal of cyanobacterial toxins is adequate and (2) salinity is acceptable for drinking water supplies. Adsorption and biodegradation characteristics of a cyanobacterial hepatotoxin in porous media under a range of environmental conditions were quantified in a related study and these results were summarised in the current analysis. It was found that riverbank filtration schemes could meet both criteria in a limited range of conditions, excluding locations where saline groundwater discharged to the river.

1 RIVERBANK FILTRATION FOR WATER TREATMENT

Riverbank filtration describes the process of extracting groundwater from wells adjacent to a river, or from horizontal collector wells beneath a river bed or within the banks in order to induce infiltration from the river. This is generally performed where the quality of water in the river is not suitable for water supplies due to intermittent or chronic pollution. The streambed sediments provide 'slow-rate filtration' and the recovered water is of higher and more consistent quality than water drawn directly from the river.

It has been known for many years that nearby streams improve the productivity of well fields (Theis, 1941, Kazmann, 1947; Rorabaugh, 1948), and analytical solutions for streamflow depletion as a result of pumping from adjacent wells have been derived for a number of configurations (eg Theis, 1948; Hantush, 1965; Jenkins, 1968; Glover, 1974; Zlotnik *et al*, 1999). Subsequently more attention has been given to the water quality improvement aspects with respect to metals and natural and synthetic organic substances due to the effects of industry on rivers in central Europe.

For example, Sontheimer (1980) found that sustained heavy metal removal in the lower River Rhine exceeded 90% for chromium and arsenic, and was greater than 50% for cadmium, zinc, lead, copper and nickel. The primary removal mechanism was adsorption, and without an ongoing fixing process within the sediments, this treatment has a finite lifetime before breakthrough or desorption occur. Schwarzenbach et al. (1983) studied attenuation of organics in gravel-sand alluvium with low organic carbon and found that about 60% of dissolved organic carbon and purgable organochlorine compounds were removed within a few weeks of travel time at two sites within the aquifer, but that the remaining fraction was persistent. Some specific compounds such as 1,4-dichlorobenzene had greater than 90% removal and were not detectable in recovered water. Others, such as chloroform, trichloroethylene, tetrachloroethylene had removals of less than 20%. In that study, biodegradation, principally within the first few metres of infiltration was found to be the primary removal process, and the recalcitrant species were those unable to be metabolised by anaerobic bacteria which colonised this zone.

Miettienen et al. (1994) observed that bank filtration of humus-rich lake water changed the quality of organic matter rather than its total amount, and the ratio of chemical oxygen demand to total organic carbon decreased when the filtration distance increased. They also noted that organic matter was retained within sediments during winter months and then liberated during spring when water temperature increased. It is becoming increasingly clear that the functional processes of the streambed sediment ecosystem need to be understood in order to predict sustainable pollutant removal rates. A recent conference in USA on riverbank filtration (National Water Research Institute, 1999) reflects a rekindling of interest in this field.

In many countries, toxic cyanobacteria (blue green algae) such as *Microcystis spp*, are becoming more abundant due to higher nutrient loads from agricultural catchments, and slower flow rates caused by river regulation and increased abstraction. Studies of adsorption and biodegradation processes associated with the removal in riverbank filtration of one cyanobacterial hepatotoxin, microcystin, have been performed (Miller, 2000) to determine the potential role of riverbank filtration in protecting threatened water supplies for small communities.

Alongside the River Murray in Australia, the ambient groundwater is of variable quality, and the design of a riverbank filtration system for removal of cyanobacterial toxins needs to be considered in relation to constraining the proportion of recovered water derived from a brackish aquifer.

2 RIVERBANK FILTRATION HYDRAULICS

Riverbank filtration schemes can consist of horizontal collector wells underlying the river bed (often radiating from a central pumping sump, eg Spiridonoff, 1963), horizontal collectors parallel to the bank, or most commonly, a normal vertical well, at some distance, a, from the riverbank (Figure 1). For simplicity, only the latter was considered in this investigation.

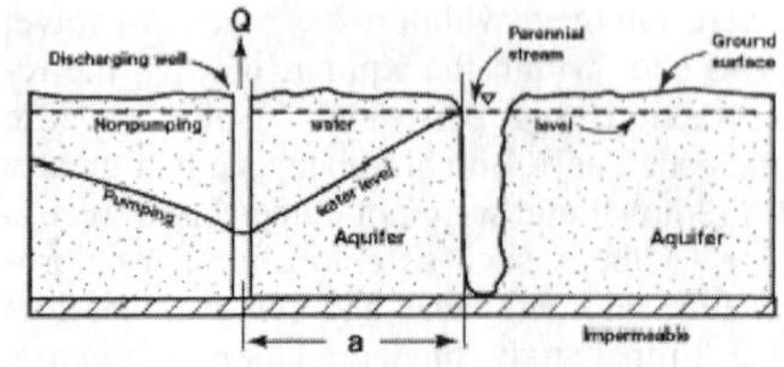

Figure 1. Cross-section though river and adjacent well.

The two factors considered most important for the quality of recovered water are the minimum travel time from the river to the well, t_{min}, and the proportion of recovered water derived from the river (q/Q). The first allows an estimate of contaminant (toxin) attenuation from adsorption and biodegradation, where these parameters can be derived from laboratory studies. The second provides for further dilution, but more importantly is used to estimate the contribution of groundwater to the salinity (assumed to be conservative) of the recovered water. For wells close to the river, the t_{min} is small and q/Q large, and with increasing distance from the river t_{min} increases and q/Q may decline. The intention therefore is to define circumstances, and design distances from streams, under which adequate algal toxin removal would occur in a product water which is sufficiently fresh for potable supplies.

Hydraulic solutions for Figure 1 for steady state and non-steady state conditions, and for water tables which are initially horizontal and sloping, and with and without the presence of a semi-permeable liner on the streambed were reported by Rhebergen and Dillon (1999) using analytical and numerical methods. Arising from that work robust approximate design parameters were found to be:

$$t_{\min} = \frac{2\pi D n_e a^2}{3Q} \qquad (1)$$

where t_{min} = minimum travel time [d]
D = aquifer saturated thickness [m]
n_e = effective porosity of aquifer [-]
a = distance of well from stream bank [m]
and Q = discharge rate of pumping well [m^3d^{-1}]

$$\frac{q}{Q} = erfc(\frac{a}{\sqrt{4\alpha\, t}}) \qquad (2)$$

after Glover and Balmer (1954) where
q = rate of induced infiltration from the river [m^3d^{-1}]
α = aquifer diffusivity = transmissivity/storage coefficient, which in an unconfined aquifer of average saturated thickness, D, and hydraulic conductivity, K, is KD/n_e and
t = time since commencement of pumping [d].

Here t_{min} is the travel time under steady state pumping, which underestimates travel time for intermittent pumping, while q/Q is the proportion of water derived from the stream for transient pumping. (Under steady state conditions q/Q is one for a semi-infinite aquifer with an initially horizontal free surface). Equations (1) and (2) therefore provide lower bounds, or worst case values, for these factors for the case of an initially horizontal watertable and no streambed liner.

Various analytical models of bank filtration were reviewed and used to verify a numerical model, based on Visual MODFLOW. The numerical model has been shown to perform with satisfactory accuracy, and has pointed out a weakness where some analytical solutions have over-specified stream boundary conditions (where both head and hydraulic gradient are fixed). This appears to have little or no effect on the quantity of water derived from a stream by a pumping well, however it has a significant effect on minimum travel times predicted by analytical models. The MODFLOW model was subsequently applied to a range of problems.

3 POLLUTANT ATTENUATION AND DILUTION OF SALINE GROUNDWATER

The maximum concentration of pollutant in the well can be defined as follows, assuming that pollutant concentrations in river infiltrate are attenuated by adsorption with a linear isotherm, that the degradation rate of the pollutant can be defined by exponential decay, and ignoring mixing from water travelling along longer flow paths from the river:

$$\frac{c_w}{c_r} = \frac{q}{Q}\exp[-\lambda R t_{\min}] \tag{3}$$

where

c_w = concentration of pollutant in the well [$\mu g.L^{-1}$]
c_r = concentration of pollutant in the river [$\mu g.L^{-1}$]
λ = exponential rate of pollutant degradation [d^{-1}] = $0.693/t_{1/2}$ ($t_{1/2}$ = half life [d]), and
R = retardation factor for pollutant as a result of adsorption [-].

Consequently, the design criteria to be met to ensure that c_w remains less than an allowable concentration, c* in recovered water is:

$$\lambda R t_{\min} - \ln(q/Q) \geq \ln(c_r/c^*) \tag{4}$$

Similarly the concentration of conservative solute in the well, s_w resulting from the mixture of ambient groundwater (at concentration s_g) and river water (at concentration s_r) is given by:

$$s_w = \left(1-\frac{q}{Q}\right)s_g + \frac{q}{Q}s_r \tag{5}$$

Therefore, in the event of brackish groundwater, where the salinity $s_g = g\ s^*$, and s* is the allowable salinity in the recovered water, and defining $s_r = r\ s^*$ (fresher than groundwater, ie r<g), the minimum acceptable value of q/Q for recovered water to be of acceptable salinity is:

$$(q/Q)_{\min} = \frac{g-1}{g-r} \tag{6}$$

The WHO guideline for microcystin in drinking water (c*) is 1 $\mu g.L^{-1}$. Concentrations of microcystin in river water during periods of toxic algal blooms are of course variable but $2 mg.L^{-1}$ is considered a likely upper value for c_r. The WHO guideline for salinity (total dissolved solids, s*) is 1000 mg L^{-1} and concentrations in groundwater (s_g) may vary up to 20,000 mg L^{-1} or more in semi-arid and arid areas adjacent streams.

4 ATTENUATION OF CYANOBACTERIAL TOXINS

Laboratory batch and sand column adsorption and biodegradation experiments were undertaken under abiotic conditions with several types of porous media including sand and clay loam, at various concentrations of microcystin, and for ranges of environmental variables such as pH, temperature and dissolved organic carbon and phosphate concentrations of the infiltrate. This work is documented by Miller (2000) and it is intended that these investigations will be extended.

For the purposes of this paper, using the soil most representative of River Murray alluvium, under environmental conditions to replicate field conditions at prospective riverbank filtration sites, the following results were obtained for adsorption and biodegradation. Although a linear adsorption isotherm is not the only interpretation of batch experimental data, a retardation factor, R of 2.5 is suggested. The rate of degradation of microcystin may decline with time, for as yet unknown reasons, from a half-life of 1-2 days to 8-16days. Choosing an intermediate value of 8 days gives λ a value of 0.087.

5 APPLICATION TO RIVERBANK FILTRATION DESIGN

Cross-sections on and adjacent a meander on the River Murray at Paringa, South Australia were used to test the potential for riverbank filtration to provide water of suitable quality for a small town water supply ($Q=150\ m^3d^{-1}$). Site characteristics are defined as D=8m, n_e=0.3, K=15 md^{-1}, s_g = 20,000 $mg.L^{-1}$, s_r = 500 $mg.L^{-1}$, with an initially horizontal watertable and no streambed liner. Substituting into equation (4) and neglecting dilution by groundwater defines a minimum travel time of 35 days, and from equation (1) a minimum distance from the river of 32 m, in order to meet microcystin concentrations in drinking water.

From equation (2) the salinity requirements for the recovered water (q/Q >0.974) are met after 1210 days of pumping, but there would need to be a means of discharging the initial saline water. Depending on mixing within the aquifer and fluctuations in relative levels of the river and adjacent watertable the potential to establish and retain a plume of fresh water from induced recharge within the vicinity of the recovery well would be marginal.

Where the groundwater gradient results in natural discharge from the aquifer to the stream, (a gaining stream) Rhebergen and Dillon (1999) showed that the very limited ability to reduce the salinity of brackish waters to acceptable levels in the recovered water renders riverbank filtration projects nonviable. This is analogous to salinity interception schemes that currently protect the River Murray from high salt loads in groundwater.

In reaches where groundwater gradients allow recharge from the stream (a losing stream) the local groundwater will be fresher, and the salinity constraint will diminish. It can be shown that ultimately all of the water extracted from the well may be derived from the stream (Rhebergen and Dillon, 1999).

A lock which was established many years ago on a meander of the river near Paringa had the effect of producing a 2m difference between water levels in parallel reaches 800m apart. Subsequently river water has almost completely flushed the saline groundwater from the meander. Table 1 shows values of t_{min} and q/Q derived from MODFLOW simulations, where q represents the steady state flux derived from the closest bank, for wells 50 and 100m from the upstream and downstream banks, and also for the theoretical case of an initially horizontal watertable at the same site. Aquifer characteristics are as detailed previously.

Table 1. Effect of groundwater slope and distance of collector well from the stream

Slope	Distance (m)	t_{min} (d)	q/Q
+0.0025	50	67	1
+0.0025	100	224	1
0	50	84	0.94
0	100	335	0.88
-0.0025	50	99	0.36
-0.0025	100	432	0.14

To maximise travel time and toxin degradation the downstream wells would be preferred, although all wells have sufficient residence time to remove microcystin. However for the down-gradient wells at 50 and 100m to produce water of acceptable salinity the highest ambient groundwater salinities would be 1280 and 1080 mg/L respectively. The upstream wells would produce the freshest water, particularly under intermittent operation if there is any residual salinity from remnant groundwater. Under continuous pumping salinity is not a constraint for the upstream wells. Hence in this instance, on balance, the preferred well location is 50 m from the upstream bank, where Rhebergen and Dillon (1999) showed that this produces the minimum drawdown and meets both water quality criteria.

6 CONCLUSIONS

Bank filtration is a useful pre-treatment technique for town water supplies and, like slow sand filtration, has proven successful in filtering surface water especially for removal of sediments, metals, and some organic contaminants. It also smoothes out variations in temperature and chemical composition of water, which can improve the performance of treatment plants for which this is a source of water.

For the field site with a sloping initial piezometric surface in a meander near Paringa the model was used to determine the effect of well location and pump rate on q/Q and t_{min}. For a large t_{min} the best location for a well is further away from the river and on the downstream side of a meander. This allows the largest travel times and opportunities for biodegradation of algal toxins in the river water. For a q/Q approaching 1, the best place for the well is close to the river on the upstream side of a meander. This gives minimum drawdown and if the groundwater has a higher salinity than river water, it produces fresher supplies when operated on an intermittent basis. For steady state withdrawal, all water reaching the well will originate from the river.

This paper has considered a variety of conditions affecting bank filtration design which extend the range previously rerported. This is intended to be used by those considering bank filtration in arid and semi arid environments, as a starting point to assess feasibility of alternative designs before commencing field studies.

7 ACKNOWLEDGMENTS

This study was in part supported by SA Water, and the contribution of Bruce Roberts (SA Water), in identifying potential field sites adjacent the River Murray and Steve Barnett and Bob Newman (Dept Water Resources) for hydrogeological information is gratefully acknowledged. The authors thank Prof Ian Simmers for his encouragement of this visiting student traineeship activity. The authors record their thanks to Paul Pavelic for his tuition in MODFLOW modelling and Abdi Ismail who initiated the literature review and undertook some piezometric investigations adjacent the River Murray related to this study.

REFERENCES

Glover, R.E., (1974) Transient ground water hydraulics. Water Resources Publications. 413 pp.

Glover, R.E., and Balmer, G.G., (1954) River depletion resulting from a well pumping near a river. Transactions American Geophysical Union. 35. pp. 468 - 470

Hantush, M.S., (1959) Analysis of data from pumping wells near a river. Journal of Geophysical Research 64. pp. 1921 – 1932

Hantush, M.S., (1964) Drawdown around wells of variable discharge. Journal of Geophysical Research, 69 pp 4221 - 4253

Hantush, M.S., (1965) Wells near a stream with semi-pervious beds. Journal of Geophysical Research, 70. pp. 2829 - 2838

Jenkins, C.Y., (1968) Techniques for computing rate and volume of stream depletion by wells. Groundwater 6. pp. 37 - 46

Kazmann, R.G., (1946) Induced infiltration supplies most productive well field. Civil Engineering. Vol. 16. pp. 544 - 545

Kazmann, R.G., (1947) River infiltration as a source of ground water supply. Transactions American Society of Civil Engineers. Paper No. 2339. pp. 404 - 424

Miettienen, I.T., Martikainen, P.J., and Vartiainen, T (1994) Humus transformation at the bank filtration water plant. Water Science Technology. 30:10, pp. 179-187

Miller, M. (2000) Removal of cyanobacterial toxins by aquifer filtration. PhD Thesis, Dept of Environmental Health, Faculty of Health Sciences, Flinders University of South Australia.

National Water Research Institute (1999) International Riverbank Filtration Conference, Abstracts, Louisville, Kentucky, 4-6 Nov, 1999, 55p.

Rhebergen, W. and Dillon, P. (1999) Riverbank filtration models for assessing viability of water quality improvement. Centre for Groundwater Studies Report No. 90.

Rorabaugh, M.I., (1948) Induced infiltration of river water to wells. Transactions, American Geophysical Union. 29:1, pp. 85 - 92

Rorabaugh, M.I., (1956) Ground water in North-eastern Louisville, Kentucky, with reference to induced infiltration. U.S. Geological Survey Water-Supply Pater 1360-B. 68 pp.

Schwarzenbach, R.P., et al (1983) Behavior of organic compounds during infiltration of river water to groundwater: Field studies. Environmental Scientific Technology. 17, pp 472-479.

Sontheimer, H., (1980) Experience with riverbank filtration along the Rhine river. Journal of American Water Works Association. pp. 386 - 390

Spiridonoff, S.V. (1963) Design and use of radial collector wells. Journal of American Water Works Association. pp. 689 - 702

Theis, C.V., (1935) The relation between the lowering of the piezometric surface and the rate and duration of discharge of a well using groundwater storage. Transactions American Geophysical Union. 16, pp. 519 - 524

Theis, C.V., (1941) The effect of a well on the flow of a nearby stream. Transactions American Geophysical Union. 22, pp. 734 - 737

Todd, D.K., (1959) Ground water hydrology. Wiley & sons. 336 pp.

Wilderer, P.A. Förstner, U, and Kuntschik, O.R. (1985) The role of riverbank filtration along the Rhine river for municipal and industrial water supply pp. 509-528 in: Artificial Recharge of Groundwater. T. Asano (ed). Butterworth Publ., Boston.

Zlotnik, V.A., Huang, H. and Butler, J.J. (1999) Evaluation of stream depletion considering finite stream width, shallow penetration, and properties of streambed sediments. Proc. Water 99 Congress, Brisbane July 1999, p221-226, The Inst. of Engineers, Aust.

Groundwater: Past Achievements and Future Challenges, Sililo et al. (eds) © 2000 Balkema, Rotterdam, ISBN 90 5809 159 7

Small scale aquifer storage and recovery – A case study from England and implications for wider development

J.E. Dodds
Newton Burgoland, UK

ABSTRACT: Aquifer Storage and Recovery (ASR) is a water resource management technique that is growing in popularity around the world, particularly for municipal or public water supply. A number of schemes have been developed in the UK for agricultural irrigation. The schemes demonstrate that the technique is feasible for smaller scale operations supplying 150 000 - 200 000 m^3/year. The success of these smaller scale schemes has implications for the use of ASR techniques in the management of water resources in other situations such as tourist centres, mineral exploitation sites and remote villages or towns. In order to develop small scale schemes, the health, economic and development benefits must be judged against over cautious aquifer protection.

1 INTRODUCTION

Aquifer Storage and Recovery (ASR) was defined by the England and Wales Environment Agency (Jones et al. 1998) as "involving recharge of water into relatively deep, possibly poor quality and little used, aquifers to create a bubble of fresh groundwater for subsequent re-abstraction". The use of borehole recharge to inject water into good quality, generally well utilised aquifers, to enable better use of water resources was referred to as artificial recharge. Payne 1995 has a wider definition in which he states that the injection and re-abstraction of water from the same borehole, is the critical aspect of an ASR scheme.

ASR has traditionally been used as a water management technique for public or municipal water supply, where it can be demonstrated to have major cost benefits over surface storage systems. There is little documented information on the use of ASR techniques being used in local or small scale systems.

As water resources become scarce, and demand increases due to increased domestic, industrial, agricultural and environmental need, better use of the storage capabilities of aquifers will be required, whether the storage exists in depleted fresh water aquifers or in poor quality aquifers.

This case study shows that the development of small scale ASR schemes is viable, feasible and cost effective and could form an appropriate part of a water resources strategy in many locations.

The ongoing development of ASR schemes is to a greater or lesser extent constrained by regulation in Europe and the USA. Over cautious regulation regarding recharge quality, could result in ASR being under utilised.

2 CASE STUDY

2.1 *Project background*

In order to grow root crops in the sandy soils of the Sherwood Sandstone outcrop in Nottinghamshire, England, irrigation is required. In the order of 150 - 200 mm of water is applied during the growing season to help control disease, maintain a uniform shape and increase the size and yield of the crop.

Over exploitation of the Sherwood Sandstone aquifer as a result of intensive public water supply abstraction has led to declining groundwater levels and the concomitant environmental impacts. As a result, the Environment Agency have ceased issuing new licences to allow abstraction to take place and are actively trying to reduce the quantities of water abstracted.

In order to increase the certainty of water supply from river abstractions in the summer, the area of land under irrigation or the amount of water applied, farmers were advised to construct surface water storage reservoirs which could be filled during periods of peak surface water flows in the winter. Due to a lack of naturally occurring clays in the area under discussion, such reservoirs had to be artificially lined thereby resulting in high prices when compared to naturally sealed reservoirs.

The cost of reservoir construction led one farmer to explore for water in a deeper aquifer. Although this aquifer provided water of suitable quality, direct abstraction yields were insufficient for direct irrigation. Thus storage was still required. Inspired by the use of artificial recharge in and around London, the farmer decided to develop a scheme in Nottinghamshire.

2.2 *Geology*

The area comprises the full Permo-Triassic sequence of the English Midlands (Figure 1), which lies unconformably on the Upper Carboniferous. The Permo-Triassic sequence is made up from basal breccias overlain by the Lower Permian Marl, a grey limey mudstone which passes upwards into the Lower Magnesian Limestone. The Lower Magnesian Limestone is a variable group consisting of dense grey limestone, thinly interbedded with grey limey mudstones. Towards the top of the group, cream, buff or pink crystalline granular or oolitic dolomites or dolomitic limestones are common. Various portions of the Lower Magnesian Limestone are in places sandy limestones or calcareous/dolomitic sandstones. The Middle Permian Marl separates the Lower and the Upper Magnesian Limestone and comprises, for the most part, a red-brown mudstone, which is often plastic in nature (Edwards 1967).
The Upper Magnesian Limestone is generally a pale grey, cream, buff or pink dolomitic limestone which again is sandy in places. In some areas, particularly in the south of the district, the Upper Magnesian Limestone and Upper Permian Marl form one continuous unit comprising, clays with interbedded sandstones, sandy limestones/dolomites and calcareous sandstones. In these areas the upper part of the Permian sequence is almost identical to the overlying Sherwood Sandstone. The Sherwood Sandstone comprises a uniform medium to coarse grained red-brown sandstone, with occasional pebble beds and clay or marl bands.

The Permo-Triassic sequence dips at between 2° and 3° to the east and is affected by faulting. The faulting strikes in an approximately east west direction, and is well mapped in the Carboniferous strata to the west but generally ill-defined in the Permo-Triassic strata. Faulting and jointing may also be affected by deep coal mining.

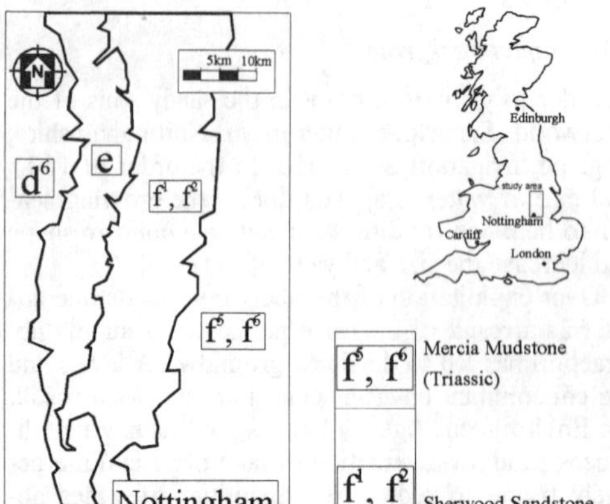

Figure 1 Geology

2.3 *Hydrogeology*

The principal water bearing zones are the Lower Magnesian Limestone and the Sherwood Sandstone. The Sherwood Sandstone is a major English aquifer and is the source of the order of 300 - 350 Ml/day of public water supply. The aquifer also supports important agricultural demand of approximately 30 Ml/day. The Sherwood Sandstone is approximately 100 - 200 m thick in the area of interest and is a high storage/high transmissivity aquifer with values in the region of 15% and 250 m^2/day respectively.

The Lower Magnesian Limestone is typically 15 m thick in the south and over 90 m thick in the north and behaves in a typical fractured carbonate aquifer manner. Yields are low or non existent unless solution widened fractures are intersected. Yields are often higher in areas affected by deep mining subsidence and are significantly enhanced by acidisation. Due to the traditionally poor yield of the aquifer, little exploitation has been carried out in the confined area and therefore little reliable hydraulic data are available. However, storage and transmissivity values obtained during this work were typically 0.01% (confined) and 20 m^2/day (post acidisation). It is considered that the Lower Magnesian Limestone is hydraulically separate from the Sherwood Sandstone by virtue of the Upper Permian Marl and/or the Middle Permian Marl.

2.4 *Water quality*

The Sherwood Sandstone is characterised by good quality water, albeit susceptible to surface contamination, particularly from agricultural activities. The water is generally of a calcium bicarbonate type with a positive oxidation reduction potential. The limestone water is enriched with magnesium due to the presence of dolomite and may be saturated with respect to calcium and dolomite. The water from the limestone usually has a very low positive or negative oxidation reduction potential. In the southern part of the study area where the development sites are close to the outcrop of the limestone, the chemistries of the two waters are very similar.

2.5 *Design concept*

The principal of the system is based on Inter-Aquifer Transfer. That is, water is pumped from one aquifer and stored in another. This allows the resources from the lower yielding aquifer to be exploited together with the high storage and transmissivity features of the over exploited aquifer. Figure 2 shows the system diagrammatically.

The boreholes are located close together and the water from the limestone is pumped directly into the sandstone injection pipes. Two injection pipes are used in order to reduce pipe diameter and increase head losses. The use of an orifice plate or other restriction at the discharge end of the injection pipe further restricts flow. The total head losses and the placement of the end of the injection tube well below the pumping water level in the borehole ensures that the pipe remains filled with water. A full pipe minimises the risk of the introduction of air bubbles

Table 1. Chemistry of the water

Determinand	unit	South (limestone)	North (limestone)
pH		8.8	7.8
Conductivity	µS/cm	448	795
Tot' Alkalinity	$CaCO_3$	137	120
Chloride	mg/l	23	159
Sulphate	SO_4	36.1	33.3
Calcium	mg/l	30.1	30.7
Magnesium	mg/l	30.7	22.8
Sodium	mg/l	8.68	83
Potassium	mg/l	2.92	9.53
Iron (Dissolved)	mg/l	0.05	0.05
Zinc (Dissolved)	mg/l	0.24	0.01

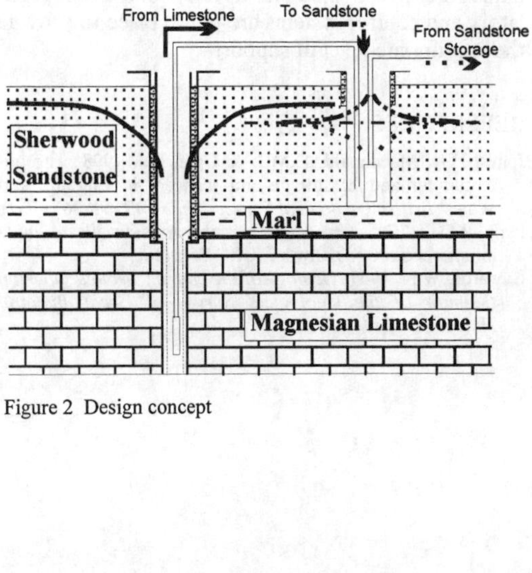

Figure 2 Design concept

by the recharged water cascading. The discharge from the injector is passed through a section of well screen acting as a diffuser to reduce the flow velocity and prevent erosion of the borehole wall.

Recharge rates are limited by the performance of the limestone borehole and are typically 15 - 20 m^3/hour. Pumping is essentially continuous giving a total annual recharge of the order of 100 to 150 Ml. Abstraction from the sandstone is often at 110 m^3/hour.

2.6 *System performance*

The systems that have been developed have not been supported by intensive research. The developers have simply implemented the techniques and hoped that they will work. The schemes have not had specialist input at all stages and therefore data on the performance of the systems are patchy.

The data that are available indicates that after one or two recharge/re-abstraction cycles, no long term deterioration in borehole performance was identified. A temporary deterioration in performance was identified after prolonged recharge. The deterioration was reversed during the first stages of step testing, or high rate re-abstraction as a result of development at the higher pumping rates, relative to the recharge rate.

2.7 *Regulatory Framework*

Within England and Wales abstraction of water is regulated and most abstractions must be licensed by the Environment Agency. Similarly discharges are also regulated via both UK regulations and European Union directives. In essence the ASR schemes described in this paper are not allowed to abstract a greater volume of water than is recharged and the recharge water must be of a potable quality. In addition, the recharge process must not cause damage to the receiving aquifer.

3 OTHER APPLICATIONS

The work that has been presented in this paper demonstrates that small scale ASR schemes are viable. The hydrogeological setting in Nottinghamshire means that the management of the schemes within the UK and European regulatory system is relatively straightforward. There are several controls on the feasibility of the systems:

- Lack of easily accessible water resources
- Skewed supply/demand curve
- Cost of surface storage
- Access to water of suitable quality for direct recharge with no, or minimal treatment

It is considered that these criteria form the basic requirements of any small scale ASR scheme. The principal driver for any scheme will be a problem relating to available water resources. If water resources are abundant and easily accessible then management will be straightforward and an ASR scheme will not be required. In the case of Nottinghamshire, the resource control is due to low river baseflows in the summer, causing the regulatory authorities to prohibit surface water abstraction or to threaten to stop abstraction and therefore make planning difficult. Available resources are limited further due to the closure of the principal aquifer to new abstraction licences.

The second driver is the supply/demand relationship as this defines the amount of storage required. In the cases discussed in this paper there is a requirement for water, delivered at a high rate over about a five month period. The accessible water resources either cannot be relied upon during the demand period or is not capable of supplying the demand.

The lack of natural reservoir sealing materials means that the cost of constructing surface storage is high when compared to an ASR scheme and the availability of a good quality, albeit low yielding aquifer, means a source of water is readily available.

The source of water and the recharge are the trickiest part of the process. European Community law/regulation and similar laws and regulations in the USA means that the quality of the water that is recharged must be very high and often of drinking water quality. For this reason the public water supply schemes that have been investigated in the UK have tended to follow the US model, whereby drinking quality water is stored in poor quality or brackish aquifers. Much effort has been expended on research into the effects of dual porosity systems, particularly in the English Chalk aquifer and issues relating to diffusion of contaminants out of primary porosity and into fractures.

It can be argued that such an approach is over stringent and can make small scale schemes impractical. Although it is important that aquifers are protected and it would be foolish to recharge with water that is of worse quality than that in the aquifer, this must be balanced with the opportunity to use available storage.

Small scale systems, if funded independently, cannot justify intensive research. The risks associated with the storage of water must be balanced against the benefits of local water supply and the economics of underground, rather than above ground, storage.

In many rural areas the water being used is of poor quality due to surface or shallow groundwater contamination as a result of human and animal waste disposal and in certain circumstances agricultural practices and leaching. The development of small scale ASR schemes in such situations may provide a much improved supply.

On commercial or commercial/domestic sites (eg. tourist accommodation), the development of an ASR scheme which uses good, but not necessarily water of an internationally recognised drinking water standard, for non potable uses such as manufacturing and ablutions, may have major economic benefits in terms of the provision of local and regional infrastructure and the support of the local economy, by enabling development to take place.

Small scale ASR schemes provide an opportunity to help manage moderate scale water supplies in remote areas. ASR schemes would be particularly appropriate in both hot and cold arid situations where evaporative losses, or non accessibility due to freezing, cause significant water management problems. Mineral development (oil, gas, mining) in remote arid areas involves the construction of townships and processing facilities which are fixed by the location of the mineral deposit. Identification of suitable aquifers and the use of rainfall runoff and possibly treated effluent, may make ASR an attractive proposition. In this situation it is probable that due to the nature of the activities on site, a high level of scientific and engineering skills will be available, thereby allowing a greater level of system management and possible water treatment. The higher level of management/design input would allow more complicated aquifers, whether they be deeper or saline, to be considered and developed, as well as extending the range of options in terms of source water.

4 CONCLUSIONS

The work carried out in many parts of the world demonstrates that the ASR technique is highly effective and that carried out in Nottinghamshire shows that the technique is appropriate on the small local scale.

A number of factors are required to make ASR schemes feasible, which relate to a need to store water and the high cost or impracticality of surface storage. Where these factors are present then small scale ASR provides a useful water resource management technique.

It is important that as water resources become more stretched, that innovative management techniques are investigated and developed and that regulatory and control systems are put in place to provide the appropriate level of support.

REFERENCES

Jones, H.K. Macdonald, D.M.J. & Gale, I.N. 1998. The potential for aquifer storage and recovery in England and Wales. *British Geological Survey Report WD/98/26.*

Pyne, R.D.G. 1995. *Groundwater recharge and wells: A guide to aquifer storage recovery.* CRC Press Inc.

Edwards, W.N. 1967. *Geology of the country around Ollerton (Memoir of the Geological Survey of Great Britain). HMSO London.*

Groundwater: Past Achievements and Future Challenges, Sililo et al. (eds) © *2000 Balkema, Rotterdam, ISBN 90 5809 159 7*

Development scenarios for sustainable urban water systems

M. Eiswirth & H. Hötzl
Department of Applied Geology, Karlsruhe, Germany

L.S. Burn
CSIRO Building, Construction and Engineering, Melbourne, Vic., Australia

ABSTRACT: Current urban water design in most countries has evolved from 19th century European experiences to control water borne disease transmission. This generally involves the large-scale transport of water into and out of major urban areas. This approach has effectively eliminated water borne disease in nearly all communities; however, it also has a number of problems such as a heavy demand on source water catchments, escalating costs for infrastructure investment and environmental problems caused by large-scale discharges of wastewater and contaminants back to the environment. Alternative approaches to the provision of urban water services do exist, but they have not been thoroughly investigated because of a reluctance to experiment with any system, which could in any way compromise human health. With population densities increasing (particularly in urban areas) and constraints on funding, it is becoming increasingly evident that present urban water systems will not be suitable models for service provision into the 21st century.

1 INTRODUCTION

The growth of cities has required advances in the sciences of civil engineering and hydrology, as water supplies have brought from increasingly distant sources. There have been three major drivers behind the enormous expansion of water resources infrastructure in the past century: (1) population growth; (2) industrial development and (3) expansion of irrigated agriculture. All three factors have increased dramatically. As a result, traditional water planning regularly concludes that future water demands will inexorably rise and eventually exceed water supplies available in any given location (Gleick 2000).

However, the increasing concerns about the environmental impacts of water projects and their increasing economic costs mean that traditional planning concepts, which assume unlimited supplies of potable water, must be questioned. This includes the source of the water supplies and its appropriate use. Alternative sources of water supply can replace traditional dams as well as large- and small-scale desalination technology, water reclamation and reuse. In addition to this, other non-traditional sources of supply will play an increasing role, including reclaimed or recycled water and, in some limited circumstances, desalinated brackish water or seawater (Gleick 2000).

A recent study of the role of water in the Australian economy (ATSE 1999) has shown that future economic growth will be limited by the availability of water. Efficient water use and sustainable urban water systems are the key to the future of Australia and other countries. Management of complex urban water system interactions, contained within the concept of sustainability and inter-generational equity is necessary (Harris 2000).

This paper discusses twentieth century urban water management approaches and looks at the new paths being explored for the coming decades.

2 STRATEGIES FOR THE DEVELOPMENT OF SUSTAINABLE URBAN WATER SYSTEMS

Urban transport systems and the provision of water have been identified as the most critical factors that determine the future of cities in this century. The effectiveness of conventional technical urban water concepts has reached a limit, and the sustainability of such practice is in question. Urban water systems fail more frequently as a consequence of a lack of appropriate maintenance procedures rather than a result of poor engineering design (Butler & Parkinson 1997).

A pilot study in the Netherlands has been carried out investigating the feasibility of a sustainable urban-water chain (Van der Graaf et al. 1997). In this study unsustainability factors have been developed and weighted. Important factors of this analysis have been the emission of pollutants into surface water, groundwater management, and sludge production and sewer exfiltration. The main topic of the research was to answer the question how unsustainability or unsustainability factors could be reduced. Therefore, different ideas of sustainable urban development have been applied to the model (Table 1).

Although the zero emission model as discussed by Van der Graaf et al. (1997) has been offered as the most feasible way to achieve a sustainable urban water system, such systems are not practicable in society where cost has to be questioned. In this scenario the urban-water chain is virtually closed and all used water is collected and transported to the wastewater treatment plant. This also means that there are no more overflows or leakage from the sewerage system. Whilst such a system would be ideal, growing cities are in a period of ageing infrastructure and limited resources. However, improvements are needed over the current systems, where wet weather flows in sewers often exceed 10 times dry weather flows. Consequently leakage detection and improved rehabilitation of damaged sewers are important to sustainable urban water systems.

2.1 Sustainable urban drainage systems

Butler & Parkinson (1997) have proposed three strategies for the development of sustainable urban drainage systems:

1 Reduction of reliance on water as transport medium for waste
2 Elimination of the mixing of industrial wastewater with domestic wastewater
3 Separation of storm runoff from flows of polluted wastewater.

These strategies may potentially be realised through the adoption of some techniques listed in Table 1.

2.1.1 Water transport

Conventional drainage systems utilise water extremely inefficiently. Installing low-flush toilets, dry sanitation systems, vacuum sewerage or possibly greywater recycling systems can reduce household water consumption. However it must be ensured that existing systems will operate effectively with much lower dry weather flows.

Table 1. Sustainable strategies for the development of different components of urban water systems.

Components of the urban water system	Proposed sustainable development	References
Wastewater drainage	Sustainable leak detection Improved separated sewer system Decision support systems for rehabilitation	Burn et al. 1999, Burton 1998, Hedberg 1999, Eiswirth 2000a
Stormwater drainage	Utilization of natural drainage patterns Using extended detention basins and porous paving systems Re-use of stormwater Creation of temporary catchments Urban watercourse corridors	Butler & Parkinson 1997, Ellis 1995
Water supply & distribution	Sustainable leak detection and rehabilitation Application of inert pipe materials Decision support systems for rehabilitation	Burn et al. 1999, Eiswirth 2000a
Wastewater treatment	Separation of faeces and/or urine from domestic wastewater Waste separation and re-use Advanced wastewater treatment Sludge processing Local sanitation technology Use of on-site treated greywater as fertilisers Incorporation of sanitary system in agricultural system	Chen & Beck 1997, Jeffrey et al. 1997, Otterpohl et al. 1997 & 1999, Van der Graaf et al. 1997
Groundwater management	Use of locally discharged groundwater Artificial groundwater recharge	Tredoux et al. 1999
Water consumption	Water demand management Small-scale recycling of greywater & rainwater On-site storage or infiltration of stormwater Use of rainwater in and around house Installation of water-saving provisions & second water system Sanitation systems without water consumption (dry toilets)	Jong et al. 1998, Terpstra 1999
Economics	Cost-sustainability analysis Life-cycle analysis	Icke et al. 1999, Varis & Somlyódy 1997
Environmental policies	Promotion of transition scenarios Active approach instead of reactive Holistic views and integrated approaches Multicriteria decision analysis Social-cost benefit analysis	Dirkzwager 1997, Larsen & Gujer 1997, Rauch 1998

2.1.2 Separation of wastewater

Removing the industrial waste component of municipal sewage is critically important to sustainable water systems (Butler & Parkinson 1997). Therefore dealing with waste as close to its source as possible reduces the problems associated with treating, and increases the re-use potential of the more valuable components of waste. Additionally it can produce major reductions in the cost of the transport system. Where the complete isolation of an industrial waste stream is impractical, pre-treatment using the best available technology is required prior to discharge into the sewer.

2.1.3 Separation of stormwater

Whilst separate sewage and stormwater systems are still the current practice in most developed countries, but this is not recommended as a sustainable approach (Butler & Parkinson 1997). Combined systems are more sustainable, if the storm water and wastewater components can be separated. This is possible by the utilization of smart control systems that can release sewage on a timed basis, in conjunction with short transport systems, which additionally means that the lowly diluted sewage can be treated as a resource, rather than a waste. In conjunction with smart treatment systems, the polluted first flush of storm water could be treated as sewage and then the later flows discharged with minimal treatment. This would significantly reduce the environmental discharge to the environment from polluted storm water. Other alternative approaches recommend utilising the natural drainage patterns of the catchment for on-site infiltration into the soil. Where the soil and the quality of the runoff permits direct infiltration into the ground is preferable in order to recharge groundwater reserves. Where infiltration is not possible, the development of natural drainage patterns is advocated (Ellis 1995).

2.2 Sustainable urban water and waste management

The traditional sanitation concept of industrialised countries produces material flows that result in accumulated mixing of different water flows. Reuse of material and directing flows to advantageous places is limited. The separation of faeces and /or urine from the domestic wastewater can be considered as one important step towards sustainable water concepts. Results of resent studies indicate the strong advantages of a decentralised system (vacuum - biogas system) on energy consumption, pollutant load and nutrients emitted to the receiving waters. This includes a decentralised aerobic treatment of greywater in constructed wetland, and infiltration of stormwater to avoid a centralised treatment system for these components (Otterpohl et al. 1997, 1999).

3 PIPE LEAKAGE AS ONE MAIN UNSUSTAINIBILITY FACTOR FOR URBAN WATER SYSTEMS

Influences on the sustainability of urban groundwater (quality and natural groundwater flow) include surface sealing, water leakage from mains sewers, and storage tanks, leakage from artificial lakes and reservoirs, water withdrawal, urban land over-irrigation, etc. Anthropogenic pollutants from different sources are also potential risks for the groundwater quality in urban areas (Eiswirth 2000b) (Fig. 1).

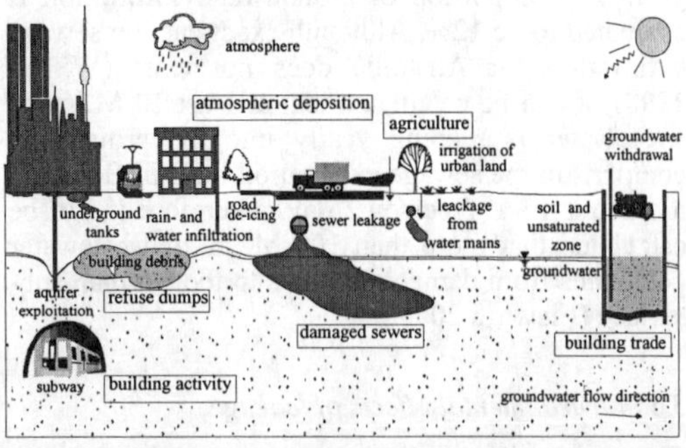

Figure 1. Unsustainability factors for the urban groundwater (Eiswirth 2000b).

Pipe leakage is perceived to be a major problem for the sustainable management of urban water systems by many water authorities worldwide. Associated costs are incurred on condition monitoring of sewerage systems, the over-design of water systems and the treatment of additional potable water that is lost due to leakage.

In many water distribution systems, a significant percentage of water is lost in transit from treatment plants to consumers. The amount of lost or "unaccounted for water" (UFW) is typically between 20–30% of production (Cheong 1991). In the case of some systems, mostly older ones, the percentage of lost water could be as high as 50%. Water authorities in Australia report UFW levels on average being 15% in 1997/98 (WSAA 1998). As pipes age, the problems of infiltration and exfiltration increase as well as the failure levels (WSAA 1998).

3.1 Water supply systems

Supply systems in most capital cities typically consist of large supply lines, referred to as trunk mains, delivering water from a catchment or treatment plant to a network of distribution mains that deliver to reticulation mains.

The failure rate for water mains in Australia is about 44 bursts / 100 km / year and for Germany 18 bursts / 100 km / year. However, an additional 63 bursts / 100 km / year on property connections have been reported in Germany (Eiswirth 2000a). In

Australia the annual water consumption is 2.15 Million. ML and the calculated water loss amounts 142 l / household / day. The current value of the annual water loss in Australia is about $90 M (US) (WSAA 1998).

3.2 *Sewer system*

One predominant problem with sewer systems is thought to be infiltration / exfiltration through joints. In Australia about 1.3 billion m^3 of wastewater and 0.12 billion m^3 of tradewater are collected every year. The proportion of groundwater infiltration is estimated to be 12%. Although exact data on sewage exfiltration for Australia does not exist (WSAA 1998), it can be calculated that at least 10 M. m^3 of wastewater is seeping yearly into the ground. In comparison the sewage exfiltration rates in Germany are about 15 L / person / day. Therefore it can be calculated that more than 100 M. m^3 of wastewater percolates from damaged sewers to the German subsurface (Eiswirth 2000a).

3.3 *Environmental effects of leakage*

Characterization of sewerage system performance as a main unsustainability factor requires recognition of the impacts of leakage into and out of the pipe network. The results of detailed investigations in an in German city show that damaged sewerage systems are the main source for groundwater contamination of sodium, chloride and nitrogen sulphate compounds (Fig. 2).

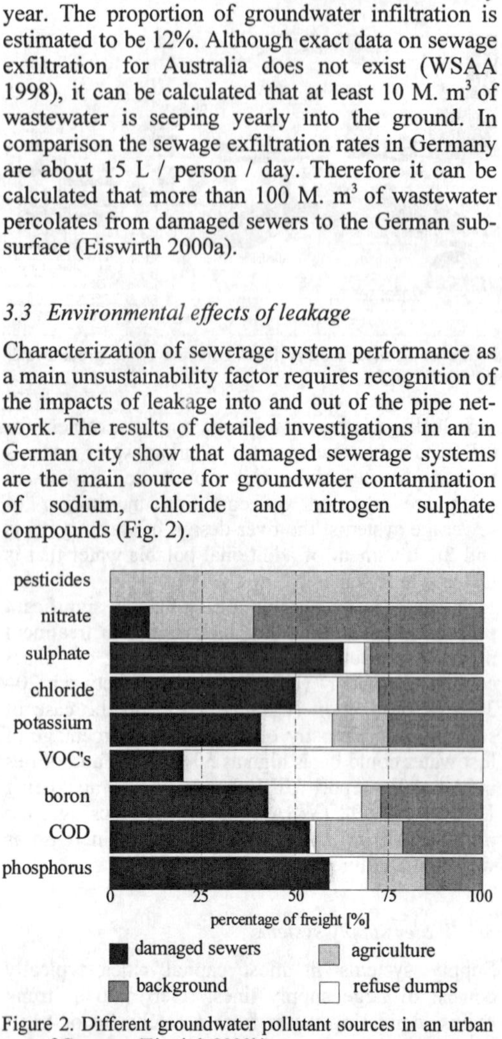

Figure 2. Different groundwater pollutant sources in an urban area of Germany (Eiswirth 2000b).

To solve leakage in both sewer and water reticulation systems in order to provide urban water system sustainability, sophisticated leak detection techniques are required (Burn et al. 1999).

3.4 *Rehabilitation costs and leakage control solutions*

It is recognised that leakage of sewer and water reticulation pipes is a problem worldwide. Although the problem in Australia is not as great as countries with older systems such as Germany, as the Australian infrastructure continues to age the problems will also increase.

The choice of which rehabilitation technique or whether rehabilitation needs to be undertaken at all is currently difficult, as little information is available to aid selection of the appropriate technology. Asset management systems need to be developed to allow technology selection, taking into account lifecycle costing methodologies as well as externalities, such as customer disruption and health risks (Burn et al. 1999).

The Australian water authorities spend about $140 Mio. (US) every year for the rehabilitation of water mains (WSAA 1998). In Germany the total annual infrastructure costs for water mains are about $2.5 billion (US) and for the rehabilitation of mains at least $300 M (US) (Eiswirth 2000). For damaged property connections the necessary rehabilitation costs have been calculated to $20 B (US) for the period from 1995 to 2015. In Germany for rehabilitation of the damaged public sewer systems about $2 B (US) have been spent every year. For the restoration of the complete German sewer system approximately $50 B (US) is required (Eiswirth 2000a).

4 AUSTRALIA'S URBAN WATER PROGRAM

4.1 *Objectives and aims*

The Objective of the Australia's CSIRO urban water program (UWP) is to identify and develop systems and technologies, integrative processes and tools of analysis, which are commercially valuable, scientifically robust and which improve the cost effectiveness of urban water services, in line with the program's vision of ecological sustainability. The main tasks of this program are given in Table 2.

Table 2. Main tasks of the UWP (Speers 1998).

Task 1 Understanding the Urban Water System	Domestic water use study Water balance study Contaminant flows Lifecycle costs Sustainability indicators
Task 2 Future Needs & Demands	Building blocks of alternatives Social context
Task 3 Scenario Development & Analysis	Scenario development & analysis based on a test case site in Perth
Task 4 Algorithm Development	Algorithm development

4.2 *Urban water and contaminant balance*

As part of this program, a tool has been developed by CSIRO to estimate the water flows and contaminant loads within the urban water system. This conceptual urban water and contaminant balance model is called UVQ (Urban Volume and Quality) and represents water and contaminant flows through the existing urban water, wastewater and stormwater systems, from source to discharge point (Mitchell et al. 2000) (Fig. 3).

As shown in figure 3 the actual model includes the leakage of water mains as well as in/exfiltration from the wastewater drainage. The contaminant balance of this model has the advantage of directly representing the way in which alterations in the water delivery and disposal routes affect the movement and distribution of contaminants in the urban environment. Consequently this model will enable analysis of different scenarios and view effects on ground contamination (Speers 1998).

4.3 *Decision support system for selection of operation or rehabilitation technique*

If the state of the urban water system is known, e.g. by advanced leakage detection, decisions on how to operate asset management schemes and whether to undertake maintenance, repair or replacement can be made. Existing decision support systems to evaluate replacement or preventive maintenance priorities for sewers and water mains do not necessarily address the issue of selection of the repair/replacement technology to combat the problem of pipe leakage.

It is widely recognised that true asset management requires a balance between risk and planning and it is essential to develop a decision support system for rehabilitation that will support rational and objective comparisons between the engineering, cost and social benefits of each technique.

For this process to occur, a decision-support system to allow selection of rehabilitation techniques incorporating lifecycle-costing methodologies, to assess the viability of each selection decision, needs to be developed and to be implemented into the UWP model. The existing model is called TAWS and has been implemented already as a decision support system (DSS). The actual modeling approach integrates water supply, wastewater and stormwater services into a single framework. However, this system needs to be expended to allow operational and rehabilitation procedures to be analysed (Maheepala & Zoppou 2000).

4.4 *Scenario development & analysis*

A number of procedures can be undertaken to reduce leakage in both sewers and water supply systems. Within the UWP for example, an analysis of alternative urban water systems and related costs has been carried out for different scenarios. One alternate scenario analysed in the UWP is shown in Figure 4.

For water reticulations pipelines, either rehabilitation can be undertaken or pressure management techniques instigated. Pressure management can be a very cost-effective means of reducing losses or controlling demand for many utilities, with a wide diversity of conditions. For example, regulation of pressure supplied to households offers water authorities a saving of 9.5% and the householder a saving of 7.7%. However, peak levelling methods are currently considered controversial and are in need of further research (DeSilva et al. 2000).

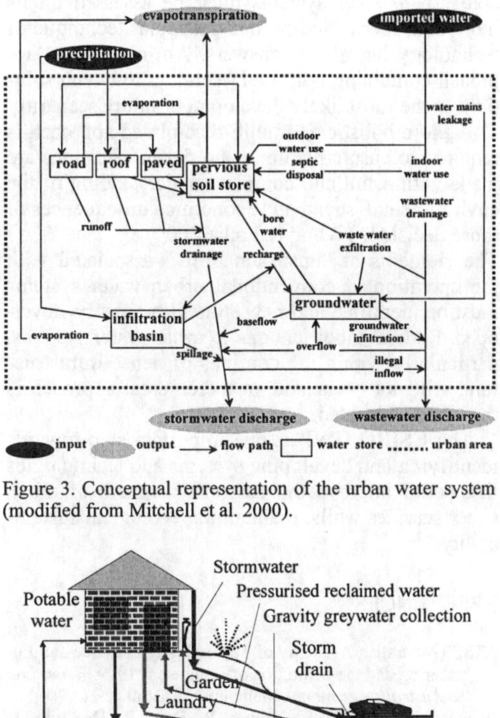

Figure 3. Conceptual representation of the urban water system (modified from Mitchell et al. 2000).

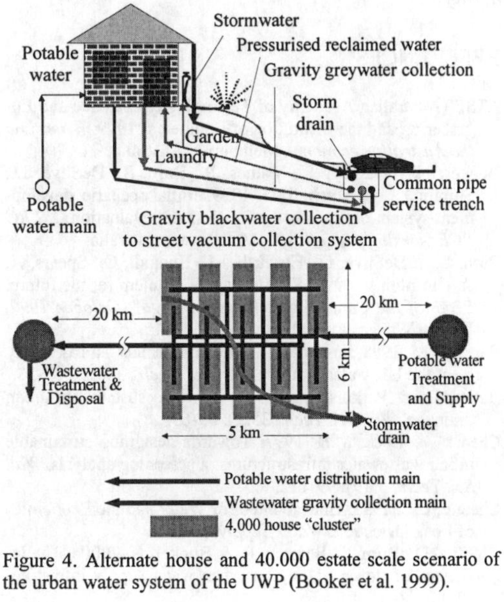

Figure 4. Alternate house and 40.000 estate scale scenario of the urban water system of the UWP (Booker et al. 1999).

5 CONCLUSIONS

Recent research identifies two possible future development scenarios for urban water system infrastructure. Either, a continuation and further refinement of the present approach of high-tech, centralised, end-of-pipe drainage approaches or the widespread provision of low-cost, low-energy solutions based upon maximisation of source control procedures and the application of alternative decentralised biological systems (Butler & Parkinson 1997).

To fulfil the criteria of sustainability, urban water systems need to be adaptable to meet future environmental quality objectives and demands. Each approach or technique for the development of sustainable urban water systems must be assessed on its individual merits and no one particular technique or technology has all the answers. An incremental approach containing both high-tech and low-tech solutions is the most likely development future scenario.

A more holistic and multi-disciplined approach is required to ensure planning and design decisions are made with a full and complete understanding of the environmental, social and economic consequences of those decisions (Wong & Eadie 2000).

The elements of "unsustainability" associated with the operation of conventional urban water systems must be identified and new strategies must be developed. Radical modifications to urban water systems, particularly within the confines of dense infrastructure, will not occur and an incremental approach is therefore advocated.

The CSIRO UWP is one important step towards identifying and developing systems and technologies that could improve the cost effectiveness of urban water services whilst maintaining ecological sustainability.

REFERENCES

ATSE (Australian Academy of Technological Science and Engineering and the Institution of Engineers) 1999. *Water and the Australian economy*. Melbourne, Australia.

Booker, N., Priestley, T., Mathes, P., Burn, S., DeSilva, D., Shipton, R. & Mitchell, G. 1999. Initial scenario development system design, cost estimation and evaluation. *CSIRO BCE working paper T3-05*, Melbourne, Australia.

Burn, L.S., DeSilva, D., Eiswirth, M., Hunaidi, O., Speers, A. & Thornton, J. 1999. Pipe leakage a problem for the future. *Proc. of the Conf. "Pipes Wagga Wagga", October 1999*, Wagga Wagga, Australia.

Burton, J.M. 1998. Sustainable waste management - a case study from the UK water industry. *Wat Sci. Tech.*, 38(11): 51-58.

Butler, D. & Parkinson, J. 1997. Towards sustainable urban drainage. *Wat. Sci. Tech.*, 35(9): 53-63.

Chen, J. & Beck, M.B. 1997. Towards designing sustainable urban wastewater infrastructures: a screening analysis. *Wat. Sci. Tech.*, 35(9): 99-112

Cheong, L.C. 1991. Unaccounted for Water and the Economics of Leak Detection. *Water Supply*, 9(3&4).

DeSilva, D., Burn, S., Shipton, R. & Booker, N. 2000a. The influence of flow levelling on urban water supply infrastructure. *Proc. of the Conf. Water 2000, 19-23 March 2000, Auckland, New Zealand.*

Dirkzwager, A.H. 1997. Sustainable development: new ways of thinking about water in urban areas. *European Water Pollution Control*, 7(1): 28-40.

Eiswirth, M. & Hötzl, H. 1997. The impact of leaking sewers on urban groundwater. In Chilton, J. et al. (Eds.), *Groundwater in the urban environment. Vol. 1, Problems, processes and management*, p. 399-404.

Eiswirth, M. 2000a. Leak detection – future challenges and solutions. *EntsorgungsPraxis, 6* (in press) (in German).

Eiswirth, M. 2000b. Balancing the contaminant fluxes in the urban groundwater. *Urban Water, 4* (in press).

Ellis, J.B. 1995. Integrated approaches for achieving sustainable development of urban storm drainage. *Wat. Sci. Tech.*, 32(1): 1-6.

Gleick, P.H. 2000. Meeting basic water needs as a human right. *Proc. of the Xth World Water Congress, 12-17 March* 2000, Melbourne, Australia.

Harris, G. 2000. Water, science and ecology in an age of cynicism. *Proc. of the Xth World Water Congress, 12-17 March* 2000, Melbourne, Australia.

Hedberg, T. 1999. Attitudes to traditional and alternative sustainable sanitary systems. *Wat. Sci. Tech.*, 39(5): 9-16.

Icke, J., Van den Boomen, R.M. & Aalderink, R.H. 1999. A cost-sustainability analysis of urban water management. *Wat. Sci. Tech.*, 39(5): 211-218.

Jeffrey, P., Seaton, R., Parson, S. & Stephenson, T. 1997. Evaluation methods for the design of adaptive water supply systems in urban environments. *Wat. Sci. Tech.*, 35(9): 45-51.

Larsen, T A. & Gujer, W. 1997. The concept of sustainable urban water management. *Wat Sci. & Tech.*, 35(9): 3-10.

Maheepala, S. & Zoppou, C. 2000. Decision support system for assessing integrated urban water supply systems. *Proceeding of the Xth World Water Congress, 12-17 March* 2000, Melbourne, Australia.

Mitchell, V.G., Gray, S. & Farley, T. 2000. Accounting for Water and Contaminants in Urban Areas. *Proc. of the Xth World Water Congress, 12-17 March 2000*, Melbourne, Australia.

Otterpohl, R., Albold, A. & Oldenburg, M. 1999. Source control in urban sanitation and waste management: ten systems with reuse of resources. *Wat. Sci. Tech.*, 39(5): 153-160.

Otterpohl, R., Grottker, M. & Lange, J. 1997. Sustainable water and waste management in urban areas. *Wat. Sci. Tech.*, 35(9): 121-133.

Rauch, W. 1998. Problems of decision making for a sustainable development. *Wat. Sci. Tech.*, 38(11): 31-39.

Speers, A. 1998. CSIRO Urban Water Program. *Internet info, http://www.dbce.csiro.au/urbanwater/index.htm.*

Terpstra, P.M.J. 1999. Sustainable water usage systems: models for the sustainable utilization of domestic water in urban areas. *Wat. Sci. Tech.*, 39(5): 65-72.

Tredoux, G., King, P. & Cavé, L. 1999. Managing urban wastewater for maximising water resource utilization. *Wat. Sci. Tech.*, 39(10/11): 353-356.

Van der Graaf, J.H.J.M., Meester-Broertjes, H.A., Bruggeman, W.A. & Vles, E.J. 1997. Sustainable technological development for urban-water cycles. *Wat. Sci. Tech.*, 35(10): 213-220.

Varis, O. & Somlyódy, L. 1997. Global urbanization and urban water: can sustainability be afforded? *Wat Sci. & Tech.*, 35(9): 21-32.

Wong, T.H.F. & Eadie, M.L. 2000. Water sensitive urban design - a paradigm shift in urban design. *Proc. of the Xth World Water Congress, 12-17 March* 2000, Melbourne, Australia.

WSSA 1998. Facts 98: The Australian Urban Water Industry. *Water Services Association of Australia.*

Groundwater: Past Achievements and Future Challenges, Sililo et al. (eds) © *2000 Balkema, Rotterdam, ISBN 90 5809 159 7*

Hydrogeological decision analysis for selection of alternative road stretches and designs

H.S.Eklund & L.Rosén
Department of Geology, Chalmers University of Technology, Göteborg, Sweden

ABSTRACT: Road projects can result in contamination of ground- and surface waters during construction and operation due to stormwater production and accidental spills. Protection measures are costly and from a societal economical point of view it is important to implement measures where they are most useful. Whether the costs for protection measures are acceptable to the society depends mainly on (1) the probability for contaminant release, (2) possibilities for remediation in case of a release, (3) the efficiency of possible protection measures, and (4) the protection value of the water environment. This paper describes a method using risk-cost decision analysis, taking these factors into consideration, for evaluating the need for protective measures along roads. The method can be used as a decision support in selection between alternative road stretches and alternative designs to protect the surrounding water environment.

1. INTRODUCTION

An extension of the trans-European highway E6 through the northern Bohuslän in southwestern Sweden is planned. Road projects impact the natural environment. They can result in contamination of ground- and surfacewater both during time of construction and during operation due to stormwater production and accidental spills. It is important to make the impact as small as possible to preserve the water environment and protection measures are thus sometimes needed. Protection measures are costly and it is therefore important to implement them where they are most useful. Whether the costs for protection measures are acceptable depends on several factors of uncertainty, such as:

- Will a contaminant release occur that will impact the water environment?
- Will remediation be successful in case of a release?
- How effective are protection measures compared to their costs?
- What is the protection value of the water?

This paper describes an extension of a method for risk management of petroleum spills close to water supplies using risk-cost-benefit decision analysis (Swedish Road Administration, 1998). The purpose of the extended method described here is for decision support in the planning and design stages of new roads.

Risk is defined as the product of the probability of affecting water resources and the economical consequence of the damage. To decrease the risk of damaging water resources, protection measures can be implemented either to reduce the probability of contaminant release or to reduce its consequences. The reduction in economical risk can be compared to the costs and benefits associated with protection measures and the cost-efficiency of measures can be assessed.

The application described here is on a more general level than the original method. For the overview type of application described in this paper, the concept of hydrogeological type settings for estimations of the transport of contaminants in ground and groundwater is of primary importance.

2. DECISION ANALYSIS

For cost-effective risk management, cost-benefit-risk decision analysis can be used. A short background to this approach to decision analysis in a hydrogeological context is given below, primarily based on the review paper by Freeze *et al* (1990).

In cost-benefit-risk decision analysis the trade-offs between a given set of alternatives are evaluated in order to select the most cost-efficient risk reduction alternative. The benefits, costs, and risks of each alternative are taken into account by defining an *objective function*, ϕ_i, for each alternative i=1,...,n. The objective function reflects the specific problem *and* the preferences of the decision-maker, and thus varies according to the key variables involved. The objective function has the general form:

$$\Phi_i = \sum_{t=0}^{T} \frac{1}{(1+r)^t} [B_i(t) - C_i(t) - R_i(t)] \qquad (2.1)$$

where B_i are the benefits of alternative i in year t [Skr], C_i are the costs of alternative i in year t [Skr], R_i are the risks, or probabilistic costs, of alternative i in year t [Skr], r is the discount rate [decimal fraction], and T is the time horizon [years]. (The notation Skr represents the Swedish *kronor* currency; 1 Skr ≈ 8.50 US$.) The objective function represents the net present value of the alternative i. A simplified type of the objective function, a *risk-cost minimization* objective function, was used in this paper, as the benefits were assumed to be independent on the costs and risks. The risk-cost objective function is:

$$\Phi_i = \sum_{t=0}^{T} \frac{1}{(1+r)^t} [C_i(t) + R_i(t)] \qquad (2.2)$$

Risk, R, is in this paper defined as:

$$R = P_f C_f \qquad (2.3)$$

where P_f is the probability of failure and C_f denotes the consequences in case of failure. For risk assessment concerning water pollution, *failure* may be defined with respect to e.g. compliance levels for water quality in streams or groundwater or retention time of contaminants in groundwater between the point of release and a receptor. The consequences of failure involve e.g. costs for remedial actions and loss of environmental value of the water resource.

The costs term includes all costs associated with efforts made to reduce the risk, e.g. from implementing protective measures. It was assumed in this study that a reasonable decision criterion is to minimize the risk-cost objective function, given that this risk level is socially accepted. The risk level associated with the minimized risk-cost objective function is referred to as the *optimal risk*. Using this decision criterion, actions that are more costly than the risk-reduction they provide cannot be justified. If a socially acceptable risk has been defined, then the objective of the decision-maker is to reach that risk level to the lowest posssible cost. Figure 1 gives a schematic description of the risk-cost objective function, optimal risk and acceptable risk.

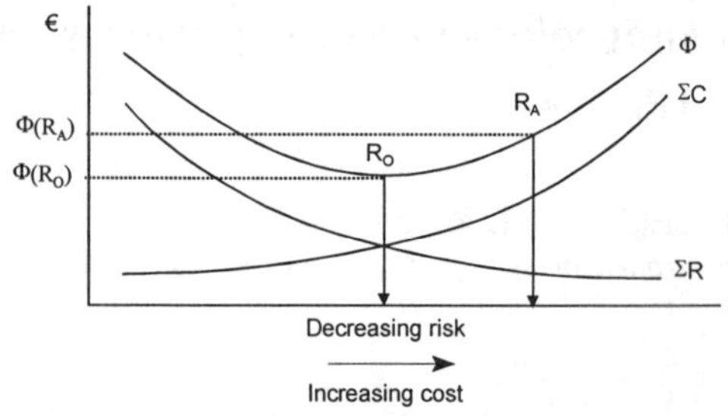

Figure 1. Risk-cost minimization. The concepts of optimal risk (R_O) and acceptable risk (R_A) and how they produce different outcomes of the objective function, Φ (after Freeze et al, 1990).

3. DECISION MODEL

The present study was performed with respect to the risk of a transport of hazardous goods, mainly petroleum products, to contaminate the surrounding surface water environment. *Failure* was defined as contamination of surface water streams to such an extent that existing biota being affected. The groundwater in the area is not used for water supply and it was assumed that the groundwater exhibits no specific protection values. The event tree shown in Figure 2 was used to structure the risk assessment. Similar trees are used by the Swedish Road Administration (1998).

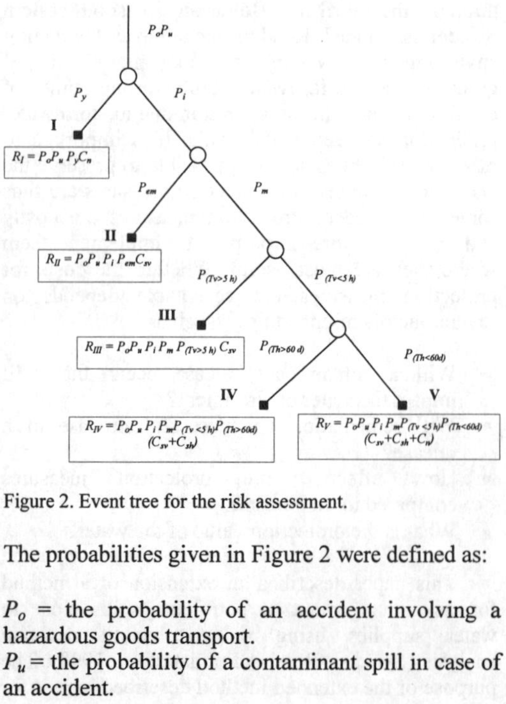

Figure 2. Event tree for the risk assessment.

The probabilities given in Figure 2 were defined as:

P_o = the probability of an accident involving a hazardous goods transport.
P_u = the probability of a contaminant spill in case of an accident.

P_y = the probability of surface run-off of the contaminant in case of an accident with a spill.
P_i = the probability of infiltration of contaminants into the ground in case of an accident with a spill.
P_{em} = the probability of the contaminant to be retained in the unsaturated zone in case of an accident with a spill and infiltration.
P_m = the probability of the contaminant to percolate through the unsaturated zone in case of an accident with a spill and infiltration.
$P_{(Tv>5h)}$ = the probability of the retention time of the contaminant being more than 5 hours in the unsaturated zone (5 hours being the estimated time required for a successful remedial action to prevent the contaminant to reach the groundwater) in case of an accident with a spill, infiltration and percolation.
$P_{(Tv<5h)}$ = the probability of the retention time of the contaminant to be less than 5 hours in the unsaturated zone, i.e. the probability for the contaminant to reach the groundwater in case of a spill, in case of an accident with a spill, infiltration and percolation.
$P_{(Tv>60\ days)}$ = the probability of the retention time of the contaminant being more than 60 days between the point of release and the receptor of the contaminant (60 days being the estimated retention time required in order to perform a successful remedial action before the contaminant reaches the receptor).
$P_{(Tv<60\ days)}$ = the probability of the retention time of the contaminant being less than 60 days between the point of release and the receptor of the contaminant.

The consequence variables given in Figure 2 were defined as:
C_n = loss of environmental values [Skr].
C_{sv} = costs for remediation in the unsaturated zone [Skr].
C_{sh} = cost for remediation in the groundwater zone [Skr].

The risk was calculated to for each outcome marked I-V in Figure 2. A total risk for each road alternative *i* and was calculated as:

$$R_i = \sum_{t=j}^{J} \sum_{t=1}^{T} \frac{1}{(1+r)^t} [R_{Ij}(t) + R_{IIj}(t) + R_{IIIj}(t) + R_{IVj}(t) + R_{Vj}(t)] \tag{3.1}$$

for *j=1...J* road sections and *t=1...T* years.

Annual probabilities of accidents involving hazardous goods transports and probabilities for contaminant spills in case of an accident were made according to Swedish Road Administration (1998). The probabilities regarding infiltration, percolation and retention times were calculated using stochastic representations of infiltration capacity, retention capacity of the unsaturated zone, and time of travel of water through the unsaturated and saturated zones. It was assumed that the contaminants will be transported by advective flow. It was also assumed that the hydrogeological properties are homogeneous but uncertain. Statistical properties of uncertain parameters were estimated based on literature data, existing information, subjective assessments and the guidelines given by the Swedish Road Administration (1998). Probabilities were estimated through Monte Carlo statistical simulation.

Consequence costs were estimated based on the guidelines for remedial cost estimations given by the Swedish Road Administration (1998) and an environmental economical valuation of the surrounding water resources, see below.

All calculations were made using a time horizon, *T*=40 years and a discount rate, *i*=4%.

4. HYDROGEOLOGICAL TYPE SETTINGS

Hydrogeological type settings are used to describe the hydrogeological conditions and the vulnerability of groundwater in an area. The concept using hydrogeological type settings allows extraction of information from a well-known area to another area where the information is limited.

Descriptions of hydrogeological type settings are a kind of comparative hydrogeology, which represent quick and useful means of discerning the significance of processes and principles in particular environments. The development of hydrogeological setting descriptions has its origin in work by LeGrand (1962, 1970, 1983 and 1988) and Heath (1984). Hydrogeological setting description allows inference of information from a well-known area to another area with limited information. The concept of hydrogeologic type settings is used in groundwater vulnerability classification, e.g. the DRASTIC-system (Aller et al. 1987). The National Ground Water Association (NGWA), USA, also uses hydrogeological type settings to develop a hydrogeological database for groundwater modelling (Newell et al 1990).

A hydrogeologic type setting is defined as a mapable unit with common hydrogeologic characteristics. It is a composite description of all the major geologic and hydrologic factors, which affect and control groundwater movement into, through and out of an area. (Aller et al 1987). In Sweden the geology is varied and the settings often constitute small units. At the Department of Geology, Chalmers University of Technology, hydrogeological settings have been used in different projects, both in vulnerability classifications (Rosén 1991, Stejmar 1995 and Bengtsson 1996) and in assessments of groundwater quality (Bengtsson and Gustafson 1996 and Stejmar 1996).

Five different hydrogeological type settings were identified in the study area:

1. Aquifers in crystalline bedrock
2. Unconfined aquifers in glacial till
3. Confined aquifers in glacial till or glaciofluvial deposits
4. Unconfined aquifers in beach deposits
5. Wetland areas

The type setting *Unconfined aquifers in glacial till* is presented in Figure 3 as an example of the graphical representation of the hydrogeological settings in the studied area.

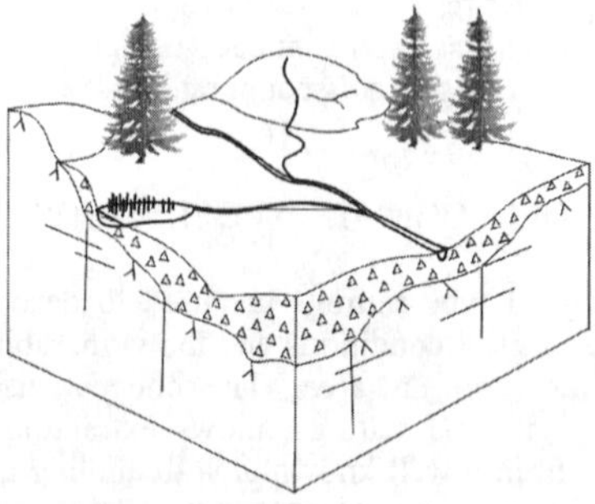

Figure 3. Unconfined aquifers in glacial till; an example of a hydrogeological setting in the study area.

5. VALUATION OF THE WATER

In a decision analysis from a water management perspective, the cost of failure, C_f, includes costs of remedial actions, charges, fines, and values from services provided by the water that are restricted or canceled due to contamination. The value of water resources typically fall in two different categories, related to the services provided, based upon the physical state of groundwater (NRC, 1997): extractive and *in situ* values. The first category includes municipal, agricultural and industrial uses of water. *In situ* values include ecological values, buffer values, recreational values, existence values, and bequest values. The values of these two categories make up the total economic value (TEV) of groundwater.

Difficulties in valuing water, especially *in situ* values, compared to many other assets are stemming from (1) economic data unavailability, i.e. many services provided by water are not traded in regular commodity markets (NRC, 1997), and (2) uncertainty related to the biological and ecological effects of elevated contaminant concentrations. However, as noted by NRC (1997) even incomplete or partial estimations of the TEV may often provide substantial information and support decision-making. The impact of incomplete knowledge of the TEV can be studied by applying a range of different values in the decision analysis (Massmann *et al.*, 1991). A concept of using alternative approaches of valuing in decision analysis regarding water protection measures was discussed by Wladis *et al.* (1999)

Several both direct and indirect approaches to valuing water can be distinguished. Direct approaches to non-market valuation use different types of survey techniques. This type of valuation requires the construction of hypothetical markets in which a set of changes are valued. The most common approach to this type of valuing non-market goods and services is the contingent valuation method (CVM), which is a survey-based procedure to investigate peoples willingness to pay (WTP) or willingness to accept (WTA) compensation (expressed in monetary units) for the goods or service. Indirect methods include the travel cost method, the averting behavior method, and methods based on market prices. Indirect methods do not measure *in situ* values. The contingent valuation method (CVM) provides a means to estimate the TEV, including extractive and *in situ* values. However, it should be emphasized that there are numerous methodological controversies associated with the application of CVM, as described by NRC (1997).

An accident with a contaminant spill was assumed to primarily affect the angling fish habitat of adjacent streams and the nearby coastal area. This means that only the surface water was valued in this paper. Groundwater was regarded as a contaminant transport media to the streams and coastal area. Two approaches for valuation of the water resources with respect to the fish habitat were applied:

Direct valuation (*Method 1*): Valuation based on the value of the angler fish, including fishing equipment and travel costs and accommodation for anglers. It was assumed that the streams are affected by the contamination for three years. The value of each fish was estimated to be approximately Skr 800. Valuation was performed an agreement with the county authorities.

Indirect valuation (*Method 2*): Valuation based on the willingness of the inhabitants in the region to pay for keeping the angling fish habitat free from contamination. The valuation was based on Silvander (1991). It was estimated that each of the 11 000 inhabitants in the region would be willing to pay about Skr 160 each year to safeguard the angling fish habitat.

6. DECISION ALTERNATIVES

According to the decision criterion in this study, protection measures that are more costly than the risk-reduction they provide are not cost-effective and cannot be justified. Infiltration barriers of bentonite or plastic liners can be regarded as a standard type of protection measure along road stretches near valuable water resources in Sweden. The investment cost for this type of measure is approximately Skr 5000 per meter road. An analysis of the cost-efficiency of infiltration barriers at each section of the two road alternatives was made, using the decision model described by equation 2.2.

7. RESULTS

Estimation of risk was accomplished for the two different alternative road stretches using both methods described above for valuation of the water environment. The results from the analysis are shown in Figures 4 and 5.

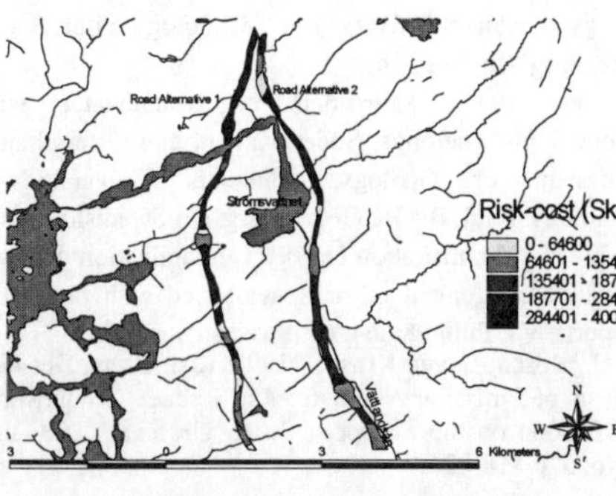

Figure 4. Estimated risk for the two alternative road stretches; valuing of the water environment according to Method 1.

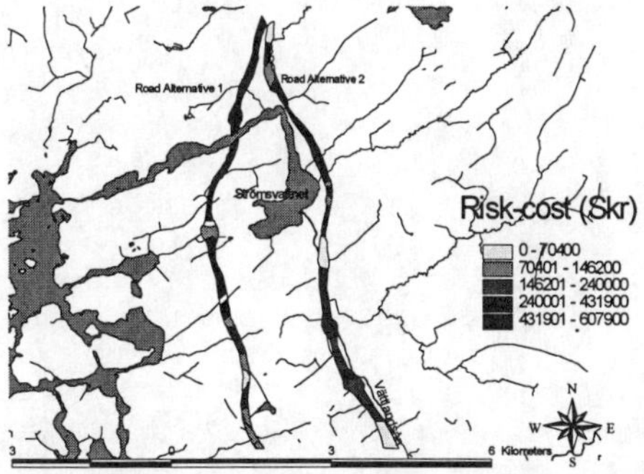

Figure 5. Estimated risk for the two alternative road stretches; valuing of the water environment according to method 2.

Each road alternative was divided into several sections with respect to hydrogeological type settings and the water catchment areas. For each section a specific risk was estimated as shown in the figures.

The total risk for the two road stretches calculated with the two different valuation methods are presented in Table 1.

Table 1. Risk (Skr) for the two road alternatives (Alt. 1 and 2, see Figures 4 and 5) calculated with the two different valuation methods.

	Valuing method 1	*Valuing method 2*
Alt. 1	3 900 000	5 700 000
Alt. 2	5 000 000	6 400 000

The risk is 15-25 % higher for the eastern road alternative compared to the western alternative. Valuation method 2 gave a higher risk for both alternative road stretches.

The total objective function for each road alternative and for the two valuation methods described in Section 5 are shown in Table 2. It can be seen that the studied protection measures are not cost-effective given the used decision model and valuing of the water resources.

Table 2. Objective function (Skr) for the two road alternatives (Alt. 1 and 2, see Figures 4 and 5) with and without protection measures.

	Valuation method 1		*Valuation method 2*	
	Protection	*No Protection*	*Protection*	*No Protection*
Alt. 1	25 400 000	3 900 000	26 400 000	5 700 000
Alt. 2	24 100 000	5 000 000	25 300 000	6 400 000

To make this type of protective measures defensible, an annual WTP value for protecting the water environment of approximately Skr 5000 per person would be required. Given existing information regarding WTP valuing of water resources in Sweden and the United States (NRC, 1997; Silvander, 1991; Wladis *et al.*, 1999), such a valuing appears to be unrealistically high.

8. CONCLUSIONS

The following conclusions were made from the present study:

- The method presumes that the water environment can be valued in economical terms. However, valuing of non-market goods, such as water, is still controversial; different stakeholders have different valuing perspectives and unambiguous valuing may often be difficult to obtain.
- Even though distinct valuing of water is difficult, the method is considered useful for comparison purposes and for identifying reasonable road deign alternatives with respect to the protection value of the water environment.
- The results show that protection measures of the type considered in this analysis should not be implemented. However, given the present probabilities of accidental spills from transports of hazardous goods, the existing hydrogeological and hydrological conditions, and the applied

failure criteria, an unreasonably high valuing would have been required to motivate this type of protection measures.

- The presented approach is thought to assist in considering uncertainties, structuring complex problems, identifying cost-efficient alternatives, and achieving a basis for communication between different stakeholders.
- The described method is suitable for overview studies in the planning and design stages of road projects. It should be considered as one of several important tools in the decision-making process of road projects.

ACKNOWLEDGEMENT

This work was kindly sponsored by the National Swedish Road Administration, Western Region.

REFERENCES

Aller, L. T. Bennet, T. Lehr, J. H. Petty, R. J. and Hackett, G. (1987) DRASTIC: A Standardised System for Evaluating Groundwater Pollution Potential Using Hydrogeological Settings. U.S. Environmental Protection Agency/600/2-87/035. Washington D.C.

Bengtsson, M-L. (1996) Hydrogeological vulnerability classification as a tool in municipal planning. Department of Geology, Chalmers University of Technology. Publ. A 81. Gothenburg. (In Swedish)

Bengtsson, M-L. and Gustafson, G. (1996) Assessment of groundwater using hydro-geological settings. Pilot study in Western Sweden. Department of Geology, Chalmers University of Technology. Publ. B 426. Gothenburg. (In Swedish).

Freeze, R.A., Massmann, J., Smith, J.L., Sperling, T., and James, B.R., 1990: Hydrogeological Decision Analysis. 1. A Framework. *Ground Water*, Vol. 28, No 5, p. 738-766.

Heath, R (1984) Ground-water Regions of the United States. US Geological Survey, *Water Supply Paper* 2242.

LeGrand, H (1962) Perspectives on Problems of Hydrogelogy. *Geological Society of American Bulletin*, Vol. 73, p 1147-1152.

LeGrand, H. (1970) Comparative hydrogeology; An example of its use. *Geological Society of American Bulletin*, Vol. 81, p 1243-1248.

LeGrand, H (1983) A Standardized System for Evaluating Waste-Disposal Sites. Second Edition. National Water Well Association. Worthington, Ohio.

LeGrand, H (1988) Region 21. Peidmont and Blue Ridge. In Hydrogeology. Edited by W. Back, J.S. Rosensheim, and P.R. Seaber. Geological Society of America. Geology of North America. Vol. O-2, p 201-208.

Massmann, J., Freeze, R.A., Smith, L., Sperling, T., and James, B., 1991: Hydrogeological decision analysis: 2. Applications to ground-water contamination. *Ground Water*, Vol. 29, No. 4, p. 536-548.

Newell, C. J. Hopkins, L. P. and Bedient, P. B. (1990) A Hydrogeologic Database for Ground-Water Modelling. *Ground Water* Vol. 28, no 5 p 703-714.

NRC (National Research Council), 1997: Valuing Ground Water - Economics Concepts and Approaches. National Academy Press. Washington D.C. 189 pp.

Rosén, L., (1991) Classification of groundwater vulnerability. Applications of standardized classification systems in Sweden. Geohydrological Research Group, Chalmers University of Technology. Gothenburg Publ No. 92. (In Swedish).

Silvander, U., 1991: Willingness to pay studies for angling fishing and groundwater in Sweden. Swedish Agricultural University, Uppsala. Department of Economics. Dissertations No 2. (In Swedish).

Stejmar, H. (1995) Hydrogeological vulnerability mapping of Save river valley in municipality Lerum. Department of Geology, Chalmers University of Technology. Publ. B 415. Gothenburg, (In Swedish).

Stejmar, H. (1996) Assessment of groundwater using hydrogeological settings. Södermanland and Östergötland. Department of Geology, Chalmers University of Technology. Publ. B 435. Gothenburg, (In Swedish).

Swedish Road Administration (1998): Contamination of water supplies. Management of risks associated with petroleum transports. VV Publ 98:064. (In Swedish).

Wladis, D., Rosén, L. and Kros, J., 1999: Risk-Based Decision Analysis of Emission Alternatives to Reduce Groundwater Acidification on the European Scale. *Ground Water*, Vol. 37, No. 6, p 818-826.

Groundwater: Past Achievements and Future Challenges, Sililo et al. (eds) © 2000 Balkema, Rotterdam, ISBN 90 5809 159 7

Investigation of trace elements for the assessment of sustainable water supply

B. Fritz, C. Sommer-von Jarmersted & A. Pekdeger
Department of Economic and Environmental Geology, Free University of Berlin, Germany
S. Wischnack & M. Jekel
Department of Water Quality Control, Technical University of Berlin, Germany

ABSTRACT: The drinking water supply in the Berlin metropolitan region is based on bank infiltration of approximately 70 %. Drinking water treatment consists only of aeration and subsequent rapid sand filtration in order to remove iron and manganese. With regard to the increasing amounts of persistent organic and inorganic components in the surface water, there is a significant need in surveying the infiltration processes in order to achieve a sustainable drinking water quality. Hence, several investigations were done on instructions of the Administration of Urban Development, Environmental Protection and Technology to predict changes in drinking water quality because of the expected decreasing run-off of the river spree. All investigations showed that most sewage water compounds reach the drinking water wells. Some of them change their chemical behavior caused by the fluctuation of water table.

1 INTRODUCTION

The inhabitants of Berlin, about 3.4 million, are supplied with drinking water exclusively from groundwater resources. 100% of public water supply in the Berlin metropolitan region comes from groundwater with a contribution of approximately 70% from bank filtration and artificial recharge and that depend on the condition of rivers and groundwater quality (SOMMER-VON JARMERSTED 2000).

The 12 Water works of Berlin are located nearby the surface water system. Their discharging wells are drilled mostly in a short distance (1-600 meter) around the rivers and lakes, which are also affected by the clear water input of the water treatment plants. In low-water times the share of clear water may be even equal or higher than the surface water influent. Because of the predicted decreasing run-off of the river Spree the surface water quality in future will change for the worse.

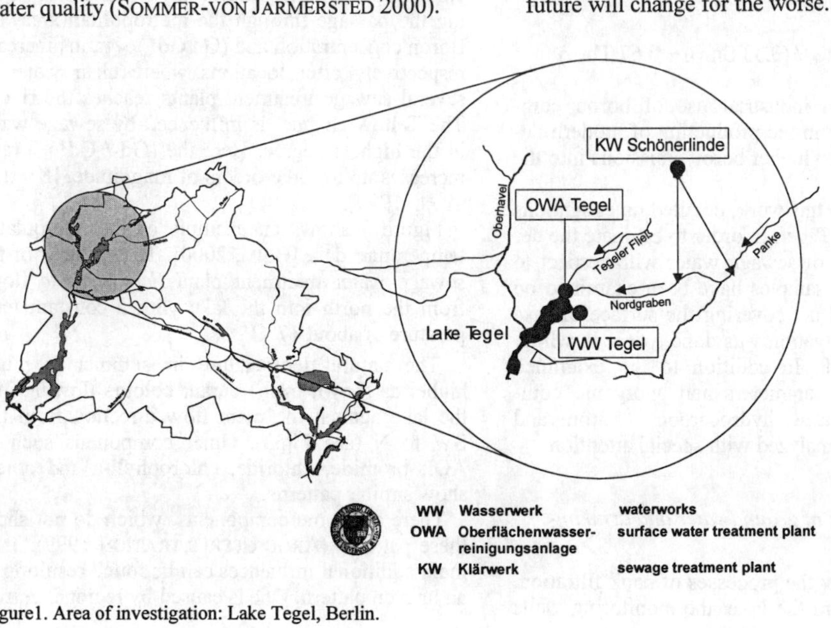

Figure1. Area of investigation: Lake Tegel, Berlin.

Lake Tegel in Berlin was chosen for special study because it represents an area with high amounts of bank filtration (today about 60%) and is the location of Berlin's biggest water. The lake would naturally be fed by two streams with low influents (Nordgraben and Tegeler Fließ, see Fig.1). In addition, the sewage treatment plant *Schönerlinde* lead clear water into the river Nordgraben. Yet, the water quantity and quality is strongly influenced by sewage discharge, which will rise in the future.

2 METHODS

2.1 *Investigations of surface water*

Firstly, in summer 1997 surface water samples were taken of the entire river and lake system of Berlin and examined systematically to get an idea of distribution and quantity of sewage water components in the river system of Berlin. Particularly Boron and Gadolinium Gd (SEE) are used to trace the influence of treated waste water on surface water quality (KNAPPE ET AL, 1999). The positive Gd anomalies are of anthropogenic origin and result from the application of gadopentetic acid Gd (DTPA)2-, used in magnetic resonance imaging (MRI) as a paramagnetic contrast agent. The organic aqueous Gd(III) complex is very stable and is excreted very fast by the human body. In order to get the anomaly factors, it was necessary to analyse the whole group of rare earth elements (REE) (KNAPPE ET AL. 1999). By interpolation of Gd^*_{SN} with its neighbour elements Sm and Tb the size of anomaly can be quantified by:

$$(Gd / Gd^*)_{SN} = Gd_{SN} / (0.33\ Sm_{SN} + 0.67\ Tb_{SN})$$

The increase in industrial use of boron compounds, especially in the production of modern detergents, has lead to higher boron emissions into the main outfalls.

Subsequently to this work, detailed investigations were done at *lake Tegel*. In order to estimate the detailed distribution of sewage water with respect to the current, water samples have been examined on the base of a grid net covering the surface of *lake Tegel*. This investigation was done twice, in winter and summer 1998. In addition to the examined physico-chemical parameters and anorganic components, halogenated hydrocarbons, boron and chlorophyll were analyzed with special attention.

2.2 *Investigations of groundwater and drinking water*

In order to study the processes of bank filtration, water samples from the lake, the monitoring wells and the discharge wells are being investigated every month over a long period of time (1997-recent). Physico-chemical parameters, anorganic and organic compounds, drugs, halogenated hydrocarbons and algae toxins are determined. A scheme of the study site is shown in Fig.2. The distance from the shore to the discharging well is about 160 meter and travel times estimated with different tracer (e.g. radon, temperature) vary depending on the discharging rate from 5 to 6 months.

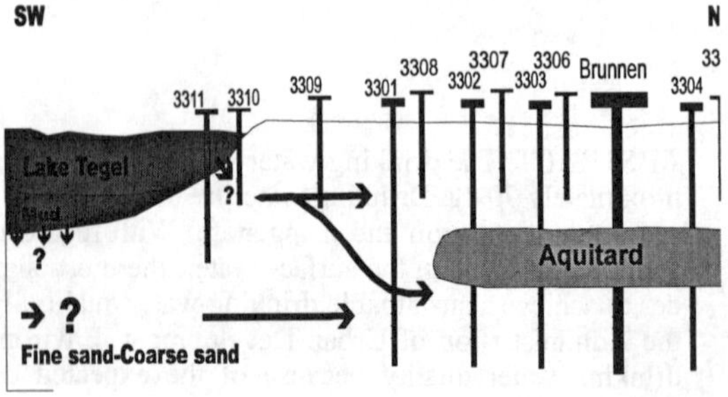

Figure 2. Investigation area.

3 RESULTS/DISCUSSION

3.1 *Surface water*

The main rivers in Berlin do not show clear **Gd_{SN}** anomalies and less amounts of boron before they reach the metropolitan area. Here the $(Gd/Gd^*)_{SN}$ ratio varies from 1.3 to 1.8 and in this study these values represent the 'natural' input into Berlin. During the passage through the metropolitan area, the Boron concentration and $(Gd/Gd^*)_{SN}$ ratios increase respectively after locations where clear water of several sewage treatment plants reaches the rivers. The Teltow channel is influenced by sewage water in the highest degree. Here the $(Gd / Gd^*)_{SN}$ ratio increases up to three orders of magnitude. [KNAPPE ET AL. 1999].

Figure 3 shows an example of the interpolated temperature data [HEIM, 2000]. The effluent of the sewage water treatment plant *Schönerlinde* flows from the north into the lake with a constant temperature of about 17°C.

The natural temperature in summer is much higher as shown in the darker colours flowing into the lake against the main flow direction from the SW to N (see Fig.3). Other compounds such as AOI, bromide, chloride, chlorophyll-a and others show similar patterns.

There are some components, which do not show these patterns (ARGE UFERFILTRATION 1999). Two main additional influences can be found resulting in an uneven pattern. One is caused by regional entries e.g. boron from households on the islands (see Fig.4).

The other one - the main source for another additional input - are pollutants caused by cleaning agents and antifouling paints from yachting ports (e.g. AOCl). The influent of organic chlorine compounds over the Nordgraben is about 2 - 3 µg/L. In September 1998, a point concentration of 50 µg/L AOCl was measured next to the yachting port. Therefore it can be concluded that AOX in the lake only partly originates from wastewater influent from the Nordgraben. Therefore, single wells can present different analyses depending on their position near the lake.

Surprisingly, the organo bromines in the lake show a distinct variation in summer and winter. The minimum concentration was observed in winter with 10 µg/L and the maximum of over 40 µg/L in mid-summer. Throughout the year the concentration is above the values which are initiated by the waste water influent into the lake. The organic bromine compounds are evenly distributed over the entire lake. No other input source was detected. The correlation with the high organo bromine values suggests that blue algae bloom could be responsible for the formation of the AOBr (WISCHNACK ET AL 2000).

3.2 *Groundwater*

In general, most parameters will be filtered through the bank filtration process. No problems exist at lake Tegel with bacteria and resulting from the long residence time of about half a year. Only during summer, problems with bacteria arise in the areas of the lake used for recreational activities, e.g. swimming. The technical standard of Berlin's sewage treatment plants is very high causing less problems with most parameters such as phosphate, sulfate or nitrogen. Table 1 contains a summary of the previous water quality data. DOC shows a decrease especially in the first few meters of soil passage. Background groundwater shows lower levels of DOC (about 3.9 mg/l). Remobilization of Phos-

Table 1. Water quality data (Average of one year: March 1998- March 19996 standard deviation).

	Lake	Observation well (3301-03)	B-13 (well)	Groundwater (3304)
DOC	9.560.8	6.360.6	5.360.9	3.960.9
PO_4 (mg/l)	0.08 60.06	0.260.1	0.160.1	0.0560.05
SO_4 (mg/l)	118612.1	122.4610	152.4612	183.4611.3
NO_3 (mg/l)	14.566.3	4.564	2.462	1.762.5

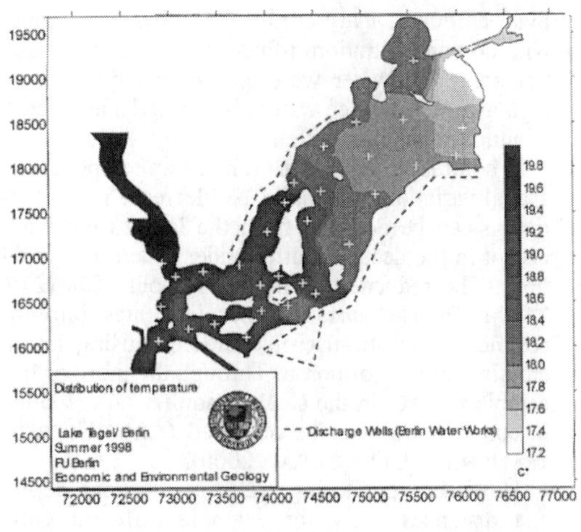

Figure 3. Distribution of Temperature in Lake Tegel, Berlin.

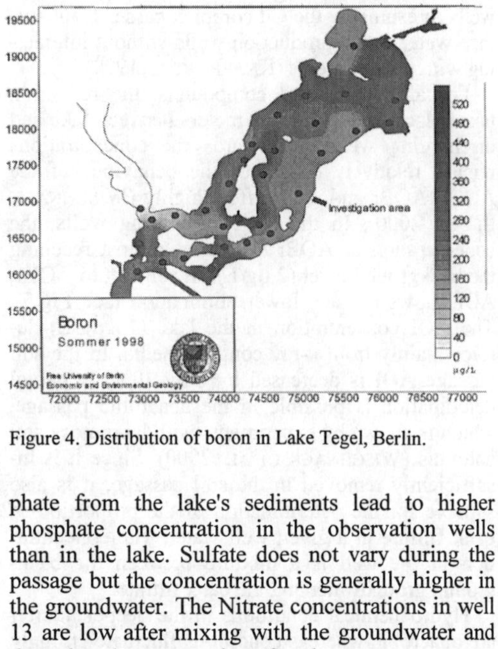

Figure 4. Distribution of boron in Lake Tegel, Berlin.

phate from the lake's sediments lead to higher phosphate concentration in the observation wells than in the lake. Sulfate does not vary during the passage but the concentration is generally higher in the groundwater. The Nitrate concentrations in well 13 are low after mixing with the groundwater and the reduced concentration after the passage.
Yet, this study concentrates on the sewage water components, which are investigated with regard to their unknown behavior during bank filtration. Some results are presented in this paper.

Ethylenediaminetetraacetic acid (EDTA) is one of the most conservative tracers we tested. The concentrations of EDTA in the surface water are about 5-7 µg/l and show only little temporal variation and no variation during the passage from the lake to the wells (ARGE UFERFILTRATION 1999). Compared to the surface water, high concentrations of EDTA measured in the discharging wells are caused by high amounts analysed in groundwater samples with no bank filtration influence. EDTA concentration in groundwater were approximately twice as high as in the surface water (10-39 µg/l) as a result of anthropogenic pollution.

In contrast to EDTA, Boron shows temporal and spatial variation in the lake. No decrease in concentration can be detected from the lake to the wells except in the deeper aquifer under anaerobic conditions. The reduction of boron is about 10% (279-248 µg/l). The surface water infiltrates into the sediment with an anaerobic milieu causing an increasing amount of boron. Through the more or less aerobic passage in the shallow aquifer no deviation in concentration can be observed (ARGE UFERFILTRATION 1999, FRITZ ET AL. 2000).

However, monitoring wells in different depths and distances show for example different Gd_{SN} anomalies and drug concentrations. The investigation proved that under influent conditions Gd_{SN} anomalies will reach groundwater and production wells, presumably the Gd complex passes from surface water to the production wells without interacting with the sediments (KNAPPE ET AL 1999).

For all halogenated compounds, the reduction takes place in the first few meters between lake and discharging wells. Afterwards the concentrations remain relatively constant. The behaviour of the AOCl, AOBr and AOI differs highly (WISCHNACK ET AL. 2000). In the deep monitoring wells, the concentration of AOBr decreases, almost reaching the background level (2 µg/L). In contrast to AOBr, AOI shows a much lower elimination (see Fig.5). The AOI concentration in the lake (15 µg/L) derives mainly from x-ray contrast media. In the soil passage AOI is decreased by only 30%. A partial deiodination is possible in the anaerobic passage, which is shown by experiments with laboratory soil columns (WISCHNACK ET AL. 2000). Since it is insufficiently removed in the soil passage, it is also suitable for the determination of the proportion of bank filtrate in a given well water. The concentration in the well is a mixture between the background groundwater and the bank filtrate.

Hydrochemical conditions in the deeper aquifer are anaerobic whereas conditions fluctuate strongly in the shallow levels. This is the reason why no elimination of AOI and only a small elimination of AOBr in the shallow levels can be observed.

4 CONCLUSIONS

In order to investigate bank filtration processes in Berlin, it is inevitably to know the distribution of the sewage water components in the surface water. These components are particularly suitable for such investigations in urban areas. It has been proved that most tested organic substances being put into

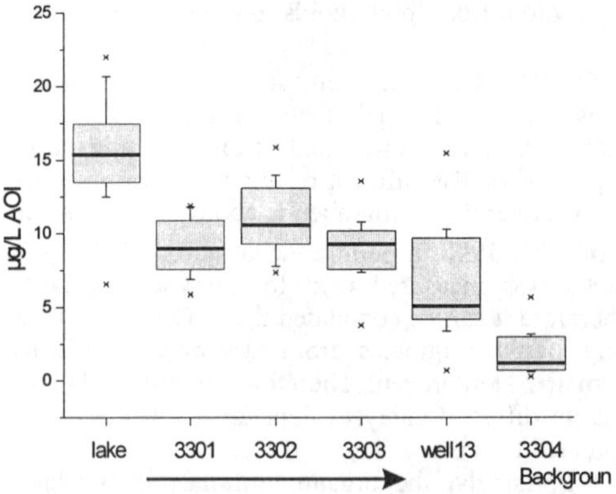

Figure 5. AOI in groundwater monitoring wells between lake and discharging wells, Lake Tegel, Berlin.

the lake reach the discharging wells and are not completely filtered during the passage. In the future, the influence of the treated waste water on the drinking water quality in Berlin could gain a higher significance due to increasing amounts of sewage water in the surface water.

Fluctuation of the groundwater table resulting from high rates of discharge produces extreme changes in the hydrochemical system. Seasonal input variations must also be considered (ARGE UFERFILTRATION 1999). Thus, further detailed investigation must be done in the future. Water samples from observation wells drilled between Lake Tegel and the discharging wells have shown no decrease especially for Boron, EDTA and AOI. They can be used as ideal tracer for bank filtration calculations.

In Berlin, AOI, Gd and boron are possible tools for quantifying the proportions of wastewater in surface water and groundwater (ARGE UFERFILTRATION 1999).

The AOCl concentration in the lake derives only partly from the wastewater AOCl. The main source for the AOCl contamination are the yachting ports.

The high AOBr concentration in the lake with the variation over the year is surprising. The AOBr is obviously formed in the lake. In the soil passage the contamination increases in the first few meters, thus an actual contamination danger for the groundwater does not exist.

5 ACKNOWLEDGEMENTS

The financial support of the Administration of Urban Development, Environmental Protection and Technology is gratefully acknowledged. For cooperation and financial report (halogenated hydrocarbons) we would like to thank the Berlin Water Works.

REFERENCES

Arge Uferfiltration (1999).Abschlußbericht Projekt Uferfiltration Berlin.- Senatsverwaltung für Stadtentwicklung, Umweltschutz und Technologie. Berlin.

Fritz B. et al. (2000). Sewage water components used as trace elements to investigate water quality in Berlin/ Germany: Part 1: Detailed investigations of surface water and groundwater. 1st World Congress of the International Water Association, 3-7. July 2000, Paris, France.

Heim, B. (2000). Untersuchungen zur Bestimmung der Wassergüte im Tegeler See anhand von in-situ-Daten und Flugzeugimager Daten (CASI).- Thesis, Department of Econmic and Environmental Geology, Free University of Berlin.

Knappe, A.; Sommer - von Jarmersted, C.; Bau, M. & Dulski, P. (1999). Gadolinium in aquatic systems as indicator for sewage water contamination.- Proceedings of the 5th International Symposium on Geochemistry of the Earth's surface, Reykjavik, Island, 187-190.

Sommer-von Jarmersted, C., Fritz, B. & Pekdeger, A. (2000). Groundwater resources in Berlin (Germany): Will there be enough water in the future?.- XXXth Congress 2000. Groundwater: Past achievements and future challenges. Cape town, 26.11-01.12.2000.

Wischnack, S. & Jekel, M. (2000). Sewage water components used as trace elements to investigate water quality in Berlin/Germany: Part 2. Behaviour of organo halogens in surface water and groundwater - 1st World Congess of the International Water Association, 3-7. Juli 2000, Paris, France.

Groundwater: Past Achievements and Future Challenges, Sililo et al. (eds) © 2000 Balkema, Rotterdam, ISBN 90 5809 159 7

Managing groundwater resources with increasing sulphate concentration: A case study from Germany

T.Grischek, V.Schmalz & E.Worch
Institute for Water Chemistry, Dresden University of Technology, Germany

J.Herlitzius
Water and Soil Consultants Limited, Bannewitz, Germany

W.Nestler
Institute for Geotechnics and Water Sciences, University of Applied Sciences, Dresden, Germany

ABSTRACT: Long-term application of fertilizers in agriculture, autotrophic denitrification and direct inputs have led to increased sulphate concentrations in an alluvial aquifer. An intensive case study for the catchment area of a water works in Germany was started in 1999. Options for groundwater management are discussed based on the initial identification and quantification of the main sources for sulphate in the aquifer.

1 INTRODUCTION

Whereas the nitrate problem as a result of application of fertilizers is well known and investigated, increasing sulphate concentrations are often neglected in Europe. High sulphate concentrations are also a sign of increasing salinisation of groundwater resources. For sustainable management of such groundwater resources, methods for the quantification of the main sources and reactions of sulphur species, for the prognosis and modelling of future sulphate concentrations, and monitoring strategies are needed.

Limits for sulphate concentration in drinking water were mainly set to prevent corrosion of materials. Sulphate concentrations of more than 500 mg L^{-1} can also cause objectionable taste and odour in drinking water and can result in diarrhoea. Furthermore, sulphate plays an important role in acidification processes in groundwater and in the deterioration of foundations and underground structures.

The investigated aquifer is used for drinking water production for more than one million people and has experienced an increase in sulphate concentration by up to 200 mg L^{-1} within 20 years. In some production boreholes the sulphate concentration increased to 240 mg L^{-1}, which is the limit fixed in German guidelines for drinking water quality (TrinkwV 1990). In the new European guideline for drinking water quality, sulphate is fixed as an indicator compound and proposed not to exceed 250 mg L^{-1} (EC 1998). The water works needs to check different management measures for its effects on the future sulphate level in the pumped raw water. Therefore, a prognosis has to be given based on investigations into the sulphur sources, and the reactions in the soil and aquifer (Fig. 1).

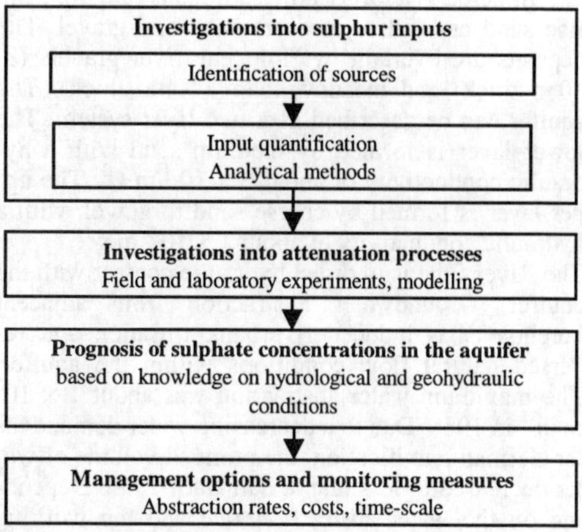

Figure 1. Concept of investigation.

The observed sulphate concentrations are low compared with other reported studies, e.g. in Australia or Kuwait where high sulphate concentrations limit drinking water consumption. But, experience gained in the multiple application of hydrogeological, hydrogeochemical and isotopic methods allows conclusions to be drawn from the investigated aquifer that should be considered for monitoring programmes in other regions.

2 SITE HYDROGEOLOGY

The water works is situated within the Torgau basin of the River Elbe between the cities of Dresden and Hamburg (Fig. 2). The basin is filled with Pleistocene deposits to a depth of 50-60 m that comprise

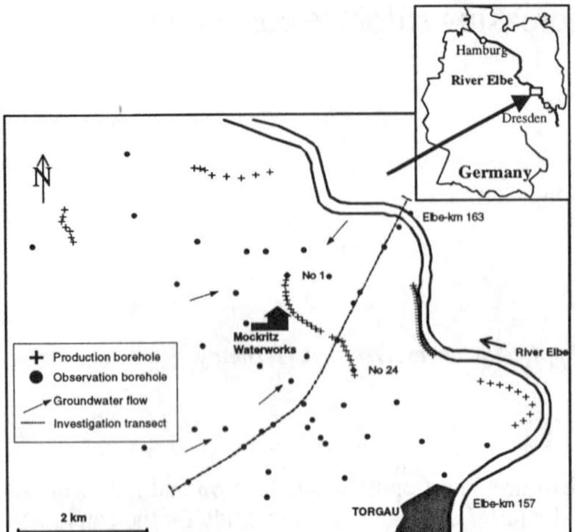

Figure 2. Location map.

interfingered glaciofluviatile sediments ranging from fine sand and silt to medium sand and gravel. The deposits are overlain by Holocene river gravels (8-10 m thick) and meadow loam (1-4 m thick). The aquifer can be described as a two-layer system. The lower layer is formed by medium sand with a hydraulic conductivity of about 6×10^{-4} m s^{-1}. The upper layer is formed by coarse sand to gravel, with a hydraulic conductivity of about 1×10^{-3} m s^{-1}.
The River Elbe is in direct hydraulic contact with the aquifer. Groundwater abstraction from adjacent boreholes has induced riverbank filtration and reversed natural flow conditions within the aquifer. The maximum water abstraction was about 1×10^5 m^3 d^{-1} in 1988. Due to a decreasing water demand after German reunification, the pumped volume of water decreased to less than a half until 1999. Depending on the water abstraction rate and the working production boreholes, the proportion of pumped riverbank filtrate varies between 50 and 80 %.

Land use in the catchment area of the production boreholes is predominantly arable farming, recently dominated by growing rape and corn. The crops grown depend to a high extent on the actual agricultural policy of the European Union. This can cause significant changes from growing traditional crops to specific ones, e.g. rape, including the application of specific fertilizers, and limits the prediction of changes in land use.

3 METHODOLOGY

3.1 Sampling

Sampling of water from 75 production boreholes has been undertaken since 1974. Groundwater samples have been taken from about 69 observation boreholes and two production boreholes (Fig. 2, production boreholes No. 1 and No. 24) at a half-yearly interval since 1995. An additional sampling campaign in April and May 2000 included more than 300 observation boreholes and surface water measuring points in the catchment area of the Mockritz waterworks. Groundwater was obtained using a mobile submersible pump (Grundfos Inc.).

Soil samples were taken in November 1999 and March 2000 using hammer probes and from cores obtained during drilling the observation boreholes.

3.2 Analytical Methods

In the field, well-head measurements included temperature, pH, Eh, electrical conductivity, alkalinity and O_2. In the laboratory, analyses included DOC, SO_4^{2-}, NO_3^-, Cl^-, Ca^{2+}, Mg^{2+}, K^+, Na^+, Fe^{2+}, Fe_{total}, Mn^{2+}, NH_4^+. In general, analyses were carried out according to German guidelines (DIN 1986). Prior to analysis, water samples were filtered through a 0.45 µm filter. Anions were determined using ion-chromatography (Dionex) or capillary electrophoresis (Hewlett Packard).

Soil samples were dried, sieved to a less than 2 mm fraction and ground to a less than 100 µm fraction. The following parameters were determined: water-soluble sulphate by extraction with deionised water, total inorganic sulphate by extraction with a 0.016 M KH_2PO_4 solution, disulphidic sulphur by reduction with Zn/Cr(II), organic sulphur as C-O-S species by reduction with HI and total sulphur by combustion at 1350°C.

Data loggers for measurement of electrical conductivity, temperature and hydraulic head were installed in six boreholes along the investigation transect (Fig. 2), at places where dynamic behaviour was expected.

3.3 Laboratory Experiments

Laboratory batch and column experiments were designed to model exchange processes between inorganic and organic sulphur compounds in the dynamic water level zone, the effects of ion exchange and sorption processes during sulphate transport in the anoxic aquifer and to estimate the rates of autotrophic denitrification. Columns were constructed from liners, 1 m in length (i.d. = 100 mm), and packed with material taken during the installation of the observation boreholes, and closed with specific plugs to prevent oxidation of the aquifer material. The plugs include a filter plate and connectors for tubes.

3.4 Data Handling

All available data on hydrogeology and water quality are stored in a MS ACCESS based data bank using GCI-GMS software (Nillert et al. 1996). The programme offers a wide range of useful options for handling and presentation of geological, hydrological and hydrochemical data as well as options for simple groundwater flow pathline calculations. In the future, it will be used by the water company for interpretation of monitoring results and for groundwater management.

3.5 Modelling

For the catchment area of the water works a quasi-3D groundwater flow and transport model has been created using Processing Modflow (Chiang & Kinzelbach 1997). To understand the controls on the concentration of sulphates in the aquifer, the geochemical modelling software tools PHREEQC (Appelo & Parkhurst 1998) and TBC (Schäfer et al. 1998) are to be used.

4 RESULTS

4.1 Recent sulphur pool in soils

Water-soluble sulphate concentrations and pH were found normally between 15 and 160 mg kg^{-1} and pH 6.3 but at some locations were between 600 and 1000 mg kg^{-1} and pH 4.2, respectively. High concentrations of water-soluble calcium, magnesium and chloride occurring together with high sulphate concentrations indicated an input from agriculture.

The total sulphur content ranged between 100 and 2000 mg kg^{-1}. In meadow loam, organic sulphur amounts to 90 % of total sulphur. Therefore, organic sulphur compounds and their transformation to inorganic sulphur compounds have to be determined to allow a prognosis of future sulphate concentrations.

For some samples, discrepancies between total sulphur content and the sum of determined sulphur fractions were observed. Further studies are needed to clarify uncertainties in the determination of sulphur fractions in soils.

4.2 Sulphate concentration in groundwater

Before 1975, sulphate concentrations in groundwater were found between 40 and 160 mg L^{-1}. Hydrogen-sulphide smell indicated sulphate reduction especially in the lower layer of the aquifer (depth 3) which has low sulphate concentrations. Between 1974 and 1995 a steep increase in the mean sulphate concentration of about 200 mg L^{-1} was observed for the upper layer of the aquifer (depth 1). In the middle layer of the aquifer (depth 2) the mean increase was about 100 mg L^{-1} and started at a later time. Figure 3 shows the trends in sulphate concentration for three different depths (compare Fig. 5). Dotted lines indicate a small data set, full lines are based on the mean value from more than 20 observation boreholes at each depth.

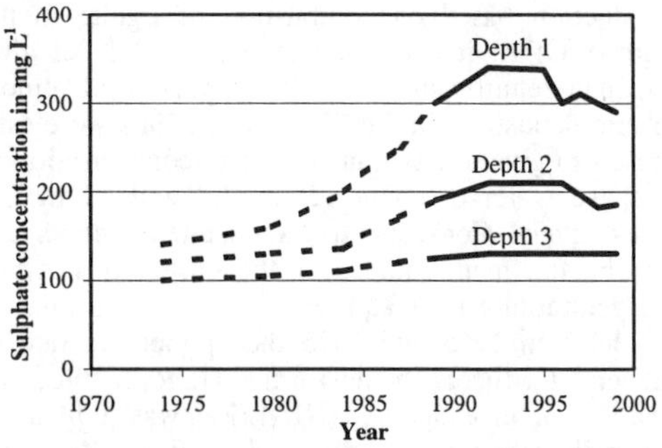

Figure 3. Mean sulphate concentrations in groundwater in all observation boreholes at depth 1, 2 and 3 in the river basin.

However, looking only at recent results from those observation boreholes near the production boreholes it can be seen that the highest sulphate concentrations are found at depth 2. Furthermore, the sulphate concentration seems to go down with increasing distance from production boreholes and it can be noted that the mass of nitrate diminishes consistently with increasing sulphate concentrations. The highest sulphate concentration was found in production boreholes with the lowest proportion of bank filtrate (Fig. 4).

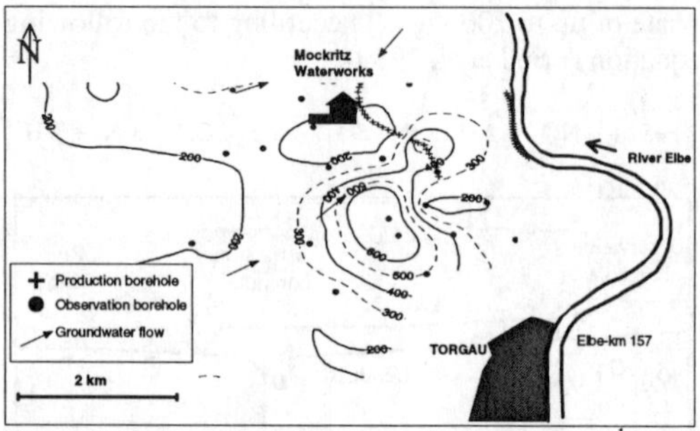

Figure 4. Contoured sulphate concentrations in mg L^{-1} in the upper layer of the aquifer, November 1999 – May 2000.

4.3 Identification of main sources of sulphate

The input from potential sources of sulphate to groundwater in a catchment area is site-specific and depends also on hydrogeological conditions and water abstraction management. Main sources for direct inputs can be atmospheric deposition, fertilizer application in agriculture and forestry, waste deposits, waste water or surface water infiltration and geogenic sources e.g. upwelling deep groundwaters with high salinity. Indirect inputs can come through autotrophic denitrification and sulphide oxidation during water level fluctuations.

Whereas the atmospheric input of sulphur (Fig. 5, A) was relatively high in the region due to energy production based on combustion of lignite in the former GDR, it decreased to a normal level after German reunification in 1989. Inputs from atmospheric deposition in the `70s and the `90s were estimated to give a maximum sulphate concentration in seepage water of about 120 mg L^{-1} and 60 mg L^{-1}, respectively. Consequently, atmospheric inputs can not be the main source for the observed sulphate concentrations in the aquifer.

Between 1955 and 1975 the application rate of mineral fertilizers in the former GDR doubled. At the site of investigation $(NH_4)_2SO_4$ was applied as the main nitrogen fertilizer until 1989, resulting in a yearly input of sulphate of about 19 g m^{-2} (Mallen, 1999). A mean groundwater recharge of 100 mm per year could result in a maximum sulphate concentration in the seepage water of 190 mg L^{-1} from this fertilizer application only (Fig. 5, B). This value is a theoretical value because a large portion of the applied sulphur was withdrawn by the harvested plants.

In the catchment area outside the river basin nitrate concentrations of up to 200 mg L^{-1} were determined in groundwater. In the river basin no or only low nitrate contents of up to 10 mg L^{-1} were found. Nitrate concentrations in raw water from the production boreholes were generally < 5 mg L^{-1}, in most cases < 1 mg L^{-1}. If we assume a total autotrophic denitrification (Fig. 5, C) with pyrite present in the aquifer, this would give an additional source of sulphate of up to 206 mg L^{-1} according to the following equation (Frind et al. 1989):

$$2\ FeS_2 + 6\ NO_3^- + 2\ H_2O \rightarrow 4\ SO_4^{2-} + 2\ FeOOH + 3\ N_2 + 2\ H^+$$

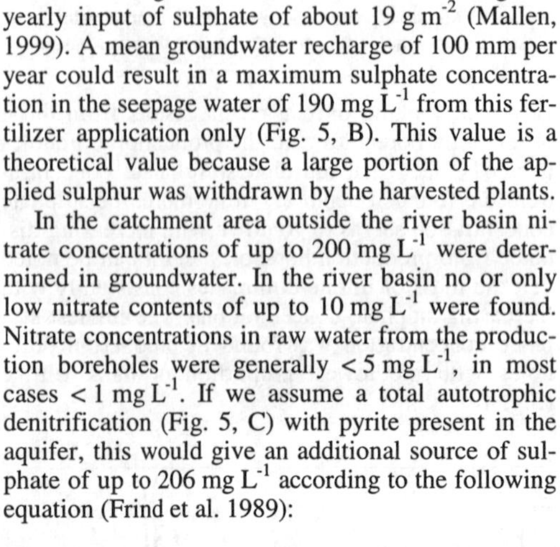

Figure 5. Sources and processes affecting sulphate concentrations in the aquifer.

It was assumed that the high abstraction rate during the 80`s caused a wide cone of depression resulting in oxidation of sulphides to sulphates in the aquifer (Fig. 5, D). A decreasing abstraction rate and an increasing groundwater level by about 2 m near the production boreholes could have caused a leaching of these sulphates. Additionally, fluctuations of the groundwater level of more than 3 m due to the nearby river result in changing redox conditions in the upper layer of the aquifer promoting sulphide oxidation and sulphate leaching. High sulphate concentrations in the upper layer of the aquifer and the significant increase since 1985 are evidence for these processes at the investigation site.

Drainage water from the catchment area outside the basin flows to the creeks which are partly in hydraulic contact with the aquifer. Infiltrating surface waters showed sulphate concentrations between 100 and 200 mg L^{-1} and nitrate concentrations mainly between 100 and 300 mg L^{-1} in the year 2000. In areas where infiltration from creeks occurs, the nitrate reacts with pyrite and an increase in sulphate concentration is observed. The estimated portion of infiltrate from the creeks is only 1-5 % of the abstracted water volume and therefore is not seen as a major source for sulphate but, however, cannot be neglected (Fig. 5, E).

Evaporation of water from old branches of the river, which are filled during flooding, can also cause an increase in sulphate concentration in the infiltrate (Fig. 5, F). But this effect is assumed to have little importance for the system.

Local inputs from small-size manure deposits or a recycling site for building materials were found to have a significant effect on water quality in nearby observation wells but no importance for the whole system (Fig. 5, G).

4.4 Options for groundwater quality management

The specifics of river bank filtration and groundwater abstraction at the investigated site allow effective management options to ensure a high water quality of the raw water and sulphate concentrations below 250 mg L^{-1}. Sulphate concentrations in River Elbe water varied from 76 to 108 mg L^{-1} with a median concentration of 85 mg L^{-1} in 1995-97 (Grischek et al. 1998).

Direct, short-term options are:

- Mixing of raw water from different production boreholes using on-line measurements of electrical conductivity.
- Continuous pumping of boreholes abstracting a high proportion of riverbank filtrate.

Indirect, long-term options are:

- Changes in application of fertilizers containing sulphur and nitrogen in parts of the catchment area which show a low storage potential for sulphur in its organic forms.
- Management of water abstraction to prevent high water table fluctuations.

Whereas short-term options are already practiced, a prognosis has to be made on the effects of the long-term options. It is especially important to evaluate the impact of the fluctuating groundwater table due to the dynamic stage of the nearby river and the operation of different production boreholes.

5 CONCLUSIONS

Redox reactions in the soil and aquifer, infiltration of surface water, changes in land use and in the application of fertilizers, decreasing atmospheric deposition of S-emissions after German reunification and local contamination demand a complex investigation to allow the prognosis of future sulphate concentrations in the groundwater and pumped raw water. A sufficient prognosis demands further expenditures for investigations into the processes of sulphate formation and reduction in the aquifer, field work, laboratory experiments, geohydraulic and geochemical modelling.

ACKNOWLEDGEMENTS

This work was supported financially by the BMBF-grants 02WT9928/29, LfUG Saxony, DVGW and FWV Elbaue-Ostharz GmbH. The presentation was supported by the Society of Friends of Dresden University of Technology.

REFERENCES

Appelo, C.A.J. & Parkhurst, D.L. 1998. Enhancements to the geochemical model PHREEQC - 1D transport and reaction kinetics. In G.B. Arehart & J.R. Hulston (eds), Proc. 9th Int. Symp. On Water-Rock Interaction, Taupo, New Zealand, 30 March - 3 April 1998: 873-876. Rotterdam: Balkema.

Chiang, W.-H. & Kinzelbach, W. 1997. Processing Modflow. User`s handbook.

DIN (1986) German standard methods for the examination of water, waste water and sludge. Beuth, Berlin.

EC (1998) Council Directive 98/83/EC of 3 November 1998 on the quality of water intended for human consumption. OJ L 330/32, 5.12.1998.

Frind, E.O., Duynisveld, W.H.M., Strebel, O., Boettcher, J. 1989. Simulation of nitrate and sulfate transport and transformation in the Fuhrberger Feld aquifer, Hannover, Germany. In Kobus & W. Kinzelbach (eds) Contaminant Transport in Groundwater: 97-104. Rotterdam: Balkema.

Grischek, T., Hiscock, K., Metschies, T., Dennis, P.F., Nestler, W. 1998. Factors affecting denitrification during infiltration of river water into a sand and gravel aquifer in Saxony, Germany. Wat. Res. 32(2): 450-460.

Mallen, G. 1999. Isotopic characterization of hydrodynamic and biogeochemical processes in bank filtrate and groundwater in the quarternary aquifer in the Elbe river basin near Torgau. PhD thesis, Institute for Geophysics and Geology, Univ. Leipzig, Germany.

Nillert, P., Hoffknecht, A., Schäfer, D., Ziesche, M. 1996. Groundwater monitoring and model prognosis. Geowissenschaften 14(3/4): 129-134

Schäfer, D., Schäfer, W., Kinzelbach, W. 1998. Simulation of reactive processes related to biodegradation in aquifers, 1. Structure of the 3D reactive transport model. J. Contam. Hydrol. 31: 167-186

TrinkwV (1990) Verordnung über Trinkwasser und über Wasser für Lebensmittelbetriebe. 05.12.1990, BGBl Teil 1, Nr. 66, 2613-2628.

Groundwater: Past Achievements and Future Challenges, Sililo et al. (eds) © 2000 Balkema, Rotterdam, ISBN 90 5809 159 7

Salinity and floodplain tree health in the lower River Murray, Australia

K.L. Holland – *Flinders University of South Australia, Centre for Groundwater Studies, CSIRO Land and Water, Adelaide, S.A., Australia*

A.J. Telfer – *Australian Water Environments Pty Limited, Australia*

G.R. Walker – *CSIRO Land and Water, Centre for Groundwater Studies, Glen Osmond, S.A., Australia*

S.D. Tyerman – *Flinders University of South Australia, Adelaide, S.A., Australia*

L.J. Mensforth – *River Murray Catchment Water Management Board, Australia*

ABSTRACT: Salinity has attracted much attention in the Murray Darling Basin of south-eastern Australia, with current predictions of instream salinities exceeding the desirable limit for drinking water (0.8 dSm^{-1}) within 100 years. Much of this salt comes from the large regional groundwater systems that discharge to the river via the floodplains. Approximately 26,000 ha of a total 100,000 ha of floodplain have been identified as being salt affected in the lower River Murray. Salt accumulation is the major cause of floodplain tree dieback, reducing the environmental, ecological and social values of the floodplain environment. Six key processes leading to salt accumulation and floodplain tree death (reduced flooding frequency and duration; irrigation; regional groundwater discharge; river regulation; blocked flow paths and grazing) have been identified. An understanding of these complex interactions is essential in order to manage the floodplains at both a regional and more localised scale.

1 INTRODUCTION

The Murray Darling Basin covers an area of 1.06 million square kilometres in south-eastern Australia. It acts as a salt trap, storing salt in the unsaturated and saturated zones, restricting salt export to surface water flows (Simpson and Herczeg, 1994), resulting in high groundwater salinities in parts of the basin. The regional groundwater system discharges this salty groundwater into the lower River Murray via the floodplain aquifer. This saline discharge affects both instream and floodplain soil salinities. The lower River Murray is an important water supply for irrigation, urban users and industry in the state of South Australia. Recently, predictions have been made that instream salinities will exceed the desirable limit for drinking water (0.8 dSm^{-1}, NHMRC/ARMCANZ, 1996) within 100 years (MDBMC, 1999).

In addition to the increase in instream salinities, one of the largest environmental impacts in the lower River Murray is the decline in riparian vegetation health. Salt accumulation has been identified as the major cause of floodplain tree dieback, principally from water stress (Margules and Partners *et al.*, 1990). Recent estimates (Herbert, 1999) based on a study of River Murray floodplains by Nichols, 1998) have indicated that a quarter (26,000 ha of a total 100,000 ha) of the floodplain in the lower River Murray is affected by salt. Salinisation threatens not only the quality of water supplies, but also the health of the flora and fauna of the floodplain environment.

1.1 *Hydrogeology*

The Murray Basin groundwater system consists of three aquifers separated by low permeability confining layers. The Pliocene Sands aquifer consists of sediments deposited between 2 to 6 million years ago, as layers of sands and gravel over most of the Murray Basin. Underlying the Pliocene Sands is the Murray Group aquifer, a limestone unit deposited during sea incursion between 12 and 32 million years ago. The Renmark Group aquifer which underlies the Murray Group aquifer, is an accumulation of riverine sand, silt and clay deposited 30 to 50 million years ago (Evans and Kellett, 1989). It is approximately 200 m to 400 m thick, extending to a maximum depth of 500-600 m. The water table lies within the Pliocene Sands to the east of Overland Corner, and within the Murray Group limestone aquifer to the west. The Pliocene Sands and Murray Group aquifers discharge directly into the river (Barnett and SA Department of Mines and Energy, 1991).

Within the river valley, the Monoman Sands aquifer (approximately 30 m thick) underlies a semi-confining layer of grey brown silty clays, the Coonambidgal Formation (typically 4-5 m thick)

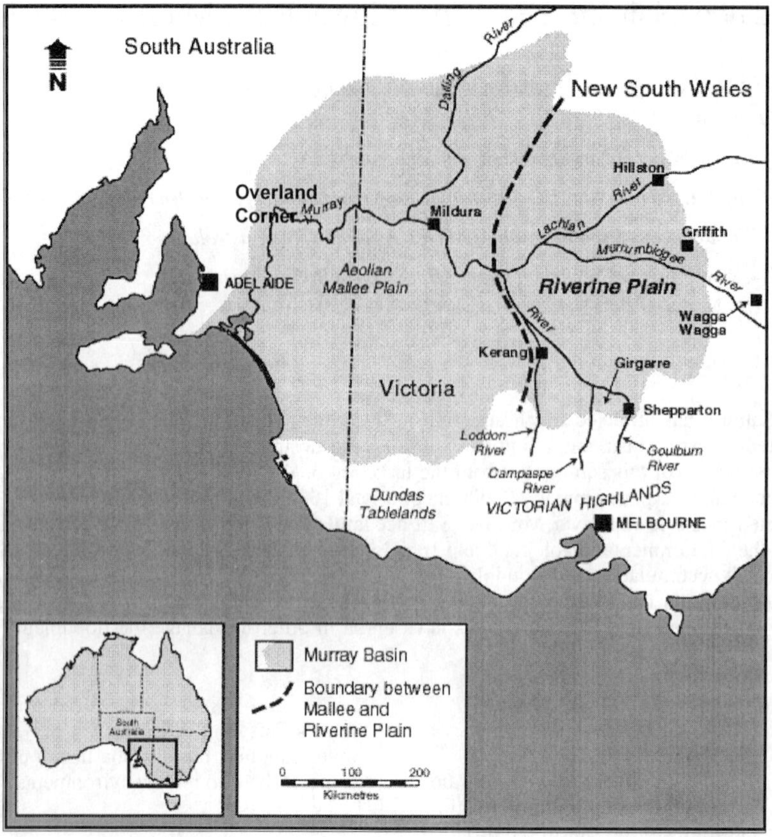

Figure 1. Lower River Murray, south-eastern Australia

(Hollingsworth, 1989). The Monoman Sands aquifer is in good hydraulic connection with the Pliocene Sands or Murray Group Formations and with the floodplain creeks and main river channel (Jolly *et al.*, 1994). Where the water table is in the Pliocene Sands, there is some upward leakage from the underlying Murray Group aquifer. To the west of Overland Corner there is significant upward leakage from the Renmark Group aquifer into the Murray Group aquifer and hence into the river (Barnett and SA Department of Mines and Energy, 1991). The significant salt loads transported in the regional groundwater system are further concentrated by evapotranspiration as the groundwater passes through the floodplain to the river. Groundwater salinities in the lower River Murray floodplain aquifers range from 20 dSm^{-1} to over 100 dSm^{-1}.

1.2 *Floodplain vegetation*

Three principal overstorey species occur on the floodplains of the lower River Murray. Blackbox (*Eucalyptus largiflorens*) is a salt and drought tolerant eucalypt, which inhabits the more elevated parts of the floodplain Slavich *et al.*, 1999a. River red gums (*E. camaldulensis*) are a less salt and drought tolerant tree that inhabit the river banks and more frequently flooded areas. Lignum (*Muehlenbeckia florulenta*), a woody perennial shrub occurs in frequently flooded parts of the floodplain on heavy clay soils (Craig *et al.*, 1991).

The wetland and floodplain vegetation of the lower River Murray have important ecological, environmental and recreational values. They contain a high biodiversity of both flora and fauna, and provide an important native fish nursery and breeding area for water birds and aquatic invertebrates (O'Malley and Sheldon, 1990; Cunningham *et al.*, 1981). The Chowilla floodplain which lies near the New South Wales, Victorian and South Australian borders supports 200 km^2 of native woodland and has an UNESCO Ramsar Convention listing as a Wetland of International Importance and as a habitat for migratory birds (National Environmental Consultancy, 1988). This woodland is

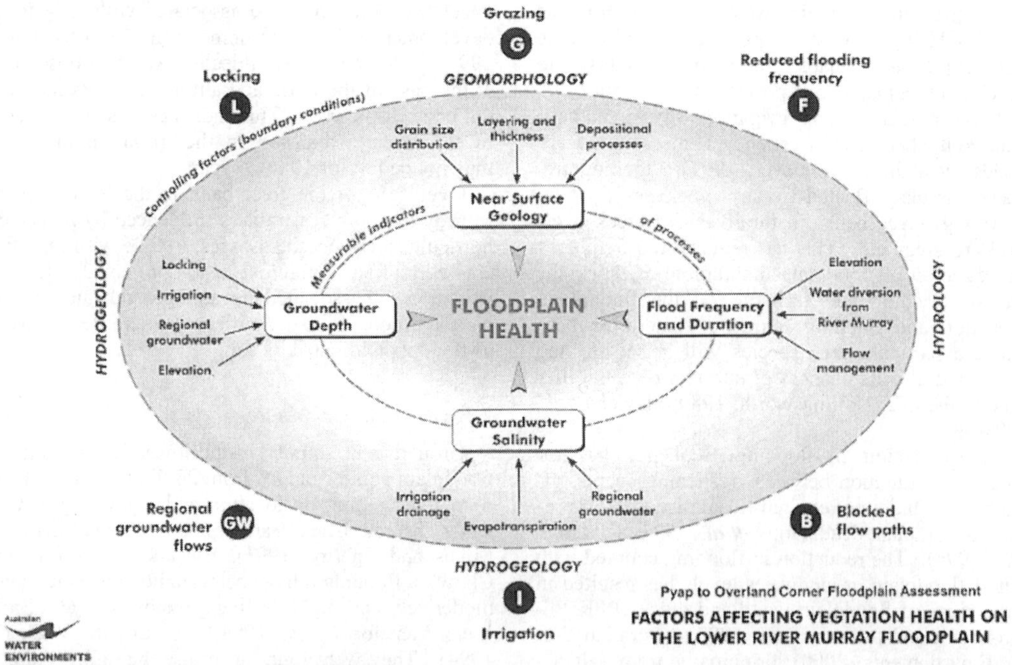

Figure 2. Factors affecting vegetation health on the lower River Murray floodplain (Australian Water Environments, 2000)

typical of the floodplain vegetation in the lower River Murray, of which 25% is estimated to be affected by salt (Herbert, 1999).

This paper presents the current understanding of how four key processes affect floodplain vegetation health interacts at a range of spatial and temporal scales, with particular emphasis on the hydrogeology. The focus of this paper will be on how groundwater processes affect blackbox (*Eucalyptus largiflorens*) tree health on the floodplains of the lower River Murray in South Australia.

2 KEY INDICATORS OF FLOODPLAIN TREE HEALTH

Previous studies have identified most of the likely potential causes of floodplain vegetation decline (e.g. Akeroyd *et al.*, 1998; Craig *et al.*, 1991; Jolly *et al.*, 1996; Slavich *et al.*, 1999b; Telfer and Overton, 1999). The notion of a critical depth to groundwater has been raised, however the studies into floodplain vegetation health have found that it is dependent on a combination of depth to groundwater, groundwater salinity, near surface geology and flooding frequency. Due to the complex interactions between each of these processes at a range of spatial and temporal scales, it is very difficult to identify which processes are having the greatest impact on a particular floodplain.

The work described in this paper is aimed at discerning which of these four key processes are having the strongest effect on vegetation health at several different sites along the river. Armed with this knowledge for each floodplain or portion of each floodplain, management options can then be implemented which will have the greatest benefit to floodplain vegetation health. The four key measurable indicators that have the greatest role in determining floodplain vegetation health, and also ease of measurement or data availability are groundwater depth, groundwater salinity, near surface geology and flooding frequency.

Figure 2 below is a schematic model of all the factors which affect the four indicators of floodplain tree health. It is the first step in developing a large spatial model to determine which factors are having the greatest impact on tree health at various floodplain locations. This large spatial model will incorporate a hydrogeological model of floodplain groundwater level and salinities, an elevation flooding frequency model and near surface geological information based on geomorphological information and field sampling.

2.1 *Flood frequency and duration*

Ten weirs and locks were constructed on the lower River Murray and four major reservoirs in the headwaters between 1920 and 1940 to aid river boat navigation and provide water during droughts (Jacobs, 1990). The stable pool levels and reliable water supply have enabled the irrigation industry to develop into a major food provider for the nation.

River regulation and irrigation diversions have reduced the height and frequency of medium to large floods by a third (Ohlmeyer, 1991). Flooding frequency is also affected by any blockages to flood flows e.g. levee banks, naturally silted creek beds, road crossings etc. This has reduced the frequency and duration of floodplain inundation, affecting the rate of leaching and recharge on the floodplain. Elevation and flooding frequency determine both where a particular tree species will germinate and grow, and also its chances of survival over the lifetime of the tree (Hollingsworth, 1989; Slavich *et al.*, 1999b).

The reduction in flooding frequency has increased the duration between freshening events, often beyond the time required to salinise the soil profile causing plant death (Jolly *et al.*, 1993, Slavich *et al.*, 1999b). The reduction in flooding coupled with raised floodplain groundwater levels has resulted in large areas of floodplain woodland dying (PPK Environment and Infrastructure, 1997, Australian Water Environments, 2000), principally from salt accumulation.

2.2 *Groundwater depth*

River regulation and the construction of 10 weirs and locks in the lower River Murray has raised the minimum river level upstream of the weir by approximately three metres. Prior to river regulation the floodplain water table was typically in the Monoman Sands aquifer, however the increased pressure from higher river levels has raised the water table into the overlying Coonambidgal clay (Slavich *et al.*, 1999b), increasing evaporation and salt accumulation rates (Jolly, 1996).

When the water table is raised into the overlying Coonambidgal Clay, the rate of capillary rise through the clay soil increased, compared to the sand. This has increased saline groundwater discharge and hence salt accumulation rates in the floodplain soils. Prior to river regulation there was a balance between salt accumulation and leaching by rainfall and floodwaters to which blackbox was adapted (Jolly and Walker, 1995). However, the raised water levels and reduced flooding frequency increased the rate of salt accumulation beyond the tolerance levels of the blackbox woodland, resulting in their slow death (Jolly and Walker, 1995).

Increased recharge beneath irrigation areas has also caused water tables to rise, e.g. by 18 m above river level beneath the Loxton Irrigation Area (PPK Environment and Infrastructure, 1997) and consequently beneath the adjacent floodplain. Floodplain vegetation mapping at Loxton revealed trends in vegetation health decline associated with irrigation development (PPK Environment and Infrastructure, 1997). Vegetation health decline was first observed at the base of the cliffs and almost 20 years later at the edge of the river. This indicates that the impact of the recent expansion in the irrigation industry may not be felt for 10 to 20 years.

Over 43% of the river bank in the lower River Murray in South Australia is influenced by irrigated horticulture (Bone and Davies, 1992), with significant current and future expansion projected. Hence, irrigation and associated elevated groundwater levels have the potential to significantly increase the area of floodplain affected by salt.

2.3 *Groundwater salinity*

Groundwater salinity in the lower River Murray floodplain aquifer ranges from 20 dSm^{-1} to over 100 dSm^{-1}, depending on location and regional groundwater characteristics (Barnett and SA Department of Mines and Energy, 1991). Blackbox trees on the Chowilla floodplain have been studied growing over moderately and highly saline groundwater (60 dSm^{-1}, e.g. Mensforth *et al.*, 1994; Thorburn and Walker, 1994). They were found to utilise the highly saline groundwater for transpiration, despite the availability of fresh creek water (<1 dSm^{-1}) within ~15 m for transpiration (Mensforth *et al.*, 1994). The use of highly saline groundwater enables the trees to survive long dry periods between floods, but reduces the transpiration rate to below 0.1-0.3 $mmday^{-1}$ depending on soil and groundwater salinity (Thorburn *et al.*, 1993). A reduced transpiration rate also reduces the rate of salt accumulation in the root zone from transpiration of water and the exclusion of salts, allowing the blackbox to survive for longer periods of time on highly saline groundwater (Jolly, 1996).

An analytical model to predict salinity limited transpiration rates, based on saturated zone water and salt balance during dry periods (Thorburn *et al.*, 1995) showed that groundwater depth and salinity were the main controls on groundwater uptake, while soil properties appeared to have a lesser effect. The model predictions were compared to 15 months of field measurements; the predicted values were close to the measured values. The model showed that the soil profile would reach complete salinisation within 4 to 30 years without leaching by rainfall or floodwaters (Thorburn *et al.*, 1995). This again indicates the importance played by the four key processes identified in this paper on blackbox tree health. Groundwater salinity has a very strong influence on the rate of soil salinisation and also potential transpiration rates and hence tree growth rates.

2.4 *Near surface geology*

Floodplains soils in the lower River Murray are typically composed of a surface layer of 2-5 m thick grey clay, the Coonambidgal Formation, (Barnett and SA Department of Mines and Energy, 1991). The Coonambidgal Formation is underlain by the Monoman Formation, a fine to coarse river sand. Erosional and sedimentation processes controlled by floodplain geomorphology affect the distribution of these two formations and their sand, silt and clay percentages. The floodplains in the lower River Murray are characterised by the presence of low-lying depressions, creek beds and sand dunes. This results in a largely heterogeneous distribution of floodplain soils, further enhancing the patchy nature of the floodplain vegetation.

The heavy, grey clay / sandy clay found on the Chowilla floodplain has a low hydraulic conductivity, particularly during inundation as a result of its sodic and swelling nature, which limits floodwater recharge (Jolly *et al.*, 1994). The studies of salt accumulation rates (Jolly *et al.*, 1993, Thorburn *et al.*, 1995) and flooding frequency requirements assumed that the soil profile was leached of salts following floods. Work following flooding showed that only the surface 0.7 m was leached (McEwan *et al.*, 1994), and that the accumulated salts were unchanged in the remaining profile, despite up to 78 days of flooding. As expected, the clay content of the soils strongly affected the degree of leaching (Akeroyd *et al.*, 1998.

Near surface geology can reduce transmission between the river and the floodplain aquifer if a "skin" or veneer of low hydraulic conductivity clays and silts impedes groundwater discharge into the river. This raises groundwater levels beneath that area of the floodplain and can be compounded by raised groundwater levels associated with irrigation mounds or floodplain disposal basins.

Recent fieldwork on other lower River Murray floodplains has shown that this heavy clay is not ubiquitous, but includes coarse sands and clayey sands, depending on geomorphology. Hence the distribution of soil types across individual floodplains will strongly affect tree health for the same condition of depth to groundwater, flooding frequency and groundwater salinity.

3 DISCUSSION

The areas of floodplain vegetation that have been most affected are those that naturally had a very tight balance between flooding frequency and salt accumulation rates (Slavich *et al.*, 1999b). Reduced flooding and raised groundwater levels are the most likely indicators to have changed, whereas, groundwater salinity and near surface geology are more stable. It is noted that groundwater salinity may change, e.g. when lower salinity drainage water displaces highly saline native groundwater from irrigation mounds.

The difficulty lies not in identifying the controlling factors, but interpreting how they interact to affect floodplain tree health. A study by Zubrinich *et al.*, in review) found a planar relationship between predawn xylem water potentials (a measure of tree water stress), groundwater depth and salinity on a three-dimensional plot. This work was done on the Chowilla floodplain where the soils were the typical heavy, grey, sodic, swelling clay. A similar two-dimensional analysis calculated the substrate severity index (SSI = groundwater salinity / groundwater depth * 100). The plot of SSI and predawn xylem water potential also found a good relationship, with R^2 values > 0.9 (Zubrinich, 1996).

The groundwater system in the lower River Murray is highly saline and management cannot change this, however it is important to understand the interactions between groundwater salinity, groundwater depth, near surface geology and flooding frequency and how they affect blackbox health.

The difficulty lies in incorporating the other two factors (near surface geology and flooding frequency), knowing that the factors are not independent, as for example flooding frequency and groundwater depth are related to elevation, as well as location within the floodplain. Current research aims to combine field data with modelled data to determine the relationship between the four key indicators. When this relationship is determined and compared to field data, large scale management decisions can be evaluated and implemented based on knowledge of the likely outcomes of the management scenarios.

ACKNOWLEDGEMENTS

The first author acknowledges the support of an Australian Postgraduate Award and a Centre for Groundwater Studies Bursary. This work was supported by research grants from the Land and Water Resources Research and Development Corporation and the River Murray Catchment Water Management Board.

REFERENCES

Akeroyd, M.D., Tyerman, S.D., Walker, G.R. and Jolly, I.D. (1998). Impact of flooding on the water use of semi-arid riparian eucalypts. *Journal of Hydrology* 206, 104-17.

Australian Water Environments (2000). Pyap to Overland Corner Floodplain Assessment Report. Prepared for the Loxton to Bookpurnong Local Action Planning Committee. No. 98027.

Barnett, S.R. and SA Department of Mines and Energy (1991). 'Renmark Hydrogeological Map (1:250,000 Scale).' (Bureau of Mineral Resources, Geology and Geophysics: Canberra, Australia.)

Bone, B. and Davies, G. (1992). Effects of irrigation on riparian vegetation in the South Australian Riverland of Agriculture. South Australian Department of Agriculture.

Craig, A.E., Walker, K.F. and Boulton, A.J. (1991). Effects of edaphic factors and flood frequency on the abundance of lignum (*Muehlenbeckia florulenta* Meissner) (Polygonaceae) on the River Murray floodplain, South Australia. *Australian Journal of Botany* 39, 431-43.

Cunningham, G.M., Mulham, W.E., Milthorpe, P.L. and Leigh, J.H. (1981). 'Plants of Western New South Wales. ' (Government Printer: Sydney.)

Evans, W.R. and Kellett, J.R. (1989). The hydrogeology of the Murray Basin, southeastern Australia. *BMR Journal of Australian Geology and Geophysics. BMR* 11, 147-66.

Herbert, T. (1999, unpublished). Current and future floodplain area affected by salt.

Hollingsworth, I.D. (1989). A reconnaissance soil survey of the Chowilla Anabranch System of the River Murray in South Australia and New South Wales. Murray-Darling Basin Commission by the Loxton Research Centre, South Australian Department of Agriculture.

Jacobs, T. (1990). Regulation and Management of the River Murray. *In* 'The Murray.' (Eds McKay, N. and Eastburn, D.) (Murray-Darling Basin Commission: Canberra.)

Jolly, I.D. (1996). The Effects of River Management on the Hydrology and Hydroecology of Arid and Semi-Arid Floodplains. *In* 'Floodplain Processes.' (Eds Anderson, M.G., Walling, D.E. and Bates, P.D.) pp. 577-609. (John Wiley & Sons Ltd.: Chichester, New York.)

Jolly, I.D. and Walker, G.R. (1995). A sketch of salt and water movement in the Chowilla floodplain. CSIRO Division of Water Resources, Adelaide.

Jolly, I.D., Walker, G.R., Hollingsworth, I.D., Eldridge, S.R., Thorburn, P.J., McEwan, K.L. and Hatton, T.J. (1996). Chapter 10. The Causes of Decline in Eucalypt Communities and Possible Ameliorative Approaches. *In* 'Salt and Water Movement in a Saline Semi Arid Floodplain of the River Murray.' (Eds Walker, G.R., Jolly, I.D. and Jarwal, S.D.) pp. 119-38. (CSIRO Division of Water Resources.)

Jolly, I.D., Walker, G.R. and Narayan, K.A. (1994). Floodwater recharge processes in the Chowilla Anabranch system. *Australian Journal of Soil Research* 32, 417-35.

Jolly, I.D., Walker, G.R. and Thorburn, P.J. (1993). Salt accumulation in semi-arid floodplain soils with implications for forest health. *Journal of Hydrology* 150, 589-614.

Margules and Partners, P and J Smith Ecological Consultants and Victorian Department of Conservation, Forests and Lands (1990). River Murray riparian vegetation study for the Murray Darling Basin Commission. Canberra.

McEwan, K.L., Jolly, I.D. and Holub, A.N. (1994). Effects of flooding on the water use of *Eucalyptus largiflorens*: Compilation of results from the 1992/93 flood. CSIRO Division of Water Resources, Technical Memorandum No. 5/1994, Adelaide, SA.

MDBMC (1999). The Salinity Audit of the Murray-Darling Basin. A 100-year perspective, 1999. Murray-Darling Basin Ministerial Council, Canberra, Australia.

Mensforth, L.J., Thorburn, P.J., Tyerman, S.D. and Walker, G.R. (1994). Sources of water used by riparian *Eucalyptus camaldulensis* overlying highly saline groundwater. *Oecologia* 100, 21-28.

National Environmental Consultancy (1988). Chowilla Salinity Mitigation Scheme, Draft Environmental Impact Statement. Engineering and Water Supply Department, Adelaide.

NHMRC/ARMCANZ (1996). Australian Drinking Water Guidelines. National Health and Medical Research Council/Agriculture and Resource Management Council of Australia and New Zealand, Canberra.

Nichols, S. (1998). Riverine Local Action Planning Associations, Wetland Management Study. Wetland Care Australia.

O'Malley, C. and Sheldon, F. (1990). '*Chowilla Floodplain Biological Study*.' (Nature Conservation Society of South Australia: Adelaide.)

Ohlmeyer, R.G. (1991). Report on implications of manipulating water levels in the River Murray between Lock 1 and 10. South Australian River Murray Wetlands Management Program. Murray-Darling Basin Natural Resources Management Program. Engineering and Water Supply Department.

PPK Environment and Infrastructure (1997). Assessment of the impact of the Loxton Irrigation District on Floodplain Health and implications for future options.

Simpson, H.J. and Herczeg, A.L. (1994). Delivery of marine chloride in precipitation and removal by rivers in the Murray-Darling Basin, Australia. *Journal of Hydrology* 154, 323-50.

Slavich, P.G., Walker, G.R. and Jolly, I.D. (1999a). A flood history weighted index of average root-zone salinity for assessing flood impacts on health of vegetation on a saline floodplain. *Agricultural Water Management* 39, 135-51.

Slavich, P.G., Walker, G.R., Jolly, I.D., Hatton, T.J. and Dawes, W.R.J. (1999b). Dynamics of *Eucalyptus largiflorens* growth and water use in response to modified watertable and flooding regimes on a saline floodplain. *Agricultural Water Management* 39, 245-64.

Telfer, A. and Overton, I.C. (1999). The impact of irrigation on floodplain vegetation health adjacent the River Murray. *In* 'Proceedings of the Second Australian Stream Management Conference, Adelaide, (Eds Rutherford, I.D. and Bartley, R.)

Thorburn, P.J., Hatton, T.J. and Walker, G.R. (1993). Combining measurements of transpiration and stable isotopes of water to determine groundwater discharge from forests. *Journal of Hydrology* 150, 563-87.

Thorburn, P.J. and Walker, G.R. (1994). Variations in stream water uptake by *Eucalyptus camaldulensis* with differing access to stream water. *Oecologia* 100, 293-301.

Thorburn, P.J., Walker, G.R. and Jolly, I.D. (1995). Uptake of saline groundwater by plants: an analytical model for semi-arid and arid areas. *Plant and Soil* 175, 1-11.

Zubrinich, T.M. (1996). 'An Investigation of the Ecophysiological, Morphological and Genetic Characteristics of *Eucalyptus Largiflorens* F. Muell. and *Eucalyptus gracilis* F. Muell; in Relation to Soil Salinity and Groundwater Conditions Throughout the Chowilla Anabranch. PhD Thesis.' (Flinders University of South Australia: School of Biological Sciences.)

Zubrinich, T.M., Byrne, B., Gallasch, S. and Tyerman, S.D. (in review). Tolerance of salinised floodplain conditions in a naturally occurring Eucalyptus hybrid related to lowered plant water potential. *Tree Physiology*

Groundwater: Past Achievements and Future Challenges, Sililo et al. (eds) © 2000 Balkema, Rotterdam, ISBN 90 5809 159 7

Advancing technologies in the hydrogeological design of urban subdivisions

K.W.F. Howard
University of Toronto, Scarborough, Ont., Canada

B. Beatty & J. Thompson
W.B. Beatty and Associates Limited, Bolton, Ont., Canada

S. Motkaluk
SDM Incorporated, Waterloo, Ont., Canada

ABSTRACT: A study is underway which aims to improve the hydrogeological design of urban subdivisions by developing and testing novel technologies ("best management practices" - BMPs) which would alleviate the potentially detrimental impacts of urban development on the groundwater. The work is being carried out along the Oak Ridges Moraine, a ridge of sand north of Toronto, Canada, featuring kettle lakes and numerous closed depressions, both of which provide foci for indirect recharge. Field studies have established recharge conditions under natural, "undeveloped conditions". In ongoing work, alternative urban sub-division plans are being evaluated in order to select a design that will not only maintain the quality and quantity of recharge at the site but will preserve the hydrogeological integrity of the system. An important component of this work is the testing of sites appropriate for the installation of infiltration ponds and rapid infiltration basins (RIB's). The intention is not to maximise recharge but to maintain it at pre-development levels

1 STUDY OBJECTIVE

The overall, long-term goal of the study is to improve the hydrogeological design of urban subdivisions by developing and testing novel technologies (best management practices – BMP's) which would minimise, if not eliminate potentially detrimental impacts of urban development on the quality and quantity of Ontario's groundwater. Since this work requires a detailed understanding of the local water balance, a secondary objective of a more "fundamental" or "scientific" nature is to develop a quantitative understanding of the surface and sub-surface flow dynamics under pre- and post-development conditions.

The study has been initiated at sites on the Oak Ridges Moraine (Fig. 1A), a 20-km wide ridge of till-draped sand and gravel extending eastwards from the Niagara Escarpment for over 150 km. The Oak Ridges Moraine is a significant recharge area for one of Ontario's more important aquifers. It is also recognized by many as prime land for real estate development, notably in the towns of Richmond Hill and Uxbridge (Fig. 1B).

The hydrogeology of the moraine has been the subject of considerable study during the past fifteen years, notably by the University of Toronto (Howard *et al.* 1995; Howard and Gerber 1997; Howard 1996, 1997; Gerber & Howard 1996, 1997; Gerber 1994, 1999), the University of Waterloo (Romano 1998) and the Geological Survey of Canada (GSC 1997, for example). While these studies have provided considerable information on the moraine's three-dimensional geological framework, its multi-layer aquifer system and the overall water balance, the level of detail has, until recently, proved insufficient to enable site specific decisions to be made in the context of land-use planning and urban development. For example, the typical kettle relief of moraines and the presence of numerous internal drainage basins (Fig. 2) exerts a significant influence on the nature of aquifer recharge. In addition, there has been some debate on the degree to which tills that drape the moraine limit natural infiltration. An important challenge is to maintain pre-development recharge rates, wetland and kettle lake water levels, and baseflow to tributary streams during the post-development phase. With appropriate study, this can be achieved through the design and installation of infiltration ponds and rapid infiltration basins (RIB's). To date, two infiltration basins have been proposed for the site and pilot studies are underway. Initial test results demonstrate that the moraine sediments readily accommodate the runoff water and are ideally suited for artificial recharge technology.

2 OAK RIDGES MORAINE AQUIFER

To illustrate the water balance approach being adopted in the Oak Ridges Moraine area, reference is made here to the Yonge East study area (Fig. 2) in

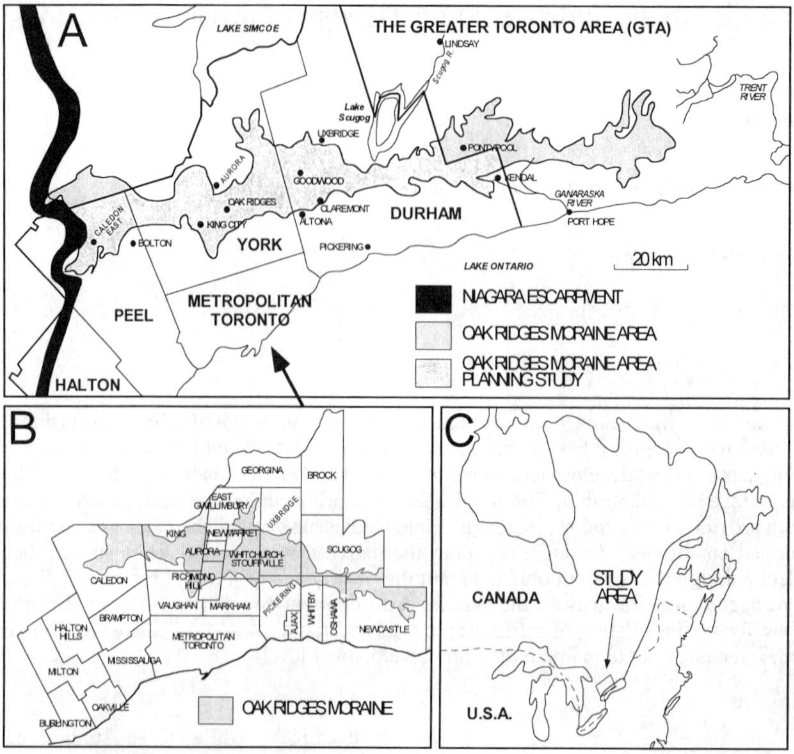

Figure 1. A) Location map showing the Oak Ridges Moraine (ORM) in relation to Metropolitan Toronto and the four Regions that together comprise the Greater Toronto Area (GTA). The Oak Ridges Moraine Planning study area refers to that part of the moraine contained within the GTA; B) Local Townships; and C) Regional location (after Howard *et al.* 1995).

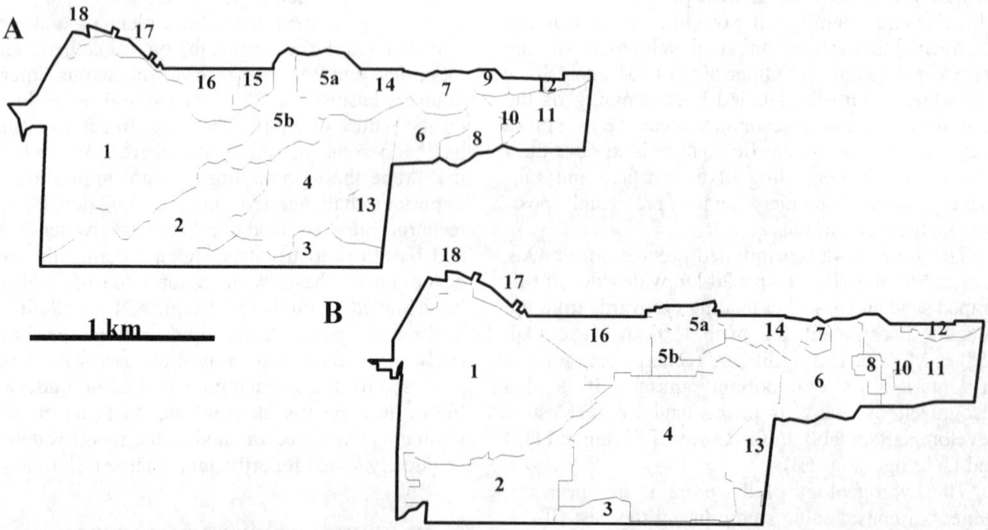

Figure 2. Kettle relief in the Yonge East study area features numerous local drainage basins (numbered 1 to 18): A) before development and B) post-development. Drainage Basins 1, 5a, 5b, 7, 8 and 10 contain significant kettle lake features.

the town of Richmond Hill (Fig. 1B). In this area a three-phase work program is underway:

1. To determine recharge characteristics under pre-development conditions;
2. To determine recharge characteristics under post-development conditions;
3. To select, develop, test, and adapt appropriate methodologies that will ensure protection of the area's hydrogeological function.

The Yonge East study area is situated on the Oak Ridges Moraine to the south of an east-west regional water divide (Fig. 3). The primary aquifer in the study area is the Upper Aquifer, or "Oak Ridges Aquifer Complex". This aquifer is normally unconfined and is the preferred source of water for most domestic users. It also feeds headwater springs of numerous rivers which pass through the city of Toronto and enter Lake Ontario.

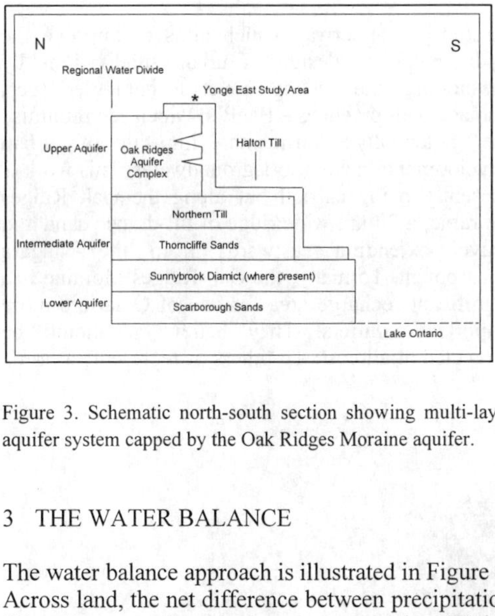

Figure 3. Schematic north-south section showing multi-layer aquifer system capped by the Oak Ridges Moraine aquifer.

3 THE WATER BALANCE

The water balance approach is illustrated in Figure 4. Across land, the net difference between precipitation (P) and evaporation plus evapotranspiration (E) i.e. P-E either infiltrates through the soil zone as direct groundwater recharge or runs-off to enter surface water bodies (wetlands, streams and lakes). For closed basins lacking a surface water outlet, all of P-E will ultimately enter the subsurface as groundwater recharge. As indicated by Figure 3, direct recharge via the soil zone is termed RLAND, while runoff is termed either REX where the run-off enters streams and leaves the catchment, or RIN, where run-off enters wetlands and lakes and stays within the study area. In urbanised areas where runoff from impermeable areas is controlled and directed into storm water management ponds, this quantity is referred to as SWM.

While the lakes and wetlands will vary in size and capacity over the course of year, it can be assumed

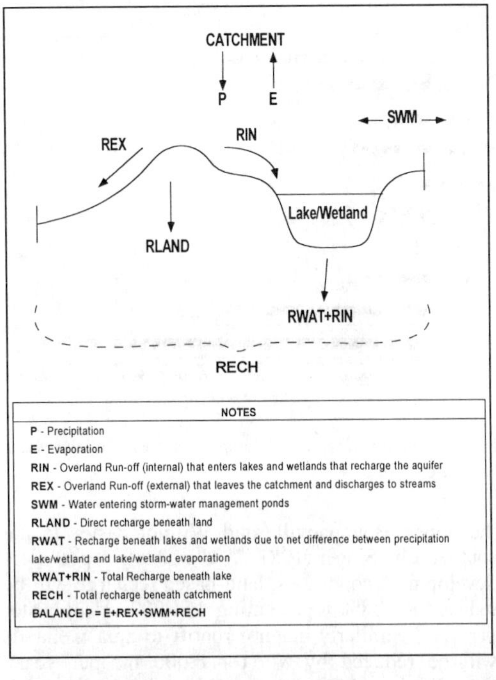

Figure 4. Components of the water balance assessment.

that most lake and wetland water levels fluctuate about a steady-state long-term mean. Thus a balance exists between inflow to and outflow from the lake/wetland system. Inflow to this system includes two components: the excess of P-E as determined immediately over the lake/wetland area (a volume termed RWAT), together with internal surface water run-off to the lake/wetland (defined above as RIN). There is no evidence of a groundwater contribution to lakes and wetlands in the study area.

None of the lakes or wetlands in the study area displays any form of surface water outlet, thus all outflow occurs as groundwater recharge through the bottom of the lake or wetland (i.e. RIN + RWAT). If this quantity is not maintained, the wetland or lake will experience a decline in water level over time, and inevitably become dry. Total groundwater recharge across the catchment (termed RECH) will include both the land and water based components i.e. RECH = RIN + RWAT + RLAND.

Figure 5 shows preliminary annual estimates for each of the water balance components under pre- and post- development conditions.

Under pre-development conditions, annual recharge to the underlying aquifer varies between between 100 and 300mm (Fig. 6) with an average of 201 mm. The average value compares well with the range of 150 to 250 mm given for recharge via the Halton Till by Gerber (1994), and the range of 150 to 200 mm for recharge via the Halton Till for

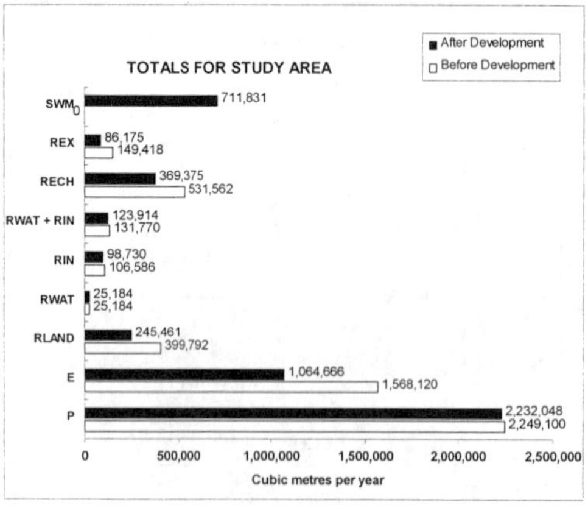

Figure 5. Preliminary annual estimates for each of the water balance components.

the Soper, Bowmanville and Wilmot basins to the southeast by Singer (1981). Under post-development conditions, land-based recharge will be reduced by 38% representing 154,000 m^3 of water per year. Similarly, natural runoff to area wetlands will be reduced by 8% (or 8,000 m^3 per year). Overall it is estimated that recharge to the underlying aquifer will be reduced by 30% or 162,000 m^3 per year.

On the positive side, development will generate 700,000 m^3 of storm water each year of which less than 25% is required to restore losses to wetlands and the sub-surface. This water is sourced entirely from residential areas and is generally of very good quality. Strategic design of the sub-division will allow significant additional water to be directed to the wetlands in catchments 5a, 5b, 7, 8 and 10, thus allowing their recharge function to be maintained (Fig. 7). In selecting the best approach for additional resource augmentation, it should also be recognised that approximately 95% of the recharge shortfall is land based recharge (RLAND) and that artificial recharge technologies are required. To this end, it is proposed to infiltrate an additional 74,000 m^3 of water annually using a rapid infiltration basin (RIB) that straddles the boundary between catchments 4 and 13 (Fig. 2). With the RIB in place, the overall effect on the distribution of recharge across the area is shown on Figure 8.

4 CONCLUSION

A study is underway which aims to improve the hydrogeological design of urban subdivisions by developing and testing novel technologies (best management practices – BMP's) which can minimise the potentially detrimental impacts of urban development on Ontario's groundwater. This work is presently being carried out along the Oak Ridges Moraine, a 20-km wide ridge of till-draped sand and gravel extending eastwards from the Niagara Escarpment. To many, the Oak Ridges Moraine is a significant recharge area for one of Ontario's more important aquifers. They believe it should be protected at all cost. To others, it represents a prime

Figure 6. Annual distribution of recharge under pre-development conditions (preliminary estimates).

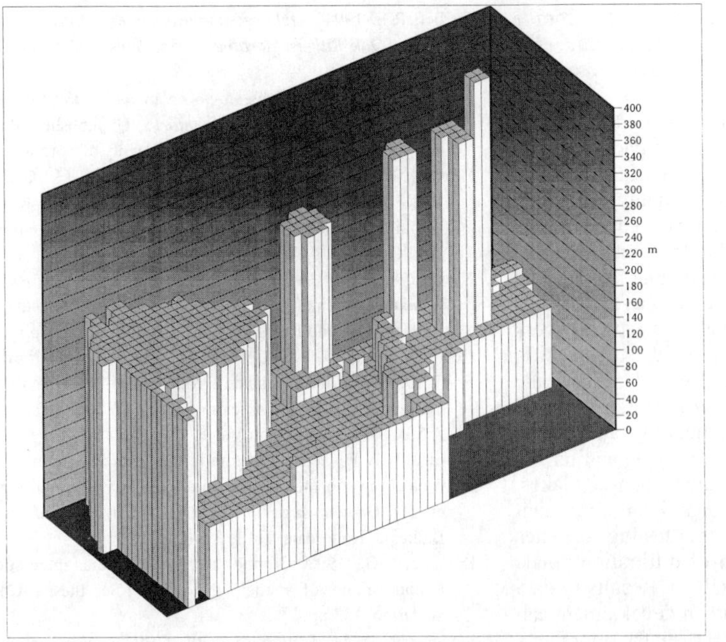

Figure 7. Annual distribution of recharge under post-development conditions (without RIB) (preliminary estimates).

Figure 8. Annual distribution of recharge under post-development conditions (with RIB in place) (preliminary estimates).

opportunity for real estate development. From a groundwater perspective, the argument is immaterial providing the hydrogeological integrity of the moraine can be preserved.

Water balance studies along the Oak Ridges Moraine have allowed the potential impacts of urban development on the recharge function of the moraine and its lakes and wetlands to be reliably evaluated. Results indicate that urbanization will reduce natural recharge to the underlying aquifer, but will also generate large volumes of water suitable for artificial replenishment. This water is obtained exclusively from residential areas and its quality is generally very good. In ongoing work, alternative urban subdivision plans are being evaluated and refined, the primary objective being to select a design that will not only maintain the quality, quantity and spatial distribution of recharge to the underlying aquifer but will preserve hydrologic features such as lakes, wetlands and emerging springs. An important component of this work is the testing of sites appropriate for the installation of infiltration ponds and rapid infiltration basins (RIB's). Results to date provide every indication that urban development can proceed with negligible impacts on the moraine's important hydrogeological function.

ACKNOWLEDGMENTS

The authors gratefully acknowledge Tecia White, Massimiliano Pizzato, Patrizia Gatti and Karina Gelo for their help in collecting and analyzing the data, preparing figures and reviewing the manuscript. Initial work was sponsored by the Ontario Ministry of the Environment and Energy. More recently, significant financial support has been provided by the Natural Sciences and Engineering Research Council (NSERC) (Strategic Grant to Frind *et al.*), and the Centre for Research in Earth and Space Technology (CRESTech).

The views presented here are those of the authors and are not necessarily endorsed by colleagues or by the funding agencies.

REFERENCES

Geological Survey of Canada,1997. Water Resource Investigations of the Oak Ridges Moraine, Ontario. GSC Open File Report 3540, 60pp.

Gerber, R.E. & K.W.F. Howard 1996. Evidence for recent groundwater flow through Late Wisconsinan till near Toronto, Canada. *Canadian Geotechnical Journal* 33: 538-555.

Gerber, R.E. & K.W.F. Howard 1997. Ground-water recharge to the Oak Ridges Moraine. In: Eyles, E. (ed.) *Environmental Geology of Urban Areas*. Special Publication of the Geol. Association of Canada. Geotext #3: 173 - 192.

Gerber, R.E. 1994. *Recharge analysis for the central portion of the Oak Ridges Moraine.* M.Sc. Thesis: University of Toronto.

Gerber, R.E. 1999. *Hydrogeologic behaviour of the Northern till aquitard near Toronto, Ontario.* Unpublished Ph.D. Thesis: Department of Geology, University of Toronto.

Howard, K.W.F., Eyles, N., Smart, P.J., Boyce, J.I., Gerber, R.E., Salvatori, S. & M. Doughty 1995. The Oak Ridges Moraine of southern Ontario: A groundwater resource at risk. *Geoscience Canada* 22: 101-120.

Howard, K.W.F. , 1996. *The Hydrogeology of the Oak Ridges Moraine* - Final report to the Ontario Ministry of Environment and Energy (MOEE). June, 1996. 450pp.

Howard, K.W.F. & R.E. Gerber 1997. Do tills beneath urban Toronto provide adequate groundwater protection? In: Chilton J. et al., *Groundwater in the Urban Environment.* AA Balkema, Rotterdam Vol 1: 433-438.

Howard, K.W.F. 1997. Incorporating policies for groundwater protection into the urban planning process. In: Chilton J. et al., *Groundwater in the Urban Environment.* A.A. Balkema, Rotterdam Vol 1: 31-40.

Romano, C.G., 1998. Groundwater-surface water interaction: a comparison of modelling tools. M.Sc. thesis. Univ. Waterloo. 133 pp.

Singer, S.N. 1981. *Evaluation of the Groundwater Resources Applied to the Bowmanville, Soper and Wilmot Creeks.* International Hydrological Decade (IHD) Representative Drainage Basin: Ontario Ministry of the Environment, Water Resources Report 9b, 153 pp.

Groundwater: Past Achievements and Future Challenges, Sililo et al. (eds) © 2000 Balkema, Rotterdam, ISBN 90 5809 159 7

A framework for assessing water resources and abstraction sustainability

S. Kirk
Environment Agency, National Groundwater and Contaminated Land Centre, Solihull, UK

R. Soley
Entec UK Limited, Shrewsbury, UK

ABSTRACT: The Environment Agency is seeking to regularise the way in which water resources are assessed throughout England and Wales. A framework methodology has been developed which considers the interaction of groundwater and surface water and regards these components as a single integrated resource. It draws on other Environment Agency led research, most notably an on-going programme of work on the estimation of impacts of groundwater abstraction on river flows. The framework methodology requires: catchment delineation, the recording of a simple conceptual understanding of the catchment's hydrology and hydrogeology, estimation of artificial impacts, calculation of aquifer water balance, and comparison of existing and minimum acceptable flows to help derive a sustainability status for a catchment. It has been developed for use within the regulatory community in helping achieve a consistent approach to the sustainable management of water resources.

1 INTRODUCTION

The Environment Agency has a statutory duty under the Water Resources Act 1991 and the Environment Act 1995 to protect and improve the water environment in England and Wales through effective management of water resources and by reductions in pollution. Much of the current legislation on water abstraction in England and Wales remains essentially unchanged from the Water Resources Act 1963, which first introduced a unified system of water resource management and abstraction licensing. There currently exist over 48,000 licences to abstract water from 'sources of supply' throughout England and Wales. In all, some 24,500 million cubic metres of water per year are authorised for abstraction, of which approximately one third is from groundwater and the remaining from non-tidal surface waters. In recent years there has been growing concern that many rivers and valued wetlands have been degraded by over-abstraction.

Public awareness of water resources issues in the UK has been particularly heightened by a recent series of unusually dry summers. In response to these and related concerns, the UK Government intends to make a number of changes to the abstraction licensing system of England and Wales. The changes aim to "ensure that abstraction licensing provides full protection to the water environment while enabling fair and flexible arrangements for meeting properly managed demand for water resources".

The proposed changes include a shift towards time-limited abstraction licences and the production, by the Environment Agency, of new 'Catchment Abstraction Management Strategies' for all of England and Wales. These new strategies may involve the re-assessment of the available water resources of many catchments.

These developments should help the Environment Agency achieve and demonstrate 'sustainable' water resource management. However, it remains for the Agency to determine its technical approach to the re-assessment and reporting of the water resources of each catchment.

An additional force for change in water resource management in the UK and elsewhere in Europe, is the impending European Union Water Framework Directive (WFD). This is the first time that European Community legislation has addressed the management of water quantity. The European Directive is designed to prevent further deterioration of, and to protect and enhance, the quality and quantity of aquatic ecosystems. The Directive proposes a river basin approach to water management and embodies a philosophy of highly integrated management of surface water, groundwater and aquatic ecology, involving both quantity and quality considerations. In order to meet the requirements of the Directive the

Agency is moving towards increasingly integrated and holistic approaches to resource management.

This paper describes a new water resource assessment framework, the 'Available Resource Methodology'(ARM), that embraces the integrated approach (to surface and groundwater management) of the Water Framework Directive, and may be used to help re-assess water resources as part of the new 'Catchment Abstraction Management Strategies'. The ARM has been developed and trialled over an eighteen month period by the Environment Agency and environmental consultants, Entec UK Ltd. (Environment Agency 2000).

2 ARM FRAMEWORK

The ARM sets out a series of steps to assess the sustainability status of a catchment given environmental water needs. The main output from the assessment is usually a statement of the surplus or deficit of resources in the area derived by consideration of predefined minimum acceptable flows. Results from different areas can be compared to set priorities for remedial action and to help develop strategies for future management. A key feature of the ARM is that it is very flexible, such that each step of the assessment can incorporate a range of simple to more sophisticated approaches depending upon the users' needs and catchment characteristics.

Unlike many existing groundwater or surface water yield methodologies, the ARM does not separate the management of groundwater and surface water, as this can be unhelpful when seeking to protect and manage river flows that integrate both groundwater base-flow and surface run-off. Instead the resource assessment is based on river flows or groundwater outflows from a catchment and the artificial impacts on these. These flows are then compared with minimum acceptable flows for the area under study to help derive an abstraction sustainability status. Gauged river flows are the preferred focus for the method although flow estimation techniques can be used where no gauged data are available, albeit with higher uncertainties. In catchments where there is no or negligible hydraulic connection between surface drainage and the groundwater, then groundwater-only assessments are possible (e.g. for confined or coastal aquifers) but (as with any approach) will be subject to much greater uncertainties.

The ARM seeks to improve on the more traditional methods of groundwater resource assessment by explicitly addressing seasonal variations in the availability of groundwater resources. Whereas more traditional methods typically produce a single annual figure of groundwater resource availability, the ARM produces resource estimates for individual months. Furthermore, the ARM allows a tiered approach to water resource assessment by accommodating both simple techniques, that produce a relatively rapid overview of resource availability, and more sophisticated techniques such as numerical groundwater modelling that can deliver a more detailed assessment.

3 APPLICATION OF THE ARM

The principal steps of the ARM, as they would apply to a river that has a significant base-flow component and for which gauged flow data are available, are set out below. For this illustration, an individual year has been selected on which to base the assessment.

- Define assessment area, delineate catchment and sub-catchments/river reaches
- Record conceptual understanding of catchment hydrology and key processes
- Specify assessment year e.g. 1992
- Plot gauged river flows for selected year, 1992
- Estimate artificial impacts on river flows (abstractions and discharges)
- Estimate and plot naturalised river flows for 1992
- Estimate and plot natural base-flows for 1992 and estimate groundwater recharge
- Define and plot minimum acceptable river flows for 1992
- Calculate and plot surplus or deficit resource for 1992 by subtracting gauged flows from minimum acceptable flows
- Check groundwater abstractions are acceptable relative to recharge/baseflow
- Include uncertainty analysis for input data and determine abstraction sustainability status as shown in Figure 1.

Note that the analysis and plots can be produced for several points on the river to help determine the relative sustainability status for various sub-catchments/river reaches. Other, optional, steps include:

- Calculate the maximum impact by abstraction that can be permitted at each gauge site by subtracting the minimum acceptable flows from the naturalised flows. However, care must be taken in interpreting this figure due to the inherent time lags and seasonality associated with the impact of many groundwater abstractions.
- Plot river flows or base-flows and artificial influences as compound bar charts in Ml/d. These plots illustrate the relative importance of artificial influences at the various assessment points on the river and the pattern of flow accretion along the river.

For rivers or reaches for which flow data are unavailable, the above process is varied to include flow

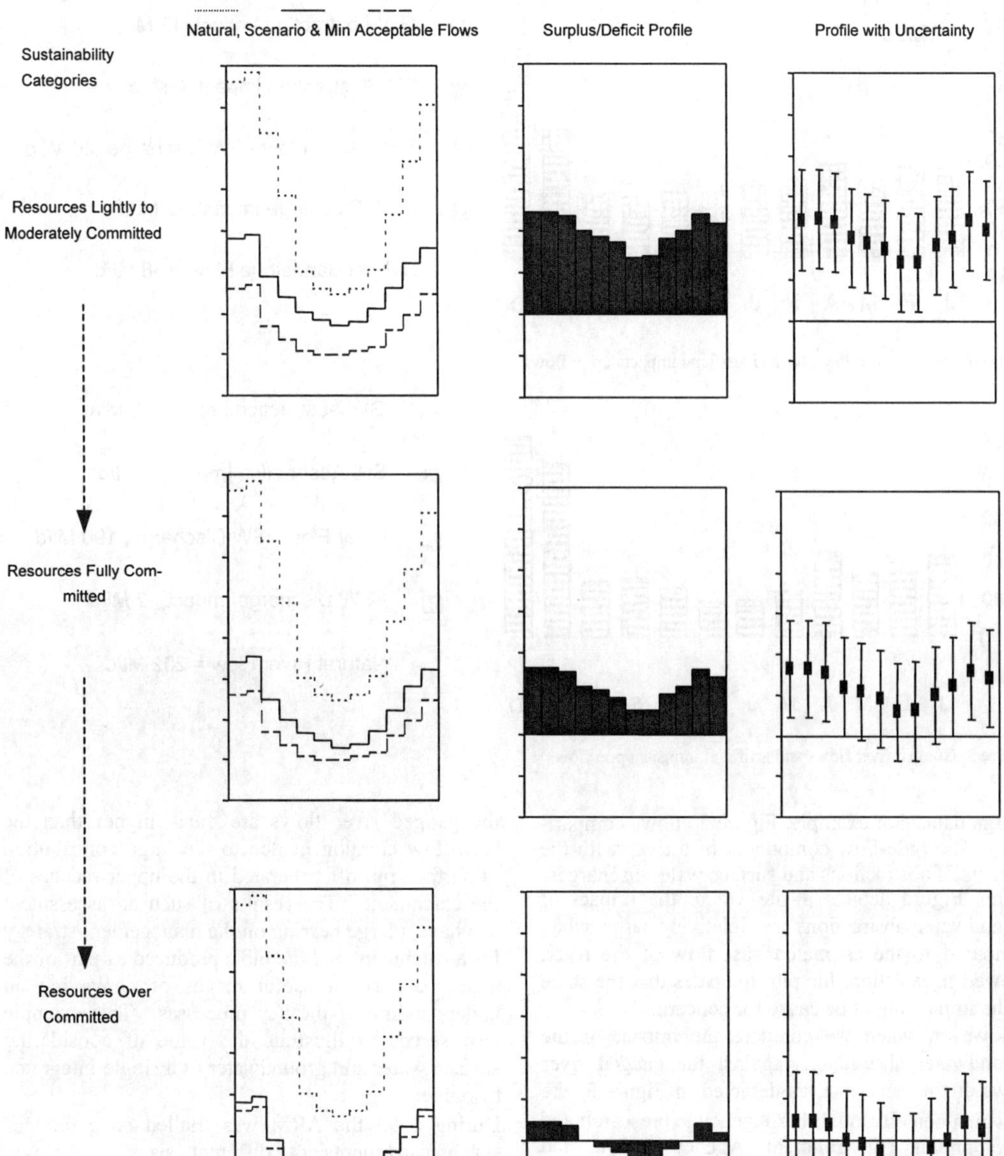

Figure 1. ARM hydrographs and sustainability status.

estimation techniques to first determine a naturalised flow. Thereafter, estimation of the impacts of artificial influences (abstractions and discharges) can be used to derive an estimate of actual flow. In the case of an aquifer that is not in hydraulic connection with a river, groundwater outflows are the focus of the assessment rather than river flows.

A series of spreadsheets support the application of the ARM and provide an audit trail for the assessment. The spreadsheets record a simple conceptual understanding and all hydrological and artificial impact data used in the ARM. They incorporate procedures for flow naturalisation, impact estimation and means of setting minimum acceptable flows, all

of the resource assessment. Many items are available to plot as compound bar charts. These include: river flows, base-flows, abstraction data (surface water and groundwater sources), and surface water dis-

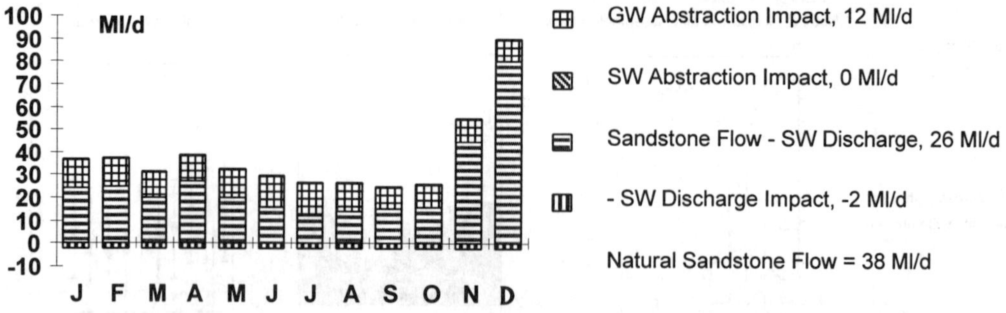

Figure 2. Base-flow of the river and artificial impacts upon flows.

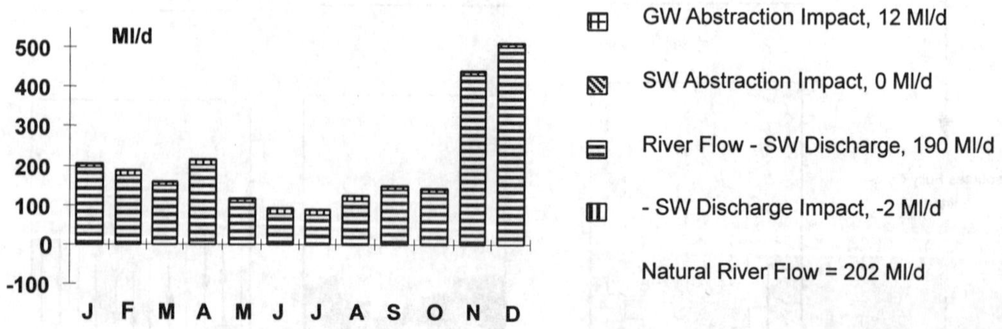

Figure 3. Guaged river flows and artificial impacts upon flows.

charge data. For example, Figure 2 allows comparison of the base-flow component of a river with the impacts of abstractions and surface water discharges.

This Figure depicts a site where the impact of groundwater abstractions are relatively large when compared to the estimated base-flow of the river. Viewed in isolation this plot indicates that the scale of the impact might be cause for concern.

However, when we compare the impact of the groundwater abstractions against the gauged river flows at the same site, as depicted in Figure 3, the of which benefit from other Agency led research and development (Environment Agency 1999a and 1999b, National Rivers Authority 1996). This complementary research is required to provide standard procedures and guidance on best practice to help ensure a high level of consistency in the application of the ARM or indeed other methods of water resource or impact assessment. The spreadsheets also incorporate a wide variety of plots designed to aid analysis and reporting, this includes those presented in this paper.

The various plots produced by the ARM spreadsheet are designed to convey a clear picture of the relative importance of each of the main components impact of the groundwater abstractions are clearly only a small proportion of the total flow and hence are acceptable. This reflects the fact that at this site the gauged river flows are much higher than the base-flow component due to the large contribution of surface run-off generated in the upper reaches of the catchment. The results of such an assessment can have a large bearing on the management strategy for a catchment and the plots produced as part of the assessment are a useful means of conveying an understanding of the key processes. This example also serves to illustrate the value of considering surface water and groundwater as a single integrated resource.

During 1999 the ARM was trialled on a total of seven catchments of different sizes, hydrology, geology, and level of artificial influence. The results of the trials are encouraging, with resource and abstraction sustainability assessments being successfully delivered in each case. A series of project reports, user guide and spreadsheet tools have been produced by the National Groundwater and Contaminated Land Centre of the Environment Agency to support the future application of the methodology. The implementation of the ARM is currently being considered by a multi-functional working group of the Environment Agency.

4 CONCLUSIONS

1 The ARM provides a consistent but flexible framework to help in the assessment and sustainable management of water resources across a wide variety of catchment types.
2 There are clear benefits in treating groundwater and surface water as an integrated resource from the outset when carrying out resource assessments that focus on river flows.
3 The ARM's supporting spreadsheets allow simple graphical reporting of river flows (or groundwater outflows), minimum acceptable flows, and artificial impacts, that convey a clear picture of the key processes involved in the resource and sustainability assessment.
4 Further improvements in the consistency of results from the ARM are expected from the continuing development of standard techniques and best practice guidance for items such as: flow naturalisation, the estimation of impacts of groundwater abstractions on river flows, and procedures for the determination of minimum acceptable river flows.

REFERENCES

Environment Agency, 2000. The available resource methodology: a framework for assessing water resources and abstraction sustainability. *National Groundwater and Contaminated Land Centre project report NC/99/68, Solihull, UK.*Environment Agency, 1999a. Good practice guideline for flow naturalisation. *Internal report in preparation, Bristol, UK.*

Environment Agency, 1999b. Impact of groundwater abstractions on river flows. *National Groundwater and Contaminated Land Centre project report NC/06/28, Solihull, UK.*

National Rivers Authority, 1996. Determination of Minimum Flows. *R&D Note 449, Bristol, UK.*

DISCLAIMER

Any opinions expressed herein are those of the authors and do not necessarily reflect the views of their respective organisations.

Groundwater: Past Achievements and Future Challenges, Sililo et al. (eds) © *2000 Balkema, Rotterdam, ISBN 90 5809 159 7*

Information on interactions between groundwater and vegetation relevant to South African conditions: A review

D.C. Le Maitre, D.F. Scott & C. Colvin
Division of Water, Environment and Forestry Technology, Stellenbosch, South Africa

ABSTRACT: Sustainable management of groundwater resources requires understanding both how vegetation influences recharge and how groundwater discharges influence ecosystems. Changes in vegetation cover and structure can have a significant impact on groundwater recharge by altering components of the hydrological cycle such as interception and transpiration. Groundwater discharge can also have an important impact on ecosystems, through, for example, baseflow in rivers, dry season discharge into wetlands or springs and in special situations such as cave ecosystems. Interactions between groundwater and vegetation appear to be generally more pervasive and important than was believed in the past. Research is needed to develop our understanding of these ecosystems and their groundwater dependence and to provide guidelines aimed at ensuring that these ecologically and often economically important ecosystems are conserved.

1 INTRODUCTION

There is increasing interest in the interactions between vegetation and groundwater, both in the impacts of vegetation on recharge and in the impacts of discharge on ecosystems. There is also growing recognition of the ecological importance of groundwater-dependent ecosystems and that they are remarkably diverse, complex, and fragile. The sustainability of the world's freshwater resources depends strongly on the maintenance of the ecological systems and processes that regulate the flows, stabilize soils and maintain water quality (Pimentel et al. 1997, Chichilnisky & Heal 1998, Gleick 1998).

Traditionally, research has tended to focus on surface water hydrology, and the associated aquatic ecosystems, or on groundwater itself and not on the interfaces between these water systems. In Australia and South Africa recent changes in the Water Law and the approach to the management of water resources to embrace the entire water cycle have highlighted the lack of studies of the interactions between groundwater and vegetation (Hatton et al. 1998, Xu et al. 2000). This paper summarises some of the main findings of a recent review of groundwater-vegetation interactions (Le Maitre et al. 1999) and subsequent work in applying the approach developed by Hatton et al. (1998) for interpreting groundwater dependence in South African ecosystems.

2 IMPACTS OF VEGETATION ON RECHARGE

Vegetation affects the proportion of the rainfall that reaches and infiltrates the soil, percolates through the soil and recharges aquifers. This occurs at several stages:

- A proportion of the rainfall is intercepted and redirected by the plant canopy and branches. Some of the water evaporates directly from the plant surface (interception) and some is redirected as stem flow or canopy drip. The proportion lost through interception depends on the surface area of the plant canopy and on the drop size, intensity, and duration of the rainfall (Larcher 1983). The losses vary from about 5% to 20% of the rainfall under South African conditions but may be considerably higher in situations with light rains and dense canopies (Le Maitre et al. 1999). Stemflow concentrates the water entering the soil and may influence infiltration rates and efficiencies (Navar & Ryan 1990).
- Organic matter from plants influences recharge in two ways: Litter layers protect the soil but also intercept rainfall which then evaporates, and organic matter improves soil structure and promotes infiltration (Dawson 1993). This is most evident on degraded land where infiltration may be 2-10 times less than under good vegetation cover (Macdonald 1978, O'Connor 1985).
- Root systems of plants may provide pathways for preferential flow (Allison & Hughes 1983, Sharma et al. 1987; Sandström 1996) and roots

extract soil and tap the capillary fringe to extract groundwater for the plants to transpire. Transpiration accounts for 45-80% of the rainfall, but in dry years, or areas with less than about 400 mm of rainfall per year, can account for all the rainfall (Larcher 1983, Zhang et al. 1999). Replacement of grassland with trees (e.g. commercial plantations) can significantly reduce streamflow or even dry up streams completely (Dye 1996). The drier soil profile under the plantations can absorb more of the rainfall, decreasing or preventing groundwater recharge – the converse occurs when trees are cleared (Salama et al. 1993, Tyler & Walker 1994, Zhang et al. 1999).

The net effects on recharge with a conversion from grassland to forest vary from 2-10 fold and even tc complete cessation of recharge (Le Maitre et al. 1999).

3 DEPTHS OF ROOT SYSTEMS

One of the reasons why the effects of vegetation on groundwater and vice-versa have tended to be overlooked is a lack of information on plant root systems, particularly the depths they can reach. Research has tended to focus on the shallow roots (<0.5 m deep) because these comprise most of the volume or weight of the root system (Stone & Kalisz 1991). Sinker roots may only be a small fraction of the root system, but they are highly specialized to transport water (Dawson & Pate 1996). The typical depths of plant root systems are much greater than was generally appreciated (Figure 1); some of the deepest known root systems (>60 m) have been recorded in the Kalahari sands (Jackson et al. 1996, Canadell et al. 1996).

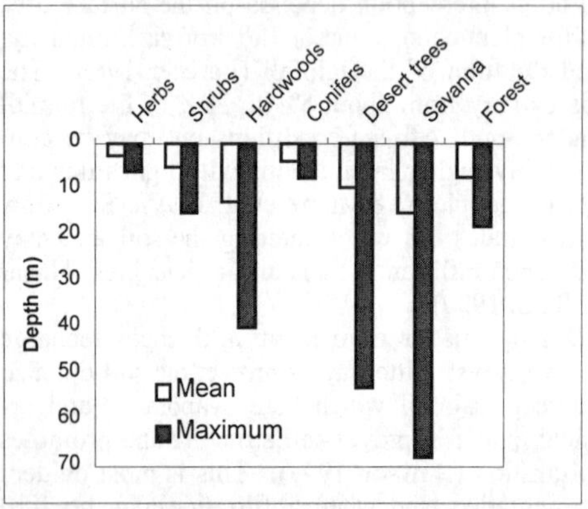

Figure 1. Depths of root systems for different plant growth forms and types of trees (after Canadell et al. 1996).

4 IMPACTS OF GROUNDWATER DISCHARGE ON ECOSYSTEMS

The main source of information on this topic comes from the limited research into the impacts of groundwater abstraction on ecosystems, particularly wetlands and riparian vegetation because there have been few studies of ecosystem dependence as such (Hatton et al. 1998, Le Maitre et al. 1999).

4.1 *Ecological background*

Some basic tenets or principles in ecology need to be dealt with before groundwater dependence can be defined:

- Ecosystems will generally utilise resources in proportion to their availability - whether the resource be water, light, nitrogen or anything else - and the availability of different resources will be a significant determinant of their structure, composition and dynamics (Sousa 1984). Thus, where groundwater is accessible ecosystems will develop some degree of dependence on it and that dependence is likely to increase with increasing aridity of the associated environment (Hatton et al. 1998).
- Ecosystems exist in a dynamic state and are continuously being altered by disturbances or perturbations; and the nature, timing and magnitude of those disturbances can have profound effects on the future state of those ecosystems (Sousa 1984). Changes in groundwater availability will, therefore, alter the affected ecosystems, but the nature of the responses is essentially unknown. The changes in ecosystem structure or functioning may be directly proportional to the changes in groundwater availability, linear or non-linear, or they may take the form of abrupt changes as critical thresholds are exceeded. The responses may also be similar to those described for state and transition models where ecosystems persist in a particular state or condition and then abruptly switch to a new condition (Westoby et al. 1989). Once the switch has occurred a high degree of intervention may be needed to return the system to its previous state.

4.2 *Groundwater dependence*

The next step is to define groundwater from the perspective of ecosystem dependence. Defining groundwater as all underground water or saturated soil, regolith or rock aquifers and including periodic saturation (e.g. perching) is not useful for understanding groundwater dependence. A more useful definition is: *that water in the system which would be unavailable to plants and animals were it to be extracted by pumping* (Hatton et al. 1998). This includes both lowering of water tables and decreasing discharge rates and has to take account of variability (e.g. inter-annual) and seasonality (e.g. ecosystems dependent on base flows). Groundwater dependence can now be defined as: *the degree of dependence is proportional to the fraction of the water budget that the ecosystem derives from groundwater* (Hatton et al. 1998).

The review by Hatton et al. (1998) of groundwater dependent ecosystems in Australia identified four categories of groundwater dependence (Table 1).

Table 1. A classification of the groundwater dependence of ecosystems (after Hatton et al. 1998).

Australian classification	Groundwater dependence	Type of ecosystem
Aquifer & cave	Entirely to high	Typical wetlands (where above ground level), cave ecosystems
Wetlands	Variable but generally high, sometimes entirely	This category includes typical wetlands, river floodplains, endorhoeic systems, estuaries and vegetation dependent on spring discharge
River base flow	Variable, high to proportional	Riverine (riparian vegetation but may include floodplain)
Terrestrial	Variable, high to none apparent	Other forms of discharge areas -near surface water tables (<10-20 m deep?), cryptic discharge

This approach must still be evaluated and tested under South African conditions but it looks useful. In terms of the South African National Water Act (with its focus on aquatic ecosystems) riverine, estuarine and wetland ecosystems form part of the Ecological Reserve but terrestrial ecosystems do not. Protection for these ecosystems needs to be sought in terms of the Resource Quality Objectives (Xu et al. 2000).

Most of South Africa is classified as semi-arid or arid and a prominent feature of many of the seasonal or ephemeral river systems in this environment is the riverine vegetation. These ecosystems, like those elsewhere in the world are critical habitats for biodiversity, offering refuges and habitat for a variety of organisms (Naiman et al. 1993, Milton 1990, Milton & Dean 1999). The convergence of surface and groundwaters in floodplains and hyporhoeic zones is an important factor determining landscape morphologies and their biodiversity and productivity (Stanford & Ward 1993). This is partly a consequence of groundwater carrying higher quantities of nutrients than surface waters, with the latter usually being better oxygenated. The upwelling creates patches of high productivity in the hyporhoeic zones and aquatic systems which support greater animal densities and diversity compared with non-upwelling situations.

4.3 *Evidence of groundwater dependence*

The groundwater dependence of river floodplain and wetland ecosystems has been reasonably well documented (Le Maitre et al. 1999). Studies in Australia have shown that floodplain trees may largely or exclusively tap groundwater even where surface water is available (Dawson & Ehleringer 1991). Many riparian species, particularly phreatophytes, which have deep roots and tap the capillary fringe, are highly sensitive to manipulations of the water table and reductions in base flow (Smith et al. 1991, Mahoney & Rood 1992, Bacon et al. 1993). Anecdotal observations suggest that invading *Prosopis* trees may outcompete *Acacia erioloba* for groundwater on ephemeral rivers in the Kalahari (H. Zimmerman pers. comm. 1999).

One documented case is based on observations in the Kuiseb River in Namibia from 1979/80 to 1982/83. During this period the flows in the river did not reach the delta, and the piezometric surface in the alluvial aquifer dropped by 3 m (Ward & Breen 1983, Bate & Walker 1993). A number of large *Faidherbia (Acacia) albida* trees (riparian fringe woodland) died and the growth and vitality of riverine vegetation declined. Localised stands of young *F. albida* (established in 1974 and 1976) did survive suggesting that the large dead trees had lost their ability to adjust to the lowering of the water table. *Acacia erioloba*, a non-riparian species, did not show any signs of mortality, presumably because it is better able to track changing water tables. *Faidherbia albida* seedlings apparently were able to keep their roots in contact with the falling water table unlike the unlike the mature trees.

4.4 *Viewing things in perspective*

Although it is clear that groundwater dependent ecosystems exist and are important, there is also a need to find a compromise between total environment protection and economic development. This will require ecologists to determine the ecological significance of these ecosystems. Some criteria could include: uniqueness (e.g. biodiversity, cultural, aesthetic and socio-economic value); and the expected or likely vulnerability (sensitivity) to change.

5 CONCLUSIONS

Interactions between groundwater and vegetation are clearly important. Vegetation changes can have significant impacts on recharge and groundwater discharge, whether obvious as springs or baseflow or cryptic through roots tapping capillary fringes, and this, in turn has significant effects on ecosystems. Vegetation can decrease recharge by depleting soil moisture throughout the root zone or increase it by promoting infiltration and preferential flow. Deep root systems are pervasive and play an important role in groundwater vegetation interactions as they conduct the water lost through transpiration which, in turn accounts for 45-80% or even more of the rainfall. The effects of deep root systems will be greatest in vegetation types with woody shrubs and trees such as forest, savanna and plantations.

The requirements of ecosystems that depend on groundwater are poorly understood, yet this is an issue that we shall increasingly be called on to address in South Africa. Surface and groundwater hydrologists and ecologists need to actively seek ways of enhancing their understanding of the interactions between groundwater and vegetation with the aim of supporting the sustainable utilisation of South Africa's limited but vital water resources and the ecosystems that provide key economic benefits to society.

REFERENCES

Allison, G.B. & M.W. Hughes 1983. The use of natural tracers as indicators of soil-water movement in a temperate semi-arid region. *Journal of Hydrology* 60: 157-173.

Bacon, P.E., C. Stone, D.L. Binns, D.J. Leslie & D.W. Edwards 1993. Relationship between water availability and *Eucalyptus camaldulensis* growth in a riparian forest. *Journal of Hydrology* 150: 541-561.

Bate, G.C. & B.H. Walker 1993. Water relations of the vegetation along the Kuiseb River, Namibia. *Madoqua* 18: 85-91.

Canadell, J., R.B. Jackson, J.R. Ehleringer, H.A. Mooney, O.E. Sala, & E.-D. Schulze 1996. Maximum rooting depth of vegetation types at the global scale. *Oecologia* 108: 583-595.

Chichilnisky, G. & G. Heal 1998. Economic returns from the biosphere. *Nature* 391: 629-630.

Dawson, T.E. 1993. Woodland water balance. *Trends in Ecology and Evolution* 8: 120-121.

Dawson, T.E. & J.R. Ehleringer 1991. Streamside trees that do not use stream water. *Nature* 350: 335-337.

Dawson, T.E. & J.S. Pate 1996. Seasonal water uptake and movement in root systems of Australian phreatophytic plants of dimorphic morphology: a stable isotope investigation. *Oecologia* 107: 13-20.

Dye, P.J. 1996. Climate, forest and streamflow relationships in South African afforested catchments. *Commonwealth Forestry Review* 75: 31-38.

Gleick, P.H. 1998. The human right to water. *Water Policy* 1: 487-503.

Hatton, T., R. Evans & S.K. Merz 1998. *Dependence of ecosystems on groundwater and its significance to Australia.* Occasional Paper No 12/98, Land and Water Resources Research and Development Corporation, CSIRO Australia.

Jackson, R.B., J. Canadell, J.R. Ehleringer, H.A. Mooney, O.E. Sala & E.-D. Schulze 1996. A global analysis of root distributions for terrestrial biomes. *Oecologia* 108: 389-411.

Larcher, W. 1983. *Physiological plant ecology.* Springer-Verlag, Berlin.

Le Maitre, D.C., D.F. Scott & C. Colvin 1999. A review of information on interactions between vegetation and groundwater. *Water SA* 25: 137-152.

Macdonald, I.A.W. 1978. Pattern and process in semi-arid grassveld in Rhodesia. *Proceedings of the Grassland Society of South Africa* 13: 103-109.

Mahoney, J.M. & S.B. Rood 1992. Response of a hybrid poplar to water table decline in different substrates. *Forest Ecology and Management* 54: 141-156.

Milton, S.J. 1990. Life styles of plants in four habitats in an arid Karoo shrubland. *South African Journal of Ecology* 1: 63-72.

Milton, S. & R. Dean 1999. Big old camelthorn trees. Key resource for wildlife. *African Wildlife* 53(5): 10-12.

Naiman, R.J., H. Decamps & M. Pollock 1993. The role of riparian corridors in maintaining regional biodiversity. *Ecological Applications* 3: 209-212.

Navar, J. & R. Ryan 1990. Interception loss and rainfall redistribution by three semi-arid growing shrubs in northeastern Mexico. *Journal of Hydrology* 115: 51-63.

O'Connor, T.G. 1985. *A synthesis of field experiments concerning the grass layer in the savanna regions of southern Africa.* Report No 114, South African National Scientific Programmes, Foundation for Research Development, Pretoria.

Pimentel, D., J. Houser, Preiss, O. White, H. Fang, L. Mesnick, T. Barsky, S. Tariche, J. Schreck & S. Alpert 1997. Water resources: agriculture, the environment and society. *BioScience* 47: 97-106.

Salama, R.B., P. Farrington, G.A. Bartle & G.D. Watson 1993. Salinity trends in the wheatbelt of Western Australia: results of water and salt balance studies from Cuballing catchment. *Journal of Hydrology* 145: 41-63.

Sandström, K. 1996. Hydrochemical deciphering of streamflow generation in semi-arid East Africa. *Hydrological Processes* 10: 703-720.

Sharma, M.L., R.J.W. Barron & D.R. Williamson 1987. Soil water dynamics of lateritic catchments as affected by forest clearing for pasture. *Journal of Hydrology* 94: 29-46.

Smith, S.D., A.B. Wellington, J.L. Nachlinger & C.A. Fox 1991. Functional responses of riparian vegetation to streamflow diversion in the eastern Sierra Nevada. *Ecological Applications* 1: 89-97.

Sousa, W.P. 1984. The role of disturbance in natural communities. *Annual Review of Ecology and Systematics* 15: 353-391.

Stanford, J.A. & J.V. Ward 1993. An ecosystem perspective of alluvial rivers: connectivity and the hyporhoeic zone. *Journal of the North American Benthological Society* 12: 48-60.

Stone, E.L. & P.J. Kalisz 1991. On the maximum extent of roots. *Forest Ecology and Management* 46: 59-102.

Tyler, S.W. & G.R. Walker 1994. Root zone effects on tracer migration in arid zones. *Soil Science Society of America Journal* 58: 25-31.

Ward, J.D. & C.M. Breen 1983. Drought stress and the demise of *Faidherbia albida* along the lower Kuiseb River, central Namib Desert: preliminary findings. *South African Journal of Science* 79: 444-447.

Westoby, M., B. Walker & I. Noy-Meir 1989. Opportunistic management for rangelands not at equilibrium. *Journal of Range Management* 42: 266-274.

Xu, Y., E. Braune, C. Colvin, D. Le Maitre, K. Pietersen, T. Hatton 2000. Comprehensive determination of the Resource Directed Measures for groundwater in South Africa. *Proceedings IAH XXX Congress.* Cape Town: Balkema.

Zhang, L., W.R. Dawes & G.R. Walker 1999. Predicting the effect of vegetation changes on catchment average water balance. Technical Report No. 99/12, Cooperative Research Centre for Catchment Hydrology, CSIRO Land and Water, CSIRO, Australia.

Groundwater: Past Achievements and Future Challenges, Sililo et al. (eds) © 2000 Balkema, Rotterdam, ISBN 90 5809 159 7

Failure of rural schemes in South Africa to provide potable water

G. Mackintosh & C. Colvin
Cape Water Programme, Council for Scientific and Industrial Research, Stellenbosch, South Africa

ABSTRACT: The impact of water borne disease in South Africa is significant. An estimated 43 000 deaths per annum, including 20% of deaths in the one to five age group, are directly attributable to diarrhoeal diseases. Drinking water quality provision in many rural areas is substandard. This paper describes the results of sampling drinking water supplies in rural communities in the Western and Eastern Cape, South Africa. The majority of samples collected failed microbial drinking water quality standards. Overall, schemes dependent on groundwater provided a worse quality water at point of use than surface water dependant schemes. This is thought to be the result of pump breakdown, deterioration of the storage and reticulation system, and insufficient monitoring and management of the schemes. Importantly, it is shown that the implementation of well considered, community accepted drinking-water quality management procedures can effectively change an unacceptable water quality to one which satisfies drinking-water specifications.

1 INTRODUCTION

In almost all South African metropolitan areas the consumer is provided with a high quality drinking-water. However, in many rural communities, the situation is very different. In 1996 an estimated 12 million South Africans were without access to adequate water supply (DWAF 1996) and, even where water supply exists, quality is often poor (Mackintosh & Linde 1997) and the impact of water-borne disease is significant. In South Africa, diarrhoea is responsible for some 20% of all deaths in the one to five age group (Bourne & Coetzee 1996) and an annual estimated 43 000 deaths and 3 million incidences of illness, with an associated treatment cost of some R3.4 billion (approximately US$ 523 million, April 2000) (Pegrum et al. 1998).

Groundwater plays an important role in meeting basic human needs for drinking water. Well-managed groundwater supply schemes reliably provide water of high microbiological and physico-chemical quality. Typically, groundwater is advantageous for rural water supply in that it is protected from surface contamination and abstraction can be maintained during (short-term) periods of drought. However, where poorly designed groundwater schemes are installed, users are often supplied with water of very poor quality. The challenge exists for hydrogeologists and water utilisation engineers to ensure sustainable high-quality drinking-water provision.

2 THE PURPOSE OF THIS PAPER

This paper has three main objectives:

- to document typical drinking-water microbial quality in existing rural water supply schemes in the Eastern and Western Cape, South Africa;
- to draw attention to the very poor drinking-water quality that results from a high proportion of the groundwater supply schemes assessed,
- to introduce the concept of "drinking-water quality management" procedures for implementation in rural communities and show that their introduction can change an unacceptable drinking-water quality to one which satisfies drinking-water specifications.

3 RURAL DRINKING WATER QUALITY INVESTIGATION

3.1 *Selection of rural supply schemes*

Rural communities of both the Eastern and Western Cape of South Africa were studied. The rural communities of these two provinces differ considerably. The population of the Eastern Cape is largely non-urban, poor and with inadequate water supply infrastructure. Rural communities of the Eastern Cape comprise both scattered villages and subsistence farmers, and formalised towns serving subsistence farmers. The poverty rate in 1998 was 71%, and only 25% of the households had a tap inside the dwelling (Mey 1998 & CSS 1998). The Western Cape is largely urban, relatively wealthy, and with relatively well developed water supply infrastructure. Rural communities of the Western Cape typically are formalised towns connected to commercial agriculture and enterprise. The poverty rate in 1998 was 28%, and 76.4% of households had a tap inside the dwelling (Mey 1998 & CSS 1998).

Rural communities typical of each region were identified. Communities selected included both smaller communities of about 1 000 to 4 000 inhabitants, and larger communities of about 40 000 inhabitants. Three larger communities, and 16 smaller communities were studied in the Eastern Cape. Two larger communities, and 18 small communities were studied in the Western Cape.

3.2 *Sampling*

Once-off sampling was carried out in the selected communities. Sampling was focussed on points of use (eg. tap stands, homes, schools, municipal offices and hospitals). Storage reservoirs and the distribution system were also sampled, but no well-head samples were obtained. The number of samples varied with the size of community.

3.3 *Selection of water quality variables*

Microbiological testing was limited to SABS 241-1984 guideline indicator organisms; namely heterotrophic plate count, total coliform and faecal coliform counts (SABS 1984 & 1990). In addition, chemical and physical determinants that are linked to the survival of micro-organisms (such as pH, temperature and turbidity), and those which provide an indication as to the condition of the distribution network (calcium, total alkalinity, iron and electrical conductivity) were measured according to standard techniques (APAH, AWWA & WPCF 1995)

Table 1. South African Bureau of Standards microbial drinking-water quality limits (1984).

Determinands	Recommended Limit	Maximum allowable Limit
Heterotrophic plant count (per 1 mL)	100	None
Total coliforms (per 100 mL)	0	5
Faecal coliforms (per 100 mL)	0	0

3.4 *Further observations*

Assessment included consideration of the type of raw water source, water supply management, disinfection practises, other water treatment practices and water quality monitoring practices.

4 RESULTS

A total of 39 communities were assessed, 19 in the Eastern Cape. In each province, groundwater was the sole water source for 7 communities.

In the Eastern Cape, 5 of the 7 groundwater supply schemes were non-functional (71%) and users were forced to procure alternative untreated water supplies, whereas in the Western Cape all groundwater schemes visited were functional.

In both regions a high number of schemes have no disinfection capability. In the Eastern Cape, all surface water schemes visited practised forms of flocculation and settling as a result of the high turbidity of surface waters in the region. In the Western Cape, where mountain catchment's streams generally have low turbidity, a large number of the communities practise no additional water treatment other than disinfection (65%).

4.1 *Limitations of the data*

In addition to the microbiological results presented in this paper, physico-chemical results were obtained. However, it is the aim of this paper to indicate broadly the drinking-water quality of rural schemes, and in this regard the microbiological indicators were the critical parameters. Figures 1 and 2 present the results in a qualitative manner with respect to compliance or non-compliance with microbial guidelines. The results are not correlated to aquifer types, abstraction methods, land-use, point of use, etc. Time series monitoring would also be required to give a further indication of the mechanisms of scheme failure. In both provinces fewer samples from groundwater schemes were obtained. It is recommended that a further phase of study pursue these issues. Nevertheless, despite these quantitative limitations to the data, the results indicate the existing state of the schemes visited, their vulnerability to contamination and the general state of drinking-water quality management.

4.2 *Overall drinking water quality*

The overall efficiency of drinking-water provision in the Eastern Cape regarding microbiological water quality is very poor with some 50% of samples failing the Maximum Allowable Limit (MAL) (See Figure 1). Figure 1 testifies that of the worst cases of drinking-water quality exist in failed groundwater supply schemes where consumers were forced to make do with very poor alternatives (eg. ponds)

The overall efficiency of drinking-water provision in the Western Cape regarding microbiological water quality is very poor with some 62% of samples failing MALs (Fig. 2). Surprisingly, the percentage failure of MALs was higher in the Western Cape, despite a higher standard of living and level of supply infrastructure.

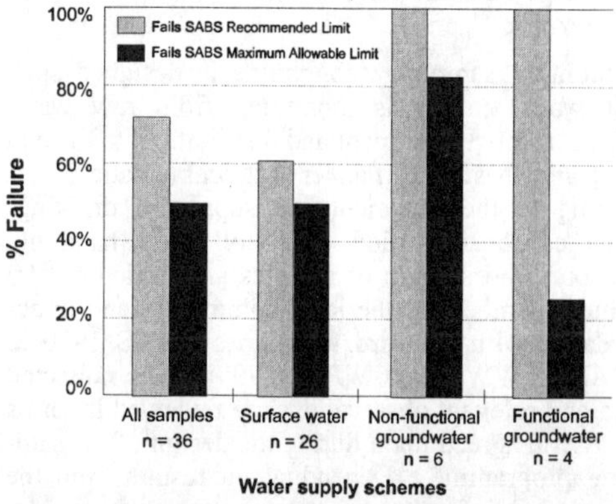

Figure 1. % Failure regarding microbiological indicator organ isms of drinking-water quality as tested at rural water supply schemes in the Eastern Cape, South Africa

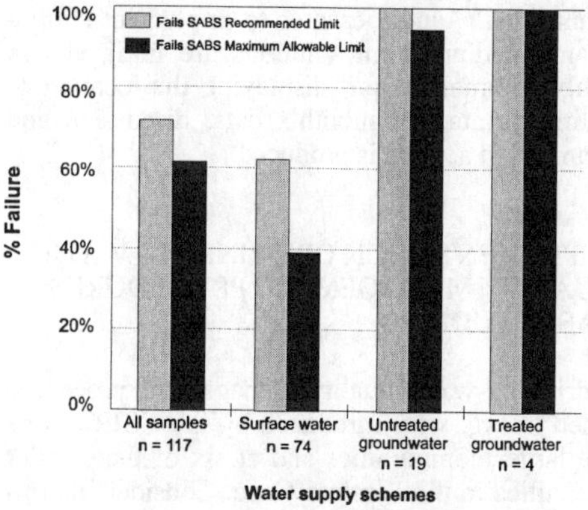

Figure 2. % Failure regarding microbiological indicator organisms of drinking-water quality as tested at rural water supply schemes in the Western Cape, South Africa.

4.2.1 *Groundwater quality*

The microbiological quality of samples from groundwater schemes was worse than surface water samples. All groundwater samples failed the recommended limits. In the Western Cape almost all groundwater samples failed the MALs. In the Eastern Cape, the majority of non-functional schemes failed MALs. Three of the four samples from the two functional schemes passed the MALs.

4.2.2 *Monitoring*

Only 26% of communities visited in the Eastern Cape monitor drinking- water quality; in nearly all cases this is limited to *ad hoc* microbiological sampling. In the Western Cape some 80% of communities monitor drinking- water quality. Again in most instances this is *ad hoc* and ineffective, as evidenced by the generally very poor microbiological quality of drinking-water.

4.3 *Summary of results*

In brief, the investigation has shown that drinking-water quality is very poor in rural communities in both the Eastern and Western Cape, with a high incidence of failure regarding MALs for microbiological indicator organisms. Additional points of interest are:

- in the more developed Western Cape, almost all groundwater samples failed the MALs;
- non-functional groundwater schemes in the Eastern Cape resulted in sub-standard supplies being used;
- the equally poor overall microbiological drinking-water quality in both regions, despite the Western Cape being more affluent, and having better services;
- in both regions drinking-water quality management is inadequate.

5 DISCUSSION

The failure of rural groundwater schemes to deliver acceptable quality drinking water to consumers has been shown in several areas of Southern Africa previously.

For example, in 1991 a survey of almost all the Transkei government water supply schemes revealed only 10% to be delivering on acceptable quality and quantity of water (Ravenscroft & Cain 1997). Observations included that almost all boreholes with windpumps were not functional; functional schemes relied on handpumps and springs. In many of the former 'homeland' areas, we are left with a legacy of inappropriate delivery systems.

Observed reasons for the failure of rural groundwater schemes to deliver adequate quality drinking water are:

- failure of the pump (particularly windpumps) forcing people to use water from contaminated sources;
- contamination via the reticulation system, particularly where corrosive waters are abstracted (such as in the Western Cape);
- contamination via the storage reservoir (many are open and others have broken roofs with nesting birds, etc);

- failure to chlorinate; and
- In particular, inadequate or non-existent water supply management practices fail to timeously prevent or remedy microbial pollution.
- Where water quality monitoring was carried out it was typically *ad hoc,* and the results had little or no impact on supply management practices.

Microbiological sampling and analyses pose particular problems for remote rural areas. Current practices are very costly and there is a high risk that quality assurance procedures are not adhered to. The current drive internationally towards developing rapid field detection methods for microbial quality indication would benefit rural water supply management tremendously.

6 DEVELOPMENT OF RURAL DRINKING-WATER QUALITY MANAGEMENT PROCEDURES

A common denominator to the poor water quality results observed above is the lack of regular assessment/management of water quality, including source protection, optimisation of water treatment, ensuring reliability of supply, and the maintenance of drinking-water quality within the distribution network. An appropriate procedure was developed to address these problems as outlined below.

6.1 *Stage One: Initial familiarisation with water supply scheme*

Basic information relating to the water scheme is collected: water sources, vulnerability of sources, water treatment requirements and existing treatment procedures, the drinking-water distribution network, drinking-water quality records and existing drinking-water quality management procedures and local capacity.

6.2 *Stage Two: Preliminary water quality investigation*

Potentially problematic issues and areas where historical data is insufficient are targeted. Special consideration is given to sampling raw water sources, post-water treatment works, network dead-ends, high occupancy buildings, hospitals and schools, areas perceived to be problematic and any regions using untreated water.

6.3 *Stage Three: Define operations, local management team and responsibilities*

A series of meetings are held with local operators responsible for drinking-water supply. The operations and management structure, and associated responsibilities, are clearly defined. Focus is given to skills training, capacity building, technical support, planning and facilitation of project implementation. Consensus on community responsibilities is negotiated with the objective of maximising community involvement and responsibility for operation of the water supply scheme.

6.4 *Stage Four: Implementation of monitoring process*

A monthly sampling programme is designed such that water quality is monitored from raw water source, through treatment and distribution, to the end user; and in such a manner that problematic issues relating to the treatment and supply of drinking-water can be identified. The collection, handling, transport and storage of samples are carried out by trained members of the local community and in accordance with standard techniques (SABS 1990 & APAH, AWWA and WPCF 1995). The collected data and relevant observations are reviewed in terms of criteria agreed upon during the design of the sampling programme. Depending on results from the review of information, actions are carried out. Where no failures occur, information is used to optimise treatment procedures. Where failures occur, a process of investigation and trouble shooting takes place. Where necessary, consumers are notified of relevant details, and measures to implement temporary and medium term solutions are initiated. A monthly summary report displaying the location of sampling points, the month's data, discussion and recommended actions is produced.

7 IMPLEMENTATION OF DRINKING-WATER QUALITY MANAGEMENT PROCEDURES: CASE STUDIES

The drinking-water quality management procedure outlined above, was introduced by the CSIR at one of the larger communities and at six of the smaller communities in the Western Cape. Considerable improvement of drinking-water quality was seen.

7.1 *Six smaller communities*

In the six smaller communities, a 100% failure regarding MALs for microbiological indicator organisms was improved to total compliance in a period of three months in five of the six communities. These improvements have been maintained. In the sixth community there were water quality improvements during the period that CSIR were able to assist with monitoring, evaluation and mentoring. Unfortunately, despite community support and enthusiasm, these improvements lapsed as a result of the lack of necessary resources to upgrade water treatment facilities.

7.2 *One larger community/town*

The original failure of 24% for MALs was improved to total compliance in a period of 3 months.

8 CONCLUSIONS AND RECOMMENDATIONS

In conclusion, this paper clearly demonstrates that:

- sub-standard quality drinking-water is supplied to the majority of rural communities sampled in the Eastern and Western Cape, South Africa;
- although groundwater resources have advantages over surface water sources for rural water supply, rural groundwater supply schemes fail more frequently;
- the most important reasons for the incidence of failure of groundwater to supply quality drinking-water are thought to be pump breakdown, deterioration of the storage and reticulation system, and, most importantly, insufficient monitoring and management;
- the implementation of well considered, community accepted drinking-water quality management procedures can effectively change an unacceptable water quality to one which satisfies drinking-water specifications;
- skills training, capacity building, technical support, and mentoring are vital to ensure the successful implementation of drinking-water quality management procedures in rural communities.

Recognising the considerable impact of poor drinking-water quality, it is recommended that Departments of Local Government, Health, and Water Affairs and Forestry assist in achieving broader implementation of drinking-water quality management procedures in rural communities. These procedures should form a blue-print for monitoring requirements of Water Service Providers under the new Water Services Act. Further study and research are required to refine implementation of drinking-water quality management procedures in rural communities, and identify supporting technologies. For example, easy tools and methodologies for field assessment of microbiological water quality. Source Directed and Resource Directed Measures under the National Water Act, should give due consideration to well-head protection and prevention of on-site sanitation pollution.

ACKNOWLEDGEMENTS

The authors are indebted to Mziwabantu Ramba of Barox Projects, who facilitated the site visits and sampling in the Eastern Cape. Finally, this work could not have been completed without the cooperation and support of the communities involved.

REFERENCES

APAH, AWWA and WPCF. 1995. *Standard methods for the examination of water and wastewater.* 19th Edition. Eds A D Eaton, L S Clesceri and A E Greenberg. American Public Association. Washington DC, USA.

Bourne, D.E. and Coetzee, N. 1996. An Atlas of Potentially Water-Related Diseases in South Africa. WRC Report No 584/1/96. Pretoria, South Africa.

CSS *Statistics in Brief.* 1998. Central Statistical Services. Pretoria, South Africa.

DWAF. 1996. Community Water Supply and Sanitation: Strategic Study -National Assessment. Pretoria, South Africa.

Kempster, P.L. 1998. *Draft Drinking Water Guideline Notes*, Water Research Commission Workshop, Pretoria, South Africa .

Mackintosh, G.S. and Linde, H. 1997. Drinking-water Quality Management: A Pressing Need for Rural Communities. Paper presented at Conference of the South African Institute of Environmental Health, Pretoria, South Africa.

Mey, J. 1998. *Poverty and Inequality in South Africa.* Report prepared for the Office of the Executive Deputy President and the Inter-Ministerial Committee for Poverty and Inequality, Durban, South Africa.

Pegrum, G.C., Rollins, N. and Espey, Q. 1998. Estimating the cost of diarrhoea and epidemic dysentery in KwaZulu-Natal and South Africa. *Water SA* Vol. 24 No. 1.

Ravenscroft, P., and Cain, J. 1997. Sustainability in operation and maintenance. In J. Pickford (ed) Proceedings of the 23rd WEDC Conference, Durban, South Africa: Water and Sanitation for All.

SABS. 1984. South African Bureau of Standards 241-1984: Specification for water for domestic supplies. Pretoria, South Africa.

SABS. 1990. South African Bureau of Standards SM 221-1990: Bacteriological Quality of Water. Pretoria, South Africa.

Groundwater: Past Achievements and Future Challenges, Sililo et al. (eds) © 2000 Balkema, Rotterdam, ISBN 90 5809 159 7

Aquifer Storage and Recovery (ASR) using water treated to irrigation standards

R.R.Martin & N.Z.Gerges
Department for Water Resources, Adelaide, S.A., Australia
P.J.Dillon
CSIRO Land and Water, Glen Osmond, S.A., Australia

ABSTRACT: Since the early 1990's Aquifer Storage and Recovery (ASR) has been increasingly used in South Australia to address specific surface or groundwater problems which have needed innovative solutions. The schemes implemented range from harvesting catchment runoff in urban areas to providing safe potable rural water supplies. In recent years, the focus for ASR in South Australia has shifted from the single well projects to much more ambitious undertakings. A pipeline has been constructed to deliver reclaimed water from the Bolivar Wastewater Treatment Plant, which processes much of metropolitan Adelaide's urban and industrial effluent, to irrigators during the summer months. The use of ASR is being trialed to determine the economic viability and environmental sustainability, as well as providing guidance for the implementation and management of a much larger scheme to store all of the winter excess (approximately 10 GL) in the aquifer. This paper describes briefly the history of ASR in South Australia and presents some of the results from the Bolivar ASR trial using reclaimed water. Issues that must be confronted when injecting water that is of a quality less than drinking water standard is also discussed.

1. INTRODUCTION

ASR as a management tool for the conjunctive use of surface and groundwater resources was not recognised in South Australia until the 1950's. Although initial attempts to use wells for storage of water during this time were very successful, South Australia was not yet ready to embrace this method of water resource management, dams were considered 'safer' and more 'visible' options (Sibenaler, 1996). ASR was first implemented in South Australia to harvest winter runoff, initially in rural then urban catchments to enhance water supplies for irrigation use.

Guidelines, for the reuse of stormwater developed by Dillon and Pavelic (1996) and adopted by statutory authorities have provided a practical basis for the storage of water within aquifers and have facilitated the development of ASR as an accepted water resources management tool within South Australia. For example, if the aquifer is brackish, the water that is proposed to be injected need not be fit for drinking, so long as all existing potential beneficial uses eg irrigation, industrial, stock use, and ecosystem maintenance are protected. The guidelines also recommend that if supplies are drawn for drinking water from within fresher parts of the same aquifer, an evaluation is required on whether contaminants would have attenuated to drinking water standards within the minimum travel time to such wells to ensure they are not compromised.

In keeping with the 1995 National Water Quality Management Strategy (NWQMS) groundwater protection guidelines, this offers a higher level of protection to aquifers containing higher quality water or discharging into sensitive environmental surface water bodies. These guidelines are quite different to those currently applying in USA where groundwater less than 10 000 mg/L total dissolved solids are considered to be potentially potable, and water needs to be treated to drinking water standards prior to injection (Dillon and Pavelic, 1996).

2. THE BOLIVAR ASR TRIAL

The Northern Adelaide Plains (NAP), centred around the township of Virginia approximately 30 kilometres to the north of Adelaide, the capital city of South Australia, (Fig. 1) is an extremely important horticultural area for the city as well as interstate. Approximately 3,500 ha of the available 12,000 ha produces a diverse range of crops including almonds, olives, tomatoes, grapes and a wide range of green-leaf vegetables yielding a farm-gate value of approximately AU$45 million. The principal source of water for irrigation is from groundwater, but the sustainability of this resource is under threat as a result of over 40 years of exploitation. These threats include:

- A decline in aquifer pressures resulting in the potentiometric surface being some 60 metres lower than when pumping first commenced
- The development of a large cone of depression within the aquifer centred around Virginia because of the intensity of pumping wells
- Salinity increases from vertical leakage as a result of the lower hydraulic heads in the aquifer.
- The possible ingress of the surrounding saline water toward the centre of the cone of depression

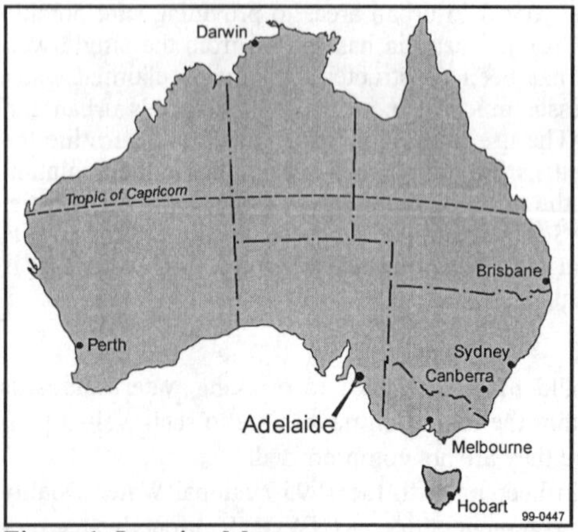

Figure 1. Location Map

A partial solution to help relieve the stresses on the aquifer and further realise the horticultural potential of the region lies in the proposal to re-use water from the Bolivar Wastewater Treatment Plant (WWTP). Bolivar processes much of Adelaide's urban and industrial effluent. After only secondary treatment this water was traditionally discharged to sea, which has resulted in impacts on the fragile coastal marine ecosystems. Significant upgrades to the wastewater treatment process have been undertaken including the installation of a Dissolved Air Flotation/Filtration (DAF/F) plant. These upgrades have resulted in a high quality water product that meets the South Australian Health Commissions (SAHC) guidelines for spray irrigation of reclaimed water onto crops including those that may be eaten raw. Parameters for the treated water quality to meet 'Class A' standards as set out in the "South Australian Reclaimed Water Guidelines (treated effluent)" include:

- Turbidity < 10 NTU (mean), 15 NTU (max)
- E. Coli < 10 organisims/100mL (median)
- Pathogens <1/50L (objective zero).

The Virginia pipeline has been built at a cost of AU$22 million to supply the treated water from Bolivar to growers in the Virginia area. The pipeline delivered the first water to customers in October 1999. At present some 22 gigalitres (GL) of the available 32 GL summer supply has been contracted to growers (Kracman et al., 1999). The pipeline scheme realises the potential for re-use of all of the available reclaimed water in summer with some minimal winter demand. However, to achieve the vision of potentially zero discharge to the marine environment a means of storing all of the winter excess (approximately 10 GL) is required. Winter storage in the aquifer combined with summer reuse of the product water from Bolivar provides a unique opportunity for improved water management. Sztajnbok and Kracman (2000) identify some of the outcomes that can be achieved from conjunctive management as:

- further reductions in discharge of nutrient rich wastewater to the marine environment
- provision of additional water to facilitate the expansion of the horticultural industry
- partial restoration of aquifer pressures especially during the winter months
- protection of the aquifer from saline water intrusion
- cost effective below ground storage in comparison to above ground storage.

In economic terms, the delivery of the reclaimed water in summer via the pipeline scheme has the potential to expand horticulture in the area by an additional 6000 ha resulting in an additional income at the farm-gate of around AU$150-$250 million/yr. An ASR scheme to store and recover the winter excess of reclaimed water from Bolivar may provide irrigation for up to a further 2400 ha of land potentially realising an additional AU$30 million/yr at the farm-gate.

3. RISKS

The reclaimed water from Bolivar has moderate nutrient and organic concentrations, elevated partial pressure of CO_2 and a slight possibility of containing pathogenic microorganisms. However, the greater control over the processing of the water considerably reduces this risk which is why highly treated reclaimed water is often preferred over stormwater. (Dillon et al 1999) summarised the risks as follows:

- growth or survival in the aquifer of injected pathogens

- growth of opportunistic pathogens indigenous to the aquifer
- injected water reaching existing water supply wells
- irreversible clogging of the well and aquifer at the location of the trial
- well or aquitard instability due to aquifer matrix dissolution or pressure changes
- recovered water containing an excessive proportion of brackish native groundwater, and
- aquifer storage capacity being inadequate for economic ASR.

4. OUTCOMES

Stage 1 of the trial has concentrated on laboratory column studies carried out on aquifer matrix material using both native groundwater taken from the site, and reclaimed water, to assess some of the risks identified above (eg Rinck-Pfeiffer et al., 2000; Toze, 1999). Additional work carried out under Stage 1 has included the development of models to assess solute transport and mixing within the aquifer (Dillon et al, 1999) and to determine the efficiency of recovery. The preliminary modelling has examined issues concerned with:

- where the reclaimed water will go in the aquifer
- the effects on nearby wells and effects of leaky wells and;
- the required design of the groundwater monitoring system for Stage 2.

Preliminary results from the laboratory studies and modelling have found that:

- bioclogging can be controlled
- recovered water will meet required standards
- some calcite dissolution will occur and be redeposited
- injected harmful organisms die out in the aquifer
- injection of effluent does not affect the growth of opportunistic pathogens already present in the aquifer
- the confining clay layer is pliable and pressures induced during injection will not breach it
- all injected water will remain within 300m of the injection well during the trial and most will be contained within 100m
- 3D mapping of the spatial distribution of the injected water within the aquifer can be accomplished using geophysics, and
- a nearby "leaky" well may have a small effect on salinity of wells very close to the injection site but no injected water would be detected.

These results provided sufficient assurances to indicate that the project should progress to the field trial stage (Stage 2) and any risks associated with the injection of reclaimed water into the aquifer could be appropriately managed. The focus for the field trial is the deep Tertiary (T2) aquifer (Fig. 2) because:

- groundwater at the trial site location is too saline (>2100mg/L TDS) for irrigation use
- the nearest operational well using groundwater from the T2 aquifer is 1.25 km to the north of the trial site
- the nearest operational well using groundwater from the upper aquifer (known as T1) is 400m from the trial site
- an eight metre thick, impervious clay separates the two aquifers so there will be no impact on existing users of the upper aquifer

Injection of the reclaimed water into the aquifer commenced in October 1999. Injection volumes of around 250 ML are needed before breakthrough of the reclaimed water into the observation wells located at a radius of 50m (Fig. 2) from the injection well occurs. So far results have been as expected with periodic redevelopment required to minimise the impacts of potential clogging. On-going evaluations are being carried out into rates of pathogen die-off using membrane cells specifically constructed for the purpose. Water quality measurements and field monitoring are also being carried out on a routine basis throughout the duration of the field trial.

5. THE FUTURE

Injection of about 250 ML of reclaimed water is expected to be completed by late 2000. Recovery of the injected water is expected to commence after a short period of storage. Based on the information gained from the field trial it is anticipated that the operational rules (Stage 3) for a fully implemented scheme to store and recover all of the available winter excess will be developed provided such a scheme proves economically viable.

Community acceptance in supporting a possible larger scheme is seen as an important prerequisite to the success of such a project. Such a project is also seen as a long-term solution to the sustainable management of water resources on the Northern Adelaide Plains. To this end, the project proponents have undertaken an extensive community liaison program designed to educate groundwater users about their aquifer, and to inform them about managing the possible risks and the benefits to be gained from such a

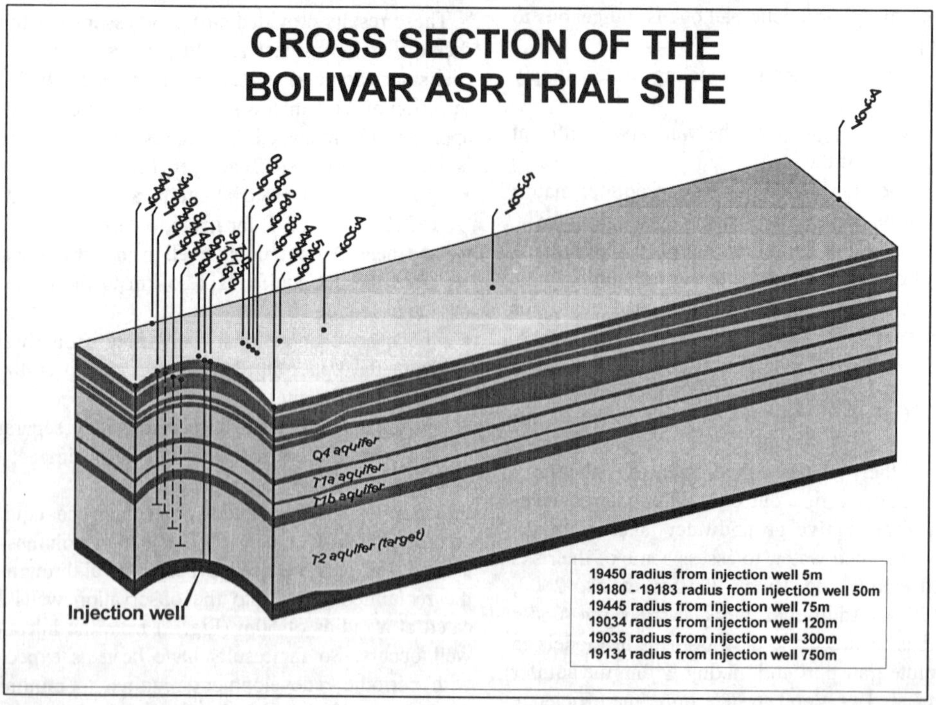

Figure 2. Cross section of Bolivar ASR trial site

large-scale ASR scheme. The community liaison has identified that misconceptions existing within the community about what an aquifer is and the movement of water within aquifers need to be addressed before a much larger scheme could be implemented. The community was concerned with health issues and would need verifiable assurances that the injected reclaimed water would not reach any wells used for domestic supply purposes.

6. CONCLUSIONS

The Bolivar ASR trial is a unique project involving the use of deep well injection for storage and subsequent retrieval of reclaimed water that has been treated to a level suitable for irrigation use. Previous studies of a similar nature have not been identified in the literature. This project provides the opportunity to realise significant environmental, economic and social benefits through the use of reclaimed water. Significant economic benefits accrue as a result of ;

- targeted upgrading of the treatment plant to only provide water that is of a quality fit for the intended end-use (irrigation)
- expansion of the existing AU$45 million/yr horticultural industry (to around AU$150-$280 million/yr inclusive of the pipeline and ASR scheme combined) previously constrained by the limited quantity of groundwater
- commercialisation of the reclaimed water to provide a return to the distributor thereby offsetting some of the cost of the treatment plant upgrade to metropolitan ratepayers.

The environmental benefits include:

- maintenance or improvement of existing pressures within the aquifer
- maintenance or improvement of the existing groundwater quality by injecting the reclaimed water along the margins of the basin where the water quality is poor
- zero discharge of nutrient rich water to the sea thereby preserving the delicate environmental balance in the coastal marine waters.

The social benefits that accrue from reusing all of the reclaimed water are inter-related with the economic and environmental benefits such as a cleaner marine environment; greater water availability when needed; increased wealth and productivity for the region which in turn creates significant flow on effects.

The aims of the trial are to address the issues that confront the proponents when injecting a quality of water that is less than potable. Further, the trial is also structured to provide the necessary information to undertake the design of a full-scale scheme capable of storing and recovering up to 10 GL/yr.

Community acceptance in supporting a possible future scheme to store all of the winter excess of reclaimed water from Bolivar is seen as an important pre-requisite to the success of such a project.
The risks and immediate benefits of the trial are small, but a successful trial and community acceptance of a full-scale scheme will lead to major benefits. If the trial proves that ASR using reclaimed water treated to a level suitable for irrigation is safe; can be well managed; is both technically and economically viable; will have no adverse environmental impacts then; the storage and subsequent retrieval of the reclaimed water will be seen as a long-term solution to the sustainable management of all of the available water resources on the NAP.

ACKNOWLEDGEMENTS

Funding for this project has been provided by the Natural Heritage Trust (NHT) and a consortium comprising United Water, SA Water, CSIRO, Department for Water Resources and Department of Administration and Information Services.

REFERENCES

Dillon, P. and Pavelic, P. 1996. Guidelines on the quality of stormwater for injection into aquifers for storage and reuse. *Urban Water Research Assoc. of Aust. Research Report No 109.*

Dillon, P., Toze, S., Pavelic, P., Ragusa, S., Wright, M., Peter, P., Martin, R., Gerges, N., Rinck-Pfeiffer, S., 1999. Storing recycled water in an aquifer at Bolivar: benefits and risks. *Australian Water and Wastewater Association Jour.* Water 26(5): 21-29.

Kracman, B., Martin, R., Sztajnbok, P., 1999. The Virginia Pipeline: Australia's largest water recycling project. *3rd International Symposium on Wastewater Reclamation, Recycling and Reuse.* Paris 3-6 July 2000.

Rinck-Pfeiffer, S. Ragusa, S., Sztajnbok, P. and Vandevelde, T. 2000. Interrelationships between biological, chemical, and physical processes as an analog to clogging in Aquifer Storage & Recovery (ASR) wells. *Water Research*, 34(7): 2110-2118.

Sibenaler, X.P., 1996. ASR experience in South Australia. *Aquifer Storage and Recovery (ASR) Short Course.* 1-2 October, Adelaide 1996.

Sztajnbok, P., and Kracman, B., 2000. Wastewater reuse and aquifer storage and recovery – key elements for a more sustainable use of water. *Water 2000 Auckland New Zealand.* March 20-22.

Toze S. 1999 PCR and the detection of microbial pathogens in water and wastewater. *Water Research* 33(17): 3545-3556.

Groundwater: Past Achievements and Future Challenges, Sililo et al. (eds) © 2000 Balkema, Rotterdam, ISBN 90 5809 159 7

Major issues in sustainable water supply in Botswana

O.A. Masedi, O. Katai, I. Muzila & L. Carlsson
Department of Water Affairs, Gaborone, Botswana

ABSTRACT: Development of sustainable water supply in Botswana is hampered by the country's climatic, topographic and geologic conditions. The low rainfall and high potential evapotranspiration, combined with the country's flat topography and deep sandy soils, result in low rates of surface runoff and low rates of recharge to its groundwater. The estimated annual average run-off of 1.2 mm is one of the lowest in the world. All streams originating in Botswana are ephemeral. The low topographic relief of Botswana limits the construction of high dams storing large volumes of water. Groundwater development is hampered by low replenishment over most of the country and also by poor groundwater quality (high salinity) in some parts.

The water scarcity indicators and the water development constraints are a harsh reality, which leaves very little room for mistakes in water resources planning in Botswana. Optimum water development in a semi-arid country will be better obtained through a conjunctive use of surface and groundwater resources.

1. INTRODUCTION

1.1 *Background*

Botswana is a landlocked country occupying an area of 582,000 km^2 between the latitudes 18°S and 27°S in the southern part of the African continent.

The country is flat with most of the land lying between 800 and 1,200 mamsl. Annual rainfall varies gradually from about 250 mm in the south-west to about 690 mm in the north-east. About 95% of the rainfall occurs between October to April; this is the period when the evaporation is at its highest. The evaporation averages about 2 000 mm annually over the country.

The total population of Botswana is currently around 1.5 million with a growth rate of around 3.4% per annum. The most populated area is in the south-east and eastern parts along a corridor comprising urban centers of Lobatse, Kanye, Molepolole, Gaborone, Mochudi, Mahalapye, Serowe, Palapye and Francistown, see Figure 1.

1.2 *Present and Future Water Demands*

Botswana has experienced a rapid development since its independence in 1966. Besides a stable democratic political system, one key to the development has been the supply of water to every sector of the community. The development of the water sector has been heavily supported by international aid in capital investment, technology transfer and human resources development. The goal has been achieved through development of groundwater supplies for the rural and major villages and surface water for the urban areas.

The Ministry of Minerals, Energy and Water Affairs, through the Department of Water Affairs (DWA) launched a comprehensive National Water Master Plan (NWMP) study during 1989 to 1992. The plan not only considered the future water demands and possible supply schemes, but also addressed water conservation, management, legislation and institutional aspects. The Master Plan projected a period of 30 years (1990-2020) but in the same time promoted flexibility allowing adjustments and amendments to be made during the course of its implementation (DWA, 1991). The water consumption and demand figures for the whole country as estimated in the NWMP are summarized in Table 1. The country has no hydropower, hence no water projections for the subsector.

2 WATER RESOURCES

2.1 *Surface Water Resources*

The low rainfall and high evapotranspiration, combined with the country's flat topography and deep sandy soils, result in low rates of surface runoff. All streams originating in Botswana are ephemeral and only flow for some 10 to 75 days a year on average. Over the whole of the west and center of the country there is no surface runoff. A comparison of surface water resources of Botswana with those of other countries emphasizes the meager resources of the country see (Figure 2).

Perennial rivers are border rivers originating from outside Botswana such as the Zambezi and Chobe rivers in the north. In the north-west the Okavango river flows into the country and dissipates in the 12,000 km^2 large Okavango inland delta. In the south, the Limpopo River is in general ephemeral and can occasionally carry large amounts of water. Table 2 shows the estimated annual runoff of these international border rivers. Any development of the water resources of these rivers must be subject to international agreement. SADC states have signed and ratified a Protocol on Shared Water Course Systems; this forms the base for the development of international water resources.

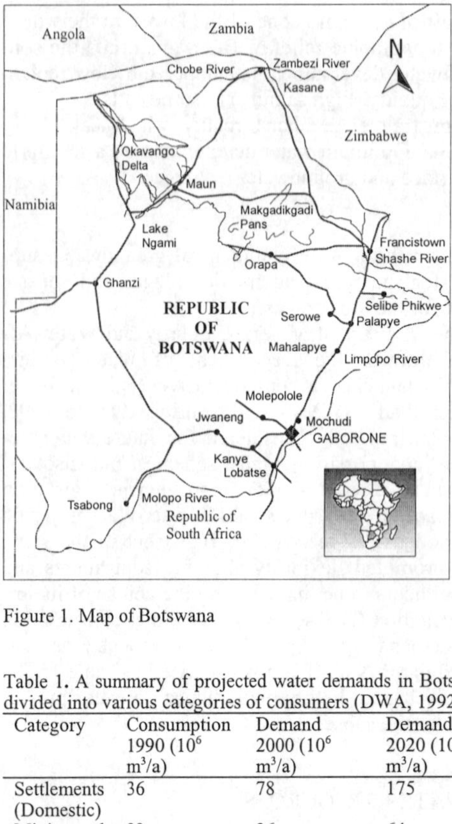

Figure 1. Map of Botswana

Table 1. A summary of projected water demands in Botswana divided into various categories of consumers (DWA, 1992)

Category	Consumption 1990 (10^6 m^3/a)	Demand 2000 (10^6 m^3/a)	Demand 2020 (10^6 m^3/a)
Settlements (Domestic)	36	78	175
Mining and Energy	23	36	64
Livestock	35	45	44
Irrigation & Forestry	19	29	47
Wildlife	6	6	6
Total	119	194	336

2.2 *Groundwater Resources*

Botswana's groundwater resources are related to the bedrock geology (see Table 3). Large parts of south-eastern and eastern Botswana comprise Archean crystalline rocks. Weathering is shallow and groundwater occurrence is mostly in fractures. Dolomite formations are found in relatively small areas but the formations are of considerable importance for water supply, particularly in areas south and south-west of Gaborone.

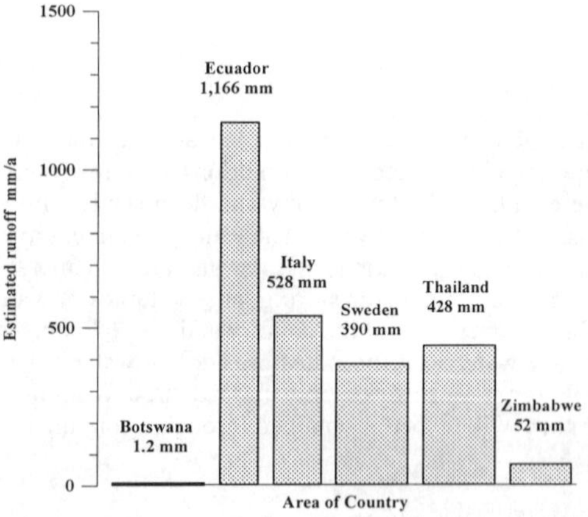

Figure 2. Average annual runoff from selected countries. The width of the bars illustrate the areas of the countries

Table 2. Estimated runoff of international border rivers (DWA, 1991)

River	Area km^2	Estimated annual mean runoff 10^6 m^3	mm
Zambezi at Victoria Falls	360,700	40,100	113
Okavango at border inflow	200,000	9,910	50
Chobe	131,000	1,310	10
Limpopo	158,000	620	4

The sandstone formations of the Karoo Supergroup constitute the most important aquifers in Botswana due to their large areal extent and favourable hydraulic properties. They are fractured porous aquifers and cover wide areas in the central, southern, western, and most eastern parts of the country. The sandstones are found mainly at two different horizons in the Supergroup. The upper horizon, Ntane sandstone comprises a sequence of about 100 m and in many places covered by the basalt. The lower sandstone sequence, the Ecca sandstone contains a number of layers with varying thickness intercalated with coal seams which give rise to complex aquifer systems. One of the country's largest wellfields located 50 km north of Jwaneng, abstracts water (7.7 10^6 m^3/a) from Ecca sandstones.

The Kalahari Beds deposited over an undulating pre-Kalahari relief during the Tertiary and Quaternary blankets almost 85% of Botswana's bedrock. The sequence can attain a thickness more than 200 m. Where the Kalahari Beds are sandy and reach below the groundwater table as along the identified pre-Kalahari valleys, moderate aquifers are developed. Calcrete and silcrete crusts intercalated in the sands give rise to localized perched aquifers.

The ephemeral rivers of eastern Botswana are infilled with sands and gravel, which can be up to 10 meters thick. The river bed sediments (Sand Rivers) are recharged annually from percolating river water and form elongated localised small aquifers.

3 MAJOR ISSUES

3.1 *Surface Waters*

Two characteristics, spatial and temporal variability of the surface water resources are of relevance. Most of the surface water resources are located in the eastern corridor of the country and some of the undeveloped streams are some 400 km from the major demand centers in the south. A feature of arid and semi-arid climate is the high variability of rainfall and streamflow leading to long sequences of dry years interspersed with a few wetter years. Accordingly, the development of a reservoir would require ample filling time before the transfer of the water is carried out; this may be worse in case the dam construction is completed during the drought spell. The low topography limits the construction of high dams to store large volumes of water.

The development of surface water resources is also hampered by difficulties in assessing the is about four times greater than rainfall and streamflow records are of relatively short duration.

Environmental considerations are also a potential impediment in some surface water development projects.

3.2 *Groundwater*

Major factors that hamper the use of groundwater for water supply are: (i) low replenishment rate, (ii) water quality, (iii) location of major resources in relation to consumers and (iv) overdevelopment.

Semi-arid climatic and geological conditions do not favor high replenishment rate. Recharge from precipitation has been assessed in various research studies (Beekman et al, 1996, Selaolo et al, 1994, Selaolo, 1998). In general, low values (0-3 mm/a) are assessed for the Kalahari area and higher values for the eastern Botswana where the bedrock is not covered by Kalahari sediments.

The sustainability of the groundwater resources is a major challenge. Abstraction for minor villages, settlement, livestock and wildlife is in general under sustainable conditions. The major concern is with large concentrated water abstractions for major villages, towns and mines. Figure 3 illustrates the two major wellfields of Jwaneng and Orapa showing the

Table 3. Main aquifer formations in Botswana

Age	Litho-stratigraphic unit	Lithology	Aquifer area km^2	Type	Regional Transmissivity m^2/d	Borehole Success rate %
Cenozoic	Sand Rivers of E. Botswana and the Okavango delta	Gravel, sand		P	400-1500	
	Kalahari Beds	Sand, sandstone, calcrete, silcrete, clay	40	P	350	100
			201,000	P	0.7-35	72-100
Mesozoic	Karoo Supergroup					
	Stormberg	Basalt	6,000	F	1.5-2.5	68-94
	Lebung	Ntane sandstone	138,000	F-P	1-40	92-100
	Ecca, Beaufort	Arkose, sandstone, siltstone, mudstone, coal	94,000	F-P	0.9-70	86-100
Proterozoic	*E and S Botswana*					
	Waterberg Supergroup	Conglomerate, siltstone, quartzite, shale	49,000	F	1-8	68-100
	Transvaal Supergroup	Dolomite	4,000	F-K	2-5	71-87
	W, NW and N Botswana					
	Olifantshoek Sequence	Quartzite, schist, porphyry	400	F	0.2	62
	Damara Sequence	Arkose, quartzite, schist, mudstone, shale	33,000	F		
		Dolomite	1,000	F-K	0.6-2	79-100
Archean	Ventersdorp Supergroup	Porphyry, felsite, greywacke	50	F	0.7	6
	Archean basement of SE and E Botswana	Granite, gneiss, migmatite, amphibolite	61,000	F	0.5-2	44-96

F=fractured K=karstified P=porous Successful borehole rate > 0.01 m^3/h

drawdown versus time for various scenarios and models (Debswana, 1997, AAC, 1994).

Parallel to research on recharge, a continuous monitoring of water abstraction at all major wellfields was launched by DWA in 1990 (Mannathoko et al, 1994). In a comprehensive updating of the monitoring system, DWA undertook a modelling exercise of all wellfields in order to assess their sustainable yields (DWA, 2000).

Of the 12 wellfields modelled, only 3 are currently working under sustainable conditions. Figure 4 illustrates the on-going lowering of the water level in the Dukwi wellfield.

Low recharge rate together with geological and topographical conditions are the background to the groundwater quality. Figure 5 shows the total dissolved solids (TDS) distribution in groundwater. The desired limit of 1,000 mg/l, inline with the WHO guidelines, is prevalent in about 25% of the country. The maximum permissible TDS of 2,000 mg/l shows that about 50% of the country has unsuitable water quality. In order to obtain water supplies for settlements and rural villages in areas with high TDS, water has to be conveyed from other places or the desalination process must be used. Rainwater harvesting is also performed in some of the areas.

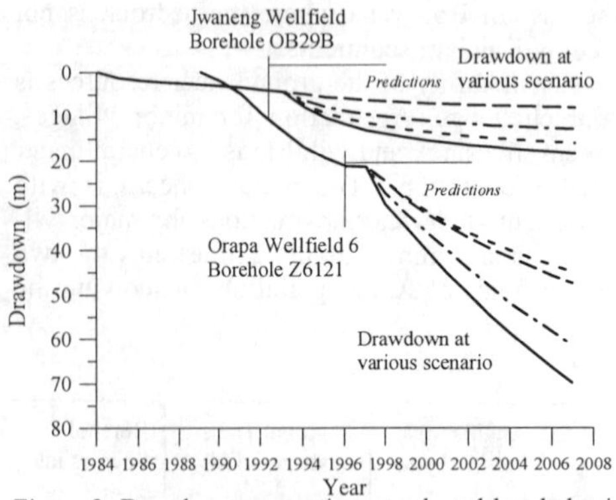

Figure 3. Drawdown versus time at selected boreholes in the Jwaneng and Orapa Wellfields. Predictions for various scenario (AAC, 1994, Debswana, 1997).

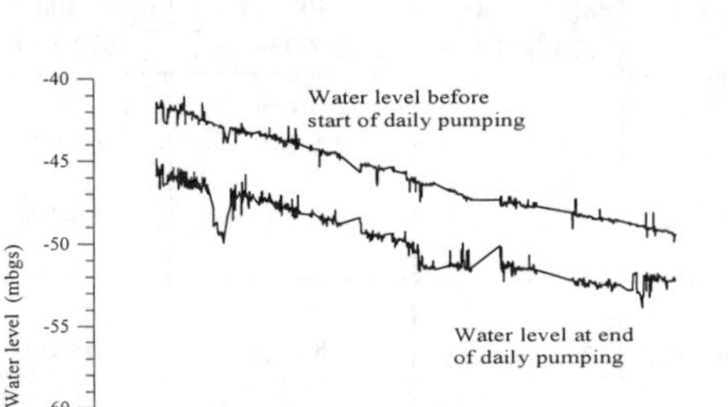

Figure 4. Monitoring of groundwater level in production borehole 4700 at Dukwi wellfield. The borehole is pumped approximately 11 hours per day at a rate of 35 m^3/h. The whole wellfield produces 2,250 m^3/day.

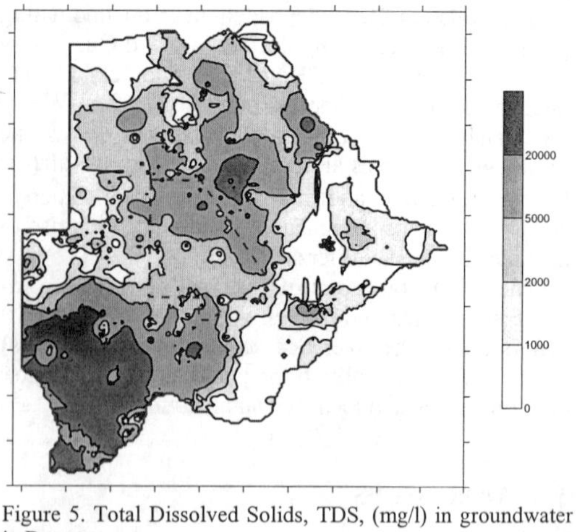

Figure 5. Total Dissolved Solids, TDS, (mg/l) in groundwater in Botswana.

High nitrate in the groundwater may locally be of concern in water development schemes.

4 CONJUNCTIVE USE OF SURFACE AND GROUNDWATER

4.1 *General*

Optimum water development in a semi-arid country will best be achieved by conjunctive use of surface and groundwater resources. In Botswana, the NWMP has used the term "conjunctive use of water resources" simply meaning use of groundwater or surface water to support each other especially during times of droughts. The term conjunctive use is understood differently; conjunctive use may also refer to the active water transfer under optimum conditions from surface to groundwater storage (Gieske, 1994).

Conjunctive use of surface and groundwater must be considered in relation to the scale or water production figures. In the local scale, it includes rainwater catchment as a support to groundwater use. In a large scale, the conjunctive use will comprise of the use of surface water from dams and groundwater supplies. In the regional scale covering urban areas and large clusters of villages, transfer of water from surface and groundwater sources is considered for the purpose of sustainable supplies.

4.2 *Rainwater harvesting*

Traditionally people in Botswana have for centuries collected rainwater run-off in excavated pits and from the eaves of thatched roofs in pots and other small containers. Two major types of catchments are ground and roof tanks.

In a large scheme of rainwater catchment, more than 700 ground tanks were constructed (Chilume, 1993). The tanks were originally designed to provide water for draught animals in the lands area; invariably people used the water for domestic purposes including drinking since family members particularly women, often spend long hours fetching water for household consumption which affects their productivity in farming activities. Bacterial quality and pollution problems of the rainwater were encountered (Jay et al, 1993)

Roof catchment tanks are a common sight at primary schools in villages throughout Botswana. About half of the over 800 primary schools have tanks with an average capacity of about 12 m^3. Rainwater tanks are also found at clinics and other government buildings in several villages. Many private households have invested in roof catchment tanks especially in rural Botswana.

While water from ground catchment is certainly not suitable for consumption without being boiled or chlorinated, water from well maintained roof catchment systems is generally potable. Rainwater catchment systems can be used almost anywhere and can potentially provide a convenient high quality water source once proper precautions are taken. The system can supplement other sources of water in various communities. In some settlement, rainwater may be the only portable source of water since other sources may be saline or located too far away.

4.3 *The North-South water carrier*

The NWMP study estimated a five-fold increase in water consumption for the major urban centers over the next 30 years. To supply Greater Gaborone, Lobatse, Francistown and Selebi-Phikwe and associated major and rural villages a transfer of water from the eastern part of Botswana was recommended and a project entitled the North-South Carrier Water project (NSCWP) commenced in 1993.

The project entails the construction of a dam along the Motloutse River (Letsibogo Dam), and a 360 km pipeline to transfer water to Greater Gaborone and major villages along the course of the pipeline (see Figure 6). An envisaged second phase project scheduled for 2008 will duplicate this project with a second dam on the Lower Shashe River. The project is massive and raises a myriad of considerations associated with institutional reorganization, funding, construction, training, public education, environmental and social issues.

The NSCWP will not deliver the total supply to the Greater Gaborone and the major villages along its route. Existing supplies for Gaborone will still be used but all increases in water demand will be taken through the NSC. Some villages like Palapye and Mahalapye cluster currently have sufficient supplies from groundwater but will be fully supplied from the NSC and the wellfields converted to strategic back-ups. With time, the dependency on surface water through the NSC will increase.

The highly dependence on a surface water storage for more than half a million consumers must however be supported by reliable back-up supplies. A conjunctive use of surface and groundwater is therefore applied and strategic backup wellfields are identified in four areas along the route of the NSC. The resources identified are in the Ntane sandstone aquifer. Considerations will also be given to the possibility of artificial recharge to this aquifer during periods of surplus surface water.

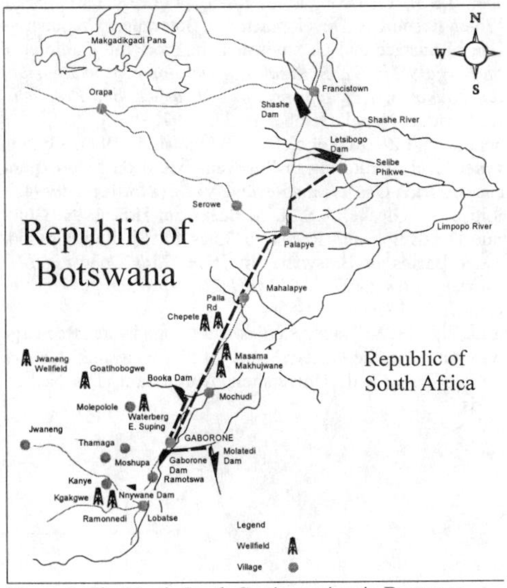

Figure 6. The North-South Carrier project in Botswana

4.4 *Water Supply for Major Villages in the neighborhood of Gaborone, including Conjunctive Use of Surface and Groundwater*

The major villages surrounding the capital Gaborone are mainly supplied by their own wellfields or connected to Gaborone water supply. Due to rapid population increase, the water abstraction in many of these villages is under strained conditions and sustainability can no longer be maintained. The two villages of Molepolole and Thamaga have recently been connected to a wellfield at the fringe of the Kalahari. The replenishment is however low and abstraction will be under 'mining' condition implying that water abstraction exceeds the sustainable yield. The villages of Kanye and Moshupa are supplied from a karstified dolomite aquifer where boreholes are in general high yielding. The aquifer is extensive and has a comparatively high replenishment through recharge (in the order of 20 mm/a). This aquifer can meet the present demand under sustainable conditions but the future abstraction will be under non-sustainable conditions.

The supply of groundwater to major villages from improved and new wellfields will add to the over-abstraction of the precious groundwater. An almost fivefold increase in groundwater utilization is envisaged. A situation of over-abstraction is in the perspective unacceptable and other sources of water supply must be introduced on sustainable basis. A second phase of the NSCWP, scheduled to be commissioned in 2012 will offer this possibility. In the mean time, over-abstraction of the groundwater resources under controlled conditions has to be accepted in order to promote continued prosperity and social development. The time period of over-abstraction must however be utilized in such a way that the sustainable yields of the resources can be established with an ever increasing reliability. Monitoring of groundwater abstraction and quality correlation are key factors in any assessment of the sustainable yield (Khupe et al, 1997)

Whatever feasible method is applied in the assessment of the sustainable use, a key issue in the concept is conjunctive use of surface and groundwater in a semi-arid environment such as Botswana (Carlsson et al, 1994).

5 CONCLUSION

Development of water supply in Botswana must continue on sustainable basis. The issues of concern in such development are related to the climatic, geological and topographical conditions of the country. Surface water is only available for part of the year and dams are needed to store the yearly inflows. The surface water is however available only in the northern and eastern part of the country and topographical sites for large dams with limited areal extent are scarce. The groundwater resources, even the ones with good yields have limited replenishment and for long term supply they must be used with careful monitoring and management. Conjunctive use is the way forward and the country adopts this concept. Until the conjunctive use is developed for the major part of the country, abstraction of groundwater on a non-sustainable basis has to be accepted as an intermediate solution.

6 ACKNOWLEDGEMENT

The authors wish to express their gratitude to the Permanent Secretary, Ministry of Minerals, Energy and Water Affairs for allowing them to publish this paper.

7 REFERENCES

Anglo American Corporation, AAC. 1994. *Jwaneng Northern Wellfield - Modelling of Groundwater Flow with Optimized Management Recommendations on Wellfield Utilization.* Report No. CED/029/94, Johannesburg, April 1994.

Beekman, H.E, Gieske, A. & Selaolo, E.T. 1996. GRES: Groundwater Recharge Studies in Botswana 1987-1996. *Botswana Journal of Earth Sciences*, Volume 3, 1996, pp 1-17.

Carlsson, L., Selaolo, E. & von Hoyer, M. 1994. Assessment of ground water resources in Botswana: experiences from the Botswana National Water Master Plan study. *Africa Geoscience Review*, Vol. 1, No. 1, pp 11-20.

Chilume, G.P.N. 1993. Lands Area Water Supply: The ALDEP Rainwater Catchment Tank Programme. In Proceedings from *Rainwater Catchment Systems Application and Technology Workshop*, 1-3 March 1993, Gaborone/Kanye, Botswana, pp 17-26.

Debswana Diamond Co. 1997. *Orapa Groundwater Model Review.* Final report to Debswana Diamond Co (Pty) Ltd, Orapa and Letlhakane Mines by Water Surveys (Botswana) (Pty) Ltd, November 1997.

Department of Water Affairs, DWA 1991. *Botswana National Water Master Plan Phase 1* Report in 12 volumes, Gaborone July 1991.

Department of Water Affairs, DWA 1992. *Botswana National Water Master Plan Phase 2.* Report in 2 volumes, Gaborone July 1992.

Department of Water Affairs, DWA 2000 (in prep.). *Groundwater Monitoring. Review of Monitoring Performed by DWA and DGS, Assessment of Water Resources and Suggestions for Improvements. Phase 1 of the National Groundwater Information Systems.* Final report prepared by Geotechnical Consulting Services.

Gieske, A. 1994. Conjunctive Use of Surface and Ground Water in Botswana - Urban and Rural Aspects. In Proceedings from *Integrated Water Resources Management Workshop 1994*, 17-18 March, Kanye-Gaborone, Botswana.

Jay, B., Gould, J.E. 1993. An Overview of Rainwater Catchment Systems in Botswana. In Proceedings from *Rainwater Catchment Systems Application and Technology Workshop*, 1-3 March 1993, Gaborone/Kanye, Botswana, pp 3-13.

Khupe, B.B.J., Carlsson, L. & Katai, O. 1997. Large Scale Water Resources Development in Botswana: Conjunctive Use of Surface and Groundwater. In Proc. Int. Conference on *Large Scale Water Resources Development in Developing Countries: New Dimensions of Prospect and Problems*, Kathmandu, Nepal, October 20-23, 1997.

Mannathoko, I.Z., Makobo, P. & Mogotsi, I. 1994. Ground water level monitoring in Botswana based on field experiences. *Africa Geoscience Review*, Vol. 1, No. 1, pp 89-94.

Selaolo, E.T., Gieske, A.S.M. & Beekman, H.E. 1994. Chloride Deposition and Recharge Rates for Shallow Groundwater Basins in Botswana. In *Proc. 25th Congress IAH "Water Down Under'94", Adelaide, Australia,* 21-25 November, Vol 2A pp 501-506.

Selaolo, E.T. 1998. Tracer Studies and Groundwater Recharge Assessment in the Eastern Fringe of the Botswana Kalahari. *Ph.D. Thesis*, Vrije Universiteit, Amsterdam, The Netherlands

Groundwater: Past Achievements and Future Challenges, Sililo et al. (eds) © 2000 Balkema, Rotterdam, ISBN 90 5809 159 7

Groundwater resources management in the SADC region: A field of future regional co-operation

P. Molapo
SADC Water Sector Co-ordination Unit, Maseru, Lesotho

S.K. Pandey
Groundwater Consultants, Maseru, Lesotho

S. Puyoô
French Technical Assistance, SADC Water Sector Co-ordination Unit, Maseru, Lesotho

ABSTRACT: The Southern African Development Community (SADC) is a regional grouping of 14 sovereign states. The SADC institutions provide an opportunity to address trans-boundary water management issues, through regional co-operation. Groundwater is the main source of water supply for about 37% of the total population in the SADC region, while surface water accounts for 23%. Remaining 40% of the population still rely on unsafe water sources, which are more often prone to drought. Groundwater has a potential role to play in future development of secured water supply facilities. The SADC Water Sector prepared a "Groundwater Management Programme for the SADC Region", focused on exchange of information, research and training, monitoring, mapping, characterisation and management of groundwater resources in aquifers of regional magnitude.

1 THE SADC INSTITUTIONS AND GROUNDWATER

The Southern African Development Community (SADC) is a regional grouping of 14 sovereign states with the main purposes of fostering co-operation and mutual benefit from the resources of this region.

The population of the SADC region is expected to double in 25 years and major water and environment crisis will occur if decisive actions are not taken towards sustainable and integrated water resources management. With the establishment of the SADC Water Sector in August 1996 the critical importance of water to regional integration and economic development was recognised by the Member States. Water was upgraded to be a separate sector from the Environment and Land Management Sector in SADC with its own Committee of Ministers. The Government of the Kingdom of Lesotho is given the responsibility to coordinate the Water Sector. The SADC Water Sector Coordination Unit is based in Maseru.

The regional activities in the field of integrated water resources management are not seen as a substitute to individual States responsibility, but rather as a support and an incentive to increase the efficiency of national programmes, with due consideration to trans-boundary issues.

Despite the present and future eminent socio-economic role of groundwater, regional water resources programmes in SADC have been and still are dominated by surface water issues. The SADC institutions i.e. the Water Resources Technical Committee (WRTC), the Sectoral Committee of Senior Officials and the Sectoral Committee of Ministers of Water have acknowledged that more attention should be given to groundwater development and protection. They identified the need for the establishment of a Sub-committee for Hydrogeology, in which each Member State is represented.

The Sub-committee for Hydrogeology serves as supervisor or steering group for the implementation of projects of regional magnitude in the field of Hydrogeology. It may call specific national and regional institutions, experts and consultants for implementation and guidance of these projects.

The Water Sector developed a "Groundwater Management Programme for the SADC Region". The Programme is part of the Regional Strategic Action Plan for Integrated Water Resources Development and Management in the SADC Region (Water Sector Coordination Unit 1998), a SADC Water Sector initiative undertaken in cooperation with the United Nations Development Programme (UNDP).

2 GROUNDWATER IN THE SADC REGION

Growing attention was given in the recent years to groundwater resources at the sub-continental or regional level (United Nations 1989, M. MacDonald et al. 1990, OACT 1991, Chenje et al. 1996, FAO 1997). Several workshops were held on water resources (Burke et al. 1993, Stanley 1995,) and more specifically on groundwater (SADCC-ELMS 1991, SARCCUS 1995, SARCCUS 1996, SARCCUS 1997). During the preparation phase of the Regional Strategic Action Plan for Integrated Water Resources Development and Management in the SADC Countries, Country Situation Reports were produced between 1996 and 1998 (Water Sector Coordination Unit 1998). In most cases they provided updated national information on groundwater occurrence, use and management.

An assessment of the groundwater resources management situation in the SADC Member States was carried out, as well as consultation with major stakeholders, as a basis for regional groundwater management programme identification, with the financial support of UNEP (Groundwater Consultants 1998).

2.1 *Groundwater Occurrence*

The SADC Region represents a broad spectrum in terms of physiography, climate and geology. Topographic differences ranges from mountainous terrain (D.R.Congo, Lesotho, Tanzania, etc) to extremely flat sandy bushveld of Botswana and Namibia. Humid equatorial to arid regions are present with annual rainfall ranging from more than 2,000 mm (D.R.Congo, Mauritius) to effectively none in the Namib Desert of Namibia.

The hydrogeology of the SADC Region, although variable when considered at local scale, consists of a few major aquifer types on a regional scale. The fractured type aquifers of the crystalline basement and metamorphic units occupy 25 to 78 % of the SADC countries. In Namibia, Swaziland, Mozambique, Zimbabwe, Tanzania and Malawi, they form more than 50% of their respective territories. Volcanic rocks are the main outcropping formation in Mauritius, Lesotho, and Swaziland. They also cover significant areas of Botswana, South Africa, Zimbabwe and Tanzania. The sedimentary formations overlie the depressed crystalline basement in the axial part of the sub-continent, including the Congo basin, the Okavango and the Kahalari. The Karoo basin is a fossil basin raised in a vast plateau. Their essentially continental formations consist of fairly coarse sandstones, which are good dual porosity aquifers. At the edge of the basement rocks, calcaro-dolomitic Lower Cambrian formations are extensive in D.R.Congo, Angola, Namibia, Tanzania and Zambia. (United Nations, 1989). Important karstic aquifers are developed in such formations, in particular in Zambia. Coastal sedimentary basins are found in D.R.Congo, Angola, Namibia, South Africa, Mozambique and Tanzania. The main rivers of the region are associated with alluvial aquifers.

Most of these aquifers are shared by several countries. In the majority of Member States, groundwater occurrence and recharge conditions are complex and highly variable. The assessment of overall renewable groundwater resources has been done in some countries at local or regional level.

2.2 *Groundwater Development and Usage*

The status of groundwater development across the region largely reflects the importance of the resource within the specific country. In countries with a strong dependence on groundwater, such as Botswana, Namibia, Mauritius, and to some extent, South Africa, the level of development in the sector is quite high. In countries where alternative sources from surface water are abundant, the level of groundwater development can be relatively limited and characterised by little or no management of the resource.

The actual format and modes of groundwater development vary at a basic level between those countries where government is the main implementing agency (with varying degrees of cost recovery) and those where significant degree of implementation is carried out by the private sector, including NGOs.

In the SADC region, excluding D.R. Congo and Seychelles for which global data is not available, the total water use is estimated to nearly 35,000 millions cubic meters per year, of which 83% is used for irrigation and livestock, 11.5 % for domestic water supply facilities, and 5.5% for mining and industries.

Groundwater accounts for 11.6% of the total quantity of water used with 4,000 millions cubic meters and 20% of the water used for domestic water supply (Table 1). As far as total water use is concerned, groundwater is quite important in Botswana, Namibia, Lesotho, and very significant in Mauritius, South Africa, Zimbabwe and Zambia. Groundwater provides a significant volume of water in water supply facilities in all the SADC Members States, varying from 10 to 58% of the total domestic water demand.

Despite their moderate quantitative contribution to the satisfaction of the global water demand compared to surface water, groundwater is the main

Table 1. Water use and part of groundwater in the SADC countries

SADC Member State	Domestic water supply: Surface and Groundwater $10^6 m^3$/year	Domestic water supply: Groundwater only $10^6 m^3$/year	Part of Groundwater in domestic water supply %	Total water use: Surface and Groundwater $10^6 m^3$/year	Total water use: Groundwater only $10^6 m^3$/year	Part of Groundwater in total water use %
ANGOLA	130	*28*	*22*	2,474	*35*	1
BOTSWANA	36	15	41	119	76	64
D.R.CONGO	-	-	-	-	-	-
LESOTHO	19	11	58	*37*	*15*	*41*
MALAWI	120	35	29	1,161	35	3
MAURITIUS	170	80	47	620	101	16
MOZAMBIQUE	107	36	34	630	36	6
NAMIBIA	144	53	37	278	140	50
SEYCHELLES	-	-	-	-	-	-
SOUTH AFRICA	2,128	*319*	*15*	18,965	2,844	15
SWAZILAND	24	8	33	1,716	40	*2*
TANZANIA	*263*	*66*	*25*	*2,423*	*108*	*4*
ZAMBIA	271	75	28	2,221	189	9
ZIMBABWE	410	40	10	*3,930*	390	10
SADC	3,823	766	20.0	34,574	4,009	11.6

Source: Statistics established by the SADC Water Sector Coordination Unit. Figures in regular characters are abstracted from country situation reports (1996-1998). Figures in italic are rough estimates.

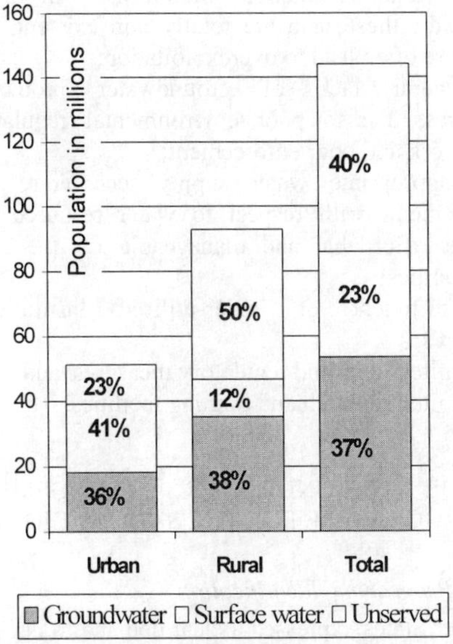

Figure 1 Source of domestic water supply in the SADC region (DRC and Seychelles not included)

source of the population having access to developed water supply (Table 2), either through private connection, public standpipe or hand pump.

Groundwater is particularly important in rural areas, in which the per capita consumption is much lower (10 to 30 l/day) than in major cities (up to several hundreds of l/day). The dominant role of groundwater in rural areas is a fairly consistent characteristic of SADC Member States. It is the only source of water during the dry periods in many areas.

As a whole, groundwater is the main source for about 37% of the total population in the SADC region, while surface water accounts for 23%. Remaining 40% of the population still rely on unsafe water sources, including open pits, shallow wells and sand river underflows, which are more often prone to drought (Figure 1). Groundwater is expected to be a major component of increased water supply coverage particularly in rural areas, and of drought management strategies in the future, provided that it is adequately developed.

2.3 *Groundwater Management*

SADC Member States that are largely dependent on groundwater are initiating the sustainable development concepts. They have effective

Table 2. Access to developed water supply in the SADC region

SADC Member State	Total population	Population Ratio Urban	Population Ratio Rural	Part of the population having access to developed water supply Urban	Part of the population having access to developed water supply Rural	Part of the population having access to developed water supply Total	Part of the population having access to developed Groundwater Supply Urban	Part of the population having access to developed Groundwater Supply Rural	Part of the population having access to developed Groundwater Supply Total
	Millions	%	%	%	%	%	%	%	%
ANGOLA	12.6	37	63	60	20	35	*9*	*18*	*15*
BOTSWANA	1.6	43	57	*98*	50	71	66	50	57
D.R.CONGO	49.0	20	80						
LESOTHO	2.1	27	73	*65*	56	58	*14*	56	*45*
MALAWI	11.2	13	87	82	49	53	25	29	28
MAURITIUS	1.2	50	50	*98*	*98*	*98*	*46*	*46*	46
MOZAMBIQUE	16.5	15	85	34	37	37	7	37	33
NAMIBIA	1.6	32	68	99	62	74	27	43	38
SEYCHELLES	0.08								
SOUTH AFRICA	42.3	59	41	80	60	72	*56*	*41*	*50*
SWAZILAND	1.0	25	75	80	45	54	4	40	31
TANZANIA	32.0	22	78	68	48	52	*20*	*29*	*26*
ZAMBIA	10.1	43	57	85	58	70	20	52	38
ZIMBABWE	11.9	31	69	*98*	80	*86*	1	72	50
SADC	193.1	32	68	77	50	60	35	38	37

Source: Statistics established by the SADC Water Sector Coordination Unit. Figures in regular characters are abstracted from country situation reports (1996-1998). Figures in italic are rough estimates.

management planning and programmes, notably Mauritius, Botswana, Namibia and South Africa. In such cases, groundwater development is based on preliminary assessment of groundwater resources at the local and national level, recharge monitoring and assessment programmes, prediction of abstraction, quality and water level monitoring. Artificial recharge is now being considered at an experimental level in Namibia, South Africa, Mauritius and Botswana.

All SADC countries have national hydrogeological maps, edited from 1980 to present (Namibia on going), but they show considerable variation in level of accuracy, aquifer classification and legend.

Storage and management of groundwater related information reflect the various perception of the importance of groundwater of individual countries. They range from basic borehole record archive to fully developed databases linked to Geographic Information Systems.

In some SADC Member States, there have been very limited efforts and funding for management of the resource to parallel development. The main drawbacks in terms of groundwater resources management are:

- less understanding of the significance of groundwater resources, as a result of poor knowledge of natural recharge and long term viability of withdrawals;
- undervaluing of the groundwater potential, especially in areas where surface water is abundant;
- depletion of aquifers due to mine dewatering and pollution from mining activities;
- limited monitoring of groundwater level, quality and abstraction monitoring. In some wellfields, these data are totally non existent, and this more often lead to overexploitation;
- general lack of groundwater protection measures, due to poor environmental regulatory framework and poor enforcement;
- inappropriate water supply technology and management, with respect to water resource and what is affordable and manageable for the local communities;
- insufficient or under-utilised institutional structures;
- limited legal and regulatory measures and
- limited research and training facilities.

3 THE GROUNDWATER MANAGEMENT PROGRAMME FOR THE SADC REGION

3.1 *Programme Identification*

Member States expressed a clear demand to develop the exchange of information, combine efforts and improve wise development and management of groundwater resources at both national and regional/SADC level.

The SADC Water Sector developed in 1998 and

Table 3. Groundwater Management Programme for the SADC Region

Project Title ⟺	Objectives	Project Title ⟺	Objectives
1- Capacity Building within the Context of Regional Groundwater Management Programme	Address the various issues on capacity of SADC Member States for the successful programme implementation	2- Develop Minimum Common Standards for Groundwater development in the SADC Region	Develop Minimum Common Standards and Guidelines for Groundwater Development.
3- Development of a Regional Groundwater Information System	Management of monitoring and groundwater data of regional significance.	4- Establishment of a Regional Groundwater Monitoring Network	Assess regional impact of recharge and drought. Joint management of river basins.
5- Compilation of a Regional Hydro-geological Map and Atlas for the SADC Region	Support to regional integrated water resource development and management.	6- Establish a Regional Groundwater Research Institute/Commission	Co-ordination of research and training activities at regional level.
7- Construct a Website on Internet and publish quarterly Newsletters	Facilitate access and exchange of groundwater related information.	8- Regional Groundwater Resource Assessment of Karoo Aquifers	Resource assessment for joint management of a major aquifer system at regional scale.
9- Regional Groundwater Resource Assessment of Precambrian Basement Aquifers	Characterisation of groundwater occurrence and assessment of the potential development of Precambrian aquifers.	10- Groundwater Resource Assessment of Limpopo/Save Basin	Develop a pilot study on surface and groundwater interaction at basin level, including drought management.

formally adopted in 1999, a Groundwater Management Programme for the SADC Region. The main objectives of the programme are:

- promote sustainable development of groundwater resources at regional scale, in terms of research, assessment, exploitation and protection;
- assess groundwater management issues in member states and Provide technical support for their assessment and management programmes (including capacity building, institutional and legal framework, policies, socio-economics and financing);
- develop the groundwater component in the regional integrated water resources development and management approach, with a particular emphasis on the role of groundwater in drought management issues and
- intensify links between national and regional levels of activities within a general framework of regional economical integration.

The Groundwater Management Programme for the SADC region includes 10 projects listed in order of priority (Table 3).

3.2 *Programme Implementation*

The basis for programme implementation is the willingness of Member States and their optimal contribution, through public sector, private sector and users participation. External funds are also requested and various cooperating partners have been approached.

At the date of March 2000, 3 projects are being implemented, or have been initiated: The projects "Establishment of Minimum Common Standards for Groundwater Development", "Compilation of a Regional Hydrogeological Map and Atlas" and "Construct a Website and publish quarterly News Letters". The Establishment of Minimum Common Standards is based on collection of data through country visits. It includes all aspects of groundwater exploration, from borehole siting techniques to contract documents and data management. It intends to provide guidelines based on regional experience and expertise. The Hydrogeological Map is developed in conjunction with the SADC Mining Sector, which is depository of many existing regional geological documents. The Website is developed in the Water Sector and covers all aspects of regional water resources.

4 CONCLUSION

The establishment of the SADC Water Sector has created a forum to enhance regional cooperation and exchange information on groundwater resources management. A regional programme for groundwater management was prepared. It reflects the commitment of SADC Member States to the regional integration. Several projects have been initiated, aiming at creating a framework for exchange of information, developing common standards for groundwater development and unifying the hydrogeological mapping in the SADC region.

Implementation of some of the programme components will be considered in the context of drought mitigation strategy, in order to plan groundwater development and move from crisis management to preparedness. They will include, in addition to the regional hydrogeological mapping, the set up of a regional monitoring network, linked to a regional information system, and on the ground demonstrations in high priority drought prone basin(s), the establishment of relevant early warning systems, the mapping of groundwater drought vulnerability, and a menu of proactive mitigation measures to combat the impact of drought.

Regional permanent institutions need to be secured to address the management of common groundwater resources at the regional level.

REFERENCES

Burke, Jacob J. and Patorni, Francois-Marie, 1993. Workshop on Water Resources Management in Southern Africa, *Proceedings of the Workshop. Victoria Falls, Zimbabwe, July 5-9, 1993, World Bank (EDI), UNDSED, University of Zimbabwe, Ministry of Lands, Agriculture and Water Development, Government of Zimbabwe.*

Chenje M., Chivasa M. Z., Serra King A.,Laisi E., 1996. Water in Southern Africa. *SADC/IUCN/SARDC, Maseru/Harare. Chenje, M. And Johnson, P., (eds.), Harare, Zimbabwe.*

FAO Land and Water Development Division, 1997. Irrigation potential in Africa, a basin approach. *FAO land and water bulletin, no 4, Roma.*

Groundwater Consultants Bee Pee (Pty) Ltd, 1998. Draft Groundwater Management Programme for SADC Region, Final Report. *SADC Water Sector Co-ordination Unit (WSCU), United Nations Environment Programme (UNEP), Maseru, Lesotho.*

M.Macdonald and Partners Limited, Hidroprojecto Consultores de Hidraulica e Salubridade, S.A., 1990. Sub Saharan Africa Hydrological Assessment SADCC Countries, Regional Report. *The World Bank, The United Nations Development Programme, Consulting Services.*

OACT, 1991. Hydrogeological Map of Africa at the scale of 1/5,000,000 . Programme of the International Hydrogeological Map of Africa, *Organisation of African Unity and African Organization of Cartography and Remote sensing, Algiers.*

SADCC-ELMS, 1991. Draft Report of the Hydrogeological Mapping Workshop, 28 January – 8 February 1991. *Southern African Development Coordination Conference, Environment and Land Management Sector .*

SARCCUS, 1995. Proceedings of the 1st Meeting of the SARCCUS Sub Committee for hydrogeology: Gross Barmen, Namibia, 20 July 1995. *Southern African Regional Commission for the Conservation and Utilisation of the Soil, SADC-WSCU archive.*

SARCCUS, 1996. Proceedings of the 2nd Meeting of the SARCCUS Sub Committee for hydrogeology: Maputo, Moçambique, 3-5 July 1996. *SADC-WSCU archive.*

SARCCUS, 1997. Proceedings of the 3rd Meeting of the SARCCUS Sub Committee for Hydrogeology: Thaba Bosiu, Lesotho, 12 – 15 May 1997. *SADC-WSCU archive.*

Stanley Consultants, 1995. *SARP* – Southern Africa Regional Water Sector Assessment, *Project no 690-0280, July 1995, Agency for International Development, USAID, Southern Africa Regional Center, Gaborone, Botswana.*

United Nations, Department of Technical Co-operation for Development, 1989. Groundwater in Eastern, Central and Southern Africa. *Natural Resources/Water Series No 19, ST/TCD/6, United Nations Publication, New York.*

Water Sector Co-ordination Unit, SADC, Sept. 1998. Regional Strategic Action Plan for integrated Water Resources Development and Management in the SADC Countries (1999-2004). Summary Report and country situation reports, *SADC-WSCU, Maseru (Lesotho), with the support of UNDP.*

Groundwater: Past Achievements and Future Challenges, Sililo et al. (eds) © 2000 Balkema, Rotterdam, ISBN 90 5809 159 7

Testing borehole injection for artificially recharging secondary aquifers in Southern Africa

E.C. Murray & G. Tredoux
Groundwater Group, Environmentek, Council for Scientific and Industrial Research, Stellenbosch, South Africa

ABSTRACT: This paper presents results from borehole injection tests on two very different aquifer systems. One is the Windhoek (Namibia) aquifer, which is highly permeable and has the ability to store tens of millions of cubic metres of water; and the other is the Karkams (South Africa) aquifer, which is a low permeability and low porosity fractured aquifer. In the case of the Windhoek quartzitic aquifer, four injection tests were done using good quality drinking water from the town's supply system. The longest test continued for 195 days at an injection rate of 17.1 L/s. In Karkams the maximum injection rate achieved was 1.5 L/s using filtered surface runoff. This paper presents the results from two borehole injection tests, one in Windhoek and the other in Karkams. Both tests indicate that artificial recharge to fractured aquifers is feasible, and that it should be encouraged for augmenting water supplies.

1 INTRODUCTION

The concept of artificially recharging Southern Africa's hard-rock, secondary aquifers is being tested at four sites that were selected after a feasibility evaluation in 1998 (Murray and Tredoux, 1998). The sites are Windhoek's quartzitic aquifer, Karkams' granitic aquifer, a cylindrical, brecciated and metasomatised sedimentary rock aquifer near Calvinia and Pietersburg's gneissic aquifer (Figure 1). This paper presents the results from the two most advanced pilot sites, namely Windhoek and Karkams.

2 WINDHOEK ARTIFICIAL RECHARGE TESTS

On average about 10% of Windhoek's water is obtained from the Windhoek aquifer. During periods of drought, this figure rises to about 25% (Van der Merwe, 1999). The rest of the city's water is obtained from three dams in ephemeral rivers and a reclamation plant. However, the aquifer is vulnerable to over-pumping. Historical data shows that after years of high abstraction the water level in the aquifer drops by tens of metres, and does not fully recover after years of low abstraction (Figure 2).

Since1950, when abstraction records were kept, the total volume of water pumped from all municipal boreholes was approximately 75 Mm^3. This water is relatively "old" water, with ^{14}C-determined ages ranging between 3 100 to 24 400 years before present. It would appear that since large scale abstraction began, most of the groundwater has come from storage, and that abstraction rates have exceeded natural recharge.

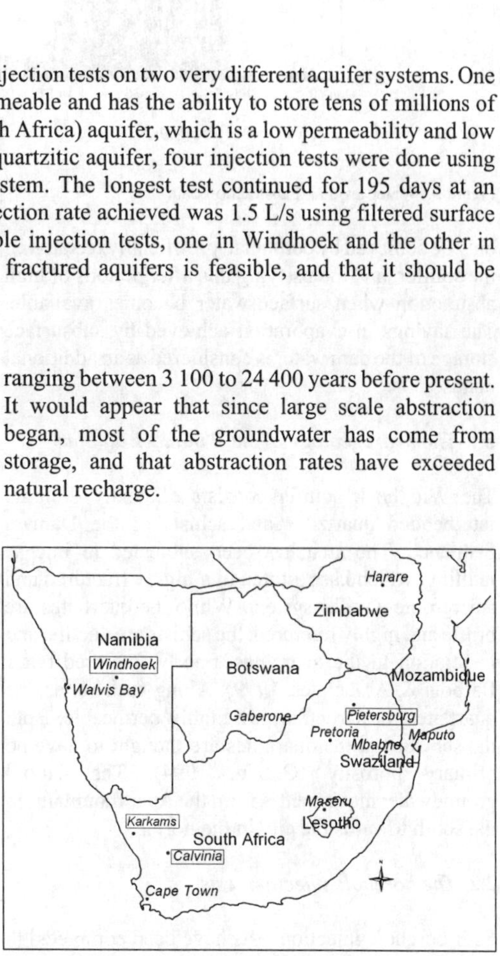

Figure 1. Location of pilot artificial recharge schemes

One aim of artificially recharging the Windhoek aquifer is to prevent the further over-abstraction of this valuable resource. The overall objective, however, is to ensure a minimum level of supply from the Windhoek aquifer during periods of drought. This can

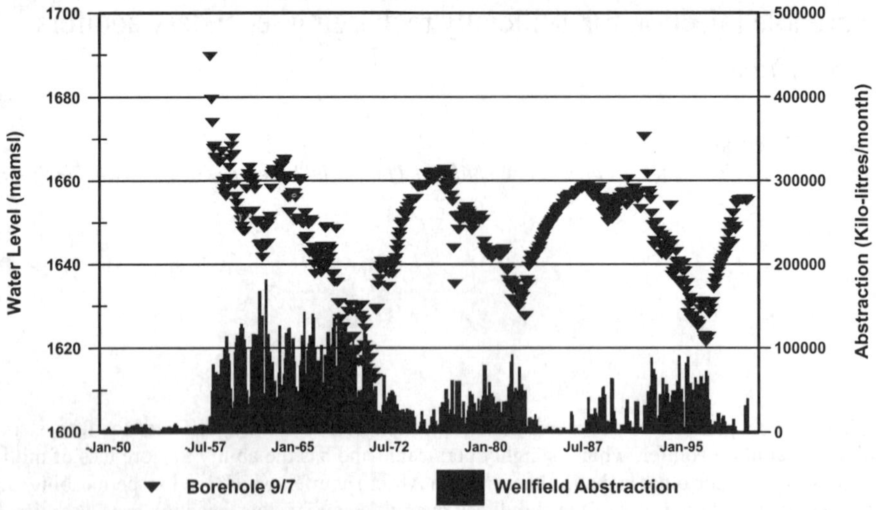

Figure 2. Water level and abstraction data

only be achieved by continually artificially recharging the aquifer, and intensifying this after periods of high abstraction when surface water becomes available. The savings in evaporation achieved by subsurface storage of the dam water is considered as an additional bonus.

2.1 *The hydrogeology of the Windhoek aquifer*

The Windhoek aquifer consists of gently dipping, interbedded quartzites and schists of the Damara Sequence. The area has been subjected to intense faulting, which has resulted in a highly fractured and anisotropic aquifer system. While the quartzites are brittle and highly fractured, the schists are ductile, and in relation to the quartzites, poorly fractured (Carr Barbour & Associates, 1999). Along fault zones, the quartzites, in particular, are highly permeable. Both the schists and the quartzites are thought to have no primary porosity (Colvin, 1994). The natural groundwater movement is from the Auas mountains in the south towards the city in the north.

2.2 *The borehole injection tests*

Four borehole injection tests have been completed to date (Table 1). This paper describes the results of one of the tests, namely at borehole12/3 (Figure 3).

2.3 *The injection test on borehole 12/3*

Borehole 12/3 was drilled into quartzites and schists to a depth of 180 m. Water was struck at 108 m, 145 m and 165 m, and the "blow"yield was 20 L/s. This part of the aquifer was deemed suitable for artificial recharge in terms of available storage and aquifer permeability.

Table 1. Pilot borehole injection tests

Injection borehole	Period (days)	Injection rate (L/s)	Total volume injected (kL)
9/5	21	2.8	5 000
9/8A	195	17.2	289 000
9/16	45	6.2	24 000
12/3	35	32.7	99 000

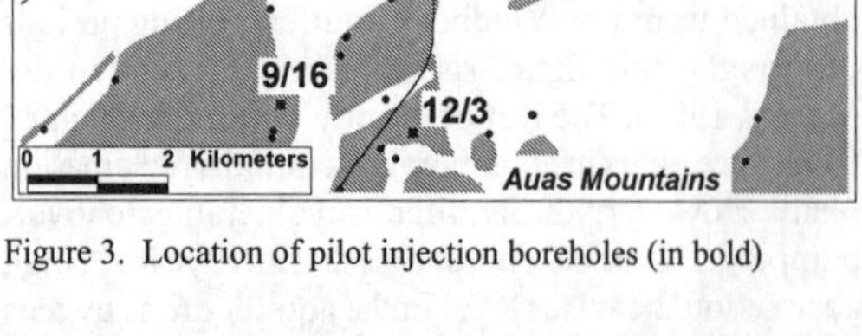

Figure 3. Location of pilot injection boreholes (in bold)

Decades of pumping from boreholes to the north of this area has induced groundwater flow from this area towards the City of Windhoek, thereby creating space in this part of the aquifer. Water injected here will thus occupy local storage and it will flow down gradient towards the hydraulic depression near the city.

This area was also seen as suitable for artificial recharge in terms of aquifer permeability. Virtually all the boreholes in this area are high-yielding, with fracture transmissivity values being in the order of several thousand m^2/day.

A step injection test was conducted at rates that ranged from 14.4 L/s to 59.4 L/s. During all the steps, the water level rapidly stabilized, thereby confirming the high permeability in this area. It was not possible to do the 35-day constant injection test at 59.4 L/s since this would have required booster pumping at a high rate for the duration of this test. Rather, gravity flow from the supply reservoir was used for the test. This gave an average rate of 32.7 L/s.

2.4 *The regional effect of injection*

Figure 4 shows the water level response in boreholes 12/1A and 12/2 which are 1 348 m and 459 m south-west of borehole 12/3, and 12/4 which is 1 133 m east of borehole 12/3. Like the previous injection tests, the response in these observation boreholes was very rapid - within hours of injection. Like all other boreholes within the wellfield which have been monitored with pressure transducers, these boreholes also show the sinusoidal water level fluctuations, which, together with the rapid water level response, indicate that the aquifer behaves in a confined manner.

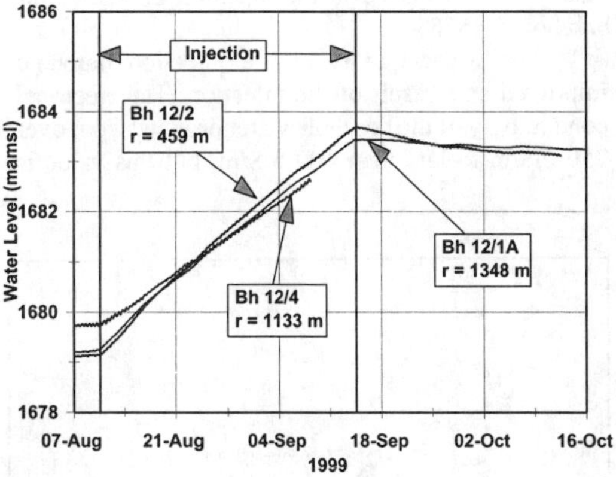

Figure 4. Water level response to injection

2.5 *Proposed artificial recharge plan*

Existing information points to three pertinent observations:

i. The groundwater has, in all probability, been mined over the past few decades;

ii. In the aquifer south of Windhoek, a high proportion of storage occurs in the quartzites of the Auas Mountains;

iii. Abstraction from the Group 9 boreholes induces groundwater flow from the Auas Mountains towards the boreholes, thereby draining the water that is held in storage below the mountains.

Should these observations be correct, then the main area of artificial recharge should be as close to the Auas Mountains as possible (Figure 5). By recharging this area, the depleted storage under the mountain will gradually be replenished, as the artificially created hydraulic mound will slow down the natural groundwater flow towards the town by "damming" the stored water below the mountain.

3 KARKAMS ARTIFICIAL RECHARGE TEST

The village of Karkams, with a mean annual rainfall of 250 mm, and a population of 2000, depends solely on groundwater. The sustainable yield of borehole G39002, which was used for trial injection purposes, is approximately 10 m^3/day, even though its drilling "blow" yield was 25.9 m^3/h. Although apparently high yields can be obtained during drilling by siting boreholes on faults, the low natural recharge in the area limits the sustainable supply of groundwater.

The aims of artificial recharge to the Karkams aquifer is to rapidly replenish the aquifer when surface runoff is available, and to improve the groundwater quality.

3.1 *The hydrogeology of the Karkams aquifer*

The Karkams aquifer consists of porphyritic granites, on which borehole G39002 is located, and older, fine-grained granitic gneisses. Most of the high yielding boreholes have been sited on the north-south striking faults that cut through this mountainous area.

Groundwater levels are generally a subdued replica of topography, with borehole G39002 being located in a high lying area close to the natural recharge area.

The groundwater salinity in the region generally increases from east to west and is directly related to elevation above sea level, which in turn, controls precipitation (Visser, et al, 1995). The groundwater quality also deteriorates with depth. Salinity profiles in the three boreholes located within 30 m from G39002, show that the electrical conductivity values increase on average from 111 mS/m at 40 mbgl to 256 mS/m at 80 mbgl.

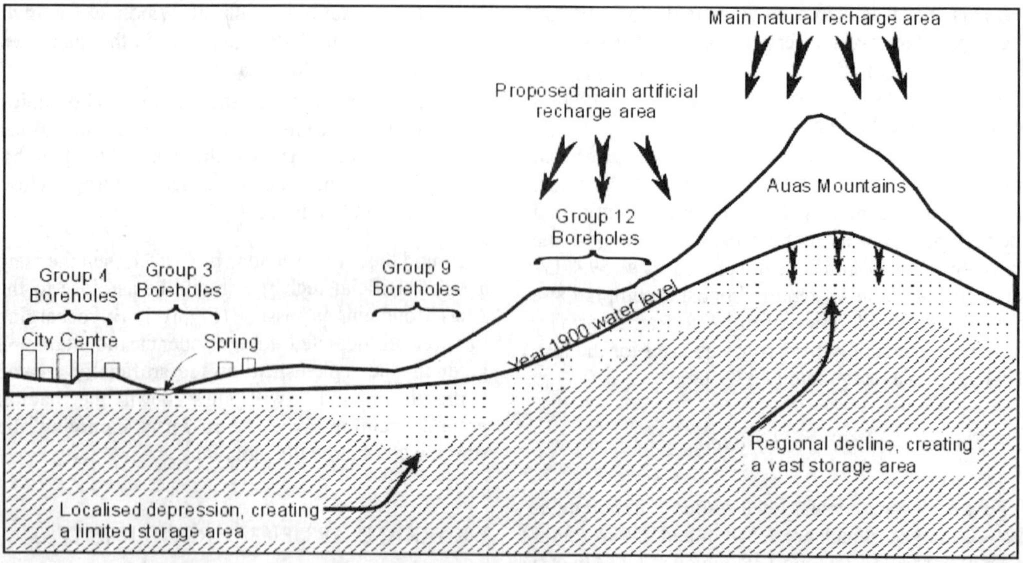

Figure 5. Hypothetical diagram showing postulated groundwater level changes since the commencement of abstraction, and the vast storage area under the Auas Mountains which has been reduced as a result of pumping (not to scale)

The fluoride content of the groundwater in the area is generally high which is typical of granitic and gneissic terrains. The trend generally follows that of salinity, with the highest values found in the low lying areas in the west (Visser, et al, 1995). The fluoride concentration of 2.8 mg/L at borehole G39002 is relatively low in comparison to some boreholes in the west which have values up to 5.3 mg/L (Visser, et al, 1995).

3.2 *The borehole injection test*

In order to test the feasibility of artificially recharging this fractured granitic aquifer, runoff was gravitated into borehole G39002 during the 28 days of streamflow in 1999. A sand filter was constructed upstream of the borehole on this ephemeral river for removing turbidity.

At first the flow through the filter was very weak, but with time it improved and gave a maximum flow of 1.5 L/s, or 130 m^3/day. This was the maximum injection rate the borehole received for this test. A total of 1042 m^3 was injected at an average rate of 37 m^3/day.

Before the test, monitoring boreholes were drilled near G39002 for assessing the aquifer response. The layout of the site is shown schematically in Figure 6.

3.3 *The effect of injection on observation boreholes*

The water level in all observation boreholes responded immediately to injection into borehole G39002. Figure 7 shows the response in the injection borehole and borehole G45758.

The water quality at the injection borehole improved as a result of the injection. The electrical conductivity of the borehole water dropped from over 250 mS/m to less than 100 mS/m, but this value is

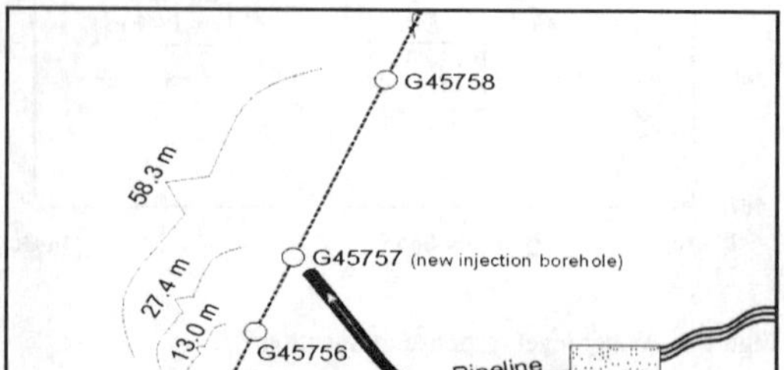

Figure 6. Layout of the borehole injection site (the injection line to G45757 is a recent addition)

increasing rapidly as a result of pumping and blending with the natural groundwater (Figure 8).

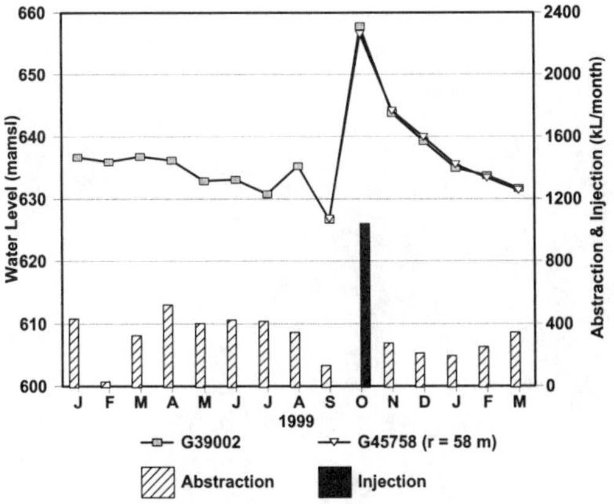

Figure 7. Water level response to injection

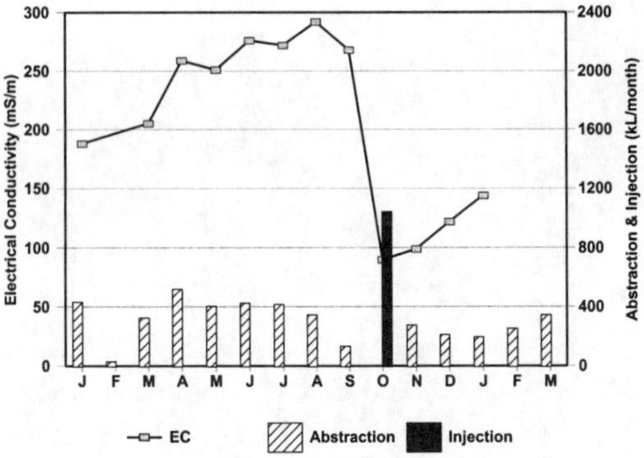

Figure 8. Water quality response to injection

3.4 *Conclusions from the Karkams injection test*

This test demonstrated that the borehole can receive the maximum flow rate that the sand filter can deliver, which is 130 m^3/day. Considering that the ephemeral river that supplies this borehole usually flows for one to three months each year, artificially recharged water should be able to supply 3 900 m^3/a to 11 700 m^3/a, which is 1.1 to 3.2 times the borehole's sustainable yield.

An additional benefit of introducing this fresh water to the aquifer, is that it will improve the groundwater quality for a few months after the injection has stopped.

4 CONCLUSIONS

The feasibility of artificially recharging secondary aquifers has been demonstrated both on a large scale with the Windhoek aquifer, and on a small scale, with the Karkams aquifer. In the case of Windhoek, the volume of water that could be injection into a single borehole in a month is in the order of 150 000 m^3, whereas in Karkams, it is about 4 000 m^3. In both cases artificially recharged water will contribute significantly towards the water needs of the towns, particularly if a number of injection boreholes are used.

Artificial recharge and the principle of conjunctive use offer valuable tools for augmenting Southern Africa's rather limited natural recharge. It is, however, essential to investigate the hydrogeology of the aquifer and the characteristics of the available water sources for providing recharge water. The design must be such that it can make full use of the available surface water.

ACKNOWLEDGMENTS

The following organisations have contributed towards this study on artificial recharge:
The Water Research Commission; the Department of Water Affairs and Forestry (RSA); the City of Windhoek; the Department of Water Affairs (Namibia); NamWater (Namibia); Carr Barbour & Associates (Namibia); the Leliefontein Transitional Local Council; and Toens and Partners.

REFERENCES

Carr Barbour and Associates. 1999. *Windhoek Wellfield Extension Project. Part 1: Final Report.* City of Windhoek.

Colvin, C . 1994. *The hydrogeology of the Windhoek Aquifer, Namibia.* Unpublished M.Sc, University College, London.

Murray, E.C. & Tredoux, G. 1998. *Artificial Recharge: A technology for sustainable water resource development.* Water Research Commission Report No. 842/1/98. Pretoria South Africa.

Van der Merwe, B. 1999. *Investigation into the viability of artificial recharge of the aquifer in Windhoek, Namibia.* Draft report to the City of Windhoek.

Visser, D., Toens, P.D., Du Toit, A. & King, J. 1995. *The importance of establishing an adequate water supply prior to approving rural community development: Karkams, Leliefontein Rural Area, Namaqualand, South Africa - A case study.* Conf. Proc. Ground Water Recharge and Rural Water Supply. Sroundwater Division of the GSSA, September 1995, Midrand, South Africa. Paper No. 7.

Groundwater: Past Achievements and Future Challenges, Sililo et al. (eds) © 2000 Balkema, Rotterdam, ISBN 90 5809 159 7

Management of groundwater resources in Lusaka, Zambia, and expectations for the future

D.C.W. Nkhuwa
School of Mines, University of Zambia, Lusaka, Zambia

ABSTRACT: Rapid growth of population in Lusaka has overwhelmed the provision of basic needs and services by the local authority, resulting, among others, in inadequate collection and disposal of waste. Because the city is underlain by karstified marbles that form the unconfined aquifer from which the city draws most of its potable water, such practices form potential sources of groundwater pollution. Further, uncontrolled drilling of boreholes to satisfy increasing demands for water is being done without due consideration to the amount of recharge to the groundwater store. Consequently, this activity appears to have contributed to dwindling water levels in the aquifer.

1 INTRODUCTION

Although people in Lusaka have lived in relative harmony with their surroundings for many years on account of their small populations, the current progressive reduction of the city's resources has been closely linked to the rise in human population. Problems of over-crowding have continued to multiply as an ever-increasing number of people is added to the city's population each year. This has mainly resulted from the rural-urban migration in search of a better life. In the last five years, the change of ownership of the mines has resulted in job losses, which has forced most of them to migrate to Lusaka.

Consequently, the city has experienced rapid rates of urbanisation, which have led to uncontrolled patterns of settlement. In many cases, this has also heightened varying forms of anthropogenic activities in the city, thereby negatively impacting on its environment.

Because of these activities, groundwater resources have increasingly become susceptible to quality and quantity depletion. In the long-term, this scenario may cause rapid escalation of water supply costs that may, in turn, seriously affect public health and the sustenance of human life. Three major factors appear to be responsible for this quality and quantity depletion of groundwater in Lusaka:

a) The nature of bedrock on which the city is built.
b) Increased use of the ground to dispose of various forms of liquid and solid wastes. Because of the unconfined nature of the aquifer, these activities make it very vulnerable to pollution.
c) Aquifer over-abstraction to satisfy rising demands for potable water.

This paper discusses these anthropogenic activities in Lusaka and their implications on the city's environment, particularly with regard to management of groundwater resources.

2 PHYSIOGRAPHY AND HYDROGEOLOGY

The plateau on which Lusaka is built stands at 1,200 m above sea level. However, to the north of the city are flat-topped hills that stand at 1,300 metres, while the land gently drops off towards the south, east, and west. The most conspicuous feature of the plateau is the lack of surface drainage, displaying only minor seasonal streams along which surface water in excess of infiltration capacity flows.

The main lithologies that underlie Lusaka comprise schists and quartzites dominated by thick and extensive sequences of marbles. Differential dissolution of the marbles has developed a system of conduits and solution channels, which occur as deep as 100 m below ground surface (Fig. 1). The presence of these features has transformed these rocks into a favourable and comparatively cheap source of water supply to the city.

The flow of groundwater in the Lusaka marbles appears to be controlled by a network of fissures/fractures along which a water circulation system has been initiated.

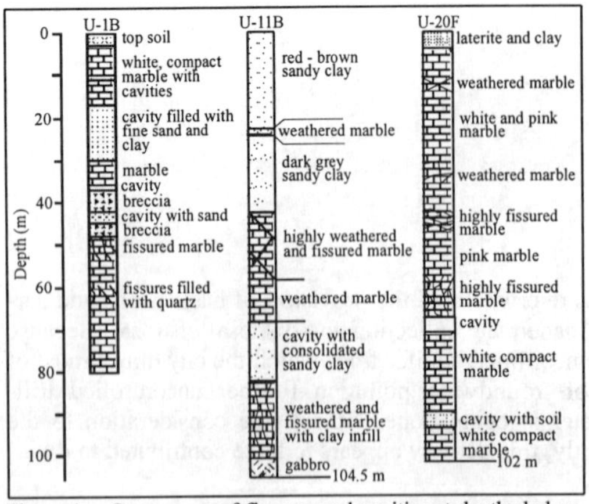

Figure 1: Occurrence of fissures and cavities at depths below 50 m in the Lusaka marbles (see Figure 2 for borehole location)

Water circulation in the marbles has resulted in enlargement of the fissures/fractures by dissolution as depicted in Figure 1. Evidence for the presence of major discontinuities in these rocks is exhibited by some recognisable surficial features, such as lineaments and the association of sinkhole-distribution with zones of high lineament density (Fig. 2). Although no studies have been conducted to determine the area of recharge of the aquifer, postulations from earlier works have indicated this area to lie to the south-eastern part of the city. However, the presence of well-developed karstic features over the entire marble terrain (Fig. 2) would indicate that recharge occurs over the whole city area underlain by marble.

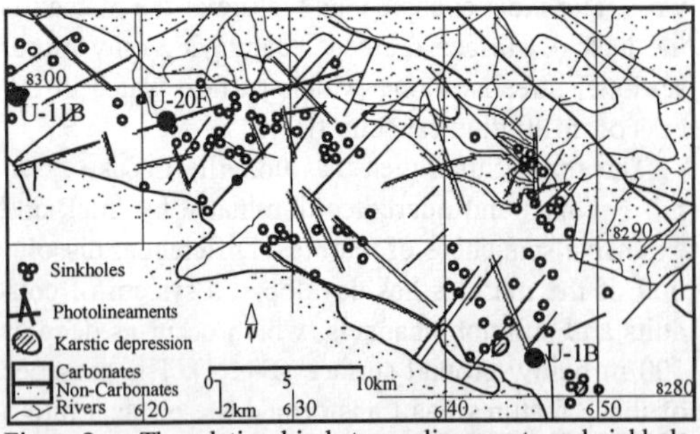

Figure 2: The relationship between lineaments and sinkhole manifestations.

In addition, shearing along the marble-schist lithologic contact must also have given rise to open conduits, which together with fracture zones, formed essential channel-ways for water into the rock mass.

Evidence for dissolution widening along lithologic contacts has been confirmed by Mdala (pers. comm.), who observed in the Chalimbana area 20 km east of Lusaka that boreholes located on the lithologic contact proved very successful and highly productive. Even some of the Lusaka Water and Sewerage Company (LWSC) borehole-drilling records indicate an evident relationship between water strikes and large scale fracturing and/or faulting/shearing (Table 1).

Table 1. Influence of geologic features on borehole yield

Borehole	Associated geologic feature	Pumping rate (ls^{-1})
Leopards Hill	breccia	360
Roadside	breccia	89
Shaft 5	breccia	100
U-5	fracturing and solution channels	50
International School	fractured marble-schist contact	50
Old Pump Stn.	breccia	83
Mass Media	fractured marble-schist contact	50
Water Works	breccia	42

Source: Drilling records from Lusaka Water and Sewerage Company (LWSC).

3 DISPOSAL OF LIQUID AND SOLID WASTE

Lusaka has one of the rapidly growing populations among the African cities. At independence in 1964, the city had a population of 195,753, which increased to 535,850 in 1980, rising to 769,353 in 1990 (Central Statistics Office, CSO 1990). By 1996, the population was envisaged to have reached one-and-half (1½) million. Currently, the population is estimated to have reached two million.

Undoubtedly, a city without a commensurate increase in the provision of social services cannot cope with such a high rate of population increase. Such is the situation for Lusaka, which has had no formal plan or strategy for water supply, as well as liquid and solid waste management. As such, current disposal practices have increased use of the ground to dispose of various forms of waste with no due consideration for the underlying geology.

Currently, for instance, about 25 % of the city's two million inhabitants are serviced by a sewer system and about 20 % by septic tanks to dispose of their sewage, while 55 % rely on pit latrines. The 55% consist of high-density residential townships that have generally developed in close proximity to areas of natural groundwater discharge or springs. Unfortunately, the drying-up of most of these springs as a result of lowering of the water table arising from aquifer over-abstraction (section 5) to satisfy rising demand for water has reversed their role to that of local recharge zones.

On the other hand, of the estimated 765,000 kg of solid waste generated each day [Lusaka City Council (LCC) & Environmental Council of Zambia (ECZ) 1997], only about 10 % are freighted to the dumpsite per day. Consequently, the larger part of the remaining waste is randomly dumped into *unofficial* dumps scattered over the city, most of which are sinkholes and surface depressions. Even the dumping of wastes at the official dumpsite at Libala (Fig. 3) is done in dugout karst conduits and solution features from which laterite was quarried for construction purposes. As such, it comprises numerous scattered excavations, which fill with water during the rainy season.

This site is estimated to be in excess of 25 hectares in size. It was licensed and approved as an official dumpsite in December 1993 as a temporary location to receive only domestic waste. However, due to delays in securing land for a new site, the Libala dumpsite is still in use today, almost seven years later.

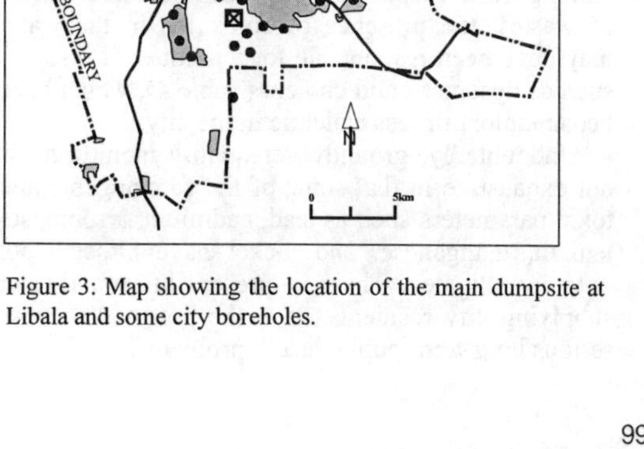

Figure 3: Map showing the location of the main dumpsite at Libala and some city boreholes.

However, instead of only domestic waste, the dumpsite currently also receives commercial and industrial wastes. A study (*unpub.*) conducted by the Lusaka City Council (LCC) in 1996 determined this waste to comprise 20 % each of plastics and metal cans, 15 % oil sludge, 10 % each of vegetable remains, building debris, paper, and glass, 3 % clinical waste, and 2% constituting other forms of waste.

With lack of monitoring at the site and laxy enforcement of existing environmental laws, the situation appears to have progressively deteriorated in the context that all types of waste are randomly disposed of throughout the area of the dumpsite. The lack of control on the type of waste dumped at the site has led to hazardous liquid and clinical wastes being frequently encountered at the site.

The site is also inadequately equipped comprising only one D4 and one D6 caterpillar wheel dozer compactors. Because of lack of repair services, these machines are rarely available at the site. In addition, the site has no reliable services of truck-type tractors, mechanical loaders, and tipper trucks. The absence of these pieces of equipment has affected effective operations at the site to compact the waste and to bring in soil to cover the compacted waste.

Unfortunately, the Libala dumpsite is located over the recharge zone of the city aquifer. As such, the presence of moisture at the site, particularly during the rainy season, constitutes the most important factor that facilitates the breakdown of the different forms of waste as it provides the basic medium for microbial activity. Further, the nature of waste-bedrock boundary and the degree of saturation of the waste constitute important hydrogeological environments in the control processes that affect attenuation of the leachate. With the envisaged subterranean channel-network in the marbles underlying the site, there is little control, if at all, at the waste-bedrock boundary to the outward migration of leachate into the surrounding strata (Fig. 4). Therefore, lack of consideration of (i) the nature of the geology and its permeability to water and leachate, (ii) the nature of groundwater regime, its gradient and seasonal fluctuations, and (iii) effective rainfall and groundwater recharge, among others, poses great potential for groundwater pollution.

In addition, the Libala dumpsite is located in close proximity to some boreholes (Fig. 3) supplying water to the city. Undoubtedly, cones of depression of some of the surrounding boreholes tap some of their recharge from seepage through the waste, thereby encouraging the unconstrained interaction of migrating leachate with groundwater. In which case, the presence of hazardous and clinical wastes in the dumpsite has the potential to produce one or more soluble elements and mobile fractions that have the potential to present some serious long-term threat to the environment and to human health.

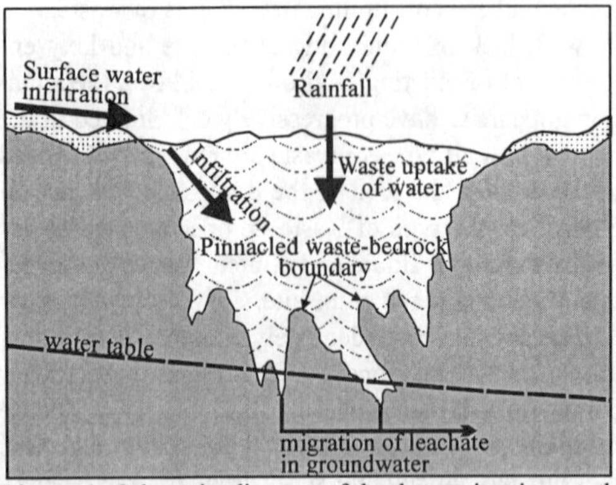

Figure 4: Schematic diagram of leachate migration at the waste-bedrock boundary into the surrounding strata.

4 GROUNDWATER QUALITY MONITORING

Most of the routine water quality monitoring is done by the Lusaka Water and Sewerage Company (LWSC) on samples obtained from its boreholes scattered over the city. Physico-chemical parameters analyzed for are given in Table 2. As indicated from this Table, some parameters such as nitrates, nitrites, conductivity and ammonia, do not meet the acceptable WHO levels of concentration for drinking water at all times.

Table 2: Chemical analyses (mean values of 42, 21, 22 and 18, respectively) of groundwater samples from some boreholes in different townships of Lusaka.

	Buckley	J/L	Kamanga	WL
Alkalinity	361	211	564	124
T-hardness	399	261	265	172
Ca-hardness	275	186	211	102
Chloride	52	102	42.3	90.1
Nitrite	0.4	1.9	0.002	41.6
Nitrate	13	39.5	10.3	0.03
Ammonia	0.08	59.7	0.08	0.9

All measurements are in mg/L; T-, & Ca-hardness = Total & Calcium hardness, respestively; J/L = John Laing; WL = Woodlands.

Source: *Unpublished data from Lusaka Water and Sewerage Company (LWSC) Limited*

Bacteriological analysis results also reveal problems due to the presence of coliforms in the groundwater for some high-density residential townships as exemplified by Buckley, Chawama, and John_Laing townships (Table 3). The conspicuous use of pit latrines in the affected areas would appear to be responsible for the bacteriological contamination of water from boreholes. Although little is known about how much treatment is done to water supplied by private boreholes, a large proportion of this water is untreated, more so in high-density areas, which generally rely on shallow hand-dug wells.

Table 3: Bacteriological analyses of groundwater from some boreholes in Chawama (September 1995 – June 1997) and Buckley (September 1995 – March 1996)

Date	Total coliforms	Faecal coliforms
CHAWAMA		
26.09.95	21	
16.10.95	1	
14.11.95	25	21
23.11.95	17	
26.12.95	TNTC	
14.01.96	124	
07.03.96	28	
12.09.96	16	
18.10.96	13	
20.12.96	3	
04.02.97	1	
04.04.97	TNTC	
04.06.97	TNTC	
23.06.97	TNTC	TNTC
BUCKLEY		
19.09.95	6	
01.11.95	3	
29.11.95	TNTC	20
09.01.96	3	
16.01.96	10	
08.02.96	3	
20.02.96	3	
01.03.96	1	
19.03.96	3	

All readings are in counts per 100 milliltres.
TNTC = Too Numerous To Count
Source: Unpublished data from Lusaka Water and Sewerage Company (LWSC) Limited.

Among other causes resulting from mismanagement of wastes, the presence of coliforms in the water may have been responsible for a number of diseases such as dysentery and cholera (Table 4), which have become more or less endemic in the city.

Undoubtedly, groundwater quality monitoring is not exhaustive in that some of the carcinogenic and toxic parameters such as lead, cadmium, arsenic, selenium, manganese, and nickel have hitherto not been investigated. These, if present in groundwater supplying city residents, have the potential to pose serious long-term public health problems.

Table 4: Incidence of cholera cases reported to Health Centres for high-, medium-, and low-density residential areas.

High-density areas	Cases reported
Bauleni	24
Chainda	2
Chaisa	173
Chawama	246
Chipata	123
Chunga	87
Garden	233
George	300
Kalikiliki	7
Kamanga	61
Kanyama	383
Kaunda Square	17
Mtendere	40
Misisi	109
Medium-density areas	
Chilenje	11
Kabwata	10
Libala	5
Matero	51
Low-density areas	
Avondale	3
Bimbe	5
Total	1,948

Source: Public Health Department, Lusaka City Council

5 EFFECTS OF GROUNDWATER OVER-ABSTRACTION

The above scenario notwithstanding, rapid growth of population in Lusaka has overwhelmed the rate of provision of water supplies. For instance, the current population in the city estimated at two million, at a per capita consumption of 200 litres per day, requires about 4×10^5 m^3 d^{-1} (146×10^6 m^3 per annum) to satisfy its domestic requirements. The current optimistic daily supply of about 2.5×10^5 m^3 (91.25×10^6 m^3 per annum) comprises 100,000 m^3 per day from a surface source situated 50 km south of the city and 105,000 m^3 from the Lusaka aquifer. Undoubtedly, the conspicuous deficit in supply must come from ground sources because of, among others, its comparatively low capital investment. In addition, its exploitation has the added advantage that as *a hidden resource*, it does not portray direct competition and conflicts among different user-interests that the surface source may create from different users along its course.

However, hydrologic studies conducted by Nkhuwa (1996) reveal that there is an average of about 200 mm of replenishment to the groundwater store per annum. Taking a total surface area of 680 km^2 for Lusaka and its environs, an average of about 1.4×10^8 m^3 of recharge is envisaged per annum. Therefore, to fulfil the shortfall in supply to the domestic requirements, alongside fulfilling demands for agriculture and industry from ground sources within the city aquifer may invariably *suck-up* the annual replenishment from the groundwater store.

In spite of the above circumstances, the city aquifer has progressively experienced indiscriminate drilling of boreholes in an effort to augment the shortfall in supply. In places, this has caused aquifer over-abstraction, which has initiated lowering of the water table (Fig. 5). Although this represents too short a period to portray very meaningful long-term behaviour of the aquifer, it provides the only longest and continuous monitoring data available. If this should portray the true long-term trend, natural recharge may, sooner than later, take so long to replenish the aquifer that future generations may essentially lose this vast source of groundwater supplies.

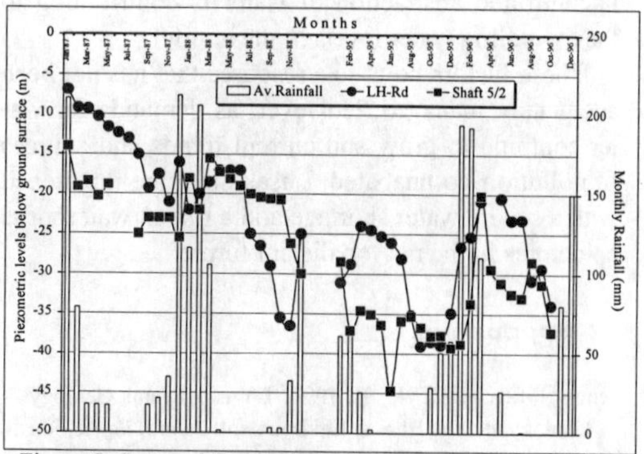

Figure 5: Seasonal fluctuations of water levels in the Leopards Hill Road and Shaft 5 boreholes.

Lowering of the water table generally has the consequence of withdrawing the water pressure in the pores of the aquifer that helps support the weight of the overlying rocks and keeping soil grains from collapsing upon each other. Consequently, soil particles of the aquifer settle, resulting in either compaction or stoping and collapse of the ground. Among the long-term consequences of these processes, will be reductions in borehole-yields and subsequent drying-up of some boreholes, signifying a reduction or complete sealing of waterways in the aquifer. This has led to the drying-up of most springs and some boreholes.

Further, progressive lowering of the water table will, among others, increase production costs and initiate the development of ground instabilities. Consequently, there will be:

- Disruptions in the land surface
- Structural damage to buildings and roads.
- Damage to buried cables, pipes and drains. In addition, this may also constitute an important factor in increasing the amount of land susceptible to flooding as is already becoming increasingly frequent in Lusaka.

6 CONCLUSION

Rapid growth of population and lack of forward planning by city authorities to provide residents with basic needs and services has led to inadequate collection and disposal of wastes and the provision potable water.

The siting of the official and other unofficial dumpsites over the aquifer recharge area raise serious risks of groundwater pollution and far-reaching public health implications for its residence. Further, uncontrolled abstraction of water is contributing to the dwindling water levels in the aquifer.

These factors point to a resource that has not been adequately managed. However, as demands for water continue to grow and current trends and sources of pollution go unabated, Lusaka may be threatened with a severe water shortage and a rise in waterborne epidemics in the not too distant future.

7 REFERENCES

Bennet, M.R. & Doyle, P. 1997: Environmental Geology – Geology and the Human Environment. *John Wiley & Sons.*

Central Statistical Office. 1990: 1990 Census of population, Housing and Agriculture – Preliminary report, 17pp, Lusaka.

Lusaka City Council (LCC) & Environmental Council of Zambia (ECZ), 1997: Solid waste management master plan project for the city of Lusaka. Phase 1 – Diagnosis Final Report

Muck, B.W., Skinner, B.J. & Porter, S.C. 1996: Environmental Geology. *John Wiley & Sons, Inc.*

Nkhuwa, D.C.W. 1996: Hydrogeological and Engineering Geological problems of urban development over karstified marble in Lusaka. *Mitt.Ing. –u. Hydrogeol., Heft 63, 251S, Aachen*

Nkhuwa, D.C.W. 1997: Waste disposal practices and the state of environmental health in Lusaka. Paper presented at the National Conference on Science and Technology on 'The Role of Science and Technology in socio-economic development in Zambia', 26-30 August 1997, Lake Kariba Inns, Siavonga

Groundwater: Past Achievements and Future Challenges, Sililo et al. (eds) © 2000 Balkema, Rotterdam, ISBN 90 5809 159 7

Determination of the groundwater component of the reserve using the intermediate reserve determination method

Roger Parsons
Parsons and Associates Specialist Groundwater Consultants, Somerset West, South Africa
Heather MacKay
Department of Water Affairs and Forestry, Pretoria, South Africa

ABSTRACT: The recently promulgated National Water Act of South Africa requires water for basic human needs and aquatic ecosystems be reserved before allocation to other water users. Determination of the Reserve can be done at four levels of detail and confidence. This paper describes the method to set the groundwater component of an intermediate Reserve determination. The method centres around determination of recharge, adjustments for basic human water needs and baseflow and ongoing monitoring.

1 INTRODUCTION

In response to the democratisation of South Africa in 1994, outdated water legislation was revised. The National Water Act, promulgated in 1998, is based on guiding principles of equity and sustainability. The new Act recognises a unitary hydrological cycle, thus requiring adoption of integrated water resource management in the country. This is considered a major advancement on earlier legislation which regarded surface water as public water and groundwater as a private water resource.

2 THE RESERVE

The National Water Act also recognises basic human water needs of present and future generations as well as the legitimate right of the environment to an adequate supply of water. Therefore, a volume of water has to be set aside or reserved before allocating water for other uses. This is referred to as the Reserve and is defined by the Act as 'the quantity and quality of water required to satisfy basic human needs by securing a basic water supply (as prescribed under the Water Services Act 1997) for people who are or who will in the reasonably near future be relying upon, taking water from, or being supplied from the relevant water resource; and to protect aquatic ecosystems in order to secure ecologically sustainable development and use of water resources.'

At present basic human need is set at 25 L/p/d.

The Act requires the Reserve be determined before any water is allocated for other uses. It is the Minister of Water Affairs and Forestry's responsibility to determine the Reserve for significant water resources.

3 LEVELS OF RESERVE DETERMINATION

Methods and tools needed to determine the Reserve are currently being developed (DWAF 1999). It was recognised it would take a number of years for these to be put in place. It was also recognided it would take years or decades for Reserve determinations for all 1 946 quaternary catchments to be completed. The Act therefore allows for preliminary determinations to be made. Four levels of Reserve determination are envisaged, the first three of which are considered preliminary determinations while comprehensive determinations will only be accepted as Reserve determinations once the methods are accepted in the Government Gazette:

- desktop estimate - very low confidence and should take about 2 hours to complete;
- rapid - low confidence and should taken about 2 days to complete, allows for a quick field assessment;
- intermediate - medium confidence and should take about 2 months to complete, allows for specialist field studies;
- comprehensive - high confidence assessment to be completed within 8 to 12 months, allows for extensive field data collection.

This paper focuses on the method to determine the groundwater component of an intermediate Reserve determination (IRD). Methods to determine the groundwater component of desktop and rapid level determinations are described by Haupt (1999) and DWAF (1999) while comprehensive tools are being developed separately by Braune et al. (1999).

4 ENVIRONMENT WATER REQUIREMENTS

Increasing demand on water resources has resulted in more and more attention has being given to water requirements of the environment (Tharme 1996; O'Keefe et al. 1998; Hatton & Evans 1998; Petts et al. 1999). Methods to determine instream flow requirements (IFR) have been developed in South Africa by Tharme & King (1998) and others. These methods form the basis of the IRD method. However, IFR methods essentially focussed on rivers. To ensure the entire hydrological cycle (and functioning thereof) was properly considered, components addressing groundwater, wetlands and estuaries have been included in the Reserve determination methods.

It is not readily possible to calculate the volume of water required by the environment. It is therefore assumed environmental water requirements are met if a certain depth of water is left in place. For ease of implementation, the required depth is converted to a flow rate by means of hydraulic modelling.

Whereas surface water components revolve around depth of water to be left in the river, such an approach is inappropriate for groundwater. Rather, it is depth to groundwater which is important as this provides various parts of the environment access to subsurface water. Lowering the water table or piezometric surface reduces or removes such access. It was therefore decided the groundwater component of the IRD method be set such that groundwater abstraction would not induce a lowering of the regional depth to groundwater. In essence, this requires the sustainable volume of abstractable groundwater be calculated. This is referred to as a groundwater allocation. Similar approaches were proposed independently by O'Keefe et al. (1998) and Petts et al. (1999).

5 METHODOLOGY

5.1 *Generic process*

Based on experience from setting IFR's, a generic six step process was initially proposed. This was later modified to a seven step process when it became apparent monitoring was an important means of assessing the validity of Reserve assessments. The seven step process has been found to be robust and applicable to the surface water, groundwater, wetland and estuary components. Adoption of the generic process by the four components has resulted in a high level of integration during Reserve determinations.

Generally the need for a Reserve determination is triggered by the Minister of Water Affairs and Forestry. Typical triggers include license applications, protection of high priority areas and management of areas with high water demand and hydrological stress.

5.2 *Delineate boundaries*

The first step in the process is to delineate boundaries of significant water resources. Quaternary catchment boundaries have been selected as default boundaries of a significant water resources. However, a significant water resource could comprise parts of or more than one quaternary catchment. The selected boundary is generally applicable to the four components of the resource.

5.3 *Delineate geohydrological response units*

The surface water component delineates areas within a significant water resource with similar or characteristic ecological functioning (Kleynhans et al. 1998). Similarly, the groundwater component defines geohydrological response units in which the geohydrological characteristics and functioning are similar. A two-step process is used, the first of which requires determination of homogenous response unit. This is usually based on geology or aquifer type. A dolomitic aquifer and a fractured and weathered sandstone aquifer within a significant water resource, for example, would be delineated as separate homogenous response units. National and regional scale geohydrological maps prepared by the Water Research Commission and Department of Water Affairs and Forestry have proved to be invaluable support tools for this task..

The second step requires delineation of geohydrological region types. This concept was based on earlier work by Braune & Dziembowski (1997). The geohydrological region type classification is based on the role groundwater plays in supporting the environment (Figure 1). The types are grouped according to whether they maintain system integrity, discharge integrity or ecological integrity. Based on the role played, an area is excluded from the recharge calculation, the groundwater allocation reduced by the required base flow or local drawdown limitations set as part of Resource Quality Objectives (RQOs). RQOs

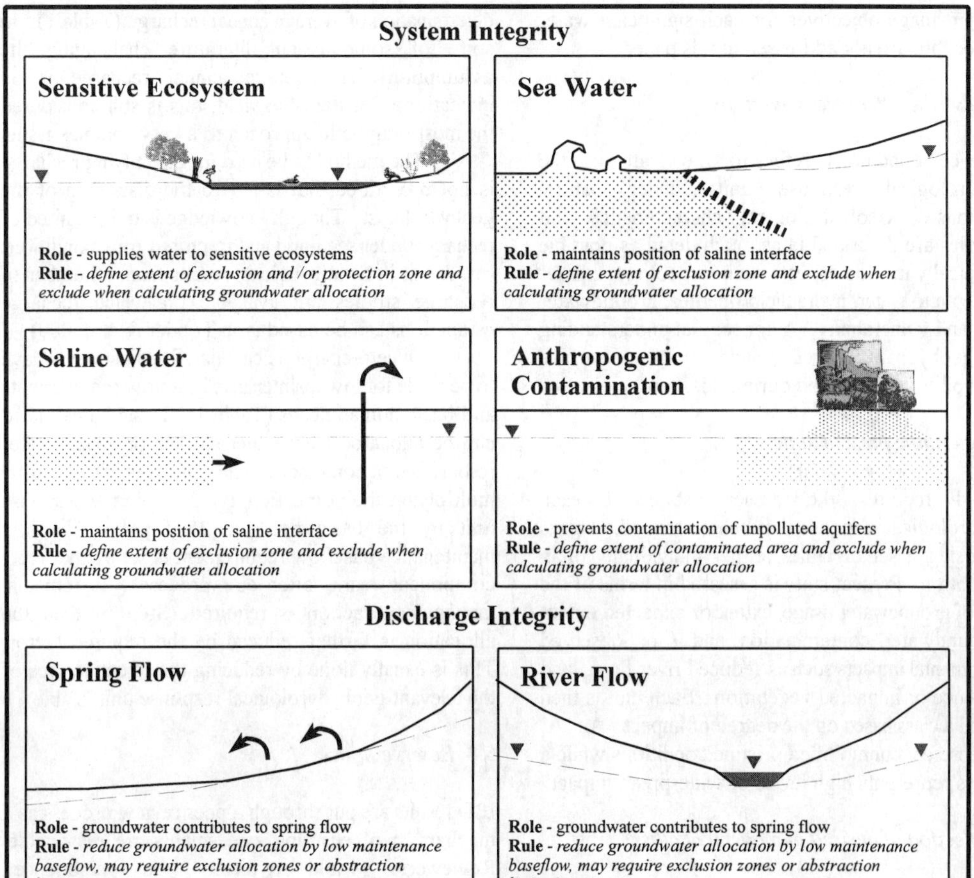

Figure 1: Geohydrological region type classification (DWAF, 1999)

specify manage objectives for each significant water resource, but are not addressed in this paper.

5.4 *Describe reference conditions*

Reference conditions refer to the state of the geohydrological response units under natural conditions. Geohydrological characteristics and properties are described in as much detail as possible and typically include aspects such as geology, aquifer type, depth to water, hydraulic properties, groundwater quality and vulnerability. A conceptual understanding of the geohydrological regime and linkages to surface water bodies is developed during this step.

5.5 *Determine present state*

Step 4 requires the present state of each geohydrological response unit be determined. This is done using a set of tables, either individually or in combination. Present state is assessed in terms of the extent of groundwater usage, extent or expected extent of groundwater contamination and / or observed environmental impacts such as reduced river flow, land subsidence or impacted vegetation. Each unit is then assigned a class based on the degree of impact. An 'A' class represents unmodified pristine conditions while a 'F' class represents high levels of widespread impact.

5.6 *Selection of desired management class*

A degree of environmental impact is inevitable whenever water is abstracted. During this step, the acceptability of the impact determined in Step 4 is assessed. If the level is considered acceptable, the resource can continue to be used and managed in it's present state. However, if the impact is considered unacceptable, mitigation measures have to be implemented so the state of the resource is improved.

During comprehensive Reserve determinations, this step requires extensive input from the public. Because of the 2 month time constraint of the IRD method, the desired management class is usually set by the Minister based on input from the IRD team.

The selected desired management class is then used to define monitoring and frequency of review requirements as well as protection and management actions. Actions include reduction of the groundwater allocation by a prescribed factor and exclusion of contaminated areas from recharge calculations.

5.7 *Quantify the groundwater allocation*

The basis for quantifying the groundwater allocation is determination of average annual recharge (Table 1). In spite of some recent literature challenging the assumption that average annual recharge is an indication of sustainable yield, this is still considered the most pragmatic approach to a very complex issue.

The method to be used for quantifying recharge is not prescribed, but is left to the discretion of the geohydrologist. Though knowledge is quite limited of recharge under semi-arid and fractured rock conditions (prevalent in South Africa), results from a number of recharge studies are available on which recharge estimations can be based (van Tonder & Xu 2000).

Once recharge is calculated, adjustments have to be made for low maintenance baseflow requirements and basic human needs (Table 1). Basic human need can be allocated from either surface or groundwater resources or a combination thereof. It is calculated by multiplying the population living within the response unit by the basic need of 25 L/p/d. The low maintenance baseflow requirement is set by the rivers component team, after due process. If remedial management action is required (Step 5) then the allocation is further reduced by the required factor. This is usually done by reducing the effective area of the relevant geohydyrological response unit (Table 1).

5.8 *Reserve notice*

IRD results are put through a peer review process as a mechanism of ensuring assessments are reasonable. Reviewers would typically be experienced geohydrologists familiar with the geohydrological regimes being assessment. Once published in the gazette, the determination can be used as the basis for allocating water. If water users in a catchment disagree with the outcome of the assessment, they can petition the Minister by submitting an independent Reserve determination for his consideration.

5.9 *Monitoring*

By definition, the IRD method provides an assessment of the Reserve at a medium level of confidence. The only means of checking the validity of assessments is by means of ongoing monitoring. Groundwater monitoring (both level and quality) in South Africa is still in its infancy and lags behind comprehensive surface water monitoring networks (flow, quality, river health etc.). The ability of the Minister to set the groundwater component of the Reserve will ultimately depend on ability to check the validity of assessments.

Table 1. Data sheet used when assessing the groundwater component of the Reserve using the IRD method

BOUNDARIES AND TYPING		MANAGEMENT CLASS		RECHARGE				ADJUSTMENTS		ALLOCATION	
Homogeneous Response Unit	Geohydrological Region Type	Current Status	Desired Management Class	Total Area	Effective Area	Recharge Method	Annual Recharge	Low Maintenance Baseflow Adjustment	BHN Adjustment	GRU Groundwater Allocation	Confidence
				(km^2)	(km^2)		(10^6 m^3)	(10^6 m^3)	(10^6 m^3)	(10^6 m^3)	
Dolomitic aquifer	Terrestrial Vegetation	A	a	50	50	20% MAP	7.00	0.00	0.00	8.89	very high
	River Flow	A	a	30	30	20% MAP	4.20	3.50	0.01		
	High Volume Abstraction	D	b	20	15	20% MAP	2.10	0.00	0.90		
Sandstone aquifer	Terrestrial Vegetation	A	a	80	80	5% MAP	2.80	0.00	0.00	1.85	medium
	River Flow	A	a	30	30	5% MAP	1.05	2.00	0.00		
	Anthrop. Contamination	E	c	25	0	exclud.	0.00	0.00	0.00		
				Total for Significant Water Resource			17.15	5.50	0.91	10.74	

Notes

1. Theoretical example for illustration purposes.
2. Effective area of dolomite aquifer high abstraction zone reduced because of 'c' management class

6 PILOT TESTING

The IRD method was tested on the Crocodile and Pienaars Rivers. The method was found to work well while geohydrological input into the IRD process provided valuable information regarding hydrological functioning. During both tests integration of surface and groundwater proved problematic. Though the conceptual understanding of interaction is considered to be quite good, quantification of baseflow and recharge did not always match. Close collaboration by the responsible hydrologist and geohydrologist throughout the process provided a mechanism for overcoming this problem. However, fundamental research of surface-groundwater interaction and quantification thereof is urgently required.

7 DISCUSSION

Although considered visionary, the new Act fails to take account of unique characteristics of groundwater. Large parts of the geohydrological regime are, to an extent, excluded from the Reserve as groundwater supports both aquatic and terrestrial ecosystems. Whereas the Reserve is an effective protection mechanism for rivers, estuaries and wetlands, this is not true for the geohydrological component. A separate set of resource directed measures is required to address groundwater protection needs outside the Reserve. Careful formulation of groundwater protection measures under Chapter 3 of the National Water Act will embrace the principal of an integrated water resource on which the Act is based.

8 CONCLUSIONS

A method to determine the groundwater component of the Reserve within a 2 month period has been developed. The method is based on determining average annual recharge of geohydrological response units. Determined recharge is then adjusted by taking into account basic human needs and low maintenance baseflow requirements as well as remedial management requirements. Peer review, regular reassessment and ongoing monitoring form important mechanisms for assessing the validity of determinations. The developed method is a first attempt at quantifying the groundwater component of the Reserve. A number of shortcomings in our geohydrolgical knowledge and the impact of groundwater abstraction on the environment, however, still need to be addressed.

REFERENCES

Braune, E. & Dziembowski, Z.M. 1997. Towards integrated groundwater and surface water management in South Africa. Paper presented at 8th South African National Hydrological Symposium, Pretoria.

Braune, E., Xu, Y. & van der Voort, I. 1999. WRC programme progress report - Groundwater Reserve Determination. unpubl. report K5/1091 - Doc 3/99, Water Research Commission, Pretoria.

Department of Water Affairs and Forestry 1999. Resource directed measures for protection of water resources - Version 1.0. Department of Water Affairs and Forestry, Pretoia.

Hatton, T. & Evans, R. 1998. Dependance of ecosystems on groundwater and its significance to Australia. Occasional paper No. 12/98, Land and Water Resources Research and Development Corporation, CSIRO, Australia.

Haupt, C. 1999. Water resources situation assessment study of South Africa - groundwater; report in preparation, Water Systems Management, Pietersburg.

Kleynhans, C.J., Sibberbauer, M. & Kemper, N. 1998. Preliminary ecoregion 1 classification for South Africa. unpubl. internal report, Institute for Water Quality Studies, Department of Water Affairs and Forestry, Pretoria.

O'Keefe, V., Gates, G. & Kalaitzis, P. 1998. Developing groundwater policy for sustainable groundwater management in NSW, Australia. IAH Int. Conf. Proc., Melbourne, February 1998, pp 857 -862.

Petts, G.E, Bickerton, M.A & Crawford, C. 1999. Flow management to sustain groundwater-dominated stream ecosystems. *Hydrol.Process.* Vol. 13, 497 - 513.

Tharme, R.E. 1996. Review of international methodologies for the quantification of the instream flow requirements of Rivers. unpubl. report prepared for DWAF, Freshwater Research Unit, University of Cape Town.

Tharme, R.E. & King. J.M. 1998. Development of the Building Block Methodology for instream flow assessments and supporting research on the effects of different magnitude flows on riverine ecosystems. WRC Report 576/1/98, Water Research Commission, Pretoria.

Groundwater: Past Achievements and Future Challenges, Sililo et al. (eds) © 2000 Balkema, Rotterdam, ISBN 90 5809 159 7

Challenges in evaluating solute transport from a long-term ASR trial in a heterogeneous carbonate aquifer

P. Pavelic & P.J. Dillon
CSIRO Land and Water and the Centre for Groundwater Studies, Glen Osmond, S.A., Australia

N.Z. Gerges
Department for Water Resources, South Australia & Centre for Groundwater Studies, Adelaide, S.A., Australia

ABSTRACT: A five-year Aquifer Storage and Recovery (ASR) trial at Andrews Farm, South Australia has provided the opportunity to study advection and dispersion processes in a deep heterogeneous aquifer. Three observation wells were drilled to monitor the fate of fresh stormwater injected into the brackish ambient groundwater. Over four successive seasons, over 250 ML of stormwater was injected into the aquifer and 60% of the injected volume was recovered in the fifth year. Using chloride as a tracer, breakthrough responses at observation wells were measured as well as the proportion of injectant in recovered water. Temperature profiles clearly showed the heterogeneous nature of the aquifer. A direct hydraulic connection between the injection well and nearest observation well, together with the incomplete penetration of the remaining observation wells denied the opportunity to develop an acceptable conceptualisation of the aquifer, and to develop a model which could be useful in a predictive sense. However, this study may offer some insight into the design of future solute transport studies involving ASR.

1 INTRODUCTION

Aquifer Storage and Recovery (ASR) is a form of artificial recharge of groundwater that makes use of a single well to inject and recover water (Pyne, 1995). It is particularly appealing in regions where there is a seasonal shortfall between water availability and supply. Such is the case for much of Australia, and as a result, aquifers which have little or no beneficial use due to their high salinities are becoming targets for ASR using large quantities of urban stormwater and treated sewage effluent for subsequent non-potable reuse (Dillon *et al.* 1997,1999).

Since the ambient groundwater is more saline than the injected water the usefulness of the recovered blend is defined by its intended use. The salinity of the recovered water depends on the degree of mixing which occurs during aquifer storage. The quantity of water that can be usefully recovered, expressed as the proportion of the injected volume in any injection/recovery cycle is the so called "recovery efficiency" (Merritt, 1986). Recovery efficiency at any given site is not constant, and in general, tends to increase in successive injection/recovery cycles as the proportion of ambient groundwater in the zone occupied by the injected water decreases.

The intrinsic nature of the subsurface as well as the way in which a site is operated make recovery efficiency difficult to predict with reasonable confidence without conducting field trials. Recent trials in South Australia have shown that recovery efficiency in the first year may vary from a few percent in fissured limestone to over 100% in lower permeability limestones containing brackish groundwater (Gerges and Howles, 1998).

Whilst information of this type is emerging, there are still knowledge gaps, and particularly on the role of heterogeneity on solute transport and recovery efficiencies in aquifers. Whilst tracer tests are the most commonly applied approach, we have selected to examine transport within the context of a long-term, large-scale ASR field trial. It will be made apparent that transport processes in heterogeneous aquifers are complex; that differences between tracer tests and ASR are not subtle, and they create challenges in the interpretation of the trial. In documenting these results, we have attempted to highlight the main issues and lessons learnt.

2 DESCRIPTION OF SITE AND EXPERIMENTAL PROGRAM

The study was conducted at the Andrews Farm site, situated in the northern metropolitan area of Adelaide in South Australia. Over its five year duration this site was the focus for research on well clogging, geochemical reactions, and attenuation of chemical contaminants and pathogenic microorganisms (Dillon *et al.* 1997).

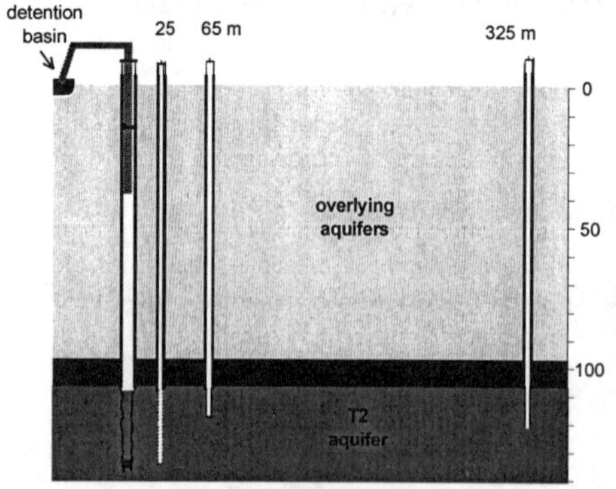

Figure 1. Vertical cross section.

The aquifer targeted for ASR, known as the "T2" aquifer, is the second of a series of confined Tertiary marine sediments underlying the 60 m thick surficial Plio-Pleistocene Hindmarsh Clay formation. Intersected at a depth of 105 metres and confined beneath 9 m of highly plastic Munno Para Clay, it consists of interbedded sequences of variably cemented carbonate (calcarenite) and sand (sandstone).

Although the T2 extends to a greater depth, the injection/recovery well was completed as an "open hole" in the upper sequence (108 to 127 m) which was less susceptible to collapse since the sediments were better consolidated than at greater depths (Fig. 1). A bulk aquifer transmissivity of 180 $m^2 day^{-1}$ and storage coefficient is $5x10^{-4}$ have been measured through aquifer testing (Gerges *et al.,* 1995). Three observation wells were drilled along a transect at distances of 25, 65 and 325 metres down-gradient of the ASR well to trace the movement of the injected fluid (Fig. 1).

The source of recharge water is a stormwater collection system comprising of three inter-connected detention basins. During the trial there were four years of stormwater injection between November 1993 and October 1996, with an earlier mains water injection test in August 1993, which is referred to as "1993a" in Figure 2. These injections took place during the winter to spring period when excess runoff was available for recharge.

Note that the volumes of stormwater injected roughly doubled in each successive year due to a combination of improved operational procedures and higher rainfall. A net total of 256 ML of water was added to groundwater storage with little abstraction up to the end of 1996 apart from a small recovery to refill the wetland during the summer of 1993/94. In all cases, the average rates of injection varied between around 1,300 to 1700 $m^3 day^{-1}$.

The rationale for allowing multiple years of injection prior to recovery was for full breakthrough to occur in at least one of the observation wells. Since this did not occur (in the strictest sense) after the four years, for reasons which become clear later, a major recovery phase was conducted in the following year (July 1997 to July 1998). The recovered volume was 150.8 ML, 60% of the total volume injected.

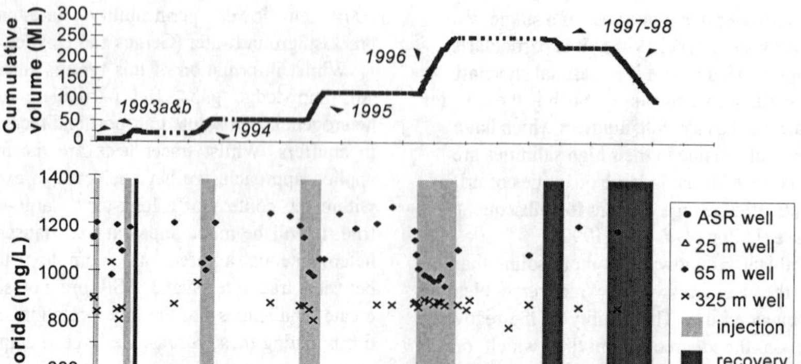

Figure 2. Cumulative volume of water stored in aquifer and chloride concentration in injection and observation wells over the duration of the experiment. *In the upper plot, periods of pumping are inferred by a decrease in volume whilst periods of storage by no change.*

Table 1. Injectant, ambient groundwater and recovered water concentrations for five environmental tracers.

	Injectant [1]					Ambient groundwater [4]	Recovered water [5]
	1993a [2]	*1993b* [3]	*1994*	*1995*	*1996*		
Volume injected (ML)	*6.5*	*19*	*32.7*	*64.9*	*132.8*		
EC (μS cm^{-1})	490±10	274±13	296±77	238±82	263±46	3400 - 4300	320 - 1920
Chloride (mg L^{-1})	93±3	27±8	40±15	25±18	26±7	850 - 1200	30 - 473
Temperature (°C)	14±<1	20±1.2	10±2.4	13±1.0	17±2.1	24 - 25	21.5 - 18.3
Deuterium (% SMOW)	<--------------------		-26.3 - 1.9		-------------------->	-25.0 - -22.4	n/a
Oxygen-18 (% SMOW)	<--------------------		-4.6 - -1.3		-------------------->	-4.7 - -3.9	n/a

1 means and standard deviations in each year (ranges over 4 years only for isotopes)
2 mains water injection 3 stormwater injection 4 mean concentration at each well 5 values at beginning and end

There were several main factors which made evaluating the trial more difficult:

(1) methods employed to develop the ASR well substantially altered the characteristics of the aquifer in the immediate vicinity of the well
(2) in this same vicinity, aquifer properties varied with time through clogging and unclogging processes upon commencing injection. Many tonnes of sediment were introduced and subsequently recovered each year (Pavelic *et al.* 1998), whilst calcite dissolution mobilised sand grains, altering the total porosity and its distribution in the aquifer (Rattray *et al.* 1996).
(3) injection events were intermittent due to lack of available water and or well clogging, which lasted up to several weeks in the worst case
(4) the slotted intervals of the 65 and 325 m observation wells did not cover the entire depth subject to injection and recovery (Fig. 1)

Many of these issues are related to the very nature of ASR, and will be explored in further detail below.

3 RESULTS OF THE FIELD TEST

3.1 Selection of tracers

Since the temporal variability associated with stormwater quality was unknown, several types of tracers were monitored. The parameters that were routinely monitored which could usefully identify the presence of the injected water included electrical conductivity, temperature, chloride, fluoride, deuterium and oxygen-18. Groundwater samples were collected from observation wells by pumping at rates of 0.3-0.4 Ls^{-1} and would be indicative of the flow-weighted concentration in the open interval.

The data for most of these tracers are shown in Table 1. The most useful parameter over the entire duration of the experiment was found to be chloride due to the strong distinction in salinity between the two water types. The chloride concentration was 93 mg L^{-1} for the mainswater injected first off, whilst for the subsequent stormwater injections it was relatively consistent at 25 to 40 mg L^{-1} and significantly lower than for the ambient groundwater, which varied from around 850 to 1200 mg L^{-1} across the site.

For other tracers, the range in injectant values was large with respect to the differences between the injectant and the ambient groundwater. This diminished their value in discriminating between the two water types in a quantitative sense, but still proved useful in supporting the chloride data.

The history of chloride concentrations in groundwater at the injection and observation wells over the duration of the experiment are shown in Figure 2. Over the four years of injection the response at each of the wells are distinctly different, apart from the injection and 25 m wells which are similar.

3.2 Response at the 0 and 25 m wells

The behaviour of the injection and 25 m wells, both in terms of piezometric pressure and water quality, are very similar. Since the aquifer is granular and not fractured in nature, this can only be explained by a direct hydraulic connection between these wells.

After drilling the 150 mm diameter injection well it was extensively airlifted to enhance its recharge capacity. This resulted in significant quantities of sand being recovered (Gerges *et al.* 1995). Comparison of the caliper logs before and after well-development, show the development of large cavities extending beyond the 340 mm span of the tool below 120 m. This has the effect of enhancing the secondary porosity of the aquifer through de-sanding of the least cemented parts of the formation. The hydraulic head and water quality data suggest that the disturbance extends at least to the closest observation well at 25 metres. This is supported by downhole TV camera images showing colloidal-rich waters characteristic of the injectant in the 25 m well almost immediately after injection starts. This effectively negated the benefits of having this closest observation well in terms of representing flow of injected water through the aquifer to this point.

Therefore we have established that permeable flow paths allow water to flow at higher velocities away from the well along discreet pathways and breakthrough to the 25 m well is almost "instantaneous". Mixing occurs during the intervals of storage between each annual injection period, as indicated for the ASR well (Fig. 3). The relative chloride concentrations (C/Co) tended to decline during the storage period. The rate of decline in C/Co in successive years was from 1.7×10^{-4} per day after injecting mainswater (1993a) to $<5 \times 10^{-5}$ per day in 1996, where effectively all of the ambient fluid in the 0 to 25 m interval was flushed. The mechanism for mixing involves exchange between the injected fluid which occupies the most permeable zones, and any ambient groundwater remaining in the less permeable parts of the aquifer.

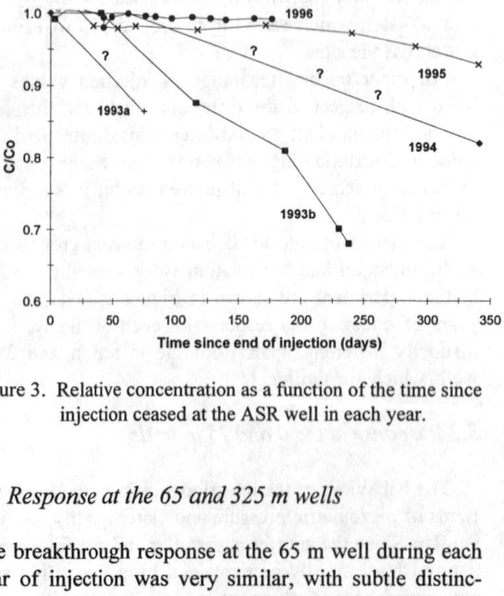

Figure 3. Relative concentration as a function of the time since injection ceased at the ASR well in each year.

3.3 *Response at the 65 and 325 m wells*

The breakthrough response at the 65 m well during each year of injection was very similar, with subtle distinctions probably due to variations in flow rate (Fig. 4). In all years injectant is first detected within a few days, although only in the fourth year was the response more fully developed, with peak C/Co values less than 0.25. Assuming this as an upper limit, then the average advective travel time to 65 m is around 7 days, at a mean velocity of ~ 9 m/day.

After injection has ceased, the solute concentration tended towards its ambient level, which was achieved in all years, except for possibly the last. As with the ASR well, the rate at which this occurs was most rapid in the earliest years and declined in subsequent years as a result of the larger injection volumes or accumulated effects of previous injection cycles.

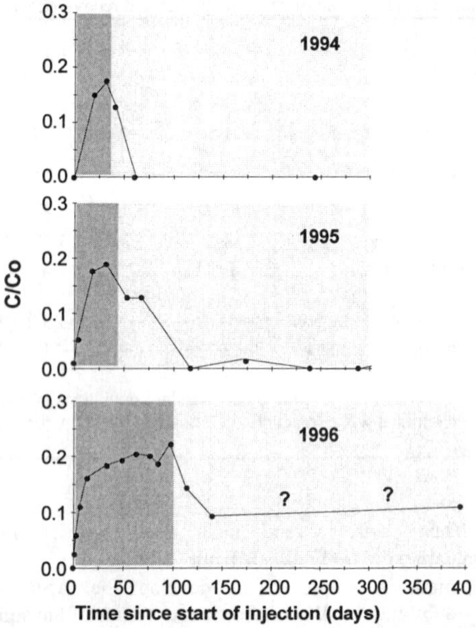

Figure 4. Relative concentration as a function of the time since injection started at 65 m well during last three years. *Periods of injection are indicated by the darker background.*

Examination of the 325 m well data did not reveal breakthrough of injectant. Perturbations on several occasions, both above and below the mean (Fig. 2), are most likely the result of measurement errors as there is no significant difference in chloride between the injection and other periods.

3.4 *Evidence of heterogeneity*

The clearest evidence of heterogeneity comes from the downhole temperature profiles collected before and after mainswater injection (EC logging was not performed in this instance). Figure 5 shows the impact of the 14°C injectant on the ambient groundwater at 24-25°C. Since colder water is indicative of higher rates of flow, most flow occurs deeper in the aquifer, and peaks at a depth of 127 m. Temperatures fail to decline to that of the injected water due to heat exchange with the solid phase. Responses for injection and 25 m wells are similar, as expected, however its rather unusual that the temperature response in the top part of the aquifer is greater at 25 m than at the injection well. More recent work in the T2 aquifer at a nearby ASR site has clearly shown that hydraulic conductivity may vary by several orders of magnitude (Pavelic *et al.* 1999).

A temperature decline in the order of ~0.5°C is observed at 65 m (supporting chloride data shown in Figs. 2&4). Based on this chloride data, the C/Co expected after injection of 6.5 ML would be 0.09. The corresponding value based on temperature is ~0.05 lower as expected, as a result of an effective retardation factor of approximately 2 caused by the thermal mass of the aquifer matrix (Stuyfzand, 1998). The observation well at 65 m only penetrates the aquifer to ~118 m, with Figure 5 suggesting that the majority of flow may pass undetected beneath this well.

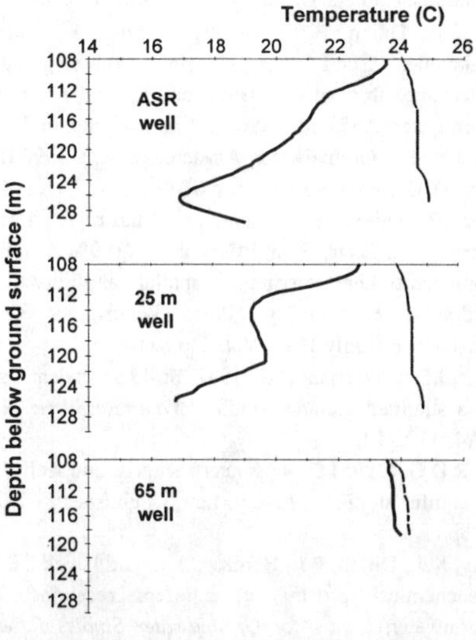

Figure 5. Temperature profile in wells before (-----) and after (-----) 6.5 ML of mains water injection (1993a)

3.5 *Recovery phase*

The chloride concentration of the withdrawn water increases (and C/Co declines) during recovery (Fig. 6). Whilst the initial relative concentration starts off at a value corresponding to that of the injectant, it declines as pumping progresses to a minimum value of 0.60. The step-like decrease at ~20 ML coincides with a period where extended airlifting was performed to control sand ingress after the breakdown of a submersible pump. A five-month delay between airlifting and the recommencement of recovery with a second pump saw only a slight increase in chloride. From ~25 ML onwards, two types of responses are evident, a non-linear change to 80 ML followed by a slower and approximately linear response to the end of pumping. The reason behind the change in response is unclear.

4 INTERPRETATION

4.1 *Towards a robust conceptual model ?*

The solute data we have presented provides firm evidence that flow around the injection well is complex. There seems little doubt that this has been exacerbated through the construction and development of the well itself. Preferential flow paths appear to dominate the solute responses measured at two of the observation wells, and no breakthrough was observed at the third.

The measured responses differ vastly from that of the idealised case of a homogeneous/isotropic aquifer in the absence of a regional hydraulic gradient where the advective front of the injected water plume after the fourth year is estimated to have been at ~115 m; well in excess of all but the furthest observation well. Under this scenario, breakthrough at 25 m would have occurred in the first year, whilst breakthrough at 65 m would have occurred in the third.

Our ability to validate any form of numerical model with these data is hampered by our limited understanding of the flow system since much of the injected water went undetected through the aquifer beyond 25 m. A number of different modelling approaches were trialled before coming to this conclusion. Whilst it is not our intention to provide a detailed account of these here, some explanation is necessary.

Different models could be applied with mixed success to particular aspects of the data set but no single model could be applied to the piezometric heads and solute breakthrough at observation wells and in recovery as a whole.

The obvious starting point is to establish the flow regime in operation across the site. Attempts to replicate piezometric head gradients by applying something conceptually "reasonable", such as by specifying different transmissivity zones around the point of injection failed. The problem is intractable for several reasons. The size and orientation of the cavity in the aquifer which controls the effective radius of the well remains unknown. The assumption that the strata are horizontally layered may be invalid, and the number of unknowns greatly exceed the number of observation points. Although there is little doubt that a good fit between measured and modelled heads could have been produced, we consider that this would be at the cost of the integrity of the model.

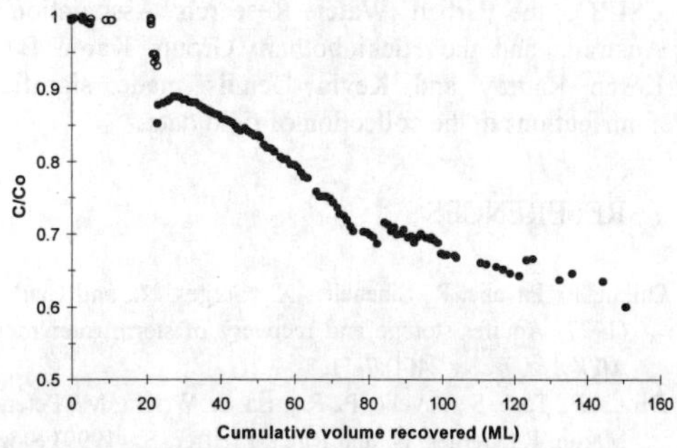

Figure 6. Relative concentration response during recovery stage. *Measurements made during airlifting are shown as open circles.*

Flow and transport models were developed that characterise the aquifer as homogeneous, layered, heterogeneous in plan, and horizontally fractured in a 2D-axisymmetric sense. In order to achieve an integrated calibration to all data, it was clear that a 3D model was required, but that there was inadequate data to warrant the effort required to develop such a model.

5 CONCLUSIONS

These results indicate that it is possible to conceptualise mechanisms to account for observed hydraulic heads, solute breakthrough and recovery efficiency in ASR operations without knowing the details of aquifer heterogeneity. Groundwater chloride data helped to explain aggregated effects of mixing mechanisms, but temperature data taken on vertical profiles in wells provided the clearest picture of these aquifer heterogeneity and mixing processes. However, in spite of these data, it was not possible to find a unique model to simulate processes when there are insufficient observations. This is especially so where well development and geochemical dissolution of aquifer matrix have created preferential flow paths of unknown geometry and dimension, and the system can only be explained with a 3-D model.

In ASR studies where there is a desire to develop a predictive ability, significantly more effort will need to be focussed on characterising the nature of the aquifer; coming to terms with spatial and temporal variability of aquifer properties, and locating observation wells at appropriate radial distances and depths. Whilst some of these observations are not new (eg. Pickens, and Grisak, 1981), this study demonstrates some of the shortfalls in not adequately addressing these issues. The current research work being undertaken by the authors (Dillon *et al.* 1999) has benefited from the experience gained at Andrews Farm.

6 ACKNOWLEDGEMENTS

Financial support for the project was provided by CSIRO, the Urban Water Research Association of Australia and the Hickinbotham Group. Karen Barry, Karen Rattray and Kevin Dennis made significant contributions in the collection of field data.

7 REFERENCES

Dillon, P., Pavelic, P., Sibenaler, X., Gerges, N., and Clark, R. (1997) Aquifer storage and recovery of stormwater runoff. *AWWA J. Water* 24(4):7-11.

Dillon, P., Toze, S., Pavelic, P., Ragusa, S., Wright, M., Peter, P., Martin, R., Gerges, N. and Rinck-Pfeiffer, S. (1999) Storing recycled water in an aquifer: benefits and risks. *AWWA J. Water* 26(5):21-29.

Gerges, N.Z., Sibenaler, X.P. and Armstrong, D. (1995) Experience in injecting stormwater into aquifers to enhance irrigation water supplies in South Australia. *Proc. 2nd Int. Symp. on Artificial Recharge of Groundwater*, Orlando, Florida, July 1994, (Eds. A.I. Johnson & R.D.G. Pyne), p.436-445.

Merritt, M.L. (1986) Recovering fresh water stored in saline limestone aquifers. *Ground Water* 24(4):516-529.

Pavelic, P., Dillon, P., Barry, K., Herczeg, A., Rattray, K., Hekmeijer, P, and Gerges, N. (1998) Well clogging effects determined from mass balances and hydraulic response at a stormwater ASR site. Proc 3rd Intl. Symp. on Artificial Recharge of Groundwater, Amsterdam, Sept. 1998 (Ed. J.H. Peters), Balkema, Rotterdam, p.61-66.

Pavelic, P., Dillon, P., Wright, M., Martin, R., Lau, C-S., Gerges, N., Peter, P. and Correll, R. (1999) A multi-scale approach to characterising the spatial variability of aquifer hydraulic conductivity. Proc. Water '99 Congress, Brisbane, 6-8 July 1999, Vol. 2, p.871-876.

Pickens, J.F. and Grisak, G.E. (1981) Scale-dependent dispersion in a stratified granular aquifer. *Water Resources Research* 17(4):1191-1211.

Pyne, R.D.G. (1995) Groundwater recharge and wells: a guide to aquifer storage recovery. Lewis Publishers, CRC Press, 376p.

Rattray, K.J., Dillon, P.J., Herczeg, A.L. and Pavelic, P. (1996) Geochemical process in aquifers receiving injected stormwater. *Centre for Groundwater Studies Report.* No. 65.

Stuyfzand, P.J. (1998) Quality changes upon injection into anoxic aquifers in the Netherlands: Evaluation of 11 experiments. Proc 3rd Intl. Symp. on Artificial Recharge of Groundwater, Amsterdam, Sept. 1998 (Ed. J.H. Peters), Balkema, Rotterdam, p.283-291.

Groundwater: Past Achievements and Future Challenges, Sililo et al. (eds) © 2000 Balkema, Rotterdam, ISBN 90 5809 159 7

Sustainable management of groundwater in Neapolitan volcanic areas, Italy

V. Piscopo, F. Formica & V. Allocca
Dipartimento di Geofisica e Vulcanologia of the 'Federico II' University of Naples, Italy

ABSTRACT: The data from recent hydrogeological investigations have been used to examine the present use of groundwater resources of the Neapolitan volcanic areas (Southern Italy). Significant differences emerge in the areas studied concerning groundwater flow rate and quality. The shallow aquifer of the Isle of Ischia appears to be over-exploited in the summer, causing provokes the intrusion of cold salt-water in some portions of the coastal aquifer. The shallow aquifer of the Phlegraean Fields appears to be under-exploited in spite of hydrothermal circulation in the southern part of the volcanic aquifer. The present abstraction of drinking water from the Vesuvian aquifer actually exceeds groundwater resources of suitable quality. New strategies in groundwater management are possible with present hydrogeological knowledge, however it is necessary to modify the quotas of the various users of fresh, mineralised and thermal groundwater.

1 INTRODUCTION

While in the last ten years our knowledge of volcanic aquifers around Naples has evolved considerably, the same cannot be said for the management of these groundwater resources. It is in this context that the present study has been undertaken, aimed at examining current use of volcanic aquifer resources in relation to the present state of hydrogeological knowledge. This has been made possible by studying both the results of recent research conducted by the Dipartimento di Geofisica e Vulcanologia of the "Federico II" University of Naples (Celico et al., 1991, 1992, 1994, 1998, 1999) and new data on the area in question.

2 HYDROGEOLOGICAL KNOWLEDGE OF NEAPOLITAN VOLCANIC AREAS

The Neapolitan volcanic area occupies the central-western sector of the Campana Plain (Fig. 1), a wide structural depression arising from the dislocation of Mesozoic carbonate sequences of the Southern Apennines. The site is intersected by important faults in carbonate bedrock and is close to points of maximum depth of volcanic deposits. The potassic volcanism of the Naples area originated in the Pleistocene and is still active today (Fig. 1). The products erupted from the various volcanic sites (Isle of Ischia, Phlegraean Fields and Somma-Vesuvius), together with marine and alluvial deposits covered the lowered area of the plain to a depth of some thousands of metres.

The most important hydrogeological structures are the carbonate reliefs bordering the plain (average yield from 0.02 to 0.03 $m^3/s \times km^2$), feeding numerous basal springs (discharge approximately 12 m^3/s) and the aquifers of the plain (discharge approximately 5 m^3/s). The latter also are recharged

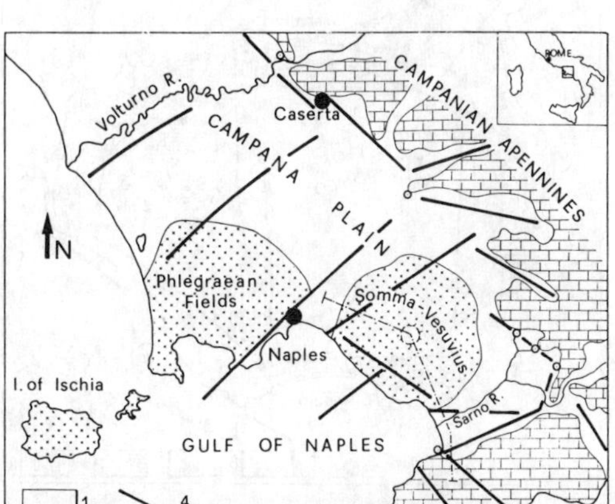

Figure 1. Map of Campana Plain: 1. Pyroclastic and alluvial deposits; 2. Volcanic rocks; 3. Carbonate rocks; 4. Main faults; 5. Main springs; 6. Trace of section of figure 5.

by groundwater from the volcanic centres of Somma-Vesuvius and the Phlegraean Fields, which, together with the Isle of Ischia, constitute further hydrogeological structures (Fig. 1).

2.1 *The Isle of Ischia*

The Isle of Ischia is made up of both effusive products (domes and dome-lava flow) and explosive products (Green Tuff, Citara Tuff, scoriae, etc.) originating from several craters, interposed with and covered by detrital deposits and sedimentary marine formations. This has given rise to a compound aquifer consisting of a disordered series of permeable beds interposed with poorly permeable or even impermeable beds. It is possible however to identify two distinct different areas from a hydrogeological point of view, coinciding with the main volcano-tectonic structures of the island: the "graben" of Ischia and the "horst" of Mount Epomeo (Fig. 2).

In the area of Ischia graben the shallow aquifer (the first 100-200 metres of saturated thickness), made up of prevalently pyroclastic deposits and lavas, is characterised by a relatively high transmissivity (from 10^{-2} to 10^{-4} m^2/s) and hosts a single water-table recharged by both infiltration and salt-water intrusion (Celico et al.,1999). Under normal conditions (in the absence of pumping) the aquifer discharges mainly into the sea (Fig. 2).

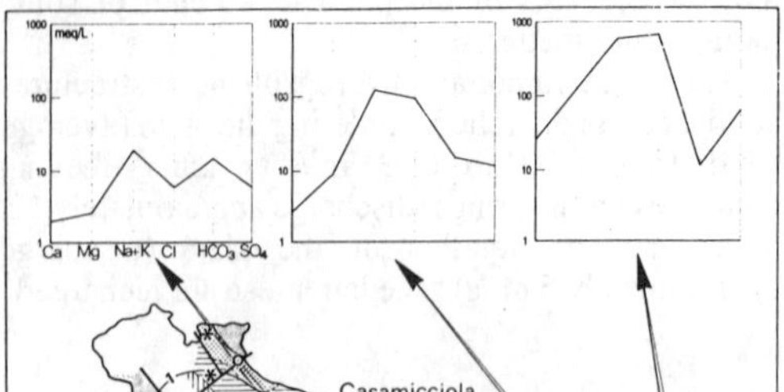

Figure 2. Hydrogeological scheme of Isle of Ischia (adapted from Celico et al., 1999): 1. Faults and fractures (dashed if supposed); 2. Piezometric contour lines of basal water-table in 1996 (m on sl); 3. Springs; 4. Fumaroles; 5. Wells 1 and 2 in figure 6; 6. Groundwater temperature from 20 to 40 °C; 7. Groundwater temperature from 40 to 60 °C; 8. Groundwater temperature from 60 to 80 °C; 9. Groundwater temperature from 80 to 90 °C.

Groundwater temperature in this zone is relatively low compared with other areas of the island (Fig. 2). Proceeding from the recharge area to the discharge area, that is, towards the sea, groundwater temperature falls while salinity increases (Electrical Conductivity EC from 5 to 40 mS/cm) along with chloride (Fig. 2). Groundwater composition increasingly comes to resemble that of seawater.

In the area of Mt. Epomeo (787 m) and its margins the aquifer, made up of unconsolidated pyroclastic deposits, debris, tuffs and marine sediments, is, on the whole, less transmissive (from 10^{-4} to 10^{-6} m^2/s), more heterogeneous and anisotropic than the former. This explains the presence of a basal water-table (Fig. 2) and several perched water-tables, which discharge mainly into the sea (approximately 0.2 m^3/s) and partly to the springs (totalling 0.02 m^3/s). Furthermore, the numerous faults and fractures of this sector of the island, which represent the preferential axes of vertical groundwater flow, may enable several overlapping aquifer layers to communicate. Salt-water recharge is present also in this area.

Groundwater temperature ranges from 20 to 90° C with the highest temperatures found at wells or springs sited near the most recent faults, where fumaroles are also present (Fig. 2). The perched aquifers are generally characterised by relatively less mineralised waters (EC from 5 to 10 mS/cm) of a bicarbonate-alkaline type, while those of the basal aquifer are generally of a sulphate-chloride-alkaline type with a high salinity (EC from 10 to 50 mS/cm). Intermediate chemical compositions are found where water from different overlapping aquifers may mix.

In general, the heat flow which affects the island is the cause of salt-water recharging of the aquifer (infiltration recharge amounts to approximately 8×10^6 m^3/year, groundwater discharge amounts to approximately 13×10^6 m^3/year).

2.2 *The Phlegraean Fields*

Intense volcanic activity provided by numerous craters concentrated in such a small area has principally given rise to tuffs, pyroclastic fall deposits and dome-lavas. Thus, as with the Isle of Ischia, medium-high enthalpy hydrothermal manifestations are present, concentrated mainly in the southern coast sector, where regional tectonic lineaments and local volcano-tectonic fractures meet (Figs 1, 3).

In the first hundred metres below the surface a complex geometry of both aquifers (unconsolidated coarse pyroclastitic deposits, debris, tuffs and fissured lavas) and aquitard beds (poorly fissured tuffs and fine grain pyroclastic deposits) may be distinguished. Together they constitute an aquifer which is at the same time both highly heterogeneous and also anisotropic, where fissured tuffs prevail. The highest transmissivity values belong to unconsolidated pyroclastic deposits (from 10^{-3} to 10^{-4} m^2/s), while the lowest belong to the tuff beds (less or equal to 10^{-4} m^2/s).

On a large scale groundwater circulation may be recognised with a pseudo-radial flow and with the main flow directions orientated towards the sea (Fig. 3). Infiltration recharge for this area amounts to approximately 16×10^6 m^3/year, comparable with that of groundwater discharge into the sea and the springs on land. The springs are located along the southern coastal zone and their total discharge amounts to less than 0.01 m^3/s.

The chemical composition of the groundwater (Fig. 3) makes it possible to distinguish the northern sector, with bicarbonate-alkaline and bicarbonate-alkaline-earthy waters poorly mineralised (EC = 0.8 – 1.3 mS/cm) and cold (15-20°C), from the southern sector, characterised by chloride-alkaline and sulphate-alkaline waters, more mineralised (EC up to 50 mS/cm) and at a higher temperature (up to 80°C). The chemical nature of the water of the latter may be traced to the mixing of groundwater from the shallow aquifer with deep fluids rising due to the heat flow present in this area (geothermal gradient 15°C/100 m). The rise of gas of magma origin (H_2S and CO_2) and of salt-water, which boils at depth due to heating from below, is greater in fractured tuff zones, where preferential flow is vertical and in the coastal zone, where the freshwater lens is less pronounced.

2.3 *Somma-Vesuvius*

The Somma-Vesuvius (1281 m) is a strato-volcano (formed by two calderas, Mount Somma and Vesuvius) whose eruptive activity has alternated between explosive and effusive phases.

In this area it is possible to identify both a shallow and a deep aquifer (Figs 4, 5).

The shallow aquifer is limited to the volcanic complex and is made up of extremely permeable rocks (scoriae, pumice, coarse sands and fissured lavas) alternating with less permeable rocks (tuffs, low fissured lavas, ash and paleosols). It is recharged mainly by precipitation on the volcano

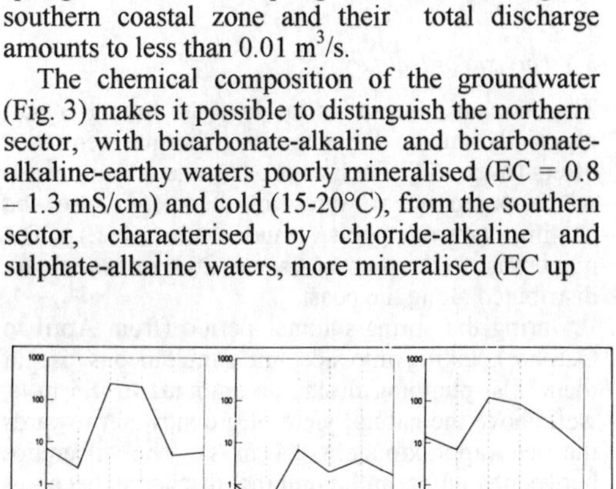

Figure 3. Hydrogeological scheme of Phlegraean Fields : 1. Faults and fractures (dashed if supposed); 2. Caldera rims; 3. Springs; 4. Fumaroles; 5. Piezometric contour lines in 1997 (m on sl); 6. Groundwater temperature from 30 to 50 °C; 7. Groundwater temperature >50 °C; 8. Wells 1 and 2 in figure 7).

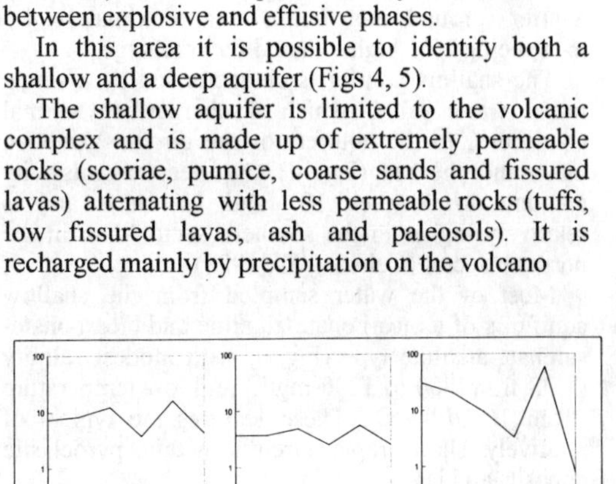

Figure 4. Hydrogeological scheme of Somma-Vesuvius (adapted from Celico et al., 1998): 1. Pyroclastic and alluvial deposits; 2. Pyroclastic deposits and lavas; 3. Caldera rims; 4. Piezometric contour lines in 1994 (m on sl); 5. Well-fields.

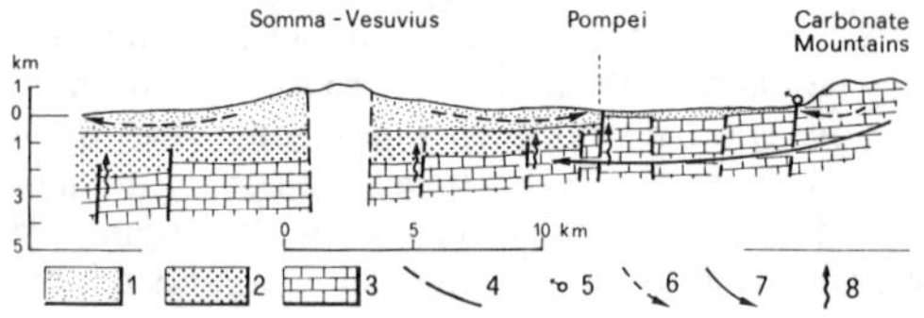

Figure 5. Schematic hydrogeological section through Somma-Vesuvius (adapted from Celico et al., 1998): 1. Shallow aquifer; 2. Intermediate leaky aquifer; 3. Carbonate aquifer; 4. Main faults (dashed if supposed); 5. Spring; 6. Local groundwater flow; 7. Regional groundwater flow; 8. Deep fluids rising.

itself (infiltration amounts to 37 × 10^6 m^3/year), and to a lesser extent by rising water from the deep aquifer, groundwater outflow (approximately 1.6 m^3/s) being 25% higher than direct recharge.

The shallow aquifer is characterised by a single radial water table, which discharges, in normal conditions, to the aquifers of the circum-Vesuvian plain and the sea (Fig. 4). Aquifer transmissivity amounts to between 10^{-1} and 10^{-4} m^2/s and is relatively higher in the southern sector than in the northern sector (Celico et al., 1998).

Most of the water sampled from the shallow aquifer is of a bicarbonate-alkaline and bicarbonate-sulphate-alkaline type (Fig.4), with modest salinity (TDS from 800 to 1200 mg/L) and low temperature (from 16 to 18°C). These features are typical of relatively short, rapid circuits within pyroclastic deposits and lavas.

The deep aquifer corresponds to the carbonate rocks of the substratum of the Campana Plain (Fig. 5). It is larger than the former and is affected by regional groundwater flow, relatively slower and longer, recharged by the carbonate reliefs which border the Campana Plain. The rising of fluids from depth towards the shallow aquifer is possible through the more fractured zones of the carbonate substratum, especially where the covering of volcanic, alluvial and marine deposits is less thick. This occurs in the southern sector of the Vesuvius area, between the towns of Torre A. and Pompei (Fig.5). In this area, water sampled from wells drawing from the shallow aquifer are of a bicarbonate-alkaline-earthy type (Fig.4), with a relatively high saline content (TDS up to 8800 mg/L), temperature (up to 25°C) and carbon dioxide content (from 300 to 2000 mg/L). This may be accounted for precisely by the mixing of water circulating in the shallow aquifer with water of a deep origin flowing upwards.

3 PRESENT GROUNDWATER USE AND FUTURE OPTIONS

The water supply for the Campana Plain (approximately 4 million inhabitants) is provided almost entirely by groundwater. Most of the water destined for drinking is tapped from the carbonate reliefs. The resources of the plain aquifers are used for irrigation and local industry. Groundwater resources below the Neapolitan volcanoes have until now been utilised partly for drinking water and partly for therapeutic purposes (thermal and mineral waters).

Such use, in the light of the hydrogeological schemes outlined, does not always seem to be consistent with the qualitative and quantitative potential of the resources of these aquifers.

3.1 *The Isle of Ischia*

The Isle of Ischia is a famous tourist resort especially due to the numerous thermal centres it offers (over 150). Each centre has one or more wells which pump warm water for therapeutic use and heating (warm pools, mud baths, etc.). The numerous wells present (some 200 recorded) are distributed along the coast.

During the spring-summer period (from April to October), taking into account simultaneous use of the wells, pumping discharge amounts to 0.6 m^3/s, well above the natural yield of groundwater towards the sea (approximately 0.4 m^3/s). The difference between aquifer and pumping discharge becomes even more evident when the high concentration of wells in some areas and the absence in others is taken into account.

Thus in periods of peak activity natural salt-water intrusion into the aquifer is induced. This has been confirmed by detailed data from 61 wells, where upconing of the fresh-salt water interface occurs in 65% of cases during pumping.

Salt-water intrusion due to the effect of a single well or a group of wells may trigger, at least locally, a fall in temperature of the waters of the shallow aquifer. This has been detected in many wells along the coast when aquifer transmissivity is high, as shown, for example, in Figure 6 (well 1).

By contrast, in other cases, particularly near the faulted zones of Mt. Epomeo, as pumping proceeds

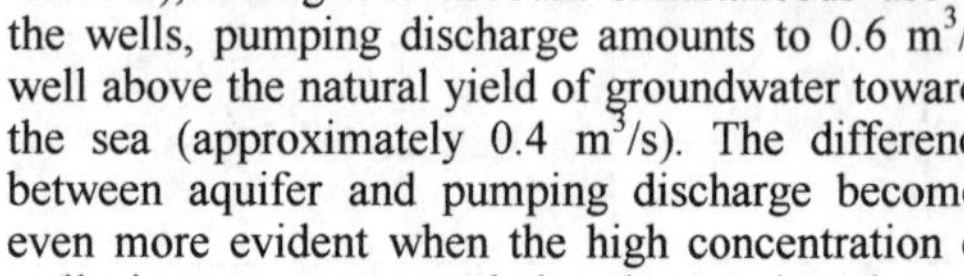

Figure 6. Drawdown (d) and temperature (T) response of wells 1 and 2 of the Isle of Ischia.

water temperature actually increases (up to over 90°C) which may be explained by the drawing up of warmer deep level fluids. Associated with the drawing up of steam in particular is the rise of the water level during pumping of some of these wells (well 2 in Fig. 6), those wells that reveal the highest increases in temperature during pumping. In this case prolonged pumping could trigger an explosion of the well, or in other words, produce a geyser.

For a more rational and effective use of the groundwater resources of the island, it would be perhaps opportune to regulate the combined action of pumping, creating for example pumping schedule with assigned duration and discharge of the pumping. Through the use of numeric modelling of "coupled phenomena", the adjustment of the technical characteristics of the wells and their redistribution across the territory, it would certainly be possible to improve the management of those processes which lead to high temperatures of groundwater and to control colder salt-water intrusion in the coastal sector.

3.2 *The Phlegraean Fields*

Groundwater resources of the Phlegraean Fields are used for irrigation, industry and therapeutic purposes.

In the northern sector, where the waters are poorly mineralised and at a low temperature, numerous wells are present (some hundreds) which utilise the resource at a reduced discharge and only occasionally throughout the year, with a total volume of approximately 2×10^6 m^3/year, considerably less than the yield of the aquifer of this sector (approximately 9×10^6 m^3/year).

In the southern sector, where the main hydrothermal activity of the area occur, some thermal centres are present (at present 4, compared with 13 active forty years ago). These centres use the warm mineralised waters of the coastal zone tapping either through wells or directly from the springs. In this case too the annual volume returned to the springs and pumped from the wells (approximately 1×10^6 m^3/year) is considerably lower than that of the recharge of this underground basin (approximately 7×10^6 m^3/year).

In the Phlegraean Fields groundwater abstraction has a minimal effect on the resource. This is confirmed by the substantial stability of the piezometric levels over the last ten years.

Unlike the Isle of Ischia, the utilisation of groundwater in the southern sector of Phlegraean Fields should be maximised. However, this shallow aquifer is vulnerable to surface pollutants. In the Agnano basin (Fig.3), for example, where the water-table is shallow and where the main thermal centres currently active are present, there are risks from urban areas and the training grounds of the racecourse nearby. Groundwater is already being contaminated in the northern sector of the Phlegraean Fields near the town of Quarto where the concentration of groundwater nitrates (NO_3) lies between 50 and 90 mg/L (Regione Campania, 1998).

A further aspect regarding the use of groundwater is linked to the surveillance of volcanic activity in the area, already studied during the last bradyseismic event (in the eighties). In that case, though correlations between some chemical parameters of the waters of the wells and seismic activity in the area were revealed (Fig. 7), the correlations were too weak to allow a single interpretation to be made. This is probably linked to the weakening of the "message from the deep" by the groundwater circuits closest to the surface, which contribute to feeding the wells sampled. With current hydrogeological knowledge it should be possible to create special efficient monitoring wells, taking into account the extreme heterogeneous and anisotropic nature of the aquifer.

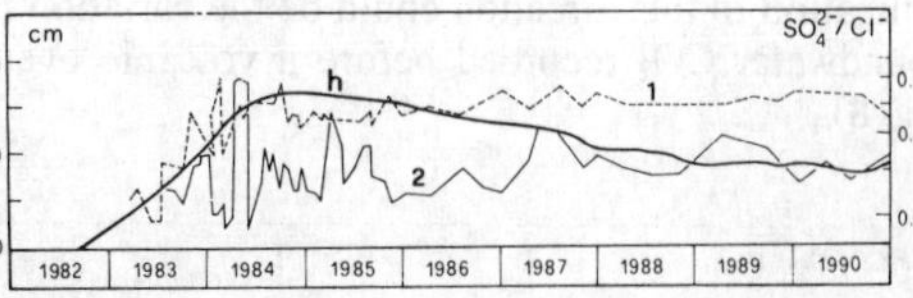

Figure 7. Soil height (h in cm) at port of Pozzuoli and variation of SO_4^{2+}/Cl^- ratio of wells 1 and 2 of Phlegraean Fields during bradyseismic crisis (adapted from Celico et al., 1992).

3.3 *Somma-Vesuvius*

The Somma-Vesuvius aquifer is, among the three Neapolitan volcanic aquifers, characterised by the most substantial groundwater resources (37×10^6 m^3/year of infiltration recharge) and by relatively cold groundwater, at least in the area nearest the surface.

At present, groundwater resources of the Vesuvian aquifer are used mainly for the local drinking water supply, irrigation and industry. While irrigation and industrial use is limited over time and spread across the territory, drinking water accounts for the largest use in terms of volume continuously drawn from below ground. This occurs on the northern and western slopes of the volcano where groundwater is less mineralised, through several well fields of the local water company (Fig. 4). In this area an estimated 19×10^6 m^3/year has been abstracted during the last twenty years. These amounts would not seem proportionate to the aquifer yield in the sectors concerned, considering that in these zones the aquifer is less transmissive and that 61% of groundwater outflow of the whole aquifer has been found to involve the southern sector. It is likely therefore that should such use be prolonged in the future, more heavily mineralised waters, unsuitable for drinking water purposes, will be drawn from both deeper water circuits and those of the southern sector. Some clues as to the considerable impact of this pumping are already evident around the towns of Somma V., where piezometric contour lines reconstructed during pumping shift away from the general radial pattern which characterises the volcano (Fig. 4).

It is thus necessary that the northwestern sector regulates and perhaps resites its groundwater pumping stations used for drinking water supplies.

Groundwater of the southern sector of the volcanic aquifer is, at present, little used because of its high salinity, though its future use for therapeutic purposes could be encouraged. Between Torre A. and Pompei there are some extraction systems of CO_2 from groundwater abstracted from the shallow aquifer. This area seems particularly interesting for the monitoring of groundwater in future volcanic surveillance, at one of the deep fractured zones of the carbonate substratum of the Campana Plain. One application in this direction could be the variation in groundwater CO_2 recorded before a volcanic event (Fig. 8).

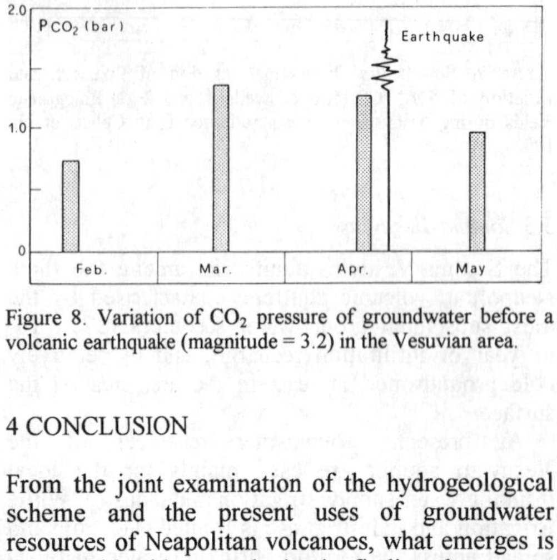

Figure 8. Variation of CO_2 pressure of groundwater before a volcanic earthquake (magnitude = 3.2) in the Vesuvian area.

4 CONCLUSION

From the joint examination of the hydrogeological scheme and the present uses of groundwater resources of Neapolitan volcanoes, what emerges is the considerable complexity in finding the correct criteria to define the "optimal yield" (Domenico, 1972). This is due to the co-existence within the volcanic aquifers of resources which differ in both quality and quantity. These resources have so far allowed only an inconsistent use of the effective availability.

Current hydrogeological knowledge does constitute a starting point from which to set up short-term measures aimed at improving the future management of groundwater resources. However it is necessary to reassess current laws in order to identify quotas according to criteria based on a beneficial use; which is strictly linked to the hydrogeological and hydrogeochemical processes highlighted above.

ACKNOWLEDGEMENTS

The authors wish to thank Professor Pietro Celico of the Dipartimento di Geofisica e Vulcanologia and Professor Damiano Stanzione of the Dipartimento di Scienze della Terra of the "Federico II" University of Naples, for their valuable comments and suggestions. This research was supported by the Consiglio Nazionale delle Ricerche – Gruppo Nazionale Difesa Catastrofi Idrogeologiche (Operative Unit 4.21N).

REFERENCES

Celico, N., P. Celico, M.R. Ghiara, V. Piscopo, D. Stanzione & S. Aquino, 1994. Caratteristiche geochimiche delle acque sotterranea dell'area del Somma-Vesuvio. *Geologica Romana* 30: 709-724.

Celico, P., M. Dell'Aglio, M.R. Ghiara, D. Stanzione, M. Brondi & M. Prosperi, 1992. Geochemical monitoring of the thermal fluids in the Phlegraean Fields from 1970 to 1990. *Boll. Soc. Geol. It.* 111: 409-422.

Celico, P., P. De Vita, F. Nikzad, D. Stanzione & A. Vallario, 1991. Schema idrogeologico e idrogeochimico dei Campi Flegrei. In Andrea Cancelli (ed.), *Ricerca scientifica ed educazione permanente* 93: 287-296.

Celico, P., D. Stanzione, L. Esposito, F. Formica, V. Piscopo & B. M. De Rosa, 1999. La complessità idrogeologica di un'area vulcanica attiva: l'Isola d'Ischia (Napoli-Campania). *Boll. Soc. Geol. It.* 118: 485-504.

Celico, P., D. Stanzione, L. Esposito, M.R. Ghiara, V. Piscopo, S. Caliro & P. La Gioia, 1998. Caratterizzazione idrogeologica e idrogeochimica dell'area vesuviana. *Boll. Soc. Geol. It.* 117: 3-20.

Domenico, P.A. 1972. *Concepts and models in groundwater hydrology*. New York: Mc Graw Hill.

Regione Campania, 1998. *Sviluppo e valorizzazione delle risorse termali dei Campi Flegrei.* Consulting report for Assessorato Industria ed Artigianato, Regione Campania, Napoli (unpubl.).

Groundwater: Past Achievements and Future Challenges, Sililo et al. (eds) © 2000 Balkema, Rotterdam, ISBN 90 5809 159 7

Support for integrated groundwater/surface water monitoring and assessment for sustainable catchment management

D.W. Peach, B. Adams & J.P. Bloomfield
British Geological Survey, Wallingford, UK

H. Wheater
Imperial College of Science, Technology and Medicine, London, UK

ABSTRACT: Sustainable management of groundwater at the catchment scale requires conceptual and possibly subsequent numerical and/or analytical models to be developed. These depend on an understanding of the key hydrological, hydrogeological and geomorphological process interactions. Recent initiatives have indicated: 1) a shortage of the basic data relating to these processes, 2) uncertainty as to what additional data may be required, how and where to collect them. A new research initiative will develop and test methodologies to produce guidelines for monitoring and assessment at the catchment scale. This will be achieved by addressing key scientific and technical issues in a series of carefully instrumented lowland permeable catchments.

1 INTRODUCTION

Lowland catchments (<300m AOD) are subject to a complex set of environmental pressures and associated management problems, and yet approaches to monitoring have generally been fragmented and undertaken on a reactive basis to operational problems. A strategic approach to monitoring is not yet the norm in most European countries at the river basin or even the catchment scale. As environmental awareness has grown, monitoring has increased in scope and breadth but has remained centred on fixed sampling-point and fixed time-interval sampling, often with little interpretation of data. However, this is not just a European issue.

Worldwide, past monitoring efforts were not generally able to provide early warning of:

- The wide effects of eutrophication in the 1960s;
- The effects of acidic deposition on water quality in the 1970s & 1980s;
- The movement through soils and occurrence in groundwater and surface water of a range of pesticides in the 1970s & 1980s
- The widespread occurrence of selenium in irrigation return waters in the western USA;
- The impact of water abstractions on wetland habitats through into the 1990s;

Furthermore, vast sums of money have been spent worldwide on treatment of municipal and industrial wastewaters, reclamation of degraded rivers and wetlands and re-establishment of characteristic floral and faunal communities. However, in many cases the existing monitoring systems have not been able to provide the scientific evidence to judge whether the required or expected improvements in environmental status have been achieved. Whilst large sums of money have been spent on monitoring, policy and management results are often lacking.

This situation illustrates the difficulty of collecting, interpreting and using data adequately to address a range of objectives and issues which may as yet not be apparent. A truly comprehensive assessment system must address the current, known water issues and provide background information necessary for the early detection of future changes in surface and groundwater status (and the associated impacts on the environment) from whatever causes.

This paper describes a unique initiative being undertaken in the UK to gain improved understanding of the hydrological cycle at the meso-catchment scale. The Lowland Catchment Research programme (LOCAR) will focus on two pairs of catchments on the Chalk (England's prime aquifer in terms of public water supply) and one catchment on Permo-Triassic sandstone (England's second most important aquifer), see Figure 1.

Proposed European legislation (The Water Framework Directive) requires establishment and maintenance of good ecological, quantitative and qualitative status of surface waters and good quantitative and qualitative status of groundwaters. It is evident that any human intervention in the hydrological cycle (e.g. groundwater abstraction, changes of land use, introduction of treated sewage into surface waters) will have an impact on the extant status of surface and/or groundwaters. To be able to predict the impact that such interventions will have and also to be able to plan measures to recover "good status" has implications for data requirements and also the degree to which the processes involved are understood.

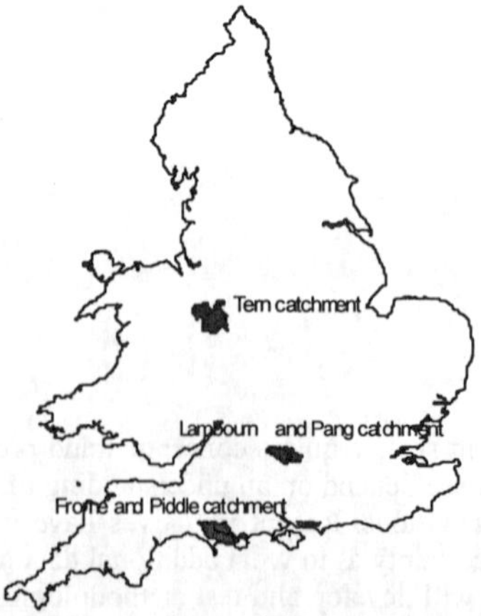

Figure 1. Locations of the LOCAR research catchments

In addition, the perceived impacts of future climate change have important implications for water resources management. In the south and east of the United Kingdom winters are expected to be wetter (with generally more intense storm events) and summers drier whilst in the north-west rainfall is expected to increase throughout the year. There is evidence to suggest an amplification of the current monthly distribution of both rainfall and runoff through the year (Marsh 1996). However, the precise response of catchments depends on the current climate, catchment geology and land use.

Recent plans announced by the UK's Department of the Environment, Transport and the Regions (DETR) indicate that some 4 million new homes will be required by 2016 in the South of England to meet the predicted future demand for housing. This will inevitably put a much greater demand on water resources in this part of the country – an area where groundwater currently provides up to 80% of public water supply.

2 THE LOCAR PROGRAMME

The lack of an integrated science base for lowland catchments has been recognised in a number of recent national and European reviews within the water sector (e.g. Acreman & Adams 1998, European Commission 1998, Dept. of Environment 1996, Grey et al. 1995). There is limited understanding of the underlying surface and subsurface hydrological processes and their interactions, the effects of land-use change on water, sediment and solute fluxes, the associated transport and degradation of contaminants, the hydrological controls on ecological systems (in particular with respect to aquatic, riparian and wetland habitats), and the effects of climatic variability. Integrated models are required to provide a focus for integrated science and to combine understanding of process interactions for predictive studies. The LOCAR programme will address the following issues:

1. What are the key hydrological processes controlling surface water-groundwater interactions and the movement of groundwater in lowland catchments?
2. What are the key physical, chemical and biological processes operating within the valley floor corridor which affect the surface water and groundwater?
3. How do the varying flow regimes control in-stream, riparian and wetland habitats?
4. How does land use management impact on lowland catchment hydrology, including both water quantity and quality?
5. How can the hydrological, hydrogeological, geomorphological and ecological interactions resulting from natural or anthropogenic changes be predicted using integrated mathematical models?

The long term results from the five-year research programme will include improved characterisation of subsurface flow and transport processes. It will also provide improved knowledge of the impacts of land use, better understanding of the hydrological controls on stream and wetland ecology, and new holistic methods to evaluate the impacts of natural and anthropogenic change on lowland catchment systems.

The meso-scale (200-500 km^2) catchments were chosen to encompass the two main aquifers in UK, the Cretaceous Chalk and Permo-Triassic sandstones and on the basis of availability of data.

A review of existing hydrological, hydrogeological and ecological data for the three catchments was carried out (Peach 2000). In spite of the relatively large amount of previous work that had been carried out in these catchments there was still a significant lack of knowledge on various aspects of, for example, the hydrogeology. In the Pang catchment there is uncertainty over both the location of a groundwater divide and the importance of karstic groundwater flow in the Chalk below Palaeogene cover. Whilst the local geology is fundamental to understanding the hydrogeology, some parts of these catchments were last mapped some 70 or more years ago and geological models and mapping techniques have changed significantly since then (Bristow et al. 1998).

In summary, results of the review indicated a great variability of data availability, both in type and space, e.g. good monitoring of groundwater in part

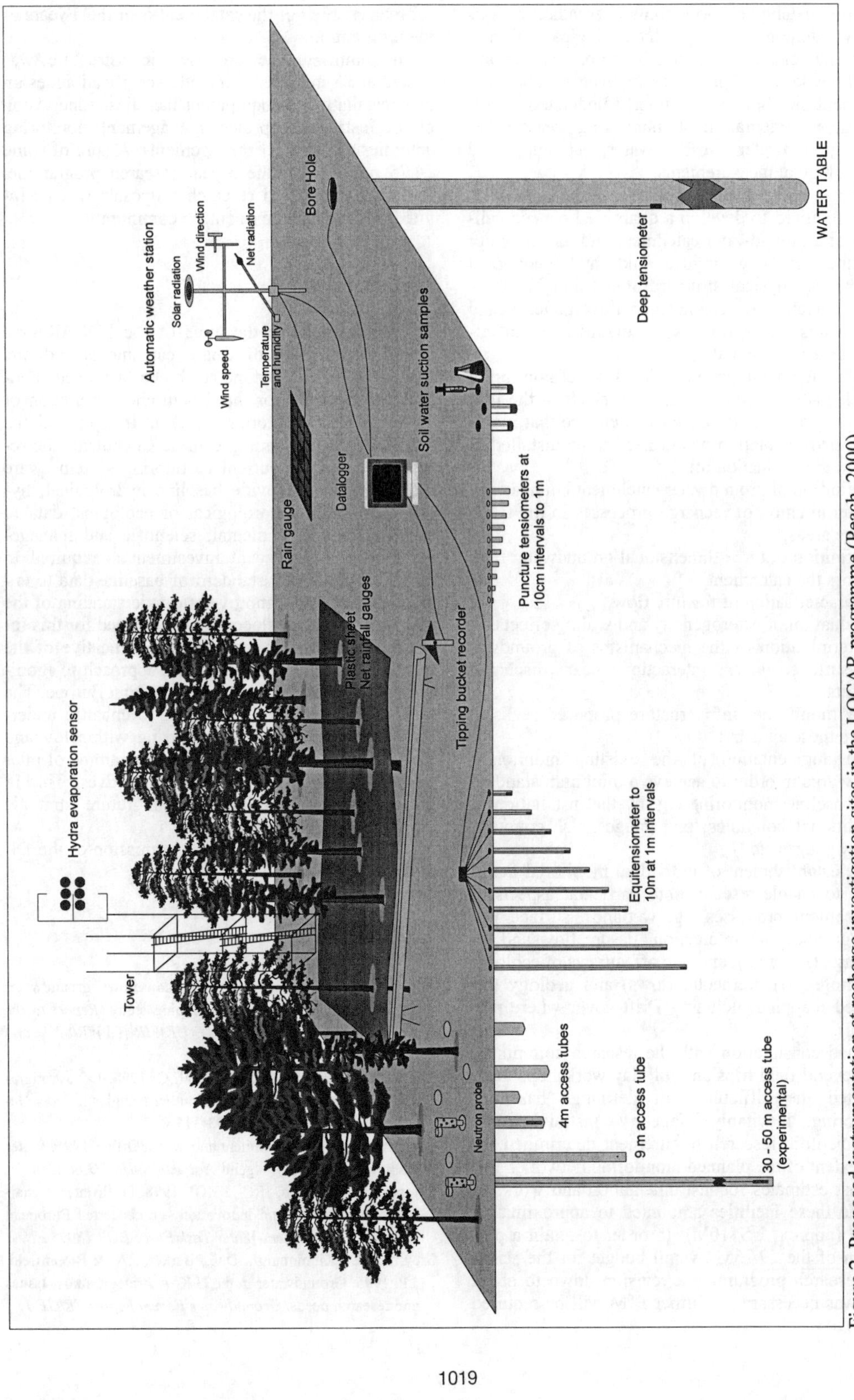

Figure 2. Proposed instrumentation at recharge investigation sites in the LOCAR programme (Peach, 2000)

of one catchment, no automatic weather stations on any catchment, very patchy ecological information, basic conceptual models based on very incomplete knowledge of processes or even groundwater catchment size/boundaries, poorly understood geology and no integration of monitoring systems between groundwater, surface water, ecological and meteorological measurements.

From a groundwater perspective the knowledge that is required to develop a quantified conceptualisation of a groundwater catchment and its behaviour encompasses: - geological and hydrogeological boundaries, physical structure of the aquifer(s), recharge, discharge, groundwater flow patterns and mechanisms, rock-groundwater interactions and effects of the unsaturated zone.

Both catchment scale and local scale monitoring is inadequate to serve the research needs of the LOCAR programme. It was clear therefore that monitoring instrumentation would need to be installed to assist in the evaluation of:

1 Definition of groundwater catchment boundaries,
2 Determination of recharge processes in the interfluve areas,
3 Definition of three-dimensional groundwater flow across the catchment,
4 Characterisation of fracture flow,
5 Evaluation of heterogeneity and scaling effects,
6 Determination of the mechanisms of groundwater-surface water interaction near discharge points.

The monitoring infrastructure proposed seeks to achieve these aims by:

1 The augmentation of the existing monitoring networks in order to achieve a minimum standard of baseline monitoring e.g. by the installation of additional boreholes, rain gauges, fish counters, flow gauges etc.
2 The establishment of individual monitored facilities to enable research into particular aspects of catchment processes e.g. wetlands, surface water/groundwater interaction, fissure flow and recharge (Figure 2), and one off surveys of ecology, hydrology (incremental flows) and geology (revised mapping including Drift cover where present).

Broad consultation with the research community over several field trips and full day workshops demonstrated the difficulties of defining "baseline" monitoring. Inevitably, budgetary constraints rather than scientific research requirement determined the final extent of the planned monitoring networks. Initial cost estimates for instrumentation and works to provide these facilities amounted to approximately £6.7M (approx. US$10M). In order to retain a proportion of the £9.75M overall budget for the scientific research programme a revision down to about £4M was necessary. A further £1M will be required for management of the catchment over the 5 years of the programme.

In summary, the proposed infrastructure was costed at £5 million (to include specification/design and installation of equipment/facilities, supervision of the installation, project management monitoring activities and project management). A sum of some £4.75 remains for the 5-year research programme, for which integrated research proposals will be invited from within the scientific community.

3 CONCLUSIONS

In order to achieve the aims of the LOCAR programme two pairs of Chalk catchments and one Permo-Triassic sandstone catchment have been chosen and reviewed for the design and installation of an upgraded monitoring system. In spite of the knowledge base relating to these catchments the review showed that current monitoring systems were not adequate to provide baseline hydrological, hydrogeological, meteorological or ecological data to address the environmental, scientific and management issues. Significant investment is required to provide a minimum standard of baseline data to enable researchers to improve our understanding of the catchment processes occurring. The need for this investment in basic infrastructure is indicative of the responsive, rather than strategic, approach to monitoring that is not uncommon throughout Europe. The LOCAR programme is designed to enhance understanding of the processes occurring within lowland permeable catchments through a programme of integrated monitoring and research initiatives. This in turn will enable optimisation of future strategic monitoring programmes.

This paper is published by permission of the Director BGS (NERC).

4 REFERENCES

Acreman, M.C. & Adams, B. 1998. Low flows, groundwater and wetland interactions – a scoping study. *Report to the Environment Agency (W6-013), UKWIR (98/WR/09/1) and NERC (BGS WD/98/11)*

Bristow, B., Mortimore, R. & Wood, C. 1998. Lithostratigraphy for mapping the Chalk of southern England. *Proc. Geologists' Association.* 109: 293-315.

Department of the Environment and Welsh Office. 1996. Water resources and supply: Agenda for action. *HMSO London.*

European Commission, JRC-DGXII. 1998. Freshwater, a challenge for research and innovation – a concerted European response. *Environment-Water Task Force EUR 18098 EN.*

Grey, D.R.C., Kinninburgh, D.G., Barker, J.A. & Bloomfield, J.P. 1995. Groundwater in the U.K. A strategic study. Issues and research needs. *Groundwater Forum Report FR/GF 1.*

Marsh, T.J. 1996. The 1995 U.K. drought – a signal of climate instability? *Proc. Instn. Civ. Engrs. Wat., Marit., & Energy.* 118, pp 189-195.

Peach, D. (Editor) 2000. Proposals for infrastructure and monitoring on the LOCAR catchments. *Unpublished report to Natural Environment Research Council.* pp114.

Groundwater: Past Achievements and Future Challenges, Sililo et al. (eds) © 2000 Balkema, Rotterdam, ISBN 90 5809 159 7

Semiarid regions being the millstone of new developed countries: Case study Rioverde basin, Mexico

B. Planer-Friedrich & B.J. Merkel
Freiburg University of Mining and Technology, Germany

ABSTRACT: Water balance, groundwater flow and groundwater hydrogeochemistry of a semiarid basin in the central northern part of Mexico were investigated. The discharge of a karst spring with 4.35 m^3/s into the basin turned out to be a key factor for the basin´s water and constituents balance due to high sulfate concentrations (900 mg/L). Intensive irrigation, mostly with this spring water, leads to an increase of evaporation and in consequence an increase of sulfate in the shallow Quarternary aquifer, while the deeper, confined Cretaceous aquifer still is little affected. Without changes in irrigation techniques, gypsum equilibrium will be achieved within 60 years in the Quarternary aquifer. Besides this chlorinated pesticides (e.g. DDT) were found in the groundwater at considerably high levels as well as some trace elements (e.g. Al, As, U). Sustainable groundwater management with restrictions of irrigation, certain crops or implementation of better irrigation techniques has to be established.

1 CLIMATE AND LANDUSE

The study area is part of a 4500 km^2 large sedimentary basin in the central-northern part of Mexico surrounded by the volcanics of the Mesa Central to the east and the Sierra Madre Occidental to the west.

Annual temperatures in the basin are approximately 21°C, precipitation is about 500 mm/a. Typical for a semiarid climate rainfall occurs in few, heavy rain events during the rainy season around September while especially in February and March there may be no rainfall at all. Potential evaporation with about 1600 mm/a exceeds precipitation by far.

The biggest town, Rioverde, is located in the southern part of the basin, on the southern riverside of the rio verde („green river"), that crosses the basin from the west to the east. Westsouthwest of the town Rioverde the area around the village El Refugio is intensively used for irrigation agriculture.

The irrigation water is provided by a karst spring with a yield of about 4.35 m^3/s, the Media Luna, and approximately 600 wells, about 10 of them guaranteeing also Rioverde´s drinking water supply. North and east of Rioverde agriculture is limited due to poor soil properties and water quality. Drinking water in that area is supplied by trucks.

2 GEOLOGY

The most important structure for the geological development in the area is the platform Valles - San Luis Potosi that formed during the Appalachian and remained emerged during the entire Triassic and Jurassic. At the beginning of Cretaceous the platform subsided. The oldest sediments found at the basin´s margins and underneath the basin fill are Cretaceous limestones. At the transition Cretaceous to Tertiary compressive tectonical forces caused the formation of a basin and range province today bordering the basin to the north and northeast. Subsequent extension enabled acid volcanic extrusions, mainly rhyoliths, forming the basin´s western border.

Tectonically the basin is supposed to be a graben with it´s main thrust probably being an elongation of the Meozoic platform that was activated again in early Quarternary. The sudden subsidence of the graben created a drainageless depression, filled by shallow lakes which were subject to intensive evaporation in the semiarid environment with the deposition of chalk and gypsum. It is assumed that the rio verde filled the depression rather fast with debris from the nearby Sierra creating a deltaic fan in the southwestern part of the basin. Later on basalts and tuffs intruded in these Quarternary sediments. The whole depth of the Quarternary basin fill is around 200 m in average, 450 m at the most.

3 HYDROLOGY

The basin is build up by two main aquifers, a confined Cretaceous and an unconfined Quarternary one both interconnected without distinct aquiclud in between. The main recharge area for the Cretaceous aquifer is the Sierra west of the basin. Within the basin groundwater recharge is very low (~5 mm/a). The calculated amount of recharged groundwater for the basin´s southwestern part equals the yield of the karst spring Media Luna (4.35 m^3/s), while in the northwestern part it is only one quarter of that. The main drain in the southwestern part is the river rio verde causing a general groundwater flow for both aquifers from the west to the east.

4 HYDROGEOCHEMISTRY

Chemical analysis conducted in June and October 1999 and interpreted by cluster analysis based on the geological and hydrological background came up with 3 groups of different water types figured out in table 1.

Table 1. Chemical analysis of representative samples for each of the 3 groups of different water types; concentrations in mg/L

group sample	group 1 San Diego well	group 2 Chilera well	group 3 Media Luna karst spring
water type	Ca-HCO_3	Ca-SO_4	Ca-SO_4
temperature [°C]	24.6	25.0	29.0
pH	7.17	6.93	7.03
conductivity [µS/cm]	449	1242	1782
Ca	62.65	218.79	352.78
Mg	7.84	35.58	64.18
Na	28.58	37.88	18.25
K	6.24	4.76	3.4
Cl	11.5	18	9
SO_4	13.964	514.13	941.3
NO_3	35.42	29.06	1.6
HCO_3	216.55	245.07	304.54
As	0.0062	0.0118	0.0053
Ba	0.170	0.0323	0.0434
Br	0.072	0.215	0.073
Ce	0.018	0.017	0.024
F	0.45	0.90	1.4
I	0.040	0.036	0.036
Li	0.014	0.015	0.023
Pb	0.0011	0.0049	0.0003
Sc	0.027	0.018	0.017
Sr	0.413	3.350	4.850
Ti	0.030	0.021	0.026
V	0.0067	0.0070	0.0017
U	0.436	0.0585	0.0019
Zn	0.0322	0.119	0.0353

Figure1 displays the scheme of groundwater flow in the basin: a deep karst aquifer in contact with gypsum from the Lower Cretaceous Guaxcama formation with sulfate concentrations of approximately 800-1000 mg/L, a second shallower karst system without sulfate impact and the quarternary shallow aquifer of the basin sediments.

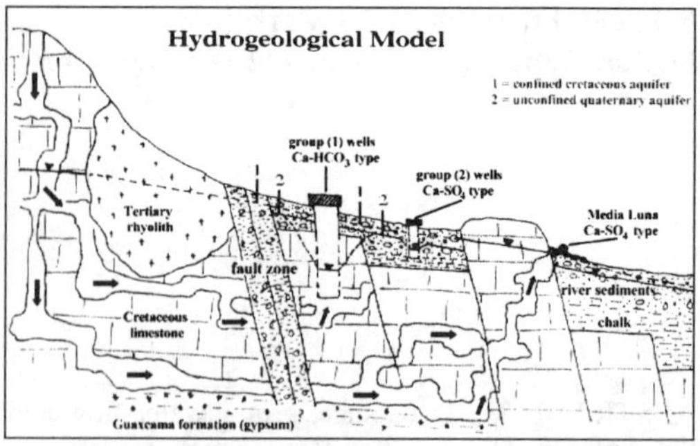

Figure 1. Hydrogeological model explaining the origin of the 3 groups of water types in the SW´ part of the Rioverde basin

5 CONSEQUENCE OF CHANGES IN LANDUSE

In order to keep up with growing population and agriculture being the only chance for increasing income the cultivated land in the study area was enlarged 5 times the original size from 1972-1998. A large net of irrigation and drainage channels from the karst spring Media Luna helped to extend cultivation both to former dry wastelands and areas were there have been wetlands and small lakes.

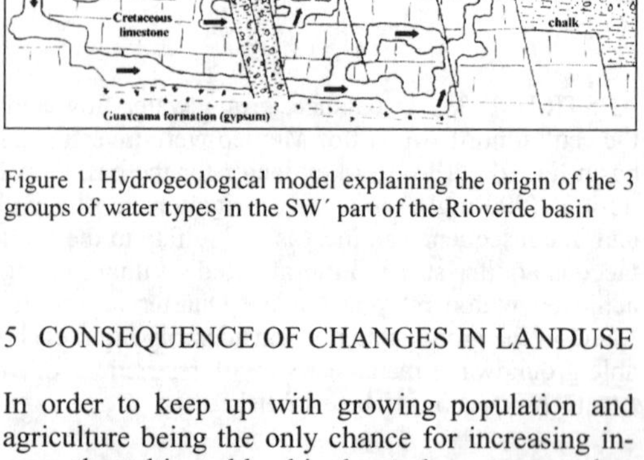

Figure 2. Variations and changes (1965-1992) of water quantities flowing into (rio verde flow gauge Nogal Obscuro) and out of the basin (rio verde flow gauge Vigas) as well as the yield of the karst spring Media Luna and the precipitation in the basin (Pastora).

Figure 2 shows variations and changes of water quantities flowing into and out of the basin as well as the yield of the karst spring Media Luna and the precipitation in the basin. The inflow to the basin varies between 0 and 280 million m^3 while the output ranges from 60 to 400 million m^3. The Media Luna spring serves the basin considerably constant with 100 to 180 million m^3 per year.

There is no correlation to be seen between precipitation and inflow to the basin. However unfortunately no rainfall gauge is available in the Sierra, the mountain area west to the basin, thus the precipitation data were taken from a station within the basin which might be only a rough estimation for the real situation in the recharge area of the river rio verde.

Before the intensive use through irrigation agriculture evaporation in the area of 432 km^2 was about 500 mm/a (precipitation) and only at the small lakes and wetland areas (about 2.9 km^2) actual evaporation was similar to potential evaporation of 1600 mm (4.64 million m^3). Most of the Media Luna spring water was feeding the shallow groundwater and finally discharging to the river rio verde.

With the construction of the irrigation and drainage system the wetland areas decreased in size (about 0.7 km^2 today) with less evaporation on the one hand. On the other hand large areas (71.4 km^2) were irrigated and cultivated increasing total evatranspiration of the basin by 75 million m^3. This number was calculated assuming 500 mm actual evapotranspiration for non-irrigated and 1600 mm (potential evapotranspiration) for irrigated farmland. Since only a small part of the irrigation water is recharging the shallow groundwater by irrigation return flow, less water (125 million m^3) is flowing out of the basin than before (200 million m^3) as can be seen as "delta Vigas+ML-Nogal" in figure 2.

The combination of high evaporation from large areas, less outflow from the basin (about 62.5%) and the considerably high sulfate concentrations of the Media Luna irrigation water lead to an enrichment of sulfate in the Quarternary groundwater and the rio verde. On the long term the basin presents a sink for sulfate and other elements.

This effect already becomes obvious by the steady increase of sulfate concentrations in well P12 tapping groundwater from the Quarternary aquifer:

1993	9.6 mg/L
1994	216 mg/L
1995	330 mg/L
1999	762 mg/L

Calculating the annual sulfate load from the Media Luna with an average 900 mg/L SO_4^{2-} and a yield of 4.35 m^3/s comes up to 123,463 tons/a. Referencing this load on the volume of the Quarternary aquifer with $7.78*10^9$ m^3 (area 432 km^2, thickness ca.180m, effective porosity 0.1) results in a load of approximately 16 mg/(a*L). Assuming an average sulfate concentration of 500 mg/L for the Quarternary aquifer nowdays it can be stated from the presented rough estimations that in about 60 years the sulfate concentrations will reach gypsum equilibrium (15 mmol/L). Na, Cl, Br and I also already increased nowadays, may show comparable effects.

The deeper wells, which are interconnecting the Cretaceous and the Quarternary aquifer by one screen, also show variations in the sulfate concentrations, yet not as large and not with a general trend over time:

P3	17.2 - 98.58 mg/L
P9	6.0 - 96.08 mg/L
P16	69.95 - 235.0 mg/L
P17	4.0 - 93.0 mg/L
PSD	17.28 - 48.0 mg/L

The enrichment of solution load by changed evaporation conditions in the Quarternary aquifer is further engraved by drawdown of -10 to -15 m from 1972-1997. In the southern part several former dug wells with depths of about 10 m are dry nowadays and new wells have been drilled with watertables 4-15 m below the bottom of the old dug wells.

Groundwater quality is further affected by uncontrolled, widespread use of pesticides and fertilizers as may be seen from table 2.

Table 2. Pesticide analysis of selected drinking water wells analysed by GC-ECD and GC-MS; units [ng/L], "-" stands for not detected by the method used (PLANER-FRIEDRICH 2000)

sample	P3	P9	P12	Chilera	standards
α HCH	39	-	-	9	200 EPA
β HCH	-	-	-	-	
γ HCH	610	-	-	30	2000 WHO
Heptachlor	46	-	-	-	400 EPA 30 WHO
cis Heptachlorepoxide	161	103	52	90	200 EPA 30 WHO
Aldrine	505	480	145	458	30 WHO
Dieldrine	528	666	506	751	30 WHO
Endrine	814	1199	1115	1048	2000 EPA
2, 4, DDD	450	641	469	700	
4, 4, DDD	817	778	661	952	
2, 4, DDT	617	1322	604	844	2000 WHO
4, 4, DDT	1349	2007	1504	1610	(all isomers)
2, 4, DDE	-	-	-	-	
4, 4, DDE	574	749	464	356	
α Endosulfane	1536	524	703	-	
β Endosulfane	16	210	-	-	
Hexachlorbenzol	-	34	-	-	1000 WHO

Maximum contamination levels (MCL) of WHO (1996) and EPA (2000) are in some cases violated by far. It should be pointed out that these pesticide screening was the very first performed in the area and not covering the entire spectrum of possible pesticides. Further investigations are ongoing especially focussing on phosphorous organics like Parathion-Methyl, Fenchlorphos and Chlorphyriphos, already detected qualitatively but not quantitatively yet, and Triazines.

Due to the fact that some trace elements are increased at some wells compared to other wells as well as in comparison to international standards (table 3) the water problem in the Rioverde basins turns out to be of multiple character.

Table 3. Critical trace element concentrations at selected wells (TW = tap water, IW = irrigation water, units in [µg/L]

element	TW wells	IW wells	EPA	WHO
Al	137 - 898	20 - 650	50 - 200	200
As	3.8 - 11.8	5.0 - 18.6	50	10
Fe	292 - 1420	934 - 1600	300	300
U	2.9 - 436	1.9 - 698	20	---

6 SUSTAINABLE GROUNDWATER MANAGEMENT

Drilling campaigns in the past suffered from poor core sampling techniques thus leading to a poor knowledge of the basins geology frame work and poor documentation of well construction. Thus there is an urgent need for ruling any further drilling actions effectively.

Drinking water wells nowadays are monitored for groundwater tables every 3-8 days (depending on the drought potential) by the local water authority SASAR. Chemistry is checked about every month in the local university. However those analysis cover only major ions as well as pH, conductivity and oxygen demand, determined in the laboratory 2-3 days after sampling. A well head protection zone of about 10 x 10 m exists for the drinking water wells maintained by SASAR.

In order to improve the groundwater management observation wells have to be drilled and a groundwater monitoring program has to be established. All future drillings should include detailed recording of geology and the well construction. Latitude, longitude and especially altitude of observation wells have to be determined exactly.

Groundwater tables and chemical composition, including trace elements and organic contaminants, have to be checked on a regular schedule, e.g. weekly for the depth to the groundwater and every 6 months for the chemical parameters.

The drinking water wells should be reconstructed using screens only in the Cretaceous formation and casings in the Quarternary sediments in order to tap only the shallow karst due to it´s better quality.

The numerous irrigation wells should get a consistent numbering and flow gauges, in order to determine the amount of water extracted.

River flow gauge stations should record not only the yield of the river, but also the chemistry of important elements. The information have to be stored in a database, freely accessible for future investigations.

Age determination of groundwater from the Cretaceous aquifer would be helpfull in order to estimate mean residence times of the karst water being recharged in the Sierra.

To avoid further negative impact on groundwater quantity and quality irrigation must be restricted depending on actual evaporation and the basin´s water and constituents balance. Cropping has to be focused on plants with low water demand. Drip irrigation and greenhouses reducing evaporation could be an alternative technique to be used in the future.

7 CONCLUSION

This semiarid region nowadays being an important economical factor of the new developed country Mexico concerning agriculture especially citrus fruit may soon present a problem with increasing element concentrations in groundwater. Thus without a water management in the near future the water might be no longer suitable for drinking and irrigation purposes. A basin that has been in equilibrium for over thousands of years can turn into an inhospitable habitat in a period as short as some tens of years.

REFERENCES

EPA (2000): http://www.pura.com/contamin.htm

PLANER-FRIEDRICH, B. (2000): Hydrogeological and hydrochemical investigations in the Rioverde basin, Mexico.- Freiberg Online Geoscience (FOG), Vol. III; http://www.geo.tu-freiberg.de/fog/issues.phtml.

WHO (1996): WHO Guidelines for Drinking Water Quality, Vol.2, 2nd edition, Health criteria and other supporting information. WHO, Geneva, Switzerland.

Groundwater: Past Achievements and Future Challenges, Sililo et al. (eds) © 2000 Balkema, Rotterdam, ISBN 90 5809 159 7

Towards a groundwater management plan for the Taaibos Spruit and Leeu Spruit catchments, South Africa: A situation analysis

J.A.Pretorius
Department of Water Affairs and Forestry, Bloemfontein, South Africa
B.H.Usher
Institute for Groundwater Studies, University of the Orange Free State, South Africa

ABSTRACT: A Groundwater Situation Analysis on two catchments in the Free State, South Africa, was done to provide a framework on which a sustainable Groundwater Management Plan could be based. All available data was reviewed and integrated. Fieldwork to verify the available data and confirm the conceptual model of the area was done. This included groundwater sampling and airborne geophysics. Conventional laboratory analysis and toxicity testing on selected samples was done to assess the threat associated with different levels of contamination. From this data, catchment groundwater quality was obtained and used to define catchment quality guidelines. Numerical modelling was done to identify trends and areas of vulnerability and a simplified prioritization spreadsheet (Threat-Action Guide) was developed to be used by regulators and impactors in the area. Based on the information different groundwater management zones were defined and a management plan suggested.

1 INTRODUCTION

The South African Department of Water Affairs and Forestry (DWA&F) has recently moved towards the management of regions on a catchment and sub-catchment scale. Early in 1998 the Taaibos Spruit and Leeu Spruit catchments were identified, as catchments where investigation into water related issues needed to be done. The final objective of this work would be the implementation of a Catchment Management Plan (CMP) pertinent to all water related issues. This plan would from part of the Integrated Catchment Management (ICM) strategy currently favoured by the department. This study was targeted towards the groundwater in these catchments.

These catchments house several possible influences on groundwater quality. These include the urban area of Sasolburg and Zamdela, several chemical industries, mining activities, power stations and agriculture throughout the catchments.

It was therefore considered necessary to do a groundwater situation analysis in these catchments to highlight all relevant groundwater components in the area.

2 AIM OF THE INVESTIGATION

The aim of the situation analysis was to provide a comprehensive and accurate information base on which a successful and sustainable Groundwater Management Plan (GMP) may be developed. To achieve this aim the objectives of the situation analysis were:

- To determine the status quo of the study area with regard to all groundwater quality and quantity issues
- To identify all land and water users and establish an amiable working relationship with and between them
- To identify emerging water quality and quantity issues and fill any information gaps needed to design the GMP
- To make recommendations towards future management of the groundwater in the study area, identifying the aspects that need to dealt with for the eventual implementation of a GMP

It can be seen that the principle aim of the situation analysis is to give a complete overview of the study area and make recommendations based on this overview so that a groundwater management plan can be developed and implemented.

3 METHOD OF INVESTIGATION

This investigation has used existing information from published and unpublished reports, EMPR reports as submitted to the DWA&F, and published and unpublished maps and plans. In addition site specific water quality (borehole chemical analyses) and in certain cases, water levels and borehole logs (in excess of 200 different sites) were obtained from several parties in the catchment. This data has been generated through monitoring programs and investigations done by the industries, mines and regulatory bodies, which have been in place, in some cases, for several years.

Detailed study and analysis of this data was performed and problem areas in terms of impact and in terms of data deficiency were identified.

To supplement this information a field investigation was conducted, where all possible impactors in the catchment were visited, 101 sites were sampled and a geophysical investigation was undertaken.

Using all the data, numerical modelling was used to simulate the interactions in the catchment. This included a stochastic risk assessment at selected points based on the data available.

Based on this, management strategies including different groundwater protection zones, groundwater standards and immediate management options using a developed spreadsheet, have been developed.

The project was divided into four main sections:

1. A description of the study area including theoretical information regarding the conditions described and present data and observations for the area.
2. Conceptual and numeric modelling and calculations of impacts.
3. Present techniques to identify and prioritise problem areas and broad management techniques for the catchment
4. An in-depth look at the steps that need to be taken regionally and on each site for the development of a Groundwater Management Plan

This approach satisfies the aims set out and provides an extensive synopsis of the study area, with several conclusions that could be drawn and consequent recommendations made to develop and implement a CMP.

4 THE TAAIBOS SPRUIT AND LEEU SPRUIT CATCHMENTS

4.1 *Description of Study Area*

Use was made of available data to describe the system. From this information the principal conclusions that could be drawn were:

- This is a large study area (over 1000km^2) with typical highveld type climate, warm summers and cold frosty winters and rainfall (MAP 644 mm/a). The topography is very flat with consequent slow flowing river systems. Land uses consists of dry land and cattle farming agriculture over the majority of the catchment with several industries, mines and power stations as major influences on the groundwater.

The major industries include: A phosphoric catalyst factory, a company producing diverse chemicals from xanthates and alkylates to agricultural chemicals, a petroleum refinery, two fertilizers and explosives plants, a company responsible for pumping petroleum and crude oil across the country, a manufacturer of PVC, polyethylene, sodium and calcium cyanides, organic peroxides, arctons/frezones, carbon tetrachloride (CTC), and perchloroethylene (PCE), a plant making high density polyethylene (PE-HD) and polypropylene (PP),and a very large company which manufactures a host of diverse inorganic and organic chemicals.

In addition there are several transport companies handling a variety of substances including different types of chemicals from the industries mentioned above and three collieries of concern in the area:

1. An underground colliery, in the process of closure
2. An opencast mine, still operating in the northern part of the area
3. A defunct and refilling colliery to the south

There is also a decommissioned power station within the study area.

There are various other diverse activities which could impact on the groundwater in the area; the most important of these are the abattoir, leather tannery and feedlots.

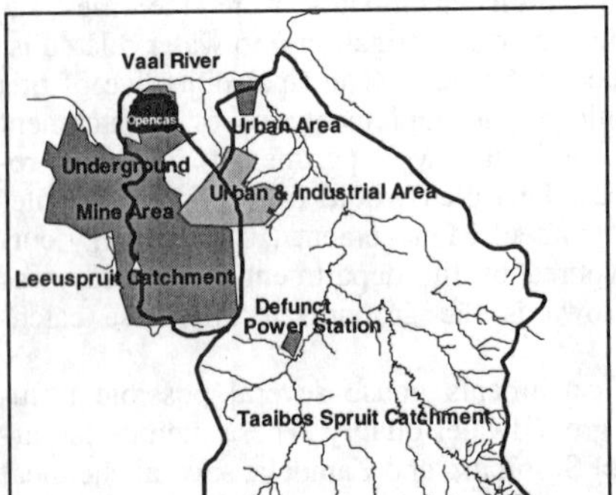

Figure 1: Study Area with Salient Features Shown

The results of previous studies allowed broad definition of the geohydrological regime in the area. From previous investigations it appears as if the ma jor control on groundwater movement, particularly in the industrial areas, is a dolerite dyke of varying depth and thickness. Four different aquifers were defined in the area:

1. The shallow aquifer system, which was defined as all the unconsolidated and fractured material lying above and including the dolerite dyke
2. The alluvial aquifer, which consists of unconsolidated material associated with the rivers and stream to the north of the area
3. A sandstone aquifer represented by clean, white, arenaceous sandstone horizon, located some distance above the coal seams and below the dolerite sill.
4. A karst aquifer represented by the dolomitic formations that partly underlie the Karoo sediments.

The effect of the underground workings adds another dimension to the system. The influence of the latter has as yet been poorly defined.

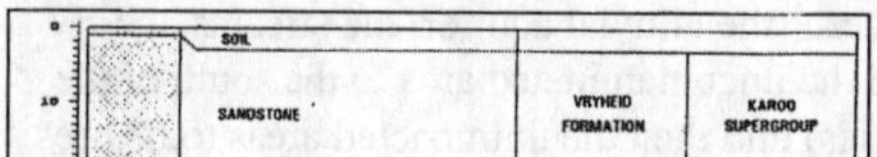

Figure 2: Generalised lithology (after VSA GeoConsultants, 1993)

In order to set up a proper management strategy for the aquifers the aquifer vulnerability also needs to be considered. Thus the shallow aquifer is a minor aquifer system with a high vulnerability while the other three are major aquifers. The aquifer vulnerability of the sandstone and dolomite aquifers is low, thus management strategies must be implemented so as to protect the alluvial aquifer system as priority followed by the shallow and then sandstone and dolomite aquifers.

4.2 *Results of Field Work*

One hundred and one samples were taken from different localities using the same sampling protocol (conductivity depth profiling and specific depth bailer) throughout. This facilitated comparison of results from different areas. The sampling protocol used was verified by results obtained by monitoring done in other investigations in the area. The results of the inorganic analysis showed the extreme variability in quality and a good correlation to land-use activities. Thus for example, high nitrate and phosphates were found in areas of fertilizer manufacture and sewerage works, high sulphates found at industrial effluent areas and in the opencast pit area, elevated sodium and sulphate values downgradient of the treatment and holding ponds at the refinery and elevated chemical oxygen demand in areas suspected of organic contamination. A preliminary indication of ambient catchment quality was also obtained suggesting a good quality (conductivity approximately 60 mS/m) of typical calcium/magnesium or sodium bicarbonate nature.

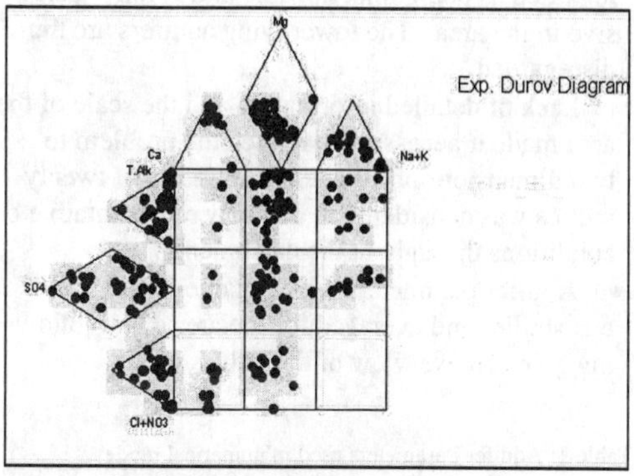

Figure3: Expanded Durov Plot showing variability of quality in area

Several samples were simultaneously collected for toxicity testing. The standard 24-hour and 48-hour *Daphnia pulex* (waterflea) and the 96 hour *Poecilia reticula* (Guppy fish) acute lethality tests were used. Oxygen and pH levels were adjusted to > 40% saturation and the range of 6-9 respectively. This testing yielded good results, as the correlation with chemical analyses was good. They also indicated the extreme risk associated with certain groundwaters, should these waters find their way into the aquatic ecosystem.

From the literature study and sampling the princi pal pollution sources in the area could be identified and discussed for each individual site in the area.

A geophysical investigation was also undertaken by DWA&F. The aims were to verify the current understanding of the geology surrounding the industrial area and using airborne geophysical techniques to locate any major geological structures to the north of the industries. The results of these investigations provided verification of the proposed conceptual model of the geology and that there was very little evidence of a major geological structure, which could act as a preferential pollution path from the industries to the Taaibos Spruit in the north.

4.3 *Numerical Modelling*

The results of previous investigations and the observations and deductions made from the fieldwork were integrated to construct a conceptual model for to be used for simplified numerical modelling of the area. Broadly this conceptual model included the following:

- The shallow aquifer was regarded as being the most vulnerable and important in terms of contaminant transport.
- For the purposes of modelling the dolerite sill is regarded as being impermeable and is fairly pervasive in the area. The lower lying aquifers are thus disregarded.
- Lack of detailed information and the scale of the area made it necessary to reduce the problem to two dimensions and a one layer system of twenty metres was considered to be fairly representative of conditions throughout the catchment.
- Aquifer parameters were obtained from previous studies and averaged for the area. The following gives an overview of the values used:

Table 1: Aquifer Parameters used in numerical model

Zone	Transmisivity (m^2/d)	Storativity
Shallow Aquifer	3	0.001
Alluvial Aquifer	25	0.07
Subsidence Zone	15	0.1
Opencast Pit	50	0.2

The aim of this simplified modelling was to simulate the system so that generalised indications could be obtained as to the sources, magnitude and direction of pollution plume spreading. The modelling suite, AquaMod for Windows (van Tonder &Buys,1995) was used. At this scale the results were very generalised but still compared favourably with several site-specific modelling studies done in the area.

The models highlighted areas of concern, where pollution was coming from, which areas needed protection and where plumes flowed into each other. As a tool for identifying trends and highlighting the most important issues the numerical modelling was very successful. The vulnerability of the Leeu Spruit to the west and the Taaibos Spruit tributary which lies to the east of the industrial area was shown as was the vulnerability of the alluvial aquifer to the north of the catchment to the proximity of the opencast mine.

4.4 *Groundwater Management Techniques Prioritizing Areas of Concern*

To allow the groundwater issues to be easily managed in the region, three different types of protection zones were defined. The first zone encompasses all the industries and is also used around the current opencast mine. Surrounding these areas a second "buffer" zone which separates zone 1 from the areas of concern (the alluvial aquifer, the streams and rivers and the uncontaminated area to the south of the industries) and then the unimpacted areas to be protected which all fall into the third zone. Different preliminary groundwater objectives have been set in the different zones, thus allowing for realistic goals for the industries and mine to work towards in zone 1, a more stringent objective for the buffer zone and very strict values which should be striven towards, in the protected areas. As an example the maximum value for sulphate is 1200 mg/l in zone 1, 600 mg/l in zone 2 and an ideal value of 60 mg/l to be striven to in zone 3.

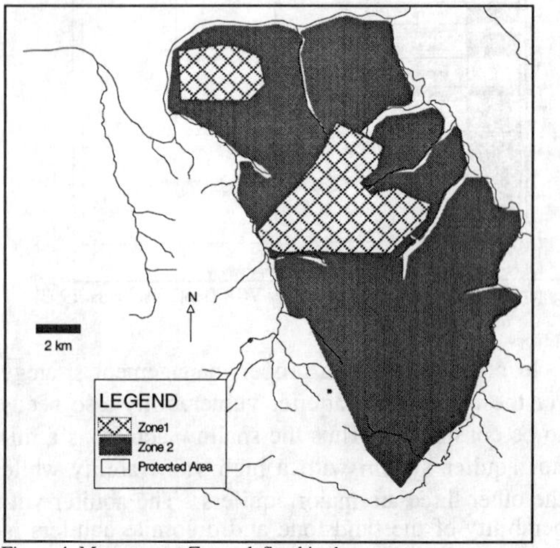

Figure 4: Management Zones defined in the area

Areas that have already been affected therefore have realistic objectives to attain in the short term, while the buffer zones objectives are strict enough to prevent degradation of the water quality in the protected areas.

In addition to using all the results from the literature study and fieldwork, two other techniques were used to highlight and prioritise problem areas.

As part of the numerical modelling, stochastic risk assessment was done. Specific observation points in the system were used in the model and randomly generated aquifer parameters, within the limits set by the modellers, are used to simulate a range of possibilities. The effect is to obtain values at these points for the different contaminants, with a certain degree of confidence. This Monte Carlo Risk Analysis highlighted several areas of concern, principally around the Leeu and Taaibos Spruit in the industrial areas.

A second technique was the development of an Excel based Threat-Action Guide (TAG). This guide uses the current concentrations of different species as measured during sampling and combines this with the distance the point is from an area of concern. Using these values and defined groundwater standards, the aquifer classification and the groundwater management zone within which the point falls, a set of empirical equations is used to give a threat index for the observation point. Based on this threat index the guide suggests one of three management options viz. immediate remedial action, intensive risk assessment or continued monitoring.
By applying these two techniques the areas of greatest concern could be identified and prioritised.

5 RECOMMENDATIONS ARISING FROM THE PROJECT

Situation analyses such as this form an excellent basis from which an integrated CMP can be developed.

Based on what has been learned in the course of this study the following actions are recommended:

- Site specific monitoring must continue in the interim until the monitoring positions for regional monitoring have been finalised,
- A technical sub-committee should be formed to address the issue of organic contamination in the industrial area. The function of this committee would be to quantify the problem, set appropriate groundwater standards, define key parameters which should be analysed for and decide on management plans to limit the long-term impact of these chemicals.
- Before the final integrated management plan for the catchments is developed, full integration of surface water and groundwater data must be done.
- The results of the numerical modelling and the TAG program must be used to identify areas where priority management is needed. The appropriate level of risk assessment and remedial action required at each site must then be implemented as part of the management plan for the two catchments.
- Areas where pollution plumes from different sources converge must be investigated. This is very important to determine synergistic or antagonistic interactions. These interactions are difficult to predict without detailed information and where the secondary mixtures hold severe risks to the receiving environment, measures will need to be implemented to limit the long-term impact.
- It is suggested that toxicity testing, as done in this project, be done as part of the regional monitoring plan. The results from this study have shown that these tests give a very good indication of the true risk associated with each water sample. The frequency of this testing should be at least once a year for the short term until there is enough data to make further recommendations, when the frequency and location of the sampling can be modified according to what emerges.
- Detailed information on the flooding of the underground mines will need to be included in the final management plan. Although it appears that this will not effect the shallow aquifer system, the positions and time to decant will be very important to the water management of the catchments.
- The interaction between the deep, refilling workings and the shallow system must also be quantified.
- More sampling in areas unaffected by potential impactors must be done. This will allow refinement of the suggested interim objectives.
- New monitoring boreholes may need to be drilled north of the industries to form part of the regional monitoring plan. Information from these new boreholes could provide confirmation of the geophysical investigation. This data will also provide valuable insight into the conceptual model of the area.
- All the industries and mines should submit the results of the monitoring to DWA&F in an electronic format that is WISH (Lukas & Hodgson, unpublished software) compatible.
- In phase two of the CMP the interaction between surface and groundwater must be looked at in more detail. Evidence collected in this phase

suggested that groundwater pollution could impact on the surface water quality in the streams and rivers in these catchments. This will need quantification.

- The interim objectives suggested and the catchment management zones defined must be used to manage groundwater within these catchments. In this way realistic goals are set for the industries and mines and a good measure of protection is given to areas regarded as needing protection
- Groundwater resource/quality guidelines for the catchment should be set as soon as possible. The input from the organic pollution sub-committee will be needed to make these final catchment groundwater quality standards as inclusive as possible.
- In areas where severe contamination is in publicly accessible areas, measures must be put into to place to prevent this water from being accessed or used by any persons or animals.
- In several areas there is more than one industry responsible for pollution. It is proposed that there needs to be co-operation between these industries to solve the problem.
- The parameters that need to analysed for in the regional monitoring also need to be determined. It is suggested that as a minimum the following parameters are included: water level, pH, electrical conductivity, sulphate, nitrate, sodium, and chemical oxygen demand/ total organic carbon. Once the technical subcommittee has been established the recommended parameters for organic contaminants can be included.

6 CONCLUSIONS

This was one of the first situation analyses in South Africa, which regarded all the influences on groundwater on a catchment scale, with the objective of setting up an integrated catchment management plan and catchment-scale groundwater management plan.

Some of the of the most important aspects which arose from this study and can be applied to similar studies are the following:

- Co-operation between all the role players in a catchment needs to be established as the first step to undertaking such a study. The current structure of catchment forums provides an excellent vehicle to achieve this goal.
- Through this co-operation a large amount of diverse data (GIS, water quality, water levels etc) was integrated into a single database to form an overall portrayal of all pertinent issues.
- From this it is possible to select appropriate sites to used to evaluate the catchment groundwater quality and resources.
- By using a combination of diverse techniques (integration of available data, once-off sampling to verify and broaden current water quality data, geophysical techniques, simple numerical models and definition of tiered management zones) a cost-effective overview of the system can be obtained.

REFERENCES

Lukas, E and Hodgson, F.D.I. 1999 *Windows Interpretation System for Hydrogeologists*, Software developed at the Institute for Groundwater Studies , University of the Free State

Van Tonder, G.J. and Cogho, V.E. (1987). AQUAMOD: A two dimensional Galerkin finite element microcomputer program for the simulation of groundwater flow and mass transport. Water SA,13 (3), p.175-180.

VSA GeoConsultants, 1993, Definition of the Hydrogeological regime at Sigma Colliery with special reference to the mutual impact between underground mining and the environment. Confidential report to Sigma Mine.

Groundwater: Past Achievements and Future Challenges, Sililo et al. (eds) © *2000 Balkema, Rotterdam, ISBN 90 5809 159 7*

Sustainable water resources management in an agricultural environment

M. Razack, M. Larminach & O. Gaillard
Department of Hydrogeology, University of Poitiers, France

ABSTRACT: Poitou-Charentes is a region in Western France where agriculture is a major activity. Irrigation water is supplied from groundwater and surface water and amounts up to 90% of the whole demand. Water supply to all users has led to overexploitation the resources. Sustainable conjunctive management is at present a crucial need and aims at the following main objectives: to maintain the crop production; and to preserve water resources. To this end, an experimental four year study (1993-1996) has been carried out in river Charente basin. Water resources were assessed with the use of numerical modeling (Modflow™). Irrigation demands were controlled in real-time with the use of software (Irritel™) which computes a daily soil water budget. Experimental corn sowed sites were selected, in which non-controlled and controlled irrigation were performed. The final results are very encouraging and reveal that water resources problems can be solved through proper conjunctive management.

1 INTRODUCTION

In agricultural regions, which usually come under a strong water deficit during the summer season, demands for irrigation water can be very important and subsequently can have significant and negative impacts on water resource sustainability. The Poitou-Charentes region, in Western France, spreads over 25 810 km² and has a population of 1 600 000 inhabitants. Irrigated agriculture is a major activity and irrigation is extensively practiced during the summer season. Irrigation sources are groundwater as well as surface water. During this period, water supplies to all users (agriculture, municipalities, and industries) has led to overexploitation of the resources and induced critical water level decline in aquifers and streamflow depletion in related rivers. In order to set up sustainable conjunctive management of available water resources, a pilot study was undertaken in the river Charente basin (Fig.1) aiming at optimizing the agricultural water demands while maintaining corn crop productivity.

2 STUDY AREA AND METHOD

2.1. *Water resources availability*

The pilot study is related to 7000 ha of agricultural lands located upstream of Angoulème city (Fig.2).

The lands are mainly sowed with corn. The main river is La Charente. The aquifer system consists of finely fissured limestones (middle Jurassic to lower Cretaceous) and river alluvium. The limestones con-

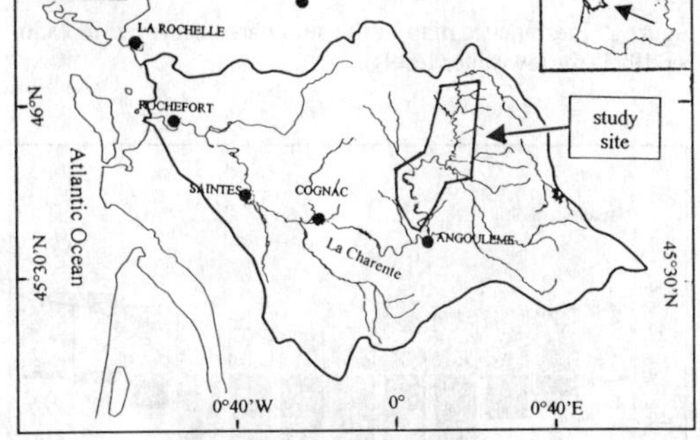

Figure 1. River La Charente Basin and study area

tain an unconfined groundwater. The study extended from 1993 to 1996. All available data were collected and reviewed (thickness of aquifer, physical properties). New water level data were collected twice each year for low and high water level. The piezometric map in October 1995 is reported on figure 2 and shows that the rivers drain the groundwater.

On figure 3 is reported the average monthly water balance of the Charente basin estimated over 30 years (from 1961 to 1990). Such a diagram clearly shows that the summer period, from June to September, undergoes a significant water deficit.

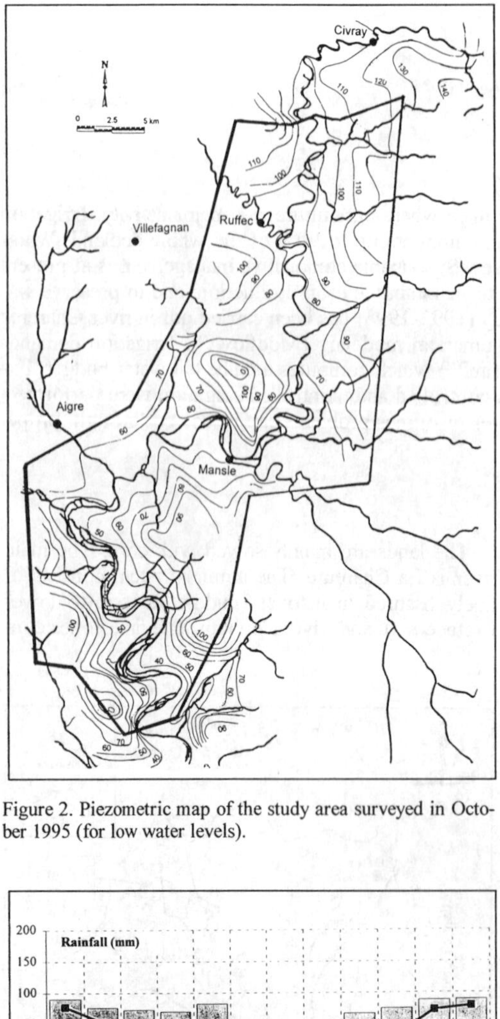

Figure 2. Piezometric map of the study area surveyed in October 1995 (for low water levels).

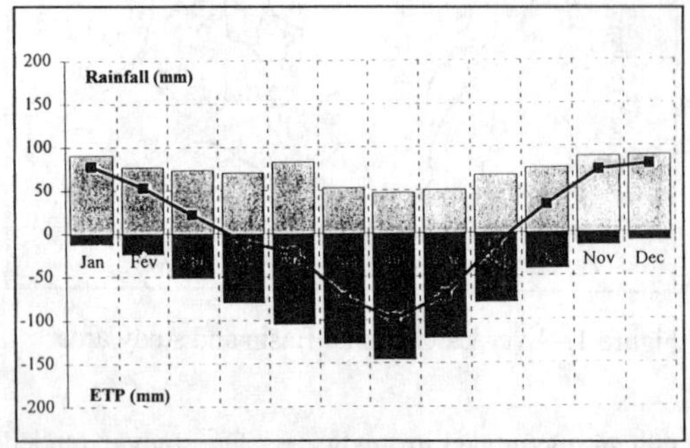

Figure 3. Average monthly water balance of La Charente basin computed over 30 years (1961-1990)

A finite difference numerical model was developed to simulate the river-aquifer system under consideration in order to evaluate the available water resources with the use of MODFLOW™ software (McDonald & Harbaugh, 1984). The average yearly water budget calculated by the numerical model is shown on the figure 4. It can be seen that the agricultural withdrawals for irrigation are extremely important and amount up to 90% of the whole water demand. It should be kept in mind that the irrigation water demands are concentrated during the summer season and thus lead to critical declines in water levels in the rivers and the groundwater (as summer is a period of water shortage).

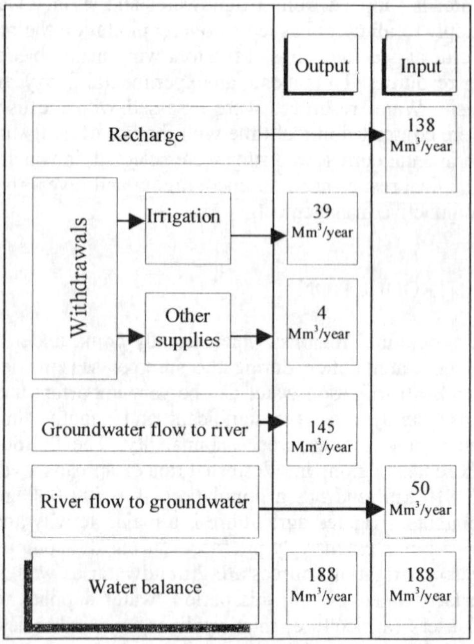

Figure 4. Diagram showing the average yearly water budget of the study area derived from the numerical model of the river-aquifer system.

2.2 *Soil water balance and corn water demands*

The general expression of the soil water balance (W) is the following (Hillel, 1974) :

$$\Delta W \text{ (mm)} = W_{input} - W_{output} = P-(R+I+ET) \qquad (1)$$

where P (mm) = rainfall; R (mm) = runoff; I (mm) = drainage in depth; ET (mm) = evapotranspiration .

The corn water demand can be expressed as follows :

$$ETM = ETP*K \quad (2)$$

where ETM = corn water demand (mm) ; ETP = potential evapotranspiration (mm); K= cultural coefficient which depends on the climatic conditions.

Thus the water balance of a soil planted with corn can be written as follows :

$$\Delta W\,(mm) = (P + SM) - ETM \quad (3)$$

where P = rainfall (mm); and SM = soil moisture (mm). It should be noted that the soil moisture (SM) is variable according to the soil nature and varies here from 30 mm to 100 mm (Thinon et al.,1995).

During the summer period, the soil water balance can show a deficit ($\Delta W<0$) . The cumulative water deficit in summer season in the study area is reported on figure 5 for the four years (1993 to 1996) during which the pilot study took place. The cumulative deficit for 1990 is also shown, as in 1990 the region faced a severe drought. The climatic conditions were quite different during the four years. 1995 was the most dry year while 1994 was the most rainy year.

However this figure clearly reveals, that whatever the climatic conditions, the region undergoes a chronic water deficit in summer. By the end of July, the soil moisture is almost completely emptied whatever its initial value (note that the cumulative water deficit is evaluated from 1st July; earlier deficit is not taken into account).

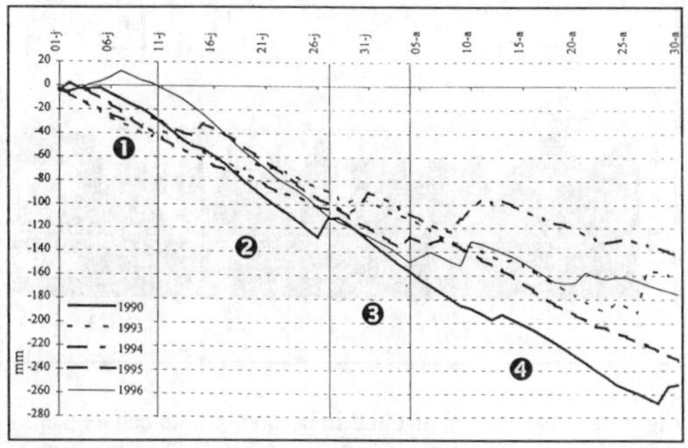

Figure 5. Cumulative water deficit in the study area from 1st July to 30th August

Consequently water supplies are needed and are provided through irrigation :

$$ETM = (P + Ir) - SM \quad (4)$$

where Ir (mm) = irrigation water.

Irrigation's objective is thus to supply to the crops the needed amounts of water during periods of rainfall shortage. An insufficient irrigation would of course reduce the crop production. But on the other hand, an excessive irrigation would produce negative effects (water resource wastage, and fertilizer leaching to the groundwater). It is thus highly recommendable to compute the soil water balance terms at a relatively short time step in order to evaluate optimal irrigation demands with time (water fluxes and periodicity). A daily time step was selected and appeared suitable to this end.

2.3. *Operating procedure of the pilot study*

As stated above, the pilot study extended over four years, from 1993 to 1996. Each year several experimental sites were selected (5 in 1993, 15 in 1994, 21 in 1995, and 22 in 1996). In each site, two corn-cultivated parcels were considered. On the first land parcel, the farmer performed irrigation according to his usual practice. On the second parcel (belonging to the same farmer) experimental and controlled irrigation was organized. This controlled irrigation aimed at providing to the crops, the exact amount of water when needed. The controlled irrigation was performed as shown in figure 6.

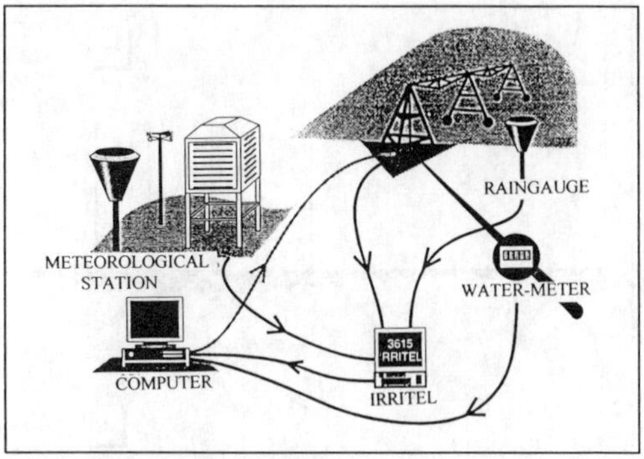

Figure 6. Controlled experimental irrigation organization.

The controlled irrigation was under the supervision of a software (IRRITEL™, Specty, 1988). This software computed the daily soil water budget, according to data provided by the meteorological service (rainfall, evapotranspiration) and the farmer himself (soil parameters, physiological state of the plant, and water supplied by irrigation). Using these data, Irritel evaluated the daily soil water budget. Irritel uses a two reservoirs conceptual model. The soil moisture (SM) is divided into a first reservoir (SM1) representing water that can easily be taken up by the plant and a second reservoir (SM2) representing water available with difficulty to the plant.

The controlled irrigation is performed under the following constraints :

-no irrigation before corn sowing;

-irrigation is started the day after when the reservoir SM1 (easily usable soil moisture) is half-emptied;

-the daily water supplied by irrigation is given by :

$$Irr_j \text{ (mm)} = SM_{max} - SM_j - 10 \qquad (5)$$

where Irr_j = irrigation for the jth day ; SM_{max} = maximum value of the soil moisture; SM_j = soil moisture value for the jth day ; 10 mm represents average daily rainfall.
-no irrigation is supplied during rainy days;
-minimal irrigation supply is equal to 10 mm.

This procedure appeared to be the most adapted one in order to take into account the climatic conditions that prevail in the region.

3 RESULTS AND DISCUSSION

Controlled and non-controlled irrigation consequences are shown on figures 7 and 8. Note that discussion on the role of the irrigation equipment is beyond the scope of this paper and will not be approached.

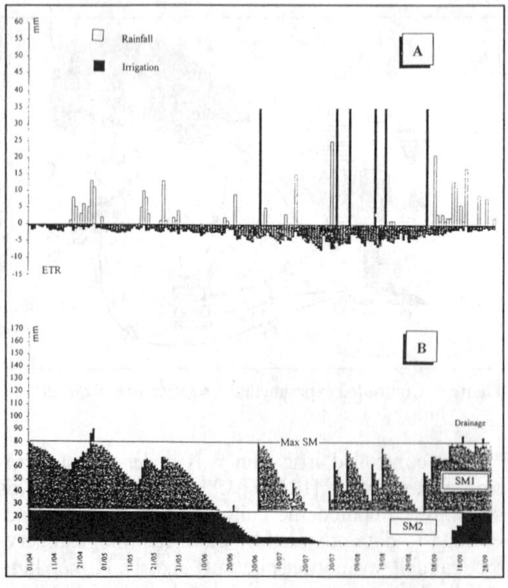

Figure 7. Non controlled irrigation and its consequences. Example taken in 1995. A. Climatic conditions and usual irrigation supply. B. Evolution of the soil reserve. P: rainfall (mm); ETR: actual evapotranspiration (mm); Irr: irrigation (mm); SM: soil moisture (mm); SM1: 1st reservoir representing water easily available to plants; SM2: 2nd reservoir representing water available with difficulty to plants. X axis is the same for both figures A and B.

Usually, farmers did not start irrigation until beginning July. They also met some difficulty to determine the frequency and the quantity of water to supply through irrigation. We can see from figure 7, that the 1st reservoir of the soil moisture (SM1) can be completely emptied, as well as the second one (SM2). Irrigation supplies are systematically equal to 30 mm without any consideration of the soil moisture. Now, during rainy episodes, the farmers had no means to estimate the infiltrated water. Thus this component is also not taken into account. In all cases, the result is a loss for the farmer. When the soil moisture SM1 is emptied, the plant undergoes a stress and the crop productivity is less. If more water than needed were supplied through irrigation, this would induce water wasting and would thus affect water resource sustainability. How controlled irrigation contributes to a better water resource management is pointed out on figure 8.

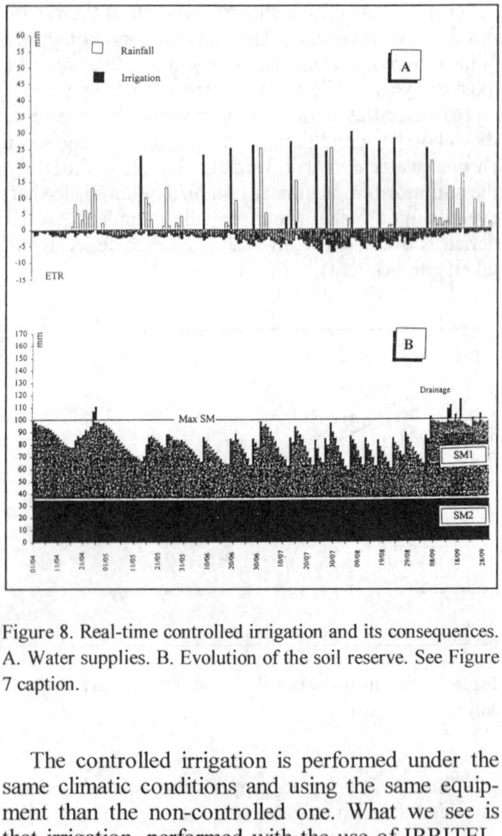

Figure 8. Real-time controlled irrigation and its consequences. A. Water supplies. B. Evolution of the soil reserve. See Figure 7 caption.

The controlled irrigation is performed under the same climatic conditions and using the same equipment than the non-controlled one. What we see is that irrigation, performed with the use of IRRITEL, is started much earlier (here 2nd period of ten days of May) and water is supplied according to the constraints explained above. This way the soil moisture in the 1st reservoir SM1 is never emptied more than half of its maximum value. The second reservoir (SM2) is never used. This point is very significant as the plant never undergo any stress and the agricultural productivity is much better. Controlled irrigation thus respects very closely the plant water demand as well as the availability of water resources.

A comparison of the water supplied by both kinds of irrigation over the four years points out that on average 25% of the water usually used for irrigation during summer season can be saved up with the use of such a controlled irrigation. Let's recall the study area yearly water budget, in which 39 Mm^3 of water is used on average for irrigation. According to the pilot study, almost #10 Mm^3 of water could thus be saved annually while keeping the same corn production in this area.

The results acquired in the study area can be extended to the whole Poitou-Charentes region. A sustainable use of the available surface and ground water resources in this region can then be envisaged.

4 CONCLUSIONS

This four years pilot study has put emphasis on the following points :

-Present irrigation according to the farmers usual practice does not take into account the available water resources;

-This usual practice may produce significant water wasting;

-Farmers in these agricultural areas facing water deficit in summer season, at present are aware that surface and underground water resources should cunjunctively be managed in a proper way;

-A controlled irrigation constitutes a powerful tool to manage water resources and preserve the crop production.

Ongoing works on the sustainable use of cunjunctive surface and underground water resources in agricultural environments are now focused on the application of decision aiding methods. These methods (Kholghi et al., 1996; Razack et al., 1996) are capable of defining the best water management policy which takes into account several criteria and may be effective in the overall preservation of the water resources. Such best management policy is the one which optimizes some objective function subject to a set of physical and economical constraints.

ACKNOWLEDGEMENTS

The authors would like to acknowledge the cooperation of the farmers in the study area who actively participated to the pilot study. Funds were provided by the Conseil Général de la Charente and the Syndicat des Irrigants de Nord-Charente. Help for soil data collection by the Chambre d'Agriculture de la Charente is gratefully acknowledged. Reviewers comments and corrections helped improving the paper and are gratefully acknowledged.

REFERENCES

Hillel D. 1974. L'eau et le sol. Principes et processus physiques. *Ed. Vander, Bruxelles, 288 pages.*

Kholghi M., M.Razack & W.Treichel 1996. Modélisation et gestion quantitative des systèmes nappe-rivières. *Hydrogéologie, no.4, pp.11-19.*

McDonald M.G. & A.W.Harbaugh 1988. A modular three dimensional finite difference groundwater flow model. *U.S.Geol.Survey Paper, vol.16, 548 pages.*

Razack M., M.Kholghi & W.Treichel 1996. Multi-Criterion decision making approach for stream-aquifer conjunctive use management. *Suppl. EOS Trans. AGU, vol.77. no.46, p.F184 (abstract).*

Specty R. 1988. Irritel ou l'avertissement par télématique en Alsace. *BTI, no.426/427, pp.69-82.*

Thinon M., M.Razack, JP. Arrondeau, B.de Grissac & M.Sicard. 1995. Gestion active des prélèvements d'eau pour l'irrigation. *Hydrogéologie, no.1, pp.27-36*

Groundwater: Past Achievements and Future Challenges, Sililo et al. (eds) © 2000 Balkema, Rotterdam, ISBN 90 5809 159 7

Sustainable groundwater supplies for rural communities: A common approach to project audit

N.S. Robins & R.C. Calow
British Geological Survey, Wallingford, UK

K. Pietersen
Water Research Commission, Pretoria, South Africa

R. Titus
University of Western Cape, Bellville, South Africa

ABSTRACT: Groundwater development programmes are normally deemed successful to the benefit of the implementer rather than appropriateness to the user. A system of common audit is proposed whereby all projects are measured against a uniform set of criteria which include not only the immediate success rate of borehole or well yields, but also the projected sustainablity of those yields within the prevailing physical, social and economic environment. An example from the Northern Cape Province of South Africa highlights some of the criteria that need to be assessed as part of an audit system.

1 INTRODUCTION

The need for sustainable water points for rural communities in sub-Saharan Africa and elsewhere is widely advocated, but measurement of sustainability and success is poorly defined. Physical sustainability is achieved by careful selection of hydrogeologically favourable sites, and the appropriateness of technical choice, including well or borehole design, and the selection of optimum pump type for the proposed normal head, yield and use. Socio-economic and institutional determinants include community involvement, commencing with site selection and technical choice, ownership, and aspects of operation, maintenance and monitoring.

The concept of success in rural community water point development is less well defined. A *successful borehole* may provide a sustainable yield of only 0.3 $l\ s^{-1}$ in, for example, the escarpment area of Malawi, whereas a yield of 5.0 $l\ s^{-1}$ may be required for a village supply in South Africa. More importantly, success measured from a yield-drawdown test of only a few hours duration says little about likely sustainability of that yield at times of water stress.

A common approach to defining long-term or sustainable success is desirable, not least so that implementers can report statistics on a like-for-like basis. A complex array of factors contributing to success in a sustainable sense can be identified as the basis for common audit. These factors can be reduced to a simpler index based on aquifer type, effective rainfall, site selection process and water demand supported by limited documentation for quality assurance. Obviously, a source with unacceptable water quality is automatically deemed unsuccessful.

Project audit and project monitoring are powerful tools. They can provide continuous feedback to enable assessment, and ensure that problems are identified at source so that appropriate action can be taken. In addition, they may help achieve sustainable groundwater development, identifying threats to the long-term provision of water supply. This does however, depend on appropriate programme objectives for sustainability being in place from the outset.

2 CONCEPT OF SUCCESS

The criteria for success can be defined in a variety of ways, but the definition given by Van Dongen and Woodhouse (1994) is hard to better: "A borehole is deemed successful when the yield and water quality satisfy the needs of a particular household or community". In terms of quantity, a range of 0.6 $l\ s^{-1}$ from a shallow groundwater body, to only 0.1 $l\ s^{-1}$ or less for deeper aquifers requiring higher lift pumps, may be adequate. Thus, if they are sustainable supplies they are considered successful. The criteria defining success may vary between projects. There may, for example, be a minimum yield, or in a coastal project, a maximum salinity. The success of a borehole siting programme may be judged more readily. In financial terms, siting techniques are justified if they increase the chances of subsequent boreholes or wells being successful, such that the saving in drilling costs is greater than the costs of the siting (Farr et al, 1982).

Physical sustainability can be described in broad terms from the evaluation of test pumping data to predict long term yield, and the assessment of the long term water balance to determine renewable yield. It must also include examination of the human and social aspects of both the physical engineering sustainability of the water point and of environmental sustainability, both in terms of recharge and groundwater quality. It cannot be viewed in isolation and must encompass sustainability of livelihood as well. Climate change is yet another factor that may need to be taken into account.

The international attitude to water supply development has now evolved into demand responsive policies in many countries. This has significant implications for the way in which rural water supply projects are planned, managed and evaluated. Of relevance here, local institutions tasked with facilitating projects on behalf of communities need performance benchmarks and indicators for measuring and comparing outcomes. This is particularly important where siting and drilling has to be contracted out to the private sector, as private contractors are likely to be driven by commercial interest rather than community benefit.

3 UNIFORM APPROACH

Because of the diversity of both physical and human frameworks into which any rural community water development programme might impact, the key to uniform audit of these programmes has to come from a set of common criteria for success. These derive in turn from the establishment of achievable objectives for the promotion of success. The common objectives for water supply (and these would normally parallel those for sanitation) are summarised in Table 1 (after Carter et al, 1999).

These objectives assume in the first place that the physical water supply is sustainable and that it is neither liable to long term exhaustion nor long term contamination. They also assume an adequate level of water resource management based on data collection and monitoring, evaluation of levels of existing and future projected demand, and the sustainable and equitable use of the resource.

Table 1. Proposed aims and objectives of rural water supply projects and umbrella programmes

Objectives in relation to impact:
Achieve minimum daily consumption of 25 l/d per person.
Reduce time taken in water carrying to say one hour per person per day.
Achieve water quality targets, perhaps 10 faecal coliforms per 100 ml at the point of use.
Effect a water supply downtime of no more than say 7 days per year.
Ensure costs of supply are attainable.
Objectives in relation to sustainability:
Caretakers, committees and revenue collection in place.
Regular contact with implementing agency (possibly NGO).
Functional infrastructure.

Given an appropriate set of objectives it is feasible to establish a parallel set of criteria with which to judge the success of a rural water supply project and at the larger scale, of a water supply programme. Secondary indicators can also be ascribed to the evaluation, including health and welfare as well as poverty alleviation. The broad overall criteria are:

1. The water resource is capable of sustaining adequate quantity and quality to satisfy current and future demand.
2. The water resource is monitored and managed and exploited in a sustainable and equitable manner.
3. The community is involved with planning and implementation of the water supply.
4. There is adequate provision for operation and maintenance at community level, including financing and monitoring of water point status.
5. There is adequate provision for institution building at all levels necessary for the maintenance of the water supply system.

Adequacy is not easy to quantify and will vary from project to project depending on the prevailing conditions. However, these criteria provide a framework into which more detailed criteria can be added to provide an overall project score of sustainability.

4 CASE STUDY – BUFFELS RIVER CATCHMENT

This case study highlights some of the problems and failings common to rural water supply projects. The Buffels River catchment is located in the semi-arid Northern Cape Province of South Africa (Figure 1). Aquifers within the catchment are classified as poor, negligible to low yielding with variable water quality, or at best minor, moderate yielding with variable water quality (Parsons and Conrad, 1998). However, all the rural communities are reliant on groundwater, but the inadequacy of much of the water in supply impacts adversely on quality of life and health.

The weathered basement rocks of the Namaqua Metamorphic Province form the main aquifer. There is a strong structural influence on groundwater occurrence from features such as faults and shear zones. The strata are otherwise a complexly deformed heterogeneous group of gneisses and intrusions of regional medium- to high-grade metamorphism, collectively known as Namaqua Metamorphic Complex.

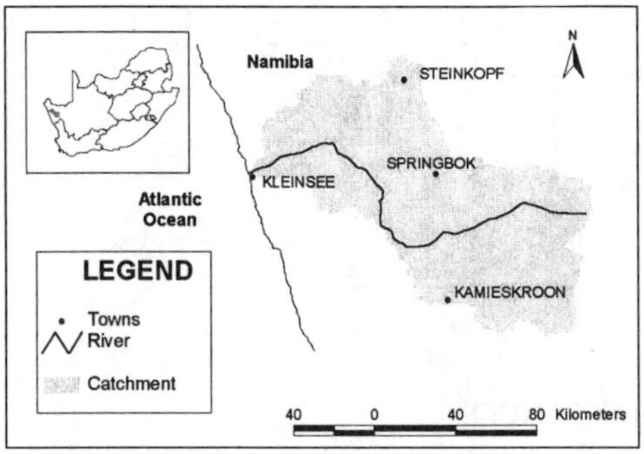

Figure 1 The Buffels River catchment

Borehole yields commonly range between 0.2 and 2.0 l s^{-1}, with 37% of the boreholes capable of yields between 0.1 to 0.5 l s^{-1}, and only 5% of the boreholes capable of a sustained yield greater than 1 l s^{-1} (Toens et al, 1996). The degree of over-exploitation of a production borehole used for community water supply is illustrated in Figure 2. This is typical for many communities in the region, where heavy demands are made at the water point for stock watering as well as agricultural purposes.

There are significant water quality problems (Table 2). Some sources have a high fluoride concentration and this poses a serious health risk to long term users. In addition, elevated levels of trace constituents such as uranium and arsenic in part of the region have a positive correlation with typical lymphocytes (as a proxy of haematological abnormality), the exact cause of which has not yet been established (Toens et al, 1998).

The key groundwater management issues in the regions are (Pietersen et al, 1999):

1. Low rainfall and high evaporation rates inhibit recharge to the crystalline aquifers.
2. The low storage and hydraulic conductivity of the crystalline aquifers.
3. Recurrent droughts occur resulting in water stress.
4. Lack of suitable institutional arrangements at local and catchment scale.
5. Groundwater protection measures are inadequate.

5 DEVELOPMENT OF PROJECT AUDIT

Agencies in South Africa have been accustomed to designing technical solutions with little responsibility beyond implementation. The Community Water Supply and Sanitation Programme was intended to bring safe water to millions of South Africans, but focus on delivery (construction of water systems) has been at the expense of equity and sustainability (Schmitz, 1999). A fundamental flaw in the process was the lack of accountability of the role players, be they the government or consultants.

It is apparent that implementing agents in Namaqualand have not adhered to a uniform approach in water supply development. A study conducted in two communities (Williams, 1998) showed that 80% of people indicated that improved water supply was a priority, albeit partly for stock watering purposes. However, the coverage remained inadequate (more than 300 people per borehole) and equipment was li-

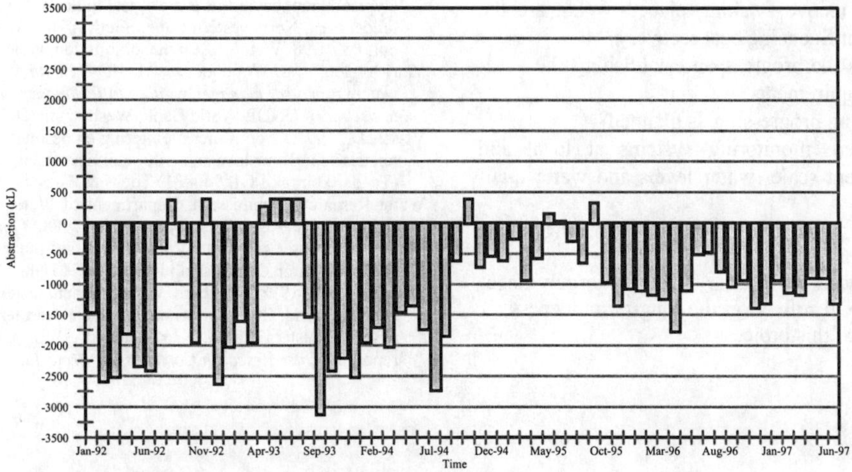

Figure 2. Over-abstraction (estimated sustainable yield minus actual abstraction) for a borehole at Komaggas (Visser, 1997)

Table 2. Groundwater quality in the Buffels River catchment

	Conductivity (mS m^{-1})	Cl (mg l^{-1})	F (mg l^{-1})	NO_3 (m l^{-1})
Rooifontein	208	467	5.4	36
Bulletrap	184	587	1.8	8
Nourivier	358	1058	1.1	7
Leliefontein	83	174	2.8	0.4
Kamassies	142	270	5.7	39
Komaggas	287	684	0.5	16
Guideline level*	370-520	200-400	1.5-3.5	20-40

*WRC et al, 1998

able to misuse. Some management recommendations had been ignored, and cost recovery considerations were not adequate. There was limited community involvement in the provision of water supplies and resultant unhappiness about imposed rules and regulations.

Project audit asks the following questions:

1. Can the resource meet needs - now and in the future (quantity and quality)?
2. Has an operation and maintenance plan been developed and agreed, detailing roles, responsibilities and training requirements?
3. Have cost recovery mechanisms been established?
4. Have the necessary institutional arrangements been established?
5. What is the impact of the project on the environment?
6. What are the current water usage/practices?
7. Has the community been educated about the groundwater system (yield & quality aspects)?
8. To what extent are the institutions functioning (addressing issues representative of the various sectors in the community)?
9. Is cost recovery taking place? What are the factors influencing cost recovery?
10. What is the progress on establishing the hardware requirements?
11. Is training progressing as planned?
12. Are there monitoring systems at local and catchment scale (water levels and water quality)?
13. Are systems and institutional arrangements functional?

The answers at Buffels River are largely negative. This will reflect on the long-term success, or otherwise, of this project.

REFERENCES

Carter, R.C., Tyrrel, S.F. & Howsam, P. 1999. The impact and sustainability of community water supply and sanitation programmes in developing countries. *J. Instn. Wat and Envir. Mangt.* 13, 292-296.

Farr, J.L., Spray, P.R. & Foster, S.S.D. 1982. Groundwater supply exploration in semi-arid regions for livestock extension - a technical and economic appraisal. *Wat. Supply and Mangt.* 6: 4, 343-353.

Parsons, R. & Conrad, J. 1998. Explanatory notes for the aquifer classification map of South Africa (map included). WRC Report, KV 116/98. Water Research Commission, Pretoria.

Pietersen, K.C., Williams, M., Taylor, V. & Titus, R.A. 1999. The impact of inadequate groundwater supply and sanitation facilities on rural communities and the implications for sustainable livelihoods - the Namaqualand experience. *IAIsa Conference,* Bloemfontein.

Schmitz, T. 1999. Rethinking delivery? A review of the efforts of the Department of Water Affairs, 1994-9. *CPS Policy Brief 16.* Center for Policy Studies, November 1999.

Toens, P.D. Visser, D. & Esterhuyse, C. 1996. Northern Cape community water supplies and sanitation: a preliminary assessment of the hydrogeology of the Northern Cape. Toens & Partners, Report Number 1.

Toens, P.D., Stadler, W. & Wullscheger, N.J. 1998. The association of groundwater chemistry and geology with atypical lymphocytes (as a biological indicator) in the Pofadder area, Northwestern Cape, South Africa. WRC Report 839/1/98. Water Research Commission, Pretoria.

Van Dongen, P. & Woodhoouse, M. 1994. *Finding groundwater: a project manager's guide to techniques and how to use them.* UNDP-World Bank, Washington, DC.

Visser, D., 1997. Grondwater monitering en bestuur van waterstelsels in Namakwaland. Vorederingsverslag No 199. (Tender kontrak LGB 940001) Toens & Partners.

Water Research Commission, Department of Water Affairs & Forestry and Department of Health. 1998. Quality of domestic water supplies. Volume 1: Assessment Guide. Water Research Commission Report TT 101/98.

Williams, M., 1998. Houbare watervoorsieningstelsels in Namakwaland. Groundwater supply and assessment strategy for Western Karoo, Namaqualand - WRC progress report 3. *Water Research Commission, Pretoria.*

Groundwater: Past Achievements and Future Challenges, Sililo et al. (eds) © 2000 Balkema, Rotterdam, ISBN 90 5809 159 7

Groundwater resources in Berlin, Germany: Will there be enough fresh water in the future?

C. Sommer-von Jarmersted, B. Fritz & A. Pekdeger
Department of Economic and Environmental Geology, Free University of Berlin, Germany

ABSTRACT: At the Senate's Administration for the Urban Development, Environmental Protection and Technology (SenSUT), different data stocks of the total area of Berlin were pooled and evaluated in order to define the hydrological and hydrochemical situation of Berlin's fresh-water resources. This provided a basis for a chemical assessment of the groundwater reserves and the spatial distribution of the hydrochemical composition of the aquifers.

The groundwater reserves potentially available in this urban area are sufficient. A gradual shrinkage of the potential yield is not expected, provided the balance components do not significantly change in the future. Berlin's potable water supply is mainly based on bank filtration. Large quantities of sewage contribute considerably to the total runoff of the surface waters. The nature of the groundwater reservoirs has already changed to a large extent. Taking all these factors into account, the long-term potable water supply cannot be viewed without certain reservations.

1 INTRODUCTION

Since 1991, Berlin's drinking water supply is based exclusively on groundwater. In order to conserve the natural available groundwater, it is enriched intensively by bank filtration from surface waters and artificial recharge. Hence, the volume of the groundwater storage, its quality and the proportion of bank filtration are of significant importance for a sustainable water management in Berlin. Due to the highly complicated hydrogeological conditions, the quantification of the groundwater storage and the proportion of bank filtration has not been exact yet.

However, the impact of infiltrating surface water on the groundwater-quality and the amount of bank filtration were recently subject to examinations on an interdisciplinary level.

In this paper a first assessment of the groundwater reserves and the spatial distribution of the hydrochemical composition of the aquifers are presented.

2 HYDROGEOLOGY

Berlin is located in a lowland region between two plateaus, Barnim and Teltow, in a glacial valley with many rivers and lakes. The area is formed by Quaternary and Tertiary glacial sequences with heterogeneous series of aquifers and aquitards. The main part of the glacial sediments found in Berlin and its surrounding represent excellent aquifers and the vulnerability of the groundwater is obviously high.

In the Berlin area four freshwater aquifers form the multiple aquifer which is separated by a confining bed from deeper saltwater aquifers. Elsterian channels of coarse sand form windows in this impermeable layer and therefore saltwater can ascend as a result of groundwater drawdown.

About 26,000 drillings are distributed all over the area of the city (Fig.1). Groundwater levels were measured by data logger in more than 2000 observation wells.

3 VOLUME OF GROUNDWATER STORAGE

The calculation of the complete sustainable water volume were done with geostatistical methods. At first, the volume of the water filled underground were detected on the basis of 900 geological drillings in the freshwater aquifers. All geological sequences were allocated to aquifers or aquicludes. To get an idea of the realistic groundwater volume of the Berlin underground the second step of calculation considers the proportion of the effective porosity from 15% up to 30% (sand and gravel). The total volume of groundwater can be indicated from 15 to 30 * 10^9 m^3 (Fig.2). In the years 1995 and 1996 the groundwater withdrawal

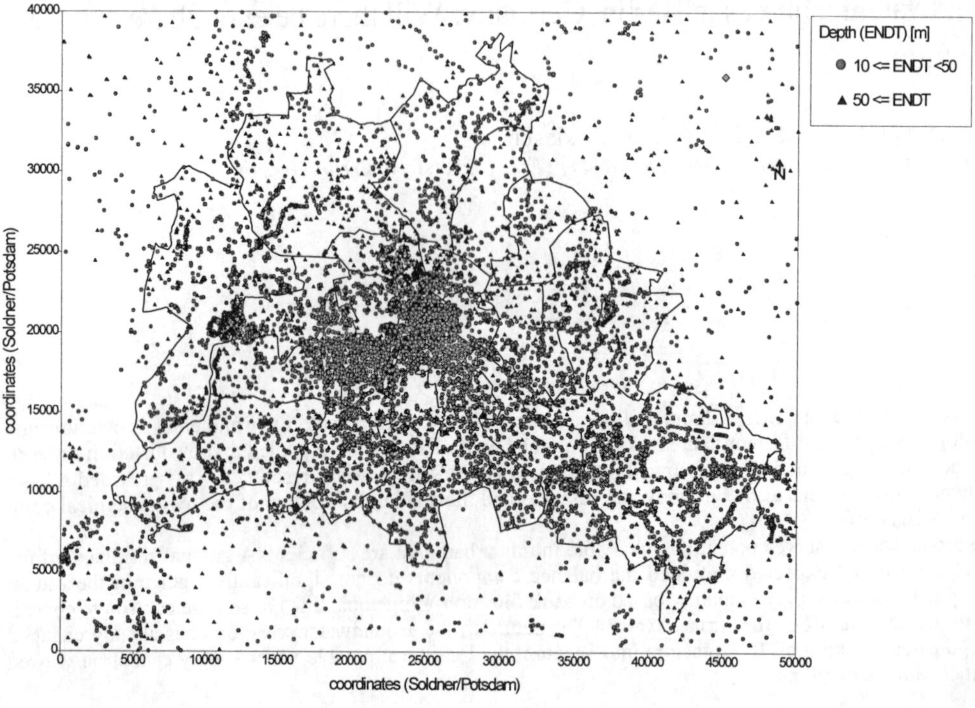

Figure 1. Spatial distribution of boreholes.

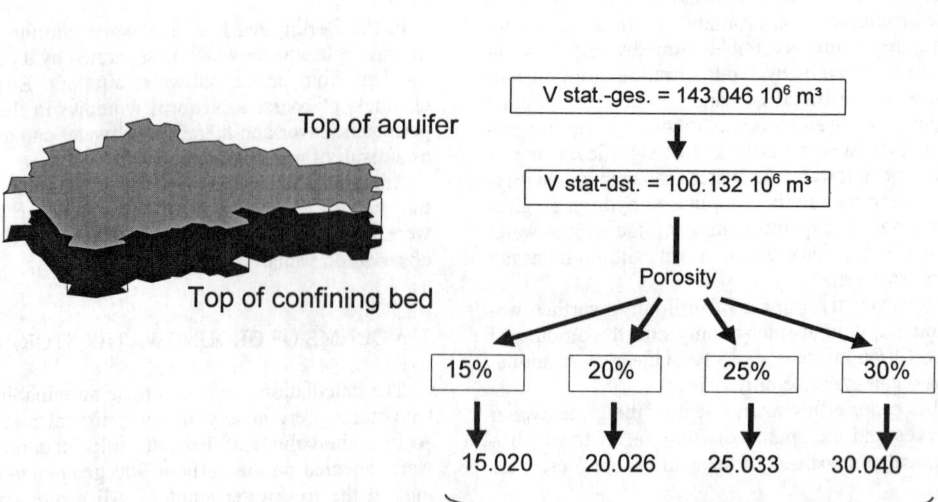

Figure 2. Calculation of the potential water volume.

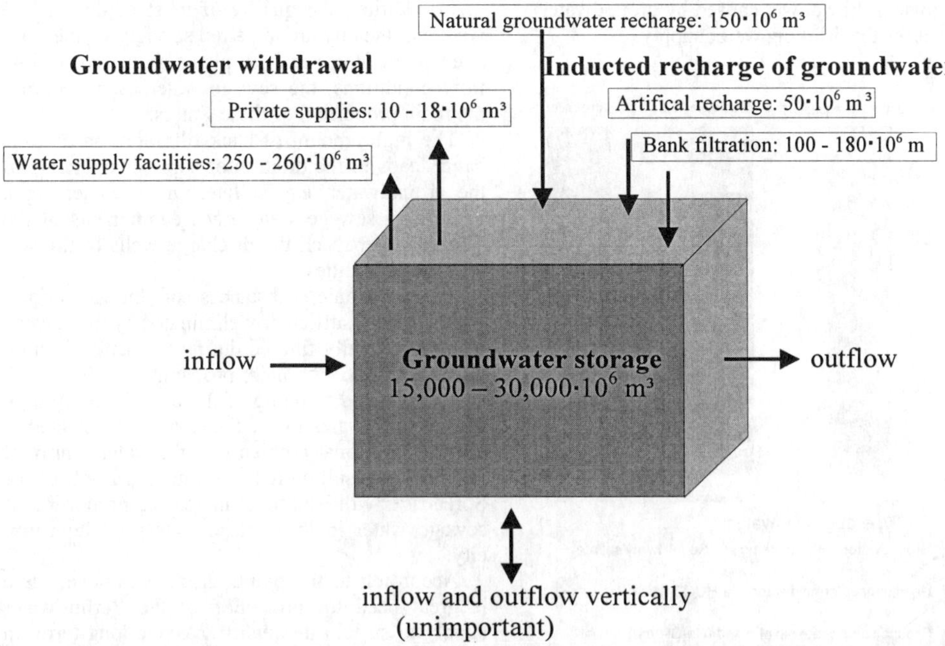

Figure 3. Summary of the balance components (modified after Sommer-von Jarmersted et al. 2000)

of the Berlin water supply amounted to approximately 250-260 $*10^6$ m³. Due to the variable proportions of bank filtered water (40-70%), depending on the extremely different hydrological systems, it is not possible to fix the amount of river bank filtered water in the total water withdrawal. Hence, a range from 100 to 180 $*10^6$ m³ can be calculated. Additionally, artificial recharge produces a possible yield of 50 $*10^6$ m³ groundwater. As another element of the water balance natural groundwater recharge supplies in Berlin ca. 150 $*10^6$ m³.

The summary of the balance components (see Fig.3) show a positive balance result. There are no groundwater outflow out of the Berlin city area. Unfortunately, the amount of recent groundwater inflow can not be determined currently. Hence, all calculations show that there is enough water in Berlin, and without changes of the balance components, there might be enough in future. However, continued utilization depends on the water quality, which will be the primary problem in future.

4 GROUNDWATER-QUALITY

For the assessment of the recent groundwater-quality, hydrochemical data from about 1100 observation wells have been analysed. Water quality of the different aquifers was differentiated (Fig. 4). By means of the hydrochemical data basis the spatial distribution of the main cations and anions in the groundwater has been determined. Special local situations, such as areas with abandoned waste sites and contaminated sites including the special observation wells are not taken into consideration.

Especially sulfate and chloride indicate impairment of the groundwater.

Concentrations of sulfate exceeding the maximum limit of the national drinking water regulation (240 mg/l) are determined in the first and second aquifer throughout the city, sometimes even reaching more than 1000 mg/l. These high concentrations are explained predominantly by human impact, in Berlin especially by the leaching of deposited rubble from the Second World War. So far, the deeper aquifers do not show any sulfate impact, possibly due to reducing conditions with increasing depth.

The concentrations of chloride show an opposite trend with low concentrations in the shallow aquifers, whereas the highest concentrations are found in the third and fourth aquifers. The fourth, i.e. the Tertiary aquifer even reaches a maximum value of 11,000 mg/l. These results partly correlate with missing parts in the confining bed between the saltwater and freshwater aquifers. Hence, locally indicated high concentrations of chloride are directly related to the geological background of the area. Additionally, the rise of saltwater is enhanced by groundwater drawdown, caused by groundwater withdrawal for the drinking water supply.

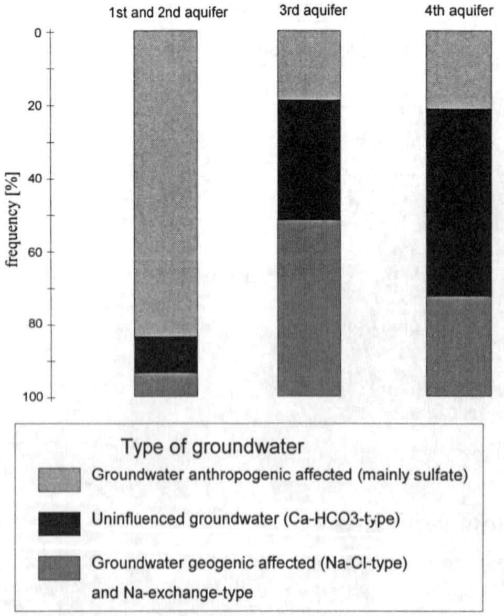

Figure 4. Anthropogenic and geogenic influence on the different aquifers.

The surface water in the Berlin area is highly influenced by sewage water (Fritz et al. 2000). During the soil passage via bank filtration some proportion of sewage water reaches the discharging wells. Pathogens and most of the organics and nutrients were eliminated in the first few meters between the lake and the wells. Nevertheless, some trace organic substances like Gd-complex, EDTA, AOI, AOBr, AOCl, etc., Boron etc. can be found in the drinking water. These components were used by us to estimate the amounts of sewage water in the surface and groundwater. Additionally they are ideal tracers for bank filtration investigations (Fritz et al. 2000, Sommer-von Jarmersted et al. 2000).

5 CONCLUSION

The groundwater storage in the freshwater aquifer of the Berlin area consists of a volume between 15 and $30*10^9$ m^3. Groundwater withdrawal and groundwater recharge, including artificial recharge are approximately balanced, implying that the groundwater storage is currently not limited.

However, high concentrations of sulfate and chloride indicate a significant natural and human impact on the groundwater quality.

In addition, the quality of most of the surface waters in the city are impaired strongly by the sewage water of the treatment plants. In periods of low flow conditions, the sewage water even becomes the main proportion of these waters.

The high amount of bank filtration, on the one hand, leads to the dilution of some components of the groundwater, e.g. sulfate. On the other hand, infiltrating sewage water trace components of the surface water reach the discharge wells of the water supply facilities.

Trace organic substances of human origin, which are not sufficiently eliminated by the sewage plants nor totally filtered during the bank filtration process could cause major problems in future.

The planned flooding of the open lignite pits in the catchment area of the Spree river is expected to impose additional problems on the water supply of Berlin. The resulting reduction in the run-off of the Spree river will lead to an increasing proportion of sewage water in the surface waters of the entire city.

According to the points discussed above, there is great need for protection of the Berlin water supply to sustain its quality over the long-term, in order to assure that the potentially available volume of groundwater is fit for water supply.

6 ACKNOWLEDGEMENTS

The financial support of the Berlin Administration of Urban Development, Environmental Protection and Technology is gratefully acknowledged.

REFERENCES

Sommer-von Jarmersted, C., Kösters, E., Fritz, B., Knappe, A. & A. Pekdeger 2000. Grundwassermenge und –güte in Berlin. *Wasserwirtschaft* 90(2): 74-80.

Fritz, B., Sommer-von Jarmersted, C. & Pekdeger, A. (2000). Investigation of trace elements for the assessment of sustainable water supply.- IAH XXXth Congress 2000. Groundwater: Past achievements and future challenges. Cape Town, 26 Nov-1 Dec 2000.

Groundwater: Past Achievements and Future Challenges, Sililo et al. (eds) © 2000 Balkema, Rotterdam, ISBN 90 5809 159 7

Sustaining ecosystems and cultures dependent on springs of the Grand Canyon, USA

Abraham E. Springer
Northern Arizona University, Flagstaff, Ariz., USA

Eric S. Wilson
Maxim Technologies Incorporated, Missoula, Mont., USA

ABSTRACT: Ecosystems and cultures at seeps and springs on the South Rim of the Grand Canyon of the Colorado River in Arizona, USA, are supplied with ground water from a large regional carbonate aquifer. Approximately 2.0 m^3/sec of water discharges into three major and 15 minor springs and numerous seeps. The bedrock aquifer in this arid region covers approximately 13,000 km^2 and receives only as much recharge as the volume of discharge from the seeps and springs. These seeps and springs located in the middle of a desert support a diversity of species, sites which are culturally significant to Native Americans, and wilderness areas. A numerical ground-water flow model was constructed from a digital geologic framework model to delineate capture zones for the major springs. This information was combined with conceptual ecosystem and cultural information to predict the impacts on the spring-dependent systems. The sustainable yield of groundwater from this aquifer is capable of supporting little new human development in the region.

1 INTRODUCTION

When managing water resources in arid regions, it is difficult to balance water requirements for humans with *in situ* water requirements. Providing water for *in situ* values to be met allows a water source to be considered sustainable. Gleick (1998) defines a yield of water from a source as sustainable if it (1) guarantees all humans a basic water requirement, (2) guarantees all dependent ecosystems a basic water requirement, (3) maintains minimum water quality standards, and (4) doesn't disrupt the natural hydrological cycle. Therefore, a sustainable ground-water supply can't just support humans, it must support all dependent Earth systems. To manage an aquifer in an arid region as a sustainable supply, therefore, requires a holistic approach (Sophocleous 1998).

One such aquifer needing sustainable management is the Redwall-Muav aquifer of the South Rim of the Grand Canyon, Arizona, USA. This deep, large, regional bedrock aquifer was not exploited as a ground-water supply until 1989. Until then, all water which recharged the 13,000 km^2 area of the Coconino Plateau discharged from 3 major springs and about 15 minor springs and numerous seeps. A proposal in 1993 by a private developer to exchange private in holdings for federal land to build a new community at the South Rim has renewed a discussion of how to manage these types of aquifers. The Havasupai Indians have a community with strong cultural and religious ties to the largest of the major springs discharging from the aquifer. All of the other springs and seeps occur in Grand Canyon National Park and have value for the ecosystems they support and the aesthetic and wilderness values they provide. Grand Canyon National Park is a world heritage site and one of the most popular tourist destinations in the USA. It receives nearly 5 million visitors a year, but has no dependable, reliable water supply for visitors and employees.

To characterize the impacts of ground-water withdrawals from this regional aquifer, a geologic framework model and a numerical ground-water flow model were constructed (Wilson 2000) and coupled with conceptual ecosystem and cultural information. Capture zones of the major springs were delineated to determine which portions of the aquifer were influenced by which proposed wells. The conceptual ecosystem and cultural information were used to assess impacts of the changes in spring discharge on ecosystems and significant cultures.

1.1 *Site Description*

The Coconino Plateau, located just south of the Grand Canyon (Fig. 1), consists of relatively horizontal layers of Paleozoic strata underlain by Precambrian basement rock. The stratigraphy of the Grand Canyon has been studied in great detail. The

Precambrian rocks which crop out along the bottom of the Grand Canyon can be divided into two groups. The first group includes the Vishnu Schist, the Trinity and Elves Chasm Gneiss, and the Zoroaster plutonic complex. The Grand Canyon Series comprises the second group of Precambrian rocks and can be further divided into the Unkar and Chuar Groups. These are predominantly sedimentary rocks that exhibit little metamorphism.

The Paleozoic formations, listed in ascending order, are as follows: the Tapeats Sandstone, Bright Angel Shale, Muav Limestone, Temple Butte Limestone, Redwall Limestone, the Supai Group, Hermit Shale, Coconino Sandstone, Toroweap Formation, and the Kaibab Limestone.

The hydrogeologic characteristics of these strata are directly tied to their geologic characteristics. None of the Precambrian basement rocks are considered ground-water sources significant for water supply.

The Muav and Redwall limestones together make up the most significant aquifer in the Grand Canyon and the Coconino Plateau. The combination of the excellent permeability and the confining effect of the underlying Bright Angel Shale account for the number of springs that issue from the Muav Limestone. Most of the springs on the south side of the river flow from this layer. Intermittently sandwiched between the Muav and the Redwall is the Temple Butte Limestone. No springs flow from the Temple Butte Limestone and it is not considered of importance in the flow of ground water in the Grand Canyon (Metzger 1961). The Redwall is an aquiclude where not fractured. However, faults and joints which cut through the entire unit have extremely high hydraulic conductivity values and act as vertical and lateral conduits for ground-water movement and recharge.

The Colorado Plateau was formed as a result of tectonic uplift followed by various erosional events (Huntoon & Sears 1975). Extensive geologic mapping and hydrologic analysis of the Coconino Plateau and Grand Canyon (Huntoon et al. 1996; Billingsley et al. 1985; Billingsley & Huntoon, 1983; Huntoon 1982, 1994, and 1995; Montgomery & Associates 1985; Huntoon & Billingsley 1981; Shoemaker et al. 1978; Harshbarger & Associates, 1973; Metzger 1961; Moore et al. 1960) describe the location and interconnection of structural features.

Two of the most important controls on the movement of groundwater are the fractures due to faulting and the regional dip of the Paleozoic strata in the Grand Canyon area. The initial deformation of the basement in the Proterozoic has controlled subsequent faulting, folding, and karst formation (Huntoon & Sears 1975; Wenrich 1986). The Precambrian crystalline basement was first deformed during the late Proterozoic, when tectonic forces shortened the lithosphere and northeast trending thrust faults were formed by the compressional forces (Huntoon & Sears 1975). Subsequent crustal extension facilitated the formation of northwest trending transverse normal faults, which are conjugate to the northeast oriented thrust faults (Huntoon 1974).

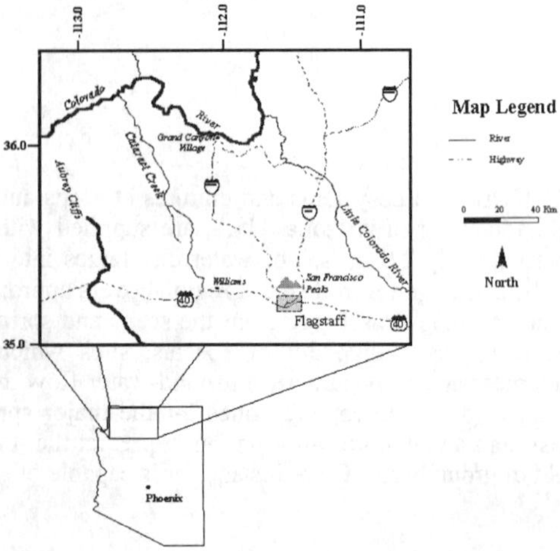

Figure 1. Map showing location of Grand Canyon study area.

2 METHODS

2.1 *Geologic framework model*

The purpose of building a framework model is to create a three dimensional, representation of geologic features to be used for visualization and as the basis for a ground-water flow model. If digitized, the framework model can be an extremely useful foundation for future modeling efforts for a particular geologic region and provides a detailed characterization of the hydrologic system. The objective of building a framework model for the South Rim of the Grand Canyon and the entire Coconino Plateau groundwater sub-basin was to provide the database for a new generation of more accurate flow models of the area.

In this study, the framework model is described as a three-dimensional computer generated model including geologic structure and stratigraphy as well as hydrogeologic information. For this project, we used Stratamodel (v4.000)(Landmark Graphics Corporation 1998) to create the framework model. Stratamodel is a three dimensional stratigraphic geocellular modeling package that integrates raw data and knowledge of the local geology, geophysics, and engineering to create an intricate three dimensional visualization of the area using combinations of cross sections, stratigraphic slices, and cross-sequence connectivity displays.

2.2 *Numerical ground-water model*

Ground-water flow through the Coconino Plateau groundwater sub-basin and Redwall-Muav aquifer on the South Rim of the Grand Canyon was simulated with a three-dimensional numerical, ground-water flow model, MODFLOW (Harbaugh & McDonald 1996). MODFLOW was selected because it is relatively well suited to simulate flow in heterogeneous and anisotropic systems such as the Redwall-Muav aquifer and because it is well documented and widely used in the public domain. Although it was originally developed to simulate flow through porous media, it has been applied successfully to similar regional fractured aquifer systems.

The purpose of the ground-water flow model was to characterize the impacts of ground-water pumping on the three major springs draining this aquifer in the Grand Canyon. The flow model had two main objectives. The first was to calibrate a steady-state ground-water flow model to pre-pumping (pre-1989) hydrogeologic data. This included eight wells, three springs, and limited measurements of recharge, and specific capacity. The second was to delineate capture zones for the three major springs in the region. A numerical postprocessor to the numerical flow model (Pollock 1994) was used to track water particles and delineate capture zones for each of the major springs. This provided insight into which regions are most likely to impact on specific springs if more wells were to be drilled. The approach used for building the numerical ground-water flow model included building a conceptual model, selecting a numerical model, constructing the model, calibrating to steady-state conditions, and conducting a sensitivity analysis.

The simulated area of the Redwall-Muav aquifer is approximately 9,500 km^2. A two-layer, three-dimensional grid was constructed to simulate flow and delineate capture zones. The top layer represented the Supai Formation and the bottom layer represented the Redwall and Muav Limestones. The grid was oriented 60 degrees west of north to parallel the principle direction of groundwater flow toward Havasu Springs (Fig. 2). Data from the digital geologic framework model were imported for the top and bottom elevations of both layers.

The boundaries of the model included no-flow boundaries for ground-water divides, specified fluxes for recharge boundaries, and head-dependent fluxes for spring discharges. Because of the lack of measurements of aquifer properties (hydraulic conductivity and storativity), aquifer properties were determined through calibration and sensitivity analyses. The model was calibrated to pre-pumping (pre-1989) hydraulic head measurements from eight wells and spring discharge measurements from three springs. Calibration was considered acceptable when the root mean squared error (Anderson and Woessner 1992) was approximately 15 % of total hydraulic head change across the region.

2.3 *Conceptual ecosystem and cultural information*

Information about the diversity of life supported by spring and seeps in the Grand Canyon was gathered from existing reports of vegetation and wildlife surveys. Springs typically support a very narrow corridor of riparian vegetation in the reach of the perennial channel supported by the spring. Seeps typically support vegetation in hanging gardens. Seeps typically do not support a corridor of riparian vegetation downstream of the hanging garden.

The Havasupai are a tribe with significant cultural and religious ties to the largest spring discharging from this regional aquifer. The Havasupai refer to the water as the "lifeblood of the Earth and the Havasupai". Over 97 % (approximately 157,000 m^3/day) of the total discharge from the aquifer discharges into Havasu Springs.

3 RESULTS

3.1 *Geologic framework model*

Construction of a digital geologic framework model resulted in a model which could be viewed in any three-dimensional perspective as a block diagram (Wilson, 2000). These digital views of the system aided in the development of the conceptual model of the groundwater flow system. Digital data from the geologic framework model were imported directly into the numerical groundwater flow model. The digital framework model is a working document which will be revised in the future as new information is collected about the aquifer system.

3.2 *Numerical groundwater model*

Aquifer properties and boundary conditions were constrained by field measurements where available, but were adjusted through calibration of the model and sensitivity analyses. Different hydraulic conductivity values were assigned for matrix and fractured portions of the Redwall-Muav and Supai. Fractured hydraulic conductivity values were either 10 or 100 m/d. Matrix hydraulic conductivity values in the x-axis direction were 0.4 m/d for the Redwall-

Muav and 0.006 m/d for the Supai. Anisotropy was built in for both layers with matrix hydraulic conductivity values in the y-axis direction of 0.1 m/d for the Redwall-Muav and 0.0001 m/d for the Supai.

A total recharge of 160,079 m^3/d was applied to the model. Simulated (and target values in parentheses) steady-state, pre-development spring discharges for the Redwall-Muav aquifer were 157,090 (157,000) m^3/d for Havasu Spring, 1,644 (1,640) m^3/d for Hermit Spring, and 1,640 (1,640) m^3/d for Indian Garden Spring. The root mean squared error for hydraulic head values from the steady-state simulation was 41.2 m, or 16 % of the total head change across the region.

A capture zone analysis with hydraulic heads from the steady-state pre-development numerical model indicated that capture zones for Hermit and Indian Garden Springs do not extend far from the South Rim (Fig. 2). The model indicates that water is actually drawn to those springs from a distance of about only 5 km south of the South Rim. Havasu Springs captures most of the groundwater in the aquifer.

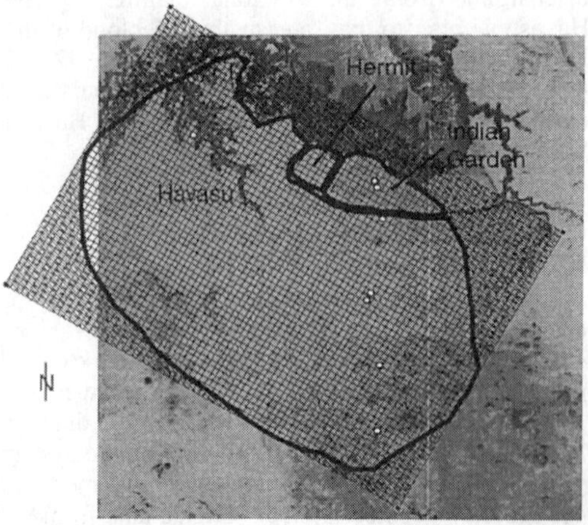

Figure 2. Steady-state, pre-development capture zones for Havasu, Hermit and Indian Garden Springs on shaded relief map of the Coconino Plateau and Grand Canyon. Dots show locations of wells with water levels used for calibration. Each grid cell is 1,750 m^2.

3.3 *Conceptual ecosystem and cultural information*

Surveys of seeps indicates that there are 47 native species of vegetation and 30 species of wildlife (J. Spence, pers. comm.). There are at least 2 non-native vegetation species. Surprisingly, through this study, it was determined that there were no detailed surveys of vegetation supported by springs and no accurate measures of seep discharge. These surveys are now in progress and will be incorporated in the process when they are complete (L. Stevens, pers. comm.).

The Havasupai have indicated that they are unwilling to accept any diminishment in spring discharge from Havasu Spring. It has been estimated that spring discharge may decrease under some groundwater withdrawal proposals of some potential development scenarios along the South Rim (Montgomery and Associates 1999). Although the developer of the land exchange discussed in this study found an alternative water supply which doesn't rely on groundwater, other developments may not be held to this environmental standard. These developments may use water from the Redwall-Muav aquifer as a water supply, and may lead to eventual decreases in spring and seep discharge.

4 CONCLUSIONS

The creation of a three dimensional geologic framework model can prove to be useful. The creation of a single hydrogeologic database is a data presentation method allowing an area to be viewed in a way that can be visually revealing, interpretively significant, and can aid in developing the conceptual model of the flow system. Secondly, a digital framework can easily be updated and modified as new data are collected. As more information on the system becomes available, it can be added to the database and model. This is especially true for large regional aquifers such as this one on the Coconino Plateau, where there currently is a paucity of data. Growth and development will increase the knowledge base through new wells, new site investigations, and more monitoring of groundwater conditions.

The numerical ground-water flow model when coupled with the conceptual model were useful to improve the knowledge of flow in the aquifer. Some of the insight gained includes distribution of hydraulic conductivity, distribution of recharge, and connectivities of the fracture systems.

Particle tracking allowed capture zones for the three major springs to be defined. The model showed that the capture area for Havasu Spring includes most of the basin. Capture areas for the two other relatively major springs were very small. Because of continued future demands for human growth within the Indian Garden Spring capture zone, it is anticipated that future groundwater extraction will impact discharge of this spring.

REFERENCES

Anderson, M.P. and Woessner, W.W., 1992. *Applied Groundwater Modeling, Simulation of Flow and Advective Transport*. Academic Press, Inc., San Diego, CA, 381 p.

Billingsley, G.H., Barnes, C.W., and Ulrich, G.E., 1985. Geologic map of the Coconino Point and Grandview Point quadrangles, Coconino County, Arizona: United States Geological Survey. MAP I-1644. 1:62,500 scale.

Billingsley, G.H. and Huntoon, P.W., 1983. Geologic map of the western Grand Canyon Formation, Arizona. Grand Canyon Natural History Association, Grand Canyon, Arizona.

Gleick, P.H. 1998. *The World's Water: The Biennial Report on Freshwater Resources 1998-1999*. Washington, D.C.: Island Press.

Harbaugh, A.W., and McDonald, M.G., 1996, User's documentation for MODFLOW-96, an update to the U.S. Geological Survey modular finite-difference ground-water flow model: U.S. Geological Survey Open-File Report 96-485, 56 p.

Harshbarger & Associates, Inc., 1973. Groundwater conditions and potential development on Babbitt Ranch area, Coconino County, Arizona. Unpublished.

Huntoon, P.W., 1974a. The Karstic Groundwater Basins of the Kaibab Plateau, Arizona. Water Resources Research. Vol. 10, No. 3, p. 579-590.

Huntoon, P.W., 1974b. The post-Paleozoic structural geology of the eastern Grand Canyon, Arizona, in Breed, W.J. and Roat, E.C. (editors), Geology of the Grand Canyon. Flagstaff, Arizona, Museum of Northern Arizona and Grand Canyon Natural History Association, p. 82-115.

Huntoon, P.W., 1982. The ground-water systems that drain to the Grand Canyon of Arizona. Laramie, WY: Department of Geology and Water Resources Institute, University of Wyoming, 25 p.

Huntoon, P.W., 1994. Is it appropriate to apply porous media ground water circulation models to karstic aquifers?, in Proceedings from assessment of models for groundwater resources analysis and management, University of Hawaii at Manoa, March, 1994.

Huntoon, P.W., 1995. Unpublished photogeologic mapping. May 1995.

Huntoon, P.W., and Billingsley, G.H., 1981. Geologic Map of the Hurricane fault zone and vicinity, western Grand Canyon, Arizona.

Huntoon, P.W., Billingsley, G.H., and others, 1996. Geologic Map of the Eastern Part of the Grand Canyon National Park, Arizona: Grand Canyon Association. 1:62,500 scale.

Huntoon, P.W., and Sears, J.W., 1975. Bright Angel and Eminence Faults. Geological Society of America Bulletin, v. 86, pp. 465-472.

Landmark Graphics Corporation, 1998. Stratamodel, version 4.000, Houston, Texas.

Metzger, D.G., 1961. Geology in Relation to Availability of water along the South Rim, Grand Canyon National Park, Arizona. U.S. Geological Survey Water Supply Paper 1475-C, 138 p.

Montgomery & Associates, Inc., 1985. Groundwater conditions, Canyon Mine region, Coconino County, Arizona. Appendix F of Draft Environmental Impact Statement, Canyon Uranium Mine, December 1985.

Montgomery & Associates, Inc., 1999. Supplemental assessment of the hydrologic conditions and potential effects of proposed groundwater withdrawal, Coconino Plateau Groundwater Sub-Basin, Coconino County, Arizona, in Appendix of the Final Environmental Impact Statement of Tusayan Growth, Kaibab National Forest, Williams, Arizona, 85 p.

Moore, R.T., Wilson, E.D., and O'Haire, R.T, 1960. Geologic map of Coconino County, Arizona: Arizona Bureau of Mines. 1:375,000 scale.

Pollack, D.W., 1994. User's guide for MODPATH/MODPATH-PLOT, Version 3: A particle tracking post-processing package for MODFLOW, U.S. Geological Survey Open-File Report 94-464, 234 p.

Shoemaker, E.M., Squires, R.L., and Abrams, M.J., 1978. Bright Angel and Mesa Butte fault systems of northern Arizona. Geologic Society of America, Memoir 152.

Sophocleous, M. ed. 1998. *Perspectives on Sustainable Development of Water Resources in Kansas*. Bulletin 239. Lawrence, Kansas: Kansas Geological Survey.

Wenrich, K.J., 1986. Geochemical exploration for mineralized breccia pipes in northern Arizona. Applied Geochemistry, v. 1, pp. 469-485.

Wilson, E. 2000. Geologic framework and numerical groundwater models of the South Rim of the Grand Canyon, Arizona. Unpublished M.S. thesis. Northern Arizona University: Flagstaff, Arizona, 72 p.

Groundwater: Past Achievements and Future Challenges, Sililo et al. (eds) © 2000 Balkema, Rotterdam, ISBN 90 5809 159 7

Managing mine de-watering for bulk groundwater supply in the Kalahari region, South Africa

G.S.du Toit van Dyk & L.D.Ferris
Department of Water Affairs and Forestry, Kimberley, South Africa

J.M.Nel
Department of Water Affairs and Forestry, Pretoria, South Africa

ABSTRACT: The Vaal Gamagara Government Water scheme was constructed to deliver surface water from the Vaal river for mining, industrial, domestic and agricultural use in the Kalahari region, South Africa. Water extracted and purified at the river is delivered via a 280 km pipeline at an average annual demand of 9.3 million m^3. Groundwater abstracted from the Sishen, Kathu, Postmasburg and Beeshoek aquifers at 31.6 million m^3/a is the major source. Groundwater delivery cost of less than 30 % of surface water and recently confirmed mine de-watering schedules, favors increased utilization. These aquifers are fractured sediments and karstic dolomites developed along dolerite /fault zone weathering, lithological contacts, folding and mineralization. The water is of exceptable quality for all use. The aim of this study is to compile the hydrogeological characteristics of the area, determine possible long-term impact from water utilization, evaluate current water monitoring systems, ensure reliable future monitoring systems and ensure that local management structures are capacitated to manage the resource.

1 INTRODUCTION

The Kalahari area of South Africa is semi-arid with no perennial water resources available. Initially only primary use could be satisfied from groundwater in this mineral rich area. Exploration revealed economical deposits of asbestos, iron and manganese. To satisfy the increased demand for the mines and related industries surface water was imported from the confirmed sustainable resource of the Vaalriver.

Development of the mines revealed aquifers of enormous proportions associated with the ore. These aquifers are de-watered for mining purposes with the water used in the mining process and to sustain the mining communities. Long-term use of these resources confirmed their sustainability.

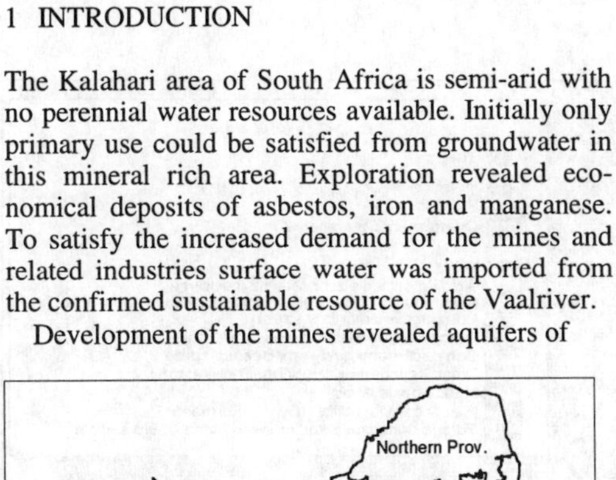

Figure 1. Locality of the D41K, D41J and D73A catchments in South Africa.

2 DELINEATION OF THE WATER AREA

The area is delineated by the surface quartenery catchment boundaries of the 3 catchments depicted in Figure 1. These catchments are bound by the Kuruman Hills in the east and the Lange and Koranna Hills to the west. The area is open to the south and North with both surface and groundwater leaving the area.

The surface water divide of hills east and west is mirrored by groundwater. (Leskiewicz 1981, Willemink 1983). A surface and groundwater divide in the Glossam area separates north and south draining water. The Gamogara River drains toward the north with the Groenwaterspruit draining to the south. These rivers are non perennial and flow only with extreme precipitation.

The area is generally sparsely populated with only the mining communities at Hotazel, Kathu, Postmasburg and agricultural centers at Olifantshoek and Dibeng.

3 CLIMATE AND HYDROLOGY

The rainfall ranges from 480 mm in the north east to 320 mm in the southwest.
Precipitation occur during summer as high intensity thunderstorms. The long-term precipitation trend is erratic as evident from Figure 2.

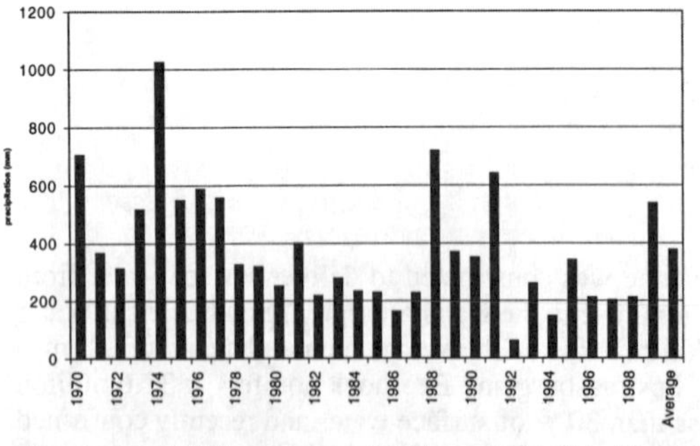

Figure 2. Rainfall histogram over time (mm/annum) (S.A. Weather Buro).

A semi-arid Acacia and grass vegetation is sustained, of which the Kathu Forest Natural Heritage Site formed by the *Acacia Ericloba* population is well known. Groundwater is sustaining the older trees. Mortality of the trees is from old age and not as a result of water fluctuations. (Laan 1998)

4 HYDROGEOLOGY

4.1 *General hydrogeology*

The area is characterized by 6 diverse water-bearing rocks.

The different hydrogeological units (Vegter 1995) are indicated in Figure 3 with the Vaal Gamagara pipeline and major aquifers in the area. The groundwater occurrence is complicated by shallow dipping strata resulting in multi-layer aquifers of different lithology. Most major aquifers are developed in karstic dolomite and fractured compact sediments associated with folding and mineralization.

Two distinctive karsts are developed. The shallow karst associated with surface lineaments and limited weathering covers most of the area. Limited aquifer thickness and erratic recharge limit the usefulness of this aquifer. The deeper karst associated with mineralization, folding and lithological contacts is targeted for long-term utilization.

Away from the folding zone the compact sediments are dominated by banded ironstone formation and jaspelite with limited exploitation potential due to low yields. The compact arenaceous strata, dominated by quartzite's are also poor yielding. Towards the north the mafic lavas are low yielding with higher potential in the south.

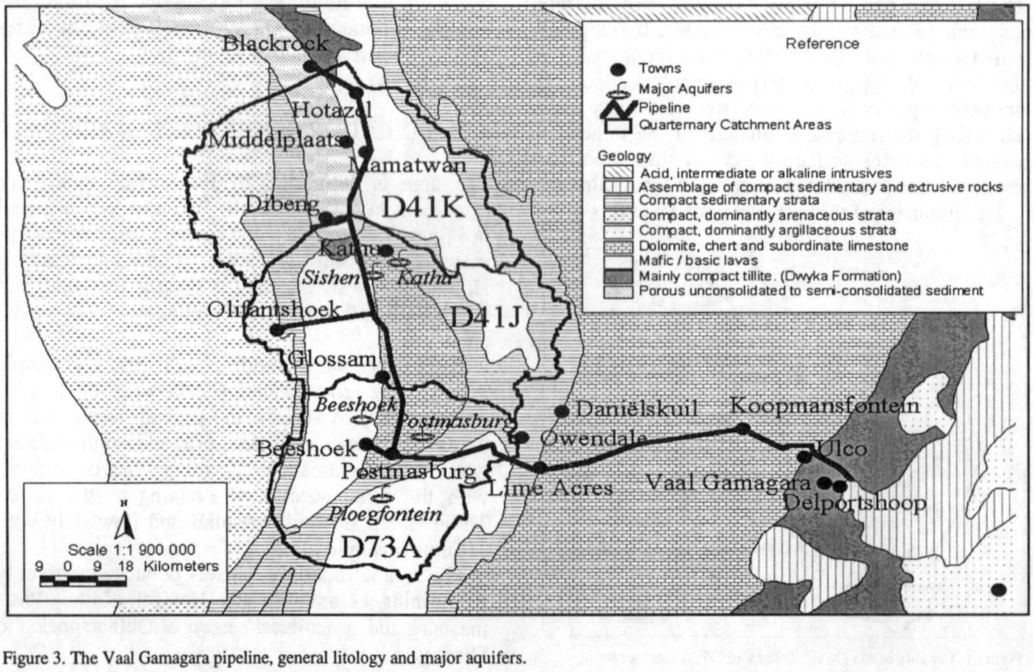

Figure 3. The Vaal Gamagara pipeline, general litology and major aquifers.

The porous sediments form well developed primary aquifers where course material is saturated along the Gamogara river. The tillite practically is an aquitard forming a barrier between the dolomites and porous sediments.

4.2 *Major aquifer recharge and potential.*

The major aquifers have been or are in process of quantification. Table 1 is a summary from local studies.

Table 1. Major aquifer potential.

Aquifer	Potenial m^3/a	Local demand m^3/a	Current unused m^3/a	Recharge % MAP
Sishen	24	15	9	15 to 30
Kathu	12.2	7.8	3.1	2 to 4
Beeshoek	10	7.8	2.2	12
Postmasb	1.6	1.5	0.1	2 to 5
Ploegftn	3	?	?	?

The calculated recharge varies considerably between these aquifers. Early modeling revealed conservative recharge of between 2 and 5 % for most aquifers. Long-term abstraction and mine de-watering led to the re-evaluation of recharge with major adjustment to deep aquifer systems. (Sishen, beeshoek and Ploegfontein). From the Shishen aquifer 0.5 million m^3 per annum is already used with capacity being built to distribute the remaining unused 9 million m^3 water per annum. This could sustain the total pipeline area at an average demand. Peak daily demand could be 150% of average demand. The Beeshoek aquifer is earmarked for this purpose. The water, a by-product of mining must be utilized optimally whilst all aquifers are monitored to assess the impact.

4.3 *Groundwater quality*

The quality of water in all rocks is good for human consumption except for portions of the porous sediments. Areas along the hills form topographic highs with rock outcrops. The electrical conductivity (EC) here is lower than 70 mS/m which conform with probable recharge. The areas with groundwater EC in excess of 300 mS/m conform to groundwater evaporation areas along topographic depressions. Generally water from these areas do not contribute to recharge of aquifers. (Ver Hagen and Dziembowski, 1979)

4.4 *Waterlevel dynamics.*

The waterlevel in the catchment is highly dependent on recharge with approximately 5 m recovery in response to high recharge events in the Sishen area. In the Postmasburg area recovery of 3 to 4 m has been recorded. These recoveries occur in unaffected and de-watered aquifers.

The cone of influence from the mine de-watering has been determined with local studies. At Sishen de-watering in excess of 100 m below the 1180-m (pre-de-watering waterlevel) elevation has taken place since 1977. The radius of influence from abstraction extends for approximately 3-km south and 2 km north and west. Dolerite dykes and impermeable formations form groundwater boundaries towards the east. Proposed pumping in the west would enlarge the cone of depression in that direction. The total depth of de-watering would be between 300 and 500 m below the 1180-m elevation within 30 years. At Beeshoek approximately 100 m de-watering would take place within 15 years. The cone of influence is unknown.

4.5 *The relative cost of groundwater*

The energy cost to deliver the groundwater by the mine is R0.366. This cost does not include developing wellfields, personnel or maintenance.

Vaalriver water is supplied to the northern section of the pipeline at R1.14 (personnel and maintenance included). Addition of the Trans Caledon Transfer (TCTO) and Vaalriver levies of R0.82 and R0.13 respectively result in a total cost of R2.09.

Comparatively the groundwater can be supplied at 20 to 30 % the cost of surface water.

5 CATCHMENT POTENTIAL

The impact of de-watering is calculated by aquifer modeling exercises. Total abstraction of 24.5-million m^3/ annum the past 30 years proofed sustainable with only localized de-watering. The impact of the proposed 37 million-m3/ annum abstraction for the next 30 years could have a wider impact. The Modeling of fractured aquifers on regional scale has not been attempted yet.

To obtain a measure of volumes of water available from the total catchment a water balance (WB) based on general parameters and the groundwater harvest potential (HP) map of South Africa (Seymore, Seward, 1996) can be utilized. The water balance relies on average general recharge to the catchment whilst the HP relies on the aquifer storage coefficient and aquifer dimensions. By applying both methods the results in Table 2 were calculate.

The lateral baseflow from a catchment is not accounted for by both these methods. The expected 9 million m^3 / annum water export from the D41J to D41K catchment is still well within the potential after the local need is satisfied. The total annual abstraction of 37 million m^3/ annum is well with in the estimated of 84 million m^3/ annum available in the 3 catchments. (considering the conservative method)

Table 2. The catchment potential as determined by the WB and HP methods. Volume given in million m^3 per annum.
WB :Area*Avg. Precip*%MAP Rech=Annual vol
HP :Area*max annual volume absract (from map)

Catch ment	Area Km^2	Annual precip mm	2.5% MAP Re-charge	Annual Volume	Local need
D41K	3850	380	2.5WB HP	36 23	3
D41J	3550	400	2.5WB HP	36 53	17
D73A HP	2980	340	2.5WB HP	25 45	12
Total				84	32

6 MONITORING SYSTEMS

In this area where de-watering is the major goal, declining waterlevels indicative of over exploited aquifers would have to be managed. The current monitoring networks are aimed at local impact evaluation.

The Sishen mine is operating a waterlevel, water quality and abstraction network on mine property for the purposes of their de-watering management.

DWAF is operating a waterlevel, water quality and abstraction-monitoring network in the Kathu aquifer to qualify local abstraction for Bestwood and Kathu.

Limited monitoring is done on the Postmasburg and Beeshoek aquifers. All networks are localized with no networks aimed at a regional level.

For the purpose of managing the area a regional waterlevel-monitoring network is proposed to determine the water level dynamics in the recharge, intermediate and discharge zones.

7 GROUNDWATER MANAGEMENT FORUMS

Since 1991 partial participatory management of groundwater was done through the quarterly groundwater provision meeting between DWAF, Sishen mine, Kathu municipality, Bestwood farmers and the Kalahari East stockwatering scheme.

Current liaison between the Sishen mine and DWAF is aimed at substitution of surface water by groundwater. The National Water Act (36 of 1998) enforces communication between the Sishen mine and interested and effected parties. A forum has been established to meet this need with the mine committed to provide access to water or alternative water resources where proof of de-watering impact can be provided.

Open discussions in the D73A catchment have been initiated, which would be stimulated with availability of additional information.

For the water management of both the Sishen and Postmasburg areas, existing forums should be transformed to Water User Associations with representation of all users for co-operative water management.

8 CONCLUSION

The long-term availability of groundwater through de-watering has been proven after more that 20 years of use from these resources. De-watering schedules recently announced, confirmed the availability of this resource for the next 30 years.

The direct economic cost of less than 30 % of Vaalriver water and ensured beneficial use, justify the small capital layout to distribute this groundwater.

The necessary systems to monitor the effect of de-watering should be established to ensure evaluation of impact.

The necessary forums for open information flow and discussion should be established to ensure benefit to all water users from the groundwater resources.

9 REFERENCES

Dziembowski, Z.M. & Verhagen, B. Th., 1979. *Groundwater studies in the Gamogara Cathment.* Joint report: DWAF and WRC.

Laan, R.W. 1998. *The population structure of the Camel thorn trees at different sites.* Honours study University of Stellenbosch.

Leskiewicz, A.F. 1981. *Geohydrology of Dimoten Syncline.* Internal report GH3930. DWAF, Pta.

Nel, J.M. & Van Wyk, E, 1999. *Groundwater resource evaluation for the Kathu and Khai-Appel aquifers.* Internal report GH3930. DWAF, Pta.

Steenekamp, G. 1998. *Geohydrological Modeling of the Sishen Aquifer: Sishen Iron Ore Mine.*

Stroebel,L.E. & Rudolph,D.C., 1999. *Postmasburg: Estimation of the groundwater potential.* Geohydro Technologies, Bloemfontein.

Van Wyk, E. 1999. *Preliminary Evaluation of Groundwater resources: D41J and D73A catchments.* DWAF internal report.

Vegter, J.R. 1995. *Map Groundwater Resources of South Africa,* Water Research Commission and DWAF.

Willemink, J.1983. *Evaluasie van die grondwater bronne tussen die Korannaberg en die Noordelike gedeelte van die Langeberg.* Internal report GH3930. DWAF, Pta.

Groundwater: Past Achievements and Future Challenges, Sililo et al. (eds) © 2000 Balkema, Rotterdam, ISBN 90 5809 159 7

The role of groundwater quality monitoring in the strategy of groundwater protection

J.Vrba
DHV CR Consultancy Company, Prague, Czech Republic

ABSTRACT: Groundwater quality monitoring plays an important role in the strategy of groundwater protection and quality conservation and in enforcement of an anti-pollution policy. Groundwater monitoring programmes supply data about the current state and long-term trends of groundwater quality and help to clarify and analyse the extent of natural processes and human impacts on groundwater systems. International, national, regional and local groundwater quality monitoring programmes are operating all over the world and frequently are correlated or integrated with other environmental monitoring programmes. The objectives of monitoring programmes govern the extent of monitoring activities. Establishment and operation of monitoring networks and standardisation of monitoring methods and procedures is discussed particularly with regard to national groundwater quality monitoring programmes.

1 INTRODUCTION

Groundwater quality monitoring supports the policy and the strategy of groundwater protection and quality conservation on national, regional or local level.

Groundwater quality monitoring data provides a valuable base for assessment of the current state and forecasting trends in groundwater quality. Analysis of groundwater quality monitoring data support sustainable management of groundwater resources development and protection.

A holistic approach should be implemented in the policy, strategy and management of groundwater resource protection, reflecting the social, economic and ecological value of groundwater, considering the close connection between groundwater and surface water. The same attention should be given to both quantitative and qualitative aspects, while dealing simultaneously with sustainable development and environmentally sound protection of water resources.

The holistic concept appears in the co-ordination or even the integration of national groundwater monitoring programmes with surface water, precipitation, climatic and soil monitoring networks and in the establishment of pilot monitoring stations with complex observation of all components of the hydrological cycle and soil. Such a concept also helps to reduce the costs of design and operation of environmental monitoring networks.

2 DEFINITION OF GROUNDWATER QUALITY MONITORING

Groundwater quality monitoring can be understood as a continuous, methodically and technically standardised programme of observations, measurements and analysis of selected physical, chemical and biological variables of groundwater with the objectives:

1. To collect, process and analyse the data on groundwater quality as a baseline for assessing the current state, anticipating changes and forecasting trends in groundwater quality due to natural processes and human impacts in time and space;
2. To provide information for improvements in the planning, policy, strategy and management of groundwater protection and quality conservation.

3 GROUNDWATER QUALITY MONITORING PROGRAMMES

Groundwater quality monitoring programmes can be used to evaluate the chemical composition and evolution of natural groundwater, to address groundwater quality problems in relation to the economic development and social and ecological needs, as well as to define the priorities in groundwater resource protection and the potential risks and conflicts. Legislative measures and institutional structures on all governmental levels support groundwater quality monitoring activities.

Groundwater quality monitoring programmes operate at the international, national, regional and local levels. The objectives of each of the above programmes govern the extent of monitoring activities, such as design of monitoring networks, frequency and methods of sampling and measurements and number of variables to be analysed.

International and national groundwater quality monitoring programmes are typically background monitoring activities, whereas regional and local monitoring programmes are oriented to the specific monitoring purposes (van Lanen 1998). This paper is directed particularly at groundwater background monitoring on a national level.

The groundwater quality monitoring programme is composed of a monitoring system, information system and management system (Table 1).

Clearly defined objectives of monitoring are essential to achieve the expected results. The requested output information cannot be defined precisely when monitoring objectives are not stated clearly. The objectives required by the user (policy and decision making sphere, scientific community, public) have to be respected, but it should be realistic. The objective and definition of requested output information is the first critical point of the groundwater monitoring system and have to be defined before the monitoring network is established and the first sample of water is taken (Table 1).

Prior to outlining the strategy for the monitoring system, an initial inventory of accessible data and analyses is made to gain a basic knowledge about the composition and contemporary state of the groundwater system and to define the extent of the required complementary hydrogeological investigation.

The strategy for the monitoring system includes:

1. Delimitation of the monitoring area;
2. Identification of the geological framework, the flow system and the groundwater quality aspects;
3. Identification and inventory of potential and existing pollution sources;
4. Selection of sites and density of monitoring stations and establishment of monitoring network;
5. Selection of field and laboratory methods and devices;
6. Determination of sampling frequency and selection of variables to be analysed, including field and trip blank samples as part of the quality assurance (QA) and quality control (QC) process;
7. Database management, especially the methods of data handling and processing (statistical, visual and/or modelling methods).

Data analysis and transformation of analysed monitoring data into a user tailored information product is an other critical point of the groundwater quality monitoring programme (Table 1). It is being emphasized that however high the quality of monitoring data, is as long they are not transmitted in an intelligible form to the user, the groundwater quality monitoring programme is not effectively utilised and the expenditures on its design and operation are not vindicated. The information on groundwater quality can be used for technical, socio-economic, ecological or scientific purposes and must prove that the problem has been understood correctly and resolved.

A groundwater quality management system as the final stage of the groundwater quality monitoring programme (Table 1) is based on monitoring results and aspires to implement groundwater quality protective measures in the frame of the existing legal basis and institutional structure.

3.1 *International Groundwater Quality Monitoring Programme*

An international groundwater quality monitoring programme was initiated by the United Nations Environment Programme and operates over the world within the Global Environment Monitoring System (GEMS) since 1974.

3.2 *National Groundwater Quality Monitoring Programme*

National groundwater quality monitoring programmes are concerned with data collection of the natural background level and current state of groundwater quality. These can serve as a baseline for national policy of groundwater preventive protection and quality conservation. Monitoring of groundwater quality on a national level is an organisationally and technically demanding and expensive process in terms of both capital and operational costs. It is, therefore, generally accepted that a national monitoring programme is beneficial, if the results it yields are applied.

Baseline and trend stations, according to the classification of Maybeck (1985), play dominant roles in national groundwater monitoring networks (Table 2). They have to represent natural environmental and hydrogeological conditions and are located outside the influence of pollution sources and groundwater abstraction sites.

National groundwater quality monitoring networks are in operation in a few countries only. Monitoring methods and procedures are not standardized and vary in each country. This causes difficulties in mutual comparison of monitoring data especially in transboundary aquifers.

TABLE 1. Groundwater quality monitoring programme

Groundwater quality monitoring programme			
Groundwater quality monitoring system		Objective	
	Critical point	Definition of requested output information	
		Strategy	
		Delimitation of monitoring area	
	Identification of the groundwater system geometry and parameters, the groundwater flow system and evaluation of relevant data of groundwater quality		Identification and inventory of potential and existing pollution sources
		Design and establishment of monitoring network and designation of monitoring methods	Feedback
		Data acquisition	
	Samples collection		Field and laboratory analysis
		Data handling	
	Transmission	Processing	Storage
		Database management system	
		Data evaluation	
Groundwater quality information system	Critical point	Information products	
		Information communication	
Groundwater quality management system		Information utilization	
		Legislative and institutional implementation of protective measures	

Importance of groundwater as a drinking water source and vulnerability of aquifers control the density of monitoring stations (wells, springs) and sampling frequency which both affect the unit monitoring costs.

A group of experts of EU member states has been charged with the study of the costs needed on establishing and operating groundwater monitoring networks in the member states (1996). In a questionnaire sent to the different organisations carrying out the groundwater monitoring in EU countries, the following information was asked: number of monitoring sites, parameters monitored, frequency, and the total and unit monitoring costs. Unfortunately, information about groundwater sampling methods, which affect the sample representativity and costs, was not requested. Noted differences were recognised in technical and cost parameters of national groundwater quality monitoring programmes in individual EU countries. Some examples are given in the table 3.

The highest expenditures on the operation of national groundwater quality monitoring networks are reported from Austria, where drinking water supplies depend 99% on groundwater and Denmark, where 98% of drinking water is obtained from groundwater. In Sweden and Finland the distribution of monitoring stations in shallow aquifers over a large land area, long travel distances and high staff costs result in high unit operation costs. In Greece, where the number of variables analysed is limited, operating unit costs are very low (179 ECU/year), compared with Scandinavian countries. Significant differences exist in sampling frequency and in number of variables determined (Table 3).

TABLE 3. Technical and cost parameters of national groundwater quality monitoring networks of some European countries

Country	A	B	C	D
Austria	1884	4	50	858
Denmark	1100	1-4	84	800
Sweden	76	1-8	19	3125
Finland	45	6	28	2395

A - Number of sampling sites
B - Sampling frequency per year
C - Number of variables analysed
D - Unit operating costs/ annum (ECU)

Also the costs of drilling monitoring boreholes in porous aquifers differ: in Austria 1 m borehole (depth < 50 m) costs 151-265 ECU, in Portugal 41 ECU.

In the Czech Republic the national groundwater monitoring network is divided into three parts: shallow aquifers (network established from 1956-1962), deep aquifers (1986-1990) and springs (1955-1980). The groundwater quality monitoring network at present includes 492 stations, sampling frequency is two times a year and more than 50 variables are analysed. Samples are taken after well purging, when temperature, pH and electrical conductivity have stabilized.

The objectives of national groundwater quality monitoring programmes and methods of monitoring are discussed in chapter 4.

3.3 *Regional Groundwater Quality Monitoring Programmes*

Regional groundwater quality monitoring programmes are concerned on specific purposes and complement the national monitoring programme.

The main objective of regional programmes is to acquire statistically significant sets of data as a basis for creating the regional policy of groundwater resources protection and management of their quality and for supporting decision making in integrated land use planning. Monitoring activities are concerned with the specific potential or existing diffuse pollution problems (particularly on the impact of agriculture on groundwater quality) and on public groundwater supply sources and their catchment areas.

The trend stations dominate in the design of regional groundwater quality monitoring networks (Table 2). However through selected baseline stations of the national monitoring network, both monitoring programmes are interconnected.

Frequency of sampling in regional monitoring programmes concerned with diffuse pollution problems is higher compared with national monitoring programme (up to 12 times per annum). Variables analysed are selected according to the objective of monitoring. Design of monitoring wells should permit separate measurement and testing of individual aquifers and their vertical hydrochemical profiling. Monitoring of vertical movement of pollutants in the unsaturated zone (lysimeters, centrifugal extraction of interstitial water from core samples, sampling by means of compressed gas injection or other methods) should be recommended especially in areas with a high potential for aquifer pollution.

Groundwater quality monitoring programmes operating on a regional scale should be established in catchment areas of groundwater supply sources.

The objects of monitoring are water supply wells or springs and monitoring wells located in designated protection areas of the groundwater source. By reason of preventive protection of groundwater resource quality, monitoring networks are established in outer or second degree protection areas encompassing usually the whole source catchment area or zone of contribution.

The number of monitoring stations and frequency of monitoring strongly depends on aquifer vulnerability and related risk to pollution. Drinking water standards valid in individual countries control the extent of variables analysed. Sampling frequency is high; some variables from water supply wells sensitive to human impacts are analysed daily. Analysis of some organic chemicals due to their low mobility is less frequent (as little as once per annum).

3.4 *Local or Site Specific Groundwater Quality Monitoring Programmes*

Local or site specific groundwater quality monitoring programmes are established around:

1. Potential point pollution sources, with the objective to give an early warning about groundwater quality deterioration;
2. Existing point pollution sources with the purpose to observe and control movement of pollution plume or effects of pollution remediation.

Good knowledge of the hydrogeological properties of the unsaturated zone and aquifer, groundwater flow nets and pollutant behaviour is desirable before designing a local monitoring programme.

Local monitoring programmes operate independently of national or regional programmes. A great density of monitoring stations, high sampling frequency and field and laboratory analyses of variables selected according to the type and properties of pollutants are typical for local monitoring programmes.

TABLE 2. Categories of groundwater monitoring stations operating in groundwater quality monitoring programmes

Monitoring programme	Category and importance of monitoring station Baseline	Trend	Impact	Station density	Sampling frequency	Variables analysed	Monitoring stations characteristics
International	D	C	LS	VL	L	B + O -	Baseline station: natural background groundwater quality
National	D	C	LS	L	L	B + O -	Trend station: trends in groundwater quality due to natural and human impacts
Regional	C	D	LS	M	M	B + O +	
Local	LS	C	D	H	H	O +	Impact station: changes of groundwater quality due to human impacts

Station's significance: D Dominant, C Complementary, LS Low significance

Station density: Station per km^2: H High – m^2 to 10, M Medium – 10 to 100, L Low – 100 to 1000, VL Very low – 1000 and more

Sampling frequency: H High – more than 12 times a year, M Medium – 2 to 12 times a year, L Low – 1 to 4 times a year

List of variables analysed: B Basic – physical, chemical and biological variables included into the drinking water standards
O Optional – heavy metals, organochlorine compounds, oil hydrocarbons and other variables depending on monitoring program objectives
+ Regular analysis
- Occasional analysis

The impact monitoring stations (Maybeck 1985) dominate in local monitoring networks (Table 2). Monitoring of the unsaturated zone and remote sensing methods applied with respect to pollutant properties supplement groundwater monitoring activities. Some of monitoring wells are located close to the potential pollution source and should be used as recovery wells when pollution occurs.

The refinement of the local groundwater quality monitoring network based on pollution control and the effect of the remediation procedure is frequent. Fast and simple methods of data recording and retrieval should be used to submit the monitoring results and information to the individual users in time and in a intelligible form. Operational costs of local monitoring programmes are high, however they are covered by the polluters (polluter pays principle). Operation of site specific groundwater quality monitoring networks is mostly limited for a certain period.

4 DISCUSSION

Several experiences have been acquired about groundwater quality monitoring in countries where national groundwater quality monitoring programme were designed and carried out. Even though natural, social and economic conditions and land use vary from country to country it is possible to give a generalization about national monitoring programmes. The following suggestions are based on the experience with the establishment and long term operation of a national groundwater quality monitoring programme in the Czech Republic and on available information from other countries (Vrba 1987, Duijvenbooden 1987, European Commission - DG XI 1996).

An elaboration of conceptual models of main country aquifers based on aquifer boundaries and the recharge and discharge areas delineation, aquifer structure, geometry and hydrodynamic parameters and on groundwater flow system is essential before the initial design of a groundwater monitoring programme is proposed.

Existing or abandoned wells drilled for other

purposes of which the casing, screening and geological and hydrogeological data are not relevant to the objectives of the national groundwater quality monitoring programme, should not be incorporated into the monitoring network.

When designing a monitoring network in an area with fissured rock permeability the orientation and geometry of the fracture system (preferred flow path) should be carefully studied. However the location of monitoring wells in such complicated hydrogeological conditions is always difficult. In karst areas where groundwater flow is in conduits or in large open fissures, springs allow more representative observation of groundwater quality than monitoring wells. Sampling of groundwater in karst regions must be frequent, because groundwater is moving through the karst system quickly and water quality changes as a consequence of pollution are usually fast.

In the monitoring network formed by monitoring wells with standing water and water supply or irrigation wells which are in a dynamic stage groundwater samples represent a different hydraulic and hydrochemical situation of the aquifers and they are not mutually comparable.

Monitoring wells of national monitoring programme should be located beyond the influence of pollution sources or groundwater abstraction sites.

In case of a groundwater system formed by two or three aquifers, it is preferable to have separate monitoring of each aquifer by individual small diameter wells. Employment of the packer technique on the separation of aquifers in the monitoring well is less suitable because when packers operate in the long term, the monitoring well should be affected by natural or human impacts and aquifers should be interconnected. However in shallow anisotropic aquifers intercalated by low permeable sediments two or three small diameter piezometers of different depths should be placed into a large diameter borehole. Water samples are taken from the bottom part of the piezometer. Such a national groundwater monitoring system successfully operates in the Netherlands and facilitates vertical hydrochemical profiling of the first shallow vulnerable aquifer (Duijvenbooden 1987).

The method of taking representative samples of groundwater is the most problematic part of the sampling procedure. The water is standing in the well several weeks or even months in the column above the screen without direct contact with the groundwater body and under the impact of atmosphere and casing material which all affect its chemical composition. Purging the well is, therefore, desirable to take water samples at a dynamic stage. The main sampling methods implemented according to the properties of aquifers are 1) sampling after removing the volume of water accumulated above the well screen, 2) in low yielding wells sampling by suction or sample collection after pumping the well dry and the recuperation of water column, 3) purging the well until selected field measured parameters (pH, Eh, temperature, electrical conductivity) are stabilised. In the case of deeper, unconfined and high yielding aquifers with lateral groundwater flow, simple or special multistage samplers are used to take water samples from a defined depth. A mixed water sample from the entire profile of the aquifer is not considered satisfactory.

Sampling collection and field and laboratory analysis are usually carried out by various companies. The State agency responsible for the operation and management of groundwater monitoring programmes has the responsibility to coordinate and control observance of monitoring methods and procedures.

Groundwater quality monitoring of transboundary aquifers requires close cooperation among neighbouring countries. Sampling collection methods have to be unified as much as possible to minimise potential problems with data evaluation and implementation.

After some period of operation of the monitoring program, feedback is an essential feature of the monitoring process (Table 1). Modification of the monitoring system (number and location of monitoring wells, sampling frequency, extent of chemical analyses) is based on a data evaluation. Adaptation of the monitoring process should be repeated when a long term data set is available and statistically evaluated.

5 CONCLUSIONS

The objective of groundwater monitoring programmes is to provide data and information about the current state and trends in groundwater quality, and to support the policy and strategy of groundwater protection and quality conservation on national, regional and local levels. National groundwater quality monitoring programmes are an quality, and to support the policy and strategy of groundwater protection and quality conservation on national, regional and local levels. National groundwater quality monitoring programmes are an important early warning system, that facilitate timely identification of changes in groundwater quality due to the influence of natural processes and human impacts. However groundwater quality monitoring programmes on a national level exist in a few countries only. Establishment and operation of national monitoring programmes is a technically and financially demanding process in terms of both capital and operational costs. A cost-benefit analysis is, therefore, required by governmental authorities to compare the value of the data and information with the financial expenditures involved.

REFERENCES

European Commission – DG XI 1996. Evaluation of the Costs of Groundwater Inspection in the Member States. *Final report*: 1-79

Maybeck, M. 1985. The Gems Water Programme 1978-1983. *Water quality bull., volume 10, no.4*, Canada: 167-173

Van Lanen, H. 1998. Monitoring for Groundwater Management in (Semi-) Arid Regions. *Studies and reports in hydrology 57,* UNESCO

Vrba, J. 1987. Groundwater Quality Monitoring-Strategy, Objectives, Programmes and Methods, *TNO-RIVM, Proceedings and information no. 38*, The Hague: 167-177

Groundwater: Past Achievements and Future Challenges, Sililo et al. (eds) © 2000 Balkema, Rotterdam, ISBN 90 5809 159 7

The impacts of dual-porosity aquifers on Aquifer Storage Recovery (ASR) schemes

A.T.Williams, I.Gaus & I.N.Gale
British Geological Survey, Wallingford, UK

ABSTRACT: Dual porosity formations, such as the Cretaceous Chalk, can be prolific aquifers and have the potential to store large volumes of artificially recharged potable water. The unique characteristic of these aquifers is the large volume of effectively immobile pore water stored in the rock matrix (up to 40% porosity) and the relatively small volume of mobile water stored in fractures (1 to 2% porosity). Diffusive mixing of the fracture water and pore water does occur however, so the quality of the native pore water impacts on the quality of the water recovered subsequent to injection and storage. The SWIFT/486 model was used to investigate the relative significance of diffusive and dispersive mixing. The diffusive mixing process was shown to have a significant deleterious effect on the quality of the recovered water after a period of storage. This would not happen if only dispersive processes are significant. A good fit between measured and modelled data for non-reactive solutes gave confidence that the diffusive model adequately describes the significant transport properties of the chalk aquifer. If the native groundwater contains solutes at unacceptable concentrations then large volumes of injected water need to be invested to create a buffer zone within which an operational scheme can be used. Water-water and water-rock geochemical reactions, both detrimental and beneficial, are not considered here and need to be monitored and modelled separately and, if possible, linked to the predictive operational scenarios.

Aquifers are natural, underground water storage reservoirs. They are extensively used, either as a sole source of supply or in conjunction with surface water sources. Where the limits of sustainable use of these resources are approached, imaginative management of groundwater, including artificially replenishment of aquifers, can optimise their use with minimum environmental impacts.

Artificial recharge is a means of supplementing or replacing natural infiltration into an aquifer; it involves storing water in an aquifer at times when there is a surplus, for use at times of scarcity. Traditionally, the technique has utilised aquifers containing potable water and water has either been infiltrated through basins or injected into wells or boreholes.

Where treated, potable water is stored in the aquifers local to boreholes that are used for both injection and abstraction the schemes are referred to as Aquifer Storage Recovery (ASR) schemes (Pyne 1995). A high percentage of the water injected is abstracted at a later date and the schemes may utilise aquifer containing poor quality or brackish water, although this does not exclude the use of aquifers containing potable water. ASR schemes enable maximum use to be made of existing licensed resources.

Around 50 years ago, the first investigations of injecting potable water into saline aquifers were carried out (Cederstrom 1957). This involved recharge of freshwater into the brackish aquifers of the coastal plains of Virginia in 1946. The water injected formed a lens of potable water within the saline body of groundwater. The borehole clogged during injection; however the withdrawal phase gave encouraging results. Investigations have been carried out since, mainly in the United States (Pyne 1998) but also in other countries (Connorton & McIntosh 1994. Jones et al., 1999).

The main mechanisms for loss of injected fresh water are dispersion (hydrodynamic mixing and molecular diffusion), and density stratification (gravitational segregation). Mixing occurs along the contact of injected and native waters, and a dispersion zone, or buffer zone, develops rather than a sharply defined interface. The degree of mixing is partly dependent on the type of flow in the aquifer: Harpaz (1971) showed that sandstones are likely to yield higher recovery efficiencies than limestones. This is due to the relatively slow groundwater movement in sandstones: the recharging water follows a radial expansion pattern and tends to retain the shape of a coherent water body (except in highly stratified formations). Harpaz calculated the thickness of the dispersion zone to be of the order of tens of metres.

In some cases, the effects of molecular diffusion may be significant. This is considered to be the dominant mixing process in the Chalk aquifer in the UK (Jones et al., 1998). The Chalk is unusual in that it has a high porosity (up to 40%) and may contain ancient saline or brackish water in the pore spaces. Due to the small pore throat diameters, groundwater flow is dominantly through fractures (1-2% porosity) and not through the pores. Saline water in the pores can therefore diffuse across a relatively large surface area.

If a Chalk aquifer contains poor quality water then injection of potable water will mix with, and displace the water in the fissures. However, diffusive mixing will be the dominant process between the fissure and the matrix waters. This is a slow process and the rate at which it occurs depends upon the surface area of contact between the two waters, their relative volumes and concentrations and the ionic species under consideration. Understanding the scale and rate of this process is therefore key to the success of ASR schemes in aquifers containing non-potable native groundwater.

It is worthwhile examining the processes that occur during an injection-recovery cycle by modelling the changes in water quality. The modelling package chosen as the most suitable is SWIFT/486. The

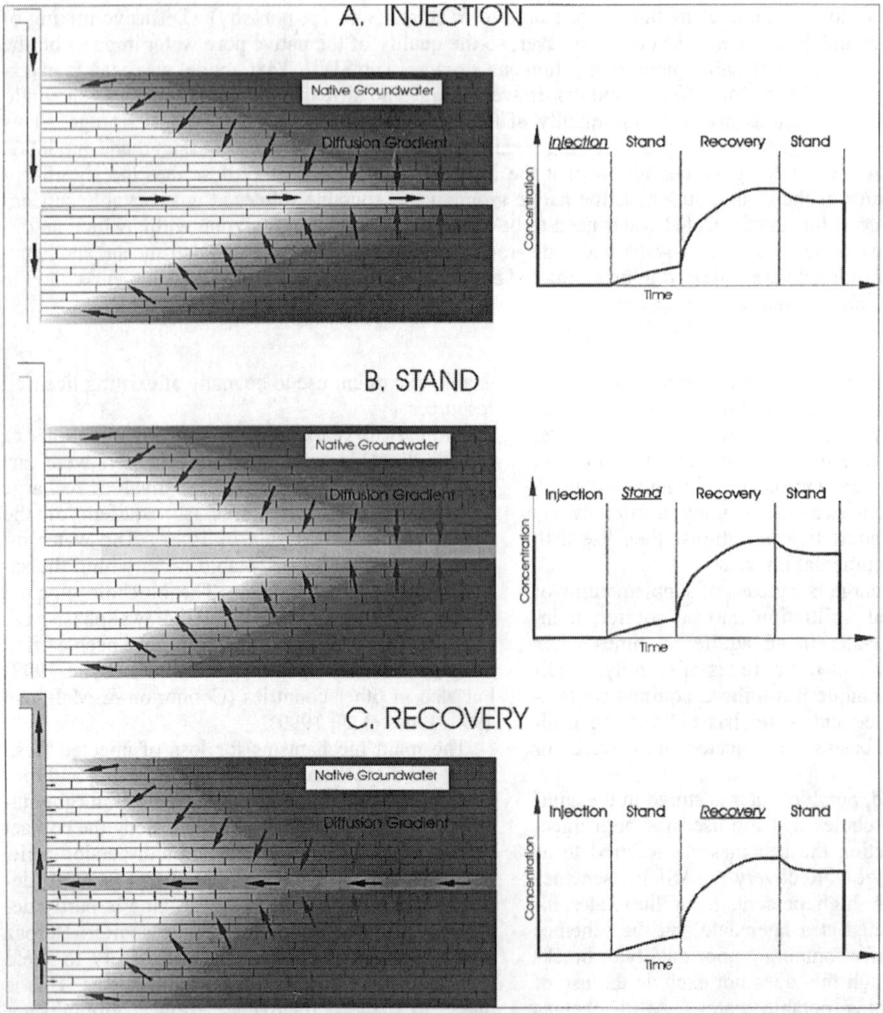

Figure 1. Modelled response of a dual-porosity aquifer to a typical injection-stand-recovery cycle

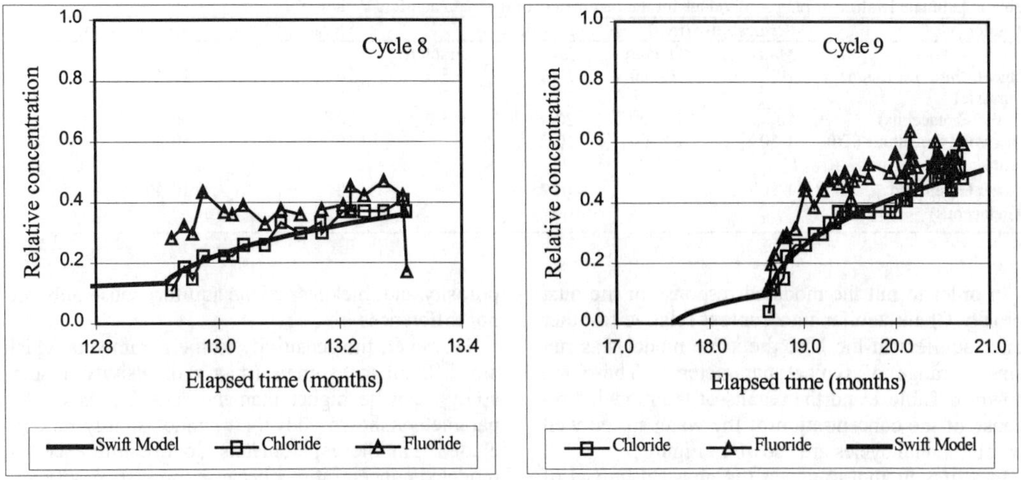

Figure 2. Data of dual porosity modelling against data from cycle testing in the Chalk aquifer (cycles 8 and 9)

model was developed for use in the analysis of deep geological facilities for nuclear waste disposal but can be used for numerous applications including groundwater transport in porous or fractured media. The versatility of the modelling package means that it has a large number of input parameters, many of which may not be utilised. The processes to be incorporated into a model are the hydraulic properties of the fractured aquifer and the dual-porosity transport properties, particularly diffusive exchange of solutes between the mobile fracture water and the immobile matrix water.

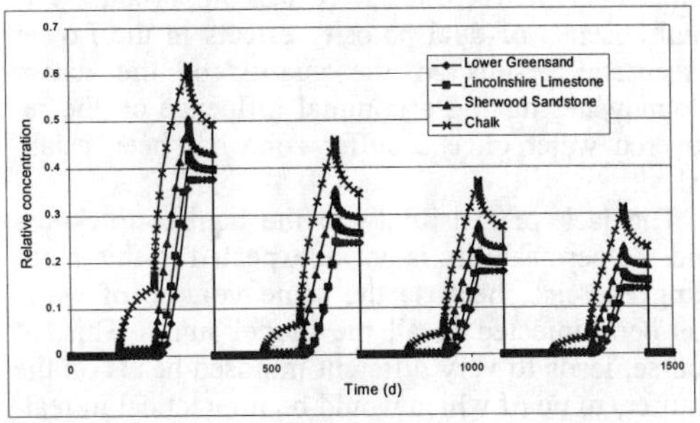

Figure 3. Predicted responses of the four major aquifers i the UK to ASR over four annual cycles

The system that is being modeled is a cylinder of fractured aquifer centered around a well. Initially, the aquifer is full of water with a nominal salinity of 1.0 i.e. native groundwater. This water is in the fractures and the matrix. The responses of the model to a typical annual cycle are shown in Figure 1A, B, C. The relative concentration modeled is that at the well. During the injection phase the concentration in the well is 0.0, that is, the injected water concentration. If the well is left standing after injection stops the concentration in the well gradually increases, due to the effect of the diffusive exchange between the low salinity injection water in the fractures and the higher salinity water in the matrix. When the recovery phase starts the concentration in the well increases as higher salinity water from further in the aquifer is drawn into the well. This is the key phase of the cycle as predictions can be compared to measured data.

When recovery stops, diffusive mixing causes a decrease in salinity as contaminants from the fracture water are diluted by fresher water in the matrix. As soon as injection starts again the salinity in the well drops to 0.0 i.e. the concentration of the injected water. Changing the parameters used to describe the aquifer, and the previous abstraction-injection history, will affect the relative importance of the different features.

The modelled results were calibrated against data collected during an extensive cycle-testing programme at a site in the confined chalk aquifer in southern England. The model was matched to the non-reactive solute, chloride. The match between modelled and measured data are shown in for cycles 8 and 9 of an 18 month programme (Figure 2), a good match being obtained for all the preceding seven cycles. Also shown in Figure 2 are the relative concentrations for a reactive solute, fluoride. The measured relative concentration of fluoride is higher than the modeled concentration indicating that additional solute is being brought into solution in excess of that resulting from either dispersive or diffusive mixing.

Table 1. Estimated values or ranges of values for the parameters of the main aquifers in the UK

Aquifer	Permeability (m/d)		Porosity (%)		Storativity (m^{-1})	Thickness (m)
	Matrix	Fracture	Matrix	Fracture		
Lincolnshire Limestone (Jurassic)	10^{-3}	10-100	1-10	0.5-5	10^{-6}	10-40
Chalk (Cretaceous)	10^{-3}	1-100	20-40	0.5-2	10^{-6}	50-200
Sherwood Sandstone (Triassic)	1-50	10-500	10-25	1	10^{-5}	50-300
Lower Greensand (Cretaceous)	1-100		10-25		10^{-5}	10-40

In order to put the modeled response of the dual porosity Chalk aquifer into context relative to other major aquifers in the UK, the same model was run using a range of typical parameters. These are shown in Table 1 and the results of the modeled response of the concentration of the water in the well for four annual cycles are shown in Figure 3.

In a first attempt to assess the physical impact of ASR-schemes on the recovered water quality in the British aquifers a sensitivity analysis was carried out with the SWIFT-model. Based on estimates of aquifer parameters the impact of operational and test cycles on the recovered water quality were modelled. No chemical reactions between the injected water and the native groundwater and the aquifer material are included in the model. The changes occurring reflect solely mixing processes.

The Lower Greensand is a single porosity aquifer and thus has no fracture porosity or fracture permeability. The Chalk and the Lincolnshire Limestone aquifers are both dual-porosity aquifers, the significant differences being that the fracture porosity in the limestone is higher than in the chalk and the matrix porosity in the chalk is considerably higher than it is in the limestone.

The behaviour of the Sherwood Sandstone lies somewhere between the dual-porosity and single-porosity conceptualisations and is probably best described as a dual-permeability medium. This is because the matrix is permeable enough to transmit water at the low rates which are relevant in regional flow systems, but probably not in the near-well situation, where fracture flow is likely to dominate. The SWIFT model cannot simulate this type of aquifer, and as such it may not be an appropriate model for simulating ASR behaviour in the Sherwood Sandstone.

In general, varying the parameter values within the estimated ranges for the Chalk, the Sherwood Sandstone and the Lincolnshire Limestone shows that their sensitivity is low and the changes in relative concentration of non-reactive solutes are small. Because of these low sensitivities, the conclusions drawn for the short-term test cycles can be extrapolated to long term operational cycles. From the estimated parameters only the matrix porosity has an influence on the modelled relative concentrations. The other parameters (fracture permeability, fracture porosity and thickness of the aquifer) cause only minor differences

However, the sensitivity to the parameters which are difficult to estimate (matrix diffusivity, dispersivity), can be higher than the sensitivities for the parameters shown above. It cannot simply be concluded that these parameters do not influence the modeled concentrations because the dispersivity and matrix diffusivity depend on the values of the parameters such as fracture permeability, rock porosity and matrix porosity.

For the Lower Greensand, which is a single-porosity aquifer, none of the tested parameters have a significant influence on the modeled concentrations when chosen within the estimated ranges.

The modeled base case for each aquifer is presented in Figure 2. Based on the parameter estimates the Chalk seems to be the least suitable for ASR-purposes in terms of recovered water quality based on physical processes. This is due to its high matrix porosity in comparison with the other aquifers leading to a high degree of mixing of injected and native water. Due to the low matrix porosity, for the Sherwood Sandstone and the Lincolnshire Limestone, matrix effects are less important. The total absence of dual porosity effects in the Lower Greensand results in the quality of the native groundwater having a minimal influence on the recovered water once a buffer zone has been established.

The lack of sensitivity to the aquifer thickness and the permeability is not unexpected in this modeling exercise, because the same volume of water has been injected in all the model runs. This, of course, leads to very different imposed heads on the aquifer, some of which would be impractical in reality. However the volume of the aquifer (and thus the volume of the native groundwater) which comes in contact with the injected water is the same in all the cases where only the permeability or the thickness is changed. For the dual-porosity aquifers, the matrix permeability has little effect on predicted concentrations because of the assumption in SWIFT that there is no flow in the matrix, except perpendicular to the fractures. This is a reasonable assumption for the Chalk and the Lincolnshire Limestone but not for the Sherwood Sandstone, for the reasons described earlier.

The difference between the single and dual-porosity response of the aquifers is significant and so it is important to be certain that the conceptual model used for predictions is appropriate to the aquifer being tested.

The main conclusion of this modeling work is that dual-porosity aquifers will need more conditioning (injection of non-recoverable water to build up a buffer zone of mixed waters) than single porosity aquifers before they are suitable for use for ASR. However the suitability of any aquifer depends on the dispersive and diffusive parameters which can generally only be obtained from field trials.

ACKNOWLEDGEMENTS

This work has been funded jointly by a Foresight LINK award and UK Water Industry Research Ltd as a project called ASR-UK. Wessex Water Services are gratefully acknowledged for permission to use data from its trial site in the chalk aquifer to calibrate the model.

This paper is published with the permission of the Director of the British Geological Survey.

REFERENCES

Cederstrom, D. J. 1957. Geology and ground-water resources of the York-James peninsula, Virginia, *US Geological Survey*: 237.

Connorton, B. J. & P. McIntosh. 1994. EURAU survey on artifical recharge. *Artifical Recharge of Groundwater, II, Walt Disney World, Florida.*

Harpaz, Y. 1971. Artificial groundwater recharge by means of wells in Israel. *Proc. of American Society Civil Engineers, Journal of Hydraulics Division*: 1947-1964.

Jones, H. K., D. M. J. Macdonald and I. N. Gale. 1998. The potential for Aquifer Storage and Recovery in England and Wales. *British Geological Survey Report WD/98/26.*

Jones, H K., Gaus, I., Williams, A T. and Gale, I N. 1999. ASR-UK. A review of the status of research and investigations. *British Geological Survey Report WD/99/54.*

Pyne, R. D. G. 1995. *Groundwater recharge and wells: A guide to Aquifer Storage Recovery. Florida, USA.*, CRC Press.

Pyne, R. D. G. 1998. Aquifer storage recovery: Recent developments in the United States. *Artificial recharge of groundwater*, Amsterdam, Netherlands, A A Balkema.

[illegible]

[illegible]

ACKNOWLEDGEMENTS

[illegible]

REFERENCES

[illegible]

Groundwater: Past Achievements and Future Challenges, Sililo et al. (eds) © 2000 Balkema, Rotterdam, ISBN 90 5809 159 7

Comprehensive determination of the resource directed measures for groundwater in South Africa

Y.Xu & E.Braune – *Department of Water Affairs and Forestry, Pretoria, South Africa*
C.Colvin & D.Le Maitre – *Council for Scientific and Industrial Research, Stellenbosch, South Africa*
K.Pietersen – *Water Research Commission, Pretoria, South Africa*
T.Hatton – *CSIRO, Australia*

ABSTRACT: Resource Directed Measures (RDM) are a series of measures under the South African National Water Act (1998) which aim to ensure the protection of all water resources. The RDM include classification, setting Resource Quality Objectives and setting the Reserve. This paper describes the *comprehensive* determination of the RDM for groundwater. This is the most detailed level of RDM determination with the highest level of confidence. The paper discusses the progress of the comprehensive RDM team, in terms of outlining tools and methodologies to be used, and discusses some of the limitations of the National Water Act with respect to groundwater. Comparisons to approaches to groundwater protection in other countries are made.

1 INTRODUCTION

1.1 *Resource Directed Measures*

The National Water Act No. 36 of 1998 (NWA) (DWAF 1998) outlines a series of measures to be progressively developed to protect water resources in South Africa. These Resource Directed Measures (RDM) consist of three core concepts: (1) Classification, (2) the Reserve and (3) Resource Quality Objectives (DWAF 1999).

The Water Research Commission (WRC) and the Department of Water Affairs and Forestry (DWAF) established a research programme in 1999 to support the development of tools required to comprehensively quantify RDM for groundwater (Braune et al. 1999). Guideline criteria for a comprehensive determination have been put forward as:

- Taking between 8 and 12 months to complete;
- Being of high to very high confidence; and
- Being based on site-specific data, adequate to meet the assessment confidence criteria

The result is to be used when considering licence applications where impacts on a catchment are likely to be considerable and where abstraction is to take place in a sensitive catchment.

Classification

The classification system entails adaptation of existing management classes and establishment of sets of classification rules for ecosystems, basic human needs and water users. A principal criterion for classification is the degree of modification of the system from unimpacted conditions. The classification system should also address vulnerabilities of the system to stresses, e.g. sinkhole formulation, land subsidence, important vegetation die-off and seawater intrusion. More information on the classification system is given by van der Voort and Conrad (2000).

Resource Quality Objectives (RQOs)

Resource Quality Objectives (RQOs) are to be established for each geohydrological response unit. There are clear goals to quantify all aspects of the resource. For groundwater this includes: the pattern and timing of water levels; the quantity of flow; the physical, chemical and biological quality of the water; the character, condition and distribution of dependent biota. A method to comprehensively determine groundwater RQOs is yet to be established.

The Reserve

According to the National Water Act of 1998, "Reserve", in terms of geohydrology, means the quantity and quality of groundwater required:

- to satisfy basic human needs by securing a basic water supply for people who are now or who will, in the foreseeable future, be (1) relying on, (2) taking water from, (3) being supplied by, the relevant water resource; and
- to protect aquatic ecosystems, in order to secure ecologically sustainable development and use of the relevant water resource.

The former provides for the minimum living requirements of individuals served by the water resource in question and includes water for essential uses, such as drinking, food preparation and personal hygiene (set, for the time being, at 25 l/d/person). The latter is largely reliant upon expert judgement and set at a value so as to ensure that the natural ecology of the identified area is maintained or improved, depending on the Resource Quality Objectives (RQOs) set. The overall Reserve (i.e. basic human needs and ecological) is defined as both a quantity and quality of the whole water resource, and will vary depending on the class of the resource. The Minister is required to determine the Reserve for all or part of any significant water resource.

A Reserve determination of the water resource has to be made if a licence application for water allocation is processed. Depending upon significance and sensitivity of the resources in question, the Reserve determination required may be an intermediate estimate or comprehensive evaluation. The latter may be achieved through improvement and upgrading of the former into a more comprehensive and accurate Reserve assessment as more information becomes available. Before the methodology for determination of the RDM is formally gazetted in 2004, all the proposed methods are preliminary.

1.2 *Purpose of this Paper*

This paper provides an overview of the on-going research programme and shares experience of multidisciplinary approaches to water resource management with the international community.

It is also hoped to generate discussion on:

- Identifying knowledge gaps, validation or recommendations regarding current projects and methods, in support of R&D planning.
- Improving conceptual and methodological thinking regarding the comprehensive RDM determination for groundwater and dependent ecosystems.
- Linking South African and international experts in this area.

2 INTERNATIONAL COMPARISONS

2.1 *Australia*

The national principles for the provision of water for ecosystems were outlined by the Subcommittee on Water Resources of the Sustainable Land and Water Resources Management Committee (1996). Hatton and Evans (1998) assessed the significance of groundwater dependent ecosystems in Australia.

Similarities between Australia and South Africa are as follows:

- National Guidelines or Principles reserve groundwater water for the environment.
- Both countries have processes in which a first, conservative general allocation is identified with subsequent allocation requiring more substantial analysis and investigation.
- The treatment of groundwater in this sense lags behind surface water allocation; both countries have more experience in setting instream flow requirements for aquatic systems than in identifying groundwater requirements.
- Both involve stakeholders drawn from the community, with much power vested in local catchment management authorities.
- Groundwater is increasingly seen as the basis for further water resource development.
- The two countries share the highest temporal rainfall variability in the world; this has risk implications associated with Reserves.
- There is not a complete consensus among the hydrologists in either country regarding the conceptualisation of flow systems.

Contrasts between Australia and South Africa can be summarised as follows:

- Australia has national *principles* that guarantee allocation of groundwater to the environment, but *policy* is not national.
- In South Africa, water allocation over the Reserve is based on the greatest beneficial use as assessed by catchment management agencies. Australia is moving toward a market-based approach, with water trading among users.
- The Australian approach does not include Basic Human Needs as a water right.
- Australian guidelines require that water allocations meet the needs of all ecosystems, while the South African National Water Act explicitly identifies only aquatic systems.
- South African procedures define acceptable impacts through prioritising process.
- South Africa aspires to establish comprehensive determinations for the entire country, with the catchments assessed in order of technical priorities but also in response to licence applications. Australia (or its respective states) tends to react to the conjunction of sensitive, highly significant ecosystems and proposals for local development.
- Australia does not share the vast challenge of the state delivering basic water supplies to rural and remote communities.
- Australia has a national assessment of ecosystem dependence on groundwater that identifies key ecosystems, their significance, and nature of dependency.

2.2 *UK*

In the UK, groundwater protection is predominantly reactive (Herbertson 1994). The basic process entails

(1) obtain estimate for long-term average groundwater recharge; (2) then subtract percentage that accounts for unreliability of recharge; (3) further subtract percentage that satisfies river and environmental requirement and (4) allocate the remaining percentage for human use. Approaches vary from the empirical to detailed modelling and monitoring. The Precautionary Principle applies, therefore empirically deduced allocations are likely to leave more water in aquifers than necessary. Consequently, more detailed analyses may then allow further exploitation.

3 METHODOLOGY

3.1 *Objectives of the comprehensive RDM determination*

The method for comprehensive determination of the groundwater component must specify a groundwater allocation according to section 23 of the NWA. The following aspects must be considered in determining the groundwater allocation:

- water quantity and quality to meet the requirements of the Reserve for aquatic ecosystems;
- water quantity and quality for basic human need;
- the integrity of the groundwater resource; and
- water to supply needs beyond the boundaries of the delimited area ('downstream' requirements).

Additionally, whilst adhering to the criteria put forward in Section 1 of this paper, the procedure used must:

- be based on sound hydrogeological principles;
- fulfil DWAF's administrative requirements; and
- be fully integrated with the other components of RDM (rivers, wetlands, estuaries), *via* procedures in the Integrated Manual (DWAF 1999).

Where specific knowledge, information or expertise is lacking or cannot be obtained within reasonable cost, a determination can be based on expert judgement, applied within a structured and consistent process. However, all assumptions must be properly documented. A formal, independent review process will be used to ensure quality and consistency. Finally, the method used must accommodate the need for appropriate consultation with stakeholders, as specified by the NWA.

3.2 *Determination of the Reserve*

The Reserve concept has its origin in river ecology thinking. A critical assumption in the interpretation of the NWA is that groundwater only plays a role in the ecological Reserve determination if there is hydraulic connection between groundwater and surface water bodies that sustains aquatic ecosystems. In such situations the often complex role of groundwater in supporting the aquatic ecosystem and human population, has to be ascertained. Where aquifers have minimal connection with aquatic ecosystems, the groundwater Reserve would refer to the Basic Human Needs component (25 l/d/person using groundwater for supply in the area).

The crux of a comprehensive determination of the groundwater component is an adequate understanding of the affected hydrological system, the water balance (in particular groundwater recharge) and the critical factors that influence the availability of groundwater to aquatic ecosytems and people (Wright and Xu 2000).

A key surface water expression of the Reserve is Instream Flow Requirements (IFRs). To create the groundwater linkage to this important measure, it is critical that geohydrologists become part of hydrograph analysis and flow duration studies. The Reserve includes the quality of water required and this must also be determined. These requirements impose constraints on the aquifer in the vicinity of the aquatic ecosystems. Such constraints should be translated into groundwater terms like water levels, gradient and flux towards a surface water body. An objective function may be formulated to account for such constraints in abstraction modelling.

Due to the fact that all aquifers are heterogeneous (e.g. hydraulic conductivity and aquifer storativity) and that critical ecosystem linkages mostly occur in specific parts of the landscape, distributed models are required to determine the best strategy for the allocation and protection component. In setting the Reserve, for instance, the catchment of interest will be treated as a distributed model.

The link to surface water is critical; clearly the groundwater component of the Reserve cannot be treated as independent of the total Reserve. Modelling in support of the comprehensive determination must treat, often simultaneously, all components of the hydrological cycle. The feasibility and methodology for developing a coupled surface water / groundwater flow and management model from existing codes is being considered (Chiang 2000). It is proposed that MODFLOW is used for groundwater flow simulation while BRANCH is selected for surface water flow simulation. These two models are connected with MODBRANCH for simulating the water exchange between aquifers and surface water bodies. The computer code MODOFC is used to find the best combination of pumping rates and borehole locations to achieve specified constraints, for example draw-down, pumping rates or costs. The software structure of a graphical user-interface, referred as MODGSM, for linking MODFLOW, BRANCH and MODOFC is also introduced.

Determination of the reserve requires certain assumptions. The accuracy of the result is largely dependent on whether or not the assumptions are adequately met. Improvement in our conceptual understanding and data, will reduce uncertainty.

3.3 *Determination of RQOs*

Through the setting of Resource Quality Objectives (RQOs) any requirements or conditions that may need to ensure that the water resource is maintained in a desired state or condition can be achieved. While the focus of the Reserve is on the protection of aquatic ecosystems, the RQOs can also address additional considerations relating to physical and chemical degradation of the aquifer itself. To set appropriate RQOs will require a good understanding of the critical system relationships, e.g. between abstraction rates and changes in water quality, between groundwater level and physical integrity of the aquifers and between hydraulic gradients in the aquifers and spring flow and river base flow discharge.

In the drier two thirds of the country where groundwater does not support aquatic ecosystems, Basic Human Needs as a right must be safeguarded. While the norm of 25 l/d/person is minimal, to locally guarantee its quantity and quality may entail strict proactive protection measures (Xu and Braune 1995). These measures can be regarded as sets of RQOs and include delineation of borehole protection areas if necessary.

3.4 *Procedures for comprehensive RDM determination*

A generic RDM procedure has been designed to accommodate the needs of all the different components of the RDM and sets out all the steps needed to carry out the determination of the Reserve, whilst outlining links to the licence application process (MacKay 1998; DWAF 1999). Such generic procedures will be adapted to reflect the unique requirements for the protection of groundwater resources. A flow chart shows the steps for comprehensive RDM determination (Fig. 1).

The requirement for delimitation of groundwater management units (Step 1) involves a change of scale from the generic methodology, which calls for an eco- or geohydrological-region typing (i.e. having defined geographical, regional boundaries). For groundwater, the management unit is a sub-division of a geohydrological-region type in the overall geographical hierarchy. As a part of a significant water resource linked to other ecosystems, the management unit has its own particular management requirements. A system for delimiting the groundwater system and the methodology for determining boundaries from the national level to the management unit is currently being researched (Braune et al. 1999).

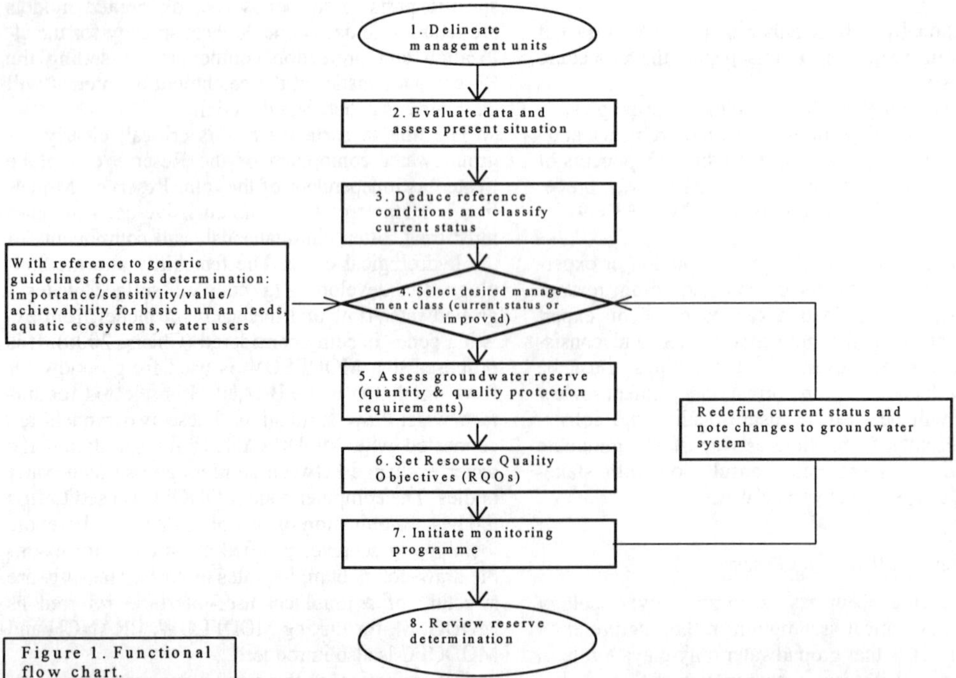

Figure 1. Functional flow chart.

The comprehensive determination should be based on available hydrogeological information and that anticipated within the 12 months. Basic conceptual modelling scenarios will ascertain the additional data required. Further work should be performed where data is needed for the high confidence of a comprehensive RDM determination.

Where sufficient data exists, a conceptual model of the hydrological system and its ecosystem link census of the chosen area is important. Sound judgement regarding the type of model and data requirements is essential.

National databases, such as the National Groundwater Data Base (NGDB), can be used to address resource classification and determine aquifer management units at large scales. For example, GIS can be used to produce regional vulnerability maps. Information systems involving hydrological characteristics, such as Water Resources 90 (WR90), are useful for a conceptual understanding of aquifer-ecosystem linkage. One of the least understood and least quantified aspects of the hydrological balance needed for a comprehensive Reserve determination is the groundwater - vegetation interaction. A particular issue here is the dependence of terrestrial vegetation on groundwater. Until more comprehensive methodology has been developed specific to different environments, reliance on expert knowledge will be essential (Braune et al. 1999).

Once the boundaries of aquifer units and management units within a significant water resource have been decided (Step 1), the resource must be characterised in terms of its intrinsic properties and recharge, as well as stress (Step 2).

In practice, reference conditions (Step 3) are deduced from an understanding of present conditions (Step 2). Current status must be established and classified.

As mentioned previously, RQOs include objectives for the maintenance of the Reserve as set out in Step 5 and also are dependent on the determination of the management class. In practice, Step 4 sets the basis for the remaining steps and Steps 4-7 form an iterative loop because the information gathered may require revision. In essence, the procedures must provide input to meet the objectives in section 3.1.

Since characteristics of natural flow systems, such as water levels, are dynamic, follow up monitoring in Step 7 is critical to the improvement of the Reserve determination process. Groundwater levels tend to fluctuate naturally, sometimes significantly, over time. This is especially true for catchments with weak lateral hydraulic connection and where groundwater recharge depends largely on episodic rainfall. Furthermore, natural fluctuations are, in most areas, modified to a certain degree by human impacts. Monitoring is required to verify and improve the baseline conditions and the assessment of groundwater resources. Such monitoring has not been undertaken at a regional scale in South Africa and will only be meaningful if it is based on an understanding of the flow system.

Interaction between water and ecosystems may not be revealed until systems are stressed. A full understanding of ecosystem impacts may only be revealed by monitoring following groundwater development.

4 LIMITATIONS OF COMPREHENSIVE RDM DETERMINATION

The NWA is based on principles which include water as an indivisible national asset, the interdependence of all elements of the water cycle, the importance of the ecological functions of all water, and the need to treat all water consistently in law (DWAF 1996). The generic approach to establishing the Resource Directed Measures is robust and, on the whole, valid for all elements of the water cycle. However, there are some aspects that are more difficult to transpose onto groundwater and need careful treatment. Some of these are discussed below, with recommendations for implementation.

The RDM represent a shift in thinking from water quality to water *resource* quality. The resource base includes ecological integrity and functioning. In recognition of this the Reserve sets aside the quality and quantity of water required by aquatic ecosystems. The ecological support functioning of groundwater is therefore considered only in terms of surface water resources. Groundwater's role in supporting important terrestrial ecosystems and stygofauna, in particular in the large semi-arid parts of South Africa is not accommodated in the Reserve. In the long term a wiser definition of aquatic ecosystems may be required while in the short term maximum use should be made of the other RDM measures, in particular the classification of certain areas requiring special management.

It is not yet clear how the Reserve for groundwater will be specified. For surface water the quantity and quality may be specified and the quantity of the total resource remaining may then be allocated to other users. A simple bucket analogy is valid because whatever is 'left over' in the bottom of the bucket can be used by the dependent people and biota. However, to ensure that sufficient volumes of groundwater are available for baseflow and basic human needs (where shallow abstraction points are used) a minimum *water level* should be maintained. Groundwater at the base of the aquifer is not as accessible from the surface and does not fulfil the same ecological functions, therefore a volume set aside here will not fulfil the requirements of the Reserve. It is recommended that the comprehensive groundwater Reserve be specified in terms of quality parameters and spatially referenced water levels or discharge requirements.

5 WAY FORWARD

One of most significant features of the NWA is the acceptance of groundwater as a component of the hydrologic cycle. Consequently, determination of groundwater resources entails apportionment of the Basic Human Needs, Ecological Reserve and resource allocation components of the entire water resource (i.e. surface and groundwater, etc.) and subsequent quantification of the groundwater contribution to each of the components.

Groundwater systems and their linkage to surface water and related ecosystems are still poorly understood. The guidelines presented in this paper require ongoing improvement and verification. A systematic process of monitoring and research is required. The existing legislation itself, which is largely based on river ecology thinking, will also have to be improved as our experience grows. In the meantime the full potential of the Resource Directed Measures should be utilised to ensure the protection of groundwater resources and groundwater-dependent ecosystems.

Issues that need to be addressed to improve the comprehensive RDM determination methodology to groundwater systems are, *inter alia*:

- Integrated classification to link delineated geohydrological systems with the eco-region classification systems;
- Approaches to establish reference conditions;
- Approaches to identify and map groundwater-dependent ecosystems;
- Appropriate models that include groundwater/surface water linkages and
- Appropriate Resource Quality Objectives and adequate monitoring systems, addressing ecosystem and groundwater resource integrity.

ACKNOWLEDGEMENTS

The following team members are thanked for their valuable comment on a draft of the paper: J Conrad, I van der Voort, WH Chiang, GJ van Tonder and K Wright. We thank H MacKay and N Kleynhans who have positively influenced our thinking. The Department of Water Affairs and Forestry (DWAF) and Water Research Commission (WRC) are thanked for supporting this study and allowing publication of the ongoing work.

REFERENCES

Braune E, Xu Y, Van der Voort I, Conrad J, Colvin C, Le Maitre D, Van Tonder GJ, Chiang WH, Zhang J, Wright K, Hughes S & Pietersen K. 1999. *Comprehensive Determination of Water Resource Reserve: Groundwater Component (Zero Version)*. WRC Report, Programme K5/1007/0/1. Water Research Commission, Pretoria

Chiang, WH 2000. Personal communication. University of Free Stste, Bloemfontein, South Africa

DWAF 1996. Discussion document: Water Law Principles, Pretoria.

DWAF 1998. *National Water Act*, Government Gazette of the Republic of South Africa, Vol. 398 No. 19182, 26 August 1998, Cape Town.

DWAF 1999. *Resource Directed Measures for Protection of Water Resources*, Volume 2: Integrated Manual Version 1.0, Pretoria

DWAF 1999. *Resource Directed Measures for Protection of Water Resources*, Volume 3: River Component Version 1.0, Pretoria

DWAF 1999. *Resource Directed Measures for Protection of Water Resources*, Volume 6: Groundwater Component Version 1.0, Pretoria

Hatton, T. and Evans, R. 1998. *Dependence of Ecosystems on Groundwater and its Significance to Australia*, Occasional Paper no. 12/8, Land & Water Resources Research & Development Corporation, Australia.

Herbertson, PW. 1994. Utilising licensing practice to protect groundwater resources, NRA Southern Region, UK.

MacKay, H.M. 1998. *Part 3. Development of a methodology for determination of the ecological Reserve*. Draft Docu-ment for discussion dated 25 October 1998.

Subcommittee on Water Resources of the Sustainable Land and Water Resources Management Committee 1996. *National principles for the provision of water for ecosystems*, ISBN: 0-7310-2382 x, Agriculture and Resource Management Council of Australia and New Zealand and the Australian and New Zealand Environment and Conservation Council, Canberra.

Van der Voort, I and Conrad, J. 2000. Personal communication. CSIR, Stellenbosch, South Africa

Wright, KA and Xu, Y 2000. A water balance approach to the sustainable management of groundwater in South Africa, ISSN 0378-4738, *Water SA*, Vol. 26 No. 2, April 2000.

Xu, Y and Braune, E. 1995. *A guideline for groundwater protection for the community water supply and sanitation programme*, ISBN: 0-621-16787-8, DWAF, Pretoria.

Groundwater: Past Achievements and Future Challenges, Sililo et al. (eds) © 2000 Balkema, Rotterdam, ISBN 90 5809 159 7

Groundwater resources protection and aquifer recovery in China

Han Zaisheng
China Geological Survey, Beijing, People's Republic of China

ABSTRACT: The groundwater resources in China and especially the north part represent a vital water resource. Both the shallow and deep aquifers are highly over-exploited in a large area of north China. The heavy over-exploitation of the groundwater resources is causing huge environmental damage. For protecting groundwater resource, several technical feasibility studies were taken. Artificial recharge using floodwater and wastewater were tried and performed. Surface spreading systems are applicable in many areas of Yellow River basin and Hai River basin. Deep aquifer injections are undertaken in the urban area. A far better strategy is to reduce the extraction of groundwater, especially for stopping or slowing land subsidence and seawater intrusion. To address the problem of falling groundwater levels and aquifer recovery, there is needed to incorporate both a reduction in groundwater extractions and artificial recharge

1 INTRODUCTION

Groundwater has played a key, albeit largely unknown, role in the history and development of China. As early as 1835 the Chinese had mastered the cable tool drilling method and had drilled to over a kilometer deep, while the rest of the world was only able to tap relatively shallow aquifers. Groundwater is now a major water resource providing immense volumes for agriculture, industry, and urban and domestic needs. This historical and current demand on the groundwater resources of China has been at a price – falling groundwater levels, subsidence, eco-system destruction, seawater intrusion, pollution and water quality deterioration.

For groundwater resources protection and aquifer recovery, the potential of artificial recharge to address these problems is assessed, especially considering the possible use of wastewater and floodwater.

2 GROUNDWATER RESOURCES OF CHINA AND 3-H BASINS

2.1 Groundwater resources of China

The groundwater resources of China were calculated on the basis of the mean annual precipitation for China from 1956 to 1979, and presented in surface water basins (Table 1). (MWR China 1992)

The assessment of the groundwater resources of China is based upon an understanding of the recharge over a region and then considering technical, economic and environmental factors. This typically results in the exploitable yield being less than the recharge volume.

Table 1 Groundwater Resources of Basins in China (Billions of cubic meters per year)

	River Basins	Mt. area	Plain area	Overlap	Total
1	Songliao	31.9	33.0	2.4	62.5
2	Hai	12.5	17.8	3.8	26.5
3	Huai	10.7	29.7	1.1	39.3
4	Huang	29.2	15.7	4.3	40.6
5	Yangtze	221.8	26.1	1.5	246.4
6	Pearl	102.8	9.3	0.5	111.6
7	South-east	56.2	5.2	0.1	61.3
8	South-west	154.4			154.4
9	Inland	56.7	50.6	21.1	86.2
Total		676.2	187.4	34.8	828.8

Overlap = groundwater flow from mountainous area to plain area (i.e. Double counted)

It is noted that the Hai River, Huai River and Huang River (Yellow River) basins (3-H basins) have moderate sized groundwater resources in comparison with the rest of China. If the same data is compared with available surface water resources, then in proportion the groundwater resources of the 3-H basins is a much more important component of the total available water resources (Table 2). (Wu Yiao et al .1989)

Table 2 Groundwater, surface runoff and their Ratios of major basins (in range with Ratio)
(Billions of cubic meters per year)

Basins	Surface runoff	Ground-Water	Ratio (G/S)
1 Hai	28.8	26.5	0.92
2 Inland	116.4	86.2	0.74
3 Huang	66.2	40.6	0.61
4 Huai	74.1	39.3	0.53
5 Songliao	165.0	62.5	0.38
6 Yangtze	951.3	246.4	0.26
7 SW. China	585.3	154.4	0.26
8 Pearl	468.5	111.6	0.24
9 SE. China	255.7	61.3	0.24
Total(Average)	2712.0	829.0	0.31

2.2 Exploitable groundwater resources of the 3-H basins

Considering the 3-H basins, the groundwater resources can be divided into two major hydrogeological units – the mountains and the plains. The mountain groundwater resources are generally in fractured rock and karst (limestone) formations, of Cambrian to Ordovician age. The plains, of which the North China plain is the most significant, consist of Quaternary alluvial sediments up to almost 1000m in thickness.

The exploitable fresh groundwater resources in the 3-H basins are presented in Table 3.

Table 3 Groundwater Resources in 3-H basins
(Billions of cubic meters per year)

River Basin	Area (km^2)	Annual Precipitation (mm)	Re-charge	Exploitable fresh GW TDS<2g/L
Hai	318160	549	27.5	17.3
Huai	331620	859	34.5	24.0
Huang	799680	454	40.4	18.2
Total	1449660	578	102.4	59.5

2.3 Groundwater use in the 3-H basins

The water supplied (taken as meaning the use) from groundwater and surface water in the 3-H basins is, for 1997, as follows:

Table 4 Surface water and groundwater use in 3-H basins in 1997
(Billions of cubic meters per year)

Basin	Surface water	Ground-water	Total
Hai	16.158	27.368	43.775
Huai	55.488	18.438	74.208
Huang	27.781	13.874	41.802
Total	99.428	59.681	159.785

There are reportedly 2.8 million wells in all of China. Groundwater use is very important in all three basins, but especially important in the Hai River basin. In terms of total water use in all of China, groundwater plays a much more important role in northern China than in southern China.

A comparison between the amount of groundwater available (the exploitable fresh groundwater) and the groundwater use in 1997 is provided in Table 5. Those regions where the use exceeds the exploitable resource are shaded. Even though on average over all of the 3-H basins, the use is almost equal to the available groundwater resource, this is actually misleading. If data were available at a smaller scale, the extent of over development would be much more extreme.

Table 5 Groundwater Resources and Use in 3H Basins (Billions of cubic meters per year)

River Basin	Exploitable fresh groundwater	Use of groundwater in 1997	Use/ exploitable
Hai	17.3	27.4	1.58
Huai	24.0	18.4	0.77
Huang	18.2	13.9	0.76
Total	59.5	59.7	1.003

3 GROUNDWATER ISSUES

3.1 Resource over-exploitation

Few regions in the world would have such vast areas of over-exploited groundwater resources. The large scale dropping in water table level in the shallow unconfined aquifers and pressure levels in the deeper confined aquifers not only causes many bores to go dry but also causes many other environmental problems. For example, in Hebei province, the shallow water table level has dropped 5m, to 12m depth, from 1983 to 1999 over large areas. The deep aquifer pressures have dropped up to 10m below sea level in some areas. The rate of water level decline from 1990 to 1995 was 1m/year. There are many smaller areas, which still represent huge cities and regions, where the groundwater resources are grossly over-exploited. This especially applies to the deep confined aquifers where the recharge rate is really very small, but approximately 25% of all groundwater use is from the deep aquifers.

3.2 Pollution

Unpolluted groundwater is usually of better quality than surface water, especially from the deep confined aquifers. Extensive areas of China have considerable groundwater pollution. The shallow unconfined aquifer, and in some cases the deep confined aquifers, are extensively polluted. The assessment indicates that groundwater pollution is widespread. The major pollution sources are primarily wastewater discharges from the cities and towns, which are used for irrigation, and also include pesticides, nutrients and heavy metals. The rapidly expanding sources of pollution, especially the large increase in untreated sewage disposal to land, tend to be the major problems. Cities within the 3-H basins focus prominently as having the greatest pollution problem (Duan et al. 1993).

3.3 Quality degradation

Quality degradation is completely different from pollution in the sense that pollution refers to the introduction of external pollutants into aquifers, whereas quality degradation refers to the migration of poorer quality groundwater into good quality aquifers as a result of the decrease in pressure/water level within the exploited aquifer. Hence, for example, where extraction occurs from the intermediate depth confined aquifer; there is both horizontal and vertical migration of saline water into the good quality aquifer being pumped. The relative significance of horizontal to vertical leakage is very important from the point of view of developing management strategies, especially artificial recharge, to counter this aquifer salinisation mechanism. It is believed that vertical leakage is far more important than horizontal migration.

3.4 Seawater intrusion

Seawater intrusion occurs in many coastal regions of China where the coastal groundwater resources are over-exploited. Table 6 indicates major areas where seawater intrusion is well recognized.

Table 6 Seawater intrusion in China

Location	Area (km^2)	Velocity (M/year)	Period
Dalian city	223.5		1975-1998
Qinhuangdao city	50	22	1970-1995
Laizhou gulf	202	177	
Longkou city	78		
Yantai city	33.5	170	1970-now
Qingdao city	39		
Beihai city	4		1980- now

Seawater intrusion has reportedly caused 8000 wells to be abandoned, resulting in over one million people having problems accessing drinking water. Perhaps the key consequence of seawater intrusion is that it is virtually an irreversible process, in the sense that for practical timeframes once intruded, it is difficult to restore the original aquifer water quality (Han 1987).

3.5 Subsidence

There are 30 examples of land subsidence in China resulting from groundwater resource over-exploitation. 20 of these examples are due to extraction from deep confined aquifers. In Tianjin Municipalities by 1994, subsidence reached 1.523m, with an average rate from 1991 to 1994 of 48mm/year. This has caused settlement of structures, bridges to collapse, significantly aggravated storm water drainage problems, and a reduction in the capacity of the river levees to resist high tides and floods.

In Shanghai, major groundwater usage from 1921 to 1986 resulted in a water table drawdown of 35m and resulting subsidence of 2.63m (Liu et al. 1987). The introduction of an artificial recharge scheme has largely controlled the subsidence. In other locations, reducing extractions has only controlled ongoing subsidence over large areas. Still there are many areas, mostly associated with urban groundwater use, where serious subsidence is continuing.

3.6 Land Salinisation

Land salinisation is due to excessive use of surface water in irrigation areas causing the groundwater level to rise, a build up of salt (NaCl predominantly) in the soil and unsaturated zone due to high evapotranspiration, and ultimately reduced yields or even death of vegetation. Salinisation is a significant problem in sub areas 3 and 6 of the Huang River basin – between Lanzhou and Hekou, and between Sanmen Gorge and Huayuankou. Solving land salinisation problems involves a multitude of land, vegetation, and water management strategies. However it is relevant in the sense that increasing groundwater usage is one of the key components of a salinity management plan. Integrated with increased groundwater usage is the conjunctive use of surface water and groundwater resources.

3.7 Overview

The groundwater problems described above (with the exception of the last topic – land salinisation) do not occur as individual issues. They are frequently interrelated. The ongoing drawdown in groundwater levels results in reduced yields of wells, sharp increases in operating costs, frequent need for replacement of pumping facilities, and the abandoning of wells. Subsidence, salinity degradation, migration of polluted groundwater into good quality aquifers, and seawater intrusion (for coastal areas) are all consequential problems of over-exploitation.

4 ARTIFICIAL RECHARGE AND AQUIFER RECOVERY

4.1 Principles and experience in China

The artificial recharge of water into aquifers and subsequent extraction is gradually becoming a well accepted technique to enhance the availability of groundwater and to solve many associated groundwater related problems – for example, subsidence, sea water intrusion, ecosystem damage and water quality deterioration. This is a method of storing surplus water underground in order to meet needs in times of shortage. The potential widespread adoption of the method to address over development of groundwater resources in the 3-H basins is considered. The different techniques, which are available to be used, are evaluated, as well as the problems - principally clogging.

The methods of artificial recharge fall into two groups: Surface systems – spreading basins and trenches and Deep systems – deep well injection. Surface spreading of effluent with periodic infiltration and drying to avoid clogging and to promote nitrification, denitrification, and other pollutant removal or retardation, should be sustainable indefinitely for a wide range of non-potable uses. Primary treated sewage effluent (i.e. settled effluent) has been successfully used for recharging groundwater at land-spreading sites (Pyne 1995).

Clogging is a difficulty for artificial recharge managers but it can be properly treated by means of good design of facilities and preventive measures. Current knowledge, complemented with pilot tests, is enough to assure the efficiency of a cheap and very useful technique.

Six existing artificial recharge schemes in China are shown in Table 7. Artificial recharge in Shanghai to control land subsidence, is well known and has been very successful. Good quality potable water has been used for the deep well injection since 1965. The second example is really a very impressive scheme. The artificial recharge fields are located in the middle reaches of Chaobai River, downstream of Miyun reservoir. After the reservoir was constructed in 1958, the surface water flow down the Chaobai River was significantly reduced. The adjacent groundwater level declined 6m - 9m from the 1970's to the 1980's. Some of the adjacent shallow wells became dry. A groundwater well field for urban water supply in this area had to reduce its water yield. To enhance groundwater resources nine weirs were constructed along the river channel from 1984 to 1998. The first one was concrete, while rubber was used for the later eight. The length of these dams is 300m – 400m; their height is 3m – 3.5m. The artificial recharge project on the Yongding River, at the edge of Beijing City, involved extensive testing prior to the implementation of the scheme.

The Tianjin city deep well injection artificial recharge scheme is interesting because the fluoride content of the injected water is 3mg/L, which makes it unsuitable for potable purposes. Considering all the schemes in Table 7, the deep well injection is always using excellent quality water, while the use of surface systems is restricted to high hydraulic conductivity surface soils (i.e. coarse sands and gravel) generally associated with river deposits and readily available surface waters.

Table 7 Existing artificial recharge systems in China

Location	Type of Scheme	Year	Recharge Volume	Reason	Depth of Aquifer	Comment
Urban area of Shanghai city	Deep well injection	1965		For land subsidence control and energy storage in aquifers.	60-260m	Source of water is water supply from Huangpu River. Land subsidence has been controlled.
Chaobai River, Beijing	Spreading Basin	1984	40-80 mcm/year	To enhance groundwater resources.	0-30m	Groundwater level has risen 2 – 3 m since 1990 in this area.
Yongding River, Beijing city	Trench and deep well injection	1981	200000 m^3/day	To enhance groundwater resources.	Surface and 10-50 m	Source is from Yongding River. The trend of groundwater level decline was slowed.
Tanggu section, Tianjin city	Deep well injection	1980	Unknown	To protect geothermal water resources.	160 – 600 m	Geothermal water is source of recharge. AR slowed the speed of geothermal water level decline in deep aquifers.
Dagu River, Qingdao	Trench	1990	Unknown	To enhance groundwater resources.	0-30 m	AR ensured the groundwater supply for Qingdao city.
Middle reach of Luo River	Trench	1994	50000 m^3/d	To enhance groundwater resources.	Surface	Luo River is the source of water. Groundwater is in dynamic balance in the recharge area.

4.2 Artificial recharge using wastewater in 3-H Basins

Treated wastewater from urban areas is currently available and will increasingly be available in the future. It can be used for a variety of purposes, the most likely being: increased irrigation, as a substitute for untreated wastewater, as a substitute for groundwater or as artificial recharge.

To address the groundwater problems, three options are available: 1.Reduce the volume of groundwater being extracted. This can be achieved by a variety of methods, including substitution with wastewater, inter-basin transfers, or more efficient irrigation practices. 2. Artificial recharge, using either deep injection or surface systems as appropriate. 3. Both 1 and 2.

Where relatively coarse sediments outcrop at or near the surface, then wastewater treatment to relatively low standards only is necessary. Coarse sands and gravel's outcrop or subcrop in a wide belt around the western margin of the North China Plain. Artificial recharge using surface systems in these environments would be relatively simple, and high infiltration rates would be very achievable. Infiltration rates of 5m/day would be feasible. In the central and eastern part of the North China Plain, finer grained surface sediments are present. There are however localized coarser areas (clayey sands and silty sands), associated with ancestral river courses (sometimes called "prior streams"), where infiltration rates of at least in the order of 0.2 m/day would be expected. These ancestral river courses have been mapped.

Tertiary level treatment of wastewater is required for deep well injection. Secondary level treatment of wastewater is satisfactory for surface artificial recharge systems. The actual volumes of treated wastewater available for artificial recharge may not be great because of competition from other uses, principally irrigation.

4.3 Artificial recharge using floodwaters in 3-H Basins

Even though long dry periods are common, major floods are also a characteristic of the highly variable hydrology of the 3-H basins. Flooding is generally contained within the levees surrounding all major rivers. However for major floods, detention basins are used. These detention basins operate with either gated structures or simply by natural or man-made breaching or overtopping of the levees during major floods. The frequency of inundation of the detention basins varies considerably, as does the duration of inundation. Nonetheless, the option exists to use this excess floodwater to help alleviate problems associated with over-exploited aquifers. As well as detention basins, the huge areas between the levees of all the major rivers offer the potential for artificial recharge during floods. There are detention basins and regions where artificial recharge may have a role. Meeting the water quality requirements, especially turbidity for floodwaters would require treatment.

There are some areas beneath detention basins (and along the levees) where the groundwater is over-exploited and it would be advantageous to facilitate recharge using untreated floodwaters. It is recognized that due to the high turbidity of the floodwaters in some cases, the enhanced recharge areas would clog within several weeks and the infiltration rate would decrease significantly. Prior to the next flood, it would be necessary to clean these enhanced recharge areas. Nonetheless, considering the large areas involved, the useful addition to the groundwater resource would be significant. Assuming a (short-term) average infiltration rate of 1m/d over a ten-day flood period, this would produce a shallow aquifer recharge of 10 million m^3/km^2. Some of this recharge would be currently occurring.

Considering the Yellow River basin even though the groundwater use is high or over-exploited, the artificial recharge potential is only ranked as medium because the surface soils are very silty and hence the ability to achieve reasonable infiltration rates is only moderate.

The four situations in the Hai River basin are all geologically more suitable and some "enhancement" works as described above for detention basins are worthy of further evaluation. Even though the flood frequency is less in the Hai River basin, the turbidity is also less, hence this region is much more attractive from the point of view of artificial recharge. In the Hai River basin, in both cases the deep confined aquifers are overdeveloped. However, it is understood that there are some lakes in this region where the turbidity would be expected to be much lower and hence the use of deep drainage bores may be feasible, provided there is a downward hydraulic gradient.

4.4 Overall groundwater management approach

The serious water shortage in North China, whereby both the shallow and deep aquifers are frequently over-exploited and polluted. It can be addressed in the mid to long term by: increased reuse of wastewater, including artificial recharge, increased conjunctive use of groundwater and surface water, reduced groundwater extraction, especially and primarily from the deep confined aquifers, or inter-basin water transfer.

The deep confined aquifers represent an especially vulnerable resource which has very limited recharge and an exploitable yield possibly far smaller than officially recognized. Problems of subsidence, seawater intrusion, and salinity degradation are primarily associated with over-exploitation of the deep confined aquifers.

Hence a clear groundwater management program focussed toward the above objectives is required in over-exploited areas. Nonetheless, engineering measures (e.g. artificial recharge, inter-basin transfers) will not address the huge existing problems. Correctly managing the existing groundwater resource to prevent pollution, inter-aquifer contamination, salinity degradation, and other problems must be the primary objective. Hence groundwater management is essential to ensure the long-term sustainability of all groundwater resources in the 3-H basins. Responsibility for groundwater resource investigations is with the China Geological Survey. Increased groundwater level and quality monitoring is required in some parts of the 3-H basins to provide the basic data necessary for water resources management. The over-exploitation should be reduced to a sustainable level within a time frame agreed to in the ground water management program.

5 CONCLUSIONS

The huge extent of over-exploitation of the groundwater of the 3-H basins of China has created a broad range of largely irreversible environmental problems including subsidence, seawater intrusion, salinity degradation, pollution, and falling groundwater levels. Artificial recharge is the most important engineering measure for groundwater resources protection and aquifer recovery. The use of treated wastewater for artificial recharge requires a higher standard of treatment and greater operation and maintenance effort. Nonetheless, with care, both surface and deep systems can operate successfully. There are large areas on the water flank of the North China plain where coarse sands and gravels outcrop and where surface artificial recharge systems would be efficient. Reducing groundwater extractions, by a variety of means, will generally be the most cost-effective method of solving groundwater problems. Problems caused by over-exploitation of groundwater in the North China plain should be solved by evaluation, in an integrated manner, of all available water management options.

REFERENCES

Beijing Company of Hydrogeology and Engineering Geology, 1995. Test Report on Groundwater Artificial Recharge in West Suburb of Beijing.

Committee on Mineral Reserves of Tianjin, 1995. An example of approximate volumes of water required addressing groundwater problems in urban area.

Department of Hydrology, Ministry of Water Resources, PR China, 1992. Water Resources Assessment for China. China Water and Power Press.

Duan Yonghou, Luo Yuanhua, Liu Yuan, etc., 1993. The Geological Hazard in China, China Architecture and Building Press.

Han Zaisheng, 1998. Groundwater for Urban Water Supplies in Northern China. Hydrogeology Journal, 1998, No. 6, 416-420.

Liu Tiazhou and Sun Shusuang, 1987. Case Study on the Laws of Land Subsidence in Shanghai. Urban Geology of Coastal Lowlands in China, UN, ESCAP, Volume 3, 3-11.

Pyne, R.D.G., 1995, Groundwater Recharge and Wells. A Guide to Aquifer Storage and Recovery. Lewis Publishers.

Institute of Hydrogeology and Engineering Geology, 1979, Atlas Publishing Company, The Hydrogeological Atlas of PR China

Institute of Hydrogeology and Engineering Geology, 1995, Research on rational utilization of groundwater resources and the techniques for interception of floodwater to artificial recharge to groundwater in the Yellow River basin.

Groundwater: Past Achievements and Future Challenges, Sililo et al. (eds) © 2000 Balkema, Rotterdam, ISBN 90 5809 159 7

Geostatistical approach to groundwater quality assessment in a rural zone of the Tejo River basin border, Portugal

J. Duque
Centro de Geologia da Faculdade de Ciências da Universidade de Lisboa, Portugal

A. Chambel
Universidade de Évora, Portugal

ABSTRACT: The Chança region is a rural zone in the border of Tejo river sediments, near the contact with the igneous and metamorphic Paleozoic Massif. In some places there are groundwater quality problems, due to high mineralization associated with natural contamination. The groundwater in the igneous rocks of the northern part of the region are scarce but has lower mineralization rates. The water is mainly HCO_3^--Cl^- Ca^{2+}-Mg^{2+} type. The aim of this paper is to apply a non parametric geostatistical methodology which allows the construction of robust probabilistic risk maps, showing the probability of the concentration of any parameter exceeding an indicator threshold value. These maps are very important tools for decision makers, resulting in the integration of water quality and potential water use, which is very important in zones of scarce groundwater resources.

1 INTRODUCTION

The hydrogeological studies in order to define potential drilling zones for human, industrial and agricultural supplies, usually plays with the uncertainty of the quantity and quality of groundwater. An alternative approach is to map this variables by interpolation from sample locations. Geostatistical techniques such as ordinary kriging, indicator kriging, multigaussian kriging, conditional simulation (Journel 1983, Journel & Isaaks 1984, Johnson & Dreiss 1989, Deutsch & Journel 1992, Wingle & Poeter 1996), are a host of spatial data exploratory methods and robust tools that can be used for this purpose.

In this study we are interested in creating risk maps, with low costs, depicting the current distribution of electrical conductivity, as a groundwater quality indicator of mineralization, in a rural region of Chança, near the border of Tejo river basin. Here the grade of groundwater mineralization has a major influence on groundwater and land use (Chambel & Duque 1999). This maps can show a quantifiable risk assessment, through probabilities of this hydrogeological variable, which is very important for groundwater resources management.

2 GEOLOGICAL FRAMEWORK

This is a very flat region controlled by the Tejo river. Geologically the study area are in the East side of the Tejo river basin, which contacts with diverse Hercynian rocks (Fig. 1). These rocks belong to the geostructural Ossa-Morena Zone (OMZ), and are represented by some granitic rocks and volcano-sedimentary rocks of Precambrian to Ordovician age (granodiorites, gneissic-granites, granites, micaschists, amphibolites, marbles, schists and greywackes). These rocks were subject to intense metamorphic, igneous and tectonic activity.

The Tertiary sediments of Tejo river are discordant above the hard rock Hercynian substrate and become very thin to the exterior of the basin: Miocene to Pleistocene - sandstones, sands, and conglomerates; Paleocene - conglomerates, limestones and clays.

3 HYDROGEOLOGICAL FEATURES

The study area is located in a typical rural zone where the main human activities are related to agriculture, namely cattle breeding, small irrigation farms and production of cork trees. The region has a typical Mediterranean climate and the annual precipitation is around 650 mm/year (Chambel et al. 1998).

The hydrogeological knowledge about this region is very sparse and are related to some local studies for implementation of groundwater public supplies. Chambel & Duque 1997 gave the first approach to the hydrogeological characterization.

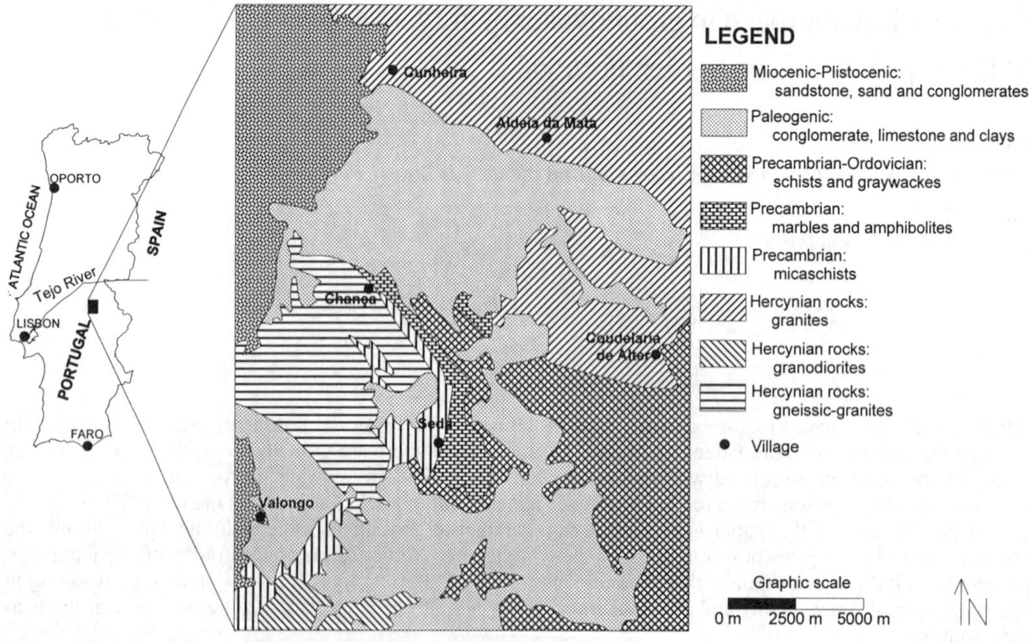

Figure 1. Map of the region under research and their geological features.

There are in the study area two kinds of aquifers: sedimentary and hard rock aquifers. Unfortunately all of them have low productivity, usually less than 1 L/s per production well.

Sometimes in the sedimentary aquifers is possible to get yields about 3 L/s from gravel and carbonate levels. This is what happen in the wells around Chança. The same yields are possible in hard rocks aquifers, specially when the geostructural context is favourable, with high grade of fracturing and large deepness of rock weathering. It is possible to increase these well yields when they cut a sedimentary level with a gravel base plus hard rock with high grade of weathering and deep fractures, such as the south of Aldeia da Mata wells, where the yields achieve about 5-6 L/s and transmissivity are around 50 m^2/day.

In all the region the groundwater is very hard and highly mineralized, except in the north and northeast in granitic rocks. There are some wells in the sedimentary rocks with excessive chloride (sometimes more than 200 mg/L). The major groundwater type in the region is HCO_3^--Cl^- Ca^{2+}-Mg^{2+} (Chambel & Duque 1997).

4 GEOSTATISTICAL SCOPE

4.1 *Introduction*

The praxis in Geostatistics encompasses in general two interrelated steps: the structural analysis and kriging simulation (Ribeiro et al., 1997a).

The structural analysis of the hydrogeologic variables are defined by a function $\gamma(h)$, the variogram (Davis 1986). A variogram is a graph describing the expected difference in value between pairs of samples with a given relative orientation (Clarke, 1979). The variograms are a useful means of summarizing the spatial variability of a variable (Coombes 1997). The basic philosophy behind a variogram is that, at a particular orientation within a given domain, they measure the difference between concentration values relative to the distance separating the concentration sample locations, this procedure can be used with any variable.

Deterministic interpolation techniques do not provide any measure of the reliability of the estimate (Journel 1983).

Kriging is a powerful tool which can be used to determine the concentration of a contaminant or any chemical parameter based on sampling from the surface. The procedures are designed to characterize and model the spatial relationships from sample data, then use the model to estimate values between sample observations. It also provides information on the uncertainty associated with each estimate through an error variance, which is minimized and attached to the estimate (Tarboton et al. 1995).

The uncertainty about any unknown value is displayed by a probability distribution estimated using indicator kriging (Tarboton et al. 1995). This nonparametric method is not dependent of any statistical distribution of the variable and can be applied in contaminant problems (Journel 1983, Bilonick 1984, Ribeiro et al. 1997a, Ribeiro et al. 1997b, Armstrong 1998, Atkinson et al. 1998).

So, indicator kriging determines, by using the samples in the neighborhood, the probability of data values in a given area being grater than a defined threshold value (Isaaks & Srivastava, 1989). Here the data values are transformed into indicator values: original values, which exceed the chosen threshold value, are coded 1, and those below the threshold value are coded 0. These indicators are then analyzed to determine their spatial directional variability with a series of experimental variograms.

4.2 *Data and univariate analysis*

In 1997 it was carried on a general hydrogeological inventory in 366 wells (84 deep wells, 61 springs and 221 large wells). Some field parameters were measured: pH, electrical conductivity, redox potential and temperature. Here, we consider further the electrical conductivity as an mineralization indicator. In this area it seems that deepness don't affect the mineralization content in the same kind of rock. The differences in electrical conductivity in a large well near a deep well only have a standard difference of 20 μS/cm. So, all the inventory data was used in the geostatistical approach methodology.

The 366 data points are distributed randomly in the study area of 18 per 24 km (432 km^2), which corresponds almost to 1 well per 1 km^2 (Fig. 2). The univariate statistics for electrical conductivity are represented in the Table 1. The electrical conductivity histogram in the figure 3 is very asymmetric. It has an asymmetry coefficient of 3.54 μS/cm, and have approximately a log-normally distribution. The principal data class are between 0 and 1000 μS/cm.

Table 1. Univariate statistics of electrical conductivity.

Parameter	Value
Samples	366
Mean	636 μS/cm
Median	627 μS/cm
Minimum	30 μS/cm
Maximum	5823 μS/cm
1. Quartile	233 μS/cm
3. Quartile	835 μS/cm
Variance	321530 μS/cm
Asymmetry	3.54 μS/cm

4.3 *Indicator structural and kriging analysis*

The risk maps were developed considering two different indicator threshold values of electrical conductivity. The chosen threshold values should take in account some statistical significance for all data

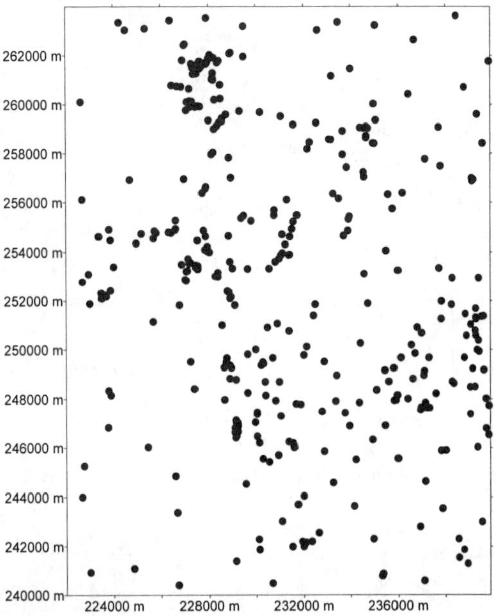

Figure 2. Distribution of the sample data set.

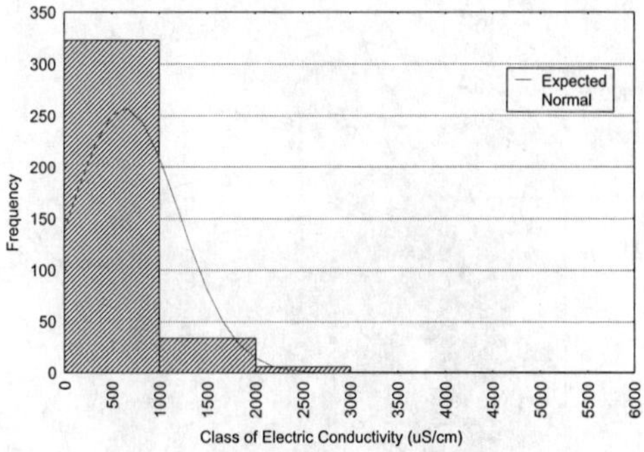

Figure 3. Histogram of the data set.

set. The first risk map have a indicator threshold value which corresponds to the sample median (627 μS/cm), and the indicator threshold for the second risk map corresponds to the third quartile which is 835 μS/cm. All the samples then were coded 1 when exceed these threshold values and 0 when they are below them resulting in two data sets.

A variogram structural analysis was performed, considering a hide range of orientations for the two data sets. The Figure 4 shows the chosen variograms.

Considering the first risk map the variogram analysis revels no significant geometric anisotropy for electrical conductivity. The omnidirectional variogram were modelled by a spherical function

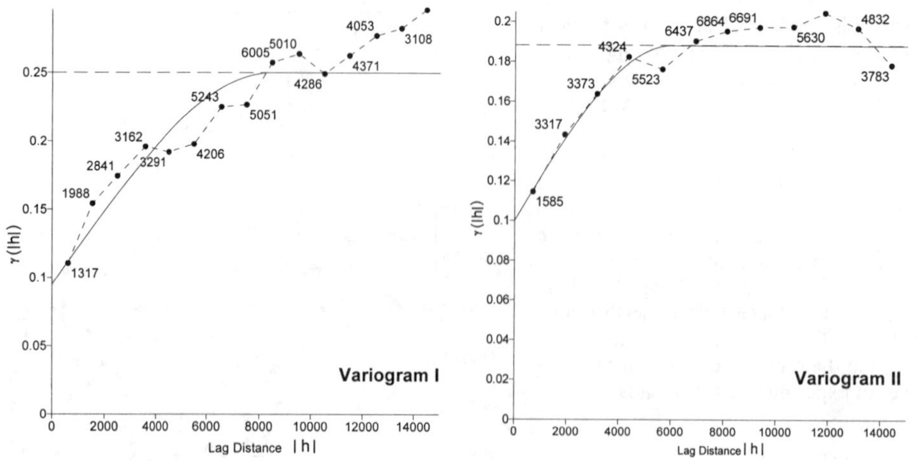

Figure 4. Indicator variograms I and II and fitted models use in kriging.

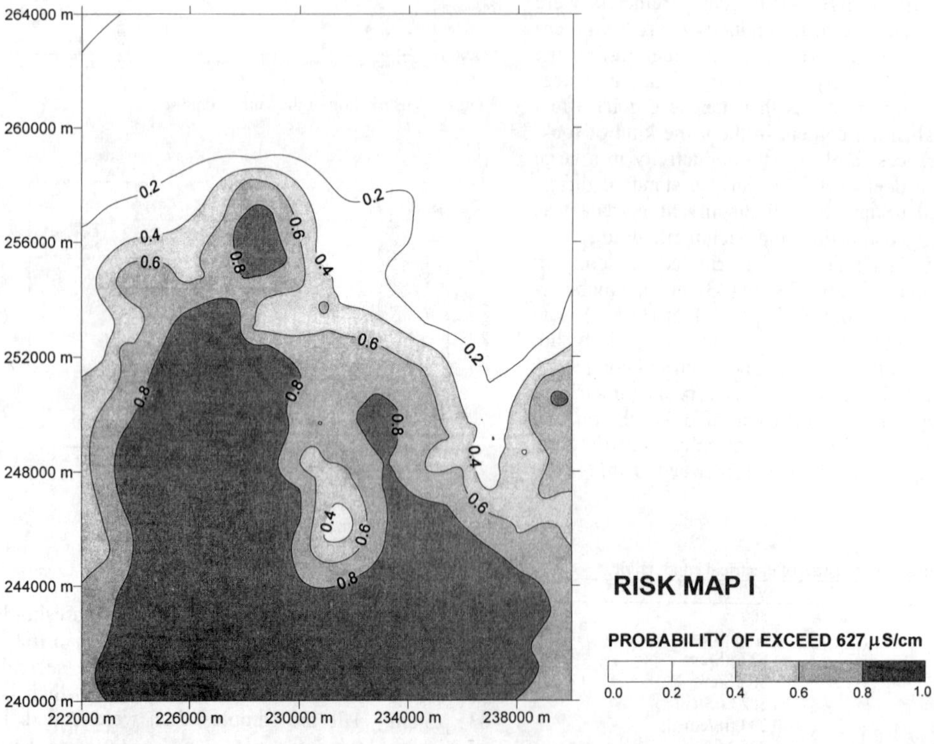

Figure 5. Risk map I which represents the probability of the electrical conductivity on groundwater exceed the indicator threshold value of 627 μS/cm.

considering 17 lags. The lag width was 1000 m, which corresponds to the sample density (1 sample per 1 km^2). The nugget effect is 9.5% of the total variance.

Also the structural analysis for the second risk map shows no significant geometric anisotropy. The omnidirectional variogram was modeled by a spherical function, considering 15 lags with 1000 m width. The nugget effect is 8% of the total variance.

The indicator kriging 2D simulations used for both risk maps were performed through a 100x100 m grid.

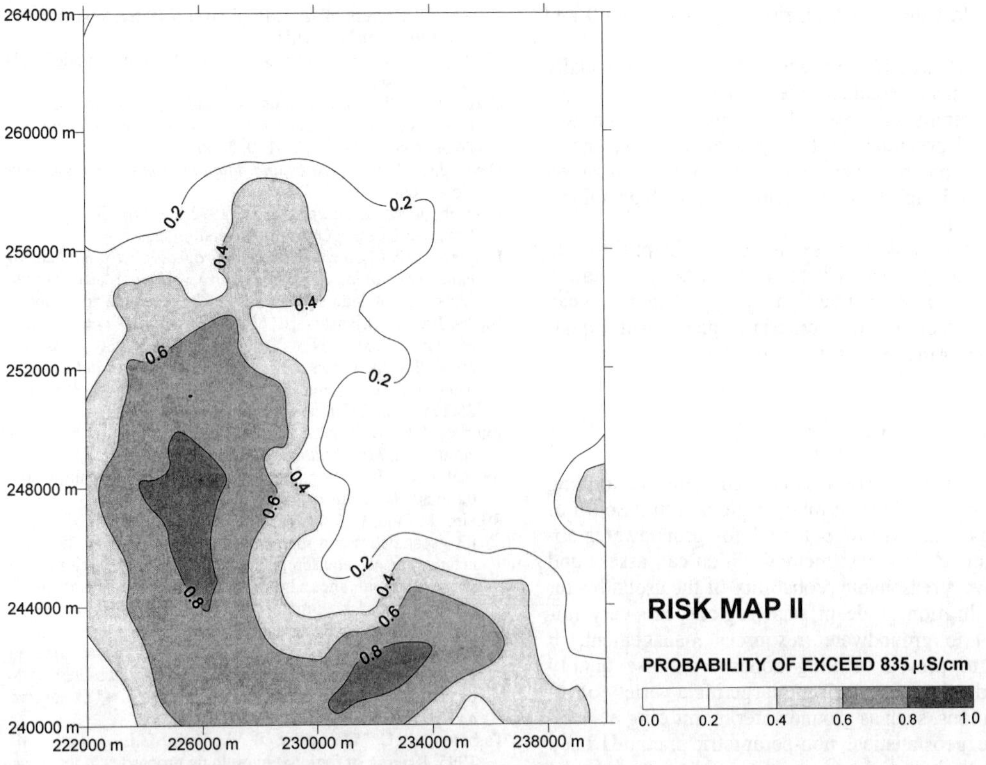

Figure 6. Risk map II which represents the probability of the electrical conductivity on groundwater exceed the indicator threshold value of 627 μS/cm.

The Figure 5 and 6 represents the risk maps I and II. The Table 2 shows the summary of the two fitted model parameters used in the kriging simulations which were used respectively in the risk map I and II. The risk maps I and II give two levels of information related to the probability of occurrence of certain mineralization grade in groundwater.

Table 2. Fitted models parameters.

Parameters	Risk map I	Risk map II
Nugget effect - c_o	0.095	0.1
Sill – c_1	0.155	0.08
Range - a_1 (m)	8500	6000

Combining the geological information with the risk maps is possible to understand that high values of electrical conductivity are related with the Palaeogenic sedimentary rocks, namely conglomerate, limestone and clays. These rocks are formally named as Sedimentary rocks of Vale do Guiso, which belongs to the Tejo river basin.

The signature of these highly mineralized waters is visible in the risk maps. It is almost possible to depict the border limit of the Tejo river basin. Recent studies shows that mineralization are highly related with chloride concentrations which probably coming from natural contamination of Tejo river in the past (Chambel & Duque 1997).

Moreover natural contamination, groundwater in sedimentary rocks, seems to show high residence times. The occurrence of large thickness of clay levels and the influence of low slops of topography reduces the groundwater flow, allowing more time for water-rock reactions, which will increment of groundwater mineralization.

The use of these highly mineralized waters in human supply have an health negative impact through renal diseases, which are very common. However, this groundwater is plenty used for cattle drinking and irrigation.

Rural people usually drink waters collected in several springs in the northern granitic massif around Cunheira and Aldeia da Mata, in the Miocenic sandstones near Valongo and in the western area of Cunheira which have very low mineralization content and are far from contamination sources.

This human water supply, only for drinking, is very common in almost rural zones of south Portugal (Duque 1997).

Almost all granitic and Miocenic rocks on the northern and western part of the map of Figure 1 have low probability of exceeding the indicator threshold values for both risk maps (between 0 and 20%).

The schists and greywackes (East of Seda) usually have electrical conductivities above 800 μS/cm.

Combining the probability maps with groundwater yield potential of the regional aquifers, shows that low potential aquifers have low mineralization levels and high potential aquifers have high mineralization levels.

However, risk maps seems to be a helpful tool to define quality zones where groundwater exploitation with low quality can be firstly avoided, and also can be used to define some criteria of groundwater quality monitoring networks.

5 CONCLUSIONS

The financial costs associated with abandoning wells are very high in many regions in the South of Portugal, due to low potential for groundwater abstraction. A low cost method which can assess and can give a reasonable probability of the groundwater mineralization grade in such regions, are very important to groundwater resources management, in order to prevent drilling zones with low quality groundwater, or give rise to targeting a variety of interventions, such as groundwater monitoring plans.

The geostatistical non-parametric methods used in this study in order to produce risk maps, gives the level of probability of the hydrogeological variable concentration in a certain place of the aquifers, i.e. it is possible to have a quick groundwater quality probability, which will allow the decision maker to have a certain reliability in decisions taken.

REFERENCES

Armstrong, M. 1998. *Basic linear geostatistics*. Berlin: Springer Verlag.

Atkinson, P.M. & Lloyd, C. D. 1998. Mapping precipitation in Switzerland with ordinary and indicator kriging. *Journal of Geographic Information and Decision Analysis* 2(2): 72-86.

Bilonick, R.A. 1984. Risk qualified maps for a long term sparse network – mapping the USGS New York acid precipitation data. In G. Verley (ed.), *Geostatistics for natural resources characterization*: Part 2, 851-862. D. Reidel Publishing Company.

Chambel, A. & Duque, J. 1997. Caracterização hidroquímica das águas subterrâneas do sector oeste do concelho de Alter do Chão. *4° Congresso da Água*: CD-74. Lisboa: APRH.

Chambel, A. Duque, J. & Fialho, A. 1998. Groundwater in a semi arid area of south Portugal. In J. Brahana, Y. Eckstein, L. Ongley, R. Schneider, & J. Moore (eds.) Gambling with Groundwater-Physical, Chemical and Biological Aspects of Stream Relations - Proc. XXVIII Congress of IAH, Las Vegas, 5-9 September 1998: 75-80. Las Vegas: IAH.

Chambel, A. & Duque, J. 1999. Hard rock aquifers of Alentejo Region (South Portugal) – Contribution to the water and land use management. In M. Fendeková & M. Fendek (eds), *Hydrogeology and land use management* - Proc. XXIX Congress of IAH, Bratislava, 6-10 September 1999: 171-176. Bratislava: IAH.

Clarke, I. 1979. *Practical Geostatistics*. London: Applied Science Publishers.

Coombs, J. 1997. Handy hints for variography. *Proc. AusIMM Ironmaking Resources and Reserves Conference, Perth, 25-26 September 1997:* 127-130. Perth: AIMM.

Davis, J.C. 1986. *Statistics and data analysis in geology*. New York: John Wiley & Sons.

Deutsch, C.V. & Journel, A.G. 1992. *GSLIB-Geostatistical Software Library*. Oxford: University Press

Duque, J. 1997. *Caracterização hidrogeológica e modelação matemática do aquífero dos gabros de Beja*. Master Thesis. Lisboa. Faculdade de Ciências da Universidade de Lisboa.

Isaaks, E.H. & Srivastava, R.M. 1989. *An introduction to applied geostatistics*. New York: Oxford University Press.

Johnson, N.M. & Dreiss, S. J. 1989. Hydrostatic interpretation using indicator geostatistics. *Water Resources Research* 25(12): 2501-2510.

Journel, A.G. & Isaaks, E. H. 1984. Conditional indicator simulation. *Mathematical Geology* 16(7): 685-718.

Journel, A.G. 1983. Nonparametric estimation of spatial distributions. *Mathematical Geology* 15(3): 445-468.

Ribeiro, L. Pina, P. & Muge, F. 1997a. Contribution of indicator geostatistics and mathematical morphology to the characterization of aquifer heterogeneities in the vicinities of waste disposal sites. In Marinos, Koukis, Tsiambaos & Stournaras (eds), *Engineering geology and the environment*: 2127-2132. Rotterdam: Balkema.

Ribeiro, L. Santos, A.C. & Silva, J.D. 1997b. Caracterização geoestatística da contaminação salina por dissolução de evaporitos – Um caso na zona da Maragota. *4° Congresso da Água*: CD-69. Lisboa: APRH.

Tarboton, K. C., Wallender, W.W., Fogg, G.E. & Belitz, K. 1995. Kriging of regional hydrologic properties in the western San Joaquin Valley. *Hydrogeology Journal* 3(1): 5-23.

Wingle, W.L. & Poeter E.P. 1996. Evaluating subsurface uncertainty using zonal kriging. *Uncertainty 96 (ASCE). University of Wisconsin, Madison-Wisconsin, 1-3 August 1996*: 1-15. Madison: ASCE

6 Groundwater and mining

Topic editor:

Kim Morton
KLM Consultant, South Africa

Groundwater: Past Achievements and Future Challenges, Sililo et al. (eds) © 2000 Balkema, Rotterdam, ISBN 90 5809 159 7

Hydrogeochemical tools to define sources of ground-water inflow to surface and underground mines

D.A. Bird & J.J. Mahoney
Hydrologic Consultants Incorporated, Lakewood, Colo., USA

ABSTRACT: Chemical analyses of ground water, surface water, and precipitation at surface and underground mining sites often will reveal spatial or temporal patterns that can help determine, and sometimes quantify, the sources of ground-water inflows to the mine. Methods of data evaluation range from very simple analyses of major ion ratios, using graphical analytical tools such as trilinear or fingerprint diagrams, up to more complex analyses such as isotope geochemistry. Waters derived due to contact with sulphide-bearing orebodies typically have a chemical signature that distinguishes them from local background waters. Concentrations of sulphate, iron, aluminium, and other trace elements are typically anomalous compared to background. Mass balance analyses using such ions can help quantify contributions from different water sources. Waters derived from other types of mining environments can also show anomalous chemistry, because mineral deposits are, in themselves, geologic anomalies, and waters derived from them should also be anomalous compared to background. Stable isotopic data from oxygen, deuterium, tritium, sulphur, nitrogen, and carbon can prove useful in segregating waters of different origin. Analyses of major ions or trace elements over time can indicate the time at which mine-water inflow due to ground-water storage gives way to inflow due to recharge, which proves valuable in dewatering plans. Case histories are presented in which analyses of major ions, trace elements, stable isotopes and other tools have been applied to help differentiate sources of ground-water inflows to the mines, and help design cost-effective mine dewatering systems.

1 INTRODUCTION

Hydrogeochemistry is often overlooked as a tool in hydrologic investigations at mine sites, such as mine dewatering programs. Ground water entering surface or underground mines often can be divided into different chemically distinct types depending on various factors. Such factors include: 1) the lithology/mineralogy through which the ground water has flowed and reacted before entering the mine, 2) the residence time of the water on ground surface before infiltration to the subsurface, 3) elevation of recharge of the ground water, and 4) age of the ground water. This paper presents various hydrogeochemical analytical methods by which these factors can be evaluated for the purpose of supporting hydrogeologic investigations. The operational benefits, such as cost savings for mine dewatering, can be substantial.

Table 1 shows three chemical analyses from waters at actual mining sites. Water A is a mountain spring less than a kilometer up-gradient from a waste disposal area. Water B is surface water draining an acid-rock drainage environment, and Water C is a shallow ground water from a desert environment with high evaporation relative to precipitation. Table 1 demonstrates the diversity of water chemistry that can exist at mining sites. This chemical diversity is what allows hydrogeochemists, using the tools described here, to determine the sources of various waters entering the mining facility.

Table 1. Selected analytical data from three chemically distinct waters (mg/L).

Sample	Water A	Water B	Water C
Constituent			
pH	7.4	3.6	8.2
Calcium	26	1131	480
Magnesium	2	474	660
Sodium	0.1	30	3080
Potassium	0.04	26	80
Alkalinity	74	U	320
Sulphate	4	9495	9300
Chloride	U	4	1200
Aluminum	0.03	869	2
Iron	0.003	363	U

U = undetected

2 METHODS OF DATA EVALUATION

2.1 *Trilinear (Piper) Diagrams*

Trilinear diagrams (Figure 1), also known as Piper diagrams, are probably the most commonly used graphical presentation tool used in water chemistry investigations. The relative amounts, as converted to equivalents, of the major cations (Na^+, K^+, Ca^{2+}, Mg^{2+}) and anions (HCO_3^-, CO_3^{2-}, Cl^-, SO_4^{2-}) in a given water sample are plotted in their respective triangular field and projected to the upper diamond-shaped (quadrilateral) field. Different water types will plot in different areas, and both temporal and spatial trends can be identified with the plots. For example, Figure 1 shows three different groups of water samples from three distinct water-bearing formations. The waters depicted by triangles were collected from siltstones; the waters display approximately equal proportions of the different cations and anions. The waters depicted by the squares are from carbonate rocks, and contain no-

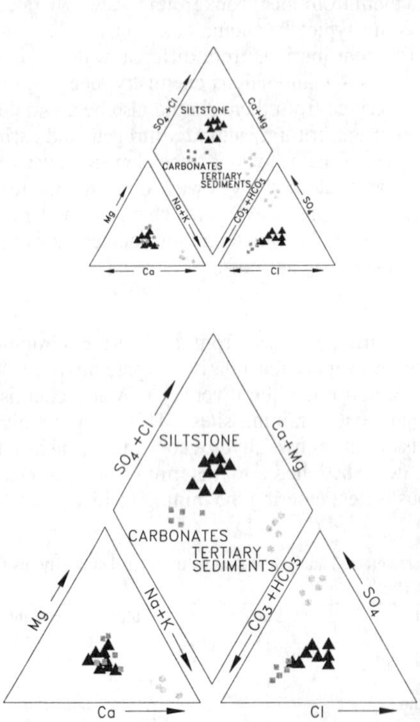

Figure 1. Trilinear Diagram

Piper diagrams provide more than a simple method to classify water types. They can be used to identify changes in composition as a function of time. Changes can be detected as the compositions of selected samples "march" through the diamond field of the diagram. Changing compositions may reflect pumping of waters from different zones, or the appearance of another water source. These sources may include water from tailings and waste rock piles.

For waste rock piles, changes in water chemistry can often be identified by increases in sulphate concentrations. Waters with increasing sulphate concentrations will "migrate" to the top of the quadrilateral. Furthermore, simple mixing calculations can be performed through use of the lever rule. Piper diagrams can also be used to identify the early release of tailings waters. At a mine in central Nevada (USA), the tailings contained cyanide that had been neutralized using an alkaline chlorination process (Filipek 1999). The resulting tailings water had elevated chloride concentrations. Slight increases in chloride concentrations provided early identification of a potential problem.

Piper diagrams can also help identify the presence of ion exchange reactions. In the case of a cation exchange process, for example in bentonite rich soils, it may happen that even though the cation and anion concentrations increase significantly, the proportions of cations may remain the same. All of the cation concentrations will remain on the same point in the cation triangle, whereas data points in the anion triangle may show a high degree of variability.

2.2 *Stiff Diagrams*

Stiff diagrams are another commonly used graphical presentation of the major cations and anions in a water sample. Stiff diagrams plot major cations and anion concentrations in terms of equivalents, or more usually milliequivalents, per liter. Cations are plotted on the left side, and anions are plotted on the right hand side of the diagram. Because the diagrams are based upon equivalents, the figures provide a quick check on the quality of the analyses. Charge balance requires that the solution must be electrically neutral. Consequently, the area of the left of the centerline (cations) should be approximately equal to the area to the right of the centerline (anions).

The utility of Stiff diagrams lies in the characteristic shape formed by the chemical composition of a given water. Figure 2 shows Stiff diagrams for waters from four monitoring wells and one pumping well. The diagrams suggest that water from monitoring well MW-161, and to a lesser extent MW-630, has a similar origin as pumping well PW-1.

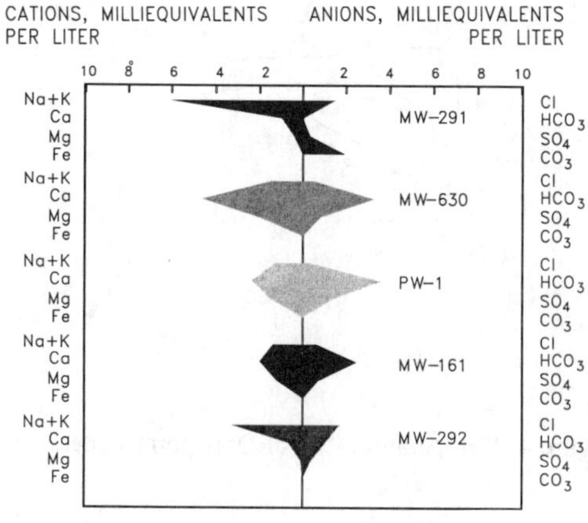

Figure 2. Stiff Diagrams

2.3 *Fingerprint Diagrams*

Fingerprint diagrams present the absolute concentrations of a suite of analytes. Again, the shapes of the plots reveal the similarities and differences of various water samples. Figure 3 plots analytical data from a mining site. The concentrations of cations, and the ratios of cations to each other, are similar among different waters, but the anions and trace elements show wide variability.

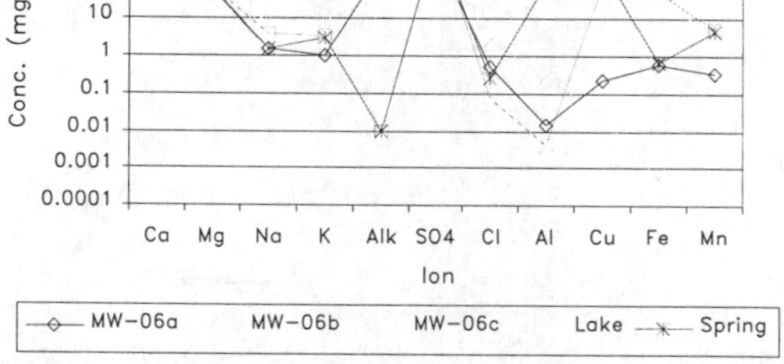

Figure 3. Fingerprint Diagram

The high sulphate and calcium concentrations are consistent with an origin form an acid-rock drainage environment. Because of the dominance that calcium and sulphate would have over their respective counterparts, methods depending upon ratios, such as Trilinear diagrams are often less useful for analyses of mine waters than methods such as fingerprint diagrams that plot absolute concentrations. This particular diagram illustrates the variability often seen in trace metals such Al, Cu, and Fe. Attenuation processes (e.g., precipitation and adsorption) downgradient from the source of the metals can significantly reduce the concentrations of the metals in solution.

2.4 *Isotopic Analyses*

The isotopic composition of oxygen and hydrogen in ground water varies depending on such factors as elevation of recharge, temperature at the recharge site, and distance from the ocean. The proportions of heavy isotopes of hydrogen (deuterium or D), and oxygen (^{18}O) tend to decrease at higher elevations. Therefore, precipitation that falls in the mountain ranges will produce waters with lighter isotopic signatures than precipitation that fell at a lower elevation. In the Basin and Range of Nevada, isotopic signatures are commonly used to distinguish waters from the near surface local system from the deeper regional aquifers that commonly are recharged in the mountains.

Because in most near surface environments, isotopic exchange reactions involving water molecules in minerals do not occur, the stable isotope ratios (represented by the symbol δ or del) for deuterium and ^{18}O will only change as a result of mixing with waters of a different isotopic signature. Barring evaporation, the isotopic signatures in near surface ground waters are essentially conserved. Therefore, hydrogen and oxygen stable isotopic ratios provide a reliable means to calculate extent of mixing.

Similarly, the isotopic compositions of sulfur, carbon, and nitrogen can evolve as ground water moves, reacts with rock, and acquires dissolved solids and other elements dissolved in the water. The isotopic composition, as well as the major and trace element chemistry, will evolve to reflect the geochemical environment through which the water moves. Differences in isotopic composition allow waters to be separated by hydrogeologic origin.

2.5 *Geochemical Models*

Geochemical models provide another set of quantitative tools to evaluate the sources of waters. Mass balance models such as *NETPATH* (Plummer et al. 1991) provide a means to calculate water proportions using a variety of possible mixing models. In a typical calculation, waters are selected as possible end members and mixtures. If a water is formed by the mixing of two other end member waters, the mixing proportions, based upon the behavior of conservative components, should be similar for all of the conservative components. *NETPATH* calculates possible mixtures and allows for mineral precipitation and dissolution. Usually, mixing proportions are calculated based on the conservative components in the solution. Chloride is the most common conservative component, although the program can use ratios of deuterium and oxygen isotopes to calculate mixture proportions. The effects of evaporation and dilution can also be included in these models.

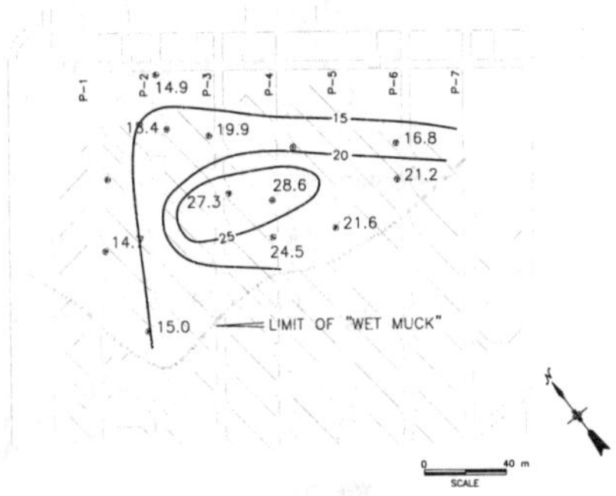

Figure 4. Temperature (°C) of Drawpoint Waters

3 CASE HISTORIES

3.1 *IOZ Block Cave Mine*

At PT Freeport Indonesia's Intermediate Ore Zone (IOZ) block cave mine in Irian Jaya, Indonesia, hydrogeochemical data were used to confirm that ground water was entering the cave not only via direct precipitation above the cave, but also via lateral ground-water inflow from adjacent in-situ country rock (Bird et al. 1998). Excess water in the block cave was believed responsible for the unplanned release of fine-grained material from certain drawpoints. These muck rushes were considered a hazard, and control measures were required to reduce the incidence of these muck rushes. Samples of water discharging from seeps at individual drawpoints on the extraction level of the mine revealed graded magnitudes of various parameters. Figures 4 through 6 show how temperature and the concentrations of sulphate and dissolved oxygen vary with distance from the northwestern edge of the mine workings.

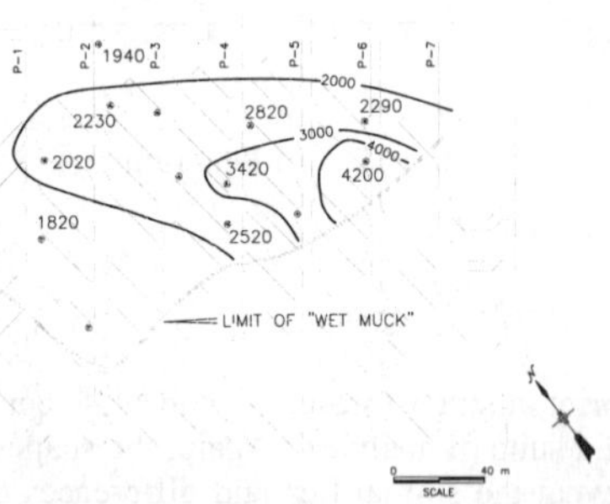

Figure 5. Sulphate Concentrations (mg/L) in Drawpoint Waters

A trilinear diagram (Figure 7) indicated a bimodal signature of ground waters entering the system through the dewatering drillholes. The data point labeled W on Figure 7 represents the average of a group of 13 westerly directed drillholes, and the data point labeled E represents the average of 7 easterly directed drillholes. The data points R and D are from rainfall and the extraction level drawpoints, respectively. The drawpoint water more closely resembles the western drillhole group, which was interpreted to be less evolved ground water than the E group. The E group is presumed to represent more geochemically evolved ground water, due to more rock-water interaction. This graph helped add to the body of evidence that two distinct water sources were entering the cave. Geochemical modeling using NETPATH (Plummer et al., 1991) indicated that lateral ground-water inflows were as much as 50% of total mine water inflows.

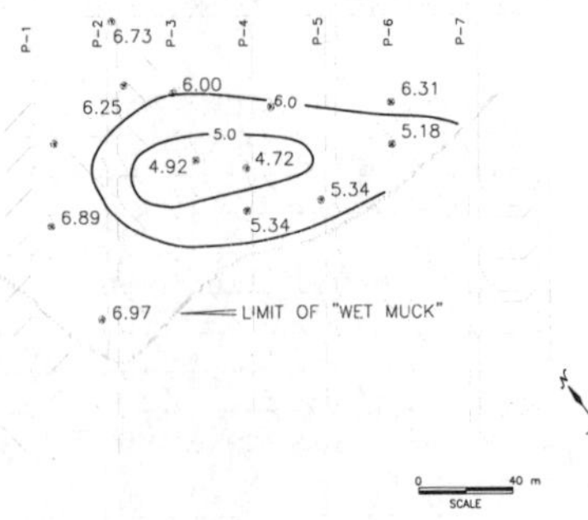

Figure 6. Dissolved Oxygen Concentrations (mg/L) in Drawpoint Waters

Armed with the knowledge that lateral ground water inflow was a significant source of the total inflow, PTFI undertook efforts to reduce that component of water through increased dewatering efforts. A geologic mapping program ultimately identified several structures (faults) to the north as new targets for dewatering drilling. The new program intercepted a source that produced water at an initial flow rate of 900 liters per minute (equivalent to 15 L/s) in one drillhole, which at the time was the largest flow rate in the IOZ mine dewatering drilling program to date. Intercepting this water in the dewatering program essentially removed it from the reservoir of ground water that could potentially enter the cave.

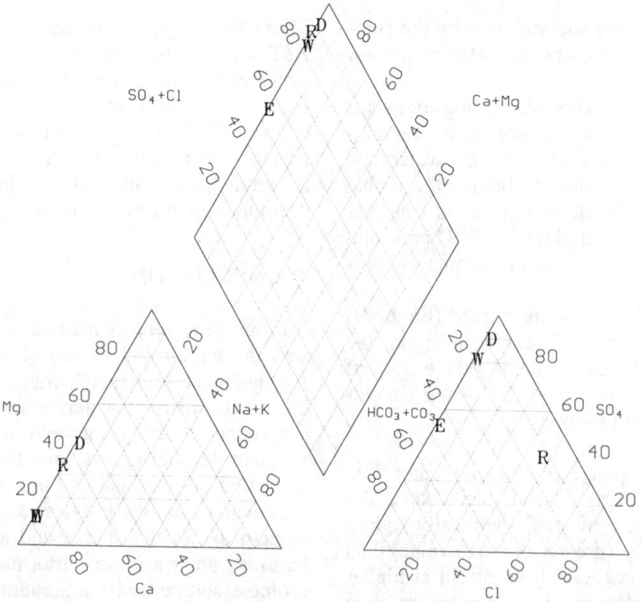

Figure 7. Piper Diagram of Waters Near the IOZ

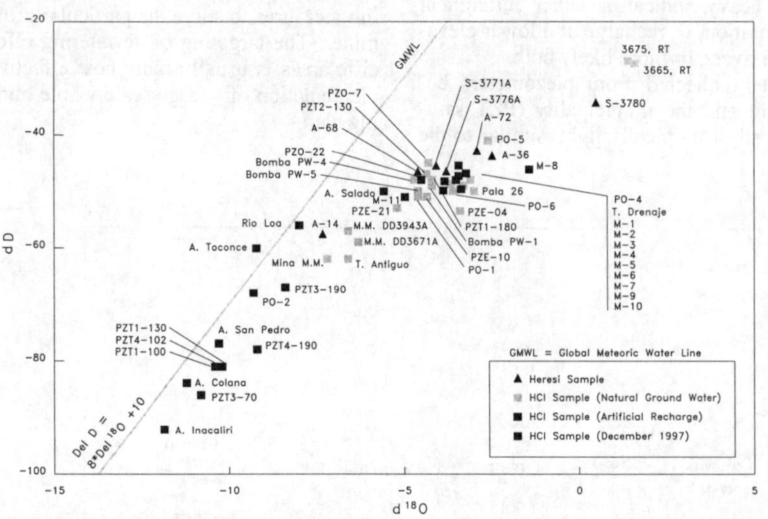

Figure 8. Isotopic Composition of Surface and Ground-Water Samples

3.2 *Chuquicamata open pit copper mine*

Despite its location in one of the driest regions on the planet, the Chuquicamata copper mine in Chile requires dewatering due to ground-water seepage into the pit highwalls (Bird et al. 1999). A hydrogeochemical sampling program was conducted to determine the source of the water so that it could be diverted before entering the pit. Common sense suggested that the water source was from the ore processing facilities and municipality located adjacent to the pit. The seeps are perched well above the regional water table, so the regional ground water was ruled out. The nearest natural surface water is from a group of rivers 80 to 120 km away, eliminating them as a source. However, water from these rivers is piped to the mine site and used by the company in mineral processing and as water supply for the encampment.

Figure 8 plots the $\delta^{18}O$ vs δD compositions for various ground-water and surface-water samples from the site. The figure shows that the waters are generally parallel to, but slightly below, the global meteoric water line. The deviation due to evaporation increases with distance along the $\delta^{18}O$ axis, and becomes significant as the waters become isotopically heavy.

When only the $\delta^{18}O$ values are plotted (Figure 9) as a function of the samples' relative location, the data reveal a pattern that helps identify the origins of the different ground waters. The following conclusions can be drawn from Figure 9:

1) The source waters (i.e., rivers) for all water used at the mine (Inacaliri, Colana, San Pedro, Toconce, and Loa) are mostly isotopically lighter with respect to $\delta^{18}O$ (greater negative value) than the other waters, suggesting an origin at higher elevations, and relatively less evaporation than the other waters.
2) Samples from observation (PO) and pumping wells (PW), which tap deep ground water, are isotopically heavy, indicating either enrichment due to evaporation, or recharge at a lower elevation than the rivers, or more likely both.
3) Ground water collected from piezometers between the pit and the municipality (PZT samples) are mostly isotopically light, similar to the rivers.

These data suggest that the source of water in the PZT wells is the same source as the municipality's water supply, namely the aqueducts conveying water from the rivers Inacaliri, Colana, San Pedro, Toconce, and Loa. We concluded that the shallow ground water adjacent to the pit was derived from the municipality, and a dewatering program has been developed to intercept this water.

4 CONCLUSIONS

Various geochemical methods are available to assess the nature and source of waters near mines. An understanding that a mine may have several chemically distinct waters entering the pit or underground workings can help in designing a more efficient dewatering system. The two case studies presented in this paper have demonstrated that chemical variations in water samples can be used to help assess the source and amounts of waters that may enter a mine. Information about the hydrologic source and the amount of water is commonly required because the additional water may contribute to geotechnical problems such as slope failure or muck rushes. Identification of the source area permits implementation of specific dewatering measures to solve the particular problem at the mine. The targeting of dewatering efforts to specific areas is usually more cost effective than implementation of a large scale whole mine dewatering plan.

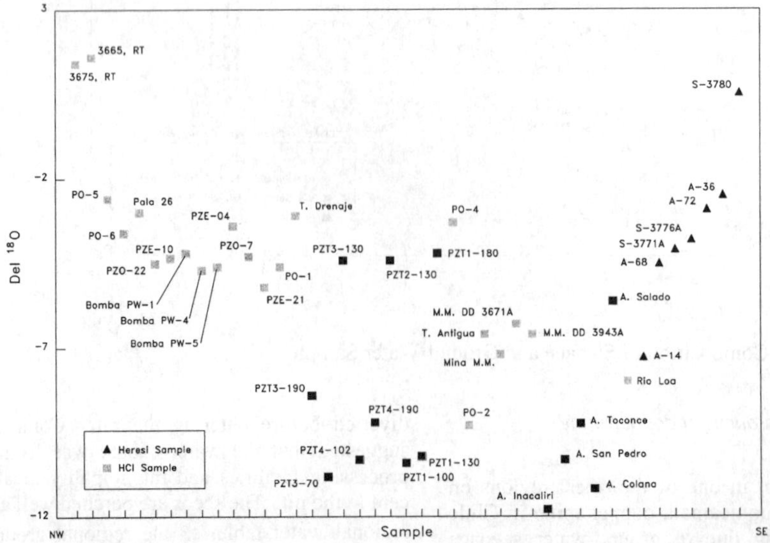

Figure 9. $\delta^{18}O$ in Water Samples

REFERENCES

Bird, D.A., D.C. Flint, and N. Pearson 1998. Identification of sources and interception of water inflow at P.T. Freeport Indonesia's IOZ block cave mine. Presentation at 1998 Annual Meeting of the Society for Mining, Metallurgy, and Exploration, Orlando, Florida, USA, March 9-11, 1998, SME Preprint No. 98-81.

Bird, D.A., G. Leone, and G. Flores 1999. Verification of sources of ground-water inflows at Chuquicamata mine using major ion and stable isotope data. Presentation at 1999 Annual Meeting of the Society for Mining, Metallurgy, and Exploration, Denver, Colorado, USA, March 1-3, 1999.

Filipek L.H., 1999, Determination of the source and pathway of cyanide-bearing mine water seepage, *in* Filipek, L.H., and Plumlee G.S. (eds.) T*he Environmental Geochemistry of Mineral Deposits Part B: Case Studies and Research Topics. Reviews in Economic Geology Volume 6B*. p549-565. Society of Economic Geologists, Inc. Littleton, CO.

Plummer, L.N., E.C, Prestemon and D.L. Parkhurst 1991. An interactive code (NETPATH) for modeling NET geochemical reactions along a flow PATH. *U.S. Geological Survey Water-Resources Investigations Report 91-4078*. 227 p.

Groundwater: Past Achievements and Future Challenges, Sililo et al. (eds) © 2000 Balkema, Rotterdam, ISBN 90 5809 159 7

Hydrogeological and geochemical investigations of an acidic mining lake as a basis for a biotechnological remediation

E. Bozau, K. Knöller & G. Strauch
Department of Hydrogeology, UFZ Centre for Environmental Research, Leipzig-Halle, Germany

ABSTRACT: As a part of the research project "Remediation of organically and inorganically polluted ground water and surface water" the acidic mining lake 111 (situated in Eastern Germany) has been investigated to assess the possibilities of a successful biotechnological remediation. These investigations include a detailed hydrogeological study of the whole catchment area. Ground, seepage and lake water, as well as the surrounding sediments have been examined to model the hydrodynamic processes and the geochemical development of the lake. The isotope signatures of the various waters have also been analysed.

1 INTRODUCTION

Coal mining has caused significant changes to the landscape of the Lower Lusatia (Germany). After the coal exploitation ended, many mining lakes were formed by ascending ground water. The oxidation of pyrite occurring in dump sediments exposed to atmospheric oxygen is catalysed by bacteria (*Thiobacillus thiooxidans*) and can be described by the following simplified reaction:

$$FeS_2 + 3\frac{1}{2}\,O_2 + H_2O \rightarrow Fe^{2+} + 2\,SO_4^{2-} + 2\,H^+$$

The contact of ground water with these sediments leads to the development of acidic mining lakes. These lakes with pH-values of about 3 can be used neither as a drinking water reservoir nor for recreation activities. The biological diversity is limited due to the unfavourable living conditions. More than 100 lakes are affected by acidification in Lusatia (Geller et al. 1998). Several research projects deal with the chemical and ecological state of such mining lakes to predict the possibilities for a natural or an active initiated restoration.

The lake chosen for our investigations (Acidic Mining Lake 111, figure 1) has existed for more than 40 years. It is characterised by a pH-value of 2.6. A biotechnological remediation which includes the control of microbial sulphur conversions could improve the situation of the lake. Former investigations of the natural development of acidic mining lakes showed that the addition of organic matter (for instance leaves) could reduce the acidity of such lakes (Campbell & Lind 1969).

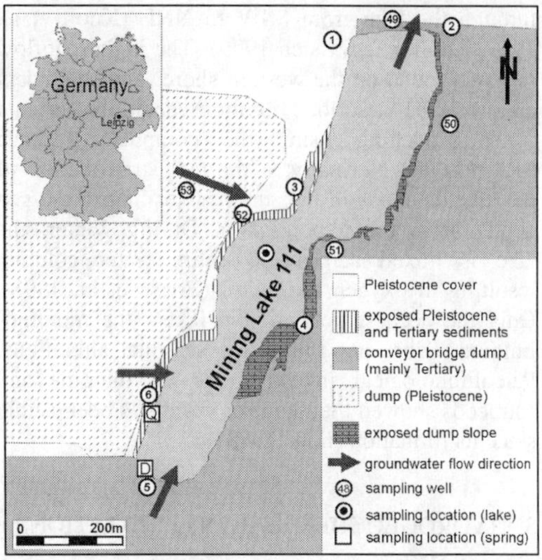

Figure 1. Regional setting and groundwater flow directions of ML 111 (Knöller & Strauch 1999).

The living conditions of the microbial biocoenosis can be improved by the addition of organic matter to accelerate the sulphate reduction, which leads to an increase in pH. Successful biological laboratory experiments are tested in enclosures placed in the lake. However, the success of the biotechnological remediation of the whole lake depends on the hydrogeological and geochemical conditions in the catchment area of the lake.

2 REGIONAL SETTING

The lake 111 - situated in the former lignite mining pit Plessa - is bordered by the Lusatian glacial marginal valley in the south and a disturbed moraine in the north. Figure 1 shows the geological situation of the catchment area. Lignite mining took place between 1925 and 1956. The world´s first overburden conveyor bridge was set up in the lignite mine of Plessa in 1926. This technology formed the heterogeneous structure of the dump.

The lake is about 900 m long and consists of three basins separated by narrow and shallow zones. The basins are between 120 and 140 m wide. The deepest point of the lake (10.2 m) lies in the middle basin. The north and south basin reach depths of about 7 m. The lake has a surface area of 110,000 m^2 and a volume of 500,000 m^3.

The water level of the lake is controlled by the in- and outflow of ground water, the surface runoff from the banks of the lake, as well as by the precipitation and evaporation rates. The ground water passes through the lake from SSW to NNE (Knoll et al. 1999, Knöller & Strauch 1999). The highest inflow rate was found on the western shore. The catchment area will be described in more detail by ground water modelling using the computer package MODFLOW. Normally, a thermal stratification of the lake develops in the summer, leading to oxygen depletion in the hypolimnion. During winter, the lake is mixed across the entire water column, resulting in oxygen saturation (Friese et al. 1998). Only the deepest point of the lake - the monimolimnion - does not follow the seasonal alterations. Paleolimnological investigations on diatoms and ostracods showed that the lake water has been acidic since its formation (BTU 1998).

3 HYDROGEOCHEMICAL INVESTIGATIONS

3.1 *Lake water*

The lake water is characterised by a pH-value of 2.6. Mean sulphate concentrations of 1300 mg/l and iron concentrations of 150 mg/l are typical. The $\delta^{34}S(SO_4)$-values of the water profile vary from 3.2 ‰ (CDT) for the surface water to 5.4 ‰ (CDT) in the deepest part of the lake. The amount of evaporation and the relation between lake and ground water were investigated by $\delta^{18}O$ and δ^2H values. The average annual ground water inflow and outflow were estimated at 24,000 m^3 and 16,000 m^3, respectively (Knöller 1999). More lake water data (profile of the deepest part from June 1999) are given in the Table 1.

3.2 *Ground water*

The ground water from the Quaternary and Tertiary aquifers and from the dump aquifer monitored for 4 years is characterised by high sulphate, iron and aluminium concentrations typical for pyrite oxidation controlled mine waters. Table 1 a - c reports hydrogeochemical and isotopic analyses for ground and lake water sampled in June 1999.

Table 1a. pH-values and sulphate concentrations of the ground and lake water (June 1999).

	pH	SO_4^{2-} (mg/l)
Lake (0 - 10 m)	2.6 - 3.1	950 - 1800
Dump aquifer	4.0 - 4.1	2300 - 4200
Quaternary/Tertiary aquifer	4.0 - 5.0	580 - 670
Lake outflow	4.2	760

Table 1b. Iron (total) and aluminium concentrations of the ground water and lake water (June 1999).

	Fe (mg/l)	Al (mg/l)
Lake (0 - 10 m)	130 - 480	32 - 38
Dump aquifer	510 - 1200	57 - 60
Quaternary/Tertiary aquifer	40 - 76	4 - 23
Lake outflow	190	11

Table 1c. $\delta^{18}O$- and δ^2H-values (VSMOW) of the ground water and lake water (sampling June 1999).

	$\delta^{18}O$ (‰)	δ^2H (‰)
Lake (0 - 10 m)	-2.4 - -4.2	-33 - -40
Dump aquifer	-9.2 - -9.4	-65 - -66
Quaternary/Tertiary aquifer	-8.9 - -9.1	-64 - -66
Lake outflow	-7.8	-61

A decrease or seasonal changes of the mineralisation in ground water and lake water could not be observed during the last 4 years. Calculations of the sulphur input using sulphur isotopes have been described by Knöller (1999).

3.3 *Seepage water*

Twelve seepage meters are installed on the lake bottom to estimate the ground water flow into the lake. The flow rates range from <1 to 10 l/m^2 d due to the heterogeneous structure of the aquifers and the lake bottom. Chemical analyses of the seepage water indicate ion exchange between ground water, lake water and the argillaceous lake sediments. Hence, the seepage water shows almost the same chemistry as the lake water. However, the isotope signatures ($\delta^{18}O$ and $\delta^{2}H$) of the seepage water from the Tertiary aquifer of the western shore are typical for ground water. This supports the hypothesis of a higher ground water inflow from the western shore and a stronger interaction between the lake water and the ground water of the dump aquifer on the eastern shore.

3.4 *Precipitation*

Precipitation water was collected by bulk samplers to assess the deposition of atmospheric sulphate, as well as to determine the elution potential of the rain water. The pH-values of precipitation range from 4.4 to 7.8 and are typical for European open field sampling sites (Krieter 1991). The average sulphate concentration of rain water (collected monthly from May to December 1999) is 8.9 mg/l, the average yearly precipitation is 617 mm in this area. That yields an annual input to the lake of about 600 kg atmospheric sulphate.

4 INVESTIGATIONS OF THE SEDIMENTS

4.1 *Surface sediments*

The dump sediments cover the main part of the lake shore (15,000 m^2). Spare vegetation on the dump area causes rather high erosion. Werner (1999) calculated that 10 cm soil from uncovered dump sediments are yearly removed by erosion. Therefore, about 3000 t/a of the dump sediments are washed into the mining lake. The average amount of total sulphur in the surface sediments which consists mainly of sulphuric acid, gypsum, jarosite and low amounts of pyrite and organic sulphur is 0.7 wt.% (table 2). Pyrite amounts of about 0.02 wt.% were measured indicating a high oxidation rate.

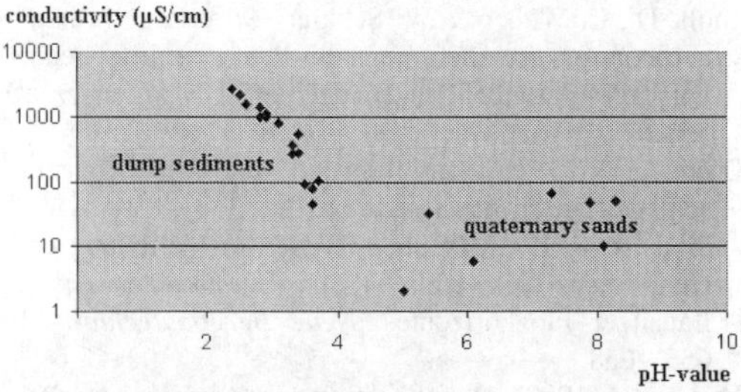

Figure 2. Acidification potential of the surface sediments.

Assuming all this sedimentary sulphur is converted to sulphate, the yearly sulphate input to the lake could reach 63 t. Elution tests confirmed the high acidification potential of the surface sediments. Depending on the mineral composition all sulphur can be washed out as sulphate. Low pH-values (minimum 2.3) correlate with high conductivity values (figure 2) in the investigated surface sediments of the dump. The Quaternary sands do not influence the acidification process.

4.2 *Aquifer sediments*

Three boreholes up to 30 m were drilled to get sediment material of the dump aquifer (mainly silty sand) and the Quaternary/Tertiary aquifer (mainly medium grained sand). Some chemical data are given in table 2. Lusatian lignite has an average total sulphur content of 0.6 wt.% (Vulpius & Neubert 1982) varying from 0.2 to more than 3 wt.%. FeS_2-contents lie between 0.1 up to 1.6 wt.%. The total sulphur consists of organic sulphur, sulphate and sulphide (analysed as acid volatile and chromium reducible sulphide). The relation between these components varies. The lignite has a higher content of organic sulphur than the surrounding sediments. But often there are higher FeS_2-concentrations in the Tertiary sediments than in the lignite. Correlations between the sediment species, the total sulphur contents, the pyrite contents and $\delta^{34}S$-values of the pyrite could not be found.

Table 2. Sulphur and pyrite content of the investi-gated sediments.

Sediment	S_{total} (wt.%)	FeS_2 (wt.%)
Surface sediments	0.13 - 2.14	<DL - 0.09
Dump aquifer sediments	0.65 - 2.61	0.08 - 0.92
Quaternary/Tertiary aquifer sediments	0.27 - 0.58	0.25 - 0.49

*Detection Limit = DL

4.3 *Lake sediments*

In June 1999, four sediment cores from several parts of the lake bottom were sampled. Depending on the oxygen content of the water at the sampling point the geochemistry and mineralogy of the lake sediments varies.

Table 3. pH- and Eh-values of the lake sediments at the sediment-water interface (June 1999).

Core	Location/depth	pH	Eh (mV)
1	North Basin/ 6 m	3.5	450
2	Shoal between north and middle basin/ 1.2 m	2.6	730
3	South basin/ 6 m	2.6	650
4	Middle Basin/ 10 m	5.0	-70

The pH- and Eh-values measured at the sediment-water interface represent these conditions (table 3). Chromium reducible sulphur (CRS) which indicates pyrite was only found in the sediments from the deepest part of the lake (table 4).

Table 4. Sulphur and pyrite contents of the investigated lake sediments.

Lake sediment	S_{total} **(wt.%)**	FeS_2 **(wt.%)**
Core 1 - 3 North/south basin	0.30 - 5.00	-
Core 4 Middle basin	<DL - 1.90	<DL - 0.06

*Detection Limit = DL

The lake bottom can function as a primary source of biologically produced alkalinity which until now occurs only at the deepest part of the lake in the middle basin.

If biological reduction is successful in the deeper parts of the lake, the bottom sediments of these parts will act as a sink for reduced sulphate and iron. However, the bottom near the shore, where anoxic conditions cannot be established, will still react as a source of acid, sulphate and iron.

5 CONCLUSIONS

The acidity of the lake water is higher than the acidity of the ground water as a result by changes from anoxic conditions in the aquifer to oxygen enriched parts of the dump and the lake and by evaporation. Another source of acidity is the erosion of dump sediments and its elution by lake water. Ground water supplies annually about 36 t sulphate (Knöller 1999), erosion of the dump sediments supplies about 63 t/a. The atmospheric sulphate deposition (about 0.5 t/a) can be neglected in this case.

A high amount of water soluble acid remains in the surface sediments, as well as in the deeper sediment layers and the aquifers. The relocation of the pyrite oxidation zone will progress to the deeper sediment layers.

Even if all pyrite is oxidised, the stored sulphate in the sediments will contribute to the lake acidification for a long time.

A biotechnological remediation of the lake water does not consider the influence of ground water and the surrounding sediments. If the biotechnological remediation is to succeed in the whole lake, remediation of the surface sediments, in order to decrease the erosion rate (by planting), as well as remediation of the ground water and the dump aquifer sediments (for instance the installation of neutralisation zones) have to be considered. An investigation of the lake bottom sediments as a source of sulphate and iron during the remediation must be included, too.

ACKNOWLEDGEMENTS

This research was performed within the HGF-Project "Systemintegrated Environmental Biotechnology - Remediation of Organically and Inorganically Polluted Groundwater and Surface Water".

Obtaining all the data of this study without the help of our colleagues from the Department of Hydrogeology and Analytics would have been impossible. Special thanks are due to the laboratory staff.

REFERENCES

BTU 1998. Niederlausitzer Bergbaufolgelandschaft - Erarbeitung von Leitbildern und Handlungskonzepten für die verantwortliche Gestaltung und nachhaltige Entwicklung ihrer naturnahen Bereiche. *Abschlußbericht, BMBF-Förderkennzeichen. 0339648.*

Campell, R. S. & O. T. Lind 1969. Water quality and aging of strip-mine lake. *Journal WPCF*, Vol. 41, No 11: 1943 - 1955.

Friese, K., K. Wendt-Potthoff, D. W. Zachmann, A. Fauville, B. Mayer & J. Veizer 1998. Biogeochemistry of iron and sulfur in sediments of an acidic mining lake in Lusatia, Germany. *Water, Air, and Soil Pollution* 108: 231 - 247.

Geller, W., H. Klapper & W. Salomons 1998: *Acidic Mining Lakes.* Berlin-Heidelberg-New York: Springer.

Knöller, K. 1999. *Anwendung stabiler Umweltisotope zur Bewertung hydrochemischer Zustände und Prozesse in Folgelandschaften des Braunkohlebergbaus.* UFZ - Sektion Hydrogeologie. Dissertation (in press).

Knöller, K. & G. Strauch 1999. Assessment of the flow dynamic of a mining lake by stable isotope investigations. *Isotopes Environ. Health Stud.* 35: 75 - 83.

Knoll, D., L. Weber & W. Schäfer 1999. Grundwasseranbindung von alten Tagebau-Restseen im Niederlausitzer Braunkohletagebaugebiet. *Grundwasser* 2: 55 - 61.

Krieter, M. 1991. Wasserkreislauf und saure Niederschläge. *Geogr. Rundschau* 6: 326 - 333.

Vulpius, R. & K.-H. Neubert 1982. Zur Verteilung und Genese des Schwefels in den Braunkohlen des 2. Lausitzer Flözhorizontes. *Neue Bergbautechnik* 11: 653 - 658.

Werner, F. 1999. *Modellgestützte Untersuchungen zur Genese der Wasserbeschaffenheit in Folgelandschaften des Braunkohlebergbaus.* Dresdener GFZ e.V., Heft 14.

Groundwater: Past Achievements and Future Challenges, Sililo et al. (eds) © 2000 Balkema, Rotterdam, ISBN 90 5809 159 7

Investigation on deep water resources by tunnels within the "Piton des Neiges" volcano, Reunion Island

L. Bret, J.-L. Join & J. Coudray
Laboratoire des Sciences de la Terre, Université de la Réunion, Saint Denis, France

ABSTRACT: Reunion Island deep volcanic aquifers are a recent field of investigation. Various author's analysis based on others islands and volcanic archipelagos (Hawaii, Canarias, West Indies) highlights the difficulty in defining a general representation of the deep groundwater circulation conditions in such media. In Reunion Island, the coastal water table is usually separated from the high altitude water tables.
So, the geological survey of an eighteen kilometres long tunnel one kilometre below the ground surface, and of other tunnels drilled in the Piton des Neiges volcano flanks, gives precious information regarding the deep hydrogeological structures of a shield volcano.

1 INTRODUCTION

Tunnels have been used for groundwater supply since antiquity. This extraction method is especially well adapted to mountainous area and fractured aquifers. In the high volcanic islands, these conditions are often fulfilled. In the Canarias (Custodio, 1978) this kind of work has been achieved successfully in spite of average annual rainfall equivalent to 500 mm. In Reunion Island, with a 3000 mm annual rainfall, this method should be a solution for water supply.

A feasibility study has been undertaken for the Piton des Neiges massif, thanks to the project of drill tunnel, part of a big irrigation network. The geological monitoring of this tunnel will provide us with important information concerning the hydrogeological conditions of the deep volcanic aquifers.

2 GENERAL BACKGROUND

Reunion Island is a volcanic one characterized by a hot rainfall season from December to April and a rather temperate dry season from May to December. It is located in a humid tropical area in the southwestern Indian Ocean (21°S/55°E).

The island is formed of two shield volcanoes (Fig. 1) :

- the Piton des Neiges (3069m), covering the northwestern two thirds of the island, has been extincted for 20 000 years (P.Y. Gillot & P. Nativel, 1982),

-the Piton de la Fournaise (2632m), a still active volcano covering the remaining southeastern third of the island.

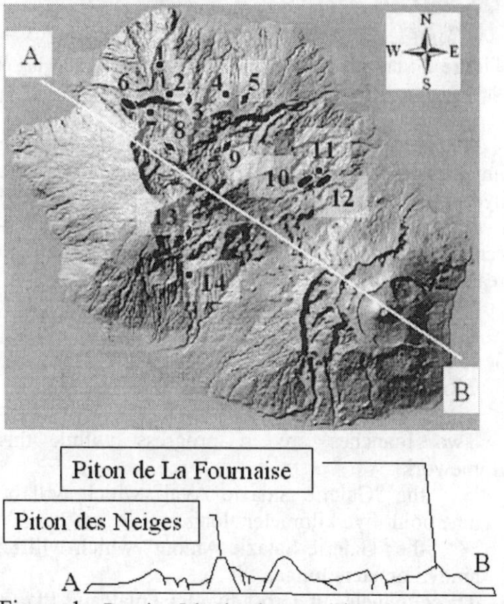

Figure 1 : Reunion Island volcanoes (Piton des Neiges and Piton de la Fournaise), waters transfer site and other galleries. 1 : Bras Guillaume ; 2 : Galerie Salazie Amont ; 3 : Plaine des Chicots plateau ; 4 : Galerie Salazie Aval ; 5 : Plaine des Fougères plateau ; 6 : Rivière des Galets ; 7 : Bras Sainte Suzanne ; 8 : Mafate ; 9 : Salazie ; 10 : Bras Patience ; 11 : Takamaka II ; 12: Rivière des Marsouins ; 13 : Cilaos ; 14 : Bras de Cilaos.

Water resources are important compared with the needs, but the high topography of the island brings about an important disparity of rainfall between the eastern region (subdued to trade-winds) and the western leeward region. The eastern region has an average annual pluviometry of 5000 mm per year, whereas the western region (the most developed area) has one of 1300 mm and suffers from a chronic drought six months per year.

Many solutions have been considered in order to solve this lack. Eventually the chosen solution was to transfer water from the east to the west through pipeline. Surface water has to be caught in the rivers of Mafate and Salazie "cirques" and transferred to the western region through a network of thirty kilometers long tunnels, drilled through the mountain (Fig. 2).

Figure 2 : the tunnel boring machine used for the digging of the leading galleries of the east to west transfer.

Given the tunnels low altitude (350m) and the importance of the covering (more than 1000m), the hydrogeological interest is unquestionable. Deep hydrogeological structures of the phase II old volcanic sequence of the Piton des Neiges, can be reached.

3 HYDROGEOLOGICAL SURVEY OF THE TRANS-PITON DES NEIGES TUNNEL.

Two branches are in progress within this framework :

✓ the "Galerie Salazie Aval" which will be nine point five kilometers long,

✓ the "Galerie Salazie Amont" which will be nine kilometers long.

These tunnels cut across two lava plateau : Plaine des Chicots plateau and Plaine des Fougères plateau respectively (Fig. 1).

The identification of the geological conditions in the core of the shield volcano will be deduced from observations in "Galerie Salazie Aval" and "Galerie Salazie Amont". It starts to bring important information to us, insofar as after only two hundred meters of digging it has reached a saturated level of extended size.

The work consists now in :

- a geological survey of tunnel parements and sampling on mine face,
- analyses and hydrodynamic tests on water inflows,
- interpretation of cuttings delivered by the tunnel boring machine.

Moreover, the information gathered in tunnel are interpreted in their regional context. This approach is based on a geological study and a hydrological analysis set on shield volcano scale. It is directed towards the analysis of the transfer of fluid in a volcanic environment.

The digging of the first 198 meters of "Gallerie Salazie Amont" site has been achieved in intensely zeolite basalts of Piton des Neiges phase II. Water inflows where scares but some inrushes happen when crossing some dykes (Fig. 3).

Eg. The dyke crossed at the metric point (mp) 198 produced a discharge of about 35 $l.s^{-1}$, with water pouring out of fractures.

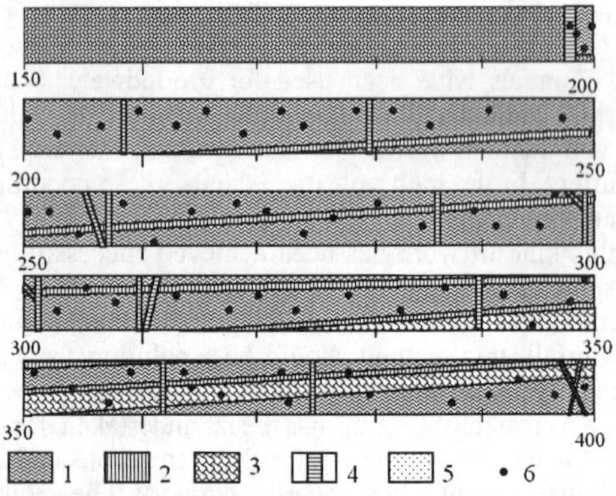

Figure 3 : geological environment of "Galerie Salazie Amont" Inflows. 1 : "pahoehoe" lava flows of ± olivine basalt ; 2 : "aa" lava flows ; 3 : «pahoehoe » lava flows with a low cohesion ; 4 : dyke ; 5 : highly zeolitised "pahoehoe" lava flows ; 6 : inflows.

This discharge continuously increased with the progress of the tunnel. It has gone through most vesicular hence most permeable "pahoehoe" flow draining water to the tunnel, so the discharge has risen to more than 60 $l.s^{-1}$. Then the digging has crossed a new dyke and has kept progressing through most permeable "pahoehoe" flows.

Three dykes divided this aquifer into four compartments (Fig. 4).

The discharge has increased until pm 439 to 500 $l.s^{-1}$. These water inflows have stabilized sometimes in a compartment, but they have increased again after the crossing of some dykes.

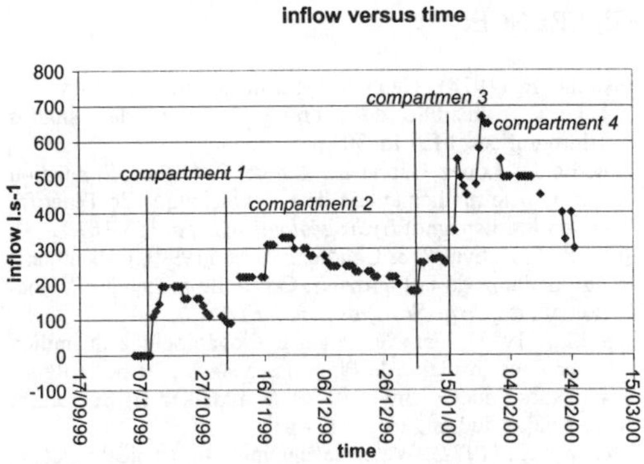

Figure 4 : inflow versus time in the Galerie Salazie Amont highlight the dyke influence on the water flow.

It appears clearly that the dykes have a considerable influence on the water flow in these aquifers. This phenomenon of groundwater impounded by dyke is described by Peterson (Peterson 1972). This is true in particular for the dykes at mp 198, 283, 375 and 425.

4 PATTERN OF GROUNDWATER FLOW DEDUCED FROM TUNNELS IN REUNION ISLAND.

Tunnels that have already been drilled in Reunion Island are under distinct geological conditions. They have been concerned by important water inflows.

The achieved four tunnels geological sections allow us to have a condition of inflows sketch view in such tunnels.

4.1 *The Bras Guillaume tunnel.*

The Bras Guillaume gallery is a drainage tunnel located at an altitude of 800 meters in the Piton des Neiges basalt formations (Figs 1-5). Its whole length is 310 meters, and its discharge of 44 l.s-1. Inflows have appeared in favorable structures (rubefied beds, thick lava flows with unaltered "aa").

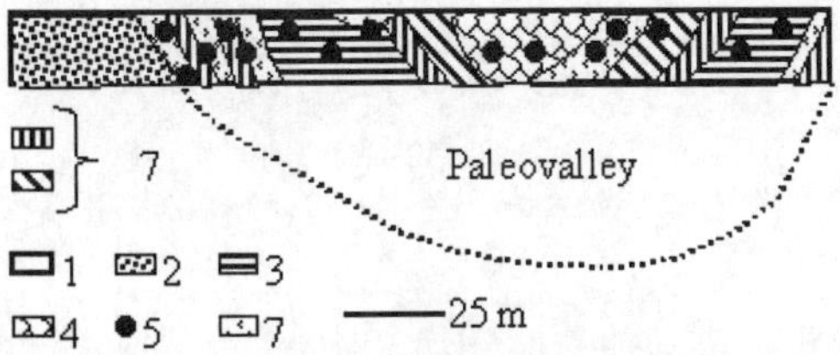

Figure 5 : the Bras Guillaume tunnel section. 1 : tunnel ; 2 : highly fractured basalt ; 3 :phase IIa (olivine basalt, altered basalt) ; 4 : scoria and "aa" ; 5 : breccia and scree ; 6 : tuffs et ashes ; 7 : rubefied bed

The water flows conditions – inflows on both parements and through the wall – are well representative of a saturated heterogeneous environment context (Join 1991).

4.2 *.The Takamaka II tunnel.*

The tunnel is located under a plateau between the Bras Patience and the Rivière des Marsouins, made up of a filling of recent volcanic formations coming from the Piton des Neiges and in an ancient valley located in the phase II older formations (Figs 1-6). The discharges in this part of the tunnel are in the order of 300 $l.s^{-1}$.

Recent formations are constituted of aphyric basalt lava flows, generally thick for well channelled parallel to the river present axis. Pyroclastic layers forming impermeable layer and favouring the outlet of many springs along the wall, come in between the lava series. In these pyroclastic facies, erosion has easily dug narrow valleys where upper columnar basalt structures often show a strong "v" morphology.

4.3 *The Bras de Cilaos leading tunnel.*

The first five hundred meters are characterized by highly altered formations with clayey minerals in the fractures. Next, formations become less altered. The layers seems to be comparatively homogeneous on the whole section. Its like a regular alternation of vesicular basalts more or less rich in olivine crystals with scoriaceous inter-beds. The thickness of basalt lava flows varies from metric to decametric, and as well as the degree of fracturation (Fig. 7).

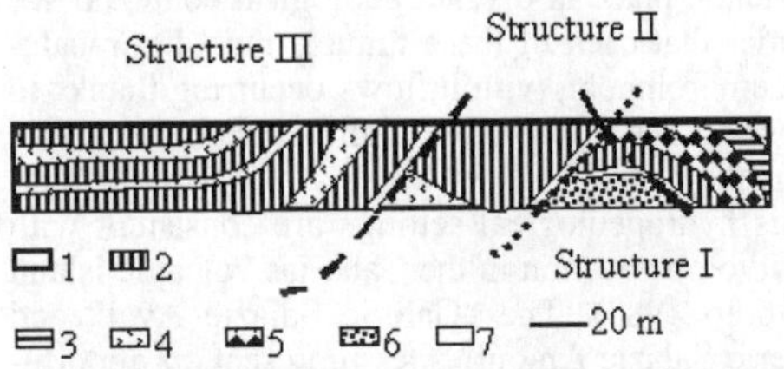

Figure 6 : geological environment of inflows on the Takamaka tunnel. 1 : tunnel ; 2 : tuffs and ashes ; 3 : very fractured basalt ; 4 : fractured basalt and prisms ; 5 : inrushes ; 6 fractured basalt ; 7 : scoria and "aa".

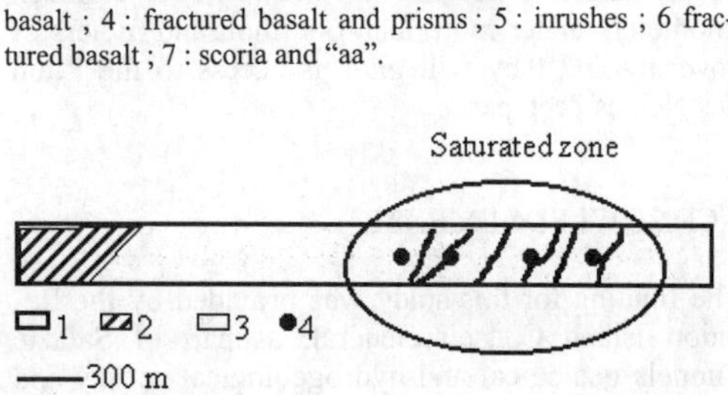

Figure 7 : geological environment of inflows on the Bras de Cilaos tunnel. 1 : tunnel ; 2 : the Piton des Neiges phase II altered basalt ; 3 : basalt II / scoria alternation ; 4 : inflows.

The inflows take place in this tunnel thanks to the fracturation with discharges of about 20 l.s^{-1}. This low productivity is explained by the very small size of the watershed basin (Fig.1).

4.4 *The Bras de Sainte Suzanne tunnel.*

This digging, located in the outpouring basin of the Rivière des galets, goes under the River bed (Fig. 1).

One can notice that this 2500 meters long tunnel goes through four highly productive areas, two of which are very fractured and the other two are paleovalleys (Fig. 8). The cumulative disflow of this tunnel is 90 l.s^{-1}.

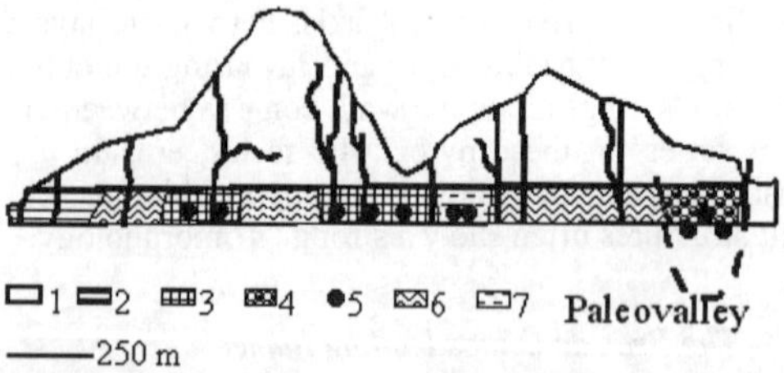

Figure 8 : geological environment of inflows on the Bras de Sainte Suzanne tunnel. 1 : tunnel ; 2 : fractured basalt ; 3 : very fractured basalt; 4 : alluvium (rock, blocks) ; 5 : inflows ; 6 : scoria and "aa" ; 7 : mud and debris.

5 CONCLUSION

In all the tunnels dug in Reunion Island, we have been confronted with important inflows. Yet they have taken place in distinct geological contexts. We consider that each of these tunnels must be in saturated environment, with inflows occurring thanks to discontinuities : fractures, highly permeable areas, paleovalley, etc.

This hydrogeological settings are consistent with Custodio observation in the Canarias volcanic island (Custodio 1988). The "Galerie Salazie Aval" and "Galerie Salazie Amont" sites represent an opportunity to improve our knowledge of these media. Indeed, these tunnels will be followed over eighteen kilometers under more than one thousand meters of covering and they will give us access to the Piton des Neiges deep parts.

ACKNOWLEDGEMENTS

The funding for this study was provided by the Reunion Island Conseil Général, as part of Salazie Tunnels geological and hydrogeological survey and synthesis research program, in collaboration with the "G.I.E. Salazie" company (SPIE Batignolles, Bouygues, Razel).

REFERENCES

Custodio E. (1978). Geohidrologia de terrenos en islas volcanicas – Instituto de Hidrologia, Centro de Estudios Hidrograficos, Madrid, 303 p.

Custodio E., Lopez Garcia L., Amigo E. (1988). Simulation par modèle mathématique de l'île volcanique de Ténériffe (Canaries, Espagne) *Hydrogéologie* n¡2, pp. 153-167.

Gillot P-Y., Nativel P. & Condomines M. (1990). Geochronology of Piton de la Fournaise, Centre de recherches Volcanologiques, *Jean-François Lénat, ED.*

Join J.L., (1991). Caractérisation hydrogéologique du milieu volcanique insulaire, le Piton des Neiges , Ile de la Réunion. thèse doct., Université de Montpellier II- Science et Technique du Languedoc, 179 p.

Peterson F.L. (1972). Water development on tropical volcanic islands, type example : Hawaii- *Ground Water*, vol. n¡5, pp. 18-23.

Groundwater: Past Achievements and Future Challenges, Sililo et al. (eds) © 2000 Balkema, Rotterdam, ISBN 90 5809 159 7

Dewatering and the tourist potential of hydrogeological resources of the Gavorrano mine, Tuscany, Italy

C.A.Garzonio
Dipartimento di Urbanistica e Pianificazione del Territorio, Firenze, Italy

ABSTRACT: The paper describes the hydrogeological problems linked to the dewatering of the groundwater contained in the decommissioned mine of Gavorrano. The study of mine water quality, discharge and re-utilisation is important because it can lead to an economically profitable rehabilitation of mining areas and their water resources. The geological and structural settings which control the water quality, chemistry and discharge, as well as the water rebound, are analysed. In particular the main data regarding the evolution and the quantity of the water resource, both fresh and thermal waters, are shown the aim being to highlight the aspects connected with the possibility of creating a tourist mining park.

1 INTRODUCTION

The Gavorrano mine was one of the largest pyrite mines in Europe last century. Production ceased in 1981 and since then the mine has been under maintenance. A multi-purpose study was set up in 1995 to evaluate the possibilities of creating a natural reserve and mining park, by recovering the mining area and all its historical mining structures. The environmental rehabilitation of the area involves quarries (with trekking and rock climbing tracks and the construction of an open theatre), tailing ponds and spoil dump areas, and consists in reforesting, stabilization procedures and restoration of significant old mining structures. A major aspect of the rehabilitation involves the evaluation of the possibility of increasing the recovery of water resources, and planning the correct exploitation of the waters. In particular the paper describes the dewatering operations, the effects of the water level rebound and the evaluation of the groundwater volume storable within the mine in safe conditions.

The water drainage problem was analysed starting with the hydrogeological scheme of the area, using historical data concerning both water springs and water level recovery, and then moved on to hydrochemical data. The analysis of this case history, as the ones in literature (Civita et al. 1994, Banks et al. 1996; Sammarco 1993), show the importance of ground water level monitoring both during and after mining activity, in order to estimate rebound potential. It is also necessary to keep records of water hydrochemistry in order to assess the potential for acid generation on closure. Attention is given to water characteristics, water circulation, mixing and chemical changes within the excavated rock masses and their possible relationships with superficial aquifers. The waters pumped from the tunnels have a complex origin. A surface circulation is present, due to the infiltration in fractured and karstified carbonatic rocks, and there is a deep circulation of thermal water. The latter, sulphate-calcic water, with a temperature of almost 40°, was cut off and deviated during the pyrite mining activity, and mixed with the other groundwater pumped out. The mining complex was the most important one in Europe for the extraction of pyrite (used mainly in the chemical (over 400 m. from the surface). All other entrances and wells have been closed for the dewatering and safety operations. The mining system is over 5 km long, over 400 m. deep, with an exploitation thickness of over 500 m. The study area corresponds to the ridge of Mt. Calvo, which is composed of fractured and karstic limestones. The reliefs are crossed by more than one fault system, with throws > 600 m. The masses of pyrite which were mined coincided with the contact with faults between the sedimentary units and intrusive granite, quartz-monzonitic masses.

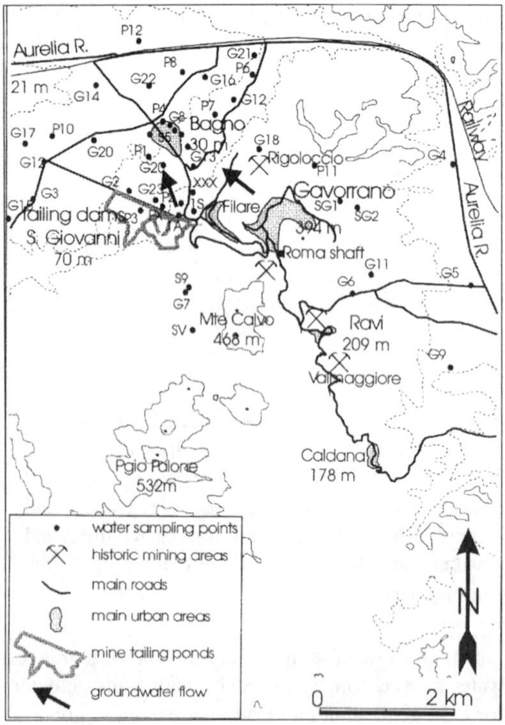

Figure1. Gavorrano mine area with location of superficial water sampling points.

2 HYDROGEOLOGICAL FEATURES

The hydrogeological circuit of the Gavorrano area is characterized by the presence of three systems: a superficial alluvial system, a karstic system and a deep hydrothermal system (Crosta & Garzonio 1998) Waters from the two systems have been forcefully mixed by human intervention during mining. In fact, 500 m of production levels were excavated in more than a century depressing the pre-existing groundwater level up to -250 m b.s.l.. Old thermal springs were drained during this lowering, and hot water springs (up to 47°C) were found during mining. Figure 2 is a geo-hydrogeological map representing lithotypes with different permeabilities through different shading. The permeability classes were attributed by considering: the lithology, the degree of fracturing, the degree of weathering and alteration, the presence of karstic structures, observed both at the surface and within mine drifts. Alluvial deposits were separated from the other materials, because of their characteristics and their natural northward groundwater flow direction. Karstic features and the fracturing degree were essential for distinguishing the determinant to differentiate among carbonate rocks. Due to their very low permeabilities flysch and shales work as a permeability threshold controlling the location of the main springs around the carbonate massif. Intrusive

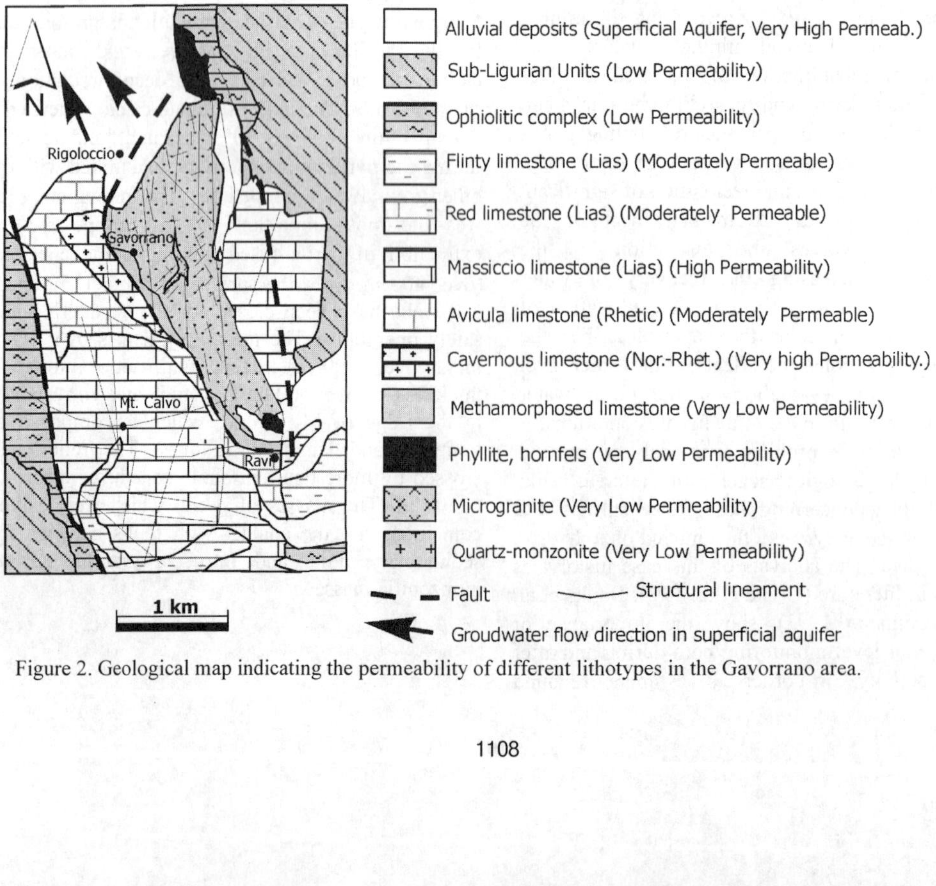

Figure 2. Geological map indicating the permeability of different lithotypes in the Gavorrano area.

rocks are characterised by a relatively low permeability because of their superficial weathering and their low fracture density. This unit is more permeable in places due to the presence of the fault system. A groundwater balance was performed by attributing different coefficients of potential infiltration to the different lithotypes and by calculating the contributing areas for each lithotype. Average annual rainfall in the area ranges from 750 mm y^{-1} to 800 mm y^{-1}, with an average evapo-transpiration of 430 mm y^{-1} . The rainy season lasts from October to February, with the maximum average monthly rainfall in November (120 mm) and very low precipitation in summer (20 mm in July). According to these data, the calculated evapo-transpiration and a series of infiltration coefficients, it results that 1.1 Mm3 of water form the annual groundwater recharge. However, it must be stressed that rock mass properties, especially hydraulic conductivity, have been strongly and permanently influenced by mining and subsequent induced processes (e.g.: the presence of tunnels, the increase in fracturing degree and the enlargement of existing fractures by acid water circulation and water level lowering which increased karst solution in places). In fact, tunnels and drifts form a drainage network characterised both by voids and refilled spaces.

3 THE DEWATERING STOPPAGE

Groundwater level rebound after the closure of mine activities is a common process that can be understood starting from the data recorded during the water level lowering, and some occasional water rising, phase. Groundwater rebound at the Gavorrano mine is regulated by a submersible pumping system. This is adopted to avoid too fast a groundwater level rising which could induce turbulent flow and internal erosion of the back-filling; and excessive hydraulic gradients and groundwater re-emergence at the surface within inhabited areas. Finally, the controlled rebound allows the evaluation of the volume of water storable within the mine voids system, as well as the changes in the chemical composition of drainage waters. The passage of the Second World War front between 1944 and 1945 caused a forced water level rising (Figures 3), and together with the data collected since 1995, it constitutes the only information available for the mine. The data for the 1944-45 forced rising are available for the main mining centres of that time (Rigoloccio, Boccheggiano and Valmaggiore from N to S). The analysis of the groundwater level recovery for the 1944-45 and the 1995-1999 periods has allowed us to observe the non linearity of the process. In particular, the rising is slower at the beginning and faster in the final stage. This is imputable to the change in the fracturing degree of the rock mass, the opening of main joint sets and the possible decrease in volume of voids and of the area of the main aquifer. The pump discharge has remained almost constant since 1995, at 60 l s^{-1} with the exception of two periods of heavy rainfall. This allowed the storage of about $300*10^3$ m^3 of water, excluding the initial uncontrolled water level rising between -236 m b.s.l. and -197 m b.s.l. According to the recession analysis of the increased discharge the average water resources that are renewed yearly amount to 2.1 Mm^3 or 66 l sec^{-1}. These values are quite close to the average yearly pump discharge (60 l/s or 2.08 Mm^3). Finally, by comparing these data with the ones obtained by the effective infiltration analysis (1.1 Mm^3) it results that every year almost 1 Mm^3 of hot water regularly flows from the deep circulation system into the mine. Furthermore, because of the almost constant temperature of the pumped water, it can be assumed that the hot water discharge increases with the fresh karstic water discharge during wet periods, maintaining an almost constant ratio between the two. Geophysical and geomechanic soundings were carried out in order to identity the geometry of the aquifer above the built up area of Bagno di Gavorrano, above all to contact the hazards. The results of the mechanical soundings, for a depth of over 250 m, which went through a debris and "Renone" cover (granite alteration) for less than 30 m, then a crystalline calcareous level and finally the cavernous limestone, are in contrast with previous geo-structural analyses and the geophysical results. In other words the cavernous limestone is less permeable than elsewhere, with few and closed fissures, few cavities, even though scarce in gypsum

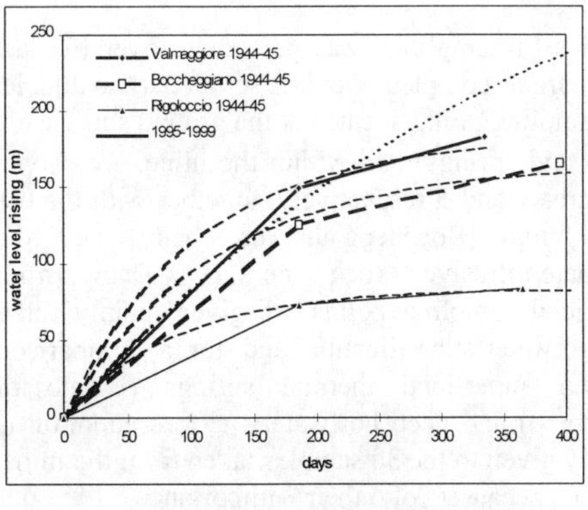

Figure 3. Groudwater level rising during the force 1944-45 and controlled 1995-1999 rebounds. T compare the gradients, data have been referred to common zero.

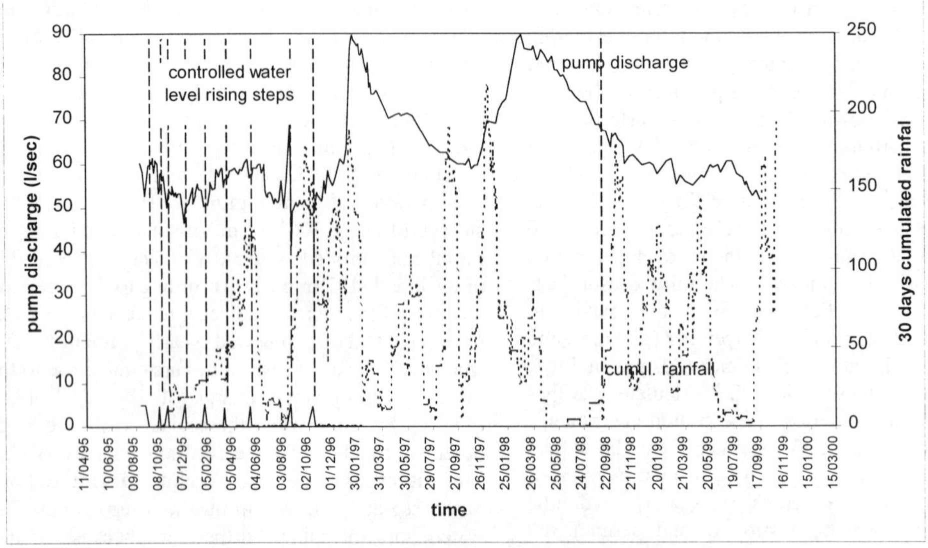

Figure 4. Plot showing the relationships between the pump discharge, from the groundwater level controllin system, and the cumulated 30 days rainfall during the 1995-1999 period. It is also shown the sequence of risin steps adopted during the controlled flooding of the mine.

levels. For this reason too, before carrying out further drillings, it was decided to re-open a shaft near Rigoloccio in order to reach the tunnel (up to – 200 m.b.s.l., which connected the Rigoloccio mine with the Gavorrano one (Roma shaft), and then intercept the waters, analyse them and install a new pumping system. According to an initial analysis, this operation could produce good results, not only for the control and the safety of the water rising operations, but also for the separation albeit not total, of the hot and cold waters infiltrated in Mt.Calvo.

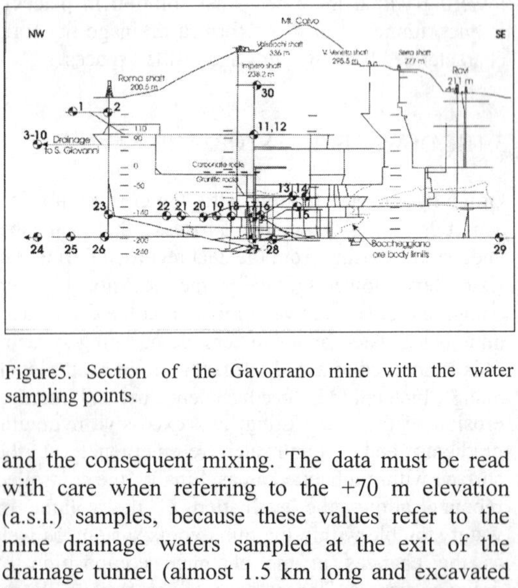

Figure5. Section of the Gavorrano mine with the water sampling points.

4 WATER CHEMISTRY

A series of samplings was performed over the last four years to complete the data set already available. The sampling points located at the ground surface (wells and springs) and within the mine, are shown in figures 1 and 5 respectively, together with the ore body limits (Boccheggiano ore body) and the carbonate/intrusive rock contact. Data from superficial sampling points are given mainly for a general water classification and for a comparison between superficial thermal springs (Bagno di Gavorrano) and deep hot waters. More attention is actually given to the 30 samples taken from the mine (fig.5) because of their importance for our understanding of the main aquifer hydro-dynamics, the hydro-chemical composition of waters and their possible changes with the proceeding of water rising and the consequent mixing. The data must be read with care when referring to the +70 m elevation (a.s.l.) samples, because these values refer to the mine drainage waters sampled at the exit of the drainage tunnel (almost 1.5 km long and excavated in flyschoid rocks).

At a regional scale it has been observed that sulphate waters commonly spring from the Cavernoso limestone and from intrusive rocks. Calcium bicarbonate waters are mainly associated with carbonate formations and small depths and are considered the less mature water class. Seawater influence is absent, with the exception of very few points where wells probably intercepted small remnants of old entrapped seawater. Sea water

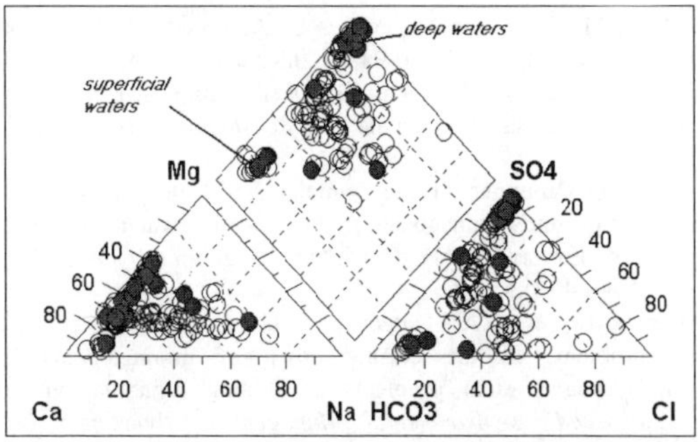

Figure 6. Piper plot of all the analysed waters, showing the differentiation between karst waters and deep mine waters.

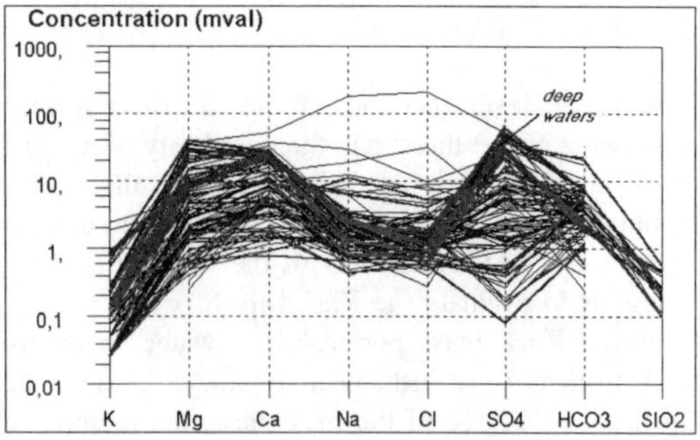

Figure 7. Schoeller diagram for some of the deep and fresh water samples.

intrusion is becoming problematic to the north of the area (Avio *et al*., 1995).

Mine water samples can be clearly differentiated in two main groups: "superficial" bicarbonate waters (from levels +240 m a.s.l., + 155m: Mg-Ca, +90 m: Ca-Mg-HCO3) and deep sulphate waters (Ca-Mg-SO4, -80 m b.s.l., - 110 m, -140 m, -200 m) as suggested by Piper and by Schoeller plots (figs. 6 and 7). This grouping can be done on the basis of the sulphate (from sulphide oxidation and evaporites solution), iron and silica contents and it is also suggested by the few temperature data even if more qualitative observations have been made. In fact, by fitting temperature measurements, excluding samples taken at the outlet of the drainage tunnel (1.5 km long), a geothermal gradient of about 75°C km^{-1} has been determined. This geothermal gradient is in agreement with the data published by Baldi *et al.* (1995) concerning southern Tuscany and indicating, for the marginal areas of the geothermal fields (Larderello, Amiata, Travele, Radicofani), a gradient no lower than 70°C km^{-1}.

A water type sub-group, with peculiar characteristics, is composed by four samples (13 to 16) all collected within the intrusion in a relatively localised area (fig. 5), and characterised by very low pH values and high TDS contents. The area was characterised by a conspicuous air flow and abundant water presence, with clear and cold waters. The pH values strongly characterise this minor group. All the other sampled waters generally presented values just a little over neutral (7-8). Mine drainage is characterized by an average temperature of 32°C, a 7 to 8 pH value, and it can be classified as a Ca-Mg-SO4 or a Ca-Mg-SO4-HCO3 water, probably as a result of temporal changes and the mixing action caused by pumping. In comparison, the water of the old spring at Bagno di Gavorrano was characterised by a constantly higher Na and Cl content, with a minor SO4 content.

4.1 *Isotopic analyses*

Three samples of mine waters were analysed to determine the ^{2}H, ^{18}O, ^{3}H and ^{13}C content. These samples were collected from a dripping stalactite (-140 m b.s.l., n. 19), from the drainage pipe system at -140 b.s.l. (sample n. 17) and at the drainage tunnel outlet (+70 m a.s.l., sample n. 7). From the ^{2}H vs ^{18}O analysis (Zuppi 1986) the waters do not show any fractionation of isotopes as occurs with geothermal fluids, and they are comparable with the ones of the regional recharge. This suggests homogenisation due to the length of the flow path in the fractured rocks. The sample taken from the stalactite shows a slightly higher ^{18}O content and for this reason it could be considered a deep water, unlike the other two samples. The Tritium content is quite constant and this could be the result of a long circulation time or industry). It had its major development period in the 1940-50's and closed down in 1982. At present, some of the drifts and lift systems are in perfect working order, in the central part of the mine (Roma and Impero shafts), up to a depth of 143 m b.s.l. of a very short one. Previous considerations concerning ^{2}H and ^{18}O contents, water chemistry and some published research works suggest that a residence time of about 15 years or more is probable. The contribution of more superficial waters can then be read in the increase in tritium content in the drainage tunnel sample. The two samples from the tunnel are characterised by a different trend suggesting mixing of different waters (drainage water at -140 m) and a superficial contamination (drainage tunnel). The equilibrium of deep waters is also suggested by the over-saturation both in the carbonate mineralogical phases and gypsum.

5 CONCLUSIONS

The water level rebound of the Gavorrano mine represents both a threat and the possibility of making a valuable resource accessible. The results of the continuous monitoring of the water level underway since 1995, and the analysis of the water chemistry, allow us to evaluate the feasibility of water reusage projects. Rock mass permeability, water chemistry and discharge are influenced by the lithology and fracturation degree of the rock mass, the method of excavation and the consequent volume of mining voids, the depth of excavation, the composition of back-filling material and the stage of the mine activity. The drainage waters of the Gavorrano mine show an almost constant temperature (34°C at the discharge gallery outlet, almost 250 m above the actual groundwater level) and an almost neutral pH (7-8), with high TDS content, but very low quantities (below the limit of the law) of polluting elements (As, Cr, Cu, Zn, etc). This may be due to the alkaline minerals within the carbonate rocks and the back-filling material made of limestone mixed with clay and cement.

According to the groundwater balance performed it is possible to estimate that 1 to 1.5 Mm^3 of water could be stored within the voids of the decommissioned mine. This groundwater storage would be extremely useful and valuable, given the dry climatic conditions which characterize the area. In fact, due to the average quality of the drainage water it would be possible to use it both for irrigation and in a dual distribution network. If the hot waters were separated, they could be used as thermal spa waters. This enhancement of the waters would be another tourist attractions for the project underway on the Gavorrano mining park.

REFERENCES

Avio C.M., Bianchi S. & M. Marchetti 1995. Prima valutazione del chimismo delle acque di Pianura di Follonica e Scarlino e dei rilievi contermini (prov. Di Grosseto). *Geol.Tecnica & Ambientale*, 3:45-56.

Baldi P., Bellini S., Ceccarelli A., Fiordalisi A., Rocchi G., Squarci P. & L. Taffi 1995. Geothermal anomalies and structural features of southern Tuscany (Italy). *Int. Inst. Geotherm. Res. Papers, World Geothermal Congr.*, Florence, May 18-31/1995 : 7-11

Civita M., Cocozza T., Forti P., Perna G., & Turi B. 1983. Idrogeologia del bacino minerario dell'Iglesiente (Sardegna sud occidentale). SAMIM, *Mem. Istituto It. di Speleologia*, Roma, II (2), 137 pp.

Civita M., Guarascio M., Pelizza S., Antonietti M., Fioravanti E., & Lipari D. 1994. L'interruzione, in condizioni controllate, dell'eduzione nel bacino minerario di Iglesias (Sardegna sud-occidentale). *De Re Metallica*, Torino, 289-297

Banks D., Younger P.L., & Dumpleton S. 1996. The historical use of mine-drainage and pyrite-oxidation waters in Central and Eastern England, United Kingdom. *Hydrogeology Jour.*, 4 (4):55-68

Crosta G. & C.A. Garzonio 1998. Engineering geology problems in the closing down of mining areas II: Hydrogeological problems of mining rehabilitation. In *Proceed. Environmental Management*, Sivakumar M., Chowdhury R.N. eds., Pergamon, 2:1153-1162.

Sammarco O. 1993. Karst and thermal water in an underground mine during and after exploitation. Proc. 1st African Symp. *On Mine drainage and Env. Protection from Mine Waste Water Disposal*:671-682.

Zuppi G.M. 1986. Introduzione all'idrologia isotopica. In: Zavatti A. ed. *Proc. Conf. Ambiente: Protezione e Risanamento*, Pitagora Ed., 1:281-335.

Groundwater: Past Achievements and Future Challenges, Sililo et al. (eds) © 2000 Balkema, Rotterdam, ISBN 90 5809 159 7

Salinity issues relating to decommissioned open pits in Western Australia

J.W. Hall
Aquaterra, Perth, W.A., Australia

ABSTRACT: There is increasing concern among the community and the regulators in Western Australia about the potential impacts of final mine voids, on groundwater salinity and downstream beneficial use, as a result of evaporative concentration of pit waters and the down-gradient migration of saline plumes. The initial focus of this concern is in the Pilbara region, Australia's principal iron ore mining province, where most of the groundwater is fresh to brackish. However, the same issues can affect all mines that extend below the water table and where evaporation significantly exceeds rainfall. That is, everywhere in Western Australia and in many of the world's other mining provinces. This paper presents a broad outline of the hydrogeological processes controlling pit void salinity, the critical environmental and regulatory issues and how these are being dealt with in the Pilbara, and how these issues might be dealt with elsewhere.

1 INTRODUCTION

A final void is often the end result of open pit mining. While some pits have been used for the deposition of mine tailings or waste rock (i.e. backfilled), most have been decommissioned leaving a final void.

If the pit extended below the water table, water levels in the final voids will recover to some equilibrium level. In regions where annual evaporation significantly exceeds annual rainfall, as is the case for all of Western Australia's mining provinces, evaporation from the pit water surface will lead to an increase in pit water salinity.

The potential for pit water salinisation and downstream impacts has become a major concern of the community and regulators in recent years. The initial focus of this concern is in the Pilbara region of Western Australia, where most of the groundwater is fresh to marginally brackish and has a high beneficial use. The regulators are taking a cautious approach to new mine approvals and a joint industry/community/government committee, the Pilbara Iron Ore Environmental Committee, has been established to assist government in developing new policy and guidelines.

However, there have been issues raised in a recent discussion document prepared for the Committee which, if applied elsewhere, could put constraints on, or set costly conditions, on the development of some new projects in other mining provinces, even where the background water quality is saline. In assessing the potential for, and possible consequences of, pit void salinisation, an understanding of the hydrogeological issues and processes involved is critical. An outline of these processes is presented, followed by a discussion of the environmental and regulatory issues and how these are being dealt with in the Pilbara region of Western Australia. The paper concludes with how these issues might be dealt with elsewhere, the possible impacts on new projects and how we might overcome them.

2 HYDROGEOLOGICAL ISSUES

2.1 *General hydrogeological processes*

Once pit dewatering (by perimeter bores or in-pit bores or sumps) ceases, a pit lake will generally develop as water levels recover. The water levels will recover to an equilibrium level or balance between inflows to and outflows from the pit.

Where annual evaporation is much higher than annual rainfall, evaporative losses from the pit water surface will become a major component of the pit water balance and the pit will, to some degree become a groundwater sink. Evaporative concentration will also result in increasing pit water salinity. The process is very slow, and can take in the order of hundreds of years before significant salinity increases are observed. There is also the potential for saline water to move out of the pit into surrounding aquifers and impact on local and regional water quality.

2.2 *Pit-aquifer models*

The relationships between pits and aquifers are dependent on numerous site specific characteristics and no two pit-aquifer systems are exactly the same. However, there are three generic pit-aquifer models which cover most existing and potential pit salinisation occurrences. These are known as the groundwater sink, the throughflow pit and the recharge pit, although the latter is rare.

2.2.1 *Groundwater sink*

This occurs where evaporative losses are equivalent to, or exceed groundwater inflows, and the pit becomes a permanent groundwater sink (Fig. 1).This most commonly occurs where the final void is surrounded by an aquiclude, where groundwater flow is restricted by low permeability.

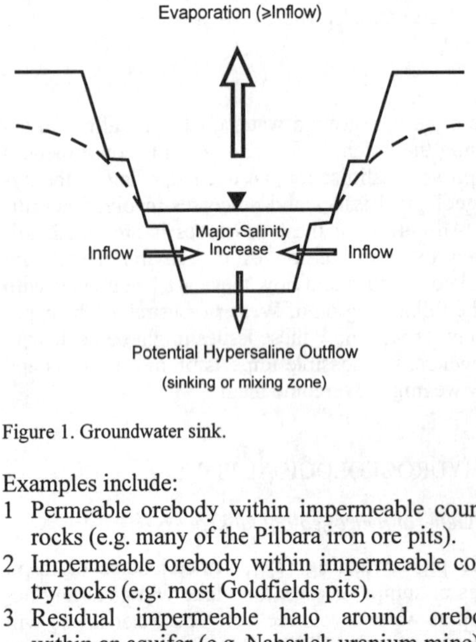

Figure 1. Groundwater sink.

Examples include:
1 Permeable orebody within impermeable country rocks (e.g. many of the Pilbara iron ore pits).
2 Impermeable orebody within impermeable country rocks (e.g. most Goldfields pits).
3 Residual impermeable halo around orebody within an aquifer (e.g. Nabarlek uranium mine).

In the initial stages of pit salinisation, the only loss of water from the pit will be by evaporation and the increasing salinity will be constrained to the pit. With time, however, the pit water salinity may increase significantly above that of the surrounding groundwaters and a plume of the more saline pit water may sink into the surrounding aquiclude or aquifer due to density difference.

This plume may continue to sink through the surrounding less saline groundwater until it reaches some barrier (e.g. more impermeable strata), after which it may move along the surface of the barrier under the influence of gravity (i.e. downhill). Alternately, if there is a significant regional groundwater throughflow around or beneath the pit, the saline plume will mix with the throughflow and move (diluted) down hydraulic gradient. Downgradient salinity will depend on the relative contributions of, and salinities of, pit outflow and groundwater throughflow.

Groundwater sinks can also develop where pits are in good hydraulic connection with aquifers if the surface area of the final void is large enough for evaporative loss to exceed groundwater inflow (e.g. some mine closure options for the Marillana Creek channel iron deposit pits in the Pilbara).

2.2.2 *Throughflow pit*

This occurs where groundwater inflows exceed evaporative losses, and there will be some groundwater outflow (Fig. 2). This usually occurs where the orebodies and pits are within, and in good hydraulic connection with, surrounding aquifer systems.

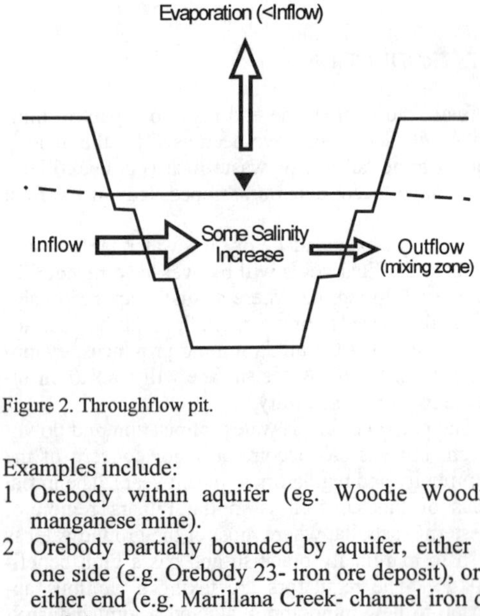

Figure 2. Throughflow pit.

Examples include:
1 Orebody within aquifer (eg. Woodie Woodie-manganese mine).
2 Orebody partially bounded by aquifer, either on one side (e.g. Orebody 23- iron ore deposit), or at either end (e.g. Marillana Creek- channel iron deposits).
3 Orebodies bounded by multiple aquifers and aquicludes, such as permeable shallow horizons over impermeable deep material (e.g. St Ives-goldmine) or impermeable shallow horizons over permeable deep aquifer (e.g. Mt Pleasant- goldmine).

In these cases, there will still be evaporative losses from the pit water surface and this will result in some salinity increase, although the throughflow of water in the pit will tend to restrict the magnitude of the salinity. The longer term pit water salinity will depend on the relative magnitudes of evaporative losses and the pit throughflow. Downgradient salinity will depend on the relative contributions of, and salinities of, pit outflow and groundwater flow around or beneath the pit.

2.2.3 *Recharge Pit*

This occurs when pit inflows exceed evaporation and the final void becomes a source of groundwater recharge. However, this would require significant imputs from incident rainfall and/or rainfall runoff, and although this situation rarely exists as a long term condition. In the tropical north of Australia and the cooler southwest parts of Western Australia, final voids, which are normally a sink or throughflow pit, can become short–term temporary recharge pits during the wetter months.

3 ENVIRONMENTAL & REGULATORY ISSUES

The main issue being addressed in the Pilbara at the moment, is the potential for increases in final void water salinity to impact on other water resources. However, there are two sub-issues that complicate the overall approach to potential salinity problems and management options. These are the concept of beneficial use, and the lack of empirical data on the pit water salinity trends with time.

3.1 *The concept of beneficial use*

This term is being used more and more in relation to a whole range of issues, including water quality management. There are a number of wordy and often confusing definitions for beneficial use (EPA, Water and Rivers Commission etc). There is also confusion because there could be more than one beneficial use. However, generally, the highest potential value use of a water body has been considered its priority beneficial use. That is, if the water body is of potable water quality then its priority beneficial use is as a potable supply source. Similarly if the water body is of stock water quality (up to 10,000 mg/L) then this is its beneficial use. Higher quality water resources have the greatest range of beneficial uses. As water quality deteriorates, the range of beneficial uses is reduced.

However, in the Goldfields mining province, the issue is becoming less clear. The bulk of mineral processing has been by CIL/CIP gold processing plants which have been using both saline and hypersaline groundwater supplies. Historically, all water over 10,000mg/L TDS has been assigned a beneficial use as undifferentiated saline process water. At the present time, however, it could be argued that water at less than around 30,000 to 40,000 mg/L (saline) has a higher beneficial use than water exceeding 50,000 mg/L (hypersaline) in terms of process water supplies.

With recent improvements to reverse osmosis desalination technology, raw water feeds at less than 40,000 mg/L can now be economically desalinated to provide the "boiler feed" quality process water requirements for the pressure acid leach technologies being used in the new laterite nickel projects.

Also, there is an incremental cost of using hypersaline water in processing in terms of reagant use, scaling and corrosion. A recent study indicated that, for example, the cost of using a 100,000 mg/L raw water feed can be as much as 50% higher than using 40,000 mg/L water.

However, having defined the beneficial use, how do you apply the concept? If it were applied rigidly, then it would be difficult to manage any project where there would be some change (however small) in water quality. Happily, the regulatory framework allows for a practical and balanced approach.

The EPA's draft Western Australian Water Quality Guidelines also state that the aim of the National Water Quality Management Strategy is to "achieve the sustainable use of the nation's water resources by protecting and enhancing their quality while maintaining economic and social development". That is, there is recognition that there needs to be a balance between the beneficial use of the water resource and the beneficial outcomes of development.

The current regulatory and approvals processes take the concept of balancing needs and values into account, and is by no means prescriptive (ie. project and site specific assessments are preferred to setting rigid guidelines). However, it is the proponent's responsibility to identify the beneficial uses of water resources that might be impacted, assess the severity of any impacts and to present the case for balancing any reduction in beneficial use options against the overall beneficial value of the project. That is, the overall economic and social value of the project may be such that the downgrading of a part of a water resource from say "stock use" to "process use", or from "saline process use" to "hypersaline process use" may be acceptable.

However, during the review process of a project Environmental Impact Assessment there may well be many different interpretations of the balance between beneficial use and beneficial outcome. These different interpretations may be as a result of different levels of knowledge and understanding of the project, different vested interests and different political agendas.

3.2 *Lack of Empirical Data*

One of the problems facing proponents and regulators, in predicting long-term impacts and developing appropriate management plans, is the lack of relevant empirical data. There just haven't been that many final voids left open to the water table for long enough to provide actual performance data. Also, most of the longer term work that has been done (UK and US) relates to coal and base metals mines, and has been largely focussed on acid mine drainage issues, rather than salinity.

The monitoring of abandoned pits has commenced in Australia and good post mine closure data is being collected. However, the available data only covers tens of years (at best) and only covers certain pit-aquifer types.

For example, the old Goldsworthy pit, in the northern Pilbara, is being used as a living model for pit water salinisation and has generated an extensive monitoring database. The 15 or so years of data do allow for some calibration of predictive salinity models, however, the reliability of the predictions over hundreds and thousands of years remains largely unknown. Also, the empirical model provided by the Goldsworthy pit covers only one pit-aquifer type (groundwater sink).

The application of sound engineering and scientific principles and a detailed understanding of local/regional hydrogeological conditions allows for the best possible prediction of long term salinity increases and resulting impacts. However, the lack of actual performance or impact data means that there will always remain some doubt. It is this doubt, or risk, that causes the problem. There are often and will continue to be differing interpretations on what is acceptable risk.

4 HOW THESE ISSUES ARE BEING DEALT WITH IN THE PILBARA

4.1 *General*

There are numerous projects in the Pilbara where mining has gone, or is planned to extend below the water table. As outlined previously, there are many different types of pit-aquifer interaction, and most of these are represented somewhere in the Pilbara. The general approach has included the following:

1. Extensive baseline data collection and hydrogeological/environmental investigation.
2. Prediction of impacts using modelling techniques- ranging from simple analytical approaches to complex numerical models.
3. Acceptance that there will be some impacts on water resources. Some mines have been approved, where some reduction in the beneficial uses of an aquifer were accepted (e.g. Marillana Creek).
4. Development of management strategies based both on environmental and economic considerations- in some cases, backfilling of pits has been a viable option (e.g. Mining Area C, Paraburdoo 4 ENW).
5. Initiation of detailed monitoring programmes to provide real data for checking of predictions.
6. Development of, and commitment to "fallback" management strategies- if the results of ongoing monitoring confirm that the longer term impacts will be unacceptable and/or cannot be managed.

Fallback strategies that have been considered and investigated include:

- using the pits as water supply sources (short to medium term solution only);
- backfill/infill using waste material from future pit developments;
- isolating the pit (hydraulically) from the aquifer using artificial barriers (eg. neutral barrier technology, grout curtains etc);
- promoting throughflow and rapid turnover of pit water by diversion of creek flows into (and out of) the pit; and
- other, as yet unspecified, techniques and technologies that may be developed in the future.

4.2 *Specific examples*

The following specific examples provide a guide to the level of detail (and consequent cost) that has been required for the assessment of impacts and the planning required for the approval to mine for several recent new projects assessed by the regulators.

4.2.1 *Marillana Creek*

The Marillana Creek channel iron deposit orebodies are part of an extensive paleochannel aquifer system. However, the surface areas of the final voids are very large and the voids have the potential to become groundwater sinks. In approving mine development, the regulators acknowledged that there would be some irreversible affects on the groundwater system in the mining areas.

Since this time, extensive investigation has been directed at developing mine closure plans to restrict impacts to the local groundwater system, and reduce the impacts on the local surface water flow system and reliant vegetation. This work has involved many years of data collection and predictive modelling of future impacts, using both regional scale numerical flow models (which incorporate state of the art stream-aquifer interaction simulation) and minesite specific analytical models to assess the interaction between the final pits and the local aquifers. This work to date has taken over 10 years.

Approval to mine has been granted on the basis that pits will be backfilled if no other long-term management measures can be proven.

4.2.2 *Orebody 18*

Orebody 18 will be a groundwater sink. It was demonstrated, by a brief investigation (less than 3 months) that any saline water emerging (sinking) from the final void would be restricted to a low permeability aquifer which was not being used, and would not likely ever be used for water supply purposes.

However, much follow-up work was requested, as part of the project approval process, to define outflow rates of saline water from the pit, the salinity of the outflow and the impact on the aquifer as a whole. For this investigation, a simple mass balance model was developed.

4.2.3 *Orebody 23*

The Orebody 23 pit is located on the upstream edge of a gorge through which a major river flows.

The orebody is a major aquifer bounded to the north and west by aquicludes and to the south and east by a valley fill alluvial aquifer, and located downstream of an artificial recharge scheme, designed to enhance recharge to the valley fill aquifer. The pit will be a throughflow pit.

As part of dewatering design and environmental investigations, a multi layered groundwater flow model was developed and calibrated against drilling and test pumping data from several phases of work stretching over 3 years. The model was used to predict dewatering requirements, the impacts of mine dewatering and the impacts of the final void after mine closure. Solute transport modelling, using both simple one dimensional analytical models and the multi layered flow model, was also undertaken as part of the prediction of downstream impacts. The predictive modelling and assessment took some 6 months to complete.

The predicted downstream impacts were such that the proponent committed to ongoing research and monitoring programme with a fallback position of infilling the pit with waste from other future developments.

4.2.4 *Mining Area C*

A major hydrogeological investigation programme was undertaken as part of the EIS for the future development of the Marra Mamba orebodies in the central Pilbara region, located upstream of the environmentally and culturally sensitive Weeli Wolli Spring. This investigation, which took almost a year to complete, included the drilling and testing of numerous bores (to supplement existing bores from previous investigations) and the development of a regional scale, multi layered groundwater flow model.

Preliminary model predictions indicated that the final voids would have little measurable impact on the downstream aquifers and spring. However, modifications to the mine plan and scheduling meant that infilling of the pits became an economically viable option and was incorporated into the mine plan.

5 IMPLICATION FOR OTHER MINING PROVINCES (PARTICULARLY THE GOLDFIELDS)

The current focus of the regulators in this regard is the Pilbara, where most of the groundwater resources are of potable or near potable quality. Indications are that mining operations in the southwest part of Western Australia (mostly mineral sands) will probably be the next area of focus. Again much of the groundwater resources in this region are fresh.

Is it likely to happen in the Goldfields? While much of the groundwater resources are saline, there are pockets of brackish and fresh groundwater. Also, as outlined above, some of the saline groundwaters do have a beneficial use as selective process water as distinct from the hypersaline sources.

As such, existing and new projects will eventually have to deal with the issue. That is, where the brackish and fresh waterbodies may be impacted by final voids, the potential beneficial uses of these water bodies will need to be taken into account during the planning and approvals processes. This will require some investigation and assessment of the hydrogeology of the pit and local regional aquifers as outlined above. However, where any pit is planned to go below the water table some investigation will be necessary in any case to assess potential pit inflows and determine mine dewatering strategies.

In many cases the work required to assess the likely impacts of the final pit, may be only marginally more than a specific dewatering investigation. In other cases, a great deal of additional investigative effort may be required to address the final void issue. However, in either case, much of the work required to address dewatering and longer term impacts of the final pit will be similar. If the work is done concurrently, duplication can be avoided and costs reduced.

Planning then, is the key. If there are water bodies near the final voids that might be impacted, these should be identified and investigated (along with dewatering) during mine feasibility studies. In this way, the proponent can develop realistic and economically viable management plans and/or prepare a case for balancing loss of beneficial use of a (or part of a) water body with the economic and social benefits of the mine as a whole.

REFERENCES

BHP Iron Ore Pty Ltd, 1995. *Yandi Mine Expansion Central Mesa 1 and* 2. Consultative Environmental Review.

BHP Iron Ore Pty Ltd, 1996. *Orebody 18 Consultative Environmental Review.*

BHP Iron Ore Pty Ltd, 1997. *Newman Satellite Development Mining of Orebody 23 Below the Water Table. Consultative Environmental Review.*

AusIMM, IE Aust & Goldfields Esperance Development Commission, 1997. *Hypersaline Water Seminar, Kalgoorlie, WA.*

BHP Iron Ore Pty Ltd, 1997. *Multiple Iron Ore Development Project Mining Area C. Public Environmental Review.*

Environmental Protection Authority, 1993. *Draft Western Australian Water Quality Guidelines for Fresh and Marine Waters.* EPA Bulletin 711.

Environmental Protection Authority, 1995. *Duplication of Iron Ore Mining Operation, Yandi Mine ML270SA, Hamersley Range, 90km North-west of Newman.* EPA Bulletin 80.

Golder Associates, 1997. *Draft Discussion Paper on Mining Below the Water Table in the Pilbara.* Prepared for the Department of Resources Development and the Pilbara Iron Ore Environmental Committee.

Water and Rivers Commission, 1997. *Draft Overview-Mining and Mineral Processing-Protection of Water from Pollution in Western Australia.*

Groundwater: Past Achievements and Future Challenges, Sililo et al. (eds) © 2000 Balkema, Rotterdam, ISBN 90 5809 159 7

The impact of mining on groundwater quality in the Tarkwa area, Ghana

B.K. Kortatsi
Water Research Institute, Council for Scientific and Industrial Research, Accra, Ghana

ABSTRACT: The emissions of SO_2 due to roasting of sulphide ores for gold processing at Prestea-Bogoso has led to mild acid ($5.0 \leq pH \leq 6.5$) rain production in Prestea-Tarkwa area. The infiltrating water acquires further acidity from the soil zone due to the abundance of CO_2 resulting from decaying forest vegetation and runoff over loosened sulphide minerals. Groundwaters are thus acidic ($3.9 \leq pH \leq 6.9$). Due to the acidity of the groundwaters, trace metals (iron, manganese, cadmium and nickel) are released from the rock matrix into the groundwater system in higher concentrations than the WHO (1993) guideline limits. The concentrations of Fe^{++}, Mn^{++}, Cd and Ni are respectively in the ranges 0.01-18.0 mg l^{-1}, 0.01- 2.0.8mg l^{-1}, 0.007-0.084 mg l^{-1} and 0.009 - 0.126mg l^{-1}. These preliminary results suggest that mining has negative impact on groundwater quality. However, results from identical hydrogeological environment without mining are being processed for comparison to confirm the preliminary findings.

1 INTRODUCTION

The Prestea–Tarkwa area is one of the major gold mining areas of Ghana. Gold mining in this area started about five hundred years ago however written records have restricted the period to only 120 years (Marston et al. 1993). Intense mining activities however began in 1930s and peaked in 1960. The Ghana government's policy of trade liberalisation and amendment of mining laws to attract foreign investments into the mining industry has yielded dividend and accounted for the gold rush from1980s todate. Both small and large-scale mining activities have increased about five fold; new mining concessions are acquired, old and abandoned mines have been rehabilitated and mineral prospecting has been intensified. Gold is not the only mineral mined in the area, the principal manganese mines in the country occurs at Nsuta, which is also situated within the area. The intense mining activities in the area have resulted in environmental degradation including pollution of water resources particularly surface water resources (Acquah 1993). River Ankobra, which is the major river draining area, is regarded dead due to its high level of pollution (WRRI 1986). Though surface water pollution or contamination is well documented as stated above, very little knowledge exits on the level of groundwater pollution or contamination as a result of mining operations. The only data available on groundwater quality are those generated by the individual mines. These are often extremely difficult if not impossible to come by. In spite of lack of knowledge on groundwater quality, the drinking water supply of this area is increasingly being based on groundwater due to the assumption that surface water resources are polluted while the groundwater excluded from the surface are less polluted. Recent complains as the Friday 21st January 2000 issue of the Ghanaian Times (Newspaper) of some rashes, respiratory, urinal and eye problems associated with drinking waters from some boreholes suggest that heavy metal related health problems may be associated with the groundwater.

In order to understand the hydrochemistry, water quality and determine the effect of mining on groundwater quality and pollution levels of the groundwater, the University of Ghana (Geology Department), DANIDA-ENRECA hydrogeological project has instituted a Ph.D programme in groundwater monitoring in the in the area. The data for this paper is taken from this programme.

2 BACKGROUND OF THE STUDY AREA

The study area lies within a 30 kilometres radius of Tarkwa and has an area of approximately 2824km^2. It extends from 1° 54′ to 2° 11′ W and from 5° 06′ to 5° 35′ N (Fig. 1).

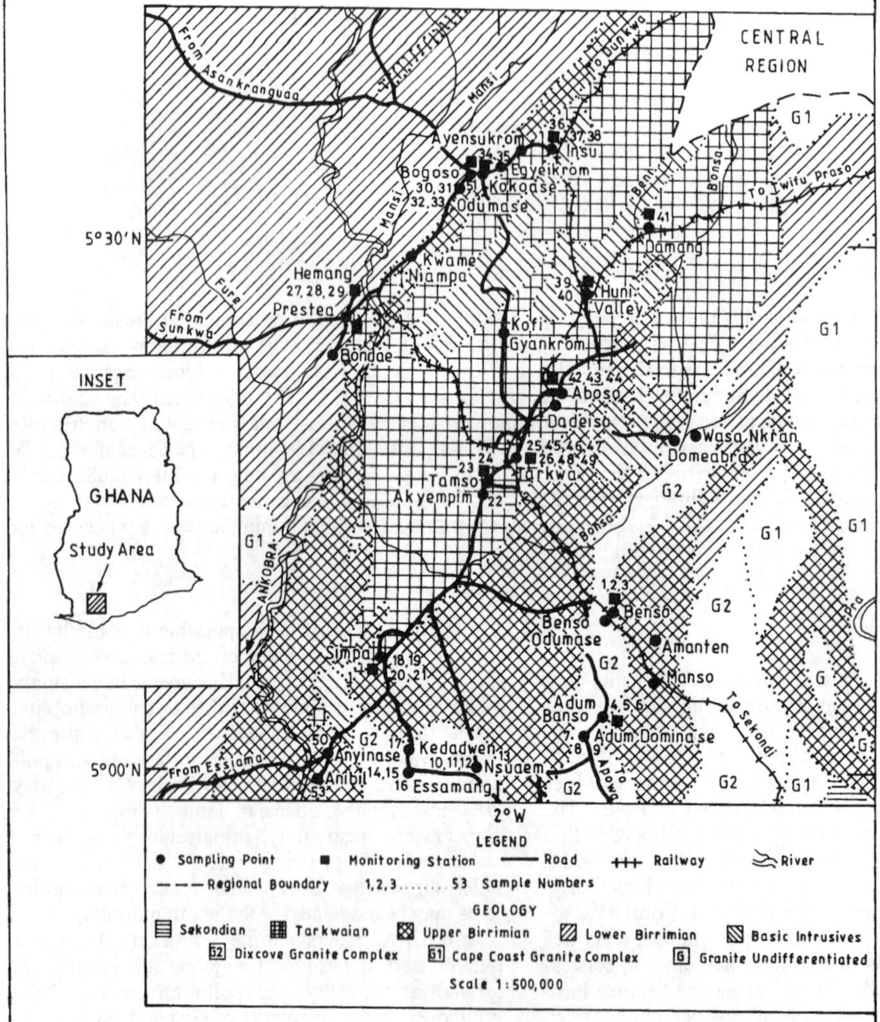

Fig. 1 : Geological map of Tarkwa area showing multi-sampling points and monitoring stations

2.1 *Climate and vegetation*

The climate of the study area falls partly under the wet semi-equatorial and partly under south- western equatorial climatic zones of Ghana (Fig. 2). It is characterised by double rainfall maxima with a mean annual rainfall between 1450 and 2600mm.The highest mean monthly temperature of the study area is approximately 30°C and occurs between March and April. The lowest temperature is approximately 26°C and occurs in August (Dickson & Benneh 1980). The average monthly humidities are in the range (75-80%), (Dickson & Benneh 1980).

The vegetation of the study area consists partly of the tropical rain forest and partly of the moist semi-deciduous forest. The tropical rain forest occurs in the south-western portion of the study area while the moist semi-deciduous forest covers the remaining portion. The combined effect of high temperatures and heavy rainfall that is well distributed through out the year has given rise to rapid and luxurious forest growth in the south- western portion of the study area. The forest is always evergreen with trees that are between 15 metres and 45 metres tall (Dickson & Benneh 1980). The trees of the lower layers stay evergreen all the year round due to the cover provided by the upper and middle tree layers. Because of farming and other land use practices, most of the original vegetation has been reduced to secondary forest.

2.2 *Hydrogeological setting*

The geology of the study area is mainly of the lower Proterozoic divided into the Birimian system, unconformably overlain by the Tarkwaian system. Sills and dykes of igneous rocks ranging from felsite and quartz porphyry to meta-dolerite, gabbro and norite intrude into the Birimian and the Tarkwaian system at several places.

The Tarkwaian system occupies an elongated and narrow geosyncline about 250km long and 16km wide that stretches in a northeastern-southwestern direction. It consists of an overall thick clastic sequence of argillaceous and arenaceous sediments (mainly arenaceous) with two well-defined zones of pebbly beds and conglomerate in the lower members of the system. The Tarkwaian system is subdivided into four main groups, namely Kawere group, Banket series, Tarkwa Phyllite and the Huni sandstone (Junner et al 1942).

The Birimian (Upper and Lower) rocks are largely folded, intruded by granitic rocks and have undergone a low-grade metamorphism. The Upper Birimian system is dominantly of volcanic and pyroclastic origin. The rocks consist of a bedded group of green lavas (greenstones), tuffs and sediments with minor bands of Phyllite that include a zone of manganiferous phyllites containing manganese ore. The Lower Birimian is mainly pelitic in origin and consists of great thicknesses of alternating shales, phyllites, greywacke and argillaceous beds with tuffs and lavas. The stratigraphy of the study area is presented in Table 1.

Table 1. Stratigraphy of the Tarkwa-Prestea area.

Formation	Thickness [m]	Composite lithology
Quartz/pyrites/tourmaline veins		
Intrusive dolerite, felsites, porphyry and granitoids		
TARKWAIAN SYSTEM		
Huni Sandstone & Dompim Phyllite	1370	Sandstones, grits feldspathic quartzites with bands of phyllites.
Tarkwa Phyllites	120-400	Green and greenish grey chloritic and sericitic phyllites and schists,
Banket Series	120-600	Quartzite, Staurolitic quartzite, sandstones, grits, breccias and quartz-pebble conglomerates.
Kawere Group	250-700	Quartzites, grits and phyllites and conglomerates.
Angular unconformity		
Birimian rocks		

Groundwater in the study area is associated with the development of secondary porosities as a result of fissuring and weathering. Tectonic lines of weakness created by the orogenic regimes have facilitated extensive and deep weathering/thick regolith development under the highly humid tropical climate with distinct wet and dry seasons. The weathering depth is greatest in the Birimian rock areas where it has reached between 90m and 120m. The weathering in the granites, porphyries, felsites and other intrusive igneous rocks is deep but less than in the Birimian rocks. However, in the Banket Series quartzites, grits, and conglomerates and the Tarkwa phyllites, the weathered depth rarely exceeds 20m (Junner et al. 1942). Aquifers are mainly semi-confined or confined in the weathered zones just above the transition zone between the weathered and the fresh aquifer or in the fractured /fissured zone normally at some depth below the weathered zone. Recharge to the aquifer system is mainly through seepage of rainwater along fissures and the weathered portions at rock outcrops where the argillite content of the formation is low.

3 METHODOLOGY

Water samples were collected from all wells (both hand-dug wells and boreholes) and major rivers in the study area. In all 56 samples were collected. All samples for major ions and trace elements analyses were collected in high-density linear polyethylene (HPDE) 1-litre bottles. The bottles were cleaned at the Environmental Chemistry Laboratory of the Water Research Institute of the Council for Scientific and Industrial Research of Ghana by filling them with a 50:50 solution of concentrated reagent grade hydrochloric acid and allowing them to stand for at least 24 hours. The bottles were then triple rinsed with distilled and de-ionised water. Glass bottles (30-ml glass bottles with poly- sealed lids) were used for the collection of samples for stable isotope analyses.

Water samples were collected using sampling protocol described by (Claasen 1982, Barcelona et al. 1985, Gale & Robins 1989). To remove particulate matter from samples for major and minor ion analysis, filtering was performed using a Sartorius polycarbonate filtering apparatus and a 0.45 μm. cellulose acetate filter membrane. The samples meant for heavy metals analysis were acidified to a pH < 2 after filtration (Appelo & Postma 1999). Unfiltered ground and surface water samples were collected in 30-ml glass bottles with poly-sealed lids for stable isotopic analyses.

On-site analyses of temperature, redox (Eh), pH and electrical conductivity were conducted using WTW-Multiline P4 Universal Meter. Alkalinity titration was carried out at the wellhead using HACH Digital Titrator Model 16900. All major ions (Sodium (Na^+), Potassium (K^+), Calcium (Ca^{++}), Magnesium (Mg^{++}), Bicarbonate (HCO_3^-), Chloride (Cl^-), Sulphate (SO_4^{--}) as well as some trace elements such as Nitrate (NO_3^-) Fluoride (F^-) were analysed using Standard Methods for the Examination of Water and Wastewater (APHA-WWA-WPCF 1994). The analyses of heavy metals were carried out using the Unicam 969 Atomic Adsorption Spectrometer (AAS). The stable Isotopes of deuterium and ^{18}O analyses were performed at the Geological Institute of the University of Copenhagen.

4 RESULTS AND DISCUSSIONS

4.1 *Geochemical data*

Geochemical data on the study area is scanty. However, according to Marston et al. 1993, the chemistry of the Banket ore (the Tarkwaian system) is very matured and free of significant quantity of all but silica and iron oxides and trace elements content is mainly a function of the opaque mineral content. Hematite, magnetite, pyrite, chalcopyrite, and trace amounts of arsenopyrite are also found associated with the ore. The analysis of representative of the oxidised mineralisation in the sandstones and dolerite at Damang in the northeastern part of the study area does not indicate concentrations of unusual trace elements and arsenic is virtually absent. High iron, vanadium and chromium contents have been observed in the samples. These have been attributed to the dolerite host rock. Other elements with significant concentration are manganese, barium and aluminum.

4.2 *Hydrochemical data*

Representative chemical data from boreholes, hand-dug wells and major streams (Rivers) are presented in Table 2. Approximately 32% of the groundwaters are (Mg + Ca + Na) $HC0_3$ water type where no particular cat-ion dominates while Bicarbonate is the major anion. Other water types of significance that occur in the study area are the Ca $(HC0_3)_2$ and Na Cl water types. Each occurs in 15% of the wells (hand dug wells and boreholes).

The waters are generally poorly buffered and acidic ($3.9 \leq pH \leq 6.9$) with low electrical conductivity values, which are in the range of 37 – 670 μS cm^{-1} with a mean value of 246.4 μS cm^{-1}. Eh values vary from 24.0mV to 186 mV suggesting aerobic (Oxidising) environment. The low electrical conductivity or total dissolved solids (TDS) values of the waters suggest short residence time or very little rock/water interaction. Since groundwater recharge is mainly by rainfall, the low pH of the groundwater could be attributed partly to the emissions of SO_2 due to roasting of sulphide ores for gold processing at both Prestea and Bogoso (Acquah 1993). The mean pH of rainwater in the Prestea-Tarkwa area is 6.0 (Osafo et al. 1995). Although acid mine drainage have not been seriously considered in the study area (the dominant gold ore in the study area is non-sulphidic paleoplacer or free milling ore), pyrite and chalcopyrite are known to be associated with the dolerites and quartz veins that have intruded the Tarkwaian (Junner et al. 1942). Furthermore, the Birimian ore that occurs mainly at the fringes of the study area is refractory quartz-Fe/As sulphide lode gold (Marston et al. 1993).

Acid mine drainage is bound to occur when recharging water comes into contact with pyrite, chalcopyrite and pyrrhotite present in the rock wastes. Thus the acidity of the waters can also be partly attributed to acid mine drainage. The infiltrating water may acquire further acidity from the soil zone due to the abundance of CO_2 resulting from decaying forest vegetation.

Due to the acidity of the infiltrating water, some trace metals notably iron, manganese, cadmium and nickel that are present in the matrix of the rock loosened by mining (waste rocks) are easily leached into the groundwater system in higher concentrations than the (WHO 1993) guideline limits. The release of the trace metals from the host rock into the groundwater system also occurs in the saturated zone due to the aggressiveness of the groundwater. Table 2 contains some trace metals analyses of water samples collected from the study area. The concentrations of Fe^{++}, Mn^{++}, Cd and Ni, which are respec-

Table 2. Representative chemical data from boreholes, hand-dug wells and major streams (rivers) within the study area.

Sample No	Temp °C	pH	EC µS/cm	Ca Mg/l	Mg mg/l	Na mg/l	K mg/l	HCO3 mg/l	Cl mg/l	SO4 mg/l	Fe mg/l	Mn mg/l	Cu mg/l	Zn mg/l	Cd mg/l	Pb mg/l	NI mg/l	As mg/l	Hg mg/l
1	26.3	5.7	313	19.9	9.9	20.7	6.2	63.4	40.0	22.1	0.12	0.27	0.001	0.104	<0.001	<0.001	<0.001	<0.001	<0.001
2	26.4	5.3	105	4.0	1.5	13.6	2.5	35.5	11.2	0.5	0.58	0.10	0.044	0.027	0.027	<0.001	0.041	<0.001	<0.001
3	26.2	5.2	85	5.1	3.9	8.0	2.4	45.1	4.0	0.5	0.09	0.13	0.040	0.116	0.011	<0.001	0.055	<0.001	<0.001
4	26.2	5.6	357	35.9	9.0	18.5	0.8	87.8	46.0	30.9	0.13	0.42	0.020	0.042	0.028	<0.001	0.001	<0.001	<0.001
5	27.2	5.6	281	22.5	9.8	16.5	0.4	64.7	32.8	21.0	0.87	0.58	0.001	0.070	<0.001	<0.001	<0.001	<0.001	<0.001
6	26.6	6.2	362	29.6	12.2	14.5	2.0	129.0	23.6	15.4	0.16	0.41	0.046	0.053	0.061	<0.001	<0.001	<0.001	<0.001
7	26.3	5.6	263	16.7	5.7	20.8	2.5	53.7	33.2	6.3	0.01	0.12	0.001	0.042	0.013	<0.001	0.009	<0.001	<0.001
8	27.5	5.5	239	15.5	5.5	20.3	1.9	51.2	33.0	0.5	0.08	<0.001	0.046	0.035	0.029	<0.001	<0.001	<0.001	<0.001
9	26.5	5.5	377	26.3	9.2	22.4	4.4	56.1	43.0	6.1	0.01	<0.001	0.054	0.048	0.075	<0.001	<0.001	<0.001	<0.001
11	27.7	6.0	152	13.6	6.4	12.5	0.5	76.9	4.8	11.2	0.11	0.03	0.049	0.174	0.040	<0.001	0.001	<0.001	<0.001
12	27.3	5.9	189	13.1	6.5	16.0	0.7	85.4	13.6	10.8	0.12	<0.001	0.010	0.165	0.051	<0.001	0.045	<0.001	<0.001
13	27.9	6.0	268	22.8	8.3	17.5	0.8	98.8	43.8	8.4	0.11	0.05	0.067	0.117	0.072	<0.001	0.022	<0.001	<0.001
14	27.2	5.8	118	10.6	3.2	10.7	0.5	63.4	5.0	10.3	0.12	<0.001	0.071	0.367	0.048	<0.001	0.001	<0.001	<0.001
15	27.1	5.8	239	13.1	10.6	17.9	1.0	83.0	24.8	10.2	0.09	0.02	0.026	0.137	0.022	<0.001	0.001	<0.001	<0.001
16	26.6	5.9	158	17.2	3.5	11.9	0.5	84.2	6.8	9.9	0.02	0.15	0.067	0.189	0.011	<0.001	0.001	<0.001	<0.001
17	27.6	5.7	266	52.0	5.9	19.8	0.8	129.3	31.2	8.8	0.21	0.27	0.006	0.183	0.031	<0.001	0.001	<0.001	<0.001
18	26.8	6.3	237	29.6	3.1	15.8	0.8	129.0	4.2	15.6	4.17	0.30	0.001	0.084	0.053	<0.001	0.001	<0.001	<0.001
19	27.5	4.8	92	4.2	1.3	11.7	1.7	12.2	11.6	12.0	0.11	0.18	0.001	0.117	0.062	<0.001	0.001	<0.001	<0.001
20	27.6	6.2	236	30.2	1.1	17.0	1.1	90.3	16.6	14.1	0.26	0.18	0.012	0.352	0.011	<0.001	0.016	0.001	<0.001
21	26.9	6.1	621	15.2	8.0	37.6	2.3	13.5	104.8	29.9	11.58	1.18	0.001	0.147	<0.001	<0.001	0.001	0.002	<0.001
22	27.6	5.9	248	20.3	3.1	19.1	6.0	64.7	27.4	20.8	1.42	0.45	0.001	0.215	0.018	<0.001	0.001	0.002	<0.001
23	26.5	4.4	87	1.6	1.5	9.6	0.4	7.3	8.0	10.5	0.01	0.33	0.001	0.036	0.011	<0.001	<0.001	<0.001	<0.001
24	26.5	4.1	201	2.6	1.7	19.5	2.1	10.0	23.8	13.7	0.17	0.50	0.001	0.068	0.075	<0.001	<0.001	<0.001	<0.001
25	26.2	5.9	263	17.2	2.7	15.5	3.1	35.4	10.6	42.5	0.34	0.37	0.001	0.072	0.028	<0.001	<0.001	<0.001	<0.001
26	26.0	5.6	180	28.4	2.1	11.3	5.5	57.3	13.4	45.5	0.20	0.13	0.001	0.053	0.030	<0.001	<0.001	<0.001	<0.001
27	26.9	5.6	233	4.2	13.1	19.0	1.9	64.7	29.2	25.5	0.14	0.24	0.001	0.086	0.019	<0.001	<0.001	<0.001	<0.001
28	26.5	5.7	104	7.9	4.5	8.9	0.5	67.1	1.0	11.4	2.66	0.20	0.001	0.107	0.001	<0.001	<0.001	<0.001	<0.001
29	26.2	6.2	226	16.0	10.5	11.4	0.6	124.0	2.5	12.5	8.93	0.17	0.001	0.078	0.047	<0.001	<0.001	<0.001	<0.001
30	26.9	5.9	144	0.3	8.7	15.3	0.7	56.1	9.6	15.8	3.53	0.18	0.001	0.110	0.067	<0.001	<0.001	<0.001	<0.001
31	27.5	4.0	522	11.5	12.8	39.8	24.1	85.0	82.0	30.6	6.40	0.23	0.073	0.371	0.035	<0.001	<0.001	<0.001	<0.001
32	26.1	5.7	91	1.0	4.6	6.5	0.4	37.9	5.4	15.5	9.81	0.14	0.001	0.064	0.021	<0.001	<0.001	0.037	<0.001
33	25.9	5.8	113	4.6	4.1	7.7	0.4	47.6	3.8	18.4	11.30	0.25	0.001	0.045	0.045	<0.001	<0.001	0.001	<0.001
34	27.0	5.4	105	5.9	4.1	11.8	0.7	28.1	10.0	14.2	0.12	0.00	0.001	0.108	0.070	<0.001	0.119	<0.001	<0.001
35	26.7	5.4	278	19.4	4.4	21.3	1.7	40.5	41.5	12.1	0.07	0.18	<0.001	0.035	0.033	<0.001	0.016	<0.001	<0.001
WHO 1993 guideline values:						200			250	250	0.3	0.5(P)	2.0(P)	3.0	0.003	0.01	0.02	0.01(P)	0.001

* (P): provisional guideline value

tively in the ranges 0.01-18.0mg l^{-1}, 0.01-2.08 mg l^{-1}, 0.007-0.084 mg l^{-1} and 0.009-0.126 mg l^{-1} are above (WHO 1993) guideline limits in several wells (Table 2). The concentrations of several other trace metals are below detection as result of the impoverishment of the parent rocks of those trace metals.

The chemical reactions that produce acid mine drainage can occur naturally without mining, when sulphide rocks, particularly pyrites and chalcopyrite are exposed to the air and rain by weathering processes. The acidity and the heavy metal levels of the groundwater cannot be attributable to mining alone without comparison from similar unmined hydrogeological environments. Results from similar hydrogeological environment without mining are being processed for comparison.

5 CONCLUSIONS

From the present study following conclusions could be made:

- Groundwaters in the Prestea-Tarkwa area are generally acidic probably due to mild acid rain production as a result of sulphide ore smelting at both Prestea and Bogoso and acid mine drainage as recharge waters infiltrate through mine waste rocks.
- The natural acidity of the groundwater allows trace metals (iron, manganese, cadmium and nickel) to be released from the rock matrix into the groundwater system in greater concentrations than the WHO (1993) guideline limits.
- The concentrations of several trace metals are below detection due to the impoverishment of the host rocks of these trace metals.
- The preliminary results suggest that mining has a negative impact on groundwater quality. However, since the reactions that produce acid mine drainage can take naturally take place without mining, results from identical hydrogeological environment without mining are needed for comparison to confirm the preliminary findings.

REFERENCES

Acquah, P.C. 1992. Emerging Trends in gold are processing and some related environmental issues in Ghana. Regional Trends in African Geology. Proceedings of the 9th International Geological Conference (Accra, 2nd to 7th November 1992) Geological Society of Africa.

APHA-WWA-WPCF 1994. Standard methods for the examination of water and wastewater. American Public Health Association (APHA) Washington DC, USA.

Appelo, C.A.J. & D. Postma 1999. Chemical analysis of groundwater. Geochemistry, groundwater and pollution. AA Balkema/Rotterdam./Brookfield.

Barcelona, M., J. P Gibb, J. A. Helfrich & E. E Garske. 1985. Practical guide for groundwater sampling. Illinois State Water Survey ISWS Contract Report 374.

Claasen, H. C. 1982. Guidelines and techniques for obtaining water samples that accurately represent the water quality for an aquifer. U.S. Geological Survey Open File Report 82-1024. 49 pp.

Dickson, K. B. & G. Benneh. 1980. A New Geography of Ghana, Longmans Group Limited, London.

Gale, I. N. & N. S. Robins. 1989. The sampling and monitoring of groundwater quality. British Geological Survey. Hydrogeology Report, No 89/37 .

Junner, N. R., T. Hirst & H. Service. 1942. Tarkwa Goldfield. Memoir No.6. Gold Coast Geological Survey.

Marston, R. J., P. Woolrich & J. Kwesi. 1992. Closely associated stock work and Paleoplacer gold mineralisation in the early Proterozoic Tarkwaian System of Ghana. Regional Trends in African Geology. Proceedings of the 9th International Geological Conference, Accra. 2nd-7th November 1992. Geological Society of Africa/ Geological Society of Ghana.

Osafo R. A & Paintsil D. A. 1995. Pollution and water quality studies. Effect of mining activities on River Ankobra and its tributaries. Unpublished Water Resources Research Institute's Technical publication.

WHO 1993. Guidelines for drinking water quality. Geneva.

Groundwater: Past Achievements and Future Challenges, Sililo et al. (eds) © *2000 Balkema, Rotterdam, ISBN 90 5809 159 7*

Witwatersrand deep Au mines: Windows into deep subsurface microbial communities

T.C.Onstott & D.P.Moser
Geosciences Department, Princeton University, N.J., USA
R.J.Wilson
Turgis Technology (Pty) Limited, Northcliff, South Africa

ABSTRACT: A major challenge to studying microbial life at great depths is the limited access to this environment. The deep (>3 km) Au mines of the Witwatersrand in South Africa, offer a unique opportunity to examine microbial diversity and biogeochemical processes within the continental crust. We recently collected rock, water and biofilm samples from several of these mines. Rock sampling focused on an actively-mined, Au, UO_2 and sulfide-rich organic layer at 3.2 km beneath the surface. Water samples were collected from boreholes at 0.9 to 3.3 km below surface. Rock and water enrichments yielded aerobic, anaerobic, mesophilic, thermophilic, radiation- and saline- tolerant bacteria that utilize organic and inorganic compounds, including H_2 and CH_4. The "service" water microbial communities are distinguishable from the "fissure". Isotopic analyses indicate that the borehole water is of meteoric origin, but exhibits extensive rock/water interaction in some cases. Open boreholes are rich in metal-oxidizing and -reducing bacteria.

1 INTRODUCTION

Borehole water samples collected in 1996 from 3.3 km below surface. at level 109 of shaft #1 at Western Deep Levels South Africa (Onstott et al., 1997) yielded a novel *Thermus* sp., designated strain SA-01 (Kieft et al. 1999). This bacterium is radiation resistant (D_{37} = 200 Grays) and grows anaerobically with Fe(III) as the terminal electron acceptor, optimally at 65°C. SA-01 can also reduce S° and other metals including Cr(VI), Co(III)-EDTA, and U(VI). Although the physiology and genetics of the genus *Thermus* have been studied for nearly three decades this metabolic versatility has not previously been reported. This strain was enriched initially on an Fe(III) medium with H_2 as the electron donor. Samples of the Carbon Leader exhibited response to the same medium, although attempts at subsequent transfer and purification were not successful. Based upon these results, a field campaign was undertaken in late 1998 to determine whether these mines offer ready access microorganisms with unique capabilities, some of which may be indigenous to the subsurface crustal environment. The following questions were addressed:

To what extent are rock and water samples microbially contaminated by mining activity? Are the microbial contaminants distinct from potential indigenous microorganisms? What sampling strategies must be developed to reduce and quantify this contamination?

Does the radiation emanating from the UO_2-rich, organic layer provide a significant source of energy for indigenous microbial communities by radiolytic generation of H_2 and O_2 (Hofmann, 1992)?

2 PROCEDURES

An expedition of scientists from USA, Canadian and Swedish universities and U.S. DOE National labs investigated microorganisms associated with deeply-buried rocks and groundwater from East Driefontein, West Driefontein, Kloof and Beatrix mines South Africa. Rock sampling focused on a Au, UO_2 and sulfide-enriched organic layer at 3.2 km below surface. in East Driefontein Mine. The groundwater samples were collected from boreholes and dripping fissures at depths ranging 0.87 to 3.3 km below surface and temperatures ranging from 20° to 60°C from all four mines.

Twenty water samples, fourteen Carbon Leader and quartzite rock samples and three biofilm samples were collected. Samples were collected of the "service" water used in mining, from the "dolomite" water and from several boreholes emanating "fissure" water at varying rates (Moser et al. 1999). Microbial water samples were collected in sterile serum bottles. Microbial rock samples were collected after applying rhodamine dye and fluorescent microspheres to track contamination.

Over 20 types of microbial media were inoculated on site in an anaerobic glove bag with rock samples, groundwater and biofilms to culture aerobic and anaerobic, mesophilic, thermophilic and halotolerant bacteria capable of utilizing a variety of organic and inorganic compounds as energy sources. H_2 was injected into the headspace of most media to determine if the bacteria were capable of utilizing it as an electron donor. CH_4 was injected into the headspace of select media to determine if autotrophic methanotrophy was present (Kotelnikova and Pedersen, 1997). Additional water samples were collected in 12 liter, autoclaved metal canisters, carried to the surface, filtered. The filters were frozen immediately to preserve the microorganisms for DNA and phospholipid fatty acid (PLFA) analyses (Pfiffner et al., 1999).

3 RESULTS

Some of the more significant results are as follows: PLFA analyses revealed the presence of a significant "epoxide" fatty acid in the "service" water at East Driefontein and an acid mine water at West Driefontein that is not in "fissure" water samples. The epoxide probably reflects "disinfection" of the "service" water with chlorine and bromine (Pfiffner et al., 1999). This compound provides a valuable marker for bacterial contamination from "service" water. The PLFA data also indicate that the bacterial biomass varies from 200 to 10^5 cells/ml in the "fissure" water.

Methanotrophy in the presence of Fe(III) and Mn(IV) was observed for 60°C incubations of some "fissure" water collected at 3.2 kmbls. These may represent consortia of formate producing methanotrophs and anaerobic respirers (Kotelnikova et al., 1999).

An extremely large, 4µm diameter, aerobic microorganism was grown at 60°C in the presence of aromatic organic compounds from "fissure" water samples collected at 3.2 kmbls. The 16S rDNA identity of this isolate as belonging to the genus *Bacillus* (DeFlaun et al., 1999).

The $\delta^{13}C$ of the C1-4 compounds present in "fissure" water exhibits a trend similar to those previously reported for the Canadian and Scandinavian shield groundwater (Sherwood-Lollar et al. 1993).

The $\delta^{18}O$ and chemistry of the "fissure" water samples are distinct from that of the "dolomite" and "service" water (Onstott et al. 1999) and modern precipitation (Duane et al., 1997) suggesting at least a Pleistocene age (Fig. 1).

The $\delta^{13}C$ (-1 to -3‰ PDB) and total DIC of the "fissure" water is less than that of the "dolomite" water ($\delta^{13}C$ = -10‰ PDB) suggesting that CO_2 reduction or fixation may be occurring.

"Service" water and air samples yielded low levels of aerobic, mesophilic heterotrophs, some sulfur oxidizers and one positive thermophilic heterotroph enrichment. The "dolomite" water also appears to contain some sulfur and methane oxidizers based primarily upon the DNA data.

The "fissure" water typically contains mesophilic to thermophilic anaerobes. The type of bacteria that were enriched appears to correlate with the geochemistry of the "fissure" water. For example, the 60°C, "fissure" water from Kloof contains 1 ppm of dissolved O_2 with trace sulfate and inorganic carbon. Positive enrichments for hydrogen-oxidizing or "Knallgas" bacteria were obtained from these water samples. Many hydrogen-oxidizing bacteria are autotrophic, coupling the energy of this reaction with growth by using the Calvin cycle to fix CO_2.

Sulfate and sulfide-rich, 45°C, "fissure" water from 2.7 km below surface in West Driefontein yielded H_2 utilizing, sulfate reducing bacteria (Baker et al. 2000). Saline tolerant, sulfate reducing bacteria are also present in a FeS biofilm associated with a dripping borehole in East Driefontein where the salinity is 5 wt.% and the temperature is 37°C (Pfiffner et al. 1999). Mesophilic Fe-oxidizers also form a component in this biofilm community (Ghiorse et al. 1999) as has been reported for other Au mine biofilms (Doig et al. 1995).

4 CONCLUSIONS

The "fissure" water encountered in the deep Au mines of South Africa contain populations of bacteria that appear to be indigenous to the subsurface environment and are distinguishable from bacteria associated with the mining water. The isotopic data suggest that these bacteria have been in residence over geologically significant periods of time and that they may be influencing the geochemistry of the "fissure" water and gases. Future research will focus upon determining if autolithotrophic microbial communities dominate in this environment as has been found to be the case in the Scandinavian shield (Kotelnikova and Pedersen,

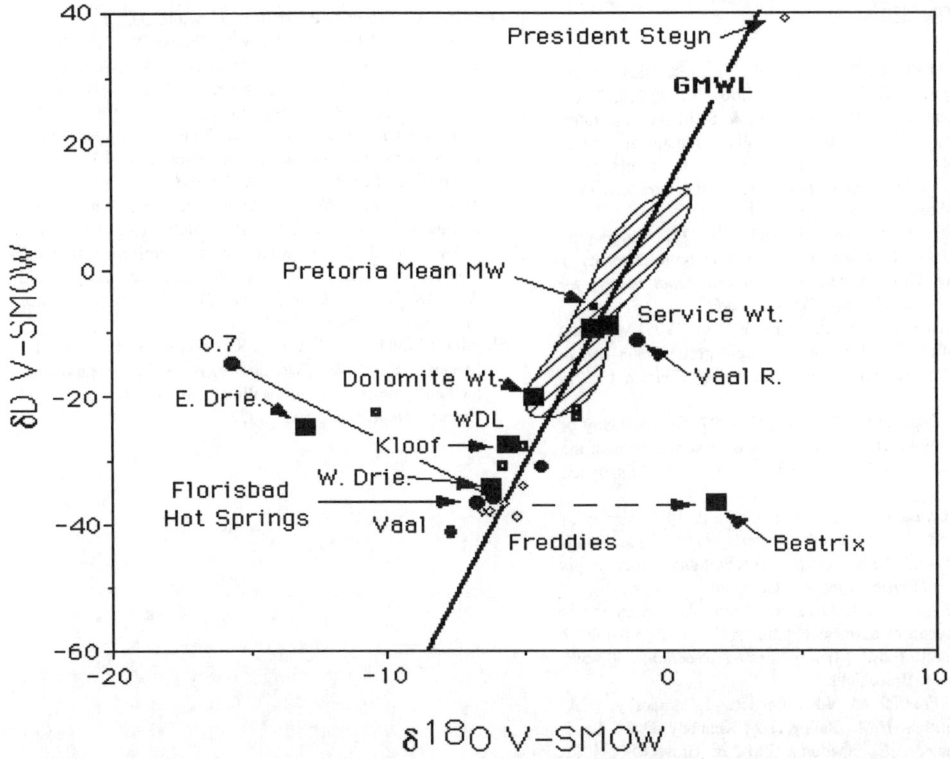

Figure 1. δD versus $\delta^{18}O$ plot for "service", "dolomite" and "fissure" water from East Driefontein (E. Drie), West Driefontein (W. Drie), Kloof and Beatrix. Data from other mines were reported in Duane et al. (1997). Hatched field represents seasonal isotopic variation in precipitation at Pretoria from IAEA database. The data for the Vaal River and for Florisbad springs were reported in Mazor and Verhagen (1983). GMWL represents the global meteoric water line.

1997) and to determine if these communities have also affected the geochemistry of the rocks and could be supported by radiolysis.

5 ACKNOWLEDGEMENTS

We are extremely grateful to Dr. J.J. Alexander of the Dept. of Microbiology, Witwatersrand University, South Africa for her logistical support, the equipment she loaned to us and for granting us access to her department's laboratories.

We would like to thank the following collaborators:
J. Hoek and Dr. G.I. Omar of the Univ. of Pennsylvania, Drs. K. Takai, J.K. Fredrickson and F.J. Brockman of PNNL, Dr. T.J. Phelps of ORNL, Dr. M.F. DeFlaun of Envirogen, Inc., Drs. S.M. Pfiffner and D.C. White and A. Peacock of the Center for Environmental Biotechnology, Univ. of Tennessee, Dr. W.C. Ghiorse of Cornell University, Drs. S. Kotelnikova and K. Pedersen of the Univ. of Göteborg, Göteborg, G. Slater and Dr. B. Sherwood-Lollar of the Univ. of Toronto, Dr. G. Southam of Northern Arizona Univ., Dr. T. Kieft of the New Mexico Institute of Mining and Technology, Dr. F.S. Colwell of the INEEL, Drs. L. Pratt and J. Fong of Indiana Univ., Dr. A.L. Reysenbach, Portland State Univ., B. Baker and C. Wimpee of Univ. of Wisconsin-Milwaukee, Drs. D.A. Stahl, B. MacGregor and S. Fishbain of Northwestern Univ., and D.A. Elias, Dr. L. Krumholz, and Dr. J. Suflita of the Environment, Univ. of Oklahoma.

Finally, we like to express our sincere appreciation to all of the staff of East Driefontein, West Driefontein, Kloof and Beatrix mines South Africa who assisted us in gathering our samples, secured safe passage into and out of the mines and provided crucial information about the environments from which the samples were collected.

6 REFERENCES

Baker, B.J., Moser, D. P.. Fishbain, S., Takai, K., Jackson, B., MacGregor, B. J., Speolstra, N., Loos, S., Onstott, T. C., Fredrickson, J.K., Wimpee, C. F. & Stahl , D. A. 2000. Phylogeny of a geochemically functional gene, dissimilatory sulfite reductase (DSR) in communities associated with dripping boreholes in the ultradeep gold mines of South Africa. *in preparation.*

Deflaun, M.F., Kotelnikova, S., Balkwill, D.L., Streger, S., Onstott, T.C., & Moser, D.P. 1999. Metabolic Diversity in Ultra-Deep Groundwater. *Abstr. International Society for Subsurface Microbiology, Vail Colorado.*

Doig F., Sherwood-Lollar, B., & Ferris, F.G. 1995. Microbial communities in deep Canadian Shield groundwater--An in situ biofilm experiment. *Geomicrobiology Journal* 13: 91-102.

Duane, M.J., Pigozzi, G., & Harris, C. 1997. Geochemistry of some deep gold mine waters from the western portion of the Witwatersrand Basin, South Africa. *Jour. African Earth Sci.* 24: 105-123.

Ghiorse, W. C., Eaglesham, B. S., & Garen R. 1999. Survey of Presumptive Iron-oxidizing Bacteria in a Deep South African Gold Mine. *Abstr. International Society for Subsurface Microbiology, Vail Colorado.*

Hofmann, B.A. 1992. Isolated reduction phenomena in red beds: A result of porewater radiolysis? In Y.K. Kharaka & A.S. Maest (eds.), *Water-Rock Interaction*: 503-506. Rotterdam & Brookfield.

Kieft, T.L., Fredrickson, J.K., Onstott, T.C., Gorby, Y.A., Kostandarithes, H.M., Bailey, T.J., Kennedy, D.W., Li, S-M, Plymale, A.E., Spadoni, C.M., & Gray, M.S. 1999. Dissimilatory Reduction of Fe(III) and other electron acceptors by a Thermus Isolate, *Applied Environmental Microbiology* 65: 1214-1221.

Kotelnikova S. & Pedersen K. 1997. Evidence for methanogenic *Archaea* and homoacetogenic *Bacteria* in deep granitic rock aquifers. *FEMS Microbiology Reviews* 20: 339-349.

Kotelnikova S., Pedersen K., Moser D.P., & Onstott, T.C. 1999. The Witwatersrand Deep Microbiology Project: Methane and Hydrogen Dependent Metal Reduction. *Abstr. International Society for Subsurface Microbiology, Vail Colorado.*

Mazor, E. & Verhagen, B.Th. 1983. Dissolved ions, stable and radioactive isotopes and noble gases in thermal waters of South Africa. *Jour. Hydrology* 63: 315-329.

Moser D.P., Onstott, T.C., Pfiffner S., White D.C., Peacock A., Phelps T.J., Deflaun M., Hoek J., Ghiorse W.C., Colwell F.S., Kieft T.L., Reysenbach A-L, Fredrickson J.K., Southam G., Kotelnikova S., Slater G., Omar, G., Pratt L., Boone D., Pedersen K., & Sherwood-Lollar B. 1999. The Witwatersrand Deep Microbiology Project: A Window Into The Extreme Environment Of Deep Subsurface Microbial Communities. *Abstr. International Society for Subsurface Microbiology, Vail Colorado.*

Onstott, T.C., Tobin, K., Dong, H., DeFlaun, M.F., Fredrickson, J.K., Bailey, T., Brockman, F.J., Kieft, T., Peacock, A., White, D.C., & Balkwill, D.L. 1997. The deep gold mines of South Africa: Window into the subsurface biosphere. *SPIE 42nd Annual Mtg. Proceedings* 3111: 344-357.

Onstott, T.C., Moser D.P., Pfiffner S., White D.C., Peacock A., Phelps T.J., DeFlaun M., Hoek J., Ghiorse W., Colwell F.S., Kieft T.L., Reysenbach A-L, Fredrickson J.K., Southam G., Kotelnikova S., Slater G., Omar, G., Pratt L., Boone D., Pedersen K., & Sherwood-Lollar B. 1999b. The Witwatersrand Basin, S. Africa: Potential biogeochemical pathways and reaction rates. *Abstr. International Society for Subsurface Microbiology, Vail Colorado.*

Pfiffner S.M., Peacock A., Macnaughton S., White D.C., Phelps T.J., Moser, D.P., & Onstott, T.C. 1999. The Witwatersrand Deep Microbiology Project: Microbial Community Assessment by Phospholipid Fatty Acid Analysis. *Abstr. International Society for Subsurface Microbiology, Vail Colorado.*

Sherwood-Lollar, B., Frape, S.K., Weise, S.M., Fritz, P., Macko, S.A. & Welhan, J.A. 1993. Abiogenic methanogenesis in crystalline rocks. *Geochim. et Cosmochim. Acta* 57: 5087-5097.

Groundwater: Past Achievements and Future Challenges, Sililo et al. (eds) © *2000 Balkema, Rotterdam, ISBN 90 5809 159 7*

Assessment and rehabilitation of reclaimed gold tailings dams areas in South Africa

T. Rösner – *Radian Environmental GmbH (URS Corporation), Kaiserslautern, Germany*
J.J.G. Vermaak – *Wates, Meiring and Barnard (Pty) Limited, South Africa*
A. van Schalkwyk – *Department of Earth Sciences, University of Pretoria, South Africa*
C.J. Viljoen – *Envirogreen Consulting (Pty) Limited, South Africa*

ABSTRACT: Mining in South Africa resulted in vast volumes of residues deposited at hundreds of sites. Some gold mine tailings dams, in particular, have been partially or completely reclaimed and left a contaminated subsurface which could pollute underlying aquifers. This paper describes methods for assessing reclaimed tailings sites and remediation measures to prevent contaminant migration into the groundwater. It compares various remediation approaches e.g. free flow drainage, natural attenuation and paddock systems. Two field experiments, in which the free flow drainage approach and engineering soil covers have been used to reduce the impacts on the soil and groundwater regime and also to enable safe land use, are briefly discussed.

1 INTRODUCTION

Since mining operations started more than a century ago, South Africa has been the largest producer of gold in the world (Department of Minerals and Energy 1998). In 1996 alone, 377 million tons of mine waste was produced, accounting for 81 per cent of the total waste stream in South Africa (Anon. 1997). Rösner et al. (1998) estimated that gold mining operations in South Africa resulted in hundreds of tailings dams, covering a total area of about 18 000 ha. These tailings dams contain substantial amounts (10-30 kg/ton) of sulphide minerals which are prone to generate acid mine drainage (AMD) in the presence of oxygen, water and catalysing bacteria. The reaction, with iron oxide as the end product, may be expressed as follows (Appelo & Postma, 1994):

$$4FeS_2 + 15O_2 + 8H_2O \rightarrow 2Fe_2O_3 + 8SO_4^{2-} + 16H^+ \quad (1)$$

Sulphide oxidation generally results in salinification and acidification of soil and water. AMD is often associated with significant concentrations of heavy metals such as Co, Ni and Zn, as well as radioactive elements (Rösner et al. 1998). These contaminants can dissolve in the acid range and limit growth conditions for plants or migrate through the soil into the groundwater system.

Since the 1970's, the high operating costs of deep underground gold mining operations encouraged some companies to focus on reprocessing (termed reclamation) of existing tailings for the recovery of gold still present in economically viable quantities. After reclamation has been completed, a contaminated subsurface or footprint of the former tailings material remains. Such footprints from gold tailings dams only, cover about 1 300 ha of land (Rösner et al. 1998).

The vadose zone in tailing footprints will remain contaminated for a long period of time inducing particular risks to human health and the environment. The aims of remediation measures at these footprint areas are to reduce these risks to acceptable limits and to enable land use after reclamation.

The objective of this paper is to present concept approaches in assessing footprint areas of reclaimed tailings dams. It then discusses various rehabilitation approaches that can be considered. Special emphasis is given to the free flow drainage methods and engineered soil covers.

2 ASSESSMENT CONCEPTS

2.1 *Closure objectives*

The first stage in the rehabilitation of tailings footprints is to set clearly defined closure objectives which should be well formulated and should be realistic and cost effective (Smithen, 1998). The closure objectives could imply that the residual impacts of the tailings footprints should be acceptable in terms of the ultimate product (post-mining land use). Residual impacts often comprise:

- Windblown tailings material which results in dust problems
- Contamination of surface water caused by contaminated run-off
- Contamination of groundwater caused by leaching of contaminants into the groundwater regime
- Contamination of the soil zone caused by leaching of contaminants into the soil zone

It is necessary to have a series of guidelines against which residual impacts could be measured. Although guidelines on the assessment and classification of hazardous waste (Department of Water Affairs and Forestry, 1998) and for water quality (Department of Water Affairs and Forestry, 1996) are available in South Africa, no regulations dealing with soil quality criteria for former mining sites are in place.

2.2 *Risk Assessments*

The term risk, as used in this paper, may be defined as the identification of potential adverse effects to humans or ecosystems resulting from exposure to environmental hazards (United States EPA, 1989). Risk Assessment is useful in determining the level of remediation most appropriate at any particular site.

Several Risk Assessment approaches have been developed world-wide. In the USA, authorities require site-specific risk analyses, involving the identification of relevant and appropriate requirements, site specific baseline risk assessment and the determination of site specific remediation measures, if necessary (United States EPA, 1989). In contrast, national risk protocol standardises methods for the assessment of risks and specifies actual numerical values (i.e. the Dutch or German soil and groundwater quality standards) for national clean-up or action standards.

In South Africa, few guidelines and regulations in terms of Risk Assessment and soil quality exist and a flexible risk-based approach is preferred. The Environmental Risk Assessment and Management approach (ERAM) has been applied in mining related investigations in South Africa (Swart *et al*, 1998). ERAM has the advantage of not relying on stringent guidelines (e.g. the USA), which could prove too expensive and complex. It also does not require national standards (e.g. the Netherlands and Germany), which currently do not exist in South Africa. However, the absence of guidelines implies that the hazards may be overlooked while the subjective nature of the evaluation could lead to inconsistent results.

2.3 *Assessment of contaminant mobilisation*

In order to conduct a Risk Assessment, it is important to assess the potential for contaminants to be mobilised from the vadose zone (the pollution source) to the groundwater and surface water regimes. A good understanding of the mechanisms controlling the mobility of contaminants and the capacity of the vadose zone to retain and attenuate contaminants in the long-term is required.

Soil extraction tests are used to estimate the mobile portion of contaminants in a soil. The levels at which extractable contaminants exceed the threshold value indicate the potential for contaminants to be mobilised from the soil zone resulting in an increased plant uptake and groundwater contamination.

The mineralogical properties of the soil play an important role to determine if the acid neutralisation potential is sufficient to maintain a neutral soil pH, in which most metals are not mobile. Changes in the soil pH or Eh conditions can cause significant increases in contaminant loads.

Sililo *et al* (1999) showed that batch experiments can be used to measure the adsorptive capacity of soil materials at different pH values. Batch experiments can also be used to determine the distribution coefficient, necessary in solute transport modelling of contaminants in the vadose zone.

The load calculation of contaminants, as introduced by Müller (1979), can be considered to rank the degree of soil contamination. The load calculation entails comparison of the total contaminant concentration with a background value of soils originating from similar geological conditions after which a load class is then determined. This index represents the long-term capacity of the soil to retain contaminants. High load indices indicate that the soil is contaminated. This method can only be used if background values of the different soil types are known.

3 REMEDIATION MEASURES

The remediation strategies should ensure that risk is reduced to an acceptable level and that closure objectives are achieved. In addition, remediation strategies should be cost effective. Currently, the following remediation measures are considered for reclaimed tailings sites:

- Paddock systems
- Natural attenuation
- Soil amendment methods
- Free flow drainage
- Construction of engineered soil covers
- In situ soil clean-up methods
- Removal of contaminated soils

3.1 Paddock systems

Paddock systems are used at reclaimed sites to prevent surface run-off from entering surface water regimes. Paddocks consist of low dam walls (approximately 1-2 m high), which are build in order to capture rainwater on site to prevent surface-run-off.

Paddock systems should be considered with caution since many risks are associated with this remediation measure. Ponding of water will result in increased infiltration contributing to groundwater contamination. Maintenance that could have significant cost implications is required. In addition, safety risks with respect to the possibility of drowning are associated with paddock systems.

3.2 Natural attenuation

Natural attenuation is the process where the contaminant load to the groundwater regime is reduced by sorption, precipitation, bio-degradation and other processes. The latter process applies only to degradable organic pollutants (e.g. petroleum). Natural attenuation implies that no remediation measures in terms of preventing groundwater and surface water contamination are necessary at reclaimed sites.

Natural attenuation is not appropriate at all sites, since the rate of natural processes is often slow and the attenuation capacity of South African soils is generally small (Sililo *et al* 1999).

3.3 Soil amendment methods

The most common amendment method applied on affected soils in South Africa, is the addition of lime and fertilisers (e.g. superphosphate). Lime is used in the case of acidic soils to increase the soil pH range so that most of the dissolved metals would precipitate. However, if the soil is acidic at depths of more than 10 cm below surface, then the only practical option is to grow acid-tolerant plants (Fenton, 1997). Fertilisers are introduced to such soils to reduce limitations on crop or plant growth that would be caused by an inadequate supply of nutrients in the soil (Wild, 1988).

The success of soil amendment methods depends on the degree to which the reclaimed sites have been contaminated and to what extent the soils can sustain vegetation. Where degradation of the vegetation cover can be expected in the long-term, alternative measures should be considered.

3.4 Free flow drainage

The focal point of free flow drainage systems is to increase the organic material content of the soil. This will have an effect on the physical and chemical properties of the soils, which will assist in their attenuation potential. In addition, the evapotranspiration potential is also increased by the well-established vegetation and greater water-holding capacity of the soils and this will result in decreased seepage towards the groundwater regime.

The free flow drainage approach may not be effective in cases where contaminants can readily be mobilised into the groundwater regime. Free flow drainage may also not be applicable in cases where significant concentrations of bio-available phytotoxic elements such as Co, Ni and Zn are present, or where soils may turn acidic, resulting in the degradation of the vegetation cover (Rösner et al., 1998).

3.5 Construction of engineered soil covers

Engineered soil covers (or final covers) should be considered where significant risks resulting from contamination of the groundwater, exist. Generally, that implies that the vadose zone is severely contaminated at depths greater than 0.5 metres and that contaminants can easily be mobilised and enter the groundwater regime. Engineered soil covers should also be considered if the soil contains phytotoxic elements such as Co, Ni and Zn in excessive concentrations, or could become acidic, which could result in the degradation of established vegetation. Engineered soil covers are an expensive option compared to many other remediation measures.

3.6 In situ soil treatment

Soil treatment techniques (e.g. soil flushing, solidification/stabilisation) have not been applied on reclaimed tailings sites in South Africa. Various in situ techniques have been developed to treat metal contaminated soils (United States EPA, 1997). These methods are likely to be too costly, since the quantities of contaminated material to be treated are very large. In addition, these techniques address only a portion of the problem and will, for instance, not impact on high salt loads that reach surface water and groundwater regimes.

3.7 Removal of contaminated soils

In the case of heavily contaminated soils, their removal should be considered. This approach is likely to be the most expensive of all remediation measures and should be considered in cases of significant health or ecological risks. Significant benefits are associated with this approach since the contaminant source is removed. However, this aspect in itself presents a problem since the contaminated material needs to be disposed of elsewhere (e.g. a well-designed landfill site).

4 PRELIMANARY EXPERIMENTAL FINDINGS

Two field experiments are currently being conducted in which remediation measures have been applied to contaminated sites. In one experiment, the free flow drainage approach has been applied to a section of the reclaimed tailings site. In the other, an engineered soil cover has been placed on top of fine coal discard material as part of an experiment to monitor infiltration and oxygen ingress into coal discard. Although not a reclaimed tailings site, it is argued that the results of the latter experiment can be extrapolated to reclaimed tailings since the processes involved, i.e. sulphide oxidation and the mobilisation of contaminants to the groundwater regime, are similar.

4.1 Free flow drainage

The experiment comprises seven test sections constructed on reclaimed footprint areas in which rehabilitation to different degrees was conducted. One section was not treated (Section 1), one section is located outside the footprint area in order to assess movement of contaminants to surrounding areas, if any (Section 2), and the remaining sections were treated with increasing volumes of organic material. A monitoring programme was initiated in which soil samples were collected and tested for contamination. The results for the period from March 1998 till June 1999 are indicated in Table 1.

Table 1. Results of footprint 6L20 experimental area.

Section no.	pH	EC mS/cm	SO4 mg/kg	C (%)
1 (No treatment)	4.03	1 460	8 210	0.56
2 (Adjacent area)	6.89	89	37	0.94
3 (40 t/ha OM)	6.04	293	679	1.02
4 (60 t/ha OM)	6.96	256	610	1.09
5 (80 t/ha OM)	6.72	201	490	1.13
6 (100 t/ha OM)	6.85	143	439	1.19
7 (200 t/ha OM)	6.92	127	398	1.24

OM = Organic material

The preliminary results indicate that, compared to the untreated section, pH has been stabilised (pH of between 6 and 7 for treated sections against 4 of the untreated section). In addition, while soil electrical conductivity and sulphate concentrations have been significantly reduced, carbon content has increased. No significant changes have been recorded within the adjacent test section, which implies that the contaminants have not moved to the surrounding areas.

The long-term trends and sustainability of the vegetation cover will only be known once an adequate monitoring database has been established.

4.2 Engineered soil covers

This experiment comprises a seven-year field pilot study to investigate the performance of engineered soil covers on rehabilitated opencast coal mines and coal discard dumps. (See also Wates *et al* 1999 and Vermaak & Bezuidenhout 2000). The soil covers were provided to limit water infiltration and to restrict oxygen ingress into the waste material.

The ten test sections were provided with different soil cover configurations (Table 2). The cover material comprised clayey sand (Avalon soils) used in single covers and as surface layers in the complex covers. Estcourt soils, comprising clayey sand, were used in the complex soil covers.

Some results are indicated in Table 2, which shows outflow expressed as a percentage of the Mean Annual Precipitation (MAP) and oxygen concentrations expressed as a percentage. (Atmosphere = 21%)

Table 2. Outflow and oxygen concentration results.

Cell No.	Outflow (% of MAP)	$[O_2]$ (% O_2)
No 1 Uncovered fine coal discard	35.6%	18.1%
No 4 0.3m uncompacted Avalon	21.5%	15.5%
No 5 0.5m compacted Avalon	16.7%	2.9%
No 6 0.3m uncompacted Avalon + 0.7m compacted Estcourt	13.6%	1.6%
No 8 0.7m uncompacted Avalon + 0.3m compacted Estcourt	13.0	1.4%

The results show that outflow (recharge) is significantly reduced, even through the simple soil covers. Little difference in outflow rate was recorded for the complex soil covers and the 0.5m compacted Avalon cover. Other experiments have shown that the outflow rate can be reduced to approximately 5% of MAP if a properly installed compacted clay liner is provided and the cover is protected against degradation (Bennet *et al*, 1988)

Oxygen measurements reveal that the 0.3m simple soil cover is not adequate in limiting oxygen ingress. In the case of Cell 5 and the complex soil covers, oxygen concentrations of 0% were recorded most of the year, with the exception of the latter part of dry seasons where preferential oxygen ingress through desiccation cracks resulted in higher oxygen concentrations (Vermaak & Bezuidenhout, 2000).

4.2.1 Relationship between oxygen ingress and leachate quality

The leachate of all cells is of poor quality as characterised by high total dissolved solids and heavy metal concentration. The poor quality leachate can be attributed to leaching of contaminants from the fine coal discard and sulphide oxidation processes. The sulphate concentrations are an indication of the rate of sulphide oxidation processes. Results of sulphate concentrations are indicated in Table 3.

Table 3. Measured sulphate concentrations.

Cell No	[SO4] (mg/l) February 1993	[SO4] (mg/l) October 1998	Change %
No 1	2 803	5253	87.4
Uncovered fine coal discard			
No 4	2 779	4 212	51.6
0.3m uncompacted Avalon			
No 5	2 716	3 022	11.3
0.5m compacted Avalon			
No 6	3 019	2 736	-9.4
0.3m uncompacted Avalon + 0.7m compacted Estcourt			
No 8	3 390	2 249	-33.7
0.7m uncompacted Avalon + 0.3m compacted Estcourt			

Sulphate concentrations in the uncovered cell increased significantly after acid breakthrough occurred in 1997, which mobilised contaminants from the fine coal discard. Positive sulphate concentration trends for the single layered cells are indicative of high sulphide oxidation rates and the leachate quality is expected to decrease with time.

Negative sulphate concentration trends indicate that sulphates are leached from the coal discard. It is also indicative of low sulphide oxidation rates, caused by low oxygen ingress rates into the waste material. The quality of the leachate is expected to improve up to point where the leachate quality will reflect sulphide oxidation rates.

The relationship between leachate quality and oxygen ingress for Cell 8, provided with complex soil covers, is shown in Figure 1. The negative sulphate trend indicates ongoing leaching of sulphates from the fine coal discard at times when little oxygen ingress takes place. The exception is the latter part of 1996, when high oxygen ingress rates resulted in a temporarily increase in sulphate concentration, caused by higher sulphate oxidation rates.

5 DISCUSSION

Reclaimed tailings sites pose significant challenges for rehabilitation. Remediation measures should be implemented to attain well-defined closure objectives in terms of intended land use. These objectives also entail compliance with respect to surface water, groundwater and soil quality.

A Risk Assessment should form the basis on which remediation measures are to be implemented. The Risk Assessment should quantify the risks associated with the final land use. The absence of South African guidelines may be problematic as some hazards may be overlooked and the subjective evaluation could result in inconsistent results.

The absence of soil quality standards in South Africa complicates rehabilitation in that it is not possible to measure the effectiveness of soil remediation and it is recommended that such guidelines be established as a matter of priority.

In terms of the determination of the attenuation capacity of soils, Sililo *et al* (1999) proved the effectiveness of batch experiments for South African conditions and recommended that these should be used to determine the mobilisation potential of contaminants. Long-term risks are influenced by dynamic geochemical processes in the soil/sediment system and these should be understood in order to predict the mobilisation potential of contaminants (Salomons & Stigliani, 1995).

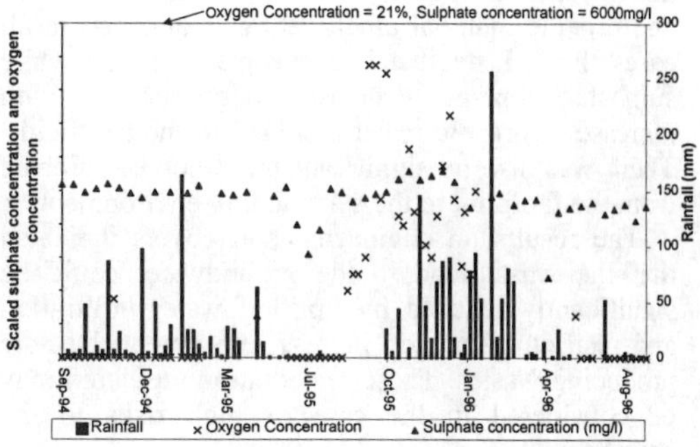

Figure 1. Relationship between oxygen ingress and leachate quality for a complex cover.

No single remediation measure can be uniformly imposed on all reclaimed tailings sites, and the type and degree of remediation will be site-specific. The different remediation strategies are compared in Table 4. Paddock systems can be used to prevent contaminated surface run-off, but considerable risks, in terms of drowning and groundwater contamination, associated with this remediation measure. Soil amendment and free flow drainage measures can be considered as an alternative to paddock systems, but may not be appropriate in cases where degradation of the vegetation cover (caused by acidic soil conditions or the presence of phytotoxic elements) is possible in the long term. In cases where sites are severely contaminated and pose significant risks, more expensive remediation measures such as engineered soil covers may be considered. All remediation measures should include a monitoring period in which the success of the remediation measure can be quantified. In most cases, maintenance will be required for some period after remediation

Table 4. Comparison of some remediation measures.

	Paddock system	Natural attenuation	Soil amendment	Free flow Drainage	Construction of soil covers	Removal of contaminated soils
Prevent contaminated runoff	Yes	No	No	No[2]	Yes	Yes
Groundwater contamination	Severe	Moderate to severe[1]	Limited to severe[1]	Limited to severe[1]	Limited	None
Soil contamination	Severe	Moderate to severe[1]	Limited to severe[1]	Limited to moderate[1]	None	None
Costs	Expensive	Not expensive	Not expensive	Moderately expensive	Expensive	Very expensive
Land use	None	Limited	Limited	Limited	Limited	Unlimited
Applicability	High risk	Low risk	Low risk	Medium risk	High risk	Very high risk

[1] Depending on extent of contamination
[2] Preliminary experimental results suggest otherwise

The preliminary experiments on free flow drainage have shown significant correlation between the increase in organic material content in the soil and organic material amelioration treatments. In all cases the pH stabilised to acceptable limits while sulphate content decreased significantly with increase in organic material added to the test fields. There was also no significant migration of sulphate from the footprint to the surrounding environment.

The results on engineered soil covers indicated that the waste load to the groundwater could be significantly reduced by limiting water infiltration and restricting oxygen ingress into potential acidic producing wastes. These remediation measured may be considered in the case of high risks to the environment

REFERENCES

Anon. 1997. Engineering News (South Africa), Vol. 17 no 6.

Appelo, C.A.J. & Postma, D. (1994): Geochemistry, groundwater and pollution. Rotterdam: Balkema.

Bennet, JW, Harries, JR and Ritchie, AIM. 1988. Rehabilitation of Waste Rock Dumps at the Rum Jungle Mine Site. Proceedings. *American Society of Surface Mining and Reclamation and the U.S. Department of the Interior Conference of Mine Drainage and Surface Mine Reclamation*, Pittsburg, PA, April 17-22, 1988. pp 104-108.

Department Of Minerals And Energy. 1998. *South Africa's mineral industry 1996/97*. Johannesburg, pp. 216.

Department Of Water Affairs And Forestry (DWAF). 1996. *Volumes 1-6: South African water quality guidelines*. Pretoria.

Department Of Water Affairs And Forestry (DWAF). 1998. *Minimum requirements for water monitoring at waste management facilities. Documents 1-3, second edition*. Pretoria.

Fenton, G. 1997. Answers to questions on soil acidity. *Agricultural Notes 5/114,* electronic edition (www.agric.nsw.gov.au/arm/soil_pub).

Müller, G. 1979. Schwermetalle in den Sedimenten des Rheins – Veränderungen seit 1971. Umschau, 79: 778-783.

Rösner, T., Boer, R.H., Reyneke, R., Aucamp, P. & Vermaak, J.J.G. 1998. *A preliminary assessment of pollution contained in the unsaturated and saturated zone beneath reclaimed gold-mine residue deposits*. Pretoria: Water Research Commission (Rep. No. K5/797/0/1).

Salomons, W. & Stigliani, W.M. 1995. Biogeodynamics of pollutants in soils and sediments – Risk assessment of delayed and non-linear responses. Berlin: Springer.

Sililo, O.T.N., Conrad, J., Murphy, K.O.H., Tredoux, G., Eigenhuis, B., Ferguson, M.C.D. & Moolman, J.H. 1999. Investigation of the contaminant attenuation characteristics of the soil aquifer system with special emphasis on the vadose zone. Draft Final Report, WRC Report No. 572/1/99, Pretoria

Smithen, 1998. Setting rehabilitation objectives. In *Environmental Rehabilitation in Mining*. CSIR, Pretoria

Swart, S.J., Pulles, W., Boer, R.H., Kirkaldy, J. & Petit, C. 1998. *Environmental risk assessment as basis for mine closure at Iscor Mining*. Journal of the South African Institute of Mining and Metallurgy, January/February 1998

United States Environmental Protection Agency. 1989. *Comparing Risks and Setting Environmental Priorities, Overview of Three Regional Projects*. Washington DC.

United States Environmental Protection Agency. 1997. *Recent developments for in-situ treatment of metal contaminated soils*. Washington DC: EPA doc. 011059.

Vermaak, J.J.G & Bezuidenhout, N. 2000. *Design of soil covers as barriers for the South African climate*. In Young Geotechnical Engineers Conference – Geotechnics into Africa, Stellenbosh, South Africa

Wates J.A., Vermaak, J.J.G. & Bezuidenhout, N. 1999 *Design of soil covers to reduce infiltration on rehabilitated opencast coal mines* In Wardle, Blight & Fourie Ed. Geotechnics for developing Africa, Balkema publishers, Rotterdam

Groundwater: Past Achievements and Future Challenges, Sililo et al. (eds) © 2000 Balkema, Rotterdam, ISBN 90 5809 159 7

Management of coal mining impacts on water resources in South Africa

D.A. Salmon
Anglo Coal Environmental Services, Witbank, South Africa

ABSTRACT: Acid and neutralised acid mine drainage are arguably the most serious threats posed to the environment by coal mining activities. Changes in South African legislation have focused attention onto water management in the coal mining industry. Water management strategies employed by coal mines may incorporate the water management principles of prevention of water pollution, the separation of clean and dirty water, maximizing water reclamation and the reuse of water, holistic mine planning, and the treatment of polluted water. The relative importance of these principles and the management practices to achieve them varies according to the mining method and the phase in the life cycle of the mine. Implementing preventative measures is the ideal management option as this is more cost effective than installing water treatment systems.

1 INTRODUCTION

Over the past decade, South African coal mines have become increasingly aware of their impact on water resources due to factors such as: the promulgation of the new National Water Act (Act No 36 of 1998), the Minerals Act (Act No 50 of 1991) requiring an Environmental Management Programme Report (EMPR), National Environmental Management Act (Act No 107 of 1998), increased awareness of economic liabilities associated with environmental risks, implementation of Environmental Management Systems such as ISO 14001, the development of catchment management plans, the cost savings achievable by water conservation measures, a need to display environmental responsibility, avoidance of bad publicity associated with pollution incidents, and overseas and local market pressures.

Coal mining impacts on water resources vary according to the life cycle of the mine. The life cycle of a colliery consists of pre-mining (Exploration and Planning), Operational, Decommissioning and Closure phases.

1.1 *Location*

Coal mines are located in the north and western parts of South Africa in three major drainage regions, the Vaal River, the Olifants River, and the north eastern escarpment within which four main rivers, the Tugela, the Mkuzi, the Umfolozi, and the Pongola rivers flow. The position of the coalfields within these regions is shown in figure 1. The environmental impacts of polluted water from the coal mining areas can have international consequences. The Olifants River crosses the international boundary into Mozambique, and the Vaal is a tributary of the Orange River which forms the boundary between South Africa and Namibia.

1.2 *Geology*

The South African coalfields are found in the Karoo Supergroup. The Ecca Group coal bearing sediments consist of formations of sandstone, shale, mudstone and coal. This sequence overlies the glacial deposits of the Dwyka, which in turn lies on basement rocks

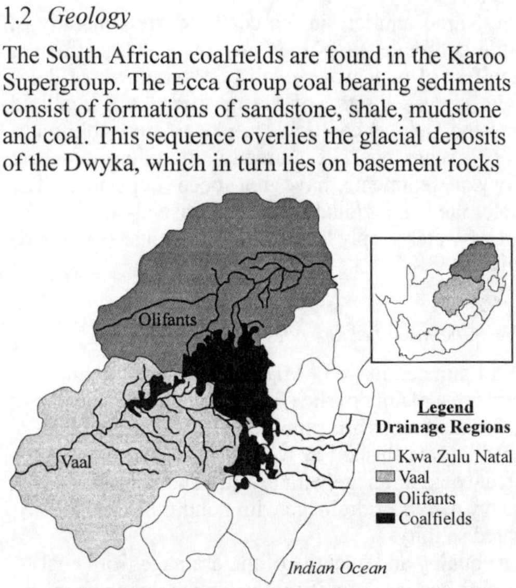

Figure 1. Location of drainage regions and coalfields (After Middleton et al. 1995a, b and Pitman et al. 1995)

of the Transvaal sequence. With the exception of the KwaZulu Natal (KZN) coalfield, the basal seams (the number 1 and number 2 seams) are associated with fluvio-glacial outwash deposits, where as the upper seams, number 3, 4 and 5 seams, were deposited in fluvio-deltaic environments. Acidic mine drainage is more often associated with the upper coal seams deposited in marine-brackish waters. Dolerite sills and dykes occur throughout the coalfields.

1.3 *Aquifers*

Four distinct groundwater systems are recognized in South African coalfields. A shallow, perched, upper aquifer in the weathered Ecca Sediments and soil horizons 5 to 20 metres (m) below the surface. Yields are low (100 to 2000 litres per hour (l/hr) and permeabilities range from 0,01 to 1,0 metres per day (m/d). It occurs throughout the coalfield.

Recent alluvial deposits up to 30m thick and overlying the Ecca strata are developed in parts of the northern Orange Free State (OFS) coalfield. These are an unconfined aquifer with permeabilities of up to 33 m/d.

A secondary aquifer occurs in the fractured Ecca sandstone and coal sediments. This aquifer is generally shallow, however in the deeper parts of the Mpumalanga region of the coalfield it may be associated with coal operations at 150 to 200 m depth.

Yields can vary from between 100 to 6000 l/hr. Water in the Ecca sequence also occurs along the contact zones and fractures adjacent to and within dolerite dykes and sills intruded into the Ecca formation.

Pre Karoo aquifers in the coalfield are generally of little consequence as they are low yielding fracture aquifers of poor water quality. A basement of dolomite underlies parts of the northern OFS coalfield and the Delmas coalfield. In places, mineable number 1 seam coal lies directly on dolomite where Dwyka sediments have not been deposited. The dolomite is a regional aquifer extensively utilised for major water supply to urban areas, industry and agriculture.

1.4 *Chemical background*

Acid mine drainage (AMD) arises when sulphide minerals, mainly pyrite (FeS_2), contained in coal and associated rocks are exposed to oxidation and the products are flushed by water. These processes can create acid drainage with low pH and dissolved heavy metals or drainage with neutral pH and elevated salinity.

The quality and quantity of the drainage from coal mining can vary significantly. The drainage characteristics are determined by the coal seam mined, the mining method employed, the area of mining, the chemistry of the rock sequence in which the coal occurs, the quality and quantity of the natural groundwater, and the location, size and design of coal discard dumps and slimes.

2 POTENTIAL IMPACTS OF COAL MINING ON WATER RESOURCES

The impacts of coal mining on water resources depend on how mining effects the exposure of rock material to oxygen and water. The various mining methods of opencast strip mining, shallow or deep underground bord and pillar mining and total extraction mining, and coal waste handling techniques disturbs natural water regimes, permitting water and oxygen ingress and contact with different sulphide bearing rocks. Mining changes land surface features producing changes in evaporative capacity, infiltration, and runoff characteristics. It can completely disrupt portions of aquifers and change their characteristics.

2.1 *Opencast Mining*

2.1.1 *Operational phase impacts*

Opencast mining totally disrupts the geology, land surface, surface and ground water regimes. Breakage of the rock overburden creates numerous faces that are exposed to oxygen and water. Sources of water include direct precipitation, surface runoff water into the open pit, seepage water from the spoil piles and the rehabilitated spoils areas, and flood water from nearby streams or rivers. As material is removed pressure on any underlying aquifer is released increasing the risk of water inflow from this source. Water movements during this phase of mining are shown in figure 2.

2.1.2 *Operational phase management principles*

Preventing water from entering the mining area can be achieved by digging cutoff trenches in front and behind the mining areas, constructing stream diversions to divert surface water courses and constructing flood protection levees.

Rapid removal of water arriving in pit decreases re-

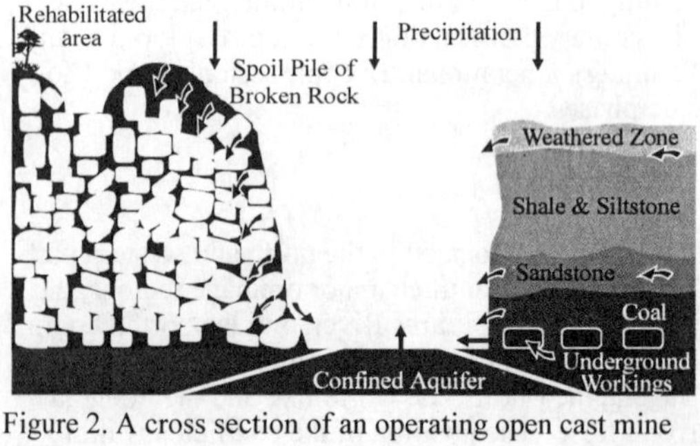

Figure 2. A cross section of an operating open cast mine

tention time and ARD reactions. Polluted water can be stored in evaporation dams. Storage of polluted water in natural pans, a method used in the past, is no longer an accepted practice.
Controlled (seasonal) discharge of polluted water to surface streams during the rainy season is practiced in the Olifants River Catchment. The regulators rarely give licenses for release of polluted water throughout the year.
Discharge to underground mine workings storage areas, and reuse of water, for example dust suppression on haul roads and in coal washing plants, is commonly practiced.
Timeous rehabilitation of spoil heaps reduces recharge by decreasing the size of the operating window of the mine where infiltration is greatest. Recharge rates vary in different parts of the operating window as shown on table 1.

Table 1. Recharge values for an opencast strip mine as a percentage of Mean Annual Precipitation (MAP). (After Hodgson and Krantz, 1995)

Recharge source	Average recharge value
Rain onto ramps and voids	70%
Rain onto rehabilitated spoils	60%
Runoff onto levelled spoils	5%
Seepage into levelled spoils	20%
Runoff onto rehabilitated spoils	10%
Seepage into rehabilitated spoils	8%
Surface runoff from surrounding areas	6% total pit water
Ground water seepage	10% total pit water

Minimising fractures caused by blasting operations, the movement of heavy equipment on the mining floor and accidental overdrilling of boreholes can prevent the development of conduits to underlying aquifers.
Removal of water contained in underground bord and pillar workings being remined by opencast methods has to stored or treated.

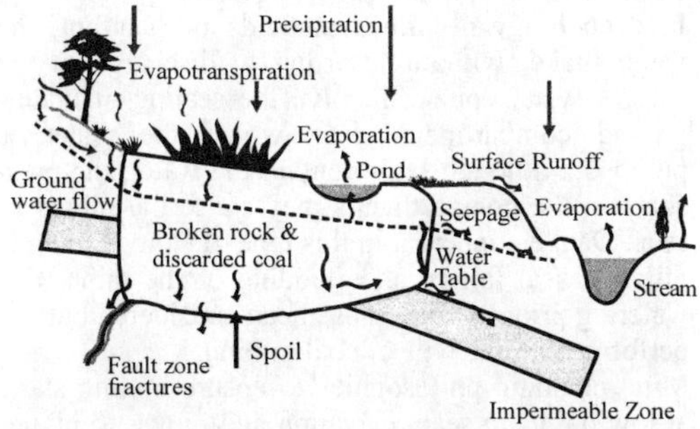

Figure 3. A cross-section through a closed opencast pit

2.1.3 Closure phase impacts
As the water table re-establishes, spoil water may migrate out of the pit into the upper aquifer or decant at surface from where it flows to surface streams. Piezometric heads in underlying aquifers drive water into the spoils. However, if the underlying aquifer is utilized in adjacent areas the hydraulic gradient slopes away from the mine and a flow of spoils water into the aquifer is possible.

2.1.4 Closure phase management practices
Selective spoil can reduce the potential for AMD generation. Rock material with high acid generating potential is deposited at the base of a pit where it is flooded as the water table re-establishes. Anaerobic conditions are created which prevents the development of AMD. However, selective spoil handling is often impractical in large opencast strip mines using draglines that mix the strata.
Cut off trenches are required upslope of rehabilitated pits, to reduce the water entering into the rehabilitated ground.
Dewatering wells can be drilled up gradient to reduce groundwater inflow.
Possible self-sustainable solutions are the use of water consuming trees such as bluegums, which can consume between 75 and 240 litres per day (per mature four-year-old tree). Trees can be planted on the unmined ground upslope of the rehabilitated pit to intercept clean water that would flow into the pit.
Opencast strip mining lowers portions of the land-surface. Often final cuts and ramps are left as voids in which water collects. This allows evaporation of the affected spoil water which can be balanced against recharge thereby keeping spoils water levels lower than seepage or decant points.

2.2 Underground bord and pillar mining

2.2.1 Operational phase impacts
The impacts on surface runoff are usually limited to the effects of surface infrastructure and collapse of bords as shown in figure 4.
Dewatering of the Ecca aquifer will occur but the dewatering cone is limited to no more than a few hundred metres. Occasionally fractures in the Ecca issue notable but discontinuous volumes of water.
Aquifers underlying bord and pillar mines can be intersected if the barrier between the workings and the aquifer is thin or absent. Large inflows of water of up to 8 Megalitres per day were experienced at the now closed Cornelia Colliery, located in the OFS, when the number 1 coal seam was worked directly on top of the dolomite.
Boreholes used by farmers for domestic supply or livestock watering may be affected if mining occurs within 500 metres.

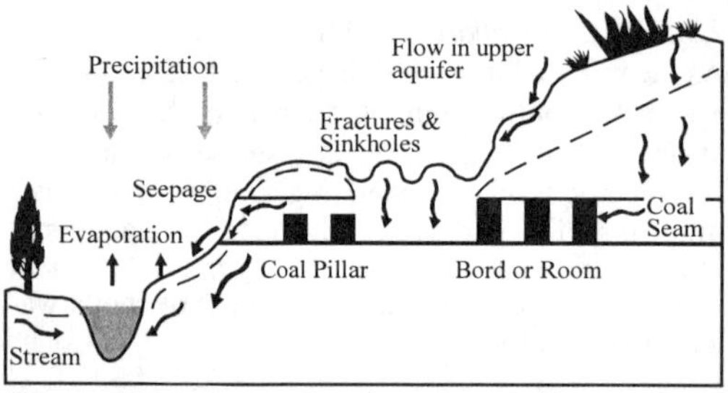

Figure 4. A cross section through a shallow bord and pillar mine.

Average recharge of some 0,5 Megalitres per day (Ml/d) are masked in the operational phase due to evaporative losses by mine ventilation systems. Rainfall recharge may be 3% for an average rainfall of 700 millimetres per annum (mm/ann.), but this may increase to 8% during above average MAP of 1000 mm/ann.

2.2.2 Operational phase management options

Underground storage dams should be flooded to the roof of the workings to reduce oxidation of the coal. Leaving enlarged pillars as barriers around farmers boreholes can prevent water loss. If this method is unsuccessful the mine may be required to supply the farmer with water.

The surface facilities of bord and pillar mines require clean water cut off drains to prevent excess runoff moving through facilities and contaminating the upper aquifer.

2.2.3 Closure phase impacts

Bord and pillar mines, such as Cornelia Colliery and Vierfontein Colliery in the OFS, and Ingagane Colliery in KZN have filled within 25 years of closure. Vierfontein and Ingagane Collieries are shallow and have commenced decanting. In bord and pillar mines where conditions allow only limited or no decant of water, the workings fill, creating anaerobic conditions, the water table re-establishes and the water held in the workings void improves in quality to the extent that the water is often considered a resource.

When the bord and pillar mine is shallow and collapses have occurred, surface water runoff and water from the upper aquifer flows into the workings filling them more rapidly. Where the coal suboutcrops in valley sides, the workings water seeps through the coal barrier pillars and out of unsealed or poorly sealed adits and thence into surface water courses.

2.2.4 Closure phase management options

As the water table re-bounds the water contained in the workings may become a resource that requires management. This management is not the normal business of a mining house. Farmers may wish to abstract this water. Abstraction can lower water levels such that oxidation can recommence and AMD produced. Abstraction must be done on a controlled basis to minimise the risk of AMD generation. Furthermore, repeated wetting and drying, and acidification and neutralization, will use up the available calcite in the pillars enhancing pillar spalling, instability and collapse. Windmills which pump between 2 to 8 cubic metres per day (m^3/d) are an option for limiting and controlling water abstraction under these conditions.

To close all fractures and sinkholes adequately is often economically impossible in old closed, shallow mines. Sealing of adits normally allows the head of water in the workings to increase and seepage will occur along lines of geological weakness at various levels. Collection of the seepage water to allow controlled discharged during the rainy season can only partially address the problem. Water treatment systems have to be considered and these may involve liming plants with the associated problem of removal and storage of the iron precipitates (yellow boy). Collection for treatment may require piping and pumpage of water. The problem with these types of options is that they are not self sustaining. Passive water treatment systems require less operational management than conventional treatment systems, however the area required to handle the volumes of water generated are normally too large to be considered a practical option. Other options include infilling the mine with power station ash, or remining the reserves by opencast methods.

2.3 Underground total extraction mining

2.3.1 Operational phase impacts

Total extraction methods include longwall, shortwall and pillar extraction (stopping). Goafing (collapse and breakage) of the strata may cause fractures through to surface, enhancing recharge from rainfall and from the upper aquifer, the Ecca aquifer, water associated with dolerite sills and dykes, due to fracturing. Water make can be 6 ML/d on average.

2.3.2 *Operational phase management options*

In deep longwall mines methods of handling the water make, without resorting to discharge to the surface water course, involve the sealing of underground compartments into which the water is pumped. Although seals contain the water, the pressures in the compartments may cause the pillars to seep. Dangers inherent in this type of storage include pillar or seal failure and flooding of the mine. Dewatering prior to mining has been considered but not performed in this type of coal mining.

Mine planning philosophies to ensure mining starts at low points in seam elevation and moves to higher ground so that water is left behind in the collapsed

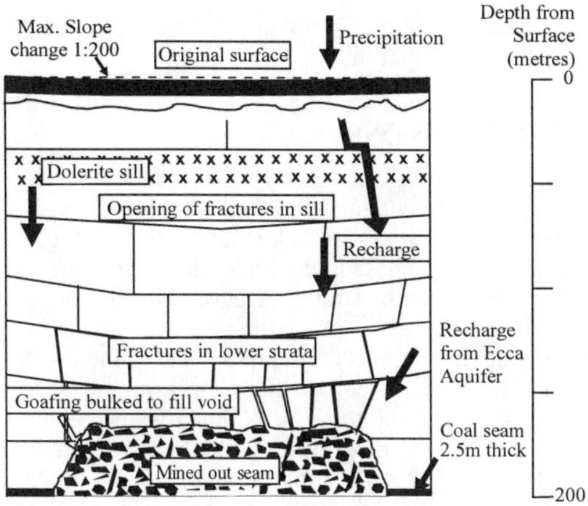

Figure 5. A cross section of total extraction in a deep mine

area, and to ensure that surface water courses are not undermined, have only recently been employed.

Water adits were used where pillar extraction was employed in collieries located in mountainous areas of KZN. These adits are developed in low lying areas of the mine where water accumulates. Box holes are dug at, or close to the centre of these areas and an adit, normally sloping at 1:200, is driven to the valley side. Although this removes water the retention time can be long enough to allow acidification. Liming plants have been used to neutralize the water prior to it reaching the surface watercourses.

2.3.3 *Closure phase impacts*

The individual aquifers present prior to longwall mining become interlinked, as a result of goafing, to form a single aquifer. Water quality stratification occurs in this ground water body, as clean, unaffected recharge drains down from the surface to lie on top of the polluted water.

Water adits may be left open and flushing out of water may occur. Sealing adits allows a build up of the water table that may force water out around sealed mine access adits or at higher levels in the stratigraphic sequence. As these mines frequently stretch from one side of a mountain to the other discharge from a mine can be made to occur into different surface water catchments.

2.3.4 *Closure phase management options*

Under normal circumstances the compartments created in deep, longwall areas will require no management. As the water resources created will be useable there may by a demand to abstract it. If abstraction wells are drilled deep enough and abstraction of the water is uncontrolled then dewatering of the poorer quality water at depth may occur. Water management of the created resource should ensure over utilization does not occur.

Where acid water emanates from mountainsides the only options are treatment by passive systems in very low flows or conventional treatment in high flows.

3. COAL WASTE FACILITIES

There are two major types of coal waste facilities, Waste Rock or Discard Dumps consisting of reject material, and Co-Disposal Facilities where the discard is co-deposited with the slurry from the coal washing plant. These are illustrated in figure 9 and 10 respectively.

3.1 *Waste rock or discard dumps*

These can consist of a mixture of various lithological units although commonly they are composed of coal parting material, and coal seam roof and floor material. The layers of coarse material provide oxygen pathways as well as preferential water conduits. Water will be stored in the fine layers during dry spells. Sampling dump material for acid base accounting (ABA) is more difficult in old and established dumps. The surface rock is not representative of the total dump material and specifically not of the material within the interior of the dump. Drilling is possible but difficult to perform.

3.1.1 *Impacts of waste and discard dumps*

The older coal waste dumps have a large impact upon all water resources. Old dumps have frequently been located in valleys and close to or within stream courses. Cognizance of the underlying geology which controls groundwater movement was rarely taken. Polluted water moves down beneath the dump and along any permeable layer.

Old dumps have been placed directly on topsoil. This has destroyed a source material used in dump rehabilitation and means polluted water from the dump can enter into the upper aquifer.

Many old dumps have not been rehabilitated allowing maximum infiltration of water and oxygen and ensuring that a high head of water exists to drive

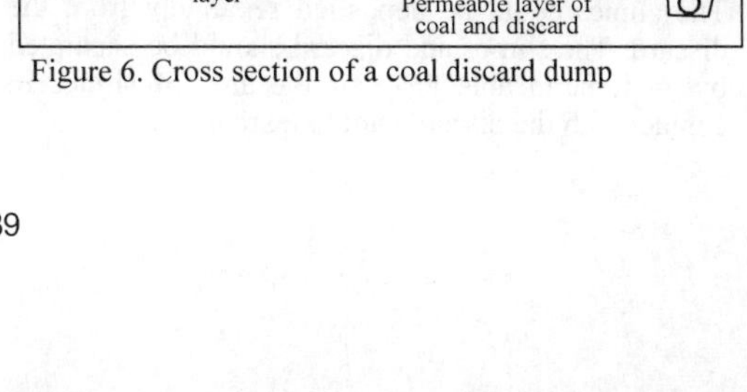

Figure 6. Cross section of a coal discard dump

polluted water into and through the upper aquifer.

3.1.2 *Management options*

Dumps have to be designed to limit ingress of oxygen and water and to keep all water away from the discard material. Both smooth "whaleback" and terrace designs are used in rehabilitating these dumps. Rainfall infiltration reduction and the maximization of runoff are achieved by using various cappings over the dump. Dump slopes are designed to limit infiltration, maximize runoff while minimizing soil and dump side erosion.

Runoff from upslope of the dump should be intercepted and directed around the feature. This is to prevent clean water from entering the dump area and becoming contaminated.

In some instances it may be possible to rework a dump. Reworking has the advantage of removing part of the pollution problem while making a profit and reducing the size of the pollution source.

3.2 *Co-disposal facilities*

Co-disposal of tailings (slimes) and coarse reject (discard) has involved placing of the slimes within a dam constructed of discards, as shown in figure 7, rather than the mixing of these two wastes.

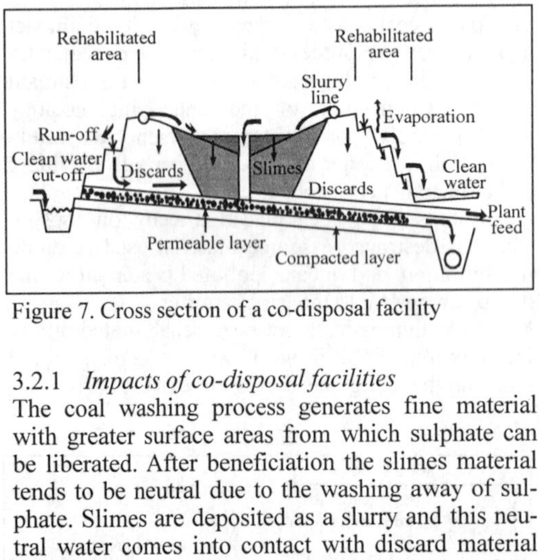

Figure 7. Cross section of a co-disposal facility

3.2.1 *Impacts of co-disposal facilities*

The coal washing process generates fine material with greater surface areas from which sulphate can be liberated. After beneficiation the slimes material tends to be neutral due to the washing away of sulphate. Slimes are deposited as a slurry and this neutral water comes into contact with discard material which contains sulphide minerals. The water from the slurry migrates through the discard effectively creating a heap leach system.

3.2.2 *Management options for co-disposal facilities*

The slimes could be deposited separately from the discard. The slimes and discard should be separated by an impermeable layer or the area of slimes in contact with the discard should be reduced.

Dewatering of slimes prior to disposal is an area of ongoing investigation.

4. CONCLUSIONS

Coal mine drainage chemistry is dependant upon the acid generating potential and the base neutralizing potential of the coal and rock material, the mining method and the phase in the life cycle of the mine. Water management strategies have been developed and the principles of prevention of water pollution managing the impacts of mining and mine waste disposal on all water resources, maximizing water reclamation and the reuse of polluted water, separation of clean and dirty water systems and the treatment of polluted water implemented in the South African coal mining industry to reduce impacts on water resources.

REFERENCES

Middleton, B.J. et al. 1995a. Surface water resources of South Africa Volume 1. Drainage Region A and B, The Limpopo - Olifants System. Water Research Commission and University of Witwatersrand *Hydrological Research Unit Report* No 101.

Middleton, B.J. et al. 1995b. Surface water resources of South Africa Volume II. Drainage Region C, the Vaal Basin. Water Research Commission and University of Witwatersrand *Hydrological Research Unit Report* No 8/81.

Pitman, B.V., et al. 1995. Surface water resource of South Africa Vol V1. Drainage Regions U, V, W, X The Eastern Escarpment. Water Research Commission and University of Witwatersrand *Hydrological Research Unit Report* No 9/81.

Hodgson, F.D.I. & R.O. Krantz, 1995. Investigation into groundwater quality deterioration in the Olifants River catchment above the Loskop Dam with specialized investigation in the Witbank dam sub-catchment, *Report to the Water Research Commission,* Pretoria, South Africa.

Groundwater: Past Achievements and Future Challenges, Sililo et al. (eds) © 2000 Balkema, Rotterdam, ISBN 90 5809 159 7

Author index